W9-BVB-450

LEWIS' DICTIONARY OF TOXICOLOGY

Robert A. Lewis

Library of Congress Cataloging-in-Publication Data

Lewis' dictionary of toxicology / edited by Robert A. Lewis.
p. cm.
Includes bibliographical references.
ISBN 1-56670-223-2 (alk. paper)
1. Toxicology--Dictionaries. I. Lewis, Robert A. (Robert Alan)
RA1193.L48 1996
615.9'003—dc20 96-35759
CIP

International Standard Book Number 1-56670-223-2
Library of Congress Card Number 96-35759
Printed in the United States of America 1 2 3 4 5 6 7 8 9 0
Printed on acid-free paper

ABOUT THE AUTHOR

Professor Lewis is the author or editor of numerous scientific papers and books written in English and German. Contributions include basic research in physiology, endocrinology, and animal behavior; applied research on air pollution and toxic chemicals; and chemical, biological, and environmental monitoring and specimen banking as applied to human health and environmental effects.

In addition to a long record of basic and applied research, the author has made a number of contributions to contemporary society through outstanding national and international achievements as an inventor and as an academic and government administrator in both the United States and Germany. He developed and coordinated several programs of research that were national or international in scope, and has served as an administrator of governmental organizations and national programs in the United States, and also as an academic administrator, director of a research institute, and leader of national programs of research in the Federal Republic of Germany. He led the team that selected the National Environmental Research and Assessment Parks of that country and later developed sampling designs and protocols for chemical and biological monitoring and environmental specimen banking to be conducted on a continuing basis.

The author has made a number of technological innovations including the invention and codesign of the Zonal Air Pollution System (ZAPS), a research tool that is able to continuously deliver and monitor the direct and indirect effects of gaseous pollutants such as SO_2 on biota throughout entire growing seasons. This system has been used in a number of major studies in the United States and Europe. See, for example, R. Guderian, D. Tingey, and R. Rabe (1985, in "Air Pollution by Photochemical Oxidants," R. Guderian, ed., Springer-Verlag, NY). Other examples of technological innovation include the design, construction, and application of an air quality monitoring and micrometeorological network in Montana; cooperating in this project were scientists from Battelle Pacific NW Laboratories. While a graduate student, he codesigned, tested and applied a state-of-the-art controlled environment system used in comparative physiological research. Numerous studies have been conducted with the use of this system.

The author recently served as a rapporteur at the "White House Conference on Technology for a Sustainable Future." He has also served on the Governor's Committee for Disease Control, New Jersey; The President's Task Force on Water Quality; the American-German Specimen Banking Program; Board of Examiners, Banares Hindu Univ., Varanassi, India; and the Health, Safety and Environmental Task Force for the U. S. Strategic Petroleum Reserve.

Prof. Lewis holds degrees from the Ohio State University, Rutgers University, and the University of Washington. His biography appears in current and recent editions of "Who's Who in America," "Who's Who in the World," "Who's Who in Science and Technology," and "Who's Who in Medicine and Health Care."

ACKNOWLEDGMENTS

Toxicology, as any other science, depends on the work of diverse scholars, practicing scientists, and other professionals, past and present, known and unknown. While these people cannot, for the most part, be acknowledged individually, they have this author's deepest respect and heartfelt thanks for their contributions to the science and language of toxicology. The body of knowledge that they built collectively, brick by brick, is impressive. In preparing this text, I consulted the original work of many such persons as well as a large number of standard references in biology, chemistry, physics, law, toxicology, and other fields of study and practice that enabled the emergence of toxicology as a mature science.

While only one name appears on the cover many people worked to make possible the production of this book. My thanks to the executives and staff of CRC Press/Lewis Publishers. All are thorough, even-handed professionals. They gave unstintingly of their time to assure the quality and success of this work. I worked most closely with Joel Stein, Susan Alfieri, Suzanne Lassandro, Carole Sweatman, and Mimi Williams. Their advice, assistance, editorial input, patience, encouragement, and many kindnesses are much appreciated.

In my own court, thanks and appreciation are due to Terri Brust for aid in research and word processing; to Cynthia Lewis for word processing, editing, and critical commentary; and to Carolyn W. Lewis who provided technical, editorial, and linguistic criticism.

Larry Barnthouse of Oak Ridge National Laboratory provided invaluable input on risk assessment and related subjects.

Thanks are due also to the reference librarians of the National Institutes of Health, the National Library of Medicine, the Library of Congress, and in Maryland, to those of the Hood College Library and the library systems of Frederick and Montgomery counties; they helped in pursuit of a number of areas of research.

I trace my introduction to and love of science to Charles and Marjorie Pensyl, dedicated and brilliant teachers.

Robert A. Lewis
January, 1998

DEDICATION

There are many persons who have contributed substantially to my development and to the development of this book. I would like to dedicate my work to each and every one of them. Nevertheless, years ago, when my daughter was born, it motivated me to make a major contribution (major for me) to the health and environmental sciences with the hope that the work would somehow help to make the world a bit better for her and her generation. But how could I omit my wife who is the most important supporter and professional critic of my work; and how could I omit my deceased mother who sacrificed much to nurture me, my dreams, and my personal and intellectual development.

this work is thus Dedicated with thanks and with love to the most important women in my life -

to my mother, my wife

Ethel (nee Hamm) Lewis, Carolyn (nee Weber) Lewis,

and especially to Cynthia Lewis, my daughter, and to her generation with the deep hope that, however indirectly, this work will contribute to a better future for them.

Robert A. Lewis
January, 1998

α. See alpha.

Α. See alpha.

a posteriori. pertaining to a manner or system of reasoning whereby one attempts to determine cause from knowledge of the effect or outcome. *Cf. a priori*.

a priori. pertaining to a manner or system of reasoning whereby one infers an effect or outcome from a notion or theory regarding cause. *Cf. a posteriori*.

A. absorbance.

A_r. relative atomic mass.

A 66. phenmetrazine.

a-. a prefix meaning **1:** on, in, at. **2:** lacking.

AA. amino acid; aminoacyl.

AAALAC. American Association for Accreditation of Laboratory Animal Care.

AAF. *N*-2-fluorenylacetamide.

2-AAF. *N*-2-fluorenylacetamide.

AAG. α-acid glycoprotein.

AAL. Ambient Air Level.

AAS. atomic absorption spectrophotometry.

ab. antibody; aberratio.

Ab. antibody.

ab-. a prefix indicting from, away from, off, derived from.

ABA. **1:** autonomic blocking agent. **2:** abscisic acid.

abalone. common name for certain low, flattened forms of gastropod mollusks of the genus *Haliotis*, *q.v.* Some are considered delicacies, others may contain a toxin concentrated in the digestive gland. Symptoms of poisoning include erythema, edema, and pain about the face, neck, and sometimes the extremities. A fulminating dermatitis is present in severe cases.

abamectin (ABM; avermectin B_1). an avermectin, *q.v.*

abate. **1:** to diminish or decrease; often applied to the amelioration of an endogenous or environmental condition (e.g., the intensity of pollution or the severity of a symptom). **2a:** to bring about a lessening or decrease; to arrest. **2b:** to cease; be arrested. **3:** temephos; a colorless, crystalline, pesticidal cholinesterase inhibitor, $[(CH_3O_2)PSOC_6H_4]S$. It is toxic by ingestion and inhalation.

abatement. **1:** a decrease in severity or elimination of a condition such as that of a symptom or the intensity of pollution. **2:** the act or means of abating. **pollution abatement**. the reduction of pollution by technological means, by the enactment and application of legislative measures, or by both. **smoke abatement**. the reduction of smoke by technological means, by the enactment and application of legislative measures, or both.

abbreviation. **1:** shortening, abridgement, curtailment, or reduction. **2:** the successive shortening or abridgement of ontogeny through loss of developmental stages.

ABCW. a method of extrapolation, based on published information on the biological effects of ionizing radiation. It is used to extrapolate results of germ cell mutation tests from one species to another. In this method, the mutation rate is normalized to the amount of DNA in the haploid genome of the species of concern.

abdomen. **1:** the part of the coelom (body cavity) of vertebrates in which viscera such as the stomach, intestines, and liver are suspended. In most vertebrates it is continuous with the thoracic cavity, but in mammals the abdomen is separated from the thoracic cavity by the diaphragm. **2:** the posterior body region of arthropods; often a series of similar segments.

abdomin-. See abdomino-.

abdominal. pertaining to the abdomen.

abdominal dropsy. ascites.

abdomino-, abdomin-. prefixes that denote a relationship to the abdomen.

aberrant. **1:** pertaining to an abnormal subject or process or to something that deviates substantially from that which is considered normal or typical. **2:** (of an anatomic structure) not found in the usual or expected place.

aberratio (ab). Latin, meaning aberration or an aberrant individual.

aberration. **1:** abnormality; anomaly; a departure from that which is considered normal or expected. See also chromosomal aberration.

abience. **1:** the tendency to avoid or withdraw from the source of a stimulus. **2:** withdrawal from a stimulus; an avoidance response. *Cf.* adience.

abient. of or pertaining to the tendency to withdraw from or avoid the source of a stimulus. *Cf.* adient.

Abies balsamea (balsam fir; Canada balsam; eastern fir). the only fir tree (Family Pinaceae) that is native to northeastern North America where it occurs mostly in coniferous forests, often in pure stands. The foliage, as browse, is important to deer and moose during the winter. It contains juvabione, *q.v.*

abiological. not biological; not pertaining to, involving, or produced by living organisms.

abiosis. **1:** nonviability. **2:** incompatibility with or absence of life. **3:** abiotrophy.

abiotic. **1a:** said of physical and chemical factors only: nonliving, devoid of life; devoid of living organisms, their products, and effects; nonbiological. **1b:** pertaining to nonliving components of the environment. **2:** incompatible with or antagonistic to life. **3:** disposed to prevent, inhibit, or destroy life.

abiotic environment. **1:** the nonliving components, collectively, of an ecosystem; the spectrum of physical and chemical factors in any system that do not arise from biological activity, but which may affect a living system. **2:** that part of the environment, in reference to an individual organism, taxon, or living system, that consists of nonbiological factors. *Cf.* biotic environment, abiotic, factor (abiotic).

abiotrophic. of or pertaining to abiotrophy.

abiotrophy. a loss of vitality by an organism or deterioration, usually of specific cells and tissues, in the absence of apparent injury.

abirritant. capable of soothing or of relieving irritation.

abirritation. **1a:** weakness. **1b:** lacking tone or strength. **2:** diminished responsiveness to stimuli. *Cf.* asthenia; atony.

abluent. a cleansing agent (e.g., a detergent).

ablution. a cleansing or washing.

ABM. abamectin. See avermectin.

abnormal. **1:** not normal; unusual; departing significantly from that which is usual, expected, or considered to be normal. **2:** outside the range of conditions considered to be healthful.

abnormal base analog. any xenobiotic analog of the normal DNA bases (e.g., 5-bromouracil, 5-fluorouridine, 6-mercaptopurine). Such compounds are extremely toxic and potent mutagens. Many are used as antineoplastic agents. They replace a natural base in DNA, thereby altering base-pair sequencing.

abnormality. **1a:** deviation from normality; the

fact or condition of being abnormal. **1b:** a structure, condition, or process that deviates from the normal state. **2:** an aberration, deformity, or anomaly. *Cf.* abnormity.

abnormity. **1:** abnormality (esp. def. 2). **2:** deformity. **3:** a monstrosity. **4:** a deformed neonate or juvenile.

abort. **1:** to terminate an activity or procedure before completion; to terminate an activity in progress. **2:** to arrest development. **3:** to give birth to a nonviable embryo or fetus. **4:** to arrest a disease in its early stages. **5:** to arrest the growth or development of an organ of a plant or animal; to cause an organ to remain rudimentary. **6:** to remove the developing organism from the uterus at an early stage.

aborticide. abortifacient (def. 2).

abortient. abortifacient (def. 1).

abortifacient. **1:** abortient, abortigenic, abortive, capable of causing or inducing abortion. **2:** an agent (e.g., mifepristone) that induces abortion.

abortigenic. abortifacient (def. 1).

abortion (miscarriage). **1a:** the termination of an activity or procedure prior to completion. **1b:** the termination or arrest of development of an organ or tissue in a plant or animal. **2:** the expulsion of an embryo or fetus from the prospective mother prior to the stage of viability (often given as ca. 20-28 wks of gestation in *Homo sapiens*). **3:** the premature birth of a dead fetus. **4:** the product of an abortion. **artificial abortion**. induced abortion. **induced abortion**. **1:** an intentionally caused (e.g., therapeutic) abortion. **2:** an abortion provoked by an abortifacient. **necrogenic abortion**. rapid necrosis of the tissues of a plant that lie immediately adjacent to the point of contact with a pathogen. This tends to prevent expansion of the disease process. **spontaneous abortion**. a naturally occurring abortion or miscarriage.

abortionist. **1a:** a person who performs abortions. **1b:** one who illegally induces abortions in humans.

abortive. See abortifacient.

abortus. an aborted conceptus.

abric acid. a component of the seeds of *Abrus precatorius* that contains a tetanic glycoside. Abric acid inhibits digestion; autopsy may thus show undigested food. It is used in the tropics as an arrow poison. Symptoms of intoxication include diarrhea, nausea, vomiting, tachycardia, convulsions, diffuse hemorrhages, coma, and death from heart failure. Ulcers occur in the mouth. See also abrin.

abrin (toxalbumin; abrus agglutinin). one of the most toxic naturally occurring substances known. It is an intensely irritant, extremely toxic, proteinaceous phytotoxin that is inactivated by heating. It is the chief active principle of *Abrus precatorius*. Abrin is actually any of five nearly identical lectins (abrins a-d; mol. wt. 63,000 to 67,000 Da) composed of 2 disulfide-linked polypeptide chains. The A-chain inhibits ribosomal protein synthesis causing cell death; the B-chain binds to the plasma membrane of the cells of the intestinal wall, permitting entry of the A-chain into the cytoplasm. Initial symptoms of intoxication are those of intense gastrointestinal irritation such as vomiting, and a bloody diarrhea that develops within 1-3 days following consumption of *Abrus* seeds. As a result, changes in electrolyte balance and plasma composition occur, causing cardiac arrhythmia and cerebral edema with drowsiness, convulsions, coma, and cardiovascular collapse. Further symptoms may include diffuse hemorrhages, hemolytic anemia, oliguria, and fatal uremia with death from heart failure. Digestive processes are inhibited such that autopsy may reveal undigested food. Ulcers occur in the mouth. LD_{50} *i.p.* in laboratory mice is reported at 0.02 mg/kg. Abrin is chemically similar to botulinum toxin. See abric acid, *Abrus precatorius*, phytotoxin.

abrin poisoning. abrism.

abrism (abrin poisoning). poisoning by abrin or by the abrin-containing seeds of *Abrus precatorius*. See abrin.

Abrus. a genus of climbing or creeping tropical leguminous shrubs (Family Fabaceae, formerly Leguminosae) that are indigenous to tropical Asia. The seeds are extremely toxic when thoroughly chewed. ***A. precatorius*** (bead vine; black-eyed Susan; crab's eyes; crabs-eye vine;

crabseye; Indian bean; Indian licorice; jequirity bean; love bean; lucky bean; mienie-mienie; Paternoster pea; prayer bead; prayer bean; precatory bean; redbead vine; rosary bead; rosary pea; Seminole bead; weather plant; wild licorice). a tall, twining, perennial vine of the tropics common in Africa and Asia, also occurring in central and southern Florida, Hawaii, Guam, Central America, southern Europe, and India. The root is sometimes used as a substitute for licorice, but the distinctive beans are extremely poisonous. One bean thoroughly chewed (or with a broken seed coat) can prove fatal to an adult human. Symptoms appear within 1 to 3 days and include those of intense gastroenteritis, ulcers in the mouth, tachycardia, drowsiness, convulsions, coma, circulatory collapse, hemolytic anemia, oliguria, and fatal uremia. Digestion is inhibited, thus autopsy may reveal undigested food in the stomach. The active principle is abrin, although the seeds contain other toxic agents such as abric acid and hypaphorine (a convulsant). See also abrin, abrus agglutinin.

abrus agglutinin. abrin.

abscess. a localized injury or infection that contains pus and involves the displacement or disintegration of tissue with or without necrosis. An abscess may form at the site of an envenomation due to the trauma, chemical erosion of tissue, or infection.

abscisic acid (abscisic acid; [S-(Z,E)]-5-(1-hydroxy-2,6,6-trimethyl-4-oxo-2-cyclohexen-1-yl)-3-methyl-2,4-pentadienoic acid; dormin; abscisin II; ABA). a naturally occurring sesquiterpenoid, epoxide auxin, $C_{15}H_{20}O_4$, that exerts numerous, mainly inhibitory, effects on the growth and development of many plant species. It promotes leaf and fruit abscission (leaf fall and fruit drop), leaf senescence, and is active in the control of seed and bud dormancy, apical dominance, and inhibition of flowering of long-day plants maintained on short days. Under stress, especially due to a water deficit, the rate of secretion of ABA by the plant increases greatly, causing stomatal closure during the day, thus decreasing water loss. Plants normally regulate abscisic acid at nontoxic levels by converting it to an abscisyl glucoside.

abscisin II. abscisic acid.

abscissic acid. abscisic acid.

abscopal. refers to effects on nonirradiated tissue due to irradiation of parts of a living organism.

absinthe. **1:** a neurotoxic alcoholic beverage made from *Artemisia absinthium* and other bitter herbs. The alcohol content is 60% or more; it also contains thujone extracted from *A. absinthium*. Regular use causes CNS manifestations such as anxiety and tension, insomnia, nightmares, trismus, amblyopia, optic neuritis, vertigo, tremors, convulsions, and mental deterioration. **2:** *Artemisia absinthium*.

absinthium. **1:** an intensely bitter, toxic, essential oil, $C_{30}H_{40}O_7$, used as a flavoring in liqueurs, vermouth. **2:** common name of *Artemisia absinthium*.

absinthol. thujone.

absolute alcohol. See alcohol.

absolute environmental persistence. See persistence.

absolute temperature scale. Kelvin temperature scale.

absorb. **1:** to take up or take in a substance, especially (in biology) across membranes that bound tissues, organs, or cells. **2:** to interact with incident radiant energy such that the energy is attenuated. **3:** to immerse or engage completely, as the mind.

absorbance (A). **1:** the amount of incident light of a particular wavelength that is absorbed by a substance in solution. It is measured spectrophotometrically and can be used, e.g., to determine the concentration of a substance in solution. **2:** a measure of the capacity of a medium to absorb ionizing radiation. It is expressed as the logarithm of the ratio of the intensity of the radiation that enters the medium to that which is transmitted.

absorbed dose. See dose.

absorbent. **1:** absorptive; having the capacity to absorb. **2:** an absorbefacient (def. 3). **3:** a biological structure involved in absorption.

absorbifacient. **1:** referring to the facilitation of absorption. **2:** causing or facilitating absorption. **3:** an agent that facilitates absorption.

absorption (uptake). **1a:** the penetration of one substance (the absorbate) into or through the inner structure of another (the absorbent). **1b:** the assimilation or incorporation of one substance into another by capillary action, chemical processes, osmosis, or solvation. **1c:** the transport of molecules across tissues or cell membranes into a living system or system component. **1d:** the process (active or passive) of taking in. **1e:** the uptake into a living system of water, solutes, or other substances by active or passive mechanisms. **2:** the attenuation of energy (e.g., light, sound waves) impinging on a substance or structure. Such diminution indicates that some portion of the energy striking the object is neither transmitted nor reflected. **3:** in toxicology and the health sciences, the process of movement of a xenobiotic from the site of application or exposure to the systemic circulation. *Cf.* absorption of xenobiotics. **4:** the concentration of one's thought upon a single subject, object, or activity to the exclusion of others. **cellular absorption**. the absorption of a substance through a cell membrane into the interior of a cell. Cellular absorption may occur via passive diffusion, filtration, carrier-mediated transport (active transport and facilitated diffusion), or by endocytosis (phagocytosis and pinocytosis). **cutaneous absorption** (percutaneous absorption). absorption through the skin. **dermal absorption**. the passage of a substance into the interior of a vertebrate animal via the skin. **external absorption**. absorption of substances by the skin and mucous membranes. **mouth (oral) absorption**. absorption through the oral mucosa. Certain xenobiotics, especially alkaloids, can be so absorbed. **parenteral absorption**. absorption other than from the gastrointestinal tract. **pathologic absorption**. the absorption of substances, such as the products of disease, into blood or lymph that would otherwise normally be excreted. See also route (route of absorption), absorption of xenobiotics. **percutaneous absorption**. cutaneous absorption. **route of absorption**. **1:** route of exposure. See exposure (route of exposure). **2:** occasionally taken to mean the "intestinal route" only. This usage is not recommended.

absorption spectrum. the differential pattern of absorption of light of different wavelengths that characterizes many organic substances. It is usually presented as a plot of absorbance against wavelength. The resulting curve is characteristic of a given substance.

absorption of xenobiotics. movement of the molecules of a xenobiotic across cell membranes or tissues (e.g., via skin, gills or pulmonary surfaces, stomata, or the gastrointestinal tract) into a living system or system component (e.g., the cytoplasm of protozoa, the blood or lymphatic system of mammals). Physiological and biochemical transport mechanisms are generally unavailable to xenobiotics as they are structurally unrelated to normal metabolites. Most are absorbed by passive diffusion; lipophilic compounds are typically most readily absorbed, while ionic or highly polar substances are not easily absorbed.

absorptive. pertaining to or capable of absorbing. See absorbent (def. 1).

abundance. the total number of individuals of a given taxon or specified type of organism (usually a species) per unit area, volume, population, or community. Abundance may be measured directly or indirectly (inferred) and is often measured as cover in terrestrial plants; relative abundance. **absolute abundance**. the precise number of individuals of a taxon or specified type of organism in a given area, volume, population, or community. **relative abundance**. the abundance of one taxon or specified type of organism in a given area, volume, population, or community relative to specified others within the same boundaries.

abuse. as used in toxicology: misuse, excessive use, or physiologic dependence on the use of a drug or other substance. **alcohol abuse**. excessive use of alcoholic beverages to the point of developing the symptoms of alcoholism. **drug abuse**. substance abuse. **substance abuse** (drug abuse). **1a:** the use of substances that are generally disapproved by a society, especially if such are proscribed by law. Such abuse commonly involves any of a number of different types of behavior, such as experimental and recreational use; use of psychoactive drugs to relieve problems or symptoms; use of drugs at first for the above reasons but with later development of dependence and continuation at least partly to prevent the discomfort of withdrawal. Common substances of abuse in-clude narcotics, alcohol, and psychoactive agents. **1b:** excessive use of a drug or use of a drug for purposes other than that approved (as defined by a particular society). **1c:** the use of a drug by other than those for whom use of the drug is legitimately intended or accepted. **2:** the use of a drug to the point of developing physiological or behavioral dependence. **3:** use of a drug in amounts that are seriously detrimental to a person's health or well being.

Ac. **1:** the symbol for actinium. **2:** acetyl.

acacia. **1:** common name for plants of the genus *Acacia*. **2:** a commonly used vehicle for insoluble toxicants.

Acacia (acacia). a genus of some 50 species of warm-climate leguminous shrubs and trees (Family Fabaceae, formerly Leguminosae, several of which contain toxic or medically important principals. ***A. berlandieri*** (guajillo). a deep-rooted, often spiny, perennial shrub or small tree of the southwestern United States that contains a toxic amine, *N*-methyl-*beta*-phenylethylamine. Consumption of large amounts of the leaves or fruit over a period of months by sheep and goats produces ataxia ("limber leg," "guajillo wobbles") and sometimes death. ***A. catechu***. a source of catechin. ***A. georginae***. a tree of northern Australia, the leaves of which contain concentrations of fluoroacetate that are lethal to cattle and sheep. ***A. greggii*** (cat's claw, catclaw, catclaw acacia). a cyanogenic shrub. ***A. senegal***. an African tree and a source of gum arabic.

Acalyptophis peronii (Peron's sea snake). a large, rare species of sea snake (Family Hydrophidae) known largely from the Gulf of Siam, Arafura Sea, Gulf of Carpentaria, and the Coral Sea. The scales about the eye have spiny projections and the body scales have pointed keels. As with most hydrophids, this snake is considered dangerous.

Acanthina (unicorns; drills). a genus of marine snails; the shells bear knobs and ridges. ***A. spirata*** (angular unicorn). a muricacean gastropod mollusk that secretes a paralytic toxin produced in the hypobranchial gland and salivary-accessory gland complex; it is rich in acetylcholine.

Acanthophis (death adders). a genus of venomous snakes (Family Elapidae) indigenous to Australia and nearby islands. They are remarkably viper-like in appearance, with a broad, flattened head that is distinct from the neck; the canthus rostralis is distinct. The body is thick, depressed, with a short tail bearing a long terminal spine. ***A. antarcticus*** (common death adder; death adder; New Guinea death adder; bartezgooremil; burrek; dudbe; pomnica). a widely distributed, deadly, largely nocturnal, ovoviviparous snake of Australia and New Guinea; it is absent from the central desert regions of Australia. This snake is brown, gray, reddish, or yellowish in color. Less than 1 m long, it is one of the most deadly snakes of the region and is as greatly feared as *Notechis scutatus*. The victim may faint almost at once; cardiac and respiratory distress are common. The bite is fatal in half of untreated cases. ***A. pyrrhus*** (death adder; Centralian death adder; desert death adder). a poorly known desert snake of central and western Australia.

Acanthuridae (surgeonfishes; doctorfishes; tangs). a family of brightly colored, tropical marine bony fishes (Class Osteichthyes) characterized by a hinged, movable, scalpel-like spine (a modified scale) that folds into a groove on either side of the caudal peduncle. This spine may be used by the moving fish to slash or warn away other fishes. The body is deeply compressed, with eyes high on the head. The small terminal mouth bears finely serrated spatula-shaped teeth.

Acanthurus (surgeonfishes). a genus of bony marine fishes (Family Acanthuridae), certain populations of which are transvectors of ciguatera. The teeth are generally adapted to scrape algae from hard substrates such as rocks and coral. See ciguatera. ***A. glaucopareius***. a fish of the tropical Pacific Ocean, Indonesia, and the Philippine Islands. ***A. triostegus***. a fish of the Hawaiian and Johnston islands.

acapnia. a state of reduced carbon dioxide in the circulating blood.

Acari (Acarina). a large order of small, even microscopic, arachnids that includes ticks and mites. The body is characteristically small and round and the abdomen is fused with the cephalothorax, concealing the segmentation.

acaricidal. lethal to mites.

acaricide (miticide). **1:** an agent (usually a chemical) lethal to mites. **2:** an agent specifically lethal to phytophagous, as opposed to parasitic, mites. **3:** a pesticide lethal to mites that may be applied to domestic animals, crops, or humans. Some insecticides are also acaricidal. **4:** also in common usage: an agent lethal to mites and ticks (acarines).

acarin. See dicofol.

Acarina. Acari.

acarine. **1:** pertaining to mites. **2:** an arachnid of the order Acarina.

acarology. the study of mites and ticks.

acarotoxic. poisonous to mites and ticks.

acathisia. akathisia. See *Acacia* (*A. berlandieri*).

ACC. actinomycin C. See cactinomycin.

accelerator. an agent, organ, or apparatus that increases the rate of any process (e.g., a reaction, absorption, toxic effect).

acceptable daily intake (ADI; allowable human daily intake). **1:** the amount of a substance (usually a pesticide or food additive) that can be indefinitely consumed daily in one's diet without significant risk. It is usually expressed as mg/kg body weight/day. **2:** a subjective term, sometimes used for contaminants of food, air, and water. It is derived from the No-Observed-Effect-Level (NOEL) by applying a subjective safety factor (usually of 10, 100, or even 1000) intended to allow for the variability in sensitivity among species and the amount and quality of data supporting the NOEL. Aside from the subjective judgment, it is equivalent to the term "reference dose."

accident. **1:** a happening, incident, occurrence. **2:** an unexpected or unplanned occurrence, especially during the course of a usual or necessary act, event, or process. **3:** a mishap; the occurrence of a sudden, unexpected event or result; misfortune; misadventure. In toxicology this term implies an event that results from negligence and/or has harmful, unexplained results (e.g., injury, damage, or death). The cause may remain unexplained, and a victim may or may not be the cause.

accident proneness. the tendency of some individuals to have more accidents than normal. The causes may relate to ill health, but are often obscure and may depend upon unconscious factors.

accident site. **1a:** the location of an unexpected occurrence or accident; the scene of an accident. **1b:** in toxicology and the health and environmental sciences generally, an accident site is the location of an accidental release of toxic or hazardous materials, whether from a stationary or mobile source.

acclimatation. acclimatization.

acclimate. acclimatize.

acclimation. **1:** a normal phenotypic adjustment of an organism to environmental change, including behavioral, physiological, and morphological changes. **2a:** a gradual and reversible physiological or morphological adjustment of physiology or morphology in response to changing environmental conditions. **2b:** habituation or increased tolerance of an organism to an altered environment. **3:** acclimatization. *Cf.* acclimatization, accommodation, adaptation, adjustment.

acclimatization (acclimatation). **1:** the process or the result of acclimatizing. **2:** increased tolerance of a species to a pronounced environmental change over several generations; adaptation. *Cf.* acclimation, accommodation, adaptation, adjustment.

acclimatize (acclimate; harden). **1:** to adapt to a new environment. **2:** to develop increased tolerance to new environmental conditions.

accommodation. **1:** the state or condition of adapting. **2:** the capacity of a living system to adapt to changes in the environment. **3:** a decrease in the threshold of response due to repetitive stimulation. See also acclimation, adaptation, habituation. **4:** adjustment of the lens of the eye to enable close vision. See adaptation (def. 3c). *Cf.* acclimation, acclimatization, adaptation, adjustment, habituation.

accumulation. **1:** the process of concentrating, amassing, or collecting together, or the state of having concentrated or amassed an amount of something. **2:** in toxicology, the state whereby the concentration of a substance within a specified system is greater than that in the immediate environment. See also bioaccumulation, specificity, variation. **degree of accumulation**. the ratio of the concentration of a substance within the target system to that component of the environment from which accumulation is considered to take place (e.g., plant tissue, soil). The degree of accumulation of a toxicant is often system specific (e.g., tissue specific, organ specific, species specific). **tissue-specific accumulation**. the preferential accumulation of a substance in a specific tissue. See, for example, cephaloridine, paraquat. See also bioaccumulation, bioconcentration factor, specificity, variation.

accumulator (concentrator; accumulator organism). **1a:** any living organism or species that actively concentrates a particular exogenous element or compound to levels above that in the food or environment to which it is exposed. **1b:** accumulator plant (See plant). **facultative accumulator**. an organism in which an environmental chemical can accumulate under some circumstances. **obligate accumulator**. an organism that selectively accumulates a specific environmental chemical (element or compound), one of its derivatives, or a class of compounds. Mechanisms of accumulation vary but are often due to an inability to translocate or further metabolize the substance once it enters the organism.

accumulator organism. accumulator.

accumulator plant. See plant (accumulator).

accuracy. **1:** the quality or degree of freedom from error. **2a:** the degree of agreement between a calculated or measured value and the true value of a variable. **2b:** the degree of agreement between the true value and the mean of the measured values obtained by applying a particular procedure a large number of times. **3:** the reciprocal of the degree of uncertainty of a measured value of a given quantity. **4:** the degree of agreement between a measured value and the value of a reference, certified standard, or a value accepted as true. This practice is necessary and generally accepted because the "true" value is an ideal that cannot be known. *Cf.* precision.

accurate. **1a:** free from error. **1b:** quantitatively equivalent to or in conformity with a standard, reference, or accepted true value.

Accutane®. a trademark name for the 13-*cis*-form of retinoic acid. It is fetotoxic. Pregnant women who use Accutane to treat acne (even for a few days) are risking malformations of the fetus that can include severe facial malformations, heart defects, and mental retardation. See also retinoic acid.

ACD. dactinomycin (actinomycin D).

-aceae. the suffix of a family name in botanical nomenclature. *Cf.* -idae.

aceclidine. the acetate ester of quinuclidinol, *q. v.*

acenocoumarin (3-(α-acetonyl-4-nitrobenzyl)-4-hydroxycoumarin). a white, tasteless, odorless, crystalline powder that is slightly soluble in water and organic solvents. It is an anticoagulant.

acephalous. having no head.

acephate (acetylphosphoramidothioic acid *O,S*-dimethyl ester; Ortho 12420; Orthene). a white, water soluble, crystalline solid, $C_4H_{10}NO_3PS$, with a melting point of 72 to 80°C, used as an insecticide against aphids and other foliage pests. It is moderately toxic to humans and laboratory animals by ingestion and is carcinogenic to some laboratory animals. Signs of acute oral intoxication of mallards include ataxia, imbalance, jerkiness, hopping and falling, mild spasms in the legs and feet, immobility, spread wings, and intermittent tremors.

acephatemet. a white, slightly water-soluble, crystalline solid, $CH_3OCH_3SPONH_2$, with a melting point of 39 to 41°C, used as an insecticide to control cutworms and borers.

acet-, aceto-. prefixes that denote the two-carbon skeleton of acetic acid or any moiety of a chemical comprised of such a fragment.

acetal (diethylacetal; 1,1-diethoxyethane; ethylidenediethyl ether). a colorless volatile liquid, $CH_3CH(OC_2H_5)_2$, with an agreeable odor. It is used in organic synthesis, in cosmetics, as a solvent, and in perfumes and flavorings. Acetal is narcotic and moderately toxic at high concentrations.

acetaldehyde (ethanal; ethyl aldehyde; acetic aldehyde; "aldehyde"). a colorless, highly flammable, irritant, organic liquid (b.p. 21°C) with a strong, harsh, fruity odor. It is produced commercially and used in the preparation of various aromas and flavorings and in the manufacture of acetic acid, aniline dyes, butanol, various drugs, paraldehyde, synthetic rubber, and the silvering of mirrors. Acetaldehyde is also used in small amounts as a flavor enhancer in various foods. It is synthesized metabolically in the vertebrate liver. It is moderately toxic to laboratory rats by both the oral and respiratory routes of absorption. Contact can cause severe irritation and injury of the skin, eye, and mucous membranes. Inhalation depresses the CNS, causes headaches, nausea, vomiting, runny nose, and sometimes severe pulmonary distress. Ingestion produces symptoms similar to alcoholic intoxication. It is carcinogenic.

acetaldehyde *N*-methyl-*N*-formylhydrazone. gyromitrin.

acetamide (acetic acid amine; ethanamide). a solid, CH_3CONH_2, with colorless, long deliquescent crystals, and a characteristic mouse-like odor; it is soluble in water and ethanol and is slightly soluble in ether. Acetamide is used chiefly as an experimental carcinogen and as a solvent, wetting agent, and penetrating agent.

8-acetamido-5-amino-2-naphthalenesulfonic acid (acetyl-1,4-naphthalenediamine-6-sulfonic acid; acetylamino-1,7-Cleve's acid). a toxic paste, $C_{10}H_5(NHCOCH)(NH_2)(SO_3H)$, used as a chemical intermediate and in dyes.

2-acetamidofluorene. *N*-2-fluorenylacetamide.

L-α-acetamido-β-mercaptopropionic acid. acetylcysteine.

6-(acetamido)-4-methyl-1,2-dithiolo-[4,3-b]pyrrol-5(4H)-one. thiolutin.

3-acetamido-5-methylpyrrolin-4-one[4,3-d]-1,2-dithiole. thiolutin.

***p*-acetamidophenol**. acetaminophen.

2-acetaminofluorene. *N*-2-fluorenylacetmide.

acetaminophen (APAP; *N*-(4-hydroxyphenyl) acetamide; 4′-hydroxyacetanilide; *p*-hydroxyacetanilide; *p*-acetamidophenol; *p*-acetaminophenol; *p*-acetylaminophenol; *N*-acetyl-*p*-aminophenol; paracetamol; APAP). a simple nonprescription oral analgesic and antipyretic that occurs as a white, odorless, crystalline powder, $CH_3CONHC_6H_4OH$. It relieves pain without altering consciousness. Although hepatotoxic and nephrotoxic, acetaminophen is generally considered safe when the recommended dosage is used for no longer than 10 days. It has a lower incidence of hypersensitivity than salicylates. The actual hepatotoxicant is a toxic intermediate of acetaminophen metabolism. This intermediate is normally detoxified by glutathione conjugation. When this detoxication system is saturated by a large overdose, the concentration of the toxic intermediate increases and binds to various liver constituents. *N*-acetylcysteine will usually prevent hepatotoxic effects if given within 12 hours following ingestion of the drug. Acetaminophen is very toxic to laboratory mice, causing acute renal tubular necrosis and renal failure in susceptible strains (e.g., Fischer-344) of laboratory rats. Sprague-Dawley rats are much less likely to develop serious symptoms. See also receptor (Ah receptor), Cimetidine.

acetaminophen poisoning. poisoning generally due to ingestion of large overdoses of acetaminophen. Nausea and vomiting mark the first 2-3 days following an acute dose; additional associated symptoms in humans are pallor, sweating, dyspnea, jaundice, delirium, and unconsciousness. Signs of hepatic damage may appear within 3-4 days, sometimes accompanied by signs of CNS, renal, and myocardial damage. The condition may be fatal as a result of hepatic necrosis. Treatment may include the use of *N*-acetylcysteine or cysteamine.

***p*-acetaminophenol**. acetaminophen.

acetanilide, acetanilid (*N*-phenylacetamide; acetylaminobenzene; acetylaniline; antifebrin). a toxic white powder or crystalline solid obtained by the reaction of acetic acid with aniline. It is an analgesic, an antipyretic, and an antiinflammatory agent that is sometimes used in the treatment of neuralgia and rheumatism. Chronic or acute acetanilide poisoning can be quite serious. It is moderately toxic to laboratory rats.

acetanilide poisoning. acute or chronic poisoning, the latter being the result, usually, of prolonged use of acetanilide. Symptoms may include chills and sweating, nausea, vomiting, tinnitus, an irregular pulse, dyspnea, hypotension, cyanosis due to the formation of methemoglobin, unconsciousness, coma, convulsions, and death from sudden respiratory or cardiovascular collapse. Lesions include methemoglobinemia and kidney and liver injury.

acetate. **1a**: the acetate ion, CH_3COO^-. **1b:** any salt of acetic acid. **1c:** any compound (e.g., ethyl acetate) that contains the acetate group. **2:** a textile fiber (formerly acetate rayon) produced from partially hydrolyzed cellulose acetate.

(acetato)phenylmercury. phenylmercuric acetate.

(acetato)trimetaarsenitodicopper. cupric acetoarsenite.

acetene. ethylene.

acetic acid (ethanoic acid; vinegar acid; methanecarboxylic acid). a weak, caustic, organic acid, CH_3COOH, with a pungent odor, used as a reagent. It is the distinctive component (4-6%) of vinegar. The reagent grade is an aqueous solution containing not less than 36% acetic acid by weight. Acetic acid occurs in the defensive fluid of the arachnid, *Mastigoproctus giganteus*, and is an active component of certain ant venoms (e.g., that of *Myrmicaria natalensis*). It is the sole (e.g., *Chlamydomonas spp.*) or an alternative (e.g., *Chlorella*) source of carbon in some algae. **glacial acetic acid**. an acid con- [illegible] is an extremely toxic vesicant and corrosive agent produced by destructive distillation of wood or from acetylene and water. Ingestion can cause severe damage to the lining of the gastrointestinal tract. Effects of acute exposure can include vomiting, hematemesis, uremia, diarrhea, and death from cardiovascular collapse. Chronic exposure can cause bronchitis, irritation of the eye, and erosion of tooth enamel.

acetic acid allyl ester. allyl acetate.

acetic acid amine. acetamide.

acetic acid benzyl ester. benzyl acetate.

acetic acid-2-butoxy ester. *sec*-butyl acetate.

acetic acid butyl ester. *n*-butyl acetate.

acetic acid-*tert*-butyl ester. *tert*-butyl acetate.

acetic acid-1,1-dimethylethyl ester. *tert*-butyl acetate.

acetic acid-1-methylpropyl ester. *sec*-butyl acetate.

acetic acid phenylmethyl ester. benzyl acetate.

acetic acid-2-propenyl ester. allyl acetate.

acetic aldehyde. acetaldehyde.

acetic oxide. acetic anhydride. See anhydride (acetic).

acetoacet-*p*-chloranilide. a white, combustible, crystalline powder, $CH_3COCH_2CONHC_6H_4Cl$, used as an intermediate in the synthesis of azo dyes. It is toxic by ingestion.

acetoacetic acid (acetone carboxylic acid; acetyl acetic acid; diacetic acid). a colorless, oily liquid, CH_3COCH_2COOH, used in organic syntheses, that is soluble in water, ethanol, and ether. It is irritant to the eyes and mucous membranes.

acetoaminofluorene. *N*-2-fluorenylacetamide.

7-(2′-aceto-6′, 8′-dimethoxy-3′-methyl-1′- hy- [illegible] **3,8,9,-trihydroxy-1(2*H*)-anthracenone** (T-516). a buck-thorn toxin.

acetone (2-propanone; dimethylketone; *β*-ketopropane; pyroacetic ether; ketone propane; propanone; pyroacetic ether). a colorless, pleasant-smelling, flammable liquid solvent,

C_3H_6O, with numerous industrial and commercial uses. It is extremely toxic to laboratory rats. Acetone is an irritant to the eyes and respiratory tract and can cause a contact dermatitis by dissolving fats in the skin. Inhalation can cause headaches, excitement, bronchial irritation, faintness, and fatigue. It has narcotic effects and large amounts can cause unconsciousness. Severe poisonings are rare.

acetone carboxylic acid. acetoacetic acid.

acetone cyanohydrin (2-hydroxy-2-methyl-propanenitrile; 2-methyllactonitrile; α-hydroxyisobutyronitrile). a colorless, oxygen-containing liquid nitrile, $(CH_3)_2C(OH)CN$, produced by the condensation of acetone with hydrocyanic acid. It has numerous industrial applications (e.g., in polymerization reactions and in the synthesis of foaming agents, insecticides, and pharmaceuticals). It is extremely toxic by ingestion and is also readily absorbed through the skin, decomposing in the body with the release of acetone and HCN. The latter is the primary proximate toxicant. See also nitrile.

acetonemia. high concentrations of acetone in circulating blood.

acetonitrile (methyl cyanide; cyanomethane; ethanenitrile). a flammable, moderately toxic, colorless liquid, CH_3CN, with a mild, ether-like odor. A by-product of the manufacture of acrylonitrile, it is an important industrial solvent for many organic and inorganic compounds because of its relatively low boiling point (m.p. -45°C; b.p. 81°C). It is especially important as an easily recovered reaction medium. It is readily absorbed by all routes of exposure and can damage the respiratory tract if the vapor is inhaled. Human deaths are rare. See also nitriles.

acetonylacetone (1,2-diacetylethane; hexanedione-2,5; 2,5-diketohexane; 2,5-hexanedone; α,β-diacetylethane). a clear, colorless, liquid γ-diketone, $CH_3COCH_2CH_2COCH_3$, that gradually turns yellow. It is water soluble and is used as a solvent and intermediate in the preparation of pharmaceuticals and photographic chemicals. Acetonylacetone is an irritant to the skin and mucous membranes and is a moderately toxic neurotoxicant, with narcotic effects at high concentrations. It forms protein adducts (pyrroles) with consequent crosslinking of neurofilaments and is marked by specific behavioral, pathological, and biochemical effects. See γ-diketone neuropathy.

3-(α-acetonylbenzyl)-4-hydroxycoumarin. warfarin.

acetonylchloride. chloroacetone.

3-(α-acetonylfurfuryl)-4-hydroxycoumarin. coumafuryl.

2-acetonyl-1-methylpyrrolidine. hygrine.

3-(α-acetonyl-4-nitrobenzyl)-4-hydroxycoumarin. acenocoumarin.

***p*-acetophenetide**. acetophenetidin.

acetophenetidin (phenacetin; *N*-(4-ethoxyphenyl)acetamide; *p*-ethoxyacetanilide; *p*-acetphenetidin; acet-*p*-phenetidin; *p*-acetophenetide). an odorless white powder or crystalline substance, $CH_3CONHC_6H_4OC_2H_5$, it is a *p*-aminophenol derivative used as an antipyretic and oral analgesic. It is a CNS stimulant. Symptoms of intoxication may include sweating, nausea, vomiting, tinnitus, hypotension, and chills. Cyanosis, coma, convulsions, and death from cardiopulmonary collapse may ensue. Lesions may include methemoglobinemia, and kidney and liver damage. In the United States, containers of analgesics that include this substance must bear a warning label stating that kidney damage may occur with prolonged use. Unlike the salicylates, however, *p*-aminophenol derivatives do not cause pronounced disruption of acid/base balance. Acetophenetidin may impair male reproductive function. Chronic use of high doses of this drug can cause bladder cancer in humans.

acetopyrrothine. thiolutin.

***o*-acetoxybenzoic acid**. acetylsalicylic acid.

2-acetoxybenzoic acid. aspirin.

acetoxyphenylmercury. phenylmercuric acetate.

(2-acetoxypropyl)trimethylammonium chloride. methacholine chloride.

3-acetoxyquinuclidine. the acetate ester of quinuclidinol, *q.v.*

α-acetoxytoluene. benzyl acetate.

acetozone. acetylbenzoyl peroxide.

acet-*p*-phenetidin. acetophenetidin.

***p*-acetphenetidin**. acetophenetidin.

acetyl (Ac). the univalent radical, CH_3CO.

acetyl carbromal (*N*-acetyl-*N*-bromodiethyl acetylurea). a slightly water-soluble, crystalline compound, $(C_2H_5)_2CBrCONHCOCH_3$, with a slightly bitter taste; it is completely soluble in ethanol and ethyl acetate. It is a sedative; overdosage can prove fatal.

acetyl chloride (ethanoyl chloride). an acrid, fuming, corrosive, flammable liquid, CH_3COCl. It is an acetylating agent used in organic preparations, in dyestuffs, and pharmaceuticals. It causes severe burns on contact and is extremely irritant to the eyes.

acetyl CoA. acetylcoenzyme A.

acetyl ethylene. methyl vinyl ketone.

acetyl nitrate. a colorless, fuming, hygroscopic liquid, CH_3COONO_2, that is corrosive to the skin and mucous membranes.

acetyl oxide. acetic anhydride. See anhydride (acetic anhydride).

acetyl peroxide (diacetyl peroxide). a colorless crystalline compound, $(CH_3CO)_2O_2$, that is slightly soluble in cold water, ethanol, and ether. It is used as an initiator and catalyst for resins. The pure compound is a strong oxidizer, is violently explosive, and should not be heated above 30°C nor stored following preparation. The commercially available preparation (a 25% solution in dimethyl phthalate) is a strong irritant to the skin, eyes, and mucous membranes.

***N*-acetyl-Wieland-Gumlich aldehyde**. diaboline.

acetylacetic acid. acetoacetic acid.

acetylacetone (2,4-pentanedione; diacetylmethane). a colorless or slightly yellow, pleasant-smelling, water-soluble, flammable liquid, $CH_3COCH_2OCCH_2$; used as a solvent, a paint and varnish drier, lubricant additive, insecticide, and fungicide. It is a mild irritant to the skin of higher vertebrates.

acetyladonitoxin. a constituent of *Adonis vernalis*, *q.v.*

acetylaminobenzene. acetanilide.

***o*-acetylaminobenzoic acid** (*N*-acetylanthranilic acid). a flammable, crystalline compound, $CH_3CONHC_6H_4COOH$, that is soluble in hot ethanol, ether, and benzene; slightly soluble in water. It is the detoxified form of *p*-aminobenzoic acid.

acetylamino-1,7-Cleve's acid. 8-acetamido-5-amino-2-naphthalenesulfonic acid.

2-acetylaminofluorene. *N*-2-fluorenylacetamide.

***N*-acetyl-*p*-aminophenol**. acetaminophen.

***p*-acetylaminophenol**. acetaminophen.

8-acetyl-10-[(3-amino-2,3,6-trideoxy-α-L-lyxo-hexopyranosyl)oxy]-7,8,9,10-tetrahydro-1,6,8,11-tetrahydroxy-5,12-naphthacenedione. carubicin.

(8*S*-*cis*)-8-acetyl-10-[(3-amino-2,3,6-trideoxy-α-L-lyxo-hexopyranosyl)oxy]-7,8,9,10-tetrahydro-6,8,11-trihydroxy-1-methoxy-5,12-naphthacenedione. daunorubicin.

acetylaniline. acetanilide.

***N*-acetylanthranilic acid**. *o*-acetylaminobenzoic acid.

acetylation. the introduction of one or more acetyl radicals (CH_3CO) into an organic molecule that has OH or NH_2 groups. Metabolic acetylations often render xenobiotic compounds less toxic. In systems where acetylation predominates over deacetylation, there is usually a net excretion of the acetylated substances. Domestic rabbits have high acetyltransferase activity and low deacetylase activity. They therefore excrete significant amounts of acetylated amines. See also deacetylation.

acetylbenzoyl aconine. aconitine.

acetylbenzoyl peroxide (acetozone; benzozone; benzoyl acetyl peroxide). a white, toxic, crystalline compound, $C_6H_5CO{\bullet}O_2{\bullet}OCCH_3$, that decomposes in water, alkaloids, some organic solvents, and other organic matter. It evaporates when ground, compressed, or heated gently; there is some risk of explosion. It is a strong oxidizing agent and strong irritant used chiefly as a germicide and disinfectant. It can seriously damage skin and mucous membranes of humans and other vertebrates.

acetylbenzoylaconine. aconitine.

***N*-acetyl-*N*-bromodiethylacetylurea**. acetyl carbromal.

acetylcholine (ACh; acetylethanoltrimethylammonium hydroxide). an acetyl ester, $(CH_3)_3N$-$(CH_2)_2COOCH_3$, it is a reversible product of choline and is the normal substrate of acetylcholinesterase. It is a neurotransmitter that acts across synapses between adjacent nerve endings in the PNS and CNS, and also across neuromuscular junctions in vertebrates and in many invertebrate animals. It is released from the nerve terminal in response to a neuronal action potential, transmits the stimulus across the junctional gap to the adjacent neuron or muscle cell, and then rapidly decomposes in a hydrolytic reaction catalyzed by acetylcholinesterase. Acetylcholine is a potent parasympathomimetic agent, and is therefore highly toxic. Certain agents (e.g., β-bungarotoxin and *Latrodectus* venom) induce hyperexcitability of cholinergic (acetylcholine secreting) pathways by stimulating neuronally induced release of acetylcholine. Ergot contains ACh, as does the venom of hornets. When used as a drug it is administered by injection.

acetylcholine acetylhydrolase. acetylcholinesterase.

acetylcholine chloride (2-(acetyloxy)-*N,N,N*-trimethylethanaminium chloride). a very toxic, deliquescent crystalline powder. It is a miotic agent and a very toxic cholinergic drug.

acetylcholine receptor. See receptor.

acetylcholinesterase (AChE; acetylhydrolase; specific cholinesterase; true cholinesterase). **1:** a carboxylic esterase that is widely distributed throughout the vertebrate nervous system and in many nonnervous tissues such as muscle and erythrocytes. It specifically inactivates acetylcholine by catalyzing its conversion to choline plus acetate in the body. Its presence in the vertebrate body plays an essential role in nerve action and prevents cholinesterase poisoning. AChE also occurs in many venoms of snakes of the family Elapidae and those of the Hydrophidae. It is absent in venoms of the Viperidae and Crotalidae. Its function in other snake venoms is unclear. The toxicity of some large classes of pesticides (e.g., carbamates, organophosphates) is due to inhibition of this enzyme. Furthermore, acetylcholinesterase is the only hydrolytic enzyme that is a major target of acutely toxic substances. See also acetylcholine crisis (cholinergic crisis), enzymes of snake venoms, snakebite, venom (snake venom). **2:** pseudocholinesterase.

acetylcholinesterase inhibitor. anticholinesterase.

acetylcholinesterase reactivation. acetylcholinesterase (AChE) inhibition by carbamates and organophosphates is spontaneously reversible. AChE activity is restored by hydrolysis that removes the carbamylating or phosphorylating moiety from the enzyme.

acetylcholinesterase-inhibiting pesticides. See anticholinesterase.

acetylcoenzyme A (acetyl CoA). a coenzyme, $C_{23}H_{39}O_{17}N_7P_3S$, derived mainly from the metabolism of glucose and fatty acids. It takes part in many biological acetylation reactions. For example, it is the acetyl donor in the metabolic acetylation (catalyzed by N acetyltransferase) of xenobiotic amines. See also coenzyme A.

acetylcysteine (*N*-acetyl-L-cysteine; L-α-acetamido-β-mercaptopropionic acid; *N*-acetyl-3-mercaptoalanine). an organic chemical used to treat acetaminophen poisoning, *q.v.*

***N*-acetylcysteine conjugate** (mercapturic acid). any of a class of compounds, R-CH_2-(CHCOOH)NH(COCH3), that is the terminal receptor in the detoxication of potentially harmful electrophiles (R).

***N*-acetyl-L-cysteine**. acetylcysteine.

3β-[(2-*o*-acetyl-6-deoxy-3-*o*-methyl-α-L-glucopyranosyl)oxy]-7β,8-epoxy-14-hydroxy-5β-card-20(22)-enolide. tanghinin.

8-acetyldiacetoxyscirpenol (acetylneosolaniol). a trichothecene first isolated from *Fusarium*.

1-acetyl-19,20-didehydro-17,18-epoxycuran-17-ol. diaboline.

acetyldigitoxin. the α-acetyl ester (acetyldigitoxin-α) and the β-acetyl ester (acetyldigitoxin-β) of digitoxin. It is produced by enzymatic hydrolysis from lanatoside A. The molecule, $C_{43}H_{66}O_{14}$, consists of digitoxigenin (the aglycone) and 3 molecules of digitoxose, one of which bears an acetyl group. It has the same cardiotonic and cardiotoxic activity and applications as digitoxin, but with more rapid onset and a shorter duration of action.

acetyldigoxin. a cardiotoxic digitalis glycoside obtained by enzymatic hydrolysis of lanatoside C. Its action is similar to that of digoxin.

acetylene (ethyne; ethine). **1:** a slightly toxic, asphyxiant, narcotic, highly flammable, colorless gas, HC≡CH, with a garlicky odor. It has been used for anesthesia. Exposure can cause headache, dizziness, and gastric disturbances. Some adverse effects may be due to impurities in the commercial product. It is used widely as a chemical raw material and fuel for oxyacetylene torches. **2:** an alkyne.

acetylene hydrocarbon. alkyne.

acetylene tetrachloride. *sym*-tetrachloroethane.

acetylethanoltrimethylammonium hydroxide. acetylcholine.

acetylethyltetramethyltetralin (1,1,4,4-tetramethyl-6-ethyl-7-acetyl-1,2,3,4-tetrahydronaphthalene; polycyclic musk; musk tetralin; [illegible] pound formerly used in perfumes. Injection, *i.p.*, into laboratory rats and rabbits, but not monkeys, causes a blue discoloration of tissue. Acute intoxication may cause hyperexcitability, depression, and intensifying tremors. Less severe exposures of laboratory rats cause hyperirritability, ataxia, muscle weakness, foot drop with eversion of the hind feet, and occasional arching of the back. The urine becomes green. Lesions include widespread pigmentation of neurons followed by scattered degeneration, edema, and segmental demyelination.

(1S,3S)-3-acetyl-1,2-3,4,6,11-hexahydro-3,5,10,12-tetrahydroxy-6,11-dioxo-1-naphthacenyl 3-amino-2,3,6-trideoxy-α-L-lyxo-hexopyranoside. carubicin.

***N'*-[5-[[4-[[5-(acetylhydroxyamino)pentyl]-amino]-1,4-dioxobutyl]hydroxy-amino]-pentyl]-*N*-(5-aminopentyl)-*N*-hydroxybutanediamide**. deferoxamine.

***N*-acetyl-3-mercaptoalanine**. acetylcysteine.

3-acetyl-6-methoxybenzaldehyde. 5-acetyl-2-methoxybenzaldehyde.

5-acetyl-2-methoxybenzaldehyde (3-acetyl-6-methoxybenzaldehyde). a plant growth inhibitor that occurs in the leaves of *Encelia farinosa* (Family Compositae). See also allelopathy.

acetyl-β-methylcholine chloride. methacholine chloride.

***o*-acetyl-β-methylcholine chloride**. methacholine chloride.

acetyl-1,4-naphthalenediamine-6-sulfonic acid. 8-acetamido-5-amino-2-naphthalenesulfonic acid.

acetylneosolaniol. 8-acetyldiacetoxyscirpenol.

acetylnitro peroxide (PAN; peroxyacetyl nitrate). an organic peroxide, CH_3 C(O)OONO$_2$, often referred to as PAN. It is partially responsible for the eye irritation caused by photochemical smog.

16-(acetyloxy)-3[[8-[[4-[(aminoiminomethyl)-[illegible] octyl]oxy]-14-hydroxybufa-20,22-dienolide. bufotoxin (def. 3).

2-(acetyloxy)benzoic acid. aspirin.

16β-(acetyloxy)-3β-[(2,6-dideoxy-3-*o*-methyl-L-arabino-hexopyranosyl)oxy]-14-hydroxycard-20-(22)-enolide. oleandrin.

16-(acetyloxy)-3,14-dihydroxybufa-20,22-dienolide. bufotalin.

(3α,4α,8α,9β,11α,13α,14β,16β,17Z)-16-(acetyloxy)-3,11-dihydroxy-29-nordammara-1,7-(20),24-dien-21-oic acid. fusidic acid.

25-(acetyloxy-2-(β- D-glucopyranosyloxy)-16,20-dihydroxy-9-methyl-19-norlanosta-1,5,23-triene-3,11,22-trione. colocynthin.

(25-acetyloxy)-2β,16α,20-trihydroxy-9β-methyl-19-*nor*-10α-lanosta-1,5,23E-triene-3,11,22-trione, 1,2-dihydro-α-elaterin). cucurbitacin E.

2-(acetyloxy)-*N, N,N*-trimethylethanaminium chloride. acetylcholine chloride.

2-(acetyloxy)-*N,N,N*-trimethyl-1-propanaminium chloride. methacholine chloride.

acetylphosphoramidothioic acid *O,S*-dimethyl ester. acephate.

β-acetylpyridine. methyl pyridyl ketone.

3-acetylpyridine. methyl pyridyl ketone.

acetylsalicylic acid (HAsc; ASA; A.S.A.; *o*-acetoxybenzoic acid). the acidic form of aspirin, $CH_3COOCC_6H_4COOH$. At a pH substantially below the pKa of 3.2 (as in the stomach of most mammals) most of this acid remains in a neutral form and is thus rapidly absorbed. See aspirin.

***N*-acetyltransferase**. a cytosolic enzyme isolated from rat liver that catalyses the acetylation (with acetyl CoA as the specific acetyl donor) of xenobiotic (and certain endogenous) amines; any of a number of isoenzymes that occur in liver, kidney, and certain other organs.

ACGIH. American Conference of Governmental and Industrial Hygienists.

ACh. acetylcholine.

ACh receptor. See receptor (acetylcholine receptor).

achalasia. **1:** failure of the smooth muscles of the alimentary canal to relax at junctures between parts, especially failure of the esophogastric sphincter to relax on swallowing. **2: sphincteral achalasia**. failure of a sphincter of any tubular organ to relax in response to a normal physiologic stimulus.

ache. **1:** a persistent, usually somewhat diffuse, pain of moderate severity. **2:** to suffer continuing pain of moderate severity.

AChE. acetylcholinesterase.

achil. *Naja nigricollis*.

achipeten. *Walterinnesia aegyptia*.

acholia. a decrease or absence of bile or a condition in which bile fails to reach the duodenum.

acholic. pertaining to acholia.

acid. **1:** any substance that releases hydrogen ions (protons) or hydroxonium ions (H_3O^+) in aqueous solution. The pH is thus below 7. **2:** a Lowry-Brønsted acid (Brønsted acid) is a substance that tends to release a proton. **3:** a Lewis acid is a substance (e.g., $AlCl_3$, BF_3, SO_3) that accepts an electron pair from a base. In substances whose toxicity is due primarily to their acidity (defs. 1, 2, 3), symptoms of acute oral toxicity may include immediate pain and corrosion of the mucous membranes of the mouth, throat, and esophagus; difficulty in swallowing; abdominal pain; thirst; nausea; coffee-ground vomitus; shock syndrome; circulatory collapse; and death. Death may also result from asphyxia due to glottic edema. **4:** vernacular usage: **a:** any substance that has a sour taste. **b:** LSD. **c:** sour tasting. **5:** acidic. See also acid poisoning. **corrosive acid**. a strong acid, especially a mineral acid (e.g., hydrochloric, sulfuric), which, because of its high degree of ionization and solubility in water, causes rapid, deep, and painful destruction of tissues. See also acid poisoning, corrosive, corrosive poisoning.

acid anhydride. See anhydride.

acid-base balance. **1:** the ratio of acids to bases in a solution. **2:** the normal ratio of acids to bases maintained by a healthy organism with body fluids of normal pH. This ratio is usually maintained within a very narrow range. Disturbances of acid-base balance (e.g., by the action of toxicants) can have extremely serious impacts on the affected organism. **3:** the maintenance of the acid:base ratio in blood and other body fluids.

acid-base metabolism. the metabolic processes that maintain acids and bases within a living organism at normal (viable) levels.

acid burn. See burn.

acid deposition. acidic air pollutants (chiefly nitrogen oxides and sulfur oxides) deposited in the biogeosphere primarily as nitric acid or sulfuric acid. These materials may be deposited as dry particulates or acidic gases (**dry deposition**) or in rain, snow, or fog (**wet deposition** or acid precipitation). *Cf.* acid precipitation.

acid dissociation constant. See pKa.

acid ester sulfate. See ester sulfate.

acid ethyl sulfate. ethylsulfuric acid.

acid gland. **1:** one of the complex of honeybee poison glands. **2:** one of the lateral glands of the trunk segments of a millipede. It secretes hydrogen cyanide and/or other substances.

α1-acid glycoprotein (acute-phase reaction protein; orosomucoid; AAG). an anionic polymorphic conjugated protein that is chiefly hepatic in origin. The carbohydrate moiety, made up of equal parts of hexosamine, hexose, and sialic acid, comprises 45% of the molecule. Plasma levels may rise during chronic inflammation or conditions of acute physiological stress; levels may fall as a consequence of severe hepatic damage, certain serious gastroenteropathies, and severe malnutrition. It combines electrostatically with many basic xenobiotics.

acid methyl sulfate. See methyl sulfate.

acid mine drainage. leachate from pyritic coal mines and spoils. It is often acid due to acidic sulfates (mainly ferrous sulfate) and bacterial and chemical oxidation of reduced iron to ferric hydroxide precipitates. Acid mine drainage can kill fishes and other aquatic organisms or render them infertile. It can also introduce heavy metals into the water supply and otherwise otherwise lower the quality of drinking water. *Cf.* acid mine water.

acid mine water. mine water that contains free sulfuric acid due to the weathering and leaching of iron pyrites. *Cf.* acid mine drainage.

acid mist. See sulfur trioxide, sulfuric acid.

acid phosphatase. a lysosomal acid hydrolase that enables the hydrolysis of phosphoric acid esters and facilitates the absorption and metabolism of carbohydrates, nucleotides, and phospholipids. Serum concentrations of acid phosphatase are elevated in cases of metastatic prostatic carcinoma, benign prostatic hypertrophy, prostatitis, Paget's disease, and metastases to bone and liver from breast carcinoma.

acid poisoning. poisoning by ingestion of a toxic acid. Symptoms of acute poisoning include those of acute gastroenteritis. They may include severe and immediate pain in the affected tissues, destruction of the lining of the mouth and upper gastrointestinal tract, thirst, abdominal pain, nausea, vomiting (with coffee-ground colored vomitus), shock, circulatory collapse, and death. A possible complication is asphyxia from glottic edema. See also acid, acidism, alkali poisoning.

acid precipitation. rainfall and other forms of precipitation that contain high concentrations of nitric or sulfuric acids. It results mainly from the reaction of atmospheric sulfur oxides and nitrogen oxides with moisture. These gases arise mainly from the combustion of fossil fuels and often disperse over long distances, producing acidic precipitation that may damage a whole ecosystem in regions that are remote from the source. Water bodies with poor buffering capacity may become acidified to the point where fishes and other aquatic biota are killed or population growth is impaired. The acidic waters may release toxic metals (especially aluminum) from bottom sediments, adding significantly to the toxicity of the water. Metal ions may also leach from the soils into affected water bodies. Acid precipitation may also damage terrestrial vegetation by direct effects and indirectly, for example, by leaching nutrients from the soil. *Cf.* acid deposition.

acid rain. rainfall with a pH of about 4 or less; a form of acid precipitation, *q.v.* Prior to the

Industrial Revolution, rain had a pH of about 5.6. Acid rain and other forms of acid precipitation are formed from atmospheric sulfur oxides and nitrogen oxides that are released into the atmosphere chiefly from fossil fuel-burning systems, most importantly coal-fired power plants, metal-processing factories, and motor vehicles. These gases undergo complex reactions in the atmosphere and combine with moisture to form sulfuric or nitric acids. Acid rain is especially injurious to aquatic life, but can damage entire ecosystems. *Cf.* acid deposition.

acidic. **1:** having the properties of an acid. **2:** pertaining to or denoting an acid; acid forming; having the properties of an acid. **3:** pertaining to a substance, material, or system (e.g., a habitat) with a pH below 7. *Cf.* alkaline.

acidic habitat. a habitat of low pH and usually poor nutrient content (e.g., a cranberry bog).

acidism (acidismus). poisoning by exogenous acids. See acid poisoning.

acidismus. acidism.

acidity. **1:** the state of being acid. **2:** the acid content of a solution or extent to which a solution is acid. **3:** the concentration of any chemical species that is titratable by strong base in a solution.

acidobiontic. living in an acidic habitat.

acidophile. an organism that lives successfully in acidic habitats and environments.

acidophilic (acidophilous; aciduric; oxyphilous). **1:** tolerant of or thriving in acidic habitats or environments. **2:** staining readily with acid stains (said of a cell or tissue).

acidophilous. acidophilic.

acidophily. the condition of being acidophilic, tolerant of, or successful in acidic habitats or environments.

acidophobe. an organism that is intolerant of acidic environments.

acidophobic. intolerant of acidic environments.

acidophoby. the condition of being acidophobic, intolerant of, or unsuccessful in acidic habitats or environments.

acidosic. acidotic.

acidosis. a serious reduction of pH within a living organism from excessive acid or a decreased alkali reserve. Any number of conditions can cause acidosis. It usually results from diabetes, starvation, liver impairment, respiratory blockage or respiratory failure, or failure of the kidneys to excrete hydrogen or to use bicarbonate efficiently. Other causes are the use of narcotics, sedatives, tranquilizers, and anesthetics that inhibit breathing reflexes. Symptoms can include dyspnea, tremors, headache, tachycardia, and high blood pressure. Acidosis can be fatal. **metabolic acidosis**. acidosis caused by excess acid in, or loss of bicarbonate from, body fluids. Metabolic acidosis may be caused by shock or heart failure. **renal tubular acidosis**. acidosis caused either by failure of the kidney to eliminate hydrogen or to effectively use bicarbonate. **respiratory acidosis**. that which is due to a low breathing rate with consequent high levels of carbon dioxide in the blood and the formation of large amounts of carbonic acid. This form of acidosis has numerous causes which may include the use of narcotics, sedatives, tranquilizers, anesthetics, and other substances that inhibit breathing reflexes.

acidotic (acidosic). of, pertaining to, or characterized by acidosis.

acidotrophic. feeding on acidic food or acidic substrates.

acidum acetylsalicylicum. aspirin.

aciduric. tolerant of acid media. See acidophilic.

Acipenser. a genus of sturgeons. See Acipenseridae, acipenserin.

Acipenseridae (the sturgeon family). a family of bony fishes that includes the genus *Acipenser*. They are large, ranging in size to about 3.5 m; they occur in large rivers, lakes, and in marine waters where they feed on bottom vegetation. They are characterized by a heterocercal tail, 5 longitudinal rows of keeled bony plates, a long snout, ventral mouth, and a lack of teeth.

acipenserin. a toxic substance isolated from the gonads of sturgeons of the genus *Acipenser* (Family Acipenseridae).

ackee. See ***Blighia sapida***.

ackee poisoning (Jamaican vomiting sickness). an acute and often fatal reaction, mainly among children, to eating the unripe fruit of *Blighia sapida* and perhaps sometimes the ripe aril. The ripe fruit readily becomes rancid and may also be toxic. The ripe aril, properly prepared, apparently is safe. This condition is marked by sudden onset of nausea and violent vomiting with periods of relief that last no more than a few hours, followed by convulsions, coma, and death in most cases. The condition is marked clinically by severe hypoglycemia.

aclacinomycin. any of a family of some 20 toxic, antineoplastic anthracycline antibiotics isolated from the fungus *Streptomyces galilaeus*. The major components are aclacinomycin A and aclacinomycin B, *q.v.* Both are extremely toxic to laboratory mice. **aclacinomycin A**. (2-ethyl-1,2,3,4,6,11-hexahydro-2,5,7-trihydroxy-6,11-dioxo-4-[[2,3,6-trideoxy-4-*O*-[2,6-dideoxy-4-*O*[(2R-*trans*)-tetrahydro-6-methyl-5-oxo-2H-pyran-2-yl]-α-L-lyxohexopyranosyl]-3-(dimethylamino)-α-L-lyxohexopyranosyl]oxy]-1-naphthacenecarboxylic acid methyl ester; antibiotic MA 144A1; NSC-208734; aclarubicin). an extremely toxic antineoplastic anthracycline antibiotic. See aclacinomycins. **aclacinomycin B** (4-[[[2''',3''-anhydro]-*O*-3,6-dideoxy-α-L-erythro-hexopyranos-4-ulos-1-yl-(1→4)-*O*-2,6-dideoxy-α-L-lyxohexopyrano-syl-(1→4)-2,3,6-trideoxy-3-(dimethylamino)-α-L-lyxohexopyranosyl]oxy]-2-ethyl-1,2,3,-4,6,11-hexahydro-2,5,7-trihydroxy-6,11-dioxo-1-naphthacenecarboxylic acid methyl ester; antibiotic MA 144B1). an extremely toxic antineoplastic anthracycline antibiotic.

aclarubicin. aclacinomycin A.

acme. **1:** the highest point attained; a period of maximum vigor; **2:** the stage of peak intensity of any process, disease, sign, or symptom.

acmic. of or pertaining to the acme of a system or process or to periods of change.

acne. an inflammatory condition of the sebaceous glands and hair follicles marked by papular and pustular eruptions; scarring is common. **acne artificialis**. that produced by exogenous irritants or drugs. **acne vulgaris**. common acne. **asbestos acne**. that produced by occupational exposure to asbestos. **bromide acne**. that caused by ingestion of bromides; the face, trunk, and extremities may be affected. **chlorine acne**. chloracne. **halogen acne**. bromide acne and iodide acne. **iodide acne**. that caused by ingestion of iodides; the face, trunk, and extremities may be affected. **petroleum acne**. occupational acne in petroleum workers. **steroid acne**. that caused by the use of steroid drugs. **tar acne**. chloracne.

acneform. resembling acne.

acnegenic. **1:** having the capacity to cause acne. **2:** of or pertaining to substances that may aggravate the lesions of acne.

acneiform. acneform.

Acocanthera (*Akocanthera*). a genus of plants (Family Apocynaceae), the stems and leaves of which yield cardiotoxic substances used by various tribes to make arrow poisons and sometimes in trial by ordeal. Thus, poison makers of the Giriama tribe of the seacoast of Kenya trim and slice the branches of the "muriju" (*Acocanthera spp.*) and boil them with reptile intestines. The product is a sticky cardiotoxic mixture sold to the bowmen of the Wakamba tribe. The latter have often used arrows tipped with this poison to poach elephants. The active principle is ouabain, *q.v.*

acocantherin. ouabain.

acomia. alopecia.

aconine. a monobasic diterpenoid alkaloid, $C_{25}H_{41}NO_9$, derived from and much less toxic than aconitine, that occurs in various species of *Aconitum*. It initially stimulates and then [illegible]

aconitase. the enzyme that catalyzes the reversible interconversion of citric acid, *cis*-aconitic acid, and isocitric acid in the tricarboxylic acid cycle (citric acid cycle; Krebs cycle) by dehydration and hydration reactions. Aconitase is the target of fluoroacetate, a potent rodenticide that condenses with oxaloacetate to yield an aconitase inhibitor, with resultant accumulation of citric acid and consequent inhibition of aerobic energy production.

aconite (monkshood; wolfsbane; "queen mother of poisons"). **1:** any plant of the genus *Aconitum*. **2a:** the dried tuberous root of *Aconitum*, especially *A. napellus* and *A. lycoctonum*, the favored poison of the legendary Greek sorceress, Medea. It is extremely toxic and is the presumed arrow poison of ancient China and possibly ancient Gaul. **2b:** the dried root of *A. ferox* (Indian aconite). **3:** an extremely poisonous and rapid-acting mixture of alkaloids extracted from the dried root of *A. napellus*. It contains aconitine, aconine, isoaconitine, napelline, picraconitine, pseudoaconitine, and aconitic acid. Aconite was at one time used as an anodyne, antipyretic, diaphoretic, diuretic, topical analgesic, and a cardiac and respiratory depressant. Symptoms of intoxication include initial burning and tingling sensations and numbness of the tongue, mouth, throat, and face; this may be followed by nausea, vomiting, blurred vision, low blood pressure, a slowing of the pulse, chest pain, giddiness, sweating, convulsions, respiratory paralysis, and collapse. Recovery from non-lethal exposures are rapid, usually less than a day.

aconite poisoning. poisoning by plants of the genus *Aconitum*. The condition in humans and livestock usually runs a severe course; death may occur within a few hours. Domestic cattle salivate heavily, frothing at the mouth, belching, and bloating; they become weak and restless, with an irregular heartbeat. Aconite poisoning of cattle has been confused with larkspur (delphinium) poisoning. The presenting symptoms in humans are usually a tingling and burning sensation about the mouth and throat, followed by numbness of the tongue, throat, and face, and constriction of the throat. Common also are nausea, vertigo, vomiting, giddiness, muscular weakness, severe thoracic pain, hypotension, impaired vision and speech, paralysis of the respiratory system, convulsions, and syncope. There is a progressive burning, tingling, and numbness of the entire body with CNS depression and finally respiratory arrest and death due to cardiac failure. Terminal stages are marked by severe pain and paralysis of facial muscles. Many victims remain conscious until near death; some complain of yellow-green vision and tinnitus.

aconitic acid (propene-1,2,3-tricarboxylic acid). a white to yellowish, crystalline alkaloid, $H(COOH)C{:}C(COOH)CH_2(COOH)$. It is toxic and flammable. See aconite, *Aconitum*, *Adonis vernalis*.

aconitine (acetyl benzoyl aconine; (1α,3α,6α,-14α15α,16β)-20-ethyl-1,6,16-trimethoxy-4-(methoxymethyl)aconitane-3,8,13,14,15-pentol-8-acetate-14-benzoate; acetylbenzoylaconine; 16-ethyl-1,16,19-trimethoxy-4-(methoxymethyl)aconitane-3,8,10,11,18-pentol-8-acetate 10 benzoate; monkshood). a dangerous and highly toxic cardiotoxic and neurotoxic agent, $C_{34}H_{49}NO_{11}$, obtained from plants of the genus *Aconitum* and other aconites. It is a white, crystalline, polycyclic diterpenoid alkaloid that is readily absorbed through the skin. It is nearly identical in structure to delphinine. As little as 4 mg is lethal to *Homo sapiens*. Aconitine acts on nerve axons by opening sodium channels, inhibiting repolarization of the excitable membrane, and causing repetitive firing. It slows cardiac conduction and causes reflex bradycardia, cardiac arrhythmias, a weak pulse, and hypotension. The first indication of intoxication is a tingling, burning sensation about the lips, mouth, gums, and throat, followed by numbness and constriction of the throat. The victim frequently experiences nausea, vertigo, vomiting, muscular weakness, impaired vision and speech, and syncope. There is a progressive burning, tingling, and numbness of the body, with depression and finally cardiac failure. There is no reliable antidote, but procaine sometimes reverses the effects of aconitine; atropine has also been recommended. See also *Aconitum*. **amorphous aconitine**. a mixture of amorphous alkaloids from *A. napellus*, including aconitine, hypaconitine, l-ephedrine, mesaconitine, napelline, neoline, and neopelline. It is extremely toxic and hazardous as it is readily absorbed through the skin. **mild aconitine**. amorphous aconitine.

Aconitum (aconite; friar's cowl; monkshood; mousebane; wolfbane; wolf's-bane). a genus of more than 100 species of poisonous North Temperate Zone flowering plants (Family Ranunculaceae). Some are violently poisonous. Aconitum was the queen mother of poisons and was the poison of choice by Medea, the fabled Greek sorceress. Several species occur in North America, most notably *A. napellus*. *Aconitum* is closely related to, and toxicologically similar to, *Delphinium* (larkspurs). The entire plant is toxic; even small amounts of leaves or roots can be fatal if ingested. There is no dependable antidote. A number of monobasic diterpenoid alkaloids have been isolated from various species of *Aconitum*. The most abundant, aconitine, is nearly identical to delphinine; other common examples are picratonitine, aconine, benzoylamine, and neopelline. These alkaloids act rapidly to stimulate and then depress the CNS. See aconite poisoning. ***A. columbianum*** (monkshood; wolfbane; aconite; western monkshood). a species that occurs at higher elevations in the northern Rocky Mountain and Pacific Coast states of the United States. ***A. ferox*** (Indian aconite; bish; visha; bishma; bikhroot). This species is violently poisonous. It is cardiotoxic and was placed on arrow points by natives to hunt big game. A very small dose can be fatal. ***A. lutescens*** (yellow monkshood). a native North American species found from Idaho to New Mexico. ***A. napellus*** (cultivated aconite; garden monkshood; garden wolfbane). an extremely dangerous plant with attractive violet-blue (sometimes white or mauve) hooded flowers. Widely cultivated in gardens, they readily escape and are often abundant on roadsides, in old fields, and other abandoned sites. This plant contains a number of toxic alkaloids, the chief of which is aconitine. ***A. uncinatum*** (wild monkshood). a species found in the United States from Pennsylvania to Georgia. All parts are toxic, producing symptoms similar to those of *A. napellus*, usually including dysphagia and paresthesia of the lips, mouth and throat, a colicky abdominal pain, nausea, vomiting, bloody diarrhea, ataxia, convulsions, coma, and respiratory collapse.

acorn poisoning. a syndrome of livestock that have been eating green acorns. The condition is more common in Europe than in North America. It occurs only when animals are allowed to forage in oak stands in years when acorns are abundant. Symptoms and lesions are very similar to those of oak poisoning. Milk letdown decreases and the milk may be bitter. See also *Quercus*.

Acorus calamus (sweet flag; flagroot; calmus; calamus; sweet cane; sweet grass). an iris-like herb (Family Araceae). The oil is carcinogenic, cardiotoxic, and hepatotoxic. See also asarone, oil of calamus, Araceae.

acoustic. pertaining to sound or hearing.

ACPC. cycloleucine.

acquinite. chloropicrin.

acquired. **1:** pertaining to a nonheritable property or propensity; not inherited; not genetic. **2:** obtained or received from materials or factors originating in the external environment. **3:** pertaining to or denoting a disease that occurs following birth in response to environmental factors.

ACR. acute-to-chronic ratio.

acraconitine. pseudoaconitine.

acral. **1:** of or pertaining to an apex or extremity. **2:** affecting the extremities.

acraldehyde. acrolein.

acrasin. cyclic AMP.

Acremonium. a genus of fungi (Family Moniliaceae, Order Moniliales) that causes eumycotic mycetoma. ***A. coenophialum***. an endophytic fungus that infests *Festuca arundinacea*, *q.v.* It produces toxic alkaloids that are responsible for the plant's toxicity. See also fescue foot.

acrid. **1:** unpleasantly sharp or strong tasting, biting, irritating; pungent. **2:** irritant, causing irritation; caustic; intensely or violently bitter.

9-acridinamine. aminacrine.

acridine (dibenzo[*b,e*]pyridine; 10-azaanthracene). a toxic, colorless, solid component, $C_{13}H_9N$, of coal tar. It is comprised of small monoclinic crystals, is soluble in ethanol, ether, or carbon disulfide, and slightly soluble [illegible] strong irritant of the skin and mucous membranes. It is used in the manunfacture of dyes.

acrisia. uncertainty of prognosis or diagnosis of a disease or condition.

acritical. not marked by or having a crisis.

acrodynia. **1:** pain in the extremities or peripheral parts of the body. **2:** erythredema, acrodynic erythema, pink disease, Swift's disease, dermatopolyneuritis; a syndrome of humans (primarily of infants) associated with the ingestion of mercury. Symptoms in infants and children include erythema of the nose, chest, and extremities; polyneuritis; gastrointestinal problems; and failure to thrive. The disease in adults is marked by anorexia, photophobia, tachycardia, and excessive sweating. **3:** a condition observed in rats, dogs, and swine due to a deficiency of pyridoxine, in which the tips of the ears, nose, lips, and feet become red, swollen, and eventually necrotic.

acrodynic erythema. See acrodynia.

acrolein (2-propenal; acrylic aldehyde; acrylaldehyde; acraldehyde; aqualin; allyl aldehyde). a volatile, highly toxic, alkenic aldehyde. It is a colorless to light yellow, asphyxiating, irritant, highly reactive liquid, $CH_2{=}CHCHO$, that is soluble in water, ethanol, and ether. It is produced by the dry distillation of glycerin. Acrolein is a herbicide and has been used in poisonous war gases. It is a component of photochemical smog and an environmental pollutant that enters the environment from numerous industrial sources, automotive exhausts, and smoke (e.g., from tobacco, forest fires). Acrolein is extremely toxic to laboratory mice and rats. Poisoning can result from contact, inhalation or ingestion. It is especially irritating to the skin and mucous membranes; it can produce tissue necrosis on contact and easily damages the eye. The vapor is an intense lacrimator. The fumes are extremely irritant to the lungs and upper respiratory tract and when inhaled may produce an asthmatic reaction, bronchitis, tracheobronchitis, pulmonary edema, pneumonia, emphysema, and nephritis. Symptoms of ingestion are mainly those of severe gastroenteritis.

acronarcotic. having both irritant (acrid) and narcotic properties.

acrorhagus. a tubercle near the margin of the body in certain sea anemones that contains specialized nematocysts.

acrotism. lack of a discernible pulse.

acrylaldehyde. acrolein.

acrylamide (2-propene amide; acrylamide monomer; acrylic acid amide; propenamide; vinylformic acid). the crystalline amide of acrylic acid, it is a toxic, carcinogenic, irritant substance, $CH_2CHCONH_2$, that is readily absorbed through the skin. A contact dermatitis is common among those working with acrylamide. It is very toxic orally to laboratory rats, guinea pigs, and rabbits. Acrylamide monomer is neurotoxic to humans and acute exposures can produce a seemingly reversible encephalopathy. It is a cumulative poison and repeated exposure to relatively low doses affects neuronal transmission and can produce degenerative distal sensorimotor axonopathies characterized by the accumulation of neurofilaments in the axonal terminal. The large-diameter peripheral nerves are most affected, although axonal degeneration is also seen to some extent in the CNS. The presenting symptoms are incoordination, difficulty in walking, the loss of distal deep tendon reflexes, and numbness of the extremities. Excessive sweating is common.

acrylamide monomer. acrylamide.

acrylate. a salt or ester of acrylic acid.

acrylic acid. a colorless, corrosive liquid, $CH_2{:}CH{\bullet}COOH$, used to produce acrylic polymers and resins. See also acrylamide, acrylate.

acrylic acid amide. acrylamide.

acrylic aldehyde. acrolein.

acrylic resin. **1:** any of various polymers of acrylic acid and their esters. See acrylic plastic.

acrylonitrile (vinyl cyanide; cyanoethylene; 2-propenenitrile; propene nitrile; AN). a clear, colorless, flammable, highly reactive, and extremely toxic synthetic liquid cyanide, $CH_2CH{\cdot}CN$. The fumes are toxic and highly explosive in air. On combustion, acrylonitrile releases extremely poisonous cyanide gases. It is manufactured in large quantitities and is used chiefly in the manufacture of plastics, acrylic fibers, synthetic rubber, dyes, pharmaceuticals, and pesticidal fumigants of grain. Substantial quantities are released to the atmosphere during manufacture and usage. It is readily absorbed via contact, inhalation, or ingestion. The vapor is damaging to the eye and pulmonary tissues. Symptoms of intoxication in humans include eye irritation, headache, dizziness, dyspnea, nausea, vomiting, diarrhea, muscular weakness, fatigue, sleeplessness, mild jaundice, diarrhea, and irritation of the skin (with blistering and erythema), eyes, and respiratory tract. In severe cases, victims may experience profound weakness, convulsions, asphyxia, coma, and death. It causes hemorrhagic necrosis of the adrenal glands with chronic or acute adrenocortical insufficiency. It is embryotoxic to laboratory mammals at doses that are near toxic to the dam. Acrylonitrile is carcinogenic to some laboratory animals and increases the incidence of a variety of cancers in exposed humans.

acrylonitrile/butadiene/styrene. ABS plastic.

acrylylcholine. a choline ester toxin produced by the gastropod mollusk, *Buccinum undatum*, with an action similar to that of muscarine and nicotine. It is cardiotoxic, hypotensive, stimulates gastric motility and secretion, and increases respiration.

ACS. American Chemical Society.

Actaea (*Actea*; baneberry; cohosh; necklaceweed; coralberry; herb-Christopher). a genus of mildly to highly toxic, perennial herbs (Family Ranunculaceae) with white flowers and white or colored berries; native to the woodlands of eastern North America or to forests in western North America from the Rocky Mountains to the Pacific Coast. There is one European species. Prolonged contact with a baneberry plant produces a contact dermatitis. Symptoms following ingestion are similar to those of digitalis. Symptoms may not present for several hours to days, averaging about 48 hours, following ingestion. Small quantities of roots, sap or berries can cause burning in the stomach, dizziness, and increased pulse rate. Larger amounts can induce nausea, vomiting, bloody diarrhea, tachycardia, shock, and convulsions. Human fatalities sometimes result. The toxic principle is an irritant oil, protoanemonin. Descriptions of some of these plants and the various scientific names applied by different authors can cause some confusion to one inexperienced with this genus. ***A. alba*** (white baneberry; doll's eyes; white cohosh). the fruit is white. ***A. odorata*** (bitter weed). a species that causes heavy losses of sheep and goats in the southwestern United States. ***A. richardsoni*** (rubber weed). known to be poisonous to sheep. ***A. rubra*** (red baneberry; snakeberry; red cohosh). the fruit is cherry red and mildly poisonous, but with a disagreeable taste. ***A. spicata*** (black cohosh; black baneberry; red cohosh). a species with white or bluish flowers and purple-black berries.

Actea. *Actaea*.

ACTH (corticotrop(h)in; adrenocorticotrop(h)in; adrenocorticotrop(h)ic hormone; adrenotropic hormone). a straight-chain polypeptide hormone comprised of 39 amino acid residues. It is synthesized by the pars distalis of the mammalian pituitary gland and is released partly in response to hypothalamic corticotropin-releasing hormone (CRH) and partly (as part of a negative feedback loop) by the level of hydrocortisone in the circulating blood. ACTH promotes growth and maintenance of the adrenal cortex and stimulates the synthesis and release of various adrenal corticosteroid hormones (e.g., hydrocortisone, aldosterone) and androgens. It affects other tissues also; it promotes, for example, the breakdown and release of fatty acids from fat cells. ACTH secretion increases during most conditions of stress including trauma, burns, infection, emotional stress, and hypotension. A form of ACTH ("big ACTH") is produced by certain tumors. It lacks the biological activity characteristic of ACTH. See also hormone, pituitary gland. ACTH is used chiefly in medicine to treat inflammatory disorders. Its therapeutic action is due to overproduction of hydrocortisone by the adrenal gland. See hydrocortisone.

Actinea odorata. See *Hymenoxys odorata*.

Actiniaria (true sea anemones). an order of the coelenterate class Anthozoa comprising the zoantherian anthozoans or true sea anemones. They are venomous and poisonous; have no skeleton; the polyp is columnar and sessile, but is able to "creep" across the substrate. The larger species are attached to the substrate by a pedal disk; septa are paired. They are solitary or gregarious. Typical genera are *Adamsia*, *Gonactinia*, and *Metridium*. Poisoning of humans from the ingestion of sea anemones is extremely rare. A few species, however, are potentially dangerous. See *Actinodendron plumosum*, *Anemonia sulcata*, and *Condylactis* *gigantea*. Poisonings of humans have occurred in parts of the Indo-Pacific region, especially Samoa where selected sea anemones are eaten after cooking. See *Rhopdactis howesi*, *Physobrachia douglasi*, and *Radianthus paumotensis*. See also sea anemone.

actinium (Ac)[Z = 89; most stable isotope = ^{227}Ac, half-life, 21.6 yr]. a rare, toxic, radioactive, metallic element that occurs in minute amounts in uranium ore, but is normally obtained by neutron bombardment of radium. It is the first member of the actinoid series of the periodic table.

actinobiology (radiation biology; actinology). the science that deals with the effects and mechanisms of action of potentially damaging radiation on living organisms.

Actinodendron plumosum (Hell's fire sea anemone). a dangerous anthozoan coelenterate (Order Actiniaria) typically found on the shady side of rocks or under coral ledges. It can inflict extremely painful and serious wounds. The wound may ulcerate and is slow to heal. Severe systemic effects such as chills, fever, gastrointestinal distress, and extreme thirst; prostration may develop. See also Anthozoa.

actinogen. a radiogenic substance, one that produces radiation.

actinogenesis. radiogenesis.

actinogenic. radiogenic.

actinolite. **1:** any substance that is chemically altered by exposure to light. **2:** a greenish crystalline or fibrous variety of asbestos.

actinology. actinobiology.

actinomyces. a common name of microorganisms of the genus *Actinomyces*.

Actinomyces (*Ascomyces*; ray fungi). a genus of microorganisms that includes *A. antibioticus*, *A. bovis*, *A. israelii* (Order Actomycetales, Family Actinomycetaceae). All are pathogenic and contain Gram-positive staining filaments. *A. micropolyspora faeni* is the causative agent of farmer's lung. See aleukia (alimentary toxic aleukia).

Actinomycetales. an order of bacteria comprised of the families Mycobacteriaceae, Actinomycetaceae, Actinoplanaceae, Dermatophilaceae, Micromonosporaceae, Nocardiaceae, and Streptomycetaceae.

actinomycete. **1:** any of a group of branched, filamentous, Gram-positive, mostly anaerobic, nonmotile, fungus-like bacteria (Family Actinomycetaceae; Order Actinomycetales) found in soil and river and lake bottom sediments. Included are bacteria of the genera *Actinomyces* and *Streptomyces*. Many species are pathogenic and many produce antibiotics. **2:** any bacterium of the order Actinomycetales.

actinomycin. any of a large family of antibacterial, antifungal, and cytotoxic antibiotics produced by *Streptomyces spp*. They have cytostatic and radiomimetic activity. Among the more important actinomycins are cactinomycin and dactinomycin.

actinomycin A. an orange-red, highly toxic, heat-stable, antibacterial agent that is soluble in ethanol and ethyl ether. It is produced by *Actinomyces antibioticus* and some species of *Streptomyces*. It blocks transcription in gram-positive bacteria and in eukaryotic cells.

actinomycin A_{IV}. dactinomycin.

actinomycin B. an extremely toxic, ethanol-insoluble, antibacterial agent, with properties otherwise similar to those of actinomycin A. It is not used clinically.

actinomycin C. cactinomycin.

actinomycin C_1. dactinomycin.

actinomycin D. dactinomycin. See also cactinomycin.

Actinomycin D®. dactinomycin.

actinomycin I_1. dactinomycin.

actinomycin X_1. dactinomycin.

actinonarian. any coelenterate of the subclass Actinonaria.

actinotoxemia. radiation sickness (obs.).

actinotoxin. a toxic principle from the tentacles of sea anemones (actinonarians) isolated by alcoholic extraction. *Cf.* actinotoxemia.

Actinozoa. Anthozoa.

action. **1a:** performance; the process of doing. **1b:** an act; a thing done; a step taken; deed. **2:** a morbid process (pathology). **antagonistic action**. the opposition of a chemical or muscle to the action, respectively, of another chemical or muscle. **bacteriocidal action**. an action that is lethal to bacteria. **bacteriostatic action**. an action that retards growth of bacteria. **biologic action**. an action that produces a biological effect. **cumulative action**. the production (usually sudden) of a biologic, pharmacologic, or toxic effect by a chemical following several subthreshold exposures. **drug action**. the action of a drug on the whole organism or on various body systems. **specific action**. any unique action produced by a substance on a particular organism or part of such organism. **synergistic action**. the production, by a substance or stimulus, of an enhanced effect (i.e., greater than additive) when used together with another substance or stimulus.

action level. **1:** in general, the concentration of a chemical residue in food or feed above which adverse effects are possible and corrective action is indicated. **2:** any concentration recommended by EPA for enforcement by the FDA and USDA of pesticide residues that accidentally or inadvertently occur in food or animal feeds. **3:** a concentration in the environment of a toxic or hazardous substance that is high enough to warrant action.

action system. a behavior pattern in an organism.

activate. **1:** to render active; trigger, actuate, stimulate, start, mobilize. **2:** to induce radioactivity by bombardment with neutrons or other types of radiation. See activation.

activated carbon (activated charcoal; active carbon). a highly adsorbent form of carbon produced by heating wood or other organic matter (e.g., coconut shells, bones) to a temperature of 800-900°C in the absence of air (and in the presence of steam) in order to drive off adsorbed gases and to produce a highly porous material. Activated carbon is very effective in decolorizing and deodorizing materials, and in removing impurities and toxic substances from liquids or gases (e.g., as in gas masks). It is used to remove odors and toxic substances from gaseous emissions, and in waste treatment to remove dissolved organic matter from wastewater.

activated charcoal. activated carbon.

activated ergosterol. vitamin D_2.

activation. **1:** the process of rendering active. **bioactivation**. biological activation. **biochemical activation**. **1:** activation by means of a biochemical process. **2:** metabolic activation. **biological activation**. **1:** activation by means of a biological process. **2:** metabolic activation. **enzyme activation**. the conversion of a preenzyme into an active enzyme, usually catalyzed by a kinase or another preenzyme. **metabolic activation** (biochemical activation; biological activation; physiological activation) the enzymatic conversion of a foreign or endogenous substance to a more reactive (or in toxicology to a more toxic) form. Activation reactions of toxicological significance are usually monooxygenations that yield electrophiles that may react with nucleophilic moieties on endogenous proteins or DNA. **multistep metabolic activation.** carcinogens and certain other toxic chemicals undergo a series of transformations before generating the ultimate toxic (or carcinogenic) metabolites. **physiological activation**. metabolic activation.

activator. **1a:** something that activates; stimulus, trigger. **1b:** any substance that renders another substance active (i.e., capable of entering into its normal reactions). **2:** a substance that provokes a specific activity (e.g., an inducer of specific hormone secretion). See also activation.

active. **1a:** energetic; productive of, or capable of action or movement. **1b:** in action, in motion or performing [illegible] tendency to act. **3:** capable of acting in a specific manner. **biologically active** (bioactive). able to elicit a response from living tissues, organs, or systems.

active avoidance. See conditioned avoidance.

active carbon. activated carbon.

active immunity. resistance to disease effected by the production of antibodies by an organism in response to antigens from a pathogen or disease organism.

active metabolite. See metabolite.

active methionine. *S*-adenosylmethionine.

active process. any process that requires metabolic energy; *cf.* passive process.

active serum. See serum.

active site (catalytic site). the region on the surface of an enzyme to which a substrate attaches.

active transport. See transport.

active uptake. the absorption of ions with the expenditure of metabolic energy. *Cf.* passive uptake.

activity. **1a:** motion or any process associated with the flow of energy. **1b:** the state or quality of being active. **2:** the effective concentration of a solute in a given solution. It is often expressed as the product of the activity coefficient, *q.v.*, and the concentration of the solute. The activity of an ideal or pure substance is unity. **3:** the behavior of a cell, tissue, organ, or living organism as reflected in action or movement. **biological activity** (bioactivity). **1:** the activity of a biological system. See activity (def. 3). **2a:** the capacity of an exogenous or endogenous agent (e.g., a toxicant) to directly affect one or more biological processes. **2b:** the specific action of any agent on a biological system or process.

activity coefficient. activity (def. 2).

Aculeata. a division of the insect order Hymenoptera that includes bees, ants, and true wasps; the ovipositor is modified as a sting.

aculeate. **1:** having sharp points, prickles, or a sting; pointed, stinging. **2:** denoting, having the characteristics of, or belonging to the Aculeata.

acute. **1:** sharp, pointed. **2:** transient or of short duration; **3:** intense; marked by sharpness or severity; critical. **4a:** of or pertaining to a single, relatively brief, usually intense, exposure to a toxicant or other biologically active agent. **4b:** the development of symptoms following an acute exposure (see 4a) to a toxicant or other biologically active agent. **5:** of or pertaining to a stimulus intense enough to induce a rapid response. In aquatic toxicity tests, for example, an effect (not necessarily a lethal effect) that presents within 96 hours is considered acute. *Cf.* chronic, exposure, toxicity.

acute alcoholic intoxication. acute alcoholism.

acute-to-chronic ratio (ACR). the ratio of the acute toxicity of a toxicant or other biologically active substance to its chronic toxicity. While sometimes used as an estimate of chronic toxicity by extrapolation from data on acute toxicity (or conversely), the significance of this ratio beyond a narrow range where both toxicities are established is usually unclear.

acute condition (acute disease; acute disorder). any condition characterized by rapid onset, sharp rise, short duration, pronounced or severe symptoms (often including pain, high fever, dyspnea). *Cf.* chronic condition.

acute delusional state. See delirium.

acute disease. acute condition.

acute disorder. acute condition.

acute effect. See effect.

acute exposure. See exposure.

acute hallucinosis. hallucinosis.

acute lethal injury. See injury.

acute myelogenous leukemia. See leukemia.

acute parenchymatous hepatitis. acute yellow atrophy (See atrophy).

acute poisoning. See poisoning.

acute radiation syndrome. the acute form of radiation sickness, it may be manifested clinically in humans as any of three syndromes: the cerebral syndrome, the gastrointestinal syndrome, and the hematopoietic syndrome, *q.v.*

acute toxicity. See toxicity.

acute toxicity end point. See end point.

acute yellow atrophy. See atrophy.

acute-phase reaction protein. α-acid glycoprotein.

acutely toxic condition. any ambient condition that makes an area toxic to an organism following a brief exposure.

acyanotic. pertaining to or characterized by the absence of cyanosis.

acyclic terpene. a class of aliphatic hydrocarbons. See terpene.

acyl anhydride. acid anhydride. See anhydride.

acyl CoA:amino acid *N*-acyltransferase. See conjugation (amino acid conjugation).

acyl CoA synthetase. See conjugation (amino acid conjugation).

acyl halide. See halide.

acylating agent. a carrier of activated acyl groups (e.g., acid anhydrides, coenzyme A) that introduces RCO group into a substance.

acylation. the incorporation of an acyl radical, RCO—, into a compound. Acid anhydrides, acid halides, and coenzyme A are acylating agents. See acid anhydride, acyl halide, coenzyme A.

***N,O*-acyltransferase**. an enzyme activator of carcinogenic arylamines.

ADAM. See MDMA.

Adam and Eve. *Arum maculatum*.

adamsite. phenarsazine chloride

adaptability. the capacity to adapt.

adaptation (adaption; adaptive adaption). **1:** any process by which an organism adjusts to environmental stress. **2a:** a process of natural selection that results in improved fitness of an organism or population to a particular environment; **2b:** the fitness of an organism to a particular environment. **3:** an inherited phenotype (structure or trait), resulting from natural selection, that enhances survival or reproductive success of an organism or population in a specific environment, environmental variable, or activity. **4a:** a process by which an organism or sense organ ceases to respond to a continuing or repeated stimulus of a particular type when the intensity of the stimulus remains constant. **4b:** a reduction in the frequency of impulse generation in a receptor following a repeated or continuing stimulus of constant intensity; sensory adaptation; habituation. **4c:** adjustment of the pupil and retina to changes in illumination. *Cf.* accommodation. **5:** the dynamic adjustment of an organ or organism to a constantly changing environment. Such adaptation depends upon biophysical, biochemical, physiological, and behavioral responses. *Cf.* acclimation, acclimatization, habituation.

adaptation autoinhibitor. a substance secreted and released by individuals that limits the population to its equilibrium level.

adaptation to a toxicant. 1: habituation to a toxicant. **2:** reduced sensitivity to a toxicant through a process of natural selection that results in improved fitness. Thus, particular herbivores may have evolved systems that detoxify particular phytotoxins, whereas other organisms have not. *Cf.* resistance, tolerance.

adaptedness. the degree of adaptation to an environment.

adaption. adaptation.

adaptive adaption. adaptation.

adaptive behavior. behavior of an animal that improves its fitness under a given set of environmental circumstances. Thus, animals structurally adapted to fly do so (under appropriate circumstances) and most animals perceive, discriminate, and avoid many toxic chemicals such that exposure is limited. See adaptation.

adaptive enzyme. inducible enzyme.

adaptive value. the comparative fitness, *q.v.*, of different genotypes in a given environment; the survival and reproductive value of one genotype relative to other genotypes in the population. One can give numerous examples. As one example, numerous plants have developed toxins, presumably in some cases in response to herbivory. Such toxins are often much less toxic to herbivores that have co-adapted to the evolving plant toxins than to other animals.

ADCC. antibody-dependent cellular cytotoxicity.

ADD. Acceptable Daily Dose.

adder. **1a:** part of the common name of any of a variety of venomous snakes, especially those of the family Viperidae. **1b:** *Vipera berus*. **2:** true adders are snakes of the genus *Vipera*. **3:** any of several harmless North American snakes such as the hognose snake ("puff adder") and milksnake.

addict. See drug addict.

addiction. **1:** craving, compulsion, dependence, dependency. **2:** the behavioral syndrome or life style of one who is addicted to, dependent on, or obsessed with some aspect of his or her environment, especially dependence on the use of a drug (usually alcohol or a narcotic). **drug addiction**. the compulsive, continuing, uncontrolled use of a habit-forming substance (substance of abuse), especially alcohol or a drug, under conditions that are harmful to the individual or to society. Addiction can be due to psychological or physiological dependency. The subject may suffer symptoms of withdrawal when attempting to break the addiction. *Cf.* substance abuse, dependence, alcoholism.

addictive drug. any chemical substance (usually a narcotic) that causes chronic or periodic intoxication by repeated usage and fosters dependence.

addition product. a compound formed by the chemical addition of two species. *Cf.* adduct.

additive. **1:** of or pertaining to addition. **2:** referring to the ability to add or be added; cumulative. **3:** any substance not a normal part of a material (e.g., food) which is added to it in order to alter or enhance its natural properties, to preserve or extend its life, or to confer new properties. **4:** a substance which, when added to a petroleum product, confers special properties or enhances the natural properties of the product. *Cf.* synergistic, antagonistic. **food additive**. any substance added to a food prior to sale, usually to enhance flavor or appearance or to preserve or extend its useful life. See, for example, nitrate, nitrite, sulfite, color (artificial color), butylated hydroxytoluene (BHT), butylated hydroxyanisole (BHA), and flavor (artificial flavor). See also contaminant, Delaney Clause, sulfite.

additivity. the quality or condition of being additive. **causal additivity**. the phenomenon whereby two or more causal agents (e.g., biologically or chemically active agents or stimuli) have a combined effect that is the algebraic sum of their individual effects.

adduct. **1:** to draw toward the main axis or median plane of an animal's body or limb. **2a:** short for addition product, *q.v.* **2b:** any simple combination of chemical groups joined by a coordinate bond.

ademetionine. *S*-adenosylmethionine.

Adenia (passion flower). a genus of toxic, flowering African plants. ***A. digitata***. a flowering plant of southern Africa, sometimes called passion flower. It is extremely toxic to humans. Symptoms of intoxication include drowsiness and weakness, worsening to paralysis followed by death. Material from this plant and that from *A. volkensii* are cortical depressants. Ingestion of minute quantities of *A. digitata* are sufficient to cause death. ***A. volkensii.*** a flowering plant considered to be a serious human poison in eastern Africa; it contains both a cyanogenic glycoside and a phytotoxin, like that of *A. digitata*. Material from this species acts as a cortical depressant in vertebrate animals, affecting the higher centers of the brain. Symptoms in humans include drowsiness and weakness worsening to paralysis, and subsequent death.

adenine-D-ribosephosphate-phosphate-D-ribose-nicotinamide. nadide.

adenitis. glandular inflammation.

adeno-. a prefix denoting a gland or indicating a glandular state.

adenocarcinoma. any of a large number of malignant tumors that arise from epithelium. They contain malignant cells characteristic of, and named for, the tissue from which they arise. The cancer cells often become so profoundly degenerate that specific tissue characteristics are unrecognizable, and the tissue of origin of a disseminated adenocarcinoma may thus be indeterminant.

adenohypophyseal growth hormone. somatotrophin.

adenohypophysis. the anterior lobe of the pituitary gland. See pituitary gland.

adenoma. a gland-like, usually benign tumor that arises usually from glandular epithelium or mucosal epithelium. The development of an adenoma in a laboratory rat that has received a test chemical is considered evidence that the substance might be carcinogenic in humans. *Cf.* adenocarcinoma.

adenomatosis. a condition characterized by numerous glandular tissue enlargements or overgrowths. See silo-filler's disease.

Adenorhinos barbouri (worm-eating viper). formerly *Atheris barbouri*, it is a small moderately slender viper (Family Viperidae), with a head that is distinct from the neck and a moderately slender tail. It occurs in Tanzania and is probably not dangerous to humans. The snout is short and rounded and the canthus rostralis is obtuse; the tail is moderately slender. The eyes are very large (1½ times the distance from eye to lip) with vertically elliptical pupils. *Cf. Vipera supercilliaris*.

adenosine cyclic 3′,5′-hydrogen phosphate. cyclic AMP.

adenosine 5′-dihydrogen monophosphate. adenosine monophosphate.

adenosine diphosphate (adenosine 5′-trihydrogen diphosphate; ADP). a nucleotide comprised of the [illegible] ribose to which two phosphate groups are attached, one of which is a high-energy group. ADP is a component of all living cells and is an important coenzyme in numerous biochemical reactions. It undergoes oxidative phosphorylation in mitochondria to yield ATP and is reconstituted by the condensation of adenosine and phosphoric acid. High concentrations of ADP stimulate (and low concentrations inhibit) oxidative phosphorylation. See also adenosine monophosphate, adenosine triphosphate, and adenosine triphosphatase.

adenosine monophosphate (adenosine 5′-dihydrogen monophosphate; AMP; 5′-AMP; adenylic acid). a mononucleotide ester of phosphoric acid and adenosine. Oxidative phosphorylation of AMP in mitochondria (catalyzed by adenylate kinase) can add more phosphate groups to yield the diphosphate (adenosine diphosphate) or the triphosphate (adenosine triphosphate). AMP is reconstituted by hydrolysis of ATP → ADP → AMP. See adenosine triphosphate for more information. See also adenosine diphosphate, adenosine triphosphatase, cyclic AMP.

adenosine 3′,5 ′-monophosphate. cyclic AMP.

adenosine 3′,5′-phosphate. cyclic AMP.

adenosine 5′-phosphosulfate kinase. See sulfate conjugation.

adenosine phosphosulfate kinase. See sulfate conjugation.

adenosine 5′-tetrahydrogen triphosphate. adenosine triphosphate.

adenosine 5′-trihydrogen diphosphate. adenosine diphosphate.

adenosine 5′-(trihydrogen diphosphate) 5′→5′-ester with 3-(aminocarbonyl)-1-β-D-ribofuranosylpyridinium, hydroxide, inner salt. nadide.

adenosine triphosphatase (adenosine 5′-triphosphatase; ATPase). an enzyme that catalyzes the hydrolysis of ATP with the release of phosphate and free energy. The latter drives numerous metabolic processes (e.g., active transport, muscular contraction). ATPase also catalyzes [illegible] ADP → P_i → ATP. Certain other proteins (e.g., myosin) have ATPase activity. See adenosine triphosphate.

adenosine 5′-triphosphatase. adenosinetriphosphatase.

adenosine triphosphate (adenosine 5′-(tetrahydrogen triphosphate); ATP). a nucleotide comprised of the purine base, adenosine, and D-ribose attached to three phosphate groups, two of which are high-energy groups. It is an essential energy-storage and energy-transfer component of all living cells. ATP is hydrolyzed to adenosine diphosphate (ADP) with the release of free energy and pyrophosphate (PP_i); ADP may then form adenosine monophosphate (AMP) with the release of additional free energy and PP_i. The energy so released, drives numerous endergonic metabolic processes such as active transport and muscle contraction. ATP is reconstituted chiefly from ADP during photosynthesis and oxidative phosphorylation. See also adenosine diphosphate, adenosine monophosphate, adenosine triphosphatase. Some poisons block ATP synthesis by uncoupling mitochondrial oxidative phosphorylation (e.g., dinitrophenol) or by inhibiting mitochondrial electron transport (e.g., cyanide).

adenosine-3′,5 ′-monophosphate. cyclic AMP.

Adenostoma fasciculatum. a small California evergreen shrub (Family Rosaceae) with small heath-like flowers found in the hard chaparral community of California. Together with a number of other species (e.g., *Arctostaphylos spp.*) that make up the hard chaparral community, these plants secrete phenolic compounds such as arbutin and *p*-hydroxy cinnamic acid, which dissolve in rainfall, wash to the ground, and inhibit the growth of grasses (Gramineae) near the plant. Substantial bare areas thus develop within and somewhat beyond the edge of the community. When the chaparral is destroyed by fire, the growth of grasses is initially vigorous, but slows as the chaparral community is reestablished.

***S*-adenosylmethionine** (5′-[(3-amino-3-carboxypropyl)methylsulfono]-5′-deoxyadenosine hydroxide, inner salt; S-(5′-desoxyadenosin-5′-yl)-L-methionine; active methionine; ademetionine). a physiologically active methionine derivative found in all living organisms. It donates the methyl group in the enzymatic transmethylation of xenobiotic substances.

adenyl cyclase. adenylate cyclase.

adenylate cyclase (adenyl cyclase). the enzyme that catalyzes the conversion of ATP to cyclic AMP. It is sometimes a site of toxicant action (e.g., cholera toxin, pertussis toxin) that can have grave consequences. The active site of this enzyme is part of a complex polymer that includes hormone or neurotransmitter receptors, and guanine nucleotide-binding proteins (G proteins, *q.v.*).

adenylic acid. **1:** adenosine monophosphate. **2:** a generic term applied to a group of isomeric nucleotides that contain the adenyl radical, $C_5H_4N_5$·.

adermine hydrochloride. pyridoxine hydrochloride.

ADH. **1:** vasopressin. **2:** alcohol dehydrogenase.

adherence. the amount of sediment or soil adhering to the skin that could be absorbed into the body, usually expressed as surface density in mg/cm^2.

adhesion. **1:** a molecular attraction that holds the surfaces of two bodies together. **2:** a firm solid attachment. **3a:** the union of two body parts (organs or tissues) that are not normally joined, by the formation of fibrous tissue; the site of union of such tissue. **3b:** the tissue that joins or binds two organs or tissues that are not normally joined. **4:** the attraction of molecules in the walls of the interstices of a sediment to water molecules.

ADI. Acceptable Daily Intake.

adience. **1:** the tendency to approach the source of a stimulus. **2:** movement toward a stimulus; an approach response. See also abience.

adient. positive; having the tendency to approach the source of stimulus.

adipose. fat, fatty, pertaining to fat.

adipose tissue (fatty tissue). a type of connective tissue comprised of fat cells (adipocytes). In mammals, this tissue occurs chiefly in subcutaneous tissue, in the mesenteries, around the kidneys, and in the mediastinum. An adipocyte typically contains one large fat droplet that restricts the cytoplasm to a narrow zone near the cell periphery. Adipose tissue serves as a reserve (sometimes the main) energy source of animals; it also provides insulation against heat loss, and can act as a shock absorber. Many lipophilic chemicals (e.g., chlorinated hydrocarbons) accumulate to levels in adipose tissue that are several orders of magnitude higher than that in blood. Polychlorinated biphenyls, when absorbed by higher vertebrates, occur at first in highly perfused organs such as liver and muscle, but later accumulate in poorly perfused organs such as adipose tissue and skin. Most animals living at mid or high latitudes undergo annual cycles of lipid deposition and mobilization. Thus, actual concentrations, total burden, distribution of xenobiotics throughout the body, and their effects may strongly depend upon the time of year and other factors (such as dieting) that affect lipid storage, mobilization, and metabolism. See lipophilic.

adipsia. **1a:** absence of thirst. **1b:** complete failure to drink.

adjustment. **1:** acclimation; a functional, often transitory, response to stimuli by which an organism adapts to its immediate environment. **2:** the process by which an organism becomes adjusted to its environment. *Cf.* acclimation, acclimatization, accommodation, adaptation.

adjuvant. an excipient that may enhance the toxicity, antigenicity, or pharmacologic action of a chemical formulation. Thus, the activities of dithiocarbamate fungicides may be increased by the addition of 2-mercaptothiazole. **Freund's adjuvant**. an emulsion that contains detergent and dead mycobacteria (usually *Mycobacterium tuberculosis*). It enhances antibody response apparently by delaying antigen clearance from the site of inoculation. It also activates the lymphoid system by promoting the influx of leukocytes. Incomplete Freund's adjuvant contains no dead bacteria or bacterial products.

administration. **1:** management, direction, supervision. **2:** giving, dispensing, delivery; as in the administration of a drug to a patient. **route of administration**. the path of entry into the body of a drug or poison administered directly to the recipient organism by human agency (e.g., a human health care professional, a veterinarian). Possible routes are essentially the same as those noted under the term, route of exposure and are treated there. Some terms used only, or most often in the administration of xenobiotics are best mentioned here, although they apply also to "route of exposure." All of the following routes of administration are used in research and in the health-related fields: intracerebral, intracervical, intradermal, intraduodenal, inhalation, intramuscular, intraplacental, intrapleural, intraperitoneal, intrarenal, intratracheal, intravaginal, intravenous, ocular, oral, rectal, subcutaneous, and skin contact. *Cf.* exposure (route of exposure). **anal route of administration**. drugs are often administered *per anus* as a suppository. **extravascular route of administration**. any route in which the drug is not introduced directly into the cardiovascular system. **parenteral route of administration**. any route other than via the gastrointestinal tract. **rectal route of administration**. the introduction of a drug into the rectum. This route is used only to test or use pharmaceuticals that are normally administered as suppositories (See, for example, chlorpromazine).

admissibility. the state of being admissible. Thus, toxicological data that are relevant and proper to be considered in reaching a decision concerning issues raised in a judicial proceeding are admissible. Data are admissible by precedent or mutual agreement and not require defense prior to acceptance. *Cf.* defensibility.

admissible. a term used in toxicology regarding data or physical evidence deemed relevant and proper for consideration in arriving at a decision in a judicial proceeding.

adolescence. the period of life from onset of puberty until maturity; used mainly in reference to *Homo sapiens*.

adonidin. any of the cardiotoxic phenanthrene glycosides found in various species of *Adonis*. Effects are similar to those of digitalin.

Adonis (pheasant's eye; bird's eye; false hellebore). a genus of toxic herbs (Family Ranunculaceae). They are native to Europe, Asia, and Africa and have been introduced elsewhere, including the Australian region and North America. They [illegible] under field conditions in Australia, New Zealand, and Europe, but incidents are unknown in North America, where they have been introduced as garden plants. The toxic principles are phenanthrene glycosides such as adonidin. ***A. vernalis*** (false hellebore, vernal pheasant's eye, bird's eye). Native to Europe, Asian parts of the former U.S.S.R., and Labrador. Constituents include adonitoxin, acetyladonitoxin, adonivernoside, adonitol, adoniverhith, convallatoxin (strophanthidin), and vernadin. See adonidin.

adonitol. a constituent of *Adonis vernalis*.

adonitoxin (3-[(6-deoxy-α-L-mannopyranosyl)-oxy]-14,16-dihydroxy-19-oxocard-20(22)-enolide). a toxic phenanthrene glycoside isolated from *Adonis vernalis*. See also adonidin.

adoniverhith. a constituent of *Adonis vernalis*.

adonivernoside. a constituent of *Adonis vernalis*.

ADP. adenosine diphosphate.

ADR. adverse drug reaction.

adrenal. **1:** toward the kidney. **2:** an adrenal gland.

adrenal cortex. See adrenal gland.

adrenal gland (suprarenal gland; adrenal, def. 2). either of a pair of glandular bodies in vertebrates. They are situated near the kidneys in mammals and are superior to the kidney in primates. Each is comprised of an inner medulla that secretes adrenalin and noradrenalin, and an outer cortex that secretes steroid hormones (corticosteroids). In some vertebrates the two types of adrenosecretory tissue form separate glands.

adrenal medulla. See adrenal gland.

adrenalin. **1:** epinephrine. **2:** a frequently used spelling in the United States for adrenaline. **3:** often the preferred term for epinephrine in physiology and allied sciences such as toxicology. *Cf.* adrenaline.

adrenalin oxidase (adrenaline oxidase). monoamine oxidase.

adrenaline. **1:** epinephrine. **2:** the official British Pharmacopeia name for epinephrine and the preferred usage in the United Kingdom. *Cf.* adrenalin.

adrenergic. **1:** pertaining to or denoting nerve cells, fibers, or nerve terminals of the autonomic nervous system that release adrenalin or noradrenalin as a neurotransmitter from the presynaptic membrane of the neuron. The postganglionic sympathetic neurons of vertebrates are adrenergic. **2:** pertaining to drug actions similar to those of the sympathetic nervous system.

β-adrenergic antagonist. any of a class of drugs that competitively inhibit the binding of catecholamines to β-adrenergic receptors. They are used to reduce cardiovascular overactivity, improve blood circulation, and to prevent vascular headaches, hypertension, cardiac arrhythmias, and angina. Possible adverse effects include bronchospasm, fatigue, and hypoglycemia.

α-adrenergic receptor. See receptor. See also adrenergic.

β-adrenergic receptor. See receptor. See also adrenergic.

adrenocortical. pertaining to or denoting the cortex of the adrenal gland or its secretions. See also ACTH.

adrenocorticotrophin. ACTH.

adrenocorticotrophic hormone. ACTH.

adrenocorticotropic hormone. ACTH.

adrenocorticotropin. ACTH.

adrenogenital syndrome. a condition caused by hypersecretion of, or excessive treatment with adrenal androgens. This brings about premature puberty of the male, penile enlargement, hirsutism in adult females, and shortened stature. In congenital cases, the female may be misidentified as a male.

adrenolytic. **1:** antagonistic to or inhibiting the action of adrenalin, noradrenalin, and related sympathomimetics. **2:** antagonistic to or inhibiting the activity of adrenergic nerves.

adrenomimetic. having an action similar to that of adrenalin and noradrenalin.

adrenotoxin. any substance, especially a toxin (defs. 1a and 1b), that is poisonous to the adrenal glands.

adrenotropic hormone. ACTH.

Adriamycin® (adriamycin). a trademark name and former generic name of doxorubicin.

adsorb. **1:** to adhere as an ultrathin layer on the surface of a solid or fluid; said of molecules. **2:** to attract and retain other substances on a surface by means of chemical or physical forces.

adsorbate. a solid, liquid, or gas that is adsorbed, as to the surface of a solid.

adsorbent. **1:** pertaining to or denoting adsorption or the capacity to adsorb. **2:** a solid or liquid (e.g., charcoal, silica, metals, water, mercury) that adsorbs other materials.

adsorption. **1:** the adherence of a substance (liquid, gas, or solid) to the surface of another as an ultrathin layer. **2:** a process of attracting and retaining other substances on a surface by chemical or physical forces. **3:** an advanced treatment method in which activated carbon is used to remove organic matter from wastewater. **4:** in spectroscopy, the absorption of light energy by a substance.

adsorption assay. See assay.

adulterant. any chemical impurity or substance that is not allowed by law in a particular food, pesticide, or other formulation. **food adulterant**. any impurity that impairs the quality or bulk of a foodstuff, including toxic substances, poisonous organisms, and radioactive fallout.

adulterated. said of a food, plant, animal, pesticide formulation, or any substance that contains an adulterant; one from which an active or significant constituent has been wholly or partly removed; or one, the purity of which falls below a legal standard.

adulteration. a reduction in purity of any substance by the deliberate addition of an ingredient not ordinarily part of that substance, especially where its presence is restricted by law. Use of the term usually implies that the quality of the substance is diminished.

advection. **1:** the process of transfer of air mass properties by virtue of the motion (the velocity field) of the atmosphere. **2:** the process of transfer of heat or matter by horizontal movement of a body of water.

adventive. **1a:** newly arrived; appearing. **1b:** of or pertaining to an organism that is accidentally introduced to a new region and is imperfectly adapted; not native. Such an organism may be more susceptible than native organisms to certain poisonous substances in the region due to a lack of physiological or genetic adaptation.

adverse. in toxicology, adverse usually means detrimental, harmful, damaging, or unfavorable to a living system. **adverse reaction**. **1:** adverse effect. See effect. **2:** in medicine and pharmacology, the development of a deleterious, usually unintended effect, caused by the administration of a medication. **adverse drug reaction** (ADR). any of a great variety of dose-dependent and dose-independent drug interactions (*q.v.*) that increase toxicity or result in therapeutic failure of a drug.

advisory. **1:** having or exercising the authority or capacity to advise. **2:** intended to advise. **3:** a nonregulatory document that communicates risk information to persons who may be responsible for making risk management decisions.

aer-, aero-. prefix meaning air, atmosphere.

aerial. referring to air; occurring in the air; airborne.

aerobe. an aerobic organism.

aerobic (oxybiotic). **1a:** of or pertaining to an aerobe or to processes that function in or tolerate the presence of molecular oxygen. **1b:** growing or occurring only in the presence of molecular oxygen; oxybiotic. **1c:** pertaining to growth in the presence of molecular oxygen. **2:** pertaining to environments that contain molecular oxygen; oxygenated.

aerobic cellular respiration. cellular respiration in which oxygen is the final acceptor of hydrogen atoms from a series of reactions that produce ATP.

aerobic treatment. the decomposition of complex organic compounds in the presence of oxygen by the use of microbes.

aerobiology. a branch of science that treats the

occurrence, transport, and effects of atmospheric components that are of biological origin or significance.

aerohygrophobe. an organism that is intolerant of high atmospheric humidity.

aerohygrophobous. intolerant of high atmospheric humidity.

aerohygrophoby. the condition of an organism such that it is intolerant of high atmospheric humidity.

aerometric. pertaining to measurement of the properties or contaminants of air.

aerosol. a "colloidal" suspension of ultramicroscopic liquid or solid particles dispersed in a gas, the particles of which have a negligible falling velocity. The size of the particles often exceeds that of the normal colloidal limits (ca. 1 nm to 1 μm in diameter). **2:** a product, packaged under pressure, that contains materials intended for inhalation, topical application, or introduction into body orifices (e.g., pesticides, various therapeutic drugs, shaving cream). Many of these products are eye or skin irritants, some are seriously toxic, and inhalation is difficult to avoid.

aerosolization. the dispersion of a liquid as a fine mist.

aerothionin A. a bromine-containing antibacterial fungal metabolite.

Aerotrol™. isoproteronol hydrochloride.

aesculin. esculin.

Aesculus (buckeye; horsechestnut). a genus of ornamental deciduous shrubs and trees (Family Hippocastanaceae) with compound leaves composed of 5 to 7 palmately arranged leaflets. *A. glabra* (Ohio buckeye) and *A. hippocastanum* (horsechestnut) are the best known species. The seed is a distinctive, shiny brown nut about 2.5 to 3.0 cm in diameter and similar in appearance to that of the edible chestnut (*Castanea spp.*). The nuts and probably the young growth and sprouts are neurotoxic due to the presence of esculin. These plants can be lethal to humans, livestock, and laboratory animals. The main symptoms are those of gastroenteritis, with inflammation of mucous membranes, depression, debility, and paralysis.

aesthesia. esthesia.

aesthesis. esthesis.

Aethusa (formerly *Cynapium*). a genus of herbs (Family Apiaceae). ***A. cynapium*** (fool's parsley, dog parsley, "wild parsley," fool's cicely). a fetid, poisonous, annual European herb, naturalized in America, where it occurs in fields and waste places in the northeastern United States and eastern Canada. It looks very much like poison hemlock, but lacks the purple spotting. All parts of the plant are toxic to humans when ingested. Mild intoxication causes nausea and vomiting. The fruit contains cynapine, a volatile alkaloid.

aethylis chloridium. ethyl chloride.

aetiologic. etiologic.

aetiologic agent. etiologic agent.

aetiological. etiological.

aetiology. etiology.

AETT. acetylethyltetramethyltetralin.

AFDW. free dry weight.

afebrile (apyretic). without fever.

afee. *Echis carinatus*.

affect. **1:** to alter, influence, modify; to feign, pretend, simulate. **2:** mood; conscious awareness of feelings. **3:** in medicine, an outward manifestation or sign of a person's mood, feelings, or emotions.

affection. **1:** any pathology or diseased state. **2:** the awareness or affective aspect of consciousness.

affective. pertaining to the expression of feelings or emotions.

affective disorder. any disorder marked by inappropriate emotional responses or disturbances in mood (e.g., depression, mania).

affinin (*N*-(2-methylpropyl)-2,6,8-decatriene-amide); *N*-isobutyl-2,6,8-decatrieneamide; *N*-isobutyldeca-*trans*-2-*cis*-6-*trans*-8-trienamide; spilanthol). an insecticidal analgesic lipid amide from *Heliopsis longipes*.

affinitin. lectin.

affinity. the propensity of a toxicant or other pharmacologically active substance to bind to a specific receptor (ligand). **affinity constant** (binding constant). a quantitative measure of the strength of the interaction in ligand binding. The constant is derived from the equation:

$$K_a = [TP] / [T] [P] = 1/K_d$$

where K_a is the affinity (or binding) constant, K_d is the dissociation constant, [TP] is the concentration of bound ligand, [T] is the concentration of free toxicant, and [P] is the concentration of free protein. See also dissociation constant.

affliction. ailment, infirmity. In toxicology, this term is roughly synonymous with disease, *q.v.*

affluent. **1a:** flowing toward. **1b:** any fluid that flows toward. **2a:** a body of water (e.g., river, stream, canal, sewage) that flows into a larger body of water (e.g., river, lake). **2b:** the raw sewage that flows into a waste treatment plant.

aflatoxicol (aflatoxin P_1). a mutagenic bisfuranoid mycotoxin.

aflatoxicosis (aflatoxin poisoning; turkey X disease). a mycotoxicosis acquired by eating peanuts, groundnut (peanut) meal, peanut products, soybeans, maize, and other cereal grains including feeds such as groundnut (peanut) meal, contaminated by aflatoxin produced by secreting strains of *Aspergillus flavus* and *A. parasiticus*. All domestic animals are vulnerable, but growing fowl (e.g., ducklings, turkey poults, pheasant chicks), young swine, pregnant sows, calves, and dogs are most commonly affected. Suckling animals are also highly susceptible and may be exposed via milk while nursing. Adult cattle, sheep, and goats are fairly resistant to the more acute forms of the disease, but may respond to long continued toxic diets. Effects and symptoms are varied and diverse as a function of sex, age, species, nutritional status, amount and duration of intake, the strain involved, the level of contamination of the rations, and perhaps other factors. Ataxia, convulsions, and emprosthotonos are common, and a catarrhal enteritis, especially of the duodenum, is usually evident. Common symptoms are depression; inappetence; reduced growth rate; loss of condition; bruising; decreased egg production, fertility, and hatchability; and increased mortality. Lesions may include membranous glomerulonephritis and hyaline droplet nephrosis together with some amount of ascites and visceral edema; the liver is pale, mottled, with extensive necrosis. Excessive bile production is common. Most disease outbreaks formerly identified as "moldy corn toxicosis," "hemorrhagic syndrome of poultry," and "*Aspergillus toxicosis*" were caused by aflatoxins in the feed.

aflatoxin. any of a family of polynuclear bisfuranoid mycotoxins produced by toxigenic strains of *Aspergillus flavus*, *A. parasiticus*, and related species. All contain the coumarin ring and an α,β-unsaturated lactone moiety. The aflatoxins may infest, for example, milk, peanuts, peanut meal, peanut butter, corn, cottonseed meal, and dried chili peppers. Aflatoxins are potent toxins, carcinogens, teratogens, and mutagens. They have a pronounced effect on cell-mediated immunity in *Homo sapiens* and the domestic fowl (*Gallus*). The presence of *Aspergillus* in grains, meals, or other food materials does not always imply the presence of aflatoxins because their production depends upon factors such as temperature, moisture, aeration, and substrate. Most important are the aflatoxins B, G, and M. Others are known (See aflatoxicol). **aflatoxins B**. the B aflatoxins exhibit blue fluorescence. **aflatoxin B_1** (2,3,-6aα,9aα-tetrahydro-4-methoxycyclopenta-[c]-furo-[3′,2′:4,5]furo[2,3-h][1]benzopyran-1,11-dione). the most extensively studied of the aflatoxins, it is an extremely powerful hepatotoxicant and hepatocarcinogen in many animal species. Species sensitivity and target organ specificity vary greatly, however. It is extremely toxic to trout, day-old ducklings, turkeys, cattle, rabbits, cats, some laboratory rats, and swine. Dogs, sheep, and guinea pigs are somewhat less sensitive; hamsters, some laboratory rats, and the domestic fowl are more resistant; and some strains of laboratory mice appear to be completely resistant. Liver lesions differ among susceptible species, being, for example, centrilobar in swine and guinea pigs, periportal in laboratory rats and ducklings, and midzonal in rabbits. Aflatoxin B_1 is an extremely potent environmental mutagen and carcinogen and contributes to the etiology of primary cancer of the liver in humans. It is metabolized by the cytochrome P-450 dependent monooxygenase system in the liver, yielding a highly reactive electrophilic epoxide. The latter may bond covalently to protein, DNA, and RNA. Reaction with protein probably accounts for its hepatotoxicity, while reaction with DNA probably initiates carcinogenesis. **aflatoxin B_2** (2,3,6aα,8,9,9aα-hexahydro-4-methoxycyclopenta[c]furo-[3′,-2′:4,5]furo[2,3-h][1]benzopyran-1,11-dione). The less toxic 8,9-dihydro derivative of aflatoxin B_1 is extremely toxic to 1-day-old ducklings and is mutagenic and carcinogenic. **aflatoxins G**. the G aflatoxins exhibit green fluorescence. **aflatoxin G_1** (3,4,7aα,10aα-tetrahydro-5-methoxy-1H,12H-furo[3′,2′:4,5]-furo[2,3-h]pyrano[3,4-c][1]-benzopyran-1,12-dione). a mycotoxin secreted by *Aspergillosus flavus* and *A. parasiticus*, it is extremely toxic to 1-day-old ducklings and is mutagenic and carcinogenic. **aflatoxin G_2** (3,4,7aα,9,10,10aα-hexahydro-5-methoxy-1H,12H-furo[3′,2′:4,5]-furo[2,3h]-pyrano[3,4-c][1]benzopyran-1,12-dione). the less toxic 9,10-dihydro derivative of aflatoxin G_1, it is extremely toxic to ducklings and is mutagenic and carcinogenic. **aflatoxins M** (milk toxins). extremely toxic 4-hydroxylated derivatives of aflatoxins B. They occur in the milk of cattle fed infested meal. **aflatoxin M_1** (2,3,6a,9a-tetrahydro-9a-hydroxy-4-methoxycyclopenta[c]furo[2,3-h][1]benzopyran-1,11-dione). a compound that is mutagenic, carcinogenic, and extremely toxic to day-old Pekin ducklings. **aflatoxin** M_2 (2,3,6a,8,9,9a-hexahydro-9a-hydroxy-4-methoxycyclopenta[c]furo-[3′,2′:4,5]-furo[2,3-h][1]benzopyran-1,11-dione). effects are similar to those of aflatoxin M_1. **aflatoxin P_1**. aflatoxicol. See Reye's syndrome.

aflatoxin poisoning. aflatoxicosis.

African black cobra. *Naja melanoleuca*.

African brown snake. *Dispholidus typus*.

African bush viper. a snake of the genus *Atheris*.

African coffee tree. *Ricinus communis*.

African coral snake. *Aspidelaps lubricus*.

African desert horned viper. *Cerastes cerastes*.

African dwarf gartersnakes. snakes of the genus *Elaps*.

African garter snake. *Elapsoidea sundevallii*.

African lowland viper. *Vipera superciliaris*.

African milk plant. *Euphorbia lactea*.

African necklace snake. *Elapsoidea sundevallii*.

African night adder. *Causus rhombeatus*.

African puff adder. *Bitis arietans*.

African rue. *Paganum*.

African sand viper. *Cerastes vipera*.

African shield-nosed snake. *Aspidelaps scutatus*.

African viper. a snake of the genus *Bitis*. See also Viperidae.

Afrikanische Korallenschlange. *Elaps lacteus*.

after damp. See damp.

afterbirth. a common name for the placenta and associated membranes expelled from the mammalian uterus, through the vagina, soon after giving birth. In humans, it is usually expelled within 8-10 minutes following birth.

afterdamp. the carbon monoxide produced from methane (firedamp) explosions in coal mines.

aftershave lotion. a product, usually a soothing or bracing liquid, applied to the face after shaving. The alcohol content is high. Ingestion of about a tablespoon of such a product by a small child may cause intoxication or even death. This condition is marked by reduced blood sugar of the child who may become unconscious. The same is true of perfumes.

aftertaste. a taste that persists when the substance producing it is no longer in the mouth.

Ag. **1:** antigen. **2:** the symbol for silver.

agaric. **1:** a common name applied to mushrooms that bear gills on the undersurface of the cap.

agaricoid. resembling an agaric.

agaricoid agent. **1:** agaric acid. **2:** the dried fruiting body of *Polyporus officinalis* (Family Polyporaceae). It contains agaric acid. See also *Agaricus*.

agaricus. a common name of mushrooms of the genus *Agaricus*.

Agaricus (agaric; agaricus). a large genus of mushrooms (Family Agaricaceae) that includes both edible and poisonous species. Species which are foul-smelling should be considered poisonous. Poisoning by *Agaricus* species typically produces mild to severe symptoms of gastrointestinal distress which may appear within ½ to a few hours following ingestion. Recovery can take from 1 hour to 2 days. ***A. californicus*** (California agaricus). a poisonous mushroom usually found in urban lawns and parks in California. Symptoms are mostly those of gastric distress. It is nearly indistinguishable from *A. campestis*, the edible meadow mushroom. ***A. hondensis*** (felt-ringed agaricus). a poisonous, foul-smelling mushroom usually found scattered in woodlands from British Columbia to California. ***A. meleagris*** (Western flat-topped agaricus). foul-smelling and apparently poisonous, it grows in lawns, meadows, and along woodland paths in western North America and sometimes in the east. ***A. xanthodermus*** (yellow-foot agaricus). a poisonous species with an unpleasant odor that grows in lawns, along paths and under hedges in urban areas. Symptoms, which include vomiting and/or diarrhea, appear within 2 hours of ingestion and may persist for a day or two.

agarin. muscimol.

Agave. a genus of attractive evergreen plants (Family Agavaceae) that usually have long suc- [illegible] ginal spines. ***A. lecheguilla*** (lechuguilla). a hepatotoxic plant that also causes hepatogenic photosensitivity in grazing animals. Sheep and goats are most frequently affected; cattle are rarely affected; and horses are apparently never poisoned naturally. Grazing can cause sluggishness, progressively intense jaundice with a yellowish discharge from the eyes and nostrils, eventual coma, and death. The urine is often dark and a purplish area may appear beneath the coronary band of the hoof. Symptoms of photosensitization may or may not appear. Lesions are those of nephritis and severe toxic hepatitis. See also phylloerythrin.

age class. a category of individuals within a population of biota that fall within a specified age range (i.e., are of about the same age); a cohort.

age distribution (age structure, age composition). the number or proportion of individuals within each age class of a population.

age specific death rate. the death rate for a specified age class or cohort of a population. It is calculated as the number of individuals that die within a specified age class divided by the number of individuals that reach the age class.

age specific fecundity rate. the average number of offspring produced per unit time by individuals in a specified age class or cohort.

age specific survival rate. the number or proportion of individuals that survive to a specified age from an initial age class or cohort.

age structure (age distribution, age composition). the number or proportion of individuals within each age class of a population.

agent. means, method, instrument. See also entries for specific agents of interest. **1a:** anything that produces an effect; the means, vehicle, or instrument used to produce an effect. **1b:** an active force or substance capable of producing an effect. **1c:** that which acts upon a target. See also stimulus. **aetiologic agent**. etiologic agent. **biological agent**. an agent (e.g., chemicals, heat, electricity, mechanical devices) that is biologically active; one that causes a change in or a response by a sensitive biologic system; one that acts upon a biologic target. **damaging agent**. an agent, the predominant action of which is injurious and without immediate, direct biologic value to the affected living system. **etiologic agent**. an agent that acts alone or in conjuction with other such agents to cause disease. **nonselective agent**. nonspecific agent. **nonspecific agent** (nonselective agent). an agent that acts on numerous targets (e.g., hypoxia, cold, a pyrogen). All are selective to some degree. **oxidizing agent**. an electron recipient in a coupled oxidation-reduction reaction. **reducing agent**. an electron donor in a coupled oxidation-reduction reaction. **selective agent** (specific agent). an agent that acts on one or a few targets or on one type of target. **specific agent**. selective agent. **toxic agent**. a chemical substance or physical agent that acts directly to poison the system at concentrations or in amounts that are not normally encountered.

Agent Blue (herbicide blue). a solution of cacodylic acid (0.37 kg/l) used as a selective herbicide in the short-term control of rice and other food crops. It was employed by the United States army during the Vietnam war.

Agent Orange (herbicide orange). a defoliant herbicide formulation sprayed by the US military forces in the Vietnam war to remove forest cover and destroy the crops of the Viet Cong (1964 to 1970). It contained equal parts of 2,4-D and 2,4,5-T and was contaminated with trace amounts of TCDD (a dioxin). A large number of birth defects among infants was soon associated with exposure of Vietnamese families to Agent Orange. Subsequent to their return to the United States, American veterans who were potentially exposed to Agent Orange reported an unusually high incidence of cancer and birth defects among their children. See Agent Purple; Agent White; dioxin; TCDD; 2,4-D; 2,4,5-T.

Agent Purple (herbicide purple). a herbicide used by United States military forces to control vegetation early in the Vietnam conflict; it was replaced by Agent Orange in 1964. It is a mixture of the *n*-butyl esters of 2,4-D and 2,4,5-T and the isobutyl ester of 2,4,5-T, at respective concentrations of 0.50, 0.31, and 0.22 kg/l). See Agent Orange; Agent White; 2,4-D; 2,4,5-T.

Agent White (herbicide white). a herbicide used by the United States military forces to control vegetation in the Vietnam conflict. It is a mixture of the triisopropanolamine salt of 2,4-D and picloram, at respective concentrations of 0.24 and 0.07 kg/l. See Agent Orange; Agent Purple; 2,4-D; 2,4,5-T.

agglutinable. able to agglutinate; able to respond to an agglutinant by agglutinating.

agglutinant. **1a:** a substance that effects adhesion. **1b:** able to effect adhesion. **2:** aglutinin.

agglutinate. **1a:** to clump together as the result of an antigen-antibody reaction; used mainly in reference to cells (e.g., erythrocytes, pollen) or bacteria. **1b:** to cause to agglutinate (def. 1a).

agglutination. **1:** the clumping of cells (e.g., erythrocytes or bacteria) due to the interaction of antigens on the cell surface with certain antibodies (agglutinins). *Cf.* lectin. **2:** the clumping of solid particles coated with a thin adhesive layer or the arresting of solid particles by impact on a surface coated with adhesive material.

agglutinative. able to cause agglutination. *Cf.* agglutinant.

agglutinin. **1a:** an antibody that causes agglutination. **1b:** a substance in normal or immune serum that can cause cross-linking or attachment of specific antigens on the surface of cells or bacteria. See also snakebite, venom (snake venom). **1c:** any substance capable of agglutinating particles (e.g., a lectin). See also lectin, snakebite, venom (snake venom). **2:** abrin. **abrus agglutinin**. abrin. **plant agglutinin** (phytoagglutinin). a lectin of plant origin. See also phytohemagglutinin.

agglutinogen. an antigen that resides on the surface of particles such as erythrocytes and reacts with agglutinin to produce agglutination.

Agkistrodon (syn: *Ancistrodon*; moccasins; Asian pit vipers; Drieckskopfen; Mokassinschlangen). a genus of venomous snakes (Family Crotalidae) with 12 recognized species, three from North and Central America and nine from Asia with the range of *A. halys* extending into southeastern Europe. The head is broad and compressed, with a distinct canthus; and the head is distinct from the neck. They lack rattles and are sometimes mistaken for harmless species. The body is often stout, tapered and cylindrical, or depressed, with a short to moderately long tail. Loreal scales may be present or absent and the loreal pit is bordered anteriorly by a second supralabial. Signs and symptoms of envenomation by North American moccasins (*A. spp.*) may include pain, discoloration, swelling and edema about the bite, ecchymosis, thirst, nausea, weakness, some hemorrhage, some degree of shock, respiratory distress, and swelling of regional lymph nodes. The species most dangerous to humans are *A. acutus*, *A. piscivorus*, and *A. rhodostoma*. ***A. acutus*** (sharp-nosed pit viper; long-nosed pit viper; one-hundred-pace snake; hundred-pace snake; hundred pacer; five pacer; snake of hundred design; hyappoda). a dangerous species that reaches a length of up to 1.6 m. Found in South Vietnam, Taiwan, and China, it is the most venomous Asian pit viper. The bite is characterized by very rapid onset of edema and tissue damage, internal hemorrhage. Antivenin is available. ***A. bilineatus*** (cantil; Mexican moccasin; Mexikanische mokassinschlange). an extremely dangerous, brightly colored, chocolate brown to black aquatic pit viper with one thin white line along the canthus rostralis which continues behind the eye and another such line that runs along the superior part of the supralabials to the corner of the mouth. The colors and pattern are distinctive throughout its range. Juveniles are marked by light-edged crossbands on a lighter ground color; the venter is dark brown with white markings edged with black. Adults may reach a maximum length greater than 1.2 m. This snake is often aggressive and the bite may cause serious local lesions, although human fatalities are uncommon. Its range extends from the low regions of the Rio Grande to Nicaragua. The cantil occurs along stream banks and in swampy areas on either coast of Mexico, from Sonora and Nuevo Leon, southward to the east coast of Nicaragua and the west coast of Guatemala. ***A. contortrix*** (*A. mokasen*; American copperhead; copperhead; highland moccasin). one of the most common venomous snakes throughout its range from Massachusetts to northern Florida, west through Nebraska, Oklahoma, and west Texas. The head is triangular with a facial pit; the pupil is elliptical; the top of the head is yellowish to coppery red and paler laterally; the body is relatively stout and most subcaudal scales are undivided. The ground color is russet, orange-brown, or pinkish with dark brown or reddish crossbands; the belly is pale pinkish with large black spots or mottling. The tip of the tail is yellow in young snakes, but is dark green, brown, or black in adults. In the eastern United States it occurs most often in rocky sites in low wooded mountains, wet meadows and the edges of swamps; in the southern US it also occurs in lowland swamps; and along the Rio Grande it occurs in thick stands of cane. It is nocturnal in warm weather; diurnal in cool weather. The copperhead often basks on rocks or mudflats and often hibernates in rocky areas on ledges together with various venomous and nonvenomous snakes. Copperheads usually strike reluctantly. Although the bite is painful, it is seldom fatal to humans and antivenin is rarely needed. When the snake is angry, the tail vibrates. There are several races: *A. c. contortrix* (Southern copperhead), *A. c. laticinctus* (broad-banded copperhead), *A. c. mokeson* (Northern copperhead), *A. c. phaeogaster* (Osage copperhead), *A. c. pictigaster* (trans-pecos copperhead). ***A. halys*** (mamushi; Pallas' pit viper; Pallas' viper; Korean pit viper; Mongolian pit viper; Japanese mamushi; Korean mamushi; chi-tsun-tsze; fei-shang-ts'ao; halysotter; hami; hiraguchi; kuchibami; shchitomordnik; schitomordnik). a day-active snake that occurs in a wide variety of habitats including the grasslands and desert of the vast central Asian steppe. It occurs also in a small portion of Europe between the Volga River and Ural mountains; from there it ranges eastward to southern Siberia and Mongolia, eastern and northern China, Korea, Japan and the Bonin and Pescadores Islands. Additional habitats that often support this species include low marshy river valleys and mountains at elevations up to nearly 4000 m; it is sometimes reported in the outskirts of Tokyo and other large cities. The mamushi is the only Eurasian snake with a loreal pit, and is distinguished from other vipers within most of its range by the nine large shields on the crown and the abutment of at least one supralabial with each eye. Over much of its range, the facial pit alone is distinguishing. In most populations, the ground color is yellowish, reddish-brown, tan, or grayish with many wide, dark brown or gray crossbands that alternate with spots on the sides or with crossbands and wide, dark brown, irregularly shaped spots with black margins. The spots fuse, forming an irregular network. The belly is cream to yellow with fine black punctuations, especially toward the tail. There is a dark spot above each eye and at the nape; the tip of the tail is yellowish. Adults average 56-72 cm. and rare individuals may attain a length of nearly 90 cm. Bites are not infrequent (a few thousand per year in Japan), but are seldom serious, and rarely fatal. When agitated, the mamushi withdraws if possible, and is usually considered to be inoffensive. Human fatalities are rare. ***A. himalayanus*** (Himalayan pit viper). a species that occurs at high elevations throughout much of Nepal, Sikkim, Bhutan, Kashmir, Tibet, Sinkiang, Tsinghai, western Pakistan, and northwestern India. ***A. hypnale*** (hump-nosed viper; kunn katuva; polon thelissa). a snake of southern India and Ceylon. The head is broad, with large frontal and parietal shields and a pointed, upturned snout; the body is stout. The ground color is grayish, either heavily powdered or blotched with brown; belly yellowish or brownish with dark mottling; the tip of the tail is reddish or brownish. It occurs in dense jungle or on coffee plantations in hilly terrain. It is easily irritated, usually signaling annoyance by vibrating its tail; it readily strikes humans, but usually without serious consequence. ***A. mokasen***. *A. contortrix*. ***A. nepa*** (Ceylon hump-nosed viper). a species found only in Ceylon. ***A. piscivorus*** (cottonmouth; cottonmouth moccasin; American water moccasin; water moccasin; cottonmouth water moccasin; Wassermokassinschlange). a dangerous semiaquatic, heavy-bodied snake, up to 1.8 m long. The body is brown or olive colored with very dark crossbands that surround lighter centers; large specimens are sometimes uniformly black above. The underside is yellow, densely marbled with dark gray or black. The cottonmouth is often mistaken for nonvenomous watersnakes of the genus *Natrix*. The latter are not usually pugnaceous and, as in the case of other nonvenomous snakes within the range of *A. piscivorus*, do not have the distinctive elliptical pupil and facial pit. It inhabits swamps, shallow ponds and lakes, lagoons, and sluggish waters of the southern and central United States from Dismal Swamp in southeastern Virginia to Illinois, southward to southern Florida, and west to eastern Oklahoma and central Texas. Cottonmouths are pugnaceous and often threaten or strike when approached. The bite is rarely fatal to humans, but tissue destruction can be severe. ***A. rhodostoma*** (Malayan pit viper; oraj lemah; ular bandotan bedor; ular biludak; ular gebuk; ular kepac daun; ular tanah). a pugnaceous and dangerous crotalid of southeast Asia. It is the only Asian pit viper with large scales on the crown and smooth body scales. It occurs in forests and rubber plantations at low elevations in Thailand, northern Malaysia, Cambodia, Laos, Vietnam, Java, and Sumatra. It has a triangular head, pointed snout, and facial pits. It is a source of ancrod. Adults average about 0.7 m, a few may reach a length of 1 m or more.

agkistrodon rhodostoma venom protease. ancrod.

agkistrodon serine proteinase. ancrod.

aglucon. aglycone.

aglycone (aglucone; genin). The noncarbohydrate moiety of a glycoside molecule. It is largely or wholly responsible for the action of a toxic glycoside. Operationally, it is any nonsugar resulting from hydrolysis of a glycoside.

agmatine (4-(aminobutyl)guanidine; 1-amino-4-guanidobutane). a substance that occurs in ergot, sponges, the salivary glands of cephalopods, the pollen of *Ambrosia artemisfolia*, herring sperm, and in octopus muscle.

agnogenic. of unknown origin or etiology.

agonal. **1:** of or pertaining to death, agony, the time immediately preceding death, or to the moment of death. **2:** pertaining generally to death or agony. **3:** pertaining to terminal infection (e.g., agonal leukemia).

agonist. **1:** a person engaged in a struggle. **2:** a muscle that is automatically opposed and controlled in its action by the (appropriately graded) contraction of an opposing muscle (an antagonist). **3a:** a substance that interacts with a receptor to initiate a biological response that increases or decreases cellular activity. *Cf.* antagonist. **3b:** a xenobiotic substance whose toxic or pharmacologic activity derives from its interaction with a receptor in the same way as the endogenous ligand for the receptor. **direct-acting agonist**. one that interacts directly with the receptor. **indirect-acting agonist**. one that induces an increase in concentration of the en dogenous compound that activates the receptor, thereby increasing the rate of activation. **partial agonist**. any structural analog of an agonist that exhibits both agonist and antagonist activities. Receptor activation usually dominates at low concentrations, but receptor blockade dominates at high concentrations. Dichloroisoproteronol is, for example, a partial agonist that induces β-receptor-mediated effects similar to those of isoproterenol at low doses but blocks the interaction of isoproterenol with β-receptor sites at high doses.

agony. **1:** extreme pain, suffering, or anguish; severe distress, misery, or grief. **2:** death throes.

agranulocyte. a type of leucocyte (e.g., lymphocytes and monocytes) that contains no distinct cytoplasmic granules. *Cf.* granulocyte.

agranulocytic. of or pertaining to agranulocytosis; able to effect agranulocytosis.

agranulocytosis (granulocytopenia). **1a:** an acute, extremely low leukocyte count with pronounced neutropenia. **1b:** the condition associated with 1a, above, which is marked by high fever; necrotic lesions of the oral, rectal, and vaginal mucosae; prostration. The condition may be idiopathic in origin or induced by drugs (e.g., aminopyrine) or ionizing radiation. **2:** a system complex from such a decrease in granulocytes (def. 1). Effects include lesions of the skin and mucous membranes of the gastrointestinal tract, including the lining of the throat. Risk of rapid, massive infection is considerable. Agranulocytosis often results from leukemia or radiation or chemically induced bone marrow toxicity, and occasionally from administration of a therapeutic agent.

agricultural chemical (agrochemical). any chemical commonly used for agricultural purposes, usually to enhance crop productivity or quality. Included are fertilizers, soil conditioners, fungicides, insecticides, weed killers, and other chemicals used to increase farm crop productivity and quality. These chemicals or intermediates produced during decomposition in the environment are often toxic, sometimes dangerously so, and may accumulate in soils, water bodies, or through food chains.

agricultural ecosystem (agroecosystem). any of a variety of types of ecosystems produced through human agricultural activities. The nature of such an ecosystem depends upon agricultural practices; on the particular plants and animals that are introduced and managed; and upon geographic, physiographic, climatological, and other factors that also affect natural ecosystems. Agricultural chemicals (e.g., fertilizers, pesticides) are employed to sustain productivity and biotic composition of the system or to improve or maintain the commercial value of the system and its products. Many of these chemicals are toxic to humans and other living organisms. Such may be exposed, for example, due to resuspension of chemicals in air or surface waters; by percolation through soil into groundwater; and/or by transport through food chains. Individuals living outside the ecosystem may thus also be exposed.

Agrimet. phorate.

Agrimonia eupatoria (meadow grass). an herb (Family Rosaceae), it causes a phototoxic contact dermatitis (meadow grass dermatitis). Sunbathers are most often affected.

agrochemical. jargon for agricultural chemical.

agroecosystem. jargon for agricultural ecosystem.

Agropyron (wheat grasses, quack grass). a genus of wild North American grasses (Family gramineae) that may become heavily ergotized and therefore hazardous. See *Claviceps*.

Agrostemma (corn cockle). a genus of tall, silky, annual or biennial, toxic European herbs (Family Caryophyllaceae) that contain saponins. ***A. githago*** (corn cockle; purple cockle). a common noxious weed of grainfields (especially corn and winter wheat) and sometimes roadsides. It is sometimes cultivated as an ornamental annual. The flower petals are purplish red with black spots. The whole plant is toxic due to the presence of githagin and saponin glycosides. The seeds are especially toxic and are sometimes unintentionally harvested with winter wheat or rye; unless carefully separated from the grain, intoxication of livestock may occur, sometimes with deadly effects, Horses, cattle, and poultry have been poisoned, but swine rarely suffer acute poisoning because they expel the material by vomiting. Milled or ground seeds are moderately toxic to livestock and to humans who have been poisoned by using flour made from contaminated grain. Symptoms usually appear within ½ to 1 hour following ingestion and may include rawness of the throat, nausea, acute gastroenteritis with severe stomach pains, dizziness, fever, weakness, delirium, and depression of respiratory rate, sharp pains along the spine, coma, and death from respiratory arrest.

Agrostis alba (redtop). a wild North American grass (Family Gramineae) that may become heavily ergotized and therefore hazardous. See *Claviceps*.

aguason. *Naja naja*.

Ägyptische Sandrasselotter. *Echis carinatus*.

***Ah* locus**. a gene locus that regulates the induction of various enzymes, such as the monooxygenases, by aromatic hydrocarbons. Variations in this locus among strains of laboratory mice may account for differing sensitivities to certain toxicants such as TCDD. See receptor (Ah receptor).

ailment. illness (def. 2); a condition of mild illness, physical disorder, or chronic disease. This term is roughly synonymous with disease, *q.v.*

aipomiro. *Bitis arietans*.

Aipysurus (brown sea snakes). a genus of moderately large to large sea snakes (Family Hydrophidae) that have a spotty occurrence in the region from the Gulf of Siam to the Coral Sea; but most are restricted to Australian and New Guinean waters. They are characterized by a smoothly rounded snout; anterior fangs are followed by several small maxillary teeth. The nostrils are dorsal in position and the crown shields are small. The ventrals are well developed, extending at least one-third the width of the body. While little information is available on these snakes, all, especially the larger species, should be considered dangerous. ***A. apraefrontalis***. a species found in the Timor sea. ***A. duboisii*** (Dubois sea snake). a species found in the Timor sea. ***A. eydouxi*** (spine-tailed sea snake). a species found in the China Sea, the Gulf of Siam (Gulf of Thailand), the Timor Sea, the Java and Flores seas, the Arafura Sea, and the Gulf of Carpentaria; specimens are also known from the Banda Sea and the Celebes Sea. ***A. foliosquama***. a species known from the Timor Sea. ***A. fuscus***. a species found chiefly in the Timor Sea; known also from the Java, Flores, and Banda seas, and possibly from the Arafura Sea. ***A. laevis*** (olive-brown sea snake). a stout, very heavy snake (Family Hydrophidae) found in the Timor Sea, the Arafura Sea, the Gulf of Carpentaria, and the Coral Sea. The body is somewhat compressed. Adults are uniformly olive brown, sometimes with a row of dark spots on the flanks and belly. The head is large, somewhat wider than the neck; the nostrils are dorsal; the crown shields are small; the tip of the tail is typically ragged; ventral scales are well developed. Adults may reach a length of nearly 2 m. Although sometimes found stranded on beaches, this species does not normally leave the water; it is awkward on land. See also *Astrotia stokesii*.

air. the mixture of gases and their proportions in the earth's atmosphere. Dry air today contains by volume, about 78% nitrogen, 20.95% oxygen, 0.93% argon, and 0.03% carbon dioxide, plus amounts of many other components. Living organisms and all ecosystems are strongly influenced by atmospheric properties, processes, and pollutants. But air is an exceedingly variable, dynamic, and complex environmental compartment that offers difficulties for monitoring and evaluation. The amount of water vapor is highly variable, depending on other atmospheric conditions. The composition of air has changed over the course of the earth's history and continues to change, especially with respect to CO_2 concentration. Since the industrial revolution, the air environment, indoors and outdoors, has become a major transporter and source of exposure to toxic substances.

air changes per hour (ACH). a representation of the movement of a volume of air in a specified period of time. Thus, a building has one air change per hour if all of the air in the building is replaced in each hour.

air pollution. See pollution.

air pollution index. an arbitrary index that is a function of the concentration of one or more pollutants that are taken as a measure of pollution intensity. This index has, for example, been used in the United States: 10 times the SO_2 concentration plus the CO concentration (both in ppm by volume) plus twice the coefficient of haze. It was considered to be a cause for alarm with reference to human health when this index rose from its average value of about 12 to 50 or more. Several such indices are used locally. See air quality index. See also pollution.

air quality index. a standardized indication of pollution intensity formulated by the United States Environmental Protection Agency (EPA) for public use. The index is based on a set of numbers based on a combination of the environmental levels of ozone, suspended particulates, sulfur dioxide, and carbon dioxide. In this index, 0 indicates no pollution; 100 is standard; 200, alert; 300, warning; 400, emergency; 500, significant harm can be expected. See air pollution index.

air quality standard. the pollutant concentration of a particular air pollutant that may not be exceeded during a specified time in a defined geographic area as prescribed by governmental regulations.

air toxics. See toxics.

airborne release. the release of any chemical into the air.

airway. **1:** a tubular passage through which air passes to the respiratory tissues of terrestrial animals. **2:** a passage for air in a mine.

Aizen Food Blue. FD&C Blue No. 1.

ajacine(4-[[[2-acetylamino(benzoyl]oxy]methyl]-20-ethyl-1,6,14,16-tetramethoxyaconitane-7,8-diol; *N*-acetylanthranilic acid ester of lycoctonine). a toxic alkaloid, $C_{34}H_{48}N_2O_9$ isolated from the seeds of *Delphinium ajacis* (Family Ranunculaceae).

ajmaline. an amber, crystalline indole alkaloid found in plants of the genus *Rauwolfia*, especially *R. serpentina*.

ajowan oil (ptychotis oil). an oil distilled from [illegible] of *Carum copticum* (*Ptychotis ajowan*) that contains dipentene, *p*-cymene, α-pinene, τ-terpinene, and thymol. The oil is toxic to earthworms; alcoholic extracts of the oil are toxic to *Escherichia coli* and staphylococci.

Ajuga remota (bugle weed). a European herba ceous plant (Family Laminaceae), the leaves of which contain ajugarins.

ajugarin. any of a class of diterpenes isolated from the leaves of *Ajuga remota* and related plants. Ajugarins I, II, and III are effective antifeedants against the African army worm, a moth larva (Family Noctuidae) that feeds on alfalfa and various grains, sometimes in huge numbers. Ajugarin IV is insecticidal, and ajugarin V appears to be biologically inactive.

akathisia (acathisia; kathisophobia). a condition of fitfulness caused by phenothiazines in which one may be unable to sleep or even to sit or lie quietly. See *Acacia* (*A. berlandieri*), extra-pyramidal side effects.

akee, Akee. **1:** *Blighia sapida*. **2:** the native Jamaican name for the fruit of *B. sapida*.

Aki. *Blighia sapida*.

akinesis. the absence or cessation of movement.

akipom. *Bitis arietans*.

Akocanthera. *Acocanthera*.

Al. the symbol for aluminum.

alachlor (2-chloro-*N*-(2,6-diethylphenyl)-*N*-(methoxymethyl)acetamide; 2-chloro-2′,6′-diethyl-*N*-(methoxymethyl)acetanilide; metachlor). a preemergence selective anilide herbicide used chiefly to control weeds in corn and soybean fields. It is absorbed through the roots and inhibits root elongation. Alachlor persists in the soil for up to 3 months. It is a common groundwater pollutant in some agricultural areas. Propachlor, diphenamid, and naptalam are herbicides with similar action. Alachlor is moderately toxic to laboratory rats and is carcinogenic to laboratory mice and rats. It is readily absorbed through the skin of humans and may produce allergic skin reactions, eye irritation, and liver damage. Some effects associated with alachlor may be largely due to contamination by polychlorinated dib[illegible]

ALAD. porphobilinogen synthetase (aminolevulinic acid dehydratase).

alanine nitrogen mustard. melphalan.

ALARA. as low as reasonably achievable.

alarm reaction. the sum of all nonspecific responses to the sudden exposure to stimuli or to a biologically active agent to which the organism is not adapted.

alarm stimulus. a stressor; a stimulus that causes an alarm reaction.

Alaskan butter clam. *Saxidomus giganteus*.

Albizia anthelmintica. a small tropical and subtropical deciduous tree (Family Fabaceae, formerly Leguminosae, the bark of which is used as an anthelmintic and purgative, but a small overdose is said to be lethal. It occurs in southeast Africa and Ethiopia.

Albula vulpes (ladyfish; bonefish). a silvery, bottom-feeding marine bony fish (Family Albulidae) that occurs in all tropical, subtropical, and warm temperate seas. Certain populations are transvectors of ciguatera.

Albulidae (ladyfishes, bonefishes). a family of tropical, silvery, marine bony fishes (Class Osteichthyes), with a ventral mouth. They are typically bottom-feeders in shallow inlets. See *Albula vulpes*, ciguatera.

albumen (egg white; dried egg white). commercially prepared egg white. It comprises about 60% of the fresh egg of domestic fowl. Albumen has numerous uses and is an antidote for mercury poisoning. The pure protein is "albumin," *q.v.*

albumin. **1:** any of a class of proteins that occur in nearly all animal tissues and in many plants. Albumins coagulate on heating and dissolve in dilute salt solutions. The best characterized and most interesting toxicologically are the serum albumins of mammalian blood. **2:** the pure protein of egg white (albumin, ovalbumin). See also albumen. **serum albumin**. any of the largest class of blood proteins, they are importantly involved in the transport of many (nonlipophilic) toxicants. The toxicants are bound to specific sites on the protein by hydrophilic bonds and are thus temporarily inactive. *Cf.* albumen.

albuminuretic. referring to or causing albuminuria.

albuminuria. the presence of easily detected amounts of serum protein (especially serum albumin) in urine. It may be due to poisonous substances and the presence of albumin may indicate renal impairment. **renal albuminuria**. albuminuria due to increased permeability of renal epithelial cells to proteins in blood. It occurs in all types of nephritis. **toxic albuminuria**. that caused by xenobiotic substances or by toxins generated within the body.

alcohol. **1:** any of a class of oxygenated organic compounds that have one or more functional hydroxyl groups bonded to a hydrocarbon chain but not to carbon atoms of an aromatic ring. Phenols are thus excluded from the family of alcohols even though a few phenols (e.g., phenyl methanol have the properties of an alcohol. Most alcohols are toxic. See alcohol poisoning. **2:** a vernacular term for beverages that contain ethanol (e.g., beers, wines, and spirits). **3:** a vernacular term for ethanol, *q.v.* **absolute alcohol**. a liquid that contains 99% ethanol and not more than 1% water. **aliphatic alcohol** (alkyl alcohol). an alcohol in which the hydroxyl groups (1 or more) are attached to a carbon atom of a branched or straight-chain aliphatic hydrocarbon. They are classified as primary, secondary, and tertiary based on the location of the substituted carbon atom in the hydrocarbon chain (thus there is primary, secondary, and tertiary butanol). Aliphatic alcohols are widely used as industrial solvents. Ethanol is toxic, but used extensively in beverages; the other aliphatic alcohols are more toxic than ethanol. **alkyl alcohol**. aliphatic alcohol. **allyl alcohol** (2-propene-1-ol; 1-propenol-3; vinyl carbinol; olefinic alcohol). a colorless, toxic, strongly irritant liquid, $CH_2{=}CHCH_2OH$, with a pungent, mustard like odor. It is used in the manufacture of war gas, resins, and plasticizers. It is strongly irritating to eyes, mouth, and lungs and very toxic to laboratory rats. **amino alcohol**. alkamine. **α-(aminomethyl)-3,4-dihydroxybenzyl alcohol**. norepinephrine. **α-(aminomethyl)-*p*-hydroxybenzyl alcohol**. octopamine. **amyl alcohol**. **1:** amyl hydrate; a compound, $C_5H_{11}OH$, with eight possible isomers (discounting optically active isomers); six are commercially available (*n*-amyl alcohol, primary; isoamyl alcohol, primary; 2-methyl-1-butanol; 2-pentanol; 3-pentanol; *tert*-pentyl alcohol), all of which are toxic. In addition to the pure isomers, definite mixtures are sold under a variety of names, some of which are identical to that of a pure isomer. **2:** *n*-amyl alcohol, primary. ***dl-sec*-amyl alcohol**. 2-butanol. **amyl alcohol, commercial**. a highly flammable aromatic liquid used in low concentrations as a flavoring agent in foods and liquor. It is an irritant of the eyes and upper respiratory tract. Inhalation of the vapor causes violent coughing, and ingestion of even small amounts can cause headache and vertigo. Ingestion of large amounts can be fatal. **amyl alcohol normal**. *n*-amyl alcohol, primary. **amyl alcohol, primary**. **1:** *n*-amyl alcohol, pri-mary. **2:** a flammable, commercial mixture of primary amyl alcohols that is sold under this name. It contains 60% primary amyl alcohol, 35% 2-methyl-1-butanol, and 5% 3-methyl-1-butanol. ***n*-amyl alcohol, primary** (amyl alcohol normal; amyl alcohol, primary; 1-pentanol; *n*-butyl carbinol; *n*-pentanol; pentyl alcohol; primary amyl alcohol). a commercially available straight-chain primary alcohol and isomer of amyl alcohol. It is a colorless liquid with a mild odor and is moderately toxic by ingestion and skin contact, a moderate fire hazard and explosive in air if exposed to flames. It is a severe eye, skin, and upper respiratory tract irritant. Symptoms following ingestion of a toxic dose may include headache, nausea, vomiting, delerium, and methemoglobinemia. See amyl alcohol. ***sec*-amyl alcohol**. 2-pentanol. ***sec*-amyl alcohol, active**. 2-pentanol. **amyl alcohol, primary active**. 2-methyl-1-butanol. ***sec-n*-amyl alcohol**. 2-pentanol, 3-pentanol. **anhydrous alcohol**. absolute alcohol. **anise alcohol**(4-methoxybenzenemethanol;methanol; *p*-methoxybenzyl alcohol; anisyl alcohol). an extremely toxic organic liquid that is nearly insoluble in water, but freely soluble in ethanol and ethyl ether. **anisyl alcohol**. anise alcohol. **aromatic alcohol** (aryl alcohol). any alcohol in which the hydroxyl group is a substituent of an alkyl side chain of an aromatic compound. **aryl alcohol**. an aromatic alcohol. **benzal alcohol**. benzyl alcohol. **benzoyl alcohol**. benzyl alcohol. **benzyl alcohol** (benzene-methanol; benzene carbinol; phenylcarbinol; phenylmethanol; α-hydroxytoluene; α-hydroxytoluene; benzal alcohol; benzoyl alcohol; phenol carbinol; phenylmethyl alcohol; α-toluenol). a flammable, white, liquid hydrocarbon, $C_6H_5CH_2OH$, with a slightly sweetish odor, and a sharp, burning taste. It is used as a solvent in perfumes, as a local anesthetic, and as a synthetic fruit flavoring for beverages, sweets, and chewing gum. Benzyl alcohol occurs in *Jasminum* (jasmine), *Hyacinthus* (hyacinth), ilangilang oils, Peru and Tolubalsams, and many other plant-derived oils in which it may also occur as the ester. It is moderately toxic to laboratory mice when ingested, causing respiratory and muscular paralysis, convulsions, and narcosis. It is moderately irritant and corrosive to the skin and severely irritant to the eyes and mucous membranes of humans. Contact may cause inflammation and blistering of the skin and mucous membranes. Ingestion of excessive amounts can cause a burning sensation in the mouth and stomach, vomiting, diarrhea, headache, confusion, CNS depression, and coma. It can cause adverse reactions in persons sensitive to petrochemical derivatives. *Cf.* sodium benzoate. **butyl alcohol**. 1-butanol. **butyl alcohol, secondary**. 2-butanol. **1-butyl alcohol**. 1-butanol. **2-butyl alcohol**. 2-butanol. ***n*-butyl alcohol**. 1-butanol. ***sec*-butyl alcohol**. 2-butanol. ***tert*-butyl alcohol**. *tert*-butanol. ***n*-butyl carbinol**. *n*-amyl alcohol, primary. **butyric alcohol**. 1-butanol. **caprylic alcohol**. 1-octanol. **caprylic alcohol, secondary**. 2-octanol. **2-chloroethyl alcohol**. ethylene chlorohydrin. **crotonyl alcohol**. crotyl alcohol. **crotyl alcohol** (2-buten-1-ol; crotonyl alcohol). an extremely toxic, moderately water-soluble liquid, $CH_3CH{=}CHCH_2OH$, that is miscible with ethanol. It is produced by the reduction of crotonaldehyde. **dehydrated alcohol**. absolute alcohol. **denatured alcohol**. alcohol (ethanol) rendered unfit for consumption by the addition of toxic substances. It is used primarily as a solvent. **dihydric alcohol**. dihydroxy alcohol. **dihydroxy alcohol**. an alcohol that contains two functional hydroxyl groups. **3,4-dihydroxy-α-[(methylamino)methyl]benzyl alcohol**. epinephrine. **diluted alcohol** (diluted ethanol). USP alcohol that contains not less than 41% nor more than 42% by weight of ethanol. It is used as a solvent. **ethyl alcohol**. ethanol. **ethyl dimethyl carbinol**. *tert*-pentyl alcohol. **fermentation butyl alcohol**. isobutyl alcohol. **grain alcohol**. ethanol. ***n*-hexyl alcohol**. 1-hexanol. **isoamyl alcohol, primary** (3-methyl-1-butanol; isopentyl alcohol; isobutyl carbinol; 2-methyl-4-butanol). a moderately toxic, commercially available, branched-chain primary alcohol and isomer of amyl alcohol. It is a colorless liquid, $(CH_3)_2CHCH_2CH_2OH$, with a pungent taste and a disagreeable odor. It is poisonous by all routes of exposure; the vapor is toxic and irritant. It is a fire and explosion hazard; when heated to decomposi tion, it releases acrid smoke and fumes. ***sec*-isoamyl alcohol**. a branched chain secondary alcohol and isomer of amyl alcohol; it is not commercially available. **isobutyl alcohol** (2-methyl-1-propanol; isopropylcarbinol; 1-hydroxymethylpropane; isobutanol; fermentation butyl alcohol). a colorless, flammable, moderately toxic liquid. It is a mild irritant and a narcotic at high concentrations. **isopentyl alcohol**. isoamyl alcohol, primary. **isopropyl alcohol** (IPA; dimethylcarbinol; *sec*-propyl alcohol; isopropanol; 2-propanol). a highly toxic, colorless liquid, $(CH_3)_2CHOH$, with a pleasant odor; it is an isomer of propyl alcohol, and is more toxic than ethanol. It is used in high volume industrially. Isopropyl alcohol is a CNS depressant with effects when ingested that are similar to those of ethanol intoxication: the appearance of being very drunk, but with more extreme effects. Even when used as rubbing alcohol to sponge parts of the body, there is some danger of coma. *Cf.* isopropanol. See also isopropyl rubbing alcohol, isopropyl alcohol poisoning. **isopropyl rubbing alcohol**. an aqueous solution, for external use only, that contains 68 to 72%, by volume, isopropyl alcohol. It is flammable, toxic by ingestion or inhalation, and is used as a rubefacient; it is suitable only for external use. *Cf.* rubbing alcohol. ***p*-methoxy-benzyl alcohol**. anise alcohol. **methyl alcohol**. methanol. **α-[1-(methylamino)ethyl]benzyl alcohol**. ephedrine. **1-methylbutyl alcohol**. 2-pentanol. **2-methylpropyl alcohol**. isobutanol. **monohydric alcohol**. monohydroxy alcohol. **monohydroxy alcohol**. an alcohol that contains only one functional hydroxyl group. ***n*-octyl alcohol, primary**. 1-*n*-octanol. **olefinic alcohol**. **1:** an alcohol in which the hydroxyl group is attached to an unsaturated hydrocarbon skeleton. **2:** allyl alcohol. ***tert*-pentyl alcohol** (*tert*-amyl alcohol; 2-methyl-2-butanol; dimethyl ethyl carbinol; ethyl dimethyl carbinol; *tert*-pentanol; amylene hydrate). a commercially available branched-chained tertiary alcohol and isomer of amyl alcohol. It is a volatile, moderately toxic liquid, $(CH_3)_2C(OH)CH_2CH_3$, with a characteristic odor. It has a burning taste, irritates mucous membranes, and has a narcotic action at high concentrations. It is also a serious fire hazard. **pentyl alcohol**. *n*-amyl alcohol, primary. ***sec*-pentyl alcohol**. 2-pentanol. **phenyl alcohol**. phenol (def. 1). **phenylmethyl alcohol**. benzyl alcohol. **polyhydric alcohol**. polyhydroxy alcohol. **polyhydroxy alcohol**. any alcohol that contains 2 or more functional hydroxyl groups. Such alcohols are widely utilized as heat exchangers, in hydraulic fluids and in antifreeze. They are toxic to laboratory animals and humans. **primary alcohol**. an alcohol in which the carbon atom that bears the hydroxyl group is also attached to two hydrogen atoms. **primary amyl alcohol**. *n*-amyl alcohol, primary. **propyl alcohol, secondary**. 2-propanol. ***n*-propyl alcohol**. 1-propanol. ***sec*-propyl alcohol**. isopropyl alcohol. **propylic alcohol**. 1-propanol. **rubbing alcohol**. a 70% aqueous solution of ethanol by volume plus denaturants, or an aqueous solution of isopropyl alcohol (isopropyl rubbing alcohol). It is flammable, toxic, and suitable only for external use. **secondary alcohol**. an alcohol in which the carbon atom that bears the hydroxyl group is bonded also to one hydrogen atom and two other groups. **secondary propyl alcohol**. 2-propanol. **sugar alcohol**. an alcohol produced by reducing an aldehyde or ketone of a sugar. ***tert*-amyl alcohol**. *tert*-pentyl alcohol. **tertiary alcohol**. an alcohol in which the carbon atom that bears the hydroxyl group is bonded to three other groups, none of which is hydrogen. **thiobutyl alcohol**. *n*-butyl mercaptan. **trihydric alcohol**. trihydroxy alcohol. **trihydroxy alcohol**. an alcohol that contain three functional hydroxyl groups. **wood alcohol**. methanol.

alcohol amnestic syndrome. an amnestic syndrome due to chronic alcoholism and vitamin deficiency. It may be accompanied by neurological complications (e.g., peripheral neuropathy, cerebellar ataxia) and malnutrition. *Cf.* Korsakoff's syndrome.

alcohol C-8. 1-*n*-octanol.

alcohol dehydrogenase (ADH). any of several isomeric enzymes that occur, for example, in yeast and in the soluble fraction of the liver, kidney, and lung. They catalyze the conversion of ketones or aldehydes to alcohol as in alcoholic fermentation.

$$RCH_2OH + NAD^+ \leftrightarrows RCHO + NADH + H^+$$

The reverse reaction may be regarded as an activation reaction if the product is an aldehyde. Aldehydes are toxic, lipophilic, and may remain in the body for some time. Detoxication is usually by oxidation of the aldehyde.

alcohol inky. *Coprinus atramentarius*.

alcohol poisoning. poisoning, acute or chronic, caused by drinking ethyl alcohol, isopropal alcohol, methyl alcohol, or beverages that contain any of these alcohols. Of these substances, methyl alcohol is most toxic and ethyl alcohol is least toxic. See alcoholism.

alcohol-related birth defect (ARBD). any birth defect in the newborn due to the use of alcohol by the mother during pregnancy. See fetal alcohol syndrome.

alcohol sensitization. See sensitization.

alcohol syndrome. See fetal alcohol syndrome.

alcohol withdrawal syndrome. the set of symptoms observed in an alcoholic who has suddenly stopped using alcohol. These may include tremor, delirium tremens, various neurologic effects, and seizures.

alcoholic. **1:** referring to or denoting alcohol, usually ethanol. **2:** a person afflicted by alcoholism. **acute alcoholic**. a person suffering from acute alcoholism. **chronic alcoholic**. a person suffering from chronic alcoholism.

alcoholic beverage. any beverage that contains ethanol. Such beverages are addictive in some people. They are teratogenic and consumption by pregnant women can produce fetal alcohol syndrome, *q.v.*

alcoholic cirrhosis. See cirrhosis.

alcoholic hallucinosis. hallucinosis.

alcoholic neuritis. alcoholic neuropathy (See neuropathy).

alcoholic-nutritional cerebellar degeneration. a condition of poorly nourished alcoholics typified by a sudden loss of the ability to bring the legs together. [illegible] remain far apart.

alcoholic psychosis. See psychosis.

alcoholic tremor. See tremor.

alcoholism. **1:** a progressive and sometimes fatal disease caused by excessive consumption of alcohol (usually in alcoholic beverages). It is characterized by physiologic addiction and tolerance to alcoholic intoxication and by characteristic pathologies that are the direct or indirect result of alcohol consumption. Ethanol is responsible for more deaths than any other chemical. The deaths most commonly result from the effects of chronic alcoholism, *q.v.*, vehicular fatalities and alcohol-related homicides. **2:** addiction to alcoholic beverages. **acute alcoholism** (acute alcoholic intoxication). a condition due to excessive consumption (usually over a short period of time) of beverages with a high concentration of ethanol. It is characterized by symptoms of central nervous system depression, lowered inhibitions, slowed reaction times, muscular incoordination, and mental disturbances; the victim may become stuporous, or even comatose. Effects may be transient or may lead to death, sometimes rapidly. Certain other conditions such as insulin shock and certain cases of intracranial disease mimic acute alcoholism. See alcoholic psychosis. **chronic alcoholism**. a condition that results from long-term habitual consumption of alcohol in toxic amounts. It is marked by malnutrition, pharmacodynamic tolerance, vitamin deficiency, gastritis, alcoholic cirrhosis of the liver, and neuropathies which may include tremulousness, hallucinosis, delirium tremens, coma, and death. Hepatic cirrhosis alone may cause death.

alcoholomania. an abnormal craving for alcoholic beverages. *Cf.* alcoholophilia.

alcoholophilia. a morbid craving for alcoholic beverages. *Cf.* alcoholomania.

alcoholuria. the presence of ethanol in urine.

Alcyonacea (soft corals). an order of anthozoans (Subclass Alcyonaria), the soft corals, certain species of which are extremely poisonous. The polyps are characterized by an endoskeleton [illegible] of separate calcareous spicules, but form a gelatinous mass from which only the oral ends of the polyps protrude. See *Palythoa*, palytoxin.

alcyonacian. a member of the order Alcyonacea of coelenterate subclass Alcyonaria (Class Anthozoa). See *Palythoa*.

Alcyonaria. one of two subclasses (Class Anthozoa) of coelenterates (formerly Octocorallia), the individuals of which have an endoskeleton, eight complete septa, one siphonoglyph, and bear eight pinnately branched tentacles; most are colonial. Included are the soft corals which secrete a gelatinous matrix about the polyp and the horny corals which have a horny skeleton. Some soft corals (Subclass Alcyonaria) are extremely poisonous (see *Palythoa*, palytoxin). See also Anthozoa.

alcyonarian. any member of the subclass Alcyonaria, phylum Coelenterata.

ALD. approximate lethal dose. See dose.

aldehyde. 1: any partially oxidized hydrocarbon compound that contains a terminal carbonyl group, C=O. They have the general formula RHC=O, in which R is a hydrocarbon group that may consist of saturated or unsaturated straight chains, branched chains, or rings. Many aldehydes are poisonous and most are intense irritants. Because they readily dissolve in water, the lower aldehydes are damaging to the eyes and mucous membranes, especially of the upper respiratory tract. The higher aldehydes are less water soluble and so tend to penetrate deeper into the respiratory tract and may seriously damage the lungs. **2:** acetaldehyde. *Cf.* ketone.

aldehyde dehydrogenase. an enzyme that catalyzes the formation of acids from aldehydes; the acids are then accessible as substrates of phase II conjugating enzymes.

aldehyde oxidase. a flavoprotein in the soluble fraction of hepatic cells. It is very similar to xanthine oxidase and acts chiefly in the oxidation of endogenous aldehydes. *Cf.* aldehyde dehydrogenase, xanthine oxidase.

aldehyde reduction. See reduction reaction.

alder buckthorn. *Rhamnus frangula*.

aldicarb (2-methyl-2-(methylthio)propanal *O*-[(methylamino)carbonyl]oxime; 2-methyl-2-(methylthio)propionaldehyde *O*-(methylcarbamoyl)oxime; Temik). a colorless, crystalline carbamate, $CH_3SC(CH_3)_2CH=NOCONHCH_3$, m.p. 100°C. It is a highly dangerous, highly toxic (anticholinesterase) systemic carbamate insecticide, acaricide, and nematocide, synthesized from ethyl isocyanate and sold under the name Temik®. It is used also in some household products to eradicate pests. See also carbamate poisoning.

aldosterone (electrocortin). a corticosteroid hormone (a mineralocorticoid), $C_{21}H_{28}O_5$, secreted by the *zona glomerulosa* of the outer mammalian adrenal cortex and by the adrenal tissue of many other vertebrates. It is a potent regulator of electrolyte (sodium and potassium) balance. Excessive secretion produces hypertension due to salt retention (Conn's syndrome in humans). Aldosterone secretion is normally regulated by angiotensin, but can also be stimulated by ACTH, lowered blood volume, or by insufficiency of sodium in the diet.

aldrin (1,2,3,4,10,10-hexachloro-1,4,4a,5,8,8a-hexahydro-1,4:5,8-dimethanonaphthalene; HHDN; compound 118). the common name of a colorless, water-insoluble, crystalline chlorinated diene insecticide, $C_{12}H_8Cl_6$. It is a derivative of chlorinated naphthalene and is closely related chemically and toxicologically to dieldrin. Its manufacture and use are now banned in the United States. It is highly toxic to insects and to most vertebrate animals, including humans, and is carcinogenic in some laboratory animals. Intoxication of most vertebrate animals can result from absorption via skin, respiratory organs, or by ingestion. In humans, aldrin causes headaches, dizziness, nausea, and vomiting. Acute exposures can also cause excessive sweating, ataxia, tremors, convulsions, and kidney damage, and subsequent CNS depression, respiratory collapse, and death. Prolonged chronic exposure can cause liver damage. See also DDT.

Aldrin®. aldrin.

alert level. a concentration of gaseous pollutants that has been defined by a competent authority as indicating an approaching, a potential, or an actual hazard to human health.

alertness. awareness or attentiveness to aspects of the environment or to changes in one's surroundings. Drowsiness, stupor, torpor, or even unconsciousness are states of limited alertness.

aleukia. the absence of leukocytes in circulating blood. *Cf.* leukopenia, aleukocytic, aleukocytosis. alimentary toxic aleukia (ATA). A mycotoxicosis of livestock associated with the ingestion of old grain infested with fungi of the genera *Actinomyces, Alternaria, Cladosporium, Fusarium, Mucor, Penicillium, Piptocephalis, Rhizopus, Trichoderma, Trichothecium,* and *Verticillium*. The toxicants responsible for the condition appear to be trichothecenes such as acetyl T-2 toxin and monoacetylnivalenol (fusarenon-X).

aleukocytic. having no leukocytes.

aleukocytosis. a low proportion of leukocytes in the circulating blood.

Aleurites (tung tree; tung oil tree; tung nut). a genus of small trees (Family Euphorbaceae) indigenous to China and Japan but widely cultivated in the United States, especially in states that border the Gulf of Mexico. Leaves and seeds contain toxic saponins and a phytotoxin (sapotoxin). The toxicity of foliage, ranked in descending order, is *A. fordii*, *A. montana*, *A. moluccana*, and *A. trisperma*. *A. trisperma* is about half as toxic as *A. fordii* to domestic fowl. Ingestion of foliage or untreated tung meal by cattle causes hemorrhagic inflammation with necrotic lesions of the gastrointestinal tract and consequent hemorrhagic diarrhea. This may be accompanied by anorexia, listlessness, and emaciation if exposure is prolonged. A diet containing as little as 3.5 g/kg of fresh macerated leaves can produce these symptoms. Symptoms in humans appear in less than 1 hour following ingestion of as little as a single nut and include nausea, severe vomiting, abdominal cramps, weakness, and exhaustion. In more acute cases, shock, dehydration, cyanosis, diminished reflexes, and respiratory depression may also be observed and death may result. Gastrointestinal lesions in humans are essentially the same as those of poisoned livestock. ***A. fordii*** (tung nut). ingestion of seeds causes nausea, vomiting, abdominal pain, weakness, hypotension, and shallow respiration. [illegible] in sapotoxin.

alexeteric. protective or effective against poison or infection.

alexic. **1:** protective, as an alexin. **2:** pertaining to alexia (loss of ability to read; word blindness).

alexin. nonspecific bactericidal substances (as distinct from antibodies) that are present in blood plasma and serum.

alexipharmac. alexipharmic. *Cf.* alexiteric.

alexipharmic, alexipharmical (alexipharmac). **1:** antidotal; effective against or counteracting the harmful effects of a poison. **2:** an antidote.

alexiteric. protective against poison, venom, or infection. *Cf.* alexic, alexipharmic.

alfalfa. *Medicago sativa*.

alfombrilla. *Drymaria arenarioides*.

ALG. antilymphocyte globulin. See antilymphocyte serum.

alga (plural, algae). a vernacular term for individual organisms of three phyla of the Kingdom Plantae: Phaeophyta (brown algae), Rhodophyta (red algae), and Chlorophyta (green algae). The term formerly included the blue-green bacteria (Cyanobacteria). Algae are primitive eukaryotic, often microscopic, photosynthetic, nonflowering plants. Included are unicellular, multicellular and colonial organisms. They primarily inhabit natural bodies of water and sometimes soil. In large numbers, they can adversely affect water quality by reducing or even exhausting the amount of dissolved oxygen in the water. Some are poisonous. See bloom, *Gonyaulax catenella*, algal poisoning, red tides, shellfish poisoning, Phaeophyta.

algae. plural form of alga.

algal. of, referring to, resembling, or caused by algae.

algal bloom. See bloom.

algal poisoning. poisoning of wildlife, livestock, pets, or humans who drink the water from water bodies that have algal blooms. The onset of symptoms is usually rapid (15-45 minutes) following ingestion of a toxic dose. The course of poisoning is rapid, often culminating in death within 24 hours. Symptoms are usually progressive and include nausea, vomiting, abdominal pain, diarrhea, prostration, muscle tremors, dyspnea, cyanosis, paralysis, and convulsions. In cases that do not progress rapidly, bloody stools, jaundice, and photosensitization are common findings. See also bloom, alga, fast death factor, and *Microcystis aeruginosa*.

algelasine. a toxin isolated from *Algelas dispar*.

algesia (algesthesia). unusual sensitivity or hypersensitivity to pain. *Cf.* hyperesthesia.

algesic (algetic). painful.

algesthesia. algesia.

algetic. algesic.

-algia. a suffix denoting pain.

algicide. a substance (e.g., copper sulfate) used to destroy algae.

algogenesia (algogenesis). the induction of pain.

algogenesis. algogenesia.

algogenic. **1:** causing or able to cause pain. **2:** lowering or able to lower body temperature, especially below normal.

algorithm. a finite series of logical steps used to solve a specific type of problem.

algos. pain.

algospasm. a painful spasm or cramp.

alicyclic (cycloaliphatic; cycloalkane). **1:** cycloalkane; having the structure and properties of both aliphatic and cyclic compounds. **2:** of or pertaining to a class of saturated hydrocarbon compounds that combine the properties of straight- or branch-chain aliphatic and cyclic nonaromatic hydrocarbons. **3:** having the structure and properties of both aliphatic and cyclic hydrocarbons. See also hydrocarbon. **4:** an alicyclic hydrocarbon.

alicyclic compound. alicyclic hydrocarbon. See hydrocarbon.

alimentary canal. the tubular portion of an animal's digestive tract through which foods and other ingested substances pass. There are usually two orifices, a mouth through which food enters and an anus through which wastes are egested. The Platyhelminthes and Coelenterata have a single opening. The vertebrate canal usually includes a mouth, pharynx, esophagus, stomach, and intestines. Food is propelled through the canal either by cilia or muscular contractions from mouth to anus. Reverse movements may occur, as in the case of ruminant animals or animals that vomit or regurgitate food.

alimentary toxic aleukia. See aleukia.

aliphatic chlorinated hydrocarbons. See hydrocarbons.

aliphatic epoxidation. See epoxidation.

aliphatic esterase. See esterase.

alkadiene. any unsaturated aliphatic hydrocarbon with two double bonds. See 1,3-butadiene.

alkalemia. excessive alkalinity of blood, as reflected in a pH of arterial blood that lies above the normal range (7.35 to 7.45 in humans).

alkali. **1:** any strong water-soluble base, especially a hydroxide of an alkali metal (group I) (e.g., potassium hydroxide, sodium hydroxide). They neutralize acids, react with acids to yield salts, react with fatty acids to yield soap, and turn red litmus paper blue. Because of their strong dissociation in water and their affinity for lipids, alkalis are highly corrosive and can cause rapid, deep erosion of tissue. See alkali poisoning, corrosion, corrosive poisoning. **2:** alkali soil. **corrosive alkali**. any strongly corrosive base, especially a metallic hydroxide (e.g., sodium, ammonium, and potassium hydroxides) or a carbonate. Because of their high degree of ionization, solubility in water, and their action on the fatty tissues, they cause rapid, deep, and very painful destruction of tissues. They tend to gelatinize tissue, giving it a grayish color and a slippery or soapy texture. See also caustic, corrosion, alkali poisoning, corrosive poisoning.

alkali burn. See burn.

alkali chlorosis. See chlorosis.

alkali disease. **1:** (Western duck sickness; botulism of ducks). a form of botulism causing death chiefly of ducks (Family Anatidae) and, to a lesser extent, loons, mergansers, gees, and gulls, in North America. The disease is caused by ingestion of vegetation contaminated with the toxin produced by *Clostridium botulinum*, type C. (type C botulinum toxin). Tens of thousands of birds die in most years, but losses in the western United States may reach a million or more during great outbreaks. **2:** a chronic disease of livestock marked by emaciation, stiffness, and anemia due to grazing on plants with a high selenium content. Humans have been affected on occasion by eating grain grown on soils rich in selenium. **3:** blind staggers, an acute form of selenium poisoning.

alkali grass. a common name of *Zigadenus elegans* and *Z. venenosus*.

alkali metal. any of the generally univalent elements of group IA of the periodic table, lithium, sodium, potassium, rubidium, cesium, and francium. They are essential micronutrients that are easily absorbed and excreted by living organisms, thus limiting their toxicity. They are the most strongly electropositive of metals, often reacting vigorously or even violently with water. The salts, especially in diseased organisms, are toxic.

alkali poisoning. a serious, often life-threatening type of corrosive poisoning due to ingestion of a strong water-soluble corrosive alkali (e.g., ammonia, lye, and certain soap powders or other cleaning agents); tissues of the gastrointestinal tract may be rapidly and deeply eroded. The victim may suffer excruciating pain and death may come swiftly as a result of edema of the throat and pharynx with closure of the respiratory tract, choking, and asphyxia. See also alkali disease, acid poisoning.

alkali reserve. 1: the relative amount of buffering substances in circulating blood (e.g., sodium bicarbonate, dipotassium phosphate, and proteins) that can neutralize acid. **2:** the relative amount of bicarbonates in blood. **3:** the set of dissolved substances in natural water bodies that tends to maintain constant pH.

alkali soil. any soil with a pH of 8.5 or more that contains salts injurious to plant life.

alkaline. **1:** pertaining to or having properties or reactions of an alkali. **2:** having a pH above 7.

alkaline earth (alkaline earth oxide). an oxide of an element of group IIA in the periodic table (Be, Mg, Ca, Sr, and Ba). All are essentially nontoxic except beryllium, which is highly toxic when inhaled. Magnesium and calcium are important constituents of living tissue and are toxic only in high dosages.

alkaline earth metal. any of the heaviest elements of group IIA of the periodic table; these are Mg, Ca, Sr, Ba, and Ra.

alkaline earth oxide. alkaline earth.

alkaline hypochlorite solution. See hypochlorite solution, alkaline.

alkaline phosphatase. any of a class of enzymes that catalyze the hydrolysis of phosphate monoesters. Elevated activity in blood serum usually indicates obstructive jaundice, Paget's disease, or bone carcinoma.

alkaline salt. any salt of a strong base. They are extremely toxic. In acute cases, immediate collapse and unconsciousness with death from respiratory paralysis are possible.

alkaline sodium hypochlorite solution. See sodium hypochlorite solution, alkaline.

alkalinity. **1:** the state of being alkaline. **2a:** the degree of alkalinity. **2b:** the extent to which water, an aqueous solution, or a water body is alkaline or exceeds a pH of 7.0. **3:** the number of milliequivalents of H^+ neutralized by 1 liter of seawater at 20°C.

alkalinization. alkalization.

alkalinize. alkalize.

alkaliphile (alkalophile). an alkaliphilic organism is one that occurs in, is tolerant of, or is successful in an alkaline habitat or environment.

alkaliphilic (alkalophilic). occurring in, tolerant of, or flourishing in alkaline habitats or environments. *Cf.* alkaloduric.

alkaliphily (alkalophily). the condition or state of being highly tolerant of alkaline conditions in the environment.

alkaliphobe. an alkaliphobic organism.

alkaliphobic. avoiding or intolerant of alkaline habitats or conditions. *Cf.* alkaloduric.

alkaliphoby. the condition or state of avoiding or being highly intolerant of alkaline habitats or environments.

alkalization (alkalinization). the process of making alkaline.

alkalize (alkalinize). to make alkaline.

alkalizer. an agent that neutralizes acids or effects alkalization; alkaline.

alkaloduric. extremely tolerant of high pH (alkaline) conditions. *Cf.* alkaliphilic.

alkaloid. **1:** any of a large heterogeneous group of alkaline, bitter-tasting, biologically active, usually water-insoluble, nitrogenous organic compounds produced by plants. They are usually derivatives of the nitrogen ring compounds quinoline, isoquinoline, pyridine, and pyrrole. Most are crystalline solids; a few are liquids or gums. Many flowering (dicotyledonous) plants produce alkaloids, especially those of the families Solanaceae, Fabaceae, and Papaveraceae. Monocots rarely produce alkaloids, although many groups of lower plants (but not algae) contain species that do. Alkaloids yield salts on reaction with acids and most occur in plants as soluble organic acid-alkaloid salts which can be extracted with organic solvents. The nitrogen usually occurs as part of a heterocyclic and/or aromatic ring. Most are potent, physiologically active agents. They are usually neurotoxic or cardiotoxic and many have therapeutic value; some are hepatotoxic or cytotoxic. They generally exhibit pharmacologic specificity, e.g., as analgesics (morphine), tranquilizers (reserpine), respiratory stimulants (nicotine), vasoconstrictors (scopolamine), local anesthetics (cocaine), muscle relaxants (strychnine), and psychedelic agents (LSD, psilocybin). Effects are usually also strongly species specific. Alkaloids of similar structure usually occur in closely related plants. Their trivial names usually end in -ine (e.g., atropine aconitine, brucine, caffeine, cocaine, coniine, morphine, nicotine, quinine, scopolamine, strychnine). Lysergic acids and lysergic acid amides are alkaloids as are most nitrogen-containing phytotoxins. **2:** the term alkaloid is often restricted to nitrogen-containing heterocyclic or aromatic compounds with pharmacological activity and the general properties of amines. **3:** any of certain synthetic compounds (e.g., procaine) with structures similar to naturally occurring alkaloids. **amine alkaloid**. usually an ergot alkaloid such as ergonovine, lergotrile, D-lysergic acid, methylergonovine, or methylsergide. **amino acid alkaloid**. usually an ergot alkaloid such as bromocriptine or ergotamine. **animal alkaloid**. **1:** a ptomaine. **2:** a leukomaine. **artificial alkaloid**. a synthetic alkaloid. **belladonna alkaloid**. any organic ester compounded from tropic acid and an organic base such as tropine or scopine (e.g., atropine and scopolamine). These alkaloids are produced by solanaceous plants such as *Atropa belladonna*, *Datura stramonium*, *Hyoscyamus niger*, and *Scopolia carniolica*. *A. belladonna* produces mostly atropine, whereas *H. niger* and *S. carniolica* produce scopolamine. **clavine alkaloid.** a type of ergot alkaloid (e.g., elymoclavine, ergoclavine). **ergot alkaloid**. any of a large group of alkaloids that are constituents of *Claviceps* (ergot); they contain lysergic acid in combination with various amine moieties. The four main types are clavine alkaloids, lysergic acids, lysergic acid amides, and ergot peptide alkaloids. They are very toxic and may produce nausea, vomiting, a weak rapid pulse, coma, and death. They have various complex effects on cardiovascular performance, increase uterine motility, and suppress prolactin secretion. Ergot alkaloids have been used therapeutically to treat postpartum hemorrhage, migraine, and Parkinson's syndrome. **ergot peptide alkaloid**. any of a type of alkaloid isolated from *Claviceps* (ergot). There are ten of these: ergotamine, ergosine, ergocristine, ergocryptine, ergocornine, and the isomers ergotaminine, ergosinine, ergocristinine, ergocryptinine, and ergocortinine. They are made up of lysergic acid, dimethylpyruvic acid, proline, and phenylalanine moieties linked by amide bridges. **ergotoxine alkaloid**. any ergot alkaloid having oxytocic activity. See ergotoxine. **indole alkaloid**. any of a group of alkaloids (e.g., strychnine, yohimbine), that contain an indole nucleus. They are derivatives of tryptophan or phenylalanine. **isoquinoline alkaloid.** any of a group of alkaloids derived from tyrosine and phenylalanine that contain an isoquinoline nucleus (e.g., curare, morphine). They occur chiefly in plants of the family Papaveraceae(poppies) and related plants. **mydriatic alkaloid**. an alkaloid that dilates the pupil of the eye (e.g., atropine, homatropine, cocaine). **papaver alkaloid**. isoquinoline alkaloid. See also alkaloid nitrogen oxides. **piperidine alkaloid**. any alkaloid that contains the piperidine ring (e.g., nicotine). **plant alkaloid**. any naturally occurring alkaloid. **protoalkaloid**. any alkaloid that lacks a heterocyclic ring (e.g., mescaline, ephedrine). Most are simple amines, some of which may be precursors of true alkaloids. **pseudoalkaloid**. a substance (e.g., theobromine, caffeine) that is derived from a compound such as a purine, a sterol, or a terpene. True alkaloids and protoalkaloids are, however, derivatives of amino acids. **pyridine alkaloid.** any alkaloid that contains a pyridine nucleus (e.g., coniine). **pyrrolizidine alkaloid**. a type of hepatotoxic alkaloid produced by various plants, notably *Senecio* and *Crotalaria*. Toxicity varies considerably among these alkaloids. Acute intoxication characteristically produces a focal hemorrhagic, centrilobar necrosis in the liver of susceptible taxa, such as the laboratory rat or mouse, horses, and cattle. Challenge with repeated small doses produces a chronic liver lesion characterized in part by megalocytosis in some species. Pulmonary lesions also occur in some taxa. **spiropiperidine alkaloid**. any alkaloid with two piperidine rings. They occur in the poisonous skin secretions of frogs of the genus *Dendrobates*. **steroid alkaloid**. an alkaloid that includes a steroid nucleus (the perhydrocyclopentanophenanthrene ring), e.g., solanidine, veratramine. **synthetic alkaloid**. any synthetic substance that is structurally similar to plant alkaloids. **thiooxazolidone alkaloid**. See L-5-vinyl-2-thiooxazolidone. **tropane alkaloid.** any of a group of tropane-like alkaloids (e.g., hyoscyamine, atropine, cocaine alkaloids) derived from the amino acid ornithine. **true alkaloid.** any alkaloid (e.g., the isoquinoline alkaloids) that includes a nitrogen-containing heterocyclic nucleus.

alkaloid nitrogen oxide. any alkaloid with a nitrogen oxide group; they occur in some plants (e.g., *Senecio, Lupinus*). See nitrogen oxide.

alkalophile. alkaliphile.

alkalophilic. alkaliphilic.

alkalophily. alkaliphily.

alkalosis. a pathophysiological condition characterized by an increase in pH of body fluids, usually including circulating arterial blood, above the normal range, by an excess of base (metabolic alkalosis), or by the loss of CO_2 due to hyperventilation (respiratory alkalosis). **acapnial alkalosis**. respiratory alkylosis. **compensated alkalosis**. alkylosis in which blood bicarbonate levels have changed but the pH of body fluids is near normal. Thus, for example, respiratory alkylosis may be compensated by increased metabolic production of acids or by increased excretion of bicarbonate. **hypochloremic alkalosis**. metabolic alkalosis that results from a loss of chloride due to severe vomiting. **metabolic alkalosis**. that which may be caused by loss of acid (HCl) through severe, persistent vomiting or from excessive acid in the urine, excessive intake of alkaline substances, or altered potassium ion concentration. It is associated with excess base and increased arterial concentrations of bicarbonate. Many poisons can induce metabolic alkalosis by one or more of these mechanisms. **respiratory alkalosis** (acapnial alkalosis). that resulting from abnormal loss of CO_2 due either to hyperventilation or poisoning (e.g., by salicylates). **uncompensated alkalosis**. that in which the pH of body fluids is elevated because of a lack of countering or compensating mechanisms.

alkamine (amino alcohol). an alcohol that contains an amine group (e.g., solanidine).

alkane (paraffin). the systematic name for any straight-chain, branched-chain, or cyclic (e.g., hexane) aliphatic hydrocarbon in which adjacent carbon atoms are joined by a single covalent (sigma) bond; they contain no unsaturated (double) carbon-carbon bonds. They are the primary constituents of petroleum products. Straight-chain or branched-chain alkanes have the general molecular formula C_nH_{2n+2}; that of cyclic alkanes is C_nH_{2n}. Alkanes are flammable, burning readily and sometimes violently in air. Incomplete oxidation, as in an oxygen-deficient atmosphere, may produce quantities of carbon monoxide. Alkanes are neurotoxic, cytotoxic, and irritant compounds that chiefly affect the central nervous system and skin. Alkanes have a narcotic effect, causing unconsciousness at low concentrations. Acute exposures can be fatal. **higher alkane**. any alkane higher than C8. They occur in fuel oil, diesel fuel, jet fuel, kerosene, and mineral oil. Most are not very toxic. Occupational exposure is mostly by inhalation. Symptoms may include dizziness, headache, and stupor. Acute exposure to high concentrations in air can result in coma and death. Inhalation of mists or aspiration of vomitus containing the liquid can cause aspiration pneumonia. Mineral oils may be carcinogenic to the skin and scrotum. See also aliphatic hydrocarbon. **lower alkane**. any alkane with 1-8 carbon atoms. The C1-C4 alkanes are those with one to four carbon atoms per molecule: methane, ethane, propane, and butane. All are gases under ambient conditions. Exposure is by inhalation; they are simple asphyxiants. C5-C8 alkanes are those with 5-8 carbon atoms per molecule: pentane, hexane, heptane, octane. All are volatile liquids under ambient conditions. They have numerous commercial and industrial uses in addition to their use in gasoline and other petroleum products. Exposure is chiefly by inhalation of the vapor. Humans and laboratory animals suffer central nervous system depression and loss of coordination on exposure to the fumes. High concentrations can be fatal. Contact with the liquid (an occupational risk) can produce a dermatitis in which the skin becomes dry, inflamed, and scaly due partly to defatting of the skin.

alkene (olefin). the systematic chemical name for any aliphatic hydrocarbon comprised of straight or branched chains of carbon atoms that contain one or more carbon-carbon double bonds (e.g., ethene, propene). An alkene that contains two double bonds is termed a diene, those with three are trienes, etc. Alkenes are produced in large amounts during the cracking of crude oils and are common components of gasoline and other refined petroleum products. They are usually more toxic than alkanes of similar molecular weight, and less toxic than aromatic hydrocarbons. See alkadiene.

alkenyl halide. See halide.

alkenyl mercaptan. alkenyl thiol. See thiol.

alkenyl thiol. See thiol.

alkoxy. **1:** pertaining to, denoting, or containing an alkoxyl group. **2:** an alkyl radical connected to a molecule by oxygen.

alkoxy-alkylated mercury compound. an organomercurial compound that contains an alkoxy radical; one of three types of organomercurials used as pesticides. See also organomercurial pesticides.

alkoxyl. a univalent radical (e.g., methoxyl) composed of an alkyl group linked to an oxygen.

alkyl. **1:** pertaining to, denoting, or containing an alkyl radical. **2a:** a monovalent aliphatic radical C_nH_{2n+1} (e.g., methyl, ethyl) which can be thought of as an alkane moiety that has lost one hydrogen atom. **2b:** a monovalent aliphatic, alicyclic, or aliphatic-aromatic hydrocarbon radical. **3:** a compound comprised of one or more alkyl radicals joined to a metal (e.g., sodium alkyls).

alkyl carbonyl. a carbonyl in which a metal atom is bonded to an aromatic group and coordinated with several carbon monoxide molecules.

alkyl ester hydrolase. See esterase.

alkyl esterase. See esterase.

alkyl group. a substituent of a compound with one less hydrogen than the corresponding alkane or aliphatic hydrocarbon.

alkyl halide. See halide.

alkyl mercaptan. alkyl thiol. See thiol.

alkyl mercury. See alkylated mercury compound.

alkyl polyamine. any amine in which 2 or more amino groups are bonded to alkane moieties. Such amines are chelating agents (e.g., triethylenetetramine) and are also used industrially as solvents, emulsifiers, epoxy resin hardeners, stabilizers, and starting materials for dye synthesis. They are strongly alkaline and most are thus skin, eye, and respiratory tract irritants, especially those of lower molecular weight.

alkyl sulfonate (linear alkylate sulfonate; LAS). any straight-chain alkylbenzene sulfonate (or branched only at the end) having 10 or more carbons. They are cytotoxic alkylating agents that were developed as biodegradable detergents.

alkyl thiol. See thiol.

alkylamine. a compound comprised of an alkyl group attached to the nitrogen of an amine. Ethylamine, $C_2H_5NH_2$, is an example.

alkylated mercury compound (alkylmercury compound; alkylmercury; alkyl mercury). any mercury-containing alkane. See also organomercurial pesticides.

alkylated organotin compound (alkyltin compound; alkyltin). any tin-containing alkane. A number of these compounds are immunotoxic and induce thymus atrophy without apparent effects on other organ systems. See also dialkylated organotin compounds, trialkylated organotin compounds, organotin compounds.

alkylating agent. any substance that is able to form covalent bonds and attach alkyl groups (e.g., methyl, ethyl) to nitrogen and oxygen atoms of the DNA bases and to other macromolecules. They are highly reactive and often strongly carcinogenic and/or mutagenic, causing errors in replication (e.g., chromosome breaks, mispairing of bases). Alkylating agents can produce serious cytotoxic, immunotoxic, fetotoxic, teratogenic, and abortifacient effects, and can cause widespread damage to tissues. The action of alkylating agents on proliferating lymphocytes and myeloid tissues can cause permanent damage or death to humans and other animals. See, for example, *N*-dimethylnitrosamine, gas (mustard gas), alkyl sulfonate.

alkylation. the addition of an alkyl group (e.g., methyl) to a molecule. See alkylating agent.

alkylene. an organic radical such as the ethylene radical, $C_2H_3\cdot$, that is formed from an unsaturated aliphatic hydrocarbon.

alkylmercury, **alkylmercury compound**, **alkyl mercury**. alkylated mercury compound.

alkyltin. alkylated organotin compound.

alkyltin compound. alkylated organotin compound.

alkyne (acetylene hydrocarbon; acetylene). any of a class of unsaturated straight-chain or branched-chain hydrocarbons that contain at least one triple bond between adjacent carbon atoms (e.g., acetylene, propylene). They have the generic formula, C_nH_{2n-2}. See hydrocarbon.

all-or-none (all-or-none response; all-or-none principle). a term that indicates or refers to a characteristic of an irritable cell or tissue or other biological system whereby it either fails to respond to a stimulus or responds maximally regardless of the stimulus intensity above that which induces a response at all (the threshold value). The initiation of a nervous impulse is a typical all-or-none response. Similarly, the muscle fibers of a motor unit contract maximally when exposed to a stimulus at or above a threshold intensity.

all-or-none principle. See all-or-none.

all-or-none response. See all-or-none.

allantiasis (sausage poisoning). obsolete for botulism acquired from eating improperly preserved sausage.

allantotoxicon. botulinic acid.

allelochemic (allelochemical). **1:** pertaining to interspecific chemical interactions. **2a:** of or pertaining to a semiochemical agent of interspecies communication. **2b:** a waste product or metabolite produced by an organism that influences the growth, behavior, or population dynamics of another species.

allelochemical. **1:** allelochemic. **2:** any substance produced by a living organism that affects individuals of another species.

allelochemical interaction. any interspecific interaction that involves biologically active chemicals. See also allelopathy, allomone, kairomone, signal.

allelochemics. **1:** plural form of allelochemic. **2:** intraspecific chemical interactions that involve the release of biologically active substances.

allelochemistry. the science of allelochemical compounds.

allelopathic. of, or pertaining to allelopathy or to an allelopathic substance. See also antibiotic.

allelopathic chemical. See allelopathic substance.

allelopathic substance (allelopathic chemical). **1:** an allelochemic; a waste product, excretory product, or metabolite having a detrimental, usually inhibitory or regulatory, effect on organisms of species other than that producing the substance. **2:** any allelochemic used by humans as herbicides. Included are any of a large number of naturally occurring herbicides produced by plants or microorganisms (mostly soil bacteria and fungi). Toxicities and specificities vary greatly. These substances are usually obtained by extraction from the producing plants by the leaves or root exudates, or from decomposed plant tissue. By this definition, amygdalin, arbutin, caffeine, and gallic acid are, for example, all considered allelopathic.

allelopathy (biogenic toxicity). **1:** the production of interspecific effects at a distance, usually by chemical means. See allelochemical. **2:** the chemical inhibition of one organism by another regardless of species; antibiosis. **3:** an allelochemical effect. See, for example, 3-acetyl-6-methoxy benzaldehyde, arbutin, *Adenostoma fasciculatum*, *Arctostaphylos*, *Juglans nigra*. **indirect allelopathy**. the release of a chemical by one plant that kills or inhibits the growth of a second plant or microorganism that is essential to the growth or viability of a third plant. Allelopathic substances sometimes play an important role during plant succession.

allelotoxin. a toxic substance released by an allelopathic process.

allenthesis (obs.). the introduction of a foreign substance into an organism. *Cf.* allergization.

allergen. an antigen (e.g., a foreign protein, microorganism, pollen) that induces a response in the immune system such that subsequent exposure to the same allergen produces an allergic reaction or specific hypersensitivity. The most common reactions are localized inflammation, watery eyes, and runny nose. Sometimes, a serious, often fatal systemic condition (anaphylaxis; anaphylactic shock) presents. *Cf.* allergin.

allergenic. pertaining to or acting as an allergen; capable of producing allergy; antigenic. *Cf.* allerginic, antigenic, immunogenic.

allergenicity. the capacity of a substance to provoke an allergic reaction.

allergic. pertaining to allergy or to any response to an allergen.

allergic arteritis. inflammation of the arterial walls due to an allergic state.

allergic asthma (bronchial asthma). a common form of asthma due to hypersensitivity to an allergen.

allergic conjunctivitis (atopic conjunctivitis). **1:** hay fever. **2:** an allergic reaction of the conjunctiva to a specific substance.

allergic dermatitis (allergodermia). See dermatitis.

allergic reaction. See reaction (allergic reaction).

allergic response. allergic reaction. See reaction.

allergic shellfish poisoning. See shellfish poisoning.

allergic shock. anaphylaxis.

allergic vasculitis syndrome. a possibly immunologic skin disease, marked by ulcers due to destructive inflammation of underlying blood vessels.

allergie. allergy.

allergin. **1:** an antigen that triggers anaphylaxis. **2:** allergen.

allerginic. **1:** anaphylaxis-inducing. **2:** allergenic.

allerginicity. **1:** the capacity of a substance to induce anaphylaxis. **2:** allergenicity.

allergization. active sensitization or the induction of allergy by natural or artificial contact of foreign matter (allergen) with susceptible tissues, including the introduction of a specific allergen into the body in order to effect sensitization. *Cf.* allenthesis.

allergize. to sensitize or to make allergic by contact with or the introduction of foreign matter (allergen) into the body.

allergized. made specifically sensitive or reactive by contact with an allergen.

allergodermania. allergodermia.

allergodermia (allergodermania). an allergic dermatitis.

allergology. the science of allergy.

allergosis. any allergic disease.

allergy (allergie). **1a:** the immunologic state of a sensitive individual induced by an allergen. It is marked by the liberation of histamine and other molecules with histamine-like effects. An allergic reaction (*q.v.*) is not a poisoning in the narrow sense because the response depends upon a preexisting sensitivity of an individual to a specific chemical. **1b:** sensitivity to an allergen. **2:** commonly used synonymously with hypersensitivity, especially delayed hypersensitivity and class I (complement independent) hypersensitivity. **3:** that branch of medicine concerned with the study, diagnosis, and treatment of allergic manifestations. *Cf.* hypersensitivity, anaphylaxis, atopy. See also allergic reaction. **atopic allergy**. atopy. **bacterial allergy**. hypersensitivity to a specific bacterial allergin as a result of prior infection. **cell-mediated allergy**. delayed hypersensitivity. **chemical allergy**. unusual sensitivity to a chemical due to prior exposure to that chemical or to one that is structurally similar. *Cf.* drug allergy. **contact allergy**. an eczematous, cutaneous reaction to direct contact with an allergen to which the individual is hypersensitive. **delayed allergy**. delayed hypersensitivity. **drug allergy**. unusual sensitivity to a drug or other chemical or the allergic reaction that occurs when exposed to the drug; roughly synonymous with chemical allergy. **endocrine allergy**. hypersensitivity to an endogenous hormone. **food allergy**. that in which the antigens include ingested foods, drugs, or other materials to which one has been sensitized. See sulfite. **immediate allergy**. immediate hypersensitivity. See also immune response, immune reaction.

allethrin (allyl cinerin). **1:** the generic name for 2-allyl-4-hydroxy-3-methyl-2-cyclopenten-1-one ester of chrysanthemummonocarboxylic acid, $C_{19}H_{26}O_3$, used as an insecticide or synergist. **2:** any of a class of synthetic analogs (pyrethroids) of naturally occurring pyrethrins such as cinerin, jasmolin, and pyrethrin. Toxicities are similar to those of pyrethrins. Allethrins are allethrolone esters and pyrethrolone esters of chrysanthemummonocarboxylic acids such as chrysanthemic acid, barthrin, and cyclethrin. They are viscous, water-insoluble, hepatotoxic and nephrotoxic liquid insecticides that are readily absorbed by the lungs, skin, and mucous membranes. **allethrin I** (2,2-dimethyl-3-(2-methyl-1-propenyl)cyclopropanecarboxylic acid 2-methyl-4-oxo-3-(2-propenyl)-2-cyclopenten-1-yl ester; allethrone ester of chrysanthemummonocarboxylic acid). the commercial product is very toxic to laboratory mice. **allethrin II** (3-(3-methoxy-2-methyl-3-oxo-1-propenyl)-2,2-dimethylcyclopropane-carboxylic acid 2-methyl-4-oxo-3-(2-propenyl)-2-cyclopenten-1-yl ester; allethrolone ester of chrysanthemumdicarboxylic acid monomethyl ester). a toxic, oily, pale yellow insecticidal mixture. It is less toxic than allethrin I.

allethrolone (2-methyl-4-oxo-3-(2-propenyl)-2-cyclopentenol). an analog of pyrethrolone in which 2-propanyl replaces the 2,4-pentadienyl group. It is used in the synthesis of allethrins.

allethrolone ester of chrysanthemumdicarboxylic acid monomethyl ester. allethrin II.

allethrone ester of chrysanthemummonocarboxylic acid. allethrin I.

allidochlor (2-chloro-*N,N*-di-2-propenylacetamide; *α*-chloro-*N,N*-diallylacetamide; *N,N*-diallyl-2-chloroacetamide; CDAA). an amide used as a preemergence herbicide with a persistence in soil of up to two months. It modifies RNA and protein biosynthesis and inhibits cell division in primary meristems. Effects vary, but usually include inhibition of photosynthesis. [illegible] moderately toxic to laboratory rats.

alliette. fosetyl aluminum.

allium. common name for herbs of the genus *Allium*.

Allium. a genus of some 300 species of bulbous herbs (Family Liliaceae) with strong, often distinctive, odors. Both wild and domestic forms are included. About 70 species are cultivated; those commonly grown for the table include garlic, leeks, onions, and chives. ***A. cepa***. a wild biennial herb (Family Liliaceae) with firm, white bulbs; also cultivated as a common root vegetable. Ingestion by humans of large amounts of the raw bulb may induce severe and ultimately fatal anemia.

allo-. a prefix **1:** indicating divergence from or opposition to the normal. **2:** denoting the more stable form of two isomers.

alloantibody. an antibody that reacts with an alloantigen.

alloantigen. a class of isoantigens present in some members of a strain or species, but not in others. Examples are blood group antigens on erythrocytes and histocompatability antigens in transplanted tissues that induce an alloimmune response in a recipient who does not have them. *Cf.* isoantigen.

allobiosis. altered responses of an organism following physiological changes or exposure to environmental changes.

allochemic. **1:** any secondary compound produced by plants that reduces or inhibits the activity of herbivores (e.g., a toxin, inhibitor of digestion). **2:** having the properties of an allochemic agent.

allochthonous (allocthonous). exogenous; of foreign origin, as in the case of living organisms, soils, rocks, or sediments in a given ecosystem that originated elsewhere. *Cf.* enthetic, exogenous, xenogenous.

allocthonous. allochthonous.

allogeneic. pertaining to or descriptive of genetic differences among individuals of the same strain or species. The term is used chiefly in reference to tissue or organ grafts between individuals of the same species. *Cf.* allogenic.

allogenic (allothogenic). **1:** pertaining to an ecologic succession. See succession (allogenic succession, autogenic succession). **2:** allogeneic.

allogenic succession. See succession.

allolactose. a disaccharide isomer of lactose that occurs in milk. It is an effector molecule that permits activation of the *lac* operon by binding to the repressor molecule.

allomone. **1a:** an allelochemical that confers an advantage on the organism that produces it by its effects on other species. Allomones include repellents, poisons, venoms, attractants, counteractants, distractants, inductants, suppressants (e.g., antibiotics), screens (e.g., inks of cephalopods), etc. **1b:** any substance secreted by an organism that confers an adaptive advantage on that organism. **2:** any coactone that serves to defend the producing organism. See also allelochemical, kairomone, signal.

allorphine. nalorphine.

allosteric. pertaining to or involving proteins that exhibit allostery.

allostery. a property of many proteins that have two or more receptor sites, such that the binding of a small molecule at a particular site induces a change in the properties or specificity of distant sites. Allostery apparently depends on changes in the tertiary structure shape of the protein due to binding of the first molecule.

allothogenic. allogenic.

allotoxin. any endogenous substance produced within an organism that neutralizes the action of a specific toxin.

allowable human daily intake. acceptable daily intake.

alloxan (2,4,5,6(1H,3H)-pyrimidinetetrone; 2,4,5,6-tetraoxohexahydropyrimidine; mesoxalylurea; mesoxalylcarbamide; mesoxalyl urea; mesoxalylcarbamide). a crystalline diabetogenic oxidation product, $C_4H_2N_2O_4$, of uric acid that causes selective necrosis of the β islet cells of the pancreas. It is sometimes used as an antineoplastic agent. It also has antibacterial and antifungal activity.

allozyme. a type of enzyme that has the same activity but differs somewhat in the amino acid sequence produced by different alleles at a single gene locus.

allyl-. a prefix indicating the presence of the allyl.

allyl acetate (acetic acid allyl ester; acetic acid-2-propenyl ester). a flammable, water-insoluble liquid, $C_5H_8O_2$. It is a skin and eye irritant and is poisonous by ingestion, inhalation, and skin contact.

allyl aldehyde. acrolein.

allyl bromide. a colorless to light yellow, flammable liquid, $H_2C{=}CHCH_2Br$, that is soluble in organic solvents; boiling point, 71.3°C. It is toxic and a strong irritant to the skin and eyes. Allyl bromide is used in the synthesis of organic compounds and in the manufacture of synthetic perfumes.

allyl catechol methylene ether. safrole.

allyl chloride (3-chloro-1-propene; 3-chloropropylene). a colorless, volatile, flammable, moderately toxic liquid, chlorinated propene, $CH_2{=}CHCH_2CL$, with an unpleasant pungent odor. Allyl chloride is used as an intermediate in the manufacture of glycerol, various allyl compounds allyl alcohol and a number of pharmaceuticals, insecticides, plastic resins, and thermosetting varnish. It is an irritant. Contact with the skin may produce burns and numbness and the vapor is a strong irritant of the eyes, nose, and throat. Exposure to the fumes can cause headaches, dizziness, and at high concentrations, unconsciousness. Allyl chloride is readily absorbed through the skin as well as the respiratory passages and can produce systemic effects. Chronic exposure is marked by deep muscle pain and injury to the kidneys, liver, and respiratory tissues (with pulmonary edema).

allyl cinerin. allethrin.

1-*N*-allyl-7,8-dihydro-14-hydroxynormorphinone. naloxone.

17-allyl-4,5α-epoxy-3,14-dihydroxymorphinan-6-one hydrochloride. naloxone hydrochloride.

allyl glucosinolate. sinigrin.

2-allyl-4-hydroxy-3-methyl-2-cyclopenten-1-one ester of chrysanthemummonocarboxylic acid. See allethrin.

1-*N*-allyl-14-hydroxynordihydromorphinone. naloxone.

allyl isosulfocyanate. allyl isothiocyanate.

allyl isothiocyanate (3-isothiocyanato-1-propene, isothiocyanic acid allyl ether, allyl isosulfocyanate, 2-propenyl isothiocyanate; volatile oil of mustard; mustard oil; Redskin). a common vesicating, volatile, colorless to pale yellow, mustard oil, $CH_2CH{:}CH_2NCS$, with a pungent, irritating odor. It is slightly soluble in water and in alcohol, with a boiling point of 152°C. Found in *Brassica nigra* (black mustard seed), *q.v.*, and many other plants, it repels many insects, but attracts butterflies of the genus *Pieris*. It is extremely toxic and is used as a fumigant, a war gas, and therapeutically as a counterirritant. The lethal oral dose for domestic cattle is about 10^{-3}% body weight. See also sinigrin.

allyl isovalerianate. allylisovalerate.

allyl 3-methylbutyrate. allylisovalerate.

allyl sulfide (3,3′-thio-bis-1-propene; diallyl sulfide; thioallyl ether; oil garlic; garlic oil; "oil garlic"). a colorless, irritant, nearly water-insoluble, irritant, hepatotoxic, and nephrotoxic liquid, $(CH_2{=}CHCH_2)_2S$, with a garlic odor. It is used to dehydrate onions, and in the manufacture of fruit and garlic flavorings. It is an irritant to the eyes and respiratory tract and high concentrations can cause unconsciousness. Long-term chronic exposure can damage liver and kidneys.

12-allyl-7,7a,8,9-tetra hydro-3,7a-dihydroxy-4a*H*-8,9c-iminoethanophenanthro[4,5-bcd]furan-5(6*H*)-one. naloxone.

allylacetone. a liquid, $CH_2CHCH_2CH_2COCH_3$, that is colorless and soluble in water and organic solvents. It is used in the synthesis of perfumes, pharmaceuticals, fungicides, and insecticides.

allylamine (2-propenylamine; 2-propen-1-amine; 3-aminopropylene). a colorless to light yellow toxic irritant oil, $C_3H_5NH_2$, with a strong ammoniacal odor and a burning taste. It is soluble in water, ethanol, ether, and chloroform. It is used in organic synthesis and is an intermediate in the manufacture of mercurial diuretics. It attacks rubber and cork, is extremely toxic to laboratory mice, and the vapors cause sneezing and tearing at low concentrations.

allyldioxybenzene methylene ether. safrole.

1-*N*-allyl-14-hydroxynordihydromorphinone. naloxone.

allylisopropylacetamide. an amide that initiates autooxidation, destroying microsomal cytochromes, especially cytochrome P-450.

allylisovalerate (isovaleric acid, allyl ester; allyl isovalerianate; allyl 3-methylbutyrate; 3-methyl-butanoic acid, 2-propanyl ester; 3-methylbutyric acid, allyl ester; 2-propenyl isovalerate; 2-propanyl 3-methylbutanoate). an organic compound, $C_8H_{14}O_2$, that is toxic by ingestion and to a lesser extent by contact. It is a skin irritant and is carcinogenic to some laboratory animals (e.g., mouse, rat). It is a probable human carcinogen.

5-allyl-5-(1-methylbutyl)-2-thiobarbituric acid. thiamylal.

4-allyl-1,2-methylenedioxybenzene. safrole.

***N*-allylnormorphine**. nalorphine.

***N*-allylnormorphine bis(pyridine-3-carboxylic acid) ester**. nalorphine dinicotinate.

***N*-allylnormorphine dinicotinate**. nalorphine dinicotinate.

***N*-allylnoroxymorphone**. naloxone.

1-(*o*-allylphenoxy)-3-(isopropylamino)-2-propanol. alprenolol.

12-allyl-7,7a,8,9-tetra hydro-3,7a-dihydroxy-4a*H*-8,9c-iminoethanophenanthro[4,5-bcd]furan-5(6*H*)-one. naloxone.

almitrine (6-[4-[bis(4-fluorophenyl)methyl]-1-piperazinyl]-N,N′-di-2-propenyl-1,3,5-triazine-2,4-diamine; 2,4-bis[allylamino]-6-[4-[bis(p-fluorophenyl)methyl]-1-piperazinyl]-*s*-triazine). a respiratory stimulant, $C_{26}H_{29}F_2N_7$. It is very toxic.

almond. See *Prunus amygdalus*.

alnawana. *Bitis arietans*.

alocasia. common name of plants of the genus *Alocasia*.

Alocasia (alocasia). a genus of some 70 species of tropical Asiatic herbs (Family Araceae) with numerous hybrids. These plants are characterized by basal, often showy leaves, with long petioles, a glaucous boat-shaped spathe, and reddish berries; some are evergreen, others are herbaceous. All parts of the plant are toxic and, when chewed, can produce an intense burning pain in the mouth with swelling of tongue and throat. Additional symptoms include nausea, salivation, vomiting, diarrhea, and occasional direct systemic effects. The effects are due chiefly to the presence of calcium oxalate crystals in the plant tissue; other as yet unidentified active principles may also be involved. *Cf. Dieffenbachia*. See also Araceae.

aloe. the thickened juice of various species of plants of the genus *Aloe* (Family Lilaceae). Formerly used as a cathartic, its use was largely discontinued because it can cause severe abdominal cramps. The active principle is aloe-emodin, *q.v.*

Aloe. a genus of perennial succulent herbs (Family Lilaceae) that are native chiefly to southern Africa, but also to tropical Africa and Madagascar. Various species contain emodin, *q.v. Cf.* aloe.

aloe-emodin (1,8-dihydroxy-3-(hydroxymethyl)-9,10-anthracenedione;1,8-dihydroxy-3-(hydroxymethyl)anthraquinone; 3-hydroxymethylchrysazin; emodin; rhabarberone). a cathartic anthraquinoid that occurs free and as a glycoside in *Rheum* (rhubarb), *Cassia* (senna) leaves, and in various species of *Aloe* (Lilaceae). *Cf.* emodin.

aloin. a yellow, bitter, cathartic substance isolated from *Aloe*. It contains the pentose sugar, *d*-arabinose, and one or more glycosides. See arabinose.

alopecia (acomia; baldness; calvities; pelade; baldness). absence, loss, or deficiency of hair. Causes vary. **alopecia medicamentosa**. alopecia due to the administration of a medication, especially one containing cytotoxic agents. **alopecia toxica**. alopecia presumed to be caused by toxins or infectious disease.

Alopex lagopus (blue fox). See Ronnel.

alpha (α, A). **1:** first letter of the Greek alphabet, it has many applications in science. For terms beginning with alpha, see the root word. **2:** the first; first in order of magnitude. **3:** the beginning; the first in a series or classification. **4:** the closest to or first in position from a part of an organic molecule (e.g., the position adjacent to a phenyl group). **5:** an aromatic substituent linked to an aliphatic chain. **6:** the direction of a chemical bond opposite the plane of view.

alpha emission. the ejection of alpha particles from an atom's nucleus.

alpha irradiation. subjecting a substance to a flow of alpha particles.

alpha particle ("alpha ray"). a positively charged particle (atomic mass, 4, Z = 2) consisting of two protons and two neutrons and identical to the helium nucleus, He^{2+}. It is emitted at high speed in certain radioactive transformations and is the largest, and, when emitted within the organism, the most damaging type of radioactivity. Within the lung they can cause cell and tissue damage and initiate processes that induce cancer. *Cf.* helium.

alpha-particle detector. an instrument used to detect the presence of alpha particles.

alpha ray. alpha particle.

α-alphahydroxybenzeneacetonitrile. mandelonitrile.

alphaprodine (1,3-dimethyl-4-phenyl-4-piperidinol propanoate; α-1,3-dimethyl-4-phenyl-4-propionoxypiperidine; α-prodine). a narcotic analgesic, it is very toxic to laboratory mice and rats. It may be habit forming and is a controlled substance listed in the U.S. Code of Federal Regulations (CFR).

alprenolol(1-[(1-methylethyl)amino]-3-[2-(2-propenyl)phenoxyl]-2-propanol; 1-(*o*-allylphenoxy)-3-(isopropylamino)-2-propanol; H56/28). a β-adrenergic blocker, used therapeutically as an antihypertensive, antianginal, and antiarrhythmic. It is very toxic to laboratory mice and rabbits; somewhat less so in laboratory rats.

ALS. antilymphocyte serum.

alsike clover. *Trifolium hybridum*.

alsike poisoning. trifoliosis.

Alsophis (West Indian racers). a genus of venomous snakes (Family Colubridae). Envenomation produces effects essentially similar to those of other venomous colubrids. See colubrid venom poisoning.

Alternaria solani. a fungus pathogenic to plants of the family Solanaceae, it produces alternaric acid, a withering agent. See also tentoxin.

alternaric acid. isolated from *Alternaria solani*, a pathogen of the Solanaceae, it is a withering agent that acts on leaves at concentrations as low as 5 μg/ml.

alternative test method. any method or technique used to test products or other substances (e.g., for toxicity) that does not depend upon the use of animals. Such approaches include the use of cell cultures, bacteria, living tissue *in vitro* (e.g., cloned skin cultures). Such tests are often used in conjunction with computer simulations. While computer simulations alone are sometimes used, the output should be considered hypothetical without alternative substantiation.

altretamine (*N,N,N',N',N'',N''*-hexamethyl-1,-3,5-triazine-2,4,6-triamine; 2,4,6-tris-(dimethylamino)-*s*-triazine; heme; hexamethylmelamine; HMM). an antineoplastic alkylating agent and insect chemosterilant. It is very toxic to laboratory rats and guinea pigs.

alumina. the native form of aluminum oxide, Al_2O_3. It occurs as corundum, a powder or crystalline material. See aluminosis.

aluminium. aluminum.

aluminosis (aluminosis pulmonum). a chronic (occupational) pulmonary inflammation (a pneumoconiosis) due to aluminum-bearing dust (e.g., alumina) in the lungs.

aluminosis pulmonum. aluminosis.

aluminum (Al; aluminium (British, Canadian))[Z = 13; A_r = 26.98]. a soft, silvery white, moderately reactive, toxic, low-density metallic element with many commercial and industrial uses. It occurs naturally in shale, igneous rocks, clay, and most soils. Humans are rarely poisoned by aluminum except as indicated below. It is readily soluble in water of low pH and is thus sometimes a major environmental poison in highly acidic water bodies where it leaches into the water from aluminum-containing bottom sediments. In such cases, it not only poisons fishes and other aquatic biota, but may threaten humans by entering water supplies. It can damage the skin, eyes, and mucous membranes of the upper respiratory tract. Soils with a pH greater than 5.5 may also contain soluble aluminum in dangerous quantities. Workers in the smelting and refining industries who inhale large amounts of aluminum dust incur Shaver's disease, and have sometimes been fatally poisoned. High concentrations of aluminum in renal dialyzing solutions can cause a condition known as dialysis encephalopathy or dialysis dementia. Symptoms include memory loss, dementia, aphasia, ataxia, convulsions, and altered electroencephalograms. Airborne particulates bearing the metal may cause aluminosis. Effects typically include kidney and liver necrosis and severe brain damage. Aluminum has also been associated with various brain disorders, including Alzheimer's disease. See also aluminosis, anhidrotic.

aluminum chloride (anhydrous aluminum chloride). a colorless to white hexagonal deliquescent anhydrous crystalline compound, $AlCl_3$. It has a number of uses, including that of a catalyst, a dyestuff intermediate, and a detergent alkylate. It fumes in air and reacts explosively with water, emitting hydrogen chloride gas. The vapor, Al_2Cl_6, is composed of double molecules and is soluble in water. Aluminum chloride is a powerful irritant and is moderately toxic when ingested.

aluminum chloride, anhydrous. See aluminum chloride.

aluminum chlorohydrate (aluminum hydroxy chloride; basic aluminum chloride; aluminum chlorohydroxide). a compound with the generally accepted formula, $Al_2(OH)_5Cl{\cdot}2H_2O$. It is used in commercial preparations of antiperspirants and deodorants and also in water purification and the treatment of sewage and factory effluents. As a component of antiperspirants and deodorants, it sometimes causes infection of hair follicles of the armpit and skin irritations that are severe enough to require medical attention. There is a possibility that the aluminum salts in antiperspirants may contribute to a buildup of aluminum in the body, with a consequent contribution to certain brain disorders, including Alzheimer's disease.

aluminum chlorohydroxide. aluminum chlorohydrate.

aluminum hydroxychloride. aluminum chlorohydrate.

aluminum oxide. See alumina.

aluminum tris(ethyl phosphite). fosetyl aluminum.

aluminum tris-(*O*-ethylphosphonate). fosetyl aluminum.

alupong. *Naja naja*.

Alutera scripta (filefish). a widely distributed marine bony fish (Family Monacanthidae) of warm waters, certain populations of which are transvectors of ciguatera.

alveolar clearance. the clearance of particles that reach the respiratory surfaces of the lung. This is usually effected either by (1) phagocytosis and removal by a mucociliary process (with or without coughing) or by the lymphatic system; or by (2) dissolution of the particles, passage in solution either to the cardiovascular or lymphatic system.

alveolar ventilation. the volume of fresh air that reaches the alveolar surface. It is equal to the volume inhaled minus the volume that is exhaled during a breathing cycle.

alveolitis. inflammation of the alveoli.

alveolus (plural, alveoli). **1:** a small hollow, cavity, or blind sac. **2:** any of exceedingly numerous blind sacs (ca. 100 to 500 million in humans) at the terminus of bronchioles in the lungs of mammals and reptiles. They are thin-walled hollow structures, ca. 250-350 μ in diameter and form the respiratory surface of the lung. They have a surface area in humans of up to 100 m^2 at maximum inspiration, thereby providing not only an excellent respiratory exchange surface, serving the exchange of respiratory gases (carbon dioxide and oxygen), but also as an important portal of entry and/or elimination for volatile xenobiotics and for volatile products of toxicant metabolism.

Alytes obstetricans (European midwife toad). a species of terrestrial toad (Class Amphibia; Subclass Anura). The male gathers the strings of eggs laid by the female around his hind legs and carries them thus until about to hatch. He then returns to water, where the tadpoles spawn and swim away. Adult midwife toads have few predators due to a potent poison released by the dermal glands ("warts").

Am. the symbol for americium.

amac-asa-hebi. *Bungarus multicinctus*.

amalgam (amalgamation). a liquid or solid alloy of mercury with one or more other metals.

amalgamate. to make an amalgam.

amalgamation. **1:** the process of making an amalgam. **2:** an amalgam.

amanin. an amanitin. It is extremely toxic.

Amanita (amanita). a genus of fairly large, gilled, mostly toxic mushrooms (Family Amanitaceae). Most species occur on the ground in forested areas. Some designated species are actually complexes of closely related species in which the toxicity of some forms is not established. Toxic species of *Amanita* cause about 90% of fatal intoxications by mushrooms in the United States. Toxicity varies among poisonous species. They contain a number of toxic substances, including amanitahemolysin, amanitatoxin, amanitin, amanitine, choline, ibotenic acid, muscaridine, muscarine, muscimol, and phalloidin. See also amatoxin, phallotoxin. The few edible species should be collected only by an expert on the genus. These fungi are plentiful in America and Europe. In the United States, they occur mainly in the mid-Atlantic states southward to Florida and west to Texas. They flourish between October and December. Most are encountered in drypine woods, but *A. phalloides* occurs on damp, sandy soils at mid to low elevations. See thioctic acid. ***A. aspera*** (western yellow veil). a striking species found on the ground under conifers along the Pacific coast of N. America. The cap is brownish with yellowish-gray patches, a ringed stalk, and a bulbous base bounded by loose fragments. It is not known to be edible. ***A. bisporiger*** (smaller death angel). similar in appearance to *A. phalloides*, it can be found growing from May through October in mixed woods or sometimes wooded lawns, especially near oaks. ***A. brunnescens*** (cleft-foot amanita). the cap is brownish with white patches; the stalk is large with a split basal bulb. It grows in dry, deciduous woods, especially under oaks in Quebec and in the eastern United States from the Atlantic coastal states west to Michigan and to eastern Texas. It is possibly poisonous and easily confused with the deadly *A. phalloides*. ***A. citrina*** (citron amanita). a deadly species with a yellowish cap marked by light buffy patches; the stalk is ringed; the basal bulb is large with a saclike cup. It grows in oak and pine woods in eastern North America. ***A. cokeri*** (Coker's amanita). a large white mushroom found in oak-pine woods in the eastern United States from New York to North Carolina, west to Indiana and Texas. It should not be eaten. ***A. cothurnata*** (booted amanita). a poisonous species found in oak, pine or mixed oak-pine woods. The cap is white, often with a yellow center with white patches; the stalk is ringed and the upper margin of the bulb is rimmed. It is possibly a variety of *A. panthera*. ***A. gemmata*** (gemmed amanita). possibly poisonous, it occurs in oak and pine woods and in urban areas of the eastern United States from New York to Florida, west to Michigan. ***A. muscaria*** (fly agaric; fly fungus; fly mushroom). a toxic species with a broad, flat or convex, yellow to blood-red cap and white gills; the stalk is ringed and has a bulbous base. All parts are poisonous and contain muscimol, muscarine, muscarin, muscazone, choline, muscaridine, ibotenic acid, and isoxazoles. It occurs under both evergreen (pine, spruce, madrone) and deciduous (oak, birch) trees in the Rocky Mountains and Pacific Coast of North America; rarely seen in the eastern United States. Symptoms of intoxication usually begin within about 3 hours after ingestion, and may initially include those of severe gastroenteritis with lacrimation, sweating, salivation, nausea, vomiting, diarrhea, abdominal cramps, excessive thirst, profuse sweating, vertigo, delirium, jaundice, convulsions, and possibly coma. Other symptoms that may occur are drowsiness and dizziness, followed by increased motor activity, elation, and even tremors, illusions, and manic excitement. The symptoms are seldom severe in adult humans and fatalities are rare, although convulsions and coma may persist for as long as 12 hours and may require respiratory maintenance. ***A. muscaria* var. *formosa*** (yellow-orange fly agaric). the cap is yellow to orange-red with buff or pinkish patches. It occurs mainly under pine, spruce, and eastern hemlock. Widely distributed in the United States, it is most common in late summer and fall in the northeast. It is poisonous, but not deadly. Symptoms of intoxication may include disorientation, sweating, deep sleep. ***A. mutabilis***. a species marked by an anise-like odor and reddish granules on the cap. It is not as deadly as others of this genus, but effects can be damaging if not treated in time. The two chief toxic principles are slow-acting amanitin and phalloidin. The former toxin produces hypoglycemia and the major symptoms of intoxication; phalloidin acts much more quickly, causing degenerative changes in liver, kidney, and cardiac muscle. Symptoms usually appear 6-15 (occasionally as much as 48) hours following ingestion. The longer a victim remains without symptoms the greater the danger, since the liver is assaulted immediately after digestion. Lacking symptoms, medical attention will not be sought and additional toadstools may be eaten. The first symptoms are usually nausea, vomiting, discomfort, and bloody diarrhea, followed by a sudden onset of extreme stomach pains, violent vomiting, intense thirst, and cyanosis of extremities. In fatal cases, the subject usually remains conscious until near death. Long intervals when the subject is alert, may be interrupted by brief periods of unconsciousness with eventual coma and death. Because of severe dehydration, it's only a matter of time before the potassium concentration cause cardiac arrest. ***A. pantherina*** (panther mushroom). the cap is brownish with whitish patches and a bulbous stalk, the margin of which may be free, rolled, or with rows of cottony scales. It is poisonous and can prove lethal. It contains muscarin, muscimole, ibotenic acid, and isoxazoles. It grows under conifers, especially Douglas fir, from the Rocky Mountains to the Pacific Coast of North America; rarely seen in the east. Symptoms of intoxication include delirium and coma or a coma-like sleep, and resemble the CNS effects of atropine. ***A. phalloides*** (death cap; deadly amanita; deadly agaric; green death cap). the cap is smooth and greenish; the top of the stalk has a skirt-like ring and the base bears a sac-like cup. It grows in evergreen and deciduous forests, especially on damp, sandy soils and in urban areas. It is abundant in Europe and North America and occurs in the eastern United States from Massachusetts to Virginia and Ohio, the Pacific Northwest, and California. It is deadly and often mistaken for edible mushrooms if the ring on the stalk is ignored. It closely resembles *A. verna*, *A. virosa*, and *A. bisporiger*. All parts are toxic due to the presence of principles such as phalloidin, amanitin, and antamanide. It is the most dangerous amanita. Symptoms may not appear until 10 to 15 hours after ingestion. They include nausea, severe intermittent vomiting, profuse diarrhea, thirst, weakness, extreme abdominal pain, restlessness, delirium, hallucinations, coma, and late-developing jaundice. Remissions alternating with these symptoms occur during a 6 to 8 day period, with rapidly developing renal failure and a hepatic insufficiency that is indistinguishable from acute viral hepatitis. This may be followed by a slow recovery or death due to cardiac, renal, hepatic, or CNS lesions. Lacking proper treatment, death may ensue in about 5-10 days. One or two mushrooms, even though cooked can be fatal to an adult; less for children. Mortality among affected humans ranges from 50 to 90%. See also lectin. ***A. porphyria*** (gray veil amanita). a brown-capped mushroom with a ringed, bulbous, grayish-patched stalk. It grows in evergreen and mixed woods of the eastern United States and the Pacific Northwest. It is considered poisonous. ***A. verna*** (destroying angel; fool's mushroom). a deadly mushroom of the Pacific Northwest, with a ringed stalk and a smooth, unlined white cap. ***A. virosa*** (destroying angel; death angel). a beautiful and deadly white mushroom with a flaring to ragged ring on the stalk and a cupped base. It is widely distributed throughout North America, growing alone or in small scattered groups in the grass under or near trees in mixed woods. It resembles *A. phalloides*, and the symptoms of intoxication are similar. Lacking proper medical care, Death may result.

amanita muscaria poisoning. poisoning due to ingestion of the mushroom, *Amanita muscaria*. Symptoms usually appear within 2-3 hours following ingestion and are largely those of parasympathetic stimulation with salivation, sweating, vomiting, abdominal pain, diarrhea, vertigo, coma, and convulsions. With proper therapy, complete recovery within 24 hours is possible.

amanita phalloides poisoning. poisoning due to the ingestion of the mushroom, *Amanita phalloides*, *q.v.* Symptoms are in some respects similar to those of Amanita muscaria poisoning and usually appear within 6-15 hours following ingestion. Effects may include weakness, abdominal pain, vomiting, purging, thirst, shock syndrome, restlessness, delirium, hallucinations, late icterus, acute renal failure, and coma. Lesions revealed on autopsy include acute yellow liver atrophy, acute renal failure, cardiopathy. Death may result from lesions of the CNS, kidney, liver, or heart. See also amanitine.

amanita poisoning. poisoning by ingestion of mushrooms of the genus *Amanita*, *q.v.* See also amanita muscaria poisoning, amanita phalloides poisoning.

amanitahemolysin. a hemolytic substance isolated from *Amanita phalloides*. It is heat labile and is destroyed during digestion. Thus, it does not account for the major symptoms of poisoning by *A. phalloides*.

amanitatoxin (*Amanita* toxin, amanitotoxin). a toxic principle isolated from *Amanita phalloides*. In animals it causes symptoms and lesions that are similar to the poisoning of humans by *A. phalloides*. It contains amanitine, α-amanitin, β-amanitin, and phalloidin.

amanitin. a group (or any of a group) of highly toxic, heat-stable, cyclic octapeptides (α-, β-, τ-amanitin, and amanin) isolated from various species of mushrooms. Of these, [illegible] ($C_{39}H_{54}N_{10}O_{13}S$) is the most toxic and is 10 to 20 times as toxic as phalloidin; β-amanitin ($C_{39}H_{53}N_9O_{14}S$) has been isolated as acicular crystals that are soluble in water, methanol, and ethanol. North American mushrooms that contain amanitin are *Amanita*, especially *A. phalloides*, *Galerina* spp., *Lepiota* spp., and *Conocybe filaris*. Amanitins are potent inhibitors of RNA polymerase II and III and thus of RNA synthesis. The toxic group from *A. phalloides* is comprised of α-, β-, and τ-amanitin, and amanin. Of these, α-amanitin is the major constituent. The group inhibits protein synthesis of mammalian cells. Amanitin causes salivation, vomiting, bloody stools, cyanosis, muscular twitching, convulsions, and can be fatal. In contrast to phalloidin, *q.v.*, amanitin has a delayed action; even at high doses, the lethal interval is more than 15 hours. *Cf.* amatoxin.

amanitine. a polypeptide, $C_{30}H_{45-47}O_{12}N_7S$, that contains an indole ring. A toxin, it is a component of amanitatoxin and is responsible for the major symptoms of poisoning from ingestion of *Amanita phalloides*. It also produces hypoglycemia.

amanitotoxin. amanitatoxin.

amaranth (C.I. Acid Red 27; C. I. 16185; E123; FD&C Red No. 2; Red Dye No. 2; Red No. 2). a dark reddish-brown to purple azo dye, banned in 1976 by the US Federal Drug Administration for use in foods, drugs and cosmetics. It is still used for these purposes in several countries, including those of the EEC. It is a tumorigen in female laboratory rats and is used as an experimental carcinogen.

Amaranthus (pigweed; carelessweed; redroot). a widely distributed genus of weedy, mostly annual herbs of North America (Family Amaranthaceae). Plants of this genus sometimes accumulate toxic concentrations of nitrate and some species (e.g., *A. palmeri* and *A. retroflexus*) are believed to poison livestock at times.

amaril. a hypothetical poison, once thought to cause yellow fever.

amarillic. pertaining to amaril.

[illegible] 1. 2,4,5-triphenylimidazoline. 2. various bitter plant-derived principles.

amaryllid. a plant of the family Amaryllidaceae.

Amaryllidaceae (amaryllis family). a family of perennial plants with bulbs, corms or rhizomes. Many contain toxic, usually emetic, alkaloids (e.g., lycorine) and some are toxic to livestock.

The bulbous members of the family (e.g., daffodils, narcissus, spider lilies, snowdrops, and snowflakes) can be especially dangerous to small children who may mistake them for food. Ingestion may induce severe and repeated episodes of vomiting with little or no diarrhea. Recovery is usually complete within 24 hours. See *Narcissus*.

amaryllis. a collective common name for the numerous hybrids of *Amaryllis belladonna*.

Amaryllis (amaryllis). a genus of South African bulbous plants (Family Amaryllidaceae) that is toxic to livestock. There are many cultivated hybrids, but only one true species, *A. belladonna* (belladonna-lily, naked lady). In the United States it grows well outdoors only in California.

amatoxin. one of two groups of thermostable toxins isolated from poisonous species of *Amanita*. An example is α-amanitin. They are extremely toxic, bicyclic octapeptides that act on the RNA-polymerase II system of eukaryotic cells, blocking RNA transcription and thus inhibiting protein biosynthesis and cell maintenance. They occur in many species and genera of mushrooms, most notably in species of *Amanita* and *Galerina*. The symptoms of amatoxin intoxication usually resemble acute hepatitis, with renal damage observed after a week or more. See amanitin, phallotoxin.

amaurosis. a complete loss of vision, especially where there is no lesion or pathologic condition of the eye in evidence. See also blindness. **intoxication amaurosis**. that caused by a systemic poison (e.g., ethyl alcohol, tobacco). **lead amaurosis**. amaurosis due to lead poisoning. **toxic amaurosis**. amaurosis due to neuritis of the orbital portion of the optic nerve. It is caused by methanol, lead, arsenic, quinine, and certain other systemic xenobiotics or endogenous substances.

amaurotic. pertaining to amaurosis or to a person suffering from amaurosis; afflicted with amaurosis.

Amazon slender coral snake. *Leptomicrurus narducci*.

Amazonian coral snake. *Micrurus spixii*.

Amazonian tree viper. *Bothrops bilineatus*.

ambient. **1:** surrounding. **2:** pertaining to or denoting the surrounding (background, or prevailing) environment, components, or aspects thereof (as ambient environment, ambient air, ambient temperature).

ambient air. **1:** the air surrounding or in the immediate vicinity of a living system. **2:** usually, in air pollution science, unconfined outdoor air.

ambient air level (AAL). any of a variety of state and local health-based guidelines for noncriteria air pollutants (toxics) in ambient air. They are based variously on risk assessments, NOAELs or LOAELs, OELs (RELs, PELs, TLVs). AALs for a given chemical vary greatly among jurisdictions.

ambient air standard. the highest concentration of a given air pollutant that is permitted under law within a given jurisdiction. See also ambient air.

ambient temperature. the temperature of the surrounding medium in the immediate vicinity of a living system or object.

ambiguous. **1:** vague, indefinite, unclear, open to alternative interpretations or misinterpretation. **2:** of uncertain origin.

ambivore. an animal that feeds on grasses and broadleaved plants.

ambivorous. feeding on grasses and broadleaved plants.

ambivory. the selective feeding of an animal on grasses and broadleaved plants.

Amblyomma. a genus of ticks. Some North American species cause tick paralysis, *q.v.* See also *Dermacentor*.

amblyopia. unilateral reduced visual acuity with no detectable change in the eye itself. **toxic amblyopia**. that due to optic neuritis involving the orbital portion of the optic nerve. It is caused by methyl alcohol, lead, arsenic, quinine, or other toxicants.

amblyopic. pertaining to, or afflicted with, amblyopia.

amebacidal. amebicidal.

amebacide. amebicide.

amebicidal (amebacidal). destructive to amebae.

amebicide (amebacide). any agent that destroys amebas.

amelia. the congenital absence of one or more limbs. See thalidomide.

ameliorate. to ease, improve (e.g., the condition of); to alleviate.

amelioration. moderation or improvement of an illness or condition.

ameliorative. able or tending to ameliorate.

American arborvitae. *Thuja occidentalis*.

American aspidium. *Dryopteris marginalis*.

American Association for Accreditation of Laboratory Animal Care (AAALAC). a nonprofit corporation, founded in 1965, that inspects by invitation and accredits laboratory animal care facilities and programs. The association encourages optimal care for laboratory animals through peer review, site visits, and specific recommendations to the responsible authority for each facility reviewed.

American blacksnake. See blacksnake.

American Chemical Society. See grade.

American chestnut. *Castanea dentata*.

American Conference of Governmental and Industrial Hygienists (ACGIH). a professional organization in the United States that develops [illegible] (TLV) for airborne toxicants in the workplace.

American copperhead. *Agkistrodon contortrix*.

American coral snake. any snake of the genus *Micrurus*. Sometimes used loosely to include *Micruroides*.

American hellebore. *Veratrum viride*.

American ivy. *Parthenocissus quinquefolia*.

American lance-headed viper. See *Bothrops*.

American lanceheads. See *Bothrops*.

American laural. plants of the genus *Kalmia*.

American mistletoe. *Phoradendron serotinum*.

American mole shrew. *Blarina brevicauda*.

American musquash root. *Cicuta maculata*.

American nightshade. common name of *Solanum americanum* and sometimes of *Phytolacca americana*.

American red elder. *Sambucus pubens*.

American toad. common name of *Bufo americanus*.

American veratrum. *Veratrum viride*.

American water moccasin. *Agkistrodon piscivorus*.

American white hellebore. *Veratrum viride*.

American white ipecac. *Euphorbia ipecacuanha*.

American wormseed. *Chenopodium ambrosioides*.

American yew. *Taxus canadensis*.

americium (Am). a highly toxic, silvery, metallic radioactive element of the actinoid series. It does not occur naturally, but is obtained from the bombardment of uranium with neutrons or is synthesized from plutonium 241 and 243. The half-life of the most stable isotope, ^{243}Am, is 7.4 x 10^3 years.

Amerikanische Lanzenottern. See *Bothrops*.

Ames assay. See test (Ames test).

amethopterin. methotrexate.

ametryn (2-ethamino-4-isopropylamino-6-methylmercapto-*s*-triazine. ametryn; *N*-ethyl-*N'*-(1-

methylethyl)-6-(methylthio)-1,3,5-triazine-2,4-diamine; ametryne). a methylthiotriazine similar in biological properties to atrazine and simazine, but absorbed through foliage as well as the roots of green plants. Ametryne and related methylthiotriazines are used both as preemergence and postemergence herbicides.

ametryne. ametryn.

Amianthium muscitoxicum (synonym, *A. muscaetoxicum*; fly-poison; staggergrass; cowpoison). a highly toxic perennial herb (Family Liliaceae) that contains the steroid alkaloid, veratramine. It is highly toxic either on contact or by ingestion. The bulb is especially toxic.

amianthosis. asbestosis.

amidase. any deamidizing hydrolase that enables hydrolysis of a monocarboxylic acid amide, yielding a monocarboxylic acid and ammonia or one that hydrolyzes a nonpeptide C-N bond of an amide. *Cf.* esterase, amidinase.

amide. **1a:** any organic compound that contains the $CONH_2$ radical; a primary amide. **1b:** any organic compound with the general formula $RCONH_2$ (primary amide), $(RCO)_2NH$ (secondary amide), or $(RCO)_3N$ (tertiary amide). All are basic, white, crystalline solids. **2:** any inorganic salt that contains NH_{2^-}. They are formed by the reaction of ammonia with metals such as sodium and potassium. See also herbicide (amide herbicide).

amidinase. a deamidinizing enzyme that enables the hydrolysis of a nonpeptide C-N bond of an amidine. *Cf.* amidase.

amidine. a strong monobasic organic compound with the general formula, $RC(=NH)NH_2$. Such are formed by the reaction of ammonia with nitriles or with imido esters.

***N*-amidinosarcosine**. creatine.

amidocyanogen. cyanamide.

amikacin. an antibacterial aminoglycoside.

aminacrine (9-acridinamine; 5-aminoacridine; 9-aminoacridine). an antiinfective agent and local antiseptic used in veterinary practice. It is very toxic to laboratory mice.

amine. any of a class of nitrogenous organic compounds with the general formula R_3N where R can be hydrogen or a hydrocarbon group. Amines can be thought of as derivatives of ammonia, NH_3, in which one or more of the hydrogen atoms has been replaced by organic radicals. Synthetic amines are prepared by reducing amides or nitro compounds. Naturally produced toxic amines are uncommon, but include the toxic principles of some mistletoes (Family Loranthaceae) and some wild peas (*Lathyrus*), and they also contribute to the toxicity of a number of mushrooms. Amines are rare among phytotoxins, most of which are alkaloids (but See *Acacia berlanieri*). See also *Claviceps*. **aliphatic amine**. any amine in which the substituted hydrocarbon groups are aliphatic hydrocarbon groups. All have strong odors. **aromatic amine** (arylamine). any amine derived from an aromatic hydrocarbon by the replacement of at least one hydrogen on the benzene ring by an amino group ($—NH_2$). Some bioaccumulate through food chains. The most common target organs among laboratory animals are the kidney, liver, gastrointestinal tract, and urinary bladder. Many are carcinogenic to laboratory animals and to humans; others are suspected human carcinogens. The most commonly induced cancers are cancers of the liver, breast (chiefly in women), and urinary bladder (chiefly in males). See aromatic hydrocarbon, benzidene, 4-biphenylamine, 2-naphthylamine; *o*-toluidine. **biogenic amine**. any of a large group of naturally occurring, biologically active compounds that contain a primary amine group. Most are neurotransmitters, such as the catecholamines (adrenalin, noradrenalin, and dopamine) and the indoleamine, serotonin. See also biogenic. Other examples are histamine and serotonin. **carbocyclic aromatic amine**. an amine in which at least one substituent group is an aromatic ring containing only C atoms as part of the ring structure, and with one of the C atoms in the ring bonded directly to the amino group. Several of these compounds cause cancer in the human bladder, ureter, and pelvis, and are suspected of causing lung, liver, and prostate cancer. **cyclic amine**. an amine in which nitrogen atoms form part of the ring structure. **fatty amine**. an amine that contains alkyl groups having more than 6 carbon atoms. The commercially useful fatty amines are synthesized from naturally occurring fatty acids and are used, for example, as fabric softeners, chemical intermediates, emulsifiers of petroleum and asphalt, and flotation agents (e.g., of ores). **higher aliphatic amine**. any amine in which one or more aliphatic hydrocarbon groups contains seven or more carbon atoms. Most are less hazardous toxicologically than lower aliphatic amines. **lower aliphatic amine**. an amine in which none of the aliphatic hydrocarbon groups contains more than 6 carbon atoms; all are highly flammable. They are used chiefly as intermediates in the manufacture of other chemicals, including polymers, agricultural chemicals, and medicinal chemicals. Lower aliphatic amines are among the more toxic substances that are used routinely on a large scale. Toxicity is due in part to their high pH. They are also rapidly and easily absorbed; are corrosive to tissue and can cause tissue necrosis at the point of contact; eyes are especially vulnerable. Systemic effects may be seen in many organs. Necrosis of the liver and kidneys is common, as is hemorrhage and edema of the lungs. The immune system can become sensitized to amines. **primary amine**. an amine in which one hydrogen on NH_3 is substituted by a hydrocarbon group. **secondary amine**. an amine in which 2 hydrogens are each replaced by a hydrocarbon group. **tertiary amine**. an amine in which 3 hydrogens are replaced by hydrocarbon groups.

amine oxidase. **1:** monoamine oxidase. **2:** any enzyme that oxidizes amines. See diamine oxidase.

amino-. a prefix indicating the presence of an amino group, NH_2.

amino acid (aminoalkanoic acid; AA). any of a class of organic acids with the general formula R-CH(NH_2)-COOH (α-amino acids) where R is a distinguishing group. Amino acids are the basic structural units of polypeptides and proteins. They occur as optically active D- and L-isomers, the latter predominating in living organisms. Some 24 distinct amino acids occur in proteins. They are synthesized by autotrophic organisms. Heterotrophs synthesize some amino acids, but acquire others from dietary proteins. Amino acids are also biosynthetic building blocks of many important compounds. Those that are not synthesized by a particular species of animal are termed essential amino acids. For humans these are arginine, histidine, isoleucine, leucine, lysine, methionine, phenylalanine, threonine, tryptophan, and valine. See antagonist (amino acid antagonist), conjugation (amino acid conjugation), L-amino acid oxidase.

amino acid antagonist. See antagonist.

amino acid conjugation. See conjugation.

L-amino acid oxidase. a group of homologous enzymes that occur in all snake venoms. They catalyze the oxidation of L-α-amino and α-hydroxy acids. That from *Crotalus adamanteus* (Eastern diamondback rattlesnake) is extremely toxic. See also enzymes of snake venoms.

amino acid transpsort. See transport.

amino compound. any organic compound that contains an amino group, *q.v.*

amino group. NH_2, the essential component of an amino acid. Compounds with amino groups occur in numerous important or essential biological compounds. Numerous amino compounds are important toxicants (e.g., many polypeptides, proteins).

4-amino pyridine. a compound that inhibits the conduction of impulses in a neuron by blocking potassium conductance channels.

aminoacetic acid. glycine.

aminoacidemia. excessive amounts of amino acids in the blood.

5-aminoacridine. aminacrine.

9-aminoacridine. aminacrine.

aminoalkanoic acid. amino acid.

5-amino-6-(7-amino-5,8-dihydro-6-methoxy-5,8-dioxo-2-quinolyl)-4-(2-hydroxy-3,4-dimethoxyphenyl)-3-methylpicolilic acid. streptonigrin.

5-amino-6-(7-amino-5,8-dihydro-6-methoxy-5,8-dioxo-2-quinolyl)-4-(2-hydroxy-3,4-dimethoxyphenyl)-3-methyl-2-pyridinecarboxylic acid. streptonigrin.

***o*-aminoanisole**. See anisidine.

_p_-aminoanisole. See anisidine.

aminoanthroquinone. either of two tricyclic compounds, 1-aminoanthroquinone and 2-aminoanthroquinone, $C_6H_4(CO)_2C_6H_3NH_2$. They are used as dyes and pharmaceutical intermediates. 2-aminoanthraquinone is carcinogenic in some laboratory animals. It is a crystalline substance with red or orange-brown needles that are soluble in ethanol, chloroform, benzene, and acetone; insoluble in water.

aminoazo compound. any amino compound with aromatic rings linked to two nitrogen atoms by an azo bond (—N=N—); all are carcinogenic. See aminoazobenzene, *o*-aminoazotoluene.

aminoazo dye. azo dye.

aminoazobenzene (*p*-phenylazo)aniline; C.I. 11000; C.I Solvent Blue 7). a carcinogenic aromatic amine, $C_{12}H_{11}N_3$, with relatively high lipophilicity and a strong tendency to bioaccumulate. It is toxic by intraperitoneal route and is a probable mutagen. It is used as a dye in lacquers, varnishes, stains, and resins.

_o_-aminoazotoluene. a red crystalline solid; it is a known carcinogen. See aminoazo compound.

aminobenzene. aniline.

_p_-aminobenzenearsonic acid. arsanilic acid.

4-aminobenzenesulfonamide. sulfanilamide.

_p_-aminobenzenesulfonamide. sulfanilamide

2-aminobenzoic acid. anthranilic acid.

4-aminobenzoic acid. *p*-aminobenzoic acid.

_p_-aminobenzoic acid (4-aminobenzoic acid; vitamin B_x; bacterial vitamin H^1; chromotrichia factor; antichromotrichia factor; trichochromogenic factor; anticanitic vitamin; PABA). a compound that occurs widely among biota, it is generally considered to be a growth factor in laboratory rats, the domestic fowl, and yeast, and is the anti-gray-hair factor in laboratory rats. PABA is moderately toxic to laboratory mice and rabbits; less so to rats. It is used as a topical sunscreen, in research as a sulfonamide antagonist, and formerly as an antirickettsial. See also acetylamino-benzoic acid.

2-aminobenzoic acid-3-phenyl-2-propenyl ester. cinnamyl anthranilate.

4-aminobiphenyl (*p*-aminobiphenyl; *p*-biphenylamine). a colorless or purplish, crystalline aromatic hydrocarbon, $C_6H_5C_6H_4NH_2$, which, on decomposition by heating, emits toxic nitrogen oxide gases. It is carcinogenic to experimental animals, producing neoplasms in various tissues. When introduced by gavage in rabbits, rats, and dogs, it produces cancer of the urinary bladder. It also causes cancer of the urinary bladder in humans. This compound is no longer produced commercially in the United States.

_p_-aminobiphenyl. 4-aminobiphenyl.

3-aminobutanoic acid. β-aminobutyric acid.

4-aminobutanoic acid. γ-aminobutyric acid

3-(2-aminobutyl) indole acetate. etryptamine acetate.

4-(aminobutyl)guanidine. agmatine.

γ-aminobutyrate-α-ketoglutarate aminotransferase. τ-aminobutyric acid transaminase.

β-aminobutyric acid (3-aminobutanoic acid; β-amino-*n*-butyric acid). a highly toxic, nearly tasteless, water-soluble compound, $CH_3CH(NH_2)\text{-}CH_2COOH$, that can induce a profound coma-like narcosis.

γ-aminobutyric acid (4-aminobutanoic acid; γ-amino-*n*-butyric acid; piperidic acid; GABA). an amino acid, $NH_3CH_2CH_2CH_2COOH$, that is not a component of protein. It occurs chiefly in nervous tissue as a hyperpolarizing agent. It is an inhibitory neurotransmitter of the CNS and may mediate the inhibitory actions of local interneurons. Inhibition of the synthesis or transmission of γ-aminobutyric acid can cause convulsions. It is used therapeutically as a hypertensive agent. See also bicuculline, 3-mercaptopropionic acid, picrotoxin, bicuculline.

β-amino-_n_-butyric acid. β-aminobutyric acid.

τ-amino-_n_-butyric acid. τ-aminobutyric acid.

aminocarb (4-(dimethylamino)-3-methylphenol-methylcarbamate ester; methylcarbamic acid 4-

(dimethylamino)-*m*-tolyl ester; 4-dimethylamino-*m*-tolyl methylcarbamate). a tan, crystalline, slightly water-soluble, highly poisonous compound with a melting point of 93 to 94°C; used as an insecticide. It is extremely toxic to laboratory rats.

4-[(aminocarbonyl)aminophenyl]arsonic acid. carbarsone.

***N*-(aminocarbonyl)aspartic acid calcium salt (1:1)**. calcium *N*-carbamoylaspartate.

2-[2-[(aminocarbonyl)oxy]-1-methoxyethyl]-3,6-bis(1-aziridinyl)-5-methyl-2,5-cyclohexadiene-1,4-dione. carboquone.

3-amino-*N*-(α-carboxyphenethyl)succinamic acid *N*-methyl ester. aspartame.

5′-[(3-amino-3-carboxypropyl)methylsulfonio]-5′-deoxyadenosine hydroxide, inner salt. *S*-adenosylmethionine.

2-[(aminocarbonyl)oxy]-*N,N,N*-trimethyl-1-propanaminium chloride. bethanechol chloride.

1-amino-2-chlorobenzene. *o*-chloroaniline (See chloroaniline).

1-amino-3-chlorobenzene. *m*-chloroaniline (See chloroaniline).

1-amino-4-chlorobenzene. *p*-chloroaniline.

***m*-aminochlorobenzene**. *m*-chloroaniline (See chloroaniline).

2-amino-5-chlorobenzoxazole. zoxazolamine.

1-amino-3-chloro-6-methylbenzene.5-chloro-*o*-toluidine.

2-amino-4-chlorotoluene. 5-chloro-*o*-toluidine.

2-amino-5-chlorotoluene. 4-chloro-*o*-toluidine.

2-amino-5-chlorotoluene hydrochloride. 4-chloro-2-toluidine hydrochloride.

***m*-amino-*p*-cresol methyl ester**. *p*-cresol.

3-amino-*p*-cresol methyl ester. *p*-cresol.

aminocyclohexane. cyclohexylamine.

1-aminocyclopentanecarboxylic acid. cycloleucine.

3-amino-2,5-dichlorobenzoic acid. chloramben.

α-amino-2,3-dihydro-3-oxo-5-isoxazoleacetic acid. ibotenic acid.

α-amino-2,3-dihydro-2-oxo-5-oxazoleacetic acid. muscazone.

2-amino-1-(3,4-dihydroxyphenyl)ethanol. norepinephrine.

2-amino-3-(3,4-dihydroxyphenyl)propanoic acid. levodopa.

1-amino-6,17-dihydroxy-7,10,18,21-tetraoxo-2,7-(*N*-acetylhydroxylamino)-6,11,17,22-tetraazaheptaeicosane. deferoxamine.

2-amino-4,6-dinitrophenol. picramic acid.

4-aminodiphenyl. *p*-biphenylamine.

***p*-aminodiphenyl**. benzidine.

aminoethane. ethylamine.

2-aminoethanethiol. cysteamine.

2-aminoethyl mercaptan. cysteamine.

β-aminoethylbenzene. phenethylamine.

4-(2-aminoethyl)-1,2-benzenediol. dopamine.

3-amino-9-ethylcarbazole (3-amino-*N*-ethylcarbazole hydrochloride). the hydrochloride salt of this substance and the free amine are carcinogenic to laboratory animals. It is also a poison by ingestion and by *i.p.* routes of administration. If heated to decomposition it emits NO_x.

3-amino-*N*-ethylcarbazole hydrochloride. 3-amino-9-ethylcarbazole.

β-aminoethylglyoxaline. histamine.

4-aminoethylglyoxaline. histamine.

3-(β-aminoethyl)-5-hydroxyindole. serotonin.

β-aminoethylimidazole. histamine.

3-(2-aminoethyl)-1*H*-indol-5-ol. serotonin.

2-aminoethyl mercaptan. cysteamine.

4-(2-aminoethyl)phenol. tyramine.

***p*-β-aminoethylphenol**. tyramine.

4-(2-aminoethyl)pyrocatechol. dopamine.

2-aminofluorene. a carcinogenic polycyclic arylamine.

4-aminofolic acid. aminopterin.

α-amino-3-hydroxy-5-isoazoleacetic acid. ibotenic acid.

aminomercuric chloride. ammoniated mercuric chloride (See mercuric chloride).

4-amino-10-methylfolic acid. methotrexate.

5-(aminomethyl)-3-isoxazol. muscimol.

3-(4-amino-2-methylpyrimidyl-5-methyl)-4-methyl-5,β-hydroxy-ethylthiazolium chloride. thiamine.

1-aminonaphthalene. 1-naphthylamine.

2-aminonaphthalene. 2-naphthylamine.

***p*-aminonitrobenzene**. *p*-nitroaniline.

4-amino-PGA. aminopterin.

aminophen. aniline.

4-aminophenylarsonic acid. arsanilic acid.

(4-aminophenyl)arsonic acid sodium salt. sodium arsanilate.

1-amino-2-phenylethane. phenethylamine.

1-(4-aminophenyl)-1-propanone. *p*-aminopropiophenone.

aminophylline (theophylline compounded with ethylenediamine; 3,7-dihydro-1,3-dimethyl-1*H*-purine-2,6-dione compounded with 1,2-ethanediamine(2:1); theophylline ethylenediamine). a very toxic, bitter-tasting, white or yellowish, alkaloid mixture of theophylline and aqueous ethylenediamine, $C_{16}H_{24}N_{10}O_4$, with a mild ammoniacal odor. It is used chiefly to control asthma, as a diuretic, and a stimulant of cardiac muscle and the respiratory center. See aminophylline poisoning.

aminophylline poisoning. similar to caffeine poisoning, effects include restlessness with intervals of drowsiness; rapid pulse; tinnitus; vomiting; diuresis; fever; dehydration; thirst; delerium; tremor, convulsions; coma; and death from cardiovascular or pulmonary collapse. Lesions include gastric ulceration and CNS stimulation.

β-aminopropionitrile. See β-(γ-L-glutamyl)-aminopropionitrile.

***p*-aminopropiophenone** (1-(4-aminophenyl)-1-propanone; ethyl *p*-aminophenyl ketone; PAPP). an extremely toxic organic compound and antidote to cyanide poisoning.

β-aminopropylbenzene. amphetamine.

3-aminopropylene. allylamine.

4-(2-aminopropyl)phenol hydrobromide. hydroxyamphetamine hydrobromide.

***p*-(2-aminopropyl)phenol hydrobromide**. hydroxyamphetamine hydrobromide.

aminopterin (*N*-[4-[[(2,4-diamino-6-pteridinyl)-methyl]amino]benzoyl]-L-glutamic acid; 4-aminofolic acid; 4-aminopteroylglutamic acid; 4-amino-PGA). a rodenticide, abortifacient, and antineoplastic agent, it is a folic acid antagonist that inhibits purine synthesis. It is teratogenic causing nervous system defects, abnormalities of the extremities, micrognathia, hydrocephalus, and related abnormalities in the fetus.

4-aminopteroylglutamic acid. aminopterin.

aminopurine (2-aminopurine). an analog of adenine that pairs with cytosine (not thymine) and is thus mutagenic.

2-aminopurine. See aminopurine.

aminopyrine. a very hazardous pyrazolon derivative, $C_{13}H_{17}N_3O$, used as an antipyretic, analgesic, and antiinflammatory agent. It may induce agranulocytosis. See also apazone, antipyrine, phenylbutazone, salicylate.

4-amino-*N*-2-quinoxalinylbenzenesulfonamide. sulfaquinoxaline.

4-amino-1-β-D-ribofuranosyl-1,3,5-triazine-2(1H)-one. azacitidine.

aminosalicylate. any salt of aminosalicylic acid. See 4-aminosalicylic acid.

4-aminosalicylic acid (PAS; PASA; *p*-aminosalicylic acid; 4-amino-2-hydroxybenzoic acid). a moderately toxic, white or near-white, dense, odorless or nearly odorless powder, $NH_2C_6H_3$-(OH)COOH. Its salts are used in medicine and as an industrial preservative; it is nephrotoxic.

***p*-aminosalicylic acid**. 4-aminosalicylic acid.

5-(aminosulfonyl)-4-chloro-2-[(2-furanylmethyl)amino]benzoic acid. furosemide.

5-amino-1,2,3,4-tetrahydroacridine. tacrine.

9-amino-1,2,3,4-tetrahydroacridine. tacrine.

2-aminothiazole (2-thiazolamine). a thyroid inhibitor, derived from thiazole synthesized from vinyl acetate. It is very toxic to laboratory rats.

4-amino-*N*-2-thiazolylbenzenesulfonamide. sulfathiazole.

aminotoluene. benzylamine.

***o*-aminotoluine**. *o*-toluidine.

***p*-aminotoluene**. *p*-toluidine.

3-amino-*p*-toluidene. toluene-2,4-diamine.

5-amino-*o*-toluidene. toluene-2,4-diamine.

aminotriazole. 3-amino-1,2,4-triazole.

3-amino-1*H*-1,2,4-triazole. 3-amino-1,2,4-triazole.

3-amino-1,2,4-triazole (1*H*-1,2,4-triazol-3-amine; 3-amino-1*H*-1,2,4-triazole; 1,2,4-triazol-3-ylamine; aminotriazole; ATA; amitrole). a colorless, crystalline solid, soluble in water and ethanol and used chiefly as the active component of a selective pre- and postemergence herbicide spray. It is a powerful plant growth suppressant and cotton defoliant that is readily absorbed by leaves and roots. It causes albinism in developing leaves which is irreversible if the application rate is high. It inhibits cell division in primary root meristems of some plants (e.g., *Pisum, Linum*). It is extremely toxic to laboratory mice and rats, is carcinogenic to a number of laboratory animals, and a suspected human carcinogen. All uses were banned in the United States in 1971. *Cf.* amitrole.

4-amino-3,5,6-trichloropicolinic acid. picloram.

4-amino-3,5,6-trichloro-2-pyridinecarboxylic acid. picloram.

(8S-*cis*)-10-[(3-amino-2,3,6-trideoxy-α-L-lyxo-hexopyranosyl)oxy]-7,8,9,10-tetrahydro-6,-8,11-trihydroxy-8-(hydroxyacetyl)-1-methoxy-5,12-naphthacenedione. doxorubicin.

amiton (S-[2-(diethylamino)ethyl]phosphorothioic acid *O,O*-diethyl ester; *O,O*-diethyl *S*-(β-diethylamino)ethylphosphorothiolate). an extremely toxic cholinesterase inhibitor used as a contact insecticide and miticide.

amitriptyline (3-(10-,11-dihydro-5H-dibenzo-[a,d]cyclohepten-5-ylidene)-*N,N*-dimethyl-1-propanamine). a tricyclic antidepressant usually given as the hydrochloride. It is very toxic to laboratory mice and rats.

amitrole. the generic name for 3-amino-1,2,4-triazole.

Ammi. a genus of branched annual herbs (Family Apiaceae) of the North Atlantic islands and the Mediterranean region. *Ammi majus* (bishop's weed) and *A. visnaga* seeds eaten by poultry produce severe photosensitization. Toxic concentrations of nitrate in the foliage have also been reported.

ammoidin. methoxsalen.

ammonia. 1: true ammonia is a colorless gas, NH_3, with a characteristic pungent odor. It is highly water soluble (a saturated solution at 0°C contains 36.9% ammonia). The aqueous solution is alkaline and caustic, with ammonium hydroxide (NH_4OH) predominating. Ammonia has many commercial and industrial uses, e.g., as a refrigerant, a pesticide, in the manufacture of nitric acid, dyes, fertilizers (e.g., ammonium nitrate, ammonium sulfate, urea), explosives, and synthetic fibers. Sources of occupational exposure also include the silvering of mirrors, glue-making, and tanning of leather. It is formed biologically by the decomposition and deamination of nitrogenous substances (e.g., proteins and amino acids). Ammonia is highly toxic to terrestrial and aquatic animals, less so to plants. It is important in intermediary metabolism and is the principal excretory product from protein metabolism in many bacteria, fungi, seeds, invertebrate animals, and aquatic vertebrates. In mammals, most endogenous ammonia is detoxified by conversion to urea which is temporarily stored in the urinary bladder. In terrestrial gastropods, insects other than Diptera, most reptiles, and birds, the chief nitrogenous excretory product is uric acid. Ammonia is caustic and damaging to the skin, mucous membranes, and eyes. Inhalation of concentrated ammonia can be life threatening, causing severe respiratory distress with glottic spasm, laryngeal edema, and asphyxiation. If such exposure is not fatal, full recovery is usual. See also ammonium hydroxide, ammonia poisoning. **2:** ammonia water. **3:** often loosely applied to compounds such as ammonium chloride, ammonium hydroxide, benzalkonium chloride, and quarternary ammonium salts, all of which are toxic and irritant to skin, eyes, and mucous membranes. Depending upon the concentration, site, and type of exposure, any of these compounds may cause conjunctivitis, laryngitis, tracheitis, pulmonary edema, pneumonitis, and chemical burns on contact with tissue.

ammonia intoxication. ammonia poisoning.

ammonia poisoning (ammonia intoxication; am monia toxicity). poisoning from true ammonia, NH_3, of endogenous or exogenous origin. In many vertebrate animals ammonia is detoxified in the liver; but in certain disorders, it may be shunted past the liver. In such cases, ammonia may accumulate in blood with serious consequences that may compound the deleterious effects of the precipitating condition. Effects in humans that are due at least in part to ammonia include various neurologic effects, altered consciousness, abnormalities in the electroencephalogram, coma, and a flapping tremor of the liver (See asterixis). See also ammonia toxicity. **ammonia poisoning, acute**. a life-threatening condition caused by inhalation of concentrated ammonia in animals with lungs or by absorption of dissolved ammonia through gills in fish and other aquatic organisms. Concentrated ammonia gas is an asphyxiant causing spasm of the glottis and edema of the upper respiratory tract if inhaled. Death may swiftly intervene in the absence of prompt treatment. Major signs in humans include severe respiratory distress with glottic spasm, laryngeal edema and asphyxiation. Additional signs may include headaches, salivation, burning throat, nausea, and vomiting. Lesions may include iritis, opacity of the eye lens, perforation of the cornea, corrosive inflammation of the esophagus and stomach, and laryngeal and pulmonary edema. Convulsions, coma, and death may result if poisoning is severe, otherwise full recovery is usual. Signs in fish may include hyperexcitability, disequilibrium, elevated cardiac and respiratory rates with increased oxygen uptake, and death. **ammonia poisoning, chronic, of fish**. signs may include lower rates of growth and development and impaired hatchability of eggs. Lesions of gills, liver, and kidney are also common. Percidae and Salmonidae appear to be the most sensitive families of fish. Aquatic plants appear to be less sensitive to ammonia than aquatic animals. See also ammonia toxicity.

ammonia solution. ammonia water.

"**ammonia toxicity**." sometimes used as a synonym for ammonia poisoning. This usage is not recommended.

ammonia water (ammonia; ammonia solution; household ammonia; aqua ammonia). a highly volatile, extremely irritant liquid; the fumes are extremely irritant to the eyes and lungs. The "ammonia" commonly used in dry cleaning, it is an aqueous solution of ammonium hydroxide. It is also used alone and in various formulations, as a general or all-purpose cleaner in the home where it is a common source of poisoning. It can damage the skin, causing rashes, erythema, and chemical burns. The fumes are extremely irritating to the eyes and lungs and can be seriously harmful to individuals with respiratory problems such as colds, asthma, or bronchitis. Warning labels for household ammonia vary greatly; an example: "POISON:

May cause burns. Call a physician. Keep out of reach of children." *Cf.* ammonium hydroxide.

ammoniac. ammoniacal.

ammoniacal (ammoniac). **1a:** pertaining to ammonia or its properties. **1b:** having the characteristics of ammonia. **2:** pertaining to a solution in aqueous ammonia.

ammoniated. containing or combined with ammonia; ammonia added.

ammoniated mercuric chloride. See mercuric chloride.

ammoniated mercury. ammoniated mercuric chloride. See mercuric chloride.

ammoniemia (ammonemia). the presence of ammonia or ammonia-containing compounds in blood, probably formed from the decomposition of urea. Symptoms typically include a weak pulse, lowered body temperature, gastrointestinal pain or discomfort, and coma.

ammonification (ammonization). addition of, or treatment of, a substance with ammonia.

ammonium chloride (sal ammoniac). a white, crystalline, somewhat hygroscopic substance, NH_4Cl, with a characteristic salty taste; very soluble in water and glycerol; slightly soluble in ethanol. It sublimes on heating; the fumes are toxic. Ammonium chloride is used as a soldering flux, as a pickling agent in galvanizing and electroplating, as a mordant in dyeing and printing, and in the manufacture of various ammonium compounds and of dry cell batteries. See also ammonia (def. 3).

ammonium fluoride (neutral ammonium fluoride). a granular powder, FH_4N, that yields ammonia and HF gas on heating. It is used to etch and frost glass, as a wood preservative, moth-proofing agent, and antiseptic in brewing [illegible] gestion are those of severe gastrointestinal distress with hemorrhagic gastroenteritis. Included are nausea, vomiting, abdominal pain, diarrhea, muscular weakness, tremors, and convulsions. In severe cases, hyperpnea, depression, cardiovascular collapse, and death may ensue. Chronic poisoning mottles dental enamel and produces a generalized osteosclerosis with synostoses and calcification of ligaments and tendons.

ammonium fluosilicate. ammonium hexafluorosilicate.

ammonium hexafluorosilicate (cryptohalite; diammonium hexafluorosilicate; ammonium fluosilicate; ammonium silicofluoride). a very toxic, odorless, crystalline powder used in pesticides. Symptoms are chiefly those of severe gastroenteritis. In fatal cases, symptoms may include dyspnea, tremors, convulsions, muscular weakness, and death from cardiovascular collapse.

ammonium hydrate. ammonium hydroxide.

ammonium hydroxide (ammonia solution; aqua ammonia; ammonium hydrate). a colorless, toxic, corrosive, strongly alkaline liquid, NH_4OH, with a strong pungent odor. It is an aqueous solution of up to about 30% ammonia. Both the liquid and vapor are highly irritant to the skin, eyes, and mucous membranes. When mixed with chlorine-containing products, it evolves deadly chloramine, and, on heating, releases toxic and flammable fumes. Many cleaning products contain ammonium hydroxide, but do not include it on the label. Ammonium hydroxide is the "ammonia" commonly used for dry cleaning and alone or in other products as a general-purpose household cleaner. If swallowed, it produces symptoms of gastrointestinal distress similar to that of alkali poisoning, including extreme pain, coughing, vomiting, a weak rapid pulse, and collapse. Later, the stomach and esophagus may perforate, with fever and intensification of abdominal pain with rigidity. Pulmonary edema, irritation, and even pneumonia may appear within 12 to 24 hours. Inhalation of highly concentrated fumes cause swelling of the lips and eyelids, with temporary blindness; the victim may foam at the mouth and the skin appears reddish. It is a frequent source of poisoning in households. See also ammonia (def. 3), ammonia water, ammonium hydroxide poisoning.

ammonium hydroxide poisoning. a common source of poisoning in the home, ammonium hydroxide produces symptoms following ingestion that may include those of extreme gastrointestinal distress with extreme pain in the mouth, chest, and abdomen; coughing; vomiting; and collapse. Perforations of the esophagus and stomach may occur with fever, and abdominal rigidity. At 12-24 hours after acute exposure, lung irritation and edema may appear. Inhalation of fumes may cause edema of the lips and eyelids, temporary blindness, restlessness, tightness in the chest, foaming at the mouth, reddening of the skin, the pulse is weak but rapid. See also alkali poisoning.

ammonium ion. a positively charged univalent cation, NH_4^+. The herbicidal activity of ammonium salts such as ammonium sulfate and sodium sulfamate is due chiefly to the ammonium ion. Ammonia is absorbed very rapidly into the plant cell, quickly rendering the normally acid and highly buffered sap alkaline. The increased alkalinity alone quickly kills the cell.

ammonium silicofluoride. ammonium hexafluorosilicate.

ammonium sulfate (ammonium sulphate). a toxic, white to brownish-gray crystalline substance, $(NH_4)_2SO_4$. It is highly soluble in water, forming a corrosive, acidic solution; insoluble in ethanol and acetone. It is used as a rapid-acting herbicide, in fertilizers, in water treatment, fermentation, in fire-proofing formulations, and as a food additive. Toxicity is due chiefly to the usually rapid absorption and toxicity of the ammonium ion and to complexes formed with plasma proteins. Ammonia is very rapidly absorbed by plant cells and the cytoplasm, which is normally heavily buffered at an acid pH, may become alkaline. The increased alkalinity and the toxic action of ammonia on protoplasm causes rapid death of the cell.

ammonium sulphate. ammonium sulfate.

ammonium thiocyanate. a colorless, crystalline, water-soluble solid, NH_4SCN. It is a fast-acting, extremely toxic plant poison. See also ammonia, ammonia sulfate, ammonium salt.

ammonium thioglycolate. a colorless, flammable liquid, $HSCH_2COONH_4$, with an offensive odor due partly to the evolution of hydrogen sulfide. It is used in various permanent wave solutions and preparations for hair removal. It is the most hazardous component of permanent wave solutions and can cause dermatoses on hands and scalp, erythema, edema, and subdermal hemorrhage, and is readily absorbed through the scalp. Permanent waves using preparations that contain this substance are probably contraindicated during pregnancy.

ammoniuria. excessive ammonia in urine.

ammonization. ammonification.

amnestic psychosis. Korsakoff's syndrome.

amnestic syndrome. Korsakoff's syndrome.

amnioinfusion. the introduction of solutions into the amniotic fluid, usually to induce abortion.

amnion (amniotic sac). one of the extraembryonic membranes of many vertebrate animals. It contains the fluid surrounding the fetus. It is a thin, tough, transparent membrane which, together with the placenta of mammals, forms part of the afterbirth.

amniotic sac. See amnion.

Amo-1618. a plant growth retardant.

amobarbital (5-ethyl-5-(3-methylbutyl)-2,4,6-(1H,3H,5H)-pyrimidinetrione; 5-ethyl-5-isopentylbarbituric acid; 5-ethyl-5-isoamylbarbituric acid; 5-isoamyl-5-barbituric acid; amylobarbitone; barbamil; amylobarbitone; pentymal). a barbiturate, it is a white, odorless, slightly bitter, crystalline powder, $C_{11}N_{18}N_2O_3$, used as a hypnotic and sedative. Amobarbital is a controlled substance listed in the U.S. Code of Federal Regulations (CFR). It is very toxic to laboratory mice.

amobarbital sodium. the sodium salt of amobarbital, *q.v.* it is a white, odorless, granular powder used as a short-acting sedative.

amole. *Chlorogalum pomeridianum*.

Amoracia lapathifolia (*A. rusticana*; *Cochlearia amoracia*; horseradish). a coarse perennial herb with a large taproot, the source of a condiment (horseradish). *A. lapathifolia* escaped from cultivation in the United States and now occurs widely in moist situations throughout the eastern and midwestern states. Ingestion of tops or roots, which contain mustard oil, has produced acute gastroenteritis in cattle, sheep, and swine that can result in collapse and death.

Amoracia rusticana. *Amoracia lapathifolia*.

Amorphophallis (krubi). a genus of large Asiatic herbs (Family Araceae) with immense bulb-like roots. They are sometimes grown as greenhouse curiosities in temperate climates. All parts of the plant are toxic and, when chewed, produce symptoms similar to those of *Dieffenbachia*, *q.v.* See also Araceae.

amorphous aconitine. See aconitine.

amosite, amosite asbestos. See asbestos.

AMP. adenosine monophosphate.

3′,5′-AMP. cyclic AMP.

5′-AMP. adenosine monophosphate.

amperometric method. any electroanalytic technique based on the measurement of current or current difference. Depending on the technique used, such methods are termed either "amperometry" or "amperometric titration." Amperometric methods are widely used for air pollutant measurement and are the basis of most of the instruments that are used for the continuous monitoring of sulfur dioxide concentrations in the atmosphere; such instruments are often incorrectly termed "coulometric".

amphetamine (Aktedron, Benzedrine, Elastonon, Orthedrine, Phenamine, Phenedrine). **1:** ((±)-α-methylbenzeneethanamine; DL-α-methylphenethylamine; 1-phenyl-2-aminopropane; (phenylisopropyl)amine; β-aminopropylbenzene; racemic desoxynorephedrine). a volatile, colorless, synthetic organic liquid, $C_6H_5CH_2CH(NH_2)CH_3$, used as a CNS stimulant. A sympathomimetic drug, chemically related to adrenalin, amphetamine acts centrally on dopamine-containing and norepinephrine-containing nerve terminals. It also stimulates the release of neurotransmitters from adrenergic nerve terminals in the peripheral nervous system. Amphetamine somewhat inhibits the uptake of catecholamines into presynaptic terminals. Adrenergic antagonists can be used to counteract acute CNS effects of this drug; β-adrenergic antagonists can counteract acute peripheral effects. In humans, ***d*-amphetamine** may cause restlessness, hyperactivity, dizziness, nausea, vomiting, diarrhea, dry mouth, dehydration, hyperactive reflexes, talkativeness, tremors, tachycardia, palpitations, mydriasis, fever, and hallucinations. In severe cases, delirium, mania, convulsions, coma, and death from circulatory collapse may occur. Lesions may include petechial hemorrhages in the brain. **2:** any of a number of chemically related, habit-forming, controlled substances that contain a nucleus of 1-phenyl-2-aminopropane (e.g., amphetamine sulfate (benzedrine), dexedrine (disomer), methamphetamine). All are CNS stimulants that increase heart rate, elevate blood pressure, relieve nasal congestion, and suppress appetite. Effects of intoxication may include tachycardia, cardiac arrhythmias, elevated blood pressure, and hyperpnea. These effects can be relieved by adrenergic antagonists. The production and use of these substances are limited by law. **3:** any of a number of unrelated drugs that produce CNS effects similar to those of amphetamine (e.g., diethylpropion, phenmetrazine). **hallucinogenic amphetamine**. any strongly hallucinogenic amphetamine.

amphetamine drug. an amphetamine (def. 2).

amphetamine hydrochloride (DL-α-methylphenethylamine hydrochloride; DL-β-phenyl-isopropylamine hydrochloride). a poisonous substance by intravenous, intraperitoneal and subcutaneous routes of administration. See also amphetamine, amphetamine poisoning.

amphetamine poisoning. poisoning due to overdosage of *d*-amphetamine. Symptoms are largely those of CNS stimulation, and include talkativeness, restlessness, excitement, dizziness, tremors, dry mouth, nausea, vomiting, diarrhea, dehydration, fever, heart palpitations, tachycardia, mydriasis, delirium, hallucinations, mania, convulsions, coma, cardiovascular collapse, and death. Autopsy reveals petechial hemorrhages in the brain.

amphetamine sulfate (benzedrine). a bitter-tasting, crystalline substance that is very or extremely toxic to laboratory mice and rats. ***d*-amphetamine sulfate**. dextroamphetamine sulfate. See amphetamine.

Amphibia. a class of anamniote tetrapods (Subphylum Vertebrata), comprised of some 2600 known species of toads, frogs, salamanders, and newts. A number of species are poisonous, though few are dangerous to humans. The toxins are chemically diverse. Modern amphibians are characterized by skin that is usually moist and lacking scales, feathers, or hair; typically pentadactyl; hind feet often webbed; two nares that communicate with the mouth cavity; the mouth is often lined with fine teeth; usually a protrusible tongue; a bony skeleton; two occipital condyles; a pelvic girdle that articulates with the sacrum; ribs, when present, not attached to the sternum; a middle ear and tympanum that detect sound waves; a three-chambered heart; nucleated erythrocytes; respiration via skin, the mucous membrane that lines the mouth, and gills or lungs; usually carnivorous, feeding on insects and other small invertebrates; typically oviparous; eggs with a gelatinous covering; usually a gill-breathing aquatic larva (tadpole); predominantly fresh-water, brackish-water, arboreal, and terrestrial forms occur. See, for example, Anura, Urodela, atelopidtoxin, terms with the prefix bufo-.

amphibian. **1:** pertaining to any life form defined in 2a and 2b. **2a:** any animal of the class Amphibia. **2b:** any life form adapted to or able to live in both aquatic and terrestrial environments.

amphibious sea snake. *Laticauda colubrina*.

amphibole. See asbestos.

amphioecious. a term denoting a population or species that has a wide and variable tolerance to habitat and environmental conditions as reflected in the presence of clines and subspecies. *Homo sapiens* can be regarded as amphioecious in many respects. However, humans are more importantly adaptable by virtue of an ability to modify the environment. Thus no human truly lives in the Antarctic; to a high degree the temperate zone is imported to make human life possible in the Antarctic.

amphisbaenian. a reptile of the family Amphisbaenidae

Amphisbaenidae (amphisbaenians; worm lizards). a family of burrowing, usually limbless, superficially worm-like reptiles (Order Squamata) of Africa, tropical America, and the West Indies. They are characterized by the presence of epidermal scales or plates that form segmental rings; a double, eversible, male copulatory organ; a transverse anal slit; and procoelous vertebrae. The eyes are inconspicuous, the tail is rounded.

amphotericin B. a very toxic amphoteric polyene antibiotic and antifungal agent, $C_{47}H_{73}NO_{17}$, isolated from a strain of *Streptomyces nodosus*. It is available as the sodium deoxycholate complex which is used widely to treat systemic mycoses. It has nephrotoxic, abortifacient, fetotoxic, and teratogenic activities.

ampicillin (6,Δ,α-aminophenyl-acetamido penicillanic acid). the USAN name of a semisynthetic, broad-spectrum penicillin, $C_{16}H_{19}N_3O_4S$, used to treat enteric fevers, gonorrhea, and infections of the respiratory, urinary, and gastrointestinal tracts, ear, nose, and throat. It can cause nausea, vomiting, diarrhea, a specific dermatitis, and occasionally anemia, hepatitis, and allergic reactions; a person allergic to penicillin is probably also allergic to ampicillin.

ampul. ampule.

ampule (ampul). a small, sealed glass container used to keep substances sterile.

amsinckia poisoning. poisoning of herbivores by *Amsinckia intermedia*. Symptoms in domestic cattle, horses, and swine may appear after long-continued feeding on small amounts of the seeds. Affected animals are sometimes weak, moderately icteric, and usually fail to gain weight. Horses also exhibit neurologic effects. See "walking disease." Symptoms in cattle are generally mild; swine fail to gain weight; and sheep are relatively resistant, while domestic fowl are very resistant to amsinckia poisoning. Hepatic lesions are produced that are similar to those caused by *Echium* and *Heliotropium*. Diagnosis may be made complicated by the occasional presence of toxic concentrations of nitrate in *Amsinckia*. See hard liver disease.

Amsinckia intermedia (tarweed; fiddleneck). an annual herb up to 1 m in height (Family Boraginaceae) that is native to California, Oregon, Washington, and Idaho. It is hepatotoxic and is especially dangerous when present as a weed in wheat fields. The hepatic lesions resemble those produced by *Echium* and *Heliotropium*. Symptoms in domestic cattle, horses, and swine, may appear after long-continued feeding on small amounts of the seeds. Horses also exhibit neurologic effects. See "walking disease." Lethal concentrations of nitrate have also been reported in *Amsinckia*. See amsinckia poisoning.

amudane. griseofulvum.

amygdalin (mandelonitrile-β-gentiobioside; amygdaloside). a bitter, white, crystalline substance, $C_6H_5CHCNOC_{12}H_{21}O_{10}$. It is a cyanogenic glycoside that is soluble in water and ethanol; insoluble in ether. It occurs chiefly in seeds and other parts of various plants of the family Rosaceae, most notably in the seeds of species of *Prunus* (e.g., bitter almond, apricots, black cherries, peaches, plums) and certain other human food sources such as cassava. See also benzaldehyde, mandelonitrile.

amygdaloside. amygdalin.

Amygdalus communus. *Prunus amygdalus*. ***A. communis amara***. *Prunus amygdalus amara*.

amyl (pentyl). any of eight isomers of the radical, C_5H_{11}, or a mixture of such.

amyl acetate (banana oil; amylacetic ester; isoamyl acetate; pear oil). a colorless, toxic, flammable liquid, $CH_3COO(CH_2)_2CH(CH_3)_2$, soluble in ethanol and ethyl ether and having a characteristic odor of bananas and pears. It is used in flavors and perfumes; as a warning odor; as a solvent for paints and lacquers and for phosphors in fluorescent lamps; in the extraction of penicillin; in photographic films; nail and leather polish; and in printing and finishing fabrics. Commercial preparations are mixtures of isomers, the chief of which are isoamyl, normal, and secondary amyl acetates.

amyl hydrate. amyl alcohol.

amyl hydrosulfide. 1-pentanethiol.

amyl mercaptan. 1-pentanethiol.

***n*-amyl mercaptan**. 1-pentanethiol.

amyl nitrite (isoamyl nitrite). a poisonous, strongly oxidizing, highly flammable, yellowish liquid, $(CH_3)_2C_3H_5NO_2$, with a distinctive fruity odor and a pungent aromatic taste. The vapor may explode if ignited. It is nearly insoluble in water; soluble in ethanol. It is used in the preparation of perfumes and diazonium compounds. It is an antidote for cyanide poisoning, but is itself poisonous as are all cyanide antidotes.

amyl sulfhydrate. 1-pentanethiol.

amyl thioalcohol. 1-pentanethiol.

amylacetic ester. amyl acetate.

***n*-amylamine** (pentylamine; 1-aminopentane). a colorless, flammable liquid, $C_5H_{11}NH_2$, that is soluble in water, ethanol, and ethyl ether. It is a strong irritant and fire hazard. It reacts with ammonia to yield a mixture of mono-, di-, and triamylamines.

α-amylase. an enzyme secreted by the salivary glands and pancreas that catalyzes the hydrolysis of 1,4-glucoside linkages of carbohydrates to yield a mixture of dextrins, di- and trisaccharides, plus glucose. The level of α-amylase may be elevated during renal failure and depressed following hepatobiliary toxicity.

amylene hydrate. *tert*-pentyl alcohol.

amylobarbitone. amobarbital.

amyosthenia. weakness of muscles.

amyosthenic. of or pertaining to amyosthenia.

amyotonia. a lack of muscle tone.

An. **1:** chemical symbol for actinon. **2:** anisometropia. **3:** anode. **4:** antigen.

AN. acrylonitrile.

ANA test. autoimmune assay. See assay.

Anabaena flos-aquae. a poisonous species of [illegible].

anabascin. a paralytic substance stored in the proboscis of *Paranemertes peregrina*, a venomous marine nemertine (Phylum Nemertea) worm. Do not confuse with anabasine.

anabasine (3-(2-piperidinyl)pyridine; 2-(3-pyridyl)piperidine; neonicotine). a nicotine-like insecticidal alkaloid, $C_{10}H_{14}N_2$, from the herb *Anabasis aphylla* and from *Tobaccum*. It is a colorless liquid that darkens on exposure to air; it is miscible with water and soluble in ethanol and ether. Symptoms of acute or subacute intoxication include confusion, vertigo, syncope, visual and auditory disturbances, nausea, excessive salivation, vomiting, photophobia, cold extremities, diarrhea, and clonic spasms. Do not confuse with anabascin.

anabasis. a period of increasing severity of a disease.

Anabasis aphylla. a woody perennial herb (Family Chenopodiaceae) that secretes the insecticide anabasine. It is native to the Caucasus and nearby areas about the Black and Caspian Seas.

anabatic. denoting or pertaining to anabasis.

anabiosis. reanimation or resuscitation following apparent death.

anabiotic. an agent that restores, reanimates, or resuscitates. See anabiosis.

anabolic. pertaining to or denoting anabolism. See also catabolic.

anabolic steroid. See steroid.

anabolin. anabolite.

anabolite (anabolin). any product of anabolism.

anabolism. those metabolic processes that produce complex substances from simpler substances (e.g., the synthesis of proteins from amino acids) with increased structural and organizational complexity. During growth, anabolic activities exceed catabolic activities. Anabolism is promoted by certain substances, most notably hormones such as the anabolic steroids. *Cf.* catabolism.

Anacardaceae. a family of plants native to South America and southeastern Asia. A number of genera produce poisons similar to those of *Anacardium occidentale*.

Anacardium occidentale (cashew). an evergreen tree (Family Anacardaceae) that produces an edible nut. The shell, however, contains an irritant oil that causes a dermatitis indistinguishable from that of poison ivy. The active principles are mono- and polyhydric phenols. The juice is used to make an insecticidal varnish.

anacatharsis. severe, protracted vomiting.

anacathartic. an emetic; an agent that causes vomiting.

Anacyclus pyrethrum. a prostrate perennial shrub (Family Asteraceae) native to Morocco, the flowers of which are a source of pyrethrin.

anaerobe. a microorganism that can live and multiply in the absence of molecular oxygen.

anaerobic. **1a:** anoxybiotic, anoxybiontic; able to live without oxygen. **1b:** pertaining to or denoting an anaerobe. **2:** pertaining to a process that occurs in the absence of molecular oxygen.

anaerobic respiration. cellular respiration in which glucose is converted, in the absence of molecular oxygen, to lactic acid with the release of energy to other metabolic processes.

anaerobiosis (anoxybiosis). **1:** the process of living in an oxygen-free environment. **2:** The functioning of a cell, tissue, or organ in the absence of free oxygen.

Anagallis arvensis (scarlet pimpernel; poor-man's weatherglass). an annual herb (Family Primulaceae) native to Europe and Asia and introduced in North America. All parts of the plant are toxic to humans if ingested.

Anagyris foetida (Mediterranean stinkbush). a plant that contains the cardiotoxic alkaloid, sparteine, *q.v.*

analepsis. regaining strength or returning to health following a disease or disorder.

analeptic. **1:** able to block the effects of depressant drugs. **2:** any drug that blocks the effects of depressant drugs such as ethanol, sedatives, and anesthetics. No drug is known to be broadly analeptic and many are highly toxic. They are used in medicine chiefly to treat poisoning by CNS depressants (e.g., barbiturates). Analeptics such as caffeine and amphetamines sometimes help to restore consciousness. *Cf.* stimulant. See also amphetamine, caffeine, picrotoxin, strychnine. **3:** a restorative.

analgesia. absence of or insensitivity to pain; absence of or reduced response to painful stimuli.

analgesic. **1:** analgetic; pain-relieving; able to reduce or relieve pain; able to induce analgesia, *q.v.* **2:** a substance that relieves pain or is able to produce analgesia without loss of sensation or consciousness. Aspirin and other salicylates are the most widely used analgesics. Nonsteroidal antiinflammatory drugs are also used as analgesics. Toxic doses depress respiratory rate and produce symptoms specific to the particular analgesic. While analgesics that contain acetophenetidin (phenacetin) and other 4-aminophenol derivatives are less likely to induce hypersensitivity than salicylates, they should be used with caution, since prolonged use may damage the kidneys and chronic use of high doses can cause bladder cancer. Nearly all analgesics are fetotoxic and can cause fetal or neonatal tremors, convulsions, hemorrhage, and death. **narcotic analgesic**. any narcotic used to relieve pain. The first to be used were alkaloids from the opium poppy. Narcotic analgesics include codeine, opium, morphine, heroin (diacetylmorphine), and the synthetics, meperidine and methadone. **opioid analgesic**. any opium alkaloid (e.g., morphine or codeine) used to relieve pain.

analgesic nephropathy. renal damage in susceptible individuals resulting from ingestion of excessive amounts of analgesic mixtures or of salicylate analgesics.

Analgesine. antipyrine.

analgetic. analgesic.

analgia. an illness or condition without pain.

analgic. without pain.

analog (analogue). **1a:** that which is analogous to another entity, property, or event. **1b:** any organ in one species that is similar in function to an organ of another species, but differing in structure and origin. **2:** in chemistry, a compound which is structurally similar to another.

analogous. having a similar function but a dissimilar structure and evolutionary origin.

analogue. analog.

analysis of variance (ANOVA). a statistical technique used to define and isolate the sources of variability within a set of observations. It is especially useful for examining sources of uncontrolled variation in experimental situations. Thus, ANOVA is often used in toxicology to test for mean differences among two or more experimental conditions.

Anamirta cocculus (*A. paniculata*). an East Indies woody vine (Family Menispermaceae) with bitter, shiny rhomboid leaflets. The leaflets, berries, and seeds are poisonous. The berry is called "cocculus," *q.v.* The seeds within the berry contain picrotoxin and anamirtin. It is used locally to stupefy fishes and in an ointment to control pests. ***A. paniculata***. *Anamirta cocculus*.

anamnesis. **1:** recall, recollection, remembering; the ability to remember. **2:** that which is remembered. **3:** the past medical history of a subject. **4:** immunologic memory; the condition of elevated immune response following exposures of immunocompetent cells subsequent to an initial exposure. Such responses are referred to as anamnestic or secondary immune responses.

anamnestic. **1:** aiding the memory. **2:** pertaining to one's medical history.

anamnestic immune response (anamnestic response; anamnestic reaction). the immune response to subsequent challenge, which is more rapid and more intense than the response to the primary immunizing dose of the antigen. See anamnesis, def. 4.

anamnestic response. anamnestic immune response.

anaphylactic. pertaining to or denoting anaphylaxis; displaying extreme sensitivity to an antigen.

anaphylactic conjunctivitis. hay fever.

anaphylactic intoxication. anaphylactic shock.

anaphylactic reaction. anaphylaxis

anaphylactic shock (anaphylactic intoxication; allergic shock; anaphylaxis (in part)). a severe allergic reaction with sudden onset by a hypersensitive (sensitized) individual as the result of exposure to a specific antigen, as by a bee sting, injection of a foreign serum, or blood transfusion. In humans, the condition is usually marked by increased irritability; violent circulatory activity; sometimes skin eruptions; respiratory symptoms including dyspnea, chest constriction, a violent cough; and cyanosis. Severe cases may result in convulsions, loss of consciousness, and death. Reactions are due chiefly to increased permeability of capillary endothelium and contraction of smooth muscle. *Cf.* anaphylaxis, allergic reaction, hypersensitivity.

anaphylactin. the specific antibody in anaphylaxis. It is formed following the initial encounter with the antigen and interacts with it during the second challenge.

anaphylactogen. **1a:** a substance (antigen) that can sensitize an individual such that subsequent challenge may induce anaphylaxis. **1b:** a substance (antigen) that can induce an anaphylactic reaction in a sensitized individual.

anaphylactogenesis. the induction or initiation of anaphylaxis.

anaphylactogenic. pertaining to the induction or initiation of anaphylaxis or to antigens that provoke susceptibility of an individual to anaphylaxis.

anaphylactoid (pseudoanaphylactic). resembling or similar to anaphylaxis.

anaphylactotoxin. anaphylatoxin.

anaphylatoxin (anaphylactotoxin; anaphylotoxin). **1:** a substance formed in blood serum during complement fixation and presumed to be the direct cause of anaphylactic shock. **2:** small polypeptide (C3a and C5a) fragments of the C3 and C5 components of complement. The former is split off by C3 convertase and the latter by the EAC1243 complex.

anaphylatoxin inactivator. an α-globulin (MW 300,000) that suppresses the activity of anaphylatoxins

anaphylaxis (anaphylactic reaction). **1:** originally synonymous with hypersensitivity in the sense of being a condition of abnormally heightened sensitivity to a toxin due to prior exposure to the same substance. **2a:** allergy, especially if severe, and marked by immediate and transient hypersensitivity to a foreign protein (allergin, antigen) following a second or later exposure to the same protein (allergen, antigen) to which the subject has produced specific antibodies. **2b:** a sensitivity to horse serum or other antitoxic serum. See serum sickness. **2c:** induced sensitivity to any foreign substance. **3:** anaphylactic shock. **active anaphylaxis**. that due to direct innoculation of an antigen. **active cutaneous anaphylaxis**. a localized skin reaction at the site of injection of an antigen (as by a sting, tooth, spine, or needle). **cutaneous anaphylaxis**. a specific reaction in the skin of a sensitized animal caused, for example, by insect stings or bites, or contact with any of a number of allergens (e.g., various foods, fish, eggs; dander). Such a reaction may result from direct contact of the antigen with the skin or the skin may be a target of a systemic reaction with resulting pruritis, urticaria, and angioedema. **cytotoxic anaphylaxis**. that which results from the injection of antibodies that are specific to the normal cell-surface antigens of the recipient. **local anaphylaxis**. a localized reaction. **passive anaphylaxis**. that induced in a normal individual by injecting serum (antiserum) from a sensitized individual. **passive cutaneous anaphylaxis**. that induced by intradermal injection of a specific antibody followed, within ca. 24 to 72 hours, by *i.v.* injection of the homologous antigen. The reaction is verified by injection with the antigen of Evans Blue dye, which gives a bluish cast to the skin if the reaction is positive. *Cf.* allergy, hypersensitivity, hypersensibility.

anaphylotoxin. anaphylatoxin.

anaplasia. a characteristic dedifferentiation of tumor cells with a loss of orientation to one another, to their axial framework, and to adjacent tissues. See also atypia, dysplasia.

anaplastic. **1:** characteristic of anaplasia. **2:** restoring a lost part as in plastic surgery.

anastatic. restorative.

anatomic, anatomical. **1:** of or pertaining to the structure (anatomy) of an organism. **2:** structural.

anatomy. **1:** the structure of an organism. **2:** the branch of science that deals with the structure of living organisms. **3:** dissection.

anatoxic. having or pertaining to the toxicological properties of anatoxin. See toxoid.

anatoxin. a toxoid prepared by adding a small amount of formaldehyde to a bacterial toxin.

anavenin, anavenom. a toxoid of snake venom, i.e., a venom that retains its antigenic activity following detoxication by formaldehyde.

anchovy (plural, anchovies). any small bony fish (Osteichthyes) of the Family Engraulidae, especially *Engraulis japonicus*. See also ciguatera.

Ancistrodon. *Agkistrodon*.

ancrod (agkistrodon serine proteinase; agkistrodon rhodostoma venom proteinase). a thrombin-like defibrinating enzyme isolated from the venom of the Malaysian pit viper (*Agkistrodon rhodostoma*) that causes hypofibrinogenemia and a diminution in the viscosities of whole blood and plasma. It is used therapeutically in treating chronic peripheral vascular disease. See thrombin-like enzymes.

Anderson's mole viper. *Atractaspis microlepidota andersonii*.

Andira araroba (araroba). a very large tree indigenous to Brazil (Family Fabaceae, formerly Leguminosae. It is the source of Goa powder. See chrysarobin.

Androctonus australis. a deadly venomous scorpion of arid regions in North Africa, India and Pakistan. The mortality rate of adult humans stung by this species is about 2%, and of children, 8%. *Cf. Buthus occitanus*.

androgen (androgenic hormone). any substance that stimulates the development of masculine characteristics by an individual organism. Most are 19-C steroid hormones produced from acetate and cholesterol, and can be synthesized chemically. Androgens are secreted by the testis in all vertebrates, and to some extent by the ovaries and adrenal cortex of mammals. The most common or important androgens are testosterone, androsterone, androstenedione, and dihydrotestosterone; and many other weakly androgenic steroids are known. Dihydrotestosterone, which occurs in many androgen-responsive tissues, is more potent than testosterone. Androgens promote the development and function of male sexual organs, regulate spermatogenesis, stimulate the development and maintenance of male secondary sex characteristics, and play a strong role in the development and maintenance of male sexual behavior. Androgens also act metabolically in a variety of tissues in the retention of nitrogen, the synthesis of tissue-specific proteins, and in protein anabolism. They promote masculinization and are teratogenic. When administered to pregnant women they can accelerate bone development in the fetus and masculinization of the female fetus. Disorders of androgen secretion or improper administration in humans may also cause acne, menstrual irregularities, disturbances of bone growth in children, edema, jaundice, impotence, infertility due to reduced spermatogenesis, and hepatic carcinoma.

androgenic hormone. androgen.

andromeda. common name for shrubs of the genus *Andromeda*.

Andromeda (andromeda; bog rosemary). a genus of attractive, widely distributed and cultivated shrubs (Family Ericaceae), these plants contain grayanotoxins, *q.v.* Other plants that contain grayanotoxins include *Kalmia latifolia* (mountain laurel), *Ledum* (Labrador tea), *Lyonia* (e.g., stagger bush), *Pieris* (pieris), and *Rhododendron* (rhododendrons, azaleas). See also andromedotoxin.

andromedotoxin (grayanotoxin I). **1:** a sometimes amorphous, sometimes crystalline, strongly emetic, cardiotoxic compound that occurs in the leaves of all toxic heaths (Family Ericaceae), including *Andromeda* (andromeda; e.g., bog rosemary), *Kalmia latifolia* (mountain laurel), *Ledum* (Labrador tea), *Lyonia* (e.g., stagger bush), *Pieris* (pieris), and *Rhododendron* (rhododendrons, azaleas). In humans and laboratory mammals, andromedotoxin paralyzes the motor nerve terminals in skeletal muscle. It also initially stimulates and then paralyzes the vagus nerve. See grayanotoxin, *Rhododendron*. **2:** any grayanotoxin.

androstenedione. See androgen.

androsterone. See androgen.

Anectine chloride. succinylcholine chloride.

anellated coral snake. *Micrurus anellatus*.

anemia. **1:** any condition characterized by a less than normal number of erythrocytes or a reduced concentration of hemoglobin in blood. **2:** any condition whereby the oxygen-carrying capacity of blood is reduced. Anemia can result from any of numerous causes, including extravasation of blood, increased rate of destruction or reduced rate of production of erythrocytes, irregularities in the synthesis of hemoglobin, or hemolysis. Nearly all of the proximate causes of anemia can be the result of toxic action. The condition is usually characterized by hypoxia, pallor of the skin and mucous membranes, lethargy, fatigability, shortness of breath, palpitations of the heart, and soft systolic heart murmurs. **aplastic anemia.** hypoplastic anemia. **hemolytic anemia**. anemia due to a reduced life span of mature erythrocytes in circulating blood and an inability of bone marrow to compensate. The term applies especially to anemia resulting from hemolysis, either drug-induced or congenital in origin. Oxidant stress by substances (e.g., nitrites) that oxidize hemoglobin to methemoglobin (e.g., nitrites, phenylhydroxylamine) may cause a hemolytic anemia marked by the presence of Heinz bodies within the erythrocytes. **hemotoxic anemia.** toxic anemia. **hypoplastic anemia** (aplastic anemia). anemia due to the suppression of the bone marrow's erythrocyte-producing capacity with consequent low erythrocyte concentrations in circulating blood. The condition is often due to ionizing radiation or chemicals. **lead anemia.** anemia associated with lead poisoning. It is apparently due to defective hemoglobin synthesis, whereby iron is poorly, if at all, incorporated into the porphyrin ring during the metabolic synthesis of hemoglobin. **megaloblastic anemia**. a condition in which blast cells are prematurely released from bone marrow due to excessive stimulation of their production by folic acid antagonists (e.g., methotrexate) or antimalarials (e.g., pyrimethamine), or to vitamin B_{12} or folic acid deficiency. **toxic anemia** (hemotoxic anemia). anemia due to the destructive effects of any endogenous or exogenous toxic substance.

anemic. pertaining to or characterized by anemia.

anemone. any herb of the genus *Anemone*. Sometimes used in popular speech for a sea anemone, *q.v.*

Anemone. (anemone; windflower; pasqueflower; thimbleweed). a large genus of herbs (Family Ranunculaceae) that produces ranunculin (a glycoside), protoanemonin, an irritant oil and decomposition product of ranunculin, anemonol, anemonin, and anemonine. Do not confuse with sea anemone. *Cf.* anemonin, anemonine. See also anemonism.

anemone camphor. anemonin.

Anemonia sulcata. a sea anemone (Phylum Coelenterata) with unusually long tentacles that frequently sting bathers in the Adriatic Sea. The venom is moderately toxic and contains three toxic polypeptides. Toxins I and II have similar toxicologic characteristics, but toxin II is considerably more toxic and is positively inotropic. When toxins I and II are injected into crustacea, fish, or mammals, convulsions, paralysis, and death may ensue.

anemonin (1,7-dioxadispiro[4.0.4.2]dodeca-3,9-diene-2,8-dione; 1,2-dihydroxy-1,2-cyclobutanediacrylic acid di-τ-lactone; anemone camphor; pulsatilla camphor). a colorless, crystalline, acrid, highly toxic, substituted diacrylic dilactone found in *Anemone pulsatilla*, *Caltha palustris*, *Ranunculus acer*, and other plants of the family Ranunculaceae. Anemonin has been used therapeutically as a sedative and antispasmodic. *Cf.* anemonine, anemonol.

anemonine (3-(carboxymethyl)-1,1-dimethylimidazolium hydroxide inner salt). a nontoxic constituent of *Caltha palustris* (*q.v.*) and *Ranunculus*. *Cf.* anemonin, anemonol.

anemonism. poisoning by ingestion of plants of the genus *Anemone* (Family Ranunculaceae).

anemonol. a highly toxic volatile oil found in various plants of the genus *Anemone* and other plants of the family Ranunculaceae. *Cf.* anemonin, anemonine.

anergan 25. promethazine.

anesthesia. **1:** the absence of or reduced response to external stumuli by an animal or part of an animal. **2:** a loss of sensation by an animal or part of an animal (e.g., as produced by an anesthetic drug or certain venoms). **general anesthesia**. that which causes temporary loss of consciousness and a complete inability to sense normally painful stimuli. It is normally effected by inhalation or injection. **inhalation anesthesia**. general anesthesia produced by the inhalation of anesthetic vapors or gases. See anesthetic gas. **local anesthesia**. the temporary prevention or inhibition of local pain by use of a local anesthetic injected into the concerned tissue or site, by topical application, or by injection near the nerves that serve the field to be anesthetized (nerve block). **regional anesthesia**. the insensibility of a body part produced by interrupting afferent (sensory) nerve conduction from that part.

anesthetic. **1:** pertaining to anesthesia or to the capacity to produce anesthesia. **2:** an agent that produces anesthesia (e.g., an anesthetic drug, hypnosis). **general anesthetic**. an anesthetic that causes temporary unconsciousness and a complete inability to sense normally painful stimuli. Substances that can produce general anesthesia under suitable conditions include hydrocarbons such as ethylene (flammable), cyclopropane (flammable); halogenated hydrocarbons such as chloroform (highly toxic), ethyl chloride (flammable), and trichloroethylene (toxic and flammable); ethers such as ethyl ether (highly flammable and explosive) and vinyl ether (highly flammable); and nitrous oxide (nonflammable, used with oxygen, mostly for dental surgery). **local anesthetic**. an anesthetic that acts locally (See anesthesia). Examples are certain alkaloids such as cocaine, alkyl esters of aromatic acids, synthetics such as "Novocain," and quinine hydrochloride.

anesthetic ether. ethyl ether.

anesthetic gas. a vapor or gas such as ethyl ether, nitrous oxide, or methoxyflurane used to induce general anesthesia in air-breathing animals by inhalation. Repeated exposure of hospital operating room staff to such anesthetics can reduce alertness, slow reaction times, cause miscarriages in pregnant women, and defects in the newborn that have been exposed *in utero*. Some of these substances are probably carcinogenic. See also abortifacient.

anesthetics. **1:** the art or science of anesthesia. **2:** those chemicals, collectively, that produce anesthesia. *Cf.* anesthetic.

anesthetized. absence of or diminished response to external stimuli with loss of righting reflex; treated by, or under the influence of, an anesthetic.

anestrous. pertaining to anestrus.

anestrus (anoestrous, anoestrum). **1:** in mammals, a state of nonbreeding; a nonbreeding period. **2:** the normal period of reproductive quiescence between two breeding periods in mammals that reproduce periodically.

aneuploidy. a chromosomal aberration in which a cell has either more or fewer chromosomes than normal. See also chromosome aberrations.

Anfamon. diethylpropion.

angel dust. phencyclidine.

angel trumpet. See *Brugmansia*.

angelica. **1:** *Angelica*. **2:** *Aralia spinosa* (angelica tree).

Angelica. a genus of herbs (Family Apiaceae) that is native to temperate parts of North America, Eurasia, and New Zealand. Some species are poisonous; some therapeutic. Furthermore, some species of *Angelica* are sometimes mistaken for the extremely toxic water hemlock, *Cicuta maculata*. ***A. archangelica*** (wild celery). a mildly or moderately poisonous species suspected also of being carcinogenic. Women have been poisoned by using it in vain attempts to induce abortion. ***A. venosa***. an extremely poisonous species.

angelica lactone (5-methyl-2-furanone). a lactone that exists in three forms, α-, β-, and α'-angelica. **α-angelica lactone** (5-methyl-2(3H)-furanone; Δ^2-angelica lactone; γ-methyl-β,γ-crotonolactone; 4-hydroxy-3-pentenoic acid γ-lactone). a moderately toxic, nontumorigenic lactone. **α′-angelica lactone** (dihydro-5-methylene-2(3H)-furanone; γ-methylene-γ-butyrolactone). a solid below 80°C. ***β*-angelica lactone**(5-methyl-2(5H)-furanone; Δ^1-angelica lactone; γ-methyl-α,β-crotonolactone; 4-hydroxy-2-pentenoic acid γ-lactone). more stable than the α-lactone; a known carcinogen. **Δ^1-angelica lactone**. β-angelica lactone. See angelica lactone. **Δ^2-angelica lactone**. α-angelica lactone. See angelica lactone.

angina. **1:** angina pectoris. **2a:** acute sore throat. **2b:** any severe, spasmodic pain.

angina pectoris. a condition marked by severe thoracic pain, often referred to the left pectoral area and arm; to the abdomen (rarely); or sometimes to the dorsum or jaw. It is caused by myocardial ischemia due to an insufficient blood supply to the heart.

angiogram (angiograph). a roentgenogram (X-ray record) that displays blood vessels containing a contrast medium.

angiograph. angiogram.

angiography. **1:** the preparation and analysis of angiograms. **2:** the study of blood vessels.

angioneuropathic. angioneurotic.

angioneuropathy. a neuropathy that chiefly affects blood vessels; a vasomotor disorder (e.g., vasomotor paralysis), angiospasm, or angioparalysis.

angioneurotic (angioneuropathic). pertaining to or having the characteristics of an angioneuropathy.

angiosarcoma (hemangiosarcoma; malignant hemangioendothelioma). a malignant tumor of the vascular endothelium. The cells are embryonal (presumably mesenchymal) in origin. See also sarcoma.

angiosperm. a flowering plant, one that has seeds enclosed in an ovary.

angiotensin II. a polypeptide with eight aminoacid residues, it is a powerful vasoconstrictor and stimulator of aldosterone secretion from the adrenal cortex. It is formed from angiotensin I, which was produced by cleavage of the protein angiotensinogen in a reaction catalyzed by renin (secreted by the kidneys). See also aldosterone.

Ångstrom (symbol, Å; angstrom, Ångstrom unit; Å unit; tenthmeter). a unit of length, used chiefly to express wavelengths of optical spectra. It is defined by the wavelength of the red line of cadmium, 10^{-10}) m.

angular unicorn. common name of *Acanthina spirata*.

Anguuina agrostis. a telenchid nematode, it is a plant parasite that infests *Festuca rubra*. It is responsible for the neurotoxic effects of this grass on livestock.

anhidrotic (anidrotic; antihidrotic; antiperspirant; antisudorific). **1:** preventing or inhibiting perspiration. **2:** an agent that prevents or inhibits perspiration.

anhydride. a compound formed by removal of water from an acid or occasionally a base. Thus SO_3 is the anhydride of H_2SO_4, and CO_2 is the anhydride of H_2CO_3. **acetic anhydride** (acetyl oxide; acetic oxide; acyl anhydride). a colorless, mobile, highly refractive, liquid organic anhydride, $(CH_3CO)_2O$. It is corrosive and a strong irritant that can damage eyes, skin and mucous membranes. **acid anhydride** (acyl anhydride). an organic anhydride (e.g., acetic anhydride) formed by the removal of water from two carboxyl groups, and thus having the functional group, —COOCO—, seen in the general formula RCOOCOR′, where R and R′ are alkyl or aryl groups. Acid anhydrides are highly reactive, especially with strong nucleophiles such as H_2O, NH_3 and ROH. They yield carboxylic acids on hydrolysis. See also acetic anhydride, acylation, acyl halide. **acyl anhydride**. acid anhydride. **arsenic acid anhydride**. arsenic pentoxide. **arsenic anhydride**. arsenic pentoxide. **arsenous acid anhydride**. arsenic trioxide. **arsenous anhydride**. arsenic trioxide**bis-[bisdimethylaminophosphonous]-anhydride**. schradan. **bis-*O,o*-diethyl-phosphoric anhydride**. tetraethyl pyrophosphate. **bis-*N,N,N′,N′*-tetra-methylphosphorodiamidic anhydride**. schradan. **carbonic anhydride**. carbon dioxide. **chromic anhydride**. chromium trioxide. **coumarinic anhydride**. coumarin. **exo-1,2-*cis*-dimethyl-3,6-epoxyhexahydrophthalic anhydride**. cantharidin.

2,3-dimethyl-7-oxabicyclo-[2.2.1]-heptane-2,3-dicarboxylic anhydride. cantharidin. **dithiocarbonic anhydride**. carbon disulfide. **hypochlorous anhydride**. chlorine monoxide. **perchloric anhydride**. chlorine heptoxide. **sulfocarbonic anhydride**. carbon disulfide. **trifluoroacetic acid anhydride**. See dimethyl sulfoxide. **trimethylglycocoll anhydride**. betaine.

4-[[[2‴,3″-anhydro]-*O*-3,6-dideoxy-α-L-erythrohexopyranos-4-ulos-1-yl-(1→4)-*O*-2,6-dideoxy-α-L-lyxo-hexopyranosyl-(1→4)-2,3,6-trideoxy-3-(dimethylamino)-α-L-lyxo-hexopyranosyl]oxy]-2ethyl-1,2,3,4,6,11-hexahydro-2,5,7-trihydroxy-6,11-dioxi-1-naphthacenecarboxylic acid methyl ester. aclacinomycin B.

anhydrogitalin. gitoxin.

anhydrous. lacking water; descriptive of an inorganic compound that lacks combined or adsorbed water. *Cf.* anhydride.

anhydrous alcohol. See alcohol.

anhydrous chloral. trichloroacetaldehyde. *Cf.* chloral hydrate.

anhydrous ferric chloride. ferric chloride.

anhydrous sodium hypochlorite. See sodium hypochlorite.

anidrotic. anhidrotic.

anilid. anilide.

anilide (anilid). a type of amide ($RNHC_6H_5$) in which the hydrogen of the amido group is replaced by a phenyl group. Anilides (e.g., alachlor, acetanilide) are *N*-acyl derivatives of aniline. See also herbicides (anilide herbicides).

aniline (benzenamine; aniline oil; phenylamine; aminobenzene; aminophen; kyanol). **1:** a highly toxic, colorless or brownish, oily liquid, $C_6H_5NH_2$, with a distinctive odor and acrid taste; the fumes are poisonous if inhaled. It is used industrially in numerous organic syntheses, as in the manufacture of antioxidants, antidegradants, dyes, fungicides, herbicides and defoliants, insecticides, pharmaceuticals, polyurethanes and rubber, resins, and varnishes. It is a common ingredient of some household products such as stove and shoe polishes, paints, varnishes, and inks. Aniline is readily absorbed through the skin, digestive tract, or respiratory system. It is an allergen and a toxicant, commonly causing methemoglobinemia by oxidizing the iron in hemoglobin. Moderate exposure to aniline can produce anoxia with cyanosis. Unlike carbon monoxide poisoning, oxygen therapy does not reverse this process, but it can be reversed by the action of methemoglobin reductase or of methylene blue. Infants who develop methemoglobinemia from exposure to aniline incur a 5-10% mortality. Aniline is also a known carcinogen. The toxic action of aniline is similar to that of nitrobenzene. See also acetanilid. **2:** any of a class of monocyclic aromatic amines that have a single amino group. **aniline inks, oils, and polishes**. toxic mixtures that produce symptoms such as sweating, nausea, vomiting, tinnitus, chills, hypotension, cyanosis, convulsions, coma, circulatory collapse, and death from respiratory failure. Lesions include methemoglobinemia, kidney and liver damage.

aniline dye. See exterior color (D&C color).

aniline oil. aniline.

aniline poisoning, chronic. anilinism.

***p*-anilinesulfonamide**. sulfanilamide.

anilinism (anilism; chronic aniline poisoning). chronic poisoning by aniline is characterized by methemoglobinemia, aplastic anemia, vertigo, muscular depression, cardiac weakness, intermittent pulse, and cyanosis.

anilinobenzene. *p*-biphenylamine.

***p*-anilinobenzenesulfonic acid**. *N*-phenylsulfanilic acid.

anilism. anilinism.

animal. **1:** any member of the Kingdom Animalia. **2:** any multicellular eukaryotic organism with holozoic nutrition, a capacity for spontaneous movement (locomotion), and usually rapid motor responses to stimulation. **3:** under law, any animal organism other than *Homo sapiens*. **4:** often improperly used in reference to mammals only. **domestic animal**. any tame animal that can live and breed normally in close association with and under the control of humans (e.g., horses, cattle, dogs, domestic cats). Such animals may experience a toxic environment that more closely resembles that of humans than of their feral counterparts. Livestock may ingest deposited air pollutants, pesticides, and even materials from fertilizers in their forage or supplemental feeds. Animals kept as household pets may breathe air and eat food that is similar to that of their owners. **experimental animal**. any animal used for experimental purposes. It may be a feral animal or a domestic animal held captive in the laboratory for some time prior to utilization to allow acclimatization to laboratory conditions. More often, it is a laboratory animal. **feral animal**. **1:** an animal that was formerly tamed, but has since become wild. **2:** an animal that lives under natural conditions and has never been domesticated. **laboratory animal**. any animal used primarily for research and testing that is bred, born, reared, and dies in the laboratory or a supply house. Such animals typically represent strains or populations that have no genetic counterparts among wild animals of the same species. They usually depart substantially, both phenotypically and genetically, from the wild stock from which they derive. In some instances, there are no wild or nonlaboratory representatives of a species. Many are highly inbred and few, if any, can be said to be representative of the species from which the original stock was drawn. In some cases, the actual species of the animal is questionable. For purposes of scientific discourse, such animals should be identified by an accepted or accredited common name, with a breed or strain designation if known, and their origin if they come from a long-established population of a particular laboratory, institute, or supply house. In general, a scientific name should not be used, although the use of a genus and family name is desirable. Laboratory animals are typically housed, fed, and studied under conditions that wild populations of the same species (if such exist) do not experience and often do not tolerate. They are usually characterized in part by their small size and the ability to survive and reproduce under laboratory conditions. The vast majority of animal species do not reproduce under laboratory conditions and many cannot survive. **poisonous animal**. any animal whose tissues, in whole or in part, contain poisons in sufficient quantity or concentration to injure or kill specified organisms that might feed upon it or otherwise conceivably be exposed to it such that sufficient poison is absorbed to cause injury or death. Nearly all major classes of animals contain poisonous species or populations. See also poisonous, venomous, atoxic. **venomous animal**. **1:** an animal that produces a venom. **2a:** an animal that can deliver a poison (called a venom, q.c.) to another animal by biting (usually by fangs), stinging, or in some cases by spraying or "spitting." **2b:** this definition extends def. 2a to include animals that deliver complex toxicants or venoms by contact (e.g., through breaks in the integument of the victim. See also poisonous, venomous. **wild animal**. a feral animal (def. 2).

animal coniine. cadaverine.

animal toxin. zootoxin.

anion. any negatively charged ion or group of ions (e.g., chloride, bicarbonate, phosphate). *Cf.* cation.

anisidine. either of two isomers, *o*-anisidine (*o*-methoxyaniline; *o*-aminoanisole) and *p*-anisidine (*p*-methoxyaniline; *p*-aminoanisole) with the general formula, $CH_3OC_6H_4NH_2$. Both are strong irritants and are toxic by inhalation, ingestion, and may be absorbed through the skin. Both are toxic to the blood and cardiovascular system, liver, and kidneys; symptoms of exposure may include headache, dizziness, and cyanosis. They are known carcinogens.

***o*-anisidine**. See anisidine.

***p*-anisidine**. See anisidine.

anisohydric. hydrolabile.

anisomycin (1,4,5-trideoxy-1,4-imino-5-(4-methoxyphenyl)-D-xylo-pentitol 3-acetate; [2R-(2α,3α,4β)]-2-[(4-methoxyphenyl)methyl]-3,4-pyrrolidinediol 3-acetate; 2-*p*-methoxyphenylmethyl-3-acetoxy-4-hydroxypyrrolidine). a crystalline compound, $C_{14}H_{19}NO_4$, with monoclinic needles. It is slightly soluble in water and most organic solvents, but soluble in chloroform, ketones, and low molecular weight alcohols. It is a fungicide and antiprotozoal antibiotic produced by *Streptomyces griseolus* and *S. chromogenes* and used chiefly as a fungus inhibitor and mildew preventive in vegetables.

annelated coral snake. *Micrurus annelatus*.

annelid (annelid worm). common name of any of the segmented worms (Phylum Annelida) such as an earthworm or clamworm. Examples of genera with venomous species are *Eurythoe*, *Glycera*, *Hermodice*, and *Eunice*.

annelid worm. See annelid.

Annelida (segmented worms). a phylum of coelomate invertebrate worms having elongate, soft-bodied, metamerically segmented bodies, often with paired setae ("bristles") on each segment. They have a soft, muscular, cylindrical body, a thin cuticle, a well-developed closed circulatory system, paired nephridia (usually metanephridia), a well-developed nervous system with a brain and ventral nerve cord and a well-developed blood vascular system. The setae of some species form a stinging mechanism (e.g., *Eurythöe complanata*, *Hermodice carunculata*). Certain other species have hard chitinous biting jaws (e.g., *Glycera dibranchiata*, *Eunice aphroditois*).

Annie, Fannie, and Mike. most species of blue-green bacteria that are toxic to animals belong to three genera: *Anabaena*, *Aphanizomenon* and *Microcystis*, known as "Annie, Fannie, and Mike" by water treatment professionals. See also Cyanobacteria.

annual ring. See growth ring.

annual sage. *Salvia reflexa*.

annulated sea snake. *Hydrophis cyanocinctus*.

anomaly. a process, entity, organism or part of an organism that deviates from normal.

anophelicide. **1:** destructive to mosquitoes of the genus *Anopheles*. **2:** an agent that destroys anopheline mosquitoes.

anorectic (anoretic; anorexic; appetite suppressant). **1:** pertaining to, or suffering from, anorexia, especially anorexia nervosa. **2:** an agent that suppresses appetite. Most are amphetamines or related compounds. They are generally addictive and are thus used only under some circumstances to treat life-endangering obesity. Some can elevate blood pressure, disturb cardiac rhythm, cause insomnia, restlessness, hyperactivity, headache, euphoria, tremor, dryness of mouth, unpleasant tastes, diarrhea, stomach upset, altered sex drive, and impotence, as well as depression and psychosis. Overdoses cause hallucinations, aggressiveness, and even panic. If MAO inhibi-tors are also taken, blood pressure can increase dramatically. See amphetamines, benzphet-amine, diethylpropion, phentermine, phenmetrazine.

anoretic. anorectic.

Anorex. diethylpropion.

anorexia. loss of appetite or aversion to food with a greatly reduced consumption of food and consequent weight loss which may be dramatic and life-threatening.

anorexiant. an agent or phenomenon that provokes anorexia.

anorexic. anorectic.

anorexigenic. producing or promoting anorexia.

anosmia (anosphrasia; olfactory anesthesia). the absence of a sense of smell.

anosmic. **1:** of or pertaining to anosmia. **2:** aosmic, odorless.

ANOVA. analysis of variance.

anoxemic. denoting, due to, or characterized by a serious lack of oxygen in circulating blood.

anoxia. **1a:** an absence or extreme reduction of oxygen in the surrounding medium, habitat (anoxicity), arterial blood, interstitial fluid, or tissues of an animal. Anoxia may result from low atmospheric pressure, anemia, decreased or interrupted blood flow, or a decreased capacity of cells to utilize available oxygen. **1b:** a seriously reduced oxygen supply to cells or tissues of a living organism. **2:** formerly synonymous with hypoxia. **histotoxic anoxia**. anoxia due to the failure of cells to utilize oxygen because of the poisoning of respiratory enzymes, as in cyanide poisoning. This condition is usually marked by higher than normal oxygen tension in arterial and capillary blood. *Cf.* hypoxia, oligoxia.

anoxiate. to induce a state of anoxia.

anoxic. **1:** of, pertaining to, or characteristic of anoxia. This term may be used in reference to any type of living system. **2:** said of a habitat that is devoid of molecular oxygen; anoxia, anoxicity; See also normoxic, oligoxic.

anoxicity. anoxia, *q.v.*, especially as applied to a habitat.

anoxybiontic. See anaerobic.

anoxybiosis. anaerobiosis.

anoxybiotic. See anaerobic.

ANS. autonomic nervous system.

ant. See Formicidae.

Antabuse®. disulfiram.

antacid. **1a:** having the capacity to reduce or neutralize stomach acid. **1b:** a medicine taken orally that reduces or neutralizes stomach acid.

antagonism. **1:** mutual resistance or opposition of force, principle, or action between organisms, structures, agents, diseases, biologic processes, populations, or species. *Cf.* antipathy. **2a:** the interaction of two or more biologically active agents such that the effects of one or more of them are reduced or eliminated. **2b:** the condition whereby the combined effect of two or more chemicals to which a living system is simultaneously or sequentially exposed is less than additive. **2c:** reduction of the toxicity of a substance by the presence of another. **3:** the inhibition of one species or population by the action of another, especially inhibition of growth. Any of a number of mechanisms may be involved. Sometimes antagonism is manifested by the passive creation of unfavorable conditions by the presence and normal activities of one species or sometimes by specific behaviors (e.g., competition for limited resources such as food or nesting sites, or by the secretion of inhibitory substances such as antibiotics or other toxicants). **4:** the opposing action of two muscles or sets of muscles such that the contraction of one is accompanied by relaxation of the other. **bacterial antagonism**. the destruction, injury, or inhibition of one species or strain of bacterium by the products of another. **biochemical antagonism**. functional antagonism in which two chemicals produce opposing biochemical effects that alter the process. **chemical antagonism** (inactivation). chemical interaction between two or more substances that yields products which are less toxic than the reactants. See dispositional antagonism, functional antagonism, inactivation. **competitive antagonism** (receptor antagonism). antagonism resulting from the competition of two pharmacologically active agents for the same receptor site, thus producing a less than additive effect with simultaneous exposure. Carbon monoxide poisoning results from competitive antagonism. **dispositional antagonism**. decreased toxicity of one substance due to the influence of another on its disposition (absorption, transformation, distribution, or excretion) and consequent reduction in availability to the target organ. Thus, in this type of antagonism, a less toxic compound may reach the target organ or the time at the target organ is reduced. Ipecac is a dispositional antagonist that reduces or prevents absorption of a toxicant. See chemical antagonism, functional antagonism, inactivation. **functional antagonism**. that due to the interaction of two chemicals that produce opposing biochemical or physiological effects on the same system or process and thus effectively counteract each other. See also chemical antagonism, dispositional antagonism, inactivation. **induction antagonism**. the reduction in toxicity of one substance due to prior treatment with another substance that induces detoxifying enzymes. Thus, the duration of zoxazolamine paralysis is reduced by prior treatment with phenobarbital. **noncompetitive antagonism**. the phenomenon whereby an antagonist alters the receptor or otherwise prevents an agonist from exerting an effect. In this case, the efficacy of an agonist (e.g., a toxicant) is reduced as the concentration of a second agonist increases. A noncompetitive antagonist may act at some distance from the receptor site, in which case the affinity of the agonist for the receptor is not altered. In noncompetitive antagonism the log dose-response curve shifts to the right with a reduced maximum effect. **physiological antagonism**. functional antagonism that affects a physiological process. **receptor antagonism**. competitive antagonism.

antagonist. **1:** an agent that opposes, resists, or neutralizes the action or effects of another. *Cf.* agonist. **2:** a substance that interacts with or occupies a receptor but does not initiate a response. **3:** a substance that reduces the toxicity of a toxicant. **4:** a muscle that opposes the action of another. **β-adrenergic antagonist**. any drug that competitively inhibits the binding of catecholamines to β-adrenergic receptors. They are used to reduce cardiovascular activity, improve blood circulation, and prevent vascular headaches, hypertension, cardiac arrhythmias, and angina. Possible adverse effects include bronchospasm, fatigue, and hypoglycemia. **amino acid antagonist**. a chemical (e.g., an amino acid analog) that hinders the absorption, metabolism or biological activity of an amino acid. Such chemicals may also be incorporated into proteins, greatly affecting their behavior with significant, even disastrous biochemical and physiological consequences (e.g., teratology). Examples are ethionine, azaserine and asparagine. **competitive antagonist** (receptor antagonist). a pharmacologically active agonist that competes with another for the same receptor sites in the same living system. Thus a given dose of agonist will induce a less intense response when the competitive molecule is also present. Such antagonism may be partial. **dispositional antagonist**. a substance that hinders or prevents an active toxicant from reaching the target organ or that reduces the duration of its presence at the site of action. Ipecac is an example. **functional antagonist**. a substance that produces biochemical or physiological effects that reverse or oppose those produced by another (e.g., a toxicant). **narcotic antagonist** (opiate antagonist). any of a group of drugs that can bind to opioid receptors, but with the activity of an agonist. These substances are used to reverse the effects of narcotic overdose. **noncompetitive antagonist**. a substance that either 1) blocks the action of toxicant by altering the receptor or binding irreversibly to a receptor site, thereby preventing the agonist (e.g., a toxicant) from exerting an effect, or 2) acts at a distance from the receptor site without altering the affinity of the agonist for the receptor. **opiate antagonist**. narcotic antagonist. **opioid antagonist.** any substance that binds to opiate receptors but produces little or no agonist activity. **pharmacological antagonist**. a substance that inhibits the action of an agonist at the site of action of the agonist. Such antagonism may be competitive or noncompetitive. **physiological antagonist**. functional antagonist. See agonist, antagonism. **receptor antagonist**. competetive antagonist. **structural antagonist**. antimetabolite. See also agonist, antagonism.

antagonistic (antipathic). having the action of an antagonist.

antamanide. an antitoxin isolated from *Amanita phalloides*. If large quantities of this substance are ingested just before or simultaneously with the *Amanita* toxins, the consumer is completely protected. Antamanide would seem to be of no practical use prophylactically or therapeutically because it could rarely be taken in time to be effective.

antemortem. prior to death. *Cf.* postmortem.

antenatal. prenatal; prior to birth; during gestation.

anterior pituitary gland, anterior pituitary. anterior lobe of the pituitary gland. See pituitary gland.

anterior pituitary growth hormone. somatotrophin.

anthelminthic. anthelmintic (def.1).

anthelmintic. **1:** an agent that destroys or eliminates intestinal worms; anthelminthic; anthelmintic drug; helminthagogue; helminthic; helmintic, vermifuge. **2.** able to destroy or eliminate intestinal worms; helmintic; vermifugal.

anthelmintic drug. any medication used to treat infestations of intestinal worms; anthelmintic (def. 1).

Anthemis (chamomile). a genus of perennial

(Family Asteraceae) with many solitary flowers and finely dissected, often heavily scented, foliage. See also chamomile. They are native to Eurasia and northern Africa. ***A. cotula*** (mayweed; dog-fennel; stinking chamollymile). Often found along roadsides and waste places in North America. All parts of the plant are toxic to humans.

anthophyllite. See asbestos.

anthopleurin-A. a toxin isolated from the sea anemone *Anthopleura xanthogrammica*. It is structurally similar to toxin II from *Anemonia sulcata*. See also anthopleurin-B.

anthopleurin-B. a toxin isolated from the sea anemone *Anthopleura xanthogrammica*. See also anthopleurin-A.

Anthozoa (Actinozoa). the class of coelenterates comprised of the corals, sea anemones, and alcyonarians. All are sessile, cylindrical polyps that are structurally more complex than those of the Hydrozoa; the medusa stage of the life cycle is absent. They are marine, colonial, or solitary forms that are characterized by an oral disc that bears hollow tentacles; a stomodaeum and usually one or two siphonoglyphs; and a gastrovascular cavity subdivided by partial and complete radially arranged septa. A skeleton may be present. Some soft corals (Subclass Alcyonaria) are extremely poisonous (e.g., *Palythoa*). While rare, poisonings of humans from ingestion of sea anemones has occurred in parts of the Indo-Pacific region, especially Samoa, where selected sea anemones are eaten only after cooking. See *Rhopdactis howesi*, *Physobrachia douglasi*, and *Radianthus paumotensis*. See also Alcyonaria, Zoantharia.

anthracene. the pure compound is comprised of colorless crystals with violet fluorescence, but when contaminated (usually with tetracene and naphthacene), the crystals are yellow with green fluorescence. Anthracene is a component of coal tar and is used to produce dyes. It causes skin damage and perhaps cancer of the nasal cavity, larynx, lungs, skin, and scrotum. A number of its derivatives are also toxic.

9,10-anthracenedione. anthraquinone.

anthracosilicosis (black lung; coal worker's pneumoconiosis; miner's phthisis). a pneumonoconiosis due to deposits of carbon and silica in the lungs from chronic inhalation of coal dust. Fibrous nodules are induced by the silica. *Cf.* anthracosis, black lung, silicosis.

anthracosis (collier's lung; miner's lung; melanedema). a pneumonoconiosis resulting from an accumulation in the lungs of carbon from inhaled smoke or coal dust. The disease is aggravated by cigarette smoking. *Cf.* black lung. See also pneumomelanosis.

anthracotic. pertaining to or characterized by anthracosis.

anthracotic tuberculosis. pneumoconiosis.

anthracycline. any of a class of fermentation products (e.g., daunomycin, doxorubicin) of *Streptomyces peucetius*.

anthranilic acid (2-aminobenzoic acid; *o*-aminobenzoic acid). a white or pale yellow, crystalline acid, $NH_2C_6H_4{\cdot}COOH$, used as an intermediate in the synthesis of dyes, perfumes, and pharmaceuticals. The cadmium salt of this acid is used as an ascaricide in swine. See also cinnamyl anthranilate.

anthranilic acid, cinnamyl ester. cinnamyl anthranilate.

anthraquinoid. pertaining to or denoting derivatives of anthraquinone. See mycotoxin (anthraquinoid mycotoxins).

anthraquinoid pigment. See anthraquinone (def.2).

anthraquinone (9,10-anthracenedione; 9,10-anthraquinone; 9,10-dioxoanthracene; AQ). **1:** a derivative of anthracene, it is a yellow, crystalline irritant, $C_6H_4(CO)_2C_6H_4$, that is insoluble in water. It has low systemic toxicity but may cause sensitization. **2:** any of numerous, functionally diverse, derivatives of anthraquinone (anthraquinone pigments or anthraquinoid pigments). They occur as orange or red pigments in aloes, cascara sagrade, senn, rhubarb, fungi (See, for example, luteoskyrin, rubroskyrin, rugulosin), lichens, and insects such as the cochineal beetle and lac insect; some are synthesized. They have numerous applications and many are used to color food, drugs, cosmetics, hair dyes, and textiles. They are used in medicine as cathartics and as antimicrobial and antitumor drugs. They are used by the military as ground smoke screens. Some are quite toxic or carcinogenic to laboratory animals. The risk to humans is, in many cases, poorly established. See also mycotoxin. **amino-anthraquinone**. a class of anthraquinone derivatives that are toxic and carcinogenic. All are hepatotoxic; lesions in mammals exposed to these compounds may include uterine atrophy, thymic atrophy, renal lesions (renal tubular metaplasia, glomerulonephritis), bone marrow hypoplasia, and anemia. Signs and symptoms of chronic exposure may also include ataxia, convulsions, decreased hematocrit and hemoglobin, and uremia. These substances can produce transitional cell papillomas, carcinomas, and other neoplasms including mammary gland adenomas and renal and hepatocellular neoplasms. **nitroanthraquinone**. a class of hepatotoxic anthraquinone derivatives, exposure to which may cause hepatocellular neoplasms, subcutaneous fibromas, subcutaneous hemangiosarcomas, and type II pneumocyte proliferation. **phenolic anthraquinone**. any of a class of anthraquinones that can cause adenomas and adenocarcinomas of the colon and cecum.

9,10-anthraquinone. anthraquinone.

anthraquinone derivative. See anthraquinone (def. 2).

anthraquinone pigment. See anthraquinone (def. 2).

anthropeic. anthropic.

anthropic (anthropeic; anthropical). pertaining to human influence.

anthropical. anthropic.

anthropocentric. interpreting the properties and activities of other organisms in terms of human behavior and values.

anthropogenic (anthropogenous). produced or caused directly or indirectly by humans.

anthropogenous. anthropogenic.

anthropoid ape. See ape.

Anthurium (tailflower). a genus of toxic, commonly cultivated tropical plants (Family Araceae) with heart-shaped or strap-like leaves. See Araceae.

anti-. prefix meaning against.

antiadrenergic (sympatholytic; sympathicolytic; sympathoparalytic). **1:** pertaining to the blocking of transmission by adrenergic nerves. **2:** antagonistic to, opposing, or inhibiting the action of adrenergic neurons. **3:** an antiadrenergic agent. *Cf.* anticholinergic.

antiagglutinin. a specific antibody that acts against the action of agglutinin.

antialexin. anticomplement.

antiallergic. pertaining to or denoting an agent or treatment that inhibits, prevents, or ameliorates an allergic reaction.

antianaphylaxis. acting against or preventing anaphylaxis.

antianginal. **1:** pertaining to or denoting the capacity to allay or prevent the symptoms of angina pectoris. **2:** antianginal drug; any substance that dilates the coronary arteries, thus increasing blood flow and acting to prevent or allay the symptoms of angina pectoris.

antianopheline. detrimental to, destructive to, or used against mosquitoes of the genus *Anopheles* or their larvae.

antiantibody. an immunoglobin that acts as an immunogen when innoculated into another species, or sometimes as an autoantigen. Produced following the introduction of an immunogen into the body; it interacts with the immunogen.

antiantidote. a substance that counteracts or neutralizes the action of an antidote.

antiantitoxin. an antiantibody that counteracts the effects of an antitoxin.

antianxiety drug. See anxiolytic.

antiar. **1:** a poisonous gum resin of *Antiaris toxicaria*, a very large evergreen tree. **2:** an arrow and dart poison made from antiar.

antiarin either of two potent crystalline cardiac glycosides, $C_{29}H_{42}O_{11}$, contained in antiar.

Antiaris toxicaria (upas; upas tree). a very large lowland evergreen tree (Family Moraceae) of southeastern Asia including the Philippines. It produces the extremely toxic gum resin, antiar. Fine droplets from the leaves form a mist that contains cardiac glycosides (e.g., convallatoxin, antiarin). See also upas.

antiarrhythmic. **1:** pertaining to or having the capacity to correct cardiac arrhythmias. **2:** any medication used to treat cardiac arrhythmias.

antiarsenin. a nonarsenical substance thought to be secreted by the body in response to the administration of doses of arsenic trioxide (as arsenic acid) that are protective against toxic responses to a higher dosage.

antibacterial. **1:** destructive to or inhibiting the multiplication of bacteria. **2:** an antibacterial agent; a chemical or physical agent that kills (bacteriocidal) or reversibly suppresses (bacteriostatic) the multiplication of bacteria. Common antibacterials are antibiotics, antiseptics, and disinfectants. Some are selectively toxic, others are not. *Cf.* antimicrobial.

antibacterial agent. antibacterial (def. 2).

antibactericidin. a component of certain snake venoms that inhibits the bactericidal action of phagocytes. See also snakebite, venom (snake venom).

antibiont. a living organism involved in antibiosis.

antibiosis. **1:** an association of two species of organisms to the detriment of one, or between one species and a metabolic product of another (e.g., a toxin or antibiotic). **2**. the production of a metabolite (antibiotic) by a microorganism (e.g., a bacterium or fungus) that is toxic or inhibitory to certain other organisms. See allelopathy (def. 2), antibiotic, antibiont, probiosis.

antibiotic. **1:** pertaining to antibiosis. **2:** detrimental or destructive to life. **3:** of, with, or pertaining to an antibiotic (def. 4) or to the action of an antibiotic. **4:** any of a chemically diverse group of low molecular weight organic compounds with selective antimicrobial activity. They are extremely varied in chemical structure. Most antibiotics act by inhibiting protein synthesis on the 70S (but not the 80S) ribosomes of prokaryotes, while a few (e.g., cycloheximide) inhibit protein synthesis only on the 80S ribosomes. Many are used therapeutically in human and veterinary medicine. They are effective at low concentrations. Many antibiotics also produce toxic effects or allergic reactions in animals. Some (e.g., tetracycline) are teratogenic. Many commonly used antibiotics are highly toxic to guinea pigs and hamsters. In these animals a syndrome rapidly leading to death can result from excessive growth of *Clostridium difficile* populations in the intestines with ensuing production of toxins. This causes a fatal enterocolitis, with diarrhea and death in 3-7 days. Antibiotics with an action directed chiefly against gram-positive organisms (e.g., penicillin, lincomycin, erythromycin, tylosin) should not be used in guinea pigs and hamsters. Broad-spectrum antibiotics should not be used orally because of the direct effect on intestinal flora; they may be used parenterally with caution. *Cf.* allelochemical. See also allelopathy, *Pennicillium*, *Streptomyces*.

antibiotic A 523. natamycin.

antibiotic MA 144A1. aclacinomycin A.

antibiotic-resistant. pertaining to microorganisms that continue to multiply when exposed to antibiotic substances.

antibody (antisubstance; immune body; sensitizer). **1:** a protein elaborated into the blood or body tissues in response to the presence of an antigen or to a hapten associated with a carrier. It combines specifically with either the antigen or the hapten that activated it, inhibiting or neutralizing its action. **2:** a specific serum globulin, produced endogenously in response to the presence of an antigen. It binds with the specific antigen or sometimes certain structurally similar antigens if such are subsequently encountered. Antibodies are produced as part of the development of natural immunity or artificially by immunization. Some antibodies presumably exist naturally in the vertebrate body prior to or independent of the stimulus of an antigen. Most, however, are produced by B lymphocytes that have matured into plasma cells under the influence of a specific antigen. Antibodies confer humoral immunity for a period of time, in some cases providing lifetime immunity. *Cf.* sensitinogen, sensitizer. See also allergy, antigen, antigen-antibody reaction, immune response, immune reaction.

antibody-dependent cellular cytotoxicity (ADCC). the nonspecific lysis of antibody-covered cells by cells that have not been sensitized to that particular antigen or antibody.

anticancer agent. an agent (e.g., ionizing radiation, medications) that weakens, slows the proliferation of, or destroys cancerous tissues without undue harm to healthy tissues. See antineoplastic agent.

anticancer drug (anticancer medication). a medication that weakens, slows the proliferation of, or preferentially destroys cancerous tissues.

anticanitic vitamin. *p*-aminobenzoic acid.

antichlor. sodium thiosulfate.

anticholinergic. **1:** pertaining to the blocking of transmission by cholinergic nerves. **2:** antagonistic to the action of cholinergic neurons. **3:** an anticholinergic substance; one that blocks or reduces nerve impulses in the parasympathetic nervous system (e.g., atropine; belladonna; henbane; scopolamine; henxtropine mesylate; procyclidine; propantheline bromide; jimson weed seed). Anticholinergics control some activities of the heart, gastrointestinal system, bladder, and other viscera.

anticholinergic substance. See anticholinergic.

anticholinesterase (acetylcholinesterase inhibitor). any substance that inhibits or inactivates acetylcholinesterase, thereby preventing the hydrolysis of acetylcholine. Their action leads to an accumulation of endogenous acetylcholine with a resultant hyperactivity of cholinergic neurons which can prove lethal. A number of organophosphorus or carbamate esters have anticholinesterase activity. Symptoms of acute poisoning can include irritability, tremors, and convulsions, with death usually from respiratory failure. Acetylcholinesterase-inhibiting pesticides include aldicarb, carbaryl, ethion, fenthion, malathion, methomyl, and parathion. Further examples of anticholinesterases are physostigmine, tetraethyl pyrophosphate, snake venoms, and nerve gases. Their effects may be countered by the administration of atropine sulfate or pralidoxime iodide, which reactivate acetylcholinesterase.

anticipation. **1:** expectation, hope, apprehension. **2:** subintrance (def. 1c).

anticoagulant. **1:** preventing or delaying coagulation. **2:** of or pertaining to the capacity of a substance to prevent coagulation, especially of blood. **3:** an agent that interferes with or prevents coagulation, used mainly in connection with the clotting of blood. Examples are dicoumarol, warfarin, heparin, phenindione, prothromadin, and many venoms. See also anticoagulant poison, snakebite, venom (snake venom).

anticoagulant poison. an anticoagulant used generally to control small mammal pests, of which commensal rodents in cities are the most important. They are the most widely used agents for this purpose. Many small non-pest animals are unintended victims of anticoagulant poisons.

anticodon. a sequence of three nucleotides (triplet) in transfer RNA that pairs with a complementary triplet (codon) in messenger RNA.

anticomplement (antialexin). a substance that abolishes the action of complement by combining with it, thus preventing the union of complement and antibody.

anticomplementary. pertaining to or denoting a substance that can reduce or abolish the action of a complement. See anticomplement.

anticonvulsant (anticonvulsive). **1:** an agent used to prevent, control, or to relieve convulsive states (e.g., epileptic seizures, seizures caused by toxic substances, nonspecific insults to the CNS). They can cause osteomalacia. **2:** pertaining to the capacity to prevent, control, or relieve convulsions. See also antiepileptic, convulsant, convulsions, barbiturates, benzodiazepines, diazepam, phenytoin.

anticonvulsive. anticonvulsant.

anticrotin. the antitoxin of croton.

anticurare. an agent that neutralizes the myotoxic action of curare.

anticytotoxin. a substance (usually a specific antibody) that inhibits the action of a cytotoxin.

antidepressant. any agent that prevents, eases, or cures mental depression. There are two main types: monoamine oxidase inhibitors, *q.v.*, and tricyclic antidepressants. Both are toxic by virtue of their capacity to potentiate the action of endogenous and, if ingested, exogenous monoamines. Lithium is sometimes used prophylactically against depression. Amphetamines are also antidepressants, but are rarely used today. **tricyclic antidepressant**. any antidepressant with a basic three-ring structure. They are often as effective as monoamine oxidase inhibitors and appear to be somewhat safer, although they can cause dry mouth, dilated pupils, vomiting, retention of urine, agitation, hypertension, cardiac arrhythmias, exaggerated reflexes, incoordination, hallucinations, convulsions, and loss of consciousness. See amitriptyline, imipramine.

Antideprin. imipramine.

antidiuretic hormone. vasopressin.

antidotal (antidotic). serving as or having the properties of an antidote.

antidote (counterpoison; alexipharmic (def.2)). **1:** a substance that acts specifically to prevent, inhibit, inactivate, prevent, counteract, reverse, or relieve the action or poisonous effects of a toxicant. Many supposed antidotes are themselves toxic or otherwise potentially harmful and should not be used without special training or, with few exceptions, under appropriate medical supervision. *Cf.* antitoxin, antivenin. See also antiantidote, dimercaprol (British Antilewisite), atropine, cupric sulfate, 2,3-dimercapto-1-propanesulfonic acid, dithiocarb sodium, edrophonium chloride, ferric ferrocyanide, folinic acid, iodine, methylene blue, potassium ferrocyanide, scopolamine, and tiopronin. **2:** in vernacular usage, any substance that prevents or relieves the effects of a poison is called an antidote. Definition 1 above is the only accepted professional usage. *Cf.* true antidote. **chemical antidote**. a substance that unites with a poison to form a nontoxic compound. **mechanical antidote**. a substance that prevents the absorption of a toxicant. See true antidote. **physiologic antidote**. an agent that produces systemic effects that counter those of the specific toxicant under treatment. **true antidote**. a substance that acts by a specific mechanism related to that of the toxicant (e.g., *N*-methylpyridinium-2-aldoxime). A chemical that dilutes a toxicant (e.g., water, milk) or otherwise acts nonspecifically as a barrier to contact or absorption or one that acts to alleviate symptoms are not true antidotes. This term is sometimes used to emphasize that the vernacular usage of "antidote" (def. 2, above) is not intended; "specific antidote," though not recommended, is sometimes used. **universal antidote**. a hypothetical antidote that would be therapeutically effective against all poisons. No such antidote is known to exist. The use and effectiveness of any antidote depend on the type and amount of the toxicant encountered and the route of exposure or administration. Their use ordinarily depends also on the time delay between exposure and treatment, the condition (sometimes including age and sex) and history of the victim, and usually other factors.

antidotic. antidotal.

antiemesis. the prevention or relief of nausea and vomiting.

antiemetic. **1:** preventing, suppressing, or relieving nausea and vomiting. **2:** an agent having antiemetic action (e.g., chlorpromazine, naloxone, perphenazine, prochlorperazine, promethazinc). Chlorpromazine and other phenothiazines (e.g., chlorpromazine) can block nausea and vomiting induced by a variety of causes such as vestibular stimulation, gastroenteritis, carcinoma, chemotherapeutic drugs, gastroenteritis, radiation sickness, and opioids. The nausea and emesis of motion sickness can be controlled by a number of drugs (e.g., diphenhydramine, Dramamine, promethazine, and scopolamine). Most neuroleptics also have some antiemetic activity and a few can block nausea induced by vestibular stimulation.

antiendotoxic. relieving or counteracting the effects of endotoxins.

antiendotoxin. an antibody that counteracts the effects of an endotoxin.

antienzyme. an agent that prevents, retards, or abolishes the activity of an enzyme.

antiepileptic. having the capacity to control epileptic seizures; an anticonvulsant that acts to control epileptic seizures. See anticonvulsant.

antiestrogen. **1:** inhibiting or modifying the biological action of estrogens. **2:** any substance that inhibits or modifies the biological action of an estrogen by stimulating antagonistic effects on the target tissue (e.g., androgens and progestogens) or by competing with estrogen for receptor sites.

antiestrogenic. having the capacity to inhibit or modify the biological action of estrogens.

antifebrile. antipyretic.

antifebrin. acetanilide.

antifeedant. an allelochemic produced by one organism that deters another organism from feeding upon it.

antiformin. alkaline sodium hypochlorite solution. See sodium hypochlorite solution.

antifungal. antimycotic.

antifungal agent. antimycotic (def. 2). *Cf.* antifungal drug.

antifungal drug. a medication used to treat fungal diseases. *Cf.* antimycotic.

antigen (Ag; allergen; immunogen). any substance, usually a foreign protein (e.g., an enzyme, a toxin) or polysaccharide with a molecular mass greater than 5000 to 10,000 daltons), a parasite, or a microorganism that, when introduced into the body of an animal, provokes the formation of a specific antibody. Subsequent exposure to the antigen (following a latent period of 1 to 2 weeks) results in a reaction specifically with that antibody (See antigen-antibody reaction), the outcome of which may be greatly increased specific resistance to an infection (immunity) or to a toxic substance. *Cf.* allergen, anaphylaxis, antibody, hapten, hypersensitivity, immune response. **rhus venenata antigen**. an extract of fresh leaves of *Rhus venenata* (poison sumac) that is used to determine one's sensitivity to this plant or to alleviate the dermatitis caused by contact with its leaves. **sensitized antigen**. the complex formed when an antigen combines with specific antibody. The antigen then becomes sensitive to the action of complement. **shock antigen**. an antigen that can cause anaphylactic shock in a sensitized animal.

antigen-antibody complex. See reaction (antigen-antibody reaction).

antigen-antibody reaction. See reaction (antigen-antibody reaction).

antigenic (immunogenic). having the capacity to act as an antigen; capable of inducing the production of an antibody. *Cf.* allergenic, immunogenic.

antigenic drift. a gradual change in the antigenic makeup of some viruses.

antigenicity (immunogenicity). **1a:** the state or condition of being antigenic. **1b:** the capacity of an antigen to induce a specific immune response. Antigenicity usually depends on the size and chemical configuration of a molecule. See antigen, hapten.

Antigestil. diethylstilbestrol.

antiglobulin. a substance blocks the action of globulin.

antihelminthic. anthelmintic.

antihelmintic. anthelmintic.

antihemolysin. an agent that inhibits the action or effects of a hemolysin.

antihemolytic (hemosozic). **1:** preventing or inhibiting hemolysis; able to inhibit hemolysis. **2:** a substance that inhibits hemolysis.

antihemorrhagic. hemostatic; arresting hemorrhage; an agent that is able to prevent or arrest hemorrhage.

antiheterolysin. a substance that inhibits the action of heterolysin.

antiheterolytic. pertaining to, denoting, or having the properties of an antiheterolysin.

antihidrotic. anhidrotic.

antihistamine. any of a number of synthetic antimetabolites (mostly complex amines), minute amounts of which reduce the effects of excess histamine produced by body tissues. It acts by blocking the H_1 and/or H_2 histamine receptors. Examples of antihistamines are diphenhydramine hydrochloride, chloropheniramine maleate, cimetidine, dimenhydrinate, imidazole, tripelennamine hydrochloride, and promethazine. Antihistamines are used to treat allergies (e.g., hay fever), motion sickness, and stomach ulcers. They help to counteract some allergic manifestations of snakebite or horse serum, but play no role in the initial (acute) stages of snake venom poisoning. Some unintended effects of normal usage are nausea, vomiting, fatigue, sedation, constipation, dryness of the mouth and upper respiratory tract, dysuria, and hypotension. Overuse should be avoided, and antihistamines should not be kept within reach of children. Abuse can prove fatal. Symptoms of intoxication vary, but often include drowsiness, excitement or depression, nervousness, ataxia, disorientation, fatigue, apathy, miosis, and coma. Another pattern that sometimes occurs includes anxiety, tremors, nausea, vomiting, diarrhea, hallucinations, delirium, hyperpyrexia, convulsions, and coma. Death from cardiovascular collapse or respiratory arrest sometimes ensues. Cerebral edema may be the only evident lesion. See unintended effect.

antihistaminic. **1:** pertaining to antihistamines or the action of antihistamines. **2:** an agent with the action or properties of an antihistamine.

antihormone. any substance that prevents or interferes with the action of a hormone.

antihydropic. **1a:** relieving or able to relieve generalized edema. **1b:** an agent that relieves generalized edema.

antihyperlipidemic drug. any drug that reduces lipids or lipoproteins in the blood with a view to preventing or arresting vascular disease. Examples are clofibrate, nicotinic acid, and cholestyramine. All of these drugs produce adverse effects in some individuals.

antihyperlipoproteinemic drug. See antihyperlipidemic drug.

antihypertensive. **1:** acting to counteract hypertension; pertaining to the action of an antihypertensive agent. **2:** an antihypertensive agent, usually a drug.

antihypertensive drug (antihypertensive). a substance used therapeutically to lower blood pressure. An overdose can dangerously lower blood pressure. Examples are guanethidine sulfate, α-methyl dopa, hydralazine, reserpine, and syrosingopine. See also antihypertensive (def. 2).

antihypotensive. **1:** acting to counteract hypotension; pertaining to the action of a antihypotensive agent. **2:** a drug or other agent that counteracts hypotension.

antihypotensive drug. a substance used therapeutically to increase blood pressure. See also antihypotensive (def. 2).

antiicteric. able to relieve or prevent icterus.

antiinflammatory. **1:** counteracting or having the capacity to counteract or reduce inflammation. **2:** any antioxidant agent or procedure that controls, counteracts, or reduces inflammation. Antiinflammatory agents comprise a diverse class of pharmaceuticals used to control inflammation that is not caused by infection. Examples include betamethasone, prednisolone, prednisone, and other glucocorticoids; colchicine; gold salts; and nonsteroidal antiinflammatory drugs (NSAID) such as phenylbutazone and indomethacin. The latter include 6 types of drugs: fenamates, indenes, oxicams, propionates, pyrazolones, and salicylates. Salicylates and other NSAID appear to control inflammation by inhibiting prostaglandin and/or leukotriene synthesis.

antiinflammatory agent. antiinflammatory (def. 2).

antiinflammatory drug. an antiinflammatory used to control inflammation not caused by infection. Many of the therapeutic effects of nonsteroidal antiinflammatory drugs such as acetaminophen, aspirin, and ibuprofen inhibit cyclooxygenase, thus reducing prostaglandin synthesis.

antiinflammatory hormone. hydrocortisone.

antiinvasin. an enzyme in normal blood plasma that is antagonistic to hyaluronidase. **antiinvasin I**. an enzyme, normally present in blood, that is antagonistic to hyaluronidase. **antiinvasin II**. an enzyme, normally present in blood plasma that is antagonistic to proinvasin I.

antiisolysin. any substance that inhibits, neutralizes, or counteracts the effects of an isolysin.

antiknock agent. an organometallic compound added to fuels of internal-combustion, spark-ignition engines in very small amounts to increase their resistance to knocking. In the past, tetraethyllead and/or tetramethyllead were most commonly used for this purpose. Halogenated compounds are also added to the fuel to facilitate the removal of lead after combustion. See tetraethyllead, tetramethyllead. Several nations have restricted the amount of lead agents allowed in gasoline.

antileukocidin (antileukotoxin). **1:** a substance that opposes the action of or inhibits the effects of leukocidin. **2:** a leukocidin-specific antibody.

antileukotoxin. antileukocidin.

antilirium. a parasympathomimetic agent given by injection or orally.

antilymphocyte globulin (ALG). **1a:** the globulin fraction of antilymphocyte serum. **1b:** antilymphocyte serum, *q.v.*

antilymphocyte serum (antilymphocyte globulin (def. 1b); antilymphocytic serum; ALS). a serum that contains antibodies produced by one species of animal (e.g., a rabbit or horse) against those of another such as a human. It is produced by inoculating animals with tissue from another species. The globulin fraction (ALG) is as effective at immunosuppression as whole antiserum. **antihuman antilymphocyte serum** (antihuman antilymphocytic serum;). a serum produced by an animal against human antibodies. It is usually prepared by immunizing horses with human thoracic duct lymphocytes. It is used to suppress the rejection of transplanted organs and tissues by the human host.

antilymphocytic serum. antilymphocyte serum. **antihuman antilymphocytic serum**. antihuman antilymphocyte serum.

antilysin. a substance that opposes the action of a lysin or prevents the process of lysis.

antilysis. the prevention, inhibition, or suppression of lysis.

antilytic. pertaining to, marked by, or capable of antilysis; inhibiting or counteracting lysis.

antimalarial. **1:** having the capacity to prevent or treat malaria; an antimalarial drug. **2:** an antimalarial drug.

antimalarial drug. a substance used to prevent or treat malaria; an antimalarial (def. 2).

antimetabolite (structural antagonist). **1:** a metabolically inactive organic compound (e.g., sulfanilamide) that structurally resembles a biologically active metabolite (e.g., an enzyme, amino acid, nucleic acid) which it replaces, antagonizes, or competes with, thus blocking a particular metabolic reaction sequence. Antimetabolites are particularly toxic to proliferating lymphocytes and myeloid tissues. They can be chronically disabling or lethal; they may induce thrombocytopenia, granulocytopenia, and in severe cases, aplastic anemia. Antimetabolites also have important medical uses, as in the treatment of allergies. **2:** an antimetabolite drug. See, for example, sulfa drug, antihistamine. See also imidazole.

antimetabolite drug. any antimetabolite used to treat certain cancers or autoimmune diseases.

antimethemoglobinemic. **1:** reducing methemoglobin levels in circulating blood. **2:** an agent that reduces methemoglobin levels in the circulating blood.

antimicrobial. **1:** lethal to or inhibiting the multiplication of microorganisms. **2:** an antimicrobial agent; a lethal agent or one that reversibly suppresses the multiplication of microorganisms. Physical antimicrobials include extremes of temperature, ultraviolet radiation, and ultrasound. Chemical antimicrobials include sulfonamides, numerous disinfectants, antiseptics, and chemotherapeutic agents.

antimicrobial drug. an antibiotic; an antimicrobial (def. 2).

antimicrobic. antimicrobial.

antimitotic agent. See spindle poison.

antimonial poisoning (stibialism). poisoning by salts of antimony, most of which are extreme irritants and are highly toxic. Effects are similar to those of arsenic poisoning. The fumes or dust of many antimonial compounds are extreme irritants, causing conjunctivitis, keratitis, dermatitis, and ulceration of the nasal septum. Onset of the effects of acute poisoning by ingestion of antimonial salts may be delayed for several hours. There may be a metallic taste in the mouth, and effects usually include those of severe gastrointestinal pain, vomiting and purging, shock, dehydration, convulsions, severe bloody diarrhea, coma, and possibly death. See especially stibine.

antimonium. antimony.

antimonous chloride. antimony trichloride.

antimonous fluoride. antimony trifluoride.

antimonous sulfate. antimony sulfate.

antimony sodium gluconate. See stibogluconate sodium.

antimony sodium oxide L(+)-tartrate. antimony sodium tartrate.

antimony sodium tartrate (antimony sodium oxide L(+)-tartrate; sodium antimonyl tartrate). an anthelmintic used against schistosoma. An extremely toxic, transparent or whitish, hygroscopic powder or scales, $(SbO)NaC_4H_4O_6$, used as an anthelmintic against *Schistosoma* infestations; used also as a leishmanicide in veterinary practice.

antimony sulfate (antimonous sulfate; antimony trisulfate). a poisonous, flammable, crystalline powder or lumps, $Sb_2(SO_4)_3$, that deliquesces in air. It is used in matches and pyrotechnics. It should be kept in a well-sealed container.

antimony trichloride (antimonous chloride; trichlorostibine; butter of antimony; caustic antimony). antimony trichloride, $SbCl_3$, is used commercially as a mordant, in the synthesis of other antimony salts, and as a catalyst in organic syntheses. It is corrosive to living tissue and the fumes are poisonous.

antimony trifluoride (antimony fluoride; antimonous fluoride). a poisonous substance, SbF_3, and skin irritant.

antimony trifluorodichloride. antimony dichlorotrifluoride.

antimony trisulfate. antimony sulfate.

antimony yellow. lead antimonate.

antimuscarinic. 1: acting to inhibit or block the toxic action of muscarine and muscarine-like agents. **2:** a muscarinic antagonist. **3:** inhibiting the effects of parasympathetic stimulation at the neuroeffector junction.

antimutagen. any agent that slows the rate of mutation or inhibits or reverses the action of a mutagen.

antimutagenic. relating to or having the action of an antimutagen.

antimycobacterial. antagonistic to or destructive of mycobacteria.

antimycotic (antifungal). antagonistic to, destructive to, or inhibiting the growth of fungi.

antinarcotic. 1: antagonistic to or inhibiting narcotic depresson. **2:** an agent that inhibits the action of a narcotic.

antinauseant. 1: able to inhibit or alleviate nausea. **2:** an antinauseant agent.

antineoplastic. 1: destructive to neoplasms; acting against neoplastic disease processes. **2:** an antineoplastic agent or antineoplastic drug.

antineoplastic agent. any biologically active agent that is destructive to neoplasms. See antineoplastic drug.

antineoplastic drug. any of a number of pharmacologically and structurally diverse therapeutic drugs used in chemotherapy to treat neoplastic diseases by killing neoplastic cells. All are notoriously toxic, and those used clinically have numerous toxic effects on a variety of tissues and organs. They are often abortifacient, fetotoxic, and teratogenic, causing multiple fetal or neonatal anomalies or abortion. Examples are busulfan, carmustine, cyclophosphamide, ethinylestradiol, metho-trexate, tamoxifen, thiotepa, vinblastine, and vincristine. *Cf.* chemotherapeutic agent, antineoplastic (def. 2).

antinephritic. **1:** able to relieve or prevent inflammation of the kidneys. **2:** an agent capable of preventing or relieving inflammation of the kidneys.

antineuralgic. **1:** able to relieve neuralgia. **2:** an agent that relieves neuralgia.

antineuritic. **1:** able to prevent or relieve inflammation of a nerve. **2:** an agent that prevents or relieves inflammation of a nerve.

antineuritic vitamin. thiamine.

antineurotoxin. any substance, especially an antibody, that inhibits or counteracts the action of a neurotoxin.

antiophidic. **1:** able to inhibit or alleviate the effects of snakebite. **2:** any agent that opposes or relieves the effects of snakebite. *Cf.* antiophidica.

antiophidica. **1:** drugs that oppose or relieve the effects of snakebite. **2:** plural form of antiophidic (def. 2).

antioxidant (antoxidant). **1:** any substance that inhibits or prevents oxidation by molecular oxygen, peroxides, or any active oxygen species (e.g., the superoxide radical). Examples are superoxide dismutases, reduced glutathione (GSH), alphatocopherol, ascorbic acid, betacarotene, catalase, glutathione peroxidase, glutathione reductase, histidine, methionine, and polyunsaturated fatty acids. Antioxidants are used, for example, as food preservatives to prevent or inhibit spoilage or the development of other undesirable changes such as discoloration or offensive odors in foods and other organic materials. **2:** a substance that helps to prevent foods and other materials from becoming discolored, rancid, spoiled, or otherwise offensive. **3:** of or pertaining to an antioxidant.

antioxidation. the inhibition or prevention of oxidation.

antiparalytic. **1:** relieving paralysis; able to resist, inhibit, or relieve paralysis. **2:** any drug that relieves paralysis.

antiparasitic. antagonistic to or destructive to parasites.

antiparasympathomimetic. **1:** able to produce effects similar to those of an interrupted parasympathetic innervation. **2:** a substance that produces antiparasympathomimetic effects.

antiparkinsonism drug. any medication used to treat Parkinsonism.

antipathic. antagonistic.

antipathy. **1:** antagonism. **2:** a feeling of strong aversion or dislike.

antiperspirant. a commercial product in the form of a solid stick, a liquid, or an aerosol spray that is used to control underarm wetness. Aerosol propellants often include ammonia, alcohol, formaldehyde, and fragrance, but aluminum chlorhydrate, the active ingredient that helps stop wetness, offers the greatest danger. It can promote infections in the hair follicles of the armpit and skin irritations that sometimes require medical attention. If absorbed through the skin, or if inhaled, it may have systemic effects as well. Aluminum has been associated with various brain disorders, including Alzheimer's disease. Other long-term health effects remain unknown.

antiphrynolysin (toad antivenin). an antivenin specific for the toxin of toad venom.

antipneumococcic. **1:** antagonistic to or destructive of pneumococci. **2:** an agent that inhibits or destroys pneumococci.

antiprotozoal (antiprotozoan). **1:** antagonistic to or destructive to protozoa. **2:** an agent that destroys protozoa or inhibits their growth or reproduction.

antipruritic. **1:** able to prevent or relieve itching. **2:** an agent that is able to prevent or relieve itching.

antipsychotic. neuroleptic.

antipsychotic drug. See neuroleptic drug.

antipyresis. the use or application of antipyretics.

antipyretic (antifebrile; antithermetic). **1:** acting to relieve or reduce a fever. **2:** an agent (antifebrile, antithermetic, febrifuge), such as aspirin, that allays fever.

antipyrine (1,2-dihydro-1,5-dimethyl-2-phenyl-3H-pyrazol-3-one; 1,5-dimethyl-2-phenyl-3-pyrazolone; phenazone; 2,3-dimethyl-1-phenyl-3-pyrazolin-5-one; phenyldimethylpyrazolone(e); dimethyloxychinizin; dimethyloxyquinazine). a crystalline substance or powder, $(CH_3)_2(C_6H_5)C_3HN_2O$, that is colorless, odorless, slightly bitter, and is soluble in water, ethanol, and chloroform. A pyrazolon derivative, it is produced by the condensation of methylphenylhydrazine and ethyl acetoacetate, and is used medically as a weak analgesic, potent antiinflammatory agent, and antirheumatic. It is generally moderately toxic *per os*, but serious side effects may occur, especially serious blood disorders. See also aminopyrine, apazone, antipyrine, phenylbutazone.

antiricin. an antibody or antitoxin that counters or interferes with the action of ricin.

antirobin. the antitoxin of robin.

antiseptic. any nonselective antimicrobial that is applied topically to kill or inhibit the growth of pathogenic microorganisms in living material.

antisera. plural form of antiserum.

antiserum (plural, antisera). a serum or body fluid that contains high levels of antibodies against a specific antigen. Antisera result naturally (through disease) or through immunization. Antisera are injected to give passive immunity against specific diseases and are used experimentally to identify unknown pathogens or antigenic substances. *Cf.* antitoxin, antivenin.

antisnakebite serum. snake antivenin. See antivenin.

antispasmodic. **1:** preventive or ameliorative of convulsions or spasms, usually of smooth muscle. **2:** an agent that allays spasms. *Cf.* antispastic. See also spasmolytic. **biliary antispasmodic.** an agent that alleviates spasm of the biliary duct and sphincter. **bronchial antispasmodic.** an agent that eases bronchial spasm.

antispasmodic drug. antispasmodic (def. 2).

antispastic. preventing or alleviating spasms of skeletal muscle. *Cf.* antispasmodic.

antispermotoxin. a substance that is antagonistic to the action of a spermotoxin.

antistaphylococcic. antagonistic to or destructive of staphylococci or their toxins.

antistaphylohemolysin. antistaphylolysin.

antistaphylolysin. a substance that inhibits the action of staphylolysin.

antisterility vitamin. vitamin E.

antistreptococcic. **1:** antagonistic to or destructive of streptococci or their toxins. **2:** a substance with antistreptococcic action.

antistreptolysin. an antibody that is active against streptolysin.

antisubstance. antibody.

antisudorific. anhidrotic.

antisympathomimetic. **1:** producing or able to produce effects similar to those of an interrupted sympathetic nerve supply. **2:** an agent that produces antisympathetic effects.

antisympathomimetic drug. antisympathomimetic (def. 2).

antitetanic. **1:** preventing, relieving, or curing tetanus. **2:** an agent that prevents, relieves, or cures tetanus or tetanic spasms.

antithermetic. antipyretic.

antithrombin. former name of heparin, *q.v.*

antithrombin III. a plasma protein that inactivates activated blood clotting factors, most importantly thrombin.

antithyroid drug. an antithyrotoxic medication used to counter the effects of hyperactive thyroid gland. *Cf.* antithyrotoxic.

antithyrotoxic. antagonistic to, inhibiting, preventing, or counteracting the toxic effects of thyroid gland tissue, its products, or extracts. See also antithyroid drug.

antitonic. able to diminish tonicity.

antitoxic. **1:** active against a poison. **2:** of, or pertaining to the action of an antitoxin. *Cf.* antidotal.

antitoxigen. antitoxinogen.

antitoxin (antitoxinum; immunotoxin; toxolysin). **1a:** an antibody induced by the presence of poisonous xenobiotic substances, usually by a disease-causing organism or poisonous substances that it secretes (e.g., phytotoxins, zootoxins, bacterial exotoxins). **1b:** whole serum or the globulin fraction of serum from animals (e.g., horses, rabbits) that have been immunized by injections of a specific toxoid. Examples are diphtheria antitoxin, bothropic antitoxin (antivenin), botulism antitoxin, and Crotalus antitoxin (antivenin). **diphtheria antitoxin**. that produced in the blood of horses that have been injected with diphtheria toxoid. The serum of such horses, when injected into humans, confers immunity to diphtheria. **bothropic antitoxin** (Bothrops antitoxin). an antitoxin specific for the venom of pit vipers of the genus *Bothrops* (Family Crotalidae). *Cf.* serum (antibothropic serum). **bothrops antitoxin**. bothropic antitoxin. **botulism antitoxin** (botulinum). an antitoxin specific for a toxin of one of the strains of *Clostridium botulinum*. **crotalus antitoxin**. an antitoxin specific for rattlesnake venoms (i.e., of *Crotalus*). *Cf.* serum (anticrotalus serum), antiantitoxin, antivenin. See also deuterotoxin, endoantitoxin, epitoxoid, hemoantitoxin, prototoxin, prototoxoid, toxoid, tritoxin.

antitoxinogen (antitoxigen). any substance that induces or stimulates the formation of antitoxin by an animal *in vivo*. *Cf.* antitoxin.

antitoxinum. antitoxin.

antitrichomonal. **1:** lethal to or resistant to trichomonads. **2:** a drug that is used to treat trichomonal infection.

antitussive. **1:** relieving or preventing cough. **2:** an agent that prevents or relieves cough.

antivenene. antivenin.

antivenin (antivenom; antivenene; antivenine). **1:** a specific antitoxin that acts against, inhibits, or reverses the effects of a venom. The antibodies in such immune serum preparations neutralize the venom of the taxon for which it was prepared. **2:** any serum (antiserum; antivenomous serum) that contains a venom-specific (species-specific, taxon-specific) antitoxin used in the treatment of poisoning by envenomation, as in snakebite or bee stings. *Cf.* antidote. **antivenin therapy**. the treatment of snakebite and other envenomations by the use of antivenin. *Cf.* venom extract therapy. **antivenin (Crotalidae) polyvalent [USP]**. a sterile antitoxin that contains the specific venom-neutralizing globulins produced by horses that have been immunized against the venom of four crotalid snakes: *Crotalus adamanteus*, *C. atrox*, *C. durissus terrificus*, and *Bothrops atrox*. It is used in the passive immunization of envenomations by most crotalid snakes and against that of *Agkistrodon spp*. **antivenin (*Micrurus fulvius*) [USP]**. a sterile antivenin used in the treatment of bites by *Micrurus fulvius*, the Eastern coral snake. **black widow spider antivenin**. *Latrodectus mactans* antivenin. ***Latrodectus mactans* antivenin**. any antivenin used in the treatment of bites by *Latrodectus mactans*. **Lyovac® Antivenin**. a *Latrodectus mactans* antivenin; normally used only for children. **polyvalent (Crotalidae) antivenin**. antivenin (Crotalidae) polyvalent [USP]. **polyvalent antivenin**. any antivenin used as an antidote or neutralizing agent against the venom of more than one species of animal. **polyvalent crotalid antivenin** (Wyeth, Inc., Philadelphia). an antivenin specific for the venoms of *Crotalus adamanteus* and *C. atrox*. It is to some degree effective against all crotalid venoms. **polyvalent crotaline antivenin**. antivenin (Crotalidae) polyvalent [USP]. **snake antivenin** (antisnakebite serum; antivenomous serum; Calmette's serum). any antiserum used against a snake venom. There are no other specific antidotes for snakebites. Antivenins are available for most snake venoms. Their use usually requires correct identification of the snake and their effectiveness is further affected by the site of the bite and the amount of time elapsed between bite and administration of the antivenin. Most fatalities in treated individuals are due to administration of the wrong antivenin or of too small a dose. Snake antivenins can also cause heart or respiratory problems, and their administration to persons who have not been bitten can be as dangerous as an actual envenomation. **stonefish antivenin**. an antivenin specific to stonefishes, genus *Synanceja*. It is produced by hyperimmunizing horses with stonefish venom. **toad antivenin**. See antiphrynolysin.

antivenin therapy. the treatment of snakebite and envenomations by other animals by the use of antivenin. *Cf.* venom extract therapy.

antivenine. antivenin.

antivenom. antivenin.

antivenomous. **1a:** antagonistic to or counteracting the action of a venom. **1b:** having the capacity to alleviate or counteract the effects of venom. **2:** pertaining to antivenin.

antivenomous serum. snake antivenin.

antiviral. **1:** destructive to or inhibiting the replication of a virus. **2a:** any agent that destroys viruses or inhibits their replication. **2b:** an antiviral drug.

antiviral drug (antiviral, def. 2b). a medication used to treat infection by a virus.

antivitamin (vitamin antagonist). any of a group of substances that oppose the action of certain vitamins.

antixenic. of or pertaining to the reactions of living tissue that counteract the action or effects of a foreign substance.

antizymotic. **1:** of or pertaining to the capacity to inhibit enzymatic activity. **2:** any agent that inhibits the action of enzymes.

antorphine. nalorphine.

antoxidant. antioxidant.

antrum. any nearly closed chamber in an animal body, especially in bone. See also chamber.

ants. See Formicidae.

ANTU (**1-naphthalenylthiourea; 1-(1-naphthyl)-2-thiourea; α-naphthylthiourea; *N*-1-naphthylthiourea; α-naphthylcarbamide; krysid; chemical 109**). a thiourea derivative used to control *Rattus norvegicus* (Norway rat). It is an antithyroid agent and induces massive pulmonary edema with effusion into the pleural cavity, fatty degeneration of the liver, and hypothermia with a high rate of mortality in this species. It is highly toxic to *R. norvegicus*, is less toxic to other species of *Rattus*, and is relatively nontoxic to most domesticated animals with the exception of dogs, which are quite susceptible to ANTU poisoning. It may be moderately toxic to adult humans. The lethal dose to monkeys is about 4000 mg/kg. A suicidal adult male human ingested about 80 g of 30% ANTU rat poison together with a considerable amount of alcohol. He soon vomited, suffering no significant ill effects.

ANTU poisoning. highly selective poisoning by the rodenticide, ANTU. Poisoning of the Norway rat, *Rattus norvegicus*, is marked by massive pulmonary edema with effusion into the pleural cavity, fatty degeneration of the liver, and hypothermia with a high rate of mortality. Most other animals are much less sensitive. Dogs, however, are quite susceptible to ANTU poisoning.

Anura (Salientia). the subclass of Amphibia (sometimes classified as an order of the subclass Lissamphibia) comprised of frogs and toads. They have a short vertebral column, lack tails as adults and are highly adapted to jumping and usually swimming by virtue of powerful hind legs, a strong pectoral girdle that withstands the force of landing, and webbed feet. The larvae (tadpoles) are aquatic with gill slits and a tail; they undergo a usually rapid metamorphosis in which the tail is resorbed and the gill slits are replaced by lungs. Some, representatives of the families Bufonidae, Atelopodidae, Dendrobatidae, Discoglossidae, Hylidae, Phylomedusae, Pipidae, and Ranidae, are poisonous. Anuran toxins are chemically diverse and act chiefly as repellants and/or bactericides.

anuresis. the absence of urination. *Cf.* anuria.

anuretic. of or pertaining to anuresis.

anuria. an absence of or sharp decline in the secretion of urine. *Cf.* anuresis.

anus. the posterior or distal opening of the digestive tract present in all metazoans, except the Coelenterata and Platyhelminthes, through which unabsorbed food or food residues and waste products are eliminated. It originates at or near the blastopore of the embryonic archenteron and, in some groups, opens into a cloaca. It is often controlled by a sphincter that relaxes to allow the passage of feces. The anus is an infrequent route of exposure to drugs and poisonous substances, although many drugs are easily administered *per anus*.

anxiety. a feeling of apprehension, uncertainty, and fear that something unpleasant or dangerous will occur. Such a state is usually a normal or exaggerated psychophysiological reaction to potentially threatening environmental stimuli. Anxiety (marked by sweating, tachycardia, tremor) can also indicate poisoning (e.g., by narcotics, psychoactive drugs, alcohol) and certain disease states. **free-floating anxiety**. a state of anxiety for which there is no known cause. **neurotic anxiety**. that which is disproportionate to any discernible cause. See anxiolytic.

anxiolytic. **1:** having the capacity to alleviate anxiety. **2:** any agent that alleviates the symptoms of anxiety, where anxiety is a psychophysiological reaction to potentially threatening environmental stimuli. See drug (anxiolytic drug).

anxiolytic drug. See drug.

aosmic. See anosmic.

apaconitine. a poisonous base derived from aconitine.

apamin. a small (mol wt 2027.38) highly potent neurotoxic polypeptide, $C_{79}H_{131}N_{31}O_{24}S_4$, that occurs in the venom of *Apis mellifera (mellifica)*. A CNS poison, it is able to cross the blood-brain barrier.

APAP. acetaminophen.

aparalytic. characterized by a lack of paralysis.

Aparasin®. lindane.

apazone (5-(dimethylamino)-9-methyl-2-propyl-1*H*-pyrazolo[l,2-a][1,2,4]benzotriazine-1,3(2H)-dione; 3-dimethylamino-7-methyl-1,2-(*n*-propylmalonyl)-1,2-dihydro-1,2,4-benzotriazine; azapropazone). a nearly colorless, crystalline solid used as an antiinflammatory agent, analgesic, and antipyretic. It has also been used in the treatment of rheumatoid disorders such as osteoarthritis, rheumatoid arthritis, and gout. Its action is similar to that of phenylbutazone, but it is substantially less toxic. Symptoms of intoxication most commonly include nausea, vomiting, abdominal pain, and sometimes gastric ulcers. *Cf.* aminopyrine.

ape. **1:** true apes are the large, tailless, Old World monkeys (primates) of the family Pongidae. Also called anthropoid ape or great ape, they include the chimpanzees, gibbons, gorillas, and orangutan. **2:** often applied to any monkey.

aphagia. complete failure to eat.

Aphanizomenon flos-aquae. a toxic species of Cyanobacteria. See also Annie, Fannie, and Mike.

Apheloria corrugata. a millipede known to release up to 645 μg of hydrogen cyanide when threatened. This is many times the amount that is lethal to a mouse. The hydrogen cyanide is produced in a two-chambered venom gland. The first chamber contains mandelonitrile, which passes to the second chamber where it dissociates into benzoic aldehyde and hydrogen cyanide. A muscle-controlled valve regulates the passage between the two chambers.

aphidicolin (tetradecahydro-3,9-dihydroxy-4,11b-dimethyl-8,11-methano-11aH-cyclohepta[a]-naphthalene-4,9-dimethanol). a white, tetracyclic, diterpene, fungal antibiotic that inhibits DNA replication in eukaryotes and DNA α-polymerase *in vitro*. Aphidicolin is used in research on cell multiplication and differentiation.

Aphonopelma. *Eurypelma*.

aphoxide. triethylenephosphoramide.

aphrenia. dementia.

aphrodine. yohimbine.

aphrodisiac. **1:** able to stimulate or increase sexual arousal or pleasure. **2:** any substance claimed to increase sexual desire or pleasure. Many substances are said to have this capacity, but most are ineffectual and, while many are harmless, some are poisonous. Cantharides is highly dangerous and can prove lethal when used as an aphrodisiac.

aphronesia. dementia.

aphthous. blister-producing.

Aphtiria®. lindane.

aphylactic. lacking immunity.

aphylaxis. the absence of immunity.

Apiaceae (the carrot, celery, or parsnip family). a family of hollow-stemmed herbaceous plants, occasionally of great size. They are widely distributed throughout temperate and subtropical regions. Some are weeds, others are important agricultural, ornamental, medicinal, garden, and household plants. Some are poisonous. See *Aethusa*, *Ammi*, *Angelica*, *Carum copticum*, *Apium graveolens*, *Conium maculatum*, *Cicuta*, *Cymopterus watsonii*, *Daucus carota*, *Heracleum*. See also methoxsalen, furocuomarin, psoralen.

Apidae. a family of social bees (Order Hymenoptera) that includes venomous and nonvenomous species. Some classifications include certain solitary bees. They typically build extensive combs of wax cells. Individual colonies may contain 10,000 to 40,000 individuals. There are three castes: a queen (the sole egg-laying female of a colony), drones (fertile males), and workers (sterile females). Bees are valuable to humans as pollinators and honey producers. See also *Apis*, bee.

apidosamine. a strong base and toxic irritant, $C_{22}H_{28}N_2O_2$, isolated from quebracho.

apiotherapy. therapy using bee venom.

Apis (honey bees). a genus of venomous bees; the type genus of family Apidae. Honey bees are distinguished from other genera of bees largely by wing venation. The venom is acidic and contains hemolytic enzymes. The sting causes local pain and a burning sensation with swelling, erythema, and whitishness at the wound site, respiratory distress; and shock. Multiple stings may require hospitalization. Fatalities are rare and are usually due to anaphylactic shock. ***A. mellifera*** (*A. mellifica*; common honey bee). a small venomous bee that occurs worldwide in hollow trees or in hives kept by beekeepers. They were first introduced to North America by settlers in the 17th Century. The honey is not sterile and has been known to cause botulism when used in infant formula. ***A. mellifica***. *A. mellifera*. See apamin, apitoxin.

apisination. envenomation by the sting of a bee.

Apistocalamus. a genus of small, venomous, burrowing snakes (Family Elapidae), all five species of which occur only in New Guinea. None appear to be highly dangerous to humans. They are distinguished by the poorly defined fangs and the presence of only six lower labials. ***A. grandis*** (giant apistocalamus). a species in which some individuals reach a length of nearly 1 m; it is the only species in the genus that exceeds 0.65 m in length. The remaining species are *A. lamingtoni* (Mount Lamington snake); *A. loennbergii*; *A. loriae*; and *A. pratti* (Pratt's snake).

Apistus carinatus (bullrout, sulky). a venomous marine fish (Family Scorpaenidae) found in coastal marine waters of India, the Netherlands Indies, Australia, Philippine islands, China, and Japan. The venom apparatus is of the *Scorpaena* type (See Scorpaenidae).

apitoxin. the toxic protein component of bee venom.

Apium graveolens (celery). a common table vegetable (Family Apiaceae). Toxic concentrations of nitrate that might be hazardous to foraging animals have been reported.

aplasia. arrested or defective development of an organ or tissue, or of its cellular products. *Cf.* hypoplasia, hyperplasia.

aplastic. **1:** pertaining to or characterized by aplasia. **2a:** exhibiting arrested, deficient, or defective development of an organ or tissue. **2b:** anatomically undeveloped.

aplastic anemia. See anemia.

Aplopappus. a synonym of *Haplopappus*.

Aplysia (sea hares). a genus of massive, slug-like, hermaphroditic, marine gastropod mollusks (up to 45 cm long) with a rudimentary, internal shell. They can concentrate toxic bromoterpenes from algae on which they feed. These mollusks can then poison animals (e.g., humans) higher in the food chain. The bromoterpenes aplysin and aplysinol have both been isolated from *Aplysia kurodai* and from red algae of the genus *Laurencia* on which it feeds. ***A. californica***. (California sea hare). a very large, unusually venomous and poisonous gastropod (ca. 41 cm long). It occurs from the low tide line to a depth of 18 m, from northern California to Baja California. It contains at least two extremely toxic principles: murexine in the midgut and aplysin in the digestive glands. Aplysinol, debromoaplysin, and diterpene aplysin-20 may also occur. See also *Laurencia*.

aplysin. an extremely toxic, hypotensive, cardiotoxic, brominated sesquiterpene (bromoterpene) isolated from the digestive glands of *Aplysia* (especially *Aplysia kurodai*) and from red algae of the genus *Laurencia*. Aplysin causes hypersalivation, ataxia, respiratory paralysis, and death in laboratory mice. See also *Aplysia californica*.

aplysinol. a bromoterpene isolated from *Aplysia kurodai*, *A. californica*, and from red algae of the genus *Laurencia*.

apnea. **1:** absence or temporary cessation of breathing. **2:** asphyxia.

apneic. of, pertaining to, or affected by apnea.

apo-, ap-. a prefix indicating separation, development, or derivation from.

APO. triethylenephosphoramide.

apoatropine ([illegible] acid 8-methyl-8-azabicyclo[3.2.1]oct-3-yl ester; 1αH,5αH-tropan-3α-ol atropate; atropamine; atropyltropeine). a toxic principle found in roots of *Atropa belladonna*. It is highly toxic to small mammals, and fairly small doses can cause death from respiratory collapse.

Apocynaceae (dogbanes). a family of widely distributed trees, shrubs, and vines that are most abundant in the tropics. Many are grown as ornamentals and occasionally for pharmaceuticals or their edible fruit. They have a poorly developed corona, a well-developed latex system, granular pollen, and carpels that are often joined by the style and stigma. Some are poisonous. Well-known genera with poisonous species are *Apocynum* (dogbane, Indian hemp), *Nerium* (oleander), and *Vinca* (periwinkle).

apocynin (acetovanillone). a cardiac glycoside. See *Apocynum*.

Apocynum (dogbane; Indian hemp). a genus of poisonous, chiefly North American herbs (Family Apocynaceae) with opposite leaves and small white or pink flowers; some species are cultivated. They contain toxic digitalis-like cardioactive glycosides such as apocynin. ***A. androsaemifolium*** (dogbane; wild ipecac; milk ipecac; rheumatism weed). a dogbane found in dry thickets, and woodland borders throughout much of the United States and in Canada from Nova Scotia to Newfoundland. It contains apocynin. ***A. cannabinum*** (Canadian hemp; black Indian hemp). a plant of open ground, thickets and woodland borders in southern Canada and the northern tier of states in the United States, extending from New England south to Florida in the eastern United States. It contains apocynin.

apogenic. sterile.

apogeny. sterility.

apomorphine (5,6,6a,7-tetrahydro-6-methyl-4H-dibenzo[de,g]quinoline-10,11-diol; 6aβ-aporphine-10,11-diol). a very toxic, synthetic, crystalline opiate alkaloid, $C_{17}H_{17}NO_2$, obtained by the removal of one molecule of water from morphine. Usually given intravenously as the hydrochloride, $C_{17}H_{17}NO_2 \cdot HCL \cdot ½H_2O$, it is a powerful emetic that promptly induces vomiting.

aponia. **1:** absence of pain. **2:** [illegible] exertion

aponic. **1:** able to relieve pain or fatigue. **2:** of or pertaining to aponia.

apoplectic stroke. See apoplexy.

apoplexy. **1:** being suddenly struck down, as in the case of a stroke. **2a:** apoplectic stroke; sudden neurologic impairment due to a cerebrovascular accident or disorder (e.g., intracranial hemorrhage, an embolus in a cerebral artery, occlusive cerebrovascular lesions, heat stroke). **2b:** usage restricted to sudden neurologic impairment due only to intracranial hemorrhage.

6aβ-aporphine-10,11-diol. apomorphine.

aposematic coloration. warning coloration.

apothecary. pharmacist.

3α,16α-apovincaminic acid ethyl ester. vinpocetine.

appetite suppressant. anorectic (def. 2).

appetitive. characterized by approach toward a stimulus; promoting or stimulating approach as opposed to withdrawal. Said of a stimulus or a behavior. *Cf.* aversive. See also approach-withdrawal.

apple. *Pyrus malus*.

apple of Peru. a common name of plants of the genus *Datura*, especially *D. stramonium*.

apple of Sodom. *Solanum sodomeum*.

approach-withdrawal. a highly adaptive general phenomenon whereby a motile organism tends to approach weak stimuli (e.g., as those emanating from food molecules) and to withdraw from strong stimuli (e.g., high temperature, loud noise, noxious substances). Underlying mechanisms are extremely diverse and this type of behavior in many forms of animals can be modified by experience. Numerous mechanisms have thus independently evolved whereby animals are attracted to and approach nutrient materials and avoid those which are irritant or toxic. Thus, insects, via diverse and often complex mechanisms, are attracted to their host plants and repelled by others, thereby avoiding toxic situations, plants, or even lures. Such mechanisms sometimes fail, as in the case of inexperienced animals or during unfavorable conditions when animals eat alternative materials. Thus, for example, herbivores generally graze on suitably nutritious species of plants, but during conditions of drought or overgrazing often fall victim to toxic species. See also aversive.

approximate. **1:** nearly correct, nearly accurate, close to a desired result. **2:** to calculate or evaluate by hurried or crude methods; to come close. **3:** approximation.

approximation. **1:** an estimate that is nearly correct. **2:** an estimate arrived at by the use of hurried or crude methods. **3:** the process of estimating by hurried, crude, or inaccurate methods.

apricot. *Prunus armenica*.

apricot kernel. apricot pit.

apricot-kernel oil. either of two oils extracted from the pits (seeds) of *Prunus armenica* (apricot). **1a:** persic oil. a colorless or straw-colored oil expressed from the pit. **1b:** bitter almond oil. a toxic, colorless or yellow aromatic oil obtained from the pit by steam distillation. See also benzaldehyde.

apricot pit (apricot kernel). the large, hard seed of the apricot fruit. See also amygdalin, plant poisoning, apricot-kernel oil.

Aprion virescens (snapper). a marine bony fish (Family Lutjanidae) of the tropical Indo-Pacific region, certain populations of which are transvectors of ciguatera.

aprocarb. propoxur.

apyretic. afebrile.

apyretic tetanus. See tetany.

apyrexia. absence of fever.

apyrogenetic. apyrogenic.

apyrogenic (apyrogenetic). not producing fever.

AQ. anthraquinone.

aqua. water.

aqua ammonia. See ammonia water, ammonium hydroxide.

aqua fortis. nitric acid.

aqua regia. nitrohydrochloric acid.

aqua toffana (aquetta di Napoli). an extremely toxic potion, presumably created by an Italian countess, Toffana. Its use was favored by and is associated with the Medicis. It probably contained arsenic and cantharides. Four to six drops in water or wine were usually fatal within a few hours.

Aquacide. diquat.

aqualin. acrolein.

aquatic. **1:** of or pertaining to water, especially in the environment. **2a:** living entirely or mainly in water. **2b:** living in, on, or near water; applied to plants that are adapted to partial or complete submergence. **3:** an aquatic organism. **4:** used in or conducted in or on water.

aquatic bioassay. See bioassay.

aquatic transport. See transport.

aqueous. pertaining to, denoting, or having the properties of water; watery.

aqueous humor. the transparent, thin liquid that occupies the space between the cornea and the lens of the vertebrate eye.

aquetta di Napoli. aqua toffana.

A_r. relative atomic mass.

Ar. the symbol for argon.

AR. acceptable risk. See risk.

Arabian mole viper. *Atractaspis microlepidota andersoni*.

arabinose (gum sugar). a nephrotoxic crystalline aldopentose, [illegible] from vegetable gums by acid hydrolysis and sometimes found in urine. It is nephrotoxic. The D- form of arabinose is a constituent of aloin; l-arabinose is the gum sugar.

arabinosis. poisoning by arabinose. Effects may include nephrosis.

arabuumihebi. the common name (Japanese) of the sea snake, *Laticauda semifasciata*.

Ara-C. cytosine arabinoside.

Araceae (arum family). a family of toxic plants all of which produce symptoms similar to those of *Dieffenbachia*. Many are common house plants. Important genera are *Acorus, Alocasia, Amorphophallis, Anthurium, Caladium, Calla, Colocasia, Dieffenbachia, Dracunculus, Philodendron,* and *Zantedeschia*. All parts of the plant are toxic and can cause severe irritation of mucous membranes, with swelling of the tongue and throat, salivation, nausea, vomiting, and diarrhea. Direct systemic effects occur rarely. Active principles vary among species but always include calcium oxalate.

arachidonic acid. See basophil.

arachnid. any arachnoid animal of the subclass Arachnida, *q.v.*

Arachnida. **1:** a large class of invertebrates (Phylum Arthropoda, Subphylum Chelicerata) comprised chiefly of predaceous terrestrial forms such as scorpions, spiders, harvestmen, mites, ticks, and related forms. Arachnids are characterized by a cephalothorax (a fused head and thorax, the prosoma) that bears four pairs of walking appendages; a pair of usually sensory pedipalps; and a pair of prehensile buccal chelicerae that are often served by poison glands; the abdomen (opisthosoma) bears no appendages. The sexes are separate; development is direct (except for ticks and mites); the eyes are simple; most species breathe via tracheae, book lungs, or gill lungs; they eat only fluids; and excretion is by coxal glands and Malpighian tubules. **2:** formerly a subclass of the arthropod class Arachnoidea.

arachnidism (arachnoidism). **1:** systemic poisoning due to envenomation by a spider or other arachnid; applied especially to that produced by *Latrodectus mactans*. It is a condition characterized by intense pain and muscular rigidity. **2:** any condition that results from envenomation by a spider. *Cf.* araneism.

arachnoid. **1:** of, pertaining to, or characterizing the Arachnida. **2:** any of various invertebrates that are related to or resemble arachnids. **3:** resembling a spider's web; membraneous.

Arachnoidea. Arachnida (def. 1).

arachnoidism. arachnidism.

arachnolysin. the active hemolytic principle of the venom of certain spiders.

Aralia spinosa (devils walking stick; angelica tree). a poisonous shrub or small tree with distinctive thick, club-like, spiny branches tipped with very large doubly pinnately compound leaves. It occurs from Pennsylvania to Florida, and west to Illinois and Iowa.

Araneae (Araneida). the largest order (nearly 50,000 species are known) of the class Arachnida, containing true spiders (e.g., species of *Araneus*, *Latrodectus*, *Loxosceles*, *Steatoda*, *Phidippus*, and *Chiricanthium*). Nearly all are terrestrial and venomous, but only about 200 species are harmful to *Homo sapiens* and other large animals. The body is divided into a cephalothorax and a short, usually unsegmented abdomen which are greatly constricted at their juncture; spinnerets that extrude silk are located on the ventral surface of the abdomen in females of many species; the chelicerae are modified as short, pointed fangs with a venom duct in the terminal claw and used to kill or immobilize prey; the pedipalps are short and leg-like with crushing bases and are modified as accessory sex organs in the male; external respiration is by two or four book lungs; tracheae are sometimes present; and digestion is external. Species that are dangerous to humans include the widow spiders, *Latrodectus mactans*, and related species; the brown or violin spider, *Loxosceles reclusa* (sometimes called the brown recluse), and related species; the jumping spiders, *Phidippus*; the tarantulas, *Aphonopelma* and *Pamphobeteus* species; the trapdoor spiders, *Bothriocyrtum* and *Ummidia* species; the so-called banana spiders, *Phoneutria* and *Cupiennius sallei, Lycosa* (wolf spider), and *Heteropoda*; the crab spider, *Misumenoides aleatorius*; the running spiders, *Liocranoides* and *Chiracanthium*; the orbweavers, *Neoscona vertebrata, Araneus* species, and *Argiope aurantia* (orange argiope); the running or gnaphosid spiders, *Drassodes*; the green lynx spider, *Peucetia viridans*; and the comb-footed or false black widow, *Steatoda grossa*.

araneid. pertaining to or denoting spiders; any spider.

Araneida. Araneae.

Araneidae. a family of orb-weaving spiders (Order Araneae) that usually build large, complex webs of radial threads crossed by spiral strands. See *Araneus*, *Argiope*.

araneism. poisoning by the bite of a venomous spider. *Cf.* arachnidism.

Araneus (orbweavers). a genus of true spiders (Family Araneidae) that are dangerous to humans as allergic reactions are possible. *A. marmoreus* (marbled spider; false shamrock). a common spider of Europe and North America that builds in shrubs and trees. See Araneae.

araroba. the common name of *Andira araroba*. **purified araroba**. chrysarobin.

ARBD. alcohol-related birth defect.

arboricide. a chemical that destroys trees.

arbovirus. See virus.

arbre. *Blighia sapida*.

arbutin (4-hydroxyphenyl-β-D-glucopyranoside; hydroquinone-β-D-glucopyranoside, arbutoside, ursin). a hydroquinone glycoside, $C_{12}H_{16}O_7$, and precursor of benzoquinone. It is a plant growth inhibitor that is toxic to grasses (Family Gramineae). Arbutin occurs in the leaves of a variety of green plants, especially those of the family Ericaceae. Examples are *Adenostoma spp.*, *Arctostaphylos spp.*, *Bergenia crassifolia*, *Pyrus spp.*, and various species of *Vaccinium* (e.g., blueberries, cowberries, cranberries). See especially *Adenostoma* and *Arctostaphylos*. Arbutin is sometimes used therapeutically as a diuretic and urinary antiinfective agent. See also allelopathy.

arbutoside. arbutin.

Arceuthobium pusillum (dwarf mistletoe; petit gui (Que.)). a very small mistletoe (Family Loranthaceae) with leaves reduced to tiny connate thin brown scales; the "berry" is dry and compressed. It is a parasitic epiphyte on coniferous trees, chiefly on the branches of *Picea* and rarely on *Larix* and *Pinus strobus* in North America from Newfoundland to Ontario, southward to the mid-Atlantic States, west to Michigan, Wisconsin, and Minnesota. *Cf. Phoradendron serotinum*. See also *Viscum album*.

archin. emodin.

Arctiidae (tiger moths). a family of stout, furry, small- to medium-sized moths. Many species are poisonous, containing toxic substances such as histamine. They usually have conspicuous patterns or bold colors.

Arctostaphylos (bearberry). a genus of American and circumpolar woody evergreen shrubs (Family Aricaceae) that are characterized by alternate leaves, nodding flowers, and drupaceous fruit. Certain species such as *A. glauca, A. glandulosa*, and *A. uva-ursi*, together with other shrubs such as *Adenostoma fasciculatum* that make up the hard chaparral community, secrete phenolic compounds such as arbutin and *p*-hydroxy cinnamic acid. These wash to the ground in rainfall and inhibit the growth of grasses (Gramineae) near the plant, leaving substantial bare areas within and somewhat beyond the edge of the community. When the chaparral is destroyed by fire, the regrowth of grasses is initially vigorous, but slows as the chaparral community is reestablished.

arc-welders' disease. siderosis (def. 1).

area postrema. a narrow band of nervous tissue on the caudal floor of the cisterna magna (fourth ventrical) of the mammalian brain, dorsal to the nucleus tractus solitarius and immediately lateral to the trigonum nervi vagi and delimited from it by the sulcus limitans. This region lacks the blood-brain barrier, and substances can be absorbed from the capillary bed. It is thus a possible receptor site for various xenobiotics. In particular, receptors for emetics such as apomorphine are thought to reside in this area. It also mediates pressor effects of angiotensin II.

Areca catechu (betel palm; betel nut palm; areca nut; betel nut; betel nut seed). a palm tree (Family Palmaceae) with pinnate leaves and a slender ringed trunk, found in central southwest Asia and South America. The fruit is an orange-colored, astringent drupe with a fibrous husk and a pungent odor. The seed (betel nut) is very toxic, producing gastroenteritis, miosis, dyspnea, and convulsions. The chief toxicant is arecoline, although the nut contains other poisons including the alkaloid, arecain. The nut is chewed in the east as an anthelmintic and miotic. See arecoline.

areca nut. **1:** *Areca catechu*. **2:** the seed of *Areca catechu*.

Arecaceae. Palmaceae.

arecaidine. arecain.

arecain (arecaidine; methylguvacine; 1,2,5,6-tetrahydro-1-methyl-3-pyridinecarboxylic acid; 1,2,5,6-tetrahydro-1-methylnicotinic acid). a poisonous substance found in *Areca catechu*, *q.v.* See also arecoline.

arecaline. arecoline.

arecoline (1,2,5,6-tetrahydro-1-methyl-3-pyridinecarboxylic acid methyl ester; methyl 1,2,5,6-tetrahydro-1-methylnicotinate; methyl 1-methyl-$\Delta^{3,4}$-tetrahydro-3-pyridinecarboxylate; methyl*N*-methyltetrahydronicotinate; arecaline; arecholine; methylarecaidin). a colorless, oily, very toxic, anthelmintic alkaloid, $C_8H_{13}NO_2$, isolated from seeds of the *Areca catechu*, the betel nut palm. It is a derivative of nicotinic acid and is the active principle of betel nuts.

arene. aromatic hydrocarbon.

arene carbonyl. a carbonyl in which a metal atom is bonded to an aromatic group and coordinated with several carbon monoxide molecules.

arene oxide. See epoxidation, epoxide.

ARF. acute respiratory failure. See respiratory failure.

arg. the IUPAC abbreviation for arginine.

Argemone (prickly poppy; thornapple). hardy, mostly annual herbs (Family Papaveraceae). All plant parts are toxic and contain the isoquinoline alkaloids, berberine and protopine. The seeds also contain sanguinarine and hydrosanguinarine. ***A. intermedia***. similar to *A. mexi-*

cana, *q.v.* Considered toxic. ***A. mexicana*** (prickly poppy; Mexican prickle-poppy; Mexican poppy). a toxic annual or perennial herb (Family Papaveraceae) that occurs from tropical America northward to Texas and Arizona. It also occurs in scattered locations from Florida to Pennsylvania and is widely distributed globally. The seeds and expressed oil are toxic. This plant is not usually eaten by animals and poisonings are rare. Effects of intoxication include gastroenteritis, myocardiopathy, visual disturbances, and edema.

argemone oil. a toxic oil from *Argemone mexicana* (Family Papaveraceae). It causes epidemic dropsy, and sometimes gastrointestinal and visual disturbances.

Argentine bahia grass. *Paspalum notatum*. See *Claviceps*.

argentum. silver.

argeria. silver poisoning.

arginine (arg; 2-amino-5-guanidinovaleric acid). an essential amino acid in laboratory rats. The L(+) form is physiologically active and it is used therapeutically as an ammonia detoxicant in hepatic failure and also as a diagnostic aid in the assessment of pituitary function.

arginine ester hydrolase. any of a class of enzymes that catalyze the hydrolysis of the ester or peptide linkage of an arginine residue. They are constituents of many crotalid and viperid venoms and some hydrophid venoms. See also enzymes of snake venoms.

arginine glutamate. the L(+)-arginine salt of glutamic acid, $[H_2NC(NH)]HNCH_2CH_2CH_2CH(NH_2)COOH \cdot HOOCCH_2CH_2CH(NH_2)COOH$, used as an intravenous ammonia detoxicant in cases of hepatic failure.

Argiope (garden spiders). a genus of large, predominantly black and yellow garden spiders (Family Araneidae). Two species are common throughout much of North America. ***A. aurantia*** (black and yellow garden spider; orange argiope). garden spiders with black, yellow, and silver markings. The pain from the bite is slight and localized. The venom is neurotoxic and usually produces no more than mild systemic symptoms in humans, although an allergic reaction is possible. ***A. trifasciata*** (banded garden spider). toxicologically similar to *A. aurantia*.

argon (symbol, Ar)[Z = 18; A_r = 39.95]. a noble gas. It is a colorless, odorless, monatomic, inert element comprising 0.93% by volume of the atmosphere.

argyremia. the presence of silver or silver salts in circulating blood.

argyria (argyriasis; argyrism; argyrosis). a permanent bluish-black discoloration of the skin, conjunctiva, and internal organs due to repeated application of small amounts of silver nitrate.

argyriasis. argyria.

argyric. pertaining to or caused by silver.

argyrism. argyria.

argyrosis. argyria.

Ariidae (sea catfishes). a family of venomous marine bony fishes that inhabit shallow, usually warm, coastal seas. The parents are buccal incubators and carry eggs and young in the mouth for as long as 3 months. The sharp dorsal and pectoral spines can inflict painful wounds. Typical genera are *Arius* and *Galeichthys*.

aril. a pulpy or fleshy outer covering of a seed; a fleshy appendage of a seed.

Arisaema (Indian turnip; dragon arum). a genus of about 60 species of tuberous, usually monoecious, plants (Family Araceae) that are nearly worldwide in distribution; some thrive as house and garden plants. The flowers are often distinctive. Most species have an acrid, pungent sap; some hybridize; and all should be considered toxic, although the best known for its toxicity is the familiar *A. triphyllum*. An active principle in all cases is calcium oxalate. ***A. atrórubens*** (Jack-in-the-pulpit; Indian turnip; Oignon sauvage; Petit Prêcheur). a species that occurs in rich woods and thickets from southern Canada to New England, South Carolina, Tennessee, Missouri, and eastern Kansas. The hood is purple or purple to bronze, with pale longitudinal stripes; the leaflets are often pale or glaucous beneath. It is generally, but not always. taller than *A. triphyllum*. ***A. dracontium*** (green dragon; dragon root). a tall perennial herb with fleeting blossoms found on rich or alluvial woods, thickets, and swales from Florida to Texas and northward locally to southern New Hampshire, Vermont, southwestern Quebec, southern Ontario, Michigan, and Wisconsin. The toxic effects are similar to those of *A. triphyllum*. ***A. stewardsonii*** (Jack-in-the-pulpit; Indian turnip; bog onion). found in wet or swampy woods and thickets from Nova Scotia to Minnesota, northern New Jersey, and Pennsylvania, and along the mountains to North Carolina. ***A. triphyllum*** (*A. atrou-bens*; Jack-in-the-pulpit; small Jack-in-the-pulpit; Indian turnip). a woodland herb with tripartite leaves and mottled flowers, it is one of the most familiar species of arum in eastern North America. It is usually smaller than *A. atrórubens* and is otherwise similar to it in appearance, but occurs in wet woods, swamps, and peat bogs of the Coastal Plain and Piedmont of the eastern United States, from Georgia to Kentucky, northward to southeastern Massachusetts and southeastern New York. Plants range in height from about 0.3 to nearly 1 m; the flower is about 25 cm long, green, and often spotted with purple. The fruits are a brilliant red. The entire plant, but especially the rhizome, is poisonous. The sweet-tasting berries, sour leaves, flowers, and roots are most deadly. The poisons are aroin (related to coniine) and calcium oxalate. Ingestion can cause an intense burning sensation, irritation, blistering, and edema in the mouth and pharynx when plant material is chewed. This is probably due to the presence of calcium oxalate crystals. If material is actually swallowed, symptoms may also include severe gastroenteritis, hemorrhage, convulsions, mydriasis, coma, and death. See also *Arum maculatum*.

Arisan habu. *Trimeresurus monticola*.

Aristolochia (birthwort). a genus of nearly [illegible] species of mostly tropical evergreen and deciduous shrubs, vines, and herbs (Family Aristolochiaceae). ***A. reticulata*** (Texas snakeroot). a toxic herb that is therapeutically and toxicologically similar to *A. serpentaria*. ***A. serpentaria*** (Virginia snakeroot). an herb sometimes used as a medicinal, although it is seldom recommended because of its toxicity. When ingested, it can cause severe gastrointestinal pain, dizziness, respiratory paralysis, cardiac arrest, and death in experimental animals.

aristolochic acid (8-methoxy-6-nitrophenanthro-[3,4-d]-1,3-dioxole-5-carboxylic acid; 3,4-methylenedioxy-8-methoxy-10-nitro-1-phenanthrenecarboxylic acid; aristolochic acid-I; aristolochine). a toxic, bitter, crystalline substance with shiny brown leaflets, isolated from serpentaria, *q.v.* It is one of 14 substituted derivatives of 1-phenanthrenecarboxylic acid known to occur in shrubs of the genus *Aristolochia*, *q.v.* It occurs also in the butterfly, *Pachlioptera aristolochiae*, via the caterpillars which feed on these shrubs. It is very to extremely toxic to laboratory mice and rats.

aristolochic acid-I. aristolochic acid.

aristolochine. aristocholic acid.

Arius felis (*Galeichthys felis*; hardhead catfish). a fairly small (up to 61 cm in length) venomous marine catfish (Family Ariidae). It inhabits shallow coastal waters of the eastern coast of the United States from Cape Cod, Massachusetts to the Gulf of Mexico and occasionally occurs in nearby fresh or brackish waters. The males are buccal incubators. Their dorsal and pectoral spines can cause severe and painful wounds. See catfishes.

Arizona black rattlesnake. *Crotalus viridis cerberus*.

Arizona coral snake. *Micruroides euryxanthus*.

Arizona korallenottern. *Micruroides euryxanthus*.

Arizona ridge-nosed rattlesnake. *Crotalus willardi*.

Armenian mountain adder. *Vipera xanthina*.

Armenian sand viper. *Vipera ammodytes*.

Armillariella. a genus of mushrooms (Family Tricholomataceae). ***A. mellea*** (honey mushroom). a normally edible, but sometimes toxic mushroom, that causes gastrointestinal distress; complete recovery is usual within a few hours. ***A. tabescens*** (ringless honey mushroom). Effects are similar to those of *A. mellea*.

Armoracia lapathifolia (horseradish). a tall, coarse, white-flowered perennial herb (Family Cruciferae) that is indigenous to Europe, but widely cultivated elsewhere. The grated root is used as a condiment. It is toxic and can cause severe gastroenteritis in livestock, presumably because of the mustard oil content.

arnica. **1:** plants of the genus *Arnica*. **2:** the dried flowerheads of *Arnica montana* used externally as a counterirritant ointment or liniment to reduce inflammation due to sprains and bruises. It contains helenalin, *q.v.*, which causes a dermatitis in some people; it is also a potent cardiotoxic agent if taken internally. Arnica can also produce violent gastroenteritis, neuropathies, extreme muscular weakness, collapse, and death. **3:** a tincture of arnica.

Aroclor. any of a variety of commercial mixtures of polychlorinated biphenyls (PCB) and/or triphenyls. Each mixture is identified by a four-digit number, the first of which indicates the presence of biphenyls (12), triphenyls (54), or both (25, 44). The last two digits represent the average percentage by weight of chlorine in the mixture. Thus, Aroclor 1242 is a mixture of PCB that contains about 42% chlorine. Aroclors were originally used chiefly in heat exchangers, transformers, capacitors, and other types of electrical equipment. They are immunotoxic, carcinogenic to laboratory animals, and probably carcinogenic to humans. Symptoms of intoxication in humans include gastrointestinal distress, chloracne, excessive discharge from the eyes, edematous eyelids, pigmentation of nails and skin, and characteristic changes in hair follicles. **Aroclor 1242**. a clear, mobile liquid with approximately 3.1 Cl atoms per molecule. **Aroclor 1254**. a moderately toxic, light yellow viscous liquid with approximately 4.96 Cl atoms per molecule. It is a confirmed carcinogen in laboratory animals. **Aroclor 1260**. a soft, sticky, moderately toxic, light yellow resin with approximately 6.3 Cl atoms per molecule. See polychlorinated biphenyl.

aromatic. See hydrocarbon (aromatic hydrocarbon).

aromatic amine. See amine.

aromatic compound. **1:** aromatic hydrocarbon. See also hydrocarbon (aromatic hydrocarbon). **2:** any substituted aromatic hydrocarbon.

aromatic ring (benzene ring). a six-membered planar ring of carbon atoms in which the π electrons are delocalized and the ring is sometimes said to have three conjugated double bonds. Such a ring is exemplified by benzine, which is comprised of a single aromatic ring in which a hydrogen atom is bonded to each carbon atom. In all other aromatic compounds, the aromatic ring is retained, but one or more hydrogens are replaced by other atoms or chemical groups. See also hydrocarbon (aromatic hydrocarbon), amine (aromatic amine), purine, pyrimidine.

Arothron. a genus of tropical marine bony fishes (Family Tetraodontidae) that number among the most poisonous of the family. ***A. hispidus*** (maki-maki, deadly death puffer). a species that occurs in the tropical reaches of the Pacific Ocean, to Japan, South Africa, and the Red Sea. ***A. meleagris*** (white-spotted puffer). a species that ranges from the west coast of Central America to Indonesia. ***A. nigropunctatus*** (black-spotted puffer). a Polynesian species that occurs in the tropical Indo-Pacific to Japan, to eastern Africa and the Red Sea. See puffer poisoning.

ARPPRN. nadide.

arrest. to stop, cease; the act of stopping. **cardiac arrest**. a sudden cessation of cardiac function, with complete loss of blood pressure and circulation. This condition may result from ventricular fibrillation or ventricular standstill. Cardiac arrest may be caused by a large variety of toxicants, from general anesthesia, or from asphyxia. **heart arrest**. cardiac arrest.

arrhythmia. a lack of rhythm or divergence from the normal rhythm; applied most commonly in toxicology to heart beat or breathing. **cardiac arrhythmia**. irregular heart beat or any variation from normal heart rhythm, including an irregular pulse, atrial fibrillation, atrial flutter, heart block, paroxysmal tachycardia, premature beat, and sinus arrhythmia. **pulmonary arrhythmia**. irregular breathing.

arrhythmic. aperiodic; lacking rhythmicity; characterized by an absence of rhythm.

arrow arum. *Peltandra virginica*.

arrow poison. **1:** any of a number of toxic substances prepared from various plant and/or animal extracts or mixtures by primitive peoples in various parts of the world and used to prepare poisonous arrows. See, for example, batrachotoxin, calotropin, strophanthin, *Cerbera tanghin*, *Derris*, onobaio. **2:** a common name for curare.

arrowgrass. plants of the genus *Triglochin*.

arsenate. any salt or ester of arsenic acid. They are distributed widely in nature and contaminate coal and metal ores. Arsenates are used in pesticides, herbicides, fungicides, and algicides. They are carcinogenic and less toxic than arsenites. Arsenates uncouple oxidative phosphorylation by replacing inorganic phosphorus in the ATP complex. They have a metallic taste, and the breath and feces have a sweet garlicky odor. Symptoms of acute poisoning are mainly those of acute gastroenteritis with a burning pain throughout the digestive tract, nausea and violent vomiting, intense (sometimes bloody) diarrhea, and dehydration. Severe poisoning may also include shock, convulsions, coma, paralysis, and death. See also sodium arsenate.

arseniasis (arsenicalism, arsenism). chronic poisoning due to slow accumulation of arsenic in the body. It is characterized by gastrointestinal symptoms with severe gastric distress, burning esophageal pain, nausea, vomiting and bloody diarrhea, together with dehydration, neuritis, and paralysis of muscles of the wrist and ankle. The victim may experience burning pains in the hands and feet, a numbing sensation throughout the body, skin irritations and eventually exfoliative dermatitis, cold clammy skin, localized edema, alopecia, cirrhosis of the liver (eventually icterus), weight loss, visual impairment, hypotension, and ultimately, convulsions, coma, cardiovascular collapse, and death. Skin cancer may develop. Chronic exposure of hamsters during early pregnancy to nonmaternotoxic levels of arsenic is teratogenic.

arsenic (As; arsenium; ratsbane)[Z = 33; A_r = 74.9]. a grayish-white, toxic, metalloid element that exists in several allotropic forms. It is widely distributed in soil, water, air, and a number of human foods, especially shellfish. Arsenic and many of its compounds are poisonous at low concentrations. Arsenic causes cardiac arrhythmias, toxic amaurosis, and cancer, and is a well-known teratogen to a variety of vertebrate animals. It can be absorbed via intestinal and pulmonary routes and sometimes via skin. Arsenic is a byproduct of lead, copper, and zinc smelting, and elemental arsenic is used to make alloys with lead and copper. Exposed workers in the smelting industries named above are those most commonly affected by arsenic poisoning. These workers have exceptionally high rates of lung and liver cancer. While used chiefly in herbicides and pesticides, arsenic and its compounds have a wide range of applications, being used variously as pharmaceuticals, in catalysts, bactericides, fungicides, additives in animal feed, corrosion inhibitors, tanning agents, and wood preservatives. Human exposure to arsenic occurs chiefly in the manufacture of herbicides, insecticides, wood preservatives, and during the smelting of lead, copper, and zinc ores. Humans may also be exposed to arsenic through contaminated drinking water, wood products impregnated with preservatives, and air emissions from smelting, pesticide, and glass-manufacturing plants. Arsenic acts at the cellular level by reacting with SH-containing mitochondrial enzymes, thereby impairing respiration; by coagulating protein and forming complexes with coenzymes; or by inhibiting the biosynthesis of ATP. It also interferes with oxidative phosphorylation by substituting for phosphorus. Arsenic inhibits various enzymes (e.g., choline oxidase, pyruvate oxidase, and transaminase). Chronic exposure to arsenic and accumulation in humans causes anorexia, weakness, and general ill health, gastrointestinal distress, diarrhea, discoloration and exfoliation of the skin, mental disorders, and symmetric distal neuropathy with anesthesia in the wrists and ankles. Liver and kidney degeneration and altered hematopoiesis may also occur. Effects of acute arsenic poisoning may also include anorexia, pyrexia, hepatomegaly, symptoms of gastrointestinal [illegible] [illegible] [illegible], respiratory effects, transient encephalopathy, shock that is secondary to hemorrhagic gastritis, and death. Severe exposures are also associated with lung, liver, and skin cancer. Exposure to fine arsenical powder while handling arsenic compounds produces warts on the skin that may become malignant. Edema of the face and eyelids is peculiar to arsenic poisoning. Survivors may suffer peripheral neuropathy. Arsenic is metabolically detoxified by reductive methylation, with excretion in urine of methylarsonic and dimethylarsinic acids. Antidotes to arsenic poisoning (e.g., British Antilewisite) contain sulfhydryl groups, taking advantage of its affinity for sulfur. See also amaurosis, amblyopia, arseniasis, arsenic poisoning.

arsenic acid. the white poisonous crystalline hydrate, H_3AsO_4, of arsenic trioxide or arsenic pentoxide. It is a defoliant and is used in the manufacture of insecticides, glass, and arsenates.

arsenic acid anhydride. arsenic pentoxide.

arsenic anhydride. arsenic pentoxide.

arsenic chloride. arsenic trichloride.

arsenic disulfide (arsenic sulfide; arsenic mono sulfide; red arsenic sulfide; realgar; red orpiment; red arsenic; ruby arsenic; red arsenic glass; C.I. Pigment Yellow 39; C.I. 77085). a toxic, water-insoluble, orange-red powder or deep red, orange, or black monoclinic crystals, As_2S_2 or AsS, that occurs naturally as the mineral realgar. It is used as a rodenticide, in fireworks to give an intense white flame, in pigments, and in depilating and tanning hides.

arsenic-fast. resistant to poisoning by arsenic, used mainly in reference to acquired resistance in spirochetes and other protozoan parasites following repeated exposure to arsenic. *Cf.* arsenoresistant.

arsenic hydride. arsine.

arsenic monosulfide. arsenic disulfide.

arsenic oxide. arsenic trioxide; arsenic pentoxide.

arsenic pentoxide (arsenic oxide; arsenic anhydride; arsenic acid anhydride; arsenic acid). a white amorphous, deliquescent solid, As_2O_5, that forms arsenic acid in aqueous solution and is soluble also in ethanol. It is used, for example, as an insecticide and herbicide. *Cf.* arsenic acid, arsenic trioxide. Both arsenic trioxide and arsenic pentoxide are often referred to as arsenic oxide.

arsenic poisoning. poisoning by arsenic or by an arsenical. *Cf.* arsenical poisoning. **acute arsenic poisoning**. poisoning by acute exposure to arsenic. Death may be nearly immediate or may follow exposure within a few hours or to 1 day. The only lesion is inflammation of the stomach. Some arsenic may remain in the gastrointestinal tract. The victim may be anemic and the skin may appear slightly icteric. Poisoning may result in death from circulatory failure. If death is not immediate, the victim becomes restless, headachy, has dizzy spells, and cannot void; the skin is jaundiced due to destruction erythrocytes; there may be intermittant paralysis. If death is immediate, the only lesion to be found on autopsy is an inflammation of the stomach; some arsenic may remain in the digestive tract. **arsenic poisoning, chronic**. arseniasis.

arsenic reduction. the reduction of arsenic in organic arsenicals from the pentavalent to the trivalent state. The reduced compounds are more toxic and are often used as antiparasitic or antiprotozoan agents.

arsenic sulfide. arsenic disulfide.

arsenic trichloride (arsenic chloride; arsenious chloride; arsenous chloride; butter of arsenic; caustic arsenic chloride; fuming liquid arsenic; trichloride of arsenic). a highly toxic, colorless or pale yellow, nonflammable, fuming oil, $AsCl_3$, that readily releases hydrochloric acid and dissolves readily in most organic solvents but decomposes in water. It is a systemic poison and vesicant that readily penetrates skin, causing serious damage at the point of contact. It is used as a war gas and as an intermediate in the manufacture of organic arsenical pharmaceuticals and insecticides. *Cf.* arsenic acid, arsenic pentoxide.

arsenic trifluoride (arsenious fluoride). an extremely toxic, fuming, mobile liquid, AsF_3, used as a fluorinating reagent, catalyst, a dopant, and an ion implantation source.

arsenic trihydride. arsine.

arsenic trioxide (arsenous acid; arsenous acid anhydride; arsenious acid; arsenic acid; arsenic oxide; arsenic sesquioxide; arsenous anhydride; arsenous oxide; arsenous powder; crude arsenic; flowers of arsenic; white arsenic). a sweet-tasting toxic compound of white or clear and glassy amorphous lumps or crystals, As_2O_3. It is used in the synthesis of numerous arsenic compounds and, in small amounts, is used in some therapeutic drugs. It is toxic and carcinogenic to humans. Fine particles of arsenic trioxide are absorbed through the skin and intestines; large particles pose a lesser hazard as they tend to pass through the intestinal tract and are egested. Repeated small doses taken by individuals who live at high altitudes may increase their work capacity by elevating the level of circulating hemoglobin. Both arsenic trioxide and arsenic pentoxide are often referred to as arsenic oxide. See antiarsenin, arsenite.

arsenic trisulfide (arsenic yellow; yellow arsenic sulfide; auripigment; orpiment; king's yellow; king's gold). a toxic compound, As_2S_3, used, for example, in the manufacture of glass, linoleum, oil cloth, photoconductors, and semiconductors; as a pigment; in pyrotechnics; in depilating animal hides; and sometimes therapeutically.

arsenic yellow. arsenic trisulfide.

arsenical. 1: a compound or formulation that contains arsenic. **2:** any biologically active substance that depends on the arsenic content for its action or effects. **3:** referring to or denoting arsenic or one of its compounds; containing arsenic. *Cf.* herbicide (organoarsenical herbicide); wood preservatives.

arsenical poisoning. poisoning by an arsenical. See arsenic poisoning.

arsenical solution. potassium arsenite solution.

arsenicalism. arseniasis.

arsenicophagy (arsenophagy). habitual eating of arsenic.

arsenide (arseniuret). **1a:** any compound formed from the union of arsenic with a metal. **1b:** any compound of arsenic in which the arsenic is negatively charged and is not bound to oxygen.

arsenilic acid (4-aminophenylarsonic acid; *p*-aminobenzenearsonic acid; atoxylic acid). a poisonous substance synthesized from aniline and arsenic acid. It is used in the manufacture of arsenicals, as a feed additive to promote growth and the efficient utilization of feed by poultry and swine, and to control dysentery in swine. Swine have been poisoned by feed rations containing excessive amounts of organic arsenical growth promoters. See arsenilic acid poisoning.

arsenilic acid poisoning. poisoning by arsenilic acid, mostly of domestic swine due to overdosing feed rations with organic arsenical growth promoters. Clinical signs in swine may progress from pelvic limb ataxia and paresis to tetraparesis. Affected animals may assume a dog-like sitting position. Myelin and axonal degeneration in peripheral nerves, optic nerves, and optic tracts is common. Early withdrawal of the arsenical may result in complete spontaneous recovery. Early withdrawal of the arsenical may result in spontaneous complete recovery.

arsenious. arsenous (def. 2).

arsenious acid. arsenic trioxide.

arsenious acid copper(2+) salt (1:1). cupric arsenite.

arsenious chloride. arsenic trichloride.

arsenious fluoride. arsenic trifluoride.

arsenism. arseniasis.

arsenite. 1: a trivalent negative ion, AsO_3^{3-}, in aqueous solutions of arsenic oxide, As_4O_6. **2:** any salt or ester of arsenous acid; any salt that contains trivalent arsenic. Arsenites occur widely in nature; they are common components of pesticides, and many arsenites are antiparasitic or antiprotozoal agents. They are corrosive to tissues and are carcinogenic. Arsenites are poisonous, chiefly because they react easily with the sulfhydryl groups of essential enzymes; they are generally more toxic than arsenates. Symptoms of acute exposure of humans include intense nausea and retching, abdominal pain, severe diarrhea, and dehydration. The breath and feces have a garlicky odor. See also cupric arsenite, sodium arsenite, reduced arsenic.

arsenium. elemental arsenic.

arseniuret. arsenide.

arseniureted. chemically combined with arsenic, forming an arsenide.

arseniureted hydrogen. arsine.

arseno-. a prefix indicating the inclusion of the element arsenic.

arsenoactivation. an increase or heightening of the symptoms of syphilis under treatment with arsenic.

arsenoceptor. a presumed cellular receptor of arsenic.

arsenophagy. arsenicophagy.

arsenopyrite. an arsenic-containing mineral that can be smelted to produce elemental arsenic. See also loellingite.

arsenoresistant. resistant to the toxic action of arsenic and its compounds. *Cf.* arsenic-fast.

arsenous (arsenious). **1:** pertaining to arsenic. **2:** denoting a compound that contains trivalent arsenic; arsenious, arsenic.

arsenous acid. arsenic trioxide.

arsenous acid anhydride. arsenic trioxide.

arsenous anhydride. arsenic trioxide.

arsenous chloride. arsenic trichloride.

arsenous hydride. arsine.

arsenous oxide. arsenic trioxide.

arsenous powder. arsenic trioxide.

arsenous sesquioxide. arsenic trioxide.

arsenoxide. **1:** oxophenarsine. **2:** any metabolic oxidation product of an arsphenamine. arsenoxides appear to be active against spirochetes.

arsenum. elemental arsenic.

arsine (arsenic trihydride; arsenic hydride; arsenous hydride; hydrogen arsenide; arseniureted hydrogen; AsH). **1:** a colorless, volatile, flammable, extremely poisonous (cytotoxic and hemotoxic), neutral gas, AsH_3, with a disagreeable garlicky odor. It is soluble in water and slightly soluble in ethanol and alkalies. Arsine is a hydride formed from the decomposition of arsenides (especially sodium or zinc arsenide) in water or dilute acids. When moist, it decomposes readily on exposure to light. Arsine is highly toxic by inhalation. Exposure, even to small amounts, can cause skin and pulmonary cancer, mood swings, and paranoia. Symptoms of acute exposure usually appear within a few hours and are similar to those produced by stibine. Included are headache, nausea, vomiting, hematemesis, abdominal colic, backache, shortness of breath, paresthesia, hemolysis, hematuria, anuria, icterus, and death from pulmonary edema or renal failure. See also arsines, arsonic acid, hemolytic gas. **2:** any of a class of very toxic organic derivatives of arsine in which one or more of the three hydrogen atoms are replaced by an alkyl radical; the remaining hydrogen atoms may be replaced by halogens or other substances. They are analogs of amines and phosphines. Many are chemical warfare agents.

arsinic acid. any of a class of disubstituted organic acids (e.g., cacodylic acid) derived from arsines by replacing two hydrogen atoms with organic radicals; they have the general formula RR'AsOOH. Arsinic acids are obtained by oxidizing disubstituted arsines.

Arsobal (Mel B). an organic arsenical used in the treatment of the neurological stage of African trypanosomiasis, the vectors of which are *Trypanosoma gambiense* or *T. rhodesiense*.

arsonate. **1:** any salt or ester of an arsonic acid. **2:** to introduce an arsono group into a substance; to convert into an arsonic acid or arsonic acid derivative.

arsonic acid copper(2+) salt (1:1). cupric arsenite.

arsonic acid. **1:** any of a class of monosubstituted organic acids derived from arsenic acid by replacement of a hydroxyl group by an organic radical. The general formula is $RAsO(OH)_2$. Arsonic acids are obtained by oxidizing monosubstituted arsines. **2:** arsinic acid.

arsonium. a positively charged univalent radical or ion, AsH_4^+; it behaves chemically very much like the ammonium ion, NH_4^+.

***p*-arsonophenylurea**. carbarsone.

arsphenamine. salvarsan.

arsthinol (*N*-[2-hydroxy-5-[4-(hydroxymethyl)-1,3,2-dithiarsolan-2-yl]phenyl]acetamide; 3-acetamido-4-hydroxydithiobenzenearsonous acid, cyclic (hydroxymethyl)-ethylene ester; 2-(3′-acetamido-4′-hydroxyphenyl)-1,3-dithia-2-arsacyclopentane-4-methanol; 3-acetamido-4-hydroxydithiobenzenearsonous acid cyclic 3-hydroxypropylene ester; 2-acetylamino-4-(methylolcycloethylene-dimercaptoarsine)phenol). an antiamebic substance.

artefact, artifact. an apparent structure, aspect of organization, behavior, or any experimental result that is due to the methods employed.

Artemisia (sage; sagebrush; wormwood, mugwort). a genus of plants (Family Asteraceae) comprised of more than 200 species. They produce commercially important volatile oils (absinthe, wormseed, santonica) and are toxic if ingested in large amounts. The sages of the rangelands of the western United States are, however, generally considered good forage for livestock. See sage sickness. ***A. absinthium*** (wormwood, absinthium, absinthe). a coarse perennial herb used formerly as an ingredient of absinthe. It contains a volatile narcotic oil, thujone. ***A. californica***. a sagebrush which, together with certain others such as *Salvia leucophylla*, make up the soft chaparral communities of southern California. It produces volatile terpenes that are apparently adsorbed by the soil and inhibit the growth of grasses, sometimes (with other members of the sagebrush community) producing bare areas up to a few meters wide around the community.

arterial hypoxia. See hypoxia.

arterial tension (intraarterial pressure). the blood pressure within an artery.

arteriogram. a roentgenogram (X-ray record) that displays arteries containing a contrast medium.

arteriograph. arteriogram.

arteriography. **1:** the preparation and analysis of arteriograms. **2:** the study of arteries.

arteriolar necrosis. arteriolonecrosis.

arteriole. the smallest branch of an artery; it carries blood directly into a capillary. Histologically, it resembles an artery. The tunica media, however, except in the pulmonary arterioles, is composed entirely of smooth muscle cells that are 15 to 20 μm long.

arteriolonecrosis (arteriolar necrosis). necrosis of arterioles.

arteriosclerosis (hardening of the arteries). any disease marked by a thickening with loss of elasticity of the walls of arteries.

arteriospasm. a spasm of an artery.

arteriotony. arterial blood pressure.

artery. a blood vessel with characteristic elastic walls that conveys blood away from the heart to the capillary beds. It has an inner coat (tunica intima) of endothelial cells that rests on a thin elastic membrane (lamina elastica interna). The surrounding wall of the artery is comprised of strong elastic fibromuscular tissue that comprised of a middle layer (tunica media) and an outer coat (tunica adventitia).

arthrogryposis. **1:** a persistent flexure or contracture of a joint in terrestrial vertebrates. **2:** a tetanoid spasm. **congenital arthrogryposis** (congenital articular rigidity, CAR). a myopathy characterized by congenital articular rigidity with other anomalies. It occurs in cattle, sheep, horses, and pigs. The condition is sometimes lethal, but recovery may be complete. Plant teratogens such as the following are important etiologic factors: *Datura* (thorn apple), *Lupinus* (lupine), *Nicotiana* (tobacco), *Oxytropis* (locoweed), *Prunus serotina* (black cherry), *Sorghum vulgare* (sorghum or broom corn), and *Tsuga* (hemlock).

arthropod. any animal of the phylum Arthropoda.

arthropod venom. See venom.

Arthropoda. the largest animal phylum. Arthropods are distinguished by the chitinous exoskeleton; a segmented body with segmented appendages; a reduced coelom; an open circulatory system with an extensive hemocoel; a complete digestive system; and a nervous system comprised of a dorsal anterior brain and a ventral nerve cord with fused ganglia in each segment. Many are venomous. Bites or stings can anesthetize or kill prey or even predators. Human medical emergencies and fatalities may result directly from the effects of the venom, from anaphylactic shock, or from other autopharmacologic reactions.

arthropodiasis. any condition of a vertebrate animal that is caused directly by an arthropod, to include effects of poisoning or envenomation, acariasis, allergy, dermatoses, etc.

arthropodic. arthropodous.

arthropodous (arthropodic). pertaining to arthropods.

Arthus reaction (Arthus phenomenon). any of a number of antibody-mediated reactions following intradermal injection of an antigen. Such challenge usually produces local inflammatory lesions, with vasculitis, edema, hemorrhage, and necrosis. See immediate hypersensitivity.

artificial cinnabar. red mercuric sulfide (See mercuric sulfide).

artificial essential oil of almond. benzaldehyde.

artificial flavor. food flavoring.

aruba. *Crotalus unicolor*.

Aruba rattlesnake. *Crotalus unicolor*.

arum. See *Arum*, Araceae.

Arum (arum; wild ginger). a genus of tuberous plants (Family Aracaceae) indigenous to Eurasia. The various species resemble and are often called calas. The roots are acrid-tasting and are toxic in a number of species. ***Arum maculatum***. (cuckoopint; Adam and Eve, lords-and-ladies; wild arum; wake-robin). a very toxic tuberous herb native to southern Europe and northern Africa, but now occurring as far north as England. The leaves and spathe (flower) are each about 25 cm long; the flower is green, often spotted with purple. The fruits are a brilliant red. The entire plant is highly poisonous, the sweet-tasting berries, sour leaves, flowers, and roots being most deadly. The poisons are aroin (related to coniine) and calcium oxylate. Symptoms in humans include blistering, severe gastroenteritis, hemorrhages, convulsions, mydriasis, coma, and death. See also *Arisaema*.

aryl-. a prefix indicating an aromatic radical.

aryl carbamate. See carbamate.

aryl carbamic ester. See herbicide (aryl carbamate herbicide).

aryl ester hydrolase. See esterase.

aryl halide. See halide.

aryl hydrocarbon hydroxylase (AHH). an enzyme that catalyzes the hydroxylation of aromatic hydrocarbons. Many polycyclic aromatic hydrocarbons (PAH) are carcinogenic, and AHH activity plays a role in the sequence of reactions that lead from a PAH procarcinogen to the ultimate carcinogen. Aryl hydrocarbon hydroxylase activity resides in isozymes of cytochrome P-450.

aryl mercaptan. aryl thiol (See thiol).

aryl sulfotransferase. See sulfate conjugation.

aryl thiol. See thiol.

arylamidase, aryl amidase. a hydrolase that hydrolyzes amides. See hydrolysis.

arylamine. aromatic amine (See amine).

arylarsonic acid. an arsonic acid, such as arsenilic acid, that contains an aryl radical.

arylated mercury compound. any aromatic mercury compound. All are toxic.

arylated organotin compound (aryltin compounds, aryltins) any aromatic compound of tin; they are immunotoxic.

arylating agent. **1:** any arylating compound; a compound able to give up an aryl group to another compound. **2:** a chemical carcinogen that attaches aryl moieties to N and O atoms of the nitrogenous bases of DNA. The DNA is altered such that growth and replication of neoplastic (cancerous) cells are enabled.

arylation. the chemical addition of an aryl group to a molecule. Many arylated compounds are more toxic or act in a different mode than the original molecule.

arylesterase. any enzyme that enables the hydrolysis of arylesters. See hydrolysis.

aryltin. any arylated organotin compound.

aryltin compound. any arylated organotin compound.

as low as reasonably achievable (ALARA). a subjective radiation exposure guideline intended to encourage protection practices that are better than any prescribed standard. It is a basic criterion used to control exposure in very low threshold or assumed nonthreshold dose-effect relationships.

As. the symbol for arsenic.

ASA (A.S.A.). acetylsalicylic acid.

Asagraea officinalis. *Schoenocaulon officinale*.

asarabacca. *Asarum canadense*.

asarabacca camphor. asarone.

asarin. asarone.

asaron. asarone.

asarone (1,2,4-trimethyoxy-5-(1-propenyl)benzene; 2,4,5-trimethoxy-1-propenylbenzene; asarin; asaron; asarum camphor; asarabacca camphor). **1:** a highly toxic and carcinogenic crystalline, water-insoluble phenolic ether, $(CH_3O)_3C_6H_2CH{:}CHCH_3$. It is a mixture of two isomers (α-asarone and β-asarone) extracted from the root or rhizome of certain [illegible] *Asarum*, especially *A. europaeum* (Family Aristolochiaceae), by distillation with water. It also occurs in ethereal oils of *A. europaeum* and *A. arifolium* and that of *Acorus calamus* (Family Araceae). It is used as a grain fumigant and insect sterilant. **2:** α-asarone. See also oil of calamus.

α-asarone. See asarone (def. 2).

β-asarone. See asarone.

Asarum (wild ginger). a genus of toxic, perennial, North Temperate Zone herbs (Family Aristolochiaceae) with dull-colored flowers and pungent aromatic rhizomes. Many species contain asarone. ***A. canadense*** (wild ginger; Canada snake-root; Indian ginger; asarabacca). a low, shade-tolerant, perennial herb, common to moist woods of the eastern United States and into Canada. The flowers are chocolate-colored; the leaves are more or less kidney-shaped. The roots contain asarone, methyl ergenol, and an aromatic oil. ***A. europaeum***. a species indigenous to Europe, it is a major source of asarone. ***A. arifolium*** (arum-leaved ginger). the leaves are triangular, or more or less heart-shaped or spear-shaped, and ca. 3-16 cm long and 4-15 cm wide. They occur in woodlands from southern Virginia and Tennessee to the coastal plains of Florida and Louisiana.

asarum camphor. asarone.

asbestos. a generic term for a class of natural fibrous silicates that are widely used in thermal and electric insulation, the manufacture of building materials and brake linings, and many other purposes. There are two main types: chrysotile (white asbestos), which is by far the most abundant form, and the amphiboles, of which the most important are crocidolite (blue asbestos), anthophyllite, amosite, various commercially processed fibrous amphibole (amosite, anthrophyllite, and crocidolite) and serpentine (chrysotile) minerals. All are hydrated silicates that are virtually insoluble, conduct heat poorly, have exceptional tensile strength, are moldable, and relatively inert chemically. Asbestos is or has been used in applications where thermal insulation, resistance to high temperature or to fire, resistance to saltwater, or resistance to corrosive chemicals are desired. Inhalation of asbestos fibers of any type can cause asbestosis, [illegible] characteristic pleural and peritoneal mesotheliomas. The level of risk depends on many factors, including the intensity and duration of exposure, the type of asbestos, and the length of the fibers (longer, thin fibers offer the greatest risk). If cancer develops, it does so 3 decades or more after exposure.

asbestosis (diffuse pulmonary fibrosis). a progressive, incurable, and sometimes fatal pneumoconiosis associated with chronic exposure to and inhalation of airborne asbestos fibers. The disease is marked by shortness of breath, pain in the upper chest or back, and rales. As a result of the difficulty in breathing, the victim's fingers and toes become rounded with flattened nails ("clubbed"). The lungs become progressively fibrotic and calcified, often with the development of bronchogenic carcinoma and mesothelioma.

ascaricidal. 1a: destructive to, or having the capacity to destroy, ascarid worms, particularly those of the genus *Ascaris*. **1b:** pertaining to or denoting an agent that destroys ascarids.

ascaricide. 1: causing the destruction of ascarids. **2:** an agent that destroys ascarids, especially those of the genus *Ascaris*.

ascarid. 1: the common name of any nematode worm of the family Ascarididae (Ascaridae). **2:** the common name of any species of the superfamily Ascaridoidea (Ascaroidea).

Ascaridae. Ascarididae.

Ascaridata. Ascaridida.

Ascaridida (Ascaridorida; Ascarididea; Ascaridata). an order of parasitic nematode worms (Subclass Phasmidia, Class Rhabditida). Important genera include *Ascaridia, Ascaris, Subuluris, Heterakis,* and *Anisakis*.

Ascarididae (Ascaridae). a family of large, intestinal, nematode worms that includes species such as *Ascaris lumbricoides*, that are important to humans. There are species that occur, for example, in humans, swine, sheep, cattle, horses, dogs, cats, and other mammals.

Ascarididea. Ascaridida.

ascaridole (1-methyl-4-(1-methylethyl)2,3-dioxabicyclo[2.2.2]oct-5-ene; 1,4-peroxido-*p*-menthene-2). a toxic, unstable, organic liquid peroxide, a terpene, used as an anthelmintic against nematodes. It is the active ingredient in wormseed oil. Ascaridole can be fatal to humans and animals. See *Chenopodium* (*C. ambrosioides*).

ascaron. a toxic peptone of certain helminths, especially ascarids. Symptoms of intoxication are comparable to those of anaphylactic shock.

ascites (abdominal dropsy; hydroperitonia; hydrops abdominis; peritoneal dropsy). the accumulation of serous fluid in the abdominal cavity.

ascitic. referring to, characterized by, or denoting ascites.

ascitogenous. causing ascites.

Asclepiadaceae (milkweed family). a family of widely distributed plants (herbs, shrubs, and a few vines), marked by an abundant, milky sap and a pod that opens on one side when ripe to discharge small, flat seeds with a tuft of long silky down that favors dispersal by light winds. Many have medicinal and/or toxic properties. See, for example, *Asclepias*.

Asclepias (milkweeds; silkweeds; swallow worts; mw). a genus of perennial herbs (Family Asclepiadaceae) with a poisonous, milky sap. Species frequently hybridize; many are highly poisonous; some are medicinal. The more toxic species contain galitoxin or related resinoids. North American species have poisoned livestock, usually during drought or under conditions of poor management. The minimum lethal dose of the fresh green vegetation ranges from ca. 0.2 to 2000 g/kg body weight or more. Symptoms produced by the various species differ only in intensity. See galitoxin. See asclepias poisoning of livestock. ***A. asperula*** (= *Asclepiadora decumbens*). found on open, dry soils from Kansas and Arkansas, and south into Mexico. Toxicity is similar to that of *A. latifolia*. ***A. curassavica***. a species whose sap is a source of calotropin. ***A. eriocarpa*** (woolypod milkweed). found on dry soils in California; moderately toxic. See milkweed poisoning of rabbits. ***A. incarnata*** (swamp milkweed). a species suspected of fatally poisoning sheep. ***A. labriformis*** (labriform milkweed). a very toxic plant confined to old stream beds in sandy soils in Utah. ***A. latifolia*** (broadleaf milkweed). less toxic than *labriformis*, this species occurs in dry prairies and plains from Missouri, Kansas, and Colorado to Texas and Arizona. ***A. mexicana*** (Mexican whorled milkweed). a species found in scattered locations on the western slopes from southern Washington and eastern Idaho to California, western Nevada, and Mexico. It is among the least toxic of the plants treated herein. ***A. pumila*** (low whorled milkweed; plains whorled milkweed). low to the ground (10 to 15 cm tall) with slender stems, it occurs in dry plains east of the Rocky Mountains from Montana and southwestern North Dakota to Texas and New Mexico. The toxicity is similar to that of *latifolia*. ***A. speciosa*** (showy milkweed). an herb that may attain a height of 2 m. It occurs in prairies and openings in Canada and the United States from Minnesota and British Columbia southward. Large amounts given experimentally have produced symptoms. ***A. subverticillata*** (= *A. verticillata*, formerly *A. galioides*; whorled milkweed). less toxic than *labriformis*, this species occurs mainly on dry plains and foothills in the central and southern Great Plains into Mexico. ***A. syriaca*** (common milkweed). up to 2 m tall with coarse stems, it occurs in thickets, roadsides, and old fields in much of eastern North America. The young shoots and firm, ripe pods are cooked and eaten; they are suspected of some poisonings. ***A. tuberosa*** (butterfly weed; pleurisy root; chigger flower). a rough-haired plant with brilliant orange or yellow flowers, the stems, leaves and roots of which are toxic. It occurs on dry sites in the eastern United States It has been used as an emetic, diaphoretic, expectorant, and laxative. ***A. verticillata*** (eastern whorled milkweed). less than 1 m tall, it occurs in dry woods and open sterile soil in the southern to northeastern United States, Ontario, southern Manitoba and southern Saskatchewan. It is probably somewhat less toxic than *latifolia*.

asclepias poisoning. a sometimes fatal poisoning of herbivores by ingestion of plants of the genus *Asclepias*. Early symptoms among livestock are extreme depression, weakness, and staggering. Frequently repeated seizures are common in downed animals as are pyrexia, miosis, and labored respiration. Gastrointestinal irritation may be mild to severe. [illegible] accompanies depression in the domestic fowl. The sole or primary toxicant is galitoxin.

Ascomyces. *Actinomyces*.

Ascomycetes. a large class of terrestrial, mycelial fungi (Eumycophyta) characterized by the formation (via sexual reproduction) of sac-like structures termed asci. Included are the yeasts, leaf-curl fungi, powdery mildews, black and green molds, cup fungi, morels, and truffles. Certain genera produce antibiotics and/or mycotoxins.

ascorbate-cyanide test. See test.

ascorbic acid. vitamin C.

-ase. a suffix denoting an enzyme.

ash. the solid mineral residue of effectively complete combustion of a substance. **fly ash**. ash from combustion of coal or other fuels that is entrained by combustion gases. In the absence of dust separators, such ash is emitted from the stack. **total ash.** the residue of the mineral matter obtained by controlled incineration, e.g., of coal.

AsH. arsine.

ashy pit viper. *Trimeresurus puniceus*.

Asian cobra. a common name applied to *Naja naja* or to *Naja n. naja*.

Asian coral snake. *Calliophis japonicus*, *C. sauteri*.

Asian lance-headed vipers. See *Trimeresurus*.

Asian lanceheads. See *Trimeresurus*.

Asian pit viper. See *Agkistrodon*.

Asian sand viper. *Eristicophis macmahonii*.

Asiatic horseshoe crab. *Carcinoscorpius rotundicauda*.

Asiatische Lanzenottern. See *Trimeresurus*.

asigirikolongo. *Bitis arietans*, *Naja haje*.

Asimina triloba (papaw). a tree, the fruit of which is commonly eaten by humans, a small percentage of whom develop a contact dermatitis. Some individuals experience severe gastrointestinal symptoms.

asp. *Cerastes vipera* and sometimes *C. cerastes*. See also Cleopatra's asp, *Vipera aspis*.

asp viper. *Vipera aspis*.

asparaginase (L-asparaginase; L-asparagine amidohydrolase; colaspase). an enzyme that catalyzes the hydrolysis of L-asparagine to L-aspartate and ammonia. It is an antineoplastic agent in the treatment of all acute lymphocytic leukemias. Although widely distributed in nature, it is usually derived from *Escherichia coli*. Asparaginase is toxic, and may cause anaphylaxis, nausea, vomiting, anorexia, and inhibition of protein synthesis. Target organs are primarily the brain and pancreas.

L-asparaginase. asparaginase.

L-asparagine amidohydrolase. asparaginase.

Aspartame™ (*N*-4-α-aspartyl-L-phenylalanine 1-methyl ester; 3-amino-*N*-(α-carboxyphenethyl)succinamic acid *N*-methyl ester; L-aspartyl-L-phenylalanine methyl ester; Equal™; NutraSweet™). a nonnutritive sweetener, $C_{14}H_{18}N_2O_5$, used in a large variety of foods, and said to be some 160 times sweeter to human taste than sucrose in aqueous solution. It is a dipeptide ester comprised of two amino acids, L-phenylalanine and aspartic acid. Acceptable daily intake (ADI) is set at 50 mg/kg body weight/day. While some individuals ingest substantially larger amounts, there is no evidence of widespread adverse health effects. Unconfirmed reports, however, of headaches, behavioral disturbances, diminution of IQ, and epileptic seizures have been made. A very small number of urticarial reactions occur. Homozygous phenylketonuric individuals are at risk.

L-aspartyl-L-phenylalanine methyl ester. aspartame.

N-4-α-aspartyl-L-phenylalanine 1-methyl ester. aspartame.

aspergillic acid (2-hydroxy-3-isobutyl-6(methylpropyl)pyrazine-1-oxide). a yellow, crystalline, diketopiperazine-like mycotoxin, antifungal, and antibiotic agent, $C_{12}H_{20}N_2O_2$, produced by certain strains of *Aspergillus flavus*. It is very toxic to laboratory mice (*i.p.*).

aspergillin. **1:** an amorphous black pigment that occurs in spores of various molds of the genus *Aspergillus*. **2:** a broad-spectrum antibacterial antibiotic obtained from cultures of the molds *Aspergillus flavus* and *A. fumigatus*.

aspergillomarasmin A. a marasmin.

aspergillomarasmin B. a marasmin.

aspergillomycosis. aspergillosis.

aspergillosis (aspergillomycosis). a disease condition caused by any of several species of *Aspergillus* and characterized by inflammatory granulomatous lesions of the skin and mucous membranes, including especially the ear, orbit of the eye, urethra, nasal cavities, and lungs. In severe cases, the bones, meninges, liver, kidneys, and other organs may be affected.

aspergillotoxicosis. aspergillustoxicosis.

Aspergillus. a large genus of filamentous fungi (Family Moniliaceae, Class Ascomycetes), a number of which have black, brown, or green spores. The conidiophores are elongate, bearing many chains of basipetally formed conidia. Included are common molds and a few species that are pathogenic to *Homo sapiens*, other animals, or plants. Some species, e.g., *A. flavus*, produce penicillin. See also aspergillustoxicosis. ***A. caespitosus***. a species that produces the indole mycotoxin, verruculogen. ***A. candidus***. a species that produces the mycotoxins, terphenyllin and xanthoascin. ***A. chevalieri***. a species that produces the mycotoxin, xanthocillin-X. ***A. clavatus***. a species isolated from soil and feces. It produces patulin, a carcinogenic mycotoxin, and the indole mycotoxins, tryptoquivaline and tryptoquivalone. ***A. flavus***. a species with yellow-green conidia that infests peanuts and grain. It produces aflatoxins B_1, B_2, G_1, G_2, (see also aflatoxins M) and other mycotoxins such as aspergillic acid, kojic acid, palmotoxin B_0, palmotoxin G_0, and β-nitropropionic acid. ***A. fumigatus*** (formerly *A. gliocladium*, *Eurotium malignum*). this species is thermotolerant, growing in soil and manure. It produces various mycotoxins and antibiotics (e.g., aspergillin, fumagilin, fumigacin, fumigation, fumigatoxin, fumitremorgin A and B, and helvolic acid). It is the common cause of aspergillosis in humans and birds, and is considered the primary pathogen in the case of birds. It sometimes occurs, for example, in infections of the ears, lungs, and upper respiratory tract of humans and other vertebrates. ***A. giganteus***. a species that produces gigantic acid. ***A. nidulans***. a species that commonly occurs in soil; it produces nidultoxin. It sometimes causes aspergillosis in humans and has been isolated from patients with onychomycosis and occasionally from those with maduromycosis. ***A. niger***. a pathogenic species that occurs commonly in soil. It produces a mycotoxin, malformin A, which causes severe, persistent infections in humans and laboratory mice. The spores are black and though not necessarily pathogenic, are often present in the external auditory meatus of humans with otomycosis. ***A. ochraceus***. this species ferments the coffee berry, producing the characteristic aroma. It produces the mycotoxins ochratoxin A and ochratoxin B. ***A. oryzae* var *microsporum 3***. a species that produces the mycotoxin maltoryzine. ***A. parasiticus***. a species that produces aflatoxins. ***A. pictor***. formerly used, probably for *A. versicolor*. ***A. terreus***. a species that produces the epoxide mycotoxins, terreic acid and citrinin, and antibiotics. It may sometimes infect the upper respiratory tract and has been isolated from the external auditory meatus in patients with otomycosis (mostly in Japan and Taiwan). It occasionally causes aspergillosis in humans and animals. ***A. ustus***. a species that produces the indole mycotoxin, austamide. ***A. versicolor***. a species commonly found in soil and often in dried salted beef. It produces the bisfuranoid mycotoxin, sterigmatocystin. It is probably identical with *A. pictor*, a name no longer in use. ***A. viridinutans***. a species that produces viriditoxin.

aspergillustoxicosis, aspergillus toxicosis (aspergillotoxicosis). **1:** a mycotoxicosis caused by a member of the genus *Aspergillus*. **2:** an acute lethal disease of livestock caused by feed contaminated by *Aspergillus spp*. The disease is marked by widespread internal hemorrhage; death usually intervenes within a week. *Cf.* aflatoxicosus.

asphalt (bitumen; asphaltum, pitch; [illegible] pitch; Judean pitch; mineral pitch; petroleum pitch; road asphalt; road tar). a mixture of bitumen and mineral matter. In North America this term is also used for bitumen alone. Asphalt is a source of intense local pollution at road-building sites and plants where the artificial product is manufactured. It is a suspected carcinogen and a moderate irritant.

asphaltum. asphalt.

asphyxia. **1a:** a condition of imminent death due to lack of oxygen in respired air. **1b:** suffocation from insufficient intake of air. **2:** severely impaired oxygen exchange (external or internal) by a living animal. **3:** a condition marked by hypercapnia and extreme hypoxia or anoxia. See also asphyxiant.

asphyxial (asphyctic). pertaining to asphyxia.

asphyxiant. **1:** asphyxiating; causing asphyxia. **2:** any agent with the capacity to asphyxiate; one that deprives an organism, tissue, or cell of oxygen. **chemical asphyxiant**. an asphyxiant that acts chemically to inhibit oxygen transport to (e.g., carbon monoxide), or oxygen utilization by (e.g., cyanide) cells or tissues. **simple asphyxiant**. an asphyxiant, a gas, or an aerosol that dilutes the air or oxygen that reaches a respiratory surface (e.g., nitrogen, hydrogen, helium). Respirable air containing high levels of simple asphyxiants lacks sufficient oxygen to support respiration.

asphyxiate. to suffocate; to cause or induce asphyxia.

asphyxiating. **1:** asphyxiant (def. 1). **2:** the induction of asphyxia. **3:** having the ability to suffocate or asphyxiate.

asphyxiation. **1:** the act of asphyxiating. **2:** the state of asphyxia.

Aspidelaps (shield-nose snakes). a genus of two species of small, semiburrowing, venomous snakes (Family Elapidae) of southern Africa. The snout is broad and adapted to burrowing. Neither species attains a length of over 0.8 m, and although the fangs are relatively large, these snakes are not considered dangerous. The head is short with an indistinct canthus rostralis and is only slightly distinct from neck. The maxilla bears two rather large tubular fangs with external grooves. The body is stout and cylindrical or somewhat depressed with a short, obtuse tail. The eyes are moderate in size with round or vertically elliptical pupils. The crown bears the usual 9 scales; the rostral is very large, concave below, recurved over the snout, and separated from other scales on sides; prefrontals are very short; the nasals are in broad contact with the single preocular. ***A. lubricus***

(African coral snake). a very small elapid snake of southern Mozambique, southern Angola, and southwest Africa. It has black and white bands over the bright orange ground color. ***A. scutatus*** (African shield-nosed snake; shield snake). a very small elapid snake of southern Mozambique, Botswana, Namibia, Zambia, and the Republic of South Africa. Dorsal body scales are smooth or faintly keeled posteriorly.

aspidin. a toxic principle, $C_{25}H_{32}O_8$, of aspidium.

aspidium. the dried rhizomes and stems of the ferns, *Dryoptens filix-mas* (European aspidium) or of *D. marginalis*. It is toxic and sometimes used to treat otherwise unmanageable tapeworm infestations.

Aspidomorphus (crowned snakes). a genus of small venomous snakes (Family Elapidae), none achieving a length greater than about 0.8 m. Two species occur in New Guinea and nearby islands; the others are restricted to Australia. None appear dangerous to humans. Included are *A. christieanus* (yellow-naped snake) of northern Australia; *A. diadema* (diadem snake; red-naped snake) of southern Australia; *A. harriettae* (white crowned snake) of northeastern Australia; *Aspidomorphus krefftii* (dwarf crowned snake; dwarf snake) of eastern Australia; *A. muellerii* (Muller's snake) of New Guinea and the islands of Ceram and Schouten; *A. schlegelii* (Schlegel's snake) of New Guinea and the island of Schouten; *A. squamulosus* (golden crowned snake; red-bellied snake) of eastern Australia.

aspidosperma. See quebracho.

Aspidosperma (quebracho). a genus of tropical American trees or occasional shrubs (Family Apocynaceae) with small flowers, alternate leaves, compressed peltate seeds with a flat papery wing, and follicular fruit. ***A. quebrachoblanco*** (quebracho; white quebracho) a tree and important source of quebrachine, adiposamine, and aspidospermine. See also quebracho.

aspidospermine. a white to brownish, crystalline irritant, alkaloid, $C_{22}H_{30}N_2O_2$, from the dried bark (quebracho, def. 2) of *Aspidosperma quebrachoblanco* (Family Apocynaceae).

aspiration. **1:** removal of fluid from a cavity or space with a needle. **2:** accidental inhalation of fluids or objects into the lungs.

aspiration pneumonia. pneumonia resulting from the inspiration of foreign matter such as food particles into the bronchial tree.

aspirin (2-(acetyloxy)benzoic acid; salicylic acid acetate; 2-acetoxybenzoic acid; acidum acetylsalicylicum; acetylsalicylic acid; sodium acetylsalicylate). an over-the-counter drug, aspirin is the most widely used analgesic. It is also used as an antipyritic, anticoagulant, antiinflammatory, and antirheumatic agent. It is considered safe for most individuals with normal usage. Its antiinflammatory activity is due to its inhibition of the synthesis of prostaglandins and related autocoids by damaged tissues. Aspirin is also one of the most common poisons, often from overdosage. It stimulates the central nervous system with subsequent CNS depression. In severe cases, this may present as respiratory failure, circulatory collapse, coma, and death. Early toxic effects include tinnitus, hearing loss, headache, dizziness, vomiting, hyperpnea, intense irritability, and peculiar behavior. Aspirin sometimes interferes with blood clotting and, at high doses, is hepatotoxic. Chronic use can induce or aggravate peptic ulcer and cause gastric distress, hemorrhage, heartburn, and iron-deficiency anemia. Some individuals are hypersensitive to aspirin and may have anaphylactic reactions to the drug even at very low doses. Allergic reactions range from rashes, hives, and edema, to severe, even fatal, asthma attacks. Children are at special risk from the unsupervised use of colored and flavored aspirin. If given aspirin, individuals under 18 years of age with chicken pox or flu are at risk of developing Reye's syndrome, which is often fatal. Aspirin is moderately toxic to most laboratory rodents and is fetotoxic to laboratory rats, but evidently not to humans. *Cf.* acetylsalicylic acid.

assay. **1a:** analysis; test of purity; trial. **1b:** the chemical analysis of a material to determine the proportion of constituents present. **2:** to subject to analysis; to perform an assay. **adsorption assay** (ELISA test; enzyme-linked immunosorbent assay; enzyme-linked immunoassay; enzyme-linked assay; ELISA test). an assay, usually conducted in the solid phase, using a ligand and a binding protein (usually an antibody), one of which is conjugated to an enzyme. A substrate is added following separation of the antibody-ligand complex from free ligand and free binding protein. The activity of the enzyme on the substrate yields products in amounts that are proportional to the amount of bound ligand. See also immunoassay. **Ames assay**. Ames test. **autoimmune assay** (ANA test). a blood test to identify autoimmune disease. **biological assay** (bioassay). the establishment of the relative strength of a biologically active substance by comparing measured effects in test organisms to those of a reference standard. **competitive adsorption assay**. a competitive assay in which antibody is adsorbed to a surface, and the test ligand allowed to compete with a known amount of labeled ligand. **competitive binding assay** (displacement analysis; saturation analysis). a type of assay in which a binder competes for a labeled versus an unlabeled ligand. The free and bound ligands are separated, and the amount of ligand is determined by comparing bound and unbound ratios with reference standards. **four-point assay**. an assay that employs a mixture of two doses of test material and two doses of a standard. **immunochemical assay**. immunoassay. **immunoenzymatic adsorption assay**. an assay in which unlabeled ligand is adsorbed to a surface and exposed to labeled antibody and test ligand in solution. The amount of bound, labeled antibody is inversely proportional to amount of free test ligand. **microbiological assay**. a biological assay that uses microorganisms as the test system. **sandwich adsorption assay**. an assay in which cold antibody is adsorbed to a surface to which the test ligand and cold ligand are added, followed by the addition of labeled antibody, thus the term "sandwich." The enzyme product is proportional to the concentration of bound ligand. **stem cell assay**. the determination of the effectiveness of a given drug against human cancer by the use of human tumor cell suspensions that are incubated with a number of drugs, suspended in agar, and plated over a layer of agar. The number of treated colonies that grow compared to the number of untreated (control) colonies that grow provides a measure of a drug's effectiveness. **toxicity assay**. toxicity test (See test). See also test.

assessment. a critical, often official, appraisal or evaluation of the nature, status, significance or importance of an entity or process. See dose effects assessment, dose-response assessment, dose scaling, endangerment assessment, exposure assessment, hazard assessment, liability assessment, risk assessment, screening-level assessment, toxicity assessment.

assimilate. **1:** to absorb digested food or nutrients into the system (i.e., into protoplasm). **2:** that which is assimilated. *Cf.* dissimilate.

assimilation. the conversion of food or nutrients into a cellular substance by a living system. *Cf.* dissimilate.

assimilation capacity (assimilative capacity). the capacity of a biotic system (most commonly an ecosystem) to receive a certain amount or to sustain certain concentrations of a xenobiotic chemical or waste discharges without suffering significant or permanent harm or degradation.

assimilative capacity. assimilation capacity.

assumption. **1:** that which is taken for granted or supposed to be true in the absence of proof. **2:** the act of taking for granted.

astasia. muscular incoordination with the inability to stand or walk.

astatic. referring to or denoting astasia.

astatine (symbol, At)[Z = 85; most stable isotope, ^{210}At; half-life 8.3 hours]. an artificial radioactive element of the halogen series with 20 known, short-lived isotopes.

aster. common name of plants of the genus *Aster* (Family Asteraceae). This term is sometimes used for specific plants or genera of other families, e.g., *Callistephus chinensis* (China aster), *Xylorrhiza* (woody asters), and *Gutierrezia* (tansy asters).

Aster (asters, true asters, true hardy asters, starworts, Michaelmas daisies). a genus of perennial herbs (Family Asteraceae, formerly Compositae), many of which are indigenous to the temperate zone of North America. Some are facultative accumulators of selenium from soil. Examples are *A. adscendens*, *A. coerulescens*, *A. commutatus*, *A. ericoides*, *A. glaucoides*, and *A. laevis*. See also *Xylorrhiza*.

Asteraceae. the aster, composite, or sunflower family (formerly Compositae).

asterixis (flapping tremor; hepatic tremor; liver flap; liver tremor). a coarse, abnormal muscle tremor marked by jerking or flapping movements, especially of the arm and hands, sometimes the tongue, or muscle groups in the feet. See also ammonia intoxication. When associated with hepatic coma, asterixis is often referred to as hepatic (or liver) tremor or liver flap. See also alcoholism, ammonia intoxication, tremor.

asteroid. any animal of the class Asteroidea (Phylum Echinodermata).

Asteroidea (starfishes; sea stars; asteroids). a class of the phylum Echinodermata; the true starfishes. They have radial (pentagonal) symmetry usually characterized as star-shaped and having from 5 to 50 laterally projecting rays or arms that are usually not sharply delimited from a central disc that is covered by a calcite test. The mouth is situated on the ventral surface. Many hard spines of calcite and associated organic materials project from the upper surface of the rays. These are sheathed by a thin integument comprised of an epidermis and dermis. Acidophilic cells within the epidermis presumably secrete a toxin into the surrounding water or discharge it on direct contact (as when handled by humans). Starfishes also have pedicellariae (tube feet) that serve in feeding and locomotion. They contain poison glands in the concave cavity of their valves. Some species are poisonous when ingested. See *Asterias*.

asterosaponin. any of a number of toxic saponin-like glycosides found in starfish (Asteroidea). Similar compounds occur also in sea cucumbers (Holothurans). **Asterosaponin A** has anti-hormonal activity.

asthenia. muscular weakness or debility; loss of muscular strength. *Cf.* abirritation, hyposthenia.

astheniant. **1:** weakening, debilitant. **2:** a debilitating agent. See also hyposthenia nt.

asthenic. weak, debilitated; pertaining to or characterized by asthenia; one who suffers from asthenia. See also hyposthenic.

Asthenosoma. a genus of venomous sea urchins (Class Echinoidea, Phylum Echinodermata). ***A. ijimai***. a dangerous sea urchin with well-developed, sharp, venomous aboral spines. ***A. jimoni***. a moderately venomous red sea urchin. See sea urchin.

asthma. a recurrent condition characterized by paroxysmal dyspnea with bronchospasm. Some cases of asthma are allergic reactions of sensitized individuals, others are provoked by vigorous exercise or various sources of physical or psychological stress. See steam fitter's asthma.

Astragalus (milk vetches; poisonvetches; locoweeds). the largest genus of plants of the Family Fabaceae. Some species are harmless and afford good forage for livestock, and certain oriental species (e.g., *A. gummifer*) have medicinal value. A number of species are, however, poisonous. There are three distinct categories of poisonous *Astragalus*: (1) The poisonvetches are obligate selenophiles and the selenium content accounts for their chief toxic effects. Many of these species are distasteful to livestock and do not usually pose a hazard on the range if other forage is available. The species that most often poison livestock are *A. bisulcatus*, *A. pattersonii*, *A. pectinatus*, and *A. racemosus*. (2) Some species are true locoweeds (see loco). They contain the mydriatic alkaloid, indolizidine-1,2,8-triol. Locoine was the name given to early toxic extracts from locoweeds. See also *Oxytropis*. (3) Certain species produce a distinctive type of paralytic poisoning of livestock with respiratory involvement. Symptoms in range cattle include poor condition, a dry or peeling muzzle, chronic cough, loss or huskiness of voice, anorexia, and incoordination, especially of the hind legs which tend to knock together (hence, "cracker-heel" disease). Forced exercise causes a wheezing or even a roaring exhalation. Paralysis may develop, and sudden death from asphyxia is common in animals that are driven. Mortality is otherwise unpredictable. The syndrome is more severe in sheep and especially so in horses. Species responsible for this third type of poisoning include *A. emoryanus*, *A. michauxii*, *A. pubentissimus*, *A. sabulosus*, and *A. tetrapterus*.

astrogliosis. See gliosis.

Astrotia stokesii (Stokes's sea snake). a rather rare

sea snake of open, moderately deep, marine waters. The ground color is light brown, yellowish or orange with broad black rings or bars and spots above. The belly is lighter; the head is olive to yellowish. As in the case of *Aipysurus laevis*, the body is massive for a sea snake and the head is large. It differs from *A. laevis* by the larger head shields and the presence of fragmented and poorly differentiated ventrals. The body scales are large, keeled, and strongly imbricate. The length of adults averages about 1.5 m and the body of large individuals may have a circumference of about 25 cm. While considered dangerous, it is a snake of moderately deep marine waters, seldom encountering humans; bites are unknown as is the toxicity of the venom.

asymmetric, asymmetrical. lacking symmetry; uneven in size, shape, or position.

asymptomatic. lacking or without symptoms.

At. the symbol for astatine.

ATA. **1:** amitrole, the generic name for 3-amino-1,2,4-triazole, *q.v.* **2:** alimentary toxic aleukia (see aleukia).

Atabrine. quinine.

atactic. referring to or characterized by ataxia or incoordination; uncoordinated.

atactostele. See stele.

atamasco lily. *Zephyranthes atamasco*.

atamasco poisoning. poisoning by foliage and/or bulbs of *Zephyranthes atamasco*, *q.v.* See also staggers.

ataxia (ataxy; dyssynergia; incoordination). incoordination or failure of coordination of voluntary muscular action; often manifested by unsteady gait. **partial ataxia**. dystaxia.

ataxia telangiectasia. a rare inherited disease of *Homo sapiens* in which affected individuals develop tumors following X-irradiation. The condition is apparently due to certain defective DNA polymerases that repair DNA.

ataxiadynamia. muscular weakness with incoordination. See also ataxia, incoordination.

ataxic. pertaining to, characterized by, or afflicted by ataxia.

ataxy. ataxia.

ATE. acute toxicity end point. See end point.

atelopidtoxin. an extremely toxic, dialyzable toxin isolated from the skin of frogs of the genus *Atelopus*, notably *A. zeteki*. Its chemical and pharmacological properties are not fully established.

Atelopodidae. a family of frogs that includes *Atelopus zeteki*.

Atelopus zeteki (golden arrow frog). a small, extremely poisonous frog (Family Atelopodidae) of Central and South America. It is the source of atelopidtoxin, secreted by its skin.

Atergatus floridus (reef crab). a poisonous tropical marine crustacean (Order Decapoda) of the Indo-Pacific region. The poisonous principle appears to be structurally identical to tetrodotoxin. Symptoms of intoxication in humans may include nausea, vomiting, numbness, tingling sensations beginning about the lips and tongue, muscular paralysis, and collapse. Severe intoxications of humans sometimes terminate fatally.

Atheris (African bush vipers; bush vipers). a genus of venomous, arboreal, tropical, African snakes (Family Viperidae). Bites are rare but considered to be very dangerous to humans. Bush vipers are rather small, attaining a maximum length of less than 1 m. The head is very broad and distinct from the narrow neck; the crown is covered by small imbricate scales; the snout is broad; the canthus rostralis is sharply defined; there are 2-3 flat scales between the nasal and each eye; and 1-3 rows of small scales separate each eye from the labials; the eyes are moderately large with vertically elliptical pupils. The body is rather slender, tapering, and somewhat compressed with a moderately long prehensile tail. Dorsal body scales are strongly keeled and bear apical pits; lateral scales are smaller than the midline dorsals and are not serrated. ***A. barbouri***. the former scientific name of *Adenorhinos barbouri*. ***A. ceratophorus*** (horned tree viper). an arboreal viper of Togo and Tanzania. ***A. chloroechis*** (West African tree viper; galobou). a small viper (very similar to *A. squamigera*) that inhabits the forests of west Africa. ***A. nitschei*** (sedge viper; bush viper; Great Lakes bush viper; green viper; Nitsche's tree viper; rugando). a small, rather stout, semiterrestrial snake that is very common in montane areas of Zaire and Uganda (at elevations of 2000 m or so) southward to Zambia. It usually occurs in reeds and papyrus along lake margins, in upland swamps, or up to a height of 3 m or so in the elephant grass of humid valleys; it is occasionally seen on the ground. The dorsal scales are small and numerous. Adults average about ½ m in length, rarely more. Normally, the crown is green with a V-shaped or A-shaped mark; the crown of occasional individuals is almost entirely black. The body is usually bright to olive green with irregular black markings due to black or black-tipped scales. Occasional individuals are almost uniformly black with a lighter tail. The belly is yellowish or very pale green. ***A. squamigera*** (green bush viper; bush viper; leaf viper; tree viper; Rauhschuppige bush viper; kisigosogo; kyozima; nalukonge). a green (sometimes yellow) viper usually with yellow crossbands or paired yellow spots; sometimes nearly unicolor; black markings are entirely lacking. The crown is uniformly green; chin yellow; labials light yellow or cream. The body is usually slender, but may be stout in large individuals. Adults average about 5 cm shorter than *A. nitschei*. Dorsal scales are larger and fewer than in *A. nitschei*. It occurs in rain forests from western Kenya and the Cameroons to Angola and on the island of Fernando Po.

athymia. dementia.

Atlantic coral snake. *Micrurus diastema*.

Atlantic Jackknife. *Ensis directus*.

atlas. a book, map or other format that presents information by means of illustrations rather than by description.

Atlas adder. *Vipera lebetina*.

atmosphere. **1:** the gaseous envelope that surrounds the earth. **2:** a standard unit of pressure representing that exerted by a column of mercury at sea level at 45° latitude. It is equal to 1000 gm per square centimeter or 760 mm of mercury. In the English system, this is equivalent to a force of 14.695916 pounds per square inch.

atmosphere generation. the production of reproducibly accurate mixtures of gases or vapors in the field of inhalation toxicology to test airborne toxicants.

atomic. pertaining to or denoting an atom.

atomic absorption spectrophotometry. See spectrometry, atomic absorption.

atomic fallout. See fallout.

atomic mass. See relative atomic mass.

atomic mass unit. Dalton.

atomic number. proton number.

atomic pile. a little-used synonym for nuclear reactor, *q.v.*

atomic weight. former term for relative atomic mass, *q.v.*

atomization. **1:** the production of an atomic vapor, one comprised of free atoms of a given substance as in analytical flame spectroscopy. **2:** nebulization. **3:** the nebulization of a liquid into a spray comprised of very fine particles.

atomize. **1:** to produce an atomic vapor. See atomization. **2:** to nebulize; to divide a liquid into very fine particles by impact with a jet of steam or compressed air, or by passage through a porous mechanical device.

atonic. referring to atony.

atony. **1:** lack of tonus. **2:** weakness, lack of energy, debility. See abirritation.

atopic conjunctivitis. allergic conjunctivitis.

atopic dermatitis. an allergic skin reaction to agents such as poison ivy and poison oak; a kind of delayed hypersensitivity.

atopy (atopic allergy). a genetic predisposition to develop a particular type of allergy.

atoxic. **1:** not toxic at concentrations or in amounts to which a specified living system may conceivably be exposed. **2:** not due to a poison. *Cf.* nontoxic.

atoxigenic. not producing or releasing toxins.

Atoxyl. sodium arsenilate.

atoxylic acid. arsanilic acid.

ATP. adenosine triphosphate.

ATP-dependent acid:CoA ligase (AMP). See conjugation (amino acid conjugation).

ATPase. adenosine triphosphatase.

Atractaspis (burrowing vipers; mole vipers). a genus of venomous, burrowing snakes (Family Viperidae). All are limited to Africa except for *A. engaddensis* (the oasis mole viper), which ranges from Egypt to Israel, and *A. microlepidota andersonii* (Arabian mole viper; Anderson's mole viper), which occurs in the southern Arabian Peninsula. All are less than 1 m in length. They have large fangs, however, and can inflict serious bites to persons who pick them up or step on them with bare feet. There are no external characteristics that distinguish mole vipers from nonvenomous snakes. Mole vipers are the only viperid snakes that have divided anal plates. The head is short, conical, and indistinct from the neck; there is no canthus; and the snout is broad, depressed, and often pointed. The eyes are very small with round pupils. The body is cylindrical, slender in small individuals, stout in large ones; the tail is short, ending in a distinct spine. Scales on the head include the usual 9 crown scales; an enlarged rostral that extends somewhat between the internasals; the loreal is absent and so the nasal adjoins the single preocular. Species not described herein include *A. atterima, A. battersbyi, A. boulengeri, A. coallescens, A. congica* (Congo burrowing adder), *A. dahomeyensis, A duerdeni, A. engdahli, A. irregularis, A. leucomelas, A. reticulata,* and *A.* [illegible] Bibron's burrowing adder; burrowing adder; Bibron's viper; black burrowing viper; Southern mole viper). a slender viper that is usually uniformly dark above (purplish brown or black) with a creamy, yellowish, or white venter. The head is small with a strongly projecting snout. The average length of adults is about 40-45 cm. The range stretches from Angola and southern Rhodesia to southern Southwest Africa and Natal. It lives under rocks and in burrows and usually appears on the surface only following heavy rains. It strikes with little provocation; experienced herpetologists have often been bitten by this snake. The bite of even a small snake is painful, causing edema and serious systemic effect. ***A. corpulenta*** (western mole viper). a rather stout, poorly known, slate-colored viper that occurs in the tropical rain forest region from the Ivory Coast to the eastern Congo. The head is small with a strongly projecting snout. Adults average about 48 cm in length but some individuals reach a length of about 60 cm. It is slate gray or slate blue above, lighter beneath; the terminal portion of the tail is often white. There are five supralabials and a single anterior temporal; the second infralabial is much enlarged, fused with chin shields and in contact with its fellow. There are 23-29 dorsals at midbody and 178-208 ventrals. The anal plate is entire and all or nearly all of the subcaudals are single. ***A. microlepidota*** (Arabian mole viper; Anderson's mole viper; Northern mole viper; balaiwona; balewol; himmin; pagha wubré). a mole viper indigenous to the southern Arabian Peninsula and throughout much of northern and central Africa, southern Mauritania, southern Mali, southern Niger, and possibly Chad. It is one of the most common venomous snakes in the Sudan, and a frequent cause of death; 25% or more of untreated bites of adult humans apparently result in death.

atractylic acid. See atraxtyloside.

atractylin (C_{30} glucoside). atraxtyloside.

atractyloside ((2β,4α,15α)-15-hydroxy-2-[[2-*O*-(3-methyl-1-oxobutyl)3,4-di-*O*-sulfo-β-D-glucopyranosyl]-oxy]-19-norkaur-16-en-18-oic acid dipotassium salt; potassium atractylate; atractylin, C_{30} glucoside). a very toxic glycoside from the thistle, *Actractylis gummifer* (Family Compositae).

atracurium (Tracium). an extremely toxic neuromuscular blocking agent.

Atrax (funnel-web spiders; Atrax spider; funnel whip spider). a genus of Australian tarantula-like spiders (Family Dipluridae); many are large. The bite often results in the slow onset of dyspnea and cardiovascular difficulties. Mortality is generally low; antivenin is available. The venom of eight species, including *A. robustus* and *A. formidabisis* (tree funnel-web spider), is known to be injurious to humans, and that of *A. robustus*, at least, is sometimes lethal.

atrazine (2-chloro-4-ethylamino-6-isopropylamino-1,3,5-triazine; 6-chloro-*N*-ethyl-*N*′-(1-methylethyl)-1,3,5-triazine-2,4-diamine). a selective triazine derivative used as a pre- and post-emergence herbicide, characterized by high toxicity to a diversity of monocotyledonous and dicotyledonous plants when applied to the soil. It is slightly toxic to fish and slightly to moderately toxic to laboratory mice. Acute poisoning of domestic cattle and sheep can cause muscular spasms, fasciculations, unsteady gait, pulmonary congestion, and hyperpnea. Lesions may include degenerative changes in the adrenal glands, kidneys, and liver. Atrazine is a possible mutagen, carcinogen, and teratogen. It is degraded non-enzymically in maize to the inactive hydroxy derivative and is thus used extensively in controlling annual grasses and broadleaved weeds in maize.

atrial fibrillation. See fibrillation.

atrichous isorhiza. a small isorhiza nematocyst with a smooth tubule. See nematocyst.

atrioventricular bundle (bundle of His; AV bundle). a small band of specialized, rapidlyconducting cardiac fibers originating in the AV node and extending into the interatrial septum to the ventricular myocardium. Interruption of this pathway produces heart block.

atrioventricular heart block. See heart block.

atrioventricular node (AV node; nodus atrioventricularis). a microscopic set of specialized conducting fibers situated in the interatrial septum of the heart. They are continuous with the atrioventricular bundle and the atrial muscle fibers. See atrioentricular bundle.

Atriplex (saltbushes). a genus of plants (Family Chenopodiaceae) that are facultative accumulators of selenium from soil.

atriplex poisoning. atriplicism.

atriplicism (atriplex poisoning). poisoning due to ingestion of the spinach-like herb, *Atriplex littoralis* (Family Chenopodiaceae).

Atropa belladonna (deadly nightshade; belladonna; death's herb). a shrublike, perennial European herb (Family Solanaceae), the source of belladonna. It is the most notorious and probably the most dangerous of the solanaceous plants, which also include *Cestrum, Datura, Hyoscyamus, Mandragora, Physalis,* and *Solanum*. It contains atropine and various related alkaloids (e.g., 1-hyoscyamine, scopolamine, and tropane); the roots contain scopoletin and scopolin. All parts of the plant are toxic and toxicity is greatest at maturity. Validated poisonings are rare. Children may be attracted to the black berries, and consumption of as few as three of these can be fatal to a child. See also belladonna poisoning.

atropamine. apoatropine.

atrophia. atrophy (def. 1).

atrophic. pertaining to organisms or stages in the life cycle of an organism that do not feed.

atrophie. atrophy (def. 1).

atrophy (atrophia; atrophie). **1:** a wasting away or decrease in the size of a cell, tissue, organ, or part. Atrophy may result, for example, from loss of cells through death and resorption, reduced rates of cellular proliferation, ischemia, malnutrition, physical pressure, decreased activity, hormonal changes, or directly or indirectly from empoisonment. **2:** to undergo or to cause atrophy. **acute yellow atrophy** (acute parenchymatous hepatitis; Budd's jaundice; malignant jaundice; massive hepatic necrosis; Rokitansky's disease) atrophy of the liver marked by wasting, massive necrosis, and jaundice; it a complication of fulminant hepatitis. It is an infrequent and usually fatal complication of viral hepatitis and may result from toxic hepatitis due to poisoning, for example, by carbon tetrachloride, fluorinated hydrocarbons, or *Amanita phalloides*.

atropia. atropine.

atropine (*dl*-hyoscyamine; tropine tropate; tropic acid ester with tropine; *dl*-tropyl tropate; α-(hydroxymethyl)benzeneacetic acid 8-methyl-8-

azabicyclo[3.2.l]oct-3-yl ester; 1α*H*,5α*H*-tropan-3α-ol (+)-tropate; atropia). an alkaloid, $C_{17}H_{23}NO_3$, that occurs naturally in *Atropa belladonna*, and related plants (*Datura*, *Hyoscyamus*); it is also prepared synthetically. It is used in medicine mainly as a muscle relaxant (e.g., to dilate the pupil of the eye or to inhibit movements of the intestine), as an antispasmodic, antisudorific, and anticholinergic. Atropine (as atropine sulfate) is an antidote for anticholinesterases and various other toxic substances such as the organophosphorous pesticides (e.g., parathion); artificial respiration may also be required. It is moderately toxic to adult humans. Infants and young children are, however, very vulnerable to atropine intoxication, which may result from its use in ophthalmologic treatments. See also atropinic effects, atropine poisoning.

atropine poisoning (atropinic poisoning). poisoning from exposure to atropine. Signs and symptoms may appear within a few hours or a few days following exposure. The condition is marked by intense CNS excitation and parasympathetic paralysis. Mania, delirium, psychotic behavior, hallucination, disorientation, and aggressive behavior are usually observed. Symptoms in humans may also include suppressed salivation and an immediate sensation of dryness and burning of the mouth with intense thirst; inhibition of sweating; dry, hot, flushed skin, especially of the face; mild to severe hyperthermia; difficulty or inability to swallow; dysphagia; and blurred vision with impaired accommodation to changing light intensity due to pupillary dilation. Victims may also exhibit rapid pulse and respiration, muscle rigidity, urinary retention, convulsions, and coma. A rash may appear with subsequent exfoliation of skin, chiefly about the face, neck, and upper trunk. Death may result from cardiovascular and respiratory collapse. See also atropinism. *Cf.* atropinism.

atropine sulfate. See atropine.

atropinic. of or pertaining to atropine or to atropine-like properties or activity. See effect (atropinic effect).

atropinic poisoning. atropine poisoning.

atropinism (atropism). poisoning by atropine, belladonna, or other preparations containing atropine. See also atropine, *Atropa belladonna*, belladonna poisoning, atropine poisoning.

atropinization. exposure to or exposing to atropine.

atropinize. to expose to atropine.

atropism. atropinism.

atropyltropeine. apoatropine.

atroscine. racemic scopolamine.

atrotoxin. a component of *Crotalus atrox* (diamondback rattlesnake) venom. It reversibly elevates voltage-dependent calcium ion currents in isolated myocytes.

attack. an episode or sudden onset of illness; a spasm, fit, seizure, or paroxism.

attention deficit disorder. See hyperactivity (def. 2).

attenuation. **1:** a weakening or reduction in strength or intensity, e.g., of a stimulus or a toxicant. **2:** a process whereby a compound is reduced in concentration with time through adsorption, degradation, dilution, and/or transformation. **3:** a loss of virulence of a pathogenic microorganism, often by preparation of successive subcultures under controlled conditions of medium and temperature. Suspensions of attenuated cultures may bring about immunity to the disease caused by the virulent form.

attractant. **1:** a substance that attracts. **2:** a substance produced by one organism or species that attracts another, usually conferring an advantage on the producing organism. Examples are sexual attractants and baits used by humans and other species to lure prey. See allomone.

atypia. a condition or disorder of nonconformance to type (said, for example, of cells or tissues). See anaplasia, dysplasia.

atypical. abnormal, unusual.

Au. the symbol for gold (aurum).

auditory startle. an unconditioned, whole-body startle reflex resulting from presentation with a standard sound stimulus greater than 90 dB. Prior presentation of a low-intensity sound stimulus inhibits the auditory startle (prepulse inhibition). Tests of auditory startle using such inhibition offers one method of testing drugs for sensory effects.

auger shells. See Terebridae.

augers. See Terebridae.

auramine (4,4′-(imidocarbonyl)bis(*N*,*N*-dimethylaniline)). a yellow, crystalline or powdery solid, $C_{17}H_{22}ClN_3 \cdot HCl$, that is soluble in water, ethanol, and ethyl ether. Usually as the hydrochloride, it is used mainly as a dye for paper, leather, and other products. It is also used as an antiseptic and fungicide. Auramine is an occupational carcinogen, causing bladder cancer in humans. Rubber workers, tile dyers, textile dyers, paint and dye manufacturers, and workers in the manufacture of auramine are at greatest risk.

auramine hydrochloride (yellow pyoktanin). a yellow, water-soluble, crystalline compound, also soluble in ethanol, $C_{17}H_{22}ClN_3 \cdot H_2O$. It is used chiefly as a commercial dye and antiseptic. Workers in the manufacture of auramine are at risk of developing bladder cancer, as are others who work with auramine (e.g., rubber workers, paint and dye workers).

aureine. senecionine.

***Aurelia*.** a genus of large scyphozoan jellyfish (Order Discomedusae) of the North Atlantic, Indian, and Pacific oceans. Stings by many larger forms are intensely painful to humans.

aureomycin. former name of chlorotetracycline hydrochloride. See chlorotetracycline. See also *Streptomyces*.

Aureomycin®. trademark for a crystalline preparation of chlorotetracycline hydrochloride. See chlorotetracycline.

Aurora Yellow. cadmium sulfide.

auric cyanide. gold tricyanide.

auripigment. arsenic trisulfide.

1-aurothio-*d*-glucopyranose. gold thioglucose.

aurothioglucose. gold thioglucose.

aurothiosulfate natrium. gold sodium thiosulfate.

aurous cyanide. gold monocyanide.

aurum. gold.

aurum paradoxum. tellurium.

austamide. an indole mycotoxin produced by *Aspergillus ustus*.

Australian black snake. *Pseudechis porphyriacus*.

Australian blue-ringed octopus. *Octopus maculosus*.

Australian broad-headed snake. any snake of the genus *Hoplocephalus*.

Australian brown snake. See *Demansia textilis*.

Australian collared snakes. snakes of the genus *Glyphodon*.

Australian copperheads. See *Denisonia*.

Australian coral snake. *Brachyeurophis australis*

Australian mulga snake. *Pseudechis australis*.

Australian necklace snake. *Brachyurophis australis*.

Australian nettle. *Dendrocnide moroides*.

Australian sea lion. *Neophoca cinerea*.

Australian tigersnake. *Notechis scutatus*.

Australian yellow-spotted snake. *Hoplocephalus bungaroides*.

autoallergy. autoimmunity.

autoantibody. any antibody that is produced in response to, and reacts with, molecules that are part of one's own body.

autoantigen (autoimmunogen). any normally endogenous substance against which the body elaborated specific antibodies (autoantibodies).

autochthonous (autologous). occurring in the same individual.

autocoid. any of a variety of structurally and functionally diverse endogenous substances that appear to be importantly involved in keeping the body free of disease and maintaining a state of health. Included are angiotensin, bradykinin, kallidin, histamine, serotonin, and the prostaglandins.

autocytolysis. autolysis.

autocytolytic. autolytic.

autocytotoxin. a cytotoxin that has the capacity to poison the organism that produces it.

autogenic. resulting from or involving biotic activity.

autogenic succession. See succession.

autoimmune assay. See assay (autoimmune assay).

autoimmune disorder. a disease state in which the immune system produces antibodies that attack the individual's own tissues.

autoimmunity (autoallergy). a condition or process of inappropriate immunological reaction by an organism to components of its own tissues (autoantigens), usually with detrimental outcomes. Hypersensitivity may result or, in severe cases, autoimmune disease.

autoimmunogen. autoantigen.

auto-inhibitor. See adaptation auto-inhibitor.

autointoxicant. **1a:** any endogenous substance that poisons the producing organism or the progeny of the producing organism. **1b:** any metabolic waste or toxicant in the excreta of an organism to which it or its progeny may be exposed. *Cf.* autotoxin. See also autointoxication.

autointoxication (enterotoxism; enterotoxication; endogenic toxicosis; intestinal intoxication; self-poisoning; self-empoisonment). **1a:** autotoxico-sis; intoxication due to a toxicant produced within the affected organism. **1b:** intoxication of an individual due to excreted toxicants. A few plants (e.g., certain tropical forest trees) inhibit sprouting of their own seeds, which can thus develop only in areas occupied by other species. This may sometimes confer an adaptive advantage. **2:** enterotoxism; poisoning from the absorption of metabolic wastes, decomposed matter in the intestines, or the products of necrotic or infected tissue. **3:** symptoms of constipation (obs.).

autologous. autochthonous.

autolysis (autocytolysis). lytic disintegration of a cell, part of a cell, or tissue by the action of cellular enzymes. Autolysis occurs either within a lysosome or upon the release of enzymes from lysosomes into the cytoplasm. It occurs in damaged tissues, immediately prior to and after the death of a cell, or in developmental processes such as metamorphosis. *Cf.* cytolysis.

autolytic (autocytolytic). pertaining to autolysis or the capacity to autolyze.

autolyze. to undergo or to cause autolysis.

Automeris io (io moth). a reddish-brown moth (Family Saturniide), the hind wing of which has a characteristic black eyespot. The caterpillar is green with reddish-pink and white lateral stripes. See io-moth dermatitis.

autonomic. **1:** of, or pertaining to the autonomic nervous system (ANS) or a part of the ANS. **2:** involuntary; reflex. **3:** pertaining to a chemical substance that affects tissues served by the ANS. **4:** the autonomic nervous system. **5:** due to endogenous influences (in botany). Thus, autonomic movement of a plant or plant part is that which results from internal growth changes.

autonomic agent. a substance that inhibits or intensifies the rate of nerve impulse transmission across synaptic junctions, especially those of the autonomic nervous system.

autonomic nervous system (ANS). the division of the vertebrate peripheral nervous system that controls, but does not usually trigger, glandular secretion and the action of viscera and smooth (involuntary), cardiac, and some striated muscle. The ANS is controlled by the CNS mainly from centers in the hypothalamus and medulla oblongata. It is thus thought by many scientists to respond only indirectly to external stimuli. There are two anatomically and functionally distinct divisions of this system, the sympathetic and parasympathetic nervous systems, *q.v.* Most autonomic effectors receive dual innervation from these divisions, although only sympathetic neurons serve the blood vessels of the skin and limbs. With the exception of the innervation of the adrenal medulla, the autonomic system consists of two-neuron pathways. A preganglionic myelinated neuron arises in the CNS and synapses in a ganglion with numerous postganglionic unmyelinated neurons that terminate at the effectors. The visceral nervous system of arthropods is structurally and functionally analogous to the ANS. Poisons that act entirely or in part on the ANS can produce any of a number of effects, such as miosis, bronchoconstriction, salivation, lacrimation, respiratory failure, cardiovascular collapse, and death. See atropine, blocking agent, carbachol, curare, muscarine, nervous system, parasympathetic nervous system, sympathetic nervous system, peripheral neuropathy.

autonomic poison. a poisonous autonomic agent.

autopharmacologic. pertaining to or denoting substances (e.g., hormones, neurotransmitters) produced by a living organism that are pharmacologically active.

autopharmacologic reaction. any response by the cells or tissues of a living organism to an endogenously produced pharmacologic agent.

autopoisonous. autotoxic (def. 2).

autopsy. 1a: (postmortem examination; necropsy; necroscopy; thanatopsy; postmortem examination). the postmortem examination of an animal or human by gross dissection (accompanied as needed by histological, chemical, analytic, and other methods) to determine circumstances accompanying or causing the death, including the pathology of disease. This is usually required under law for an unexplained, unnatural, or violent human fatality. **1b:** permission to perform 1a. **2:** to perform a postmortem examination. **3:** a critical analysis of a past event or process, sometimes by reproducing the effect or the circumstances surrounding the event or process. For example, the duplication of an event, action, or circumstance that occurred during the course of a disease and observing its relationship to the aggravation or amelioration symptoms.

autoradiography. the science and method of producing an image of a radioactive specimen by the effect of the ionizing rays or particles emitted by the specimen on a photographic emulsion or other radiation-sensitive medium. Autoradiography can be used to demonstrate the distribution of a selected substance in tissues and organs of an individual organism.

autosepticemia. septicemia caused by toxic substances produced within the organism.

autotoxaemia. autotoxicosis.

autotoxemia. autotoxicosis.

autotoxic. **1:** pertaining to autointoxication. **2:** autopoisonous; poisonous to the producing organism. See also autointoxication, autopharmacologic, autotoxicosis.

autotoxic reaction. a toxic autopharmacologic reaction.

autotoxicity. the capacity of an endogenous substance or excreta to harm the producing organism or its progeny.

autotoxicosis (autotoxemia; autotoxaemia; autotoxis; autoxemia). intoxication (poisoning) by substances produced within the affected organism; autointoxication (def. 1a) *Cf.* autointoxication.

autotoxin. **1:** an autotoxic toxin. **2:** an autointoxicant.

autotoxis. autotoxicosis.

autotroph. autotrophic organism.

autotrophic (holophytic; lithotrophic). of or pertaining to autotrophic organisms. *Cf.* holophytic, holotrophic, holozoic, lithotrophic.

autotrophic organism (autotroph; holophytic organism; lithotrophic organism; producer). producer (in ecology); an organism (e.g., a green plant) capable of growth and maintenance on purely inorganic media; one that produces food from inorganic substances. Autotrophs utilize carbon dioxide and nitrates or ammonium salts as sources of carbon and nitrogen, respectively. The energy source for these syntheses is either light via photosynthesis (photoautotrophic organisms) or inorganic chemical reactions independent of light energy via chemosynthesis *(chemoautotrophic* or *chemosynthetic* organisms). Photoautotrophic organisms include green plants and some purple bacteria; all contain chlorophyll. Chemoautotrophic organisms include certain bacteria that derive the energy that drives carbon dioxide assimilation and biosynthesis from the oxidation of reduced inorganic compounds. As examples, nitrifying bacteria oxidize ammonia to nitrate or nitrite; iron bacteria convert ferrous compounds to ferric; and colorless sulfur bacteria convert hydrogen sulphide to sulfur.

autoxemia. autotoxicosis.

autoxidation. the spontaneous combination of a substance directly with oxygen at ordinary or moderate temperatures, without catalysis.

autoxidative. of or pertaining to autoxidation.

autoxidative injury. an injury that results from membrane disruption, lipid peroxidation, mutation, DNA damage; or direct tissue or organ injury which, if acute, may be lethal. Autoxidative injury can also manifest itself as cardiovascular disease and attendant complications, diabetes, cataracts, male sterility, nephrotoxicity, arthropathies, and carcinoma. Autoxidative injury is thus difficult to distinguish from degenerative diseases.

autumn crocus. *Colchicum autumnale*.

autumn galerina. *Galerina autumnalis*.

autumn squill. *Scilla autumnalis*.

auxin (growth substance; plant hormone). any organic substance produced in growing shoot and root apices of plants that promotes or regulates growth of the producing plant when present in low concentrations. This is effected by the stimulation of cell elongation in the region just behind the apex. Auxins are very toxic to plants, inhibiting growth at higher concentrations. Consequently, both natural auxins and synthetic compounds that are closely related to auxins are used as herbicides (e.g., 2,4-D). Auxins also promote cell division in plant tissues and regulate phototropic and geotropic movements of plants. Auxins are antagonists of abscissic acid, retarding leaf abscission. The most widely distributed natural auxin is indole-acetic acid. Other natural auxins (e.g., indole-acetonitrile and indolepyruvic acid) also contain an indole ring. See herbicide (auxin herbicide).

auxin 1AA. 3-indolacetic acid.

auxotox. an atomic group which, when present in the molecules of a compound, confers or intensifies the toxicity of the compound.

AV bundle. atrioventricular bundle.

AV node. atrioventricular node.

Avena sativa (cultivated oats). a cereal grain (Family Gramineae) variously involved in four types of livestock poisoning: nitrate poisoning, grass tetany, photo-sensitization, and smutty oat poisoning. There are reports of the smutty oat disease of horses and cattle from the 19th century. See also sunscald.

avermectin (AVM). any of a group of derivatives of pentacyclic 16-membered lactones that are used as broad-spectrum antiparasitic antibiotics. They are produced by the actinomycete, *Streptomyces avermitilis*. Oral toxicity of avermectins to terrestrial vertebrates appears to be moderate at most; fishes and aquatic invertebrates are much more susceptible to these substances. **avermectin B_1**. abamectin.

aversive. **1:** avoiding, avoidance; characterized by avoidance (e.g., aversive behavior). **2:** noxious; promoting or stimulating avoidance. *Cf.* appetitive. See also approach-withdrawal.

Aves (birds). until recently thought to be the only class of vertebrate animals (birds) [illegible] no naturally poisonous or venomous species. The skin, muscles, and feathers of *Pitohui dichrous* (*q.v.*), (the hooded pitohui), are now known, however, to be highly poisonous.

avian. pertaining to or denoting birds or the vertebrate class Aves.

Avicenna viper. *Cerastes vipera*.

Avicenna's sand viper. *Cerastes vipera*.

avicide. any substance lethal to birds, especially those intended for use in killing birds. Avicides include organophosphates such as monocrotophos. See 3-chloro-*p*-toluidine.

avidin. a protein component of the white of raw eggs of birds and amphibians and in the genital tract of many animals. It induces biotin deficiency in laboratory rats and chicks of the domestic fowl. If cooked, egg white does not lead to biotin deficiency.

AVM. avermectin.

avocado. **1:** *Persea americana*. **2:** the fruit of *Persea americana*.

AW. atomic weight. See relative atomic mass.

awnless brome grass. *Bromus inermis*. See *Claviceps*.

awnless grass. *Bromus inermis*.

axenic. free of microorganisms.

axon. the elongated, unbranched process of a neuron, the cytoplasm of which is continuous with that of the cell body or perikaryon. It conducts nerve impulses away from the cell body. Its volume is generally much greater than that of the cell body. See neuron.

axonal. of or pertaining to an axon or axons.

axonopathy. axonal functional disturbances, degeneration, or pathological changes in axons where the primary injury is to the axon itself. The term is not applied to indirect injury, as when changes in the nerve cell body (perikaryon) indirectly injure axonal structure or function. Since the normal activity and viability of an axon depend on continuous transport of substrate and macromolecules from the perikaryon, the death or gross impairment of the cell body may, of itself, cause degenerative changes or even death of the axon. Examples of neurotoxicants that directly injure the axon are tri-*o*-cresyl phosphate (TOCP) and the γ-diketones.

10-azaanthracene. acridine.

5-aza-10-arsenaanthracene chloride. phenarsazine chloride.

1-azabicyclo[2.2.2]octan-3-ol. 3-quinuclidinol.

azacitidine (4-amino-1-β-D-ribofuranosyl-1,3,5-triazine-2(1H)-one; 5-azacytidine; ladakamycin). a moderately toxic antibiotic glycosyl derivative of 5-azocytosine. It is carcinogenic to some laboratory animals.

5-azacytidine. azacitidine.

9-azafluorene. carbazole.

azaguanine. a mitotic poison, structurally similar to guanine. It is a competitive inhibitor of nucleic acid synthesis.

azalea. *Rhododendron*.

azapropazone. apazone.

azaserine (L-serine diazoacetate(ester), *O*-diazoacetyl-L-serine, azaserin, Cl 337, CN 1575-7, P 165). an antibiotic, antifungal, and antineoplastic agent produced by *Streptomyces* and synthetically. It is very toxic to laboratory mice and rats.

azathioprine (Imuran). a cytotoxic substance used for immunosuppression, especially in the case of kidney and liver transplants.

Azemiops feae (Fea's viper). the sole species in the genus (Family Viperidae), it is known only from the mountains of southeastern Asia. Individuals are small, less than 1 m in length and the danger they may pose to humans is unknown. The head is somewhat depressed and distinct from the neck; the snout is broad and short, and the canthus is obtuse. The body is cylindrical and moderately slender with a short tail. The eyes are moderate in size, with vertically elliptical pupils.

azid. azide.

azide (azid). **1:** any inorganic compound that contains the ion N_3^-. **2:** any organic compound that contains an azido group, N_3. They are cytotoxic and neurotoxic, attacking various parts of the central nervous system. Azides inhibit cytochrome oxidase, affecting the electron transport system and producing anoxia by an action similar to that of cyanide. They also inhibit catalase and peroxidase. They are respiratory irritants and may cause bronchitis and pulmonary edema. Symptoms of acute intoxication in humans may include dyspnea, headache, nausea, vomiting, hypotension, tachycardia, and leukocytosis. **halogen azide**. any azide that contains a halogen. All are extremely reactive and can explode spontaneously. Halogen azides emit irritant vapors. They can react with water to yield toxic fumes that contain the elemental halogen and an acid such as hydrochloric acid.

azido-. in chemistry, indicates a compound that contains the azido group, N_3. See azide.

azido group. —N_3, a derivative of hydrazoic acid. See azide.

azinphos ethyl. See azinphos methyl.

azinphos methyl (*o,o*-dimethyl-S-4-oxo-1,2,3-benzotriazin-3(4H)-yl methyl phosphorodithioate; "Guthion"). a toxic, brown, waxy solid, $_{12}N_3O_3PS_2$, which is soluble in water and in many organic solvents. It is a cholinesterase inhibitor that is readily absorbed by the skin of higher vertebrates; its toxicity and toxicological behavior are similar to that of azinphos ethyl. The maximum nontoxic oral dose for calves in one study was 0.44 mg/kg body weight; for cattle and goats, 2.2 mg/kg; and for sheep 4.8 mg/kg.

aziridine. ethylenimine.

azo compound. any of a broad class of organic compounds with the forumla, RN=NR′, where R and R′ are aromatic groups. These compounds are formed by coupling an aromatic phenol or aromatic amine with a diazonium compound. Most are colored because of the azo group, —N=N—. Included are the azo dyes, *q.v.* Most azo compounds are carcinogenic, most often producing tumors of the urinary bladder and liver in mammals. The azo group does not necessarily determine the carcinogenicity of a particular compound.

azo dye (aminoazo dye). any of a very large class of azo compounds derived from coal tar that contain -N=N- as a chromophore group. They are synthesized from amino compounds by linking an aryldiazonium group with a coupler such as an aromatic amine, phenol, or pyrazolone. Most azo dyes are sodium salts of sulfonic acids. They are often intense yellows, reds, violets, or blues and are used as dyes and food colorings. They often undergo metabolic reactions that produce toxic intermediates.

azo group. a nitrogen group, -N=N-, that occurs in a variety of organic compounds that are often brightly colored because of its presence. See also azo-.

azo reduction. the reduction of azo compounds. This process presumably involves cytochrome P-450; *in vitro*, it requires NADPH and proceeds only under anaerobic conditions. The process is inhibited by carbon monoxide. Azo reduction by mammals *in vivo* is somewhat poor and may be aided by intestinal microflora.

azo-. a prefix indicating the presence of an azo group, -N=N-, united to carbon atoms at each end. The carbons are usually part of univalent organic radicals, most often aryl radicals. Azo compounds are solids of varying color from yellow through red and violet to blue. See azo dyes.

azobenzene. a component of FD&C Yellow No. 5. It is carcinogenic to some laboratory animals.

Azodrin. an organophosphate cholinesterase inhibitor that is structurally and toxicologically very similar to Bidrin, *q.v.*

azoic. lacking animal life; without life.

azoic red 36. *p*-cresol.

azotic acid. nitric acid.

azuncena de Mejico. *Atropa belladonna*.

B

β. See beta.

B. See beta.

B. the symbol for boron.

B_{max}. the theoretical number of binding sites in a radioreceptor assay, usually inferred from a Scatchard, Eadie Hofstee, or Woolf plot.

B cell. B lymphocyte.

B lymphocyte (B cell). any of a type of lymphocyte that provides humoral immunity by the secretion of antibodies. Such cells are derived from the bursa of Fabricius of birds and the bone marrow of mammals. Mature B cells, usually in the presence of helper T cells, differentiate into plasma cells. This process results in a high serum concentration of specific antibody and the establishment of immunological memory. *Cf.* plasma cell, T lymphocyte.

b.p. boiling point.

B.P. before present.

Ba. the symbol for barium.

BA. naphth-2-yl-oxyacetic acid.

Baam. amitraz.

babies' slippers. *Lotus corniculatus*.

Babylonia japonica (ivory shell). a gastropod mollusk that contains surugatoxin.

bachelor's buttons. *Nux vomica*.

bacillus. a general term for any rod-shaped bacterium of the class Schizomycetes, and for bacteria of the genus *Bacillus*.

Bacillus. a genus of bacteria (Class Schizomycetes) that synthesizes numerous polypeptide antibiotics such as enniatin and gramicidins. ***B. subtilis***. a species used in prokaryote mutagenicity tests.

bacillus anthracis toxin. anthrax toxin. See toxin.

back mutation. the mutation of a mutant gene back to its original state.

back-fanged. rear-fanged.

back-fanged snake. *Dispholidus typus*.

background concentration. background level.

background level (background concentration). the normal environmental concentration of a substance. This term is often applied to ionizing radiation and to air pollutants that have both natural and anthropogenic sources.

background radiation. **1:** the sum of all natural radioactive (ionizing) radiation at a given location, including elementary particles from outer space (cosmic radiation) and terrestrial radiation from naturally occurring isotopes. **2:** the level of radioactive radiation (natural and anthropogenic) that would occur at a given location if the level were not influenced by a specified source.

bacteremia. the presence of bacteria in circulating blood. **Gram-negative bacteremia**. the presence of Gram-negative bacteria in circulating blood. **Gram-positive bacteremia**. the presence of Gram-positive bacteria in circulating blood.

bacteria (plural of bacterium). common name of a large, extremely diverse assemblage of unicellular prokaryotic microorganisms, usually with cell walls. Reproduction is by binary fission or asexual spores. The transfer of genetic material may be effected by conjugation (a sexual process), or mediated by virus or bacteriophage (transduction). Included are the actinomycetes, Cyanobacteria (blue-green bacteria), eubacteria, iron bacteria, mycoplasmas, rickettsiae, spirochaetes, and sulfur bacteria. In terms of numbers and diversity of habitats, bacteria must be considered one of the most successful of life forms. Many are useful to humans (e.g., in food processing, waste treatment, genetic engineering). Many are, however, parasitic or otherwise harmful, and some secrete powerful toxins (e.g., botulinum toxin). *Cf.* microbe. See also exotoxin, endotoxin. **blue-green bacteria**. cyanobacteria. **denitrifying bacteria**. See denitrification.

bacterial mutagenesis. See mutagenesis.

bacterial pyrogen. endotoxin (See toxin).

bacterial toxin. See toxin.

bacterial vitamin H[1]. *p*-aminobenzoic acid.

bactericidal (bacteriodical). **1:** pertaining to a bactericide. **2:** lethal to bacteria.

bactericide (bacteriocide). **1:** an agent that kills bacteria. Bactericides in common use are hypochlorite, various phenolics, mercuric chloride, and organic mercurials. **2:** bactericidal. *Cf.* disinfectant, bacteriostat. **specific bactericide**. a bacteriolytic immune serum destructive to a specific species or genus of bacteria.

bactericidin. any substance in blood plasma that kills bacteria.

bacteriocidal. bactericidal.

bacteriocidal agent. bactericide.

bacteriocide. bactericide.

bacteriocin. any of a group of antibacterial proteins produced by various bacteria that selectively kill or inhibit, but do not lyse, other taxa of bacteria. They act by complexing with specific receptors on the cell wall and inducing a specific metabolic block. The different [illegible] group of bacteria that produce them. Thus those produced by *Escherichia coli* are termed colicins, those by *Staphylococcus* are staphylococcins, etc. Bacteriocins are coded by genes on particular plasmids and not by those that are part of the main genome. Certain bacteriocins tend to form large aggregates.

bacteriocinogenic. producing bacteriocins.

bacteriolysant. **1a:** having the capacity to lyse bacteria. **1b:** an agent that lyses bacteria. See bacteriolysis. *Cf.* bacteriolysin.

bacteriolysin. an antibacterial antibody that lyses bacteria. See bacteriolysis. *Cf.* bacteriolysant.

bacteriolysis. destruction of a bacterial cell by a specific bacteriolysin, with the release of the cell contents.

bacteriolytic. pertaining to, denoting, promoting, or having the capacity to destroy bacteria by lytic action.

bacteriolyze. to effect bacteriolysis.

bacteriostat (bacteriostatic agent). any agent (e.g., low temperature, dyes, food preservatives, some antibiotics, weak antiseptics) that reversibly inhibits the multiplication of bacteria.

bacteriostatic. **1:** reversibly inhibiting the multiplication of bacteria. **2:** pertaining to agents that reversibly inhibit the multiplication of bacteria. *Cf.* bactericidal.

bacteriostatic agent. bacteriostat.

bacteriotoxic. **1:** toxic to bacteria. **2:** caused by or pertaining to bacterial toxins.

bacteriotoxin. **1:** any toxin produced by bacteria. **2:** any toxin that kills or inhibits the growth of bacteria.

bacterium. singular form of bacteria, *q.v.*

badge. See film badge.

BAF. bioaccumulation factor. See factor.

bagasscosis. bagassosis.

bagasse. See bagassosis.

bagassosis (bagasse; bagasscosis). a respiratory disorder resulting from inhalation of bagasse, the residue of sugarcane that remains following the extraction of sugar. The proximate cause of the disease is the inhalation of the spores of a fungus that infects sugarcane. An asthma-like attack usually ensues several hours after contact in an individual who has worked with cane for several months. With repeated attacks, the lungs become scarred with development of a pneumoconiosis.

bagpod. *Sesbania*.

Bagre marina (gafftopsail catfish). a species of venomous sea catfish (Family Ariidae) that commonly occurs in shallow waters and embayments along the Atlantic Coast of the Americas from Cape Cod to Rio de Janeiro. Individuals have long ribbon-like dorsal and pectoral fins. The male incubates the eggs in his mouth. The pectoral and dorsal spines can cause painful wounds. See catfish.

bagre sapo. *Thalassophryne reticulata*.

Baileya multiradiata (desert baileya; cloth of gold). a small poisonous showy annual or perennial plant. It is native to the southwestern desert region of the United States and is locally common on dry sandy and graveled soils from Mexico north into Texas and California (Family Compositae). It is lethal to sheep when relatively large amounts are eaten. Signs and symptoms include sluggishness, anorexia, depression, and the drooling of green saliva. Red urine and a tendency to develop pneumonia have also been reported. If unmanaged, death ensues within several days or weeks following the appearance of symptoms.

bait. an attractant used to lure prey. The bait may be used to attract specific types of prey into a trap or snare (e.g., a spider web), or in some cases may be poisoned (ant and roach baits), thus serving to attract and kill or immobilize the prey. **snail bait**. until recently, baits used to attract and kill snails contained inert ingredients that were attractive to most dogs. These have been reformulated in the United States with a dramatic decrease in the number of metaldehyde poisonings of dogs. Some bait poisonings are, however, misdiagnosed as strychnine poisoning due to the similarity of signs. See metaldehyde, metaldehyde poisoning.

baking soda. sodium bicarbonate.

BAL. dimercaprol.

balaiwona. *Atractaspis microlepidota*.

balanticidal. **1:** destructive to protozoa of the genus *Balantidium*, many species of which occur in the intestinal tracts of vertebrates and many invertebrates. *B. coli* sometimes causes diarrhea and dysentery, with ulceration of the colonic mucosa of primates, a condition known as balantiasis. **2:** an agent that destroys protozoa of the genus *Balantidium*. See carbarsone.

BALB/3T3. a type of cultured mouse embryo fibroblast used to test the capacity of a chemical *in vitro* to induce the transformation of mammalian cells. See cell transformation, mammalian cell transformation tests.

baldness. alopecia.

balewol. *Atractaspis microlepidota*.

Balistidae (leatherjackets). a family of bony fish in which the pelvic fins are absent or reduced to small spine-like projections of the pelvic bone. The triggerfishes are included in this family; some are palatable, and others are toxic. See *Balistoides conspicillum*, ciguatera.

Balistoides conspicillum (triggerfish). a species of marine bony fish (Family Balistidae) of the tropical Pacific Ocean, from Polynesia to Madagascar, China, and Japan. See ciguatera.

Balkan cross adder. *Vipera berus*.

ball nettle. *Solanum carolinense*.

balloonfish. one of numerous common names of fishes of the family Tetraodontidae.

ball-point ink. any ink used in a ball-point pen. The amount in one pen is usually not toxic if ingested (as by a child).

balor. *Naja naja*.

balsam apple. See *Momordica charantia*.

balsam pear. *Momordica charantia*.

bamboo curare. See curare.

bamboo pit viper. *Trimeresurus gramineus*.

bamboo snake. See *Trimeresurus*.

bamboo viper. *Trimeresurus albolabris*, *T. graminnneus*, or *T. stejnegeri*.

banana oil. amyl acetate.

banana spider. See *Cupiennius*.

band application. the spreading of pesticides or other chemicals over, or next to, each row of plants in a field.

banded cobra. See *Naja haje*.

banded coral snake. *Calliophis macclellandii*.

banded krait. *Bungrarus fasciatus*.

banded Malaysian coral snake. *Maticora intestinalis*.

banded sea snake. *Hydrophis caerulescens*, *Laticauda colubrina*.

banded small-headed sea snake. *Hydrophis fasciatus*.

banded water cobra. *Boulengerina annulata*.

banded yellow-lipped sea snake. *Laticauda colubrina*.

bandy-bandy. *Rhynchoelaps bertholdi*; *Vermicella spp.*

bane. **1a:** (obs.) to kill, especially by poisoning. **1b:** (archaic) to afflict, injure or harm. **1c:** death, destruction. **2a:** a poisonous plant. **2b:** any poison of plant origin, but used mainly in reference to specific plant poisons such as those produced by *Aconitum* and *Arnica*. **3:** any poison; not a recommended usage in toxicology.

baneberry. **1:** the poisonous fruit of any plant of the genus *Actaea*. **2:** a common name for plants of the genus *Actaea*.

banewort. *Atropa belladonna*.

banisterine. harmine.

BaP, B[*a*]P. benzo[*a*]pyrene.

BAP. 6-benzylaminopurine.

Baptisia (wild indigo; false indigo). a genus of leguminous herbs (Family Fabaceae, formerly Leguminosae. ***B. leucantha*** (wild indigo). a species that has poisoned livestock. It contains quinolizidine alkaloids. ***B. tinctoria*** (wild indigo, rattle-weed, horsefly weed). a hardy perennial herb with bright yellow flowers. The leaves and shoots are toxic; they contain cytisine and the alkaloid, quinolizidine. It causes severe diarrhea and anorexia in cattle. The dried root has been used as an emetic and cathartic.

baptitoxin, baptitoxine. cytisine.

barba amarillo. the Spanish name ("yellow chin") of *Bothrops atrox*.

Barbados lily. *Atropa belladonna*.

barbadosnut. *Jatropha curcas*.

barbamil. amobarbital.

barban (4-chlorobut-2-ynyl-*N*-(3-chlorophenyl)-carbamate). a highly selective carbamate herbicide used to control *Avena fatua* (wild oats). Applications up to 0.75 kg/ha to crops at the 2-3 leaf stage kill this grass without significant damage to cereal grains and broadleaved crops such as *Vicia faba* (broad beans) and *Beta vulgaris* (sugarbeet). It causes the plant chromosomes to contract. It is a powerful inhibitor of gibberellin-induced α-amylase synthesis in barley endosperm, and is thus thought to inhibit gene derepression (i.e., the onset of metabolic changes) but not processes that are already in progress. Barban is a skin irritant and is moderately toxic to laboratory rats.

barbasco. a vernacular term applied by Spanish-speaking people in the Americas to a variety of plants used to paralyze fish. In Mexico, this term usually refers to the roots of *Dioscorea composita* or *D. tepinapensis* (Family Dioscoreaceae), which contain up to 5% diosgenin (*q.v.*) on a dry weight basis. It is also the common name of *Jacquinia paramensis* and *J. pinnata*. See also *Lonchocarpus*.

barbasco poison. a substance prepared from barbasco that is sometimes used by the natives of eastern Peru to paralyze piranhas before catching them for food. See also *Jacquinia*.

barberry family. Berberidaceae.

barbital (5,5-diethylbarbituric acid; diethylmalonylurea; barbitone; Veronal). an odorless, bitter white powder or crystalline substance, $C_8H_{12}N_2O_3$. It is soluble in hot water, ethanol, ether, acetone, and ethyl acetate. Barbital is a long-acting barbiturate used as an orally administered sedative and hypnotic; it is also used as a stabilizer of hydrogen peroxide. As with other barbiturates, barbital is derived from barbituric acid. See barbituism.

barbital sodium (barbital soluble). the soluble monosodium salt of barbital, $C_8H_{11}N_2NaO_3$, with the same action and applications as the base. See barbituric acid, barbituism.

barbital soluble. barbital sodium.

barbitalism. barbituism.

barbitone. barbital (British).

barbituism (barbitalism; barbiturism). chronic poisoning by barbital or its derivatives. The signs and symptoms are relatively nonspecific and include headache, chills, fever, and cutaneous eruptions. See also barbiturate.

barbiturate. **1:** any of a class of organic compounds derived from barbituric acid (e.g., amobarbital, barbital, nembutal, pentobarbital, phenobarbital, secobarbital, secanol). All are CNS depressants used therapeutically as tranquilizers, hypnotics, anticonvulsants, and as an aid (given *i.v.*) to the induction of general anesthesia. Their activity is due mainly to the presence of substituted alkyl and acyl groups at position five. Barbiturates have a low therapeutic index and care is required in their use; overdosage is hazardous. They are mildly addictive, and abrupt withdrawal can lower the seizure threshold with fatal consequences. Barbiturates depress breathing rate, heart rate, blood pressure, and body temperature. Symptoms of intoxication may also include drowsiness, headache, confusion, ataxia, vertigo, slurred speech, impaired reflexes, stupor, cyanosis, coma; pulmonary edema is seen with prolonged coma. Death may result from cardiovascular collapse or respiratory arrest. **2:** denoting or having the properties of a barbiturate. See oxybarbiturate.

barbiturate poisoning. See barbituism.

barbituric acid (2,4,6-(1*H*,3*H*,5*H*)-pyrimidinetrione; 2,4,6-trioxohexahydropyrimidine; pyrimidinetrione; malonylurea). a colorless, slightly water-soluble, crystalline dibasic acid, with no hypnotic or sedative activity, but many of its derivatives (barbiturates) do possess such activity.

barbiturism. barbituism.

Barchatus. a genus of toadfish (Family Batrachoididae, *q.v.*). ***B. cirrhosus***. a toadfish (Family Batrachoididae) that inhabits the Red Sea.

bardick. *Brachyaspis curta*.

bargil. *Walterinnesia aegyptia*.

barilla. any plant of the genus *Halogeton*, especially *H. glomaeratus*.

baritosis. a type of pneumoconiosis caused by chronic or repeated inhalation of barite or barium dust.

barium (Ba)[Z = 56; A_r = 137.36]. a toxic, reactive divalent alkaline earth element; it is a silver-white powder or sometimes malleable metal. The powder is flammable at room temperature and should be stored under an inert gas. It is extremely reactive with ammonia, water, halogens, and most acids. Even partially soluble salts are highly toxic (See barium carbonate). The myotoxic Ba^{2+} ion is readily absorbed from the gastrointestinal tract. The rate of absorption can be reduced by the administration of soluble sulfate immediately after ingestion of a toxic barium salt. The barium ion causes muscular contraction by blocking the K+ channels of the sodium-potassium pump in the cell membranes and is thus violently cathartic. See barium poisoning.

barium binoxide. barium peroxide.

barium carbonate. a white salt, $BaCO_3$, that occurs in nature as the mineral witherite. Although water insoluble, it is poisonous (See barium) and used as a rodenticide.

barium chloride. a white powder or flat, colorless prisms, $BaCl_2{\cdot}2H_2O$, used for example, as a rodenticide. It is extremely toxic. Ingestion of less than one gram by an adult human can be fatal. It was formerly used as a heart tonic and to control varicose veins.

barium chromate (VI). C.I.Pigment Yellow 31.

barium dioxide. barium peroxide.

barium hydrate. **1:** barium hydroxide, octahydrate. **2:** any of the hydrates of barium hydroxide. See barium hydroxide.

barium hydroxide (barium hydrate; caustic baryta). a white or colorless, poisonous powder or crystalline substance, $Ba(OH)_2$, used for example, in the saponification of fats and the fusing of silicates. Barium hydroxide forms a strong caustic base in aqueous solution. It has many uses, e.g., as a test for sulfides; in pesticides; in the manufacture of alkali and glass. **barium hydroxide, anhydrous.** the commercially available preparation usually contains 92-95% of the monohydrate and is poisonous. **barium hydroxide, monohydrate** (barium monohydrate). a white, poisonous, slightly water-soluble, powder, $Ba(OH)_2{\cdot}H_2O$; it is soluble in dilute acids. **barium hydroxide, octahydrate** (barium hydrate; barium octahydrate; caustic baryta). a poisonous white powder, $Ba(OH)_2{\cdot}8H_2O$, that is soluble in water, ethanol, and ether; it absorbs CO_2 from the atmosphere. It is used chiefly in organic preparations and analytical chemistry. **barium hydroxide, pentahydrate** (barium pentahydrate). toxic white flakes, $Ba(OH)_2{\cdot}5H_2O$; uses are the same as those of the octahydrate.

barium monohydrate. barium hydroxide, monohydrate.

barium monosulfide. barium sulfide.

barium monoxide. barium oxide.

barium nitrate (nitrobarite). a lustrous white, crystalline, water-soluble salt, $Ba(NO_3)_2$. It is toxic and a strong oxidizing agent used in explosives, pyrotechnics, and as a rodenticide. See barium.

barium octahydrate. barium hydroxide, octahydrate.

barium oxide (barium monoxide; barium protoxide; calcined baryta). a toxic, caustic, white or yellowish-white powder, BaO, that is soluble in water and acids. It reacts violently with water, yielding barium hydroxide.

barium pentahydrate. barium hydroxide, pentahydrate.

barium permanganate. a toxic, brownish-violet, crystalline compound, $Ba(MnO_4)_2$. dissolved in water, it is used as a strong disinfectant and a fire and explosion hazard.

barium peroxide (barium binoxide; barium dioxide; barium superoxide). a white, or grayish-white, flammable, water-insoluble powder, BaO_2, that is used chiefly as a bleach in the glass manufacturing industry. It is an oxidizing substance and is a fire and explosion hazard, if allowed to come into contact with organic material. It is a skin irritant, and is toxic by ingestion.

barium poisoning. poisoning by barium and its salts. Even slightly water-soluble salts of barium are highly toxic, causing severe hypokalemia which is reversed by potassium infusion. Common effects include salivation, vertigo, nausea, vomiting, abdominal distress, violent and bloody diarrhea, bradycardia, cardiac arrhythmias, tinnitus, convulsions, and sometimes paralysis. Lesions typically include violent peristalsis, atrial hypertension, late kidney damage, and muscle contraction. Death may ensue from respiratory failure and cardiovascular collapse. See also barium.

barium protoxide. barium oxide.

barium salt. See barium.

barium sulfate (barytes; blanc fixe; basofor; C.I. Pigment White 21). a white or yellowish, odorless, tasteless, nonflammable powder, $BaSO_4$, that is soluble in concentrated sulfuric acid. It is usually not toxic if ingested, but should not be inhaled. It is given as a suspension orally or rectally for radiographic examination of the gastrointestinal tract. See barium sulfate method, barium sulfide.

barium sulfate method. a turbidimetric method used to determine the presence and concentration of sulfur dioxide in air. Air is passed through a solution of hydrogen peroxide, with which sulfur dioxide reacts to form sulfuric acid. Barium chloride is added, yielding an insoluble sulfate; the turbidity of the suspension is determined with a spectrophotometer.

barium sulfide (barium monosulfide; black ash). a poisonous, flammable, grayish-yellowish-green powder or lumps, BaS. It is used as a depilatory to remove hair from hides, as a flame retardant, to generate pure hydrogen sulfide, and in luminous paints.

barium sulfite. a toxic, white powder, BaO_3S, that is soluble in dilute hydrochloric acid, and decomposes on heating. It is used in the manufacture of paper.

barium superoxide. barium peroxide.

barium tungstate (barium white; barium wolframate; tungstate white; wolfram white). a poisonous white pigment used in roentgenography.

barium white. barium tungstate.

barium wolframate. barium tungstate.

bark. all tissues, collectively, that lie outside the vascular cambium of the stems and roots of plants. The term is sometimes applied only to the tissues that arise from and lie external to the phellogen when such tissue is exposed by sloughing off of the epidermis.

bark scorpion. any scorpion of the genus *Centruroides*.

barley. *Hordeum vulgare*.

barn rat. *Rattus norvegicus*. See also laboratory rat.

Barnard snake. *Glyphodon barnardi*.

barnyard grass. *Echinochloa crusgalli*

baroceptor. baroreceptor.

baroreceptor (baroceptor). any receptor in the walls of the vertebrate cardiovascular system that is sensitive to the pressure produced by distension of the vessel wall. Such distension is due to an increase in blood pressure. Stimulation of baroreceptors causes a reflex vasodilatation, a reduction in heart rate, decreased cardiac output, and depressed breathing. The most important are those in the carotid sinus, left atrium, and the systemic arch.

barracuda. common name of fish of the Family Sphyraenidae, especially *Sphyraena barracuda*.

barrier. **1:** an obstruction or impediment (e.g., stratum corneum, blood-brain barrier). **2:** an obstacle (biological or nonbiological), within a living system, to the passage of substances. **blood-brain barrier** (BBB; blood-cerebral barrier; blood-cerebrospinal fluid barrier, hematoencephalic barrier). a barrier that separates systemic blood from the parenchyma of the central nervous system (CNS). It consists of endothelial cells of the capillaries and surrounding glial membranes (glial endfeet) to which the capillaries are closely joined. It is impermeable to most ions and protein-bound chemicals, but allows passage of lipid-soluble molecules. The BBB thus effectively protects tissues of the brain and spinal cord from the intrusion of most toxicants. The CNS of the embryo, fetus, and the very young mammal is considerably less well protected from the influx of toxicants than that of the adult. The passage of some toxicants through the blood-brain barrier is facilitated by a specific transport system; in other cases, the toxicant may damage or otherwise reduce the effectiveness of the barrier. **blood-cerebral barrier**. blood-brain barrier. **blood-cerebrospinal fluid barrier**. blood-brain barrier. **blood-testis barrier**. a permeability barrier to the passage of blood-borne materials into the seminiferous tubules of the testis. **filtration barrier**. a barrier that separates the blood in the glomerular capillaries of the kidney from that of the capsular space of the renal corpuscle. **gastric mucosal barrier**. a physiologic barrier that limits the diffusion of nonionic materials through the gastric mucosa. Certain agents (e.g., aspirin, bile salts, organic acids) impair the effectiveness of this barrier. **hematoencephalic barrier**. blood-brain barrier. **placental barrier** (placental membrane). the mammalian placenta physically separates fetal from maternal blood and limits the passage of large blood-borne molecules between the two. In at least some species, this barrier impedes the passage of many toxicants to some extent. Its effectiveness, however, probably varies considerably among placental mammals as a function of the number of cell layers (one to six). **primary radiation barrier**. a barrier that blocks a primary beam of radiation (e.g., an X-ray beam), reducing it to a permissible exposure rate. **protective barrier**. radiation barrier. **radiation barrier**. any barrier or device of sufficient density and thickness to protect an organism from a particular type of radiation. **secondary radiation barrier**. a barrier such as a wall or lead apron, that is capable of blocking scattered radiation such that it is reduced to a permissible exposure rate.

barrier substance. any substance applied to the skin as a barrier to irritants (e.g., a water- or oil-resistant cream or lotion).

barthrin (2,2-dimethyl-3-(2-methyl-1-propenyl)-cyclopropanecarboxylic acid (6-chloro-1,3-benzodioxol-5-yl)methyl ester; 2,2-dimethyl-3-(2-methylpropanyl)cyclopropanecarboxylic acid 6-chloropiperonyl ester; 6-chloropiperonyl-2,2-dimethyl-3(2-methylpropenyl)-cyclopropanecarboxylate; 6-chloropiperonyl chrysanthemumate; chrysanthemummonocarboxylic acid 6-chloropiperonyl ester). an insecticide with low toxicity to mammals.

bartezgooremil. *Acanthophis antarcticus*.

Bartlett's test. See test.

baryta. barium oxide.

baryta yellow. C.I. Pigment Yellow 31.

barytes. naturally occurring barium sulfate.

basal application. the application of a pesticide or other chemical just above the soil line on plant stems or tree trunks.

basal ganglion. See ganglion.

basal metabolic rate (BMR). the resting rate of metabolism of an animal, as the rate of oxygen consumption or of heat production per unit time per square meter of surface area. BMR is a measure of the energy required to maintain the basic life processes of the organism. It is influenced by age, hormonal state, and ambient temperature.

basal metabolism. the normal metabolic rate of an organism at rest.

base. **1a:** basis, core, essence. **1b:** foundation, support, footing; the base or lower part; that part of an object which supports it; the part of a structure situated opposite the apex. **1c:** the part of an organ, extremity, or other body part that lies nearest to its point of origin. **2a:** any substance with a pH greater than 7. **2b:** any substance that releases hydroxyl ions, OH^-, or other ions that combine with hydrogen ions in aqueous solution. Such solutions have a pH greater than 7. **2c:** any electropositive substance (a cation) that tends to unite with electronegative substances (an anion) to form a salt. **3:** in biochemistry, any nitrogenous organic compound such as an amine, an alkaloid, a ptomaine, or any of the purine and pyrimidine constituents of nucleotides that act as a Lowry-Brønsted base. **4:** in pharmacy, the main ingredient of a mixture. **Brønsted base**. Lowry-Brønsted base. **Lewis base**. any substance that tends to donate an electron pair. Examples are ammonia, the hydroxyl ion, trimethylamine, and pyridine. **Lowry-Brønsted base** (Brønsted base). any substance that tends to accept a proton. This definition extends that of 2a to include, for example, water (which can accept a proton to form H_3O^+) and the negative ions of mineral acids (e.g., NO_3^-, SO_4^{2-}) which can be considered as weak conjugate bases of their respective acids.

base analog. See abnormal base analog.

base pair. a pair of bases (a purine and a pyrimidine) cross-linked by hydrogen bonds; adenine always pairs with thymine, and guanine with cytosine. Such pairs hold the two helical chains of the Watson-Crick model together, conferring the helical configuration to the molecule. Since each chain in the DNA molecule serves as a template for the other, any change in a base pair produces a genetic mutation.

base-pair substitution. See mutation (point mutation).

base-pair transformation. See mutation (point mutation).

base-pair transition. See mutation (point mutation).

base-pair transversion. See mutation (point mutation).

base unit. any of the fundamental units of the International System of Units, *q.v.*

baseline monitoring. See monitoring.

basement membrane. a thin layer at the base of epithelial cells in many types of epithelia. It is made up of mucopolysaccharides and a meshwork of fibrous material. The basement membrane supports the epithelium, separates it from underlying tissues, and acts as a barrier to the exchange of molecules between the surface epithelium and the underlying tissue.

basic aluminum chloride. aluminum chlorohydrate.

basic lead carbonate (lead carbonate hydroxide; lead subcarbonate; white lead; flake lead; ceruse; cerussa). a heavy white, poisonous acid-soluble white powder or crystalline substance (rhombic crystals), $PbCO_3$, that is insoluble in water and alcohol. It occurs naturally as the mineral cerussite. It is used chiefly as a pigment in paints.

basic lead chromate. lead chromate, basic.

basic lead hydroxide. lead hydroxide.

basic mercuric sulfate. mercuric subsulfate.

basic violet 10. FD&C Red No. 19.

basic zinc chromate. zinc chromate.

basidiocarp. a fruiting body of the higher basidiomycete fungi. They are the most prominent features of these fungi and include mushrooms, toadstools, and related structures. Some are edible, others are poisonous, some violently so.

basofor. barium sulfate.

basophil. **1:** staining strongly with basic dye. **2:** certain cells in the circulating blood that contain water-soluble granules that stain strongly with basic dyes. **3:** a type of granulocyte that releases histamine and serotonin from the granules in certain immune reactions. The release of the granule contents and the synthesis of secondary biological mediators (e.g., leukotrienes C4, D4, E4 which are responsible for major changes in the lungs) are the proximate causes of an anaphylactic reaction. Such a reaction may be local (e.g., when inhalation of allergen produces an asthmatic response) or systemic (e.g., in response to envenomation or from reaction to an injected drug). Degranulation is triggered by the cross-linking of surface-bound IgE. Antigen triggering (a) releases biological mediators stored in the granules and (b) effects production of secondary biological mediators, mostly via the breakdown of arachidonic acid. **4:** a type of secretory cell of the anterior pituitary.

basophilopenia. an absence of or pronounced decrease in basophils in circulating blood.

bastard acacia. *Robinia pseudoacacia*.

bastard toadflax. *Comandra pallida*.

bat ray. a stingray, *q.v.*, of the family Myliobatidae.

batoid. any of the skates and rays; pertaining to skates and rays.

Batoidea (Raji; Rajida). a suborder (Batoidei in some classifications) of cartilaginous fish of the order Selachii. Included are the sawfishes, skates, and rays. The body is depressed (flattened), the eyes are dorsal and lack an upper eyelid, and there are 5 ventral gill slits. The pectoral fins are greatly enlarged (up to 6 m across), fused to the head, and are the principal organs of locomotion. The tail is small, pointed, and serves as a ruddor. Many species bear venomous spines; a few are poisonous, and the electric rays can produce a powerful electric charge which they use to stun prey; some may be dangerous to humans.

Batoidei. See Batoidea.

Batrachia. **1:** a synonym used in some classifications for the amphibian subclass Anura (frogs and toads). See also Salientia. **2:** occasionally encountered as a synonym for the vertebrate class Amphibia.

batrachian. **1a:** a frog or toad. **1b:** pertaining to or having the characteristics of a frog or toad.

batrachoid. **1a:** any fish of the family Batrachoididae. **1b:** of, pertaining to, or resembling fish of the family Batrachoididae.

Batrachoides. a genus of toadfish (Family Batrachoididae, *q.v.*). ***B. grunniens***. a toadfish of the coastal waters of Ceylon, India, Burma, and Malaya. ***B. didactylus***. a species that inhabits the Mediterranean Sea and nearby waters of the Atlantic Coast. ***B. grunniens***. a species that occurs in the coastal waters of Ceylon, India, Burma, and Malaya.

Batrachoididae (toadfishes). a family of small, slow-moving, venomous bony fishes with broad, depressed heads. They are shoaling bottom fishes of the warmer coastal waters of the Americas, Europe, Africa, and India. It is the only family of the order Batrachoidiformes (synonym, Haplodoci), *q.v.* They are voracious, preying on mollusks. Most are marine, but a few occur in estuarine or fresh waters. The venom apparatus consists of two slender, hollow, needle-sharp spines of the dorsal fin, two hollow bones of the operculum that terminate in a sharp protruding tip, and their associated venom glands. They also have powerful jaws and should be handled cautiously. The sting is extremely painful. The pain radiates from the wound and is quickly accompanied by swelling, redness and heat. Fatalities are unknown. Genera include *Barchatus*, *Batrachoides*, *Opsanus*, *Porichthys*, and *Thalassophryne*.

Batrachoidiformes (Haplodoci). an order of venomous bony fishes that contains only the family Batrachoididae, *q.v.* The head is large and depressed, bearing a very wide mouth; the body is slimy; scales are absent; the pectoral fins are jugular in position. See Batrachoididae.

batrachotoxin (batrachotoxinin A 20-(2,4-dimethyl-1*H*-pyrrole-3-carboxylate); 3α,9α-epoxy-14β,18β-(epoxyethano-*N*-methylimino)-5β-pregna-7,16-diene-3β,11α,20α-triol, 20α-ester with 2,4-dimethylpyrrole-3-carboxylic acid; BTX). **1:** the active principle (comprised of four main toxic alkaloids) of poison-dart (poison-arrow) frogs of the genus *Phyllobates*. These are batrachotoxin (def. 2), batrachotoxinin A, isobatrachotoxin, and pseudobatrachotoxin. **2:** the most toxic of the four major steroidal alkaloids isolated from poison-dart (poison-arrow) frogs of the genus *Phyllobates*. It is an extremely toxic steroidal alkaloid with neurotoxic and cardiotoxic activity. The lethal dose in laboratory mice is ca. 100 ng (*s.c.*) and in humans less than 200 μg. The toxin does not affect intact skin, but if the skin is broken or scratched, it causes pain of similar intensity to that of a bee sting. Oral toxicity is quite low if the mucosa is intact. It acts by producing a selective and irreversible increase in permeability of membranes to Na^+. Batrachotoxin blocks neuromuscular transmission. Effects include contracture of skeletal muscles, cardiac arrhythmias, tachycardia, and fibrillation. Death usually results from respiratory paralysis. Saxitoxin and tetrodotoxin are antagonistic to the peripheral nerve activity of batrachotoxin, as are certain anesthetics.

batrachotoxinin A (1,2,3,4,7*a*,10,11,11*a*,12,13-decahydro-14-(1-hydroxyethyl)-2,11*a*-dimethyl-7*H*-9,11b-epoxy-13a,5a-propenophenanthro[2,1-f][1,4]oxazepine-9,12(8*H*)-diol; 3α,9α-epoxy-14β,18β-(epoxyethano-*N*-methylimino)-5β-pregna-7,16-diene-3β,11α,20α-triol; BTX-A). the least toxic of the four major steroidal alkaloids isolated from skin extracts of poison-dart (poison-arrow) frogs of the genus *Phyllobates*. It is, nevertheless, an exceptionally toxic alkaloid. The LD_{50} in laboratory mice is 1 mg/kg. Do not confuse with batrachotoxinin A 20-(2,4-dimethyl-1*H*-pyrrole-3-carboxylate). See batrachotoxin.

batrachotoxinin A 20-(2,4-dimethyl-1*H*-pyrrole-3-carboxylate). batrachotoxin. Do not confuse with batrachotoxinin A, *q.v.*

batracin. an arrow poison prepared from the skin [illegible] American tree toad, *Phyllobates chocoensis*.

batroxobin (*Bothrops atrox* serine proteinase; *Bothrops* venom proteinase; reptilase R). a thrombin-like enzyme from the viper, *Bothrops atrox*. The concentrated enzyme is hemostatic at low concentrations and anticoagulant at higher levels. It is used as a hemostatic agent in certain peripheral arterial circulatory conditions.

battery acid. sulfuric acid.

battery manganese. manganese dioxide.

bauxite pneumoconiosis. See pneumoconiosis.

bay mussel. *Mytilus edulis*.

BBB. blood-brain barrier.

BBC. α-bromobenzyl cyanide.

BCF. bioconcentration factor. See factor.

BCGF. B-cell replacing factor. See factor.

BCME. *sym*-dichloromethyl ether.

BCNU. carmustine.

Be. the symbol for beryllium.

be-still tree. *Thevetia peruviana*.

bead snake. *Micrurus fulvius*.

bead vine. *Abrus precatorius*.

beaded lizard (escorpion). See *Helodermatidae, Heloderma horridum*.

beak. a protruding, usually sharp, pointed mouth-part or jaw that lies distal to other mouthparts.

beaked sea snake. *Enhydrina schistosa*.

bean family. Fabaceae.

bean-hull poisoning. intoxication of livestock by *Fusarium* infestations of bean hulls used as feed. It is characterized chiefly by signs of nervous dysfunction. Symptoms include staggering, various psychic disturbances, impaired reflexes, decreased heart rate, meningism, signs of encephalitis, dermographism, hyperesthesia, convulsion, and respiratory failure. Symptoms are due mainly to trichothecene mycotoxins.

beans. See Fabaceae.

bear corn. *Veratrum viride*.

bearberry. *Arctostaphylos*.

beard tongue. *Penstemon*.

bearded darnel. *Lolium temulentum*.

bearded seal. See *Erignathus barbatus*.

beargrass. *Nolina texana*.

bear's foot. *Helleborus foetidus*.

bearwood. *Rhamnus purshiana*.

beautiful coral snake. *Calliophis calligaster*.

beaver poison. *Cicuta maculata*.

Becquerel (Bq). the SI unit of radioactivity defined as the quantity of a given radioisotope in which 1 atom is transformed in 1 second (one disintegration per second). 1 Bq = 27 picocuries. This unit is thus a measure of the strength of a particular source, but does not take into account the differing effects of different types of radiation on living matter. The sievert is generally more useful in toxicology, the health sciences, and the biological sciences.

bee. any of numerous species of stout-bodied hymenopterous insects, chiefly of the families Apidae, Bombidae, and Xylocopidae. Most are venomous, flying, social (colonial) insects with hairy bodies. They collect pollen, and have both biting and sucking mouth parts. Some species are solitary. Bee stings usually cause temporary local pain and edema, with no systemic complications. Bee stings sometimes induce anaphylaxis in sensitive individuals. See *Apis*, *Bombus*, Hymenoptera, apamin, kinin peptide, mellitin, venom (bee venom).

bee sting. See bee.

beech. any tree of the genus *Fagus*, especially *F. sylvatica*.

beech family. Fagaceae.

beechnut. **1:** *Fagus sylvatica*. **2:** the seed of *F. sylvatica*.

beefsteak morel. *Gyromitra (Helvella) esculenta.*

beer heart. See cardiomyopathy.

beer-drinker's cardiomyopathy. See cardiomyopathy (beer-drinker's cardiomyopathy).

beet. *Beta vulgaris.*

beetroot. *Beta vulgaris.*

beggar tick. *Bidens frondosa.*

behavior. **1:** any overt action or activity of a defined individual organism, population, species, or type of organism, whether apparently spontaneous or in response to internal or external stimuli. **2:** the battery of activities that comprises the responses of a living system to the entire range of factors that constitute its environment. **3a:** a complex set of activities that is associated with a particular stimulus, type of stimulus, or constellation of stimuli (internal or external). **3b:** any activity or set of activities associated with or stimulated by a type of life process or adaptation. Thus, reproductive behavior includes those (usually adaptive) behaviors that attend any or all phases of reproduction. Other examples are aversive behavior, migratory behavior, foraging behavior. In toxicology, behavioral responses are, together with certain physiological and morphological responses, the directly observable signs of poisoning. Behavioral changes in an individual or population are often among the early or warning signs of intoxication.

behavioral response. any overt activity that occurs in response to a stimulus that is usually external in origin. In toxicology, a behavioral response is an activity that is either discrete or can be measured quantitatively or semiquantitatively as a function of exposure to a chemical stimulus. The latter is represented either by a dosage delivered under specified circumstances and routes of exposure or a known concentration of a xenobiotic that is present in some compartment of the environment (e.g., the surrounding medium, food).

behavioral toxicity test. See toxicity test.

BEI. Biological Exposure Index.

belalang. *Ophiophagus hannah.*

belladonna ("fair lady"). **1:** *Atropa belladonna.* **2:** an anticholinergic extract of *Atropa belladonna* (Family Solanaceae) used by ladies during the Renaissance as a beauty aid to enlarge the pupils of their eyes. It contains atropine and other alkaloids and is used as a powder (0.3% belladonna alkaloids) or a tincture. Effects and symptoms of poisoning are very similar to those of atropine. Belladonna is now used mainly to control spasms of coughing, asthma and cramps, colic, and hyperacidity. It is also used in eye examinations to dilate the pupils and has been used in the treatment of Parkinson's syndrome. An overdose may induce an excited, confused mental state and can prove fatal. *Cf.* atropine. See belladonna poisoning.

belladonna alkaloid. See alkaloid (belladonna alkaloid).

belladonna lily. *Amaryllis belladonna.*

belladonna poisoning. a rare poisoning by ingestion of parts of *Atropa belladonna* (deadly nightshade), usually the berries. Symptoms appear within 15 minutes and include trembling, dizziness, dryness of mouth, burning throat, intense thirst, mydriasis, double vision, nausea, excitement, hallucinations, rambling speech or even loss of speech, tachycardia, a weak, rapid pulse, delirium, and paralysis. Prostration may follow in about 10 hours; death may intervene in as little as ½ hour following ingestion. See also atropinism, apoatropine.

belladonnine (1,2,3,4-tetrahydro-1-phenyl-1,4-naphthalenedicarboxylic acid bis(8-methyl-8-azabicyclo-[3.2.1]oct-3-yl) ester; isatropylditropeine; tropyl isatropate; ditropyl isatropate). an alkaloid produced by warming atropine with hydrochloric acid; found also in extracts of *Atropa belladonna* and other solanaceous plants, but it is probably an artifact of ex-[illegible]

bellaradine. cuscohygrine.

bellyache bush. *Jatropha gossypifolia.*

belmark. See fenvalerate.

Belvedere. *Kochia scoparia*.

bemegride (4-ethyl-4-methyl-2,6-piperidinedione; 3-ethyl-3-methylglutarimide; 3-methyl-3-ethylglutarimide; 4-ethyl-4-methyl-2,6-di-oxo-piperidine; 2,6-dioxo-4-methyl-4-ethyl-piperidine; methetharimide; β,β-methylethyl-glutarimide). an imide, $C_8H_{13}NO_2$, that is soluble in water and ethanol. It is used as an analeptic in barbiturate poisoning of humans and as a central stimulant in animals. Bemegride is a poison by ingestion and is extremely toxic to laboratory mice and rats. Human systemic effects, *per os*, include sleeplessness, distorted perceptions, hallucinations, and toxic psychosis. When heated to decomposition, it releases toxic fumes of NO_x.

benactyzine (α-hydroxy-α-phenylbenzeneacetic acid 2-(diethylamino)ethyl ester; benzylic acid β-diethylaminoethyl ester; β-diethylaminoethyl benzylate; 2-diethylaminoethyl diphenylglycolate). a muscarinic antagonist, $C_{20}H_{25}NO_3$, that acts on both the central and peripheral nervous systems. The hydrochloride is used therapeutically as an antidepressant and anticholinergic.

Benadryl®. diphenhydramine hydrochloride.

bendiocarb (2,2-dimethyl-1,3-benzodioxol-4-ol-methylcarbamate; methylcarbamic acid 2,3-(iso-propylidene-dioxy)phenyl ester). a water-soluble contact insecticide, it is also very to extremely toxic to a number of laboratory mammals.

bends (decompression sickness; caisson disease). a painful, sometimes fatal condition caused by nitrogen or helium bubbles in blood and other tissues due to a rapid reduction of atmospheric pressure (as in caisson workers) or a too rapid ascent by deep-sea divers or scuba divers. Signs and symptoms include painful joints, tightness in the chest, vomiting, giddiness, abdominal pain, and visual disturbances. The condition is sometimes marked by suffocation (See chokes). The victim occasionally has convulsions that may be followed by paralysis and even death.

Benedictin. See doxylamine.

benign. **1a:** mild, harmless; of no danger to life or health. **1b:** neither recurrent nor tending to progress; not malignant; **2:** pertaining to a tumor, growth, or condition that is localized, noncancerous, does not tend to progress, is usually nonrecurrent, and does not impair normal functions.

benign tetanus. See tetany.

Benlate®. benomyl.

benomyl ([1-[(butylamino)carbonyl]-1*H*-benzimidazol-2-yl]carbamic acid methyl ester; 1-(butylcarbamoyl)-2-benzimidazolecarbamic acid methyl ester; methyl 1-(butylcarbamoyl)-2-benzimidazolecarbamate; Benlate). a colorless, crystalline solid, $C_{14}H_{18}N_4O_3$, used as a ascaricide and fungicide. It is toxic to laboratory animals and humans orally or when inhaled. Benomyl is a mutagenic, teratogenic, and reproductive poison. The active agent is thought to be a degradation product, methyl-2-benzimidazolecarbamate. When heated to decomposition, benomyl emits toxic NO_x fumes.

bensulide (*N*-(2-mercaptoethyl)benzenesulfonamide). a herbicide that, like chlorthaldimethyl, inhibits cell division in the tips of root initials on *Cynodon dactylon* stolons. As in the case of bromacil, it inhibits cell division in the meristem of oat roots in association with the appearance of binucleate cells and differentiation of tissues nearer the tip.

bensylyt. phenoxybenzamine.

benthic (benthonic). **1a:** of, or pertaining to, the benthos. **1b:** of, or pertaining to, life or existence on or near the bottom of a water body. **1c:** of or pertaining to life or existence at great depths in a very large water body (a major river, lake, sea, or ocean). **1d:** sometimes used in reference to any organism or biotic community found exclusively on the bottom of any body of water.

benthic organism. any bottom-dwelling plant or animal, one that occurs on or near the bottom of a water body or at great depths in a large water body (a major lake, sea or ocean). See also benthos.

benthiocarb. an amber-colored, slightly water-soluble, liquid, $C_{12}H_{16}NOCl$, that is moderately toxic to humans and other mammals *per os*. It is a thiocarbamate herbicide, used to control sedges and grasses in rice paddies. Benthiocarb is rapidly detoxified in the liver by oxidation to the sulfoxide, the sulfur-carbon bond of which is then split by the soluble glutathione system. Effects of intoxication may suggest cholinesterase inhibition.

benthonic. benthic.

benthos. **1:** the floor of a sea or ocean; depths. **2:** a collective term for benthic organisms; applied most frequently to marine organisms that live at great depths.

1,2-benzacenaphthene. fluoranthene.

Benzahex®. lindane.

benzaldehyde (benzoic aldehyde; bitter almond oil). a flammable liquid, C_7H_6O, that readily oxidizes in air to benzoic acid. It is a decomposition product of the glycoside amygdaline, and is the source of the odor of bitter almonds that sometimes accompanies cyanide poisoning. Because benzaldehyde evaporates rapidly, the odor is detectable in poisoned animals only immediately after death. Benzaldehyde is used in small amounts to flavor beverages, chewing gum, and various foods. It is allergenic and a CNS depressant, causing convulsions when taken in large amounts. Two ounces may be fatal to humans. It is moderately toxic to laboratory rats and guinea pigs. See also apricot-kernel oil.

benzaldehyde cyanohydrin. mandelonitrile.

benzalkonium chloride. a highly toxic quaternary ammonium salt made up of bitter-tasting, white or yellowish-white powder or gelatinous pieces with an aromatic odor; it is highly alkaline in aqueous solution. It is a mixture of alkyl dimethylbenzylammonium chlorides; the general formula is $C_6H_5CH_2N(CH_3)_2RCl$, where R is a mixture of alkyl groups (C_8H_{17} to $C_{18}H_{37}$). It is used as a cationic detergent, a surface antiseptic, and fungicide. See detergent.

benzaminoacetic acid. hippuric acid.

benzamizole. isoxaben.

benzanthracene. 1,2-benzanthracene.

1,2-benzanthracene (benz[*a*]anthracene; benzanthracene; 1,2-benz[*a*]anthracene; 1,2-benzanthrene; benzanthrene; benzoanthracene; 1,2-benzoanthracene; benzo[*a*]anthracene; benzo-[*b*]phenanthrene; 2,3-benzphenanthrene; naphthanthracene; tetraphene). a colorless crystalline (leaflets or plates) carcinogenic hydrocarbon, $C_{18}H_{12}$. An isomer of naphthacene, it occurs in oils, coal tar, waxes, smoke, foods, and drugs. Benzanthracene is a confirmed carcinogen and probable human mutagen, with tumorigenic and neoplastigenic activity on contact with skin of humans and laboratory animals, and by other routes.

benz[*a*]anthracene. 1,2-benzanthracene.

9-10-benz[*a*]anthracene. 9,10-dimethyl-1,2-benzanthracene.

benzanthrene. 1,2-benzanthracene.

1,2-benzanthrene. 1,2-benzanthracene.

benzedrine. amphetamine, amphetamaine sulfate.

benzenamine. aniline.

benzene (benzol; cyclohexatriene; coal tar naphtha). the simplest aromatic hydrocarbon, consisting of a planar 6-membered equilateral ring. It is a flammable, volatile, highly toxic, transparent, colorless liquid derived from light coal tar oil. More than 7 million tons are produced annually in the United States. It is used mainly as a solvent and in the manufacture of organic chemicals. Benzene is a minor constituent (an octane booster) of gasoline (a possible source of exposure) and other fuels. It is highly toxic to terrestrial vertebrates, including humans, and to some fish and other aquatic organisms. Human exposure is chiefly via the respiratory route and uptake is partially reversible. Benzene is metabolically activated to quinones and active oxygen in mammals. It is myelotoxic and leukemogenic. Even a light exposure to benzene can cause weakness, dizziness, disorientation, euphoria, headache, nausea, vomiting, and loss of appetite. These symptoms can mimic drunken behavior. Long-term occupational exposure may cause aplastic anemia, preleukemia, leukemia, or cancer. It is an irritant of the skin, eyes, and upper respiratory tract, producing erythema (by defatting the skin) and dermatitis or blistering and edema. Acute exposures to the vapor can bring death within a few minutes. Acute systemic effects in humans from inhalation or ingestion include a burning sensation in the mouth and stomach, nausea, vomiting, chest pains, coughing, and CNS depression (sometimes with initial excitation) with headache, dizziness, vertigo, ataxia, confusion, hysteria, stupor, and fitful coma. Lesions may include late, severe blood dyscrasias, respiratory collapse, ventricular fibrillation, severe aplastic anemia with bone marrow depression, a reduced white cell count with elevated blood lymphocytes, and thrombocytopenia. Death from respiratory failure or ventricular fibrillation may result. Many chronic effects are nonspecific, e.g., fatigue, headache, dyspnea, irritability, ataxia, anorexia, confusion, disorientation, and sometimes hysteria. The victim can appear to be drunk. *Cf.* benzole.

benzene bromide. a lacrimator used as a tear gas.

benzene carbinol. benzyl alcohol.

benzene carboxylic acid. benzoic acid.

benzene hexachloride (BHC; hexachlorocyclohexane). **1:** a toxic mixture of the isomers of 1,2,3,4,5,6-hexachlorocyclohexane, $C_6H_6Cl_6$. No longer produced in the United States, it was once widely used as an insecticide, mainly for the control of cotton pests, but was later replaced by lindane. It has also been used as an ectoparasiticide for dogs and large mammals, but is highly toxic to domestic cats in concentrations that are effective against parasites. Lindane is also preferred for this usage. BHC is stored in body fat and excreted in milk. The alpha and gamma isomers are depressants of the CNS. *Cf.* hexachlorocyclohexane. **2:** hexachlorocyclohexane.

γ-benzene hexachloride. γ-hexachlorocyclohexane (See hexachlorocyclohexane).

benzene orthocarboxylic acid. orthophthalic acid.

benzene-1,2-oxide. an epoxide, it is an intermediate in the metabolic oxidation of benzene and probably accounts for the toxicity of the latter.

benzene poisoning. benzolism.

benzene ring. aromatic ring.

benzeneamine. aniline.

benzenecarboxylate. benzoate.

benzenecarboxylic acid. benzoic acid.

1,2-benzenedicarboxylic acid. orthophthalic acid.

1,2-benzenedicarboxylic acid bis(2-ethylhexyl) ester. bis(2-ethylhexyl) phthalate.

1,2-benzenedicarboxylic acid, butyl phenylmethyl ester. butyl benzyl phthalate.

1,2-benzenedicarboxylic acid dibutyl ester. *n*-butyl phthalate.

1,2-benzenedicarboxylic acid dimethyl ester. dimethyl phthalate.

1,2-benzenediol. pyrocatechol.

1,3-benzenediol. resorcinol.

1,4-benzenediol. hydroquinone.

***m*-benzenediol**. resorcinol.

benzeneethanamine. phenethylamine.

benzenemethanol. benzyl alcohol.

1,2,3-benzenetriol. pyrogallol.

benzenyl trichloride. benzotrichloride.

benzethethonium chloride. a toxic (*per os*), colorless, odorless, very bitter, synthetic solid (plates), $C_{27}H_{42}ClNO_2$, that is soluble in water, ethanol, and acetone. It is a quaternary ammonium compound used as an antiseptic and cationic detergent.

benzhydramine. diphenhydramine.

***O*-benzhydryldimethylaminoethanol**. diphenhydramine.

2-(benzhydryloxy)-*N*,*N*-dimethylethylamine. diphenhydramine.

benzidine. **1:** (benzidine base; 1,1′-biphenyl-4,-4′-diamine; *p*-aminodiphenyl; *p*-diamino-diphenyl; 4,4-diaminobiphenyl; 4,4′-di-phenylenediamine). a highly toxic, white, grayish-yellow, or reddish-gray crystalline powder. A biphenyl aromatic amine, $NH_2(C_6H_4)_2NH_2$, used chiefly in organic synthesis and dye manufacture, especially of benzine dyes, *q.v.* Most other uses have been discontinued. Benzidine and its salts are carcinogenic, causing cancer of the urinary bladder and pancreas in humans and some laboratory animals. It causes bone marrow depression, hemolysis, and kidney and liver damage. It is highly toxic by all routes of exposure, but exposure is chiefly by percutaneous absorption and inhalation. Great care should be taken in the use, storage, and disposal. It is a hazardous waste. *Cf. p*-biphenylamine. See also aromatic amine. **2:** any of a class of highly carcinogenic chemicals that contain the benzine group (e.g., dichlorobenzidine).

benzidine base. benzidine.

benzidine dye. any of a class of highly toxic, carcinogenic azo dyes derived from 3,3′-dichlorobenzidine. Included are orange and yellow, light-fast, alkali-resistant dyes and Congo Red. They may be absorbed through the skin.

4-[2-benzimidazolyl]thiazole. thiabendazole.

benzin. naphtha.

benzine. petroleum ether.

1,2-benzisothiazol-3(2H)-one 1,1-dioxide. saccharin.

benzoanthracene. 1,2-benzanthracene.

1,2-benzoanthracene. 1,2-benzanthracene.

benzo[*a*]anthracene, benzoanthracene. 1,2-benzanthracene.

benzoaric acid. ellagic acid.

benzoate (benzenecarboxylate). a salt or ester of benzoic acid.

benzodiazepine. any of a class of chemically related psychotropic drugs that includes chlordiazepoxide, clonazepam, diazepam, and oxazepam. They are potent hypnotics and sedatives that are used to treat anxiety, nervous tension, insomnia, seizures and muscle spasms, and acute alcoholic withdrawal syndrome. They have largely replaced barbiturates in a number of applications because their toxicity is generally low and the therapeutic index is high. Benzodiazepines can cause drowsiness, dizziness, incoordination, and occasionally coma. They are to some extent habit forming and occasionally contribute to the development of palatal deformity. See depressant, drug abuse.

1-[5-(1,3-benzodioxol-5-yl)-1-oxo-2,4-pentadienyl]piperidine. piperine.

benzo[*jk*]fluorine. fluoranthene.

benzoic acid (benzenecarboxylic acid; benzoyl hydrate; carboxybenzene; dracylic acid; flowers of benzoin; phenylformic acid). a white crystalline compound, $C_6H_4(OH)_2$, with the odor of benzoin or benzaldehyde. The carboxyl group is bound directly to the benzene ring. It occurs naturally in some plants, as in benzoin resin (gum benzoin). Benzoic acid is moderately toxic, but has many highly toxic derivative compounds (e.g., benzidine, 3,3′-dichlorobenzidine). It is used as a food preservative and fungistat. It is also used as an oral antiseptic, a diuretic, and expectorant. It is rapidly excreted as hippuric acid. Benzoylglucuronic acid is the detoxified form. See also bopindolol, hippuric acid.

(benzoic acid [1-[4-[(3-amino-2,3,6-trideo xy-α-L-lyxo-hexopyranosyl)-oxy]-1,2,3,4,6,11-hexahydro-2,5,12-trihydroxy-7-methoxy-6,11-dioxo-2-napthacenyl]ethylidene]hydrazide. zorubicin.

benzoic acid benzyl ester. benzyl benzoate

benzoic acid hydrazide 3-hydrazone with daunorubicin. zorubicin.

benzoic acid phenylmethyl ester. benzyl benzoate.

benzoic aldehyde. benzaldehyde.

benzol. benzene. *Cf.* benzole.

benzole. an industrial term applied to any of a number of aromatic hydrocarbons (benzene and its homologs) in pure form or as mixtures in which aromatic hydrocarbons predominate. Benzole is applied largely to products that contain benzene, methylbenzene (toluene), dimethylbenzenes (xylenes), and various other aromatics. *Cf.* benzol.

benzolism (benzene poisoning). poisoning by benzene or its fumes. See benzene.

1,2-benzophenanthrene. chrysene.

benzo[*a*]phenanthrene. chrysene.

benzo[*b*]phenanthrene. 1,2-benzanthracene.
benzophenol. phenol (def. 1).

2*H*-1-benzopyran-2-one. coumarin.

benzopyrene. See benzo[*a*]pyrene.

benzo[*a*]pyrene (benzopyrene; BaP; B[*a*]P; 3,4-benzpyrene; formerly 1,2-benzpyrene). a crystalline, polynuclear (5-ring) aromatic hydrocarbon, $C_{20}H_{12}$, that is insoluble in water, slightly soluble in ethanol, and soluble in benzene, toluene, and xylene. It is a ubiquitous environmental pollutant that occurs in coal tar, soot, tobacco smoke, and the atmosphere as a product of incomplete combustion of hydrocarbons. It is toxic and a potent local and systemic carcinogen (a procarcinogen) by inhalation. Benzo[*a*]pyrene is metabolically activated in several steps to the proximate carcinogen, the 7,8-diol-9,10-epoxide (See below). Humans are usually exposed through cigarette smoke, cooked food, or through environmental sources including the atmosphere, water, and soil. See also aryl hydrocarbon hydroxylase.

benzo[*a*]pyrene-7,8-diol-9,10-epoxide. any of four mutagenic stereoisomers resulting from the metabolism of benzo[*a*]pyrene. Of these, the (+)-benzo[*a*]pyrene-7,8-diol-2-epoxide is the most toxic.

benzo[*a*]pyrene hydroxylase. certain isozymes of cytochrome P-450 with aryl hydrocarbon hydroxylase activity. See aryl hydrocarbon hydroxylase.

1,2-benzopyrone. coumarin.

2,3-benzopyrrole. indole.

benzoquinone. **1:** any benzene derivative in which 2 hydrogen atoms are replaced by 2 oxygen atoms, thus yielding the basic quinone structure. They are usually yellow to orange or orange to red in color. Millipedes, when disturbed, release secretions from glands in each segment that contain high concentrations of benzoquinones, such as 2-methoxy benzoquinone, 3-methyl benzoquinone, and toluquinone. These are irritant substances and appear to protect against predation, especially by ants. Contact with human skin, which most frequently occurs when sleeping outdoors, causes burns, rashes, and even blisters. Certain beetles, grasshoppers, gonyleptid spiders, and termites also secrete benzoquinones. See also quinone. See Diplopoda. **2:** often used as a synonym of quinone or *p*-quinone,*q.v.*

1,4-benzoquinone. *p*-quinone. See quinone.

***p*-benzoquinone**. *p*-quinone. See quinone.

benzothiazole. a very toxic liquid derivative of thiazole synthesized from *N,N*-dimethylaniline and sulfur. It is used in organic syntheses.

benzotrichloride ((trichloromethyl)benzene; α,α,α-trichlorotoluene; ω,ω,ω-trichlorotoluene; benzenyl trichloride; toluene trichloride). a colorless, oily, unstable, fuming liquid. Commercial grades may contain benzylidene chloride, benzyl chloride, and hydrochloric acid. The pure chemical readily hydrolyzes on standing in moist air to benzoic and hydrochloric acids. It is used in the production of dyes (e.g., Malachite Green, Rosamine, Alizarin Yellow) and as an intermediate in the synthesis of ethyl benzoate. Large doses produce CNS depression in some laboratory mammals. Oral toxicity is slight to moderate, but the fumes are highly irritant to the skin and mucous membranes, and are carcinogenic and leukemogenic in laboratory mice.

benzoyl acetyl peroxide. acetylbenzoyl peroxide.

benzoyl hydrate. benzoic acid.

benzoylamine. a toxic alkaloid present in various species of *Aconitum*. It initially stimulates and then depresses the central nervous system. See also aconitine.

benzoylaminoacetic acid. hippuric acid.

benzoylglucuronic acid. the conjugation product of benzoic and glucuronic acids. It is the detoxified form of benzoic acid.

benzoylglycin. hippuric acid.

benzoylglycocoll. hippuric acid.

benzoylmethylecgonine. cocaine.

benzozone. acetylbenzoyl peroxide.

1,2-benzphenanthrene. chrysene.

2,3-benzphenanthrene. 1,2-benzanthracene.

benz[*a*]phenanthrene. chrysene.

benzphetamine (*N*,α-dimethyl-*N*-(phenylmethyl)-benzeneethanamine; *N*-benzyl-*N*,α-dimethylphenethylamine; *d*-*N*-methyl-*N*-benzyl-β-phenylisopropylamine; Didrex). an anorectic used also to assay certain isozymes of cytochrome P-450. Among numerous effects, it can elevate blood pressure, disturb heart rhythm, cause restlessness, insomnia, hyperactivity, headache, euphoria, depression, psychosis, tremor, dryness of mouth, unpleasant tastes, diarrhea, stomach upset, altered sex drive, and impotence. Overdosage can induce aggressiveness, hallucinations, and panic. Tolerance and dependence may develop from excessive use. It is a controlled substance listed in the U.S. Code of Federal Regulations (CFR). *Cf.* diethylpropion, phentermine, phenmetrazine; See also appetite suppressant.

1,2-benzpyrene. benzo[*a*]pyrene.

3,4-benzpyrene. benzo[*a*]pyrene.

benzyl acetate (acetic acid phenylmethyl ester; acetic acid benzyl ester; benzyl ethanoate; α-acetoxytoluene). a moderately toxic, colorless, flammable, liquid, $C_6H_5CH_2OOCCH_3$, that occurs in a number of plants, especially those of the genus *Jasminum* (jasmines). It has a sweet, fruity, floral odor and is used as a synthetic raspberry, strawberry, cherry, banana, and plum flavoring agent for beverages, sweets, and chewing gum; also used in perfumery. If ingested, it can cause gastrointestinal irritation with vomiting and diarrhea. It is an irritant to the skin, eyes, and respiratory tract and is carcinogenic to some laboratory animals.

benzyl benzoate (benzoic acid phenylmethyl ester; benzoic acid benzyl ester; benzylbenzenecarboxylate). an oily colorless liquid ester, $C_6H_5COOCH_2C_6H_5$. An irritant contained in Peru and Tolu balsams, it is used as an antispasmodic and parasiticide (scabicide). It is moderately toxic orally to laboratory mice, rats, guinea pigs, and rabbits. Acute effects on these animals proceed from incoordination, excitation, convulsions, to death.

benzyl butyl phthalate. butyl benzyl phthalate.

benzyl ethanoate. benzyl acetate.

benzyl mercaptan. α-toluenethiol.

6-benzyladenine. 6-benzylaminopurine.

benzylamine (aminotoluene). a toxic liquid that is soluble in water, ethanol, and ethyl ether. It is used as a chemical intermediate in the manufacture of dyes.

benzylamine oxidase. histaminase.

6-benzylaminopurine (6-benzyladenine; BAP). a synthetic cytokinin, it is one of the most active of a number of substituted adenines that show cytokinin activity.

benzylbenzenecarboxylate. benzyl benzoate.

1-(2-benzylcarbamyl)ethyl-2-iso-nicotinoylhydrazine. nialamide.

2-[benzyl(2-dimethylaminoethyl)amino]-pyridine. tripelennamine.

2-[benzyl-(2-dimethylaminoethyl)amino]pyridine dihydrogen citrate. tripelennamine citrate.

***N*-benzyl-*N*,α-dimethylphenethylamine**. benzphetamine.

N-benzyl-N',N'-dimethyl-N-(2-pyridyl)-ethylenediamine. 2,4,6-trinitrotoluene.

(E,E)-16-benzyl-6,7,8,9,10,12a,13,14,15,-15a,16,16-dodecahydro-5,13-dihydroxy-9,15-dimethyl-14-methylene-2H-oxycyclotetradec-[2,3-d]isoindole-2,18(5H)-dione. cytochalasin B.

benzyl[(dodecylcarbamoylmethyl)dimethyl]-ammonium chloride. dodecarbonium chloride.

(1-benzylethyl)hydrazine. pheniprazine.

benzylic acid β-diethylaminoethyl ester. benactyzine.

benzylisoquinoline. tetrahydroisoquinoline.

benzyltetrahydroisoquinoline. any alkaloid (e.g., papaverine) that occurs naturally in the opium poppy. See tetrahydroisoquinoline.

Berberidaceae (the barberry family). a family of hardy Northern Hemisphere herbs and shrubs, many of which are ornamental. They have spiny stems, small attractive leaves, and red, black, or yellow berries. Most are deciduous, but retain their leaves well into winter; some are evergreen. See podophyllin, podophyllotoxin, *Podophyllum peltatum*.

berberine (5,6-dihydro-9,10-dimethoxybenzo[*g*]-1,3-benzodioxolo[5,6-a]quinolizinium; berbinium; umbellatine). an isoquinoline (papaver) alkaloid, $C_{20}H_{19}NO_5$, isolated from *Hydrastis canadensis* and many other plants such as *Argemone mexicana*, *Chelidonium majus*, and *Papaver spp.*, *q.v.* The crystals are white to yellow and are insoluble in water, but soluble in ethanol and ethyl ether. Its salts are slightly soluble in water. Berberine is extremely toxic by all routes of exposure, but has been used therapeutically as an antimalarial, antipyretic, a carminative, and topically for indolent ulcers.

berbinium. berberine.

Berg adder. *Bitis atropos*.

Bergenia crassifolia. a hardy, perennial garden herb (Family Saxifragaceae) with large, fleshy evergreen leaves that contain arbutin.

bergotter. *Vipera xanthina*.

berkelium (symbol, Bk)[Z = 97; the most stable isotope = ^{247}Bk, half-life, 1400 yr]. an artificial radioactive transuranic element of the actinoid series of metals. Several radioisotopes have been synthesized.

Berlin blue. ferric ferrocyanide.

bermat. chlordimeform.

Bermuda buttercup. *Oxalis pes-caprae*.

Bermuda grass. *Cynodon dactylon*.

Bermuda oxalis. *Oxalis pes-caprae*.

berry. a fruit which is generally more or less fleshy throughout.

berylliosis (beryllium poisoning). a severe disease of the respiratory tract caused by acute or chronic exposure of beryllium salts by inhalation. Diseases associated with beryllium, *q.v.*, usually have a long latency period that hinders prognosis. Symptoms (e.g., of pulmonary granulomatous disease) may not appear until 3 months to 15 years following exposure. **acute berylliosis**. a severe pneumoconiosis due to acute exposure to beryllium salts. **chronic berylliosis**. a progressive disease caused by chronic inhalation of beryllium. It has a latent period of 5-20 years. The most damaging effect is granulomatous pulmonary fibrosis and pneumonitis. In addition to coughing and chest pain, the subject suffers from fatigue, weakness, loss of weight, and dyspnea. The impaired lungs do not transfer oxygen well. The liver, kidneys, heart, spleen, and striated muscles may also be adversely affected. Mortality in advanced cases of beryllium poisoning is high.

beryllium (Be)[Z = 4; A_r = 9.013] a light (rel density 1.85), white or steel-gray, extremely toxic metallic element of the alkaline earth series. Beryllium in powder form and its compounds are highly poisonous. It is the first metal in the periodic table to be distinctly toxic. It has major applications in nuclear reactors, automobiles, computers, business machines, and aerospace products. Air pollution by dust containing beryllium may occur in the vicinity of factories where the metal is produced or used. Not only are workers and their families at risk, but dust that contains beryllium may occur in the vicinity of factories where the beryllium is produced or used; even people using the same laundries as beryllium workers are at risk. Inhalation of dust or fumes in the workplace can cause coughing, dyspnea, pain and tightness in the chest, muscular weakness and fatigue, anorexia, and weight loss. A few minutes of exposure to high levels of beryllium dust or fumes can produce severe poisoning that affects the eyes, skin, and respiratory system. Pneumonitis may result from a single acute exposure and can be fatal. Direct contact can cause nonhealing ulceration of the skin at the site of contact and granulomas; hypersensitive individuals may develop a severe dermatitis, acute chemical conjunctivitis, and corneal burns. Death sometimes results from short exposures to very low concentrations of beryllium or its salts. Victims may recover from mild poisoning within 1 to 6 weeks; more severe cases may require 6 months or more to recover. Chronic poisoning may permanently affect the lungs, kidneys, liver, or bones. Beryllium is also carcinogenic to several laboratory animals and possibly to humans; it may cause cancers of the liver, gall bladder, and bile ducts. See also berylliosis, organo-beryllium compound.

beryllium diethyl. diethylberyllium.

beryllium poisoning. berylliosis.

bespectacled snake. *Naje naje*.

beta (β, B). **1:** the second letter of the Greek alphabet. Its applications as a symbol or prefix in scientific terms are analogous to those of alpha. For terms beginning with beta, See the root word. **2:** the second. **3:** second in order of magnitude. **4:** it indicates, for example, the second position in a series or classification; the [illegible] position of a chemical substituent or side chain of an aromatic compound. **5:** beta is used in reference to a secondary allotropic modification of a metal or compound. **6:** beta denotes a type of radioactive decay. See beta particle.

Beta (beet). a genus of some 15 species of widely cultivated, mostly biennial herbs and root crops (Family Chenopodiaceae) that are indigenous to Europe and the Middle East. ***Beta vulgaris*** (beet; beetroot; sugarbeet; foliage beet; fodder beet; mangel; Mangelwurzel; mangold; chard; Swiss chard). this species is the progenitor of several cultivated varieties that have been cultivated by humans for several centuries. Two botanical varieties, *B.v.* var. *crassa* (beet, beetroot) and *B.v.* var. *macrorhiza* (mangold), sometimes contain concentrations of nitrate that are toxic to livestock. ***Beta vulgaris* var. *cicla*** (chard) is goitrogenic and contains hazardous levels of soluble oxalates. Sugarbeet tops are sometimes used as forage by livestock. Nitrate poisoning of livestock has occurred following ingestion of sugarbeet tops. This has been associated with metabolic derangements within the plant from 2,4-D used as a weed killer. Death of livestock from consumption of the tops has also been attributed to the presence of soluble oxalates. Beetroot crops grown for fodder are also toxic under certain conditions, producing hypocalcemia and sometimes acid indigestion.

beta decay. a radioactive disintegration in which a radionuclide emits an electron or a positron (See beta particle).

beta particle (beta ray). a positively (positron) or negatively (negatron) charged electron emitted during disintegration (beta decay) of a radionuclide. Such rays can burn the skin, but have little penetrating power and can be blocked by a thin sheet of paper. Beta particles are used in scintillation counting and autoradiography.

beta ray. beta particle.

betaine (1-carboxy-*N,N,N*-trimethylmethanaminium hydroxide inner salt; (carboxymethyl)trimethylammonium hydroxide inner salt; glycine betaine; glycocoll betaine; lycine; oxyneurine; trimethylglycine hydroxide inner salt; [illegible] anhydride); an alkaloid that is widely distributed in living tissues. It is very soluble in water and soluble in ethanol and methanol. Betaine is used in organic syntheses, in soldering and resin-curing fluxes, and therapeutically as a hepatoprotectant.

betaine hydrochloride. an alkaloid used in medicine as a source of hydrogen chloride.

betel nut. **1:** *Areca catechu*. **2:** the seed of *Areca catechu*.

betel nut palm. *Areca catechu*.

betel nut seed. *Areca catechu*.

betel palm. *Areca catechu*.

bethanechol chloride (2-[(aminocarbonyl)oxy]-*N,N,N*-trimethyl-1-propanaminium chloride; Duboid®; Urecholine®). a colorless or white crystalline substance or crystalline powder, $C_7H_{17}ClN_2O_2$. It is a parasympathomimetic agent, given *s.c.* or *per os*, to treat a number of conditions in which stimulation of the parasympathetic nervous system is desirable. It is used, for example in the treatment of paralytic ileus following surgery and postoperative urinary retention. It is also used as a cholinergic in veterinary medicine; it has been used to treat urolithiasis of cats and for atonic conditions of the gastrointestinal tract. Adverse effects of therapeutic doses may include headache, flushing, gastrointestinal distress, diarrhea, hypotension, excessive salivation, and sweating; occasionally hypersensitivity may develeop. It is not given during pregnancy.

Bettendorff's test. See test.

Bettlach May disease. a fatal disease of adult *Apis mellifera* (honeybees) known mainly in Switzerland. It results from the ingestion of pollen of certain types of *Ranunculus* (buttercups). The bee is paralyzed, unable to fly, and eventually dies.

betula oil. methyl salicylate.

α-Bgt. α-bungarotoxin.

BHA. butylated hydroxyanisole.

bhang. cannabis.

BHC. benzene hexachloride.

γ-BHC. lindane. See also hexachlorocyclobenzene.

Bhopal. a city of about 1 million inhabitants in central India where large amounts of very toxic methyl isocyanate (MIC) fumes leaked from a Union Carbide factory on December 3, 1984. Some 2000 persons were killed outright and 200,000 or more were injured, in many cases with persistent sequalae. It is considered the worst chemical accident on record.

BHT. butylated hydroxytoluene.

Bi. the symbol for bismuth.

bialamicol (3,3′-bis[(diethylamino)methyl]-5,5′-diallyl-α,α′-bis(diethylamino)-*m,m*′-bitolyl-4,4′-diol; 6,6′-diallyl-α,α′-bis(diethylamino)-4,4′-bi-*o*-cresyl; biallylamicol). bialamicol and its dihydrochloride are antiamebic compounds.

bialamicol hydrochloride. See bialamicol.

biallylamicol. bialamicol.

***N,N*′-bianiline**. 1,2-diphenylhydrazine.

bias. **1:** the error caused by nonrandom deviation of an estimate or measurement from the true value. The latter cannot usually be known with certainty. **2a:** a systematic error that results from nonrandom sampling, and/or nonrandom errors in measurement. **2b:** systematic errors that are inherent in the method used to draw samples from a population. **3:** the situation whereby the probability of rejecting a null hypothesis is not a minimum when the null hypothesis is true. **investigator bias**. preferences, disposition, inclinations, or prejudices of an investigator that may influence outcomes. Such are always present, consciously or unconsciously in any investigation. It is the obligation of the investigator in structuring and conducting a test, experiment, or set of measurements to rule out the potential effects of such biases on data and outcomes.

bibenzene. biphenyl.

Bibron's burrowing adder. *Atractaspis bibronii*.

Bibron's mole viper. *Atractaspis bibronii*.

Bibron's viper. *Atractaspis bibronii*.

bicarburetted hydrogen. ethylene.

bichemorphic. pertaining to biochemomorphology or to the relationship between chemical structure and biological activity.

bichloride of mercury. mercuric chloride.

bicolorin. esculin.

bicuculline ([*R*-(*R**,*S**)]-6-(5,6,7,8-tetrahydro-6-methyl-1,3-dioxolo[4,5-g]isoquinolin-5-yl)-furo[3,4-e]-1,3-benzodioxol-8(6*H*)-one). a naturally occurring alkaloid found in *Dicentra cucullaria, Adlumia fungosa*, and *Corydalis*. It is a γ-aminobutyric acid (GABA) antagonist.

Bidens frondosa (beggar-tick). a weedy North American plant (Family Asteraceae) that sometimes contains concentrations of nitrate that are toxic to livestock.

Bidrin (3-(dimethoxyphosphinylloxy)-*N,N*-dimethyl-*cis*-crotonamide). an extremely toxic organophosphate insecticide; it is a cholinesterase inhibitor. *Cf.* Bidrin [Shell].

Bidrin [Shell]®. the dimethyl phosphate of 3-hydroxyl-*N,N*-dimethyl-*cis*-crotonamide. See dicrotophos. *Cf.* Bidrin. See also dicrotophos.

biennial. **1:** occurring every two years; of or pertaining to a cycle of two years. **2:** biennial plant. a plant that completes its life cycle in 2 years. The first year's growth is usually vegetative only, which includes a large underground food storage organ. The plant produces flowers, fruits, and seeds during the second year.

biflorine. protopine.

bighead. a photosensitization disease of sheep caused by grazing on plants such as *Tetradymia*, *Tribulus terrestris*, and *Nolina texana* with subsequent exposure to sunlight. The term "big head" is sometimes used in reference to nutritional hyperparathyroidism in horses and sometimes in goats. The latter condition is rare in sheep.

bigitalin. gitoxin.

bikhroot. *Aconitum ferox*.

bile (gall). a dark yellow or brown-green alkaline secretion produced by the vertebrate liver that contains cholesterol, bile salts, bile pigments (a reddish pigment, bilirubin, and a green pigment, biliverdin), lecithin, phospholipids, electrolytes, urea, and various xenobiotics (including toxicants) or their metabolites. Bile is stored temporarily in the gall bladder (a structure lacking in many birds) and is conducted from the gall bladder to the duodenum of the small intestine via the bile duct under pressure by contractions (stimulated by cholecystokinin) of the gall bladder. The components of bile may be excreted or recycled; lecithin and the bile salts emulsify dietary lipids, thus facilitating digestion.

bile acid. any of the steroid acids (derivatives of cholesterol) produced by the vertebrate liver, such as taurocholic and glycocholic acids. They appear in bile as sodium salts that facilitate the emulsification of dietary fats. A large portion of the bile acids are recycled by enterohepatic circulation. Certain of the bile salts can cause cholestasis or hepatic necrosis and can be metabolized by the intestinal microflora into secondary bile acids, e.g., lithocholic acid, which are promoters of colonic carcinogenesis.

bile duct. the duct in vertebrates that conducts bile from the bile tubules of the liver to the small intestine.

bile salt. See bile acid.

biliary. of bile, pertaining to bile.

bilineurine. choline.

bilirubin. a yellowish bile pigment, it is a catabolic product of hemoglobin and other heme pigments during the destruction of erythrocytes, a waste product of the normal destruction of erythrocytes. It contributes to the yellowish color of urine and can cause jaundice if it accumulates in the blood. See bile.

biliverdin (biliverdinic acid; dehydrobilirubin). a green bile pigment, $C_{33}H_{34}O_6N_4$, it is formed from the catabolism of hemoglobin during the normal destruction of erythrocytes. It is one precursor of bilirubin. See bile.

biliverdinic acid. biliverdin.

binapacryl (3-methyl-2-butenoic acid 2-(1-methylpropyl)-4,6-dinitrophenyl ester; 3-methylcrotonic acid 2-*sec*-butyl-4,6-dinitrophenyl ester; 2-*sec*-butyl-4,6-dinitrophenyl 3-methyl-2-butenoate; 2-*sec*-butyl-4-6-dinitrophenyl 3-methylcrotonate; 2-*sec*-butyl-4,6-

dinitrophenyl senecioate; 3,3-dimethylacrylic acid 2-*sec*-butyl-4,6-dinitrophenyl ester; 4,6-dinitro-2-*sec*-butylphenyl β,β-dimethylacrylate; 3-methyl-2-butanoic acid 2-*sec*-butyl-4,6-dinitrophenyl ester; dinoseb methacrylate; senecoic acid 2-*sec*-butyl-4,6-dinitrophenyl ester). a very toxic substance, $C_{15}H_8O_6N_2$, used as a fungicide and mitocide. See also dinoseb.

binary gas. See gas.

binder. an excipient that binds a mixture of solid particles together. Binders used in drug or toxicant formulations include gums and waxes, starch, gelatin, and sucrose.

binding. the attachment of one atom or molecule to another (e.g., via van der Waal's forces) without undergoing a chemical reaction. A toxicant or potential toxicant (ligand) may bind with a macromolecule (e.g., protein, nucleic acid) in a tissue, producing a higher than expected concentration in that tissue. Binding may be of an irreversible covalent type, which is generally associated with significant toxic effects, or noncovalent and reversible. Reversible binding is importantly involved in receptor binding and in the transport of toxicants by albumins and lipoproteins. **covalent binding**. irreversible binding by the sharing of a pair of electrons between two atoms or molecules such that they occupy two stable orbitals, one in each of the two atoms. Covalent binding is characteristic of many instances of chronic toxicity or toxic action that involve long-term effects, including carcinogenicity and immune responses. In the case of carcinogenicity, the toxic metabolite often forms a stable adduct with intracellular molecules, especially proteins or DNA. **hydrophobic binding**. binding between two nonpolar groups with the exclusion of water. The hydrophobic binding of two molecules forms a complex that is often highly stable in aqueous solution.

binding constant. affinity constant.

binding site. **1:** a receptor to which a drug, toxicant, or other pharmacologically active substance binds. Examples are cell membranes of various types of cells, enzymes, and blood proteins. While the binding sites of many drugs seem to be highly specific, most toxicants seem less selective and binding affinity may depend to a large extent on their lipophilicity. **2:** active site.

bindweed. *Convolvulus*.

bioaccumulate. to accumulate within an organism to a concentration above that of the exposure medium, or to accumulate progressively through a food chain (said of an exogenous chemical). See bioaccumulation.

bioaccumulation. **1:** the tendency or process of accumulating or concentrating a xenobiotic substance from all sources in the environment and by all routes of entry. **2a:** the net accumulation of a substance from the environment of a living system by all routes of entry. **2b:** state of having bioaccumulated a substance. **3:** operationally, bioaccumulation is the state whereby the concentration in an organism as a whole, or in selected tissues, is greater than the exposure concentration. Chemicals that bioaccumulate are often lipohilic and have a long biological half-life, usually because they are sequestered or accumulate in target organs within affected organisms and are not readily excreted. Heavy metals accumulate, for example, in target organs and produce a toxic response when the concentration exceeds the threshold of response.

bioaccumulation factor. See factor (bioaccumulation factor).

bioaccumulative. denoting substances that increase in concentration in exposed living organisms. Such substances are usually very slowly metabolized and excreted.

bioactivation. biological activation. See activation (metabolic activation).

bioactive. biologically active, *q.v.*

bioactivity. biological activity. See activity.

bioassay (biological assay). any controlled, reproducible test to quantitatively determine the identity (presence), character, specificity, or strength (potency or concentration) of a biological agent (e.g., a medication or toxicant) by measuring specified effects (end points) upon a living organism (usually one adapted to laboratory conditions), or on isolated tissue. The term also properly applies to any test of the effects of environmental variables on a biological system. A toxicity test is a form of bioassay. **aquatic bioassay**. any bioassay that uses aquatic systems (species, populations, communities, microcosms) to assess the biological or ecological effects of potentially toxic or hazardous materials, or environmental variables. Common test organisms used in aquatic toxicology are *Cyprinodon variegatus* (sheepshead minnow), *Daphnia spp.* (daphnids), *Lepomis macrochirus* (bluegill sunfish), *Mysidopsis spp.* (mysid shrimp), *Pimephales promelas* (fathead minnow), and *Salmo gairdneri* (rainbow trout).

bioavailability. **1a:** the potential or tendency of a xenobiotic substance to enter and to interact physiologically with a living system as distinct from its presence or concentration in the immediate environment. This is reflected in the rate and extent of intake under specified physiologic and environmental conditions for a given concentration of the substance. **1b:** in vertebrate animals, bioavailability is taken as the rate and extent of absorption from the site of administration into the systemic circulation.

biocenology (biocoenology; biocenotics; biocoenotics; cenobiology; coenobiology). the study of biotic communities. This subsumes a significant portion of ecotoxicological investigation.

biocenosis (biocoenosis; biological community; biotic community; ecological community). **1:** a well-defined natural assemblage or community of living organisms that occurs in a particular biotope and is distinguishable from other such assemblages. This term is sometimes improperly used as a synonym for ecosystem, but this concept, unlike that of ecosystem, excludes reference to the physical or nonliving aspects of the environment. **2:** the smallest naturally occurring interactive assemblage of biotic populations that is reasonably self-sufficient and mutually sustaining. *Cf.* ecosystem.

biocenotics. biocenology.

biochemical. of or pertaining to biochemistry, to biochemicals or their properties, or to chemical processes that take place within a living organism.

biochemical activation. See activation (metabolic activation).

biochemical oxygen demand. biological oxygen demand.

biochemical toxicity test. See toxicity test.

biochemistry (biological chemistry; physiological chemistry). **1:** the chemistry of living organisms. **2:** the science that treats the chemistry of living organisms and their products. See also chemistry.

biochemorphic. of or pertaining to the relationship between biological activity and chemical structure.

biochemorphology. **1:** the study of relationships between chemical structure and biological activity. **2:** the use of biochemical techniques (e.g., selective staining) in the study of gross morphology.

biocidal. destructive to life; of, pertaining to, or denoting that which is able to kill living organisms.

biocide. a substance that is destructive to living organisms.

biocoenology. biocenology.

biocoenosis. biocenosis.

biocoenotics. biocenology.

bioconcentration. biological concentration (See concentration). See also factor (bioconcentration factor).

biocontrol. biological control.

biodegradable. 1a: denoting a substance that breaks down or decomposes readily under natural conditions as a consequence of biological processes (e.g., microbial oxidation, enzymic catalysis). **1b:** susceptible to biodegradation. **2:** pertaining to the tendency of a substance to rapidly decompose under natural conditions; capable of being decomposed by natural processes.

biodegradation (biological degradation). **1:** the partial or complete decomposition of a chemical into more elementary substances by means of biological agents or processes (e.g., enzymatic action). **2:** any biological process that degrades substances (e.g., microbial oxidation, enzyme catalysis).

biodiversity. biological diversity.

bioecology. ecology.

bioenergetics. the study of energy flow through living systems.

Biofanal®. nystatin.

biogenic (biogenetic). **1:** produced by living organisms (e.g., as biogenic amines). **2:** essential to the maintenance of life.

biogenic amine. See amine.

biogenic toxicity. See toxicity.

biogeochemical cycle. the repeated cycle through which a specific chemical element passes between biotic and abiotic components of the biosphere. Examples are the carbon cycle and the nitrogen cycle. Persistent xenobiotics may also cycle through the biosphere.

biogeocoenosis. ecosystem; the biocenosis, together with its biotope.

biogeography. the science of the geography of living species, populations, and communities in relation to their origins and the influence of past and present geographic and environmental factors (including chemical substances) that have shaped their distributions, habitats, and relationships at present or historically, in an evolutionary framework.

bioindication. indication (*q.v.*) by the use of bioindicators.

bioindicator. an organism (most commonly a plant or microorganism) that is used as a qualitative indicator of specific chemical activity in, or an aspect of the chemical composition of, a natural system. *Cf.* biomonitor. See also indication, indicator, indicator species.

biolarvicide. an endotoxin of *Bacillus thuningiensis israelensis*. When the bacterium is ingested by mosquito larvae, the endotoxin is freed by stomach secretions, and disrupts the ability of the gut to regulate electrolytes, resulting in the death of the larvae.

biologic. of or pertaining to living organisms or to life processes. *Cf.* biological.

biological. **1:** biologic. **2:** any toxicologically or therapeutically active chemical produced by, or made from, a living organism.

biological activation. See activation (metabolic activation).

biological amplification. biomagnification.

biological assay. bioassay.

biological chemistry. biochemistry; formerly physiological chemistry.

biological clock (endogenous clock). See circadian rhythm, diurnal rhythm.

biological community. biocenosis.

biological concentration. See concentration.

biological control (biocontrol). the control of a pest by the use of living organisms or natural biological processes. Biological control methods include, for example, the introduction or use of natural predators, parasites, pathogens, or competitors; plant or animal hormones; toxins; the introduction of abnormal (e.g., sterilized) organisms into a population.

biological degradation. See biodegradation.

Biological Exposure Index (BEI). a measure of the amount of chemical absorbed by an organism as indicated by the concentration of a substance in body parts considered to be indicators of exposure (e.g., blood, hair). BEIs are sometimes useful supplements to TLVs.

biological half-life. See half-life.

biological indicator. See indicator.

biological integrity. **1:** the ability of a living system to carry out all vital functions and to respond adaptively to external and internal stimuli. **2:** the maintenance of normal or characteristic structure and dynamics by a living system.

biological limit value (BLV). See limit value.

biological magnification. biomagnification.

biological marker (biomarker). any alteration in the structure or function of a given organ, tissue, or cell of an organism that is induced exclusively (or nearly so) by exposure to a particular xenobiotic chemical or class of xenobiotics. In the broadest sense, any toxic end point or clinical sign that indicates exposure to a specific biologically active agent could be considered a biomarker. In practice, the term is usually applied to distinct alterations (lesions) within cells, tissues, or organs of the affected organism that are specific to a given chemical. Operationally (and ideally), a biomarker is any lesion that can be used in bioindication or biological monitoring to detect or predict a toxic event in the affected species with a potential for extrapolation to other species. *Cf.* sign, symptom, indication, monitoring, lesion, adduct.

biological mineralization. the biological decomposition of organic compounds with the liberation of inorganic minerals; immobilization. This is an important detoxication process within biotic communities and ecosystems.

biological oxygen demand (biochemical oxygen demand; BOD). a measure of the oxygen consumed in the microbial oxidation of decomposable organic matter and the oxidation of reduced substances in water. It is thus a measure of the organic pollution of a water body. In practice, the BOD is the amount of oxygen (as mg O_2/liter water) taken up from a water sample that contains a known amount of dissolved oxygen kept at 20°C for (usually) 5 days (BOD5). The higher the BOD, the greater the organic pollution. Values above 20 for an [illegible] clean river are generally considered polluting. BOD, together with the actual oxygen concentration of the water, can also be taken as a measure of the biomass that a body of water can support. See also BOD5, CBOD5. *Cf.* chemical oxygen demand.

biological remediation. See bioremediation.

biological rhythm. any of numerous periodic (e.g., diel or circadian rhythms, annual cycles, migratory cycles, molt cycles, hibernation, tidal rhythms) phenomena associated with living organisms. Such phenomena are partly endogenous, requiring only an external signal (zeitgeber) to bring the cycle into phase with environmental cycles (day, year, tidal cycles). The toxicity of a given xenobiotic chemical may vary substantially as a function of the phase or stage of a rhythm during which exposure occurs. Thus, the effect of a drug administered in the morning can differ from that given in the evening. *Cf.* biorhythm.

biological system. any living entity (e.g., cell, tissue, organ, organism) or any interacting set of such entities (population, species, colony, community, ecosystem). Such systems contain nonliving parts.

biological threshold limit value (TLV-BLV). threshold limit-biological. See limit value.

biological transport. See transport.

biological treatment (biological waste treatment). a waste treatment technology that uses bacteria to decompose organic wastes.

biological warfare (BW). the deliberate use of biological warfare agents to kill, incapacitate, or otherwise reduce the military effectiveness of enemy troops or to deprive them of meat or crops.

biological warfare agent. See warfare agent.

biological waste treatment. biological treatment.

biologically active (bioactive). pertaining to or denoting the capacity to act on and induce a response from living organisms. See active.

biologist. a scientist who studies living systems; a student of biology. Many, but not all, toxicologists are biologists by training or expertise.

biology (life science). the branch of natural science that deals with living systems (e.g., cells, tissues, organs, organisms, populations, and biocenoses), life processes, interactions among living systems, and interactions of living systems with nonliving systems and environmental variables.

biolysis. death and/or subsequent decomposition of living (organic) matter by the chemical action of living organisms.

biolytic. **1:** pertaining to or denoting biolysis. **2:** destructive to life. **3:** denoting living systems that are capable of destroying living material.

biomagnification (biological magnification; biological amplification). **1a:** the phenomenon whereby the concentration of an environmental chemical increases from the original concentration at the point of entry into the environment (usually in the media — soil, sediment, water, or air — or at a relatively low trophic level) upward through one or more steps in a food chain. **1b:** operationally, biomagnification is the increasing concentration of an environmental chemical at higher trophic levels. **1c:** the tendency of a chemical to increasing concentrations at higher trophic levels through by dietary accumulation. **2:** the increasing concentration of an environmental chemical (e.g., a pesticide, heavy metal, or other xenobiotics) within organisms as it progresses through a food chain. Substances that biomagnify (e.g., heavy metals, chlorinated hydrocarbons) are usually environmentally persistent and always have relatively long biological half-lives. See also factor (biomagnification factor).

biomarker. biological marker.

biomass. the total weight of living material of a specified type in a given area or for a given population or community. Operationally, biomass is any quantitative estimate of the total mass of specified organisms that comprise part or all of a population, area, habitat, or other defined unit. Frequently used in reference to vegetation. Biomass is usually expressed as volume or weight of live or dry material, ash-free weight, standing crop, standing stock, or as energy in calories.

biomedical. usage and definitions of this term are varied and broadly overlapping because medicine, as a science, is not a unified field in itself, but draws upon most of the natural sciences for its body of knowledge and applications. **1:** biological and medical; pertaining to the purely biological aspects of medicine, especially with respect to research. **2:** pertaining to those biological sciences or components of a biological science that contribute to, support, or are directly involved in the study or practice of medicine. **3:** pertaining to those aspects of the science and the practice of medicine that draw primarily on the biological knowledge. **4:** pertaining to or including medicine and the biological and physical sciences. General toxicology is in large measure a biomedical field, although individual investigators and practitioners usually bring a background, interests, and skills that strongly orient them toward one or more of the contributing sciences to the various fields of medicine. Application of those sciences to the field of toxicology. Even environmental toxicology and ecotoxicology draw from and contribute to all of the various biomedical sciences. **5:** of or pertaining to the application of biological knowledge or disciplines to the study or practice of medicine.

Biomet TBTO. tributyltin.

biomethylated arsenic. methanearsonic acid and cacodylic acid are biomethylated compounds. They were responsible for numerous cases of arsenic poisoning in Europe during the 19th century. The arsenic in plaster and wallpaper was converted, under humid conditions, to these biomethylated forms. Inhalation of the released fumes caused persons sleeping and working in affected rooms to become ill. See biomethylation.

biomethylation. the introduction of one or more methyl groups, CH_3, to a toxic element (e.g., mercury, arsenic, lead) through microbial (often anaerobic) activity. Methylation may greatly increase the toxicity of a particular element, as in the case of mercury. Methylated heavy metals (e.g., lead, mercury) are generally more readily absorbed by biological membranes than are inorganic species. *Cf.* organometals.

biometrician. a specialist in the field of biometry.

biometrics. biometry.

biometry (biometrics). the science that treats the statistical analysis of biological data, especially those based on the measurement of variation in living organisms.

biominification. the maintenance by a living organism of tissue concentrations of a xenobiotic chemical below that of the environment or of ingested food.

biomonitor. if the effect registered by a bioindicator is sufficiently quantitative, the indicator may be termed a biomonitor.

biomonitoring. See monitoring.

bion. a living entity; the singular form of biota.

bionecrosis. necrobiosis.

bionomics. **1:** bionomy. **2:** ecology. See bionomy.

bionomy (bionomics). the laws and principles that describe or regulate life processes; the science concerned with such laws. In application, bionomy is sometimes nearly synonymous with ecology.

biopharmaceutics. the study of the properties of drugs, especially as they relate to the factors that affect drug action and determine dosage.

biophylactic. pertaining to biophylaxis.

biophylaxis. **1:** self protection of an organism by means of nonspecific reactions to pathogens, xenobiotics, or other threats to the integrity of the organism. **2:** the reactions themselves (e.g., inflammatory processes, phagocytosis).

biophysics. application of the concepts, assumptions, theories, and methods of physics to the study of living systems, including both life processes and physical processes (e.g., electrical phenomena, mechanics, luminescence) that occur in living systems.

biopsy. 1: the removal of a small amount of tissue or fluid for examination as an aid to diagnosis. **2:** a specimen obtained by biopsy.

biorestoration. a term generally applied to a class of methods for managing hazardous wastes in subsurface (e.g., soil) environments. Such methods are designed to convert organic wastes into microbial biomass and harmless by-products. Biorestoration involves stimulating either indigenous or introduced subsurface microflora to degrade xenobiotics in place. See also restoration, remediation.

biorhythm. a popular and commercial term, the definition of which varies with the commercial application or the state of ignorance of the user. Its use is science is discouraged. *Cf.* biological rhythm.

biosis. life in general; living organisms in general.

biosphere (ecosphere). sometimes referred to as the global ecosystem, it is that part of the Earth and its atmosphere that contains and supports life. Poisoning of major portions of the biosphere has occurred (e.g., from acid rain) and must be regarded a clear possibility now and in the future by new or as yet unidentified chemicals or constellations of chemicals.

biostasis. the capacity of living organisms to maintain their status and integrity in the face of environmental changes.

biostatistics. **1:** the science that treats relationships, variations, and differences among biological data. **2:** vital statistics.

biosynthesis. 1: the synthesis of organic compounds by a living cell, utilizing simple precursors. **2:** the synthesis of organic compounds facilitated by enzymes, cell fragments, or cell extracts *in vitro*.

biosynthetic. pertaining to, denoting, or produced by biosynthesis.

biosystem. **1:** biological system. **2:** ecosystem.

biosystematics. experimental taxonomy. See taxonomy.

biota. **1:** living organisms in general. **2:** flora and fauna collectively, especially that of a particular region.

biotechnology. a relatively new and rapidly-growing family of technologies, nearly all of which have raised new questions and new challenges for toxicology. Subdisciplines of many sciences contribute to biotechnology. **1:** the use of natural and genetically engineered biological systems (e.g., cells, cell fragments, organelles, isolated enzymes, and enzyme systems) in manufacturing or in the provision of services to humans. Included are numerous biosynthetic and biodegradative processes. Examples of biotechnologies are the fermentation of sugar by yeast to produce alcohol or alcoholic beverages; the use of *Lactobacillus* to produce yoghurt; the large-scale microbial production of drugs such as antibiotics, insulin, interferon, and serum proteins; microbial breakdown of toxic and hazardous wastes. **2:** any use of biological systems to convert raw materials to marketable products. **3:** the infrastructure that supports industrial uses of biological systems, including the design and operation of equipment (e.g., fermenters, bioreactors, analytical equipment, downstream processing).

biotic. living; of or pertaining to living systems or to life processes, especially to populations, biocenoses, or ecosystems. *Cf.* biologic, biological.

biotic community. biocenosis.

biotic environment. **1:** the array of living organisms, their products, and activities that may affect a living system (i.e., an individual organism, taxon, population, community, or ecosystem). **2:** the living components, collectively, of an ecosystem.

biotic potential. the maximum population growth rate of a population under ideal conditions. See also reproductive potential, environmental resistance.

biotics. the science that treats life functions and vital activity.

biotope. a particular type of habitat or area that is defined by the particular association of organisms that typically occur in it. Examples are grassland, woodland, a southfacing slope of a hill. *Cf.* also ecosystem.

biotoxication. poisoning (intoxication) caused by a poison or venom produced by or transmitted by a plant or animal. See biotoxin.

biotoxicologist. a scientist who studies living organisms that produce, contain, secrete, or transmit poisons; the poisons concerned, their effects, biological origins, and significance. *Cf.* phytotoxicologist, zootoxicologist.

biotoxicology. See toxicology.

biotoxin. See toxin.

biotransformation. the metabolic transformation of one substance to another within a living system (e.g., within a cell, organism, community, or ecosystem). Usually used in reference to a foreign compound referred to as the parent compound. Such transformations can increase or decrease a chemical's toxicity or alter other pharmacologic, biochemical, or physiological properties and effects. See also biodegradation, environmental transformation. **hepatic biotransformation**. biotransformation that occurs in the liver. These biotransformations (metabolism, detoxication) usually take place in two stages in the mammalian liver: cytochrome P-450-dependent oxidation and subsequent conjugation with endogenous material. The conjugates are usually more polar than the original compound and thus more readily excreted. This is an essential process in the elimination of most toxicants from higher animals. See also metabolic reactions, environmental transformation.

bioturbated. disturbed or mixed by the activity, usually burrowing, of living organisms, as a bioturbated sediment. Such activity can spread a pollutant or toxic substance to levels or areas of a sediment that have not been directly exposed. See also bioturbation.

bioturbation. the mixing of materials (e.g., pollutants, toxic substances) in a sediment by burrowing, or other activities of living organisms. Bioturbation of bottom sediments increases the difficulty of determining the times of intrusion of environmental chemicals.

biphenyl (phenylbenzene; 1,1′-biphenyl; diphenyl; bibenzene). a combustible, very toxic, pleasant-smelling aromatic hydrocarbon, C_6H_5-C_6H_5. It is a powerful irritant to the human pulmonary system. Inhalation of very small amounts can cause nausea, vomiting, flaccid paralysis, and other nonspecific gastrointestinal effects.

1,1′-biphenyl. biphenyl.

biphenylamine. See *p*-biphenylamine.

[1,1′-biphenyl]-4-amine. *p*-biphenylamine.

4-biphenylamine. *p*-biphenylamine.

***p*-biphenylamine** (*p*-diphenylamine; 4-biphenylamine; [1,1′-biphenyl]-4-amine; *p*-aminobiphenyl; *p*-aminodiphenyl; 4-aminobiphenyl; 4-aminodiphenyl; anilinobenzene; biphenylamine; *p*-phenylaniline; *p*-xenylamine). a yellowish-brown or colorless, irritant, crystalline, biphenyl aromatic amine, $(C_6H_5)_2NH$, with effects similar to those of benzidine. It is a confirmed carcinogen of humans and some laboratory animals. Several analogs are also carcinogenic. It is used to detect sulfates, as a tool in cancer research, and formerly as an antioxidant in the manufacture of rubber. A relatively brief exposure (as little as 133 days) can cause bladder cancer with a latency of 15-35 years. *Cf.* benzidine. See also diphenylamine, aromatic amine.

2-biphenylamine hydrochloride. carcinogenic to some laboratory animals.

[1,1′-biphenyl]-4,4′-diamine. benzidine.

bipyridilium compound. See bipyridylium quaternary ammonium salt.

bipyridyl herbicide. See bipyridilium quaternary ammonium salt.

bipyridylium quaternary ammonium salt (bipyridilium compound; bipyridyl herbicide). any heterocyclic ammonium salt with a bipyridine nucleus. They are potent and rapid-acting herbicides. They rapidly and strongly adsorb to soil and are rapidly absorbed by leaves. As a result, rain falling even within ½ hour of spraying has no effect on the potency of these compounds. The best known of these salts (diquat, paraquat, and morfamquat) differ in phytotoxic action from conventional quaternary ammonium compounds because they are able to form free radicals by reduction in photosystem 1 of the chloroplast. They are more toxic to animals than most other herbicides.

bird. common name of any animal of the vertebrate class Aves, *q.v.* **cage bird**. any species or variety of bird that is adapted to confinement in a cage, usually small birds kept as pets or songsters in the home. Lead poisoning is the commonest poisoning of cage birds. When allowed to fly in the home, they may consume poisonous house plants. See also pitohui.

bird of paradise. See *Caesalpinia*, *Strelitzia*. See also poinciana.

bird snake. *Thelotornis kirtlandii*.

bird spider. tarantula. See Theraphosidae.

bird's eye. common name for plants of the genus *Adonis*.

birdsfoot trefoil. *Lotus corniculatus*.

birdshot. See lead shot.

Birgus latro (coconut crab; palm crab; purse crab). a large terrestrial crustacean (Order Decapoda) that weighs up to 9.1 kg. It is a reef crab found in moist forests on tropical islands of the Indo-Pacific region, returning to the ocean to breed. It can be poisonous as a result of feeding on toxic plants. Symptoms of intoxication include nausea, vomiting, headache, fever with chills, joint aches, muscular weakness, exhaustion, and occasionally death.

birit tiu. *Thelotornis kirtlandi*.

birth defect. any inherited disorder or abnormality that affects newborn mammals. Some are chiefly genetic in origin, and others are primarily the effect of environmental factors, including direct effects of xenobiotic chemicals *in utero* or indirectly via effects on the pregnant female. See also teratogenesis, teratology, placenta, placental barrier.

birth rate. natality.

birthwort. **1:** any herb of the genus *Arisaema* and *Aristolochia*. **2:** serpentaria.

6,16-bis(acetyloxy)-3,7-dioxo-29-nordammara-1,17(20),24-trien-21-oic acid. helvolic acid.

2,4-bis[allylamino]-6-[4-[bis(*p*-fluorophenyl)-methyl]-1-piperazinyl]-s-triazine. almitrine.

***N,N′*-bis-(2-aminoethyl)-1,2ethanediamine**. trientine.

2,5-bis(1-aziridinyl)-3-(2-carbamoyloxy-1-methoxyethyl)-6-methyl-1,4-benzoquinone. carboquone.

2,5-bis(1-aziridinyl)-3-(2-hydroxy-1-methoxy ethyl)-6-methyl-*p*-benzoquinone carbamate (ester). carboquone.

bis[bisdimethylaminophosphonous]anhydride. schradan.

1,3-bis(carbamoylthio)-2-*N,N*-(dimethylamino)-propane. cartap.

3,6-bis(carboxymethyl)-3,5-diazooctanedioic acid. edetic acid.

(*p*-(bis(2-chloroethyl)amino)phenyl)acetic acid cholesterol ester. fenesterine.

(4-(bis(2-chloroethyl)amino)phenyl)acetic acid cholesteryl ester. fenesterine.

5-[bis(2-chloroethyl)amino]-2,4-(1*H*,3*H*)-pyrimidinedione. uracil mustard.

5-[bis(2-chloroethyl)amino]-uracil. uracil mustard.

bis(β-chloroethyl) ether. *sym*-dichloroethyl ether.

bis(2-chloroethyl) ether. *sym*-dichloroethyl ether.

9-(4-(bis-β-chloroethylamino)-1-methylbutylamino)-6-chloro-2-methoxyacridine. quinacrine mustard.

1-bis(2-chloroethyl)amino-1-oxo-2-aza-5-oxaphosphoridin. cyclophosphamide.

1-3-[*p*-(bis{2-chloroethyl}amino)phenyl]alanine. melphalan.

3-[*p*-(*p*-{bis(2-chloroethyl)amino}phenyl]-1-alanine. melphalan.

4-[bis(2-chloroethyl)amino]-L-phenylalanine. melphalan.

5-[bis(2-chloroethyl)amino]-2,4-(1*H*,3*H*)-pyrimidinedione. uracil mustard.

4-[bis(2-chloroethyl)amino]benzenebutanoic acid. chlorambucil.

4-[*p*-[bis(2-chloroethyl)amino]aminophenyl]-butyric acid. chlorambucil.

4-[bis(2-chloroethyl)amino]-L-phenylalanine mustard. melphalan.

2-[bis(2-chloroethyl)amino]tetrahydro-2*H*-1,3,2-oxazophosphorine 2-oxide. cyclophosphamide.

***p*-*N*-bis(2-chloroethyl)amino-1-phenylalanine**. melphalan.

5-[bis(2-chloroethyl)amino]uracil. uracil mustard.

bis(2-chloroethyl)ethylamine. HN1.

***N,N*-bis(2-chloroethyl)-2-naphthylamine**. chlornaphazine.

1,3-bis(2-chloroethyl)-1-nitrosourea (BCNU). carmustine.

***N,N′*-bis(2-chloroethyl)-*N*-nitrosourea**. carmustine.

bis(2-chloroethyl)phosphamide cyclic propanolamide ester. cyclophosphamide.

bis(2-chloroethyl)phosphoramide cyclic propanolamide ester. cyclophosphamide.

***N,N*-bis(2-chloroethyl)phosphorodiamidic acid, cyclohexyl ammonium salt**. phosphoramide mustard cyclohexylamine salt.

***N,N*-bis(β-chloroethyl)-*N′,o*-propylene-phosphoric acid ester diamide**. cyclophosphamide.

bis(β-chloroethyl)sulfide. mustard gas (See gas).

bis(2-chloroethyl)sulfide. mustard gas (See gas).

***N,N*-bis(2-chloroethyl)tetrahydro-2*H*-1,3,2-oxazaphosphorin-2-amine 2-oxide**. cyclophosphamide.

1,2-bis(2-chloroethylthio)ethane. sesquimustard.

bis(2-chloroethylthioethyl)ether. *o*-mustard.

***N,N*-bis(β-chloroethyl)-*N′,o*-trimethylenephosphoric acid ester diamide**. cyclophosphamide.

bis(chloromethyl) ether. *sym*-dichloromethyl ether.

1,1-bis(*p*-chlorophenyl)ethanol. chlorfenethol.

1,1-bis-(*p*-chlorophenyl)methyl carbinol. chlorfenethol.

1,1-bis(*p*-chlorophenyl)-2-nitrobutane. Bulan.

β,β-bis(*p*-chlorophenyl)-β,β,β-trichlorethane. DDT.

1,1-bis(*p*-chlorophenyl)-2,2,2-trichloroethanol. dicofol.

3,3′-bis[(diethylamino)methyl]-5,5′-diallyl-α,α′-bis(diethylamino)-m,m′-bitolyl-4,4′-diol. bialamicol.

bis-*O,O*-diethylphosphoric anhydride. tetraethyl pyrophosphate.

bis(diethylthiocarbamoyl) disulfide. disulfiram.

bis(diethylthiocarbamyl) disulfide. disulfiram.

bis[μ-[2,3-dihydroxybutanedioato(4—)01,02:03,04]]-diantimonate dipotassium trihydrate (*stereoisomer*). antimony potassium tartrate.

bis[2-dimethylaminoethyl]succinate bis-[methochloride. succinylcholine chloride.

3,7-bis(dimethylamino)phenothiazin-5-ium chloride. methylene blue.

bis[*p*-(dimethylamino)phenyl]phenylmethylium chloride. Malachite Green; both the chloride and the oxalate are termed Malachite Green.

bis(dimethylthiocarbamyl)disulfide. thiram.

bis(2,3-epoxypropyl)ether. diglycidyl ether.

S-1,2-bis(ethoxycarbonyl)ethyl *O,O*-dimethyl-[illegible]dithioate. malathion.

bis(2,3-epoxypropyl)ether. diglycidyl ether.

bis(2-ethylhexyl) phthalate (1,2-benzenedicarboxylic acid bis(2-ethylhexyl) ester; di-*n*-octyl phthalate (DOC); di(2-ethylhexyl) phthalate; dioctyl phthalate; DEHP). a clear, colorless, odorless, oily, irritant liquid, $C_{24}H_{38}O_4$, used large quantities as a plasticizer. It is a persistent and widespread environmental pollutant. Acute oral toxicity to most laboratory animals is moderate; however, laboratory rats exposed for extended periods develop hepatomegaly with fatty degeneration, atrophy, and focal degeneration of the seminiferous tubules of the testis. DEHP has little effect on the liver of guinea pigs and dogs and little effect on the hamster testis. Cytotoxicity, pulmonary hemorrhage, teratogenicity, and narcosis (at high concentrations) have been reported. It is carcinogenic to some laboratory animals. Ingestion by humans may cause a mild gastritis. It is rapidly detoxified and excreted in the bile and urine. Do not confuse with di-*n*-octyl phthalate.

bisethylxanthogen. dixanthogen.

6-[4-[bis(4-fluorophenyl)methyl]-1-piperazinyl]-*N,N′*-di-2-propenyl-1,3,5-triazine-2,4-diamine. almitrine.

2,2′-bis[8-formyl-1,6,7-trihydroxy-5-isopropyl-3-methylnaphthalene]. gossypol.

bisfuranoid mycotoxin. any mycotoxin that contains the dihydrofurobenzofuran system. All metabolites of these mycotoxins characteristically include either the unsaturated 7,8-dihydrofurano-(2,3-b)furan or the more reduced 2,3,7,8-tetra-hydrofuro(2,3-b)furan. They are hepatotoxic and tumorigenic to various animals. Examples are aflatoxins and sterigmatocystin.

bish. *Aconitum ferox*.

bishma. *Aconitum ferox*.

bishop's weed. *Ammi majus*.

bishydroxycoumarin. dicumarol.

bis[2-hydroxy-3,5-dichlorophenyl]sulfide. bithionol.

β-bis(hydroxyethyl)[illegible]

2,2-bis(hydroxymethyl)propionic acid. dimethylolpropionic acid.

bis(6-hydroxy-2-naphthyl) disulfide. DDD, analytical.

3,4-bis(*p*-hydroxyphenyl)-3-hexene. diethylstilbestrol.

bismethomyl thioether. thiodicarb.

2,2-bis(*p*-methoxyphenol)-1,1,1-trichloroethane. methoxychlor.

α-[2-[bis(1-methylethyl)amino]-ethyl]-α-phenyl-2-pyridineacetamide. disopyramide.

bis(1-methylethyl)carbamothioic acid *S*-(2,3-dichloro-2-propenyl) ester. diallate.

bis(1-methylethyl)carbamothioic acid *S*-(2,3,3-trichloro-2-propenyl) ester. triallate.

bis(2-methylpropyl)carbamothioic acid *S*-ethyl ester. butylate.

1,3-bis(1-methyl-2-pyrrolidinyl)-2-propanone. cuscohygrine.

***N,N'*-bis[1-methylthioacetaldehyde *o*-(*N*-methyl-carbamoyl)oxime]sulfide**. thiodicarb.

bis-[*o*-(1-methylthioethylimino)-*N*-methylcarbamic acid]-*N,N'*-sulfide. thiodicarb.

bis(monothiosulfato)aurate(3—) trisodium. gold sodium thiosulfate.

bis[monothiosulfato(2—)-*O*-*S*]aurate(3—) trisodium. gold sodium thiosulfate.

bismuth (Bi)[Z = 83; A_r = 209]. a toxic, brittle, pinkish, trivalent metallic element that serves no known essential life functions. A number of its compounds are used in cosmetics and medicines. Bismuth is and most of its compounds are not very toxic. See bismuthosis.

bismuth poisoning. See bismuth, bismuthosis.

bismuthism. bismuthosis.

bismuthosis (bismuthism). chronic bismuth poisoning; signs include dermatitis, anuria, stomatitis, and diarrhea; renal damage may result.

bis(nicotinic acid) diester of *N*-allylnormorphine. nalorphine dinicotinate.

bis-*N,N,N',N'*-tetramethylphosphorodiamidic anhydride. schradan.

bis(tributyltin) oxide. tributyltin.

bis(3,5,6-trichloro-2-hydroxyphenyl)methane. hexachlorophene.

2,4-bis(isopropylamino)-6-chloro-*s*-triazine. propazine.

2,2'-bis[1,6,7-trihydroxy-3-methyl-5-isopropyl-8-aldehydonaphthalene]. gossypol.

bite. **1a:** to seize, wound, abrade, or puncture with the teeth or mouth. **1b:** an injury produced by biting. The trauma associated with a bite may be slight to severe, but the bites of most animals are not serious to humans. Poisoning or serious infection (e.g., septicemia, tetanus) is always possible, however, and appropriate preventive measures should be taken. Bites that are seriously poisonous to humans are relatively rare, but the bites of some animals can be seriously injurious or even fatal. See also sting.

bithionol (bis[2-hydroxy-3,5-dichlorophenyl]-sulfide; 2,2'-thiobis[4,6-dichlorophenol]). a whitish water-insoluble, crystalline powder, $HOCl_2C_6H_2SC_6H_2Cl_2OH$, used as an antiparasitic against human lung worm, *Paragonimus westermani* (human lung worm) and *Clonorchis sinensis* (Oriental liver fluke); also as a bacteriostat in soaps and detergents, as a deodorant, and in pharmaceuticals.

biting reef worm. *Eunice aphroditois*.

Bitis (African vipers). a genus of venomous African snakes (Family Viperidae) with broad, heart-shaped heads; a narrow neck; a short snout; and a distinct canthus. The genus includes the largest true vipers. All are dangerous, some extremely so. The body is moderately to unusually stout and moderately depressed with a short tail. The eyes are small with vertically elliptical pupils. The crown is sheathed in small scales (there are no enlarged plates). Some species have large, erect scales on the snout or above the eye. Dorsal body scales are keeled with apical pits; ventrals are rounded or have indistinct lateral keels; subcaudals are paired and laterally keeled in some species. ***B. arietans*** (*B. lachesis*; African puff adder; puff adder; aipomiro; alnawana; asigirikolongo; bulabundoo; bululu; chikorviri; chiva; choichodo; dagar; essalambwa; ibiboboca; iBululu; kipili; kipiri; kassa; kiiri; lebolobolo; lerabe; liboma; lipiri; marabe; mhiri; moma; mpoma; nawama akipom; olwero; piri; pofadder; puff otter; thamaha; thamaha-dinkotsane; thiéby; trigonocéphale; vipère du Cap; vipère hébraique; vipère heurtante; vuluvulu). a thick-bodied, extremely dangerous viper with an extremely toxic venom that occurs in Africa south of the Sahara and throughout the Middle East. It attains a length of 1.5 m. and is noted for its violent hissing. The body is gray to dark brown with thin yellow chevrons on the back. Envenomation may produce severe pain, rapidly developing, heavy edema, and intense discoloration of the skin. Otherwise the spectrum of signs and symptoms are similar to those of *Vipera berus* including dizziness. This snake probably kills more humans than any other snake. This is reflected in the many colloquial names given to this viper. Antivenin is available. ***B. atropos*** (mountain adder; Berg adder;). a South African viper that is similar in appearance to *B. caudalis*, but with no spine over the eye. ***B. caudalis*** (horned puff adder; horned viper; horned adder; Buschel-brauen viper; chipukupaku; horingadder; in-Dlonlo; shaushawane; sheushewane). a pale, light-colored, desert species of southern and southwestern Africa. It is characterized by a short snout, a raised supraorbital ridge, and a horn-like spine (usually) above the eye. Adults average about 0.4 m in length, but the venom is highly toxic; it often strikes with little provocation, and deaths of humans sometimes result from its bite. See *B. cornuta*, *B. atropos*. ***B. cornuta*** (horned adder; hornsman). a South African viper that is similar in appearance to *B. caudalis*, but with multiple spines over the eye. ***B. gabonica*** (Gaboon viper, king puff adder; Gabun; butterfly adder; h'ion; moma; mônemé; koningpofadder; vipére à cornes). the largest, heaviest, deadliest and most dangerous of the vipers, it is indigenous to the rain forests of tropical west Africa. It is a huge, brightly marked, extremely thick-bodied viper that sometimes reaches a length of more than 2 m. The color pattern is distinctive; the crown is [illegible] with a narrow brown median stripe, and the body has a complex pattern of black, tan, and blue markings, some with white edging. The fangs may be 5 cm long in large individuals. Because it is nocturnal and difficult to arouse during the daytime, envenomations of humans are common. The bite may cause internal hemorrhage and without prompt, appropriate treatment is usually fatal. ***B. inornata***. Cape puff adder. ***B. lachesis***. *B. arietans*. ***B. nasicornis*** (rhinoceros horned viper; rhinoceros viper; nose-horned viper; horned puff adder; river jack; horo; kissadi; nashornviper; nonemé; pegali; toulou; toumou; vipère à cornes; vipère nasicornis; vipère rhinocéros). a large, brightly-colored, extremely thick-bodied snake with complex markings. The head is small and the snout bears 2 or 3 characteristic horn-like growths. It is easily distinguished from *B. gabonica* by the large, dark, arrow-shaped mark on the crown. The rhinoceros viper occurs in moist situations (e.g., in swamps, along river banks) in the tropical rain forests of central and southern Africa. The venom is highly toxic, but this snake is not especially pugnacious and inflicts few bites on humans. ***B. peringueyi*** (Peringuey's adder). a species that occurs throughout the Republic of Angola, portions of southwest Africa, and southern portions of the Republic of Botswana. ***B. worthingtonii*** (Kenya horned viper). found throughout the Republic of Kenya, it is characterized by rostrals that abut the nasals.

bitter alkaloid. See *Menispermum*.

bitter almond. *Prunus amygdalus amara*.

bitter almond oil. **1:** benzaldehyde. **2:** apricot-kernal oil.

bitter apple. *Citrullus colocynthus*, colocynth.

bitter cucumber. *Citrullus colocynthus*, colocynth.

bitter gourd. **1:** *Citrullus colocynthus*. **2:** *Momordica charantia*.

bitter principles of cucurbits. common name for cucurbitacins.

bitter rubberweed. *Hymenoxys odorata*.

bitter salts. the heptahydrate of magnesium sulfate.

bitter tonic. serpentaria.

bitter weed. *Actaea odorata*. See also bitterweed.

bittersweet, European bittersweet. *Solanum dulcamara*.

bitterweed. **1:** *Hymenoxys odorata*. **2:** *Senecio glabellus*. See also bitter weed.

bitumen. See asphalt.

bituminosis. a pneumoconiosis caused by the inhalation of dust from bituminous (soft) coal.

bivinyl. 1,3-butadiene.

Bk. the symbol for berkelium.

black acacia. *Robinia pseudoacacia*.

black alder. See ***Ilex aquifolium, opaca, verticillata***. See houseplant.

black and blue. livid.

black and scarlet elder. See ***Sambucus canadensis, pubens.***

black and white-lipped cobra. *Naja melanoleuca*.

black and yellow garden spider. *Argiope aurantia*.

black ash. barium sulfide.

black-banded coral snake. *Micrurus nigrocinctus*.

black-banded sea snake. *Laticauda laticauda*.

black baneberry. *Actaea (= Actea) spicata*.

black-bellied snake. *Denisonia pallidiceps, D. signata*.

black burrowing viper. *Atractaspis bibronii*.

black cherry. *Prunus serotina*; *P. demissa*; *P. melanocarpa*.

black cobra. See *Naja naja*, *Naja melanoleuca*, *Pseudohaje goldii*.

black cohosh. *Actaea (= Actea) spicata*.

black-collared cobra. *Naja nigricollis*.

black damp. See damp.

black elder. *Sambucus canadensis*.

black elderberry. See *Sambucus canadensis, pubens*.

black-eyed Susan. **1:** the vernacular name of a common wildflower, *Rudbeckia hirta*. The central disk of the flower is large and dark; the petals that radiate from the disk are yellow. **2:** an occasional colloquial name for *Abrus precatorius*, *q.v.* The name is presumably due to coloration and pattern of the seeds which are a brilliant red with black spots.

black forest cobra. *Pseudohaje goldii*.

black garter snake. *Elapsoidea sundevallii*.

black greasewood. *Sarcobatus vermiculatus*.

black-headed snake. *Denisonia gouldii*.

black hellebore. See *Helleborus niger*.

black henbane. **1:** *Hyoscyamus niger*. **2:** See hyoscyamus.

black Indian hemp. *Apocynum cannabinum*.

black krait. *Bungarus niger*.

Black Leaf 40. a 40% solution of nicotine sulfate, formerly used as an insecticide.

black-lipped cobra. *Naja melanoleuca*.

black locust. *Robinia pseudoacacia*.

black long-spined sea urchin. *Diadema setosum*.

black lung. a common term used chiefly as a synonym of anthracosilicosis (coal worker's pneumoconiosis), but also of anthracosis and pneumomelanosis.

black lupin bean. *Lupinus niger*. See sparteine.

black mamba. *Dendroaspis polylepis*.

black manganese dioxide. manganese dioxide.

black mercuric sulfide. See mercuric sulfide.

black mercury sulfide. black mercuric sulfide. See mercuric sulfide.

black morel. *Morchella elata*.

black-mouthed mamba. *Dendroaspis polylepis*.

black mustard. *Brassica nigra*.

black-naped snake. *Denisonia gouldii*, *Vermicella bimaculata*.

black-necked cobra. *Naja nigricollis*.

black-necked spitting cobra. *Naja nigricollis*.

black nightshade. *Solanum nigrum*. *Solanum americanum* is sometimes also referred to as "black nightshade."

black oxide of iron. iron oxide.

black pepper. *Piper nigrum*.

black phthisis. coal worker's pneumoconiosis.

black poison. **1:** the toxic juice of *Metopium brownei*. **2:** a common name of *Metopium brownei*.

black poisonwood. *Metopium brownei*.

black racer. See blacksnake.

black-ringed coral snake. *Micrurus mipartitus*.

black rat. *Rattus rattus*. See also laboratory rat.

black rockfish. *Sebastes mystinus*.

black sea urchin. See *Diadema*.

black snake, blacksnake. **1:** a common name most often and most properly applied to the venomous *Pseudechis porphyriacus*, the Australian black snake. **2a:** *Coluber constrictor*, a nonvenomous snake more properly termed the racer (additional common names for this snake abound in the literature, e.g., blue racer, black racer, American blacksnake). **2b:** sometimes applied to colubrids generally.

black snakeroot. See *Zigadenus densus*.

black-spotted palm viper. *Bothrops nigroviridis*.

black-spotted puffer. *Arothron nigropunctatus*.

black-striped snake. *Denisonia nigrostriata*, *Vermicella calonota*.

black-tailed rattlesnake. *Crotalus molossus*.

black tigersnake. *Notechis scutatus*.

black-tipped reef shark. *Carcharhinus melanoperus*.

black tree snake. *Dispholidus typus*.

black walnut. See *Juglans nigra*.

black watersnake. *Elapsoidea sundevalli*.

black whip snake. *Demansia olivacea*.

black widow, black widow spider. *Latrodectus mactans* and *L. curacaviensis*.

black widow spider venom poisoning. poisoning by *Latrodectus mactans*. Symptoms of envenomation usually include intense pain, restlessness, irritability, tremors, shallow breathing, tachycardia and hypertension. Death may intervene, mostly in the case of children. An antivenin, Lyovac® Antivenin, is available but is normally used only for children. See also latrotoxin.

blackboard chalk. markers made of calcium carbonate and used to write or draw on blackboards. It is usually not toxic if ingested.

blackdamp. See firedamp.

blackjack pine. *Pinus ponderosa*.

blacksnake. See *Dispholidus typus*, *Pseudechis*, *Walterinnesia*.

blackwater. water that contains animal, human, or food wastes.

bladder. any fluid-filled or inflatable sac such as the swim bladder of bony fish, the urinary bladder of certain animals, or the air bladder of many algae of the order Fucales (e.g., *Fucus vesiculosus* (the bladder wrack). **urinary bladder**. a muscular, membranous sac that temporarily stores liquid or semiliquid wastes in certain invertebrate animals and in terrestrial vertebrates other than birds. See also 2-naphthylamine. **gallbladder**. a muscular, sac-like diverticulum of the bile duct in which bile is stored and concentrated. It contracts and empties the bile into the intestines under the influence of the hormone, cholecystokinin. The latter is released into the duodenum in response to the presence of food. A gallbladder is lacking in many birds and mammals.

blanc fixe. artificial (precipitated) barium sulfate.

Blarina brevicauda. (short-tailed shrew; American mole shrew). a shrew (Family Soricidae) up to 10 cm long that occurs in forests and grasslands of the eastern United States. One of extremely few venomous mammals. It secretes a neurotoxic saliva that flows into wounds made by the lower incisors, causing respiratory distress and bradycardia in the prey.

blasting gelatin. nitroglycerin.

blasting oil. nitroglycerin.

blastocyte. an undifferentiated embryonic cell. *Cf.* blastocyst.

blastogenic. originating in the germ plasm.

blastomogen. an agent that produces neoplasms; a carcinogen. *Cf.* oncogen, tumorigen.

blastomogenesis. the production or causation of tumors. *Cf.* oncogenesis, carcinogenesis.

blastomogenic. pertaining to or denoting the capacity to produce neoplasms; carcinogenic. *Cf.* oncogenetic, tumorigenic.

blastomogenous. blastomogenic.

blausäure. hydrogen cyanide.

bleach. See hypochlorite, hypochlorite poisoning.

bleach, chlorine. See bleach, household.

bleach, household. any product used in the home to bleach materials, usually fabrics. The chief active ingredients are usually hypochlorite salts, especially sodium hypochlorite which is produced by adding chlorine to a liquid solution of lye. Household bleaches contain other harmful ingredients such as chlorine, lye, artificial dyes, detergents, fluorescent brighteners, and synthetic fragrances. In the United States, household bleaches carry warning labels such as "CAUTION: keep out of reach of children. May be harmful if swallowed or may cause severe eye irritation. Never mix chlorine bleach with cleaning products containing ammonia, or with vinegar. The resulting chloramine fumes are deadly. It also should not be used on silk, wool, mohair, leather, spandex, or on any natural fiber that is not colorfast as it can damage or discolor the fabric, and cause colors to run." Note that such labels only warn against drinking liquid bleach. It is a strong irritant, however. Contact with the skin and inhalation of fumes should be avoided also. See hypochlorite, hypochlorite poisoning.

bleaching powder. chlorinated lime.

bleeding. the discharge of blood as from a punctured, incised, or damaged blood vessel. **arterial bleeding**. bleeding from an artery, characterized by spurts of bright red blood. **capillary bleeding**. bleeding from capillaries. Capillary bleeding is diffuse, often persistent, but easily controlled, and occurs to some extent in all bleeding. **external bleeding**. the discharge of blood from the body's exterior. **gastrointestinal bleeding**. bleeding from the gastrointestinal tract. This should always be treated as a serious condition and the cause promptly determined. **internal bleeding**. bleeding into internal body spaces. **occult bleeding**. inapparent bleeding. **venous bleeding**. the discharge of blood from a vein, characterized by continuous flow of dark red blood. *Cf.* hemorrhage.

bleeding disease. a disease of cattle caused by feeding on moldy sweetclover hay. The effects are similar to that of bracken poisoning and is characterized by uncontrollable internal bleeding, especially within the intestinal tract and under the skin from bruised muscles. The toxic principle is dicoumarin, a breakdown product of the glycoside, coumarin.

bleeding heart. *Dicentra.*

blennophthalmia. conjunctivitis.

bleomycin (BLM). any of a group of related glycopeptide antibiotics produced by *Streptomyces verticillus*. Bleomycin has antineoplastic and suspected carcinogenic and mutagenic activity. It is especially toxic when administered intravenously. It is moderately toxic by oral and intramuscular routes of administration, causing dyspnea and fibrosing pulmonary alveolitis. Toxic fumes, NO_x, are released on decomposition by heating.

bleomycin A2. a species of bleomycin noted for its adverse pulmonary effects in humans.

blepharism. spasm of the eyelid with continuous, involuntary blinking.

blepharitis. inflammation of the eyelid.

blepharospasm. twitching due to spasmodic contraction of the orbicularis occuli muscle.

Blighia sapida (akee; common names in Haiti: Akee; Aki; ackee; arbre; fricasse; a common name in Cuba and Puerto Rico is vegetal). a small tree native to western Africa (Family Sapindaceae), but cultivated in southern Florida, the West Indies, and tropical America. The seeds, cotyledons, and unripe fruit contain thepolypeptides hypoglycine A and B, which are deadly hypoglycemic agents. The aril of the unripe fruit contains hypoglycine A, which is the more poisonous of the two toxicants. The whitish, ripe aril is cooked and consumed as a delicacy, although the water in which the fruit is cooked is toxic. Overripe (rancid) fruit is also highly toxic. Ingestion of the unripe fruit causes Jamaican vomiting sickness, which is characterized by severe emesis, drowsiness, stupor, coma and convulsions, with conspicuous hypoglycemia. *Blighia* is also thought to cause defects in calves of domestic cattle similar to those of crooked calf disease.

blind staggers. an acute or subacute form of selenium poisoning in grazing animals. See also alkali disease, moldy corn poisoning, staggers.

blindness. the state of being blind; lack of, or severe impairment of vision or the ability to perceive visual stimuli. In medicine, usually taken as having less than 1/10 of normal vision in the more effective eye. See also amaurosis,

blister. a small, liquid-containing sac or vesicle, usually less than 1 or 2 cm in diameter. **blood blister**. a blood-colored blister with some blood in the contents. **water blister**. a blister with nearly clear watery contents.

blister beetle. See *Meliodae.*

blister beetle poisoning. cantharidin poisoning of livestock, especially horses, although cattle and sheep are sometimes affected. Most reports of poisoning originate in the central and southwestern United States. The 2-striped blister beetle (*Epicauta spp.*) sometimes swarms alfalfa hay while drying prior to baling. These beetles contain cantharidin. Ingestion of beetle-infested hay results in nephritis, cystitis, and hyperemia or ulceration of the oral, esophageal, and gastric mucosa. Clinical manifestations include colic, salivation, shock, frequent urination, and occasionally, synchronous diaphragmatic flutter. Hematuria, hemoconcentration, or neutrophilic leukocytosis may be observed, but are not diagnostic. Diagnosis is confirmed by detection of blister beetles in the hay and of cantharidin in the urine or stomach contents of affected animals. There is no certain treatment. See also cantharidin poisoning.

blister bug. *Lytta (Cantheris) vesicatoria.*

blistering beetle. cantharides.

blistering fly. cantharides.

blistering gas. vesicant gas (See gas).

BLM. bleomycin.

bloat. **1a:** to cause to swell (e.g., with water or air) or to make turgid (as a cell). **1b:** to cause or to result in the accumulation of gases in the digestive tract. **2a:** to fill to capacity or to overfill (stuff). **2b:** to swell, to become turgid. **3a:** the overextension of the first stomach (rumen), or the condition that results from rupture of the rumen of domestic cattle. Fermentation of ingested forage and release of the gases produced take place in the rumen. tthese gases are normally expelled by belching. When this is not possible, the expanding gases may rupture the rumen. See *Delphinium*. **3b:** any flatulent digestive disturbance of livestock. **4:** bloat colic of horses due to distension of the gastrointestinal lumen. Causes vary.

block. **1a:** to obstruct, arrest, or stop. **1b:** to arrest passage through. **2a:** to cause to swell (e.g., with water or air) or to make turgid (as a cell). **2b:** to swell, to become turgid. **3:** to stuff; to fill to capacity or to overfill. **4:** an obstacle or impediment to free passage. **5:** a stoppage, obstruction, arrest, or the cause of such. **6:** the interruption of the normal physiologic functioning of an organ or tissue (e.g., heart block, nerve block), often due to the presence of a toxicant or an abnormal chemical environment. **7:** the partial or complete, temporary or permanent arrest of an impulse through a conductive tissue (e.g., nerve, muscle). **8:** regional anesthesia, i.e., that which produces insensibility of a body part by interrupting afferent (sensory) nerve conduction to that part. **9:** to cause or to result in the accumulation of gases in the digestive tract. **10:** any flatulent digestive disturbance of livestock. **11:** the overextension of the first stomach (rumen), or the condition that results from rupture of the rumen in ruminants such as domestic cattle. Fermentation of ingested forage and release of the gases produced take place in the rumen. These gases are normally expelled by belching. When this is not possible, the expanding gases may rupture the rumen. See *Delphinium*. **12:** bloat colic of horses due to distension of the gastrointestinal lumen. Causes vary. **atrioventricular heart block**. See main entry for heart block. **depolarizing block**. paralysis of skeletal muscle with loss of polarity of the motor endplate (e.g., due to the administration of succinylcholine). **heart block**. See main entry for heart block. **nondepolarizing block**. paralysis of skeletal muscle with no change in polarity of the motor endplate (e.g., with the administration of tubocurarine). **nerve block**. the interruption of nerve impulses. This is accomplished clinically by local extraneural or paraneural injection of an anesthetic drug. See also anesthesia (local and regional), 4-amino pyridine, atracurium, antiadrenergic, anticholinergic. **phase I nerve block**. interruption of the conduction of nerve impulses across the myoneural junction with an associated depolarization of the motor endplate, as in muscle paralysis as caused by succinylcholine. **phase II nerve block**. interruption of the conduction of nerve impulses across the myoneural junction with no depolarization of the motor endplate (e.g., muscle paralysis due to the administration of tubocurarine).

blockade. strong inhibition, interruption, or prevention of an action (e.g., of a drug or toxicant) or of an activity (e.g., nerve transmission). **ganglionic blockade**. blockade of neurotransmission in autonomic ganglia, as by the use of drugs that occupy receptor sites for acetylcholine or by shielding postsynaptic membranes against the action of acetylcholine elaborated from the presynaptic nerve terminals. Pharmacologically produced ganglionic blockade usually results in arteriolar vasodilatation with hypotension with increased peripheral circulation; dilatation of veins with pool ing of blood within tissues; decreased venous return with a consequent decrease in cardiac output. Symptoms also include cycloplegia, tachycardia, mydriasis, reduced gastrointestinal tone and motility with constipation, dry mouth, decreased sweating, and urinary retention. *Cf.* blocking agent, blocker, blocking.

blocker. **1a:** something that obstructs or blocks the movement or passage as through a passage or conduit. **1b:** something that prevents or inhibits a process or activity. **2:** a blocking agent. **α-blocker**. alpha-adrenergic blocking agent. See blocking agent. **β-blocker**. beta-adrenergic blocking agent. *Cf.* blocking agent. See also antagonist, blockade.

blocking. obstructing, suppressing, inhibiting. *Cf.* blockade.

blocking agent. as used in medicine and toxicology, a blocking agent is an agent (usually a chemical substance) that: **1a:** blocks, as defined above (in particular, def. 6 in part; def. 7). **1b:** strongly inhibits, prevents, or interrupts the action of a drug or toxicant on an activity or process. **1c:** suppresses or blocks the transmission of electrochemical signals within nervous tissues or the neuromuscular system of an animal. *Cf.* blocker. See also antagonist. **autonomic blocking agent** (ABA). an agent (e.g., alpha-adrenergic and beta-adrenergic blocking agents) that selectively inhibits neuronal transmission within the autonomic nervous system. See also autonomic agent, autonomic poison. **alpha-adrenergic blocking agent** (α-blocker). a substance that induces adrenergic blockade at α-adrenergic receptors. **beta-adrenergic blocking agent** (β-blocker). a substance that induces adrenergic blockade at β_1- and/or β_2-adrenergic receptors. They are used to reduce blood pressure; all have similar toxicities and produce similar symptoms. See also alprenolol. **neuromuscular blocking agent**. an agent that blocks neuromuscular transmissions. All act very rapidly, are extremely toxic, depress respiration, and are potentially lethal if respiration is not artificially maintained. Symptoms of intoxication vary, but usually include heaviness of eyelids; diplopia; difficulty in swallowing and talking; and rapid paralysis of the extremities, neck, intercostal muscles, and diaphragm. Additional symptoms may include contracture of skeletal muscles, cardiac arrhythmias, tachycardia, and fibrillation. Death usually results from respiratory paralysis or cardiovascular collapse. See, for example, batrachotoxin.

blood. any fluid (often a complex tissue) that circulates in the vascular system of an animal. Blood typically distributes nutrients, oxygen, hormones, etc. to respiring tissues and absorbs waste products that are transported to the organs of elimination (e.g., lungs, gills, kidneys). Blood may also play specific roles in protection against infection and may scavenge dead cells. The blood often contains respiratory pigment, which is free in the plasma in most invertebrate animals, but is contained in cells in vertebrates and certain invertebrates. The blood of vertebrate animals is considered a connective tissue. Systemic poisons are potentially transported from the site of entry via the blood to all parts of the animal's body, including target organs, detoxifying tissues, and organs of elimination.

blood-brain barrier. See barrier.

blood cell. blood corpuscle.

blood-cerebral barrier. See barrier.

blood-cerebrospinal fluid barrier. See barrier.

blood circulatory system. See circulatory system.

blood clot. See clot (blood clot).

blood clot formation. See coagulation (blood coagulation).

blood clotting. See coagulation (blood coagulation).

blood coagulant. See coagulant.

blood coagulation. See coagulation.

blood corpuscle (blood cell). a nonspecific term referring to any of the various cells normally found in the blood of a particular type of organism (e.g., erythrocytes, leukocytes).

blood plasma (plasma). **1:** the liquid fraction of blood, including fibrinogen, in which the formed elements are dispersed. **2:** that portion of blood (a liquid) that which remains when blood cells are removed. *Cf.* serum (blood serum).

blood platelet. See platelet.

blood poisoning. a common nonmedical term that refers to the presence of infecting agents (e.g., bacteria) or their toxins in circulating blood. *Cf.* septicemia, pyemia, toxemia.

blood pressure (BP). **1:** a measure of the pressure exerted on the wall of any blood vessel of an animal by the blood within. In vertebrate animals blood pressure is generally measured for an artery (arterial tension) in vertebrate animals, hence def. 2. **2:** the pressure (tension) exerted by blood circulating in the main arteries of a human or other vertebrate. This pressure depends upon the force of cardiac contractions, the elasticity of the vascular wall, and the viscosity of the blood. The maximum pressure (See systolic pressure) occurs in mammals near the end of the contraction of the left ventricle (systole); minimum pressure (See diastolic pressure) occurs late during the relaxation phase of the ventricular cycle (diastole). Numerous drugs and toxicants affect blood pressure. Thus, exposure to toxic amounts of amphetamines, cocaine, tricyclic antidepressants, phenylcyclidines, and belladonna alkaloids elevate blood pressure. Toxic doses of barbiturates, cyanide, iron, nitrite, opiates, and mushroom toxins lower blood pressure, as do excessive doses of antihypertensive agents. **high blood pressure**. hypertension. **low blood pressure**. hypotension. See also arterial tension, venous tension.

blood root. See *Sanguinaria canadensis*.

blood serum. See serum.

blood-testis barrier. See barrier.

blood urea nitrogen (BUN) The amount of nitrogen in circulating urea. The normal amount in human blood is 9-25 mg/ml. Higher than normal levels of urea nitrogen in blood usually indicate diminished renal function.

blood-vascular. of, pertaining to, or involving blood vessels. See blood-vascular system.

blood-vascular system. a network or a continuous series of vessels and/or spaces within a multicellular animal (metazoan) that conducts blood through the body in response to a pressure gradient generated by one or more hearts. Such a system occurs in all but a few (all small) multicellular animals (metazoans). Such a system transports respiratory gases, nutrients, and wastes to and/or from the tissues of the body. See also vascular system, cardiovascular system, circulatory system. **open blood-vascular system**. the type of system found in most invertebrates in which blood flows partly or wholly through tissue spaces, perfusing the cells directly. **closed blood-vascular system**. the circulating blood remains within a system of closed blood vessels (e.g., the arteries, veins and capillaries of vertebrates). This permits pumping at a higher pressure, with movement of greater volumes of blood per unit time.

blood vessel. one of the closed tubular vessels that contain the circulating blood of an animal (e.g., artery, vein, venule, arteriole, and capillary).

bloodworm. **1:** the common name of the red-colored freshwater larva of many species of dipterous insects (flies and related forms) of the family Tendipedidae. The color is due to the presence of invertebrate hemoglobin dissolved in the circulating body fluid. **2:** any of many bright red marine polychaete worms. **3:** *Glycera dibranchiata*.

bloody tears (bloody weeping). an ocular exudate that contains blood due to hemorrhage from the conjunctiva.

bloody weeping. bloody tears.

bloom. **1:** an individual blossom or flower. **2:** the state or condition of blossoming or flowering. **3:** a waxy layer on the surface of certain fruits (e.g., apples). **4:** a dense growth of algae, Cyanobacteria (blue-green bacteria), and/or other plankton (usually phytpoplankton) in or on a surface water body such that the organisms color the water, sometimes forming a surface scum. Such blooms may be highly toxic (See Cyanobacteria, *Gonyaulax catenella*, red tide). **algal bloom**. a water bloom comprised mainly of algae or the Cyanobacteria (blue-green bacteria). Algal blooms have caused severe illness and extensive mortality among livestock, fish and wildlife, pets, and even humans in many areas of the world, including the northern United States and southern Canada. Decaying algal blooms can be especially toxic; they produce compounds such as hydroxylamine, hydrogen sulfide, and other compounds that are toxic to fish or other organisms that live in or drink the water. Algal blooms also deplete the oxygen in the water body at night through respiration and the process of decay, killing aquatic organisms by suffocation. Sustained algal blooms may lead to eutrophication of the water body. In some cases, poisonings attributed to algal blooms may be due to the presence of other organisms, most importantly the bacterium, *Claustridium botulinum*. See also alga, algal poisoning. **aquatic bloom**. water bloom. **dinoflagellate bloom**. red tide. **gonyaulax bloom**. red tide. **gymnodinium bloom**. red tide. **phytoplankton bloom**. a type of water bloom in which a dense growth of algae and/ or other phytoplankton occurs in or on a water body (e.g., a lake or pond). The term is usually taken to include the Cyanobacteria (blue-green bacteria). An algal bloom is a phytoplankton bloom. **red dinoflagellate bloom.** red tide. **toxic bloom**. **1a:** a toxic water bloom comprised of algae, the Cyanobacteria, and/or other aquatic microorganisms. **1b:** algal bloom. **water bloom** (aquatic bloom). **1a:** the accumulation of microorganisms at or near the surface of a water body. **1b:** the accumulation, mainly of blue-green bacteria, at the surface of a water body. Algal blooms, dinoflagellate blooms, and toxic blooms are types of water blooms. **2:** the microbiota (especially algae or the Cyanobacteria) that make up a bloom. **3:** the scum formed by a water bloom as previously defined. **4:** the proliferation of higher aquatic plants.

blow gas. producer gas.

blowfish. one of many common names of fishes of the family Tetraodontidae. See also puffer.

Blue 1. FD&C Blue No. 1.

Blue 2. FD&C Blue 2.

1206 Blue. FD&C Blue No. 1.

1311 Blue. FD&C Blue 2.

12070 Blue. FD&C Blue 2.

blue asbestos. crocidolite asbestos. See asbestos.

blue-banded sea snake. *Hydrophis cyanocinctus*.

blue bottle. *Physalia pelagica*.

blue cohosh. *Caulophyllum thalictroides*.

blue copper. the pentahydrate salt of cupric sulfate, *q.v.*

blue flag. *Iris versicolor*.

blue fox. *Alopex lagopus*. See Ronnel.

blue-green algae. former common name of Cyanobacteria (blue-green bacteria).

blue-green bacteria. Cyanobacteria.

blue krait. *Bungarus caeruleus*.

blue loco. *Astragalus diphysus* (= *A. lentiginosus* var. *diphysus*), a true locoweed.

blue Malaysian coral snake. *Matacora bivirgata*.

blue mass. mercury mass.

blue pill. mercury mass.

blue vitriol. the pentahydrate salt of cupric sulfate, *q.v.*

bluebell. *Scilla*.

blueberry. See *Vaccinium*.

bluebonnet. *Lupinus*.

bluegill sunfish, bluegill. *Lepomis macrochirus*.

bluegrass. See *Poa*.

bluejoint grass. See *Calamagrostis*.

bluestone. the pentahydrate salt of cupric sulfate, *q.v.*

bluish small-headed sea snake. *Hydrophis caerulescens*.

blurred vision. a visual deficit in which objects are not brought into sharp focus. Such may be due to any of a number of causes (e.g., a film of tears over the surface of the eyes) and may be temporary or easily corrected. Many toxicants (e.g., certain snake venoms, atropine, camphor, cannabis, carbon monoxide, cocaine, methanol, phosphate ester insecticides, quinine) can cause the condition, which is sometimes persistent or part of a serious syndrome, and usually the result of CNS effects.

BLV. biological limit value. See limit value.

BMF. biomagnification factor. See factor.

BNPD. 2-bromo-2-nitro-1,3-propanediol.

bobwhite, bobwhite quail. *Colinus virginianus*.

bocaraca. *Bothrops schlegelii*.

bococcio. *Sebastes paucispinis*.

BOD. biological oxygen demand.

BOD5. five-day biological oxygen demand: the amount of oxygen (as mg O_2/liter of water) taken up from a water sample that contains a known amount of dissolved oxygen kept at 20°C for 5 days. See also biological oxygen demand, CBOD5.

body weight. the total live weight of a living organism.

bog conocybe. *Conocybe smithii*.

bog laurel. *Kalmia polifolia*.

bog manganese. manganese dioxide.

bog onion. *Arisaema stewardsonii*.

bog rosemary. *Andromeda*.

boiciuinga. *Crotalus durissus*.

Boie's sea snake. *Enhydrina valakadyn*.

Boiga dendrophila (mangrove snake). a venomous snake (Family Colubridae) of southeast Asia. A serious bite causes effects typical of colubrid venom poisoning.

boipeva. *Bothrops itapetiningae*.

bolaffinine. a toxic protein isolated from the mushroom, *Boletus affinis* (Family Boletaceae). It is very toxic to laboratory mice.

bolesatine. a toxic protein isolated from *Boletus satanas* that inhibits *in vitro* protein synthesis.

bolete. a fleshy mushroom; any fungus of the genus *Boletus*.

boletic acid. fumaric acid.

Boletus. a genus of soft pore fungi (Family Boletaceae). Many species of this mushroom are edible; others are indigestible or strongly emetic when eaten raw. Symptoms of poisoning are frequently those of mild to severe gastroenteritis. Some boletes (e.g., *B. affinis*) may be lethal to humans and to grazing animals (e.g., *Bos indicus*, the zebu in Madagascar). ***B. affinis***. a possibly lethal species known from eastern North America, but more widely distributed. See bolaffinine. ***B. luridus***. a mushroom that produces disulfiram-like effects when alcohol is taken within 3 days following ingestion. ***B. satanas***. a European species that contains bolesatine (a toxic protein), and is considered by some to be toxic. ***B. subvelutipes*** (red-mouth bolete). a probably poisonous mushroom which, at least at some times in some places, should not be eaten. It is ground-dwelling in deciduous and mixed woods, usually near beech, balsam fir, eastern hemlock, or white spruce from Quebec to Virginia, west to Michigan.

Bolivian coca. *Erythroxylon coca*.

bombesin(2-L-glutamine-6-L-asparaginealytesin). a toxic tetradecapeptide from the skin secretions of disc-tongued frogs (Family Discoglossidae), mainly *Bombina bombina* and *B. variegata*. Also found in human small-cell pulmonary carcinoma.

Bombidae (bumble bees). a family of medium to large social bees with a long tongue. They sting viciously. The queen hibernates during winter and initiates a new underground colony each spring. They are important flower pollinators, but many species have become rare because of sensitivity to pollutants. See *Bombus*.

Bombina. a genus of venomous Eurasian disc-tongued frogs (Family Discoglossidae). ***B. bombina***. the skin secretions are toxic and contain bombesin and other peptides, free amino acids, and large amounts of serotonin. ***B. variegata***. the skin secretions are toxic and contain at least twelve α-amino acids, τ-aminobutyric acid, serotonin, two nonapeptides, and a hemolytic polypeptide.

Bombus (bumble bees). a genus of large bees (Family Bombidae). The venom is acidic and contains hemolytic enzymes. The sting causes local pain and a burning sensation, with swelling, erythema, and whitishness at the wound site. Common systemic effects are respiratory distress and shock. Multiple stings may require hospitalization. Fatalities are rare and are usually due to anaphylactic shock.

bonavist. *Dolichos lablab*.

bone. a hard, rigid connective tissue that forms the bulk of the skeleton of bony fish and the higher vertebrate classes. It is comprised of osteocytes, collagen fibers, and calcium salts (hydroxyapatite). Bones are usually hollow and in most vertebrates contains marrow, a blood-forming tissue. Bones form a framework of the body, make locomotion possible, and shield vital parts of the body (e.g., brain) from trauma. The calcium of bone is important in most species of bony vertebrates as a calcium reserve in the body. Calcium from bones forms the shell of bird eggs. Bone is the target tissue of a number of xenobiotics, some of which are simply sequestered in bone and others that are toxic (e.g., lathyrogens, excessive fluoride, lead, and strontium).

bone marrow. See marrow.

bone-marrow-derived lymphocyte. B lymphocyte.

bone marrow hypoplasia. See hypoplasia.

bonefish. a common name of fish of the Family

Albulidae, especially *Albula vulpes*.

boneset. *Eupatorium perfoliatum*.

bonetail. *Bothrops atrox*.

bongkrek intoxication. bongkrek poisoning.

bongkrek poisoning (bongkrek intoxication; tempeh poisoning). that caused by ingestion of improperly-prepared bongkrek, a Javanese food made from fermented copra press cake. Symptoms are nonspecific and include nausea, vomiting, diaphoresis, muscle cramps, and coma.

bonito. common name of any of several mackerel-like marine fishes (Family Scombridae). See *Euthynnus pelamis*, scombroid poisoning.

bony fish. fish of the vertebrate class, Osteichthyes.

boomslang (plural, boomslangs or boomslange). *Dispholidus typus*.

bopindolol ((±)1-[(1,1-dimethylethyl)amino]-3-[(2-methyl-1H-indol-4-yl)oxy]-2-propanol benzoate ester; (±)-1-(*tert*-butylamino)-3-[(2-methylindol-4-yl)oxy]-2-propanol benzoate (ester)). a β-adrenergic blocker used as an antihypertensive. It is extremely toxic to laboratory mice.

boracic acid. boric acid.

borane (boron hydride). **1:** any of a class of hazardous binary compounds of boron and hydrogen (hydrides). The simplest borane that is stable (as a gas) at normal ambient temperature and pressure is diborane, B_2H_6. Most are highly toxic, relatively unstable, and react with water to release hydrogen; many react violently in air. They are used as fuels. **2:** any derivative of a boron-hydrogen compound such as $B_{10}H_{12}I_2$ or BCl_3. [illegible] borane, decaborane, organoborane.

borate. **1:** any basic salt or ester of boric acid. Very little is known of chronic intoxication by borates at low environmental concentrations. **2a:** related to the salts of orthoboric acid, H_3BO_3. **2b:** less commonly, related to boric oxide, B_2O_3. See borism, borate poisoning. **anhydrous sodium borate**. anhydrous borax (See borax). **sodium borate**. sodium tetraborate; sodium pyroborate; borax decahydrate; tincal). a naturally occurring substance found in salt lakes, alkali lakes, and alkali soil. It is an odorless, white crystalline substance or powder, $Na_2B_4O_7 \cdot 10H_2O$. It loses water of crystallization when heated to melting. Used as a detergent, in cleaning solutions, herbicides, fertilizers, rust inhibitors, bleaches, a flame-retardant fungicide for wood, and a laboratory reagent, it should be considered hazardous as it is toxic by inhalation or ingestion. See also borax, borism.

borate ester. See borate.

borate poisoning. poisoning by any basic salt or ester of boric acid. Symptoms in humans may include headache, nausea, vomiting, diarrhea, gastrointestinal distress, and symptoms of shock such as weakness, restlessness, lethargy, tremors, convulsions, coma, vascular collapse, and death. Kidney and liver damage sometimes occur. Chronic borate poisoning of humans (following at least several weeks of exposure) usually produces a distinctive bright red dermatitis, conjunctivitis, a red tongue, cracked lips, gastrointestinal distress, and alopecia. Symptoms may also include periorbital edema, headache, irritability, persistent nausea, vomiting, diarrhea, anorexia and weight loss, stomach pain, weakness and lethargy, and restlessness. Kidney and liver damage are occasionally seen. Tremor, convulsions, and coma are fairly common, especially in infants. See also borism. Death from cardiovascular collapse may result.

borate salt. See borate.

borax. the commercial name for sodium borate. See also borism. **anhydrous borax** (dehydrated Borax; anhydrous sodium borate; anhydrous sodium borax) a toxic, white, hygroscopic, nonflammable, free-flowing crystalline [illegible] that forms a partial hydrate in humid air. It is used as a herbicide and in the manufacture of glass, enamels, and ceramics. **dehydrated borax**. anhydrous borax. *Cf.* Borax.

Borax. a laundry agent that contains boric acid. *Cf.* borax. See also borax (anhydrous borax).

borax decahydrate. sodium borate (See borate).

Bordeaux mixture. a mixture of cupric sulfate and calcium hydroxide applied to plants as a fungicide. It was originally used in the grape vineyards near Bordeaux, France. This mixture can injure certain crops. See also Burgundy mixture, cupric sulfate.

boric acid (boracic acid). a very toxic white, weakly acidic, crystalline substance, H_3BO_3. It is a nonirritating, weakly bacteriostatic substance, and has been used widely in optic irrigants, as an antiinfective agent, and as a disinfectant. The effects of poisoning are usually reversible, but this substance can prove lethal. It can be absorbed through abrasions, burns, or breaks in the skin. In some countries, it is still a common cause of poisoning in infants. The oral MLD is about 2-3 g for human infants, about 5-6 g for children, and about 15-20 g for adults. Systemic poisoning may result from doses as low as 0.17-0.2g/kg body weight. Boric acid is a cumulative poison. Poisonings also arise from agricultural and industrial use and from careless usage of household products that contain boric acid. Unfortunately, it is not regarded by most people as poisonous; it resembles sugar in appearance and has been consumed by children and accidentally used on foods. Boric acid is toxic to all cells. Symptoms of chronic poisoning may appear following several weeks of exposure. See boric acid poisoning.

boric acid poisoning. poisoning by boric acid. The effects are usually reversible but can prove fatal. **boric acid poisoning, chronic**. symptoms may appear following several weeks of exposure and usually include a distinctive bright red dermatitis, conjunctivitis, a red tongue, cracked lips, and alopecia. Symptoms may also include periorbital edema, headache, irritability, persistent nausea, vomiting, diarrhea, anorexia and weight loss, stomach pain, weakness and lethargy, restlessness. Tremor, convulsions, and coma are not infrequent, especially in infants. In serious cases, the victim may enter shock and die in a state of cardiovascular collapse. See borate poisoning.

borism. poisoning by ingestion of sodium borate (borax) or a compound of boron. Symptoms may include dry skin with a distinctive dermatitis and gastric distress. Additional symptoms are common. Very little is known of chronic intoxication by borates at low environmental concentrations. See also borate, borate poisoning.

(+)-2-bornanone. (1*R*,4*R*)-(+)-camphor. See camphor.

2-bornanone. camphor.

D-2-bornanone. camphor.

Borneo cobra. *Naja n. miolepis*.

borneol. a white amorphous substance with a camphor-like odor isolated from serpentaria. It is used in medicine, perfumes, and chemical syntheses.

boroethane. diborane.

boron (B)[Z = 5; A_r = 10.81]. a hard crystalline mass or brown powder, it is a trivalent metalloid element that forms borates and boric acid. The most common source of boron is naturally occurring disodium decahydrate tetraborate (borax). See also boric acid, boric acid poisoning, borism.

boron hydride. borane.

borquira. *Crotalus durissus*.

Bos indicus (zebu). a species of domestic cattle (Family Bovidae) of Asia with large ears, a large dewlap, short horns, and a prominent hump between the shoulders. See *Boletus*.

bosré. *Elapsoidea sundevallii*.

botanic, botanical. **1:** of, pertaining to, or derived from plants or plant culture. **2a:** a biologically active (usually therapeutic) substance derived from plant material. **2b:** any insecticide made from plant material, e.g., pyrethrum, rotenone, ryania, sabadilla. **3:** concerning the science of botany.

botete. one of numerous common names of fishes of the family Tetraodontidae.

Bothriocyrtum. a common genus of trapdoor spiders (Family Ctenizidae) that feed on other arthropods. They are dangerous to humans. ***B. californicum*** (California trapdoor spider). a spider of dry, sunny, hillsides or sloping ground with dry soil; the tunnel is as much as 20 cm deep. The female averages about 30 mm in length. The cephalothorax is nearly as wide as long, is blackish brown and darkest in the male; the abdomen brown to yellow-gray (palest in the male), is ovate, and lacks longitudinal grooves. See trapdoor spider.

bothropic. pertaining to, denoting, or produced by snakes of the genus *Bothrops*.

bothropic antitoxin. See antitoxin.

Bothrops (American lance-headed vipers; American lanceheads; lanceheads; hog-nosed vipers; Amerikanische Lanzenottern; palm vipers). a genus of venomous snakes (Family Crotalidae) of tropical and southern South America and the West Indies. The head is broad, depressed, and distinct from the narrow neck; the canthus rostralis is well defined, supraoculars are usually present, and the internasals are usually distinct or sometimes separated by small scales. The body is relatively slender to stout and cylindrical or slightly compressed. Dorsal body scales are keeled, forming 19-35 non-oblique rows at the midpoint of the body. Individual species fall into one of three groups: 1. large, long-tailed terrestrial species, usually with paired subcaudal scales; 2. small, short-tailed terrestrial species with unpaired subcaudal scales; 3. small-to-medium-sized arboreal species with a prehensile tail and unpaired anterior subcaudal scales. Large snakes of this genus, especially the bush and tree vipers, are very dangerous and are responsible for most snakebites in tropical Central America. The venom is chiefly hemolytic and cytolytic, causing interstitial bleeding. The blood does not coagulate and there may be prolonged bleeding from the wound and into tissues. Effects of envenomation may include shock, local pain, edema, skin discoloration, ecchymosis, necrosis and sloughing of tissue, weakness, a weak rapid pulse, blurring of [illegible], respiratory distress, hypotension, destruction of erythrocytes, and increased coagulation time sometimes with prolonged bleeding from the wound and into tissues. The EKG of the victim may be abnormal. Coma occurs in serious cases, but death is uncommon except in the case of bites by *B. atrox* and *B. jajaraca*. Species of *Bothrops* not defined herein are *B. albocarinatus*, *B. alticola*, and *B. lojanus*, which are widely distributed in Ecuador; *B. ammodytoides* (Patagonian pit viper; yarara nata) widely distributed in Argentina; *B. andianus* widely distributed in Peru; *B. barbouri* in southwestern Mexico; *B. barnetti* in northern Peru; *B. bicolor* in southern Mexico and southern Guatemala; *B. castelnaudi* (cuatro narices; macabrel; macaurel) in Ecuador southeastward to eastern Peru and northern Brazil; *B. cotiara* in southern Brazil and northeastern Argentina; *B. dunni*, *B. melanurus*, and *B. sphenophrys* of southern Mexico; *B. erythromelas* and *B. iglesiasi* limited to eastern Brazil; *B. fonsecai* and *B. itapetiningae* (boipeva; cotiarinha) in southern Brazil; *B. godmanni* (Godmann's pit viper) found in southern Mexico and much of Central America; *B. hyoprorus* widespread in Peru, ranging northward into southern Columbia and eastern Ecuador and eastward into western Brazil; *B. insularis* (island viper) found only on the island of Queimada Grande in southeastern Brazil; *B. lateralis* (yellow-lined palm viper) in Costa Rica and eastern Panama; *B. lichenosus*, *B. venezuelae* (tigra mariposa), *B. nasutus* (hognosed viper; horned hog-nosed viper; nahuyaka), and *B. medusa* are widely distributed in Venezuela; *B. microphthalmus* found throughout much of Venezuela, Peru, Ecuador, and may occur locally in Bolivia; *B. neglectus* (rabo de raton) widely distributed in Columbia, Venezuela, and Surinam; *B. nigroviridis* (black-spotted palm viper; yellow-spotted palm viper) in southern Mexico and Central America exclusive of Honduras and possibly El Salvador and Nicaragua; *B. peruvianus* occurs in southeastern Chile; *B. picadoi*, a snake of Costa Rica; *B. pictus* occurs throughout Chile; *B. pifanoi*. a snake of northern Venezuela; *B. pirajai* confined to eastern French Guiana; *B. pulcher*. found in eastern Peru and eastern Ecuador. *B. punctatus* (rabo de chuncha) found in Panama; *B. undulatus* and *B. yucatanicus* are indigenous to eastern [illegible]; *B.* [illegible], widespread in Ecuador and may occur in Columbia. ***B. alternatus*** (crossed pit viper; urutu; wutu). a dark brown snake with about 20 pairs of distinctive rounded lateral darker brown splotches narrowly outlined in yellow; the latter are shaped rather like semicircles or telephone handsets, the apices of which nearly meet on the dorsal midline. The markings on the crown are distinctive. The belly is white, spotted with brown or black. Dorsals strongly keeled, in 29-35 rows at midbody; ventrals 167-181; the 34-51 subcaudals are paired. Adults rarely exceed 1.5 m in length. It occurs along watercourses throughout southern Brazil, Uruguay, Paraguay, and northern Argentina. This snake bites a large number of humans each year. The bite is not usually lethal, but causes severe local effects. ***B. asper***. former scientific name of *B. atrox*. ***B. atrox*** (formerly *B. asper*; Fer-de-lance; barba amarillo; bonetail; caicaca; cuatro narices; equis; manapare; tomigoff; torciopelo). the Fer-de-lance (French meaning lancehead) is called "barba amarillo" (yellow chin) in Spanish. It is a large venomous crotalid that is widely distributed in forested areas of Mexico, Central America, and South America as far as Peru and northern Brazil; also in part of the West Indies where it is common and greatly feared on sugar plantations. On the mainland it occurs not only in native forests but is also common in banana, cocoa, and coffee plantations and along streams. The tail is tipped by a horny spine and it is further distinguished by an olive-green, gray, or brownish ground color with a lateral pattern of dark, usually black-bordered triangles whose apices meet (or nearly meet) at the dorsal midline. The dark border may have a light edging. The belly is light cream to yellow with dark blotches. It reaches a length of up to about 2.4 m. The fangs are long. This snake is often irritable, readily strikes (sometimes repeatedly), and is able to deliver a large amount of a powerful venom. It is responsible for more human deaths in tropical Central America (and probably in the Americas altogether) than any other snake. Other species of tropical *Bothrops* and some species of *Trimeresurus* are sometimes called Fer-de-lance. *B. atrox* is most often confused with *B. jararaca*, *q.v.* Envenomation of humans causes local pain and bleeding from the bite, gums, nose, mouth, and rectum. Blood does not coagulate and seeps into muscles and the nervous system. Shock and respiratory distress precede death. ***B. bilineatus*** (Amazonian tree viper; surucucu patiabo). a tree viper that is usually uniformly bright green above (some are speckled with black) with a narrow lateral yellow stripe. The tail is usually tipped with red or red-brown; belly white, without markings; snout rounded; canthus rostralis sharp and slightly raised. Internasals are large and in contact with one another; canthals large; 5-8 rows of scales between large supraoculars. Adults average about 0.7 m. in length, growing to a maximum of about 1 m. One of the most widely distributed of the prehensile-tailed tree vipers of South America, it is indigenous to the Amazon Basin in Brazil, British Guiana, Bolivia, Peru, Ecuador, and Venezuela. It is not regarded as dangerous anywhere within its range. ***B. caribbaeus*** (Saint Lucia serpent; Saint Lucia viper). the only venomous snake on the West Indian island of Santa Lucia where it occurs in cacao and coconut plantations and in damp forest. Adults average 1-1.3 m in length, occasionally reaching a length of about 2.2 m. The head is dark gray with a postorbital band that extends across the upper edge of the supraorbitals. The body is light gray or yellowish (sometimes with a suffusion of rusty red) with obscure gray blotches that are only a little darker than the ground color; the belly is yellowish with a few gray markings. The chin is white or cream. Dorsal scales are strongly keeled. The bite of this dangerous snake causes severe focal tissue damage and is sometimes lethal to humans. Care should be taken with the earlier literature as this snake was at times mistaken for *B. atrox* and *B. lanceolatus*. ***B. jajaraca*** (jararaca; yarara). a deadly and dangerous species of tropical and southern South America found most commonly in the grassy regions of Brazil. It is one of the most common venomous snakes within its range and is second in the region only to *Crotalus durissus* as a cause of human deaths from snakebite. Adults average somewhat more than 1 m in length, but may attain a maximum length of nearly 2 m. The head is rather long with a short snout; a dark olive crown, usually with dark brown irregular patches that often have light edging; laterally, the head is light, with a well-defined dark brown postorbital stripe. The body is olive or gray brown with about 25 pairs of darker brown blotches which are triangular anteriorly becoming progressively rounded and irregular in shape posteriorly. The belly is yellowish, spotted with gray, and may be completely gray posteriorly. This snake is often confused with *B. atrox* and *B. jararacussu*, but one can distinguish among them by color patterns and scales of the snout. ***B. jararacussu*** (jararacussú; yararaguassu). a dull black and yellowish species with a broad, lance-shaped head. It is semiaquatic, occurring near rivers and lakes in southern Brazil, eastern Bolivia, Paraguay, and northern Argentina. Adults reach an average length of somewhat more than 1 m. The crown is uniformly black and dark brown with dark, yellowish lines above the temporal areas; laterally the head is mostly yellowish. There are some 15 sets of black, lateral, inverted U-shaped body blotches that may alternate with one another or may sit opposite and join over the dorsum. The back is often covered with irregular dark patches which are edged in dark yellow laterally. The belly is yellow with dark brown or black blotches. Prefrontals (canthals) are broader than long and separated from one another by 1-2 rows of small scales. The dorsal scales are strongly keeled, and form 23-27 rows at midbody, fewer posteriorly. It is not very common, but is unusually dangerous. The venom is very toxic and is produced in large amounts; 100 mg or more is usually obtained by a single milking. The bite is often fatal; blindness is a common early effect. ***B. lanceolatus*** (Fer-de-lance). a large lance-headed viper distinguished by the dark, truncated lateral blotches and high numbers of dorsal (31-33 rows at midbody) and ventral (215-230) scales; the dorsal scales are strongly keeled. It is the only venomous snake on the island of Martinique in the Caribbean Sea. Adults average somewhat more than 1 m in length, some individuals reaching lengths of more than 2 m. The head is brown with a sharply defined darker postorbital band that extends to the corner of the mouth. The ground color is gray, olive, or brown with a series of 22-27 indistinct hourglass-shaped blotches along the dorsum. The belly is white or cream with some grayish or brown stippling anteriorly and posteriorly. Dorsals are strongly keeled and in 31-33 rows at midbody. ***B. lansbergii*** (hognosed viper; Lansberg's hog-nosed viper; patoca; Lansberg's pit viper; chatilla; tamaga; Western hog-nosed viper). a small brownish, [illegible] than 0.7 m long and marked by the pointed upturned snout, the raised canthus, and a series of dark brown angular blotches along the back that are separated by a thin light line. They occur in brushy areas and semiarid forests in southern Mexico and throughout much of Central America. The head is broad, depressed; the body is moderately stout and short. It resembles *B. nasutus* and *B. nummifer*. ***B. nasutus*** (nahuyaca). a viper that resembles *B. lansbergii*. The ranges broadly overlap, but *lansbergii* usually occurs in drier situations. *B. nasutus* is also sometimes mistaken for a young bushmaster, *Lachesis mutus*. ***B. neuwiedi*** (jararaca pintada; Maximilian's viper; Maximilian's pit viper; palm viper; urutu; Weid's lancehead; jararaca). a tan or grayish, rather small, pit viper with a distinctive pattern on the crown. Adults average somewhat less than 1 m in length. It occurs widely in grasslands and open country on the plateau of southern Brazil, eastern Bolivia, Paraguay, and northern Argentina. It is a major cause of snakebite in Argentina. The crown is light tan or brown with a series of distinct spots and often with a distinctive U-shaped mark posteriorly; the arms of the "U" are directed posteriorly and sometimes abut the body pattern. The body pattern varies geographically but is basically a series of small, paired triangular or rhomboidal black or dark brown dorsal blotches. They alternate or fuse across the back, forming small X-shaped markings. Dark, rounded spots may lie on the midline between the main series; a lateral series of small spots alternate with the dorsal markings. All of these markings are bordered with bright yellow. The belly is yellowish with some ventrals with gray edging. ***B. nummifer*** (jumping viper; Central African jumping viper; jumping snake; mano de piedra; timba; toboba chingu tommygoff). an aggressive, small, light brown or gray terrestrial viper, with diamond-shaped cross-markings, a short, extremely thick body, and short tail. The crown is dark with a light-colored oblique postorbital band. The snout is rounded, the canthus rostralis sharp, and the eye is separated from the labials by three or four rows of small scales. The fangs are relatively short and the venom is not very toxic. About 20 dark brown or black saddle-shaped blotches run along the dorsum. These often connect with lateral spots, forming narrow crossbands. The body is whitish beneath, sometimes with dark brown blotches. The dorsal scales are keeled or even tubercular in large specimens. It occurs on plantations and in low hilly rain forests from eastern Mexico, throughout much of Central America to Panama, exclusive of Honduras and possibly El Salvador. It is larger, but not much longer (usually less than 0.6 m) than *B. atrox* and *B.*

lansbergii. It resembles *B. lansbergii* and while the ranges of the two species broadly overlap, *lansbergii* is usually found in drier situations. *B. nummifer* is also sometimes mistaken for young specimens of *Lachesis mutus*. It is an aggressive species and the strong, stout body allows this unusual snake to strike at a distance further than the length of its body. It often attacks so vigorously that it lifts itself ("jumps") completely off the ground. ***B. schlegelii*** (eyelash viper; eyelash snake; horned palm viper; bocaraca; colgadora; Griefschwanz Lanzenotter; Schlegel's palm viper; sleeping gough; manapare cejuda; nahuyaka; oropel; toboba de pestana). a small, moderately stout, tree viper with a prehensile tail. The head is broad and distinct from the body, with a row of small, elevated, pointed scales above the eye. Few individuals are more than 0.6 m (24 inches) long. The ground color is tan, green, olive-green, or yellow with small black spots which are often dispersed, but sometimes form irregular crossbands. Tan and green specimens commonly have narrow brown and reddish crossbands or a reticulated pattern in red. The belly is green or yellow with black spots. The eyelash viper is found chiefly in trees and shrubs in rain forest and cacao plantations in southern Mexico and throughout much of Central America. The most common of the green "palm vipers," and the only one that bears raised scales above the eyes. It is not considered dangerous.

bothrops antitoxin. bothropic antitoxin. See antitoxin.

bothrops atrox serine proteinase. batroxobin.

bothrops venom proteinase. batroxobin.

Bothryocyrtum. a genus of common trapdoor spiders (Family Ctenizidae) indigenous to California. The burrow is cylindrical and closed by a wafer-like lid. They are dangerous to humans.

Botropase. batroxobin.

bottom disease. crotalism.

bottom sediment. See sediment.

botulin. botulinum toxin. See toxin.

botulinal. pertaining to *Clostridium botulinum* or to botulinum toxin.

botulinic acid (allantotoxicon). a toxin in putrescent sausage. It is probably a mixture that includes allantotoxicon.

botulinogenic. botulogenic.

botulinum antitoxin. botulism antitoxin. See antitoxin.

botulinum toxin. See toxin (botulinum toxin).

botulinum poisoning. botulism.

botulinus toxin. botulinum toxin (See toxin).

botulism (botulinum poisoning; claustridium botulinum poisoning; Lamziekte). a rare but often rapidly fatal poisoning by ingesting botulinum toxin in food. The animals most frequently known to be affected are humans, domestic fowl, waterfowl, most livestock, and fish. Botulism is a serious and often fatal disease in humans, wildfowl, and other affected animals, but cats, dogs, and swine are fairly resistant to all types of botulinum toxin when it is administered by mouth. Species other than humans may acquire the toxin from carrion, contaminated vegetation, wounds, or directly from soil, alkaline muds, or bottom sediments where anaerobic conditions favor the proliferation of the causative bacterium, *Claustridium botulinum*. Paralysis is a feature of the disease in all species affected. Humans usually become intoxicated by consuming inadequately sterilized canned or bottled food that has become putrescent. Symptoms in humans usually appear within 12 to 36 hours following consumption of contaminated food. The first symptoms to present are disturbances of vision general weakness, and the victim is easily fatigued. The condition progresses to include nausea, vomiting, gastrointestinal distress, difficulty in swallowing, dysarthria, pronounced muscular weakness and paralysis, ptosis, loss of visual acuity, double vision, loss of pupillary reflexes, paralysis of the respiratory muscles, profuse sweating, a weak, rapid pulse, and sometimes diarrhea, with death from cardiovascular or pulmonary collapse. Congestion and hemorrhage may be noted in all organs, especially the CNS. Suspected botulism requires prompt medical aid. Botulism in humans may be treated with type ABE botulinus antitoxin. See also alkali disease (Western duck sickness), allantiasis.

botulism antitoxin. See antitoxin.

botulism of ducks. See alkali disease.

botulismotoxin. botulinum toxin (See toxin).

botulogenic (botulinogenic). productive of botulism.

Boulengerina (water cobras; wasserkobras). a genus of large, venomous snakes (Family Elapidae) of central Africa. While not usually aggressive, they are regarded as dangerous. The head is short with a broad base and is distinct from neck, which can form a hood; the canthus rostralis is indistinct; the crown is usually covered by 9 scales, the frontal is small; the nasal abuts a single preocular; eyes are small, pupils are round. The maxilla bears two large tubular fangs with external grooves followed by a space and 3-4 small teeth. The body is cylindrical, rather stout, with a tapering, moderately long tail. Dorsal scales are smooth; the anal plate is entire; subcaudals are paired. ***B. annulata*** (banded water cobra; Storm's water cobra). a large aquatic, piscivorous cobra with a hood found in Africa from Lake Tanganyika and the Republic of Zaire westward through the rain forests to the Congo and Camaroon. It is quite common along some shores of Lake Tanganyika. Adults average ca. 2 m or less; occasionally reaching a length of nearly 3 m.

bouncing bet, bouncing betty. common name of *Saponaria officinalis* and some related species.

bovine hemorrhagic syndrome (hemorrhagic syndrome of cattle). a disease of cattle placed on cornstalk pasturage or moldy corn, sometimes on moldy peanuts or oats, and sometimes on moldy commercial feed concentrates. The fungus that causes the disease is usually *Aspergillus flavus*. Symptoms include excitement, incoordination, coma and death. Nearly always fatal. Postmortem examination reveals petechial and ecchymotic hemorrhages of tissues throughout the body. Young cattle are most susceptible, but swine, other livestock, and dogs are also susceptible.

box. common name of *Buxus sempervirens* and *Pachysandra procumbens*.

box jelly. *Chironex fleckeri*.

boxwood. *Buxus sempervirens*.

boxwood family. Buxaceae.

Boyle's disease. dementia paralytica.

boys and girls. *Mercurialis annua*.

BP. blood pressure.

Bq. becquerel.

Br. the symbol for bromine.

Brachiaria brizantha. an annual grass (Family Gramineae) that sometimes causes hepatogenic photosensitization in livestock. See also phylloerythrin.

Brachyaspis. a genus of venomous Australian snakes (Family Elapidae) that is represented by a single species. ***B. curta*** (desert snake; bardick). a venomous snake of southwestern Australia (Family Elapidae). It is a small, short, rather stout snake (adults reach a length of about 0.65 m), with a short tail. The bite can be very painful, but is not lethal to humans. The head is large, borne on a distinct neck; the canthus rostrallis is obtuse; the eyes are small with vertically elliptical pupils.

Brachyurophis (girdled snakes). a genus of venomous snakes (Family Elapidae) found throughout most of Australia with the exception of the southeastern coastal regions. All are small sand-dwelling, burrowing species and are not believed to be dangerous. The body is moderately slender, tapering little; the tail is short. The head is short, narrow, and not distinct from the neck; the snout is pointed; the rostral is large and shovel-shaped; there is no canthus rostralis. Examples are *B. australis* (Australian necklace snake; Australian coral snake), of estern Australia; *B. fasciolatus* (narrow-banded snake), *B. roperi* (burrowing snake), indigenous to northern Australia; and *B. semifasciatus* (half-banded snake; half-girdled snake; half-ringed snake).

bracken (bracken fern). the preferred common name for any fern of the genus *Pteridium*, especially *Pteridium acquilinum*.

bracken fern. improper common name of *Pteridium* (bracken).

bracken poisoning. an acute fatal hemorrhagic disease of cattle and horses caused by eating large amounts of bracken (*Pteridium*) or hay that contains 20% or more of bracken; the toxic principle is thiaminase. The condition is marked by uncontrollable internal bleeding; in horses the disease is manifested by necrologic signs; in cattle by pancytopenia. *Cf.* bleeding disease.

bracken staggers. a condition of horses resulting from the ingestion of *Pteridium* (bracken), which contains a thiaminase. Signs and symptoms include incoordination due to thiaminase-induced thiamine deficiency.

brackish water. a mixture of fresh and salt water.

bradyarrhythmia. any disturbance of the heart's rhythm resulting in bradycardia, *q.v.*

bradycardia (bradyrhythmia; bradysystole). an unusually slow heart rate, usually taken as that below 60 beats per minute in resting humans. *Cf.* tachycardia.

bradykinin. a polypeptide kinin found in blood, $C_{50}H_{73}N_{15}O_{11}$. It is a powerful vasodilator and also causes increased capillary permeability. It stimulates pain receptors and the contraction of smooth muscle. Tests have shown it to be teratogenic and to have reproductive effects. Data on mutagenicity have also been reported. See kinin.

bradypnea. an abnormally low breathing rate.

bradyrhythmia. bradycardia.

bradysystole. bradycardia.

brain. the enlarged, usually anteriorportion of the central nervous system (CNS) of most bilaterally symmetric animals. It is the chief coordinating center of the CNS. In vertebrates the brain and spinal cord constitute the CNS. Many lower animals with discrete nervous systems (e.g., coelenterates) have no distinct coordinating center. In most other invertebrates that lack a distinct brain, supraesophageal or suprapharyngeal ganglia serve similar functions. The brain of vertebrates is the largest portion of the vertebrate central nervous system situated at the anterior (or as in humans, superior) end of — and continuous with — the spinal cord. It is housed in a cartilaginous or bony cranium. The vertebrate brain is the organ of perception and the chief organ of interpretation of peripheral input and neural coordination. It includes all of the neural centers that receive and interpret input from the special senses (olfaction, taste, vision, hearing), to which it normally responds with appropriate voluntary or involuntary motor output. See also brain death.

brain death (cerebral death). many governments have enacted statutes that define human death based in part on brain-related criteria. Many have adopted the criteria of the Uniform Determination of Death Act: "An individual who has sustained either (1) irreversible cessation of circulatory and respiratory function, or (2) irreversible cessation of all functions of the entire brain, including the brain stem, is dead. A determination of death must be made in accordance with accepted medical standards."

brain mushroom. *Gyromitra (Helvella) esculenta*.

branchia (plural, branchiae). a gill or gill-like structure.

branchiae. plural of branchia.

branchial. pertaining to, denoting, or resembling gills.

branchial route. See route of exposure.

brand name. the name under which a product, class of products, or a service is produced and/or marketed by a particular firm. Numerous brand names exist and offer more confusion than help in differentiating products that are, in many cases, significantly different. Many new names for old drugs are added to the inventory of brand names daily. Practicing physicians sometimes run serious risks of failing to identify potential drug interactions that may affect patients who take more than one drug. *Cf.* generic name, trade name, trademark.

Brassica. a genus of plants (Family Brassicaceae, formerly Cruciferae) that includes a wide variety of crop species, forage plants, and weeds (e.g., mustards, rape, kale, cabbage, brussels sprouts, broccoli, cauliflower, rutabaga, turnip). The classification of taxa within *Brassica* is complicated, with numerous synonyms and much confusion in the literature per-taining to some species or varieties. A number of *Brassica* are toxic. The active principles in such cases are often components of mustard oils (mixtures of isothiocyanates in glysidic combination) and/or a thyrotoxic (goitrogenic) glycoside, the aglycone of which is L-5-vinyl-2-thiooxazolidone. Some also accumulate concentrations of nitrate under certain growing conditions that are toxic to livestock, and some have been implicated in trefoil dermatitis and rape scald. See also *Erodium*, *Medicago*, *Trifolium*. ***B. hirta*** (*B. alba*; *Sinapis alba*; white mustard). a somewhat hairy annual herb with oval leaves that grows to a height of ca. 1.3 m. The seeds are a commercial source of mustard and mustard oil, and the foliage is used as greens. The seeds are toxic and goitrogenic. Seeds or seed meal fed to livestock in small quantities cause no problems. Cattle and sheep fed on mustard seed meal, pod-bearing plants, or stubble are sometimes poisoned. Symptoms of poisoning are those of severe, sometimes fatal, gastroenteritis. ***B. juncea*** (Indian mustard, mustard greens). an annual plant with narrow lobed leaves that grows to a height of ca. 1.3 m. It is used for spring greens. It can cause severe gastroenteritis in livestock. ***B. kaber*** (*B. arvensis*; charlock; wild mustard). a common annual weed of cereal crops throughout much of North America; introduced from Europe. Seeds or pod-bearing plants can produce a severe gastroenteritis when ingested by cattle, swine, or sheep. ***B. napus*** (rape; cultivated rape). a widely cultivated annual plant that is toxic to livestock. It contains a thyrotoxic glycoside, the aglycone of which is L-5-vinyl-2-thiooxazolidone. This compound can be extracted from the seed meal with hot water. The main effects of rape seed meal can be counteracted by the addition of iodinated protein to the diet. The seed also contains at least three glycosidic isothiocyanates (mustard oils). It can cause hepatogenic photosensitivity in grazing animals. Toxic concentrations of nitrate have also been reported. ***B. napus* var. *napobrassica*** (*B. napobrassica*; *B. oleracea* var. *napobrassica*; rutabaga; yellow turnip; swedish turnip). a hardy, highly goitrogenic biennial plant. The active principle is L-5-vinyl-2-thiooxazolidone. Toxic concentrations of nitrate have also been reported. ***B. nigra*** (black mustard). a branching annual with feathery leaves that grows to a height of 2 m. It is a source of mustard oils and condiments. The seeds are toxic and goitrogenic. They contain sinigrin, a glycoside that releases the mustard oil, allyl isothiocyanate. The latter is a potent vesicant. The seeds also contain potassium hydrogen sulfate produced by a reaction catalyzed by myrosinase. ***B. oleracea*** (broccoli, kale, wild cabbage, etc.). an annual or perennial species that includes the largest number of varieties within the genus. Some may contain concentrations of nitrates that are toxic to grazing animals, and many varieties have thyrotoxic or vesicant activity. ***B. oleracea* var. *capitata*** (cabbage). a variety that is goitrogenic and produces symptoms of hypothyroidism. ***B. oleracea* var. *acephala*** (common kale). grown for its foliage (used by humans as greens), it is toxic to livestock. It is goitrogenic, producing symptoms of hypothyroidism. The active principle is a thyrotoxic glycoside, the aglycone of which is L-5-vinyl-2-thiooxazolidone. ***B. oleracea* var. *botrytis*** (broccoli). toxic to livestock. It is goitrogenic, producing symptoms of hypothyroidism. The active principle is a thyrotoxic glycoside, the aglycone of which is L-5-vinyl-2-thiooxazolidone. ***B. oleracea* var. *gemmifera*** (brussels sprouts). a vegetable developed from wild cabbage, it is toxic to livestock. It is goitrogenic, producing symptoms of hypothyroidism by means of a thyrotoxic glycoside, the aglycone of which is L-5-vinyl-2-thiooxazolidone. ***B. oleracea* var. *gongylodes*** (synonym, *B. caulorapa*) (kohlrabi, stem turnip). toxic to livestock, it is goitrogenic, producing symptoms of hypothyroidism. The active principle is a thyrotoxic glycoside, the aglycone of which is L-5-vinyl-2-thiooxazolidone. ***B. rapa*** (field mustard). an annual or biennial mustard with long, lobed leaves and flat or globular pods. ***B. rapa* var. *pekinensis*** (Chinese cabbage, pakchoi). it is goitrogenic, producing symptoms of hypothyroidism. ***B. rapa* var. *rapifera*** (turnip). a goitrogenic root vegetable that sometimes accumulates toxic concentrations of nitrate. It can also cause pulmonary emphysema in cattle. Toxic concentrations of nitrate have also been reported. See emphysema (pulmonary emphysema of cattle).

Brassicaceae (mustard family). a large family (formerly Cruciferae) comprised of at least 3000 species of dicotyledonous plants. Wild forms occur chiefly in the north temperate latitudes. Most are herbs with alternate leaves and flowers, the four petals of which are inserted into a raceme or corymb, in the shape of a cross. Many are cultivated and important as crops. See *Brassica*.

braza de piedra. *Bothrops nummifer*.

bread. a food staple made from flour or meal, usually from cereal grains. Breads often contain potentially harmful ingredients, many of which may not be on the label. Examples are butylated hydroxyanisole (BHA), butylated hydroxytoluene (BHT), colors, flavors, hydrogenated shortening, mineral oil, and pesticide residues. French breads are usually free of additives. See also bread poisoning.

bread poisoning. a local name for poisoning by seeds of plants of the genus *Senecio* (groundsel). This plant has been an important source of human mortality in parts of Africa where the seeds sometimes contaminate grain crops. The condition is not rapidly fatal. The hepatotoxic alkaloids contained in the seed cause necrosis and enlargement of liver cells and rupturing of veins.

breathing (pneusis). the entrance and exit of air or gaseous mixtures into and out of the respiratory system by inhalation and exhalation. **irregular breathing**. aperiodic, variable, or erratic breathing. See also dyspnea. **labored breathing**. dyspnea. **shallow breathing**. breathing with abnormally low intake (low tidal volume).

breathing rate. respiratory rate.

breeches flower. *Dicentra cucullaria*.

breeding success. reproductive success.

brei. a finely divided, nearly homogeneous tissue suspension.

brevetoxin (BTX). any of three closely related, chemically unique neurotoxins, brevetoxins A, B, C. They are produced by the dinoflagellate, *Gymnodinium brevis* (= *G. breve*, *Ptychodiscus brevis*) which has caused massive fish kills, and poisoning of mollusks and humans in the Gulf of Mexico and along the west Florida coast. All of these toxins are rapidly and acutely lethal to small fish at concentrations of ca. 4-30 ng/l. Unlike other dinoflagellate toxins (e.g., saxitoxin) that are water-soluble sodium channel blockers, the brevetoxins are lipid-soluble sodium channel activators. See also bloom, red tide.

brevianamide. an indole mycotoxin produced by *Penicillium brevicompactum*.

Brillenschlange. *Naja naja*.

brimstone. sulfur.

brinolase. a fibrinolytic enzyme secreted by the fungus, *Aspergillus oryzae*.

bristleworm. See *Eurythoe*, *Hermodice*.

British AntiLewisite. dimercaprol.

British mandrake. See *Bryonia*.

broad bean. *Vicia faba*.

broad-banded blue sea snake. *Laticauda semifasciata*.

broad-banded copperhead. See *Agkistrodon contortrix*.

broad-banded coral snake. *Micrurus distans*.

broad-banded sea snake. *Hydrophis mamillaris*.

broad-headed snake. *Hoplocephalus bungaroides*.

broadcast application. the spreading of pesticides over an entire area.

broadleaf milkweed. *Asclepias latifolia*.

broccoli. *Brassica oleracea* var. *botrytis*.

brodifacoum (3-[3-(4′-bromo[1,1′-biphenyl]-4-yl)-1,2,3,4-tetrahydro-1-naphthalenyl]-4-hydroxy-2H-1-benzopyran-2-one). a whitish, poisonous, water-insoluble powder that is exceptionally toxic to rats. It is used as a rodenticide.

bromacil (5-bromo-3-*sec*-butyl-6-methyluracil; 5-bromo-6-methyl-3-(1-methylpropyl)-2,4-(1*H*,3*H*)-pyrimidinedione;5-bromo-6-methyl-3-(1-methyl-propyl)uracil). a nonselective herbicide. It is a uracil, $C_9H_{13}BrN_2O_2$, recommended for use as a weedkiller on non-crop lands. It strongly inhibits root growth of oats by blocking cell division in the meristem; it also affects the integrity of cell walls and membranes. Chloroplast development in oat leaves is partially inhibited. Bromacil is moderately toxic by ingestion to humans and laboratory animals. Very toxic fumes, NO_x and Br^-, are released on decomposition by heating.

bromadiolone (3-[3-(4′-bromo[1,1′-biphenyl]-4-yl)-3-hydroxy-1-phenylpropyl]-4-hydroxy-2*H*-1-benzopyran-2-one; 3-[α-[p-(p-bromophenyl)-β-hydroxyphenethyl]-benzyl]-4-hydroxycoumarin). a whitish, slightly water-soluble powder used as an anticoagulant rodenticide. It is highly toxic to small rodents.

bromate. **1:** the anion, BrO_3^- (e.g., of bromic acid, $HBrO_3$). It is extremely toxic and has caused death in children due to acute renal failure. **2:** a salt of bromic acid. Bromates are flammable and are generally more toxic than chlorates. They are neurotoxic, causing CNS paralysis. When heated to decomposition, they release toxic fumes of Br^-. **3:** brominate. See potassium bromate, sodium bromate.

bromated. brominated.

bromatotoxin (bromatotoxismus). a toxicant in food that results, for example, from fermentation.

bromatotoxismus. bromatotoxin.

bromatoxism. a food poisoning. See bromatotoxin.

brome grass. grasses of the genus *Bromus*.

bromethalin (*N*-methyl-2,4-dinitro-*N*-(2,4,6-tribromophenyl)-6-(trifluoromethyl)benzenamine). a pale yellow water-insoluble crystalline substance, used as a rodenticide. It is an uncoupler of oxidative phosphorylation. It is highly poisonous not only to rats and mice, but to cats, dogs, and presumably other mammals. Acute exposure of humans causes headache, confusion, personality changes, seizures, coma, and death. Warfarin-resistant strains of laboratory mice and rodents succumb readily to bromethalin.

bromic. of or pertaining to bromine, especially bromic acid, $HBrO_3$.

bromic acid. a colorless or slightly yellow liquid, $HBrO_3$, that turns yellow on exposure to air. It exists only in aqueous solution and is unstable except in dilute solutions. A strong oxidizing agent and irritant, it is poisonous if ingested or inhaled.

bromic acid, sodium salt. sodium bromate.

bromide. **1:** the bromine anion, Br^-. **2:** any salt or ester of hydrobromic acid, or any binary compound of bromine in which bromine has a valence of -1. Formerly used as sedatives, bromides are CNS depressants that can cause serious psychic and sensory disturbances with overdosage. As a consequence, they are rarely used. Symptoms of intoxication may include immediate nausea and vomiting, drowsiness, irritability, ataxia, vertigo, confusion, mania, hallucinations, and coma. Death is rare. Lesions may include dermatoses and acneform skin eruptions, inflammation of mucous membranes, CNS depression, and elevated spinal fluid pressure. Death is rare. See also bromism.

brominate (bromate). to treat with bromine or a bromate (def. 3).

brominated (bromated). treated with, combined with, or containing bromine or a compound of bromine.

brominated biphenyl. polybrominated biphenyl.

bromindione (2-(4-bromophenyl)-*H*-indene-1,3-[illegible] dioxohydrindene; 2-(*p*-bromophenyl)-1,3-indandione). an oral anticoagulant.

bromine (symbol, Br; bromum)[Z = 35; A_r = 79.9]. a deep red, corrosive, moderately reactive, nonmetallic, volatile liquid element, Br_2, of the halogen series with valences of 1 to

7 inclusive. Bromine is used chiefly in the production of organobromine compounds. It reacts with most metals. It reacts with hydrogen to yield hydrobromic acid, which can react with many metals to form bromides. Bromine is toxic when inhaled or ingested. The fumes are noxious and both vapor and liquid are strong irritants, attacking especially the mucous membranes of the eyes and respiratory tract. Severe bromine poisoning can cause pulmonary edema. See also bromism, bromide, bromate. Species of *Aplysia* concentrate, in their tissues, bromoterpenes from the algae on which they feed.

bromine cyanide. cyanogen bromide.

bromine poisoning. poisoning by bromine or its fumes. Exposed tissues may be eroded and severe poisoning may cause pulmonary edema.
bromine poisoning, chronic. bromism.

brominism. bromism.

bromism (brominism). chronic poisoning from excessive prolonged use of bromine or bromides. Symptoms of intoxication include headache, mental dullness and confusion, drowsiness, memory loss, slurred speech, muscular weakness, unsteady gait, irritability, mania, cardiac depression, acneform eruptions, fetid breath, anorexia, gastrointestinal distress, and skin eruptions. Death is rare. See also bromine poisoning.

bromoacetone (1-bromo-2-propanone). a slightly water-soluble, colorless liquid, CH_3COCH_2Br; m.p. -36.5°C, b.p. 137°C, when pure. It rapidly turns violet even in the absence of air. Bromacetone is a violent lacrimator (tear gas) used as a chemical warfare agent. It is a strong irritant and toxic by inhalation and skin contact.

bromobenzene (monobromobenzene, phenylbromide). a liquid aryl halide, C_6H_5Br, used as an intermediate in organic syntheses, as a solvent, and as an additive to motor oils. It is a skin irritant, and readily enters the body via the respiratory tract, gastrointestinal tract, or skin. Bromobenzene is hepatoxic to laboratory rats.

α-bromobenzeneacetonitrile. α-bromobenzyl cyanide.

α-bromobenzyl cyanide (α-bromobenzeneacetonitrile; α-bromophenylacetonitrile; α-bromo-α-tolunitrile; BBC). a light yellow, oily substance, $C_6H_5CHBrCN$. It is a highly toxic chemical warfare gas and a tear gas used for riot control; it is a potent lacrimator. It is only slightly water soluble, but is freely soluble in organic solvents such as acetone, ethanol, ethyl ether, and chloroform.

3-[3-(4′-bromo[1,1′-biphenyl]-4-yl)-3-hydroxy-1-phenylpropyl]-4-hydroxy-2*H*-1-benzopyran-2-one. bromadiolone.

3-[3-(4′-bromo[1,1′-biphenyl]-4-yl)-1,2,3,4-tetrahydro-1-naphthalenyl]-4-hydroxy-2*H*-1-benzopyran-2-one. brodifacoum.

5-bromo-3-*sec*-butyl-6-methyluracil. bromacil.

2-bromo-2-chloro-1,1,1-trifluoroethane. halothane.

bromocriptine ((5′α)2-bromo-12′-hydroxy-2′-(1-methylethyl)-5′-(2-methylpropyl)ergotaman-3′,6′,18-trione; 2-bromoergocryptine; 2-bromo-α-ergokryptine). an ergot alkaloid derivative and powerful dopamine agonist. It inhibits prolactin secretion and release from the pituitary and retards tumor growth and is thus used in the treatment of endocrine disorders such as hyperprolactinemia. It is also used as an antiparkinsonian. Bromocriptine is moderately toxic to laboratory rabbits. Acute adverse reactions in humans include nausea, vomiting, and orthostatic hypotension. Chronic effects include constipation, dyskinesias, psychoses, digital spasm, and erythromelalgia. See also ergot alkaloids, dopamine.

***O*-(4-bromo-2,5-dichlorophenyl)-*O*-methyl phenylthiophosphonate**. leptophos.

bromoderma. an acneform or granulomatous skin eruption due to hypersensitivity to, or prolonged use of, bromides. See bromism.

bromo-DMA. DOB.

2-bromoergocryptine. bromocriptine.

2-bromo-α-ergokryptine. bromocriptine.

bromofenoxim. a nitrophenol herbicide used for selective weed control in cereal crops. It has a potent contact action against dicotyledons.

(5′α)2-bromo-12′-hydroxy-2′-(1-methylethyl)-5′-(2-methylpropyl)ergotaman-3′,6′,18-trione. bromocriptine.

bromoiodism. poisoning with bromine and iodine or their compounds.

bromomania. a mental disorder induced by prolonged, excessive use of bromides. See bromism.

bromomethane. methyl bromide.

5-bromo-6-methyl-3-(1-methylpropyl)uracil. bromacil.

5-bromo-6-methyl-3-(1-methylpropyl)-2,4-(1*H*,3*H*)-pyrimidinedione. bromacil.

2-bromo-2-nitropane-1,3-diol. 2-bromo-2-nitro-1,3-propanediol.

2-bromo-2-nitro-1,3-propanediol (BNPD; 2-bromo-2-nitropane-1,3-diol; 2-bromo-2-nitropropan-1,3-diol; β-bromo-β-nitrotrimethyleneglycol). a diol that is toxic by all routes of exposure and an irritant of the eye and human skin. It is moderately toxic by skin contact. BNPD is used as an antiseptic. It is potentially carcinogenic in the sense that when mixed with certain compounds such as amines on the skin or in the body it can produce nitrosoamines. It emits very toxic fumes (NO_x and Br^-) when heated to decomposition.

β-bromo-β-nitrotrimethyleneglycol. 2-bromo-2-nitro-1,3-propanediol.

α-bromophenylacetonitrile. α-bromobenzyl cyanide.

2-(4-bromophenyl)-1,3-dioxohydrindene. bromindione.

3-[α-[*p*-(*p*-bromophenyl)-β-hydroxyphenethyl]benzyl]-4-hydroxycoumarin. bromadiolone.

2-(*p*-bromophenyl)-1,3-indandione. bromindione.

2-(4-bromophenyl)-*H*-indene-1,3-(2*H*)-dione. bromindione.

3-(*p*-bromophenyl)-1-methoxy-1-methylurea. metobromuron.

1-bromo-2-propanone. bromoacetone.

5-bromo-2,4(1*H*,3*H*)-pyrimidinedione. 5-bromouracil.

bromoterpene. any brominated terpene. See terpene, aplysin, plysinol.

α-bromo-α-tolunitrile. α-bromobenzyl cyanide.

bromotoxism. See food poisoning.

5-bromouracil (5-bromo-2,4(1*H*,3*H*)-pyrimidinedione; BrU). a potent mutagenic pyrimidine analog. It modifies the base-pair sequencing of DNA by replacing thymine. It causes mutations because it can pair with either guanine or adenine. It is used as an experimental mutagen.

bromoxynil (3,5-dibromo-4-hydroxybenzonitrile; 3,5-dibromo-4-hydroxyphenyl cyanide; 2,6-dibromo-4-cyanophenol; broxynil). a postemergence herbicide that is less active than ioxynil. The first visual effects are the appearance within 24 hours of necrotic spots on leaves; tissue damage spreads and the plant eventually dies. It is very toxic to laboratory mice.

bromum. bromine.

Bromus (brome grass). a genus of annual, biennial, and perennial temperate zone grasses (Family Graminae) with few, many-flowered spikelets. ***B. catharticus*** (rescue grass). concentrations of nitrates toxic to livestock have been reported. ***B. inermis*** (awnless, smooth, or Hungarian brome grass). Sometimes becomes heavily ergotized and therefore hazardous. See *Claviceps*.

bronchi. plural of bronchus.

bronchial. pertaining to or denoting bronchi or bronchae.

bronchial asthma. allergic asthma.

bronchial tube. bronchus.

bronchiectasis. chronic bronchodilatation with paroxysmal coughing, foul breath, and expectoration of mucopurulent matter.

bronchiole (bronchiolus). any of many small air passages (less than 1 mm in diameter) that branch from the bronchi of the mammalian lung. The wall is elastic with smooth muscles but no cartilage. Terminal bronchioles communicate with respiratory bronchioles, which terminate in the alveolar ducts.

bronchioli. plural of bronchiolus (See bronchiole).

bronchiolitis. bronchial pneumonia.

bronchiolitis fibrosa obliterans. bronchial pneumonia (bronchiolitis) with invasive growth of connective tissue from the walls of the terminal bronchi that obliterates the bronchial lumina. See also nitrogen dioxide poisoning.

bronchiolus. bronchiole.

bronchitis. an inflammation of the bronchi. It may be acute or chronic, and can be due to an irritant xenobiotic substance.

bronchoconstriction. a reduction in luminal caliber of the bronchi and bronchioles due to autonomic nervous stimulation or in response to a pharmacologically active agent.

bronchoconstrictor. **1:** causing a reduction in luminal caliber of bronchi and bronchioles. **2:** an agent with bronchoconstrictor activity. See def. 1.

bronchodilatation (bronchodilation). an increase in luminal caliber of the bronchi and bronchioles due to autonomic nervous stimulation or in response to a pharmacologically active agent (bronchodilator).

bronchodilation. **1:** bronchodilatation. **2:** bronchiectasis (rarely).

bronchodilator. **1:** causing bronchodilatation. **2:** an agent that causes bronchodilatation.

bronchodilator drug. any bronchodilator (def. 2) used as a therapeutic agent to treat asthma and other diseases of the bronchi that cause shortness of breath.

bronchogenic. originating in the bronchus.

bronchospasm. spasmodic contraction of smooth muscle in the walls of the bronchi and bronchioles causing a narrowing of the lumen. Such occurs in asthma and certain other respiratory disorders.

bronchus (plural, bronchi). either of two tubes in air-breathing vertebrates that conduct air between the trachea and the lungs. They branch from the lower end of the trachea, each communicating with a lung. They subdivide further into lobar, segmental, and subsegmental bronchi.

Brønsted base. See base.

brood survival. the proportion of newly hatched birds that survive until fledging.

Brook's smallheaded sea snake. *Hydrophis brookei*.

broom. common name of two closely related genera of shrubs, *Sparteum* (*Cytisus*) and *Genista*, although it is sometimes applied to other leguminous shrubs with long, slender branches.

broom corn. *Sorghum vulgare*.

broom groundsel. *Senecio spartoides*.

broom poisoning. poisoning caused by eating *Cytisus scoparius*. Poisoning is due to the presence of sparteine and cytisine within the plant.

broom-snakeroot. *Gutierrezia*.

broomweed. *Gutierrezia microcephala*.

brown algae. See Phaeophyta.

brown asbestos. amosite asbestos. See asbestos.

brown-banded snake. *Notechis scutatus*.

brown canesnake. See *Oxyuranus scutellatus*.

brown cobra. *Naje haje*.

brown-headed snake. *Glyphodon tristis*.

brown lung. byssinosis.

brown lung disease. byssinosis.

brown mamba. *Dendroaspis polylepis*.

brown rat. *Rattus norvegicus*. See also laboratory rat.

brown recluse, brown recluse spider. See *Loxosceles*.

brown sea snake. See *Aipysurus*.

brown spider. See *Loxosceles*.

brown snake. See *Demansia textilis*, *Dispholidus typus*.

brown-tail moth. *Euproctis chrysorrhoea*.

brown-tail moth dermatitis. See dermatitis (brown-tail moth dermatitis).

brown-tail rash. brown-tail moth dermatitis.

brown widow. See *Latrodectus*.

brownsnake. See *Demansia textilis*, *Oxyuranus scutellatus*, *Pseudechis australis*, *Elapognathus minor*.

broxynil. bromoxynil.

BrU. 5-bromouracil.

Bruce Ames procedure. See test (Ames test).

brucia. brucine.

brucine (2,3-dimethoxystrychnidin-10-one; 10,-11-dimethoxystrychnine; dimethoxystrychnine; [illegible]); a bitter alkaloid found in the seeds and bark of *Strychnos ignatii* and *S. nux-vomica* (Family Loganaceae). It is structurally related to strychnine and produces similar effects on humans and laboratory mammals. It is extremely toxic, but less so than strychnine. See also ignatia, nux vomica, strychnine.

Brugmansia (angel trumpets). a genus of herbs and occasional shrubs or small trees (Family Solanaceae) that is closely related to *Datura*. All species contain belladonna alkaloids in toxic concentrations. All plant parts are toxic, including the nectar, but most poisonings are due to the ingestion of seeds. Physostigmine opposes the atropinic effects of ingestion.

bruinkapel. *Naja nivea*.

bruise. **1:** bruising; an injury to skin and sometimes to the superficial muscles, characterized by discoloration under the skin and usually accompanied by discomfort and a dull, persistent ache. **2:** to acquire a bruise, to injure by bruising.

bruneomycin. streptonigrin.

brussels sprout. *B. oleracea* var. *gemmifera*.

bryonia. the dried roots of *Bryonia spp.*, especially *B. alba*. It contains several toxic glycosides, most importantly bryonin and bryonidin. It is a virulent purgative, formerly used in medicine.

Bryonia (bryony; devil's tulip; British mandrake). a genus of climbing plants with greenish-yellow flowers (Family Cucurbitaceae), *B. dioica*, *B. alba* (white bryony), and *B. cretica*. They occur most commonly in Wales and England where they are often seen in public gardens. The berries and roots are very toxic if ingested, the chief toxic principles being the glycosides bryonin and bryonidin. Symptoms of intoxication include a burning sensation in the mouth, nausea, and vomiting. Symptoms of serious poisoning also include violent diarrhea, convulsive coma, paralysis, and death from respiratory arrest. The juice is an irritant, blistering the skin on contact.

bryonidin. a toxic glycoside of bryony (See *Bryonia*).

bryonin. a toxic glycoside of bryony (See *Bryonia*).

bryony. *Bryonia spp*.

BTX. **1:** brevetoxins. **2:** batrachotoxin.

BTX-A. batrachotoxinin A.

buccal. pertaining to the mouth.

buccal cavity. the mouth cavity in vertebrate animals and certain advanced invertebrates. In vertebrate animals it communicates with the pharynx.

buccal mass. a thick, dense mass of tissue in the head of many mollusks (gastropods, amphineurans, scaphopods, and cephalopods). It surrounds the anterior part of the digestive tract and contains the radula with supporting musculature and connective tissue. See also *Octopus*.

buckeye. See *Aesculus*.

buckeye family. Hippocastanaceae.

buckshot. See lead shot.

buckthorn. **1:** *Rhamnus*. **2:** sometimes erroneously applied to *Karwinskia humboldtiana*. See toxin (buckthorn toxin).

buckthorn family. Rhamnaceae

buckthorn toxins (Coytillo; Tullidora). a mixture of anthracenones in the fruit of *Karwinskia humboldtiana*. Chief among them are 7-[3′,4′-dihydro-7′,9′-dimethoxy-l′,3′-dimethyl-10′-1′*H*-naphtho(2′,3′-c′)pyran-5′-yl]-3,4-dihydro-3-methyl-3,8,9-trihydroxy-1(2*H*)-anthracenone (T-544); 3,4-dihydro-3,3′-dimethyl-1′,3,-8,8′,9-pentahydroxy(7.lO′-bianthracene)-l,9′-(2*H*,l0′*H*)-dione (T-496); 7-(2′-aceto-6′,8′-dimethoxy-3′-methyl-1′-hydroxynaphth-4′yl)-3,4-dihydro-3-methyl-3,8,9,-trihydroxy-1(2*H*)-anthracenone (T-516); and 3,3′-dimethyl-3,3′,8,8′,9,9′-hexahydroxy-3,3′,4,-4′-tetrahydro-(7,10′-bianthracene)-1,1′(2*H*,2′*H*)-dione (T-514). These toxins produce an often fatal demyelinating neuropathy of peripheral nerves in humans and all forms of livestock. See *Karwinskia humboltiana*.

buckwheat. *Fagopyrum sagittatum*.

buckwheat family (knotweed family). Polygonaceae.

buckwheat poisoning (fagopyrism). **1:** an allergic reaction of grazing animals due to ingestion of *Fagopyrum sagittatum*. **2:** fagopyrism; a rare primary photosensitivity in animals that consume *F. sagittatum*.

Budd's jaundice. acute yellow atrophy (See atrophy).

bufa-, bufo-. prefixes that indicate a relationship to, or origin from toads (Family Bufonidae). These prefixes are used in both systematic and trivial names of toxic substances of plant and animal origin that contain a bufanolide moiety.

bufadienolide. a derivative of bufanolide, *q.v.*

bufagenin. bufagin.

bufagin (bufagenin). any of a group of cardiotoxic steroid toxins (bufanolides) found in toads of the family Bufonidae. Their action is similar to that of digitalis. See also bufotoxin.

bufalin (3,14-dihydroxybufa-20,22-dienolide) **1:** a cardiotonic steroid component of Ch'an Su (Senso), the dried venom of the toad, *Bufo b. gargarizans*. **2:** sodium bufalin-3-sulfate. Isolated from the Japanese toad, *Bufo vulgaris formosus*.

bufanolide. any of a group of basic steroid lactones of certain squill-toad (Family Bufonidae) toxins and plant glycosides such as digitalis. Bufanolide derivatives are usually classified according to the number of unsaturated bonds: **bufenolides** with one double bond, **bufadienolides** with two double bonds, and **bufatrienolides** with three, etc. The number of hydroxyl groups at positions 3, 5, 14, and 16 may vary and can also be substituted.

bufadienolide. a derivative of bufanolide, *q.v.*

bufatrienolide. a derivative of bufanolide, *q.v.*

bufencarb (3-(1-ethylpropyl)phenol methylcarbamate mixture with 3-(1-methylbutyl)-phenyl methylcarbamate (1:3); methylcarbamic acid *m*-(1-ethylpropyl)phenyl ester mixture with *m*-(1-methylbutyl)phenyl ester; metalkamate). a very toxic, very dangerous cholinesterase inhibitor used as an insecticide; it is readily absorbed through the skin.

buffer. **1:** a substance or mixture that, in solution, resists large changes in pH when acidic or alkaline material is added. A buffer is usually a mixture of a weak acid and its conjugate base (e.g., acetic acid and sodium acetate), or a weak base and its conjugate acid (e.g., ammonium hydroxide and ammonium chloride). Bicarbonates and proteins, for example, act as buffers in biological materials. **2:** to treat with a buffer. **3:** to insulate or protecta system from change due to external factors. **4:** to lessen the effect or impact of. **5a:** something that separates two entities. **5b:** anything that insulates or protects a system from external factors that may cause an impact.

bufin. a white cardiotoxic secretion isolated from the parotid gland of certain toads.

Bufo. a genus of some 100 species of toads (Family Bufonidae), about 18 of which occur in North America. Nearly all are terrestrial and active at night, or at twilight, remaining in burrows or otherwise hidden and protected from high temperature and dehydration during the day. A few are diurnal. They typically enter water only to spawn. Some common or biologically important species that occur in North America are *B. americanus* (American toad), *B. marinus* (giant toad), *B. terrestris* (southern toad), *B. cognatus* (Great Plains toad), *B. boreas* (Western toad), *B. punctatus* (red-spotted toad), and *B. woodhousi* (Woodhouse's toad). ***B. marinus*** (giant toad). a giant, semiaquatic, venomous toad indigenous to Central and South America and the extreme southern tip of Texas; it has been widely introduced elsewhere, including southern Florida. It attains a length of up to 22 cm and a body weight of 450-700 g. The venom, a milky fluid secreted by the parotid gland, contains the toxins, marinobufagin and marinobufotoxin. It is a strong irritant of the skin and mucous membranes; it is neurotoxic and often lethal to small animals. Symptoms in humans include nausea, vomiting, hypertension, severe headache, tachycardia, and paralysis. Cats and dogs that bite this toad are sometimes fatally poisoned. ***B. vulgaris*** (common European toad); the skin and saliva contain bufotalin. ***B. vulgaris formosus*** (gama). a poisonous Asiatic (Chinese) toad.

bufo-. See bufa-.

bufogenin (bufogenine). any of a class of toxic principles in toad secretions (e.g., bufogenin B, bufotalin, resibufogenin). In general, their toxicities are poorly characterized, but they have marked effects on involuntary muscle, including the heart. See bufogenin B.

bufogenin B (3,14,16-trihydroxybufa-20,22-dienolide; desacetylbufotalin). a toxicant isolated from the Chinese drug, Ch'an Su, *q.v.*; also derived from bufotalin, *q.v.*

Bufonidae. a family of anuran amphibians (Order Anomocoela) that includes many of the true toads. They are nearly cosmopolitan in distribution, although they are absent from the Madagascar and Australian regions. Maxillary teeth are usually absent and the parotid gland is conspicuous. The dermal glands secrete a number of pharmacologically active agents, some of which are toxic. Bufonid toxins are chemically diverse and include biogenic amines, bufogenins, and indolalkylamines. Several of these toxins have an action similar to that of digitalis. *Bufo* is the only New World genus. See also bufalin, bufanolide, bufagin, bufin, bufogenin, bufonin, bufotalin, bufotenin, bufotenine, bufotenidin, bufotoxin, bufoviridin, bufotalin.

bufonin. a toxin of toads of the family Bufonidae.

bufotalin (16-(acetyloxy)-3,14-dihydroxybufa-20,22-dienolide**;** 3β,14,16β-trihydroxy-5β-bufa-20,22-dienolide-16-acetate). a crystalline genin (a bufagen) found mainly in the skin and saliva of the common European toad, *Bufo vulgaris*. See also bufotoxin.

bufotalin 3-suberoylarginine ester. bufotoxin.

bufotenidin. a vasoactive and apparently hallucinogenic indolalkylamine isolated from toad secretions.

bufotenin (3-[2-(dimethylamino)ethyl]-1*H*-indol-5-ol; 3-(2-dimethylaminoethyl)-5-indolol**;** *N,N*-dimethyl-5-hydroxytryptamine; 5-hydroxy-*N,N*-dimethyltryptamine; *N,N*-dimethylserotonin; 3-(β-dimethylaminoethyl)-5-hydroxyindole; bufotenine; mappine). a very toxic, water-insoluble, genin and serotonin derivative, $C_{12}H_{16}N_2O$, produced synthetically and found in certain toads (Family Bufonidae), a number of green plants, including *Piptadenia peregrina* (Family Mimosaceae), and in very small amounts in the usually edible mushroom, *Ama-*

nita citrina. It is a pressor and psychotomimetic genin.

bufotenine. bufotenin.

bufotoxin. **1:** any toxin present in or derived from the skin of a toad. **2:** any of a group of steroid lactones secreted by the skin of toads (Family Bufonidae). They are conjugates of bufagins with suberylarinine at C-3. Their action is similar to but weaker than that of the bufagins. **3:** 16-(acetyloxy)-3[[8-[[4-[(amino-iminomethyl)-amino]-1-carboxybutyl]amino]-1,8-dioxo-octyl]-oxy]-14-hydroxybufa-20,22-dienolide; vulgaro-bufotoxin; bufotalin 3-suberoyl-arginine ester). the principal toxin of the common European toad, *Bufo vulgaris*. See also bufotalin, *Bufo*.

bufoviridin. a vasoactive and apparently hallucinogenic indolalkylamine isolated from toad secretions.

bukizi. *Dendroaspis jaesoni*.

bulabundoo. *Bitis arietans*.

Bulan (1,1′-(2-nitrobutylidene)bis[4-chlorobenzene]; 1,1-bis(*p*-chlorophenyl)-2-nitrobutane; 2-nitro-1,1-bis(*p*-chlorophenyl)-butane). an insecticide that is very toxic, *per os*, to laboratory rats.

bulb. a fleshy underground organ of perennation and usually of vegetative propagation in many plants of the Liliaceae (e.g., *Convallaria*, *Gloriosa*, *Hyacinthus*, *Ornithogalum*, *Veratrum*) and Amaryllidaceae (e.g., *Galanthus*, *Hymenocallis*, *Leucojum*, *Narcissus*), spider lilies, snowdrops, and snowflakes. It consists of a highly modified shoot with a very short stem surmounted by swollen scale leaves or thickened leaf bases that store food. Growth begins with the elongation of the adventitious roots, followed by sprouting of the apical bud, usually followed by the growth of one or more additional buds. Growth is initially supported by nutrients stored in the bulb. The bulbs of many plants (as in the genera named above) are poisonous and may resemble edible bulbs such as onion. *Cf.* corm.

bulbous water hemlock. *Cicuta bulbifera*.

bulk transport. See transport.

bull nettle. a common name of *Cnidoscolus stimulosus*, *Jatropha stimulosa*, and *Solanum carolinense*.

bull pine. *Pinus taeda*.

bullrout. a common name of *Apistus carinatus* and *Notesthes robusta*.

bululu. *Bitis arietans*.

bumble bee. a common name applied to any of a wide variety of large bees, especially *Bombus* (Family Bombidae).

BUN. blood urea nitrogen.

bundle of His. atrioventricular bundle.

bungarotoxin. any of a class of proteins contained in the venom of the southeast Asian banded krait, *Bungarus multicinctus* (Family Elapidae), and other snakes of this genus, especially *B. caeruleus* (Indian krait; kauryala). The crude venom is highly toxic. **α-bungarotoxin** (α-Bgt). a potent neurotoxic protein comprised of a single polypeptide chain with a mol. wt. of about 8000. It binds specifically to acetylcholine receptors. **β-bungarotoxin** (β-BuTX). an extremely potent neurotoxic material that contains several components, the most important being β_1-bungarotoxin. It is comprised of two subunits with mol. wts. of about 13,000 and 7,000, connected by sulfide bonds. It is a presynaptic neurotoxin that inhibits acetylcholine release at myoneural junctions of voluntary muscles without affecting the postsynaptic membrane. The LD_{50} in laboratory mice is ca. 0.02 mg/kg.

Bungarus (kraits). a genus of venomous snakes (Family Elapidae) of southeastern Asia, chiefly northern India. Kraits resemble many nonpoisonous snakes in general appearance. The head is short, slightly depressed, and noticeably distinct from the neck; there is no hood. The eyes are small and dark, with pupils that are nearly invisible in live specimens. The jaws bear 2 large tubular (maxillary) fangs with external grooves followed by an interspace and 1-4 small, slightly grooved teeth. The body of most species has a vivid pattern of crossbands; scales are smooth and glossy. Kraits are alert and active at night, but are quiet and sluggish at best during the day. While they cause few snakebites, the case fatality rate is very high. They prey largely on other snakes. Adults of most species average about 1.2-1.5 m in length, although *B. fasciatus* may exceed 2 m. All species are extremely dangerous. The venom is extremely virulent and contains the neurotoxic bungarotoxins, *q.v.*, and the bite produces generalized anesthetic and paralytic effects. The neurotoxic effects are similar to those induced by cobra venom. Symptoms of envenomation include headache, dizziness, disorders of speech and vision, unconsciousness, convulsions, gastrointestinal pain, and muscular paralysis lasting as long as four days. Death from respiratory paralysis is frequent in untreated cases. Kraits not described below include *B. javonicus* (Javan krait), *B. lividus* (lesser black krait), *B. niger* (black krait), and *B. walli* (Wall's krait). ***B. caeruleus*** (blue krait; common krait; Indian krait; chitti; pee-un; sangchul; yennai viriyan; yettadi viriyan). a jet black to dark brown krait with a series of narrow, white or yellow, usually paired, crossbands that often fade out or break up on the anterior quarter of the body. The upper lip is white or yellow; belly an immaculate white. All subcaudals are undivided. The length of adults averages about 1.1 m with occasional individuals reaching a length of about 1.6 m. It occurs in a variety of habitats at low and moderate elevations in dry open country from western Pakistan and India to China. It also occurs near human habitation, usually occupying poorly constructed or dilapidated buildings. This agile snake is usually active on hot humid nights. When disturbed they coil loosely with the body slightly flattened and the head concealed; they make jerky movements and sometimes elevate the tail. They do not strike but often make a quick snapping bite. During the day they are passive and sluggish. They rarely strike humans, but are considered to be highly dangerous because the venom is extremely virulent. Some 50-80% of victims who do not receive antivenin die. ***B. candidus*** (Malayan krait; blue krait; common krait; ular weling). a species that is very similar in appearance to *B. caeruleus*, but with fewer crossbands (15-25 versus 35-55). *Cf. Bungarus ceylonicus*. ***B. ceylonicus*** (Ceylon krait; dunu karawala; karawala; tel-karawala). a species that is very similar in appearance to *B. caeruleus*, but with fewer crossbands (15-25 versus 35-55). Furthermore, the bands are narrow and often broken in *B. ceylonicus*. ***B. fasciatus*** (banded krait; pama; oraj welang; ngu sam liem; oraj weling; pama; ular katam tabu; ular welang). a species that is indigenous to eastern India to southern China and south through much of Malaysia and Indonesia where it occurs in more or less open country to elevations of about 1500 m, often near surface water. It is slightly larger than the *B. caeruleus* and is characterized by a pronounced vertebral ridge causing this snake to appear emaciated; a blunt tail; and a pattern of bright, alternating, light (usually bright yellow) and black bands encircling the body that are almost identical in width. In occasional individuals the light bands are white, tan, or orange. The average length of adults is about 1.3 m with occasional individuals attaining a length of 2 m or more. The venom is lethal to humans, although it is not as virulent as that of most other kraits. Nevertheless, this snake is usually regarded as nearly harmless since it appears docile and bites of humans are almost unknown. When disturbed, it coils up, hides its head beneath its coils, and makes jerky flinching movements but rarely bites. ***B. flaviceps*** (red-headed krait; yellow-headed krait; ular tandjon api; yellow-headed snake). a rare snake that occurs in jungles, chiefly in hilly or mountainous areas, from southern Burma to Viet Nam and southward through Malaysia and the larger Indonesian islands. It is similar in appearance to *B. fasciatus* but the tail is not so blunt and while the anterior subcaudals are entire, the posterior ones are divided. The markings are bright and distinctive, the head and tail being bright red and the body black with a narrow bluish white stripe low on side. In some individuals there is a narrow orange stripe or row of dots down middle of back. ***B. multicinctus*** (many-banded krait; Southeast Asian banded krait; hundred segment snake, Taiwan krait; Taiwan banded krait; umbrella snake). a krait found from Burma through southern China to Hainan and Taiwan. It commonly occurs in wooded or grassy areas near surface water and may be found in villages and suburban areas. It is common in rice paddies. It is very similar in appearance to *B. caeruleus* but the light crossbands are unpaired and the underside may show dark mottling. This snake is nocturnal and is most active on damp or rainy nights. It is usually inoffensive and is not easily disturbed. Nevertheless, the venom is extremely toxic to laboratory animals, with an LD_{50} of about 0.1 mg/kg. Bites by this krait are fairly frequent in Taiwan, but the case fatality rates are much less than those reported for India.

Buphane disticha. a plant used widely in Africa by medicine men and to prepare arrow poisons. All parts of the plant are poisonous and contain buphanie, a toxin with effects very similar to those of scopolamine.

buphanie. See *Buphane disticha*.

burden. hardship, ordeal; strain, stress; load, weight. **body burden. 1:** the total activity of a radiopharmaceutical retained by an organism at a designated time following administration. **2:** the total amount of a xenobiotic chemical that occurs in a living organism at a specified time. **chemical burden**. the total amount of a chemical above background or baseline that occurs in any component of a living organism or ecological system (e.g., soil burden). **environmental burden**. the total amount of a chemical above background or baseline that occurs in a specified region or environment.

burdening. **1:** establishing, imposing, or sustaining a burden. **2:** the imposition or addition of a xenobiotic chemical to any natural entity, biotic or abiotic, over and above that which is normal.

Burgundy mixture. a copper-containing fungicide that is more potent than Bordeaux mixture and is used primarily against rusts. It contains 60 g sodium carbonate and 50 g cupric sulfate in 51 g water. See Bordeaux mixture, cupric sulfate.

Burma bean. *Phaseolus limensis*.

burn. **1:** a lesion (e.g., irritation, inflammation, erosion, or destruction of tissue) due to undue contact with thermal energy sources (e.g., flame, searing heat from any source, steam); irritant or caustic chemicals (e.g., strong acids, strong bases), electrical energy (lightning, electricity), radiant energy (e.g., intense light, ionizing radiation), or even friction. Regardless of cause, the appearance and effects of most burns are similar. Shock is always, and pain is nearly always, a feature of severe burns and may cause death. The risk of infection is high in the case of severe, extensive burns. **acid burn**. a burn resulting from exposure to corrosive acids (e.g., hydrochloric, nitric, sulfuric). An acid burn is a chemical burn. **alkali burn**. a burn resulting from exposure to strong alkalies (e.g., potassium hydroxide, sodium hydroxide). An alkali burn is a chemical burn. **chemical burn**. **1:** any irritation, inflammation, or erosion of tissue (usually of the skin or mucous linings of the eye, gastrointestinal tract, or upper respiratory tract) due to contact with an irritant or caustic substance (e.g., a strong base or acid). **2:** severe primary irritant dermatitis, dermatitis escharotica. See dermatitis. **first-degree burn**. that which involves only the epidermis and characterized by erythema and edema without vesiculation. **inhalation burn**. a burn resulting from the inhalation of hot gases. Such are often associated with elevated carboxyhemoglobin levels in circulating blood. **radiation burn**. a burn caused by excessive exposure to radiant energy (e.g., sunlight, ionizing radiation). **respiratory burn**. inhalation burn. **second-degree burn**. that which involves the epidermis and dermis, usually with blistering, with or without deep dermal necrosis. The epithelium can regenerate. **third-degree burn**. that in which the entire skin is destroyed, often with much scarring. The injury may extend into subcutaneous fat, muscle and other deep tissues.

burning bean. *Sophora secundiflora*.

burning bush. *Euonymus atropurpurea*.

burnt lime. calcium oxide.

burrek. *Acanthophis antarcticus*.

burrfish. a common name of fish of the family Diondontidae.

burrow weed. See *Haplopappus*.

burrowing adder. *Atractaspis bibronii*.

burrowing cobra. *Paranaja multifasciata*.

burrowing snake. *Brachyurophis roperi*.

burrowing viper. See *Atractaspis*.

Burton's carpet viper. *Echis coloratus*.

Buschelbrauen viper. *Bitis caudalis*.

bush bean. See *Phaseolus*.

bush vipers. vipers of the genus *Atheris*.

bushmaster. *Lachesis mutus*.

Bushnell's wasp. See *Odynerus*.

busulfan (1,4-butanediol dimethanesulfonate esters; 1,4-bis(methanesulfonoxy)butane; 1,4-di(methanesulfonyloxy)butane; 1,4-di(methylsulfonoxy)butane; methanesulfonic acid tetramethylene ester; tetramethylene bis-(methanesulfonate); busulfan; BSF). a crystalline, acetone-soluble alkylsulfonate. It is an extremely toxic antineoplastic (pulmonary) alkylating agent and insect sterilant. Busulfan is a primary carcinogen, and induces clastogenic, teratogenic, and immunosuppressive effects. It causes delayed bone marrow aplasia, cataracts, pigmentation, pulmonary thrombosis, cardiotoxic effects, and thrombocytopenia in humans.

Busycon. a genus of large, carnivorous, marine gastropods (whelks or conchs) that produce tetramethyl ammonium. The shell is up to 24 cm in length.

butachlor ((*N*-(butoxymethyl)-2-chloro-*N*-(2,6-diethylphenyl)acetamide; *N*-(butoxymethyl)-2-chloro-2′,6′-diethylacetanilide; 2-chloro-2,6-diethyl-*N*-(butoxymethyl)acetanilide). a pre-emergence selective anilide herbicide that is absorbed through the roots. Its herbicidal properties and applications are similar to other anilide herbicides, but it is also used to control grass and broad-leaved weeds in seeded and transplanted rice paddies. It is moderately toxic to laboratory rats.

1,3-butadiene (vinylethylene; erythrene; bivinyl; divinyl). a colorless liquid (b.p. -4.5°C) or gas, $H_2C{=}CHHC{=}CH_2$, it is a dialkene with a mildly aromatic odor. The gas is highly flammable and may form explosive peroxides on exposure to air; the liquid phase can be safely managed and stored. It polymerizes readily and is widely used in the manufacture of various polymers, including synthetic rubber and elastomers such as neoprene, nitriles, polybutadiene, and styrenebutadiene. It is also used as a chemical intermediate in the production of rocket fuels. Butadiene is manufactured in huge quantities and is a toxic and hazardous occupational chemical. It is an irritant to the eyes, skin, and mucous membranes (especially of the lungs). The liquid causes a frostbite-like contact dermatitis. Narcotic effects are seen at relatively high concentrations and can cause drowsiness, fatigue, unconsciousness, respiratory paralysis, and death. Chronic occupational exposure can cause CNS and liver damage, and anemia. It is also carcinogenic to some laboratory animals and to humans, causing cancer of the heart, lungs, spleen, liver, kidneys, and stomach.

1,2,3,4-butadiene epoxide. the oxidation product of 1,3-butadiene; both *cis*- and *trans*- forms are known. It is a direct-acting (primary) carcinogen.

butane (trimethylmethane). an extremely stable but highly flammable, colorless gas at room temperature and standard atmospheric pressure, $CH_3CH_2CH_2CH_3$, with a slight odor. It is very soluble in water and soluble in ethanol and chloroform, with no corrosive action on metals. There are two isomers of butane (*n*-butane and isobutane (trimethylmethane). Both are simple asphyxiants with no known toxic systemic effects. isobutane (2-methylpropane). An important component natural gas.

***n*-butane**. See butane.

butanedioic acid mono(2,2-dimethylhydrazide). daminozide.

1,4-butanediol dimethane sulfonate. bisulfan.

butanethiol. *n*-butyl mercaptan.

1-butanethiol. *n*-butyl mercaptan.

2-butanethiol. *sec*-butyl mercaptan.

***n*-butanethiol**. *n*-butyl mercaptan.

butanoic acid 2,2,2-trichloro-1-(dimethoxyphosphinyl)ethyl ester. butonate.

butanol-1. 1-butanol.

1-butanol (*n*-butyl alcohol; butanol-1; butyl alcohol; 1-butyl alcohol; butyric alcohol; propyl carbinol). a colorless, flammable liquid solvent and food flavoring, $CH_3(CH_2)_2CH_2OH$. It easily biodegrades and does not bioaccumulate. It is an irritant of the skin and mucous membranes, and can induce dizziness, headaches, and drowsiness in humans. Exposure may also affect hearing and light adaptation of the eye. In severe cases, liver damage may result. It is slightly to moderately toxic to laboratory animals, not directly toxic to aquatic animals, and practically nontoxic to algae. Acute effects in humans are substantially those of acute alcoholic intoxication and narcosis. Prolonged, severe exposure to vapor mixtures of 1-butanol and isobutanol has resulted in vertigo. See also *tert*-butanol.

2-butanol (*sec*-butanol; *sec*-butyl alcohol; 2-butyl alcohol; butylene hydrate; secondary butyl alcohol; ethyl methyl carbinol; methyl ethyl carbinol; 2-hydroxybutane). a colorless, flammable liquid that readily decomposes in the environment, possibly contributing to oxygen depletion of aquatic habitats. It does not bioaccumulate. 2-Butanol is used in the extraction of fish meal to produce fish protein concentrate. It is used as a flavoring agent and also occurs naturally in food and beverages. 2-Butanol is essentially nontoxic to aquatic organisms and is only slightly toxic to laboratory animals and man, but may have mutagenic activity. Probable acute effects on humans are those of acute alcoholic intoxication.

***n*-butanol**. 1-butanol.

***sec*-butanol**. 2-butanol.

***tert*-butanol** (*tert*-butyl alcohol; 2-methyl-2-propanol; tertiary butanol; *tert*-butyl hydroxide; trimethyl carbonol; trimethyl methanol). a slightly to moderately toxic colorless liquid or white crystalline solid, $(CH_3)_3COH$, soluble in water and ethanol. It is used as a solvent, an octane booster, a moderately toxic denaturant of ethanol, in flavors and perfumes, and in several manufacturing processes. At normal ambient levels it is not toxic to fish, amphibia, crustacea, algae, or bacteria. Acute effects in laboratory animals are those of alcoholic intoxication. It causes physical dependence and postnatal effects from exposure *in utero*. In humans, *tert*-butanol is also a skin irritant. It has a number of uses, including those of a denaturant of ethanol, a solvent for pharmaceuticals, and an octane booster in unleaded gasoline.

2-butanol acetate. *sec*-butyl acetate.

butanone. methyl ethyl ketone.

2-butanone. methyl ethyl ketone.

Butazolidin. phenylbutazone.

2-butenal. crotonaldehyde.

1-butene. See butylene.

2-butene. See butylene.

2-buten-1-ol. crotyl alcohol.

butenolide. a mycotoxin produced by *Fusarium sporotrichioides*. It contains an α,β-unsaturated lactone moiety and is extremely toxic (*i.p.*) or very toxic (*p.o.*) to laboratory mice.

Δ^3-2-butenone. methyl vinyl ketone.

3-buten-3-one. methyl vinyl ketone.

3-butenyl isothiocyanate. a mustard oil.

buthid (buthid scorpion). any scorpion of the family Buthidae.

buthid scorpion. a buthid. See Buthidae.

Buthidae (buthids; buthid scorpions). the largest family of scorpions (Order Scorpionida), with some 700 species worldwide. Five species occur in North America. They range in length from 50-70 mm. The sternum is triangular and each side of the distal tarsal segment of the last pair of legs bears one or two spurs.

Buthus occitanus. a deadly scorpion (Family Buthidae) found in arid regions of North Africa, India, and Pakistan. The venom contains four basic neurotoxins, each comprised of a single polypeptide chain with four disulfide bridges; the average molecular weight is about 7000. All are extremely toxic. The venom also contains 5-hydroxytryptamine, a compound that probably does not contribute significantly to toxicity, but may cause intense local pain. See also *Androctonus australis*.

butilate. butylate.

butinol-*N*-(3-chlor ophenyl)carbamate. a highly selective aryl carbamate herbicide. It is used to control *Avena fatua* (wild oats) in cereal crops.

Butinox. See tributyltin.

butonate (butanoic acid 2,2,2-trichloro-1-(dimethoxyphosphinyl)ethyl ester; butyric acid ester with dimethyl (2,2,2-trichloro-1-hydroxyethylphosphonate; *O,O*-dimethyl-2,2,2-trichloro-1-*n*-butyryloxyethylphosphonate;*O,O*-dimethyl 2,2,2-trichloro-1-phosphonoethyl butyrate). a cholinesterase inhibitor used as an insecticide and as an anthelmintic in horses.

butopyronoxyl (3,4-dihydro-2,2-dimethyl-4-oxo-2*H*-pyran-6-carboxylic acid butyl ester; butyl 3,4-dihydro-2,2-dimethyl-4-oxo-2*H*-pyran-6-carboxylate; butyl mesityl oxide oxalate; α,α-dimethyl-α'-carboxydihydro-γ-pyrone butyl ester). a yellow to amber, essentially water-insoluble, aromatic liquid used as an insect repellant on skin or clothing. It occurs chiefly as two species in equilibrium, the dihydropyrone and the enol. Butopyronoxyl is extremely toxic orally to laboratory mice and rats and has caused liver necroses in experimental animals.

α-[2-(2-butoxyethoxy)ethoxy]-4,5-methylenedioxy-2-propyltoluene. piperonyl butoxide.

5-[[2-(2-butoxyethoxy)ethoxy]methyl]-6-propyl-1,3-benzodioxole. piperonyl butoxide

***N*-(butoxymethyl)-2-chloro-2′,6′-diethylacetanilide**. butachlor.

***N*-(butoxymethyl)-2-chloro-*N*-(2,6-diethylphenyl)acetamide**. butachlor.

butter clam. a common name of *Saxidomus giganteus* and *S. nuttalli*.

butter of antimony. antimony trichloride.

butter of arsenic. arsenic trichloride.

butter-yellow (*p*-dimethylaminoazobenzene). a banned, carcinogenic food coloring. See also food colors.

buttercup family. Ranunculaceae.

butterfly. an insect of the order Lepidoptera.

butterfly adder. *Bitis gabonica*.

butterfly ray. a stingray, *q.v.*, of the family Gymnuridae.

butterfly root. *Asclepias tuberosa*.

butterfly weed. *Asclepias tuberosa*.

β-BuTX. β-bungarotoxin.

butyl acetate. a colorless, flammable, irritant, toxic, liquid, $C_6H_2O_2$, with three isomers.

1-butyl acetate. *n*-butyl acetate.

2-butyl acetate. *sec*-butyl acetate.

***n*-butyl acetate** (acetic acid *n*-butyl ester; butyl acetate; 1-butyl acetate; butyl ethanoate). an irritant liquid, $CH_3COOCH_2CH_2CH_2CH_2$, with a strong fruity odor. It is slightly water soluble and is miscible with ethanol, ethyl ether, and propylene glycol. It is used in solvents, lacquer thinners, and as a synthetic flavoring agent in foods. It is also used as an experimental teratogen. Butyl acetate is moderately toxic, *i.p.*, but only slightly toxic orally and by inhalation. It is also a mild allergen and a strong irritant of the skin and especially the eye; it can irritate the conjunctiva and can cause pinkeye and other conditions in the vertebrate eye. High doses have narcotic effects. ***sec*-butyl acetate** (acetic acid-2-butoxy ester; acetic acid-1-methylpropyl ester; 2-butanol acetate; 2-butyl acetate; *sec*-butyl alcohol acetate). an allergen and irritant liquid, $CH_3COOCH(CH_3)(C_2H_5)$, with a mild odor. ***tert*-butyl acetate** (aceticacid-*tert*-butyl ester; acetic acid-1,1-dimethylethyl ester). a liquid, $CH_3COOC(CH_3)_2$, used as a solvent and a gasoline additive.

***sec*-butyl alcohol acetate**. *sec*-butyl acetate.

butyl benzyl phthalate (benzyl butyl phthalate; 1,2-benzenedicarboxylic acid, butyl phenylmethyl ester; *n*-butyl benzyl phthalate). a clear, oily, moderately toxic, liquid plasticizer used in large quantities. It is a persistent and widespread environmental pollutant. Experimentally induced reproductive effects have been reported and it is carcinogenic to some laboratory animals. ***n*-butyl benzyl phthalate**. butyl benzyl phthalate.

***n*-butyl-*n*-butanoate**. butyl butyrate.

butyl butyrate (*n*-butyl-*n*-butanoate; *n*-butyl butyrate; *n*-butyl-*n*-butyrate). a colorless, transparent, flammable, toxic liquid with a pineapple-like odor. It is used as a flavoring in beverages, ice cream, and other sweets to give a taste of butter, apple, banana, peach, or nut. It is moderately toxic *i.p.*, but mildly toxic orally. It is moderately irritant to the skin, eyes, and mucous membranes and, in high concentrations, may have a narcotic effect.

***n*-butyl butyrate**. butyl butyrate.

***n*-butyl carbinol**. 1-pentanol.

***N*-butyl-*N'*-(3,4-dichlorophenyl)-*N*-methylurea**. neburon.

butyl-3,4-dihydro-2,2-dimethyl-4-oxo-2H-pyran-6-carboxylate. butopyronoxyl.

4-*n*-butyl-1,2-diphenyl-3,5-pyrazolidinedione. phenylbutazone.

butyl ethanoate. *n*-butyl acetate.

***tert*-butyl hydroxide**. *tert*-butanol.

6-*tert*-butyl-3-(2-imidazolin-2-ylmethyl)-2,4-dimethylphenol. oxymetazoline.

***n*-butyl mercaptan** (1-butanethiol; normal butyl thioalcohol; thiobutyl alcohol; *n*-butanethiol; butanethiol; butyl mercaptan). a colorless to light yellow, mobile, flammable liquid, $CH_3(CH_2)_2CH_2SH$, and a characteristic, odorous, component of skunk secretion. It is a powerful oxidizer and an irritant poison that is moderately toxic by inhalation and ingestion. This compound is used commercially as an odorant to detect leaks in gas lines and storage tanks. Do not confuse with *n*-propyl mercaptan. **butyl mercaptan**. *n*-butyl mercaptan.

***sec*-butyl mercaptan** (2-butanethiol; *sec*-butyl thioalcohol). a colorless, flammable liquid, $C_2H_5CH(SH)CH_3$, with an offensive odor. Toxicity and usage are similar to those of *n*-butyl mercaptan.

***tert*-butyl mercaptan** (2-methyl-2-butanethiol). a compound, $(CH_3)_2CSH(C_2H)_5$, with properties, usage, and probably toxicity that are similar to those of *n*-butyl mercaptan.

butyl mesityl oxide oxalate. butopyronoxyl.

***n*-butyl nitrite** (nitrous acid butyl ester). an oily organic liquid, $CH_3(CH_2)_3ONO$, with a characteristic odor. It is a hypotensive agent (a vasodilator); inhaling the vapor can cause headaches and weakness.

***n*-butyl phthalate** (1,2-benzenedicarboxylic acid dibutyl ester; di-*n*-butylphthalate; phthalic acid dibutyl ester; dibutyl phthalate). a toxic oily liquid impregnated in clothing as an insect repellant. At fairly high doses, it causes rapid and severe seminiferous tubular atrophy in laboratory rats and guinea pigs, focal atrophy only in mice, and no such effects in hamsters.

5-butylpicolinic acid. fusaric acid.

5-butyl-2-pyridinecarboxylic acid. fusaric acid.

***sec*-butyl thioalcohol**. *sec*-butyl mercaptan.

[1-[(butylamino)carbonyl]-1*H*-benzimidazol-2-yl]carbamic acid methyl ester. benomyl.

(±)-1-(*tert*-butylamino)-3-[(2-methylindol-4-yl)oxy]-2-propanol benzoate (ester). bopindolol.

butylate. **1:** to introduce a butyl group into a substance. **2:** (bis(2-methylpropyl)carbamothioic acid *S*-ethyl ester; diisobutylthiocarbamic acid *S*-ethyl ester; *S*-ethyl-*N,N*-diisobutylthiocarbamate; butilate; diisocarb). a selective, preplanting thiocarbamate herbicide that is moderately toxic to laboratory mammals.

butylated hydroxyanisole (BHA). a toxic mixture of 2- and 3-*tert*-4-methoxyphenol, it is an antioxidant used for fats and oils and is sometimes used as a preservative in drug or toxicant formulations. Its use in foods is restricted in the United States but may be used in breads. BHA is synergistic with acid and is moderately toxic to laboratory mice and rats. BHA can cause depression of growth rate, weight loss, liver damage, baldness, and fetal abnormalities. It has been linked to hyperactivity and behavioral disturbances in children. See also bread, butylated hydroxytoluene.

butylated hydroxytoluene (2,6-di-*tert*-butyl-*p*-cresol; BHT). a toxic, flammable, white, crystalline solid, $[C(CH_3)_3]_2CH_3C_6H_2OH$, soluble in organic solvents. It is used as an antioxidant in various products, including food products and animal feeds, although its use in foods in the United States is restricted, but may be used in breads. BHT causes depression of growth rate, loss of weight, damage to the liver, baldness, and fetal abnormalities in laboratory animals. It is carcinogenic to mice, and also appears to promote existing tumors. It is a suspected human carcinogen. BHT has been linked to hyperactivity and behavioral disturbances in children. See also butylated hydroxyanisole.

(5-butyl-1*H*-benzimidazol-2-yl)carbamic acid methyl ester. parabendazole.

1-(butylcarbamoyl)-2-benzimidazolecarbamic acid methyl ester. benomyl.

di-*n*-butylcarbamylcholine sulfate. dibutoline sulfate.

***sec*-butylcarbinol**. 2-methyl-1-butanol.

5-butyl-2-(carbomethoxyamino)benzimidizole. parabendazole.

***N*-butyl-*N'*-(3,4-dichlorophenyl)-*N*-methylurea**. neburon.

2-(4-*tert*-butyl-2,6-dimethyl-3-hydroxybenzyl)-2-imidazoline. oxymetazoline.

2-*sec*-butyl-4,6-dinitrophenol. dinoseb.

2-*sec*-butyl-4,6-dinitrophenyl isopropyl carbonate. dinobuton.

2-*sec*-butyl-4,6-dinitrophenyl 3-methyl-2-butenoate. binapacryl.

2-*sec*-butyl-4-6-dinitrophenyl 3-methylcrotonate. binapacryl.

2-*sec*-butyl-4,6-dinitrophenyl senecioate. binapacryl.

butylene. any of three extremely flammable isomeric monoalkenes with the formula C_4H_8 (1-butene {= α-butylene}, 2-butene, {= β-butylene}, isobutene {= γ-butylene). All are easily liquified gases at ambient conditions. They readily isomerize and undergo addition reactions to form polymers. They are simple asphyxiants and mildly narcotic when inhaled, but are otherwise not regarded as especially toxic.

α-butylene. See butylene.

β-butylene. See butylene.

γ-butylene. See butylene.

butylene hydrate. 2-butanol.

butylene oxide (2,3-epoxybutane). a liquid (b.p. 63°C) epoxide with toxicity and effects similar to those of ethylene oxide. It may form toxic vapors at room temperature. It is highly flammable and a serious fire hazard.

butyric acid. lead butyrate.

butyric acid ester with dimethyl (2,2,2-trichloro-1-hydroxyethylphosphonate. butonate.

β-butyrolactone (3-hydroxybutanoid acid β-lactone; hydroxybutyric acid lactone; 3-hydroxybutyric acid lactone; 4-methyl-2-oxetanone). a suspected carcinogen and mutagen. It is moderately irritant, but only slightly toxic when ingested. *Cf.* γ-butyrolactone.

γ-butyrolactone. a nontumorigenic lactone. *Cf.* β-butyrolactone.

butyronitrile. a colorless, flammable liquid, $CH_3(CH_2)_2CN$, that is slightly soluble in water and soluble in ethanol and ethyl ether. It has various uses, chiefly in chemical and pharmaceutical products. It is a skin irritant and is [illegible] tion, and by intraperitoneal and subcutaneous routes. On decomposition by heating it releases toxic fumes of NO_x and CN.

butyrophenone. any of a class of antipsychotic drugs that includes haloperidol.

Buxaceae (the boxwood family). a small family

of chiefly tropical or subtropical trees, shrubs, and perennial herbs comprised of about six genera. See *Buxus*.

buxine. a neurotoxin isolated from the leaves of *Buxus sempervirens*, *q.v.*

Buxus sempervirens (box; common box; boxwood). a slow-growing, highly toxic evergreen hedge plant or tree (Family Buxaceae) of southern Europe, northern Africa, western Asia, and widely cultivated elsewhere. The sickly looking, yellowish flowers grow in small clusters and produce seed capsules containing three seeds. All parts of the plant are poisonous. The principle is buxine, an alkaloid. Despite a presumably repellant odor and bitter taste, grazing animals sometimes die from eating clippings from box hedges. The symptoms are mainly those of gastrointestinal distress; convulsions may occur.

BW. biological warfare.

by-product (co-product). a substance produced incidental to a process such as the synthesis, manufacture, or processing of another substance.

byssinosis (brown lung; brown lung disease; Monday fever; Monday dyspnea; cotton-dust asthma; stripper's asthma; cotton-mill fever). a pneumoconiosis due to inhalation of cotton dust by textile workers and workers who process soft hemp and flax. Symptoms of acute exposure include tightness of the chest, dyspnea, and coughing on return to work following an absence of a few days. Long-continued exposure characteristically causes a permanent, crippling dyspnea and eventual death. Byssinosis is reversible if detected early. See also dust (cotton dust, proliferative dust).

byssinotic. **1:** of or pertaining to byssinosis. **2:** a victim of byssinosis.

byssophthisis. obsolete term for byssinosis.

BZ. 3-quinuclidinol.

C

c. centi-.

C. the symbol for carbon.

^{12}**C.** the symbol for carbon-12.

C-283. See nitracrine.

C.I. Colour Index.

C.I. 749. FD&C Red No. 19.

C.I. 11000. aminoazobenze.

C.I. 11020. *p*-dimethylaminoazobenzene.

C.I. 11380. FD&C Yellow No. 3.

C.I. 12100. 1-(*o*-tolylazo)-2-naphthol.

C.I. 16035. FD&C Red No. 40.

C.I. 16155. FD&C Red No. 1.

C.I. 16185. amaranth.

C.I. 19140. tartrazine.

C.I. 24110. dianisidine.

C.I. 37035. *p*-nitroanaline.

C.I. 42053. FD&C Green No. 3.

C.I. 42095. FD&C Green No. 2.

C.I. 42640. FD&C Violet No. 1.

C.I. 45170. rhodamine B.

C.I. 45365. C.I. Solvent Orange 32, the disodium salt of 4',5'-dichlorofluorescein, *q.v.*

C.I. 52015. methylene blue.

C.I. 73015. FD&C Blue No. 2.

C.I. 75500. walnut extract.

C.I. 75670. quercetin.

C.I. 76035. toluene-2,4-diamine.

C.I. 76050. 2,4-diaminoanisole.

C.I. 76051. 2,4-diaminoanisole sulfate.

C.I. 76505. resorcinol.

C.I. 76515. pyrogallol.

C.I. 76605. 1-naphthol.

C.I. 77085. arsenic disulfide.

C.I. 77103. C.I. pigment yellow 31.

C.I. 77199. cadmium sulfide.

C.I. 77491. iron oxide.

C.I. 77577. lead monoxide.

C.I. 77578. lead oxide red.

C.I. 77600. lead chromate.

C.I. 77601. lead chromate, basic.

C.I. 77610. lead cyanide.

C.I. 77727. manganese(III) oxide.

C.I. 77728. manganese dioxide.

C.I. 77945. zinc.

C.I. 77947. zinc oxide.

C.I. 77955. zinc chromate.

C.I. acid blue 1. FD&C Blue No. 2.

C.I. acid green. FD&C Green No. 2.

C.I. acid red 27. amaranth.

C.I. acid yellow 23. tartrazine.

C.I. azoic coupling component. 2-naphthol.

C.I. azoic diazo component 37. *p*-nitroanaline.

C.I. azoic diazo component 48. dianisidine.

C.I. azoid red 83. *p*-cresol.

C.I. basic blue 9. methylene blue.

C.I. basic red 9 monohydrochloride. a commercial coloring agent that is carcinogenic to some laboratory animals.

C.I. basic violet 10. rhodamine B.

C.I. developer 4. resorcinol.

C.I. developer 5. 2-naphtol. 2-naphthol.

C.I. developer 17. *p*-nitroanaline.

C.I. direct black 38. a commercially prepared coloring agent that is carcinogenic to some laboratory animals.

C.I. direct blue 6. a commercially prepared coloring agent that is carcinogenic to some laboratory animals.

C.I. direct brown 95. a commercially prepared coloring agent that is carcinogenic to some laboratory animals.

C.I. disperse black. dianisidine.

C.I. disperse blue 1. a commercially prepared coloring agent that is carcinogenic to some laboratory animals.

C.I. disperse orange 11. a confirmed carcinogen.

C.I. disperse yellow 3. a commercially prepared coloring agent, it is an allergen, tumorigen, and is carcinogenic to some laboratory animals.

C.I. food blue. FD&C Blue No. 1.

C.I. food blue 1. FD&C Blue No. 2.

C.I. food green 2. FD&C Green No. 2.

C.I. food green 3. FD&C Green No. 3.

C.I. food red. FD&C Food No. 1.

C.I. food red 6. FD&C Food No. 1.

C.I. food red 15. rhodamine B.

C.I. food violet 2. FD&C Violet No. 1.

C.I. food yellow 10. FD&C Yellow No. 3.

C.I. natural brown. manganese(III) oxide.

C.I. natural brown 7. walnut extract.

C.I. natural red 1. quercetin.

C.I. natural yellow 10. quercetin.

C.I. oxidation base. toluene-2,4-diamine.

C.I. oxidation base 12. 2,4-diaminoanisole.

C.I. oxidation base 12A. 2,4-diaminoanisole sulfate.

C.I. oxidation base 31. resorcinol.

C.I. oxidation base 32. pyrogallol.

C.I. pigment black 14. manganese dioxide.

C.I. pigment black 16. zinc.

C.I. pigment brown 8. manganese dioxide.

C.I. pigment metal 6. zinc.

C.I. pigment orange 20. cadmium sulfide.

C.I. pigment orange 21. lead chromate, basic.

C.I. pigment red. lead chromate, basic.

C.I. pigment red 101. iron oxide.

C.I. pigment red 104. lead-molybdenum chromate.

C.I. pigment red 106. black mercuric sulfide. See mercuric sulfide.

C.I. pigment white 4. zinc oxide.

C.I. pigment white 21. barium sulfide.

C.I. pigment yellow 31 (barium chromate (VI)). a heavy, yellow, poisonous, flammable, water-insoluble, crystalline powder, $BaCrO_4$, that is soluble in acids. It is used as a reducing agent, an explosion initiator, and in pyrotechnics. It is a confirmed human carcinogen.

C.I. pigment yellow 32. strontium chromate.

C.I. pigment yellow 34. lead chromate.

C.I. pigment yellow 36. zinc chromate.

C.I. pigment yellow 37. cadmium sulfide. cadmium sulfide.

C.I. pigment yellow 39. arsenic disulfide.

C.I. pigment yellow 46. lead monoxide.

C.I. pigment yellow 48. lead cyanide.

C.I. solvent blue 7. aminoazobenze.

C.I. solvent orange 2. 1-(*o*-tolylazo)-2-naphthol.

C.I. solvent orange 32. the disodium salt of 4',5'-dichlorofluorescein.

C.I. solvent yellow 2. *p*-dimethylamino-azobenzene.

C.I. solvent yellow 5. FD&C Yellow No. 3.

C.I. solvent yellow 14. a commercially prepared coloring agent that is carcinogenic to some laboratory animals.

C.I. vat orange No. 1. contains C.I. vat yellow No. 4, which is carcinogenic to some laboratory animals.

C.I. vat yellow No. 4. a commercially prepared coloring agent that is carcinogenic to some laboratory animals; a component of C.I. vat orange No. 1.

c-onc. cancer cell oncogene.

C1-C4 alkanes. See alkane.

C3 proactivator. properdin factor B. See factor.

C3H10T1/2 cell. See cell.

Ca. the symbol for calcium.

cabbage. *Brassica oleracea* var. *capitata*.

cabbage butterfly. *Pieris rapae* (See *Pieris*, def.1).

cabbage goiter. goiter resulting from the ingestion of goitrogenic foods such as cabbage.

cachectic. referring to or characterized by cachexia.

cachectin. tumor necrosis factor α. See factor.

cachexia (cachexy). a general and profound state of constitutional illness, wasting, and malnutrition due to chronic disease, poisoning, emotional disturbance. **cachexia mercurialis.** that resulting from chronic mercurial poisoning. **saturnine cachexia.** that resulting from chronic lead poisoning.

cachexy. cachexia.

cacodyl (tetramethyldiarsine; tetramethylbiarsine; dicacodyl). a colorless, flammable, oily liquid with an extremely foul garlicky odor. It is flammable and emits a toxic vapor when exposed to air. Symptoms are those of arsenic poisoning.

cacodyl cyanide. a flammable white powder that emits a highly toxic vapor when exposed to air.

cacodyl hydride. a colorless liquid with a garlicky odor. It emits a toxic vapor on exposure to air and combusts spontaneously.

cacodylic acid (dimethylarsinic acid; hydroxydimethylarsine oxide; Phytar). a moderately to very toxic pentavalent organic arsenical, $(CH_3)_2As(O)OH$, used as a nonselective postemergence herbicide to control weeds in non-crop areas and as a desiccant. The sodium salt was formerly used in treating leukemia and human skin disease. It is moderately toxic to laboratory rats. Toxicity is greatest when ingested because the acidic gastric juices change this compound to inorganic (trivalent) arsenic. Some inorganic arsenic in the body is converted to cacodylic acid. The latter is exhaled, and excreted in urine and sweat, giving a strong garlicky odor. Cacodylic acid is the major metabolite that is excreted in urine of humans and dogs following ingestion of trivalent or pentavalent inorganic arsenic. See sodium cacodylate. See also Agent Blue, biomethylated arsenic.

CACP. cisplatin.

Cactaceae (cactus family). a family of perennial, succulant plants native to the Western Hemisphere. They have reduced leaves, showy flowers, and often spiny stems. See also *Lophophora williamsii*.

cactinomycin (actinomycin C). an antineoplastic agent, it is an extremely toxic mixture of actinomycins C_2, C_3, and actinomycin D (dactinomycin) produced by *Streptomyces chrysomallus*. Formerly used as an antineoplastic agent.

cactus family. Cactaceae.

cadaver. a dead human or animal body used for autopsy or study by dissection.

cadaveric rigidity. rigor mortis.

cadaverine(1,5-pentanediamine;pentamethylenediamine; animal coniine). a rather viscous liquid, $H_2N(CH_2)_5NH_2$. It is a biogenic amine and homolog of putrescine, produced in flesh or meat by the decarboxylation of lysine by putrefying bacteria. It is noxious and toxic, occurring also in ergot.

cadaverous. corpse-like; having the pallor and appearance of a corpse.

cadmiosis. a pneumoconiosis due to inhalation of cadmium dust.

cadmium (Cd)[$Z = 48$; $A_r = 112.40$]. a silvery white transition metal obtained as a by-product of the commercial extraction of zinc. It is used chiefly in metal plating and the production of certain alloys. It occurs in compounds only in the +2 oxidation state. Cadmium is a widespread air pollutant, and a common component of acid rain. It is a frequent contaminant in soil, groundwater (drinking water), sediments, and biota. Cadmium and its compounds are extremely toxic with immunotoxic, nephrotoxic, hepatotoxic, teratogenic, and carcinogenic activities. Acute oral poisoning of humans produces a severe enteritis with nausea and vomiting, diarrhea, headache, muscular aches, salivation, abdominal pain, and shock. The vomiting and diarrhea are severe, causing dehydration that can result in venous thrombosis, embolism, coma, and death. Cadmium is a cumulative poison with a half-life in humans of 20-30 years. Acute inhalation exposure produces severe lung damage with acute pulmonary edema and death in the acute phase. See also poisoning (cadmium poisoning), cadmium halides, organocadmium compounds, itai-itai.

cadmium bromide. a yellow, highly hygroscopic, crystalline powder, $CdBr_2$, that is soluble in water and ethanol. It is used in lithography, process engraving, and photography. See cadmium.

cadmium chloride (cadmium dichloride). toxic, colorless, hexagonal crystals, $CdCl_2$. It is soluble in water, methanol, and ethanol. It has a number of industrial and commercial uses, e.g., as a lubricant, a fungicide, in photography, the dyeing and printing of calico, vacuum tube manufacture, and as an ice nucleating agent. Cadmium chloride is a confirmed human carcinogen and teratogen; reproductive effects have also been reported. It is very toxic to laboratory rats. Cadmium chloride may occur as the hydrate, $CdCl_2 \cdot 5/2H_2O$, the monohydrate, $CdCl_2 \cdot H_2O$, or the dihydrate, $CdCl_2 \cdot 2H_2O$. All are toxic, carcinogenic, and release toxic fumes containing cadmium and Cl^- on decomposition by heating. See cadmium.

cadmium chloride, dihydrate. See cadmium chloride.

cadmium chloride, hydrate. See cadmium chloride.

cadmium chloride, monohydrate. See cadmium chloride.

cadmium compounds. See cadmium.

cadmium cyanide. a crystalline substance or white powder, $Cd(CN)_2$, that is soluble in water and turns brown on standing. It resists the action of most organic acids, but decomposes (with the evolution of hydrogen cyanide) when treated with mineral acids. It is very toxic and readily forms complex cyanides. It is used in copper bright electroplating.

cadmium dichloride. cadmium chloride.

cadmium diethyl. diethylcadmium.

cadmium dimethyl. dimethyl cadmium.

cadmium dipropyl. dipropyl cadmium.

cadmium fluoride. a poisonous, water-, and acid-soluble compound, CdF_2, m.p. 1110°C, with cubic crystals. It is a confirmed human carcinogen. Cadmium fluoride has a number of applications in optics and electronics, and is used in the manufacture of glass and phosphors and in nuclear reactor controls.

cadmium golden 366. cadmium sulfide.

cadmium halide. See halide.

cadmium iodide. a halide, CdI_2, that forms white, lustrous, hexagonal, flake-like scales, comprised of two water-soluble allotropes. It is used as a nematocide, a lubricant, in the electrodeposition of cadmium, in photography, lithography, process engraving, in analytic chemistry, and formerly as an antiseptic.

cadmium lemon yellow 527. cadmium sulfide.

cadmium metal fume fever (cadmium pneumonitis). ore-processing workers or those who weld or cut cadmium alloys or cadmium-plated metals and certain other workers may be exposed to cadmium oxides, dusts, or fumes. Such exposure can produce a disabling emphysema. *Cf.* magnesium metal fume fever, zinc metal fume fever. See also cadmium oxide.

cadmium orange. cadmium sulfide.

cadmium oxide. a water-insoluble brown amorphous powder in the cubic form, CdO. Inhalation of the fumes or dusts produces cadmium metal fume fever that resembles that caused by zinc oxide, *q.v.* Acute exposure causes a pneumonitis characterized by edema and pulmonary epithelium necrosis. The edema may prove fatal. Chronic exposure sometimes produces a disabling emphysema. See also cadmium metal fume fever.

cadmium pneumonitis. cadmium metal fume fever. See also cadmium oxide.

cadmium poisoning. poisoning either from inhaling cadmium fumes or ingesting soluble salts. Effects of mild poisoning in humans are coughing, sweating, and chest pain. More acute intoxication is marked by damage to lungs, kidneys, and bones. Cadmium frequently impairs renal tubular function with resultant proteinuria (including albuminuria). Because cadmium is such a powerful irritant and violent emetic, acute, fatal poisonings by ingestion are rare. Even brief exposure to high concentrations of cadmium-containing dusts or fumes, however, may cause pulmonary edema and death with little or no discomfort to warn the victim of danger. Survivors suffer severe, crippling lung disease. Chronic cadmium poisoning may produce a disabling emphysema ending in death from acute renal failure. Symptoms include diarrhea, nausea, and abdominal pain. A prominent and characteristic symptom is the formation of a yellow ring around the gums at the tooth line. Cadmium poisoning may cause birth defects as well as pulmonary and prostate cancer. See also itai, itai disease.

cadmium primrose 819. cadmium sulfide.

cadmium sulfate (1:1) hydrate (3:8) (cadmium sulfate octahydrate; sulfuric acid, cadmium salt, hydrate). a confirmed human carcinogen, $SO_4 \cdot Cd \cdot 8/3H_2O$.

cadmium sulfate octahydrate. cadmium sulfate (1:1) hydrate (3:8).

cadmium sulfate tetrahydrate (sulfuric acid, cadmium salt, tetrahydrate). a confirmed human carcinogen, $SO_4Cd \cdot 4H_2O$. When decomposed by heating it releases toxic fumes, Cd and SO_x.

cadmium sulfide (Cadmium Golden 366; Cadmium Lemon Yellow 527; Cadmium Orange; Cadmium Primrose 819; cadmium sulphide; C.I. 77199; C.I. Pigment Orange 20; C.I. Pigment Yellow 37). a compound, CdS, with yellow-orange crystals. It is moderately toxic by ingestion and inhalation, and is a confirmed human carcinogen.

cadmium sulphide. cadmium sulfide.

Caesalpinia (poinciana; bird of paradise). a genus of mostly showy tropical evergreen shrubs and small trees (Family Fabaceae, formerly Leguminosae, often cultivated in warm climates. Cultivated poincianas have caused serious poisoning in children. Signs and symptoms include severe gastrointestinal distress, vomiting, and profuse diarrhea. ***C. pulcherrima***. indigenous to Guatemala and Panama, the leaves are used to poison fish; the seeds have been used to poison criminals.

caesium. cesium.

caffeic acid 3-methyl ether. ferulic acid.

caffeine (3,7-dihydro-1,3,7-trimethyl-1*H*-purine-2,6-dione; 1,3,7-trimethylxanthine; 1,3,7-trimethyl-2,6-dioxopurine; coffeine; thein; theine; guaranine; methyltheobromine). an odorless, bitter-tasting, white powder, $C_8H_{10}N_4O_2 \cdot H_2O$. It is a pseudoalkaloid that occurs in a wide variety of plants and in beverages prepared from certain of these plants or those to which caffeine has been added: coffee (ca. 125 mg/cup), tea (ca. 60 mg/cup), cocoa (ca. 50 mg/cup), many soft drinks (colas contain ca. 40-75 mg/serving), and numerous drugs (e.g., Aminophylline, Cafergot, dyphylline, No-Doz, pentoxifylline, Trental). Caffeinated drinks relieve fatigue and drowsiness. Caffeine is a CNS stimulant, a heart and respiratory stimulant, and diuretic, and is used in the treatment of headache, shock, asthma, and heart disease. Even moderate amounts can produce any of the following symptoms in some individuals: excitement, restlessness, insomnia, frequent urination, nausea, gastric irritation, sweating, vomiting, rapid pulse, palpitations, cardiac arrhythmias, tinnitis, fever, tremors, and sometimes delirium, convulsions, and coma. Large amounts can cause fever, tremors, photophobia, inability to walk, delirium, convulsions, and coma. Allergic reactions and [illegible] depression are also known. Caffeine convulsions are less often fatal than theophylline convulsions. Caffeine is very toxic to laboratory mice, rats, and rabbits. The lethal dose for humans is about 10 g. See theophylline.

caffeine poisoning (obs, caffeinism). poisoning by ingestion of caffeine. Symptoms of intoxication may include excitement and restlessness with intervening periods of drowsiness; a rapid pulse; tinnitus; nausea and vomiting; diuresis; fever; dehydration; thirst; delirium; tremor; convulsions; coma. Death from cardiovascular or pulmonary collapse may result. Lesions may include gastric ulceration and CNS stimulation.

caffeinism. an obsolete term for caffeine poisoning.

3-caffeoylquinic acid. chlorogenic acid.

caicaca. See *Bothrops atrox*.

caisson disease. bends.

cake urchin. See Echinoidea.

cal. calorie.

cala. See *Arum*.

calabar bean. **1:** common name of physostigma and *Physostigma venosum*. **2:** the seed of the *P. venenosum*. The principal toxicant is physostigmine. See *Physostigma*.

calabash curare. See curare.

calabucab. See *Hydrophis ornatus*.

calactin. a toxic cardiac glycoside isolated from milkweeds of the genus *Asclepias*. See also *Danaus*.

caladium. a common name of herbs of the genera *Caladium* and *Xanthosoma*.

Caladium (caladium; elephant's ear). a genus of attractive, large-leaved herbs (Family Araceae). All parts contain calcium oxalate and other toxic principles. When plant material is chewed, the needle-sharp crystals of calcium oxalate are released by ruptured plant cells and [illegible] the tissues of the mouth, tongue, and throat, quickly producing a severe burning sensation and irritation of the mucous membranes with swelling of the affected tissues; nausea; vomiting; diarrhea; salivation. Occasionally direct systemic effects are seen. On occasion, the swollen tongue may become immobile, interfering with speech, swallowing, and breathing; obstruction of the airway is a rare complication. *Cf. Colocasia*, which is commonly referred to as elephant ear.

Calamagrostis (reed bentgrass; bluejoint grass). a genus of perennial grasses (Family Graminae) that occurs in cool and temperate regions. Most species have simple, erect culms and multiflowered panicles. They sometimes become heavily ergotized. See *Claviceps*.

calamus (sweet flag; calmus; sweet cane; sweet grass). **1:** the dried rhizome of *Acorus calamus*. It is used as a carminative and anthelmintic. **2:** *Acorus calamus*. See also oil of calamus.

calcemia. hypercalcemia; sometimes synonymous with hypocalcemia.

calcicosilicosis. a pneumconiosis produced by the inhalation of dusts (chiefly marble dusts) that contain silica and lime (calcium carbonate).

calcicosis. a morbid condition of the lung due to chronic inhalation of marble dust.

calciferol. vitamin D_2.

calcined baryta. barium oxide.

calcined soda. soda ash.

calcinosis (exudative calcifying fasciitis; calcium gout; hypercalcification). a condition in which abnormal amounts of calcium salts are deposited in various tissues of the animal body. **enzootic calcinosis** (hypercalcification). a disease of herbivores (known especially in ruminants and horses) characterized by extensive calcification and degeneration of soft tissues (especially the heart, lungs, and tendons) with weight loss, emaciation, and attendant muscular, postural, locomotor, respiratory, and cardiovascular problems. It is a chronic, progressive disease with a variable, but often high mortality rate among livestock (apparently more than 50% in some areas). The disease may be accompanied by osteoporosis and osteopetrosis (See *Trisetum flavescens*). The victim becomes weak and listless, breathing becomes shallow and diaphragmatic, and the pulse rate increases with mild exercise. The etiology is not always clear, but can be the result of poisoning by certain plants that contain 1α,25-dihydroxycholecalciferol(1α,25-[OH],-D_3) glycoside, a derivative of calcitriol) or other biologically active substances that imitate the calcinogenic activity of calcitriol. Such plants include *Cestrum diurnum*, *Trisetum flavescens*, Nierembergia vietehii, *Solanum esuriale*, *S. torvum*, and *S. malacoxylon*. All of these plants contain calcitriolglycoside or a substance that imitates its calcinogenic activity. This disease sometimes results from mineral imbalances in the soil. See also osteopetrosis.

calcitonin (thyrocalcitonin). a highly active calcium-regulating hormone secreted by the thyroid gland in mammals or from the ultimobranchial gland of nonmammalian species. It suppresses the osteolytic effect of toxic amounts of vitamins A and D.

calcitriol ((1α,3β,5Z,7E)-9,10-secocholesta-5,7,10(19)-triene1,3,25-triol; 1,25-DHCC; 1α,25-dihydroxycholecalciferol; 1α,25-dihydroxyvitamin D_3). a calcium regulator, it is the active form of vitamin D_3. See also calcinosis (enzootic calcinosis).

calcium (Ca) [Z = 20; A_r = 40.08]. a moderately soft, low-melting reactive metal. It is an essential component of and the most abundant mineral in living systems. Calcium is the fifth most abundant element, occurring naturally only in combined form (e.g., as calcium carbonate, calcium phosphate), mostly in large deposits of chalk, limestone, or marble and in seawater at a concentration of about 400 g/ton. Because of its relatively large atomic radius and low ionization potential, calcium is a very electropositive element, valence 2. A number of its compounds are toxic. See also calcicosilicosis, calcicosis, calcinosis, calcinosis (enzootic calcinosis), hypercalcemia.

calcium acetate (acetate of lime; brown acetate of lime; gray acetate or lime; lime acetate; lime pyrolignite; vinegar salts). a substance, $Ca(CH_3COO)_2$, that occurs naturally as a fine, white, bulky powder or very hygroscopic, rod-shaped crystals or granules that are soluble in water, slightly soluble in methanol, and practically insoluble in ethanol. It occurs also as a monohydrate in the form of monoclinic crystals, granules, or powder, and as the dihydrate with long, transparent monoclinic crystals. It is moderately toxic, orally, to laboratory rats and possibly mutagenic.

calcium arsenate (tricalcium arsenate). a poisonous substance, $Ca_3(AsO_4)_2$, used as an insecticide in the control of cotton pests and as a molluscicide. It is very toxic to female laboratory rats.

calcium arsenite. a poisonous, white, granular substance used as an insecticide. It is a mixture of variable composition, prepared by passing steam over a dry mixture of calcium oxide and arsenic trioxide.

calcium *N*-carbamoylaspartate (*N*-(aminocarbonyl)aspartic acid calcium salt (1:1); *N*-carbamoylaspartic acid calcium salt; calcium ureidosuccinate). a moderately toxic psychostimulant.

calcium carbimide. calcium cyanamide.

calcium chlorate. a white, water-soluble, heat-labile, crystalline (monoclinic) substance, $Ca(ClO_3)_2 \cdot 2H_2O$. It is used as a herbicide, insecticide, and seed disinfectant. It is moderately toxic to laboratory rats.

calcium chloride. a colorless, crystalline substance or deliquescent powder, $CaCl_2$, that is soluble in water and ethanol. It is used as an antifreeze, a wood preservative, an antidust agent, a firming agent for sliced apples and other fruits, an ingredient in jelly and canned tomatoes. It is also used in highway deicing and in the fireproofing of materials. It is extremely toxic to laboratory mice, *i.v.*, and moderately toxic when ingested. Calcium chloride is strongly irritating to the eyes and, if ingested, to the gastrointestinal tract. Symptoms of intoxication include stomach discomfort or pain and cardiac arrhythmias.

calcium cyanamid. calcium cyanamide.

calcium cyanamide (calcium carbimide; calcium cyanamid; nitrolime; CCC). a colorless phytotoxic substance with hexagonal, rhombohedral crystals, $CaNC=NC$, used as a fertilizer, defoliant, herbicide, pesticide, and as an anthelmintic in veterinary practice. It is poisonous by all routes of exposure and is moderately toxic to humans by ingestion, 20-30 g being usually fatal to an adult. Toxic fumes, NO_x and CN^-, are released on decomposition by heating. See also cyanamide (def. 2).

calcium cyanide (cyanogas). an extremely toxic white, crystalline (rhombohedral) substance or white powder, $Ca(CN)_2$, used as a fumigant and rodenticide. Pure calcium cyanide gives off hydrogen cyanide in air at normal temperatures and humidity, and when heated to decomposition, also releases NO_x. It is a deadly poison by ingestion and probably via all routes of exposure.

calcium disodium edatate. edetate calcium disodium.

calcium disodium ethylenediaminetetraacetate. edetate calcium disodium.

calcium disodium (ethylenedinitrilo)-tetraacetate. edetate calcium disodium.

calcium folinate (leucovorin calcium). **1:** the pentahydrate calcium salt of folinic acid. It is a water-soluble, nearly ethanol-insoluble, odorless, off-white to light beige, powder, $C_{20}H_{21}CaN_7O_7 \cdot 5H_2O$. It is used to counteract the toxic effects of folic acid antagonists, and as an antianemic in folic acid deficiency. **2:** a citrovorum factor. See factor.

calcium gluconate. a tasteless, odorless, white, crystalline powder or granules, $Ca[HOCH_2(CHOH)_4COO]_2$, it is a moderately toxic salt of gluconic acid, $C_{12},H_{22}CaO_{14}2$. It is given *per os* as an antidote to fluoride and oxalate poisoning, and is administered intravenously or orally as a source of calcium and to decrease capillary permeability in certain conditions. It is also used medically as a foaming agent and as a buffer in foods, as well as a foaming agent in the purification of sewage.

calcium gout. calcinosis in humans.

calcium hypochlorite. a white powder with a strong chlorine odor, $Ca(OCl)_2 \cdot 4H_2O$, it is used in commercial and household algicides, bactericides, bleaches, deodorants, fungicides, and oxidizing agents. It is commonly used as a bleaching agent and disinfectant for swimming pools. The pure compound is unavailable. Solid calcium hypochlorite is 90-95% pure. Commercial products usually contain 50% or more of the compound. A 50% solution is used to sterilize fruit and vegetables. Calcium hypochlorite is also used in sugar refining. It is the active ingredient of chlorinated lime, *q.v.* Laundry bleaches and household bleaches also contain dilute calcium hypochlorite. It is a strong irritant to eyes, skin, and mucous membranes. Symptoms produced by ingestion may include pain and inflammation of the stomach and upper respiratory tract. *Cf.* sodium hypochlorite. See also hypochlorite, poisoning (hypochlorite poisoning).

calcium hyposulfite. calcium thiosulfate.

calcium oxalate (ethanedioic acid calcium salt). a white crystalline salt, CaC_2O_4, of oxalic acid. It is insoluble in water or body fluids. The crystals occur in large quantities in the tissues of certain plants, especially in those of the Family Araceae (arums such as *Caladium*, *Colocasia*, *Dieffenbachia*, *Dracunculus*, *Philodendron*). The crystals probably cause the intense burning sensation and oral edema that results from chewing the plant. This is partly effected by penetration of the oral mucosa by the crystals and partly by the chemical properties of the salt. See also oxalic acid.

calcium oxide (lime; burnt lime; calx; caustic lime; quicklime; unslaked lime; fluxing lime). a strongly caustic, odorless, white or grayish-white, crystalline substance, powder or lumps, CaO. It sometimes has a yellowish or brownish tint due to the presence of iron. It is used, for example, in insecticides and fungicides, in the manufacture of bricks, mortar, plaster and stucco, as an alkali in the production of dairy products such as ice cream, and as a yeast nutrient and dough conditioner in baked goods. In high concentrations it can damage the skin, eyes, mucous membranes, and upper respiratory tract.

calcium thiosulfate (calcium hyposulfite). a crystalline compound, CaS_2O_3, used to treat dermatitis and jaundice caused by arsphenamine.

calcium ureidosuccinate. calcium *N*-carbamoylaspartate.

caldoriomycin. a chlorine-containing antibacterial fungal metabolite.

calfkill. *Kalmia angustifolia*.

caliche. impure industrial sodium nitrate.

calico bush. *Kalmia latifolia*.

California agaricus. *Agaricus californicus*.

California chicory. *Rafinesquia californica*.

California fern. *Conium maculatum*.

California sea hare. *Aplysia californica*.

California soap plant. *Chlorogalum pomeridianum*.

California trapdoor spider. *B. californicum*.

California water hemlock. *Cicuta californica*.

californium (Cf)[Z = 98; most stable isotope = Cf^{251}; half-life, 800 yr]. an artificial transuranium element of the actinoid series, first discovered by bombardment of curium-242 with alpha particles. ^{252}Cf is used as an antineoplastic source of radiation.

calisaya bark. cinchona.

calla, calla lily. See *Zantedeschia aethiopica*. Do not confuse with wild calla.

Calla palustris (wild calla; water arum). a toxic herb (Family Araceae) of cold bogs of the North Temperate Zone. Biting into or chewing plant parts can produce an intense burning sensation, and severe irritation and edema of the tongue and mouth. The symptoms are essentially similar to those of *Dieffenbachia*, *q.v.*, and are caused by calcium oxalate crystals. See also Araceae.

Callimorpha jacobea. a moth that contains large amounts of histamine (ca. 750 $\mu g/g$ body wt.). It also accumulates the hepatotoxic pyrrolizidine alkaloids of *Senecio*.

Calliophis (*Callophis*; oriental coral snakes; schmuckottern), a genus of generally small snakes (Family Elapidae), only a few of which reach a length of 1 m. All have a small head which is indistinct from the neck or nearly so. The body is long, slim, and cylindrical, with a short, smooth, barely tapered tail. There are usually 13 to 15 scale rows at midbody. They are difficult to distinguish from some nonvenomous snakes. All occur throughout southeast Asia, including the Philippines, Malaysia, and Taiwan. They inhabit forested lands, ranging well into the mountains except in dry regions. They are sometimes found under logs or ground litter, occasionally on open ground at night, and have been known to enter towns and the suburbs of large cities. They are secretive, rarely encountered, and seemingly reluctant to bite humans. Even so, the larger species are considered dangerous. Species are not individually treated: *Calliophis calligaster* (McClung coral snake); *C. gracilis* (graceful coral snake; spotted coral snake); *C. japonicus* (striped coral snake; Asian coral snake); *C. kelloggi* (Kellogg's coral snake); *C. macclellandii* (red-ringed snake; Macclelland's coral snake; wamon-beni-hebi); *C. maculiceps* (slender coral snake; small-spotted coral snake); and *C. sauteri* (Asian coral snake; hatori-hai; Taiwan-hai).

Callista. a genus of marine clams that occur in the south Atlantic and Gulf coasts of North America. At least some species are poisonous. ***C. brevisiphonata***. ingestion produces a rapid onset of an allergic-like reaction, possibly due to choline in the ovaries. Symptoms typically include wheezing, flushing, urticaria, and gastrointestinal distress. ***C. brevisphonata*** (Japanese callista). ingestion of the ovaries frequently produces mild illness in humans that is typified by urticaria.

Callophis. *Calliophis*.

Calluna (heather). a genus of hardy flowering shrubs (Family Ericaceae). They secrete chemicals that inhibit the growth of mycorrhiza, thereby retarding the development of forests. See mycorrhiza.

calmus. calamus; *Acorus calamus*.

calomel. mercurous chloride.

calonectrin. a trichothecene toxin first isolated from *Fusarium*.

calor. one of the four principal signs of inflammation asserted by Celsus: calor, dolor, rubor, and tumor.

calotropin ([2α(2S,3S,4S,6R),3β,5α]-14-hydroxy-19-oxo-3,2-[(tetrahydro-3,4-dihydroxy-6-methyl-2H-pyran-2,3-diyl)bis(oxy)]card-20(22)-enolide). an extremely poisonous cardiac glycoside, $C_{29}H_{40}O_9$, used as an African arrow poison. Originally isolated from the sap of milkweeds, *Asclepias curassavica*, it is also the chief toxic component of the milk-weeds, *Calotropis spp.* (Family Asclepiadaceae). A reported lethal dose in cats is 0.12 mg/kg. See also *Danaus*.

Calotropis. a genus of Asiatic and African shrubs and trees (Family Asclepiadaceae). The latex sap of these milkweeds contains the cardiac glycoside, calotropin. ***C. procera***. the sap contains calotropin. ***C. gigantea*** (giant milkweed). the sap contains calotropin and is extremely toxic; it is used to make arrow poison in Africa. Minute amounts can cause death. It is also a popular drug for suicides and murders in some parts of its range.

Caltha palustris (marsh marigold; cowslip). a small succulent perennial herb (Family Ranunculaceae) native to North America from the Carolinas to Canada and westward, but now also found widely throughout the Northern Hemisphere. The active principle is a volatile, irritant oil, protoanemonin, which is a breakdown product of the glycocide ranunculin. Symptoms of intoxication are very similar to those produced by *Ranunculus*. Ingestion of any part of the raw plant can produce paresthesia and a burning sensation in and about the mouth, nausea, vomiting, hypotension, and convulsions. See anemonin, protoanemonin.

caltrop. *Tribulus terrestris*.

Calvatia gigantea (giant puffball). a normally edible mushroom (Family Lycoperdaceae), but sometimes toxic, producing gastrointestinal distress. The victim usually recovers within a few hours.

calvities. alopecia.

calx. calcium oxide.

calycanthine. a bitter, extremely toxic crystalline alkaloid, $C_{22}H_{26}N_4$, obtained from the seeds of the shrubs *Calycanthus floridus*, *C. glaucus*, and *Chimonanthus praecox* (Family Calycanthaceae). It can cause violent convulsions, cardiac depression, and paralysis.

Calycanthus. a genus of American shrubs with aromatic bark and seeds that contain a highly toxic alkaloid, calycanthine, *q.v.*

camamala. See *Naja*.

camas. a common name of plants of the genus *Zigadenus*.

cAMP. cyclic AMP.

2-camphanone. camphor.

D-2-camphanone. (1*R*,4*R*)-(+)-camphor. See camphor.

camphene (3,3-dimethyl-2-methylenenorbornane). **1a:** a colorless, crystalline, bicyclic terpene, $C_{10}H_{16}$, with three optical isomers. It is synthesized from pinene as a step in the synthesis of camphor and is used as a raw material in the synthesis of insecticides such as toxaphene. It is mutagenic in some systems. **1b:** any of several terpenes related to camphor. **2:** camphine (oil of terpentine or a mixture of oil of terpentine and ethanol, used as an illuminant).

camphenechlor. toxaphene.

camphor (1,7,7-trimethylbicyclo[2.2.1]-heptane-2-one; 1,7,7-trimethylbicyclo(2.2.1)-2-heptanone; 2-keto-1,7,7-trimethylnorcamphane; 1,7,-7-trimethylnorcamphor; 2-bornanone; 2-camphanone; 2-oxobornane; gum camphor; laurel camphor; Formosa camphor; Japan camphor). **1:** a saturated bicyclic terpene ketone, $C_{10}H_{16}O$, with a characteristic, fragrant, and penetrating odor, and comprised of colorless or white, transparent rhombohedral (from ethanol) or cubic (made by melting then chilling) crystals, crystalline masses, or granules. It occurs naturally as optically active levo and dextro forms and as a racemate. The dextro form occurs in all parts of *Cinnamomum camphora* (camphor tree), but is usually obtained from the bark and wood. The levo form occurs in certain essential oils, and the racemate is obtained from an Asiatic chrysanthemum. Optically inactive camphor can also be synthesized from certain terpenes. Camphor is used as a plasticizer of cellulose esters and ethers, a moth repellent, in embalming fluid, as a preservative in pharmaceuticals and cosmetics, in the manufacture of explosives, and in lacquers. Camphor also occurs in certain spice flavorings of food and beverages. It is used therapeutically in liniments, topical anesthetics, and as a topical antiinfective and antipruritic. Camphor is an irritant, but poisoning is usually by ingestion. It is readily absorbed into the body, causing any of a number of effects associated with intense CNS excitation. Ingestion or injection may cause any or all of the following effects: nausea, vomiting, a feeling of warmth, headache, blurred vision, vertigo, excitement, restlessness, mental confusion, delirium, hallucinations, tremors, clonic convulsions, depression, coma, and death due to respiratory failure. Ingestion by a pregnant woman can be fatal to the fetus. **2:** any of a number of compounds that have characteristics similar to those of camphor (def.1). **(+)-camphor**. (1*R*,4*R*)-(+)-camphor. **(1*R*,4*R*)-(+)-camphor** (1*R*)-1,7,7-trimethylbicyclo(2.2.1)heptan-2-one; (+)-2-bornanone; D-2-bornanone; D-2-camphanone; (+)-camphor; D-camphor; D-(+)-camphor). a moderately toxic isomer of camphor. **2:** any of a number of compounds that have characteristics similar to those of camphor (def.1). **D-camphor**. (1*R*,4*R*)-(+)-camphor. **D-(+)-camphor**. (1*R*,4*R*)-(+)-camphor.

camphor tree. *Cinnamomum camphora*.

camphorism. poisoning by camphor. It is characterized by convulsions, coma, and gastritis.

Campsis radicans (trumpet creeper; trumpet vine). a tall ornamental deciduous shrub or weed (Family Bignoniaceae) that may climb to 10 m. The leaves and flowers are toxic. They contain a contact irritant to which people have differing degrees of sensitivity.

Canada snakeroot. *Asarum canadense*.

Canada thistle. *Cirsium arvense*.

Canadian hemp. *Apocynum cannabinum*

Canadian yew. *Taxus canadensis*.

canadine. a toxic crystalline alkaloid, $C_{20}H_{21}NO_4$, isolated from the root of *Hydrastis canadensis*. *Cf.* berberine, hydrastine.

canary chrome yellow 40-2250. lead chromate.

Canavalia (jack bean). a small genus of twining or bushy, tropical annual herbs (Family Fabaceae, formerly Leguminosae) with large seeds in long, tough pods. They are occasionally used as food, but more commonly as forage, or adulterants of coffee. They contain the amino acid canavanine. ***C. ensiformis*** (jack bean). a bushy tropical American plant with long pods and white seeds, grown especially as forage. It is the source of concanavalin A. ***C. gladiata*** (sword bean). a twining, tropical Old World plant with red or pink seeds in long pods. They have long been cultivated in the Orient; both pods and seeds are used as food. Even so, the pollen and beans are toxic to some individuals. Symptoms of intoxication may appear 2-3 hours. following inhalation of pollen or 2 to 3 days following ingestion of beans. They may include headache, GI distress, and severe hemolytic anemia may occur. This reaction is seen in families of Italian and Mediterranean origin with G6PD deficiency. Lectins may be responsible.

canavanine. an amino acid that occurs in *Canavalia*.

cancellation. **1:** the act or process of canceling, nullifying, or deleting. **2:** the process of, or means used to nullify, cancel, or otherwise destroy the force, effectiveness, or validity of a contract, ruling, or legal instrument. **3:** a cancellation order under Section 6(b) of the U.S. Federal Insecticide, Fungicide and Rodenticide Act (FIFRA) authorizes the review and cancellation of a pesticide registration if unreasonable adverse effects to humans or the environment may be expected to develop when the product is used according to widespread and commonly accepted practice, or if the material (e.g., labeling) submitted to the U.S. EPA does not comply with the provisions of FIFRA. The review process may be protracted, sometimes continuing for several years. During this time, the manufacturing, marketing, and distribution of the product is permitted.

cancer (malignant tumor; malignant neoplasm) **1:** any of various diseases characterized by malignant, uncontrolled proliferation of cells resulting in a tumor or other abnormal condition. **2:** any of various types of malignant neoplasms, regardless of the tissues or body parts involved. Most are able to invade adjacent tissues, and have a tendency toward metastasis. Most are likely to recur following attempted removal, and are usually fatal if untreated. Many cancers are initiated by xenobiotic substances, both natural and anthropogenic, or by viruses. Five to 20% of deaths from cancer may be caused by exposure to toxic substances. **3:** any tumor characterized by abnormally rapid cell division and dedifferentiation of tissue. See also cancerous growths.

cancericidal (cancerocidal; carcinolytic). destructive to cancerous tissue.

cancerigenic. carcinogenic.

cancerocidal. cancericidal.

cancerous. **1a:** pertaining to previously differentiated normal cells of a metazoan that have undergone changes that enable them to divide indefinitely and to invade adjoining tissue. **1b:** pertaining to or of the nature of a cancer. **2:** being afflicted with a cancer, invaded by cancer cells, or afflicted with the process of cancer formation. **3:** productive of cancer; malignant.

candelabra cactus. See Euphorbia.

Candex®. nystatin.

candicidin. a macrolide antifungal antibiotic complex (candicidins A, B, C, D) in which candicidin D predominates. It is produced by a strain of the actinomycete, *Streptomyces griseus*. Toxic fumes, NO_x, are released on decomposition by heating. Candicidin is moderately toxic by ingestion and extremely toxic to laboratory mice, *i.p.*

candidin. a golden-yellow crystalline substance, it is the chief component of an antifungal antibiotic complex produced by the soil actinomycete, *Streptomyces viridoflavus*. It is nearly insoluble in water and most organic solvents, but is moderately soluble in dimethylformamide (DMF), glacial acetic acid, and pyridine. It is extremely toxic, *i.v.*, *i.p.*, and *s.c.*, in laboratory mice.

Candio-Hermal®. nystatin.

canesnake, brown. See *Oxyuranus scutellatus*.

Cannabaceae. Moraceae.

cannabic. pertaining to or obtained from *Cannabis*.

cannabidiol. a physiologically inactive crystalline diphenol, $C_{21}H_{28}(OH)_2$, obtained from the resin of *Cannabis sativa*. It is activated metabolically to tetrahydrocannabinol which is highly psychoactive. See also cannabinoid.

cannabinoid. any of a class of C_{21} compounds and their analogs thus far found only in *Cannabis sativa*. Only a few of these (some 60 are known) have been characterized pharmacologically (e.g., cannabinol, ([-])-Δ_9-*trans*-tetrahydrocannabinol (THC)). Of these, THC is the major source of psychoactivity in cannabis. See also cannabidiol, cannabinol.

cannabinol. a psychoactive cannabinoid with a potency only about 5% that of ([-])-Δ_9-*trans*-tetrahydrocannabinol (THC). It is produced spontaneously by dehydrogenation of tetrahydrocannabinol from cannabis.

cannabis (bhang; ganja; hashish; marihuana; marijuana; hemp; Indian hemp; charas; pot; maryjane). **1:** the dried flowering tops of pistillate plants of *Cannabis sativa*. It contains isomeric tetrahydrocannabinols, cannabinol, and cannabidiol. Preparations of cannabis are used in many cultures to induce psychomimetic effects. The effects of the inhaled smoke or ingesta are variable, but usually include euphoria, hallucinations, delirium, weakness, drowsiness, and hyporeflexia. Cannabis was formerly used in Western countries as a sedative and analgesic, but legal usage has recently been restricted to the management of iatrogenic anorexia. See also hashish. **2:** *Cannabis sativa*. **3:** any of the many psychoactive preparations derived from *Cannabis sativa*.

Cannabis sativa (marihuana; marijuana; bhang; charas; dagga; ganja; hemp; hemp plant; Indian hemp; pot; maryjane). a plant (Family Moraceae) named and described by Linnaeus in 1753. It is an annual herb indigenous to central Asia and the northwestern Himalayas and has been naturalized in parts of North America. It is cultivated widely throughout the world for the fiber contained in its inner bark. It can be legally cultivated in the United States only with a federal permit. The leaves, flowers, and sap are toxic, producing blurred vision, exhilaration, hallucinations, delusions, ataxia, stupor, and coma. The active principles are cannabinoids. The chief toxic, psychoactive agent, tetrahydrocannabinol, can be synthesized. See also cannabis. It contains psychoactive narcotic resinoids. Cannabis has been cultivated for at least 5000 years, originating in central Asia and spreading to all but the polar regions of the world. In addition to hemp which is used to produce twine, rope, and cloth, *Cannabis sativa* an excellent source of paper pulp, produces up to five times as much cellulose per acre-year as pulp trees. Varieties grown for legitimate purposes do not contain high concentrations of psychoactive cannabinoids. Thus plants grown for fiber contain less than 0.5% ([-])-Δ_9-*trans*-tetrahydrocannabinol (THC), whereas psychoactively potent varieties contain more than 1% and one type (sinsemilla) may contain as much as 15%. *Cannabis* seeds are used as food by humans, livestock, poultry, and other birds. It is the source of hemp seed oil used in paint and soap. The dried flowering tops (cannabis, *q.v.*) have been smoked or even ingested by humans since antiquity; such use in India dates back at least to ca. 1500 BC. The various preparations from this plant form the most widely used group of illicit drugs. Ingestion of *Cannabis* seed press cake has caused fatalities among livestock. See also cannabism, hashish, hemp. The leaves, flowers, and sap are toxic, producing excitement or exhilaration; hallucinations; delusions; blurred vision; ataxia; stupor and eventually coma. The active principle is a mixture of cannabinol and related compounds.

cannabism. **1:** habituation to the use of cannabis. **2:** chronic poisoning or a morbid state caused by habitual use of cannabis.

canned food. See food.

cannula. a plastic or metal tube with a smooth, unsharpened tip, used for insertion into a blood vessel, duct, or into a body cavity to deliver or withdraw fluids (e.g., in blood transfusions or removal of blood for testing). It can remain in place for several days.

cannulate. to introduce a cannula, usually for the infusion or withdrawal of liquids. The cannula is often left in place for some time to allow for continuous or intermittent use.

cantarella. an arsenic-containing poison. Its use

is associated with Catherine and Cesare Borgia in the 15th Century.

cantharidal. pertaining to or containing cantharides.

cantharidate. to saturate, impregnate, or treat with cantharides. *Cf.* canthharidize.

cantharidean. cantharidian.

cantharides (Spanish fly). **1:** plural of cantharis. **2:** a preparation of dried blister beetles (Spanish flies), most importantly *Lytta (Cantharis) vesicatoria*. An important source of cantharidin, it is used as a counterirritant and formerly administered internally as a diuretic and aphrodisiac. It is a powerful irritant, vesicant, and rubefacient. It is highly toxic to humans when ingested or absorbed through the skin or mucous membranes. Symptoms of poisoning are those of severe gastroenteritis and nephritis. Collapse and death may occur in acute cases. The active principle is cantharidin, *q.v.*

cantharides camphor. cantharidin.

cantharidian (cantharidean). containing or composed of cantharides.

cantharidic acid. a dibasic acid, $C_{10}H_{14}O_5$, produced by the reaction of cantharidin with water. See cantharidin.

cantharidin (hexahydro-3a,7a-dimethyl-4,7-epoxyisobenzofuran-1,3-dione; 2,3-dimethyl-7-oxabicyclo-[2.2.1]heptane-2,3-dicarboxylic anhydride; exo-1,2-*cis*-dimethyl-3,6-epoxy-hexahydrophthalic anhydride; cantharides camphor). a white powder or colorless, crystalline terpene derivative, $C_{10}H_{12}O_4$, that is slightly soluble in water, acetone, chloroform, and ethanol. It is secreted by certain beetles of the family Meloidae of which the Spanish fly, *Lytta vesicatoria*, is an important source. Canth[illegible] cantharidic acid. It is the highly toxic active principle of cantharides, *q.v.*, and is used in medicine as a counterirritant and formerly used internally as a diuretic and aphrodisiac. Cantharidin is used also in veterinary medicine as a vesicant, rubefacient, and counterirritant. It is a powerful irritant to all cells and tissues, and (as Spanish fly) has an undeserved reputation as an aphrodisiac. See also cantharidin poisoning.

cantharidin poisoning (cantharidism; blister beetle poisoning). **1:** poisoning of humans by the misuse of cantharides. Symptoms appear immediately following ingestion and may include severe skin irritation, blistering of mucous membranes, abdominal pain, nausea, diarrhea, vomiting of blood, severe fall in blood pressure, hematuria, coma, and death due to respiratory failure. Lesions commonly include necrosis of esophageal and gastric mucous membranes, intense congestion of blood in the genitourinary tract, damaged cells of the renal tubules, and hemorrhagic changes in the ovaries. **2:** blister beetle poisoning of livestock. *Cf.* See blister beetle poisoning.

cantharidism. cantharidin poisoning.

cantharidize. to treat with cantharides. *Cf.* cantharidate.

cantharis (Spanish fly; blistering fly; blistering beetle; blister bug). **1:** the Spanish fly, *Lytta (Cantharis) vesicatoria*, a beetle that is indigenous to southern and central Europe. **2:** singular of cantharides (def. 2).

canthus. **1:** the angle where the eyelids meet on either side of the eye. **2:** canthus rostralis.

canthus rostralis. the angle between the flat crown of the head of a snake and the side, between the snout and head. *Cf.* canthus.

cantil. *Agkistrodon bilineatus*.

Cantor's sea snake. *Microcephalophis cantoris*.

canutillo. *Ephedra nevadensis*.

cap. **1:** the top, crown. **2:** an overlying or covering structure. **3:** the enlarged top of a mushroom. **4:** a layer of substantially impermeable material (e.g., clay) placed over the top of a landfill at closure to prevent passage of rainwater into the fill, thereby limiting the leaching of materials from the site.

Cape belladonna. *Atropa belladonna*.

Cape cobra. Naja nivea. See also *Micrurus coallinus*, *Micrurus frontalis*.

Cape night adder. *Causus rhombeatus*.

Cape puff adder. *Bitis inornata*.

Cape snake. *Dispholidus typus*.

Cape viper. *Causus rhombeatus*.

caper spurge. *Euphorbia lathyris*.

capillary. any of numerous minute blood vessels (5-20 μ in diameter) that conduct blood from an arteriole to a venule. They are endothelial tubes comprised of flat pavement cells, which are separated from cells of the surrounding tissue cells by a layer of connective tissue. They form extensive intercommunicating networks (capillary beds) in nearly all vertebrate tissues. Dissolved gases, inorganic ions, water, and proteins are exchanged primarily between the capillary blood and intercellular fluid by hydrostatic and osmotic pressure.

capillary bed. See capillary.

Caprifoliaceae (honeysuckle family). a family of widely distributed, chiefly woody, often medicinal shrubs with large showy flowers. Many are cultivated. See also *Sambucus pubens*.

caprylic acid. octanoic acid.

capsaicin (*N*-[(4-hydroxy-3-methoxyphenyl)-methyl]-8-methyl-6-nonenamide; *trans*-8-methyl-*N*-vanillyl-6-nonenamide). a compound found in hot peppers that has sensory neurotoxic effects. Capsaicin causes severe pain in humans and experimental animals upon administration. Capsaicin appears to act selectively on primary afferent neurons, depleting these processes of the peptide substance P. As a result, chronic administration causes an insensitivity to some types of nociceptive stimuli. Neonatal rodents are particularly sensitive to capsaicin, which causes a selective, profound and permanent degeneration of C and A fibers. The sensory deficits induced by capsaicin appear to be due to plasma membranes that are in a chronic depolarized state. Its use appears to be limited to a neurobiological research tool.

capsicin. an extract of cayenne pepper (*Capsicum spp.*) and of paprika.

capsicum. the dried fruit of any of various species of *Capsicum*. It is used as an irritant and carminative.

Capsicum. a rather diverse, mainly tropical genus (Family Solanaceae) of shrubby, perennial plants that includes the cayanne (red) pepper. Many are grown for their fruit. See capsaicin, capsicin, capsicum.

captafol (3α,4,7,7a-tetrahydro-2-[(1,1,2,2-tetrachloroethyl)thio]-1*H*-isoindole-1,3(2*H*)-dione; *N*-(1,1,2,2-tetrachloroethylthio)-4-cyclohexene-1,2-dicarboximide; *N*-(1,1,2,2-tetrachloroethylmercapto)-4-cyclohexene-1,2-dicarboximide; *N*-(1,1,2,2-tetrachloroethylthio)Δ^4-tetrahydrophthalimide; *N*-(1,1,2,2-tetrachloroethylsulfenyl)-*cis*-4-cyclohexene-1,2-dicarboximide). a fungicide used mainly on potatoes. It is slightly to moderately toxic to laboratory animals and is carcinogenic to laboratory mice and rats.

captan (3a,4,7,7a-tetrahydro-2-[(trichloromethyl)-thio]-1*H*-isoindole-1,3-(2*H*)-dione; *N*-(trichloromethylthio)-4-cyclohexene-1,2-dicarboximide; *N*-(trichloromethylthio)cyclohex-4-ene-1,2-dicarboximide; *N*-trichloromethylthio-3a,4,7,7a-tetrahydrothalimide; 1,2,3,6-tetrahydro-*N*-(trichloromethylthio)phthalimide; *N*-trichloromethylmercapto-4-cyclohexene-1,2-dicarboximide; *N*-trichloromethylmercapto)-Δ^4-terahydrophthalimide). an odorless, crystalline substance, $C_9H_8Cl_3NO_2S$, used as an agricultural fungicide and a bacteriostat in soap. It is insoluble in water, but soluble in benzene and chloroform. The acute oral LD_{50} in laboratory rats is 9 g/kg. In 2-yr trials, the no-effect level, *q.v.*, for laboratory rats was 1 g/kg body weight in the diet. Ingestion of large quantities can cause vomiting and diarrhea in humans; it is carcinogenic and teratogenic to some laboratory animals and is a suspected human carcinogen. It is also a skin irritant.

CAR. congenital articular rigidity. See arthrogryposis.

Carangidae (jacks; scad; pompanos; leatherbacks). an extremely diverse family of fast-swimming marine bony fish characterized by forked caudal fins and the presence of two spines anterior to the anal fin. Body shape is variable, though most are fast swimmers. North

American waters house 38 species. A few are venomous and some are linked to ciguatera poisoning. See *Caranx*, *Scomberoides sanctipetri*.

Caranx. a genus of bony fish (Family Carangidae; jacks), some of which are linked to ciguatera poisoning, *q.v.* ***C. hippos*** (common jack, crevalle). a greenish to gold-colored jack common to the tropical Atlantic coasts of the Americas. Large jacks of this species may be extremely toxic and should be eaten with caution. ***C. melampygus*** (jack). a jack of the tropical Pacific from Japan southward. Large specimens may be extremely toxic and should be eaten with caution.

carbachol (2-[(aminocarbonyl)oxy]-*N,N,N*-trimethylethanaminium chloride; (2-hydroxyethyl)trimethyl ammonium chloride carbamate; choline chloride carbamate; carbamylcholine chloride; carbocholine). an odorless toxic substance with hard, hygroscopic, prismatic crystals that are soluble in water, ethanol, and methanol, $[NH_2COOCH_2CH_2N(CH_3)_3]^+Cl^-$. It is used as a cholinergic and miotic in humans and on larger animals as a parasympathomimetic drug. Carbachol has substantial effects on the gastrointestinal and urinary tracts, but little effect on the cardiovascular system. It is also dipsogenic. Carbachol is extremely toxic to laboratory mice. It acts mainly at muscarinic receptors and nicotinic cholinergic receptors, especially at autonomic ganglia.

carbamate (carbamoate). any salt or ester of carbamic acid. Some occur naturally in the blood and urine of mammals; others are manufactured xenobiotics with pesticidal activity. Those used as pesticides (e.g., carbaryl, carbofuran, pirimicarb, propoxur) generally have lower dermal toxicities than most organophosphate pesticides and are less persistent environmentally than the organochlorine insecticides. Carbamates vary greatly in toxicity from essentially nontoxic to lethal in trace amounts. Some are immunotoxic. The insecticidal carbamates kill animals mainly through direct, relatively reversible inhibition of anticholinesterase, although they also affect other enzymes. See carbamate poisoning.

carbamate insecticide. See insecticide (carbamate insecticide).

carbamate poisoning. poisoning by toxic carbamates, most often by carbamate insecticides. Some carbamates are lethal to humans in trace amounts. Some are immunotoxic. The insecticidal carbamates kill animals mainly through direct, relatively reversible inhibition anticholinesterase, although they affect other enzymes also. They may cause convulsions and even death through contact or ingestion in humans and other animals. Symptoms in mammals include excessive salivation, lacrimation, miosis, bradycardia. The animal either dies, usually from bronchoconstriction with consequent respiratory failure, or recovers completely. Atropine is a useful antidote.

carbamazepine (5*H*-dibenz[*b,f*]azepine-5-carboxamide; Tegretol). a white to off-white powder, it is an oral antiepileptic and an analgesic used specifically to control the pain of trigeminal neuralgia. Its use must be carefully monitored and controlled since it is nephrotoxic, may seriously affect the hemopoietic system, and can cause aplastic anemia. Usage should be discontinued at the first indication of a drop in the erythrocyte count.

carbamic acid. a hypothetical or postulated acid, NH_2-COOH, known only by its derivatives (salts or esters) such as urea.

carbamic acid ethyl ester. urethane.

carbamic acid 2-methyl-2-propyltrimethylene ester. meprobamate.

carbamide. urea.

***p*-carbamidobenzenearsonic acid**. carbarsone.

carbaminohemoglobin. the form of hemoglobin that predominates in venous blood. See hemoglobin.

carbamoate. carbamate.

carbamothioc acid, S,S′-[2-(dimethylamino)-1,3-propanediyl] ester. cartap.

_N_-carbamoylarsanilic acid. carbarsone.

_N_-carbamoylaspartic acid calcium salt. calcium *N*-carbamoylaspartate.

3-carbamoyl-1-β-D-ribofuranosylpyridinium hydroxide 5'-ester with adenosine 5'-pyrophosphate inner salt. nadide.

Carbamult. promecarb.

4-carbamylaminophenylarsonic acid. carbarsone.

_N_-carbamylarsanilic acid. carbarsone.

carbamylated. pertaining to the chemical product of the covalent addition of a carbamyl moiety (e.g., from a carbamate insecticide) to a macromolecule such as acetylcholinesterase.

carbanilic acid isopropyl ester. propham.

carbaril. carbaryl.

carbarsone ([4-[(aminocarbonyl)aminophenyl]-arsonic acid; *N*-carbamoylarsanilic acid; *p*-ureidobenzenearsonicacid; *N*-carbamylarsanilic acid; *p*-carbamidobenzenearsonic acid; 4-ureido-1-phenylarsonic acid; 4-carbamylaminophenylarsonic acid; *p*-arsonophenylurea). a pentavalent organic arsenical occurring as a white powder. It is an antiamebic that is administered orally or in a retention enema. It is an oral balanticidal and is administered vaginally to control vaginitis due to *Trichomonas vaginalis*. Carbarsone is used as an antihistomonad in domestic fowl. It is moderately to very toxic in laboratory rats.

carbaryl (1-naphthalenol methylcarbamate; methyl carbamic acid 1-naphthyl ester; 1-naphthyl *N*-methylcarbamate). a skin irritant and poisonous white powder that is used chiefly as a broad-spectrum contact insecticide and pediculicide. It reversibly inhibits cholinesterase. It is generally regarded as moderately toxic to mammals. Symptoms of intoxication in humans include dizziness, nausea, excessive salivation, vomiting, abdominal cramps, diarrhea, blurred vision, headaches, bronchoconstriction and dyspnea, muscle twitching, incoordination, slurred speech, general weakness. In severe poisoning, symptoms can progress to convulsions, cyanosis, coma, respiratory failure, and death. Carbaryl is carcinogenic to humans and may cause birth defects as well. See also anticholinesterase.

carbazilquinone. carboquone.

carbazotic acid. picric acid.

Carbicron®. dicrotophos.

carbimide. cyanamide.

carbinol. methanol.

carbocycle. a molecular ring comprised only of carbon atoms. Usually used in reference to macrocyclic ring. See also lactone.

carbocyclic. of or pertaining to a compound, the structure of which includes one or more rings of carbon atoms.

carbocyclic aromatic amine. any amine in which at least one substituent group is an aromatic ring containing only carbon atoms as part of the ring structure, and with one of the C atoms in the ring bonded directly to the amino group. There are numerous compounds with many industrial uses in this class of amines. A number of these compounds are known to cause cancer of the bladder, ureter, and pelvis in humans and are suspected of causing lung, liver, and prostate cancer.

carbodiimide. cyanamide.

carbofos. malathion.

carbofuran (2,3-dihydro-2,2-dimethyl-7-benzofuranol methylcarbamate; methyl carbamic acid 2,3-dihydro-2,2-dimethyl-7-benzofuranyl ester; 2,2-dimethyl-2,3-dihydro-7-benzofuranyl-*N*-methylcarbamate; 2,2-dimethyl-7-coumaranyl *N*-methylcarbamate). a white, highly toxic, water-soluble, solid carbamate. It is a cholinesterase inhibitor used as a plant systemic insecticide and also as an acaricide and nematocide. The inhibition may be reversed by injection of atropine sulfate. Carbofuran is absorbed by the roots and leaves of plants and by insects feeding on the plant material. It is used in some household products to eradicate pests and should be considered highly dangerous, especially in households with children. The oral LD_{50} in the laboratory rat is

8 mg/kg body weight. The minimum toxic dose in cattle and sheep is 4.5 mg/kg body weight, becoming lethal at 18 mg/kg rat and 9 mg/kg in the latter two animals, respectively. The oral LD_{50} in the dog is 19 mg/kg body weight. Swine have been poisoned after drinking water contaminated by this compound. See insecticide (carbamate insecticide).

carbohydrate. a compound with the approximate general formula, $(CH_2O)_n$. Included are a wide range of substances from simple sugars (e.g., glucose) to polymers of high molecular mass (e.g., starch, cellulose, glycogen). Most carbohydrates play a role in storing and transferring energy. Some have structural or additional physiologic functions.

carbolic acid. phenol (def. 1).

carbolic acid poisoning (phenol poisoning). carbolic acid (See phenol) is a corrosive, protoplasmic poison that is damaging to all types of cells. Properly diluted it is an effective bacteriostat. Acute exposures of vertebrates affect mainly the CNS. In the case of humans, death, usually resulting from respiratory failure, may intervene as early as 1 hour following an acute exposure. Accidental ingestion causes corrosive burns in the mouth, esophagus, and stomach. Further symptoms may include abdominal pain, dyspnea, headache, pallor, sweating, dizziness, vomiting, weakness, stomach, abdominal pain, bloody diarrhea, hypotension, lowered body temperature, cyanosis, and coma. Complete renal failure sometimes occurs. Lesions most commonly seen at autopsy are corrosive burns of the skin and mucous membranes, and gastric perforation. Esophageal stricture and renal damage are less common. Lesions produced most often by chronic exposure to phenol are the spleen, pancreas, and kidneys. Pulmonary edema may also occur. Skin contact produces pain and subsequent numbness and corrosive burn.

β-carboline. any of a number of derivatives of tryptoline or 1,2,3,4-tetrahydro-β-carboline produced by the Pictet-Spengler isoquinoline synthesis. All have powerful effects in/on the nervous system (e.g., as serotonin agonists), and some occur in the nervous system where they may act as neuromodulators or neurotransmitters. They occasionally occur in foods at concentrations that make them of some toxicological concern. *Cf.* tetrahydroisoquinoline. See also Pictet-Spengler isoquinoline synthesis.

carbomethane. ethenone.

carbomethene. ketene.

2β-carbomethoxy-3β-benzoxytropane. cocaine.

α-2-carbomethoxy-1-methylvinyl dimethyl phosphate. Mevinphos.

2-carbomethoxy-1-methylvinyl dimethyl phosphate. Mevinphos.

2-carbomethoxy-1-propen-2-yl dimethyl phosphate. Mevinphos.

carbomonoxyhaemoglobin. carboxyhaemoglobin. See carboxyhemoglobin.

carbomonoxyhemoglobin. carboxyhemoglobin.

carbomycin (magnamycin). an antibacterial, antibiotic, monobasic macrolide synthesized by *Streptomyces halstedii*. It is structurally similar to erythromycin and leucomycin and is bacteriostatic to Gram-positive bacteria. **carbomycin A** ((12S,13S)-9-deoxy-12,13-epoxy-12,13-dihydro-9-oxoleucomycin V 3-acetate 4^B-(3-methylbutanoate); M-4209; magnamycin A; deltamycin A_4). a crystalline compound, $C_{42}H_{67}NO_{16}$, with blunt needles, used as an antibacterial. It is moderately toxic to laboratory mice. **carbomycin B** (9-deoxy-9-oxoleucomycin V 3-acetate 4^B-(3-methylbutanoate); magnamycin B). a colorless crystalline compound, $C_{42}H_{67}NO_{15}$, with anisotropic plates. It is used as an antibacterial and antimicrobial in veterinary medicine.

carbon (C)[Z = 6; A_r = 12.01]. a nonmetallic tetravalent element with two natural isotopes, ^{12}C and ^{13}C, and two artificial, radioactive isotopes of interest, TIC and ^{14}C. The relative atomic mass of carbon-12 is set at 12.00000 as the standard for all molecular weights. The principal sources of carbon are biological in origin, e.g., coal, oil, gas, carbonates, and atmospheric CO_2; it is also present in the mineral dolomite, and in wood, charcoal, coke, and soot. Minute amounts of pure carbon occur as the allotropes diamond and graphite. Carbon combines to form a vast number of compounds. Carbon and certain classes of its compounds are universal constituents of living organisms. Organic chemistry and biochemistry are the sciences that investigate the structure and behavior of these substances. Numerous carbon compounds and classes of carbon compounds are toxic, many extremely so. Of particular note are organic acids and many compounds that also contain nitrogen, heavy metals, or halogens. **carbon-12** (symbol, ^{12}C). the standard of atomic mass comprising 98.89% of natural carbon. **carbon-13** (symbol, ^{13}C). a natural isotope comprising 1.11% of naturally occurring carbon. **carbon-14** (symbol, ^{14}C; radiocarbon). a naturally occurring radioactive isotope of carbon, having a mass number of 14. It is a pure beta-emitter with a half-life of 5580 years that is used as a tracer in estimating the time of death of organisms, or of artifacts that contain carbon of natural origin, within the past 50,000 years and in studying various aspects of metabolism. ^{14}C arises naturally from bombardment of carbon by cosmic rays in the upper atmosphere. It is used to date biological materials and artifacts that contain carbon of natural origin.

carbon bichloride. perchloroethylene.

carbon bisulfide. carbon disulfide.

carbon black. a substantially pure, finely divided form of carbon. It is produced mainly from gaseous or liquid hydrocarbons by controlled combustion with a restricted air supply to obtain incomplete combustion. It is carcinogenic to some laboratory animals. Complete removal of carbon black from the gases that leave the stack is extremely difficult; consequently, the production of carbon black often produces intense air pollution.

carbon dichloride. perchloroethylene.

carbon dioxide (carbonic acid gas; carbonic anhydride). a colorless, odorless, tasteless gas, CO_2, with a density ca. 1.5 times that of air. The concentration of carbon dioxide in the atmosphere is 0.034% by volume. It is formed by natural and anthropogenic processes. Enormous amounts enter the atmosphere from the combustion of fossil fuels such that the concentration of atmospheric CO_2 is increasing by about 0.27% annually. Local concentrations in air rarely rise to levels that are dangerous to life or health. In Albert Park in Africa, however, there are several areas where gases vented from the earth contain CO_2 concentrations of 40% or more. Various animals (and occasional humans) attracted to the lush vegetation die from anoxia. Scavengers, recruited by the resulting carrion may suffer the same fate. Carbon dioxide is a toxic product of aerobic respiration in living organisms. In most metazoans, it is dissolved in blood and excreted through the skin, lungs, and/or gills. At high ambient levels, CO_2 is an irritant and simple asphyxiant. In humans, it can cause headache, drowsiness, dyspnea, dizziness, muscular weakness, and tinnitis. It can cause hypoxia in the fetus, thereby producing teratogenic effects. The solid form (dry ice) can cause severe frostbite. The gas is used medically as a respiratory stimulant.

carbon disulfide (carbon bisulfide; dithiocarbonic anhydride; sulfocarbonic anhydride). a toxic, colorless, slightly etherial smelling, highly refractile, mobile, extremely flammable liquid, CS_2. The vapor is 2.67 times denser than air. Carbon disulfide is widely used in the manufacture of cellophane, rayon, and textiles and as an industrial solvent. It has also been used as a fumigant in grain elevators and as a parasiticide and insecticide. Its use as a fumigant was banned in the United States in 1985. It lacks a disulfide (-SS-) group, but rather has two sulfur atoms, each separately bonded to a carbon atom. Carbon disulfide is narcotic, neurotoxic, hepatotoxic, and immunotoxic. It destroys microsomal cytochromes by initiating autoxidation. In humans and laboratory animals, it is metabolically activated to carbonyl sulfide and the sulphene derivative of cytochrome P-450, a neurotoxicant and hepatotoxicant. Poisonings are usually due to inhalation of vapor, but can occur by any route of exposure. Acute intoxication of humans is often marked initially by euphoria, followed by restlessness, depression, and lethargy with severe inflammation of mucous membranes, nausea, vomiting, unconsciousness, and sometimes death with terminal convulsions. Contact with concentrated solutions can cause erythema, exfoliation, and intense pain. Repeated inhalation of the fumes or prolonged contact with the skin can also damage the heart, the CNS, and both male and female reproductive systems. Chronic exposure may also produce anemia, moderate to severe psychic disorders including extreme irritability and mania with hallucinations, tremors, seizures, weight loss, and blood dyscrasias. The victim may suffer auditory and visual disturbances (e.g., blurred vision), neuritis, irritability, insomnia, depression, toxic psychoses, seizures, and delirium. A parkinsonian-like paralysis is sometimes seen in younger victims. Encephalopathies and associated behavior are varied, severe, and may be permanent.

carbon disulfide poisoning. intoxication by carbon disulfide. It occurs most commonly among rubber workers and manufacturers of rayon by the viscose process. The condition is marked by listlessness, breathlessness on mild exertion, nausea and vomiting, irritability, and insomnia, followed by paralysis, impaired vision, peptic ulcer, confusion, ataxia, vertigo, mania. Lesions are chiefly those of CNS depression, sometimes with permanent neurological sequalae.

carbon monoxide (CO). a colorless, almost odorless, flammable, highly poisonous gas, CO, produced mainly by the incomplete combustion of organic materials without sufficient air. A major air pollutant, CO occurs in the exhaust fumes from all internal combustion engines, fumes from household kerosene-burning heaters, in the gas from coke and charcoal ovens, electric furnaces, blast furnaces, gas manufacturing plants, oil distilleries, refuse processing plants, kilns, coal mines, and sewers, charcoal grills, and kerosene heaters. It is a common cause of accidental and deliberate poisoning. Poisoning is by inhalation. CO is highly toxic at concentrations exceeding about 100 cm^3/m^3 (0.01%). Tobacco smoking in unventilated rooms can also produce concentrations of CO that are sufficiently high to be of concern. The toxicity of carbon monoxide resides in its strong affinity for hemoglobin (200-300 times greater than that of oxygen) in circulating blood and strong inhibition of the cytochrome oxidase system. By complexing with these substances, it reduces the oxygen transport capacity of hemoglobin and blocks oxygen utilization with consequent asphyxiation and frequently the death of victims. See also poisoning (carbon monoxide), carboxyhemoglobin, combustion, hemoglobin.

carbon monoxide hemoglobin. carboxyhemoglobin.

carbon monoxide poisoning, a potentially fatal acute or chronic intoxication caused by inhalation of carbon monoxide gas. The condition is essentially one of asphyxiation due to the irreversible combination of carbon monoxide with hemoglobin to form carboxyhemoglobin. Surviving individuals may exhibit brain damage. Most instances of carbon monoxide poisoning in the home involve gas appliances such as kitchen ranges, space heaters, wall heaters, central heating systems, and clothes dryers. See also antagonism (receptor antagonism). Early symptoms or those of mild poisoning are dizziness, headache, sleepiness, irritability, and an inability to concentrate. In more severe poisoning, symptoms progress to include nausea and vomiting, blurred vision, shortness of breath, convulsions, unconsciousness, and, finally, death.

carbon monoxide poisoning, acute. symptoms of acute intoxication may include initial hyperactivity, mild headache, irritability, breathlessness on moderate exertion, fatigue, vomiting, confusion, incoordination, ataxia, transient syncope, convulsions, and incontinence. In some cases the skin is a characteristic cherry red. **carbon monoxide poisoning, chronic**. chronic exposure is not well characterized, but causes disorders of the heart and respiratory system. Lesions revealed on autopsy are those of asphyxial death.

carbon oxychloride. phosgene.

carbon oxysulfide (COS). a toxic volatile organic liquid by-product of petroleum or natural gas refining that boils at 50°C. It is structurally related to carbon disulfide. The vapor is a strongly narcotic toxicant. Carbon oxysulfide vapor is a toxic irritant. On decomposition, carbon oxysulfide liberates hydrogen sulfide.

carbon pernitride. cyanogen azide.

carbon tetrachloride (tetrachloromethane; perchloromethane). a toxic, colorless, nonflammable liquid, CCl_4, with a distinctive chloroform-like odor and a density 5.3 times that of air. It was formerly used as a degreasing solvent, fire extinguisher, dry cleaning agent, fumigant, and anthelmintic (used mainly against hookworms), but is now used almost entirely as a chemical intermediate in the manufacture of chlorofluorocarbon refrigerants. Sale for home use was banned by the U.S. Food and Drug Administration (FDA) in 1970; its use as a fumigant was banned in 1985; and it is a designated hazardous substance and hazardous waste by the U.S. EPA. It is an anesthetic and narcotic but is too toxic for such use; it is also a CNS depressant. Systemic poisoning may result from inhalation, ingestion, or contact. Serious neurotoxic effects usually result from inhalation, whereas gastrointestinal, liver, and kidney damage follow ingestion. Death may result from respiratory failure, cardiovascular collapse, or ventricular fibrillation, or respiratory failure or in some cases, renal failure may result. Common lesions are those of hepatic central lobular necrosis and necrosis of renal tubular epithelium. Effects are potentiated by the use of ethanol. Chronic intoxication by carbon tetrachloride may cause headaches, dizziness, nausea, and vomiting. Symptoms of severe kidney and liver damage may appear within a few weeks. Carbon tetrachloride is activated in the liver by cytochrome P-450-mediated formation of trichloromethyl radical ($CCl_3\cdot$) and active oxygen. Both metabolites destroy microsomal cytochromes by initiating autoxidation. The trichloromethyl radical also binds to nucleophilic substituents of proteins. Carbon tetrachloride is a confirmed carcinogen in humans and a number of laboratory animals. It also causes hepatogenic photosensitization in livestock. The laboratory rat and many species of birds are highly resistant. Humans have been fatally poisoned when using fumigants containing a mixture of acrylonitrile, carbon tetrachloride, and methylene chloride.

carbon tetrachloride poisoning. the lesions and symptoms of poisoning by carbon tetrachloride are chiefly those of CNS depression, with renal and hepatic failure as common complications. Symptoms may include nausea, vomiting, intense abdominal pain, headache, confusion, and coma, with late hepatic and renal complications with the possibility of renal failure. Death may result from cardiovascular or respiratory collapse, or ventricular fibrillation. Common lesions are central lobular necrosis of the liver and necrosis of the renal tubular epithelium.

carbon tetrafluoride (tetrafluoromethane). a colorless, odorless, thermally stable, nearly inert, gas, CF_4, used as a refrigerant. It acts as a simple asphyxiant, and is especially dangerous because its density is substantially greater than that of air. At high concentrations it has a narcotic effect.

carbon-12 (symbol, ^{12}C). the standard of atomic mass comprising 98.89% of natural carbon.

carbon-13 (^{13}C). a natural isotope comprising 1.11% of natural carbon.

carbon-14 (^{14}C). a beta-emitter, half-life of 5730 years, used as a tracer in studying various aspects of metabolism. Naturally occurring ^{14}C arising from cosmic rays, is used to date biological materials and artifacts that contain carbon of natural origin.

carbon oxychloride. phosgene.

carbonate. a salt formed by the reaction of carbonic acid with a metal or an ester an aliphatic or aromatic hydrocarbon (e.g., diphenyl carbonate). Carbonate esters are liquids used as solvents and in the synthesis of polycarbonate resins. See carbonate poisoning.

carbonate crystal. sodium carbonate, monohydrate.

carbonate poisoning. poisoning by salts (or esters) that contain the carbonate ion, CO_3^{2-}. Prominent signs are tachycardia, late esophageal stenosis, and death in shock or asphyxia from glottic edema.

carbonic acid. a weak organic acid, H_2CO_3, produced by the reaction of carbon dioxide and water. Organic compounds or metals react with carbonic acid to yield carbonates.

carbonic acid 2-*sec*-butyl-4,6-dinitro-phenyl isopropyl ester. dinobuton.

carbonic acid gas. carbon dioxide.

carbonic acid 1-methylethyl 2-(1-methylpropyl)-4,6-dinitrophenyl ester. dinobuton.

carbonic anhydrase. an enzyme that catalyzes the reversible formation or decomposition of carbonic acid from water and carbon dioxide. When carbon dioxide concentrations are relatively high, there is a net formation of carbonic acid from CO_2 and H_2O; decomposition is favored under conditions of low CO_2 concentration. This reaction that takes place chiefly in erythrocytes aids the transport of carbon dioxide from tissues to the respiratory surfaces and also increases the buffering capacity of blood. Interference with carbonic anhydrase activity by organochlorine insecticides is associated with eggshell thinning in birds.

carbonic anhydride. carbon dioxide.

carbonic dichloride. phosgene.

carbonic difluoride. carbonyl fluoride.

carbonization. the treatment of coal by heat in a closed vessel usually to produce coke or gas. High-temperature carbonization is that conducted generally at temperatures above 900°C and medium-temperature carbonization, between 600 and 900°C. This process is used to produce coal gas and, at a temperature of 1300°C), blast furnace coke. Low-temperature carbonization, that generally below 600°C, yields reactive cokes for use as "smokeless" fuels. Carbonization also gives many by-products (e.g., coal tar) of great importance to chemical industries and is often highly polluting.

carbonochloridic acid ethyl ester. ethyl chloroformate.

carbonochloridic acid trichloromethyl ester. diphosgene.

carbonuria. a rarely used term indicating the presence of carbon dioxide or other carbon compounds in urine.

carbonyl. **1:** pertaining to the carbonyl group or to any substance containing a carbonyl group. **2:** a complex, such as tetracarbonyl nickel, in which carbon monoxide ligands coordinate with a metal atom. Toxicologically, the most important carbonyl is probably nickel carbonyl. See for example, carbonyl compound, carbonyl group, chromium carbonyl, iron pentacarbonyl, lithium carbonyl, nickel carbonyl, phosphorus carbonyl, sodium carbonyl.

carbonyl chloride. phosgene.

carbonyl compound. any compound that contains the carbonyl group, -C=O. Most contain several carbon monoxide molecules bonded to a metal. Many transition metal carbonyl compounds are known. Of these, the most significant toxicologically is nickel carbonyl, $Ni(CO)_4$), *q.v.* Perhaps the next most abundant carbonyl compound is $Fe(CO)_5$. Other examples are $V(CO)_6$ and $Cr(CO)_6$. In some configurations, bonding favors compounds with two metal atoms per molecule, e.g., $(CO)_5Mn-Mn(CO)_5$, $(CO)_4Co-Co(CO)_4$.

carbonyl fluoride (carbonic difluoride; fluophosgene). a colorless, highly irritant, water-soluble, very hygroscopic gas, CF_2O, used in organic syntheses.

carbonyl group. a organic group, -C=O, that occurs in aldehydes, ketones, carboxylic acids, and complexes of transition metals.

carbophenothion (phosphorodithioic acid *S*-[[(4-chlorophenyl)thio]methyl] *O,o*-diethyl ester; *S*-[[(*p*-chlorophenyl)thio]methyl] *O,o*-diethyl phosphorodithioate; *O,o*-diethyl *S*-(*p*-chlorophenylthio)methyl phosphorodithioate). a light amber, nearly water-insoluble liquid pesticide used on fruits, nuts, vegetables, and fiber crops. It is also used in certain household products to eliminate pests. It is extremely toxic to laboratory rats, dairy cattle, sheep, and goats and should be considered dangerous to humans.

carboquone (2-[2-[(aminocarbonyl)oxy]-1-methoxyethyl]-3,6-bis(1-aziridinyl)-5-methyl-2,5-cyclohexadiene-1,4-dione;2,5-bis(1-aziridinyl)-3-(2-hydroxy-1-methoxyethyl)-6-methyl-*p*-benzoquinone carbamate (ester); 2,5-bis(1-aziridinyl)-3-(2-carbamoyloxy-1-methoxyethyl)-6-methyl-1,4-benzoquinone; carbazilquinone). an extremely toxic, red to reddish-brown crystalline compound used as an antineoplastic alkylating agent. It is nearly insoluble in water and only slightly soluble in most organic solvents.

carboxybenzene. benzoic acid.

2-carboxy-3-carboxymethyl-4-isopropenylpyrrolidine. kainic acid.

(*S*)-α-carboxy-2,3-dihydro-*N*,*N*,*N*-trimethyl-2-thioxo-1*H*-imidazole-4-ethanaminium hydroxide inner salt. ergothioneine.

2-carboxy-1,1-dimethylpyrrolidinium hydroxide inner salt. stachydrine.

carboxyethane. propionic acid.

carboxyhaemoglobin. carboxyhemoglobin.

carboxyhemoglobin (HbCO; carboxyhaemoglobin; carbon monoxide hemoglobin). carbon monoxide-containing hemoglobin. It is a much more stable complex than oxyhemoglobin. It is formed in the presence of carbon monoxide, which binds to hemoglobin much more readily than oxygen. The affinity of hemoglobin for CO is ca. 200-300 times greater than that of O_2. In carbon monoxide poisoning, CO displaces much of the oxygen complexed with hemoglobin (oxyhemoglobin), preventing the exchange of carbon dioxide and oxygen during the circulation of blood, with increasingly severe asphyxiation and even death.

carboxyhemoglobinemia. the presence of carboxyhemoglobin in circulating blood as in carbon monoxide poisoning.

3-carboxy-5-hydroxy-1-*p*-sulfophenyl-4-*p*-sulfophenylpyrazole trisodium salt. tartrazine.

carboxyl group. an organic chemical group, -COOH that is present in and defines carboxylic acids.

[3-[[2-(carboxylatomethoxy)benzoyl]amino]-2-methoxypropyl]hydroxymercurate(1—) sodium. Mersalyl.

carboxylesterase. carboxylic esterase (See esterase).

carboxylic acid. an organic compound with the general formula RCOOH, where R is a hydrocarbon, or other organic moiety. Examples are formic acid, acetic acid, and trichloroacetic acid. Many carboxylic acids occur in plants and as esters in fats and oils. Carboxylic acids are termed **monobasic, dibasic**, or **tribasic**, depending upon the number of carboxyl groups (COOH), 1, 2, or 3, respectively. The acidity of carboxylic acids is due mainly to the carbonyl group, C=O. Carboxylic acids are the oxidation products of aldehydes and are often synthesized by that route. Applications of carboxylic acids are extremely diverse. All are toxic or irritant; some serve as defense mechanisms, as in the case of formic acid (e.g., in the venom of ants), *q.v.*

carboxylic ester hydrolase. carboxylic esterase (See esterase).

carboxylic esterase. See esterase.

[1-carboxy-2-[2-mercaptoimidazol-4(or 5)yl]ethyl]trimethylammonium hydroxide, inner salt. ergothioneine.

carboxymethyl ether cellulose sodium salt. carboxymethylcellulose sodium.

carboxymethylcellulose. dehydrating material packed with drugs, film, etc.

carboxymethylcellulose sodium (carboxymethyl methyl ether cellulose sodium salt; CMC; sodium carboxymethylcellulose; sodium cellulose glycolate). a white granular colloid with variable solubility in water depending upon the degree of substitution. It is commonly used as a vehicle for insoluble toxicants or drugs.

5-(carboxymethyl)-1,1-dimethylimidazolium hydroxide inner salt. anemonine.

2-carboxy-4-(1-methylethenyl)-3-pyrrolidineacetic acid. kainic acid.

(carboxymethyl)trimethylammonium hydroxide inner salt. betaine.

20β-carboxy-11-oxo-30-norolean-12-en-3β-yl-2-*o*-β-D-glucopyranuronosyl-α-D-glucopyranosiduronic acid. glycyrrhizic acid.

1-[[2-carboxy-8-oxo-7-[(thienylacetyl)amino]-5-thia-1-azabicyclo[4.2.0]oct-2-en-3-yl]-methyl]pyridinium hydroxide inner salt. cephaloridine.

***N*-[9-(2-carboxyphenyl)-6-(diethylamino)-3*H*-xanthen-3-ylidene]-*N*-ethylethanaminium chloride**. Rhodamine B.

α-carboxy-*N*,*N*,*N*-trimethyl-1*H*-indole-3-ethanaminium hydroxide inner salt. hypaphorine.

1-carboxy-*N*,*N*,*N*-trimethylmethanaminium hydroxide inner salt. betaine.

carbyloxime. fulminic acid.

carcass. **1:** the body of a dead animal. **2:** in the food industry, the animal body after the head, hide, tail, extremities, and viscera are removed.

Carcharhinus melanoperus (black-tipped reef shark). a poisonous shark of the Indo-Pacific region; South Africa to the East Indies, Hawaii, Tuomotu, and the Marianas. See elasmobranch poisoning.

carcino-, carcin-. a prefix referring to or indicating a carcinoma.

carcinogen (oncogen; tumorigen). any cancer-inducing agent; an agent that increases the incidence of malignant neoplasms above that of a comparable control group. Examples are asbestos, polycyclic aromatic hydrocarbons, aflatoxin B1, mustard gas (See gas), tobacco smoke, ionizing radiation, ultraviolet light. Most carcinogens are also mutagens. In vertebrate animals, carcinogens generally act as inducers of hepatic microsomal enzymes. **chemical carcinogen**. a carcinogen that has the ability to form covalent bonds with macromolecular molecules, especially DNA. Examples are alkylating agents, arylating agents. **direct-acting carcinogen**. primary carcinogen. **epigenetic carcinogen** (nongenotoxic carcinogen). a carcinogen that acts without interacting directly with DNA. Such agents may alter methylation patterns or the tertiary structure of DNA. **genotoxic carcinogen**. a carcinogenic agent that produces its effects by interacting directly with DNA. Such an agent may cause gene mutations, or it may alter chromosome structure or the number of chromosomes. Carcinogens that produce a consistent positive response in short-term mutagenicity tests are defined as acting by genetic mechanisms. This type of carcinogen often produces cancers at doses below that at which they produce other toxic effects. They usually produce cancers in more than one target organ, sometimes with a single exposure. The latent period is usually short. **initiator**. genotoxic carcinogen. **inorganic carcinogen**. any inorganic carcinogen (e.g., heavy metals and their salts, asbestos). Most chemical carcinogens are organic compounds. Inorganic carcinogens may not act directly on DNA but may alter DNA indirectly via their effects on the behavior of DNA polymerases and perhaps other indirect mechanisms. **mutagenic carcinogen**. a carcinogen that is also a mutagen. **nongenotoxic carcinogen**. epigenetic carcinogen. **nonmutagenic carcinogen**. a carcinogen that has not given positive results in mutagenicity tests. **precarcinogen**. the initial compound to which the organism is exposed, prior to metabolic activation. **primary carcinogen** (direct-acting carcinogen). a carcinogen that does not require metabolic activation; one that is reactive enough to act, in the unmodified or unmetabolized form to cause cancer. Some examples are alkylating agents, dimethyl sulfate, epoxides, nitrogen and sulfur mutards, peroxides, sulfones, sulfonic esters, and ionizing radiation. **procarcinogen**. proximate carcinogen. **proximate carcinogen** (procarcinogen). **1:** a metabolic product that is more active than the precarcinogen; a chemical whose carcinogenic action is expressed only after metabolic activation. Examples include nitrosamines, nitrosoureas, polyaromatic hydrocarbons, aromatic and heterocyclic amines, azo dyes, chlorinated hydrocarbons, aflatoxin and mycotoxins. **2:** an intermediate metabolite between the precarcinogen and the ultimate carcinogen. See metabolic activation. **solid-state carcinogen**. a solid compound whose physical structure is essential to carcinogenesis. Examples are asbestos, metal foils, and plastics. **ultimate carcinogen**. the metabolic species that directly interacts with a cell constituent, usually DNA, to initiate carcinogenesis. See also alkylating agent, arylating agent.

carcinogenesis. any process that produces malignant neoplasms. See carcinoma. **chemical carcinogenesis**. the production of malignant neoplasms by chemicals. **radiation carcinogenesis**. the production of malignant neoplasms by ionizing radiation. The cytopathological changes preceding the full-blown development of neoplasms are diverse. Malignant tumors are late developments in individuals that have survived the acute phases of radiation poisoning.

carcinogenic (cancerigenic). causing or able to cause cancer; cancer-causing; productive of malignant tumors. See carcinoma.

carcinogenicity. the capacity to cause cancer. See carcinoma. See carcinogenic.

carcinolytic. cancericidal.

carcinoma (plural, carcinomata). **1:** any malignant tumor arising in epithelial tissue; any malignant neoplasm of any tissue arising from embryonic ectoderm and endoderm. They occur most commonly in the skin and gastrointestinal tract of vertebrates. In humans, the bronchi and prostate gland in men, and the breast and cervix in women are often affected. Carcinomas are invasive. Carcinomatous tissue may be undifferentiated or show varying degrees of anaplasia. **2:** sometimes used in reference to any malignant neoplasm.

carcinomata. pl. of carcinoma.

carcinomatosis (carcinosis). a condition in which cancer is widely disseminated throughout the body.

carcinomatous. pertaining to or manifesting the properties of cancer.

Carcinoscorpius rotundicauda (Asiatic horseshoe crab). a species of arthropod (Class Merostomata) that occurs in southeast Asia. The unlaid green eggs and viscera are sometimes poisonous during the reproductive season. In spite of this, the eggs are highly regarded as food. See also Xiphosura.

carcinosis. carcinomatosis.

carcinostatic. **1:** having the capacity to check the growth of a carcinoma. **2:** a carcinostatic agent.

cardenolide. a type of cardiac glycoside found in milkweeds (Family Asclepiadaceae) that causes emesis in vertebrates that ingest the plant. As a rule, this causes them to avoid the plant in the future.

cardiac. **1:** of or pertaining to the heart. **2:** a cardiac patient or a person with a heart disorder. **3:** a restorative heart medicine. **4:** pertaining to the part of the stomach that lies near the esophagus or to the esophageal opening of the stomach.

cardiac arrest. See arrest.

cardiac arrhythmia. arrhythmia.

cardiac dyspnea. See dyspnea.

cardiac muscle. See muscle.

cardiac output. **1:** the total volume of blood pumped by the heart of an animal during a specified period of time. **2:** the total volume of blood pumped by either the right or left ventricle of the mammalian heart per minute. The minute-volume of the heart.

cardialgia. cardiodynia.

cardiant. an agent that stimulates heart action. *Cf.* cardiotonic.

cardiataxia. incoordination (extreme irregularity) of heart movements.

cardinal. preeminent; of primary importance.

cardinal flower. *Lobelia cardinalis.*

cardinophyllin. carzinophilin.

cardioaccelerator. **1:** pertaining to cardioacceleration, the speeding up of heart action. **2:** an agent that accelerates heart action.

cardioactive (cardiokinetic). capable of affecting the heart, especially the action of the heart. *Cf.* cardioaccelerator, cardioinhibitory, cardiokinetic, cardiotonic, cardiotoxic.

cardiodynia (cardialgia). heart pain, pain in the heart.

cardiogram. the record or tracing made by a cardiograph.

cardiograph. an instrument that records some cardiac event.

cardioinhibitor. an agent that slows or arrests the movements of the heart.

cardioinhibitory. having the effect of slowing or arresting the movements of the heart.

cardiokinetic. cardioactive.

cardiolysin. a lysin that acts on cardiac muscle.

cardiolytic. having the properties of a cardiolysin.

cardiomuscular. relating to the heart and blood vessels.

cardiomyopathy (myocardiopathy). **1:** a disease of the myocardium, often of unkown etiology. **2:** a primary disease process of heart muscle in the absence of a known underlying etiology (World Health Organization). *Cf.* cardiopathy. **alcoholic cardiomyopathy**. a disease of chronic alcoholics, it is a congestive cardiomyopathy accompanied by enlargement of the heart and reduced cardiac output. **beer-drinker's cardiomyopathy** (beer heart). a myocardial myopathy with pericardial effusion, associated with chronic heavy consumption of beer. **congestive cardiomyopthy**. a syndrome marked by enlargement of the heart, especially of the left ventricle, disfunction of the myocardium, ultimately with reduced cardiac output and congestive heart failure. *Cf.* cardiopathy.

cardiopathy. any disorder of the heart. *Cf.* cardiomyopathy.

cardiopulmonary (pneumocardial). pertaining to the heart and lungs.

cardiothyrotoxicosis. thyrotoxicosis with involvement of the heart.

cardiotonic. **1:** exerting a favorable or stimulatory effect on the action of the heart. **2:** pertaining to, or denoting a chemical agent that tends to increase the efficiency of contraction of cardiac muscle. **3:** an agent that has a favorable effect on the heart, especially on the efficiency or force of contraction of cardiac muscle.

cardiotoxic. **1:** denoting the capacity of an agent to impair heart function; toxic to heart tissue. **2:** a cardiotoxic agent.

cardiotoxicology. See toxicology.

cardiotoxin. **1:** any biological substance that produces low blood pressure and heart failure. Such are components of some snake venoms. **2:** a cytolytic and cardiotoxic polypeptide component of cobra venom comprised of a single chain of 60 amino acid residues cross-linked by disulfide bridges. The *N*- and *C*- terminal residues of the chain are leucine and asparagine, respectively. It irreversibly depolarizes the nerve cell membrane, with contraction of both striated and smooth muscle. The toxicity of cardiotoxin is potentiated by phospholipase A and is inhibited by the action of heparin and RNA gangliosides. See also snakebite, venom (snake venom).

cardiotropic. pertaining to, denoting, or characterized by cardiotropism.

cardiotropism. the affinity of a chemical substance or of a pathogenic agent for heart tissue.

cardiovascular. pertaining to the heart and blood-vascular system. See also circulatory system.

cardiovascular system. a circulatory system that distributes blood throughout the body of higher metazoans. It consists generally of one or more hearts and blood vessels or sinuses. The functional properties of the cardiovascular system of vertebrates, such as blood pressure and pulse rate, are often affected directly or indirectly by a toxicant, and its behavior often provides clues regarding the nature of the toxicant. *Cf.* circulatory system, blood-vascular system.

cardiovasculorenal. pertaining to the heart, vascular system, and kidneys.

Cardium edule (common cockle). a pelecypod that inhabits European seas. It is often a source of paralytic shellfish poisoning of humans.

Carduus (nodding thistle; plumeless thistle). a genus of hardy, robust, spiny-leaved annual or perennial herbs with purple, tubular flowers (Family Compositae) that are indigenous to southeast Asia. Toxic concentrations of [illegible] have been reported.

carelessweed. *Amaranthus*.

Carissa ovata stylonifera. a spiny shrub (Family Apocynaceae) of Australia from which carissin, a cardiotoxic glucoside, is obtained.

carissin. a highly toxic cardioactive glucoside isolated from a spiny shrub, *Carissa ovata stylonifera*, of Australia.

carminomycin, carminomycin I. carubicin.

carmustine (*N,N'*-bis(2-chloroethyl)-*N*-nitrosourea; BCNU; BiCNU). a chloroethylnitrosourea derivative and alkylating agent. It is used as an antineoplastic agent in the treatment of Hodgkin's disease, other lymphomas, meningeal leukemia, and metastatic brain tumors. It is extremely toxic to laboratory mice and is a probable carcinogen. Toxic effects include CNS depression, pulmonary fibrosis, and renal and hepatic damage. It is cytotoxic, immunosuppressive, and is probably a carcinogen itself since it alkylates DNA and RNA. It causes delayed thrombocytopenia in humans.

Carnivora. a large order of eutherian mammals that are usually characterized by powerful jaws, carnassial premolar teeth specialized to shear and tear, and canine teeth that are to a greater or lesser extent modified as fangs. Most are carnivorous, but bears are, for example, omnivorous and pandas are herbivorous. They are small to large in size, usually with five clawed toes (sometimes four) on all four feet. The claws are usually long, sharp, and sometimes retractile. Some examples are dogs and related forms (Family Canidae); cats (Family Felidae); weasels, ferrets, otters, skunks, and related forms (Family Mustelidae); bears (Ursidae); harbor seals (Family Phocidae); sea lions (Family Otariidae), and walruses (Family Odobaenidae).

carnivore. **1:** a mammal of the order Carnivora. **2:** used loosely in reference to any carnivorous animal (especially a terrestrial mammal).

carnivorous. pertaining to the eating or consumption of animal flesh.

Carolina ipecac. *Euphorbia ipecacuanha*.

Carolina jessamine. *Gelsemium sempervirens*.

Carolina yellow jessamine. See *Gelsemium sempervirens*.

carp. See Cyprinidae, *Cyprinus carpio*.

Carpentaria whip snake. *Denisonia carpentariae*.

carpet viper. *Echis carinatus*.

Carpilius maculatus (red spotted crab). a tropical, poisonous marine crustacean (Order Decapoda) that occurs in the Indo-Pacific region. The poisonous principle of this reef crab seems to be structurally identical to tetrodotoxin. Symptoms of intoxication in humans include nausea, vomition, numbness, tingling sensations beginning about the lips and tongue, muscular paralysis, and collapse. Severe intoxications may terminate fatally.

Carrel-Dakin solution. sodium hypochlorite solution, dilute.

carriage. bearing or posture, especially during locomotion. **low carriage**. a condition whereby an animal's torso is carried very close to the ground during locomotion.

carrier. **1:** a molecule that combines with a substance and actively transports it across the cell membrane. See also facilitated diffusion, transport (active transport, carrier-mediated transport). **2:** an individual that transmits an infectious or hereditary disease.

carrier-mediated transport. See transport.

carrier protein. *q.v.* (membrane transport protein). a protein in the plasmalemma (cell membrane) that binds to a specific type of solute molecule and carries it across the plasmalemma which is otherwise the molecule thus transported. See also transport protein.

carrot. See *Daucus carota*.

carrot family. Apiaceae.

carrying capacity. **1:** the maximum number of individuals of a particular species that is indefinitely self-sustaining in a given ecosystem. Usually denoted by K, it is the upper asymptote of the logistic equation. **2:** in recreation management, the amount of use a recreation area can sustain without a loss of environmental quality. **3:** in wildlife management, the maximum number of animals an area or ecosystem can sustain during a given period of the year.

cartap (carbamothioc acid, *S,S'*-[2-(dimethylamino)-1,3-propanediyl] ester; 1,3-bis-(carbamoylthio)-2-*N,N*-(dimethylamino)propane). a colorless, crystalline, synthetic insecticide. It is very toxic to laboratory mice and rats.

carubicin (8-acetyl-10-[(3-amino-2,3,6-trideoxy-α-L-lyxo-hexopyranosyl)oxy]-7,8,9,10-tetrahydro-1,6,8,11-tetrahydroxy-5,12-naphthacenedione; (1*S*,3*S*)-3-acetyl-1,2,-3,4,6,11-hexahydro-3,5,10,12-tetrahydroxy-6,11-dioxo-1-naphthacenyl 3-amino-2,3,6-trideoxy-α-L-lyxohexopyranoside; 4-*o*-dimethyldaunorubicin; carminomycin; carminomycin I; karminomycin). a highly toxic anthracycline antineoplastic antibiotic related to daunomycin and doxorubicin. It is produced by *Actinomadura carminata* (Family Actinomycetaceae, Order Actinomycetales).

Carum copticum. an aromatic herb (Family Apiaceae) the seeds of which yield ajowan oil (*q.v.*) on distillation.

Caryophyllaceae (the pink family). a family of widely distributed, annual or perennial, temperate or cold-climate herbs and occasionally shrublike. The various species have brightly colored, mostly regular flowers and stems with characteristically swollen joints. See *Agrostemma*, *Drymaria*, *Saponaria*.

carzinophilin (cardinophyllin; CZ). an antineoplastic antibiotic isolated from *Streptomyces sahachiroi* (Family Streptomycetaceae, Order Actinomycetales).

Carzinophilin A. an extremely toxic, slightly water-soluble crystalline substance. LD_{50} *i.v.* for laboratory mice is ca. 150 γ/kg body weight.

cascabel, cascabela. *Crotalus durissus*, *C. terrificus*, *Lachesis mutus*.

[illegible]

cascabela muda. See *Lachesis mutus*.

cascara sagrada. the dried bark of *Rhamnus purshiana* (cascara buckthorn). It contains emodin and is used as a mild laxative.

cascavel. a common name of *Crotalus durissus* and *Lachesis mutus*.

case-control study. an epidemiological study in which the exposure histories of humans who show a particular toxic effect are compared with unexposed individuals.

case fatality rate. **1a:** the percentage of individuals diagnosed as having a specified disease who die from that disease. **1b:** the percentage of individuals exposed to a specified disease who die following such exposure (e.g., the percentage who die due to the bite of an eastern diamondback rattlesnake).

cashew, cashew nut. See *Anacardium occidentale.*

cask. a thick-walled, usually leaden container used to transport radioactive material.

cassava (casava, manioc, tapioca plant). **1:** any of several widely cultivated plants of the genus *Manihot* (e.g., *M. esculenta*, *M. utilissima*). The tubers are often a local staple food source and are the source of tapioca. The raw tubers often contain lethal concentrations of cyanide, but are generally rendered harmless either by heating the peeled tubers or by repeated mashing and soaking in water. **2:** the rootstock of *M. esculenta*. See also *Manihot*.

Castanea (chestnut). a genus of rough-barked, temperate-zone trees or shrubs (Family Fagaceae), the chestnuts. ***C. dentata*** (American chestnut). a nearly extinct species due to chestnut blight. Its wood and leaves contain 6-11% tannin. The nut is edible, but the leaves and buds when ingested in large quantities are poisonous to livestock.

Castilleja (paint brushes; Indian paint brush; painted cup). a genus of herbs (Family Scrophulariaceae) with showy red, orange, or yellow bracts that encompass the true flowers. They are partially parasitic on the roots of nearby plants. All species are facultative accumulators of selenium from soil.

castor bean. **1:** *Ricinus communis*. **2:** the toxic seed of *Ricinus communis*.

castor bean poisoning. poisoning by the seed of

Ricinus communis, the castor bean. The active principle is ricin. Symptoms are chiefly those of violent gastrointestinal distress. Although reported in the literature, hemolysis is not a major effect of castor bean poisoning. See ricin, ricin agglutinin, ricinism.

castor oil. usually not toxic if ingested, but see *Ricinus communis*.

castor oil fish. *Ruvettus pretiosus*.

castration. the surgical removal of, interference with, or chemical inactivation of the generative organs. The term is commonly used to indicate the removal of the testes (orchidectomy), although it applies also to removal of the ovaries (oophorectomy).

casualty gas. choking gas.

catabolic. pertaining to or denoting catabolism. See also anabolic.

catabolic steroid. See steroid.

catabolism (katabolism). degradative metabolism; that which involve in the decomposition of complex substances into less complex substances with the transfer of energy to ATP, which drives the anabolic processes. See anabolism. Some of these processes detoxify certain substances and some produce more toxic species (activation). *Cf.* anabolism.

catabolite. any substance that is produced by catabolism.

catalase. a type of enzyme found in a diversity of living organisms that catalyzes the decomposition of hydrogen peroxide to oxygen and water. They are best known from studies of mammalian catalase. Catalase occurs in most mammalian tissues. It prevents accumulation of (and tissue damage by) hydrogen peroxide produced during metabolism. Commercial preparations of catalase are obtained from mammalian liver, bacteria (*Micrococcus lysodeikticus*), or fungi (*Aspergillus niger*).

catalepsy. a seizure characterized by a waxy rigidity of the muscles such that the animal is unable to initiate movement. If standing, it will fall down. Otherwise, the body tends to remain in any position in which it is placed. The condition can be induced by antipsychotics and certain other centrally acting agents. *Cf.* catatonia.

catalysis. a process whereby the presence of a small amount of a substance (a catalyst) alters the rate of a reaction without itself being modified chemically. **enzyme catalysis**. any biochemical reaction mediated by an enzyme.

catalyst. a substance that alters the rate of a chemical or physical reaction without itself being modified chemically or consumed during the reaction. Catalysts act by favoring an alternate pathway in which the rate-limiting step has a lower energy of activation than in the uncatalyzed reaction. Biological reactions are catalyzed by enzymes. **negative catalyst**. a catalyst that decreases the rate of a reaction. **positive catalyst**. a catalyst that increases the rate of a reaction.

catalytic converter. a device that removes the polluting high-boiling hydrocarbons from motor vehicle exhaust by oxidizing them catalytically to carbon dioxide and water or reducing them to nitrogen and oxygen.

catalytic incinerator. a device that oxidizes volatile organic compounds (VOCs) using a catalyst to promote combustion. Catalytic incinerators require lower temperatures than conventional thermal incinerators, with savings in fuel and costs.

catalytic site. active site.

cataract. an opacity of the crystalline lens of the eye, of its capsule, or both.

cataractogenic. able to induce the formation of cataracts.

catatonia. a psychomotor disturbance characteristic of certain schizophrenic patients who are unresponsive, mute, and who tend to assume and remain in an abnormal position. Symptoms may also include negativism, stupor, and excitement. *Cf.* catalepsy.

catatonic. pertaining to catatonia; stuporous.

catclaw, catclaw acacia. *Acacia greggii*.

catechin (*trans*-2-(3,4-dihydroxyphenyl)-3,4-

4-dihydro-2*H*-1-benzopyran-3,5,7-triol; catechol; 3,3',4',5,7-flavanpentol; catechinic acid; catechuic acid; cyanidol; flavan). a crystalline flavonoid found chiefly in higher woody plants such as *Acacia catechu* (Family Fabaceae). It is used therapeutically as an astringent and the d-form is used in the treatment of hepatic disease. Do not confuse with pyrocatechol or pyrocatechuic acid, both of which are often called catechol. See catechol.

catechinic acid. catechin.

catechol. any of three diphenols in which the two hydroxyl groups are ortho in position to each other. **a:** catechin. **b:** pyrocatechol. **c:** pyrocatechuic acid.

catechol *O*-methyltransferase. a type of enzyme, typically found in the soluble fraction of a number of tissues, that catalyzes the methylation of catecholamines (e.g., epinephrine, norepinephrine, dopamine) and other derivatives of pyrocatechol (catechol). See also methyltransferases, *o*-methylation.

catecholamine. any amine derivative of pyrocatechol (e.g., L-dopa, dopamine, epinephrine, norepinephrine, octopamine) that contains a 3,4-dihydroxyphenylethylamine nucleus. These compounds are extremely important biologically, serving as neurotransmitters and/or hormones. Toxicologically, they produce effects similar to those induced by stimulating the sympathetic nervous system and are responsible for a number of CNS disturbances such as schizophrenia and depression. They play important roles in sensory-motor integration, locomotion, memory, and response to stress. The effects of manganese and tetrahydropyridine (MPTP) are probably due mainly to their effects on catecholamines.

catechuic acid. catechin.

catenarin. an anthraquinoid mycotoxin that is very toxic [illegible] mutants.

caterpillar. **1:** the wormlike larva of a sawfly (Order Hymenoptera). **2:** the terrestrial wormlike larva of a butterfly or moth. It bears a biting mouth, six pairs of thoracic legs, and two to four pairs of prolegs. Numerous species are venomous and/or poisonous. Some poisons result from the bioconcentration of compounds of food plants. Venom is secreted by glands that open through pores near the tip of each cheliceral fang and/or via stinging bristles. See dermatitis (caterpillar dermatitis), urticaria endemica, urticaria epidemica, Lepidoptera, lepidopterism, *Thametopoea pityocampa*.

caterpillar dermatitis. See dermatitis.

caterpillar hair poisoning (flannel moth dermatitis). that caused by *Megalopyge caterpillars*, *q.v.*

caterpillar poisoning. urticaria endemica and urticaria epidemica.

caterpillar rash. caterpillar dermatitis (See dermatitis).

catfish. a common name applied to bony fish (Class Osteichthyes) of several familes. They vary considerably in size and shape; the lips usually carry long fleshy tactile processes (barbels). The skin is thick and slimy, or covered with bony plates; true scales are absent. Most of the approximately 1000 species occur in tropical freshwater streams. They are benthic feeders, have a relatively large amount of subcutaneous fat, and are accumulators of lipophilic hydrocarbons, including PCB. Many are consumed by humans. Certain, mostly marine and brackish water species such as *Arius felis*, *Clarias batrachus*, *Plotosus lineatus*, and *Bagre marinus*, are of further toxicological interest. The venom apparatus or sting consists of a single sharp, strong spine that lies immediately cephalad to the soft-rayed portions of the dorsal and pectoral fins. The spine is covered by a thin integumentary sheath. The epidermal layer of the latter contains a series of glandular cells that secrete the venom. The sting in some species has recurved teeth and can thus produce severe penetrating lacerations that are subject to secondary infection and favor absorption of the venom. The catfish can voluntarily lock the spine into the extended position, thus increasing the effectiveness of the sting. Immediately after a sting, the wound becomes pale, and the victim typically suffers a sharp stinging, throbbing, or even violent pain or scalding sensation which may remain localized or spread up the affected limb. The wound soon becomes cyanotic, erythematous, and edematous. The edema is sometimes extreme, and the area about the wound may become numb and gangrenous. Systemic effects include shock. Occasional deaths have been reported.

Catharanthus roseus. the currently accepted name of *Vinca rosea*. See *Vinca*.

catharsis (purgation). **1:** the loosening of the bowels or the induction of defecation. Some poisons have a violent cathartic action. The primary danger of such an effect is that large amounts of fluid are lost from the body. This may result not only in dehydration, but also in a life-threatening loss of electrolytes. Catharsis is sometimes used to remove unabsorbed poisons from the digestive tract. **2:** a freudian term (psychocatharsis) relating to the method of recalling past (especially repressed) experiences with a view to releasing emotional tension or anxiety.

cathartic (purgative; eccoprotic). **1:** pertaining to catharsis. **2:** an agent that stimulates bowel movements; a powerful laxative. Care should be taken in their use and all should be considered poisonous, some violently so.

cathepsin (catheptic enzyme). a lysosomal proteolytic enzyme responsible for autolysis following the death of a cell. Such enzymes are active at the low pH characteristic of lysosomes. Cathepsins A and C are exopeptidases; B and D are endopeptidases.

catheter. a hollow tube used to introduce or to remove fluids from the body of an animal.

cation. a positively charged ion (e.g., potassium, sodium, calcium, magnesium). *Cf.* anion.

cat's claw, catsclaw, catclaw acacia. *Acacia greggii*.

Caucasus adder. *Vipera kaznakovi*.

Caudata. Urodela.

caulophylline (1,2,3,4,5,6-hexahydro-3-methyl-1,5-methano-8*H*-pyrido[1,2-a][1,5]diazocin-8-one; 12-methylcytisine; *N*-methylcytisine). an extremely toxic alkaloid isolated from *Caulophyllum thalictroides*.

caulophyllum. the dried roots and rhizomes of the herb, *Caulophyllum thalictroides*.

Caulophyllum thalictroides (blue cohosh; papoose root; squaw root). a cardiotoxic perennial herb (Family Berberidaceae) native to the eastern United States. All parts are toxic. It contains an alkaloid, caulophylline, and a glycoside, caulosaponin.

caulosaponin. a toxic glycoside isolated from an herb, *Caulophyllum thalictroides*.

causal (causative). **1:** indicating or expressing cause. **2:** pertaining to or denoting a cause. **3:** acting as a cause or involving cause. **4:** resulting from a cause. See cause.

causal agent (causal organism, causative agent, etiologic agent). the agent (e.g., organism, chemical, radiation) that is the proximate or primary cause of a given disease state.

causal organism. See causal agent.

causative. causal.

causative agent. causal agent.

cause. any condition or activity that consistently and invariably precedes a particular result or effect is regarded as a cause. This is primarily a layman's term. The natural sciences are essentially empirical in nature and results of studies are contingencies, correlations, and the falsification of hypotheses, all of which vary quantitatively upon replication and which therefore bear certain probabilities regarding their association with a given effect or result. **proximate cause**. the immediate, direct cause of a phenomenon; the last known element in a chain of phenomena ("chain of causation") that are associated with and precede the phenomenon of interest (effect). It is both a legal and a scientific concept (insofar as cause can be scientifically established). **ultimate cause**. that process thought to have initiated or conditioned the sequence, progression, or train of events preceding the proximate cause of a phenomenon; the cause of a cause. It is both a legal and a scientific concept (insofar as cause can be scientifically established). Thus, an ultimate cause of unusual sensitivity to a toxicant might, for example, be the presence of mutant gene or prior exposure to the toxicant.

caustic. **1:** corrosive; producing damage to tissues resembling that of a burn. **2:** a substance that produces caustic effects. Common caustics in the home are drain and toilet cleaners and electric dishwasher detergents. **3:** pertaining to or denoting a corrosive alkali (e.g., caustic soda, sodium hydroxide). **4:** a hydroxide of a light metal. See also corrosion, corrosive alkali, poisoning (corrosive poisoning).

caustic antimony. antimony trichloride.

caustic arsenic chloride. arsenic trichloride

caustic barley. See sabadilla.

caustic baryta. barium hydroxide octahydrate.

caustic lime. calcium oxide.

caustic potash. potassium hydroxide.

caustic soda. sodium hydroxide.

Causus (night adders; chinigani; krotenvipern). a small genus of nocturnal African venemous snakes (Family Viperidae) that occurs south of the Sahara. All are indigenous to tropical and southern Africa: *C. defilippi* (night adder; snouted night adder; kinangananga; fuko), *C. lichtensteinii* (Lichtenstein night adder; olive-green viper), *C. lineatus* (lined night adder), *C. resimus* (night adder; green night adder; velvety green night adder; green viper), and *C. rhombeatus* (Cape night adder; African night adder; common night adder; rhombic night adder; Cape viper; demon night adder; night adder; choichodo; embalasasa; fonfoni; kalelea; kasambwe; pwéré; rhombic night adder; saindé; zakra). None attains a length of more than 1 m. The venom is only mildly toxic and, although the bite is painful, these snakes are not considered a threat to human life. They look very much like nonpoisonous snakes. The head is of medium size; the neck is distinct; the canthus is obtuse. These are rather slender snakes with a cylindrical (or slightly depressed) body, and a short tail. The eyes are moderate in size, the pupils are [illegible], and the [illegible] appearance is very much like that of non-venomous snakes.

cauterant. an acid, caustic, or toxic chemical used (in chemocautery) to destroy abnormal or diseased cells on the skin.

CAUTION. See label (warning label).

Cavia. a genus of rodents (Family Caviidae) that includes the guinea pig, a domesticated form of unknown origin. The guinea pig is a small, stout-bodied, short-eared, very short-tailed (the tail cannot be discerned externally) hystricomorph rodent. It is raised for food in South America, as a pet, and as a laboratory animal in biological research. There are numerous variants. While it is sometimes considered to be a distinct species, *C. cobaya*, it is more probably a variety of a feral South American species (possibly *C. porcellanus* or *C. cutleri*). It is especially unwise to assign a scientific name to domestic variants and lines used in laboratories. See also cavy.

cavies. plural of cavy.

Caviidae. a family of rodents that includes the domesticated guinea pig, *Cavia*, and cavies.

cavy (pl. cavies). a common name of any of several South American rodents (Family Caviidae), especially *Cavia porcellus*. All are short-tailed, rough-coated herbivores. The domesticated guinea pig, *Cavia*, is a cavy.

cay note. *Ephedra nevadensis*.

CBOD. chemical and biological oxygen demand. See CBOD5. See also biological oxygen demand, chemical oxygen demand.

CBOD5. five-day chemical and biological oxygen demand: the amount of dissolved oxygen taken up in 5 days from the carbonaceous portion of a water sample that contains a known amount of dissolved oxygen kept at 20°C for 5 days. The procedure is the same as that for BOD5, except that nitrogen demand is suppressed. See also, biological oxygen demand, BOD5.

CCA. chromated copper arsenate.

CCC. **1:** calcium cyanamide. **2:** chlorcholine chloride.

CCNU. 1-(2-chloroethyl)-3-cyclohexyl-1-nitrosourea. See lomustine.

Cd. the symbol for cadmium.

CD. **1:** chlordecone. **2:** curative dose.

CD4$^+$ helper/inducer T lymphocyte. helper T lymphocyte (See T lymphocyte).

CD$_{50}$. median curative dose. See dose.

CDAA. allidochlor.

CDEC. sulfallate.

CDM. chlordimeform.

Ce. the symbol for cerium.

cecropia moth. *Hyalophora cecropia*.

cecropin. a protein antibacterial isolated from immunized cecropia moths, *Hyalophora cecropia*. Cecropin has been found also in other lepidopterous insects (moths and butterflies). It is a basic protein that rapidly lyses selected Gram-negative and Gram-positive bacteria.

cedar leaf oil. an essential oil distilled from the leaves of *Juniperus virginiana*. Other common sources include *Artemisia absinthium*, *Tanacetum*, and *Thuja occidentalis*. It has been used internally as an expectorant, emmenagogue, and anthelmintic, and externally as a mild counterirritant. It is nevertheless toxic and can cause violent reactions and death if consumed in large or even in moderate amounts. The toxic component is thujone, *q.v.* Symptoms of intoxication include a weak, rapid pulse, severe gastritis, foaming at the mouth, violent siezures, convulsions, and death.

cefaloridin. cephaloridine.

ceiling effect. See maximal response.

ceiling level. the maximum allowable concentration or level of an airborne chemical in the workplace. It is not to be exceeded even for an instant.

ceiling value (DL). the concentration of a substance that should not be exceeded even momentarily.

celandine. common name of plants of the genus *Chelidonium*, especially *C. majus*.

Celastraceae (the stafftree family). a widely distributed family of trees and often climbing shrubs. See *Euonymus*.

celery. *Apium graveolens*.

celery family. Apiaceae.

celiac. pertaining to the abdomen.

celiac sprue. See gluten-induced enteropathy.

cell. **1:** the structural and functional unit of a living organism. It is the smallest structure capable of performing all of the life functions and is, in many cases, capable of independent existence. All cells have a highly organized structure and contain protoplasm enclosed by a membrane, the plasmalemma, through which controlled exchanges of materials with the environment occur. They also contain nucleic acids. Many living organisms are unicellular, others are multicellular. **C3H10T1/2 cell**. any of a cell line originating from mouse embryo fibroblasts. They are easily cultured and are used in cell transformation tests for carcinogenicity or promotion. They lack activating capacity and are aneuploid in the untransformed state. See cell transformation.

cell-mediated hypersensitivity. hypersensitivity (delayed hypersensitivity).

cell-mediated immunity. immunity in which a specific antigen is bound to receptor sites on the surface of sensitized T-lymphocytes. No antibodies are involved. The T-lymphocytes may kill infected cells or they may secrete soluble substances that facilitate or augment inflammation and elimination of the antigen. Subsequent exposure to the same antigen will produce a more rapid and powerful response. *Cf.* humoral immunity.

cell-mediated transport. See transport.

cell membrane. plasmalemma.

cell nucleus. See nucleus.

cell transformation. a phase of carcinogenesis whereby a cell acquires the capacity of uncontrolled growth. *Cf.* test (mammalian cell transformation tests).

cell transport. See transport.

cell wall. the outer covering or wall of most prokaryotes and most plant cells. The primary wall is comprised of cellulose microfibrils within a matrix of complex polysaccharides with some protein and water-filled spaces. Pores may perforate the wall at intervals, permitting cytoplasmic interchanges between neighboring cells. In some cells, a secondary wall develops internal to the primary wall. The collective mechanical strength of the cell walls can often support the weight of a large plant. The tensile strength of cellulose fibrils is nearly as great as that of steel. Furthermore, support of especially large structures is possible because the cell wall is elastic but relatively inextensible, permitting the development of high turgor pressures within the plant tissues. Cell walls of fungi contain chitin, but not always cellulose and the cell walls of prokaryotes are reinforced by mucopeptides rather than by cellulose. Penicillin inhibits cross-linking in the mucopeptide of Gram-positive bacteria, thereby preventing cell wall formation and killing the organism.

cellular absorption. See absorption.

cellular respiration. the energy-releasing metabolic pathways in a cell (glycolysis, citric acid cycle, and electron transport system) that oxidize organic molecules such as glucose, fatty acids, and others and release energy to ATP.

cellulicidal. destructive to cells.

cellulotoxic. **1:** toxic to cells. **2:** caused by cytotoxins.

Celsius degree. a temperature interval that is identical to that of the kelvin (absolute) scale. See Celsius temperature scale.

Celsius temperature scale (formerly centigrade temperature scale). a temperature scale in which 0 degrees is the freezing point of water at standard atmospheric pressure, almost exactly zero degrees, and the boiling point is almost 100 degrees. Temperature, Θ_c in degrees Celsius (°C) is related to the Kelvin scale by the equation:

$$\Theta_c = T_k - 273.15.$$

cenobiology. biocenology.

centi- (c) a prefix to units of measurement that indicates a quantity of 0.01 (10^{-2}) times the value indicated by the root. See International System of Units.

centigrade temperature scale. former name of the Celsius temperature scale.

centipede. common name of any arthropod of the order Chilopoda. They have a venom apparatus that communicates with hollow, piercing claws or fangs that inject a relatively weak venom elaborated in the head. The bite can be painful to humans and commonly produces a local inflammatory response, with mild lymphangitis about the bite and some necrosis. Bites are rarely dangerous to adults. See Chilopoda, *Scolopendra morsitans*.

central. pertaining to or located at or near the center. **2:** located nearer the center than some designated point or object of interest.

Central African jumping viper. *Bothrops nummifer*.

Central Asiatic cobra. Naje naja.

central canal. tube within the spinal cord that contains cerebrospinal fluid. It is continuous with the ventricles of the brain.

central nervous system (CNS). the chief neurobehavioral coordinating mechanism of vertebrates and higher invertebrates, it is characterized by a high concentration of nerve cell bodies and synapses and is often closely associated with primary or special sense organs (olfactory, optic, auditory). The cerebral region of the CNS (the cerebral ganglia of invertebrates and the brain of vertebrates) coordinates and integrates the activities of the whole nervous system. The CNS communicates with various receptors and effectors throughout the animal body by means of a system of nerves (the peripheral nervous system). The CNS of metazoan invertebrates, except those having only a nerve net, consists of paired or single ventral or dorsal nerve cords and sometimes a brain, which is a collection of single segmental ganglia connected by commissures. The CNS of vertebrates consists of a brain and dorsal spinal cord derived from the embryonic neural tube. The vertebrate brain lies within a bony or cartilaginous skull, and the spinal cord lies within a vertebral column. The central nervous system is seriously affected by many toxic substances. Common responses to poisoning by the CNS are convulsions, paralysis, hallucinations, and ataxia, either alone or in combination. Behavioral or psychic effects are common and may include headache, agitation or excitement, hyperactivity, drowsiness, depression, disorientation, confusion, delirium, twitching, convulsions, and coma. The dynamics of interactions between the toxicant and the nervous system are diverse and often complex, supporting the major subdisciplines of neurotoxicology and behavioral toxicology. See also barrier (blood-brain barrier), peripheral nervous system.

Centralian death adder. *Acanthophis pyrrhus*.

Centropogon australis (waspfish; fortescue). a venomous fish (Family Scorpaenidae) found in the waters of New South Wales and Queensland, Australia. The venom apparatus is of the *Scorpaena* type (see Scorpaenidae).

Centruroides (bark scorpions). a genus of about 30 species of New World scorpions; the most common are *C. gracilis* (Margarita scorpion), *C. vittatus* (stripe-back scorpion), and *C. sculpturatus* (deadly sculptured scorpion). See also Scorpionida. They are crab-like with long, segmented tails that end in a bulbous sac and stinger. The venoms are chiefly neurotoxic. An allergic reaction is possible. Seven species are of interest toxicologically. Most of these occur in Mexico where fatalities, especially in young children, are very common. Toxicities and effects of envenomations vary with the species implicated. The sting of *C. exilicauda* (= *sculpturatus*), for example, usually causes little pain. Symptoms in children may include restlessness, abnormal head and neck movements, nystagmus, and often roving eye movements. Hypertension and tachycardia often appear within the hour. Within 1½ hr the child may appear seriously ill with symptoms that may include fasciculations over the face or large muscle masses; general malaise; some muscular weakness and incoordination; respiratory distress which may progress to paralysis, excessive salivation which may further interfere with breathing; slurred speech; convulsions. Nonfatal cases generally become asymptomatic within 36 hr. The course in adults is rather different and generally less severe; victims usually becoming asymptomatic within 12 hr. *C. suffusus* is an especially dangerous scorpion of Mexico; it may reach a length of 9 cm. *C. noxious* rarely exceeds 5 cm but is quite dangerous.

cephaeline (7',10,11-trimethoxyemetan-6'-ol; desmethylemetine; dihydropsychotrine). an emetic and antiamebic, it is the second most important alkaloid, $C_{28}H_{38}N_2O_4$, of ipecac.

cephaeline methyl ether. emetine.

Cephaelis ipecacuanha (ipecac). a tropical South American plant (Family Euphorbiaceae) with drooping flowers, sometimes called Brazilian ipecac. See ipecac.

cephalalgia. headache.

cephalea. headache.

cephalodynia. headache.

cephalopathy. encephalopathy.

Cephalopholis argus (sea bass; grouper). a marine bony fish (Family Serranidae) of the tropical Indo-Pacific region. It is a common transvector of ciguatera.

cephalopod. any animal of the class Cephalopoda.

Cephalopoda. a class of living mollusks, comprised of animals characterized by two horny jaws surrounded by a ring of eight or ten arms or numerous prehensile tentacles. They have a highly developed nervous system and sensory organs. Rapid locomotion is effected by contractions of the muscular mantle that propel water through a siphon, which is a modified portion of the foot. All but Nautilus have an ink gland that secretes a dark fluid into the surrounding water when the animal is threatened. The shell may be external (as in Nautilus, chambered nautilus, or pearly nautilus), internal (as in *Sepia*, cuttlefish), or entirely absent (as in *Octopus*). The salivary glands of various cephalopods secrete poisonous substances, although these are not necessarily used in defense or to capture prey. Some octopods are venomous and certain squids are venomous or poisonous. The toxins of some species can be rapidly fatal to humans.

cephaloridine (1-[[2carboxy-8-oxo-7-[(thienylacetyl)amino]-5-thia-1-azabicyclo[4.2.0]oct-2-

en-3-yl]methyl]pyridinium hydroxide inner salt; 1-[7'-β-[2-(2-thienyl)acetamido-8'-oxo-1'-aza-5'-thiabicyclo[4.2.0]oct-2'-en-3'-yl)-methyl]-pyridinium-2'-carboxylate; *N*-[7-[(2-thienyl)-acetamido]ceph-3-em-3-ylmethyl]pyridinium-4-carboxylate; *N*-[7-(2'-thienylacetamidoceph-3-yl-methyl)]pyridinium-2-carboxylate; cefaloridin). an antibacterial agent. It is moderately to very toxic to laboratory animals. It is selectively taken up by the anion transport system of the kidney, where it accumulates due to a cationic moiety in the molecule that prevents its excretion by the anion transport system.

cephalosporin. any of several antibiotics related to penicillin and originally isolated from the fungus, *Cephalosporium acremonium*. They inhibit development of the cell wall. They are nephrotoxic, may cause anaphylaxis in humans, and often crossreact with penicillin. Cephalosporins are contraindicated for penicillin-sensitive patients.

Cephalosporium. a genus of fungi that secrete epoxide mycotoxins. See cephalosporin.

cephalotoxin. a proteinaceous poison isolated from the posterior salivary glands of the decapod mollusks, *Sepia officinalis* and *Octopus vulgaris*. See also eledoisin.

Cerastes (horned vipers). a genus of two small venomous North African desert snakes (Family Viperidae). Their ranges are confined to the desert regions of northern Africa and western Asia. Although the bite is painful, they are not usually considered dangerous to humans. The head is broad, compressed, with a very short, broad snout, and an indistinct canthus rostralis. It is covered by small irregular scales with tubercular keels; there is a large, keeled, horn-like scale above the eye. Each nasal is separated from the rostral by 1-3 rows of small scales; 3-5 rows of small scales separate the eye from the supralabials. The neck is distinct. The eyes are small to moderate in size with vertically elliptical pupils. The body is depressed, tapered, and has a short tail. The dorsal scales are large and heavily keeled on the back; they become progressively smaller laterally with serrated keels. ***C. cerastes*** (African desert horned viper; fiyah; lefa bin kurum; métyi; vipére à cornes). a small, yellowish, pale gray, pinkish, or pale brown, viperine snake with dark brown, blackish, or bluish spots in rows; they may fuse, forming crossbars; the underside is whitish and the tip of the tail is black. There is a long, distinguishing, spine-like horn above the eye in most individuals, but is short or absent in some; the crown is covered by small scales. The maximum length is about ¾ m, with adults averaging little more than ½ m. It has a wide triangular head, a distinct neck, thick body, and a short tail that tapers abruptly behind the anus. The subcaudal scales are paired and the ventrals are weakly keeled. *C. cerastes* differs from *Pseudocerastes persicus*, chiefly in that the ventral body scales are weakly keeled. The range of this snake includes Saharan Africa, the Arabian Peninsula, and parts of the Middle East. It occurs chiefly in arid deserts that have fine sand and rock outcroppings, and also frequents oases. It is mostly nocturnal and remains concealed during the day under stones or in rodent burrows. It often uses sidewinding locomotion. When annoyed, it makes a rasping sound by expanding the loops of its body and rubbing them together in a fashion like that of *Echis*. It is not considered to be an especially aggressive or dangerous snake, although it does not usually retreat if disturbed. This snake occasionally bites humans. It may produce swelling and tissue damage, but fatalities are rare. Antivenin is available. ***C. vipera*** (African sand viper; Avicenna viper; Avicenna's sand viper; Sahara sand viper; "asp"). a small, generally harmless, viper that averages ca. 0.4 m in length; occasionally reaching a length of nearly 0.6 m. It lacks horns, but is otherwise very similar in color and appearance to *C. cerastes*. There are 9-13 scales across the crown and less than 130 ventrals. The ground color tends to be more faded and the spots less well defined than in *C. cerastes*. Females are somewhat larger than males; the tip of the tail is black in the female, light in the male. It occurs in sandy desert in the eastern and central Sahara to Israel. It is nocturnal and normally spends the day buried in sand at the base of a shrub; if alarmed, it will usually bury itself in loose sand. Care should be taken to differentiate this snake from the much more dangerous *Echis carinatus*.

Cerbera. a genus of trees (Family Apocynaceae) related to *Acocanthera* and *Strophanthus*. ***C. odallam***. an Asiatic tree; the source of cerberin and neriifolin. ***C. tanghin*** (tanghin; tanghinia).

an extract of the extremely toxic seeds is used as an arrow poison and sometimes in trial by ordeal.

cerberigenin. digitoxigenin.

cerberin (monoacetylneriifolin; cerberine; veneniferin). the 2′ acetate of neriifolin. See *Cerbera*, neriifolin.

cerberine. cerberine.

cerberoside (3-[(*O*-β-D-glucopyranosyl-1(1→6-*O*-D-glucopyranosyl-(1→4)-6-deoxy-3-*O*-methyl-α-L-glucopyranosyl]-14-hydroxycard-20(22)-enolide; thevetin B). a cardiotoxic glycoside from *Cerbera odollam* and *Thevetia neriifolia*.

Cercocarpus (mountain mahogany). a genus of small semideciduous shrubs and small evergreen trees of western North America (Family Rosaceae) that are often browsed by animals. They are cyanogenic.

cerebellar degeneration. See alcoholic-nutritional cerebellar degeneration.

cerebellum. a part of the brain that lies dorsally and posterior to the cerebral hemispheres of vertebrate animals. It controls the coordination of voluntary movements and balance, functions that are affected by many toxicants. See alcohol amnestic syndrome, Minamata disease, mercury methylazoxymethanol.

cerebral. **1:** pertaining to the cerebrum. **2:** often used loosely in reference to the vertebrate brain.

cerebral cortex. the rigid "gray matter" of the brain that forms the outer layer of each hemisphere of the cerebrum. It receives and interprets nerve impulses from the sense organs. It is concerned also with higher mental functions, such as perception, intelligence, memory, and consciousness.

cerebral death. brain death.

cerebral radiation syndrome. cerebral syndrome.

cerebral syndrome (cerebral radiation syndrome). one of three manifestations of acute radiation syndrome in humans. This condition, always fatal, results from extremely high dose rates of ionizing radiation (> 3000 rads). It is always fatal. The presenting symptoms (prodromal phase) are nausea and vomiting followed by listlessness, drowsiness, apathy, and prostration. The terminal phase is marked by tremors, convulsions, and ataxia with death intervening within a few hours.

cerebralgia. headache.

cerebritis. **1:** inflammation of the cerebrum. **2:** poisoning (moldy corn poisoning).

cerebropathia. encephalopathy.

cerebropathy. encephalopathy.

cerebrosis. and disease of the cerebrum.

cerebrospinal. pertaining to the brain and spinal cord.

cerebrospinal meningitis. **1:** inflammation of the membranes of the brain and spinal cord. It is caused by any of numerous pathogens. **2:** moldy corn poisoning.

cerebrum. a main portion of the brain of most vertebrates, it coordinates and integrates activities of the entire nervous system. In humans and probably certain other primates, it is responsible for consciousness. The cerebrum is the largest part of the brain in humans and is situated beneath the roof of the skull. It consists of two hemispheres that are separated by a furrowed division. The outer layer is the cerebral cortex ("gray matter"). White matter lies beneath the cortex and consists of three kinds of fibers that connect the hemispheres, convey impulses to and from the cortex and spinal cord, and interconnect different areas of the cortex. The effects of poisons on the cerebrum can be quite serious and, in many cases, life threatening. See, for example, abrin, antihistamine, methylergonovine.

Cereus pecten. a Mexican cactus (Family Cactaceae) that contains pectenine, a toxic alkaloid.

ceriman. See *Monstera deliciosa*.

cerium (Ce)[Z = 58; A_r = 140.12]. a silvery metallic element of the lanthanoid series.

ceroid. a waxy, golden or yellow-brown substance that occurs in some cirrhotic livers and certain other tissues of mammals.

cerubidin. daunorubicin.

cerulenin (3-(1-oxo-4,7-nonadienyl)oxirane-carboxamide; (2R,3S)-2,3-epoxy-4-oxo-7E,10E-dodecadienamide; 2,3-epoxy-4-oxo-7,10-dodecadienoylamide; helicocerin). an epoxide antifungal antibiotic isolated from *Cephalosporium caerulens*, *Acryocylindrium oryzae*, and *Helicoceras oryzae*. It is very toxic *i.v.* or *i.p.* and moderately toxic *per os* to laboratory mice.

ceruloplasmin. a blue glycoprotein in vertebrates to which most copper in blood is attached. Its level in blood is depressed in hypocupremia and Wilson's disease.

cerumen. See sebum.

ceruse. basic lead carbonate.

cerussa. basic lead carbonate.

cervical. refers to the neck of an organism, to the uterine cervix, or to any neck-like structure. Thus, bones of the vertebral column that are situated in the neck region and support the head are termed the cervical vertebrae.

cervicarcin (1,2,3-tetrahydro-1,3,4,5,10-pentahydroxy-2-methyl-3-[(3-methyloxiranyl)carbonyl]-4a,9a-epoxyanthracen-9(10H)-one; 4a,-9a-epoxy-3-(2,3-epoxybutyryl)-1,2,3,4,4a,9a,-hexahydro-1,3,4,5,10-pentahydroxy-2-methyl-anthrone). an antineoplastic epoxide antibiotic isolated from *Streptomyces ogaensis*; it is extremely toxic, *i.p.*, to laboratory mice.

cervicitis. inflammation of the neck of a structure, such as that of the uterine cervix.

cervix. the neck of an organism or of any organ; any structure that resembles a neck. See cervix uteri, cervical.

cervix uteri. the lower portion, the neck, of the mammalian uterus, which protrudes slightly into the vagina.

cesium (caesium; Cs)[Z = 55; A_r = 132.91]. a soft, silvery white, ductile, highly reactive, alkali metal comprised of hexagonal crystals or (above 28.5°C) a silvery liquid. In addition to being a radiation hazard, cesium is the most electropositive element known. **Cesium-137** (radiocesium). a relatively strong emitter of gamma radiation with a half-life of 30 years. Because it is a radionuclide, the chemical toxicity of cesium has hardly been studied. It is known, however, to produce hyperirritability with spasms in laboratory animals when administered orally in amounts equal to the potassium content of the diet. When dietary potassium is replaced by cesium, laboratory rats may die within 10-17 days.

cestrum. *Cestrum*.

Cestrum (jessamine; cestrum). a genus of large, handsome shrubs or small trees (Family Solanaceae) with trumpet-shaped flowers that are native to the New World tropics. All parts of the plant are poisonous. ***C. diurnum*** (day-blooming jessamine; wild jasmine; king of the day). a common ornamental shrub in the southern United States, it also occurs in waste areas in south Texas and the Florida Keys. Toxicity and history of poisonings is similar to that of *C. nocturnum*. It can induce calcinosis accompanied by osteopetrosis in herbivores. ***C. nocturnum*** (night-blooming jessamine). a common ornamental shrub in the southern United States. Neurotoxic; poisonings of humans and pets have been reported. Symptoms are similar to those of atropine poisoning. ***C. parqui*** (green cestrum; willow-leaved jessamine). poisoning of cattle, hogs, horses, sheep, and domestic fowl have been reported. All parts of the plant are toxic, but the berries are most toxic. Signs and symptoms include pyrexia and those of gastroenteritis, including bloody feces. Death occurs within 10 hours following the appearance of symptoms.

cevadilla. sabadilla.

cevadilline. a toxic alkaloid and component of veratrine mixture. See veratrine

cevadine ((*Z*)-4α,9-epoxycevane-3β,4,12,14,-16β,17,20-heptol 3-(methyl-2-butenoate; veratrine; crystalline veratrine). an extremely toxic alkaloid isolated from seeds of *Schoenocaulon officinale* (sabadilla) and *Veratrum*. When isolated from *Veratrum*, it is often referred to as veratrine. Even local application can produce serious poisoning. It is extremely irritating, especially to mucous membranes. The cevadine of *Veratrum* is commonly termed veratrine. See veratrine.

(3β,4α,16β)-cevane-3,4,12,14,16,17,20-heptol 3-acetate. sabadine.

cevine(4,9-epoxycevane-3α,4α,12,14,16β,17,20-heptol). a very toxic alkaloid and component of veratrine mixture. It can be prepared by hydrolysis of cevadine. See cevadine, veratrine.

Ceylon hump-nosed viper. *Agkistrodon nepa*.

Ceylon krait. *Bungarus ceylonicus*.

Cf. the symbol for californium.

Cf. compare.

CF. **1:** chronicity factor. See factor. **2:** folinic acid.

CFC. chlorofluorocarbon.

CFR. Code of Federal Regulations, a publication that contains regulations promulgated under U.S. federal laws.

cGMP. cyclic GMP.

cgs system. a system of scientific units based on the centimeter-gram-second.

CH. acronym for cyclohexanone.

chaeta (plural, chaetae). seta.

chaetae. plural of cheta.

chaetoglobosin. any of a variety of mycotoxins produced by various fungi. They are structurally and toxicologically related to the cytochalasins and phomins. **chaetoglobosin A**. a mycotoxin from *Chaetomium globusum*. It is extremely toxic to laboratory mice. See cytochalasins.

Chaetomium. a genus of ascomycetous fungi (Family Chaetomiaceae), the species of which produce various mycotoxins including cochliodinol. ***C. globusum***. a species that produces chaetoglobusin A. ***C. trilaterale***. a species that produces oosporein.

chain reaction. See radical (free radical).

chain snake. *Vipera russelii*.

chain viper. *Vipera russelii*.

chalcanthite. the naturally occurring pentahydrate of cupric sulfate.

chalcosis (chalkitis). **1:** chronic poisoning by copper. It causes a severe colic similar to that of lead colic. **2:** copper deposits in tissue, especially the lungs. Do not confuse with chalicosis. See also chalcosis lentis.

chalcosis lentis (copper cataract). a cataract caused by unduly high concentrations of intraocular copper. See also chalcosis.

chalice vine. *Solandra*. any plant of the genus *Solandra*.

chalicosis (flint disease). an occupational pneumoconiosis caused by inhaling dust produced by stone cutting.

chalkitis. chalcosis.

challenge. **1:** demand, test, trial. **2:** exposure to an antigen subsequent to specific immunization against that antigen.

chalone. any of a number of substances found in mammalian tissue homogenates that inhibit mitosis in intact cells, especially in the presence of adrenaline and corticosteroids. They may play a role in cancer and aging. The chalone of epithelial tissue appears to be a glycoprotein of high molecular weight.

Chamaelirium luteum (devil's bit). a genus of plants (Family Melanthaceae) of eastern North America. All parts are toxic.

chamber. a compartment, an enclosed space, or antrum. **anterior chamber**. the space between the iris and the cornea of the eye. **controlled environment chamber**. a chamber in which ambient conditions of temperature, lighting, and humidity can be maintained within narrow limits or changing (usually cycling, as between periods of light and dark) according to the requirements of the experiment. **experimental chamber**. a chamber in which animals or plants can be kept in comparative comfort and isolation for purposes of experimentation. **exposure chamber**. a sealed experimental chamber for controlled exposures of plants or animals to dispersed agents (e.g., gases, aerosols, sprays, dusts). **inhalation chamber**. an exposure chamber into which aerosols, gases or dusts of controlled composition can be introduced. See also inhalation. **lethal chamber**. a chamber that can be filled with a toxic gas to kill small animals.

chandra bora. *Vipera russelii*.

channelization. a straightening and deepening stream so water will move faster. It is usually employed as a flood-reduction or marsh-drainage tactic. It can directly disturb fish and wildlife habitats and can disrupt waste assimilation capabilities with consequent accumulation of toxicants in the system.

Ch'an Su (Ch'an-Su; Senso). a Chinese drug, it is a dry mixture of the venom of the Chinese toad *Bufo b. gargarizans* = *B. gargarizans* = *B. asiaticus* (Family Bufonidae). See also bufalin and bufogenin B.

Ch'an-Su. Ch'an Su.

charas. cannabis.

charcoal. See activated carbon.

chard. *Beta vulgaris* var. *cicla*.

charlock. *Brassica kaber*.

Chasen's pit viper. *Trimeresurus chaseni*.

chatilla. *Bothrops lansbergii*.

cheese poisoning. tyrotoxism.

cheese reaction. the symptoms that appear in patients who eat certain cheeses while taking mono amine oxidase inhibitors.

cheeseweed. *Malva parviflora*.

cheilitis. inflammation of the lips.

cheirospasm. spasm of the hand muscles.

chelate. **1:** claw-like, pincer-like. **2:** descriptive of an appendage with a pincer-like organ or claw. **3:** to combine with a metal ion to form a stable compound. **4:** a molecular complex that forms stable (heterocyclic) rings with the unshared electrons of neighboring atoms, especially a complex of organic and metal ions formed by chelation. It is a coordination complex in which one ligand coordinates to a metal ion at two or more points. Such complexes contain rings of atoms that include the metal atom. This arrangement helps to keep the metal in solution and greatly reduces its toxicity.

chelating agent. a metal-complexing agent. An organic compound (e.g., edetic acid, penicillamine) that forms multiple coordinate covalent bonds with a metal ion, such as lead and mercury, to form a stable complex. Some complex with a number of metals, others show some degree of specificity. They are used in medicine and the biological sciences largely to limit or control poisoning by heavy metals. Chelating agents may exhibit some degree of toxicity by removing certain trace elements from the body. *Cf.* sequestrant. See also chelation therapy.

chelation. the formation of a chelate by organic and metal ions. See chelate.

chelation therapy. the use of chelating agents to complex free metal ions, greatly reducing their toxicity until the metal chelate complex can be excreted. Chelation therapy may sometimes aid the elimination of stored metals from the body. In the treatment of lead poisoning by chelation therapy, for example, the lead is solubilized and eliminated by a chelating agent such as the calcium chelate of edetic acid which binds strongly to most +2 and +3 cations. Dimercaprol (British Antilewisite) is another chelating agent used to treat lead poisoning. It chelates lead through its sulfhydryl groups, and the lead chelate is excreted via the kidneys and bile.

chelerythrine (1,2-dimethoxy-12-methyl[1,3]-benzodioxolo[5,6-c]phenanthridinium; toddaline). a toxic, slightly ethanol-soluble, crystalline, papaver alkaloid isolated from the roots of *Chelodonium majus* and from *Papaver spp*. It has a narcotic action.

chelicera (plural, chelicerae). a pair of anterior appendages beneath the cephalothorax of arachnids. In spiders they comprise a heavy first segment and a sharp, pointed second segment or fang through which venom is introduced into the prey.

cheliceral fang. See fang.

chelidonine. a crystalline papaver alkaloid. It is an active principle of *Chelidonium majus* and other papaveraceous plants.

Chelidonium (celandine; felonwart; rock poppy; swallow wort; wort weed). a genus of toxic herbs (Family Papaveraceae), the species of which contain isoquinoline alkaloids. They have a conspicuous yellow or yellowish-orange, acrid irritant sap, brittle stems, pinnately divided leaves, and small yellow flowers in pedunculate umbels. ***C. majus*** (celandine, swallow wort). a toxic herbaceous biennial or perennial weed (Family Papaveraceae) that was introduced from Europe and is now widely distributed in the eastern United States. All parts are deadly, but the leaves, stems, and roots are especially toxic. The sap is a strong irritant and produces severe stomatitis, gastroenteritis, fainting, and coma when ingested. *C. majus* also contains several papaver alkaloids such as chelidonine, chelerythrine, sanguinarine, berberine, protopine, and tetrahydrocoptisine. Chelidonine and isoquinoline are the chief active principles. Occasional poisonings, some fatal, of humans and animals have been reported in Europe. See chelidonine.

Chelonia. an order of reptiles comprised of the turtles, terrapins, and tortoises. Chelonians are nearly cosmopolitan in distribution; they are absent from New Zealand and western South America; included are freshwater, marine, and terrestrial species. They are characterized by a broad body encased in a usually rigid shell that consists of a bony arched dorsal carapace and a flat ventral plastron, which are connected laterally and covered with polygonal scutes or leathery skin. The mandibles are often sharp and powerful, but teeth are absent; thoracic vertebrae and ribs are usually fused to shell. *Cf. Chelonia*.

Chelonia (green turtles). a genus of warm-water marine turtles, prized as food and weighing up to 180 kg. The extremities are modified as flippers. *Cf.* Chelonia. ***C. mydas*** (green sea turtle). a marine turtle (Family Chelonidae) with a broad, oval, unkeeled carapace and paddle-like limbs that occurs in all tropical and subtropical seas, ranging into temperate waters during the summer. The meat and eggs are normally considered a gourmet treat, and it produces other popular products (flipper leather, cosmetic and cooking oils) such that it is rapidly approaching extinction. In some locations, however, the meat and especially the liver are extremely toxic. Nearly half of those persons who are poisoned die.

Chelonidae (sea turtles). a family of reptiles comprised of marine-dwelling turtles that are characterized by heart-shaped, scuted carapaces and paddle-like limbs with one or two claws. The eggs are laid on shore. Some are seriously poisonous under certain circumstances. See poisoning (turtle), *Chelonia mydas*, *Dermochelys coriacea*, and *Eretomochelys imbricata*.

Chemhex®. lindane.

chemical. 1: any substance; an element, compound, or a mixture of two or more substances. **2:** pertaining to a substance or to a process that changes the structure of one or more substances.

Chemical 109. ANTU.

chemical agent. any substance or active principle that can produce an effect on a living organism by interacting with endogenous substances. Included are an extremely diverse array of drugs, poisons, and substances with other types of action.

chemical antagonism. See antagonism (chemical).

chemical burn. See burn.

chemical carcinogen. See carcinogen.

chemical carcinogenesis. See carcinogenesis.

chemical concentration. See concentration.

chemical conjunctivitis. inflammation of the conjunctiva by chemical irritants.

chemical disposition. the absorption, distribution, metabolism, and excretion of toxicants or of a class of toxicants.

chemical ecology. the science of chemical relionships within and among biotic populations.

chemical emergency. See emergency.

chemical equilibrium. a condition in which a chemical reaction occurs at equal rates in its forward and reverse directions such that the concentrations of the reactants do not change with time.

chemical fallout. See fallout.

chemical idiosyncracy. See idiosyncrasy.

chemical intermediate. **1a:** a substance formed in the course of a chemical process that is essential to the formation of the end product. **1b:** a substance formed during the decomposition of a substance.

chemical lesion. See lesion.

chemical lure. a chemical substance secreted by one organism that attracts another. See kairomone.

chemical oxygen demand (COD). the amount of oxygen depleted in a water sample as a result of chemical oxidation by sulfuric acid-potassium dichromate solution. The result is a measure of total oxidizable compounds in a sample. See also biological oxygen demand.

Chemical Mace. chloropicrin.

chemical treatment. any of a variety of technologies or processes that use chemicals to treat industrial, municipal or agricultural wastes.

chemical warfare (CW). the deliberate development and use of toxic and other hazardous agents to kill or incapacitate enemy troops or to deprive them of meat, crops, or plant cover. Such chemicals have also been employed in the deliberate destruction of civilian populations. The use of poisons (chemicals) in warfare have been condemned or banned by any number of civilizations. For example, the Greeks and Romans denounced such use as a violation of the law of nations (*ius gentium*). Poisons and other inhumane weapons of war were banned by the Manu Law of India in the 5th or 6th century, B.C. and by the Saracens in the 6th century A.D. When not "banned" outright, the use of chemicals (and now biological agents) in warfare has been broadly condemned by humanity. The nature of war, however, is such that international condemnation or law holds little sway in many cases. Warfare agents were used in the 1st World War, and while apparently not used as such in the 2nd World War against combatants, such chemicals were used against civilian enemies of the Nazi state. The 1925 Geneva Protocol forbids the use of chemical and bacteriological agents in war. Nevertheless, chemical warfare was engaged by Iraq during its war with Iran during the 1980s and possibly also during the war in the Persian Gulf War of 1991. Iraq has also used such agents to slaughter internal "enemies" such as the Kurds in northern Iraq. The above is a small sampling of chemical warfare. See warfare agent.

chemical warfare agent. See warfare agent.

chemical warfare gas. a chemical warfare agent used in gaseous form. See warfare agent.

chemically pure (CP). an accepted grade of fine chemicals and drugs that contain a minimum of impurities.

cheminosis. any disease caused by chemical substances.

chemist. a student or practitioner of chemistry, *q.v.*

chemistry. the science of matter. There are numerous subdisciplines within this branch of science; a few are: **analytical chemistry**. the development and application of qualitative and quantitative techniques that yield information of a quantitative nature, when possible, regarding composition, structure, and/or energy changes of chemical systems. It exists as a major subdiscipline in its own right and also as a subdiscipline of each of the subdisciplines defined here. **biological chemistry**. biochemistry. **ecological chemistry**. the study of naturally occurring chemicals and their interactions in the biosphere. **environmental chemistry**. the science that deals with chemicals and chemical phenomena in the environment. Of special concern is (1) the in-introduction of chemicals into the environment, their movement, chemical reactions, transformations, and fates under varying circumstances and within diverse media, and (2) the relationship between the properties of environmental chemicals and their environmental effects. **inorganic chemistry**. the science that deals with elements other than carbon and their compounds. A few carbon compounds in which carbon is part of a simple ion are treated as inorganic. These include carbon disulfide, hydrogen cyanide, halides, and simple carbon salts such as cyanates, cyanides, and carbonates. **organic chemistry**. the science of the vast number of compounds that contain covalently bonded carbon. Relatively few carbon-containing compounds are classified as inorganic. **physiological chemistry**. biochemistry, *q.v.* **toxicological chemistry**. the science that deals with the chemical properties and behavior of toxic substances, including, but not restricted to reactions that occur within exposed organisms (e.g., toxification, detoxication reactions, transformations, fate). Also included is the study of the synthesis or origins of toxic chemicals, their applications, and the characteristics and behavior of reactants and products. See also biochemistry.

chemoautotroph. chemolithotroph.

chemoautotrophic. chemolithotrophic.

chemoautotrophic organism. See autotrophic organism.

chemobiodynamics. the study of relationships between structural properties of various chemicals and their biological activity or effects.

chemocautery. the destruction of abnormal tissue by means of acids, caustics, or poisons (cauterants).

Chemocorticoid. any of a number of substances presumed to originate in the adrenal cortex and determined by chemical methods.

chemolithotroph. an autotrophic organism that utilizes CO_2 as its principle source of carbon and obtains its energy by the oxidation of inorganic compounds.

chemolithotrophic. referring to or denoting the mode of nutrition of a chemolithotroph; pertaining to a chemolithotroph.

chemolysis. decomposition by the action of chemicals.

chemopharmacodynamic. of or pertaining to the relationship between chemical structure and pharmacologic activity.

chemoreceptor. a neuroreceptor (e.g., taste buds, the carotid body) that is activated or stimulated by alterations in the chemical composition of the surrounding medium.

chemoreflex. a chemically induced reflex.

chemosis. edema of the conjunctiva(e), usually in response to noxious stimuli.

chemostat. an apparatus in which microorganisms are maintained in continuous steady-state growth through continuous input of a growth-limiting nutrient and removal of microorganisms and their products.

chemosterilant. a substance that causes reproductive sterility. Such chemicals are sometimes used to control a pest population by sterilizing a portion of the males. This is often accomplished by releasing large numbers of laboratory-reared sterile males (e.g., as with *Drosophila*, fruit flies) into a pest population at appropriate times. To the extent that the population is limited by reproductive rate and the sterile males compete successfully with normal males, the number of offspring in the next generation will be reduced. See also sterilant (chemical sterilant).

chemosynthetic organism. See autotrophic organism.

chemotherapeutic. **1:** pertaining to chemotherapy or the capacity of a drug for use in chemotherapy. **2:** an agent used for chemotherapy. **3:** antineoplastic.

chemotherapeutic agent. **1:** a chemical agent used in the treatment of disease. Such agents are selected largely on the basis of their selective toxicity to a particular type of disease-causing microorganism. A fully effective chemotherapeutic agent thus destroys or inhibits proliferation of the microorganism at dosages below that which is toxic or injurious to the host. Most such agents are, however, extremely toxic and can be fatal or chronically disabling to humans, for example, through their cytotoxic effects on proliferating lymphocytes and myeloid tissues. Examples are antibiotics, antimalarials, antifungals, antivirals, antiseptics, and cytotoxic drugs. **2:** an antineoplastic agent.

chemotherapeutic index. **1:** the ratio of the minimum effective dose of a chemotherapeutic agent to the maximum tolerated dose. It is thus a measure of relative toxicity. **2:** in practice, usually the ratio of LD_{50} to ED_{50}.

chemotherapy. **1:** the treatment of infectious disease by the use of chemical agents (e.g., bactericidal or bacteriostatic antibiotics) that have a selective toxic effect on the targetted pathogen, but are not unacceptably harmful to the victim at the doses used. Many microorganisms have become resistant to commonly used chemotherapeutic agents. **2:** the treatment of malignancies by the use of chemical agents that destroy cancer cells without, ideally, harming healthy tissues or organs. See chemotherapeutics, antineoplastics.

chemozoophobe. a chemozoophobous plant. A plant that produces noxious chemicals (allelochemics) that inhibit consumption by herbivorous animals.

chemozoophobous. pertaining to or denoting plants that inhibit consumption by herbivorous animals by producing noxious chemical substances (allelochemics).

Chenopodiaceae (goosefoot; pigweed). a cosmopolitan family of some 75 genera of weedy herbs and shrubs, many members of which contain high concentrations of oxalate salts. See *Chenopodium*.

Chenopodium (goosefoot; pigweed). a nearly cosmopolitan genus of toxic or potentially toxic, mainly weedy plants (Family Chenopodiaceae). Concentrations of nitrate toxic to livestock have also been reported in some species. ***C. ambrosioides*** (American wormseed, Mexican tea, Spanish tea). a strong-smelling annual herb of tropical America, now widely distributed throughout most of eastern North America. The raw leaves are toxic, but have been used in medicinal teas. It is also cultivated for wormseed oil, an irritant oil used as an anthelmintic. The active principle is the terpene, ascaridole. Concentrations of nitrate toxic to livestock have been reported. ***C. botrys*** (Jerusalem oak, feather geranium). a strong-smelling annual herb with toxicological properties similar to those of *C. ambrosioides*. The leaves are toxic. Concentrations of nitrate that are toxic to livestock have also been reported.

chernetid. any member of the family Chernetidae.

Chernetidae (chernetids). a family of small (up to 4 mm long), common pseudoscorpions (Order Pseudoscorpionida). Each pedipalp contains a well-developed venom gland that communicates by means of a duct to the movable segment of the pincer. Most species are predaceous, feeding on small insects and other small terrestrial invertebrates.

cherry. a common name of various species of *Prunus* (e.g., black cherry, cherry laurel) and a number of unrelated plants.

cherry, black. a common name for any of these species: *Prunus serotina*, *P. demissa*, *P. melanocarpa*, etc.

cherry laurel. *Prunus lawrocerasus*.

cherry laurel oil. oil of cherry laurel.

chestnut. common name of trees of the genus *Castanea*.

Chevron. See Diazinon.

chewing fescue. *Festuca rubra*.

chewing tobacco. a form of tobacco that is chewed. It is a cause of tongue cancer. See also nicotine, snuff, tobacco.

Cheyne-Stokes breathing (Cheyne-Stokes respiration). breathing characterized by rhythmic changes in the amplitude of breathing, and periodically recurring periods of apnea. It occurs especially in coma resulting from affected nerve centers.

chi-square test. X-square test. See test.

chick. **1:** a newborn or young bird of any flying species prior to the onset of flight. **2:** a nestling. **3:** the still downy young of precocious species (e.g., waterfowl and gallinaceous birds). **4:** in the medical literature the term is often loosely applied specifically to the chick of *Gallus domesticus*, the domestic fowl.

chick embryo. **1:** the embryo of any bird, but most commonly applied to gallinaceous birds and waterfowl. **2:** in the biomedical sciences, this term often refers only to the embryo of *Gallus domesticus*. See test (chick embryo test).

chick embryo test. See test.

chicken mushroom. *Laetiporus sulfureaus*.

chickweed. *Stellaria media*.

chicona. *Chondodendron tomentosum*.

chigger flower. *Asclepias tuberosa*.

chikorviri. *Bitis arietans*.

child. a young person past infancy but not yet at the age of puberty; often regarded as a person in the first 10 years of life. Sometimes because of behavior and sometimes because of physiological and metabolic differences with adults, a child may be at substantially greater or substantially less risk from a given toxicant than an adult.

Chile saltpeter. natural, impure, sodium nitrate. It occurs in large quantities in South America.

chillipepper. See *Sebastes goodei*.

Chilopoda (centipedes). a class of about 2000 species of nocturnal, predatory, terrestrial arthropods that includes centipedes. They superficially resemble millipedes (See Diplopoda). The body is elongate, depressed, with 15 to 173 body segments and one pair of legs per segment except the last two. The head bears a pair of antennae, a labrum, mandibles, and two pairs of maxillae. The claws (maxillipeds) on the first body segment are modified hollow fangs that inject a relatively weak venom originating in the head. The venom of some of the larger species of *Scolopendra* is painful to humans, producing an inflammatory response, often with mild lymphangitis and lymphadenitis; symptoms rarely persist more than 2 days in adult humans.

chimaera. **1:** any cartilaginous fish of the subclass Holocephali. **2:** an organism whose tissues are of two more genetic origins. **3:** a mythical two-headed animal with the head and body of a lion plus the head of a goat.

Chimaera monstrosa (European ratfish). a venomous marine cartilaginous fish (Family Chimaeridae) that occurs in the north Atlantic Ocean from Norway and Iceland to the Mediterranean Sea, Morocco, the Azores, South Africa, and Cuba. See chimaera, Chimaeridae, chimaeroids, *Hydrolagus colliei*.

Chimaeridae (ratfishes). marine, usually pelagic, cartilaginous fish (Order Chimaeriformes, Subclass Holocephali) with a short, rounded snout and a long, whip-like caudal fin. They are the only cartilaginous fishes to have a single gill opening on either side of the body. The second dorsal fin extends from a point just anterior to the pelvic fin almost to the origin of the caudal fin. See chimaera, chimaeroid, *Chimaera monstrosa*, *Hydrolagus colliei*.

chimaeroid. **1:** relating to or resembling a chimaera. **2:** common term for any cartilaginous fish of the subclass Holocephali.

chimaeroid poisoning. empoisonment due to envenomation by chimaerods or by ingestion of the flesh of these fish which has been shown to be toxic to laboratory animals. See also Holocephali.

chimney-bellflower. *Chimonanthus*.

chimney sweeps' cancer. scrotal cancer.

Chimonanthus (chimney-bellflower; wintersweet). a genus of evergreen and deciduous shrubs (Family Calycanthaceae) that are native to China. ***C. praecox***. a bush or shrub with fragrant, yellow flowers that is grown horticulturally in mild climates. It is a source of a highly toxic alkaloid, calycanthine.

China berry. See *Melia azedarach*.

China tree. *Melia azedarach*.

Chinaball tree. *Melia azedarach*.

Chinaberry tree. *Melia azedarach*.

Chinaman fish. *Lutjanus nematophorus*.

Chinawood oil. tung oil.

Chinese anise. any tree of the genus *illicium*. The fruit is also called Chinese anise.

Chinese bamboo viper. *Trimeresurus stejnegeri*.

Chinese blistering flies. mylabris.

Chinese blue. ferric ferrocyanide.

Chinese cabbage. the common name of oriental varieties of *Brassica rapa* (field mustard), in particular *B. r. pekinensis* (*q.v.*) and *B. r. Chinensis* (pak choi).

Chinese cantharides. mylabris.

Chinese cobra. *Naja n. atra*.

Chinese green tree viper. *Trimeresurus stejnegeri*.

Chinese habu. *Trimeresurus mucrosquamatus*.

Chinese mountain viper. *Trimeresurus monticola*.

Chinese pit viper. *Trimeresurus monticola*.

Chinese restaurant syndrome (CRS). a syndrome that presents soon after the ingestion of food that contains monosodium glutamate (MSG). The effects are somewhat variable, but by generally include numbness at the nape of the neck, radiating gradually to the arms and back. This is accompanied by generalized muscular weakness and palpitations. Susceptibility varies considerably among individuals. Clinical data have thus far not demonstrated MSG to be the sole factor in CRS.

Chinese vermilion. red mercuric sulfide. See mercuric sulfide.

Chinese white. zinc oxide.

Chinese wisteria. *Wisteria sinensis*.

chinigani. See *Causus*.

chinone. quinone or *p*-quinone. See quinone.

chipukupaku. *Bitis caudalis*.

Chiricanthium (running spiders). a genus of tiny, greenish-white or pale yellow spiders with neurotoxic venom that cling tenaciously when biting, causing an intensely painful wound. Only four of the approximately 160 species are known to bite humans. Systemic effects in humans and other large animals are usually mild, although an allergic reaction is possible. ***C. diversum***. the species that most often bites humans, it ranges in length from 7-16 mm. The abdomen is ovoid and the color varies from yellow, green or greenish white to red-brown. The chelicerae are strong and the legs are long and hairy. They must often be forcefully removed from the site of the wound. Persistent pain about the wound, restlessness, lymphadenitis, and lymphadenopathy are characteristic.

chiriquitoxin. a toxin isolated from the frog, *Atelopus chiriquensis*.

Chironex fleckeri. an extremely dangerous venomous jellyfish or medusa. It is one of the deadliest stinging marine animals; the crude venom is supertoxic to humans. Victims may die within a few minutes of envenomation. Extracts of tentacle preparations or of nematocysts have exhibited neurotoxic, hemolytic, dermonecrotizing, cardiotoxic, hypotensive, and cytolytic effects. Many of the effects of *Chironex* toxin are due to protein fractions. See also cubomedusae.

Chironex toxins. See *Chironex fleckeri*, cubomedusae.

chitti. *Bungarus caeruleus*.

chittul. *Hydrophis caerulescens*.

chiva. *Bitis arietans*.

chloracetone. chloroacetone.

chloracne (tar acne). a generalized acneiform dermatitis induced by prolonged contact with chlorine-containing compounds such as naphthalenes, biphenyls, and TCDD. The condition was first observed in workers who inhaled chlorobiphenyls. Symptoms include cutaneous eruptions, systemic manifestations, and sometimes death. Fatal cases are known from exposure to mixtures of chlorinated hydrocarbons. Chloracne is one of the major manifestations of TCDD poisoning in humans. It does not occur in most laboratory animals. Exceptions are monkeys and rabbits.

chloral. **1:** trichloroacetaldehyde. **2:** sometimes used as a synonym for chloral hydrate.

chloral alcoholate. See chloral derivative.

chloral, anhydrous. trichloroacetaldehyde.

chloral betaine. See chloral derivative.

chloral derivative. any of a number of sedative hypnotics derived from chloral (trichloroacetaldehyde). Examples are chloral hydrate, trichlofos, chloral betaine, β-chloralose, dichloral phenazone, chloral alcoholatc, and the hemiacetals. As in the case of barbiturates, chloral derivatives induce drug-metabolizing enzymes, tolerance, and physical dependence, and a physiological withdrawal syndrome characterized by convulsions and delirium. All are presumably metabolically converted to the active species, 2,2,2-trichloroethanol.

chloral hydrate (2,2,2-trichloro-1,1-ethanediol; trichloroacetaldehyde monohydrate; "knockout drops;" trichloroacetaldehyde, hydrates; trichloroethylidene glycol). a very toxic monohydrate of chloral, used mainly as a hypnotic, sedative, and anticonvulsant. Chloral hydrate is a CNS depressant, may be habit-forming, and is a controlled substance in the United States. An overdose produces neurological effects similar to those of the barbiturates. Symptoms of intoxication in humans may include vomiting, drowsiness, confusion, respiratory depression, shock, pinpoint pupils, hypothermia, hypotension, coma and death. Liver and kidney injury with acute renal and hepatic failure may be noted. It is very toxic to laboratory rats; a skin irritant. See also chloralism, chloral derivative, drug abuse, mickey finn.

chloral hydrate poisoning. chloralism (def. 2).

chloralism. **1:** chronic consumption of chloral compounds. **2:** a morbid condition produced by chronic use of chloral hydrate (chloral hydrate poisoning). The condition usually presents as CNS depression, progressing to eventual paralysis. Symptoms of gastrointestinal irritation are common. Symptoms may also include shallow and irregular respiration, a weak pulse, dizziness, nausea, vomiting, lassitude, weakness, and prolonged sleep. See also chloralization.

chloralization. **1:** chloralism. **2:** chloral-induced anesthesia (obs.).

chloralkyl ethers. carcinogenic in some laboratory animals.

β-chloralose. See chloral derivative.

chloramben (3-amino-2,5-dichlorobenzoic acid). a white, solid substituted benzoic acid used as a selective preemergence herbicide. It is carcinogenic to some laboratory animals.

chlorambucil (4-[*p*-[bis(2-chloroethyl)amino]-benzenebutanoic acid; 4-[*p*-[bis(2-chloroethyl)-amino]amino]phenyl]butyric acid; γ-[*p*-di(2-chloroethyl)aminophenyl]butyric acid; *N,N*-di-2-chloroethyl-γ-*p*-aminophenylbutyric acid; chloraminophene; chloroambucil; CAB). an extremely poisonous nitrogen mustard derivative. It inhibits lymphocytic maturation and proliferation and is used as an antineoplastic agent in the treatment of lymphocytic leukemia. Chlorambucil is an alkylating agent with both cytotoxic and immunosuppressive effects and is a primary carcinogen. It causes delayed liver damage and delayed reproductive effects in humans.

chloramine. **1:** a colorless, unstable, pungent, extremely hazardous, water-soluble liquid, NH_2Cl; the fumes can be deadly. It yields nitrogen, hydrochloric acid, and ammonium chloride on decomposition. It is used in the manufacture of hydrazine and is a hazard in households, where it is produced when chlorine or chlorine-containing products are mixed with ammonium hydroxide. This is especially dangerous because ammonia is a component of many cleaning products (See household cleaning product) without being named on the label. **2:** any of a class of monochloramine (See chloramine B, chloramine T) and dichloramine compounds (e.g., dichloramine T) that form during the purification of drinking water to provide combined available chlorine. See chlorine (combined available chlorine).

chloramine B (sodium benzenesulfochloramide). a monochloramine with a faint chlorine odor, $C_6H_5SO_2NClNa$. It is used as a water-soluble disinfectant and medicinally as an antiseptic. See chloramine.

chloramine T (sodium *p*-toluenesulfochloramine). a white or yellowish crystalline or powdery water-soluble monochloramine with a slight odor of chlorine $CH_3C_6H_4SO_2NaCl{\cdot}3H_2O$. It is soluble in water, insoluble in most organic solvents. It decomposes in ethanol and also slowly decomposes in air, releasing chlorine. Chloramine T contains 11.5 to less than 13% active chlorine. It is used as a antiseptic and disinfectant. Do not confuse with chloramine (def. 1); See also chloramine (def. 2).

chloraminophene. chlorambucil.

chloramphenicol (2,2-dichloro-*N*[2-hydroxy-1-(hydroxymethyl)-2-(4-nitrophenyl)ethyl]-acetamide; D-threo-*N*-dichloroacetyl-1-*p*-nitrophenyl-2-amino-1,3-propanediol; D(—)-threo-2-dichloroacetamido-1-*p*-nitrophenyl-1,3-propanediol; D-threo-*N*-(1,1'-dihydroxy-1-*p*-nitrophenylisopropyl)dichloroacetamide; numerous other synonyms). a very toxic broad-spectrum antibiotic originally isolated from the bacterium *Streptomyces venezuelae*, later from other spirochaetes and from *Lunatia heros* (moon snail), and now produced synthetically. It is chiefly used in the treatment of typhus and other rickettsial infections, but is also effective against a broad range of bacteria, including *Brucella abortus*, *Staphylococcus aureus*, and Friedlander's bacillus. It is no longer favored for therapeutic use in most cases because it can cause blood dyscrasias in humans, including granulocytopenia, thrombocytopenia, and severe, even fatal, aplastic anemia. It can cause acute myelogenous leukemia. Chloramphenicol is, however, widely used experimentally as a powerful inhibitor of bacterial protein synthesis. See also chloromycetin.

chloranil (2,3,5,6-tetrachloro-2,5-cyclohexadiene-1,4-dione; 2,3,5,6-tetrachloro-*p*-benzoquinone; tetrachloroquinone). a yellow crystalline substance used as a fungicide and as a reagent in the assay of pamaquine in urine. It is irritating to the skin and mucous membranes and is carcinogenic to some laboratory animals.

chlorate. **1:** a negative ion, ClO_3^-, derived from chloric acid. **2:** any salt of chloric acid. Chlorate salts are generally less toxic than bromates. They are moderately toxic orally, with symptoms that may include diarrhea, vomiting, abdominal pain, methemoglobinemia, intravascular hemolysis, cyanosis, jaundice, delirium, convulsions, and coma. Death may ensue from respiratory arrest of acute renal failure. Chronic poisoning may produce kidney and heart damage. Chlorates are powerful oxidizing agents. They present a fire hazard and are explosive when exposed to heat, friction, or shock, especially when contaminated with organic materials or various reducing agents. **3:** to chlorinate.

chlorate salt. See chlorate.

chlorated. chlorinated.

chlorazine (2-chloro-4,6-bis(diethylamino)-1,3,5-triazine). a solid triazine herbicide that is insoluble in water but soluble in alcohols, hydrocarbons, and ketones.

chlorcholine chloride (CCC). a plant growth retardant. It may act by inhibiting gibberellin synthesis. See also mevalonic acid.

chlordane (1,2,4,5,6,7,8,8-octachloro-2,3,-3a,4,7,7a-hexahydro-4,7-methano-1H-indene; 1,2,4,5,6,7,8,8-octachloro-3a,4,7,7a-tetrahydro-4,7-methanoindan; 1,2,4,5,6,7,8,8-octachloro-4,7-methane-3a,4,7,7a-tetrahydroindane). a colorless or pale yellow, viscous, flammable, liquid chlorinated diene hydrocarbon (a cyclodiene), $C_{10}H_6Cl_8$. It is a contact and stomach poison used as an insecticide and fumigant. Its use was discontinued in the United States in 1987. Chlordane does not readily degrade in the environment. It is lipophilic and accumulates in body fat and bioaccumulates in food chains. Poisoning of mammals and birds can occur via ingestion, inhalation, or contact. Chlordane is mildly to very toxic to laboratory animals and is slightly more toxic to humans than DDT. Large doses are carcinogenic to some laboratory animals and probably to humans. It is a cumulative poison, and repeated or chronic exposure may damage the liver, kidneys, and spleen. Symptoms of acute intoxication in vertebrate animals include irritability, hyperexcitability, tremors, and convulsions punctuated by deep depression. Even relatively small doses produce respiratory, gastrointestinal, and central nervous system effects with coughing, nausea, vomiting, blurred vision, confusion, delirium, abdominal pain, convulsions, and perhaps blindness and paralysis.

chlordecone (1,1a,3,3a,4,5,5a,5b,6-decachloro-octahydro-1,3,4-metheno-2*H*-cyclobuta[*cd*]-pentalen-2-one; Kepone; CD). a hepatotoxicant, nephrotoxicant, and neurotoxicant, chlordecone is used as an insecticide to control ants and roaches and as a fungicide. Its use was largely terminated following contamination (first noted in bald eagle and osprey eggs) of the James River in the United States. Chlordecone is very toxic to laboratory rats and causes hepatocellular carcinoma in laboratory rats and mice. Its neurotoxic effects are due to blocking of Na^+/K^+-ATPase. Cholestyramine, a nonabsorbed anion exchange resin, increases the rate of elimination of chlordecone. See also poisoning (chlordecone poisoning).

chlordecone poisoning (Kepone poisoning). poisoning by chlordecone; intoxication of animals is characterized by neurological signs and symptoms such as tremors and exaggerated startle response. Factory workers exposed to this substance experienced chest pain, weight loss, skin rash, muscular weakness, incoordination, arthralgia, slurred speech, nervousness, visual and auditory hallucinations, anxiety, irritability, and short-term memory loss. See also chlordecone.

chlordiazepoxide (7-chloro-*N*-methyl-5-phenyl-3*H*-1,4-benzodiazepin-2-amine 4-oxide; 7-chloro-2-methylamino-5-phenyl-3*H*-1,4-benzodiazepine 4-oxide; metaminodiazepoxide; methaminodiazepoxide; clopoxide). a benzodiazepine, *q.v.*, used as an anxiolytic and a veterinary tranquilizer. It is a controlled substance (a depressant) listed in the U.S. Code of Federal Regulations (CFR).

chlordimeform (*N*'-(4-chloro-2-methylphenyl)-*N*,*N*-dimethylmethanimidamide; *N*'-(4-chloro-*o*-tolyl)-*N*,*N*-dimethylformamidine; chlorophenamidine; chlorphenamidine; chlorophedine; bermat; spanon; CDM). a formamidine neurotoxicant, insecticide, and acaricide that is effective against a wide range of pests that are resistant to carbamates and organophosphates. It is an effective ovicide against a variety of lepidopterous insects. It is an octopaminergic agonist in insects, reversibly inhibiting monoamine oxidase. Chlordimeform is an adrenergic agonist/antagonist (low/high doses) in mammals. It is very toxic to male laboratory rats, somewhat less toxic to female rats, and is carcinogenic to some laboratory animals. Symptoms of acute intoxication in humans include dermatitis, hemorrhagic cystitis, abdominal pain, dysuria, nocturia, hematuria, and proteinuria. Death from respiratory collapse may ensue.

chlorendic acid. a white crystalline substance, $C_9H_4Cl_6O_4$, used in fire-resistant polyester resins and as an intermediate in the synthesis of dyes, fungicides, and insecticides. It is carcinogenic to some laboratory animals.

chlorethene. vinyl chloride.

chlorethene homopolymer. polyvinyl chloride.

chlorethyl. ethyl chloride.

chlorethylene. vinyl chloride.

chlorfenac (2,3,6-trichlorobenzeneacetic acid; 2,3,6-trichlorophenylacetic acid; fenac). a herbicide used to control annual weeds, *Agropyron repens*, and seedling perennials in maize and *Beta vulgaris*. It is moderately toxic to laboratory rats.

chlorfenethol (4-chloro-α-(4-chlorophenyl)-α-methylbenzenemethanol; 4,4'-dichloro-α-methylbenzhydrol; *p,p*'-dichlorodiphenylmethyl carbinol; 1,1-bis(*p*-chlorophenyl)ethanol; 1,1-bis-(*p*-chlorophenyl)methyl carbinol; DMC). a colorless, crystalline, water-insoluble compound used as a miticide on ornamental plants with effects similar to those of DDT. It is very toxic to laboratory rats.

chlorfenvinphos (phosphoric acid 2-chloro-1-(2,4-dichlorophenyl diethyl ester; *O,O*-diethyl-*O*-[2-chloro-1-(2,4-dichlorophenyl)vinyl] phosphate; 2,4-dichloro-α-(chloromethylene)benzyl alcohol diethyl phosphate; CVP). an amber, liquid insecticide and acaricide used mainly to control flies, lice, mites, and ticks on cattle. The minimum oral toxic dose for cattle is about 22 mg/kg body weight. The acute oral LD_{50} for laboratory rats is 10-39 mg/kg body weight. It is also toxic to fish.

chlorhydrin. chlorohydrin.

α-chlorhydrin. α-chlorohydrin. See chlorohydrin.

chloric acid. **1:** a colorless liquid, $HClO_3$, with a pungent odor. A strong acid formed by the action of dilute sulfuric acid on barium chlorate, it is a strong oxidizing agent and bleach. Concentrated solutions ignite paper and other organic materials; it decomposes at 40°C. **2:** also properly referred to as chloric acids: hypochlorous acid, chlorous acid, perchloric acid.

chloric (I) acid. hypochlorous acid.

chloride (chloride salt). any compound that contains chlorine at a valence of -1 together with an electropositive element such as sodium or potassium. Transition metal chlorides have some covalent bonding. Many chloride salts are toxic. Inorganic metal chlorides, however, are toxic only if the metal ion is toxic. Chlorides that release hydrogen (e.g., HCl) are generally toxic. Symptoms of intoxication in humans may include nausea, vomiting, intense abdominal pain, violent hematemesis, a slow irregular pulse, tinnitis, dizziness, convulsions, or paralysis. Death may follow from respiratory or cardiovascular collapse. Lesions may include violent peristalsis, cardiac disturbances, atrial hypertension, and kidney damage.

chloride of copper. cuprous chloride.

chloride of lime. chlorinated lime.

chloride salt. See chloride.

chloridimetry. the process of determining the amount of chlorides in blood, urine, or other fluids.

chloridometer. an apparatus used for chlorodimetry.

chloriduria. chloruresis.

chlorinate (chlorate). to treat with chlorine or a chlorate.

chlorinated (chlorated). treated with, combined with, or containing chlorine or a compound of chlorine.

chlorinated aromatic hydrocarbon. See hydrocarbon.

chlorinated biphenyl. polychlorinated biphenyl.

chlorinated camphene. toxaphene.

chlorinated camphene compound. toxaphene.

chlorinated cyclodiene insecticide. See cyclodiene insecticide.

chlorinated diphenyl. polychlorinated biphenyl.

chlorinated diphenylene. polychlorinated biphenyl.

chlorinated ethane. any compound with an ethane nucleus in which chlorine is substituted for one or more hydrogen atoms. They are carcinogenic in some laboratory animals.

chlorinated hydrocarbon. any hydrocarbon in which chlorine is substituted for one or more hydrogen atoms. Many are very to highly toxic and environmentally persistent, cumulative poisons; some are carcinogenic. See, for example, hexachlorophene, methoxychlor, pentachlorophenol, organochlorine insecticide, herbicide.

chlorinated hydrocarbon insecticide. organochlorine insecticide.

chlorinated lime (bleaching powder, chloride of lime). a complex mixture of calcium calcium hypochlorite, $Ca(OCl)_2 \cdot H_2O$, calcium hydroxide, $Ca(OH)_2$, chloride, $CaCl_2$, and water, H_2O, combined in varying proportions. It is a variable and relatively unstable carrier of chlorine and decomposes rapidly on exposure to moist air. Commercial powders do not contain more than about 39% and the content is usually much lower. It is used as a bleaching agent and disinfectant of drinking water and sewage. Chlorinated lime is also used to decontaminate mustard gas (See gas) and similar substances. Inhalation of the fumes can damage the lungs and upper respiratory tract, causing pulmonary edema and death. Strong solutions can damage the skin and ingestion can cause serious injury to the mucous membranes of the mouth, esophagus, and stomach.

chlorinated naphthalene. See naphthalene.

chlorinated naphthalene poisoning. See hyperkeratosis.

chlorinated organic insecticide. See organochlorine insecticide.

chlorinated phenol. See phenol.

chlorinated propane. See propane.

chlorinated propene. a class of irritant organic compounds with a pungent, repulsive odor; avoidance is thus a typical response of exposed subjects. All are irritants to the eyes, skin, and respiratory tract. Contact with the skin can produce a dermatosis which can be severe with blistering and burns.

chlorination. **1:** the introduction of chlorine into a compound. **2:** to disinfect, oxidize, or destroy undesirable compounds and microorganisms in drinking water, sewage, or industrial waste by adding chlorine. Chlorination can effectively disinfect drinking water. However, water so treated may contain a number of potentially carcinogenic halogenated hydrocarbons such as bromochloromethane, bromoform, chloroform, carbon tetrachloride, dibromochloromethane, and 1,2-dichloroethane.

chlorindanol (7-chloro-4-indanol). a spermicide.

chlorine (Cl)[Z = 17; A_r = 35.453]. a reactive, highly toxic, greenish-yellow, gas (m.p. 101°C; b.p. -34.5°C), Cl_2, that is manufactured in large quantities. A halogen, it occurs naturally in seawater, salt lakes, and in underground halite deposits (salt domes). Chlorine is a strong oxidizing agent, disinfectant and bleaching agent, used mainly as hypochlorite or chlorinated water. It is used also in the detinning and dezincing of iron, in the manufacture of chlorinated lime, and the synthesis of organochlorine solvents and many other organic and inorganic compounds. Large amounts of liquified Cl_2 are transported by rail, and exposure of humans to chlorine from accidents is not uncommon. Chlorine is highly corrosive and very damaging to exposed tissue, including pulmonary tissues. It is also asphyxiant, causing choking and edema of the mucous membranes of the respiratory tree. Subacute and chronic exposures are characterized by coughing, dyspnea, chest pain, cyanosis, pulmonary edema, rales, and pneumonia. Systemic effects commonly include nausea, vomiting, anxiety, and syncope. Chlorine was used as a chemical warfare agent early in World War I. Exposure to concentrations of 10-20 ppm Cl_2 is instantly irritating; even brief exposures to 1000 ppm can be fatal. The 1 hr LC_{50} values for inhalation are 137 and 293 ppm (by volume) for laboratory mice and rats, respectively. Chlorides that release hydrogen (e.g., HCl) are generally toxic, but inorganic metal chlorides are toxic only if the metal ion is toxic. Chlorinated hydrocarbons are among the most toxic and environmentally persistent of organic compounds. Many chlorine-containing compounds are hepatotoxic and/or hepatocarcinogenic. See also chlorides, chlorinated hydrocarbons, chlorinated lime, DDT, TCDD. **combined available chlorine**. a relatively weak disinfectant that persists longer in water than stronger disinfectants such as Cl_2, HOCl, or OCl^-. See chloramine.

chlorine-containing peptide. cyclochlorotine.

chlorine dioxide (chlorine oxide; chlorine peroxide; chloroperoxyl; chloryl radical). an orange gas or orange-red crystals, ClO_2, it is a powerful oxidizing agent and is explosive in the presence of reducing agents. Concentrations in air above 10% are explosive. Chlorine dioxide is commercially the most important of the halogen oxides. It is used chiefly as a disinfectant, in the bleaching of wood pulp, and in the control of odors. It is an important halogen oxide, used chiefly as a disinfectant and to bleach cellulose (wood pulp), to purify water, and to control odors. Because of its extreme instability, it is prepared (from chlorine and sodium chlorite) at the location where it is to be used. Cl^- is released on decomposition by heating. Chlorine dioxide is moderately toxic when inhaled; it is an irritant of the eye and mucous membranes, may cause reproductive effects, and may be mutagenic.

chlorine fluoride. chlorine monofluoride.

chlorine group. the halogens.

chlorine heptoxide (dichlorine heptoxide; perchloric anhydride). an important halogen oxide, Cl_2O_7, it is a colorless, very volatile irritant oil. It is the most stable chlorine oxide but explodes violently on contact with iodine, a flame, or upon concussion.

chlorine monofluoride (chlorine fluoride). a colorless, corrosive gas, ClF, that reacts violently with water, rapidly erodes glass, and in the presence of moisture, quartz. It ignites organic matter on contact and is an extremely hazardous, highly irritant substance that can damage the eyes, skin, and mucous membranes on contact, and the respiratory tract if inhaled.

chlorine monoxide (dichlorine monoxide; dichloromonoxide; dichloroxide; hypochlorous anhydride). an extremely hazardous yellowish-brown gas, Cl_2O, with a disagreeable odor that is used as a chlorinating agent. It explodes on contact with a spark, flame, or organic matter. It is a highly irritant substance that attacks the eyes, skin, mucous membranes, and respiratory tract.

chlorine oxide. chlorine dioxide.

chlorine peroxide. chlorine dioxide.

chlorine tetroxyfluoride. fluorine perchlorate.

chlorine trifluoride. an extremely reactive, corrosive, colorless gas, ClF_3, with a sweetish suffocating odor. It is used as a fluorinating agent. Chlorine trifluoride is strongly irritating to the eyes, skin, mucous membranes, and upper respiratory tract.

chlorine war gas, chlorine warfare gas. chlorine gas packaged for release against enemy troops in combat. See warfare agent (chemical warfare agent).

chlorinity. **1a:** a measure of the amount or concentration of dissolved chlorine or other halogen in water, especially salt water. **1b:** total weight of dissolved chlorine per unit volume of solvent. *Cf.* salinity.

chlormequat chloride (2-chloro-*N,N,N*-trimethylethanaminium chloride; (2-chloroethyl)trimethylammonium chloride; chlorocholine chloride; choline dichloride). a phytocidal bipyridilium quaternary ammonium salt used as a plant growth regulator. It is a white, water-soluble, very hygroscopic, crystalline solid, with a fishlike odor, $[ClCH_2CH_2N^+\text{-}(CH_3)_3]Cl$. It is extremely toxic to humans and to laboratory mice and is corrosive in aqueous solution.

chlormethine. mechlorethamine.

chlornaphazine (*N,N*-bis(2-chloroethyl)-2-naphthylamine; dichloroethyl-β-naphthylamine; β-naphthyldi(2-chloroethyl)amine; β-naphthylbis(β-chloroethyl)amine; di(2-chloroethyl)-β-naphthylamine). a compound that is carcinogenic to laboratory animals and humans.

chloroacetic acid (chloroethanoic acid; monochloroacetic acid; MCA). a herbicide used also in the manufacture of various organic chemicals. Chloroacetic acid and its salts (chloroacetates) are very toxic to laboratory animals and presumably to humans. It is irritating to the skin and mucous membranes. It has been used primarily as a contact preemergent herbicide and defoliant to control weeds in *Brassica oleracea* (kale), *B. oleracea* var. *capitata* (cabbage), *Allium ampeloprasum* var. *porrum* (leeks), and sprouts. Effects of intoxication on mammals include anuria, clonic and tonic convulsions, and respiratory depression.

chloroacetone (1-chloro-2-propanone; chloracetone; monochloroacetone; monochloracetone; acetonylchloride; chloropropanone; 1-chloro-2-ketopropane; 1-chloro-2-oxopropane). a liquid lacrimator with a pungent odor. Chloroacetone is extremely irritating to the eyes and mucous membranes. It is extremely toxic to laboratory mice and rats orally and by inhalation. It is used industrially as an enzyme inactivator and in the synthesis of various drugs and perfumes. It is also used in insecticide formulations and may also be a component of certain military gases.

α-chloroacetophenone. ω-chloroacetophenone.

ω-chloroacetophenone (2-chloro-1-phenylethanone; 2-chloroacetophenone; α-chloroacetophenone; phenacyl chloride). a nearly water-insoluble crystalline solid used as an intermediate in organic syntheses, in riot control, and as a chemical warfare agent. It is a lacrimator and sensitizing agent to humans. It is a potent irritant of the skin, eye, and mucous membranes and may cause corneal damage. *Cf.* *o*-chlorobenzylidenemalononitrile.

2-chloroacetophenone. ω-chloroacetophenone.

chloroacetylchloride. a corrosive and toxic acid.

γ-chloroallyl chloride. 1,3-dichloropropene.

2-chloroallyldiethyldithiocarbamate. sulfallate.

chloroalonil. chlorothalonil.

chloroambucil. chlorambucil.

4-chloro-2-aminotoluene. 5-chloro-*o*-toluidine.

5-chloro-2-aminotoluene hydrochloride. 4-chloro-2-toluidine hydrochloride.

chloroaniline. a chlorinated aromatic amine, C_6H_6ClN, with three isomers, all of which are toxic. These compounds are highly toxic and are thought to be more toxic than aniline. They are readily absorbed through the skin of humans and laboratory animals. Major systemic effects include sometimes severe methemoglobinuria and occasionally hematuria. The latter may be due to hemorrhagic cystitis. **2-chloroaniline**. *o*-chloraniline. **3-chloroaniline**. *m*-chloraniline. **4-chloroaniline**. *p*-chloraniline. ***m*-chloroaniline** (3-chloroaniline; 3-chlorobenzenamine; 3-chlorophenylamine; *m*-aminochlorobenzene; 1-amino-3-chlorobenzene; *m*-chlorophenylamine). a nearly water-insoluble liquid, but soluble in many organic solvents. It is toxic by all routes of exposure and is probably mutagenic. ***o*-chloroaniline** (2-chloroaniline; 1-amino-2-chlorobenzene). a nearly water-insoluble liquid, but soluble in acid and most organic solvents. It is toxic by all routes of exposure; it is probably mutagenic. ***p*-chloroaniline** (4-chloroaniline; 1-amino-4-chlorobenzene; 4-chlorobenzenamine; 4-chlorophenylamine). a crystalline solid that is soluble in hot water and freely soluble in ethanol, acetone, and carbon disulfide. It is a skin and eye irritant and a systemic poison by all routes of exposure. It is very toxic, orally, to laboratory rats. *Cf.* chloronitrobenzene.

chloroazotic acid. nitrohydrochloric acid.

***o*-chlorobenzalmalononitrile**. *o*-chlorobenzylidenemalononitrile.

3-chlorobenzenamine. *m*-chloroaniline (See chloroaniline).

4-chlorobenzenamine. *p*-chloroaniline.

6-chloro-1,3-benzodioxol-5-yl)methyl ester. barthrin.

chlorobenzylate (4,4'-dichlorobenzilic acid ethyl ester; 4,4'-dichlorobenzilate; ethyl-4-chloro-α-(4-chlorophenyl)-α-hydroxybenzeneacetate; ethyl-*p,p'*-dichlorobenzilate; ethyl-4,4'-dichlorodiphenyl glycollate; ethyl-4,4'-dichlorophenyl glycollate; ethyl ester of 4,4'-dichlorobenzilic acid; ethyl-2-hydroxy-2,2-bis(4-chlorophenyl)acetate; benz-*o*-chlor; chlorbenzilate). a viscous, slightly water-soluble, liquid pesticide, $C_{16}H_{14}Cl_2O_3$. It is a skin and eye irritant and is moderately toxic orally to humans and laboratory animals. Chlorobenzylate is nephrotoxic, and is carcinogenic to some laboratory animals. Toxic fumes (Cl^-) are released on decomposition by heating.

***o*-chlorobenzylidenemalononitrile** (([(2-chlorophenyl)methylene]propanedinitrile; *o*-chlorobenzalmalononitrile; β,β-dicyano-*o*-chlorostyrene; CS). an extremely toxic, white, slightly water-soluble crystalline solid used as a riot-control agent. It is a more potent irritant than ω-chloroacetophenone, but less disabling.

7-chlorobicyclo-[3.2.0]hepta-2,6-dien-6-yl dimethylphosphate. heptenophos.

chlorobiphenyl. polychlorinated biphenyl.

2-chlorobutadiene-1,3. chloroprene.

4-chlorobut-2-ynyl-*N*-(3-chlorophenyl)-carbamate. barban.

***m*-chlorocarbanilic acid isopropyl ester**. chlorpropham.

chlorocarbonate. See ethyl chlorformate.

1-chloro-2-(β-chloroethoxy)ethane. *sym*-dichloroethyl ether.

2-chloro-*N*-(2-chloroethyl)-*N*-ethylethanamine. HN1.

2-chloro-*N*-(2-chloroethyl)-*N*-methylethanamine. mechlorethamine.

1-chloro-2-(β-chloroethylthio)ethane. mustard gas (See gas).

4-chloro-α-(4-chlorophenyl)-α-methylbenzene methanol. chlorfenethol.

4-chloro-α-(4-chlorophenyl)-α-trichloromethyl)-benzenemethanol. dicofol.

chlorocholine chloride. chlormequatchloride.

chlorochromic anhydride. chromyl chloride.

4-chloro-*o*-cresoxyacetic acid. See MCPA.

chlorodane. chlordane.

α-chloro-*N*,*N*-diallylacetamide. allidochlor.

3-chloro-1,2-dibromopropane. dibromochloropane.

5-chloro-2-(2,4-dichlorophenoxy)phenol. triclosan.

7-chloro-4-(4-diethylamino-1-methylbutylamino)quinoline. chloroquine.

2-chloro-2,6-diethyl-*N*-(butoxymethyl)-acetanilide. butachlor.

2-chloro-2-diethylcarbamoyl-1-methylvinyldimethylphosphate. phosphamidon.

6-chloro-3-(*O*,*O*-diethyldithiophosphorylmethyl)benzoxazolone. phosalone.

2-chloro-2',6'-diethyl-*N*-(methoxymethyl)-acetanilide. alachlor.

2-chloro-*N*-(2,6-diethylphenyl)-*N*-(methoxymethyl)acetamide. alachlor.

10-chloro-5,10-dihydroarsacridine. phenarsazine chloride.

7-chloro-1,3-dihydro-1-methyl-5-phenyl-2*H*-1,4-benzodiazepin-2-one. diazepam.

10-chloro-5,10-dihydrophenarsazine. phenarsazine chloride.

3-chloro-1,2-dihydroxypropane. α-chlorohydrin (See chlorohydrin).

7-chloro-4,6-dimethoxycoumaran-3-one-2-spiro-1'-(2'-methoxy-6'-methylcyclohex-2'-en-4'-one). griseofulvum.

2-[*p*-chloro-α-(2-dimethylaminoethyl)benzyl]-pyridine. chlorpheniramine.

7-chloro-4-dimethylamino-1,4,4a,5,5a,6,11,12a-octahydro-3,6,10,12,12a-pentahydroxy-6-methyl-1,11-dioxo-2-naphthacenecarboxamide. chlorotetracycline.

2-chloro-10-(3-dimethylaminopropyl)phenothiazine. chlorpromazine.

3-chloro-10-(3-dimethylaminopropyl)phenothiazine. chlorpromazine.

chlorodimethyl ether. chloromethyl methyl ether.

2-chloro-*N*,*N*-dimethyl-10*H*-phenothiazine-10-propanamine. chlorpromazine.

2-chloro-*N*,*N*-di-2-propenylacetamide. allidochlor.

1-chloro-2,3-epoxypropane. epichlorhydrin.

chloroethane. ethyl chloride.

chloroethanoic acid. chloroacetic acid.

2-chloroethanol. ethylene chlorohydrin.

chloroethene. vinyl chloride.

(2-chloroethenyl)arsenous dichloride. lewisite.

chloroethyl ether. *sym*-dichloroethyl ether.

2-[[4-chloro-6-(ethylamino)-1,3,5-triazin-2-yl]-amino]-2-methylpropanenitrile. cyanazine.

2-[[4-chloro-6-(ethylamino)-s-triazin-2-yl]-amino]-2-methylpropionitrile. cyanazine.

1-(2-chloroethyl)-3-cyclohexyl-1-nitrosourea. lomustine.

***N*-(2-chloroethyl)-*N*'-cyclohexyl-*N*-nitrosourea**. lomustine.

chloroethylene. vinyl chloride.

***N*-(2-chloroethyl)-*N*-(1-methyl-2-phenoxyethyl)-benzenemethanamine**. phenoxybenzamine.

***N*-(2-chloroethyl)-*N*-(1-methyl-2-phenoxyethyl)-benzylamine**. phenoxybenzamine.

di(2-chloroethyl)-β-naphthylamine. chlornaphazine

1-chloro-3-ethyl-1-penten-4yn-3-ol. ethchlorvynol.

(2-chloroethyl)trimethylammonium chloride. chlormequatchloride.

chloroflavonin. a chlorine-containing antifungal fungal metabolite.

chlorofluorocarbon (CFC). any of a number of easily liquefied compounds formed by the replacement some hydrogen atoms in hydrocarbons by chlorine and some by fluorine. Most are chemically inert and considered nontoxic; they may, however, act as simple asphyxiants. All are manufactured and used chiefly as refrigerants, aerosol propellants, solvents, insulation, and in packaging and the production of plastic foams. These compounds, which are used in large quantities, drift into the upper atmosphere, releasing chlorine, which attacks the ozone layer. For this reason, many (but by no means all) countries have banned or restricted their use. For example, their use in aerosol products was banned in the United States by the Environmental Protection Agency and the Food and Drug Administration in 1977. They are still used widely in the United States, however, as refrigerants, coolants, and industrial solvents. Fatalities from exposure to CFC have been recorded. See, for example, Freon.

chloroform (trichloromethane; methane trichloride; methenyl trichloride; methyl trichloride; TCM). **1:** a heavy, clear, colorless, volatile, nonflammable, toxic, liquid trihalomethane, and common water pollutant, $CHCl_3$, with a sweetish taste and a heavy, ethereal odor. Chloroform is usually produced by the chlorination and oxidation of acetone, or by the chlorination of methane or methyl chloride. Chloroform is a CNS depressant and was the first widely used general anesthetic, but has been gradually replaced by safer, less toxic drugs, since the 1950s. It is now used chiefly as a solvent. A possible source of exposure is drinking water. Chloroform is hepatotoxic, nephrotoxic, and can cause fatal cardiac arrhythmias. Acute exposures of humans to the vapor may cause hypotension, and respiratory or myocardial depression with fatal damage to the liver or heart. Moderate exposures may cause dilation of the pupils of the eyes (from inhalation), irritation of the mucous membranes, nausea, vomiting and other gastrointestinal effects, and may distort perception and produce hallucinations. High levels of chloroform cause cancer in laboratory mice and rats and, when ingested, can cause cancer in humans. Experimental data indicates that chloroform may be teratogenic, mutagenic, and may cause reproductive effects in humans. **2a:** to anesthetize or to make a human or other organism tractable or insensible by the use of chloroform. **2b:** to kill using chloroform.

chloroformic acid ethyl ester. ethyl chloroformate.

chloroformic acid trichloromethylcarbonochloridate. diphosgene.

chloroformic acid trichloromethyl ester. diphosgene.

chloroformism. the habitual inhalation of chloroform or the resulting effects (See chloroform).

chloroformyl chloride. phosgene.

4-chloro-*N*-(2-furfurylmethyl)-5-sulfamoylanthranilic acid. furosemide.

4-chloro-*N*-furfuryl-5-sulfamoylanthranilic acid. furosemide.

Chlorogalum pomeridianum (California soap plant; amole). a plant (Family Lilaceae), the bulb of which is the source of chlorogenin.

chlorogenic acid (3-[[3-(3,4-dihydroxyphenyl)-1-oxo-2-propanyl]oxy]-1,4,5-trihydroxycyclohexanecarboxylic acid; 1,3,4,5-tetrahydroxycyclohexanecarboxylic acid, 3-(3,4-dihydroxycinnamate, 3-caffeoylquinic acid, 3-(3,4-dihydroxycinnamoyl)quinic acid). a plant growth inhibitor that occurs in green coffee beans and in the leaves, fruit, and other tissues of dicotyledenous plants. See also *Helianthus annus*.

chlorogenin. a saponin isolated from the bulbs of *Chlorogalum pomeridianum*, it is used to kill or stun fish without affecting palatability.

chlorohydrin (chlorhydrin). any of a number of compounds obtained by substitution of chlorine for some hydroxyl groups of a group of glycols or polyhydroxy alcohols. **α-chlorohydrin** (3-chloro-1,2-propanediol; 3-chloro-1,2-dihydroxypropane; α-monochlorohydrin; β,β'-dihydroxyisopropyl chloride; glycerol α-monochlorohydrin; 3-chloropropylene glycol; α-chlorhydrin). a clear to straw-colored, sweet liquid, $CH_2ClCHOHCH_2OH$, used as a rodent sterilant. It is a very toxic systemic poison. *Cf.* epichlorohydrin, ethylene chlorohydrin.

5-chloro-8-hydroxy-7-iodoquinoline. iodochlorhydroxyquin.

chloroidoquin. iodochlorhydroxyquin.

7-chloro-4-indanol. chlorindanol.

chloroiodoquin. iodochlorhydroxyquin.

5-chloro-7-iodo-8-quinolinol. iodochlorhydroxyquin.

chloro-IPC. chlorpropham.

α-chloroisobutylene. 1-chloro-2-methylpropene.

τ-chloroisobutylene. 3-chloro-2-methylpropene.

2-chloro-*N*-isopropylacetanilide. propachlor.

1-chloro-2-ketopropane. chloroacetone.

chloromethane (methyl chloride). a CNS depressant and known carcinogen. It was once widely used as a refrigerant and an aerosol propellant. More recent usage is in the manufacture of silicones.

chloromethoxymethane. chloromethyl methyl ether.

chloromethyl methyl ether (chloromethoxymethane; methyl chloromethyl ether; monochloromethyl ether; chlorodimethyl ether; CMME). a colorless, volatile, flammable liquid, CH_3OCH_2Cl. It is highly irritant to the skin and mucous membranes. Exposure is usually via inhalation of the vapor. The technical grade is a known carcinogen to some laboratory animals and may cause lung cancer in exposed human workers (e.g., as in plastic manufacturing plants); it often contains *sym*-dichloromethyl ether, a confirmed human carcinogen. Chemical workers in plastic manufacturing plants are at risk.

4-chloro-2-methylalanine. 4-chloro-*o*-toluidine.

4-chloro-6-methylalanine. 4-chloro-*o*-toluidine.

7-chloro-2-methylamino-5-phenyl-3*H*-1,4-benzodiazepine 4-oxide. chlordiazepoxide.

3-chloro-6-methylaniline. 5-chloro-*o*-toluidine.

5-chloro-2-methylaniline. 5-chloro-*o*-toluidine.

4-chloro-2-methylaniline hydrochloride. 4-chloro-2-toluidine hydrochloride.

4-chloro-6-methylaniline hydrochloride. 4-chloro-2-toluidine hydrochloride.

4-chloro-2-methylbenzeneamine. 4-chloro-*o*-toluidine.

4-chloro-2-methylbenzeneamine hydrochloride. 4-chloro-2-toluidine hydrochloride.

4-chloro-α-(1-methylethyl)benzeneacetic acid cyano(3-phenoxyphenyl)methyl ester. fenvalerate.

2-chloro-*N*-(1-methylethyl)-*N*-phenylacetamide. propachlor.

chloromethyloxirane. epichlorhydrin.

(4-chloro-2-methylphenoxy)acetic acid. MCPA.

7-chloro-1-methyl-5-phenyl-3*H*-1,4-benzodiazepin-2-one. diazepam.

7-chloro-*N*-methyl-5-phenyl-3*H*-1,4-benzodiazepin-2-amine 4-oxide. chlordiazepoxide.

chloro-1-methyl-5-phenyl-3*H*-1,4-benzodiazepin-2(1*H*)-one-7-chloro-1-methyl-5-phenyl-3*H*-1,4-benzodiazepin-2(1*H*)-one. diazepam.

***N'*-(4-chloro-2-methylphenyl)-*N*,*N*-dimethylmethanimidamide.** chlordimeform.

1-chloro-2-methylpropene(α-chloroisobutylene; β,β-dimethylvinyl chloride; isocrotyl chloride). a colorless, volatile liquid, C_4H_7Cl, used in organic syntheses. It is an irritant of the eyes and mucous membranes, has anesthetic activity, and is carcinogenic to some animals. Toxic fumes, Cl^-, evolve on heating to decomposition.

3-chloro-2-methyl-1-propene. 3-chloro-2-methylpropene.

3-chloro-2-methylpropene (CMP; 3-chloro-2-methyl-1-propene; τ-chloroisobutylene; methallylchloride; β-methallylchloride; β-methallyl chloride; 2-methylallyl chloride; isobutanyl chloride). a colorless, volatile liquid with a disagreeable odor, C_4H_7Cl, used as an insecticide, fumigant, and reagent. CMP is an irritant to the eyes and respiratory tract and is mildly toxic by inhalation. It is carcinogenic and probably mutagenic to some laboratory animals. It is a serious fire hazard and emits very toxic Cl^- fumes when heated to decomposition.

3-chloromethylpyridine. a poisonous substance, *per os*, it is also carcinogenic and probably mutagenic to the laboratory mouse and certain other animals. It releases very toxic fumes, NO_x and Cl^-, when heated to decomposition.

chloromycetin. an antibiotic isolated from *Streptomyces venezuele*. It is chemically identical to chloramphenicol, *q.v.*

chloronaphthalene. chlorinated naphthalene. See naphthalene.

chloronitrobenzene. a crystalline aromatic compound, $C_6H_4ClNO_2$, used in the dye industry. There are three isomers: ***m*-chloronitrobenzene** (1-chloro-3-nitrobenzene; *m*-nitrochlorobenzene), ***o*-chloronitrobenzene**, and ***p*-chloronitrobenzene**. All are very toxic by inhalation or ingestion, although the *para*-isomer is probably somewhat less toxic than the *ortho*-isomer. They form methemoglobin, produce cyanosis, and other changes in the blood. They are probably metabolically converted to the corresponding chloroaniline isomers. Serious occupational exposure is usually via inhalation of chloronitrobenzene-containing dusts.

1-chloro-3-nitrobenzene. *m*-chloronitrobenzene (See chloronitrobenzene).

chloronitrous acid. nitrohydrochloric acid.

***S*-(6-chloro-2-oxobenzoxazolin-3-yl)methyl diethyl phosphorothiolothionate.** phosalone.

1-chloro-2-oxopropane. chloroacetone.

chloroperoxyl. chlorine dioxide.

chlorophedine. chlordimeform.

chlorophenamidine. chlordimeform.

chlorophenol. 1: any chlorinated derivative of phenol. All are toxic, and most are potent contact irritants of the eye, skin, and mucous membranes; many are systemic irritants if ingested, inhaled, or contact the skin. Many are suspected carcinogens. See pentachlorophenol, trichlorophenol. **2:** any of three monochlorinated derivatives of phenol (*m*-chlorophenol; *o*-chlorophenol; *p*-chlorophenol) used mainly as intermediates in the synthesis of dyes.

***N*-[[(4-chlorophenol)amino]carbonyl]-2,6-difluorobenzamide.** diflubenzuron.

chlorophenothane. DDT.

chlorophenoxy acid. See, for example, chlorophenoxy compound, chlorophenoxyacetic acid, and chlorophenoxypropionic acid.

chlorophenoxy compound. any of a number of noninsecticidal organohalide pesticides and growth regulators that profoundly affect the growth and structure of plants. They are widely employed in herbicide formulations. Examples are 2,4-D, 2,4,5-T (See also Agent Orange), chlorophenoxyacetic acid, and chlorophenoxypropionic acid. They produce intermediates that cause downward (epinastic) bending of the plant, cessation of growth, tumor formation, and secondary root induction. Meristematic cells fail to divide, and cells which normally elongate enlarge only radially. Parenchyma cells of mature plants swell, divide, and produce callus tissue (i.e., tissue that usually develops only at an injured surface) and expanding root primordia. Roots no longer elongate and root tips swell. Young leaves fail to expand further; roots fail to absorb water and salts; photosynthesis is inhibited and phloem becomes plugged. These compounds modify nucleic acid metabolism in plants and alter numerous enzyme systems in plants. They are generally moderately or very toxic to animals. All should be considered hazardous to humans, some extremely so.

chlorophenoxyacetic acid (*p*-chlorophenoxyacetic acid; 4-chlorophenoxyacetic acid; 4-CP; CPA; PCPA). as with other chlorophenoxy compounds, *q.v.*, chlorophenoxyacetic acid, $C_8H_7O_3Cl$, has profound effects on the growth and structure of plants. It is used as a herbicide and as a growth regulator for tomatoes and peaches. Chlorophenoxyacetic acid is moderately toxic and probably mutagenic, *per os*, to humans. See also chlorophenoxy compound.

4-chlorophenoxyacetic acid. chlorophenoxyacetic acid.

***p*-chlorophenoxyacetic acid**. chlorophenoxyacetic acid.

2-(4-chlorophenoxy)-2-methylpropanoic acid ethyl ester. clofibrate.

chlorophenoxypropionic acid. as with other chlorophenoxy compounds, *q.v.*, chlorophenoxypropionic acid, $C_9H_9O_3Cl$, has profound effects on the growth and structure of plants. It is used as a growth regulator and fruit thinner for plums and prunes.

4-chlorophenylamine. *p*-chloroaniline.

***m*-chlorophenylamine**. *m*-chloroaniline (See chloroaniline).

3-chlorophenylamine. *m*-chloroaniline (See chloroaniline).

(3-chlorophenyl)carbamic acid 1-methylethyl ester. chlorpropham.

(*N*-4-chlorophenyl)-*N*-(3,4-dichlorophenyl)urea. triclocarban.

τ-(4-chlorophenyl-*N*,*N*-dimethyl-2-pyridinepropanamine. chlorpheniramine.

3-(*p*-chlorophenyl)-1,1-dimethylurea.monuron.

***N*'-(4-chlorophenyl)-*N*,*N*-dimethylurea**. monuron.

2-chloro-1-phenylethanone. ω-chloroacetophenone.

4-[4-(4-chlorophenyl)-4-hydroxy-1-piperidinyl]-1-(4-fluorophenyl)-1-butanone. haloperidol.

2-(2-chlorophenyl)-2-(methylamino)cyclohexanone. ketamine.

[(2-chlorophenyl)methylene]propanedinitrile. *o*-chlorobenzylidenemalononitrile.

(4-chlorophenyl)-τ-(2-pyridyl)propyldimethylamine. chlorpheniramine.

1-(*p*-chlorophenyl)-1-(2-pyridyl)-3-*N*,*N*-dimethylpropylamine. chlorpheniramine.

3(*p*-chlorophenyl)-3-(2-pyridyl)-*N*,*N*-dimethylpropylamine. chlorpheniramine.

S-[[(*p*-chlorophenyl)thio]methyl] *O*,*O*-diethyl phosphorodithioate. carbophenothion.

Chlorophyllum molybdites (green-spored lepiota). a very common poisonous mushroom (Family Agaricaceae) that grows in lawns, meadows, and pastures from New York to Florida west to Michigan, Colorado, and California. It is a potent purgative, causing one to several days of violent purging. Symptoms appear within ½-3 hours following ingestion and include severe nausea, vomiting, diarrhea, and abdominal pain. Hospitalization may be necessary.

chloropicrin (nitrochloroform; trichloronitromethane; acquinite; vomiting gas; Chemical Mace). a dense, colorless, acrid, slightly oily, highly toxic liquid, CCl_3NO_2. It is a potent, potentially lethal, skin and pulmonary irritant and lacrymator. It is the active ingredient in tear gas, and is used in grenades and personal protective devices to repel prospective assailants. It is also used as a fungicide, larvacide, soil insecticide, fumigant, war gas, tracer gas, and a disinfectant of cereal grains. A 10 minute exposure to a concentration of 0.85 mg/l is probably lethal to humans.

6-chloropiperonylchrysanthemumate. barthrin.

6-chloropiperonyl-2,2-dimethyl-3(2-methylpropenyl)cyclopropanecarboxylate. barthrin.

chloroprene (2-chlorobutadiene-1,3). a colorless, slightly water-soluble, liquid, chlorinated propene, $H_2C{:}CHCCL{:}CH_2$, produced in large amounts for use in the manufacture of neoprene. It is a colorless, flammable liquid which is strongly irritant to the skin and mucous membranes of the eye and pulmonary system. It can cause a contact dermatitis in exposed workers with alopecia in the affected areas. Inhalation of small amounts can cause CNS depression and irreversible, sometimes lethal damage to vital tissues. Symptoms of intoxication may include nervousness and irritability. Chloroprene is also immunosuppressive and is carcinogenic to humans.

3-chloro-1,2-propanediol. α-chlorohydrin (See chlorohydrin).

3-chloropropanenitrile. β-chloropropionitrile.

chloropropanone. chloroacetone.

1-chloro-2-propanone. chloroacetone.

3-chloropropanonitrile. β-chloropropionitrile.

3-chloro-1-propene. allyl chloride.

chloropropham. chlorpropham.

chloroprophenpyridamine. chlorpheniramine.

β-chloropropionitrile (3-chloropropanenitrile; 3-chloropropanonitrile). a liquid, $ClCH_2CH_2CN$, with a characteristic acrid odor, and miscible with a variety of organic solvents. It is used in the synthesis of pharmaceuticals and polymers. It is highly toxic by all routes of exposure, readily penetrating the skin and causing systemic cyanide poisoning and death. On heating to decomposition, it releases toxic fumes (HCl). It is somewhat less hazardous than acrylonitrile only because it is somewhat less volatile.

3-chloropropylene. allyl chloride.

3-chloropropylene glycol. α-chlorohydrin (See chlorohydrin).

γ-chloropropylene oxide. epichlorhydrin.

chloropsia (green vision). a condition in which objects are perceived as green in color, as in digitalis intoxication.

chloroquine (N^4-(7-chloro-4-quinolinyl)-N^1,N^1-diethyl-1,4-pentanediamine; 7-chloro-4-(4-diethylamino-1-methylbutylamino)quinoline). a white or yellowish crystalline powder used (chiefly as the diphosphate) as an antimalarial. Its use is associated with an especially high occurrence of injuries to the visual sytem. Dose-related retinopathies and damage to hearing may result. It concentrates in the liver; its use is thus contraindicated in persons with liver disease. Chloroquine is used also as an antiamebic in cases of extraintestinal amebiasis, as an antirheumatic, and as a suppressant of lupus erythematosus.

chloroquinol (Enterovioform). a toxic antidiarrheal that has been used widely in Japan. It is a cause of subacute myelo-optic neuropathy characterized by optic nerve damage and an associated stiffness in the joints.

N^4-(7-chloro-4-quinolinyl)-N^1,N^1-diethyl-1,4-pentanediamine. chloroquine.

chlorosilane. See silane.

chlorosis. **1:** a condition of unhealthy green plants whereby chlorophyll levels drop and normally green structures such as leaves become pale, yellow, or even white in color. The condition may be due to pollutants, toxic substances, disease (esp. certain virus infections), mineral deficiencies (esp. of iron, magnesium, or copper), or to poor lighting. **2:** a disorder mainly of adolescent women of the 19th Century marked by hypochromic erythrocytes and a greenish-yellow skin discoloration.

chlorosity. the chlorine content per liter of seawater. See also chlorinity.

chlorosulfuric acid. sulfuryl chloride.

chlorotetracycline (7-chloro-4-dimethylamino-1,4,4a,5,5a,6,11,12a-octahydro-3,6,10,12,12a-pentahydroxy-6-methyl-1,11-dioxo-2-naphthacenecarboxamide; 7-chlorotetracycline). a golden-yellow, crystalline, chlorine-containing antimicrobial, antiamebic fungal metabolite. It is hepatotoxic and nephrotoxic, but only slightly toxic to laboratory rats. *Cf.* Aureomycin.

7-chlorotetracycline. chlorotetracycline.

chlorothalonil (2,4,5,6-tetrachloro-1,3-benzenedicarbonitrile; 2,4,5,6-tetrachloro-3-cyanobenzonitrile; tetrachloroisophthalonitrile; *m*-tetrachlorophthalodinitrile; *m*-TCPN; 2,4,-5,6-tetrachloro-1,3-dicyanobenzene; 1,3-(4-dicyano-2,4,5,6-tetrachlorobenzene; 1,3-dicyanotetrachlorobenzene; chloroalonil; chlorthalonil). a nearly water-insoluble crystalline substance used as a bactericide, nematocide, and as an agricultural horticultural fungicide. It is slightly toxic orally to laboratory rats, and moderately toxic, *i.p.*, but is carcinogenic to some laboratory animals. Very toxic fumes, NO_x, Cl^-, and CN, are released when it is heated to decomposition.

(4-chloro-*o*-toloxy)acetic acid. MCPA.

3-chloro-*p*-toluidine (CPT). a selective nephrotoxic (sometimes hepatotoxic) avicide. It is extremely toxic to most species of birds but is only slightly toxic to the laboratory rat. Acute exposure produces renal necrosis in all sensitive species and, in quail, a focal hepatic necrosis also occurs. CPT causes methemoglobinemia in laboratory rats and rabbits. Death in some species appears to result from a complex of causes and may sometimes be due to cardiovascular and respiratory effects. Renal deacetylase activity is high in the very sensitive starling and domestic fowl, but is undetectable in the nonsusceptible *Buteo jamaicensis* (red-tailed hawk) and laboratory rat. **4-chloro-2-toluidine**. 4-chloro-*o*-toluidine.

4-chloro-2-toluidine hydrochloride (2-amino-5-chlorotoluene hydrochloride; 5-chloro-2-aminotoluene hydrochloride; 4-chloro-2-methylaniline hydrochloride; 4-chloro-6-methylaniline hydrochloride; 4-chloro-2-methyl-benzeneamine hydrochloride; 4-chloro-*o*-toluidine hydrochloride; 2-methyl-4-chloro-aniline hydrochloride). a substance, $C_7H_8ClN{\cdot}ClH$, that is moderately toxic, *i.p.*, and carcinogenic to some laboratory animals.

4-chloro-*o*-toluidine (2-amino-5-chlorotoluene; 4-chloro-2-methylalanine; 4-chloro-6-methylalanine; 4-chloro-2-methylbenzeneamine; 4-chloro-2-toluidine). a moderately toxic substance, C_7H_8ClN, when ingested. It is carcinogenic to some laboratory animals and is a confirmed human carcinogen and probable mutagen. *Cf.* 4-chloro-2-toluidine hydrochloride.

5-chloro-*o*-toluidine (1-amino-3-chloro-6-methylbenzene; 2-amino-4-chlorotoluene; 4-chloro-2-aminotoluene; 3-chloro-6-methylaniline; 5-chloro-2-methylaniline). a moderately toxic substance when ingested; carcinogenic to some laboratory animals.

4-chloro-*o*-toluidine hydrochloride. 4-chloro-2-toluidine hydrochloride.

***N*'-(4-chloro-*o*-tolyl)-*N*,*N*-dimethylformamidine**. chlordimeform.

2-chloro-1-(2,4,5-trichlorophenyl)vinyl dimethyl phosphate. stirofos.

2-chloro-1,1,2-trifluoroethyl difluoromethyl ether. enflurane.

5-[2-chloro-4-(trifluoromethyl)-phenoxy]-2-nitrobenzoic acid). acifluorfen.

7-chloro-2',4,6-trimethoxy-6'-methylspiro-[benzofuran-2(3*H*),1'-[2]cyclohexene]-3,4'-dione. griseofulvum.

2-chloro-*N*,*N*,*N*-trimethylethanaminium chloride. chlormequatchloride.

chlorovinylarsine dichloride. lewisite.

2-chlorovinyldichloroarsine. lewisite.

β-chlorovinyldichloroarsine. lewisite, the legal label name is in more common usage. See Lewisite.

chlorphenamidine. chlordimeform.

chlorphenamine. chlorpheniramine.

chlorpheniramine (τ-(4-chlorophenyl-*N*,*N*-dimethyl-2-pyridinepropanamine; 2-[*p*-chloro-α-(2-dimethylaminoethyl)benzyl]pyridine; 1-(*p*-chlorophenyl)-1-(2-pyridyl)-3-*N*,*N*-dimethylpropylamine; 3(*p*-chlorophenyl)-3-(2-pyridyl)-*N*,*N*-dimethylpropylamine; τ-(4-chlorophenyl)-

γ-(2-pyridil)propyldimethylamine; chloroprophenpyridamine; chlorphenamine). an antihistaminic, it is very toxic to laboratory mice.

chlorphenothane. DDT.

chlorpromazine (2-chloro-*N,N*-dimethyl-10*H* phenothiazine-10-propanamine;2-chloro-10-(3-dimethylaminopropyl)phenothiazine; 3-chloro-10-(3-dimethylaminopropyl)phenothiazine; *N*-(3-dimethylaminopropyl)-3-chlorophenothiazine; Thorazine). a very toxic phenothiazine derivative used medically as an antipsychotic, especially to control manic-depressive individuals and certain severe behavior problems; used also to treat hiccups, tetanus, and as an antiemetic drug. It is often given by rectal suppository, although the hydrochloride (very toxic to laboratory rats) is administered orally, *i.m.*, or *i.v.* It is used in veterinary practice as a sedative and peripheral vasodilator. Overdosage in humans may cause drowsiness, fainting, hypotension, tachycardia, hypothermia, tremor, dizziness, convulsions, stupor, coma, areflexia. Extrapyramidal effects such as restlessness, slobbering, unsteady gait, stuttering, rigid back muscles, contraction of face and neck muscles, and hand tremors commonly occur. Damage to the liver is common. Sudden death from respiratory or cardiovascular collapse may result from large doses of chlorpromazine or other phenothiazines (See phenothiazine sudden death).

chlorpropham ((3-chlorophenyl)carbamic acid 1-methylethyl ester; *m*-chlorocarbanilic acid isopropyl ester; isopropyl-*m*-chlorocarbanilate; isopropyl*N*-(3-chlorophenyl)carbamate; chloro-IPC; chloropropham; CIPC). a carbamate herbicide and plant growth regulator, it is the 3-chloro derivative of propham. It inhibits gibberellin-induced α-amylase synthesis in barley endosperm, and is thus thought to inhibit gene derepression (i.e., the onset of metabolic changes) but not processes that are already in progress. Chlorpropham is generally effective against monocotyledons, but not dicotyledons, and is used to control *Digitaria* (crabgrass). Chlorpropham is moderately toxic to laboratory rats. See also propham.

chlorprophenpyridamine. chlorpheniramine.

chlorpyrifos (phosphorothioic acid *O,O*-diethyl *O*-(3,5,6-trichloro-2-pyridinyl) ester; *O,O*-diethyl *O*-3,5,6-trichloro-2-pyridyl phosphorothioate; chlorpyrifosethyl). chlorpyrifos and its *O,O*-dimethyl analog, chlorpyrifos methyl) are used as insecticides and acaricides; it is a popular termiticide. The LD_{50} of chlorpyrifos in goats has been reported as 500 mg/kg body weight orally and 97 mg/kg in the laboratory rat. Bulls in comparison to calves, steers, and cows are highly susceptible to a single dose. This agent in pour-on preparations has poisoned bulls, particularly of exotic breeds. Chlorpyrifos, is extremely toxic to fish, birds, and aquatic invertebrates; and is very toxic to laboratory rats. See also termiticide.

chlorpyrifos ethyl. chlorpyrifos.

chlorpyrifos methyl. the *O,O*-dimethyl analog of chlorpyrifos.

chlortetracycline (7-chloro-4-dimethylamino-1,4,4a,5,5a,6,11,12a-octahydro-3,6,10,12,12a-pentahydroxy-6-methyl-1,11-dioxo-2-naphthacenecarboxamide; 7-chlorotetracycline). one of the best known tetracyclines (from *Streptomyces aureofaciens*), it is a golden-yellow, crystalline, chlorine-containing antimicrobial, antiamebic, fungal metabolite. It is hepatotoxic; only slightly toxic to laboratory rats. *Cf.* Aureomycin.

chlorthal-dimethyl (DCPA; 2,3,5,6-tetrachloro-1,4-benzenedicarboxylic acid dimethyl ester; 2,3,5,6-tetrachlorotere-phthalic acid dimethyl ester; dimethyl 2,3,5,6-tetrachloroterephthalate; chlorthal methyl). a toxic aromatic compound, $C_{10}H_6Cl_4O_4$, used as a preemergent herbicide. It inhibits the rooting of stolons in *Cynodon dactylon* by completely inhibiting cell division in the root primordia. Cell enlargement may continue for a while, however. Chlorthal-dimethyl blocks mitosis in maize and onion roots at metaphase and the meristematic tissues become disorganized with the development of many hypertrophied cells.

chlorthal-methyl. chlorthal-dimethyl.

chlorthalonil. chlorothalonil.

chlorthiepin. endosulfan.

chloruresis (chloruria; chloriduria). the excretion of chloride in the urine.

chloruretic. **1a:** of or pertaining to an agent that increases the excretion of chloride in the urine. **1b:** a chloruretic agent. **2:** pertaining or denoting chloruresis or an agent with the capacity to induce chloruresis.

chloruria. chloruresis.

β-chlorvinyl ethylethynylcarbinol. ethchlorvynol.

chloryl radical. chlorine dioxide.

CHO. Chinese hamster ovary cells.

choichodo. See *Bitis arietans*, *Causus rhombeatus*.

chokes, the. the bends when marked by severe dyspnea or suffocation.

choking gas (casualty gas). any war gas that irritates and inflames the bronchial tubes and lungs (e.g., phosgene).

choking staggers. See moldy corn poisoning.

cholangitis. inflammation of bile vessels.

cholanthrene (1,2-dihydrobenz[j]aceanthrylene). a pentacyclic, carcinogenic hydrocarbon, it is the structural precursor of 3-methylcholanthrene, a potent carcinogen.

cholera. **1:** any condition characterized by profuse vomiting and diarrhea (seldom used) **2:** (Asiatic cholera). an acute epidemic infectious disease of humans caused by *Vibrio cholerae* (*V. comma*), a comma-shaped bacterium. Symptoms appear suddenly one to five days following infection. These include vomiting; profuse, watery diarrhea; and dehydration. The loss of fluids and electrolytes may be intense and untreated patients may suffer extreme dehydration, shock, rapid collapse, and death with no lesions of the intestinal mucosa. The fluid loss results from the action of a toxin, choleragen, that promotes the passage of fluid from circulating blood into the large and small intestines. Infection is always from consumption of food or water that is contaminated by the vibrio. See choleragen, the cholera toxin. Originally confined to Asia, cases are now seen worldwide. A few cases occur annually in the United States. See also choleragen, *Vibrio cholerae*.

cholera vibrio. the bacterium, *Vibrio cholerae* (*V. comma*) that causes cholera. *Cf.* vibrio.

choleragen (cholera toxin). the soluble exotoxin secreted in the intestinal tract by the cholera vibrio. It is an 87,000-Dalton protein with an A_1 peptide subunit coupled to an A_2 subunit and five B subunits. Choleragen apparently alters the permeability of the mucosa of the small intestine by increasing the adenylate cyclase activity of the mucosa of the small intestine. The cyclic AMP concentration within the cells is thus increased and active transport of ions by these epithelial cells is stimulated, with a consequent increase in the efflux of Na^+, other electrolytes, and water into the gut. This produces a profuse watery diarrhea. The process is self-sustaining because choleragen also blocks the GTPase activity of the cell, effectively destroying the adenylate deactivation mechanism, and allowing continual production of cyclic AMP. The loss of fluids and electrolytes may be intense and the untreated patient may suffer extreme dehydration, shock, and collapse with no lesions of the intestinal mucosa. See cholera, *Vibrio cholerae*.

choleresis. the secretion of bile (by the liver).

cholestasis. the suppression or cessation of bile flow with consequent hyperbilirubinemia and retention of sulfobromophthalein. The causes of cholestasis are numerous and include induction by any of a wide variety of chemicals, hepatic and extrahepatic, such as lithocholate, manganese, and α-naphthylisothiocyanate.

5-cholesten-3-β-ol 3-(*p*-(bis(2-chloroethyl)amino)phenyl)acetate. fenesterine.

cholest-5-en-3β-ol. cholesterol.

cholesterin. cholesterol.

cholesterol (cholest-5-en-3β-ol; cholesterin). a monohydric secondary alcohol, of the cyclopentenophenanthrene system, with the formula, $C_{27}H_{45}OH$, it is comprised of 4 fused rings with one double bond. It occurs widely among animals and is the principal sterol of higher animals, found in all body tissues, associated mainly with lipids. In vertebrate animals, including humans, it occurs especially in the brain and spinal cord, fats and oils, and plasmalemmas. It is the main constituent of gallstones. In humans and other higher animals some occurs as the sterol and some as a lipid (i.e., esterified with higher fatty acids) in blood serum. Most cholesterol in higher animals is synthesized in the liver; some is dietary. It is the parent compound of bile acids and many steroids (e.g., the steroid hormones) and its presence in excess in the blood of humans and certain other animals contributes importantly to cardiovascular disease. See also cholesterosis.

cholesterolosis. cholesterosis.

cholesterosis (cholesterolosis). a condition in which elevated amounts of cholesterol accumulate in tissues. **extracellular cholesterosis**. a condition in which cholesterol accumulates in extracellular sites. It is characterized by the development of erythematous nodules of the extremities, especially the hands.

cholesteryl-*p*-bis(2-chloroethyl)amino phenylacetate. fenesterine.

cholestyramine resin. a synthetic, strongly basic anion exchange resin that contains quaternary ammonium functional groups joined to a styrene-divinylbenzene copolymer. The main component is polystyrene trimethylbenzylammonium as Cl^- anion). Administered orally, it facilitates detoxification in chlordecone (Kepone) poisoning. Its use greatly reduces the half-life of Kepone (chlordecone) in the body by increasing the rate of excretion and hastening. It is also used therapeutically as an antihyperlipoproteinemic, to limit the absorption and enterohepatic circulation of cardiac glycosides. Its use can cause gastrointestinal discomfort or distress and can retard fat absorption.

choline (2-hydroxy-*N*,*N*,*N*-trimethylethanaminium; (*β*-hydroxyethyl)trimethylammonium; bilineurine; hydroxyethyl trimethyl ammonium hydroxide; lipotropic factor; transmethylation factor; trimethylammonium ion). one of the B complex vitamins, it is a basic constituent of lecithin (as phosphatidyl choline) and other pholipids and is essential to normal fat and carbohydrate metabolism. Choline is a precursor of acetylcholine. It occurs in most animal and many plant tissues including, for example, *Amanita muscaria*, belladonna, bile, brain tissue, ergot, hops, strophanthus, and egg yolk. It is also produced synthetically. Choline is an essential nutrient from some animals and microorganisms; humans, however, synthesize adequate amounts. Hepatic lipoidosis results from an insufficiency of choline. Choline uptake by nerve terminals, a rate-limiting step in the synthesis of acetylcholinesterase, is the target of certain toxicants. See choline acetyltransferase, hemicholinium-3.

choline acetyltransferase (CAT; ChAT). a mitochondrial enzyme that catalyzes the biosynthesis of acetylcholine from choline and acetyl CoA. Choline absorption by nerve terminals is the rate-limiting step in this process. See choline, hemicholinium-3.

choline chloride succinate (2:1). succinylcholine chloride.

choline dichloride. chlormequatchloride.

choline succinate dichloride. succinylcholine chloride.

cholinergic. **1a:** stimulated, activated or conveyed by choline (acetylcholine). **1b:** pertaining to or denoting a nerve fiber that liberates the neurotransmitter, acetylcholine (ACh), from the presynaptic terminal of its synapse. The voluntary motor neurones of vertebrates, parasympathetic and preganglionic sympathetic neurones, and some invertebrate neurones are cholinergic. **2:** a parasympathomimetic agent. *Cf.* adrenergic.

cholinergic agent. parasympathomimetic agent.

cholinergic crisis. See crisis.

cholinergic nervous system. See parasympathetic nervous system.

cholinergic receptor. See receptor (cholinergic receptor).

cholinesterase. **1:** any enzyme such as acetylcholinesterase that catalyzes the hydrolysis of choline esters. They are inhibited by physostigmine (eserine). See also pseudocholinesterase. **2:** sometimes used in text as a synonym for acetylcholinesterase.

cholinesterase inhibitor. See anticholinesterase.

cholinesterase, nonspecific. See also pseudocholinesterase.

cholinesterase-inhibiting pesticide. any of a group of anticholinesterases used as pesticides (e.g., carbaryl, ethion, fenthion, malathion, methomyl, parathion). See anticholinesterase.

cholinoceptor. a cholinergic receptor (See receptor).

cholinolytic. **1:** blocking the action, or having the ability to block the action, of acetylcholine, or of cholinergic agents. **2a:** a cholinolytic agent, one that blocks the action of cholinergics in skeletal muscle, and in organs supplied by parasympathetic system nerves. **2b:** an agent that blocks the action of acetylcholine.

cholinomimetic. **1:** having an action, or producing effects similar to that of acetylcholine. **2:** an agent that has an action or produces effects similar to that of acetylcholine. *Cf.* parasympathomimetic.

cholinoreceptor. a cholinergic receptor (See receptors).

Chondodendron tomentosum (liana; chicona). a tree of Central America and northern South America (Family Menispermaceae) related to the North American moonseed. All parts of the plant are highly toxic and lethal; the sap contains curare. It is a source of chicona bark and curare. See also tubocurarine chloride.

Chondrichthyes. the vertebrate class comprised of cartilaginous fishes, including sharks, rays, skates, and chimaeras. Most are marine preda- [illegible] dermal placoid scales, a cartilaginous endoskeleton, numerous and complete vertebrae, median and paired pelvic and pectoral fins, the absence of lungs and swim bladder, and working jaws that bear numerous vitreodentine-covered teeth. The body is depressed, the mouth ventral, the intestine has a spiral valve, there are one or two olfactory sacs that do not open into the mouth, the heart has a sinus venosus, one auricle, one ventricle, and a conus arteriosus. Five to seven pairs of gills are present except in chimaeras (three), each with a separate gill slit. There are about 300 species; some are venomous.

chop nut. common name of physostigma and *Physostigma venosum*.

Chordata. a phylum of coelomate animals that includes a number of small, primitive marine forms and the subphylum Vertebrata. Members are marked by the presence of a dorsal tubular nerve cord; a notochord and usually pharyngeal gill slits at some period of the life cycle; a postanal tail. The body is metamerically segmented and a complete digestive tract is present.

chordate. 1: any member of the phylum Chordata. The vertebrates are included in this group. **2:** having a notochord.

choreoathetoid. pertaining to or denoting choreoathetosis.

choreoathetosis. a nervous condition characterized by choreic (jerky, rapid) and athetoid (relatively slow, writhing) movements of the limbs, face, and trunk that are common in cerebral palsy. The movements are involuntary, uncontrollable, and unceasing, with fluctuations in muscle tone from hypotonia to hypertonia. Pyrethroids can produce a similar condition.

Choridactylus multibarbis (stonefish). a venomous fish (Family Scorpaenidae) that occurs in coastal areas of India, China, the Phillipine Islands, and Polynesia. The venom apparatus is of the *Synanceja* type (see Scorpaenidae).

choroiditis. inflammation of the choroid coat of the eye, that part which supplies blood to the retina.

chorus frogs. See Hylidae.

CHRIS. Chemical Hazards Response Information System, a U.S. Coast Guard information system pertaining to water transport of hazardous chemicals. Included are: the Condensed Guide to Chemical Hazards (handbook), the Hazardous Chemical Data Manual, the Hazard-Assessment Handbook, the Response Methods Handbook, Data Bases for Regional Contingency Plans, and the Hazard-Assessment Computer System (HACS).

Christmas rose. *Helleborus niger*.

chromaffin (chromaphil). **1a:** having an affinity for and readily and strongly stained by chromium salts. **1b:** of or pertaining to cells of the adrenal medulla, to certain cells of the carotid and coccygeal glands, to the paraganglia or sympathetic nerves; and organs with cytoplasmic granules that turn brown when exposed to chromium salts. Such cells are stained green by ferric chloride, yellow by iodine, and brown with osmium tetroxide (osmic acid). See also chromaffin reaction.

chromaffin cell. a cell that contains granules that are stained brown by stains that contain chromium salts, especially potassium bichromate.

chromaffin reaction. the reaction of certain organs in which the cytoplasmic granules that contain adrenalin turn brown when exposed to stains that contain chromium salts. See also chromaffin.

chromaffinoma (paraganglioma). a chromaffin cell tumor.

chromaffinopathy. any disease of chromaffin tissue.

chromaphil. chromaffin.

chromate. **1:** any chromium salt with hexavalent chromium; any salt of chromic acid. **2:** any oxygen-containing derivative of chromium (e.g., sodium dichromate, Na_2CrO_4). Chromates are water soluble and are readily absorbed into the bloodstream via the lungs. Chromates are teratogenic to mammals and birds. They are mutagenic to bacteria, yeast, and cultured mammalian cells, and most are carcinogenic to humans and laboratory animals. Exposure of human workers to atmospheric chromates may cause bronchogenic carcinoma with a latent period of 10-15 years. Hexavalent chromium is readily reduced to the less toxic trivalent form in the body, but the reverse reaction does not occur. Chromates are mutagenic in bacteria, yeasts, and cultured mammalian cells.

chromated copper arsenate (CCA). a toxic wood preservative that is widely used in the building industry. CCA poses a health hazard during manufacture and in the handling and burning of treated wood. Risk is incurred through prolonged skin contact, and inhalation of sawdust or ash from burning treated wood. The worker may be exposed to chromium and arsenic salts with possible damage to the proximal tubular cells of the kidney.

chromatin. the diffuse, thread-like material of the cell nucleus during interphase. It condenses into chromosomes during mitosis. The main constituents are DNA and proteins (mostly histones).

chromatogram. the pattern formed by zones or spots of separated solute compounds resulting from a chromatographic procedure.

chromatographic adsorption. differential adsorption of gases or liquids in order of ascending weight onto a solid adsorbent material (e.g., activated carbon, alumina, silica gel). The process is used to separate and chemically analyze mixtures. *Cf.* chromatography.

chromatography. any of several analytic techniques used mainly to separate the components of a sample mixture or solution by distributing them between a stationary and a mobile phase by chromatographic adsorption or differential solubility. Retention times or rates of migration (and hence the degree of physical separation) of the solute compounds generally depend on their relative solubilities in the mobile and stationary phases. The stationary phase may be a liquid, gel, or solid in the form of a column, layer, or film; the mobile phase is either a liquid or a gas. **adsorption chromatography**. separation of a mixture of gaseous or liquid solutes by conducting it over a bed that adsorbs compounds at differential rates. **adsorption phase chromatography**. normal phase chromatography. **affinity chromatography**. a method that takes advantage of the ability of biological macromolecules to specifically and irreversibly turn to ligands. **bonded phase chromatography**. a high-pressure liquid chromatography technique that uses a stable, chemically bonded stationary phase. **column chromatography**. the simplest and oldest chromatographic separation technique. There are two general types: (1) Packed column chromatography, in which the stationary phase is a column of powdered inactive adsorbent (e.g., aluminum oxide or kieselguhr) confined within a vertical tube. The sample is delivered into the top of the tube and percolates slowly through the stationary bed. The different solute compounds are absorbed in varying degrees and thus move down the column at different rates, those that are most strongly absorbed moving most slowly. The degree or completeness of separation depends upon the length of the tube, the adsorbent used, and the composition of the sample. If the components are pigmented, colored bands develop down the tube. (2) Open-tube chromatography, in which there is a thin layer of partitioning liquid on the column walls leaving the center of the column open so that gas can pass through the column. In column chromatography, the solutes pass through the column and are collected and analyzed. Column chromatography also serves to concentrate the solutes, favoring chemical analysis. **gas chromatography** (GC). the mobile phase in gas chromatography is a gas and the stationary phase is either a liquid (gas-liquid chromatography) or a solid (gas-solid chromatography). This method, especially when used in conjunction with electron capture detection, can be used for the separation and quantitative analysis of volatile organic air pollutants and certain chlorinated hydrocarbons. Used in conjunction with flame ionization detection, gas chromatography can be used to detect or to determine carbon monoxide levels in air. **gas-liquid chromatography** (GLC; gas-liquid partition chromatography). a gas chromatographic method used to separate volatile solutes in which the mobile phase is an inert gas and the stationary phase is a nonvolatile liquid distributed on the surface of an inert porous support. The solute compounds to be separated are partitioned between a carrier gas and the stationary phase. These compounds are retained by the stationary phase for a period of time that is a function of their individual solubilities. **gas-liquid partition chromatography**. gas-liquid chromatography. **gas-solid chromatography** (GSC). gas chromatography in which the mobile phase is a gas and the stationary phase is a surface-active sorbent such as charcoal, silica gel, or activated alumina. **gel permeation chromatography**. a method in which the stationary phase is a porous polymer (e.g., a cross-linked dextran carbohydrate derivative such as Sephadex) and the mobile phase is a liquid. **high performance liquid chromatography** (HPLC; high-pressure liquid chromatography; liquid chromatography). a technique in which the mobile phase is a liquid and the stationary phase is typically a polar absorbent (e.g., microparticulate silica) to which an array of functional groups may be bonded. **high-pressure liquid chromatography.** an unacceptable synonym for high-performance liquid chromatography. **ionexchangechromatography**. a method in which the stationary phase consists of absorbents to which anionic or cationic groups have been bonded. **liquid chromatography**. those chromatographic techniques that use a liquid as the mobile phase and a solid or a liquid as the stationary phase. **normal phase chromatography**. high performance liquid chromatography in which solute retention is due to the interaction of a nonpolar mobile phase with the surface of a polar absorbent stationary phase. The mobile phase i scomprised of variable-length acyl carbon chains that bond chemically to the polar absorbent. *Cf.* reversed-phasematography. **open-tube chromatography**. See column chromatography. **packed column chromatography**. See column chromatography. **paper chromatography**. a method used to separate and analyze complex chemical mixtures using a sheet of purified cellulose paper of varying fiber coarseness and packing density as the stationary phase. The mobile phase is a solution of varying proportions of polar and nonpolar solvents that contains the solute compounds to be assayed. A small amount of the sample is spotted near one edge and eluted by allowing the solvent to descend or ascend the sheet of paper by capillary action. The solutes are separated by a combination of differential adsorption (e.g., via hydrogen bonding, van der Waals' forces) to the cellulose fibers; ion exchange with the carboxy groups of the cellulose; and partitioning between the mobile phase and the water content of the stationary phase. Colorless compounds are observed by viewing in ultraviolet light or by chemical staining. The relative migration of solutes is expressed as *Rf* values, *q.v.* This method is often used to analyze mixtures of amino acids (e.g., as from the hydrolysis of proteins). The acid is identified by the distance that it moves relative to that moved by the solvent under standard conditions. **partition chromatography**. a technique in which the stationary phase is a high boiling-point liquid, dispersed as a thin film on an inert support, and the mobile phase is a vaporous mixture of the analytes in a carrier gas. **reversed-phase chromatography**. a bonded-phase paper chromatographic method that utilizes a polar mobile phase (e.g., of water, acid, organic solvent) and a nonpolar stationary phase (e.g., paraffin, paraffin jelly, grease). *Cf.* normal phase chromatography. **thin-layer chromatography** (TLC). a technique that uses a liquid mobile phase which moves by capillary action through the stationary phase. The latter is a thin coating of adsorbent(e.g., aluminum oxide, cellulose, charcoals, ion exchangers, polyamides, silica gel, silicates) applied to an inert, rigid plate (usually of glass). The mobile phase or solvent system consists of a single solvent or a mixture of solvents of varying polarity. The separation mechanisms (e.g., hydrogen bonding, ion exchange, phase partitioning) depend on the sorbent used and the physicochemical properties of the solutes. Typically, the sample to be separated is applied to the edge of the sorbent layer as a spot, and the plate is developed in an enclosed chamber by contacting the solvent system. Additional separation power can be obtained by two-dimensional development in which a developed plate is rotated through 90° and redeveloped using a different solvent system. The relative migration of a solute compound is expressed as an *Rf* value, *q.v.*

chrome green. lead chromate.

chrome lemon. lead chromate.

chrome yellow. lead chromate.

chromic acid. **1:** an aqueous solution, H_2CrO_4, of chromium trioxide. It is a strong oxidizing agent. **2:** chromium trioxide is often referred to as chromic acid. The latter occurs only in solution, however.

chromic acid, lead and molybdenum salt. lead-molybdenum chromate.

chromic acid lead salt with lead molybdate. lead-molybdenum chromate.

chromic acid, zinc salt. zinc chromate.

chromium (Cr)[Z = 24; A_r = 52.01]. a metallic transition element that occurs naturally as chromite. It has many uses; for example, in the manufacture of alloys such as stainless steel, and to electroplate other metals to increase hardness and corrosion resistance. Certain of its compounds, especially those with hexavalent chromium (chromates), are toxic. Chemically combined chromium exists in oxidation states +2 through +6. In the +3 oxidation state (trivalent chromium), it is essential to glucose and lipid metabolism in mammals. Although cases of chromium deficiency are unknown, experimental Cr^{3+} deficiency has produced the symptoms of diabetes mellitus. Chromium and most of its compounds, especially those with hexavalent chromium (chromates), are, however, nephrotoxic and most chromates are also carcinogenic to humans and laboratory animals. They are also irritants that can cause serious damage to the skin and mucous membranes, especially of the lungs and gastrointestinal tract. Workers who have been exposed to chromium fumes experience an increased incidence of pulmonary cancer and cancer of the larynx, nasal cavity, and sinuses with a latent period of about 20 years. Chromium is toxic to vertebrate animals other than humans, to at least some marine invertebrates, and to plants. It bioaccumulates in various freshwater fish.

chromium acid, strontium salt (1:1). strontium chromate.

chromium carbonyl (chromium hexacarbonyl; hexacarbonyl chromium). a crystalline substance, $Cr(CO)_6$, that sublimes at room temperature. It is very toxic to laboratory mice. and rats. See carbonyl.

chromium hexacarbonyl. chromium carbonyl.

chromium lead oxide. lead chromate, basic.

chromium oxychloride. chromyl chloride.

chromium poisoning. poisoning by chromium and its compounds, especially those with hexavalent chromium (chromates). These substances are irritants, are nephrotoxic, and most chromates are also carcinogenic to humans and laboratory animals. Symptoms of intoxication may include an unpleasant taste in the mouth, pain, diarrhea, cramps, collapse, and sometimes death from uremia.

chromium trioxide (chromic anhydride). a dark purplish-red crystalline or powdery substance, CrO_3. It is very soluble in water in which it forms a dibasic acid, H_2CrO_4. CrO_3 is a powerful oxidizing agent, reacting (often violently) with most organic materials. Exposures are mostly occupational as this substance is used primarily in chromium plating, aluminum anodizing, copper stripping, and as a corrosion inhibitor. It is also used in photography, as an oxidizing agent, and in the preparation of microscopic specimens. It is a fire hazard if stored or used where it may come in contact with combustible materials; it may explode on contact with reducing agents. It is a powerful irritant and can cause ulceration and allergic eczema on contact. Ingestion can cause violent gastrointestinal irritation, and inhalation can cause damage to the nasal septum or to the lungs. Inhalation of chromate dust can cause bronchial carcinoma as well. Nephrotoxicity has been reported for several types of laboratory animals. *Cf.* chromic acid.

chromium zinc oxide. zinc chromate.

chromodacryorrhea. the presence of reddish conjunctival exudate in the eyes; no blood cells are present in exudate. It is not true "bloody tears."

chromosomal aberration (chromosomal mutation). any change (mutation) in the number, structure, or sequence of genes as reflected in gross chromosomal structure or number. That is, any modification (detectable by light microscopy) of the normal chromosome complement as a result of deletion (gaps, breaks), duplication, inversions, or translocations of chromosomal material. Many toxicants as well as ionizing radiation can cause chromosomal aberrations. Alkylating agents can cause chromosome breakage by cross-linking with DNA. A chromosomal mutation.

chromosomal mutation. See chromosomal aberration.

chromosome (karyomite). **1a:** any of a number of deeply staining rod-like or thread-like bodies that occur in the nucleus of eukaryotic cells. Such are composed largely of DNA in association with RNA and protein (mostly histones), and comprising a linear sequence of hundreds of genes. The number of chromosomes is usually constant in any given species. **1b:** the condensed form of chromatin (DNA plus nucleoproteins). It is visible during mitosis. **2:** sometimes used in reference to the genetic material of prokaryotes and viruses that consist exclusively of DNA (RNA in certain viruses) with no associated protein. It is never visible as discrete thread-like or rod-like structures, as in eukaryotes.

chromosome aberration. chromosomal aberration.

chromosome aberration test. See test.

chromosome deletion. a chromosome mutation with a loss of genes.

chromosome insertion. a chromosome mutation with the addition of duplicate genes.

chromosome inversion. a chromosome mutation in which the sequence of some genes are reversed.

chromosome translocation. a chromosome mutation in which a segment of a chromosome is transferred to the homologous chromosome.

chromotoxic. **1:** poisonous to hemoglobin. **2:** poisonous to an organism by virtue of the ability to destroy hemoglobin.

chromotoxicity. **1:** the capacity of a substance to poison hemoglobin. **2:** the capacity of a substance to poison an organism by destroying hemoglobin.

chromyl chloride (chromium oxychloride; chlorochromic anhydride). a mobile, dark red, fuming, poisonous liquid, CrO_2Cl_2. It is a strong oxidizer and corrosive agent that is highly destructive to tissue and other organic materials.

chronic. **1:** extended in time, long-term, continuing, of long duration; mild; of frequent occurrence. **2a:** of, pertaining to, or describing a disorder of long duration or one that frequently recurs (See chronic condition). **2b:** the persistence of signs or symptoms (due, for example, to poisoning, infection, or disability) for an extended period of time. **3:** of or pertaining to continuous or frequently repeated exposure (usually at a low level) over an extended period of time, to a hazardous substance or other stressor.

chronic condition (chronic disease; chronic disorder). any condition that has at least one of these characteristics: **1.** it is marked by slow onset and is of relatively long duration; **2.** it is permanent or nearly so; **3.** it leaves disabling sequalae; **4.** it causes an irreversible lesion at some level of organization; **5.** special training is required for rehabilitation.

chronic disease. See chronic condition.

chronic disorder. See chronic condition.

chronic exposure. See exposure.

chronic obstructive pulmonary disease. See emphysema.

chronic poisoning. See poisoning.

chronic toxicity. See toxicity (chronic toxicity). chrysanthemummonocarboxylic acid; chrysanthemumic acid). an acid found in pyrethrum flowers as esters (See pyrethrin) and forms toxic synthetic esters (See allethrin).

chronicity. the state of being chronic.

chronicity factor (CF). See factor (chronicity factor).

chronicity index. the application of the therapeutic index or the margin of safety to chronic exposure to a substance. The chronicity index (c.i.) is a numerical expression of a specific cumulative toxic effect of a chemical. A conventional toxicity index is the ratio of the single dose (LD_{50}(1)) to the 90 day dose (LD_{50}(90)). The latter is that dose which, administered daily for a period of 90 days, will cause 50% mortality in the test population. Thus, for this example:

$$\text{c.i.} = LD_{50}\ (1\ \text{day})/LD_{50}(90)$$

In the case of air pollutants or aquatic organisms, a 90-day chronicity test using a constant concentration might be used. For relatively short-lived organisms, shorter test periods or life cycle tests might provide the basis for a chronicity index. Even so, the determination of an accurate chronicity index is rarely, if ever, warranted because of the time and resources required, the number of animals sacrificed, and the uncertain relationship of an index based on mortality to the relative or comparative toxicity of any toxicant.

chrono-. a prefix denoting or relating to time.

chronobiology. the biolological science that treats the temporal characteristics of living systems and system components. The primary focus is on the exogenous and endogenous factors that influence biological rhythms and the mechanisms of control. Included is the development of knowledge regarding temporal variation in the effects of toxicants on living systems. See also circadian rhythm, circalunadian rhythm, circamonthly rhythm, circannual rhythm, circasemiannual rhythm, diurnal rhythm.

chronotropic. able to influence the rate of occurrence of an event such as the beating of a heart.

chrysaniline yellow. a yellow crystalline basic dye, $C_{19}H_{15}N_3$, used as a stain. It is a by-product of fuchsine manufacture; sometimes called phosphine.

chrysanthemic acid (2,2-dimethyl-3-(2-methyl-1-propanyl)cyclopropanecarboxylic acid; chrysanthemummonocarboxylic acid; chrysanthemumic acid). an acid found in pyrethrum flowers as esters (See pyrethrin) and forms toxic synthetic esters (See allethrin).

chrysanthemum. a common name of many of the plants of the genus *Chrysanthemum*.

Chrysanthemum (chrysanthemum). a large genus of widely cultivated annual and perennial Old World herbs (Family Asteraceae). Many are toxic, producing a contact dermatitis with erythema and vesiculation in sensitive individuals. The active principle is a resin. ***C. (Pyrethrum) cinariifolium*** (= *Pyrethrum cinaraefolium*). a perennial herb, indigenous to Dalmatia and Montenegro. It grows to 30-40 cm, with many stems and white flower heads. The dried, ground flowers are the major source of pyrethrum insecticides. Cultivated pyrethrums are derived form *C. coccineum*. See pyrethrum flowers. ***C. coccineum*** (= *C. roseum*). a hardy perennial herb, indigenous to western Asia, which grows to about 1 m in height; the flowers, which bloom in late spring, are various shades of red, pink, and white. It isa source of pyrethrum flowers, *q.v.* ***C. marschalli***. a hardy perennial indigenous to western Asia. It is also a source of pyrethrum flowers.

chrysanthemumate. barthrin.

chrysanthemumic acid. chrysanthemic acid.

chrysanthemummonocarboxylic acid. chrysanthemic acid.

chrysanthemummonocarboxylic acid 6-chloropiperonyl ester. barthrin.

chrysanthemummonocarboxylic acid ester with 3-(2-cyclopenten-1-yl)-2-methyl-4-oxo-2-cyclopenten-1-ol). cyclethrin.

chrysanthemumdicarboxylic acid monomethyl ester pyrethrolone ester. pyrethrin II.

chrysanthemummonocarboxylic acid pyrethrolone ester. pyrethrin I.

Chrysaora quinquecirrha (sea nettle). a large marine venomous jellyfish (Class Scyphozoa) that occurs from Cape Cod to Florida and Texas. It is pink with radiating red stripes, 40 tentacles, and a bell covered with small warts and a finely scalloped margin. Contact often results in a mild, itchy dermatitis; severe stings may require hospitalization. The venom is extremely toxic with cardiotoxic, dermonecrotizing, musculotoxic, neurotoxic, antigenic, and collagenase activity. Toxicity resides in two proteins. A related species of the Pacific Ocean, *C. melanaster* (lined sea nettle) is often found [illegible] from Alaska to southern California. It has similar toxicologic properties.

chrysarobin (purified Goa powder; purified araroba). **1:** 1,8-dihydroxy-3-methyl-9-anthrone (3-methyl-1,8,9-anthracenetriol). a flammable, poisonous, crystalline compound, $C_{15}H_{12}O_3$, a reduction product of chrysophanic acid. It is an irritant and an allergen. **2:** a commercial mixture consisting of ca. 15-30% pure chrysarobin and 70-85% various neutral anthraquinone derivatives of Goa powder. It is an odorless, tasteless, slightly water-soluble, brownish to yellow-orange microcrystalline powder used topically to treat psoriasis and other chronic, noninfectious skin diseases. It is a strong irritant; the eyes can become seriously inflamed on contact with this substance. Chrysarobin is nephrotoxic to humans and, if applied to large areas of skin can cause kidney damage via percutaneous absorption. See Goa powder.

chrysatropic acid. scopoletin.

chrysene (1,2-benzphenanthrene; 1,2-benzophenanthrene; benz(*a*)phenanthrene; benzo(*a*)phenanthrene; 1,2,5,6-dibenzonaphthalene). a toxic, water-insoluble, crystalline, polynuclear hydrocarbon, $C_{18}H_{12}$, found in coal tar. It gives a strong, blue fluorescence under ultraviolet light when pure, and easily sublimes *in vacuo*. It is only slightly soluble in common organic solvents at room temperature, but moderately so on heating. Chrysene forms during the distillation of coal and produced in very small amounts during the distillation or pyrolysis of many fats and oils. Chrysene is tumorigenic, carcinogenic, and neoplastigenic on contact with the skin of some laboratory animals and humans and is a probable human mutagen. Acrid fumes and smoke are released on heating to decomposition.

chrysotile. See asbestos.

chrysotile asbestos (white asbestos). See asbestos.

chub. See Cyprinidae.

chummar. *Naja*.

churus. hashish.

CI-395. phencyclidine.

cicatrix. a scar.

cichorin (6,7-dihydroxycoumarin-7-glucoside). a glucoside in flowers of *Cichorium intybus*; it is isomeric with esculin.

ciclosporin. cyclosporin A.

Cicuta (water hemlock). a genus of native North American herbs (Family Apiaceae) that inhabit moist situations. Included are a number of species, of which the most important toxicologically is *C. maculata*. All contain convulsant toxins, most notably cicutoxin (found in *C. maculata* and *C. virosa*), which is the most violent poison produced by plants in North America. Numerous poisonings have occurred because these plants were mistaken for parsnips, artichokes, or other rots. The odor is similar to that of wild parsnip. Most poisonings occur during early growth in the spring. Children have been poisoned by making peashooters and whistles from the hollow stems. ***C. bolanderi***. *C. californica*. ***C. bulbifera*** (bulbous water hemlock). a water hemlock of the northern United States. ***C. californica*** (= *C. bolanderi*; California water hemlock). a species that occurs in middle western California. ***C. douglassi*** (Douglas water hemlock). a species found along the Pacific Coast states and British Columbia. ***C. maculata*** (water hemlock; cowbane; musquash root; beaver poison). a perennial herb of bogs, marshes, and swampy areas in eastern North America, and west to the Great Plains. It has small white flowers and may grow to 2 m. All parts can poison fatally, especially the root. Symptoms may include those of gastroenteritis, convulsions (may be persistent), and respiratory depression. The active principle is the resin, cicutoxin. ***C. occidentalis*** (western hemlock). a plant of the Rocky Mountain states west to the Pacific Coast. ***C. vagans*** (tuber water hemlock; Oregon water hemlock). a water hemlock of the Pacific Northwest region of the United States.

cicutine. See coniine.

cicutoxin ((E,E,E)-(—)-8,10,12-heptadecatriene-4,6-diyne-1,14-diol). a very toxic, highly unsaturated higher alcohol component, $C_{17}H_{22}O_2$, of *Cicuta maculata* and *C. virosa*. It is poisonous by ingestion, by intravenous injection, and probably other routes. Its action is similar to that of picrotoxin.

cigar. a tubular roll of tobacco leaf designed for smoking. Cigars burn at a lower temperature than cigarettes, producing far fewer carcinogenic hydrocarbons. See tobacco, cigarette smoke.

cigarette. **1:** a roll of finely cut tobacco wrapped in paper. Combustion produces a smoke that is toxic, dangerously carcinogenic, and addictive when inhaled. Cigarette smoking increases the risk of heart disease and lung cancer. Mortality in persons also exposed to asbestos is greater than in nonsmokers. Cigarettes contain cadmium, cacollidin, nicotine, pyridine, picoline, large numbers of polycyclic aromatic hydrocarbons, cocarcinogenic phenols, fatty acids, carbon monoxide, hydrogen sulfide, hydrocyanic acid, nitrogen oxides, and various irritants that suppress protease inhibition and alveolar macrophage activity. See Nicotiana, tobacco smoke. **2:** a cigarette containing or made from marijuana. Such cigarettes have a large number of street names (reefers, doobies, numbers, smoke, joints, etc.).

cigarette smoke. tobacco smoke, especially from cigarette smoking, is a major indoor pollutant, that affects not only smokers but those who share their environment. Tobacco contains a number of known carcinogens such as arsenic, formaldehyde, nitrites, vinyl chloride, and polycyclic aromatic hydrocarbons as well as other toxic chemicals such as cadmium, nitropropanes (toxic and possibly carcinogenic), and radon particles (radon daughters). Thus, for example, the body burden of cadmium in cigarette smokers is nearly 2 times greater than in nonsmoking subjects of the same age. The first substantive data that linked cigarette smoke with lung cancer were published in 1950. There is little doubt today that cigarette smoke is responsible for about one-third of all cancer deaths. Since 1970, United States federal law requires that each pack of cigarettes bears the message: "Warning: The Surgeon General has determined that cigarette smoking is dangerous to your health." In closed rooms (e.g., in the home, in bars, in restaurants) the gaseous toxicants and the lighter particulates (including radon daughters) remain in the air for substantial periods of time. These materials may be inhaled by smokers and nonsmokers (passive smokers) alike. Passive smoking substantially increases one's risk of cancer. As many as 5% of those who die from cigarette smoke-induced lung cancer in the United States are passive smokers. See also Nicotiana, nicotine, TCDD, tobacco.

cigarette smoking. See cigarette smoke.

cigarette tar. the relatively non-volatile fraction of hydrocarbons in the smoke of burning cigarette tobacco. A prominent component of this mixture is benzop[a]pyrene. Cigarette tar is carcinogenic due to the presence of this compound and other aromatic hydrocarbons.

cigua. *Livona pica*.

ciguatera (ciguatoxism, ciguatera poisoning). **1:** a common and serious type of ichthyosarcotoxism in certain subtropical and tropical coastal areas that is acquired by ingesting certain tropical and subtropical reef and semipelagic marine fishes, dolphins, and some gastropod mollusks that have accumulated ciguatoxin. For example, fish that ingest toxic algae, invertebrates, or contaminated fish transmit the toxic principle (ciguatoxin) to higher trophic levels, e.g., *Homo sapiens*. The primary or ultimate source of ciguatoxin is *Gambierdiscus toxicus* and possibly other dinoflagellates. The toxin is heat stable and remains active in preserved flesh or viscera. More than 400 species of fish, mostly of tropical and subtropical shores (especially the Pacific Ocean, the Caribbean Archipelago, and the Indian Ocean) have been implicated. Many of these fish are normally palatable in some parts of their range and some are valued food fish. They may become toxic within a few hours and remain so for some years. Effects are gastrointestinal and neurological in nature. Those in laboratory animals include miosis, dyspnea, cyanosis, hypothermia, increased salivation and lacrimation, ataxia, hypoactive or abolished reflexes, and exhaustion. Signs and symptoms in humans usually appear within 30 minutes following ingestion of contaminated flesh or viscera. They include, in varying degrees, most of the above symptoms as well as itching, nausea, dizziness, vomiting, diarrhea, headache, abdominal pain, joint aches, myalgia, extreme muscle weakness, dysesthesia and paresthesia of the lips, tongue and limbs, pruritis, and bradycardia. Paradoxical sensory disturbances (e.g., perceiving hot objects as cold) are also common. Neurologic effects may include dysphagia, generalized motor incoordination, ataxia, diminished reflexes, muscular tics and tremors, clonic and tonic convulsions, paralysis, and coma; and in severe poisoning, paresis, especially of the legs. Death occurs in about 12% of cases, normally from respiratory collapse. Among survivors, complete recovery may require years. There is no specific treatment. More than 300 varieties of fish from the Caribbean and South Pacific are implicated. Bony fish with populations of known transvectors of ciguatoxin to humans are: *Acanthurus* (surgeonfish), *Albula vulpes* (ladyfish), *Alutera scripta*, (filefish), *Aprion virescens* (snapper), *Balistoides conspicillum* (triggerfish), *Caranx* (jacks), *Cephalopholis argus* (seabass, grouper), *Clupanodon thrissa* (herring), *Coris gaimardi* (wrasse), *Engraulis japonicus* (anchovy), *Epibulus insidiator* (wrasse), *Epinephelus fuscoguttatus* (sea bass; grouper), *Gambierdiscus toxicus*, *Gnathodentex aureolineatus* (snapper), *Gymnothorax* (moray eels), *Euthynnus pelamis* (oceanic bonito), *Lactophrys trigonus* (trunkfish), *Lactoria cornutus* (trunkfish), *Lethrinus miniatus* (snapper), *Lutjanus nematophorus* (Chinaman fish), *Lutjanus* (red snapper), *Monotaxis grandoculis* (snapper), *Mycteroperca venenosa* (seabass), *Myripristis murdjan* (squirrelfish), *Pagellus erythrinus* (porgie), *Pagrus* (porgie), *Parupeneus chryserydros* (surmullet; goatfish), *Plectropomus* (seabass), *Scarus* (parrotfish), *Sphyraena barracuda* (barracuda), *Tetragonurus cuvieri* (squaretail), *Upeneus arge* (surmullet, goatfish), and *Variola louti* (seabass). See ciguatera poison, ciguatoxin, maitotoxin, *Sphyraena barracuda*, *Gambierdiscus toxicus*. **2:** (obs.) poisoning caused by ingestion of the "cigua" or *Livona pica*, a common marine snail of the Caribbean.

ciguatera poison. a mixture that includes ciguatoxin and sometimes maitotoxin, scaritoxin, and perhaps others.

ciguatera poisoning. ciguatera.

ciguatoxic. **1:** of or pertaining to the toxicity of ciguatoxin. **2:** said of a transvector of ciguatoxin.

ciguatoxin. a colorless, neurotoxic, cardiotoxic, heat-stable, lipid- and water-soluble, saponin, $C_{28}H_{52}NO_5Cl$, MW ca. 1100. It is a chemically complex quaternary ammonium compound and the principal toxin that causes ciguatera. The primary or ultimate source is *Gambierdis-*

cus toxicus and certain other dinoflagellates. Toxicity is unaffected by most methods of preservation or of preparing the flesh or viscera for ingestion by humans. It causes cell depolarization by increasing membrane permeability to sodium. Heart rate and the force of cardiac contraction may be altered. Ciguatoxin is sometimes represented as anticholinergic. Physostigmine is, however, an antagonist. Do not confuse with ciguatera poison, *q.v.*

ciguatoxism. ciguatera.

cilia (pl. of cilium). fine, mobile, hair-like, projecting organelles of many cells. They are comprised of microtubule bundles with associated motor proteins. They are locomotor structures in many unicellular organisms and certain small, aquatic metazoans. Their functions vary in higher organisms. In general, the movements of the cilia either propel the cell through an aqueous medium or the cell remains fixed in position and the cilia sweep fluids or particles past the cell. Thus, in the bronchi, ciliated tracheobronchial cells sweep particles (embedded in mucus) from the respiratory tract. These might otherwise accumulate, impairing respiratory function, causing pulmonary edema, or in some cases entering the system with toxic effects.

ciliotoxicity. the capacity of an agent (e.g., a drug) or condition to interfere with or impair ciliary function, usually in reference to tracheobronchial cilia.

cilium. See cilia.

cimetidine (*N*-cyano-*N'*-methyl-*N''*-[2-[[(5-methyl-1*H*-imidazol-4-yl)methyl]thio]ethyl]-guanidine). a drug related to antihistamines; it is a histamine H_2 receptor antagonist. It inhibits gastric secretions and the hepatic microsomal oxidative metabolism of several drugs, sometimes reducing their toxicity. It is, for example, protective against acetaminophen-induced hepatotoxicity. It reversibly inhibits gastric secretion of hydrochloric acid which is normally stimulated by endogenous histamine and is thus used to promote healing of gastric and duodenal ulcers and to reduce esophagitis. Cimetidine usually relieves symptoms within 1-2 weeks. There is a slight risk that its use may temporarily mask the symptoms of stomach cancer. Some patients experience headache, dizziness, nausea, and loss of libido. Rare effects are slurred speech and delirium; these are most often seen in older individuals. It is slightly toxic to laboratory mice and rats.

CIN. cyanogen iodide.

cinchona (cinchona bark; calisaya bark; Jesuits' bark; Peruvian bark; quina; quinaquina; quinquina). the dried bark of the stem or root of various species of South American trees of the genus *Cinchona* (Family Rubiaceae). It contains some 35 alkaloids, the specific composition varying according to the source of the bark. Of these, two sets of isomers are most important (quinine and quinidine; cinchonidine and cinchonine). Bark from cultivated trees contains about 5-7% quinine. See cinchonism.

Cinchona. a genus of evergreen trees (Family Rubiaceae), native of South America but cultivated in various tropical regions. Bark of the various species (See cinchona) contains a large number of alkaloids, most notably the isomers quinine and quinidine and the isomers cinchonidine and cinchonine. ***C. ledgeriana*** (cinchona bark). this plant is indigenous to the South American Andes, but occurs also in Java, India, East Africa, as well as Australia and the Caucasus. The bark is very toxic. Symptoms of intoxication may include nausea, vomiting, hemorrhage, tinnitus, giddiness, collapse, visual disturbances, coma, and death from respiratory arrest due to paralysis. Ingestion or injection of large doses causes sudden onset of cardiac depression or heart failure.

cinchona bark. cinchona.

cinchona-bark tree. *Cinchona pubescens*.

cinchonan-9-ol. cinchonidine, cinchonine.

cinchonic. pertaining to cinchona.

cinchonidine (cinchonan-9-ol; cinchovatine; α-quinidine). an antimalarial and isomer of chinonine, it occurs in most species of *Cinchona*, but especially in *C. pubescens* (common cinchona; cinchona-bark tree). It is usually given orally as the sulfate. Cinchonidine is very toxic to laboratory rats. See cinchona.

cinchonine (cinchonan-9-ol). an antimalarial and isomer of cinchonidine, it occurs in most species of *Cinchona*, but especially in *C. micrantha*. It is very toxic to laboratory rats. See cinchona.

cinchonism (quininism). poisoning by cinchona or its alkaloids (quinine and quinidine). Symptoms include headache, tinnitus, deafness, and sometimes anaphylactoid shock.

cinchophen (2-phenyl-4-quinolinecarboxylic acid; 2-phenylquinoline-4-carboxylic acid; 2-phenylcinchoninic acid). a slightly bitter-tasting, light yellow powder. It is a quinoline derivative formerly used as an analgesic, antipyretic, and uricosuric agent. Cinchophen often produces serious toxic effects including a fatal type of hepatitis. It is now used experimentally to induce gastric ulcers in experimental animals.

cinchovatine. cinchonidine.

cinclisis. rapid spasmodic movement of any part of the body.

cinerin. either of two naturally occurring insecticidal pyrethrins, cinerin I and II, isolated from pyrethrum flowers. Both substances are water-insoluble, viscous liquids that oxidize in air, rapidly becoming inactive; they are hepatotoxic and nephrotoxic. Symptoms of acute intoxication in laboratory mammals include diarrhea, convulsions, prostration, and death due to respiratory paralysis. See also pyrethrin (def. 2).

cinerin I. See cinerin.

cinerin II. See cinerin.

cinmethylin(*exo*-(±)-methyl-4-(1-methylethyl)-2-[(2-methylphenyl)methoxy]-7-oxabicyclo-[2.2.1]heptane; (±)-[illegible] oxy)-1-methyl-4-isopropyl-7-oxabicyclo[2.2.1]-heptane). a preemergence grass herbicide. It is moderately toxic orally to laboratory rats.

cinnabar (cinnabarite). the only important ore of mercury, HgS, it is a granular (rarely crystalline) vermillion-red mineral.

cinnabarite. cinnabar.

Cinnamomum camphora (camphor tree). an aromatic evergreen tree (Family Lauraceae) that occurs in Brazil and eastern Asia, including Java, Sumatra, the central provinces of China, Formosa, and Japan. All parts of the tree contain the dextro form of camphor. See camphor.

cinnamyl alcohol anthranilate. cinnamyl anthranilate.

cinnamyl-2-aminobenzoate. cinnamyl anthranilate.

cinnamyl-*o*-aminobenzoate. cinnamyl anthranilate.

cinnamyl anthranilate (anthranilic acid, cinnamyl ester; 2-aminobenzoic acid-3-phenyl-2-propenyl ester; cinnamyl alcohol anthranilate; cinnamyl-2-aminobenzoate; cinnamyl-*o*-aminobenzoate; 3-phenyl-2-propenylanthranilate; 3-phenyl-2-propen-1-ylanthranylate). a water-insoluble, reddish-yellow powder, $C_{16}H_{15}NO_2$, that is soluble in ethanol, ethyl ether, and chloroform. It is carcinogenic to some laboratory animals.

cinobufagin. a toad venom with anesthetic properties.

Ciodrin (α-methylbenzyl-3-(dimethoxyphosphinyloxy)-*cis*-crotonate; dimethyl phosphate of α-methylbenzyl 3-hydroxy-*cis*-crotonate). an organophosphate cholinesterase inhibitor that is structurally and toxicologically similar to, but slightly less toxic than Bidrin, *q.v.*

CIPC. chlorpropham.

circadian rhythm. an endogenous biological rhythm having a periodicity of about 1 day (24 hours). Such rhythms occur in all living systems and subsystems including cells, tissues, and organs. Such rhythms appear to be of multiple evolutionary origin and operate independently of daily environmental rhythmical changes. Under natural conditions, they are synchronized anew each day by a signal from the environment (a zeitgeber) to the 24-hr light/dark cycle. They are thought to be driven by an endogenous "biological clock." There is often a pronounced daily rhythm in the susceptibility of an organism and of target organs and tissues to a particular toxic substance. *Cf.* diurnal rhythm.

circalunadian rhythm. a biological rhythm having a persistent periodicity of about 1 lunar day (24.8 hours). Such rhythms are usually bimodal. As in the case of circadian rhythms, *q.v.*, important toxicological processes and tests or evaluations of hazard, toxicity, or risk may be affected by such rhythms.

circamonthly rhythm. a biological rhythm that has a periodicity of about 29.5 days. As in the case of circadian rhythms, *q.v.*, important toxicological processes and tests or evaluations of toxicity may be affected by such rhythms.

circannian rhythm. circannual rhythm.

circannual rhythm (circannian rhythm). a biological rhythm having a periodicity of about 1 year (365.25 days). As in the case of circadian rhythms, *q.v.*, important biological processes (e.g., metabolic cycles) are affected by such rhythms. Thus, toxicity tests and evaluations of toxicity may be affected by such rhythms, sometimes greatly so.

circasemiannual rhythm. a biological rhythm that completes two cycles within a period of approximately 1 year. As in the case of circadian rhythms, *q.v.*, important toxicological processes and tests or evaluations of toxicity may be affected by such rhythms.

circulation. in toxicology, usually means the forced flow of blood through the circulatory system (*q.v.*); the flow of cerebrospinal fluid around the brain and spinal cord of vertebrate animals; the aqueous circulation of the eye; the flow of lymph through the lymphatic system of vertebrate animals.

circulatory failure. failure of the cardiovascular system to supply tissues with an amount of blood sufficient to support normal functioning.

circulatory system. any system that circulates materials (e.g., nutrients, respiratory gases, endogenous fluids) throughout the body of an organism. In higher plants and animals, circulatory systems consist of a series of vessels and/or spaces through which fluid flows, returning potentially to the point of origin. Circulatory systems are important toxicologically as a means of transport of toxicants and as a site of toxic action. Examples are the blood-vascular system and lymphatic system of vertebrates. **blood circulatory system**. a circulatory system in which blood is the principal fluid circulated. In vertebrate animals, it is the system of heart, blood vessels, lymph vessels, and the fluids (blood and lymph) contained within. See also blood-vascular system, cardiovascular system. *Cf.* vascular system.

circumneutral. pertaining to an organism that thrives in habitats of about neutral pH.

circumneutrophile. a living organism that thrives in conditions of near neutral pH.

circumneutrophilous. thriving in conditions of near neutral pH.

circumneutrophily. the condition or state of thriving in conditions of near neutral pH.

cirrhosis. **1:** hepatocirrhosis; a group of chronic, progressive inflammatory diseases of the liver characterzied by diffuse degeneration of parenchymal cells, infiltration of fats with single-cell necrosis, loss of the normal lobular architecture of the liver, nodular regeneration, and the development of dense perilobar connective tissue. It is associated with functional failure of hepatic cells and interference with hepatic circulation, often with jaundice, portal hypertension, ascites, and ultimately hepatic failure. Cirrhosis is sometimes caused by dietary deficiencies or by specific xenobiotics such as aflatoxins, carbon tetrachloride, or lead, but the most common causes are chronic alcoholism and hepatitis. **2a:** a progressive inflammatory disease of any tissue or organ; seldom used. **2b:** any condition affecting organs other than the liver that resembles cirrhosis of the liver; used primarily in veterinary medicine. **alcoholic cirrhosis**. that which results from chronic alcoholism. It is marked in early stages by hepatomegally due to fatty degeneration with mild fibrosis. It is later comparable to Laennec's cirrhosis with a reduction in size of the liver. **atrophic cirrhosis**. that in which the size of the liver decreases. **fatty cirrhosis**. early nutritional cirrhosis, seen mainly in alcoholics. The liver is enlarged by fatty infiltration, with mild fibrosis. **Laennec's cirrhosis** (portal cirrhosis). healthy liver lobules are replaced by small regeneration nodules separated by a rather regular meshwork of fine fibrous strands. It is usually due to chronic alcoholism. **portal cirrhosis**. Laennec's cirrhosis. **toxic cirrhosis**. hepatic cirrhosis due to chronic poisoning by substances such as carbon tetrachloride or phosphorus.

cirrhotic. pertaining to, characterized by, or affected with cirrhosis.

Cirsium arvense (Canada thistle). a spiny-leafed plant (Family Asteraceae) introduced to North America from Eurasia. Toxic concentrations of nitrate toxic to livestock have been reported.

cisplatin ((SP-4-2)-diamminedichloroplatinum; *cis*-diamminedichloroplatinum; *cis*-platinum II; *cis*-Pt II; *cis*-DDP; CACP; CPDC; DDP). a cytotoxic, nephrotoxic, neurotoxic, ototoxic, and carcinogenic antineoplastic agent. Cisplatin induces anaphylaxis in sensitive individuals. It is trypanocidal and is extremely toxic to laboratory mice and guinea pigs. It binds to DNA, thus inhibiting replication. Acute treatment of humans causes nausea, vomiting, delayed bone marrow depression, and renal damage.

citrate. **1:** any salt of citric acid and a base. They are naturally occurring chelating agents and are added to foods, pharmaceuticals, and household detergents. Citrates have lethally poisoned humans when given blood transfusions with blood containing citrate anticoagulants. When citrate accumulates in blood, the plasma concentration of ionic calcium falls, with consequent cardiac arrhythmias, reduced cardiac output, and death. **2:** citric acid. See citric acid.

citrate intoxication (citrate poisoning). a toxic condition in which the citrate combines with [illegible] in the blood and may produce tetany. It is an occasional consequence of massive replacement therapy with transfused blood that contains citrate as an anticoagulant.

citrate poisoning. citrate intoxication.

citreoviridin. a neurotoxic mycotoxin isolated from *Penicillium citreoviride*, *P. ochra-salmoneum*, *P. fellatum*, and *P. pulvillorum*. It is comprised of three moieties: α-pyrone chromophore, conjugated polyene, and a hydrofuran ring. It is extremely toxic to laboratory rodents. Acute poisoning of vertebrates by citreoviridin is marked by ascending paralysis and respiratory arrest. In endotherms, this is associated with hypothermia. Early symptoms of poisoning in various mammals are a progressive paralysis of the posterior extremities, impaired voluntary movement, tremors, vomiting, and convulsions. Symptoms of respiratory distress gradually develop. Hypotension and marked hypothermia are observed in more advanced stages of intoxication. Eventually dyspnea, gasping, and Cheyne-Stokes respiration appear. Death may follow from respiratory arrest due to inhibition of the respiratory center. Nervous symptoms are pronounced in larger laboratory mammals (monkeys, cats, and dogs), but are obscure in fish, amphibians, and reptiles.

citric acid (1,2,3-tricarboxylic acid). a slightly (laboratory rats) to moderately (laboratory mice) toxic, moderately strong, tribasic carboxylic acid, $C_6H_8O_7$, with colorless, translucent prisms or powder. It is odorless, with a strongly acidic taste, soluble in water and ethanol; and decomposes before boiling. It occurs in the juice of many fruits, especially citrus fruits. Lethal intoxication of laboratory rats is marked by signs of metabolic acidosis and calcium deficiency. It plays an important role in cellular respiration in both plant and animals cells as a component of the tricarboxylic acid cycle (citric acid cycle; Krebs cycle) wherein it is converted to isocitrate by aconitase, *q.v.* See citrate.

citric acid cycle. tricarboxylic acid cycle.

citrinin ((3*R-trans*)-4,6-dihydro-8-hydroxy-3,4,5-trimethyl-6-oxo-3*H*-2-benzopyran-7-carboxylic acid). a nephrotoxic mycotoxin and antibiotic produced by *Aspergillus niveus* and in small quantities by [illegible]. It is extremely toxic to laboratory rodents and rabbits.

citron amanita. *Amanita citrina*.

citrullinemia. a type of aminoaciduria due to

autosomal recessive inheritance and marked by elevated concentrations of citrulline in blood, urine, and cerebrospinal fluid. Symptoms include vomiting, ammonia intoxication, *q.v.*, and mental retardation from infancy.

citrullinuria. heightened urinary excretion of citrulline due to citrullinemia.

Citrullus colocynthus (colocynth; colocynthis; bitter apple; bitter cucumber). a pale yellowish-green plant (Family Cucurbitaceae) with a bitter-tasting, lethally toxic fruit. It is indigenous to the Mediterranean region of Africa, and Syria, but now occurs also in Central America. The dried unripe fruit is the source of colocynthin.

citrus red (1-[(2,5-dimethoxyphenyl)azo]-2-naphthol). a food color used to color the skin of mature oranges. It may be carcinogenic.

Cl. **1:** the symbol for chlorine. **2:** clearance.

Cl-PEST. chlorinated pesticides. Examples are DDT, dieldrin.

Cladosporium. a genus of imperfect fungi (Order Moniales; Family Dematiaceae). Some species are important parasites of plants and animals. Mycotoxicoses caused by these fungi include alimentary toxic aleukia and black degeneration of the brain. ***Cladosporium epiphyllum***. a species that produces the mycotoxins epicladosporic acid and fagicladosporic acid. See aleukia (alimentary toxic aleukia).

clam poisoning. paralytic shellfish poisoning (See shellfish poisoning).

Clara cell. a nonciliated, metabolically active, bronchiolar cell that contains a monooxygenase system capable of activating or detoxifying contaminants that enter the lungs. The Clara cells of some animals are characterized by the presence of a basal granular endoplasmic reticulum (ER) and an apical smooth ER.

Clarence river snake. *Tropidechis carinatus*.

Clarias batrachus (walking catfish). a venomous labyrinth catfish (Family Clariidae) that occurs from India to the Netherlands and has been introduced into the United States. It inhabits nearly stagnant water bodies (lakes, canals, and swamps) with muddy bottoms that support plants. See catfishes.

Clariidae (labyrinth catfishes). a family of bony marine fish native to India, Africa, Syria, and the region that includes the Philippines, Borneo, and Java. These fish have a unique respiratory organ arising from the gills that allows them to breathe air. Some are able to move about on land. *Clarias batrachus*, *q.v.*, has been introduced into the United States. See also catfishes.

Clark I. diphenylchlorarsine.

Clark II. diphenylcyanarsine.

class. **1:** a taxonomic group within the hierarchy of taxonomic classification that ranks between phylum or division and order. A class consists of a number of closely related orders; it is usually a large and easily recognized group of organisms. Several related classes form a phylum or subphylum. A class of plants takes the ending -opsida or -phyceae, whereas animal classes typically end with -a (some exceptions are the vertebrate classes Aves, Osteichthyes, Chrondrichthyes). **2:** in statistics, any category or group of like observations; a set.

classes of toxicants. See toxicants, types of.

classification of poisons. See toxicants, types of.

classification of toxicants. See toxicants, types of.

clastogen. any agent (e.g., chemicals, ionizing radiation) that has the capacity to alter the structure of chromosomes.

clastogenic. pertaining to the ability of an agent to alter the structure of chromosomes.

clastogenicity. pertaining to or denoting the capacity to alter the structure of chromosomes.

claustridium botulinum poisoning. botulism.

Claustridium botulinum. the causative bacterium of botulsim, *q.v.*

clavacin. patulin.

Clavaria formosa. former name of *Ramaria formosa*.

clavatin. patulin.

Claviceps (ergot). a genus of ascomycetous fungi(Family Clavicipitaceae) that parasitizes various grasses. A characteristic sclerotium develops in the ovary of the parasitized, developing flower. These contain highly toxic ergot alkaloids and amines which are the causative agents of ergotism. Cultivated plants that are parasitized include *Triticum aestivum* (wheat), *Secale cereale* (cultivated rye), and *Hordeum vulgare* (cultivated barley). Wild grasses (Family Gramineae) in North America that may become heavily ergotized and thus hazardous include *Bromus inermis* (awnless, smooth or Hungarian brome grass), *Poa spp.* (meadow grasses, speargrasses, bluegrasses), *Agropyron spp.* (wheat grasses, quack grass), *Elymus spp.* (wild rye), *Calamagrostis spp.* (reed bentgrass, bluejoint grass), *Agrostis alba* (redtop), *Phalaris arundinacea* (reed canarygrass), *Paspalum dilatatum* (Dallis grass), and *P. notatum* (Argentine bahia grass). ***C. cinerea***. a parasite of *Hilaria mutica* (tobosagrass) and *H. jamesii* (galletagrass). It occasionally causes neurotropic ergotism of cattle. ***C. paspali***. parasitizes *Paspalum dilatatum, P. notatum*, and certain other grasses. It is not known to cause ergotism in humans, but has been a fairly frequent source of neurotropic ergotism in livestock. ***C. purpurea*** (ergot; ergot of rye; sansert; St. Anthony's fire). a parasite of *Triticum aestivum*, *Secale cereale*, and *Hilaria vulgare*, as well as a number of wild grasses. It originated in Europe, but is now widely distributed throughout the world. It is extremely toxic and the most common cause of ergotism in humans. Poisoning stimulates involuntary muscles and paralyzes the sympathetic system. It is sometimes used as an abortifacient, but the effective dose is often fatal. Ergot is often found in obstetrical departments and is given to women after birth to help the uterus return to normal, 17th-century midwives used ergot for the same purposes. In the Middle Ages, ergotism caused by this species often reached epidemic proportions. In A.D. 944, 40,000 people in France died from ergotism.

Clavicipitaceae. a family of fungi (Order Clavicipitales) that includes the genera Claviceps and *Cordyceps*. All species have long, cylindrical asci and long filiform ascospores.

claviformin. patulin.

clavine alkaloid. See alkaloid.

clay pigeon poisoning. pitch poisoning.

clean room. a room in which elaborate safeguards are employed to limit dust particles and other contaminants in the air and on surfaces as required to protect against the contamination of delicate instruments or of substances being processed or under investigation.

cleaner (cleaning agent). any formulation designed for use in cleaning materials, objects, or surfaces. Numerous such products are available commercially; some are designed for very special purposes in the home, industry, and in hospitals. A few of the many types of cleaners are defined below. See also household cleaning product, cleaning product, potassium hydroxide, sodium hydroxide, hydrochloric acid, phosphoric acid, caustic. **abrasive cleaner**. a cleaner (e.g., chrome cleaner, copper cleaner) whose effectiveness depends partially or chiefly on the inclusion of an abrasive (e.g., pumice, silica). Such cleaners generally range from slightly to very toxic, but should be treated as dangerous. **acid cleaner**. a cleaner (e.g., certain individual metal cleaners, and some toilet bowl and dairy cleaners) whose action depends on the presence of caustic mineral acids (e.g., hydrochloric acid, phosphoric acid, oxalic acid). Examples are boiler cleaners. **alkaline cleaner**. a cleaner (e.g., some toilet bowl and dairy cleaners) whose action depends upon the presence of a caustic alkali, usually ammonium hydroxide or sodium hydroxide. **all-purpose cleaner**. any of numerous irritant formulations used in the home or the work place for general cleaning of surfaces. Such cleaners contain toxic ingredients that usually include ammonia water, artificial dyes, detergents, and fragrances. They are usually slightly to moderately toxic. See also household cleaning product, household product. Such cleaners are a frequent cause of household poisonings. **aluminum cleaner**. a cleaner that contains aluminum salts (e.g., a metal cleaner). These are generally considered moderately toxic. **basin, tub,**

and tile cleaner. any cleaner used to clean ceramic surfaces in the bathroom. They usually contain potentially toxic ingredients such as aerosol propellants, ammonia, detergents, ethanol, fragrances. All should be stored securely away from children. These products bear no warning labels. **drain cleaner**. a product used to chemically clean drains, usually of sinks. Lye, an extremely corrosive substance, is the chief component of such cleaners. A drop spilled on the skin, or a dry crystal in contact with wet skin can cause significant damage. If swallowed, the lye in a drain cleaner quickly erodes tissues of the gastrointestinal tract. The internal damage can be irreparable and fatal in such cases. Other potentially harmful ingredients often include petroleum distillates and sulfur compounds. Drano is a widely used drain cleaner. It contains a number of corrosive alkalies. If ingested, it rapidly liquefies and erodes deeply into the tissues lining the gastrointestinal tract, immediately causing intense pain that is promptly followed by other severe gastrointestinal symptoms. Vomiting and diarrhea precede collapse and death. The vomit of survivors is tinged with blood. Following initial symptoms, the victim may show improvement that may persist for one to four days. This period may terminate with the sudden onset of stomach pain, abdominal rigidity, rapidly falling blood pressure, and associated dizziness, blurred vision, headache, and fainting. These last symptoms indicate perforation of the esophagus or stomach wall. Necrotic tissue is shed in strips through vomiting; death is slow and painful, usually occuring by the third day. Lesions revealed on autopsy are gelatinous necrotic areas where the caustics penetrated the wall of the gut. Most drain cleaners, if mixed with ammonia, toilet bowl cleaners, household cleaners, or even other drain cleaners, release toxic fumes and may cause a violent eruption from the drain. **general purpose cleaner**. all-purpose cleaner. **glass cleaner**. a mixture of ammonia with water and a colorant, used to clean glass. Such cleaners can liberate highly irritant fumes which can damage the eye. The spray or mist, when aerosol spray containers are used, can be readily inhaled or get into the eyes. They are usually slightly or moderately toxic by ingestion. **metal cleaner** (metal polish). any paste or liquid used to clean and/or polish metal surfaces. They usually contain ammonium hydroxide and a variety of unknown petroleum distillates of varying toxicity. Other components that may be present and toxic include ethanol, synthetic fragrance, sulfur compounds. A typical warning label: "Danger: Harmful or fatal if swallowed. Combustible. Irritating to eyes. Contains petroleum distillates and ammonium hydroxide. If swallowed, do not induce vomiting. Call a physician immediately. Keep out of reach of children. Keep away from heat or flame." **mold and mildew cleaner**. any cleaner used to control mold and mildew. They often contain phenol, kerosene, or pentachlorophenol; formerly, they often also contained formaldehyde in the United States. All are toxic to humans by all routes of exposure. Death may result from ingestion. They are especially dangerous because they are packaged either as pump or aerosol sprays. **nonsoap cleaner**. any synthetic detergent. The term detergent originally meant cleaner. Such cleaners may be, in general, slightly to very toxic. **oven cleaner**. any of a variety of formulations used to clean baking ovens. They contain a variety of toxic substances, the most toxic of which are lye and other strong bases. Ingredients may also include ammonia, detergents, synthetic fragrances, aerosol propellants. Oven cleaners are powerful irritants to the skin, eyes, and mucous membranes. One should avoid contact with any part of the body and with clothing. Effects can be life-threatening if inhaled or ingested. Rubber gloves should be worn while using. Oven cleaners sprayed on by aerosol propellants are particularly hazardous because the spray broadcasts tiny, easily inhaled droplets of lye and ammonia into the air in the vicinity of the user; the droplets can also enter the eyes or come into contact with the skin. Symptoms of poisoning are similar to those of alkali poisoning. **rug and carpet cleaner**. any formulation used to clean carpeting. They are generally regarded as slightly to very toxic when ingested. Such products often contain perchloroethylene, the fumes of which are highly toxic and carcinogenic to humans. **toilet bowl cleaner**. any of a number of caustic formulations of varying strength designed to use on toilet bowls and other surfaces that are hard to clean. All are toxic (some are very toxic) and should be considered hazardous and kept secure from children. One should never mix them with other cleaning solutions. See drain cleaner. **upholstery cleaner** (upholstery shampoo). any product designed for use in cleaning upholstery. The active ingredient is usually perchloroethylene. Naphthalene, ethanol, ammonia, and detergents are also common components. They may vary from slightly to very toxic when ingested.

cleaning agent. See cleaner.

cleaning product. See household cleaning product. See also cleaner.

clearance (Cl). **1:** the rate of loss of a substance from an organism or part of an organism. It is a measure of the capacity to eliminate a substance metabolically (e.g., by decomposition, transformation, or excretion). Clearance is usually expressed as the volume of tissue from which the substance is removed in a given period of time. See factor (clearance factor). **renal clearance**. the rate at which a substance is removed from circulating blood by the kidney through secretion, tubular absorption (reabsorption), and filtration. **2:** removal of material from a site, as in the clearance of mucus from the respiratory tract. **systemic clearance**. clearance from all compartments of the body. Clearance (Cl) is related to the biological half-life ($t_{1/2}$) and volume of distribution (V_d) of a xenobiotic chemical by the equation:

$$Cl = 0.693V_d / t_{1/2}$$

See volume of distribution.

cleft-foot amanita. *Amanita brunnescens*.

clematis. a common name of plants of the genus *Clematis*.

Clematis (clematis; virgin's bower). a large genus of perennial, widely distributed temperate herbs (Family Ranunculaceae). All parts are toxic.

Cleome serrulata (stinking clover; Rocky Mountain bee plant). an annual herb (Family Capparidaceae) of prairies and waste areas of Manitoba and the western United States, east to Illinois, Missouri, and Kansas. Concentrations of nitrate toxic to grazing animals have been reported for this species.

Cleopatra's asp. a small venomous Egyptian snake to which Cleopatra's death is attributed. It is generally thought to be *Cerastes cerastes*, although the bite of this viper is not usually fatal to humans, or to Naja haje. See *Homo sapiens*, suicide.

cleroden (vermifuge). a naturally occurring epoxide used as a vermifuge.

climbing lily. *Gloriosa superba*.

climbing nightshade. *Solanum dulcamara*.

clinical. **1:** of or pertaining to activities based on actual observation or measurements of patients. **2:** pertaining to a clinic.

clinical effect. See effect (clinical effect).

clinical sign. See sign.

clinical toxicology. See toxicology.

clinotropism (klinotropism). an orientation response to a gradient of stimulation.

clioquinol. iodochlorhydroxyquin.

clitocybe. a mushroom of the genus *Clitocybe*.

Clitocybe (clitocybes). a genus of mushrooms (Family Tricholomataceae), some species of which contain muscarine. ***C. clavipes*** (fat-footed clitocybe). a mushroom (Family Tricholomataceae) that occurs under conifers, especially white pine, and sometimes under deciduous trees throughout much of North America. It is not toxic in the usual sense, but contains an enzyme, coprine, that inhibits the detoxification of alcohol when the latter is taken within 3 days following ingestion of the mushroom. If alcoholic beverages are consumed with this mushroom or during the following day or two, a transient condition lasting 2-4 hours may occur. Symptoms include peripheral vasodilatation with hypotension, flushing of the face, neck, and upper body, a tingling sensation in the extremities, headache, and sometimes nausea, vomiting and diaphoresis. See disulfiram. ***Clitocybe illudens***. former name of the eastern form of *Omphalotus olearius*.

cloak fern. *Notholaena sinuata*.

clofenotane. DDT.

clofibrate (2-(4-chlorophenoxy)-2-methylpropanoic acid ethyl ester). a nearly water-insoluble oil that is miscible with acetone, chloroform, ethanol, and ether. Clofibrate is used therapeutically as an antihyperlipoproteinemic drug (one that lowers the concentration in blood of very low-density plasma lipoproteins (and plasma triglycerides). It accelerates excretion of neutral steroids and inhibits cholesterol synthesis. It is moderately toxic to laboratory mice and rats. It may cause nausea, vomiting, diarrhea, weakness, stiffness, cramps, and muscle tenderness in some humans at therapeutic doses.

clonazepam. a moderately toxic benzodiazepine, *q.v.*, with anticonvulsant, antieptileptic, anxiolytic, and antimanic properties. It is a controlled substance listed in the U.S. Code of Federal Regulations (CFR).

clonic. pertaining to, characterized by, or of the nature of clonus; descriptive of rapid, pronounced alternate contraction and relaxation of muscles, especially that associated with seizure disorders. In tetrapods, such contractions are often seen as a "paddling" motion of the forelegs. See also convulsion.

clonic convulsion. See convulsion.

clonicity. the state or condition of being clonic.

clonicotonic. of or pertaining to spasms or convulsions that are both clonic and tonic.

clonicotonicity. the state of being clonicotonic.

clonidine (2,6-dichloro-*N*-2-imidazolidinylidenebenzenamine; 2-(2,6-dichloroanilino)-2-imidazoline; 2-(2,6-dichloroanilino)-1,-3-diazacyclopentene-(2); 2-[(2,6-dichlorophenyl)-amino]-2-imidazoline). an oral antihypertensive agent that lowers body temperature and pulse rate. It is an adrenergic that is usually administered as the hydrochloride (2,6-dichloro-*N*-2-imidazolidinylidenebenzenamine monohydrochloride), $C_9H_9Cl_2N_3 \cdot HCl$. It is very toxic to laboratory mice and rats.

clonidine hydrochloride. See clonidine.

clonism (clonismus). a condition of repeated clonic spasms.

clonismus. clonism.

clonus. a rapid or spasmodic alternation of muscular contraction and relaxation. *Cf.* tonus.

clophen. polychlorinated biphenyl(s).

clopoxide. chlordiazepoxide.

cloroformic acid trichloromethylcarbonochloridate. diphosgene.

clorox. a 5.25% solution of sodium hypochlorite, *q.v.*

clorox 2. a solution of sodium carbonate and sodium perborate.

closed segment technique. a method used to estimate intestinal absorption of a substance. A section of exposed intestine is ligated at each end and the substance introduced by a hypodermic syringe through one of the ligatures. The amount absorbed per unit time is determined by removing and analyzing the isolated segment for residual (unabsorbed) chemical. The amount absorbed is calculated as the amount injected less the amount still remaining in the preparation.

clostridium, pl. clostridia. common name of any member of the genus *Clostridium*.

Clostridium. a large genus of anaerobic (some are aerotolerant), mostly motile, spore-forming soil bacteria (Family Bacillaceae) that contain Gram-positive rods. Some species produce extremely potent exotoxins. ***C. botulinum***. the causal agent of botulism. This bacterium occurs naturally under anaerobic or slightly anaerobic conditions in soil, bottom sediments, and organic materials, especially vegetables. It is a common cause of food poisoning, especially from improperly preserved foods. There are six main types of *C. botulinum* that are differentiated by their secretion of antigenically distinct but toxicologically similar neurotoxins. Human botulism is caused mainly from types A, B, E, and F; type Cα causes botulism in birds; types Cβ and D cause botulism in cattle. Botulinum toxin binds irreversibly to nerve terminals, preventing the release of acetylcholine. The toxin is inactivated by heating at 80-100°C for a sufficient time. The toxin is one of the most potent known and is chemically similar to that of *Abrus precatorius*, the precatory bean. The toxin has been extensively investigated as a chemical warfare agent. ***C. histolyticum***. the causal agent of gas gangrene. ***C. perfringens***. the most widely distributed disease-causing organism known, occurring in soil, dust, on food, and in the intestinal tracts of humans and other warm-blooded animals. Ingestion of contaminated food results in diarrhea and abdominal pain within about 12 hours following ingestion. ***C. tetani***. a common anaerobic soil bacterium, it is the causal agent of tetanus (def.3). This bacterium forms spores that can remain alive in soil for many years. It most commonly enters the animal body via puncture wounds, growing in dead or damaged tissue that has an inadequate blood supply and thus low oxygen tension. The toxin (tetanospasmin) secreted *B. tetani* is a potent neurotropic agent that interferes with neurotransmitters, such as acetylcholine, causing tetanus (lockjaw).

clot (coagulum). a semisolid mass, as of blood or lymph. **blood clot**. a jelly-like clot formed within the body or externally by the agglutination of erythrocytes. **external clot**. a blood clot formed outside a blood vessel. **internal clot**. a blood clot formed within a blood vessel.

clot formation. See coagulation, blood.

clotbur. a plant of the genus *Xanthium*.

cloth of gold. *Baileya multiradiata*.

clotting. the activity of the blood and blood vessels that causes blood to form a jelly-like clot, usually near an injury that helps to slow or stop bleeding. The body's clotting mechanism is slowed or reduced ("thinning the blood") with anticoagulants used to treat certain diseases or by anticoagulant poisons or venoms.

clotting agent. coagulant.

clover. a common name of plants of the genus *Trifolium*. See also *Medicago*. **white clover**. *Trifolium repens*.

clover disease. trifoliosis.

Clupanodon thrissa (herring). a bony fish (Family Clupeidae) of the tropical Pacific, Indonesia, India, Japan, China, Formosa, and Korea. Certain populations are transvectors of ciguatera.

Clupeidae (shad, herring). a family mainly of small, long-bodied marine bony fishes (Class Osteicthyes) with cycloid scales that breed in inshore waters. They commonly form huge schools. Members of this family have long been known to cause sporadic, violent poisonings of humans in tropical seas, mostly near islands during the summer months. These have sometimes been clearly associated with ciguatoxin, but it is by no means clear that this is the sole source of clupeotoxism, *q.v.* See *Clupanodon thrissa*, ciguatera.

clupeoid. of or pertaining to fishes of the Family Clupeidae.

clupeoid fishes (anchovies, herrings, etc.). fishes of the family Clupeidae, *q.v.*

clupeoid poisoning. clupeotoxism.

clupeotoxin. possibly a form of ciguatoxin. See also Clupeidae, clupeotoxism. See clupeotoxism.

clupeotoxism (clupeoid poisoning). intoxication by ingestion of clupeoid fishes. In some cases, clupeoid poisoning may be due to ciguatoxin. The symptoms of clupeotoxism are not always identical to those of ciguatera, hence the relationship is not clear. In humans, the first sign of clupeotoxism is a sharp metallic taste which may present immediately following ingestion of the fish. A number of neurologic symptoms quickly follow and death may occur within 15 minutes. See also Clupeidae, *Clupanodon thrissa*, ciguatera.

C_m. clearance factor. See factor.

Cm. the symbol for curium.

CMC. carboxymethylcellulose sodium.

CMDP. See Mevinphos.

CMME. chloromethyl methyl ether.

CMP. 3-chloro-2-methylpropene.

CMU. monuron.

CNI. cyanogen iodide.

cnida (plural, cnidae). nematocyst.

Cnidaria. generally regarded as a subphylum of Coelenterata, but sometimes given phylum rank. Regardless, it is the more narrow term as it excludes the Ctenophora (comb jellies; See gooseberries) which are the non-nematocyst-bearing (nonstinging) coelenterates.

cnidarian. **1a:** pertaining to a nematocyst (= cnida). **1b:** pertaining to the Cnidaria. **2a:** a member of the subphylum (phylum) Cnidaria. **2b:** any stinging coelenterate.

cnidoblast. a type of specialized epidermal cell that houses a nematocyst. When it is stimulated by dissolved substances associated with prey organisms, the nematocyst is triggered.

cnidocell. a term used variously to mean nematocyst or cnidoblast.

cnidocil. **1:** nematocyst. **2:** the sensory bristle of a nematocyst that causes release of the stinging thread.

cnidocyst. nematocyst.

cnidophore. a structure (a modified zooid) that bears nematocysts.

cnidopod. the stalk-like "foot" of a nematocyst.

Cnidoscolus. a genus of perennial herbs and shrubs (Family Euphorbaceae) with alternate, palmately veined, usually long-petioled, lobed, or divided leaves, and stipules with stinging bristles. ***C. stimulosus*** (**spurge nettle; stinging spurge; bull nettle; tread-softly**). a short herb with a stout stem and deep tap-root. It is covered by bristly, stinging hairs (2-6 mm long). The leaves are alternate with 3 to 5 deeply cut lobes and white, showy flowers.

cnidosis. See urticaria.

CNS. **1:** central nervous system. **2:** the thiocyanate radical (CNS, —CNS).

—CNS. thiocyanate radical.

CNS depressant. See depressant.

CNS stimulant. See stimulant.

Co. the symbol for cobalt.

CoA. coenzyme A.

Co I. nadide.

^{57}Co. cobalt-57. See cobalt.

^{58}Co. cobalt-58. See cobalt.

^{60}Co. cobalt-60. See cobalt.

co-actone. **1:** allelochemical (def. 2) **2:** any chemical that transfers information between living organisms.

co-carcinogen. a chemical that enhances the action of a complete carcinogen when administered simultaneous with or before that of the carcinogen. Such an agent acts by increasing absorption or bioactivation, or by inhibiting detoxification of the carcinogen, *q.v.*

co-carcinogenesis. the enhancement of chemical carcinogenesis by administration of a co-carcinogen at about the same time as that of the carcinogen. *Cf.* promotion, co-carcinogen.

co-evolution. herbivores, granivores, and frugivores accumulate energy and biomass by feeding directly upon plants. Certain metabolic and behavioral attributes of plants and herbivores result from of co-evolution. Most plant species deficient in some chemicals (e.g., essential amino acids) that are essential components of a herbivore's diet. Most herbivores thus feed upon a diversity of plant material. Many plants also produce toxic chemicals, many of which are detoxified by, and harmless to, associated herbivores. One would expect herbivores to be preadapted to at least certain poisons of human origin. Many herbivorous insects are resistant to pesticides.

co-product. by-product.

coagulant (clotting agent). **1:** an agent that causes a liquid such as blood or lymph to coagulate. **2:** able to cause coagulation. **blood coagulant**. an agent that causes blood to coagulate following, for example, damage to a blood vessel. See also coagulation, snakebite, venom (snake venoms).

coagulase. coagulating enzyme.

coagulate. **1:** to clot, to form a clot (e.g., a blood clot). **2:** to congeal, curdle, solidify, thicken. See also coagulation. **3:** to form a coagulum (See coagulation, def. 2).

coagulated. clotted or curdled.

coagulation. **1a:** the process of clotting. **1b:** the alteration of a dispersed system into a liquid phase and an insoluble mass (the clot or curd). **2:** the disruption of tissue with the formation of amorphous mass (coagulum) as by photocoagulation or electrocoagulation. **blood coagulation** (clot formation). the process of blood clot formation by conversion of whole blood or plasma into a gel. Blood coagulation is a complicated process involving the sequential activation of a number of soluble proteins (clotting factors) in blood plasma with consequent production of a fibrin network. Coagulation can be initiated by various blood disturbances, by injury to the blood vessel walls, or the introduction of coagulants in the cardiovascular system. If the blood coagulation system is deficient, uncontrolled bleeding may result from even slight trauma, external or internal. Anticoagulants (e.g., warfarin, hydroxycoumarins, many snake venoms) that impair the blood coagulation system can produce severe, even fatal hemorrhage. See, for example, coumarin, cytochalasins, fibrin.

coagulum. See clot.

coal dust. See dust.

coal gas. See gas.

coal miner's lung. anthracosilicosis.

coal tar. a black viscous, flammable liquid or semisolid byproduct of the destructive distillation of bituminous coal. It is a complex mixture of hydrocarbons including anthracene, benzene, cresol, naphthalene, phenol, toluene, and xylene. Other constituents are ammonia, pyridine, and certain other organic bases, thiophene, and various impurities. Coal tar is an irritant to the skin and damaging to the eye; it is toxic by inhalation, and is linked to cancer of the lung, larynx, scrotum, and urinary bladder of humans.

coal tar naphtha. See naphtha.

coal tar pitch poisoning. pitch poisoning.

coal tar poisoning. pitch poisoning.

coal worker's pneumoconiosis. anthracosilicosis.

coast goldenbush. *Haplopappus venetus*.

coating agent. an excipient that forms a coating about a drug or toxicant formulation. Commonly used coating agents are carnauba wax, gelatin, and shellac.

cobalt (Co)[Z = 27, A_r = 58.9]. a silvery white transition element with one naturally occurring, widely distributed isotope, ^{59}Co. Several artificial, radioactive isotopes are known (See below). It is used in the manufacture of extremely hard steel and cutting tools. As a component of vitamin B_{12}, it is essential to life. The soluble salts of cobalt are toxic and irritant. Cobalt powder can cause a contact dermatitis and inhalation of cobalt-containing dust can cause a pneumoconiosis; chronic exposure can cause hard-metals disease, *q.v.* Cobalt causes polycythemia by stimulating erythropoiesis. Long exposure to repeated high doses of cobalt may be carcinogenic or mutagenic. It also causes an erythematous papular type of contact dermatitis. **cobalt-57** (^{57}Co). a radioactive isotope, $t_{1/2}$ = 272 days, that emits medium-energy gamma rays. **cobalt-58** (^{58}Co). a radioactive isotope, $t_{1/2}$ = 72 days, that emits positrons. **cobalt-60** (^{60}Co). a radioactive isotope, $t_{1/2}$ = 5.26 years. It emits beta particles and high-energy gamma rays. It is used in radiation therapy and diagnosis.

cobaltosis. a pneumoconiosis due to the inhalation of cobalt dust.

cobra. **1:** a true cobra is any of the highly venomous snakes of the genus *Naja* (Family Elapidae). With the exception of *Naja naja*, they occur only in Africa; none, however,occur in the [illegible] The body is moderately slender, slightly depressed, and tapered; the tail is of moderate length. The eyes are of moderate size with round pupils. All cobras have two rather large tubular fangs with external grooves followed by a space and 0-3 small teeth. The neck typically spreads into a "hood" and one-third of the body rears off the ground if the snake is irritated; some species can spit venom when in this posture. Cobras range in size from about 1.3 to 2.6 m. The lethal dose of the venom for humans is only about 20 mg, yet the venom glands of large cobras may contain as much as 500 to 600 mg venom. Even so, recovery from cobra bites without effective treatment is not uncommon. **2:** *Ophiophagus* (the king cobra, Family Elapidae). **3:** certain other cobra-like elapid snakes (in particular those that lack a loreal shield and often having a hood, or at least the suggestion of a hood) are also termed cobras; for example, *Oxyuranus scutellatus* (taipan), *Hemachatus haemachatus* (ringhals cobra, spitting cobra), *Paranaja spp.* (burrowing cobras), *Pseudohaje* (tree cobra), *Dendroaspis*, *Walterinnesia aegyptia* (desert cobra). **water cobra**. See *Boulengerina*.

cobra hemotoxin. a component of cobra venom which is able to lyse human erythrocytes in the absence of blood serum.

cobra lecithinase. See phospholipase A.

cobra phospholipase. See phospholipase A.

cobra venom cofactor. See factor (cobra venom factor).

cobra venom poisoning (cobraism). envenomation of humans by cobras is generally marked by pain within a few minutes of being bitten, followed by slowly developing edema that is initially localized about the wound, weakness, salivation, and paresis of the facial muscles, lips, tongue, and larynx. Additional effects may include hypotension, a weak pulse, dyspnea. More severe cases may be further marked by a generalized muscular weakness or paralysis. Serious bites are often fatal, but recovery without medical treatment is not uncommon.

cobra venom solution. minute amounts of cobra venom in sterile physiological salt solution.

cobraism. cobra venom poisoning.

cobralysin. a hemolytic substance isolated from cobra venom.

cobratoxin (cobratoxin). **1:** the principal toxic protein of cobra venom. Its cytolytic activity is similar to that of melittin. It is a single-chain polypeptide with 62 amino acid residues that are cross-linked within the molecule by four disulfide bonds. **2:** sometimes used as a synonym for cobra venom.

cobriform. resembling or related to the cobras.

cobrotoxin. cobratoxin.

cobweb spider. a spider of the genus *Steatoda*.

coca. **1:** common name of any of several South American shrubs of the genus *Erythroxylon*, especially *E. coca*. **2:** the dried leaves of the two coca plants (*E. coca* and *E. novagranatense*) which contain several alkaloids, most importantly cocaine. See *Erythroxylon coca*.

cocaine ([1*R*-(*exo-exo*)]-3-(benzoyloxy)-8-methyl-8-azabicyclo[3.2.1]octane-2-carboxylic acid methylester; 3β-hydroxy-1αH,5αH-tropane-2β-carboxylic acid methyl ester benzoate; 2β-carbomethoxy-3β-benzoxytropane; ecgonine methyl ester benzoate; *l*-cocaine; benzoylmethylecgonine; crack). an extremely toxic alkaloid, cocaine is a white powder or colorless crystalls, $C_{17}H_{21}NO_4$, extracted from the leaves of *Erythroxylon coca* and *E. novagranatense* or synthesized from ecgonine or its derivatives. Cocaine initially stimulates, then depresses the central nervous system. Toxic doses can elevate blood pressure, cause tachypnea, palpatations, and tachycardia or cardiac arrhythmia. Symptoms of acute intoxication may initially include stimulation with euphoria (or dysphoria at high doses), restlessness, anxiety, hallucinations, nausea, vomiting, convulsions, with subsequent depression, loss of reflexes, coma, and death. Cocaine has pronounced psychotropic activity. Sensitive individuals may experience an immediate allergic response. Chronic usage can cause a condition that is nearly identical to schizophrenia. Cocaine and its derivatives are addictive and are controlled substances under the U.S. Code of Federal Regulations. See also coca, crack, erythroxyline, free basing.

***l*-cocaine**. cocaine.

cocaine hydrochloride. the hydrochloride salt of cocaine. It is a topical anesthetic, applied to mucous membranes. It is frequently used in extremely low concentrations as a topical anesthetic in nasal surgery to prevent excessive bleeding by vasoconstriction. See also poisoning (cocaine hydrochloride poisoning).

cocaine hydrochloride poisoning. poisoning by overdosage or abuse of cocaine hydrochloride. The nervous system is initially stimulated as evidenced by excitement, incoherence, restlessness, and possibly hallucinations. This phase is followed by nausea, profound depression, dizziness, a tingling sensation in the extremities, fever, hypertension, cardiac arrhythmias, hyperpnea, and occasionally convulsions, collapse, and death from respiratory arrest. In acute cases, with proper treatment, survival for 3 hours offers a good prognosis. **cocaine hydrochloride poisoning, chronic**. cocainism (def. 2).

cocainism. **1:** habitual use of cocaine. **2:** (chronic cocaine hydrochloride poisoning). the morbid condition caused by prolonged use of cocaine.

cocculin. See picrotoxin.

cocculus (cocculus indicus; fish berry; Indian berry; oriental berry). the dried, very poisonous bean-shaped fruit of the woody vine, *Anamirta cocculus*. The dried berry contains about 50% fat, cocculine alkaloid, menispermine, *para*-menispermine, picrotoxin, and picrotoxic acid. In humans, intoxication produces gastrointestinal symptoms, with restlessness, pyrexia, convulsions, and coma. The active principle is picrotoxin. Dried cocculus is used locally to stupefy fish and in an ointment to control pests. It is used therapeutically as a central and respiratory stimulant.

cocculus indicus. cocculus.

Cochlearia amoracia. *Amoracia lapathifolia*.

Cochliobolus sativus. a phytotoxic "fungus" that infects cereal grains.

cochliodinol. an extremely toxic mycotoxin from the ascomycetous fungus, *Chaetomium*.

cockerell. *Hymenoxys lemmoni*.

cockle. See *Cardium edule*.

cockle, corn. any plant of the genus *Agrostemma*, especially *A. githago*.

cocklebur. common name of *Xanthium* and of *Agrimonia*.

coconut crab. *Birgus latro*.

COD. chemical oxygen demand.

codehydrogenase. nadide.

codeine ((5α,6α)-7,8-didehydro-4,5-epoxy-3-methoxy-17-methylmorphinan-6-ol; methylmorphine; morphine monomethyl ether; morphine monomethyl ether; morphine 3-methyl ether). an alkaloid that comprises from 0.7 to 2.5% of the content of *Papaver somniferum*. It is usually obtained, however, by methylation of morphine. Codeine finds use as a mild oral analgesic, a hypnotic sedative, and antitussive. Because it acts on the CNS, codeine complements aspirin, acetaminophen, and other peripherally acting analgesics. Its effects are similar to those of morphine; it is probably less addictive. It is analgesic *per os*, 65 mg being equianalgesic to about 600 mg of aspirin. Codeine is more effective orally as an antitussive than morphine and is also useful in treating diarrhea. deleterious effects (respiratory depression, nausea, vomiting, and constipation) are similar to those of other opioids. Codeine hydrochloride and codeine phosphate are very toxic to laboratory mice and rabbits. See also *Papaver*, codeine poisoning.

codeine hydrochloride. See codeine.

codeine phosphate. See codeine.

codeine poisoning. poisoning by an acute overdose of codeine. The chief symptoms are those of CNS depression, including the centers that control breathing and heart rate. Death may result from cardiopulmonary failure.

codon. the basic unit of the genetic code comprising a sequence of three nucleotides in messenger RNA which codes for one amino acid in protein synthesis. See anticodon.

coefficient. **1a:** acting jointly to produce an effect. **1a:** something that acts jointly with another to produce an effect. **2a:** a numerical measure of the amount, degree, or intensity of some property of a substance. **2b:** any factor (as a constant) in a product considered in relation to another factor (a variable). See absorption coefficient, activity coefficient, correlation coefficient, temperature coefficient.

coefficient of activity (activity coefficient). a term that relates the actual concentration of a substance to its effective concentration (= activity). In the case of solutions, it is a measure of the deviation of a more or less concentrated solution from the ideal and is derived from theoretical considerations of ion-ion and ion-solvent interactions. The activity coefficient of an ideal or pure substance is unity.

coefficient of variation (CV). standard deviation expressed as the percentage of a mean value.

Coelenterata. a major phylum of aquatic, mostly marine, radially symmetric, acoelomate invertebrates with a diffuse nerve network. There are four classes (and some 10,000 living species): Hydrozoa, Scyphozoa, Anthozoa, and Ctenophora. The body is sac-like, with a body wall comprised of two cellular layers (epidermis and gastrodermis) separated by a noncellular jelly-like layer (mesoglea) that encloses a central cavity (coelenteron) with a single opening (proctostome) to the exterior that functions as both mouth and anus. A ring of nematocyst-bearing tentacles surrounds the mouth. The nematocysts can in some cases penetrate human skin. Venomous species occur in all classes except the Ctenophora; the toxins are unusual. At least 80 species are known to injure humans. See also Cnidaria.

coelenterate. common name of any animal of the phylum Coelenterata, including the Ctenophora (sea combs, sea gooseberries).

coenobiology. biocenology.

coenzyme. an organic molecule that acts as an enzyme cofactor (e.g., coenzyme A, NAD, FAD, pyridoxal phosphate. Some coenzymes form a tightly bound prosthetic group (e.g., NAD and NADP) that acts as an acceptor of electrons or functional groups. Such coenzymes are not easily removed.

coenzyme A (CoA). a coenzyme comprized of adenine, ribose, three phosphate groups, pantothenic acid, and mercaptoethylamine. It plays a key role in acetylation.

coenzyme I. nadide.

cofactor. a non-protein substance that is indispensable to the activity of certain enzymes. Cofactors may be metal ions or organic molecules (e.g., coenzyme A, NAD, pyridoxal phosphate). A cofactor imparts activity by forming a bridge between the substrate and the enzyme or by participating in the reaction.

coffee. **1:** the seed of the coffee tree, genus *Coffea*. These beans when roasted are pulverized and brewed with hot water to form a widely used beverage. Coffee is grown chiefly in countries where highly toxic pesticides are still used. **2:** the beverage prepared from coffee beans. Many scientists say that drinking coffee is not harmful. Nevertheless, it contains a number of harmful ingredients that may include not only caffeine, but also flavorings, hexane, methylene chloride, pesticide residues, sucrose, trichloroethylene. The scientific literature on the possible toxic effects of coffee, especially through long-term usage, is confusing. Nevertheless there are clear indications that the caffeine in coffee produces ill effects (alone, or in association with other stressors) in many people, especially habitual drinkers. These include anxiety, insomnia, depression, stomach upset, ulcers, headache, and heart attack. Pregnant women should limit their intake of caffeine; in large quantities it promotes miscarriage, premature birth, and birth defects. Coffee drinking is statistically linked to cancer of the pancreas.

coffeebean. See *Sesbania*.

coffeeweed. *Sesbania*.

coffeine. caffeine.

coffin snake. *Vipera lebetina*.

cohoba. **1:** *Piptadenia peregrina*. **2:** a narcotic snuff made from seeds of *Piptadenia peregrina*. It contains bufotenine.

cohort. **1:** an age class; a group of like individuals that are recruited into a population at the same time. **2:** a taxonomic category comprising a group of related families.

cohort study. an epidemiological study in which a cohort of individuals is followed into the future in order to elucidate factors which may, for example, affect toxicity.

cohosh. See *Actaea*.

coke. the solid residue remaining from the distillation of coal at temperatures above 800°C). The sulfur content, and hence sulfur emissions from the combustion of coke, is about the same as that of the coal from which it was produced. Coke combustion, however, produces very little particulate pollution.

coke-oven emissions (coke-oven fumes). emissions from massive industrial ovens that produce coke by heating coal in the absence of air. These emissions contain solid particles, benzene, and traces of arsenic and cadmium. Coke-oven workers incur the risk of genetic damage, bronchitis, and emphysema. They are also at risk for lung, prostate, and kidney cancer. See also coke. Coke is used in steel-producing blast furnaces and in the extraction of metals from ore.

coke-oven fumes. See coke-oven emissions.

Coker's amanita. *Amanita cokeri*.

COL. cortisol. See hydrocortisone.

COLA. hydrocortisone acetate (cortisol acetate).

colaspase. asparaginase.

colcemid. colchicine.

colchamine. demecolcine.

colchicine (*N*-(5,6,7,9-tetrahydro-1,2,3,10-tetramethoxy-9-oxobenzo[*a*]heptalen-7-yl)acetamide; colcemid). a fine, yellow, very bitter, heat-stable powder, $C_{22}H_{25}NO_6$, first obtained from the corms of *Colchicum autumnale*. It is an extremely toxic, neurotoxic alkaloid that inhibits axonal transport. Colchicine is also a mitotic poison (specifically a spindle poison) and mutagen. It promotes disassembly of microtubules by binding to dimers of the protein, tubulin, inhibiting its polymerization. This interrupts mitosis at metaphase, resulting in polyploidy, a usually lethal mutation in animals. Symptoms of intoxication in humans include nausea, vomiting, abdominal pain, purging, extreme thirst, and cold and painful extremities. A heavy overdose can cause muscular weakness, hypothermia, bone marrow depression, vascular damage, thrombocytopenia, and alopecia in survivors.

colchicine-binding protein. tubulin.

Colchicum (autumn crocus; meadow saffron). a genus of poisonous cormaceous plants (Family Lilaceae). They are unrelated to the true crocus (*Crocus spp.*), although the flowers are very similar. Poisoning by these colchicine-containing plants occurs after a partially dose-dependent latent period of several hours. Such poisoning may result from ingestion of leaves, seeds, bulbs, or flowers. Certain of these flowering plants have been used successfully to commit suicide. They are believed to cause defects in calves similar to crooked calf disease. Poisonings have occasionally resulted when persons ate the tubers of the glory lily (*Gloriosa superba* and *G. rothschildiana*), thinking them to be sweet potatoes. ***C. autumnale*** (autumn crocus; fall crocus; meadow saffron; naked ladies). a small (ca. 1/3 m tall), frequently cultivated, poisonous perennial herb (Family Liliaceae). This extremely dangerous plant is native to the United Kingdom and mainland Europe, occurring also throughout much of Eurasia. It frequents damp meadows and woodsy areas of England and Wales and some parts of Scotland. Colchicine, the chief active principle, is concentrated in the corms (solid bulbs at the base of the stem). Seeds, leaves, and flowers are also toxic, causing a burning pain in the mouth and stomach, with symptoms of gastroenteritis, anuria, hematurea, and prostration. The corm has sometimes been fatally mistaken for an onion. Livestock are usually poisoned by consuming the plant, but humans and suckling animals may be poisoned by drinking the milk of an affected animal. Symptoms in most animals are mainly those of gastrointestinal irritation, including a burning pain in the mouth and stomach, prostration, hematuria, and anuria.

cold damp. See damp.

colgadora. *Bothrops schlegelii*.

colic 1: pertaining to the colon. **2:** acute, usually intermittent, painful spasms of any visceral organ. **3:** in young infants, spasmodic gastrointestinal pain evidenced by crying and irritability. **copper colic**. severe colic due to chronic poisoning by copper; it is similar to lead colic. **Devonshire colic**. lead colic. **lead colic** (Devonshire, painter's, or saturnine colic). colic with severe abdominal pain and constipation due to lead poisoning. **milk colic**. enterotoxemia. **painter's colic**. lead colic. **Poitou colic**. lead colic. **saturnine colic**. lead colic. **zinc colic**. colic due to zinc poisoning.

colicin. a bacteriocin produced by coliform bacteria.

colicinogenic. producing colicins.

colicky. denoting, resembling, or affected by colic or the symptoms of colic.

coliform bacteria. Gram-negative, peritrichous, rod-shaped eubacteria found in the intestinal tract of higher animals. They obtain energy by fermentation or aerobic respiration. They may contaminate water via fecal pollution and are a potential source of intoxication or disease.

Coliform Index. a rating of the purity of water based on a count of fecal bacteria in a sample.

Colinus virginianus (bobwhite, common bobwhite, bobwhite quail). a galinaceous bird (Family Phasianidae), about the size of a meadowlark, that occurs from the central and eastern United States southward to Guatemala and to Cuba. It is a standard test species in wildlife toxicology.

colitis. inflammation of the colon.

colitoxemia. toxemia caused by *Escherichia coli*, the colon bacillus. *Cf.* colitoxicosis.

colitoxicosis. systemic poisoning by colitoxin.

colitoxin. a toxin produced by *Escherichia coli*, the colon bacillus. It is a systemic poison and the causative agent of colitoxicosis.

collagen. a fibrous insoluble protein of connective tissue. It comprises, in humans, approximately 30% of total body protein, from which body tissues are formed. See also collagenase.

collagenase. a rare proteinase that catalyzes the hydrolysis of collagen. Collagenase activity has been demonstrated in several venoms of crotalid and viperid snakes; it also occurs in certain species of *Clostridium*.

collapse. **1:** to break down completely. **2:** any sudden condition of extreme weakness, exhaustion, or cessation of activity due to hypovolemic shock. **3:** a state of extreme physical depression or prostration with failure of circulation. **4:** an abnormal caving in of the walls of any part of a hollow organ. **5:** in toxicology (usually): any sudden weakness or cessation of normal activity as in exhaustion, prostration, or cardiovascular or respiratory failure or insufficiency. There are a range of symptoms similar to those of hemorrhage or shock. They include apathy or loss of consciousness, hypotension, extreme pallor due to loss of blood in the peripheral circulation, a weak rapid pulse, and venous congestion especially in the splanchnic region. *Cf.* failure. **cardiovascular collapse**. complete failure of the cardiovascular system. **circulatory collapse**. shock; severe insufficiency or complete failure of the peripheral or cardiac circulation other than that due to congestive heart failure. **respiratory collapse**. extreme insufficiency of external respiration.

collapse. **1:** failure (e.g., of a vital structure, system, or of the entire organism), breakdown; extreme, sudden prostration; disintegration. **2:** seizure, paroxysm.

collapsing. **1:** falling into extreme and sudden prostration that resembles shock; failing, breaking down. **2:** experiencing a seizure.

collared brownsnake. *Demansia textilis*.

collared snake, Australian. See *Glyphodon*.

collidine. any of a number of usually oily, acrid, toxic organic bases, $C_8H_{11}N$. They are the methyl, ethyl, propyl, and trimethyl homologs of pyridine and are by-products of the coking process, or are synthesized. **2,4,6-collidine**. a toxic liquid, slightly water-soluble and soluble in ethanol, used as a chemical intermediate.

collier's lung. anthracosis; sometimes applied to anthracosilicosis.

colloid. any material (gel, sol, aerosol) comprised of particles with diameters less than 10 μm.

colloidal ferric oxide. iron oxide.

Colocasia (elephant-ear). a genus of large-leaved tropical herbs (Family Araceae) commonly cultivated for their ornamental foliage. All parts of the plant contain calcium oxalate and other toxic principles. When plant material is chewed, the needle-sharp crystals of calcium oxalate are released by ruptured plant cells and penetrate the tissues of the mouth and tongue. This quickly produces a severe burning sensation and irritation of the mucous membranes of the mouth, lips, and tongue, with swelling of the tongue and throat and nausea, vomiting, diarrhea, and salivation. *Cf. Caladium*, commonly referred to as elephant's ear. ***C. esculenta*** (taro; eddo). a plant widely cultivated in warm climates, especially on islands of the Pacific Ocean. The tubers are starchy and cooked like potatoes; the sprouts (often called dasheen) are edible.

Colocephali. an order of mostly tropical and subtropical marine bony fishes that includes moray eels (Family Muraenidae). They are brightly colored and voracious. *Muraenai* typical.

colocynth (colocynthis; bitter apple; bitter cucumber). **1:** dried pulp of fullgrown but unripe fruit of *Citrullus colocynthus* (Family Cucurbitaceae). It is powerfully cathartic. See also colocynthin, colocynthidism. **2:** *Citrullus colocynthus*.

colocynth poisoning. colocynthidism.

colocynthidism (colocynthpoisoning). poisoning by the unripe fruit of *Citrullus colocynthus*, by colocynth, or by colocynthin. It is characterized in humans by blood-tinged diarrhea, cramps, headache, oliguria, kidney failure, and death on the day of poisoning. See colocynthin.

colocynthin (25-(acetyloxy-2-(β-D-glucopyranosyloxy)-16,20-dihydroxy-9-methyl-19-norlanosta-1,5,23-triene-3,11,22-trione; 2-*O*-β-D-glucopyranosylcucurbitacin E). a bitter, highly toxic, purgative glycoside, $C_{38}H_{54}O_{13}$, isolated from the dried unripe fruit (colocynth) of *Citrullus colocynthis*. It is used as an insecticide and sometimes as a purgative and as an abortifacient. *Cf.* cucurbitacin.

colocynthis. **1:** colocynth. **2:** *Citrullus colocynthus*.

Cologne yellow. lead chromate.

colon. **1:** the portion of the large intestine of vertebrates that extends from the cecum to the rectum. It is the site of much reabsorption of water and electrolytes and a frequent site of cancer from chemical carcinogens. **2:** the second portion of the intestine of insects.

colon bacillus. *Escherichia coli*.

color. **1:** the characteristic sensations produced by light waves that strike the retina of the eye of vertebrates such as birds and a few mammals such as humans and other primates. **2:** pigment, tint, hue. **3:** dye, stain, pigment, tint. **4:** to dye, paint, stain, or tint. **5:** a substance that is added to food or other materials to confer a specific, desirable color to the substance than is natural or normal. *Cf.* colorant. **artificial color**. any synthetic or manufactured color that does not also occur naturally. Such colors are nearly always coal-tar derivatives. Almost all are carcinogenic. Some have been linked to hyperactivity and behavioral disturbances in children. *Cf.* butylated hydroxyanisole, butylated hydroxytoluene. **D&C color**. any synthetic dye or pigment derived from coal tar that can be used in the United States (in certified batches) in drugs and cosmetics, but not in foods. **exterior D&C color**. any D&C color that is approved for exterior use only; such colors should not be applied to the lips or mucous membranes. "EXT. D&C" appears on the label in front of the color listing. "Traces of D&C" appearing on a label indicates that a compound of aluminum, calcium, barium, potassium, strontium, or zirconium has been added to the dye. Types of D&C colors include azo dyes which contain phenol and are easily absorbed through the skin; anthraquinone dyes; and aniline dyes. The latter, if ingested, can cause intoxication, a reduction in blood oxygen levels, dizziness, headaches, and mental confusion. **FD&C color**. any color that can be used in foods, drugs, and cosmetics (FD&C) in certified batches in the United States. Included are many artificial colors and U.S. certified colors. Animal studies have shown almost all FD&C colors to be carcinogenic. In general, the safety of coal-tar colors is assessed by the U.S. Food and Drug Administration (FDA) to be safe/unsafe on the basis of results from tests of acute oral toxicity, primary irritation, sensitization, subacute skin toxicity, and carcinogenicity by skin application. Only six such colors have been permanently listed as being "safe," although most are known animal carcinogens. See also E number. **HC color**. any color approved only for coloring hair. Colors thus classified include aniline, azo, and peroxide dyes. Effects of peroxide dyes may include skin rash, eczema, bronchial asthma, gastritis, occasional complications, and death.

color index. see Colour Index.

colorant. any substance that imparts color to another substance, material, or formulation. Dyes, pigments and colors (defs. 3 and 5). Naturally occurring colorants include chlorophyll. *Cf.* color, pigment, dye.

coloration. See warning coloration.

colorimeter. an instrument that measures the absorption of light by a colored solution, thus enabling the determination of the colored constituent in the solution. Light (often of a particular wavelength, or group of wavelengths, obtained by means of a filter) is passed through a sample of standard width, the amount that is transmitted being measured either visually or photoelectrically.

coloring. See color.

Colour Index (C.I.). a British system of classification of colors according to hue, brightness, and saturation as an aid to identification or reproduction. See also The Colour Index.

colubrid. **1a:** pertaining to or denoting the family Colubridae. **1b:** any snake of the family Colubridae. See colubrine, colubriform.

colubrid snake. any snake of the family Colubridae. *Cf.* colubrid.

colubrid venom poisoning. poisoning by venomous colubrid snakes varies in severity. Bites of the boomslang and bird snake are often severe and sometimes fatal. Effects of the bites of a number of colubrids are poorly known, but are usually associated with pain and sometimes discoloration about the bite, ecchymosis. In severe cases, there is swelling and edema of the entire affected appendage, fever, and malaise. Colubrid venom poisoning is generally less serious than that of vipers and elapids.

Colubridae. the largest and most cosmopolitan family of snakes, comprising more than 3/4 of the known species of snakes. The family is morphologically, behaviorally, and ecologically diverse. Representative genera include *Coluber, Natrix, Thamnophis, Masticophis, Pituophis, Lampropeltis, Bioga, and Tantilla*. Both jaws hold solid or grooved teeth, but no enlarged hollow fangs. In most cases, the head is wider than the neck. Most species are nonvenomous. Africa is the only continent that has dangerously venomous colubrid snakes and two of these, *Dispholidus typus* (boomslang) and *Thelotornis kirtlandi* (the bird snake), can inflict fatal bites.

colubriform. **1a:** a colubrid. **1b:** colubroid; resembling a colubrid.

Colubrinae. in earlier classifications, a subfamily of the Colubridae that is nearly coextensive with the family.

colubrine (colubroid, in part). **1a:** of or pertaining to subfamily Colubrinae. **1b:** relating to or similar to a snake. **2:** either of two colorless crystalline alkaloids (α-colubrine and β-colubrine) that occur naturally with strychnine; the general formula is $C_{22}H_{24}N_2O_3$.

colubroid. **1a:** colubrine (def. 1a, 1b). **1b:** colubriform. **1c:** a colubrid.

columbium. the former name of niobium.

Columbus grass. See Sorghum.

coma (exanimation). a state of deep, sometimes prolonged, unconsciousness or deep stupor. Five levels of coma are discernible in humans: level 0, the subject can be roused and will respond to questions; level 1, not easily roused, but all reflexes function; level 2, most reflexes function, but the subject does not withdraw from painful stimuli; level 3, all reflexes are abolished; level 4, breathing is depressed; cardiovascular collapse and death are imminent. In accord with the Glasgow Coma Scale, the absence of eye opening, verbal and motor responses supports a diagnosis of coma. Common causes are trauma, stroke, epileptic seizure, disease (e.g., diabetes, uremia), and asphyxia. Many toxicants (endogenous and exogenous) can cause coma, e.g., alcohols, benzodiazepines, carbamates, hydrocarbons, hydrogen sulfide, isoniazid, lead, narcotic analgesics, opiates, organophosphates, phenothiazines, and tricyclic antidepressants. **alcoholic coma**. coma due to overdosage of alcohol (ethanol). **barbiturate coma**. that due to ingestion or injection of barbiturates. **hepatic coma** (hepatic encephalopathy). that resulting from severe liver disease (e.g., advanced cirrhosis, hepatitis, poisoning). It may be associated with severe icterus and acute metabolic disorders (e.g., of ammonia, nitrogen, amino acids) and may be anticipated by necrologic abnormalities such as flapping tremor, delirium, or mental confusion. **irreversible coma**. that from which recovery under any circumstance is not expected. **uremic coma**. that due to autointoxication as a result of disturbed renal function. The proximate agents are metabolic end products, normally excreted by the kidneys, that accumulate to levels that interferes with acid-base balance.

coma scale. See Glasgow Coma Scale.

Comandra pallida (bastard toadflax). a North American herb with creeping stems and whitish flowers (Family Santalaceae). It is a facultative accumulator of selenium from soil.

comatose. of or in a state of coma or near coma.

comb-footed black widow. *Steatoda grossa*.

comb-footed spider. any spider of the family Theridiidae.

[illegible]

combined sewers. a sewer system that carries both sewage and storm-water runoff. The entire volume of such a system normally goes to a waste treatment plant. Heavy or prolonged rains may, however, exceed the capacity of the system and overflow. Under these conditions, mixtures of untreated sewage and storm-water may enter receiving waters. This mixture may include toxic chemicals, e.g., from industrial areas or streets.

combustion. burning, or rapid oxidation, with the evolution of energy in the form of heat and light. Many toxic and some otherwise nontoxic chemicals release toxic fumes on combustion or on decomposition by heating. The combustion of fossil fuels is a major cause of air pollution. **complete combustion**. regarding fuels that contain carbon and hydrogen, combustion is considered complete when these two elements are completely oxidized to carbon dioxide and water. **incomplete combustion**. combustion of fuels that contain carbon and hydrogen is incomplete if either of these elements is incompletely oxidized. Some consequences of incomplete combustion are that some of the carbon may be released as carbon monoxide, appreciable amounts of carbon may remain in the ash, and fuel molecules may react to yield a variety of products of greater complexity than fuels themselves. These may leave the site of combustion as smoke. Some, even many, of the compounds so formed are likely to be toxic or carcinogenic.

combustion product. a chemical produced during the burning or oxidation of a substance.

comfrey. *Symphytum*.

commercial name™. trade name.

common adder. *Vipera berus*.

common bile duct. a tube formed by the union of the hepatic duct and cystic duct that conveys bile from the liver to the small intestine of vertebrates.

common bobwhite. *Colinus virginianus*.

common box. *Buxus sempervirens*.

common broom. *Cytisus scoparius*.

common brownsnake. *Demansia textilis*.

common carp. *Cyprinus carpio*.

common cinchona. *Cinchona pubescens*.

common cobra. Naja naja.

common comfrey. *Symphytum officinale*.

common death adder. *Acanthophis antarcticus*.

common duck-billed platypus. *Ornithorhynchus anatinae*.

common mamba. *Dendroaspis polylepis*

common elder. *Sambucus canadensis*.

common European adder. *Vipera berus*.

common European toad. *Bufo vulgaris*.

common exposure route. the likely pathway whereby a pesticide may enter an organism. This most commonly via contact, the mouth, or inhalation. See also route of exposure.

common honey bee. *Apis mellifera*.

common Indian krait. *Bungarus caeruleus*.

common jack. *Caranx hippos*.

common kale. *Brassuca oleracea* var. *acephala*.

common krait. *Bungarus caeruleus*.

common laburnum. *Laburnum anagyroides*.

common law. a law, or body of laws of general application as opposed to statutory laws that apply to a specific sector of society or to specific activities. Such law derives from the rules and principles that receive their authority entirely from customs and usages or from judicial decisions that recognize and confirm such usage. Common law relates mainly to the security of persons and property. Compliance with special laws enacted by a legislature does not necessarily free one from obligations under common law. Thus, a polluter or a company that manufactures, uses, or disposes of toxic materials may meet existing standards and regulations, but may still be responsible under common law for damages incurred by others.

common long-glanded snake. *Maticora intesti nalis*.

common milkweed. *Asclepias syriaca*.

common mussel. *Mytilus californianus*.

common night adder. *Causus rhombeatus*.

common nightshade. *Solanum nigrum*.

common privet. *Ligustrum vulgare*.

common razor clam. *Ensis directus*.

common salt. sodium chloride.

common sea serpent. *Hydrophis cyanocinctus, laticauda semifasciata*.

common sea snake. *Enhydrina schistosa*.

common soft-shell clam. *Mya arenaria*.

common sunflower. *Helianthus annus*.

common Surinam toad. *Pipa*.

common tansy. *Tanacetum vulgare*.

common toadfish. *Opsanus tau*.

common violet-latex milky. *Lactarius uvidus*.

common viper. *Vipera berus*.

common Washington clam. *Saxidomus nuttalli*.

communication. any action of an individual organism that modifies the behavior or other biological functions of another organism. Communication by chemical means is common.

community. **1:** a group of interacting individuals that resides in a defined area or that have characteristics in common that are not shared by the general population. **3:** society as a whole. **4:** a biocenosis. **5:** population. This usage is not recommended. **biotic community**. **1:** biocenosis. **2:** in paleontology, a group of species that frequently occurs together.

compartment. **1a:** any convenient portion of an organism or ecosystem that can be defined and quantified for purposes of study, especially one that is semiautonomous in some respects. **1b:** in toxicology, usually, a compartment is a hypothetical volume in which the behavior of a given toxicant is homogeneous with respect to selected properties, but most especially transport, transportation, and fate (e.g., a soil layer, blood, the cardiopulmonary system).

complaint. illness, ailment, disorder; roughly synonymous with disease, *q.v.*

complement. a complex of thermolabile, cytotoxic proteins, lymphotoxin and tumor necrosis factor (cachectin) that combines with antibody to lyse bacterial and other cells. They kill cells by forming transmembrane channels. The complex consists of 20 serum proteins, normally present in blood, and 9 membrane proteins classified as components C1-C9. Activation of the C1 component initiates the sequential activation of remaining components (complement cascade). This cascade of reactions facilitates the clearance of antigen-antibody complexes and the destruction of foreign microbes. Intermediate stages in the activation sequence initiate other biological processes such as the chemotaxis of polymorphonuclear leucocytes, the action of opsonin, and the lysis of erythrocytes. This activation process is called complement fixation, *q.v.*

complement fixation. **1:** the binding (fixing) of complement to an antigen-antibody complex, whereby the complement is unavailable to react. **2:** the capacity of certain antigen-antibody complexes to bind the components of complement.

complete combustion. See combustion.

complete Freund's adjuvant. See Freund's adjuvant.

complex mixture. See mixture.

complication. in toxicology and medicine, usually **1:** an event that intervenes or is noted during the course of a disease or treatment that causes further symptoms and often delayed recovery or **2:** an added obstacle, difficulty, or complexity (e.g., due to an additional disease, drug interaction) that may affect prognosis and or treatment.

Compositae. Asteraceae.

Compound 42. warfarin.

Compound 118. aldrin.

Compound 269. endrin.

Compound 497. dieldrin.

Compound 711. isodrin.

Compound 1080 (1080). sodium fluoroacetate.

Compound B. corticosterone.

compound F-2. zearalenone.

CON. cortisone.

Con A. concanavalin A.

CONA. cortisone acetate.

concanavalin A (Con A). a lectin mitogen isolated from *Canavalia ensiformis* (Jack bean) that binds specifically to and agglutinates transformed (tumor) cells. It binds to glucose and mannose residues in the cell surface. Concanavalin A stimulates proliferation of T-cell lymphocytes, but not B-cells.

concatenation. a set of phenomena, events, or effects that act in concert or that occur at the same time.

concentrated. referring to or denoting a solution that contains a relatively high concentration of solute. *Cf.* dilute.

concentration. **1a:** the process or result of increasing the strength or relative amount of a substance within a given volume or mass, as of a solute in a solution. **1b:** the amount of a given substance in a specified unit volume or mass of a solution, mixture, or ore. Concentration is usually expressed as percentage, by molarity (molar concentration), molality (molal concentration), normality, or as mass per unit volume (mass concentraton). The concentration of radioactivity of a substance in solution is usually presented as millicuries per milliliter (mCi/ml) or millicuries per millimole (mCi/mmol). **biological concentration** (bioconcentration). **1:** trophic concentration; the accumulation of an environmental chemical in biota from a lower trophic level (usually by ingestion). **2a:** the net accumulation in aquatic biota of a xenobiotic substance in aqueous solution. **2b:** net accumulation in aquatic biota that results from or is associated with direct absorption or filtering as opposed to ingestion (See trophic concentration). **chemical concentration**. the amount of a substance (solute) per unit volume or mass in a solution. The relationship between chemical concentration within any matrix, and the toxicity of a given substance depends upon the chemical, the matrix, and environmental factors that may vary in importance. Such relationships are determined empirically. **critical concentration**. the mean concentration of a toxicant in an organ at the time any cell reaches a concentration at which adverse functional changes occur in the cell. See also effect (critical effect), critical organ. **effective concentration** (EC). that concentration of a substance that causes a response of a given extent or intensity (usually as a percentage of the maximal response) in an *in vitro* system. EC is used to compare the potencies of toxicants that produce the same effect. EC may be accompanied by a number that indicates the extent or intensity of response. **effective concentration/50% of maximal response** (EC_{50}). the concentration of a substance that causes a response that is 50% of the maximum. **lethal concentration** (LC). a concentration of a toxic chemical in the surrounding medium that kills a living organism or that kills an estimated percentage of test organisms during a specified period of exposure under narrowly defined conditions of testing. This term is sometimes taken to be synonymous with median lethal concentration, *q.v.* **lethal concentration, 50% response**. median lethal concentration. **maximum sublethal concentration** (LC_0). the highest concentration of a toxic substance (e.g., in food, air, water) to which individuals in a test population can be exposed with no mortality. **mass concentration**. the mass of a solute per unit volume of solution. **median lethal concentration** (lethal concentration, 50%; LC_{50}; LCt_{50}; MLC). that concentration of a toxic substance in the surrounding medium (e.g., air, water, sediment, culture medium) which is estimated to kill half the organisms of a test population within a designated time following a predetermined period of exposure. The estimated median lethal concentration is derived statistically from the results of a set of suitably designed toxicity tests. It is used mainly as an estimator of acute lethality of environmental chemicals in air or water. **median lethal concentration, inhalation** (inhalation LC_{50}). that concentration of a toxic substance expressed as mg per liter or parts per million of air that is lethal to 50% of a test population under specified conditions. **molal concentration**. the amount, in moles, of a substance per kilogram of solute. **molar concentration**. the amount, in moles, of a substance per liter of a solution. **trophic concentration**. See biological concentration.

concentrator. accumulator.

conceptus. the early embryo; in *Homo sapiens* the conceptus is designated as an embryo during the first 8 weeks following fertilization; thereafter and until birth, it is termed a fetus. *Cf.* abortus, embryo, fetus.

concomitant. accessory, accompanying; taking place at the same time.

concomitant mortality. See mortality.

condensation. **1:** the change of a substance from the vapor phase to the liquid phase, usually as a result of cooling. **2:** a type of chemical reaction in which two substances combine with the loss of water or a simple alcohol.

condition. **1:** to train. **2:** to subject to conditioning. **3a:** circumstance, situation, **3b:** state, as in state of health. **4:** stipulation, requirement, prerequisite, qualification. **5a:** a state of ill health, especially one that is chronic, due to injury, or heredity. **5b:** any illness, ailment, disease, or malady. **abnormal condition**. a state of ill health; roughly synonymous with disease.

conditioning. **1:** improving the condition or physical ability of a subject organism. **2:** a method to modify behavior by applying a formerly neutral stimulus that has been applied many times in conjunction with the stimulus that normally elicits the desired response. Conditioning can also be applied to test the nature and intensity of a stimulus (e.g., a potential toxicant). **avoidance conditioning**. behavioral modification by use of a stimulus that provides a negative reward (sometimes called punishment) or produces consequences that are negatively rewarding to the subject. **instrumental conditioning**. operant conditioning. **op-**

erant conditioning. behavioral modification by use of a stimulus that is considered rewarding to the subject, or produces consequences that are rewarding to the subject. See toxicity testing (behavioral).

condom. a thin sheath worn over the penis during sexual intercourse to reduce the probably of insemination or the transmission of venereal disease between partners. Allergic reactions can occur from contact with the latex rubber of which most condoms are made, or with the chemicals used to lubricate many condoms.

conductimetric. conductometric.

conductometric (conductimetric). pertaining to or denoting a general electroanalytic method based on measurement of the conductance (or sometimes the admittance or susceptance) of a solution that contains an electrolyte as the concentration of the electrolyte or the volume of the system is varied. Conductometric methods are widely used, for example, in continuous sulfur dioxide recorders.

Condylactis gigantea (pink-tipped anemone). a large (15 cm tall, 30 cm wide), impressive showy, venomous sea anemone. It secretes an extremely potent paralytic protein toxin. Sometimes called the passion flower of the Caribbean, it is the largest and most impressive anemone of the region.

cone (cone shell). common name of any gastropod of the family Conidae, especially of the genus *Conus*. They have an attractive cone-shaped shell; most occur in the South Pacific and Indian oceans. The cone uses its barbed sting to kill other gastropods and secretes a toxin which competes for acetylcholine at the motor end plate. Symptoms of envenomation in humans may include weakness, ataxia, depressed respiration, coma, and eventual death, although the case mortality rate is low. The shells of various species of cones in the Indo-Pacific region are highly prized by collectors, and a number of human deaths due to stings on the hand have been reported.

cone shell. **1:** the shell of a cone, *q.v.* **2:** a cone, *q.v.*

confusion. disorientation, bewilderment; lack of awareness of or orientation with respect to time, place, or self. *Cf.* See consciousness.

congener. **1:** a chemical compound that is structurally related to another or an element belonging to the same family of the periodic table as another. **2:** a plant or animal species that is assigned to the same genus as another. The inference is that congeners are more closely related to each other than to species outside the genus. Thus, for example, if one congener is found to be toxic or is otherwise of toxicologic interest, others may have similar properties. This often holds true.

congenital. present at birth; pertaining to or denoting any condition **1:** present at or from birth regardless of cause. **2:** present at or before birth and acquired *in utero*; not inherited.

congenital articular rigidity (CAR). congenital arthrogryposis (See arthrogryposis).

congenital hypoplastic anemia. hypoplastic anemia.

congested. said of a tissue or organ that contains excessive amounts of blood or tissue fluid.

congestin. a toxin found in the tentacles of certain sea anemones. It was originally thought to be a venom, but does not appear to occur in the nematocysts.

congestion. **1a:** the accumulation of blood, tissue fluid, or lymph in a tissue or organ. **1b:** the swelling of an organ associated with congestion (def. 1a). Congestion often arises (as in inflammation) from increased blood flow into a body part as by dilatation of the blood vessels supplying the part (active congestion). Congestion may also arise from reduced drainage of blood from the part (passive congestion). The latter may be due to vasoconstriction, to heart failure, or various other disorders. See also hyperemia.

congestive. pertaining to congestion.

congestive heart failure. See failure (congestive heart failure), cardiomyopathy (congestive cardiomyopathy).

Congo blue. C.I. Direct Blue 14, tetrasodium salt.

Congo burrowing adder. *Atractaspis congica.*

conicine. coniine.

Conidae (cone or cone shell family). a large family of venomous marine gastropods of the Indo-Pacific region. The shell is cone-shaped and often brightly colored with yellows and browns predominant, up to 10 cm long. Some species are valued collectors items and deaths of humans are known from stings on the hand. They are carnivorous, using their sting to kill other gastropods. The toxin competes for acetylcholine at the motor end plate. Symptoms of envenomation in humans include weakness, ataxia, depressed respiration, coma, and sometimes death. See *Conus*.

conifer false morel. *Gyromitra (Helvella) esculenta*.

coniine (2-propylpiperidine; cicutine; conicine). a highly toxic liquid alkaloid and derivative of pyridine, it is the chief toxic agent of poison hemlock, *Conium maculatum* (Family Apiaceae). The fetid odor of the plant is partly due to coniine. Symptoms appear within about 15 minutes following ingestion and include weakness, nervousness, drowsiness, nausea, vomiting, trembling, dyspnea, cardiac arrhythmia, and bradycardia. Body temperature may decrease and a fatal paralysis can occur. Death due to asphyxia (usually as a result of paralysis of the respiratory muscles) may occur within 3 hours. Coniine affects skeletal muscle tone. A narcosis may follow prolonged inhalation of the coniine-containing effluvia of the plant. See also conium.

coniofibrosis. fibrosis caused usually by inhalation of dust.

coniosis. any disease or morbid condition produced by dust, especially via inhalation. *Cf.* pneumoconiosis.

coniotoxicosis. a pneumoconiosis due to the direct action of the irritants on pulmonary tissue.

conium. the dried unripe fruit of *Conium maculatum*. It has been used as a sedative, antispasmodic, and anodyne.

Conium maculatum (poison hemlock; deadly hemlock; water hemlock; spotted cow bane; spotted parsley; poison fool's parsley; California fern; Nebraska fern; winter fern). an extremely poisonous neurotoxic herb that grows to a height of about 1 m before flowering, it resembles Queen Anne's lace (wild carrot). Poison hemlock is easily distinguishable from Queen Anne's lace if the leaves are directly compared; also, while the main leaf stalks of Queen Anne's lace are "hairy," those of poison hemlock are smooth. While indigenous to Europe (Family Apiaceae), it has become widely distributed in North and South America. It is the hemlock reputedly taken by Socrates. All parts of the plant are toxic and ingestion of even small amounts of seeds, leaves, or roots can prove fatal. It is rarely eaten accidently because of its noxious odor and taste. Poison hemlock contains a number of alkaloids related to nicotine; of these, coniine (a derivative of pyridine) is the chief toxicant; pyridine is also present. Symptoms of intoxication are similar to those of acute nicotine poisoning. Toxic concentrations of nitrate have also been reported. The death rate in humans is high and results from respiratory collapse due to CNS depression. *C. maculatum* is fetotoxic in domestic cattle, causing deformities in the calves that are similar to those of crooked calf disease.

conjugate (conjugated product; conjugation product). **1:** paired or joined. **2:** the compound formed by a conjugation reaction, *q.v.*

conjugated product. conjugate.

conjugating agent. a substance, usually a xenobiotic compound that reacts (usually after activation) with a substrate compound such as an amino acid, glutathione, a sugar, or sulfate to yield a new compound, the conjugation product. See conjugation reactions, glucuronides, glutathione, sulfate.

conjugation. **1:** a coupling or joining together **2a:** the union of two microorganisms with an exchange of nuclear material (e.g., as in various bacteria, algae, fungi, and protozoans). **2b:** the fusion of male and female gametes with formation of a zygote. **3:** in chemistry, a reaction in which two compounds unite, forming a third. See conjugation reaction. **4:** in toxicology, the linkage of a toxicant with a natural metabolite in the organism, forming a product that can be easily excreted or otherwise eliminated. Conjugation plays an important role in the transport and detoxification of xenobiotics and of toxic intermediates formed during metabolism. **amino acid conjugation**. a type of acylation reaction in which ATP and CoA interact to activate exogenous carboxylic acids to form *S*-CoA derivatives. The *S*-CoA derivatives thus formed, acylate the amino group of an amino acid in reactions catalyzed by ATP-dependent acid:CoA ligases as represented by the equations:

$$RCOO^- + ATP + CoA\text{-}SH \rightarrow RCOS\text{-}CoA + PP_i + AMP$$

$$OS\text{-}CoA + R'NH_2 \rightarrow CoA\text{-}SH + RCONHR'$$

The amino acids most commonly involved are taurine in fish, ornithine in reptiles and birds, and glycine and glutamine in mammals. See conjugation reaction (type II). **glutamine conjugation**. a form of amino acid conjugation that occurs in some mammals (e.g., humans and certain other primates). **glutathione conjugation**. reactions, usually catalyzed by glutathione transferases, in which xenobiotics with electrophilic constituents are conjugated with glutathione. This protects nucleophilic groups in macromolecules by effectively eliminating reactive electrophiles. **glycine conjugation**. the most common type of amino acid conjugation reaction in mammals. **glycoside conjugation**. the production of a water-soluble conjugate from a sugar and a xenobiotic. This type of reaction requires an active intermediate, either uridine diphosphate glucose (UDPG) or uridine diphosphate glucuronic acid (UDPGA). See also glucuronides, glycosides. **ornithine conjugation**. an amino acid conjugation seen most commonly in reptiles and birds. Ornithine is a nonprotein amino acid. **sulfate conjugation**. the formation of water-soluble, easily excreted, sulfate esters from xenobiotics (e.g., alcohols, arylamines, phenols) and activated sulfate ions (PAP (3'-phosphoadenosine-5'-phosphosufate)). **taurine conjugation**. a minor route of amino acid [illegible] and indolylacetic acids, observed in a wide variety of animals. It is a major route, however, in a few species such as the rock dove (*Columba livia*) and ferret. Taurine is a nonprotein amino acid.

conjugation product. conjugate (def. 2).

conjugation reaction. See reaction (conjugation reaction).

conjunctiva. the mucous membrane in the eyes. It covers the front part of the eyeball and the lining of the eyelid. It is subject to irritation and damage by irritant substances.

conjunctival. pertaining to the conjunctiva.

conjunctivitis (blennophthalmia). inflammation of the conjunctiva, causing erythema, discomfort, and a discharge from the eye. Conjunctivitis is a common disorder and can be caused by any of a very large number of irritant substances, by pathogens, or abrasion. Most conjunctival infections are caused by bacteria. **allergic conjunctivitis** (atopic conjunctivitis). hay fever. **anaphylactic conjunctivitis**. hay fever. **atopic conjunctivitis**. allergic conjunctivitis. **atropine conjunctivitis**. a follicular conjunctivitis due to repeated use of atropine. **catarrhal conjunctivitis**. that due to any of a variety of causes including irritation from toxins or other chemicals. **chemical conjunctivitis**. that due to exposure to chemical irritants.

connective tissue. any of diverse tissues that connect or support other tissues or organs. Such tissue is frequently fibrous (fascia, ligamentous tissue, scar tissue), dense, or hard (e.g., cartilage and bone). Blood may be considered a connective tissue. Fibrous connective tissue is characterized by widely spaced cells and extensive interstitial material, most importantly collagen. Chemicals that inhibit collagen synthesis (e.g., lathyrogens) adversely affect connective tissue.

Conocybe smithii (bog conocybe). an hallucinogenic mushroom (Family Bolbitaceae) that occurs in moss in bogs in Michigan and the Pacific Northwest of the United States. It contains amanitin, psilocin and psilocybin. Reactions are variable, and this mushroom should be considered

conscious. **1:** perceiving, aware, or noticing aspects of one's environment with a degree of understanding and controlled attention. See unconscious.

consciousness. a state of awareness or orientation with respect to time, space, and self; a state of awareness of one's surroundings that implies a degree of discernment, understanding, and controlled attention. *Cf.* confusion, unconsciousness. **levels of consciousness**: **1:** *alert wakefulness* in which an individual is clearly aware of the environment and responds quickly and appropriately to sensory stimuli. **2:** *drowsiness*. a state in which an individual is not fully aware or does not fully appreciate aspects of environment. Responses to sensory stimuli are appropriate but may be slow or delayed; some relatively weak or repetitious stimuli may be ignored. **3:** *stupor*. the individual is aroused only by strong stimuli and responses to such stimuli (e.g., to loud noise) may be nonspecific. In an otherwise normal individual motor responses and reflex reactions are usually normal. **4:** *coma*. See coma.

consequential mortality. See mortality.

constant. **1:** unchanging. **2:** a value that remains unchanged throughout a process.

consumer. **1a:** one who consumes or uses. **1b:** one who purchases, uses, maintains, or disposes of a product or service as opposed to one who manufactures or sells such. **2:** a heterotrophic organism.

consumer product. any tangible goods or personal property that enters into commerce and that is normally used for personal, family, or household purposes. Included is any such product intended to be attached to or installed in or on any real property.

consumer product safety commission (CPSC). a United States government agency that sets safety standards for consumer products. Products for which safe limits cannot be devised may be banned.

contact allergy. See allergy.

contact dermatitis. See dermatitis (contact dermatitis)

contact herbicide. See herbicide.

contact insecticide. See insecticide.

contact pesticide. See pesticide.

contact poison. See poison.

contact urticaria. See urticaria.

contactant. an allergen that is able to induce a delayed contact-type epidermal hypersensitivity following one or more incidents of contact with the skin.

contaminant. **1:** an impurity. **2a:** any substance that soils, stains or pollutes another. **2b:** a foreign substance that enters, mixes with, or adheres to a system in which it does not normally occur or from which it was thought to be free. **3:** radioactive material residing in any unauthorized location. **4:** an agent of contamination (e.g., a substance, an organism). *Homo sapiens* is an important agent of contamination. **food contaminant**. any substance accidentally or inadvertently introduced into or adhering to food. Pathogens, pesticides, toxins, and other potentially poisonous materials are common contaminants of food.

contaminate. **1:** to make impure, or unclean. **2:** to soil, stain, or pollute. **2:** to pollute or render impure by contact with, by mixing with, or introducing into it, a foreign substance. **3:** to render unfit for use by introduction of a foreign substance that is harmful or injurious. **4:** to deposit a radioactive substance in any place where such deposition is not allowed.

contaminated. **1:** impure or polluted, as by a pathogen, a toxicant, or other substance.

contamination. **1:** the introduction of an impurity or undesirable agent, such as a pollutant, pest, or pathogen, into a substance or system. **2:** the quality of being contaminated; a condition of impurity resulting from contact of mixing with a foreign substance.

content. that which is included or contained in something.

continuous variable. See variable (continuous variable).

contraceptive. **1:** contraceptive device, method, or process that prevents, or helps to avoid, conception. **2:** having the ability to prevent conception. **oral contraceptive** ("the pill"). any synthetic steroid that is similar to one of the natural hormones, estrogen or progesterone. They act by preventing ovulation. Additional effects are numerous and can include weight gain, acne, eczema, nausea, gum inflammation, inflammation of the optic nerve (sometimes with loss of vision, double vision, eye pains and swelling, and an inability to wear contact lenses), headaches, promotion of vaginal yeast infections, depression, loss of sex drive, formation of blood clots, heart attacks, high blood pressure, stroke, gall bladder disease, liver tumors, birth defects, breast tumors, menstrual irregularities, post-pill infertility, and infections. According to the U.S. Food and Drug Administration (FDA), oral contraceptives should not be used by women who have or have had blood clots in their lungs or legs; have heart pains; have had a heart attack or stroke; who may have ovarian; cervical; or breast cancer; who have unusual undiagnosed vaginal bleeding; or who may be pregnant. In the United States, detailed warnings on possible side effects are required with every prescription of oral contraceptives.

contraceptive sponge. a contraceptive device made of polyurethane foam and used with a spermaticide. Risks of usage may in some cases include local irritation, birth defects, and spontaneous abortion. Polyurethane has been linked to cancer in laboratory mice. In the event of accidental ingestion of a sponge, contact a poison control center, a doctor, or an emergency medical facility. See polyurethane, spermaticide.

contracted pupil. See pupil.

contraction. a shortening or reduction in size due usually to internal structural changes. **clonic contraction**. contraction of a muscle alternating with relaxation. **tonic contraction**. sustained contraction of a muscle, as in standing or pushing.

Contrathion. pralidoxime mesylate.

control. 1: to regulate, constrain, or influence a process or [illegible] trolling. **2b:** the means or method of controlling. **3:** restraint, self-restraint. **3:** the regulation, governing, or limitation of certain substances, objects, processes, or phenomena. **4:** a regulation. **5:** a standard against which experimental observations can be compared and evaluated. See treatment (control treatment).

control treatment. See treatment.

controlled substance. See substance.

Conus (cones, cone shells). a genus of some 400 tropical and subtropical species of small, venomous, marine gastropod mollusks (Family Conidae). They commonly occur on sand or under rocks or coral. All are predatory, preying on worms, other mollusks and even fish. Some piscivorous species can cause serious injury or even death to humans. The venom apparatus is highly developed and includes a muscular venom bulb, a long coiled venom duct, and a toothed radula with a sheath. The venom seems to be produced in the venom duct and pumped by contractions of the venom bulb and duct into the radular teeth which are thrust under pressure through the proboscis into the skin of the prey. *Conus* venoms are generally thought to be polypeptides or proteins. The venom of some species is only mildly toxic to humans; that of others is highly potent. Envenomation by some species (e.g., *C. geographus*) can be fatal to humans. Common early symptoms of envenomation in humans include ischemia, cyanosis, and a sharp burning sensation or numbness about the wound. The numbness and tingling may spread throughout the entire body, becoming especially pronounced about the mouth. In severe cases, paralysis, coma, and finally death from cardiac failure may intervene. ***C. californicus***. the only dangerous cone found in North American waters. Its sting produces localized pain, swelling, redness, and numbness.

convacard. convallaria (def. 2).

convalescence. the phase of recovery from an illness or injury.

convalescent. recovering, recuperating, convalescing, mending, getting well; in a state of, or undergoing convalescence.

[illegible]

convallamarin. a cardiotoxic glycoside from the root of *Convallaria majalis*. The aglycone is convallamarogenin.

convallamarogenin. the aglycone of the glycoside, convallamarin, it occurs in the root of *Convallaria majalis*.

convallan. convallaria (def. 2).

convallaria. **1:** the dried flowers, rhizome and roots of *Convallaria majalis*, *q.v.* The flowers contain convallatoxin, and convallarin; the roots and rhizome contain glycosides of convallamarin, convallatoxin, and convallarin. **2:** an aqueous extract that contains the total glycosides of *C. majalis* (synonyms, Convallaria glycosides, Convacard, Convallan, Convallen, Convalyt, Convasid).

Convallaria (lily of the valley). a genus of perennial herbs (Family Liliaceae) indigenous to Europe, northern Asia, and Alleghanian North America. ***C. majalis*** (lily of the valley, May lily, park lily, May blossom). a small, graceful, low-growing colonial herb with fragrant, white, nodding, bell-shaped flowers (Family Liliaceae). It is native to Europe and Asia; naturalized elsewhere. It is a common garden and potted plant that often escapes from cultivation, forming dense carpets along roadsides, thickets, and open woods. The literature is rife with supposedly poisonous compounds that have been isolated from this plant, especially the berries. The flowers contain convallatoxin, and convallarin; the roots and rhizomes contain glycosides of convallamarin, convallatoxin, and convallarin. These three glycosides are cardiotoxic to laboratory animals. Ingestion of any plant part may cause cardiac arrhythmia, nausea, vomiting, and dizziness. Confirmed reports of human poisonings by this plant are, however, extremely rare. See also convallaria.

convallarin. a cardiotoxic glycoside of *Convallaria majalis*. See also convallatoxin.

convallatoxigenin. the aglycone of convallatoxin.

convallatoxin (3β-[(6-deoxy-α-L-mannopyranosyl)oxy]-5,14-dihydroxy-19-oxo-5β-card-20-(22)-enolide; strophanthidin α-L-rhamnoside). an extremely potent cardiotoxic phytotoxin from the flowers of *Adonis vernalis*, *Convallaria majalis*, *Ornithogalum umbellatum* and *Antiaris toxicaria*. It is used therapeutically as a cardiotonic. Convallatoxin produces digitalis-like effects in laboratory animals. The aglycone is convallatoxigenin and the sugar is a rhamnose.

Convallen®. convallaria (def. 2).

Convalyt®. convallaria (def. 2).

Convasid®. convallaria (def. 2).

Convention on the Prohibition of Biological and Toxin Weapons. an agreement promulgated by the Conference of the Committee on Disarmament (CCD) of the United Nations and signed by representatives of more than 100 nations. The agreement effectively bans all offensive chemical warfare research, testing, development, or stockpiling of biological warfare agents.

Convolvulus (bindweed). a genus of cultivated annual or perennial herbs or shrubby twining or horizontal plants (Family Convolvulaceae) in which toxic concentrations of nitrate have been reported. Roots contain scopoletin.

convulsant (convulsivant). **1:** of or pertaining to the ability to induce convulsions. **2a:** any agent that causes or precipitates convulsions. **2b:** a convulsant drug or poison such as strychnine, brucine, and other alkaloids from nux vomica.

convulsion (seizure; fit). an involuntary, usually violent, generalized (whole-body) spasm or series of spasms of voluntary muscle. Convulsions are activated by the brain and are usually accompanied by a period of unconsciousness. Convulsions are symptomatic of such disorders as epilepsy, diabetes, hypoxia, pyrexia, certain disease states of the brain (notably encephalitis or meningitis), brain tumors (in children), head injury, arteriosclerosis (in elderly persons), toxemia of pregnancy, drug or alcohol withdrawal, and numerous toxicants. Convulsions due to toxic substances, drug overdose, diabetes, or head injury can be life threatening. **central convulsion** (essential convulsion, spontaneous convulsion). a convulsion not induced by any external cause, but rather by a CNS lesion. **clonic convulsion**. a convulsion with intermittent muscular contractions alternating with periods of relaxation. **coordinate convulsion**. a convulsion with clonic contractions that are similar to normal voluntary movements. **crowing convulsion**. a convulsion of the larynx. **epileptiform**. any convulsion with loss of consciousness. **essential convulsion**. central convulsion. **ether convulsion**.

a convulsion sometimes attending the induction of ethyl ether anesthesia. **febrile convulsion**. a convulsion, usually occurring in infants and children, associated with pyrexia. **hysterical convulsion.** any spasmodic, convulsive action that is associated with an hysterical disorder. **local convulsion**. any localized spasm (e.g., of a single muscle or body part. **mimetic convulsion** (mimic convulsion). a facial spasm; a tic. **mimic convulsion**. mimetic convulsion. **puerperal convulsion**. an involuntary spasm in perinatal women. **spontaneous convulsion**. central convulsion. **tetanic convulsion**. tonic convulsion. **tonic convulsion** (tetanic convulsion). a convulsion with sustained muscular contraction and rigidity of the body. The limbs of are fixed in position. **toxic convulsion**. that caused by the action of a toxin on the nervous system. **uremic convulsion**. that due to failure of the kidneys to excrete products of protein digestion (e.g., urea) normally.

convulsivant. convulsant.

convulsive. of, pertaining to, producing, or characterized by convulsions.

convulsive ergotism. See ergotism (neurotropic ergotism).

convulsive poisoning. poisoning by a convulsant such as strychnine or picrotoxin.

coontie. a common name of certain species of *Zamia* (e.g., *Z. integrifolia*).

copious. in a large amount.

copper (cuprum; Cu)[Z = 29; Ar = 63.55]. a golden-red metallic transition element that occurs in nature mainly as the sulfide. Copper compounds are white, except for the oxide which is red, and form blue solutions. Elemental copper is an excellent conductor of heat and electricity, with low toxicity. Copper is a constituent of cytochrome *c* oxidase, which [illegible] of [illegible] from cytochrome a_3 to oxygen in respiration. Copper is also a component of the protein, plastocyanin, thus also playing a role in photosynthetic electron transport. Soluble copper salts are strong irritants and hepatotoxicants. As a component of phylloerythrin, copper causes hepatogenic photosensitization in livestock.

copper-64 (^{64}Cu). a radioactive isotope, $t_{1/2}$ = 12.82 hours, that emits beta particles and positrons.

copper-67 (^{67}Cu). a radioactive isotope, $t_{1/2}$ = 59 hours, that emits beta particles and gamma rays. See poisoning (copper poisoning).

copper acetate arsenite. cupric acetoarsenite.

copper cataract. chalcosis lentis.

copper chloride. **1:** cupric chloride. **2:** cuprous chloride. **3:** any of various basic chlorides (e.g., $CuCl_2 \cdot 3CuO$) or mixture of chlorides formed by exposure of cupric oxide to the atmosphere. Such substances are also termed copper oxychloride.

copper (I) chloride. cuprous chloride.

copper (II) chloride. cupric chloride.

copper colic. See colic.

copper cyanide. cuprous cyanide.

copper ethanoatoarsenate. cupric acetoarsenite.

copper fungicide. any inorganic fungicide that contains copper (e.g., Bordeaux mixture, Burgundy mixture). Copper fungicides are often based on copper oxychlorides (See copper chloride, def. 3) or copper oxide; they may also be blended with other types of fungicide.

copper monosulfate. cupric sulfate.

copper oxide. cuprous oxide.

copper oxychloride. copper chloride (def. 3).

copper poisoning (toxemic jaundice). poisoning, acute or chronic, due to soluble copper salts, all of which are strong irritants and hepatotoxicants. Poisoning is accompanied by pain in mouth, esophagus, and stomach; vomiting; diarrhea with abdominal pain; a metallic taste, shock; convulsions; paralysis; coma; death. Lesions include widespread capillary damage, kidney injury, liver damage. Copper poisoning, acute or chronic, is encountered in most parts of the world. Among livestock, sheep are affected most often. Acute poisoning is usually observed after accidental administration of excessive amounts of soluble copper salts, which may be present in anthelmintic drenches, mineral mixes, or improperly formulated rations. Many factors that alter copper metabolism influence chronic copper poisoning by enhancing the absorption or retention of copper. Low levels of molybdenum or sulfate in the diet are important examples. Primary chronic poisoning is seen when excessive amounts of copper are ingested over a prolonged period. The toxicosis remains subclinical until the copper that is stored in the liver is released in massive amounts. Blood copper concentrations increase suddenly, which causes severe intravascular hemolysis. The hemolytic crisis may be precipitated by many factors, including transportation, lactation, strenuous exercise, or a deteriorating plane of nutrition. See also chalcosis. **copper poisoning, hepatogenous chronic**. a form of chronic phytogenous copper poisoning due to the repeated ingestion of plants such as *Heliotropium europaeum* or *Senecio spp*. for a period of several months. These plants contain hepatotoxic alkaloids, which cause excessive copper retention by the liver. **copper poisoning, phytogenous chronic**. a form of chronic phytogenous copper poisoning that results from long-term repeated ingestion of plants such as subterranean clover (*Trifolium subterraneum*). In this case, a mineral imbalance is incurred that results in excessive copper retention. Plants that are not hepatotoxic contain normal amounts of copper and low levels of molybdenum. Grazing animals are most commonly affected.

copper(2+) salt (1:1). cupric sulfate.

copper snake. *Pseudoechis australis*.

copper sulfate. a term commonly applied to cupric sulfate, *q.v.* or to any of the variable basic sulfate salts of copper. Such salts are phytotoxic. See cupric sulfate, Bordeaux mixture, Burgundy mixture.

copper sulfate poisoning. cupric sulfate poisoning.

copperas. ferrous sulfate.

copperhead. the common name of several species of venomous snakes, mostly of the genus *Agkistrodon*. **American copperhead**. *A. contortrix*. **Australian copperhead**. *Denisonia superba*. **Northern copperhead**. *A. contortrix mokeson*. **Solomons copperhead**. *D*. **Southern copperhead**. *A. contortrix*. **Trans-Pecos copperhead**. *A. contortrix pictigaster*.

copperon. cupferron.

coprine (1-cyclopropanol-1-N^5-glutamine; N^5-(1-hydroxycyclopropyl)-*L*-glutamine). a mycotoxin of the mushrooms *Coprinus atramentarius*, *Clitocybe clavipes*, and a few other congeners. It is a water-soluble γ-glutamyl conjugate of 1-aminocyclopropanol. Coprine is hydrolyzed to 1-amino-cyclopropanol when ingested by humans and laboratory mammals. The latter molecule or a product of further hydrolysis, cyclopropanone (or cyclopropanone hydrate) is the proximate toxicant that inhibits acetaldehyde dehydroenase in liver. If even small amounts of ethanol is ingested by a subject during this inhibition, acetaldehyde accumulates in blood plasma with an associated illness that is indistinguishable from the effects of Disulfiram in combination with ethanol. If alcohol is not consumed, *C. atramenarius* is nontoxic and edible.

Coprinus atramentarius (alcohol inky; smooth ink cap mushroom). an edible mushroom (Family Coprinaceae) that occurs in grass, woody rubble, and on soil near buried wood throughout much of North America. It is not toxic in the usual sense, but contains the mycotoxin, coprine. The latter yields a metabolite, cyclopropane hydrate, following ingestion, that mimics the action of Antabuse in that it inhibits the detoxification of alcohol. If alcoholic beverages are consumed with this mushroom or during the following 2 or 3 days, a transient condition lasting 2 to 4 hours may occur. Symptoms may be severe, but usually persist for no more than an hour or two. They may include peripheral vasodilatation with flushing of the face and neck, hypotension, tachycardia, palpitations, hyperventilation, a tingling sensation in the extremities, headache, and sometimes nausea, vomiting, and diaphoresis.

copse laurel. *Daphne mezereum*.

coral. **1:** See Anthozoa. **2:** any mushroom of the genus *Ramaria*. **3:** any snake of the genus *Micrurus*.

coral cobra. *Micrurus corallinus*, *M. frontalis*.

coral reef. a large, calcareous reef formed from the skeletons of corals joined by limestone secretions. Coral reefs vary in size, forming atolls, fringing reefs or barrier reefs in tropical seas, usually near shore. The surface is hard, razor sharp, and able to cause serious lacerations. The wounds are complex and require medical attention. They usually contain foreign matter, are subject to secondary bacterial infection and envenomation by nematocysts if the wound is produced by live coral.

coral snake. the common name of any of a number of brightly colored venomous snakes (Family Elapidae) of the Americas; they are especially well represented in the tropics. The term applies to snakes of several genera: *Aspidelaps*, *Brachyurophis*, *Calliophis* (= *Callophis*), *Leptomicrurus*, *Maticora*, *Micruroides*, and *Micrurus*, *q.v.* Coral snakes are highly venomous and the case fatality rate is very high. Nevertheless, few people are bitten because these snakes are secretive and seldom encountered. They have a relatively narrow, often flattened head that appears nearly continuous with the slender cylindrical body. The eye is small with a round pupil. A number of harmless species of snakes resemble coral snakes and are not easily distinguished from them. Coral snakes are thus often mistaken for harmless species and nearly all persons bitten were attempting to catch or kill the snake at the time. All brightly colored ringed snakes should be regarded as dangerous until a positive identification has been made.

coral sumac. *Metopium taxiferum*.

coralberry. *Actaea*.

coralillo. See *Micrurus*.

Corchorus (**jute**). a genus of tropical herbs and small shrubs (Family Tiliaceae). The fibrous [illegible] *olitorius* and *C. capsularis*) are toxic, producing allergic reactions (e.g., asthma, hay fever, contact dermatitis) in sensitive individuals.

Coriaria (tutu). a small genus of herbs shrubs or small trees (Family Coriariaceae) that are grown largely for the ornamental value of their berries. The seeds of two species in particular, *C. ruscifolia* and *C. sarmentosa* of New Zealand, are poisonous. They contain toot poison.

corilagin (gallotannine). a slightly toxic tannin.

Coris gaimardi (wrasse). a bony fish, certain populations of which are transvectors of ciguatera.

corium. dermis. See skin.

corm. a short, fleshy, erect, underground stem, usually broader than high and covered by membranous scale-like leaves. It closely resembles a bulb, but is more stemlike. *Caladium* and *Colchicum* grow from corms. Corms are often poisonous as in the above-named genera. *Cf.* bulb.

corn. *Zea mays*.

corn cockle. any plant of the genus *Agrostemma*, especially *A. githago*.

corn lily. *Veratrum californicum*.

corn oil (maize oil). a pale yellow, water-insoluble liquid with a distinctive taste and odor. It is obtained by pressing the germ of cultivated corn, *Zea mays*, and contains saturated (mainly palmitic and stearic acids) and unsaturated (mainly linoleic and oleic acids) fatty acids. Corn oil is a common ingredient of foods, soaps, margarine, salad oil, hair dressing, lubricants, and solvents. In foods it serves as an energy source and can thus affect food consumption and/or body weight. Prolonged oral administration has been associated with increased rates of carcinogenesis.

corn root-rot organism. See *Giberella zeae*.

cornea. the clear, transparent structure forming the anterior portion of the fibrous tunic of the eye. Because its curvature is greater than the remainder of the bulb, it can function as an important refractive medium. Its periphery is continuous with the sclera. See corneal reflex.

corneal reflex. closure of the eyelids in response to a light touch of the cornea (as with a soft bristle).

corneitis (keratitis). inflammation of the cornea of the eye due usually to infection or to an irritant substance.

cornstalk disease. **1:** a disease syndrome in cattle in which nitrosohemoglobin in blood is formed due to the ingestion of corn of high nitrogen content. **2:** moldy corn poisoning, a rare disease of livestock (primarily cattle) caused by the ingestion of moldy corn forage.

Cornus (dogwood). a genus of ornamental shrubs, trees and occasional herbs (Family Cornaceae) that are native to north temperate regions. The fruit is slightly toxic.

Cornybacterium diphtheriae. the species of bacterium that causes diphtheria. It does so through the agency of a toxin. See toxin (diphtheria toxin).

coronary. **1:** encircling; used most often in medicine and toxicology in reference to the blood vessels that "encircle" and supply blood directly to the vertebrate heart. **2:** a heart attack resulting from obstruction of a coronary artery. **3:** used loosely in reference to the heart and to coronary heart disease.

coronated snake. *Denisonia coronata*.

coroner. a public official whose duty is to inquire into the causes and circumstances of any suspicious (i.e., unnatural) or violent human death. The functions and responsibilities have been curtailed in many jurisdictions and replaced by those of a medical examiner.

corpse. a dead human body; sometimes used also for the body of a higher (vertebrate) animal.

corpus luteum hormone. progesterone.

corrode. to cause, or to be affected by, corrosion.

corrosion. **1:** the dissolving and erosion of one material by another. **2:** the products of corrosion.

corrosive (caustic, escharotic). **1a:** having the capacity to corrode; to be corrosive. **1b:** having the capacity to chemically dissolve or erode a material. **2:** a chemical agent, such as an acid or strong alkali, that erodes the surface of a material or substance (e.g., a tissue) by chemical action. **3:** in toxicology, any agent (usually a strong base or acid) that causes visible destruction or irreversible alterations of dermal tissues or mucous membranes by chemical action at the site of contact.

corrosive acid. See acid.

corrosive alkali. See alkali.

corrosive mercury chloride. mercuric chloride.

corrosive poisoning. poisoning by a corrosive agent such as a strong acid or base, or by strong antiseptics such as bichloride of mercury, carbolic acid (phenol), cresols, and lysol. Such agents cause tissue damage similar to that caused by burns. Ingestion of such agents may cause erosive damage to any or all parts of the gastrointestinal (GI) tract. Symptoms may include an intense burning sensation about the mouth, pharynx and throat; abdominal pain which may be severe; nausea, retching or vomiting which may be bloody; a bloody diarrhea. There may also be stains about the mouth or in the stool that are characteristic of the specific corrosive agent involved. Corrosives may perforate the GI tract and sometimes other organs and may thereby produce severe shock that can quickly lead to death. Similarly, the victim may suffer a rapid death from asphyxiation due to closure of the respiratory tract by rapidly developing edema of the throat and pharynx. The esophagus may close off, sometimes causing death by starvation.

corrosive sublimate. mercuric chloride.

cort. any mushroom of the family Cortinariaceae, especially those of the genus *Cortinarius*.

cortexone. a toxic pregnene-derived steroid secreted from the prothoracic gland of certain water beetles such as those of the genera *Cybister* and *Dysticus*.

corticoid. **1:** any substance that simulates the activity of the adrenal cortex. **2:** a corticosteroid.

corticosteroid. any of various steroid hormones produced in the cortex of the adrenal glands and synthetic derivatives with similar properties. They are classified as glucocorticoids (e.g., hydrocortisone) and mineralocorticoids (e.g., aldosterone). *Cf.* corticoid.

corticosterone ((11β)-11,21-dihydroxypregn-4-ene-3,20-dione; 4-pregnene-11β,21-diol-3,20-dione; 11-21-dihydroxy-progesterone; Compound B; Kendall's compound B; Reichstein's substance H). a natural mineralocorticoid secreted by most mammals in smaller quantities than cortisol. It is, however, the chief glucocorticoid of laboratory rodents. Corticosterone has nominal glucocorticoid and mineralocorticoid activity, but is able to maintain life in adrenalectomized mammals.

corticotrophin. ACTH.

corticotropin. ACTH.

cortinarious. *Cortinarius orellanus*.

Cortinarius (corts). a genus of mushrooms (Family Cortinariaceae), some of which are poisonous, causing polydipsia and polyuria. Species of *Cortinarius* known to be toxic in Europe, also occur in California, although poisonings have not been reported. Symptoms of intoxication appear in three to 17 days following ingestion of poisonous corts. Polydipsia appears first, followed by headaches, nausea, muscular pain, and chills. In serious cases, the polyuria may be succeeded by oliguria or anuria. Pathology includes renal tubular necrosis, fatty degeneration of the liver, and severe inflammatory changes of the intestinal wall. ***C. orellanus*** (Poznan cort). a highly toxic, deadly Central European species that occurs largely in Poland and nearby areas. It is known to have caused fatalities. It produces the mycotoxin orellanine. Symptoms usually appear 3-14 days after ingestion. The liver and kidneys become damaged to the extent that death may intervene regardless of therapy. Symptoms include nausea, vomiting, weakness, [illegible] urinary retention, somewhat bloody urine, convulsions, coma, and consequent death. ***C. gentilis*** (deadly cort; gentle cort). a deadly species that occurs commonly throughout much of North America; it also occurs in Europe.

cortisol (COL). hydrocortisone.

cortisol acetate. hydrocortisone acetate.

cortisone (17α,21-dihydroxy-4-pregnene-3,11,20-trione; 17-hydroxy-11-dehydrocorticosterone; 11-dehydro-17-hydroxycorticosterone; Kendall's compound E; Reichstein's compound Fa 6; Wintersteiner's compound F). a corticosteroid isolated from the mammalian adrenal cortex in the early 1930s and the first to be successful in the treatment of rheumatoid arthritis. It has glucocorticoid and weak mineralocorticoid activity. Often used as an anti-inflammatory.

cortisone acetate (CONA). the monoacetate of cortisone.

corydalis (squirrel corn; turkey corn). **1:** common name of any plant of the genus *Corydalis*. **2:** the dried tuber of *Dicentra (Bicuculla) cucullaria* or *D. canadensis* (Family Fumariaceae). It contains corydaline, bulbocapnine, isocorydine, corytuberine, corycavine, corybulbine, corydine, fumaric acid, acrid resin, and protopine. *Cf. Corydalis*.

Corydalis (corydalis). a genus of toxic, usually biennial, leafy-stemmed, North American herbs (Family Fumaraceae; alternatively Family Paparivaceae, Subfamily Fumarioideae) with unusual, irregular blue, purple, yellow, or rose flowers similar to those of *Dicentra*. A number of species can poison grazing livestock. The alkaloids isoquinoline, aporphine, protoberberine, protopine, and others have been isolated from various species. *C. caseana* (synonym, *Capnoides caseana*) has caused the most livestock fatalities (sheep and cattle). Symptoms may appear within a few minutes to a few hours following ingestion. The symptoms of sheep and cattle are similar to those due to ingestion of *Dicentra*. They include restlessness, patent depression, followed by twitches especially of the facial muscles, convulsive clonic spasms of 5 to 15 minutes in duration. During the spasms, muscular rigidity is readily induced by any strong stimulus such as a loud sound. Biting or snapping at objects is common as is scouring, bleating, or bawling. In severe poisoning, the animal soon weakens, becomes quiescent, and dies following several seizures. Otherwise, recovery is rapid and uneventful. *Cf.* corydalis.

corynine. yohimbine.

coryza. a feature of iodism, it is an acneform or follicular eruption or granulomatous lesion caused by a reaction to systemic iodine or iodide exposure.

COS. carbon oxysulfide.

cosmetic. **1a:** beautifying (especially of the skin). **1b:** pertaining to beautifying or beautification. **2:** cosmetics; any of a variety of products (other than soap) applied to parts of the body that are designed to, intended to, or advertised as accomplishing one or more of the following without seriously affecting the body in general: **a:** cleansing, beautifying, promoting attractiveness, or preserving or altering one's appearence; **b:** cleansing, coloring, conditioning, or protecting one's skin, hair, eyes, lips, nails, or teeth. In the United States, the law does not require that cosmetics be tested for safety prior to sale. Indeed, the U.S. Food and Drug Administration (FDA) can take action against the continued sale of a cosmetic preparation only after harm by a particular (already marketed) cosmetic has been demonstrated. **hypoallergenic cosmetic**. any cosmetic that carries the label "hypoallergenic." Numerous ingredients in ordinary cosmetics are well-established allergens (e.g., many of those derived from petrochemicals or natural animal and vegetable sources). There is thus a significant market for cosmetics labeled and sold as "hypoallergenic." The consumer should be aware that such a label indicates only that most commonly encountered allergens have not been added to (nor have they been removed from) the preparation (e.g., fragrances, lanolin, cocoa butter, cornstarch, cottonseed oils). Some individuals are likely to be allergic to such a product or to become sensitized after some period of usage.

Cosmetic Blue Lake. FD&C Blue No. 1.

cosmopolitan (cosmopolite; ecumenical; pandemic; ubiquitous). worldwide in extent, distribution or influence; pandemic.

cosmopolite. cosmopolitan.

cotiarinha. *Bothrops itapetiningae*.

cotinine. the chief metabolite of nicotine. Its detection in urine indicates that an individual has recently smoked tobacco.

cotton dust. See dust.

cotton-dust asthma. byssinosis.

cotton-mill fever. byssinosis.

cotton rat. See *Sigmodon*.

cottonmouth. *Agkistrodon piscivorus*.

cottonmouth moccasin. *Agkistrodon piscivorus*.

cottonmouth water moccasin. *Agkistrodon piscivorus*.

cottonseed. seed of the cultivated cotton plant, *Gossypium herbaceum*. It contains about 20% protein and 20% oil. The oil contains gossypol, *q.v.*, and a number of other toxic pigments, some of which are more toxic than gossypol. Meal produced from cottonseed may contain ca. 40% protein and a very low percentage of oil if solvent extraction is employed in its manufacture. It is thus usually a good food supplement for livestock. Nevertheless, occasional intoxications of livestock occur, either from poorly controlled manufacture of the meal or when the meal comprises too high a proportion of the diet. The gossypol content of the swine diet must be less than 100 mg/kg body weight to be considered safe. See gossypol poisoning.

cottonseed meal. See cottonseed.

cottonseed oil. See cottonseed.

coturnism. food poisoning caused by the common European quail, *Coturnix*, and characterized by impaired speech, dyspnea, weakness, nausea, paresthesia of the lower legs, partial paralysis, and occasionally death.

cotyledontoxin. a toxic, nonnitrogenous, neutral nonalkaloidal, nonglucosidal, amorphous material isolated from the small, succulant, evergreen, herbaceous plants *Cotyledon ventricosa* and *C. wallchii* (Family Crassulaceae).

coumafuryl (3-(α-acetonylfurfuryl)-4-hydroxycoumarin). a white, nearly water-insoluble powder, $C_{17}H_{14}O_5$. It is a coumarin anticoagulant. This compound or its sodium salt is used as a rodenticide. It is highly toxic by ingestion and inhalation.

coumaphos (O,O-diethyl-O-(3-chloro-4-methyl-2-oxo-2*H*-1-benzopyran-7-yl)-phosphorothioate). a crystalline, water-insoluble substance, $C_9H_3O_2(CH_3)ClOPS(OC_2H_5)_2$. It is an anticholinesterase used as an insecticide and anthelmintic. It is used, in particular, against cattle grubs and a number of other ectoparasites, and for treatment of areas holding cattle. The maximum concentration that may be safely used on adult cattle, horses, and hogs is 0.5%.

***p*-coumaric acid** (3-(4-hydroxyphenyl)-2-propenoic acid; *p*-hydroxycinnamic acid; β-[4-hydroxyphenyl]-acrylic acid). a crystalline substance that is only slightly soluble in cold water, but soluble in hot water, ethanol, and ethyl ether; nearly insoluble in benzene and ligroin. It is a plant growth inhibitor structurally related to ferulic acid. It is the aglycon of coumarin glycoside. Coumarin is the anhydride of *o*-coumaric acid.

coumarin (2*H*-1-benzopyran-2-one; 1,2-benzopyrone; *cis-o*-coumarinic acid lactone; cumarin; coumarinic anhydride; tonka bean camphor). a very toxic, white crystalline substance, $C_9H_6O_2$, with an odor of new-mown hay. It occurs naturally in many plants (e.g., clovers), especially in the fruit or test where it inhibits germination. The seeds of such plants germinate only when the coumarin is removed or destroyed, e.g., by photolysis. It is used in perfumes and in the production of the anticoagulant, dicoumarin (dicumarol). Coumarin is very toxic to laboratory rats and guinea pigs, but is nontumorigenic to laboratory animals. Coumarin anticoagulants, if given to pregnant women, can cause fetal hemorrhage and death. *Cf.* warfarin.

coumarin anticoagulant. any anticoagulant (e.g., warfarin, coumafuryl) that contains coumarin as the active component. These tend to be less toxic than indandione anticoagulants, *q.v.* Effects of acute exposures include hemorrhage and sudden death with no warning. In subacute [illegible] [illegible] include anemia, weakness, pale mucous membranes, dyspnea, moist rales, bloody feces, and scleral, conjunctival, and intraocular hemorrhage. Staggering, ataxia, blood-tinged froth around mouth and nose, and CNS signs may appear when hemorrhage occurs in the brain or spinal cord.

coumarin glycoside. See glycoside.

counteract. to act against or in opposition to.

counteractant. a chemical substance that opposes or acts against the action of another chemical. See also allomone.

counteracting. acting against or in opposition to. Said, for example, of the action or effects of a drug or other xenobiotic in relation to the action of another substance.

counteraction. that action of a drug or other biologically active substance which opposes that of another drug.

counterpoison. antidote.

covalent binding. See binding.

cow cockle. *Saponaria spp*.

cow killer. a large whitish, yellow, and red wasp (Family Mutillidae) covered with long hairs. The female is wingless and delivers a powerful sting.

cow-nosed ray. a stingray, *q.v.*, of the family Rhinopteridae.

cow parsnip. See *Heracleum sphondylium* var. *montanum*.

cow-poison. *Amianthium muscitoxicum*.

cowbane. *Cicuta maculata*.

cowberry. See *Vaccinium*.

cowitch. *Rhus radicans*.

cowslip. *Caltha palustris*.

coyotillo. *Karwinskia humboldtiana*.

coytillo. buckthorn toxin.

cozymase. nadide.

CP. chemically pure.

CP 4742. sulfallate.

4-CP. chlorophenoxyacetic acid.

CP 50144. alachlor.

CPA. **1:** chlorophenoxyacetic acid. **2:** cyclophosphamide.

CPDC. cisplatin.

CPT. 3-chloro-*p*-toluidine.

Cr. the symbol for chromium.

CR. conditioned reflex (See reflex).

crab. **1:** a common term that usually refers to any of the numerous species of crustaceans comprising the order Decapoda, especially those with a broad cephalothorax and a small abdomen that lies against the ventral surface of the cephalothorax. *Birgus latro* (the coconut crab) and several species of tropical reef crabs (e.g., *Atergatus floridus*, *Carpilius maculatus*, and *Demania toxica*) are poisonous. The antennae are small and the walking legs are well-developed. **2:** sometimes used in reference to crayfish or any large decapod crustacean.

crab's eye, crab's-eye vine, crab's eyes, crabseye. *Abrus precatorius*.

cracheur. Naja nigricollis.

crack, crack cocaine. a street name for an almost pure form of cocaine that is smoked. See freebasing.

cracker-heel disease. See *Astragalus*.

cramp. a painful, often spasmodic, especially tonic muscular contraction. The affected individual may experience excruciating pain and hard, contracted masses of muscle.

cranberry. See *Vaccinium*.

crapulent, crapulous. pertaining to excessive eating and drinking; intoxicated as by alcoholic beverages.

Crassostrea gigas (giant Pacific oyster; Japanese oyster). a toxic oyster of Japanese coastal waters. Ingestion by humans causes venerupin poisoning which is often fatal. This bivalve was introduced from Japan to the west coast of North America around 1900 and now occurs from British Columbia to southern California. Venerupin poisoning is not known from North American populations of *Crassostrea*. They are cultured in Puget Sound and certain sheltered bays in northern California and are marketed mainly for restaurants and as packaged frozen food. See poisoning (venerupin poisoning), venerupin.

creatine (*N*-aminoiminomethyl)-*N*-methylglycine; *N*-amidinosarcosine; (α-methylguanidino)acetic acid; α-methylguanidineacetic acid; *N*-methyl-*N*-guanylglycine; methylglycocyamine). a compound present in muscle tissue of numerous vertebrates; small amounts occur in blood. It is present in muscle, mostly in the form of phosphocreatine, which contains a store of high-energy phosphoryl groups that can be transferred to ADP, yielding ATP which supplies the energy of muscular contraction. The phosphate transfer is catalyzed by creatine phosphokinase in the reaction:

phosphocreatine + ADP $\rightleftarrows$ ATP + creatine

creatinine. a degradation product of creatine. Serum creatinine is elevated following ingestion of meat. High serum creatinine may also be an indication of renal failure.

creatoxin. kreotoxin.

creeping Charlie. *Glechoma hederacea*.

creeping fescue. *Festuca rubra*.

creeping indigo. *Indigofera endecaphylla*.

creeping lady's sorrel. *Oxalis corniculata*.

creosote. an oily, colorless or yellowish to dark green-brown toxic and flammable mixture of phenols. It is oily with a characteristic, pungent odor and is obtained by distilling coal tar or wood tar. Creosote is used as a wood preservative and disinfectant, has been used externally by humans as an antiseptic, and taken internally as an expectorant in chronic bronchitis. It is a skin irritant and carcinogen. Exposed workers may develop pulmonary cancer, or cancer of the skin, larynx, or nasal cavity after 12-30 years of exposure.

creotoxin. kreotoxin.

creotoxism. kreotoxism.

crepitation. **1:** a rattling or crackling noise or sensation as heard in the lungs following exposure to mustard gas (See gas). It is produced by rapidly spreading decay of pulmonary tissue (gas gangrene) due to damage by the gas.

cresidine, *m*-cresidine (2-methyl-*p*-anisidine; 4-methoxy-2-methylaniline; 4-methoxy-2-methylbenzenamine; 2-methyl-4-methoxy-aniline). a suspected carcinogen, $C_8H_{11}NO$, and probable mutagen. Toxic fumes, NO_x, are released on decomposition by heating. ***p*-cresidine** (*m*-amino-*p*-cresol, methyl ester; 5-methyl-*o*-anisidine; 3-amino-*p*-cresol methyl ester; 1-amino-2-methoxy-5-methylbenzene; 3-amino-4-methoxytoluene; 2-amino-4-methyl-anisole; 2-methoxy-5-methyl-aniline; 2-methoxy-5-methylbenzamine (9CI); 4-methoxy-*m*-toluidine; 4-methyl-2-aminoanisole; azoic red 36; C.I. azoid red 83). a moderately toxic compound, orally, it is carcinogenic in some laboratory animals and a confirmed human carcinogen. It is moderately toxic orally and is probably mutagenic. Toxic fumes, NO_x, are released on decomposition by heating.

cresol (tricresol; hydroxymethylbenzene; methyl phenol). a mixture of the three isomeric cresols, *m*-, *o*-, and *p*-cresol, obtained from coal tar. It is a common component of disinfectants and is used in some dandruff shampoos. It is readily absorbed through the skin and the membranes of the pulmonary system. Intoxication may result from ingestion, inhalation, or absorption through the skin. Symptoms of cresol intoxication may include dyspnea, irritability, depression, hyperactivity, and unconsciousness. Nausea, vomiting, and a burning sensation of the mouth and throat are also common. Cresols can damage the liver, kidney, spleen, pancreas, and central nervous system. See cresols, cresylic acid. The three isomers have properties similar to those of phenol, but are less toxic. ***m*-cresol** (3-methylphenol; metacresol). a moderately toxic phenolic compound used as a local antiseptic, in disinfectants, fumigants, and explosives. It often occurs mixed with *ortho*- and *para*- analogs as cresol or cresylic acid, light yellow liquid, m.p. 11°C, b.p. 203°C. ***o*-cresol** (2-methylphenol; orthocresol). a moderately toxic solid (mp 31°C, b.p. 191°C) phenolic compound used as a disinfectant. ***p*-cresol** (4-methylphenol; paracresol). a crystalline solid (mp 36°C, b.p. 202°C) with a phenolic odor that is used as a disinfectant.

cresylate. any salt of a cresol or cresylic acid.

cresylic acid. **1:** a cresol. **2:** a mixture of phenols that contains varying amounts of cresols, xylenols, and other high-boiling fractions.

crevalle. *Caranx hippos*.

Cricelulus. See hamster.

Cricetus. See hamster.

cricket frog. See Hylidae.

crinotoxic (ichthyocrinotoxic). **1:** pertaining to those fish which produce a poison (not a venom) by glandular activity (e.g., lamprey eels, hagfishes, soapfishes, some gobies, boxfishes, and toadfishes). See crinotoxic fish. **2:** pertaining to the activity or effects of an ichthyocrinotoxin. See also crinotoxic fish.

crinotoxic fish. more than 60 species of fish are known to be crinotoxic. They release a toxic substance from epidermal glands that is lethal to other fish and possibly other marine animals. See crinotoxic, ostracitoxin.

Crinum (crinum lily). a genus of toxic bulbous plants (Family Amaryllidaceae). ***C. americum*** (swamp lily). the only species native to the United States. They are toxic to livestock. ***C. asiaticum*** (St.-John's-lily; poison bulb). a commonly cultured potted plant.

crinum lily. *Crinum*.

crisis. **1:** a turning point in the course of an acute disease generally marked by an intensification of symptoms, followed by recovery, stabilization, or a worsening of condition. **2:** an emergency. **abdominal crisis**. severe abdominal pain. **cholinergic crisis**. a condition characterized by hyperactivity of cholinergic nerve pathways. The proximate cause is an excessive accumulation of acetylcholine. This may result from poisoning by organophosphate, carbamate, or other anticholinesterases, or from overdosage of myesthenics with neostigmine or other anticholinesterases. **salt-losing crisis**. acute loss of sodium with dehydration, hypotension, and sudden death due to acute vomiting. **thyroid crisis** (thyroid storm). a sudden increase in severity of the symptoms of thyrotoxicosis, characterized by extreme tachycardia and fever; sometimes fatal. **true crisis**. a fall in temperature and pulse rate.

crisis intervention. a problem-solving activity aimed at preventing or correcting the continuation of a crisis (e.g., as by poison control centers, suicide hotlines). Such activities are usually carried out by means of a telephone service.

criteria. pl. of criterion.

criteria air pollutant. any of the seven pollutants (sulfur dioxide, carbon monoxide, nitrogen dioxide, hydrocarbons, oxidants (e.g., ozone), particulate matter (PM_{10}), and lead) that are regulated by the U.S. EPA based on national ambient air quality standards (NAAQS). The term, "criteria pollutant" stems from the requirement that EPA must describe the characteristics and potential health and welfare effects of these pollutants. This information provides the criteria by which standards are set or revised.

criteria pollutant. See criteria air pollutants.

criterion (plural, criteria) **1:** a distinguishing mark or trait. **2:** a predetermined standard on which a decision or judgment may be based.

critical. **1:** dangerous; pertaining to or denoting a state of crisis. **2:** crucial, decisive. **3a:** of or pertaining to a turning point. **3b:** of sufficient size, quantity, or intensity to effect a major change or turning point (e.g., a critical concentration). **4a:** of doubtful or uncertain outcome; characterized or attended by uncertainty. **4b:** in danger of death.

critical concentration. See concentration (critical concentration).

critical effect. See effect (critical effect).

critical organ. See organ (critical organ).

crocidolite asbestos (blue asbestos). See asbestos.

Crocker's sea krait. *Laticauda crockeri*.

Crocker's sea snake. *Laticauda crockeri*.

crocus. **1:** any plant of the genus *Crocus* (Family Iridaceae), the true crocusses. **2:** a plant of the genus *Colchicum* (Family Lilaceae), more often referred to as autumn crocus or meadow saffron.

crooked calf disease. a syndrome of cattle characterized by varying degrees of cleft palate, joint contractures, torticollis, and scoliosis or kyphosis. It results from ingesting any of various species of *Lupinus* (e.g., *L. caudatus*, *L. laxiflorus*, *L. sericeus*, and *L. nootkatensis*) by pregnant cows fed between days 40 and 70 of gestation. The plants are most teratogenic early in the growing season and during seed formation. Anagyrine, an alkaloid, is the active principle.

crosier cycas. *Cycas circinalis*.

cross adder. *Vipera berus*.

cross-eye. esotropia, a form of strabismus.

cross-eyed. strabismic. See strabismus.

cross-reaction. See reaction (cross-reaction).

cross-resistance (cross-tolerance). the condition whereby an organism is resistant (or tolerant) to a substance other than the one that originally induced resistance in the population. Insect populations, for example, often become resistant to a given insecticide. These organisms may then show resistance to certain other insecticides to which they have not before been exposed. Such recruitment of resistance (tolerance) to a spectrum of chemicals is commonly due to adaptive selection with an increased incidence of high levels of nonspecific xenobiotic-metabolizing enzymes in the population.

cross-tolerance. cross-resistance.

crossed eyes. strabismus.

crossed pit viper. *Bothrops alternatus*.

crotactin. a toxic component of crotoxin with a higher index of lethality than crotoxin.

crotalaria. a common name of herbs of the genus *Crotalaria*.

Crotalaria (rattlebox; crotalaria). a genus of small herbs (Family Fabaceae, formerly Leguminosae) found in warm areas, often on sandy soils of fields and along roadsides. Several species contain pyrrolizidine alkaloids and are known to be toxic; some can prove lethal. The raw seeds are especially toxic. They cause liver cell necrosis, enlargement of certain hepatic cells, and rupture of veins. Crotalaria are involved mostly in the poisoning of livestock. ***C. retusa***. all parts contain monocrotaline. Consumption of seeds has caused major losses of domestic fowl with effects similar to those of *C. spectabilis* but less acute. Cattle rarely consume *C. retusa*. ***C. spectabilis***. a coarse annual herb of fields and roadsides of the southern United States. The toxic principle is monocrotaline. This plant has caused severe losses among domestic cattle, horses, swine, and domestic fowl and lesser losses among goats, sheep, and mules. In chronic poisoning of the domestic fowl (80-100 seeds), signs and lesions may not appear until a month or more following consumption. In acute poisoning (ca. 320 seeds or more), fowl die within 1-10 days. Symptoms include darkening, scaling, and congestion of the comb, depression, severe diarrhea with green feces, and retention of seeds in the crop. Lesions include petechial hemorrhages of the abdominal and thoracic serosa; a smaller than normal, often ruptured, marbled liver; and petachiae and suffusions of visceral fat, heart, and skeletal musculature. In less acute cases, massive hemorrhage and ascites may present. In chronic poisoning, development and progression of the above symptoms are slowed and the birds may become severely anemic with a paling of the comb. Lesions of chronic poisoning include necrotic enteritis, ascites, partial splenic atrophy, and cirrhosis of the liver. Symptoms of chronic poisoning of cattle may include nervousness or excitement, bloody feces, prostration, and death. Lesions commonly include profuse hemorrhage of the thoracic and abdominal serosa. In long-lasting cases, ascites and jaundice may be observed and hemorrhage is less severe. Symptoms of cattle sometimes include weakness, emaciation, incoordination, and constipation or diarrhea. See also crotalism.

crotalaria poisoning. crotalism.

crotalic. of or pertaining to rattlesnakes (crotalids).

crotalid (crotaline). **1:** pertaining to or characteristic of a pit viper. **2:** any snake of the family Crotalidae.

crotalid snake. any snake of the family Crotalidae.

crotalid venom poisoning. rattlesnake venoms and those of many other pit vipers produce local pain, edema, hemorrhagic vesiculation, petachiae, tissue damage, and necrosis with sloughing of the skin. In the absence of necrosis, the skin appears tense and shiny. Numbness or a tingling sensation of the tongue, mouth, or scalp is occasionally reported. Systemic effects may include blood cell changes, coagulation defects, blood vessel injury, changes in vascular resistance, hypotension, shock, and neurologic defects. The hematocrit may fall rapidly, although hemoconcentration may occur during very early stages. Thrombocytopenia is common, and coagulation is often abnormal. Pulmonary edema may result from severe poisoning, and bleeding may occur in the lungs, peritoneum, kidneys, and heart. A critical deficit in glomerular filtration secondary to hypotension may lead to renal failure as may hemolysis. Poisoning by some crotalid species commonly produce proteinuria, hemoglobinuria, or myoglobinuria. Early cardiovascular collapse occasionally occurs due to a sharp fall in circulating blood volume. The latter is probably due to a loss of blood plasma and protein through the vessel walls, and by laking of blood. Most North American crotalid venoms produce relatively minor changes in neuromuscular transmission, but the venom of *Crotalus scutulatus* may cause serious neurologic deficits. Specific diagnosis usually requires the presence of one or more fang marks, rapid onset of swelling, edema, and usually pain. See also snake venom poisoning.

Crotalidae (pit vipers and rattlesnakes). a widely distributed family of nearly 300 species of venomous snakes, exclusive of Africa and Australia. Common genera are *Agkistrodon, Bothrops, Lachesis, Sistrurus,* and *Crotalus*. Crotalids are characterized by the presence of paired, heat-sensitive, pit-like loreal depressions located between each eye and nostril; a broad head that is distinct from the narrow neck; and eyes with a vertically elliptical pupils. They have long, recurved, paired, hollow, erectile fangs situated anteriorly in the upper jaw. These remain folded back (recessed) except when striking. Crotalidae is sometimes regarded as a subfamily of Viperidae.

crotaliform. of the type of or resembling a rattlesnake.

crotalin. **1:** a protein contained in rattlesnake venom. **2:** rattlesnake venom.

crotaline. **1:** crotalid. **2:** monocrotaline.

crotalism (bottom disease; crotalaria poisoning). a venoocclusive disease of grazing animals (sometimes humans) caused by ingestion of plants of the genus *Crotalaria* (either by grazing or by ingesting hay that contains *Crotalaria*) and sometimes *Senecio* or *Heliotropum*). The condition is marked by weakness, emaciation, stupor, congestion, and hemorrhage of the liver and spleen. **2:** crotalaria poisoning (See poisoning).

crotaloid. resembling the Crotalidae (pit vipers).

crotalotoxin. a poisonous substance isolated from rattlesnake venom.

Crotalus (rattlesnakes; klapperschlangen). a genus of New World, live-bearing, venomous snakes (Family Crotalidae). The eyes are small with vertically elliptical pupils and the head is broad and distinct from the narrow neck. They have large recurved fangs that are periodically replaced throughout life. The body is moderately slender to stout and more or less cylindrical. A jointed rattle comprised of cornified, button-like structures forms the tip of the tail in nearly all species. The larger species prey mainly on small mammals and the smaller species prey chiefly on lizards. Effects of envenomation may include pain, swelling, and edema about the bite, and thirst, nausea, vomiting, weakness, a weak rapid pulse, hypotension or shock, tingling or numbness (especially about the tongue and mouth), respiratory distress, and blurred vision. In serious cases, coma and death may ensue. *C. adamanteus* and *C. atrox* cause most of the snakebite fatalities in the United States. See also crotalid venom poisoning, rattlesnake.

C. ***adamanteus*** (eastern diamondback rattlesnake). the only large rattlesnake in the southeastern United States, it reaches a maximum length of about 2.4 m. It occurs in old fields, dry pine flatwoods, sandhill or longleaf pine and turkey oak country, and palmetto thickets in coastal lowlands from North Carolina through Florida and eastward to Louisiana, from sea level to ca. 150 m. The side of the head is characterized by the presence of distinct, diagonal, whitish stripes. The fangs are long and the venom produced is considerable. These snakes feed chiefly on rabbits, squirrels, and birds. They are extremely dangerous, often irritable, and sometimes strike without warning, and the venom is highly destructive to tissue and the bite is sometimes fatal. See L-amino acid oxidase. *C.* ***atrox*** (western diamond-back rattlesnake). a snake of arid and semiarid open or sparsely wooded terrain, including rocky hills, deserts and coastal sand dunes. It often occurs also in cultivated areas and in or near farm buildings. It reaches a length of about 2.15 m. The ground color may be buff, brown, reddish, or gray and there are characteristic lateral diagonal stripes on the head; the posteriormost stripe extends to the angle of the mouth; the tail is ringed with broad black and grey or white stripes. The range extends from central Arkansas to southeastern California southward through most of Texas into northern Vera Cruz and southern Sonora with a relict population in Oaxaca, Mexico. The fangs are long and the venom produced is considerable. This species is extremely dangerous, often irritable, and sometimes strikes without warning; the bite can seriously damage tissue and is sometimes fatal. See also *C. ruber*. *C.* ***basiliscus*** (Mexican West Coast rattlesnake). a rather stout rattlesnake, nearly triangular in cross-section; some adulsts reach a maximum length of nearly 2 m. It occurs on the mountain slopes and coastal plains of western Mexico from southern Sonora to central Oaxaca. It is the only rattlesnake in this region with diamond-shaped dorsal markings. The highly toxic venom is produced in large amounts. A polyvalent antivenin is available from the Instituto Nacional de Higiene, Mexico. Large specimens are extremely dangerous; only *C. durissus* kills more humans yearly in South America than *C. basiliscus*. *C.* ***catalensis***. a rattlesnake indigenous to Catalina Island in northwestern Mexico. *C.* ***cerastes*** (sidewinder; horned rattle-snake). a small, largely nocturnal rattlesnake, reaching a maximum length of about 30 inches or 0.7 m; females are somewhat larger than males. It is characterized by the presence of a conspicuous horn-like scale over the eye and its unmistakable form of sidewinding locomotion (seen also in certain sand vipers of Africa and Asia, and occasionally in a few other desert snakes). The body is light brown, tan, cream, gray, or pinkish with rows of darker spots and a ringed tail. It occurs in desert areas, mainly on sandy flats and dunes with meager vegetation from southeastern California and southern Nevada southward through western Arizona into Mexico (Sonora and Baja California). The sidewinder is not especially irritable, the venom is produced in small quantities, and fatalities from the bite of a sidewinder are rare. It feeds largely on pocket mice, kangaroo rats, and lizards. ***C. durissus*** (cascabel; cascabela; cascabel rattlesnake; cascavel; neotropical rattlesnake; neotropical rattlesnake; tropical rattlesnake; South American rattlesnake; boiciuinga; borquira). a large, tropical rattlesnake with a stout, slightly compressed body. It occurs in grasslands, other open warm semiarid areas, and in thorny scrub from southern and eastern coastal Mexico, Central America, and in eastern South America from northern Columbia to northern Argentina. With the exception of Mexico, the cascabel is the only true rattlesnake throughout this range. It is distinguished by a series of large rhomboid blotches or diamonds along the dorsum, cervical stripes in some individuals, and the large rattle. Adults average about 1.2 m long, with occasional individuals reaching a length of nearly 2 m. The ground color is brown or olive. The tail is usually dark brown or black above and white or cream colored below. The cascabel can deliver a large volume of a very to highly toxic venom. Toxicity of the venom varies considerably throughout the range; in Brazil, where the *C. durissus terrificus* is the main cause of death from snakebite, it is extremely toxic. The venom produces minor local effects but the systemic effects are very grave. Envenomation produces more severe changes in nerve conduction and neuromuscular transmission than does that of other crotalids. Effects may include blindness, paralysis of the neck muscles, and death from pulmonary and cardiovascular collapse. Specific antivenin is not very effective and tremendous amounts are needed to counteract the effects of the bite of a snake of average size. The cascabel rattlesnake is thus one of the most dangerous snakes in the world. ***C. enyo***. a rattlesnake of northwestern Mexico. ***C. exsul***. a rattlesnake that occurs on Cedros Island in northwestern Mexico. ***C. horridus*** (timber rattlesnake; velvet-tail rattlesnake). an irritable, but not especially aggressive rattlesnake that often rattles loudly and feints repeatedly before striking. Fatalities among humans are rare. Adult specimens of *C. horridus* are the only rattlesnakes in the eastern United States that combine small interocular scales, absence of light lateral stripes on the head, and a black tail. The ground color is yellow, buff, pale brown, or gray with sooty black crossbands or chevrons with narrow white or pale yellow margins. The belly is pinkish white or cream, with dark gray scattered throughout. Some individuals from upland areas in the eastern United States are uniformly black above. There is often a pinkish, rusty, or amber stripe along the dorsal midline. Maximum length is about 1.9 m. The range extends from New England to the Florida panhandle, westward to central Texas and northward through the Mississippi Valley to southeastern Minnesota. ***C. intermedius*** a rattlesnake of southern Mexico. ***C. lepidus*** (rock rattlesnake). a small, slender rattlesnake, averaging about 0.6 m in length. It occurs largely in rocky mountainous areas (e.g., rimrock, talus slopes, gorges, and rocky streambeds) from northern Mexico to southwestern Arizona, west central and southeastern New Mexico, through the trans Pecos region. The ground color is greenish-gray, bluish-gray, or pinkish-tan in color with widely spaced, narrow, black or brown crossbands with irregular borders. ***C. mitchellii*** (speckled rattlesnake; Mitchell's rattlesnake). Adults average less than 1 m in length, although some individuals attain a length of more than 1.3 m. Ground color and pattern vary greatly, although most individuals have a sandy, speckled appearance. The dorsum is marked by crossbands made up of lightly stippled diamond-shaped markings. The large scale above the eye is pitted, creased, or has rough edges. The rostral is separated from the preanals by a row of tiny scales. The range of this species includes extreme southwestern Utah, southern Nevada, and southern California southward into northwestern Mexico (Sonora and Baja California). ***C. molosus*** (black-tailed rattlesnake). a greenish, yellowish, or grayish rattlesnake with lighter irregular crossbands. The tail is black. Found mostly in rocky mountainous areas (rimrock and limestone outcrops) throughout central and northern Mexico and in the United States from southern and central Arizona eastward to central Texas, it is not usually aggressive. ***C. polystictus*** (Mexican blotched rattlesnake). a rattlesnake of west central Mexico. ***C. pricei*** (twin-spotted rattlesnake). a small, slender rattlesnake (usually less than 0.5m long) of high, arid rocky areas of northwestern Mexico and extreme southeastern Arizona. It is characterized by the two rows of paired or slightly alternating, brown, scalloped, spots along the dorsum, and by the light bordered strip from the eye to the angle of the jaw. The rattle is buzzy and can be heard only at a short distance. ***C. pusillus*** a rattlesnake of southwestern Mexico. ***C. ruber*** (red diamond rattlesnake; red rattlesnake). a fairly large rattlesnake, reaching a maximum length of about 1.5 m. Its range encompasses southwestern California and Baja California of northwestern Mexico. This rattlesnake is more reddish in color, but is otherwise structurally very similar to *C. atrox*. It usually has 29 rather than 25 scale rows at midlength and the first lower labial is usually divided in *ruber*, but is undivided in *atrox*. However, the distribution of the two species overlaps only in a narrow band along the border between California and extreme northeastern Baja California; and red diamond rattlesnakes are largely restricted to rocky hillsides at lower elevations. *C. ruber* is generally less aggressive and more active diurnally than either *C. adamanteus* or *C. atrox* and the venom is less toxic. Fatalities are rare. ***C. scutulatus*** (Mojave rattlesnake). of moderate size, this snake attains a maximum length of about 1.2 m. It is similar in appearance to the western diamondback and prairie rattlesnakes. Usually greenish or olive in color with dark rings on the tail that are much narrower than the intervening light spaces. The range extends from western Texas northwestward to the Mohave Desert and southeastward into the arid highlands of Mexico. The venom is the most toxic of North American rattlesnakes and has a pronounced effect on respiration; it may cause a temporary paralysis. It is less irritable and more reluctant to strike than *C. adamanteus* and *C. atrox*. ***C. stejnegeri***. a rattlesnake of northwestern Mexico. ***C. tigris*** (tiger rattlesnake). a rattlesnake characterized by the presence of many gray or brownish rossbands (comprised of tiny dots) over a ground color of pale gray, buff, lavender, or pinkish gray. Scales of the body are keeled and in 23 rows. They occur from central Arizona southward into northern Mexico. ***C. tortugensis*** a rattlesnake of Tortuga Island in northwestern Mexico. ***C. transversus*** a rattlesnake of central Mexico. ***C. triseriatus***. a species of rattlesnake that is widely distributed in Mexico. ***C. unicolor*** (Aruba rattlesnake; aruba). it occurs only on the island of Aruba off the coast of Venezuela, where it is the only venomous snake. The body is usually uniformly gray or light gray-brown above and white or cream-colored ventrally. There is sometimes a faint dorsal pattern of rhomboidal blotches. There may be a series of lateral obsolete blotches that alternate with or lie opposite the dorsal series. The body is stout and somewhat depressed. Adults average about 0.7 m in length; occasional individuals measure nearly 1 m. The dorsal scales are strongly keeled and occur in 25-27 rows at midbody, fewer posteriorly. There is usually a distinct pair of posterior parallel lateral stripes on the head that may project onto the neck. It is not aggressive but readily strikes when threatened. Its close kinship with *C. durissus* suggests that the venom might be highly toxic and it may be capable of a serious bite. ***C. viridis*** (western rattlesnake). a complex of rattlesnakes given generic status by some authorities but most of the races are considered separate species by others. The size, color, and other aspects of appearance vary considerably among the races. When present, the light lateral diagonal stripes on the head extend behind the angle of the mouth; the tail is ringed or dark; the body is spotted or marked by crossbands rather than diamonds; maximum length is about 1.6 m. The defining characteristic is the presence of 2 or more internal scales that abut the rostral. Snakes of this complex are widely distributed, are largely diurnal, and are usually less irritable than diamondbacks, although bites are fairly common. The effects of envenomation are not usually severe and are often marked by prickling sensations and numbness about the mouth. ***C. v. abyssus*** (Grand Canyon rattlesnake). his snake is reddish above with obscure dorsal blotches. It occurs in the Grand Canyon. ***C. v. cerberus*** (Arizona black rattlesnake). The ground color of this crotalid is dark gray, brown, and occasionally nearly black. The large, dark splotches are partially bordered by white. It occurs from Arizona (south of the Colorado Plateau) southward into extreme western New Mexico. ***C. v. concolor*** (midget faded rattlesnake). a pale yellow snake with obscure or absent blotches that occurs in extreme southwestern Wyoming, Colorado, and eastern Utah. ***C. v. helleri*** (Southern Pacific rattlesnake). a rattlesnake that closely resembles *C. v. oreganus*, *q.v.* It occurs in San Luis Obispo and Kern counties of California, southward to southern Baja California. ***C. v. lutosus*** (Great Basin rattlesnake). a small rattlesnake that reaches a maximum length of about 1.3 m. The light stripe behind the eye is wider than in other members of the *C. viridis* complex; the rectangular blotches on the body usually lack light margins; and the ground color is buff or dull yellow. It occurs mainly in arid and semiarid rocky areas in western Utah, southern Idaho, Nevada, and southeastern Oregon. ***C. v. nuntius*** (Hopi rattlesnake). a pinkish to reddish-brown rattlesnake with well-defined blotches. ***C. v. oreganus*** (Pacific rattlesnake; Northern Pacific rattlesnake). a rattlesnake of the Pacific Coast of North America from British Columbia to lower California and east into Idaho with a disjunct population in Arizona. In Washington and Oregon it is largely confined to semiarid areas. It usually contrasts with other members of the *C. viridis* complex in that the light stripe behind the eye is wide and often indistinct; the body pattern consists of diamonds or hexagonal blotches; and the ground color is brown, olive, or dark gray. It is substantially larger than *C. v. lutosus* and is nearly indistinguishable from the more southern race, *C. v. helleri*. ***C. v. viridis*** (prairie rattlesnake). a relatively nonaggressive rattlesnake characterized by the brownish, greenish-gray, or olive ground color; the narrow, light, diagonal stripe behind the eye; and the well-defined, brown, nearly rectangular, blotches with usually narrow light-colored borders that occur on the body. It [illegible] of the Great Plains of the western United States, extreme northern Mexico, and extreme southwestern Canada, where it typically occupies dry grassland, rocky hills, and open rocky mountain slopes below timberline. It is not especially aggressive and usually avoids humans and livestock. Adults sometimes attain a length of about 1.6 m. ***C. willardi*** (ridge-nosed rattlesnake; Arizona ridge-nosed rattlesnake). a small, retiring, rattlesnake that averages about 0.5 m in length, and is characterized by the presence of a prominent ridge along the upper edge of the snout. The ground color is generally light gray, brown, or reddish, marked with dark-bordered, narrow, white crossbands. It occupies a small range in northern Mexico with extensions into southeastern Arizona and southwestern New Mexico.

crotalus antitoxin. See antitoxin.

crotocin. a naturally occurring epoxide trichothecene mycotoxin and antibiotic.

croton. a mixture of toxic proteins contained in the seeds of *Croton tiglium*, a small Asiatic tree (Family Euphorbaceae). See also crotonism.

Croton (hogwort; croton). a genus of small poisonous trees and shrubs (Family Euphorbaceae). ***C. tiglium*** (croton oil; mayapple; gamboge; purging croton). a small tropical East Indian tree (now found in the southwestern United States and elsewhere), it is the commercial source of the highly toxic croton oil. The seed is fatally toxic when ingested, producing a painful burning sensation in the mouth and stomach, tachycardia, a bloody diarrhea, coma, and death. The active principle is croton oil. The oil will blister the skin on contact. Other species such as *C. texensis* (skunkweed, Texas croton) and *C. capitatus* (hogwort, wooly croton) also contain toxic amounts of the oil. These plants are generally distasteful to animals and poisonings are rare. See croton, crotonism, croton oil.

croton oil. **1:** a thick, extremely toxic, irritant, vesicant, and violently purgative oil, isolated from a small East Indian tree, *Croton tiglium* (Family Euphorbaceae), *q.v.* It has been used therapeutically as a cathartic and counter-irritant. A few drops of the pure oil can be fatal to a human. Symptoms are those of severe gastroenteritis. Croton oil is used in alcoholic beverages as a form of "mickey finn." See also crotonism, phorbol, tetradecanoylphorbol 13-acetate. **2:** *Croton tiglium*.

croton-oil poisoning. crotonism.

crotonaldehyde (2-butenal; crotonic aldehyde; β-methylacrolein). a flammable, explosive liquid, $CH_3CH{=}CHCHO$; it is an intense irritant and lacrymator. The toxicity and effects are similar to those of acrolein. The fumes are highly irritating to the skin and mucous membranes, readily damaging the eyes.

crotonic aldehyde. crotonaldehyde.

crotonism (croton-oil poisoning). poisoning by croton oil. It is characterized by a burning sensation in the mouth, colic, and severe watery diarrhea which may be followed by vertigo, somnolence, prostration, coma, and death from circulatory or respiratory collapse.

crotonyl alcohol. crotyl alcohol.

crotoxin. **1:** a polypeptide complex of *Crotalus* venom, comprised of one acidic and one basic subunit. **2:** the original crystalline protein isolated from the venom of the rattlesnake, *C. durissus terrificus* (cascabel; maracaboia). This isolate contained crotoxin (def. 1) and hyaluronidase, phospholipase, and perhaps other enzymes. It accounted for most of the toxicity of the crude venom. It exhibited neurolytic, indirect hemolytic, and smooth muscle-stimulating activities. **3:** variously and improperly, and often ambiguously used throughout the literature in reference to any of 17 different components of the venom of *C. durissus terrificus*. See also crotamine.

crowfoot family. Ranunculaceae.

crown of thorns. *Euphorbia millii*.

crownbeard. *Verbesina encelioides*.

crowned snake. See *Aspidomorphus*.

CRS. Chinese restaurant syndrome.

crucial. critical; decisive; determining; of supremely important.

crucifer. **1:** any plant of the mustard family, Brassicaceae (formerly Cruciferae). See especially *Brassica*. **2:** pertaining to a crucifer (def. 1).

Cruciferae. Brassicaceae.

crude. **1:** unprocessed, unrefined, raw, in a natural state. **2:** crude oil (plural, crudes). See petroleum.

crude arsenic. arsenic trioxide.

crude oil. petroleum, *q.v.*, in its natural state prior to refining.

crude opium. opium.

crustacean. any animal of the class Crustacea (e.g., crabs, crayfish, lobsters, shrimps).

crustecdysone. β-ecdysone.

cryptic damage. damage, as by a toxicant, of which there is no visible, palpable, or otherwise evident sign.

cryptogenic. of uncertain, indeterminant, or obscure origin.

cryptohalite. ammonium hexafluorosilicate.

Cryptostegia (allamanda). a genus of tropical vining shrubs (Family Asclepiadceae) that are often grown as ornamentals. ***C. grandiflora*** (rubber vine, purple allamanda, pink allamanda). a thick-stemmed toxic woody vine with thick glossy leaves. It is widely cultivated in India for rubber from the sap. It has become naturalized on roadsides and waste areas in southern Florida. It has caused fatalities in humans from severe gastroenteritis.

cryptotoxic. having unknown or normally masked toxic properties as in the case of food fish that are normally nontoxic, but become toxic under certain conditions; or in the case of a liquid that is normally nontoxic, but becomes toxic if certain physical properties (e.g., solubility, colloidal state) are altered.

cryptotoxic. having unknown or normally masked toxic properties.

crystal-containing body. microbody.

crystalline digitalin. a synonym for digitoxin. Do not confuse with digitalin.

crystalline veratrine. cevadine.

Cs. the symbol for cesium (caesium).

CS, CS gas. *o*-chlorobenzylidenemalononitrile. The acronym stands for the compound's inventors, Corson and Stoughton.

Ctenizidae. a family of trapdoor spiders of the southern and western United States that build silk-lined burrows with a "trapdoor" lid. See *Bothryocyrtum*.

Ctenizidae. See trapdoor spiders.

CTG. Control Technique Guidelines.

CTL. cytotoxic T-cell (See T lymphocyte).

cu. cubic.

Cu. the symbol for copper (cuprum).

^{64}Cu. copper-64. See copper.

cuatro narices. *Bothrops atrox, B. castelnaudi*.

Cuban lily. *Scilla peruviana*.

cubeb. the dried, unripe but nearly grown fruit of *Piper cubeba* (Family Piperaceae). It is pulverized and smoked in cigarettes to reduce catarrh. See *Piper cubeba*, cubebism.

cubeb poisoning. cubebism.

cubebism (cubeb poisoning). poisoning by cubeb. Symptoms commonly include nausea, vomiting, diarrhea, fever, arthralgia, inflamed kidneys, a weak pulse, prostration, loss of consciousness, miosis, delirium, and coma. In severe cases, death from respiratory failure may intervene.

cubomedusae (sea wasps). extremely dangerous cnidarians, especially *Chironex fleckeri* and *Chiropsalmus quadrigatus* of Australia. Contact causes a stinging or burning sensation; an erythematous wheal may develop. The wheal may enlarge and vesiculate, becoming pustular, and the skin may become necrotic.

cuckoopint. *Arum maculatum*.

cucumber. *Cucumis sativus*.

cucumber family. Cucurbitaceae.

Cucumis sativus (cucumber). the common garden cucumber (Family Cucurbitaceae). Concentrations of nitrate in the foliage toxic to livestock have been reported.

Cucurbita maxima (squash). a vine-like herb with several varieties; they are common garden vegetables (Family Cucurbitaceae). Concentrations of nitrate in the foliage that are toxic to cattle have been reported.

Cucurbitaceae (cucurbits; cucumbers; gourds). a family that includes many agriculturally and commercially important plants (e.g., cucumbers, watermelon, squashes, gourds, and various ornamentals). The sap of most species contains cucurbitacins, a class of highly poisonous triterpenes. See cucurbitacin.

cucurbitacin. any of a group of 17 toxic antineoplastic tetracyclic triterpenes isolated from the sap of most species of plants of the family, Cucurbitaceae (cucumbers, gourds, squashes) and various species of Brassicaceae (formerly Cruciferae), Datiscaceae, Euphorbiaceae, and Scrophulariaceae. They have been used since antiquity as antimalarials, emetics, vermifuges, and narcotics. Cucurbitacins have sometimes been implicated in the poisoning of livestock. *Cf.* colocynth, colocynthin. **cucurbitacin B** (25-acetyloxy)-2β,16α,20-trihydroxy-9β-methyl-19-*nor*-10α-lanosta-5,23E-diene-3,11,22-trione, 1,2-dihydro-α-elaterin). this form is extremely toxic to laboratory mice. **cucurbitacin E** ((25-acetyl-oxy)-2β,16α,20-trihydroxy-9β-methyl-19-*nor*-10α-lanosta-1,5,23E-triene-3,11,22-trione, 1,2-dihydro-α-elaterin). this form is very toxic to laboratory mice.

cucurbit. See Cucurbitaceae.

Culex. a large common genus of pest mosquitos (Family Culicidae). The saliva contains irritants and anticoagulant; the bite causes local swelling and itching. Culex mosquitos may transmit filariasis.

culicicide. culicide.

Culicidae (mosquitoes). a large family of insects (Order Diptera) characterized by a slender, delicate, humped body. Females have long, piercing mouthparts used in most species to feed on the blood of mammals or birds. The saliva contains irritants and anticoagulant. Many species are vectors of disease; some are virulent (e.g., malaria, yellow fever). Males are smaller and more delicate than females and may feed on plant juices. The larvae and pupae are aquatic. Important genera are *Aedes*, *Anopheles*, and *Culex*.

culicidal. 1: destructive to mosquitoes (Family Culicidae) and to dipterous gnats. **2:** a culicidal agent.

culicide. a substance that destroys mosquitoes (Family Culicidae) and dipterous gnats.

cultivar (CV). **1:** a variety of a plant that is produced and maintained under cultivation. **2:** a taxon below the subspecies level. See variety.

cultivated aconite. *Aconitum napelius*.

cultivated carrot. *Daucus carota* var. *sativus*.

cultivated oats. See *Avena sativa*.

cultivated rape. *Brassica napus*.

cultivated rye. *Secale cereale*.

culture. **1:** a population of cells, microorganisms, or tissue maintained and grown in a nutrient medium (culture medium) in the laboratory. **2:** to maintain and grow cells, microorganisms, or tissues in a nutrient medium. **3:** the identification of microorganisms or viruses by collecting material from an infected site or tissue and maintaining it on nutrient material until the infecting agent has grown to a point where it can be examined and identified microscopically. **axenic culture** (pure culture). a culture that contains only one type or species of microorganism. **batch culture**. a culture in which the medium is not renewed. **continuous culture**: a culture in which populations of microorganisms or cells can be maintained in a state of exponential growth for long periods of time. **mixed culture**. a culture that contains more than one type of microorganism. **pure culture**. axenic culture.

cumarin. coumarin.

***cis*-cumarinic acid lactone**. coumarin.

cumulative. increasing in strength or intensity of effect with successive additions or applications. Thus, a cumulative poison may produce no observed effect on a particular organism at a given concentration or with a given dose, but the effect of subsequent exposure will be stronger than otherwise expected. The cumulative effects are often more than additive.

cumulative action. cumulative effect (See effect).

cumulative drug action. the cumulative effect of small, repeated doses of a drug that is not rapidly eliminated from the body. Cumulative drugs include lead, mercurials, silver.

cupferron (copperon; *N*-hydroxy-*N*-nitrosobenzenamine ammonium salt; *N*-nitrosophenylhydroxylamine ammonium salt). a colorless crystalline ammonium salt. It is used in acid solution as a precipitating reagent to separate copper and iron from other metals, and in the colorimetric analysis of aluminum. It is carcinogenic to some laboratory animals.

Cupiennius. the so-called banana spider. It sometimes occurs in produce or other materials imported into the United States.

cupric. pertaining to copper, especially the divalent ion.

cupric acetoarsenite. ((acetato)trimetaarsenitodicopper; copper ethanoacetoarsenate; Paris green; copper acetate arsenite; emerald green; French green; imperial green; mineral green; Mitis green; parrot green; Schweinfurt green; Vienna green). a bright blue-green, crystalline powder, $Cu(C_2H_3O_2)_2 \cdot 3Cu(AsO_2)_2$, used as an insecticide, in antifouling compositions and wood preservatives. It is very toxic to laboratory rats. Toxicity is due mainly to the arsenic content. Symptoms of intoxication in humans include gastrointestinal disturbances, muscular cramps, tremors, and peripheral neuritis.

cupric arsenite (arsonic acid copper(2+) salt (1:1); arsenious acid copper(2+) salt (1:1); Scheele's green). a poisonous, yellowish-green, highly water-insoluble, crystalline powder, ($CuHAsO_3$ or $Cu(AsO_2)_2$) of varying composition. Formerly used in wallpapers, it is now used chiefly as an insecticide, fungicide, and wood preservative.

cupric chloride (copper chloride; copper (II) chloride). **1:** a yellowish-brown, deliquescent powder, $CuCl_2$, which is soluble in water, ethanol and ammonium chloride. **2:** a green, crystalline, water-soluble, dihydrate, $CuCl_2 \cdot 2H_2O$, of cupric chloride. It is used in the refining of copper, gold, and silver, and as a mordant in the textile industry. It is poisonous by ingestion and by *i.v.*, *i.p.*, and *s.c.* routes.

cupric cyanamide. cupric cyanide.

cupric cyanide (cupric cyanamide). sometimes called copper cyanide, but see cuprous cyanide. A highly toxic yellowish-green powder, C_2CuN_2, that reacts violently with magnesium. Toxic fumes, NO_x and CN, are released on heating to decomposition.

cupric sulfate (copper sulfate; copper monosulfate; hydrous copper sulfate; sulfuric acid, copper(2+) salt (1:1)). a water-soluble, hygroscopic salt, $CuSO_4$, which occurs as grayish-white to greenish-white crystals or an amorphous powder; hydrous copper sulfate is a blue crystalline substance. Cupric sulfate is used in copper-plating baths and has various other commercial applications. It is used therapeutically as an antidote to phosphorus poisoning and as a topical antifungal. It is also used in veterinary practice as an emetic, anthelmintic, and as a feed additive in copper deficiency of ruminants. Copper sulfate is a powerful algicide that acts by inhibiting photosynthesis. Toxic fumes of SO_x are released on decomposition by heating. **anhydrous cupric sulfate** is a toxic, highly irritant, hygroscopic, grayish-white substance, $CuSO_4$, consisting of rhombic crystals or powder; it is water soluble but nearly insoluble in ethanol. It is used as a fungicide and to detect and remove water from organic compounds, especially alcohols. The **monohydrate salt** (dried cupric sulfate), $CuSO_4 \cdot H_2O$, is a hygroscopic, water-soluble, off-white powder; it is nearly insoluble in ethanol. The **pentahydrate salt** (blue copper; blue vitriol; bluestone; Roman vitriol; Salzburg vitriol), $CuSO_4 \cdot 5H_2O$, is moderately toxic by ingestion and consists of large blue triclinic crystals, blue granules, or a light blue powder, all of which are soluble in water, methanol, and glycerol; and slightly soluble in ethanol. The pentahydrate has numerous uses, e.g., as an agricultural algicide, bactericide, fungicide (in Bordeaux mixture), and herbicide; it is also used in insecticidal mixtures, as a wood preservative, food and fertilizer additive, in paints and dyes, as a precursor of other copper salts, and in electroplating solutions. It is a strong irritant and systemic poison in animals. Ingestion by humans leaves a metallic taste and can cause pain in the mouth, esophagus, and stomach, vomiting, diarrhea, shock convulsions, paralysis, coma, and death. Lesions may include extensive capillary damage, and kidney and liver damage. Toxic fumes of SO_x are released on decomposition by heating. See also poisoning (cupric sulfate poisoning).

cupric sulfate poisoning (copper sulfate poisoning). acute poisoning of humans causes pain in the upper parts of the gastrointestinal tract, vomiting, diarrhea, abdominal cramps, shock, convulsions, paralysis, coma, and death. Lesions may include gastrointestinal injury, kidney injury, liver damage, and extensive capillary damage. See also potassium ferrocyanide.

cupricin. cuprous cyanide.

Cuprid. trientine.

cuprite. the naturally occurring mineral form of cuprous oxide.

cuprophyte. a plant adapted to, or having a high tolerance to copper in the soil. Such plants are often used as an indicator of copper-containing soils.

cuprous chloride (copper chloride; copper (I) chloride; chloride of copper). a poisonous, slightly water-soluble, white powder, CuCl or Cu_2Cl_2, that turns green on exposure to air and brown on exposure to light. Prepared by reducing cupric chloride, it is used mainly as an absorbent of carbon monoxide and as a catalyst.

cuprous cyanide (cupricin; copper cyanide). a white crystalline substance of monoclinic prisms, CuCN. It is extremely toxic, reacts violently with magnesium, and releases very toxic fumes, CN^- and NO_x, on decomposition by heating. It is used as a fungicide, insecticide, in antifouling marine paints, and as a catalyst in polymerization.

cuprous oxide (red copper oxide; copper oxide). an oxide of copper, Cu_2O, that occurs in natureas the mineral, cuprite. It is formed on copper by heat and is used mainly as a pigment and fungicide. Cuprous oxide is used, for example, as a fungicide, in marine antifouling paints, as an antiseptic for fish nets, as a catalyst, as a pigment in glass and ceramics, and in photoelectric cells. It is very toxic by ingestion to laboratory rats. See also metal fume fever.

cuprum. copper.

curare (curari; arrow poison; ourari; urari; woorari; woorali; wourara). **1:** a standardized toxic alkaloidal extract of *Chondodendron tomentosum* (chicona bark). A refined preparation (either D-tubocurarine chloride or metocurine iodide) may be used as a muscle relaxant during general anesthesia. **2:** the generic name for black, sticky, paralytic substances isolated by indians from various South American trees, mainly from the bark of *C. tomentosum* (chicona bark) and from certain species of *Strychnos*, such as *S. castelnaei*, *S. crevauxii*, *S. guianensis*, and *S. toxifera* in the Amazon and Orinoco valleys. There are three types named for the type of container used in their preparation: tube (bamboo) curare, pot curare, and calabash (gourd) curare (See c-curarine I). The mixtures are not standardized and constituents vary, but tubocurarine is the principal active ingredient. Curare was originally used to poison the tips of arrows. It produces a flaccid, ascending paralysis of the voluntary musculature, with death from respiratory failure. Meanwhile, the victim remains conscious. Curare acts by blocking ACh receptor proteins at the myoneural junction, producing a nondepolarizing paralysis.

curaremimetic (curarimimetic). having an action or producing effects similar to that of curare.

curari. curare.

curariform. pertaining to or resembling curare, especially with regard to its action.

curarimimetic. curaremimetic.

c-curarine I. the alkaloid component of calabash curare. See curare.

curarization. **1:** the (usually slow, often intravenous) administration of purified curare until the desired physiologic effect is produced (a lessening of the severity of convulsions, the relaxation of skeletal muscles in tetanus or prior to anesthesia). **2:** the condition resulting from the introduction of purified curare: heavy eyelids, nystagmus, husky voice, weakness of jaw and throat muscles, inability to raise the head and extremities.

curative. healing; having the ability to cure or heal.

curcas bean. *Jatropha curcas*.

cure. **1:** the therapy or course of treatment used in any disease to restore health **2:** successful treatment; restoration to health. **3:** a system of treating diseases. **4:** a drug that is able to restore health (usually by itself) in treating a specific disease or type of disease. *Cf.* antidote.

curie. a level of radioactivity equal to 3.7×10^{10} disintegrations per second.

curium (Cm)[Z = 96; most stable isotope = ^{247}Cm, $t_{1/2}$ = ca. 1.6×10^7 yr]. a highly toxic, intensely radioactive man-made transuranic element of the actinoid series. It was initially identified by G.T. Seaborg and associates in 1944 while bombarding plutonium-239 with alpha particles and was later produced by L. B. Werner and L. Perlman (1947) by bombarding americium-241 with neutrons. There are nine known isotopes. The maximum allowable concentration of sol ^{244}Cm in air in the United States is 3×10^{-12} μCi/cc and of insol ^{244}Cm in air is 3×10^{-11} μCi/cc (National Bureau of Standards Handbook 69, p. 89, 1959).

curve. **1:** to cause to curve. **2a:** a continuous nonangular deviation from a straight course. **2b:** a line or surface that curves. **3a:** any representation or display of a curve (def. 2). **3b:** a chart or graph representing, by a continuous line or surface, the changing relationships between two or more variables. See dose-response curve, survivorship curve.

curvularin. a mycotoxin produced by *Penicillium expansum*.

cusco-bark tree. *Cinchona pubescens*.

cuscohygrine(1,3-bis(1-methyl-2-pyrrolidinyl)-2-propanone; cuskhygrine; cuskohygrine; belladine). an oily alkaloid, $C_{13}H_{24}ON_2$, isolated with crude hygrine from leaves of the common cinchona or cusco-bark tree, *Cinchona pubescens*, and from those of *Erythroxylum coca*. See hygrine.

Cushing's syndrome. a condition resulting from hypersecretion of glucocorticoids. The secretory activity is due to tumors of the adrenal cortex or to ACTH-secreting tumors of the anterior pituitary body. The condition is characterized by weight gain, excess body hair, thin arms and legs, a "moon face," altered protein and carbohydrate metabolism, hyperglycemia, hypernatremia, hypertension, and muscular weakness. See also adrenogenital syndrome.

cuskhygrine. cuscohygrine.

cuskohygrine. cuscohygrine.

cutaneous. pertaining to the skin.

cuticle. **1:** an outer protective layer, pellicle or skin, especially a layer of secreted material that covers the free surface of epithelial cells and is impervious to water. In many plants and animals, it serves as a barrier to the penetration of water and hence many chemicals. **2a:** an outer layer of variable composition, secreted by epidermal cells, that covers the body of many invertebrates. The invertebrate cuticle generally protects against mechanical damage and often against chemical damage. Its composition and structure vary greatly among invertebrates. It may be reinforced by elastic fibers (nematodes), calcium carbonate (mollusks, crustaceans), or chitin (insects). The cuticle of nematodes and arthropods forms a rigid skeleton which is molted at intervals, allowing growth of the animal. **2b:** the cuticle of plants consists of cutin, a waxy material that is impervious to water. It serves to restrict water loss and may thus be nearly continuous in leaves and other aerial organs of higher plants, being interrupted only by stomata and lenticels. The cuticle is also impervious to many chemicals. **3:** the thin outer layer of the skin, as that adjacent to a fingernail.

cuticula. a secreted horny layer.

Cutie-Pie. an instrument used to measure radiation levels.

cutis. epidermis, skin. See skin.

cutis vera. dermis.

cutleaf nightshade. *Solanum triflorum*.

Cuvierian organ. See Cuvierian tubules.

Cuvierian tubules. organs found in certain members of the Holothuroidea, *q.v.* When such animals are irritated, these organs are discharged through the anus. Once in the water, the tubules may become greatly elongated by hydrostatic pressure and the outer covering may rupture. In this way, some species release a proteinaceous adhesive mixture; others release toxins (e.g., holothurin).

CV. **1:** coefficient of variation. **2:** cultivar.

CVF. cobra venom factor. See factor.

CVP. chlorfenvinphos.

CW. chemical warfare.

CWA. Federal Water Pollution Control Act ("Clean Water Act").

Cy. cyanogen.

cyan radical. cyanogen (def. 1a).

cyano-. a prefix meaning dark blue.

cyanamide (carbodiimide; hydrogen cyanamide; carbimide; cyanogenamide; amidocyanogen). **1:** cyanamide, $H_2NC{\equiv}N$, is prepared from calcium cyanamide, $CaNC{\equiv}N$. Both of these compounds are very toxic, caustic, and are used mainly as fertilizers and raw materials. Cyanamide (a solid) produces a severe dermatitis on moist skin. Inhalation may irritate the mucous membranes of the respiratory [illegible] Inhalation or ingestion causes a transitory redness of the face, headache, vertigo, hypotension, and elevated pulse and respiration rates. **2:** "cyanamide" is sometimes used specifically for calcium cyanamide, *q.v.*

cyanate. **1:** any salt or ester of cyanic acid that

contains the CNO radical. **2:** pertains to or denotes the -CNO radical or compounds that include it.

cyanazine (2-[[4-chloro-6-(ethylamino)-1,3,5-triazin-2-yl]amino]-2-methylpropanenitrile; 2-[[4-chloro-6-(ethylamino)-*s*-triazin-2-yl]amino]-2-methylpropionitrile). a white solid used as a selective pre- and postemergence herbicide to control weeds in crops such as alfalfa, corn, cotton, sorghum, soybeans, and wheat. It is very toxic orally to laboratory rats and mice.

Cyanea. a genus of scyphozoan jellyfish (Order Discomedusae). ***C. capillata*** (lion's mane; pink jellyfish). a common jellyfish, and the largest. It floats near the surface, occuring along the Atlantic Coast of North America from the Arctic into the Gulf of Mexico, and on the Pacific coast from Alaska to southern California. It is the largest jellyfish with a disc (or bell) that is saucer-shaped with a smooth upper surface. The disc may reach a diameter of up to 2.4 m and the dangling tentacles may reach a length of up to 2.4 m. The color of the bell varies with age (and size): pink and yellowish (to ca. 1.3 cm), reddish to yellowish-brown (to ca. 46 cm), and darker red-brown in larger specimens. An extremely venomous species. Any contact with the tentacles by humans will produce severe burns and blistering. Extensive or prolonged contact can cause dyspnea, muscle cramps, and occasionally death. See cyanea principle.

cyanea principle. a toxic principle originally thought to be a jellyfish venom, but it does not appear to occur in the nematocysts. See also congestin. See *Cyanea*.

cyanemia. an obsolete term for cyanosis.

cyanhydric acid. hydrocyanic acid.

cyanic acid (hydrogen cyanate). a colorless, toxic, explosive liquid (b.p. 23.3°C, m.p. - 86°C), $N{\equiv}COH$, with an acrid odor. It is a strong irritant, lacrimator, and vesicant and easily damages skin, eyes, mucous membranes and respiratory tract. Cyanic acid forms cyanate salts such as NaOCN and KOCN. It polymerizes, forming cyamelide and fulminic acid and it readily decomposes on heating or on contact with strong acid, yielding very toxic fumes. *Cf.* isocyanic acid.

cyanic acid sodium salt. sodium cyanate.

cyanidation. the coupling of cyanide to an atom or molecule.

cyanide. **1:** any of numerous, highly toxic, compounds (e.g., hydrogen cyanide (HCN), potassium cyanide (KCN), and sodium cyanide (NaCN)) derived from hydrogen cyanide that contains the CN group and forms hydrocyanic acid in water. Cyanide, in its various forms, is toxic by all routes of exposure. They are used in fumigants, soil sterilizers, fertilizers, and rodenticides. Livestock losses sometimes result from improper or malicious use of these compounds. **2:** the -CN radical or the CN^- ion. It is a cellular asphyxiant, inhibiting cytochrome oxidase, thus blocking cellular respiration. Cyanides occur naturally in many plants, usually in the form of glycosides. Cyanide is odorless, but reputedly has the odor of bitter almonds. Benzaldehyde, a breakdown product of amygdalin, is the source of this odor which may be present briefly when amygdalin is the source of the cyanide. All cyanide antidotes (e.g., amyl nitrite) are themselves poisonous. See also *p*-aminopropiophenone, cyanide poisoning, cyanide poisoning of herbivores, cyanide poisoning of livestock, *Manihot*.

cyanide methemoglobin. cyanmethemoglobin.

cyanide poisoning. poisoning by a cyanide compound, the most important of which in the case of humans are hydrogen cyanide, potassium cyanide, and sodium cyanide. Exposure can occur by ingestion, inhalation, or absorption through the skin. Swallowing or smelling a lethal dose of cyanide as a gas or a salt can cause immediate unconsciousness, convulsions, and death within 1-15 minutes or longer. Cyanide is toxic to a wide range of living organisms, acts rapidly, and is lethal at low doses. The lethal oral dose in humans is thought to be 60-90 mg. Exposure to a near lethal dose produces effects that usually include hyperpnea, rapid pulse, gasping, dizziness, flushing, headache, nausea, vomiting, and hypotension with fainting. Smaller doses may leave an acid taste which precedes numbness in the throat, anxiety, confusion, vertigo, and hyperpnea with ensuing dyspnea. Workers chronically exposed to cyanide-containing solutions may develop a skin a "cyanide rash"; at somewhat higher concentrations, they may develop a sore throat, headache, dyspnea, and weakness; and those exposed to fairly high concentrations may further suffer lightheadedness, headache, unconsciousness, and convulsions. Convulsions usually precede death by less than 4 hours except in the case of sodium nitroprusside, where death is does not occur for as long as 12 hours after ingestion. The most important cause of cyanide poisoning among domestic animals is ingestion of such plants as arrow grass (*Triglochin spp.*), Johnson grass (*Sorghum halepense*), Sudan grass (*S. sudanense*), common sorghum (*S. bicolor [vulgare]*), wild black cherry (*Prunus serotina*), chokecherry (*P. virginiana*), pincherry (*P. pensylvanica*), and flax (*Linum usitatissimum*). These plants contain cyanogenetic glycosides that release hydrocyanic (prussic) acid when attacked by acids such as the hydrochloric acid found in the stomach of many animals. The cyanogenetic glycoside content of the plant is increased by heavy nitrate fertilization, wilting, trampling, and plant disease. Very young, rapidly growing plants contain greater quantities of the glycoside. Spraying of cyanogenetic plants with herbicides may enhance the toxic hazard. Ruminants are more susceptible than monogastric animals, and cattle more so than sheep. All antidotes for cyanide poisoning are themselves poisonous.

cyanide poisoning of herbivores. a fairly common disease of herbivorous animals, caused by ingestion of cyanogenic plants or plant parts. The cyanide is usually present in glucosides which are hydrolyzed in the alimentary canal, yielding hydrocyanic acid. Some farm chemicals such as fungicides or insecticides may cause cyanide poisoning. Hydrogen cyanide and its salts are extremely poisonous to *Homo sapiens* and a wide array of terrestrial and aquatic animals.

cyanide poisoning of livestock. cyanides are used in fumigants, soil sterilizers, fertilizers, and rodenticides. Livestock losses sometimes [illegible] malicious use of these products. The most important cause of cyanide poisoning among domestic animals is ingestion of such plants as arrow grass (*Triglochin spp.*), Johnson grass (*Sorghum halepense*), Sudan grass (*S. sudanense*), common sorghum (*S. bicolor [vulgare]*), wild black cherry (*Prunus serotina*), chokecherry (*P. virginiana*), pincherry (*P. pensylvanica*), and flax (*Linum usitatissimum*). These plants contain cyanogenetic glycosides that yield hydrocyanic acid when hydrolyzed by enzymes during digestion. The cyanogenetic glycoside content of these plants is increased by heavy nitrate fertilization, wilting, trampling, and plant disease. Very young, rapidly growing plants contain greater quantities of the glycoside. Spraying of cyanogenetic plants with herbicide may increase the toxic hazard. Cattle are more susceptible than sheep, and ruminants are generally more susceptible than monogastric animals.

cyanide rash. an itchy rash developed by workers who are exposed daily to cyanides (e.g., in electroplating and pickling processes). See cyanide poisoning.

cyanidenon. luteolin.

cyanidol. catechin.

cyanihydric acid. hydrogen cyanide.

cyanmethemoglobin (cyanide methemoglobin). a somewhat toxic compound of cyanide and methemoglobin that results from the administration of Methylene Blue in cyanide poisoning.

cyano pyrethroid. See pyrethroid.

cyano (type II) pyrethroid. a pyrethroid that contains an alpha-cyano group in the alcohol moiety. They cause profuse salivation and choreoathetosis in mammals. **non-cyano (type I) pyrethroid**. a pyrethroid that lacks the alpha-cyano group. They cause hyperexcitability, repetitive nerve firing, incoordination, and whole-body tremors in mammals.

cyanoacrylate adhesive (superglue). any of a number of monomers of *N*-alkylcyanoacrylate, available commercially as "superglue." Such compounds have wide application as strong adhesives and have been used as tissue adhesives. Usage is, however, limited by toxicity. Contact with the [illegible] can cause adhesions that must be removed surgically.

β-cyano-L-alanine. a precursor of the toxicants, β-(γ-L-glutamyl)-aminopropionitrile and L-α,γ-diaminobutyric acid.

cyanoauric acid. gold tricyanide

cyanobacteria (singular, cyanobacterium). a common collective name for members of the phylum Cyanobacteria.

Cyanobacteria (blue-green bacteria; formerly blue-green algae; Cyanophyceae; Cyanophyta; Myxophyta). a homogeneous group of closely related unicellular, prokaryotic, photosynthetic, colonial or filamentous nonmotile microorganisms. Formerly considered algae, they are now usually considered a phylum of the Kingdom Monera. Most occur in surface water bodies or in moist terrestrial situations as free-living or symbiotic associates of lichens or fungi. Some freshwater forms such as *Anabaena flos-aquae*, *Aphanizomenon flos-aquae*, and *Microcystis aeruginosa* ("Annie, Fannie and Mike") are extremely poisonous. These species occur worldwide in warm, calm, nutrient-rich waters where they sometimes form toxic blooms, *q.v.* Human intoxication is rare, but dogs, cats, domestic fowl, livestock, fish, and various wildlife species have sometimes been killed in large numbers by drinking water heavily populated with toxic blue-green bacteria. Dangerously toxic waters usually look like a thin blue or grayish paint. Livestock have been photosensitized from drinking water containing high concentrations of cyanobacteria. The toxic principles of this phlum have not been identified in natural populations (but See *Microcystis aeruginosa*). Toxic concentrations of nitrate have been reported in certain forms.

cyanochroic. See cyanotic.

cyanochrous. See cyanotic.

cyanoderma. cyanosis.

1-(3-cyano-3,3-diphenylpropyl)-4-phenyl-4-piperidinecarboxylic acid ester. diphenoxylate.

1-(3-cyano-3,3-diphenylpropyl)-4-phenyl-isonipecotic acid ethyl ester. diphenoxylate.

cyanoethane. propionitrile.

cyanoethylene. acrylonitrile.

cyanogas. calcium cyanide.

cyanogen (Cy). **1a:** the extremely toxic cyan radical, $C \equiv N$. **1b:** a loose designation for any compound that contains the cyan radical. **2:** (ethane dinitrile; dicyan; dicyanogen; oxalic acid dinitrile; NCCN). a clear, colorless, oily, highly toxic, violently flammable gas with an acrid odor. It is produced by heating mercuric cyanide. Cyanogen is extremely toxic, with effects similar to those of hydrogen cyanide and may cause permanent injury or even death in exposed individuals. It reacts with water and acids to yield highly toxic fumes. Used mainly in organic syntheses, NCCN is a compound comprised of two cyan radicals, $N \equiv C\text{-}C \equiv N$, and is readily halogenated to form highly toxic compounds with the formula XCN, where X represents a halogen atom.

cyanogen azide (carbon pernitride). a colorless, oily liquid, $N^{-}=N^{+}=N\text{-}C \equiv N$, that explodes violently on concussion or thermal or electric shock. It is extremely reactive and can be managed safely only in solution. It is a versatile chemical reagent and produces, for example, alkylcyanamides on reaction with alkanes.

cyanogen bromide (bromine cyanide). a neurotoxic, white, crystalline substance, $BrC \equiv N$, with very toxic and highly irritant fumes. It is a potent lacrimator and an extremely toxic warfare gas. Effects are similar to those of hydrogen cyanide.

cyanogen chloride. a colorless water-soluble gas or volatile liquid, $ClC \equiv N$, with a boiling point of 13.8°C. It is an extremely toxic systemic poison and lacrymator. The vapors are very poisonous and highly irritant. It is used in organic syntheses and as a pesticidal fumigant and military poisonous gas. It is extremely toxic to Norway rats and other vermin, but is less hazardous to humans. Effects are similar to those of hydrogen cyanide.

cyanogen fluoride. a toxic, colorless gas and lacrymator, $FC \equiv N$, used as a tear gas.

cyanogen iodide (iodine cyanide; CNI; ICN). a colorless, crystalline substance, $IC \equiv N$, with a very pungent odor and acrid taste. It is readily soluble in water, alcohol and ether; m.p. 146.5°C, d 1.84. CNI is a strong irritant of the eyes, skin, and mucous membranes. It appears to be lethal to all nonvertebrate animals and is very or extremely toxic to vertebrates. Effects of acute oral intoxication include convulsions, paralysis, and death from respiratory failure. It is used as a preservative, mainly of insects, in systematics and in taxidermy.

cyanogenamide. cyanamide.

cyanogenesis. the production of or elaboration of cyanogen or hydrocyanic acid as in the synthesis of cyanogen glycosides by plants such as *Prunus lawrocerasus* (cherry laurel), *Lotus corniculatus* (birdsfoot trefoil), and *Trifolium repens* (white clover).

cyanogenetic (cyanogenic; cyanogenous). **1a:** pertaining to cyanogenesis. **1b:** capable of producing cyanogen or hydrocyanic acid. **2:** pertaining to or denoting living organisms that produce hydrocyanic acid or its immediate precursors.

cyanogenetic glycoside. See glycoside (cyanogenetic glycoside).

cyanogenic. cyanogenetic.

cyanogenous. cyanogenetic.

cyanogran. sodium cyanide.

cyanomethane. acetonitrile.

cyanopathy. cyanosis.

α-cyano-3-phenoxybenzyl α-(4-chlorophenyl)-isovalerate. fenvalerate.

α-cyano-3-phenoxybenzyl-2-(4-chlorophenyl)-3-methylbutyrate. fenvalerate.

(*S*)-α-cyano-3-phenoxybenzyl-(1*R*)-*cis*-3-(2,2-dibromovinyl)-2,2-dimethylcyclopropane carboxylate. deltamethrin.

(±)-α-cyano-3-phenoxybenzyl-(±)-*cis,trans*-3-[illegible] carboxylate. cypermethrin.

cyano-(3-phenoxyphenyl)methyl 4-chloro-α-(1-methylethyl)benzeneacetate. fenvalerate.

(*R,S*)-cyano(3-phenoxyphenyl)methyl(*R,S*)-cis-trans-3-(2,2-dichloroethenyl)-2,2-dimethylcyclopropane carboxylate. cypermethrin.

cyanophoric. **1:** of or pertaining to glycosides (e.g., amygdalin) that yield glucose, hydrocyanic acid, and benzaldehyde on hydrolysis. **2:** a cyanophoric glycoside. See glycoside (cyanophoric glycoside).

Cyanophyceae. Cyanobacteria.

Cyanophyta. Cyanobacteria.

cyanophyte. blue-green algae (obs.).

cyanosed. cyanotic.

cyanosis (cyanoderma; cyanopathy). a pale to dark (bluish, purple, or grayish-blue) discoloration of skin and mucous membranes produced by hypoxia. In simple hypoxic states, it becomes evident in humans when the concentration of reduced hemoglobin in circulating blood exceeds about 5 g per 100 ml. Cyanosis may be due to any of a number of factors and occurs, for example, in mountain sickness, pulmonary edema, in some types of heart disease, and with abnormally sluggish capillary circulation. Cyanosis is also seen as a result of exposure to certain toxic chemicals such as carbon monoxide or asphyxiant gases. **apparent cyanosis**. enterogenous cyanosis. **enterogenous cyanosis** (apparent cyanosis; false cyanosis). cyanosis due to methemoglobin or sulfhemoglobin formation. It may result from exposure to nitrites and may be seen in carbon monoxide poisoning. The skin color is due to the chocolate color of methemoglobin. **false cyanosis**. enterogenous cyanosis. **toxic cyanosis**. enterogenous cyanosis due to the action of certain drugs.

cyanotic (cyanochroic; cyanochrous; cyanosed). pertaining to or characterized by cyanosis.

cyanurotriamide. melamine.

Cybister. a genus of water beetles, the prothoracic glands of which secrete a number of poisons (including several pregnene-derived steroids such as cortexone) that are very toxic to vertebrates, especially fish, which are their chief predators.

cycad. any plant of the family Cycadaceae, especially those of the genus *Cycas*.

Cycadaceae (cycads; the cycad family). a family of palm-like, but nonflowering plants, that bear cones. See Cycadales, *Cycas*.

Cycadales. an order of gymnospermous plants that includes the Cycadaceae. Plants of this order are thought to cause defects in calves of domestic cattle that are similar to those of crooked calf disease. See also *Cycas*.

Cycas (cycad; sago palm). a genus of palm-like tropical plants (Family Cycadaceae) the trunk of which may reach a height of more than 6 m and diameter of 75 cm. The seeds (nuts) of the following two species contain a toxic glucoside, cycasin, *q.v.* ***C. circinalis*** (queen-sago; fern palm; false sago palm; crosier cycas). a sago palm with leaves up to 2.6 m. ***C. revoluta*** (king-sago). a sago palm with leaves that are usually less than 1.3 m long. Flour made from the nuts of these plants are associated with hepatotoxicity, carcinogenicity, teratogenesis, and neurotoxicity. See also cycasin.

cycasin (methyl-*ONN*-azoxy)methyl β-D-glucopyranoside; methylazoxymethanol β-D-glucoside; β-D-glucosyloxyazoxymethane). a naturally occurring alkylating agent that is produced by sago palms, *Cycas*, and is found in flour made from the sago nut. Cycasin is very toxic and carcinogenic to laboratory rats, causing cancer of the liver, kidney, and digestive tract, together with hepatotoxic, carcinogenic, teratogenic and neurotoxic effects. The aglycone is methylazoxymethanol. Cycasin is probably carcinogenic to humans. See also *Cycas*.

cyclamate. **1a:** any salt of cyclamic acid. **1b:** any of a class of low-calorie, nonnutritive sweetening agents, about 30 times as sweet as refined sugar. Included are sodium, potassium, and calcium cyclamates. Early studies (1969-1970) on laboratory animals indicated that cyclamates cause bladder cancer in rats at high dosage and genetic damage to chick embryos. Many recent studies have failed to confirm the carcinogenicity of the early studies, even at extremely high dosages. There is still much uncertainty concerning the danger to humans, partly due to reliance on flawed experiments and partly due to the long latency period for cancers. Information now emerging may provide an impetus for new studies and a reassessment of the potential for cyclamates to harm humans.

cyclane. an alicyclic hydrocarbon (See hydrocarbon).

cycle. **1a:** a single sequence in a regularly recurring sequence of events. **1b:** a single period in any continuing process that returns periodically to the original state. **2:** a recurrent sequence of events such that the last event in the sequence immediately precedes the recurrence of the designated first event. **3:** the duration of a single cycle (def. 1, 2). **4:** any periodic process.

cycle-specific drug. any immunosuppressive or cytotoxic substance that destroys mitotic and resting cells.

cyclethrin(2,2-dimethyl-3-(2-methyl-1-propenyl)-cyclopropanecarboxylic acid 3-(2-cyclopenten-1-yl)-2-methyl-4-oxo-2-cyclopenten-1-yl ester; 3-(2-cyclopentenyl)-2-methyl-4-oxo-2-cyclopentenyl ester of chrysanthemummonocarboxylic acid; chrysanthemummonocarboxylic acid ester with 3-(2-cyclopenten-1-yl)-2-methyl-4-oxo-2-cyclopenten-1-ol). an insecticide which is only moderately toxic to laboratory mice and rats.

cyclic adenosine-3′,5′-monophosphate. cyclic AMP.

cyclic amine. See amine.

cyclic AMP (adenosine cyclic 3′,5′-(hydrogen phosphate); adenosine 3′,5′-cyclic monophosphate; adenosine 3′,5′-monophosphate; adenosine 3′,5′-phosphate; cyclic adenosine 3′,5′-monophosphate; acrasin; 3′,5′-AMP; cAMP). a substance that is widely distributed among living organisms at concentrations from 10^{-7}-10^{-6} moles/kg body weight, CAMP regulates a number of intracellular processes and mediates the action of a number of hormones such as ACTH, adrenaline, and glucagon. It is the conversion product of ATP catalyzed by adenylate cyclase. Some toxicants cause physiological effects by interfering with CAMP metabolism. See, for example, choleragen, pertussis toxin.

cyclic GMP (guanosine cyclic 3′,5′-(hydrogen phosphate); cyclic guanosine-3′,5′-monophosphate; guanosine 3′,5′-monophosphate; guanosine 3′,5′-cyclic monophosphate; guanosine 3′,5′-cyclic phosphate; guanosine 3′,5′-GMP; cGMP). the conversion product of GTP catalyzed by the enzyme guanylate cyclase. It is a cellular regulatory agent, sometimes referred to as a "second messenger." It is a regulatory agent of many intracellular processes. Some toxicants such as nitroprusside may cause physiological effects by disrupting cGMP metabolism.

cyclic guanosine-3′,5′-monophosphate. cyclic GMP.

cyclic nucleotide phosphodiesterase. See phosphodiesterase.

cyclical lactone. See cytochalasin.

Cyclo-Werrol. hypericin.

cycloaliphatic. alicyclic.

cycloalkane (cycloparaffin). a cyclic alkane. See alkane.

cycloate. a thiocarbamate herbicide.

cyclochlorotine (chlorine-containing peptide). a hepatotoxic, carcinogenic chlorinated cyclic peptapeptide mycotoxin isolated from *Penicillium islandicum*. It is extremely toxic to laboratory mice. See also luteoskyrin.

cyclodiene insecticide (chlorinated cyclodiene insecticide). any of a class of chlorinated insecticides based on a cyclodiene ring structure. Included are aldrin, dieldrin, endrin, chlordane, heptachlor, endosulfan, and isodrin. All are highly toxic neurotoxicants with similar effects. They appear to act on the brain, releasing betaine esters and causing headaches, dizziness, nausea, vomiting, muscle twitches, and convulsions. Some are fetotoxic, some are carcinogenic, and their use has thus been banned in some countries. The United States has banned the use of aldrin and dieldrin

1,4-cyclohexadienedione. *p*-quinone. See quinone.

2,5-cyclohexadiene-1,4-dione. *p*-quinone. See quinone.

cyclohexamide. a glutaric imide derivative produced by *Streptomyces griseus*. It is a very effective agricultural fungicide, but is too toxic to be used clinically. Cyclohexamide is strongly repellent to rodents in concentrations as low as 5 ppm in drinking water. Laboratory rats exposed to cyclohexamide may die from dehydration.

cyclohexane (hexamethylene; hexanaphthene; hexahydrobenzene). a moderately toxic, colorless, volatile, highly flammable, 6-carbon ring liquid, C_6H_{12}, with a pungent odor. It is produced by hydrogenating benzene and is used industrially mainly as a substitute for benzene and toluene and in the synthesis of cyclohexanol and cyclohexanone. Cyclohexane is weakly anesthetic. Symptoms of acute poisoning are usually temporary and may include giddiness, disequilibrium, stupor, and ultimately paralysis of the respiratory center of the brain. Severe poisoning may cause kidney and liver damage.

cyclohexanone (CH; pimelic ketone; ketohexamethene). a clear to pale yellow, slightly water-soluble, carcinogenic, liquid ketone, $C_6H_{10}O$, with a characteristic acetone and peppermint-like odor. Exposure is normally via contact or inhalation of the vapor.

cyclohexatriene. benzene.

cyclohexenamine. cyclohexylamine.

5-cyclohexenyl-3,5-dimethylbarbituric acid. hexobarbital.

5-(1-cyclohexen-1-yl)-1,5-dimethylbarbituric acid. hexobarbital.

5-(1-cyclohexen-1-yl)-1,5-dimethyl-2,4,6-(1H,3H,5H)-pyrimidinetrione. hexobarbital.

cycloheximide (4-[2-(3,5-dimethyl-2-oxocyclohexyl)2-hydroxyethyl]-2,6-piperidinedione; 3-[2(3,5-dimethyl-2-oxocyclohexyl)-2-hydroxyethyl]glutarimide; naramycin A). an antibiotic isolated from streptomycin-producing strains of *Streptomyces griseus*. It inhibits protein synthesis in eukaryotic cells by inhibiting translation. Cycloheximide is very toxic, *i.v.*, to laboratory mice and is extremely repellant to rats. It is used chiefly as a plant growth regulator and fungicide.

cyclohexylamine (cyclohexenamine; aminocyclohexane; hexahydroaniline). a poisonous liquid with a strong, fishy odor. It is strongly irritant and caustic to the eyes, mucous membranes, and skin, as well as a systemic poison. The oral LD_{50} in laboratory rats is 0.71 ml/kg body weight. Signs and symptoms of systemic cyclohexylamine poisoning include nausea, vomiting, anxiety, restlessness, and drowsiness. It adversely affects the female reproductive system. *Cf.* dicyclohexylamine.

***N*-cyclohexylcyclohexanamine**. dicyclohexylamine.

3-cyclohexyl-6,7-dihydro-1*H*-cyclopentapyrimidine-2,4-(3*H*,5*H*)-dione. lenacil.

α-cyclohexyl-α-phenyl-1-pyrrolidinepropanol. procyclidine.

1-cyclohexyl-1-phenyl-3-pyrrolidinohydrochloride. procyclidine.

1-cyclohexyl-1-phenyl-3-pyrrolidino-1-propanol. procyclidine.

1-cyclohexyl-1-phenyl-3-(1-pyrrolidinyl)-1-propanol. procyclidine.

3-cyclohexyl-5,6-trimeth yleneuracil. lenacil.

cyclol. a cyclic dipeptide found in some of the ergot alkaloids.

cycloleucine (1-aminocyclopentanecarboxylic acid; ACPC). a synthetic amino acid which acts as an immunosuppresive agent and apparently as a valine antagonist. It is very toxic orally to laboratory rats and mice. Cycloleucine is used therapeutically as an antineoplastic agent.

cyclooxygenase. an enzyme that catalyzes the conversion of arachidonic acid to one of a number of prostaglandins.

cycloparaffin. a cycloalkane. See alkane.

cyclopentadiene (1,3-cyclopentadiene). a colorless, cyclic, liquid dialkene produced by distillation during the carbonization of coal or the cracking of petroleum hydrocarbons. It is used in the manufacture of resins and in the synthesis of synthetic alkaloids, camphors, and sesquiterpenes. On standing, it polymerizes to the dimer, dicyclopentadiene, *q.v.*

1,3-cyclopentadiene. cyclopentadiene.

cyclopenta[*cd*]pyrene (CPP). a common mutagenic polycyclic aromatic hydrocarbon (PAH); a component of fossil fuel combustion and carbon black.

3-(2-cyclopentenyl)-2-methyl-4-oxo-2-cyclopentenyl ester of chrysanthemummonocarboxylic acid. cyclethrin.

cyclophospham. cyclophosphamide.

cyclophosphamide (*N*,*N*-bis(2-chloroethyl)tetrahydro-2*H*-1,3,2-oxazaphosphorin-2-amine 2-oxide; *N*,*N*-bis-(2-chloroethyl)*N'*-(3-hydroxypropyl2-[bis(2-chloroethyl)amino]tetrahydro-2*H*-1,3,2-oxazophosphorine 2-oxide; 1-bis(2-chloroethyl)amino-1-oxo-2-aza-5-oxaphosphoridin; bis(2-chloroethyl)phosphoramide cyclic propanolamide ester; *N*,*N*-bis-(β-chloroethyl)-*N'*,*O*-propylenephosphoric acid ester diamide; phosphordiamidic acid cyclic ester monohydrate; cyclophospham; cyclophosphane; CPA). a toxic compound, $C_7H_{15}Cl_2N_2O_2P$, the crystals of which very slightly soluble in ether and acetone and slightly soluble in benzene and carbon tetrachloride. A derivative of nitrogen mustard, it is an alkylating agent used as an antineoplastic agent. Target organs are the brain, testis, and ovary. It also suppresses antibody formation and B-cell activity, and has thus been used in the treatment of autoimmune disease. It is activated to phosphoramide mustard, which strongly suppresses DNA synthesis. Effects include delayed alopecia, pulmonary fibrosis, renal damage, and urinary and reproductive effects in humans. It is abortifacient, fetotoxic, teratogenic, and carcinogenic. Cyclophosphamide is very toxic to laboratory rats and mice.

cyclophosphamide hydrate. the monohydrate of cyclophosphamide. An antineoplastic, it is also a confirmed human carcinogen and is probably mutagenic.

cyclophosphane. cyclophosphamide.

cyclopiazonic acid. an indole mycotoxin produced by *Penicillium cyclopium* and related species and by several species of *Aspergillus*. It is a common contaminant of cheese, corn, and peanuts and occurs widely in mammalian tissues (notably voluntary muscle). It is used as an antifungal agent. It is extremely toxic to laboratory mice and rats, ducklings, and calves of domestic cattle. Effects vary among mammalian species. It is myotoxic, hepatotoxic, nephrotoxic, and cytotoxic to laboratory mammals. Symptoms of intoxication can include, for example, catalepsy, hypothermia, hypokinesia, incoordination, ptosis, sedation without loss of righting reflex, tremor, and convulsions.

cycloplegia. paralysis of the ciliary muscle (of the eye); paralysis of visual accommodation.

cycloplegic. **1:** of or pertaining to cycloplegia; causing cycloplegia. **2:** an agent that causes cycloplegia.

cyclopropane (trimethylene). a colorless, water-insoluble, explosive gas, C_3H_6. It was formerly used as an inhalation anesthetic.

cyclopropanone. a metabolite of coprine that inhibits the metabolic detoxification of alcohol and accounts for the toxic effects of alcoholic beverages when consumed with the mushrooms *Coprinus atramentarius* and *Clitocybe clavipes*. See coprine.

17-(cyclopropylmethyl)-4,5-epoxy-3,14-dihydroxymorphinan-6-one. naltrexone.

***N*-cyclopropylmethyl-14-hydroxydihydromorphinone**. naltrexone.

Cyclosan. hypericin.

cyclosporin. any of a group of cyclic, nonpolar, oligopeptides produced by *Tolypocladium inflatum* and certain other soil fungi. **cyclosporin A** (ciclosporin; cyclosporine; Sandimmun(e)). a highly toxic, cyclosporin, $C_{62}H_{111}N_{11}O_{12}$. It has antifungal action and inhibits T-cell activity. It is used as an [illegible] organ transplants. It is also nephrotoxic.

cyclosporine. cyclosporin A.

cymarin. a cardiac glycoside, $C_{30}H_{44}O_9$, that occurs chiefly in plants of the genus *apocynum* and in *Strophanthus kombe*. The aglycone is strophanthin.

cymarose (2,6-deoxy-3-methoxyaldohexose; 2,6-dideoxy-3-*O*-methylribohexose; 3-methyl digitoxose). a hexose, $C_7H_{14}O_4$, derived from the hydrolysis of cardiac glycosides (e.g., cymarin) that occur in members of the family Apocynaceae (e.g., *Apocynum cannabinum*, *A. androsaemifolium*, *A. venetum*). It is the 3-methyl ester of digitoxose.

Cymopterus watsonii (spring parsley). a low, glabrous, North American herb (Family Apiaceae) with white flowers and pinnately compound leaves. It may cause primary photosensitivity in livestock. See photosensitivity.

***m*-cym-5-yl methylcarbamate**. promecarb.

cynapine. a toxic, volatile alkaloid, it is the active principle of *Aethusa cynapium*. Ingestion of small amounts causes nausea and vomiting.

Cynapium. the former generic name of *Aethusa*.

Cynodon dactylon (Bermuda grass). a grass (Family Graminae) native to southern Europe that is trailing and stoloniferous in habit. It is now widely used as a lawn and pasture grass in warm climates. See bensulide.

cypermethrin (3-(2,2-dichloroethenyl)-2,2-dimethylcyclopropanecarboxylic acid cyano-(3-phenoxyphenyl)methyl ester; (±)-α-cyano-3-phenoxybenzyl-(±)-*cis,trans*-3-(2,2-dichlorovinyl)-2,2-dimethylcyclopropane carboxylate; (*R,S*)-cyano(3-phenoxyphenyl)methyl(*R,S*)-*cis,trans*-3-(2,2-dichloroethenyl)-2,2-dimethylcyclopropane carboxylate). a viscous, semisolid, synthetic type II pyrethroid, used effectively as a stomach and contact insecticide on a wide variety of insects and agricultural pests. It is insoluble in water, but soluble in acetone, methanol, methylene dichloride, and xylene. It is used in veterinary medicine as an ectoparasiticide. Cypermethrin is neurotoxic, [illegible] by increasing the permeability of the neuronal membrane to sodium. It is not environmentally persistent, however, and is no more than moderately toxic to most vertebrate animals tested. Effects of intoxication on humans include gastroenteritis, nausea, vomiting, diarrhea, paresthesia of the tongue and lips, hyperexcitability, incoordination, syncope, convulsions, muscular paralysis, collapse, and death due to respiratory paralysis.

cypress spurge. *Euphorbia cyparissias*.

Cyprinidae (true minnows). a large family of freshwater fish with pharyngeal teeth. Included are carp, chub, dace, goldfish, shiners, etc. See *Cyprinus carpio*.

Cyprinodon variegatus (sheepshead minnow). a killifish (Family Cyprinodontidae), not a true minnow, it is a standard fish species in freshwater toxicity tests. It is a small (up to 7.5 cm in length), robust, moderately compressed fish. The male is olive, iridescent blue, or greenish-blue above with poorly defined bars; the belly is yellowish; the edge and base of the caudal fin is black. It occurs in nature along the coast of North America from Massachusetts to North Mexico, where it inhabits shallow coastal marshes and tide pools; it enters fresh water. See also bioassay (aquatic bioassay).

cyprinin. a toxicant isolated from the milt of the common carp, *Cyprinus carpio* (Family Cyprinidae).

Cyprinus carpio (common carp). a medium to large, robust, moderately depressed, bony fish (Family Cyprinidae) native to Asia, but now introduced and even cultured in warm and temperate waters of the world. The milt of this fish is the source of a toxic substance, cyprinin.

Cypripedium hirsutum (lady's slipper; moccasin flower). a genus of plants (Family Orchidaceae) indigenous to temperate portions of Asia, Europe, and North America. The hairs of stems and leaves are toxic, producing a contact dermatitis. The active principles are fatty acids.

cypromid. an amide herbicide. Applied to the foliage, amide damage takes the form of localized or general necrosis, depending on the dose applied. Cypromid also inhibits growth, probably as a secondary effect.

cyst. **1:** an abnormal swelling or closed sac that usually contains fluid and may occur in nearly any tissue. In vertebrate animals, they most often occur in the skin and ovaries, where they may grow to a large size. There are several kinds of such cysts: fluid-filled nonmalignant tumors; cysts that contain cells of tissues normally found elsewhere in the body; and cysts caused by parasitic infection. **2:** a protective coat that surrounds a resting cell. **3:** a bladder or air vesicle that occurs in certain seaweeds.

cysteamine (2-aminoethanethiol; mercaptamine; β-mercaptoethylamine; 2-aminoethyl mercaptan; thioethanolamine; decarboxycysteine; MEA; mercamine). a water-soluble, crystalline, malodorous, biologically active, sulfhydryl compound, C_2H_7NS, with various effects. It oxidizes to cystamine on standing in air. It is extremely toxic orally to laboratory mice and is an antidote for acetaminophen intoxication. **cysteamine hydrochloride**. a water- and ethanol-soluble crystalline compound. It is very toxic, *i.p.*, to laboratory rats and *i.v.* to laboratory rabbits.

cysteine conjugate β-lyase. a type of enzyme that metabolizes cysteine conjugates, producing the thiol derivative of pyruvic acid and ammonia. Following methylation of the thiol, the methylthio derivative forms. The enzyme from rat liver is a soluble pyridoxal phosphate-requiring protein of about 175 kilo-Daltons. The best substrates are the cysteine conjugates of aromatic compounds.

L-cysteine thioacetal of formaldehyde. djenkolic acid.

cystitis. inflammation of the bladder.

cytarabine. cytosine arabinoside.

cytarbine. cytosine arabinoside.

Cythion. malathion.

cytisine (1,2,3,4,5,6-hexahydro-1,5-methano-8*H*-pyrido-[1,2-a][1,5]diazocin-9-one; baptitoxine; sophorine; ulexine). a bitter, highly toxic, crystalline alkaloid originally isolated from *Laburnum anagyroides*, it occurs in many other plants of the family Fabaceae. Formerly used as a cathartic and diuretic. See also *Gymnocladus dioica*.

cytisism. poisoning by *Laburnum anagyroides*, *Baptisia*, or cytisine. Symptoms include a burning in the mouth and pharynx, thirst, nausea, vomiting, diarrhea, frothing at the mouth, convulsions, extreme drowsiness, irregular pulse, and coma. Muscle tone is also affected and the victim may be aphasic, with visual disturbances and delirium. Death from respiratory paralysis may follow.

cytisus. any plant of the genus *Cytisus*.

Cytisus (broom; cytisus). a genus of stiff or spiny deciduous and evergreen shrubs (Family Fabaceae, formerly Leguminosae) that are native to Europe, western Asia, and northern Africa, but some are widely cultivated elsewhere. They have showy racemose flowers, a two-lipped calyx, and grow to a height of 3 m. ***C. scoparius*** (*Sarothamnus scoparius*; broom; common broom; Scotch broom). an attractive, rather hardy European shrub; naturalized in parts of North America. It is often planted as cover along roadsides (Family Fabaceae, formerly Leguminosae) or other dry gravelly embankments. The raw seedpods are toxic; they contain the alkaloids quinolizidine and sparteine. Symptoms of intoxication are similar to those of nicotine and include nausea, diarrhea, headache, vertigo, paralysis of the ileus, tachycardia, and circulatory collapse. ***C. laburnum*** (laburnum, golden chain). the leaves and seeds are toxic, producing effects similar to nicotine: nausea, intense vomiting, dysphagia, severe vomiting, diarrhea, prostration, irregular pulse and respiration, delirium, twitching and coma; renal damage may occur. The active principle is cytisine.

cyto-. a prefix indicating relationship to, or denoting a cell.

cytobiology. the biology of cells.

cytobiotaxis. cytoclesis.

cytochalasin (cyclical lactone). any of a large class of cytotoxic macrolide mycotoxins with a highly substituted hydrogenated isoindole ring fused to a macrocyclic ring (a carbocyclic or a lactone) comprised of 11-14 atoms and is a car- [illegible] iginally isolated from *Helminthosporium dematioideum*, *Metarrhizium anisopliae*, and *Rosellina necatrix*, but are now known from a wide variety of *Phoma*-type fungi. Cytochalasins have various unusual effects on animal cells. They interfere with cytokinesis, for example, by inhibiting the formation of contractile microfilaments. The resultant cell is multinucleate. They also induce extrusion of cell nuclei, and reversibly inhibit cell movement. They reportedly inhibit such processes as glucose transport, phagocytosis, aggregation of platelets and clot contraction, release of growth hormone, and thyroid secretion. See also *Phoma*. Cytochalasins are used in medical research to elucidate cellular mechanisms, especially that of microfilaments.The toxicology of most of the more than 20 known cytochalasins is poorly known and most information is based on *in vitro* studies. See also macrolide mycotoxins. **cytochalasin B** (7,20-dihydroxy-16-methyl-10-phenyl-24-oxo-[14]cytochalasa-6(12),13,21-triene-1,23-dione; (E,E)-16-benzyl-6,7,8,9,10,12a,13,14,15,15a,-16,16-dodecahydro-5,13-dihydroxy-9,15-dimethyl-14-methylene-2*H*-oxycyclotetradec-[2,3-d]isoindole-2,18(5H)-dione; phomin). initially obtained from *Phoma spp*. as phomin, it accounts for the phytotoxicity of *P. exigua* and others. See also *Phoma*. It has been extensively studied biologically and is used mainly as a tool in cytological research. It affects a number of cellular processes *in vitro*, including cytokinesis, single-cell movement, morphogenesis of epithelia, and contractile processes such as that of mammalian smooth muscle and cardiac cells. **cytochalasin D**. a mycotoxin from *Chaetomium globusum*. It is extremely toxic (*s.c.*) to laboratory mice. **cytochalasin E**. extremely toxic to laboratory rats and guinea pigs. Acute intoxication (*i.p* or *p.o.*) of laboratory rats may cause ataxia, drowsiness, cyanosis, and coma. Death occurs within a few hours from circulatory collapse due to rapid, massive extravasation of plasma. Brain edema, pulmonary hemorrhage, injury to the vascular wall, and congestive degenerative changes and necrosis of liver, kidney, spleen, and small intestines are sometimes reported. **kodo-cytochalasin-1**. a species that is extremely toxic to laboratory mice.

cytochrome. any of a class of heme-containing conjugated protein pigments involved in electron and/or hydrogen transport in cells due to a reversible change in valency of heme iron. They are widely distributed in plant and animal tissues. The four main types are cytochromes a, b, c, and d. They transfer electrons from flavoproteins to molecular oxygen. This is effected by the porphyrin component that contains a chelated iron ion which reversibly changes state between the oxidized (ferric, +3) and reduced (ferrous, +2) oxidation states.

cytochrome *a*. cytochrome *c* oxidase.

cytochrome a_3. cytochrome oxidase.

cytochrome b_5. a microsomal cytochrome that participates in the desaturation of fatty acids and perhaps usually in the reduction of cytochrome P-450.

cytochrome *c* (myohematin; hematin protein). any hemoprotein enzyme derived from Protoporphyrin IX and present in the cells of aerobic organisms. They play an important role in the electron transport system of plants and animals. Several are known.

cytochrome oxidase (cytochrome a_3; Warburg's respiratory enzyme). any of a family of heme- and copper-containing cytochromes in the electron transfer chain (respiratory chain) that react directly with reduced oxygen. They accept electrons from cytochrome *c* oxidase (cytochrome *a*) transferring them to oxygen with the production of water and the release of sufficient free energy to support the formation of ATP. Cytochrome oxidase is inhibited by cyanide or carbon monoxide which effectively halts electron transport and prevents the formation of ATP.

cytochrome *c* oxidase (cytochrome *a*). a cytochrome that catalyzes the transfer of electrons from cytochrome oxidase (cytochrome a_3) to oxygen in respiration. See cytochrome oxidase.

cytochrome P-420. a degradation product of cytochrome P-450, often seen in microsomal preparations, that binds carbon monoxide and a number of type II ligands, but lacks oxidative activity.

cytochrome P-448. a cytochrome complex that often activates carcinogens and other toxic chemicals. It is induced, for example, by cigarette smoking, cooked red meats, and by any of a number of halogenated hydrocarbons.

cytochrome P-450. any of a number of cytochromes in the electron transport chain that are active in enzymatic hydroxylation, demethylation, and *N*-oxidation reactions in liver microsomes and adrenal mitochondria. They are most abundant in the liver and adrenal microsomes of vertebrates, but occur also in many other vertebrate tissues as well as in bacteria, plants, yeasts, and insects. The active site contains an iron atom that can change between the +2 and +3 oxidation states. P-450s act as a monooxygenases in catalyzing metabolic functions; they affect the rate of metabolism of drugs and other xenobiotics, especially lipophilic substances. The name reflects the characteristic absorption at 450 nm.

cytochrome P-450-dependent monooxygenase system (P-450 monooxygenase enzyme system). a system comprised of a large number of distinct enzymes ("isozymes") in the mitochondrial endoplasmic reticulum. They catalyze the dehydrogenation of diverse endogenous and exogenous compounds. Substrate specificities vary among these enzymes, but are usually broad and overlapping. This system plays a crucial role in the metabolic transformation and deactivation of many lipophilic xenobiotic compounds. In some cases, toxic (e.g., hepatotoxic, carcinogenic, mutagenic) intermediates result. Many of these enzymes are inducible by many substances, especially lipophilic xenobiotics.

cytochrome P-450 isozyme. See cytochrome P-450-dependent monooxygenase system.

cytocidal. lethal to cells, cell-destroying.

cytocide. an agent that destroys cells.

cytoclastic. destructive to cells.

cytoclesis (cytobiotaxis). the influence of living cells on other cells.

cytocide. an agent that destroys cells.

cytogenesis. the formation or development of cells.

cytogenetic. 1: pertaining to cytogenetics. **2:** pertaining to cytogenesis.

cytogenetical. pertaining to cytogenetics.

cytogenetics. the biological science that treats the

relationships between genetics and cell structure. It deals with the microscopic structure and behavior of chromosome inheritance (i.e., phenotypic expression). An important focus of cytogenetics is the correlation of chromosomal defects with specific inherited diseases.

cytogenic. cell-forming; forming a cell.

cytokinesis. division of the cytoplasm of a cell into two parts, as in the latter phases of mitosis.

cytokinin (phytokinin, kinin). **1:** any of a naturally occurring class of plant growth promoters. They are derivatives of the base adenine. They stimulate cell division in the presence of auxin, induce flowering in some plants, delay senescence, and induce the breaking of dormancy in axillary buds and some seeds. Unusually high endogenous concentrations of cytokinins occur in plant diseases such as crown gall and witches broom. **2:** synthetic cytokinins, some of which are substituted phenylureas. See also growth inhibitor, mevalonic acid, tumor-inducing principle.

cytologist. a scientist whose field of study is cytology.

cytology. the study of the structure, physiology, development, reproduction, and life history of cells, especially by means of light or electron microscopy.

cytolysin. any substance that lyses cells (e.g., hemolysins). Cytolysins generally exhibit a specificity for particular types of cells and are thus usually named accordingly (e.g., hemolysin, neurocytolysin). See also snakebite, venom (snake venom).

cytolysis. the lysis, disruption, or dissolution (partial or complete) of a cell, especially by destruction of the plasmalemma. **immune cytolysis**. that which is caused by antibody together with complement.

cytolytic. pertaining to, caused by, or having the capacity to cytolyse; pertaining to cytolysis.

cytophosphane. cyclophosphamide.

cytoplasm. 1: the living contents of a cell that lie within the cell, excluding the nucleus. It contains an aqueous ground substance or cell sap (hyaloplasm), vacuoles, and a variety of organelles and inclusions. **2:** the hyaloplasm or ground substance exclusive of all organelles except ribosomes. It is partly colloidal in nature, being comprised of a wide variety of macromolecules and in part a true solution of small molecules and ions.

cytoplasmic. pertaining to or found in the cytoplasm.

cytoplasmic membrane. plasmalemma.

cytosine arabinoside (cytarbine; Ara-C). an antineoplastic agent originally used as an antileukemic, but now used to treat herpes-virus hominis infections. It can cause thrombocytopenia and delayed CNS damage to humans.

cytosol. cytoplasm, exclusive of the various membrane-bound organelles and larger insoluble inclusions.

cytosolic. pertaining to or contained in the cytosol, or pertaining to the properties or contents of a cytosol.

cytosome. microbody.

cytostatic. **1:** pertaining to or denoting the suppression of cell growth and multiplication. **2:** an agent that suppresses cell growth and multiplication.

cytotoxic. **1a:** damaging to or destructive to cells. **1b:** pertaining to a substance or other agent that destroys or damages cells. **2:** referring to the effect of noncytophilic antibody on specific antigen.

cytotoxic anaphylaxis. an exaggerated hypersensitivity to an injection of antibodies that are specific for antigenic substances that occur normally on the cell surface.

cytotoxic cell. See natural killer cell.

cytotoxic drug. any pharmacologically active substance that destroys cells or inhibits cell proliferation. They are used mainly in cancer chemotherapy. Included are alkylating agents and antimetabolites used to selectively destroy abnormal cells. The use of such drugs must be carefully controlled as they may cause hair loss, serious blood dyscrasias, or reduced resistance to infections. They are often teratogenic, mutagenic, and carcinogenic. *Cf.* antineoplastic.

cytotoxic hypersensitivity. cytotoxic reaction. See reaction (cytotoxic reaction).

cytotoxic reaction. See reaction (cytotoxic reaction).

cytotoxic T-cell. cytotoxic T lymphocyte (See T lymphocyte).

cytotoxicity. **1:** the capacity of a toxicant to damage or destroy cells generally or specifically, or to produce a structural or functional change in the target cell that is potentially harmful to the cell or to its descendants. Harmful effects may result when a toxicant binds to and acts at specific loci in or on the cell or by directly interfering with any of various physiologic processes. **2:** the state of being cytotoxic.

cytotoxicosis. **1:** any morbid state produced by a cytotoxin. **2:** any condition produced by the poisoning of cells.

cytotoxin. **1:** any substance produced by a living organism that is toxic to living cells. **2:** any substance that inhibits cellular processes or is otherwise toxic to or destroys cells. See also toxin.

CZ. carzinophilin.

δ. See delta.

Δ. See delta.

d. deci-. Do not confuse with da.

2,4-D ((2,4-dichlorophenoxy)acetic acid). a synthetic auxin used as a herbicide It selectively poisons broad-leaved plants, upsetting their growth and metabolism causing them to overgrow and die. It also interferes with biochemical pathways that process nitrates in some plants with consequent accumulation of nitrate. 2,4-D is a constituent of Agent Orange. It is very toxic and an irritant to the eye and gastrointestinal tract. symptoms of intoxication in humans typically include weakness, lethargy; diarrhea; spastic myotonia; ventricular fibrillation and cardiac arrest; and sometimes hypermetabolism, hyperpyrexia, convulsions and coma. Large doses or long-continued exposure can produce severe, protracted peripheral neuropathy and sometimes brain damage in humans. See also chlorophenoxy compounds, nitrate poisoning.

da. deca-. Do not confuse with d.

Da. Dalton.

DA. diphenylchlorarsine.

2,4-DAA. 2,4-diaminoanisole.

2,4-DAA sulfate. 2,4-diaminoanisole sulfate.

2,4-DAA sulphate. 2,4-diaminoanisole sulfate.

dab kwingu. *Naja nigricollis*.

daboia. a common name in India for *Vipera russelli*.

daboya. a common name in India for *Vipera russelii*.

dace. See Cyprinidae.

dacryogenic. inducing or promoting the flow of tears.

dactinomycin (actinomycin D; actinomycin-[thr-val-pro-sar-meval]; meractinomycin; actinomycin A_{IV}; actinomycin C_1; actinomycin I_1; actinomycin X_1). an extremely toxic antibiotic of the actinomycin family produced by *Streptomyces chyrosomallus* and several other members of this genus. It is used as an antineoplastic agent. Effects on humans include alopecia, bone marrow suppression, gastroenteritis, and possibly anaphylaxis. Dactinomycin is a potent inhibitor of DNA-dependent RNA synthesis and is much used experimentally in cell biology. See cactinomycin.

daffodil. common name of a number of hardy species of *Narcissus* (*q.v.*) that bear a single, large, trumpet-shaped flower. This term is often used incorrectly in reference to any species of hardy narcissus.

dagar. *Bitis arietans*.

dalapon (2,2-dichloropropionic acid; α,α-dichloropropionic acid). a chlorinated aliphatic acid resulting from the substitution of one methyl group by a chlorine atom of TCA. The sodium salt is used as a herbicide to selectively control to annual and perennial weed grasses on arable land. It is absorbed through roots and leaves. Dalapon is slightly toxic to laboratory mice. Dalapon and other halogenated aliphatic acids inhibit plant growth, induce leaf chlorosis, and at higher concentrations cause leaf necrosis. They affect many plant processes.

dallis grass. *Paspalum dilatatum*.

Dalmatian insect powder. pyrethrin, pyrethrum flowers.

Dalton (atomic mass unit). an arbitrary unit of mass equal to 1/12 the mass of ^{12}C or $1.657 X 10^{-24}$.

damage. **1:** harm or injury with impairment; the harm, detriment or loss sustained as the result of an injury. **2:** the loss or temporary or permanent impairment of an organ or bodily part. *Cf.* harm, injury, trauma. **genotoxic damage**. chromosomal damage induced by a toxic chemical. Such damage is marked by the presence (following cell division) of chromosomal fragments in the cytoplasm. **3:** to harm, disfigure, or injure with impairment.

damaging agent. an agent whose principal action is injurious and devoid of any immediate, direct value to the injured system. See damage.

daminozide (butanedioic acid mono(2,2-dimethylhydrazide); *N*-(dimethylamino)succinamic acid; succinic acid 2,2-dimethylhydrazide). a pesticide used to delay ripening of apples, Concord grapes, other fruit and sometimes peanuts. It reddens and firms apples, making them less likely to drop before growers are ready to pick them. Daminozide is carcinogenic to laboratory mice and rats and perhaps to humans. Residues of this compound remain on the treated produce until purchase.

damp. **1:** moist, wet, humid. **2:** a noxious gas that occurs in a mine. **after damp**. a toxic mixture that contains large amounts of carbon dioxide, carbon monoxide, and nitrogen. It results from the explosion of methane and air in a mine. **black damp**. the atmosphere in a coal mine that results from the slow absorption of oxygen and release of carbon dioxide from coal. **cold damp**. vapor in a mine with high levels of carbon dioxide. **fire damp**. methane, a gas which is commonly present in a coal mine. **white damp**. carbon monoxide.

Danaidae. a small family of large, brightly colored butterflies; most are dark orange with bold black and white markings. See *Danaus*.

Danaus. a genus of toxic, boldly patterned butterflies (Family Danaidae). The sources of toxicity are the milkweeds and other poisonous plants on which the larvae feed. Species such as *D. plexippus* and *D. chrysippus*, for example, accumulate the toxic cardiac glycosides, calactin and calotropine, from the milkweeds (*Asclepias*) on which they feed. These substances are bitter, indigestible, and toxic to birds and other vertebrate animals that may prey on them. They are also protected from herbivorous animals by their association with *Asclepias*.

D&C. drug and cosmetic. *Cf.* D&C color, FD&C.

D&C Blue No. 4. FD&C Blue No. 1.

D&C color. See color.

D&C Orange No. 2. 1-(*o*-tolylazo)-2-naphthol.

D&C Orange No. 17. a coloring agent that is carcinogenic in some laboratory animals.

D&C Orange No. 8. 4′,5′-dichlorofluorescein.

D&C Red No. 8. a coloring agent that is carcinogenic in some laboratory animals.

D&C Red No. 9. a coloring agent that is carcinogenic in some laboratory animals.

D&C Red No. 19. rhodamine B.

D&C Red No. 37. a coloring agent that is carcinogenic in some laboratory animals.

D&C, Traces of. See color.

DANGER. **1:** a hazard or threat; a potential risk. **2:** a signal word which, when displayed on a product label, indicates that ingestion of a tiny pinch can kill an adult human. See U.S. Federal Hazardous Substances Labeling Act; warning. "POISON" in this context is equivalent.

DANGER POISON (with skull and crossbones). See label (warning label).

Danilone®. phenindione.

[illegible]

daphne. *Daphne*.

Daphne (daphne; mezereum; mezereon; spurge laurel; spurge). a genus of low, woody, ornamental shrubs (Family Thymelaeceae) with acrid, tough bark that are native to Eurasia and Africa. All parts of the plants of all species are highly toxic. They are violent purgatives. Ingestion of berries or other plant parts can cause severe abdominal pain, vomiting, weakness, stupor, convulsions, and bloody diarrhea. Poisoning may produce corrosive lesions of the oral mucosa and upper gastrointestinal tract and nephritis. The chief active principle is the glycoside, daphnin, the aglycone of which is daphnetin. The berries may attract children and can prove fatal when ingested. Human fatalities are now rare. Principal species include *D. cneorum*, *D. laureola*, *D. genkwa*, and *D. mezereum*. ***D. cneorum*** (garland flower, rose daphne). an odorous plant, common to rock gardens, with rose-colored flowers. It grows to a height of less than 30 cm. ***D. laureola*** (spurge laurel). a plant with yellowish-green trumpet-shaped, unpleasant-smelling flowers that grows to nearly 1 m. The new berries are green at first, then bluish, and when ripe are purplish black. All parts of the plant are poisonous, but the berries are especially so; they are violently toxic. Do not confuse this plant with the true laurel (*Laurus nobilis*). ***D. mezereum*** (daphne; mezereum; mezereon; spurge olive; February daphne; dwarf bay; flax olive; spurge flax; wild pepper; wood laurel; copse laurel). a species indigenous to Europe and parts of Asia, it has become naturalized in southeastern Canada and in northeastern United States, where it occurs casually along roadsides, in hedges and thickets, and in abandoned limestone quarries. It grows to a height of ca. 1½ m. The flowers bloom in the spring, are fragrant, pink to purple (rarely white), and trumpet-shaped with the flare of the bell consisting of four parts. All parts are poisonous, but the berries are especially deadly. The berries are usually bright red or scarlet and are attractive to children. Ingestion of just a few berries can produce a burning sensation in the mouth and severe gastroenteritis. Symptoms of intoxication may include a burning sensation in the mouth, abdominal cramps, hematemesis, diarrhea, weakness, stupor, and convulsions. Lesions can include nephritis, and corrosive lesions of the oral mucosa and upper gastrointestinal tract.

daphnetin (7,8-dihydroxy-2*H*-1-benzopyran-2-one; 7,8-dihydroxycoumarin). the highly toxic aglucon of daphnin, it is soluble in hot dilute ethanol, hot glacial acetic acid, and boiling water; it is sparingly soluble in benzene, carbon bisulfide, chloroform, and ethyl ether. Extract from plant material (See *Daphne*) is by boiling with dilute mineral acids. Do not confuse with 7,8-dihydroxycoumarin 7-β-D-glucoside. See *Daphne*.

daphnetin 7-β-D-glucoside. daphnin.

Daphnia (water fleas). a cosmopolitan genus of freshwater crustaceans (Order Cladocera). Prominent species are *D. pulex, D. galeata,* and *D. magna*. ***D. magna.*** a standard species used in aquatic toxicity tests. It is widely used in biological research.

daphnid. 1. common name of any member of the cladoceran genus *Daphnia*. 2. pertaining to or denoting any cladoceran.

daphnin (7-β-D-glucopyranosyloxy-8-hydroxy-2*H*-1-benzopyran-2-one; 7,8-dihydroxy coumarin 7-β-D-glucoside; daphnetin 7-β-D-glucoside; 7-glucosido-8-hydroxycoumarin). a toxic coumarin glycoside isolated from the bark and flowers of *Daphne spp.*, *q.v.* It has been used in folk medicine as an abortifacient. The aglucon is daphnetin.

daphnism. poisoning by plants of the genus *Daphne*.

dapsone (4,4′-sulfonyldianiline). a nearly water-insoluble, poisonous, antibacterial, antiprotozoal sulfone, $C_{12}H_{12}N_2O_2S$, used in the treatment of Hanson's disease and in veterinary practice. It is carcinogenic to some laboratory animals.

Darier's sign. dermatitis herpetiformis.

darnel. *Lolium temulentum*.

darnel poisoning. loliism.

darpa. *Pseudechis australis*.

Darvon. propoxyphene.

Darwin's sea snake. *Hydrelaps darwiniensis*.

dasyatid. **1a:** of or pertaining to true stingrays (Family Dasyatidae). **1b:** of or pertaining to stingrays of the genus *Dasyatis*. **2:** a true stingray, especially one of the genus *Dasyatis*. **3:** of or pertaining to the dasyatid type of sting which occurs also in the Potamotrygonidae.

Dasyatidae (stingrays, whiprays). a family of venomous cartilaginous fish, the true stingrays. The sting is well developed and lies distal to the midpoint of the whiplike tail. These stingrays are among the most dangerous. Some produce deep penetrating wounds and can introduce tetanus bacilli in addition to the venom. See also stingray, dasyatid, *Dasyatis*.

Dasyatis. a common genus of stingrays or whiprays (Family Dasyatidae). All are venomous and dangerous to humans. The sting is well developed and lies distal to the midpoint of the whiplike tail. See dasyatid, stingrays. ***D. brevicaudata*** (giant stingray or giant stingray of Australia). venomous, it is possibly the largest of stingrays, reaching a length of 4.5 m, a width of 2.2 m, and a weight of more than 325 kg.

data. plural of datum, *q.v.*; facts, information, measurements, results.

data bank. a large data base such as a set of computerized files, a set of computer disks, or a library.

data call-in. part of the process of developing important test data required by the Office of Pesticide Programs (OPP) of the U.S. Environmental Protection Agency. The focus is on the acquisition of long-term, chronic effects of existing pesticides, in advance of scheduled Registration Standard reviews. "Data call-in" is intended to expedite preregistration by permitting manufacturers to "call in" (to telephone) data.

database, data base. a nonredundant collection of related data, stored in machine-readable form and available for use. See also data bank.

Datiscaceae. a small family of herbs and trees characterized by regular, apetalous and often dioecious flowers born on spikes or racemes, and a capsule with a single seed. The sap of most species contains cucurbitacin, *q.v.*

dative bond. coordinate bond.

datum (plural, data, *q.v.*). **1:** a quantitative or qualitative piece of information that is either granted as true or derived from direct observation of phenomena or from experiments. **2:** any number, symbol, or analog quantity that can be entered into a computer for processing.

datura. **1a:** a plant of the genus *Datura*. **1b:** jimsonweed seed. It is a toxic anticholinergic. See poisoning (datura poisoning)

Datura (datura; thornapple; jimsonweed; apple of Peru; Jamestown weed; tolguacha). a genus of narcotic, poisonous, warm-climate annual or perennial plants (Family Solanaceae) with large, showy flowers. They contain several tropane alkaloids in high concentration (ca. 0.25-0.7%), the most important being atropine, hyoscyamine, and hyoscine (scopolamine). Toxic concentrations of nitrate have also been reported. All classes of livestock have been poisoned, and human poisoning is not uncommon. Most accidental poisonings are due to ingestion of seeds. Approximately 4-5 g of seed or leaves may fatal to a child. Effects vary greatly. Initial symptoms include a dry mouth and intense thirst, disturbed vision, flushed skin, and hyperexcitability. Pyrexia, tachycardia, incoherence, delirium, and even violent behavior may follow. In fatal poisoning, convulsions, coma, and death ensue, and early stages are sometimes bypassed. See also poisoning (datura poisoning). ***D. metel***. a source of scopolamine. ***D. stramonium*** (thornapple; jimsonweed; stramonium; Jamestown weed; stinkweed; devil's apple; devil's trumpet; mad apple). an extremely dangerous, subcosmopolitan annual weed; height to about 2 ms. It is distributed widely throughout the United States on gravelly or rich alluvial soils. This plant probably causes more poisonings of humans than any other plant. The fruit is extremely poisonous and hallucinogenic when ingested. Some individuals suffer an allergic contact dermatitis. It is used by drug cultists, sometimes with lethal results. Children have been poisoned by the nectar, seeds, or the fruits containing the seeds. Animals sometimes eat the foliage. Early symptoms of intoxication in humans are a dry throat, gaiety, laughter, and mild delirium. Pupillary enlargement may appear early and persist for several days in recovering individuals. Further symptoms may include dimness of vision, pupillary dilation, giddiness, delusions, delirium and sometimes mania, unconsciousness, and even death. This plant is teratogenic and a cause of arthrogryposis in livestock. ***D. suaveolens***. Effects similar to those of *D. stramonium*, *q.v.* Used by drug cultists, sometimes with lethal results.

datura poisoning (daturism). poisoning due to ingestion of plants of the genus *Datura*. All classes of livestock have been poisoned and human poisoning is not uncommon. Most accidental poisonings of humans are due to ingestion of seeds. Ca. 4-5 g of seed or leaves may prove fatal to a child. Effects are largely parasympathoylytic and in severe cases include CNS depression, circulatory failure, and respiratory depression. Nevertheless, symptoms vary greatly among victims. Initial symptoms include a dry mouth, intense thirst, disturbances of vision, flushed skin, and hyperexcitability. Fever, tachycardia, incoherence, delirium, and even violent behavior may follow. Fatal poisonings are marked by convulsions, coma, and death; the early stages of poisoning are sometimes bypassed.

daturine. hyoscyamine.

daturism. datura poisoning.

Daubentonia drummondi. *Sesbania drummondi*.

Daubentonia punicea. *Sesbania punicea*.

Daucus carota* var. *sativus (cultivated carrot). a biennial herb and common garden vegetable and food crop (Family Apiaceae). The root is normally edible, but may contain toxic concentrations of nitrates when growth has been especially rapid. This sometimes occurs also in the above-ground portions of the plant. Livestock have been poisoned on occasion, apparently by nitrates, when feeding on the leaves.

daunomycin. daunorubicin.

daunorubicin ((8*S*-*cis*)-8-acetyl-10-[(3-amino-2,3,6-trideoxy-α-L-lyxo-hexopyranosyl)oxy]-7,8,9,10-tetrahydro-6,8,11-trihydroxy-1-methoxy-5,12-naphthacenedione; cerubicin; daunomycin; DMC; leukaemomycin C; rubidomycin; DRC). an extremely toxic, anthrcycline, antibiotic glycoside with antineoplastic, immunosuppressive, and antibacterial activity. It is produced by *Streptomyces peucetius* and is related to the rhodomycins. Daunorubicin is used in the treatment of solid tumors, but is cardiotoxic, nephrotoxic, and carcinogenic to laboratory animals. It inhibits DNA and RNA synthesis. Symptoms of acute intoxication are nausea and vomiting with subsequent bone marrow suppression and congestive heart failure.

daunorubicin hydrochloride. the hydrochloride salt of daunorubicin. Its biologic actions are the same as those of the base.

day-blooming jessamine. See *Cestrum*.

DBA. **1:** 1,2:5,6-dibenzanthracene. **2:** 9,10 dimethyl-1,2-benzanthracene.

1,2,5,6-DBA. 1,2:5,6-dibenzanthracene.

DB[a,h]A. 1,2:5,6-dibenzanthracene.

DBCP. dibromochloropropane.

DBDO. decabromodiphenyl oxide.

DBH. dopamine hydroxylase.

DBL. desbromoleptophos.

DBP. *n*-butyl phthalate.

DCA, DCa. desoxycorticosterone acetate. See desoxycorticosterone acetate.

DCB. 3,3′-dichlorobenzidine.

1,2-DCE. ethylene dichloride.

DCPA. chlorthal-dimethyl.

DDC. ditiocarb sodium.

DDD. **1:** DDD, analytical. **2:** 1,1-dichloro-2,2-bis(*p*-chlorophenyl)ethane.

***p,p′*-DDD**. 1,1-dichloro-2,2-bis(*p*-chlorophenyl)-ethane.

DDD, analytical (6,6′-dithiobis-2-naphthalenol;

6,6′-dithiodi-2-naphthol; bis(6-hydroxy-2-naphthyl) disulfide; 2,2′-dihydroxy-6,6′-dinaphthyl disulfide; 6,6′-dithiobis(2-naphthol)). analytical DDD, a reagent used in the determination of protein-bound sulfhydryl groups. Do not confuse with the pesticide, *p,p′*-DDD (See 1,1-dichloro-2,2-bis(*p*-chlorophenyl)ethane).

DDE. dieldrin.

DDP. cisplatin.

***cis*-DDP**. cisplatin.

DDT (1,1′-(2,2,2-trichloroethylidene)bis[4-chlorobenzene]; 1,1,1-trichloro-2,2-bis(*p*-chlorophenyl)ethane; *β,β*-bis(*p*-chlorophenyl)-*β,β,β*-trichlorethane; chlorophenothane, clofenotane, dichlorodiphenyltrichloroethane, dicophane, gesarol, pentachlorin, *p,p′*-DDT, etc.). a white, colorless, odorless, water-insoluble, crystalline solid, DDT is a polychlorinated, lipophilic, environmentally persistent contact pesticide, larvicide, ectoparasiticide, and topical pediculicide that is toxic to many vertebrate animals. The *p,p′*-isomer is weakly estrogenic and has the greatest toxicity in insects; the *o,p′*-isomer is less toxic and more estrogenic. DDT is neurotoxic and can cause serious damage to the CNS. It has been a serious cause of reproductive failure in birds and fish. It is not generally considered to be highly toxic to mammals, but note that DDT is very toxic to adult laboratory rats, while moderately toxic to newborns. Humans can be exposed by ingestion or by absorption through skin or respiratory tract. Acute toxicity is due mainly to the direct action of DDT on axonal membranes to increase sodium conductance with a consequent increase in excitability that can result in multiple impulses, tremors and tetanus. Acute effects in humans include vomiting (which may be delayed); numbness about the lips; tongue and face; headache; sore throat; weakness; tremors of the head and neck muscles; ataxia; tonic and clonic convulsions; coma; and death from respiratory or cardiac collapse within 2-24 hours. There are no lesions. Chronic exposure can cause muscular weakness, dermatitis, liver damage, CNS degeneration, agranulocytosis, convulsions, coma, and eventual death. It is carcinogenic in a number of animals; it produces liver tumors in laboratory mice, but apparently not in humans. The acute dermal toxicity to humans is low. Largely because of its persistence in the environment, accumulation in fat and induction of eggshell thinning in predatory birds, the U.S. Environmental Protection Agency cancelled all registrations and interstate sales of DDT in 1973. See also DDD, DDT dehydrochlorinase, dieldrin, eggshell thinning.

***p,p′*-DDT**. DDT.

DDT dehydrochlorinase. an enzyme (usually a tetramer with a molecular mass of about 144 Da), found in a wide variety of animals that catalyzes the dehydrochlorination of DDT and a number of DDT analogs. Most importantly, in the presence of glutathione, it catalyzes the dehydrochlorination of DDT to DDE and of *p,p′*-DDD to TDEE (2,2-bis(*p*-chlorophenyl)-l-chloroethylene).

DDTC. ditiocarb sodium.

DDVP. dichlorvos.

de Capello cobra. See *Naja naja*.

de-lead. delead.

de minimus (*diminimis non curat lex*). the doctrine of law which states that the law does not deal with, or take notice of insignificant matters. It is applied in toxicology such that chemicals that are of very low risk may be dismissed from consideration or liability.

deacetylation. removal of one or more acetyl radicals, CH_3CO—, from an organic compound (usually from a nitrogen atom) that contains one or more OH or NH_2 groups. Metabolic deacetylations often render a previous acetylated xenobiotic more toxic. Acetanilide (acetylaniline), for example, is deacetylated by an aromatic deacetylase, producing the more toxic aniline.

deacetylcalonectrin. a trichothecene first isolated from *Fusarium*.

deacetyllanatoside C. deslanoside.

***N*-deacetyl-*N*-methylcolchicine**. demecolcine.

dead. **1:** deceased, numb, lifeless, inert, inanimate. **2:** pertaining to an organism or part of an organism, having ended a state of life; no longer living. **3:** having the appearance of death. **4:** devoid of life. **5:** the dead; that which is dead (def. 1).

dead nettle. *Lamium amplexicaule*.

dead on arrival (DOA). a phrase used mainly in reference to humans that were not dead (or not known to be dead) when dispatched to a hospital or other treatment facility, but whose death was confirmed on or shortly after arrival.

deadly. productive of death, or tending to cause death; lethal; virulent. In the context of toxicology, deadly can usually be taken to mean that direct contact with or proximity to a deadly substance or organism incurs substantial risk of a fatal outcome.

deadly agaric. *Amanita phalloides*.

deadly amanita. *Amanita phalloides*.

deadly cort. *Cortinarius gentilis*.

deadly death puffer. *Arothron hispidus*.

deadly galerina. *Galerina autumnalis*.

deadly hemlock. *Conium maculatum*.

deadly lawn galerina. *Galerina venenata*.

deadly lepiota. *Lepiota josserandii*.

deadly nightshade. *Atropa belladonna*; *Solanum nigrum*.

deadly sculptured scorpion. *Centruroides sculpturatus*.

deadly sea wasp. *Chironex fleckeri*.

deadly stonefish. *Synanceja horrida*.

dealkylate. to remove alkyl groups from a compound. See dealkylation.

dealkylation. the loss or removal of alkyl groups from a compound. This process involves the removal of alkyl substituents (usually ethyl or methyl groups) from an organic compound by hydrogen atoms. Dealkylation is an important step in the metabolism of many xenobiotic compounds, many of which contain alkyl groups attached to atoms of O, N, and S. Metabolic dealkylation usually involves the replacement of alkyl substituents (usually methyl or ethyl groups) attached to nitrogen, oxygen, or sulfur atoms by hydrogen. Alkyl groups attached directly to carbon atoms are rarely removed. Dealkylations that are of most interest toxicologically are catalyzed by isozymes of the cytochrome P-450 system. Methyl and ethyl groups are most commonly involved in metabolic dealkylations. ***O*-dealkylation**. the loss or removal of an alkyl group from an oxygen atom of an organic compound. ***N*-dealkylation**. the loss or removal of an alkyl group from a nitrogen atom of an organic compound. ***S*-dealkylation**. the loss or removal of an alkyl group from a sulfur atom of an organic compound. Examples are the *O*-dealkylation of methoxychlor insecticides, the *N*-dealkylation of carbaryl insecticide, and the *S*-dealkylation of dimethyl mercaptan. *N*-dealkylation is important in the metabolism of many drugs, insecticides, and other xenobiotics.

***N*-dealkylation**. See dealkylation.

***O*-dealkylation**. See dealkylation.

deaminate. to remove an amino group from an amino acid or other amine.

deamination. the removal of an amino group, $-NH_2$, from an amino acid or other amine. Several mechanisms whereby organic xenobiotics are deaminated. Transaminations are most common and are catalyzed by several different enzymes. For example, in this reaction catalyzed by glutamate dehydrogenase, transamination is coupled with the oxidative deamination of glutamic acid:

$$\text{glutamic acid} + \text{NAD} \rightleftarrows \alpha\text{-ketoglutaric acid} + \text{NADH} + NH_3$$

The NH_3 thus produced is converted to urea in the liver and excreted.

dean pitch. asphalt.

death. **1:** demise; extinction; the cessation of life of a living system or a part thereof (e.g., an organism, appendage, population, community, species). Death may be a gradual process, not all parts of a system dying at the same time. The death of an organism is thus usually taken either as a point where the integrity of the individual fails or at a point where the process of dying is thought to be irreversible. The death of a cell or microorganism is usually marked by rapid dissolution. **2:** the permanent loss of the normal state of dynamic equilibrium typical of a live organism. **3:** the permanent cessation of all vital functions of a living organism or part of a living organism. **4:** that point or phase in the process of dying where the integrity of an individual fails or at a point where the process of dying is thought to be irreversible. **5:** the cause of dying. **apparent death**. the absence of a heartbeat or breathing. **brain death** (cerebral death). **1a:** many governments have enacted statutes that define human death based in part on brain-related criteria. Many such statutes adhere to the criteria of the Uniform Determination of Death Act: "An individual who has sustained either (1) irreversible cessation of circulatory and respiratory function, or (2) irreversible cessation of all functions of the entire brain, including the brain stem, is dead. A determination of death must be made in accordance with accepted medical standards." The guidelines state that there must be clear evidence of irreversible damage to the brain; persistent, deep coma; no attempts at breathing when the individual is removed from the ventilator; and absence of brain stem function (e.g., nonresponsiveness of the pupils to light, grimacing in response to pain, no blinking if the eye is touched). Even with great care in applying these criteria, the assessment of death is sometimes unreliable (e.g., in the case of intoxication, hypothermia, or shock). Consult, e.g., Com. versus Golston, 373 Mass. 249, 252, 366 N.E. 2d 744. **1b:** the loss of brainstem and spinal reflexes, with a flat EEG for at least 24 hours. **cell death** (molecular death). **1:** irreversible failure of a cell to maintain life functions. **2:** the point in the process of dying where loss of life functions has encompassed the cellular level. **cerebral death**. brain death. **early fetal death**. in humans, death of the fetus during the first 20 weeks of gestation. **fetal death**. **1:** the death of a fetal mammal. **2:** the death in utero (as judged by accepted medical/legal standards) of a human fetus at a gestational age where it would normally be considered viable; in practice, a fetus that weighs 500 grams or more. **functional death**. loss of CNS activity, with other vital functions being artificially supported. **genetic death**. **1:** the state of an individual who has become irreversibly unable to produce viable offspring. **2:** genetic extinction, *q.v.* **legal death**. definitions of legal death vary with legal jurisdiction. Most rely in part on brain-related criteria and typically require examination and certification by a licensed physician of the absence of all CNS, cardiovascular, and respiratory activity. **molecular death**. cell death.

death adder. a snake of the genus *Acanthophis* especially *A. antarcticus*.

death angel. See *Amanita phalloides*, *A. virosa*.

death camas. See *Zigadenus*.

death camass. *Zigadenus nuttallii*.

death cap. *Amanita phalloides*.

death point. the limit of tolerance or limit of viability (e.g., degrees of temperature, concentration of a substance) for any endogenous or environmental factor, at or beyond which death occurs.

death rate. See rate (death rate).

death seaweed of Hana. *Palythoa toxica*.

death's cup. the volva of *Amanita* and certain other mushrooms.

death's herb. *Atropa belladonna*.

death's stool. toadstool.

debilitating. weakening, enfeebling, devitalizing, crippling. **1a:** of a disease: enfeebling, crippling, damaging; often implying a degree of irreversibility. **1b:** of an agent such as a poison or pathogen: producing or [illegible] enfeeblement, or a serious decrement in health or performance, often to the point of helplessness.

debilitant. **1:** able to weaken or debilitate. **2:** a medicine used to reduce excitement. **3:** that which weakens or debilitates.

debilitate. to cause debility.

debility. **1:** disability, frailty, impairment, infirmity, sickness. **2:** weakness or loss of tonicity of organs or of bodily functions.

debromoaplysin. a debrominated derivative of aplysin.

deca- (da) a prefix to units of measurement that indicates a quantity of ten (10^{12}) times the value indicated by the root. See International System of Units.

decaborane (decaboron tetradecahydride). a colorless, crystalline, binary compound of boron and hydrogen, $B_{10}H_{14}$. It is stable at room temperature and slightly soluble in cold water, but hydrolyzes in hot water; it is soluble in organic solvents such as benzene, ethanol, hexane, carbon tetrachloride, and toluene. It is used as a catalyst in olefin polymerizations, a corrosion inhibitor, a gasoline additive, and in the manufacture of rocket propellants. Decaborane is a CNS poison, causing neurological and behavioral disturbances. It is toxic by inhalation, ingestion, cutaneous absorption, and by intraperitoneal routes. Effects are similar to those of pentaborane; symptoms of intoxication in humans may include dizziness, nausea, vomiting, and muscular tremors. Liver damage is known. It emits toxic boron oxides when heated to decomposition. It is flammable, and forms impact-sensitive explosive mixtures with ethers such as dioxane and with halocarbons such as carbon tetrachloride.

decabromodiphenyl oxide (DBDO). a highly toxic and environmentally persistent compound related to PCBs and PBBs. It is used as a flame retardant in plastics.

decaboron tetradecahydride. decaborane.

decabromodiphenyl oxide (DBDO). a highly toxic, environmentally persistent substance used as a flame retardant in plastics. It is structurally related to PCBs and PBBs. It is carcinogenic to some laboratory animals.

1,1a,3,3a,4,5,5a,5b,6-decachlorooctahydro-1,3,4-metheno-2*H*-cyclobuta[cd]pentalen-2-one. chlordecone.

1,2,3,4,7*a*,10,11,11*a*,1 2,13-decahydro-14-(1-hydroxyethyl)-2,11*a*-dimeth yl-7*H*-9,12(8*H*)-diol. batrachotoxinin A.

1,2,3,4,7*a*,10,11,11*a*,1 2,13-decahydro-14-(1-hydroxyethyl)-2,11*a*-dimethyl-7*H*-9,11b-epoxy-13a,5a-propenophenanthro[2,1-f][1,4]oxazepine-9,12(8H)-diol. batrachotoxinin A.

1,1a,1b,4,4a,7a,7b,8,9,9a-decahydro-4a,7b,9,-9a-tetrahydroxy-3-(hydroxymethyl)-1,1,6,8-tetramethyl-5*H*-cyclopropa[3,4]-benz[1,2-e]azulen-5-one. phorbol.

decamethrin. deltamethrin.

1-decanol (*n*-decyl alcohol; nonylcarbinol). a moderately viscous liquid of low acute toxicity, $CH_3(CH_2)_8CH_2$0H.

decarboxycysteine. cysteamine.

decay (organic decay). **1:** to decompose, molder, putefry. **2:** (moldering, putrefaction) the decomposition of organic matter, with nearly total oxidation. **3:** in atmospheric chemistry, the disappearance of a pollutant with time (e.g., as the result of absorption or chemical reaction at the earth's surface, removal by precipitation, or chemical transformation into another substance). See also decompose.

dechlorination. the removal of chlorine from a compound by replacing it with hydrogen or hydroxyl ions usually as a means of detoxification.

dechloro-penitrem A. penitrem b.

deci- (d). a prefix to units of measurement that indicates a quantity of one-tenth (10^{-1}) the value indicated by the root term. See International System of Units.

Declaration of Helsinki. a declaration with guidelines that apply expressly to experimental studies conducted on humans. General guidelines stipulate that such studies be undertaken only when the results cannot be otherwise obtained, when the benefits to be derived are considerable, and when the risks under the conditions of the test can be estimated as nearly nonexistent. Furthermore such studies are to be conducted only on informed volunteers and under skilled medical supervision.

decoction. **1:** the process of boiling, usually in order to isolate a flavor or the active principle of a substance. **2:** any extract (e.g., a drug) prepared by boiling. **3:** a liquid obtained by boiling a toxic or medicinal plant in water.

decoctum. a decoction.

decompensation. failure of the heart to maintain adequate circulation.

decompose. **1:** to separate or resolve a compound into simpler compounds and/or elements. **2:** to bring about chemical disintegration of organic matter. **3:** to disintegrate or decay.

decomposer. a heterotrophic organism (e.g., a bacterium, fungus) that metabolically decomposes nonliving organic matter into simpler organic and inorganic compounds and/or elements. See decomposition.

decomposition. **1:** the act or process of decomposing. **2:** the state or condition of being decomposed. **3:** the separation or resolution of a compound into simpler compounds and/or elements. **4:** decay, disintegration, putrefaction; the metabolic breakdown of organic matter into simple organic and inorganic compounds, usually by the agency of bacteria or fungi and with the liberation of energy.

decompression sickness. bends.

decontaminate. **1:** to free an entity of a contaminating (usually toxic or otherwise hazardous) substance by physical or chemical means. **2:** to render an area or environment harmless to humans or other living systems by removing, destroying, or neutralizing hazardous contaminants such as toxic substances, biologicals, or radioactive materials.

decontamination. **1:** the removal or neutralization of chemical, biological, or radiological contaminants from a person, animal, object, location, etc. **2:** the rendering of an area or environment harmless to humans or other living systems by removing, destroying, or neutralizing toxic or otherwise hazardous materials. **radiation decontamination**. the removal of unwanted radioactive contamination.

decrudescence. a decline or subsidence of a process or condition such as the symptoms of intoxication.

decumbin. a mycotoxin produced by *Penicillium decumbens*. It is very toxic to laboratory rats, causing respiratory distress and hemorrhage. The oral LD_{50} for rats is about 275 mg/kg body weight.

DEDC. ditiocarb sodium.

DeDTC. ditiocarb sodium.

DEET. See insect repellent.

deethylation. chemical or metabolic dealkylation in which an ethyl group is removed from a compound. See also dealkylation.

deferoxamine. (*N*′-[5-[[4-[[5-(acetylhydroxyamino)pentyl]amino]-1,4-dioxobutyl]hydroxyamino]pentyl]-*N*-(5-aminopentyl)-*N*-hydroxybutanediamide; *N*-[5-[3-[(5-aminopentyl)hydroxycarbamoyl]propionamido]pentyl-3-[[5-(*N*-hydroxyacetamido)pentyl]carbamoyl]propionohydroxamic acid; 1-amino-6,17-dihydroxy-7,10,18,21-tetraoxo-27-(*N*-acetylhydroxylamino)-6,11,17,22-tetraazaheptaeicosane; desferrioxamine B). a naturally occurring substance that readily complexes with iron. It is used therapeutically as a parenteral chelating agent for iron and aluminum. It has been used successfully to treat dialysis encephalopathy and to promote excretion of iron in patients with a secondary iron overload from multiple transfusions. The methanesulfonate, desferrioxamine mesylate, $C_{26}H_{52}N_6O_{11}S$, is an antidote to iron poisoning.

deferrioxamine mesylate (DFOM). See deferoxamine.

defibrase. batroxobin.

defibrillation, cardiac. the application of an electric current to the chest over the heart to interrupt fibrillation.

definitive. clear-cut, without question, conclusive, clear and final.

defoliant. an agent (e.g., the herbicides 2,4-D and 2,4,5-T) that causes leaves to fall from a plant.

DEG. diethylene glycol.

degenerate. **1a:** to decline, worsen, deteriorate; to undergo degeneration. **1b:** to lose functional integrity. **2:** debased; dissolute; characterized by dissolution or degeneration.

degeneration. deterioration, dissolution, decline, regression. **1:** a loss of integrity or a breakdown in structure of a living system or of a component of such. Operationally, degeneration is reflected in chemical and/or physical changes that take place in one or more system components. Degeneration may result from any of a number of causes (e.g., aging, a disease process, poisoning, reduced blood supply, autolysis, nutrient deficiencies). **2:** the change of an entity to a less specialized or less active form. **3:** with respect to humans and higher animals, a deterioration in physical, mental, or behavioral state.

degenerative. marked by, or accompanied by, decline or degeneration.

degenerative disorder. any of a wide variety of conditions reflected in progressive structural and functional impairment of one or more body parts. Only organic disease or accelerated aging is encompassed by this term. Degeneration due to specific infections and poisoning by xenobiotics are thus excluded. Many degenerative disorders may eventually be reclassified as external causes are discovered.

deglutition. the act or the action of swallowing.

degradation. **1:** deterioration, degeneration. **2:** the reduction of a chemical to less complex forms. *Cf.* decomposition.

degree of accumulation. the ratio of the concentration of a substance within the target system to that component of the environment from which accumulation is considered to take place (e.g., [plant tissue]/[soil]). The degree of accumulation of a toxicant is often system specific (e.g., tissue specific, organ specific, species specific).

deguelia root. derris root.

dehalogenation. the removal of halogen atoms from an organic compound. This is a critical step in the metabolism of the many xenobiotic compounds that contain covalently bound halogen substituents. Dehalogenation may be enzymatically controlled. **reductive dehalogenation**. that resulting from the substitution of a hydrogen atom for a halogen atom or the loss of a halogen atom from each of two adjacent carbon atoms, with the formation of a carbon-carbon double bond. **oxidative dehalogenation**. that in which oxygen is substituted for a halogen atom.

DEHP. See bis(2-ethylhexyl) phthalate.

dehydrated borax. See borax (anhydrous borax).

dehydration. **1:** the loss of water, as from a substance, cells, or tissues. **2:** the state of having a lower than normal water content. **3:** a procedure used to reduce the water content of tissues.

dehydroacetic acid (DHA). a carcinogenic, water-insoluble, lactone, $C_8H_8O_4$, used as a fungicide and bactericide.

dehydrobilirubin. biliverdin.

dehydrocollinusin. justicidin B.

dehydrocorticosterone (21-hydroxypregn-4-ene-3,11,20-trione; Δ^4-pregnen-21-ol-3,11,20-trione; 17-(1-keto-2-hydroxyethyl)-Δ^4-androsten-3,11-dione; Kendall's compound A, 17-desoxycortisone). a steroid hormone, $C_{21}H_{28}O_4$, secreted by the adrenal cortex and produced synthetically for use as an antiallergic agent. *Cf.* cortisone.

dehydrogenase. an enzyme that catalyzes the passage of hydrogen from the substrate to a hydrogen receptor. The various dehydrogenases are named according to their specific activity or the substrate acted upon (e.g., alcohol dehydrogenase). See also oxidase.

dehydrogenation. any chemical process in which hydrogen is removed from a compound; it is a type of oxidation. See dehydrogenase, oxidation.

11-dehydro-17-hydroxycorticosterone. cortisone.

dehydroretinol. vitamin A.

delacrimation. an excessive flow of tears.

Delaney amendment. See Delaney clause.

Delaney clause. a clause of the Food Additive Amendment of the U.S. Food, Drugs and Cosmetic Act that stipulates that no substance known to induce cancer in any animal may be a component of any food.

delayed contact sensitivity. allergic contact dermatitis (See dermatitis).

delayed hypersensitivity. See hypersensitivity.

delayed mortality. See mortality.

delayed response. delayed effect. (See effect).

delayed toxicity. delayed effect (See effect).

delead (de-lead). to mobilize and remove lead from the body or from a tissue by administering a chelate. Bone is deleaded in lead poisoning by administering edetate disodium calcium. See also chelate, chelating agent, chelation therapy.

deleterious. **1a:** harmful, injurious, having an adverse effect. See also effect (adverse effect). **1b:** pertaining to any harmful, injurious, or damaging agent. **2:** pertaining to a trait or a mutation that impairs survival or reduces fitness.

deletion. a chromosomal aberration in which a segment of a chromosome is lost.

Delfen®. a product that contains nonoxynol-9.

delimitation. to establish or set the boundary or limits of; preventing the spread of a disease or intoxicating process (as in the case of snakebite) within an organism or a community.

deliquesce. **1a:** to dissolve gradually and liquefy by absorbing moisture from the air. **1b:** to become soft or to liquefy with time as in the case of some mushrooms. **2:** to disappear as if by melting. **3:** to divide repeatedly resulting in finely divided parts, as in the venation of certain leaves.

deliquescence. the process of deliquescing or the condition of readily absorbing moisture from air, thereby liquefying or becoming moist.

deliquescent. pertaining to or exhibiting deliquescence; tending to liquefy or to form an aqueous solution by absorbing moisture from air.

deliriant, delirifacient. **1:** causing delirium. **2:** an agent that produces delirium (e.g. atropine, hyoscine).

delirious. incoherent, deranged, delusional, in a state of delirium.

delirium (acute delusional state). a temporary state confusion, disorientation, and often frenzied excitement with delusions, illusions, and hallucinations. Such a condition may result from pyrexia, intoxication, shock, alcohol withdrawal, and metabolic disorders. Symptoms may include confusion, hallucinations, incoherency, and uncontrolled shaking. **alcoholic delirium**. delirium tremens. **toxic delirium** (toxic psychosis). that caused by the action of a poison.

delirium tremens. delirium tremens. See also psychosis (alcoholic psychosis).

delirium tremens (DT; alcoholic delirium). acute delirium due to alcoholic withdrawal and characterized by anxiety, restlessness, confusion, hallucinations, tremor, sweating, atonic dyspepsia, and precordial distress. See also ethanol, delirium, psychosis (alcoholic psychosis).

delist. the use of a formal petition to have a facility's toxic designation rescinded by the U.S. Environmental Protection Agency.

delitescence. **1:** the latent period prior to the development of symptoms following exposure to a toxicant or poisonous principle. **2:** the incubation period of a pathogenic organism. **3a:** an unusually complete and rapid resolution of an inflammation. **3b:** a sudden subsidence of symptoms or disappearance [illegible] cutaneous lesion.

Delnav®. dioxathion.

delphinine. a toxic polycyclic diterpene alkaloid produced by plants of the genera *Aconitum* and *Delphinium* (larkspur). It is nearly identical in structure and activity to aconitine.

delphinium. a common name of most perennial forms of the plant genus *Delphinium*, *q.v.* See larkspur poisoning.

Delphinium (larkspur; delphinium; staggerweed). a genus of hardy toxic annual and perennial erect herbs (Family Ranunculaceae) commonly known as larkspur or delphinium. They are native to Asia Minor, Europe, and the United States. The genus can be easily identified by the distinctive flowers which have a spur that projects to the rear. The crude extract of many species contains a number of complex neurotoxic diterpenoid alkaloids that are similar to those found in *Aconitum*. All parts of the plant are toxic, but the various species and hybrids are not uniformly toxic. Young leaves and seeds are most deadly. Accurate species identification is not always possible, especially because poisoning of livestock usually occurs before flowering. Within species, toxicity varies with season, the part eaten, the species of animal affected, individual susceptibility, and other factors. Larkspur is a great killer of livestock. New growth of certain species is lethal to cattle when it makes up less than 1% of the forage. ***D. barbeyi*** (staggerweed). The seeds are highly toxic. Most livestock losses to this plant occur among cattle throughout the western United States. Horses will ingest larkspur, but rarely if ever consume a lethal dose; sheep are highly resistant and losses to larkspur poisoning are rare. See also delphinine, *aconitum*, larkspur poisoning.

delphinium poisoning. larkspur poisoning.

delphinoidine. a diterpenoid alkaloid from the seeds of *Delphinium staphisagria*.

delphisine. an alkaloid isomer of delphinine isolated from seeds of *Delphinium staphisagria*.

delta (δ, Δ). fourth letter of the Greek alphabet. It has a number of applications in mathematics and science. **1a:** Δ signifies any of a number of things that resemble it (e.g., a river delta). **1b:** Δ indicates change. **2:** δ is used somewhat arbitrarily to denote or refer to one or more related chemical substances. **3:** denoting the fourth position from a particular atom or group in an organic molecule as in δ-diketones, δ-lactones, and δ-hydroxy acids.

deltamethrin (3-(2,2-dibromoethenyl)-2,2-dimethylcyclopropanecarboxylic acid cyano(3-phenoxyphenyl)methyl ester; (*S*)-α-cyano-3-phenoxybenzyl-(1R)-*cis*-3-(2,2-dibromovinyl)-2,2-dimethylcyclopropane carboxylate; decamethrin; esbecythrin). an extremely toxic, synthetic, broad-spectrum contact and stomach pyrethroid insecticide, and agricultural pesticide. Acute oral toxicity to vertebrates appears to be generally low. It is a neurotoxicant, blocking nerve transmission by increasing sodium permeability of the neuron. It may also act at the GABA receptor/ionophore complex. Effects on humans are those of acute gastrointestinal irritation, with numbness of the lips and tongue, hyperexcitability, syncope, incoordination, muscular paralysis, convulsions, and death from respiratory collapse.

deltamycin A_4. carbomycin A.

Delvinal sodium. vinbarbital sodium.

Demania toxica. (reef crab). a tropical, poisonous marine crustacean (Order Decapoda) that occurs in the Indo-Pacific region. The poisonous principle appears to be structurally identical to tetrodotoxin. Symptoms of intoxication in humans include nausea, vomition, numbness, tingling sensations beginning about the lips and tongue, muscular paralysis, and collapse. Severe intoxications may terminate fatally.

Demansia (brown snakes; whip snakes). a genus of venomous snakes (Family Elapidae) that are indigenous to Australia and New Guinea. Included are *D. acutirostris* of southern Australia; *D. guttata* of northeastern Australia; *D. modesta* of western Australia; *D. psammophis* (grey snake; Percy Island snake; salt-bush snake; whip snake; wyree), which is widely distributed in Australia. *D typus* (green snake), *D. olivacea*, and *D. textilis*. The latter two species are highly dangerous to humans: ***D. olivacea*** (black whip snake; spotted-headed snake). a diurnal snake of open sandy areas in northern Australia, southeastern New Guinea, and Melville Island. At first glance, this snake resembles the inoffensive racers and whip snakes of North America and Eurasia. It is distinguished from them, however, by the presence of only two scales between the nares, the short snout, and the dark round eye. The average length of adults is 1.3-1.6 m; occasional individuals may exceed 2 m. The ground color is rich brown above, changing to greenish-blue beneath. Each body scale is edged in black. It moves rapidly but is usually inoffensive but defends itself fiercely when agitated and may bite several times in rapid succession. Large individuals are considered dangerous. ***D. textilis*** (Australian brown snake; collared brownsnake; common brownsnake; spotted brownsnake; malle snake; dugite; gwardae). a slender, small-headed, snake that is indigenous to eastern New Guinea and the drier areas of Australia. It is often found in wheat fields, rice fields, and certain irrigated areas. This brown snake has a fierce temper, is alert, agile, fast-moving, and often rears up to strike; it will strike repeatedly if agitated. The head is narrow, deep, and only slightly distinct from neck. It adapts to the background color of its environment, the ground color varying from dark brown, reddish brown, or light brown to dull green. Juveniles may have a series of distinct narrow crossbands and a dark collar. Most adults have a nearly uniform color above. The average length of adults is about 1.7 m; occasional individuals may reach a length of about 2.2 m. The venom is extremely lethal. Systemic effects (cardiac or respiratory distress) are marked by slow onset. Mortality is moderate, but death may occur suddenly. Antivenin is available. Although the case mortality rate is moderate, it has caused more deaths in Australia than any other snake.

demecolcine (6,7-dihydro-1,2,3,10-tetramethoxy-7-(methylamino)benzo[a]heptalen-9(5*H*)-one; *N*-Deacetyl-*N*-methylcolchicine; *N*-desacetyl-*N*-methylcolchicine; *N*-methyl-*N*-desacetylcolchicine; colchamine). an alkaloid from *Colchicum autumnale* (Family Liliaceae) used as an anti-neoplastic agent and to control gout. It is structurally similar to colchicine except that a methyl group substitutes for an acetyl group. It has an action upon mitosis similar to that of colchicine but is probably less toxic.

dementia (amentia). a general decline or loss of intellectual ability; general mental deterioration of varied organic or psychological origin. It is marked by disorientation, impaired judgment and memory, loss of intellectual capacity, and an unstable emotional state. Dementia is usually progressive. **dialysis dementia** (dialysis encephalopathy). a neurologic condition that occurs in patients that have been on dialysis for a number of years. Symptoms include speech difficulties, impaired memory, changes in the EEG, aphasia, ataxia, dementia, convulsions, and eventual death. The condition is associated with elevated levels of aluminum in the brain and is probably caused by the absorption of aluminum from the dialyzing solution together with a reduced ability to excrete this element. **toxic dementia**. that due to poisoning, as in severe forms of drug addiction, or patients that have been maintained on a heavy regimen of narcotics for some time.

demethylation. chemical or metabolic removal of a methyl group from a compound. Metabolic demethylation is catalyzed by isozymes of the cytochrome P-450-dependent monooxygenase system and usually proceeds more rapidly than the dealkylation of higher homologous groups. *N*-, *O*-, and *S*-demethylations are common reactions by xenobiotics. *Cf.* dealkylation.

4-*O*-demethyldaunorubicin. carubicin.

5-*O*-demethyl-25-de(1-methylpropyl)-25-(1-methylethyl)avermectin A_{1a} (4:1). abamectin.

demethyldopan. uracil mustard.

5-*O*-demethyllavermectin A_{1a}. abamectin.

demeton (demeton-*O* + demeton-*S*; phosphorothioic acid *O,O*-diethyl *O*-[2-(ethylthio)ethyl] ester mixture with *O,O*-diethyl *S*-[2-(ethylthio)ethyl]phosphorothioate; diethoxy thiophosphoric acid ester of 2-ethylmercaptoethanol; *O,O*-diethyl 2-ethylmercaptoethyl thiophosphate; *O,O*-diethyl *O*(+*S*)-2-(ethylthio)ethyl phosphorothioate mixture; mercaptophos; Systox). a pale yellow to light brown liquid systemic insecticide, $C_8H_{19}O_3PS_2 \cdot C_8H_{19}O_3PS_2$, with a faint odor of sulfur. It is a mixture, of demeton-*O* and demeton-*S*. Demeton is slightly soluble in water, but soluble in most organic solvents. It is a powerful anticholinesterase that is extremely toxic to a wide variety of animals and a deadly poison to humans by all routes of administration; it readily penetrates the skin. It is a cumulative poison; only acute intoxication is known. It is easily taken up by plants, rendering the leaves and fluids toxic to insects and probably other herbivores. Signs and symptoms in humans include blurred vision, giddiness, headache, nausea, diarrhea, weakness and discomfort in the chest. It is teratogenic and a probable human mutagen. Very toxic fumes, PO_x and SO_x, are released on decomposition by heating. This highly dangerous pesticide, which is readily absorbed through the skin, is used in some household products to eradicate pests.

demeton-*O*. See demeton.

demeton-*S* (diethyl-*S*-(2-ethioethyl)thiophosphate; *O,O*-diethyl-*S*-(2-ethioethyl)phosphorothioate; *O,O*-diethyl-*S*-ethyl-2-ethylmercaptophosphorothiolate; *O,O*-diethyl-*S*-2-(ethylthio)ethyl phosphorothiolate; 2-(ethylthio)-ethanethiol *S*-ester + *O,O*-diethyl phosphorothioate). an insecticide, $C_8H_{19}O_3PS_2$, it is toxic to humans and a variety of animals by ingestion, intraperitoneal, and subcutaneous routes. Very toxic fumes, PO_x and SO_x, are released on decomposition by heating. See demeton.

demeton-*O* + demeton-*S*. demeton.

demeton-methyl. demeton-*S*-methyl.

demeton-*O*-methyl (*O,O*-dimethyl-*O*-ethylmercaptoethyl thiophosphate; *O,O*-dimethyl-2-ethylmercaptoethyl thiophosphate, thiono isomer; *O,O*-dimethyl-*O*-2-(ethylthio)ethyl phosphorothioate). an insecticide, $C_6H_{15}O_3PS_2$, it is toxic to humans, many higher animals, and invertebrates probably by all routes of exposure. It is also teratogenic and has reproductive effects. Very toxic fumes, PO_x and SO_x, are released on decomposition by heating. See demeton.

demeton-*O*-methyl sulfoxide (demeton-*S*-methyl sulfoxide; demeton-methyl sulphoxide; *O,O*-dimethyl-*S*-(2-ethionylethyl)phosphorothioate; dimethyl-*S*-(2-ethionylethyl) thiophosphate; *O,O*-dimethyl-*S*-[2-(ethylsulfinyl)ethyl] phosphorothioate; *O,O*-dimethyl-*S*-(2-ethylsulfinyl)-ethyl thiophosphate; *O,O*-dimethyl-*S*-ethylsulfinylethyl phosphorothiolate; *S*-[2-(ethylsulfinyl)ethyl]-*O,O*-dimethylphosphorothioate). an insecticide. It is toxic to humans, many other higher animals and invertebrates by most or all routes of exposure. It is a probable human mutagen. Very toxic fumes (phosphates and sulfates) are released on decomposition by heating. See demeton.

demeton-*S*-methyl (methyl demeton; phosphorothioic acid *O*-[2-(ethylthio)ethyl] *O,O*-dimethylester mixed with *S*-[2-(ethylthio)ethyl]; *O,O*-dimethylphosphorothioate; demetonmethyl; methylmercaptophos; methyl systox). an oily liquid, $C_6H_{15}O_3PS_2$, that is slightly water soluble. It is used as an insecticide to control aphids. It is toxic to humans and many higher animals and invertebrates probably by all routes of exposure and is probably mutagenic. Very toxic fumes, PO_x and SO_x, are released on decomposition by heating. See demeton.

demeton-*S*-methyl-sulfone (demeton-*S*-methyl-sulphone; *O,O*-dimethyl-*S*-(2-ethsulfonylethyl)-phosphorothioate; dimethyl-*S*-(2-ethsulfonyl-ethyl)thiophosphate; *O,O*-dimethyl-*S*-ethyl-2-sulfonylethyl phosphorothiolate; *O,O*-dimethyl-*S*-ethylsulfonylethyl phosphorothiolate; dioxy-demeton-*S*-methyl). Very toxic fumes, PO_x and SO_x, are released on decomposition by heating. See demeton.

demeton-*S*-methyl sulfoxide. demeton-*O*-methyl sulfoxide.

demeton-*S*-methyl-sulphone. demeton-*S*-methyl-sulfone.

demeton-methyl sulphoxide. demeton-*O*-methyl sulfoxide.

demise. death or destruction.

demographic variable. See variable (population variable).

demon night adder. *Causus rhombeatus*.

demorphinization. the gradually increasing deprivation of a morphine addict until the physical (physiological) addiction is cured.

demulcent. an agent that soothes the skin or a body part to which it is applied; used chiefly with respect to agents that act to soothe mucous membranes.

demustardization. 1: the removal of mustard gas from a person or occasionally an animal. **2:** treatment of a person exposed to mustard gas.

demyelination. destruction of myelin; the loss of myelin from the sheaths of nerves or nerve tracts within the central or peripheral nervous systems (CNS, PNS), with consequent impairment of nerve conduction. Demyelination in the PNS is potentially reversible, but is irreversible in the CNS. Certain toxicants such as hexachlorophene and triethyltin can produce segmental demyelination (i.e., a loss of myelin from the internode) as a consequence of direct injury to the sheath; other toxicants (e.g., lead) can cause segmental demyelination by injuring the myelinating cell (the Schwann cell in the PNS; the oligodenrocyte in the CNS). Demyelination may also result from certain diseases (e.g., multiple sclerosis) or mechanical injury.

DEN. *N*-nitrosodiethylamine.

DENA. *N*-nitrosodiethylamine.

denarcotize. to deprive an addict of a narcotic drug until the addiction is abolished. It is a part of treatment for addiction.

denaturant. a denaturing agent.

denaturation. **1:** a change in the normal structural properties or nature of a substance. **2:** a change in the structural properties of a protein or nucleic acid on exposure to a chemical agent (denaturant) or to extremes of temperature or pH. These changes may or may not be reversible and are often accompanied by a loss of biological activity. **3:** the addition of a toxic denaturant (e.g., benzene, methyl alcohol) to ethanol, thereby rendering it unfit to drink.

denature. **1:** to change the normal structural properties or nature of a substance (e.g., a protein or nucleic acid) by heating or by the addition of acid or alkali. **2:** to add a small amounts of toxic denaturants such as benzene or methyl alcohol to ethyl alcohol, thereby rendering it unfit to drink.

Dendroaspis (mambas; tree cobras; namahamba). a small genus of largely arboreal snakes (Family Elapidae) that occur throughout much of equatorial Africa. They are large, fast moving, have highly toxic venom, and are among the most dangerous of snakes. The head is elongate, narrow, and nearly indistinct from the neck; there is a distinct canthus; the eyes are moderate in size with round pupils. The body is slender, tapering, and slightly compressed with a long, tapering tail. The neck may be flattened when the snake is aroused, but there is no actual hood. There is a pair of large tubular fangs that lack external grooves; these are followed only by maxillary teeth. Effects of envenomation may include pain, localized edema, numbness of the affected area, weakness, drowsiness, slurred speech, difficulty in swallowing, ptosis, blurred vision, headache, dyspnea, excessive salivation, nausea, vomiting, abdominal pain, pain in regional lymph nodes, weak pulse, hypotension, shock, paresis or paralysis, and muscle fasciculations. See also elapid venom poisoning. ***D. angusticeps*** (eastern green mamba; pale-mouthed mamba; fune; songo). a long, very slender, bright green, arboreal snake that occurs chiefly in forested and brushy areas in sub-Saharan east Africa from Kenya southward to northeastern Cape Province and occurs also on the island of Zanzibar. Adults average about 2.5 m in length, occasional individuals reaching a length of nearly 3 m. The only other mamba within its range is *D. polylepis* (black mamba), from which it is easily distinguished by its smaller size, bright green body color, and the white to bluish-white color of the oral mucosa. It is much more arboreal than the black mamba and is rarely seen on the ground. The eastern green mamba is less dangerous than the black mamba. It usually avoids humans and is less aggressive and much more retiring than the black mamba; its venom is much less toxic. It is easily confused with *Dispholidus* (boomslang) as well as *Philothamnus spp*. (the inoffensive green bush-snakes). Its eyes are smaller than individuals of either of the above taxa and there is not loreal scale; the ventral plates of bush-snakes also keels and notches. ***D. jamesoni*** (Jameson's Mamba; East African Jameson's mamba; Traill's green mamba; bukizi; serpent du bananier; toka). a mamba of the tropical rain forest region from western Kenyua and Tanganyika to Guinea and Angola, where it occurs both on the ground and in bushes. The ground color is usually green with scales edgedin black, the color becoming darker posteriorly, the tail of some specimens being entirely black. It resembles the nonvenomous bush-snakes (*Philothamnus*) and the boomslang (*Dispholidus*), but it has smaller eyes and lacks the loreal scale. It is also darker than *Philothamnus* and lacks lateral keels and notches on the ventrals. The black edging on the scales and the presence of fewer dorsals and ventrals distinguishes it from *D. polylepis*. Adults attain an average length of about 2.1 m, occasional individuals reaching a length of about 1.6 m. There are 15-19 rows of dorsals at midbody, about the same number on the neck, and fewer (11-13) posteriorly; ventrals range from 210-239. ***D. polylepis*** (black mamba; black-mouthed mamba; brown mamba; common mamba; Schwarzen Mamba; ginyambila; songo; songwe). one of the most dangerous snakes known, it is a large snake of sub-Saharan Africa, growing to a length of about 4.3 m. It occurs chiefly at altitudes below 1200 m in open bush country from Ethiopia and Somalia, southward to eastern South Africa. It is the most terrestrial of the mambas, although it sometimes hunts in trees. As an adult it is a slender, dark gray or greenish brown, fast-moving snake (clocked at speeds up to 7 miles per hour) with large scales, and very long fangs; the inside of the mouth is bluish-gray or blackish; it has a well-developed canthus rostralis. It is not easily confused with any nonvenomous snake. Hatchlings are grayish-green or olive colored. The adult is larger, darker, and has more ventral scutes (242-282) and more dorsal scale rows (21-25) than other mamba. It is extremely aggressive and rears up, often striking on a human's head or trunk. The venom is extremely toxic and the bite appears to be 100% fatal in the absence of antivenom. Systemic effects are marked by rapid onset and may include dizziness, dyspnea, and a wildly erratic heart beat, cardiorespiratory collapse, and death. The case mortality rate is high, approaching 100% unless antivenom is administered. ***D. viridis*** (western green mamba; green tree mamba; Hallowell's mamba; West African mamba; Grüne Mamba; mamba vert; serpent du bananier). an arboreal species of the tropical rain forests of the western bulge of Africa, from the Senegal to the Niger and on the island of Sao Tome. The ground color is green or yellowish, although the dorsal scales and head scales are edged in black. Dorsal scales are quite large and narrow. The loreal scale is absent and it has fewer rows (13 at midbody) of dorsal scales than any snake with which it might be confused. No other mamba occurs within its range. Adults average about 2.2 m in length.

Dendrobates. a genus of small, toothless, usually brightly colored poisonous frogs (Family Dendrobatidae, formerly Ranidae) of the American tropics. Numerous toxins have been isolated from the skin of various species. The toxic skin secretions, which contain chiefly spiropiperidine alkaloids, are used by certain Central American ndians as arrow poisons. See pumiliotoxins, spiropiperidine alkaloids. ***D. histrionicus***. a Columbian frog, the skin of which secretes a number of neurotoxic alkaloids and is used to make arrow poisons. The toxins secreted by the skin include histrionicotoxin, spiropiperidin, and derivatives of the latter. See also pitohui.

Dendrobatidae (poison dart frogs; poison arrow frogs). a family of frogs, from members of which (mainly *Phyllobates spp*. and *Dendrobates spp*.) large numbers of toxic skin secretions (mostly alkaloids) are known. See also batrachotoxin, batrachotoxin A, homobatrachotoxin, dihydrobatrachotoxin, and 3-*O*-methylbatrachotoxin), pumiliotoxins, spiropiperidine alkaloids. See also pitohui.

Dendrocnide moroides (Australian nettle). plants (Family Urticaceae) that induce a nonimmunologic contact urticaria by means of hollow stinging hairs that penetrate the skin and inject irritants that produces an almost immediate burning sensation and pruritis. The toxin of *D. moroides*, unlike that of the North American nettles (*Urtica*, *q.v.*), is a thermostable, nondialyzable carbohydrate.

dendrolasin. a component of ant venom, it also occurs some plants.

Denisonia (Australian copperheads; ornamental snakes). a genus of venomous snakes (Family Elapidae). All but one species is restricted to Australia. Those species not treated specifically below include *D. carpentariae* (carpentaria whip snake), *D. coronata* (coronated snake; werr), *D. coronoides* (white-lipped snake), *D. fasciata* (Rosen's snake), *D. flagellum*. little whip snake, *D. gouldii* (black headed snake; black-naped snake; little whip snake), *D. maculata* (ornamental snake), *D. nigrostriata* (black-striped snake), *D. pallidiceps* (black-bellied snake), *D. psammophis* (spotted-headed snake), *D. punctata* (little spotted snake; spotted snake), *D. signata* (black-bellied snake; march snake). ***D. par*** (Solomons copperhead; Guppy's snake). a moderately slender, slightly compressed snake that rarely reaches a length of 1 m. The head is broad and moderately depressed; the pupils are round or subelliptical. The ground color varies considerably among individuals, and may be uniformly gray, sand-colored, reddish, pink, or almost black. Faint irregular cross-bars are sometimes present. The scales are lustrous with edging that is darker than the body of the scale. It occurs in rain forest, grasslands, and coconut plantations on many of the Solomon Islands. It is considered dangerous. ***D. superba*** (Australian copperhead; diamond snake; superb snake). a fairly stout, short-tailed snake that averages about 1.7 m. in length. It is indigenous to Tasmania and the southern coastal region of Australia, where it occurs chiefly in coastal mountain swamps. The head is depressed and rather broad, but only slightly distinct from neck; the pupil is round. The ground color varies greatly among individuals, ranging from reddish- or copper-brown, to black or nearly black or reddish with an indistinct dark stripe along the dorsum in some parts of its range. In the Bowral region, most specimens are black dorsally with yellowish or whitish sides. The labials are distinctive in Alpine specimens, the dorsal and caudal portions are dark, while the ventral and anterior parts are edged with an oblique cream-colored streak. Reports of bites are infrequent, but serious.

denitrification. the anaerobic biological reduction of nitrate nitrogen to gaseous nitrogen, nitrous oxide, or nitrites and ammonia; also the conversion of ammonia to molecular nitrogen. This is usually accomplished by facultative aerobic soil bacteria (denitrifying bacteria) under conditions of low environmental oxygen tension.

denitrify. to effect denitrification.

denitrifying. pertaining to conditions favorable to, or organisms that can accomplish denitrification.

density. 1: the weight per unit volume of a substance. **2:** the number of individuals in a population or set per unit of space occupied.

deodorant. a substance used to mask or absorb foul or unpleasant odors. **personal deodorant**. a commercial product used to control underarm odor. Nonantiperspirant deodorants may contain the bacteriocide, triclosan, *q.v.* This substance might cause liver damage if absorbed through the skin. **underarm deodorant**. any of a variety of commercial products used to limit underarm bacterial activity and thus control unpleasant underarm odors. The potential problems are essentially similar to those due to use of an antiperspirant, *q.v.*

deoxydihydrothebacodine. racemethorphan.

25-deoxyecdysterone. ponasterone A.

11-deoxycorticosteroid. desoxycorticosterone.

***d*-deoxyephedrine**. methamphetamine.

(12S,13S)-9-deoxy-12,13-epoxy-12,13-dihydro-9-oxoleucomycin V 3-acetate 4^{B}-(3-methylbutanoate). carbomycin A.

deoxygenated. pertaining to a substance or an environment that is depleted of free oxygen; anaerobic.

deoxygenation. the removal of free oxygen from a substance or an environment.

(3β,5β,16β)-3-[(6-deoxy-4-*O*-β-D-glucopyranosyl-3-*O*-methyl-β-D-galactopyranosyl)oxy]-14,16-dihydroxy-card-20(22)-enolide. digitalin.

deoxyhemoglobin. reduced hemoglobin (See hemoglobin).

3-[(6-deoxy-α-L-mannopyranosyl)oxy]-14-hydroxybufa-4,20,22-trienolide. proscillaridin.

3-[(6-deoxy-α-L-mannopyranosyl)oxy]-14,16-dihydroxy-19-oxocard-20(22)-enolide. adonitoxin.

3-[(6-deoxy-α-L-mannopyranosyl)oxy]-1,5,11α,14,19-pentahydroxycard-20(22)-enolide. ouabain.

3β-[(6-deoxy-α-L-mannopyranosyl)oxy]-5,14-dihydroxy-19-oxo-5β-card-20(22)-enolide. convallatoxin.

3-[(6-deoxy-α-L-1-mannopyranosyl)oxy]-14-hydroxybufa-4,20,22-trienolide. proscillaridin.

2,6-deoxy-3-methoxyaldohexose. cymarose.

***O*-2-deoxy-2-(methylamino)-α-L-glucopyranosyl-(1→2)-*O*-5-deoxy-3-*C*-formyl-α-L-lyxofuranosyl-(1→4)-*N,N'*-bis(aminoiminomethyl)-D-streptamine**. streptomycin.

(3β,5β)-3-[(6-deoxy-3-*O*-methyl-α-L-glucopyranosyl)oxy]-14-hydroxycard-20(22)-enolide. neriifolin.

deoxynivalenol (Rd-toxin). a trichothecene first isolated from *Fusarium*.

deoxynivalenolmonoacetate (Rc-toxin). a trichothecene first isolated from *Fusarium*.

9-deoxy-9-oxoleucomycin V 3-acetate 4^B^-(3-methylbutanoate). carbomycin B.

deoxyribonuclease. DNAase.

deoxyribonucleic acid. DNA.

depauperate. **1:** inferior in natural size or development. **2a:** impoverished in numbers of individuals (said of a population). **2b:** impoverished with respect to numbers of kinds (taxa) of organisms present (said of a flora, a fauna, or a biotic community).

dependence (dependency). **1a:** the state or condition of being contingent upon, determined by, or conditioned by something (e.g., another person, a narcotic drug). **1b:** a condition in which a person is psychologically or physically addicted to a substance such as a narcotic or alcohol. If the dosage of the substance is reduced or use of the drug is discontinued, symptoms of withdrawal develop. In some cases, dosage must be increased to avoid withdrawal. See also addiction, tolerance. **2:** the status of a variable whereby it is influenced or controlled by another variable. **drug dependence** (substance use disorders; drug addiction; drug abuse; drug habituation). often used synonymously with "drug addiction," although this is not recommended. A single definition for drug dependence is neither desirable nor possible. different drugs have different effects, including the basis of, type, and hazard of the dependence produced. In some cases, drug dependence is due largely to habituation; in other cases, dependence is largely psychic in nature; in yet other cases, there is a strong physiological component to dependence. Thus, alcohol dependence and heroin dependence are physiologically and behaviorally different entities. *Cf.* addiction (drug addiction).

dependency. dependence.

depilatory (hair-removal product). the most toxic constituent of a hair-removal product is ammonium thioglycolate, which can cause severe skin rashes, swelling, redness, and the breaking of small blood vessels under the skin.

depletion curve. in hydraulics, a graphical representation of water depletion from storage-stream channels, surface soil, and ground water. A depletion curve can be drawn for base flow, direct runoff, or total flow.

deposition. **1:** the amount of matter (e.g., a pollutant) that settles out of the air onto a surface (usually taken as the amount per unit area during a specified period of time). **2:** the amount of an aerosol or particulate matter that remains in the lung following exhalation.

depressant. **1a:** pertaining to or able to diminish the activity of a vital function. **1b:** an agent that reduces neuromuscular or other functional activities, tone, or vitality (e.g., an anesthetic or sedative). **2a:** a substance that alters mood (depression, defs. 4 and 5). **2b:** a depressor, *q.v.* See also coactone. **cardiac depressant**. an agent that reduces the rate or force of cardiac contractions. **central nervous system depressant**. CNS depressant. **cerebral depressant**. any drug that lowers brain activity. The affected individual becomes dull and less active. **CNS depressant**. any drug that depresses CNS function. Examples are barbiturates, benzodiazepines, ethanol, glutethimide, meprobamate, methaqualone, ethchlorvynol, chloral hydrate, paraldehyde. All such drugs potentiate the action of others and tolerance to one indicates tolerance to others. Symptoms of intoxication or overdosage of CNS depressants generally, may include drowsiness, confusion, nystagmus, slurred speech, depressed respiration, hypotension, shock, depressed tendon reflexes, ataxia, delirium or coma; in severe cases, miosis. Drug-specific symptoms may also be seen. Withdrawal is often marked by insomnia, nervousness, anxiety, sweating, fever, shock, clonic blink reflex, agitation, delirium, hallucinations, and disorientation. Convulsions and cardiovascular collapse may ensue. **motor depressant**. a depressant that lessens skeletal muscle contractions. **respiratory depressant**. a drug that lowers the rate and depth of breathing. **secretory depressant**. an agent that reduces glandular secretions.

depression. **1:** a hollow or depressed area such as a fossa. **2:** dorsoventral flattening as seen, for example, in many bottom-dwelling fish (e.g., skates and rays). **3:** a decrease in activity of a vital function (e.g., depressed respiration). **4:** a mood marked by feelings of sadness, gloom, discouragement, or despair. It is considered normal unless persistent, exaggerated, or unrelated to environmental stimuli or conditions that might be expected to cause depression. **5:** an altered mood characterized by dejection, a lack of motivation, and disinterest in normally pleasurable activities (e.g., food, sex, entertainment), or in friends and loved ones. Symptoms vary greatly and may be mild or extreme, with feelings of gloom, melancholy, hopelessness, and despair that are inconsistent with the subject's situation. Severe physiological effects may also appear (e.g., weight loss, a reduction in venous pulse, psychomotor retardation). Any of a number of factors may contribute to the origin of depression in an individual (e.g., genetic, toxicologic, nutritional, neurologic, endocrine, neoplastic, or infectious components). See depressant.

depressive. causing, or having the capacity to cause depression; dampening.

depressor. **1:** that which depresses; a depressant, *q.v.* **2:** an agent that slows or prevents a process or chemical reaction. **3:** a waste product of one species that inhibits or poisons both the producer and one or more other exposed species. **4:** a waste product of one species that inhibits or poisons recipients of one or more other species, without being useful or advantageous to the producer. See also allelochemical, autotoxin. See also co-actone.

depressurization. in the context of toxicology, this term denotes a condition whereby the air pressure inside a dwelling or other building is lower than the air pressure outside. This can occur when appliances that consume or exhaust air (furnaces, fireplaces) are not supplied with enough makeup air from the outside. Under such conditions, radon-containing "soil gas" may be drawn into the building.

depuration. **1:** cleansing, purification, removal of impurities. **1a:** a self-cleansing or purging activity of shellfish when immersed in clean water (e.g., prior to harvesting or in estuaries when the water freshens) that removes pathogenic organisms and pollutants from the tissues.

derivation. extraction; origin, source.

derivative. **1:** comprised of or characterized by components or attributes arising from other material or another entity. **2a:** a chemical substance that is so similar to, or so closely related to, another substance that it can be theoretically or actually derived from that substance. **2b:** a substance that can be produced in one or more steps from another substance by modification or partial substitution. **3:** an agent that removes blood from the body part in which a disease is centered.

derived unit. units of the International System of Units, *q.v.*, which derive from the base units.

Dermacentor. a genus of wood ticks (Family Ixodidae) that transmit Rocky Mountain spotted fever; some species cause tick paralysis. They are parasitic on numerous mammals, including humans. See tick, Ixodidae, Ixodoidea, *Amblyomma*. ***D. andersoni*** (Rocky Mountain wood tick). a tick that introduces a venomous saliva on biting which can produce a flaccid ascending motor paralysis that progresses from the lower extremities. If the tick is promptly removed, rapid and complete recovery is usual. See also Ixodoidea.

dermal. pertaining to vertebrate skin, especially the inner living layer (dermis).

dermal absorption. See absorption.

dermal irritation test. See test.

dermal sensitization test. See test.

dermal test. See test.

dermatitis. inflammation of the skin. **actinic dermatitis**. any reaction of the skin to any photochemically active type of radiation (e.g., sunlight, roentgen rays, ultraviolet light). **allergic contact dermatitis** (delayed contact sensitivity). a delayed type of induced sensitivity (allergy) arising from cutaneous reexposure to a specific sensitizing substance (allergen). The condition is marked by varying degrees of erythema, edema, and vesiculation. **allergic dermatitis**. any dermatitis due to allergy (e.g., allergic contact dermatitis, contact-type dermatitis, cosmetic dermatitis, systemic lupus erythematosus). *Cf.* erythema toxicum. **atopic dermatitis**. a dermatitis of unknown origin in a human individual with inherently irritable skin. It may occur at any time after the age of 2 months. **berlock (berloque) dermatitis**. a phototoxic dermatitis, mainly of the face, neck, and chest, which is induced by sunlight following exposure to perfume or other toiletries. **brown-tail moth dermatitis** (brown-tail rash). a caterpillar dermatitis resulting from contact with setae or cocoon of the brown-tail moth, *Euproctis chrysorrhoea*. **bubble gum dermatitis**. an allergic contact dermatitis about the lips of some children who chew bubble gum. It is caused by plastics in the gum. **caterpillar dermatitis** (caterpillar rash). a transient, sometimes painful allergic contact dermatitis from contact with caterpillar setae that penetrate the skin. **cercarial dermatitis**. schistosome dermatitis. **chemical dermatitis**. an allergic contact dermatitis or a primary irritation dermatitis due to contact with chemicals. It is usually characterized by erythema, edema, and vesiculation of the exposed area. **contact dermatitis**. an acute dermatitis induced by contact with any of a variety of substances. The proximate cause may be primary chemical irritation, mechanical injury, toxicity, phototoxicity, allergy, photoallergy, or a combination. **contact-type dermatitis**. any dermatitis that resembles contact dermatitis but is caused by an ingested or injected allergen, usually a medication. It is characterized by widespread cutaneous distribution. **cosmetic dermatitis**. any contact dermatitis caused by the application of a cosmetic preparation. It may be due to primary irritation or to allergic sensitization. **exfoliative dermatitis** (dermatitis exfoliativa; pityriasis rubra; Wilson's disease). an erythematous dermatitis with generalized exfoliation and scaling of the skin. It is sometimes due to a severe drug reaction or in other cases associated with lymphomas or any of various benign dermatoses. **grass dermatitis**. meadow-grass dermatitis. **industrial dermatitis**. occupational dermatitis. **insect dermatitis**. dermatitis caused by contact with insects, usually moths or caterpillars. **marine dermatitis**. schistosome dermatitis. **meadow dermatitis**. meadow grass dermatitis. **meadow-grass dermatitis** (meadow dermatitis, grass dermatitis, dermatitis striata pratensis bullosa; phytophlyctodermatitis). a phototoxic dermatitis caused by contact with *Agrimonia eupatoria* and subsequent exposure to sunlight. There is an eruption of vesicles and bullae that occur in strips and unusual groupings reflecting the pattern of plant contact. It is often contracted during sunbathing. **nickel dermatitis**. an allergic dermatitis due to nickel or metals such as stainless steel that contain nickel. **occupational dermatitis** (industrial dermatitis). any contact dermatitis contracted as a result of one's employment. **paraffin dermatitis**. a chronic dermatitis caused by exposure to paraffin. With high exposure rates there is risk of scrotal cancer. **photoallergic dermatitis**. See photoallergic photodermatitis. **photocontact dermatitis**. an allergic contact dermatitis caused by the action of sunlight on skin sensitized by contact with an allergen such as sandalwood oil, hexachlorophene, or a halogenated salicylanide. *Cf.* phototoxic contact dermatitis. **phototoxic contact dermatitis**. a rapidly developing sunburnlike dermatitis of skin areas exposed to sun following cutaneous contact with a photosensitizing substance. *Cf.* photocontact dermatitis. **phototoxic dermatitis**. See phototoxic photodermatitis, phototoxic contact dermatitis. **plant dermatitis**. see d. venenata. **poison ivy dermatitis**. *Rhus* dermatitis. **poison oak dermatitis**. *Rhus* dermatitis. **poison sumac dermatitis**. *Rhus* dermatitis. **primary contact dermatitis.** (primary irritant dermatitis; primary chemical dermatitis). a chemical dermatitis resembling that from contact with an irritant or corrosive agent. It results from contact with substances that are toxic to cutaneous tissues. The degree of injury varies with the potency of the irritant, its concentration, and duration of contact. The lesions are usually erythematous and papular but, in more severe cases, may be purulent and necrotic. The eye is especially subject to injury, and keratoconjunctivitis and temporary blindness may result if the eye is involved. Primary contact dermatitis is frequently caused are spurges of the family Euphorbaceae, especially *Ranunculus spp*. (buttercups), *Daphne mezereum* (daphne) and *Capsicum frutescens* (wild pepper). **primary dermatitis**. a dermatitis that is induced directly by some agent; not an allergic response. **primary irritant dermatitis**. primary contact dermatitis. **radiation dermatitis**. See radiodermatitis. **rhus dermatitis**. *Rhus* dermatitis. ***Rhus* dermatitis**. an allergic contact dermatitis due to cutaneous exposure to urushiol from certain plants of the genus *Rhus* (synonym, *Toxicodendron*), e.g., poison ivy, poison oak, poison sumac. **roentgen ray dermatitis**. See radiodermatitis. **schistosome dermatitis** (cercarial dermatitis; marine dermatitis; seabathers eruption; swimmer's itch; swimmer's dermatitis). an itching dermatitis caused by infestation of the skin by blood flukes of the genus *Schistosoma* that are not normally parasites of *Homo sapiens*. The condition occurs in persons who wade or swim in saltwater. **swimmer's dermatitis**. schistosome dermatitis. ***Toxicodendron* dermatitis**. *Rhus* dermatitis. **traumatic dermatitis**. any dermatitis caused by an irritant or by a physical agent. **trefoil dermatitis**. trifoliosis. **X-ray dermatitis**. radiodermatitis. See also phytodermatitis, phytophotodermatitis, photodermatitis, photoallergic photodermatitis, radiodermatitis,

dermatitis escharotica. (chemical burn). a severe primary irritant dermatitis.

dermatitis exfoliativa. exfoliative dermatitis. See Wilson's disease (def. 2).

dermatitis herpetiformis (Darier's sign). a chronic inflammatory condition marked by intensely itching and burning erythematous, bullous, papular, or pustular lesions with a tendency to clump.

dermatitis medicamentosa. **1:** any dermatitis induced by a medication. **2:** drug eruption (See eruption).

dermatitis simplex. erythema.

dermatitis skiagraphica. radiodermatitis.

dermatitis striata pratensis bullosa. meadow-grass dermatitis.

dermatitis venenata. any severe contact dermatitis marked by edema, erythema, and vesiculation.

dermato-, derma-, dermat-, dermo-. combining forms denoting the skin or indicating a relationship to the skin.

dermatopolyneuritis. See acrodynia.

dermatosis. any skin disease. See also dermatitis.

dermatotoxic. toxic to the skin.

dermis (cutis vera; corium). the skin; true skin; the thick, deep, living layer of the skin of vertebrate animals. It lies beneath the epidermis and sometimes also beneath an outer cuticle or horny layer. See also exposure (dermal exposure), (toxicity) dermal toxicity, dermatitis. It consists of a dense connective tissue matrix, fat, nerves, sensory organs, glands (usually), a rich circulation; and in most mammals, hair follicles; and in birds, feather papillae. The dermis is more or less impervious to many xenobiotics; others are readily absorbed. The rate of penetration varies greatly among taxa. The dermis is bounded superficially by the epidermis (stratum papillare) and beneath by a thick, dense layer of connective tissue. See also skin.

Dermochelidae. a family of large, marine turtles that contains a single living species, *Dermochelys coriacea*, *q.v.* The skin is leathery and the carapace is comprised of numerous small bony platelets embedded in the skin.

Dermochelys coriacea (leatherback turtle). an endangered, circumtropical, carnivorous pelagic turtle (Family Dermochelidae) covered with smooth leathery skin; sometimes found in bays and estuaries and sometimes in temperate waters. It is the largest living turtle and the only living representative of its family. While normally safe for eating, it is not generally consumed by humans. In some locations, these turtles are extremely toxic. The liver is especially poisonous. Nearly half of those persons who are poisoned die. See turtle poisoning.

dermolysin. a substance in the peripheral circulation that is able to dissolve the skin.

dermotoxin (dermonecrotic toxin). **1:** any toxin capable of producing a localized dermal necrosis. Coagulase-positive staphylococci, especially *Staphylococcus aureus*, secrete several. They produce a focal necrosis when injected into the skin. **2a:** any of various toxins that can cause necrotic lesions of the skin. **2b:** infrequently taken to be any substance secreted by a living organism that is able to produce pathologic changes in skin.

dermotropic. affecting especially the skin.

Derrin. rotenone.

derris. **1:** common name of any plant of the genus *Derris*. **2:** a preparation of derris roots and stems that contains insecticidal rotenoids.

Derris. a genus of more than 80 species of Old World tropical woody vines and shrubs with long climbing branches (Family Fabaceae, formerly Leguminosae). The roots and stems of some species contain rotenone and other insecticidal rotenoids. Resin from the roots of a few species are used by natives as arrow poisons or to temporarily paralyze fish in streams to ease the harvest. They are commercial sources of rotenone and other insecticidal rotenoids. ***D. elliptica***. cultivated in the Dutch East Indies and British Malaya. The roots are dried and powdered or the resin is extracted, and marketed under various brand names as contact insecticides. ***D. malaccensis***. cultivated in the Dutch East Indies and British Malaya for use as noted for *D. elliptica*.

derris root (tuba root; deguelia root). the dried resinous root of plants of the genus *Derris*, especially of *D. elliptica*. and *D. mallaccensis*. It contains rotenone and other insecticidal rotenoids. The dried powdered root is very toxic *per os*.

DES. diethylstilbestrol

desacetylbufotalin. bufogenin B.

desacetyldigilanide C. deslanoside.

desacetyllanatosides. desacetyldigilanides (See lanatosides).

***N*-desacetyl-*N*-methylcolchicine**. demecolcine.

desbromoleptophos (DBL). a delayed neurotoxicant that inhibits both acetylcholinesterase and brain neurotoxic esterase. Partial or complete blindness is the first symptom to present. Subsequently, the affected animal loses the use of its tongue and cannot swallow. The affliction is thus referred to as "paralyzed tongue," *q.v.*

Descurainia pinnata (tansy mustard). a mustard (Family Brassicaceae) that is widely distributed throughout the southern United States. Range cattle have been poisoned from long-continued consumption of large quantities of this plant. *q.v.* Partial or complete blindness is the first symptom to present. Subsequently, the affected animal loses the use of its tongue and cannot swallow. The affliction is thus referred to as "paralyzed tongue."

desensitization. hyposensitization.

desert adder. *Vipera lebetina*.

desert baileya. *Baileya multiradiata*.

desert banded snake. snakes of the genus, *Rhynchoelaps*. Two species are recognized, *R. aproximans* and *R. bertholdi*, both inhabit the dry regions of Australia. Neither reaches a length of more than 0.4 m; although venomous, they are not considered dangerous to humans.

desert blacksnake. *Walterinnesia aegyptia*.

desert cobra. *Walterinnesia aegyptia*.

desert death adder. *Acanthophis pyrrhus*.

desert horned viper. See *C. cerastes*.

desert snake. *Brachyaspis curta*. See also *Elapognathus minor*.

desferrioxamine B. deferoxamine.

desglucotransvaaline. proscillaridin.

desiccant. a chemical agent that absorbs moisture (e.g., activated alumina, calcium chloride, silica gel, zinc chloride); some are able to fatally desiccate plants or invertebrate animals.

designer bugs. a popular term for microbes developed through biotechnology that can degrade specific toxic or otherwise hazardous chemicals at their source in the environment (e.g., in toxic waste dumps, in groundwater).

desipramine (10,11-dihydro-*N*-methyl-5*H*-dibenz-[b,f]azepine-5-propanamine; 10,-11-dihydro-5-[3-(methylamino)propyl]-5*H*-dibenz[b,f]-azepine; 5-(γ-methylaminopropyl)imino-dibenzyl; *N*-(3-methylaminopropyl)imino-bibenzyl; desmethylimipramine; norimipramine). a white to off-white crystalline powder. A metabolite of imipramine, it is a tricyclic oral antidepressant, usually given as the hydrochloride. Desipramine is very toxic orally and subcutaneously to laboratory mice and rats, but very or extremely toxic *i.p.*

deslanoside (3β,5β,12β)-3-[(*O*-β-D-glucopyranosyl-(1→4)-*O*-2,6-dideoxy-β-D-ribohexopyranosyl-(1→4)-*O*-2,6-dideoxy-β-D-ribohexopyranosyl-(1→4)-*O* -2,6-dideoxy-β-D-ribohexopyranosyl)oxy]-12,14-dihydroxycard-20(22)-enolide; deacetyllanatoside C; desacetyldigilanide C; purpurea glycoside C). a deacetyllanatoside (See lanatosides) isolated from the leaves of *Digitalis lanata*.

desmethoxyviridiol. a mycotoxin from *Nodulisporium hinnuleum*. It is extremely toxic orally to the cockerel of the domestic fowl.

desmethyldopan. uracil mustard.

desmethylemetine. cephaeline.

desmethylimipramine. desipramine.

desmetryne. an atrazine herbicide (a methylthiotriazine) that is similar in biological activity to atrazine and simazine, but is absorbed through foliage as well as the roots. It is used both as a preemergence and postemergence herbicide.

desmoneme. a nematocyst that discharges a thread or coiled tube, the distal end of which coils about the prey.

***S*-(5′-desoxyadenosin-5′-yl)-L-methionine**. *S*-adenosylmethionine.

2-desoxy-D-altromethylose. digitoxose.

11-desoxycorticosteroid. desoxycorticosterone.

desoxycorticosterone (4-pregnen-21-ol-3,-20-dione; 11-deoxycorticosteroid; 11-desoxycorticosteroid). a phlogistic adrenal corticoid (*q.v.*), $C_{21}H_{30}O_3$, that causes retention of salt and water by the kidney of vertebrates. It is used in medicine usually as the acetate.

desoxycorticosterone acetate (DCA; DCa; DOCA). See desoxycorticosterone.

***d*-desoxyephedrine**. methamphetamine.

desoxypatulinic acid. an analog of patulin.

desoxyribose nucleic acid. DNA (deoxyribonucleic acid).

desquamation. the shedding of skin in sheets or scales.

destroying angel. the common name of many amanitas, especially *Amanita verna* and *A. virosa*.

destructive. damaging, harmful, injurious, ruinous, lethal.

destructive lesion. See lesion.

destruxin. any of a class of insecticidal cyclodepsipeptides marked by the presence of a number of N-CH_3 groups. See also *Metarrhizium anisophae*.

desulfurization. the removal of sulfur from fossil fuels or their combustion products (e.g., from stack gases, coal, oil, or other hydrocarbons) to reduce pollution.

detection. **1a:** the observation, recognition, discovery, or discernment of the presence of a defined object, material, activity, or stimulus. **1b:** the discovery of a certain condition (e.g., sign or symptom). In toxicology, detection often refers to the observation or measurement and identification of a toxicant, qualitatively or quantitatively.

detection limit. limit of detection.

detergent. a product used in conjunction with water to clean dirt from surfaces. All are surface active agents that reduce the surface tension of water and emulsify fats and oils in contact with the water. Modern detergents are synthetic and are classified as anionic, cationic or nonionic, depending upon the ionization of the surface active moiety of the compound and hence their mode of action. Within each of these classes, there is a wide range of chemical structures and toxicities. **anionic detergent**. the surface active moiety of the detergent bears a negative charge at neutral pH. **cationic detergent**. the surface active moiety bears a positive charge at neutral Ph. This type of detergent occurs chiefly in dishwashing detergents and fabric softeners. **dishwashing detergent**. any detergent used to wash dishes and other kitchen utensils in a dishwasher. Most contain chlorine in a dry form that becomes active when wet, with the release of chlorine fumes which may seep out of the dishwasher into the kitchen. Accidental ingestion and, to a lesser degree, inhalation can be harmful. With reasonable care, the use of dishwashing detergent is considered safe for most people. Many users, however, are affected by headache, fatigue, burning eyes, and difficult breathing when exposed to the small amounts of chlorine released during normal dishwashing. Dishwashing detergent may contain other toxic materials. See dicyclohexylamine and ethylene oxide. **nonionic detergent**. the surface active moiety is nonionic at neutral pH. This type of detergent acts by hydrogen-bonding.

deteriorate. **1:** to impair or make to grow worse or to become inferior. **2:** to decline (gradually) in quality, condition or functional capacity; to worsen; decline; regress; become impaired; degenerate, decay.

deterioration. **1a:** a gradual decline in quality, condition or functional capacity as in a diseased condition; the process or action of worsening; degeneration; decay. **1b:** the state of having deteriorated.

determination. resolution, decision; establishing, or reaching a conclusion regarding the exact nature or identity of an event, process, substance, or other entity.

detersive. cleansing; detergent; purging.

detoxicant. an agent that decreases or eliminates the toxicity of a substance.

detoxicate. **1:** to detoxify a substance by detoxicaton. **2:** to detoxify a xenobiotic by metabolic conversion to a less toxic form.

detoxication. reduction or elimination of a toxic xenobiotic substance from a living system by metabolic conversion to less toxic and usually more water-soluble substance(s) and thus more readily excreted than the original substance. *Cf.* detoxication, toxification. See also environmental detoxification.

detoxication reaction. See reaction (detoxication reaction)

detoxification (detoxication). **1:** recovery from the toxic effects of a drug. **2:** the process of eliminating effects of toxicants from an intoxicated individual, especially from an alcoholic or other drug-addicted person. **3:** reduction or removal of the toxic properties of a poison or of the virulence of a pathogenic organism. **environmental (ecological) detoxification**. the reduction or elimination of a toxic substance from a biotic community or ecosystem by any of numerous processes (e.g., photolysis, biological mineralization, oxygenation, leaching, acidification, immobilization, export). See also biological mineralization. **metabolic detoxification**. See detoxication.

detoxification reaction. See reaction (detoxification reaction).

detoxify (detoxicate). **1:** to decrease or remove the toxic capacity of a substance or to diminish the virulence of a pathogenic organism through treatment. **2:** to treat the toxic state produced by an overdose or any drug or medication; used most often in reference to alcoholism and drugs of abuse. *Cf.* detoxification.

deuterotoxin. an obsolete term for one of three groups of toxins in bacterial cultures that are classified according to their affinity for antitoxin: prototoxin > deuterotoxin > tritotoxin.

devasation. the destruction of blood vessels.

development. **1:** ontogeny; regulated growth and differentiation of an individual, including cellular differentiation, histogenesis and/or anagenesis. **2:** the orderly sequence of progressive changes that result in generally increasing complexity of a living system.

deviation. departure of an observation or measurement from a targeted or expected value.

devil ray (devilfish). any of the large (up to 6m across), harmless, warm water cartilaginous fishes (Family Mobulidae, Suborder Batoidea). The body is greatly depressed, as are the huge pectoral fins which are fused to the head. Two ear-like fins project forward from the head. These rays are harmless and feed on zooplankton. See also stingray, Batoidea.

devil scorpions. See *Vejovis*.

devil's apple. *Datura stramonium*.

devil's bit. *Veratrum viride*, *Chamaelirium luteum*.

devil's tobacco. *Veratrum viride*.

devil's trumpet. *Datura stramonium*.

devil's tulip. See *Bryonia*.

devil's walking stick. *Aralia spinosa*.

devilfish. 1: devil ray. **2:** octopus.

Devonshire colic. lead colic.

dew poisoning. trifoliosis.

Dexedrine (disomer). See amphetamine (def. 2).

Dexedrine sulfate®. dextroamphetamine sulfate.

dextroamphetamine. a CNS stimulant. See also dextroamphetamine sulfate, drug abuse.

dextroamphetamine sulfate ((*S*)-α-methylbenzeneethanamine sulfate; Dexedrine sulfate; d-amphetamine sulfate; speed). an amphetamine, it is a CNS stimulant and anorexic. It is used in the treatment of exceptionally hyperactive children, but continued use inhibits growth. It also causes insomnia, tremors, exaggerated reflexes, and bad breath. Prolonged use can cause psychological dependence. It is a controlled substance in the United States; the street name is speed.

dextropropoxyphene. propoxyphene.

DF (methylphosphonyldifluoride). an organophosphate chemical warfare agent.

DFDD (1,1′-(2,2-dichloroethylidene)bis[4-fluorobenzene]; 1,1-dichloro-2,2-bis(*p*-fluorophenyl)ethane; difluorodiphenyldichloroethane). a crystalline contact insecticide. *Cf.* DFDT.

DFDT (1,1′-(2,2,2-trichloroethylidene)bis[4-fluorobenzene; 1,1,1-trichloro-2,2-bis(*p*-fluorophenyl)ethane; difluorodiphenyltrichloroethane; fluorogesarol). a crystalline, nearly water-insoluble contact insecticide. Effects are similar to those of DDT. *Cf.* DFDD.

DFOM. deferrioxamine mesylate. See deferoxamine.

DFP. diisopropyl fluorophosphate.

DGE. diglycidyl ether.

DHA. dehydroacetic acid.

1,25-DHCC. calcitriol.

DHT. dihydrotestosterone. See testosterone.

dhurrin ((*S*)-α-(β-D-glucopyranosyloxy)-4-hydroxybenzeneacetonitrile; *p*-hydroxymandelonitrile-β-D-glucoside; β-D-glucopyranosyloxy-L-*p*-hydroxymandelonitrile). a cyanogenetic nitrile glycoside, $C_{14}H_{17}NO_7$, isolated from various sorghums (e.g., *Sorghum almum*, *S. halepense*, and *S. vulgare* (var. *sudanense*)). It accumulates in the plant especially under conditions of drought, high soil nitrogen, and low phosphorous. The highest concentrations occur in young growth. When the plant is ingested (e.g., by livestock), dhurrin hydrolyzes to yield parahydroxy benzaldehyde, glucose, and hydrocyanic acid.

diaboline (1-acetyl-19,20-didehydro-17,18-epoxycuran-17-ol; *N*-acetyl-Wieland-Gumlich aldehyde). a toxic aromatic aldehyde isolated from *Strychnos diaboli* (Family Loganaceae).

diabrosis. perforation of a vessel or organ by a corrosive agent; a perforating ulceration.

diabrotic. **1:** corrosive, caustic, ulcerative. **2:** a corrosive agent or escharotic.

diacepin. diazepam.

diacetic acid. acetoacetic acid.

diacetin (1,2,3-propanetriol diacetate; glycerol diacetate; glyceryl diacetate). a water-soluble liquid plasticizer and softening agent; also soluble in ethanol, ether, and benzene; commercial preparations are probably a mixture of 1,2- and 1,3-diacetate. It is a cellular narcotic.

diacetoxyscirpenol. the first phytotoxic trichothecene to be isolated (from *Fusarium*).

diacetyl peroxide. acetyl peroxide.

diacetylcalonectrin. a trichothecene mycotoxin first isolated from *Fusarium*.

diacetylcholine dichloride. succinylcholine chloride.

α,β-diacetylethane. acetonylacetone.

1,2-diacetylethane. acetonylacetone.

diacetylmorphine. heroin.

diacetylnivalenol. a tricothecene mycotoxin.

diacritic, diacritical. **1a:** serving to separate or distinguish. distinctive; distinguishing; diagnostic; discernable. **1b:** capable of discerning or distinguishing.

diadem snake. *Aspidomorphus diadema*.

Diadema (long-spined sea urchins). a genus of large, long-spined tropical sea urchins. The spines are poisonous and can be up to 41 cm. long. They occur in tide pools, coral reefs, rocks, rubble, or reef flats. ***D. setosum*** (long-spined urchin; black sea urchin). an urchin of the Indo-Pacific region from East Africa to Polynesia, China, and Japan. It is the most venomous and dangerous of sea urchins. The spines easily penetrate soft tissues, producing erythema, edema, and an intense burning pain. They can also penetrate wetsuits, causing severe injury. The venomous spines or pedicellariae are brittle and bear three-pronged, tooth-like biting tips that easily break off, remaining in the skin. They must be removed (surgically, if necessary). ***D. antillarum*** (long-spined urchin). a fairly large (ca. 10 cm wide; 4.4 cm high), dark purple or black venomous urchin with spines as long as 41 cm. They are indigenous to the American tropics from Florida, Bermuda, the Bahamas, and West Indies to Mexico and Surinam. The spines can penetrate wetsuits, causing severe injury.

diadermic. percutaneous.

diagnose. **1:** detect, determine, ascertain. **2:** to determine the causes or nature of a disorder or disease state; to make a diagnosis.

diagnosis. **1:** determination, identification, detection; analysis, examination, scrutiny. **2:** analysis of the specific causes or nature of any state, process, situation, or problem. **3:** the art of identifying or determining the nature of diseased states. **4:** the determination of a physiological or diseased state or process and, as possible, the proximate causes of such condition. **5:** a brief technical description of a species or other taxonomic unit, giving the complete set of distinctive features. **clinical diagnosis**. that determined by history, laboratory evaluation, signs, and symptoms. **differential diagnosis**. the clinical determination of which of two or more disease states is responsible for, or conforms best with a patient's symptoms. **medical diagnosis**. that entire process and end result of assessing a patient's illness. **pathological diagnosis**. a diagnosis based on pathological data indicating the presence of lesions. **physical diagnosis**. that based on external examination only.

diagnostic. **1:** relating to, adapted to, or used to aid in diagnosis. **2:** distinctive, distinguishing; serving to identify, determine, or distinguish. **2a:** distinctive of a disease. **2b:** distinctive of a species or other taxonomic unit. **3:** the art of diagnosis.

diagnostic procedure. a process conducted to determine the cause or nature of condition, disorder, or disease.

dialifor (phosphorodithioic acid *S*-[2-chloro-1-(1,3-dihydro-1,3-dioxo-2*H*-isoindol-2-yl)ethyl] *O,O*-diethyl ester; phosphorodithioic acid *O,O*-diethyl ester *S*-ester with *N*-(2-chloro-1-mercaptoethyl)phthalimide; *O,O*-diethyl *S*-(2-chloro-1-phthalimidoethyl)phosphorodithioate; dialifos). an extremely toxic, water-insoluble, white crystalline compound. It is an insecticide and acaricide used to control pests of citrus fruits, grapes, and pecans.

dialifos. dialifor.

dialkylated organotin compound. See alkylated organotin compound, trialkylated tin compound.

dialkylcadmium compound. a compound with the general formula $(R)_2Cd$, where R is an alkyl group. All are distillable, but decompose above about 150°C, releasing toxic cadmium fumes.

dialkylhydrazine. any of several derivatives of hydrazine, RNHNHR′, in which an alkyl group (R or R′) is bonded to each of the nitrogens of the parent molecule. These compounds are carcinogenic to laboratory animals, chiefly inducing gastrointestinal, hepatic, and pulmonary tumors. Examples of dialkylhydrazines are 1,2-diethylhydrazine, 1,2-dimethylhydrazine, and procarbazine.

dialkyltin. any dialkylated organotin compound.

diallate (bis(1-methylethyl)carbamothioic acid S(2,3-dichloro-2-propenyl) ester; diisopropylcarbamic acid S-2,3-dichloroallyl ester; S-2,3-dichloroallyldiisopropylthiocarbamate). a selective thiocarbamate herbicide used mainly to control *Avena fatua* (wild oats) in cereal crops. It inhibits shoot growth largely by interfering with cell elongation. It is also a mitotic poison at higher concenrations. It is incorporated into the top 10 cm of soil just prior to sowing the crop. Diallate is very toxic to laboratory rats.

diallyl sulfide. allyl sulfide.

6,6′-diallyl-α,α′-bis(diethylamino)-4,4′-bi-*o*-cresyl. bialamicol.

***N,N*-diallyl-2-chloroacetamide**. allidochlor.

dialysis. a process in which blood is diffused across a semipermeable membrane to separate low molecular weight substances from high molecular weight substances (e.g., proteins and polysaccharides). Dialysis is used to maintain the electrolyte, fluid, and acid-base balance of circulating blood in the event of renal impairment or failure. Dialysis is also used in certain cases of poisoning (e.g., meprobamate or phencyclidine poisoning).

dialysis dementia (dialysis encephalopathy). a neurologic condition that occurs in patients that have been on dialysis for a number of years. Symptoms include speech difficulties, impaired memory, changes in the EEG, aphasia, ataxia, dementia, convulsions, and eventual death. The condition is associated with elevated levels of aluminum in the brain and is probably caused by the absorption of aluminum from the dialyzing solution, together with a reduced ability to excrete this element.

dialysis encephalopathy. dialysis dementia.

4,4′-diamidino-α,ω-diphenoxypentane. pentamidine.

diamine. any compound that contains two amino groups.

diamine oxidase. any of a class of flavoprotein enzymes that catalyze the aerobic oxidation of amines to the corresponding aldehyde and ammonia. Diamine oxidases are most effective with aliphatic diamines having a chain length of four (putrescine) or five (cadaverine) carbon atoms. They are not effective with diamines that have carbon chains of more than nine atoms; monoamine oxidases will catalyze the deamination of longer chains. histaminase is a diamine oxidase. *Cf.* monoamine oxidases.

diamino toluene. toluene-2,4-diamine.

2,4-diaminoanisol. 2,4-diaminoanisole.

2,4-diaminoanisol sulfate. 2,4-diaminoanisole sulfate.

2,4-diaminoanisole (C.I. 76050; C.I. oxidation base 12; 2,4-DAA; 2,4-diamineanisole; 2,4-diaminoanisol). a skin irritant and moderately toxic compound, $C_7H_{10}N_2O$, by ingestion and intraperitoneal routes. It is a confirmed carcinogen.

2,4-diaminoanisole sulfate (2,4-diaminoanisole sulphate; C.I. 76051; C.I. oxidation base 12A;

2,4-DAA sulfate; 2,4-diamineanisole sulfate; 2,4-diaminoanisol sulfate). a poisonous compound, $C_7H_{10}N_2O$, by intraperitoneal routes. It is a confirmed carcinogen.

2,4-diaminoanisole sulphate. 2,4-diaminoanisole sulfate.

4,4-diaminobiphenyl. benzidine.

L-α,γ-diaminobutyric acid. the toxic principle of *Lathyrus sylvestris* and *L. latifolius* that causes hyperexcitability, convulsions and death without skeletal lesions. See *Lathyrus*, lathyrism.

di-(4-amino-3-chlor ophenyl)methane. 4,-4′-methylenebis[2-chloroaniline].

diaminodiatolyl. *o*-tolidine.

4,4′-diamino-3,3′-dichlorodiphenylmethane. 4,4′-methylenebis[2-chloroaniline].

3,3′-diamino-4,4′-dihydroxyarsenobenzene dihydrochloride. arsphenamine.

2,5-diamino-4,6-diketopyrimidine-3-β-D-glucoside. vicine.

***p*-diaminodiphenyl**. benzidine.

1,2-diaminoethane. ethylenediamine.

1,3-diamino-4-methylbenzene. toluene-2,4-diamine.

2,4-diamino-1-methylbenzene. toluene-2,4-diamine.

***N*-[*p*-[[2,4-diamino-6-pteridinylmethyl]-methylamino]benzoyl]glutamic acid**. methotrexate.

diamino toluene. toluene-2,4-diamine.

2,4-diaminotoluene. toluene-2,4-diamine.

***S*-[(4,6-diamino-1,3,5-triazin-2-yl)-methyl]-phosphorodithioic acid *O,o*-dimethyl ester**. menazon.

***S*-[(4,6-diamino-s-triazin-2-yl)meth yl] *O,o*-dimethyl phosphorodithioate**. menazon.

diammonium hexafluorosilicate. ammonium hexafluorosilicate.

diamond snake. *Denisonia superba*.

dianat. dicamba.

dianisidine (3,3′-dimethoxy-[1,1′-biphenyl]-4,4′-diamine; 3,3′-dimethoxybenzidine; 3,3′-dimethoxy-4,4′-diaminobiphenyl; *o*-dianisidine). a flammable, moderately toxic, mutagenic, biphenylamine, $C_{14}H_{16}N_2O_2$. It is a confirmed human carcinogen. It is used in the manufacture of azo dyes. See also benzidine, dichlorobenzidine.

***o*-dianisidine**. See dianisidine.

2-di-*p*-anisyl-1,1,1-trichloroethane. methoxychlor.

diaphoresis. artificially induced profuse sweating.

diaphoretic. See sudorific.

diaphragm. **1:** any relatively thin sheet of tissue that separates one structure from another. The term is most commonly applied to the large sheet of muscle between the abdominal cavity and the chest cavity in mammals. **2:** a contraceptive device used with a foam, cream, jelly, suppository, or foaming tablet that contains a spermatocide. Such devices sometimes cause an allergic reaction with local irritation, edema, or blistering due to the spermatocide or to the latex from which they are made. Their use is sometimes associated with birth defects and spontaneous abortion. See spermatocide.

diarrhea. **1a:** abnormally frequent passage of soft or watery fecal discharges. In some cases, the stool may contain pus, blood, or mucus. It is a common symptom of gastrointestinal disturbances including that caused by any of numerous poisons. Diarrhea can be acute or chronic; it can limit the uptake of water and salts by the body with resultant dehydration and/or electrolyte imbalances. **1b:** the condition characterized by 1a. **cachectic diarrhea**. that associated with cachexia; **choleraic diarrhea**. acute diarrhea that resembles that of cholera. The stools are serous and attended by circulatory collapse. **critical diarrhea**. that which occurs at the crisis of a disease or which causes a crisis. **dysenteric diarrhea**. that in which the stools are bloody with mucous. **irritative diarrhea.** that due to irritation of the intestine by toxicants, unsuitable food, purgatives, various drugs, or other ingested substances. **osmotic diarrhea**. that due to the presence of osmotically active nonabsorbable solutes in the intestine. **serous diarrhea** (watery diarrhea). that which produces soft feces with large amounts of serous fluid. **traveler's diarrhea**. that which occurs among travelers, especially those who visit tropical or subtropical areas that have poor sanitation. It probably most often results from infection with enteropathogenic *Escherichia coli*. **watery diarrhea**. serous diarrhea.

diarrheogenic. able to cause diarrhea.

Diastatin®. nystatin.

diastole. the dilatation or period of dilatation due to relaxation of cardiac muscle with filling of the atria (atrial diastole) or ventricles (ventricular diastole). In humans, ventricular diastole coincides with the interval between the second and first heart sounds. See also blood pressure.

diastolic. of or pertaining to diastole.

diastolic pressure (minimum blood pressure; minimum pressure). the minimum blood pressure during a cardiac cycle, i.e., between contraction of the vertebrate ventricle(s). See also blood pressure.

diataxia. ataxia that affects both sides of the body.

diathermy. heating through; the heating of body tissues due to their resistance to the passage of high-frequency electromagnetic radiation, ultrasonic waves, or electric current. **medical diathermy** (thermopenetration). any therapeutic application of diathermy in which tissues are heated but not damaged. **surgical diathermy** (electrocoagulation). any surgical application of diathermy in which tissue is selectively destroyed.

diatomaceous earth (diatomite). a chalk-like material (fossilized diatoms) used to filter out solid waste in waste-water treatment plants. It is also used as an active ingredient in some powdered pesticides.

diatomic. descriptive of a gas whose molecules (e.g., oxygen, O_2, hydrogen, H_2) contain two atoms. *Cf.* monoatomic.

diatomite. diatomaceous earth.

diazepam (7-chloro-1,3-dihydro-1-methyl-5-phenyl-2*H*-1,4-benzodiazepin-2-one; 7-chloro-1-methyl-5-phenyl-3*H*-1,4-benzodiazepin-2(1*H*)-one; methyl diazepinone; diacepin; Valium). a toxic benzodiazepine with numerous applications as an antianxiety, anticonvulsant drug. It may be given in a tablet or by intravenous or intramuscular injection. Diazepam may be habit-forming and is a controlled substance listed in the U.S. Code of Federal Regulations. It is effective in the control of status epilepticus and in relieving convulsions induced by sarin, soman, strychnine, and tabun. Diazepam is an antidote for picrotoxin poisoning. It is used therapeutically in a variety of anxiety states, can be effective in the treatment of extreme anxiety, and potentiates other antidepressants. Traces are found in the urine. Symptoms of intoxication may include drowsiness, ataxia, muscle weakness, tinnitus, excitability, rage, hallucinations, coma, and death by cardiac arrest. *Cf.* LSD, mescaline, psilocybin.

Diazinon [Chevron]®. *O,O*-diethyl-*O*-(2-isopropyl-4-methyl-6-pyrimidinal)phosphorothioate. a colorless, slightly water-soluble compound, $[C(CH_3)_2CHC_4N_2H(CH_3)O]PS(OC_2H_5)_2$, used as an insecticide and parasiticide on open areas such as sod farms and golf courses was banned by the U.S. Environmental Protection Agency in 1986 because it posed a danger to migratory birds who gather on them in large numbers. The ban did not apply to its use in agriculture, or on lawns of homes and commercial establishments. Young calves, sprayed with Diazinon at a concentration of 0.1%, are poisoned, but seem to tolerate 0.05%. Adult cattle may safely be sprayed repeatedly at weekly intervals. Orally, young calves are poisoned at a concentration of 0.88 mg/kg body weight, but seem to tolerate 0.44 mg/kg. Adult cattle tolerate 8.8 mg/kg body weight orally, but are poisoned by 22 mg/kg. Sheep tolerate 17.6 mg/kg body weight, but are poisoned by 26 mg/kg.

Diazo Fast Red GG. *p*-nitroanaline.

dibasic acid. an acid that contains two ionizable atoms of hydrogen per molecule. See carboxylic acid.

Dibenyline. phenoxybenzamine hydrochloride.

dibenz-dibutyl anthraquinol (1,2,5,6-dibenz-9,10,di-n-butyl anthraquinol). an estrogenic compound; it is carcinogenic.

1,2,5,6-dibenz-9,10,di-*n*-butyl anthraquinol. dibenz-dibutyl anthraquinol.

1,2:5,6-dibenzanthracene (dibenz[a,h]-anthracene; DBA; DB[a,h]A; 1,2,5,6-DBA). a toxic, carcinogenic (possibly mutagenic), orange-brown crystalline aromatic polycyclic hydrocarbon, $C_{22}H_{14}$, present in trace amounts in coal tar. When introduced into the body, it can cause epithelial tumors.

1,2,5,6-dibenzonaphthalene. chrysene.

dibenzo[*b,e*]pyridine. acridine.

dibenzothiazine. phenothiazine.

3-dibenz[b,e]oxepin-11(6*H*)-ylidene-*N,N*-dimethyl-1-propanamine. doxepin.

Dibenzyline. phenoxybenzamine hydrochloride.

Dibenzyran. phenoxybenzamine hydrochloride.

diborane (diborane hexahydride; boroethane). the simplest stable borane at normal ambient temprature and pressure, it is a toxic, highly-reactive, highly flammable, colorless gas, B_2H_6, with a disgusting, sickly-sweet odor. It is soluble in carbon disulfide, but decomposes in water. It is a strong irritant and inhalation may result in pulmonary edema. Diborane has many uses, including that of an insect repellent.

diborane hexahydride. diborane.

dibromochloropropane (1,2-dibromo-3-chloropropane; 3-chloro-1,2-dibromopropane; DBCP). a dense, light yellow, or brown, environmentally persistent liquid with a pungent odor, $ClCH_2CHBrCH_2Br$, used as a soil fumigant and agricultural nematocide. an irritant and CNS depressant, it is moderately toxic by ingestion and possibly other routes of exposure. High rates of sterility or low sperm counts and liver and kidney damage among male workers exposed to dibromochloropropane have been reported. It is possibly carcinogenic and mutagenic. On decomposition by heating, it releases toxic Cl^- fumes.

1,2-dibromo-3-chloropropane. dibromochloropropane.

2,6-dibromo-4-cyanophenol. bromoxynil.

l,l-dibromoethane. a heavy liquid that was formerly used widely as a grain and soil fumigant for insect control and as a component of leaded gasoline for scavenging lead from engine cylinders.

1,2-dibromoethane. ethylene dibromide.

3-(2,2-dibromoethenyl)-2,2-dimethylcyclopropanecarboxylic acid cyano(3-phenoxyphenyl)methyl ester. deltamethrin.

3,5-dibromo-4-hydroxybenzonitrile. bromoxynil.

3,5-dibromo-4-hydroxyphenyl cyanide. bromoxynil.

dibuline sulfate. dibutoline sulfate.

dibulinsulfat. dibutoline sulfate.

dibutoline sulfate (di-*n*-butylcarbamylcholine sulfate; dibuline sulfate; dibulinsulfat; 1-(((dibutylamino)carbonyl)oxy)-*n*-ethyl-*n,n*-dimethylethanaminium sulfate(2:1)). a toxic organic sulfate, $C_{30}H_{66}N_4O_4{\cdot}O_4S$. It is an anticholinergic used in medicine as a mydriatic, a cycloplegic, and a gastrointestinal antispasmodic.

dibutyl phthalate. *n*-butyl phthalate.

di-*n*-butylphthalate (DBP). *n*-butyl phthalate.

1-(((dibutylamino)carbonyl)oxy)-*n*-ethyl-*n,n*-dimethylethanaminiumsulfate(2:1). dibutoline sulfate.

***N,N′*-dibutyl-*N,N′*-dicarboxyethylenediaminemorpholide**. dimorpholamine.

***N,N'*-dibutyl-*N,N'*-dicarboxymorpholideethylenediamine**. dimorpholamine.

DIC. disseminated intravascular coagulation.

dicamba (3,6-dichloro-2-methoxybenzoic acid; 3,6-dichloro-*o*-anisic acid; 2-methoxy-3,6-dichlorobenzoic acid; dianat). a translocated postemergence herbicide, $C_8H_6Cl_2O_3$, used to control dock (genus *Rumex*., subgenus *Lapathum*; Family Polygonaceae) in established grassland and to control bracken, *Pteridium spp.*, Family Polypodiaceae). Dicamba is poisonous to cattle and sheep and is moderately toxic to laboratory rats.

2,2-di(carbamoyloxymethyl)pentane. meprobamate.

***S*-(1,2-dicarbethoxyethyl) *O,O*-dimethyldithiophosphate**. malathion.

[*S*-(1,2-dicarbethoxyethyl)*O,O*-dimethyldithiophosphate]. malathion.

dicarbosulf. thiodicarb.

dicentra. any plant of the genus *Dicentra*.

Dicentra (dicentra, bleeding heart, Dutchman's breeches, squirrel corn, turkey corn). a genus of low, often stemless North American and Asian perennial herbs (Family Papaveraceae). All parts of the plant are toxic. More than 20 isoquinoline alkaloids have been isolated from various species. These include aporphine, protoberberine, and protopine. The latter appears to be present in all species. ***D. eximia*** (turkey corn; staggerweed). potentially toxic to livestock. ***D. cucullaria***. (Dutchman's breeches, breeches flower). toxic to cattle. Symptoms are similar to those caused by ingestion of *Corydalis*. The animals run back and forth, however, with the head held unusually high and the animals fall in a characteristic position of emprosthotonos. ***D. canadensis*** (squirrel corn). a North American woodland plant, it is potentially toxic to livestock.

dicephalous. having two heads.

Dichapetalum toxicarium. an African shrub (Family Dichapetalaceae) that accumulates toxic levels of fluorine stored as fluorooleic acid.

dichlobenil (2,6-dichlorobenzonitrile). a selective pre- and postemergence, amide herbicide applied to the soil that disrupts many aspects of growth and differentiation. It destroys phloem, cambium and parenchyma tissue in alligator weed *(Alternanthera philoxeroides)*. It prevents cell nuclei in oat (*Avena*) root tips, for example, from entering mitosis. Such blocking of cell division results in a blackening and death of the growing points in many plant species; this resembles the effects of boron deficiency. Its chief preemergence effect is the disruption of cell division in growing primary meristems. Dichlobenil is slightly toxic to laboratory mice.

dichloral. a herbicide. Although a chloral derivative, it is usually not toxic if ingested.

dichloramine. See chloramine.

dichloramine T. a dichloramine. See chloramine.

dichloricide. *p*-dichlorobenzene.

dichlorine heptoxide. chlorine heptoxide.

dichlorine monoxide. chlorine monoxide.

1,4-dichlorobenzene. paradichlorobenzene.

1,1-dichloro-2,2-bis(*p*-fluorophenyl)ethane. DFDD.

2,4-dichloro-α-(chloromethylene)benzylalcohol diethyl phosphate. chlorfenvinphos.

α,β-dichloro-β-formylacrylic acid. mucochloric acid.

4,5-dichloro-3,6-fluorandiol. 4′,5′-dichlorofluorescein.

4′,5′-dichloro-3′,6′-di-hydroxyspiro-[illegible]. 4′,5′-dichlorofluorescein.

2,2′-dichloro-*N*-methyldiethylamine. mechlorethamine.

***S*-2,3-dichloroallyldiisopropylthiocarbamate**. diallate.

2-(2,6-dichloroanilino)-1,3-diazacyclopentene-(2). clonidine.

2-(2,6-dichloroanilino)-2-imidazoline. clonidine.

3,6-dichloro-*o*-anisic acid. dicamba.

1,4-dichlorobenzene. *p*-dichlorobenzene.

***p*-dichlorobenzene** (1,4-dichlorobenzene; dichloricide; PDB). a white, crystalline, aryl halide insecticide, $C_6H_4Cl_2$, with a penetrating odor. It is moderately toxic to humans by all routes of exposure. Long-term chronic or intermittant exposure can cause liver and kidney damage. It is moderately toxic to laboratory animals by most portals of entry, but mildly toxic by the subcutaneous route. It is an eye irritant, a probable teratogen, a reproductive poison, a suspected carcinogen, and depressant. Severe (acute) clinical poisonings are unkown. Sources of exposure include mothballs and household deodorizers.

3,3′-dichlorobenzidine (3,3-dichloro-(1,1′-biphenyl)-4,4′-diamine; 3,3′-dichloro-4,4′-biphenyldiamine; DCB). an aromatic amine, $C_{12}H_{10}Cl_2N_2{\cdot}ClH$, used chiefly in the manufacture of pigments for printing inks, cloth, and plastics (and as a curing agent for urethane polymers. It is carcinogenic to laboratory mice and rats and may cause allergic dermatoses. It readily enters the human body percutaneously or via the respiratory tract. It is moderately toxic by ingestion and can also cause anemia, icterus, damage to the CNS, kidneys, and liver, and even death. See dye (direct dye).

2,6-dichlorobenzonitrile. dichlobenil.

3,3-dichloro-(1,1′-biphenyl)-4,4′-diamine.3,3′-dichlorobenzidine.

3,3′-dichloro-4,4′-biphenyldiamine. 3,3′-dichlorobenzidine.

1,1-dichloro-2,2-bis(*p*-chlorophenyl)ethane (1,1′-(2,2-dichloroethylidene)bis[4-chlorobenzene]; tetrachlorodiphenylethane; TDE; dichlorodiphenyldichloroethane; *p,p*′-DDD; *p,p*′-TDE). a nonbiodegradable pesticide and a metabolite of DDT in mammals. It was once used on many agricultural crops, but all applications were halted in the United States in 1972, largely due to its probable role in eggshell thinning among raptorial birds. DDD is slightly to moderately toxic to mammals; carcinogenic to some. Lethargy is a prominent sign of acute toxicity in humans. Chronic exposure may cause liver damage and atrophy of the adrenal cortex with symptoms similar to those of DDT exposure. Do not confuse *p,p*′-DDD with the reagent "DDD, analytical". See also DDT, DDT dehydrochlorinase, eggshell thinning.

***O*-(2,5-dichloro-4-bromophenyl)-*O*-methyl phenylthiophosphonate**. leptophos.

dichloro(2-chlorovinyl)arsine. lewisite.

dichlorodiethyl sulfide. mustard gas (See gas).

2,2-dichloro-1,1-difluoroethyl methyl ether. methoxyflurane.

4′,5′-dichloro-3′,6′-dihydroxyspiro[isobenzofuran-1(3H),9′-[9H]xanthen]-3-one. 4′,5′-dichlorofluorescein.

dichlorodiphenyldichloroethane. 1,1-dichloro-2,2-bis(*p*-chlorophenyl)ethane.

***p,p*′-dichlorodiphenylmethyl carbinol**. chlorfenethol.

dichlorodiphenyltrichloroethane. See DDT.

dichloroethane. ethylene dichloride.

1,2-dichloroethane. ethylene dichloride.

α,β-dichloroethane. ethylene dichloride.

***sym*-dichloroethane**. ethylene dichloride.

3-(2,2-dichloroethenyl)-2,2-dimethylcyclopropanecarboxylic acid cyano(3-phenoxyphenyl)methyl ester. cypermethrin.

dichloroether. *sym*-dichloroethyl ether.

dichloroethyl ether. *sym*-dichloroethyl ether.

2,2′-dichloroethyl ether. *sym*-dichloroethyl ether.

β,β-dichloroethyl ether. *sym*-dichloroethyl ether.

***sym*-dichloroethyl ether** (1,1′-oxybis[2-chloro ethane]; bis(2-chloroethyl) ether; β,β-dichloroethyl ether; β,β′-dichloroethyl ether; bis(β-chloroethyl) ether; 1-chloro-2-(β-chloroethoxy)ethane; chloroethyl ether; DCEE; 2,2′-dichloroethyl ether; dichloroethyl oxide; 1,1′-oxybis(2-chloro)ethane; dichloroether; di(β-chloroethyl)ether; dichloroethyl ether). a colorless, stable liquid used as a fumigant and as a solvent of paints, lacquers, and varnishes. It is an irritant of the eyes, skin, and mucous membranes, is flammable, and a dangerous explosion hazard. It reacts vigorously with oxidants and while insoluble in water, it reacts explosively with water or steam, releasing poisonous and corrosive fumes. Dichloroethyl ether may be carcinogenic and/or mutagenic.

di(chloroethyl)methylamine. mechlorethamine.

dichloroethyl-β-naphthylamine. chlornaphazine.

di(2-chloroethyl)-β-naphthylamine. chlornaphazine

dichloroethyl oxide. *sym*-dichloroethyl ether.

di(2-chloroethyl)sulfide. mustard gas (See gas).

dichloroethyl sulfide. mustard gas (See gas).

3-*p*-[di(2-chloroethyl)amino]-phenyl-1-alanine. melphalan.

***p*-di(2-chloroethyl)amino-L-phenylalanine**. melphalan.

***p*-di-2-(chloroethyl)amino-1-phenylalanine**. melphalan.

***p*-*N*-di(chloroethyl)aminophenylalanine**. melphalan.

γ-[*p*-di(2-chloroethyl)aminophenyl]butyric acid. chlorambucil.

***N*,*N*-di-2-chloroethyl-γ-*p*-aminophenylbutyric acid**. chlorambucil.

5-[di-(β-chloroethyl)amino]uracil. uracil mustard.

dichloroethylene. ethylene dichloride.

di(β-chloroethyl)ether. *sym*-dichloroethyl ether.

1,1′-(2,2-dichloroethylidene)bis[4-chloro-benzene]. 1,1-dichloro-2,2-bis(*p*-chloro-phenyl)-ethane.

1,1′-(2,2-dichloro ethylidene)bis[4-fluorobenzene]. DFDD.

4,5-dichloro-3,6-fluorandiol. 4′,5′-dichlorofluorescein.

4′,5′-dichlorofluorescein (4′,5′-dichloro-3′,6′-dihydroxyspiro[isobenzofuran-1(3H),9′-[9H]-xanthen]-3-one; D&C Orange No.8; 4,5-dichloro-3,6-fluorandiol). an orange powder that is insoluble in water and dilute acids; soluble in ethanol and dilute alkali; slightly soluble in glycerol and glycols. It is carcinogenic to some laboratory animals. The disodium salt is C.I. Solvent Orange 32, C.I. 45365.

dichloroformoxime. phosgene oxime.

2,6-dichloro-*N*-2-imidazolidinylidenebenzenamine. clonidine.

dichloromalealdehydic acid. mucochloric acid.

2,3-dichloromaleic aldehyde acid. mucochloric acid.

3,6-dichloro-2-methoxybenzoic acid. dicamba.

***sym*-dichloromethyl ether** (oxybis-[chloromethane]; bis(chloromethyl) ether; BCME; dichloromethyl ether). a colorless, volatile liquid, $(CH_2Cl)_2O$, with a stifling odor that is used in the manufacture of polymers. It is a strong irritant to the eyes and respiratory tract and a known carcinogen with a latent period of 10-15 years; it probably causes pulmonary cancer in workers producing certain resins. *Sym*-dichloromethyl ether is a designated priority toxic pollutant and hazardous waste of EPA.

3′,4′-dichloro-2-methylacrylanilide. dicryl.

4,4′-dichloro-α-methylbenzhydrol. chlorfenethol.

2,2′-dichloro-*N*-methyldiethylamine. mechlorethamine.

dichloromonoxide. chlorine monoxide.

2,4-dichloro-1-(4-nitrophenoxy)benzene. nitrofen.

2,3-dichloro-4-oxo-2-butenoic acid. mucochloric acid.

(2,4-dichlorophenoxy) acetic acid. 2,4-D.

2,4-dichlorophenoxyacetic acid. 2,4-D.

***o*-(2,4-dichlorophenyl) *o*-methyl isopropylphosphoramidothioate**. DMPA.

2,4-dichlorophenyl *p*-nitrophenyl ether. nitrofen.

2-[(2,6-dichlorophenyl)amino]-2-imidazoline. clonidine.

1-(3′,4′-dichlorophenyl)-3-(4′-chlorophenyl)urea. triclocarban.

***N*′-(3,4-dichlorophenyl)-*N,N*-dimethylurea**. diuron.

***N*′-(3,4-dichlorophenyl)-*N*-methoxy-*N*-methylurea**. linuron.

3-(3,4-dichlorophenyl)-1-methyl-1-*n*-butylurea. neburon.

***N*-(3,4-dichlorophenyl)methylacrylamide**. dicryl.

***N*-(3,4-dichlorophenyl)-2-methyl-2-propenamide**. dicryl.

***N*-(3,4-dichlorophenyl)propanamide**. propanil.

di(*p*-chlorophenyl)trichloromethylcarbinol. dicofol.

dichlorophos. dichlorvos.

1,3-dichloropropene (1,3-dichloropropylene; α,-γ-dichloropropylene; γ-chloroallyl chloride). a toxic, flammable liquid, $ClCH_2{=}CHCl$, with a chloroform-like odor. The technical grade is a mixture of *cis*- and *trans*- isomers. It is used as a soil fumigant, nematocide, a solvent for oils and fats, in drycleaning, and in metal degreasing. It is moderately toxic orally to laboratory rodents; dermal toxicity to laboratory rabbits is moderate; it may be a human mutagen.

3′,4′-dichloropropionanilide. propanil.

2,2-dichloropropionic acid. dalapon.

α,α-dichloropropionic acid. dalapon.

1,3-dichloropropylene. 1,3-dichloropropene.

α,γ-dichloropropylene. 1,3-dichloropropene.

***trans*-1,3-dichloropropylene**. *trans*-1,3-dichloropropene.

dichlorosilane. See silane.

4,4′-dichloro-α-trichloromethylbenzhydrol. the chemical name of docofol.

2,2′-dichlorotriethylamine. HN1.

dichlorovos. dichlorvos.

dichloroxide. chlorine monoxide.

dichlorvos (phosphoric acid 2,2-dichloroethenyl dimethyl ester; phosphoric acid 2,2-dichlorovinyl dimethyl ester; O,O-dimethyl O-(2,2-dichlorovinyl) phosphate; dichlorophos; dichlorovos; DDVP). an organophosphate, $(CH_3O)_2P(O)CH{:}CCl_2$, it is an amber, liquid, anticholinesterase that is used as a broad-spectrum insecticide and miticide especially to control house flies and mosquitoes, and to control both ecto- and endoparasites of domestic animals. Dichlorvos is moderately toxic to humans, and extremely toxic to some animals. The minimum toxic dose is 10 mg/kg body weight in young calves, and 25 mg/kg in horses and sheep. The oral LD_{50} in laboratory rats has been reported as 25-80 mg/kg. This compound is rapidly metabolized and excreted.

dichromate. any salt of dichromic acid; such are usually orange or red in color. See sodium dichromate.

dicobalt edetate (Kelocyanor). a deadly poison and antidote for cyanide poisoning. It is unobtainable in the United States.

dicofol (4-chloro-α-(4-chlorophenyl)-α-trichloromethyl)benzenemethanol; 4,4′-dichloro-α-(trichloromethyl)benzhydrol; 1,-1-bis(*p*-chlorophenyl)-2,2,2-trichloroethanol; di(*p*-chlorophenyl)trichloromethylcarbinol; DTMC).a neurotoxic organochlorine pesticide used mainly as a nonsystemic miticide on cotton and citrus fruits. The chemical name is 4,4′-dichloro-α-(trichloromethyl)benzhydrol. It interferes with the $Ca_2{}^+$, K^+, Na^+, and ATP-ases in insects and mammals. It is moderately toxic to laboratory rats and is carcinogenic to some laboratory animals. Symptoms of intoxication in humans may include nausea, vomiting, restlessness, fearfulness, tremor, convulsions, coma, respiratory collapse, and death.

dicophane. DDT.

dicoumarin. dicumarol.

dicoumarol. dicumarol.

dicrotophos (phosphoric acid 3-(dimethylamino)-methyl-3-oxo-1-propenyl dimethyl ester; phosphoric acid dimethyl ester, ester with *cis*-3-hydroxy-*N,N*-dimethylcrotonamide; 3-(dimethoxyphosphinyloxy)-*N,N*-dimethyl-*cis*-crotonamide; dimethyl 2-dimethylcarbamoyl-1-methylvinyl phosphate; dimethyl 1-dimethylcarbamoyl-1-propen-2-yl phosphate; Bidrin; Carbicron; Ektafos). the dimethyl phosphate of 3-hydroxy-*N,N*-dimethylcrotonamide,dicrotophos is an extremely toxic and hazardous liquid pesticide; it is fully miscible with water, ethanol, and xylene and somewhat soluble in kerosene. The commercial grade is brown. It is an anticholinesterase used as an insecticide and in some household products used to control pests.

dicryl (*N*-(3,4-dichlorophenyl)-2-methyl-2-propenamide; 3′,4′-dichloro-2-methylacrylanilide; *N*-(3,4-dichlorophenyl)-methylacrylamide). a postemergent amide herbicide which, when applied to foliage, produces a localized or general necrosis that is dose dependent. It also inhibits growth. Dicryl is slightly toxic to laboratory rats.

dicumarol (3,3′-methylenebis[4-hydroxy-2H-1-benzopyran-2-one]; bishydroxycoumarin; Dicoumarol; Dicoumarin; Dicumol; Dufalone; Melitoxin). a vitamin K antagonist, it decreases the activity of vitamin K-dependent clotting factors in blood plasma. It inhibits the production of prothrombin in the liver, and increases prothrombin time. Dicumarol is used therapeutically as a long-acting, oral anticoagulant with a half-life in humans of 4 days. Single-dose toxicity is probably low, but dicumarol is a cumulative poison and long-continued use at any dosage may be risky. Dicumarol occurs in *Melilotus* (sweet clover) hay and is also prepared synthetically. It is the proximate cause of bleeding disease in cattle and can affect stock feeding on fresh clover or on hay. It also affects humans, inhibiting clotting, producing pallor, bruising, nosebleeds, and bleeding gums. It is used therapeutically to prevent or to manage intravascular clotting (e.g., in acute coronary thrombosis). Dicumarol therapy may also cause nausea, abdominal pain, and diarrhea. Breast-fed infants of mothers under dicumarol therapy may have a tendency to hemorrhage and should thus be carefully monitored.

Dicumol. dicumarol.

dicyan. cyanogen.

dicyanide. a salt with two cyanide groups.

β,β-dicyan-*o*-*o*-chlorostyrene. *o*-chlorobenzylidenemalononitrile.

dicyanogen. cyanogen.

1,3-dicyanotetrachlorobenzene. chlorothalonil.

1,3-dicyano-2,4,5,6-tetrachlorobenzene. chlorothalonil.

3-dicyanotetrachlorobenzene. chlorothalonil.

3-dicyano-2,4,5,6-tetrachlorobenzene. chlorothalonil.

dicyclohexylamine (*N*-cyclohexylcyclohexanamine; dodecahydrodiphenylamine). a clear, colorless, liquid dicyclic aliphatic secondary amine used as a plasticizer; catalyst; and in insecticides, corrosion inhibitors, antioxidants, and detergents. It produces effects similar to those of cyclohexylamine. Dicyclohexylamine is, however, less polar, more lipid soluble, more toxic, and much more likely to be absorbed in toxic amounts through the skin.

dicyclopentadiene. the dimer of cyclopentadiene, it is a water-insoluble liquid used in the manufacture of polychlorinated pesticides, elastomers and polyhalogenated flame retardants. It is an irritant and narcotic; it is moderately toxic to laboratory rats via dermal absorption, but is highly toxic if ingested. See also cyclopentadiene.

12,13-didehydro-13,14-dihydro-α erythroidine. β-erythroidine.

(5α,6α)-7,8-didehydro-4,5-epoxy-3-methoxy-17-methylmorphinan-6-ol. codeine.

(5α,6α)-7,8-didehydro-4,5-epoxy-17-methylmorphinan-3,6-diol. morphine.

(5α,6α)-7,8-didehydro-4,5-epoxy-17-methylmorphinan-3,6-diol diacetate (ester). heroin.

7,8-didehydro-4,5-epoxy-17-(2-propenyl) morphinan-3,6-diol bis(3-pyridinecarboxylate) (ester). nalorphine dinicotinate.

[8β(*S*)]-9,10-didehydro-*N*-(2-hydroxy-1-methylethyl)-6-methylergoline-8-carboxamide. ergonovine.

13,19-didehydro-12-hydroxysenecionan-11,16-dione. seneciphylline.

8,9-didehydro-6-methylergoline-8-methanol. elymoclavine.

2,6-dideoxy-3-*O*-methyl-ribohexose. cymarose.

3β,5β,12β)-3-[(O-2,6-dideoxy-β-D-ribohexopyranosyl-(1→4)-O-2,6-dideoxy-β-D-ribohexopyranosyl-(1→4)-2,6-dideoxy-β-D-ribohexopyranosyl)oxy]-12,14-dihydroxycard-20(22)-enolide. digoxin.

(3β,5β)-3-[(*O*-2,6-dideoxy-β-D-ribohexopyranosyl-(1→4)-*O*-2,6-dideoxy-β-D-ribohexopyranosyl-(1→4)-2,6-dideoxy-β-D-ribohexopyranosyl)oxy]-14-hydroxycard-20(22)-enolide. digitoxin.

3-[(*O*-2,6-dideoxy-β-D-ribo-hexopyranosyl-(1→4)-*O*-2,6-dideoxy-β-D-ribohexopyranosyl)oxy]-14,16-dihydroxycard-20(22)-enolide. gitoxin.

2,6-dideoxy-D-ribohexose. digitoxose.

2,6-didesoxy-D-allose. digitoxose.

Didrex. benzphetamine.

die. cease to live. *Cf.* death, dying.

dieffenbachia. *Dieffenbachia*.

Dieffenbachia (dumbcane; dumb cane; dieffenbachia). a genus of erect, shrubby, tropical American evergreen plants (Family Araceae). They are common houseplants. Biting into a leaf or other plant part quickly causes an intense burning sensation with severe irritation of the lining of the mouth, lips, and tongue. Symptoms include salivation, edema of the affected tissues including the tongue which may become immobile as a result, interfering with swallowing, speech, and breathing. Airway obstruction is a rare complication. These effects are due to the presence in plant tissues of needle-sharp crystals of calcium oxalate that are released by ruptured plant cells and penetrate the tissues of the mouth and tongue. Symptoms may persist for a week or more. Principle species/varieties are *D.* x *bausei*, *D. imperialis*, *D. maculata* (formerly *D. picta*) (spotted dumb cane), and *D. seguine* (mother-in-law plant). *Cf. Arisaema*, See also Araceae.

dieldrin (DDE; (1aα,2β,2aα,3β,6β,6aα,7β,7a)-3,4,5,6,9,9-hexachloro-1a,2,2a,3,6,6a,7,7a-octahydro-2,7:3,6-dimethanonaphth[2,3-b]-oxirene; 1,2,3,4,10,10-hexachloro-6,7-epoxy-1,4,4a,5,6,7,8,8a-octahydro-endo,exo-1,4:5,8-dimethanonaphthalene; Compound 497; Insecticide No. 497; HEOD). a light tan, nonflammable, environmentally persistent, chlorinated diene insecticide which is no longer manufactured or used in the United States. It is a stereoisomer of endrin. Dieldrin is toxic to an extremely wide spectrum of organisms and accumulates especially in aquatic organisms and is associated with reproductive failure in birds and fish. No other compound is known to produce such a degree of eggshell thinning in birds as DDE. Birds of the orders Falconiformes, Pelecaniformes, Ciconiiformes, and Strigiformes are highly sensitive to eggshell thinning by DDE; Columbiformes, Anseriformes, and Charadriiformes, are moderately sensitive; and Galliformes and Passeriformes are relatively insensitive. DDE is extremely toxic to laboratory rats and to humans. Exposures with toxic effects readily occur through skin contact, inhalation, or ingestion of contaminated food. It is neurotoxic, with symptoms in humans that include headache, weakness, nausea, vomiting, dizziness, tremors, clonic and tonic convulsions, coma, respiratory failure, and death. Dieldrin also affects the liver, kidneys, and skin. It is a chronic immunosuppressant and carcinogen, producing liver cancer in experimental animals.

diene. See alkene.

1,2,5,6-diepoxyhexane. a carcinogenic bifunctional epoxide.

di(2,3-epoxypropyl)ether. diglycidyl ether.

diestrus. See estrus cycle.

diethadione (5,5-diethyldihydro-OH-1,3-oxazine-2,4(OH)-diane). an analeptic. It is extremely toxic to laboratory mice and rats, *i.p.*, and very toxic by ingestion.

diethoxy thiophosphoric acid ester of 2-ethylmercaptoethanol. demeton.

1,1-diethoxyethane. acetal.

2-(diethoxyphosphinylimino)-1,3-dithiolane. phosfolan.

diethylcarbamodithioic acid 2-chloro-2-propenyl ester. sulfallate.

diethyl carbinol. 3-pentanol.

***O,O*-diethyl *O*-[2-chloro-1-(2,4-dichlorophenyl)vinyl] phosphate**. chlorfenvinphos.

***O,O*-diethyl *S*-(*p*-chlorophenylthio)methyl phosphorodithioate**. carbophenothion.

***O,O*-diethyl *S*-(2-chloro-1-phthalimidoethyl)-phosphorodithioate**. dialifor.

***O,O*-diethyl *o*-[2-(diethylamino)-6-methyl-4-pyrimidinyl]-phosphorothioate**. pirimiphosethyl.

***O,O*-diethyl dithiobis[thioformate]**. dixanthogen.

diethyl 1,3-dithiolan-2-ylidenephosphoramidate. phosfolan.

diethyl ester sulfuric acid. diethyl sulfate.

diethyl ether. **1:** any ether that contains two ethoxy groups. **2:** ethyl ether.

***O,O*-diethyl 2-ethylmercaptoethyl thiophosphate**. demeton.

***O,O*-diethyl *S*-ethylmercaptomethyl dithiophosphate**. phorate.

***O,O*-diethyl *O*(+*S*)-2-(ethylthio)ethyl phosphorothioate mixture**. demeton.

***O,O*-diethyl S-[2-(ethylthio)ethyl]phosphorodithioate**. disulfoton.

***O,O*-diethyl *S*-(ethylthio)methyl phosphorodithioate**. phorate.

***N,N*-diethyl-2-(1-naphthalenyloxy)-propanamide**. napropamide.

***N,N*-diethyl-2-(1-naphthyloxy)propionamide**. napropamide.

diethyl-*p*-nitrophenyl monothiophosphate. parathion.

diethyl-*p*-nitrophenyl phosphate. paraoxon.

***O,O*-diethyl-*o,p*-nitrophenyl phosphorothioate**. parathion.

diethyl oxide. ethyl ether.

***O,O*-diethyl phosphorothioate mucous membranes**. See also ester sulfate.

diethyl phthalate. ethyl phthalate.

diethyl sulfate (diethyl ester sulfuric acid; sulfuric acid diethyl ester; ethyl sulfate). a colorless, oily liquid, $C_4H_{10}O_4S$, with a faint ethereal odor that reacts with water to yield sulfuric acid. diethyl sulfate is used in the sulfation of ethylene, and as an ethylating and sulfonating agent. It is a strong irritant with low acute toxicity by inhalation and subcutaneous injection, moderately toxic by ingestion and skin contact. Diethyl sulfate is carcinogenic to experimental animals.

***O,O*-diethyl *O*-3,5,6-trichloro-2-pyridyl phosphorothioate**. chlorpyrifos.

diethylacetal. acetal.

β-diethylaminoethyl benzylate. benactyzine.

2-diethylaminoethyl diphenylglycolate. benactyzine.

S-[2-(diethylamino)ethyl]phosphorothioic acid *O,O*-diethyl ester. amiton.

o-[2-(diethylamino)-6-methyl-4-pyrimidinyl]phosphorothioic acid *O,O*-diethyl ester. pirimiphosethyl.

5,5-diethylbarbituric acid. barbital.

diethylberyllium (beryllium diethyl). an extremely hazardous, volatile, colorless oily liquid at room temperature, $Be(C_2H_5)_2$; b.p. 63°C. It combusts spontaneously in air, releasing dense white fumes of BeO, and reacts violently with water, HCl and alcohols, releasing ethane.

diethylcadmium (cadmium diethyl). an extremely hazardous oily liquid, $(C_2H_5)_2Cd$, (mp 21°C; b.p. 64°C). It reacts explosively with atmospheric oxygen and decomposes above about 150°C, releasing toxic cadmium oxide fumes. It decomposes when moist. It is a confirmed human carcinogen.

diethylcarbamodithioic acid 2-chloro-2-propenyl ester. sulfallate.

diethylcarbamodithioic acid sodium salt. ditiocarb sodium.

O,O-diethyl-O-(3-chloro-4-methyl-2-oxo-2*H*-1-benzopyran-7-yl)-phosphorothioate. coumaphos.

diethyldithiocarbamate sodium. ditiocarb sodium.

diethyldithiocarbamic acid 2-chloroallyl ester. sulfallate.

diethyldithiocarbamic acid sodium salt. ditiocarb sodium.

3-(*O,o*-diethyldithiophosphorylmethyl)-6-chlorobenzoxazolinone. phosalone.

1,4-diethylene dioxide. dioxane.

diethylene glycol (dihydroxydiethyl ether; diglycol; DEG). a toxic, colorless, nearly odorless, syrupy, extremely hygroscopic, noncorrosive liquid with a sweetish taste. It has numerous applications including use in the production of a number of derivative compounds, and of polyurethane and unsaturated polyester resins; in solvent extraction of petroleum; a solvent; a humectant for tobacco; in cosmetics, and in antifreeze solutions. As used in the processing of color film, it offers the risk of ethylene glycol poisoning. DEG is nephrotoxic and hazardous to use in household products at concentrations of 10% or more. See also ethylene glycol.

diethylenediamine. piperazine.

4,4′-(1,2-diethyl-1,2-ethenediyl)bisphenol. diethylstilbestrol.

***O,O*-diethyl-*S*-(2-ethioethyl)phosphorothioate**. demeton-*S*.

diethyl-*S*-(2-ethioethyl)thiophosphate. demeton-*S*.

***O,O*-diethyl-*S*-ethyl-2-ethylmercaptophosphorothiolate**. demeton-*S*.

***O,o*-diethyl-*S*-ethylmercaptoethyl dithiophosphate**. disulfoton.

***O,O*-diethyl-*S*-2-(ethylthio)ethyl phosphorothiolate**. demeton-*S*.

di(2-ethylhexyl) adipate. carcinogenic to some laboratory animals.

di-(2-ethylhexyl) phthalate (DEHP). bis(2-ethylhexyl) phthalate.

1,2-diethylhydrazine (*sym*-diethylhydrazine; *N,N′*-diethylhydrazine; hydrazoethane; SDEH). a dialkyl hydrazine, C_2H_5-NHN-C_2H_5, that is soluble in ethanol and ethyl ether. It is a suspected carcinogen and exhibits tumorigenic and teratogenic activities. It readily passes through the placenta. Toxic fumes, NO_x, are released on decomposition by heating. See dialkylhydrazine.

***N,N′*-diethylhydrazine**. 1,2-diethylhydrazine.

***sym*-diethylhydrazine**. 1,2-diethylhydrazine.

***O,O*-diethyl-*O*-(2-isopropyl-4-methyl-6-pyrimidinal)phosphorothioate**. See Diazinon.

diethylmagnesium. a poisonous pyrophoric compound, $Mg(C_2H_5)_2$, that reacts violently with water and steam and combusts spontaneously in air or even in carbon dioxide. See also dimethylmagnesium, diphenylmagnesium.

diethylmalonylurea. barbital.

***O,O*-diethyl-O-[p-(methylsulfinyl)phenyl]-phosphorothioate**. fensulfothion.

diethyl-*p*-nitrophenyl monothiophosphate. parathion.

***O,O*-diethyl-*O*-(p-nitrophenyl) phosphate**. paraoxon.

***O,O*-diethyl-*O*-*p*-nitrophenylphosphorothioate**. parathion.

diethylnitrosamine. *N*-nitrosodiethylamine.

diethylpropion (2-(diethylamino)-1-phenyl-1-propanone; 2-diethylaminopropiophenone; α-benzoyltriethylamine; amfepramone; Tenuate). an amphetamine (def. 2), *q.v.* It is a controlled substance in the United States and sold as the hydrochloride under several brand names (e.g., Anfamon, Anorex, Danylen, Dobesin, Frekentine, Keramik, Keramin, Magrene, Modulor, Moderatan, Parabolin, Prefamone, Regenon, Tenuate, Tenuate Dospan, Tepanil, Tylinal). It is an appetite suppressant that also elevates blood pressure, disturbs heart rhythm, causes overactivity, restlessness, insomnia, euphoria, tremor, headache, dryness of mouth, unpleasant tastes, diarrhea, upset stomach, changes in sex drive, and impotence, as well as depression and psychosis. Overdosage causes hallucinations, aggressiveness, and panic that may result in an "accidental" death. It should not be taken by persons suffering from heart disease.

α,α'-diethylstilbenediol. diethylstilbestrol.

diethylstilbestrol (4,4'-(1,2-diethyl-1,2-ethenediyl)bisphenol; α,α'-diethylstilbenediol; stilbestrol; stilboestrol; 3,4-bis(*p*-hydroxyphenyl)-3-hexene; 4,4'-dihydroxy-α,β-diethylstilbene; DES; Antigestil). a synthetic, nonsteroidal, orally active estrogen that is several times more potent biologically than endogenous estrogens. It has been used therapeutically to treat disorders in women resulting from estrogen deficiency, in the treatment of carcinomas that have specific estrogen-binding capacity and as an oral "morning after" contraceptive, especially in cases of rape or incest. It also is given to livestock, chiefly poultry and feedlot cattle, to promote growth. Residues in meat are considered carcinogenic and its use for such purposes in the United States was banned by the Food and Drug Administration in 1979. Dietary DES causes mammary carcinomas and adenomas in laboratory mice, pituitary tumors in male laboratory rats, bladder cancer in male and female laboratory rats, renal cancers in hamsters, and uterine cancers in rhesus monkeys. Products that contain diethylstilbestrol cause reproductive failure and a high incidence of urinary tract infections in mink. DES can cause vaginal and uterine cancer in women, and if treated with DES during the first trimester of pregnancy, vaginal cancer (clear cell adenocarcinoma) may develop in female progeny. Effects of acute exposure initially include nausea and vomiting, with subsequent fluid retention, feminization and uterine bleeding. DES is moderately toxic to laboratory mice.

***N,N'*-diethylthiocarbamide**. *N,N'*-diethylthiourea.

1,3-diethylthiourea. *N,N'*-diethylthiourea.

1,3-diethyl-2-thiourea. *N,N'*-diethylthiourea.

***N,N'*-diethylthiourea** (1,3-diethylthiourea; *N,N*-diethylthiocarbamide; 1,3-diethyl-2-thiourea). a substitution product of thiourea that is moderately toxic by oral and intraperitoneal routes of administration. It is carcinogenic in some laboratory animals. Toxic fumes, NO_x and SO_x, are released on heating to decomposition.

diethylxanthogenate. dixanthogen.

diethylzinc (zinc diethyl; zinc ethide). a mobile, volatile liquid, $Zn(C_2H_5)_2$, at room temperature. A hazardous chemical that ignites spontaneously in air or in the presence of other oxidizing substances (e.g., hydrazine, methanol, ozone). It burns with a blue flame, giving off a garlicky odor. Diethylzinc is a dangerous explosion hazard, reacting violently or explosively with sulfur oxides, alkenes, water, bromine, and nitro compounds, nonmetallic halides, for example, usually releasing poisonous fumes. Properties are similar to those of dimethylzinc. Its toxicity is not well established, but see organozinc compounds. See also dimethylzinc, diphenylzinc.

dietotoxic. having the property of dietotoxicity.

dietotoxicity. the capacity of certain food substances to be toxic when part of an unbalanced diet.

difenzoquat. a bipyridilium quaternary ammonium salt, *q.v.*

difference spectrum, optical. See optical difference spectrum.

differential diagnosis. See diagnosis.

differentiation. the process and developmental stages by which cells, tissues and organs become structurally distinct and specialized.

diffuse. **1a:** to scatter, disperse. **1b:** to move passively or by thermal agitation from a region of higher concentration to a region of lower concentration within a medium (e.g., a solvent); said of an atom, ion, or molecule. **2:** to pass through a membrane, tissue or structure. **3:** not localized or focused; not limited; widely spread, dispersed, or scattered; extended, lacking a well-defined margin.

diffuse pulmonary fibrosis. uncomplicated asbestosis. See also fibrosis.

diffusion. **1:** a spreading, scattering, or dispersion, e.g., of matter, energy, or ideas. **2:** the net movement of very small particles within a solution as a result of diffusion (defs. 3 and 4). The net movement is from a region of high concentration to a region of low concentration of each specific type of molecule or particle. In a closed system, a net movement will continue until the diffusible components are evenly dispersed. Diffusive movement within solids at normal temperatures is extremely slow. **3:** the process whereby ions, molecules, and other very small particles within a gas, liquid, or solid disperse or intermingle as a result of random molecular (thermal) motion. **4:** the process whereby molecules move down a gradient of chemical potential, $\partial c/\partial x$, until the gradient is abolished. Do not confuse molecular diffusion with turbulent diffusion. The latter effects a much more rapid mixing through bulk movement of the fluid. Most toxicants enter cells via passive diffusion across the cell membrane, the rate of penetration depending upon the concentration gradient across the membrane and the lipid solubility of the toxicant (but see also Fick's Law). **facilitated diffusion** (facilitated transport). a type of carrier-mediated transport of molecules across a biological membrane down a concentration gradient. It is similar to active transport in that it proceeds more rapidly than passive diffusion, but it does not work against a concentration gradient, is not dependent on metabolic energy, and is not inhibited by metabolic poisons. See also transport, diffusion, osmosis. **passive diffusion**. movement across a cell membrane in which the rate of diffusion of the solute is directly proportional to its concentration gradient across the membrane. This gradient is the driving force for such movement. Unless qualified, the term diffusion usually refers to passive diffusion.

diffusion capacity. See pulmonary diffusion capacity.

diffusion, coefficient of. the coefficient of proportionality in Fick's law. It is the number of particles that cross a unit area in unit time per unit negative concentration gradient.

diffusion constant. diffusion, coefficient of.

diffusion hypoxia. when normal room air is inhaled following nitrogen oxide anesthesia, there is an abrupt transient decrease in alveolar oxygen tension (hypoxia) due to dilution by nitrous oxide which diffuses from the capillary blood.

diflubenzuron (*N*-[[(4-chlorophenol)amino]carbonyl]-2,6-difluorobenzamide). a stomach and contact insecticide, larvicide, and ovicide that acts as a growth regulator, blocks DNA synthesis, and inhibits the synthesis and incorporation of chitin into the insect cuticle. It is slightly or moderately toxic by ingestion to laboratory mice and rats; moderately toxic to laboratory mice, *i.p.*; and moderately toxic to laboratory rabbits via the percutaneous route of administration. The no-effect dietary dose for the laboratory rat has been reported as 40 mg/kg body weight. Diflubenzuron is, at most, moderately toxic to *Anas platyrhynchos* (mallard) and *Colinus virginianus* (bobwhite). The median lethal (96-hour) dose for *Lepomis macrochirus* (blue-gill sunfish). See dose (median lethal dose), no-effect level.

difluorodiphenyldichloroethane. DFDD.

difluorodiphenyltrichloroethane. DFDT.

digallic acid (3,4-dihydroxy-5-[(3,4,5-trihydroxybenzoyl)oxy]benzoic acid; gallic acid 5,6-dihydroxy-3-carboxyphenyl ester; 4,5-dihydroxybenzoic acid monogallate; gallic acid 3-mono-gallate; *m*-digallic acid; *m*-galloylgallic acid; gallotannin). **1:** a decomposition product of tannins, it is a crystalline phenolic ester; on hydrolysis it yields two molecules of gallic acid, *q.v.* **2:** sometimes incorrectly used as a synonym for tannin, especially in the pharmacologic literature. **3:** referred to as gallotannin in pharmaceuticals.

***m*-digallic acid**. digallic acid.

Digenea simplex. a red alga (Family Rhodomelaceae) from which kainic acid, a neuroexcitant and anthelmintic, is isolated.

digenic acid. kainic acid.

digestion. **1:** in biology, digestion is the breakdown by a living organism of complex molecules (typically food materials) into smaller, usually less complex molecules by mechanical and/or chemical processes. The process may be external as in the digestion of organic wastes by bacteria, or internal as in the digestion of food by most animals. In the latter case digested material can then usually be absorbed, metabolized, and often assimilated by the organism's cells. Many poisons that are toxic by other routes of exposure may be detoxified through digestion if ingested. **chemical digestion**. that accomplished by digestive enzymes. **extracellular digestion**. digestion whereby digestive enzymes are released from cells, usually specialized cells, onto food that is either in an alimentary canal or cavity (e.g., in vertebrate animals) or onto or into external materials (as in flies, mosquitos and other dipterous insects). **intracellular digestion**. occurs when food particles are taken into the cells of a living organism (as by phagocytosis) and digested (e.g., as in protozoans and coelenterates). **2:** in chemistry and chemical engineering, digestion refers to various processes of relevance here: **2a:** the biochemical decomposition of organic matter, resulting in partial gasification, liquefaction, and mineralization of pollutants. **2b:** the liquefaction of organic wastes by microbial action as in activated sludge. This is, of course, an application of biological digestion to engineering problems.

digestive gland. See liver (def. 1).

digestive system. the organs, collectively, that process and convert food into simpler substances that are absorbed into the circulating blood. In vertebrate animals, the major digestive organs are generally the mouth, esophagus, stomach, duodenum, small intestine, large intestine (or colon), and rectum. Organs of mastication (e.g., teeth), and digestive glands such as the liver, gallbladder, and pancreas of vertebrates may also be considered parts of the digestive system.

digestive tract. the entire route through which food physically passes prior to egestion. In vertebrate animals, this usually includes the mouth, buccal cavity, esophagus, stomach or stomachs, intestines, anus, and associated digestive glands and accessory structures. See alimentary canal, gastrointestinal tract.

Digibund. digoxin immune FAB (ovine) for injection.

digilanide C. lanatoside C (see lanatosides).

digitalin ((3β,5β,16β)-3-[(6-deoxy-4-*O*-β-D-glucopyranosyl-3-*O*-methyl-β-D-galactopyranosyl)oxy]-14,16-dihydroxy-card-20(22)-enolide; digitalinum verum; digitalinum true, Schmiedeberg's digitalin). a white, crystalline, cardioactive, steroid alkaloid isolated from the seeds and leaves of *Digitalis purpurea*. It is a component of digitalis. Do not confuse digitalin with crystalline digitalin; the latter is synonymous with digitoxin. **crystalline digitalin**. a synonym for digitoxin. Do not confuse with digitalin.

digitalinum true. digitalin.

digitalinum verum. digitalin.

digitalis (foxglove; purple foxglove; fairy gloves; digifortis; digitora). **1:** the dried leaves of *Digitalis purpurea*. **2:** a mixture extracted from the dried leaves of *Digitalis purpurea* that contains digitoxin, digitonin, digitalin, digoxin, gitoxin, antirhinic acid, digitalosmin, digitoflavone, inositol, and pectin. It is standardized for therapeutic use and is prescribed as tablets, capsules or liquid. Digitalis is diuretic and cardiotonic, but the toxic dose is close to the therapeutic dose. It is used chiefly to treat congestive heart failure and some types of palpitation. Digitalis increases the efficiency of a failing heart by strengthening contractions and lowering heart rate. Toxic effects in humans may include nausea, salivation, vomiting, gastritis, diarrhea, anorexia, severe headache, weakness, fatigue, drowsiness, a slowing irregular pulse, confusion, disorientation, visual disturbances, delirium, hallucinations, sometimes tremors or convulsions, unconsciousness, and death from ventricular fibrillation. Digitalis is a cumulative poison. It is eliminated very slowly from the body. Consequently, small frequently repeated doses may eventually produce symptoms. **prepared digitalis**. the dried powdered leaves of *D. purpurea* dispensed as a cardiotherapeutic agent. See also alkaloid, glycoside, digitoxin, digoxin, gitoxin.

Digitalis (foxglove). a genus of biennial and perennial herbs (Family Scrophulariaceae) with tall, handsome, bell-shaped flowers. They are indigenous to Europe and western Asia. ***D. purpurea*** (foxglove; purple foxglove; fairy gloves; fairy finger; fairy thimbles; fairy bells). widely grown and distributed in the United States and elsewhere. Widely cultivated in gardens in North America, foxglove frequently grows wild in the northcentral and northeastern United States and along the Pacific Coast, and in Hawaii. All parts of the plant are highly toxic. It contains about a dozen chemically related, physiologically active cardiac or steroid glycosides. Used therapeutically, they strengthen the force of cardiac contractions and stimulate the vagus nerve, prolonging diastole and slowing the rate of contractions. Numerous human fatalities have resulted from ingestion of foxglove or from overdoses of the drugs. Ingestion of a few leaves can cause nausea, vomiting, slowing of the pulse, and possibly cardiac arrest. A wide range of animals have been poisoned. As a house or garden plant, foxglove should be considered hazardous. The entire plant is toxic as is smoke from burning foliage and the water in which the flowers have been placed. Potency survives drying or boiling of plant material. Thus livestock fatalities have resulted from ingestion of cured hay infested with *D. purpurea*. See also digitalis.

digitalis poisoning. digitalism; acute or cumulative chronic poisoning by digitalis. Symptoms may include gastrointestinal disturbances and pain, severe headache, nausea, vomiting, diarrhea, irregular pulse, and yellow vision. Cardiac irregularities such as bradycardia, partial heart block, ventricular extrasystoles.

digitalism. **1:** digitalis poisoning. **2:** the symptoms resulting from digitalis poisoning.

digitalization. the subjection of a living organism to digitalis to the extent necessary to produce a desired therapeutic result.

digitaloid. chemically related to or similar in action to digitalis; any substance resembling digitalis.

digitalose. a rare sugar found only in certain digitalis glycosides.

digitin. digitonin.

digitogenin((25R)-5α-spirostan-2α,3β,15β-triol). a crystalline steroid sapogenin produced by the hydrolysis of digitonin.

digitonin (digitin). a crystalline cardiotoxic saponin isolated from the seeds and leaves of *Digitalis purpurea*. It is a component of digitalis. It is a clinical reagent used in the determination of cholesterol. See also digitogenin.

digitophyllin. digitoxin.

digitoxigen. the aglycon of digitonin.

digitoxigenin ((3,14-dihydrocard-20(22)-enolide; $\Delta^{20:22}$-3,14,21-trihydroxynorcholenic acid lactone; cerberigenin; echujetin; evonogenin). a crystalline steroid lactone. It is the aglycone of digitoxin.

digitoxin ((3β,5β)-3-[(*O*-2,6-dideoxy-β-D-ribo-hexopyranosyl-(1→4)-*O*-2,6-dideoxy-β-D-ribo-hexopyranosyl-(1→4)-2,6-dideoxy-β-D-ribo-hexopyranosyl)oxy]-14-hydroxycard-20(22)-enolide; crystalline digitalin; digitophyllin). a secondary cardiac glycoside extracted from dried leaves of *Digitalis purpurea* (Family Scrophulariaceae). A highly poisonous cardioactive crystalline steroid glycoside and the most active principle of digitalis. It is cardiotonic and used in the treatment of heart failure to increase cardiac contractility. Its cardiotonic and cardiotoxic actions and effects are essentially similar to those of digitalis and digoxin. It is very toxic orally to guinea pigs and extremely toxic to domestic cats. The therapeutic dose is close to the toxic dose. digitoxin alters cardiac membrane potential by inhibiting Na+ and K+ ATPase associated with the sodium-potassium pump; the result is premature atrial contraction, fibrillation, and atrioventricular block, with ventricular tachycardia and fibrillation. Symptoms include cardiac arrhythmia, anorexia, salivation, nausea, vomiting, diarrhea, headache, drowsiness, and disorientation. In severe cases of intoxication, delirium, hallucinations and death may result.

digitoxose (2,6-dideoxy-D-ribo-hexose; 2-desoxy-D-altromethylose; 2,6-didesoxy-D-allose). a rare water-soluble, crystalline sugar isolated by hydrolysis of digitoxin, gitoxin, and digoxin. See cymarose.

diglycidyl ether (bis(2,3-epoxypropyl)ether; di-(2,3-epoxypropyl) ether; DGE). an extremely hazardous liquid, suspected carcinogen and mutagen, $C_6H_{10}O_3$. DGE is toxic by all routes [illegible] including ingestion, inhalation, and skin contact. It is a systemic poison and is highly irritant to the eye, skin, and mucous membranes. Chronic exposure can cause bone marrow depression.

diglycol. diethylene glycol.

digoxigenin (3,12,14-trihydroxycard-20-(22) enolide; $\Delta^{20:22}$-3β,12β,14,21-tetrahydroxy norcholenic acid lactone; lanadigenin). the aglycone of digoxin, formed by hydrolysis of digoxin.

digoxin (3β,5β,12β)-3-[(O-2,6-dideoxy-β-D-ribo-hexopyranosyl-(1→4)-O-2,6-dideoxy-β-D-ribo-hexopyranosyl-(1→4)-2,6-dideoxy-β-D-ribo-hexopyranosyl)oxy]-12,14-dihydroxycard-20-(22)-enolide). a secondary glycoside extracted from the leaves of *Digitalis purpurea*. a cardiotoxic digitalis glycoside, $C_{41}H_{64}O_{14}$. Its action is similar to that of digitoxin.

digoxin immune FAB (ovine) for injection (Digibund). a biological used to treat severe overdosage of digoxin or digitoxin. This ovine FAB (fragment antigen binding) combines with digoxin or digitoxin molecules which are transported to, and excreted by the kidneys.

1,2-dihydrobenz[j]aceanthrylene. cholanthrene.

3,7-dihydro-3,7-dimethyl-1*H*-purine-2,6-dione. theobromine.

3,7-dihydro-1,3-dimethyl-1*H*-purine-2,6-dione compounded with 1,2-ethanediamine(2:1). aminophylline.

3,14-dihydrocard-20(22)-enolide. digitoxigenin.

dihydrochalcone disaccharides. See sweetener, nonnutritive.

dihydrocollidine. an oily ptomaine, $C_8H_{11}NH_2$, from decaying meat and fish.

3-(10-,11-dihydro-5H-dibenzo-[a,d]cyclohepten-5-ylidene)-*N*,*N*-dimethyl-1-propanamine. amitriptyuline.

14,19-dihydro-12,13-dihydroxy-20-norcrotolanan-11,15-dione. monocrotaline.

[illegible]-7',9'-dimethoxy-1',3'-dimethyl-10'-1'*H*-naphtho(2',3'-c')pyran-5'-yl]-3,4-dihydro-3-methyl-3,8,9-trihydroxy-1(2*H*)-anthracenone (T-544). a buckthorn toxin.

2,3-dihydro-2,2-dimethyl-7-benzofuranol methylcarbamate. carbofuran.

10,11-dihydro-*N,N*-dimethyl-5*H*-dibenz[b,f]-azepine-5-propanamine. imipramine.

3,4-dihydro-2,2-dimethyl-4-oxo-2H-pyran-6-carboxylic acid butyl ester. butopyronoxyl.

3,4-dihydro-3,3′dimethyl-1′,3,8,8′,9-pentahydroxy(7.lU′-bianthracene)-l,9′(2*H*,l0′*H*)-dione (T-496). a buckthorn toxin.

1,2-dihydro-1,5-dimethyl-2-phenyl-3H-pyrazol-3-one. antipyrine.

dihydrodiol. any diol in which both hydroxy groups are on adjacent ring carbon atoms and the bond between the hydroxyl-substituted carbon atoms is saturated. The biosynthesis of dihydrodiols from arene oxides is catalyzed by epoxide hydrolase. See 7,8-dihydrodiol-9,10-epoxide, catechol.

7,8-dihydrodiol-9,10-epoxide. a metabolically active, carcinogenic form of benzo(a)pyrene. See dihydrodiol.

6,7-dihydrodipyrido[1,2-a:2′,1′-c]pyrazinediium dibromide. diquat.

dihydroergotamine. hydrogenated ergotamine, it is less toxic than the parent compound.

7,8-dihydrofurano(2,3-b)furan. See mycotoxin (bisfuranoid mycotoxin).

dihydrofurobenzofuran. See mycotoxin (bisfuranoid mycotoxin).

3,4-dihydroharmine. harmaline.

dihydrohengol. 3-pentadecylcatechol.

(3α,14β,16α)-14,15-dihydro-14-hydroxy-eburnamenine-14-carboxylic acid methyl ester. vincamine.

(3R-*trans*)-4,6-dihydro-8-hydroxy-3,4,5-trimethyl-6-oxo-3H-2-benzopyran-7-carboxylic acid. citrinin.

3-[(4,5-dihydro-1*H*-imidazol-2-yl)methyl]-6-(1,1-dimethylethyl)-2,4-dimethylphenol. oxymetazoline.

1,2-dihydro-4-methoxy-1-methyl-2-oxo-nicotinonitrile. ricinine.

1,2-dihydro-4-methoxy-1-methyl-2-oxo-3-pyridinecarbonitrile. ricinine.

4,9-dihydro-7-methoxy-1-methyl-3H-pyrido[3,4-*b*]indole. harmaline.

10,11-dihydro-5-[3-(methylamino)propyl]-5*H*-dibenz[b,f]azepine. desipramine.

1,2-dihydro-3-methylbenz[j]aceanthrylene. 3-methylcholanthrene.

dihydro-5-(1-methylbutyl)-5-(2-propenyl)-2-thioxo-4,6(l*H*,5*H*)-pyrimidinedione. thiamylal.

10,11-dihydro-*N*-methyl-5*H*-dibenz[b,f]azepine-5-propanamine. desipramine.

dihydro-5-methylene-2(3H)-furanone. α'-angelica lactone. See angelica.

***N*-(4,5-dihydro-4-methyl-5-oxo-1,2-dithiolo[4,3-b]pyrrol-6-yl)acetamide**. thiolutin.

1,3-dihydro-7-nitro-5-phenyl-2H-1,4-benzodiazepin-2-one. nitrazepam.

6,11-dihydro-11-oxodibenz[b,e]oxepin-2-acetic acid. isoxepac.

4,5-dihydro-5-oxo-1-(4-sulfophenyl)-4-[(4-sulfophenyl)azo]-12*H*-pyrazole-3-carboxylic acid trisodium salt. tartrazine.

dihydrophenytoin. phenytoin.

2,3-dihydro-6-propyl-2-thioxo-4(1H)-pyridinone. propylthiouracil.

dihydropsychotrine. cephaeline.

1,7-dihydro-6*H*-purine-6-thione. 6-mercaptopurine.

dihydrorhengol. 3-pentadecylcatechol.

dihydrosanguinarine. See *Argemone*.

dihydrotestosterone (DHT). the active form of testosterone, *q.v.*

dihydrostreptomycin. an ototoxic hydrogenated derivative of streptomycin with similar activity to that of the parent compound.

dihydrotestosterone (7β-hydroxy-5α-androstan-3-one; DHT). a metabolite of testosterone, $C_{19}H_{30}O_2$, formed in a reaction catalyzed by 5α-reductase in many androgen-responsive tissues. It is more potent than testosterone and stimulates the development of most male secondary characteristics as well as the endocrine-based adult male sexual functions. It may also induce virilization during embryogenesis.

6,7-dihydro-1,2,3,10-tetramethoxy-7-(methylamino)benzo[a]heptalen-9(5*H*)-one. demecolcine.

3,4-dihydro-2,5,7,8-tetramethyl-2-(4,8,12-trimethyltridecyl)-2*H*-1-benzopyran-6-ol. vitamin E.

dihydrotrichothecene. a trichothecene produced by *Trichothecium roseum*.

3,7-dihydro-1,3,7-trimethyl-1*H*-purine-2,6-dione. caffeine.

1,3-dihydroxy-2-amino-4-octadecene. sphingosine.

1,2-dihydroxybenzene. pyrocatechol.

***m*-dihydroxybenzene**. resorcinol.

***o*-dihydroxybenzene**. pyrocatechol.

***p*-dihydroxybenzene**. hydroquinone.

4,5-dihydroxybenzoic acid monogallate. digallic acid.

7,8-dihydroxy-2*H*-1-benzopyran-2-one. daphnetin.

2,6-dihydroxy-5-bis[2-chloroethyl]aminopyr-amidine. uracil mustard.

3,14-dihydroxybufa-20,22-dienolide. bufalin.

1α,25-dihydroxycholecalciferol. calcitriol.

1α,25-dihydroxycholecalciferol(1α,25-[OH],-D$_3$) glycoside. a derivative of calcitriol that may be responsible for some cases of enzootic calcinosis (See calcinosis) in herbivores.

3-(3,4-dihydroxycinnamate. chlorogenic acid.

3-(3,4-dihydroxycinnamoyl)quinic acid. chlorogenic acid.

6,7-dihydroxycoumarin. **1:** See esculin. **2:** the aglucon of cichorin, *q.v.*

7,8-dihydroxycoumarin. daphnetin. *C.f.* 7,8-dihydroxycoumarin 7-β-D-glucoside.

dihydroxycoumarin glucoside. daphnin.

6,7-dihydroxycoumarin 6-glucoside. esculin.

6,7-dihydroxycoumarin-7-glucoside. cichorin.

7,8-dihydroxycoumarin 7-β-D-glucoside. daphnin. Note: do not confuse with 7,8-dihydroxycoumarin.

1,2-dihydroxy-1,2-cyclobutanediacrylic acid di-τ-lactone. anemonin.

7,8-dihydroxydiacetoxyscirpenol. a trichothecene first isolated from *Fusarium*.

dihydroxydiaminoarsenobenzene dihydrochloride. salvarsan.

dihydroxydiethyl ether. diethylene glycol.

4,4′-dihydroxy-α,β-diethylstilbene. diethylstilbestrol.

7′,12′-dihydroxy-6,6′-dimethoxy-2,2′,2′-trimethyltubocuraranium chloride hydrochloride. tubocurarine chloride.

2,2′-dihydroxy-6,6′-dinaphthyldisulfide. DDD, analytical. Do not confuse with DDD.

[illegible] 29-nor-8α,9β,-13α,14β-dammara-1,17-(20),24-trien-21-oic acid diacetate. helvolic acid.

dihydroxyethyl sulfide. thiodiglycol.

2,2′-dihydroxy-3,3′,5,5′,6,6′-hexachlorodiphenylmethane. hexachlorophene.

dihydroxy-3-(hydroxymethyl)-9,10-anthracenedione. aloe-emodin.

1,8-dihydroxy-3-(hydroxymethyl)anthraquinone. aloe-emodin.

3,4-dihydroxy-1-[1-hydroxy-2-(methylamino)-ethyl]benzene. epinephrine.

β,β′-dihydroxyisopropyl chloride. α-chlorohydrin (See chlorohydrin).

1,8-dihydroxy-3-methyl-9-anthrone. See chrysarobin (def. 1).

3,5-dihydroxy-3-methylpentanoic acid. mevalonic acid.

7,20-dihydroxy-16-methyl-10-phenyl-24-oxo-[14]cytochalasa-6(12),13,21-triene-1,23-dione. cytochalasin B.

3,5-dihydroxy-3-methylvaleric acid. mevalonic acid.

β,δ-dihydroxy-β-methylvaleric acid. mevalonic acid.

dihydroxymurexine. a compound isolated from the hypobranchial gland of the gastropod mollusc, *Thais haemastoma*.

6α,8β-dihydroxy-4-oxoambrosa-2,11(13)-dien-12-oic acid 12,8-lactone. helenalin.

3-(3,4-dihydroxyphenalalanine. dopa.

3,4-dihydroxyphenethylamine. dopamine.

L-3,4-dihydroxyphenethylamine 2-(L-3,4-dihydroxyphenyl)aminoethane. dopamine.

(—)-3-(3,4-dihydroxyphenyl)-L-alanine. levodopa.

β-(3,4 dihydroxyphenyl) α alanine. levodopa.

dihydroxyphenylalanine. dopa.

L-3,4-dihydroxyphenylalanine. See L-dopa.

α-(3,4-dihydroxyphenyl)-β-aminoethane. dopamine.

1-(3,4-dihydroxyphenyl)-2-aminoethanol. norepinephrine.

2-(3,4-dihydroxyphenyl)-3,4-dihydro-2H-1-benzopyran-3,5,7-triol. catechin.

1-(3,4-dihydroxyphenyl)-2-(methylamino)-ethanol. epinephrine.

3-[[3-(3,4-dihydroxyphenyl)-1-oxo-2-propanyl]oxy]-1,4,5-trihydroxycyclohexanecarboxylic acid. chlorogenic acid.

(11β)-11,21-dihydroxypregn-4-ene-3,20-dione. corticosterone.

17α,21-dihydroxy-4-pregnene-3,11,20-trione. cortisone.

11-21-dihydroxy-progesterone. corticosterone.

17,21-dihydroxy-4-propylajmalanium. prajmaline.

17R,21-α-dihydroxy-4-propylajmalanium hydrogen tartrate. prajmaline tartrate.

12,18-dihydroxysenecionan-11,16-dione. retrorsine

Δ^5-3β,12α-dihydroxysolanidene. rubijervine.

3,4-dihydroxy-5-[(3,4,5-trihydroxybenzoyl)oxy]-benzoic acid. digallic acid.

1α,25-dihydroxyvitamin D_3. calcitriol.

2,6-diiodo-4-cyanophenol. ioxynil.

3,5-diiodo-4-hydroxybenzonitrile. ioxynil.

3,5-diiodo-4-hydroxyphenyl cyanide. ioxynil.

diiodohydroxyquin. See iodoquinol.

diisobutylamine(2-methyl-*N*-(2-methylpropyl)-1-propanamine). a clear, flammable liquid, $C_8H_{19}N$, with an amine odor. It is toxic and a dangerous fire hazard. Toxic fumes, NO_x, are released on heating to decomposition. See also amine (lower aliphatic amine).

diisobutylthiocarbamic acid *S*-ethyl ester. butylate.

diisocarb. butylate.

diisopropyl fluorophosphonate. isoflurophate.

diisopropyl fluorophosphate (DFP; DIFP; diisopropyl phosphorofluoridate; diisopropylfluorophosphonate; diisopropylfluorophosphate; fluostigmine; isofluorphate; isoflurophate; isoflurophosphate; isopropylfluophosphate; phosphorofluoridic acid bis(1-methylethyl) ester; phosphorofluoridic acid diisopropyl ester; diisopropylphosphofluoridate). an extremely toxic fluorophosphate alkyl ester, $[(CH_3)_2CHO]_2P(O)F$, it is an oily, liquid, cholinergic that irreversibly inhibits cholinesterase. It inhibits proteolytic enzymes and acetylcholinesterase by reacting with serine residues. It is used in insecticides and nerve gas and as a miotic. DFP has pronounced miotic effects even at concentrations that are chemically undetectable. It forms hydrogen fluoride in the presence of moisture.

α-[2-(diisopropylamino)ethyl]-α-phenyl-2-pyridineacetamide. disopyramide.

4-(diisopropylamino)-2-phenyl-2-(2-pyridyl)-butyramide. disopyramide.

diisopropylcarbamic acid *S*-2,3-dichloroallyl ester. diallate.

diisopropylphosphofluoridate. diisopropyl fluorophosphate.

diisopropylthiocarbamic acid *S*-(2,3,3-trichloroallyl) ester. triallate.

2,5-diketohexane. acetonylacetone.

diketone. any ketone that contains two carbonyl groups (e.g., acetonylacetone, 2,5-heptanedione, 2,5-hexanedione, and 3,6-octanedione). the γ-diketones are neurotoxic; the α-, β-, and δ-diketones are not. See also γ-diketone neuropathy; axonopathy.

γ-diketone neuropathy (hexacarbon neuropathy). acetonylacetone, a γ-diketone, is the ultimate neurotoxic metabolite of *n*-hexane and methyl *n*-butyl ketone. Its toxicity is probably due to its reaction with lysyl amino groups of proteins with the production of 2,5-dimethylpyrrolyl derivatives. Other γ-diketones (e.g., 2,5-heptanedione and 3,6-octanedione) and hydrocarbons that can be metabolized to γ-diketones may have a similar action; the α-, β- and δ-diketones are not neurotoxic.

Dilantin. phenytoin.

dilatator. dilator.

dilated pupil. enlarged pupil of the eye.

dilation of pupils. mydriasis.

dilator (dilatator). in toxicology, a substance that enlarges an opening (e.g., the pupil of the eye) or the lumen of a hollow structure.

diluent. **1:** attenuating, diluting, having the capacity to dilute. **2:** a diluting agent; an agent that dilutes, attenuates or renders less potent. **3:** a dilutent (def. 2).

dilute. **1:** of or pertaining to a solution in which the concentration of solute is low relative to that of the solvent. **2a:** to weaken, attenuate, thin, water down. **2b:** to reduce the concentration of solute in a solution, usually by adding solvent or a more dilute solution. **2c:** to make any mixture or a powder less concentrated by adding another substance (a diluent or dilutent). *Cf.* concentrated.

dilutent. **1:** having the capacity to dilute. **2:** an excipient that increases bulk and density of a chemical formulation. Calcium carbonate, dicalcium phosphate, kaolin, milk solids, starch, and sucrose are often used for this purpose. *Cf.* diluent.

dilution. **1:** diluting or being diluted; attenuation, weakening, thinning, watering down. **2:** the state of being diluted. **3:** a diluted substance, mixture, or solution. **4:** a method of counting the number of viable microbes or cells in a suspension in which a sample is diluted to the extent that an aliquot, when plated, yields a countable number of distinct colonies.

dilution rate. the amount of diluent per unit chemical needed to obtain a desired concentration or dosage.

dilution ratio. the ratio of the volume of water in a stream to the volume of incoming water. It is a measure of the ability of the stream to assimilate toxic materials or wastes.

dimanganese trioxide. manganese(III) oxide.

dimercaprol (2,3-dimercapto-1-propanol; 1,2-dithioglycerol; British Antilewisite; 2,3-mercaptopropanol; BAL; antilewisite; dimercaptopropanol). a thick, oily, almost colorless compound, $CH_2SHCHSHCH_2OH$, with the odor of a mercaptan. It is a very toxic chelating agent, developed to treat poisoning by lewisite and other arsenicals. It is also effective as an antidote in the treatment of antimony, bismuth, chromium, gold, mercury, nickel, and lead poisonings; it is administered *i.m.* Metal complexes of dimercaprol decompose readily in acid solution. See also chelation.

2,3-dimercapto-1-propanesulfonic acid (2,3-dithiolpropanesulfonic acid). a toxic chelating agent used as an antidote for heavy metal poisoning. It is structurally related to dimercaprol.

(R*,S*)-2,3-dimercaptobutanedioic acid. succimer.

1,3-dimercaptopropane. 1,3-propanedithiol.

2,3-dimercapto-1-propanol. dimercaprol.

dimercaptopropanol. See dimercaprol.

***meso*-2,3-dimercaptosuccinic acid**. succimer.

dimetan. Dimetan®.

Dimetan® (dimethylcarbamic acid 5,5-dimethyl-3-oxo-1-cyclohexen-1-yl ester; 5,5-dimethyldihydroresorcinol dimethylcarbamate; dimetan). a synthetic neurotoxic carbamate insecticide. It is a cholinesterase inhibitor with an action similar to that of physostigmine.

dimethoate (phosphorodithioic acid *O,O*-dimethyl *S*-[2-(methylamino)-2-oxoethyl] ester; phosphorodithioic acid *O,O*-dimethyl ester, ester with 2-mercapto-*N*-methylacetamide). a systemic and contact insecticide and acaricide. A cholinesterase inhibitor, it is very toxic to laboratory rats.

2,3-dimethoxy-benzoic acid. orthoveratric acid.

3,3′-dimethoxybenzidine. dianisidine.

3,4-dimethoxybenzoic acid. veratric acid.

3,3′-dimethoxy-[1,1′-biphenyl]-4,4′ diamine. dianisidine.

1-(2,5-dimethoxy-4-bromophenyl)-2-aminopropane. DOB.

3,3′-dimethoxy-4,4′-diaminobiphenyl. dianisidine.

2,5-dimethoxy-α,4-dimethylphenethylamine hydrochloride. 2,5-dimethoxy-4-methylamphetamine hydrochloride.

2,5-dimethoxy-4-methylamphetamine. 2,5-dimethoxy-4-methylamphetamine hydrochloride.

2,5-dimethoxy-4-methylamphetamine hydrochloride(1-(2,5-dimethoxy-4-methyl-phenyl)-2-aminopropane; 2,5-dimethoxy-4-methylamphetamine; 2,5-dimethoxy-α,4-dimethylphenethylamine hydrochloride; DOM; STP). a habit-forming CNS stimulant and hallucinogen. It is biologically active and toxic when ingested or administered *i.v.* or *i.p.* It modifies thought processes and perception involving any of the senses and may promote or precipitate psychosis or emotional problems in some persons. Further effects may include hyperreflexia and restlessness. High doses produce hallucinations and loss of touch with reality which may persist as long as 24 hours. Fatalities may occur. See amphetamine.

1,2-dimethoxy-12-methyl[1,3]benzodioxolo[5,6-c]phenanthridinium. chelerythrine.

1-(2,5-dimethoxy-4-methylphenyl)-2-aminopropane. 2,5-dimethoxy-4-methylamphetamine hydrochloride.

1-[(3,4-dimethoxyphenyl)methyl]6,7dimethoxyisoquinoline. papaverine.

2-dimethoxyphosphinothioylthiomethyl-4,6-diamino-s-triazine. menazon.

[(dimethoxyphosphinothioyl)thio]butanedioic acid diethyl ester. malathion.

3-(dimethoxyphosphinyloxy)-*N,N*-dimethyl-*cis*-crotonamide. dicrotophos.

3-[(dimethoxyphosphinyl)oxy]-2-butenoic acid methyl ester. mevinphos.

N-[[6-(2,3-dimethoxypropyl)tetrahydro-4-hydroxy-5,5-dimethyl-2*H*-pyran-2-yl]-methoxymethyl]tetrahydro-α-hydroxy-2-methoxy-5,6-dimethyl-4-methylene-2*H*-pyran-2-glycolamide. pederin.

N-[[6-(2,3-dimethoxypropyl)tetrahydro-4-hydroxy-5,5-dimethyl-2*H*-pyran-2-yl]-methoxymethyl]tetrahydro-α-hydroxy-2-methoxy-5,6-dimethyl-4-methylene-2*H*-pyran-2-acetamide. pederin.

2,3-dimethoxystrychnidin-10-one. brucine.

10,11-dimethoxystrychnine. brucine.

dimethoxystrychnine. brucine.

((3β,16β,17α,18β,20α)-11,17-dimethoxy-18-[(3,4,5-trimethoxybenzoyl)oxy]yohimban-16-carboxylic acid methyl ester. reserpine.

6,7-dimethoxy-l-veratrylisoquinoline. papaverine.

dimethyl. ethane.

3,3-dimethylacrylic acid. senecioic acid.

2-dimethylallyl-5,9-dimethyl-2′-hydroxybenzomorphan. pentazocine.

3-[[(dimethylamino)carbonyl]oxy]-*N,N,N*-trimethylbenzenaminium. neostigmine.

β-dimethylaminoethanol diphenylmethyl ether. diphenhydramine.

dimethylaminoethodycyanophosphine oxide. tabun.

α-(2-dimethylaminoethoxy)diphenylmethane. diphenhydramine.

2-dimethylaminoethoxyphenylmethyl-2-picoline. doxylamine.

β-dimethylaminoethyl benzhydryl ether. diphenhydramine.

3-(β-dimethylaminoethyl)-5-hydroxyindole. bufotenine.

3-(2-(dimethylamino)ethyl)indole. DMT.

3-[2-(dimethylamino)ethyl]-1*H*-indole-4-ol. psilocin.

3-[2-(dimethylamino)ethyl]-1*H*-indol-5-ol. bufotenine.

3-(2-dimethylaminoethyl)-5-indolol. bufotenine.

3-[2-(dimethylamino)ethyl]-1H-indol-4-ol dihydrogen phosphate ester. psilocin.

3-2′-dimethylaminoethyl)indol-4-phosphate. psilocin.

3-[2-(dimethylamino)ethyl]-1*H*-indol-4-ol dihydrogen phosphate ester. psilocybin.

3-[2-(dimethylamino)ethyl]-1*H*-indol-4-ol dihydrogen phosphate ester. psilocin.

3-2′-(dimethylaminoethyl)indol-4-phosphate. psilocin.

β-dimethylaminoethyl-2-pyridylaminotoluene. 2,4,6-trinitrotoluene.

β-dimethylaminoethyl-2-pyridylbenzylamine. 2,4,6-trinitrotoluene.

2-dimethylaminoethyl succinate dimethochloride. succinylcholine chloride.

2-(dimethylamino)-*N*-[[(methylamino)-carbonyl]oxy]-2-oxoethanimidothioic acid methyl ester. oxamyl.

α-d-4-dimethylamino-3-methyl-1,2-diphenyl-2-butanol propionate. propoxyphene.

α-[2-(dimethylamino)-1-methylethyl]-α-phenylbenzeneethanol propanoate. propoxyphene.

[*S*-(*R,*S**)]-α-[2-(dimethylamino)-1-methylethyl]-α-phenylbenzeneethanol propanoate** (ester). propoxyphene.

4-(dimethylamino)-3-methylphenolmethylcarbamate ester. aminocarb.

3-dimethylamino-7-methyl-1,2-(*n*-propylmalonyl)-1,2-dihydro-1,2,4-benzotriazine. apazone.

5-(dimethylamino)-9-methyl-2-propyl-1*H*-pyrazolo[1,2-a][1,2,4]benz otriazine-1,3(2*H*)-dione. apazone.

(11β,17β)-11-[4-(dimethylamino)-phenyl]-17-hydroxy-17-(1-propynyl)estra-4,9-dien-3-one. mifepristone.

***N,N*-dimethylamino-*N'*-(phenylmethyl)-*N'*-2-pyridinyl-1,2-ethanediamine**. tripelennamine.

***N*-[4[[4-(dimethylamino)phenyl]phenylmethylene]-2,5-cyclohexadien-1-ylidene]-*N*-methylmethanaminium chloride**. Malachite Green. (Note: both the chloride and the oxalate are termed Malachite Green.)

9-[[3-(dimethylamino)propyl]amino]-1-nitroacridine. nitracrine.

***N*-(3-dimethylaminopropyl)-3-chlorophenothiazine**. chlorpromazine.

5-(3-dimethylaminopropyl)-10,11-dihydro-5*H*-dibenz[b,f]azepine. imipramine.

***N*-(γ-dimethylaminopropyl)iminodibenzyl**. imipramine.

11-(3-dimethylaminopropylidene)-6,11-dihydrodibenz[b,e]oxepin. doxepin.

***N*-(dimethylamino)succinamic acid**. daminozide.

4-dimethylamino-*m*-tolyl methylcarbamate. aminocarb.

4-dimethylamino-3,5-xylyl methylcarbamate. mexacarbate.

dimethylarsine. an extremely poisonous arsenical, often formed by microbial activity from arsenic compounds introduced into an aquatic environment. It accumulates through aquatic food chains to piscivorous animals including humans, sometimes producing serious impacts.

dimethylarsinic acid. cacodylic acid.

[(dimethylarsino)oxy]sodium arsenic oxide. sodium cacodylate.

[(dimethylarsino)oxy]sodium As oxide. sodium cacodylate.

dimethylbenz[a]anthracene. 9,10 dimethyl-1,2-benzanthracene.

7,12-dimethylbenz[a]anthracene. 9,10-dimethyl-1,2-benzanthracene.

9,10-dimethylbenzanthracene. 9,10-dimethyl-1,2-benzanthracene.

9,10-dimethyl-1,2-benzanthracene (DBA; dimethylbenz[a]anthracene; 7,12-dimethylbenz-[a]anthracene; 9,10-dimethylbenzanthracene; 9-10-benz[a]anthracene; dimethylbenzanthrene; 7,12-dimethylbenzo[a]anthracene; 1,4-dimethyl-2,3-benzphenanthrene; DMBA; 7,12-DMBA). a toxic, polycyclic aromatic hydrocarbon comprised of water-insoluble plates or leaflets that are slightly soluble in ethanol and freely soluble in benzene. DBA is a teratogen, a human mutagen, and tumorigen. It is a transplacental carcinogen and neoplastigen in laboratory mammals and is a suspected human carcinogen. DBA is a skin irritant and is probably poisonous by all routes of exposure or administration.

dimethylbenzanthrene. 9,10-dimethyl-1,2-benzanthracene.

dimethylbenzene. xylene.

dimethyl 1,2-benzenedicarboxylate. dimethyl phthalate.

α,α-dimethylbenzeneethanamine. phentermine.

***N*,α-dimethylbenzeneethanamine**. methamphetamine.

3,3'-dimethylbenzidine. *o*-tolidine.

2,2-dimethyl-1,3-benzodioxol-4-ol-methylcarbamate. bendiocarb.

1,4-dimethyl-2,3-benzphenanthrene. 9,10 dimethyl-1,2-benzanthracene.

***N,N*-dimethyl-*N*-benzyl-*N*-chloro-*N'*-dodecylglycinamide**. dodecarbonium chloride.

***N,N*-dimethyl-*N'*-benzyl-*N'*-(α-pyridyl)ethylenediamine**. 2,4,6-trinitrotoluene.

dimethylberyllium (dimethyl beryllium). a white solid with needlelike crystals. It is poisonous and a confirmed human carcinogen. When heated to decomposition, it emits highly toxic beryllium oxide fumes. It combusts spontaneously on contact with moist air or carbon dioxide and reacts explosively with water.

1,1'-dimethyl-4,4'-bipyridinium. paraquat.

dimethylcadmium (cadmium dimethyl). a hazardous, oily, explosive liquid at room temperatures, $Cd(CH_3)_2$, m.p. -4.5°C, b.p. 106°C. It is prepared from a Grignard reagent and a cadmium halide. It decomposes on contact with water and is distillable but explodes at temperatures above 150°C, releasing toxic cadmium oxide fumes. On contact with air, it produces a friction-sensitive explosive, dimethylcadmium peroxide; but if the surface area is large, it may ignite on contact with air. Dimethyl cadmium is a confirmed human carcinogen. See also cadmium oxide.

dimethylcadmium peroxide. See dimethylcadmium.

dimethylcarbamic acid 1-[(dimethylamino)-carbonyl]-5-methyl-1*H*-pyrazol-3-yl ester. dimetilan.

dimethylcarbamic acid 2-(dimethylamino)-5,6-dimethyl-4-pyrimidinyl ester. pirimicarb.

dimethylcarbamic acid 5,5-dimethyl-3-oxo-1-cyclohexen-1-yl ester. Dimetan®.

***N,N*-dimethylcarbamic acid-3-dimethylaminophenyl ester methosulfate**. neostigmine methyl sulfate.

dimethylcarbamic acid ester with (*m*-hydroxyphenyl)trimethylammonium methyl sulfate. neostigmine methyl sulfate.

dimethylcarbamic acid ester with 3-hydroxy-*N,N*,5-trimethylpyrazole-1-carboxamide. dimetilan.

dimethylcarbamic acid 3-methyl-1-(1-methylethyl)-1*H*-pyrazol-5-yl ester. primin.

dimethylcarbamic acid 3-methyl-1-phenyl-1*H*-pyrazol-5-yl ester. Pyrolan®

(3-dimethylcarbamoxyphenyl) trimethylammonium. neostigmine.

dimethyl carbamoyl chloride (chloroformic acid dimethylamide; DDC; (dimethylamino)carbonyl chloride; dimethylcarbamic acid chloride; dimethylcarbamic chloride; dimethylcarbamidoyl chloride; dimethylcarbamoyl chloride; *N,N*-dimethylcarbamoyl chloride; dimethylcarbamyl chloride; *N,N*-dimethylcarbamyl chloride; DMCC). an extremely toxic liquid and potent lacrimator that reacts with water and steam to produce corrosive fumes. It is a confirmed animal tumorigen, carcinogen and neoplastigen, causing skin and papillary tumors from direct contact, and squamous cell carcinoma from inhalation. It is a suspected human carcinogen and probable human mutagen. DDC is moderately to very toxic when inhaled or ingested, but is extremely toxic when administered *i.p*.

2-dimethylcarbamoyl-3-methyl-5-pyrazolyl dimethylcarbamate. dimetilan.

dimethylcarbinol. isopropyl alcohol.

***O,O*-dimethyl-1-carbomethoxy-1-propen-2-yl phosphate**. mevinphos.

α,α-dimethyl-α'-carboxydihydro-γ-pyrone butyl ester. butopyronoxyl.

1,1-dimethyl-3-(*p*-chlorophenyl)urea. monuron.

(+)-*O,O'*-dimethylchondrocurarine diiodide. metocurine iodide.

2,2-dimethyl-7-coumaranyl *N*-methylcarbamate. carbofuran.

***O,O*-dimethyl *S*-[(4,6-diamino-s-triazin-[illegible] methyl] phosphorodithioate**. menazon.

***N,N*-dimethyldibenz[b,e]oxepin-$\Delta^{11(6H)}$,γ-propylamine**. doxepin.

1,1-dimethyl-3-(3,4-dichlorophenyl)urea. diuron.

_O,O_-dimethyl _O_-(2,2-dichlorovinyl) phosphate. dichlorvos.

2,2-dimethyl-2,3-dihydro-7-benzofuranyl-_N_-methylcarbamate. carbofuran.

5,5-dimethyldihydroresorcinol dimethylcarbamate. Dimetan®.

5,6-dimethyl-2-dimethylamino-4-dimethylcarbamoyloxypyrimidine. pirimicarb.

dimethyl 2-dimethylcarbamoyl-1-methylvinyl phosphate. dicrotophos.

dimethyl 1-dimethylcarbamoyl-1-propen-2-yl phosphate. dicrotophos.

_O,O_-dimethyl-O-[_p_-(dimethylsulfamoyl)-phenyl]phosphorothioate (generic). famphur.

1,6-dimethyl-5,7-dioxo-1,5,6,7-tetrahydropyrimido[5,4-e]-_as_-triazine. toxoflavin.

_N,N_-dimethyl-α,α-diphenyl acetamide. diphenamid.

_N,N_-dimethyl-2,2-diphenylacetamide. diphenamid.

_N,N'_-dimethyl-γ,γ'-dipyridylium. paraquat.

dimethyldiselenide. See dimethyl selenide.

_O,O_-dimethyl dithiophosphate of diethyl mercaptosuccinate. malathion.

_O,O_-dimethyl-_S_-(2-ethionylethyl)phosphorothioate. demeton-*O*-methyl sulfoxide.

dimethyl-_S_-(2-ethionylethyl) thiophosphate. demeton-*O*-methyl sulfoxide.

_O,O_-dimethyl-_S_-(2-ethsulfonylethyl)phosphorothioate. demeton-*S*-methylsulfone.

dimethyl-_S_-(2-ethsulfonylethyl)thiophosphate. demeton-*S*-methylsulfone.

(±)1-[(1,1-dimethylethyl)amino]-3-[(2-methyl-1_H_-indol-4-yl)oxy]-2-propanol benzoate ester. bopindolol.

dimethyl ethyl carbinol. *tert*-pentyl alcohol.

_O,O_-dimethyl_O_-(2-ethyl-4-ethoxypyrimidinyl)-6-thionophosphate. etrimphos.

_O,O_-dimethyl-_O_-ethylmercaptoethyl thiophosphate. demeton-*O*-methyl.

_O,O_-dimethyl 2-ethylmercaptoethyl thiophosphate, thiono isomer. demeton-*O*-methyl.

_O,O_-dimethyl-_S_-[2-(ethylsulfinyl)ethyl]-phosphorothioate. demeton-*O*-methyl sulfoxide.

_O,O_-dimethyl-_S_-ethylsulfinylethyl phosphorothiolate. demeton-*O*-methyl sulfoxide.

_O,O_-dimethyl-_S_-(2-ethylsulfinyl)ethyl thiophosphate. demeton-*O*-methyl sulfoxide.

_O,O_-dimethyl-_S_-ethylsulfonylethyl phosphorothiolate. demeton-*S*-methylsulfone.

_O,O_-dimethyl-_S_-ethyl-2-sulfonylethyl phosphorothiolate. demeton-*S*-methylsulfone.

_O,O_-dimethyl-_O_-2-(ethylthio)ethyl phosphorothioate. demeton-*O*-methyl.

dimethylformamide. *N,N*-dimethylformamide.

_N,N_-dimethylformamide (DMF; DMFA; *N*-formyldimethylamine; dimethylformamide). the "universal organic solvent," it is a colorless, mobile liquid, $HCON(CH_3)_2$, widely used as an industrial solvent. DMF is hepatotoxic, probably carcinogenic, and the liquid or vapor is an irritant to the eyes, skin, and mucous membranes. It is a fire and explosion hazard.

3,3'-dimethyl-3,3',8,8',9,9'-hexahydroxy-3,3',4,4'-tetrahydro-(7,10'bianthracene)-1,1'(2_H_,2'_H_)-dione (T-514). a buckthorn toxin.

1,1-dimethylhydrazine (*unsym*-dimethylhydrazine; *asym*-dimethylhydrazine; *N,N* dimethylhydrazine; UDMH). a colorless, highly flammable, highly toxic, hygroscopic, water-miscible liquid, $(CH_3)_2NNH_2$, and a powerful reducing agent. UDMH is a plant growth regulator, a constituent of rocket and jet fuels, and a stabilizer for organic peroxide fuel additives. It is very toxic, orally, to laboratory rats and mice and moderately toxic by inhalation and skin contact. It is an irritant to the eyes, skin, and mucous membranes and a CNS stimulant, sometimes causing convulsions. UDMH is teratogenic and a confirmed carcinogen. Very toxic fumes, NO_x, are released on decomposition by heating. Do not confuse with 1,2-dimethylhydrazine. See also hydrazines, dialkylhydrazine.

1,2-dimethylhydrazine (*N,N'*-dimethylhydrazine; *sym*-dimethylhydrazine; SDMH; hydrazomethane). a clear, colorless (gradually turning yellow), flammable, mobile, fuming (in air), hygroscopic liquid, $CH_3NHNHCH_3$, with a characteristic fishy ammoniacal odor. It is miscible with water (evolving much heat), ethanol, ethyl ether, dimethylformamide, and hydrocarbons generally. It is used as a high-energy liquid rocket propellant. It is corrosive to the skin and other tissues. Data indicate that it is extremely toxic orally to laboratory mice and very toxic to laboratory rats. It is moderately toxic on inhalation. Effects of systemic poisoning are similar to those of 1,1-dimethylhydrazine, *q.v.* It is a confirmed carcinogen and probable human mutagen. See also hydrazines, dialkylhydrazines. Do not confuse with 1,1-dimethylhydrazine. See dialkylhydrazine.

***asym*-dimethylhydrazine**. 1,1-dimethylhydrazine.

***N,N'*-dimethylhydrazine**. 1,2-dimethylhydrazine.

***N,N*-dimethylhydrazine**. 1,1-dimethylhydrazine.

***sym*-dimethylhydrazine**. 1,2-dimethylhydrazine.

***unsym*-dimethylhydrazine**. 1,1 dimethylhydrazine.

dimethyl-5-(1-isopropyl-3-methylpyrazolyl)-carbamate. primin.

dimethyl ketone. acetone.

dimethylketone. acetone.

[illegible]. A poisonous pyrophoric liquid, $Mg(CH_3)_2$, that melts at 0^oC. It reacts violently with water or steam and combusts spontaneously in air.

dimethylmercury (mercury, dimethyl; methyl mercury). **1:** an organic compound, $(CH_3)2Hg$, that is synthesized from inorganic mercury by anaerobic bacteria in neutral or slightly acid sediments and soils through the action of methylcobalamin. Production of the latter compound is favored under such conditions. Dimethylmercury is soluble and volatile and hence may escape to the atmosphere. It is lipid soluble and bioaccumulates or biomagnifies in aquatic organisms. Fish tissues frequently have concentrations of mercury that are more than 1000 times greater than that of the surrounding water. This compound and its sister, monomethylmercury, have similar chemical and toxicological properties. See also mercury, methylmercury, Minamata disease, monomethylmercury.

***N,N*-dimethyl-α-methylcarbamoyloxyimino-α-(methylthio)acetamide**. oxamyl.

***N',N'*-dimethyl-N-[(methylcarbamoyl)oxy]-1-thiooxamimidic acid methyl ester**. oxamyl.

***O,O*-dimethyl-*S*-(*N*-methyl-*N*-formylcarbamoylmethyl)phosphorodithiolate**. formothion.

***O,O*-dimethyl *O*-(4-methylmercapto-3-methylphenyl) thionophosphate**. fenthion.

2,2-dimethyl-3-(2-methylpropanyl)cyclopropanecarboxylic acid 6-chloropiperonyl ester. barthrin.

2,2-dimethyl-3-(2-methyl-1-propanyl)cyclopropanecarboxylic acid. chrysanthemic acid.

2,2-dimethyl-3-(2-methyl-1-propenyl)cyclopropanecarboxylic acid 3-(2-cyclopenten-1-yl)-2-methyl-4-oxo-2-cyclopenten-1-yl ester. cyclethrin.

2,2-dimethyl-3-(2-methyl-1-propenyl)cyclopropanecarboxylic acid 2-methyl-4-oxo-3-(2-propenyl)-2-cyclopenten-1-yl ester. allethrin I.

2,2-dimethyl-3-(2-methyl-1-propenyl)-cyclopropanecarboxylic acid (6-chloro-1,3-benzodioxol-5-yl)methyl ester. barthrin.

2,2-dimethyl-3-(2-methyl-1-propenyl)cyclopropanecarboxylic acid 2-methyl-4-oxo-3-(2,4-pentadienyl)-2-cyclopenten-1-yl ester. pyrethrin I.

***O,O*-dimethyl*O*-(4-methylthio-3-methylphenyl) thiophosphate**. fenthion.

***O,O*-dimethyl *O*-[4-(methylthio-*m*-tolyl]-phosphorothiolate**. fenthion.

3,5-dimethyl-4-(methylthio)phenol methylcarbamate. methiocarb.

3,3-dimethyl-2-methylene-norbornane. camphene.

dimethyl monosulfate. dimethyl sulfate.

dimethyl morphine. thebaine.

***N,N*-dimethyl-*N'*-(1-nitro-9-acridinyl)-1,3-propanediamine**. nitracrine.

***O,O*-dimethyl-*O-p*-nitrophenylphosphorothioate)**. methyl parathion.

***O,O*-dimethyl *o-p*-nitrophenyl phosphorothioate**. methyl parathion.

***O,O*-dimethyl *o-p*-nitrophenyl thiophosphate**. methyl parathion.

dimethylnitrosamine. *N*-nitrosodimethylamine.

dimethylol propionic acid (2,2-bis(hydroxymethyl)propionic acid; DMPA). an off-white granular powder that is soluble in water and methanol, slightly soluble in acetone, and insoluble in benzene. It is used in the preparation of water-soluble alkyl resins, cosmetics, plasticizers, and in textile finishing. Do not confuse with the herbicide, DMPA.

***O,O*-dimethyl-*S*-4-oxo-1,2,3-benzotriazin-3-(4*H*)-yl methyl phosphorodithioate**. azinphos methyl.

3-[2(3,5-dimethyl-2-oxocyclohexyl)-2-hydroxyethyl]glutarimide. cycloheximide.

4-[2-(3,5-dimethyl-2-oxocyclohexyl)2-hydroxyethyl]-2,6-piperidinedione. cycloheximide.

dimethyloxychinizin. antipyrine.

dimethyl parathion. methyl parathion.

dimethylphenanthrene. a weakly estrogenic, carcinogenic hydrocarbon compound.

***d-N*,α-dimethylphenethylamine**. methamphetamine.

***N,N*-dimethyl-4-(phenylazo)benzenamine**. *p*-dimethylaminoazobenzene.

***N,N*-dimethyl-α-phenylbenzenacetamide**. diphenamid.

***N*,α-dimethyl-*N*-(phenylmethyl)-benzeneethanamine**. benzphetamine.

***N,N*-dimethyl-*N'*-(phenylmethyl)-*N'*-2-pyridinyl-1,2-ethanediamine**. tripelennamine.

1,3-dimethyl-4-phenyl-4-piperidinol propanoate. alphaprodine.

α-1,3-dimethyl-4-phenyl-4-piperidinyl propanoate. alphaprodine.

α-1,3-dimethyl-4-phenyl-4-propionoxypiperidine. alphaprodine.

2,3-dimethyl-1-phenyl-3-pyrazolin-5-one. antipyrine.

1,5-dimethyl-2-phenyl-3-pyrazolone. antipyrine.

***N,N*-dimethyl-2-[1-phenyl-1-(2-pyridinyl)-ethoxy]ethanamine**. doxylamine.

***N,N*-dimethyl-*N*-phenylurea**. fenuron.

dimethyl phosphate of 3-hydroxyl-*N,N*-dimethyl-*cis*-crotnamide. dicrotophos.

dimethylphosphine. a reactive, colorless, volatile liquid, $(CH_3)_2PH$, that boils at 25°C. It is poisonous by ingestion or by inhalation of the fumes. It is a pulmonary tract irritant and CNS depressant; effects similar to those of phosphine. It is also carcinogenic, weakly estrogenic, and is extremely destructive to protozoans and microscopic metazoans. Symptoms in humans include nausea, vomiting, fatigue, dyspnea, and sometimes convulsions, coma, and death. *Cf.* methylphosphine, phosphine.

***O,S*-dimethyl phosphoramidothioate**. methamidophos.

***O,O*-dimethyl phosphorothioate**. demeton-*S*-methyl.

dimethyl phthalate (1,2-benzenedicarboxylic acid dimethyl ester; phthalic acid dimethyl ester; methyl phthalate; dimethyl 1,2-benzenedicarboxylate; DMP). a highly toxic, colorless, slightly aromatic, synthetic ester that is slightly water soluble but soluble in organic solvents. It is used as a solvent and plasticizer for cellulose acetate and cellulose acetate-butyrate compositions, in lacquers, resins, perfumes, and as an insect repellant. It is a persistent and widespread environmental pollutant. Dimethyl phthalate is an irritant to the eyes and mucous membranes, but not to the skin. It is a CNS depressant.

***O,O*-dimethyl *S*-phthalimidomethyl phosphorothionate**. phosmet.

2,2-dimethyl-1-propanol. a branched-chain primary alcohol and isomer of amyl alcohol; not commercially available.

dimethyl proto catechuic acid. veratric acid.

1,6-dimethylpyrimido[5,4-e]-1,2,4-triazine-5,7-(1H,6H)-dione. toxoflavin.

dimethyl selenide. a volatile, toxic, organoselenium compound, $(CH_3)_2Se$, produced by the biomethylation of inorganic selenium by certain organisms, especially certain fungi. Its offensive odor is assumed by the plants that accumulate selenium. See organoselenium compound.

dimethylselenone. a toxic organoselenium compound, $(CH_3)_2SeO_2$, resulting from the biomethylation of inorganic selenium by certain microorganisms.

***N,N*-dimethylserotonin**. bufotenine.

dimethyl sulfate (sulfuric acid dimethyl ester; dimethyl monosulfate; DMS; DMS(methyl sulfate)). an extremely toxic and hazardous compound; one may be exposed to lethal amounts with little or no warning; it is colorless and has no odor; there is no initial irritation and other symptoms are also delayed. DMS is often referred to as methyl sulfate in the literature and may even carry the label name, methyl sulfate, *q.v.* Dimethyl sulfate. It is a highly corrosive liquid used as an industrial poison and chemical warfare gas, $(CH_3)_2SO_4$; b.p. 188°C, fp -32°C. It is also used as a methylating agent in the synthesis of numerous organic chemicals. DMS is very toxic to laboratory rats and is a primary carcinogen to humans and various laboratory animals. It appears to be poisonous by all routes of exposure. Symptoms of intoxication due to inhalation or contact are delayed and include nystagmus, convulsions, and death (usually in 3-4 days) from respiratory complications or, with a delay of several weeks, from kidney or liver damage. Brief or mild exposures may produce conjunctivitis, inflammation of the mucous lining of the nares, mouth, and upper respiratory tract, and eventual reddening of the skin. With more intense or longer exposures, symptoms may become more pronounced, the cornea may cloud, and pulmonary edema may develop after several hours. Liver and kidney damage are likely to result and are marked in survivors by suppression of urine, jaundice, hematuria, and albuminuria. Contact with the skin may cause a necrotic ulcerative dermatitis. Dimethyl sulfate can react vigorously or even violently with oxidizing agents. In some texts, dimethyl sulfate is erroneously termed methyl sulfate and may carry the label name, methyl sulfate, *q.v.*

dimethyl sulfide (methyl sulfide). a colorless, volatile, water-insoluble liquid, $(CH_3)_2S$, with a disagreeable odor; it is soluble in ethanol and ether. It is flammable, moderately toxic by ingestion, and releases sulfur dioxide when heated.

dimethyl sulfoxide (sulfinylbismethane; methyl sulfoxide; methylsulfinylmethane; dimethylsulfoxide; DMSO). a clear, colorless, nearly odorless, flammable, hygroscopic liquid, $(CH_3)_2SO$, that is soluble in water, acetone, benzene, chloroform, ethanol, and ethyl ether. It is used as an industrial solvent (e.g., for acetylene, sulfur dioxide, and other gases), paint and varnish remover, antifreeze, hydraulic fluid, and as a medium for carrying out chemical reactions. DMSO is used in medicine mainly as a topical analgesic and antiinflammatory and in veterinary medicine also as a penetrant carrier to facilitate absorption of pharmacologic agents. DMSO can carry solutes into the skin's stratum corneum where they are slowly released into the blood and lymph systems. The acute oral and subcu- taneous toxicity of DMSO is slight in a wide range of experimental animals; usually it is moderately toxic when given *i.v.* or i.p. DMSO is an irritant, producing on contact a primary inflammation of human skin, with itching, occasional urticarial wheals, and sometimes scaling. Applied to the skin, it is rapidly absorbed and spreads throughout the body; the subject experiences a garlicky taste and the breath takes on the odor of garlic. Systemic effects in humans include nausea, vomiting, and icterus. DMSO can produce anaphylactic reactions and corneal opacities. It is tumorigenic to humans, and is apparently mutagenic and possibly carcinogenic. Teratogenic and reproductive effects have been demonstrated experimentally. It reacts violently or explosively with many acyl-, aryl- and non-metal halides, boron compounds, metal alkoxides, and trifluoroacetic acid anhydride; and it forms explosive mixtures with metal salts of oxoacids. Toxic fumes (SO_x) are released on decomposition by heating.

dimethyl sulphoxide. dimethyl sulfoxide.

dimethyltelluride. a substance, $Te(CH_3)_2$, produced from inorganic tellurium compounds by the action of fungi. Because tellurium is relatively rare in nature, such biomethylation is unlikely to be a significant environmental problem. See tellurism.

dimethyl-(2,3,5,6-tetrachloroterephthalate). chlorthal-dimethyl.

dimethyl tin. See organotin compound.

dimethyltin compound. See organotin compound.

***O,O*-dimethyl 2,2,2-trichloro-1-*n*-butyryl-oxyethylphosphonate**. butonate.

dimethyltrichlorophenyl thiophosphate. Ronnel.

***O,O*-dimethyl O-(2,4,5-trichlorophenyl)ester**. Ronnel.

***O,O*-dimethyl-2,2,2-trichloro-1-phosphonoethyl butyrate**. butonate.

***N,N*-dimethyl-*N'*[3-(trifluoromethyl)phenyl]-urea**. fluometuron.

1,1-dimethyl-3-(α,α, α-trifluoro-*m*-tolyl)urea. fluometuron.

3,7-dimethyl-9-(2,6,6-trimethyl-1-cyclohexen-1-yl)-2,4,6,8-nontetraenoic acid. retinoic acid.

dimethyltryptamine. See DMT.

***N,N*-dimethyltryptamine**. DMT.

dimethyl tubocurarine iodide. obsolete name of metocurine iodide.

β,β-dimethylvinyl chloride. 1-chloro-2-methylpropene.

3,7-dimethylxanthine. theobromine.

dimethylzinc (zinc dimethyl). a white, crystalline, solid organozinc, $Zn(CH_3)_2$, with rather low melting (-40°C) and boiling (46°C) points; it is soluble in ethyl ether and miscible with hydrocarbons. While much less reactive than diethyl and dimethylzincs, it is a volatile mobile liquid at room temperature that undergoes self-ignition in air and reacts violently with water, releasing toxic fumes. The toxicity of dimethylzinc is not well established, but it is hazardous and presumed toxic. It sometimes releases phenol during decomposition.

3,3-dimethylacrylic acid 2-*sec*-butyl-4,6-dinitrophenyl ester. binapacryl.

dimethylamine, anhydrous. a colorless, irritant, flammable, and highly reactive gas. It is a very dangerous fire hazard if exposed to open flames or heat. It is moderately explosive when exposed to flame. See also amine (lower aliphatic amine).

4-dimethylaminoazobenzene. *p*-dimethylamino-azobenzene.

***p*-dimethylaminoazobenzene** (*N,N*-dimethyl-4-(phenylazo)benzenamine; 4-dimethylamino-azobenzene; butter yellow; Methyl Yellow; C.I. Solvent Yellow 2; C.I. 11020). a yellow, crystalline, water-insoluble azo compound. An ultimate carcinogen, the metabolically active form of which is *N*-hydroxymethyl-*N*-methylaminoazobenzene.

3-[[(dimethylamino)carbonyl]oxy]-*N,N,N*-trimethylbenzenaminium. neostigmine.

2-(dimethylamino)-5,6-dimethyl-4-pyrimidinyl dimethylcarbamate. pirimicarb.

4-(dimethylamino)-3,5-dimethylphenolmethylcarbamate (ester). mexacarbate.

6-dimethylamino-4,4-diphenyl-3-heptanonehydrochloride. methadone hydrochloride.

α-(+)-4-dimethylamino-1,2-diphenyl-3-methyl-2-butanolpropionate ester. propoxyphene.

(+)-4-dimethylamino-1,2-diphenyl-3-methyl-2-propionyloxybutane. propoxyphene.

dimethyltrichlorophenylthiophosphate.Ronnel.

dimetilan (dimethylcarbamic acid 1-[(dimethylamino)carbonyl]-5-methyl-1*H*-pyrazol-3-yl ester; dimethylcarbamic acid ester with 3-hydroxy-*N,N*,5-trimethylpyrazole-1-carboxamide; 2-dimethylcarbamoyl-3-methyl-5-pyrazolyl dimethylcarbamate). a carbamate insecticide. It is extremely toxic orally to laboratory rats and moderately toxic dermally.

dimexan. a preemergent contact thiocarbonate herbicide used to control seedling dicotyledonous weeds before the crops emerge. It is also used as a preharvest desiccant for onions and peas. It is non-persistent in soil.

dimorpholamine (*N,N'*-1,2-ethanediylbis[*N*-butyl-4-morpholinecarboxamide]; *N,N'*-dibutyl-*N,N'*-dicarboxymorpholideethylenediamine; *N,N'*-dibutyl-*N,N'*-dicarboxyethylene diamine-morpholide). a crystalline substance used therapeutically as a stimulant. It is very toxic orally to laboratory mice and very toxic *i.p.*, *i.v.*, and *s.c.*

Dindevan®. phenindione.

dinitramine. a yellow solid used as a pre-emergence herbicide for annual grass and broad-leaved weeds in cotton and soybeans. It is carcinogenic to some laboratory animals.

4,6-dinitro-2-aminophenol. picramic acid.

2,4-dinitro-6-*sec*-butylphenol. dinoseb.

4,6-dinitro-2*s*-butylphenol. dinoseb.

4,6-dinitro-2-*sec*-butylphenyl β,β-dimethyl-acrylate. binapacryl.

dinitro-*sec*-butylphenyl isopropylcarbonate. dinobuton.

dinitrocresol (DN; DNC; DNOC; 2-methyl4,6-dinitrophenol; 4,6-dinitro-*o*-cresol; 3,5-dinitro-2-hydroxytoluene; 3,5-dinitro-*o*-cresol; dinitro-*o*-cresol; 4,6-dinitro-2-methylphenol; Sinox). a solid, yellow compound, $CH_3C_6H_2(NO_2)_2OH$, that is slightly water soluble and soluble in ethanol, ether, and acetone. It exists in nine isomeric forms of which 3,5-dinitro-*o*-cresol is the most important commercially. It is toxic to humans and other animals. It is highly phytotoxic and cannot be used safely on or near actively growing plants. DNC is used chiefly as a herbicide to control weeds in cereal crops, an insecticide, and as a dormant ovicidal pesticide spray for fruit trees. It is also used in preharvest desiccation of potatoes. It is an extremely toxic, cumulative poison that blocks oxidative phosphorylation. It readily enters the body by inhalation, ingestion, or percutaneous absorption. Contact with the skin can produce local necrosis and serious systemic effects similar to those of dinitrophenol; it is more potent than dinitrophenol. Initial symptoms of intoxication include an increase in body temperature and basal metabolic rate, fatigue, sweating, and dehydration. Tachycardia may develop and the condition may terminate in death.

4,6-dinitro-o-cresol. an isomer of dinitrocresol, *q.v.*

dinitro-*o*-cresol (DNOC). See dinitrocresol.

3,5-dinitro-*o*-cresol. dinitrocresol.

3,5-dinitro-N^4,N^4-dipropylsulfanilamide. oryzalin.

2,6-dinitro-*N,N*-dipropyl-α,α,α-trifluoro-*p*-toluidine. trifluralin.

2,6-dinitro-*N,N*-dipropyl-4-trifluoromethylaniline. trifluralin.

2,6-dinitro-*N,N*-dipropyl-4-(trifluoromethyl)-benzenamine. trifluralin.

dinitrogen monoxide. nitrous oxide.

3,5-dinitro-2-hydroxytoluene. dinitrocresol.

4,6-dinitro-*o*-cresol. dinitrocresol.

4,6-dinitro-2-methylphenol. See dinitrocresol.

2,4-dinitrophenol (α-dinitrophenol; 2,4-DNP). a substance that is extremely toxic to plants and animals. Effects are similar to those of dinitrocresol, but less toxic. It is an intermediate in numerous organic syntheses, a wood preservative, insecticide, an indicator of pH, and a reagent used to detect ammonium and potassium ions. 2,4-DNP is extremely toxic to mammals, including laboratory rats, cattle, and humans. As an uncoupler of oxidative phosphorylation, it can greatly increase basal metabolic rate. It is readily absorbed through the skin, respiratory membranes, or gastrointestinal tract. In humans, contact with the skin or inhalation of vapors can cause nausea, vomiting, abdominal pain, restlessness, excitement, dyspnea, hyperpnea, a substantial increase in metabolism, severe hyperpyrexia, profuse sweating, cyanosis, coma, and death from respiratory or circulatory collapse. Contact can also produce an exfoliative dermatitis, weight loss, cataracts, and granulocytopenia. 2,4-DNP has been designated as a hazardous substance and hazardous waste by the U.S. Environmental Protection Agency.

α-dinitrophenol. 2,4-dinitrophenol.

dinitropyrene (DNP; 1,3-dinitropyrine; 1,6-dinitropyrine; 1,8-dinitropyrine). any of three isomers, $C_{16}H_8N_2O_4$. All are respirable carcinogens and possible human mutagens. See also nitropyrene.

1,3-dinitropyrine. a dinitropyrene isomer.

1,6-dinitropyrine. a dinitropyrene isomer.

1,8-dinitropyrine. a dinitropyrene isomer.

dinitrotoluene (2,4-dinitrotoluene; DNT). a chemosterilant, it is an important intermediate in the production of polyurethane foams, coatings, elastisomers, dyes, and explosives. It is carcinogenic to some laboratory animals.

2,4-dinitrotoluene. dinitrotoluene.

dinobuton (carbonic acid 1-methylethyl 2-(1-methylpropyl)-4,6-dinitrophenyl ester; carbonic acid 2-*sec*-butyl-4,6-dinitrophenyl isopropyl ester; 2-*sec*-butyl-4,6-dinitrophenyl isopropyl carbonate; dinitro-*sec*-butylphenyl isopropyl carbonate; isopropyl 2,4-dinitro-6-*sec*-butylphenyl carbonate). a miticide shown to be very toxic to laboratory rats.

dinocap (DNOCP). an acaricide and fungicide used mainly to control powdery mildew and other summer diseases of apples. The U.S. Environmental Protection Agency determined that dinocap is teratogenic to laboratory rabbits and recommended restricted usage in 1986.

Dinoflagellata (Dinoflagellatae). an order of predominantly marine, free-swimming protists (Class Mastigophora or Phytomastigophora in animal classification; Pyrrophyta in plant classification). Individuals have two flagella, one directed forward, the other girdling the body. Most species are holophytic (autotrophic and photosynthetic) with a cellulose test; some are heterotrophic. They are a staple food of shellfish and some crabs, and are thus in the human food chain. Certain species may become extremely abundant in warm coastal waters, causing certain of the toxic "red tides." Only a few species are actually toxic, notably *Gonyaulax catenella*, *G. tamarensis*, and *Gymnodinium brevis*. Some marine species are bioluminescent. See also red tide, saxitoxin, *Gonyaulax*.

Dinoflagellatae. Dinoflagellata.

dinoflagellate. 1: pertaining to or denoting the protozoan order Dinflagellata. **2:** the common name of any protozoan of the Dinoflagellata.

dinoflagellate bloom. See red tide; See also bloom.

dinoseb (2,4 dinitro-6-sec-butylphenol; 2-(1-methylpropyl)-4,6-dinitrophenol; 2-*sec*-butyl-4,6-dinitrophenol; DNBP). a nitrophenolic formerly widely employed as a selective herbicide against some weed species in cereal crops and also as an insecticide and miticide. It is extremely toxic to laboratory rats. Dinoseb is also a suspected teratogen in women who are exposed in early pregnancy and is suspected of causing sterility in human males.

dinoseb methacrylate. binapacryl.

dinoterb. a selective nitrophenol herbicide used chiefly for postemergence weed control in *Medicago sativa* (alfalfa, lucerne) and preemergence weed control in peas and beans.

dioctyl phthalate. bis(2-ethylhexyl) phthalate.

di-*n*-octyl phthalate. bis(2-ethylhexyl) phthalate.

Diodon. a genus of bony fish (Family Diodontidae), many members of which are poisonous (especially those in the Pacific Ocean) because of the presence of tetrodotoxin in the liver and ovary. ***D. hystrix*** (porcupine fish). a circumtropical marine bony fish (Family Diodontidae) whose range occasionally enters the temperate zone. It is one of the most poisonous species of tetraodontoid fishes. See puffer poisoning.

Diodontidae (porcupinefish; globefish; burrfish). a family of marine, mostly tropical, bony fishes (closely allied to puffers) that can puff up by swallowing large amounts of water or air. See *Diodon*.

dioscin ((25-R)-spirost-5-en-3β-yl*O*-6-deoxy-α-L-mannopyranosyl-(1→2)-*O*-6-[deoxy-α-L-mannopyranosyl-(1→4)]-β-D-glucopyranoside; diosgenin bis-α-L-rhamnopyranosyl-(1→2 and 1→4)-β-D-glucopyranoside). a saponin, it is the aglycone diosgenin, *q.v.*

diosgenin ((5R)-spirost-5-en-3β-ol; nitogenin). a toxic, steroidal sapogenin; the aglycone of dioscin, *q.v.*, isolated, for example, from roots of *Dioscorea composita*, *D. tepinapensis*, *D. tokoro*, and *D. villosa* (Family Dioscoreaceae) and the rhizomes of *Trillium erectum* (Family Liliaceae). It is used to synthesize pregnenolone and progesterone. See especially [illegible].

diosgenin bis-α-L-rhamnopyranosyl-(1→2 and 1→4)-β-D-glucopyranoside. dioscin.

1,7-dioxadispiro-[4.0.4.2]dodeca-3,9-diene-2,8-dione. anemonin.

dioxane (1,4-dioxane; 1,4-diethylene dioxide). the cyclic ether of ethylene glycol, it is a colorless, volatile, flammable liquid with a faint pleasant odor. Dioxane is completely soluble in water in all proportions and has numerous applications as a wetting agent and as a solvent for numerous organic and inorganic compounds including oils, resins, and waxes. It is used for stripping paint and varnish. It is extremely toxic to laboratory mice and rats and is carcinogenic to some laboratory animals. It is an irritant, a CNS depressant and is also hepatotoxic and nephrotoxic. Symptoms of acute exposure in humans include drowsiness, nausea, vomiting and necroses of the liver and kidney. Routes of entry into the animal body includes inhalation of vapors, ingestion, and percutaneous absorption; it is toxic by all routes.

1,4-dioxane. See dioxane.

2,3-*p*-dioxanedithiol *S,S*-bis(*O,o*-diethyl)phosphorodithioate. folinic acid.

2,3-*p*-dioxanedithiol *S,S*-bis(*O,o*-diethyl) phosphorodithioate. dioxathion.

***p*-dioxane-2,3-diylethylphosphorodithioate**. dioxathion.

***S,S'*-1,4-dioxane-2,3-diyl *O,O,O',O'*-tetraethyl ester**. folinic acid.

dioxathion (*p*-dioxane-2,3-diylethylphosphorodithioate; phosphorodithioic acid *S,S'*-1,4-dioxane-2,3-diyl *O,O,O',O'*-tetraethyl ester; 2,3-p-dioxanedithiol *S,S*-bis(*O,O*-diethyl phosphorodithioate; "Delnav"). a viscous, nonvolatile, nonflammable, brown, nearly water-insoluble liquid, $C_4H_6O_2[SPS(OC_2H_5)_2]_2$. It is a dangerous organophosphate cholinesterase inhibitor used as an insecticide and miticide. It is toxic to humans and laboratory animals by all routes of exposure. Studies indicate that it is extremely toxic, orally, to laboratory rats and extremely toxic, dermally, to female rats; it is significantly less toxic to male rats. It is hydrolyzed by heating and by [illegible]

1,1-dioxide tetrahydrothiofuran. sulfolane.

"dioxin." See TCDD.

dioxin. **1:** a class of some 75 compounds that can

be considered as substituted dioxanes. They contain a ring comprised of two oxygen atoms, four CH groups, and (usually) two double bonds (See TCDD). The positions of the oxygen atoms are indicated by numeric prefixes (as in 1,4-dioxin). Dioxins are a matter of concern not only because most are extremely toxic, but because they are by-products of the manufacture of numerous chemicals (e.g., chlorinated phenols, dyes, herbicides, various pharmaceuticals, and wood preservatives) with a potential for industrial exposures and the likelihood of contaminating the commercial products as in the case of 2,4,5-T and pentachlorophenol; they can also be released into the atmosphere when waste materials are burned. Humans may be exposed to dioxins during the manufacture of the types of products mentioned above; by living near the manufacturing facilities; when using any of the various herbicides that it contaminates, breathing the air following spraying, or ingesting foods treated by such herbicides; and by the improper burning of waste products. **2:** often used specifically in reference to dibenzo[*b,e*][1,4]-dioxin. **3:** often used erroneously as a synonym for TCDD (2,3,7,8-tetrachlorodibenzo[b,e]-[1,4]dioxin), perhaps the most toxic dioxin.

9,10-dioxoanthracene. anthraquinone.

2,2′-[(1,4-dioxo-1,4-butanediyl)bis(oxy)]bis-[*N,N,N*-trimethylethanaminium] dichloride. succinylcholine chloride.

2,4-dioxo-5-fluoropyrimidine. fluorouracil.

2,6-dioxo-4-methyl-4-ethylpiperidine. bemegride.

2,6-dioxo-3-phthalimidopiperidine. thalidomide.

2-(2,6-dioxo-3-piperidinyl)-1*H*-isoindole-1,3(2*H*)-dione. thalidomide.

***N*-(2,6-dioxo-3-piperidyl)phthalimide**. thalidomide.

1,4-dioxybenzene. *p*-quinone. See quinone.

dioxydemeton-*S*-methyl. demeton-*S*-methylsulfone.

DIPF. diisopropylphosphofluordate (See diisopropyl fluorophosphate).

diphacinone. the generic name for 2-diphenylacetyl-1,3-indanedione.

diphebuzol. phenylbutazone.

diphenadione.2-diphenylacetyl-1,3-indanedione.

diphenamid (*N,N*-dimethyl-α-phenylbenzenacetamide; *N,N*-dimethyl-2,2-diphenylacetamide; *N,N*-dimethyl-α,α-diphenyl acetamide; 2,2-diphenyl-*N,N*-dimethylacetamide). a whitish crystalline slightly water-soluble organic compound used as a soil-applied, preemergence selective anilide herbicide for food crops, fruits and ornamentals. It is absorbed through the roots and inhibits root elongation. It is moderately toxic by oral, intraperitoneal and subcutaneous routes; it is probably mutagenic. Other soil-applied amides with similar action include alachlor, diphenamid, naptalam, and propachlor.

diphenhydramine (2-diphenylmethoxy-*N,N*-dimethylethanamine; 2-(benzhydryloxy)-*N,N*-dimethylethylamine; β-dimethylaminoethyl benzhydryl ether; *O*-benzhydryldimethylaminoethanol; β-dimethylaminoethanol diphenylmethyl ether; α-(2-dimethylaminoethoxy)-diphenylmethane; benzhydramine). a moderate to very toxic antihistaminic marketed under numerous brand names and trademarks. It is usually marketed as the hydrochloride (Benadryl, etc.) or the ascorbate.

diphenoxylate(1-(3-cyano-3,3-diphenylpropyl)-4-phenyl-4-piperidinecarboxylic acid ester; 1-(3-cyano-3,3-diphenylpropyl)-4-phenylisonipecotic acid ethyl ester; ethyl 1-(3-cyano-3,3-diphenylpropyl)-4-phenylisonipecotate; ethyl 1-(3-cyano-3,3-diphenylpropyl)-4-phenyl-4-piperidinecarboxylate; 2,2-diphenyl-4-(4-carmethoxy-4-phenylpiperidino)butyronitrile; R-1132). a controlled substance under the U.S. Code of Federal Regulations. The hydrochloride, $C_{30}H_{32}N_2O_2 \cdot HCl$, is a popular, but hazardous antidiarrheal drug It is an addictive narcotic substance; an overdose can be fatal. To reduce the risk of overdosage, the drug is formulated with other substances that produce unpleasant side effects such as dry skin, flushed face, and tachycardia.

diphenoxylate hydrochloride. a popular antidiarrheal drug, that is perhaps too hazardous touse. It is a narcotic that can cause addiction, and an overdose can be fatal. The possibility of overdosage is reduced, but by no means eliminated, by formulating the drug with other substances that cause dry skin, reddening of the face, tachycardia, and other unpleasant effects.

diphenyl. biphenyl.

(+)-1,2-diphenyl-2-propionoxy-3-methyl-4-dimethylaminobutane. propoxyphene.

2-diphenylacetyl-1,3-indanedione. (diphacinone; diphenadione). an indandione anticoagulant, a yellow, odorless powder or crystalline substance, $C_{23}H_{16}O_3$, that is insoluble in water. It is an indandione anticoagulant used as a rodenticide.

diphenylamine (*N*-phenylbenzeneamine; phenylaniline; DPA). a combustible, irritant, slightly water-soluble, colorless to grayish crystalline compound, $(C_6H_5)_2NH$, with a floral odor. It is used in propellants to extend storage life by neutralizing the acidic decomposition products of nitrocellulose. Action and effects are similar to those of aniline, but less severe. It is used in tests for nitrate and nitrite poisonings and is used in veterinary medicine in anti-screwworm mixtures. See also *p*-biphenylamine.

diphenylaminearsine chloride. phenarsazine chloride.

diphenylaminechlorarsine. phenarsazine chloride.

4-diphenylaminesulfonic acid. *N*-phenylsulfanilic acid.

2,2-diphenyl-4-(4-carbethoxy-4-phenylpiperidino)butyronitrile. diphenoxylate.

diphenylchlorarsine (sneezing gas; Clark I; DA). a toxic smoke and sneezing gas, $(C_6H_5)_2AsCl$, once used as a chemical warfare agent. It causes sneezing, coughing, headache, salivation, and vomiting.

diphenylcyanarsine (Clark II). a lethal war gas, $(C_6H_5)_2AsCN$.

1,1′-biphenyl-4,4′-diamine. benzidine.

2,2-diphenyl-*N,N*-dimethylacetamide. diphenamid.

4,4-diphenyl-6-dimethylamino-3-heptanone hydrochloride. methadone hydrochloride.

1,1-diphenyl-1-(2-dimethylaminopropyl)-2-butanone hydrochloride. methadone hydrochloride.

4,4′-diphenylenediamine. benzidine.

diphenylhydantoin. phenytoin.

1,2-diphenylhydrazine (hydrazobenzene; *N,N′*-bianiline; *sym*-diphenylhydrazine; hydrazobenzene). a yellowish, very slightly water-soluble, crystalline substance, $(C_6H_5)_2NNH_2$. 1,2-diphenylhydrazine is toxic if ingested. It is tumorigenic, probably mutagenic, and is carcinogenic to some laboratory animals. The hydrochloride, $C_{12}H_{12}N_2 \cdot HCl$, is a white to grayish white, slightly water-soluble, crystalline powder used as a reagent for arabinose and lactose.

***sym*-diphenylhydrazine**. 1,2-diphenylhydrazine.

diphenylhydrazine hydrochloride. See 1,2-diphenylhydrazine.

5,5-diphenyl-2,4-imidazolidinedione. phenytoin.

diphenylmagnesium. a poisonous pyrophoric solid, $Mg(C_6H_5)_2$. It reacts violently with water and combusts spontaneously in humid but not dry air. It is somewhat less hazardous than dimethyl- and diethylmagnesium, *q.v.*

diphenylmercury. a white, crystalline, water-insoluble substance with toxicological effects similar to those of inorganic mercury. It is incompatible with non-metal oxides.

diphenylmethane-4,4′-diisocyanate (methylene-di-*p*-phenylene isocyanate; methylene(bisphenyl isocyanate); MDI). a light yellow, [illegible] fused solid at room temperature (solidifying at 37°C.), $CH_2(C_6H_4NCO)_2$, that is soluble in a number of organic solvents. It is used in the manufacture of polyurethane resins and spandex fibers. It is a powerful irritant. *Cf.* toluene-2,4-diisocyanate.

2-diphenylmethoxy-*N,N*-dimethylethanamine. diphenhydramine.

(+)-1,2-diphenyl-2-propionoxy-3-methyl-4-dimethylaminobutane. propoxyphene.

diphenylthiocarbazone. dithizone.

1,3-diphenyltriazene. a component of FD&C Yellow No. 5. It is carcinogenic in some laboratory animals.

diphosgene (carbonochloridic acid trichloromethyl ester; chloroformic acid trichloromethyl ester; trichloromethylchloroformate; trichloromethylcarbonochloridate). an acutely irritant war gas, $ClCOOCCl_3$, that produces pulmonary edema when inhaled. Exposure of rabbits to 0.9mg/l of the vapor in air is invariably fatal.

diphosphopyridine nucleotide. nadide.

diphosphoric acid tetraethyl ester. tetraethyl pyrophosphate.

diphosphorus trioxide. phosphorus trioxide.

diphtheria. See *Cornybacterium diphtheriae*.

diphtheria toxin (diphtherin; diphtherotoxin). See toxin (diphtheria toxin).

diphtherin. diphtheria toxin. See toxin (diphtheria toxin).

diphtherotoxin. diphtheria toxin. See toxin (diphtheria toxin).

Diplodia maydis. a large fungus (Family Sphaeropsidaceae, Order Sphaeropsidales, Class Coelomycetes) that produces the mycotoxin diplodiatoxin.

diplodiatoxin. a mycotoxin isolated from the fungus *Diplodia maydis*.

diplopia (double vision). a visual disorder whereby two images of an object are seen due to unequal action of the eye muscles. This condition can result from toxicants such as ethanol, barbiturates, nicotine, or organophosphate insecticides.

Diplopoda. (millipedes). a class of some 7,000 species of terrestrial arthropods (e.g., *Julus, Narceus*, and *Polydesmus*) that breathe through tracheae. The body is elongated, cylindrical and covered by a calcareous cuticle; a head with one pair of short antennae, a labrum, mandibles, and one pair of maxillae (fused to form a gnathochilarium); a thorax comprised usually of four segments, each with a single pair of legs; the abdomen is comprised of 20 to more than 100 segments formed by the fusion of 2 embryonic segments, each having 2 pairs of short legs, ganglia, four spiracles, and two pairs of cardiac ostia; a midventral genital pore occurs on the 2nd or third body segment. Excretion is by Malpighian tubules. Millipedes are sluggish detritivores that superficially resemble centipedes (Chilopoda). Many secrete an irritant (usually rich in benzoquinones) that often protects them from small predatory animals and which, if the millipede is eaten may kill the predator. When handled by humans the secretion may produce necrosis in some cases. Certain species can spray an irritant which can produce conjunctivitis. *Cf.* Chilopoda. See also *Apheloria corrugata*, benzoquinone, *Pseudopolydesmus serratus*, *Rhinocrichus lethifer*.

dipping acid. sulfuric acid.

dipropyl cadmium (cadmium dipropyl). a hazardous, flammable, oily liquid, $(C_3H_7)_2Cd$, m.p. -83°C, b.p. 84°C, that reacts with water. Behavior is similar to other dialkyl cadmiums (See dimethylcadmium). It is distillable, but decomposes at temperatures above about 150°C, releasing toxic cadmium oxide fumes.

4-(dipropylamino)-3,5-dinitrobenzenesulfonamide. oryzalin.

dipropylcarbamothioic acid *S*-ethyl ester. EPTC.

***N,N*-dipropyl-2,6-dinitro-4-trifluoromethylanaline**. trifluralin.

dipsogon. an agent that induces thirst.

dipropylthiocarbamic acid *S*-ethyl ester. EPTC.

dipsogen. an agent that induces thirst.

dipsogenic. inducing thirst; having the capacity to induce thirst.

diquat (6,7-dihydrodipyrido[1,2-a:2′,1′-c]-pyrazinediium dibromide; 1,1′-ethylene-2,2′-dipyridylium dibromide; FB/2; diquat dibromide; Aquacide; Reglone). a bipyridylium, quaternary, ammonium salt used as a rapid-acting contact herbicide, desiccant, and defoliant. It is a convulsant and at high doses a gastrointestinal irritant. It is very toxic to laboratory rats, mice, and guinea pigs. Symptoms of acute diquat poisoning include acute hyperexcitability, convulsions, and gastrointestinal irritation and distension. See also paraquat.

diquat dibromide. diquat.

direct-acting carcinogen. See carcinogen.

Direct Blue 14. C.I. Direct Blue 14, tetrasodium salt.

diriora. *Oxyuranus scutellatus*.

dirty trich. *Tricholoma pardinum*.

disassimilation. dissimilation.

disaster. **1a:** a calamity, catastrophe, cataclysm; a complete failure; a grave misfortune. **1b:** an occurrence or accident that causes widespread destruction and distress or loss of life. This term generally implies great destruction, with widespread harm (especially to humans), extensive loss of property or life support systems, injury, hardship, or loss of life. **chemical disaster**. usually a chemical accident that causes multiple fatalities, disabilities, and distress. See also chemical accident, Bhopal, Minimata disaster.

Discoglossidae. a family of disc-tongued toads (Order Opisthocoela) that is indigenous to Europe, Africa, Asia, and the Philippines. Adults have ribs, a tongue, and eyelids. Included are the obstetrical toads of Europe (*Discoglossus*, *Alytes*) and *Bombina*, *q.v.*

disease **1a:** morbus; ailment; illness; sickness; affliction; disorder; abnormal condition, infirmity; malady; complaint). **1b:** deviation of a living organism or part of an organism from a healthy or normal condition. **1c:** disruption or disturbance of a life function with or without the prospect of death, but accompanied at least by recognizable signs and symptoms which distinguish it from other diseases (e.g., pain or discomfort, weakness, poor or abnormal performance of a vital function). **1d:** a physical disorder with a specific physiological basis and recognizable signs and symptoms. **1e:** any condition in which life functions are impaired, interrupted, or fail. **1f:** any morbid state. **2:** any condition or state of a living system that is inimical or disadvantageous to the individual or to its progeny. Some examples of toxicological interest are acute disease, alkali disease, arc-welder's disease (siderosis), Bettlach May disease, Grave's disease (See toxic goiter), hemorrhagic disease of poultry, itai itai disease, locoweed disease (See loco), Minimata disease, perna disease, pictou disease, Plummer's disease, Shaver's disease (See bauxite pneumoconiosis), silo-filler's disease, stonemason's disease (See silicosis), sweet clover disease, Tommaselli's disease. Illness and ailment are similar terms. See also disorder.

disease ratio. See standardized disease ratio.

diseased. **1a:** having a disease. **1b:** having an organ, tissue, or body part that is morbid or impaired.

disi. *Naja Nigricolis*.

disilane (disilicoethane; disilicane; disilicon; hexahydride; silicoethane). a toxic gas, H_3SiSiH_3, with a repulsive odor. It is a powerful irritant and is poisonous if inhaled. It combusts spontaneously in air and reacts violently with carbon tetrachloride, chloroform, oxygen, sulfur hexafluoride, and trichloromethane. It decomposes slowly in water and is soluble in benzene, carbon disulfide, ethanol, and ethyl silicate. See silane.

disilanyl. a cyclic silane. See silane.

disilicane. disilane.

disilicoethane. disilane.

disilicon. disilane.

disinfect. **1:** to clean, sterilize, sanitize, decontaminate. **2:** to kill pathogenic microorganisms, or their vectors in or on nonliving materials or to inhibit their growth or activity, rendering them inert.

disinfectant. **1:** having the capacity to disinfect. **2a:** any agent that disinfects. **2b:** any of a wide variety of nonselective antimicrobial agents or processes that kill or inhibit the growth or activity of pathogenic microorganisms in or on nonliving material. Disinfectants (often used as aerosols sprays) contain any of a variety of toxic volatile chemicals that are dangerous when inhaled. Cresol, *q.v.*, is a frequent component of disinfectants. Other toxic chemicals frequently found in disinfectants are phenol, ethanol, formaldehyde, ammonium hydroxide, and chlorine.

disinfection. the process of destroying most living organisms (usually microbial pathogens) in water or on surfaces. *Cf.* sterilization.

disintegrator. **1:** any agent that causes a substance to decompose, decay, deteriorate, or break apart. **2:** any substance, usually cellulose derivatives and starch, that facilitates disintegration of a drug or toxicant formulation.

dismutase. any of a class of enzymes that catalyze the reaction of two molecules of a single compound to yield two new compounds that have different oxidation states; one is oxidized and the other is reduced. **superoxide dismutase** (SOD). any of a type of antioxidant metalloenzymes present in aerobic and facultative bacteria and in eukaryotic cells as well. They catalyze a reaction in which two molecules of the highly toxic, highly reactive, superoxide anion is converted into one molecule each of hydrogen peroxide and molecular oxygen:

$$O_2^- + O_2^- + 2H^+ \rightarrow H_2O_2 + O_2$$

Three forms of the enzyme are known: (1) one that contains copper and zinc. It occurs chiefly in eukaryotes and is inhibited by cyanide and hydrogen peroxide. (2) one that contains iron. It occurs mainly in prokaryotes and while inhibited by hydrogen peroxide, it is resistant to cyanide. (3) one that contains manganese. It occurs in both prokaryotes and eukaryotes and is resistant to both cyanide and hydrogen peroxide.

disodium edathamil. edetate disodium.

disodium edetate. edetate disodium.

disodium ethylenebis[dithiocarbamate]. nabam.

disodium ethylenediaminetetraacetate. edetate disodium.

disodium indigo-5,5-disulfonate. FD&C Blue No. 2.

disodium monomethanearsonate (DSMA). the disodium salt of methanearsonic acid, $CH_3AsO(ONa)_2$. It is a selective post-emergence herbicide used to control grass and weeds in cotton and rubber plantations. It is moderately toxic to laboratory rats.

disomer. Dexedrine (See amphetamine, def. 2).

disopyramide (α-[2-[bis(1-methylethyl)amino]-ethyl]-α-phenyl-2-pyridineacetamide; α-[2-(diisopropylamino)ethyl]-α-phenyl-2-pyridineacetamide; 4-(diisopropylamino)-2-phenyl-2-(2-pyridyl)butyramide). a cardiac depressant (antiarrhythmic), it is very toxic, *i.p.*, to laboratory mice.

disorder. an illness, disease, or any structural or functional disturbance or malfunction of an organism generally, or of a body part, regardless of cause. This term is roughly synonymous with disease, *q.v.*

disorganization. destruction of an organ or tissue to the point of functional loss or impairment.

disorientation. **1a:** a loss of one's bearings. **1b:** mental confusion with respect to one's surroundings, including time, place, and other individuals. It is usually associated with organic brain disorders.

dispersant. **1a:** a chemical agent that promotes the dispersion and stabilization of the dispersion of one substance within another. **1b:** a chemical agent used to break up and disperse organic material such as oil or other pollutants in the environment.

dispersion. **1:** the act or process of scattering, separating, or becoming distributed in space. **2:** the pattern of spacial distribution (e.g., of a population of organisms). **3:** the volitional movement of individual organisms into or out of an area or population. Such movements are usually more or less of a regular nature or occur at a particular stage of life. **4:** in statistics, the scatter or distribution of observations or values about the mean or other central value. **5:** the spreading of chemical warfare agents by means of a bursting charge. **6:** in chemistry, a substance distributed as finely divided particles in a medium; a dispersed phase.

Dispholidus typus (boomslang; plural, boomslange, boomslangs; back-fanged snake; black tree snake; African brown snake; Cape snake; namahamba; gurukezi; iNambezulu; inhlanhlo; iNyushu; kalilelala; khangala; kokokeyamulinga; large brown tree snake; large green tree snake; legwere; lukukuru; siana; yamuhando; yangalukwe; zokalugwagu). a venomous rear-fanged tree snake (Family Colubridae). It is indigenous to southern tropical Africa, where it occurs in open savannah and brushy country. It is the only species of the genus. The boomslang is rather shy, green or brownish black in color with an oval head that is distinct from the slender neck. The crown is convex and the snout short with a distinct canthus; a single loreal scale separates the nasal from the one or two preoculars; the eyes are very large with round pupils. The body is slender, moderately compressed; averaging 1.2 to 1.5 ms long in adults; the tail is long and slender. While not usually aggressive, it will inflate the neck to more than twice its normal size, thus exposing the often brightly colored skin between the scales of the neck. If this does not deter the targeted individual, the boomslang will bite. The boomslang has relatively long fangs and its venom is more toxic than that of African cobras and vipers. Effects of envenomation on small animals include severe internal bleeding and the oozing of blood from all mucous membranes; death may result. The venom is extremely toxic. Effects of envenomation of humans include rapid onset of nausea and dizziness, often followed by a slight amelioration of symptoms, then sudden death from severe internal hemorrhage in which all mucous membranes may exude blood. Effective bites of humans and larger animals are rare because of the position of the fangs. Nevertheless, the case mortality rate for envenomations of humans is high.

displacement analysis. competitive binding assay (See assay).

disposal. **1a:** the elimination, riddance, or discarding of used or otherwise unwanted material. **1b:** the manner of eliminating, the final storage (placement, siting), or destruction (e.g., by incineration) of toxic, radioactive, or other wastes. Disposal is most often accomplished by use of approved secure landfills, impoundments, agricultural land applications, incineration, deep well injection, or ocean dumping. Safe and acceptable disposal of waste has become a serious problem for most nations and new alternatives are constantly sought, including conservation and recycling of waste materials.

disseminated intravascular coagulation (DIC). systemic coagulation in which the coagulant spreads throughout the vascular system causing clotting within the vascular system at sites remote from the source of coagulation as in the case of envenomation by the brown recluse spider, *Loxosceles reclusa*.

dissimilate. to decompose a substance into simpler compounds with resultant elimination from the system; sometimes with release of usable energy. *Cf.* assimilate. See also dissimilation.

dissimilation (disassimilation). **1:** the act or process of dissimilating. See dissimilate. **2:** often poorly or erroneously defined as the reverse of assimilation or used in reference to biological reactions in which the products do not enter biosynthetic pathways. *Cf.* assimilation.

dissociable. easily separated into component parts.

dissociation. **1:** the process of separating or being in a state of separation, detachment, or isolatin. Toxicants or their metabolites are, for example, often transported bound to or in association with blood proteins. They continue to circulate until dissociated from the protein. **2:** the resolution of a substance into two or more simpler substances due to the addition of energy (e.g., as heat) or by the action of a solvent. *Cf.* dissimilation. **3:** the state of being separated.

dissociation constant (K_d). the equilibrium constant (K) of a reversible dissociation reaction (K_d). The simplest case is the dissociation of a compound (AB) into two molecules (A and B): $AB \rightleftharpoons A + B$. Thus $K_d = [A][B]/[AB]$. The brackets indicate concentration or activity.

dissolution. **1:** death **2:** degeneration; the pathological disintegration of a body part.

dissolve. **1a:** to enter into solution or to cause a substance to enter into solution. **1b:** to disappear, disintegrate, vanish.

dissolved oxygen (DO). the unbound, freely available oxygen in water. It is vital to aquatic organisms; it also prevents odors. The concentration of dissolved oxygen in natural and man-made water bodies is usually the single most important indicator of a water body's ability to support natural or desirable aquatic life. Any anthropogenic chemical that reduces the concentration of dissolved oxygen in a water body that supports life can be said to be poisoning the system as a whole. See also BOD, COD.

dissolved solids (DS). decomposed material contained in water. Such materials may make the water toxic or otherwise unfit to drink or to use in industrial processes. Excessive amounts reduce the level of dissolved oxygen and otherwise make the water unfit to drink or to use industrially.

distal. denoting that part of an organism, organ, tissue, or process that is furthest from the center or from the point of origin. Thus the axon is distal to the cell body or nucleus. *Cf.* proximal, peripheral.

distill (distil). to subject to distillation; to extract, purify, render, or remove by distillation. In general, to volatize by heat, then condense the evaporated material by cooling. The purpose is the purification of a liquid or the separation of a more volatile substance (a compound or sometimes a mixture of known composition) from components of lower volatility.

distilland. the material which is subjected to distillation.

distillate. material (usually a liquid) that has been isolated by distillation.

distillation. **1a:** vaporization by the application of heat, sometimes with the aid of a vacuum. **2:** the separation of a volatile liquid from less volatile substances by boiling (vaporization) so that the resulting vapor condenses to a pure liquid and the less volatile contaminants remain behind as a concentrated residue.

distinctive. **1a:** characteristic, singular, distinguishing, special, uncommon. **1b:** diagnostic (def. 2)

distinguishing. diagnostic, characteristic.

distress. physical or mental suffering or mental anguish; pain; agony.

distribution. **1a:** the act or process of being distributed or apportioned. **1b:** the movement of a xenobiotic substance from site or portal of entry to other parts of the organism. **2a:** the spreading out or scattering throughout a space or system. **2b:** the ultimate arrangement of like components within a space or system. **2c:** the set of concentrations of a xenobiotic substance at different locations within a living system.

disturbance. interference with; a deviation, departure, or divergence from; or the disruption of a normal state.

disulfide. a class of compounds that contains two sulfur atoms bound to an element or a radical. See carbon disulfide, organic disulfide. **organic disulfide**. any organic compound with the general formula, R-SS-R′ (R may = R′). These compounds may produce a contact dermatitis. A number of toxic effects, including hemolytic anemia, have been reported for laboratory animals. Toxicities to humans are not well established. See also sulfide.

disulfiram (tetraethylthioperoxydicarbonic diamide; bis(diethylthiocarbamoyl); bis(diethylthiocarbamyl) disulfide; disulfide; tetraethylthiuram disulfide; teturamin; TETD; TTD; Antabuse). a light gray, slightly odorous, water-insoluble powder, $[(C_2H_5)_2NCS]_2S_2$, that is soluble in carbon disulfide, benzene, and chloroform. It has many industrial applications and is used also as a fungicide, seed disinfectant, and as a chelator in copper and nickel poisoning. Under the name Antabuse it is used in the treatment of alcohol abuse. Ethanol is normally metabolized in the liver to acetaldehyde by alcohol dehydrogenase and the acetaldehyde so produced is rapidly oxidized by aldehyde dehydrogenase. Disulfiram inhibits the activity of aldehyde dehydrogenase, resulting in the accumulation of acetaldehyde to blood concentrations five to ten times above normal when alcohol is consumed. Thus if alcoholic beverages are taken while under treatment with disulfiram, unpleasant and even severe reactions (due mainly to the effects of the accumulated acetaldehyde) may occur. These include intense vasodilation (especially of face and neck), headache, tachycardia, respiratory distress with tachypnea, nausea and vomiting, hypotension, sweating and confusion. Occasionally, in more acute cases, the victim may exhibit shock, respiratory depression, cardiac arrhythmias, acute congestive heart failure, myocardial infarction, convulsions, unconsciousness, and sudden death. Disulfiram by itself is only slightly toxic; nevertheless an overdosage can cause circulatory collapse and death. It is teratogenic to some laboratory animals. See also *Clitocybe clavipes*, thiram.

disulfoton ("Di-Syston"; phosphorodithioic acid *O,O*-diethyl *S*-[2-(ethylthio)ethyl] ester; *O,O*-diethyl-*S*-ethylmercaptoethyl dithiophosphate; thiodemeton; dithiodemeton). an organophosphate, $(C_2H_5O)_2P(S)SCH_2CH_2SCH_2CH_3$, it is a yellow liquid when pure and is soluble in water; the technical grade is brown. Disulfoton is an anticholinesterase used as a systemic insecticide and acaricide; it is also a component of some household products used to eradicate pests. It is highly poisonous to vertebrates by ingestion or inhalation, and is probably toxic by all routes of exposure. It is extremely toxic orally and dermally to laboratory rats and is a probable human mutagen.

Di-Syston [Mobay]®. disulfoton.

diterpene aplysin-20. a brominated compound related to aplysin.

diterpenoid alkaloid. an alkaloid that contains four 2-methylbutane groups, arranged according to the isoprene rule, with additional functional groups. See especially *Aconitum*, aconitine.

δ-[3-(1,2-dithiacyclopentyl)]pentanoic acid. thioctic acid.

6,6′-dithiobis-2-naphthalenol. DDD, analytical. Do not confuse with the pesticide, DDD.

6,6′-dithiobis(2-naphthol). DDD, analytical. Do not confuse with the pesticide, DDD.

dithiobis[thioformic acid] *O,O*-diethyl ester. dixanthogen.

2,4-dithiobiuret (thioimidodicarbonic diamide). a crystalline compound, $C_2H_5N_3S_2$, used in the manufacture of insecticides and rodenticides and to prevent wilting of plants. It is very or extremely toxic by all routes of exposure and is extremely hazardous. When heated to decomposition, it emits toxic fumes (SO_X, NO_X). It can cause a fatal respiratory paralysis in humans.

dithiocarb. ditiocarb sodium.

dithiocarbamate fungicide. any of a number of fungicides based on dithiocarbamic acid. They are chelates that consist of metal salts of dimethylthiocarbamate and ethylenebisdithiocarbamate anions and are named according to the metal ion present. Thus maneb is the manganese salt of dimethyldithiocarbamate anion and zineb and nabam are the zinc and sodium salts respectively, ferbam is the iron salt of the ethylenebisdithiocarbamate anion and ziram is the zinc salt of this ion. These fungicides are used commonly in agriculture because they are effective against fungi, and though they are immunotoxic, their toxicities to animals are relatively low. Certain products of the environmental degradation of these fungicides (e.g., ethylenethiourea) are a matter of concern.

dithiocarbamic acid. See dithiocarbamate fungicide.

dithiodemeton. disulfoton.

6,6′-dithiodi-2-naphthol. DDD, analytical. Do not confuse with the pesticide, DDD.

1,2-dithioglycerol. dimercaprol.

1,2-dithiolane-3-pentanoic acid. thioctic acid.

1,2-dithiolane-3 valeric acid. thioctic acid.

1,3-dithiolan-2-ylidenephosphoramidic acid diethyl ester. phosfolan.

5-[3-(1,2-dithiolanyl)]-pentanoic acid. thioctic acid.

5-(1,2-dithiolan-3-yl)pentanoic acid. thioctic acid.

5-(1,2-dithiolan-3-yl)valeric acid. thioctic acid.

2,3-dithiolpropanesulfonic acid. 2,3-dimercaptol-propanesulfonic acid.

6,8-dithiooctanoic acid. thioctic acid.

dithiotrimethyleneglycol. 1,3-propanedithiol.

dithizone (phenyldiazinecarbothioic acid 2-phenylhydrazide; (phenylazo)thioformic acid 2-phenylhydrazide; diphenylthiocarbazone). a toxic reagent used in the analysis of the heavy metals, Co, Cu, Pb, and Hg. It is used especially to detect minute amounts of Pb. It can cause glycosuria and ocular damage in laboratory animals. See dithizone method.

dithizone method. a method for the analysis of heavy metals, especially lead. The sample is collected by passing air through filter paper which is eluted with nitric acid and the residue dissolved in nitric acid. The pH adjusted and ammonium cyanide is added and the solution which is then extracted with a chloroform solution of dithizone (a toxic reagent), followed by spectrophotometric determination.

ditiglyteloidine. an alkaloid isolated from *Datura*, *q.v.*

ditiocarb sodium (diethylcarbamodithioic acid sodium salt; diethyldithiocarbamic acid sodium salt; diethyldithiocarbamate; dithiocarb; sodium; DDC; DDTC; DEDC; DeDT; DTC). a water-soluble chelating agent and antidote, $(C_2H_5)_2NNaCS_2Na$, to nickel and cadmium poisoning. It has a strong affinity, also, for copper, mercury, and zinc. It is a T-cell-specific immunostimulant and is moderately toxic to laboratory mice and rats.

ditropyl isatropate. belladonnine.

diuresis. the secretion and passage of large volumes of urine. Diuresis occurs in diabetes mellitus, diabetes insipidus, in chronic interstitial nephritis, as a result of anxiety or fear, and when stimulated by diuretics (*q.v.*).

diuretic. **1:** tending to, or having the capacity to increase the secretion of urine by the kidneys. **2:** an agent that can increase the rates of excretion of potassium, sodium and urine by the kidneys. Diuretics act either to increase the rate of glomerular filtration or to decrease reabsorption from the tubules. They increase the permeability of kidney cells and the rate of circulation to the kidneys. Diuretics include cold applications (which contract superficial blood vessels thereby increasing blood pressure) and chemically diverse substances (e.g., coffee, tea, lemonade, chlorthalidone, furosemide, and ethacrinic acid) with varying toxicities. The more potent diuretics can effect massive urine output with resultant disturbances of electrolyte balance and consequent cardiovascular and muscular effects and sometimes dermatoses and unusual sensitivity to light. **mercurial diuretic**. any of a number of powerful, seldom used, organic diuretics that must be given parenterally; all are nephrotoxic. Because they are relatively nonkaliuretic, they are sometimes used in cirrhotic patients with low potassium intake and in heart patients in whom potassium loss might trigger digitalis toxicity. The only mercurial diuretic available for therapeutic use in the United States is Mersalyl (with theophylline).

diurnal rhythm. any biological rhythm with a period of one day (24 hours). Such rhythms are driven by an endogenous clock with a period of about one day which is synchronized daily with the natural 24-hour (light-dark) cycle by an external signal (a zeitgeber). Nearly all biological processes exhibit rates or other properties that vary with the time of day. This is true also for susceptibility to toxicants. Such rhythms are usually kept in phase by the daily light cycle, but are nevertheless endogenous and continue (with a period of about a day) in the absence of external signals. Thus short-term toxicity tests conducted at one time of day may give very different results when conducted at another time of day. Diurnal variations both in cytochrome P-450 levels and susceptibility to toxicants have been described. Some rhythms may have significant secondary effects on toxicological functions or processes. Thus diurnal feeding patterns may affect the dose of a toxicant administered in food in a given study and hence the apparent susceptibility. Such rhythms, though often important in medical diagnosis and therapeutics, are usually ignored by those in the medical, dental, and veterinary health professions. They are often ignored by experimentalists and by those who conduct toxicity tests. *Cf.* circadian rhythm.

diuron (*N'*-(3,4-dichlorophenyl)-*N,N*-dimethylurea; 1,1-dimethyl-3-(3,4-dichlorophenylurea)). an extremely persistent preemergence herbicide with low water solubility. It is very toxic to laboratory rats. A cumulative poison, repeated doses can cause anemia and methemoglobinemia in laboratory rats, if hydrolyzed endogenously.

Diursal. Mersalyl with theophylline.

divicine. a toxic base found in *Lathyrus sativus*, *q.v.* See also lathyrism.

divinyl. 1,3-butadiene.

dixanthogen (thioperoxydicarbonic acid diethyl ester; dithiobis[thioformic acid] *O,O*-diethyl ester; *O,O*-diethyl dithiobis[thioformate]; bisethylxanthogen; diethylxanthogenate; ethylxanthic disulfide; preparation K; EXD). a nearly water-insoluble, non-persistent, preemergent contact thiocarbonate herbicide with an oniony odor. It is used to control seedling dicotyledonous weeds before the emergence of crops. It is also used in insecticide formulations and as an ectoparasiticide. Dixanthogen is very toxic to laboratory rats, *per os*.

dizziness. a subjective term usually applied to sensations of humans associated with disturbed states of orientation in space such as states of faintness, giddiness, light-headedness, disorientation or disequilibrium. *Cf.* vertigo.

djalimoo. *Trimeresurus wagleri*.

djenkol bean. *Pithecolobium lobatum*.

djenkol poisoning. poisoning due to ingestion of excessive amounts of beans from *Pithcolobium lobatum* (the velvet bean) of Java; symptoms are pain in the renal region, dysuria with subsequent anuria. The djenkol bean has a high vitamin B content and is used for food despite its toxic qualities. The toxic principle is djenkolic acid.

djenkolic acid (S,S'-methylenebis-L-cysteine; 3,3'-(methylenedithio)dialanine; 3,3'-methylenedithiobis (2-aminopropanoic acid); L-cysteine thioacetal of formaldehyde). a poisonous, crystalline amino acid, $CH_2[SCH_2$-$CH(NH_2)COOH]_2$, that is isolated chiefly from *Pithecolobium lobatum*. It is the active agent in djenkol poisoning.

DL. ceiling value.

DM. phenarsazine chloride.

DMAA. dimethylarsinic acid (See cacodylic acid).

DMBA. 9,10 dimethyl-1,2-benzanthracene.

7,12-DMBA. 9,10 dimethyl-1,2-benzanthracene.

DMC. chlorfenethol.

DMDT. methoxychlor.

DMF. *N,N*-dimethylformamide.

DMFA. *N,N*-dimethylformamide.

DMN. *N*-nitrosodimethylamine.

DMNA. dimethylnitrosamine (See *N*-nitrosodimethylamine).

DMP. dimethyl phthalate.

DMPA ((1-methylethyl)phosphoramidothioic acid *o*-(2,4-dichlorophenyl) *o*-methyl ester; isopropylphosphoramidothioic acid *o*-2,4-(dichlorophenyl) *o*-methyl ester; *o*-(2,4-dichlorophenyl)-*o*-methyl isopropylphosphoramidothioate). **1:** an herbicide and plant growth regulator, it is very toxic orally to laboratory rats; it is neurotoxic to chickens. It is moderately toxic dermally. **2:** DMPA (or DMPA®) is also an acronym for dimethylolpropionic acid.

DMS. dimethyl sulfate; succimer.

DMS(methyl sulfate). dimethyl sulfate.

DMSA. succimer.

DMSO. dimethyl sulfoxide.

DMT (3-(2-(dimethylamino)ethyl)indole; *N,N*-dimethyltryptamine). a very short-acting hallucinogen. It is usually effective for about one half hour. DMT is toxic by *i.v.* and *i.p.* injection. Human systemic effects include hypertension, pupillary dilation, hallucination, distorted perceptions, illusions, and loss of touch with reality. It may trigger emotional disorders or psychosis in predisposed individuals. Death by overdosage has not been reported.

DN. dinitrocresol.

DNA (deoxyribonucleic acid; desoxyribonucleic acid; genetic material). a long, linear, self-replicating macromolecule that constitutes the genetic material of all living organisms and the DNA viruses. It codes for the synthesis of proteins. The DNA molecule is right-handed double helix. It is comprised of two chains of repeating units of the sugar, 2-deoxyribose, linked by phosphodiester bridges (3′-5′ bridges). A purine or a pyrimidine base is covalently bound to each 1′-position, and the two chains are linked by hydrogen bonds between a purine base on one chain and a pyrimidine base on the other. The particular sequence of these bases in a particular DNA molecule make up the genetic code. The two chains of a strand are complementary in that a purine on one chain is always linked to a pyrimidine on the other (e.g., adenine is always linked to thymine and guanine to cytosine). This type of complementarity allows each of the two chains to act as a template for synthesis of the other during DNA replication. Toxicants can inhibit DNA-directed protein synthesis at the level of transcription or translation, or can bring about the production of altered products.

DNA adduct. a compound formed by the covalent binding of a reactive chemical species (an electrophile) with nucleophilic sites on a DNA molecule to form an addition product (adduct). This is generally regarded as the initial event in chemical carcinogenesis and mutagenesis. Most chemical carcinogens are electrophiles or form electrophilic products during metabolism.

DNA repair. DNA may be damaged by chemicals or by irradiation. In many cases, the DNA molecule repairs or reduces such damage by the action of specific enzymes (DNA polymerases and ligases). Repair may be imperfect and produce alterations of base sequence (i.e., mutations) of the DNA molecule.

DNase (deoxyribonuclease). a type of enzyme that catalyzes the hydrolysis of DNA to nucleotides; it occurs in some snake venoms.

DNBP. dinoseb.

DNC. dinitrocresol.

DNOC. the acronym used in the United States for dinitrocresol.

DNOCP. dinocap.

DNP. dinitropyrene.

2,4-DNP. 2,4-dinitrophenol.

DNT. 2,4-dinitrotoluene.

DO. dissolved oxygen.

DOA. dead on arrival.

DOB (bromo-DMA; 1-(2,5-dimethoxy-4-bromophenyl)-2-aminopropane). a dangerous and potentially lethal synthetic hallucinogenic agent. It is very toxic, *i.v.*, to laboratory mice; extremely toxic to laboratory rats, *i.p.*; and doses of about one to four mg/kg body weight, *i.v.*, may be fatal to dogs and monkeys. DOB is a powerful serotonin 5-HT_2 agonist. Symptoms of intoxication may include restlessness, hyperreflexia, perceptual alterations and illusions that involve any or all of the senses. Systematic thought may be impaired and some persons may experience profound emotional problems. High doses may produce hallucinations, loss of touch with reality, cold and painful extremities, and peripheral arterial spasm which, if untreated, may result in gangrene.

Dobesin. diethylpropion.

DOC. di-*n*-octyl phthalate (See bis(2-ethylhexyl) phthalate).

DOCA. desoxycorticosterone acetate. See desoxycorticosterone acetate.

dock. a common name for plants of the genus *Rumex*.

***cis*-13-docosenoic acid**. erucic acid.

doctor gum. *Metopium taxiferum.*

doctorfishes. See Acanthuridae.

1,1a,2,2,3,3a,4,5,5,5a,5b,6-dodecachloro-octahydro-1,3,4-metheno-1*H*-cyclobuta-[*cd*]pentalene. mirex.

dodecachloropentacyclodecane. mirex.

dodecachloropentacyclo(3,2,2,$0^{2,6}$,$0^{3,9}$,-$0^{5,10}$)decane. mirex.

dodecahydrodiphenylamine. dicyclohexyl-amine.

[7S-(7α,7aα,14aα,14aβ)]-dodecahydro-7,14-methano-2H,6H-dipyrido[1,2-a:1′,2′-e][1,5]-diazocine. sparteine.

3-(1,3,5,7,9-dodecapentaenyloxy)-1,2-pro-panediol. an unstable compound, found in human feces, that rapidly decomposes in air or acid medium. It is a powerful mutagen and a probable cause of colon cancer in humans.

dodecarbonium chloride (*N*-[2-(dodecylamino)-2-oxoethyl-]-*N,N*-dimethylbenzenemethan-aminium chloride; benzyl[(dodecylcarbamoyl-methyl)dimethyl]ammonium chloride; *N,N*-di-methyl-*N*-benzyl-*N*-chloro-*N′*-dodecylglycin-amide;(dodecylcarbamoylmethyl)benzyldimeth-ylammonium chloride). a bitter, water- and ethanol-soluble, crystalline substance used as an antiseptic and disinfectant. It is very toxic orally to laboratory rats.

***N*-[2-(dodecylamino)-2-oxoethyl-]-*N,N*-di-methylbenzenemethanaminium chloride**. dodecarbonium chloride.

(dodecylcarbamoylmethyl)benzyldimethyl-ammonium chloride. dodecarbonium chloride.

dodecyltrimethylammonium bromide. a quater-nary ammonium salt that is toxic to plants.

dog button. strychnine.

dog button plant. *Strychnos nux-vomica*

dog buttons. nux vomica.

dog-fennel. *Anthemis cotula*.

dog parsley. a common name of *Aethusa cynapium*.

dog's mercury. herbs of the genus *Mercurialis*.

dogbane. common name of any plant of the genus *Apocynum* and the collective name for plants of the family Apocynaceae.

dogbane family. Apocynaceae.

dogfish. **1:** common name of many species of sharks, especially of the family Squalidae (e.g., *S. acanthias*, the spiny dogfish), referred to collectively as the dogfish sharks. **2:** certain freshwater bony fish of the order Protospondyli (e.g., *Amia calva*).

Dolichos lablab (hyacinth bean; bonavist; lablab; papapa bean; Hawaii). a stout, but tender, tropical, twining perennial, often cultivated as an annual (Family Fabaceae, formerly Leguminosae). It is 3-6 m long, with large deltoid-ovate leaflets and reddish purple to white flowers. Introduced into the United States from India as an ornamental, it sometimes es-capes. It is sometimes cultivated in the tropics for its seeds. The latter contain cyanide, but if boiled and the water discarded, they are edible.

doll's eyes. *Actaea (= Actea) alba.*

dolor (plural, dolores). pain. Together with calor, rubor and tumor, it is one of the principal signs of inflammation as articulated by Celsus. See pain.

dolor capitis. headache.

dolor vagus. wandering pain.

dolores. plural of dolor.

dolores praesagientes. false pains; pains that indicate falsely (e.g., the pains of false labor).

dolorific (dolorogenic). able to cause or induce pain; inducing or causing pain

dolorogenic. dolorific.

dolphin. See ciguatera.

DOM. 2,5-dimethoxy-4-methylamphetamine hydrochloride.

domestic rat. *Rattus norvegicus*. See also laboratory rat.

domino viper. *Vipera superciliaris*.

Donax serra (white mussel). a bivalve mollusk (Class Bivalvia = Pelecypoda) that occurs in South African waters. It is often involved in paralytic shellfish poisoning of humans.

doobie. any cigarette made from marijuana.

dopa (3-hydroxytyrosine; 3-(3,4-dihydroxy-phenalalanine; β-(3,4-(dihydroxyphenyl)-α-alanine; dihydroxyphenylalanine). a crystalline amino acid, it is an oxidation product of tyrosine. It occurs in the posterior salivary glands of octopod mollusks (Class Cephalopoda), various fruits and vegetables, and in the seedlings, pods and beans of *Vicia faba* (broad beans) and *Stizolobium deeringianum* (velvet beans). *Cf.* levodopa; See also dopamine.

L-dopa. Levodopa.

dopamine (4-(2-aminoethyl)-1,2-benzenediol; 4-(2-aminoethyl)pyrocatechol; 3-hydroxytyramine; 3,4-dihydroxyphenethylamine; α-(3,4-dihydroxyphenyl)-β-aminoethane). a crystalline base, it is the decarboxylation product of dopa. Dopamine is an endogenous monoamine transmitter with α- and β-adrenergic activity and is an intermediate in the biosynthesis of epinephrine and norepinephrine. In mammals it occurs chiefly in the region of the corpus striatum of the brain where it is an inhibitory transmitter; it is an inhibitory transmitter in many other animal taxa also (e.g., gastropod mollusks). dopamine receptors in higher vertebrates reside on pre- and postsynaptic sites within the CNS, in the hypophysis, the area postrema (chemoreceptor trigger zone) and in the peripheral vasculature. The hydrochloride of dopamine is used as a cardiotonic and antihypotensive. It is a dangerously toxic drug by *i.p.*, *i.v.*, or intracervical routes of administration. Symptoms of intoxication i nclude nausea, vomiting, nervousness, irritability, tachycardia, cardiac arrhythmias, dilated pupils, blurred vision, chills, pallor or cyanosis, fever, suicidal behavior, spasms, convulsions, pulmonary edema, gasping, coma, and respiratory collapse. Gangrene of the extremities has developed following dopamine (Intropin) injection. Some neurotoxicants, especially among those that cause psychotic reactions in higher animals may act on the corpus striatum or other dopaminic systems (e.g., triethyl lead). Antipsychotic drugs act mainly as antagonists of dopamine receptors. See also catecholamines.

dopamine hydroxylase (DBH; EC 1.14. 17.1). a copper-containing enzyme that catalyzes the synthesis of norepinephrine from dopamine. The process of conversion requires ascorbic acid as a cofactor and molecular oxygen. Inhibitors of DBH include copper chelators, disulfiram and fusaric acid.

dormin. abscisic acid.

dorsalgia. pain in the back.

dosage. **1:** the amount of a toxicant or medication administered to or taken by an organism. Dosage is usually expressed in units of weight or volume of the chemical per unit mass of the organism during a stated time interval (e.g., mg/kg/body weight/day). **2:** the administration of medicine or other biologically active agents in prescribed or stated amounts. **3:** determination and regulation of the effective dosage (def. 1) of a biologically active substance. *Cf.* dose, dosimetry.

dose. **1:** a measured amount of a biologically active agent (e.g., medication, ionizing radiation), to which a living system is exposed, usually at one time or in fractional amounts over a specified period of time (e.g., a specified number of tablets per day make up a daily dose). **2:** the amount of energy introduced into a target in a single application of ionizing radiation (Roentgen rays or gamma rays) or during a unit period of time. **3:** the amount of a substance or of ionizing radiation absorbed by a living organism (or part of an organism), at one time or during a short period of time. **4a:** to give a prescribed or measured amount of a medicine to. **4b:** to give or prescribe in doses. **absorbed dose**. **1:** the amount of a substance that penetrates an absorption barrier in an organism by means of a physical or biological process. **2:** radiation absorbed dose. **absorbed radiation dose**. radiation absorbed dose. *Cf.* radiological dose. **acceptable daily dose** (ADD). the total amount of a drug, especially a medication, that can be taken indefinitely without significant risk. **administered dose**. the amount of a biologically active agent that is introduced into the system (e.g., via ingestion, injection, inhalation) or applied to an absorbing surface. **air dose** (exposure dose). the intensity of radiation (in Roentgens) of a roentgen-ray or gamma-ray beam as measured in free air at the target. **approximate lethal dose** (ALD). **1:** an estimate of lethal dose (usually median lethal dose) based on imprecise data or extrapolated data. **2:** any dose that kills a large, but poorly determined portion of a population. **biologic dose**. a measure of the concentration of a substance in relation to specified biological effects. **bolus dose**. a dose of a substance that is administered intravenously at a rapid but controlled rate. **booster dose**. a dose of a therapeutic agent or of antigens administered some time after an initial dose in order to enhance or prolong the effect. **critical dose**. that dose which will produce a critical effect. See concentration (critical concentration), effect (critical effect). **cumulative dose**. **1:** the total dose of radiation resulting from repeated exposures of an organism or part of an organism. **2:** the amount of a xenobiotic substance present in the body following repeated exposure, taking into account that portion which is no longer effective because of detoxification and/or excretion. **cumulative radiation dose**. See cumulative dose (def. 1). **curative dose**. the dose that will cure a disease or remedy the effects of a dietary deficiency. *Cf.* therapeutic dose. **daily dose**. the total amount of a biologically active substance taken or administered within a 24-hour period. **depth dose**. the ratio of the amount of radiation received beneath surface of the body to that recorded at the surface. **divided dose**. a dose divided into fractional portions which are administered or taken frequently throughout a specified period (e.g., one day). **effective dose** (ED). that dose of a biologically active agent which produces an intended effect in a specified sample or biologically active agent to a living organism; population of living organisms. **effective dose/1% response** (ED_{01}). that dose of a biologically active agent which produces an intended effect in only 1% of a specified sample or population of living organisms. ED01 values are imprecise and are rarely applied except in studies of carcinogenesis. **effective dose/50% response** (ED_{50}; effective $dose_{50}$). median effective dose. **effective dose/99% response** (ED_{99}). The dose that produces an intended effect in 99% of a specified sample or population of living organisms. **epilation dose**. the minimum amount of radiation that will produce hair loss in a specified period of time (usually 10 to 14 days). **erythema dose**. the amount of radiation that is just sufficient to produce erythema within two weeks following application. **estimated maximum tolerated dose** (EMTD). an approximation of the maximum dose that can be tolerated by a living organism under conditions of chronic exposure. Such estimates are almost always based on the results of relatively short-term studies or a series of such studies spanning a range of durations. See maximum tolerated dose. **estimated median lethal dose**. that median lethal dose arrived at by calculations based on results of sequential testing. This is the usual case, the result being an approximation or estimate of dosage. **exit dose**. the amount of radiation leaving a body opposite the area of entry. **exposure dose**. air dose. **external dose**. the quantity or concentration of a biologically active substance that comes into contact with the body of an organism. This may be greater than the amount actually absorbed. **fatal dose**. lethal dose. **fractional dose**. divided dose. **initial dose** (loading dose; primary dose). the large dose given at the beginning of treatment in order to rapidly elevate blood levels and thus to quickly bring the subject under the influence of the drug. **inhibitory dose**. a dose that inhibits specific enzyme activity (usually) or some other process or function. Unless otherwise specified, this term is term is usually taken as equivalent to the median inhibitory dose. **inhibitory dose, 50% response**. median inhibitory dose. **internal dose**. the amount of a biologically active substance that is absorbed by an organism and reaches metabolically active tissues, organs or organelles. **infecting dose**. see median infecting dose. **intoxicating dose**. the dose of antigen that is just sufficient to induce an allergic reaction. **introduced dose**.

the dose of antigen that is just sufficient to induce an allergic reaction. **irreversible dose.** the lowest dose that consistently produces one or more effects that remain after the substance and its active metabolites are eliminated from the system. **L_+ dose**. the smallest dose of a toxin mixed with 1 unit of antitoxin that will kill a 250 g guinea pig within 4 days when injected subcutaneously. This unit is used to standardize an antitoxin. **lethal dose** (LD). **1:** the amount of a toxicant (usually stated as mass per unit body weight, as mg/kg) which is sufficient to kill (or is likely to kill) a designated living system within a relatively short (or designated) time following exposure under a specified set of conditions (e.g., duration of exposure, route of administration, history of exposure, environmental conditions, and physiologic status of the target system). **2:** any dose that kills a large (usually specified) portion of a population or is predicted to do so. **3a:** approximate lethal dose. **3b:** sometimes considered equivalent to median lethal dose. **lethal dose, 1% response** (LD_{01}). the amount of a toxicant or an infective agent estimated to kill only 1% of a test population that is exposed to the agent for a specified period of time. This value approximates the maximum sublethal dose. It is usually a very imprecise measure of lethality. **lethal dose, 99% response** (LD_{99}). the amount of a substance, virus, or other infective agent estimated to kill 99% of a test population that is exposed to the agent for a specified period of time. It is used to estimate the lowest dose that is likely to be lethal to an entire animal population. It is usually a very imprecise measure of lethality. **lethal intragastric dose** (lethal dose, *i.g.*; LD *i.g.*). the lethal dose of a toxicant when introduced directly into the stomach. **lethal intramuscular dose** (lethal dose, i.m.4; LD i.m.). the lethal dose of a toxicant when administered intramuscularly. **lethal dose 50% response**. median lethal dose. **lethal intraperitoneal dose** (lethal dose, *i.p.*; LD *i.p.*). the lethal dose of a toxicant when administered intraperitoneally. **lethal intravenous dose** (lethal dose, *i.v.*; LD *i.v.*). the lethal dose of a toxicant when administered intravenously. **lethal dose, *per os*** (lethal dose, *p.o.*; LD *p.o.*). the lethal dose of a toxicant when ingested or taken by mouth. **lethal subcutaneous dose** (lethal dose, *s.c.*; LD *s.c.*). the lethal dose of a toxicant when administered subcutaneously. **loading dose**. initial dose. **lowest lethal dose** (LDLo). the lowest dose administered during a given period of time in one or more divided doses that is known to have caused the death humans or other animals. LDLo's are often established on the basis of accidental exposures or overdoses (chronic or acute). **maintenance dose**. the amount of a substance that is just sufficient to maintain a desired effect. **maximal dose**. maximum dose. **maximum dose** (maximal dose). the largest amount of a substance that can be safely administered to or taken by a target organism. **maximum permissible dose** (MPD). **1a:** the greatest amount of radiation permitted in accordance with radiation protection guidelines to persons who are required to work with radioactive sources. **1b:** that amount of radiation which will produce no harmful effects in the general human population. Values differ for pregnant females, for acute and chronic exposures, and for whole body exposure versus exposure of particular organs, systems, or body regions. *Cf.* permissible dose, radiation protection guide. **maximum sublethal dose** (LD_0). the highest dose of a toxic substance to which individuals of a test population can be exposed with no mortality. *Cf.* lethal dose, 1% response. **maximum tolerated dose** (MTD). In general, the highest dose of a biologically active agent given during a chronic study that will not reduce an animal's longevity from effects other than carcinogenicity. The U.S. Environmental Protection Agency defines MTD, for purposes of toxicity testing, as that dose which will cause no more than a 10% inhibition of weight gain, which does not increase mortality, and does not produce clinical evidence of toxicity or pathological lesions that would indicate a shortening of the animal's normal life span other than as a possible result of carcinogenesis. **median curative dose** (CD_{50}). that dose of a therapeutic agent or antidote which corrects or cures a condition in 50% of the test population. **median effective dose** (ED_{50}). that dose of a biologically active agent that produces an intended effect in 50% of a specified sample or population. See also therapeutic index, median infective dose. **median infective dose** (ID_{50}). that amount of a particular population of pathogenic microorganisms which produces infection in 50% of organisms randomly drawn from a defined population of susceptible subjects. The ID_{50} is calculated on the basis of a series of graduated doses given to test populations. **median inhibitory dose** (ID_{50}; MID; inhibitory dose, 50% response; inhibitory dose). the amount of asubstance estimated to be sufficient to completely inhibit the the activity of a specified enzyme, usually in a specific organ. **median lethal dose** (MLD; lethal dose; lethal dose, 50% response; LD_{50}; LDt_{50}). **1:** the amount of a substance, virus or other infective agent estimated to be sufficient to kill 50% of a test population that is exposed to the agent for a specified period of time. LD_{50} varies widely with the status of the test organism (e.g., age, sex, strain) and conditions of testing (e.g., ambient temperature, type of housing, route of exposure, time of day, time of year, and other environmental conditions). Median lethal dose (when tests are standardized and sufficiently comparable) is the most commonly used measure of chemical toxicity and is generally considered the standard for estimating relative acute toxicities of chemicals and the comparative susceptibilities of various species of organisms to acute exposures. See also therapeutic index, comparative toxicity, relative toxicity. **2:** the amount of ionizing radiation that will kill 50% of exposed individuals within a specified period of time. **median minimum lethal dose** (MLD_{50}). the smallest amount of a toxicant that will kill 50% of the organisms receiving it within a specified period of time. **median toxic dose** (TD_{50}). the amount of a substance or infective agent estimated to be just sufficient to produce a specified toxic effect (or set of effects) in 50% of a test population that is exposed to the agent for a specified period of time. Technically, this is a median effective dose applied to toxic end points. **minimal dose**. minimum dose. **minimal infecting dose** (MID). minimum infecting dose. **minimal lethal dose**. minimum lethal dose. **minimal reacting dose**. minimum reacting dose. **minimum dose**. **1:** the smallest dose of a substance that is effective in producing a particular effect. **2:** in medicine, it is usually taken to be the smallest amount of a drug that will produce a physiologic effect in an adult. **minimum effective dose** (MED; lowest effect level; LEL). the minimum dose that produces an observed effect. It is sometimes used as an alternative to the no-effect level (NEL). **minimum infecting dose** (MID). the smallest amount of infectious material that consistently produces infection. The MID is usually expressed as ID_{50}, which is the amountthat causes infection in 50% of a sample of animals or cells in culture. **minimum lethal dose** (MLD). **1:** the smallest amount of a toxicant or infectious agent that will kill the organism to which it is administered in a designated period of time. **2:** the smallest quantity of diphtheria toxin which will, on the average, kill a 250-g guinea pig within 96 hours following subcutaneous inoculation. **minimum reacting dose** (MRD). the smallest amount of a substance that will cause a minimal standard skin reaction in susceptible test animals. The standard reaction is a focal inflammation (marked by congestion, edema, degenerative changes and desquamation of epithelial cells) that appears within 18-24 hours following *i.v.* injection of the toxicant and peaking in about 96 hours. **no-effect dose** (NED). the highest dose of a drug or other xenobiotic that produces no observable effect. **no-effect dietary dose**. the oral no-effect dose. See no-effect dose, no-effect level. **nonlethal dose**. any amount of a toxicant that will not cause death in a designated living system. **optimum dose**. that amount of a biologically active agent that will produce a desired effect with the least likelihood of causing harm or producing undesirable effects. **permissible dose**. the amount of ionizing radiation that is permitted by current radiation protection guides. See also maximum permissible dose, radiation protection guide, threshold. **potential dose**. the amount of a drug or toxicant contained in material that has been ingested, inhaled, or absorbed. **prescribed dose**. the precise amount of a biologically active substance to be administered or taken at one time or during a designated period of time, as directed by one who is legally authorized to do so. **preventive dose**. the smallest amount of a substance that will prevent occurrence of the effects of a disease or of the lack of a particular dietary factor. **primary dose**. initial dose. **protective dose**. a low dose of a toxicant that protects against a higher dose, as in the case, for example, of ozone, arsenic, cadmium, and zinc. **radiation absorbed dose** (RAD; rad; absorbed radiation dose; radiological dose). the total amount of ionizing radiation (in rads) absorbed per unit mass of a tissue, material, or living system exposed to any source of radiation. It is a measure of the energy imparted per unit mass of radiated material: 1 rad = 100 ergs/g = .01 gray (Gy), *q.v.* *Cf.* absorbed dose, radiation dose. **radiation dose**. the total amount of ionizing radiation absorbed by a substance, living cells and tissues, organs, or the entire body of a living organism. *Cf.* radiation absorbed dose. **radiological dose**. radiation absorbed dose. **reacting dose** (shocking dose). the second dose of sensitizing antigen administered to an animal. It produces an immediate anaphylactic or allergic response. *Cf.* sensitizing dose. **reference dose** (RFD, RfD). a standard dose used as a basis of comparison among different treatments or different types or species of organisms to which a drug or other xenobiotic is administered. Such a dose should be established by the use of experimental data or data from standardized tests. A reference dose is intended to allow for variability in sensitivity among species. See also Acceptable Daily Intake, factor (uncertainty factor). **reversible dose**. any amount of a biologically active agent that produces no lasting effects. **sensitizing dose**. the first dose of antigen administered to an animal in the induction of an anaphylactic or allergic response. It renders the animal susceptible (sensitive) to anaphylactic shock following a subsequent dose of the same antigen. *Cf.* reacting dose. **shocking dose**. reacting dose. **skin dose**. the amount of radiation delivered to the skin surface, including that reflected from the surroundings (backscatter). **sublethal dose**. **1:** any dose of a potentially lethal substance that does not kill a specified individual, set, or class of living systems (given an otherwise favorable environment) within a relatively short (or specified) time following exposure to a given dose of a substance (as in a toxicity test). The distinction between a sublethal and a lethal dose is not precisely defined. **2:** any dose that kills only a very small portion of a population. **therapeutic dose**. that dose of a therapeutic agent required to produce the intended effect. *Cf.* curative dose. **threshold dose**. **1a:** the minimally effective dose of a biologically active chemical that evokes a stated quantal response under a stated set of variables, environmental and biological (e.g., ambient temperature, route of exposure, age, sex, body weight, diet). **1b:** the maximum dose of a biologically active substance that produces no observable effect under a stated set of variables as in definition 1a. **1c:** toxicologically, the dose of a chemical below which no injury, harm, or adverse effect occurs to the test system or to the potential progeny of the test system, if applicable. Threshold dose is often inferred by extrapolation from results of tests of systems using doses well above threshold and in many cases, threshold values are sensitive to any of a number of variables. **2:** the minimum dose of ionizing radiation that will produce a detectable specified effect. **tissue culture infectious dose** ($TCID_{50}$, TCD_{50}). that amount of a cytopathogenic agent (e.g., a virus) which will produce a cytopathic effect in 50% of the cultures inoculated. **tissue tolerance dose**. the largest amount of a substance or of radiation that does not injure tissues. **tolerance dose**. the largest quantity of a biologically active agent that is accepted by a biological system with no evidence of harm. **total dose**. the total amount of a biologically active agent administered to or taken by a living system. **toxic dose** (TD). the smallest dose of a substance that will produce an observable toxic effect. Such effect may be temporary or permanent and may occur at any time during the life of the exposed organism or may, as appropriate, include effects on the well being or viability of progeny. **toxic dose/ 1% response** (TD_{01}). The dose that is estimated to be toxic to only 1% of a test population. Regardless of the end point used, these values offer poor precision because of the large confidence limits at the extreme ends of the dose versus percent response curve. the use of TD_{01} values has generally been restricted to studies of carcinogenesis, where the biological response at extremely low doses is of interest. **toxic dose/99% response** (TD_{99}). The dose that is estimated to produce toxic effects in 99% of a test population. Regardless of the end point used, these values offer poor precision because of the large confidence limits at the extreme ends of the dose versus percent response curve. **toxicokinetically equivalent dose**. a dose based upon the toxicokinetics of a particular toxicant in a particular type of organism. As an example, one can expect the rate of metabolism of a given toxicant in mammals (and within many other groups as well) to be inversely related to the size of the species. See also dose equivalency. **virtual safe dose** (VSD). the dose that is predicted to produce an acceptable risk, *q.v.*

dose equivalency. the expression of dosage between species in ways deemed roughly equivalent and therefore suitable for studies or representations of comparative toxicities of toxicants. The most commonly used expression of equivalent dosage is the concentration of or amount of the toxicant administered per unit body weight (e.g., mg toxicant/kg body weight). This is not universally satisfactory, however, in part because the toxicity of a given chemical may vary differently as a function of body weight in different species. Even within a species or population, toxicity as a function of body weight may vary with sex, age, time of year, and even time of day. See also toxicokinetically equivalent dose, dose response.

dose equivalent. a dosage of a drug or toxicant for one species that is thought to be roughly comparable in effect to a given dosage for another species. See dose equivalency.

dose m. dosim.

dose rate. **1:** the total dose of a drug or toxicant administered per unit time. **2:** the rate at which ionizing radiation is delivered to the target. **radiation dose rate**. the radiation dose absorbed per unit of time (e.g., one hour). **H-hour plus 1 radiation dose rate**. the radiation dose absorbed during one hour.

dose scaling. **1a:** the conversion of intertaxon dose-response data to equipotent doses. **1b:** the conversion of dose-response data from tests of laboratory animals to equipotent doses for humans or for animals commonly met in veterinary practice.

dose surrogate. any indirect measure of exposure to a biologically active agent (e.g., peak plasma concentration, concentrations of a metabolite, alkylations per nucleotide).

dose-dependent. determined or conditioned by the dose, usually referring to the condition in which the intensity or quality of an effect or effects of a biologically active agent are proportional to or related to the dose or concentration of the agent.

dose-effect. dose-response.

dose-effect-duration. See dose-response-duration.

dose-independent. not dose-dependent; not determined or conditioned by the dose. See dose-dependent.

dose-response (dose-response relationship; dose effect). any correlative or quantitative relationship between dose (exposure) and the incidence or intensity of a particular response of a specified living system to a biologically active agent administered under a particular set of conditions (e.g., the route and rate of exposure). Such relationships are n-dimensional (i.e., are usually sensitive to numerous variables) but are frequently treated as two-dimensional. Dose response may be based on external dose (assuming a predictable relationship between external and internal dose) or on internal dose and any measurable (usually toxic) effect. **quantal dose response**. a value that is a measure of a quantitative dose-response relationship. Such are used extensively in toxicology. Examples are the median lethal concentration (LC_{50}), the median lethal dose (LD_{50}), and the threshold dose. *Cf.* graded response.

dose-response assessment (dose-response evaluation). a component of toxicity assessment used as part of a risk assessment whereby a quantitative evaluation of the relationship between the dose of a biologically active substance to which a population of test or experimental animals is exposed and the estimated incidence of a specified adverse effect for a human or natural population under expected exposure regimes. Standard techniques (e.g., empirical dose-response models; pharmacokinetic models) and data sets (e.g., site-specific effluent-, sediment- or leachate toxicity tests, human epidemiological data) are used. Not only may results be extrapolated from individual effects to population or ecosystem-level effects and from one population or taxon to another, but extrapolation from the relatively high doses used in the test situation to the usually low doses expected or encountered in the field situation is usual. Dose-response assessment is usually conducted in two phases: dose-range extrapolation (See extrapolation) and dose scaling, *q.v.*

dose-response curve. the relationship between varying doses or concentrations of a biologically active agent (e.g., toxicant, hormone, enzyme, virus) in a standard test- or experimental population challenged under controlled conditions and a specified quantitative effect (end point) such as mortality rate. The curve is established for a range of doses that may elicit the specified effect. A simple dose-response curve is usually sigmoidal and extrapolation beyond the range for which there is test data is difficult and usually subject to considerable error and uncertainty. See also dose-response.

dose-response evaluation. dose-response assessment.

dose-response model. a mathematical or computer model that describes, extrapolates, and/or otherwise predicts relationships between dosage of a biologically active substance and a specified effect in a given population of organisms. **gamma multi-hit dose-response model**. a mechanistic model (an extension of the one-hit model) that assumes that the DNA of a cell must be exposed to more than one molecule ("multihit") of a carcinogen to induce a tumor. *Cf.* single-hit model. **logistic (log-logit, logit) dose-response model**. a model in which the logarithm of the dose is plotted against the logit (L) of the response:

$$L = ln\ [(1 - p)/p]$$

where *ln* is the natural logarithm and *p* is the proportion of the test population that responds to a given dose. **Mantel-Bryan extrapolation model**. a model used only to extrapolate, it computes the upper 99% confidence limit on the lowest experimentally determined concentration that produces a significant effect and then extrapolates downward on a slope of one probit per order-of-magnitude of change in dose rate. **multidimensional dose-response-duration model**. any model that can be applied to more complex multidimensional dose-response relationships. **multistage dose-response model**. a complicated single-hit model (an exponential polynomial) based on the assumption that responses advance through a number of stages:

$$P = 1 - \exp[-(a_0 + a_1D + a_2D^2 + \ldots + a_nD^n)]$$

where P is the proportion of individuals that exhibit the effect and D is the dose. This model can fit complex curves and often fits experimental data quite well. **probit model**. a type of model that is based on the assumption that the probabilities of response to a given dose within a specified population are normally or log-normally distributed. **single-hit dose-response model**. a model based on the hypothesis that a single molecule of toxicant can produce an irreversible effect by reacting with a single endogenous molecule in a living system according to the equation:

$$P = 1 - e^{-aD}$$

where P is the proportion of individuals that exhibit the effect and D is the dose. As an exponential polynomial, this model can fit complex curves.

dose-response relationship. dose-response.

dose-response-duration (dose-response-duration relationships; dose effect duration). any dose-response space (relationship) that includes the time to effect (duration). See dose-response.

dose-response-duration relationship. dose-response duration.

dosim, Dosim (dose m; dosem). an instrument or device that measures x-ray output or total exposure to ionizing radiation during a given period of time.

dosimetry. the accurate measurement of dosage. **molecular dosimetry**. an approach to extrapolating dosage between species based on radioactive-labeling methods that may permit a quantitative determination of the relationship between external (e.g., mass or concentration per unit body weight) and effective target dose (e.g., alkylations per nucleotide).

dosing. the administration of the presumed appropriate dose (or dose-rate) to an organism.

dosis. dose. **dosis curativa**. the minimum dose of a therapeutic agent that will effect a cure. **dosis efficax**. dosis curativa. **dosis refracta**. fractional dose. **dosis tolerata**. the largest dose of a therapeutic agent that can be safely administered.

double blind. of or pertaining to any test or experiment in which neither the investigator nor the subject knows which treatment is administered (e.g., experimental vs. control; the administration of an active or an inert substance).

double thiosulfate of gold and sodium. gold sodium thiosulfate.

double vision. diplopia.

double-collared snake. *Hoplocephalus bitorquatus*.

douche. **1:** a stream of fluid (e.g., water, air) directed against a body part or body cavity. **2:** a bath taken by means of a douche. **3:** an implement (e.g., a syringe) for giving douches. **4:** to apply a douche. **feminine douche**. any of a variety of products used to lave the vaginal canal. They often contain harmful ingredients, e.g., ammonia, detergents, EDTA, colors, artificial fragrance, and phenol. See especially phenol. They are irritants that probably have no direct value to a woman's health and can cause allergic dermatoses and can irritate the vaginal tissues, causing vaginitis. External tissue may also be affected by dermatitis, and systemic effects are possible.

Douglas water hemlock. *Cicuta douglassi*.

doxepin (3-dibenz[b,e]oxepin-11(6*H*)-ylidene-*N,N*-dimethyl-1-propanamine; *N,N*-dimethyldibenz[b,e]oxepin-$\Delta^{11(6H)}$,γ-propylamine; 11-(3-dimethylaminopropylidene)-6,11-dihydrodibenz[b,e]oxepin). a tricyclic antidepressant, it is an oily liquid comprised of *cis*- and *trans*-isomers. The hydrochloride (a white crystalline powder), administered orally as a sedative and hypnotic, has pronounced antianxiety and significant antidepressant activity. It is used also in veterinary medicine as an antipruritic. It is toxic to humans and laboratory animals when ingested or administered *i.p., i.v.,* or subcutaneously. It is very toxic to laboratory mice and rats when ingested and extremely toxic *i.p.* and *i.v.*

doxorubicin ((8S-*cis*)-10-[(3-amino-2,3,6-trideoxy-α-L-lyxo-hexopyranosyl)oxy]-7,8,9,10-tetrahydro-6,8,11-trihydroxy-8-(hydroxyacetyl)-1-methoxy-5,12-naphthacenedione; 14-hydroxydaunomycin). a cytotoxic and cardiotoxic antineoplastic anthracycline antibiotic isolated from the bacterium *Streptomyces peucetius* var. *caesius*. It is cardinogenic and is extremely toxic to laboratory mice. Symptoms of intoxication in humans include alopecia, stomatitis, nausea, vomiting, diarrhea, stomatitis, tachycardia, and alopecia. It causes delayed cardiotoxic effects including a sometimes fatal congestive heart failure. **doxorubicin hydrochloride**. The hydrochloride salt of doxorubicin, it has the same toxicity and therapeutic uses. See also *Streptomyces*.

doxylamine (*N,N*-dimethyl-2-[1-phenyl-1-(2-pyridinyl)ethoxy]ethanamine; phenyl-2-pyridylmethyl-β-*N,N*-dimethylaminoethylether; 2-dimethylaminoethoxyphenylmethyl-2-picoline). a slightly volatile, acid-soluble liquid that darkens on exposure to light. It is an antihistaminic and hypnotic formerly prescribed for pregnant women. Doxylamine is very toxic orally to laboratory mice and rabbits and is very toxic to laboratory mice and rats when administered subcutaneously. It is probably teratogenic and causes neurobehavioral effects in humans.

DPA. diphenylamine.

DPN. nadide.

Dracunculus. a genus of herbs (Family Araceae) that are indigenous to the Mediterranean region. *D. vulgaris* is sometimes grown as greenhouse curiosities in temperate climates. It has a vile odor and all parts of the plant are toxic due to the presence of calcium oxalate crystals. Symptoms are essentially similar to those produced by *Dieffenbachia, q.v.*

dracylic acid. benzoic acid.

dragon arum. See *Arisaema*.

dragon root. *Arisaema dracontium*.

drain and toilet cleaner. See cleaner.

Draize test. See test.

Drano. See cleaner (drain cleaner).

DRC. daunorubicin.

dredging. the removal of bottom sediments ("muds") from a water body (e.g., river, stream, lake, reservoir) using a machine with a large scoop (a dredge). The process and loss of sediments disturbs the ecosystem and causes silting that can kill aquatic life. Dredging of contaminated sediments can expose aquatic life to otherwise sequestered toxicants such as heavy metals and other persistent toxicants. In the United States, dredging is regulated under Section 404 of the Clean Water Act.

drieckskopfe. *Agkistrodon*.

dried cupric sulfate. See cupric sulfate.

drill. a name applied to any of numerous small marine gastropods that feed on other mollusks by drilling through the shells. Members of the genera *Urosalpinx* and *Acanthina* are characteristic. *Cf.* cone. See also *Acanthina*.

drinking water. See water.

drip. **1:** to fall (or to let fall) in drops; to fall a drop at a time. **2:** a falling in drops. **3:** infusion of a liquid slowly, drop by drop. **intravenous drip**. the slow but steady drop by drop intravenous introduction of a solution. *Cf.* drop.

drocannabinol. (—)-Δ^1-3,4-*trans* tetrahydrocannabinol.

dronabinol. (—)-Δ^1-3,4-*trans*-tetrahydrocannabinol.

drop. **1:** to fall, to let fall or dispense in drops (def. 2). **1b:** the act of letting something fall. **2a:** a minute spherical liquid mass; a globule; a tear. **2b:** the amount of fluid in a drop (def. 2a). **3:** a volume of liquid regarded as a unit of dosage. A drop of water is approximately 1 minim. **4:** a solid confection in globular form. Usually directed to be allowed to dissolve in the mouth. Such drops may contain biologically active or therapeutic agents as in the case of cough drops or eye drops. **5:** the failure of a part (e.g., the wrist or hand) to maintain a normal position. *Cf.* drip, drops.

droplet. a liquid particle of very small mass or a globule that is able to remain suspended in a gas. In some cases, as in a cloud, droplets may reach 200 μm in diameter.

dropping. See scat.

drops. **1:** plural of drop. **2:** a medication, the dose of which is measured in drops.

dropsy (hydrops). a common term for generalized edema. **abdominal dropsy.** ascites.

dropwort. *Oenanthe crocata*.

***Drosophila melanogaster*.** a species of small fruitfly used widely in experimental genetics and embryology and in mutagenicity testing. They are easily maintained in the laboratory and the generation time is brief (10-14 days).

drug. **1a:** a substance used as a medicine; a therapeutic agent. **1b:** a substance used to aid the diagnosis, prevention, or treatment of a disease or the relief of pain and suffering. **2a:** a narcotic substance. **2b:** any substance that can be habituating or addictive. **2c:** any agent that is narcotic in effect. **3:** a poison. **4:** to administer or to take a drug (usually used in reference to narcotics). **5:** to alter a biological process by use of a substance (usually used in reference to a potentially harmful result). **6:** to poison with a drug. **antagonistic drug**. a drug that tends to counteract or neutralize the action or effect of another. **anxiolytic drug**. any of a chemically diverse group of anxiolytics (e.g., benzodiazepine derivatives) that ease or reduce symptoms of anxiety. **street drug**. any narcotic or addictive drug sold illegally to individuals. Such drugs are extremely dangerous to use because they are often impure, or actual mixtures of drugs. Such drugs are usually impure and cut with fillers. The fillers are often toxic and can be deadly. Pure or nearly pure drugs occasionally reach the street and quickly cause fatalities.

drug abuse. See abuse.

drug addiction. addiction (def. 1).

drug allergy. See allergy.

drug eruption. dermatitis medicamentosa (See dermatitis).

drug idiosyncrasy. See idiosyncrasy.

drug interaction. **1:** the phenomenon whereby the prior or nearly concurrent administration of one drug (or toxicant) increases, decreases, or otherwise alters the effect of another drug. If a drug is retained for some time in the body, interactions may occur with a drug or drugs that are administered after use of the original drug was discontinued. Drug interactions may involve xenobiotics (drugs, toxicants, or medications) alone; interactions between one or more xenobiotics and endogenous substances, or with components of one's diet. Drug interactions may be synergistic, antagonistic, or may alter the effect of one or more of the involved substances. **2:** the effects of drug interaction (def. 1). Certain combinations are lethal. See effect (unintended effect).

drug of abuse. any drug having little or no therapeutic value or which are taken at doses higher than those used therapeutically. Mood-altering drugs, CNS depressants, CNS stimulants, hallucinogens, and narcotics are common types of drugs that are abused. Some of these drugs cause physical dependence and even death with chronic use or overdosage.

drug overdose. See overdose.

drug rash. drug eruption (See eruption).

drug reaction. See adverse drug reaction, drug interaction.

drug-receptor complex. the complex formed when a biologically active agent binds to a specific receptor.

drug-resistance plasmid (drug-resistance factor; R factor). any of the plasmids of enterobacteria and other pathogenic or medically important bacteria that carry genes that confer resistance to commonly used antibiotics. Some of these plasmids are transmissible to other bacteria of the same and related species. See also plasmid.

drug tetanus. toxic tetanus.

druggist. pharmacist.

drunk. **1:** drunken, inebriated, intoxicated. **2:** an alcoholic, drunkard, dipsomaniac. **3:** bender, binge, alcoholic spree.

drunken. **1:** drunk, inebriated. **2:** chronically inebriate,

drunkenness. alcoholic intoxication (See intoxication).

dry deposition. See acid deposition.

dry gangrene. See ergotism.

dry heaves. retching.

dry vomiting. retching.

dry-cleaning solvent. any solvent used to clean clothing that cannot be laundered. Such solvents often contain perchloroethylene, the fumes of which are highly toxic and carcinogenic to humans.

Drymaria (drymary). a genus of small branched annual or perennial herbs (Family Caryophyllaceae). ***D. arenarioides*** (Alfombrilla). a small, luxuriant short-lived perennial herb that occurs on acid soils of northern and central Mexico. It is very toxic to livestock, although horses are not known to be affected. Toxicity, symptoms and lesions in cattle are similar to those of *D. pachyphylla*. ***Drymaria pachyphylla*** (inkweed, drymary). a small branching, prostrate summer annual that occurs on alkaline clay soils of west Texas, southern New Mexico to southeastern Arizona and to central Mexico. Serious cattle losses have been recorded, mostly in drought years when other forage is unavailable. In experimental animals, symptoms usually appear within 18-24 hours following ingestion of a toxic dose. In most cases death ensues within two hours following the appearance of symptoms. Symptoms include restlessness, anorexia, diarrhea, arching of the back, depression, and often coma and death. Survivors recover in about two days.

drymary. See *Drymaria*.

drymyphyte. halophyte.

Dryopteris. a large genus of robust ferns (Family Polypodiaceae), representatives of which occur in woodlands throughout the world. They readily hybridize, producing many sterile hybrids and much taxonomic confusion. Many are cultivated. ***D. felix-mas*** (= *Aspidium felix-mas*; male fern; European aspidium). a large, attractive, widely distributed, deciduous or evergreen fern. It is the source of vermifuge and also contains thiaminase. It is presumed toxic. ***D. marginalis*** (American aspidium, marginal shield fern; marginal fern). a fern, thought to be toxic, with large dark green fronds found in clumps on rocky woodland slopes in eastern North America.

DS. dissolved solids.

DSMA. disodium monomethanearsonate.

DT. delirium tremens.

DTC. dithiocarb sodium.

DTMC. dicofol.

Duboid®. bethanechol chloride.

Dubois sea snake. *Aipysurus duboisii*.

Duboisea. a genus of herbs (Family Solanaceae), the members of which contain hyoscyamine and scopolamine. See duboisine.

duboisine. **1:** an biologically active extract of *Duboisia myoporoides* that contains the alkaloids hyoscyamine and scopolamine. It is used as a mydriatic. **2:** sometimes defined as a form of hyoscyamine or used as a synonym for hyoscyamine.

duck-billed platypus. See *Ornithorhynchus anatinus*.

duck ratten. *Veratrum viride*.

duct. any narrow tube within an organism that conveys fluids (e.g., venom duct, tear duct).

ductless gland. See endocrine gland.

dudbe. *Acanthophis antarcticus*.

Dufalone. dicumarol.

dugaldin. a toxic glycoside, originally isolated from *Helenium hoopsii*. It is a strong irritant, is toxic to cattle and the chief cause of sneezeweed poisoning, *q.v.*

Dugasiella. a genus of spiders (Family Theraphosidae) found in the southwestern United States. The venom is no more poisonous than that of a bee or wasp.

dugite. *Demansia textilis*.

dukaitch. *Pseudoechis porphyriacus*.

dulcin (*p*-phenetol carbamide; (4-ethoxyphenyl)-urea; *p*-phenetylurea). a nonnutritive sugar substitute that is about 250 times as sweet as cane sugar. It readily hydrolyses to aminophenol and may thus be injurious if used extensively for extended periods of time.

dulcin(4-ethoxyphenylurea). See sweetener, nonnutritive.

dumb cane. dumbcane. See *Dieffenbachia*.

dumbcane (dumb cane). *Dieffenbachia*.

dundugu. *Naja nigricollis*.

dunu karawala. *Bungarus ceylonicus*.

duodenectomy. surgical removal of the duodenum.

duodenum. the first portion of the small intestine of certain higher vertebrates into which the pancreatic and bile ducts open. It receives partially digested food, undergoes further digestion by bile, pancreatic enzymes, and the enzymes of the succus entericus, an alkaline secretion of the glands in the duodenal wall. The duodenum of adult humans is ca. 25-30 cm long in adult humans and extends from the pyloric sphincter to the jejunum.

duplication. a chromosome aberration in which a chromosome contains two groups of identical genes.

Duragel®. a product that contains nonoxynol-11.

Duranta repens (golden dewdrop; pigeonberry; duranta skyflower). a large tropical American drooping shrub (Family Verbenaceae) that grows to 6 m in height. The flowers are lilac-colored and the branches are sometimes spiny. The yellow berries contain a saponin that causes drowsiness, pyrexia and convulsions. Fatalities among children who have eaten the berries are on record. It is a fairly common greenhouse plant.

duranta skyflower. *Duranta repens*.

Dursban [Dow]®. a brand name for insecticides that contain chlorpyrifos. They are used chiefly to control chinch bugs about the Gulf of Mexico and to control ticks on cattle and sheep in Australia.

dust. **1:** fine dry particles or powder light enough to remain suspended in air for some time. Dust particles of a respirable size, which are toxic or to which toxicants adhere can be quite hazardous. **2:** in air pollution science, dust is defined as the largest particulate matter contributing to air pollution, or those particles with a diameter of about 10μ. Exposure of air-breathing animals to dusts can induce any of a large variety of focal respiratory diseases, including pneumoconioses with pulmonary fibrosis and obstructive lung disease. Certain toxic dusts, when inhaled, may produce systemic poisoning, allergic reactions, dermatoses, and cancer. **coal dust**. miners, and other workers in the coal-producing or coal-using trades (e.g., gashouse workers, stokers) may develop chronic pneumoconioses (e.g., black lung disease, miner's asthma) as a result of long-term exposure of the lungs to coal dust. These diseases are initially characterized by coughing, shortness of breath, and chest pains; damage to the lungs and heart, accompanied by generalized weakness, develop later. **cotton dust**. a fine powder generated during the processing of raw cotton. Inhalation can cause byssinosis (brown lung disease). Symptoms include tightness of the chest and dyspnea. If detected early the condition is reversible, but long-continued exposure to cotton dust may be permanently crippling and ultimately fatal. See byssinosis. **non-proliferative dust**. dusts other than proliferative dusts. They do not induce fibrotic changes in the lung, and their effects are generally reversible; they are cleared from the lungs by phagocytosis. **organic dust**. See organic dust toxic syndrome. **proliferative dust**. any type of dust that is not readily removed from the lungs by phagocytosis or other clearance mechanisms. Such dusts accumulate in the lung and may cause irreversible lung damage via fibrotic hardening known as a pneumoconiosis (examples are asbestosis, byssinosis, silicosis). **quartz dust**. a major component of dust in quartz mines and quarries; in metal mines where the rock between veins of ore contains free silica; and in factories that use quartz sand. Long-term exposure, via inhalation, by workers may ultimately result in massive pulmonary fibroses with the development of emphysema and an associated gradual impairment of lung function. The disease develops slowly with the relatively late appearance of symptoms, mainly dyspnea and coughing. The condition progressively worsens with death intervening, usually from failure of the right heart or from pulmonary tuberculosis. **respirable dust** (respirable particles). suspended particulates that can be deposited to a significant extent in the lungs. There is little agreement on the type and size of particles involved and the term is used variously by different authors. The size range of respirable particles varies with the depth and rate of breathing and probably with any of a number of taxon-specific factors.

Dutch drops. sulfurated oil of turpentine.

Dutch liquid. ethylene dichloride.

Dutch oil. **1:** ethylene dichloride. **2:** sulfurated oil of turpentine.

Dutchman's breeches. *Dicentra cucullaria*.

DW. dry weight.

dwale. *Atropa belladonna*.

dwarf bay. *Daphne mezereum*.

dwarf laurel. *Kalmia angustifolia*.

dwarf mistletoe. *Arceuthobium pusillum*.

dwarf snake. *Aspidomorphus kreffti*.

Dy. the symbol for dysprosium.

dye. **1:** a colorant (def. 1). Under this definition there is no distinction between a dye or pigment. **2:** any colored material that, when in solution, is able to color substances to which it applied. Common usages are in dying fabric, leather, cosmetics, and various polymers. Dyes are also widely used in science to stain tissues and other materials, as test reagents, and as therapeutic agents in medicine. Most dyes now in use are synthetic and many of these are unsaturated organic compounds that contain conjugated double bonds. The bond, group of bonds, or moiety of a dye that is responsible for the color is called a chromophore. Most dyes can cause allergic reactions. *Cf.* color, colorant, pigment. See also pigment and dyes such as C.I. Direct Black 38 that are prefixed by C.I. **3:** to color a material by use of a dye or dyes. **amphoteric dye**. a dye that contains both acidic and basic chromophores. **anionic dye**. acid dye. **acid dye**, **acidic dye** (anionic dye). a dye that is acidic in reaction, usually uniting with positively charged moieties of the material to be colored. **azo dye**. any of a large series of synthetic acid dyes, derived from amino compounds, that contain one or more azo groups (—N=N—) as a chromophore. Such dyes are used extensively on wool and cotton fabrics. They are usually sodium salts of sulfonic acids. Many are mutagenic or carcinogenic. See especially direct dye, azo compound, aminoazo compound. **basic dye** (cationic dye). a dye that is basic in reaction, usually uniting with negatively charged moieties of the material to be colored. **cationic dye**. basic dye. **coal-tar dye**. **1a:** any synthetic dye derived in part from anthracene. **1b:** any synthetic dye derived from coal-tar hydrocarbons or their derivatives (e.g., aniline, anthracene, benzene, naphthalene, toluene, xylene). **direct dye**. any of numerous water-soluble azo (usually) compounds that are used in neutral or alkaline solution to dye material (usually cellulose). They are taken up by fibers directly from an aqueous solution. Many are confirmed mutagens and carcinogens (e.g., benzidines such as 3,3′-dichlorobenzidine, *q.v.*) and are no longer used in the United States. They may, however, appear on imported fabrics. **disperse dye**. any of a type of water-insoluble dye that are dispersed in water, often as a colloidal suspension, from which they are absorbed by the fibers. Further treatment is sometimes required to yield the intended color. Nearly all contain substituted amino groups but no solubilizing sulfonic acid groups. They belong to one of three classes of chemicals (anthraquinones, azo compounds, or nitro-arylamines). They are most commonly applied to cotton and synthetic fabrics (e.g., nylon, polyester) and to thermoplastics. **fabric dye**. any dye used on fabric. Some are very toxic. See azo dye, direct dye, disperse dye. **fiber-reactive dye**. any synthetic dye that contains reactive groups that can form covalent bonds with certain moieties of the molecules of the fibers to be stained. Hundreds of such dyes are in use. **leather dye**. any dye used to color leathers. Such dyes may be very toxic. Toxic components may include methanol, xylene, nitrobenzene, *o*-dichlorobenzene, and 1,1,1-trichloroethane. **natural dye**. any organic colorant found in plant or animal materials, few of which are in use today. Examples are cochineal, indigo, logwood). **solvent dye**. any of a class of water-insoluble dyes (related to disperse dyes) that are soluble in certain organic solvents. Such dyes are used, for example, to color cosmetics, polyesters, and printing inks. **synthetic dye**. any dye derived from coal-tar and petroleum-based intermediates. They are used chiefly as brightly colored permanent dyes for various fabrics. **vat dye**. a class of water-insoluble dyes that are easily reduced ("vatted") to a water-soluble form. The latter are used to impregnate a fabric and are then oxidized (e.g., by air, dichromate) to produce the water-insoluble colored form. **vital dye**. one that is absorbed by living cells and colors certain structures without apparent serious injury to the cells.

dying. **1a:** a progressive failure of vital functions leading to death. It is a natural process in which systems or functions vital to the integrity of the organism and the continuance of life fail irreversibly. **1b:** a condition in which death seems imminent and inevitable. See terminal. **2:** the process of coloring a material by use of a dye por dyes.

dynamic. active; energetic, forceful, powerful; vital.

dynamic phase of toxicity. See phase of toxicity.

dynamic chamber. a test chamber used in inhalation toxicology in which the toxicant flows continuously at a constant, metered rate in a stream of otherwise clean air at constant temperature and humidity. See dynamic exposure.

dynamic exposure. inhalation toxicity testing in which a subject is exposed (in a dynamic chamber, *q.v.*) to a continuous, metered flow of toxicant in a stream of otherwise clean air at constant temperature and humidity.

dynamic phase of toxicity. See phases of toxicity.

dynamic steady state. See steady state.

dys- prefix meaning bad, difficult, disordered, or painful.

dysaesthesia. dysesthesia.

dysarthria. difficult, defective speech due impaired action of the tongue and other muscles of speech.

dyscrasia. **1:** any morbid condition resulting from the presence of an abnormal substance in the blood; used mainly in reference to diseases that affect blood cells or platelets. **2:** formerly, a term indicating disease.

dyscrasic (dyscratic). pertaining to or affected dyscrasia.

dyscratic. dyscrasic.

dysemia. any abnormality or disease of the blood.

dysentery. any of a number of painful intestinal disorders, especially of the colon. **bacillary dysentery**. an acute infectious disorder associated with severe toxemia due probably to endo- and exotoxins produced by bacteria of the genus *Shigella*, especially *S. dysenteriae* (*q.v.*), *S. boydii*, *S. flexneri*, and *S. sonnei*.

dysesthesia (dysaesthesia). abnormal sensations from the skin such as numbness, tingling, prickling, burning, or sharp pain. See paresthesia.

dysfunction. abnormal or impaired function, as of an organ or other body part.

dysfunctional. denoting or pertaining to a state of dysfunction.

dyskinesia. impaired control of voluntary movement, such movements being incomplete. See also ataxia, dystaxia, dyssynergia, tardive dyskinesia.

dyskinesia tarda. tardive dyskinesia.

dysmnesic psychosis. Korsakoff's syndrome.

dysmnesic syndrome. Korsakoff's syndrome.

dysmorphic. marked by dysmorphism.

dysmorphism. **1:** a morphological abnormality of development such as a congenital malformation or teratism. **2:** existing in more than one form. **3:** allomorphism: a change in crystalline form with an accompanying change in chemical structure. *Cf.* teratism.

dysmorphogenesis. abnormal morphological development; the development of abnormal body structures. *Cf.* teratogenesis.

dysmorphogenic. pertaining to dysmorphogenesis or an agent of dysmorphogenesis. *Cf.* teratogenetic, teratogenic.

dysmorphology. abnormal morphology as a result of abnormal development (ontogeny). *Cf.* teratology.

dysoemia. death from obscure causes due to chronic mineral poisoning.

dysphoria. an affective disorder marked by depression and anguish.

dysplasia. disordered growth or development of any type. See Atypia, anaplasia.

dyspnea (shortness of breath; difficult or labored breathing, respiratory distress). difficult or forced and often irregular breathing. Dyspnea can arise from numerous conditions (e.g., extreme fear, asthma, cardiac insufficiency) or environmental factors. It can be induced directly or indirectly by any of numerous toxic substance include asphyxiants, pulmonary carcinogens, irritant gases, cardiotoxicants, histamine. **cardiac dyspnea**. difficult or distressful breathing due to heart disease.

dyspneic. of, pertaining to, or characterized by dyspnea.

dysprosium (Dy)[$Z = 66$; $A_r = 162.50$]. a soft silvery element of the lanthanide (rare earth) series of metals. It occurs naturally in association with other lanthanoids.

dysrhythmia. an abnormal, disordered rhythm of brain waves as represented by an electroencephalogram.

dyssynergia. disturbance of voluntary muscular coordination. See also ataxia, dystaxia, dyskinesia.

dystaxia (partial [illegible]). difficulty in controlling voluntary movement.

Dysticus. a genus of water beetles with pygidial glands on either side of the anus that secrete bacteriostatic agents which are rubbed onto their body and appendages with the hind legs. These secretions contain aromatic acids and phenols including benzoic acid, *p*-hydroxybenzoic acid, *p*-hydroxymethyl benzoate, and *p*-hydroxyquinone. The prothoracic glands secrete a number of poisons (including several of pregnene-derived steroids such as cortexone) that are very toxic to vertebrates, especially to fish which are the chief predators of these beetles.

dystocia. slow or difficult labor, delivery or birthing by a pregnant mammal. *Cf.* eutocia.

dystonia. a state of disordered or reduced tonicity of a tissue, as of muscles. See also extrapyramidal side effects.

dysuria. painful urination.

ϵ. See epsilon.

E. See epsilon.

E. the former symbol for einsteinium.

E102. tartrazine. See also E number.

E123. amaranth. See also E number.

E127. erythrosine. See also E number.

E132. indigo carmine. See also E number.

E600. paraoxon. See also E number.

E.D. 1: effective dose. See dose. **2:** erythema disease.

E number. a number given to a food additive by the European Economic Community (EEC) that is recognized as safe for use in food by the Scientific Committee for Food.

Eadie-Hofstee equation. See Eadie-Hofstee plot.

Eadie-Hofstee plot. one of a number of methods (e.g., Lineweaver-Burk Plot, Woolf Plot) based on transformations of the Michaelis-Menton equation, that are used to linearize enzyme- or receptor-binding data. For enzyme-binding, the Eadie-Hofstee equation is:

$$1/v = -K_m(v/[S]) + V_{max}$$

where v is the initial velocity of the enzyme reaction in the presence of a given concentration of substrate [S], V_{max} is the maximum theoretical velocity of the reaction, and Km is the Michaelis-Menton constant. Thus, if one plots experimentally determined 1/v against v/[kS], the line will have a slope of $-K_m$ and will intercept the ordinate Vmax. This allows an experimental determination of the apparent K_m and the apparent V_{max} of the reaction. For receptor binding, the Eadie-Hofstee equation is:

$$1/B = -K_d(vl[F]) + B_{max}$$

where B = the amount bound; [F] = the concentration of free ligand; K_d is the dissociation constant; and B_{max} is the theoretical number of sites.

eagle ray. a stingray, *q.v.*, of the family Myliobatidae.

ear. **1a:** the organ of sound collection and hearing of vertebrate animals. It is also an organ of equilibrium. **1b:** any of diverse organs that sense vibrations in the medium surrounding the organism (e.g., air, water) at frequencies greater than several per second. **2:** the external ear (e.g., of mammals).

ear shell. *Haliotis*.

eared seal. See Otariidae.

Earle loco. *Astragalus earlei*, a true locoweed.

early spring entoloma. *Entoloma vernum*.

earth gall. *Veratrum viride*.

earthworm. any burrowing terrestrial oligochaete such as *Lumbricus*. They are thought by many scientists to be good indicators of soil pollution. They are eaten by many kinds of wildlife and some chemical elements reach concentrations in earthworms that are potentially hazardous to wildlife at higher trophic levels. Earthworms bioconcentrate heavy metals to levels about one order of magnitude above that of the surrounding soil. Even in the absence of appreciable environmental contamination, some species of earthworms often contain high concentrations of Pb, Zn, and Se. Earthworms of the genus *Eisenoides*, for example, have a remarkable ability to concentrate lead. The average bioconcentration factor for lead in *E. loennbergi* in one study was 52; in other earthworms the bioconcentration factor for lead was less than one. Metal concentrations in earthworms may poorly correlate with those in soil.

East African Jameson's mamba. *Dendroaspis jamesoni*.

eastern brown snake. *Pseudoechis textilis*.

eastern coral snake. *Micrurus fulvius*.

eastern diamondback rattlesnake. *Crotalus adamanteus*.

eastern green mamba. *Dendroaspis angusticeps*.

eastern lupine. *Lupinus*.

eastern poison oak. *Rhus quercifolium*.

eastern sand adder. *Vipera ammodytes*.

eastern whorled milkweed. *Asclepias verticillata*.

EBDC. ethylene bis(dithiocarbamate).

ebrietas. inebriety, drunkenness.

ebriety. inebriety, drunkenness.

eburnamenine-14-carboxylic acid ethyl ester. vinpocetine.

EC. effective concentration.

EC_{50}. effective concentration/ 50% of maximal response.

ec-, eco-. a prefix, indicating relationship to habitat or environment.

ecboline. ergotoxine.

ecchymoses. plural of ecchymosis.

ecchymosis (plural, ecchymoses). a small, round or irregular, hemorrhagic, non-elevated area of skin or mucous membrane that is initially blue-black changing to greenish brown or yellow. Ecchymoses are formed by the extravasation of blood into subcutaneous tissues as a result of trauma. They are larger than petechiae.

eccoprotic. cathartic.

eccrine secretion. See exocytosis.

ecdysone. **1a:** an insect molting hormone that regulates pupation. Some 30 of these compounds are known. **1b:** any of a set of hormones with a similar action, including some compounds from certain plants or invertebrate animals, some ynthetics. **α-ecdysone** (2,3,14,-22,25-pentahydroxycholest-7-en-6-one). the main ecdysone isolated from insects and crustaceans. **β-ecdysone** (2,3,14,20,22,25-hexahydroxycholest-7-en-6-one; 20-hydroxyecdysone; ecdysterone; crustecdysone; isoinokosterone). the main ecdysone of plants.

ecdysterone. β-ecdysone. See ecdysone.

ECF. ethyl chloroformate.

ecg. EKG.

ECG. EKG.

ecgonine ([1*R*-(*exo,exo*)-3-hydroxy-8-methyl-8-azabicyclo[3.2.1]octane-2-carboxylic acid; 3β-hydroxy-1α*H*,5α*H*-tropane-2β-carboxylic acid). the principal part of the cocaine molecule which is obtained by hydrolysis and used as a topical anesthetic. It is a controlled substance listed in the U.S. Code of Federal Regulations.

ecgonine methyl ester benzoate. cocaine.

echide carénée. *Echis*.

echidna. See Monotremata.

echidnase. an enzyme in viperid venoms that causes inflammation.

echidnin. used loosely in medicine to mean **a:** snake venom, or **b:** the active principle of a snake venom, especially a toxic nitrogenous principle.

echidnotoxin. a poisonous principle in viperid venom. *Cf.* echidnin.

echidnovaccine. viper venom, detoxified by heating, used as a vaccine against venom.

Echinochloa crusgalli (barnyard grass). a grass (Family Gramineae) in which concentrations of nitrate toxic to [illegible]

echinoderm. **1:** of or pertaining to the phylum Echinodermata. **2:** common name of any animal of the phylum Echinodermata.

Echinodermata (starfish; sea urchins; sand dollars; sea lillies; sea cucumbers; and allies).

a phylum of marine coelomate animals in which the adults are radially or meridionally symmetric with a calcium carbonate skeleton comprised of separate plates or ossicles that are embedded in the body wall and often bear external spines (spicules). The coelom is modified as a water-vascular system with various functions, including that of locomotion, respiration, and feeding. The circulatory and nervous systems are radially arranged and poorly developed; there is no special excretory system and the sexes are separate. The larvae are bilaterally symmetrical, ciliated, and motile. Included are sea stars, brittle stars, sea urchins, sand dollars, sea lillies, sea cucumbers. They are bottom-dwellers, some in deep waters, others occur near shore. Approximately 85 of the nearly 6000 species of echinoderms are known to be venomous or poisonous. There are four classes: Asteroidea, Ophiuroidea, Echinoidea, and Holothuroidea). See also pedicellariae.

Echinoidea. a class of venomous and often poisonous animals (Phylum Echinodermata). They have globular, hemispherical, discoidal or egg-shaped bodies and skeletal plates that are firmly conjoined, forming a rigid test that bears numerous movable spines. Most dwell on the sea bottom. They have tube feet with suckers; arms are absent; Aristotle's langern is just inside the mouth in most species. Included are cake urchins, heart urchins, sand dollars, and sea urchins. The pedicellara is three-jawed and is the principal venom organ, although certain of the spines contain poison glands. The gonads of sea urchins are poisonous and perhaps lethal. See also heart urchin, sand dollar, sea urchin.

Echis (saw-scaled vipers; carpet vipers). a genus comprised of two species of venomous snakes (Family Viperidae) that are indigenous to India, the Arabian Peninsula and North Africa. While both snakes are small, the venom is highly toxic and the bite is often fatal. When alarmed, they typically inflate the body and produce a hissing sound by rubbing the saw-edged lateral scales against one another. The same behavior is shown by the nonvenomous egg-eating snakes, *Dasypeltis*. The head is broad, flattened, with a broad snout, and is quite distinct from the neck; the canthus rostralis indistinct; the nasorostral scales are greatly enlarged. Eyes are of moderate size; pupils are vertically elliptical. ***E. carinatus*** (saw-scaled viper; carpet viper; afee; Ägyptische sandrasselotter; echide carénée; efa; fossokéré; kokodiou; phoorsa; pul suratti; pwéré; sandrassel otter; Sandrasselotter; surattai pambu; tachilett; tofoni; um jenmaib; vipére des pyramides). a small viper up to ca. 0.6 m long that occurs throughout much of Afro-Asian desert belt from northern and western Africa to western Asia, southern India and Ceylon to Morocco, Ghana, and the southern portions of Asian Russia. It is abundant throughout most of this range. The ground color is pale buff or tan to olive brown, chestnut or reddish with a midline row of whitish spots; laterally the body is marked by a narrow undulating white line; the crown usually bears a light-colored marking in the shape of three-pronged trident or arrowhead (two prongs directed caudad and one anteriorly). The belly is white to pinkish brown stippled with dark gray. It inhabits barren rocky or sandy desert into dry scrub forest, from coastal areas to elevations of nearly 2000 m. This viper is usually nocturnal and while it typically retreats when approached, it is nevertheless highly irritable and aggressive and kills more humans than any other snake. The case mortality rate is quite high; antivenin is available. This snake strikes quickly and repeatedly from a characteristic figure-8 coil. Internal and external hemorrhage, sometimes including bleeding from the gums are a conspicuous effect of envenomation. Further signs and symptoms of envenomation may include pain, edema, discoloration of the skin, weakness, nausea, vomiting, diarrhea, abdominal pain, thirst, pupillary dilation, a weak, rapid pulse, hypotension, shock, albuminuria, and proteinuria. Death often intervenes within 12-16 days. ***Echis coloratus*** (saw-scaled viper; ef'eh). a species, similar in many respects to *E. carinatus*, confined to eastern Egypt, the Arabian Peninsula and Israel.

Echium (viper's bugloss). a genus of hairy or bristly Old World annual, biennial, or sometimes perennial herbs (Family Boraginaceae) introduced and naturalized in North America. Representatives contain the hepatotoxic alkaloid, pyrrolizidine. Symptoms of reported poisonings are similar to those produced by *Heliotropium*. ***E. lycopsis*** (formerly *E. plantagineum*). commonly cultivated and extensively hybridized has been implicated in poisonings. See also *Amsinckia intermedia*.

Echte Korallenotter. true coral snakes, *Micrurus spp*.

echujetin. digitoxigenin.

eclampsia. a sometimes fatal toxemia of pregnancy. See toxemia.

eco-. See ec-.

ecocide. **1a:** the destruction of an entire ecological system or of a major portion or part of the environment. **1b:** the willful destruction of a major portion of the environment. **2:** any substance that is capable of destroying an entire ecological system.

ecological. of or pertaining to ecology.

ecological chemistry. See chemistry.

ecological community. biocenosis.

ecological effects test. See toxicity test.

ecological impact. the effect of a natural or anthropogenic agent (e.g., a toxicant), process, or activity on an ecological system (e.g., a population, community, ecosystem).

ecological succession. See succession.

ecology (bioecology; bionomics). **1:** the science of relationships within and among populations of living organisms and of populations or individual living organisms with their environment. Ecology embraces all aspects of the natural sciences that help to understand or to modify ecological relationships. **2:** the relationships of living systems with each other and with their environment (e.g., the ecology of the bald eagle). **3:** frequently used erroneously as a synonym for environment (as in "protect the ecology").

economic form. any type of living system considered to be of economic importance to humans. See selective toxicity.

economic poison. any substance used to control pests or to defoliate crops such as cotton.

ecophysiology. the science of the physiological adaptations of organisms with each other and with their environment generally.

ecosphere. biosphere.

ecosystem (ecological system; biogeocoenosis; biosystem; holocoen). an interacting, partially autonomous system comprised of a biotic community (biocenosis) and its associated abiotic environment (biotope). **aquatic ecosystem** (e.g., a pond, lake or stream). an ecosystem in which water is the exclusive or dominant medium and in which no major terrestrial elements occur. **semiaquatic ecosystem** (e.g., a marsh, swamp or bog). an ecosystem comprised of both aquatic and terrestrial components and which includes both free-standing or flowing water and soil. **terrestrial ecosystem** (e.g., a grassland or forest). **1a:** an ecosystem in which soil and/or air are the principal environmental media. **1b:** any ecosystem that contains organisms that live chiefly in or on soil.

ecotoxic. of or pertaining to the capacity of a substance to damage, destroy, or disrupt plant or animal populations, biotic communities or other ecological associations or processes.

ecotoxicity. the capacity of a substance to harm, injure, or damage a population, community, ecosystem, or a significant level of ecological organization above the level of an individual organism. The ecotoxicity of a substance is a complex function of the range of its toxicities to living organisms in the environment or ecosystem of concern, the types of organisms that are most affected, its persistence in the environment, and the potential for bioaccumulation or biomagnification. Numerous other factors can affect the ecotoxicity of a substance in a particular environment. In practice, the ecotoxicity of a substance is usually based on the toxicity of a substance to one or more test species, persistence in the environment, and its tendency to bioaccumulate. The ecotoxicity of most substances is currently largely a matter of conjecture.

ecotoxicology. See toxicology (ecotoxicology).

ectasy. See MDMA.

ectoantigen (exoantigen). **1a:** any antigen (usually a toxin) that is easily separated from its source (usually a bacterium), e.g., by shaking. **1b:** any antigen formed in the ectoplasm of a bacterium.

ectocrine. **1a:** an ectohormone, a hormone which is secreted into the medium and affects the metabolism of other organisms. **1b:** pertaining to synthetic secreted substances or products of decomposition in the external medium that stimulate or inhibit plant growth. **1c:** a substance with ectocrine (defs. 1a, 1b) properties. **ecological ectocrine**. any substance synthesized by one species which, when released into the environment, influences metabolic processes of another species (e.g., vitamins produced by ruminants which are ingested by other animals). *Cf.* endocrine, exocrine.

ectocytic. external to the cell.

ectogenic. exogenous.

ectogenous. exogenous.

ectohormone. ectocrine (def. 1a).

ectoparasiticidal. lethal to parasites living on the exterior of a host's body (ectoparasites).

ectoparasiticide. an agent, applied directly, that destroys parasites living on the exterior of a host's body (ectoparasites).

ectopic pregnancy. a pregnancy in which the embryo grows outside the womb.

ectotoxemia. (obs.) toxemia produced by a xenobiotic substance.

ectotoxin. obsolete for exotoxin.

ecumenical. cosmopolitan.

eczema. a cutaneous inflammatory condition (acute or chronic) with inflammation, and some combination of noninfectious vesicles, papules, pustules, scales, crusts, and/or scabs. Causes vary, but eczema generally results from **(1)** external agents (e.g., contact with certain toxic or irritant chemicals, allergic contact, reaction to contact with certain microorganisms) or **(2)** constitutional or predisposing factors that may be genetic or psychological in origin. Whereas chronic eczema usually benefits from treatment, there is a tendency to relapse and recurrence. **facial eczema** (pithomycotoxicosis). a mycotoxic disease of grazing livestock in which toxic liver injury commonly results in photosensitization. It is seen most commonly in New Zealand, Australia, several South American countries, occasionally in Europe, and probably also in North America. Sheep, cattle, and farmed deer of all ages may acquire the disease, but it is most severe in young animals. The mycotoxin, sporidesmin, a secondary metabolite of the *Pithomyces chartarum*, is a saprophyte that grows on dead litter in pastures. Photosensitization and jaundice appear about 10-14 days following ingestion. At that time, affected animals frantically seek shade. The disease is typified by erythema and edema of photodermatitis in unpigmented skin that rapidly results from even brief exposures to sunlight. Stress is severe, with death intervening one to several weeks after appearance of the photodermatitis. Characteristic liver and bile duct lesions are seen in affected animals. In acute cases showing photodermatitis, livers are initially enlarged, icteric, and have a clear lobular pattern. There is subsequent atrophy with marked fibrosis. The shape is distorted and large modules of regenerated tissue appear on the visceral aspect. The urinary bladder usually exhibits hemorrhagic or ulcerative erosions of the mucosa stained with bile pigment and localized edema.

ED. effective dose. See dose.

ED_{01}. effective dose/1% response. See dose.

ED_{50}. median effective dose; effective dose/ 50% response.

ED_{99}. effective dose/99% response. See dose.

EDA. ethylenediamine.

edathamil. edetic acid.

edathamil calcium disodium. edetate calcium disodium.

edathamil disodium. edetate disodium.

EDB. ethylene dibromide.

EDC. ethylene dichloride.

eddo. taro, def. 2.

edema (dropsy). **1a:** the accumulation of excessive amounts of fluid in cells, body cavities, or intercellular spaces (often under the skin). The increase in interstitial fluid volume may result from any of a variety of factors such as increased venous pressure, decreased capillary permeability (which may be due to damage to the capillary by a toxicant), decreased plasma protein, a lowered gradient in osmotic pressure between the capillary and the tissue served, or insufficient lymphatic drainage. *Cf.* ascites, dropsy. **1b:** the associated swelling of tissues. See also ascites, dropsy. **conjunctival edema**. swelling of conjunctival tissue, a common response to noxious stimuli. **pulmonary edema**. the presence of excess fluid in the lungs. It hinders alveolar absorption of oxygen, is often progressive, and may be fatal. Pulmonary edema may result from any of a number of diseases or by poisoning usually by irritants that damage the pulmonary epithelium. Cholinesterase inhibitors can cause pulmonary edema by stimulating bronchial secretion. Other proximate causes include elevated venous pressure, decreased capillary permeability (which may be due to damage by a toxicant), decreased plasma protein, or insufficient lymphatic drainage. **pulmonary edema**. excessive accumulation of fluid in the lungs. It hinders alveolar absorption of oxygen, is often progressive, and may be fatal. Pulmonary edema can result from any of a number of disease states, from excess bronchial secretion as that induced by cholinesterase inhibitors, from damage to the pulmonary epithelium by chemical irritants. Pulmonary edema interferes with oxygen uptake and can be fatal.

edemagen. an irritant that injures capillaries with resultant edema; not a true inflammagen, *q.v.*

edematigenous. edematogenic.

edematization. any process of causing edema or making edematous.

edematogenic (edematigenous). able to produce or to cause edema.

edematous. characterized by, or affected by edema.

edetate. any salt or ester of edetic acid.

edetate calcium disodium ([[*N,N'*-1,2-ethanediylbis[*N*-(carboxymethyl)glycinato]](4—)-*N,N',O,O',O^N,O^N'*]calciate(2—)disodium; ethylenediaminetetraacetic acid calcium disodium chelate; calcium disodium (ethylenedinitrilo)tetraacetate; calcium disodium ethylenediaminetetraacetate; EDTA calcium; edathamil calcium disodium; calcium disodium edatate; edetic acid calcium disodium salt; sodium calciumedetate). a water-soluble chelating agent used in lead poisoning. The calcium atom is exchanged for a heavy metal ion. See edetate disodium, de-lead, chelate.

edetate disodium (*N,N'*-1,2-ethanediylbis[*N*-(carboxymethyl)glycine] disodium salt; (ethylenedinitrilo)tetraacetic acid disodium salt; ethylenediaminetetraacetic acid disodium salt; ethylenebis(iminodiacetic acid) disodium salt; edetic acid disodium salt; edathamil disodium; disodium edathamil; EDTA disodium; tetracemate disodium; disodium ethylenediaminetetraacetate; disodium edetate). a moderately toxic chelating agent used in pharmaceutics as a sequestering agent and in chelation therapy.

edetate sodium (*N,N'*-1,2-ethanediylbis[*N*-(carboxymethyl)glycine] tetrasodium salt; (ethylenedinitrilo)tetraacetic acid tetrasodium salt; sodium edetate; tetrasodium ethylenediaminetetraacetate; ethylenebis(iminodiacetic acid) tetrasodium salt; tetrasodium ethylenebis(iminodiacetate); EDTA tetrasodium; edetic acid tetrasodium salt; tetracemate tetrasodium; tetrasodium edetate; tetracemin). the tetrasodium salt, of edetic acid, $(NaOOCCH_2)_2N$-$CH_2CH_2N(CH_2COONa)_2$. It is a chelating agent that reacts with most divalent and trivalent metal ions.

edetate trisodium (*N,N'*-1,2-ethanediylbis[*N*-(carboxymethyl)glycine] trisodium salt; (ethylenedinitrilo)tetraacetic acid trisodium salt; EDTA trisodium; ethylenediaminetetraacetic acid trisodium salt; trisodium ethylenediaminetetraacetate; trisodium edetate; edetic acid trisodium salt). the trisodium salt of edetic acid, which is sometimes used as a chelating agent.

edetic acid (*N,N'*-1,2-ethanediylbis[*N*-(carboxymethyl)glycine], 3,6-bis(carboxymethyl)-3,6-diazooctanedioic acid; ethylenebisaminodiacetic acid; ethylenediamine-*N,N,N',N'*-tetraacetic acid; (ethylenedinitrilo)tetraacetic acid; ethylenediaminetetraacetic acid; ethylenediaminetetraacetate; ethylenedinitrilotetraacetic acid; ethylenediamine tartrate; edathamil; EDTA; EDTA acid). a caustic, toxic, colorless, flammable, crystalline substance, $(HOOC-CH_2)_2-NCH_2CH_2N(CH_2COOH)_2$, with an ammoniacal odor. It is slightly soluble in water and insoluble in common organic solvents. Edetic acid is a powerful chelator of +2 and +3 cations. The sodium salt binds heavy metals most strongly and can form as many as six bonds. It is used as a chelator of metals in inorganic chemistry; is administered by injection in the treatment of heavy metal poisoning (e.g., lead poisoning, especially of livestock), and in calcinosis therapy. It also has many uses in the food-processing industry (e.g., to capture and remove miniscule metal particles that enter food from the rollers, scrapers, and blenders that are used to process foods). Whereas edetic acid can complex with inorganic lead, it also causes a rapid and detrimental depletion of calcium. Thus, injections of the calcium chelate of edetic acid are used in therapy for lead poisoning, thereby avoiding a net loss of calcium that would otherwise accompany solubilization and excretion. If injected into the injured area, edetic acid may help to curb the effects of certain viper venom enzymes on tissues of the victim. Concentrations of about 0.025 to 0.05 M edetic acid seems to retard the development of tissue damage and necrosis. Edetate is teratogenic and probably mutagenic; it is also a blood coagulant. Both liquid edetic acid and the vapor are extremely caustic to the eyes, skin and mucous membranes, and other body tissues (as are concentrated solutions of its salts).

edetic acid calcium disodium salt. edetate calcium disodium.

edetic acid disodium salt. edetate disodium.

edetic acid tetrasodium salt. edetate sodium.

edetic acid trisodium salt. edetate trisodium.

edible. consumable, comestible, suitable as food, fit to eat.

edible false morel. *Gyromitra (Helvella) esculenta*.

edrophonium chloride (*N*-ethyl-3-hydroxy-*N,N*-dimethylbenzenaminium chloride; ethyl(*m*-hydroxyphenyl)dimethylammonium chloride; (3-hydroxyphenyl)dimethylethylammonium chloride; 3-hydroxy-*N,N*-dimethyl-*N*-ethylanilinium chloride). a white, crystalline, water-soluble, cholinergic powder used as an antidote to curare principles and as a diagnostic aid in myasthenia gravis and esophageal chest pain.

EDTA. edetic acid; ethylene diamine tetraacetate.

EDTA acid. edetic acid.

EDTA calcium. edetate calcium disodium.

EDTA disodium. edetate disodium.

EDTA tetrasodium. edetate sodium.

EDTA trisodium. edetate trisodium.

EE. *sym*-dichloroethyl ether.

EEG. **1:** electroencephalography. **2:** electroencephalograph. **3:** electroencephalogram.

efa. *Echis carinatus*.

ef'eh. *Echis coloratus*.

effect. consequence, outcome, result; any change, result of, or response to a stimulus, agent, action, force, or antecedent event. In toxicology, an effect is usually any response of a living system that results from exposure to a chemical substance (e.g., a venom, carcinogen, antigen) or to ionizing radiation. **additive effect**. the result of the action of two or more substances used in combination whereby the total effect is equivalent to the arithmetic sum of the individual effects of each agent experienced alone. **adverse effect** (adverse reaction; deleterious effect). **1:** adverse reaction; a harmful or injurious effect as by a drug, toxicant, or ionizing radiation. **2:** any effect (immediate or delayed) that is injurious or otherwise harmful to a living system, including progeny. See also unintended effect, critical effect. **adverse drug effect**. **1:** an unintended adverse effect of a drug administered in normal amounts. See also unintended effect. **2:** an adverse effect of a drug regardless of cause (e.g., abuse, overdosage, error). **antagonistic effect**. any change brought about through antagonism or antagonistic actions, interactions, or relationships. See antagonism. **atropinic effect**. any biological effect of atropine or of agents with similar activity. In humans, these may include dry mouth, dysphonia, dysphagia, tachycardia, dry skin, mild to severe hyperthermia, blurred vision, mydriasis, delirium, and excitement. **biochemical effect**. an effect observed at the biochemical level. Effects of toxic substances may occur in any type of cell or any part of a living organism. Effects observed at higher levels of organization (e.g., physiological, whole organism, population, ecosystem) involve biochemical effects and mechanisms. **biological effect**. any change in an organism induced by a stimulus or a biologically active agent. **cardiotoxic effect**. any toxic effect (direct or indirect) on the heart. **clinical effect**. any clinically observable effect. **chronic effect**. a permanent, or semipermanent, effect (e.g., of a drug or disease), usually of an adverse nature, that develops over a long period of time. **critical effect**. the first adverse effect observed in a particular organ in association with a known concentration of a particular toxicant under specified conditions. The critical effect defines the critical concentration. See concentration (critical concentration), critical organ. **cumulative effect** (cumulative action). any result of repeated equivalent exposures to a biologically active agent or stimulus in which the effect of any subsequent exposure is more pronounced than that of the initial exposure. **delayed effect** (delayed response, delayed toxicity). an effect that occurs or develops following a substantial interval of time following exposure to a stimulus or substance. Carcinogenic effects of toxicants may be delayed as much as 20-30 years. **deleterious effect**. adverse effect. **direct effect**. any effect that is induced directly by the action of a substance or a stimulus. **ecological effect**. any effect expressed at an ecological level of organization as an effect on a population, biotic community or ecosystem (e.g., eutrophication; a change in diversity, primary or secondary productivity, soil pH, or sedimentation rates). **environmental effect**. any effect on elements or aspects of the environment. Ecological effects are a subset of environmental effects. **excitotoxic effect**. the injury or death of a cell due to the induction of sustained, repetitive discharge. See also excitotoxic, excitotoxin, excitatory amino acid. **functional effect**. a change in the function or behavior of a target tissue or organ. **genetic effect**. any effect on the genetic material (e.g., a gene mutation, formation of a DNA adduct). **graded effect**. an effect whose magnitude or intensity is related to the level of exposure or dose. **health effect**. any effect on the health of a living system, usually an individual or a specified population. **hepatotoxic effect**. any toxic effect (direct or indirect) on the liver. **idiosyncratic effect**. any abnormal or aberrant response of a living system. Such an effect may be genetically determined or the result of an unusual and unknown history. **immediate effect** (immediate response; immediate toxicity). any effect that occurs during exposure to a substance or a stimulus or that develops rapidly and appears following a single exposure to the causative agent. **indirect effect**. any effect that is not directly induced by the action of the causative agent or stimulus, but results from internal processes initiated by the causative agent or as the result of modified responses to the environment. Thus, a partially paralyzed (direct effect) animal can no longer compete adequately and consequently dies (indirect effect). **indirect systemic effect**. any systemic effect produced by a pronounced local effect such that the damage may itself produce systemic effects. The kidneys, for example, may be damaged by a severe acid burn on the skin in which no toxicant is distributed to the kidneys. **irreversible effect**. an effect that persists indefinitely and is refractory to attempts to eliminate it. **irreversible toxic effect**. **1a:** a permanent effect; one that persists indefinitely after cessation of exposure and elimination of the toxicant from the system. **1b:** a toxic effect that does not respond to treatment. **latent effect**. an effect that does not present initially following exposure to a biologically active agent or stimulus. Thus, bronchogenic cancer induced by chronic occupational exposure to chromate may not appear until 10-15 years following initial exposure. **lethal effect**. any effect to which the death of a living system can be directly attributed. **local effect** (local injury; local toxicity). any effect that occurs at the site of first contact between the toxicant and a biological system. Local effects are usually produced by highly reactive (e.g., caustic, oxidant, irritant) substances or by radiation. **morphologic effect**. any change in the gross or microscopic structure of organs and tissues (e.g., neoplasia, necrosis, atrophy) in response to a biologically active agent. **nephrotoxic effect**. any toxic effect (direct or indirect) on the kidney. **neurotoxic effect**. any toxic effect (direct or indirect) on the nervous system of an animal. **no effect**. the failure of an active agent or stimulus (e.g., a toxicant) to elicit an observable effect. In the case of exposure to a toxicant or infective agent, the victim is said to be asymptomatic. **nonlethal effect**. any biological effect that does not impair viability or cause death. **oxygen effect**. enhancement of radiosensitivity of cells by a high ambient concentration of oxygen. **quantal effect**. an effect whose magnitude or intensity bears no relationship to the intensity of exposure or the dose. The incidence of occurrence of such an effect may, however, be dose/exposure dependent. **residual effect**. an effect that remains following withdrawal of the causative agent; one that remains after other effects are no longer observed. **reversible effect**. a temporary effect; one that ameliorates and ultimately dissipates following termination of exposure to a stimulus or biologically active agent. **reversible toxic effect**. a toxic effect in which target tissues repair or regenerate following injury. The degree of reversibility depends upon the type and extent of injury and the capacity of the tissues to regenerate. **side effect**. unintended effect. **sublethal effect**. any effect that is deleterious to an organism, but which does not impair its viability in an otherwise normal environment. The distinction between sublethal and lethal is not absolute nor always clear. **synergistic effect**. a response in which the combined effect of two or more stimuli or biologically active agents (e.g., poisons) to which a biological system has been simultaneously exposed is much greater than the sum of the effects of each agent acting alone. **systemic effect**. an effect that occurs at remote sites in the target system following absorption and transport of a biologically active agent, usually by the blood vascular system, from the point of entry. **therapeutic effect**. any biological effect of a chemical substance or stimulus that is associated with healing. **toxic effect**. any of a great variety of signs, symptoms, or complaints observed in a victim of poisoning or envenomation, including radiation poisoning. All such effects are the result of biochemical or biophysical interactions between a toxicant and/or its metabolites and the affected organism. Toxic effects have been variously classified according to the location of the effect, level of organization (e.g., biochemical level, ecosystem level), organs affected, time of onset, processes affected, destructiveness, degree of reversibility, or the mechanisms involved. **toxic systemic effect**. any toxic effect that follows absorption and distribution of the toxicant to sites within the organism that are remote from the point of entry. **unintended effect** (side effect; unintentional effect). any effect of a substance or process other than those intended on the basis of design, selection, or application. Nearly any action, product, or technology has effects other than those for which it was developed and applied; only rarely are such effects beneficial. A drug or medication, for example, typically has multiple effects on the dosed organism. The spectrum of effects usually varies with dose, other drugs administered, the individual, the species, and various environmental factors. A drug, however, is selected for medical use on the basis of one or a limited number of effects that are desired in a given instance. Other effects that are unintended or unwanted for a given application are usually referred to as side effects. For many chemicals, so-called side effects are not well documented; sometimes these unintended (and often unpredicted) effects are significant clinically and may be more pronounced than the intended effect. Technologies, industrial, and commercial processes are designed to produce certain products and effects intended to benefit humans in some way. All such activities also produce unintended effects, some of which may be harmful. Common illustrations of unintended effects are releases (e.g, into the work space or into the general environment) of toxic substances and other hazardous materials during manufacture, transport, usage, and/or disposal. **unintentional effect**. unintended effect.

effector. a cell or organ (e.g., a cilium, a gland or a muscle) that performs an action in response nerve impulses. **independent effector**. a cell (e.g., a cnidoblast), that contains both the receptor and effector components.

effector cell. a metazoan cell that kills and removes foreign cells. See also effector.

effector molecule. a compound that binds to a regulatory molecule to form a complex that can activate or inactivate a gene. See allolactose. See also effector.

effector organ. an organ, namely muscles and

glands, that produces an effect when stimulated. See also effector.

effects assessment. quantification or estimation of the health or environmental effects of exposure of humans or other biological systems to toxic or hazardous materials including microorganisms. See also exposure assessment.

efferent (centrifugal). **1:** conveying away from (e.g., away from the center of an organ or other structure. **2:** referring to or denoting a nerve or neuron (See efferent neuron) that transmits signals from the CNS to the peripheral effector organs (See motor neurone). **3:** indicating or denoting a blood vessel that collects blood from a capillary bed, used primarily in reference to arteries of the aortic arches of fish that conduct blood from the gills to the ventral aorta.

efferent neuron. a motor neuron that conducts impulses from the CNS to effector organs (e.g., muscles, glands).

effluent. **1:** any fluid discharged from a given source into the external environment. **2:** in the environmental sciences and in environmental protection, this term commonly refers to wastes discharged into surface waters or to wastewater (treated or untreated) that flows from a treatment plant, sewer, or industrial outfall into a lake or waterway.

effusion. **1:** the escape of fluid (e.g., from blood vessels or lymphatics) into a body part or cavity or into tissues as in exudation or transudation. **2:** the fluid effused.

efosite aluminum, efosite Al. fosetyl aluminum.

egesta. undigested material egested (discharged) from the body (e.g., feces, vomitus, regurgitated pellets). See also egestion, excreta.

egestion. a form of elimination, egestion is the process of discharging undigested materials from the body (e.g., in feces, vomitus, regurgitated pellets).

egg. **1a:** the female gamete or mature sex cell, especially if fertilized. **2:** a fertilized ovum, together with nutritive tissues and surrounded by protective tissues (e.g., membranes, a shell) that undergoes development without attachment to the female. It may pass from the body and hatch externally (as in birds, most reptiles, amphibians, and insects), or remain within the body of the female and hatch internally (as in certain snakes). **3:** the mammalian ovum.

egg white. albumen.

eggplant. *Solanum melongena esculentum*.

eggshell, of birds. the calcified envelope of a normal avian (bird) egg. It lies immediately external to the two membranes (composed of keratin fibers) that enclose the white and yolk of the egg. There are numerous channels through the shell (pores) that permit the exchange of respiratory and other gases between the egg and the surrounding environment. The eggshell by virtue of its composition and structure affords mechanical protection to the egg within. If the shell is too thin or otherwise weakened, the egg may be easily crushed as by the weight of the incubating adult. See eggshell thinning.

eggshell thinning. the production of a thinner than normal eggshell in birds that have been exposed to organochlorine compounds (especially DDE) and have accumulated residues. This has brought a number of large birds (e.g., peregrine falcons, pelicans) to the verge of extinction. When such large birds incubate affected eggs, their body weight may be sufficient to break such eggs, injuring or destroying the developing chicks. Following the banning of DDT in the United States and many other nations, most of the endangered bird populations have rebounded.

Egyptian blacksnake. *Walterinnesia aegypti*.

Egyptian cobra. *Naje haje*.

Egyptian henbane. *Hyoscyamus muticus*.

Egyptian millet. *Sorghum halepense*.

Egyptian saw-scaled viper. *Echis carinatus*.

ehé. *Thelotornis kirtlandi*.

Ehrlich's phenomenon. the difference between the amount of diphtheria toxin that exactly neutralizes one unit of antitoxin and that which leaves one lethal dose free. This amount is greater than one lethal dose of toxin. In other words, one must add more than one lethal dose of toxin to a neutral mixture of toxin and antitoxin to confer lethality to the mixture.

9-eicosanoic acid. gadoleic acid.

***n*-eicosanoic acid**. arachidic acid.

einsteinium (symbol Es, formerly E)[Z = 99; most stable isotope = ^{254}Es, $T_{1/2}$ = 276 days]. an artificial transuranium element of the actinoid series. Several isotopes, all of which are radioactive, have been prepared.

EIS. environmental impact statement.

ek-, ekto-. prefix meaning out of, from, outside.

EKG (ECG;; ecg). electrocardiogram, electrocardiography.

Ektafos. a tradename for dicrotophos.

elapid. **1:** of or pertaining to a cobra-like snake or to a member of the family Elapidae. **1b:** any snake of the family Elapidae (e.g., cobras, kraits, coral snakes, and mambas). *Cf.* elapine.

elapid venom poisoning. that due to envenomation by elapid snakes. a human victim may feel pain within 10 minutes; systemic effects within 30 minutes. There may be little or no tissue damage about the bite. Moderately severe poisoning is usually marked by pain (severe in cobra bites), numbness about the wounds, localized swelling of slow onset (little or none in the case of kraits), shock, muscle weakness, paresis of facial muscles, lips, tongue and pharynx, the sensation of a thickened tongue and throat, drowsiness, slurred speech, difficulty in swallowing, excessive salivation, nausea and vomiting, abdominal pain, pain about lymph nodes, paresis or paralysis, muscle fasciculations. Optic ptosis, blurred vision, and headache are common. In serious cases, blood pressure falls, cardiac and pulmonary complications may develop, and convulsions may occur. Respiratory depression, shock, and coma may develop rapidly. Unless specific antivenin is administered, death may result from respiratory in as little as two hours. See also snake venom poisoning; mamba venom poisoning.

Elapidae. a family of front-fanged, highly venomous, burrowing, terrestrial, and arboreal snakes that includes cobras, kraits, mambas, coral snakes, death adder, the Australian copperhead, and African garter snake. They are widely distributed, favoring the warmer climates of Africa and Asia, but representatives are absent only from the northern parts of North America, Europe and Asia. Elapids are characterized by a pair of comparatively short, stout, permanently erect deeply grooved fangs at the front of the mouth. The tail is cylindrical and tapered. See envenomation (elapid envenomation). See also Elapinae, *Acanthopus*, acetylcholinesterase, arginine ester hydrolases, *Bungarus*, *Calliphos*, *Dendroaspis*, *Denisonia*, *Micrurus*, *Naja*, *Notechis*, *Ophiophagus*, *Oxyuranus*, *Pseudechis*; lactate dehydrogenase.

Elapinae. the subfamily of the Colubridae to which elapid snakes are assigned in some classifications. It is coextensive with the term Elapidae as defined herein.

elapine. of or pertaining to the Elapinae; loosely equivalent to the term elapid. *Cf.* elapid.

Elapognathus minor (little brown snake). a poorly known venomous snake (Family Elapidae) of southwestern Western Australia. It probably not dangerous. The head is small and nearly indistinct from the neck; the eyes are rather large with round pupils. The body is cylindrical, rather stout, and the tail is of moderate length. an adult may reach a length of about 0.5 m.

elapoid. of, relating to, or resembling an elapid snake; a member of the family Elapidae.

Elaps (African dwarf garter snakes). a genus of two species of small, venomous, South African burrowing snakes (Family Elapidae), *E. dorsalis* and *E. lacteus*. Neither is considered dangerous to humans. The neck is indistinct from the small head; there is no canthus. The body is short (less than 0.6 m), slender, and cylindrical with a short tail. The eyes are small with round pupils. There are 9 scales on the crown. The frontal is long and narrow, the internasals short, and the rostral is broad and rounded. Laterally, the nasal narrowly contacts the single preocular. The dorsal body scales are smooth and occupy 15 rows at midbody; the ventrals number 160-239; the anal plate is divided, and the subcaudals are paired. The maxilla bears the large tubular fangs which lack external grooves.

Elapsoidea sundevallii (African garter snake; black garter snake; black watersnake; African necklace snake; bosré; kondband slang). a venomous, sluggish snake (Family Elapidae) that appears to bite only in self defense. It occurs throughout most of tropical and southern Africa, exclusive of the Cape region. Some individuals reach a length of about 1.2 m. The head is of moderate size, indistinct from the neck; indistinct canthus. There are 9 scales on the head; the rostral is broad with an obtuse point; the internasals are short; the nasals contact the single, narrow preocular. The eyes are small with round pupils. The body is cylindrical and rather slender with a very short tail. Some individuals reach a length of about 1.2 m. Dorsal body scales are smooth, rounded, with 13 rows at midbody; 138-184 ventrals; the anal plate is entire; subcaudals are usually paired. There are two large tubular fangs, each with and external groove following an interspace, and two to four small maxillary teeth.

elasmobranch. any fish the subclass Elasmobranchii (Class Chondrichthyes), *q.v.*

elasmobranch poisoning. a type of sometimes severe or even fatal ichthyosarcotoxism due the ingestion of the flesh or liver of certain toxic sharks and skates. Most commonly the flesh of the Greenland shark, *Somniosus microcephalus*, or the livers of any of a number of species of tropical sharks, including *Carcharhinus melanopterus, Heptranchius perlo, Hexanchus grisseus, Carcharodonn carcharias*, and *Sphyrna zygaena*. Symptoms may resemble those of ciguatera poisoning.

Elasmobranchii. a subclass of cartilaginous fish (Class Chondrichthyes). Fish of this subclass have five to seven pairs of gills which open directly to the exterior. They also have a spiracle, a heterocercal tail, and many rows of deciduous teeth that develop successively. Hystylic jaw suspension is usual. Included in this subclass are Cladoselachii (extinct sharks), Selachii (sharks), and Batoidea (skates and rays). *Cf.* Holocephali.

Elavil™. amitriptyline hydrochloride.

elayl. ethylene.

elder. See *Sambucus*.

elderberry. *Sambucus canadensis*.

electrocardiogram (EKG; ECG). a record of changes in electrical potential about the heart during each cycle of cardiac activity. These alterations are monitored by means of leads (electrodes) affixed to the surface of the body (most often to the arms and legs in the case of humans). These appear as a characteristic series of waves on an oscilloscope or similar device and may be recorded. See electrocardiography.

electrocardiography (EKG; ECG). a method of assessing heart function and diagnosing heart disease by measuring electrical activity of the heart with an electrocardiograph. The record produced is called an electrocardiogram, *q.v.*

electrocoagulation. See diathermy (surgical diathermy).

electrocortin. aldosterone.

electroencephalogram (EEG). a graphic recording of the rhythms and changes in electrical potential in the brain of a living vertebrate animal; most of the electrical activity recorded is produced in the cerebral cortex. The EEG is obtained via recording electrodes placed on the scalp. Variations from normal or standard patterns can often be correlated with different mental states and conditions such as neurotoxic insult. The specific loci stimulated by a neurotoxicant can often be identified. See also EEG.

electroencephalography (EEG). the study of brain function by measuring electric activity with an electroencephalograph. See also EEG.

electrolyte. **1:** a liquefied ionic substance. **2:** any solution that conducts electricity. **3:** a dissolved, ionized chemical in blood and other body fluids. In humans, the principal electrolytes are sodium, potassium, chloride, calcium, phosphorus, magnesium, and carbon dioxide. Electrolytes are acquired normally from food and beverages. They are regulated largely by the kidneys and lungs. They are essential to all life processes. **4:** in toxicology, electrochemistry, and allied fields, this term may be applied also to the conducting medium.

electrolyte acid. sulfuric acid

electrolyte measurement. laboratory test on blood or urine to identify and measure the electrolytes present.

electrolyte supplement. in medicine, any electrolytes taken to correct or to prevent body-fluid or electrolyte imbalance.

electromyogram. a recording of the electrical activity of muscles.

electromyography. the assessment of nerve and muscle disorders by the production and evaluation of an electromyogram.

electron transfer chain. See electron transport system.

electron transport system (electron transport chain; electron transfer chain; respiratory chain). a series of electron carriers within a mitochondrion or chloroplast. Elcctrons progress along this chain (sometimes together with hydrogen ions) by a series of redox reactions. The energy of oxidation is thus captured in a graded way and ATP is generated. The electron carriers include cytochromes, quinones, and ferredoxin. See especially cytochrome, cytochrome c, cytochrome oxidase, cytochrome *c* oxidase, cytochrome P-450. See also copper, mitochondria, NAD. podophyllotoxin,

electrovalent bond (ionic bond). a chemical bond formed by the complete transfer of electrons from one element or radical to another. The resulting bond derives from the net electrostatic force between the resultant ions. This force is attractive because the ions are of opposite charge. *Cf.* covalent bond.

eledoisin (formerly moschatin). an endecapeptide toxin in the posterior salivary glands (venom glands) of the cephalopods *Eledone moschata* and *E. aldrovandi*. It produces marked vasodilation in mammals, stimulates certain extravascular involuntary muscles, and stimulates salivation. It is 50 times more potent as a hypotensive agent in the dog than acetylcholine, bradykinin, or histamine. See octopamine.

elegant pit viper. *Trimeresurus elegans*.

elegant sea snake. *Hydrophis elegans*.

element. component, part, constituent, rudiment, factor. **chemical element**. a substance that cannot be chemically decomposed into structurally simpler components. They combine variously to form all molecules. All are poisonous under some circumstances and all can form naturally occurring or synthetic toxic compounds. There are 92 naturally occurring elements and 17 transuranic elements (those of atomic mass greater than uranium). Elements heavier than lead are radioactive and unstable. **essential trace element**. a trace element, *q.v.*, that is essential to critical physiological processes; a trace element that is an essential component of nutrition. Included are Zn, Mg, Cu, Ni, Mn, Co, Cr, I, Me, and Se. Their essential roles are performed as part of the structure of metabolically active molecules (e.g., enzymes, pigments), or are important cofactors or catalysts of metabolic processes. Excessive amounts are toxic. **trace element**. **1:** an element that occurs in such minute quantities in a specified type of system that they may not be quantified by conventional chemical assays. **2:** essential trace element.

elemental mercury. metallic mercury.

elephant-ear. See *Colocasia*.

elephant's ear. common name of plants of the genus *Caladium*.

elephantfish. a chimaeroid.

Eleusine indica. (goose grass). a grass (Family Gramineae) that sometimes contains concentrations of nitrate toxic to livestock.

elimination. **1:** omission or exclusion. **2:** abolishment, eradication, extermination. **3:** egestion; extrusion, expulsion; emptying, evacuation, or voiding materials (usually wastes) from the body. See also excretion, egestion, route of elimination.

ELISA, ELISA test. adsorption assay. See assay.

ellipticine. a bright yellow, crystalline, cytotoxic alkaloid isolated from dogbanes of the genus *Ochrosia* (Family Apocinaceae). It is an antitumor agent. The hydroxide, 9-hydroxyellipticine, is cytotoxic and carcinogenic.

elongate snakes. See *Toxicocalamus*.

ELV. environmental limit value. See limit value.

elymoclavine (8,9-didehydro-6-methylergoline-8-methanol). a natural ergot alkaloid isolated from *Claviceps* cultures and fungi parasitic on *Pennisetum typhoideum*.

Elymus. (wild rye; lyme grass). a genus of wild, tall, annual or perennial Northern Hemisphere grasses (Family Gramineae) that may become heavily ergotized and therefore hazardous. A few species are planted on embankments or along garden borders. See *Claviceps*.

emaciation. an extremely lean, wasted, enfeebled condition of the body. It is sometimes seen in chronic disease or chronic poisoning.

embalasasa. *Causus rhombeatus*.

embolus. a gaseous or solid mass within a cavity or vessel (e.g., a blood vessel).

embryo. **1:** the early multicellular stages in the growth and development of an organism. **2a:** the prenatal stage of a mammal (between the fertilized egg (zygote) stage and the fetal stage) during which organogenesis occurs. The embryo is more sensitive to many poisons than at later stages in development. **2b:** in *Homo sapiens* the conceptus is designated as an embryo from the 2d through the 8th week following fertilization, at which time organogenesis is essentially complete. See fetus.

embryocidal. of or pertaining to anything able to kill an embryo.

embryofetoxic. toxic to the embryo and fetus.

embryogenesis. the process of embryo formation.

embryolethal. pertaining to or denoting the capacity to kill embryos. See also embryocidal.

embryology. the science or study of the development of embryos.

embryonic. pertaining to or resembling an embryo.

embryopathy. any morbid or pathological condition in an embryo or fetus.

embryotoxic. toxic to the embryo.

embryotoxicity. **1:** the quality of being embryotoxic. **2:** any toxic effect on an animal embryo due to prenatal or prehatching exposure.

emerald green. cupric acetoarsenite.

emergency. **1:** a sudden, usually unanticipated event or situation that is potentially harmful, requiring a rapid response; an accident. *Cf.* crisis. **chemical emergency**. one that arises from an incident or sudden situation whereby the safety of workers, residents, the environment or property is seriously and immediately threatened by the presence of toxic or otherwise hazardous chemicals. Such is usually due to an accidental release or spill of hazardous chemicals and requires immediate action to contain the chemicals or otherwise minimize the hazard. **medical emergency**. **1a:** an emergency that seriously and immediately threatens the health or life of one or more humans. **1b:** a condition that requires urgent medical treatment (e.g., cardiac arrest). **1c:** a procedure that must be performed immediately (e.g., cardiopulmonary resuscitation). **toxicological emergency**. **1:** any exposure to a poison where the poison or conditions of exposure are unknown. **2:** acute exposure to a potentially lethal or damaging poison whether intentional, accidental, or natural (e.g. snake bite, possibly poisonous mushrooms) is a medical emergency, even in the absence of signs or symptoms.

emesis. vomition, vomiting, the act of vomiting.

emetamine. a constituent of ipecac.

emetic (vomitive; vomitory). **1:** pertaining to or capable of causing emesis. **2:** any agent that causes emesis (e.g., apomorphine, syrup of ipecac, strong soapy water, or a strong solution of sodium chloride or mustard).

emetic russula. *Russula emetica*.

emetine (6',7'10,11-tetramethoxyemetan; cephaeline methylether). the principal alkaloid, $C_{29}H_{40}N_2O_4$, of ipecac. Emetine and its dihydrochloride are extremely toxic to laboratory rodents. It is used therapeutically as an antiamebic and is used in veterinary medicine in lung worm infections. See also ipecac.

emetism. poisoning by an overdose of ipecac. The victim usually suffers acute inflammation of the pylorus, hyperemesis, diarrhea, and occasionally coughing and suffocation.

emetocathartic. **1:** able to produce both emesis and catharsis. **2:** an agent that produces both emesis and catharsis.

emiocytosis. See exocytosis.

emission. **1:** emanation, discharge; a release. **2:** natural or anthropogenic pollution discharged into the atmosphere, as from surfaces, vents, residential chimneys, smokestacks, and automobile, locomotive, or aircraft exhaust, or commercial or industrial facilities. **3:** a measure of the extent to which a given emission source discharges a pollutant, usually expressed as the amount per unit time (rate) or as the amount of per unit volume of gas emitted.

Emission Standard: the maximum amount of air pollutants, including toxic emissions, that are legally allowed to be discharged from a single source, mobile or stationary.

emmenagogic. inducing menstruation.

emmenagogue. an agent, method, or process that induces menstruation. **direct emmenagogue**. an agent that induces menstruation by acting directly on the reproductive organs. **indirect emmenagogue**. an agent that induces menstruation by acting on a condition that underlies amenorrhea.

emodin (1,3,8-trihydroxy-6-methyl-9-10-anthracenedione; 1,3,8-trihydroxy-6-methyl-anthraquinone; 4,5,7-trihydroxy-2-methyl-anthraquinone; archin; frangulic acid; frangula emodin; rheum emodin). a cathartic anthraquinoid that occurs mostly as the rhamnoside in roots of *Rheum rhabarbarum* (rhubarb), *Rhamnus frangula* (alder buckthorn), and *Rhamnus purshiana* (cascara buckthorn). It also occurs in *Rumex* and related genera. Emodin is a constituent of a number of laxative drugs. It is extremely toxic to the protozoan, *Tetrahymena pyriformis,* and to *Escherichia coli* mutants.

emphysema. **1:** a pathologic distension of tissue due to gas or air in the interstices. **2:** pulmonary emphysema. **pulmonary emphysema** (chronic obstructive pulmonary disease). a chronic destructive disease of the lungs. They are congested and characterized by an abnormal increase in the size of the air spaces distal to the terminal bronchioles accompanied by destructive changes in their walls; alveoli may be ruptured. **toxic pulmonary emphysema (respiratory syndrome), of cattle**. an unmistakable disease that occurs most often in animals pastured on fresh *Brassica. napus* (rape), and sometimes on fresh *B. oleracea* var. *acephala* (kale), *B. rapa*. var. *rapifera* (turnip), or rarely *Medicago sativa* (alfalfa). Its occurrence is unpredictable and the mortality rate is variable (ca. 5-35%). Symptoms usually appear in some animals within a week or ten days following pasturage on succulent rape. Affected animals stand apart from the herd, head extended, the breathing is labored typically with an expiratory grunt. Examination will reveal an elevated heart rate, paresis of stomach and intestines, and slight jaundice. The severity of dyspnea increases with exercise. subcutaneous cervical and lumbar emphysema may develop within a few days. The liver is variably necrotic, and varying portions of the gastrointestinal tract are inflamed. The pulmonary lesions are unusual. The lung is congested, distended and edematous with numerous air pockets. Histologically, most of the alveoli are ruptured.

emprosthotonos (opisthotonos; opisthotonus; tetanus dorsalis; tetans posticus). a tetanic spasm in which the body is rigid with the head and feet bent backward. May be observed in strychnine poisoning, neurotropic ergotism, tetanus, hysteria, epilepsy, severe meningitis, and convulsions of rabies.

emptysis (hemoptysis). expectoration of blood or blood-stained mucus arising from bleeding lungs, respiratory tree, or larynx.

EMTD. estimated maximum tolerated dose. See dose.

emulsin. a hydrolase that splits amygdalin into benzaldehyde, glucose, and hydrocyanic acid. It hydrolyzes other beta-glucosides.

Emydocephalus annulatus (ringed sea snake; Ijima sea snake). a venomous marine snake (Family Hydrophidae) found in the Timor, Sulu,

Arafura, and Phillipine seas. The snout has a blunt, forward-directed spine. The fangs are very small; there are no other maxillary teeth. Not generally considered dangerous.

enallochrome. esculin.

enanthotoxin (2,8,10-heptadecatriene-4,6-diyne-1,14-diol; oenanthotoxin). a convulsant, it is the toxic principle of *Oenanthe crocata*.

encephalalgia. headache.

encephalitis (plural, encephalitides). inflammation of the brain. Such a condition is most commonly caused by an arbovirus transvectored by a mosquito. Encephalitis may also result from hemorrhage, poisoning (e.g., by lead), or as a complication of another disease (e.g., chickenpox, measles, or influenza).

encephalitogen. a substance (e.g., lead) that can cause encephalitis.

encephalitogenic. causing or tending to cause encephalitis.

encephalodynia. headache.

encephalograph. a device for recording the electrical activity of the brain.

encephalopathia. encephalopathy.

encephalopathy (cephalopathy; cerebropathia; cerebropathy; encephalopathia). any dysfunctional condition, disease, or disturbance of brain function or structure regardless of etiology, especially if damaging or degenerative. *Cf.* neuropathy, myelopathy. **bilirubin encephalopathy** (kernicterus). that due to bilirubin toxicity. **dialysis encephalopathy**. dialysis dementia (See dementia). **diffuse encephalopathy**. that due to a loss of essential metabolites, often as the direct result of acute toxic exposures or indirectly from the effects of systemic poisons. **hepatic encephalopathy**. hepatic coma. **lead encephalopathy** (lead encephalitis; saturnine encephalopathy). a rapidly developing edematous, degenerative condition of neuronal tissue of the brain that occurs mainly in small children following ingestion of large amounts of lead-containing substances. Initial signs and symptoms include restlessness, headache, vertigo, clumsiness and falling, ataxia, insomnia, and irritability. These progress to include excitement, confusion, intense nausea and vomiting, delirium, hallucinations, convulsions, and coma. There is usually a marked increase in intracranial pressure, extensive cerebral edema, neurocytolysis, and pronounced vacuolization of the cerebral cortex, and some reactive inflammation. Mortality is about 25%, and permanent damage to the CNS is sustained by about one half of the survivors. Common sequalae are abnormal EEGs, mental retardation, cerebral palsy, and optic atrophy. **saturnine encephalopathy**. lead encephalopathy. **thyrotoxic encephalopathy**. a rare bulbar encephalopathy seen in severe thyrotoxicosis and characterized by disturbances of mastication, deglutition, and speech, with unconsciousness developing into deep coma.

encephalotropic (polioencephalotropic). having an affinity for brain tissue, especially the neural tissue. *Cf.* neurotropic. **polioencephalotropic**. encephalotropic.

end point. end point.

endangered. imperiled, jeopardized, threatened.

endangered species. any species of living organism faced with the likelihood of extinction, especially those that have been so declared by legal authority (e.g., the U.S. Department of Interior and/or designated state agencies). In the United States, the Endangered Species Act contains the requirements for declaring a species endangered. A number of species in this category have been threatened by toxic environmental chemicals either directly or indirectly (e.g., by effects on prey or other species with which they normal interact).

endangerment. hazard, risk, peril. **1a:** the act of placing in danger. **1b:** the state of being placed in danger.

endemic. a disease that is continuously present in a particular population, or region, but with low mortality.

endemic disease. a disease that is continuously present in a particular population or region, but with low mortality.

endemic urticaria. urticaria endemica.

Endep™. amitriptyline hydrochloride.

endo. isodrin.

endo isomer. isodrin.

endoantitoxin. **1a:** an antitoxin that remains within the cell that produces it. **1b:** sometimes applied to an antitoxin contained within a specific tissue or organ, as an antitoxin contained within the developing skull.

endocrine. **1:** secreting internally, a term applied to glands or other organs that secrete a substance (a hormone) directly into the blood or lymph that affects specific tissues or organs. **2:** hormonal; pertaining to internal secretins. *Cf.* ectocrine, exocrine. See also hormone.

endocrine gland. See gland.

endocrinism. endocrinopathy.

endocrinologist. **1a:** a scientist who studies the endocrine system from any of a number of perspectives (e.g., histology, pathologies, mechanisms that control functioning of endocrine glands and endocrine secretions, comparative studies among vertebrate taxa, role in the control of organ systems (reproductive endocrinologist), the role and action of drugs and other xenobiotics altering endocrine activity. **2:** a medical doctor (an internist) who diagnoses and treats endocrine diseases (or refers patients for treatment).

endocrinology. the study of endocrine glands, their secretions, control, and their roles and activities in the normal and diseased states of living organisms. The causes and treatment of endocrine disorders are included in this field. **comparative endocrinology**. the study of differences and similarities among various taxa of animals. **medical endocrinology**. the science and/or practice of endocrine function and the diagnosis and treatment of endocrine diseases.

endocrinopathic. of, pertaining to, or suffering from an endocrinopathy.

endocrinopathy (endocrinism). a functional disorder of an endocrine gland and the effects of such.

endocrinosis. a disordered condition due to endocrine system malfunction.

endocytosis. a mode of uptake of a foreign substance by a living cell by invagination of the cell membrane.

endoexoteric. pertaining to or due to sources or causes that are internal to the plant or animal body, together with others of external origin.

endogenetic. See endogenous.

endogenic toxicosis. autointoxication.

endogenous (endogenetic). **1:** growing from within. **2:** of or pertaining to any property, substance, or part that occurs or arises naturally within a system. Examples are endogenous rhythms and internal secretions. *Cf.* exogenous, extrinsic.

endogenous clock. biological clock. See also circadian rhythm, diurnal rhythm.

endointoxication. self empoisonment by an endogenous substance.

endoneurium (epilemma). the interstitial connective tissue of a peripheral nerve, that separates individual neurons.

endopeptidase. a proteolytic enzyme that is able to hydrolyze peptide linkages (initially internal and then terminal).

endoplasm. the innermost, more fluid portion of the cytoplasm of a cell.

endoplasmic reticulum. an extensive interconnecting network of microcanals or tubules that pass through the cytoplasm and nucleus of a cell, sometimes communicating to the cell surface. It is granular in appearance when ribosomes are present, but has a smooth surface when they are absent. These structures function in the synthesis and transport within the cell of proteins and lipids. They are vital to the metabolic activity of the cell.

endorphin. any of a group of opioid neuropolypeptides elaborated by the pituitary gland of higher animals that act at neural junctions to reduce pain. α-endorphin, β-endorphin, and γ-endorphin have been identified. The pharmacological effects are similar to those of morphine. The most potent is β-endorphin (isolated from the gastrointestinal tract); it is a powerful analgesic in humans and laboratory animals. Narcotic drugs act, in part on the endorphin receptors.

endoscopy. the visualization of the interior of organs and spaces within the body, using an endoscope. The process is sometimes used to obtain samples for cytologic or histologic examination.

endosulfan (6,7,8,9,10,10-hexachloro-1,5,5a,6,9,9a-hexahydro-6,9-methano-2,4,3-benzodioxathiepin-3-oxide; 1,4,5,6,7,7-hexachloro-5-norbornene-2,3-dimethanol cyclic sulfite; 1,2,3,4,7,7-hexachlorobicyclo[2.2.1]-2-heptene-5,6-bisoxymethylene sulfite; chlorthiepin). the commercial product is a nephrotoxic, extremely hazardous, brown crystalline or liquid chlorinated diene insecticide and miticide, $C_6H_6Cl_6O_3S$, used on ornamental flowers, vegetables, forage crops, and to control tsetse flies and termites. It is extremely toxic orally to laboratory rats. It is poisonous by all routes of exposure.

endothall (7-oxabicyclo-(2,2,1)-heptane-2,3-dicarboxylic acid). a soil and contact herbicide, $C_8H_8Na_2O_5$, it is an oxygen heterocyclic compound, forming several isomers, of which the most active is the *exocis* isomer. It is used as a pre- and postemergence herbicide and defoliant use to control weeds in *Beta vulgaris* (sugarbeet) and spinach acreage. Effects on mitosis, amylase activity, and proteolytic activity have been observed. It is not very selective, but *B. vulgaris* is tolerant to concentrations which kill a number of weeds, such as *Polygonum*, that are commonly associated with it. It has defoliant and desiccant properties and can thus be used as a preharvest treatment in a number of crops. Endothall is a strong irritant to the eyes and skin. It also affects amylase activity, proteolytic activity, and mitosis.

endotheliotoxin, a toxin that acts specifically on the endothelium of capillaries and small veins, producing hemorrhage. *Cf. hemorrhagin*.

endothelium. a type of squamous epithelium comprised of flat cells that line vessels of the blood and lymphatic systems, the heart, and many other body cavities. It is mesodermal in origin.

endothermic. of or pertaining to a process in which the temperature falls or heat is absorbed. Salt dissolving in water is often an endothermic process. *Cf.* exothermic.

endotoxemia. presence of, and poisoning by, endotoxins in the circulating blood; shock is a common feature.

endotoxic. pertaining to an endotoxin or to the activity of an endotoxin.

endotoxicosis. poisoning by an endotoxin.

endotoxin (intracellular toxin; bacterial pyrogen; esotoxin (obsolete)). **1:** bacterial pyrogen; esotoxin; intracellular toxin. any complex, heat-stable, macromolecular poison (usually a lipopolysaccharide) contained within the outer envelope of the producing (usually Gram-negative) bacterium. They are presumably released only when the bacterium dies and decomposes. They may be extracted using trichloracetic acid and glycols. The isolated endotoxins have essentially the same activity regardless of the species of bacteria that produces them. They characteristically stimulate the release of endogenous pyrogens in the host, causing fever and an inflammatory response when injected beneath the epithelium of higher animals. They are pyrogenic even in relatively small doses, increase capillary permeability, and produce leukopenia followed by leukocytosis. They also increase capillary permeability and can cause hemorrhagic shock, gastroenteritis, and tissue necrosis with accompanying, often severe diarrhea as in the case of *Salmonella*, *Shigella*, and *Vibrio cholerae*. Endotoxins do not form toxoids; they are generally less specific and less toxic than exotoxins. **2:** any normal internal proteinaceous component of an organism which provokes the production of precipitins, agglutinins, complement-fixation antibodies, etc. (but not antitoxins) when introduced beneath the epithelium of a human or other vertebrate animal. **erythrogenic toxin** (Dick toxin). a toxin that causes inflammation. See Chilopoda. **streptococcus erythrogenic toxin**. an erythrogenic exotoxin produced by certain strains of *Streptococcus pyogenes*. Intradermal inoculation of such toxin in humans and some other mammals, produces an erythematous reaction. It causes the scarlatiniform rash of scarlet fever.

endotoxoid. a toxoid prepared from endotoxin.

endotracheal tube. tube temporarily placed in the trachea (windpipe) of patients who are unable to breathe normally due to edema or damage to the upper respiratory tree.

end point (end point). **1a:** the final result of a process. Possible end points in a poisoning are death or recovery. **1b:** a specific, measurable health or environmental effect used to assess condition, hazard, or risk. **2:** the point marking the end of a process or phenomenon as, for example, when a definite effect (e.g. a color change in a titrated indicator) is observed. **acute toxicity end point**. any toxicity test result or monitoring data that indicates the presence of, or identifies or describes, a toxic stimulus that is severe enough to rapidly produce an effect in an organism (e.g., lethality, bronchospasm, incoordination). **chronic toxicity end point**. any toxicity test result or monitoring data that indicates a toxic effect of chronic exposure. The nature of such end points varies considerably, but is often more subtle than acute toxicity end points (e.g., evidence of early stages of a disease state, behavioral deficits, natality or other effects on reproductive capacity, effects on offspring, loss of productivity, population decline). **toxicological end point**. the biological effect deemed to be most specific and meaningful and specific with respect to the toxic chemical being studied and/or the biological organism or system being tested.

end point selection. in the context of toxicology, the selection of specific health (e.g., mortality, blood pressure) or ecological effects (e.g., loss of productivity, population decline or mortality) in order to assess the toxicity of a substance. See end point.

endrin(3,4,5,6,9,9-hexachloro-1a,2,2a,3,6,6a,7,7a-octahydro-2,7:3,6-dimethanonaphth[2,3-b]-oxirene; 1,2,3,4,10,10-hexachloro-6,7-epoxy-1,4,4a,5,6,7,8,8a-octahydro-*endo,endo*-1,4:5,8-dimethanonaphthalene; mendrin; nendrin; hexadrin; Compound 269; experimental insecticide No. 269). a white, flammable, highly poisonous, water-insoluble, crystalline, chlorinated, cyclodiene insecticide. It is a stereoisomer of dieldrin with effects similar to those of dieldrin and aldrin. Endrin is teratogenic, hepatotoxic, and neurotoxic, attacking the CNS. It is highly toxic (much more toxic than DDT) to humans and other vertebrates; approximately 7 g is lethal to a human. Symptoms of intoxication include nausea, mental disorientation, convulsions, and in severe cases, respiratory collapse and death. Endrin is carcinogenic to some laboratory animals and is probably mutagenic. It is no longer manufactured or used in the United States.

energy. **1:** the capacity to do work. Energy is manifested in many forms such as light and other electromagnetic waves, motion (kinetic energy), potential (energy of position; stored energy), electrical, heat, chemical, nuclear, sound waves. **2:** the ability to engage in vigorous activity.

enflurane (2-chloro-1,1,2-trifluoroethyl difluoromethyl ether; Ethrane). a clear, colorless, volatile, nonflammable liquid with a mild sweetish odor, $C_3H_2ClF_5O$, used as a clinical anaesthetic. It is only slightly water soluble, but is soluble in organic solvents. The vapor is toxic.

English aconitine. pseudoaconitine.

English bluebell. *Scilla nonscripta*.

English ivy. *Hedera helix*.

English nightshade. *Atropa belladonna*,.

English yew. *Taxus baccata*.

Engraulidae (anchovies). a family of small marine herring-like fish with a large mouth. See *Engraulis japonicus*.

Engraulis japonicus (anchovy). a marine bony fish (Family Engraulidae), certain populations of which are transvectors of ciguatera.

engraver's acid. nitric acid.

enhexymal. hexobarbital.

Enhydrina schistosa (beaked sea snake; common sea snake; hoogly patee; hook-nosed sea snake; kawon; valakachiyan; valakadyen). a widely distributed, shallow water sea snake (Family Hydrophidae) that is often very common at river mouths and the great deltas of large rivers such as the Ganges and Indus. It is sometimes found many miles from the sea in open river channels. The rostral is unusually prominent and curves downward at the tip giving this snake a characteristic beaked profile. It is further characterized by the mental shield (that at the tip of the chin), which is greatly reduced (splinter like) and buried in the cleft between the first pair of lower labials. This permits greater flexibility in the lower jaw and allows a wider gape. This is a distinctive characteristic of *Enhydrina* which can thus seize and swallow large prey. Adults are uniformly dull olive green above or pale greenish gray with dark crossbands that often fuse anteriorly. The sides and belly are cream to dirty white and the head is uniformly greenish above and the tail is usually mottled with black. Adults average 1 to 1.3 m long; occasional individuals may reach a length of about 1.6 m. Females are noticeably larger than the males. Newborn young are milky white; crossbands almost encircle the body; the crown is dark olive and the tail is black. The venom is extremely toxic; the lethal dose for laboratory animals is generally 50 to 125 μg/kg body weight. This species is considered dangerous and has probably caused more human fatalities than all other sea snakes combined. It is not usually aggressive, but will bite if restrained. *E. valakadyn* (Boie's sea snake; hoogly patee; Jew's nosed sea snake). a species that is closely related to *E. schistosa*.

enkephalin. either of two analgesic pentapeptides (methionine-enkephalin, isoleucine-enkephalin) produced in the brain, pituitary gland, GI tract. Four of the five amino acids are identical in each compound. These two neuropeptides are apparently able to depress neurons throughout the CNS. They inhibit neurotransmitters in the pathway of pain perception.

Ensis directus (common razor clam; Atlantic jackknife; razor clam). a common, narrow, convex, bivalve mollusk up to 25 cm long with gaping ends. It is indigenous to coastal bays, estuaries, and protected sites from Florida north to the Gulf of the St. Lawrence River (Labrador). *E. directus* is commonly found in sandy or muddy bottoms above the low tide line, or below in shallow waters. It was introduced and is feral in middle Europe (e.g., in the Wattenmeer, Germany). It is a common cause of paralytic shellfish poisoning, *q.v.*, of humans.

ENT 7796™. lindane.

Esoderm™. lindane.

entasia, entasis. tonic spasm. See spasm.

enteramine. serotonin.

enteric. pertaining to the small intestine.

enteritis. inflammation of the intestine.

enterohepatic. of or referring to the intestines and liver jointly.

enterohepatic circulation. a cycle in which a substance circulates from the intestines to the liver (via blood) and returns (via bile). this process whereby a parent compound (or a metabolite) is excreted via bile into the gastrointestinal tract, where it undergoes further metabolism by the intestinal microflora with reabsorption of the products, transport to the liver where they serve as substrate for further metabolism. This cycle of biliary excretion, further metabolism, and reabsorption may lead to repeated recirculation of the parent compound or active intermediates, resulting in a long biological half-life.

enterohepatic recycling. See enterohepatic circulation.

enterotoxemia (milk colic). a condition in which toxins absorbed from the intestines occur in circulating blood.

enterotoxication. enterotoxism, a form of autointoxication.

enterotoxigenic. containing, producing, secreting, or capable of producing enterotoxins (def. 2). Some strains of the bacillus *Escherichia coli* are enterotoxigenic.

enterotoxin (intestinotoxin). **1:** a toxin originating in the intestinal contents. **2:** an exotoxin that is specific for cells of the intestinal mucosa. **3:** a bacterial exotoxin that causes intestinal changes, as in food poisoning. **4:** a relatively heat-stable proteinaceous exotoxin secreted by staphylococci, chiefly *Staphylococcus pyogenes* var. *aureus*. It is a violent cathartic and emetic agent and the main cause of staphylococcal food poisoning.

enterotoxism (enterotoxication). autointoxication by toxins (enterotoxins) originating in the intestinal contents.

enthesis. the process of introducing or implanting exogenous materials into a living system as in the insertion or implantation of inert material to replace lost or damaged tissues. See also enthetic.

enthetic. **1:** referring to or denoting enthesis. **2a:** introduced from the outside. **2b:** exogenous, xenogenous.

Entoloma. a genus of mushrooms (Family Entolomatacea), several of which can cause severe gastric distress with vomiting and diarrhea that may persist for 1-2 days. ***E. sinuatum*** (lead poisoner). poisonous. It is scattered or grows in clusters under deciduous or evergreen trees throughout much of North America. ***E. strictus*** (= *Nolanea strictor*; straight-stalked entoloma). apparently poisonous. It is found on rotting logs or in moist humus from eastern Canada and the Great Lakes to Florida. ***E. vernum*** (= *Nolanea verna*; early spring entoloma). poisonous. It grows under conifers and along trails in oak-hickory communities from Canada to New York and Wisconsin.

entomopathogenic. pathogenic to insects. Referring to or denoting any agent, e.g., a substance or an organism, that is pathogenic to or causes disease states in insects.

entomophagous. insectivorous.

entomotoxic. toxic to insects; insecticidal.

entomotoxicity. the capacity to poison insects.

entry, route of. route of exposure (See exposure).

ENU. See *N*-ethyl-*N*-nitrosourea.

envenomation. **1:** a complex, multiple empoisonment in which a mixture of toxic materials or venom is introduced into an animals by means of a bite, strike, impalement, or sting. The seriousness of any envenomation usually depends upon numerous factors such as the amount and kind of venom delivered; the nature, number, location, and depth of the bites or stings; the size and age of the victim; the sensitivity of the victim to the venom; the pathogens present in or on the sting, mouth, spine. The bite or sting of a venomous animal does not always result in envenomation; even so, there is risk of tetanus or of a communicable disease. See also snake venom poisoning. **2:** the injection of venom into a living organism by a venomous animal by means of a bite, sting, spine, or other structure. See also snake venom poisoning.

environment. **1a:** the complex of all external entities, conditions, and circumstances that affect life generally. **1b:** the entire complex of chemical, physical, and biotic factors (materials, processes, circumstances, and conditions) that may act upon, be perceived by, or otherwise affect or potentially affect a living organism, a type of organism, population, community, or ecosystem. **1c:** locale milieu, the immediate surroundings. the aggregate, at a given moment, of all external conditions and influences to which a system [or organism] may be subject. **2:** a major habitat type (e.g., marine, terrestrial, rain forest, desert, lake). **abiotic environment**. the complex or aggregate of nonliving components of an environment considered broadly. **biotic environment**. the complex of living components of an environment considered broadly. **external environment**. the term environment, unless qualified, is usually taken to refer only to the external environment. This is, those entities and factors (physical, chemical, and biotic) that lie outside the body of an organism. **human environment**. collectively, everything that humans encounter (e.g., food, air, water, drugs, heat); individually, everything outside the body including other humans. **internal environment**. those components and processes within an organism that result from its own metabolism. **physical environment**. the abiotic component of the environment, including all physical and chemical environmental factors. **toxicologic environment**. those aspects of environment that are directly or indirectly toxic to a living system or that affect toxicity or the sensitivity of target and nontarget organisms to one or more toxic chemicals, or that may affect rates of exposure to toxic chemicals. See also factor (modifying factor).

environmental assessment. a written analysis prepared pursuant to the U.S. National Environmental Policy Act to determine whether a federal action would significantly affect the environment and thus require preparation of a more detailed environmental impact statement.

environmental audit. **1:** an independent assessment of the current status of an individual's or an organization's compliance with applicable environmental requirements pursuant to an ongoing activity. **2:** an independent evaluation of an individual's or organization's environmental compliance policies, practices, and controls.

environmental chemical. a chemical introduced from without; one that is either not of natural origin or is not a normal part of the environment into which it has been introduced (or of the environment generally). Most environmental chemicals of interest in toxicology and allied fields are anthropogenic in origin (e.g., most pollutants). **toxic environmental chemical**. an environmental chemical that is toxic and thought to be hazard to human health or to the environment.

environmental chemistry. See chemistry.

Environmental Impact Statement (EIS). a document, required under the National Environmental Policy Act, that is to be prepared by federal agencies prior to undertaking a major project or making legislative proposals that might significantly affect the environment. It is a decision-making tool that describes the positive and negative effects on the environment that might result from the undertaking and lists alternative actions.

environmental limit value (ELV). See limit value.

environmental media. See media.

environmental persistence. See persistence.

Environmental Protection Agency (EPA). **1:** any of many governmental agencies (e.g., in a number of countries and in the various states in the United States. **2:** the U.S. Environmental Protection Agency; an independent agency of the U.S. federal government that was established by Presidential Executive Order in 1970. This agency bears the chief responsibility at the national level for regulation of pollution, by assessing impacts; by developing and enforcing standards and regulations to control pollutants and their impacts on human health and the environment; and by establishing and conducting control programs that help to manage pollutants of air, water, and other media; radiation, solid and toxic wastes, pesticides, and noise.

environmental resistance. the complex of factors in the environment that limit the biotic potential or numerical increase of a specified population of living organisms.

Environmental Response Team. a team of experts in the U.S. Environmental Protection Agency (EPA) that is based in Edison, New Jersey, and Cincinnati, Ohio, who can provide 24-hour technical assistance to EPA regional offices and to state governments during all types of emergencies involving hazardous waste sites and spills of hazardous substances.

environmental runoff. See runoff.

environmental specimen banking (ESB). the systematic and long-term storage of selected environmental materials (including tissues) for deferred analysis and evaluation. Its applications thus far have been primarily in human toxicology, environmental chemistry and toxicology, and ecological impacts of pollutants.

environmental transformation. the conversion of a chemical in the environment from one form to another, usually with a consequent change in biological activity (e.g., toxicity). Chemical transformations in the environment are largely degradative and lead ultimately to mineralization. Transformations may produce intermediates that are less toxic or more toxic than the chemical species that enter the environment.

environmental transport. See transport (environmental transport).

environmental variance. this term generally refers to that part of total phenotypic variance due to exposure of individuals within a population to different environmental factors.

enzootic. an endemic disease that is restricted to one or a few animal populations.

enzootic calcinosis. See calcinosis.

enzymatic (enzymic). pertaining to an enzyme or to enzyme action.

enzyme. a biological catalyst that enables metabolic processes to proceed rapidly at physiological temperatures and pH. Enzymes are denatured and become ineffective at temperatures only somewhat above their optimum and are also destroyed by a pH beyond a fairly narrow range. With the exception of many xenobiotic-metabolizing enzymes, enzymes typically catalyze a specific reaction or a specific type of reaction in a living organism. There are six classes of enzymes based on function: oxidoreductases, transferases, hydrolases, lyases, isomerases, and ligases. Each is a highly specialized protein that catalyzes biochemical reactions by dramatically accelerating the conversion of one molecule (the substrate) into another (the product) by combining with the substrate and forming an intermediate enzyme-substrate complex. Upon completion of the reaction, the product and the enzyme separate and the chemically unaltered enzyme is able to complex with another substrate molecule. Enzyme catalysis is essential to nearly all cellular metabolism. Furthermore, enzymes typically act in highly integrated and regulated sequences or pathways, as in the electron transport system and pathways of protein synthesis. Many poisons act by altering or destroying enzymes such that they no longer function or do so in ways that are detrimental to the organism. Poisons that inactivate enzymes by damaging them include cyanide, heavy metals and formaldehyde. Some of these substances destroy or denature the enzyme (e.g., heavy metal ions), often at very low concentrations. A toxicant can also interfere with enzyme catalysis by binding (reversibly or irreversibly) to the enzyme and occupying or altering the active site so that its activity is abolished or altered. **activating enzyme**. an enzyme that catalyzes the attachment of an amino acid to the correct transfer ribonucleic acid (t-RNA). **adaptive enzyme**. inducible enzyme. **autolytic enzyme**. an enzyme that causes lytic disintegration of a cell (See autolysis). **catheptic enzyme**. cathepsin. **clotting enzyme**. coagulating enzyme. **coagulating enzyme** (coagulase; clotting enzyme; curdling enzyme). an enzyme (e.g., renin, fibrin ferment) that catalyzes the conversion of soluble proteins into insoluble ones. **curdling enzyme**. coagulating enzyme. **induced enzyme**. inducible enzyme. **inducible enzyme** (adaptive enzyme; induced enzyme). One whose production is provoked or greatly accelerated by a specific small molecule (inducer) which is the substrate of the enzyme or is very similar in structure to the substrate. **inhibitory enzyme**. an enzyme that blocks a chemical reaction. **repressible enzyme**. an enzyme whose rate of production is decreased as a function of the concentration of certain metabolites. **respiratory enzyme**. an intracellular enzyme that catalyzes oxidative reactions with the release of energy (e.g., cytochromes, flavoproteins). See also enzymes of snake venoms.

enzyme activation. See activation.

enzyme induction. a process whereby the production of enzymes of a particular type increases due to exposure of an organism to a specific inductor or inducing agent (usually a small molecule).

enzyme inhibition. the retardation or suppression of enzyme activity by specific small molecules. This is an important regulatory function in normal metabolism. Some toxicants act to inhibit enzymes, thus upsetting the intricate control mechanisms that regulate the activity of enzymes. Enzyme inhibition by toxicants usually shows little specificity. Enzyme inhibition may be reversible or irreversible. **irreversible enzyme inhibition**. irreversible inhibition is caused by substances reacting with groups at the active site (for example, by forming covalent bonds) in such a way as to destroy it or modify it permanently. **reversible enzyme inhibition**. inhibition is reversible in the case either of competitive or of nonoompetitive in hibition, *q.v.* **competitive enzyme inhibition**. enzyme inhibition that can be fully reversed. This can occur when molecules that are structurally similar to the substrate bind to the active site of the appropriate enzyme, but can not provoke the correct enzyme shape. Thus, the reaction cannot proceed. The inhibition is reversed if the concentration of substrate is increased. **noncompetitive enzyme inhibition**. reversible enzyme inhibition in which a compound binds to a region of the enzyme other than the active site (an allosteric site). In this case, the binding of the substrate to the active site is not prohibited, the shape of the enzyme is altered so that the catalytic groups are not exactly aligned. Inhibition is not reversed by increasing the concentration of the substrate. An activator molecule can, however, affect the active site in a way that brings about correct alignment of a substrate molecule. Noncompetitive inhibition can play a role, for example, in feedback inhibition and in enzyme activation.

enzyme inhibitor. a substance, often a xenobiotic, that adversely affects enzyme activity by slowing or abolishing the ability of an enzyme to function normally. An adverse effect usually results when a xenobiotic acts as an enzyme inhibitor. Heavy metal ions {e.g., cadmium (Cd^{2+}), lead (Pb^{2+}), mercury (Hg^{2+})} may bind tightly to sulfur-containing functional groups in the active sites of an enzyme, rendering the enzyme inactive. Organic xenobiotics to enzymes can also inhibit enzymes through covalent bonding. The latter mechanism is a major mode of acetylcholinesterase inhibition by xenobiotics. Thus, for example, diisopropylphosphorfluoridate (a nerve gas), may inhibit acetylcholinesterase through covalent bonding.

enzyme kinetics. **1:** the rates of enzyme-substrate interactions and the forces that affect them. **2:** the study of the rates of enzyme-controlled reactions and the factors that produce, control, or modify the reaction kinetics of enzymes. At low substrate concentrations, the rate of an enzymatic reaction is proportional to the substrate concentration, but at high concentrations of substrate the available enzymes become saturated and a maximum rate of reaction is reached (See Lineweaver-Burke plot). See also Michaelis-Menton equation.

enzyme-linked assay. See adsorption assay.

enzyme-linked immunoassay. See adsorption assay.

enzyme-linked immunosorbent assay (ELISA). adsorption assay. See assay.

enzyme repression. a process whereby the production of enzymes of a particular type decreases due to exposure of an organism to a specific repressor, usually a small molecule. *Cf.* enzyme induction. The addition of certain substances to a cell (repressors) can repress the synthesis of an enzyme by binding to DNA, thereby blocking the initial stages of protein synthesis.

enzymes of snake venoms. among the more important types of enzymes of snake venoms are: 5′-nucleotidase, acetylcholinesterase, arginine ester hydrolase, collagenase, DNAase, hyaluronidase, L-amino acid oxidase, lactate dehydrogenase, NAD-nucleotidase, phopholipase A_2 (A), phosphodiesterase, phospholipase C, phospholipase B, phosphomonoesterase, proteolytic enzymes, RNase, and thrombin-like enzymes. **thrombin-like enzymes of snake venoms**. enzymes that occur in significant amounts in crotalid and viperid venoms, but are generally lacking in elapid and hydrophid venoms. They act as defibrinogenating enzymes *in vivo*, but surprisingly clot plasma *in vitro*. See also ancrod, batroxobin.

enzyme-substrate complex. an intermediate complex of enzyme and substrate in an enzymatic reaction. See enzyme.

enzymic. enzymatic.

enzymolysis. **1:** the cleavage of a substance into smaller parts by the catalytic action of an enzyme. **2:** lysis due to the disintegrative action of an enzyme.

eosinophil. a polymorphic leucocyte having large cytoplasmic granules that stain brilliantly with acid dyes, e.g. eosin. Eosinophils comprise 1-3% of the white cells in human blood. During parasitic infections and some allergic states and skin diseases, the number of eosinophils in circulating blood increases, a condition termed eosinophilia. Eosinophils may play some role in antibody-antigen reactions, possibly by releasing or absorbing histamine.

eosinophilia. a pronounced increase in the numbers of eosinophils in circulating blood.

eosinophilopenia. an absence of or pronounced decrease in eosinophils in circulating blood.

EPA. See Environmental Protection Agency.

Ephalophis greyi (Grey's sea snake). a rare sea snake (Family Hydrophidae) that is apparently restricted to the Timor Sea along the northwest coast of Australia. The tail is nearly round and somewhat paddle shaped.

ephedra. any plant of the genus ephedra.

Ephedra (ephedra). a large genus of low, bushy, nearly leafless, broom-like shrubs (Family Gnetaceae) with jointed green or gray stems. They occur in various arid regions of the earth and are the original source of ephedrine. ***E. equisetina*** (ma huang). a species that is indigenous to China and India, the stems and leaves of which contain ca. 0.75-1.0% ephedrine and variable amounts pseudoephedrine. ***E. nevadensis*** (cay note; canutillo; whorehouse tea; tapopote; teamsters' tea). This species contains little or no ephedrine.

ephedrine (α-[1-(methylamino)ethyl]benzyl alcohol; 2-methylamino-1-phenyl-1-propanol; 1-phenyl-1-hydroxy-2-methylaminopropane; 1-phenyl-2-methylaminopropanol; α-hydroxy-β-methylaminopropylbenzene). a very toxic, white, water-soluble crystalline amine alkaloid, isolated from several species of *Ephedra*, especially *E. equisetina*, and from *Aconitum* and produced synthetically. It is a sympathomimetic used chiefly to relieve nasal and bronchial congestion in asthma and hay fever, as a CNS stimulant, and mydriatic. See also alkaloids (protoalkaloids) and aconitine (amorphous aconitine).

Epibulus insidiator (wrasse). a bony fish (Class Osteichthyes) of the tropical Indo-Pacific region. Certain populations are transvectors of ciguatera.

Epicanta fabricii (blister beetle). beetles (Family Meloidae) of the midwestern and eastern United States, the tissues of which contain cantharidin. Contact with their body fluids can produce a painful blister in humans. Accidental ingestion can be harmful.

epicarcinogen. any agent that augments the effect of a carcinogen.

epichlorohydrin (chloromethyloxirane; DL-α-epichlorohydrin; 1-chloro-2,3-epoxypropane; γ-chloropropylene oxide). a colorless, unstable, flammable, water-insoluble liquid with a chloroform-like, irritating odor. It is miscible with carbon tetrachloride, chloroform, ethanol, ethyl ether, and trichloroethylene, but immiscible with petroleum hydrocarbons. Epichlorohydrin is used mainly to condition starches in food processing and as a solvent for resins and a cement for celluloid. The vapors of this toxic material are irritating to the eyes, nose, and throat. It is nephrotoxic, a strong skin irritant, a sensitizer, and is carcinogenic to some laboratory animals. See also chlorohydrin, ethylene chlorohydrin. Poisoning may take place via all routes of entry. Symptoms of chronic systemic poisoning include extreme weariness, gastrointestinal disturbances, severe eye irritation, and eventually cyanosis and pneumonitis. Lesions may include lung, kidney and liver damage.

DL-α-epichlorohydrin. epichlorhydrin.

epichroic. discoloring.

epicladosporic acid. a mycotoxin of *Cladosporium epiphyllum*. See also fagicladosporic acid.

epicytic. denoting occurrence or activity on or at the surface of a cell.

epidemic. **1a:** affecting a large number of individual organisms at the same time in the same geographic region. **1b:** said of a disease or pathological condition that affects a high proportion of a population within a community, ecosystem, or a geographic region. **2a:** a widespread outbreak of disease or pathological condition. **2b:** an unusually large number of cases of a disease in a single community or relatively small area.

epidemiologic. epidemiological.

epidemiological (epidemiologic). **1:** pertaining to the study of the etiology or causes of any disease state including toxic conditions in a localized population or community (human or non-human biota). **2:** in context: pertaining to the study of localized outbreaks of disease or poisoning. *Cf.* epiphytotic, epizootic. See epidemiology.

epidemiology. **1:** the identification of factors that contribute to the origins, incidence, rates and patterns of transmission of infectious disease and other health-related states throughout a given human population, community or geographic region with a view to controlling the further spread of the disease. Outbreaks of poisoning (e.g., ergotism), especially from natural sources often come under the purview of epidemiology. **2:** the science that deals with diseases that affect large numbers of people rather than individuals and the application of epidemiological findings to the control of disease outbreaks. Many of the principles of epidemiology apply to animals, especially domestic animals, as well.

epidermides. plural of epidermis.

epidermis (plural, epidermides). **1:** the nonvascular outermost layer of cells covering the body of an animal or plant. It may consist of a single layer of cells as in invertebrate animals and may secrete a protective cuticle. In most vertebrate animals, the epidermis makes up the outer layer of the skin; it consists of several stratified layers and continually undergoes renewal. The epidermis of plants usually consists of a single layer of rectangular cells that fit closely together. The intact epidermis (together with the cuticle, if present) affords one degree or another of protection as a barrier to mechanical and chemical damage of living tissue as well as attack by pathogenic organisms. See also skin.

epigenetic carcinogen. See carcinogen.

epiglottis. a structure that covers the glottis while swallowing.

epilemma. endoneurium.

epinasty. a striking and rapid response of the shoots of many dicotyledonous plants exposed to auxins. There is a pronounced downward bending of leaves and petioles often with random bending of the younger internodes. All the auxin-type herbicides promote epinasty. It is thus a characteristic symptom of the action of auxin herbicides on susceptible plants.

Epinephelus fuscoguttatus (seabass, grouper). a marine bony fish (Family Serranidae) of the Indo-Pacific region. Large sea basses, especially, may be extremely toxic and should be eaten with caution. Certain populations of are transvectors of ciguatera.

epinephrine (4-[1-hydroxy-2-(methylamino)-ethyl]-1,2-benzenediol; 3,4-dihydroxy-α-[(methylamino)methyl]benzyl alcohol; 1-(3,4-dihydroxyphenyl)-2-(methylamino)ethanol; 3,4-dihydroxy-1-[1-hydroxy-2-(methylamino)ethyl]-benzene; methylaminoethanolcatechol; adrenalin; adrenaline). a catecholamine hormone and the chief neurotransmitter of peripheral sympathetic nerve terminals. It is formed from norepinephrine by the action of phenethylamine *N*-methyltransferase. Epinephrine is secreted by the chromaffin tissue of the adrenal medulla of mammals and by related tissue in other vertebrates. It occurs also in some invertebrate animals. Secretion by the vertebrate adrenal gland is stimulated by the sympathetic nervous system in response to stress (e.g., fear, pain, muscular activity, decreasing blood-sugar levels). Epinephrine is an extremely powerful sympathomimetic. It causes an increase in the frequency and force of cardiac contractions, vasoconstriction or vasodilation, and relaxation of bronchiolar and intestinal smooth muscle, and dilates the pupils of the eye. Epinephrine also stimulates a number of metabolic processes such as glycogenolysis and lipolysis. It is very toxic orally to laboratory mice; while poisoning is possible per os in humans, epinephrine so administered is thought to be rapidly inactivated. Iatrogenic overdosage and accidental *i.v.* administration of epinephrine have produced serious and sometimes fatal intoxication. Effects of epinephrine poisoning from overdosage may include cerebrovascular hemorrhage and pulmonary edema due to elevated arterial blood pressure, and ventricular fibrillation. See adrenalin, catecholamine, hormone.

epiphytotic. **1:** of or pertaining to a disease epidemic in plants. **2:** a disease epidemic in plants. *Cf.* epidemic, epizootic, epidemiological.

epirrheology. the science of the effects of exogenous factors on plants.

episcleritis. inflammation of the sclera of the eye.

epithelial. pertaining to the epithelium.

epithelial tissue. epithelium.

epitheliotoxin. a cytotoxin that destroys epithelial cells.

epithelium (plural, epithelia). **1:** epithelial tissue; any tissue comprised of tightly bound cells that lines cavities or covers the internal and external organs of an animal's body, including the lining of vessels. Epithelia consist of cells bound together by connective material and varies in the number of layers and the kinds of component cells. Epithelia serve protective, secretory, absorptive, and even behavioral functions. Epithelia are, in general, important in toxicology as they occur in all portals of entry and must be penetrated by toxicants if the body is to be entered. They are also targets for toxic action. See epitheliotoxin, hyaluronidase. **2a:** the layer(s) of cells that form the outer surface of the skin of an animal. In vertebrates, these epithelial cells rest on a basement membrane and may form a simple, single layer or a stratified layer several cells (5 in humans) thick. The cells may be flat (squamous), cuboidal or columnar in form and may be ciliated or glandular, depending upon the function of the membrane and the cell. **2b:** vertebrate tissue that is composed of closely apposed cells separated by very little intercellular substance and is derived from embryonic ectoderm and endoderm. In its simplest form epithelium consists of a continuous layer of cells covering internal and external surfaces such as the skin and the linings of the gastrointestinal, respiratory, and urinogenital tracts. It also forms more specialized structures that may or may not remain in contact with the epithelium (e.g., sweat glands, salivary glands, endocrine glands). **3:** in plants, the tissue that lines either resin canals in gymnosperms or the gum ducts of dicotyledons.

epitoxoid. a toxoid that has less affinity for the corresponding antitoxin than does the toxin.

epitoxonoid. a toxonoid that has the least affinity for the corresponding antitoxin.

epizoa. plural of *epizoon*.

epizoicide. an agent that destroys epizoa (See epizoon).

epizoon (plural, epizoa). an animal parasite that lives on the outer surface of the body of the host.

epizootic. **1:** of or pertaining to an epidemic disease among non-human animals. **2a:** a disease epidemic among animals other than humans. **2b:** any disease that affects a large number of animals (other than humans) in a given area or population. *Cf.* epidemic, epiphytotic.

EPN (O-ethyl O-*p*-nitrophenyl phenylphosphonothioate (phenylphosphonothioic acid O-ethyl O-(4-nitrophenyl) ester; ethyl *p*-nitrophenyl benzenethiophosphonate). a neurotoxic, liquid or light yellow, water-insoluble, crystalline compound,$C_6H_5P(C_2H_5O)(S)OC_6H_4NO_2$, m.p. 36°C, with an aromatic odor, that decomposes in alkaline solution. It is used as an insecticide against cotton pests, a miticide (sometimes together with methyl parathion) on fruit crops, and sometimes as household pesticide. An exceedingly hazardous organophosphate anticholinesterase, it is extremely toxic orally to laboratory rats. EPN is extremely hazardous as a poison. Poisoning from EPN may readily occur via ingestion, skin contact, or by injection. Extremely toxic fumes, SO_x, PO_x, NO_x, and phosphine, are released on decomposition by heating.

epoxidation. a major type of phase I reaction of xenobiotic compounds whereby an epoxide functional group is formed by addition of an oxygen atom between two carbon atoms in an unsaturated system. The process is catalyzed by cytochrome P-450 isozymes. It is an important mechanism in the decomposition of aromatic xenobiotic compounds. Many unsaturated aliphatic and alicyclic compounds are also metabolized in part by epoxidation. The resultant epoxides may be highly reactive and unstable (e.g., arene oxides) tending to undergo further reactions, especially hydroxylation. Some epoxide intermediates are more toxic than the parent compound, and many are carcinogenic.

epoxide. **1:** any of a type of organic compound (cyclic ethers) that contains the three-membered epoxy ring as a substituent. Examples are 1,2,5,6-diepoxyhexane and 7,8-dihydrodiol-9,10-epoxide. If the two carbon atoms of the epoxy ring are aromatic, the epoxide is termed an arene oxide. Epoxides, especially arene oxides, are often strongly electrophilic and are thus highly reactive and toxic metabolites. There are numerous naturally occurring epoxides, including certain plant hormones (e.g., abscissic acid), phytotoxins (e.g., prolactones), various insect hormones, and some antibiotics. Most fungal epoxides (e.g., the trichothecene mycotoxins) are antibiotic to bacteria, protozoa, and fungi. Some synthetic epoxides are carcinogenic. See epoxy resin. **2:** the epoxide functional group. **3:** ethylene oxide (epoxyethane), a three-membered cyclic ether.

epoxy. **1:** any compound that contains the epoxy group, *q.v.* Such compounds are used chiefly in adhesives, coatings (e.g., for computer circuit boards), electrical insulation, and solder flux. They are suspected carcinogens. **2:** an epoxy resin, *q.v.* **3:** epoxy group.

epoxy group. a chemical group comprised of one atom of oxygen bound to two different carbon atoms. See epoxy, epoxy ring, epoxide.

epoxy resin. any of a class of synthetic thermosetting resins produced by copolymerization of epoxides with phenols. They are characterized by adhesiveness, flexibility, and resistance to chemicals. Their reactivity is due to the presence of the epoxy group. Epoxy resins are used chiefly in adhesives, coatings, electrical insulation, solder mix, and as protection for computer circuit boards. They are suspected human carcinogens. *Cf.* epoxide. See also epoxy.

epoxy ring. a reactive, electrophilic, three-membered ring in which an oxygen atom bridges between two adjacent carbon atoms that are already bonded. It is the functional group of epoxides.

1α,4α-epoxy-3-aza-A-homoandrostan-16β-ol. samandarine.

2,3-epoxybutane. butylene oxide.

1,2-epoxybutene-3. a carcinogenic monoepoxide. See epoxide.

4,9-epoxycevane-3α,4α,12,14,16β,17,20-heptol. cevine.

4α-9-epoxycevane-3β-4,7α,14,15α,16β,20-heptol. germine.

4,9-epoxycevane-3,4,12,14,16,17,20-heptol 3-(3,4-dimethoxybenzoate). veratridene.

(Z)-4α,9-epoxycevane-3β,4,12,14,16β,17,20-heptol 3-(methyl-2-butenoate). cevadine.

4α,9-epoxycevane-3β,4,14,15α,16β-20-hexol. zygadenine.

7β,8-epoxy-3β,14-dihydroxy-5β-card-20(22)-enolide. tanghinigenin.

4,5-epoxy-3,14-dihydroxy-17-(2-propenyl)-morphinan-6-one. naloxone.

3α,9α-epoxy-14β,18β-(epoxyethano-*N*-methylimino)-5β-pregna-7,16-diene-3β,11α,20α-triol. batrachotoxinin A.

3α,9α-epoxy-14β,18β-(epoxyethano-*N*-methylimino)-5β-pregna-7,16-diene-3β,11α,20α-triol, 20α-ester with 2,4-dimethylpyrrole-3-carboxylic acid. batrachotoxin.

epoxyethane. ethylene oxide.

1,2-epoxyethane. ethylene oxide.

(3β,23β)-17,23-epoxy-3-(β-D-glucopyranosyloxy)veratraman-11-one. pseudojervine.

1,2-epoxyhexadecane. a carcinogenic monoepoxide. See epoxide.

(3β,5β,15β)-14,15-epoxy-3-hydroxy-5-bufa-20,22-dienolide. resibufogenin.

6,7-epoxy-3-hydroxytropane. scopolamine.

(3β,23β)-17,23-epoxy-3-hydroxyveratraman-11-one. jervine.

(2*R*,3*S*)-2,3-epoxy-4-oxo-7E,10E-dodecadienamide. cerulenin.

2,3-epoxy-4-oxo-7,10-dodecadienoylamide. cerulenin.

1,2-epoxypropane. propylene oxide.

12,13-epoxy-3,4,7,15-tetrahydroxytrichothec-9-en-8-one. nivalenol.

12,13-epoxytrichothec-9-ene-3,4,8,15-tetrol 4,15-diacetate 8-(3-methylbutanoate). T-2 toxin.

6β,7β-epoxy-1α-H,5αH-tropan-3α-ol (—)-tropate. scopolamine.

6a,7a-epoxy-1αH,5αH-tropan-3α-ol(-)-tropate. scopolamine.

6β,7β-epoxy-3α-tropanyl S-(-)-tropate. scopolamine.

6,7-epoxytropine. scopolamine.

6,7-epoxytropine tropate. scopolamine.

EPP. free erythrocyte protoporphyrin (See protoporphyrin).

EPS. extrapyramidal side effect.

epsilon (ϵ, Ε). **1:** fifth letter of the Greek alphabet. **2:** in mathematics, an arbitrarily small quantity.

epsom salts. the heptahydrate of magnesium sulfate.

Eptam. a thiocarbamate herbicide that is particularly useful for controlling *Agropyron repens* and *Cyperus spp.*

EPTC (dipropylcarbamothioic acid *S*-ethyl ester; dipropylthiocarbamic acid *S*-ethyl ester; *S*-ethyl dipropylthiocarbamate). a preplanting thiocarbamate herbicide that is soluble in water and miscible with benzene, ethanol, toluene, and xylene. It selectively inhibits shoot growth, while root growth remains unaffected or even enhanced. It is particularly useful in controlling *Agropyron repens* and perennial *Cyperus spp.* EPTC is moderately toxic orally to laboratory rats.

Equal. See aspartame.

Equanil™. meprobamate.

equilibrium. **1:** balance, stability, the point at which the state variables of an isolated system no longer change with time. **2:** in relation to radiation, the point at which the radioactivity of consecutive elements within a radioactive series is neither increasing nor decreasing.

equilibrium dialysis. a commonly used dialysis method to measure the binding of toxicants to macromolecules, especially plasma proteins. The unbound ligand equilibrates across the dialysis membrane, while the ligand bound to the test toxicant does not. Thus, if the amount of ligand on the side of the membrane that contains the protein in solution exceeds that on the other side of the membrane at equilibrium, the difference represents the amount of bound ligand. A radiolabeled ligand is usually employed.

equilibrium population. 1: a population in which gene frequencies are at equilibrium. **2:** a population having a stable size in which death and emigration rates are balanced by birth and immigration rates.

equinatoxin. a venom isolated from the sea anemone, *Actinia equina*. It is highly toxic and exhibits hemolytic, cardiotropic, and antigenic activities.

equine. pertaining to a member of the family Equidae, usually the modern horse, *Equus caballus* or other members of this genus.

equine nervous syndrome. See moldy corn poisoning.

equipotent. equally powerful; having an equal effect or capability.

equis. *Bothrops atrox*.

Equisetaceae. See *Equisetum*.

equisetosis (equisetum poisoning). poisoning of livestock by ingestion of *Equisetum*. In North America, equisetosis is most common in horses and is due solely to *E. arvense*. Hay containing 20% or more of *E. arvense* produces symptoms of a potentially fatal equisetosis in horses within 2-5 weeks. Poisoning of cattle is uncommon and rarely fatal because the ruminal bacteria produce an adequate internal supply of thiamine. Symptoms are similar to bracken poisoning (see *Pteridium*). Treatment with massive doses of thiamine can reverse the disease except in the terminal stages. See *Equisetum*, equisitine, *Pteridium*, thiaminase.

Equisetum (horsetail; foxtail; rush). a genus of small, flowerless, nearly cosmopolitan herbaceous wetland plants (Family Equisetaceae) with leaves reduced to scales. The chief toxic principle is thiaminase, but these plants also contain the alkaloids equisitine, palustrine, 3-methoxypyridine, and nicotine. Ingestion by nonruminant livestock (most notably horses) may result in thiamine deficiency and eventual death. Cattle and other ruminants are not susceptible to equisetosis since the ruminal bacteria produce substantial amounts of thiamine. While generally considered only moderately toxic, horsetails sometimes may accumulate toxic amounts of nitrates and sometimes selenium during periods of rapid growth. Children, using the hollow stems as blowguns, have thus occasionally been fatally poisoned. ***E. arvense***. a weedy plant common to sandy soils in moist fields, along roadside and railroad embankments throughout the United States and Canada. In North America, equisetum poisoning of horses is not uncommon and is due solely to *E. arvense*. Hay that includes 20% or more of *E. arvense* produces symptoms of equisetosis in horses within 2-5 weeks. Poisoning of cattle by *E. arvense* is seldom fatal. ***E. hyemale***. a Californian species known to poison livestock. ***E. laevigatum***. a Californian species known to poison livestock. ***E. limosum***. a European species known to poison livestock. ***E. palustre***. a plant that is locally abundant in wet meadows and along the margins of streams and lakes from Newfoundland and the northern United States into Alaska. This species sometimes poisons cattle with loss of condition, hyperexcitability, diarrhea, muscular weakness, and decreased milk production by dairy cattle. Horses, sheep, and goats are rarely poisoned. ***E. ramosissimum***. a South African species known to poison livestock, especially sheep.

equisetum poisoning. equisitosis.

equisitine. a toxic alkaloid of *Equisetum*, *q.v.*

equitoxic. equally toxic; having equivalent toxicity.

equivalent. **1:** equal, comparable, having the same value. **2:** having the ability to neutralize or counterbalance another action, process, or entity or to compensate an action or effect. **2:** chemical equivalent. **3:** in medicine, a symptom that replaces one that is usual in a given disease. **dose equivalent**. See dose equivalency, toxicokinetically equivalent dose. **toxic equivalent**. the amount of a substance necessary to poison an animal. It is usually taken as the median lethal dose per kg body weight.

equivalent dose. See dose equivalency, toxicokinetically equivalent dose.

Er. the symbol for erbium.

erabutoxin (erabutoxin A, B, and C). any of three extremely toxic neurotoxins that comprise 30% of the proteins in the venom of the sea snake, *Laticauda semifasciata*. They act on postsynaptic membranes to block neuromuscular transmission. The lethal activity of the venom resides almost exclusively in the extremely toxic erabutoxins A and B.

erabutoxins A, B, and C. See erabutoxin.

erabu-umihebi. a Japanese name for the sea snake, *Laticauda semifasciata*.

erabu-unagi. a Japanese name for the sea snake, *Laticauda semifasciata*.

eradicate. exterminate (usually willfully), e.g., a population; to destroy totally.

eradication. complete, usually willful, destruction or extermination of a population or type of organism. *Cf.* genocide.

erbium (symbol Er)[Z = 68; A_r = 167.26]. a soft silvery element of the lanthanide (rare earth) series of metals. See also erbium chloride, erbium nitrate.

erbium chloride. anhydrous erbium chloride, $ErCl_3$; it is moderately to very toxic to laboratory mice.

erbium nitrate. the hexahydrate of erbium nitrate, $Er(No_3)_3$; it is very toxic to female rats.

Erbon (2-(2,4,5-trichlorophenoxy)ethyl-2,2-dichloropropionate; Baron; Novon). a moderately toxic herbicidal formulation of isomeric trichlorophenylacetic acids. Little, if any, is absorbed percutaneously. When fed to cattle, none appeared in the milk. Erbon is carcinogenic to some laboratory animals.

erdin. a chlorine-containing mycotoxic fungal metabolite.

Eretomochelys imbricata (hawksbill turtle). an endangered species of marine turtle (Family Chelonidae) with a hooked upper jaw, a keeled shield-shaped carapace with overlapping scutes and paddle-like limbs. They occur in all warm seas, ranging into temperate waters during summer. They are normally suitable for eating, but in some locations are extremely toxic. The liver is especially poisonous. Nearly half of those persons who are poisoned die. See Chelonia.

ergobasine. ergonovine.

ergocalciferol. vitamin D_2.

ergoclavine. an equimolar mixture of ergosine and ergosinine isolated from ergot.

ergoclavinine. ergosinine.

ergocornine (1,2′-hydroxy-2′,5′α-bis(1-methylethyl)ergotaman-3′,6′,18-trione). an extremely toxic naturally occurring polypeptide ergot alkaloid. A component of ergotoxine, it is a derivative of lysergic acid.

ergocorninine (12′-hydroxy-2′,5′α-bis(1-methylethyl)-8α-ergotaman-3′,6′,18-trione). a nearly water-insoluble isomer of ergocornine, *q.v.* See also ergotinine, ergotoxine.

ergocortinine. a constituent of ergot; an alkaloid isomer, $C_{31}H_{39}N_5O_5$, of ergocornine, *q.v.*

ergocristine (12′-hydroxy-2′-(1-methylethyl)-5′-(phenylmethyl)ergotaman-3′,6′,18-trione). a polypeptide alkaloid of ergot derived from lysergic acid. See also ergotinine, ergotoxine.

ergocristinine (12′-hydroxy-2′-(1-methylethyl)-5′α-(phenylmethyl)-8α-ergotaman-3′,6′,18-trione). an isomer of ergocristine.

ergocryptine (ergokryptine). an extremely poisonous polypeptide alkaloid of ergot. See also ergotinine, ergotoxine.

ergocryptinine (ergokryptinine). **1:** an ergot alkaloid isomer of ergocryptine. **2:** prior to 1967 this term was used in reference to α-ergocryptinine.

ergokryptine. ergocryptine.

ergokryptinine. ergocryptinine.

ergometrine. ergonovine.

ergometrine maleate. ergonovine maleate.

ergometrinine. an ergot alkaloid.

ergone. any substance which, in minute amounts, promotes a physiological process.

ergonovine ([8β(S)]-9,10-didehydro-N-(2-hydroxy-1-methylethyl)-6-methylergoline-8-carboxamide; N-[α-(hydroxymethyl)ethyl]-D-lysergamide; D-lysergic acid L-2-propanolamide; ergometrine; ergobasine; ergostetrine). a crystalline ergot alkaloid that is extremely toxic, *i.v.*, to laboratory mice. It is the simplest of the ergot alkaloids structurally. The lysergic acid moiety is linked to a single amine-bearing group. Ergonovine is the most powerful oxytocic ergot alkaloid. It is also a vasoconstrictor, a CNS stimulant, and adrenergic blocking agent. It is used medically mainly as the maleate (erogonovine maleate).

ergonovine maleate (ergometrine maleate). the form of ergonovine most used medically as an oxytocic agent; other physiologic activities are less potent than its oxytocic activity.

ergosine (12′-hydroxy-2′-methyl-5′α-(2-methylpropyl)ergotaman-3′,6′,18-trione). a clavine alkaloid from ergot with actions similar to those of ergotamine.

ergosinine (ergoclavinine). a constituent of ergot, it is the 8α-isomer of ergosine.

ergosterin. ergosterol.

ergosterol (ergosterin). a water-insoluble, crystalline, unsaturated sterol produced by ergot, yeast, and certain other fungi. It is converted to vitamin D_2 when exposed to ultraviolet light or activated by electrons.

ergostetrine. ergonovine.

ergot (secale cornutum; rye ergot; rye smut). **1a:** the sclerotium, the hardened, resistant, over-wintering mycelial mass of the ascomycetous fungus *Claviceps spp.* (*q.v.*). It is a small compact spur-like mass that contains several optically isomeric pairs of alkaloids. **1b:** the dried sclerotia. **1c:** preparations from the dried sclerotia used in medicine. They are generally relatively harmless. They are used to induce or strengthen uterine contractions (during and immediately after childbirth), to control internal hemorrhage, and alleviate certain localized vascular disorders (e.g., migraine). Ergot is also an abortifacient. With careless usage, however, ergotic preparations can induce hallucinations and even death. In hypersensitive persons, especially those with beginning vascular sclerosis as little as 0.5 mg of gynergen, an ergotamine tartrate extract, i.m., can cause gangrene or necrosis. **2:** a common name for fungi of the genus *Claviceps*. See also ergotism, ergot alkaloids, ergotoxine, ergonovine, ergometrinine, ergoclavine, elymoclavine, trimethylamine, putrescine, cadaverine, agmatine, histamine, tyramine, choline, acetylcholine, ergothioneine, and ergosterol.

ergot alkaloid. See alkaloid.

ergot of rye. *Claviceps purpurea*.

ergot peptide alkaloid. See alkaloid.

ergot poisoning. ergotism.

ergotamine (12′-hydroxy-2′-methyl-5′α-(phenylmethyl)ergotaman-3′,6′,18-trione). a crystalline, highly toxic, nearly water-insoluble, tripeptide ergot alkaloid. It is a powerful stimulant of smooth muscle, especially of the blood vessels and uterus, and produces adrenergic blockade, mainly of the alpha-receptors. On heating, ergotamine is converted to lysergic acid diethylamide (LSD). Ergotamine is chiefly used therapeutically as the tartrate to treat migraine.

ergotamine tartrate. See ergotamine.

ergotaminine (12′-hydroxy-2′-methyl-5′α-(phenylmethyl)-8α-ergotaman-3′,6′,18-trione). a much less soluble and nearly nontoxic isomer of ergotamine, *q.v.*

ergothioneine ((*S*)-α-carboxy-2,3-dihydro-*N,N,-N*-trimethyl-2-thioxo-1*H*-imidazole-4-ethanaminium hydroxide inner salt; [1-carboxy-2-[2-mercaptoimidazol-4(or5)yl]ethyl]trimethylammonium hydroxide, inner salt; L(+)-ergothioneine; thioneine; thiolhistidinebetaine; thiasine; sympectothion). a crystalline betaine derived from mercaptohistidine. It is soluble in water, nearly insoluble in benzene, chloroform, and ethyl ether, and slightly soluble in acetone, hot ethanol, and hot methanol. Ergothioneine is a constituent of ergot; mammalian blood, semen, liver, kidney, various other tissues; and *Limulus polyphemus*, the king crab. **L(+)-ergothioneine**. ergothioneine.

ergotic. of, pertaining to, or produced by ergot.

ergotinic acid. a toxic acid isolated from ergot.

ergotinine. a 1:1:1 mixture of the ergot alkaloids ergocornine, ergocristine, and ergocryptine that is isomeric with ergotoxine. It is isomeric with ergotoxine. Ergotinine forms long monoclinic crystals from acetone solutions, m.p. 229°C, that are soluble in chloroform, ethanol, and absolute ether. It is relatively inactive pharmacologically.

ergotism (ergot poisoning; Saint Anthony's fire; *mal des ardents feu de Saint-Antoine*; *gangrene des Solognots*). a disease due to the ingestion of ergot. While epidemic disease is essentially nonexistent today, epidemics in past centuries have affected large numbers of people and livestock. The disease in humans is most often due to consumption of cereal grain (especially wheat or rye) and grain products (e.g., bread) that are infested with the fungus. Livestock poisoning from infected pasturage, hay, grain, or grain screenings has, at times, also been severe. Ergotism rarely affects wildlife or livestock other than cattle, though severe cases are known among newborn pigs. Sheep are very resistant. Ergot preparations used in medical practice today (see ergot) are generally very safe, clinical disease is sometimes seen from excessive use of ergotamine tartrate in the treatment of migraine. Ergot is a potent abortifacient and is used in medicine to induce or accelerate labor. Pregnant women sometimes die from ergot alkaloids administered illegally to induce abortion. Such practice is highly dangerous. Symptoms of ergotism in humans may include nausea, vomiting, diarrhea, dizziness, weak pulse, thirst, tingling in feet, numbness and coldness of extremities, variable effects on blood pressure, dyspnea, convulsions, loss of consciousness. Lesions may include congestion and inflammatory changes in the gastrointestinal tract and kidneys and gangrene of fingers and toes from persistent peripheral vasoconstriction. Effects of acute ergot poisoning of humans include thirst, nausea, vomiting, diarrhea, tachycardia, confusion, and coma.

ergotize. **1:** to infect or poison with ergot. **2:** to add or introduce ergot to a material.

ergotized. **1:** affected or poisoned by ergot. **2:** containing ergot.

ergotoxicosis. a mycotoxicosis caused by a species of *Claviceps*.

ergotoxine (ecboline). **1:** ergotoxine group, an extremely toxic mixture of isomorphous ergot alkaloids, consisting of equal portions of ergocristine, ergocornine, and ergocryptine. It forms orthorhombic crystals, m.p. 190°C, that are soluble in organic solvents such as acetone, chloroform, ethanol, and methanol. It is a powerful stimulant of smooth muscle, especially of the blood vessels and uterus, and an adrenergic blocking agent (mainly affecting the alpha-receptors). **2:** a toxic pharmacologically active stereoisomer of ergotinine. See also ergot alkaloids.

ergotoxine group. See ergotoxine (def. 1).

Ericaceae (heath family). a large, widely distributed family of deciduous and evergreen trees or shrubs that are strongly associated with acid soils. The leaves of many species are poisonous and all contain grayanotoxins, *q.v.* A variety of mammals have been fatally poisoned by eating leaves or flowers of various species of heaths. Domestic fowl (based on tests with *Kalmia latifolia*) are highly resistant. Humans have been poisoned by honey that may have been made from the nectar of several species of Ericaceae. See also arbutin, *Arctostaphylos*, *Vaccinium*. Symptoms of poisoning in livestock include anorexia, repeated swallowing or repeated eructation and swallowing of cud without mastication, profuse salivation, dullness, depression, nausea, and vomiting. The preceding may be accompanied by dyspnea, bloat, frequent defecation, evidence of abdominal pain, and nasal discharge. Autopsy usually reveals a nonspecific gastrointestinal irritation with some hemorrhage and sometimes pulmonary lesions. Animals may eventually exhibit ataxia, prostration, coma, and death. North American genera that include species that are established as poisonous to livestock include *Kalmia*, *Ledum*, *Leucothoe*, *Lyonia*, *Menziesia*, *Pieris*, and *Rhododendron*. A few species have been implicated in serious losses of livestock. Plants in any of these genera can be expected to produce symptoms when they comprise 0.2-1.3% of the diet (based on wet weight of fresh green foliage). Many other heaths may be expected to be toxic.

Erignathus (bearded seals). a genus of large, plain colored, polar seals (Family Phocidae) marked by a tuft of long bristles on each side of the muzzle. ***E. barbatus*** (bearded seal). a circumboreal marine species that lives at the edge of the pack ice. The liver contains high concentrations of vitamin A and may be as poisonous as that of *Thalarctos maritimus* (polar bear).

erinitrit. sodium nitrite.

Eristicophis macmahonii (McMahon's viper; Asian sand viper; leaf-nosed viper). a dangerous, rather stout, fairly small snake (Family Viperidae), less than 1 m long, that occurs in the deserts of southeastern Iran, Afghanistan, and west Pakistan. The head is depressed, distinctly broader than the neck, and the crown is covered by small scales. The eyes are of moderate size with vertically elliptical pupils. The snout is broad and short and the canthus is indistinct. The body is rather stout, slightly depressed, with a short tail. Effects of envenomation on humans are similar to those of *Echis spp.*; the bite is sometimes fatal.

erode. **1a:** to abrade, corrode, wear away, to cause to deteriorate. **2:** to wear away or be worn away by erosion; to wear away or gradually destroy a surface (e.g., soil, a mucosal or epidermal surface) by a physical process such as friction, dissolution, or as a result, for example, of inflammation, injury, contact with a corrosive substance. **3:** to destroy tissue by the action of a corrosive substance.

Erodium (heron's-bill; stork's-bill). a genus of annual and perennial herbs (Family Geraniaceae), some species of which have been implicated in trefoil dermatitis and rape scald, as have species of *Brassica*, *Medicago*, and *Trifolium*.

erosion. **1:** a wearing away (e.g., of a land surface, a tooth, intestinal mucosa) or gradual destruction of a surface. **2:** a state of being worn away or corroded (e.g. by friction, wind or rain, a corrosive chemical). *Cf.* corrosion. **3:** a superficial ulcer in the mucosa of the gastrointestinal tract, one that does not invade the muscularis mucosa.

erosive. **1:** able to erode, abrade, corrode, or wear away. **2:** characterized by erosion. **3:** an agent that erodes tissues (e.g., a corrosive poison) or materials. *Cf.* corrosive.

erosmic oxide. osmic oxide.

error. mistake, fault, flaw; inaccuracy, falsity; failure. Error is a widely used but often confusing term. When using this term in the sciences, one must take care to clarify the meaning: **1:** the difference between a measurement or experimental result and the "true" value. **2:** the uncertainty of an experimental result. **experimental error**. increased imprecision or inaccuracy in the results of an experiment, presumably due to chance variation or extraneous variables, as opposed to simple errors of observation or measurement. It is also sometimes taken as variation in results that is not accounted for by the hypothesis tested. **random error**. error due to chance such that any deviation in the magnitude and direction of a variable cannot be predicted. **sampling error**. **1:** in statistics, the difference between an observed value of a statistic and the variable it is intended to estimate. **2:** the variation among the observations in any one cell of an analysis of variance. **3:** any error in selection or collection of samples (sets of specimens) from a population that results in significant differences between the sample and the population that it purports to represent. Such error often results from nonrandom selection of the specimens that comprise the sample. See also experimental error. **type I error**. **1a:** in statistics and experimental research, the rejection of a null hypothesis when it is actually true. **1b:** an error in which an event is predicted which does not, in fact, materialize. **type II error**. **1a:** in statistics and experimental research, the acceptance of a null hypothesis when it is likely to be false and should be rejected. **1b:** failure to predict an event that does in fact occur. **comment:** traditionally, the two types of error have been treated differently. Thus a 5% probability of type I error in biology (or even a 1% probability) has been considered acceptable, whereas a 20% probability of type II error may be accepted. In risk assessment, however, it is the type II error that one most wishes to avoid.

erucic acid (1 3-docosenoic acid). a C_{22} solid, unsaturated, ether-soluble, fatty acid, with one double bond, $C_8H_{17}CH{:}CH(CH_2)_{11}COOH$. It is extracted chiefly from the fats and oils contained in seeds of plants of the family Brassicaceae (e.g., rape, mustard, wallflower, and crambe). Erucic acid makes up 40-50% of the total fatty acid content of the seeds of the plants just named. It occurs in various other plants, especially of the family Tropaeolaceae (*Tropaeolum*, commonly called nasturtium). Erucic acid is thought to be cardiotoxic.

eruption. **1a:** the act of breaking out; emission; explosion; appearing or becoming visible. **1b:** a breaking out of visible skin lesions. **2a:** a rapidly developing dermatosis of the skin or mucous membranes that is usually a local manifestation of systemic disease. The lesions are often marked by prominence, erythema, or both. The shape, size, and condition of the lesions may vary widely. **2b:** a rash. **drug eruption** (dermatitis medicamentosa; or dermatosis medicamentosa; drug rash; medicinal eruption). any eruption or solitary skin lesion caused by the ingestion, injection, inhalation, or insertion of a drug or other xenobiotic substance, but not by skin contact. Such eruptions are most commonly due to allergic sensitization. **fixed drug eruption**. a fixed eruption or hypersensitivity reaction following the administration of a particular drug (e.g., barbiturates, quinine, sulfonamides, tetracycline). Lesions are typically reddish-brown or purple, intensely erythematous, sharply defined macules and are associated with hydropic degeneration of the basal layer. Following involution, the affected areas may erupt again and enlarge if the drug is administered anew. The foci that are most often involved [illegible] extremities including the penis. **fixed eruption**. a circumscribed inflammatory lesion(s) that recur(s) at the same site over a period of time; each attack lasts only a few days, but leaves residual pigmentation which accumulates. **iodine eruption**. an acneform or follicular eruption or granulomatous lesion as part of a systemic reaction to iodine. **medicinal eruption**, drug eruption.

eruptive. pertaining to or characterized by eruption.

Erysimum cheiranthoides (wormseed mustard). a short-lived perennial herb (Family Brassicaceae); plants of this genus are commonly known as blistercress or wallflower. It can cause severe gastroenteritis in grazing livestock, presumably because of the mustard oil content.

erysipelas. a contagious disease of the skin and subcutaneous tissues that is characterized by redness and swelling of the affected areas. The infectious agent is *Streptococcus pyogenes*.

erysipelotoxin. the toxin of *Streptococcus pyogenes*. It is the causative agent of erysipelas.

erythema (erythema simplex; dermatitis simplex). inflammatory redness or blushing of the skin in humans due to dilation of superficial capillaries. Causes vary (e.g., a toxicant, an allergen, endogenous neurovascular processes, nervous mechanism, sunburn).

erythema multiforme. cutaneous lesions due to subcutaneous vasculitis caused by immune complexes. Such conditions are usually linked to a drug allergy or to systemic infection. See also Stevens-Johnson Syndrome.

erythema nodosum. erythema characterized by slightly elevated, painful, erythematous nodules on the shins and sometimes the head and forearms. They are due to subcutaneous vasculitis involving small arteries. The condition is usually associated with infection and is due to antigen-antibody complexes. It sometimes results from the use of certain drugs.

erythema toxicum. erythema due to an allergic reaction to a toxicant.

erythematous. pertaining to or marked by erythema; inflamed.

erythematous shellfish poisoning. See shellfish poisoning.

erythredema. See acrodynia.

erythremia. an increase in the concentration of circulating erythrocytes.

erythrene. 1,3-butadiene.

erythrocyte (red blood cell; red blood corpuscle; RBC). **1:** a hemoglobin-containing blood cell found in all vertebrate animals. It is compressed, and in cyclostomes and most mammals, it is nearly round in outline; it is oval in other vertebrates. The erythrocytes of most mammals lack a nucleus and thus take the shape of a biconcave disk. They reversibly bind oxygen and carbon dioxide and thus serve in the internal transport and exchange of these gases. **2:** a cell that contains a respiratory pigment, as in some marine worms.

erythrocyte protoporphyrin. free erythrocyte protoporphyrin (See protoporphyrin).

erythrocytolysis. hemolysis.

erythroderma. any erythematous dermatitis. See erythema.

erythrogenic. capable of reddening, as in inflammation.

erythroidine. a crystalline alkaloid with two forms, α-erythroidine and β-erythroidine, that are isolated from plants of the genus *Erythrina*. **α-erythroidine** ((3β,12β)-1,2,6,7-tetrahydro-12,17-dihydro-3-methoxy-16(15*H*)-oxaerythrinan-15-one). one of two forms of erythroidine. **β-erythroidine** ((3β)-1,2,6,7-tetradehydro-14,-17-dihydro-methoxy-16(15H)-oxaerythrinan-15-one;12,13-didehydro-13,14-dihydro-α-erythroidine). β-erythroidine has a curare-like action and is a CNS depressant.

erythrolysin. hemolysin.

erythrolysis. hemolysis.

erythromycin (erythromycin A). any of a class of antibacterial macrolide antibiotics produced by a strain of *Streptomyces erythreus*. It inhibits protein synthesis in staphylococci and pneumococci. Three erythromycins (A, B, C) are produced by *S. erythreus*. Of these, erythromycin A predominates.

erythrosine (FD&C Red No. 3; E127; disodium or dipotassium salt of 2,4,5,7-tetraiodofluoroscein). a food color and biological stain, $C_{37}H_{36}N_2O_{10}S_3 \cdot 2Na$, used widely in jams and marmalades, maraschino cherries, pickles and relishes, etc. It is moderately toxic, the LD_{50} in rats being 150 mg/kg body weight, *i.p.*, and 600 mg/kg, *p.o.* Erythrosine has a variety of actions *in vitro* due largely to its ability to affect biological membranes and lipophilic sites of enzymes. Erythrosine was thus thought to be potentially neurotoxic, but following preclinical and clinical tests, it seems unlikely to be neurotoxlc at normal dosages. Erythrosine is a questionable (possibly weak) carcinogen.

erythroskyrine. a mycotoxin from *Penicillium islandicum*. It is very toxic to laboratory mice. See also cyclochlorotine, luteoskyrin.

erythrotoxin. a toxin that is destructive to erythrocytes.

erythroxyline. the original name of cocaine, given by its discoverer, Gaedeke, in 1855.

Erythroxylon. *Erythroxylum*.

Erythroxylum (synonym, *Erythroxylon*; coca; hayo; ipado). a large genus of small trees and shrubs (Family Erythroxylaceae) that are indigenous to South America and the West Indies, but are cultivated elsewhere. The leaves of two species, ***E. coca***. (Bolivian coca; Huanaco coca) and ***E. novagranatense*** (Peruvian coca; Truxillo coca), contain cocaine and other alkaloids. The dried leaves mixed with a little alkali are chewed by Indians of the Andean uplands to increase endurance while traveling or working by preventing hunger pangs and fatigue. See also coca, cocaine, cuscohygrine, hygrine.

Es. the symbol for einsteinium.

esau. *Trimeresurus gramineus*, *T. popeorum*.

ESB. environmental specimen banking.

esbecythrin. deltamethrin.

eschar. a dry crust or slough of skin, especially one that develops following a thermal or chemical burn.

escharotic. 1. corrosive, able to produce an eschar. 2. a corrosive agent. See corrosive, erosive.

Escherichia coli (colon bacillus). a species of Gram-negative bacterium (Family Enterobacteriaceae) that are common in the alimentary canal of humans and other vertebrates. It is normally nonpathogenic when present in the gastrointestinal tract, but when free or invasive of other tracts (notably the urinary tract) they are infectious; they also cause enteritis in infants. Some strains are pathogenic and are largely responsible for traveler's diarrhea. When present in milk or water, *E. coli* is an indicator of fecal contamination. Selected strains exhibit anomalies in arginine, galactose, nicotinic acid, or tryptophan metabolism or deficiencies in DNA polymerase. Such strains are used in prokaryote mutagenicity tests. See colitoxemia, colitoxin, colitoxicosis.

escorpion. Heloderma horridum.

esculin (aesculin; 6-β-D-glucopyranosyloxy)-7-hydroxy-2*H*-1-benzopyran-2-one; 6-β-glucosido-7-hydroxycoumarin; 6,7-dihydroxycoumarin 6-glucoside; esculoside; bicolorin; enallochrome; polychrome). a coumarin glycoside contained in the nuts, leaves and bark of trees of the genus *Aesculus* (Family Hippocastanaceae). It is an antipyritic agent and skin protectant.

esculoside. esculin.

eseridine (2,3,4,4a,9-hexahydro-2,4aα,9α-trimethyl-1,2-oxazino[6,5-*b*]indol-6-ol methylcarbamate; eserine aminoxide; eserine oxide; physostigmine aminoxide). a parasympathomimetic alkaloid obtained from the seeds of *Physostigma venenosum* (Family Fabaceae, formerly Leguminosae). See also physostigmine.

eserine. physostigmine.

eserine aminoxide. eseridine.

eserine oxide. eseridine.

esfenvalerate. a white crystalline solid, the (S,S)-isomer of fenvalerate, *q.v.*

Esoderm™. lindane.

esophageal varices. enlarged veins on the lining of the esophagus. They are subject to severe bleeding and often appear in patients with severe liver disease.

esophagoscope. an optical instrument with lenses and a lighted tip used to examine the interior of the esophagus.

esophagoscopy. examination of pathologies of the esophagus by use of an esophagoscope.

esophagus (oesophagus). a muscular tube between the pharynx and stomach. Ingested materials pass through the esophagus, but few substances are absorbed to any extent in this portion of the digestive tract.

esotoxin. endotoxin.

esotropia (cross-eye; convergent strabismus; internal strabismus). a form of strabismus in which there is an obvious convergence of the visual axis of one or both eyes toward the other, resulting in diplopia (double vision).

ESR. erythrocyte sedimentation rate.

essalambwa. *Bitis arietans*.

essence. **1:** aroma, fragrance, scent. **2:** principle, extract, concentrate, tincture; elixer.

essence of mirbane. nitrobenzene.

essential. **1a:** basic, integral, primary, irreducible. **1b:** necessary, indispensable (See for example, element (essential trace element). **2a:** of or pertaining to an essence; forming, belonging to, or constituting the essence of something. **2b:** containing the essence of that part of a plant, plant principle, toxicant, etc., that is marked by a characteristic feature (e.g., odor, toxicity). **3:** having no obvious or known cause or origin (e.g., essential fever, essential hypertension). **3:** constituting an indispensable structure. **4a:** that which is basic, fundamental or necessary (e.g., an essential amino acid). **4b:** a requirement or necessity.

ester. any of a diverse group of naturally occurring and synthetic organic compounds formed by the reaction of an alcohol and an acid (usually a carboxylic or aromatic acid) with the elimination of water. They can be synthesized directly by using a dehydrating agent which shifts the equilibrium to the right; they can also be produced by the reaction of an alcohol with an acyl halide. Esters are used as industrial solvents, plasticizers, lacquers, soaps, and surfactants. Esters formed from some long-chain carboxylic acids occur naturally in fats, oils, and waxes. The numerous naturally occurring esters of low molecular weight are volatile and fragrant, giving distinctive odors and flavors of fruits, flowers, and other plant and animal products. Most naturally occurring, and many synthetic esters are essentially nontoxic, but some are highly toxic (e.g., allyl acetate, dimethyl phthalate, vinyl acetate). Toxicologically important esters include phthalic acid esters, phenoxy acid esters, pyrethroid insecticides and organophosphate ester insecticides. Toxic esters, because of their solvent action, are not only readily absorbed by the body, but also tend to dissolve body lipids. Many volatile esters are simple asphyxiants, irritants, or narcotics. Many esters hydrolyze readily in tissues, yielding products that are the proximate toxicants. See esterase.

ester sulfate. any ester of sulfuric acid in which one or both of the ionizable hydrogen atoms is replaced by hydrocarbon substituents, such as the methyl group. Replacement of one hydrogen yields an acid ester, while replacement of both yields a sulfate ester. **acid ester sulfate** (sulfuric acid ester). an ester sulfate in which one of the hydrogen atoms is replaced by a hydrocarbon. They are synthesized in phase II metabolic reactions releasing water-soluble products of xenobiotic compounds (e.g., phenols) that are readily eliminated from the body. Acid ester sulfates (ethylsulfuric acid and methyl sulfate) are used, for example, as industrial alkylating agents to attach alkyl groups to organic molecules. Products of such alkylations include various dyes, drugs, and agricultural chemicals. See also methylsulfuric acid, ethylsulfuric acid, sodium ethylsulfate, dimethylsulfate.

***S*-ester with *O,o*-dimethyl phosphorothioate**. malathion.

esterase. a hydrolase that catalyzes the hydrolysis of an ester into, or synthesis from, its constituent alcohol and acid. See amidase, hydrolase, neurotoxic esterase. **aryl esterase** (aryl ester hydrolase). an esterase that catalyzes the hydrolysis of aromatic esters. **aliphatic esterase** (alkyl ester hydrolase). an esterase that catalyzes the hydrolysis of alkyl esters. **alkyl ester esterase**. aliphatic esterase. **carboxylic esterase** (carboxylesterase; carboxylic ester hydrolase). an esterase that catalyzes the hydrolysis of esters of carboxylic acids.

esthesia (aesthesia). **1:** sensibility; feeling, perception, sensation; the capacity to feel or sense. **2:** any disease or condition that affects sensibility. *Cf.* anesthesia.

estimated maximum tolerated dose (EMTD). See dose.

Estinerval™. the acid sulfate of phenelzine.

estradiol-17-β (estra-1,3,5(10)-triene-3,17β-diol; 1,3,5-estratriene-3,17-β-diol). a white or yellow, odorless, crystalline substance or powder, $C_{18}H_{24}O_2$. The β-isomer of estradiol is the most potent naturally occurring ovarian and placental hormone in mammals. It is a steroid synthesized chiefly by the ovary but is also produced by the testis, adrenal gland and placenta, and to some extent by other organs such as liver, depot fat, and skeletal muscle. Estradiol is synthesized metabolically from androstenedione and testosterone. It stimulates and regulates the development of, and maintains the secondary female sex characteristics beginning at puberty (e.g., development of pubic hair; growth and development of the vagina, uterus, and fallopian tubes; breast enlargement and shape; roundness of hips; growth and maturation of the long bones). Estradiol-17-β is used in medicine as a steroid hormone and is administered by intramuscular injection or by use of subcutaneous implants. It is a confirmed carcinogen. Other effects, based on research and tests have been reported in the scientific literature. These include teratogenic activity in laboratory animals as well as reproductive and mutagenic effects in humans. See also estrogen.

1,3,5-estratriene-3,17-β-diol. estradiol-17-β.

estra-1,3,5(10)-triene-3,17-β-diol. estradiol-17-β.

estriol. an estrogen isolated in 1930 from the urine of pregnant women. It is excreted by the ovary, but is probably also produced by the placenta. See estrogen. Estrogens stimulate and regulate the development of, and maintain the secondary female sex characteristics beginning at puberty (e.g., development of pubic hair; growth and development of the vagina, uterus and fallopian tubes; breast shape and enlargement; roundness of hips; growth and maturation of the long bones).

estrogen. any of a class of female sex hormones (18-carbon steroids) that are elaborated by the ovaries, the adrenal glands, testes, the fetoplacental unit, and certain other peripheral organs and tissues (See estradiol-17-β). Estrogens promote the development and maintenance of the reproductive organs and, at the onset of puberty, they promote and regulate the development and maintenance of female secondary sexual characteristics (e.g., pubic hair; breast enlargement and shape; roundness of hips; growth and maturation of the long bones) and sexual behavior. Estrogens also play an essential part in the hormonal control of menstruation, being partly responsible (with progesterone) for the cyclical changes in the lining of the womb. Estrogens also have metabolic effects on many organs and tissues not directly related to reproduction. Estrogens appear to be only moderately toxic when a single dose is given. Even so, therapeutic doses sometimes cause headache, nausea, vomiting, and occasionally vaginal bleeding. Longer-term therapy can result in tumors of the breast, uterus, testis, kidney, bone, and other tissues. Chronic estrogen therapy may also cause infertility, hypertension, cholestasis, gall bladder disease, thromboembolic disease, and cardiovascular disease. Pregnant women who receive estrogens can give birth to daughters that develop vaginal adenocarcinoma later in life; estrogens may be teratogenic. Feminizing effects including gynecomastia may occur in males who are occupationally exposed (percutaneously or through the respiratory route) to estrogens. A number of potent non-steroidal synthetic estrogens are used therapeutically. See also contraceptive (oral contraceptive), estrone, estradiol-17-β. **conjugated estrogen**. any conjugation product of an estrogen, most importantly a glucuronide or a sulfate. In addition to the toxic effects of any estrogen, the use of conjugated estrogens in therapeutics increases the incidence of cardiovascular and thromboembolic disease. D-ring glucuronide conjugates of estradiol and estriol interrupt the flow of bile, resulting in the usual symptoms of cholestasis including jaundice.

estrogenic (estrogenous). **1:** of or pertaining to estrogen. **2:** having an action similar to that of an estrogen. estrogenic compounds include estradiol), chlordecone, coumestrol, diethylstilbestrol, genistein, and zearalinone. **3:** causing estrus.

estrogenism. estrogen poisoning; any toxicosis due to excessive exposure to an estrogen or to any estrogenic substance.

estrogenous. estrogenic.

estrone (3-hydroxyestra-1,3,5(10)-trien-17-one). a relatively potent estrogen and the first to be identified. It was isolated in crystalline form from the urine of pregnant women in 1929. Estrone is a metabolic oxidation product of estradiol-17-β. It is the active steroid in a number of therapeutic preparations, used especially in the treatment of irregularities of the menstrual cycle, menopausal vasomotor symptoms, and cancer of the prostate. It is also used to prevent pregnancy. Adverse effects include thrombophlebitis, embolism, and hypercalcemia. See also estrogen. Some individuals are hypersensitive to this hormone and its use is contraindicated for women who may be pregnant and individuals with thrombophlebitis.

estrone sulfotransferase. See sulfate conjugation.

estrous. pertaining to estrus.

estrual. pertaining to estrus.

estruation. estrus.

estrum. estrus.

estrus (estrum; estruation; oestrus; heat (in part)). a phase of the estrus cycle (*q.v.*), it is the recurrent, limited period of sexual activity in female mammals other than primates. It is characterized by intense sexual receptivity (heat). See also estrus cycle.

estrus cycle (oestrus cycle). the reproductive cycle that occurs in sexually mature, nonprimate, nonpregnant, female mammals either continuously or only during the breeding season. It is accompanied by intense sexual receptivity (See estrus). The length of the cycle varies among species from 4 to about 60 days. The cycle is under central control by neuroendocrine secretions periodically released from the hypothalamus which stimulate the secretion of pituitary gonadotrophins. The latter, in turn, promote the production of estrogen and progesterone which are the proximate regulators of the cycle as part of a negative feedback loop. The phases of the cycle are (l) **estrus**. the follicular phase, it is the period of sexual receptivity, during which the Graafian follicles mature, estrogen secretion is at its highest, and the endometrium (lining of the uterus) proliferates. Ovulation and the beginning of corpus luteum formation occur late in this phase in most species. See estrogenic. (2) **metestrus**. the postovulatory phase, it is the period when one or more corpora lutea develop and begin to secrete progesterone; estrogen secretion decreases. Mating is no longer sought nor usually allowed by the female of most species. (3) **diestrus**. the luteal phase, in which the corpus luteum is fully functional; it begins to regress in this phase. The growth of new Graafian follicles begins. If fertilization occurs during this phase, the cycle does not progress further until the gestation terminates. This phase is lengthened in some species, if the female does not become gravid following mating, and is termed pseudopregnancy. See especially fusarian estrogenism. (4) **proestrus**. the phase during which the corpus luteum fully regresses, progesterone secretion decreases, the next ovarian follicle(s) grows and the secretion of follicle-stimulating hormone and of estrogen begins to increase.

estuarine ecosystem. an ecosystem that lies at the mouth of a river that empties into a salt water body. The character of the system is largely dictated by the interaction between the flowing fresh water and the near-shore ocean waters, chiefly through the action of tides. The water in most parts is brackish. These ecosystems are rich in marine life, euryhaline forms, birds, and wildlife. They are, however, subject to and sensitive to domestic, agricultural, and industrial sewage from point and non-point sources.

estuary. a bay, river mouth, delta, brackish salt marsh, or lagoon that develops at the region of interaction between flowing fresh water from a river and near-shore marine waters.

ethanal. acetaldehyde.

ethanamide. acetamide.

ethane. a colorless, odorless, asphyxiant gas, C_2H_6. It is relatively inactive chemically, but is flammable and can be a serious fire risk if exposed to sparks or an open flame. It is used in the production of ethylene and halogenated ethanes. Ethane is also used as a refrigerant and a fuel. See alkane.

ethane carboxylic acid. propionic acid.

ethane dichloride. ethylene dichloride.

ethane dinitrile. cyanogen.

ethanedioic acid. oxalic acid.

ethanedioic acid calcium salt. calcium oxalate.

ethanedioic acid disodium salt. sodium oxalate.

1,2-ethanediol. ethylene glycol.

***N,N'*-1,2-ethanediylbis[*N*-butyl-4-morpholine-carboxamide]**. dimorpholamine.

[[1,2-ethanediylbis[carbamodithioato]]-(2—)]manganese. maneb.

[[1,2-ethanediylbis[carbamodithioato]]-(2—)]zinc. zineb.

1,2-ethanediylbiscarbamodithioic acid disodium salt. nabam.

[[*N,N'*-1,2-ethanediylbis[*N*-(carboxymethyl)-glycinato]](4—)-*N,N',O,O',O^N,O^N'*] calci-ate(2—)disodium. edetate calcium disodium.

***N,N'*-1,2-ethanediylbis[*N*-(carboxymethyl)gly-cine]**. edetic acid.

(*N,N'*-1,2-ethanediyl)bis[*N*-(carboxymethyl)gly-cine] disodium salt. edetate disodium.

***N,N'*-1,2-ethanediylbis[*N*-(carboxymethyl)gly-cine] tetrasodium salt**. edetate sodium.

***N,N'*-1,2-ethanediylbis[*N*-carboxymethyl)gly-cine] trisodium salt**. edetate trisodium.

ethanenitrile. acetonitrile.

ethanethiol. a volatile liquid (b.p. 35°C), used as an intermediate in pesticide synthesis and as an odorant in natural gas, butane, and propane to detect leaks in lines and storage tanks.

ethanoic acid. acetic acid.

ethanol (ethyl alcohol; absolute alcohol; anhydrous alcohol; dehydrated alcohol; ethyl hydrate; ethyl hydroxide; alcohol). a clear, colorless, volatile, flammable liquid, C_2H_5OH, m.p. is -114°C and b.p. 78°C. It is produced by the fermentation and fractional distillation of grain. Other than water, it is probably the most widely used chemical in industry and commerce, e.g., as a beverage ingredient, synthetic chemical, solvent, bacteriostat, germicide, antifreeze, preservative, and gasoline additive. Ethanol is a moderately toxic CNS depressant and anesthetic. Acute human intoxication by fairly small amounts of ethanol is marked by motor and cognitive impairment and disinhibition. Higher doses are associated with progressively increased blood concentration, excitement or depression, anesthetic effects, impaired perceptual acuity, incoordination, flushing, nausea and vomiting, hypothermia, stupor, coma and death. Long-continued chronic use is marked by dependence and pharmacokinetic and pharmacodynamic tolerance. Chronic abuse of alcohol affects most systems of the body. Ethanol can be absorbed by pulmonary epithelium, producing symptoms of intoxication in humans at atmospheric concentrations above 1000 ppm. Ethanol also has teratogenic effects in the form of fetal alcohol syndrome. See also alcoholism, delirium tremens, fetal alcohol syndrome, hangover, alcohol (rubbing), alcoholic beverage.

β-ethanolamine. ethanolamine.

ethanolamine (monoethanolamine; β-ethanolamine). **1:** a clear, colorless, somewhat viscous, hygroscopic liquid, C_2H_7NO, with an ammoniacal odor. It is used to remove H_2S and CO_2 from natural gas, as an emulsifier in polishes and hair wave solutions, and as a dispersing agent for a variety of agricultural chemicals.

It occurs in a number of consumer products, usually as a neutral salt of any of a number of acids. These salts are thought to lack significant toxicity. It is a strong base and reacts, for example, violently with acetic acid, acetic anhydride, acrolein, acrylic acid, acrylonitrile, cellulose, chlorosulfonic acid, epichlorohydrin, hydrochloric acid, hydrofluoric acid, and many organic materials. It is irritant and corrosive to the skin, eyes, and mucous membranes. The vapor is an irritant to the pulmonary system and can cause respiratory distress, lethargy, and mild degenerative changes in the liver and kidney. Ethanolamine is moderately toxic by ingestion, skin contact, and percutaneous, intravenous, and intramuscular routes. It is a possible human carcinogen. Low concentrations of ethanolamine occur normally in urine. See also ethylene oxide. **2:** any of a class of organic derivatives of ethanolamine (e.g., diethanolamine and its salts, triethanolamine and its salts, octopamine).

ethanoyl chloride. acetyl chloride.

ethchlorvynol (1-chloro-3-ethyl-1-penten-4-yl-3-ol; β-chlorvinyl ethylethynylcarbinol; Placidyl). a colorless to yellow liquid tertiary alcohol, $HC{=}CCOH(C_2H_5)CH{=}CHCl$, with a pungent aromatic odor that darkens on exposure to light and air. It is a CNS and respiratory depressant that has been used as a sedative-hypnotic. It is marked by rapid onset and short duration of action. Tolerance and dependence may develop with long-continued usage. Abrupt withdrawal may have serious consequences, including convulsions. See also depressant, drug abuse.

ethene. ethylene.

ethenone (carbomethane). the simplest ketene, sometimes referred to as "ketene." It is an extremely toxic, colorless, unpleasant-tasting, highly reactive, irritant gas, H_2CCO. It is soluble in acetone and ethyl ether, but decomposes in water and alcohol. It is used as an acetylating agent in organic syntheses. *Cf.* ketene.

***S*-5-ethenyl-2-oxazolidinethione**. L-5-vinylthiooxazolidone.

ether. **1:** any of a class of organic compounds with the general formula, R—O—R′, where R and R′ are hydrocarbon groups (R may be identical to R′) and O is oxygen. Most ethers are not very reactive (but See ethyl ether), although some form highly explosive peroxides when exposed to air. Ethyl ether and several of the more volatile ethers affect the CNS. **2:** ethyl ether.

ether chloratus. ethyl chloride.

ether convulsion. See convulsion.

ether cyanatus. propionitrile.

ether hydrochloric. ethyl chloride.

ether muriatic. ethyl chloride.

ethical. **1a:** pertaining to ethics. **1b:** just, moral, decent; consistently behaving in a moral way. **2:** in accord with, and in general conformity with generally accepted principles and standards of personal and/or professional moral conduct.

ethics. **1:** a major field of philosophy, it is the study of morality, moral thought, and moral conduct. **2:** precepts of morality; a body of beliefs, conduct, and duties that make up a person's moral character. **biomedical ethics**. the study of ethics applied to the resolution of problems associated with the practice of medicine, the conduct of medical research, and the application of moral thought and research to diverse questions of public health and morality. This field applies to all of the sciences that treat living systems, especially if the results of their research impinge upon the health and well-being of humans. In a broader sense, biomedical ethics is a concern of nearly all professions that deal with issues related to human health. Today, biomedical ethics deals with issues (e.g., research and applications of biotechnology; euthanasia) that require public debate and input. **normative ethics**. the field of ethics as a philosophical study that attempts to determine what is morally right and what is morally wrong with respect to human actions. **metaethics**. the field of ethics that treats the nature of moral judgments, their underlying justifications, and the elaboration and evaluation of theoretical systems of ethics. **profes-**

sional ethics. 1: behavior that is morally right and appropriate in the practice of a particular profession. 2: a body of ethical rules and standards of conduct that guide the behavior of a person who practices a particular profession toward colleagues, clients, and to varying degrees, the public.

ethidium. homidium bromide.

ethine. acetylene.

ethinyl trichloride. trichloroethylene.

ethinylestradiol (ethynylestradiol; 19-nor-17-α-pregna-1,3,5(1)-trien-20-yne-3,17-diol; 17α-ethynylestradiol). a fine, white, odorless, light-sensitive, crystalline powder, $C_{20}H_{24}O_2$, prepared from estrone. It is one of the most potent estrogenic compounds due in large measure to the presence of the 17α-ethynyl group. This group delays metabolic inactivation by blocking the action of 17β-dehydrogenase which normally plays a major role in the inactivation of estradiol-17β.

ethion (ethyl methylene phosphorodithioate). an organophosphate anticholinesterase used as a miticide and insecticide. It is less toxic than parathion, but more toxic than malathion. It is toxic by all routes of exposure; applied as a spray, it is an inhalation hazard.

ethionine (2-amino-4-(ethylthio)butyric acid). a carcinogenic and hepatotoxic amino acid, it is the *S*-ethyl analog and antimetabolite of methionine. It substitutes for methionine in transmethylation and inhibits the incorporation of methionine and glycine into proteins.

ethiops mineral. black mercuric sulfide (See mercuric sulfide).

ethohexadiol (2-ethylhexanediol-1,3; 2-ethyl-1,3-hexanediol; 2-ethyl-3-propyl-1,3-propanediol; octylene glycol). a colorless, oily, liquid, insect repellant, $C_3H_7CH(OH)CH(C_2H_5)CH_2OH$, that is no longer used in the United States. It acts as a CNS depressant if ingested and is also associated with serious birth defects in laboratory animals; lungs may fail to inflate at birth. It is moderately irritating to the eyes and mucous membranes, but not to the skin of humans.

***p*-ethoxyacetanilide**. acetophenetidin.

ethoxyethane. ethyl ether.

ethoxyethanol. a glycol ether, *q.v.*

***o*-(6-ethoxy-2-ethyl-4-pyrimidinyl)-phosphorothioic acid *O,o*-dimethyl ester**. etrimphos.

(4-ethoxyphenyl)urea. dulcin.

Ethrane™. enflurane.

ethyl acetone. 2-pentanone.

ethyl acrylate (ethyl propionate). a colorless liquid, $CH_2{:}CHCOOC_2H_5$, that is soluble in ethanol and ether. It is used as a monomer for acrylic resins, in the manufacture of water-resistant paints, paper coating, leather finishing, and as food flavoring. Ethyl acrylate is toxic by all routes of exposure and is a suspected human carcinogen. Ethyl acrylate has narcotic and irritant properties; chronic intake can damage the heart, kidney, and spleen. It is strongly irritant to the eyes, skin, and mucous membranes. Inhalation of the vapor can cause lethargy or convulsions.

ethyl apovincamin-22-oate. vinpocetine.

ethylbis(2-chloroethyl)amine. HN1.

ethyl carbamate. urethane.

ethyl carbanilate. phenylurethane.

ethyl chloride (chloroethane; monochlorethane; chlorethyl; aethylis chloridium; ether chloratus; ether hydrochloric; ether muriatic). a colorless, mobile, very volatile, flammable, slightly water-soluble gas with a characteristic ethereal odor and burning taste that liquefies at 12.2°C. It is used as a refrigerant, an alkylating agent, a solvent in the biological sciences, as an ethylating agent in organic syntheses, and as an intermediate in the manufacture of tetraethyl lead. It can also induce nonspecific sympathoadrenal release. Toxicological effects are similar to those of ethylene dichloride, but are much less severe. It is a CNS depressant with anesthetic and narcotic properties, formerly used as an anesthetic.

ethyl chlorocarbonate. ethyl chloroformate.

ethyl chloroformate (carbonochloridic acid ethyl ester; chloroformic acid ethyl ester; ethyl chlorocarbonate; ECF). a clear, white, corrosive, flammable liquid, $ClCOOC_2H_5$, with an irritating odor. It is nearly insoluble in water, and gradually decomposes in water releasing highly toxic chlorine fumes; it is miscible with benzene, chloroform, ethanol, and ethyl ether. It reacts vigorously with oxidizing agents and is a very dangerous fire hazard if exposed to heat or flames. The fumes are strongly irritant to the skin, eyes, and mucous membranes.

ethyl cyanide. propionitrile.

ethyl 1-(3-cyano-3,3-diphenylpropyl)-4-phenylisonipecotate. diphenoxylate.

ethyl1-(3-cyano-3,3-diphenylpropyl)-4-phenyl-4-piperidinecarboxylate. diphenoxylate.

***S*-ethyl *N,N*-diisobutylthiocarbamate**. butylate.

ethyl dimethyl carbinol. *tert*-pentyl alcohol.

***S*-ethyl dipropylthiocarbamate**. EPTC.

ethyl ether (1,1′-oxybisethane; ethoxyethane; ether; diethyl ether; ethyl oxide; diethyl ether; diethyl oxide; sulfuric ether; anesthetic ether; "ether"). a colorless, highly volatile, explosive, $C_2H_50C_2H_5$, with a sweet, distinctive, pungent odor; it is miscible in ethanol and soluble in water and chloroform. Ethyl ether is used as an anesthetic, solvent, extractant, reagent, and an intermediate in organic syntheses. Ethyl ether is a CNS depressant and was the general anesthetic of choice in human surgery for many years. It is mildly toxic by inhalation and moderately toxic by ingestion, and is at least moderately toxic by other routes of administration. When inhaled, it can cause drowsiness, intoxication, stupor, and loss of consciousness (with nausea and vomiting in recovering victims). Severe or long-continued exposures can cause death by respiratory failure. Ethyl ether is used in veterinary practice as an inhalation anesthetic, orally to treat colic, and is administered subcutaneously as a stimulant.

ethyl formate (formic ether). a colorless, unstable, flammable liquid, C_2H_5COOH, with a pleasant, distinctive odor. It is miscible with benzene, ethanol, and ether and is slightly water soluble, but slowly decomposes in water. It is used as a yeast and mold inhibitor; a fumigant for raisins and currants; a fungicide for cashew nuts, cereals, tobacco, and dried fruits; and as a food flavoring. Even a brief exposure to the vapor may irritate the skin and eyes and produce narcotic effects. Exposures to high concentrations can be fatal due to severe CNS depression.

ethyl gas. See tetraethyl lead.

ethyl hydrate. ethanol.

ethyl hydroxide. ethanol.

5-ethyl-5-(1-methyl-1-butenyl)barbituric acid sodium. vinbarbital sodium.

4-ethyl-4-methyl-2,6-dioxopiperidine. bemegride.

***N*-ethyl-*N*′-(1-methylethyl)-6-(methylthio)-1,3,5-triazine-2,4-diamine**. ametryn.

3-ethyl-3-methylglutarimide. bemegride.

ethyl methyl ketone. methyl ethyl ketone.

4-ethyl-4-methyl-2,6-piperidinedione. bemegride.

N-[3-(1-ethyl-1-methylpropyl)-5-isoxazolyl]-2,6-dimethoxybenzamide. isoxaben.

ethyl methylene phosphorodithioate. ethion.

ethyl nitrite a colorless to yellowish, volatile, highly flammable, explosive liquid, C_2HsNo_2, with an ether-like odor; it is soluble in ethanol and ether and decomposes in water. It is used in organic reactions and as a synthetic flavoring in beverages, ice cream, ices, candy, baked goods, and chewing gum. It can cause methemoglobinemia and hypotension.

ethyl *p*-nitrophenyl benzenethiophosphonate. EPN.

***o*-ethyl-*o,p*-nitrophenyl phenylphosphorothioate**. EPN.

***N*-ethyl-*N*-nitrosoethanamine**. *N*-nitroso-diethylamine.

ethylnitrosourea. *N*-ethyl-*N*-nitrosourea.

_N_-ethyl-_N_-nitrosourea (ENU; 1-ethyl-1-nitrosourea; ethylnitrosourea). a moderately toxic, pale yellow, crystalline compound, $C_3H_7N_3O_2$, it is an alkylating agent and confirmed human tumorigen and carcinogen. It is toxic by ingestion and by subcutaneous, intravenous, and intraperitoneal routes.

13a-ethyl-2,3,5,6,12,13,13a,13b-octahydro-12-hydroxy-1*H*-indolo[3,2,1-de]pyrido[3,2,1-ij]-[1,5]naphthyridine-12-carboxylicacidmethyl ester. vincamine.

ethyl oxide. ethyl ether.

ethyl *p*-aminophenyl ketone. *p*-aminopropiophenone.

ethyl phenylcarbamate. phenylurethane.

ethyl phthalate (diethyl phthalate). the ethyl ester of phthalic acid, it is a colorless, nearly odorless, water-insoluble, bitter-tasting, oily liquid. It is used in the manufacture of celluloid, as a solvent in the manufacture of varnishes and dopes, and as a denaturing agent for ethanol. It is irritating to mucous membranes and is narcotic at high concentrations. See also phthalic acid, phthalic acid ester.

ethyl propionate. ethyl acrylate.

ethyl pyrophosphate. TEPP.

ethyl sulfate. diethyl sulfate.

ethyl sulfocyanate. ethyl thiocyanate.

ethylsulfuric acid (acid ethyl sulfate). a colorless, flammable, oily, toxic, strongly irritant liquid, $C_2H_2HSO_4$, that is soluble in water, ethanol, and ether. *Cf.* methylsulfuric acid. It is toxic by inhalation and ingestion and is a strong irritant to the eyes, skin, and mucous membranes.

ethyl thiocyanate (ethyl sulfocyanate). the ethyl ester of thiocyanic acid. It is very toxic to laboratory mice.

2-(ethylthio)-ethanethiol *S*-ester + *O,O*-diethyl phosphorothioate. demeton-*S*.

16-ethyl-1,16,19-trimethoxy-4-(methoxymethyl)-aconitane-3,8,10,11,18-pentol-8-acetate 10-benzoate. aconitine.

(1α,3α,6α,14α,16β)-20-ethyl-1,6,16-trimethoxy-4-(methoxymethyl)aconitane-3,8,13,14-tetrol 8-acetate 14-(3,4-dimethoxybenzoate). pseudoaconitine.

(1α,3α,6α,14α,15α,16β)-20-ethyl-1,6,16-trimethoxy-4-(methoxymethyl)aconitane-3,8,13,14,15-pentol 8-acetate 14-benzoate. aconitine.

ethyl urethan. urethane.

ethyl urethane. urethane.

ethyl phenylurethane. phenylurethane.

ethylaldehyde. acetaldehyde.

ethylamine (monoethylamine; aminoethane). a colorless, volatile, flammable, strongly alkaline liquid or gas, $CH_3CH_2NH_2$, with the odor of ammonia. It is a ptomaine produced in decaying plant tissue and has many of the properties of ammonia. It is a strong irritant.

2-ethylamino-4-isopropylamino-6-methylmercapto-s-triazine. ametryn.

ethylbenzene. a colorless, flammable liquid, C_8H_{10}. It is an irritant to the eyes and mucous membranes; is narcotic in high concentrations; and has immunosuppressive activity. Ethylbenzene is only slightly toxic to laboratory rats. Possible sources of exposure include gasoline.

ethylene (ethene; acetene; elayl; olefiant gas). a colorless, flammable gas, $H_2C{\equiv}CH_2$, with a sweetish odor and taste. It is soluble in water, ethanol, and ethyl ether, but poorly soluble in blood. It is the most widely used organic chemical, predominantly in the manufacture of polyethylene and other organic chemicals. It is also used as an inhalation anesthetic and as an agricultural chemical. Ethylene is a simple asphyxiant and, at high concentrations, has an anesthetic effect. It is a dangerous fire hazard and moderate explosion hazard. Although ethylene is an important plant metabolite, it is phytotoxic. It can retard growth of some plant organs, e.g., pea epicotyls. Ethylene is not very toxic to animals but it is a simple asphyxiant. At high concentrations it also acts as an anesthetic to induce unconsciousness.

ethylenebisaminodiacetic acid. edetic acid.

ethylenebis(dithiocarbamate) (EBDC). any of a class of dithiocarbamate fungicides (e.g., maneb, zineb) used on many fruits and vegetables. They are suspected carcinogens.

[ethylenebis(dithiocarbamato)]manganese. maneb.

[ethylenebis(dithiocarbamato)]zinc. zineb.

ethylenebis[dithiocarbamic acid] disodium salt. nabam.

ethylenebis[dithiocarbamic acid] manganous salt. maneb.

ethylenebis(dithiocarbamic acid) zinc salt. zineb.

ethylenebisiminodiacetic acid. edetic acid.

ethylenebis(iminodiacetic acid) disodium salt. edetate disodium.

ethylene bromide. ethylene dibromide.

ethylene chloride. ethylene dichloride.

ethylene chlorohydrin (2-chloroethanol; 2-chloroethyl alcohol; glycol chlorohydrin). a colorless, highly toxic liquid, $ClCH_2CH_2OH$, with a faint ethereal odor, boiling at 129°C, and miscible with water and ethanol. It is used as a solvent and in organic synthesis, including the synthesis of insecticides. It is very toxic orally to laboratory rats. It is toxic by ingestion, skin contact, and other routes. The nervous system, spleen, liver, and lungs may be affected. Inhalation of the vapor can be fatal. It is also a mild skin and severe eye irritant and is probably mutagenic. Ethylene chlorohydrin may react violently with oxidizing agents, and reacts violently with water and steam to yield toxic and corrosive fumes. See also chlorohydrin, epichlorohydrin.

ethylene dibromide (EDB; 1,2-dibromoethane; ethylene bromide). an extremely toxic, colorless, nonflammable liquid, $BrCH_2CH_2Br$, with a sweetish odor that is miscible with most solvents and thinners. It is an agricultural fumigant and industrial chemical with various uses. It is extremely toxic, is carcinogenic at high doses, to laboratory animals, a suspected human carcinogen, and a probable mutagen and teratogen. EDB was banned for most agricultural uses in the United States in 1984.

ethylene dichloride (1,2-dichloroethane; α,β-dichloroethane; *sym*-dichloroethane; dichloroethylene; ethane dichloride; ethylene chloride; 1.2-ethylene dichloride; 1,2-DCE; EDC; Dutch liquid; Dutch oil; glycol dichloride). a toxic, carcinogenic, heavy, clear, colorless, rather viscous, volatile, sweet-tasting, flammable, and moderately explosive liquid, $ClCH_2CH_2Cl$, with a pleasant, chloroform-like odor. It is used as a fumigant of soils and foodstuffs, as an industrial solvent (e.g., of fats. oils, resins, rubber), and in the production of several organic chemicals. Ethylene dichloride is a CNS depressant with strong narcotic effects and may cause liver and kidney damage. It is a strong irritant, producing a contact dermatitis and sometimes severe damage to the eye. The fumes are an irritant to the respiratory tract and conjunctiva of the eye and may cause corneal clouding. It is moderately toxic to laboratory rats when ingested and also to humans via inhalation, skin contact, and the intraperitoneal route. It is carcinogenic to humans and a transplacental carcinogen in laboratory mammals. Teratogenic and reproductive effects have been demonstrated in laboratory animals. Chlorine and phosgene are released on decomposition by heating.

1,2-ethylene dichloride. ethylene dichloride.

ethylene dichloride poisoning. poisoning due to inhalation or ingestion of ethylene dichloride. Systemic effects in humans via inhalation include neuromuscular blockade with flaccid paralysis, coughing, nausea, vomiting, hypermotility, diarrhea, somnolence, jaundice, bradycardia, cyanosis, and coma. Ingestion may produce any of a variety of toxic effects, including adverse effects on the eye, liver, kidneys, and CNS. The CNS effects are mostly narcotic and are marked by an initial stimulation followed by depression. High doses can cause fatal kidney failure due to the metabolic oxidation of ethylene glycol to oxalic acid with subsequent formation of insoluble calcium oxalate which obstructs the kidneys. Lesions may include ulceration with bleeding from the stomach, cirrhosis of the liver, and sometimes pulmonary edema, severe corneal injury, and impaired kidney function.

ethylene glycol (1,2-ethanediol). a colorless, relatively viscous, nephrotoxic, oxalosis-inducing liquid, $HOCH_2CH_20H$, it is the main component of liquid antifreeze/antiboil formulations used in automotive cooling systems. It is also used in hydraulic fluids, condensers and heat exchangers, cosmetics, and in organic syntheses. It is moderately toxic to humans, although exposure is rather rare because of the low volatility. The lethal oral dose in humans is about 1.4 ml/kg body weight. Ingestion produces transient excitement followed by CNS depression in humans. Symptoms may include nausea, vomiting, abdominal cramps, muscle cramps, acidemia, drowsiness, ataxia, vertigo, stupor, coma, convulsions, and death from respiratory collapse or delayed renal failure with degeneration of the renal tubular epithelium and uremia. Large doses can cause kidney failure with anuria, uremia, and death due to the deposition of calcium oxalate in the tubules. Domestic cats have a voracious appetite for ethylene glycol and often experience kidney failure when they have access to antifreeze containing this compound.

ethylene glycol poisoning. poisoning due to ingestion of either of the automobile antifreeze preparations, ethylene glycol or diethylene glycol. Symptoms of mild or moderate poisoning are similar to that of alcoholic intoxication. The victim may also vomit and may experience carpodacal spasm, lumbar pain, respiratory distress, renal failure, convulsions, and coma.

ethylene lactic acid. β-hydroxypropionic acid.

ethylene monochloride. vinyl chloride.

ethylene oxide (ETO; epoxyethane; 1,2-epoxyethane; oxirane). a three-membered ring compound, it is the simplest epoxide, $O(CH_2)_2$. It is a colorless, sweet-smelling, moderately toxic, flammable gas at room temperature, liquefying at about 12^oC; it is miscible with water and soluble in organic solvents. The fumes form explosive mixtures in air at about -5° C. It is produced by oxidation of ethene over a silver catalyst. The ring is strained and the compound is thus highly reactive. ETO is used as a chemical intermediate (e.g., in the production of surfactants, ethylene glycol and higher glycols, acrylonitrile, ethanolamines), as a petroleum demulsifier, an industrial sterilant, fumigant, a fungicide, a rocket propellant, and to lower the viscosity of water (as in fire-fighting). It is moderately to very toxic and is highly irritant to the skin, eyes, and mucous membranes. Inhalation of fumes may cause drowsiness, headache, and dyspnea; in more severe cases, cyanosis, pulmonary edema, renal damage, peripheral nerve damage, and death may ensue. This compound is carcinogenic to humans and to some laboratory animals, causing leukemia and other cancers; it is also mutagenic. *Cf.* epoxide.

ethylene tetrachloride. perchloroethylene.

ethylene thiourea (ethylene-2-thiourea; 2-imidazolidinethione; ETU). a white to pale green crystalline compound that is slightly soluble in ethanol, methanol, acetic acid, and naphtha at room temperature; slightly soluble in cold water; and very soluble in hot water. It is a by-product of the manufacture of ethylenebisdithiocarbamates (EBDC) and is the major metabolite of EBDCs in mammals. ETU is used, for example, as an accelerator in the rubber industry, an intermediate in the manufacture of fungicides, insecticide, pharmaceuticals, dyes, and synthetic resins. It has also been used as a rodenticide. It is carcinogenic to some laboratory animals. ETU is mutagenic, carcinogenic, teratogenic, and nephrotoxic in laboratory rodents. See also dithiocarbamate fungicides. It is also thyrotoxic (goitrogenic), causing thyroid hyperplasia and altered thyroid hormone in laboratory rodents. ETU is thyrotoxic, nephrotoxic, and a probable carcinogen in humans. It is toxic by respiratory and oral routes of exposure and by skin contact.

ethylene-2-thiourea. See ethylene thiourea.

ethylenediamine (1,2-diaminoethane; EDA). a colorless, alkaline liquid, $NH_2CH_2CH_2NH_2$, with an ammoniacal odor, that is soluble in water and ethanol, slightly soluble in ether, and insoluble in benzene. It is widely used in the manufacture of chelating agents such as EDTA, a solvent, an intermediate in organic syntheses, fungicide, an emulsifying agent, an antifreeze inhibitor, and textile lubricant. EDT is toxic by inhalation and cutaneous absorption. Although considered only moderately toxic, EDA is a strong irritant, causing dermal sensitization and an exfoliative dermatitis in humans; it can be very damaging to the eyes. See also alkyl polyamines.

ethylenediamine tartrate. edetic acid.

ethylenediaminetetraacetate. edetic acid.

ethylenediaminetetraacetic acid. edetic acid.

ethylenediamine-*N,N,N′,N′*-tetraacetic acid. edetic acid.

ethylenediaminetetraacetic acid calcium disodium chelate. edetate calcium disodium.

ethylenediaminetetraacetic acid disodium salt. edetate disodium.

ethylenediaminetetraacetic acid trisodium salt. edetate trisodium.

(ethylenedinitrilo)tetraacetic acid. edetic acid.

(ethylenedinitrilo)tetraacetic acid disodium salt. edetate disodium.

(ethylenedinitrilo)tetraacetic acid tetrasodium salt. edetate sodium.

1,1′-ethylene-2,2′-dipyridylium dibromide. diquat.

ethyleneimine (aziridine; ethylenimine). a clear, colorless liquid, $(CH_2)_2NH$, with an amine odor, that is miscible with water and most organic solvents. It is a dangerous fire and explosion hazard. It is a monomer used for the polymerization of polyethylenimine; a flocculant used in wastewater treatment; and an additive to textile and paper to increase wet strength. Ethyleneimine is extremely toxic, and is corrosive to, and readily absorbed by the skin. It is cytotoxic (alkylating DNA) and, in humans and laboratory animals, it is cardiotoxic, hepatotoxic, and nephrotoxic. It is carcinogenic to some laboratory animals.

ethylenimine. ethyleneimine.

ethylformic acid. propionic acid.

2-ethyl-1,2,3,4,6,11-hexahydro-2,5,7-trihydroxy-6,11-dioxo-4-[[2,3,6-trideoxy-4-*O*-[2,6-dideoxy-4-*O*-[(2R-*trans*)-tetrahydro-6-methyl-5-oxo-2*H*-pyran-2-yl]-α-L-lyxo-hexopyranosyl]-3-(dimethylamino)-α-L-lyxo-hexopyranosyl]oxy]-1-naphthacenecarboxylic acid methyl ester. aclacinomycin A.

2-ethylhexanediol-1,3. ethohexadiol.

2-ethyl-1,3-hexanediol. ethohexadiol.

2-ethyl-1-hexanol (2-ethylhexyl alcohol). a colorless, toxic liquid used as an industrial solvent and in the mercerization of textiles.

2-ethylhydrazide. podophyllinic acid.

***N*-ethyl-3-hydroxy-*N,N*-dimethylbenzenaminium chloride**. edrophonium chloride.

ethyl(*m*-hydroxyphenyl)dimethylammonium chloride. edrophonium chloride.

ethylidenediethyl ether. acetal.

ethylidine gyromitrin. gyromitrin.

α-ethyl-1*H*-indole-3-ethanamine acetate. etryptamine acetate.

5-ethyl-5-isoamylbarbituric acid. amobarbital.

ethylisocyanate. a liquid, C_2H_5NCO, that is soluble in chlorinated and aromatic hydrocarbons. It is a chemical warfare agent that is powerfully irritant to skin, eyes, and mucous membranes of the respiratory tract. It is also used as an intermediate in the production of pesticides. When decomposed by heating, it releases sulfur oxides and hydrogen cyanide.

5-ethyl-5-isopentylbarbituric acid. amobarbital.

ethylmethyl carbinol. 2-butanol.

ethylphenol (*m*-ethylphenol; 3-ethylphenol). a colorless, toxic, combustible, very slightly water-soluble liquid, $HOC_6H_4C_2H_5$. Possible sources of exposure includes drinking water. See also phenol.

3-ethylphenol. See ethylphenol.

***m*-ethylphenol**. See ethylphenol.

5-ethyl-5-phenylbarbituric acid. phenobarbital.

2-ethyl-2-phenylglutarimide. glutethimide.

5-ethyl-5-phenyl-2,4,6(1*H*,3*H*,5*H*)pyrimidinetrione. phenobarbital.

ethylphenylurethane. phenylurethane.

ethylphosphonodithioic acid *O*-ethyl *S*-phenyl ester. fonofos.

1-ethyl-1-propanol. 3-pentanol.

3-(1-ethylpropyl)phenol methylcarbamate mixture with 3-(1-methylbutyl)phenyl methylcarbamate (1:3). bufencarb.

2-ethyl-3-propyl-1,3-propanediol. ethohexadiol.

***S*-[2-(ethylsulfinyl)ethyl]-*O,O*-dimethyl phosphorothioate**. demeton-*O*-methyl sulfoxide.

α-ethyltryptamine acetate. etryptamine acetate.

ethylxanthic disulfide. dixanthogen.

ethyne. acetylene.

ethynylestradiol. ethinylestradiol.

17α-ethynylestradiol. ethinylestradiol.

etiologic, etiological (aetiological). causal; pertaining to etiology.

etiologic agent, etiological agent (aetiologic agent). a causal agent; a chemical or pathogen that causes a specific disease or injury to a living organism.

etiology (aetiology). **1a:** causation. **1b:** the demonstrated causes of a trait, condition, or disease state. **1c:** the origins and course of a disease in terms of pathogens, toxicants, or other contributing causes that might be involved. **1d:** the body of knowledge concerning the causes of disease. **2a:** the study of causality. **2b:** the science of the origins and causes of disease and their mode of operation. *Cf.* pathogenesis.

ETO. ethylene oxide.

etrimfos (*o*-(6-ethoxy-2-ethyl-4-pyrimidinyl)-phosphorothioic acid *O,o*-dimethyl ester; *O,o*-dimethyl *o*-(2-ethyl-4-ethoxypyrimidinyl)-6-thionophosphate). a contact and stomach organophosphorus insecticide used against pests of fruit and vegetable crops. It is moderately to very toxic orally to laboratory rats and mice and is also toxic to fish.

etryptamine acetate (3-(2-aminobutyl) indole acetate; α-ethyl-1*H*-indole-3-ethanamine acetate; α-ethyltryptamine acetate). a substance, $C_{14}H_{20}N_2O_2$, formerly used as a CNS stimulant, but no longer marketed because of its toxicity.

ETU. See ethylene thiourea.

Eu. the symbol for europium.

Eucalyptus (gum trees, in part). a genus of rapidly growing, tropical and temperate zone, broadleaved evergreen trees (Family Myrtaceae). The flowers are attractive and gray-green or bluish in color; they are a valuable source of honey. The leaves of many, but not all species are acutely toxic to wildlife and tend to be avoided. The koala feeds obligately on eucalyptus leaves; most individuals eventually develop blood dyscrasias that are probably due to the toxicity of the diet. The entire genus tends to be free of insects due to the repellent and possibly toxic nature of the strongly aromatic oil present in the wood (See eucalyptus oil).

eucalyptus oil. an oil used as an expectorant, an intestinal dewormer, and as a local antiseptic. It is also used as a flavoring in beverages, liquors, various sweets, and baked goods. Ingestion of as little as 3 ml by some individuals can cause a burning sensation, with nausea, weakness, and delirium. Relatively small amounts have, on occasion, resulted in coma and death.

eucapnia. the condition in which the carbon dioxide tension of circulating blood is normal.

eucaryote. eukaryote.

eukaryote (eucaryote). nucleated unicellular organisms and all multicellular organisms. Eukaryotic cells are characteristic of most organisms except the bacteria. *Cf.* prokaryote.

eukaryote mutagenicity test. See test.

eukaryotic. pertaining to eukaryotes.

eukaryotic cell. that which is characteristic of eukaryotes in which the genetic material is contained within a nucleus that is separated from the cytoplasm by the membranes of the nuclear envelope. Eukaryote cells are generally much larger than prokaryote cells and are considered more complex and exhibit greater diversity.

eumycetes. the true fungi.

Eumydrin™. methylatropine nitrate.

Eunice (palolo worms). a genus of marine polychaete worms (Family Eunicidae) with hard chitinous biting jaws. ***E. aphroditois*** (biting reef worm). a large (up to 1½ m long) venomous annelid worm with hard biting chitinous jaws that are capable of producing a painful bite. The wound, only a few mm in diameter, rapidly becomes hot, edematous, inflamed, and can become infected. ***E. gigantea***. a large species of the Pacific coast of North America. that grows to a length of at least 3 m. ***E. viridis*** (Pacific palolo worm). a species indigenous to Samoa and Fiji.

eunicid. pertaining to or denoting any polychaete worm of the family Eunicidae.

Euonymus (spindle trees). a genus of deciduous or evergreen shrubs, small trees, climbing vines, and a few prostrate species (Family Celastraceae, the stafftree family). ***E. atropurpurea*** (burning bush; wahoo). a large bush or small tree, 2-8 m tall, found in both Europe and North America. It grows to about 7.6 m in height, has small purple flowers and purple fruit. All plant parts have laxative properties, thus the common names. The leaves, seeds, and bark contain **evomonoside,** a very toxic cardiac glycoside. Ingestion of any of these plant parts can cause symptoms similar to meningitis. The victim may experience a watery, blood-tinged diarrhea, colic, nausea, vomiting, and convulsions. The major lesion is hepatomegaly. Death may intervene in 8 to 10 hours. Ingestion of the purple fruit may produce nausea, vomiting, diarrhea, weakness, chills, and convulsions. Unconsciousness or coma may follow within 12 hours, but fatalities are rare or nonexistent. ***E. europaeus*** (spindle tree). a European species with toxicity and effects that are similar to those of *E. atropurpurea*. The color of the fruit varies widely among various varieties of this plant.

eupatorin. See *Eupatorium*.

Eupatorium (snakeroot; thoroughwort; eupatorin). a large genus of weedy, many-flowered, usually perennial herbs or shrubs (Family Asteraceae, formerly Compositae). They occur mainly in tropical and temperate America, with a few species in Eurasia and Africa. The literature regarding toxicity is confusing, in part because toxicities vary with time and location and some species may often be mistaken for white snakeroot (*E. rugòsum*). The latter species is clearly and seriously toxic to both humans and livestock. Toxic concentrations of nitrate have, however, been reported for several species under conditions that favor rapid growth. ***E. perfoliatum*** (boneset; thoroughwort; formerly throughwort). Toxic concentrations of nitrate have been reported. ***E. purpureum*** (Joe Pye weed; sweet Joe Pye weed; green-stemmed Joe Pye weed; thoroughwort). Toxic concentrations of nitrate have been reported. Certain other species such as *E. maculatum*, *E. fistulosum*, and *E. verticillatum* are also called Joe Pye weed. ***E. rugosum*** (= *E. ageratoides*, *E. urticaefolium*)(snakeroot; white snakeroot; white sanicle; fall poison; rich-weed). a showy perennial white-flowered bush, usually 1 m or so tall, that commonly occurs in low moist areas from eastern Canada to Saskatchewan, south to eastern Texas, Louisiana, Georgia, and Virginia. The leaves are ovate or heart-shaped and toothed, and are borne oppositely on long stalks. The numerous tiny bright white flowers form clusters at the tips of the branches. It is a widespread and very variable weedy species. It is difficult to distinguish this species from closely related species. All parts of the plant are poisonous. The poisonous principle is tremetol. The syndrome produced in cattle is usually called trembles, and in humans, milksickness. From the late 1700s to the mid 1900s, this plant was responsible for a condition marked by weakness, tremors, nausea, severe vomition, anorexia, icterus, convulsions, and prostration. The most pronounced lesion is fatty degeneration of the liver. In fatal cases, delirium and coma precede death. The odor of acetone may be noted on the breath of a victim. Kidney damage sometimes results in oliguria or anuria.

The disease was devastating in some areas, decimating local populations. Milk sickness still occurs occasionally from drinking milk from an exposed family cow. Other animals are susceptible to intoxication, and suckling young may be poisoned by tremetol in the mother's milk. Symptoms in cattle and other livestock include weakness, unwillingness to move, stupor, muscular tremors (trembling), and prostration often followed by temporary recovery. In severe cases, the animal ultimately remains prone, becomes comatose, and dies. If made to walk, it abruptly stands still with feet wide apart and trembles. See milk sickness, trembles, tremetol.

Euphorbia (spurges). a large, structurally diverse genus of shrubs and herbs (Family Euphorbiaceae) that are distributed widely in tropical and temperate regions. Most species are poisonous and/or of medicinal value. Some are used for fish and arrow poisons. *E. systyloides*, for example, is used as a remedy for hookworms, but overdosage can cause delirium, convulsions, and death within a few hours. All parts are toxic and most species have an irritant, milky sap. Toxicity varies; some species are violently toxic, others are not known to cause severe poisoning. Livestock do not readily eat spurges. The young shoots of some species may be consumed, however, when other forage is unavailable. Some species may be photosensitizing and concentrations of nitrate that are toxic to livestock have been reported. Symptoms of poisoning may include scours, collapse, and death. In less acute cases, symptoms may be limited to less severe effects such as hair loss, irritation and edema of the mouth and tongue, and symptoms of photosensitization such as dermatitis or edematous enlargement of the head. ***E. ipecacuanha*** (ipecac; ipecac spurge; America white ipecac; wild ipecac; Carolina ipecac). a spurge (Family Euphorbiaceae) with emetic properties that is indigenous to the eastern United States. ***E. lactea*** (African milk plant; spotted spurge). a spurge with a fast growing candelabrum form. The sap is toxic and is released when the spine is broken. It is reportedly used by women of certain tribes to rid themselves of evil husbands. This plant is widely cultivated and is often used for hedges in warm climates. Additional species of toxicological interest include the following: *E. candelabrum, E. grantii, E. neglecta, E. systyloides,* and *E. tirucalli* found throughout Africa. Some are used medicinally, others for fish and arrow poisons. *E. candelabrum* is used in making arrow poisons, as is *E. neglecta*. *E. systyloides* is used as a remedy for hookworms, but overdosage can cause delirium, convulsions, and death within six hours. The following North American species are among those that have been implicated in the poisoning of livestock: *E. corollata* (flowering spurge; tramp's spurge; wild hippo); *E. cyparissias* (cypress spurge; graveyard weed); *E. esula* (leafy spurge; wolf'smilk); *E. helioscopia* (sun spurge; wart-weed). *E. maculata* (milk purslane). *E. marginata* (snow-on-the-mountain). *E. prostrata*. Human intoxications, sometimes severe, have also been traced to various species of *Euphorbia* such as *E. lathyris* (caper spurge; mole plant), *E. milli* (crown-of-thorns), *E. peplus* (petty splurge), and *E. pulcherrima* (poinsettia). A number of euphorbia species are ornamentals (e.g. poinsettia, crown-of-thorns, spotted spurge) that may be attractive to children and should thus be considered dangerous.

Euphorbiaceae (the spurge family). a family of dicots that includes poisonous species such as *Hippomane mancinella*, the manchineel tree, *q.v.* The sap of most species contains cucurbitacin, *q.v.* See also *Euphorbia*.

euphoretic (euphoriant). **1:** of, pertaining to, characterized by, or producing a state of euphoria. **2:** an agent that induces euphoria.

euphoria. **1:** a feeling of well-being or of good health with no sense of pain or distress. **2:** in psychiatry, an exaggerated feeling of physical and emotional well-being that is not consistent with the individual's condition, circumstances, events, or normally encountered stimuli that might be expected to elicit such a state. It is especially common to the manic state. Euphoria occurs also in organic brain disease and can be induced by certain toxicants, including various narcotics and opioids; examples are cannabis, cocaine, gasoline, manganese, methylenedioxy amphetamine, morphine, promethazine, and scopolamine.

euphoriant. euphoretic.

euphoric. characterized by, or in a state of euphoria.

euphorigenic. **1:** tending to produce euphoria. **2:** a substance that induces euphoria.

Euphrasia officinalis (eyebright). an herb (Family Scrophulariaceae) of cool, temperate climates, with opposite toothed leaves, and small spiked flowers. Symptoms of intoxication in humans include confusion, cephalgia, lacrimation and violent pressure in the eyes, itch, redness, edematous eyelids, photophobia, dim vision, esthesia, dyspnea, coughing and expectoration, insomnia, yawning, polyuria, and diaphoresis.

eupnea. normal, unlabored breathing.

Euproctis chrysorrhoea (brown-tail moth). a pest species of moth of the northeastern United States. The caterpillar feeds on tree leaves. See dermatitis (brown-tail moth dermatitis).

eurokous. euryoecious.

euroky (euryoky). the state or condition of being euryoecious, whereby a species or population of organisms tolerate a relatively broad range of ecological conditions. See also euryoecious, stenoky.

European adder. *Vipera berus*.

European asp. *Vipera aspis*.

European aspidium. *Dryopteris felix-mas* (See aspidium).

European beech. *Fagus sylvatica*.

European bittersweet. *Solanum dulcamara*.

European cherry laurel. *Prunus laurocerasus*.

European false hellebore. *Veratrum album*.

European midwife toad. *Alytes obstetricans*.

European mistletoe. *Viscum album*.

European privet. *Ligustrum vulgare*.

European ratfish. See *Chimaera monstrosa*.

European sea urchin. *Paracentrotus lividus*.

European snakeroot. *Asarum europaeum*.

European viper. *Vipera berus*.

European white hellebore. *Veratrum album*.

europic nitrate. the hexahydrate of europic nitrate, $Eu(NO_3)_3$, is very toxic to laboratory rats, *i.p.*, but is at most slightly toxic when administered per os.

europium (Eu). an element of the lanthanide (rare earth) series with two naturally occurring isotopes, ^{151}Eu and ^{153}Eu, both of which are neutron absorbers. There are several artificial isotopes: 143-150, 152, and 154-160. [$Z = 63$; $A_r = 151.96$]. The oxide is used widely in phosphors for television monitors. See also europous chloride, europic nitrate.

europous chloride. a white amorphous powder, $EuCl_2$, that is moderately toxic to laboratory mice, i. p., but is only slightly toxic orally.

eurybiontic. pertaining to or denoting an organism that tolerates a wide range of a specific environmental factor. *Cf.* stenobiontic.

euryhaline. the ability of an individual aquatic organism, population, or species to tolerate a wide range of dissolved salts in the environment. Salmon are typically euryhaline as are a few polychaetes, and certain sharks. *Cf.* stenohaline.

euryhydric. tolerant of a wide range of moisture levels or humidity; See also stenohydric.

euryionic. having or tolerating a wide range of pH. *Cf.* stenionic.

euryoecious (euryoekous; euryokous; eurokous). of or pertaining to a species or population that is tolerant of a wide range of habitats and environmental conditions. *Cf.* amphioecious, stenoecious; one characterized by euroky. See also amphioecious, stenoecious.

euryoekous. euryoecious.

euryokous. euryoecious.

euryoky. euroky.

Eurypelma hentzii (formerly *Aphonopelma*; American tarantula; Arkansas tarantula). a genus of large spiders (Family Theraphosidae). The venom, injected into the groin of a laboratory rat, is extremely toxic and may bring death within an hour. Lesions include gastric distension and damage to the liver, kidneys, and lungs. They are dangerous to humans, though they seldom bite. See also tarantula.

euryphagous. consuming a wide variety of foods; of or pertaining to a heterotrophic organism that uses a wide variety of foods.

eurysubstratic. tolerant of a wide range of types of ecological substrata. *Cf.* stenosubstratic.

Eurythoe (bristleworms). a genus of toxic marine polychaete worms (Phylum Annelida) with stinging bristles. The stings cause local swelling, inflammation, and numbness. The bristles may be removed with adhesive tape. *Cf. Hermodice*. ***E. complanata*** (orange fireworm). a highly toxic species that occurs on and about coral reefs and flats in Florida, the West Indies, and Gulf of Mexico. Contact with the setae may cause inflammation, edema, intense itching and numbness that persists for several weeks.

eurytropic. said of organisms that respond readily or adapt substantially to changing environmental conditions. *Cf.* stenotropic.

eurytropism. the state or condition whereby an organism responds markedly or adapts to changing environmental conditions. *Cf.* stenotropism.

euthanasia (mercy killing). **1:** literally an "easy death," one that is free from suffering. **2:** the means of providing an easy death. **3:** the practice or deliberate act of painlessly terminating the life of a suffering and incurable human or a wild or domestic animal; usually of one who is suffering from intractable pain.

Euthynnus pelamis (skipjack tuna, oceanic bonito). a lively, vigorous, circumtropical and subtropical bony fish (Family Scombridae). It is dark blue above, silvery below, with four to six dark longitudinal bands on a scaleless area below the lateral line. See scombroid poisoning, ciguatera.

eutocia. normal or uneventful parturition. *Cf.* dystocia.

eutrophic. **1:** relating to the process of eutrophication. **2:** relating to a large water body that has undergone substantial eutrophication. A eutrophic water body is rich in nutrients and has a high rate of primary production reflected in well-developed littoral and sublittoral vegetation and dense growth of plankton. There are few fish and oxygen is depleted during periods of warmth and high insolation. In areas largely free from human influence, eutrophic water bodies are rather old and shallow with well developed vegetation. *Cf.* oligotrophic.

eutrophication (eutrophy). the transition (under otherwise stable conditions) of a body of surface water from an early freshwater stage through intermediate stages of nutrient enrichment to a bog, marsh, or swamp and ultimately a terrestrial system. **natural eutrophication**. the slow, natural, aging process whereby a lake, large river, estuary, or bay evolves into a bog or marsh and eventually disappears. During the later stages of eutrophication the water body is oxygen poor and choked by abundant plant life as the result of increased amounts of nutritive compounds such as nitrogen and phosphorus. Natural eutrophication occurs slowly as a consequence of the accumulation of silt. **anthropogenic eutrophication**. enrichment of large water bodies due to human agency, greatly accelerating the process of eutrophication. The most common proximate causes are due to extravagant use of nitrate fertilizers, the use of detergents that contain phosphorus, the dumping of organic wastes, and runoff containing animal excrement. Eutrophic bodies of water are characterized by large populations of flora and fauna, often leading oxygen depletion.

eutrophy. eutrophication.

evanescent brief, ephemeral, short-lived, transitory.

evaporate. vaporize, volatilize; to undergo or cause to undergo evaporation.

evaporation. vaporization, volatilization. **1:** the change of a substance from a liquid to a gaseous or vapor phase. **2:** the reduction in mass or volume of a liquid by conversion into vapor.

evening trumpet flower. *Gelsemium sempervirens*.

evergreen. a term designating those vascular plants that produce and shed leaves throughout the year, such that the branches always bear leaves. An individual leaf may persist for several years. This designation bears no relationship to taxonomic status.

everlasting pea. *Lathyrus latifolius*.

Evipal sodium. hexobarbital sodium.

evocation. See induction.

evonogenin. digitoxigenin.

ewe (yew). a reproductively adult female sheep.

exacerbation. an increase in severity of an effect or response, as of a disease, a sign, or symptom.

exanimation. unconsciousness.

exceedance. violation of environmental protection standards by exceeding allowable limits or concentrations.

excess. **1a:** surplus; that which exists, is viable, or is used over and above that which is needed, usual, or proper. **1b:** that which goes above a just or sufficient measure or amount. **2:** immoderation, intemperance.

excessive. greater than that which is usual or proper.

exchange diffusion. a type of carrier-mediated transport of molecules across a semipermeable membrane. See active transport.

excipient. any material added to a drug formulation to give a desired consistency or form. Excipients sometimes influence toxicity substantially. Types of excipients are adjuvants, diluents, preservatives, lubricants, binders, coating agents, and disintegrators.

excise. to remove by cutting out.

excitability. sensitivity to stimulation; irritability.

excitant. **1:** stimulant (adj.). **2:** a stimulant; any agent that excites the brain or a particular vital function and so named according to the organ or function affected (e.g., cerebral excitant, motor excitant). Certain toxicants such as strychnine are excitants. Some are used therapeutically.

excitation. **1:** the act of exciting or stimulating. **2:** a state of excitement or stimulation.

excitatory. **1:** tending to or disposed to excite or stimulate. **2:** tending to disassimilation. *Cf.* inhibitory.

excitatory amino acid. an amino acid that induces firing or increases the rate of firing of neurons that have receptors for the particular amino acid. Certain of these (excitotoxins) can induce sustained repetitive firing that can damage or kill the affected neuron. See excitatory effect.

excite. stimulate, arouse, induce, activate; agitate, irritate; inspire, exhilarate.

excitement. **1:** enthusiasm, exhilaration. **2:** arousal, agitation, disturbance, irritation, excitation. **3:** induction, activation. **4:** delirium.

exciting. **1:** exhilarating, inspiring, thrilling. **2:** arousing, stimulating, inducing, activating, irritating. **3:** causing excitement. **4:** causing excitation.

excitoanabolic. inducing or stimulating anabolism.

excitocatabolic. inducing or stimulating catabolism.

excitoglandular. stimulating glandular secretion.

excitometabolic. inducing metabolic changes.

excitomotor. stimulating increased speed of muscular activity.

excitomuscular. inducing or increasing the rate or intensity of muscular activity.

excitor. an agent or stimulus that greatly heightens activity of a living system.

excitosecretory. tending to induce secretion.

excitotoxic. able to produce an excitotoxic effect. See effect (excitotoxic). See also excitotoxin, excitatory amino acid.

excitotoxin. a substance that can induce cytotoxic effects; examples are glutamate and kainic acid. See also excitatory amino acid.

excitovascular. producing increased circulatory activity.

exclusion. **1:** restriction, prohibition; removal. **2:** any form of ordinance, statute, or regulation that tends to exclude specific classes of persons, businesses, or activities from a particular type of location, district or area.

excreta. **1a:** excretion; the excretory products or waste materials eliminated from the body or any cell or tissue of a living organism, often including carbon dioxide, alkaloids, organic acids, nitrogenous compounds, water, and various metabolites of xenobiotics. **1b:** that part of assimilated matter that is eliminated from the body of a living organism as excretion, secretion, or exudation. **1c:** a component of rejecta. **2:** harmful metabolic products of a plant. *Cf.* egesta, excretion, rejecta.

excrete. to separate, concentrate, and eliminate excreta. See also excretion.

excretion. **1:** excreta; the excretory products or waste materials eliminated from the body or any cell or tissue of a living organism, often including carbon dioxide, alkaloids, organic acids, nitrogenous compounds, water, and various metabolites of xenobiotics. The term is properly restricted to the elimination of those waste materials that have (in some form) been part of the living protoplasm. **2a:** any behavioral act or physiological process of eliminating waste material from the cells, tissues, or body of a living organism, especially products of metabolism. **2b:** collectively, all processes involved in separating, concentrating, and eliminating excreta. **3:** the elimination by plants of harmful metabolic products. **alimentary excretion**. excretion via the alimentary canal. This is usually a minor route of elimination of toxicants. Small lipophilic toxicants may passively enter the lumen of the alimentary canal and thus be eliminated, however, and certain toxicants such as ammonia may also enter the canal by active transport. **biliary excretion**. See hepatic excretion. **cellular excretion**. the elimination of waste, actively or passively, from a cell by passage across the cell membrane. See diffusive excretion. **diffusive excretion**. unicellular and simple multicellular animals excrete by diffusion through the cell membrane or epidermis and in higher plants, waste products diffuse out through the leaves. Animals excrete carbon dioxide during respiration, wholly or partly by diffusion. **hepatic excretion** (biliary excretion). excretion via liver and bile duct. In vertebrate animals this is the most important route of elimination other than renal excretion. Those compounds actively excreted in bile are normally polar covalent, including conjugates of xenobiotics. **organ of excretion**. excretory organ. **placental excretion**. excretion via the placenta. **pulmonary excretion**. excretion via lungs and airways. Chiefly water-soluble, volatile compounds (e.g., ethanol, many organic solvents) are excreted via this route. **renal excretion**. excretion via the kidneys. The kidneys of vertebrate animals provide the major mechanism of excretion of the common end products of metabolism as well as polar and conjugated lipophilic xenobiotics. **salivary gland excretion**. excretion via salivary glands. *Cf.* egestion. See also excretophore, exocytosis, secretion. See also elimination.

excretophore. any cell of the coelomic epithelium that accumulates waste substances from blood of coelomate invertebrates and discharges them into coelomic fluid.

excretory. pertaining to excretion, the organs of excretion, or the processes involved in excretion.

excretory organ. excretion in unicellular organisms, small multicellular animals and most plants is by diffusion. Most multicellular animals have special excretory organs that eliminate waste materials from the body. These include, for example, nephridia (many invertebrates), Malpighian tubules (arthropods), kidneys (vertebrates); the liver of vertebrates (together with the bile duct), sweat glands (many mammals); salt glands of marine birds. Lungs and gills serve both respiratory and excretory functions. The chief organs of excretion of the end products of xenobiotic metabolism in vertebrate animals are the liver and kidneys.

EXD. dixanthogen.

exfoliation. a sloughing off or falling off in scales or layers.

exfoliative. marked by exfoliation.

exhalation. the expelling of air from the lungs; expiration (def. 1).

exhale. breathe out.

exhaled smoke. See side-stream smoke.

exhaustion. **1:** extreme fatigue or weariness with a greatly reduced ability to react to stimuli. See also fatigue. **2:** the depletion or complete consumption of something; the removal of contents. **3:** extraction of the active ingredients of a drug by treatment with a suitable solvent.

exocrine. **1:** able to secrete to the outside through a duct as in the case of digestive glands. **2:** of or pertaining to glands that have secretory ducts that communicate to the outside. **3:** of or pertaining to the secretion(s) of an exocrine gland. *Cf.* ectocrine, endocrine.

exocytosis. **1:** the presence of migrating inflammatory cells in the epidermis. **2:** (emiocytosis; merocrine secretion; eccrine secretion). the most common type of secretory process, one in which an intracellular vesicle fuses with the plasmalemma (cell membrane) in such a way that the membrane-bound contents of the vesicle are released to the outside in the form or droplets or granules. Neurotransmitters are released from nerve terminals by exocytosis. *Cf.* endocytosis.

exodermis. a subepidermal protective layer (one to several cells thick) of a plant root. It occupies a position equivalent to that of the hypodermis of the stem but is structurally and physiologically more similar to the endodermis.

exogenic. exogenous.

exogenous (allochthonous; exogenetic; ectogenous; enthetic, xenogenous; exogenic). **1a:** existing, originating, or produced outside a living system; introduced from without (as an environmental chemical or xenobiotic; an infectious disease). **1b:** caused, induced, or triggered by external environmental factors (e.g., sunlight, a bee sting). **2:** growing by additions to the outside of a living system. **3:** originating in the host in response to stimuli from a parasite or foreign body, e.g., the synthesis of a toxin by host cells. *Cf.* allochthonous, endogenous, enthetic, extrinsic, exogenous.

(±)-2-*exo*-(2-methylbenzyloxy)-1-methyl-4-isopropyl-7-oxabicyclo[2.2.1]heptane. cinmethylin.

exophthalmos. an exceptional protrusion of the eyeball from the orbit.

exoteric. exogenous; external in origin; arising outside the organism.

exothermic. of or pertaining to a process (e.g., combustion, a chemical reaction) in which the temperature rises and/or heat is released. *Cf.* endothermic.

exotic. foreign; alien; not native.

exotoxic. **1:** of, pertaining to, or having the toxic properties or effects of, or produced by an exotoxin. **2:** relating to the introduction of a toxic substance into a living system.

exotoxin (ectotoxin). any proteinaceous toxin produced by bacteria (usually Gram-positive bacteria) that is found outside the bacterial shell or is released into the surrounding medium. Exotoxins are more heat labile and more toxic than endotoxins. They are detoxified (but retain their antigenicity) with treatment by agents such as formaldehyde (formol toxoid) that do not affect endotoxins. Exotoxins are among the most toxic substances known and include cardiotoxins, hemotoxins, neurotoxins, botulinum toxin, diphtheria toxin, plague toxin, and various cell-disrupting enzymes such as lecithinase and collagenase. See also endotoxin, enterotoxin, toxemia, toxoid, phytotoxin.

expansin. patulin.

expansine. patulin.

expectoration. **1:** the spitting out of saliva with or without the expulsion of mucus and other materials from the throat and upper respiratory tree. **2:** the ejection of venom from the fangs as a stream or spray by any of the so-called spitting cobras.

experiment. **1:** to explore, investigate, to conduct an experiment, to test. **2a:** a test, trial, investigation; demonstration. **2:** an empirical test of a hypothesis under controlled conditions (controlled experiment); a test or trial. See holocoenotic. **check experiment**. a crucial experiment. **control experiment**. an experiment conducted under standard conditions to test or verify the accuracy or correctness of the observations or results of another experiment, or to determine what might have occurred had the factor under study been omitted. See also treatment. **controlled experiment**. an experiment in which one group (the control group) is maintained in an environment that is "identical" to that of one or more groups (experimental or test group(s)) to examined for the effects of some variable to which the control group remains unexposed. Treatments among test groups may vary by one (usually) or more variables or one or more values of the experimental variable (different dosages of a single drug). Such experiments are valid only if the individual groups are drawn from the population about which inferences are to be drawn on the basis of a stratified random sampling design. **crucial experiment**. a mature experiment, one preceded by pilot studies and so designed that it is thought that it will conclusively test an hypothesis. Note, however, that a conclusive test is one that disproves the hypothesis tested. Failure to disprove a hypothesis, does not imply acceptance. In practice, no toxicological experiment can rule out the toxicity of a substance except under the strict conditions that obtain in the experiment employed to do so, and only for the end points measured. See also extrapolation. **double blind experiment**. an experiment or test in which neither the investigator nor the subjects know which treatment they are receiving (experimental or control). The subjects are thus not affected by knowledge, and investigator bias in recording and reporting results is minimized. **factorial experiment**. an experimental design in which two or more treatments are applied in all possible combinations, such that analysis of the results may require a multifactor analysis of variance. **pilot experiment**. an experiment conducted with fewer resources committed (e.g., on a smaller scale or in a shorter time frame) than would be demanded by an experiment (or set of experiments) deemed necessary to sufficiently test the hypothesis of concern. The purpose of such an experiment is to determine whether commitment to full-scale experimentation may be justified.

experimental animal. See animal.

experimental insecticide No. 269. endrin.

experimental treatment. See treatment.

experimental unit. the subject or grouping to which a treatment is supplied. See treatment, experiment.

expiration. **1:** exhalation; breathing out; the expelling of air from the lungs. **2:** termination. **3:** death.

expire. **1:** exhale **2:** end, conclude, terminate. **3:** die, perish, succumb.

explant. a part of a living organism that has been removed and cultured in an artificial medium.

explantation. the removal and culture in an artificial medium of part of a living organism.

expose. to subject to exposure; to endanger; to reveal. See exposure.

exposure. **1a:** the process of revealing, displaying, exhibiting, or making something available. **1b:** endangerment. **1c:** coming into contact with something (e.g., a poisonous substance) and being subjected to its influence. In toxicology, contact with an environmental agent such as a drug, poison, ionizing radiation, or any hazardous substance such that one is potentially subjected to its influence. **2:** the amount of radiation or pollutants present in a given area (e.g., a community or ecosystem) to which living systems may be potentially exposed. **3a:** contact (or a measure of contact) between a biologically active agent (e.g., ionizing radiation, a drug, a toxicant) and a living system. Exposure is usually expressed in terms of concentration (intermittent or continuous) in the medium (air, water, soil) or dosage (e.g., in food or beverage) that contacts or interfaces with some part of the system during a specified period of time. In toxicology, exposure (as opposed to dose) is always a complex function of time and intensity (e.g., concentration). Exposure may be deliberate or accidental. See route of exposure. **3b:** binding of a drug or toxicant to a receptor. **4:** a debilitated condition due to exposure to extremes of weather (heat, cold, wind) with inadequate protection. The victim may become unconscious, develop further complications, and even die. **acute exposure**. **1:** a single, brief exposure (usually less than 24 hours in duration) to a toxicant which produces death or severe harm to an organism. **2:** as in def. 1, but exposure is to repeated doses experienced or administered during a brief period of time. **acute local exposure**. acute exposure at a given site on an organism's body (e.g., the eyes, exposed skin) during a brief period of time (from a few seconds to a few hours). Such exposures may injure or damage the affected body part, such as the eyes, skin, or mucous membranes. **acute systemic exposure**. an acute exposure to a single dose of a toxicant that can enter the body and affect tissues and structures that are remote from the site of entry. **anal exposure**. exposure via the anus. It is an infrequent route of exposure to xenobiotics, although many are easily administered *per anus*. **branchial exposure**. exposure to toxicants via gills. It is a major route in diverse aquatic and semiaquatic animals and some terrestrial species that live in damp environments. There is a high degree of vulnerability to certain toxicants since, as in the pulmonary route, toxicants may bypass the organs that normally detoxify or screen xenobiotics. **chronic exposure**. **1:** discrete, frequently repeated, or continuous exposure (usually at a low level) to a chemical over a long period of time (more than 3 months in the case of human exposures). In toxicology, chronic exposure is often defined as continuous or frequently repeated exposure for at least one tenth of the normal life span of the species. **chronic local exposure**. exposure (constant or intermittent) at a given site on an organism's body during a time span of weeks, months, or even years. **chronic systemic exposure**. prolonged exposure to a toxicant that can enter the body and affect tissues and structures that are remote from the site of entry. **common route of exposure**. the likely pathway whereby a pesticide may enter an organism. This is mostly via contact with the integument, the mouth, or by inhaling. **complex exposure**. simultaneous or nearly simultaneous exposure to one or more toxicants by more than one route as may be the case when empoisonment is due to, or associated with, physical trauma, exposure to complex mixtures; exposure to volatile liquids; and in envenomation via snakebite. Exposure is also complex where both focal and systemic effects result. **cutaneous exposure**. often used interchangeably with or as a synonym of the term "dermal exposure." Due to differences in the permeability and sensitivities of the skin, the effects of cutaneous exposure to a particular irritant might be quite different in the case of a frog, a human, and an elephant. **dermal exposure**. exposure to a chemical by contact with the dermis and not just, for example, a nonliving cuticle or cornified region of the epidermis. *Cf.* cutaneous exposure. **direct exposure** (primary exposure). exposure of an organism to a chemical directly from the source. See indirect exposure. **duration of exposure**. the period of time that an organism is exposed to a substance per exposure incident. Toxicity can be greatly affected by the duration of exposure. **gastrointestinal exposure**. exposure to an agent in which the major route of entry into the body is via the stomach and intestines. **high-end exposure**. a plausible estimate of an individual's risk of exposure to a toxic chemical or other hazard that is greater than the 90th percentile. **indirect exposure** (secondary exposure). exposure of an organism through contact with exposed environmental surfaces, ingestion of or contact with a target organism or an accidentally exposed system, as opposed to direct exposure from the source. Poisoning of pests, as with pesticides, baits, or other sources of poison, is usually indirect. **integrated exposure**. an estimate of exposure in situations (often in the workplace) where exposure is from varied sources, durations, pathways or routes of exposure, and presumably concentrations. **intramuscular exposure**. exposure to an agent by direct administration into skeletal muscle. This is a common means of deliberate exposure of an organism to a drug or toxic substance. Intramuscular exposure is also common in envenomations as via snakebite. **intravenous exposure**. exposure to an agent introduced directly into the circulatory system via a vein. This is a common means of deliberate exposure of an organism to a drug or toxic substance. **lifecycle exposure**. chronic exposure, the duration of which spans all life stages of the exposed organism. **lifetime exposure**. exposure for a period of time approximating the expected lifetime of the exposed organism. **ocular exposure**. exposure to an agent through the exposed surface of the eye. **oral exposure**. exposure to an agent via entrance into the body through the mouth. **parenteral exposure**. exposure to an agent that enters the body by a route other than the alimentary canal. **percutaneous exposure**. exposure to an agent by contact and passage through the intact skin, including organs of the skin such as hair follicles and sweat glands. **primary exposure**. direct exposure. **pulmonary exposure** (inhalation exposure; respiration exposure). exposure to toxicants via the lungs. This is a major route of exposure of terrestrial organisms to aerosols, vapors, gases, and respirable particulates. There is a high degree of vulnerability to certain toxicants since, as in the branchial route, toxicants may bypass the organs that normally detoxify or screen xenobiotics. *Cf.* branchial exposure. **rate of exposure**. the frequency of exposure incidents per unit (or specified) time period. *Cf.* duration of exposure. **rectal exposure**. exposure to an agent that can enter the body through the wall of the rectum, usually one that is administered via the anus. **respiration exposure**. a term that refers either to branchial or pulmonary exposure. **route of exposure** (route of entry; route of penetration; route of absorption; route of administration). any avenue whereby a xenobiotic chemical or a microorganism comes into contact with the tissues of a living system. These avenues in vertebrate animals, for example, include the following (*q.v.*) anal exposure, branchial exposure, cutaneous exposure, dermal exposure, gastrointestinal exposure, intramuscular exposure, intravenous exposure, ocular exposure, oral (peroral) exposure, parenteral exposure, percutaneous exposure, pulmonary exposure, rectal exposure, vaginal exposure. Effective routes of exposure of animals to xenobiotics vary with species, age, sex, size, and numerous other factors such as imperviousness or thickness of the skin. The term route of administration is subsumed by the present term, except that some routes of administration are uncommon routes of exposure. In terrestrial plants there are three major routes whereby xenobiotic chemicals can reach the living tissue in leaves: 1) translocation from the root system; 2) atmospheric deposition; and 3) intake through stomata from the air. *Cf.* administration, route of. **secondary exposure**. indirect exposure. **subacute exposure**. nearly acute exposure, but with a greater number of doses or a longer duration of exposure than for acute exposure. In the case of humans, a subacute exposure is often considered as the administration of repeated doses for a period of 1 month or less. **subchronic exposure**. nearly chronic exposure with repeated doses or continuous exposure to low concentrations in food, water, or ambient air with a shorter duration of exposure than for chronic exposure. In the case of humans, subchronic exposure is often considered as the administration of repeated doses for a period of 1 to 3 months. **test exposure**. the exposure of a living system to a toxicant as part of a toxicity test or bioassay. **time-integrated exposure**. an estimate of total or average exposure (concentration or dosage) to a substance that varies in level or quantity throughout the period of exposure. This is especially important in the case of outdoor air pollutants, concentrations of which often vary drastically at a given location. **total exposure**. the total time period over which the organism is exposed. **vaginal exposure**. exposure to a biologically active agent through the walls of the vagina. See, for example, potassium permanganate. See also route of exposure.

exposure assessment. **1:** the qualitative or quantitative estimation of the magnitude, intensity, frequency, duration, and route of exposure of a living system to toxic or other hazardous materials or microorganisms. **2:** estimation of the temporal or spatial extent of present and future environmental and human exposures (for all relevant pathways and boundary conditions) using quantitative environmental fate models adapted to the location or environment under consideration. See also effects assessment. Frequently used tools and methods include field and/or laboratory monitoring, mathematical modeling, measurement of tissue concentrations of chemicals under assessment or their metabolites, and determination of the environmental concentration of the chemical. The use of environmental specimen banking has been employed where the initial times and locations of exposure were not at first known.

exposure category. categories of exposure.

exposure commitment (E_i). the integral of the concentration of a substance in a compartment over infinite time. It is a measure of the total exposure to a pollutant and is a useful way to express source-receptor relationships.

exposure commitment analysis. an approach to estimating future exposures to a toxic or hazardous environmental chemical. See exposure commitment.

exposure dose. See dose.

exposure-effect relationship. the quantitative relations between exposure to a pollutant and the risk or magnitude of an undesirable effect under specified circumstances defined by environmental variables and target variables. See also dose-response.

exposure limit. **1:** the concentration of a harmful substance (e.g., in food, drinking water, air) at which there is no significant risk of adverse (e.g., health or environmental) effects. **2:** hygienic guidelines intended to protect individuals (usually workers) from undue exposure to a toxic chemical. See also limit value (threshold limit value); maximum allowable concentration.

exposure period. the duration of time that an object or organism is exposed to a chemical or to a microorganism.

exsorption. the movement of substances from circulating blood into the lumen of the gut.

EXT. D&C. See color (exterior D&C).

exterior. external (*q.v.*); outside; surface, covering; appearance.

exterior D&C. See color.

external (externus; exterior). **1:** on the outside; outside of; outer; exoteric; exterior; farther from the center. **2:** often incorrectly used to mean lateral.

external dose. See dose.

externus. external.

extinct. **1:** extinguished, dead, perished. **2a:** no longer living (e.g., as an organism, a population, a nation); dead; deceased. **2b:** having died out altogether.

extinction. **1:** the act or process of extinguishing or making extinct; the process of becoming extinct. **2:** the death of an entire race, species, or population. **3:** the fact of being extinct.

extracorticospinal system. extrapyramidal system.

extracorticospinal tract. extrapyramidal tract.

extract. **1a:** to remove, pull out, or isolate one substance or entity from another; to perform an extraction; to prepare an extract by physical or chemical means. **1b:** to remove all or part of a mixture by use of a solvent. **2:** that which is extracted. **3:** a concentrated preparation of a drug (usually as a syrup, a compressed solid, or a powder) or other substance obtained by extraction from the original source (e.g., plant or animal parts) by the use of suitable solvents and subsequent removal of the solvents, usually by evaporation. **allergenic extract**. an extract of the protein of any substance to which a person may be allergic. The extract may then be used in diagnosis or desensitization in cases of hypersensitivity. **animal extract**. an extract that is animal in origin. **belladonna extract** [U.S.-P]. a standard preparation, as a pill or powder which contains 1.15-1.35 gm. of belladonna alkaloids per 100 grams and is used as an anticholinergic. **compound extract**. an extract of more than one biologically active agent. **henbane extract**. hyoscyamus extract. **hyoscyamus extract**. a preparation of hyoscyamus (*q.v.*) by extraction. **nux vomica extract**. an extract of nux vomica in powder form that contains 7-7.5 gm of strychnine per 100 gm of strychnine. **poison ivy extract**. an extract from fresh leaves of poison ivy (*Rhus radicans*, *q.v.*) used for desensitization. **poison oak extract**. an extract from fresh leaves of poison oak (*Rhus toxicodendron*, *q.v.*) used for desensitization.

extraction. **1:** the process of removal, pulling out, or isolating of one substance or entity from another. **2a:** the partial or complete removal of part of a mixture by use of a solvent **2b:** the preparation of an extract by chemical or physical means.

extraembryonic. apart from and external to the embryo.

extraembryonic membrane (fetal membrane; foetal membrane). one of the membranous structures that surround and serve to support and protect the vertebrate embryo. These include the yolk sac, allantois, amnion, and chorion. They develop from ectoderm, endoderm, and mesoderm. Associated with the mammalian placenta, they become part of the decidua following birth. They aid in respiration, nutrition, excretion, and protection of the embryo or fetus.

extrahepatic. external to or unrelated to the liver.

extraneous. 1a: irrelevant, superfluous, foreign. **1b:** external to an entity (e.g., an organism) and unrelated to it.

extraphysiologic. 1: of or pertaining that which lies outside of the purview or domain of physiology. **2:** pathologic.

extrapolation. 1: the estimation of a mathema tical function at a point which is larger (or smaller) than all the points at which the value of the function is known. **2a:** a process of estimating a value beyond the range of observed values, often based on an empirical equation that sufficiently and accurately describes the behavior of the set of observed values. **2b:** a process of estimation based on information drawn from one system (or one part of a system) to make predictions about another system (or another part of the original system). **2c:** a process of estimation based on the use of information obtained under one set of conditions to make predictions about the values that might have been realized under another set of conditions. **complex extrapolation**. extrapolation in toxicology usually involves a constellation of stated and unstated assumptions and inferences. Furthermore, two or more very different systems (e.g., a set of animals in an unnatural, simplified environment, exposed to a simplified, standardized exposure regime to humans with complex and variable lives, histories, heredity, environments, and exposure regimes) are usually involved. One major concern in the use of extrapolation in toxicology relates to the assumption that laboratory responses to a toxicant actually simulate responses that occur in the system of interest under authentic environmental conditions. Another major concern relates to the assumption that the laboratory system (e.g., a species sample) is sufficiently similar (e.g., behaviorally, biochemically, physiologically, genetically) to the system of interest to simulate the latter's response quantitatively or even qualitatively. See especially thalidomide. See also grand extrapolation. **dose-based extrapolation**. extrapolation, from one population (an experimental sample or a test population) to another (e.g., the parent population or an alternate population) in which the dose is usually expressed as some function of body size (e.g., mg/kg body weight per day, mg/m^2 body surface per day; mg/kg body weight/lifetime). Differences in toxicity among various populations based on such expressions of dosage are often substantial and may vary considerably among affected taxa or morphs within taxa. Such variation is (1) often due to little understood differences in physiological mechanisms involved and (2) often associated in part with differences in absolute body size among populations or taxa of interest. The way in which the dose is expressed is critical to any assessment of risk or toxicity in a specific case. There is no general or fundamental way of expressing dosage for any wide range organisms. **dose-based interspecies extrapolation**. dose-based extrapolation from a sample population drawn from one species to another species with a body size that lies partly or completely outside the size range of the test sample. **dose-based intraspecies extrapolation**. dose-based extrapolation from a sample in which the size-range is more constrained or not representative of the species as whole. **dose-range extrapolation**. the estimation of low doses and their effects from high-dose exposures using dose-response models. See also dose-response assessment. **grand extrapolation**. the extrapolation of data from another animal or laboratory preparation to *Homo sapiens*. In toxicology this usually means estimating the effect of a test chemical on humans based on the results of toxicity tests conducted on other species. **interspecies (interspecific) extrapolation** (species extrapolation). the extrapolation of observations or test results on one species of organism to another. This is a risky process, especially if the species in question are not closely related or, even if closely related, they differ significantly in ways that may affect exposure or response (differing nutritional, behavioral, or physiological patterns). Overlooked, or seemingly small differences may influence response. Thus, because laboratory rats do not vomit, they may be poisoned by much lower concentrations that have an emetic effect on humans. See also complex extrapolation, dose-based extrapolation, the grand extrapolation. **intraspecies (interspecific) extrapolation**. extrapolation from observations, anecdotes, or test results of a subset (sample) of a population or species to the species as a whole. This type of extrapolation is (perhaps surprisingly) prone to error because the set of conditions preceding or attending the observations are unlikely to be duplicated throughout the species. Genetic variation and other factors may also have a significant influence on the validity or accuracy of the extrapolation. See also factor (especially confounding factor, density-dependent factor, modifying factor, predisposing factor, uncertainty factor). **linear extrapolation**. a method that uses a simple linear projection to predict values that lie outside the range of observation. The method is based on the assumption that the proportionality between two variables (e.g., concentration and mortality) will remain constant because it has done so within the range of observations. The error in such an extrapolation tends to increase, often dramatically, as the values depart further and further from the range of observations (e.g., test results). Systems rarely maintain linearity over a wide range of values. See also factor. **low-dose extrapolation**. extrapolation of high dose-effect relationships (determined from tests of a population exposed to a high dosage or concentration) to the low dosages or concentrations expected in the environment. Such extrapolations are usually subject to much error because the dose-response curve at lower doses may differ greatly from that at the higher experimental doses. Toxic mechanisms and their effectiveness may differ at low doses from those operating at higher doses. Indeed, effects may occur at high dosage that do not occur at lower doses. The use of low-dose extrapolation in risk assessment is often not helpful. **risk extrapolation**. extrapolation of the results of a risk assessment beyond the range of data employed for that purpose. For example, estimation of the risk of toxic effects at exposures below that for which test data were available. **species extrapolation**. interspecies extrapolation.

extrapyramidal. 1: of or pertaining to nervous structures and tracts that lie outside the pyramidal tracts of the brain and are concerned with posture and movements of the body. Excluded are the motor cortex, the corticospinal tracts, and corticobulbar tracts. **2:** of or pertaining to the function of those tissues designated as extrapyramidal in the above definition.

extrapyramidal disease. any of a large variety of maladies (e.g., athetosis, chorea, tardive dyskinesia, Parkinson's disease) marked by abnormal posture, changes in muscle tone, and dystonic involuntary movements. The dystonic movements typically involve the head and neck and may include grimacing, throat spasms, protrusion of the tongue, oculogyric crisis, and scoliosis. The dystonia my result in torticollis (a turning of the head to the side) or retrocollis (a turning of the head to the rear). Symptoms often include unsteady gait, drooling, and slurred speech.

extrapyramidal reaction. any response to therapy or to overdosage of a drug that resembles signs of extrapyramidal disease. The reaction may be persistent or may disappear upon withdrawal of the drug. See extrapyramidal side effect.

extrapyramidal side effect. an extrapyramidal reaction caused by drugs that block dopamine receptors in the extrapyramidal system. See also effect (side effect). Common neurologic effects of antipsychotic drug treatment, include dystonias, agitation and motor restlessness, and pseudoparkinsonism. Anticholinergic agents such as diphenhydramine or benztropine given *i.v.* or i.m. will rapidly suppress such drug-induced reactions.

extrapyramidal system (extracorticospinal system). a functional component of the mammalian nervous system that is comprised of the extrapyramidal tracts which include the corpus striatum, the subthalamic nucleus, the substantia nigra, the red nucleus, as well as their internal connections and interconnections with the reticular formation, and the cerebellum and cerebrum. Included also are the motor neurons of the spinal cord. The system is defined functionally, not as a gross or subgross structural unit. It controls and coordinates static, postural, supporting, and locomotor activity. This term is often used interchangeably with extrapyramidal tract.

extrapyramidal tract (extracorticospinal tract). any of the tracts that make up the extrapyramidal system. They are defined functionally, and are comprised of relays of motor neurons, that communicate within the brain between motor areas of the cerebral cortex, basal ganglia, thalamus, cerebellum, and the brainstem. They extend from the brain to the anterior horns of the spinal cords in humans. These tracts control and coordination static, postural, supporting, and locomotor mechanisms. This is effected by effecting delicate and intricate sequential and/or simultaneous contractions of involved muscle groups. This term is often used interchangeably with extrapyramidal system, *q.v.*

extrasensory. beyond the normal, physiologically demonstrated senses.

extratoxicologic. **1:** lying outside the purview or domain of toxicology; of more than toxicologic concern or significance. **2:** pertaining or denoting origins, causes, or effects that are not toxicologic (applied to disease states).

extravasate. 1: to pass out of, to escape, issue, emanate, or exude from a vessel(s) into the surrounding tissues (said of blood, lymph, or urine). **2:** the material thus exuded.

extravasation. **1:** the process of extravasating; the escape of fluid(s), especially blood, lymph, or urine, into tissue or spaces surrounding a vessel(s). **2:** an extravasate (def. 2); extravasated fluid.

extravascular. **1a:** situated or occurring outside the vascular system. **1b:** not through or by way of the vascular system. See route of administration.

extrinsic. **1a:** existing external to an individual, group or system. **1b:** from the outside or coming from without; having origins external to an individual, body part, group, or system.

exudate. **1:** a substance discharged through pores, lesions, or tissue incisions. **2:** a substance accumulated in or on tissues as the result of abnormal metabolism or disease.

exudation. the diffusion or slow passage of fluid as when fluids and blood cells pass from the vascular system into tissues during inflammation. *Cf.* effusion, transudation.

exudative calcifying fasciitis. calcinosis in humans.

eye. a special organ of light reception. Eyes vary greatly in structure and function from simple light receptors (as in many invertebrates) to those that can differentiate among wavelengths and form an image (as in most insects, crustaceans, mollusks, and vertebrates). In addition to evident direct damage to the eye by irritant poisons, a number of features of the eye in humans and other vertebrates can often provide evidence of systemic poisoning. These features include the size of the pupil and its reactivity to light; voluntary and involuntary movement of the eyes; the appearance of eye structures such as the blood vessels, lens, and conjunctiva. Blurred vision, strabismus, and behavioral effects dependent on vision (e.g., light sensitivity, or withdrawal from light of normal intensity, blindness) may also provide evidence of poisoning. See miosis, mydriasis, conjunctivitis, nystagmus.

eye irritation test. See test.

eye-horned viper. *Pseudocerastes persicus*.

eyebright. *Euphrasia officinalis*.

eyelash snake. *Bothrops schlegelii*.

eyelash viper. *Bothrops schlegelii*.

f.p. freezing point.

F_2 toxin. zearalenone.

2-FAA. *N*-2-fluorenylacetamide.

FAB. fragment antigen binding. See digoxin immune FAB (ovine) for injection.

Fabaceae (pulses; beans). the bean family, a large, heterogeneous, often climbing, nearly cosmopolitan family of herbs. All are leguminous, with nitrogen-fixing microorganisms associated with the roots. They were formerly assigned to Leguminosae (the pea family).

fabism. favism.

face powder. talcum powder.

facial pit. loreal pit.

facies (plural, facies). **1:** the face, surface, aspect, or shape of a structure. **2:** the countenance of a person, especially an appearance or expression of the face that is indicative of a morbid condition. **3a:** the general appearance, aspect, or composition of a natural grouping, e.g., a population, a type of vegetation, the fauna of a region, habitat, community, ecosystem, geologic formation. **3b:** a formal class or rank of such groupings in def. 3a. **4a:** the appearance or aspect of a particular local modification of an ecological community or a biotope; **4b:** a specialized part, usually local, of a cultural community.

facilitated transport. facilitated diffusion (See diffusion).

factitious air. nitrous oxide.

factor. **1:** an agent, condition, or an endogenous or environmental variable that contributes to or modifies an action or effect. **2a:** an essential component or ingredient. **2b:** a vitamin or other essential nutritional element. **3:** a component of a number or expression that is arrived at by multiplication. **4:** a hereditary factor or gene. **abiotic factor**. **1:** any nonliving, nonbiological component of the environment that may directly affect living components (e.g., light and various other forms of energy, temperature, climatic variables, physiographic variables, carbon dioxide, oxygen, trace elements, and water). See also biotic factor, biological factor. **2:** in toxicology, an abiotic factor is usually one that influences biological response to, or properties of, a poisonous substance. See also biotic factor, biological factor. **absorption factor**. the fraction of a substance to which an organism is exposed that is absorbed. **acetate replacing factor**. thioctic acid. **additive factor**. any of a number of non-allelic genes that affect the same phenotype such that each heightens the expression of the other. **antichromotrichia factor**. *p*-aminobenzoic acid. **anti-gray-hair factor**. See *p*-aminobenzoic acid. **antihemorrhagic factor**. a factor that promotes blood clotting. See vitamin K. **physical factor**. **application factor**. a factor applied to LC_{50}, LD_{50}, and sometimes to other toxic end points to estimate the no-adverse-effects level (NOEL) for lasting exposure to a toxicant. These values are arbitrary, variously arrived at and not easily defended, if at all, on a scientific basis. See also factor. **B-cell replacing factor** (BCGF). a lymphokine that is essential to the growth and differentiation of B and T cells. **bioaccumulation factor** (BAF). the ratio of the concentration of a substance in the tissue of an organism to the concentration of the substance in the environment, surrounding medium, or that of the material to which an organism is effectively exposed. See also bioconcentration factor, biomagnification factor. **bioconcentration factor** (BCF). the concentration of an environmental chemical in an organism divided by the concentration in the water to which the organism is exposed. See also factor, *Cf.* bioaccumulation factor, biomagnification factor. **biological factor**. any living system, or a part, aspect, product, process, or property of such a system that contributes to or influences a phenomenon or process (e.g., toxicity). A biological factor may have a physical and/or chemical component or components. *Cf.* biotic factor. See also physical factor, chemical factor. **biomagnification factor** (BMF). the concentration of an environmental chemical in an organism divided by the concentration in the food of the organism. See also factor, *Cf.* bioconcentration factor, bioaccumulation factor. **biotic factor**. **1:** any environmental factor, *q.v.*, that results from the activities of living organisms. *Cf.* biological factor, abiotic factor, modifying factor. **2:** in toxicology, a biotic factor is usually one that influences biological response, or a property of a poisonous substance. See also abiotic factor, biological factor. **blood clotting factor**. any of a number of soluble proteins in vertebrate blood plasma that play essential roles in the complicated process of blood coagulation (e.g., thrombin). See dicumerol. **chemical factor**. **1:** a factor, environmental or internal to a living system, that is chemical in nature or that has a strong chemical component (e.g., pH, an enzyme, electrolyte content of a medium). **2:** in toxicology, a chemical factor is usually one that influences biological response to, or a property of a poisonous substance. **chromotrichia factor**. *p*-aminobenzoic acid. **chronicity factor** (CF). the ratio of the acute to the chronic LD_{50} dose. **citrovorum factor**. folinic acid and any of several naturally occurring or synthetic forms of folic acid that support the growth of certain bacteria (e.g., *Leuconostoc citrovorum*). **clearance factor** (C_m). the maximum rate at which any substance is cleared from the blood. See clearance. **cobra venom factor** (CVF). a protein constituent of the venom of the *Naja naja*, the Indian cobra. It renders properdin factor B responsive to properdin factor D with consequent activation of C3b (complement component 3b, a fragment of the C3 glycoprotein) and other components of complement with lysis of unsensitized erythrocytes. CVF has been injected into laboratory mammals to destroy complement activity for research purposes. See also properdin system. **conditioning factor**. a modifying factor that has little or no activity but can significantly alter a response to a stimulus. Thus, sodium can act as a conditioning factor of DOCA activity. **confounding factor** (confounding variable). a variable or factor frequently associated with a process that obscures, confounds, or confuses one's ability to make valid inferences about the variables that may be important to the process (e.g., detoxication), or to the output or effects of the process (e.g. toxicity). **density-dependent factor**. any environmental variable that influences the growth of a population (via effects on mortality or fecundity) such that its effectiveness varies with population density (the number of organisms per unit of space occupied). Examples are competition, predation, parasitism, and available food supply. Toxic substances that act to influence a density-dependent factor such as a critical resource or by altering competitive balance are themselves density-dependent factors. See also density dependence. **density-independent factor**. any environmental variable (e.g., extremes of temperature, floods) that affects population growth uniformly with respect to population density. A toxic substance that is uniformly distributed within the space occupied by a population may be a density-independent factor if it acts directly and nonselectively upon members of the population. **drug-resistance factor**. drug resistance plasmid. **edaphic factor**. any of a large number of soil properties (physical, chemical, biological) that affect the growth and health of plants rooted in the soil and indirectly the growth health and structure of the terrestrial community (and to a usually lesser degree, that of nearby aquatic communities) that develops at a given site. Edaphic factors may influence the ionic state or the persistence of an anthropogenic chemical at a given location. **elongation factor**. any of several accessory proteins that permit translation to proceed by facilitating the correct positioning of aminoacyl-tRNA on the mRNA-ribosome complex. **endogenous factor**. a factor that is naturally internal to a system. **environmental factor**. **1:** any factor or variable in the environment that arises from or constitutes a part of the environment; an environmental variable. **2:** a modifying factor (*q.v.*); any aspect or part of the environment, or any environmental process that contributes to or influences another phenomenon or process (e.g., toxicity). *Cf.* biotic factor. See also holocoenotic, modifying factor, teratogen. **exogenous factor** (external factor). a factor external to a system. an environmental factor is an exogenous factor. **external factor**. a factor that is normally external to a system. **fast death factor**. One or more "fast death factors" are responsible for the toxicity of algal blooms.

These are a variable mix of toxic principles produced by various populations of organisms associated with the bloom (notably, *Claustridium botulinum* and certain blue-green bacteria). Only certain strains of a given species may produce it. The fast death factor accumulates in the cell and is released via leakage through the cell wall or upon cellular decomposition. See also *Microcystis aeruginosa*, algal poisoning, bloom. **growth factor**. any of numerous environmental factors or endogenous principles that variously facilitate growth (e.g., temperature, citrovorum factor). **goitrogenic factor**. any of various principles and factors that contribute to the development of a goiter (e.g., L-5-vinyl-2-thiooxazolidone). See also goitrogen, goitrogenic. **hereditary factor**. a gene. **histamine-sensitizing factor**. pertussis toxin. **internal factor**. a factor, natural or introduced, that is internal to a system. **intrinsic factor**. a variable that is inborn or inherent to an organism, part of an organism, or to a material or object. **lipotropic factor**. choline. **milk factor**. a substance found in certain strains of mice that are prone to mammary cancer. This factor is transferred to suckling pups via the milk and can induce mammary cancer in the exposed mice. **modifying factor**. **1:** any of numerous environmental variables, complex happenings, and factors in the environment that affect the toxicity of chemicals to target and non-target organisms. Many of the types of factors defined here are (or include) modifying factors. *C.f.* environmental factor. **2:** any factor that influences toxicity. **3:** any factor that influences the toxicokinetics of a substance. **nutritional factor**. a principle (e.g., protein/calorie ratio, vitamins) that is essential to or important to health-sustaining or life-sustaining nutrition. Deficiencies of such factors in the diet or defective metabolism may powerfully affect the virulence of a toxicant. **1:** any factor that is environmental or internal to a living system, that is a physical variable (e.g., temperature, moisture, aeration, and pH). **2:** in toxicology, a physical factor is usually one that influences biological response to, or a property of a poisonous substance. **physico-chemical factor**. **1:** a variable that is physical and/or chemical in character as opposed to a biotic factor. **predisposing factor**. an internal or external environmental factor that predisposes a living system to a particular disease, condition or outcome. **properdin factor B** (C3 proactivator; cobra venom cofactor; glycine rich β glycoprotein II). a blood serum protein (MW 95,000) and component of the properdin system. See also cobra venom factor, properdin system. **properdin factor E**. a blood serum protein (MW 160,000) essential to the activation of the third component (C3) of complement by cobra venom factor. See also properdin system; cobra venom factor. **pyruvate oxidation factor**. thioctic acid. **R factor**. See drug-resistance plasmid. **resistance factor**. R factor. See drug-resistance plasmid. **risk factor.** single factor (endogenous or exogenous) that is statistically associated with, but not necessarily a cause of, increased risk of injury, damage, morbidity, or mortality by a stated hazard. **safety factor**. **1:** the factor by which the NOEL (no-observed-effect level) is reduced in order to give a supposedly safe dose for humans. If chronic exposure data are available for humans, the safety factor is usually set at 10; with other sources of data, a factor of 100 is often used. **2:** uncertainty factor. **transmethylation factor**. choline. **tumor necrosis factor α** (TNF-α; cachectin). a cytotoxic serum protein (a monokine) produced by numerous types of cells (e.g., macrophages, monocytes, B- and T-lymphocytes) in response to bacterial endotoxin. It kills cells by forming transmembrane channels. It contributes to inflammation; and to wound healing, and the remodeling of tissue by facilitating leucocyte recruitment and the induction of angiogenesis, and by promoting fibroblast proliferation. It can also induce septic shock and anorexia with associated cachexia which may result in death from malnutrition. **uncertainty factor** (UF). **1:** any factor that contributes to the uncertainty of a set of observations or findings. **2:** any factor, arrived at by whatever means to establish a safe dose or concentration of a xenobiotic for a given type or taxon of living organism. **3:** a factor used to determine the reference dose from experimental data. It is intended to account for significant sources of uncertainty such as varying and poorly known sensitivity among members of the target population, uncertainty of measurement, and uncertainty in extrapolating from animal data to humans. **unconscious factor**. a mental process, below the level of awareness, that contributes to a behavior. See accident proneness. **vomiting factor**. a general term for the principles (all are trichothecenes) that cause vomiting in ducklings used in bioassays of toxic corn. **Warburg's factor**. a former name for cytochrome oxidase.

facultative. **1:** capable of, but not confined to, performing a particular action or function. **2:** able to live under a certain set of environmental conditions, but not restricted to those conditions. *Cf.* obligate.

facultative accumulator. See accumulator.

fagicladosporic acid. a mycotoxin produced by *Cladosporium epiphyllum*.

fagopyrin. a primary photosensitizing agent found in *Fagopyrum sagittatum*.

fagopyrism. buckwheat poisoning. See also *Fagopyrum sagittatum*.

Fagopyrum sagittatum (buckwheat). a fast-growing annual herb (Family Polygonaceae), farmed mostly for its seed which is used to make a characteristically dark flour that is used to make pancakes. Most cases of buckwheat poisoning are rare allergic reactions to this plant. However, *Fagopyrum* is one of just a few plant species in North America known to cause primary photosensitivity in animals. Buckwheat contains a photosensitizing agent, fagopyrin. See also *Hypericum perforatum*.

Fagus sylvatica (European beech; beechnut). an ornamental deciduous tree with numerous varieties (Family Fagaceae). Humans have occasionally been poisoned by consuming the saponin-containing seeds (beechnuts) and livestock have been poisoned by eating press cakes made from seeds from which the (nontoxic) oil has been expressed. Symptoms are those of weakness and severe gastrointestinal distress.

fahaka. one of numerous common names of fish of the family Tetraodontidae.

Fahrenheit degree. See degree.

Fahrenheit temperature scale. a temperature scale on which the freezing point of water is 32° and the boiling point of water is 212°. See also Celcius temperature scale, Kelvin temperature scale.

failure. the loss or inability to function; a state of extreme insufficiency, ineffectiveness or nonperformance. Said of an organ or system. **congestive heart failure**. a condition marked by the inability of the heart to maintain adequate blood flow. It is progressive, and associated with weakness, breathlessness, abdominal discomfort, salt and water retention with edema of the lower parts of the body due to venous stasis, pulmonary congestion. See also cardiomyopathy (congestive cardiomyopathy). **heart failure**. any condition whereby the heart is unable to pump blood normally. Such conditions may be caused directly or indirectly by a poisonous substance, and may be partial or complete, focal or diffuse, temporary or permanent. See also congestive heart failure. **renal failure**. any condition whereby the kidneys are unable to perform adequately. Such conditions may be temporary or permanent; partial or complete; chronic or acute. **respiratory failure**. any condition whereby the lungs are unable to effect the exchange of respiratory gases between blood and ambient air. Proximate causes include lesions of lung tissue and weakness or paralysis of the respiratory muscles. *Cf.* collapse (respiratory collapse).

fainting. syncope.

fair lady. *Atropa belladonna*.

fairy bells. *Digitalis purpurea*.

fairy finger. *Digitalis purpurea*.

fairy gloves. *Digitalis purpurea*.

fairy thimbles. *Digitalis purpurea*.

fall crocus. *Colchicum autumnale*.

fall poison. *Eupatorium rugosum*.

fallout. **1:** particulate materials from the atmosphere that have settled on fixed surfaces in the environment. **2:** the deposition of materials, especially of fissionable materials onto surfaces from the atmosphere. **atomic fallout**. radioactive fallout. **chemical fallout**. hazardous chemicals (produced chiefly by power plant and other industrial discharges) that are discharged into the atmosphere and subsequently fall to earth. **local fallout**. fallout that is deposited within

250 km of the source (e.g., an explosion) usually within a few hours. **radioactive fallout** (fallout, atomic fallout). the particulate materials that descend to the earth following a nuclear explosion or nuclear accident. Fallout may be highly toxic, the most hazardous materials being the fission fragments, iodine^{131} and strontium^{90}. **stratospheric fallout**. fine particulates (e.g., from volcanic eruptions, from a nuclear explosion) that are deposited widely over the earth over many months, or even many years. **tropospheric fallout**. fine particles that are transported and deposited widely around the earth within about a week following the event that produced them, roughly at the same latitude of the explosion.

false acacia. *Robinia pseudoacacia.*

false black widow. a common name of *Steatoda grossa.*

false chanterelle. *Omphalotus olearius.*

false hackled band spinners. spiders of the family Loxoscelidae.

false hellebore. a name commonly applied to a number of unrelated plants (e.g., *Adonis*, *Veratrum*). See hellebore.

false horned vipers. vipers of the genus *Pseudocerastes.*

false indigo. See *Baptisia.*

false mistletoe. *Phoradendron* spp.

false morel. any mushroom of the genera *Gyromitra* or *Helvella.*

false pennyroyal. *Hedeoma pulegoides.*

false sago palm. *Cycas circinalis.*

family. a botanic and zoologic taxonomic group comprised of related genera.

famophos. famphur.

famphur (O,O-dimethyl-O-[*p*-(dimethylsulfamoyl)-phenyl]phosphorothioate (generic); famophos). a slightly water-soluble crystalline powder, $(CH_3O)_2P(S)OC_6H_4SO_2N(CH_3)_2$, that is very soluble in chloroform and carbon tetrachloride. It is a highly toxic organophosphate cholinesterase inhibitor used as an insecticide.

fang. any long, sharp tooth by which an animal seizes, holds, envenomates, or tears its prey. **cheliceral fang**. the sharp, pointed second segment of a spider's chelicera. The venom is introduced into the prey through pores in the tip of the fang. The fangs of most spiders are too short or delicate to penetrate human skin.

fanweed. *Thlaspi arvense.*

farsightedness. hypermetropia.

FAS. fetal alcohol syndrome.

fasciculation (twitching). rapid, often continuous contraction of a skeletal muscle bundle that fails to produce a purposeful movement. *Cf.* convulsion, tremor.

fasciitis. See calcinosis.

Fastin. phentermine.

fat-footed clitocybe. *Clitocybe Claviceps.*

fatal. **1:** inevitable. **2a:** of, pertaining to, or productive of death. **2b:** causing death; able to cause death. **2c:** inescapably death causing.

fatality. **1:** a disease, condition, accident or action that results in death. **2:** any instance or individual occurrence of death.

fate. possible or likely future states or ultimate future state of an entity. **chemical fate**. environmental fate. **environmental fate** (chemical fate). a term usually applied to a pollutant or to a toxic or hazardous chemical meaning the final state (or sometimes, intermediate states) of a chemical at a given location or within a given environment as a result of its movement through the environment and the chemical and physical transformations that it undergoes (e.g., export, mineralization, storage). Fate is largely governed by the equilibrium distribution of the chemical between environmental media (e.g., soil, water, air, organisms), the environmental conditions that obtain (e.g., temperature, pH), and the distribution of the media. Properties of a chemical that help to determine the rate and path of chemical change and the ultimate fate of the chemical include water solubility, vapor pressure, molecular weight, melting point, and the effects of light, heat, air, soil, and water on its behavior.

fathead minnow. *Pimephales promelas*.

fatigue. **1a:** tiredness or weariness, it is a common complaint that is usually caused by overwork, lack of sleep. Chronic fatigue is sometimes caused by depression or anxiety. See also exhaustion. Alternative causes of fatigue usually produce additional symptoms (e.g., anemia). **1b:** a generalized feeling of tiredness, listlessness, or exhaustion due to prolonged exertion or the effects of psychotropic drugs. There is usually a decrement in performance and a build up of lactic acid in skeletal muscles. **1c:** a condition of discomfort and of muscular weakness such that one works inefficiently and with difficulty. **2:** the state of an organ or tissue whereby the response to a repeated or long-continued stimulus weakens or ceases. **3:** to induce fatigue as defined above.

fatty acid. See carboxylic acid.

fatty tissue. adipose tissue.

fava bean. *Vicia faba*.

favism (fabism). an acute form of hemolytic anemia that occurs in Italy following the ingestion of fava beans (e.g., of *Vicia faba*) or inhalation of the pollen. It affects individuals that have an inherited deficiency of glucose-6-phosphate dehydrogenase. Symptoms include fever, vomiting, headache, abdominal pain, diarrhea, severe anemia, and sometimes prostration and coma.

FB/2. diquat.

FCC. Food Chemicals Codex.

FDA. U.S. Food and Drug Administration.

FD&C. food, drug, and cosmetic. *Cf*. D&C

FD&C Blue No. 1 (Blue 1; 1206 Blue; Aizen Food Blue; C.I. food blue; Cosmetic Blue Lake; D&C Blue No. 4; Food Blue 2; Food Blue Dye No. 1; etc.) a dark purple to bronze powder used as a food coloring in bottled drinks, candy, and baked goods. It can cause chromosomal damage, is possibly carcinogenic, and is banned from French and Finnish foods. Very toxic fumes, NO_x, Na_2O and SO_x, are released on decomposition by heating.

FD&C Blue No. 2 (Blue 2; 1311 Blue; 12070 Blue; C.I. 73015; C.I. acid blue 1; C.I. food blue 1; disodium indigo-5,5-disulfonate; numerous other names) a blue-brown to red-brown powder, $C_{16}H_{10}N_2O_8S_2{\cdot}2Na$, that is soluble in water and concentrated H_2SO_4; slightly soluble in ethanol. It is used as a food coloring in candy, bottled drinks, and pet foods. It is very toxic via intravenous routes; moderately toxic by oral or subcutaneous routes. It can cause brain tumors and is possibly carcinogenic. Its use is forbidden in Norwegian foods. Very toxic fumes, NO_x, SO_x and Na_2O, are released on decomposition by heating.

FD&C color. See color.

FD&C Green No. 2 (C.I. 42095; C.I. acid green; C.I. food green 2). a moderately toxic compound, $C_{37}H_{36}N_2O_9S_3{\cdot}2Na$, by ingestion and possibly carcinogenic.

FD&C Green No. 3 (C.I. 42053; C.I. food green 3). a red to brownish-violet powder, $C_{37}H_{36}N_2O_{10}S_3{\cdot}2Na$, that is soluble in water and concentrated sulfuric acid. It is a possible human carcinogen and mutagen.

FD&C Red No. 1 (disodium 3-hydroxy-4-((2,4,5-trimethylphenyl)azo)-2,7-naphthalenedisulfonate; disodium 3-hydroxy-4-((2,4,5-trimethylphenyl)azo)-2,7-naphthalenedisulfonic acid; disodium 3-hydroxy-4-((2,4,5-trimethylphenyl)azo)-2,7-naphthalenedisulphonate; C.I. 16155; C.I. food red 6; C.I. food red). a suspected carcinogen, $C_{19}H_{16}N_2O_7S_2{\cdot}2Na$.

FD&C Red No. 2. amaranth.

FD&C Red No. 3. erythrosine.

FD&C Red No. 19 (Food Red 15). rhodamine B.

FD&C Red No. 40 (C.I. 16035). a red, water-soluble powder, $C_{18}H_{14}N_2O_8S_2Na_2$, that is slightly soluble in absolute alcohol. It is a suspected carcinogen.

FD&C Violet No. 1 (C.I. 42640; C.I. food violet 2). $C_{39}H_{41}N_3O_6S_2 \cdot Na$, a suspected carcinogen.

FD&C Yellow No. 3 (C.I. 11380; C.I. food yellow 10; C.I. solvent yellow 5). a moderately toxic color by ingestion or skin contact. It is a questionable carcinogen.

FD&C Yellow No. 5. tartrazine.

FD&CA. U.S. Federal Food, Drug and Cosmetic Act.

Fe. the symbol for iron (ferrum).

Fea's viper. *Azemiops feae*.

feather. an integumental structure that is uniquely characteristic of birds (Class Aves). It is the structural unit of plumage. See also integument. Feathers accumulate lead, mercury, and cadmium for example, and have thus been used for some time as indicators of heavy metal loadings in terrestrial and marine environments. The keratin in bird feathers is not easily degradable, and mercury is probably firmly linked to the disulfide bonds of keratin. Because feathers do not readily biodegrade, retrospective analyses spanning long periods of time are possible.

feather geranium. *Chenopodium botrys*.

febricant. causing fever, febrific. *Cf.* febrifacient.

febricide. **1:** the lowering of body temperature during fever. **2:** an agent that allays fever.

febricity. feverishness; the state of being febrile.

febrifacient. **1:** febrific, febriferous; promoting, inducing, or conveying fever. *Cf.* febricant. **2:** any agent that induces or promotes fever.

febriferous. febrifacient (def. 1).

febrific. febrifacient (def. 1).

febrifugal. antipyretic (def. 1).

febrile (feverish; pyretic; pyrectic). **1:** pertaining to or denoting fever or feverishness; feverish. **2:** an agent that induces fever. *Cf.* pyretic.

febris. fever.

February daphne. a common name of plants of the genus *Daphne*.

fecal dropping. See scat.

feces. solid or semisolid material, composed of unabsorbed food residues, bacteria, and secretions that is expelled from the gastrointestinal system. Feces in vertebrate animals are stored in the colon until expelled via the anus. Many xenobiotics or their metabolites leave the animal body via urine or feces (e.g., cadmium). Feces may thus contain poisons or residues of poisons. Monitoring feces for poisons can be misleading because the contents generally reflect only very recent exposures. Furthermore, the composition of feces may be quite variable and may not reflect the actual diet. Fecal material of birds is mixed with excreta from the kidneys. *Cf.* egesta, stool.

fei-shang-ts'ao. *Agkistrodon halys*.

felonwart. See *Chelidonium*.

felt-ringed agaricus. *Agaricus hondensis*.

feminine douche. See douche.

feminine hygeine spray. See douche.

femto- (f). a prefix to units of measurement that indicates a quantity of 1 trillionth 10^{-15} the value indicated by the root. See International System of Units.

fenac. chlorfenac.

fenchlorphos. Ronnel.

Fenclor. polychlorinated biphenyl(s).

fenformin. phenformin.

fenesterin. fenesterine.

fenestrin. fenesterine.

fenolovo. triphenyltin hydroxide.

fenoprop. silvex.

fensulfothion (O,O-diethyl-O-[p-(methylsulfinyl)-phenyl]phosphorothioate). a liquid anti-cholinesterase, $C_{11}H_{17}O_4PS_2$, used as an insecticide, nematocide, and is used in some household products to eradicate pests. It is a dangerous substance and is toxic by all routes of exposure, including skin absorption.

fentanyl (phentanyl; *N*-phenyl-*N*-[1-(2-phenyl-ethyl)-4-piperidinyl]propanamide; *N*-(1-phenylethyl-4-piperidyl)propionanilide; *N*-(1-phenylethyl-4-piperidinyl)-*N*-phenylpropionamide). a toxic opioid analgesic, it is a controlled substance in the United States. The citrate (phentanyl) is extremely toxic to laboratory mice.

fenthion (phenthion; phosphorothioic acid *O,O*-dimethyl *O*-[3-methyl-4-(methylthio)phenyl] ester; *O,O*-dimethyl *O*-(4-methylmercapto-3-methylphenyl) thionophosphate; *O,O*-dimethyl *O*-(4-methylthio-3-methylphenyl) thiophosphate; *O,O*-dimethyl *O*-[4-(methylthio-*m*-tolyl] phosphorothiolate; 4-methylmercapto-3-methylphenyl dimethylthiophosphate). an amber-colored, liquid organophosphate insecticide, acaricide (mainly for use with ornamental plants) and veterinary ectoparasiticide with a slight odor of garlic. Fenthion is a cholinesterase inhibitor. It is very toxic to laboratory rats and highly toxic to fish and hymenoptera. See also parathion, anticholinesterase.

fentin hydroxide. triphenyltin hydroxide.

fenuron (*N,N*-dimethyl-*N*-phenylurea; *N*-phenyl-*N',N'*-dimethylurea). a white crystalline herbicide, $C_6H_5NHCON(CH_3)_2$ used to control deep-rooted perennial weeds and bushes, especially *Convolvulus arvensis* (field bindweed). This compound binds less strongly to soil and is thus effective to greater soil depths than, for example, monuron. It is slightly toxic to laboratory rats. See also fenuron TCA.

fenuron TCA. the trichloroacetate of fenuron, it is a white crystalline herbicide used in non-crop areas especially to control woody plants. If hydrolyzed endogenously to aniline, it can cause methemoglobinemia. *Cf.* diuron. See also fenuron.

fenvalerate (4-chloro-α-(1-methyl-ethyl)benzene-acetic acid cyano(3-phen-oxyphenyl)methyl ester; α-cyano-3-phenoxy-benzyl α-(4-chlorophenyl)isovalerate; cyano-(3-phenoxyphenyl)-methyl 4-chloro-α-(1-methylethyl)benzeneacetate; α-cyano-3-phenoxybenzyl-2-(4-chlorophenyl)-3-methylbutyrate; phenvalerate; pydrin; sumicidin). a broad-spectrum insecticide and veterinary ectoparasiticide. It is neurotoxic, acting on both the central and peripheral nervous systems and an eye and skin irritant. It is very toxic to laboratory rats and highly toxic to fish and hymenopterans. Symptoms of intoxication in humans may include excessive salivation, nausea and vomiting, tremors, incoordination, diarrhea, hypersensitivity to touch and sound, convulsions, and death.

FEP. free erythrocyte protoporphyrin (See protoporphyrin).

Fer-de-lance. the usual common name in English of *Bothrops atrox* (formerly *Bothrops asper*). It is a French term meaning "lancehead." Other species of *Bothrops* (especially *B. lanceolatus*) and even some Asian lance-headed vipers (*Trimeresurus*) are sometimes called Fer-de-lance.

feraconitine. pseudoaconitine.

feral. **1a**: living normally in the wild. **1b:** referring to plants or animals that were domesticated but have reverted to the wild state. **2:** untamed, undomesticated. **3:** wild or savage. See also animal (*Cf.* domestic animal, feral animal, laboratory animal).

ferbam. the iron salt of ethylenebisdithiocarbamate. Its toxicity to animals is low. See dithiocarbamate fungicides.

fermium (symbol, Fm)[Z = 100; most stable isotope = ^{257}Fm, $T_{1/2}$ = 80 days]. an artificial radioactive transuranic element produced by the bombardment of plutonium by neutrons.

fern palm. *Cycas circinalis*.

ferric chloride (anhydrous ferric chloride; ferric trichloride; flores martis; iron chloride). a brown, poisonous, crystalline compound, $FeCl_3$, that is soluble in water, ethanol and glycerol. It is used in copper plating, photoengraving, as a coagulant and deodorizer in the treatment of industrial and municipal wastes, as a disinfectant, a chlorinating and oxidizing agent, a mordant, and as an intermediate in the manufacture of inks, pigments, and other iron salts. Anhydrous ferric chloride is irritant and highly toxic. The toxicity in solution depends on the iron content.

ferric ferrocyanide (hexakis(cyano-C)ferrate-(4-)-iron(3+) (3:4); ferric hexacyanoferrate (II); iron ferrocyanide; Berlin blue; Chinese blue; Hamburg blue; mineral blue; Paris blue; Prussian blue). a blue pigment used in paints, printing inks, carbon papers, typewriter ribbons, linoleum, etc. It has been used as an antidote for thallium poisoning.

ferric hexacyanoferrate (II). ferric ferrocyanide.

ferric hydrate. ferric hydroxide.

ferric hydroxide (ferric hydrate; iron hydrate; iron oxide, hydrated; ferric oxide, hydrated; iron hydroxide). the hydrated oxide of iron, $Fe(OH)_3$, a brown or reddish-brown flocculent precipitate which forms the oxide on drying. It is used in the purification of water, as a catalyst, a rubber pigment, and was formerly used as an antidote in arsenic poisoning.

ferric oxide. See iron oxide, ferric hydroxide.

ferric oxide, hydrated. ferric hydroxide.

ferric trichloride. ferric chloride.

ferrisulfas. ferrous sulfate.

ferrisulphas. ferrous sulfate.

ferrohaemoglobin. hemoglobin.

ferrohemoglobin. hemoglobin.

ferrous fumarate. a toxic, reddish-brown, odorless, flammable, powdery, essentially tasteless, chemically stable, anhydrous salt of a mixture of fumaric acid and ferrous iron, $FeC_4H_2O_4$. It is a widely prescribed dietary supplement that contains 33% elemental iron by weight. It can be dangerous in the home. Up to 10 tablets may produce mild poisoning in a child; up to 20 tablets may prove moderately toxic; and >20 tablets may cause severe poisoning. See iron poisoning.

ferrous gluconate (iron gluconate). a fine yellowish-gray or pale greenish-yellow powder or granules, $Fe(C_6H_{11}O_7)\cdot 2H_2O$. It is widely used as a feed and food additive and in vitamin tablets. It can be dangerous in the home. It contains 33% elemental iron. Up to 10 tablets may produce mild poisoning in a child; up to 20 tablets may prove moderately toxic; and >20 tablets may cause severe poisoning. See iron poisoning.

ferrous sulfate (ferrisulfas; copperas; green copperas; green vitriol; iron vitriol; iron sulfate; sal chalybis). a blue-green, water-soluble, crystalline substance, $FeSO_4\cdot 7H_2O$. It is moderately toxic to laboratory mice, per os, and very toxic *i.v.* It is toxic to humans, causing gastrointestinal disturbances. Ingestion of large amounts by children can cause hematemesis, tachycardia, liver damage, and peripheral vascular collapse. Ferrous sulfate has a number of applications, including usage as a herbicide, wood preservative, and is used in some pesticides.

ferrum. iron.

fertilization. the fusion of male and female pronuclei, forming a single cell (the zygote) and triggering development.

ferulic acid (3-(4-hydroxy-3-methoxyphenyl)-2-propenoic acid; 4-hydroxy-3-methoxycinnamic acid; 3-methoxy-4-hydroxycinnamic acid; caffeic acid 3-methyl ether). a plant growth inhibitor found in low concentrations in many species of plants. It is usually nontoxic to the producing plant. There are two isomers. The *cis*-form is a yellow oil and the *trans*-form precipitates from water solutions as orthorhombic crystals.

FES. zearalenone.

fescue. any grass of the genus *Festuca*.

fescue foot (fescue lameness; gangrenous fescue poisoning). a dry gangrene of livestock (mainly cattle; sheep are rarely affected) caused by ingestion of toxic tall fescue, *Festuca arundinacea*. The symptoms and lesions are essentially similar to those of gangrenous ergot poisoning, but fungal sclerotia have not been found in poisonous fescue. Lameness of one or both hind feet usually develops within 10-14 days. In severe cases, the front feet and tips of the ears may be affected. Resistance varies among cattle. Some animals recover spontaneously, but the disease usually terminates fatally. An endophytic fungus that infests *Festuca arundinacea* (tall fescue) has been implicated in fescue foot. This fungus was first identified as *Epichloe typhina* and has recently been renamed *Acremonium coenophialum*. It is not clear whether the toxic principle is produced directly by the fungus or whether it is a product of the fungus-plant association. *Cf.* summer fescue toxicosis.

fescue lameness. fescue foot.

fescue poisoning. See fescue foot.

fescue toxicosis. See summer fescue toxicosis. See also fescue foot.

Festuca (fescue). a genus of grasses (Family Arundinaceae). ***F. arundinacea*** (*Festuca elatior* var. *arundinacea*; fescue; tall fescue). a coarse, deeply rooted perennial grass (Family Arundinaceae) that reaches a height of more than a meter. It is sometimes used in lawns and athletic fields. Fescue may be infested by a endophytic fungus (*Acremonium coenophialium*) which contains toxic alkaloids. Livestock consuming the infested grass may develop "fescue foot," *q.v.* Fescue can also accumulate toxic levels of nitrate. ***F. elatior* var. *arundinacea***. *F. arundinacea*. ***F. fallax***. *F. rubra* var. *commutata*. ***F. rubra* var. *commutata*** (*F. fallax*; chewing fescue; creeping fescue). this species is used in lawn seed mixtures and turf. The seeds and hay infested with galls of the nematode, *Anguina agrostis*, are neurotoxic and have caused fatalities in most classes of livestock. Symptoms in sheep and cattle are similar and include ataxia, tremors, and falling.

fetal (foetal). pertaining to, denoting, or having the characteristics of a fetus.

fetal alcohol syndrome (FAS). the full set of associated birth defects that may occur in infants whose mothers continue to drink alcoholic beverages throughout pregnancy. Affected infants may show signs of alcohol withdrawal soon after birth, and some will have physical abnormalities and may show insufficient weight gain and mental deficiency. See also fetal alcohol effect.

fetal death. See death (fetal death).

fetal membrane. See extraembryonic membrane.

fetal trimethadione syndrome. a fetal syndrome due to ingestion of trimethadione (an anticonvulsant) during the early stages of pregnancy. Embryonic/fetal development is delayed and the fetus develops v-shaped eyebrows, an epicanthic fold, low-set ears with anteriorly folded helices, a palatal anomaly, and irregular teeth.

fetal warfarin syndrome. a fetal bleeding syndrome due to administration of warfarin to a pregnant woman. The disease is marked by fetal bleeding, nasal hypoplasia, optic atrophy, and death.

fetation (foetation). the development of the fetus within the uterus.

feticide. the destruction of a fetus.

fetid nightshade. *Hyoscyamus niger*.

fetotoxic. toxic to a fetus; any agent that is toxic to the fetus. Fetotoxic substances are numerous and include ethanol, morphine, salicylates, tetracyclines, thiazides, tobacco smoke, and large doses of vitamin K.

fetotoxicity. **1:** the property, state, or degree of being toxic to the fetus. **2:** injury, retardation, or death of the fetus due to exposure to a toxic agent. Manifestations of fetotoxicity include: lethality, impaired growth, mental retardation, and other dysfunctional states.

fetter bush. a common name of plants of the genus *Leucothoe*, and also of *Lyonia lucida* and *Pieris floribunda*.

fetus (foetus). the latter stages of development of the unborn young of a vertebrate animal (i.e., from the completion of organogenesis) within the uterus or the egg. In *Homo sapiens* the conceptus is designated as a fetus from the 9th week following fertilization until birth. Prior to this time, it is referred to as the embryo.

feu de Saint-Antoine. St. Anthony's fire (See ergotism).

fever (febris; pyrexia). a condition marked by a body temperature elevated above that which is considered normal for a species or population.

feverish. febrile.

FFA. free fatty acid.

fibrillate. **1:** having fibrillae or hair-like structures. **2:** exhibiting spontaneous, uncoordinated, continuously changing patterns of electrical activity and associated movements of muscle fibers, especially in fibrillation of the heart muscle. The muscle is unable to contract as a unit. If it occurs in the ventricles, fibrillation of the heart can be fatal.

fibrillation. **1:** spontaneous, uncoordinated contractions of muscle fibrils as may occur in skeletal muscle following severing of the nerve supply; or especially in the heart. **2:** the formation of fibrils. **3:** any abnormal biolectric potential. **atrial fibrillation**. fibrillation of the atrial muscles of the heart; the contractions are extremely rapid, incomplete, and uncoordinated. **ventricular fibrillation**. similar to atrial fibrillation, but involves the ventricular muscle of the heart. Contractions are rapid, tremulous, and ineffectual in pumping blood. This phenomenon may result from trauma, including the trauma of surgery; electrical stimuli, or the action of certain drugs or toxicants such as digitalis or chloroform. See fibrillate.

fibrin. the fibrous, insoluble protein that establishes the gel matrix of a blood clot and determines the semisolid character of the clot. Fibrin is produced by the action of thrombin on fibrinogen.

fibrinogen. a soluble dimeric protein consisting of three pairs of polypeptide chains that is converted to fibrin by the enzyme thrombin in the presence of calcium ions. It is the inactive precursor of fibrin.

fibroblast. a type of connective tissue cell that forms fibers. They are widely employed in mutagenicity testing.

fibrosarcoma. See sarcoma.

fibrositis. inflammation of connective tissue of the muscles, joints, ligaments, and tendons.

Fick's law. the number of molecules that cross a given area per second is proportional to the area and to the concentration gradient in a direction normal to the area crossed. See also diffusion, coefficient of.

ficusin. psoralen.

fiddle-back spiders. *Loxosceles*.

fiddleneck. *Amsinckia intermedia*.

field mustard. *Brassica rapa*.

fierce snake. *Parademansia microlepidota*.

figwort family. Scrophulariacieae.

Fiji snake. *Ogmodon vitianus*.

film badge. a badge containing X-ray-sensitive photographic film. It is worn on the person to determine cumulative exposure of persons potentially exposed to radiation.

filtration barrier. See barrier.

finger cherry. See *Rhodomyrtus macrocarpa*.

fire ant. See *Solenopsis*.

fire coral. See *Millepora*.

fire-bellied toad. See Bombina.

fireball. *Kochia scoparia*.

firedamp. methane that seeps into coal mines. It sometimes accumulates, forming an explosive mixture. The degraded air remaining in the mine following such an explosion is called blackdamp, and the carbon monoxide that is generated by such an explosion is called afterdamp.

Firemaster FF-1. polybrominated biphenyl mixture (See polybrominated biphenyl).

firethorn. See *Pyracantha coccinea*.

first mediator of damage. hypothetical substance(s) that originate in the area of an injury or insult and transmit(s) a message to other parts of the body stimulating the systemic responses of damage or shock. The "first mediator of hormonal defense" may be identical.

first mediator of hormonal defense. hypothetical substance(s) that originate in the area of an injury or insult and transmit(s) a message to the pituitary that stimulates the release of ACTH. The "first mediator of damage" may be identical.

first-order kinetics. the progress of a reaction in which the rate is proportional to the amount of material present. It is described by a proportionality constant (the elimination rate constant) which relates the rate of elimination of a substance from the reacting system to the amount present.

fish berry. See *Cocculus indicus*.

fish poison. tetrodotoxin.

fish poisoning. **1:** a type of food poisoning caused by eating decomposing, inherently poisonous, or infected fish. See also ichthyotoxism.

fit. See convulsion.

Fitzinger's coral snake. *Micrurus fitzingeri*.

five pacer. *Agkistrodon acutus*.

fiyah. *Cerastes cerastes*.

FK506. a potent immunosuppressive agent produced by *Streptomyces tsukubaensis*. It has been used experimentally to prevent the rejection of transplants. It is nephrotoxic and also has neurotoxic properties and may be diabetogenic.

flaccid. relaxed, limp, not turgid, lacking tone.

flaccidity. the condition of being flaccid.

flagroot. *Acorus calamus*.

flake lead. basic lead carbonate.

flame retardant. a substance which when used to treat a flammable material will retard the spread of flames. Many such substances are toxic. TRIS (See tris(hydroxymethyl)aminomethane), a leading flame retardant, is mutagenic and carcinogenic to animals. It can be absorbed through the skin from garments that have been repeatedly washed. Materials treated with tetrakis hydroxyl-methyl phosphonium chloride (THPC), another retardant, release formaldehyde when the fabric is wet. The following are also used as flame retardants: tetrakis hydroxyl-methyl phosphonium (THP), phenol, polybrominated biphenyls (PBBs), and polychlorinated biphenyls (PCBs).

flammable. combustible, inflammable, capable of burning; incindiary, the capacity of a substance or material to ignite and burn readily. See also ignition.

flannel moth dermatitis. See *Megalopyge operculatis*.

flapping tremor. asterixis.

flare-and-wheal reaction. a skin condition characterized by local edema and a red flare. It is a reaction of the skin to injury or the administration of antigens, caused by the release of histamine and related molecules.

flat-nosed pit viper. *Trimeresurus puniceus*.

flatpod pea. *Lathyrus cicera*.

flavan. catechin.

3,3′,4′,5,7-flavanpentol. catechin.

flavin enzyme. flavoprotein.

flavoprotein (flavin enzyme; yellow enzyme). a dehydrogenase composed of protein and mononucleotide or dinucleotide coenzymes that contain riboflavin. Flavoproteins function in tissue respiration, the hydrogen atoms being taken up by the riboflavin group.

flavoskyrin. an anthraquinoid mycotoxin that is extremely toxic to cultured HeLa cells, the protozoan *Tetrahymena pyriformis*, and *Escherichia coli* mutants.

[illegible]

flax olive. *Daphne mezereum*.

flesh. soft animal tissue, especially meat, the color of which, and sometimes the odor (in mammals and birds), can sometimes offer clues to the cause of death including death by poisoning. Flesh that is very pale or dark purple, odorous, soft or inelastic to the touch, or moist may be inedible. A green or violet coloration is characteristic of the onset of toxic putrefaction; an intense, bright red may result from toxic action of bacteria; dark reddish-brown may indicate poisoning, drowning or suffocation; scarlet coloration indicates poisoning by arsenic or carbon monoxide; white flesh indicates certain diseases and is not to be ingested. *Cf.* meat.

flint disease. chalicosis.

FLIT-MLO. a mosquito larvicide composed of petroleum-derived aliphatic hydrocarbons.

floor polish. any commercially available, usually volatile liquid used to polish floors or floor coverings. Toxicological considerations regarding floor polish are essentially the same as those for furniture polish, *q.v.*

floor sweep. a vapor collector designed to capture vapors which are heavier than air and which accumulate above the floor.

flora. **1:** plant life, especially of a designated locality or region. **2:** microflora, the set of microbes that populate external and internal surfaces of a normal healthy organism. *Cf.* microflora.

flores martis. ferric chloride.

Florida arrowroot. *Zamia integrifolia*.

flowering spurge. *Euphorbia corollata*.

flowers of arsenic. arsenic trioxide.

flowers of benzoin. benzoic acid.

flowers of zinc. zinc oxide.

fluohydric acid. hydrofluoric acid.

fluometuron (*N,N*-dimethyl-*N'*[3-(trifluoromethyl)phenyl]urea; 1,1-dimethyl-3-(α,α,α-trifluoro-*m*-tolyl)urea; *N*-(3-trifluoromethylphenyl)-*N',N'*-dimethylurea). a herbicide with a half-life in soil of about 60 days, it is absorbed mainly through the roots of plants. It is used to control broadleaved and weedy grasses in cotton. The alkoxyurea herbicides are similar in action, but are more water soluble and less persistent in soil. Fluometuron is very toxic to laboratory rats.

fluophosgene. carbonyl fluoride.

fluoranthene (1,2-benzacenaphthene; benzo-(jk)-fluorine; 1,2-(1,8-naphthalenediyl)benzene; 1,2-(1,8-naphthylene)benzene). a colorless, flammable, tetracyclic hydrocarbon, $C_{10}H_{10}$, with monoclinic crystals that is soluble in common organic solvents such as benzene and ethyl ether. It occurs in petroleum and coal tar fractions. Fluoranthene is poisonous *i.v.*, and is moderately toxic when ingested or via dermal contact. It is an experimental tumorigen, a questionable carcinogen, and a probable human mutagen.

***N*-2-fluorenylacetamide** (2-acetamidofluorene; 2-AAF; 2-acetaminofluorene; 2-acetylaminofluorene; 2-FAA; AAF; acetoaminofluorene; *N*-9*H*-fluoren-2-yl-acetamide; *N*-fluoren-2-yl acetamide). a compound, $C_{15}H_{13}NO$, that is moderately toxic by ingestion and intraperitoneal routes. It is a potent carcinogen to some laboratory animals and is a confirmed, potent, human carcinogen. It is metabolically activated via *N*-hydroxylation in dogs and hamsters (but not in guinea pigs) to *N*-hydroxyacetamidofluorene, the proximate carcinogen. As a consequence, this compound has not proven to be carcinogenic in guinea pigs. 2-FAA is teratogenic in laboratory mice, rats, and chicks. It releases toxic NO_x fumes when heated to decomposition. While originally developed as an insecticide, it is now used chiefly as a research tool in studies of cytotoxicity, hepatotoxicity, genotoxicity, and carcinogenesis.

***N*-fluoren-2-yl acetamide**. *N*-2-fluorenylacetamide.

***N*-9*H*-fluoren-2-yl-acetamide** *N*-2-fluorenylacetamide.

fluoric acid. hydrofluoric acid.

fluoridation. the addition of fluoride to a substance; used especially in reference to fluoridation of drinking water.

fluoride. any salt of hydrofluoric acid, HF, in which florine has a valence of -1. Fluorides aremajor environmental pollutants released into the atmosphere from aluminum reduction, steel manufacturing, and coal-fired power production. Fluoride pollution is associated with extensive damage to livestock, agricultural crops, and timber. Fluoride has a strong tendency to bioaccumulate and can thus be hazardous at very low environmental concentrations. Some plants accumulate tissue concentrations of fluoride that are on the order of 10^6 times that of the ambient air; fluorides have high biological activity that can induce a variety of toxic effects in plants and animals. Some inorganic fluorides are highly toxic; sodium fluoride and sodium monofluorophosphate are, for example, very toxic to laboratory rats. See fluoride toxicosis.

fluoride ion (F^+). the highly toxic component of hydrogen fluoride and of soluble fluoride salts (e.g., sodium fluoride). At low concentrations (ca. 1 ppm) in drinking water, it helps to prevent tooth decay, but higher concentrations cause fluorosis, *q.v.* Livestock may be poisoned from the deposition of fluoride from industrial pollution on grazing land. As a consequence, animals become lame and occasionally die.

fluoride poisoning. See fluoride toxicosis, fluorosis.

fluoride toxicosis. systemic, usually chronic (fluorosis), fluoride poisoning. More acute poisoning by fluorides may occur as a result of exposure to hydrogen fluoride or its soluble salts. Symptoms of intoxication may include a peculiar taste, excessive salivation, thirst, nausea, vomiting, abdominal pain, diarrhea, muscular weakness, tremor, central depression, and shock. Death may ensue due to vascular collapse. Typical lesions are those of hemorrhagic gastroenteritis, hypocalcemia, and the inhibition of cellular glycolysis. See also fluorosis, fluoride.

fluorine (symbol F)[Z = 9; A_r = 18.998]. a slightly greenish, pale yellow, highly toxic, corrosive, and flammable nonmetallic gaseous or liquid element, F_2 (m.p. -218°C, b.p. -187°C), belonging to the halogen group of elements. Fluorine is the most powerful oxidizing agent and the most electronegative chemical known. It can react directly with nearly all elements including the "inert" gas xenon. Fluorine is extremely dangerous and destructive of skin, mucous membranes, and other tissues. Accidental contact or inhalation should be treated as a medical emergency. Some of its more soluble salts are highly reactive and poisonous.

fluorine perchlorate (chlorine tetroxyfluoride). an extremely poisonous, colorless, explosive gas with a pungent odor. Inhalation of trace amounts is destructive of lung tissue.

fluorite. a pure naturally occurring form of calcium fluoride, CaF_2. At one time it was widely used as a flux in metallurgical industries, polluting the local air with fluorine compounds.

fluoroacetamide (fluoroacetic acid amide; monofluoracetamide). a crystalline solid, freely soluble in water. It is used as a rodenticide and sometimes as an insecticide. It is extremely toxic to laboratory mice and rats.

fluoroacetate. any salt of fluoroacetic acid in which carbon-bound hydrogen atoms are replaced by fluorine atoms. All are rapid-acting and extremely poisonous. Fluoroacetates are probably not directly toxic but are rapidly converted *in vivo* to fluorocitrate which inhibits aconitase and hence the tricarboxylic acid (Krebs) cycle. It may used as a rodenticide in the United States only by registered pest control operators. Toxic doses in humans and other mammals increase respiratory rate, and cause vomiting, facial paresthesias, CNS stimulation which may advance to convulsions, alternating with periods of severe CNS depression, cardiac arrhythmias, and sometimes death from ventricular fibrillation. Secondary poisoning of cats, dogs, and other animals by consumption of carcasses of fluoroacetate-killed rodents is possible. See also, aconitase, *Acacia georginae*.

fluoroacetic acid. sodium monofluoroacetate.

fluoroacetic acid amide. fluoroacetamide.

fluoroacetic acid, sodium salt. sodium fluoroacetate.

fluorocarbon (fluorinated hydrocarbon). **1:** any organic compound in which one or more hydrogen atoms have been replaced by fluorine. While all are non-flammable, chemically inactive liquids or compressed gases, they react violently with substances such as barium, potassium, and sodium. They have many industrial and household applications. **1a:** sometimes used synonymously with chlorofluorocarbon, *q.v.* This usage is not recommended; the latter is, rather, a subclass of fluorocarbons.

fluorogesarol. DFDT.

5-fluoro-2,4(1*H*,3*H*)-pyrimidinedione. fluorouracil.

fluorosilic acid (hydrofluosilicic acid; fluosilic acid; hexafluorosilicic acid; hydrogen hexafluorosilicicate; hydrosilicofluoric acid; hydrofluorosilicic acid; silicofluoric acid). a transparent, colorless, fuming liquid (in aqueous solution), $F_6Si{\cdot}2H$, that attacks glass and stoneware. It is used in the fluoridation of water, in ceramics and cement to increase hardness, to disinfect copper and brass vessels, as a wood preservative, in electroplating, the production of synthetic cryolite, aluminum fluoride, and hydrogen fluoride, and to sterilize bottling and brewing equipment. It is very toxic by ingestion, inhalation, and the subcutaneous route; it is also highly corrosive to the skin, eyes, and mucous membranes. The concentrated acid produces toxic, corrosive fumes and on decomposition by heating releases toxic F^-.

fluorosilicate (fluosilicate; fluorosilicate salt). any salt of fluorosilic acid. All (e.g., salts of ammonium, barium, magnesium, sodium, and zinc) are very toxic (as toxic as the corresponding fluorides), yielding fluoride ions on solution; they act primarily by altering the calcium balance of the blood. Most fluorosilicates are more soluble than the corresponding fluoride salt; an exception is sodium fluorosilicate. See, for example, lead hexafluosilicate, magnesium hexafluorosilicate, sodium hexafluorosilicate.

fluorosilicate salt. See fluorosilicate.

fluorosis. chronic fluorine poisoning, often observed in domestic cattle. Common lesions produced by fluoride in grazing livestock include mottling, pitting, hypoplasia, and hypocalcification of the teeth; hyperosteosis, osteoporosis, and osteomalacia. Symptoms may include poor nutrition, stunted growth, lameness, and irregular tooth wear. Fluorosis occurs in many regions of the earth due to the presence of rock phosphate deposits that contain high levels of fluoride, volcanic soils, fluoride pollution (of air and water), and the addition of fluorine-containing minerals (e.g., ground rock phosphate) as supplements to livestock rations. In the United States, the use of defluorinated phosphates is generally an adequate safeguard against fluorosis in livestock. Humans are subject to fluorosis, usually from high levels in drinking water. Symptoms typically include mottling of the teeth (dental fluorosis) and increased bone density (skeletal fluorosis). These effects are not usually serious, but continued intake of excessive levels can soften teeth and produce serious bone abnormalities with crippling. Severe cases are sometimes fatal. See also fluoride, fluorine, toxicosis.

fluorouracil (5-fluoro-2,4(1H,3H)-pyrimidinedione; 2,4-dioxo-5-fluoropyrimidine; 5-FU). a crystalline substance, it is an antineoplastic and inhibitor of DNA synthesis.

fluosilic acid. fluorosilic acid.

fluosilicate. See fluosilicate.

fluostigmine. diisopropyl fluorophosphate.

flush. **1a:** to redden. **1b:** a sudden redness of the skin; blush. **2:** to irrigate a cavity with water.

flushing. **1:** rinsing out with a stream of fluid. **2:** transient erythema.

fluxing lime. calcium oxide.

fly agaric. *Amanita muscaria.*

fly fungus. *Amanita muscaria.*

fly mushroom. *Amanita muscaria.*

fly-poison. *Amianthium muxcitoxicum.*

Fm. the symbol for fermium.

focal reaction. See reaction (focal reaction)

fodder beet. *Beta vulgaris.*

foetal. fetal.

foetal membrane. See extraembryonic membrane.

foetation. fetation.

foetoprotein. fetoprotein.

foetus. fetus.

fog fever. a sometimes fatal condition of cattle marked by acute respiratory distress. It presents about 1 week following placement on pasturage that has been recently mowed. The exact cause is unknown, but it is thought to be an allergic response of sensitized animals to proteins in grass, pollen, and/or fungal spores; non-immunologic poisoning has not been ruled out.

foliage beet. *Beta vulgaris*.

folic acid (*N*-[*p*-[[(2-amino-4-hydroxy-6-pteridinyl)methyl]-amino]benzoyl] glutamic acid). a widely distributed, yellow, water-soluble, crystalline vitamin and essential growth factor in many microorganisms and animals. See folinic acid.

folinic acid. 1: (*N*-[4-[[(2-amino-5-formyl-1,4,5,6-7,8-hexahydro-4-oxo-6-pteridinyl)methyl]-amino]benzoyl]-L-glutamic acid; *N*-[*p*-[[(2-amino-5-formyl-5,6,7,8-tetrahydro-4-hydroxy-6-pteridinyl)methyl]amino]benzoyl]glutamic acid; CF; citrovorum 2,3-*p*-dioxanedithiol *S,S*-bis(*O,o*-diethyl)phosphorodithioate; 5-formyl-5,6,7,8-tetrahydrofolic acid; 5-formyl-5,6,7,8-tetrahydropteroyl-L-glutamic acid; 5-formyl-5,6,7,8-tetrahydropteroyl-L-thioic acid; S,S′-1,4-dioxane-2,3-diyl *O,O,O′,O′*-tetraethyl ester; leucovorin). an active metabolic derivtive of folic acid, it reverses the effects of folic acid antagonists. See also calcium folinate.

follicle-stimulating hormone. FSH.

folliculoid. any substance that simulates the hormonal activity of an ovarian follicle.

follitropin. FSH.

fonfoni. *Causus rhombeatus*.

fonofos (ethylphosphonodithioic acid *O*-ethyl *S*-phenyl ester). an extremely toxic phosphonate. It is a cholinesterase inhibitor used as a soil insecticide and as a pesticide in some household products. It should be considered dangerous when so used. Fonofos is activated by FAD-containing, monooxygenase, and by cytochrome P-450.

food. any substance which, when ingested by a heterotrophic organism, contributes to the maintenance of life. **canned food**. food for humans or domestic animals that is stored in sealed cans. Such food may contain toxins and other toxicants such as pesticides, residual detergent used to clean the food, traces of lead from solder used to seal the can (not allowed in the United States), and additives incorporated into the food at any stage of processing. United States law requires only that food manufacturers list on the label those ingredients and additives that are introduced or combined in the final stage of processing. Canned foods imported into the United States are often more hazardous than domestic goods. See plant (food plant).

food additive. See additive.

food chain. the transfer of energy and nutrients from one organism to the next as they feed on one another. Food chains begin with green plants. Certain non-nutrient materials are also transported through the food chain as well. Notorious examples are lipophilic xenobiotics such as PCBs and organochlorine pesticides. See also food chain transfer.

food chain transfer. the movement of xenobiotics with or without metabolic transformation from one level usually to the next higher level. This sometimes occurs with increasing concentration throughout the chain or a portion of the chain. Predators or even omnivores are thus often at far more risk from poisoning by xenobiotics than organisms at lower trophic levels. At any level, certain xenobiotics may be also be absorbed passively by contact or absorption through gills, lungs, or the body surface.

food color (food coloring). any additive used to color a food product. There are seven food colors or food dyes now generally in use. All are made from coal tar and are identified by the color name and a number. Several million pounds of these additives are consumed annually in the United States alone. See, for example, Blue 1, Red 40, Yellow 5.

food coloring. food color.

food flavoring (artificial flavor; imitation flavor). any substance added to food to enhance flavor or to imitate natural flavor. More than 1500 such substances derived from petrochemicals are in use. Most are thought to be safe, but some have been linked to hyperactivity and behavioral disturbances in children. See also additive.

food plant. See plant.

food poisoning (bromatoxism). **1:** a vague term that is often used in reference to any illness that results from the ingestion of foods that contain poisonous substances. **2:** poisoning due to ingestion of decaying food that contains toxin-producing microbes. Effects may include vomiting, diarrhea, nausea, and fever. **3:** sitotoxism. See also coturnism. **bacterial food poisoning**. enteritis or gastroenteritis due to the presence of soluble bacterial exotoxins or to multiplication of gastrointestinal bacteria *per se*. This term does not pertain to enteric fevers and dysenteries. A common cause of such poisoning is *Claustridium botulinum*. **true food poisoning**. a term that encompasses poisoning due to ingestion of poisonous organisms (e.g., mushroom poisoning, shellfish poisoning); of plant or animal products (e.g., milk poisoning); or of foods that contain toxin-producing microbes. **staphylococcal food poisoning**. food poisoning due to exotoxins produced by *Staphylococcus spp.*, usually *Staphylococcus pyogenes* var. *aureus*. It is marked by violent vomiting and diarrhea.

food web. the network of interconnected food chains of an ecosystem. See also food chain.

Food Blue 2. FD&C Blue No. 1.

Food Blue Dye No. 1. FD&C Blue No. 1.

Food Chemicals Codex. a publication that gives test methods and specifications for chemicals used in food. See grade.

Food Red 15. rhodamine B.

Food Yellow 4. tartrazine.

fool's cicely. a common name of *Aethusa cynapium*.

fool's mushroom. *Amanita verna*.

fool's parsley. a common name of *Aethusa cynapium*.

foothill death camas. *Zigadenus paniculatus*.

forage. **1a:** to search for food or provisions, applied especially to wild and domestic animals. **1b:** to secure by foraging. **2a:** vegetable feed such as hay, grain, and fodder provided to domestic animals. **2b:** the food that wild or domestic animals (especially grazing and browsing animals) secure for themselves.

forage poisoning (obsolete). **1:** any poisoning of livestock associated with the ingestion of fungus-infested forage. See, for example, aflatoxicosis. **2:** moldy corn poisoning. **3:** sleepy stage.

forage sorghums. *Sorghum vulgare*.

foreign compound. xenobiotic (defs. 1b, 2).

forensic. legal; pertaining to, denoting or belonging to law, especially that which concerns courts of justice.

forensic toxicology. See toxicology.

forest cobra. See *Naja melanoleuca*.

formaldehyde (oxymethylene; formic aldehyde; methanal). a colorless, toxic, flammable, pungent gas, CH_2O, that is easily polymerized. It is commercially available in aqueous solution as formalin, *q.v.* Formaldehyde is one of a very few compounds known to occur in space. It is toxic by all routes of exposure; when working with formalin, one should avoid breathing the fumes and should not allow contact with the skin. An important application is the use of formalin as a fixative in the preservation of biological materials or in the preparation of tissue specimens in histology and pathology. Other major applications of formaldehyde are in the manufacture of plastics, textiles, building materials, insulation products, auto parts, and fungicides. It is a lung and skin irritant that can cause permanent damage due to the binding of strands of DNA together. Formaldehyde is carcinogenic to laboratory rats and is associated with increased cancer rates among exposed workers.

formaldehyde poisoning. poisoning by exposure to formaldehyde. Symptoms may include local irritation or numbness of skin, eyes, or mucous membranes; irritation of the respiratory tract, gastrointestinal tract. The central nervous system may be involved. The victim may complain of abdominal pain and may exhibit vertigo, stupor, convulsions, and unconsciousness. Renal damage may result. See also methanol poisoning.

formaldehyde sodium sulfoxylate. sodium formaldehydesulfoxylate.

formaldehydesulfoxylic acid sodium salt. sodium formaldehydesulfoxylate.

formalin (formol). a 37-50% aqueous solution of formaldehyde, usually with small amounts of methanol added. It is used as a fumigant and disinfectant, histological fixative, preservative, and embalming agent. It is very toxic when ingested, causing severe gastrointestinal disturbances including vomiting and diarrhea. Damage can be permanent and surgical intervention may be required. Contact with even small amounts can cause permanent damage to the skin and underlying tissues. *Cf.* formol. See also formaldehyde.

formalinize. to add formalin to inactivate a vaccine without destroying its immunizing capacity.

formamidine. See chlordimeform.

formate. a salt or ester of formic acid.

formed element. a cellular component of blood.

formic. pertaining to ants or to formic acid.

formic acid. a clear, sour-tasting, colorless, moderately toxic, dangerously caustic, liquid carboxylic acid, H_2CO_2, with a pungent odor. It is the simplest and most acidic of the monobasic fatty acids. Chronic absorption via the skin in humans can cause albuminuria and hematuria. It occurs naturally as a component of ant venom (from which it was originally prepared), in a few other insects, and in certain plants. Formic acid is used in a number of organic syntheses, including the synthesis of acetic acid and certain insecticides and drugs. It is also used to dissolve bones and remove the hair from animal carcasses, and is a component of a number of food flavorings. Formic acid is a strong oxidizing agent, and concentrated solutions are corrosive to tissues; ingestion can prove fatal. Exposure may be via inhalation of the vapor, percutaneous absorption, or ingestion.

formic aldehyde. formaldehyde.

formic ether. ethyl formate.

formiciasis. irritation produced by the bites of ants.

Formicidae. a family of polymorphic colonial hymenopterous insects (ants), 2-25mm long. They are usually black, brown, or reddish in color. Most species are scavengers or predators. Most ants are venomous. The venom is used in defense and to kill insects and other small animals on which some species prey. Envenomation by ants may cause painful reactions, but is rarely fatal to humans. See also venom (ant venom), formic acid, myrmicacin.

formol. an aqueous solution of formaldehyde. *Cf.* formalin.

formonitrile. hydrogen cyanide.

Formosa camphor. camphor.

formothion (*O,O*-dimethyl-*S*-(*N*-methyl-*N*-formylcarbamoylmethyl)phosphorodithiolate; phosphorodithioic acid *S*-[2-(formylmethylamino)-2-oxyethyl] *O,O*-dimethyl ester; phosphorodithioic acid *O,O*-dimethyl ester *S*-ester with *N*-formyl-2-mercapto-*N*-methylacetamide; *S*-(*N*-formyl-*N*-methylcarbamoylmethyl) *O,O*-dimethyl phosphorodithioate). a yellow oil or crystalline cholinesterase inhibitor used as a systemic insecticide (used chiefly against flies and aphids) and miticide. It is very toxic to laboratory rats and rabbits. See also parathion.

formulate. **1:** to state as a formula. **2:** to prepare or compose by a prescribed or established method (a formula).

formulation. the process or product of formulating.

formyl. the radical of formic acid, HCO.

***N*-formyldimethylamine**. *N,N*-dimethylformamide.

***S*-(*N*-formyl-*N*-methylcarbamoylmethyl) *O,O*-dimethyl phosphorodithioate**. formothion.

2-formyl-1-methylpyridinium chloride oxime. pralidoxime chloride.

5-formyl-5,6,7,8-tetrahydrofolic acid. folinic acid.

5-formyl-5,6,7,8-tetrahydropteroyl-L-glutamic acid. folinic acid.

5-formyl-5,6,7,8-tetrahydropteroyl-L-thioic acid. folinic acid.

formyl trichloride. sometimes seen in the literature as an inappropriate designation for chloroform.

fors. See G.

fortescue. *Centropogon australis*.

fosetyl Al. fosetyl aluminum.

fosetyl aluminum (phosphonic acid monoethyl ester aluminum salt; aluminum *tris*(ethyl) phosphite); aluminum *tris*-(*O*-ethylphosphonate), efosite Al, phosethyl Al, LS 74-783, Alliette). a systemic fungicide that is slightly to moderately toxic to laboratory mammals and birds, but is carcinogenic to some.

fossokéré. *Echis carinatus*.

four o'clock. *Mirabilis jalapa*.

Fourneau 694. pamaquine.

Fourneau 710. pamaquine.

fowl mannagrass. *Glycera striata*.

fowl meadow grass. *Glycera striata*.

Fowler's solution. potassium arsenite solution.

foxglove. common name of *Digitalis spp.*, especially of *D. purpurea*.

foxtail. See *Equisetum*.

fp. freezing point.

Fr. the symbol for francium.

fraction, S-9. See S-9 fraction.

fragrance. **1:** odor, scent, bouquet, perfume. **2:** this term on a label can represent up to 4000 distinct, mostly synthetic, ingredients that need not be named. Symptoms from exposure to these products, as reported to the FDA, may include headaches, dizziness, skin discoloration, violent coughing, vomiting, and allergic dermatoses. In addition, clinical tests and observations have associated the use of fragrances with a variety of CNS effects including depression, hyperactivity, irritability, inability to cope, and various other behavioral changes.

francium (symbol, Fr)[Z = 87; most stable isotope = ^{223}Fr, $T_{1/2}$ = 21 min]. a radioactive element of the alkali metal series formed by the decay of actinium-227. It emits β-particles

frangula emodin. emodin

frangulic acid. emodin.

frank. obvious, apparent; often applied to clinical signs (e.g., visually obvious blood in the stool) or to lesions.

Franseria discolor (white ragweed). a weedy herb (Family Compositae). toxic concentrations of nitrate have been reported.

Fratol. sodium fluoroacetate.

free erythrocyte protoporphyrin. See protoporphyrin.

freebasing. the smoking of cocaine through liquid or mixed with ether. Freebasing with crack cocaine produces the fastest, most intense effects ("highs"). See cocaine, crack.

Frekentine. diethylpropion.

Fremy's salt. potassium bifluoride.

French green. cupric acetoarsenite.

Freon [Du Pont]. trademark for a series of fluorocarbons and chlorofluorocarbons used in refrigeration and air conditioning equipment, and as fire extinguishing agents, solvents, and cleaning fluids. "Freon" is a clear, water-white liquid or vapor with an etherial odor. In the gas phase is a deadly and insidious hazard and should be handled with care: It is nonirritant and noncorrosive, but is denser than air and can act as a simple asphyxiant. Fatalities from exposure to "Freon" have occurred.

freshwater stingray. a stingray, *q.v.*, of the family Potamotrygonidae.

friar's cowl. *Aconitum.*

fricasse. *Blighia sapida.*

frijolito. *Sophora.*

fructivore. an animal that feeds on fruit.

frusemide. furosemide.

FSH (follicle-stimulating hormone; follitropin; urofolitrophin). a gonadotropic glycoprotein secreted by the anterior pituitary gland that regulates the metabolic activity of the granulosa cells of the mammalian ovary. FSH stimulates maturation of the Graafian follicles of the ovary and the production of estrogens. It is also partially responsible for spermatogenesis. See hormone, pituitary gland.

5-FU. fluorouracil.

fugu. one of numerous common names of fish of the family Tetraodontidae.

fugu poison. tetrodotoxin.

fugu poisoning. tetrodotoxism.

fugutoxin. tetrodotoxin.

fuko. *Causus resimus.*

fulgurant. fulminating.

fulminate. to appear or emerge suddenly and follow a severe, extremely intense, or rapid course.

fulminating (fulgurant). occurring or developing with great rapidity.

fulminic acid. an unstable isomer of cyanic acid, *q.v.* Its salts are notoriously explosive.

fumarine. protopine.

fumbe. *Naja melanoleuca.*

fumes. vapors or fine particulates, especially those with a caustic or irritant action. See nitric acid.

fumigacin. helvolic acid.

fumigant. an irritant, usually lethal, smoke, gas, or vapor used (usually in enclosed spaces) to control pests such as rodents or insects. Substances commonly used as fumigants are acrylonitrile, carbon tetrachloride, ethylene oxide, hydrogen cyanide, and methyl bromide. **soil fumigant**. a pesticidal chemical used to destroy pests in the soil.

fumigation. the process of fumigating; the use of a fumigant

fumigatoxin. a mycotoxin from *Aspergillus fumigatus*. It is very toxic to laboratory mice.

fuming liquid arsenic. arsenic trichloride.

fuming sulfuric acid. See sulfuric acid.

fuming tin chloride. stannous chloride.

fumitory family. Fumariaceae.

fune. *Dendroaspis angusticeps.*

fungal poisoning. See mycotoxicosis.

fungi. plural of fungus.

fungicide. **1a:** a substance that destroys fungi or prevents their growth. Fungicides are secreted by some insects such as ants and occur in some plants. **1b:** a pesticide used to control or destroy fungi.

Fungicidin™. nystatin.

fungistat (fungistatic agent). any agent that inhibits the growth of fungi.

fungistatic (mycostatic). **1:** having the capacity to inhibit the growth of fungi. **2:** a fungistat.

fungistatic agent. fungistat.

fungitoxic. poisonous to, or impeding the growth of, fungi.

fungitoxicity. the property of being, or capacity to be, fungitoxic.

fungus (plural, fungi). a heterotrophic, generally non-motile, non-photosynthesizing, chiefly multicellular organism (Kingdom Fungi) that absorbs nutrients from dead or other living organisms. Formerly considered plants, they are now considered to be an independent line of evolution. Certain mushrooms (e.g., *Amanita*) and many of the lower fungi are dangerously toxic. See, for example, ergot.

funnel-web spider. a spider of the genus *Atrax*.

fur seal. See Otariidae.

furniture polish. any commercially available, usually volatile liquid or aerosol spray used to polish furniture. Such polishes often contain phenol and other poisonous ingredients such as acrylonitrile, ammonia, detergents, artificial fragrances, naphtha, and petroleum distillates. Such substances are toxic by all routes of exposure. Following application, furniture polish (and floor polish) can give off sufficient residual fumes to pose a hazard to human health. In the United States, furniture polishes carry a warning label: "Harmful or fatal if swallowed." See also floor polish.

7*H*-furo[3,2-g][1]benz opyran-7-one. psoralen.

furo[3,2-g]-coumarin. psoralen.

furocoumarin. any of a group of photoactive antifungal dyestuffs found in plants of the families Apiaceae and Rutaceae that cause primary photosensitization in domestic fowl and livestock. See photosensitivity (primary photosensitivity).

furosemide (5-(aminosulfonyl)-4-chloro-2-[(2-furanylmethyl)amino]benzoic acid; 4-chloro-*N*-furfuryl-5-sulfamoylanthranilic acid; 4-chloro-*N*-(2-furylmethyl)-5-sulfamoylanthranilic acid; frusemide; fursemide). a diuretic and antihypertensive agent used to treat cardiovascular and renal diseases. Furosemide is moderately toxic, but large doses cause extensive hepatic and renal damage. In the laboratory mouse it produces a coagulative necrosis of the inner renal cortex and subcortex and a midzonal to central lobular necrosis of the liver. Laboratory rats and hamsters are resistant to the hepatic effects of furosemide.

fursemide. furosemide.

fusaric acid (5-butyl-2-pyridinecarboxylic acid; 5 butylpicolinic acid). a mycotoxin produced by fungi of the genus *Fusarium*, especially *Fusarium moniliforme*. It is an antibiotic and wilting agent that causes yellowing of infected plants. Fusaric acid is very toxic to laboratory mice, rabbits, and dogs.

fusariotoxicosis. a condition of poultry caused by ingestion of moldy grain. It is characterized by edematous and necrotic oral lesions and is probably due to trichothecene mycotoxins. *Cf.* fusarium toxicosis.

fusariotoxin T-2. T-2 toxin (See toxin).

Fusarium. a genus of extremely common ascomycetous fungi (Order Moniales, Family Tuberculariaceae) with sickle-shaped, multicellular conidia. They produce marasmins and fusaric acid. Some are important plant pathogens; some opportunistically (accidentally) infect animals, including humans, and have been isolated from mycotic keratitis and otomycosis externa, for example. See aleukia (alimentary toxic aleukia). *F. graminarium*, produces the mycotoxin zearalenone. *F. moniliforme* produces the mycotoxins fusaric acid and moniliformin. *F. nivale* produces the trichothecene mycotoxins, monoacetylnivalenol (fusarenon-X) and nivalenol. *F. orthoceras* produces enniatin A. *F. roseum* produces the mycotoxin zearalenone. *F. sporotrichioides*, produces the mycotoxin butenolide. *F. solani* produces 4-ipomeanol and novarubin. *F. tricinctum*, produces the trichothecene mycotoxin, T-2 toxin. See toxin; poisoning (bean-hull poisoning).

fusarium estrogenism (*Fusarium* hyperestrogenism; zearalenone hyperestrogenism). a toxicosis caused by zearalenone, *q.v.*, a nonsteroidal estrogen produced by certain toxigenic strains of *Fusarium spp*. Affected animals may be poisoned by fresh or stored feeds or pasturage contaminated with *Fusarium*. The condition is indistinguishable from excessive administration of estrogen. In swine, chiefly weaned and prepubertal gilts are observed with this condition; prominent signs are hyperemia, enlargement of the vulva, and in severe cases, uterine and mammary gland hypertrophy; prolapse of the uterus is seen on occasion. Effects of zearalenone on multiparous sows include diminished fertility, reduced litter size, smaller piglets, malformations, and probably fetal resorption. Some sows exhibit constant estrus or pseudopregnancy. Lesions in swine include ovarian atrophy and follicular atresia, uterine edema, and hypertrophy, and a cystic appearance in degenerative endometrial glands. Dairy heifers may experience weight-loss, vaginal discharge, nymphomania, uterine hypertrophy. Gravid heifers may abort 1-3 months following conception, usually with multiple returns to service. Ewes receiving 25 mg/day of zearalenone for 10 days exhibited poor reproductive performance due to decrease ovulation rates. Gravid ewes fed heavily infested, ensiled corn with a high zearalenone content experienced abortions, premature live births, prostration, and occasionally died. See *Fusarium*, zearallenone.

fusarium hyperestrogenism. fusarium estrogenism.

fusarium toxicosis. poisoning by trichothecenes secreted by various species of *Fusarium*. This term refers chiefly to skin necroses so produced. A skin test is employed to screen toxic strains of *Fusarium*. *Cf*. fusariotoxicosis.

fusel oil. a volatile toxic by-product of the alcoholic fermentation of carbohydrates. It is a mixture of butyl, heptyl, isoamyl, and propyl alcohols.

fusidic acid (3α,4α,8α,9β,11α,13α,14β,16β,17-Z)-16-(acetyloxy)-3,11-dihydroxy-29-nor-dammara-17(20),24-dien-21-oic acid; 3α, 11α,-16β-trihydroxy-29-nor-8α,9β,13α, 14β-dammara-17(20),24-dien-21-oic acid 16-acetate; 3α,11α,16β-trihydroxy-4α,8,14-trimethyl-18-nor-5α,8α,9β,13α,14β-cholesta-17(20),24-dien-21-oic acid 16-acetate; 3,11, 16-trihydroxy-4,8,10,14-tetramethyl-17-(1′-carboxyisohept-4′-enylidene)cyclopentanoperhydrophenanthrene 16-acetate; ramycin). a terpene antibiotic, $C_{31}H_{48}O_6$, isolated from *Fusidium coccideum*, it is a moderately toxic inhibitor of protein synthesis and is very active against *Staphylococci*.

fusiform bacteria. slender rod-shaped bacteria.

FW 925. nitrofen.

G

γ. See gamma.

Γ. See gamma.

G (grav; fors). a unit of acceleration equal to the standard acceleration of gravity on earth, 9.80665 m/sec^2 or approximately 32.174 feet/sec^2. See, for example, S-9 fraction.

g. gram.

G. giga-.

G6P. glucose-6-phosphate.

G6P dehydrogenase. glucose-6-phosphate dehydrogenase.

G6PD. glucose-6-phosphate dehydrogenase.

Ga. the symbol for gallium.

GA. tabun.

GABA. τ-aminobutyric acid.

GABAT. γ-aminobutyric acid transaminase.

gaboon viper. *Bitis gabonica*.

Gabun. a common name of *Bitis gabonica*.

gadolinium (symbol, Gd)[Z = 64; A_r = 157.25]. a silvery metal of the lanthanoid series of elements. A number of natural isotopes are known; one (^{152}Gd) is radioactive ($T_{1/2}$ = 1.1 X 10^{14}; an α-emitter) Eleven artificial radioactive isotopes are known. See gadolinium nitrate.

gadolinium nitrate. the hexahydrate of this salt, $Gd(NO_3)_3 \cdot 6(H_2O)$, is very toxic to laboratory rats when administered *i.p.*; oral toxicity is slight.

galactogogue. causing the production of milk; an agent that induces the production of milk.

galactophorous. milk-carrying.

4-O-β-D-Galactopyranosyl-D-glucose. lactose.

4-(β-D-Galactosido)-D-glucose. lactose.

Galactorrhea. 1: continued breast-milk flow after weaning. **2:** excessive breast-milk flow during nursing.

galactoside. a glycoside, *q.v.*, that yields galactose on hydrolysis.

galactotoxin. a toxic substance in milk that is produced by bacteria. See galactotoxism.

galactotoxism (galactoxism; galactoxismus). poisoning due to the ingestion of milk that contains toxic substances, especially galactotoxin.

galactoxism. galactotoxism.

galactoxismus. galactotoxism.

Galanthus (snowdrops). a genus of hardy, bulbous, spring-blooming herbs with drooping white flowers (Family Amaryllidaceae). The bulbs are poisonous. *G. nivalis* (snowdrop). a small ornamental plant with nodding flowers introduced to North America from Europe.

Galeichthys felis. *Arius felis*.

galerina. a common name of any mushroom of the genus *Galerina*.

Galerina (galerinas). a genus of brown-spored, gilled fungi (Family Cortinariaceae); the edibility of most species is uncertain, but some are poisonous and contain amanitin or other mycotoxins, some of which have yet to be identified. ***G. autumnalis*** (deadly galerina; Autumn galerina). a small, brown, innocent-looking, but deadly, mushroom (Family Cortinariaceae). It is widely distributed throughout North America on well-decayed logs. Symptoms of intoxication, which are similar to those of poisoning by *Amanita virosa*, rarely appear earlier than 10 hours following ingestion. Early symptoms typically include those of acute gastroenteritis with nausea, vomiting, cramps, and diarrhea. Victims may also complain of headaches and pains in the back and joints. following a remission of variable duration, the victim may develop serious lesions, most notably acute or chronic kidney and liver dysfunction or complete failure, as well as damage to the intestines, genital organs, heart, and nervous system. There may be several remissions during the course of the disease with death occurring several months following exposure. The poisoning is due to the presence of orellanin and possibly other as yet unidentified toxins. ***G. tibicystis*** (sphagnum-bog galerina). this species is difficult to identify and little is known concerning its toxicity. Common to sphagnum bogs, it is one of many moss-inhabiting galerinas. ***G. venenata*** (deadly lawn galerina). a small, deadly mushroom with a moist reddish-brown to buff-colored cap that occurs on lawns and over buried, decomposed wood in Washington and Oregon. It occurs commonly throughout the United States, especially on lawns in the Northwest. Effects are similar to those of *G. autumnalis*.

galitoxin. a toxic resinoid found in the more toxic species of *Asclepias*. It is the principal active agent in asclepias poisoning, *q.v.*

gall. See bile.

gallic acid. a colorless, water-soluble, crystalline compound. It is the toxic, hydroxyphenol moiety of oak tannin. It occurs naturally on the twigs of trees, especially *Quercus* (oak) as a reaction to gall wasp eggs. It is used in the production of antioxidants, ink dyes, and in photography.

gallic acid 5,6-dihydroxy-3-carboxyphenyl ester. digallic acid.

gallic acid 3-monogallate. digallic acid.

gallium (symbol, Ga). [Z = 31; A_r = 69.72]. a toxic, silvery-gray, metallic element with a low melting point. There are two known naturally occurring isotopes of gallium (69 and 71), and 12 artificial radioactive isotopes (63-68; 70; 72-76). Administration to humans can cause skin rashes and bone marrow depression. The nitrate is used therapeutically as an antineoplastic. Some of its compounds are toxic. See gallium nitrate, gallium lactate.

gallium lactate. a white, very toxic amorphous powder.

gallium nitrate. a white, water-soluble, very toxic crystalline powder, GaN_3O_9. See also gallium.

gallotannic acid. gallotannin, tannin.

gallotannin. tannin.

gallotannine. corilagin.

***m*-galloylgallic acid**. digallic acid.

galobou. *Atheris chlorechis*.

gama. common name of *Bufo vulgaris formosus*.

gamabufagin. gamabufotalin.

gamabufogenin. gamabufotalin.

gamabufotalin (3β,11α,14-trihydroxy-5β-bufa-20,22-dienolide; gamabufogenin; gamabufagin). a genin found in the venom of the toad, *Bufo vulgaris formosus* (Family Bufonidae) and in the Chinese drug, Ch'an-Su, prepared from the poisonous Chinese toads, *B. asiaticus* and *B. gargarizans*. A trihydroxybufadienolide, it is chemically and pharmacologically similar to digitalis. It is cardiotoxic and numbs the tongue.

Gambierdiscus toxicus. a dinoflagellate, *G. toxicus* (and possibly other dinoflagellates) is the principal or ultimate source of ciguatera.

gamboge. *Croton tiglium*.

game fish. See fish.

Gamene™. lindane.

gametocide. an agent that destroys gametes.

gametocyte (gamont). a cell able to divide, producing gametes, (spermatocytes, oocytes).

Gamexan™. lindane.

Gamiso™. lindane.

gamma (γ, Γ). **1:** the third letter of the Greek alphabet. **2:** the symbol for microgram. **3:** a prefix with meanings similar to those of α, *q.v.* It is used, for example, to designate the location of substituents in a chemical compound and to designate a particular form of an organic substance, e.g., γ-globulin. **4:** gamma is the designated prefix for the most intense form of short-wave radiation, the γ-ray.

gamma benzene hexachloride. See lindane.

gamma globulin (γ-globulin). one of a group of serum proteins in blood with distinct electrophoretic mobility. They are produced by the immune system and help to destroy or neutralize infection-causing microbes. The group also includes serum immunoglobulins and other globulins. Gamma globulin, derived and concentrated from the blood of other humans, is used to help provide temporary immunity to certain diseases.

γ-globulin. gamma globulin.

gamma hexachlor. lindane.

gamma hexane. lindane.

gamma multi-hit dose-response model. See dose-response model.

γ-ray. gamma ray.

gamma ray (γ-ray). a type of high energy, ionizing, electromagnetic radiation of extremely short wavelength that can react with atoms and molecules (e.g., of tissue) to eject electrons and produce positively charged ions. See ionizing radiation.

Gammalin™. lindane.

Gammexane™. lindane.

Gamoxol™. lindane.

ganglia. plural of ganglion.

ganglion (plural, ganglia). a more or less discrete bundle of nervous tissue outside the CNS, comprised mainly of nerve cell bodies, dendrites, and synapses, that acts as a relay station in the transmission of nerve impulses. Examples are the dorsal root ganglia of spinal nerves. A basal ganglion, however, is located within the brain and spinal cord. Ganglia occur throughout the autonomic nervous system. See also blockade (ganglionic blockade).

ganglionic blockade. See blockade.

gangrene. decay and death of tissue usually due to insufficient vascular supply, followed by bacterial infection and putrefaction; a large mass of tissue may be involved. Untreated gangrene can spread rapidly and is life threatening. It is caused by disease or trauma.

gangrene des Solognots. a French name for the ergotism caused by *Clavaceps purpurea*.

gangrenous. pertaining to or affected with gangrene.

gangrenous ergotism (chronic ergotism; dry gangrene). one effect of ergot alkaloids is constriction of arterioles and capillaries which may interrupt circulation of blood in the extremities with consequent necrosis of the affected tissues, especially in the lower limbs. Symptoms in humans also typically include nausea, vomiting, abdominal pain. Symptoms in cattle are similar and very small doses of ergot may block circulation of the extremities and dispose the animal to thromboses. The hind feet are affected first and the hoofs may slough off; the early stages thus resemble foot-and-mouth disease. The front feet, legs, tips of the ears and tail, and sometimes the tongue may subsequently become necrotic. Secondary infection may occur. The comb and beak are most affected in the domestic fowl. See ergotism.

gangrenous fescue poisoning. fescue foot.

ganja. cannabis.

gaper. *Schizothaerus nuttalli*.

garden huckleberry. *Solanum intrusum*, *S. nigrum*.

garden monkshood. *Aconitum napelius*.

garden plant. See plant.

garden sorrel. *Rumex acetosa*.

garden spider. See *Argiope aurantia*.

gargantilla. See *Micrurus*.

gargantilla coral. *Micrurus mipartitus*.

garget. *Phytolacca americana*.

garland flower. *Daphne cneorum*.

garlic oil. allyl sulfide.

gartersnakes. See *Elaps*, *Elaposoidea*,

Gärtner's bacillus. *Salmonella enteritidis*.

gas. one of the three basic physical forms of matter (gas, liquid, solid) in which the molecules move freely in all directions such that it expands indefinitely, filling its container. A gas can be converted into a liquid and eventually a solid by compression at low temperatures. **binary gas**. a toxic nerve gas produced by the mixing of two relatively harmless components. Such gases are used chiefly as chemical warfare gases. **blistering gas**. vesicant gas. **chemical warfare gas**. war gas. **coal gas**. a flammable, explosive, toxic gas comprised chiefly of methane, carbon monoxide, and hydrogen. It is produced by the distillation of coal and used for heating and lighting. **fluohydric acid gas**. hydrogen fluoride. **illuminating gas**. a flammable, poisonous mixture of various gases, including hydrogen and carbon monoxide. **lewisite gas**. lewisite. **lung irritant gas**. any irritant gas (e.g., chlorine, phosgene, hydrogen chloride). **mustard gas** (1,1′-thio-bis[2-chloroethane]; bis(2-chloroethyl)sulfide; β,β′-dichloroethyl sulfide, 2,2′-dichlorodiethyl sulfide, bis(β-chloroethyl)sulfide, 1-chloro-2-(β-chloroethylthio)ethane, sulfur mustard, yellow cross, yellow cross liquid, Kampfstoff "Lost," Yperite). a deadly, oily, vesicant liquid and carcinogen, $(CH_2ClCH_2)_2S$, used only as a chemical warfare agent. It can be neutralized and inactivated by sodium hypochlorite. It is extremely toxic to a wide variety of terrestrial animals including humans, destroying tissue on contact. Initial symptoms may include conjunctivitis and blindness. Other effects that may appear within the first 12 hours include coughing, edematous eyelids, erythema and severe pruritis. The respiratory tract and exposed skin may become ulcerated, edematous, and necrotic. Damage to eyes and the pulmonary system may be severe and permanent. Ingestion may induce vomiting. See also Lewisite. **natural gas**. **1:** any gas formed naturally in the earth's crust. Many of these (e.g., hydrogen sulfide) are dangerously toxic. **2:** a naturally occurring gaseous combustible mixture used as fuel. American natural gas is comprised of ca. 85% methane, 9% ethane, 3% propane, 2% nitrogen, 1% butane, and smaller amounts of higher hydrocarbons. Unprocessed natural gas may contain natural gasoline, sometimes carbon dioxide, hydrogen sulfide, and helium. An occasional well head will yield nearly pure methane. Natural gas, when present in ambient air at high concentrations, is a deadly narcotic and asphyxiant. Carbon monoxide results from incomplete combustion of natural gas. **nerve gas**. any gas that interferes with or blocks neurotransmission. **nose irritant gas** (diphenylchloroarsine). an irritant smoke which can cause intense pain in the nose, mouth, and upper respiratory tree. Additional symptoms may include sneezing, headache, aching teeth and jaws, vomiting, and acute mental depression. **sewer gas**. toxic, usually flammable and explosive gas elaborated by decaying organic matter in sewage. **suffocating gas**. any of a number of irritant chlorinated chemical warfare gases (e.g., chlorine, phosgene, diphosgene) that attack the lungs, producing pulmonary edema. **tear gas**. an irritant gas (e.g., bromoacetone) that causes tearing. **toxic gas**. any toxic gas. **vesicant gas** (vesicating gas; blistering gas). any irritant gas that causes blistering and usually destruction of tissue on contact (e.g., lewisite, mustard gas). Symptoms may be delayed for several hours and may include intense pain in the eyes, and sometimes the mouth, throat, and upper respiratory tract; lacrimation; a diffuse erythema may appear followed by blistering and ulceration; swelling of eyelids such that vision may be blocked. **vesicating gas**. vesicant gas. **vomiting gas**. **1:** chloropicrin. **2:** any gas that induces emesis. **war gas**. any solid, liquid, or vapor used to produce toxic, irritant gases for use in warfare. A war gas may fall into one or more of these classes: lacrimator, lung irritant, sternutator, vesicant, nerve gas, or other systemic poison.

gas appliance. any piece of equipment, such as a kitchen range, oven, space heater, wall heater, central heating system, or clothes dryer, that provides heat by burning natural gas. Most cases of carbon monoxide poisoning in the home involve such appliances.

gasoline. a toxic, volatile, flammable liquid petroleum distillate, the most common components being C4-C12 aliphatic hydrocarbons. It typically boils within a range of 30 to 220°C. Applications are mainly as a fuel for internal combustion engines, as a diluent, and as an industrial solvent; it has sometimes been used in museum practice as a defatting and cleaning agent (e.g., to remove fat adhering to the skin of some animal preparations). Toxicity is related primarily to the benzene content. Gasoline is a CNS depressant and an irritant to the skin, conjunctiva, and mucous membranes. Tetraethyllead is sometimes added, which introduces the possibility of lead poisoning from repeated exposure. Combustion of gasoline releases toxic fumes that contain toxic compounds such as carbon monoxide and nitrogen oxides. The vapor is carcinogenic to some laboratory animals. Gasoline is a serious fire hazard and the vapor is highly explosive. See also TCDD.

gasoline poisoning. poisoning by exposure to gasoline, usually by contact, ingestion, or inhalation of fumes (but not the fumes released by combustion). **acute gasoline poisoning**. early symptoms are flushing, staggering, slurred speech, confusion, and sometimes vomiting. Serious cases can result in unconsciousness or coma, cyanosis, and death. Lesions may include pancreatic hemorrhage, splenic congestion, hepatic and renal degeneration, hematopoietic changes, pneumonitis, pulmonary edema. **chronic gasoline poisoning**. long-continued (chronic) exposure to gasoline or its fumes can cause dermatitis with blistering and defatting of the skin. Systemic effects may result from ingestion (as when starting a siphon by mouth), cutaneous absorption, or inhalation. Sniffing of small amounts produces euphoria.

gasping. spasmodic, or laborious paroxysmal breathing with the mouth open.

gassing. poisoning by use of toxic, asphyxiating, irritant, or noxious gases.

gastric. of or pertaining to the stomach.

gastric lavage. See lavage.

gastric mucosal barrier. See barrier.

gastritis. inflammation of the stomach, especially of the mucosa. It is marked by nausea, vomiting, pain, or tenderness in the epigastrium. In chronic cases, the symptoms are mild or often lacking. Vomiting, if persistent, may alter the electrolyte balance. Lesions may include atrophy or hypertrophy of the mucosa. Victims of acute gastritis may exhibit persistent vomiting, thirst, moderate pyrexia, intense epigastric pain, and prostration. **toxic gastritis**. gastritis due to a toxic substance. *Cf.* gastroentercolitis, gastroenteritis.

gastrocolitis. inflammation of the stomach and colon, especially of the mucosa. See gastrointestinal tract.

gastroduodenitis. inflammation of the stomach and duodenum, especially of the mucosa. See gastrointestinal tract.

gastrodynia. stomach ache; pain in the stomach and/or epigastrium.

gastroenteralgia. pain in the stomach and intestines.

gastroentercolitis. inflammation of the stomach, small intestine and colon. *Cf.* gastroenteritis, gastritis.

gastroenteric. of or pertaining to the stomach and intestines.

gastroenteritis. inflammation of stomach and intestines. There are any number of causes including caustic and irritant chemicals, certain other toxicants, viruses, bacteria, and food poisoning. Any substance that induces vomiting can cause gastroenteritis. Severe gastroenteritis is extremely painful and can be fatal. Depending on the severity of the condition, symptoms may include salivation, nausea, frequent vomiting, diarrhea with watery or bloody stools, abdominal colic, a fall in blood pressure, collapse and death.

gastroenterocolitis. inflammation of the stomach, small intestine, and colon. See gastrointestinal tract.

gastroenterologist. a medical doctor who specializes in the diagnosis and treatment of diseases of the gastrointestinal system.

gastroenteropathy. any disorder, disease, or pathological state of the gastrointestinal tract.

gastroesophagitis. inflammation of the stomach and esophagus, especially of the mucosa. See gastrointestinal tract.

gastrogavage. gavage.

gastrohepatitis. concurrent gastritis and hepatitis.

gastroileitis. inflammation of the stomach and ileum.

gastrointestinal. **1:** of, referring to, related to, or denoting the stomach and intestines of vertebrae animals. **2:** broadly used in reference to the entire alimentary canal. See also digestive tract.

gastrointestinal radiation syndrome (gastrointestinal syndrome). See acute radiation syndrome.

gastrointestinal route. oral route.

gastrointestinal (GI) syndrome, gastrointestinal radiation syndrome. one of three forms of acute radiation syndrome seen in humans. It usually results from a dose rate 400-1000 rads. Characteristic symptoms include unmanageable nausea, vomiting, and diarrhea with resultant dehydration; reduced plasma volume; cardiovascular collapse, and death. This condition is due chiefly to tissue necrosis and consequent toxemia. The condition is always fatal. See also acute radiation syndrome.

gastrolavage. lavage of the stomach.

gastropathy. any disorder, disease, or pathological state of the stomach.

gastropod. any mollusk of the class Gastropoda.

gastroscope. an optical instrument with a lighted tip used to visually examine the inside of the stomach.

gastroscopy. visual examination of the inside of the stomach by use of a gastroscope.

gastrosis. any disease of the stomach.

gastrospasm. spasm of the stomach.

gastrostolavage. the irrigation or rinsing out of the stomach through a gastric fistula.

gastrotomy. opening the stomach with an incision. It is used clinically, for example, to remove a mass of (e.g., toxic) material from the stomach that cannot be removed by other means such as gavage or induced vomiting.

gastrotoxic. toxic to the stomach.

gastrotoxin. a cytotoxin that acts specifically on the gastric mucosa.

gautheria oil. methyl salicylate.

gonyaulax bloom. See red tide. See also bloom.

gavage (gastrogavage). the provision of a liquid or semiliquid food or drugs directly by a catheter or tube passed through the nares, pharynx, and esophagus into the stomach or via a gastric fistula.

GB. sarin.

GC. gas chromatography. See chromatography.

Gd. the symbol for gadolinium.

GD. soman.

Ge. the symbol for germanium.

geeldikkop. the South African name for "bighead" in sheep. See, for example, *Tetradymia*, *Tribulus terrestris*.

geelkapel. *Naja nivea*.

gelsemia. gelsemium.

gelsemic acid. scopoletin.

gelsemicine. a phytotoxin of *Gelsemium sempervirens*. See also gelsemium.

gelsemine. a toxic, crystalline alkaloid from the roots and rhizome of *Gelsemium sempervirens*.

It is a mydriatic and CNS stimulant. Effects of intoxication include muscular weakness, double vision and, in higher doses, respiratory distress. It is used therapeutically as a CNS stimulant.

gelseminic acid. scopoletin.

gelsemium (gelsemia). the dried roots and rhizome of *Gelsemium sempervirens*. It is highly poisonous, containing gelsemine, gelsemoidine, scopoletin, gelsemic acid, and other biologically active compounds. It was formerly used therapeutically as a CNS stimulant.

Gelsemium (yellow jessamine; trumpet flower). a small genus of twining or trailing woody evergreen shrubby plants or vines (Family Loganiaceae) that occur in woodlands and thickets in Asia and the southern United States. The flowers are yellow, trumpet-shaped, and fragrant. They contain various indole alkaloids. ***G. sempervirens*** (evening trumpet flower; Carolina jessamine; Carolina yellow jessamine; yellow jessamine; yellow jasmine). a woody, twining, perennial evergreen vine that occurs in dry to wet woods, thickets, and sands of the Coastal Plain and lower Piedmont from Virginia to Florida, Arkansas, and Texas. These plants contain several indole alkaloids related to strychnine. They are poisonous to all classes of livestock. Children have been poisoned by sucking the nectar from the fragrant flowers. Symptoms are mostly those of central depression and may include headache, dizziness, visual disturbances, profuse sweating, pronounced ptosis, and dry mouth accompanied by dysphonia and dysphagia. Muscular weakness may be pronounced in serious cases, with convulsions, respiratory depression, and sometimes respiratory failure and death. Milder cases may resemble strychnine poisoning, but lack convulsions. The active principles are the three potent phytotoxic alkaloids gelsemine, gelsemoidine, and gelsemicine which occur in all parts of the plant.

gelsemoidine. one of at least 3 virulent alkaloids in *Gelsemium sempervirens*.

gemmed amanita. *Amanita gemmata*.

gene mutation. a point mutation; a change in the nucleotide sequence of a single gene.

general purpose cleaner. all-purpose cleaner. See cleaner.

generic. **1a:** general, prevalent, in general use, indefinite. **1b:** descriptive of an entire group or class. **2:** nonproprietary. **3:** pertaining to or having the status of a biological genus.

generic name. the general or non-trademark name of a product. the generic name of a drug or other substance is usually taken to be the chemical name, a name in common use, or an official name as presented in an official compendium. *Cf.* brand name, trade name, trademark, U.S.N.

genetic. **1:** pertaining to, caused by, or involving genes. **2:** pertaining to genesis.

genetic death. See death (genetic death).

genetic extinction. the selective elimination of a genotype that bears mutant alleles that reduce fitness. See death (genetic death).

genetic material. See DNA.

genin. aglycone.

-genin. a suffix that denotes the basic steroid unit of the toxic substance, especially that of a steroid glycoside.

genocide. the willful, planned extermination of a given social or ethnic group. This has been attempted on a mass scale by various means, including the use of war gases or other toxic gases. *Cf.* eradication. See also gas, tabun, sarin.

genome. the complement of genes that comprise the haploid set of chromosomes.

genotoxic. of or pertaining to a substance that is toxic to the genetic material.

genotoxic carcinogen. See carcinogen.

genotoxic damage. See damage.

genotoxicology. genetic toxicology (See toxicology).

genotype. the genetic constitution of any individual organism.

genotypic. pertaining to genotype.

gentamycin (gentamicin). an amorphous water-soluble, solid, antibacterial aminoglycoside, $C_{21}H_{43}N_5O_7$.

Gentersal™. a product that contains nonoxynol-9.

gentiotannic acid. a hepatotoxic form, $C_{14}H_{10}O_5$, of tannic acid.

gentle cort. a common name of the deadly mushroom, *Cortinarius gentilis*, which is more commonly called the deadly cort.

genus (pl. genera). a taxonomic group of closely related species of plants or animals. Closely related genera are placed into a family.

Geomet. phorate.

germ. an organism that causes infection such as pathogenic bacteria, viruses, or fungi.

germanium (symbol, Ge)[Z = 32; A_r = 72.61]. a lustrous bluish-gray, brittle metalloid element of group IV of the periodic table. It is used as an intestinal astringent and in a few cosmetics and pharmaceuticals. Some of its compounds are caustic or strongly irritant and a few of its organic compounds are markedly neurotoxic.

germanium tetrachloride. a fuming liquid, $GeCl_4$; the fumes of which are seriously irritating to the eyes, mucous membranes, and pulmonary tissues.

germanium tetrafluoride. a highly irritant, colorless gas, GeF_4, with a garlicky odor. It hydrolyzes in water to yield GeO_2 and H_2GeF_6. It should be handled with caution.

germicidal. **1:** lethal to pathogenic microorganisms. **2a:** germicide. **2b:** of or pertaining to a germicide.

germicidal agent. germicide.

germicide (germicidal (def. 1); germicidal agent). an agent that kills pathogenic microorganisms.

germidine. See *Veratrum* (*V. viride*).

germine (4α-9-epoxycevane-3β-4,7α,14,15α,16-β,20-heptol) . an alkamine of many of the polyester alkaloids that occur in various species of *Veratrum* and *Zygadenus*.

germitrine. a cardiotoxic alkaloid from *Veratrum viride*.

gesarol. DDT.

gesse. See *Lathyrus*.

gestation. synonymous with pregnancy, but usually applied to nonhuman viviparous animals.

Gexane™. lindane.

GH. somatotrophin.

GI. gastrointestinal.

giant apistocalamus. *Apistocalamus grandis*.

giant brownsnake. *Oxyuranus scutellatus*.

giant coral snake. *Micrurus spixii*.

giant desert hairy scorpion. *Hadrurus arizonensis*.

giant hairy scorpion. *Hadrurus arizonensis*.

giant helvella. *Gyromitra (Helvella) gigas*.

giant milkweed. *Calotropis gigantea*.

giant puffball. *Calvatia gigantea*.

giant stingray (or giant stingray of Australia). See *Dasyatis brevicaudata*.

giant toad. *Bufo marinus*.

Giberella zeae (corn root-rot organism). an ascomycete; corn grain on which this fungus has been grown is extremely toxic to rats.

gid. See staggers (def. 3).

giddiness. lightheadedness, dizziness, a light-headed sensation.

gifblaar. *Dichapetalum cymosum*.

gifblaar poison. fluoroacetic acid.

giga- (G). **1:** a prefix that denotes giant size. **2:** a prefix to units of measurement that indicates a quantity of 1 billion (10^9) times the value indicated by the root. See International System of Units.

gigantic acid. an antibiotic substance, $C_{14}H_{22}O_9N_2S$, produced by *Aspergillus giganteus*.

gila monster. See *Heloderma suspectum*.

gila monster family. Helodermatidae.

gill. **1:** any of numerous vertical plate-like structures (lamellae) on the undersurface of the cap of the fruiting body of agaric fungi (mushrooms and toadstools). **2:** a respiratory organ of many aquatic or semiaquatic animals such as fish, crustaceans, and certain amphibians. It is ordinarily a highly vascular outgrowth or projection from the body surface or from a portion of the digestive tract (e.g., of the pharynx in bony fish). Gills are usually thin, highly vascular, membranous, leaf-like, or filamentous organs. They are often major sites of entry or excretion of many toxicants. See also branchia, branchial, branchial route, respiratory route. **internal gill**. a gill of fish, derived from endoderm, that projects from the wall of the pharynx. They lie in gill slits and are protected by a covering, the operculum. **external gill**. **1:** a gill, ectodermal in origin and often protected by folds of the body wall as in amphibian larvae. **2:** one of the lamellae on the under surface of the cap of agaric fungi.

gill arch. See visceral arch.

gill-over-the-ground. *Glechoma hederacea*.

ginger. See *Asarum*.

ginger jake. an illegal alcoholic beverage, flavored with ginger extract, that was sold during Prohibition in the United States (a period when the sale of alcoholic beverages was illegal). Cases of delayed neuropathies, known as Jamaica ginger paralysis (See paralysis), first appeared in the early 1930s. The cause was the use of ginger extract contaminated with the cresyl phosphates that were used in extraction. See also tri-*o*-tolyl phosphate.

gingivitis. inflammation of the gums (gingiva).

ginseng family. Araliaceae.

ginyambila. *Dendroaspis polylepis*.

girdled snakes. See *Brachyurophis*.

GIT. gastrointestinal tract.

Gitaligin™. gitalin.

gitalin (Gitaligin). a stable extract of *Digitalis purpurea*, it is a cardiotoxic mixture of the cardiotoxic glycosides digitoxin, gitoxin, and gitaloxin, together with small amounts of related alkaloids and genins. Gitalin is used therapeutically as a cardiotonic.

githagenin. a sapogenin and toxic principle of *Agrostemma githago*.

githagin. the toxic principle of *Agrostemma githago*, *q.v.*

githagism. a condition similar to lathyrism, due to poisoning by the seeds of *Agrostemma githago (= Lychnis githago)*.

gitoxin (3-[(*O*-2,6-dideoxy-β-D-ribohexopyranosyl-(1→4)-*O*-2,6-dideoxy-β-D-ribohexopyranosyl)oxy]-14,16-dihydroxycard-20-(22)-enolide; anhydrogitalin; bigitalin; pseudodigitoxin). an extremely toxic cardiotoxic glycoside isolated chiefly from *Digitalis purpurea* and *D. lanata* and produced as a byproduct of digitoxin synthesis. (Family Scrophulariaceae).

gjurza. *Vipera lebetina*.

glacial acetic acid. See acetic acid.

gland. an organ or aggregation of cells that excretes materials not related to the normal maintenance metabolism of the organ. **endocrine gland**. a gland of internal secretion, one that secretes (hormones) directly into blood or lymph. See also gland, endocrine. **exocrine gland**. a gland of external secretion. See exocrine. **poison gland**. any gland that secretes (or excretes) a material which is toxic or irritating to another animal. a venom gland is thus a poison gland. **proboscis gland**. a modified salivary gland of the Toxiglossa, *q.v.*, that secretes venom. **salivary gland**. any of several ectodermal glands derived from the buccal cavity (vertebrates) or foregut (invertebrates) that secrete saliva. Functions vary, but may include roles in digestion, excretion, predation, and defense. The salivary glands of some animals are specialized and secrete venom or adhesive substances (as in some birds). See also agmatine, cephalotoxin, eledoisin, excretion (salivary gland excretion), maculotoxin, octopamine, *Octopus*, parasympathetic nervous system, saliva, tetramethylammonium hydroxide, *Thais haemastoma*. **venom gland**. any gland that secretes (or excretes) a material toxic or irritating to another animal. See also specific glands (e.g., adrenal gland, lacrimal gland, parathyroid gland, pituitary gland, salt gland, thymus gland, thyroid gland).

glandula thyroidea. thyroid gland.

Glasgow Coma Scale. a scale that quantifies and represents the stage or depth of coma based by scoring motor, verbal, and eye-opening responses to standard stimuli. The absence of eye-opening, verbal, and motor responses supports a diagnosis of coma.

glass cleaner. See cleaner.

glass viper. *Vipera russelii*.

glaucophyllum. *Solanum malacoxylon*.

GLC. gas-liquid chromatography. See chromatography.

Glechoma hederacea (= *Nepeta hederacea*; ground ivy; gill-over-the-ground; creeping Charlie). a prostrate perennial herb (Family Labiatae) that forms dense, pure stands of ground cover. It occurs throughout most of the United States and Canada. It contains irritant oils and has proven to be lethal to horses if ingested in large amounts, either fresh or in hay.

gliadin. See gluten-induced enteropathy.

gliotoxin. a mycotoxin from *Penicillium terlikowskii*. It is very toxic to laboratory mice.

globefish. one of numerous common names of fish of the families Tetraodontidae and Diodontidae.

globose. more or less spherical.

glomerin (glomerine). an extremely poisonous convulsant quinazolinone toxin of certain myriapods. It is lethal to other arthropods and small mammals. See *Glomeris marginata*.

glomerine. glomerin.

Glomeris marginata. an Old World millipede (Order Oniscomorpha) that rolls into a ball when attacked and releases the quinazolinones, glomerin and homoglomerin, from intersegmental pores. These are deadly convulsants that are lethal to arthropods and small vertebrate animals that prey upon this species.

glonoin. nitroglycerin.

gloriosa. a common name of *Gloriosa superba*.

Gloriosa superba (glory lily; climbing lily; gloriosa). a toxic plant (Family Liliaceae) native to tropical Africa and Asia. It contains incompletely characterized alkaloids, one of which is similar to colchicine. Human fatalities are known from ingestion of parts of this plant.

glory lily. *Gloriosa superba*.

glossitis. inflammation of the tongue.

glottic. of or relating to the tongue or glottis.

Glottidium vesicarium. *Sesbania vesicaria*.

glottis. the sound-producing structure of the larynx, consisting of the vocal folds and the intervening space between the vocal cords. It is protected by a fibrocartilaginous leaf-shaped cover, the epiglottis. The glottis may become inflamed and edematous from any of a number of causes, including various chemical fumes, or alcohol or tobacco abuse. Presenting symptom is usually hoarseness of voice which may give way to complete aphonia. In extreme cases, inspiratory dyspnea and subsequently expiratory dyspnea may occur. Involvement of the epiglottis may produce shrill, rasping sounds with breathing and a barking cough.

glucoproscillaridin A. scillaren.

2-*O*-β-D-glucopyranosylcucurbitacin E. colocynthin.

3β,5β,12β)-3-[(*O*-β-D-glucopyranosyl-(1→4)-*O*-2,6-dideoxy-β-D-ribo-hexopyranosyl-(1→4)-*O*-2,6-dideoxy-β-D-ribo-hexopyranosyl-(1→4)-*O*-2,6-dideoxy-β-D-ribo-hexopyrano-syl)oxy]-12,14-dihydroxycard-20(22)-enolide. deslanoside.

3-[(*O*-β-D-glucopyranosyl-1(1→6-*O*-D-glucopyranosyl-(1→4)-6-deoxy-3-*O*-methyl-α-L-glucopyranosyl]-14-hydroxycard-20(22)-enolide. cerberoside.

[6-*O*-β-D-glucopyranosyl-β-D-glucopyranosyl)-oxy]benzeneacetonitrile. amygdalin.

α-(β-D-glucopyranosyloxy)benzeneacetonitrile. mandelonitrile glucoside.

(*S*)-α-(β-D-glucopyranosyloxy)-4-hydroxybenzeneacetonitrile. dhurrin.

6-β-D-glucopyranosyloxy)-7-hydroxy-2*H*-1-benzopyran-2-one. esculin.

7-β-D-glucopyranosyloxy-8-hydroxy-2*H*-1-benzopyran-2-one. daphnin.

β-D-glucopyranosyloxy-L-*p*-hydroxymandelonitrile. dhurrin.

7-(β-D-glucopyranosyloxy)-6-methoxy-2*H*-1-benzopyran-2-one. scopolin.

(*d*-glucopyranosylthio) gold. gold thioglucose.

glucosamine. the dextrorotatory form (D-glucosamine) of this amino sugar, $CH_2OH(CHOH)_3$-$CHNH_2CHO$, is a component of heparin and a number of other polysaccharides such as hyaluronic acid, *q.v.*

D-glucosamine. See glucosamine.

glucoscillaren A. a cardiotoxic glycoside. It is a component of scillaren B.

glucoscillipheoside. a cardiotoxic glycoside. It is a component of scillaren B.

glucose. the most common simple (six-carbon) sugar, it is essential to the energy economy of living organisms. In most vertebrate animals it is stored chiefly in the liver and muscle as glycogen. Glucose metabolism is affected by certain toxicants and plays a role in the structure, metabolism, or activity of many toxicants (See, for example, concanavalin A, cyanophoric glycoside, cytochalasins, glucoside). See also sugar.

D-glucose-6-(dihydrogen phosphate). glucose-6-phosphate.

glucose-6-phosphatase. an enzyme produced by the kidney and liver of vertebrates. It catalyzes the conversion of glucose-6-phosphate into glucose, enabling release into the circulating blood.

glucose-6-phosphate (D-glucose-6-(dihydrogen phosphate; glucose-6-phosphoric acid; G6P). a normal constituent of resting muscle and an important intermediate in carbohydrate metabolism, $C_6H_{13}O_9P$. Individuals deficient in this enzyme are susceptible to the hemolytic effects of a number of substances (e.g., fava beans, phenylhydrazine, vitamin K).

glucose-6-phosphate dehydrogenase (G6P dehydrogenase; G6PD). an enzyme that catalyzes the synthesis of phosphogluconate from glucose-6-phosphate. Individuals deficient in this enzyme are susceptible to the hemolytic effects of a number of substances such as fava beans, phenylhydrazine, and vitamin K; the effects of phenazopyridine are more severe. A heritable deficiency of this enzyme results in a hemolytic anemia induced by the antimalarial pamaquine.

glucose-6-phosphoric acid. glucose-6-phosphate.

β-glucosidase. emulsin.

glucoside. **1:** any glycoside in which the sugar moiety is glucose; one that yields glucose on hydrolysis. Glucosides are *N*, *O*, and *S* derivatives of uridine diphosphate glucose. They occur commonly in plants and insects, but are rare in vertebrate animals. **2a:** the original name for a glycoside. **2b:** sometimes used erroneously as a synonym of glycoside. *Cf.* glucuronide, glycoside. See also phase II reaction.

6-β-glucosido-7-hydroxycoumarin. esculin.

7-glucosido-8-hydroxycoumarin. daphnin.

glucosin. any of a class of bases obtained from glucose by the action of ammonia; some are extremely toxic.

β-D-glucosyloxyazoxymethane. cycasin.

glucuronic acid. **D(+)-glucuronic acid**.

D(+)-glucuronic acid (glucuronic acid). a crystalline compound, $CHO(CH_2O)_4COOH$, that is soluble in both water and ethanol. It is an oxidation product of glucose and glucuronic acid that is widely distributed in both plants and animals, usually as part of a larger molecule (e.g., with alcohols or phenols; in gums). Glucuronic acid detoxicates, by conjugation in the liver, toxicants that have been absorbed from the gastrointestinal tract (e.g., salicylic acid, menthol, phenol). See also glucuronide.

β-D-glucuronidase (glusulase; glycuronidase). an enzyme that occurs in the intestinal microflora. It catalyzes the hydrolysis of various β-D-glucuronides with the liberation of free glucuronic acid. This allows the enterohepatic recirculation of xenobiotics excreted as glucuronides in bile.

glucuronidate. to form a glucuronide by the conjugation of UDP-glucuronic acid with an aglycone in a reaction catalyzed by a UDP-glucuronyltransferase (UDPGT).

glucuronide. any glycoside of glucuronic acid. Glucuronides result from the reaction of uridine diphosphate glucuronic acid (UDPGA) with an aglycone and are the most common endogenous conjugating agents in the vertebrate body. They react with xenobiotics through a nucleophilic displacement (SN_2 reaction) of the functional group of the substrate, catalyzed by a glucuronosyltransferase. Many xenobiotic chemicals, their intermediates, and certain other products of catabolism (e.g., steroids) in vertebrates are conjugated in the liver with glucuronides and excreted in the urine. *Cf.* glucoside. See also phase II reaction.

glucuronose. See glucuronic acid.

glue sniffing. the inhalation of fumes from certain glues and solvents (especially plastic cements) that contain organic solvents such as benzene, toluene, or xylene. This activity produces central nervous system stimulation followed by depression with sometimes fatal results. The practice is engaged in, chiefly by adolescents, to experience the altered state of consciousness so induced.

glusulase. β-D-glucuronidase.

glutamate. a salt of glutamic acid; often used interchangeably with glutamic acid, although glutamate technically refers to the negative ion. It is a common acceptor of amino acids in mammals. See conjugation (amino acid conjugation).

glutamine. a white, crystalline, water-soluble powder, $H_2NC(O)(CH_2)_2CH(NH_2)COOH$, that is insoluble in most organic solvents. It is a nonessential amino acid. See conjugation (glutamine conjugation).

2-L-glutamine-6-L-asparaginealytesin. bombesin.

glutamine conjugation. See conjugation.

γ-L-glutamyl-α-amino-β-(2-methylenecyclopropyl)propionic acid dipeptide. hypoglycine B.

β-(γ-L-glutamyl)-aminopropionitrile. the skeleton-deforming principle of at least some herbs of the genus *Lathyrus*, *q.v.* It causes paralysis and skeletal lesions. The nitrile group, together with the reactive amine group is responsible for toxicity. See *Lathyrus*, lathyrism.

γ-glutamylcysteinylglycine. glutathione.

γ-L-glutamylhypoglycine. hypoglycine B.

***N*-L-γ-glutamyl-3-(2-methylenecyclopropyl) alanine**. hypoglycine B.

glutathione (γ-glutamylcysteinylglycine). an odorless, white, water-soluble, crystalline powder, $C_{10}H_{17}O_6N_3S$, that is soluble also in dilute ethanol. It occurs universally in living cells and contains chemically bound glutamic acid, cysteine, and glycine which are readily separated on hydrolysis. **reduced glutathione** (GSH) is an antioxidant. See conjugation (glutathione conjugation).

glutathione conjugation. See conjugation.

gluten. the protein of wheat and other grains that gives the tough elastic character to dough. See gluten-induced enteropathy.

gluten-induced enteropathy (celiac sprue; gluten-sensitive enteropathy; nontropical sprue). a condition seen in individuals (mostly Caucasians, occasionally in African Americans, but not in Asians) that are hypersensitive to cereal grain storage proteins such as gluten or its product gliadin that occur in wheat, barley, and oats. Symptoms include diarrhea, weight loss, bloating, and steatorrhea. Lesions include villous atrophy (a flattening of the mucosal surface) and infiltration of the epithelial layer and lamina propria with inflammatory cells, leading to malabsorption in the small intestine. Gluten is thought to cause this disease via direct toxic action and by recruitment of immunological responses. The disease is usually resolved by administering a gluten-free diet.

gluten-sensitive enteropathy. See gluten-induced enteropathy.

glutethimide (2-ethyl-2-phenylglutarimide). a white, crystalline powder, $C_{13}H_{15}NO_2$. It is a CNS depressant used in medicine as a sedative. It is a drug of abuse; both manufacture and use is controlled by law.

Glycera*. 1:** a widely distributed genus of usually brightly colored, marine polychaete worms (Family Glyceridae) that occur on sand and mud flats and are important as fish bait. They have hard chitinous, hook-like, biting jaws and superficially resemble centipedes. The bite is oval and soon becomes erythematous, inflamed, and subsequently numb and itchy. See Annelida, Polychaeta. ***G. dibranchiata (bloodworm). a venomous annelid worm with hard, biting, chitinous jaws. It occurs along the Atlantic Coast of North America from North Carolina to northern Canada. The bite is painful and the area of the wound may become inflamed and edematous; numbness and itching may ensue. See also Annelida. **2:** a genus of grasses. See *Glycera striata*.

Glycera striata (fowl meadow grass; fowl mannagrass). a slender, to loosely tufted, cyanogenic grass (Family Gramineae) that may grow to a height of 1.5 m. It grows in moist ground from Newfoundland to Alberta, Canada, south in the United States to northern Florida, Alabama, and Texas.

glycerid. pertaining to or denoting any polychaete worm of the family Glyceridae.

glycerol diacetate. diacetin.

glycerol dioleic ester. a fly attractant of *Amanita muscaria*.

glycerol α-monochlorohydrin. α-chlorohydrin (See chlorohydrin).

glycerol nitric acid triester. nitroglycerin.

glyceryl diacetate. diacetin.

glyceryl trinitrate. nitroglycerin.

glycidaldehyde. a carcinogenic monoepoxide.

glycine (aminoacetic acid). a nonessential amino acid, it is a common acceptor of amino acids in mammals. See conjugation (amino acid conjugation).

glycine conjugation. See conjugation.

Glycine max (cultivated soybean). a normally nontoxic, cultivated plant (Family Fabaceae, formerly Leguminosae). It is, however, potentially goitrogenic. Toxic concentrations of nitrate have been reported. Poisoning of cattle and horses that have consumed trichloroethylene-extracted soybean-oil meal (TCESOM) is also recorded from several countries. Symptoms vary considerably among various species of livestock and laboratory animals.

glycine rich β-glycoprotein II. properdin factor B. See factor.

glycocoll betaine. betaine.

glycol chlorohydrin. ethylene chlorohydrin.

glycol dichloride. ethylene dichloride.

glycol ether. any of 4 chemical compounds: ethoxyethanol, methoxyethanol, and their acetates. They are used in numerous products including paints, stains, varnishes, and solvents. These are hazardous chemicals that retard growth, induce birth defects in laboratory animals, are also toxic to human fetuses, and cause hematologic and nervous disorders.

β-glycoprotein II. See glycine rich β-glycoprotein II.

β-glycoprotein (II). a blood serum protein and component of the properdin system. See also factor (cobra venom factor, properdin system).

glycoside. **1:** an acetal that yields a sugar (most often a pentose or hexose, but see 3-methyl digitoxose) and a nonsugar (aglycone) on hydrolysis. They are produced naturally in plants as a result of the chemical combination of diverse hydroxy compounds with various sugars. Glycosides occur more widely and more abundantly in plants than alkaloids. Toxicity varies from nontoxic to extremely toxic. The amount and type of glycosides present in many plants vary greatly due to both endogenous and exogenous modifying factors such as age, growth rate of the plant, climate, and edaphic factors. **2:** glucoside, now obsolete. **cardiac glycoside**. **1:** any of a class of cardiotonic, cardiotoxic glycosides that occur in plants of the genus *Digitalis* (Family Scrophulariaceae), *q.v.* The aglycones are derivatives of cyclopentenophenanthrene; the sugars are unique methyl pentoses. See also digitalid, 3-methyl digitoxose. **2:** any plant-derived steroid glycoside with cardiotonic activity. The cardioactivity depends on the presence in the aglycone of an unsaturated lactone ring and a hydroxyl group that have a specific spatial relationship illustrated, for example, by oleandrin. Some 400 such compounds have been identified, mostly in plants of three families: Scrophulariaceae (figworts), Liliaceae (lilies), and Apocynaceae (dogbanes). **coumarin glycoside**. any of a class of glycosides, the aglycone of which is a modification of coumarin. Examples are aesculin and daphnin. **cyanogenetic (cyanophoric, nitrile) glycoside**. a glycoside that yields hydrocyanic acid (HCN) upon hydrolysis. Ruminant animals are more susceptible to cyanide poisoning from plants than are monogastric animals. See, amygdalin, *Sambucus*. **cyanophoric glycoside** (cyanophoric, def. 2). a glycoside that yields glucose, hydrocyanic acid, and benzaldehyde on hydrolysis. **goitrogenic glycoside**. a goiter-producing thyrotoxic glycoside. **saponic glycoside**. a noncardioactive steroid glycoside; it is a saponin. See saponin. **steroid glycoside**. any glycoside in which the aglycone contains a sterol group (e.g., solanine). Their physical properties are similar to those of saponins. While the aglycones are toxic, the toxicity of the intact glycoside is at least partially determined by the sugar moiety on which the solubility of the molecule greatly depends. **tetanic glycoside**. any of a class of tetany producing glycosides. Included are cardiac and saponic glycosides. **triterpenoid glycosides**. a glycoside in which the aglycone is polycyclic and contains 30 carbon atoms. The aglycone is toxic, but the toxicity of the intact glycoside is at least partially determined by the sugar moiety on which the solubility greatly depends.

glycoside conjugation. See conjugation.

5-glycosyl-1,4,5-trihydroxy naphthalene. a labile glycoside that releases trihydroxynaphthalene which is readily oxidized to juglone.

glycuronidase. glucuronidase; β-D-glucuronidase.

glycyrrhetinic acid glycoside. glycyrrhizic acid.

glycyrrhiza (licorice; licorice root; liquorid). the dried rhizome and roots of varieties of *Glycyrrhiza glabra*. Glycyrrhizin is extracted from it.

Glycyrrhiza glabra (licorice; liquorice; sweet root). a perennial plant (Family Fabaceae, formerly Leguminosae). Licorice should not be ingested by anyone with a cardiovascular disorder. In large doses over extended periods of time it is highly toxic. Licorice contains a saponin-like glycoside, glycyrrhizic acid (*q.v.*), which is much sweeter than sucrose. It has both therapeutic and toxic effects. Licorice root and candy should be consumed with some caution, and should be avoided by hypertensive individuals, those with cardiac or renal disease, and pregnant women who are especially subject to edema. Licorice has caused high blood pressure, paralysis of the extremities, electrolyte imbalance, and shortness of breath. Some people are allergic to licorice. Intoxication has resulted from as little as one gram of glycyrrizic acid in chewing tobacco.

glycyrrhizic acid (20β-carboxy-11-oxo-30-nor-olean-12-en-3β-yl-2-*o*-β-D-glucopyranuronosyl-α-D-glucopyranosiduronic acid; glycyrrhizin; glycyrrhizinic acid; glycyrrhetinic acid glycoside). a very sweet saponin-like glycoside extracted from glycorrhiza. It is an adrenocortical stimulant used in the treatment of Addison's disease. It is toxic in large doses, producing headache, lethargy, water and salt retention, potassium secretion, and elevated blood pressure. The depletion of potassium sometimes causes renal failure, cardiac arrest, and death. See *Glycyrrhiza glabra*.

glycyrrhizin. glycyrrhizic acid.

glycyrrhizinic acid. glycyrrhizic acid.

glyoxosome. glyoxysome.

glyoxysome (glyoxosome). any of a type of microbody present in certain microorganisms and plant cells that resemble the peroxisomes of vertebrate animal cells. See also peroxisome.

Glyphodon (Australian collared snakes). a small genus of venomous snakes (Family Elapidae) of southeastern New Guinea and a few nearby islands: *G. barnardi* (Barnard snake), *G. dunmalli*, *G. tristis* (brown-headed snake). All are extremely reluctant to bite humans and none are considered dangerous.

glyphosate (*N*-(phosphonomethyl)glycine; Roundup; Kleen-up™). a herbicide of low toxicity to humans, but some domestic dogs and cats develop eye, skin, and upper respiratory effects when exposed during or following an application of this compound to weeds or grass. On occasion, these animals vomit, stagger, and develop muscular weakness of the hind legs when exposed to this chemical on freshly treated foliage.

Glyptocranium gasteracanthoides (Peruvian tarantula; pruning spider). a venomous spider indigenous to Peru. The bite causes focal gangrene as well as hematuria and neurotoxic effects.

GMP. guanosine monophosphate. See also cyclic GMP.

Gnaphalium purpureum (purple cudweed). a plant (Family Compositae) in which concentrations of nitrate toxic to livestock have been reported.

Gnathodentex aureolineatus (snapper). a marine bony fish (Family Lutjanidae) of the Tuamotu Archipelago west to East Africa. Certain populations are transvectors of ciguatera. See ciguatera.

Goa powder (araroba). **1:** the source of chrysarobin, it is a bitter, brownish-yellow to burnt orange powder that accumulates unevenly in the interspaces of the wood of a tree, *Andira araroba* (Family Fabaceae, formerly Leguminosae) that is indigenous to Brazil. **2: purified Goa powder**. chrysarobin.

goatfish. **1:** a common name of *Parupeneus chryserydros* and *Upeneus arge* (See ciguatera). **2:** See Mullidae.

goat's rue. *Tephrosia virginiana*.

goatweed. *Hypericum perforatum*.

gobori. *Oxyuranus scutellatus*.

God's chair. *Psilocybe mexicana*.

Godman's pit viper. *Bothrops godmani*.

goiter. simple goiter is a condition in which an enlarged thyroid produces low levels of thyroxin.

goitrin. See L-5-vinyl-thiooxazolidone.

goitrogen. any goiter-inducing substance.

goitrogenic. having the ability to cause goiter; goiter-inducing. Many species of plants contain goitrogenic substances, including, for example, many species and variants of *Brassica*, and food plants in several plant families. Some can be quite dangerous to livestock. See also thiocyanate, L-5-vinyl-2-thiooxazolidone, glycoside.

gold (aurum; symbol, Au)[Z = 79; A_r = 196.9665]. a nearly nonreactive, yellow transition metal. Gold is unaffected by acids and oxygen, but does react, for example, with strong oxidizing agents (especially if they contain halogens), with alkali cyanides, double cyanides, and thiocyanate solutions. Gold is toxic to marrow; it may induce thrombocytopenia and, in extremely severe cases, aplastic anemia. Some gold compounds are toxic; most gold salts are nephrotoxic.

gold cyanide. may refer to gold monocyanide or gold tricyanide.

gold monocyanide (aurous cyanide; gold cyanide). a yellow, odorless powder, AuCN, that liberates HCN under some conditions. It should be considered dangerous and highly toxic.

gold sodium thiosulfate (bis[monothiosulfato-(2—)-*O*-*S*]aurate(3—) trisodium; bis-(monothiosulfato)aurate(3—) trisodium; aurothiosulfate natrium; aurothiosulfate sodium; double thiosulfate of gold and sodium; hyposulfite of gold and sodium; sodium aurothiosulfate). used in the treatment of lupus erythematosus and some cases of rheumatoid arthritis, it is toxic by all routes of administration. Toxic systemic effects in humans include dermatitis, granulocytopenia, and thrombocytopenia.

gold thioglucose (1-aurothio-*d*-glucopyranose; aurothioglucose; (*d*-glucopyranosylthio) gold); (1-thio-*d*-glucopyranosato)gold). a toxic, watersluble, white, crystalline compound or yellow powder, $C_6H_{11}O_5S{\cdot}Au$. Symptoms may include dermatitis, stomatitis, hepatitis, nephritis, and gastrointestinal and hematologic disturbances. It is moderately toxic, *i.v.*, and by subcutaneous injection. Dimercaprol, *q.v.*, is an antidote. Gold thioglucose is also a confirmed human carcinogen.

gold tricyanide (auric cyanide; cyanoauric acid). the trihydrate is a white, deliquescent crystalline substance that is soluble in ethanol and ether, $Au(CN)_3{\cdot}3H_2O$ or $HAu(CN)_4{\cdot}3H_2O$. It decomposes at 50°C. Gold tricyanide is dangerously poisonous and should be protected from light and stored in tightly closed containers. It is used mainly as an electrolyte in the electroplating industry.

golden arrow frog. *Atelopus zeteki*.

golden chain. *Cytisus laburnum*; *Laburnum*.

golden chain tree. *Laburnum*.

golden dewdrop. *Duranta repens*.

golden false pholiota. *Phaeolepiota aurea*.

golden oats. *Trisetum flavescens*.

golden seal. *Hydrastis canadensis*.

goldenbush. *Haplopappus venetus* (Coast goldenbush).

goldenrod. *Solidago*.

goldenseal. *Hydrastis canadensis*.

goldenweed. *Oonopsis*.

goldfish. See Cyprinidae.

Gold's tree cobra. *Pseudohaje goldii*.

gonad. either of the two primary sex organs of higher animals, the ovary or testis. It produces gametes and sex hormones. Its reproductive function can be seriously impaired or destroyed by many toxicants.

gonadal. pertaining to the gonads.

gonotoxemia. a toxemia due to the effects of the endotoxin of hematogenously disseminated gonococci. See gonotoxin.

gonotoxin. the endotoxin of the gonococcus, *Neissena gonorrhoeae.*

Gonyaulax. a genus of protozoa (Order Dinoflagellata, Class Phytomastigophora) that have yellow to brown chromatophores. They are found in marine, brackish, and freshwaters. *G. catenella*. a species associated with shellfish poisoning along the Pacific coast of North America. *G. tamarensis*. the species associated with shellfish poisoning along the Atlantic Coast of North America. See also gonyaulax toxin, shellfish poisoning.

gonyaulax toxin. paralytic shellfish poison. See also saxitoxin.

good laboratory practice. as part of the common fund of knowledge that most research scientists gained and practiced during their education (undergraduate, graduate, and post-graduate) was (and hopefully continues to be) a constellation of practices that limit the introduction of errors into the results of research and testing. A few scientists, however, left to their own devices do not approach testing and research with adequate knowledge and/or rigor. Because the health and environmental effects of some chemicals can be extremely damaging, and because the cost of managing such chemicals is so high, governmental bodies (e.g., the EPA and OEC) have established a body of regulations that govern "good laboratory practice." With respect to toxicology, these regulations include all aspects of testing, from the management of test organisms to the sampling, management, analysis and reporting of data. Regulations bear also on facilities and the training of personnel; they provide also for inspections. Research conducted for the purposes of acquiring data to be used for regulatory purposes must conform to these regulations.

goose grass. *Eleusine indica.*

goosefoot (plural, goosefoots). *Chenopodium.*

gossypol (1,1′,6,6′,7,7′-hexahydroxy-3,3′-dimethyl-5,5′-bis(1-methylethyl)[2,2′-binaphthalene]-8,8′-dicarboxaldehyde; 1,1′,6,6′,-7,7′-hexahydroxy-5,5′-diisopropyl-3,3′-dimethyl[2,2′-binaphthalene]-8,8′-dicarboxaldehyde; 2,2′-bis[1,6,7-trihydroxy-3-methyl-5-isopropyl-8-aldehydonaphthalene];2,2′-bis[8-formyl-1,6,-7-trihydroxy-5-isopropyl-3-methylnaphthalene]). a toxic, polyphenolic pigment, $C_{30}H_{30}O_8$, found in the seeds of *Gossypium* (cotton). It is a gastrointestinal irritant; large doses cause pulmonary edema, dyspnea, and paralysis in laboratory animals. Some breeds of cattle and other ruminants are resistant to gossypol and may feed on raw seeds, usually with little, if any, effect. Gossypol may also produce temporary sterility in males by lowering the sperm count and has been used in China as an oral contraceptive. It is moderately toxic (e.g., to laboratory rats, domestic fowl) to very toxic (swine, laboratory rabbits, guinea pigs) to some nonruminant animals. It acts in nonruminant animals by lowering the oxygen-carrying capacity of the blood. See also cottonseed.

gossypol poisoning. symptoms of gossypol poisoning in swine, dogs, and Holstein cattle are similar and usually appear abruptly following some weeks of cottonseed meal consumption. They include dyspnea and sometimes frothing (sometimes bloody) at the mouth. Even though the appetite usually remains normal, affected animals (usually livestock or poultry) may become weak and emaciated. Death, sometimes with convulsions, usually intervenes within 2-6 days following the onset of symptoms. Cyanosis develops immediately before death. Nonlethal doses of gossypol in poultry may reduce the hatchability of eggs, depress growth rates, and discolor or alter the consistency of eggs.

gourd. **1:** See Cucurbitaceae. **2:** the biblical name of *Ricinus communis*.

gourd curare. See curare.

gourd family. Cucurbitaceae.

gousiekte. a sometimes fatal cardiac condition in sheep that have browsed on *Vangueria pygmora*, a poisonous bush. The condition is characterized by myocarditis with dilation of the cardiac muscle and heart failure.

gout plant. *Jatropha multifida*; also *J. podagrica* (which is not treated in the present work).

graceful coral snake. *Calliophis gracilis*.

graceful small-headed sea snake. *Microcephalophis gracilis*.

grade. **1:** degree, rank, stage, status, quality. **2:** any of a number of standards of chemical purity of varying specifications that have been established by various professional associations, organs, or bodies. These include ACS reagent grade (analytical reagent quality or grade); CP (chemically pure, *q.v.*); FCC, a grade that conforms to National Formulary Specifications; NF, a grade of purity conforming to National Formulary specifications; U.S.P, a grade of purity conforming to the specifications of the U.S. Pharmacopeia. A number of other grades exist. See purity.

gram (g or gm.). the basic unit of mass in the metric system; it equals 15.432 grains, or 0.035 ounces avoirdupois.

gramicidin. any of a class of polypeptide antibiotics that contains *D*-amino acids.

gramicidin S. a cyclic peptide produced by *Bacillus brevis* and other *Bacillus* species that acts on Gram-positive bacteria (staphylococci, streptococci). It is not synthesized by the mRNA/tRNA/ribosome system.

Grand Canyon rattlesnake. *Crotalus viridis abyssus*.

grand extrapolation, the. See extrapolation.

granivore. a heterotrophic organism that feeds exclusively (obligate granivore) or predominantly on seeds.

granulocyte. a type of leukocyte that contains distinctive cytoplasmic granules (basophils, eosinophils and neutrophils).

granulocytopenia. agranulocytosis.

Granutox. phorate.

grass pea. *Lathyrus sativus*.

grass staggers. grass tetany.

grass tetany (grass staggers; wheat pasture poisoning; protein poisoning; grass tetanus; lactation tetanus; hypomagnesemia). a syndrome of livestock that is apparently due to impairment of an animal's ability to absorb dietary magnesium (See hypomagnesemia). This impairment is related to elevated rumen ammonia levels and consequent conversion of dietary magnesium to the relatively insoluble hydroxide. This suggests intoxication of ruminal flora since part of the ammonia produced by the digestion of nitrogenous material is normally processed by the ruminal microflora. The disease is usually associated with lush pasturage or forage and characterized by hypomagnesemia and frequently by hypocalcemia and other ionic imbalances in serum. Chronic grass tetany may not be accompanied by gross symptoms, and is detected clinically by low serum magnesium levels to which affected animals may have adapted. The acute syndrome has a latent period of about 1 week to 6 months. Initial symptoms in both cattle and sheep are excitement, anorexia, and incoordination progressing to include muscular twitches, grinding of teeth, and salivation, to more severe symptoms that may include belligerence, staggering, and prostration. Tetanic seizures may become general and accompanied by dyspnea and tachycardia with intermittent remissions, terminating in death. *Cf.* hypomagnesemic tetany.

Gratus strophanthin. ouabain.

grav. See G.

graveyard weed. *Euphorbia cyparissias*.

gravid. synonymous with pregnant, but usually applied to nonhuman viviparous organisms.

gravida. a gravid female; a pregnant woman.

gray (Gy). the International System unit of absorbed dose. It replaces the rad in this system. 1 Gy = 1 joule/kg body weight. = 100 rads.

gray lady spider. See *Latrodectus*.

gray sea snake. *Hydrophis torquatus*.

gray-veil amanita. *Amanita porphyria*.

Gray's sea snake. *Hydrophis ornatus*.

grayanotoxin (andromedotoxin). any of a group of poisonous substances that paralyze the motor nerve terminals in skeletal muscle of humans and laboratory mammals. They also initially stimulate and then paralyze the vagus nerve. Plants that contain grayanotoxins include *Andromeda* (andromeda; bog rosemary), *Kalmia* (laurels), *Ledum* (Labrador tea), *Lyonia* (e.g., stagger bush), *Pieris* (pieris), *Rhododendron* (rhododendrons, azaleas). See *Andromeda*.

Grayia (hop sage). a genus of plants (Family Chenopodiaceae), the members which are facultative accumulators of selenium from soil.

greasewood. **1:** *Sarcobatus vermiculatus*. **2:** loosely, any plant of the family Chenopodiaceae. **3:** certain other unrelated small woody shrubs of arid lands are also sometimes termed greasewood, e.g., **a:** ***Adenostoma fasciculatum*** (chamiso, chamisa). a California shrub. **b:** ***Ramona polystachya*** (white sage). a low shrub of the southwestern United States and northern Mexico.

great ape. See ape.

great barracuda. *Sphyraena barracuda*.

Great Basin rattlesnake. *Crotalus viridis lutosus*.

Great Lakes bush viper. *Atheris nitschei*.

great laurel. *Rhododendron maximum*.

Great Plains toad. common name of *Bufo cognatus*.

great white shark. *Carcharodon carcharias*.

greater cerastes viper. *Cerastes cerastes*.

greater weever. *Trachinus draco*.

Greek alphabet. not included in this dictionary. The Greek letters that are most important in toxicology or the most frequently encountered are defined herein. See alpha, beta, gamma, delta, kappa, lambda, mu, sigma.

Green 3. a tumorigenic food coloring, the use of which has been banned in several European nations because it can cause bladder cancer. It is used in bottled drinks and candy.

green bush viper. *Atheriis squamigera*.

green cestrum. *Cestrum parqui*.

green copperas. ferrous sulfate.

green deathcap. *Amanita phalloides*.

green dragon. *Arisaema dracontium*.

green fire worm. *Hermodice carunculata*.

green hellebore. **1:** *Helleborus viridis*. **2:** See *Veratrum spp*.

green lynx spider. *Peucetia viridans*.

green mamba. See *Dendroaspis angusticeps*, *D. jamesoni*, *D. polylepis*, *D. viridis*.

green night adder. *Causus resimus*.

green pit viper. *Trimeresurus trigonocephalus*.

green sea turtle. *Chelonia mydas*.

green snake. *Demansia typus*.

green tree mamba. *Dendroaspis viridis*.

green tree viper. *Trimeresurus gramineus*.

green turtle. See *Chelonia*.

green vetch. *Lathyrus sativus*.

green viper. *Atheris nitschei*, *Causus resimus*.

green vision. chloropsia.

green vitriol. ferrous sulfate.

green-spored lepiota. *Chlorophyllum molybdites*.

greenbriar. plants of the genus *Smilax*.

greenish-black mica. adamsite.

grey partridge. *Perdix perdix*.

grey snake. *Demansia psammophis*.

Grey's sea snake. *Ephalophis greyi*.

Griefschwanz Lanzenotter. *Bothrops schlegelii*.

Grindelia (gumweeds). a genus of herbs of the family Compositae; all are facultative accumulators of selenium from soil.

grip strength (screen grip). the strength of grip of the fore or hind limbs of laboratory rats and mice. It may be rated quantitatively or by subjective estimate of the degree of impairment as by a toxicant.

griseofulvin (7-chloro-2′4,6-trimethoxy-6′-methylspiro[benzofuran-2(3*H*),1′-[2]cyclohexene]-3,4′-dione; 7-chloro-4,6-dimethoxycoumaran-3-one-2-spiro-1′-(2′-methoxy-6′-methylcyclohex-2′-en-4′-one); amudane). a water-insoluble, chlorine-containing, slightly ethanol-soluble, fungistatic antibiotic produced by *Penicillium griseofulvum*, *P. patulum*, and *P. janczewski* (= *P. nigricans*). It is used mainly as an agricultural antifungal agent. It is slightly toxic to humans, usually causing no more than headache and gastrointestinal distress. The toxicity of griseofulvin is generally low, but it is very toxic to laboratory rats, causing liver damage. It is also teratogenic and carcinogenic to laboratory rats.

Gross Alpha Particle Activity. total radioactivity from a source due to emission of alpha particles, usually measured in picocuries. It serves as the screening measurement for radioactivity generally due to naturally occurring radionuclides.

Gross Beta Particle Activity. total radioactivity from a source due to emission of beta particles, usually measured in picocuries. It serves as the screening measurement for radioactivity from man-made radionuclides since the decay products of fission are emitters of beta particles and gamma rays.

ground cherry, **groundcherry**. *Physalis*.

ground hemlock (American yew). See *Taxus canadensis*.

ground ivy. *Glechoma hederacea*.

ground lichen. *Parmelia molliuscula*.

ground rattlesnake. common name of rattlesnakes of the genus *Sistrurus*.

groundsel. *Senecio*.

groundwater. See water.

grouper. the common name of any of a number of large tropical and subtropical marine bony fish such as *Cephalopholis argus* and *Epinephelus fuscoguttatus*.

growth. **1:** increase, especially an increase in size. Growth may take place by accretion (i.e., by the addition, accumulation, or adherence of materials directly from the environment, as in the growth of the shell of some living organisms. **2:** a tumor, cancer, growth, or swelling.

growth hormone. somatotrophin.

growth rate. the rate of change in size of a living organism, a component part, or a population. **absolute growth rate**. the actual rate of growth per unit time under a given set of conditions. **relative growth rate**. the rate of growth of one living system under a given set of conditions relative to another living system or to itself at another time or under an alternative set of conditions.

growth substance. auxin.

Grüne Mamba. *Dendroaspis viridis*.

GSC. gas-solid chromatography.

GSH. reduced glutathione. See glutathione.

GTX. grayanotoxin.

guajillo. *Acacia*.

guajillo wobbles. See *Acacia berlandieri*.

guanidase. a hydrolase that catalyzes the metabolism of guanidine into urea and ammonia. It is produced by several organisms (e.g., *Aspergillus niger*).

guanidine. a strong poisonous base, $NH:C(NH_2)_2$. It is the amidine of amino carbamic acid, occurring in lower animals and in some plants as the hydrochloride. A cholinergic, it is used medically as a stimulant of striated muscle.

guanidinemia. the presence of guanidine in circulating blood.

guanosine cyclic 3′,5′-(hydrogen phosphate). cyclic GMP.

guanosine 3′,5′-cyclic monophosphate. cyclic GMP.

guanosine 3′,5′-cyclic phosphate. cyclic GMP.

guanosine 3′,5′-GMP. cyclic GMP.

guanosine 3′,5′-monophosphate. cyclic GMP.

guaranine. caffeine.

gulf puffer. *Sphaeroides annulatus*.

gum camphor. camphor.

gum opium. opium.

gum sugar. arabinose.

gum tree. See *Eucalyptus*.

gumweed. See *Grindelia*.

Günther's sea snake. *Hydrophis torquatus*.

Guppy's snake. *Denisonia par*.

gurukezi. *Dispholidus typus*.

gut. gastrointestinal tract.

"**Guthion.**" azinphos methyl.

Gutierrezia (snakeweeds; tansy aster; broom-snakeroot). a genus of glabrous, often glutinous woody shrubs (Family Asteraceae). They are facultative accumulators of selenium from soil. ***G. microcephala*** (= *G. sarothrae* var. *microcephala*; *Xanthocephalum lucidum*; *Xanthocephalum sarothrae*; broomweed; perennial snakeweed; slinkweed; turpentine weed). a densely branched perennial resinous woody shrub of dry rangelands from Colorado and Idaho, southern Texas, California, and Mexico. It contains the toxic abortifacient saponin. Losses of livestock, especially sheep and cattle, are sometimes severe in the Trans-Pecos and High Plateau regions of Texas. Adult animals are sometimes killed by ingesting the plant, but the major impact is through abortion. Symptoms of acute poisoning in adult cattle and sheep may include anorexia, sluggishness, diarrhea or constipation, rough coat, mucus in the feces, vaginal discharge, nasal discharge and peeling muzzle (in cattle), mild jaundice (mainly in sheep), and hematuria. Lesions are those of gastroenteritis, nephritis (sometimes severe), and hydropic degeneration of liver cells with eventual necrosis. The uterus is edematous and the extraembryonic membranes are hydropic.

gutta serena. amaurosis.

gwardae. *Demansia textilis*.

Gy. gray.

Gymnocladus dioica (Kentucky coffee tree). a large rough-barked deciduous tree (Family Fabaceae, formerly Leguminosae) that is native to moist woods in the eastern and central United States. The raw seeds and fruit pulp are toxic, producing gastrointestinal symptoms such as vomiting and diarrhea with cardiac arrhythmia and coma. The toxic principle is cytisine.

gymnodinium bloom. red tide.

Gymnodinium. a genus of subspherical dinoflagellates (See Dinoflagellata) with numerous variously colored chromatophores. The cell is marked by one annular and one posterior longitudinal groove. Both fresh water and marine species are known. ***G. brevis*** (= *G. breve*, *Ptychodiscus brevis*). a toxic dinoflagellate responsible for red tides, sometimes called gymnodinium blooms, with massive mortality among fish along the west coast of Florida and elsewhere. Millions of fish have been killed by these blooms during the past century and have been responsible for shellfish poisoning of humans. Onshore winds during such blooms may induce transient coughing in humans. See also bloom, brevetoxin, red tide. Related species are *G. mikimatoi*, *G. splendens*, and *G. spectabilis*, all of which contain toxins such as psilocin and psilocybin.

Gymnothorax (moray; moray eel). a genus of marine bony fish (Family Muraenidae) found in tropical and subtropical seas, usually about reefs. The sharp, spiny teeth can cause deep, severe lacerations with considerable danger of infection. These eels are dangerous in or out of the water, although some are retiring and not easily provoked. Most injuries occur when a diver's hand enters the hole in which a moray lives. These animals are powerful enough to hold a diver down long enough for drowning. Some populations of at least the following species are transvectors of ciguatera. ***G. flavimarginatus***, ***G. javanicus***, ***G. meleagris***, ***G. pictus***, and ***G. undulatus***.

Gymnura (butterfly rays). a genus of venomous cartilaginous fish (Family Gymnuridae). The sting is small, poorly developed, and lies close to the base of the short tail. See stingrays.

gymnurid. any of the butterfly rays, Gymnuridae, especially those of the genus *Gymnura*.

Gymnuridae (gymnurids). a family of venomous

cartilaginous fish, the butterfly rays. See also stingrays, *Gymnura*, gymnurid.

Gyromitra (false morels). a genus of ascomycetous, mostly toxic mushrooms (Family Helvelaceae) with a large, simple, stipitate fruiting body (ascoma) with a folded ascus-bearing cap. Members of this genus are closely related to and easily confused with those of *Helvella*. Poisonous mushrooms of both genera usually produce effects that are similar to those produced by amanitin, but are less severe. Symptoms appear (primarily gastrointestinal) in 6 to 12 or more hours following ingestion. Some species produce hemolysis and central nervous system effects. The active principles vary (See gyromitra toxin). Some humans are resistant. See also helvellic acid, gyromitrin. ***G. (Helvella) esculenta*** (conifer false morel; brain mushroom; beefsteak morel; lorchel; edible false morel; turban top). found throughout North America, but is most common in the north and at high elevations. It may be mistaken for the edible morel, but is deadly poisonous, producing serious effects and death on occasion. It is said to be edible in some locations, but doing so would be unwise. It is tumorigenic in laboratory animals. It contains gyromitrin and methylhydrazine, a volatile pyridoxine antagonist. The toxins can usually be removed (at least partially) by air drying or by boiling, rinsing, and reboiling. Symptoms appear abruptly ca. 6-7 hours following ingestion or inhalation of vapor from cooking; they include headache, malaise, abdominal fullness, and emesis without diarrhea. Recovery without treatment within 2-6 days is common. Fatal hepatic necrosis occurs in some cases. Actual poisonings are rare. See also helvellic acid. ***G. (Helvella) gigas*** (snowbank false morel; giant helvella). found on mountains in Colorado, California and the Pacific Northwest, mainly near melting snowbanks. It grows to ca. 1 kg and is eaten in many locations, but poisonings have been reported. See also gyromitrin. ***G. (Helvella) infula*** (saddle-shaped false morel; hooded helvella). a poisonous species found throughout North America on or by rotten wood or woody debris. It contains helvellic acid. ***G. brunnea*** (*Helvella underwoodii*; gabled false morel). a mushroom with a brownish, convoluted, brain-like cap (about 10 cm wide) on a whitish, branched stalk; the cap may be saddle-shaped or lobed. It occurs throughout much of the eastern United States to Michigan, south and west to Tennessee, Missouri, and Oklahoma. It occurs on humus in deciduous forests. It is poisonous; no deaths have been reported, but the effects of intoxication include severe headaches, vomiting, diarrhea, and blood poisoning. See also gyromitrin. ***G. fastigiata*** (*G. korfi*; thick-stalked morel). closely related to *G. esculenta* and should be considered poisonous. It occurs in northeastern North America, the Rocky Mountains, and locally in the Pacific Northwest on the ground in deciduous or mixed forests. ***G. hydrazine esculenta***. poisonous; contains methyl hydrazine. ***G. korfi***. *G. fastigiata*.

gyromitrin (ethylidine gyromitrin; acetaldehyde-*N*-methyl-*N*-formylhydrazone). an extremely toxic hydrazone of acetaldehyde isolated from various species of *Gyromitra* and *Helvella* (e.g., *G. (Helvella) esculenta*, *G. (Helvella) gigas* and *G. brunnea* (= *H. underwoodii*). It occurs in association with a number of other hydrozones and related compounds, which are readily hydrolyzed to the proximate toxic metabolite, methylhydrazine, *q.v.*, that appears to be responsible for most or all of the toxic effects of gyromitrin. Symptoms in humans are similar to those produced by ingestion of *Amanita phalloides*.

h. hecto-.

H. the symbol for hydrogen.

H 56/28. alprenolol.

H2S. the hydride of sulfur; a highly toxic gas.

HA. hyaluronic acid.

HA-1A. a human monoclonal IgM antibody that is specific for the lipid A domain of endotoxin. It is effective in preventing the death of laboratory animals that have Gram-negative bacteremia and endotoxemia. HA-1A is considered to be safe and effective in the treatment of sepsis and Gram-negative bacteremia in humans even when they are in shock; it is not effective against focal Gram-negative infections.

Haarlem oil. sulfurated oil of turpentine.

Haber's rule. a rule applied to gaseous toxicants, approximate at best, which states that where the concentration (C) and the time (T) are constant, the toxic effect (K) is constant.

habitat. **1:** the particular place (area) and type of environment in which a population of organisms (e.g., human, animal, plant, microorganism) naturally occurs, including the nature of the surroundings, both living and nonliving. **2:** the particular abiotic environment in which an organism lives.

habituation. the gradual reduction or loss of response to a particular stimulus following frequent or continuous stimulation of the stimulus without reinforcement. Habituation can be due to a simple learning process, or can be physiologic in nature.

habu. *Trimeresurus flavorviridis*.

Hadrurus arizonensis (giant desert hairy scorpion; giant hairy scorpion). the largest North American scorpion, up to 140 mm long. The cephalothorax is black, with each segment bordered by yellow. It occurs in the southwestern deserts of the United States. The undersurface is pale as are the abdomen, legs, and pedipalps. It preys on insects, and sometimes on snakes and small lizards.

haem. heme.

haemagglutination. See hemagglutination.

haemagglutinin. See hemagglutinin.

haematodyscrasia. See hemodyscrasia.

haematogenous. See hematogenous.

haematologist. See hematologist.

haematology. See hematology.

haematolysis. See hemolysis.

haematolytic. See hematolytic.

haematopathy. See hemopathy.

haematopoiesis. See hematopoiesis.

haematopoietic. See hematopoietic.

haematotoxic. See hemotoxic.

haematotoxin. See hemotoxin.

haematoxic. See hemotoxic.

haematoxin. See hemotoxin.

haematuria. See hematuria.

haemerythrin. See hemerythrin.

haemoantitoxin. See hemoantitoxin.

haemochromatosis. See hemochromatosis.

haemoclasis. See hemolysis.

haemoclastic. See hemoclytic.

haemocoagulin. See hemocoagulin.

haemocytolysis. See hemolysis.

haemodilution. See hemodilution.

haemodyscrasia. See hemodyscrasia.

haemoglobin. See hemoglobin.

haemoglobinuria. See hemoglobinuria.

haemolymph. See hemolymph.

haemolysin. See hemolysin.

haemolysis. See hemolysis.

haemolytic. See hemolytic.

haemolyzation. See hemolyzation.

haemolyze. See hemolyze.

haemopathic. See hemopathic.

haemopathy. See hemopathy.

haemophilia. See hemophilia.

haemoprotein. See hemoprotein.

haemoptysis. See hemoptysis.

haemotoxic. See hemotoxic.

haemotoxin. See hemotoxin.

Haff disease. a disease of humans living along the coast of the Baltic Sea. The disease is associated with the ingestion of fish or fish liver from areas supporting algal blooms.

hafnium (symbol, Hf)[[illegible]; A_r = [illegible]] a rare, hard, brittle, transition metal, with gray crystals, that occurs in all zirconium ores. An explosion hazard, it is also toxic by inhalation and forms toxic compounds. It may be found, for example, in lightbulb filaments, electrodes, and control rods in water-cooled nuclear reactors.

Hagen's pit viper. *Trimeresurus hageni*.

HAH. halogenated aromatic compound.

haia amia. *Vipera lebetina*.

haia soda. *Walterinnesia aegypta*.

hair. **1:** a keratinized, filamentous integumental structure that is uniquely characteristic of mammals (Class Mammalia). **2:** pelage; fur; the hair collectively that covers the body. A number of toxicants such as chromium, lead, mercury, selenium, arsenic, and PCB and other halogenated compounds are associated with hair. Such substances are sometimes present in higher concentrations in hair than in any of the internal organs of a poisoned mammal; hair generally contains the highest Hg level of any mammalian tissue. Hair is easily collected and relatively resistant to decay. It can be used to assess the levels and sometimes the history of exposure to halogenated compounds such as PCB, heavy metals such as chromium, lead and mercury, and the metalloid arsenic. The accumulation of heavy metals in the washed hair of mammals exposed to heavy metals can reach values one to two magnitudes greater than in control subjects. Serial segments of human hair have sometimes proven useful as in the case of the Minimata disaster in determining the history of exposure.

hair color (hair coloring). any product used to color hair whether in the home or done by a professional. A composite (generic) warning label is required in the United States: "CAUTION: This product contains ingredients that may cause skin irritation on certain individuals, and a preliminary test according to accompanying directions should first be made. This product must not be used for dyeing eyelashes or eyebrows: to do so may cause blindness." "WARNING: Contains an ingredient that can penetrate your skin and has been determined to cause cancer in laboratory animals." A substantial number of chemicals commonly used in hair coloring products are suspected human carcinogens. Toxic or potentially toxic hair coloring products cannot, however, be banned in the United States because of a 1938 law which is still in effect.

hair coloring. See color (HC color).

hair dye. any hair coloring preparation. Such preparations commonly contain 4-methoxy-*m*-phenylenediamine (4-MMPD). This compound is linked to cancer of the breast and bladder. For this reason, some manufacturers have withheld 4-MMPD from their products.

hair-removal formulation, hair-removal product. a product used to chemically remove hair from the human body. The most dangerous ingredient in hair-removal products is ammonium thioglycolate, which can cause severe skin rashes, swelling, redness, and the breaking of small blood vessels under the skin.

hair spray. any commercially available spray used to set or groom hair. Such products commonly contains ethanol, carcinogenic polyvinylpyrrolidone plastic (PVP), formaldehyde, aerosol propellants, and artificial fragrance. Regular users of hair spray risk acquiring a pulmonary form of thesaurosis caused by the PVP content. The disease is reversible if hair spray and other sources of PVP are avoided. Eye and nasal irritations and allergic reactions to hair spray are not uncommon.

hair-removal formulation (hair-removal product). See depilatory.

hairy nightshade. *Solanum villosum*.

Haldol™. haloperidol.

half-banded snake. *Brachyurophis semifasciatus*.

half-girdled snake. *Brachyurophis semifasciatus*.

half-life ($T_{1/2}$). **1:** the average period of time during which the mass, concentration, or activity of a substance decreases to one-half of the initial value in a given system (e.g., a living organism, the atmosphere, a population, community) under specified conditions. **2:** radioactive half-life; the time required for half of the atoms of a radioactive element to undergo decay. See also effective half-life, physical half-life. **atmospheric half-life**. half-life of a specified substance (e.g., a specified air pollutant, vapor, aerosol, microorganism) at a specified location or component of the air environment (e.g., an airshed, a room, a factory). **biological half-life. 1a:** the half-life of a substance within a specified living system as a result of biological processes (e.g., metabolism, excretion, inactivation). **1b:** the half-life of a total dose of a drug, toxicant, radionuclide, or test substance from a living organism. **ecotoxicological half-life**. half-life of a toxicant within a specified ecological system (e.g., a particular population of living organisms, community, or ecosystem). **effective half-life** (effective radioactive half-life). the half-life of a particular radionuclide within a biological system due to radioactive decay and other processes such as excretion. **effective radioactive half-life**. effective half-life. **environmental half-life. 1:** the half-life of a substance introduced into the environment from anthropogenic (or sometimes other) sources. Thus, the environmental half-life of DDT is about 15 years and that of radium, 1580 years. **2:** the half-life of a substance (usually a pollutant) in the environment measured by its effect on a specified component of the environment. **pharmacologic half-life**. biological half-life as measured by the rate of loss of pharmacologic activity of a substance within a living system or component of a living system. **physical half-life**. the half-life of a radioactive substance through radioactive decay alone. **radioactive half-life**. the half-life of a radioactive substance. See also effective half-life and physical half-life. **toxicological half-life**. half-life of a toxicant within a specified living system (e.g., a particular type of cell, tissue, organ, or organism).

half-ringed snake. *Brachyurophis semifasciatus*.

halfmoon loco. *Astragalus argillophilus*, a true locoweed.

haliangine. a naturally occurring epoxide auxin.

halide. a compound that contains one or more halogen molecules. **acyl halide**. any acylating agent with the general formula RCOX, where X is a halogen. All are fuming, irritant substances. They are highly reactive, especially with strong nucleophiles such as H_2O, NH_3 and ROH. See anhydride (acid anhydride), acylation. **alkenyl halide** (alkenyl organohalide). any organic compound that contains at least one halogen atom and one carbon-carbon double bond. Alkenyl halides are widespread environmental toxicants that exhibit a wide range of acute and chronic toxicities. Examples are vinyl chloride, trichloroethylene, dichloroethylenes, tetrachloroethylene, hexachlorobutadiene, and chloroprene. **alkenyl organohalide**. alkenyl halide. **alkyl halide** (alkyl organohalide). any organohalide in which one or more halogen atoms is substituted for hydrogen in an alkyl group. Most commercially important alkyl halides are derivatives of low molecular weight alkanes. Toxicities vary considerably among alkyl halides. Most, if not all, are central nervous system depressants. **alkyl organohalide**. alkyl halide. **aryl halide**. any of a class of aromatic compounds having two benzene rings that contain substituted halogens (usually chlorine). The two major classes of aryl halides are those based upon naphthalene and the other upon biphenyl. Aryl halides have been sources of substantial human exposure and environmental contamination in the past. Examples are monochlorobenzene, dichlorobenzenes, hexachlorobenzene, bromobenzene. **cadmium halide**. any inorganic compound that contains cadmium and one or more halogen atoms. See cadmium, cadmium chloride, cadmium fluoride. **hydrogen halide**. any compound with the general formula HX, where X is a halogen atom. All are toxic gases. Of these, because of their industrial applications and abundance in the environment, HF and HCl are of greatest concern toxicologically. **phosphorus halide**. any of a class of phosphorus compounds with the general formulas PX_3 and PX_5, where X represents a halide. Examples are phosphorus trifluoride and phosphorus pentachloride (the most important commercially). All are very toxic. they react violently with water to produce the corresponding hydrogen halides plus oxo phosphorus acids. They tend to form strongly acidic solutions and are strong irritants to eyes, skin, and mucous membranes. **phosphorus oxyhalide**. a class of compounds with the general formula POX_3. Only the fluoride, chloride, and bromide are known, of which, only phosphorus oxychloride, *q.v.*, has commercial value. **vinyl halide**. a halide that contains the vinyl group, $CH_2{=}CH—$. These compounds are toxic, highly reactive, and are carcinogenic to humans and some laboratory animals.

hali-ichthyotoxin. a poisonous nitrogenous base of bacterial origin in stale fish.

Haliotis (abalone; ear shell). a genus of more than 100 species of marine gastropod mollusks; some are edible. The shell is large, up to 30 cm, and ear-shaped, with an enormous aperture and a series of holes along the margin of the shell. At least some species secrete a blue-green pigmented toxin, concentrated in the digestive gland, which causes photodermatitis in cats and humans.

halite. sodium chloride.

Hallowell's mamba. *Dendroaspis viridis*.

hallucination. in psychology, a false perception that has no relation to reality and cannot be accounted for by any external stimulus. The individual affected is unable to recognize the perception as false.

hallucinative. characterized by or able to cause hallucinations.

hallucinatory. characterized by hallucination.

hallucinogen (psychomimetic agent; psychedelic agent). **1a:** any drug whose chief action is on the CNS, producing visual or auditory hallucinations (e.g., LSD (D-lysergic acid diethylamide); psilocybin; mescaline; phencyclidine). These are accompanied by perceptual disturbances, disturbances of thought, depersonalization, and sometimes persistent behavioral changes in humans (and presumably other higher animals). Acute intoxication may be marked by elevated body temperature, increased heart rate, hypertension, euphoria, anxiety or even panic, paranoia, inappropriate affect, temporal and visual distortions and illusions, visual hallucinations, depersonalization. **1b:** any drug or other agent capable of inducing hallucinations.

hallucinogenesis. the production of hallucinations.

hallucinogenetic. hallucinogenic.

hallucinogenic (hallucinogenetic; psychedelic). **1:** able to produce hallucinations; having the properties of a hallucinogen. **2:** pertaining to a hallucinogen.

hallucinogenic amphetamine. See amphetamine.

hallucinosis (alcoholic hallucinosis; acute hallucinosis). an alcoholic psychosis marked by auditory hallucinations and vague delusions of persecution.

hallucinotic. of or pertaining to hallucinosis.

halogen. any of a group of highly electronegative chemical elements of group VIIA of the periodic table: fluorine, chlorine, bromine, iodine, astatine (here listed in order of their activity). They form monobasic acids with hydrogen, and their hydroxides. Fluorine is the most active of all chemical elements. All are toxic and highly reactive, producing a large number of compounds, many of which are toxic. See halogen oxide, interhalogen compound. Fluorine and chlorine are highly corrosive gases that are very damaging to exposed tissues.

halogen azide. See azide.

halogen oxide. any halogen bound to one or more oxygen molecules such as chlorine dioxide or chlorine heptoxide. Most are relatively unstable and highly reactive. Fluorine oxides are often referred to as fluorides because fluorine is more electronegative than oxygen. All are toxic and irritating or corrosive to exposed tissues. Toxicities are similar to those of the interhalogen compounds.

halogenated. having one or more bound halogen molecules; of or pertaining to compounds or mixtures that contain halogen molecules.

halogenation. the chemical introduction of one or more halogen atoms into a compound or mixture.

halogeton. common name of any plant of the genus *Halogeton*, especially *H. glomeratus*.

Halogeton (halogeton; barilla). a small genus of herbs and shrubs (Family Chenopodiaceae) found on rangelands in the western United States and other arid regions of the world, they are indigenous to the Mediterranean region and central Asia. They are poisonous to cattle and sheep because of the high content of soluble oxalates. ***H. glomeratus*** (halogeton, barilla). a small, coarse, annual, extremely toxic herb native to arid, alkaline soils and saline clays of Russia from the vicinity of the Caspian Sea to the steppes of northwestern China and southwestern Siberia. It was introduced from Siberia and has become naturalized in the southwestern United States where it occurs as a weed on disturbed or barren soils. Soluble oxalates may reach extremely hazardous levels in these plants (up to ca. 35% of dry weight). Livestock fatalities have been confined largely to sheep and are uncommon today. Symptoms are those of oxalate poisoning.

haloid salt. a salt comprised of a base and a halogen.

halomethane. any compound comprised of one or more halogen atoms and one or more methane groups. Examples are chloromethane (methylchloride), bromomethane (methyl bromide), dichloromethane (methylene chloride), bromodichloromethane, tribromomethane (bromoform), dichlorodifluoromethane, and trichlorofluoromethane. These compounds are carcinogenic in some laboratory animals.

haloperidol (4-[4-(4-chlorophenyl)-4-hydroxy-1-piperidinyl]-1-(4-fluorophenyl)-1-butanone; Haldol; Pernox). a moderately toxic, cumulative CNS depressant. It is a highly potent antipsychotic that appears to act as a selective antagonist of D_2 dopamine receptors. Haloperidol is a major tranquilizer used chiefly in the management of psychotic states. Although there seems to be a wide margin of safety between lethal and therapeutic doses in humans, reactions to this drug can be serious and include blurred vision; extrapyramidal symptoms (unsteady walk, drooling, slurred speech); tachycardia; hypotension; muscle rigidity; coma; collapse; and death. It can also intensify a psychosis or induce psychotic reactions in previously normal individuals. Additional effects may occur such as headache, depression, and confusion; grand mal seizures; dermatitis; thoracic pain, excessively deep breathing; and sudden death. It sometimes exacerbates the psychosis under treatment and may even induce psychotic reactions in persons who were normal in this respect. Acute poisoning can bring immediate death.

halophilic. denoting organisms that grow well only in media with high concentrations of NaCl.

halophobe. a salt-intolerant plant.

halothane (2-bromo-2-chloro-1,1,1-trifluoroethane). a general anesthetic, $C_2HBrClF_3$, administered by inhalation. It is a colorless, heavy, volatile, nonflammable liquid with a sweetish odor. It is slightly water soluble and is miscible with many organic solvents. Toxic effects are uncommon if care is taken, but are often quite serious when they do present. Halothane relaxes muscle, extends the duration, and increases the magnitude of relaxation provided by other agents (e.g., d-tubocurarine). Halothane can cause pronounced hypotension via direct effects on the myocardium; tachyarrhythmias, sometimes of the reentrant type. Rare, but serious effects also include malignant hyperpyrexia and halothane hepatitis. The former is marked by a rapid rise in body temperature, and increased oxygen consumption, and often death. Onset of halothane hepatitis is marked by fever, nausea, vomiting, and anorexia beginning 2 to 5 days following surgery. The condition soon worsens, taking on the appearance of hepatitis; liver failure and death frequently ensue.

halothane hepatitis. See halothane.

haloxon. an organophosphorus compound, $C^{14}H_{14}Cl_3O_6P$, used in veterinary practice against intestinal nematodes. Clinical signs of exposure in hypersensitive sheep appear in 3-5 weeks and are characterized by symmetric spastic paraparesis and ataxia. The pelvic limbs are partially flexed, often with the dorsal surface of the hoof on the ground. Swollen axons may occur in the lumbar region of the spinal cord and in the sciatic nerve.

halvisol. a compound, $C_{21}H_{27}FN_2O_2$, that is toxic by subcutaneous and intravenous routes. Toxic fumes (F^- and NO_X) are released when heated to decomposition.

halysotter. *Agkistrodon halys*.

hamadrayad. *Ophiophagus hannah*.

Hamburg blue. ferric ferrocyanide.

hami. *Agkistrodon halys*.

hamster. any small murine rodent (Subfamily Cricetinae, Family Muridae) of the genera *Cricetus, Cricelulus, Mesocricetus, or Phodopus*. They are widely used in research and as pets. They are granivorous and phytophagous; all store food and hibernate. They are easily maintained in captivity and can breed throughout the year under laboratory conditions. *Cricetus cricetus* is the most studied species and most common as a pet. It is native to temperate Eurasia, ca. 15 cm long, gray or brown, with round ears and a broad head.

HANDLE WITH CARE. See label (warning label).

hangover. an unpleasant condition following excessive consumption of alcoholic beverages. Symptoms include irritability, headache. and nausea. The same subjective feelings and symptoms may result from using certain medications. See also ethanol.

Hapalochaena maculosa. *Octopus maculosa*.

Haplodoci. Batrachoidiformes.

Haplopappus (synonym *Aplopappus*; rayless goldenrod; jimmy weed; burrow weed). a genus of mostly herbaceous plants (Family Compositaceae) of North and South America. Most species have alternate, rigid leaves and yellow flowers. ***H. venetus*** (coast goldenbush). all parts of the plant are toxic; ingestion by cattle causes trembles. Tremetol, the toxic principle, is excreted in milk. Suckling animals occasionally become poisoned even when the parent is asymptomatic. Humans may also become secondarily poisoned by drinking the milk of affected cattle; the disease in humans is called milk sickness. Toxic concentrations of nitrate have also been reported in this plant. *H. venetus* is closely related to *E. rugosum*.

hapten (haptin; haptene). **1:** a small, non-protein molecule that acts (in part) as an antigen by combining with specific sites on an antibody. Unless bonded to a carrier protein, a hapten does not, however, induce the formation of antibodies. A number of xenobiotics are haptens that form hapten-carrier conjugates which induce antibody formation. The antibodies so induced will usually react with the complex or with the xenobiotic hapten alone. **2:** that part of an antigen that contains the grouping on which the specificity depends. See antigen.

haptene. hapten.

haptenic. pertaining to a hapten.

haptin. hapten.

haptophore. that part of a molecule (e.g., of a toxin, agglutinin, antigen, antibody) that bears the site where it binds to other free molecules or cells. *Cf.* toxophore.

harassing agent (riot-control agent). a chemical agent used to disperse crowds or suppress a riot. They temporarily distress or incapacitate the targets, but are not usually damaging. Tear gases are the most widely used harassing agents.

hard liver disease. poisoning by *Amsinckia* of domestic cattle and swine. These animals do not show clear or intense signs of Amsinckia poisoning; consequently, diagnosis is generally made at the abattoir based on the appearance of the liver. See *Amsinckia intermedia*.

hard-metals disease. a progressive disease of workers involved in drilling, sawing, or cutting metal alloys or materials made of compressed cobalt powder. Symptoms may include fatigue, dyspnea, and sometimes emphysema. The heart, kidneys, and liver may be affected. There is no known cure.

harden. acclimatize, inure.

hardening of the arteries. See arteriosclerosis.

hardhead catfish. *Arius felis*.

Hardwick's sea snake. *Lapemis hardwicki*.

harebell. *Scilla nonscripta*.

Harlekin Korallenotter. See *Micrurus fulvius*.

harlequin, harlequin snake. *Micrurus fulvius*.

harm. a temporary or permanent loss, handicap, detriment, hurt, injury, or impairment of any kind that results from any cause. *Cf.* damage, injury, trauma.

harmaline (4,9-dihydro-7-methoxy-1-methyl-3H-pyrido[3,4-*b*]indole; 1-methyl-7-methoxy-3,4-dihydro-β-carboline; 3,4-dihydroharmine; harmidine; harmalol methyl ether; *o*-methyl harmalol). an indole alkaloid, it is a central nervous system stimulant isolated from seeds of *Peganum harmalia*. See also harmine.

harmalol methyl ether. harmaline.

harmidine. harmaline.

harmine. (banisterine; telepathine; leucoharmine; 7-methoxy-l-methyl-9H-pyrido[3,4-*b*]indole; yageine). an indole alkaloid isolated from seeds of *Peganum harmala* (Family Zygophyllaceae) and *Banisteria caapi* (Family Malpighaceae). It is a CNS stimulant and powerful monoamine oxidase inhibitor. Psychic effects are similar to those of LSD, but may be masked by sedative and depressive effects of this drug. See also harmaline.

harvester ant. See *Pogonomyrmex*.

hasach. hashish.

HAsc. the neutral form of acetylsalicylic acid, *q.v.*

hashish (charas; churus; hasach). a form of cannabis, it is a prepared chiefly by scraping resin from the flower ring tops of cultivated female hemp, *Cannabis sativa*. It is smoked or chewed. Hashish contains the highest levels of cannabinols of any preparation from *Cannabis*; it is much more intoxicating than marijuana. See cannabis.

hatori-hai. *Calliophis sauteri*.

"Hawaii". *Dolichos lablab*.

hawksbill turtle. *Eretomochelys imbricata*.

hay fever (allergic conjunctivitis; anaphylactic conjunctivitis; polenosis; seasonal allergic rhinitis). an allergic disease induced by allergens in the atmosphere (e.g., pollens, fungal spores). Symptoms include inflammation of the mucous membranes of the nose and upper respiratory tract, catarrh, watery discharges from the eyes, headache, coryza, and asthmatic symptoms.

Haynon. chlorpheniramine.

hayo. erythroxylon.

hazard. **1a:** danger, peril, threat. **1b:** exposure to the possibility of loss or injury. **2:** the adverse effect (e.g., cancer, asphyxiation, cardiac arrest) that might result from a potentially harmful component of the environment or from a living system's own behavior. **3:** any environmental component, material, or energy source, (e.g., pollutant or toxic chemical, an explosive substance, ionizing radiation, open flame, heat, noise) that can adversely affect humans, other living systems, the environment, or property. **4:** the hazard of a toxic environmental chemical is a function of exposure and toxicity. This may be defined by the ratio PEC/NOEL, where PEC is the Predicted Environmental Concentration and the NOEL is the No-Observed-Effect Level (NOEL), which is the inverse of the "margin of safety." **imminent health hazard**. a significant threat or danger to health as inferred from evidence sufficient to demonstrate that a product, process, practice, circumstance, or event is of such a danger that immediate correction or cessation is necessary to prevent severe or potentially severe injuries or damage to health. **radiation hazard**. any hazard to public health or to any biotic system due to exposure or potential exposure to ionizing radiation.

hazard determination. hazard identification.

hazard identification (hazard determination). **1:** the qualitative determination of the nature of a risk, including sources, mechanisms of action and potential adverse effects. In the case of chemical substances, it is a determination of whether exposure to a chemical may cause an increased incidence of a toxic or other type of adverse effect (e.g., cancer, birth defects) in a population of test animals and an evaluation of the relevance of this information to the potential of the chemical for causing similar effects in humans or in ecological systems. **2:** the identification of the contaminants of probable health or environmental significance at a specified site, region, or environment.

hazardous. dangerous, perilous, threatening; exposed to, or concerned with danger.

hazardous material (HAZMAT). any substance capable of endangering life or causing serious injury or damage to humans or other life forms. The term applies broadly to toxic, readily flammable, or explosive materials; to those that emit ionizing radiation; and to any material that is dangerous to handle, use, store, or transport.

hazardous substance. **1:** a substance that can adversely affect living systems by causing illness, injury, malfunction, or a reduced quality of life. **2:** a substance that can cause death or irreparable harm to a living system. **2:** a substance declared hazardous by law, especially a substance that can cause death or irreparable harm. In the United States, the Environmental Protection Agency and the Occupational Safety and Health Administration have, for example, identified numerous chemicals as hazardous.

hazardous waste. waste that contains hazardous substances of such kinds and amounts that they may cause or significantly contribute to serious illness or death of humans or other living organisms. About 80 million tons of hazardous waste are generated in the United States each year.

HAZMAT. acronym for hazardous material.

Hb. the symbol for hemoglobin.

HbCO carboxyhemoglobin.

HbO_2. oxyhemoglobin.

HBr. hydrogen bromide; hydrobromic acid.

HC-3. hemicholinium-3.

HCB. hexachlorobenzene.

HCCH. hexachlorocyclohexane.

HCH. hexachlorocyclohexane. See also lindane.

α-HCH. α-hexachlorocyclohexane.

β-HCH. β-hexachlorocyclohexane.

τ-HCH. τ-hexachlorocyclohexane. See hexachlorocyclohexane. See also lindane.

δ-HCH. δ-hexachlorocyclohexane. See hexachlorocyclohexane.

HCl. hydrogen chloride; hydrochloric acid.

HCN. hydrogen cyanide; hydrocyanic acid.

HD_{50}. a dose of complement that lyses 50% of a suspension of sensitized erythrocytes.

He. the symbol for helium.

^{3}He. See helium.

^{4}He. See helium.

head down disease. milkweed poisoning of rabbits. See Asclepiadaceae

headache (cephalalgia; cephalea; cephalodynia; cerebralgia; dolor capitis; encephalalgia; encephalodynia). pain in the head, especially superficial diffuse pain not associated with the distribution of specific nerves. Occasional transient headaches may be due to any of numerous causes. **histamine headache**. that resulting from the presence of excessive amounts of histamine in circulating blood. This can arise from ingestion of histamine (as by drinking certain wines that contain large amounts of this substance) or by injection. The proximate cause is dilatation of the carotic artery. **toxic headache**. that due to systemic poisoning. Such may be exogenous (e.g., due to poisonous gases, foul or polluted air, fumes from solvents or furnaces; various drugs such as atropine, histamine, morphine, quinine, tobacco) or endogenous (due e.g., to absorption of bacterial toxins; fever; bacteremia) in origin.

heal. **1:** to return to a normal or healthy state. **2:** to cure, restore, rehabilitate, make whole or healthy.

health. **1:** fitness, vigor, well being; the condition or overall state of being of an organism when it functions normally with no evidence of disease or persistent distress. **2:** in the charter of the World Health Organization (WHO), 'health' is defined as not only an absence of infirmity or illness, but also a state of complete physical, mental, and social well-being.

healthy. in a state of health, fit, well; sound in mind and body; free from disease, pain, or injury; free from any condition of impairment of normal behavior or life functions.

heart. a muscular organ that rhythmically propels blood and maintains its flow through a circulatory system or a vascular network. The single heart of vertebrate animals is composed of cardiac muscle, which contracts rhythmically in response to an intrinsic pacemaker. Nervous control is thus indirect. In most invertebrates that have one or more hearts, cardiac contractions are initiated by the nerves of a heart ganglion.

heart block. **1:** a condition in which the conductile tissue of the heart (the sinoatrial (SA) and atrioventricular (A-V) nodes, Purkinje fibers, bundle of His) is impaired with a partial or complete impairment of signal transmission and a resultant cardiac arrhythmia. See also block. **2:** atrioventricular heart block. Causes include a tumor of, or damage to, the myocardium (as from coronary occlusion); the effects of drugs, toxins, or other xenobiotics; endocrine factors; nutritional factors. **atrioventricular heart block**. that due to impedance of impulses in the atrioventricular junctional tissue (A-V node; bundle of His or its branches). **bundle branch heart block** (interventricular heart block). that in which impulses are blocked in a branch of the bundle of His. **complete heart block**. third degree heart block. **first degree heart block**. atrioventricular heart block in which conduction time is prolonged but all atrial beats are followed by ventricular beats. **interventricular heart block**. bundle branch heart block. **partial heart block**. second degree heart block. **second degree heart block** (partial heart block). atrioventricular heart block in which some, but not all, atrial beats are conducted. **sinoatrial heart block**. that in which the transmission of impulses from the sinoatrial node is impaired or completely impeded. **third degree heart block** (complete heart block). atrioventricular heart block in which no impulses are conducted by the junctional tissues, due to pathologic factors. This condition may be paroxysmal or permanent; if syncopal attacks occur, the condition is termed Adams-Stokes disease.

heart urchin. See Spatangoida, Echinoidea.

heat. **1a:** estrus. **1b:** the usually intense sexual receptivity of the female mammal that accompanies estrus.

heath. common name of any plant of the family Ericaceae.

heath family. Ericaceae.

heavenly blue. *Ipomoea violacea.*

heaves. vomiting (often applied to domestic animals).

heavy metal. See metal.

heavy metal poisoning. poisoning by any of the heavy metals. See metal (heavy metal).

heavy-metal tolerance. the ability of certain plants to grow in situations where the concentration of heavy metals is toxic to most plants. Such plants are usually able to detoxate the metal; some, however, do not absorb the metal from the medium in which it grows. Copper tolerance in some species of *Agrostis* is due to the ability of the plant to restrict copper to the roots; it is thus unavailable to the more sensitive shoots. See also metal.

heavy oxygen. oxygen-18.

Hebeloma crustuliniforme (poison pie). a poisonous mushroom (Family Cortinariaceae) that occurs in groups or fairy rings in wooded areas and often in residential areas that have deciduous or coniferous trees. Symptoms of intoxication are those of gastrointestinal distress that usually appear within an hour or so of ingestion. Recovery is usually complete within hours. This species is easily confused with nonpoisonous species of *Hebeloma*; thus, none should be eaten.

hecto- (h) a prefix to units of measurement that indicates a quantity of 100 (10^2) times the value indicated by the root. See International System of Units.

Hedeoma pulegoides (false pennyroyal; pennyroyal of America). a small, low, odorous, purple-flowered, herbaceous plant of (Family Labiatae) of eastern North America. The emanations from the leaves are said to be repellent to mosquitos. The oil from this herb is presumably similar in toxic properties to that of *Mentha pulegium* (true pennyroyal), *q.v.*

Hedera helix (English ivy). a member of the ginseng family (Araliaceae), it is an evergreen creeping or climbing shrub with dark green leaves and black berries. It is native to Eurasia and northern Africa, but is widely used in landscaping. The leaves and berries are toxic; the toxic principle is hederin. Children have occasionally been poisoned from eating the berries. The toxic principle is hederin (a saponic glycoside), the aglycone of which is hederagenin. Symptoms seem to include excitement, dyspnea, and coma in humans. No cases are known from the United States. Allergic persons react as most people do to *Rhus toxicodendron* (poison ivy).

hederagenin. the triterpenoid aglycone of the saponin hederin. See also *Hedera helix*.

hederin (α-hederin; helixin). a toxic, crystalline, antibiotic saponin found in *Hedera helix*; the aglycone is hederagenin.

α-hederin. hederin.

hedge, hedge plant. See plant (hedge plant).

Hediger's snake. *Parapistocalamus hedigeri*.

hedonal. **1:** 2,4-D **2:** formerly a synonym for carbamic acid 1-methylbutyl ester.

hegari. *Sorghum vulgare*.

Heinz body. a dark-staining granule comprised of damaged hemoglobin. Such occur on the inner surface of the cell membrane in erythrocytes. They seem to impair membrane function and are associated with hemolytic anemia of infancy. They may occur in premature infants, in certain individuals that have abnormal hemoglobins, in certain forms of drug sensitivity, and as a result of exposure to certain toxicants (e.g., aniline, hydroxylamine, nitrobenzenes).

helenalin (3,3a,4,4a,7a,8,9,9a-octahydro-4-hydroxy-4a,8-dimethyl-3-methyleneazuleno[6,5-b]furan-2,5-dione; 6α,8β-dihydroxy-4-oxoambrosa-2,11(13)-dien-12-oic acid 12,8-lactone). an extremely poisonous, slightly water-soluble, pseudoguaianolide sesquiterpenoid lactone, $C_{15}H_{18}O_4$, isolated from any of several species of *Helenium*, especially *H. autumnale* and from *Arnica montana*. Symptoms of poisoning in humans are chiefly those of paralysis of both skeletal and cardiac muscle and gastroenteritis that can be fatal.

Helenium. a genus of annual or perennial shrubs or herbs (Family Compositaceae). A number of species are reported as toxic to livestock: *H. autumnale*, *H. hoopesii*, *H. microcephalum*, *H. nudiflorum*, *H. tenuifolium*. ***H. hoope-sii*** (= *Dugaldia hoopesii*) (sneezeweed; orange sneezeweed). A stout, erect, perennial herb, up to 1 m tall, found on mountain slopes and valleys mostly from 2200-3200 m in the Rocky Mountain and Pacific Coast states except Montana and Washington. It contains dugaldin, *q.v.* See also sneezeweed poisoning.

Helianthus annus (common sunflower; wild sunflower). a large, erect, annual herb (Family Asteraceae) that may reach a height 4-5 m with huge flower heads. It is native to and widely distributed throughout the western prairies of North America and is extensively cultivated. It contains various plant growth inhibitors such as chlorogenic acid and scopoletin. Toxic concentrations of nitrate have also been reported.

helicocerin. cerulenin.

Heliopsis longipes (oxeye). a hardy perennial herb (Family Asteraceae) that contains affinin. It is used in Mexico as a dental analgesic.

heliotrine. a slow-acting hepatotoxin, producing atrophic hepatosis following long-continued exposure. With continued exposure, liver cells increase in size, are less viable, and their regenerative ability is impaired. It occurs in *Heliotropium europaeum*.

heliotrope. common name of *Heliotropium europaeum*, not the garden heliotrope, *Valeriana officinalis*.

heliotrope poisoning. poisoning by *Heliotropium europaeum*, *q.v.*

Heliotropium europaeum (heliotrope). a poisonous annual weed (Family Boraginaceae) that contains hepatotoxic alkaloids of the senecio or pyrrolizidine type (e.g., heliotrine, lassiocarpine, and their *N*-oxides). The oxides are less toxic than the parent alkaloids, but resist digestion, are more soluble and may thus account for the severity of effects. Heliotrope poisoning decreases the metabolic activity of the liver, causing atrophic hepatosis with some fibrosis wherein liver cells increase in size, and are less viable with impaired regenerative ability. The most severe liver damage often occurs on lush, rich pastures. Death in such cases may result from ammonia intoxication due to the high nitrogen content of the forage and the reduced ability of the damaged liver to detoxicate nitrogen. Usually, however, animals may consume large amounts of heliotrope during a whole season with no signs of disease. Hepatic effects will have occurred, however, predisposing the animals to poisoning during the next season. Effects (e.g., loss of condition, changes in metabolism, changes in composition of blood, and death) are cumulative, progressive, and due entirely to liver atrophy. Sheep exhibit a distinctive syndrome with sudden hematogenous jaundice due to altered copper metabolism. It usually presents when sheep pastured for a season on heliotrope are then transferred to pasturage that supports a high intake of copper. Cattle are more sensitive than sheep, but are rarely poisoned. See also *Amsinckia intermedia*.

helium (symbol, He)[Z = 2; A_r = 4.0026]. the second most abundant element in the universe, but rare on Earth. It is a colorless, odorless, completely nonreactive gas and the lightest known element. Natural helium is comprised of ^{4}He with a trace of ^{3}He (ca 1ppm). The latter isotope is produced by the beta-decay of tritium. Helium is emitted by radium, uranium, and a wide variety of radioactive nuclides as positively charged helium ions or alpha particles, *q.v.* It is extracted on a commercial scale from natural gas. Liquid He with temperatures approaching 0°K and extremely high volatility must be handled with caution. Because of its inertness and low aqueous solubility (hence, low solubility in blood) it is supplied (mixed with air or oxygen) to persons working under conditions of high atmospheric pressure. The time needed to adjust to the increasing or decreasing pressure and the danger of bends, *q.v.*, is thus reduced. Because of its low solubility, helium is used variously as a diluent for medicinal gases, e.g., in general anesthesia and with oxygen in the treatment of certain types of respiratory obstruction.

helium-3 (^{3}He). See helium.

helium-4 (^{4}He). See helium.

helixin. hederin.

Hell's fire sea anemone. *Actinodendron plumosum*.

hellebore. **1:** common name of plants of the genus *Helleborus* (especially *H. niger*) and sometimes of the genus *Veratrum* (false hellebore). **2:** an insecticide prepared from the roots of plants of the genus *Veratrum* (false hellebore or white hellebore).

helleborin. a toxic, narcotic glycoside produced by *Veratrum viride*.

helleborism. poisoning by plants of the genera *Veratrum* and *Helleborus*.

helleborus. the dried roots and rhizome of *Hellehorus niger*. It is used as a cardiovascular tonic, cathartic, and diuretic.

Helleborus (hellebore). a genus of European perennial herbs with fibrous roots (Family Ranunculaceae). Several species have been introduced in North America. All parts of the plant are toxic. Active principles are alkaloids such as veratrin. ***H. foetidus*** (stinking hellebore, bear's foot, setterworth). a large branching ornamental plant similar to *H. niger*, but with a bad odor. The powdered leaves have been used as a vermicide for children, but may violently affect or even kill them. ***H. niger*** (Christmas rose; hellebore; black hellebore; true hellebore). a very toxic evergreen plant with white flowers. It is native to Europe and Asia and widely cultivated elsewhere. It contains two highly toxic glycosides, helleborin and helleborein, which are violent irritants and cathartics; other toxic components are helebrin, protoanemonin, and certain saponins. All parts of the plant are neurotoxic and cathartic, especially the leaves. Effects on humans and livestock are similar. Ingestion may cause a burning sensation of the mouth and skin, blistering of the lining of the mouth, paresthesia, nausea, vomiting, gastric distress, hypotension, a weak pulse, and convulsions. It is rarely fatal. ***H. viridis*** (green hellebore; winter aconite). a poisonous evergreen plant similar to *H. niger*, but with greenish flowers tinged at the edges with purple.

helminthagogue. anthelmintic (def. 1).

helminthic. See anthelmintic.

helminthicide. vermicide.

helminthosporal. a toxic aldehyde isolated from *Cochliobolus sativus*. It inhibits phosphorylating oxidations in mitochondria.

helminthosporin. a dark maroon, crystalline phenolic, $C_{15}H_{10}O_5$, pigment produced by certain molds (e.g., of the genus *Helminthosporium*.

Helminthosporium. a genus of parasitic or saprophytic imperfect fungi (Family Dematiaceae) with erect conidiophores. ***H. dematioideum***. a fungus that produces cytochalasins. ***H. oryzae***. a phytotoxic "fungus" that infects rice leaves. The toxic principle is ophiobolin A.

helmintic. anthelmintic (def. 1).

Heloderma. a genus of lizards comprised of two species of heavy-bodied lizards with short stout legs (Family Helodermatidae, *q.v.*); the body is covered by tubercular scales. They are native to Mexico and the southwestern United States and are the only known venomous lizards. ***H. horridum*** (beaded lizard, escorpion). a venomous lizard confined to the southwestern United States, Mexico, and Central America. It is yellow and black, large, slow-moving, and mostly nocturnal. A venom gland is situated in either side of the lower jaw. The bite can be lethal to small prey animals but is rarely fatal to humans. ***H. suspectum*** (gila monster). a venomous lizard confined to the southwestern United States and Mexico. It is a pink and black, large, slow-moving, mostly nocturnal animal. The bite causes intense local pain and edema, nausea, respiratory distress, heart failure, and occasionally death.

Helodermatidae (Gila monster family). a family of lizards in which the dorsal scales have been replaced with rough tuberculated ("beaded") skin. It contains only three species, one of [illegible] The two remaining species (See *Heloderma*) are venomous with a thick, cylindrical body and tail; a body length of ca. 70 cm; short, blunt, fang-like teeth with venom glands on either side of the lower jaw. The bite is often fatal to small prey animals that can be grasped tightly in the jaws, but is rarely fatal to humans.

helper T cell. helper T lymphocyte (See T lymphocyte).

Helvella (false morels). a genus of ascomycetous fungi (Family Helvellaceae) with stalked, often folded, pileate or saddle-shaped ascocarps (fruiting bodies). Members of this genus are closely related to and easily confused with those of *Gyromitra*. Poisonous mushrooms of both genera usually produce effects that are similar to those produced by amanitin but are less severe. Symptoms (primarily gastrointestinal) usually appear in 6 to 12 or more hours following ingestion. Some species are of questionable toxicity, others are deadly poisonous. Identification is difficult and best accomplished by microscopic examination. It is probably unwise to consume any mushrooms of this genus. See also *Gyromitra*, gyromitra toxins, helvellic acid. ***H. acetabulum*** (ribbed-stalked cup). found throughout North America on open or wooded ground. It may be inedible. ***H. esculenta***. *Gyromitra esculenta*. ***H. esculenta***. *Gyromitra esculenta*. ***H. gigas***. *Gyromitra gigas*. ***H. infula***. *Gyromitra infula*. ***H. underwoodii***. *Gyromitra brunnea*.

helvellic acid. a thermolabile cytotoxic hemolysin isolated originally from the mushroom *Helvella infula*.

helvolic acid (6,16-bis(acetyloxy)-3,7-dioxo-29-nordammara-1,17(20),24-trien-21-oic acid; (Z)-6β,16β-dihydroxy-3,7-dioxo-29-*nor*-8α,9β,-13α,14β-dammara-1,17-(20),24-trien-21-oic acid diacetate; fumigacin). an antibiotic and mycotoxin produced by *Aspergillus fumigatus*. It is very toxic to laboratory mice.

Hemachatus haemachatus (ring-necked spitting cobra; South African spitting cobra; spitting cobra; spitting snake; ipHimpi; ringhals; ringhals cobra; rinkals; kake; petia; spuugslang; unobhiya; unobibi). a very dangerous venomous snake (Family Elapidae) of the veldt, savannahs, and open land of southern and southeastern Africa; it is the sole species within the genus and is the only "spitting" cobra in southern Africa. Adults have an average length of about 1.1 m and occasional specimens may reach a length of about 1.5 m. The dorsal scales are strongly keeled; this distinguishes it from the closely related true cobras (*Naja*). It spits or squirts a fine spray or stream of extremely irritant venom for distances up to about 2 m. Entering the eyes, it causes intense pain, spasm of the eyelids, and sometimes permanent damage to the tissue of the eyes and even blindness in human victims. The eyes should be immediately rinsed or flushed with clean water or other harmless liquid. The stream of venom is expressed through the anterior venom orifices of the short fangs under pressure from the contraction of the strong venom gland musculature. This snake strikes on rare occasions, usually only when restrained; the bite can prove fatal. When in a defensive attitude, the anterior portions of the body are raised off the ground and the hood is spread, revealing a black throat with 1-3 narrow, light-colored bands below the hood.

hemachrosis. an intense redness of the blood.

hemagglutination (haemagglutination). the clumping together or coagulation of erythrocytes, either spontaneously or when treated with a specific antibody or other agent. Some viruses contain specific hemagglutinating proteins.

hemagglutinin (haemagglutinin). **1a:** a type of antibody that agglutinates erythrocytes. **1b:** sometimes used for any substance that agglutinates erythrocytes. **autologous hemagglutinin**. an antibody that agglutinates erythrocytes of the producing individual. **homologous hemagglutinin**. an antibody that agglutinates erythrocytes from the same species. **heterologous hemagglutinin**. an antibody that agglutinates erythrocytes of species other than the producing species.

hemangioma. See angioma.

hemangiosarcoma. angiosarcoma.

hematemesis (vomitus cruentes). the vomiting of blood.

hematin. the former name for heme.

hematin protein. cytochromes C.

hematite. the naturally occurring α-form of ferric oxide, Fe_2O_3. The chief ore of iron, it occurs as steel gray to black crystals or as red earthy material. It causes benign pneumoconiosis and possibly lung cancer in miners.

hematocrit. the percentage of whole blood comprised of erythrocytes as opposed to serum or plasma. Normal hematocrit of humans ranges from ca. 35 to 45%, but varies with sex and age. The hematocrit value is used as an index of anemia or polycythemia.

hematodyscrasia. hemodyscrasia.

hematoencephalic barrier. See barrier.

hematogenic. See hematogenous.

hematogenous (haematogenous; hematogenic). **1:** pertaining to blood formation. **2:** pertaining to or originating in the blood; produced by or derived from the blood; disseminated by the circulation or through the blood stream.

hematologist. **1:** a physiologist who specializes in the study of blood and blood-forming organs. **2:** a medical or veterinary doctor who specializes in the diagnosis and treatment of diseases of the blood and blood-forming organs.

hematology (haematology). the study of blood, its formation, and the blood-forming organs.

hematolysis. hemolysis.

hematolytic. hemolytic.

hematopathy. hemopathy.

hematopoiesis (haematopoiesis). the process of blood cell formation and development, including the maturation of stem cells into erythrocytes or one of the several types of leukocytes. A number of toxic compounds, ionizing radiation, and vitamin deficiencies can impede or arrest hematopoiesis. **extramedullary hematopoiesis**. that taking place outside the bone marrow (e.g., in spleen, lymph nodes, liver). **medullary hematopoiesis**. that taking place in the red bone marrow.

hematopoietic (haematopoietic). **1:** of, pertaining to, promoting, or effecting hematopoiesis. **2:** an agent that stimulates or promotes hematopoiesis.

hematopoietic (radiation) syndrome. one of three forms of acute radiation syndrome seen in humans. It usually results from exposures of 200-1000 rads. Characteristic symptoms include rapidly developing, transient anorexia, nausea, vomiting, apathy. These symptoms subside after 6-12 hours and the victim may be asymptomatic for a day or so. At this time, the spleen, lymph nodes, and bone marrow begin to atrophy with consequent development of pancytopenia; lymphopenia develops, usually peaking in 24-36 hours; neutropenia develops more slowly; and thrombocytopenia may become pronounced within 3-4 weeks. The victim has an increased susceptibility to infection. Whole body exposures above about 600 rads produce both hematopoietic and gastrointestinal effects and invariably proves fatal. See also acute radiation syndrome, radiation sickness.

hematopoietic syndrome. hematopoietic radiation syndrome.

hematopoietic system. the organs, collectively, that take part in hematopoiesis (e.g., red bone marrow, spleen, lymph nodes, liver).

hematosepsis. septicemia.

hematotoxic. hemotoxic.

hematotoxicosis. any diseased condition of, or damage to, the hematopoietic system due to poisoning. See also hemotoxic.

hematotoxin. hemotoxin.

hematoxic. hemotoxic.

hematoxin. hemotoxin.

hematuria (haematuria). blood in the urine, a condition that can result from chemical poisoning.

heme (protoporphyrin IX). a cyclic tetrapyrrole or porphyrin, based on protoporphyrin IX, that contains iron. It occurs as an oxygen-carrying prosthetic group in various proteins, including a number of cytochromes (e.g., cytochrome P 450), hemoglobin, and myoglobin. Heme is toxicologically significant not only as the focus of action of several toxicants such as carbon monoxide and cyanide, but also in the metabolism of toxicants, including both detoxication and activation. See cytochrome P-450, monoxygenase reaction.

hemel. altretamine.

hemerythrin (haemerythrin). a respiratory pigment found in blood corpuscles of various invertebrate animals. It contains two iron ions coupled to polypeptide chains. The ions bind molecular oxygen under conditions of high oxygen tension, releasing it to tissues of lower oxygen tension. *Cf.* hemoglobin.

hemiacetal. See chloral derivative.

Hemibungarus. formerly a genus of oriental coral snakes (Family Elapidae), the species of which are now included in *Calliophis*.

hemicholinium-3 (HC-3). a substance that blocks the synthesis of acetylcholine by blocking the sodium-dependent, high-affinity transport of choline into nerve terminals. See also choline acetyltransferase.

hemiglobin. obsolete for methemoglobin.

hemiplegia. paralysis of one side of the body.

hemitoxin. a toxin whose toxicity is reduced by one half.

hemlock. **1:** a tree of the genus *Tsuga* (hemlock fir, hemlock spruce). **2:** the wood of *Tsuga*. **3:** any of a number of poisonous herbs, especially poison hemlock (*Conium maculatum*) or a water hemlock (*Cicuta*). **4:** any of several species of the genus *Taxus* (e.g., ground hemlock).

hemlock fir. See *Tsuga*.

hemlock spruce. See *Tsuga*.

hemlock water dropwort. *Oenanthe crocata*

hemoantitoxin (haemoantitoxin). an antibody that inhibits or neutralizes the action of a hemotoxin, such as the hemolytic principle of cobra venom.

hemochromatosis (haemochromatosis). a disease, most frequent in males, in which excess iron accumulates in the liver, pancreas and skin. Symptoms include a bronzed skin color, hepatomegaly, diabetes mellitus, and cardiac failure. **exogenous hemochromatosis**. that due to multiple transfusions or excessive intake of iron.

hemoclasis. hemolysis.

hemoclastic. pertaining to, characterized by, or causing destruction or dissolution of erythrocytes. See hemoclasis.

hemocoagulin (haemocoagulin). a blood-coagulating component of certain snake venoms.

hemocytolysis. hemolysis.

hemodilution (haemodilution). an increase in fluid content of the blood with a consequent decrease in erythrocyte concentration.

hemodyscrasia (haemodyscrasia; haematodyscrasia; hematodyscrasia). any disorder of the blood or hemopoietic system, especially of the formed elements.

hemoglobin (haemoglobin; Hb; ferrohaemoglobin; ferrohemoglobin). the respiratory pigment of erythrocytes, Hb is the chief oxygen carrier in vertebrate blood. It transports oxygen via the cardiovascular system from the oxygen-rich organs of external respiration (e.g., lungs, gills, and/or skin) to the relatively oxygen-poor tissues of the body. Hb is a compact globular heme protein (MW of 68,000) comprised of four globin moieties, each consisting of two pairs of polypeptide chains. Each chain is associated with one heme group. It reversibly binds molecular oxygen to the heme moiety, forming oxyhemoglobin which is the form found in arterial blood. About 97% of the oxygen is transported in the blood as oxyhemoglobin, while only 3% is dissolved in the plasma. The globin moiety reversibly binds carbon dioxide, forming carbaminohemoglobin which predominates in venous blood. Of importance toxicologically is the high affinity of hemoglobin for carbon monoxide, which is 200 times greater than that for oxygen. See also carbon monoxide, methemoglobin. **carbon monoxide hemoglobin**. carboxyhemoglobin. **oxygenated hemoglobin,** oxyhemoglobin. **reduced hemoglobin** (deoxyhemoglobin). the form of Hb following the release of oxygen from oxyhemoglobin. The iron moiety is in the reduced ferrous state able to bond with oxygen (or, for example, carbon monoxide).

hemoglobin M. a class of hemoglobins, associated with methemoglobinuria, that have amino acid substitutions in the alpha or beta chains.

hemoglobinemia. the presence of abnormal amounts of hemoglobin in blood plasma. **postparturient hemoglobinemia** (puerperal hemoglobinuria). chiefly a disease of dairy cattle, mainly high-producing dairy cattle, that presents 2-4 weeks following parturition. It is marked by intravascular hemolysis, hemoglobinuria, and anemia. The condition is uncommon, but one half of affected animals may die. The condition is associated with hypophosphatemia and with a diet high in cruciferous plants (e.g., rape, kale) or beet pulp, or sometimes turnips. Under some conditions, dietary copper seems to be protective. It is thought the aforementioned plants contain hemolysins and that hypophysphatemia and probably hypocuprosis render the erythrocytes more susceptible to these agents.

hemoglobinolysis. the dissolution of hemoglobin.

hemoglobinuria. the presence of hemoglobin, without the presence of erythrocytes, in the urine. This may occur, for example, in hemolytic anemia, scurvy, a number of infectious disease, in response to the action of certain toxic chemicals, and in septicemia. The proximate cause is the escape of hemoglobin from disintegrating erythrocytes (usually due to hemolysis) in amounts that exceed the ability of blood proteins to bind the hemoglogin. **toxic hemoglobinuria**. that which results from xenobiotics such as arsenic, muscarine, phosphorus, snake venom, or from endogenous substances resulting from certain infectious diseases (e.g., typhoid fever, yellow fever, syphilis).

hemoglobinuric. of, pertaining to, or characterized by hemoglobinuria.

hemolymph (haemolymph). the circulatory fluid of various invertebrates.

hemolysate. the product of hemolysis.

hemolysin (erythrocytolysin; erythrolysin). **1:** any substance produced by a living system that can lyse erythrocytes with the release of hemoglobin. **2:** a sensitizing, complement-fixing antibody that combines specifically with erythrocytes of the antigenic type that stimulated its production. This results in the dissolution of cells, with the release of hemoglobin. **bacterial hemolysin**. any hemotoxin elaborated by a bacterium. **venom hemolysin**. any hemolytic agent found in a venom. *Cf.* hemolytic substance, hemotoxin.

hemolysis (haemolysis; erthrocytolysis; erythrolysis; hematolysis; haematolysis; hemoclasis; hemocytolysis; haemocytolysis; laking). the liberation of hemoglobin into the blood plasma with subsequent hemoglobinuria by either of two types of processes. Among the many agents of hemolysis are many bacterial toxins, snake venoms, and chemical poisons. **1:** the lysis of erythrocytes with rupture of the plasmalemma and loss of hemoglobin, e.g., by specific complement-fixing antibodies, toxins, various xenobiotics, and alteration of temperature or tonicity. **2:** alteration of the erythrocyte plasmalemma with consequent liberation of hemoglobin. Such may be induced by chemicals that directly attack the membrane, or any of various solvents that alter the tonicity of the blood. **biologic hemolysis**. that caused by hemolysins elaborated by various animal and plants. **drug-induced hemolysis**. hemolysis caused by drugs or other xenobiotics. Mechanisms vary but either involve oxidation of hemoglobin in intrinsically defective erythrocytes or indirect lysis by inducing direct or indirect hemolysis by elements of the immune system. **venom hemolysis**. that caused by hemolytic substances in the venom of various species of venomous animals.

hemolytic (haemolytic; hematolytic; hemotoxic (def. 2)). of, pertaining to, or causing hemolysis; destructive to erythrocytes with the liberation of hemoglobin. See also hemolytic anemia, hemolytic gas.

hemolytic gas. any gas (e.g., arsine), the inhalation of which causes hemolysis and associated hemoglobinuria, icterus, gastroenteritis, and nephritis.

hemolytic substance. any substance, chemical or biologic, that lyses erythrocytes. *Cf.* hemolytic gas. See also hemolysin, hemotoxin.

hemolytic uremic syndrome. a condition often associated with the use of oral contraceptives in adults. It may also present *post partum*, following complications of pregnancy. It is marked by hemolytic anemia, thrombocytopenia, and acute renal failure. Children also exhibit this syndrome with sudden onset of hematuria, oliguria, gastrointestinal bleeding, and microangiopathic hemolytic anemia.

hemolyzation (haemolyzation). the production or process of hemolysis.

hemolyze (haemolyze). **1:** to cause or undergo hemolysis.

hemopathic. able to affect the blood or hemopoietic system pathologically; able to cause hemopathy.

hemopathy (hematopathy; haematopathy). any disease or abnormality of the blood or hemopoietic tissues.

hemoprotein (haemoprotein). a protein that contains a heme prosthetic group.

hemoptysis (haemoptysis). emptysis.

hemorrhage (hemorrhea). **1:** abnormally high rate of discharge of blood internally or externally from blood vessels. While external hemorrhage is visible, internal hemorrhage indicated indirectly by signs and symptoms such as those of mild to severe shock, low blood pressure, fainting, and even collapse. In some cases a hematoma can be palpated or seen. **2:** to bleed, especially to bleed excessively. **arterial hemorrhage**. hemorrhage from an artery, characterized by spurts or waves of bright red blood. **capillary hemorrhage.** the oozing of blood from a capillary bed. **concealed hemorrhage**. hemorrhage into an area where the blood is not visible. **external hemorrhage**. hemorrhage with escape of blood from the body. **gastrointestinal hemorrhage**. profuse bleeding from the gastrointestinal tract. Bleeding from the gastrointestinal tract. This should be treated as a serious condition, the cause promptly determined, and treatment initiated. Gastrointestinal hemorrhage can result from ingestion of irritant, caustic, or anticoagulant substances. **internal hemorrhage**. **2:** hemorrhage into internal body spaces with the blood remaining in the body. This term is inclusive of the term concealed hemorrhage. **pulmonary hemorrhage** (pneumorrhagia). hemorrhage from the lungs. **recurring hemorrhage**. intermittent bouts of hemorrhage. **venous hemorrhage**. steady, profuse bleeding of dark red blood. *Cf.* bleeding.

hemorrhagenic (hemorrhagiparous). causing hemorrhage.

hemorrhagic. 1: of or pertaining to hemorrhage. **2:** hemotoxic.

hemorrhagic disease (syndrome) of poultry. any of several poultry diseases of differing etiology that result from the consumption of moldy grain. All are marked by extensive internal hemorrhages, erosion of the gizzard and proventriculus, and usually hypoplastic bone marrow and anemia. The causative agents are toxicants from a variable combination of fungi (e.g., *Alternaria*, *Aspergillus*, *Penicillium*). See aflatoxicosis.

hemorrhagic syndrome of cattle. bovine hemorrhagic syndrome.

hemorrhagin. any of a class of cytolytic toxins (e.g., ricin) that causes dissolution of the endothelial cells of capillaries and other small blood vessels with extravasation of blood into the tissues served. They occur chiefly in certain plants (e.g., *Ricinus communis*) and snake venoms (e.g., of crotalids). *Cf.* cytolysin, hemolysin, hemotoxin. See also ricin.

hemorrhagiparous. hemorrhagenic.

hemorrhea. hemorrhage.

hemosiderosis. discoloration of a tissue by deposition of an iron pigment.

hemosozic. antihemolytic.

hemostasis. the stanching of blood flow from a wound. Complete stoppage usually involves vascular spasm, formation of a platelet plug, and blood coagulation.

hemotoxic (hematotoxic; hematoxic; haematotoxic; haematoxic). **1a:** able to cause blood poisoning. **1b:** toxic to the blood and hematopoietic system. **1c:** pertaining to hematotoxicosis. **2:** hemolytic or hemorrhagic; having the action of a hemotoxin.

hemotoxicosis. hematotoxicosis.

hemotoxin (hematotoxin; hematoxin; haemotoxin). **1a:** a hemolytic exotoxin. **1b:** a hemolytic substance, especially one of biological origin. **1c:** often applied to toxins that cause hemorrhage. **cobra hemotoxin**. the hemolytic constituent of cobra venom that is able to lyse erythrocytes in the absence of blood serum. *Cf.* hemolysin, hemorrhagin. See also cobra venom poisoning.

hemp. **1:** the durable fibers of the woody trunk of *Cannabis sativa*. They are used to produce twine, rope and cloth. **2:** cannabis. **3:** *Cannabis sativa* is sometimes referred to has hemp or the hemp plant.

hemp family. See Moraceae.

Hemprich's coral snake. *Micrurus hemprichii*.

henbane. **1:** a common name of *Hyoscyamus spp*. **2:** hyoscyamus.

henbit. *Lamium amplexicaule*.

Henderson-Hasselbach equation. an equation used to determine the pH of a buffer when the pKa of the acid in the buffer is known. pH = pKa + log ([salt]/[acid]). Its most important application in toxicology is to help determine the uptake of ionizable toxicants. This is accomplished by relating the degree of ionization of a weak acid or base to the pH of the medium in which it is dissolved. The most suitable forms of the equation for this application are:

for acids:

log ([unionized form]/[ionized form]) = PKa - pH

for bases:

log [ionized form]/[unionized form] = PKa - pH

Henri-Michaelis-Menten equation. See Michaelis-Menten equation.

HEOD. dieldrin.

heparin. a polysaccharide (a glycosaminoglycan) anticoagulant present in all mammalian tissues, especially the lung. It is produced, stored, and secreted by mast cells and acts chiefly by increasing the rate of inactivation of thrombin by antithrombin III.

hepat-, hepatico-, hepato-. prefixes referring to or denoting the liver.

hepatargia (hepatargy). autointoxication due to impaired liver function.

hepatargy. hepatargia.

hepatatrophia (hepatatrophy). atrophy of the liver.

hepatatrophy. hepatatrophia.

hepatic. of or pertaining to the liver.

hepatic biotransformation. See biotransformation.

hepatic duct. a tubule that carries bile from the liver. It joins the cystic duct from the gallbladder, forming the common bile duct.

hepatic peroxisome. See peroxisome.

hepatic portal circulation. the return of venous blood via the hepatic portal vein from the digestive system through the liver sinusoids before returning to the heart.

hepatic portal system. the system of veins and capillaries (beginning at the villi of the small intestine) that transports blood from the digestive system through the liver before its return to the heart. Absorbed foods (except fats) are thus transported directly to the liver, where they are metabolized.

hepatic portal vein. the vein leading to the liver that collects blood from blood vessels arising in the villi of the small intestine.

hepatic tremor. asterixis.

hepatico-. See hepat-.

hepatism. an illness due to liver disease.

hepatitides. plural of hepatitis.

hepatitis (plural, hepatides). a general term for inflammation of the liver. Although the term is usually used for viral hepatitis, it is also used for instances of inflammation produced by a poison (hepatotoxicant). It is often accompanied by jaundice and in some cases liver enlargement. **drug-induced hepatitis**. toxipathic hepatitis caused by a drug or medicine. **halothane hepatitis**. that which results from hepatocellular damage due to halothane anesthesia. **toxic hepatitis**. toxipathic hepatitis. **toxipathic hepatitis** (toxic hepatitis). that due to the direct action of any of a large number of poisons, drugs, or toxins on liver cells as by ingestion of *Amanita phalloides*

hepato-. a prefix indicating the liver.

hepatocarcinogen. any agent that causes cancer of the liver.

hepatocirrhosis. cirrhosis of the liver. See cirrhosis (def. 1).

hepatocyte. a parenchymal liver cell, the most common type of cell in the liver. They are importantly involved in the metabolism of xenobiotics and the site of direct action of hepatotoxicants.

hepatogenic photosensitivity. See photosensitivity.

hepatogenous photosensitivity. See photosensitivity.

hepatolysin. a cytolysin that destroys hepatic parenchymal cells.

hepatolytic. destructive to liver tissue.

hepatomegaly (megalohepatia). pathologic enlargement of the liver.

hepatonecrosis. the death of liver cells.

hepatonephoric syndrome. hepatorenal syndrome.

hepatopathy. any disease of the liver.

hepatorenal syndrome (hepatonephoric syndrome). acute renal failure in patients with a damaged liver or biliary tract. It may sometimes result from carbon tetrachloride poisoning.

hepatosis. any functional disorder of the liver.

hepatotoxemia. blood poisoning that originates in the liver.

hepatotoxic (hepatoxic). **1:** toxic to the liver. **2a:** of or pertaining to a hepatotoxicant. **2b:** pertaining to the capacity of an agent to poison the liver, or to the action of such an agent.

hepatotoxicant. any chemical that adversely affects the liver.

hepatotoxicity. **1a:** the ability of a toxicant to poison liver cells. **1b:** injury or damage to the liver by a poison.

hepatotoxin. a toxin that is destructive to hepatic parenchymal cells, especially a toxin that is induced by injecting an animal with liver cells.

hepatoxic. hepatotoxic.

heptachlor (1*H*-1,4,5,6,7,8,8-heptachloro-3a,4,-7,7a-tetrahydro-4,7-methanoindene). a highly toxic, white to light tan, waxy, water-insoluble, solid, chlorinated, diene insecticide, $C_{10}H_7Cl_7$, making up about 8% of technical grade chlordane. It is nephrotoxic and hepatocarcinogenic to various test animals; it is very toxic to laboratory rats. Effects on humans include irritability, hyperexcitability, tremors, and convulsions with intervening periods of depression, blood dyscrasias, late liver necrosis, and possibly severe gastroenteritis. It causes liver cancer in some laboratory animals. Its use is prohibited in the United States. See also cyclodiene. See cyclodiene, heptachlor epoxide.

heptachlor epoxide. the principal metabolite of heptachlor. It is carcinogenic to some laboratory animals. See cyclodiene, heptachlor.

heptachlorepoxide. a degradation product of heptaclor, $C_{10}H_9CL_7O$, used as an insecticide. It is also the chief metabolite of heptachlor and is carcinogenic to some laboratory animals. See heptachlor, cyclodiene.

1,4,5,6,7,8,8-hept achloro-3a,4,7,7a-tetrahydro-4,7-methanoindene. heptachlor.

2,8,10-heptadecatriene-4,6-diyne-1,14-diol. enanthotoxin.

E,E,E)-(—)-8,10,12-heptadecatriene-4,6-diyne-1,14-diol. cicutoxin.

heptane. See alkane.

2,5-heptanedione. a γ-diketone. See γ-diketone neuropathy.

heptenophos (phosphoric acid 7-chlorobicyclo-[3.2.0]hepta-2,6-dien-6-yl dimethyl ester; 7-chlorobicyclo-[3.2.0]hepta-2,6-dien-6-yl dimethylphosphate). a nonpersistent contact and systemic phosphate insecticide used as an ectoparasiticide in veterinary practice.

Heptranchias perlo (seven-gilled shark). a poisonous shark of the coastal waters of the Atlantic Ocean, the Mediterranean Sea, Cape of Good Hope, and Japan. See elasmobranch poisoning.

heptyl carbinol. 1-*n*-octanol.

Heracleum. a genus of perennial or biennial herbs of the Northern Hemisphere (Family Apiaceae). ***H. lanatum***. See *Heracleum sphondylium*. ***H. maximum***. See *H. sphondylium*. ***H. sphondylium* var. *montanum*** (formerly *H. maximum*, *H. lanatum*; cow parsnip). an herb native to North America; it occurs in wet areas. The entire plant is toxic, causing a dermatitis similar to that of poison ivy. ***H. maximum***. See *H. sphondylium* var. *sphondylium*.

herb. **1:** a nonwoody (nonlignified) annual seed plant. **2:** an aromatic plant or plant part used as a food flavoring, aromatic, or medication.

herb mercury. herbs of the genus *Mercurialis*.

herb of grace. *Ruta graveolens*.

herbaceous. pertaining to or resembling an herb or a nonwoody plant stem.

herbicide (weedkiller). **1a:** any substance or formulation intended for use as a plant regulator, defoliant, or desiccant. **1b:** a chemical agent (a pesticide) used to kill or seriously inhibit the growth of noxious or otherwise undesirable plants. Herbicides are classified as contact or systemic, selective or nonselective, and preemergence or postemergence. Many herbicides are also toxic to animals if swallowed or absorbed through the skin. They are usually less toxic to animals than are insecticides and other pesticides used to control animals. Most of the livestock losses associated with herbicides are usually the result of improper or careless use. **amide herbicide**. any of a family of herbicides that inhibit seed germination and/or growth of seedlings, probably by blocking protein synthesis in the primary meristems. When applied to foliage, localized or general necrosis is seen at the loci of contact. **anilide herbicide**. any of a family of herbicides (e.g., alachlor, butachlor, propachlor) used as selective pre- and postemergence herbicides to control annual grasses and broadleaved weeds in crops such as maize, beans, cotton, radish, and brassicas. **antiauxin herbicide**. any of a number of growth-regulating compounds that are chemically related to auxins. Included are phenoxy acids such as 4-chlorophenoxy-*iso*-butyric acid (PCIB), and some benzoic acids (e.g., 2,3,5-tri-iodobenzoic acid (TIBA)). These compounds have effects on cell elongation that are generally opposite to those of auxins. They stimulate root elongation and inhibit that of coleoptile segments. Some, including PCIB, function as auxin antagonists and are thus true antiauxins. **aryl carbamate herbicide**. (aryl carbamic ester). any of a class of herbicides that are absorbed via the roots and have colchicine-like actions. Examples are chlorpropham, diallate, phenylurethan, and propham. They are generally toxic to monocotyledons, but not to dicotyledons and are thus used to control grasses in crops such as peas and beets. The most prominent effects are inhibition of oxidative phosphorylation and of RNA and protein synthesis. Photosynthesis is also inhibited and the ATP content of tissues is reduced. **auxin herbicide**. any of a number of synthetic auxins (e.g., 2,4-D) used as herbicides. While these herbicides often cause dedifferentiation and the initiation of cell division in mature cells, they usually inhibit cell division in the primary meristems of intact plants. **carbamate herbicide**. any of a number of carbamates used as herbicides (e.g., propham, chlorpropham, and barban). They act selectively on monocotyledonous seedlings in which they strongly inhibit cell division by interrupting mitotic cycles in root and shoot meristems. **contact herbicide**. one that damages or kills a plant on contact. **hormone-type herbicide**. auxin herbicide. **nitrophenol herbicide**. nitrophenols were the first organic chemicals used in weed control. All are selective herbicides. Some are active on contact, others by absorption through the roots. **nonselective herbicide**. a herbicide that acts on a wide range of plant species. **organo-arsenical herbicide**. any organoarsenical used as a herbicide. They cause chlorosis, cessation of growth, and progressive browning followed by dehydration and death. Buds fail to sprout. **phenolic herbicide**. any of a number of phenolic compounds used as herbicides. At high rates of application, they destroy plant membranes. This produces water-logging of the leaves followed by desiccation and death of the plant. **pyrimidine herbicides**. pyrimidines act almost exclusively to inhibit photosynthesis and have little effect on nonphotosynthesizing organisms. **selective herbicide**. a herbicide that acts on one or a few types of plants. **substituted benzoic acid herbicide**. a family of herbicides that includes 2,3,6-TBA, chloramben, and dicamba. Their auxin and herbicidal activity greatly resembles that of the phenoxy auxins. Some of these compounds interfere with geotropic and phototropic curvatures. They also inhibit cell elongation. **substituted urea herbicide**. any of a number of rather selective preemergence herbicides that are absorbed by foliage and roots (e.g., diuron, fenuron, fluometuron, linuron, monuron, neburon, and siduron). Some penetrate deeply into soil and are quite persistent. The half-lives of these compounds in the soil vary considerably. Some penetrate deeply into the soil and are quite persistent. Such herbicides can thus be used to control weeds for extended periods of time. **systemic herbicide**. one that is absorbed and transported internally, killing the plant by acting on tissues that may be remote from the point of contact. **thiocarbamate herbicide**. any of a small family of herbicides with the characteristic group —CS·SS·CS—. They are toxic to germinating seeds. Uptake is via roots and through underground portions of the shoot system. Most are volatile and nonpersistent. Thiocarbamate treatment of grasses causes abnormal emergence of leaves from the coleoptiles. High doses can inhibit leaf emergence completely. Broadleaved weeds become necrotic, with a consequent suppression of growth. Thio- and dithiocarbamates alter or interfere with a variety of processes such as photosynthesis, respiration, oxidative phosphorylation, protein synthesis, and nucleic acid metabolism. **translocated herbicide**. a herbicide that is distributed throughout the plant following absorption through the leaves or roots. **triazine herbicide**. **1a:** any triazine compound used to broadly inhibit plant growth. The leaves of treated plants become chlorotic, necroses develop, and both photosynthesis and respiration are inhibited with consequent retardation of growth. **1b:** any of a class of herbicides used to control annual grasses and broadleaf weeds. Included are atrazine, prometon, prometryn, propazine, simazine, triazine; the acute oral LD_{50} in laboratory rats ranges among these herbicides from less than 100 to about 7000 mg/kg body weight. Triazine herbicides have a symmetrical triazine structure. These compounds function as herbicides by inhibiting the Hill reaction in photosynthesis. Overdosage in humans may cause abdominal pain, impaired adrenal function, anemia, dermatitis, diarrhea, eye irritation, mucous membrane irritation, nausea, disturbed thiamine and riboflavin function, or vomiting.

herbicide blue. Agent Blue.

herbicide orange. See Agent Orange.

herbicide purple. See Agent Purple.

herbicide white. See Agent White.

herbicide-resistant. pertaining to plants that continue to propagate when exposed to herbicidal substances. *Cf.* insecticide-resistant, pesticide-resistant.

herbivore. a phytophagous animal; one that feeds mainly or exclusively on vegetation.

herbivority. the quality or state of being herbivorous.

herbivorous. phytophagus; feeding on vegetation.

heredity. **1:** the transmission or mechanism of transmission of specific genetically based traits from parent to offspring. **2:** the genetic constitution of an individual.

Hermodice (bristle worms; stinging hairs). a genus of venomous polychaete worms (Phylum Annelida) with stinging setate parapodia ("setae," "bristles," or stings). The stings often remain in the skin of the victim and produce swelling, inflammation, and numbness. ***H. carunculata*** (green fire worm; bristleworm). a species that occurs in water to a depth of 15 m under rocks, on coral reefs, and in turtle grass in Florida, the Bahamas, and the West Indies. When touched, this species inflicts a painful sting. The setae (bristles) pierce the skin and break off, presumably releasing a toxin. The pain, inflammation, edema, and intense itching may persist for several weeks. *Cf. Eurythoe*.

heroin ((5α,6α)-7,8-didehydro-4,5-epoxy-17-methylmorphinan-3,6-diol diacetate (ester); diacetylmorphine). a white to brown, odorless, bitter, crystal or crystalline powder, synthesized by the acetylation of morphine. A poisonous, major addictive drug, it is an alkaloid that is much more potent than opium. Addicts usually inject heroin into the veins or sometimes simply beneath the skin; it is also sniffed. The effects are essentially the same as those of morphine. Of most importance toxicologically are the effects of overdosage, which occurs not infrequently among addicts due chiefly to differences in the original purity of the preparation and wide variations in the degree to which the drug is diluted before sale on the street. The cardinal signs of overdosage are depressed respiration, coma, and miosis. Death from respiratory paralysis may intervene. Conscious victims of overdosage may show disturbed vision, restlessness, cramps in extremities, cyanosis, weak pulse, and hypotension. Narcotic antagonists such as naloxone and naltrexone are used in heroine therapy. Heroin is fetotoxic, and can produce convulsions, tremors, hemorrhage, and even death in the neonate.

heron's-bill. a common name of plants of the genus *Erodium*.

herring. any clupeoid fish, those of the family Clupeidae, especially *Clupanodon thrissa*.

Hesperocnide (western stinging nettle). a tall plant with stinging hairs. *Cf. Cnidoscolus, Laportea canadensis, Urtica dioica*.

heteroauxin. former name of auxin 1AA (See 3-indole acetic acid).

heterocyclic. pertaining to or denoting a heterocyclic molecule or compound.

heterocyclic compound. a compound with one or more rings that contain more than one type of atom. Most are organic compounds (e.g., pyridine, glucose).

heterocytolysin. See heterolysin.

heterocytotoxin. a toxin produced by one species of organism that poisons cells of another species.

heterocytotropic. having an affinity (said of an antibody) for cells of more than one species.

heterogeneous. comprised of dissimilar constituents; mixed, varied, nonuniform with respect to specified properties (e.g., structure, configuration, composition, quality). *Cf.* homogeneous.

heterolysin (heterocytolysin). **1a:** a lytic agent produced by one species of animal that lyses cells of another. **1b:** a lysin that is formed following the introduction of antigen from another species.

heterolysis. the lysis or digestion of cells of one species of animal by the action of a lytic agent (an enzyme or antigen) produced by another species.

heterolytic. of, pertaining to, or caused by heterolysis or by a heterolysin.

Heteropneustes fossilis. a venomous marine catfish found along the coasts of India, Ceylon and Viet Nam. See catfish.

Heteropoda. a genus of spiders (Family Sparassidae) that are dangerous to humans. ***H. venatoria*** (banana spider). a large rufous and yellow tropical spider that is often unintentionally imported to higher latitudes in shipments of fruit.

heterotonia. any condition of abnormal or unusual tension or tonus.

heterotroph. a heterotrophic organism.

heterotrophic (organotrophic; zootrophic). pertaining to the process of obtaining nourishment from exogenous, preformed organic material.

heterotrophic organism (heterotroph). an organism whose growth is supported by preformed organic matter; a consumer (in ecology). Such an organism is unable to synthesize organic compounds from inorganic substrates.

hexabutyldistannoxane compound. tributyl tin.

hexacarbon neuropathy. γ-diketone neuropathy.

hexacarbonyl chromium. chromium carbonyl.

1,2,3,4,7,7-hexachlorobicyclo[2.2.1]-2-heptene-5,6-bisoxymethylene sulfite. endosulfan.

1,3,4,5,5-hexachloro-1,3-cyclopentadiene dimer. mirex.

1,2,3,6,7,8-hexachlorodibenzo-p-dioxin. carcinogenic in some laboratory animals.

1,2,3,4,10,10-hexachloro-6,7-epoxy-1,4,4a,5,-6,7,8,8a-octahydro-*endo,endo*-1,4:5,8-dimethanonaphthalene. endrin.

1,2,3,4,10,10-hexachloro-6,7-epoxy-1,4,4a,5,6,-7,8,8a-octahydro-*endo,exo*-1,4:5,8-dimethanonaphthalene. dieldrin.

1,2,3,4,10,10-hexachloro-1,4,4a,5,8,8a,-hexahydro-1,4:5,8-dimethanonaphthalene. aldrin.

6,7,8,9,10,10-hexachloro-1,5,5a,6,9,9a-hexahydro-6,9-methano-2,4,3-benzodioxathiepin 3-oxide. endosulfan.

1,4,5,6,7,7-hexachloro-5-norbornene-2,3-dimethanol cyclic sulfite. endosulfan.

3,4,5,6,9,9-hexachloro-1a,2,2a,3,6,6a,7,7a-octahydro-2,7:3,6-dimethanonaphth[2,3-*b*]oxirene. endrin.

hexachlorobenzene (HCB; perchlorobenzene). an extremely toxic, high melting point aryl halide, C_6Cl_6, comprised of white needles. It is used chiefly as an intermediate in the synthesis of chlorinated organic solvents and as an agricultural fungicide; it is also a persistent, widespread environmental chemical that bioaccumulates. It is harmful to the eyes, liver, and respiratory system. The most common toxic effects are increased respiratory rate, increased body temperature, hepatomegaly, and immunosuppression (sometimes immunostimulation). Chronic ingestion may result in cutaneous porphyria. Immunotoxic effects can appear at concentrations or doses that do not produce obvious toxic effects. Do not confuse with hexachlorocyclohexane, a synonym for benzene hexachloride.

hexachlorobutadiene. a clear, colorless, nonflammable liquid, $Cl_2C{:}CClCCl{:}CCl_2$, with a mild odor. It is insoluble in water, soluble in ethanol and ether, and compatible with many resins. It is toxic by ingestion and inhalation. hexachlorobutadiene is hepatotoxic, nephrotoxic, and is a suspected human carcinogen.

hexachlorocyclohexane (1,2,3,4,5,6-hexachlorocyclohexane; benzene hexachloride; BHC; HCH; HCCH; TBH). a chlorinated benzene, $C_6H_6Cl_6$, in the form of a white or yellowish, water-insoluble powder or flakes with a musty odor. There are eight stereoisomers having different orientations of the H and Cl atoms; physical properties and toxicity vary among them. Do not confuse with hexachlorobenzene. **α-hexachlorocyclohexane** (α-HCH). an excitant; the technical grade is carcinogenic to some laboratory animals. **β-hexachlorohexane** (β-HCH). a weak or inert depressant. **δ-hexachlorohexane** (δ-HCH). a strong depressant. **τ-hexachlorocyclohexane** (1α,2α,-3β,4α,5α,6β-hexachlorocyclohexane; τ-HCH; γ-benzene hexachloride; gamma-benzene hexachloride; gamma-hexachlor; gamma-hexane). an environmentally persistent insecticide and a constituent of a wide variety of agricultural, medical, and veterinary products. It is the most toxic isomer of hexachlorocyclohexane and the chief toxic component of lindane. It is the only isomer that is insecticidal and is also highly toxic to humans and to certain livestock and laboratory animals by all routes of exposure. Symptoms of acute poisoning include dizziness, headache, weakness, nausea, vomition, dyspnea, tremors, convulsions, cyanosis, and death from cardiopulmonary collapse. The vapor may irritate the eyes, nose, and throat, and topical application may cause a localized reaction. Lesions from chronic poisoning may include liver damage associated with fatty tissue and histoplastic anemia; and kidney damage with degeneration of kidney tubules. See lindane poisoning. **1α,2α,3β,4α,5α,6β-hexachlorocyclohexane**. τ-hexachlorocyclohexane. See hexachlorocyclohexane. See also lindane. **1,2,3,4,5,6-hexachlorocyclohexane**. hexachlorocyclohexane. *Cf.* benzene hexachloride.

γ-hexachlorocyclohexane. hexachlorocyclohexane. See also lindane.

hexachlorocyclopentadiene. a very toxic, highly reactive, cyclic alkenyl halide with two double bonds that readily undergoes substitution and addition reactions. It is a light yellow liquid (fp 11°C, b.p. 239°C, density 1.7 g/cm^3) with a pungent odor. It was originally used as an agricultural fumigant and as an intermediate in the manufacture of insecticides (e.g., Mirex and Kepone). Hexachlorocyclopentadiene is an irritant of the skin, eye, and mucous membranes; is strongly lacrimating and is damaging to most vital organs such as the adrenal glands, brain, heart, kidney, and liver.

hexachloropentadiene dimer. mirex.

1aα,2β,2aα,3β,6β,6aα,7β,7a)-3,4,5,6,9,9-hexachloro-1a,2,2a,3,6,6a,7,7a-octahydro-2,7:3,6-dimethanonaphth[2,3-*b*]oxirene. dieldrin.

hexachlorophane. hexachlorophene.

hexachlorophene (2,2′-methylenebis[3,4,6-trichlorophenol]; 2,2′-dihydroxy-3,3′,5, 5′,6, 6′-hexachlorodiphenylmethane; bis(3,5,6-trichloro-2-hydroxyphenyl)methane). a polychlorinated bisphenol, it is a white, odorless, water-insoluble powder, $(C_6HCl_3OH)_2CH_2$. It has been used chiefly as a bacteriostat in cosmetics, in dermatologicals, and in anti-infective germicidal soaps. Hexachlorphene has also been used as an agricultural fungicide and bactericide with applications in veterinary medicine. It is especially effective against Gram-positive bacteria. It is a photocontact allergen and is neurotoxic to laboratory animals and to humans, causing central neurological lesions. It is toxic when applied repeatedly to the skin, and can be lethal, especially in the case of infants. Symptoms of intoxication may include confusion, listlessness, and convulsions. Diffuse status spongiosus of the reticular formation is a common lesion. Hexachlorophene is teratogenic at least in the case of pregnant women who routinely use it.

hexadimethrine bromide (*N,N,N′,N′*-tetramethyl-1,6-hexanediamine polymer with 1,3-dibromopropane; polymer of *N,N,N′,N′*-tetramethylhexamethylenediamine and trimethylene bromide; poly(*N,N,N′,N′*-tetramethyl-*N*-trimethylenehexamethylenediammonium dibromide)). an extremely toxic heparin antagonist, $(C_{13}H_{30}Br_2N_2)_x$.

Hexadon™. lindane.

hexadrin. endrin.

hexafluorosilicic acid. fluorosilic acid.

hexahydride. disilane.

3,4,5,6,9,10-hexahydro-14,16-dihydroxy-3-methyl-1*H*-2-benzoxacyclotetradecin-1,7(8H)-dione. zearalenone.

hexahydro-3a,7a-dimethyl-4,7-epoxyisobenzofuran-1,3-dione. cantharidin.

[1aR-(1aα,2aβ,3β,6β,6aβ,8aS*,8bβ,9S*)]-hexahydro-2a-hydroxy-9-(1-hydroxy-1-methylethyl)-8b-methyl-3,6-methano-8*H*-1,5,7-trioxacyclopenta[ij]cycloprop[*a*]azulene-4,8(3*H*)-dione. picrotin.

2,3,6a,8,9,9a-hexahydro-9a-hydroxy-4-methoxycyclopenta[*c*]furo[3′,2′:4]furo[2,3-h][1]benzopyran-1,11-dione. aflatoxin M_2.

[1aR-(1aα,2aβ,3β,6β,6aβ,8aS*,8bβ,9R*)]-hexahydro-2a-hydroxy-8b-methyl-9-(1-methylethenyl)-3,6-methano-8*H*-1,5,7-trioxacyclopenta[ij]cycloprop[*a*]azulene-4,8-(3*H*)-dione. picrotoxinin.

1,2,3,4,5,6-hexahydro-1,5-methano-8*H*-pyrido-[1,2-a][1,5]diazocin-9-one. cytisine.

2,3,6aα,8,9,9aα-hexahydro-4-methoxycyclopenta[*c*]furo[3′,2′:4,5]furo[2,3-h][1]-benzopyran-1,11-dione. aflatoxin B_2.

***dl-cis*-1,2,3,9,10,10a-hexahydro-6-methoxy-11-methyl-4*H*-10,4a-iminoethanophenanthrene**. racemethorphan.

***dl-cis*-1,3,4,9,10,10a-hexahydro-6-methoxy-11-methyl-2*H*-10,4a-iminoethanophenanthrene**. racemethorphan.

(3,4,7aα,9,10,10aα-hexahydro-5-methoxy-1H,12H-furo[3′,2′:4,5]furo[2,3-h]pyrano[3,4-*c*][1]benzopyran-1,12-dione. aflatoxin G_2.

1,2,3,4,5,6-hexahydro-3-methyl-1,5-methano-8*H*-pyrido[1,2-a][1,5]diazocin-8-one. caulophylline.

hexahydrothymol. menthol.

1,2,3,4,5,6-hexahydro-3,6,11-trimethyl-2,6-methano-3-benzazocin-8-ol. metazocine.

2,3,4,4a,9-hexahydro-2,4aα,9α-trimethyl-1,2-oxazino[6,5-*b*]indol-6-ol methylcarbamate. eseridine.

(3α*S*-*cis*)-1,2,3,3α,8,8α-hexahydro-1,3α,8-trimethylpyrro[2,3-*b*]indol-5-ol methylcarbamate (ester). physostigmine.

hexahydroaniline. cyclohexylamine.

hexahydropyridine. piperidine.

2,3,14,20,22,25-hexahydroxycholest-7-en-6-one. β-ecdysone (See ecdysone).

1,1′,6,6′,7,7′-hexahydroxy-5,5′-diisopropyl-3,-3′-dimethyl[2,2′-binaphthalene]-8,8′-dicarboxaldehyde. gossypol.

1,1′,6,6′,7,7′-hexahydroxy-3,3′-dimethyl-5,5′-bis(1-methylethyl)[2,2′-binaphthalene]-8,8′-dicarboxaldehyde. gossypol.

4,5,4′,5′,7′-hexahydroxy-2,2′-dimethylnaphthodianthrone. hypericin.

1,3,4,6,8,13-hexahydroxy-10,11-dimethylphenanthro[1,10,9,8-*opqra*]perylene-7,14-dione. hypericin.

hexakis(cyano-C)ferrate(4-)iron(3+) (3:4). ferric ferrocyanide.

hexalhydrobenzene. cyclohexane.

hexamethonium chloride (hexamethylene-bis-(trimethylammonium)chloride; *N,N,N,N′,N′,N′*-hexamethyl-1,6-hexanediaminium chloride). a white, crystalline, hygroscopic powder, $(CH_3)_3NCl(CH_2)_6NCl(CH_3)_3$, that is very soluble in water, soluble in ethanol, methanol, and *n*-propanol, but insoluble in ether and chloroform. It is a quaternary ammonium salt used therapeutically as an antihypertensive. See also quaternary ammonium compound.

***N,N,N,N′,N′,N′*-hexamethyl-1,6-hexanediaminium chloride**.

***N,N,N′,N′,N″,N″*-hexamethyl-1,3,5-triazine-2,4,6-triamine**. altretamine.

hexamethylene. cyclohexane.

hexamethylenebis(trimethylammonium)chloride. hexamethonium chloride.

hexamethylmelamine. altretamine.

hexan-2,5-dione. a metabolically active form of *n*-hexane; it is a neurotoxicant.

hexanaphthene. cyclohexane.

Hexanchus grisseus (six-gilled shark). a poisonous shark of the Atlantic ocean, the Pacific coast of North America, Chile, Australia, Japan, South Africa, and the southern Indian Ocean. See elasmobranch poisoning.

hexane. See hydrocarbon (aliphatic hydrocarbon).

***n*-hexane**. a colorless, volatile, flammable, liquid, with a faint odor. It is a six-carbon alkane, $CH_3(CH_2)_4CH_3$, distilled from petroleum. It is used chiefly as a solvent in the extraction of oils from seeds (e.g., of cotton and sunflower seeds), as a paint diluent, in a number of polymerization processes. Together with other more polar solvents, it is used in the separation of fatty acids. It is a moderately toxic neurotoxicant. Major symptoms of intoxication in humans with acute, very high rates of exposure are those of CNS depression. Chronic exposure of humans may result in a sensorimotor peripheral axonopathy, the initial lesion of which affects the paranodal axonal swellings with subsequent distal degeneration. The proximate toxicants are metabolites of *n*-hexane (e.g., Hexan-2,5-dione, acetonyl acetone, and methyl *n*-butyl ketone). See also hydrocarbon (aliphatic hydrocarbon), axonopathy.

2,5-hexanedione. acetonylacetone.

3-hexanoic acid lactone. a nontumorigenic lactone.

2-hexanone. methyl n-butyl ketone.

hexavalent chromium. See chromate.

2-hexenal (hex-2-enal; hex-2-en-1-al; hexylenic aldehyde; leaf aldehyde). a poisonous aldehyde,

$C_6H_{10}O$. It is toxic by intraperitoneal route and is a possible mutagen.

hex-2-enal. 2-hexenal.

hex-2-en-1-al. 2-hexenal.

4-hexenoic acid lactone. a carcinogenic lactone.

hexobarbital (5-(1-cyclohexen-1-yl)-1,5-dimethyl-2,4,6(1*H*,3*H*,5*H*)-pyrimidinetrione; 5-(1-cyclohexen-1-yl)-1,5-dimethylbarbituric acid; 5-cyclohexenyl-3,5-dimethylbarbituric acid; *n*-methyl-5-cyclohexenyl-5-methylbarbituric acid; methylhexabital; methexenyl; enhexymal; hexobarbitone). a very toxic, short-acting intravenous sedative and hypnotic. The sodium derivative (hexobarbital sodium, *q.v.*) is also used to induce surgical anesthesia.

hexobarbital sodium (sodium 5-(1-cyclohexenyl)-1,5-dimethylbarbiturate; Evipal sodium). the sodium derivative of hexobarbital (*q.v.*), it is used *i.v.* (also active per os) to induce surgical anesthesia. Its action is rapid and of ultrashort duration.

hexobarbitone. hexobarbital.

hexone. methyl isobutyl ketone.

4-hexyl-1,3-benzenediol. 4-hexylresorcinol.

hexylenic aldehyde. 2-hexenal.

4-hexylresorcinol (4-hexyl-1,3-benzenediol; 4-hexyl-1,3-dihydroxybenzene). a white to yellowish, sharp-tasting, crystalline substance that is slightly water soluble and soluble in most organic solvents. It is a moderately to very toxic topical antiseptic and anthelmintic agent used against nematodes.

Hf. the symbol for hafnium.

HF. hydrogen fluoride, hydrofluoric acid.

Hg. the symbol for the element mercury (hydrargyrum).

HGPRT locus (hypoxanthine guanine phosphoribosyltransferase locus). a gene locus that supports the incorporation of purines from the medium into cultured mammalian cells; these purines can be converted into nucleic acids. A mutation at this locus prevents such uptake of toxic (e.g., 8-azaguanine or 6-thioguanine) as well as normal purines. Consequently, the cultured cells can produce purines de novo and will grow. This phenomenon is used as the basis for mutagenicity tests in which cultured mammalian cells are exposed to toxic purines together with a test mutagen. If the cells grow, one may infer that a mutation in the HGPRT locus has occurred.

HHDN. aldrin.

HI. hydrogen iodide.

hierarchy. 1: a general integrated system of two or more levels, the higher controlling to some extent the activities of the lower levels. **2:** a series of consecutively subordinate categories; nested hierarchy. **3:** dominance hierarchy *q.v.* **4:** taxonomic hierarchy, *q.v.*

High-Level Radioactive Waste (HLW). waste generated in the fuel of a nuclear reactor and occurs also at nuclear fuel reprocessing plants. It is highly toxic and extremely dangerous radiation hazard in the absence of adequate shielding.

highland moccasin. *Agkistrodon contortrix*.

Hilaria mutica (tobosagrass). a grass (Family gramineae) that is sometimes a host to *Claviceps cinerea*. Cattle sometimes contract ergotism from grazing on this grass. See *Claviceps*, ergot, ergotism.

hills of snow. *Hydrangia macrophylla*.

Himalayan pit viper. *Agkistrodon hymalayanus*.

hime-okoze. *Minous monodactylus*.

himehabu. *Trimeresurus okinavensis*.

himmin. *Atractaspis microlepidota*.

hiochic. mevalonic acid.

h'ion. *Bitis gabonica*.

hippo. ipecac.

Hippocastanaceae (the buckeye family). a family of trees and shrubs with opposite, palmately lobed leaves, the leaflets of which have straight veins; showy white flowers in large clusters; and a nut-like fruit encased in a leathery capsule. See *Aesculus*.

Hippomane mancinella (manchineel; manchineel tree). a small (3½-7 m) poisonous tree (Family Euphorbiacieae) of Central America, the West Indies, and remote areas of peninsular Florida. It has a palatable, but poisonous, usually greenish or greenish-yellow, apple-shaped fruit (ca. 2½-4 cm in diameter) and a milky, caustic, vesicant sap. All parts of the plant are poisonous. The fruit contains indole. Contact produces severe skin irritation that is slow to heal; there is considerable danger of conjunctivitis and temporary (occasionally permanent) blindness in humans and other animals. Individual sensitivity varies greatly. Ingestion of the fruit produces vomiting, abdominal pain, a bloody stool, and sometimes death. Extracts of the fruit lower blood pressure and induce lacrimation, salivation, leukocytosis, and extensive hemorrhage in laboratory animals.

hippurate. a salt or ester of hippuric acid.

hippuria. the presence of abnormally high concentrations of hippuric acid in the urine.

hippuric acid (benzaminoacetic acid; benzoylaminoacetic acid; benzoylglycin; *N*-benzoylglycine; benzoylglycocoll). a colorless, crystalline compound, $C_6H_5CONHCH_2$, that is very water soluble and soluble in ethanol and ethyl ether. It is a detoxication and excretory product of benzoic acid, formed by the liver and kidneys. It occurs in the urine of herbivorous animals and in smaller amounts in humans. It is used in organic synthesis, and its salts and esters (hippurates) are used therapeutically.

hiraguchi. *Agkistrodon halys*.

Hirschfeld's bacillus. *Salmonella hirschfeldii*.

hirudicidal. destructive to leeches.

hirudicide. any agent that destroys leeches.

histaffine. **1:** having affinity for tissues. **2:** asubstance in the blood serum of animals with certain diseases; it combines with certain tissue components, producing complement fixation.

histaminase (diamine oxidase; benzylamine oxidase; histamine deaminase; histamine oxidase). an enzyme that deactivates histamine. It is a copper-containing diamine oxidase that catalyzes the oxidative deamination of histamine and other diamines. It occurs throughout the body, but is most abundant in renal and intestinal mucosa. It is inhibited by carbonyl reagents. *Cf.* monoamine oxidase.

histamine (1*H*-imidazole-4-ethanamine; 4-aminoethylglyoxaline; 2-(4-imidazolyl)ethylamine; 4-imidazoleethylamine; 5-imidazole-ethylamine; β-aminoethylimidazole; β-aminoethylglyoxaline). an amine derivative of histidine, it is an organic base, $C_5H_9N_3$, that occurs naturally in living tissues and also as a product of decay. In vertebrate animals, histamine concentrates especially in the lungs, in mast cells and other connective tissue cells, and basophilic blood cells. It is a decarboxylation product of histidine. Histamine is released from tissues during allergic reactions or cellular injury. It is a potent vasodilator, a powerful stimulant of the secretion of acid and pepsin by the stomach, and a constrictor of many smooth muscles (notably those of the bronchial tubes) and a dilator of others. It is moderately toxic and produces many of the symptoms of inflammation and allergy. Symptoms of histamine intoxication include dizziness, intense headaches, epigastric pain, thirst, generalized erythema and flushing of the face, severe itching, urticarial eruptions, burning of the throat, abdominal pain, bronchospasm, an inability to swallow, suffocation, and severe respiratory distress. Shock (histaminia) is a complication and death sometimes results. See also ergot, histaminemia, poisoning (scombroid poisoning).

histamine deaminase. histaminase.

histamine oxidase. histaminase.

histamine-*N*-methyltransferase. See N-methylation.

histamine oxidase. histaminase.

histaminemia. the presence of histamine in circulating blood.

histaminergic. of or pertaining to those responses to histamine by histamine receptors that are blocked by histamine antagonists such as cimetidine.

histaminia. shock due to excessive amounts of metabolically available or introduced histamine.

histaminuria. the presence of histamine in the urine.

histanoxia. tissue anoxia due to a reduced blood supply.

histidine. a common amino acid, $C_6H_9O_2N_3$, and antioxidant.

histochemical. of or pertaining to histochemistry, to chemical components of tissues, or to the chemical processes or activities of cells or tissues.

histochemistry. the science that treats the identification of chemical processes and components of cells and tissues.

histocompatibility. the compatibility of tissues (e.g., those successfully used in organ or tissue transplantations) within or between species.

histodiagnosis. diagnosis by microscopic examination of tissues.

histogram. a graph or diagram on which the value of each of a set of variables is depicted by the height of vertical bars.

histohypoxia. an abnormally low concentration of oxygen in a tissue or tissues.

histology (histomorphology; microscopic anatomy). **1:** the science that deals with the identification and study of the microscopic structure of tissues, including the study of intracellular features. It is often of critical importance in the diagnosis of disease. **2:** the organization and cellular structure of tissues.

histolysis. the dissolution or disintegration of tissue.

histolytic. of, pertaining to, or causing hystolysis.

histomorphology. histology.

histopathological. **1a:** pertaining to the science of histopathology. **1b:** of or pertaining to pathological tissues or their condition.

histopathology. the study of pathologic histology; the microscopic study of abnormal, diseased, or damaged tissue.

histotoxic. toxic to a specific tissue or to a variety of tissues, acting especially by disrupting cellular respiration.

histotoxic hypoxia. See hypoxia.

histrionicotoxin. a neurotoxic alkaloid secreted by the skin of *Dendrobates histrionicus* that affects sodium and potassium transport across nerve cell membranes. See spiropiperidin.

hives. See urticaria.

HLW. High-Level Radioactive Waste.

HMM. altretamine.

HN1 (2-chloro-*N*-(2-chloroethyl)-*N*-ethylethanamine; 2,2′-dichlorotriethylamine; bis(2-chloroethyl)ethylamine; ethylbis(2-chloroethyl)amine). a volatile, water-insoluble, liquid nitrogen mustard, $(CH_2CH_2Cl)_2$, with a slight fishy odor. A deadly vesicant and extreme irritant to the eyes, skin, and mucous mem-branes; a deadly toxic vesicant.

HN2. nitrogen mustard.

Ho. the symbol for holmium.

hoary pea. *Tephrosia*.

HOG. phencyclidine.

hog's bean. 1: a common name of *Hyoscyamus spp*. 2: hyoscyamus.

hog's potato. *Zigadenus venenosus*.

hog-nosed vipers. See *Bothrops*.

hogfish. See Labridae.

hogwort. *Croton*.

Holcus lanatus (velvet grass). a grayish, tufted, velvety pubescent grass (Family Gramineae) with flowers born on two-flowered spikelets. It is a cyanogenic plant indigenous to Europe and Africa, but has been naturalized in other continents.

holism. the conception of living organisms and other complex systems as a functioning whole.

holistic. viewing, relating to, and treating living organisms and other complex dynamic systems as an integrated functioning whole.

holly. *Ilex aquifolium, opaca, verticillata.* See houseplant.

holmium (symbol, Ho)[Z = 67; A_r = 164.93]. an element of the lanthanoid series with one naturally occurring isotope, ^{165}Ho. It has few applications. See holmium chloride.

holmium chloride. a moderately to very toxic, bright yellow, water-soluble, crystalline solid, $HOCl_3$.

Holocentridae (squirrelfish; soldierfish). a large family of carnivorous, marine percomorph fish (Class Osteichthyes) with dazzling colors, large eyes, and large terminal mouths. Most species inhabit shallow, tropical coral reefs or tropical and subtropical rocky shores to a depth of 275 m or more. They are most active at night.

Holocephali (elephantfish; ratfish). a subclass of venomous and poisonous cartilaginous fish (Class Chondrichthyes) found in cold oceans, often in deep waters. They have large pectoral fins, no scales, and a single pair of external gill openings, each covered by a fold of skin. The body is more or less laterally compressed, tapering caudally to a slender whip-like tail. The snout is rounded or cone-shaped (and superficially rat-like in lateral aspect), or supports a curiously shaped proboscis. The anterior dorsal fin is triangular, bordered anteriorly by a strong, sharp, bony spine or sting. A shallow depression, containing a strip of soft tissue (the venom gland) runs along the caudal edge of the spine. The sharp sting can produce a very painful puncture wound. Ratfish also have a large tooth plate in each jaw and can inflict a serious bite. The sting may produce severe pain that persists for several hours, subsiding into a dull ache that may last for some days. The area of the wound may become numb, cyanotic, and edematous. Aching joints and swollen lymph nodes are common. The viscera are toxic to laboratory animals. See also *Chimaera monstrosa* and *Hydrolagus colliei.*

holocoenotic. pertaining to the effect of combined or interacting environmental factors on a biological system. In the case of environmental toxicology, insults to biological systems are not often due to the action of a single toxicant. Exposure and intoxication usually occur in an environment that is dynamic with respect not only to the constellation of toxicants present, but also with respect to environmental variables that influence system susceptibility. See also hological method. See also environmental factor, biotic factor, experiment.

hological method. any method of study aimed at the properties and actions of a whole rather than on specific component processes. See also holocoenotic, merological method.

holophytic. of or pertaining to a plant that synthesizes all organic compounds from inorganic substrates, usually by photosynthesis. *Cf.* autotrophic, holotrophic, holozoic. See also autotrophic organism.

holophytic organism. autotrophic organism.

holothurin. **1:** a steroid glucoside (saponin) first isolated from the Bahamian sea cucumber, *Actinopyga agassizi*. It has antibiotic properties, has reportedly suppressed growth of tumors in laboratory mice, and is one of the most potent hemolysins known. It is neurotoxic and cytolytic and manifests some antimetabolic activity as well. If a sea cucumber has Cuvierian tubules, it emits holothurin into the water; if not, it is otherwise secreted through the body wall. Holothurin is toxic to various invertebrate taxa and also to fish, frogs, and other vertebrate animals. **2:** any toxic, sulphated steroid glucoside secreted by Cuvier glands of holothurians; some are neurotoxic.

Holothuroidea (sea cucumbers). a class of sluggish, soft-bodied Echinoderms with a body that is elongated along the oral-aboral axis and covered by a thin, leathery integument comprised only of microscopic calcareous plates or spicules; mouth surrounded by tentacles; tube feet; internal madreporite; no pedicellariae, spines, or arms. Some holothuroideans contain Cuvierian tubules, *q.v.* Some species are poisonous by ingestion. Effects are usually short lived and only moderately severe. Pruritis and mild swelling of the hands may follow the handling of certain sea cucumbers. See also holothurin.

holotonia. muscular spasm of the entire body.

holotonic. marked by, pertaining to, or causing holotonia.

holotrophic. **1:** holophytic. **2:** of or pertaining to a predator that preys only on one species of organism. *Cf.* autotrophic, holophytic, holozoic. See also autotrophic organism.

holozoic. pertaining to or denoting organisms that feed entirely on living organisms or on other complex, organic matter, following which the material is digested, absorbed, and assimilated. Most animals are holozoic (at least in part). *Cf.* autotrophic, holophytic, holotrophic. See also autotrophic organism.

homarin. a toxic principle with curare-like effects that occurs in the venoms of cones (gastropod mollusks of the family Conidae).

homatropine (mandelytropine; tropine mandelate). a white, crystalline, slightly water-soluble alkaloid, $C_{16}H_{21}NO_3$. It has anticholinergic, mydriatic, and cycloplegic actions and is toxic by ingestion and inhalation. Usually the salts (the hydrobromide and the methylbromide) are used in medicine.

homeokinetogenic activity. a rarely used term, it is: **1a:** the capacity of corticoids to restore an organism or various organs to a normal or nearly normal level of dynamic equilibrium in the face of stress (e.g., from inflammation or trauma). **1b:** the capacity of corticoids to relieve the effects of stress.

homeostasis. **1:** the dynamic steady internal state or equilibrium of the internal environment of a healthy biological system, or a part of such system. **2:** the maintenance of a relatively steady state within a living organism through regulatory mechanisms. In humans, a nearly constant body temperature is maintained by dynamic mechanisms, as are acid-base balance, blood pressure, pH, and numerous other physiological variables. **3:** the tendency to stability of mature ecological systems.

homicidal. **1a:** of, pertaining to, arousing, or provoking to homicide. **1b:** capable of killing other humans. **2:** of or pertaining to a psychological state (homicidal mania) aroused to a point such that any opportunity will provoke attempted, usually violent homicide.

homicide. the act or process of destroying or depriving a person of life; the killing of a human being (intentionally or unintentionally), directly or through the agency of others. *Cf.* killing. **criminal homicide**. unlawful homicide; that brought about knowingly, purposely, negligently, or recklessly. The main types are negligent homicide, manslaughter, and murder.

homidium bromide (ethidium). a trypanocide used in veterinary medicine.

hominid. **1:** pertaining to the family of humans (Hominidae). **2:** any human or human-like species (e.g., *Homo spp.*, *Australopithecus spp.*). All are characterized by erect posture and other features that distinguish them from the great apes (pongids). See human, man, woman, *Homo sapiens*.

Hominidae. a family of primates (Superfamily Hominoidea), that includes *Homo sapiens* (*q.v.*) and several fossil species of humans. The only other primate family, Pongidae, contains the great apes.

homininoxious. injurious to *Homo sapiens*.

Homo. the genus of primates (Family Hominidae, Superfamily Hominoides) that includes *H. sapiens* and a number of fossil human species. See also human, man, woman. ***Homo sapiens*** (modern man). a highly variable species of hominid, characterized taxonomically by its fully erect posture, bipedal locomotion, reduced dentition, and large brain. It also has an opposable thumb, the ability to talk, and to make and use complex tools. This species is toxicologically unique. It is able to produce far more poisons and to poison a far wider array of organisms (including innumerable microorganisms, plant, and animal species) than any other species. Individuals and groups of this species sometimes poison others of its kind on an individual and sometimes on a mass scale; poisons are also frequently used in suicides. Poisonous or live venomous animals and plants may also, for example, be used to murder or commit suicide. See aqua toffana, arsenic, calotropin, Cleopatra's asp, genocide, human, hydrogen cyanide, *Kalmia latifolia*, Lewisite, gas (mustard gas), sarin, strychnine, tabun.

homobatrachotoxin. an extremely toxic neurotoxic steroidal alkaloid found in the skin of poison arrow frogs (Dendrobatidae) and the skin, muscle, and feathers of the pitohui, a perching bird (Order Passeriformes).

homocyclic. pertaining to a homocyclic compound or molecule.

homocyclic compound. a compound that has a ring comprised of only one type of atom. Most are organic compounds with carbon atoms in the ring (e.g., benzene, phenol).

homocytotropic. having an affinity for cells of the same species; as of an antibody.

homogenate. a homogenized material.

homogeneity (homogenicity). the state of being homogenous.

homogeneous. comprised of similar ingredients or otherwise having a uniform quality throughout.

homogenicity. homogeneity.

homoglomerin (homoglomerine). an extremely poisonous, convulsant, quinazolinone toxin of certain myriapods. It is lethal also to other arthropods and small vertebrate animals. See *Glomeris marginata*. See also glomerin.

homoglomerine. homoglomerin.

homolysin. a lysin (e.g., isohemolysin) produced by injection into the body of antigen derived from an individual of the same species.

homolysis. lysis of cells by homolysins and complement.

honey. See *Apis mellifera*.

honey bee. See *Apis*.

honey mushroom. *Armillariella mellea*.

honeysuckle family. Caprifoliaceae.

hontipeh pura. *Naja naja*.

hooded helvella. *Gyromitra (Helvella) infula*.

hooded pitohui. *Pitohui dichrous*.

hoogly patee. *Enhydrina schistosa*, *Enhydrina valakadyn*.

hook-nosed sea snake. *Enhydrina schistosa*.

hoolamite. a detector for carbon monoxide that contains fuming sulfuric acid, iodine pentoside, and powdered pumice. The color changes from light gray to green in the presence of carbon monoxide.

hop sage. *Grayia*. See also sage.

Hopi rattlesnake. *Crotalus viridis nuntius*.

Hoplocephalus (Australian broad-headed snakes). a small genus of venomous Australian snakes (Family Elapidae). The ventral scales bear a lateral keel and notch. They are the only truly arboreal snakes in Australia. ***H. bitorquatus*** (double-collared snake). a snake of northeastern Australia. ***H. bungaroides*** (broad-headed snake; yellow-spotted snake). a snake of southeastern Australia, found in the mountains and coastal areas of southern Queensland and New South Wales. It is the only member of this genus thought to be truly dangerous. The head is broad and distinct from the neck; the pupils of the eyes are round. Adults may attain a length of about 1.3 m. ***H. stephensii*** (Stephen's banded snake; tiger snake; yellow-banded snake). a snake of southeastern Australia. See also *Notechis*.

Hordeum vulgare (cultivated barley). a cultivated grain (Family Gramineae), often infected by any of a number of fungi. Toxic concentrations of nitrate have also been reported. Grain infected with *Gibberella saubinetii* (the imperfect stage = *Fusarium graminearum*) is toxic to some animals. Infected seeds are shrunken, may have bluish-black tips, and may confer a darkened color to seed mixtures in which they occur. Even small quantities of infected seed cause intoxication and vomiting in domestic swine. Fatalities among swine are rare, however, because they do not consume heavily infested grain. Horses and mules will not normally eat scabby barley, and ruminants such as cattle can consume heavily infected grain without toxic effects. Laboratory rats are sensitive; guinea pigs are highly resistant. Toxic concentrations of nitrate have been reported. See also *Claviceps*.

horingadder. *Bitis caudalis*.

hormesis. the stimulatory effect produced by exposure of a living organism to nontoxic concentrations of a toxic substance.

hormone. **1a:** a substance that is produced by one tissue or organ (usually an endocrine gland), directly enters the cardiovascular system, through which it is transported to one or more remote tissues or organs. The hormone stimulates one or more specific physiological effects that are characteristic of the affected target tissue or organ. These effects are often regulatory in nature, and control many vital bodily functions. Hormones are of considerable toxicologic interest. (1) hormonal activity, as well as the storage, synthesis, and release of hormones, may be adversely affected by toxicants. (2) most hormones are toxic, some extremely so. Toxic effects may result from excessive secretion or from therapeutic or other usage (e.g., ACTH, ADH, androgens, estrogens, epinephrine, somatotropin, thyroid hormone, thyrotrophin). See also thyrotoxicosis, parathyrotoxicosis. (3) some are used as agents of biological control (*q.v.*). (4) some are used as herbicides (e.g., auxin herbicide). (5) some are used as insecticides (e.g., juvabione). (6) some inhibit or suppress the effects of specific poisons (e.g., calcitonin). (7) some have therapeutic value (e.g., ACTH, epinephrine, hydrocortisone, estrogens). **2:** phytohormone, *q.v.* Phytohormones include auxins, gibberellins, and cytokinins.

hormonosis. the condition of having excessive amounts of hormone(s). **endogenous hormonosis**. that due to increased rates of secretion or storage of hormones. **exogenous hormonosis**. that due to introduction of hormone(s) into the system from without, as in various types of hormone therapy.

horn viper. *Cerastes*.

horned adder. *Bitis caudalis*, *Bitis cornuta*.

horned desert vipers. See *Pseudocerastes*.

horned hog-nosed viper. *bothrops nasutus*.

horned palm viper. *Bothrops schlegelii*.

horned puff adder. *Bitis caudalis*, *Bitis nasicornis*.

horned rattlesnake. *Crotalus cerastes*.

horned shark. See *Squalus acanthias*.

horned tree viper. *Atheris ceratophorus*, *Trimeresurus cornutus*.

horned viper. *Bitis caudalis*, *Cerastes*.

hornet. See Vespidae, Hymenoptera.

hornsman. *Bitis cornuta*.

horny layer. See stratum corneum.

horo. *Bitis nasicornis*.

horror. intense fear, dread, terror; revulsion. **horror autotoxicus**. a term no longer in use and replaced by the term "self tolerance." It was introduced in 1900 by Ehrlich and Morgenroth to express the failure of a normal animal to form autoantibodies. It was thought that such might result in specific antigen-antibody reactions within the animal body that would lead to self-destruction.

horse bean. *Vicia faba*.

horse brush. *Tetradymia*.

horse chestnut. See *Aesculus*.

horse mussel. See *Modiolus modiolus*, *Volsella modiolus*.

horse nettle. *Solanum carolinense*.

horsebean. *Parkinsonia aculeata*.

horsebrush. See *Tetradymia*.

horsechestnut. See *Aesculus*.

horsefly weed. *Baptisia tinctoria*.

horseradish. *Armoracia lapathifolia*.

horseshoe crab. common name of arthropods of the Merostomata (especially *Limulus polyphemus*). Certain living species are poisonous. See *Carcinoscorpius rotundicauda*.

horsetail. *Equisetum*

hortensia. *Hydrangia macrophylla*.

host. **1:** the organism into which a gene from an other organism has been transplanted. **2:** a living organism that is infected by or parasitized by another organism.

host-mediated assay. host susceptibility assay.

host susceptibility assay (host-mediated assay). an assay used to assess the effect of chemicals on the immune system *in vivo*. In this assay, resistance of the test organism to transplanted tumor cells, to infectious bacteria (e.g., *Listena monocytogenes*, *Streptococcus pyogenes*), or to endotoxins followed by challenge by bacteria.

hourglass spider. *Latrodectus mactans* and *Latrodectus curacaviensis*.

house fly. any of about 30 species of flies of the genus *Musca*, but more strictly applies to *Musca domestica*, *q.v.*

house plant. houseplant.

house rat. *Rattus norvegicus*. See also laboratory rat.

household ammonia. ammonia water.

household cleaning product. a cleaner, *q.v.*, used to clean surfaces, materials, or objects in the home (e.g., drain cleaner, oven cleaner, metal cleaner, toilet bowl cleaner). Many such products are toxic (some extremely so) or otherwise hazardous. Accidental poisonings by cleaners are not uncommon among infants; all such products should be kept safely away from children. The fumes from many such products may accumulate to toxic levels in a closed room or a house where the indoor air is recycled without the introduction of some outside air. In the United States and many other countries, household cleaners designated as toxic or otherwise hazardous must bear labels with suitable warnings that specify the hazard. In the case of toxic cleaners, such warnings are usually incomplete and inexact. In general, the manufacturers are not required to list the exact ingredients on the label. Thus precautions for their safe usage are unclear. The New York Poison Control Center found, in one study, that "instructions" on what to do in case of poisoning are inadequate and sometimes in error. In the United States, labels of hazardous household cleaning products must prominently display one of these signal words (bold print): (1) DANGER or POISON, accompanied by a skull and crossbones. If ingested, a single drop or pinch of such a product can seriously poison or kill an adult person. (2) WARNING. very toxic - if ingested, a teaspoon full of such a product might kill an adult. (3) CAUTION. 2 tablespoons to 2 cups, if ingested, can prove lethal to an adult. Labels in the United States must include the following (bold print): (1) Toxic/Highly Toxic. With this label one must presume that the cleaner is poisonous on contact and by ingestion or inhalation. (2) Corrosive. Such a cleaner must be severely corrosive of skin, eyes, or mucous linings of the gastrointestinal tract. Strong Sensitizer. With continued usage allergic reactions are a significant risk. See also cleaner, label, signal word.

household poisoning. poisoning in the home by (1) household poisons, *q.v.* or (2) by usage of household chemicals (including fuels) that are in normal use in the home. A variety of chronic or intermittently experienced symptoms (e.g., headaches, depression) can be caused by such products. Some household poisonings are extremely serious or even fatal. Incomplete combustion of coal or other fossil fuels is a common source of carbon monoxide poisoning. See also poison (household poison).

household product. any of numerous manufactured products used primarily in the home. Many are poisonous and pose a special risk for young children.

houseplant. any ornamental plant that is commonly raised in ordinary rooms in the home, usually as an ornamental. Many are poisonous; some are deadly. Azaleas (*Rhododendron*), chrysanthemums (*Chrysanthemum*), creeping Charlie (*Glechoma hederacea*), Autumn crocus (*Colchicum*), hydrangeas (*Hydrangea*), lily-of-the-valley (*Convallaria majalis*), American mistletoe (*Phoradendron flavescens*), morning glory (*Ipomoea violacea*), oleander (*Nerium oleander*), and rhododendron (*Rhododendron*), if eaten, may all cause illness that requires medical attention. Other plants contain substances that are irritating to the skin, mouth, and tongue. Some plants such as the following may cause stomach upset and dyspnea: Boston Ivy (*Parthenocissus*), calla lily (*Zantedeschia aethiopica*), dumb cane (*Dieffenbachia*), philodendron (*Philodendron*), pothos (*Epipremnum*), tailflowers (*Anthurium*), trefoil (*Trifolium*), and peace lily (*Spathiphyllum*). Contact with the following plants can cause dermatoses and, if ingested, may cause nausea, vomiting, diarrhea, and abdominal cramps: amaryllis (*Amaryllis*), buttercup (*Ranunculus*), carnation (*Dianthus*), cyclamen (*Cyclamen*), daffodil (*Narcissus*), weeping fig (*Ficus benjamina*), geranium (*Geranium*), berries of holly (*Ilex*), iris (*Iris*), poinsettia (*Euphorbia pulcherrima*), firethorn berries (*Pyracantha*), tulip bulbs (*Tulipa*). See also plant (garden plant, food plant).

hr. hour, hours.

5-HT. 5-hydroxytryptamine. See serotonin.

Huanaco coca. *Erythroxylon coca*.

Hugy's viper. *Vipera aspis*.

human. **1a:** of, relating to, affecting, or characteristic of any member of the primate family Hominidae, especially *Homo sapiens*. **1b:** human being; any individual of the family Hominidae. **2:** modern man (*Homo sapiens*), or humankind. **3:** humane. **4:** used in the body of the present work as **a:** the common name of *Homo sapiens* (but see man, *Homo sapiens*). **b:** a collective term for modern men and women, although human is the preferred term in the present work. See also man, modern man, woman, **Hominidae**.

humane. See human.

humoral immunity. that form of acquired immunity conferred by the secretion of antibodies by plasma cells in response to stimulation by antigens. *Cf*. cell-mediated immunity.

humorsol. a parasympathomimetic agent that is administered topically.

hump-nosed viper. *Agkistrodon hypnale*.

humpback. kyphosis.

hunched posture. a posture in which the extremities and both ends of the body are drawn-in and the back is sharply arched.

hundred pacer. *Agkistrodon acutus*.

hundred segment snake. *Bungarus multicinctus*.

hundred-pace snake. *Agkistrodon acutus*.

Hungarian brome grass. *Bromus inermis*.

husk tomato. *Physalis*.

hyacinth. See *Hyacinthus orientalis*.

hyacinth bean. *Dolichos lablab*.

Hyacinthus orientalis (hyacinth). a flowering bulbous plant (Family Liliaceae) that is indigenous to Asia Minor, but is cultivated as an ornamental worldwide. The bulb is toxic, producing severe gastrointestinal distress. It is a source of benzyl alcohol.

Hyaenanche globosa. a tree (Family Euphorbiaceae) of southern Africa. The outer envelope of the fruit, used to kill hyenas (mammals of the family Hyaenidae), contains hyenanchin.

hyaline cartilage. cartilage with a clear, milky, glassy appearance. It is composed of very fine collagenous fibers with a shiny glass-like, milky appearance.

Hyalophora cecropia **(cecropia moth)**. the largest North American moth (Family Saturniidae), with an average wingspan of about 13.4 cm. It has gray-brown speckled wings with rusty shading. The caterpillar is green, shaded with blue on the sides. See also cecropin.

hyaloplasm. See cytoplasm.

hyalurate. hyaluronate.

hyaluronate (hyalurate). a salt or ester of hyaluronic acid.

hyaluronate glycanohydrolase. hyaluronidase.

hyaluronate lyase. hyaluronidase.

hyaluronic acid (HA). a gelatinous high molecular weight mucopolysaccharide polymer of alternating units of acetylglycosamine, $C_8H_{15}NBO_6$, and glucuronic acid. It occurs in vitreous humor of the eye, synovial fluid, pathological joints, and generally throughout the animal body. It acts as an intercellular cement, apparently by binding water in the interstitial spaces and forming a highly viscous gel. It is hydrolytically decomposed by hyaluronidase.

hyaluronidase. **1:** any of a class of enzymes (hyaluronate lyase; hyaluronate glycanohydrolase; hyaluronoglucuronidase) that catalyze cleavage of glycosidic bonds of particular mucopolysaccharides (hyaluronates). One or more occurs normally in many animal cells, including testes, leeches, spider and snake venoms, and in a number of virulent microorganisms such as type II pneumococci, hemolytic streptococci (groups A and C), *Staphylococcus aureus*, and *Claustridium welchii*. As a component of venoms, they reduce the viscosity of connective tissues and promote tissue penetration of the venom. **2:** a soluble enzyme product from mammalian testes that is used to test for the presence of hyaluronic acid or chondroitin sulfates in secretions of a tissue, to increase the effect of local anesthetics, to facilitate wider infiltration of subcutaneously administered fluids, to accelerate resorption of traumatic or postoperative edema and hematomas, and in combination with collagenase to dissociate organs (e.g., liver and heart) into viable cell suspensions. See also hyaluronoglucosaminidase.

hyaluronoglucosaminidase. a hyaluronidase that attacks the 1,4-linkages in hyaluronates. See also hyaluronic acid, hyaluronidase.

hyaluronoglucuronidase. hyaluronidase.

hyappoda. *Agkistrodon acutus*.

hybrid. an organism that results from the crossing of genetically different phenotypes, strains, populations, or species.

hybridomas. fused lymphocytes and cancer cells that produce monoclonal antibodies in large quantities.

hydracrylic acid. β-hydroxypropionic acid.

hydralazine. "can produce an SLE-like (systemic lupus erythematosis)" condition.

Hydrangea (hydrangea). a genus of deciduous shrubs (Family Saxifragaceae) native to Asia and North and South America. All parts of these plants are toxic, due to the presence of hydrocyanic acid in the tissues. Ingestion of any part can produce symptoms of cyanide poisoning. *H. arborescens* (hydrangea; wild hydrangea; seven bark). a North American shrub, growing to more than 3 m tall. Several varieties occur in various habitats in the United States including rich woods, banks of streams, and calcareous rocky slopes. It is widely cultivated as a hedge plant. This plant is normally nontoxic; hydrocyanic acid is sometimes present, however, and ingestion of any part can produce symptoms of cyanide poisoning. The dried rhizome was formerly used as a diuretic. *H. macrophylla* (hydrangea, hortensia, hills of snow, snowballs). a common horticultural shrub with large, showy, white, pink, or blue flowers. They are indigenous to Japan, but now occur in Europe and most of the Western Hemisphere. All parts, especially the flower buds, are deadly toxic. The toxic principle is a hydrangin, a cyanogenetic glycoside. Symptoms appear several hours following ingestion and include gastroenteritis and other symptoms of cyanide poisoning. The delay is due to time taken for the glycosides to undergo hydrolysis with release of the poison. *H. paniculata*. both ornamental and wild varieties occur in the United States. They contain hydrocyanic acid and are extremely toxic to livestock. *H. quercifolia*. both ornamental and wild varieties occur in the United States. They contain hydrocyanic acid and are extremely toxic to livestock.

hydrangea. common name of *Hydrangea spp*.

hydrangin. a highly toxic cyanogenetic glycoside. See *Hydrangea macrophylla*.

hydrargism. mercurialism.

hydrargyria. mercury poisoning.

hydrargyrum. the element mercury.

hydrastine. an active principle of *Hydrastis canadensis*, it is an isoquinoline alkaloid. The hydrochloride is used in the treatment of gastritis, as a uterine stimulant, to check uterine hemorrhage, and is used topically to treat catarrhal inflammation of the mucous membranes. See also berberine, canadine.

Hydrastis canadensis (orange root; goldenseal). a low, bitter, perennial herb (Family Ranunculaceae). All parts, including the raspberry-like fruit, are toxic. It has mildly antibiotic and astringent effects when used as an eyewash or a douche. It has also been used to stimulate uterine contractions. It cannot be considered safe, however, because effects of a given dosage are uncertain. It can cause severe ulceration internally or externally. Large quantities taken internally are neurotoxic and may produce hypotension, a weak pulse, convulsions, respiratory failure, or miscarriage. Other common symptoms include nausea, vomiting, diarrhea, and a burning sensation of the skin. In acute cases, paralysis and death may result. The toxic principles are berberine, canadine, and hydrastine, *q.v.*

hydrated gadolinium nitrate. See gadolinium nitrate.

hydrazine. **1:** a toxic, colorless, oily, fuming, hygropscopic liquid, $H_2N\text{-}NH_2$, with a pungent odor similar to that of ammonia. Hydrazine is a powerful reducing agent. It is used as rocket fuel, a corrosion inhibitor in boilers, and is used in the synthesis of explosives, antioxidants, and biologically active compounds. Hydrazine is very toxic to laboratory mice. It is a strong irritant; direct contact can cause severe burns. The fumes are irritating and can damage eyes, nose, throat, the pulmonary system, and the liver and kidneys. It is carcinogenic in some laboratory animals. See also chloramine. **2:** any organic derivative of hydrazine with the general formula RNHNHR' where R and R' are organic substituents such as alkyl, acyl, or phenyl groups. Many are potent hepatotoxicants, reproductive poisons, or carcinogens. See also dialkylhydrazines, 1,1-dimethylhydrazine, 1,2-dimethylhydrazine.

hydrazine yellow. tartrazine.

hydrazinolysis. the splitting of chemical bonds (notably of proteins and nucleic acids) by hydrazine.

hydrazobenzene. 1,2-diphenylhydrazine.

hydrazoethane. 1,2-diethylhydrazine.

hydrazoic acid (hydrogen azide; hydronitric acid; triazoic acid). a potent protoplasmic poison, it is a colorless liquid with an offensive odor that explodes in the presence of oxidizing agents. Exposure of humans can produce eye irritation, headache, coughing, weakness, falling blood pressure, and collapse. Its salts and azides are used as detonators. See also azido group.

hydrazomethane. 1,2-dimethylhydrazine.

Hydrelaps darwiniensis (Darwin's sea snake; Port Darwin sea snake). a small, brightly colored sea snake (Family Hydrophidae) of the Timor and Arafura seas. It is not generally regarded as dangerous.

hygric acid. stachydrine.

hydriodic acid. a colorless to pale yellow aqueous solution of hydrogen iodide. It is a strong irritant. anhydrous hydriodic acid. hydrogen iodide.

hydriotic acid, anhydrous. hydrogen iodide.

hydroblepharon. edematous enlargement of the eyelid.

hydrobromic acid. an aqueous solution of hydrogen bromide, usually 40%. It is a clear colorless liquid used in analytic chemistry, in organic synthesis, and in medicine. Its salts are bromides.

hydrocarbarism. hydrocarbonism.

hydrocarbon. any organic compound comprised

entirely of carbon and hydrogen atoms (but see derivative hydrocarbon). The carbon atoms may form a straight, or branched, chain or a closed ring. Hydrocarbons are the chief constituents of crude oil, natural gas, and refined petroleum products. A variety of hydrocarbons are major air pollutants, being released in large amounts into the atmosphere from combustion of fossil fuels used in industrial processes, transportation, and the production of electric power. These hydrocarbons devastate plant and animal populations worldwide due to both routine and large oil spills, especially in marine environments. Many are toxic and/or carcinogenic. See also herbicide. **acetylene hydrocarbon**. See alkyne. **alicyclic hydrocarbon** (alicyclic compound; cyclane; alicyclic). a cyclic aliphatic hydrocarbon consisting of three or more carbon atoms (e.g., cyclohexane) that form a single ring. They combine the properties of straight- or branched-chain aliphatic and cyclic nonaromatic hydrocarbons. See also alicyclic. **aliphatic aromatic hydrocarbon**. any aromatic hydrocarbon in which an aliphatic chain replaces one or more hydrogen atoms on an aromatic ring. **aliphatic hydrocarbon**. any straight- or branched-chain hydrocarbon (branches when present are essentially straight). Included are noncyclic alkanes (paraffins), alkenes (olefins), alkadienes, acetylenes, and acyclic terpenes. The C-1 to C-4 members of this group (e.g., methane and ethane) are simple asphyxiants. The C-5 to C-8 hydrocarbons (e.g., pentane, hexane, heptane, octane) are CNS depressants and *n*-hexane is an especially hazardous neurotoxicant. **aromatic chlorinated hydrocarbon**. a class of aromatic hydrocarbons in which one or two hydrogen atoms have been replaced by a chlorine atom. **aromatic hydrocarbon** (arene; aromatic compound; aromatic). any of a class of hydrocarbons (e.g., anthracene, benzene, naphthalene) that includes one or more aromatic (benzene) rings, i.e., one or more planar rings of atoms with delocalized clouds of π (pi) electrons above and below the ring. This distribution of π electrons over several carbon atoms provides a resonance stabilization of the molecule. Aromatic hydrocarbons have an agreeable odor and are further characterized by a low hydrogen to carbon atomic ratio; C-C bonds that are quite strong and intermediate in length between those in alkanes and those in alkenes; their tendency to undergo nucleophilic reactions in which hydrogen or other groups on the ring are substituted (addition reactions are uncommon). Benzene is the simplest aromatic hydrocarbon. It forms an equilateral planar six-membered ring. Other aromatic hydrocarbons derive from benzene by substitution of one or more hydrogen atoms by CH_3 or longer carbon chains. Aromatic hydrocarbons are extremely important in the chemical industries, in toxicology, and in all fields of organic and toxicological chemistry. They are the most toxic hydrocarbons found in petroleum, and occur in all crude oils and in many petroleum products. Many are water-soluble, increasing the risk of poisoning to aquatic organisms and to organisms higher in the food chain. Some are environmentally persistent and therefore especially hazardous. **atmospheric hydrocarbon**. a popular (not a chemical) term that encompasses all hydrocarbon compounds that are common air pollutants. They may enter the atmosphere as the result of evaporation, incomplete combustion of fossil fuels, or resuspension by physical means following deposition from the atmosphere. **C-1 to C-4 aliphatic hydrocarbons**. alkanes with 1 through 4 carbon atoms (e.g., methane, ethane). These compounds are simple asphyxiants. **C-5 to C-8 aliphatic hydrocarbons**. alkanes with 5 through 8 carbon atoms (e.g., pentane, hexane, heptane, octane) consist of *n*-alkanes. The branched-chain hydrocarbons of this group have many isomers (e.g., there are 9 isomers of heptane C_7H_{16}). These compounds are all volatile liquids under ambient conditions. They are used in fuels, as solvents in formulations of a number of commercial products (e.g., glues, inks, varnishes), and in the extraction of fats. They are CNS depressants and *n*-hexane is an especially hazardous neurotoxicant. Exposure to most is chiefly via inhalation. High levels have proven fatal to laboratory animals. Included are a number of environmentally persistent, broad-spectrum insecticides (e.g., DDT, aldrin, dieldrin, heptachlor, chlordane, lindane, endrin, mirex, hexachloride, and toxaphene). **chlorinated hydrocarbon** (organochlorine compound). any halogenated hydrocarbon compound in which one or more hydrogen atoms have been replaced by chlorine atoms. They are broad-spectrum poisons and are toxic to nearly all animals including many vertebrates. Many are carcinogenic, teratogenic, and neurotoxic. About 100 are or have been used as insecticides that degrade only slowly in the environment and accumulate at higher trophic levels. They are lipiphilic and accumulate in depot fat (stored fat, as that in adipose tissue). Among the most widely known chlorinated hydrocarbons are aldrin, benzenehexachloride, chlordane, chlordecone, DDT, dieldrin, endrin, heptachlor, hexachloride, PCBs, and toxaphene. Most have been banned or restricted. **cyclic hydrocarbon**. any hydrocarbon with a closed ring. See also alkane. Included are alicyclic hydrocarbons, aromatic hydrocarbons, and cyclic terpenes. **derivative hydrocarbon** (substituted hydrocarbon). any of numerous organic compounds in which one or more hydrogen atoms have been replaced by another element. **fluorinated hydrocarbon**. fluorocarbon. **halogenated hydrocarbon**. a class of derivative hydrocarbons in which one or more hydrogen atoms have been replaced by a halogen atom. Many of these compounds are highly toxic and some are carcinogenic (e.g., vinyl chloride, dichloroethane). Many are used as industrial solvents, pesticides, chemical intermediates, and anesthetics. Examples of the many highly toxic halogenated hydrocarbons are carbon tetrachloride, chloroform, polychlorinated biphenyls, and TCDD. **monocyclic aromatic hydrocarbon**. any aromatic hydrocarbon with a single 6-membered ring. All except benzene have substituent groups. **polychlorinated aromatic hydrocarbon**. any derivative aromatic hydrocarbon that contains more than one or two substituted chlorine atoms. These compounds cause a hyperplastic and hypertrophic lesion of the gastric mucosa in the rhesus monkey and in cattle, but apparently not in the laboratory mouse, rat, and guinea pig. They produce edema and ascites in the domestic fowl, testicular damage in several species, and pancytopenia, especially in monkeys. **polycyclic aromatic hydrocarbon** (PAH; polynuclear aromatic hydrocarbon). an aromatic hydrocarbon having two or more rings. They are formed by the incomplete combustion of other hydrocarbons. Of these, benzo(a)pyrene is the most studied. PAH compounds occur in engine exhausts, wood stove smoke, cigarette smoke, and charbroiled food and high concentrations of PAH occur in coal tars and petroleum residues. Some metabolites of PAH are carcinogenic (e.g., 7,8-diol-9,10 epoxide of benzo(a)pyrene). **polyhalogenated hydrocarbon**. a substituted hydrocarbon in which more than 2 hydrogens are replaced by halogen atoms. Such are highly toxic, immunotoxic, and are ubiquitous in the environment with a high potential for bioaccumulation and biomagnification. See, for example, polybrominated biphenyl, polychlorinated biphenyl, polychlorinated dibenzofuran, and polychlorinated dibenzo-*p*-dioxin (PCDD). **polynuclear aromatic hydrocarbon**. polycyclic aromatic hydrocarbon. **substituted hydrocarbon**. derivative hydrocarbon. **unsaturated hydrocarbon**. any hydrocarbon that has multiple bonds, each involving more than 2 shared electrons between carbon atoms.

hydrocarbon oxide. a group of toxic compounds characterized by the presence of an epoxide functional group (e.g., ethylene oxide; propylene oxide; 1,2,5,6-diepoxyhexane; and 7,8-dihydrodiol-9,10-epoxide.

hydrocarbon poisoning. hydrocarbonism.

hydrocarbonism (hydrocarbon poisoning; hydrocarbarism). poisoning by any hydrocarbon compound, but especially that due to ingestion of petroleum distillates such as gasoline, kerosene, and paint thinners. Symptoms and signs relate chiefly to the respiratory system, GI tract, and CNS. Initially, the victim coughs and chokes, even with a small taste. Cyanosis, breath-holding, and persistent coughing may follow. Older children may complain of a burning sensation in the stomach, and vomit spontaneously. CNS symptoms include lethargy, coma, and convulsions. These are usually dose related and are most severe with lighter fluid and mineral seal oil ingestions. See hydrocarbon.

hydrochinone. hydroquinone.

hydrochloric acid (muriatic acid). a colorless, fuming, corrosive, and highly toxic aqueous solution of hydrogen chloride, HCl. Concentrated hydrochloric acid (a saturated solution of HCl) contains 36% HCl; it forms a constant-boiling mixture at a concentration of 20% HCl. It has numerous industrial applications and is used widely as a reagent and in organic syntheses. HCl is much less toxic than HFl. Inhalation can cause spasms of the larynx, pulmonary edema, and even death. Hydrogen chloride gas, which constitutes the fumes of hydrochloric acid, has a high affinity for water and thus desiccates moist tissues (e.g., eyes and mucous tissue of the upper respiratory tract) that it contacts. If swallowed, HCl causes immediate, intense burning pain, soon followed by vomiting; dark, bloody diarrhea; and a steep drop in blood pressure. Brownish or yellowish stains form in or around the mouth. The throat can swell, suffocating the victim. Peritonitis may occur later if the acid has ruptured the stomach; this is announced by fever and a rigid abdomen. At this stage, a fatal outcome is nearly certain in the absence of immediate medical attention. HCl is much less toxic than hydrofluoric acid. See also hydrogen chloride.

hydrochloric acid, anhydrous. hydrogen chloride.

hydrocollidine. a poisonous oily ptomaine, $C_8H_{13}N$, in nicotine, decaying meat, and stale fish.

hydrocortisone (11,17,21-trihydroxypregn-4-ene-3,20-dione; cortisol; 4-pregnene-11β,-17α,21-triol-3,20-dione; 17-hydroxycorticosterone; anti-inflammatory hormone). First isolated in 1937 from extracts of bovine tissue, it is the principal glucocorticoid of humans and many other mammals. It has weak glucocorticoid and mineralocorticoid activity. It is often used as an antiinflammatory. Systemic administration may produce serious adverse effects that may disturb endocrine and neurologic activity as well as fluid and electrolyte balance. See corticosteroid.

hydrocortisone acetate (COLA; cortisol acetate). the monoacetate of hydrocortisone.

hydrocyanic acid. hydrogen cyanide.

hydrocyanic ether. propionitrile.

hydrocyanic acid poisoning. hydrocyanism.

hydrocyanism (hydrocyanic acid poisoning). poisoning by hydrogen cyanide.

hydrocyanite. the naturally occurring mineral form of hydrous cupric sulfate.

hydrofluoric acid (fluohydric acid; fluoric acid). a nearly colorless, highly poisonous, corrosive, and weakly acidic fuming aqueous solution of hydrogen fluoride (30-60% HF by mass). It dissolves glass, silica, and silicic acid. It is used to clean metals and to etch glass. Hydrogen fluoride and hydrofluoric acid are extreme irritants to any part of the body that they contact. Inhalation of fumes and contact with skin is damaging to the exposed tissues, often with a delay of about 1 day; healing is slow and gangrene may develop. Contact with eyes or eyelids can cause serious persistent or permanent tissue damage, visual impairment, and even destruction of the eyes. Symptoms of acute intoxication from ingestion can include nausea, vomiting, diarrhea, cardiovascular collapse, and death. Necrotic lesions of the upper portions of the digestive tract typically result. Inhalation can result in pulmonary inflammation and congestion with ulceration of the upper respiratory tract. Symptoms of chronic exposure via inhalation or ingestion are those of fluorosis, *q.v.* Because it dissolves glass, hydrofluoric acid should be stored in lead, paraffin paper, plastic, or wax containers. Salts of hydrofluoric acid are used as insecticides. See also hydrogen fluoride, gas (fluohydric acid gas).

hydrofluoric acid gas. hydrogen fluoride.

hydrofluorosilicic acid. fluorosilic acid.

hydrofluosilicic acid. fluosilicic acid.

hydrogen (proteum; symbol, H)[Z = 1; A_r = 1.00794]. the most abundant element (by mass) in the universe and the ninth most abundant on Earth. Hydrogen is the simplest of all atoms, consisting of one proton and one extranuclear electron. It is a colorless gaseous element assigned to no periodic group, though it has properties similar to both the alkali metals and the halogens. The pure gas is not toxic, but can act as a simple asphyxiant at relatively high concentrations.

hydrogen arsenide. arsine.

hydrogen azide. hydrazoic acid.

hydrogen bond. a weak chemical bond between a hydrogen atom carrying a partial positive charge and an atom of another molecule carrying a partial negative charge.

hydrogen bromide (HBr; hydrobromic acid). a very dense, pale yellow or colorless gas with an irritating odor that fumes in moist air. It is irritant to the skin, eyes, and mucous membranes. It forms hydrobromic acid in aqueous solution.

hydrogen chloride (anhydrous hydrochloric acid). a colorless, pungent, corrosive, highly toxic, nonflammable gas, HCl. It is soluble in water, ethanol, and ethyl ether. It has numerous industrial and scientific applications, mainly in aqueous solution. One report gives the LC_{50} (30 min) for laboratory mice and rats, respectively, as 2142 and 5666 ppm. See also hydrochloric acid.

hydrogen cyanamide. cyanamide.

hydrogen cyanate. cyanic acid.

hydrogen cyanide (HCN; hydrocyanic acid; cyanihydric acid; blausäure (German); prussic acid; formonitrile). an extremely toxic, colorless, water-soluble gas or liquid, HCN, that oxidizes readily in air; the vapor is highly flammable. It is used in the manufacture of acrylonitrile, cyanide salts, and dyes. The compressed gas is used by professionals as a fumigant to kill insects and rodents in enclosed spaces (e.g., grain storage bins, warehouses, holds of ships). HCN was used as a war gas in World War I. It is produced in gas works, coke ovens, blast furnaces, and in the laboratory by treating a cyanide salt with acid. It is also produced naturally from cyanohydrins or amino-acid derivatives by more than 800 species of trees in more than 70 families; it occurs at relatively high concentrations in a variety of plants eaten by humans and domestic animals (e.g., pits of wild black cherry, peaches, and apricots, in apple seeds, some varieties of almonds, and lima beans from the tropics). HCN is an extremely toxic cellular asphyxiant that inhibits cytochrome oxidase, with consequent suppression of cellular respiration. Toxic doses also lower blood pressure, but the oxygen-carrying capacity of circulating blood is unimpaired. Inhalation of fumes causes tachypnea (further increasing the rate of exposure to HCN) progressing to dyspnea. Symptoms of acute hydrocyanism (hydrogen cyanide poisoning) may additionally include giddiness, headache, paralysis, unconsciousness, and convulsions, with a high likelihood of respiratory arrest and death. Death of mammals from a lethal dose ensues within a few minutes to several hours. Headache, irritation of the throat, watering of the eyes, dyspnea, vertigo, nausea, and vomiting may occur in milder cases. Exposure of humans to 150 ppm for one half hour is life threatening; exposure to 300 ppm is usually lethal to mammals within a few minutes to several hours. See, for example, cyanide, cyanogenesis, amygdalin, *Hydrangea*, glycoside (cyanogenetic glycoside), *Apheloria corrugata*, *Pseudopolydesmus serratus*, *Zygaena*.

hydrogen fluoride (hydrofluoric acid gas; fluohydric acid gas; anhydrous hydrofluoric acid). a colorless, extremely dangerous, poisonous, highly irritant gas, HF. It forms corrosive fumes when exposed to the atmosphere and forms hydrofluoric acid in aqueous solution. It is very soluble in alcohol and many organic solvents. Hydrogen fluoride is used industrially as a catalyst in paraffin alkylation, fluorination, the synthesis of fluorides, and the separation of uranium isotopes. It should be used only with safe, specially built apparatus, including gloves and face shields. Accidental exposure should be treated as a medical emergency. Effects and symptoms are essentially the same as those of hydrofluoric acid, *q.v.* See also ammonium fluoride, hydrofluoric acid. **anhydrous hydrogen fluoride** (anhydrous hydrofluoric acid). a toxic, corrosive, extremely acidic substance that forms a weak acid in aqueous solution.

hydrogen halide. See halide.

hydrogen hexafluorosilicicate. fluorosilic acid.

hydrogen iodide (HI; anhydrous hydriotic acid). a powerfully irritating, very dense, colorless to pale yellow, acrid, nonflammable gas. It is a powerful irritant to the skin, eyes, and mucous membranes.

hydrogen ion (H^+). a hydrogen atom with unit positive charge due to the loss of its electron. It reacts with a water molecule to form a hydronium ion, H_30+. See acid.

hydrogen peroxide. The pure (anhydrous) compound, (H_2O_2), at room temperature is a colorless liquid that explodes violently on contact with organic matter. This compound, even when stabilized by small amounts of inhibitors such as acetanilide or sodium stannate, is a dangerous fire and explosion hazard. Concentrated solutions of hydrogen peroxide are highly irritant, highly toxic.

hydrogen phosphide. phosphine.

hydrogen selenide (selenium hydride). a colorless, water-soluble, toxic, flammable, and explosive gas, H_2Se, soluble also in carbon disulfide and phosgene. It is used in the manufacture of metallic selenides, the preparation of semiconductor materials, and the synthesis of organoselenium compounds. It is irritant to the skin, eyes, and mucous membranes. It is damaging to the lungs and liver; inhalation can destroy the sense of smell and cause an obstinate coryza.

hydrogen sulfate. sulfuric acid.

hydrogen sulfide (sulfureted hydrogen; hydrosulfuric acid; sulfhydric acid). a colorless, flammable, very poisonous gas, H_2S, with the odor of rotten eggs. Hydrogen sulfide is a natural constituent of the atmosphere, although in concentrations low enough (except near volcanoes) to be harmless. Natural sources include sulfur springs, fumarols, coal pits, gas wells, and sulfur-containing decaying matter. H_2S is sometimes produced by algal blooms in quantities that are sufficient to kill fish and other aquatic or semiaquatic biota. Accidental escapes of the gas from various industrial processes have caused serious poisoning; sometimes death. A cytochrome oxidase inhibitor, H_2S is an extremely hazardous substance. Toxic concentrations quickly fatigue the olfactory senses and thus may not be detected. Acute poisoning quickly brings about collapse, unconsciousness, and death from respiratory failure. Just one or a few inhalations can prove fatal. Victims must be immediately removed from the toxic atmosphere. Subacute challenge is debilitating, with lingering headache, dizziness, nausea, and weakness. Low concentrations irritate the conjunctiva of the eye, respiratory tract, and skin. Chronic exposure may produce pulmonary edema. Higher concentrations may cause sudden collapse, unconsciousness, and death from respiratory paralysis.

hydrogen telluride. a colorless gas, H_2Te, that is soluble (but unstable) in water. It is also soluble in ethanol and alkalies. It is highly toxic by inhalation and a strong irritant to the skin, eyes, and mucous membranes. It is a dangerous fire and explosion hazard.

hydrogen thiocyanate. thiocyanic acid.

hydrogenase. an enzyme that catalyzes the reversible dissociation of molecular hydrogen into hydrogen ions and electrons.

hydrogenation. any reaction of hydrogen with an organic compound. Such reactions are usually accomplished by the action of heat and pressure in the presence of a catalyst or, metabolically, via enzyme catalysis (See hydrogenase). Two types of reactions occur: addition hydrogenation and hydrogenolysis. **addition hydrogenation**. the direct addition of hydrogen to the double bond of an unsaturated compound. See hydrogenolysis.

hydrogenolysis. hydrogenation of an organic compound or by the rupture of bonds within the organic reactant and subsequent reaction of hydrogen with the resulting fragments. See also hydrogenase.

hydrogeology. the study of the geology of groundwater systems, with particular emphasis on the chemistry and movement of water. This science is important to the assessment of the persistence and fate of toxic substances present at a given location, or in a given aquifer. See hydrology.

hydroid. **1:** the polyp form of a coelenterate, in contrast to the medusa or jellyfish form. They may be solitary or colonial, and are usually sessile. **2:** any species belonging to the coelenterate Class Hydrozoa.

Hydrolagus colliei (spotted ratfish). a venomous, marine, cartilaginous fish (Family Chimaeridae) that grows to nearly a meter in length. Situated immediately anterior to the first dorsal fin is a long, venomous spine. The wound inflicted by the spine is very painful. The range of *H. colliei* extends from southeastern Alaska to Bahia San Sebastian Vizcaino, Baja California, with disjunct populations in the upper Gulf of California. It is the only North American representative of the family.

hydrolase (hydrolytic enzyme; hydrolyzing enzyme). any of a large number of enzymes that catalyze hydrolytic reactions. They comprise one of the six classes of enzymes. Hydrolases play a very important role in the catabolism of xenobiotic compounds in the liver of vertebrate animals and also in the intestinal mucosa, blood plasma, nervous tissue, kidney, and muscle tissue. See also hydrolysis, enzyme, esterase, amidase.

hydrology. the science that deals with the properties, distribution, and circulation of water in the environment. This science is important to the assessment of the transport, persistence, and fate (and hence the potential hazard) of toxic substances that are present in water bodies at a given location or in a given region. See hydrogeology.

hydrolysis. **1:** any reaction of water with a salt to yield an acid and a base, one of which is more dissociated than the other. **2:** a decomposition reaction in which a compound reacts with water such that its molecules are cleaved into two new species of lower molecular mass. The hydroxyl group is incorporated into one fragment, and the hydrogen atom into another fragment. Hydrolysis is an extremely common type of reaction in living systems. It is important, for example, in the digestion of many foods and in intermediary metabolism. Many hydrolytic reactions are reversible; the reverse reaction is called neutralization, esterification, or condensation. metabolic hydrolysis. the metabolic (enzymatic) hydrolysis of organic compounds. Many xenobiotics, most notably organophosphate esters and amides.

hydrolytic. pertaining to hydrolysis.

hydrolyzing enzyme. hydrolase.

hydromorphone. an opioid used with caution as a narcotic analgesic in the treatment of moderate to severe pain. The most serious adverse reactions in nonhypersensitive individuals include drowsiness, dizziness, nausea, constipation, cardiopulmonary depression, and addiction. See also drug abuse.

hydronitric acid. hydrazoic acid.

hydroperitonia. ascites.

hydroperoxide. any of a class of potent irritants and lacrimators with the general formula, ROOH, that occur in photochemical smog.

hydroperoxyl radical. superoxide.

hydrophid (hydrophid). 1: any snake of the family Hydrophidae. 2: of or pertaining to a member of the snake family Hydrophidae.

hydrophid venom poisoning. sea snake venom poisoning.

Hydrophidae (Hydrophiidae; true sea snakes; sea snakes). a family of some 50 species of venomous, mostly tropical, mostly marine snakes. They were formerly treated as a subfamily (Hydrophinae) of the Colubridae. Most species occur in south Asian and Australian coastal marine waters, usually within 20 miles of shore. Hydrophids occur, however, near most warm coasts except those of the Atlantic Ocean. Many species sometimes enter brackish or fresh water, but only two species are confined to such waters: *Hydrophis semperi*, is a freshwater species and *Laticauda crockeri* is confined to brackish waters. Only a few species occur far into the southwestern Pacific Ocean (as far as the Society and Gilbert islands). Most species attain a length of 3 m or more. All have strongly flattened, oarlike tails that are used in sculling. The body is laterally compressed and the nostrils of most species open onto the top of the head. All have large crown shields and the ventral scutes are small. The venom of all species is extremely toxic and they should be considered extremely dangerous. Nevertheless, they are usually docile. The fangs are small and of the cobra type. The most dangerous species to humans are *Enhydrina schistosa*, *Hydrophis caerulescens*, *Hydrophis nigrocinctus*, *Lapemis hardwicki* (Hardwick's sea snake), and *Pelamis platurus* (yellow-bellied sea snake). Sea snakes prey upon eels and small bottom fish and often remain submerged for hours.

hydrophilia. **1:** the state or condition of being hydrophilic (*q.v.*). **2:** hydrophilism, the tendency of a tissue to attract and hold water.

hydrophilic. water-loving; having a strong tendency to bind, absorb, or dissolve in water. This term refers, for example, to chemicals that are water soluble or to the regions of a molecule that are polar and therefore attracted to water. hydrophilic substances include many carbohydrates and complex proteins such as gelatin and collagen. Many toxicants are not hydrophilic in their native form, but become more hydrophilic through metabolism and are thus more readily excreted. Hydrophilic compounds do not diffuse easily through membranes. See also hydrophobic, hydrophobia.

hydrophilism. hydrophilia (def. 2).

hydrophilous. taking up water.

Hydrophis. a genus of some 25 widely distributed species of sea snakes (Hydrophidae), most of which may be identified only by an expert familiar with the genus. Most species are considered dangerous to humans and the bite of some is lethal (See Hydrophidae). As with other sea snakes, the body and tail are compressed body and the nostrils are located on the top of the head in nasal shields that abut one another. The eyes are small with round pupils; the loreal shield is lacking. The ventral scales are of similar size throughout the length of the animal. The head of some species is remarkably tiny and neck of the adult is long and slender. The young differ little from the young of other hydrophids. Species of *Hydrophis* in addition to those defined more more fully below, include: *H. belcheri*. a small-headed species of Australian and Pacific seas. *H. bituberculatus* (Peters sea snake). a species that occurs only along the coasts of Ceylon. *H. brookei* (Brook's small-headed sea snake). a small-headed species that occurs in Indo-Malaysia, Java, and the Flores Sea. *H. caerulescens* (bluish small-headed sea snake; Merrem's sea snake). one of the most dangerous snakes to humans, it has been observed from the Arabian Sea to the South China Sea, Java, and the Molucca Sea. *H. elegans* (elegant sea snake). a species that has been recorded from the Timor Sea, Arafura Sea, Gulf of Carpentaria, and the Coral and Tasman seas. *H. kingii* (King's sea snake). a species known from the Timor Sea, Arafura Sea, Gulf of Carpentaria. *H. klossi*. a small-headed species of Indo-Malaysian waters. *H. lapemoides*. a species restricted to the Persian Gulf and the Arabian Sea. *H. Major*. a species known only from the Timor Sea, Arafura Sea, and the Gulf of Carpentaria. *H. mamillaris* (broad-banded sea snake; thatta pam). a species indigenous to Indo-Maylasia, with possible sightings in the Gulf of Siam and the Andaman Sea. *H. melanosoma*. this species may occur only in the South China Sea, Java, and the Flores Sea. *H. mertoni*. a species known only from the Arafura Sea. *H. nigricinctus* (kerril patte). a species known only from the Bay of Bengal, it is one of the most dangerous hydrophids to humans. *H. obscurus* (Russell's sea snake; kale shootsursun; shootur sun). a species confined to brackish waters in the Bay of Bengal. *H. parviceps* (Smith's small-headed sea snake). a species known only from the coast of South Vietnam in South China Sea. *H. semperi* (Semper's sea snake). a species found only in Lake Taal on Luzon (Phillipines). *H. stricticollis*. a species known only from the Bay of Bengal. *H. torquatus* (gray sea snake; Günther's sea snake). a species indiginous to Indo-Malaysia, the South China Sea, and Java, and the Flores Sea. ***H. cyanocinctus*** (annulated sea snake; blue-banded sea snake; common sea snake; chittul). a sea snake (Family Hydrophidae) found as far east as the Persian Gulf and the Arabian Sea and as far west and north as the East China, Timor, and Sulu seas. The head is smaller, the neck is longer and narrower, and the body is more compressed than in *H. spiralis*. The head scales are similar to those of *H. spiralis* except that there are usually 2 anterior temporals. Body scales with central keel or row of tubercles. The ground color is dirty white, pale greenish, yellow or olive with blackish crossbands that may or may not encircle the body, are widest along the vertebral midline and are as wide as, or wider than the interspaces between them. The head of the adult is olive, reddish, or dull yellow; in young blackish with the yellow horseshoe mark seen in some other species. The adult length averages 1.4 to 1.7 m with record specimens somewhat longer than 2 m. It often bites if restrained; the venom is more toxic than that of *H. spiralis*. It probably causes more human deaths than any other sea snake except *Enhydrina schistosa*. ***H. fasciatus*** (banded small-headed sea snake; milagy kadiyan). a thick-bodied sea snake with a ground color predominantly of gray to dirty yellow with dark crossbands that taper laterally. The head is very small and uniformly dark; the neck is dark olive to black with yellow spots or crossbars and is long and slender. Adults average nearly 1 m in length, ranging to about 1.3 m. This snake preys upon eels and other long-bodied fish. The venom, although extremely toxic, is produced in very small quantities; less than 1 mg can be expressed from the glands of a single snake; the lethal dose for humans is much less than this amount, however. ***H. ornatus*** (Gray's sea snake; malabasahan reef sea snake; spotted sea snake; calabucab). a widely distributed species from the Central and Southwest Pacific to the Gulf of Siam, the Bay of Bengal. It is rare or accidental in the Arabian Sea and Persian Gulf. *H. ornatus* is stout-bodied with a large head. In most parts of its range, the ground color is typically pale greenish white, olive, or yellow with an olive-colored head and marked by wide, dark crossbands or rhomboid spots along the length of the body. Those in Philippine waters are uniformly grayish green above and whitish below; a race that inhabits Australian waters is spotted or bears lateral ocellate markings. The body scales are small, bearing a central tuburcle that is larger in the male. Adult specimens average about 0.8 m in length, but may occasionally reach or exceed a length of 1.1 m. ***H. spiralis*** (yellow sea snake; narrow-banded sea snake; kadel nagam; kadel pambu). a species that occurs in the Persian Gulf, Arabian Gulf, Indo-Maylasia, the South China Sea, Java, the Flores Sea, and occasionally elsewhere within this range. The ground color is golden yellow to yellowish-green that shades to pinkish white below; narrow black rings encircle the body, being widest along the dorsal midline and narrowest on the flanks. The head is uniformly yellow in adults, and dark with a yellow horseshoe shaped mark on the crown in the young. The head is of moderate size, but is distinct from the neck; head shields are large and symmetrical; a slight prolongation of the tip of the rostral fits into a notch in the tip of the lower jaw. The body is moderately slender and not greatly compressed. It is the longest of the sea snakes, reaching an average length of nearly 2 m; rare individuals may reach a nearly 3 m in length. This snake frequents deep water, often basking at the surface. The venom is less toxic than that of most sea snakes and yields on milking are only about 3-10 mg. Human fatalities are known.

hydrophobia. **1a:** inability to drink. **1b:** a morbid fear of water. **2:** hydrophobia (def. 1) is a characteristic of rabies, which is thus commonly called hydrophobia. See also hydrophilia.

hydrophobic. water-fearing; antagonistic to water; unable to dissolve in water. See also hydrophilic, hydrophobia. Fats, oils, waxes, many resins, and certain finely divided powders (e.g., carbon black, magnesium carbonate, talc) are hydrophobic.

hydropic. of or pertaining to dropsy.

hydrops. dropsy.

hydrops abdominis. ascites.

hydroquinol. hydroquinone.

hydroquinone (1,4-benzenediol; *p*-dihydroxybenzene; hydroquinol; quinol; hydrochinone). a white, moderately toxic, crystalline substance, soluble in water, ethanol, and ethyl ether. It is the toxic principle of *Xanthium spp.* (cockleburs) and is not known to occur in any other plants. It is used as a photographic developer for black-and-white films, a dye intermediate, a stabilizer in paints and varnishes, in oils and motor fuels, as an antioxidant for fats and oils, and an inhibitor of polymerization. While safe for most purposes, quantities of ca. 2-5 g may be lethal to humans. Effects following ingestion of about 1 gram or so may include tinnitis, nausea, vomiting, feelings of suffocation, dyspnea, cyanosis, delirium, convulsions, and collapse. Hydroquinone is an irritant and may cause dermatitis and opacification of the cornea in workers exposed to concentrations below that which causes systemic effects.

hydroquinone-β-D-glucopyranoside. arbutin.

hydrorrhea. a profuse watery discharge.

hydrosilicofluoric acid. fluorosilic acid.

hydrosulfuric acid. an aqueous solution of hydrogen sulfide, *q.v.*

hydrourushiol. 3-pentadecylcatechol.

hydrous copper sulfate. See cupric sulfate.

hydroxide. any compound that contains the hydroxyl ion, OH^-, or the hydroxyl group (hydroxyl radical), —OH. Hydroxides of metals are bases; those of nonmetals are usually acids. A few of the many toxicologically significant hydroxides are ammonium, barium, potassium sodium, tetramethylammonium, and thallium hydroxides; hydroxylamine; neurine; soda lime. See also alcohol, cleaner, ellipticine, zinc chromate. Toxicologic significance may reside in moieties other than the hydroxyl group (e.g., acetylcholine, ferric hydroxide, hypaphorine, lead hydroxide, lithium hydroxide).

hydroximino. nitroso.

***N*-hydroxyacetamidofluorene**. a carcinogenic, metabolically active form of 2-acetamidofluorene, *q.v.*

4'-hydroxyacetanilide. acetaminophen.

***p*-hydroxyacetanilide**. acetaminophen.

4-hydroxyaflatoxin B_1. aflatoxin M_1.

4-hydroxyaflatoxin B_2. aflatoxin M_2.

5-hydroxy-3-(β-aminoethyl)indole. serotonin.

3-hydroxy-5-aminomethylisoxazole. muscimol.

hydroxyamphetamine hydrobromide (4-(2-aminopropyl)phenol hydrobromide; α-methyltyramine hydrobromide; *p*-(2-aminopropyl)phenol hydrobromide). a white, crystalline powder, $C_9H_{13}NO{\cdot}HBr$. An adrenergic, it is used as a mydriatic, a nasal decongestant and is given orally in the management of heart block, carotid syndrome, and postural hypertension.

7β-hydroxy-5α-androstan-3-one. dihydrotestosterone.

17-β-hydroxy-androst-4-en-3-one. testosterone.

4-hydroxy-α-angelica lactone. α-angelica lactone. See angelica lactone.

hydroxybenzene. phenol.

α-hydroxybenzeneacetonitrile. mandelonitrile.

6-hydroxy-5-benzofuranacrylic acid δ-lactone. psoralen.

***p*-hydroxy benzoic acid**. a plant growth inhibitor, $C_6H_4(OH)COOH{\cdot}H_2O$, that is soluble in ethanol and ether and partially water soluble. It is a food preservative and an intermediate in the synthesis of fungicides; its butyl, methyl, and propyl esters are preservatives in cosmetics and pharmaceuticals.

12′-hydroxy-2′,5′α-bis(1-methylethyl)-8α-ergotaman-3′,6′,18-trione. ergocorninine.

2-hydroxybutane. 2-butanol.

hydroxybutanoic acid. hydroxybutyric acid.

3-hydroxybutanoic acid β-lactone. β-butyrolactone.

hydroxybutyric acid (hydroxybutanoic acid; oxybutyric acid). a toxic, viscid, yellow mass, $CH_3CHOHCH_2COOH$, that sometimes occurs in the blood and urine of diabetics and in starvation. The most common of several isomers is β-hydroxybutyric acid (3-hydroxybutanoic acid).

hydroxybutyric acid lactone. See β-butyrolactone.

3-hydroxybutyric acid lactone. β-butyrolactone.

***m*-hydroxycarbanilic acid methyl ester *m*-methylcarbanilate**. phenmedipham.

hydroxycholine. muscarine.

***p*-hydroxy cinnamic acid**. a phenolic compound secreted by *Adenostoma fasculatum* and other plants of the hard chaparral community (e.g., *Arctostaphylos*). It leaches in the ground apparently inhibiting the growth of nearby grasses.

***p*-hydroxycinnamic acid**. See coumaric acid.

17-hydroxycorticosterone. hydrocortisone.

1-(4′-hydroxy-3′-coumarinyl)-1-phenyl-3-butanone. warfarin.

3-hydroxycrotonic acid methyl ester dimethyl phosphate. mevinphos.

N^5-(1-hydroxycyclopropyl)-*L*-glutamine. coprine.

14-hydroxydaunomycin. doxorubicin.

17-hydroxy-11-dehydrocorticosterone. cortisone.

3α-hydroxy-4β,15-diacetoxy-8α-(3-methylbutyryloxy)-12,13-epoxy-Δ^9-tricothecene. T-2 toxin.

7-hydroxydiacetoxyscirpenol. a trichothecene first isolated from *Fusarium*.

8-hydroxydiacetoxyscirpenol (neosolaniol). a trichothecene first isolated from *Fusarium*.

hydroxydimethylarsine oxide. cacodylic acid.

3-hydroxy-*N*,*N*-dimethyl-*N*-ethylanilinium chloride. edrophonium chloride.

4-hydroxy-*N*,*N*-dimethyltryptamine. psilocin.

5-hydroxy-*N*,*N*-dimethyltryptamine. bufotenine.

3-hydroxy-4,5-dimethylol-α-picoline hydrochloride. pyridoxine hydrochloride.

20-hydroxyecdysone. β-ecdysone (See ecdysone).

9-hydroxyellipticine. a cytotoxic, carcinogenic substance. See ellipticine.

3-hydroxyestra-1,3,5(10)-trien-17one. estrone.

(2-hydroxyethyl)diisopropylmethylammonium bromide xanthene-9-carboxylate. propantheline bromide.

(β-hydroxyethyl)trimethylammonium. choline.

hydroxyethyl trimethyl ammonium hydroxide. choline.

[illegible] patulin.

2-[(hydroxy-imino)methyl]-1-methylpyridinium. pralidoxime.

2-[(hydroxyimino)methyl]-1-methylpyridinium chloride. pralidoxime chloride.

2-hydroxy-3-isobutyl-6(methylpropyl)pyrazine-l-oxide. aspergillic acid.

α-hydroxyisobutyronitrile. acetone cyanohydrin.

***p*-hydroxymandelonitrile-β-D-glucoside**. dhurrin.

***o*-[[3-(hydroxymercuri)-2-methoxypropyl]-carbamoyl]phenoxyacetic acid sodium salt**. Mersalyl.

***N*-(γ-hydroxymercuri-β-methoxypropyl)salicylamide-*O*-acetic acid sodium salt**. Mersalyl.

4-(hydroxymercuri)-2-nitrophenol sodium salt. mercurophen.

hydroxymethanesulfinic acid sodium salt. sodium formaldehydesulfoxylate.

6-hydroxy-7-methoxy-5-benzofuranacrylic acid δ-lactone. methoxsalen.

7-hydroxy-6-methoxy-2*H*-1-benzopyran-2-one. scopoletin.

4-hydroxy-3-methoxycinnamic acid. ferulic acid.

7-hydroxy-6-methoxycoumarin. scopoletin.

6-hydroxy-7-methoxy-1-methyl-1,2,3,4-tetrahydroisoquinoline. salsoline.

4-[1-hydroxy-2-(methylamino)ethyl]-1,2-benzenediol. epinephrine.

α-hydroxy-β-methylaminopropylbenzene. ephedrine.

[1R-(exo,exo)]-3-hydroxy-8-methyl-8-azabicyclo[3.2.1]octane-2-carboxylic acid. ecgonine.

α-(hydroxymethyl)benzeneacetic acid 8-methyl-8-azabicyclo[3.2.1]oct-3-yl ester. hyoscyamine.

3-hydroxymethylchrysazin. aloe emodin.

***N*-hydroxymethyl-*N*-ethylaminoazobenzene**. the activated form of *N*,*N*-dimethylaminoazobenzene. It is a carcinogen.

N-[α-(hydroxymethyl)ethyl]-D-lysergamide. ergonovine.

12′-hydroxy-2′-(1-methylethyl)-5′-(phenylmethyl)ergotaman-3′,6′,18-trione. ergocristine.

12′-hydroxy-2′-(1-methylethyl)-5′α-(phenylmethyl)-8α-ergotaman-3′,6′,18-trione. ergocristinine.

12′-hydroxy-2′-methyl-5′α-(2-methylpropyl)-ergotaman-3′,6′,18-trione. ergosine.

5-hydroxy-2-methyl-1,4-naphthalenedione. plumagin.

5-hydroxy-2-methyl-1,4-naphthoquinone. plumagin.

(2β,4α,15α)-15-hydroxy-2-[[2-*O*-(3-methyl-1-oxobutyl)3,4-di-*O*-sulfo-β-D-glucopyranosyl]-oxy]-19-norkaur-16-en-18-oic acid dipotassium salt. atraxtyloside.

12′-hydroxy-2′-methyl-5′α-(phenylmethyl)ergotaman-3′,6′,18-trione. ergotamine.

12′-hydroxy-2′-methyl-5′α-(phenylmethyl)-8α-ergotaman-3′,6′,18-trione. ergotaminine.

1-hydroxymethylpropane. isobutanol.

2-hydroxy-2-methylpropanenitrile. acetone cyanohydrin.

5-hydroxy-6-methyl-3,4-pyridinedimethanol. pyridoxine.

5-hydroxy-6-methyl-3,4-pyridinedimethanol hydrochloride. pyridoxine hydrochloride.

3-hydroxy-1-methylpyridinium bromide dimethylcarbamate. pyridostigmine bromide.

1-hydroxynaphthalene. 1-naphthol.

2-hydroxynaphthalene. 2-naphthol.

α-hydroxynaphthalene. 1-naphthol.

β-hydroxynaphthalene. 2-naphthol.

5-hydroxy-1,4-naphthalenedione. juglone.

5-hydroxy-1,4-naphthoquinone. juglone.

8-hydroxy-1,4-naphthoquinone. juglone.

4-hydroxy-3-nitrobenzenearsonic acid. roxarsone.

4-hydroxy-3-nitrophenylarsonic acid. roxarsone.

***N*-hydroxy-*N*-nitrosobenzenamine ammonium salt.** cupferron.

(17α)-17-hydroxy-19-norpregna-5(10),20-dien-3-one. norgesterone.

4-hydroxy-3-(3-oxo-1-phenylbutyl)-2H-1-benzopyran-2-one. warfarin.

[2α(2S,3S,4S,6R),3β,5α]-14-hydroxy-19-oxo-3,2-[(tetrahydro-3,4-dihydroxy-6-methyl-2*H*-pyran-2,3-diyl)bis(oxy)]card-20(22)-enolide. calotropin.

6-(10-hydroxy-6-oxo-*trans*-1-undecenyl)-β-resorcylic acid lactone. zearalenone.

4-hydroxy-2,4-pentadienoic acid γ-lactone. protoanemonin.

4-hydroxy-2-pentenoic acid γ-lactone. β-angelica lactone. See angelica lactone.

3-hydroxyphenol. resorcinol.

2-[6-(β-hydroxyphenethyl)-1-methyl-2-piperidyl]acetophenone. lobeline.

4-hydroxyphenethylamine. tyramine.

***N*-(4-hydroxyphenyl)acetamide.**acetaminophen.

β-[4-hydroxyphenyl]-acrylic acid. coumaric acid.

α-(4-hydroxyphenyl)-β-aminoethane. tyramine.

1-(*p*-hydroxyphenyl)-2-aminoethanol. octopamine.

α-hydroxy-α-phenylbenzeneacetic acid 2-(diethylamino)ethyl ester. benactyzine.

(3-hydroxyphenyl)dimethylethylammonium chloride. edrophonium chloride.

D-*p*-hydroxyphenylethanolamine. octopamine.

***p*-hydroxyphenylethanolamine**. octopamine.

2-[6-(2-hydroxy-2-phenylethyl)-1-methyl-2-piperidinyl]-1-phenylethanone. lobeline.

2-*p*-hydroxyphenylethylamine. tyramine.

4-hydroxyphenyl-β-D-glucopyranoside. arbutin.

3-(4-hydroxyphenyl)-2-propenoic acid. coumaric acid.

21-hydroxypregn-4-ene-3,11,20-trione. dehydrocorticosterone.

β-hydroxypropionic acid (hydracrylic acid; 3-hydroxypropanoic acid; ethylene lactic acid). a viscous, highly water-soluble, strongly acidic liquid, CH_2OHCH_2COOH. It is nontumorigenic in laboratory animals.

(2-hydroxypropyl)trimethylammonium chloride acetate. methacholine chloride.

5-(α-hydroxy-α-2-pyridylbenzyl)-7-(α-2-pyridylbenzylidene)-5-norbornene-2,3-dicarboximide. norbormide.

3-hydroxyquinuclidine. 3-quinuclidinol.

14-hydroxy-3β-(rhamnosyloxy)bufa-4,20,22-trienolide. proscillaridin.

12-hydroxysenecionan-11,16-dione. senecionine.

5-hydroxy-1-(*p*-sulfophenyl)-4-[(*p*-sulfophenyl)-azo]pyrazole-3-carboxylic acid trisodium salt. tartrazine.

hydroxytoluene. See cresol

α-hydroxytoluene. benzyl alcohol.

2′-hydroxy-2,5,9-trimethyl-6,7-benzomorphan. metazocine.

2-hydroxy-*N,N,N*-trimethylethanaminium. choline.

[S-(Z,E)]-5-(1-hydroxy-2,6,6-trimethyl-4-oxo-2-cyclohexen-1-yl)-3-methyl-2,4-pentadienoic acid. abscisic acid.

hydroxytriphenylstannane. triphenyltin hydroxide.

hydroxytriphenyltin. triphenyltin hydroxide.

3β-hydroxy-1αH,5αH-tropane-2β-carboxylic acid. ecgonine.

3β-hydroxy-1α*H*,5α*H*-tropane-2β-carboxylic acid methyl ester benzoate. cocaine.

5-hydroxytryptamine (5HT). serotonin.

3-hydroxytyramine. dopamine.

3-hydroxytyrosine. dopa.

3-hydroxy-L-tyrosine. levodopa.

17β-hydroxy-17α-vinylestr-5(10)-en-3-one. norgesterone.

(16α,17α)-17-hydroxyyohimban-16-carboxylic acid methyl ester. yohimbine.

hydroxyl group (hydroxyl radical). the univalent —OH group. It occurs in many inorganic compounds that form hydroxyl ions, OH^-, in aqueous solution. It is also the characteristic group of alcohols. See also hydroxide.

hydroxyl radical. hydroxyl group. See also hydroxide.

hydroxylamine (oxammonium). a colorless, basic, crystalline compound, NH_2OH, that is soluble in cold water, ethanol, and acids. The free base is unstable, rapidly decomposing at room temperature; explosive at higher temperatures. Hydroxylamine can be produced by algal blooms in amounts sufficient to kill fish

hydroxylamine methyl ether. methoxyamine.

hydroxylation. the addition of one or more hydroxyl groups to hydrocarbon chains or rings. Hydroxylation may follow epoxidation in the metabolism of xenobiotic chemicals and may bring about the addition of more than one epoxide group to a substance. These processes bring about the metabolic toxification of a number of xenobiotic compounds. The 7,8-diol-9,10-epoxide of benzo(a)pyrene is, for example, formed by these processes. See also epoxidation, hydroxyl group.

3-(4-hydroxy-3-methoxyphenyl)-2-propenoic acid. ferulic acid.

hydroxymethylbenzene. cresol.

2-hydroxyphenol. pyrocatechol.

Hydrozoa. one of three classes of the phylum Coelenterata (Cnidaria). Well-known genera are *Hydra*, *Obelia*, and *Gonionemus*, and *Physalia*. The stomodaeum is absent; the mesoglea has few or no cellular elements; usually metagenetic. Most species are marine. The polyp stage is small, colonial, and usually sessile, although *Physalia* forms an exceptional free-floating colony. The medusae are produced asexually; they are small, have a velum, are free-swimming, and reproduce sexually. *Hydra* is, however, solitary, lives in fresh water, has no medusa phase, and the polyps reproduce both sexually and asexually.

hyenanchin. a poison, $C_{15}H_{18}O_7$, in the outer envelope of the fruit of *Hyaenanche globosa*. The action is similar to that of strychnine, *q.v.*

hygrine ((*R*)-1-(1-methyl-2-pyrrolidinyl)-2-propanone; 2-acetonyl-1-methylpyrrolidine; *N*-methyl-2-acetonylpyrrolidine). a colorless liquid ketonic alkaloid, $C_8H_{15}NO$, derived from pyrrolidine. It occurs in leaves of *Erythroxylum coca* and various other plants of the family Erythroxylaceae. See cuscohygrine.

hygrostomia. ptyalism.

Hylesia. a genus of South American butterflies. The body is covered by fine, poisonous, hair-like bristles that resemble poison darts. These setae detach easily on contact and can cause a dermatitis and allergies. A severe dermatitis can also result indirectly from contact with clouds of "hairs" that have detached from these insects during flight.

Hylidae (tree frogs; cricket frogs; tree toads; chorus frogs). a large, mainly New World family of small frogs (usually less than 5 cm long) of the Order Procoela. Both jaws bear teeth; claw-like terminal digits with enlarged adhesive pads used in climbing. Typical genera are *Acris, Pseudacris, Gastrotheca*, and *Hyla*.

Hymenocallis (spider lily). a genus of bulbous, summer-blooming herbs (Family Amaryllidaceae) with stout, naked stems and strap-shaped leaves and unusual flowers that occur in flattened clusters. the bulbs are poisonous.

Hymenoptera (ants, bees, wasps, hornets, yellow jackets, etc.). a large order of endopterygote insects with distinct head, thorax, and abdomen; legs attached to the thorax; usually the thorax and abdomen are separated by a narrow waist. Winged individuals have two pairs of small membranous wings that are coupled by small hooks, have few veins, the anterior pair being the largest. Mouth parts are usually biting, but may be adapted to chewing, lapping, or sucking. All females have a conspicuous ovipositor which may be modified as a sting, saw, or drill. Metamorphosis is complete, the larva are usually apodous, but some are caterpillar-like with true thoracic legs and abdominal prolegs; pupae are herbivorous or parasitic and in some cases are enclosed in a cocoon. Highly developed social or colonial behavior is common; some species are solitary. The sting or bite may be painful and sometimes lethal, although death of nonsensitive individuals is rare. Common or important genera include *Apis*, *Bombus*, *Nematus*, *Formica*, *Polistes*, *Vespa*, *Vespula*. See Aculeata, Hymenoptera sting, hymenopterism, venoms (hymenopteran venoms).

hymenoptera sting. some hymenopterans are not venomous; others can be deadly to humans. The effects of a sting may be largely local and restricted to local symptoms such as slight to intense pain, tenderness, edema, erythema, and itching about the site of the bite. More general reactions are not uncommon, especially in sensitive individuals or to multiple stings. Signs and symptoms often include a generalized urticaria, sometimes with generalized edema, nausea, vomiting, dizziness, malaise, anxiety, wheezing, and thoracic constriction. In severe reactions, further systemic effects such as dyspnea, hoarseness, dysphagia, and confusion may present. Systemic shock, while rare, can be fatal. Signs and symptoms (in addition to any of those above) may include two or more of these: cyanosis, lowered blood pressure, incontinence, unconsciousness, and collapse. Delayed reactions to hymenopteran stings may also occur, and anaphyllactoid shock is a risk in subsequent stings. See Hymenoptera, venom (hymenopteran venom).

hymenopteran. any insect of the order Hymenoptera.

hymenopterism. poisoning by the stings or bites of hymenopterous insects (e.g., bees, wasps, hornets). Effects on humans may be localized about the bite, but often include a generalized urticaria, sometimes with generalized edema, nausea and vomiting, dizziness, malaise and anxiety, wheezing, and thoracic constriction. Systemic effects in severe cases may include hoarseness, dyspnea, dysphagia, and confusion. Systemic shock is rare, but can be fatal. Additional effects may include two or more of these: cyanosis, lowered blood pressure, incontinence, unconsciousness, and collapse. Delayed reactions to hymenopteran stings also occur and anaphylactoid shock is a risk in subsequent stings. See Hymenoptera, Hymenoptera sting.

hymenopterous. of or pertaining to insects of the order Hymenoptera.

Hymenoxys. a genus of small annual or perennial, often weedy herbs (Family Compositaceae). ***H. lemmoni*** (cockerell). a small herb native to California, Utah, northeastern Arizona, and Nevada. It is highly toxic; one half pound per sheep per day is invariably fatal within a few days. The plant is normally grazed by sheep only under poor range conditions. ***H. odorata*** (= *Actinea odorata*; bitterweed; bitter rubberweed). a highly toxic, poisonous, annual weed from a few centimeters to more than 1 m tall. It occurs in semiarid rangelands from southwestern Kansas, western and central Texas, into Arizona, New Mexico, and Mexico. The leaves contain a highly toxic, water-soluble cumulative poison. Sheep may die within a day or two following ingestion of ca. 1% of their body weight of fresh green leaves. The leaves retain toxicity on drying. Symptoms of acute (experimentally forced, single lethal dose) intoxication include salivation, anorexia, weakness, depression, and vomition. In natural poisoning of sheep, symptoms are similar but less pronounced. Lesions are variable and sometimes absent. They commonly include, however, pulmonary congestion, inflammation of the abomasum, and hemorrhages of the epicardium. Livestock usually eat bitterweed only under poor range conditions. *H. richardsonii* (= *Actinea richardsonii, Hymenoxys floribunda*)(pingue, Colorado rubberweed). a perennial herb with a woody root, up to ½ m tall that is widespread from Canada to Texas and west to Oregon and California. It is somewhat less toxic to livestock than *H. odorata*. Sheep are most often poisoned and have suffered extensive losses from poisoning by pingue throughout portions of the plant's range. Symptoms of acute intoxication include salivation, uneasiness, anorexia, weakness, depression, vomition, rumen stasis with signs of severe abdominal pain, and prostration.

hyoscine. scopolamine.

hyoscine poisoning. poisoning by scopolamine, *q.v.*

hyoscyamine (daturine; α-(hydroxymethyl)benzeneacetic acid 8-methyl-8-azabicyclo[3.2.1]-oct-3-yl ester; l-hyoscyamine; 1αH,5αH-tropan-3α-ol(—)-tropate; 3α-tropanyl S-(—)-tropate; 1-tropic acid ester with tropine; l-tropine tropate). an extremely toxic, white, crystalline, anticholinergic alkaloid of *Hyoscyamus niger*, *Atropa belladonna*, *Datura stramonium*, *Mandragora*, and other plants of the family Solanaceae. Hyoscyamine is freely soluble in ethanol and dilute acids. It is the levorotatory component of the racemic mixture, atropine. Hyoscyamine and hyoscyamine hydrobromide are used medically as preanesthetics, antispasmodics, analgesics, sedatives, and as antidotes to cholinesterase inhibitors. They reduce salivary and branchial secretions, relax the gastrointestinal tract in certain spastic conditions. Hyoscyamine is highly toxic and should be used with great care. It has been used for centuries as a poison (See *Hyoscyamus*). It paralyzes the parasympathetic nervous system by blocking the action of acetylcholine at nerve endings. Renal function must be normal to eliminate it.

***dl*-hyoscyamine**. atropine.

l hyoscyamine. hyoscyamus.

hyoscyamine hydrobromide. See hyoscyamine.

hyoscyamus (henbane; henbane; hog's bean; insane root; poison tobacco). the dried leaves of *Hyoscyamus niger* or *H. muticus* (Egyptian henbane), with or without the stems and flowering tops. It contains hyoscyamine and hyoscine, and was formerly used as a smooth muscle relaxant and antispasmodic. See *Hyoscyamus*.

Hyoscyamus (henbane; hog's bean; insane root; poison tobacco). a genus of poisonous plants (Family Solanaceae). ***H. niger.*** (henbane; black henbane; insane root; fetid nightshade; poison tobacco; stinking nightshade). a poisonous annual or biennial weed sometimes found on dry roadsides and waste areas. It is indigenous to Africa and Asia from Egypt to India and is grown commercially in many parts of North America, especially California. The seeds and leaves contain tropane alkaloids such as hyoscyamine, scopolamine (hyoscine), atropine, and tropane; the seeds are slimy and less toxic than the leaves, but all parts are fatally toxic CNS depressants. Symptoms are similar to those of *Datura* or *Atropa* poisoning, except that henbane also produces profuse salivation. Poisonings are rare. The *H. reticulatus* variety of *H. niger* grows only in India. ***H. muticus***. See hyoscyamus. ***H. reticulatus***. similar in most respects to *H. niger*, this henbane is often considered a variety of *H. niger*. It occurs only in India.

hypaconitine. See also *Aconitum*.

hypaphorine (α-carboxy-*N,N,N*-trimethyl-1*H*-indole-3-ethanaminium hydroxide inner salt). the betaine of tryptophan, $C_{14}H_{18}N_2O_2$, it is a convulsant isolated from the seeds of *Abrus precatorius*, *Erythrina americana*, *E. crista-galli*, and *E. sandwicensis*. See *Abrus precatorius*.

hyper-. a prefix that means above, beyond, or excessive,

hyperabsorption. an unusually high rate of intestinal absorption of a given substance.

hyperacidity. a condition of excessive acidity.

hyperactive. **1:** hyperkinetic; having an unusually high level of motor activity. **2:** suffering from attention deficit disorder with hyper activity.

hyperactivity. **1:** an abnormally high level of motor activity. Such is sometimes a response to toxic insult, to the trauma surrounding such insult (e.g., snakebite), or to associated fear of a prospective insult. Such a response may be seen at times in humans and in wild and domestic animals. **2:** attention deficit disorder with hyperactivity; xenobiotics are hypothesized to play a role at times. *Cf.* hypoactivity.

hyperacute. extremely acute or intense.

hyperalimentation. a method of meeting the complete nutritional needs of a patient who is unable to eat normally. This is usually accomplished by supplying complete nutrients intravenously or by use of a tube placed through the nose into the stomach.

hyperbaric chamber. a large, sealed room in which air pressure can be raised to several atmospheres above normal levels. It is used mainly to treat patients with decompression sickness or severe burns.

hypercalcemia (calcemia; hypercalcinemia). abnormally elevated concentrations of calcium in the blood. This condition usually results from excessive bone resorption with release of calcium. Symptoms may include anorexia, weakness, depression, confusion, nausea, abdominal pain, muscle pain, and constipation. Hypercalcemia occasionally signals a malignancy.

hypercalcification. calcinosis (enzootic calcinosis).

hypercalcinemia. hypercalcemia.

hypercapnia (hypercarbia). an abnormally high concentration of carbon dioxide in circulating blood.

hypercapnic. referring to or characterized by hypercapnia.

hypercarbia. hypercapnia.

hypercatharsis. frequent and excessive purgation.

hypercathartic. **1:** extremely cathartic; able to cause hypercatharsis. **2:** an extremely potent cathartic agent.

hyperchloremia. unusually high concentrations of chloride in circulating blood.

hyperchloremic. pertaining to or marked by hyperchloremia.

hyperchloridation. the administration of an excess of sodium chloride.

hyperchloruration. excess chloride in the body of a living organism.

hyperchloruria. abnormally high levels of chloride in urine.

hyperchromatism (hyperchromatosis). **1a:** excessive pigmentation, especially of the skin. **1b:** a type of degeneration of a cell nucleus in which it fills with particles of pigment or chromatin.

hyperchromatosis. **1:** increased staining capacity. **2:** hyperchromatism, especially of the skin.

hypercupremia. an abnormally high amount of copper in blood plasma.

hypercupriuria. abnormally high amounts of copper in urine.

hypercyanotic. extremely cyanotic.

hypereccrisia (hypereccrisis). a state marked by an extremely high rate of excretion.

hypereccrisis. hypereccrisia.

hypereccritic. pertaining to or denoting hypereccrisia.

hyperemesis. excessive vomiting.

hyperemetic. characterized by excessive vomiting.

hyperemia. an excessive amount of blood in a tissue, organ, or body part; engorgement of blood.

hyperemic. of, pertaining to, or characterized by hyperemia.

hyperesthesia. abnormally high or increased sensitivity to stimuli.

hyperesthetic. pertaining to or characteristic of hyperesthesia.

hyperestrogenism. See *Fusarium* estrogenism.

hyperexcretory. characterized by excessive excretion.

hyperfolliculinemia. excessive amounts of estrogen in circulating blood.

hyperfolliculinism. any condition due to excessive amounts of estrogen in the body.

hyperfolliculinuria. excessive amounts of estrogen in the urine.

hyperfunction. increased function of any tissue, organ, or system.

hyperglycemia. a high concentration of glucose in circulating blood. Hyperglycemia can result from diabetes mellitus or exposure to hyperglycemic chemicals.

hyperhidrosis (hyperidrosis). abnormally profuse or excessive perspiration.

hyperhidrotic. of, pertaining to, or causing hyperhidrosis.

hyperhormonic. hyperhormonal.

hyperhormonism. chronic endocrine hyperfunction.

hyperhydration. the retention of abnormally large amounts of water in the body of an organism; the state of having excessive amounts of water in the body.

hypericin (1,3,4,6,8,13-hexahydroxy-10,11-dimethylphenanthro[1,10,9,8-[*opqra*]perylene-7,14-dione; 4,5,7,4′,5′,7′-hexahydroxy-2,2′-dimethylnaphthodianthrone; hypericum red, Cyclo-Werrol; Cyclosan). a primary photosensitizing agent found in *Hypericum* spp. It is a reddish-brown fluorescent dianthrone pigment. It remains undigested and is chemically unaltered during passage through the liver. Hypericin initiates feeding in the beetle *Chrysolina brunsvicensis*, but can accumulate in mammals that graze on *Hypericum*, causing an intense photosensitization; the symptoms are similar to fagopyrism. See photosensitivity.

Hypericum (St. John's wort). a large genus of perennial herbs and shrubs (Family Guttiferae; sometimes given as Hypericaceae). Most species are ornamental (often grown in rock gardens or borders). Some feral, mostly weedy species occur in dry pastures, old fields, woodland borders and along roadsides. The leaves are characteristically dotted on either surface with translucent pinpoint marks that are black in reflected light or yellowish from transmitted light. The flowers are yellow, numerous, and about 5 cm in diameter. They contain photosensitizing pigments that are unaffected by digestion, are not eliminated by the liver and reach the capillaries of the skin unchanged. A number of species have poisoned livestock. Common or important species include *H. ascyron*, *H. calycinum*, *H. frondosum*, *H. kalmianum*, *H. patulum*, *H. perforatum* (the most important, toxicologically), *H. prolificum*, and *H. punctatum*. ***H. perforatum*** (St. John's wort; Klamath weed; goatweed). a common roadside weed, native to Europe and naturalized elsewhere. It is one of few plant species known to cause primary photosensitivity in animals. Cases are uncommon. It contains the phototoxic pigment, hypericin. Sheep are more resistant than cattle or horses. Effects on sheep include generalized skin irritation, anorexia, and sometimes loss of vision. Because hypericin is resistant to drying and is heat stable, animals can be poisoned by weedy hay. The plant is used in folk medicine for its antibacterial and astringent properties and to soothe the digestive system. Photosensitization and other toxic effects also occur in humans. While small doses may not be harmful, it is probably unsafe for human usage. See, for example, furocoumarins, *Fagopyrum sagittatum*, *Ammi*, *Cymopterus watsonii*, *Trifolium*, and *Medicago*.

hypericum red. hypericin.

hyperidrosis. hyperhidrosis.

hyperimmune. **1:** having an unusually high level of immunity to specific antigens due to repeated immunization. **2:** having exceptionally large quantities of specific antibodies in the serum.

hyperimmunity. an unusually high degree of immunity to specific antigens, due usually to repeated immunization with the production of large numbers of effective antibodies.

hyperimmunization. the establishment of a heightened state of actively acquired immunity by repeated administration (boosters) of immunogen to an animal, or of increased passively acquired immunity by the administration of hyperimmune gamma-globulin.

hyperirritability. pathologically high or intense responsiveness to slight stimuli.

hyperkalemia (hyperkaliemia; hyperpotassemia). an abnormally high concentration of potassium in the blood, often due to impaired renal excretion. See also hypokalemia.

hyperkaliemia. hyperkalemia.

hyperkeratosis. hypertrophy of the horny layer of the skin of vertebrates or any disease or condition characterized by it. **bovine hyperkeratosis** (hyperkeratosis of cattle). a disease of domestic cattle, characterized by a thickening of the skin and hair loss, especially from the neck and shoulders. The usual cause is the ingestion of feed concentrates that contain polychlorinated naphthalenes as a result of contamination by wood preservatives or lubricating oils during milling or pelleting. The disease may also result from the contamination of feedstuffs by toxic fungi such as *Aspergillus chevalieri*, *A. chevalieri*, and *A. fumigatus*.

hyperkeratosis of cattle. bovine hyperkeratosis (See hyperkeratosis).

hyperkinesia (hyperkinesis). hyperactivity (def. 1); an unusually high level of motor activity.

hyperkinesis. hyperkinesia.

hyperkinetic. having an unusually high level of motor activity.

hyperlethal. more than enough in quantity, strength, or intensity to cause death.

hyperlipidemia. abnormally high levels of lipids, usually as lipoproteins in blood.

hyperlipoproteinemia. abnormally high levels of lipoproteins (cholesterol and other fatty materials) accumulate in the blood.

hyperlithemia. abnormally high concentrations of lithium in the blood.

hypermagnesemia. unusually large concentrations of magnesium in the blood plasma. Symptoms may include weakness, lethargy, and electrocardiographic abnormalities. With increasing levels, deep tendon reflexes are lost, the victim becomes somnolent and comatose.

hypermetabolic state. a condition of greatly increased metabolism. It is observed in fever and salicylate poisoning.

hypermetropia (farsightedness). ability to see distant objects clearly while nearby objects appear blurred.

hypermutation. mutation that occurs at a rate that is higher than normal for a particular gene or species.

hypernatremia (hypernatronemia). abnormally high concentrations of sodium in the blood.

hypernatremic. of or pertaining to hypernatremia.

hypernatronemia. hypernatremia.

hypernitremia. abnormally high concentrations of nitrogen in the blood.

hyperoxemia. hyperoxia of blood marked by low pH; excessive acidity of the blood;

hyperoxia. excessive amounts of oxygen in the system due to exposure to high oxygen concentrations, especially at hyperbaric pressure.

hyperoxic. pertaining to or marked by hyperoxia.

hyperoxidation. excessive oxidation.

hyperoxide. superoxide.

hyperperoxyl radical. superoxide.

hyperplasia. **1a:** excessive development of an organ or normal tissue due to an increased rate of cell division; enlargement of the tissue or organ may occur. Hyperplasia may occur in normal fetal development, wound healing, inflammation, tumor growth, and lymph node activity. **1b:** an abnormally high rate of proliferation of normal cells. An abnormally high rate of proliferation of normal cells that are arranged normally within a given tissue or organ. *Cf.* hypoplasia, aplasia.

hyperplastic. of, pertaining to, or characterized by hyperplasia.

hyperpnea. abnormally deep, rapid breathing.

hyperpneic. of or pertaining to hyperpnea.

hyperpotassemia. hyperkalemia.

hyperptyalism. ptyalism.

hyperpyretic. pertaining to hyperpyrexia.

hyperpyrexia. a greatly elevated body temperature in homeothermic animals.

hyperpyrexial. pertaining to hyperpyrexia.

hyperreactive. exhibiting hyperreactivity.

hyperreactivity. a greater than normal responsiveness to stimuli. *Cf.* hypersensitivity.

hypersaline. of, pertaining to, or containing unusually large amounts of sodium chloride; highly saline. Most often used with respect to seawater.

hypersensibility. excessive sensibility or sensitivity to a biologically active agent.

hypersensitive. **1:** exhibiting hypersensitivity, *q.v.* **2:** used also in toxicology to denote organisms within a given population that are killed by a specific toxicant when exposed to a dose equivalent to LD_{05} or below.

hypersensitiveness. **1:** unusual or abnormal sensitivity to the action of a specific agent, as in allergy, anaphylaxis, or hay fever. **2:** obsolete: an increase in an already existing sensitivity to a toxin by repeated exposure. Such cases are, on the contrary, allergic or immunotoxic in character; they are due to a sensitivity to a substance that was formerly innocuous.

hypersensitivity. **1:** an abnormally heightened sensitivity to an external stimulus of any kind. **2a:** an exceptionally strong reaction to an external agent or stimulus. **2b:** extreme sensitivity to any agent, chemical or physical. **3:** allergy, *q.v.* See also anaphylaxis. **4:** heightened sensitivity of a plant to infection or infestation by a pathogen such that plant tissue at the site of contact dies, preventing further spread of the disease. Plants showing such sensitivity to a particular pathogen are highly resistant to infection. Symptoms are limited to minute necrotic lesions. **cell-mediated hypersensitivity**. delayed hypersensitivity. **cytotoxic hypersensitivity**. cytotoxic reaction (See antigen-antibody reaction). **delayed hypersensitivity** (cell-mediated hypersensitivity). an allergic reaction characterized by a localized inflammation which appears as much as 24-48 hours following exposure to an antigen, e.g., contact dermatitis. It is produced solely by T-lymphocytes, which secrete lymphokines that direct an assortment of cellular responses. This form of hypersensitivity is used in toxicology to assess the integrity of cell-mediated immunity following exposure to suspected immunosuppressive or immunomodulatory agents. **immediate hypersensitivity** (type I hypersensitivity). a specific reaction to an antigen that is evident within seconds, minutes, or at most, a few hours following exposure to an antigen. The reaction often results only in a focal inflammation, hemorrhage, and vasculitis at the site of contact with the antigen. There are two classes of immediate hypersensitivity reactions. **class I:** complement-dependent, i.e., induced by IgG and IgM antibodies. These reactions are equivalent to a type II, cytolytic, or a type III, immune complex-mediated allergic reaction. Severe cases may present as immune complex diseases including drug-induced systemic lupus erythematosus. **class II:** complement-independent, i.e., induced by IgE antibodies, and equivalent to a type I (anaphylactic) allergic reaction. Severe cases may present as anaphylactic shock or allergic asthma. **induced hypersensitivity**. any reaction that results from exposure to one or more doses of a particular toxicant, as in the case of a severe allergic reaction to an antibiotic. **type I hypersensitivity**. immediate hypersensitivity, class II. **type II hypersensitivity**. immediate hypersensitivity, class I (in part) which may present, for example, as drug-induced hemolytic anemia. **type III hypersensitivity**. immediate hypersensitivity, Class I (in part). **type IV hypersensitivity**. delayed hypersensitivity.

hypersensitization. **1:** producing or inducing hypersensitivity. **2:** a state of heightened sensitivity to a stimulus of any kind.

hyperserotonemia. an unusually high serum serotonin concentration.

hypersusceptibility. **1a:** a state of abnormally heightened responsiveness to foreign (usually) agents such as pathogens or chemical substances (e.g., toxicants) which do not affect most individuals within an animal population. **1b:** abnormally heightened response to a biologically active agent. *Cf.* anaphylaxis.

hypertension (high blood pressure). higher than normal blood pressure, especially diastolic pressure. In adult humans, hypertension is often clinically defined as repeated readings of systolic pressure of 140 mm Hg or more and a diastolic pressure of 90 mm Hg or more. Hypertension may result from any of a number of causes, such as unusually intense sympathetic stimulation of the heart (with increased force and rate of contractions) or blood vessels (causing peripheral vasoconstriction); atherosclerosis; arteriosclerosis. Numerous drugs, venoms, other toxicants, and even ordinary foods and condiments can cause hypertension. See, for example, barium poisoning, *Centruroides*, cocaine hydrochloride poisoning, hallucinogen, hypervitaminosis D, ibuprofen, *Juniperus virginiana*, *Latrodectus*, picrotoxin, pyrocatechol, sodium chloride. **idiopathic hypertension**. hypertension of unknown origin.

hypertensive. **1:** refers to higher than normal blood pressure. **2:** a hypertensive agent. **3:** a person or animal with high blood pressure. See also hypotensive.

hypertensive agent. an agent that elevates blood pressure.

hypertensor. pressor.

hyperthermia. hyperpyrexia, especially that which is purposely induced therapeutically.

hyperthyroidism. overactivity of the thyroid gland or the resulting condition. Effects include increased basal metabolism, goiter, exophthalmos, and disturbances of autonomic nervous system function and of creatine metabolism.

hypertonic. **1:** descriptive of a cell or tissue that has a higher osmotic pressure than the solution in which it is bathed. **2:** said of a solution that has a higher solute concentration and thus a higher osmotic pressure than another solution with which it is compared, usually cytoplasm or tissue fluids. *Cf.* hypotonic, isotonic. See also tonicity.

hypertonic solution. a solution that has higher osmotic pressure than the cell or system from which it is separated (or potentially separated) by a semipermeable membrane (e.g., the plasmalemma). *Cf.* hypotonic solution, isotonic solution.

hypertonicity. **1:** the state or quality of being hypertonic, *q.v.* **2:** spasticity.

hypertoxic. extremely or excessively toxic.

hypertoxicity. a state or quality of extreme or excessive toxicity.

hypertrophy. increase in the size of a body tissue or organ due to cell growth.

hyperventilation. an unusually high rate of ventilation of the lungs due to increased rate and depth of breathing. There is an associated decrease in blood carbon dioxide concentration to levels below normal with eventual alkalosis. Hyperventilation can result from excessive exertion, various kinds of stress (physical or psychological), from excessive concentrations of carbon dioxide in circulating blood, or increased oxygen in the blood.

hypervitaminosis. any of a number of toxic syndromes that result from ingestion of an excessive amount of vitamin preparations. Symptoms vary with the vitamin or spectrum of vitamins involved. Overdosage of fat-soluble vitamins, such as vitamins A or D, generally produce the most serious effects.

hypervitaminosis A. chronic or acute poisoning due to overdosage of vitamin A preparations. This condition is distinguished from hypervitaminosis D by the high serum levels of vitamin A and normal serum calcium levels. *Cf.* hypervitaminosis D. **acute poisoning**. has been attributed to ingestion of the livers of sharks, halibut, seals, and polar bears. Polar bear liver may contain at least as much as 180,000 units (60,000 μg) of vitamin A per gram. Symptoms primarily due to central nervous system effects include drowsiness, lethargy, irritability, severe headache, and vomiting, which appear within a few hours following ingestion of seal or polar bear liver. Subsequent peeling of the skin has been reported. Acute toxicity in children can result from doses of 300,000 IU (100,000 μg) or more. It is characterized by increased intracranial pressure and vomiting. Recovery is spontaneous, with full recovery usually within a few days to a few weeks; fatalities are unknown. **chronic poisoning**. occurs in children and adults who receive more than 100,000 units (33,000 μg) daily (sometimes as little as 50,000 units) for several months. Regular consumption of beef liver is a common cause in adults in the United States. Infants may develop evidence of toxicity within a few weeks if given 20,000 to 60,000 units of water-dispersible vitamin A per day. Fetotoxicity has been reported in the case of women receiving 13-*cis*-retinoic acid (isotretinoin) for skin conditions during pregnancy. Early signs and symptoms in adults include sparse, coarse hair, spotty alopecia, dry rough skin, and cracked lips. These effects are followed by generalized weakness, severe headache, anorexia, bone pain, spotty alopecia, skin lesions and pigmentation, and hepatomegaly. Infants exhibit some of the above, usually with vomiting, increased intracranial pressure, bulging fontanels, hemorrhagic diathesis (hypoprothrombinemia), and decalcification of bones with retardation or arrest of growth.

hypervitaminosis D. chronic or acute poisoning due to overdosage vitamin D. Tolerance to vitamin D varies greatly among human individuals. Thus a dosage of as little as 50,000 units/day for as little as one month is toxic to some individuals; others can tolerate several hundred thousand units/day indefinitely. Early symptoms typically include anorexia, nausea, and vomiting. Ensuing symptoms are often very similar to those of hyperparathyroidism with polydipsia, polyuria, weakness, itching, and nervousness. Hypercalcemia is a constant finding in hypervitaminosis D, however. Lesions may include metastatic calcification in the kidneys with consequent renal injury and hypertension; calcification of the cerebral arteries and a resulting cerebellar ataxia and other dysfunctional conditions. Most symptoms are relieved upon withdrawal of vitamin D, although renal damage may be irreversible. *Cf.* hypervitaminosis A.

hypesthesia. hypoesthesia.

hypnotic. **1:** capable of, or having the effect of, inducing sleep or hypnosis. **2:** a medication that induces sleep.

hypnotoxin. **1:** a toxic CNS depressant isolated from the nematocysts of the tentacles of *Physalia* (the Portuguese man-of-war). **2:** a hypothetical toxic substance (perhaps secreted by brain tissue) that accumulates in the blood during wakefulness until it reaches a concentration sufficient to induce sleep.

hypo. **1:** sodium thiosulfate. **2:** jargon for hypodermic; hypodermic syringe; use of a hypodermic syringe.

hypoacidity. a deficiency of acid.

hypoactivity. an abnormally low or diminished level of activity of the body as a whole or its organs.

hypoallergenic. having a relatively limited ability to induce

hypoallergenicity. the quality of having a limited ability to induce hypersensitivity.

hypocalcemia. an abnormally low concentration of calcium in circulating blood; it usually indicates abnormally low amounts of calcium in the body. *Cf.* calcemia.

hypocapnia. See acapnia.

hypochlorite. **1:** any of a class of salts derived from hypochlorous acid that contain the ion, ClO_3^-. All are highly toxic, caustic, oxidizing, and bleaching agents. Commonly used, dry or in solution as bleaches, they are dangerous household poisons. They produce active oxygen which is largely responsible for their toxicity. See also hypochlorite poisoning, hypochlorous acid, calcium hypochlorite, sodium hypochlorite. **2:** the hypochlorite ion, ClO_3^-.

hypochlorite poisoning. poisoning by solutions of hypochlorite salts (most often as household or commercial bleaches and related chlorinated products) by ingestion or skin contact. Ingestion produces immediate gastrointestinal distress, with pain and inflammation of the mucous membranes of the mouth, pharynx, esophagus, and stomach; the vomitus has the appearance of coffee grounds. Systemic effects are secondary to local injury and shock. These include confusion, delirium, edema of the glottis, coma, and death due to circulatory collapse. Lesions may include edema of the pharynx, glottis, or larynx and perforation of stomach or esophagus. Inhalation of fumes may produce pulmonary edema.

hypochlorite salt. See hypochlorite.

hypochlorite solution (household bleach). an aqueous solution of a metallic salt of hypochlorous acid. It is a strong oxidizing agent and irritant to the skin, eyes, and mucous membranes. Such solutions (e.g., household bleach) are used in the bleaching of textiles and as antiseptic agents. Ingestion of hypochlorite salts or their solutions usually produces pain and shock locally about the mouth. The human victim may suffer extreme gastrointestinal distress with pain and inflammation of mouth, pharynx, esophagus, and stomach; glottal edema; vomitus the color of coffee grounds; confusion; delirium; coma; circulatory collapse. Lesions in humans commonly include edema of the pharynx, glottis, and/or larynx, with perforation of the stomach or esophagus in severe cases. The fumes can cause pulmonary edema. The systemic effects are secondary to the local irritation and erosion of tissues caused by these substances. Hypochlorous acid and hypochlorites are used chiefly as bleaches, disinfectants, and to purify water. They produce active (nascent) oxygen which, through its oxidizing action, accounts for most of the toxicity of hypochloric acid and hypochlorites. See also hypochlorous acid.

hypochlorous acid (chloric (I) acid). an irritant, greenish-yellow, chlorinated oxyacid, HClO, produced by the action of water on chlorine. It is a weak, highly unstable, dilute, aqueous solution of chloride of lime. It is a poor proton donor and hence a weak acid. It is highly unstable, readily decomposes to yield hydrogen chloride and oxygen, and exists only in dilute solution. It is used as a bleach, an antiseptic, and to purify water. Hypochlorous acid and hypochlorites produce active (nascent) oxygen which, through its oxidizing action, accounts for most of the toxicity of hypochloric acid and hypochlorites. See also hypochlorite solution, calcium hypochlorite, sodium hypochlorite, hypochlorite poisoning.

hypochlorous anhydride. chlorine monoxide.

hypochondriac. one who suffers from hypochondriasis.

hypochondriasis. a mental condition in which a person is convinced of being afflicted with a serious illness even though thorough medical examination and judgment contradicts this belief. The symptoms seem real to the patient.

hypocinesia. hypokinesis.

hypocoagulable. characterized by abnormally decreased coagulability.

hypocupremia. unusually low copper content of the blood. It is a finding in exfoliative dermatitis in which, even though serum-albumin-attached copper increases, ceruloplasmin is depressed.

hypodermic. of or pertaining to hypodermis or anything applied or administered beneath the skin.

hypodermis. See subcutaneous tissue.

hypoergasia. abnormally diminished functional activity. See hypoergia.

hypoergia. **1:** hypoergasia. **2:** hyposensitivity to specific allergens.

hypoergic. **1:** substantially less energetic than normal. **2:** of or pertaining to hypoergy.

hypoergy. abnormally low or diminished reactivity.

hypoesthesia (hypesthesia). abnormally low sensitivity of sensory receptors in the skin to stimuli, e.g., pain, touch, heat, or cold receptors.

hypoesthetic. of or pertaining to hypoesthesia.

hypofibrinogenemia. unusually low concentration of fibrinogen in circulating blood. See also snakebite, ancrod, venom (snake venoms).

hypofunction. reduced or below normal function of any tissue, organ, or system.

hypoglossal. beneath the tongue.

hypoglycemia. a condition marked by unusually low concentrations of glucose in circulating blood. Symptoms may include headache, hunger, weakness, ataxia, visual disturbances, anxiety, and personality change. Unless carefully managed, delirium, coma, and death may ensue. The most common causes are an increase in insulin due to excessive secretion of the islet cells of the pancreas, therapeutic administration of too much insulin, and dietary deficiency. Hypoglycemia may also be induced by any of a variety of oral hypoglycemic agents, such as sulfonylureas. The hypoglycemic action (and hence the toxicity) of such agents is not very great when taken alone, but their toxicity can be substantially potentiated by antiinflammatory agents (e.g., phenylbutazone, salicylates).

hypoglycin. See hypoglycine.

hypoglycin A. hypoglycine A. See hypoglycine.

hypoglycin B. hypoglycine B. See hypoglycine.

hypoglycine (hypoglycin). either of two closely related potent hypoglycemic agents (the peptides, hypoglycine A and B) isolated from the seeds and fruit of *Blighia sapida* (the akee), a poisonous tree indiginous to tropical West Africa, but cultivated in Florida and tropical America. They inhibit gluconeogenesis. **hypoglycine A** (α-amino-2-methylenecyclopropanepropanoic acid; 2-methylenecyclopropanalanine; 2-amino-4,5-methylenehex-5-enoic acid; α-amino-β-(2-methylenecyclopropyl)propanoic acid; hypoglycin A). a more potent substance, $C_7H_{11}NO_2$, than hypoglycine B, it occurs in the aril of the unripe fruit and is the presumed cause of Jamaican vomiting sickness, *q.v.* Hypoglycine A is heat labile and the ripe aril, if cooked, is considered a delicacy. **hypoglycine B** (N-L-γ-glutamyl-3-(2-methylenecyclopropyl)alanine; γ-L-glutamyl-α-amino-β-(2-methylenecyclopropyl)propionic acid dipeptide; γ-L-glutamylhypoglycine). a hypoglycemic agent, $C_{12}H_{18}N_2O_5$, similar in action, but less potent than hypoglycine A.

hypohepatia. inadequate functioning or insufficiency of the liver. It is a rarely used term.

hypokalemia (hypokaliemia; hypopotassemia). an abnormally low concentration of potassium in the blood, usually the result of potassium loss by renal secretion, vomiting, or diarrhea. See also hyperkalemia.

hypokalemic. **1:** pertaining to or characterized by hypokalemia. **2:** an agent that promotes or induces hypokalemia.

hypokaliemia. hypokalemia.

hypokinesia. hypokinesis.

hypokinesis (hypokinesia; hypocinesis; hypocinesia; hypomotility) slow or reduced movement.

hypokinetic. pertaining to or characterized by hypokinesis.

hypolethal. of insufficient quantity, power, or intensity to cause death.

hypomagnesemia. 1: reduced serum magnesium levels. **2:** grass tetany.

hypomagnesemic tetany. a condition of domestic cattle and sheep, mostly affecting adult cows and ewes. Affected ewes are usually lactating animals feeding on lush pastures. Cows of any age or condition may be affected, but the disease is most common in beef cattle grazing on cereal crops (especially wheat) or those that are undernourished and exposed to cold, changeable weather. Signs of the disease are irritability, tetany, and convulsions. Severe cases may present abruptly and, in some cases, cattle at pasture may be found dead without prior signs of illness. The condition is similar in sheep and cattle, is strikingly similar to and may be confused with grass tetany, and may be complicated by grass tetany (hypomagnesemia). Furthermore, since it is almost always accompanied by hypocalcemia it is difficult to determine which condition is primary. Hypomagnesemic tetany of calves is clinically identical to grass tetany, *q.v.*

hypomotility. hypokinesis.

hyponatremia. an abnormally low concentration of sodium in the blood, due to inadequate excretion of water or by excessive retention of water or by excessive water in circulating blood. **severe hyponatremia**. water intoxication.

hyponitrous acid anhydride. nitrous oxide.

hypophosphorous acid. a poisonous, monobasic reducing acid, H_3PO_2.

hypophyseal growth hormone. somatotrophin.

hypophysis. pituitary gland.

hypophysis cerebri. pituitary gland.

hypoplasia (hypoplasty). underdevelopment of a tissue or organ that is less severe than aplasia. *Cf.* hyperplasia, aplasia. **bone marrow hypoplasia**. defective development of bone marrow with serious consequences such as anemia, leukopenia, and lymphocytopenia.

hypoplastic. of, pertaining to, or characterized by hypoplasia.

hypoplastic anemia (aplastic anemia). any of a broad class of anemias marked by a reduced amount of bone marrow and a consequent reduction in the production of erythrocytes. It can be life threatening.

hypoplasty. hypoplasia.

hypopnea. abnormally slow, shallow breathing.

hypopneic. pertaining to or denoting hypopnea.

hypopotassemia. hypokalemia.

hyporeactive. **1:** denoting a less than normal response to stimuli; hyposensitive. **2:** denoting abnormally low sensitivity to a noxious substance, often as a result of exposure to repeated, gradually increasing doses.

hyposensitive (insensitive). in toxicology, a term usually reserved for individuals that survive a dose of a specific toxicant equivalent to the LD_{95} dose.

hyposensitivity (insensitivity). the state of being hyposensitive to a specific toxicant. *Cf.* insensitivity. See hyposensitive.

hyposensitization (desensitization). the process of inducing hyposensitivity.

hyposthenia. weakness; a weakened or enfeebled condition. *Cf.* asthenia.

hyposthenian. **1:** weakening, enfeebling, debilitant, debilitating. **2:** an agent that enfeebles or weakens; a debilitating agent. *Cf.* astheniant.

hyposthenic. weak, enfeebled; of or pertaining to hyposthenia. *Cf.* asthenic.

hyposulfite (sodium hyposulfite). See sodium thiosulfate.

hyposulfite of gold and sodium. gold sodium thiosulfate.

hypotaxia. a state of poor coordination.

hypotension (low blood pressure). lower than normal blood pressure. In adult humans hypotension is often clinically defined as repeated readings of a systolic pressure below 100 mm Hg and a diastolic pressure below 40 mm Hg. Hypotension may result from shock (*q.v.*) loss of blood, various infectious agents, and any of numerous vasoactive drugs, venoms, or other toxicants. See, for example, *Echis*, haloperidol, halothane, *Helleborus niger*, heroin, iron poisoning, kinin peptide, larkspur poisoning, *Ligustrum vulgare*, meprobamate, morphine, nitroglycerine, paraldehyde, parasympathomimetic agent, phenytoin, potassium bromate, potassium permanganate, ricinine, sodium cyanide, tetrodotoxin, thiamylal, toxic shock syndrome.

hypotensive. **1:** characterized by, or causing lower than normal pressure, as of blood pressure. **2:** a subject with abnormally low blood pressure. **3:** a hypotensive agent or hypotensor. Toxicants that act on the autonomic nervous system to constrict blood vessels or reduce heart rate, or the force of ventricular contractions can produce hypotension.

hypotensive agent. a stimulus or substance that produces hypotension, *q.v.* *Cf.* hypotensive (def. 3), hypotensor.

hypotensor. a substance that produces hypotension, *q,v.* *Cf.* hypotensive agent.

hypothalamic-hypophyseal axis. See pituitary gland.

hypothalamic-pituitary system. See pituitary gland

hypothalamus. a region of the brain that regulates body functions such as temperature, blood pressure, appetite, and thirst.

hypothesia. abnormally diminished sensitivity to stimuli.

hypothesis. a scientifically testable inference, assertion, or working explanation of a phenomenon. A hypothesis is usually proposed as a possible explanation of existing data in order to devise a test (usually a controlled experiment) that can nullify the hypothesis, suggest other testable hypotheses, or lead to provisional acceptance.

hypothetical substance. See first mediator of damage, first mediator of defense.

hypothyroidism. a diminished activity of the thyroid gland or the resulting condition. Effects may include reduced basal metabolism, weight gain, sluggishness, somnolence, dryness of skin, constipation, arthritis, and sometimes myxedema. Overdosage of antithyroid drugs can cause the condition.

hypotonic. **1:** descriptive of a cell or tissue that has a lower osmotic pressure than the solution in which it is bathed. **2:** said of a solution that has a lower solute concentration, and thus a higher osmotic pressure than another solution with which it is compared, usually cytoplasm or tissue fluids. *Cf.* hypertonic, isotonic. See also tonicity.

hypotonic solution. a solution that has lower osmotic pressure than the cell or system from which it is separated (or potentially separated) by a semipermeable membrane (e.g., the plasmalemma). *Cf.* hypertonic solution, isotonic solution.

hypotoxic. slightly toxic.

hypotoxicity. diminished toxicity; the capacity to be only slightly poisonous.

hypoventilation. a state of reduced ventilation of the pulmonary alveoli.

hypovolemia (oligemia; oligohemia). diminished blood volume.

hypovolemic (oligemic, oligohemic). pertaining to or denoting a state of diminished blood volume.

hypovolemic shock. a rapid drop in blood pressure due to diminished blood volume.

hypoxanthine guanine phosphoribosyl transferase locus. HGPRT LOCUS.

hypoxemia. a deficiency or insufficiency of oxygen in circulating blood.

hypoxia. a term used variously in the literature of physiology and toxicology. **1a:** less than normal levels of oxygen in inspired air or gases, blood, or tissue, but not so extreme as anoxia. **1b:** a state of reduced oxygen supply to living tissues, even though they are adequately perfused by blood. **2b:** a level of oxygen within living tissue that is below that considered physiologically normal (i.e., less than adequate for normal metabolism). **2b:** a deficiency of oxygen in respiring tissues. **3. anemic hypoxia**. hypoxia due to reduced oxygen-carrying capacity of the blood due to anemia, *q.v.* It may result from hemorrhage or poisoning by carbon monoxide, chlorates, nitrites, and certain venoms. *Cf.* hypoxemia. **anoxic hypoxia**. an inadequate supply of oxygen to respiring tissues. It can be due to irritants of the respiratory epithelium, or to poisons that depress respiration. The partial pressure of oxygen in arterial blood is typically low, even though blood oxygen capacity and flow are normal. **arterial hypoxia**. anoxic hypoxia. **diffusion hypoxia**. a sudden transient decrease in alveolar oxygen tension when ambient air is inhaled following nitrous oxide anesthesia. It is due to the diffusion of nitrous oxide from the blood with consequent dilution of alveolar air. **histotoxic hypoxia**. this is not a hypoxia in the strict sense, but is that due to impaired utilization of oxygen by tissue, as in the case of poisoning by cyanide or hydrogen sulfide. *Cf.* anoxia. **stagnant hypoxia**. that due to decreased blood flow.

hypoxic. of, pertaining to, or characterized by hypoxia.

hypoxidosis. diminished cell function due to an inadequate supply of oxygen.

hystricomorph. any rodent of the suborder Hystricomorpha. Included are the guinea pig, New World porcupines, and the nutria. See guinea pig.

I

I. the symbol for iodine.

***i.d.*, I.D., ID**. intradermal, intradermic.

***i.g.*, I.G., IG**. intragastric.

***i.m.*, I.M., IM**. intramuscular.

***i.p.*, I.P., IP**. intraperitoneal.

***i.v.*, I.V., IV**. intravenous.

IA, IAA. 3-indolacetic acid.

IAA. indole acetic acid.

IAN. indoleacetonitrile.

IARC. International Agency for Research on Cancer of the World Health Organization (WHO).

iatrogenic. originating from the physician or from treatment by a physician, surgeon, dentist, or equivalent health care professional.

iatrogenic disorder. any adverse condition brought about as a result of treatment by physician, surgeon, dentist, or equivalent medical professional. Improper care, negligence, or less than judicious treatment is implied.

ibenzmethyzin. procarbazine.

Iberian cross adder. *Vipera berus*.

ibiboboca. *Bitis arietans*.

Iboboca. *Micrurus lemniscatus*.

ibotenic acid (α-amino-2,3-dihydro-3-oxo-5-isoxazoleacetic acid; α-amino-3-hydroxy-5-isoazoleacetic acid; amino-(3-hydroxy-5-isoxazolyl)acetic acid). a neurotoxic isoxazole, ibotenic acid together with muscimol is largely responsible for the toxicities of *Amanita* species such as *A. muscaria* and *A. pantherina* and certain other poisonous mushrooms. It is insecticidal and very toxic to laboratory mice and rats. Ibotenic acid is neuroexcitatory, but potentiates narcosis. It is the precursor of muscimol.

iBululu. *Bitis arietans*.

ibuprofen (α-methyl-4-(2-methylpropyl)benzeneacetic acid). an antiinflammatory, $C_{13}H_{18}O_2$, with analgesic and antipyretic action. It is administered per os, to treat rheumatoid arthritis and osteoarthritis. Reactions to its use may include nausea, vomiting, stomach pain or cramps, constipation, diarrhea, heartburn, stiff neck, headache, fever, dizziness, depression, insomnia, blurred vision, or swelling of hands and legs. Persons allergic to aspirin or have a history of dizziness, bronchospasms, liver disease, hypertension or heart disease, nasal polyps, stomach ulcer, or intestinal bleeding should not take ibuprofen. During the last three months of pregnancy it may cause problems that affect the fetus or cause complications during delivery.

ichthyism (ichthyismus). poisoning from ingestion of decomposing or otherwise toxic fish. See also ichthyismus.

ichthyismus. ichthyism.

ichthyismus exanthematicus. toxic erythematous eruption from ingesting decomposing fish.

ichthyoacanthotoxin. a toxin or venom present in the sting, spines, or teeth of certain venomous fish.

ichthyoacanthotoxism. poisoning from envenomation by the stings, spines or teeth of venomous fish.

ichthyocrinotoxic. See crinotoxic.

ichthyocrinotoxin. any toxic substance produced by fish glands other than venom glands.

ichthyohemotoxic. 1: pertaining to fish which have a toxin in their blood (e.g., eels of the genera *Anguilla* or *Muraena*, certain marine fish). **2:** pertaining to the activity or effects of an ichtyohemotoxin. See also ichthyohemotoxin, ichthyohemotoxism.

ichthyohemotoxin. a toxin in the blood of certain fish. See also ichthyohemotoxic, ichthyohemotoxism.

ichthyohemotoxism. poisoning from ingesting fresh blood of certain fish such as eels of the genera *Anguilla* or *Muraena*. Such poisonings are extremely rare. *Cf.* ichthyotoxism. See also ichthyohemotoxic, ichthyohemotoxin.

ichthyootoxic. pertaining to **1:** fish that produce toxic roe. **2:** fish that produce a toxin that appears to be restricted to their ovaries or that is related to ovarian activity. **3:** the activity or effects of an ichthyootoxin.

ichthyootoxin. a toxin present in the roe and gonads of certain fish. See also ichthyootoxism, ichthyotoxin, ichthyotoxism.

ichthyootoxism. intoxication resulting from ingestion of toxic fish roe or ovaries. It is characterized by rapid onset of gastrointestinal symptoms including nausea, vomition, epigastric distress, and sometimes diarrhea, thirst, tinnitis, and malaise. In severe cases, respiratory distress, chest pain, convulsions and syncope or coma may result. See also ichthyootoxin, ichthyotoxin.

ichthyosarcotoxic. **1:** pertaining to fish which have a toxin (e.g., ciguatoxin, saurine, tetrodotoxin) in the musculature (flesh), viscera, or skin. **2:** pertaining to the activity or effects of an ichthyosarcotoxin.

ichthyosarcotoxin. a toxin (e.g., ciguatoxin, saurine, tetrodotoxin), excluding those produced by ordinary bacterial contamination, that occurs in the musculature (flesh), viscera, or skin of fish.

ichthyosarcotoxism. poisoning caused by eating the flesh of fish containing a toxin, excluding those produced during decomposition (See ichthyism). Examples are chimaeroid poisoning, ciguatoxism, clupeoid poisoning, elasmobranch poisoning, gymnothorax poisoning, hallucinatory fish poisoning, puffer (fugu, tetraodon) poisoning, and scombroid poisoning.

ichthyotoxic. **1:** pertaining to fish which contain a toxin. **2:** pertaining to the activity or effects of an ichthyotoxin.

ichthyotoxicology. See toxicology (ichthyotoxicology).

ichthyotoxicon. a toxic principle of certain fish.

ichthyotoxicum. an obsolete term referring to a presumed toxin in eel serum.

ichthyotoxin. **1:** an obsolete term referring to a presumed hemolytic principle of eel serum. **2:** any toxic substance in fish.

ichthyotoxism. **1:** intoxication resulting from any toxic substance present in fish. It is a form of fish poisoning (*q.v.*). Symptoms vary, but the chief signs are usually vomiting and muscular paralysis that occur usually within 30 minutes to 4 hours of ingestion. These signs may be accompanied by abdominal cramps, diarrhea, shock, and convulsions. *Cf.* ichtyosarcotoxism, ichthyootoxism, ciguatera.

ICN. cyanogen iodide.

icterepatitis. icterohepatitis.

icteric (icteritious; jaundiced). pertaining to, affected with, or characterized by jaundice.

icteritious. icteric.

icterogenic. causing jaundice (icterus).

icterogenicity. ability to cause jaundice (icterus).

icterohepatitis. inflammation of the liver with marked jaundice (icterus).

icterus. jaundice.

ictrogenic. able to cause icterus (jaundice).

ictrogenic photosensitivity. See photosensitivity.

ictus. a stroke, blow, seizure, convulsion; sudden attack.

ICU. Intensive Care Unit.

ID. **1:** intradermal, intradermic. **2:** infective dose (See dose).

ID_{50}. median infective dose (See dose).

-idae. the suffix of a family name in zoological nomenclature. *Cf.* -aceae.

idiolysin. a lysin that occurs naturally in blood in the absence of active or passive immunization.

idiopathetic. idiopathic.

idiopathic (idiopathetic). pertaining to idiopathy; self-generated; of unknown cause; pertaining to a morbid state with no known cause (e.g., idiopathic hypertension).

idiopathy. a disorder or morbid condition of spontaneous or unknown origin; one that is not preceded or provoked by another, or by an established cause.

idiospasm. a spasm limited to one part of the animal body. *Cf.* convulsion.

idiosyncrasy (idiosyncracy). **1a:** a peculiarity; an unusual structural or recurring behavioral trait that distinguishes one individual or type within a set; a trait observed in occasional individuals that is not normative for the set. **1b:** any trait that is distinctive of an individual. **2:** a physical or behavioral eccentricity. **3:** idiosyncratic reaction. **chemical idiosyncrasy** (drug idiosyncracy). **1a:** a genetically based and unusual reactivity to a food, drug, or toxicant such as an extreme sensitivity to low doses or extreme insensitivity to high doses of a xenobiotic chemical. See idiosyncratic reaction, idiosyncratic effect. **1b:** an unusual, inappropriate, or toxic response to a normal therapeutic dose of a drug. **drug idiosyncracy**. approximately synonymous with chemical idiosyncracy, but applies only to medications or drugs. This term is usually used in reference to the development of an adverse reaction with the initial use of a drug. The effect(s) may be identical to toxic effect(s) expected at higher doses (more correctly termed intolerance); it may be an exaggeration of a common mild side effect, such as antihistaminic sedation; or it may be unique.

idiosyncrasy of effect. the situation whereby a chemical or stimulus has no effect on an individual living system or one that is opposite or different from that which is usual or expected. See also idiosyncratic effect, idiosyncratic reaction.

idiosyncratic. of, pertaining to, resulting from, or having the characteristics of an idiosyncrasy; having the capacity to react idiosyncratically; eccentric; peculiar to the individual.

idiosyncratic reaction (idiosyncratic response). **1:** an atypical or abnormal response to a specific stimulus or biologically active agent; one that is qualitatively different from the expected, or one that is outside the range of response for a given population or species. **2a:** a genetically determined atypical reaction to a chemical. *Cf.* idiosyncrasy (chemical idiosyncrasy). **2b:** an abnormal (usually adverse) reaction to a drug, other than an allergic reaction, that is unrelated to the drug's known pharmacologic effects.

idiosyncratic response. idiosyncratic reaction.

IDLH. immediately dangerous to life and health.

IDPN. iminodipropionitrile.

Ig. immunoglobulin.

IgA. immunoglobulin A. See immunoglobulin.

IgD. immunoglobulin D. See immunoglobulin.

IgE. immunoglobulin E. See immunoglobulin.

IgG. immunoglobulin G. See immunoglobulin.

IgM. immunoglobulin M. See immunoglobulin.

ignatia. the dried ripe seed of *Strychnos ignatii* (Family Loganaceae). It contains several toxic alkaloids including strychnine (2-3%), brucine (ca. 1%), igasuric acid, and loganin. It is used as a bitter tonic.

Ijima sea snake. *Emydocephalus annulatus*.

Ikaheka snake. *Micropechis ikaheka*.

IL1, ILI. interleukin I.

IL2. interleukin II.

ileocolitis. inflammation of the ileum and colon.

ill (sick). not well.

Illicium (star anise; Chinese anise; sikimi). a genus of small evergreen trees and shrubs (Family Magnoliaceae) with persistent, poisonous leaves and nodding yellow or purplish flowers. The fruit, Chinese anise, is the source of anise oil. ***I. religiosum*** (sikimi). the leaves of this species are poisonous and are the source of sikimin.

illness (sickness). **1a:** the state of being unwell or of having a disease, *q.v.* **1b:** any condition of suffering. **2:** ailment.

illuminating gas. See gas.

iloyi. *Naja haje*.

imagocide. an agent that destroys adult insects (imagos), especially mosquitoes.

imfezi. *Naja nigricollis*.

1*H*-imidazole-4-ethanamine. histamine.

5-imidazoleethylamine. histamine.

4-imidazoleethylamine. histamine.

2-imidazolidinethione. ethylene thiourea.

2-(4-imidazolyl)ethylamine. histamine.

4,4′-(imidocarbonyl)bis(*N,N*-dimethylaniline). auramine.

imidol. imipramine hydrochloride.

iminodipropionitrile (ββ′-iminodiproprionitrile; IDPN). a synthetic, neurotoxic nitrile that slows axonal transport, producing a neuropathy in the spinal cord and brain stem that is marked by an accumulation of neurofilaments in the proximal portion of the axon. This is followed by a secondary progressive atrophy of the distal axon, with demyelination and gliosis. Symptoms may include hyperactivity, described as a waltzing syndrome, with circling and head-rolling in laboratory mice and rats.

ββ′-iminodiproprionitrile. iminodipropionitrile.

imipramine (10,11-dihydro-*N,N*-dimethyl-5H-dibenz[b,f]azepine-5-propanamine; imizin; Antideprin). a very toxic phenothiazine-like substance in which the sulfur atom of the phenothiazine nucleus is replaced by a 2-carbon chain. Usually referred to as a tricyclic antidepressant, it is used chiefly in the long-term treatment of depression. Beneficial effects may not appear for up to six weeks. Adverse effects may include blurred vision, dry mouth, excessive sweating, dizziness, nausea, and constipation; older men may experience difficult urination. Overdosage, especially in children, can be fatal. **imipramine hydrochloride** (Imiprin, Imidol, Tofranil). action is the same as above; the hydrochloride is very toxic to laboratory mice and rats.

Imiprin. imipramine hydrochloride.

imitation flavor. food flavoring.

imizin. imipramine.

immediate hypersensitivity. See hypersensitivity.

immediate response. **1:** immediate effect. See effect. **2:** an action or set of actions that are implemented with little or no planning and little regard for alternatives to manage a rapidly worsening situation or to minimize the impacts of a critically severe hazardous situation.

immediate toxicity. immediate effect. See effect.

immediately dangerous to life and health (IDLH). **1a:** a measure of the level of concern regarding exposure to a particular chemical. **1b:** the maximum concentration or amount of a particular chemical (in air) from which a healthy individual can escape in 30 minutes without impairment of the ability to escape and without suffering irreversible impairment of health.

immedicable. **1:** pertaining to a disease or condition that is in such a state that it cannot be remedied; incurable in the medical sense. **2:**

incurable or irremedial in a social sense.

immune amplification. the maintenance of competency of the immune system by controlled amplification of one component (e.g., the humoral component) with the other component (e.g., the cellular component) becomes deficient. This process is extremely sensitive to disruption by toxic chemicals.

immune body. antibody.

immune clearance. the rapid elimination of the antigen-antibody complex resulting from the reaction of antigen-specific antibodies with antigen (e.g., a toxin) that is introduced into an immune animal.

immune complex. the product of an antigen-antibody reaction. It is a complex consisting of antibody, antigen, and complement.

immune cytolysis. the destruction of cells mediated by a particular antibody acting in concert with complement.

immune reaction. **1:** an antigenic response to a specific antibody. **2:** a specific reaction by host cells to antigenic stimulation. Such reactions include allergic and autoimmune reactions, may cause tissue damage, and are sometimes fatal. Immune reactions are the most frequent adverse effects of drugs. See immune response.

immune response. a complex selective response of the vertebrate immune system whereby specific antibodies and/or cytotoxic cells are produced in response to any of numerous substances (antigens, which are usually proteins or polysaccharides), microorganisms, parasites, cells, and tissues which are foreign to the body or interpreted as foreign. The chief components of the vertebrate immune system are immunoglogulins, lymphocytes (B and T lymphocytes), phagocytes, complement, properdin, interferon, and the migratory inhibitory factor. **antibody-mediated immune response**. humoral immune response. **cell-mediated immune response**. cellular immune response. **cellular immune response**. the process whereby cytotoxic T lymphocytes are produced in response to antigen exposure. This response is central to delayed hypersensitivity, rejection of foreign tissue, and the destruction of virus-infected cells and neoplasms. Antigens that usually induce cellular immune responses include fungi, intracellular viruses, cancer, and foreign tissue. **humoral immune response**. the process whereby antibodies that circulate in the blood are produced by B lymphocytes in response to antigen exposure. This type of response can produce either immunity or hypersensitivity. Antigens that usually induce a humoral response include toxins, extracellular viruses, and bacteria. Antibodies so produced may, for example, combine with the active sites and thus neutralize toxins or prevent mucosal attachment of gastrointestinal parasites. They can agglutinate bacteria and viruses, rendering them less effective and facilitating removal. Antibodies attached to a cell surface can activate the complement pathways, provoke inflammation, and/or facilitate phagocytic activity. **nonspecific immune response.** a response to antigen exposure that does not involve antibody formation. Included are inflammatory reactions, phagocytosis of microbiota, and complement activation. See allergen, allergy, antibody, antigen, antigen-antibody reaction, complement, immune system, immunogen, immunity, lymphocyte (B lymphocyte, T lymphocyte), phagocytosis.

immune suppression. immunosuppression.

immune serum globulin. immunoglobulin.

immune system. a complex of biochemical and cellular elements (See especially immune response) that serves to protect an organism from foreign bodies (e.g., pathogenic organisms, complex xenobiotic molecules).

immunity. **1a:** the state of being immune; a state of resistance. **1b:** quality of being impervious to or unaffected by a particular disease, condition; unresponsive to or unaffected by a drug or poison at doses well above that which affects most individuals. **1b:** a state in which a host organism is more or less resistant (immune) to a pathogen or a toxic chemical such as bee or snake venom. So defined, there are two main types of immunity: nonspecific immunity and specifically acquired immunity (which includes passive immunity and active immunity). **2:** complete resistance to a particular disease. **3:** in law, exemption from a duty or obligation that is generally required by law; exemption from prosecution for felony, usually for providing information of importance to a law enforcement agency, or for giving evidence at a criminal trial. **active immunity**. the resistance of an organism to disease due to the production of antigen-specific antibodies in response to antigens produced by or associated with a pathogen. **active induced immunity**. an immune response (resistance) due to the production of either humoral antibody or lymphocytes that contain cell-bound antibody (cell-mediated immunity) in response to antigens from a pathogen or disease organism (as by vaccination). **active natural immunity**. an immune response that develops after exposure to infection. **active immunity**. that due to the development of an immune response following stimulation with an antigen. **cell-mediated immunity**. active induced immunity marked by the production of T lymphocytes that contain cell-bound antibody; free antibodies are not involved. The lymphocytes may kill infected cells or may secrete soluble substances that facilitate or augment inflammation and elimination of the antigen. Subsequent exposure to the same antigen will provoke a more rapid and powerful response. *Cf.* humoral immunity. **concomitant immunity** (relative immunity). resistance to a transplanted tumor by a host who already bears such a tumor. This state does not inhibit the growth of the original (primary) tumor. **2:** resistance to reinfection by a host that is currently infected by the same pathogen. **nonspecific immunity**. that which does not involve the production of antibodies and includes such mechanisms as phagocytosis and the action of interferon and lysozyme. Plants have no immune system, rather entry of pathogens is restricted by physical barriers or physiological responses (e.g., the production of phytoalexins). **passive immunity**. that which results from the presence of antibodies derived from another individual; included are passive natural immunity and passive induced immunity. **passive natural immunity**. that which results from immunoglobulins transferred from the mother to the fetus through the placenta. **passive induced immunity**. that acquired from heterologous antibodies administered to protect the organism from the effects of disease. **relative immunity**. concomitant immunity. **specifically acquired immunity**. that due to the presence of antibody in an animal; it subsumes **passive immunity** and **active immunity.**

immunization. the process of increasing an animal's resistance to pathogens by conferring either active or passive immunity. Active immunity may be conferred by a vaccine (injected or given orally). The vaccine is comprised of antigens that stimulate the development of clones of specific B or T lymphocytes by the immune system, thereby providing specific long-term or permanent immunity. Alternatively, a temporary passive immunity may be conferred by the injection of antibodies produced by another organism. Passive immunity is lost when the injected antibodies are catabolized.

immunocompromised. a condition of diminished immune responsiveness due to the administration of, or exposure to immunosuppressants; or to a disease process; irradiation; or malnutrition.

immunodepressant. **1:** immunosuppressant. **2:** immunosuppressive agent.

immunodepression. immunosuppression.

immunodepressive. immunosuppressive.

immunodepressor. See immunosuppressant (def. 2).

immunogen. any substance that can induce a humoral antibody and/or a cell-mediated immune response, as opposed to a tolerogen. See also antigen.

immunogenic. of or pertaining to the capacity to induce a humoral antibody and/or a cell-mediated immune response, but not immunologic tolerance.

immunogenicity. **1:** the quality of being immunogenic; the ability of an antigen to evoke an immune response. **2:** the extent or degree to which a substance possesses this quality.

immunoglobulin (immune serum globulin). a secretion of a mature B lymphocyte, produced in response to stimulation by an antigen. The immunoglobulin monomer is comprised of four polypeptide chains, two heavy and two light, linked together by disulfide bonds. Each antibody is an immunoglobulin with a high degree of specificity. There are five structurally distinct classes of immunoglobulins. These are immunoglobulins A, D, E, G, and M, *q.v.* See also antibody, antigen, immunity, lymphocyte.

immunoglobulin A (IgA). a type of humoral antibody that comprises 5-15% of serum immunoglobulins and has a half-life of 6 days. It is a four-chain monomer with a molecular weight of 160 kD; it may also appear as a dimer, trimer, or multimer. It occurs in all bodily secretions, and is the chief antibody in the intestinal mucosa, bronchi, saliva, and tears. It protects body surfaces from infectious microorganisms. About 1 in 600 persons are seriously deficient in IgA; levels in such cases may be only 1-5% of normal, even though the B lymphocytes have IgA on the cell surface. Such individuals have a heightened susceptibility to infection by pyogenic microbes (e.g., sinopulmonary infections). Such individuals may also suffer from intestinal lymphangiectasia, arthritis, gluten-sensitive enteropathy, allergies, myotonic dystrophy, cirrhosis, and autoimmune disease. They may also produce low molecular weight IgM antibodies against foods such as milk. Blood transfusions to an IgA-deficient individual who has anti-IgA antibodies can induce anaphylactic shock or a lethal hemolytic transfusion reaction.

immunoglobulin A deficiency. See immunoglobulin A.

immunoglobulin D (IgD). this type of humoral antibody comprises about 1% of serum immunoglobulins and has a half-life of only 2-3 days. It is a four-chain monomer with a molecular weight of 185 kD. IgD attaches to the surface of the B lymphocyte and, acting as the antigen receptor, is important in the activation of B lymphocytes. Its levels increase in individuals during allergic reactions to milk, insulin, penicillin and a variety of toxins.

immunoglobulin E (IgE). this type of humoral antibody comprises only 1% of total immunoglobulins and has a half-life of about 2.5 days. It is a monomer with a molecular weight of 190 kD. IgE is concentrated in the lungs, skin, and mucous membranes. It provides the primary defense against xenobiotic antigens and is responsible for anaphylactic hypersensitivity (Type I) in humans.

immunoglobulin G (IfG). a specialized type of protein produced in response to invading bacteria, fungi, and viruses. It comprises about 85% of serum immunoglobulins in adult humans and has a half-life of 23 days. It is a four-chain monomer with a molecular weight of 154 kD. IgG crosses the placenta and protects against erythrocyte and leucocyte antigens.

immunoglobulin M (IgM). the largest of the humoral antibodies, it comprises 5-10% of serum immunoglobulins in adult humans and has a half-life of 5 days. It is a pentamer comprised of five four-chain monomers and has a molecular weight of 900 kD. It is the first immunoglobulin produced when the body is challenged by antigens and occurs in circulating blood. IgM triggers increased production of IgG and complement fixation. It is the antibody chiefly responsible for ABO blood incompatibilities.

immunologic memory. See anamnesis (def. 2).

immunology. the science that treats the study of the immune system and its behavior, especially the reaction of the immune system to antigenic stimulation. The concerns of this field overlap significantly with toxicology (See, for example, antiantitoxin, toxin, allergy).

immunosuppressant (immunodepressant). **1:** immunosuppressive. **2a:** an agent that produces immunosuppression; immunodepressor; immunodepressive agent. **2b:** any drug used in immunosuppression therapy to inhibit immune responses.

immunosuppression (immune suppression; immunodepression). **1a:** the suppression, inhibition, diminution, or interference, usually temporary, with cellular or humoral immunity in animals. It may be due to immunologic unresponsiveness (tolerance), but is usually due to disease or to the action, at sublethal doses, of biologically active agents such as ionizing radiation, anticonvulsants, chemotherapeutic (cytotoxic) drugs, corticosteroids, and various environmental chemicals (e.g., benzene, organometallic compounds, heavy metals, and halogenated aromatic compounds). The actual immune mechanisms may be impaired, or in some cases, there is an inhibition or impairment of the cellular immune response, resulting in leukemia or lymphoma. The victim of suppression is more susceptible to infection by pathogenic organisms. **1b:** the administration of an agent that impairs the ability of the immune system to respond to antigenic stimulation. **2:** an abnormal condition of the immune system, characterized by a pronounced depression in the ability to respond to antigenic stimuli.

immunosuppressive (immunodepressive). **1:** pertaining to or causing immunosuppression. **2:** an agent (e.g., X-rays, antimetabolites, antilymphocyte serum, specific antibody). that causes immunosuppression. Many of these are extremely toxic. *Cf.* immunosuppressant.

immunosuppressive agent (immunodepressant). a chemical (e.g., a toxicant or therapeutic agent) that suppresses or interferes with cellular or humoral immunity. Examples are antimetabolites, antilymphocyte serum, ionizing radiation, specific antibodies, and certain environmental chemicals. Such agents are used to treat conditions where certain antibodies must be inactivated (e.g., following organ transplants). The use of immunosuppressive therapy increases the risk of infection and secondary cancer among those treated.

immunotoxicant. a chemical that adversely affects the immune system. Included are organometallic compounds (e.g., organotins, methyl mercury), heavy metals (e.g., cadmium and lead), and halogenated aromatic compounds (e.g., PCB and TCDD). See also immunosuppression, immunosuppressant.

immunotoxicology. See toxicology.

immunotoxin (IT). **1:** a toxin formed by the linkage of a monoclonal antibody specific for target cell antigens to a plant or animal cytotoxic substance. With parenteral injection, the antibody portion of the immunotoxin binds to a specific target cell and destroys it on contact. The A chain of ricin, on contact with a tumor cell, inhibits protein synthesis by the cell. **2:** the product of the fusion of a monoclonal antibody (or a fraction thereof) to any toxic molecule. As in def. 1, the antibody portion directs the molecule to antigens on the surface of the target cell and the cell is destroyed by the toxic moiety. Such toxins find a major application in cancer therapy (immunotoxin therapy). **3:** any antitoxin or antivenin.

impervious. impenetrable, impermeable, airtight, waterproof.

Imuran. azathioprine.

In. the symbol for indium.

in extremis. at the point of death; irreversibly near death.

inactivate. **1:** to make inactive; deactivate. **2:** to cause a substance (e.g., an enzyme, a toxicant, complement) to lose a specific type of activity. **3:** to cause an infective agent (e.g., a microbe) to lose its disease-producing capability. **4:** to transform a substance (reversibly or irreversibly) into an inactive or much less active form. For example, complement is inactivated (destroyed) when serum is heated.

inactivation. **1a:** the state or process of inactivating. See inactivate. **1b:** elimination of the activity or effects of a chemical or biologically active agent by physical or chemical means. *Cf.* chemical antagonism (See antagonism).

inactive. not active, static, inert.

iNambezulu. *Dispholidus typus*.

inanition. a debilitated condition due to an insufficiency of essential nutrients, as in starvation or maladsorption syndrome.

incapacity. intolerance.

incidence. **1:** occurrence; the frequency of occurrence of any event or condition during a given span of time in a designated population. **2a:** the rate at which new instances or cases of an event, situation, or condition (e.g., a disease, of a crime) appear. **2b:** in epidemiology, the rate at which new cases are encountered in a specified population.

incomplete combustion. See combustion.

incoordination. See ataxia. **partial incoordination**. See dystaxia.

incubation period. See latent period.

indandione. any of a class of synthetic anticoagulants (e.g., diphacinone, phenindione, pindone) derived from 1,3-indanedione. All have actions similar that of the coumarins, but tend to be more toxic and to additionally cause cardiopulmonary and neurologic damage.

Inderal. a beta-adrenergic blocking agent.

Indian aconite. **1:** *Aconitum ferox*. **2:** the dried tuberous root of *Aconitum ferox*. See also aconite.

Indian apple. podophyllum.

Indian bean. *Abrus precatorius*.

Indian berry. See cocculus.

Indian cherry. *Rhamnus caroliniana*.

Indian cobra. a common name applied to *Naja naha* or to *Naha n. naha*.

Indian corn. *Zea maize*.

Indian ginger. *Asarum canadense*.

Indian green tree viper. *Trimeresurus gramineus*.

Indian hemp. *Apocynum*, cannabis, or *Cannabis sativa*.

Indian krait. *Bungarus caeruleus*.

Indian licorice. *Abrus precatorius*.

Indian mustard. *Brassica juncea*.

Indian paint brush. Castilleja.

Indian pea. *Lathyrus sativus*.

Indian poke. *Veratrum viride*.

Indian tobacco. *Lobelia inflata*.

Indian turnip. a common name of *Ariseama atrórubens* and of *A. triphyllum*.

indicant. **1:** indicating, providing evidence of, pointing to. **2:** an agent that indicates.

indication. **1:** the use of indicators or the act of indicating. **2a:** evidence (qualitative or semiquantitative; *Cf.* monitor, monitoring) of the cause of a disease, condition, or presence; a guide or warning. **2b:** a sign, symptom, clue, other evidence, or demonstration that serves as a basis for initiating an action such as the treatment of a disease or condition. **3:** a sign, symptom, or circumstance which provides evidence as in def. 2. **4:** that which indicates or serves as a warning, especially an early warning. The use of bioindicators or the act of bioindicating. See also bioindication. **causal indication**. an indication of the cause of a disease or phenomenon. **specific indication**. an indication of the nature of a disease or condition. **symptomatic indication**. an indication of the symptoms that may accompany a disease or condition.

indigo. **1:** any plant of the genus *Indigofera*; **2:** A blue dye from *Indigofera spp.*, especially *I. tinctoria* and *I. suffruticosa*, and certain other plants.

Indigofera (indigo). A genus of some 700 mostly Old World tropical and subtropical herbs and shrubs (Family Fabaceae, formerly Leguminosae) used to prepare dyes. Some species are toxic. ***I. endecaphylla*** (creeping indigo). a toxic annual or perennial herb of the Old World tropics. Beta-nitropropionic acid is the toxic principle. The unusually high nitrite content may also contribute to this plant's toxicity. Toxicity varies among the various classes of livestock; swine do not eat this plant.

indisposed. slightly ill; not feeling well; mildly debilitated; ailing.

indisposition. a slight illness or a mild (usually temporary) debility.

indium (symbol, In)[Z = 49; A_r = 114.82]. a rare, soft, ductile, silvery metallic element of group III of the periodic table found usually in zinc ores. Salts of indium are relatively nontoxic per os, but are highly toxic when administered *i.v.* or subcutaneously. They have proven to be cardiotoxic, hematoxic, hepatoxic, and nephrotoxic. *Cf.* indium sulfate, indium trichloride.

indium chloride. indium trichloride.

indium sulfate. a white or grayish hygroscopic powder that decomposes on heating. It is moderately toxic to rabbits per os, but extremely poisonous when administered *i.v.*

indium trichloride (indium chloride). a yellowish crystalline deliquescent substance. It is an extremely toxic substance when administered *i.v.* or *s.c.* to laboratory rats.

inDlonlo. *Bitis caudalis*.

Indocybin. psilocybin.

3-indolacetic acid (formerly heteroauxin; indole-acetic acid; 1*H*-indole-3-acetic acid; IA; IAA; auxin 1AA; indol-3yl-acetic acid). the most widely distributed natural auxin and the chief auxin of higher plants, $C_8H_6NCH_2COOH$. It is a phenoxyalkane carboxylic acid. It is also used as a plant growth regulator and herbicide.

indole (2,3-benzopyrrole). an ergot alkaloid (See *Claviceps*, ergot) that occurs also in feces and certain bacterial cultures as a decomposition product of tryptophan as well as certain plants (e.g., *Gelsemium*, *Hippomane mancinella*, and *Peganum*). It is obtained also from coal tar. Indole is highly toxic and carcinogenic. It is used in highly dilute solutions in perfumes.

indoleacetic acid. 3-indolacetic acid.

1*H*-indole-3-acetic acid. 3-indolacetic acid.

indolent. **1:** indisposed to action; inactive; lazy. **2:** sluggish; inactive; not developing. **3:** said of a morbid condition that is painless or nearly so.

indolepyruvic acid. a naturally occurring auxin.

indolizidine-1,2,8-triol. a mydriatic alkaloid, that occurs in species of *Astragalus*, *q.v.* See also *Oxytropis*.

indol-3yl-acetic acid. auxin 1AA.

induced. produced artificially; produced by induction.

induced tolerance. See tolerance (induced tolerance).

inducing agent (inductor; enzyme inducing agent). a substance, usually a small molecule that promotes the production of a specific enzyme in an organism. See also repressor, enzyme induction. The addition of such to the cell can effect enzyme synthesis. Not all enzymes are inducible.

induction. **1:** initiation; the process of initiating or causing to occur (e.g., the induction of labor with oxytocic drugs). **2:** evocation; the production of a specific morphogenic effect in an embryo by a chemical produced in another part of the embryo. **3a:** the production of paralysis, unconsciousness, or anesthesia by specific agents. **3b:** the stage from administration of an anesthetic until the optimal level of anesthesia is reached. **4:** inductive phase; the time from administration of an antigen until immune reactivity is detected.

industrial poisoning. poisoning that takes place in an industrial setting.

industrial toxicology. See toxicology.

inebriant. **1:** making drunk; intoxicating **2:** an intoxicant, especially ethyl alcohol. See inebriety.

inebriate. to make drunk, to intoxicate.

inebriated. drunk, drunken; intoxicated, especially by excessive intake of alcoholic beverages.

inebriation. intoxication, especially by alcoholic beverages. *Cf.* inebriety.

inebriety (ebriety; ebrietas). **1a:** drunkenness, alcoholic intoxication. **1b:** habitual consumption of excessive amounts of alcoholic beverages.

inert. **1:** inactive; nonreactive; physiologically inactive. **2:** denoting regions of hetero-chromatin with inactive genes.

infant. a child between the ages of 2 weeks and 1 year.

infanticidal. able to, or prone to commit infanticide.

infanticide. **1:** the murder or killing of an infant or small child. **2:** one who commits infanticide.

infestation. invasion of a living organism by parasites.

infiltrate. to penetrate the interstices of a substance, environmental medium, or tissue. *Cf.* perfuse.

infiltration. **1:** the penetration and accumulation of a foreign substance into cells, tissues, or environmental media, or the penetration and accumulation of a substance normally present in amounts in excess of that which is normally present (e.g., infiltration of water into soil). *Cf.* perfusion.

infirmity, bodily infirmity. roughly synonymous with disease, *q.v.*

inflammagen. an irritant that induces the complex series of cellular processes that produce inflammation. *Cf.* edemagen

inflammation. erythema, heat, pain, and swelling of living tissue that occur as a reaction to chemical or physical irritation (traumatic inflammation) or injury, or to infection. In this process, mast cells release a number of substances, including histamine and kinins. The former is thought to be responsible for most of the effects. It increases blood flow to the injured tissue, causing erythema and heat; it makes the capillaries more leaky with consequent local extravasation and edema. The pain of inflammation is due to stimulation of nerve endings by inflammatory substances released by the mast cells. Leucocytes are usually attracted to the inflamed tissue by the inflammatory chemicals. They act against invading microorganisms and facilitate repair of tissue. If inflammation is extensive, it may be accompanied by loss of function of the affected part. **acute inflammation**. inflammation marked by rapid onset and a relatively short course. **chronic inflammation**. inflammation that progresses slowly, is of long duration, and usually causes scarring. **pseudomembraneous inflammation**. that during which a pseudomembrane forms. It is due a necrotizing toxin as in diphtheria. **purulent inflammation** (suppurative inflammation). acute exudative inflammation in which polymorphonuclear leukocytes accumulate to such an extent that their enzymes locally or diffusely liquefy the inflammed tissues. The purulent exudate (pus) consists of plasma, degenerate and necrotic cells, the products of tissue digestion, and various other constituents such as leukocytes and other formed elements of blood, debris, and often the causative agent. **specific inflammation**. that due to a specific organism. **subacute inflammation**. a relatively mild inflammation. It may worsen and become acute or chronic. **suppurative inflammation**. purulent inflammation. **toxic inflammation**. that due to a toxin or other toxicant.

infusion. the introduction of a substance interstitially or directly into a vein by means of gravity flow. *Cf.* injection, instillation.

ingesta. foodstuffs or any material that is normally consumed by a given species of animal. *Cf.* ingestant, ingestion.

ingestant. any substance that enters the body via the digestive tract. *Cf.* ingesta, ingestion.

ingestion. **1:** the process of taking in food material or liquids through an oral cavity into a digestive tract or a food cavity; feeding. **2:** the consumption of food materials. **3:** commonly used for the act of taking any materials into the body of an animal through a mouth.

ingravescent. of gradually increasing severity.

INH. isoniazid.

inhalant. a substance that may be taken into the body by way of the mouth and/or nose, upper respiratory tree, and lungs. See inhalation.

inhalation (inspiration). **1:** inspiration; the act of inspiring, inhaling, or breathing in; drawing air into the lungs. Inhalation brings to the lungs and upper respiratory tree, not only air, but other gases, vapors, fumes, or mists, and small particulates (e.g., respirable dusts, ash, smoke, small aerobiota). Inhalation is a major route of entry of many microorganisms and toxicants that are borne on gases or respirable particles or that are respirable themselves (e.g., fumes, vapors). **2:** the inhaling of a medicated vapor with the breath. **3:** a medication or drug in solution, introduced by the oral or nasal respiratory route as a nebulized mist or aerosol in order to penetrate the respiratory tree and produce local or systemic effects. **solvent inhalation**. self-intoxication by inhaling the fumes of volatile organic solvents contained, for example, in cleaning fluids, gasoline, glue, lacquer thinners, lighter fluid, and nail polish remover. See also glue-sniffing.

inhalation burn. See burn.

inhalation LC_{50}. median lethal inhalation concentration. See concentration.

inhale. to take into the pulmonary system (through the mouth and/or nose) by breathing.

inhibit. to retard, arrest, block, restrain, suppress.

inhibition. the arrest, restraint, retardation, blocking, or suppression of a process. *Cf.* induction. See also enzyme inhibition.

inhibitive. inhibitory.

inhibitor. **1:** a substance or other agent that arrests, retards, blocks, or suppresses the progress of a chemical reaction, of growth, or other biological activity. **2:** a chemical substance that inhibits the action of a tissue organizer or the growth of microorganisms.

inhibitory (inhibitive). having the capacity to restrain, retard, arrest, or suppress a process; causing inhibition.

inhlanhlo. *Dispholidus typus*.

Inimicus japonicus (lupo). a venomous fish (Family Scorpaenidae) that occurs in the coastal waters of Japan. The venom apparatus is of the *Scorpaena* type.

initiation. **1:** induction; introduction, inception, beginning. **2:** in carcinogenesis, the transformation of a benign cell to one with the potential for malignant growth.

initiator. a substance that can cause the initial step or phase in a process (e.g., a chain reaction, carcinogenesis). See initiation.

inject. to introduce a fluid forcefully or under pressure into an organism or part of an organism (e.g., by using a hypodermic syringe).

injectable. **1:** able to be injected. **2:** a substance that can be injected.

injected. **1a:** introduced by injection. **1b:** of or pertaining to a fluid introduced into a body by injection; introduced by injection. **2:** congested. **3:** denoting or referring to blood vessels that are noticeably distended with blood.

injection. **1:** the introduction of a fluid into an organism or part of an organism as by a hypodermic syringe, compressed air, or other means. Except for drugs of abuse and executions in some governmental jurisdictions, exposure to poisons is rarely by injection. *Cf.* infusion, instillation. **2:** an injectable or injected preparation, or one prepared to be injected. **3:** hyperemia or congestion. **depot injection**. the injection of a substance in a vehicle that remains at the site of injection for some time, permitting slow absorption. **hypodermic injection** (subcutaneous injection). injection into subcutaneous cellular tissue. **intramuscular injection**. injection into muscular tissue. **intravenous injection**. injection into the lumen of a vein. **intraarterial injection**. injection into the lumen an artery. **rectal injection**. injection into the rectum. **vaginal injection**. injection into the vagina.

injure. to harm, afflict, hurt, impair, wound.

injurious. harmful, detrimental; not necessarily destructive or irreversible. In toxicology, injury is generally taken to be less debilitating than damage.

injury. **1a:** harm, hurt, affliction, impairment, wound. **1b:** physical trauma to some part of a living system. This may be limited to physical pain or may include structural impairment. **1c:** physical or functional impairment. **1d:** psychological trauma. **1e:** in toxicology, injury is taken to be less debilitating and less permanent than damage. *Cf.* damage, harm, trauma. **bodily injury**. any impairment of physical condition (e.g., pain, disfigurement, sensory deficits, physical dysfunction of any type). **irreparable injury**. any injury great or small that is of constant or frequent occurrence that cannot reasonably be repaired or receive reasonable compensation under law. **permanent injury**. injury that appears to be beyond the possibility of repair. **serious bodily injure**. severe injury including permanent or protracted loss or impairment of any organ or bodily member, permanent disfigurement.

inkberry. *Phytolacca americana*.

inkweed. *Drymaria pachyphylla*.

innocuous (innoxious). harmless, benign, nontoxic, not injurious, damaging or harmful.

innoxious. innocuous.

Inocybe a genus of muscarinic mushrooms (Family Cortinariaceae) that includes a large number of species and varieties. The most poisonous and dangerous are *I. napipes* and *I. fastigiata*. As with all members of the family, the spores are brownish. The various species and varieties range from moderately to very toxic. Many contain relatively large amounts of muscarine. These corts affect chiefly the autonomic nervous system and liver. Early symptoms are profuse sweating, salivation, stupor, and rapid loss of consciousness. The face becomes bluish and the lips swell, becoming pinker due to dilation of blood vessels. Muscle spasms in the extremities may occur periodically, even though the muscles are flaccid. Reflexes are suppressed and the pulse is very weak. In the absence of suitable medical treatment, the victim may die from cardiac arrest. Inocybe group, the most dangerous are *I. napipes* and *I. fastigiata*.

inotropic. of or pertaining to the force of muscular contraction, especially those of cardiac muscle.

inotropic agent. an agent that increases myocardial contractility.

INPC. propham.

insane root. *Hyoscyamus*.

insariotoxin. T-2 toxin (See toxin).

insect. any arthropod of the Class Insecta.

insect repellent. the most commonly used pesticide in insect repellents is DEET. It is neurotoxic and can cause brain disorders, slurred speech, difficulty in walking, tremors, and even death. There are documented cases of acute neurotoxicity in children who had been exposed to or accidentally swallowed DEET; some terminated fatally. It is moderately toxic to laboratory rats and rabbits, causing chromodacryorrhea, depression, loss of righting reflexes, dyspnea, tremors, coma, terminal convulsions, and death; in at least one study, survivors recovered fully with no evident sequalae.

Insecta. a very large class of invertebrate animals (Phylum Arthropoda) that represent more than 75% of known animals species. They are characterized by the presence of three main body parts (head, thorax, and abdomen), three pairs of walking legs borne on the thorax, one pair of antennae, external respiration by means of a tracheal system, Malpighian tubules, variously modified mouthparts in a association with a diversity of modes of ingestion, two pairs of wings (usually) that insert on the second and third thoracic segments.

insecticidal. capable of destroying insects. *Cf.* entomotoxic.

insecticidal synergist. See pesticidal synergist.

insecticide. a chemical agent, a pesticide, that kills insects. Solvents and emulsifiers are required in most liquid preparations of insecticides and while they are generally of very low toxicity, they must be considered as associated causes of poisoning. *Cf.* pesticide. **carbamate insecticide**. a carbamate (any salt or ester of carbamic acid) used as an insecticide (e.g., carbaryl, carbofuran, methomyl, propoxur) with an action similar to that of an organophosphate (OP) in that it inhibits cholinesterase at nerve junctions. However, the inhibiting bond is much less durable. Frequently, the inhibition cannot be seen in the laboratory because of this reversibility. Atropine sulfate injections readily reverse the effects of inhibition. The oximes (e.g., pralidoxime, or 2-PAM) are contraindicated in carbamate poisoning, at least for carbaryl poisoning since it may increase the bonding and inhibition of cholinesterase. Signs and lesions are similar to those of the OP poisonings. See also carbamate, carbamate poisoning. **contact insecticide**. a poison that is effective on contact with the outer surface of an insect. **contact insecticide.** an insecticide that kills insects or other arthropods when it comes into direct contact with the outer surface of their bodies. Such poisons may be, for example, in controlling phytophagous insects that have sucking mouth parts and cannot therefore be killed by ingestion of insecticides applied to plant surfaces. **household insecticide**. any insecticide produced for use in the home. Many are multipurpose insecticidal sprays that contain pyrethrum. Commercial formulations containing pyrethrum also contain toxic "inert" components. While pyrethrum is relatively nontoxic to humans, it is toxic to honeybees and many other beneficial insects, fish, and wildlife. **natural insecticide**. any insecticide traditionally derived from plants. These substances (e.g., rotenone, pyrethrum) have traditionally been regarded as safe for use on animals. Nicotine (as nicotine sulfate) is an exception. Unless carefully used, poisoning may result. Furthermore, piperonyl butoxide and rotenone may be highly toxic to mink and should not be used where mink under 8 weeks of age (e.g., nest boxes) can come into contact with them. **organophosphate insecticide**. organophosphate esters from a very large class of neurotoxic insecticides. Well-known examples are Malathion, parathion, and leptophos. They are powerful acetylcholinesterase inhibitors. See also carbamate insecticide, organophosphorus compound. Organophosphates are short-lived in the environment and have thus been considered less hazardous to non-target organisms than many other pesticides, such as the organochlorines. See also organophosphorus compound.

Insecticide 3960-X14. a trademark of Strobane®.

Insecticide No. 497. dieldrin.

Insecticide No. 4049. malathion.

insecticide-resistant. pertaining to insects (and related target organisms) that continue to multiply when exposed to insecticidal substances. *Cf.* herbicide-resistant, pesticide-resistant.

Insectivora. an order of small, primitive, carnivorous or omnivorous, eutherian mammals. The snout is typically long and tapering with numerous small, sharp teeth; the surface of the brain is smooth, the cerebral hemispheres are small. Locomotion is of the plantigrade type and the feet have 5 clawed digits. They are usually nocturnal. Typical genera include *Sorex* (shrew), *Erinaceus* (hedgehog), and *Talpa* (mole).

insectivore. **1:** any eutherian mammal of the Order Insectivora. **2:** less commonly, any animal that feeds on insects.

insectivorous (entomophagous). feeding on insects.

insensate. unconscious.

insensible. unconscious.

insensitive. **1:** anesthetized, numbed, deadened. *Cf.* insensate. **2:** hyposensitive.

insensitivity. **1a:** a complete lack of sensitivity. **1b:** anesthetized, numb. **2:** in toxicology, this term is usually synonymous with hyposensitivity, *q.v.*

insomnia. the inability to fall asleep or to remain asleep long enough to feel rested or refreshed.

insorption. the movement of a substance (as from the gastrointestinal tract) into the circulating blood

inspirate. inhalant; that which is inhaled; inhaled gas or air.

inspiration. inhalation.

inspiratory. of, pertaining to, or timed during inhalation.

inspire. inhale.

instillation. **1:** the slow introduction of a fluid into a cavity or passage in the body which is allowed to remain for a specified length of time and then drained or withdrawn. **2:** a solution introduced by instillation (def. 1). *Cf.* infusion, injection.

integrator. **1a:** that which consolidates, combines, or unites. **1b:** that which integrates (as in def. 1a) in such a way as to introduce or reintroduce normal interaction and functionality of the component parts. **2:** an animal that moves about a known and relatively fixed home range and that over some period of time acts as a sampler, indicator, or monitor of a particular chemical or chemicals in the area occupied.

integument. the covering of the body. In humans and other vertebrate animals it is comprised of the dermis and epidermis. The integument offers a partial barrier to absorption of xenobiotics; but its effectiveness varies greatly among vertebrates. Included also as part of the **integumentary system** are those structures such as hair, feathers, nails, dermal scales, and antlers that arise from the dermis and/or epidermis. Concentrations of environmental chemicals (most notably heavy metals) in hair, feathers, and other integumental structures such as the antlers of deer, can often be related not only to temporal and geographic gradients of exposure, but to nutritional deficiencies and disease states as well.

integumentary. pertaining to the integument or a structure arising from the integument.

interaction. the state, condition, or process whereby two or more parts of a system act on each other, directly or indirectly. Operationally, an interaction is the observed result or effect which is qualitatively or quantitatively different than that seen in the absence of a potential for interaction. See effect (e.g., additive effect, antagonistic effect, synergistic effect). **drug interaction**. **1:** the action of one drug upon that of another, altering its effects. **2:** the pharmacological interaction of two or more drugs taken concurrently. Such interactions may be antagonistic, synergistic, or may result in effects (including death) that are unknown for each drug acting independently.

interferon. a lymphokine that heightens natural killer cell activity disrupts viral replication.

interhalogen compound. any compound formed between the various halogens. Their toxicities are similar to those of the elemental halogens and to the halogen oxides.

interictal (interparoxysmal). pertaining to intervals between episodes (e.g., of attacks, spasms, or seizures). See ictus.

interleukin I (IL1). a lymphokine that chiefly affects the activity of macrophages. See lymphokine.

interleukin II (IL2). a lymphokine essential to the growth and differentiation of T cells.

intermediate. lying or placed between two extremes; somewhere between the beginning and the end; midway; intervening; transitional; partly resembling either of two extremes.

intermittant cramp. See tetany.

intermittant tetanus. See tetany.

International System of Units (SI; SI Units; Systéme International d'Unités). a system of weights and measures adopted by the International Conference of Weights and Measures in 1960. The system is comprised of (1) seven base units (coherent units), the meter, kilogram, second, ampere, kelvin, candela and mole; (2) their decimal multiples and submultiples; (3) other metric units; and (4) derived or supplementary units. One class of derived units (e.g., force, power, frequency) are those that represent the base units to the powers of 10 (e.g., kilometer, millimeter); note that the kilogram, not the gram, was selected as the base unit for mass in this system. Another class of derived units consist of powers of base units (other than 10) and of base units in algebraic relationships (e.g., square meter, velocity, or meters per second). Many derived SI units have individualized names (e.g., joule, hertz, volt). The multiple and submultiple prefixes are tera- (T, 10^{12}), giga- (G, 10^9), mega- (M, 10^6), kilo- (k, 10^3), hecto- (h, 10^2), deca- (da, 10^1); the submultiples are deci- (d, 10^{-1}), centi- (c, 10^{-2}), milli- (m, 10^{-3}), micro- (μ, 10^{-6}), nano- (n, 10^9), pico- (p, 10^{-12}), femto-(f, 10^{-15}). The ISU has not entirely replaced older systems of measurements in the natural sciences.

interparoxysmal. interictal.

intestinotoxin. enterotoxin.

intolerance. **1a:** incapacity; inability to endure, withstand, or consume. **1b:** a condition of adverse reactivity to normally nonbiologically active or toxic doses of a drug or other xenobiotic even with an initial exposure. See also intolerance, idiosyncrasy (drug idiosyncracy). **alcoholic intolerance. 1:** the inability to consume alcoholic beverages without drinking to excess. **2:** an excessive response to small amounts of ethanol. **drug intolerance**. the inability to fully tolerate the normal pharmacologic or therapeutic doses of a drug; the state of reacting to a normal dosage with the symptoms of overdosage.

intoxation. inebriation; poisoning by substances other than alcohol, especially by those of biological origin. *Cf.* intoxication.

intoxicant. **1:** able to intoxicate. **2:** an intoxicating agent; an inebriant or a poison.

intoxicated. **1a:** under the influence of (affected by) a toxicant or intoxicant; poisoned. **1b:** showing signs of poisoning (intoxication). **2:** under the influence of an intoxicating beverage. *Cf.* inebriated.

intoxicating. **1:** inebriating. **2:** producing or able to produce intoxication or poisoning.

intoxication. **1:** any condition resulting from exposure to an intoxicant. **2:** poisoning (def.2). *Cf.* intoxation. **3a:** a condition that results from exposure to toxic amounts of a drug or to a toxc substance. **3b:** a condition of disturbed or impaired mental or physical capacities resulting from the introduction of a substance into the body of an animal, especially into a human. **4a:** acute alcoholic intoxication. **4b:** a condition of excitement or delirium induced by chemicals other than ethanol. *Cf.* intoxation. **5a:** any process, reaction, or means whereby the toxicity of a substance is increased. **5b:** an increase in toxicity of a parent compound by metabolic reactions, e.g., epoxidation. See also alcoholic psychosis). **acute alcoholic intoxication**. acute alcoholism. **alcoholic intoxication**. intoxication due to excessive consumption of alcoholic beverages or ethanol; acute alcoholism; drunkenness. **anaphylactic intoxication**. the condition resulting from an anaphylactic reaction. **citrate intoxication**. the condition (tetany) that may develop from massive transfusion of blood that contains citrate as an anticoagulant. The citrate combines with calcium and may result in tetany. **intestinal intoxication**. autointoxication. **water intoxication**. that due to excessive intake or retention of water.

intoxicative. of, pertaining to, or tending to produce inebriation or intoxication.

intraabdominal. within the abdomen; within the abdominal cavity.

intraarterial. within the lumen of an artery.

intraarterial pressure. arterial tension.

intraatrial. within the atrium or atria of the heart.

intrabronchial. within a bronchus.

intrabuccal. **1:** within the mouth or buccal cavity (the part of the oral cavity toward the cheek). **2:** within the tissue of the cheek.

intracardiac. within the heart.

intracellular. within the substance of a cell or of cells collectively.

intracerebellar. within the cerebellum.

intracerebral. within the cerebrum.

intracranial. within the cranium or skull.

intractable. refractory or resistant to change, relief, cure, or control; unmanageable, uncontrollable, resistant to treatment, relief, control, or cure.

intracutaneous. within the skin, especially the dermis. *Cf.* intradermal, intradermic.

intracutaneous injection. injection into the skin as with a hypodermic syringe.

intracutaneous reaction. a reaction of the skin to material injected into the substance of the skin.

intradermal, intradermic (*i.d.*, I.D., ID). intracutaneous; within the dermis of the skin.

intradermal injection. injection within the dermis; intracutaneous injection.

intradermal reaction. intracutaneous reaction to material injected into the dermis.

intraduodenal. within the duodenum.

intragastric (*i.g.*, I.G., IG). within the stomach.

intramuscular (*i.m.*, I.M., IM). within muscle, within a muscle.

intraperitoneal (*i.p.*, I.P., IP). within the peritoneal cavity.

intraspecific. within a species. See also variation.

intratracheal. within the lumen of the trachea. **intrauterine**. within the cavity of the uterus.

intrauterine death. death of a fetus while inside the mother's uterus.

intrauterine device (IUD). a small birth-control device placed permanently in the uterus to prevent the growth of fertilized eggs.

intravascular. within the lumina of blood vessels or within vascular tissue.

intravenous (*i.v.*, I.V., IV). into or within the lumen of a vein, used especially in reference to hypodermic injection or withdrawal of blood.

intravenous pressure. venous tension.

intravenous pyelogram (IVP). See pyelogram (intravenous).

intrinsic. situated in or denoting internal origin; inborn, innate, inherent.

intrinsic activity (efficacy; potency) [Z = 53; A_r = 126.9045]. the biological effectiveness per unit of drug-receptor complex.

intrinsic clearance. See clearance.

introsusception. intussusception.

intussusception (introsusception). growth by the transformation of materials taken into an organism (e.g., food) and transformed metabolically into biomass. This is the normal mode of growth of living organisms.

invasin. hyaluronidase.

inversion. **1:** a reversing; or turning over or up side down or inside out; or a reversal of the normal relation between parts of a body. **2:** a chromosomal aberration (mutation) in which a the sequence of base pairs in chromosome segment become reversed.

invertebrate. an animal that does not possess a vertebral column or backbone.

involuntary nervous system. See visceral nervous system.

iNyushu. *Dispholidus typus*.

io-moth dermatitis. dermatitis resulting from contact with irritant setae of the caterpillar of *Automeris io*, the io moth.

ioderma. an eruptive dermatitis of follicular papules and pustules or a granulomatous lesion caused by topical use of iodine.

iodic. of, pertaining to, or caused by, iodine or a compound of iodine in its pentavalent state.

iodic acid anhydride. iodine pentoxide.

iodic anhydride. iodine pentoxide.

iodide. a compound that contains pentavalent iodine, which is usually bound ionically to electropositive atoms; transition metal iodides often have some covalent bonding. All organic iodide contrast agents used in medical diagnosis are nephrotoxic.

iodine (symbol, I; iodum). a toxic diatomic element, I_2, with lustrous deep-violet, flat or plate-like, rhombic crystals that sublimate readily. It has the lowest electronegativity of the stable halogens and is thus the least reactive, but iodides are generally more soluble than related halides. Iodine combines to form a wide variety of organic compounds, a number of which are medically, toxicologically, or biologically important. Toxic effects are similar to those of bromine and chlorine; it is a more potent pulmonary irritant than either. Respiratory exposure to iodine is, however, limited by its relatively low vapor pressure. Symptoms of iodine intoxication in humans include a burning pain in the mouth and throat, thirst, nausea, vomiting, abdominal pain, diarrhea, albuminuria, and shock due to loss of electrolytes. Severe poisoning is marked by extreme thirst, purging with a bloody stool, and death either from circulatory collapse, inspiration pneumonia, or asphyxia from glottic edema. The lips may be stained brown and the vomitus may be blue if the stomach contains starches. Consumption of 2-4 g by a human has proven fatal. A number of its salts are caustic or irritant. Iodine is an antidote for alkaloid Al poisoning.

iodine cyanide. cyanogen iodide.

iodine pentoxide (iodic anhydride; iodic acid anhydride). the acid anhydride of iodic acid, it is a colorless oil or a white, crystalline solid, I_2O_5, with monoclinic hygroscopic crystals. It is very water soluble, but insoluble in chloroform, absolute ethanol, and ethyl ether. It is a strong oxidizing agent. Because it oxidizes carbon monoxide (CO) to carbon dioxide (CO_2), major uses are in the detection of low concentrations of CO and in removing carbon monoxide from air (e.g., in respirators).

iodine poisoning. poisoning by iodine or its compounds. The condition is characterized by lips and mouth that are stained brown; a burning pain in the mouth, pharynx, and stomach; thirst; vomiting with blue vomitus (if starchy food was recently eaten); bloody diarrhea; anuria; the urine contains albumin or blood. Death may result from circulatory collapse, asphyxia from glottic edema, or aspiration pneumonia. Lesions observed on autopsy are irritation of the throat with glottic edema, shock due to loss of fluid and electrolytes; and uncommonly, late esophageal edema. **chronic iodine poisoning**. iodism.

iodism. chronic iodine poisoning due to prolonged excessive usage of iodine or its compounds. It is characterized by a severe catarrhal inflammation of the nasal mucosa attended by a profuse nasal discharge (coryza), ptyalism, frontal headache, an acneform eruption, weakness, emaciation, excessive salivation, and foul breath. See iodine poisoning.

iodoacetate (iodoacetic acid). a colorless or white, crystalline compound, $C_2H_3IO_2$, that is soluble in water and ethanol, and only very slightly soluble in ether. It is poisonous by oral, subcutaneous, and intravenous routes of exposure. Some data indicate that it is mutagenic and teratogenic.

iodoacetic acid. See iodoacetate.

iodochlorhydroxyquin (5-chloro-7-iodo-8-quinolinol; 5-chloro-8-hydroxy-7-iodoquinoline; chloroiodoquin; iodochlorohydroxyquinoline; iodochloroxyquinoline; quinoform; clioquinol). a brownish-yellow, nearly odorless powder. It is very toxic to laboratory cats and is implicated as an etiologic factor in subacute myelooptico neuropathy (SMON), a clinical syndrome seen in Japan that affects the optic nerve, spinal cord, and peripheral nerves. Early symptoms include diarrhea, disturbances of gait and paresis of the lower limbs, abnormal deep tendon reflexes, and psyschic disorders.

iodochlorohydroxyquinoline. iodochlorhydroxyquin.

iodochloroxyquinoline. iodochlorhydroxyquin.

ododerma. **1a:** a contact dermatosis caused by iodine and marked by an eruptions of follicular papules and pustules, or by granulomatous lesions. **1b:** the skin lesion or eruption due to iodism.

iodoform (triiodomethane). a very or highly toxic, hazardous, greenish-yellow powder or lustrous crystalline compound, CHI_3, with a strong, penetrating odor. It contains about 96% iodine and is soluble in acetone, benzene, chloroform, glycerol, ethanol, ether, and slightly soluble in ethanol and water. Iodoform decomposes violently at 204°C. It is an irritant, and is applied externally as a topical anti-infective agent. Poisoning most commonly occurs through wounds or breaks in the skin when iodoform dressings are applied. This may produce a dermatitis. Systemic poisoning of humans is marked by vomiting, tachycardia, a slight fever (sometimes), and cerebral depression or excitation of any degree, which may include hallucinations, delirium, coma, and death. Laboratory rats suffer severe hepatic degeneration similar to that produced by carbon tetrachloride. There is no satisfactory antidote.

iodoformism. chronic poisoning by iodoform. Effects are chiefly due to the iodine content and are thus similar to those of iodism, *q.v.*

iodomethane. See methyl iodide.

iodoquinol (formerly diiodohydroxyquin; Yodoxin). an antiamebic, $C_9H_5I_2NO$, used to treat amebiasis and *Trichomonas hominis* infections of the intestines. This drug can have serious effects on the CNS in the form of subacute myelooptic neuropathy, *q.v.*

iodum. iodine.

ion. an atom or radical that has lost or gained one or more electrons and thus bears a net positive (cation) or negative (anion) electric charge. Both the chemical and toxicological properties of an ion differ from that of the neutral atom from which it derives.

ionamin. phentermine.

ionic. pertaining to a substance with a net positive or negative charge when in solution; pertaining to an ion or ions.

ionic bond. electrovalent bond.

ionization. **1:** dissociation of a molecule into ions, as when an electrolyte is dissolved in water or another polar solvent or is exposed to an electrical discharge or to ionizing radiation. **2:** iontophoresis (a form of electroosmosis whereby ions of soluble salts are introduced into tissues, often for therapeutic purposes (ion therapy)).

ionize. to separate into ions; to undergo ionization.

ionizing radiation. **1:** electromagnetic radiation that is at least as energetic as X-rays and charged subatomic particles of like energies. While neutrons may induce ionization, they are not classified as ionizing radiation. **2:** highly energetic radiation of extremely short-wavelength (gamma rays, X-rays, subatomic charged particles) that can react with atoms and molecules to eject electrons and produce positively charged ions. Ionizing radiation generates short-lived, highly toxic hydroxyl radicals in the aqueous phase of cells via a homolytic splitting of H_2O to yield OH· and H· radicals; these are able to initiate decomposition of numerous organic compounds. The hydroxyl radicals can cause autoxidative injury to proteins, carbohydrates, nucleotides, and nucleic acids. They can thus induce gene mutations and destroy cells and tissues; they assault bone marrow and other hemopoietic tissues (and may thus induce thrombocytopenia and in severe cases, aplastic anemia); is leukemogenic; and can poison cells by peroxidizing unsaturated lipids. Exposure to even brief bursts of ionizing radiation can be hazardous.

ionizing radiation. See radiation (ionizing radiation).

iontophoresis. ionization (def. 2).

iophobia. toxicophobia.

ioxynil (3,5-diiodo-4-hydroxybenzonitrile; 3,5-diiodo-4-hydroxyphenyl cyanide; 2,6-diiodo-4-cyanophenol). a postemergence contact herbicide with little residual activity. It is used to control broadleaved seedling weeds in cereal crops and sports turf. Effects appear within 24 hours of application as necrotic foci on leaves. This damage spreads, ultimately killing the plant. Untreated leaves develop chlorosis. Ioxynil is more toxic than bromoxynil.

ioxynil octanoate. a waxy, water-insoluble solid used as an insecticide on cereal grains and sugarcane.

IPA. isopropyl alcohol.

ipado. See *Erythroxylum*.

IPC. propham.

ipecac . **1:** *Cephaelis ipecacuanha* (ipecacuanha). **2:** (ipecacuanha; hippo). the dried rhizome and roots of *C. ipecacunha*. It was formerly used as an emetic, but is now prized chiefly as a source of emetine. It contains the alkaloids cephaeline, emetamine, psychotrine, protemetine, and resin. See also ipecac syrup. Ipecac poisoning is termed emetism, *q.v.* **3:** the dried roots or rhizomes of any of several plants having uses similar to those of *C. ipecacunha*. **4:** *Euphorbia ipecacuanha*.

ipecac spurge. *Euphorbia ipecacuanha*.

ipecacuanha. See ipecac.

ipHimpi. *Hemachatus haemachatus*, *Naja nigricollis*.

ipomea. See *Ipomoea orizabensis*.

ipomeamarone. a terpenoid phytoalexin that provides the sweet potato, *Ipomoea batatas*), with resistance against attacks by the fungus, *Ceratostomella*. The degree of resistance is correlated with the concentration of ipomeamarone.

Ipomoea (morning glories). a genus of chiefly herbaceous plants (Family Convolvulaceae); included are both ornamental and weedy species. The seeds are hallucinogenic, euphoric, and toxic, due to the presence of ergot (indole) alkaloids such as lysergic acid amide, isolysergic acid amide, chanoclavine, elymoclavine. ***I. batatas*** (sweet potato). a tropical, small-flowered American plant which is now grown worldwide as a food crop. The vines may contain toxic concentrations of nitrate. ***I. orizabensis***. the dried root (ipomea, orizaba jalap root; Mexican scammony root) is used as a cathartic. ***I. tricolor*** (morning glory). a tender perennial with many varieties, grown as an annual; it reaches a height greater than 3 m. the seeds are toxic and hallucinogenic. ***I. versicolor***. the seeds are toxic and hallucinogenic. ***I. violacea*** (morning glory, heavenly blue, pearly gates, etc.). the seeds are toxic with LSD-like effects. The active principles are indole alkaloids such as ergine, isoergine, and elymoclavine.

iproniazid. a monoamine oxidase inhibitor that is no longer available in the United States because of its toxicity. It is especially dangerous when used in combination with certain other drugs such as barbiturates, antihistamines, and other antidepressants such as aminopyrine, meperidine, and morphine.

iproniazid phosphate (Marsilid phosphate; 1-isonicotinoyl-2-isopropylhydrazine phosphate). a potent CNS stimulant and monoamine oxidase inhibitor that is no longer commercially available in the United States because of its toxicity. Toxic effects are numerous and may be serious. It is especially dangerous when used in combination with certain other drugs (barbiturates, antihistamines, and other antidepressants such as meperidine, morphine, and aminopyrine).

Ir. the symbol for iridium.

iridectomy. surgery performed to treat some kinds of glaucoma.

iridemia. bleeding from the iris of the eye.

iridium (symbol Ir)[Z = 77; A_r = 192.22]. a highly corrosion-resistant, white, silvery metallic element [Z = 77; A_r = 192.22].

iris. **1:** common name for certain members of the iris family (Iridaceae), especially those of the genus *Iris*. **2:** the rhizome of *Iris versicolor*, *q.v.*, which was formerly used as a purgative, emetic, and diuretic. **3:** the most anterior part of the vascular tunic of the vertebrate eye. It is a circular pigmented membrane that lies behind the cornea; it is perforated by the pupil.

Iris (iris). a large genus of wild North American and cultivated perennial plants (Family Iridaceae) named after the Greek godess of the rainbow. Both wild and cultivated forms contain an irritant substance in the leaves and especially in the thick, fleshy underground rhizome. Ingestion of moderate amounts can produce severe gastrointestinal distress. ***Iris versicolor*** (blue flag; poison flag). the common wild iris of North America. It is beardless and rhizomatous with blue-violet or perianth white flowers with yellow markings. All parts are toxic, especially the rhizome.

Irish potato. *Solanum tuberosum*.

iron (ferrum; symbol, Fe)[Z = 26; A_r = 55.85]. a transition element that occurs in many ores mainly as an oxide or carbonate. One of the most toxic of elements, it is essential to life and a component of hemoglobin, cytochromes, ferrichromes, and certain other biological pigments. Ferrous iron (Fe^{++}) is readily oxidized to ferric iron (Fe^{+++}) which can be coupled in aqueous solution to the toxic superoxide anion and peroxide; molecular oxygen is reduced in the process. The iron can also catalyze the conversion of superoxide anion plus peroxide to the extremely toxic hydroxyl radical with release of oxygen. Symptoms of iron intoxication in humans can include nausea, vomiting, pallor, abdominal pain, headache, hypotension, confusion, convulsions, and unconsciousness. Iron deficiency in humans can cause anemia, lowered vitality, and retarded development.

iron carbonyl. iron pentacarbonyl.

iron chloride. ferric chloride.

iron ferrocyanide. ferric ferrocyanide.

iron gluconate. ferrous gluconate.

iron hydrate. ferric hydroxide.

iron hydroxide. former name of ferric hydroxide.

iron oxide, black (black oxide of iron; iron oxide, magnetic; C.I. 77491; C.I. Pigment Red 101; colloidal ferric oxide; ferric oxide). a reddish or bluish-black, amorphous powder, $FeO{\cdot}Fe_2O_3$ or Fe_3O_4, that is soluble in acids, but insoluble in water, ethanol, and ether. It is used as a catalyst, pigment, polishing compound, in coatings for magnetic tapes, and in ferrites used in the electronics industry. It is poisonous and a possible human carcinogen.

iron oxide, hydrated. ferric hydroxide.

iron oxide, magnetic. iron oxide, black.

iron pentacarbonyl (iron carbonyl; pentacarbonyliron). a yellowish or colorless pyrophoric liquid, $Fe(CO)_5$, that burns to yield Fe_2O_3 in air. It is a potent lung irritant and is neurotoxic, hepatotoxic, and nephrotoxic. See carbonyl.

iron poisoning. poisoning by elemental iron or many of its compounds. Most poisonings are due to accidental ingestion of any of more than 100 commercial, iron-containing products. Mild to moderate effects are produced if 20 to 60 mg of elemental iron/kg body weight is ingested. Ingestion of 200 to 250 mg/kg is life threatening. Moderate or severe iron poisoning of humans usually passes through some or all of four stages. **stage I.** occurs within 6 hours and is marked by vomiting, irritability, abdominal pain, explosive diarrhea, seizures, lethargy, and possibly coma which is a grave sign if it appears at this stage. Any or all of the following may be seen if iron serum titers are high: hemorrhagic gastritis, hypotension, tachypnea, tachycardia, and metabolic acidosis. **stage II.** appears within 10-14 hours following ingestion and is characterized by a subjective amelioration of symptoms. **stage III.** occurs from 12-48 hours after ingestion. Effects may include shock, hypoperfusion, and hypoglycemia even though serum iron levels may have fallen to normal. Any or all of the following may appear at this time: disorientation, restlessness, lethargy, evidence of liver damage, fever, leukocytosis, bleeding disorders, inverted T waves on the ECG, convulsions, coma, shock, acidosis, and death. **stage IV.** is marked by late complications among those who survive for 2-5 weeks beyond stage 3. The problems that arise during this stage are due to pyloric, antral, or intestinal obstruction, hepatic cirrhosis, and/or CNS damage.

iron sulfate. ferrous sulfate.

iron vitriol. ferrous sulfate.

ironweed. See *Sideranthus*.

irrbartbart. *Pseudoechis australis*.

irrespirable. **1:** the quality of being impossible to inhale (breathe), as in the case of large particulates in the atmosphere. **2:** unsafe or unfit to inhale (breathe).

irreversibility. the quality or state of being irreversible, *q.v.*

irreversible. impossible to reverse; of or pertaining to a permanent change, one that cannot be reversed or undone (e.g., an irreversible chemical reaction; an irreversible action or effect of a poison; damage that cannot be repaired; death of a cell, organism, or other living system).

irrigate. to flood, flush out, or thoroughly rinse (e.g., a cavity or wound) with a liquid.

irrigation. flooding, flushing, or rinsing with water or other liquid, as in the flushing out of a canal or cavity, or the washing of a wound. *Cf.* lavage.

irritability. **1:** the quality or state of being irritable or readily responsive to stimuli. **2:** a state of abnormally high sensitivity to stimuli; having a low threshold of response to stimuli.

irritable. **1:** able to react to a stimulus. **2:** unusually or abnormally sensitive to stimuli.

irritant. **1:** irritating; irritative; painful. **2:** an agent that irritates. **3a:** an irritant or corrosive poison; any substance that inflames or erodes living tissue. **3b:** a substance which, while not corrosive, causes a reversible nonimmunologic inflammatory response at the site of contact with tissue. **primary irritant**. an irritant that acts directly at the point of contact with the skin or other tissue to produce an irritation (e.g., inflammation, dermatitis).

irritation. **1:** reaction to an irritant. **2:** an extreme reaction to pain or to a pathological condition. **3:** a normal response to a stimulus as by a nerve or muscle. **4:** aggravation, provocation.

irritative. irritant; inducing or causing irritation; that which irritates.

isano oil. an unsaturated oil obtained from the nut of a West African tree, *Ongokea klaineana* (Family Oleoceae). It is a violent purgative; no longer used in medicine.

isatropylditropeine. belladonnine.

isikhotsholo. *Naja nivea*.

island viper. *Bothrops insularis*.

iso-BTX. isobatrachotoxin.

isoamyl acetate. amyl acetate.

isoamyl nitrite. amyl nitrite.

5-isoamyl-5-barbituric acid. amobarbital.

isoantigen. an antigen found in an individual animal that induces an immune response if injected into a genetically dissimilar member of the same species. Isoantigens carry identical determinants in a given individual (i.e., chemical groups on macromolecular antigens that provoke an immune response). Isoantigens of two individuals may or may not have identical determinants. If they do not, they are said to be allogeneic with respect to each other and are called alloantigens. *Cf.* alloantigen.

isobatrachotoxin (iso-BTX). one of the four major steroidal alkaloids isolated from the skin of poison-dart (poison-arrow) frogs of the genus *Phyllobates*. See batrachotoxin.

isobutane (2-methylpropane). See butane.

isobutanol (methyl propanol; isobutyl alcohol; 1-hydroxymethylpropane; isopropylcarbinol; 2-methyl-1-propanol; 2-methylpropyl alcohol; fermentation butyl alcohol). a colorless, flammable, partially water-soluble liquid, $(CH_3)_2CHCH_2OH$, used as a solvent and in the manufacture of esters used as fruit-flavoring concentrates. It is moderately toxic to laboratory rats. Chronic (lifetime) exposures of rats caused liver damage ranging from steatosis to cirrhosis; malignant tumors developed in some cases. Isobutanol is a strong irritant to the eyes and mucous membranes, moderately irritant to the skin of humans, and a narcotic at high dosage. Exposure is chiefly occupational. It is not directly toxic to aquatic animals or algae, biodegrades readily, and does not bioaccumulate. See also 1-butanol.

isobutanyl chloride. 3-chloro-2-methylpropene.

isobutene. See butylene.

isobutyl carbinol. isoamyl alcohol, primary.

***N*-30isobutyldeca-*trans*-2-*cis*-6-*trans*-8-trieneamide**. affinin.

***N*-isobutyl-2,6,8-decatrieneamide**. affinin.

isocarb. propoxur.

isocrotyl chloride. 1-chloro-2-methylpropene.

isocyanate. refers to the radical -N=C=O from isocyanic acid or to a compound with the general formula, R-N=C=O, where R is an alkyl or aryl group. They are used in numerous chemical syntheses, especially the manufacture of polymers. Isocyanates (e.g., methyl isocyanate, *q.v.*) are highly reactive and toxic.

isocyanatomethane. methyl isocyanate.

isocyanic acid (carbamide). a gas, HN≡C≡O, released by depolymerization of cyanuric acid at 300-400°C in a stream of carbon dioxide. It polymerizes readily and should be stored in a dilute solution of carbon tetrachloride or ether at -30°C to retard polymerization. It is a strongly acidic, an irritant to the skin, eyes, and mucous membranes, and will blister the skin on contact. Isocyanic acid is a serious explosion hazard. *Cf.* cyanic acid.

isocyanic acid methyl ester. methyl isocyanate.

isocyanide. the radical -NC or a compound with the general formula R-NC. *Cf.* isonitrile.

isodrin (endo; endo isomer; Compound 711). an isomer of aldrin, it is an extremely toxic chlorinated diene insecticide (no longer manufactured or used in the United States). Chemical and toxicological behavior is very similar to that of aldrin, *q.v.*

isoflurophate. diisopropyl fluorophosphate.

isoflurophosphate. diisopropyl fluorophosphate.

isoinokosterone. β-ecdysone.

isolate. **1:** to separate a substance or entity from other entities or materials, as the isolation (separation) of a biogenic toxicant from the producing organism. **2a:** to free from contaminants. **2b:** to separate in pure form from a mixture. To determine the properties of a toxicant or other biologically active substance, it must first be isolated from other substances which might confound the results of testes or analyses. **3:** that which is separated or isolated (e.g., a chemical isolate, a viable organism separated from its host or culture system). **4:** to separate from others physically (quarantine) or psychologically.

isolation. **1:** the process of separating and identifying a substance or entity (e.g., a chemical compound, a type of microorganism) from other entities or materials, as in the isolation of a chemical substance from a tissue or chemical mixture. **2a:** the process of freeing of contamination. **2b:** separation in pure form from a mixture. **3:** separation from others as when carrying, or when especially susceptible to, a communicable disease. **3:** a state or feeling of loneliness or solitude. **4:** the state or condition of being isolated; segregation; quarantine. **reverse isolation**. the set of procedures used to prevent spread of infection in a health care facility such as a hospital. The process is intended to protect staff, visitors, and susceptible patients (e.g., those suffering immunosuppression) from acquiring a contagious disease from a other staff, visitors, or patients.

isolysin. a lysin that acts on the cells of animals of the same species as that which produced it.

isolysis. the lysis of cells by an isolysin.

isolytic. pertaining to isolysis.

isoniazid (4-pyridinecarboxylic acid hydrazide; isonicotinic acid hydrazide; INH). a tuberculostatic antibacterial used as an antimycotic agent in veterinary practice. Adverse effects of long-term use include liver damage and peripheral neuropathy. Rashes, fever, and CNS effects are also common.

isonicotinic acid hydrazide. isoniazid.

1-isonicotinoyl-2-isopropylhydrazine phosphate. iproniazid phosphate.

isonitrile. an organic isocyanide, R-NC.

isonitrosoacetone (monoisonitrosoacetone; pyruvaldoxine; propanone 1-oxine). a cholinesterase reactivator, $CH_3CO\text{-}CH{=}NOH$, that readily penetrates the blood-brain barrier, causing significant reactivation of phosphorylated acetylcholinesterase in the CNS. It is used to protect humans and animals from otherwise lethal poisoning by organophosphorous anticholinesterases.

isophthalic acid (*meta*-phthalic acid). the *meta*-isomer of phthalic acid.

isoPPC. propham.

isopropanol. the legal commercial label name of isopropyl alcohol.

isopropanol poisoning. isopropyl alcohol poisoning.

isopropene cyanide. methacrylonitrile.

2-isopropenyl-2,3-dihydro-5-acetylbenzofuran. tremetol.

isopropenylnitrile. methacrylonitrile.

isopropoxymethylphosphoryl fluoride. sarin.

2-isopropoxyphenyl-*N*-methylcarbamate. propoxur.

***o*-isopropoxyphenyl *N*-methylcarbamate**. propoxur.

isopropyl alcohol poisoning (isopropanol poisoning). poisoning by isopropyl alcohol is similar to that of ethanol intoxication, but usually with more severe effects. As little as eight ounces can cause respiratory collapse or heart failure. Symptoms may appear within several minutes to half an hour following exposure. Symptoms may include dizziness, nausea, hematemesis, abdominal pain, incoordination, headache, excessive sweating, depressed breathing, reduced urination, confusion, stupor, coma, and death from circulatory collapse or respiratory failure. Lesions commonly include blood in the upper respiratory tree; aspiration pneumonia with edema; blood in the thoracic cavity; and severe CNS depression. Some 15 percent or more of ingested isopropanol is converted to acetone in the body and acetonuria without glycosuria is pathognomonic. See also alcohol poisoning, alcoholism, alcohol (isopropyl alcohol).

***N*-4-isopropylcarbamoylbenzyl-*N'*-methylhydrazine**. procarbazine.

isopropyl carbanilate. propham.

isopropylcarbinol. isobutyl alcohol (See alcohol).

***N*-isopropyl-α-chloroacetanilide**. propachlor.

isopropyl-*m*-chlorocarbanilate. chlorpropham.

isopropyl *N*-(3-chlorophenyl)carbamate. chlorpropham.

isopropyl 2,4-dinitro-6-sec-butylphenyl carbonate. dinobuton.

isopropylfluophosphate. diisopropyl fluorophosphate.

isopropyl methanefluorophosphonate. sarin.

***N*-isopropyl-α-(2-methylhydrazino)-*p*-toluamide**. procarbazine.

isopropylmethylphosphonofluoridate. sarin.

***O*-isopropyl *N*-phenyl carbamate**. propham.

isopropylphosphoramidothioic acid *o*-2,4-(dichlorophenyl) *o*-methyl ester. DMPA.

isoproterenol hydrochloride (Isuprel Hydrochloride; Norisodrine; Aerotrol; Vapo-Iso). a sympathomimetic amine used to alleviate bronchoconstriction in asthma; it is also used as a cardiac stimulant. Overdosage by inhalation can prove fatal.

isoquinoline. very toxic, colorless plates or a liquid alkaloid, C_9H_7N, with a pungent odor. It is nearly insoluble in water, but is miscible with many organic solvents. Isoquinoline occurs in several genera of the poppy family, Papaveraceae: *Argemone* (prickly poppy), *Chelidonium* (celandine, swallowwort [not to be confused with celandine poppy]), *Corydalis* (corydalis), *Dicentra* (bleeding-heart; Dutchman's breeches; squirrel-corn; turkey-corn or staggerweed), *Papaver* (poppy), and *Sanguinaria* (bloodroot). It is also derived from coal tar. Isoquinoline is used in the synthesis of dyes, insecticides, antimalarial drugs, and rubber accelerators.

isospermotoxin. a spermotoxin that destroys spermatozoa of the same species. The production of such a toxin by the injection of sperm of the same species into a particular animal.

isothiocyanate (sulfocarbamide). **1:** any organic compound with the general formula R—N=C=S, where R is any alkyl or aryl group. While closely related chemically to the goitrogenic thiocyanates, isothiocyanates show no goitrogenic activity. See also *Brassica*, thiocyanate, mustard oil. **2:** the radical of isothiocyanic acid, -N=C=S.

1-isothiocyanatonaphthalene. 1-naphthylisothiocyanate.

3-isothiocyanato-1-propene. allyl isothiocyanate.

isothiocyanic acid allyl ester. allyl isothiocyanate.

isotonic. **1:** descriptive of a cell or tissue with an osmotic pressure equal to that of the solution in which it is bathed. **2:** descriptive of a solution that has the same solute concentration and thus the same osmotic pressure as another solution with which it is compared, usually cytoplasm or tissue fluids. *Cf.* hypertonic, hypotonic. See also tonicity.

isotonic solution. a solution that has the same osmotic pressure as the cell or system from which it is separated (or potentially separated) by a semipermeable membrane (e.g., the plasmalemma). *Cf.* hypertonic solution, isotonic solution.

isotoxic. of or pertaining to an isotoxin.

isotoxin. a toxin that is poisonous to other animals of the same species.

isotretinoin. the 13-*cis*-form of retinoic acid, *q.v.* It is used in a number of pharmaceuticals to treat acne, e.g., Accutane™, Isotrex™, Roaccutane™, and Teriosal™. It is moderately toxic to adult humans, but is a potent teratogen. See also Accutane.

Isotrex™. the 13-*cis*-form of retinoic acid. See also Accutane™.

isovaleric acid, allyl ester. allylisovalerate.

isoxaben (*N*-[3-(1-ethyl-1-methylpropyl)-5-isoxazoleyl]-2,6-dimethoxybenzamide; benzamizole). a selective preemergent herbicide used for cereal crops. It is slightly toxic *per os* to laboratory mice and rats, but the inhaled vapor is extremely toxic to laboratory rats.

isoxazole. a class of natural insecticides produced by poisonous mushrooms such as *Amanita muscaria*, *A. pantherina*, and *Tricholoma muscarium*. These compounds also have narcotic and psychomimetic effects on humans. Examples are ibotenic acid, muscazone, muscimol, and tricholomic acid. All but ibotenic acid cause visual damage, mental confusion, spatiotemporal dislocation, and memory loss in humans.

isoxepac (6,11-dihydro-11-oxodibenz[b,e]oxepin-2-acetic acid; oxepinac). a very toxic nonsteroidal antiinflammatory agent with analgesic and antipyretic properties.

Isuprel Hydrochloride™. isoproteronol hydrochloride.

IT. immunotoxin.

itai, itai disease ("ouch, ouch"). a form of cadmium poisoning described in people living in the Jintsu River Valley near Fuchu, Japan. Itai, itai in the Jintsu Valley was ascribed to rice that had been irrigated with water from an upstream mine that produced lead, zinc, and cadmium. It is characterized by renal tubular dysfunction, painful osteomalacia, pseudofractures, and anemia, caused by the dietary intake of contaminated shellfish or other sources containing cadmium.

Itchweed. *Veratrum viride*.

IUD. intrauterine device.

iurid (iurid scorpion). any scorpion of the family Iruidae.

iurid scorpion. iurid.

Iuridae (iurid scorpions). a family of scorpions (Order Scorpionida), with about 25 species worldwide. Included is North America's largest scorpion (*Hadrurus arizonensis*). Specimens may be 37-140 mm long, with a broad pentagonal sternum. They differ from the Buthidae by having only one spur on the distal tarsal segment of the last pair of legs and no spine beneath the stinger. a conspicuous tooth is situated beneath the chelicerae.

ivory shell. *Babylonia japonica*.

IVP. intravenous pyelogram.

ivy. the common name of *Hedera spp.*, but sometimes applied to other climbing or spreading plants.

ivy bush, ivybush. *Kalmia latifolia*.

Ixodes. a genus of ticks (Family Ixodidae) that are parasitic on mammals. See tick, Ixodidae, Ixodoidea.

Ixodidae. a family of arachnids (Order Acarina, Superfamily Ixodoidea) that includes the hard ticks. They are usually parasitic on mammals. See tick, Ixodoidea, *Ixodes*, *Dermacentor*.

Ixodoidea (ticks). a superfamily of small biting arachnids (Order Acarina) comprised of species with a conspicuous toothed hypostome. Most are parasitic on mammals. Certain species are venomous (e.g., *Dermacentor andersoni*), producing local erythema, pain, edema, and muscle cramps or serious systemic effects (See tick paralysis); others are important vectors of disease (e.g., Rocky Mountain spotted fever, lyme disease); some are chiefly a nuisance to the host. In any case, the wound may become infected. The application of gasoline or heat to the tick usually causes it to loosen its jaws and it can then be removed with forceps (tweezers) and the wound carefully washed. If the jaws remain embedded in the wound, infection or other complications may result. In such cases, a physician should be notified.

J

jack. the collective common name of fishes of the family Carangidae, used especially for species of the genus *Caranx*.

jack bean. See *Canavalia*.

jack-in-the-pulpit. a common name of *Ariseama triphyllum*.

Jack o' Lantern. *Ompholatus olearius*.

Jack-o'-Lantern mushroom. *Ompholatus olearius*.

jacks. See *Caranx*.

jacobine. a poisonous alkaloid, $C_{18}H_{25}O_6N$, isolated from the *Senecio jacobea*; it may cause necrosis of the liver.

jacodine. seneciphylline.

Jacutin™. lindane.

jake. Jamaican ginger. See tri-*o*-tolyl phosphate.

jake leg. See tri-*o*-tolyl phosphate.

jake neuritis. Jamaica ginger paralysis (See paralysis).

Jamaica ginger paralysis. See paralysis.

Jamaica ginger polyneuritis. Jamaica ginger paralysis (See paralysis).

Jamaican ginger (jake). See tri-*o*-tolyl phosphate.

Jamaican vomiting sickness. A usually fatal reaction, mainly among children, to eating the unripe fruit of *Blighia sapida*, *q.v.*, or perhaps the ripe aril of an unripe berry. The ripe fruit readily becomes rancid and may also be toxic. The ripe aril, properly prepared is apparently safe. Symptoms appear abruptly and include nausea and violent vomition with quiescent intervals, for ½ day or less, followed by convulsions, coma, and death in most cases. Severe hypoglycemia is characteristic.

Jameson's mamba. *Dendroaspis jamesoni*.

Jamestown weed. a common name of *Datura stramonium*.

Jan's banded snake. *Rhynchoelaps bertholdi*.

Jan's Mexican coral snake. *Micrurus affinis*.

Japan camphor. camphor.

Japanese belladonna. *Scopolia japonica*.

Japanese callista. *Callista brevisphonata*.

Japanese mamushi. *Agkistrodon halys*.

Japanese pagoda tree. *Sophora japonica*.

Japanese pieris. *Pieris japonica*.

Japanese privet. *Ligustrum vulgare*.

jararaca. See *Bothrops jajaraca*, *Bothrops neuwiedi*.

jararaca pintada. *Bothrops neuwiedi*.

jararacussú. *Bothrops jararacussu*.

jargon (paraphasia). **1:** confused, incomprehensible language; gibberish. **2:** communication that includes unfamiliar words, abbreviations, or permutations of normally accepted terms, usually within the context of a special occupation, science, or specialty, or clique.

jarlong. *Pseudechis australis*.

jasmine. common name of plants of the genus *Jasminum*. Other showy plants with fragrant flowers are sometimes erroneously called jasmines when they are properly termed jessamines. See also *Gelsemium*.

jasmolin. either of two naturally occurring insecticidal substances (pyrethrins), jasmolins I and II, isolated from pyrethrum flowers, *q.v.* They are moderately toxic to vertebrate animals. See pyrethrin (def. 2).

jasmolin I. See jasmolin.

jasmolin II. See jasmolin.

Jatropha (coral plant; lucky leaf). a genus of poisonous, tropical, coarse annual shrubs or small trees (Family Euphorbaceae) with a long flowering season (some constantly bear flowers). They are closely related to *Croton* and *Ricinus*. ***Jatropha curcas*** (barbadosnut; physic nut; purge nut; curcas bean). the seeds contain large amounts of a useful purgative oil; they also contain a phytotoxin similar to ricin. Numerous cases of human intoxication have been reported throughout the world. The seeds have a pleasant taste; those of at least one variety are mildly toxic, if at all. In other cases, as few as three seeds may induce symptoms. Symptoms are essentially those resulting from severe gastroenteritis and include nausea, vomiting, abdominal pain, and diarrhea. ***J. gossypifolia*** (bellyache bush). a species native to Key West, Florida. ***J. multifida*** (gout plant; coral plant; physic nut). a shrub or tree with scarlet flowers; it is native to southern peninsular Florida. Symptoms and lesions are similar to those of *J. curcas*. Poisonings are not uncommon, especially of children who find the yellow fruit attractive, especially after falling to the ground. ***J. spathulata***. A small shrub of western Texas and Mexico. ***J. stimulosa*** (bull nettle). native to the southeastern United States, this species bears bristles that cause a painful irritation of the skin on contact. Some persons experience a severe reaction.

jaundice (icterus). a condition characterized by a yellowish skin, sclerae, mucous membranes, deeper tissues, and body fluids; the stool is light and the urine dark. The discoloration is due to the deposition of bilirubin, the concentration of which is increased in blood plasma. Jaundice is seen in a number of disease states and may result from obstruction of the biliary passageways, from hemolysis, or from impaired functioning of hepatic cells. See also atrophy (acute yellow atrophy). **toxic jaundice**. that caused by chronic or acute poisoning of the liver by bacterial toxins, or hepatotoxicants such as carbon tetrachloride, phosphorus, or arsphenamine. It is the most common type of jaundice.

jaundiced. icteric.

Java bean. *Phaseolus limensis*.

Java pepper. *Piper cubeba*.

Javan krait. *Bungarus javonicus*.

jellied-base coral. *Ramaria gelatinosa*.

jellyfish (true jellyfish). **1:** the venomous, free-swimming, marine umbrella-shaped (medusa) stage of the class Scyphozoa. Contact produces symptoms similar to, but less severe than, those resulting from contact with *Physalia spp.*, *q.v.* **2:** a term sometimes erroneously used in reference to the free-swimming medusa stage of certain Hydrozoa. **3:** sometimes erroneously used in reference to the Portuguese man o' war and other large floating hydrozoans.

jequirity bean. *Abrus precatorius.*

Jerdon's sea snake. *Kerilia jerdonii*.

Jerdon's viper. *Trimeresurus gramineus*.

Jerusalem cherry. *Solanum pseudocapsicum*.

Jerusalem oak. *Chenopodium botrys*.

jervine ((3β,23β)-17,23-epoxy-3-hydroxyveratraman-11-one). a toxic steroidal alkaloid isolated from the rhizomes and roots of *Veratrum grandiflora*, *V. album*, and *V. viride.* See veratrine.

jessamine. **1:** See *Cestrum*. **2:** See *Gelsemium*.

Jesuits' bark. cinchona.

jet berry bush. See *Rhodotypos.*

Jew's nosed sea snake. *Enhydrina valakadyn.*

jimmies. a syndrome of livestock, especially sheep grazing on *Notholaena sinuata* (Jimmy fern). Affected animals appear normal until exercised, when they exhibit progressive incoordination, finally stopping in spasm with the back arched. There are associated violent tremors and a pronounced increase in heart and respiratory rates; sometimes prostration. If an affected animal is unmolested, it will usually recover in less than 30 minutes. The attacks recur with repeated exercise. If the poisoning is severe, or further exercise is forced on the animal, death by respiratory failure may ensue. Pregnant animals are most susceptible. Livestock, including domestic fowl, usually recover within 1-2 weeks following removal from a range with Jimmy fern.

Jimmy fern. *Notholaena sinuata*.

jimmy weed. *Haplopappus*.

Jimsonweed. See *Datura* especially *D. stramonium*.

Joe Pye weed (joe-pye-weed). the common name of any of a number of species of *Eupatorium*. This name stems from the belief that the plant was used by an Indian physician named Joe Pye in the early days of the Massachusetts colony as a medicine.

John's sea snake. *Microcephalophis gracilis*.

Johnson grass, **Johnsongrass**. *Sorghum spp.*, especially *S. halepense*.

joint. any cigarette made from marijuana.

jointed charlock. *Raphanus raphanistrum*.

jonquil. common name for *Narcissus jonquilla*. Often incorrectly used in reference to any species of hardy narcissus.

Judean pitch. asphalt.

Juglans nigra (black walnut). a tree (Family Juglandaceae) that poisons plants situated under the canopy. The toxic principle is juglone. See also allelopathy, *Salvia leucophylla*, *Artemesia californica*.

juglone (5-hydroxy-1,4-naphthalenedione; 5-hydroxy-1,4-naphthoquinone; 8-hydroxy-1,4-naphthoquinone, nucin, regianin). a phytotoxin of *Juglans nigra*. It is a glycoside found in high concentrations in the roots and leaves. It washes to the ground in rainfall, poisoning the area under the canopy and preventing or inhibiting the growth of the propagules of numerous species of plants (e.g., Gramineae, apple trees, tomato plants) that might otherwise grow in this situation. Juglone is also a bacteriostat and fungistat and has a sedative effect on small animals. The main parts of the tree contain a precursor, 5-glycosyl-1,4,5-trihydroxy naphthalene, *q.v.* Juglone is also an inhibitor of mycelial growth of the fungus *Fusicladium effusum*. It has sedative effects on fish and mammals.

jumble bead. *Abrus precatorius*.

jumping pit viper. *Bothrops nummifer*.

jumping snake. *Bothrops nummifer*.

jumping spider. See *Phidippus*.

jumping viper. *Bothrops nummifer*, *B. nasutus*.

Juniperus (juniper). a genus of small to medium-sized evergreen trees and shrubs (Family Cupressaceae). ***J. virginiana***. a small evergreen tree with a narrow pyramidal shape that grows to nearly 10 m. It is toxic to livestock, but is extremely distasteful and usually avoided. The main active principle is terpin-4-ol. It occurs in the leaves and wood of the plant, but is most concentrated in fresh, ripe berries. The berries in small doses are used to flavor gin. Even small doses, if repeated, may cause convulsions and renal damage. Typical symptoms in humans are pain in the kidney region, sustained diuresis, albuminuria, hematuria, dark purplish urine, tachycardia, hypertension. In rare cases, convulsive apparitions, metrorrhagia, and even miscarriage are reported. ***J. sabina***. a widely distributed, low, spreading, dark green evergreen shrub. The entire plant is poisonous and irritant due to the content of savin oil, *q.v.*

Jura viper. *Vipera aspis*.

justicidin. any of several lignans isolated from various species of *Justicia* (Family Acanthaceae). **justicidin B** (dehydrocollinusin). used as a piscicidal agent.

jute. **1:** any plant of the genus *Corchorus*, *q.v.* **2:** the fibers of any plant of the genus *Corchorus*.

juvabione. an analog of insect juvenile hormones isolated from certain trees such as *Abies balsamea*. It inhibits growth and metamorphosis of larvae. See also juvenile hormone, juvenoid.

juvenile hormone. any lipid hormone secreted by the corpus allatum of insects that prevents metamorphosis to the adult form.

juvenoid. any of several analogs of insect juvenile hormones produced by plants. See, for example, juvabione.

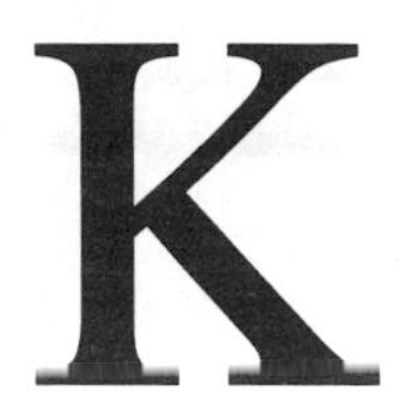

κ. See kappa.

K. See kappa.

K. the symbol for potassium (kalium).

K_a. affinity constant.

K_b. binding constant. See affinity constant.

K_d. dissociation constant.

K_m. Michaelis constant. See Michaelis-Menten equation.

^{40}K. potassium-40.

kaapse geelslang. *Naja nivea*.

kadel nagam. *Hydrophis spiralis*.

kadel pambu. *Hydrophis spiralis*.

kafir corn. *Sorghum vulgare*.

kainic acid (2-carboxy-4-(1-methylethenyl)-3-pyrrolidineacetic acid; 2-carboxy-3-carboxymethyl-4-isopropenylpyrrolidine; digenic acid; α-kainic acid; L_s-xylokainic acid). a neurotoxicant compound isolated from *Digenea simplex*, a red alga. It is used in neurobiological research and as an anthelmintic. **α-kainic acid**. kainic acid.

kairomone. a biologically active chemical produced and disseminated by living organisms of one species that is advantageous to recipient species. Examples are chemical lures (sometimes), inductants that stimulate adaptation, stimulants (e.g., growth factors), and warning substances (e.g., that signal toxicity or danger). See also allelochemical, allomone, signal.

kake. *Hemachatus haemachatus*.

kala-azar. See leishmaniasis.

kale. *Brassica oleracea* var. *acephala*.

kale shootsursun. *Hydrophis obscurus*.

kalelea. *Causus rhombeatus*.

Kalgan™. the acid sulfate of phenelzine.

kalilelala. *Dispholidus typus*.

kalium. potassium.

kallikrein. a proteolytic, shock-inducing substance found in urine, pancreas, and kidneys.

Kalmia (laurel; American laurel). a genus of mostly North American evergreen shrubs (Family Ericaceae). Poisonous species occur throughout North America, usually in moist (mesic) areas. They contain grayanotoxins, the toxic principle in most cases being andromedotoxin (grayanotoxin I). All parts of the plant are very toxic to humans, producing lacrimation, rhinorrhea, salivation, nausea, vomiting, dyspnea, hypotension, bradycardia, depression, convulsions, paralysis, and sometimes renal failure. Death may intervene in as little as 12 hours. ***K. angustifolia*** (dwarf laurel; sheep-laurel; calfkill; lambkill; sheepkill; wicky). an open woody shrub, growing to a meter or so, usually in abandoned fields in northeastern North America. It is decidedly less dangerous than *K. latifolia*. ***K. latifolia*** (mountain laurel; calico bush; poison laurel; spoonwood; mountain ivy; ivy bush). a dense woody shrub or small tree up to 3½ m tall growing in rocky wooded areas or clearings in the eastern United States. This is probably the best known broadleaf evergreen shrub in North America. The leaves and shoots are highly toxic. They were used by American Indians to commit suicide. Mountain laurel is especially dangerous to children; fatalities have occurred. It is, nevertheless, an important browse for deer during winter. ***K. polifolia*** (pale laurel, bog laurel). a small, woody bush up to 25 cm tall, of wet meadows and bogs in the Rocky Mountains from Alaska southward. At least one variety is known to be toxic.

Kalymin. pyridostigmine bromide.

Kampfstoff "Lost". mustard gas (See gas).

kanamycin. any of a group of aminoglycoside antibiotics produced by *Streptomyces kanamyceticus*; they interfere with bacterial protein synthesis.

Kanechlor. polychlorinated biphenyl(s).

kannadi virian. *Vipera russelii*.

kappa (κ, K). **1:** the tenth letter of the Greek alphabet. **2:** kappa particle. **3:** one of the two types of light chains in human antibodies.

kappa particle (kappa). any of a type of self-replicating cytoplasmic particles (the DNA of a commensal bacterium) found in certain strains. *Paramecium*. in the presence of a dominant I gene, they produce the toxin, paramecin. This confers the ability to kill other paramecia. The terms "kappa particle" and "paramecin" are sometimes erroneously used interchangeably.

karawala. *Bungarus ceylonicus*.

karminomycin. carubicin.

Karst adder. *Vipera ursinii*.

Karwinskia humboldtiana (coyotillo; tullidora; "buckthorn"). a poisonous, spineless, woody shrub or small tree (Family Rhamnaceae), 1-6½ m tall, that occurs in the deserts of northern Mexico and southwest Texas. The fruit (including seeds and pulp) is very toxic to all forms of livestock and to humans. A paralytic syndrome results from ingestion of the fruit, which contains a mixture of anthracenones ("buckthorn toxins"). Symptoms appear within a few days to several weeks following exposure (1-3 weeks in humans) and include malaise, quadriparesis, and ultimately paralysis of bulbar and respiratory muscles. In humans, the muscles become progressively weaker, beginning with the lower extremities. In livestock, weakness and incoordination first appear in the hind legs, the forelegs often being affected later. Appetite remains normal, but complete prostration can develop in about 1 month, often followed by death from respiratory failure. Recovery may take up to a year. A second syndrome has been produced experimentally by feeding the less toxic leaves to livestock. This produces nausea, wasting, progressive weakness, and death. See buckthorn toxins.

karyolysis (caryolysis). lysis of a cell nucleus.

karyolytic. denoting lysis of a cell nucleus.

karyomite. chromosome.

kasambwe. *Causus rhombeatus*.

kassa. *Bitis arietans*.

Kastenbaum-Bowman test. a test of statistical significance used to calculate mutation frequencies.

kasugamycin. an antibiotic that blocks the initiation of translation in bacteria.

katabolism. catabolism.

kathisophobia. akathisia.

kaukasus otter. *Vipera kaznakovi*.

kauryala. *Bungarus caeruleus*.

kawon. *Enhydrina schistosa*.

kawosia. *Naja nigricollis*.

kd. kilodalton (1,000 daltons).

keautia. *Naja naja*.

keeled rat snake. *Zoacys* (See *Ophiophagus hannah*).

keisau. *Trimeresurus gramineus*.

Kellogg's coral snake. *Calliophis kelloggi*.

Kelocyanor™. dicobalt edetate.

kelp. common name for marine multicellular brown algae (seaweed). See Phaeophyta.

Kelthane™. dicofol.

Kelvin absolute temperature scale. See Kelvin temperature scale.

Kelvin degree. See degree.

Kelvin temperature scale (absolute temperature scale; Kelvin absolute temperature scale). a temperature scale in which the ratios of the temperatures of two reservoirs equal the ratio of the amount of heat absorbed from one of them by a heat engine operating in a Carnot cycle to the amount of heat transferred by the engine to the other reservoir. The temperature of the triple point of water is 273.16°K and the temperature at which all molecular motion ceases and a mass would have no heat energy (absolute zero) is 0°K. See also Celsius temperature scale, Faherenheit temperature scale.

Kendall's compound A. dehydrocorticosterone.

Kendall's compound B. corticosterone.

Kendall's compound E. cortisone.

kendawang. *Maticora bivirgata*.

Kentucky coffee tree. *Gymnocladus dioica*.

Kenya horned viper. *Bitis worthingtoni*.

kepone. chlordecone.

kepone poisoning. chlordecone poisoning.

Keramik. diethylpropion.

Keramin. diethylpropion.

keratitis. corneitis.

keratoectasia. protrusion of the cornea.

Kerilia jerdonii (Jerdon's sea snake; kerril; shiddil). a venomous sea snake of the Indian Ocean.

kerosene (kerosine). a clear, pale yellow or nearly colorless mixture of liquid petroleum hydrocarbons, mainly of the methane series having 10-16 carbons per molecule. It is used as a reagent, in insecticides, and is occasionally burned for heating or light. Orally, it is very toxic to laboratory mice; inhalation of high concentrations may produce drowsiness, coma, and death. Kerosene stoves, unless very well ventilated, may release toxic levels of carbon monoxide.

kerosine. kerosene.

kerril. *Kerilia jerdoni*

kerril patte. *Hydrophis nigricinctus*.

ketene 1: any organic compound that contains the group C=C=O; general formula, RCHCO. Some ketenes occur in photochemical smog. **2:** sometimes used in reference to ethenone, the simplest ketene.

17-(1-keto-2-hydroxyethyl)-Δ^4-androsten-3,11-dione. dehydrocorticosterone.

ketoacidosis. a serious complication of diabetes mellitus in which the body produces acids that cause fluid and electrolyte disorders, dehydration, and sometimes coma.

ketohexamethane. cyclohexanone.

γ-keto-β-methoxy-δ-methyleneΔ^{α}-hexenoic acid. penicillic acid.

ketone. any of a class of partially oxidized hydrocarbons, (e.g., acetone) that are formed by the body during the catabolism of fats and fatty acids into carbon dioxide and water. Ketones contain the carbonyl group, C=O, which is bonded to two carbon atoms in the middle of a hydrocarbon chain or ring. They have the general formula, $R_1C(O)R_2$, where R_1 may be either the same as, or different from, R_2. Ketones are widely used as industrial solvents. Certain ketones occur in photochemical smog. The hydrocarbon moiety may consist of saturated or unsaturated chains (straight or branched) or of rings. *Cf.* aldehyde; see also acetone, methyl-*n*-butyl ketone and methylethyl ketone.

ketone body. any of a class of compounds (e.g., acetone, beta-hydroxybutyric acid, acetoacetic acid), produced chiefly in the liver by the condensation of acetyl CoA, and produced during fatty-acid oxidation. They enter the circulation and undergo further degradation in peripheral tissues. Examples are acetoacetate, See ketosis.

ketone propane. acetone.

β-ketopropane. acetone.

ketosis. accumulation in an individual of ketone bodies. It is frequently associated with acidosis (ketoacidosis) and is often misdiagnosed. The condition results from incomplete metabolism of fatty acids due usually to carbohydrate deficiency or the poor utilization of carbohydrates as in starvation, a high fat diet, pregnancy, after ether anesthesia, and in inadequately controlled cases of diabetes mellitus. The causes of ketosis in livestock seem to be predominantly dietary in origin (See, for example, lupinosis, trembles). In ketosis, large amounts of ketone bodies may be eliminated in the urine. If untreated, ketosis may progress to ketoacidosis, coma, and death.

ketosuria. the presence of ketone bodies in urine. The presence of ketosuria, especially the presence of acetone or diacetic acid, is an early sign of ketosis.

ketotic. pertaining to ketosis.

2-keto-1,7,7-trimethylnorcamphane. camphor.

Ketten viper. *Vipera russelii*

kg. kilogram.

khangala. *Dispholidus typus*.

khuppur. *Naja nigricollis*.

kidney. either of a pair of excretory and osmoregulatory organs of the vertebrate urinary system. They are situated dorsally near the midline on either side of the vertebral column and outside (dorsal to) the peritoneum (retroperitoneal). They filter blood, excreting water, nitrogenous wastes (e.g., urea), and certain other end-products of metabolism. They also regulate pH, the osmolarity, and the concentration of electrolytes such as sodium, phosphate, and potassium in the extracellular fluid. The kidneys play an important role in the excretion of the end products of xenobiotic metabolism. They are frequent targets of toxic action and are generally considered the chief target organ in heavy metal toxicity. In mercury poisoning of any kind, the kidney (or liver) usually accumulates the highest concentration of mercury; the kidney also accumulates the metalloid arsenic. See also nephron, nephrotoxicity, renal. See also oxalic acid.

kidney bean. See *Phaseolus*.

kidney dropsy. See renal failure.

kif. hashish.

kigau. *Naja nigricollis*.

kiiri. *Bitis arietans*.

kikanga. *Naja nigricollis*.

kill. **1:** to deprive of life, put to death, destroy, slay, murder, slaughter, exterminate. See also sacrifice, snuff. **2:** to perform the act of killing. **3:** to eliminate; nullify; to put an end to, especially abruptly. **4:** to consume totally, as a bottle of whisky. **5:** in law, to destroy the life of a person or an animal, although the term "homicide" is reserved for the killing of a human being.

killer paramecium. See *Paramecium*.

killing. the act or process of destroying or depriving an organism of life. *Cf.* homicide.

kilogram. 1000 grams. See also International System of Units.

kinangananga. *Causus defilippii*.

kinase. a biological agent, usually a protein or ion, that can activate a zymogen (the inactive form of an enzyme).

kindling. the process or phenomenon of eliciting a convulsive response by repeated application of nonconvulsive stimuli (e.g., chemical, electrical).

kinetic phase of toxicity. See phase of toxicity.

king brownsnake. *Pseudechis australis*.

king cobra. *Ophiophagus hannah*.

king crab. See *Limulus polyphemus*, Merostomata

king of the day. *Cestrum diurnum*.

king puff adder. *Bitis gabonica*.

king sago. *Cycas revoluta*.

king's gold. arsenic trisulfide.

King's sea snake. *Hydrophis kingii*.

king's yellow. **1:** arsenic trisulfide. **2:** lead chromate.

kinin peptide. a component of bee venom that can cause smooth muscle contraction and hypotension.

kinotoxin (fatigue toxin). a hypothetical toxin associated with muscular fatigue.

kipara nunga. *Naja haje*.

kipili. *Bitis arietans*.

kipiri. *Bitis arietans*.

kisigosogo. *Atheris squamigera*.

kissadi. *Bitis nasicornis*.

klamath weed. See *Hypericum perforatum*.

klapperschlangen. See *Crotalus*.

Kleen-up™. a brand name under which glyphosphate is sold.

klinotropism. clinotropism.

knight's spur. *Delphinium*.

"**knockout drops**". chloral hydrate.

knotweed family. Polygonaceae.

kobra. cobra. See *Naja*.

Kochia scoparia (Belvedere; fireball; summer cypress). a fast-growing, shrubby, annual, ornamental plant (Family Chenopodiaceae) that is used most often in borders. It causes hepatogenic photosensitization in livestock due to the presence of phylloerythrin. Toxic concentrations of nitrate have also been reported for this species.

kodo. Hindi for kodo millet.

kodo-cytochalasin-1. See cytochalasin.

kodo millet (*Paspalum scrobiculatum*, synonym, *P. commersonii*; kodo). a minor grain crop of India. The seeds, which are often contaminated by fungi (e.g., by *Phomopsis paspalli*, *q.v.*), have often been implicated in human and animal intoxications and were, in ancient times, used as a poison for tigers.

kodrava. Sanskrit for kodo millet.

kodrava poisoning. poisoning by kodo millet, *q.v.* Human symptoms include delerium, vomiting, violent tremors, difficulty in swallowing, and unconsciousness.

kohlrabi. *Brassica caulorapa*.

kojic acid. a toxic pyrone produced by certain molds (e.g., *Aspergillus fumigatus*) and aerobic bacteria. It is very toxic to laboratory mice.

kokodiou. *Echis carinatus*.

kokokeyamulinga. *Dispholidus typus*.

koli. the Hawaiian name of *Ricinus communis*.

kombi. a cardiotoxic extract from a West African vine, *Strophanthus kombé* and related species. It is prepared from the seeds of the vine and used to kill elephants.

komourtiou. *Naja nigricollis*.

kondband slang. *Elapsoidea sundevallii*.

Königskobra. *Ophiophagus hannah*.

koning-pofadder. *Bitis gabonica*.

konkati. *Thelotornis kirtlandii*.

koperkapel. *Naja nivea*.

Korallenottern. coral snakes (German).

Korallenschlange, Afrikanische. *Elaps lacteus*.

Korean mamushi. *Agkistrodon halys*.

Korean pit viper. *Agkistrodon halys*.

Korsakoff's psychosis. Korsakoff's syndrome.

Korsakoff's syndrome (Korsakoff's psychosis; amnestic psychosis; dysmnesic psychosis; polyneuritic psychosis; amnestic syndrome; dysmnesic syndrome). an amnestic syndrome usually seen in chronic alcoholics, but occasionally the result of brain damage from other sources (e.g., cerebral tumors, head injuries, minor strokes). It may be preceded by delirium tremens. The condition is marked by polyneuritis, disorientation, severe impairment of short-term memory, insomnia, muttering delirium, hallucinations, bilateral wrist or foot drop, and confusion. The individual attempts to compensate for short-term memory loss by inventing plausable accounts of what he or she has been doing during recent days or weeks. Long-term memory is usually unimpaired and skills learned in the past can be performed normally, although the ability to learn and perform a new skill is impaired. The condition probably results from nutritional deficiencies associated with the alcoholism, rather than direct toxic effects of alcohol.

koufi. *Vipera lebetina*.

Kr. the symbol for krypton.

krait. any snake of the genus *Bungarus* (Family Elapidae).

krait venom poisoning. effects of envenomation by kraits (*Bungarus spp.*) in humans usually include rapid onset of numbness, drowsiness. The clinical picture is similar to that due to cobra envenomation except there is little or no pain or swelling about the bite and the systemic effects may be more severe. Shock with respiratory depression and coma often develops rapidly. As with mambas and taipans, abdominal pain is often intense. Mortality may reach 50% even with the use of antivenin. See also cobra venom poisoning, elapid venom poisoning. See also snake venom poisoning.

kranawang. *Maticora bivirgata*.

Krebs cycle. tricarboxylic acid cycle.

kreotoxicon. any toxic substance in poisonous meat that is responsible for the symptoms of poisoning.

kreotoxin (creatoxin). a generic term for any toxin generated by a microorganism that is responsible for meat poisoning, *q.v.* See also kreotoxicon.

kreotoxism. meat poisoning.

krotenvipern. See *Causus*.

krubi. *Amorphophallis spp*.

krypton (symbol, Kr)[Z = 36; A_r = 83.80]. a colorless, odorless, inert, monatomic element of the rare-gas series. It is present in small amounts in the atmosphere. Six stable isotopes and 17 radioactive, artificial isotopes are known.

krysid. ANTU.

kuchibami. *Agkistrodon halys*.

kufah. *Trimeresurus okinavensis*.

kunn katuva. *Agkistrodon hypnale*.

kunuadi vyrien. *Vipera russelii*.

kupramite. an adsorbent for ammonia fumes that is used in gas masks.

kusari hebi. *Vipera russelii*.

kuturee pamhoo. *Vipera russelii*.

Kwell™. lindane.

kyanol. aniline.

kyozima. *Atheris squmigera*.

Kyphosidae (sea chubs). a family of tropical marine percaform bony fishes (Class Osteichthyes). They are oval in shape, compressed, with small mouths. The dorsal fin is comprised of rays and spines; the caudal fin is lunate to nearly truncate. Most feed on algae. See *Kyphosus cinerascens*.

Kyphosus cinerascens (sea chub). an oval, compressed, marine percaform fish (Family Kyphosidae) of the Indo-Pacific region. It is a common cause of hallucinatory fish poisoning in humans.

La. symbol for lanthanum.

label. **1a:** something that identifies, warns, alerts, or informs one about a specific item; an identifying mark, sign, tag, etc. **1b:** any printed or written matter (e.g., a note, notice, tag, information, title, warning) attached or appended to a larger writing or to a material or substance, a manufactured item. **1c:** a notice attached to a container that conveys information concerning its contents, intended usage, the manufacturer, and/or any notifications, concerns, or warnings relating, as appropriate or required by law concerning hazards that attend usage, transport, storage, and/or disposal. **2a:** an identifying mark, tag, marker, isotope, etc. **2b:** a substance that has a special affinity for a cell, tissue, organ, microbe such that it may be introduced and fixed, serving to identify the cell, etc. **3:** an atom or molecule attached to a ligand or to a binding protein that is able to generate a signal that can be monitored in the binding reaction. **3a:** to append or affix a written or printed label. **3b:** to add or introduce and fix a substance into a cell, tissue, organ, or microbe. **radioactive label** (radioactive tag). a radioactive isotope which, when introduced into a chemical, tissue, organ, or whole organism, etc., labels (or tags) specific portions of the recipient. Such labels are used, for example, to **a:** trace the pathways and roles of the normal isotope in intermediary metabolism. **b:** monitor the movements and fate of the labeled element or compound within a given system (an organism, the environment), or to allow the movements of an organism in the environment to be monitored. **warning label**. any label that indicates a danger. The legitimate purpose of a warning label (e.g., on a product) is to inform a party of the existence of a danger (e.g., presence of a toxicant or hazardous substance) of which he might otherwise remain unaware. Federal laws in the United States require that warning labels be affixed to potentially dangerous products such as drugs, cigarettes, household chemicals, tools, and to waste materials. Warning labels may underestimate, or be ambiguous about, the hazard involved. Consequently, items so labeled should be used with considerable caution and stored properly and safely away from access by children. See also warning. Examples of warning labels (with sample interpretations in the United States, depending on the nature of the product) that one may encounter are: **CAUTION**. (1) Harmful if swallowed. Irritant, avoid contact with eyes and prolonged contact with skin. Do not swallow; do not inhale vapors. Use in a well-ventilated area. (2) This label usually means that the contents are not lethal unless two tablespoons or more are ingested. (3) Contents are irritant or erosive to skin, eyes, and mucous membranes. **DANGER POISON** (with skull and crossbones). able to kill an adult if only a tiny pinch or a drop is ingested. See "skull and crossbones." **Extremely Flammable/Flammable/Combustible**. can ignite if exposed to a flame or spark. **HAZARDOUS WASTE - HANDLE WITH CARE**. material container or conveyance (e.g., rail car, truck) so marked serves no useful purpose and may be extremely dangerous if it escapes the container. There may be a subsidiary label that indicates the nature of the hazard (e.g., Inhalation Hazard, Marine Pollutant, corrosive). The use of such labels is regulated by law and improper disposal of such material is prohibited by law. **Strong Sensitizer**. may provoke an allergic reaction. **Toxic/Highly Toxic**. hazardous cleaning products must prominently display one of these signal words. They indicate (1) that the contents are poisonous if one drinks it, breathes the fumes, or if absorbed through the skin; (2) the product could kill an adult if about one teaspoon is ingested. Such products should be considered toxic in any amount.

labeled. affixed with a label or tag.

labeled phosphorus. radiophosphorus.

labile. **1:** not fixed, unsteady. **2a:** unstable, readily undergoing change. Applied, for example, to genes that have a relatively high mutation rate. **2b:** chemically unstable. **3:** plastic, def. 1a. **4:** psychological and psychiatric usage: characterized by rapidly changing emotions; unable to control the expression of emotions. **heat labile**. thermolabile (heat labile). **thermolabile** (heat labile). easily altered or decomposed by heat; labile when heated to a certain level.

lability. the state of condition of being labile.

Labrador tea. *Ledum*.

Labridae (wrasses; hogfishes; tautogs). a family of brightly colored marine percomorph fishes (Class Osteichthyes) of tropic and temperate seas. Many species inhabit coral reefs. The coloration may differ in the sexes, and both sex and color pattern may change as the fish matures. The mouth is small, terminal, and protractile, with conspicuously thick lips. Usually edible, sometimes toxic. See ciguatera, *Coris gaimardi*, *Epibulus insidiator*.

labriform milkweed. *Asclepias labriformis*.

laburnum. a plant of the genus *Laburnum*, especially *L. anagyroides*.

Laburnum (laburnum). a genus of poisonous shrubs and small trees (Family Fabaceae) that are native to Europe. They have bright yellow flowers, trifoliate leaves, and pendulous racemes. ***L. anagyroides*** (= *Cytisus laburnum*; common laburnum; golden chain; golden chain tree; laburnum). a large, toxic, ornamental shrub or small tree (Family Fabaceae) which grows to 10 m. It is native to the mountains of southern Germany, Switzerland, and France but is cultivated as a garden plant in most parts of the world. The leaves and seed pods are very toxic to humans and to grazing or browsing animals; the active principle is the quinolizidine alkaloid, cytisine, which produces effects similar to nicotine. The seeds are very toxic to livestock and humans; browsing or grazing animals may eat the leaves and pods with fatal results. Symptoms of intoxication include excitement, miosis, nausea, and severe vomiting, diarrhea, irregular pulse, and breathing, dysphagia, incoordination, convulsions, delirium, coma, and death by asphyxiation; miosis may be noted. Renal damage may occur. The milk of poisoned cattle may be toxic. See also cyticism.

Lachesis muta. *Lachesis mutus*.

Lachesis mutus (= *L. muta*; bushmaster; cascabel; cascabela muda; cascavel; mapana; mapipire; mapipire z'Auana; pineapple snake; sirocucu; verrugosa). a venomous snake that inhabits rain forests and tropical deciduous forests from Costa Rica, Panama and southeastern Nicaragua to the coastal lowlands of Ecuador and the Amazon basin of Peru, Brazil, Bolivia, and Paraguay. It is a huge, nocturnal, rarely encountered, venomous snake (Family Crotalidae). It has no congeners and is the longest pit viper in the New World, occasionally reaching a length of about 3.7 m. The ground color is brown or tan with large tan or brown rhomboid markings. The head is broad and distinct from the narrow neck; the snout is bluntly rounded. The internasal scales are small and the interoculars are narrow, and most other parts of the crown are covered by small scales, although the third supralabial is very large. A distinctive feature of the tip of the tail of the bushmaster is its spiny (burr-like) appearance. The dorsal body scales are weakly imbricate, and heavily keeled with bulbous tubercles. Although humans rarely encounter this snake and are infrequently bitten, it is a dangerous snake as the fangs are long and it produces large amounts of a relatively toxic venom. The case mortality rate is high.

lachrymal. lacrimal.

lachrymal gland. lacrimal gland.

lachrymation. lacrimation.

lachrymator. lacrimator.

lachrymatory. lacrimatory.

lacrimal (lachrymal). **1:** of, pertaining to, or situated near the lacrimal glands. **2:** characterized by tears or lacrimation.

lacrimal gland (lachrymal gland). one of several acinous glands in the eyelids of terrestrial vertebrates that secrete tears which clean the cornea and keep it moist.

lacrimation (lachrymation). the secretion of tears; the discharge (tearing, weeping, crying) of tears, especially in excessive amounts.

lacrimator (lachrymator). an agent (e.g., chloroacetophenone) that stimulates lacrimation on contact with the eyes of a vertebrate animal.

lacrimatory (lachrymatory). of, relating to, causing, or able to cause lacrimation.

Lactarius (milkies). a genus of common and abundant ground-dwelling gilled mushrooms (Family Russulaceae). They inhabit forests and are associated with particular trees. Most species are edible. Symptoms of intoxication by poisonous species of this genus are those of gastrointestinal distress. Symptoms usually appear within ½-3 hours following ingestion. Healthy victims will usually recover within a few hours. Among those species which may be poisonous are those treated here. ***L. piperatus*** (peppery milky). This species occurs in deciduous woods in eastern North America and is normally considered to be nontoxic. One variety, however, *glaucescens*, is probably toxic. ***L. representaneous*** (northern bearded milky). a possibly poisonous ground-dwelling species that grows under spruce in the Pacific Northwest, from Quebec to Vermont, and in the mountains of Arizona and Colorado. ***L. rufus*** (red hot milky). a rather common and widespread species that grows in sphagnum bogs and in woods under pine trees. North American varieties are considered poisonous, but those in Scandinavia are canned and marketed. ***L. scrobiculatus*** (spotted-stalked milky). This is really a complex of species that are very difficult to separate. They are reportedly poisonous. ***L. torminosus*** (pink-fringed milky). considered poisonous in North America, but are sold commercially in Finland. ***L. uvidus*** (common violet-latex milky). considered poisonous, it grows under aspen, birch, and pine trees from Maine to Tennessee, west to Colorado and Michigan. ***L. vinaceorufescens*** (yellow latex milky). possibly poisonous.

lactate dehydrogenase. an enzyme that reversibly catalyzes the conversion of lactic acid to pyruvic acid. It occurs in the venoms of some elapid snakes. See also enzymes of snake venoms.

lactation tetanus. grass tetany.

lactimorbus. milk sickness.

Lactophrys (trunkfish). a genus of boxfishes (Family Ostraciidae), some species of which are poisonous. ***L. trigonus***. a greenish, olive or nearly tan, marine fish with a fixed carapace. It occurs in seagrass beds in shallow waters and in coral reefs from the Atlantic Coast of tropical South America to Cape Cod. It may cause ciguatera in humans if improperly prepared. It is deep-bodied and triquestrous in cross section. A spine projects posteriorly from a ventrolateral ridge that runs the length of the body. The head lacks horn-like spines and is steep in profile with a blunt snout; the mouth is small and ventroterminal and the eyes are dorsolateral in position. The dorsal fin is incomplete behind. Adults may reach a length of up to 53 cm.

lactose (milk sugar; saccharum lactis; 4-O-β-D-Galactopyranosyl-D-glucose; 4-(β-D-Galactosido)-D-glucose). a sweet-tasting, odorless, white powder or hard, crystalline, water-soluble mass, $C_{12}H_{22}O_{11} \cdot H_2O$, that is insoluble in ether and chloroform; it is stable in air. It has various uses including orally as a diluent for substances that would be excessively irritant if given in pure form.

lactotoxin. a toxic substance formed in milk.

Lactuca. a genus of lettuce (Family Compositae). concentrations of nitrate that are toxic to livestock have been reported in some species such as *L. sativa* (includes salad varieties such as Romaine lettuce) and *L. scariola* (prickly lettuce).

ladies' fingers. *Lotus corniculatus*.

lady's slipper. *Cypripedium hirsutum*.

lady's sorrel. See *Oxalis corniculata*.

ladyfish. certain bony fishes of the Family Albulidae. See *Albula vulpes*, ciguatera.

lae. *Scomberoides sanctipetri*.

Laetrile™ (*l*-mandelonitrile-β-glucuronic acid). **1a:** a yellow, oily, water-soluble liquid derived from amygdalin. It is alleged antineoplastic properties remain unconfirmed by objective tests. See mandelonitrile. **1b:** sometimes used as a synonym for amygdalin.

laking. hemolysis.

lambkill. *Kalmia angustifolia*.

Lamellibranchiata. See Pelecypoda.

Lamium amplexicaule (henbit; dead nettle). a common weed (Family Labiatae) of eastern North America and the Pacific Coast. It can cause staggers (*q.v.*) in sheep, horses, and cattle.

lamium amplexicule poisoning. poisoning of livestock by *Lamium amplexicule*, a common weed of eastern North America and the Pacific Coast. See staggers.

Lampropeltis triangulum nelsoni (Nelson's milksnake). a neotropical milksnake (Family Colubridae). A relatively small nonpoisonous constrictor, it is a harmless to humans. It is a mimic of the broad-banded coral snake, *Micrurus distans*. Both snakes inhabit the same region, western Mexico from Sonora to Guerero. *L. t. nelsoni* has broad, red, crossbands and narrow black bands as in *M. distans*. As in most coral snake mimics, the black bands are paired; this is never the case in coral snakes.

lamziekte. botulism.

lance-headed viper. See lancehead.

lancehead. a common name of snakes of the genus *Bothrops*; sometimes also applied to some Asian lance-headed vipers (*Trimeresurus*) that occur in South America.

land farming (of waste). a process in which toxic or other hazardous wastes are deposited on or in the soil and are degraded naturally by microbes.

Lanolin. usually not toxic if ingested.

Lanoxin™. digoxin, USP.

Lansberg's hog-nosed viper. *Bothrops lansbergii*.

Lansberg's hognose viper. *Bothrops lansbergii*.

Lansberg's pit viper. *Bothrops lansbergii*.

lantadene. Substances isolated from *Lantana*. **lantadene A**. rehmannic acid. **lantadene B**. an apparently physiologically inactive substance related to lantadene A.

lantana. See *Lantana*.

Lantana (lantana). a genus of shrubby plants (Family Verbenaceae) that are native mainly to Central America northward to the United States regions that border the Gulf of Mexico; also widely used as ornamentals. All varieties can cause hepatogenic photosensitivity in grazing animals. The hepatotoxin is a polycyclic triterpene, rehmannic acid (=lantadene). The original toxic extract of *Lantana* was named lantanin, q.v.). The phototoxic agent is phylloerythrin. ***L. camara*** (lantana; red sage; wild sage). A perennial shrub in central and southern Florida; herbaceous further north. The native plant is usually tall and scraggly, but a number of varieties have been developed as houseplants, greenhouse, and bedding plants. All parts, especially the green berries, are toxic, producing photosensitization of increasing severity with increased exposure to sunlight. In addition to hepatogenic photosensitization, this plant produces symptoms of severe gastroenteritis in livestock. Symptoms of acute poisoning resemble those of belladonna alkaloid (atropine) poisoning. The affected animal is weak and experiences severe gastroenteritis with bloody, watery feces within a day of consumption and dies within 3-4 days. In chronic cases, the animal lives longer and mild to severe symptoms of photosensitization also develop. Complications are common, especially in chronic poisoning. These may include conjunctivitis, severe constipation, jaundice, infestations of bacteria or blow flies, corneal opacity, and blindness. Livestock fatalities are known from Africa, Australia, and the United States. The berries are believed to have fatally poisoned children in the United States.

lantanin. **1:** the original toxic extract of *Lantana*. **2:** this term has also been used as a synonym for rehmannic acid. See also lantadine, *Lantana*.

lanthanides. See lanthanoids.

lanthanoids (lanthanides; lanthanons; rare earths; rare earth metals). a closely related series of elements that include lanthanum. They are of similar size and have a characteristic oxidation state of M^{3+} and are therefore similar in chemical behavior. All but lanthanum (which has no *f*-electrons) have electronic configurations that display back-filling of the 4*f*-level.

lanthanons. See lanthanoids.

lanthanum (symbol, La)[Z = 57; A_r = 138.91]. a soft, malleable, silvery, metallic element that tarnishes in air. It is the first member of the lanthanoid series and occurs naturally in many minerals, usually in association with other lanthanoids. See also lanthanum nitrate.

lanthanum nitrate. a white, deliquescent, water-soluble, crystalline hexahydrate. It is a moderately to very toxic compound used in electron microscopy to stain extracellular mucopolysaccharides.

LAP. lyophilized anterior pituitary tissue.

Lapemis hardwicki (Hardwick's sea snake; Shaw's sea snake; Viet Nam sea snake; walo-walo; potal pambu). a rather short, stocky, venomous, sea snake (Family Hydrophidae). The average length of adults is about 0.7 m; occasional individuals may reach a length of nearly 1 m. While this snake yields very little venom, it is one of the most dangerous to humans, with a fair number of fatalities known. It is encountered usually during the rainy season, from southern Japan to the Merguri Archipelago and the coastal waters of northern Australia.

large brown tree snake. *Dispholidus typus*.

large green tree snake. *Dispholidus typus*.

larkspur. plants of the genus *Delphinium*.

larkspur poisoning (delphinium poisoning). poisoning by ingestion of plants of the genus *Delphinium*, commonly known as larkspurs. All parts of these plants are toxic, producing the same symptoms as monkshood poisoning. These may include paresthesia, a burning sensation of the mouth and skin, itching, dermatitis, nausea, vomiting, bradycardia, hypotension, weak pulse, respiratory distress, cyanosis, convulsions, and death. Symptoms appear almost immediately and death may occur within 6 hours. The active principles are delphinine and related alkaloids. Sheep are highly resistant and horses rarely eat enough to be gravely poisoned. **larkspur poisoning of cattle**. the indigenous larkspurs of the western United States are an important source of mortality in range cattle. New growth and seeds have the highest concentrations of toxic alkaloids. Both cattle and sheep readily eat larkspur; the cattle are often fatally poisoned, but the sheep rarely die from larkspur poisoning. Range cattle lose control of motor functions and are eventually unable to stand or regain their feet. Poisoned cattle may choke to death if they vomit while prone. Larkspur-poisoned cattle on hilly ranges often fall with their heads downhill and may be unable to belch and are thus susceptible to fatal bloat. Death is more often due to respiratory paralysis. In surviving animals, recovery is complete.

larvicide. an agent that destroys or incapacitates larvae, especially those of insects and other arthropods.

laryngismal. of, pertaining to, or resembling laryngismus.

laryngismus. spasm of the larynx.

larynx. the organ of voice, it is an enlarged musculocartilaginous structure at the upper end of the trachea below the root of the tongue.

LAS. alkyl sulfonate.

lassiocarpine. a slow-acting hepatotoxin, that produces atrophic hepatosis following long-continued exposure. See also *Heliotropium europaeum*. With continued exposure, liver cells increase in size, are less viable, and their regenerative ability is impaired.

late effect. any effect of acute exposure to ionizing radiation that first appears 60 or more days following the exposure.

latency period. latent period.

latent. quiescent, hidden, inactive, undeveloped, potential. See latent period.

latent period (latency period). **1a:** the time interval between the initiation of a disease process and the first indication of disease. **1b:** in practice (in toxicology and medicine), the latent period is usually taken as the interval between initial exposure to a poison or pathogen and appearance of clinical signs or symptoms. **2:** if the latent period is short it may be referred to as a reaction time, *q.v.* See, for example, neuromuscular blocking agent, potassium bromate. Do not confuse with latency period.

latentiation. any process of making latent, such as altering the chemical structure of a drug in order to affect the rate of absorption or other properties that might delay the appearance of effects.

lathyrism (sweet pea poisoning; lathyrus poisoning). poisoning by plants of the genus *Lathyrus*. Many epidemics have been recorded, although lathyrism in humans is now mainly of historical interest. In many countries the poisonous species are those used as livestock feed. It is a paralytic syndrome caused by the ingestion of large amounts of raw or cooked seeds or flour from plants of the genus *Lathyrus*, especially *L. cicera*, *L. clymenum*, *L. latifolius*, *L. odoratus*, and *L. sativus*; *Vicia sativa* may also be involved (See vicine). The condition is characterized by partial or total spastic paraplegia usually of the legs, but may also involve the arms in severe cases. Other symptoms may include prodromal pain, cramps, hypothesia or hyperesthesia, and paresthesia. See also lupinosis, osteolathyrism, *Lathyrus*.

Lathyrus (vetchling; wild pea). a cosmopolitan genus of more than 200 species of annual and perennial herbs (Family Fabaceae). The seeds of a number of species are toxic. Toxic species include *L. cicera* (flatpod pea), *L. clymenum* (Spanish vetch), *L. hirsutus*, *L. odoratus* (sweet pea), *L. latifolius* (sweet pea; perennial pea; everlasting pea; pois vivace (Quebec)), *L. pusillus*, *L. sativus* (grass pea, Indian pea, green vetch), *L. sphaericus*, *L. splendens*, *L. strictus*, *L. sylvestris* (perennial pea; everlasting pea), and *L. tingitanus*. The toxic principles are related to, or derived from, β-cyano-L-alanine. The toxic compound in at least some species of *Lathyrus* that causes paralysis and skeletal lesions is β-(γ-L-glutamyl)-aminopropionitrile. The toxic compound in *L. sylvestris* and *L. latifolius* is L-α,γ-diaminobutyric acid. It causes hyperexcitability, convulsions, and death without skeletal lesions. Both substances have a common precursor, β-cyano-L-alanine. See also lathyrism, divicine, vicine.

lathyrus poisoning. lathyrism.

latrodectism (latrodectus poisoning). poisoning by the venoms of spiders of the genus *Latrodectus*, *q.v.*

Latrodectus (widow spiders). a genus of small, to medium-sized terrestrial, highly venomous spiders that are widely distributed throughout the temperate and tropic zones; none occur in central or northern Eurasia. They have numerous common, sometimes local, names including black widow, brown widow, red-legged widow spider, hourglass, poison lady, deadly spider, T-spider, gray lady spider. They are dark brown or black, often with red markings, with a globular and often glossy abdomen. The female is much larger than the male (10-18 mm) with larger and stronger fangs and is dangerous to humans; ground color ranges from gray or brown to black in the various species. The venoms are very or extremely toxic and are chemically complex, containing several proteins, including a neurotoxin with a molecular mass of about 130,000. Unlike *Loxosceles spp.*, the bites of widow spiders, while sometimes painful, usually cause little or no apparent local injury. Symptoms in humans are varied but may include nausea, vomiting, systemic pain, cramps, sweating, headache, lightheadedness, tremor, and hypertension; the bites are rarely fatal to humans. The best known species toxicologically are the black widows, *L. mactans* and *L. curacaviensis*. In both, the bite is usually painful but sometimes may go unnoticed. Symptoms are mainly those of acute abdominal distress. Painful, localized muscular cramps are the first to appear, usually starting in the shoulders, back, and thighs and spreading to the abdomen. The abdominal muscles become rigid and breathing becomes thor acic and labored. Subsequent symptoms usually include restlessness, weakness, tremors, a weak pulse, and cold clammy skin as in shock, difficult speech, mild stupor. Delirium and convulsions, especially in children, may follow. Allergic reactions are possible. If the bite is not fatal, most symptoms disappear within 2 to 3 days, although a generalized weakness with paresthesia and transient muscle spasms may persist for as long as several months. Species of *Latrodectus*, other than the three described below, that are dangerous to humans include *L. bishopi* (red-legged widow spider); *L. hasselti* of India and Malaya; and *L. tredecimguttatus* of southern Europe and Africa. See also venom, Latrodectus venom. ***L. curacaviensis*** (black widow; black widow spider; hourglass spider). a widow spider that is similar in appearance, habits, and effects to the better known *L. mactans*, *q*.v. ***L. geometricus*** (brown widow). a New World spider common to western portions of South America, the Antilles, Central America, Mexico, much of the United States, and parts of Canada. The body is brown, but the markings are the same as those of *L. mactans*. ***L. mactans*** (black widow; black widow spider; hourglass spider). a New World spider common to western portions of South America, the Antilles, Central America, Mexico, much of the United States, and parts of Canada. In the United States it is most common in the middle and southern Atlantic states, those bordering the Gulf of Mexico, and west of the Rocky Mountains. It is one of the most feared and dangerous spiders in the New World. The adult female (slightly more than 1 cm long) has a shiny black body with small, brilliant red dumbbell- or hourglass-shaped markings on the ventral surface of the abdomen. The male is much smaller, lacks the hourglass figure, and is not dangerous. The black widow occurs in dark, sheltered locations (e.g., outdoor toilets) and is seldom seen before biting. The bite is often accompanied by sharp, pain, although many victims are initially unaware of the bite. The venom is virulent, destroying vascular tissue about the bite, with early localized and painful proximal muscle spasms. These typically begin in the shoulder, back, and thigh, spreading to the abdomen. The affected individual weakens, becomes restless and may experience shock, difficult speech, and mild stupor or delirium within the hour. Convulsions may follow. Symptoms that are easily confused with those of flu may predominate for several days (e.g., chills, urinary retention, sweating, nausea, vomiting, and intense abdominal cramps). Death is usually due to cardiac failure. Allergic reactions sometimes occur. In nonfatal cases, symptoms usually progress in severity for several hours to about a day and then gradually abate. Recovery is complete for most victims, although a generalized weakness with paresthesia and transient muscle spasms sometimes persists for several months. Antivenin (Lyovac® Antivenin) is available, but is normally used only for children or severe cases in adults.

latrodectus poisoning. latrodectism.

latrotoxin. the extremely toxic neurotoxin of *Latrodectus mactans*, the black widow spider, and others of this genus. It is a basic protein, MW 5000. Latrotoxin binds to specific receptors in the presynaptic plasma membrane of both central and peripheral nervous systems. It induces massive, nonspecific release of vesicle-bound neurotransmitters from the presynaptic nerve terminals of vertebrates with subsequent destruction of the prejunctional nerve endings. The mechanism of action is not known. The oral LD_{50} in mice is 0.55 mg/kg body weight.

laudanum. opium tincture.

laurel. **1:** laurel is included in the common name of many woody plants (e.g., *Rhododendron maximum*, *Leucothoe davisiae*). **2:** *Kalmia* species are the laurels of America.

lavage (lavation). **1:** the irrigation or washing out of a cavity or lumen (e.g., that of the stomach or intestines). **2:** to wash out or irrigate (e.g., a cavity or tissue). **blood lavage** (systemic lavage). the washing out of toxic substances from circulating blood by injecting serum into the veins. **gastric lavage**. washing out of the stomach with sterile water normal saline, 1-5% sodium bicarbonate, or activated charcoal is used as indicated. This is usually performed to dilute and remove irritants or poisons or to cleanse the stomach prior to or following surgery. In such cases, an early sample of the fluid removed from the stomach is retained for chemical analysis. If appropriate, an antidote is added to the water used to rinse the stomach. Gastric lavage is not used if a corrosive poison has been swallowed due to the risk of the lavage tube perforating tissues. Corrosive acids or alkalis may be diluted by giving large amounts of water or milk to the victim. **intestinal lavage**. dialysis by lavage using a fluid to remove, through the intestinal mucosa, materials that are not being excreted by the kidneys. **systemic l.** blood lavage.

lavation. lavage.

law. **1:** a binding rule or body of rules enacted by a governing authority such as a legislature or parliament; that which is ordained or decreed by a ruler or ruling body. **2:** a statement of a universal principle concerning natural phenomena (e.g., chemical, physical, or biological).

law of tolerance. See tolerance.

lawrencium (symbol, Lr; formerly Lw)[Z = 103]. an artificial transuranium element of the actinoid series. Several very short-lived (longest lived = 3 minutes) radioactive isotopes have been prepared.

laxation. bowel movement, with or without laxatives.

LBM. little brown mushroom.

LC. lethal concentration (See concentration). LC is usually followed by a number indicating the percentage of individuals of a test population estimated to be killed at the stated concentration.

LC_{50}. median lethal concentration (See concentration). See also LC.

LCLo. lowest lethal concentration. See concentration.

LCt_{50}. lethal concentration X time/50% response. See concentration.

LD. lethal dose. See dose.

LD *i.g.* lethal dose, *i.g.*

LD *i.m.* lethal intramuscular dose.

LD *i.p.* lethal intraperitoneal dose. See dose.

LD *i.v.* lethal intravenous dose. See dose.

LD *p.o.* lethal dose, *per os*. See dose.

LD *s.c.* lethal subcutaneous dose. See dose.

LD_0. maximum sublethal dose. See dose.

LD_{01}. lethal dose/1% response. See dose.

LD_{50}. lethal dose/50% response. See dose.

LD_{99}. lethal dose/99% response. See dose.

LDLo. lowest lethal dose. See dose.

leachate. 1a: the liquid product of leaching. **1b:** the solution formed when water (usually rain water) has filtered through soil or another solid or semisolid matrix in the environment (e.g., toxic waste). The water may be enriched or contaminated as a result of such a process.

leaching (lixiviation). the separation of soluble from insoluble matter by infiltration of a liquid solvent and drawing off the resulting solution. Examples are the percolation of rainwater through soil in the environment or through the contents of a hazardous waste site; the extraction of sugar from sugar beets; the washing of a soluble salts from an insoluble precipitate.

lead (plumbum, symbol, Pb)[Z = 82; A_r = 207.19]. a toxic, malleable, easily melted metallic element of high density (a heavy metal). A lustrous silvery metal when freshly cut, but becomes dull gray on exposure to the atmosphere. It is a member of group IV of the periodic table and shares many properties with tin. It is the end-product of a radioactive decay series. The most important lead ore is galena (PbS), which is found in Australia, Mexico, and the United States. Lead is one of the most hazardous of the toxic metals because: (1) it has been so widely used and dispersed in the environment by humans since ancient times; (2) all forms of lead are toxic; (3) it is a cumulative poison, thus relatively small exposures over a period of years may ultimately reach tissue concentrations that produce symptoms and disability. (4) many animals, including humans, accumulate lead to levels well above that found in their food or the surrounding air. Even today, lead causes untold poisonings (mainly chronic, but often fatal), in various human societies, among fish and wildlife, and in other living systems. See lead poisoning.

lead acetate (neutral lead acetate; normal lead acetate; sugar of lead; salt of Saturn). the trihydrate is a very toxic, white, water-soluble, crystalline, granular, or powdery substance, $Pb(C_3H_2O_3)_2 \cdot 3H_2O$. Lead acetate may reasonably be expected to be carcinogenic. The dust should not be inhaled. It is an analytical reagent and is used therapeutically as an astringent. See lead poisoning.

lead-acid battery. a storage battery with grids of spongy lead plates connected in series to the negative terminal. These are interleaved with lead oxide plates that are connected to the positive terminal, and the whole is immersed in dilute sulfuric acid. The lead oxides change in composition during charging and discharging. The voltage (electromotive force) of each cell of such a battery, when fully charged, is about 2 volts. Such batteries are widely used, especially in automotive vehicles. Junked batteries are substantial sources of environmental lead in the United States and certain other countries. See also lead, lead poisoning.

lead antimonate (Naples yellow; antimony yellow). a poisonous, water-insoluble, orange-yellow powder, $Pb_3(SbO_4)_2$, used as a paint pigment and in stains for glass and ceramic. It is toxic by inhalation. See also lead, lead poisoning.

lead arsenate. poisonous, water-insoluble white crystals, $Pb_3(AsO_4)2$, soluble in nitric acid. It occurs naturally as the mineral schultenite. It is moderately toxic *per os* to laboratory rats and very toxic to laboratory rabbits. Lead arsenate is a constituent of various insecticides and larvicides and is used in veterinary medicine as a teniacide. Humans may be poisoned by inhalation, ingestion, or contact. Symptoms may include inflammation of the skin, nausea, vomiting, diarrhea, abdominal pain, anorexia, weakness, and fatigue. See also lead, lead poisoning.

lead arsenite. a white, highly toxic, water-insoluble powder, $Pb(AsO_2)_2$, used as an insecticide. See also lead, lead poisoning. It is poisonous and a confirmed human carcinogen. When heated to decomposition it releases very toxic fumes of lead and arsenic.

lead borate. a water-insoluble, nonflammable, white powder, $Pb(BO_2)_2 \cdot H_2O$. It is used as a paint and varnish drier, to waterproof paints, to lead glass, and in electrically conductive ceramic coatings. It poisonous when inhaled. See also lead, lead poisoning.

lead bromate. a poisonous, slightly water-soluble compound, $Pb(BrO_3)_2$, with colorless monohydrate crystals. The pure compound is not explosive, but when prepared by precipitating lead acetate with an alkali bromate, some acetate is incorporated into the product, in which case, it may detonate when heated, jarred, rubbed, scraped, abraded, or struck. See also lead, lead poisoning.

lead bromide. an alcohol-insoluble, white crystalline powder, $PbBr_2$ that is slightly soluble in hot water. It is toxic if inhaled. See also lead, lead poisoning.

lead butyrate (butyric acid; lead salt). a poisonous, water-insoluble compound, $Pb(C_4H_7O_3)_2$, soluble in dilute nitric acid, HNO_3. See also lead, lead poisoning.

lead carbonate. See lead carbonate, basic.

lead carbonate, basic (ceruse; cerussa; flake lead; lead subcarbonate; white lead). a poisonous, heavy, white, amorphous powder, $2PbCO_3 \cdot Pb(OH)_2$, used as a pigment in oil base paints, watercolors, and cements; used also in the manufacture of lead carbonate paper and putty; and in the preparation of parchment. It is toxic by inhalation. See also lead, lead poisoning.

lead carbonate hydroxide. basic lead carbonate.

lead chloride. a moderately poisonous, white, crystalline compound, $PbCl_2$, that occurs naturally as the mineral cotunite. It is slightly soluble in hot water, and insoluble in alcohol and cold water. It is used as an analytical reagent and in the synthesis of lead salts and lead chromate pigments. See also lead, lead poisoning.

lead chromate (Canary Chrome Yellow 40-2250; Chrome Green; Chrome Lemon; Chrome Yellow; C.I. 77600; C.I. Pigment Yellow 34; lead chromate (VI); Paris Yellow; Cologne yellow; King's yellow; Leipzig Yellow). a very toxic acid-soluble, water-insoluble, yellow to orange-yellow powder, $PbCrO_4$, that occurs naturally as the minerals crocoite and phoenicochroite. It is used as a pigment in oil- and water-based paints. See also lead, lead poisoning.

lead chromate (VI). lead chromate.

lead chromate, basic (basic lead chromate; chromium lead oxide; C.I. 77601; C.I. pigment orange 21; C.I. pigment red). a red, amorphous or crystalline solid below 92°C, $CrO_4Pb{\cdot}OPb$. It is a confirmed human carcinogen and probable human mutagen. Lead fumes are released when this compound is heated to decomposition.

lead chromate, sulfate, and molybdate. lead-molybdenum chromate.

lead colic (Devonshire colic; painter's colic; saturnine colic). the severe abdominal pain, with diarrhea (or constipation) that is symptomatic of, and provides a sensitive index of, the alimentary type of chronic lead poisoning because it may appear prior to discernable elevation of blood lead levels. It is due to intestinal spasm and may be associated with rigidity of the abdominal wall. The symptoms of copper colic are similar. See also lead, lead poisoning.

lead cyanide (lead(II) cyanide; C.I. 77610; C.I. Pigment Yellow 48). a toxic, white to yellow, slightly water-soluble powder, $Pb(CN)_2$, that decomposes in acids. It is toxic by all routes of exposure. See lead, lead poisoning.

lead(II) cyanide. lead cyanide.

lead difluoride. lead fluoride.

lead encephalopathy. See encephalopathy.

lead fluoride (lead difluoride; plumbous fluoride). a poisonous, white to colorless, water-soluble crystalline substance, PbF_2. Crystals are orthorhombic, converted to cubic above 316°C. See also lead, lead poisoning.

lead fluosilicate. lead hexafluorosilicate.

lead formate. a poisonous, water-soluble, brownish-white compound, $Pb(CHO)_2$, with lustrous monoclinic or prismatic crystals. It is used as an analytical reagent. See also lead, lead poisoning.

lead hexafluorosilicate (lead fluosilicate; lead silicofluoride). a poisonous, crystalline, water-soluble dihydrate, $PbSiF_6$. It is used in the electrolytic refining of lead. See also lead, lead poisoning.

lead hydroxide (lead oxide hydrate; basic lead hydroxide). a poisonous, water-insoluble, white powder, $3PbO{\cdot}H_2O$, that absorbs atmospheric CO_2. See also lead, lead poisoning.

lead hypophosphite. a poisonous, hygroscopic, powdery crystalline substance, $Pb(H_2PO_2)_2$. It is alcohol-insoluble; only slightly soluble in cold water, but more so in hot water. See also lead, lead poisoning.

lead iodide. a poisonous, water- and alcohol-insoluble, bright golden-yellow, heavy crystalline compound, PbI_2; used in photography, printing, bronzing. See also lead, lead poisoning.

lead lactate. a poisonous, water-soluble, heavy, white crystalline powder. It slowly absorbs atmospheric CO_2, lowering its solubility. See also lead, lead poisoning.

lead line. the bluish line on the gums seen in lead poisoning.

lead metasilicate. lead silicate.

lead molybdate. a poisonous yellow powder, $PbMoO_4$, that is soluble in nitric acid and insoluble in water and ethanol. It is used in pigments, as an analytical reagent, and the crystals have applications in electronics and optics. See also lead, lead poisoning.

lead-molybdenum chromate (chromic acid, lead and molybdenum salt; chromic acid lead salt with lead molybdate; C.I. Pigment Red 104; lead chromate, sulfate, and molybdate; molybdenum lead chromate; molybdenum orange). a powerful oxidizing agent and severe irritant of the eye, skin, and mucous membranes. It is a possible human carcinogen and mutagen. See also lead, lead poisoning.

lead monosubacetate. lead subacetate.

lead monoxide (lead oxide; lead(II) oxide; lead oxide yellow; plumbous oxide; litharge; litharge yellow; massicot; lead protoxide; C.I. 77577; C.I. Pigment Yellow 46;). a very toxic dimorphic compound, PbO, of red to reddish-yellow tetragonal crystals (stable to 489°C) and yellow orthorhombic crystals (relatively stable above 489°C). Inhalation of dust that contains lead monoxide is hazardous. When working in a lead monoxide-containing atmosphere, an approved dust mask (United States Bureau of Mines) should be worn; hands should be scrubbed before handling, eating food, or smoking. See also litharge, lead, lead poisoning.

lead nitrate. a strongly oxidizing, poisonous, white, water- and alcohol-soluble, crystalline compound, $Pb(NO_3)_2$; used as a mordant, paint pigment, photographic sensitizer, in the manufacture of matches and certain explosives, and in engraving and tanning. See also lead, lead poisoning.

lead oleate (oleic acid lead salt). a water-insoluble, white, ointment-like material, $Pb(C_{18}H_{33}O_2)_2$, soluble in alcohol and other organic solvents. Slightly toxic in tests on guinea pigs. Used in varnishes, lacquers, and high-pressure lubricants. See also lead, lead poisoning.

lead orthoplumbate. lead oxide red.

lead oxalate. a poisonous, heavy, water-insoluble white powder, PbC_2O_4. See also lead, lead poisoning.

lead oxide. lead monoxide.

lead(II) oxide. lead monoxide.

lead oxide hydrate. lead hydroxide.

lead oxide red (C.I. 77578; lead orthoplumbate; lead tetraoxide; lead tetroxide; mineral orange; mineral red; minimum; orange lead; Paris red; pigment red; plumboplumbic oxide; red lead; red lead oxide; Saturn red; trilead oxide red; trilead tetroxide). a poisonous, bright red, heavy powder, Pb_3O_4, soluble in excess glacial acetic acid and dilute hydrochloric acid. The commercial product contains about 90% lead tetroxide and 10% lead monoxide. It has numerous uses, e.g., in plasters, ointments, ship paints, cements for glass and steam pipes, in the manufacture of colorless glass, in storage batteries, and the manufacture of lead peroxide. See also lead, lead poisoning.

lead oxide yellow. lead monoxide.

lead palsy. lead polyneuropathy. See palsy.

lead perchlorate. a white, crystalline compound, $Pb(ClO_4)_2 \cdot 3H_2O$, that decomposes at 100°C. It is highly soluble in cold water and in ethanol. A strong oxidizing agent, it is very toxic and dangerous. It destroys tissue and other organic materials on contact. See also lead, lead poisoning.

lead phosphate. a poisonous water- and alcohol-insoluble, white powder, Pb_3PO_4, soluble in nitric acid and in fixed alkali hydroxide; it is likely to be carcinogenic. Used as a stabilizer for plastics. See also lead, lead poisoning.

lead poisoning (plumbism; saturnism). poisoning by lead and lead compounds is one of the most common occupational diseases and a special threat to children in many communities around the world. Effects in humans may include a persistent metallic taste in the mouth, a burning abdominal pain, constipation followed by diarrhea, muscular weakness, convulsions, paralysis of the extremities, cold, cyanotic skin, and delayed severe anemia. Death may result from peripheral vascular collapse or encephalopathy. Lead in children also interferes with development of the central nervous system. Lesions may include gastrointestinal inflammation, liver and kidney injury, and often encephalopathy in children. **acute lead poisoning** (acute plumbism). poisoning from a single exposure is uncommon or rare and is usually the result of accidental or intentional ingestion of solutions that contain soluble lead salts. There is an immediate gastrointestinal reaction which may subside only to be followed within several days or a few weeks by severe, often persistent, even fatal attacks which may be indistinguishable from saturnism. If enough lead has been absorbed, the victim may experience muscle weakness, leg cramps, paresthesias, depression, coma, and possibly death within 1-2 days. Fatal acute poisonings with these symptoms have been reported in cases where the subject ingested more than 30 gm of lead acetate or lead carbonate. Acute lead poisoning in children usually produces symptoms of encephalopathy (cerebra lead poisoning). **alimentary lead poisoning.** symptoms typically include a metallic taste, lead colic (severe epigastric pain), and constipation.

cerebral lead poisoning (lead encephalopathy). the most common type of chronic lead poisoning in children. On occasion an alert, fully conscious child may lapse into coma within a few hours and die. More often the affected child will experience headache, irritability, restlessness, insomnia, persistent (sometimes projectile) vomiting, delirium, hallucinations, convulsions, coma, and often death from exhaustion and respiratory collapse. The case mortality rate is high and recovery is slow and often partial. Lead colic may or may not be observed. **chronic lead poisoning** (saturnism; plumbism; chronic plumbism). lead intoxication that usually results from the retention (accumulation) of lead from repeated ingestion of small amounts of lead. It may also result from retention of lead from acute poisoning from which a subject may have apparently recovered. On occasion chronic lead poisoning has been attributed to retention of lead bullets. There are three main types of chronic lead poisoning: alimentary, cerebral, neuromuscular. **neuromuscular lead poisoning**. this syndrome is usually seen in adults, but occasionally in children. It is characterized by lead palsy which is usually due to a painless peripheral polyneuritis that involves only extensor muscles. On occasion, muscular wasting, myalgia, and arthralgia are seen.

lead protoxide. lead monoxide.

lead salt. lead butyrate.

lead silicate (lead metasilicate). a poisonous, insoluble, white, crystalline substance, $PbSiO_3$; used in paints, enamels, ceramics, and fireproofing of fabrics. See also lead, lead poisoning.

lead silicofluoride. lead hexafluorosilicate.

lead sodium hyposulfate. lead sodium thiosulfate.

lead sodium thiosulfate (lead sodium hyposulfate, sodium lead hyposulfate). a poisonous, white, crystalline substance, $Na_4Pb(S_2O_3)_3$. It is slightly water-soluble, but soluble in thiosulfate solutions. Used in the manufacture of matches. See also lead, lead poisoning.

lead stearate (stearic acid lead salt). a poisonous, white, water-insoluble, alcohol- and ether-soluble powder, $Pb(C_{18}H_{35}O_2)_2$, that melts at 125°C. Used in high-pressure lubricants and as a drier in lacquer and varnish. It is readily absorbed through the skin. See also lead, lead poisoning.

lead subacetate (lead monosubacetate; monobasic lead acetate). a poisonous, heavy, white powder, $Pb(C_2H_3O_2)_2 \cdot 2Pb(OH)_2$. It is a decolorizing agent for sugar and other organic substances. See also lead, lead poisoning.

lead subcarbonate. basic lead carbonate.

lead sulfate. a poisonous, heavy, white, crystalline powder, $PbSO_4$, that is slightly soluble in hot water, and insoluble in alcohol. Used as a paint pigment, and in lead-acid batteries, *q.v.* See also lead, lead poisoning.

lead sulfocyanate. lead thiocyanate.

lead telluride. a silver-gray crystalline (cubic) solid, PbTe, that is very toxic if inhaled or ingested. Single crystals are used as semiconductors or photoconductors. See also lead, lead poisoning.

lead tetraethyl. tetraethyl lead.

lead tetraoxide. lead oxide red.

lead tetroxide. lead oxide red.

lead thiocyanate (lead sulfocyanate). a poisonous, yellowish, slightly water-soluble, crystalline (monoclinic) substance, $Pb(SCN)_2$. See also lead, lead poisoning.

lead tree. *Leucaena glauca*

leaded gasoline. automotive gasoline that contains small amounts of tetraethyl lead, *q.v.*, to prevent preignition ("knocking"). See also lead, lead poisoning.

leaf aldehyde. 2-hexenal.

leaf-nosed viper. *Eristicophis macmahonii.*

leaf viper. *Atheris squamigera.*

leafy spurge. *Euphorbia esula*.

least fatal dose (LFD). the smallest dose of a substance that can cause death to a specified type of organism.

leatherback. *Scomberoides sanctipetri*.

leatherback turtle. *Dermochelys coriacea*.

leatherjacket. **1:** any bony fish of the family Balistidae. **2:** any turtle of the family Dermochelidae, represented today only by the leatherback turtle, *Dermochelys coriacea*, the largest living turtle.

lebolobolo. *Bitis arietans*.

lecheguilla. lechuguilla.

lechuguilla (lecheguilla). *Agave lecheguilla* or any of several closely related species in this genus.

lechuguilla poisoning, lechuguilla fever (swellhead). photosensitization of sheep and goats in western Texas, southeastern New Mexico, and northern Mexico due to ingestion of *Agave lecheguilla* (lechuguilla) and subsequent exposure to sunlight. The condition is marked by swelling and crusting of the face and ears. Additional effects include liver damage and consequent icterus, sometimes hemoglobinuria, and often death. Lesions include liver and kidney necroses.

lecithid. any compound produced by the action of venom hemolysin on lecithin. **cobra lecithid**. a hemolytic compound formed by the action of cobra toxin on lecithin in circulating blood.

lecithinase. a class of enzymes (lecithinases A through D) now known as phospholipase (A through D). **cobra lecithinase**. phospholipase A.

lectin (agglutinin; affinitin; plant agglutinin; phytoagglutinin; phytohemaglutinin; phasein; phasin; protectin). any of a variety of toxic proteins that agglutinate erythrocytes and other types of cells (e.g., tumor cells). They are also mitogens and are apparently related only by their unusual binding specificity. They occur widely among plants, especially the seeds of legumes and also bacteria, invertebrate animals, and fish. Almost all are inhibited or inactivated by specific free monosaccharides or oligosaccharides. Important lectins are abrin, concanavalin A, ricin, soybean agglutinin (SBA), and wheat germ agglutinin (WGA).

Ledakrin. See nitracrine.

Ledum (Labrador tea). a genus of low toxic boreal shrubs (Family Ericaceae) that contain grayanotoxins. They occur in bogs, swamps, or wet meadows of North America from Greenland, Labrador, and Alaska southward to New Jersey, Pennsylvania, Ohio, Minnesota, Washington, and to the mountains of western North America from California northward. They contain grayanotoxins. See also *andromeda*.

lefa bin kurum. *Cerastes cerastes*.

Leguminosae. former name of Fabaceae, the bean family.

legwere. *Dispholidus typus*.

Leipzig yellow. lead chromate.

LEL. lowest effect level. See dose (minimum effective dose).

lemon chrome. C.I. pigment yellow 31.

lenacil (3-cyclohexyl-6,7-dihydro-1*H*-cyclopentapyrimidine-2,4-(3*H*,5*H*)-dione;3-cyclohexyl-5,-6-trimethyleneuracil; 3-cyclohexyl-1,5,6,7-tetrahydro-2*H*-cyclopentapyrimidine-2,4(3*H*)-dione). a solid compound, $C_{13}H_{18}N_2O_2$, that is slightly soluble in water and most organic solvents; it is soluble in pyrimidine. It is a uracil, used as a selective herbicide. It effectively controls many annual and some perennial weeds with little or no damage to a variety of crops. It is absorbed chiefly through the roots. Lenacil is probably only slightly toxic to humans.

Lepidoptera (moths, butterflies). a large order of insects that includes the butterflies and moths. The appendages are covered by fine, overlapping scales. They have long antennae and large eyes; a long proboscis adapted to sucking; four large wings that are usually brightly pigmented or patterned, and cross-veined; terrestrial wormlike larva (caterpillar) with a biting mouth. Metamorphosis is complete; the pupa is exposed or covered by a woven silk cocoon. A few species are poisonous or noxious as adults and the caterpillars of many are poisonous or venomous. See lepidopterism, caterpillar, *Callimorpha*, *Hylesia*, *Thametopoea pityocampa*, *Zygaena*.

lepidopterism. any of a number of reactions, usually skin reactions (e.g., dermatitis, urticaria, allergy of humans and other animals to the venoms produced by the bristles of certain caterpillars (Order Lepidoptera).

Leptomicrurus (slender coral snakes; Schlank Korallenottern). brightly colored, venomous and potentially dangerous snakes (Family Elapidae) of northern South America. The head is small and indistinct from the neck; the snout is rounded; the eyes are small and round; there is no distinct canthus. The body is extremely elongate, slender, and untapered, with a short tail. Some individuals reach a length of about 1 m (three feet). These are the only coral snakes in which the light crossbands are incomplete dorsally. The mental and anterior chin shields meet; this is also a distinguishing feature. No antivenin is produced for these rare snakes. ***L. collaris***. a rare South American coral snake that occurs in British Guiana, French Guiana, Surinam, and Venezuela. This snake is similar in many respects to *L. narducci* and should be considered dangerous. ***L. narducci*** (Amazon slender coral snake). a small, rare, very elongate, mostly black snake with a broad yellow band over the back of the head. It occurs in the upper Amazon region of South America including Columbia, northwestern Brazil, eastern Ecuador, Peru, and Bolivia. Occasional individuals reach a length of nearly 1 m. The dorsum is solid black but the belly is marked by red or yellow and black crossbands; red and yellow bands sometimes extend laterally up the sides as triangular blotches. It has fewer ventral scales than *L. collaris* but is otherwise very similar in appearance. Both species should be considered dangerous.

Leptophis (parrot snakes). a genus of venomous snakes (Family Colubridae) of western South America. The effects of envenomation on humans typically include local pain and edema, sometimes with ecchymosis, and local discoloration of the skin. In serious cases, the swelling and edema may be substantial, involving the entire affected extremity; the victim may become feverish, enfeebled, and apathetic. The acute phase may persist for as long as a week.

leptophos (*O*-(4-bromo-2,5-dichlorophenyl)-*O*-methylphenylthiophosphonate; *O*-(2,5-dichloro-4-bromophenyl)-*O*-methyl phenylthiophosphonate; *O*-methyl-*O*-(4-bromo-2,5-dichlorophenyl)phenyl thiophosphonate; MBCP). an insecticide and pesticide, $C_{13}H_{10}BrCl_2O_2PS$, that was in use in the United States during the early to mid 1970s. It was banned from further use because of toxic effects on both production workers and agricultural workers. It is toxic to humans by all routes of exposure. It is neurotoxic, causing severe harm. Symptoms of intoxication are similar to those of multiple sclerosis. Symptoms may include muscular weakness, staggering gait, tremors, paresthesia, tingling sensations, blurred vision, disorientation and temporary memory loss, auditory hallucinations, and paralysis of the legs. It may have teratogenic and mutagenic effects.

lerabe. *Bitis arietans*.

lesion. **1a:** any circumscribed area of morbid or otherwise pathologically altered tissue. **1b:** focal injury or damage, as a wound. **1c:** a single infected area in a skin disease. **2:** any breakdown or discontinuity in a process, a tissue, or the loss of function of a part, however small, that is due to injury or disease. **degenerative lesion**. one caused by or undergoing degeneration. **destructive lesion**. one that causes tissue necrosis or the death of an organ. **diffuse lesion**. one that spreads to encompass a large area. **focal lesion**. a small, well-defined lesion or one that is limited to the area of contact with the causative agent. **gross lesion**. one that is visible to the naked eye. **histologic lesion**. a microscopic lesion that appears at the cell/tissue level of organization. **irritative lesion**. one that provokes activity in the part of the body where it is located. **local lesion**. one that arises in the nervous system and produces local symptoms. **molecular lesion.** a lesion expressed at the molecular level of organization. **organic lesion**. structural lesion. **partial lesion.** one that involves only a part of a tissue, organ or other type of structure. **peripheral lesion**. a lesion of nerve endings. **precancerous lesion.** a lesion that appears unlikely to become malignant due either to its location or characteristics. **primary lesion. 1:** the first lesion to appear as a consequence of poisoning or a disease process. **2:** the lesion (chancre) that anticipates the onset of syphilis. **structural lesion** (organic lesion). one that produces distinct changes in a tissue. **systemic lesion.** one that is confined to an organ system or to a set of organs that have a common function. **total lesion.** one that involves an entire organ or the entire diameter of a conductile vessel or tract. **toxic lesion.** that due to the action of bacterial toxins or of conventional poisons.

lesser black krait. *Bungarus lividus*.

lesser cerastes viper. *Cerastes vipera*.

lesser weever. *Trachinus vipera*.

lethal. fatal, mortal, noxious, toxic, virulent. **1a:** said of any causal agent: able to bring about death; causing death; deadly, fatal. **1b:** of, pertaining to, or causing death by direct action, deadly. **2a:** said of a poison or pathogen: deadly, noxious, virulent, fatal. **2b:** deadly in relation to a particular situation or target (e.g., a particular species or type of organism).

lethal concentration. See concentration.

lethal synthesis. **1a:** the synthesis *in vivo* of a toxic metabolite from a xenobiotic that is less toxic. Many organophosphorus insecticides undergo lethal synthesis (e.g., parathion to paroxan). **1b:** the metabolic transformation of a xenobiotic that is structurally similar to an endogenous substrate, and its incorporation into the same pathway with lethal results. See, for example, fluoroacetate.

lethality. **1a:** deadliness; the property, ability, quality, or state of being lethal. **2a:** the ratio of fatalities from a given cause to that which is normal to the population of interest. **2b:** an index of mortality. **2c:** an index of toxicity; the ratio of the number of fatalities to the number of exposed individuals (or other type of living system).

lethargy. **1:** listlessness; sluggishness; indolence. **2:** a state of deep, prolonged depression or stupor from which it is possible to be aroused, but with immediate relapse.

Lethrinus (snappers). percoid bony fishes (Family Lethrinidae) of warm sunny seas. Some species are important food fishes, others are dangerously toxic. ***L. miniatus*** (snapper). a species found in Polynesian waters west to East Africa. Certain populations are transvectors of ciguatera.

lettuce. See *Lactuca*

Leucaena glauca (lead tree). a deep-rooted perennial shrub or tree (Family Fabaceae, formerly Leguminosae). In the United States, it occurs locally in wastelands of the Coastal Plain from Florida to Texas, and in arid lands in the Hawaiian Islands, where it is also cultivated. The poisonous principle is the alpha amino acid mimosine. Ingestion of large quantities of this plant reduces rates of weight gain, results in generally poor condition, and loss of hair in livestock, and laboratory rats. The disease has affected horses, donkeys, mules, rabbits, and swine in Hawaii, but cattle and domestic fowl do very well on *Leucaena* forage. The condition is reversible.

leucoharmine. harmine.

Leucoium (snowflakes). a genus of old world bulbous, amaryllidaceous plants with drooping white flowers (Family Liliaceae). *L. aestivum*, *L. autumnale*, and *L. vernum* are popular ornamentals. All are emetic and toxic.

Leucothoe (fetter bush). a genus of American woodland shrubs (Family Ericaceae). ***L. davisae*** (Sierra laurel; black laurel). a woody shrub ca. 1½-2½ m tall that grows in wet forest soils at elevations of ca. 1000-2700 m.

leucovorin calcium. calcium folinate.

leukocidin. a toxin produced by some pathogenic bacteria that destroys polymorphonuclear leukocytes with or without lysis.

leukocytolysis. lysis of leukocytes. **venom leukocytolysis**. lytic destruction of leukocytes by snake venom.

leukocytolytic. **1:** of, pertaining to, or causing leukocytolysis. **2:** an agent (e.g., a snake venom) that causes leukocytolysis.

leukocytosis. a transient increase in the concentration of leukocytes in circulating blood. Causes vary, but may include, for example, hemorrhage, fever, infection, and inflammation. **toxic leucocytosis**. leukocytosis caused by leukocytotoxin and other blood poisons.

leukocytotoxin. any toxin that destroys leukocytes.

leukoencephalitis. inflammation of the white matter of the brain. See moldy corn poisoning.

leukomaine. any one of a large group of basic substances resembling alkaloids that are normally present in living tissues. They are products of metabolism. Many are physiologically active; some are toxic.

leukomainemia. abnormally high amounts of leukomaines in circulating blood.

leukomainic. of, pertaining to, or characterized by a leukomain.

leukopenia. a condition of reduced numbers of leukocytes in circulating blood; in humans a count of 5000 cells/mm^3 or less.

leukopenic. of, pertaining to, characterized by, or able to cause leukopenia.

leukotoxic. poisonous to leukocytes.

leukotoxicity. the quality of being poisonous to leukocytes.

leukotoxin. a cytotoxin that destroys leukocytes.

Levante viper. *Vipera lebetina*.

Levantine adder. *Vipera lebetina*.

Levantine viper. *Vipera lebetina*.

levodopa (3-hydroxy-L-tyrosine; (−)-3-(3,4-dihydroxyphenyl)-L-alanine; β-(3,4-dihydroxyphenyl)-α-alanine; L-3,4-dihydroxyphenylalanine; 2-amino-3-(3,4-dihydroxyphenyl)propanoic acid; L-dopa). the naturally occurring L-form of dopa. A derivative of tyrosine, it is the amino acid precursor of the catecholamine neurotransmitters, dopamine, norepinephrine, and epinephrine. It is moderately toxic to laboratory animals.

Lewis base. See base.

lewisite (dichloro(2-chlorovinyl)arsine; (2-chloroethenyl)-arsonous dichloride; 2-chlorovinyldichloroarsine; chlorovinylarsine dichloride). the legal label name of β-chlorovinyldichloroarsine. It is an extremely toxic, oily, colorless to brown, or violet, vesicant liquid, $ClCH{=}CHAsCl_2$, with the faint odor of geraniums. It is a respiratory and systemic poison used as a warfare gas. Even small amounts, absorbed via the skin, respiratory, or gastrointestinal tracts, cause severe vesication and systemic effects. It is readily absorbed via the skin and can penetrate rubber gloves. As little as 2 ml applied to the skin can be fatal. Dimercaprol (British Antilewisite), a toxic chelating agent, was developed as an antidote to lewisite poisoning.

LFD. least fatal dose.

LH (luteinizing hormone; interstitial cell stimulating hormone; lutotropin; ICSH). a glycoprotein gonadotropin that is secreted by the anterior lobe of the pituitary gland. It stimulates the ovaries to produce progesterone, acts in concert with FSH to stimulate release of estrogens from the Graafian follicles of the ovary. It also induces extrusion of the ripe ovum from the ovarian follicle (ovulation) and subsequent formation of the corpus luteum. See also hormone, pituitary gland.

Li. the symbol for lithium.

Liability Assessment. an assessment of the legal responsibility for violations or damages related, for example, to toxic or hazardous materials. Potential liabilities may result from various federal laws, state laws, and common law.

liana. See *Chondodendron tomentosum*.

lihomo. *Bitis arietans*.

lichen. a distinctive organism with a thallus composed of symbiotically associated fungal and algal cells. The fungus (mycobiont) is usually an ascomycete and is dominant to the algal component (phycobiont). The latter is usually green algae or blue-green bacteria. Many are used as livestock and wildlife feeds, wildlife browse (e.g., *Cladonia rangiferina* reindeer moss), and as emergency foods for humans in high latitudes. Several contain mildly toxic chemicals that are removed by boiling in water. Some contain antibiotics. See *Parmelia molliuscula*.

Lichtenstein night adder. *Causus lichtensteinii*.

licorice (liquorice). *Glycyrrhiza glabra* (and other species of this genus); glycyrrhiza.

licorice root. glycyrrhiza.

lienotoxin. splenotoxin.

life cycle. **1a:** the set of stages through which an organism passes from "birth" to death. **1b:** life history; the stages through which an organism passes from "birth" to maturity, including the reproductive cycle.

life expectancy. the period during which an organism of a given class (e.g., species, age, sex, occupation) can be expected to live on the basis of statistical analysis of existing data usually in the form of actuarial or life tables. The existance of a profusion of toxic and otherwise hazardous chemicals in the environment may significantly influence life expectancy of various species including humans.

life history. life cycle (def. 1b).

life root. *Senecio aureus*.

life science. biology.

lifquorid. glycyrrhiza.

ligand. **1:** a molecule, atom, or ion (e.g., EDTA, ammonia) that complexes with another molecule by attaching to the central atom of a coordination compound such as a chelate or other complex. **2:** in toxicology and immunology, a ligand is generally defined as a small molecule bound to a protein by noncovalent forces. Examples are antigens, haptens, and molecules such as oxygen that are reversibly bound to the heme iron of heme proteins. **3:** an organic molecule attached to a tracer element.

ligand-receptor complex. the product of a reaction of a ligand (e.g., a toxicant) with a receptor from which it later dissociates, releasing an unaltered ligand and receptor. While bound to the ligand, the receptor does not function normally and a toxic effect may result.

ligustrin. the toxicologically active principle of *Ligustrum vulgare*.

Ligustrum (privet). a genus of deciduous or evergreen shrubs (Family Oleaceae) that are native to Europe, Asia, and Australia. ***L. vulgare***. (privet; European privet; prim; lovage; hedge plant; Japanese privet). a shrub native to northern Europe, but now often widely grown in parks and gardens as a shrub or as a hedge plant. It often reaches a height of nearly 5 m. The entire plant is poisonous, especially the blackish berries; the poisonous principle is ligustrin, *q.v.* Ingestion of the leaves or berries may cause severe gastroenteritis, frequent vomiting, watery stools, abdominal colic, hypotension, collapse, and kidney damage which may terminate in death. A lethal dose can cause death in as little as 2 hours.

likwétéma. *Thelotornis kirtrlandii*.

lily of the valley. *Convallaria majalis*.

lima bean. See *Phaseolus*.

limber leg. See *Acacia berlandieri*.

lime. calcium oxide. See also chlorinated lime.

lime arsenate. a poisonous aqueous solution of white arsenic and sal sod. It is used as an insecticide.

limen. Latin meaning a threshold, limit, or boundary.

limit. **1:** a boundary, border, ceiling, or limitation. **2:** that which constrains or confines. **3:** limen. See, for example, limit value, detection (detection limit), maximum residue limit, tolerance limit, exposure limit. **lower limit**. a level below which further decrease in a variable, function, or process is not possible or is not permitted. **upper limit**. a point or line above which further increase in a variable, function, or process is not possible or is not permitted.

limit of detection. **1:** the size or amount below which an object may not be detected by a specified means (e.g., direct visual observation, light microscope, electron microscope). **2:** the minimum concentration (or in some cases, the amount) of a substance that can be detected and identified 99% of the time under a given set of circumstances (e.g., method used, detection device used, matrix). **3:** in analytical chemistry, the smallest amount or the lowest concentration of a substance or material that one may observe or measure with a given tool or by applying a particular process. See also sensitivity (def. 3).

limit value. a limit is usually based on some measure of concentration, dose, or exposure to an industrial chemical and a specified health effect as indicated below in the various types of limit values. **biological limit value** (BLV). a limit based on the relationship between some measure of internal dose and either the environmental exposure in the workplace or some measure of a specified health effect. **environmental limit value** (ELV). a limit that is usually based on the relationship between the airborne concentration of an industrial chemical in the workplace environment and a specified adverse health effect based on experimental data from tests on animals or humans. **short-term exposure limit value** (STEL). See threshold limit value. **threshold limit value** (TLV). the maximum concentration in air at which it is believed that a substance will not produce adverse health effects on workers with repeated daily exposure. It can be a time-weighted average (TLV-TWA), a short-term value (TLV-STEL), or an instantaneous value (TLV-Ceiling). TLVs are based on the best available information from industrial experience, experimental studies of humans, and/or animals. A TLV is a consensus guideline developed and revised annually by the American Conference of Government Industrial Hygienists. TLVs are expressed either as ppm or mg/m^3. The original aim was to prevent acute stress, illness, and clinical manifestations of occupational disease. With time and the development of more sensitive indicators of biological response, emphasis shifted toward prevention of less obvious signs of toxicity. TLVs provide valid guidance only if (1) the assumption that there is a threshold concentration for effects is true and (2) the data can reasonably be assumed to be valid for a person's working lifetime. These conditions are rarely met. TLVs are thus usually determined by applying safety factors to the apparent thresholds for effects determined by animal studies. Even so, TLVs are intended only for use in the practice of industrial hygiene as guidelines or recommendations in the control of potential health hazards. They are not intended to be lines of demarcation between safe and unsafe concentrations. See also exposure limit, maximum allowable concentration. **threshold limit value-biological** (TLV-BLV; biological threshold limit value). a TLV based on measurements of a biological variable in exposed individuals that is known to be related to exposure levels. Data in support of such values are limited and biological TLVs are tentative. **threshold limit value-ceiling** (TLV-C; ceiling TLV). the concentration in air of a substance that should not be exceeded even momentarily. It is thus a ceiling or maximum permissible concentration. The TLV-C is an exacting guideline that is applied to a few fast-acting, highly toxic, or extremely irritating substances (e.g., acetic anhydride) that may cause serious effects even during a brief period of exposure. This value is similar to the maximum acceptable concentrations (MACs) recommended by the American National Standards Institute. The TLV-C is used in place of, rather than in conjunction with, a time-weighted average. **threshold limit value - short-term exposure limit** (TLV-STEL). the maximal concentration to which workers can be exposed for a period of up to 15 minutes continuously without suffering from **1:** irritation, **2:** chronic or irreversible tissue change, or **3:** narcosis of sufficient degree to increase accident proneness, impair self-rescue or materially reduce work efficiency, provided that no more than four excursions per day are permitted, that at least 60 minutes elapse between exposure periods, and that the daily TLV-TWA is not exceeded. **threshold limit value-time-weighted average** (TLV-TWA). the time-weighted average concentration of a substance in air for a normal 8-hour working day or a 40-hour work week to which nearly all workers may be repeatedly exposed, day after day, without evident adverse health effects. Certain excursions above this limit are allowed if compensated for by equivalent excursions below the limit during the work day (sometimes the work week). Most TLVs are based on time-weighted average concentrations. In the United States, OSHA (the Occupational Safety and Health Administration) has established similar TLV-TWAs. OSHA values are given in parts per billion (ppb), parts per million (ppm), or milligrams per cubic meter (mg/m3). The U.S. EPA and other organizations publish lists of limits for chemicals that are not on the OSHA list.

limit of visibility. the size below which an object cannot be seen with the unaided eye.

Limu-make-o-Hana. common Hawaiian name of *Palythoa toxica*.

Limulus polyphemus (horseshoe crab; king crab). a marine, arachnid-like arthropod (Family Limulidae, Class Merostomata) that occurs in the coastal waters of the Americas from Maine to Yucatan. It is greenish-tan in color, but the carapace is usually covered with algae in older individuals. There is a compound eye on either side of the carapace and 2 simple eyes on the forepart of the midline. It grows to a length of about 61 cm. For a general description, see Merostomata. See also ergothioneine.

linamarase. An enzyme of *Linum usitatissimum*. It releases cyanide from linamarin.

linamarin. a bitter-tasting cyanogenic glycoside in *Linum usitatissimum* and *Phaseolus limensis*. It is chemically identical to phaseolunatin, the cyanogenic glycoside of the lima bean.

Lindafor™. lindane.

lindane (γ-HCH; 1α2α3β4α5α6β-hexachlorocyclohexane; γ-benzene hexachloride (γ-BHC); gamma-benzene hexachloride; gamma-hexachlor; Aparasin; Aphtiria; Benzahex; Chemhex; ENT 7796; Esoderm; Gammalin; Gamene; Gamexan; Gamiso; Gammexane; Gamoxol; Gexane; Hexadon; Jacutin; Kwell; Lindafor; Lindatox; Lorexane; Pecusanol; Pultox; Quellada; Streunex; Tri-6; Vermaxan; Viton). the legal label name for γ-benzene hexachloride, *q.v.*, lindane contains 99% or more of this isomer. It is used as a commercial scabicide, pediculicide, insecticide, and herbicide, and is toxic by all routes of exposure. Lindane is a useful ectoparasiticide for large animals, and for dogs, but is highly toxic to cats in concentrations necessary for parasite control. Great care must be taken in its use; only cattle in good condition can be expected to tolerate 0.2% lindane applications. Stressed or emaciated cattle have been poisoned by spraying or dipping in 0.075% solutions. Horses and hogs appear to tolerate 0.2-0.5% sprays, whereas sheep and goats usually tolerate 0.5% applications. Very young, emaciated or lactating animals should be treated with extreme caution. The toxic effects of lindane are very similar to those of DDT. Degeneration of kidney tubules, liver damage associated with fatty tissue and histoplastic anemia have been observed in humans, and exposures to small amounts of lindane can cause headaches, nausea, depression, and respiratory difficulties. Serious acute exposures can cause convulsions, elevated white blood cell counts, and heart failure. Lindane is fetotoxic and probably carcinogenic. Children are especially susceptible to the toxic effects of lindane. See also lindane poisoning.

lindane poisoning. poisoning by lindane, the commercial grade of γ-benzene hexachloride, used chiefly to control ectoparasites of dogs and larger animals. Cats are an exception to such usage, however, since they are poisoned at concentrations below that required for parasite control. Young animals, lactating animals, and those in poor condition have heightened susceptibility to lindane poisoning and should be treated with extreme caution. Poisoning can occur by all routes of exposure. Toxic effects are very similar to those of DDT. **lindane poisoning, acute**. symptoms in humans may include dizziness, irritability, restlessness, headache, weakness, nausea, vomiting, dyspnea, tremors, convulsions, cyanosis, and death from respiratory failure or circulatory collapse. The vapor may irritate the eyes, nose, and throat, and topical application may cause a localized reaction. Lesions may include liver damage associated with fatty tissue, hyaline degeneration of the renal tubules, and histoplastic anemia. Liver damage has occurred in chronically challenged experimental animals. See also γ-benzene hexachloride, lindane.

Lindatox™. lindane.

linear alkylate sulfonate. alkyl sulfonate.

linear response. a response (e.g., of a measuring instrument) that is directly proportional, within specified limits, to the value of the quantity being measured (e.g., to the concentration of a toxicant in air).

lined night adder. *Causus lineatus*.

Lineweaver-Burk equation. See Lineweaver-Burk plot.

Lineweaver-Burk plot. one of a number of methods (e.g., Eadie-Hofstee plot, Woolf plot), based on transformations of the Michaelis-Menton equation, used to linearize enzyme- or receptor-binding data. The Lineweaver-Burk equation is:

$$1/v = K_m/v_{max}(1/[S]) + 1/V_{max}$$

where v is the initial velocity of the enzyme reaction in the presence of a given amount of substrate [S], V_{max} is the maximum theoretical velocity of the reaction, and K_m is the Michaelis-Menton constant. Thus, if one plots experimentally determined l/v against l/[S], the line will have a slope of Km/V_{max} and will intercept the ordinate at $1/V_{max}$.

lingual. of or pertaining to the tongue.

Linne's sea krait. *Laticauda laticauda*.

linotoxin. a toxin produced by *Linum neomexicanum*. It is uniformly toxic to laboratory animals.

linseed. See *Linum*.

linseed meal. See *Linum usitatissimum*.

linseed oil (flaxseed oil). an amber to brown drying oil with a characteristic odor and bland taste that is derived from *Linum usitatissimum*, *q.v.* The unboiled oil is usually not toxic if ingested. It is used, for example, in some paints, varnishes, putty, printing inks, oilcloths, soaps, and various pharmaceuticals.

Linum (flax; linseed). a genus of narrow-leaved herbs or shrubs (Family Linaceae) that is nearly worldwide in distribution. ***L. usitatissimum*** (common flax; linseed). a casual weed that is grown commercially to produce linen and linseed oil. The leaves and seeds contain a cyanogenic glycoside, linamarin, which is chemically identical to phaseolunatin. The plant is highly toxic, producing symptoms that are identical to those of cyanide poisoning. All classes of livestock have been poisoned by consuming linseed press cake, meal, and flaxseed chaff or screenings. The commercially prepared presscake in the United States rarely contains toxic levels of HCN. Sometimes small quantities of liberated HCN following consumption of linseed meal forms goitrogenic thiocyanates in the liver. Linseed meal fed to pregnant ewes has thus produced serious mortality among their lambs in iodine-deficient areas of New Zealand. See also linamarase. ***L. neomexicanum*** (yellow pine flax). an uncommon herbaceous annual or winter annual found in rangelands of New Mexico, Arizona, southward into Mexico. It contains saponins, is toxic and may have caused livestock losses. See also linotoxin. ***L. usitatissimum***. All parts of the plant contain cyanogenic nitrates and glucosides, especially linamarin. The immature seedpods are unusually toxic. Symptoms of intoxication include increased respiratory rate, gasping, excitement, weakness, staggering gait, paralysis, and convulsions. Concentrations of nitrate that are toxic to livestock are known.

linuron (*N'*-(3,4-dichlorophenyl)-*N*-methoxy-*N*-methylurea; methoxydiuron). a substituted urea used as a selective pre- and postemergence herbicide. Linuron is moderately toxic to laboratory rats.

lion's mane. *Cyanea capillata*.

lionfish. See *Pterois volitans*, *Scorpaena*, Scorpaenidae.

lipase. any enzyme that catalyzes the hydrolysis of triglyceride fats into glycerol, and fatty acids.

lipid peroxidation. a process whereby the C=C double bonds in unsaturated lipids, when attacked by free radicals in the presence of oxygen, undergo chain reactions that result in their oxidative destruction.

lipiri. *Bitis arietans*.

lipocorticoid. any corticoid that is especially effective in causing fat deposition (especially in the liver).

lipofuscinosis. any complaint due to abnormal storage of lipofuscins, which are a class of fatty pigments produced by dissolving a pigment in fat.

dl-α-lipoic acid. thioctic acid.

Lippia rehmanni. an herb (Family Verbenaceae) that causes hepatogenic photosensitivity in grazing animals. It contains the hepatotoxin rehmannic acid, a polycyclic triterpene which is also the toxic principle of *Lantana*. The phytotoxic pigment is phylloerythrin.

liquid petrolatum. mineral oil.

liquid silver. the element mercury.

liquorice. licorice.

liquorrhea. excessive discharge of a body fluid as in the case of a "runny" nose (liquorrhea nasalis).

lirio. *Atropa belladonna*.

lissive. able to relieve muscle spasm without affecting function.

listenon. succinylcholine chloride.

lisuride. a soluble ergot derivative and serotonin inhibitor with endocrine effects similar to those of bromocriptine.

liteo. *Naja melanoleuca*.

litharge. the most common form of lead monoxide, *q.v.* It results when the compound is heated to melting.

Litharge yellow. lead monoxide.

lithic acid. uric acid.

lithium (symbol, Li)[Z = 3; A_r = 6.9]. a rare, light, silvery-white, moderately reactive monovalent element of the alkali metal group. The lightest metal known, it occurs naturally in a number of complex silicates (e.g., lepidolite, spodumene, petalite) and a mixed phosphate (tryphillite). It exists in the environment only as lithium ions or salts. Lithium ion is a CNS toxicant causing dizziness, nausea, anorexia, apathy, and prostration. Large doses can cause coma and death. Lithium is nephrotoxic to humans, especially if sodium consumption is limited. Prolonged exposure to most lithium compounds may cause toxic effects; usually neurotoxic effects, except where the anion is responsible for the toxicity. Acute toxicity is uncommon. Li is used primarily as a catalyst. Lithium salts are effective against mania and depression. See also organolithium compounds.

lithium bromide. a toxic white, deliquescent, granular powder, LiBr, with a slightly bitter taste. It is used therapeutically as a sedative. Chronic absorption of LiBr can disturb blood electrolyte balance, and cause skin eruptions and CNS disturbances. The latter are due to the bromide moiety. See also lithium chloride.

lithium carbonate. a toxic, white, slightly water-soluble powder, Li_2CO_3. It is a strong irritant in solution. It is used in certain pharmaceuticals (given as a pill). The toxicity and effects are similar to those of lithium citrate, *q.v.*

lithium carbonyl. an unstable, probably very toxic carbonyl, LiCOCOLi. The two carbonyl groups, CO, form bridges between the lithium atoms.

lithium chloride. a moderately toxic, crystalline, granular or powdery, water-soluble substance, LiCl. Chronic absorption of LiCl can disturb blood electrolyte balance, cause CNS disturbances, and is nephrotoxic. It is used therapeutically as an antidepressant. See also lithium bromide.

lithium citrate. a very toxic, white powder or granular substance, $Li_3C_6H_5O_7{\cdot}4H_2O$, that is soluble in water. It is used in beverages and pharmaceuticals (given as a liquid). At toxic levels, a fatal acidosis or alkalosis may develop. Early symptoms are tremors, muscular twitchings, apathy, difficulty speaking, confusion, and ultimately coma, and death. Exaggerated reflexes and twitching or jerking in response to noise or light stimulation. *Cf.* lithium carbonate.

lithium hydrate. lithium hydroxide.

lithium hydride. the pure compound is a white, translucent crystalline mass or powder, LiH, but the commercial product is light bluish gray because of the presence of minute amounts of dispersed colloidal lithium. It is a desiccant that reacts vigorously on contact with moisture (e.g., in living tissue), producing lithium hydroxide (*q.v.*) and hydrogen gas, the latter may ignite spontaneously. The consequent damage to tissue may be greater than from other intensely corrosive alkalis such as sodium or potassium hydroxides. Lithium hydride is also highly flammable, igniting spontaneously in moist air. A dry chemical extinguisher should be used to extinguish such a fire.

lithium hydroxide (lithium hydrate). a granular, colorless, strongly alkaline, water-soluble crystalline or powdery compound, LiOH or $LiOH \cdot H_2O$, formed by the reaction of organolithium compounds, *q.v.*, with water. It is used in ceramics and lubricating greases, as an electrolyte in storage batteries and as a CO_2 absorbent. It is caustic and extremely irritating to the skin. It is also a systemic poison with effects similar to those of lithium bromide, lithium chloride, and other lithium compounds.

lithium selenate. a poisonous, water-soluble, crystalline substance, LiO_4.

little brown mushroom (LBM). certain of the brown-spored gilled mushrooms are edible, others such as *Cortinarius gentilis* and *Cortinarius orellanus* are deadly poisonous. Thus, mushroom hunters are well advised not to gather or eat LBMs.

little brown snake. *Elapognathus minor*.

little brownsnake. *Elapognathus minor*.

little desert snake. *Elapognathus minor*.

little spotted snake. *Denisonia punctata*.

little whip snake. *Denisonia flagellum*, *Denisonia gouldii*.

liver. **1:** digestive gland; the large digestive organ (or digestive gland) of many invertebrates. **2:** a major organ of vertebrate animals that plays complex roles in digestion, excretion, and detoxication. It is the principal site of metabolism and detoxication of most xenobiotics and can thus be expected to carry burdens of most chemicals that enter the body. The highest concentrations of mercury, for example, in many wild vertebrates occur in the liver. Concentrations of xenobiotics in liver, however, usually represent only recent exposures; exceptions would be in the case of birds, in which the liver is a major storage organ of lipids.

liver flap. asterixis.

liver tremor. asterixis.

livestock. animals of any kind that are kept or raised for pleasure or use. The term is most often used for domestic mammals raised for meat, meat products, or dairy products (e.g., cattle, swine, horses, goats, sheep).

livid (black and blue). discolored, as by congestion or a contusion; black and blue.

lividity. the quality or condition of being livid, as in postmortem lividity. See livor mortis.

livor. discoloration.

livor mortis (postmortem lividity). discoloration that appears in dependent parts of the body following death. It results from stagnation, settling (due to gravity), and pooling of blood following cessation of circulation.

lixiviation. leaching.

lizard. any reptile of the suborder Sauria, *q.v.* **venomous lizard**. Only two lizards, the Gila monster (*Heloderma suspectum*) found in Arizona, Sonora, and adjacent areas, and the beaded lizard (*H. horridum*) of Mexico are known to be venomous. Their venom, somewhat similar to those of some snakes, contains hyaluronidase, phospholipase A, and one or more salivary kallikreins. See *Heloderma*, Helodermatidae.

LLD. lowest lethal dose.

LOAEL. lowest-observed-adverse-effect level.

Lobelia (lobelia). a genus of mostly erect, usually unbranched, semicosmopolitan annual or perennial herbs (Family Lobeliaceae). They usually have red or blue irregular flowers and an acrid milky sap. Those species that are indigenous to North America contain a variety of alkaloids and are considered highly toxic. ***L. berlandieri***. a range plant of western Texas. ***L. cardinalis*** (cardinal flower). a perennial herb with showy red flowers. It is cultivated and occurs naturally in damp soils and along waterways throughout the eastern United States, and southern Canada. ***L. inflata*** (indian tobacco). an annual herb native to the southern United States. It is sometimes sold as a dietary aid and was formerly used medicinally, but fatalities from overdoses were not infrequent. It is extremely toxic and contains pyridine, lobeline, and a number of related pyridine alkaloids. All parts are toxic, causing vomiting, pain, weakness, stupor, tremors, miosis, a weak rapid pulse, progressive paralysis, depressed body temperature, collapse, coma, and death in humans. The plant appears to be equally toxic to other vertebrate animals. The principal toxicant is α-lobeline. See also Lobelia. ***L. siphilitica*** (great lobelia, blue cardinal flower). similar to *L. cardinalis* in structure, habitat, and distribution but occurs also in open moist woods and it is less widely distributed.

lobelia. the dried leaves and tops of *Lobelia inflata*, it is an extremely toxic and violent poison with effects similar to nicotine. Lobeline and related alkaloids may initially excite and then depress the central nervous system. If too much is consumed, respiration slows and blood pressure falls dangerously. Symptoms of acute intoxication include nausea, uncontrollable vomiting, tremors, paralysis, convulsions, coma, and death.

lobeline, α-lobeline. (2-[6-(2-hydroxy-2-phenylethyl)-1-methyl-2-piperidinyl]-1-phenylethanone; 2-[6-(β-hydroxyphenethyl)-1-methyl-2-piperidyl]acetophenone). a crystalline substance isolated from the seeds and foliage of *Lobelia inflata* (Indian tobacco), it is the principal alkaloid and chief toxicant of this plant. It is used therapeutically as a respiratory stimulant.

loblolly pine. *Pinus taeda*.

LOC. level of concern.

local. **1a:** of, pertaining to, or restricted to one locus, area, or part; not general, not systemic. **1b:** confined, isolated, limited, narrow. **2:** native, provincial, regional, territorial.

local injury. **1:** local effect. See effect (local effect) and effect (systemic effect). **2:** an injurious local effect.

local reaction. focal reaction.

local toxicity. a toxic effect that occurs at the site of initial contact between a toxicant and a living system. See effect (local effect), local injury, systemic toxicity.

lockjaw. trismus. See also tetanus (def. 3).

loco. **1:** locoweed poisoning; locoweed disease; a neuropathy of horses, sheep, and cattle caused by the ingestion of certain species of *Astragalus* and *Oxytropis*. It has caused serious losses of livestock in the western United States. Cattle initially carry their heads somewhat lower than normal; progressive locomotor ataxia soon appears. Common also are head tremors, the inability to eat or drink, and abortion. Sheep suffer extreme cerebral depression. Horses are much more susceptible than other livestock; symptoms develop with less intake of locoweed and the time to onset of symptoms and death is much shorter than in other livestock. The horse becomes listless, but in the early stages of poisoning is easily excited and dangerous to ride or manage. Goats develop an increasingly severe ascending paralysis. Head tremors and terminal emprosthotonos are usual. Histologically, transitory edematous vacuolization of the brain cells followed by eventual fragmentation of the Golgi apparatus may be seen. **2:** common name of true locoweeds, *Astragalus spp*. and *Oxytropus spp*. **3:** to poison with locoweed.

loco poisoning. See loco.

"**loco weed**." marijuana or a cigarette made from marijuana. Do not confuse with locoweed.

locoine. obsolete name of the toxic extract of a true locoweed, *Astragalus earlei*. See indolizidine-1,2,8-triol.

locoism. loco.

locoweed. loco (def. 2); any plant of the genera *Astragalus* and *Oxytropis* that cause locoweed poisoning (loco) as opposed to those that may cause selenium poisoning.

locoweed disease. See loco.

locoweed poisoning. See loco.

LOEL. lowest-observed-effect level.

loliism (darnel poisoning). poisoning by flour made from the seeds of *Lolium temulentum*, *q.v.* Symptoms include vertigo, tremor, green vision, miosis, ataxia, dysphagia, prostration, and sometimes nausea, and vomiting. The toxic principle is temuline.

Lolium temulentum (bearded darnel; poison darnel, a grass (Family Gramineae), the seeds of seeds of which are toxic; they contain a potent narcotic alkaloid, temuline. The plant is fairly rare, but hazardous because it may occur in grain fields from Quebec and northeastern Minnesota, Missouri, and Kansas, south to the Gulf of Mexico. Consumption of flour contaminated by darnel seeds can cause vertigo, nausea, vomiting, a burning pain in the mouth, dysphagia, tremors, ataxia, visual disturbances, and prostration.

lomustine (*N*-(2-chloroethyl)-*N*′-cyclohexyl-*N*-nitrosourea; 1-(2-chloroethyl)-3-cyclohexyl-1-nitrosourea; CCNU). an extremely toxic alkylating agent used as an antineoplastic in the treatment of Hodgkin's disease and other lymphomas. It can cause delayed thrombocytopenia in humans and is likely to prove carcinogenic. Toxicity is expressed clinically by delayed bone marrow suppression, nausea, and vomiting.

Lonchocarpus (cube or cubé). **1:** a genus of tropical American shrubs and vines (Family Fabaceae, formerly Leguminosae) used in the preparation of insecticides and as fish poisons. The roots contain rotenone, *q.v.* The most important species in this respect are *L. utilis* and *L. urucu*. **2:** a resinous substance (syn: barbasco, timbo) from *Lonchocarpus spp.* which is ground, purified, and used as an insecticide and fish poison.

long clam. *Mya arenaria*.

long-neck clam. *Mya arenaria*.

long-glanded coral snake. See *Maticora*.

long-glanded snakes. See *Maticora*.

long-nosed pit viper. *Agkistrodon acutus*.

long-nosed viper. *Vipera ammodytes*.

long-spined sea urchin. See *Diadema setosum*, *D. antillarum*.

longevity. the average life span of the individual members of a group (e.g., population or species) under a specified set of conditions. *Cf.* life span.

α-longilobine. seneciphylline.

β-longilobine. retrorsine.

Lophogorgia rigida (sea whip). a venomous marine hydrozoan, the toxic principle of which is lophotoxin.

Lophophora williamsii. (mescal; mescal cactus; peyote). a small cactus (Family Cactaceae) with soft, spineless, blue-gray stems and cottony tufts of wool in the areoles. The button (flowering head) is toxic, causing nausea, vomiting, headache, hallucinations, delusions, and illusions. Early vomiting usually prevents serious poisoning. It contains a number of hallucinogenic alkaloids. See lophophorine, mescaline.

lophophorine. an alkaloid, $C_{13}H_{17}NO_3$, isolated from *Lophophora williamsii*. It is poisonous with effects similar to those of mescaline.

lophotoxin. a highly toxic neurotoxin isolated from the sea whip, *Lophogorgia rigida*.

Lopressor. a beta-adrenergic blocking agent.

Loranthaceae (the mistletoe family). a widely distributed family of toxic shrubby epiphytes with jointed stems that are chiefly parasitic on trees. The flowers are small, leaves usually broad, and each species produces berries that germinate on the host. These plants are grown and gathered for use as Christmas decorations. Some mistletoes contain toxic amines, notably phenethylamine. Cattle have been seriously poisoned by *Phoradendron serotinum* (American mistletoe), *q.v.* See also *Arceuthobium pusillum* (dwarf mistleltoe), *Viscum album* (European mistletoe).

lorchel. *Gyromitra (Helvella) esculenta*.

lords-and-ladies. *Arum maculatum*. See *Arum*.

loreal pit (facial pit; pit). either of a pair of deep lateral heat-sensitive depressions in the loreal region (i.e., between the eye and nostril) of the head in pit vipers (Crotalidae). The pits aid in locating warm-blooded prey.

Lorexane™. lindane.

Lorphen™. chlorpheniramine.

Lotus corniculatus (birdsfoot trefoil; ladies' fingers; babies' slippers). a low, sprawling, hardy cyanogenic perennial herb (Family Fabaceae, formerly Leguminosae) introduced to the United States from Europe. It has many small usually red buds that blossom into yellow flowers. It is sometimes grown as a forage crop. See cyanogenesis.

lousicide. pediculicide.

lovage. *Ligustrum vulgare*.

love apple. *Mandragora officinarum*.

love bean. *Abrus precatorius*.

low blood pressure. hypotension.

low carriage. See carriage.

low lethal dose. lowest lethal dose.

low whorled milkweed. *Asclepias pumila*.

lowest effect level (LEL). See dose (minimum effective dose).

lowest lethal dose. See dose.

Lowry-Brønsted base. See base.

Loxosceles (brown spiders; brown recluse; violin spiders; fiddle-back spiders). a genus of more than 100 species of primitive, six-eyed, fawn-colored to dark brown spiders (Family Loxoscelidae). They are highly venomous and widely distributed in temperate regions, often in human dwellings. The cephalothorax is depressed and bears a characteristic dark "frying pan" or "violin-shaped" spot. The legs are long and the body of the female is about 9 mm in length; that of the male is somewhat less. Envenomation can cause severe local injury, marked by a highly ulcerative, spreading lesion in which the underlying muscle and other tissues about the bite suffer extensive necrosis. The resulting wound may be as much as 10 cm across. Systemic effects may include nausea, vomiting, and pyrexia; death is rare. ***L. arizonica*** (Arizona violin spider). a United States species that is toxicologically similar to *L. reclusa*, *q.v.* ***L. deserta*** (desert violin spider). a United States species that is toxicologically similar to *L. reclusa*, *q.v.* ***L. laeta*** (Chilean brown spider). the most dangerous species of *Loxosceles* in South America. Its venom is more toxic than that of *L. reclusa* and it causes more fatalities. ***L. reclusa*** (brown recluse spider, North American brown spider). this spider occurs in 25 of the 50 states of the United States, from California to New Jersey, Texas to Illinois, and in Hawaii. It is the most dangerous (especially the female) of the loxoscelide and is the main cause of loxoscelism in North America. Pain about the bite may be slight initially, becoming intense within 2 to 8 hours. Serious bites produce late local and severe invasive necrosis. A bleb ringed by an edematous, erythematous area develops over the bite. This eventually sloughs off to expose an area of acute ischemia, ecchymosis, and edema. This area darkens and discolors during the next day or so. A black eschar usually develops within a week and separates within 2-5 weeks. The remaining ulcer is slow to heal. Systemic effects usually include intravascular hemolysis, thrombocytopenia, weakness, and malaise. Acute cases may produce pyrexia, chills, nausea, vomiting, abdominal cramps, generalized pruritic morbilliform lesions, arthralgia, cyanosis, and delirium. Additional effects may include jaundice, splenomegaly, hematuria, and hemolytic anemia. Bites are rarely fatal, but occasionally (in the absence of suitable medical care) death due to kidney failure may result, usually within 48 hours; allergic reactions sometimes occur.

loxosceles poisoning. loxoscelism.

loxoscelism (loxosceles poisoning). **1a:** a morbid condition due to envenomation by the brown recluse spider, *Loxosceles laeta*, *L. reclusa*, and perhaps certain other members of the genus. It is characterized by a highly ulcerative, painful, spreading lesion about the site of the bite (initially an erythematous vesicle, progressing to a gangrenous slough about the wound), nausea, weakness, pyrexia, hemolysis, and thrombocytopenia. A condition encountered chiefly in South America, although North American cases are known. **1b:** the morbid condition about the site of the wound made by the bite of *Loxosceles laeta* or *Loxosceles reclusa* (See def. 1a).

Lr. the symbol for lawrencium.

LS 74-783. fosetyl aluminum

LSD (D-lysergic acid diethylamide; LSD-25; lysergide). a synthetic hallucinogenic alkaloid, $C_{20}H_{25}N_3O$, derived from lysergic acid. It has no medical usage and is a substance of abuse. See also alkaloid (ergot alkaloid). Toxicity varies widely among laboratory animals, being about 0.3 mg/kg body weight for rabbits, 17 mg /kg for rats, and 46 mg/kg for mice. LSD is considered supertoxic to humans. Profound psychologic effects can be induced by oral doses as low as 1 μg/kg; doses that are only slightly higher can produce devastating psychological effects, sometimes with permanent sequalae; doses as low as 20 μg/kg or perhaps slightly higher may prove lethal in some cases. Effects of so-called "bad trips" may include fear and even panic, accompanied by physical effects such as dizziness, nausea, and muscular weakness. In more severe cases, the affected individual may require hospitalization with sedation for several days. In acute poisonings autonomic and somatic symptoms may appear; included are salivation, vomiting, piloerection, tremor, mydriasis, fever, hypertension, hyperreflexia, ataxia, and spastic paresis. While psychosis-like manifestations may appear, there is no evidence that LSD is psychotogenic; it may, however, act as a trigger in a person disposed to mental illness. There is evidence that LSD damages chromosomes.

LSD-25. LSD.

L_s-xylokainic acid. kainic acid.

LT_{50}. lethal time 50. See time.

Lu. the symbol for lutetium.

lubricating oil. any of a variety of hydrocarbon mixtures derived from crude petroleum and generally thought to be only slightly toxic. Even so, frequent or prolonged skin contact may cause acneform lesions and sometimes dermatitis. Hypersensitivity reactions are also known. Additives to some of these oils may increase the toxicity.

lucern, lucerne. mainly British vernacular for *Medicago sativa*.

lucid. clear, bright, especially with respect to the mind; clearheaded, perceptive, rational.

lucidity. clearness, brightness; clarity of mind.

lucky leaf. a plant of the genus *Jatropha*.

lucky nut. *Thevetia peruviana*.

lukukuru. *Dispholidus typus*, *Thelotornis kirtlandii*.

luminal sodium. sodium phenobarbital.

lung irritant gas. See gas.

lupin. *Lupinus*.

lupin beans. See *Lupinus*.

lupine. *Lupinus*.

lupine poisoning. a nervous syndrome caused by alkaloids present in bitter lupines. *Cf.* lupinosus. See also *Lupinus*.

lupinidine. sparteine.

lupinine. an alkaloid isolated from the foliage and seeds of various species of *Lupinus*. See also lupinosis.

lupinosis. a mycotoxic disease of grazing animals, marked by liver damage and jaundice, and caused chiefly by feeding on sweet lupines (See *Lupinus*). The disease is common in Australia and South Africa, and has been reported in New Zealand and Europe. Sheep, and sometimes cattle and horses, are affected; pigs are susceptible. The causal fungus is *Phomopsis leptostromiformis*, a phytopathogen causing *Phomopsis* stem-blight, especially in white and yellow lupines; blue varieties are very resistant. This fungus produces hepatotoxic secondary metabolites (phomopsins, chiefly phomopsin A) on infected lupine material. Early effects in sheep and cattle are inappetence and listlessness. Complete anorexia and jaundice follow and ketosis is common. Cattle may show lacrimation and salivation. Sheep may become photosensitive. In acute outbreaks, deaths occur in 2-14 days. In acute lupinosis, icterus is prominent. Livers are enlarged, orange-yellow, and fatty. Chronic cases may show small, bronze or tan-colored, fibrous and firm, often distorted livers that are shaped rather like a boxing glove. Copious transudates may be found in the abdominal, and thoracic cavities and the pericardial sac. See *Lupinus* lupinine.

Lupinus (lupine, bluebonnet). a genus of usually low, perennial shrubs (Family Fabaceae, formerly Leguminosae). Some 100 species occur in North America. Toxic plants contain the alkaloid quinolizidine and numerous additional alkaloids (mostly quinolizidine alkaloids) and sometimes also nitrogen oxides. The alkaloids are most concentrated in the seeds. Drying does not affect the alkaloids and hay containing *Lupine* may be quite toxic. The lupine alkaloids are highly reactive and minor differences in structure are often reflected in major differences in toxicity. Whereas many lupines are toxic under certain conditions, others may normally provide acceptable forage. Many species and varieties are difficult to place taxonomically. Symptoms produced by lupines vary with species (or variety) and the class of livestock as does susceptibility. Sensitivity/toxicity also varies seasonally, with varying range conditions. Lupines have proven fatal to a large number of species; sheep are most frequently poisoned whereas cattle and horses are rarely affected. Symptoms are quite variable. Lupines are probably responsible for greater losses of sheep in the north central states than any other plant. see also crooked calf disease, piperidine, lupinine, lupinosis, lathyrism. ***L. luteus*** (yellow lupin bean). a European species with fragrant yellow flowers. It is an annual that grows to 0.7 m. See sparteine. ***L. niger*** (black lupin bean). See sparteine.

lupo. *Inimicus japonicus*.

lupus erythematosus. today, used without qualification, this term has no clear meaning. It is often applied to any inflammatory dermatitis. **drug-induced lupus erytematosus**. drug-induced systemic erythematosus. See systemic lupus erythematosus.

lupus erythematosus, systemic. See systemic lupus erythematosus.

lutecium. lutetium.

luteinizing hormone. LH.

luteohormone. progesterone.

luteoskyrin. one of a group of hepatotoxic hydroxyquinones synthesized by *Penicillia* (e.g., *P. brunneum*, *P. islandicum*, *P. rubrum*, *P. tardum*). See also rubroskyrin, rugulosin. **(-)luteoskyrin**. an anthraquinoid mycotoxin from *Penicillium islandicum*. It is extremely toxic (*i.p.* or *i.v.*) or very toxic (*s.c.* or *p.o.*) to laboratory mice; extremely toxic to cultured Ascites tumors, *Tetrahymena pyriformis* (a protozoan), and to *Escherichia coli* mutants. It is a potent carcinogen. Responses of exposed mice generally include ruffled fur, loss of skin tone, inactivity, shallow respiration, prolonged coma, and death. The liver of mice subjected to acute exposure exhibits centrolobar necrosis with coagulation necrosis of the cells, karyolysis, karyorrhexis, hyperchromatosis of the nuclear membrane, and fatty metamorphosis. Atrophy of the thymus, spleen, and lipid tissue accompany the liver necrosis. Luteoskyrin is also a powerful carcinogen. See anthraquinoid, anthrquinone, also cyclochlorotine, erythroskyrine, rubroskyrin.

lutetium (lutecium; symbol, Lu)[Z = 71; A_r = 174.97]. a silvery-white element of the lanthanoid series of metals; used as a catalyst.

lutetium chloride. a colorless, water-soluble, crystalline substance, $LuCl_3$. It is slightly toxic to laboratory mice.

Lutjanidae (snappers). a family of oblong, moderately compressed, marine, mostly carnivorous, bony percomorph fishes (Class Osteichthyes); mostly of warm seas with a large, terminal mouth that bears sharp, conical teeth. Most are reef dwellers or live on the continental slope; many are important food fishes. See *Lutjanus*, ciguatera.

Lutjanus. a genus of marine bony percomorph fish (Family Lutjanidae). The following species have certain populations which are transvectors of ciguatera, *q.v.*: *L. bohar* (red snapper), a fish of the tropical Pacific Ocean to East Africa and the Red Sea; *L. gibbus*, a species of the tropical Indo-Pacific region; *L. monostigma* (snapper), found from Polynesia, west to the Red Sea and China; *L. nematophorus* (Chinaman fish), an Australian fish; *L. vaigiensis* (red snapper), a common species found from Polynesia, west to eastern Africa and Japan.

lutotropin. LH.

luwando. *Causus rhombeatus*.

Lw. formerly the symbol for lawrencium.

lycine. betaine.

lycomarasmin. a marasmin, *q.v.*

Lycopersicon esculentum (tomato). a common garden herb and berry (usually thought of a vegetable) with numerous varieties. The vines and suckers, which contain small amounts of cyanide in the form of the steroid alkaloids, solanine and solanidine, have proven toxic to livestock.

Lycosa tarantula (European tarantula; European wolf spider). the true tarantula. Its bite was once believed to cause madness, which encouraged strenuous, frenzied dancing, and contortions to eliminate the venom from the body. The bite is actually harmless to humans.

lye. **1:** sodium hydroxide. Lye is the vernacular, and often the label term for this compound as used in various commercial (e.g., cleaning) formulations. It is the primary component of drain cleaners; it is an extremely corrosive substance that can destroy any body surface. Even a drop or a dry crystal that comes into contact with damp human skin can cause damage. When ingested, lye quickly erodes the mucosa and other internal tissues and can irreversibly damage part or all of the gastrointestinal tract, with possible lethal consequences. Symptoms are chiefly those of severe gastroenteritis. Death may result from any of a number of complications such as asphyxia due to glottic or laryngeal edema; shock with circulatory collapse; infection due, for example, to mediastinitis that follow perforation of the esophagus; inanition; and occasionally pulmonary complications. See sodium hydroxide. See cleaner. **2:** any of a number of strongly corrosive alkalis such as caustic soda, caustic potash, potassium hydroxide, sodium hydroxide, as well as sodium and potassium carbonates, oxides, and peroxides. accidental poisoning of children by these compounds, with burns of the mouth, are not uncommon, but severe, deep esophageal and gastric damage is more common among suicidal adults.

Lyell's syndrome. toxic epidermal necrolysis (See necrolysis).

Lygodesmia juncea (skeletonweed). a perennial herb (Family Compositae) found on prairies and plains of western North America; it grows to a height of a 0.2-0.4 m. Toxic concentrations of nitrate in the foliage have been reported.

lyme grass. a common name of grasses of the genus *Elymus*.

lymphangitis. an inflammation of lymph vessels. It is most commonly caused by an acute streptococcal infection of an extremity.

lymphocythemia. lymphocytosis.

lymphocytosis (lymphocythemia). an unusually elevated number of lymphocytes in circulating blood as observed, for example, in certain chronic diseases and during recovery from acute infections.

lymphocytotoxin. a toxin that damages or destroys lymphocytes. Do not confuse with lymphotoxin.

lymphokine. any of a variety of soluble, biologically active agents, released by sensitized lymphocytes, that contribute to cell-mediated immunity. They act on numerous cell types with varying effects and individual lymphokines may act variously on any of several different types of target cells. They are produced by sensitized lymphocytes usually in the presence of specific antigen, but are also stimulated by cell contact or by other lymphokines. Examples are B-cell replacing factor, interferon, interleukin I and macrophage activation factor. Complex mixtures of lymphokines, added to cell cultures, are sometimes used to determine which stages of growth or differentiation of a given cell type are most susceptible to a particular toxicant and/or to identify the stage at which the toxicity is expressed.

lymphoma. any new growth or tumor in the lymphatic system; most are malignant. Examples are Hodgkin's disease, diffuse histiocytic lymphoma, lymphatic leukemia, and reticuloses.

lymphotoxicity. the ability of a serum antibody in the recipient of an allograft to react directly with the lymphocytes or other cells of the donor to produce a hyperacute type of graft rejection.

lymphotoxin (tumor necrosis factor β; TNF-β6). a lymphokine, produced by activated T-lymphocytes, that damages or kills a variety of cell types by interfering with cell metabolism. Do not confuse with lymphocytotoxin. Lymphotoxins kill or inhibit the growth of tumor cells that are sensitive to them. They also block the transformation of cells by ultraviolet light and carcinogens.

Lyonia. a genus of North American and east Asian brushy, deciduous and evergreen shrubs (Family Ericaceae). ***L. mariana*** (staggerbush). a deciduous shrub, with nodding white or pinkish flowers, that grows to a height of 2 m. It occurs on peaty or sandy pineland soils and open woods from southern Rhode Island, New York, New Jersey, eastern Pennsylvania, western Tennessee, and Arkansas, south to Florida and eastern Texas. Poisoning of lambs and calves has been reported. All parts are toxic if ingested, causing salivation, increased tear formation, nasal discharge, vomiting, bradycardia, hypotension, convulsions, and paralysis. The active principle is andromedotoxin.

lyophilized anterior pituitary tissue (LAP). a crude preparation used in early experiments designed to demonstrate the effect of pituitary extracts upon inflammation. It has a pronounced prophlogistic effect, presumably due to the presence of somatotrophic hormone (STH).

lyre snake. *Trimorphodon boscitatus*.

lyse. to cause or to undergo lysis.

lysergic acid. any of a type of alkaloid with a structure based on a condensed 4-ring nucleus, derived by the alkaline hydrolysis of ergot alkaloids or prepared synthetically. See also alkaloid (lysergic acid), LSD.

lysergic acid amide. a type of ergot alkaloid. See also alkaloid.

D-lysergic acid diethylamide. LSD.

D-lysergic acid L-2-propanolamide. ergonovine.

lysergide. LSD.

lysin. **1:** any specific complement-fixing antibody that promotes lysis or causes the dissolution of living cells. Lysins are named after the type of antigen that stimulates their production and/or specific action (hence hemolysin, bacteriolysin, cytolysin, etc.). **2:** any substance that lyses cells.

lysinosis (lyssinosis). a lung disease due to inhaling cotton fibers, usually by workers in cotton mills.

lysis. **1a:** the dissolution of cells, usually by rupture of the cell membrane. Causative agents include phagocytes, enzymes, or other compounds within the cell itself (autolysis), hypotonic solutions, bacteriophages, and specific lysins. **1b:** the destruction or dissolution of cells by a specific lysin. Such action may be direct via a toxin or venom or indirect via an immune mechanism. **2:** gradual abatement of the symptoms of an acute illness as opposed to crisis, *q.v.* **3:** decomposition of a chemical compound or system by a specific agent.

lyso-, lys-. affixes pertaining to lysis.

lysogenesis. **1:** the promotion, initiation, or production of lysis or the production of lysins. **2:** lysogenicity.

lysogenic. **1:** having the capacity to produce lysins. **2:** lytic; causing or able to cause lysis. **3:** pertaining to lysogenicity or to bacteria in a state of lysogenicity.

lysogenicity (lysogeny). **1:** the state or quality of being lysogenic; the ability to produce lysins or to cause lysis. **2a:** the ability of bacteria to produce phage. **2b:** specific association of prophage with the bacterial genome such that few, if any, phage genes are transcribed.

lysogeny. lysogenicity.

Lysol. a toxic, corrosive, proprietary mixture of cresols.

Lysol poisoning. ingestion of Lysol can cause acute gastrointestinal distress, with or without vomiting; corrosion of tissue; pulmonary edema; immobility of the pupils; collapse; and death. Death may occur following abatement of symptoms.

lysophosphatide. a phosphatide which has lost one molecule of fatty acid, as by the action of a phospholipase A in snake venom.

lyssinosis. lysinosis.

Lytta (Russian fly). a genus of blister beetles (Family Meloidae) with larvae that feed on grasshopper eggs. *L. (Cantharis) vesicatoria* (syn. *L. vesicatoris*; Spanish fly; blister bug). a bright green, European blister beetle of southern Europe. The dried beetles (cantharides) are an important source of cantharidin. See also cantharides.

M

μ. See mu.

m. meter; milli-; minim.

M. mega; mu.

m.p. melting point.

m.u. mouse unit.

M-C. mineralocorticoid.

M phase. the phase of the cell cycle in which mitosis occurs.

M.R.D. minimum reacting dose.

M-4209. carbomycin A.

mγ. milligamma. See nanogram.

mμg. millimicrogram. See nanogram.

ma huang. *Ephedra equisetina*.

MAC. maximum allowable concentration; maximum acceptable concentration.

macabrel. *Bothrops castelnaudi*.

macaurel. *Bothrops castelnaudi*.

Macclelland's coral snake. *Calliophis macclellandii*.

Mace®. an aerosol mixture of organic lacrymators.

Machaeranthera (tansy asters). a genus of plants (Family Compositae) that are facultative accumulators of selenium from soil. ***M. ramosa*** (tansy aster). a species hazardous to livestock because the foliage may contain very high concentrations of selenium.

mAChR. muscarinic acetylcholine receptor. See muscarinic cholinergic receptor.

mackerel. a common name of marine fish of the genera *Scomber*, *Scomberomorus*, and others (Family Scombridae). See scombroid poisoning.

macleyine. protopine.

Macquer's salt. potassium arsenate.

macroazamin. See amygdalin, methylazoxymethanol.

macrobiota. those macroscopic organisms (collectively), larger than a few centimeters in length or diameter that occur in a particular habitat, ecosystem, or specified area. Most frequently applied to soil organisms and organisms of bottom sediments in freshwater systems. *Cf.* microbiota.

macrobiotic. long-lived.

macrocyclic. pertaining to a macrocyclic molecule or compound.

macrocyclic compound. a compound with at least one macrocyclic moiety (i.e., a ring with more than six atoms). See macrocyclic molecule.

macrocyclic molecule. a molecule with a ring comprised of more than six atoms.

Macrocystis. a genus of giant kelp. See Phaeophyta.

macrogol nonylphenyl ether. nonoxynol.

macrolide. a substance (e.g., erythromycin, magnamycin, nystatin, and rifamycin) that contains macrocyclic lactone. See also mycotoxin.

macromolecule. a giant molecule (e.g., a protein, nucleic acid, or polysaccharide) that is comprised of smaller subunits that are chemically linked together.

macronutrient (essential element). **1:** an element that is essential to optimal growth and development of a living organism and is required in relatively large amounts. These are generally taken to be carbon, hydrogen, oxygen, nitrogen, sulfur, phosphorus, potassium, magnesium, calcium, and iron. See also micronutrient. **2:** having the properties of a macronutrient (def. 1).

macrophage. a type of motile, mononuclear, intensely phagocytic, microbicidal, and tumoricidal cell in circulating blood. They play an important, but nonspecific role in the immune defenses of vertebrate systems. Macrophages are easily identified by their propensity to absorb and store certain vital dyes (e.g., trypan blue, Indian ink). They occur chiefly as resting cells in connective tissue in the reticuloendothelial system, in the liver, adrenal glands, the anterior pituitary, and in the blood as monocytes. **activated macrophage**. a macrophage that has developed the ability to lyse bacteria and tumor cells. See macrophage activation factor. **resting macrophage**. a macrophage that has not yet been activated. Activation is a complex process.

macrophage activation factor (MAF). a lymphokine (e.g., γ-interferon) that enhances the bactericidal, tumoricidal, and phagocytic activity of macrophages. See lymphokine.

macrophage chemotactic, and activating factor (MCAF). a chemoattractant and activator of macrophages. They are secreted by fibroblasts, monocytes, and endothelial cells in response to exogenous stimuli, and by endogenous lymphokines (cytokines). It also contributes to the activation of monocytes, and is a powerful degranulator of basophils.

macrophage chemotactic factor (MCF). a lymphokine (e.g., interleukins, interferons) that facilitates the migration of macrophages. See lymphokine.

macrophage inhibitory factor (MIF). a lymphokine that inhibits macrophage activity. See lymphokine.

maculotoxin. a potent neurotoxin isolated from the posterior salivary glands of *Octopus maculosus*, *q.v.* It is chemically indistinguishable from tetrodotoxin, *q.v.*

mad apple. *Datura stramonium*.

mad hatter's disease. mercurialism.

madtom. See *Schilbeodes*.

MAF. macrophage activation factor.

magnamycin. carbomycin.

magnamycin A. carbomycin A.

magnamycin B. carbomycin B.

magnesium (symbol, Mg)[Z = 12; A_r = 24.3]. a light, metallic, electropositive element of the alkaline earth series. It is the 8th ranking element in abundance, comprising 2.09% of the Earth's crust, occurring naturally in a wide variety of minerals. Inhalation of magnesium dust is irritating to the pulmonary system and fumes can cause metal fume fever. Particles imbedded in the skin can raise small blisters which persist for some time. Magnesium is important commercially as a component of lightweight alloys. It has many additional uses as in pyrotechnics, flash bulbs, flares, intense signal lights, the ignition of thermite, and in dry cell batteries. See also Grignard reagent.

magnesium acetate. a compound, $Mg(C_2H_3O_2)_2$ or $Mg(OOCCH_3)_2 \cdot 4H_2O$, with white or colorless crystals that readily absorbs water from the atmosphere and is soluble in water and dilute ethanol. It is used as a dye fixative in textile printing, a disinfectant, an antiseptic, and a deodorant. It is very toxic to laboratory mice.

magnesium fluosilicate. magnesium hexafluorosilicate.

magnesium hexafluorosilicate (magnesium silicofluoride; magnesium fluosilicate). a white, very toxic, efflorescent, odorless, hexahydrate crystalline compound, $MgSiF_6$; used as a mothproofing agent for fabrics.

magnesium silicofluoride. magnesium hexafluorosilicate.

magnesium sulfate. the heptahydrate (bitter salts, epsom salts) is an efflorescent crystalline or powdery substance, $MgSO_4$. It is used therapeutically as a cathartic and anticonvulsant. Magnesium intoxication may result if it is used parenterally or in cases of renal insufficiency.

Magrene. diethylpropion.

mahogany family. Meliaceae.

maize. *Zea maize*.

maize oil. corn oil.

major tranquilizer. neuroleptic drug.

maki-maki. *Arothron hispidus*.

mal des ardents feu de Saint-Antoine (St. Anthony's fire). a french name for ergotism caused by *Clavaceps purpurea*.

malabasahan. *Hydrophis ornatus*.

malachite green (*N*-[4[[4-(dimethylamino)phenyl]phenylmethylene]-2,5-cyclohexadien-1-ylidene]-*N*-methylmethanaminium chloride; bis[*p*-(dimethylamino)phenyl]phenylmethylium chloride). a green, crystalline triphenylmethane dye. It is used, e.g., as a fungicide, a parasiticide in fish, a biological stain, and a clinical reagent. It is very toxic to laboratory rodents, less so to a variety of fish species.

Malacostraca. the largest subclass of Crustacea. Both freshwater and terrestrial species are represented; the head bears compound eyes, usually on movable stalks; the thoracic appendages bear gills used in respiration, walking or swimming, and in some forms in feeding; the abdominal appendages are used in swimming and (in the females) to carry eggs. In many malacostracans, development is direct. The subclass contains five orders, including the Amphipoda (e.g., *Gammarus),* Isopoda (woodlice), and Decapoda (crabs, lobsters, etc.).

malady. illness; roughly synonymous with disease, *q.v.*

malaise. 1: weakness, infirmity; uneasiness, despondency, lassitude. **2:** a generalized or vague feeling of discomfort or uneasiness. It is sometimes an early indication of a disease.

malanga. a common name of herbs of the genus *Xanthosoma*, especially *X. violaceum*.

malaoxon ([*O,O*-dimethyl *S*-1,2-bis(ethoxycarbonyl)ethylphosphorothionate]; MO). an active oxygen analog of malathion, used as a pesticide. Malaoxon, a convulsant, is very toxic to laboratory rats and is carcinogenic, producing both benign and malignant neoplasms. Malaoxon is the active metabolite of malathion and is produced metabolically by the action of cytochrome P-450 on malathion. In higher animals it increases brain phosphoinositide signaling, probably subsequent to inhibition of acetylcholinesterase and causes convulsions. MO also brings about changes in cerebral lipid composition and may damage several brain regions.

malathion (*O,O*-dimethyl dithiophosphate of diethyl mercaptosuccinate; *O,O*-dimethyl-*S*-1,2-di(ethoxycarbonyl)ethyl phosphorothiolothionate; *S*-(1,2-dicarbethoxyethyl) *O,O*-dimethyldithiophosphate; *S*-1,2-bis(ethoxycarbonyl)ethyl *O,O*-dimethylphosphorodithioate; *S*-ester with *O,o*-dimethyl phosphorothioate; [(dimethoxyphosphinothioyl)thio]butanedioic acid diethyl ester; carbofos; insecticide no. 4049; malathon (obsolete); mercaptosuccinic acid diethyl ester; mercaptothion; phosphothion; Cythion). an organophosphorus insecticide and acaride; also used as an ectoparasiticide in veterinary practice. It is moderately toxic to adult laboratory rats, but very toxic to newborn rats. It is neurotoxic and a powerful skin sensitizer, causing an allergic dermatitis. It may irreversibly inhibit ACH. It is generally less toxic than parathion, but produces similar symptoms. It is a significant environmental poison because it is relatively persistent in soil. Malathion is carcinogenic, producing both benign and malignant neoplasms; it is mutagenic, causing chromosome damage in bone marrow cells and spermatogonia; it induces dominant lethal mutations. See also ethion, malaoxon, phosphamidon, insecticide (organophosphate insecticide).

Malay cobra. *Naja naja sputatrix*.

Malayan krait. *Bungarus candidus*.

Malayan pit viper. *Agkistrodon rhodostoma*.

maleic anhydride (2,5-furandione). a toxic, colorless substance with monoclinic crystals, $(CH)_2C_2O_3$. It is an irritant, a carcinogenic lactone, and a plant growth inhibitor. It is used, for example, in polyester resins, the manufacture of fumaric and tartaric acids, pesticides, and preservatives for fats and oils.

maleic hydrazide (MH). a pyridazine herbicide used as a growth retardant for turf grasses, hedges, and trees; to inhibit the sprouting of onions and potatoes; and to prevent the development of suckers in tobacco. It temporarily inhibits growth, often without visible damage to the plant. It interrupts cell division in both the apical, and subapical meristems of the shoot by specifically inhibiting DNA replication, but not cell extension.

malformation. an abnormal form or structure (e.g., cleft palate) often due to abnormal development. See teratogen.

malformin A. a mycotoxin from *Aspergillus niger*. It is extremely toxic to laboratory mice.

malfunction. **1a:** to function abnormally or inadequately. **1b:** a failure in performance, a breakdown; an abnormal or ineffective activity or process.

malignancy (virulence). **1:** a cancerous neoplasm or tumor, one that is poorly differentiated, unencapsulated, rapidly growing, invasive, and productive of metastases. **2:** the state or condition of being malignant. **3:** a tendency of a disease, poisoning, or pathological condition toward progressive virulence; the condition of being malignant.

malignant. **1a:** destructive; able to cause great harm, including death. **1b**. tending to become progressively worse, with death as a probable outcome; resistant to treatment. **1c:** tending to cause death. **2:** pertaining to a severe type of a normally benign condition. **3a:** of or pertaining to a condition of tumor cells marked by anaplasia, invasiveness, and metastasis; having properties similar to that of a cancerous (malignant) growth. **3b:** pertaining to, denoting, or indicating properties similar to those of a cancerous growth (e.g., invasive, spreading).

malignant hemangioendothelioma. angiosarcoma.

malignant hyperpyrexia. See halothane.

malignant jaundice. acute yellow atrophy (See atrophy).

malignant malaise. a general, often vague, feeling of physical discomfort or uneasiness. It is sometimes an early indication of disease.

malignant neoplasm. See cancer.

malignant tumor. See cancer.

mallard. *Anas platyrhynchos*.

malle snake. *Demansia textilis*.

mallow. *Malva parviflora*.

malnutrition. **1a:** undernourishment; nutrient deficiency; the lack of suitable food substances in the body due, for example, to inadequate diet, or deficient digestion, absorption, or assimilation. **1b:** nutrient excess; excess intake of nutrients or the presence in food of an excess of certain nutrients. **2:** any nutritional disorder, regardless of cause. **3:** the condition of being malnourished. Such a condition may arise from factors mentioned in defs. 1a and 1b. Malnutrition may also arise from poisons present in an otherwise normal diet (e.g., aflatoxins, ergot, pesticide residues, fertilizer, pollutants). Many poisons adversely affect nutrition, *q.v.*

malodorous lepiota. *Lepiota cristata*.

malonic acid (methanedicarbonic acid). a white, crystalline substance, $CH_2(COOH)_2$, that is soluble in water, ethanol, and ether. It is a strong irritant found as an end product of metabolism in plants such as *Aconitum* and *Rheum rhaponticum*. Malonic acid is a metabolic poison that blocks cellular respiration by reversibly inhibiting succinate dehydrogenase. It is used as an intermediate in the synthesis of barbiturates and pharmaceuticals.

malonylurea. barbituric acid.

malpractice. professional misconduct or failure to exercise reasonable skill and diligence in performing services that could be expected from the average reputable member of the profession with consequent loss, injury, or damage to the recipient of the service on which he or she was entitled to rely. Any failure to provide proper service through unreasonable ignorance, lack of skill, lack of due diligence, immoral or illegal conduct, or criminal intent. Legal malpractice applies to anyone who provides professional services for a fee, but in practice is most often applied to medical doctors (and certain other health professionals), attorneys, and accountants.

maltoryzine. a mycotoxin from *Aspergillus oryzae* var. *microsporum 3* that is extremely toxic to laboratory mice.

Malus. the botanical name for the common apple tree. See *Pyrus malus*.

Malva (mallows). a genus of mostly perennial herbs (Family Malvaceae) that have five-parted, white or rose-colored, silky or papery flowers. ***M. parviflora*** (mallow; cheeseweed). a widely distributed annual weed in North America. This plant is probably responsible for "shivers" in grazing Australian and African livestock. Toxic concentrations of nitrate have been reported.

mamba. preferred common name for the extremely venomous elapid snakes of the genus *Dendroaspis*. See also *Dispholidus typus*, *Naja melanoleuca*.

mamba venom poisoning. that due to envenomation by a mamba (*Dendroaspis*). Common early signs and symptoms are weakness, headache, nausea, vomiting, abdominal pain, slurred speech, blurred vision, excessive salivation. These are often followed by hypotension, respiratory distress, and shock. See krait venom poisoning, elapid venom poisoning.

mamba vert. *Dendroaspis jamesoni*, *D. viridis*.

mammal. common name of vertebrates of the class Mammalia.

Mammalia (mammals). a class of mostly homeothermic, viviparous, tetrapod vertebrate animals marked by the presence of hair, mammary or milk-producing glands, a secondary palate, a four-chambered heart, a muscular diaphragm, and with rare exception, seven cervical vertebrae. The teeth are in sockets and are generally differentiated into incisors, canines, and cusped molars and premolars. Many mammals accumulate lead, mercury, PCB, and a number of other environmental chemicals to levels well above that in their food or the surrounding air. It is comprised of three infraclasses - Eutheria, Prototheria, and Marsupialia - only the first two of which contain venomous species.

mammalian. of or pertaining to mammals. See Mammalia.

mammary gland. the milk-producing gland that is characteristic of all female mammals. One or more pairs of these glands develop on the ventral surface of the body and are thought to have evolved from sweat glands. Although the state of the gland fluctuates during the estrus cycle, its complete development and functioning (See lactation) generally occurs only after parturition, for the purpose of suckling the young. The mature gland consists of several lobes of multiple branching ducts embedded in fatty tissue. The ducts lead from numerous alveoli of milk-secreting cells and in most mammals drain into sinuses that open to the exterior via pores at the nipple. In the Monotremata nipples are absent and the milk is secreted directly from the sinuses onto the surface of the body. See, for example, diethylstilbestrol, factor (milk factor), fusarium estrogenism, Mammalia, melphalan, progesterone, vinyl chloride.

mamushi. *Agkistrodon halys*.

man. **1a:** a male human being. **2:** the male of any species of the family Hominidae. **3:** any human or human being. **modern man**. the currently accepted common name of *Homo sapiens* (Family Hominidae) used in the biological sciences. Its use in vernacular speech is not recommended. *Cf.* human (especially def. 4). See also person, woman.

manapare. *Bothrops atrox*.

manatee. See Sirenia.

manchineel, manchineel tree. *Hippomane mancinella*.

manda-dalag. *Trimeresurus flavomaculatus*.

mandelic acid nitrile. mandelonitrile.

mandelonitrile (α-alphahydroxybenzeneacetonitrile; mandelic acid nitrile; benzaldehyde cyanohydrin). the cyanohydrin of benzaldehyde, $C_6H_5CH(OH)CN$; it is an oily, yellow liquid that is soluble in ethanol, chloroform, and ether; nearly insoluble in water. It is the source of hydrocyanic acid produced by millipedes and the so-called active ingredient of Laetrile™.

mandelonitrile-β-gentiobioside. amygdalin.

mandelonitrile glucoside (α-(β-D-glucopyranosyloxy)benzeneacetonitrile). the d-form (prunasin) of this glucoside, $C_{14}H_{17}NO_6$, is soluble in water, ethanol, and acetone. It occurs in *Prunus serotina* (black cherry) and *Eucalyptus corynocalyx*. It can be prepared by the action of yeast on amygdalin.

***l*-mandelonitrile-β-glucuronic acid**. Laetrile™.

mandible. **1:** the lower jaw of a vertebrate animal. **2:** one of a pair of appendages in crustaceans, insects, and myriapods which is usually adapted to biting food.

Mandragora officinarum (mandrake; Devil's apple; love apple). the true European or oriental mandrake (Family Solanaceae). It grows to a height of about 30 cm. and propagates by seed or division. The leaves are large, nearly covering the single purple or yellowish bell-shaped flower. The root is thick, tuberous, and contains the highly poisonous alkaloids, mandragorine, hyoscyamine, and scopolamine, *q.v.* The root was formerly used as a narcotic and sedative. In ancient times it was used as an aphrodisiac and a poison. *Cf.* Podophyllum peltatum.

mandragorine. a toxic alkaloid found in *Mandragora officinarum*.

mandrake. **1:** *Mandragora*. **2:** *Podophyllum peltatum*. **3:** the root of *M. officinarum* or *P. peltatum* (See podophyllum).

mandrake root. podophyllum.

maneb ([[1,2-ethanediylbis[carbamodithioato]]-(2—)]manganese; [ethylenebis(dithiocarbamato)]manganese; ethylenebis[dithiocarbamic acid] manganous salt; manganous ethylenebis-[dithiocarbamate]). a dithiocarbamate fungicide used on foliage; it is the manganese salt of dimethyldithiocarbamate. Its toxicity to animals is low.

manganese (symbol, Mn; manganum)[Z = 25; A_r = 54.94]. a toxic, steel-gray, hard, brittle, lustrous, transition metal that occurs naturally as oxides. It is an essential trace element, but deficiencies have not been demonstrated in humans. See manganese poisoning.

manganese binoxide. manganese dioxide.

manganese black. manganese dioxide.

manganese difluoride (manganous fluoride; manganese fluoride). a very toxic, water-insoluble compound, produced by dissolving manganese carbonate in hydrofluoric acid.

manganese dioxide (manganese binoxide; black manganese oxide; manganese black; battery manganese; bog manganese; C.I. 77728; C.I. Pigment Black 14; C.I. Pigment Brown 8; manganese black; manganese oxide; manganese (IV) oxide; manganese peroxide; manganese superoxide). a black, flammable, crystalline compound, MnO_2, or powder that is insoluble in water but soluble in hydrochloric acid. It is a powerful oxidizer that reacts violently with aluminum, other oxidizers, potassium azide (if warmed), and diboron tetrafluoride. It is potentially explosive with hydrogen peroxide, peroxomonosulfuric acid, anilinium perchlorate, and certain other compounds. It must be secured away from heat, organic matter, and flammable materials. Manganese dioxide is destructive to tissue and is toxic by *i.v.* and subcutaneous administration.

manganese ethylenebisdithiocarbamate. maneb.

manganese fluoride. manganese difluoride.

manganese madness. See manganese poisoning.

manganese manganate. manganese(III) oxide.

manganese oxide. manganese dioxide.

manganese (III) oxide (C.I. 77727; C.I. Natural Brown; dimanganese trioxide; manganese manganate; manganese trioxide; walnut stain). a fine, water-insoluble, black powder, Mn_2O_3. It is moderately toxic to humans by the subcutaneous route. See also manganese dioxide, manganese poisoning.

manganese (IV) oxide. manganese dioxide.

manganese peroxide. manganese dioxide.

manganese poisoning (manganism). an uncommon but serious condition in workers (e.g., miners, smelters, welders, steel workers, glass, and ceramic workers, paint and varnish manufacturing workers, certain pharmaceutical workers) who are chronically exposed to manganese dust or fumes. Whereas short-term exposure to dust and fumes of manganese produces only minor irritant effects of the skin and eyes, long-continued, routine exposure has a number of systemic effects, especially on the liver, spleen, and certain neurons. Early symptoms of chronic poisoning include apathy, anorexia, irritability, headache, and muscular weakness of the legs. The victim soon develops manganese psychosis, marked by confusion, inexplicable laughter, euphoria, rash acts, aggressiveness, and hallucinations. This early phase (sometimes called manganese madness) is followed within a few weeks by the so-called Parkinsonian phase which is marked by languor, drowsiness, feebleness, salivation, a mask-like face, disturbances of speech, tremor, stooped posture, a peculiar spastic, and paralysis. Death is infrequent, but the condition unless treated early is often permanently disabling. Most of these effects are due to effects of manganese on dopamine neurotransmission in the brain.

manganese psychosis. See manganese poisoning.

manganese superoxide. manganese dioxide.

manganese trioxide. manganese(III) oxide.

manganism. manganese poisoning.

manganous ethylenebis[dithiocarbamate]. maneb.

manganous fluoride. manganese difluoride.

manganum. manganese.

mangol. *Beta vulgaris.*

mangelwurzel. *Beta vulgaris.*

Mangifera indica (mango tree). a tropical evergreen tree, the fruit of which is the common mango. The skin of fruit and the sap of the tree are irritant and toxic, producing a contact dermatitis and, if ingested, gastrointestinal distress. The peel of the fruit should not be eaten.

mango. the fruit of *Mangifera indica.*

mango tree. *Mangifera indica.*

mangold. *Beta vulgaris.*

mangrove snake. *Boiga dendrophila.*

mangrove viper. *Trimeresurus purpureomaculatus.*

manic-depressive illness. a condition in which behavior alternates between unrealistic enthusiasm and deep depression.

Manihot (cassava; manioc). a genus of herbs and shrubs (Family Euphorbaceae) that are indigenous to tropical America, but are now widespread throughout the tropics. ***M. esculenta*** (cassava; tapioca). a widely cultivated shrub of tropical America with long, thick, tuberous roots that are the source of a starchy staple (cassava) and of commercially produced tapioca. The raw root is cyanogenic, however, and potentially lethal. ***M. utilissima*** (cassava). the root is toxic, containing a soluble cyanogenetic glycoside, amygdalin, in the sap; it is present in sufficient quantity to cause cyanide poisoning unless removed. It is, nevertheless, an important food plant in certain parts of South America because the people using it have learned to remove or hydrolyze the toxins (linamarin and lotaustralin) and to destroy the β-glucosidase that is present, prior to serving.

manioc (manioca, mandioc, mandioca). cassava.

mannagrass. *Glycera striata.*

mano de piedra. *Bothrops nummifer.*

manta. See devil ray, Batoidea, Mobulidae.

mantle. **1a:** a cloak or covering. **2a:** a fold of epidermal tissue covering the dorsal and lateral surfaces of the body of the Mollusca. It secretes the shell (when present) and protects the gills (which lie in the mantle cavity between the mantle and the body); in the Cephalopoda, the mantle is muscular, and functions in locomotion and respiration. **2b:** a similar fold of tissue in the Brachiopoda. **3:** the body wall of the Urochordata, situated below the outer test (a calcareous "shell").

manufacturers' warning label. See label.

many-banded krait. *Bungarus multicinctus*.

many-ringed snake. *Vermicella multifasciata*.

MAO. monoamine oxidase.

MAO inhibitor. monoamine oxidase inhibitor.

MAO-I. monoamine oxidase inhibitor.

mapana. *Lachesis mutus*.

mapanare cejuda. *Bothrops schlegelii*.

mapipire. *Lachesis mutus*.

mapipire z'Auana. *Lachesis mutus*.

mappine. bufotenin.

marabe. *Bitis arietans*.

maracaboia. *Crotalus d. terrificus*.

marasmin. any of three toxins, lycomarasmin, and aspergillomarasmins A and B, that occur in *Fusarium*. They are amino acid derivatives that are joined by bonds other than peptide bonds. Because of the presence of nitrogen atoms on 2 adjacent carbon atoms, they readily complex with ferric ions. These ions potentiate the activity of marasmins *in vitro*. Marasmins disrupt water metabolism in many plants, causing them to wither.

marble dust. See calcinosis, calcium.

march snake. *Denisonia signata*.

Margarita scorpion. *Centruroides gracilis*.

margin of safety (MS). a measure of the relative safety of a drug, it is a function of exposure and dose, but is variously defined. **1:** traditionally: MS = ED_{01}/ED_{99}, where ED_{01} is the adverse effect dose and ED_{99} is the therapeutic dose. It is rarely used in public policy decisions regarding the therapeutic use of drugs. **2:** MS = LD_{01}/ED_{99}, where LD_{01} is the lethal dose, 1% response. **3:** the inverse of hazard, *q*.v. See also therapeutic index. **environmental margin of safety**. the margin of safety of an environmental chemical may be defined by the ratio NOEL/PEC where NOEL is the No-Observed-Effect-Level and PEC is the Predicted Environmental Concentration. This is the inverse of hazard, *q*.v.

marginal fern. *Dryopteris marginalis*.

marginal shield fern. *Dryopteris marginalis*.

margosa oil. neem oil.

marihuana. **1a:** cannabis. **1b:** *Cannabis sativa*.

marine dermatitis. schistosome dermatitis (See dermatitis).

marine protista. See protista.

marinobufagin. a cardiac poison, $C_{24}H_{32}O_5$, secreted by the skin of the toad, *Bufo marinus*.

marinobufotoxin. a poison secreted by the parotid gland of the toad, *Bufo marinus*.

markry. *Rhus radicans*.

markweed. *Rhus radicans*.

marrow (bone marrow; medulla osseum; medulla of bone). the medullary tissue of bone; the soft connective tissue that fills the cavity of bones. It is comprised chiefly of a web of branching fibers, within which one finds marrow cells (fat cells, myelocytes, and megakaryocytes). It plays a key role in the elaboration of erythrocytes (erythropoiesis) and the development and elaboration of cells of the lymphatic system in mammals and many other vertebrate animals.

marsh gas. methane produced by the decomposition of vegetation in wetlands.

marsh marigold. *Caltha palustris*.

Marsilid phosphate. iproniazid phosphate.

Marsupialia. one of three infraclasses of the class Mammalia. It contains no known poisonous or venomous species, as opposed to the infraclasses Eutheria and Prototheria (See Monotremata).

marvel of Peru. *Mirabilis jalapa*.

maryjane. cannabis.

mascara. a cosmetic used to darken and enhance the appearence of eyelashes. It often contains formaldehyde, alcohol, and plastic resins. It may thus irritate the eye, causing erythema, burning, and swelling.

mask. **1:** to conceal or obscure, as the masking of one disease state by the symptoms of another. **2:** a covering over the nose and mouth to restrict or prevent the inhalation of toxic or otherwise noxious materials or to deliver oxygen or anesthetic gas to a subject.

mass concentration. See concentration.

massasauga, massasauga rattlesnake. common names of *Sistrurus catenatus*.

massicot. lead monoxide.

massive hepatic necrosis. acute yellow atrophy (See atrophy).

mast cell. any of a type of mononuclear, strongly basophilic, granulocyte found in connective tissue. These cells bind passively acquired IgE antibody molecules. When the IgE molecules crosslink multivalent antigen or an antiIgE antibody, the mast cell degranulates and releases a number of potent biologically active agents such as heparin, histamine, and serotonin. These and apparently other substances thus released from the mast cells play a role in inflammation, anaphylaxis, and immune response.

mast cell degranulating peptide. a component of bee venom that causes mast cells to disperse and to release histamine.

master gland. pituitary gland.

masterwort. *Heracleum maximum*.

Mastigoproctus giganteus. a tropical whip scorpion (Order Pedipalpi), some 2-5 cm long, that ejects a defensive fluid to distances up to about 80 cm. Acetic acid, the chief component, makes up about 84% of the fluid, which also contains caprylic acid (5%) and water (11%). The caprylic acid presumably acts as a wetting agent, facilitating penetration of the liquid.

MATC. maximum acceptable toxicant concentration.

Maticora (long-glanded snakes; oriental coral snakes; long-glanded coral snakes; sunbeam snakes). a genus of two species of rather small, slender, venomous snakes (Family Elapidae) with a short tail. *M. bivirgata* (red-bellied long-glanded snake; kendawang; kranawang; pito; ular matahari; ular sina), however, occasionally reaches a length of nearly 1.5 m. These snakes are generally shy and rarely bite humans. The larger individuals are potentially dangerous, however, and serious bites are known including at least one death by *M. bivirgata*. Both species are indigenous to southeastern Asia from Thailand and the Philippines to Sumatra, Java, Borneo, and Celebes. The head is small and there is no distinct neck. The eyes are small to moderate in size with round pupils. There are two large, tubular maxillary fangs and no other teeth imbedded in bone. The ground color and markings vary considerably among individuals. Most specimens are dark brown to blue-black above with narrow light stripes of yellow, red, pale blue, violet, or white. The belly of *M. intestinalis* (common long-glanded snake; banded Malaysian coral snake; Philippine long-glanded snake; oro otto; tadung munggu; ular chabe; ular kapala dua) is black and white; the tail is red with black bars. In *M. bivirgata* the head, tail and belly are typically bright red. The common name of these snakes derives from the presence of distinctively elongated venom glands that extend posteriorly for about ⅓ the length of the body. The heart is displaced to the middle third of the body.

Matulane. procarbazine hydrochloride. See procarbazine.

maximal response (ceiling effect). that response or degree of effect of an organism to a biologically active agent which cannot be intensified by increased dosage.

Maximilian's pit viper, Maximilian's viper. *Bothrops neuwiedi*.

maximum acceptable concentration. See maximum allowable concentration.

maximum acceptable toxicant concentration (MATC). the geometric mean of the lowest concentration that produces a statistically significant effect on survival, growth or fecundity in any life stage in a life cycle, partial life cycle, or early life stage test and the highest concentration that produces no such effect. It is usually considered a threshold for toxic effects in exposures of indefinite duration, but is not limited to a particular concentration or type of effect on any specific life stage.

maximum allowable concentration (MAC; maximum acceptable concentration; maximum permissible concentration). the maximum exposure to a biologically active physical or chemical agent that is allowed during an 8-hour period (a workday) in a population of workers, or during a 24-hour period in the general population, which does appear to cause appreciable harm, whether immediate or delayed for any period, in the target population. In defining the term "appreciable harm" and setting an allowable concentration the relevant authority must, in each case, consider the need for including a safety factor, and the degree of risk that may or may not be acceptable to the target population or to the general population that may be affected. See also exposure limit, tolerance limit, ceiling level, ceiling value.

maximum blood pressure. systolic pressure.

maximum permissible concentration. maximum allowable concentration.

maximum pressure. maximum blood pressure (See systolic pressure).

maximum residue limit (MRL; formerly, pesticide residue tolerance). the maximum allowable concentration of a pesticide residue from the time of application to the crop, at harvest, or during post-harvest processing that is unlikely to exceed the concentration at the point of human ingestion (terminal residue) that exceeds the acceptable daily intake (ADI).

maximum tolerated dose. See dose.

May blossom. *Convallaria majalis*.

May lily. *Convallaria majalis*.

mayapple. **1:** *Podophyllum peltatum*, podophyllum. **2:** *Croton tiglium*.

Mayweed. *Anthemis cotula*.

MBCP. leptophos.

3-MC. 3-methylcholanthrene.

MC. mineralocorticoid.

MCA. chloroacetic acid.

MCAF. macrophage chemotactic, and activating factor.

McClung coral snake. *Calliophis calligaster*.

MCF. macrophage chemotactic factor.

mcg. microgram.

McGregor's pit viper. *Trimeresurus flavomaculatus*.

McMahon's viper. *Eristicophis macmahonii*.

MCP. MCPA.

MCPA ((MCP; 4-chloro-2-methylphenoxy)acetic acid; 2-methyl-4-chlorophenoxyacetic acid; (4-chloro-*o*-toloxy)acetic acid). a water-soluble, synthetic phenoxyacetic acid. It is an auxin used as a potent selective herbicide. It is moderately toxic to laboratory mice.

MCPB. γ-(4-chloro-2-methylphenoxy)butyric acid. activity and applications are similar to those of MCPA, *q.v.*

MDI. diphenylmethane-4,4′-diisocyanate.

Me. methyl.

MEA. mercaptoethylamine (See cysteamine).

meadow grass. See *Agrimonia eupatoria*, *Poa spp*.

meadow-grass dermatitis. See dermatitis.

meadow saffron. *Colchicum autumnale*.

meat poisoning (kreotoxism). a form of food poisoning (*q.v.*) caused by ingestion of meat infested with *Salmonella, Clostridium, Staphylococcus,* or other bacteria that elaborate toxins. Symptoms are those of severe gastroenteritis. See also kreotoxin, kreotoxicon.

3-MECA. 3-methylcholanthrene.

mechlorethamine (2-chloro-*N*-(2-chloroethyl)-*N*-methylethanamine; 2,2'-dichloro-*N*-methyldiethylamine; *N*-methyl-2,2'-dichlorodiethylamine; di(chloroethyl)methylamine; bis(β-chloroethyl)methylamine; methylbis(β-chloroethyl)amine; methyldi(2-chloroethyl)amine; chlormethine; nitrogen mustard). a dark, mobile, volatile, irritant, vesicant, necrotizing liquid, $CH_3N(CH_2CH_2Cl)$, with the faint odor of herring. It was developed for, and used, as a blistering warfare gas. Mechlorethamine is a deadly poison and should not be used without wearing a gas mask. It is a confirmed human carcinogen, and a powerful skin and eye irritant producing blisters and tumors on contact. The hydrochloride has been used as an antineoplastic in both human and veterinary medicine.

Mecoptera (scorpion flies). a small order of endopterygote insects. Scorpion flies have a slender body with two pairs of membranous wings and chewing mouthparts which are borne on a beak-like *rostrum*. They feed mostly on other insects. The tip of the abdomen in males is often upturned, and looks superficially like a scorpion's tail. The larvae resemble caterpillars.

MED. minimum effective dose.

median infective dose (ID_{50}). that amount of a particular population of pathogenic microorganisms that causes infection in 50% of organisms randomly drawn from a defined population of susceptible subjects. The ID_{50} is calculated on the basis of a series of graduated doses given to test (sample) populations.

median lethal concentration. See concentration.

median lethal dose (LD_{50}). See dose.

median tolerance limit. See tolerance limit.

median toxic dose (TD_{50}). See dose.

mediator. a substance that **1:** facilitates the transmission of a nerve impulse from one cell to the next. **2:** participates in the allergic process.

medicable. treatable with medication; subject to treatment with reasonable expectation of cure.

Medicago (clovers; alfalfa). a genus of herbs and a few shrubs (Family Fabaceae, formerly Leguminosae), some species of which have been implicated in trefoil dermatitis and rape scald, as have species of *Brassica*, *Erodium*, and *Trifolium*. ***M. sativa*** (alfalfa; lucerne). a herbaceous perennial legume and valuable forage and hay crop. The sprouts are popular as a salad garnish. It contains saponins, is slightly toxic, and is usually mildly photosensitizing but with no known liver damage. Alfalfa may sometimes contain toxic concentrations of nitrate during periods of rapid growth.

medical examiner. a public official who investigates the causes and circumstances of any sudden, unexplained, suspicious, or unnatural death in his or her jurisdiction. This officer performs autopsies and assists the state in cases of homicide. The medical examiner has replaced the coroner in many jurisdictions.

Medical Subject Headings. See MESH.

medicament. a medicinal substance, medicine, medication, or remedy.

medication. **1:** administration, infusion, or impregnation with a medicine. **2:** the administration of or treatment with a medicine. **3:** a medicament.

medicinal herb. an herb that is a medicinal plant.

medicinal plant. any plant that contains substances that have a therapeutic application either in native form or following modification into a partially synthetic product. Many such native substances or semisynthetic products are also potent poisons (e.g., digitalis and other cardiac glycosides)

medicinal squill. *Urginea maritima*.

medicine. **1:** any drug or remedy; a medicament. **2a:** the diagnosis and treatment of disease and the maintenance of health; the practice of medicine. **2b:** the nonsurgical treatment of disorders or disease.

Mediterranean stinkbush. *Anagyris foetida*. See sparteine.

MEDLARS-on-line. MEDLINE. See also MEDLARS.

MEDLARS. a computerized bibliographic data base (Medical Literature Analysis and Retrieval System) of the U. S. National Library of Medicine; Index Medicus is produced from this data base.

MEDLINE (MEDLARS-on-line). an on-line segment of MEDLARS that provides bibliographic information telephonically to various medical libraries and other users.

medulla. the innermost portion of a structure such as an organ.

medulla of bone. bone marrow (see marrow).

medulla osseum. bone marrow (see marrow).

medusa. **1:** the free-swimming, bell-shaped, or umbrella-shaped, jellyfish stage in the life cycle of many hydrozoans coelenterates. **2:** sometimes loosely synonymous with jellyfish. *Cf.* Scyphozoa. **3:** sometimes applied loosely to any animal with a jellyfish-like appearance.

medusoid. pertaining to or resembling a medusa.

Mees lines. transverse white lines that appear above the lunula (the crescent-shaped area) of the fingernails about 5 weeks following exposure to arsenic.

mega- (M) a prefix to units of measurement that indicates a quantity of 1 million (10^6) times the value indicated by the root. See International System of Units.

megacin. any of a group of bacteriocins isolated from strains of *Bacillus megaterium*.

megadose. a dose the greatly exceeds that which is usually prescribed or recommended.

megaloblastic anemia. See anemia.

megalohepatia. hepatomegaly.

Megalopyge. a genus of hairy moths whose larvae (caterpillars) have stinging hairs. ***M. opercularis*** (wooly worm, puss-moth caterpillar, pus caterpillar, "asp"). a brownish, fuzzy, teardrop-shaped, stinging caterpillar, ca. 20-30 mm long and 10-20 mm across, with a pointed tail. It is a public health problem in the southern United States and Mexico. The setae can puncture human skin causing large erythematous areas with wheals and pruritus. The condition, which can be severe, is called caterpillar hair poisoning or flannel moth dermatitis. Lesions include a grid-like track on the skin, papules and vesicles; desquamation occurs within a few hours or days. Frequent symptoms include local pain (which may be intense) and edema, headache, and lymphadenopathy. Symptoms are sometimes severe causing systemic reactions including shock-like manifestations and convulsions that require hospitalization and occasionally bringing the victim near to death. Allergic reactions are possible but uncommon. The toxic principle is a polypeptide.

meiosis. the type of division of a primordial germ cell, as it matures, into two, then four gametes (e.g., sperm or ova) that contain one half the number of chromosomes that are normal to the somatic cells of the species.

meiotic. pertaining to meiosis.

MEK. methyl ethyl ketone.

Mel B. Arsobal.

melamine (1,3,5-triazine-2,4,6-triamine; 2,4,6-triamino-*s*-triazine; cyanurotriamide). a white, slightly water-soluble compound with monoclinic crystals. It is a cyclic trimer of cyanamide used in tanning leather, and with formaldehyde, to form synthetic melamine resins. It is a skin and eye irritant, toxic by ingestion, and is carcinogenic to some laboratory animals.

melanedema. anthracosis.

melanodermatitis toxica lichenoides. a usually occupational photodermatitis induced when individuals are exposed to tars and subsequently to sunlight. The condition is characterized by pruritis, lichenoid papules, hyperpigmentation, and lesions of the capillaries and other small blood vessels in the skin.

melena. **1:** the passage of dark, tarry feces stained with blood altered by the action of intestinal secretions and/or blood pigments. Common in the newborn. See also mercuric chloride. **2:** black vomit (formerly termed melenemesis).

melenemesis. See melena.

melenic. characterized by melena.

melfalan. melphalan.

Melia azadirachta (neem; neem tree; nim). a subtropical shade tree (Family Meliceae) that is native to semiarid regions of India, Pakistan, and parts of Africa. It is the source of neem oil, *q.v.*; the crushed seeds are used as an insect repellent and antifeedant.

Melia azedarach (Chinaberry tree; China berry; China tree; Chinaball tree; pride of India; white cedar). a widely distributed ornamental shrub or small deciduous tree (Family Meliaceae). All parts contain highly toxic (neurotoxic and intensely irritant) resinoids, but most poisonings are due to eating the fruit. Ingestion may produce nausea, vomiting, constipation, or scours, as well as nervous symptoms such as excitement or depression and weakened heart action.

melilotoxin. dicumarol.

Melilotus (sweetclovers). a genus of biennial herbs that occur as roadside weeds throughout much of southern Canada, and the United States. The two principal species are *M. alba* (white sweet clover), and *M. officinalis* (yellow sweet clover). Both species contain coumarin. See sweetclover poisoning. Toxic concentrations of nitrate have been reported.

melitoxin. dicumarol.

melittin. a polypeptide, $C_{131}H_{229}N_{39}O_{31}$, of bee venom that consists of a single chain of 27 amino acids. It is a hemolysin that is active in humans and other vertebrates. Symptoms of intoxication in humans include bradycardia, and cardiac arrhythmia. It is also a potent antibacterial that inhibits the growth of many bacteria and has been especially useful against *Staphylococcus aureus* 80, which is resistant to penicillin. *Cf.* cobratoxin.

melleine. a probable precursor of ochratoxin isolated from *Aspergillus ochraceus*.

mellitin. a peptide hemolysin in bee venom comprised of a chain of 27 amino acids. It can produce symptoms of bradycardia and arrhythmia in humans. *Cf.* cobratoxin.

Meloidae (oil beetles; blister beetles). a family of medium-sized beetles with a narrow prothorax that feed, as adults, on foliage and flowers. Some species exhibit hypermetamorphosis with a number of specialized larval stages and a prepupa. They blister the skin of humans at the site of contact due to the presence of a vesicant; an allergic reaction is possible. Cantharidin is extracted from the dried bodies of certain species. Important genera are *Epicauta, Lytta, Meloë,* and *Sitaris.*

melphalan (4-[bis(2-chloroethyl)amino]-L-phenylalanine; 4-[bis(2-chloroethyl)amino]-L-phenylalanine mustard; *p*-di(2-chloroethyl)-amino-L-phenylalanine; L-phenylalanine mustard; alanine nitrogen mustard; L-PAM; melfalan; L-sarcolysine; 1-3-[*p*-(bis{2-chloroethyl}amino)phenyl]alanine; 3-*p*-[di(2-chloroethyl)amino]-phenyl-1-alanine; 3-[*p*-(*p*-{bis(2-chloroethyl)amino}phenyl]-1-alanine; 1-phenylalanine mustard; *p*-di-2-(chloroethyl)amino-1-phenylalanine; *p*-*N*-bis(2-chloroethyl)amino-1-phenylalanine; *p*-*N*-di(chloroethyl)aminophenylalanine; phenylalanine nitrogen mustard). a highly toxic, crystalline substance that is practically water insoluble, but soluble in ethanol and propylene glycol. Melphalan is used chiefly as an antineoplastic. It causes bone marrow depression in humans and sometimes gastrointestinal irritation, hypersensitivity, menstrual changes, infertility, and pulmonary damage. It is carcinogenic and leukemogenic, causing pulmonary cancer and lymphosarcoma in laboratory mice and local and mammary cancers in laboratory rats; it also causes Hodgkin's disease. Internal transport of melphalan is effected by a carrier-mediated, leucine-type, high-affinity amino acid transport system. Thus melphalan effects can be prevented by administering leucine or glutamine.

membrane transport protein. See carrier protein.

mendrin. endrin.

meningitides. plural of meningitis.

meningitis (plural, meningitides). any infection or inflammation of the meninges of the brain or of the brain and/or spinal cord (cerebral meningitis, spinal meningitis, cerebrospinal meningitis). Meningitis is usually due to bacterial or viral infection, but can result from any of a number of causes including septicemic blood poisoning (septicemic meningitis).

Menispermum canadense (yellow parilla; moonseed). a woody, twining vine (Family Spermaceae) similar in appearance to wild grape. It grows along fence rows, along the banks of streams, and in thickets in eastern North America. The root and bluish-black fruit are extremely toxic, producing gastrointestinal effects with bloody diarrhea, convulsions, and shock. The active principles are alkaloids with picrotoxin-like activity. The plant is uncommon and poisoning is rare. The fruit is grape-like and may be appealing to children. Fatalities have been reported.

Mentha pulegium (pennyroyal; true pennyroyal). a low, perennial herb (Family Labiatae), the true pennyroyal of Europe. Used, as are other mints, as a tea to aid digestion and soothe upset stomachs. The oil has also been taken to induce menstruation or as an abortifacient. The oil is, however, extremely toxic and the use of an effective dose can also prove fatal to the mother. As little as ½ teaspoon of the oil can produce convulsions and coma; one tablespoon has proven fatal. The oil is 85-92% pugelone, *q*.v.

menthol (hexahydrothymol; methylhydroxyisopropylcyclohexane; *p*-menthan-3-ol; 3-*p*-menthanol; peppermint camphor). a very toxic derivative of terpene, it is a white, crystalline solid, $CH_3C_6H_9(C_3H_7)OH$, with a characteristic odor, obtained from oil of peppermint, which is about 50% methanol, or synthesized. It is used in human medicine in drops, sprays, creams, and ointments, largely for colds, nasal disorders, chest rubs, and the relief of itching. It is a topical antipruretic and has a cooling and soothing effect on mucous membranes. It is used also in cough drops, cigarettes, liquors, confections, and perfumes. It is very toxic to humans. Ingestion of concentrated menthol can cause severe abdominal pain, nausea, vomiting, vertigo, ataxia, drowsiness, and coma. It is an irritant to mucous membranes on inhalation. Severe eye injury may result from contact with concentrated menthol or the fumes.

Menziesia (mock azalea; rustyleaf). a genus of low shrubs (Family Ericaceae) indigenous to North America and east Asia. All have straggling branches and alternate, usually hairy leaves. ***M. ferruginea*** (minnie bush; rustyleaf). a low, woodland shrub found along the Pacific Coast of North America from Alaska to northern California, and eastward at lower elevations in the mountains to Montana. The leaves are poisonous. See Ericaceae.

Mepergan. meperidine.

meperidine (ethyl 1-methyl-4-phenyl-4-piperidinecarboxylate; ethyl 1-methyl-4-phenylisonipecotate; Demerol; Mepergan). an analgesic opioid, often used as the hydrochloride to treat moderate to severe pain. It is also used as a preoperative drug to help relieve pain and anxiety. Its use is prohibited in the case of concomitant use of a monoamine oxidase inhibitor and of known hypersensitivity. Its use is contraindicated in the case of asthma, head injuries, impaired hepatic or renal function, unstable cardiovascular status. This drug is very toxic to nonaddicts (i.e., those with no drug tolerance). Normal therapeutic doses may cause drowsiness, dizziness, nausea, sweating, constipation, cardiopulmonary depression, and addiction. Toxic doses of meperidine cause CNS depression, anxiety, ataxia, confusion, tremor, hallucinations, and convulsions. In severe meperidine poisoning, death may result from severe respiratory depression that appears only terminally. See also drug abuse.

mephitic foul-smelling, noxious, or poisonous.

mephitis. a foul exhalation or emanation.

meprobamate (2-methyl-2-propyl-1,3-propanediol dicarbamate; carbamic acid 2-methyl-2-propyltrimethylene ester; 2,2-di(carbamoyloxymethy)pentane; 2-methyl-2-propyltrimethylene carbamate; procalmadiol; procalmidol; Equanil; Meprospan; Miltown). a widely used anxiolytic and relaxant, $H_2NCOOCH_2C(CH_3)$-$(C_3H_7)CH_2OOCNH_2$, meprobamate is a controlled substance in the United States and may be addictive. It is a moderately toxic CNS depressant that can cause drowsiness, muscular weakness, flaccidity, stupor, sleep, loss of reflexes, severe and persistent hypotension, respiratory depression, shock, coma, and heart failure; it can cause neonatal bleeding and may retard development. There are usually no significant lesions in adult humans, but pulmonary edema, hyperbilirubinemia, and skin reactions may be observed. Large doses (10-40 gm; known from attempted suicides) are extremely dangerous to adults and may cause profound, persistent hypotension with or without muscle flaccidity. Even in such cases, if hypotension is averted, the prognosis is usually good. Meprobamate has multiple sites of action including the thalamus and limbic system.

Meprospan®. See meprobamate.

mercamine. cysteamine.

mercaptamine. cysteamine.

mercaptan. a poor synonym for thiol.

2-mercaptobenzothiazole. a compound mixed, as an adjuvant, with dithiocarbamate fungicides such as maneb, nabam, zineb, ferbam, and ziram to enhance their effects.

mercaptodimethur. methiocarb.

β-mercaptoethylamine. cysteamine.

***N*-(2-mercaptoethyl)benzenesulfonamide**. bensulide.

***N*-(mercaptomethyl)phthalimide *S*-(*O*,*o*-dimethyl phosphorodithioate)**. phosmet.

mercaptophos. demeton.

2,3-mercaptopropanol. dimercaprol.

3-mercaptopropanol. 1-propanethiol.

3-mercaptopropionic acid. an inhibitor of γ-aminobutyric acid synthesis.

α-mercaptopropionylglycine. tiopronin.

***N*-(2-mercaptopropionyl)glycine**. tiopronin.

6-mercaptopurine (6-MP; 1,7-dihydro-6*H*-purine-6-thione; 6-purinethiol). a potent mutagenic base analog used as an antineoplastic agent. It becomes incorporated into DNA, modifying base pair sequencing. It is very toxic to laboratory rodents, acting to suppress DNA and RNA synthesis. Effects of poisoning in humans include extreme nausea and vomiting, with subsequent marrow depression, liver damage, reproductive effects, and urinary effects.

mercaptosuccinic acid diethyl ester. malathion.

mercaptothion. malathion.

3-mercapto-D-valine. penicillamine.

***d*,3-mercaptovaline**. penicillamine.

mercapturic acid. See *N*-acetylcysteine conjugate.

mercuramide. Mersalyl.

mercurial. **1a:** capricious, temperamental, volatile, unpredictable. **1b:** sudden, immediate. **2:** pertaining to mercury **3:** any mercury-containing substance.

mercurial diuretic. See diuretic.

mercurial palsy. paralysis produced by mercury poisoning.

mercurial rash. a rash caused by topical applications of mercury.

mercurialentis. a brown discoloration of the anterior capsule of the lens of vertebrates; it is an early sign of mercurial poisoning.

mercurialine (oil of Euphorbia). the highly toxic principle of herbs of the genus *Mercurialis*. It is a cumulative poison.

Mercurialis (mercury; dog's mercury; herb mercury). a small genus of slender herbs (Family Euphorbiaceae) native to Eurasia. The dried plants were formerly used as a purgative, diuretic, and antisyphilitic. ***M. annua*** ("boys and girls"). a weak, leafy-stemmed European annual weed (Family Euphorbaceae) introduced into North America where it occurs in many waste areas. It has poisoned most classes of livestock. It acts not only as an emetic and purgative, but also is an irritant and narcotic.

Affected livestock suffer from gastrointestinal irritation, diarrhea, and hematuria with extreme anemia; death from cardiac depression may intervene. The entire plant contains the toxic principle, mercurialine.

mercurialism (hydrargism; mad hatter's disease). chronic mercury poisoning due to repeated, usually occupational, exposure to elemental mercury or its vapor. Gastrointestinal and CNS effects usually predominate. Symptoms may include inflammation of the mouth and gums, excessive salivation, sore gums, loosening of the teeth, fetid breath, nervousness, irritability, CNS depression, personality changes, sensorial impairment, tremors, hyperflexia, and a spastic gait. See also tremor (mercurial tremor), mercury poisoning, stomatitis (mercurial stomatitis), sodium formaldehydesulfoxylate. See also mercury poisoning, micromercurialism.

mercurialized. treated with, impregnated with, or influenced by mercury.

mercuric. of or pertaining to bivalent mercury (mercury in the 2+ oxidation state).

mercuric acetate (mercury acetate). white, light-sensitive, crystals or crystalline powder, $Hg(C_2H_3O_2)_2$, with a slightly acetic odor, that is soluble in ethanol and water. It is a strong irritant and is toxic by all routes of exposure. It is used in pharmaceuticals and as intermediate in organic syntheses.

mercuric arsenate (mercury arsenate; mercury arseniate). a yellow powder, $HgHAsO_4$, that is insoluble in water, slightly soluble in nitric acid, and soluble in hydrochloric acid. It is a strong irritant and is toxic by all routes of exposure. It is used in waterproofing paints and marine antifouling paints.

mercuric bromide (mercury bromide). white, light-sensitive powder or rhombic crystals, $HgBr_2$, that is slightly water-soluble, but is sparingly soluble in ether and ethanol. It is a strong irritant and is toxic by all routes of exposure.

mercuric chloride (bichloride of mercury; mercury bichloride; mercury chloride; mercury (II) chloride; corrosive sublimate; mercury perchloride; corrosive mercury chloride). a violently poisonous, nephrotoxic, white, granular or crystalline, water-soluble compound, $HgCl_2$. The risk of exposure is high due to its high water solubility and relatively high vapor pressure. This compound is no longer used in medicine, except in dilute solution as an antiseptic. It is used in photography, the manufacture of mercurials, and as a fungicide. See mercuric chloride poisoning. **ammoniated mercuric chloride** (mercury amide chloride; mercury ammonium chloride; ammoniated mercury; white precipitate; white mercuric precipitate). a poisonous, white, odorless, water- and ethanol-insoluble powder, ClH_2HgN, used as a topical antiinfective in human and veterinary medical practice. If applied too vigorously ptyalism, an allergic dermatitis, or even systemic poisoning may result. Chronic use may cause local pigmentation of the skin or eyelids. It is almost as toxic by ingestion as mercuric bichloride; symptoms may include abdominal pain, nausea, and vomiting. It may have been responsible for many cases of acrodynia in children. Demercaprol is an antidote.

mercuric chloride, ammoniated. ammoniated mercuric chloride (See mercuric chloride).

mercuric chloride poisoning. poisoning due to ingestion of mercuric chloride or sometimes by inhalation of dust that contains this compound, or medical uses such as the application of $HgCl_2$ solutions to large areas of skin, and its use in intrauterine douches. Mercuric chloride is caustic and also immunosuppressive. See also mercurous chloride poisoning. **acute mercuric chloride poisoning**. principal effects of acute poisoning by ingestion are those of severe gastrointestinal irritation with constriction of the throat, a metallic taste in the mouth, intense nausea, vomiting (with bloody vomit), severe abdominal pain that may cause the victim to faint, diarrhea, with melena, scanty urine, convulsions, and prostration. Renal damage may occur. In the absence of immediate treatment death is usual. Dimercaprol q.v. is an antidote. **chronic mercuric chloride poisoning**. a condition, symptoms of which, in humans, may include halitosis, sore tongue, loosening of the teeth, fever, nausea, diarrhea, difficult urination, paralysis, weakness, and death. See also mercuric chloride, corrosive poisoning.

mercuric cyanate. mercury fulminate.

mercuric cyanide (mercury cyanide). an odorless, colorless, ethanol and water-soluble, light-sensitive substance, $Hg(CN)_2$, with tetragonal crystals. It is a violent poison by all routes of exposure. Mercuric cyanide is used in germicidal soaps, the manufacture of cyanogen gas, in photography, and has been used as a topical antiseptic.

mercuric dichromate (mercury dichromate (VI); mercury bichromate). a highly toxic, heavy, red, crystalline, water-insoluble powder, $HgCr_2O_7$, that is soluble in mineral acids.

mercuric dichromate (VI). See mercuric dichromate.

mercuric fluoride (mercury difluoride; mercury fluoride). a toxic compound, HgF_2, that is moderately soluble in ethanol and water. It is a a white powder or transparent crystals that decompose when heated. It is used in the fluorination of organic compounds. A strong irritant, it is highly toxic by all routes of exposure.

mercuric iodate. a highly poisonous, water-soluble, white powder, HgI_2O_6.

mercuric iodide (mercuric biniodide; mercury iodide, red; red mercuric iodide). a toxic substance, HgI_2, with heavy, odorless, nearly tasteless, scarlet-red crystals that turn yellow when heated to 150°C; they dissolve in boiling ethanol. Mercuric iodide is used as a topical vesicant in medicine and also as a counterirritant and vesicant in veterinary practice. A strong irritant, it is highly toxic by all routes of exposure.

mercuric lactate. a toxic, white, water-soluble, crystalline powder, $Hg(C_3H_5O_3)_2$, that decomposes on heating. It is toxic by all routes of exposure.

mercuric nitrate (mercury nitrate; mercury pernitrate). a hazardous and toxic crystalline compound, $Hg(NO_3)_2$. The crystals are colorless and soluble in water and nitric acid, insoluble in ethanol, and decompose on heating. It is used to destroy plant lice (*Phylloxera*), in nitrating organic aromatics, and in the manufacture of felt and mercuric fulminate.

mercuric oleate (mercury oleate; 9-octadecenoic acid mercury salt; oleate of mercury). a highly toxic, red to yellowish-brown, semi transparent, semi-solid or solid, water-insoluble mass, $C_{36}H_{66}HgO_4$, that is slightly soluble in ethanol and ether. It has been used as an ectoparasiticide, an antiseptic, and in antifouling paints.

mercuric oxide, red (mercury oxide, red; red precipitate). a toxic, heavy, odorless, bright red or orange-red, light-sensitive powder, HgO, that is usually more coarse than that of mercuric oxide, yellow). It is soluble in dilute hydrochloric, and nitric acids, but insoluble in water and ethanol. It is prepared by heating mercurous nitrate for a protracted period of time. It is highly toxic (but less so than mercuric oxide, yellow), and is a fire hazard when in close proximity to organic matter. It is used, for example, in pharmaceuticals, as a fungicide and antiseptic; in perfumery, as a paint pigment; in marine bottom antifouling paints; in dry cell batteries (especially those used in miniaturized equipment); and as an analytical reagent and catalyst in organic reactions. See also mercuric oxide, yellow.

mercuric oxide, yellow (mercury oxide, yellow). a highly toxic, fine, yellow or light orange-yellow powder, HgO, that darkens in light. It is soluble in hot water; dilute nitric and hydrochloric acids, potassium permanganate, and magnesium chloride; and in concentrated solutions of alkaline earths; it is insoluble in ethanol. It becomes a reversibly red solid on heating. It is prepared by the action of sodium hydroxide on mercuric chloride or by adding sodium carbonate to a solution of mercuric nitrate. It is extremely toxic, more so than mercuric oxide, red. It was formerly used as an antibacterial agent. See also mercuric oxide, red.

mercuric oxycyanide (mercury cyanide oxide). a violently poisonous, explosive, white, crystalline powder, $HgO{\cdot}Hg(CN)_2$, that is moderately soluble in water. The commercial product is often a mixture of the oxycyanide and cyanide; this reduces the risk of explosion. It is used in medicine as a topical antiseptic. Symptoms of intoxication are those of mercury and/or cyanide poisoning. In dogs, mercurial effects dominate. In humans, initial symptoms are often largely those of cyanide poisoning. This is especially true if the stomach is empty and free gastric acidity is high. In some cases, symptoms associated with cyanide poisoning apparently do not appear.

mercuric phosphate (trimercuric orthophosphate; mercury phosphate). a toxic, yellowish or white powder, $Hg_3(PO_4)_2$, that is soluble in acids, but not in ethanol and water.

mercuric potassium iodide solution. potassium triiodomercurate(II) solution.

mercuric salicylate (mercury subsalycilate; salycilated mercury). a white, yellow, or pinkish, odorless and tasteless, nearly water- and ethanol-insoluble powder, $Hg(C_7H_5(PO_4)_2$. an FDA-approved over-the-counter drug, it is poisonous if administered subcutaneously and is incompatible with alkali iodides.

mercuric sodium *p*-phenolsulfonate (mercuriphenoldisulfonate sodium; mercury and sodium phenolsulfonate). a colorless, water-soluble powder, $C_{12}H_8HgNa_2O_8S_2$, used as an antiseptic and germicide. It is toxic by ingestion and skin absorption.

mercuric stearate (mercury stearate; octadecanoic acid mercury salt). a yellow, granular powder, $(C_{17}H_{35}CO_2)_2Hg$, that is insoluble in water and ethanol, soluble in fatty oils. It is highly toxic by ingestion, inhalation, and skin absorption. Its toxicity is greatly favored by its solubility in lipids. It is used as a germicide.

mercuric subsulfate (mercury oxide sulfate; mercury oxonium sulfate; basic mercuric sulfate; Turpeth mineral). a heavy, highly poisonous, lemon-yellow, odorless powder, $HgSO_4 \cdot 2HgO$, that is nearly insoluble in water, but soluble in acids.

mercuric sulfate (mercury bisulfate; mercury persulfate; mercury sulfate). a white, odorless, acid-soluble, crystalline powder, $HgSO_4$, that decomposes in water and is insoluble in ethanol. It is used in galvanic batteries and as a catalyst in biology. This salt is highly toxic by ingestion, inhalation, and skin absorption.

mercuric sulfide (cinnabar). a highly toxic compound, HgS, that occurs naturally as cinnabar and locally as Ethiops mineral. **black mercuric sulfide** (black mercury sulfide; Ethiops mineral). it occurs as (1) a black or grayish-black, heavy, odorless, tasteless, amorphous powder, HgS, that is insoluble in water, ethanol, and nitric acid, but soluble in sodium sulfide solution, or as (2) black crystals. It is used as a pigment. This substance is highly toxic by skin absorption, ingestion, and inhalation. **red mercuric sulfide** (vermilion; Chinese vermilion; C.I. Pigment Red 106; quicksilver vermilion; red mercury sulfide; artificial cinnabar; red mercury sulfuret). a fine, bright, scarlet powder, HgS, that is insoluble in water and ethanol; used as a pigment. It is highly toxic by skin absorption, ingestion, and inhalation.

mercuric sulfocyanate. mercuric thiocyanate.

mercuric sulfocyanide. mercuric thiocyanate.

mercuric thiocyanate (mercuric sulfocyanate; mercuric sulfocyanide; mercury thiocyanate; mercury sulfocyanate). an odorless powder or crystalline compound, $Hg(SCN)_2$, that is slightly soluble in ethanol and cold water (decomposes in boiling water), soluble in dilute HCl and solutions of alkali cyanides and chlorides. It decomposes vigorously on heating, emitting toxic fumes. Mercury thiocyanate is used in fireworks (Pharoah's serpents) and as an intensifier in photography. It is highly toxic to humans by ingestion, inhalation, and intraperitoneal route and moderately toxic by skin absorption.

mercuriphenoldisulfonate sodium. mercuric sodium *p*-phenolsulfonate.

mercurophen (4-(hydroxymercuri)-2-nitrophenolate; sodium hydroxymercuri-*o*-nitrophenolate; 4-(hydroxymercuri)-2-nitrophenol sodium salt). a dark red, odorless powder, $C_6H_4HgNNaO_4$, that is soluble in hot water. Aqueous solutions are deep amber in color. It is poisonous and used in human and veterinary medicine as a local antiseptic and disinfectant.

mercurous. of or pertaining to monovalent mercury (mercury in the 1+ oxidation state).

mercurous acetate (mercury protoacetate; mercury acetate). a toxic substance, $HgC_2H_3O_2$, with colorless scales or plates that is slightly water-soluble, insoluble in ethanol and ether, but soluble in dilute nitric acid; it decomposes in boiling water or on exposure to light. Sometimes called mercury acetate (as is mercuric acetate). It is highly toxic by all routes of exposure. It is used in medicine as an antibacterial.

mercurous chlorate (mercury chlorate). a hazardous and toxic, white, ethanol- and water-soluble crystalline compound, $Hg_2(ClO_3)_2$, that explodes on contact with combustible substances. It decomposes at 250°C.

mercurous chloride (mercury monochloride; mercury(I) chloride; mercury protochloride; mercury subchloride; mild mercury chloride; calomel; precipité blanc). probably the best known mercurial, Hg_2Cl_2, it is a very toxic, odorless, tasteless, heavy crystalline powder or white, rhombic crystals, that blackens in air. It is nearly insoluble in water; insoluble in ethanol, ether, and cold dilute acids. Hydrochloric acid, alkaline earth chlorides, and alkalis increase solubility in water, although mercurous chloride decomposes in solutions of alkali bromides, cyanides, or iodides; it also decomposes slowly in alkali chlorides. Calomel is used in small doses as an oral laxative, but if laxation does not result, mercury poisoning is possible (See also mercurous chloride poisoning); in this case a saline laxative must be administered as a preventive measure. It is used in powder form as an application to relieve ulcers and skin rashes. Therapeutic uses have also included that of a topical antiseptic, a diuretic, and syphilitic. Additional applications are, for example, as a fungicide; in agriculture to control root maggots on cabbage and onions; in pharmaceuticals; calomel paper; and in calomel reference electrodes (e.g., the saturated calomel electrode, SCE).

mercurous chloride poisoning ("calomel sickness"). a rare poisoning, the effects of which usually include salivation, epigastric discomfort, and diarrhea. Retention of 30-40 mg./kg body weight of calomel following administration as a laxative may prove fatal. See also mercurous chloride, mercuric chloride poisoning.

mercurous sulfate (mercury (I) sulfate). a highly toxic, water-insoluble, yellow to white crystalline powder, Hg_2O_4S, that becomes gray when exposed to light. It is soluble in hot sulfuric acid and in dilute nitric acid. It is used as a catalyst and in dry-cell batteries. It is poisonous by ingestion and intraperitoneal routes and moderately toxic through skin contact.

mercury. **1:** (symbol, Hg; metallic mercury; elemental mercury; hydrargyrum; liquid silver; quicksilver)[Z = 80; A_r = 200.59]. a toxic transition metal that occurs naturally as the mercuric sulfide (cinnabar); a heavy metal. At ordinary temperatures, the pure metal is a silver, heavy, slightly volatile, non-wetting, mobile liquid. Whereas elemental mercury is usually not very toxic, salts and esters of mercury are. They are violently caustic to the skin and mucous membranes. alkyl- and phenyl mercurials in particular are readily absorbed, but also cause serious skin burns that are manifested hours later, thus giving no warning of the systemic hazard of exposure. Mercury reduces the rate of photosynthesis by phytoplankton, and produces neurotoxic effects in higher animals. It is a known mutagen, teratogen, and carcinogen. It readily crosses the mammalian placenta and accumulates in milk. Mercury, especially as a methyl mercury (*q.v.*), is a significant environmental poison. Mercury has many applications and is used, for example, in thermometers, barometers, hydrometers, fluorescent lamps, dental amalgams, in the synthesis of mercury salts. It has many scientific applications, and is a component of a number of pharmaceuticals, fungicides, and agricultural chemicals. See especially alkyl mercury, methylmercury, monomethal mercury, mercurialism, mercury poisoning, Minamata disease. **elemental mercury**. mercury, metalic mercury. **elemental mercury vapor**. the vapor phase of elemental mercury. It is readily absorbed through the skin, gastrointestinal tract, and respiratory tract, but is not unusually toxic unless heated. **metallic mercury** (mercury, elemental mercury). the only metal that is a liquid at normal ambient temperatures. The toxicity of metallic mercury (as with most of its compounds) depends upon the release of the mercury ion *in vivo*. The vapor and many of its compounds are poisonous. Its relatively high vapor pressure augments its toxicological hazard. **2:** a common name of *Rhus radicans*. **3:** a common name of *Mercurialis*.

mercury acetate. mercuric acetate or mercurous acetate.

mercury amide chloride. ammoniated mercuric chloride (See mercuric chloride).

mercury ammonium chloride. ammoniated mercuric chloride (See mercuric chloride).

mercury arsenate. mercuric arsenate.

mercury arseniate. mercuric arsenate.

mercury bichloride. mercuric chloride.

mercury bichromate. mercuric dichromate.

mercury bisulfate. mercuric sulfate.

mercury bromide. mercuric bromide or mercurous bromide.

mercury chlorate. mercurous chlorate.

mercury chloride. mercuric chloride.

mercury(I) chloride. mercurous chloride.

mercury (II) chloride. mercuric chloride.

mercury chloride, mild. mercurous chloride.

mercury cyanide. mercuric cyanide.

mercury cyanide oxide. mercuric oxycyanide.

mercury dichromate. mercuric dichromate.

mercury difluoride. mercuric fluoride.

mercury, dimethyl. dimethylmercury.

mercury fluoride. mercuric fluoride.

mercury fulminate (mercuric cyanate). a gray, crystalline powder, $Hg(CNO)_2$, that explodes when dry. It is used as an explosives' detonator. Mercury fulminate is highly toxic.

mercury iodide, red. mercuric iodide.

mercury monochloride. mercurous chloride.

mercury naphthenate. a toxic, dark amber liquid that is soluble in mineral oils. It is used as an antimildew promoter in paint and as an antiknock compound in gasoline.

mercury nitrate. mercuric nitrate.

mercury oleate. mercuric oleate.

mercury oxide, red. mercuric oxide, red.

mercury oxide, yellow. mercuric oxide, yellow.

mercury oxide sulfate. mercuric stearate.

mercury oxonium sulfate. mercuric subsulfate.

mercury perchloride. mercuric chloride.

mercury pernitrate. mercuric nitrate.

mercury persulfate. mercuric sulfate.

mercury phosphate. mercuric phosphate.

mercury poisoning. poisoning caused by any of numerous mercury compounds, most of which are intensely poisonous (See mercury) by all routes of absorption. The proximate poisons are most commonly the mercury ion or the methylmercury ion or radical. See especially mercuric chloride poisoning, mercurous chloride poisoning, Minamata disease, sodium formaldehydesulfoxylate, stomatitis (mercurial stomatitis). **acute mercury poisoning**. inhalation of heated mercury vapor may initially cause lung damage with edema and consequent shortness of breath. Contact of mercury salts with the skin, eyes, or mucosa may cause severe damage; contact with alkylmercury compounds causes severe burns which may not appear for several hours following exposure, long after systemic effects are observed. Ingestion of mercury compounds (but not elemental mercury) causes mucosal erythema, edema of the mucosa, and ulceration. Deposition of mercurial sulfide in the inflamed tissues may be discolored by the deposition of mercurial sulfide which gives an appearance similar to that seen in lead stomatitis. If swallowed, most mercury compounds can cause symptoms of severe gastroenteritis including hypersalivation, constriction of the throat, nausea, vomiting (sometimes with bloody vomit), abdominal pain, diarrhea, and sometimes melena. The damage sustained can be sufficient to cause death. Further systemic effects begin to occur after mercury (transported via the circulatory system) reaches and accumulates in various organs. Those chiefly affected are the brain and kidneys. The victim may thus experience a variety of symptoms in addition to those due to the direct caustic action of the poison. These may include fatigue, incoordination, excitability, tremor, paresthesia, sometimes impaired vision, dementia, and eventual renal failure (foreshadowed by restricted urine formation, and proteinuria). In the absence of immediate treatment, severe cases may prove rapidly fatal. **alkyl mercury poisoning**. See Minamata disease. **chronic elemental mercury poisoning**. micromercurialism. **chronic mercury poisoning**. mercurialism.

mercury protoacetate. mercurous acetate.

mercury protochloride. mercurous chloride.

mercury and sodium phenolsulfonate. mercuric sodium *p*-phenolsulfonate.

mercury stearate. mercuric stearate.

mercury subchloride. mercurous chloride.

mercury subsalycilate. mercuric salicylate.

mercury sulfate. mercuric sulfate.

mercury (I) sulfate. mercurous sulfate.

mercury sulfocyanate. mercuric thiocyanate.

mercury thiocyanate. mercuric thiocyanate.

mercy killing. a vernacular term for euthanasia applied to humans.

merocrine secretion. See exocytosis.

merological method. any method of study aimed at elucidating the properties of component parts and processes, as in earlier days individual tissues and organs were studied by anatomists and physiologists without regard for their structural and functional relationship to the whole organism. See also hological method, holocoenotic.

Merostomata (king crabs; horseshoe crabs). a class of arthropods that was previously classified as an order (Xiphosura) of the Arachnida, to which they are related. All except five species are extinct. Of these, *Limulus polyphemus* occurs on the American continent from Maine to Yucatan; three species of *Tachypleus* occur on the east coast of Asia, and *Carcinoscorpius rotundicauda* is indigenous to southeastern Asia. Some classifications place all of these in the genus *Limulus*. The extant species occur only in Asia and North America in muddy and sandy shallows along the shores of marine waters. Certain Asiatic species are poisonous. The body is heavily chitinized and may grow to a length in excess of 60 cm. The cephalothorax is massive and covered by a horseshoe-shaped, heavily armored carapace; the abdomen is broad with 6 pairs of appendages, terminating in a long spine. The mouth is situated in the center of the cephalothorax among the legs, which have crushing bases used to masticate food; it is surrounded by small pincer-like chelicerae, and masticatory pedipalps. Respiration is by five pairs of gill books. See *Carcinoscorpius rotundicauda*, *Limulus polyphemus*, ergothioneine.

merphenyl nitrate. phenylmercuric nitrate.

Merrem's sea snake. *Hydrophis caerulescens*.

merryhearts. *Zigadenus nuttallii*.

mersalyl ([3-[[2-(carboxylatomethoxy)benzoyl]-amino]-2-methoxypropyl]hydroxymercurate-(1—) sodium; *o*[[3-(hydroxymercuri)-2-methoxypropyl]carbamoyl]phenoxyacetic acid sodium salt; sodium *o*-[(3-hydroxymercuri-2-methoxypropyl)carbamoyl]phenoxyacetate; *N*-(γ-hydroxymercuri-β-methoxypropyl)salicylamide-*O*-acetic acid sodium salt; mercuramide; Diursal; Neptal). a substance with bitter, somewhat deliquescent crystals. It is a poisonous mercurial diuretic that is sometimes used in medical and veterinary practice (See diuretic).

mescal. **1:** *Lophophora williamsii*. **2:** a liquor distilled from pulque, a fermented beverage made from certain species of agave.

mescal bean. *Sophora secundiflora*.

mescal button. a flowering head of *Lophophora williamsii*, *q.v.*

mescal cactus. *Lophophora williamsii*.

mescaline (3,4,5-trimethoxyphenylethylamine). an alkaloid, $(CH_3O)_3C_6H_2CH_2CH_2NH_2$, it is a toxic, hallucinogenic, colorless, alkaline oil isolated from the flowering heads (mescal-buttons) of *Lophophora williamsii*. See also drug abuse. It produces delusions of color and music. It commonly causes pupillary dilation, hyperreflexia and restlessness, perceptual alterations, and illusions. Synethesia is common, and the thinking process is substantially altered. High doses cause hallucinations and loss of contact with reality.

mescaline oxidase. an amine oxidase of rabbit liver. *Cf.* monoamine oxidase.

mescalism. intoxication caused by mescaline, usually in the form of mescal buttons.

MESH. Medical Subject Headings, a list of words used in the storage and retrieval of medical references in the U.S. Library of Medicine.

meso-, mesi-, mes-. combining forms that: **1:** indicate middle or intermediate. **2:** pertain to a mesentery (in anatomy). **3:** secondary or partial (in medicine).

Mesocricetus. See hamster.

mesosaprobic. of or pertaining to a polluted aquatic habitat that has a below normal oxygen concentration and a moderately high level of organic decomposition.

mesothelioma. a tumor, benign or malignant, originating from the pleura, pericardium, or peritoneum. Mesotheliomas may result from exposure to asbestos. **malignant mesothelioma**. a primary, usually rapidly fatal malignant tumor. It is usually associated with a history of occupational exposure to amphibole asbestos.

mesoxalyl urea, mesoxalylurea. alloxan.

mesoxalylcarbamide. alloxan.

Mestonin Bromide. pyridostigmine bromide.

metabasis. any change in the symptoms or course of a disease.

metabolic. relating to metabolism.

metabolic dealkylation. See dealkylation.

metabolic failure. rapid failure of physical and mental abilities terminating in death.

metabolic pathway. a train of enzyme-mediated biochemical reactions within a living cell in which molecules undergo various chemical reactions and transformations.

metabolic reaction. any reaction that occurs during a metabolic process. Such reactions detoxicate many, but not all, poisonous substances; some substances are toxified and distributed to receptors; and some remain unmodified. Most xenobiotics undergo two types of reactions in mammals (phase I and phase II). **phase I reactions** are biotransformations (hydroxylation, dealkylation, dehydrogenation, reduction, and hydrolysis) of lipophilic xenobiotics in which a polar, reactive group is introduced into the molecule. The product of such a reaction is usually able to pass easily through lipid-containing cell membranes and may then bind to and be transported through the body by lipoproteins. Furthermore, because of the functional group attached in the phase I reaction, the product is usually more water-soluble (more polar) and has a site to which a substrate can attach, thus allowing elimination of the toxicant from the body. **phase II reactions** are conjugation reactions (e.g., acetylation, methylation) that bind such a substrate to the phase I endproducts. The resulting product is usually rapidly excreted. See also conjugation reactions, mixed-function oxidases, cytochromes P-450, epoxidation, hydroxylation, dealkylation, reduction, hydrolysis, dehalogenation.

metabolin. metabolite.

metabolism. the aggregate of all physical and chemical changes that occur within the living substance of an organism including all energy and material transformations that take place within a cell (cellular metabolism), organ, organ system, or whole organism. Normal metabolism is the underpinning of all life processes. *Cf.* anabolism, catabolism. **abnormal metabolism**. metabolism that is dysfunctional or maladaptive with respect to the well being of the organism. **ammonotelic metabolism**. that in which ammonia is the end product of nitrogen metabolism, as in many fresh water and marine animals. **basal metabolism**. the energy expended in simple maintenance of the organism at rest. **carbohydrate metabolism** the catabolic and anabolic processes that involve carbohydrates. Included is the metabolic oxidation of carbohydrates to CO_2 and H_2O with the release of energy. Long, complicated, reaction sequences and cycles are components of this process. Carbohydrate biosynthesis also takes place, resulting in the production of storage glycogen. Toxicants can interfere with carbohydrate catabolism and biosynthesis. **ecological metabolism**. respiration by an ecological community or ecosystem. **endogenous metabolism**. metabolism of proteins of the body itself. **energy metabolism**. the sum of metabolic processes that release energy. **intermediary metabolism**. the metabolic processes involved in the transformation of food into cellular and tissue components. **lipid metabolism**. complex catabolic and anabolic processes that involve lipids. Disruption of lipid metabolism as by a toxicant can result in pathological accumulations of lipids in the liver (cirrhosis of the liver, fatty liver). **ureotelic metabolism**. that in which urea is the end product of nitrogen metabolism, as in mammals, for example. **uricotelic metabolism**. that in which uric acid is the end product of nitrogen metabolism, as in birds, for example. **xenobiotic metabolism** (xenobiotic biotransformation). the metabolism substances that are foreign to the body of an organism. Metabolism of toxic xenobiotics may result in detoxication and elimination from the body of the end products of metabolism, toxification and distribution to receptors, or the toxicants may remain chemically unmodified and distributed to receptors. Because most xenobiotics that enter the body are lipophilic, the end result of dctoxication is increased water-solubility of the end products, thereby facilitating excretion.

metabolite (metabolin). **1:** any product of metabolism, especially a transformed chemical. **2:** any substance involved in metabolism. **active metabolite**. **1a:** a biologically active substance produced by, or modified by, a metabolic process from a precursor. **1b:** any substance that becomes more active than the parent compound as the result of metabolic processes. **reactive metabolite**. a metabolite that is more reactive than the parent compound. Such metabolites mediate many types of toxic effects (e.g., carcinogenesis, cell necrosis, mutagenesis, teratogenesis, hypersensitivity reactions). **stable metabolite**. a stable, normally non-reactive metabolite. **toxic metabolite**. a reactive metabolite that is more toxic than the parent compound. **proximate toxic metabolite**. a stable, toxic metabolite produced in a tissue remote from the target organ. **ultimate toxic metabolite**. a toxic metabolite produced within the target organ by metabolic activation, from a proximate or intermediate metabolite.

metabolizable. capable of being chemically transformed by metabolism.

metacetaldehyde. metaldehyde.

metacetonic acid. propionic acid.

metachlor. alachlor.

metacresol. *m*-cresol.

metal. **1a:** any element that is electropositive and reactive (e.g., alkali metals, alkaline earth metals), tending to lose electrons on reaction. Most are further marked by luster, malleability, ductility, and conductivity of electricity and heat. Their oxides and hydroxides are basic. **1b:** the transition elements which are less reactive, amphoteric, and tend to form complexes. The distinction between metals and non-metals is not sharp. **2:** in vernacular terms, any lustrous, malleable, ductile, solid element or alloy that readily conducts electricity and heat and which forms positive ions in aqueous solution. Not all of these properties need be present in all forms of the substance to be popularly designated as a metal (e.g., cast iron is not malleable or ductile, mercury is a liquid at normal ambient temperatures, some metals are poor conductors). **heavy metal**. a variously defined term and often used in the literature without clear definition. Definition 1b is most useful in toxicology. **1a:** a metal with a density greater than 5 g/cm^{-3}. This definition is arbitrary and includes 53 naturally occurring elements and 16 synthetic elements with widely divergent chemical properties. It includes, for example, elements of the lanthanide and actinide series which are not accepted as heavy metals by most authors because of their chemical behavior. Metals that fit this definition are antimony, manganese, arsenic, bismuth, cadmium, chromium, cobalt, copper, gallium, gold, indium, iridium, iron,lead, manganese, mercury, nickel, palladium, platinum, rhodium, silver, thallium, tin, titanium, vanadium, and zinc. **1b:** any element with a density greater than 5 g/cm^{-3} that (1) that binds preferentially to ligands that contain nitrogen or sulfur as opposed to those that preferentially bind to ligands that contain oxygen, and (2) those that are borderline with respect to ligand selection. This definition includes those metals that are of toxicological significance such as mercury, chromium, cadmium, arsenic (sometimes classed as a metalloid), and lead (See also def. 1c, below). They are extremely important environmental poisons and are the most extensively investigated chemically and toxicologically. They are environmentally persistent, are toxic even at low concentrations, and tend to bioaccumulate. See also heavy-metal tolerance. **1c:** in plant nutrition, any metal of moderate to high atomic number (e.g., Cu, Zn, Ni, Pb) present in certain soils (e.g., due to a mine spoil or outcrop) that inhibits plant growth except in the case of a few tolerant species and ecotypes. **2:** in electron microscopy, a metal of high atomic number used to introduce electron density into a biological specimen by staining, shadowing, or negative staining. **3:** often loosely applied to any metal or metalloid of relative high atomic mass, including some metals that do not fit the chemical definition of a heavy metal (e.g., copper and zinc) and some (e.g., arsenic, a metalloid) which are not metals.

metal cleaner. See cleaner.

metal fume fever. any of a number of allied occupational diseases characterized by malaria-like symptoms, that result from inhalation of particles and fumes of metallic oxides. See, for example, zinc metal fume fever, cadmium metal fume fever.

metal polish. See cleaner (metal cleaner).

metaldehyde (metacetaldehyde). a white, crystalline substance, $(CH_3CHO)_n$, it is a toxic, flammable polymer of acetaldehyde. Metaldehyde is a commonly used molluscacide (snail bait) and herbicide. It is usually blended with bran and is palatable to dogs and farm animals. Metaldehyde is hydrolyzed in stomach acid to acetaldehyde polymers that readily enter the brain. It is a major cause of poisonings in small animals. See also metaldehyde poisoning. Metaldehyde is a strong irritant.

metaldehyde poisoning. poisoning due to ingestion of metaldehyde. The effects are similar in all mammalian species. Neurologic signs dominate and initially include hyperesthesia, muscle weakness, muscle tremors, and incoordination, followed by continuous tonic convulsions and emprosthotonos. Additional signs include hypersalivation (frothing at the mouth), polypnea, dyspnea, tachycardia, and pyrexia (with profuse sweating in horses), cyanosis, and loss of consciousness. Convulsions frequently occur. In severe cases, early death from respiratory failure may intervene; survivors may suffer liver failure. Lesions include congestion and edema of liver, kidney, and lungs; intestinal hemorrhage. Opening of the stomach or rumen on autopsy may reveal a formaldehyde-like odor.

metallic. of, pertaining to, or resembling metal.

metalloid. any chemical element having properties that are intermediate between those of metals and non-metals. The metalloid elements are boron, silicon, germanium, arsenic, antimony, tellurium, and astatine. See also metal (heavy metal).

metallophyte. a plant that tolerates substrates (e.g., soils, bottom sediments) that have very high levels of heavy metals. See also pseudometallophyte. **facultative metallophyte**. pseudometallophyte. **obligate metallophyte**. a metallophyte that is restricted to substrates that contain relatively high concentrations of heavy metals.

metallothionein (MT). metallothionein is an intracellular, cytosolic, low-molecular-weight, metal-binding, inducible protein that is involved in the detoxication of heavy metals. It consists of two similar low-molecular-weight moieties (ca. 6500-7000 daltons each). The cysteine content is high, forming a large number of thiol (-SH) groups that bind tightly to heavy metals. Metallothionein may act as a sink in the removal of toxic metals from the systemic circulation. Metallothionein synthesis is induced by sublethal doses of heavy metals (e.g., Au, Cd, Cu, Hg, Pb, or Zn) and by stressors such as cold, exercise, and food restriction. Following experimental induction by cadmium, the levels increase and are protective against levels of exposure that would be fatal to a previously unexposed animal. Metallothionein has been isolated from all of the major organs of mammals.

metallum problematum. tellurium.

metaphos. methyl parathion.

metaphosphate. See phosphate.

metaplasia. **1a:** transdifferentiation. **1b:** the process of transforming certain types of mature living cells into a form that is abnormal for that tissue or locus. It is an abnormal or disease process as seen, for example, in tumor formation or in response to exposure to toxic or irritant chemicals. Thus, epithelial metaplasia in the respiratory tract may result from exposure to irritants such as smoke and sulfur dioxide. **2:** an evolutionary state marked by maximum vigor and diversification of existing organisms.

metaplastic. pertaining to metaplasia.

Metarrhizium anisophae. an entomopathogenic mushroom that contains a number of insecticidal cyclodepsipeptides (destruxins).

metastases. cancerous cells or pathogenic organisms that spread from their original location to other parts of the body.

metastasis. **1:** a change of state, form, position, or function. **2:** the transportation of pathogenic organisms around the host body. **3a:** the transfer of abnormal cells or pathogenic microrganisms from one organ to another in the body, usually through blood or lymph channels. **3b:** the spread of disease from one part of the body to another.

metastatic infection. pyemia.

metavanadic acid. vanadic acid.

Metazoa. a subkingdom of multicellular animals whose bodies in most cases are composed of many specialized cells that are organized into tissues and organs. They possess a coordinating nervous system. Included are all animals except the Protozoa and Parazoa (sponges).

metazocine (1,2,3,4,5,6-hexahydro-3,6,11-trimethyl-2,6-methano-3-benzazocin-8-ol; 2′-hydroxy-2,5,9-trimethyl-6,7-benzomorphan; methobenzmorphan). an analgesic narcotic, $C_{15}H_{21}NO$. It may be habit forming and is a controlled substance listed in the U.S. Code of Federal Regulations.

metestrus. See estrus cycle.

"meth." methamphetamine hydrochloride. See methamphetamine.

methacholine, methacholine chloride (2-(acetyloxy)-*N,N,N*-trimethyl-1-propanaminium chloride; acetyl-β-methylcholine chloride; *o*-acetyl-β-methylcholine chloride; (2-hydroxypropyl)trimethylammonium chloride acetate; (2-acetoxypropyl)trimethylammonium chloride; trimethyl-β-acetoxypropylammonium chloride). an extremely toxic parasympathomimetic agent that is rarely used because of the unpredictability of effects.

methacrylonitrile (2-methyl-2-propenenitrile; isopropenylnitrile; α-methylacrylonitrile; isopropene cyanide). a clear, colorless liquid, $CH_2C(CH_3)C{=}N$, bp 90°C, that is slightly soluble in water, soluble in acetone. It is used in the manufacture of solvent-resistant thermoplastic polymers and copolymers. It is a fire hazard and is toxic to humans by ingestion, inhalation, or cutaneous absorption.

methadone, methadone hydrochloride (6-dimethylamino-4,4-diphenyl-3-heptanone hydrochloride; 1,1-diphenyl-1-(2-dimethylaminopropyl)-2-butanone hydrochloride; 4,4-diphenyl-6-dimethylamino-3-heptanone hydrochloride). a bitter, crystalline, water-, ethanol-, and chloroform-soluble, synthetic narcotic, $(C_6H_5)_2C(COC_2H_5)CH_2CH(CH_3)N(CH_3)_2{\cdot}HCl$, that is practically insoluble in ether and glycerol. It is a toxic, addictive and has been used for maintenance in the treatment of heroin addiction. See also analgesic, opioid, propoxyphene. It is also used therapeutically as a sedative.

methallylchloride, β-methallylchloride. 3-chloro-2-methylpropene.

methamidophos (*O,S*-dimethyl phosphoramidothioate). a highly dangerous pesticide, miscible with water, that is used to protect cole, cotton, lettuce, and potato crops. It is used also in some household products to eradicate pests.

methamphetamine (*N*,α-dimethylbenzeneethanamine; d-N,α-dimethylphenethylamine; d-N-methylamphetamine; d-deoxyephedrine; d-desoxyephedrine; 1-phenyl-2-methylaminopropane; d-phenylisopropylmethylamine; methyl-β-phenylisopropylmethylamine; methedrine; "meth;" "speed"). a drug, formerly used as an anorectic in the treatment of obesity, it is a widely abused, very toxic, sympathomimetic drug. It can cause various CNS effects, hypertension, palpitations, and tachycardia, and at relatively high doses may also produce psychotic effects. Methamphetamine acts indirectly as an adrenergic and dopamine agonist by effecting the release of endogenous catecholamines. Many of its actions can be blocked by adrenergic and dopamine antagonists. It is a controlled substance that can bring about tolerance and physical dependence. Symptoms of abuse may include vomiting, diarrhea, fever, an irregular tachycardia, hallucinations, delirium, convulsions, and unconsciousness. Chronic abuse of this and related drugs, especially at high concentrations, can cause a psychotic conditon that closely resembles, and is sometimes indistinguishable from, schizophrenia.

methanal. formaldehyde.

methane (marsh gas). a colorless, odorless, nontoxic, flammable gas, CH_4, that forms explosive mixtures with air. Methane is the principal component of most natural gas and a major component of coal gas. It is formed by the anaerobic decomposition of organic matter, as in marshes. Methane is a simple asphyxiant.

methanearsonic acid (methylarsonic acid; methylarsinic acid; monomethylarsinic acid). a biomethylated arsenic herbicide, CH_5AsO_3, prepared from sodium arsenite and methyl iodide. It is a skin and eye irritant and, as the monosodium salt, is moderately toxic by ingestion to laboratory rats. Methanearsonic acid and cacodylic acid are the two organoarsenic compounds most likely to be encountered in the environment. See also biomethylated arsenic, [illegible] methanearsenate, disodium monomethanearsenate.

methanecarboxylic acid. acetic acid.

methanedicarbonic acid. malonic acid.

methanethiol. a type of gas used as an intermediate in the synthesis of pesticides and odorants placed in lines and tanks that contain natural gas, propane, and butane to warm of leaks. Human toxicity is not known, but these compounds and l-propanethiol are considered and should be treated as dangerously toxic.

methanogenic. methane producing. This is accomplished by certain autotrophic and chemolithotrophic bacteria.

methanoic acid. See formic acid.

methanol (methyl alcohol; wood alcohol; wood naphtha). the simplest of the primary aliphatic alcohols, it is a clear, colorless, mobile, highlypolar, flammable, liquid, CH_3OH, with a characteristic odor. It is miscible with ether, ethanol, and water, and is used widely in paint, varnish removers, as an industrial solvent, a synthetic intermediate, and in certain types of antifreeze fluids; mixed with ethanol and soap it is used as a solid canned fuel. It is also used as a denaturant of ethanol. Methanol is highly toxic to humans and damaging to the kidney. It is readily absorbed by the skin and gastrointestinal mucosa. The vapor is also toxic if inhaled, increasing breathing rate. It can cause blindness, and death. See methanol poisoning.

methanol poisoning (methyl alcohol poisoning). poisoning due to ingestion of methanol. The condition may be initially marked by transient exhilaration followed by drowsiness, headache, nausea, vertigo, back pain, severe abdominal pain, blurred vision, fatigue, muscular weakness, a weak, rapid pulse, dizziness, vomiting. In chronic methanol poisoning subjects may show delirium and visual disturbances which may presage irreversible blindness The blindness is caused by formalin which is added to the methanol prior to sale In acute poisoning, the pupils are usually dilated and nonreactive, respiration is rapid and shallow, blood pressure falls rapidly, the victim becomes cyanotic, comatose, and may die from respiratory failure. Formic acid, which is formed as an intermediate in the metabolism of methanol, causes a severe metabolic acidosis. Symptoms of poisoning usually appear 12-18 hours following ingestion, sometimes as much as 48 hours. During this latent period, the only clinical signs are those of mild inebriation. Massive organ damage is revealed on autopsy. Lesions may include liver, kidney, heart damage, pulmonary edema, brain swelling, and severe optic nerve damage. The condition is debilitating and recovery is protracted. See also alcohol poisoning.

methaqualone (Quaalude; Mandrax). a CNS depressant that is prescribed in the treatment of insomnia and anxiety and tension. It is extremely toxic; serious adverse effects include gastrointestinal distress, peripheral neuropathy, drug hangover, loss of inhibition and the development of drug dependence. See depressant, drug abuse. Some individuals may become hypersensitive to this drug; its use is also contraindicated for children and pregnant women. This drug has been removed from the market in the United States.

MetHb. methemoglobin.

methedrine. methamphetamine hydrochloride. See methamphetamine.

methemoglobin (MetHb; ferrihemoglobin). a form of hemoglobin that does not bind oxygen reversibly. The iron of MetHb has been oxidized to the ferric state, with essentially ionic bonds. Methemoglobin is normally present in circulating blood of vertebrates but its concentration is increased under certain circumstances as by poisoning by nitrates and certain other chemicals. Significant increases result in cyanosis. See methemoglobinemia. See also nitrite poisoning.

methemoglobinemia. the presence of unusually large amounts of methemoglobin in the blood with resulting cyanosis due to lack of oxygen in the peripheral tissues. The condition may be caused by nitrates and certain other toxic chemicals; by injury; or due to defective NADH methemoglobin reductase (a recessive autosomal trait); or an abnormality in hemoglobin M, *q.v.* See methemoglobin, nitrite poisoning.

methenyl trichloride. chloroform.

methetharimide. bemegride.

methexenyl. hexobarbital.

methiocarb (3,5-dimethyl-4-(methylthio)phenol methylcarbamate; methylcarbamic acid 4-(methylthio)-3,5-xylyl ester; 4-(methylthio)-3,5-xylyl methylcarbamate; 4-methylthio-3,5-dimethylphenyl *N*-methylcarbamate; mercaptodimethur; metmercapturon). a molluscicide and an insecticide used on vegetable and fruit crops.

methobenzmorphan. metazocine.

Methocel. methyl cellulose.

methomyl. a carbamate insecticide, it is a cholinesterase inhibitor; injection of atropine sulfate readily reverses inhibition. Symptoms of poisoning are similar to those of organophosphate poisoning. The oral LD_{50} in rats is 17 mg/kg body weight. Cattle have reportedly been poisoned by ingesting forage accidentally sprayed with this compound.

methorphan. racemethorphan.

methotrexate (amethopterin; 4-amino-10-methylfolic acid; *N*-[*p*-[[2,4-diamino-6-pteridinylmethyl]methylamino]benzoyl]glutamic acid; methylaminopterin). an orange-brown, crystalline powder that is insoluble in ethanol, water, chloroform, and ether; slightly soluble in dilute hydrochloric acid; and soluble in dilute solutions of alkali carbonates and hydroxides. Methotrexate is a very toxic folic acid antagonist that is used as a chemotherapeutic agent, especially in the treatment of subacute leukemia. This substance is used also as an antirheumatic, antipsoriatic, and occasionally as an immunosuppressive agent. It inhibits both cell-mediated and humoral immune responses. It can cause serious gastrointestinal distress and irritation with ulceration; a dose-related hepatic fibrosis; and can cause hypersensitivity pneumonitis; and megaloblastic anemia. Methotrexate is nephrotoxic, fetotoxic, and abortifacient.

methoxsalen (9-methoxy-7H-furo[3,2-g][1]-benzopyran-7-one; 6-hydroxy-7-methoxy-5-benzofuranacrylic acid δ-lactone; 9-methoxypsoralin; 8-methoxy-4′,5′:6,7-furocoumarin; 8-methoxy[furano-3′,2′:6,7-coumarin]; ammoidin; xanthotoxin; 8-methoxypsoralen; 8-MOP; 8-MP). an odorless, white to cream-colored, crystalline solid, $C_{12}H_8O_4$, that is slightly soluble in ethanol and nearly insoluble in water. It is a flammable, very toxic, phytoalexin and analog of psoralen that is present in plants of the families Fabaceae, Rutaceae, Apiaceae, and in certain fungi such as *Sclerotinia sclorotiorum* (Family Sclerotiniaceae). It is used as a suntan accelerator, a sunburn protector, and therapeutically in the treatment of psoriasis and mycosis fungoides.

methoxyamine (*o*-methylhydroxylamine; methoxylamine; α-methylhydroxylamine; hydroxylamine methyl ether). a colorless, unpleasant smelling liquid, CH_3ONH_2, that is used chiefly as an analytical reagent. It is a strong irritant to the skin, eyes, and mucous membranes.

2-methoxyaniline. *o*-anisidine hydrochloride.

***o*-methoxyaniline**. See anisidine.

***p*-methoxyaniline**. See anisidine.

4-methoxybenzenemethanol. anise alcohol.

1-methoxycarbonyl-1-propen-2-yl dimethyl phosphate. mevinphos.

8-methoxy[furano-3′,2′:6,7-coumarin]. methoxsalen.

2-*p*-methoxyphenylmethyl-3-acetoxy-4-hydroxypyrrolidine. anisomycin.

8-methoxypsoralen. methoxsalen.

9-methoxypsoralen. methoxsalen.

3-methoxypyridine. a toxic alkaloid found in *Equisetum*.

((±)-2-methoxy-5-[(1,2,3,4-tetrahydro-6,7-dimethoxy-2-methyl-1-isoquinolinyl)methyl]-phenol. laudanine.

4-methoxy-*m*-toluidine. *p*-cresol.

6-methoxyumbelliferone. scopoletin.

methoxychlor (1,1′-(2,2,2-trichloroethylidene)-bis[4-methoxybenzene]; 2,2-bis(*p*-methoxyphenol)-1,1,1-trichloroethane; methoxy-DDT; DMDT). a white, crystalline, chlorinated hydrocarbon, $Cl_3CCH(C_6H_4OCH_3)_2$, that is incompatible with alkaline materials. It is soluble in water, insoluble in ethanol. It is an insecticide used to destroy mosquito larvae and houseflies, especially in dairy barns; it is considered relatively safe to use. Methoxychlor is nephrotoxic and hepatotoxic. Dairy calves tolerate 265 mg/kg body weight; 500 mg/kg is mildly toxic; and 1 g/kg produces rather severe poisoning. Sheep are little affected by methoxychlor. One dog was given 990 mg/kg daily for 30 days with no signs of poisoning.

2-methoxy-3,6-dichlorobenzoic acid. dicamba.

methoxydiuron. linuron.

methoxyethanol. a glycol ether, *q.v.*

methoxyflurane (2,2-dichloro-1,1-difluoroethyl methyl ether). a clear, colorless, combustible liquid, $HCCl_2CF_2OCH_3$, with a fruity odor. A general anesthetic and analgesic that is administered by inhalation. Methoxyflurane is a nephrotoxic oxalosis-inducing agent.

8-methoxy-4′,5′:6,7-furocoumarin. methoxsalen.

9-methoxy-7*H*-furo[3,2-g][1]-benzopyran-7-one. methoxsalen.

3-methoxy-4-hydroxycinnamic acid. ferulic acid.

methoxylamine. methoxyamine.

2-methoxy-5-methylaniline. *p*-cresol.

4-methoxy-2-methylaniline. *m*-cresidine.

2-methoxy-5-methylbenzamine (9CI). *p*-cresol.

4-methoxy-2-methylbenzenamine. *m*-cresidine.

6-methoxy-8-(1-methyl-4 diethylamino)butylaminoquinoline. pamaquine.

(±)-3-methoxy-17-methylmorphinan. racemethorphan.

3-methoxy-5-methyl-4-oxo-2,5-hexadenoic acid. penicillic acid.

3-(3-methoxy-2-methyl-3-oxo-1-propenyl)2,2-dimethylcyclopropanecarboxylic acid 2-methyl-4-oxo-3-(2,4-pentadienyl-2-cyclopenten-1-yl ester. pyrethrin II.

7-methoxy-1-methyl-9*H*-pyrido[3,4-b]-indole. harmine.

***N*-[4-(methoxymethyl)-1-[2-(2-thien yl)ethyl]-4-piperidinyl]-*N*-phenylpropanamide.** sufentanil.

***N*-[4-(methoxymethyl)-1-[2-(2-thien yl)ethyl]-4-piperidyl]propionanilide.** sufentanil.

8-methoxy-6-nitrophenanthro-[3,4-d]-1,3-dioxole-5-carboxylic acid. aristolochic acid.

4-methoxy-*m*-phenylenediamine (4-MMPD). a carcinogen. See hair dye.

[2R-(2α,3α,4β)]-2-[(4-metho xyphenyl)methyl]-3,4-pyrrolidinediol 3-acetate. anisomycin.

(±)-2-methoxy-5-[(1,2,3,4-tetrahydro-6,7-dimethoxy-2-methyl-1-isoquinolinyl)-methyl]phenol. laudanine.

methscopolamine bromide. scopolamine methylbromide.

methyl (Me). the chemical group or radical, CH_3—.

methyldi(2-chloroethyl)amine. mechlorethamine.

methyl β-pyridyl ketone. methyl pyridyl ketone.

methyl 3-pyridyl ketone. methyl pyridyl ketone.

methyl acetate. a colorless, volatile, flammable, solvent, $CH_3CO_2CH_3$, with a pleasing odor; it is miscible with the common hydrocarbon solvents. It is a respiratory tract irritant. As an additive to foods, it has a fruity taste.

methylacetic acid. propionic acid.

methyl acetone. methyl ethyl ketone.

β-methylacrolein. crotonaldehyde.

methyl acrylate. a colorless, watery, volatile, flammable, explosive, water-soluble liquid, CH_2:$CHCOOH_3$. It is toxic by all routes of exposure and is strongly irritant to the skin, eyes, and mucosa. Exposure to high concentrations of fumes can cause convulsions. It is used as a chemical intermediate to form acrylic polymers, to package, and to coat paper and plastic film, and in amphoteric surfactants. See also acrylic plastic.

α-methylacrylonitrile. methacrylonitrile.

methyl alcohol poisoning. methanol poisoning.

β-methylallyl chloride. 3-chloro-2-methylpropene.

2-methylallyl chloride. 3-chloro-2-methylpropene.

methylaminoethanolcatechol. epinephrine.

2-methylamino-1-phenyl-1-propanol. ephedrine.

***N*-(3-methylaminopropyl)iminobibenzyl.** desipramine.

5-(γ-methylaminopropyl)iminodibenzyl. desipramine.

2-methyl-*p*-anisidine. *m*-cresidine.

5-methyl-*o*-anisidine. *p*-cresol.

3-methyl-1,8,9-anthr acenetriol. See chrysarobin (def. 1).

methyl anthraquinone. emodin.

methyl arsinic acid. methanearsonic acid.

methyl arsonic acid. methanearsonic acid.

methylate. 1: a compound of methyl alcohol combined with a base. **2:** to add a methyl group (-Ch_3) to a compound or element.

methylated. containing a methyl group; subjected to methylation.

methylation. a reaction that introduces one or more methyl groups, Ch_3, to an element or compound. Methylation facilitates environmental transport of some of the heavier metals. Elements that have methylated forms in the environment include antimony, arsenic, cobalt, germanium, lead, mercury, phosphorus, selenium, silicon, sulfur, the halogens, and tin. Enzymatic methylation is common a common mechanism in the metabolic detoxication of xenobiotics. See methyl transferase. ***O*-methylation**. chemical bonding of a methyl group to the oxygen atom of an organic hydroxyl substituent. It is catalyzed by catechol *O*-methyltransferase. ***N*-methylation**. the chemical binding of a methyl group to a nitrogen atom of an organic molecule. This type of reaction is mediated by any of a number of enzymes of varying specificity. ***S*-methylation**. the chemical binding of a methyl group to the thiol group of an organic compound, mediated by thiol *S*-methyltransferase.

methylatropine nitrate (Eumydrin). a quaternary ammonium compound. It is a derivative of atropine, with similar actions.

methylazoxymethanol. the aglycone of cycasin, *q*.v. See also amygdalin.

methylazoxymethanol β-D-glucoside. cycasin.

methyl-*O*,*N*,*N*-azoxymethyl β-D-glucopyranoside. cycasin.

methylbenzene. toluene.

***o*-methylbenzene**. toluene.

4-methyl-1,3-benzenediamine. toluene-2,4-diamine.

methyl bromide (bromomethane). a colorless, transparent, easily liquefied gas, CH_3Br, with a chloroform-like odor and burning taste. It is used chiefly in drug manufacture, as a soil fumigant, and a fumigant in grain storage warehouses. It is also used as a pesticide against fruit flies in shipments of citrus fruit. It is a strong irritant to the skin and is poisonous by all routes of exposure. Methyl bromide is carcinogenic to laboratory animals. Exposed workers may experience irritation of the eyes, skin, mucous membranes, and/or pulmonary system. Presenting symptoms are seen one-half to six hours after exposure and may include visual problems, headache, nausea, vomiting, and tremor. Coma may ensue and the victim may die from respiratory or circulatory failure. Survivors may experience permanent kidney or brain damage. Low vapor concentrations above threshold that are undetectable by taste or odor may cause chronic poisoning that affects the CNS. Symptoms include disturbances of speech and vision, confusion, lethargy, muscle pains.

***O*-methyl-*O*-(4-bromo-2,5-dichlorophenyl)-phenyl thiophosphonate**. leptophos.

3-methylbutanoic acid, 2-propanyl ester. allylisovalerate.

3-methyl-2-butanoic acid 2-*sec*-butyl-4,6-dinitrophenyl ester. binapacryl.

3-methyl-2-butanoic acid 2-(1-methylpropyl)-4,6-dinitrophenyl ester. binapacryl.

2-methylbutanol. 2-methyl-1-butanol.

2-methylbutanol-1. 2-methyl-1-butanol.

2-methyl-1-butanol (amyl alcohol, primary active; *sec*-butylcarbinol; 2-methylbutanol-1; 2-methylbutanol). a commercially available branched-chain primary alcohol and isomer of amyl alcohol. It is a slightly water-soluble, flammable, explosive, colorless liquid, $CH_3CH_2CH(CH_3)CH_2OH$. It is a moderate fire hazard, and a moderately toxic irritant by ingestion, inhalation, and skin absorption. It is the active alcohol from fusel oil. Symptoms of intoxication may include headache, nausea, vomiting, deafness, and delirium.

2-methyl-2-butanol. *tert*-pentyl alcohol.

2-methyl-4-butanol. isoamyl alcohol, primary.

3-methyl-1-butanol. isoamyl alcohol, primary.

methyl-5-butyl-2-benzimidazole carbamate. parabendazole.

methyl 1-(butylcarbamoyl)-2-benzimidazole-carbamate. benomyl.

methyl *n*-butyl ketone (propylacetone; 2-hexanone). a colorless, flammable, irritant liquid, $CH_3COC_4H_9$, used as a solvent and in the preparation of solvents, varnishes and stains, lacquers, adhesives, waxes, dyes, celluloid, and oils. It is soluble in ethanol and ether, and is a moderate fire and explosion risk. It is irritant to the eyes, skin, and respiratory tract, and is readily absorbed by the skin of humans. It is a CNS poison, having a narcotic action at high concentrations. Chronically exposed workers may develop dermatitis, headaches, muscular weakness, and drowsiness. See also *n*-hexane.

5-(1-methylbutyl)-5-(2-propenyl)-2,4,6-(1H,3H,5H)-pyrimidinetrione monosodium salt. secobarbital.

3-methylbutyric acid, allyl ester. allylisovalerate.

methyl chloride. chloromethane.

2-methyl-4-chloroaniline hydrochloride. 4-chloro-2-toluidine hydrochloride.

methylchlorocarbonate. methylchloroformate.

methylchloroformate (methylchlorocarbonate). a colorless, flammable liquid, $ClCOOH_3$. It is stable in cold water; decomposes in hot water; and is soluble in benzene, methanol, and ether. It is highly corrosive to the skin and eyes. It is a lacrimatory gas used as a military poison, as a warning agent in fumigations with hydrocyanic acid, in insecticides, and organic syntheses.

2-methyl-4-chlorophenoxyacetic acid. MCPA.

methyl chloroform. 1,1,1-trichloroethane.

methyl chloromethyl ether. chloromethyl methyl ether.

20-methylcholanthrene. 3-methylcholanthrene.

3-methylcholanthrene (1,2-dihydro-3-methylbenz[*j*]aceanthrylene; 20-methylcholanthrene; 3-MECA; 3-MC). a yellow, crystalline, polynuclear, pentacyclic hydrocarbon, $C_{21}H_{16}$, that is insoluble in water but soluble in benzene. It is derived from deoxychoic acid, cholic acid, and cholesterol and is used experimentally in cancer research. 3-MC is poisonous by intravenous and intraperitoneal routes. It is powerfully carcinogenic and probably teratogenic to humans. Induction by this compound causes a substantial increase in liver weight.

methylcobalamin. a vitamin B12 analog used as an intermediate in the synthesis of methane. It is responsible for the methylation of inorganic mercury by anaerobic bacteria in bottom sediments. Through the action of methylcobalamin in anaerobic bacteria in bottom sediments of aquatic systems, arsenic(III) is methylated to methanearsonic acid then to cacodylic acid. See also dimethyl mercury, monomethyl mercury, methanearsonic acid.

3-methylcrotonic acid 2-*sec*-butyl-4,6-dinitrophenyl ester. binapacryl.

γ-methyl-α,β-crotonolactone. β-angelica lactone. See angelica lactone.

γ-methyl-β,γ-crotonolactone. α-angelica lactone. See angelica lactone.

methyl cyanide. acetonitrile.

***n*-methyl-5-cyclohexenyl-5-methylbarbituric acid**. hexobarbital.

12-methylcytisine. caulophylline.

***N*-methylcytisine**. caulophylline.

methyl demeton. demeton-*S*-methyl.

***N*-methyl-*N*-desacetylcolchicine**. demecolcine.

methyl diazepinone. diazepam.

***N*-methyl-2,2′-dichlorodiethylamine**. mechlorethamine.

methyldi(2-chloroethyl)amine. mechlorethamine.

3-methyl digitoxose. cymarose.

β-methyldigoxin (medigoxin; metildigoxin; 4‴-*O*-methyldigoxin, 3β,12β,14β-trihydroxy-5β-card-20(22)-enolide-3-(4‴-*O*-methyltridigitoxoside). an extremely toxic derivative of digoxin.

methyl 3-(dimethoxyphosphinyloxy)crotonate. mevinphos.

methyl 1-(dimethylcarbamoyl)-*N*-(methylcarbamoyloxy)thioformimidate. oxamyl.

2-methyl-4,6-dinitrophenol. dinitrocresol.

***N*-methyl-2,4-dinitro-*N*-(2,4,6 tribromophenyl)6-(trifluoromethyl)benzenamine**. bromethalin.

methylhexabital. hexobarbital.

methylhydrazine (monomethylhydrazine; MMH). a colorless, toxic, flammable, hygroscopic liquid, CH_3NHNH_2, with an ammoniacal odor. It is used as a missile propellant, a chemical intermediate, and solvent. Methylhydrazine is also a decomposition product of gyromitrin, related hydrozones and other compounds that occur in mushrooms of the genera *Gyromitra* and *Helvella*, *q*.v. It is also a suspected human carcinogen. Poisoning of laboratory animals causes early, severe hypoglycemia, and liver and kidney damage; damage to the liver is much more severe.

methyl hydrogen sulfate. methyl sulfate.

2-methyl-3-hydroxy-4,5-bis(hydroxymethyl)-pyridine hydrochloride. pyridoxine hydrochloride.

1-methyl-3-hydroxypyridinium bromide dimethylcarbamate. pyridostigmine bromide.

methyl hygrate betaine.

β-methylindole. skatole.

3-methyl-1*H*-indole. skatole.

methyl iodide (iodomethane). a colorless, flammable, partially water-soluble, ethanol- and ether-soluble liquid, CH_3I, that turns brown when exposed to light. It is irritant, narcotic, and neurotoxic by all routes of exposure. Nervous system effects can be severe. Effects include vertigo, nausea, vomiting, slurred speech, drowsiness, dermatitis, blistering of the skin, and eye damage; acute exposures can be fatal. Methyl iodide is carcinogenic to laboratory animals and probably to humans. It is used in organic syntheses, in the manufacture of many pharmaceuticals and some pesticides, and in microscopy.

methyl isobutyl ketone (hexone; 4-methyl-2-pentanone). a colorless, stable, flammable liquid, $(CH_3)_2CHCH_2COCH_3$, that can form explosive mixtures. It is slightly soluble in water and miscible with most organic solvents. It is used as a solvent for paints, varnishes, lacquers, gums, resins, nitrocellulose, fats, oils, and waxes; in organic syntheses; as a denaturant for ethanol; in the extraction of uranium from fission products; and as a synthetic fruit flavoring. Toxicity is similar to that of methyl ethyl ketone. It is irritant to the eyes, skin, and mucous membranes. Severe, acute exposures usually cause CNS depression and gastrointestinal irritation.

methyl isocyanate (MIC; isocyanic acid, methyl ester; isocyanatomethane; methyl mustard oil). a colorless, slightly water-soluble, highly volatile liquid, CH_3NCO, used in the synthesis of pesticides such as Sevin Carbaryl; it is the active ingredient in certain soil fungicides and nematocides. MIC is highly flammable and a significant fire hazard. Both the liquid and the vapor are very toxic and are strong irritants and sensitizers of the eye, skin, and mucous membranes. Systemic effects in humans can result from all routes of exposure. Initial symptoms are usually tearing and damage to the cornea, evidenced by the opacity produced. If inhaled, MIC causes immediate constriction of the respiratory passages and lungs and the victim gasps for breath. It can inflame the pulmonary tissue (bronchitis) with accumulation of fluid, resulting in labored, painful breathing and possibly death. At high concentrations, the vapor can also cause blindness, pulmonary fibrosis, emphysema, and gynecologic effects. Severe acute exposure can result in sudden death. Chronic exposure may cause irreversible damage to the eyes, liver, and kidneys; chronic bronchitis; and may exacerbate the symptoms of asthma. MIC decomposes rapidly in the environment due to atmospheric moisture. It has been considered for use as a chemical warfare agent. See Bhopal.

***N*-methyl-2-isopropoxyphenylcarbamate**. propoxur.

3-methyl-5-isopropyl *N*-methylcarbamate. promecarb.

4-methylmercapto-3-methylphenyl dimethylthiophosphate. fenthion.

4-methyl-1-(1-methylethyl)bicyclo[3.1.0]-hexan-3-one. thujone.

***exo*-(±)-methyl-4-(1-methylethyl)-2-[(2-methylphenyl)methoxy]-7-oxabicyclo[2.2.1]heptane**. cinmethylin.

3-methyl-5-(1-methylethyl)phenol methylcarbamate. promecarb

***N*-methyl-*N*-(1-methylethyl)-*N*-[2-[(9*H*-xanthen-9-ylcarbonyl)oxy]ethyl]-2-propanaminium bromide**. propantheline bromide.

2-methyl-*N*-(2-methylpropyl)-1-propanamine. diisobutylamine.

methyl 1-methyl-$\Delta^{3,4}$-tetrahydro-3-pyridinecarboxylate'. arecoline.

2-methyl-2-(methylthio)propanal *O*-[(methylamino)carbonyl]oxime. aldicarb.

2-methyl-2-(methylthio)propionaldehyde *O*-(methylcarbamoyl)oxime. aldicarb.

methylmorphine. codeine.

methyl mustard oil. methyl isocyanate.

***N*-methyl-*N*-nitrosomethanamine**. *N*-nitrosodimethylamine.

4-methyl-2-oxetanone. *β*-butyrolactone.

methyloxirane. propylene oxide.

methyl parathion (phosphorothioic acid *O,O*-dimethyl *O*-(4-nitrophenyl) ester; *O,O*-dimethyl-*O*-*p*-nitrophenylphosphorothioate). a white, crystalline, slightly water-soluble solid when pure $(CH_3O)_2P(SO)OC_6H_4NO_2$. It is miscible with acids, alcohols, esters, and ketones, although slightly decomposed by acids. It is the methyl homolog of parathion and is a commonly used, extremely toxic organophosphate plant insecticide; a cholinesterase inhibitor. It is also used in some household products to eliminate pests. It is an explosion hazard when heated and is toxic by all routes of exposure. Symptoms of exposure in humans may include nausea and vomiting, abdominal cramps, diarrhea, involuntary defecation and urination, blurred vision, twitching, bradycardia, and dyspnea.

4-methyl-2-pentanone. methyl isobutyl ketone.

(*α*-methylphenethyl)hydrazine. pheniprazine.

methyl phenol. See cresol.

3-methylphenyl)carbamic acid 3[(methoxycarbonyl)amino]phenyl ester. phenmedipham.

***N*-methyl-*beta*-phenylethylamine**. a toxic amine isolated from *Acacia berlandieri*.

***N*-methyl-*β*-phenylethylamine**. the toxic principle of *Acacia berlandieri*.

(1-methyl-2-phenylethyl)hydrazine. pheniprazine.

3-methyl-2-phenylmorpholine. phenmetrazine.

3-methyl-1-phenyl-5-pyrazolyl dimethylcarbamate. pyrolan®.

3-methyl-2-phenyltetrahydro-2*H*-1,4-oxazine. phenmetrazine.

4-methyl-*m*-phenylenediamine. toluene-2,4-diamine.

methylphosphine. a reactive, colorless gas, CH_3PH_2, that is very toxic by inhalation, with effects similar to those of phosphine.

methylphosphonofluoride acid, isopropyl ester. sarin.

methylphosphonofluoridic acid-1,2,2-trimethylpropylester. soman.

methyl phthalate. dimethyl phthalate.

2-methylpropane. isobutane. See butane.

2-methyl-2-butanethiol. *tert*-butyl mercaptan.

methyl mustard oil. methylisocyanate.

methyl propanol. isobutanol.

2-methyl-1-propanol. isobutanol.

2-methyl-2-propanol. *tert* butanol.

2-methyl-2-propenenitrile. methacrylonitrile.

methyl-2-propylamine. a toxic secondary aliphatic amine.

methylpropylcarbinol. 2-pentanol.

2-(1-methylpropyl)-4,6-dinitrophenol. dinoseb.

methyl propyl ketone. 2-pentanone.

***N*-methylpyridinium-2-aldoxime chloride**. pralidoxime chloride.

methyl pyridyl ketone (1-(3-pyridinyl)ethanone; 3-acetylpridine; *β*-acetylpyridine; methyl 3-pyridyl ketone; methyl *β*-pyridyl ketone). a liquid, C_7H_7NO, that dissolves freely in acids. It is a nicotinic acid antagonist.

(*R*)-1-(1-methyl-2-pyrrolidinyl)-2-propanone. hygrine.

methyl salicylate (oil of wintergreen; winter-green oil; sweet-birch oil; betula oil; gautheria oil). a colorless to reddish, oily liquid, $C_6H_4OHCOOCH_3$, the odor of wintergreen. It is toxic by ingestion and its use as a food flavoring is restricted by the U.S. Food and Drug Administration. It is used also in perfumery and as a counterirritant in medicine. Severe poisoning and death may result from ingestion of small amounts. The initial symptoms are nausea, vomiting, and pulmonary edema. A salicylate, its toxicity is potentiated by certain anticoagulants, antidepressants, antineoplastics, and by drugs used to control arthritis.

methyl sulfate (sulfuric acid monomethyl ester; methylsulfuric acid; methyl hydrogen sulfate; monomethyl sulfate; acid methyl sulfate; "Methylsäure"). **1:** an oily liquid, CH_3HSO_4 or CH_3OSO_2OH, that is slightly water-soluble, very slightly ethanol-soluble, but miscible with anhydrous ether. It is used as a sulfonating agent, an industrial alkylating agent, and a specialty solvent. Do not confuse this compound with the extremely hazardous dimethyl sulfate, *q.v.*, which is often referred to as methyl sulfate in texts and which often carries the label name, methyl sulfate. See also neostigmine methyl sulfate. **2:** a legal label name for dimethyl sulfate.

methyl sulfide. dimethyl sulfide.

methyl sulfocyanate. methyl thiocyanate.

4-(methylsulfonyl)-2,6-dinitro-*N,N*-dipropylaniline. nitralin.

4-(methylsulfonyl)-2,6-dinitro-*N,N*-dipropylbenzenamine. nitralin.

methyl sulfoxide. dimethyl sulfoxide.

methyl systox. demeton-*S*-methyl.

methyl 1,2,5,6-tetrahydro-1-methylnicotinate. arecoline.

methyl thiocyanate (methyl sulfocyanate). a colorless liquid, $CH_3SC{\equiv}N$, with an onion-like odor that is nearly insoluble in water but is miscible with ethanol and ether. It is extremely toxic to laboratory cats.

methylthionine chloride. methylene blue.

methyl-3-(*m*-tolylcarbamoyloxy)phenylcarbamate. phenmedipham.

methyl trichloride. chloroform.

methyltrinitrobenzene. 2,4,6-trinitrotoluene.

methylurethane. *N*-methyl carbamate.

methyl vinyl ketone (3-buten-3-one; Δ^3-2-butenone; methylene acetone; acetyl ethylene; *δ*-oxo-*α*-butylene; vinyl methyl ketone). **1:** a liquid, $CH_3CO{=}CH_2$, that is soluble in water, ethanol, methanol, acetone, glacial acetic acid, and slightly soluble in liquid hydrocarbons. It is an irritant to the eyes, skin, and mucous membranes and an extremely toxic system poison that is readily absorbed through the skin, causing widespread systemic effects. Do not confuse with vinyl methyl ketone. **2:** the legal label name for vinyl methyl ketone.

methyl viologen (2+). paraquat.

methyl yellow. *p*-dimethylaminoazobenzene.

5-methyl-2-furanone. angelica lactone.

methyl-CCNU. See semustine.

methyl mercaptophos. demeton-*S*-methyl

***N*-methyl-2-acetonylpyrrolidine**. hygrine.

methylamine. a toxic, gaseous, primary aliphatic amine.

4-methyl-2-aminoanisole. *p*-cresol.

methylaminopterin. methotrexate.

methylate. **1:** a compound of methyl alcohol with a metal ion. **2a:** to introduce a methyl group, CH_3-, into a compound. **2b:** to mix with methyl alcohol.

methylated. containing or combined with a methyl group.

3-methyl-2-butenoic acid. senecioic acid.

methylcrotonic acid. senecioic acid.

methylene. **1a:** a radical that contains a bivalent carbon atom, $—CH_2—$. **1b:** indicating a compound that contains a methylene radical.

methylene acetone. methyl vinyl ketone.

methylenebis(*o*-chloroaniline). 4,4′-methylenebis[2-chloroaniline].

4,4′-methylenebis(2-chloroaniline) (4,4′-methylenebis[2-chlorobenzeneamine]; 4,4′-diamino-3,3′-dichlorodiphenylmethane; di-(4-amino-3-chlorophenyl)methane; methylenebis-(*o*-chloroaniline)). a moderately toxic chlorinated hydrocarbon structurally related to hexachlorophene, *q.v.* It is carcinogenic to some laboratory animals.

4,4′-methylenebis(2-chlorobenzeneamine). 4,-4′methylenebis[2-chloroaniline].

S,S′-methylenebis-L-cysteine. djenkolic acid.

3,3′-methylenebis(4-hydroxy-2H-1-benzopyran-2-one). dicumarol.

3,3′-methylenebis(4-hydroxycoumarin) (bishydroxycoumarin). dicumarol.

methylene(bisphenyl isocyanate). diphenylmethane-4,4′-diisocyanate.

2,2′-methylenebis(3,4,6-trichlorophenol). hexachlorophene.

methylene blue (methylthionine chloride; 3,7-bis(dimethylamino)phenothiazin-5-ium chloride, C.I. basic blue 9, methylthionium chloride; tetramethylthionine chloride; 3,7-bis-(dimethylamino)phenazathionium chloride; Swiss Blue; C.I. 52015; Solvent Blue 8; Urolene Blue). a dark, lustrous green, water- and ethanol soluble compound, $C_{16}H_{18}N_3SCl{\cdot}3H_2O$; toxic by ingestion. It is used as a reducing agent, a bacterial stain, and a dye $((C_{16}H_{18}N_3SCl){\cdot}\ ZnCl_2{\cdot}H_2O)$. It is used therapeutically in human medicine as an antimethemoglobinemic, and an antidote to cyanide poisoning. In veterinary practice it serves as an antiseptic, disinfectant, and an antidote to cyanide poisoning. It is sometimes effective in ameliorating the course of nitrite poisoning by reducing the methemoglobin to the normal ferrous form. Do not confuse with Methyl Blue.

γ-methylene-γ-butyrolactone. α′-angelica lactone. See angelica lactone.

methylene chloride (dichloromethane; methylene dichloride; methylene bichloride; methane dichloride). a clear, colorless, volatile, nearly nonflammable, liquid, halogenated hydrocarbon, CH_2Cl_2. It is used as a refrigerant; a solvent for organic substances; a degreasing agent; a component of many aerosol paint sprays and nonflammable paint strippers; to decaffeinate coffee and as a propellant in aerosol cans. Methylene chloride is metabolized to carbon dioxide and carbon monoxide by the P-450 cytochrome oxidase system. Carbon monoxide, *q.v.*, can significantly increase the concentration of carboxyhemoglobin in circulating blood. Methylene chloride is an irritant of the skin, eyes, and mucous membranes. Contact can produce a dry, scaly dermatitis, skin burns, and irritation to the eyes and upper respiratory tract. Human fatalities have resulted from the use of fumigant mixtures containing methylene chloride, acrylonitrile, and carbon tetrachloride. At high exposures, methylene chloride is a CNS depressant. The fumes are slightly toxic when inhaled by humans or laboratory animals; the liquid is moderately toxic by all other routes of exposure. It can cause liver and lung cancer in laboratory mice but does not produce tumors in hamsters; it is a suspected human carcinogen. Symptoms in humans by inhalation may include dizziness, nausea, fatigue, somnolence, paresthesia of the extremities, euphoria, irritability, and convulsions. Severe or prolonged exposure may result in anesthesia, hallucinations, pulmonary edema, respiratory depression, and coma. Death from respiratory insufficiency is known.

2-methylenecyclopropanealanine. hypoglycine A.

[3,4-(methylenedioxy)-6-propylbenzyl] butyl diethyleneglycol ether. piperonyl butoxide.

methylene-di-*p*-phenylene isocyanate. diphenylmethane-4,4′-diisocyanate.

3,3′-methylenedithiobis (2-aminopropanoic acid). djenkolic acid.

3,3′-(methylenedithio)dialanine. djenkolic acid.

5-methylene-2(5*H*)-furanone. protoanemonin.

5-methylene-2-oxodihydrofuran. protoanemonin.

methylhydroxyisopropylcyclohexane. menthol

methylisocyanate (methyl mustard oil). a chemical warfare agent, it is a powerful irritant to the eyes, skin, and respiratory tract.

methylmercury (II) cation. monomethylmercury.

methylmercury. **1:** either of two compounds that contain the methyl group: dimethylmercury, $(CH_3)_2Hg$, and monomethylmercury, CH_3Hg^+. Both compounds are notorious poisonous environmental pollutants. The methylmercuries are water soluble, lipid soluble, and are thus bioavailable and readily absorbed by aquatic organisms. Once formed in sediment, they may ultimately become widely distributed in both aquatic and terrestrial food webs. Fish tissues may contain more than 1000 times the concentration of mercury as the surrounding water in surface waters contaminated by these organomercurials. Their lipid solubility and high vapor pressure also favor exposure by skin absorption and the pulmonary route. Both are CNS poisons that easily pass through the blood brain barrier (See especially Minimata disease). **2:** This term is indiscriminantly, and often without clarification, used variously in the literature to indicate one or the other of these two compounds as well as common salts of monomethylmercury. Compare mercury, monomethylmercury, and dimethylmercury.

methylmercury ion (1+). monomethylmercury.

methylmercury ion. monomethylmercury.

methylmercury neurotoxicity. Minamata disease.

methylmethane. ethane.

methylmethane sulfonate (MMS). the methyl ester of methylsulfonic acid, it yields the alkylating carbonium ion in solution. MMS is a primary (direct-acting) carcinogen and is readily detoxified by nucleophiles (e.g., proteins), esterases, sulfhydryl groups, and water.

2-methyl-4-methoxyaniline. *m*-cresidine.

1-methyl-7-methoxy-3,4-dihydro-β-carboline. harmaline.

methylphenidate. a CNS stimulant (See stimulant). See also drug abuse.

methylphosphonofluoridic acid 1-methylethyl ester. sarin.

***N*-methylproline methylbetaine**. stachydrine.

1,1-methyl-2-(3-pyridyl)-pyrrolidine sulfate. nicotine sulfate.

3-(1-methyl-2-pyrrolidinyl)pyridine. nicotine.

(S)-3-(1-methyl-2-pyrrolidinyl)pyridine sulfate (2:1). nicotine sulfate.

1,3-(1-methyl-2-pyrrolidyl)pyridine sulfate. nicotine sulfate.

methyltransferase. any of a class of enzymes (*N*-, *O*-, or *S*-methyltransferases) that catalyze the transfer of a methyl group to an organic molecule. The methyl donor is *S*-adenosylmethionine, formed from methionine and ATP. See methylation.

α-methyltyramine hydrobromide. hydroxyamphetamine hydrobromide.

methylxanthine. any of a class of xanthine derivatives in which one, two, or three hydrogen atoms (in ring positions 1, 3, and/or 7) are replaced by a methyl group. A number of these compounds are CNS stimulants. Theobromine and theophylline are, for example, dimethylxanthines; caffeine is a trimethylxanthine.

metobromuron (3-(*p*-bromophenyl)-1-methoxy-1-methylurea). a herbicide used mainly to control weeds in carrots, groundnuts, peas, soyabeans, and potatoes. It is moderately toxic to laboratory rats.

metocurine iodide (dimethyl tubocurarine iodide; Metatubine Iodide). an extremely dangerous skeletal muscle relaxant used as an adjuvant in surgical anesthesia. An overdose can cause prolonged apnea and cardiac arrest with fatal consequences. It should only be used by those with a thorough knowledge of its pharmacology, and then only in facilities where emergency respiratory and cardiac resuscitation is available.

metoestrus. metestrus (See estrus cycle).

Metopium brownei. a genus of trees (Family Anacardaceae) that are related to *Rhus*. ***M. brownei*** (black poison; black poisonwood). a West Indian tree, the sap of which is poisonous, causing severe dermatitis and staining the skin black. It has alternate pinnate leaves and small greenish flowers. ***M. taxiferum*** (poison tree; poisonwood; coral sumac; doctor gum). a tree similar to *M. brownei*; the sap is toxic, causing severe dermatitis and leaving black stains.

Metrazol. pentylenetetrazole.

metridiolysin. a hemolytic toxin isolated from the sea anemone, *Metridium senile*.

métyi. *Cerastes cerastes*.

mevalonicacid(3,5-dihydroxy-3-methylpentanoic acid; 3,5-dihydroxy-3-methylvaleric acid; β,δ-dihydroxy-β-methylvaleric acid; hiochic). an organic acid, $CH_2OHCH_2C(OH)(CH_3)CH_2$-COOH, it is an intermediate in the biosynthesis of squalene, cholesterol, and coenzyme Q in plants and animals.

mevinphos (3-[(dimethoxyphosphinyl)oxy]-2-butenoic acid methyl ester; 3-hydroxycrotonic acid methyl ester dimethyl phosphate; 1-methoxycarbonyl-1-propen-2-yl dimethyl phosphate; methyl 3-(dimethoxyphosphinyloxy)-crotonate; O,O-dimethyl-1-carbomethoxy-1-propen-2-yl phosphate; α-2-carbomethoxy-1-methylvinyl dimethyl phosphate; 2-carbomethoxy-1-methylvinyl dimethyl phosphate; 2-carbomethoxy-1-propen-2-yl dimethyl phosphate; CMDP). a yellow, liquid phosphate ester, α-2-$(CH_3O)_2P(O)OC(CH_3).CHCOOH_3$. Mevinphos is an extremely poisonous, extremely dangerous systemic poison used chiefly to control pests such as mites, aphids, and houseflies. It is a cholinesterase inhibitor and is toxic by all routes of exposure. The *cis*-isomer is about 100 times more toxic than the *trans*-isomer. The commercial product is a mixture of *cis*- and *trans*-isomers. When heated to decomposition it releases toxic phosphate fumes. Mevinphos is used in some household products to eradicate pests. See also Phosdrin®.

mexacarbate (4-(dimethylamino)-3,5-dimethylphenol methylcarbamate (ester); methylcarbamic acid 4-(dimethylamino)-3,5-xylyl ester; 4-dimethylamino-3,5-xylyl methylcarbamate). a colorless, crystalline, water-insoluble substance, $C_{12}H_{18}N_2O_2$, that is soluble in ethanol, acetone, and benzene. It is a carbamate pesticide used mainly to control snails and slugs. It is extremely toxic to laboratory rats.

Mexican black-headed snake. *Tantilla atriceps*.

Mexican blotched rattlesnake. *Crotalus polystictus*.

Mexican moccasin. *Agkistrodon bilineatus*.

Mexican pigmy rattlesnake. *Sistrurus ravus*.

Mexican poppy. *Argemone mexicana*.

Mexican scammony root. See *Ipomoea*.

Mexican tea. *Chenopodium ambrosoides*.

Mexican West Coast rattlesnake. *Crotalus basiliscus*.

Mexican whorled milkweed. *Asclepias mexicana*.

Mexikanische mokassinschlange. *Agkistrodon bilineatus*.

mezereum. *Daphne*.

mg. an abbreviation of milligram.

Mg. the symbol for magnesium.

mg/m^3. milligrams per cubic meter. It is a common standard used to express the number of allowable milligrams of a substance in a cubic meter of air.

mgm. an abbreviation of milligram.

MHD. minimal hemolytic dose.

mhiri. *Bitis arietans*.

MIC. methyl isocyanate.

mica (muscovite). any of several silicate minerals, with different chemical compositions, but similar physical properties. They occur naturally (synthetic mica is also available) as thin, flexible, colorless, odorless flakes or sheets. It is used industrially as insulation in electrical equipment and in the manufacture of roof shingles, wallpaper, and paint. Mica dust is a skin and respiratory irritant. When inhaled the dust causes coughing, weakness, and weight loss. Frequent of chronic exposure may damage the lungs.

micatosis. a pneumoconiosis due to repeated inhalation of mica particles.

Michaelis constant. a constant, K_m, derived from the Michaelis-Menton equation, *q.v.* It is a measure of enzyme-substrate affinity.

Michaelis-Menten equation (Henri-Michaelis-Menten equation). one form of the equation is:

$$V = v(1 + K_m/c)$$

where V is the initial rate of an enzyme-catalyzed reaction in which the initial concentration of substrate is high enough to saturate the enzyme; v is the rate of reaction at a lower substrate concentration, c; K_m is a constant, the *Michaelis-Menten constant*, which is equal to the substrate concentration at half the maximum velocity. K_m finds particular use in toxicology as a measure of the potential effects of enzyme inhibitors, and possible modes of action, in test systems. See also Michaelis-Menton kinetics.

Michaelis-Menton kinetics. descriptive of the kinetics of an enzymatic process that can be described in terms of the rate at which the substrate is processed (dX):

$$-dX/dt = V_{max}C / (K_m + C)$$

or as the rate of change in concentration:

$$-dC/dt = V_mC / (K_m + C)$$

where dC/dt) is the rate of change in concentration, V_{max} is the maximum reaction rate, K_m is the affinity constant, C is the substrate concentration; V_m is the maximum reaction rate per unit of distribution volume (V_d). $V_m = V_{max}/V_d$. See also volume of distribution, clearance (systemic clearance).

mickey finn. an alcoholic beverage doctored with a powerful purgative or a stupefying drug (e.g., croton oil, nicotine), usually used on an unsuspecting victim to ease the commission of a crime such as robbery. In practice, any of a variety of substances have been used for such a purpose including chloral hydrate (knockout drops). Such drinks can prove extremely hazardous as the amount and type of drug may not be well controlled. See also nicotine.

micro- (μ). a prefix to units of measurement that indicates a quantity of 1 millionth (10^{-6}) the value indicated by the root. See International System of Units.

microbe. microorganism.

microbial toxin. any toxin produced by a microorganism. Pathogenic microorganisms generally normally harm of kill a host organism by the production of toxins rather than by cell invasion. The virulence of such a microorganism is thus due to the ability of the microbial toxins to damage cells that are important to the vitality or viability of the host. See endotoxin, exotoxin.

microbicidal. destructive to microbes.

microbicide. microbicidal; an agent that destroys microbes.

microbioassay. an assay to determine the potency of nutrient, a toxicant, or active substance or factor by observing its effect upon the growth of selected microorganisms. Such tests are considered fundamental by some professionals; other believe they are generally very limited in application and interpretation.

microbiota. the microscopic (unicellular and multicellular) organisms of a habitat, ecosystem or specified area. *Cf.* macrobiota, microorganism.

microcurie. one millionth of a curie.

Microcystis aeruginosa. a toxic species of Cyanobacteria. At least one toxic principle, a "fast death factor," that is algal in origin has been isolated from nearly pure laboratory cultures. This factor is highly toxic to humans and livestock. This factor is a cyclic polypeptide that contains aspartic acid, glutamic acid, D-serine, valine, ornithine, alanine, and leucine. See factor (fast death factor).

microcytotoxicity. the ability to lyse or otherwise damage or destroy cells using minute amounts of material.

microdosage. dosage in very small amounts.

microdose. a very small dose.

microflora. See flora.

microgamma (symbol, $\mu\gamma$). picogram

microgram (mcg; symbols, γ, and μg mcg). one-millionth of a gram (10^{-6} g); one-thousandth of a milligram (10^{-3} mg.).

micromercurialism (chronic elemental mercury poisoning). a syndrome in humans caused by chronic inhalation of elemental mercury and characterized by gingivitis and a number of effects due to impairment of CNS function such as excitability, neurasthenia, tremor (especially of the hands), and loss of memory. Associated psychological manifestations may include insomnia, shyness, irritability, and depression. See also mercury, mercurialism, mercury poisoning. Many of the psychopathological effects may be due to disruption of metabolic processes in the CNS and damage to the blood-brain barrier by mercury.

micrometer. See micron.

micromicrogram (symbol, $\mu\gamma$). picogram.

micromilligram. nanogram.

micron (symbol, μ; micrometer (symbol, μm)). 10,000 Å units or one-millionth (10^{-6}) meter.

micronutrient. a naturally occurring chemical in essential to normal growth and development of an organism but is required only in small amounts. Micronutrients include the trace elements and the vitamins. *Cf.* macronutrient.

microorganism (microbe). **1:** any minute one-celled organism, not distinguished as plant or animal in nature. **2:** a minute living organism (unicellular, or in some cases multicellular), especially a bacterium or protozoan, most of which are invisible to the naked eye. Many are of toxicological interest. *Cf.* bacterium, microbiota.

Micropechis (Pacific coral snakes). a genus of two fairly large species of venomous snakes (Family Elapidae). Although some individuals of both species may attain lengths somewhat greater than 1.5 m and are considered dangerous, bites of humans are rare. ***M. elapoides***. a species indigenous to the Solomon Islands, Florida, Guadalcanal, Malaita, and Ysabel. All individuals have a distinct banded pattern and are in most respects similar to *M. ikaheka*. ***M. ikaheka*** (Ikaheka snake). a nocturnal, burrowing species that is indigenous to New Guinea and the nearby islands of Aru, Batanta, Mefoor, Mios Num, Misool, Jobi, Mansinam, and Valise. Although some individuals reach a length of 1.5 m, the average length of adults is about 1.2 m. The ground color of the body is usually yellow or tan, with irregular black or brown crossbands edged in yellow. Dorsal scales are smooth and glossy, in 15 rows at midbody. The eyes are small. Extremely rare but fatal bites of humans are known.

microscopic anatomy. histology.

microsomal. pertaining to microsomes or their contents.

microsome. a type of particle (actually a fragment) derived from the endoplasmic reticulum by centrifugation of cells.

microtubule. any of a set of large protein assemblies comprised of long hollow cylinders within which 13 protofilaments of tubulin are situated parallel to the axis of the cylinder. Microtubules play an essential role in the control of form and dynamics of eukaryotic cells. They may also be involved in phagocytic motility. Colchicine and vinblastine are antimicrotubule agents. See also tubulin.

Micruroides euryxanthus (formerly *Micrurus euryxanthus*; Arizona coral snake; Arizona korallenottern; Sonoran coral snake). a small, rare snake (Family Elapidae) that inhabits semidesert areas of the southwestern United States and northwestern Mexico. It is the only species in the genus. Body length rarely exceed 50 cm. A glossy snake with a small head that is indistinct from the neck; the snout is rounded; the canthus is indistinct. The eyes are small with round pupils. The body is slender, elongate, does not taper; the tail is short. As opposed to *Micrurus*, *q.v.*, the head is solid black to the angle of the jaw, terminating in a straight edge across the parietal scales. The posterior tips of these scales mark termination of the head which is bordered by a yellow or whitish neck band succeeded by a red ring. *Micruroides* is distinguished from the nonvenomous sand snakes (*Chilomeniscus*) and shovel-nosed snakes (*Chionactis*) by the elongate body, black snout, and unmodified rostral plate. The red bands on the body of *M. euryxanthus* are bordered by yellow or white in contrast to king snakes (*Lampropeltis*), which have red bands bordered by black. The body is ringed by alternating bands of red and black rings, each separated by light rings; the tail is banded by alternating black and light rings. It is secretive and not usually regarded as dangerous by many individuals; the venom is highly toxic, however, and this snake should be approached with caution.

Micrurus (American coral snakes; coralillo; gargantilla; corals; Echte Korallenottern). a genus comprised of some 40 species of usually small, slender, brightly colored, venomous, New World snakes (Family Elapidae). They occur from North Carolina to Texas in the United States and from Coahuila and Sonora in Mexico to Argentina and Bolivia in South America. They are distinguished by a small head with small eyes, no distinct canthus, a rounded snout, and no evident neck; a long, slender untapered body with a short tail; a rounded snout, and small eyes with a round pupil. There are nine scales on the crown; the nasal abuts on the single preocular; there is no distinct canthus. The only teeth rooted in bone are the rather large maxillary tubular fangs. Nearly all species are brightly colored with complete alternate rings of black, yellow (sometimes white), and red (usually). The venom is highly toxic and all are dangerous. The bite is sometimes painless, but a bite with envenomation is typically accompanied by localized pain (which may be severe), edema, and sometimes paresthesia of the affected area. This may be followed by slow onset of paralysis and impaired breathing, that very often terminates in death if untreated. Additional effects may include excessive salivation, weakness, drowsiness, slurred speech, difficulty in swallowing, ptosis, cardiovascular effects, and sometimes hypotension, a weak pulse and other signs of shock. Death is rare in individuals receiving antivenin therapy and mechanical respiration. Species not further defined below include: *M. affinis* (Jan's Mexican coral snake); *M. albicinctus*; *M. alleni* (Allen's coral snake), a South American coral snake that is widely distributed in Nicaragua, Costa Rica and Panama; *M. ancoralis*, occurs only in southern Panama; *M. averyi*; *M. balzani*; *M. bernadi*, of western Mexico; *M. bocourti*; *M. browni*, found in southern Mexico and throughout Guatemala; *M. carinacauda*; *M. circinalis*; *M. clarki*, found throughout much of Costa Rica and Panama; *M. corallinus* (coral cobra; Rio de Janiero coral snake; Cape cobra), a South American species; *M. corallinus*; *M. decoratus*; *M. dissoleucus*, of southern Panama; *M. dumerilii*; *M. elegans*, of Guatemala and eastern Mexico; *M. ephippifer*, of western Mexico; *M. filiformis*; *M. hollandi*; *M. ibiboboca*; *M. isozonus*; *M. langsdorffi*; *M. laticollaris*, of southwestern Mexico; *M. latifasciatus*, of southern Mexico and southern Guatemala; *M. lemniscatus* (iboboca); *M. mertensi*; *M. nuchalis*, of southern Mexico; *M. ornatissimus*; *M. peruvianus*; *M. psyches*; *M. putumayensis*; *M. pyrrhocryptus*; *M. ruatanus*, of Honduras; *M. spurelli*; *M. stewarti*, of Panama; and *M. tschudii*. See also venom (neurotoxic snake venoms). ***M. annellatus*** (annelated coral snake). a small species, usually 60-75 cm long. It is typically mostly black with a narrow yellow band across the parietal scutes. Broad, red crossbands are seen in young specimens; these become nearly black in adults. It occurs at altitudes from 460 to 1800 m in river valleys in the mountains of Peru, Bolivia and Ecuador. ***M. diastema*** (Atlantic coral snake). a species found from eastern Mexico southward to western Honduras. The color patterns are rather diverse, but these snakes typically have numerous narrow and complete black rings of approximately uniform width that alternate with yellow and red rings, the red rings, unlike those of other members of the genus, contain irregular black spots and the red scales have black tips. The black rings are usually narrowly bounded by yellow; the red rings are of approximately uniform width. Individuals may reach a length of about one m. ***M. distans*** (broad-banded coral snake). a species found in western Mexico. The head (with the exception of light-colored spots on the snout) is black to the level of the eyes; The lips are yellow (sometimes white). The body is mainly red, with broad red bands and narrow black bands that encircle the body. The red scales are not red-tipped. Occasional individuals attain a length of about 1.1 m. See *Lampropeltis triangulum nelsoni*. ***M. euryxanthus*** (Arizona coral snake). a small, rare snake limited to the deserts of the southwestern United States. It rarely strikes humans. ***M. fitzingeri*** (Fitzinger's coral snake). an unusual red, yellow, and black species of eastern Mexico south into the southern portions of the Mexican Plateau. ***M. frontalis*** (southern coral snake; tropical coral snake; coral cobra; "Cape cobra"; Brazilian giant coral snake). a coral snake with 6-15 repeating sets of crossbands made up of a triad of black crossbands separated by two yellow rings bordered by broad red bands; this sequence is common to many South American coral snakes. The head is black with plates edged in red. The crown is black to the posterior end of the parietals; crown scutes are edged with red or yellow; the labials and temporals bear yellow spots. *M. frontalis* is one of the larger coral snakes, the adults averaging slightly more than 1 m in length. The range extends from southern Brazil and northern Argentina, through Bolivia, Uruguay, and Paraguay. It has been responsible for many human deaths. ***M. fulvius*** (eastern coral snake; bead snake; harlequin snake; Harlekin Korallenotter). a small, shy, secretive, brightly colored snake of the southern United States from coastal North Carolina to western Texas and southward into Mexico. Adults attain a length of about 1 m. It occurs at low elevations, usually in grassland and dry open woods and occasionally along streams. The head is small, the body is slender (about 2 cm in diameter), and only slightly tapered. The scales are smooth and glossy. The tip of the snout is black, bordering on a broad yellow band that runs across the base of the head, and a wide black neck ring. The entire length of the body is completely encircled by a repeated series of broad red and black rings separated by narrow yellow rings; the red and black rings to do not border each other. A number of nonvenomous North American snakes resemble the eastern coral snake, but the tip of the snout of such snakes is not black. This snake is not aggressive unless provoked. The mouth is small and the teeth short, producing a wound that is usually slight. The venom is highly toxic but is expressed in small amounts. Many bites do not produce symptoms of systemic envenomation, but among persons who show systemic effects, mortality is quite high. ***M. hemprichii*** (Hemprich's coral snake). a small species that occurs along the rim of the Amazon Basin in northeastern Brazil, the Guianas, Colombia, Ecuador, and Peru. It is characterized by a series of 5-10 broad black bands about the body separated by narrow yellow and red rings. The crown, snout, and tip of the chin are black, with a red collar (narrow above, broader below) behind the crown. It is the only coral snake in which the anal plate is typically entire. ***M. mipartitus*** (black-ringed coral snake; gargantilla coral). a species that occurs widely throughout the rain forests from Nicuaragua to Peru and northern Venezuela. The snout is black, bordered by a broad red band that covers the posterior portions of the head from immediately behind the eye. The body pattern consists of broad black rings with many narrow white, yellow or red rings between them. The pattern is distinctive. The maximum length reached by adults is about 0.8 m. ***M. nigrocinctus*** (black-banded coral snake). a South American species found in lowland rain forests from southern Mexico, southern Guatemala and widely throughout British Honduras, El Salvador, Honduras, Nicaragua, Costa Rica, and Panama, to northwestern Columbia. The snout is broad and black and the body is characterized by the presence of alternating broad red bands and narrower black rings (totalling 12-20, with 3-7 on the tail); these bands are separated by narrow yellow or whitish rings. There is a broad yellow band over the posterior part of the head and the red scales are often tipped with black. Occasional individuals attain a length of 1.2 m or more. ***M. spixii*** (Amazonian coral snake; giant coral snake; Riesen Korallenotter). the largest of the coral snakes, adults average more than 1 m in length; some individuals attaining a length of 1.5 m. or more. It is indigenous to the Amazon Basin of Brazil, Colombia, Venezuela, Ecuador, Peru, and Bolivia. The crown is predominantly black, with shields that are often spotted with yellow. Lateral to the crown, the head is mostly light. The black rings about the body in this species are all of approximately equal width and are narrower than the yellow and red rings. There may be a black cervical collar followed by a yellow ring. There are 4-9 repeating sets of bands along the body, each comprised of a triad of narrow black rings separated by somewhat broader bands of yellow and red. The anal plate is divided and there are 203-275 ventrals and 16-25 subcaudals. Human fatalities have been reported. ***M. surinamensis*** (Surinam coral snake). a coral snake easily distinguished by the red head and triads of black rings, the median ring of each set being clearly broader than the lateral ones. Apparently a semiaquatic snake that inhabits the borders of the Amazon Basin, including portions of the Guianas, Brazil, Venezuela, Ecuador, Peru, and Bolivia. Adults average about 1 m in length; occasional individuals attain a length of perhaps 1.3 m. The crown is red, with each plate bordered in black. There are 5-8 complete triads along the length of the body, each with a broad black band in the middle and narrow bands laterally. The yellow rings are narrow dorsally. There are 17-19 dorsal scales on the anterior portions of the body; 15 at midbody and posteriorly. Ventral scales total 162-206 and subcaudals, 30-40; the anal plate is usually divided.

midget faded rattlesnake. *Crotalus viridis concolor*.

midwife toad. See *Alytes obstetricans*.

mienie-mienie. *Abrus precatorius*.

MIF. macrophage inhibitory factor.

Mifegyne. a commercial name of mifepristone.

mifepristone ((11β,17β)-11-[4-(dimethylamino)-phenyl]-17-hydroxy-17-(1-propynyl)estra-4,9-dien-3-one; Mifegyne, RU-486, RU 38486). a progesterone antagonist used as an abortifacient, reinforced with prostaglandins; also useful in the treatment of some malignancies.

milagy kadiyan. See *Hydrophis fasciatus*, *Microcephalophis gracilis*.

mild mercury chloride. mercurous chloride.

mildew. **1:** any fungus that grows on plant material. Downy or powder mildews are ascomycetous fungi that cause plant disease. **2:** any fungal disease of plants in which the fungus is seen on the surface of the plant; any disease produced by mildew (def. 1).

mildewcide. a substance that destroys mildew.

military gas. a chemical warfare agent delivered as a gas or dispersed as droplets in the atmosphere. See warfare agent.

military poison. a toxic chemical used as, or developed for use as, a warfare agent, *q.v.*

milk colic. enterotoxemia.

milk fever (parturient apoplexy; parturient fever; parturient paresis; parturient paralysis). an endemic paralytic disease of cattle near parturition. It is a metabolic disease accompanied by hypocalcemia.

milk purslane. *Euphorbia maculata*.

milk sickness. tremetol poisoning in humans is called milk sickness. Empoisonment results from the ingestion of milk or meat from cattle intoxicated by *Eupatorium rugosum*, *q.v.* A nearly identical condition is produced by *Haplopappus*. See also tremetol, trembles.

milk sugar. lactose.

milk vetch. See *Astragalus*.

milkweed. any plant of the Family Asclepidaceae, especially those of the genus *Asclepias*.

milkweed poisoning of rabbits. this type of poisoning is caused by feeding hay that contains woolly pod milkweed. *Asclepias eriocarpa*, reported only from the Pacific southwestern USA. It sometimes is called head down disease, inasmuch as the affected rabbits develop paralysis of the neck muscles and loss of coordination. If the animal has not consumed too much of the weed and the paralysis has not progressed too far, an attempt may be made to treat it. The head of the rabbit is held so that it can drink water and consume food. Leafy greens and carrots should be fed. The poisonous principle is a resinoid; consumption of the green plant equalling approximately 0.25% of an animal's weight is lethal.

milky. See *Lactarius*.

Millepora (fire corals; stinging corals). the only genus comprising the hydrozoan suborder *Milleporina*, *q.v.* Members of this genus are a common and regular component of coral reefs. ***M. alcicornis*** (fire coral). an extremely toxic coral that occurs from Florida to Mexico, in the Bahamas, and the West Indies. It is an upright, branching, or plate-like calcareous coral that may grow to a height of more than 60 cm, and a width of more than of 46 cm. The whitish polyps occupy tiny pores in this brown to creamy yellow coral. Persons touching this coral suffer a severe burning sensation and a blistery, sometimes necrotic rash. The toxin is hemolytic, dermonecrotizing, and antigenic. The LD_{50}, *i.v.*, in mice is 0.04 mg/kg body weight.

Milleporina. a suborder of colonial hydrozoan coelenterates that contains only the genus *Millepora*. The polyps are embedded in a massive upright, leaf-like or branching calcareous exoskeleton.

milli- (m) a prefix to units of measurement that indicates a quantity of 0.001 (10^{-3}) times the value indicated by the root. See International System of Units.

milligamma. nanogram.

milligram (mg; mgm) a unit of mass in the cgs (centimeter-gram-second) system, taken as 10^{-3} gram, which equals 0.015432 grains or 0.000035 ounces.

millimicrogram. nanogram.

millimole (mM). one-thousandth (10^{-3}) mole (def.3).

millipede. common name of any member of the arthropod class Diplopoda.

milo. *Sorghum vulgare*.

Miltown®. See meprobamate.

mimosine. an alpha amino acid contained in large quantities in the seeds of *Leucaena* and *Mimosa*. It causes loss of hair and inhibits hair growth.

Minamata disaster. thousands of persons living near Minamata Bay in Japan during the period 1953-1960 were poisoned by eating fish contaminated by mercury wastes from several chemical plants that had been dumped into the bay; more than 100 persons died. The proximate poisons were alkyl mercuries, notably monomethylmercury and dimethylmercury. The family of symptoms experienced by these people became known as Minamata disease, *q.v.*

Minamata disease (methylmercury neurotoxicity). a severe toxic neuropathy first recognized as a result of the Minamata disaster, *q.v.* The syndrome is caused by alkyl mercury poisoning (chiefly by methylmercy, *q.v.*) that may result in permanent paralysis, other neurologic and mental disabilities, or death. Symptoms may include constriction of the visual field (tunnel vision), cerebellar ataxia, paresthesias, progressive paralysis, deafness, and dysarthria. Lesions include loss of neurons, especially from the cerebral cortex and cerebellar cortex. CNS development in the young may be seriously affected. At risk are those who eat seafood contaminated with organomercury compounds. See also Minamata disaster, methylmercury, dimethylmercury, monomethylmercury.

miner's lettuce. *Montia perfoliata*

miner's lung. anthracosilicosis.

miner's phthisis. anthracosilicosis.

mineral blue. ferric ferrocyanide.

mineral oil (liquid petrolatum; white mineral oil). a clear, colorless, nearly tasteless and odorless, flammable, oily liquid hydrocarbon mixture derived from petroleum. It is insoluble in water and ethanol, soluble in benzene, chloroform, and ether. It is used as a suspending agent or vehicle for medicinal agents to be applied locally; also as a laxative. It is irritant to the eyes and is toxic by all routes of exposure. Mineral oil is a human teratogen by inhalation, causing testicular tumors in the fetus. Aspiration of the vapor or droplets can cause lipoid pneumonia. Hazards of usage are not widely recognized, increasing the risk of usage.

millirem. 10^{-3} rem.

mineral orange. mineral red.

mineral pitch. asphalt.

mineral red. lead oxide red.

mineralization. See biological mineralization.

mineralo-corticotrophic action. the stimulation of mineralocorticoid production.

mineralocorticoid (MC; M-C). any corticoid hormone that is especially effective in causing sodium ion retention and potassium ion loss (e.g., DCA, desoxocortisone, aldosterone) by the kidney. An indirect effect is water retention.

minimal hemolytic dose (MHD). the smallest amount of complement that can completely lyse a designated volume of a standard suspension of erythrocytes that have been sensitized with antibody.

minimum. **1a:** the least, smallest, lowest. **1b:** the least quantity or lowest limit. See dose. **2:** lead oxide red.

minnie bush. *Menziesia ferruginea*.

minnow. See Cyprinidae.

Minous monodactylus. a venomous fish (Family Scorpaenidae) of the coastal waters of Japan, China, and the South Pacific Islands. The venom apparatus is of the *Synanceja* type (See Scorpaenidae).

mintweed. *Salvia reflexa*.

miosis. constriction of the pupil of the eye.

miotic. **1:** of, pertaining to, characterized by, or producing miosis (def.1). **2:** an agent that causes the pupil to contract.

Mirabilis. a genus of common perennial garden herbs (Family Nyctaginaceae) that are native to the American tropics. About ten species are cultivated in the warmer parts of the United states including *M. jalapa* (four-o'clock; marvel of Peru), an herb with fragrant yellow, red, or white flowers. All parts of these plants are poisonous.

mirex (1,1a,2,2,3,3a,4,5,5,5a,5b,6-dodecachlorooctahydro-1,3,4-metheno-1*H*-cyclobuta-(c,d)pentalene; dodecachloropentacyclodecane; dodecachloropentacyclo(3,2,2,$0^{2,6}$,$0^{3,9}$,$0^{5,10}$)-decane; hexachlorocyclopentadienedimer; 1,3,-4,5,5-hexachloro-1,3-cyclopentadiene dimer; perchloropentacyclo(5.2.1.$0^{2,6}$.$0^{3,9}$.$0^{5,8}$)decane). an intensely white, odorless, crystalline substance, $C_{10}Cl_{12}$, that is insoluble in water, soluble in benzene and dioxane. It is a nephrotoxic chlorinated hydrocarbon and confirmed human carcinogen. Mirex is highly toxic to humans by ingestion and moderately toxic by inhalation and skin absorption. Teratogenic and reproductive effects have been demonstrated in laboratory animals. It is an insecticide, that is especially dangerous because it is highly persistent in the environment, bioaccumulates, and is poisonous to nontarget animals including humans.

miscarriage. **1:** failure, malfunction, misapplication. **2:** a natural or accidental abortion; the premature birth of a nonviable fetus. Most miscarriages occur between the 4th and 7th months of pregnancy.

mistletoe. *Phoradendron flavescens* (American mistletoe), and *Viscum album* (European mistletoe).

mistura. a mixture of suspended, insoluble substances prepared for internal use. Because of settling, such a mixture should always be shaken before using.

Mitchell's rattlesnake. *Crotalus mitchellii*.

mite. a minute arachnid (Order Acarina). Some are parasitic and/or vectors of disease.

mithridatism. the attainment of immunity to the effects of a given poison by ingestion of gradually increasing amounts of the poison.

miticide. a pesticide used to destroys mites and other small arachnids such as the fruit-infesting European red mite and the common red spider.

mitigate. to moderate; allay, palliate, render milder.

mitigated. diminished in severity.

mitochndria. plural of mitochondrion

mitochondrion (plural, mitochondria). an organelle in the cytoplasm of eukaryote cells that. Each is bounded by a double membrane, the inner of which is invaginated. The latter is the site of citric acid cycle (Krebs cycle, tricarboxylic acid cycle) and of oxidative phosphorylation and electron transport. Mitochondria contains DNA that specifies rRNAs, tRNAs, and certain proteins (e.g., acetylcholine). The mitochondrial enzymes that facilitate the above processes are the targets of metabolic inhibitors such as cyanide, helminthosporal, podophyllotoxin, and rotenone.

mitochondrial function test. See test.

mitomycin (Mitocin-C; Mutamycin). a complex of 3 antineoplastic antibiotics (mitomycin A, B, and C) produced by *Streptomyces caespitosus*. It is used in a variety of neoplastic diseases. Common serious adverse effects include gastrointestinal disturbances, alopecia, and skin reactions. This drug can also exacerbate existing clotting deficiencies and thrombocytopenia; individuals may become hypersensitive. See *Streptomyces*.

mitotic poison. a substance that interrupts or inhibits cell division during mitosis; mechanisms vary. Some herbicides are mitotic poisons that block mitosis in the primary meristems (e.g., the carbamates, thiocarbamates, maleic hydrazide, trifluralin, and nitralin). Their action is associated with mitotic aberrations and precocious vacuolation and enlargement of the meristematic cells. See also azaguanine.

mixture. **1:** any heterogeneous combination of two or more substances without chemical union (e.g., air, alloys, blood, concrete, petroleum, sea water, many venoms). A mixture cannot, therefore, be represented by a chemical formula. The components may or may not be uniformly dispersed and in many cases can be separated into the component materials by mechanical means. **2:** mistura. **complex mixture**. a mixture comprised of numerous substances (e.g., crude petroleum). Operationally, a complex mixture is one in which the component parts cannot be separated out by convenient or cost-efficient means. Such mixtures, if toxic, (petroleum, many petroleum distillates, liquid synfuels) are extremely difficult to assess. In particular, during the course of testing or experimentation, the makeup of such a mixture may change (e.g., due to differential volatilization, rates of decay, and reactivity). Furthermore, such mixtures may produce complex and confusing combinations of signs and symptoms. The set of symptoms may vary complexly with concentration and age of the mixture due to differential stabilities, solubilities, toxicities, and mode of action and other properties of the component materials. Many animal venoms are complex mixtures.

MLC. median lethal concentration (See concentration).

MLD. median lethal dose. See dose.

mm. millimeter.

MMH. methylhydrazine.

4-MMPD. 4-methoxy-*m*-phenylenediamine.

MMS. methylmethane sulfonate.

Mn. the symbol for manganese.

MNTD. maximum non-toxic dose.

Mo. the symbol for molybdenum.

MO. malaoxon.

Mobulidae. a family of venomous cartilaginous fish, the devil rays.

moccasin. See *Agkistrodon*.

moccasin flower. *Cypripedium hirsutum*.

mock azalea. See *Menziesia*.

model, dose-response. See dose-response model.

modern man. See man.

modified Dakin's solution. sodium hypochlorite solution, dilute.

modifier. **1a:** a process or condition that alters an outcome such as the intensity, extent, or character of a disease, the toxicity of a xenobiotic, the susceptibility of an individual to a toxicant. **1b:** modifying factor (See factor).

Modiolus modiolus (horse mussel; Northern horse mussel). a heavy circumboreal species of bivalve mollusk, about 15 cm long and 9 cm high. It occurs along the Pacific Coast of North America from Baja, California to the Arctic Ocean, and on the Atlantic coast from northern Florida to the Arctic, where it is found in crevices on rocks from low tide to a depth of about 150 m. It is often involved in paralytic shellfish poisoning of humans.

Mojave rattlesnake. *Crotalus scutulatus*.

mokassinschlange. *Agkistrodon*.

mol. abbreviation for mole in chemistry.

molal. a solution that contains 1 mole of solute per kg of solvent. *Cf.* molar.

molality. the number of moles of a solute per kg of pure solvent. *Cf.* molarity.

molar. **1:** pertaining to a mass; not molecular. **2:** pertaining to a mole (def. 2). **3a:** pertaining to the most posterior teeth in mammals. Such teeth are generally used to crush or grind food; they also serve as a major jaw support in the dental arch. Humans have 12 molar teeth. **3b:** a molar tooth. See def. 3a. **4:** containing one mole of solute per liter of solution. *Cf.* molal. **5:** adapted for grinding or crushing, as a molar tooth.

molar concentration. See concentration.

molarity. the number of moles of a solute per liter of solution. *Cf.* molality.

mold and mildew cleaner. See cleaner.

moldy corn poisoning (equine nervous syndrome; choking staggers; cornstalk disease; stomach staggers). a nervous disease especially of horses due to ingestion of moldy forages and feeds, especially corn. Symptoms vary but usually include excitement and paralysis. Autopsy reveals degenerative changes (most commonly necrosis and liquefaction) in the white matter of the brain and upper spinal cord. There are three clinical types: lethargic, nervous, and paralytic moldy corn poisoning. It is likely that early outbreaks designated as moldy corn poisoning were caused chiefly by aflatoxins. *Cf.* aflatoxicosus, forage poisoning, moldy corn toxicosis. See also encephalitis (virus equine encephalitis). Poisoned cattle experience a severe, usually fatal gastroenteritis.

mole. **1:** a common name of subterranean mammals of the family Talpidae. See Insectivora. **2:** that amount of a substance considered to contain 6.023×10^{23} (Avogadro's number) elementary particles (atoms, ions, molecules, or radicals). In practice, a mole is that amount of a chemical compound whose mass in grams is equivalent to its formula mass (molecular weight).

mole death. strychnine.

mole-nots. strychnine.

mole plant. *Euphorbia lathyris*.

mole viper. a viper of the genus *Atractaspis*.

molinate. a thiocarbamate herbicide.

mollusc. mollusk.

molluscacidal. destructive to mollusks.

molluscacide (molluscicide). an agent that is destructive to mollusks.

molluscicide. molluscacide.

mollusk (mollusc). any member of the Phylum Mollusca. Most venomous mollusks are either gastropods (univalve mollusks) or cephalopods (octopuses and squids). Bivalve mollusks (clams, oysters and mussels), however, may accumulate and even concentrate in their tissues, poisons (e.g., chromium) from the medium, from contaminated sediments, or in their food. See also shellfish poisoning.

Molten's disease. Pictou disease.

molybdenum (symbol, Mo)[Z = 42; A_r = 95.94]. a transition element that occurs naturally as the sulfide in molybdenite. It occurs as a dark-gray to black powder with a metallic sheen or as a silver-white mass. It is used in special steel alloys, in glass-to-metal seals, and electric light bulbs. Molybdenum is slightly toxic to humans, but is toxicologically important because of its role as an antagonist of copper in animal nutrition. Certain species of plants, in soils with high concentrations of molybdenum, can accumulate this element to concentrations that are not toxic to the plants, but are poisonous to grazing animals. Legumes generally accumulate much higher levels of molybdenum than most other plants under a given set of conditions. Soils with normal copper content and low concentrations of molybdenum can similarly poison grazing animals by copper. On the other hand, soils with normal copper content and high levels of molybdenum (or with low copper and moderate molybdenum levels) may produce symptoms of copper deficiency in grazing animals. Molybdenum toxicity to livestock is strongly influenced by the amount of inorganic sulfate in the forage consumed. Even modest amounts of sulfates in the soil can effectively control molybdenum intoxication of livestock.

molybdenum anhydride. molybdenum trioxide.

molybdenum lead chromate. lead-molybdenum chromate.

molybdenum orange. lead-molybdenum chromate.

molybdenum trioxide (molybdenum anhydride; molybdic oxide; molybdic acid hydride). a white, yellow, or bluish, slightly water-soluble powder, MoO_3. It is very soluble in concentrated mineral acids and in excess alkali with formation of molybdates. The two known hydrates, $MoO_3{\cdot}H_2O$ and $MoO_3{\cdot}2H_2O$, readily form polymers by combining with acids and bases. A powerful irritant, it is poisonous by ingestion and by subcutaneous and intraperitoneal routes. When heated to decomposition it releases toxic fumes of elemental molybdenum.

molybdic acid hydride. molybdenum trioxide.

molybdic oxide. molybdenum trioxide.

moma. *Bitis arietans, B. gabonica.*

Momordica. a genus of plants (Family Cucurbitaceae) with gourd-like fruit and a record of toxicity in Africa, Australia, and India. With few exceptions, the foliage, leaves, and outer coat of the fruit have strong cathartic properties. ***Momordica charantia*** L. (wild balsam-apple; balsam pear; bitter gourd). a common, weedy, creeping or climbing vine of sandy soils and waste areas in the southeastern United States. All parts of the plant are toxic.

Monacanthidae (filefishes). a family of tropical to temperate marine fishes (Class Osteichthyes) with a deep, compressed body and a tough integument with tiny hard scales, a single dorsal spine, and a small terminal mouth.

monalide. a post-emergent anilide herbicide that decomposes rapidly in the soil. It is used mainly to destroy weeds of umbelliferous crops such as carrots, celery and parsley.

monamine. an amine with one amine group. Do not confuse with monoamine.

mônemé. *Bitis gabonica.*

money shell. *Saxidomus nuttallli.*

Mongolian pit viper. *Agkistrodon halys.*

monkshood. See *Aconitum.*

monkshood poisoning. poisoning due to ingestion of aconite (*Aconitum*). Symptoms in humans may include paresthesia, a burning sensation of the mouth and skin, itching, dermatitis, nausea, vomiting, bradycardia, hypotension, weak pulse, respiratory distress, cyanosis, convulsions, and death.

monoacetylneriifolin. cerberin. See *Cerbera.*

monoacetylnivalenol (fusarenon-X). a tricothecene mycotoxin.

monoamine. **1:** containing one amine radical. **2:** a compound that contains one amine radical. *Cf.* monamine.

monoamine oxidase (amine oxidase; adrenalin oxidase; tyraminase; MAO). an enzyme that is distributed widely among animals. It inactivates a number of biogenic amines (e.g., dopamine, epinephrine, norepinephrine, serotonin, tyramine) by catalyzing the oxidative deamination of these compounds. Various other amine oxidases (e.g., plant amine oxidase, histamine oxidase, mescaline oxidase, spermine oxidase) are known from a wide variety of life forms. Unlike monoamine oxidase, however, they do not catalyze *N*-substituted amines and are inhibited by carbonyl reagents. See monoamine oxidase inhibitor.

monoamine oxidase inhibitor (MAO Inhibitor; MAO-I). any of a group of psychiatric drugs that act chiefly as CNS stimulants used in the control of depression. They help to control anxiety, especially anxiety linked to abnormal fears or phobias. They are sometimes also used to treat migraine headaches and hypertension. Like other antidepressants they presumably change the chemical composition of the blood reportedly responsible for the depression. The antidepressant action of this type of drug is not fully explained, however, by the inhibition of monoamine oxidase. They are used for hypertension as well as depression. All are toxic and must be used with caution. Overdosage may cause trembling, a heightened sense of well being, or psychotic behavior. More severe intoxication may produce nausea, vomiting, lethargy, dry mouth, ataxia, stupor, rise or fall in blood pressure, fever, tachycardia, acidosis, convulsions, liver damage, and jaundice. Death is usually from cardiac or respiratory failure. MAO inhibitors potentiate the action of barbiturates, antihistamines, and other antidepressants such as meperidine, morphine, and aminopyrine among others. Even certain foods (e.g., cheeses, yogurt, red wine, beer, and smoked or pickled herring) must be avoided.

monoatomic. of or pertaining to molecule, radical, or ion that consists of a single atom. Thus, helium is a monoatomic gas and the hydrogen radical, H·, is a monoatomic radical.

mono-. a prefix denoting one.

monobasic. having one hydrogen atom that is replaceable by a metal or positive radical as in the case of hydrochloric acid, HCl.

monobasic acid. an acid such as HCl that contains one ionizable atom of hydrogen per molecule. See carboxylic acid.

monobasic lead acetate. lead subacetate.

monocellate cobra. *Naja naja*.

monochloracetone. chloroacetone.

monochloroacetic acid. chloroacetic acid.

monochloroethene. vinyl chloride.

monochloroethylene. vinyl chloride.

monocled cobra. *Naja naja*.

monocrotaline (14,19-dihydro-12,13-dihydroxy-20-norcrotolanan-11,15-dione; crotaline). a pyrrolizidine alkaloid, it is a major, very toxic component of *Crotalaria spectabilis*. Monocrotaline is a diester of monocrotalic acid and retronecine. It is hepatotoxic and acts by constricting the lumen of the medium and smaller veins of the hepatic circulation by subendothelial swelling. Sinusoidal congestion and hemorrhage follow, with subsequent fibrosis and cirrhosis. See also necine.

monocrotophos. a highly toxic organophosphate insecticide. The oral LD_{50} in male laboratory rats has been reported at 17 mg/k body weight; in female rats, 20 mg/kg; and the dermal LD_{50} in males at 126, in females at 112 mg/kg. It is used in some household products to eradicate pests.

monoethanolamine. ethanolamine.

monoethylamine. ethylamine.

monofluoracetamide. fluoroacetamide.

monofluoroacetamide. fluoroacetamide.

monoisonitrosoacetone. isonitrosoacetone.

monomethylmercury (methylmercury ion; methylmercury ion (1+); methylmercury (II) cation). **1:** a highly reactive, poisonous, mutagenic ionic compound, CH_3Hg. It rarely occurs free in the environment as it readily ionizes and usually occurs as a component of another compound (e.g., methylmercury chloride). Monomethyl mercury compounds are often found in association with dimethyl mercury in contaminated fish and marine (aquatic) birds. It is synthesized in sediments from inorganic mercury by anaerobic bacteria in neutral or slightly acid sediments through the action of methylcobalamin, *q.v.* These conditions also favor anaerobic decay and the production of methylcobalamin. It is used as a fungicide. **2:** methylmercury (def. 2).

mononuclear phagocyte system. reticuloendothelial system.

monosilane. silicon tetrahydride.

monosodium glutamate (MSG). the sodium salt of glutamic acid, $C_6H_8NNaO_4 \cdot H_2O$, it is a popular, white, crystalline powder with a meat-like flavor that is used to enhance or intensify the flavor of foods. Some persons react to MSG in what has come to be called the Chinese restaurant syndrome. Although symptoms vary among those affected, many experience paresthesia, weakness, heart palpitations, a cold sweat, and headache. Many sensitive individuals experience a burning sensation in the back of the neck and forearms, a tightness in the chest, and headaches about 30 minutes following a Chinese meal. Data indicate that MSG can cause brain damage, obesity, retarded skeletal development, and female sterility in humans. It may also be mutagenic. MSG should not be used by those known to be affected by the disease, by pregnant women, and people whose sodium intake is restricted. MSG also destroys nerve cells in the hypothalamus of certain animals.

monosodium methanearsonate (MSMA). the monosodium salt of methylarsonic acid, CH_4AsNaO_3. MSMA, a herbicide, is somewhat less toxic to laboratory rats than the disodium salt, disodium monomethanearsonate (also a herbicide).

Monotaxis grandoculis (snapper). a marine bony fish (Class Osteichthyes) of Polynesia, west to east Africa. See ciguatera.

Monotremata. the sole order of mammals in the subclass Prototheria. Living representatives of the order are characterized by the presence of a cloaca or common chamber that receives urinary and reproductive products as well as egesta. The external ear is absent; adults have a horny beak, only the young have teeth; a uterus and vagina are lacking; the penis transports sperm only; the testes remain abdominal and there is no scrotum; the mammary glands lack nipples; and the females lay fertilized eggs. All three genera are confined to the Australian region: *Ornithorhynchus* (duck-billed platypus), *Tachyglossus* (echidna; spiny anteater), and *Zaglossus* (echidna; long-billed spiny anteater). See especially *O. anatinus*. It is an aquatic mammal with webbed feet and a bill that contains horny pads that are used to crush invertebrates. *Tachyglossus* and *zaglossus* are nocturnal, terrestrial insectivores with sharp, quill-like hairs and spines that protect the body from predators, a long toothless snout, and a tongue used to catch ants, worms, termites, and other insects. The limbs and claws are powerful and are used to break into insect nests.

monotreme. any mammal of the order Monotremata.

monozygotic twins. See identical twins.

montane viper. *Vipera hindii*.

Monticelli's pit viper. *Bothrops monticelli*.

monuron (*N'*-(4-chlorophenyl)-*N*,N-dimethylurea; 1,1-dimethyl-3-(*p*-chlorophenyl)urea; 3-(*p*-chlorophenyl)-1,1-dimethylurea; CMU). an environmentally persistent substituted urea used as a somewhat selective preemergence herbicide and soil sterilant. It is used mainly to control broadleaved weeds and grasses in noncropland areas and in crops such as cotton, sugar cane, pineapple and asparagus. It is moderately toxic to laboratory rats. If hydrolyzed to the anilide, it can cause anemia and methemoglobinemia.

moonseed. *Menispermum canadense*

moray, moray eel. any fish of the family Muraenidae (e.g., *Muraena* spp., *Gymnothorax* spp.). All are considered venomous, but not deadly. The blood itself is toxic and may be injected into prey when the moray bites. Certain species cause gymnothorax poisoning, *q.v.*; other species that may be considered safe should not be eaten since they may be at times violently toxic. Together with *Sphyraena barracuda*, moray eels are the most common cause of ciguatera in the Carribean archipelago. See also ciguatera.

morbid. **1:** diseased; of, or pertaining to, disease. **2a:** unhealthy or unwholesome; gruesome. **2b:** preoccupied with unwholesome ideas and conditions.

morbidity. **1:** the state of being diseased or morbid. **2:** the illness rate; the number of sick individuals or the ratio of sick to well individuals in a defined population or community during a given period of time.

morbidity rate. the number of cases of a specific illness or disease in a defined population or community during a specified period of time (usually a month or a year) per unit (usually 1000, 10,000, or 100,000) of population.

morbific. pathogenic; pertaining to or causing disease or morbidity.

morbigenous. pathogenic.

morbus. disease.

Morfamquat. a highly selective bipyridylium quaternary salt that has been used as a contact postemergence herbicide to control a variety of annual weeds in cereals; grasses are highly resistant. *Cf.* diquat, paraquat.

Morgan's cobra. *Walterinnesia aegyptia*.

morgue. **1:** mortuary; a public mortuary; a facility used to store dead human bodies for identification and preparation for burial or other legal disposition. **2:** a facility, usually a unit of a hospital, used to store dead human bodies and to conduct autopsies.

moribund. **1:** dying; in a dying condition or state; in a condition of impending death, often characterized by a deepening stupor or coma. **2:** dormant. **3:** a dying person or animal.

morning glory. See *Ipomoea*.

morph. See polymorphism.

morphactin. any of a class of derivatives of fluorene-9-carboxylic acids that powerfully regulate plant growth and development. They inhibit shoot elongation and cause morphological abnormalities in intact plants. This is accomplished by blocking endogenous auxin transport. Morphactins also promote the cell divisions that give rise to lateral primordia in the root pericycle. The organization of such primordia is, however, greatly disturbed and further growth is inhibited.

morphine. the principal alkaloid of opium, it occurs as a bitter, colorless crystalline powder, $C_{17}H_{19}NO_3 \cdot H_2O$, that is slightly soluble in water, ethanol, and ether. Obtained from opium, morphine is the prototype of the opioid analgesics. Morphine (in the form of soluble salts such as the acetate, hydrochloride, or tartrate) is analgetic at a dose (about 10 mg *i.m.*) that does not result in severe alterations in consciousness. Adverse effects of morphine are dose related and include CNS depression, with decreased respiratory responsiveness to CO_2, diminished cough reflex, nausea, and vomiting. Morphine also produces miosis and stimulates release of ADH from the hypothalamus. Peripheral smooth muscle effects reduce the propulsive movements in the GI system, causing constipation. Venules dilate following morphine administration, and hypotension may be seen in hypovolemic individuals or those who suddenly assume an upright position. Depression of central vasomotor centers and the release of histamine may be at least partly responsible for the hypotension. The hypotensive effect of opioids is intensified by concomitant administration of phenothiazines. See also *Papaver spp*.

morphine poisoning. poisoning by excessive dosage of morphine. A large dose can cause immediate CNS depression. Smaller doses may induce a brief, initial mental exhilaration or languor succeeded by weariness, drowsiness, miosis. The pulse is rapid and forceful, becoming slow and feeble; breathing is slow and shallow; the victim becomes unconscious and can be aroused with difficulty. Muscles relax, reflexes diminish, and body temperature drops. The skin is cold, pale, and moist. Victims may also suffer loss of pain perception, nausea, vomiting, hypotension, and pulmonary congestion. At a sufficiently high dosage (depending on tolerance), coma with relaxation of muscles, circulatory insufficiency, cyanosis, and death from respiratory arrest (due to CNS depression) may ensue. *Cf.* morphinism.

morphinic. pertaining to morphine.

morphinism. **1:** a morbid state due to habitual or excessive use of morphine. *Cf.* morphine poisoning, morphinomania. **2:** morphinomania (def. 1).

morphinistic. of, pertaining to, characteristic of morphine; having the characteristics of morphine.

morphinium. morphine.

morphinization. to expose to the influence of morphine.

morphinomania (morphiomania). **1:** morphine addiction. **2:** psychosis due to the misuse of morphine. *Cf.* morphinism, morphine poisoning.

morphiomania. morphinomania.

morphogen. any substance that promotes morphogenesis or embryonic development generally.

mortality. **1:** the state or quality of being mortal (subject to death). **2:** a heavy loss of life; death in large numbers. **3:** the total number of deaths in a given population or community at a given time. **4:** death rate; **4a:** the proportion of deaths to population. **4b:** the ratio of deaths that occur to those expected. **concomitant mortality**. death during exposure to a hazardous or toxic substance or stimulus. *Cf.* consequential mortality, delayed mortality, concomitant immunity (See immunity). **consequential mortality**. death that is secondary to the action of a given agent (e.g., a toxicant). **delayed mortality**. death that occurs some time following trauma or removal of a toxic or otherwise deadly stimulus. **perinatal mortality**. mortality that occurs at about the time of birth, or during the period when life *ex utero* is considered possible; sometimes taken to include stillbirths and deaths during the first week of life. *Cf.* natimortality.

mortality ratio. See standardized mortality ratio.

mortuary. See morgue.

Morus (mulberry tree). a genus of hardy, long-lived Asiatic trees with milky sap (Family Moraceae). They are grown in the United States for their berry-like fruit. The unripe fruit and shoots are toxic. The principle species are *M. alba* (white mulberry) which is native to China and *M. nigra* (black mulberry) which is native to Asia.

moschatin. See eledoisin.

moth. See Lepidoptera.

moth flakes. See naphthalene.

mothball. a marble-sized ball, formerly made from camphor, but now made from naphthalene (usually with paradichlorbenzene). It is a compressed, usually white sphere. Both naphthalene and paradichlorobenzene are moderately toxic in the case of acute poisoning by ingestion or inhalation of the fumes. Mothballs apparently look much like candy and are very attractive to small children. A single mothball swallowed by a toddler could produce seizures in less than an hour. Chronic toxicity is poorly known, but the risk of serious effects may be substantial in some cases (See naphthalene, paradichlorobenzene). Fumes from mothballs in a bedroom or linen closet can permeate a home with the risk of nearly constant long-term exposure in a room or a home that is poorly ventilated. The vapors from mothballs are also absorbed by clothing and blankets providing another source of exposure. In the United States, mothball packages carry a warning label: "CAUTION: May be harmful if swallowed. Avoid prolonged breathing of vapor or repeated contact with skin. Keep out of reach of children."

mothball poisoning. naphthalene poisoning.

mother-in-law plant. *Dieffenbachia seguine*.

motor. **1:** of, or pertaining to, movement or to structures (e.g., nerves, muscles, CNS centers) that cause or affect movement. **2:** producing, supporting, or influencing movement.

Mount Lamington snake. *Apistocalamus lamingtoni*.

Mount Stanley snake. *Toxicocalamus stanleyanus*.

mountain adder. *Bitis atropos*, *Vipera lebetina*.

mountain ivy. *Kalmia latifolia*.

mountain laurel. *Kalmia latifolia*.

m.p. mp.

mp (m.p.). melting point.

MPC. maximum permissible concentration (See maximum allowable concentration).

MPS. mononuclear phagocyte system (See reticuloendothelial system).

MPTP (1,2,3,6-tetrahydro-1-methyl-4-phenylpyridine; 1-methyl-4-phenyl-1,2,3,6-tetrahydropyridine). a derivative of piperidine, it ($C_{12}H_{15}N$) causes irreversible parkonsonian-like symptoms in humans and laboratory monkeys, apparently by selective destruction of dopaminergic neurons.

MSMA. monosodium methanearsonate.

MTD. **1:** maximum tolerated dose (See dose). **2:** toluene-2,4-diamine.

mucochloric acid (2,3-dichloro-4-oxo-2-butenoic acid; dichloromalealdehydic acid; α,β-dichloro-β-formylacrylic acid; 2,3-dichloromaleic aldehyde acid). a compound, $C_4H_2Cl_2O_3$, the crystals of which are only slightly soluble in cold water, but soluble in hot water, hot benzene, and ethanol. It is a strong irritant to the skin and eyes and a powerful skin sensitizer.

mud clam. *Mya arenaria*.

mule killer. common name of any of a number of arthropods in the southern United States (e.g., whip scorpions, mantids, walking sticks) alleged to kill livestock by stinging, biting, or when swallowed.

multidimensional dose-response-duration model. See dose-response model.

multistage dose-response model. See dose-response model.

Muller's snake. *Rhinoplocephalus bicolor*.

murex. a common name for marine snails of the genus *Murex*.

Murex (murex; rock shells). a large genus of carnivorous marine snails (Class Gastropoda) with heavy shells that bear 3 or more rows of spines. The aperture is round, terminating in a long canal. Bivalves are common prey. The hypobranchial gland produces a poisonous secretion comprised of two pharmacologically active substances, murexine and serotonin. Serotonin from this source was formerly called enteramine.

murexine (2-[[3-(1*H*-imidazol-4-yl)-1-oxo-2-propenyl]oxy]-*N,N,N*-trimethylethanaminium; β-(4-imidizolyl)acrylcholine; urocanylcholine). a neurotoxin, $[C_{11}H_{18}N_3O_2]^+$, isolated (often in large quantities) from the median zone of the hypobranchial gland of snails of the genus *Murex trunculus*, *Purpura*, *Aplysia*, and related forms. The neurotoxin isolated from *Purpura* is called purpurine, which is identical or nearly identical to murexine. See especially *Aplysia californica*. **2:** purpurin.

muscarine (tetrahydro-4-hydroxy-*N,N,N*,5-tetramethyl-2-furanmethanaminium; hydroxycholine). an extremely toxic alkaloid from the red variety of *Amanita muscaria*, from *A. pantherina* and various species of *Inocybe* and *Clitocybe*. High concentrations occur in some species of the latter two genera. Ingestion of these mushrooms is the source of muscarinism, the first signs of which usually appear soon after ingestion. It is an extremely toxic parasympathetic poison, but does not account for the major effects of *A. muscaria* poisoning.

muscarine poisoning. muscarinism.

muscarinic. **1a:** of, pertaining to, resembling or characteristic of muscarine. **1b:** pertaining to or denoting the effects of muscarine. **1c:** pertaining to stimulation of, or the ability to stimulate, postganglionic parasympathetic receptors. See receptor (muscarinic cholinergic receptor). **1d:** pertaining to ACh receptors that are blocked by muscarine and similar drugs (as opposed to the nicotinic ACh receptors).

muscarinic cholinergic receptor. See receptor.

muscarinism (muscarine poisoning). poisoning by muscarine-containing mushrooms. Effects may include lacrymation, miosis, nausea, vomiting, abdominal colic, evacuation of bowels and urinary bladder, diaphoresis, sialorrhea, bronchoconstriction, respiratory depression, bradycardia, dilation of arterioles, and death by cardiovascular collapse. Muscarine is given orally in medicine as a parasympathomimetic agent. Effects of muscarine are countered by atropine but its use is generally contraindicated in muscarinism. See also muscarine.

muscazone (α-amino-2,3-dihydro-2-oxo-5-oxazoleacetic acid; α-amino-2-oxo-4-oxazoline-5-acetic acid). an isoxazole isolated from *Amanita muscaria* that causes visual damage, mental confusion, spatiotemporal dislocation, and memory loss in humans. See isoxazoles, muscimol, tricholomic acid.

muscimol (5-aminomethyl)-3(2H)-isoxazolone; 5-(aminomethyl)-3-isoxazolol; 5-aminomethyl-3-hydroxyisozazole; 3-hydroxy-5-aminomethyl-isoxazole; agarin; pantherine). an isoxazole isolated from the poisonous mushroom, *Amanita muscaria*. It is an extremely toxic CNS depressant and GABA agonist that causes visual damage, mental confusion, spatiotemporal dislocation, and memory loss in humans. See isoxazoles, muscazone, tricholomic acid.

mushroom. **1:** any of a large variety of fungi (mostly of the subdivision Basidiomycetes) that usually have a fleshy, usually aerial, umbrella-like, fruiting body (actually a sporophore); it arises from an underground mycelium. The mushroom is typically differentiated into a *stipe* (the stalk) and *pileus* (the cap). The spores develop in the folds (gills) or pores on the undersurface of the pileus. In some species, the edge of the cap is united with the stalk by a membrane, the *partial veil*. This ruptures in the mature mushroom, exposing the underside of the cap and leaving a ring (the annulus) about the top of the stalk. Many are edible, others are unpalatable or toxic, and some are deadly poisonous. In some cases, certain populations of a species may be edible while other populations are poisonous. **2:** any macroscopic fungal fruiting body. *Cf. toadstool.* See also mycetismus.

mussel poisoning. **1a:** paralytic shellfish poisoning (See shellfish poisoning). **1b:** this term is most frequently applied to paralytic shellfish poisoning along the Pacific coast of the United States that results from eating mussels or clams. It usually occurs from June to October. See, for example, *Modiolus modiolus, Mytilus edulis, M. californianus*.

mutafacient. mutagenic.

mutagen. an agent (chemical or physical) that induces mutagenesis; one that brings about an increase in the natural rate of mutation within a population (e.g., radioactive radiation; short-wave electromagnetic radiations such as ultraviolet light, X-rays, and cosmic rays; chemicals such as ethylmethanesulfonate, nitrogen mustard, nitrous acid, proflavin and base analogues). Mutagens are often also carcinogenic or teratogenic; they are of major toxicological concern.

mutagenesis. **1a:** the origin, development, or production of a mutation in heritable material. **1b:** the induction of heritable changes in genetic material within living cells. Whereas mutagenesis is a normal process that plays an important role in adaptive evolution of plants and animals, most mutations are harmful. Consequently xenobiotic substances (and ionizing radiation) that induce mutagenesis (mutagens) are of great toxicological importance. **1c:** any process that causes mutation.

mutagenic (mutafacient). able to induce mutations or to increase the rate of mutations in a population. See mutagen.

mutagenic carcinogen. See carcinogen.

mutagenicity. the capacity to induce mutations.

mutagenize. to treat with a mutagen.

mutant. **1a:** any organism, gene, or character that has undergone a mutational change. **1b:** of, pertaining to, or designating a gene that has undergone a mutation. **1c:** an organism or cell carrying one or more mutant genes. Such an organism or cell may differ significantly from its parent or immediate precursor cell in phenotype, including susceptibility and response to toxicants, generally or specifically. **1d:** the feature or phenotype produced as the result of gene mutation. **2:** any gene that is rare in a population, especially if harmful or produces a phenotype that differs from the wild-type.

mutate. **1:** to undergo mutation. **2:** to effect a

change in the genetic constitution of a cell by altering its DNA.

mutation. **1:** any sudden or discontinuous, potentially heritable, change in the genetic constitution of a cell or organism. Most mutations are recessive and deleterious to the affected organism. If a mutation is advantageous, it may eventually becomedominant through natural selection. Some mutations occur naturally, or stimulated by unknown mechanisms, but can be caused (or the rate of mutation accelerated) by ionizing radiation and many xenobiotic substances (See, for example, acetonylacetone, 5-bromouracil, colchicine, malathion, podophyllotoxin, ionizing radiation, mutagen). often having detrimental effects on the progeny. **2:** a point mutation. **3:** a change in the character of a gene that is perpetuated in subsequent divisions of the cell in which it occurs. **4:** a change in the sequence of base pairs in the DNA molecule by substitution, addition or deletion of nucleotides. **5:** the loss of entire chromosomes. See also base pair, genotoxic, genotoxic, chromosome deletion, chromosome insertion, genotoxic carcinogen, chromosome inversion, chromosome translocation, hypermutation, ionizing radiation, revertant, back mutation. **6:** a mutant; a cell or individual that has undergone a mutational change; mutant. **chromosomal mutation**. any change in chromosome number or structure. Such changes are usually visible on microscopic examinatin. See chromosomal aberration. **gene mutation**. See point mutation. **germinal mutation**. any mutation that takes place in germinal cells (eggs or sperm) and are potentially heritable. *Cf.* somatic mutation. **induced mutation**. a mutation that does not occur naturally, but is induced as by certain chemicals, types of radiation, etc. See mutagen. **lethal mutation**. a mutation that is so disruptive to life processes that the affected cell or organism dies. **natural mutation**. spontaneous mutation. **point mutation**. gene mutation; probably the most common type of mutation, it is a change in the nucleotide sequence of a single gene, or more accurately a change in one or more base pairs in DNA. The main effect on the mutant organism results from the incorporation of one or more incorrect amino acids being substituted in the protein being made. The resulting protein differs from the normal protein and is either ineffective in carrying out its metabolic role or has unusual and usually deleterious effects. Such mutations occasionally give rise to wild-type revertants (reverse mutation). Such changes may occur by any of several means. **reverse mutation**. a mutation that changes a mutant gene, causing it to revert to the original wild-type gene. **silent mutation**. any alteration in DNA that does not cause a detectable phenotypic change. **somatic mutation**. any mutation that occurs in a body cell of a multicellular organism. Such mutations and are only transmitted to the immediate cell line (by mitosis) within the affected organism. They are not otherwise inherited. Certain somatic mutations are, nevertheless, of toxicologic interest, as are a probable first step in cellular transformation leading to cancer. *Cf.* mutations **spontaneous mutation** (natural mutation). a mutation that occurs naturally due to errors in normal cell processes.

MW. molecular weight.

Mya arenaria (soft-shelled clam; common soft-shell clam; long clam; long-neck clam; sand clam; mud clam; steamer clam). a species of normally edible clam (Class Bivalvia, Phylum Mollusca) that ranges from Britain, Scandinavia, Greenland, southward along the Atlantic coast of North America to the Carolinas; and on the Pacific coast of North America (where it was introduced many years ago), northward from California to Alaska, and west to Japan. The shell is approximately 15 cm long and is a dull, chalky, white with a brown periostracum. The siphons form a very long tubular projection. It is often a transvector of paralytic shellfish poisoning to humans.

mycetismus (mycetism; mushroom poisoning; mushroom mycotoxicosis). poisoning due to ingestion of a fungus, especially a poisonous mushroom. In North America alone more than 70 species of mushrooms are known to be poisonous to humans. Of these, *Amanita phalloides* causes most of the confirmed fatalities. Diagnosis is often difficult due in part to (1) the time between exposure and presentation of symptoms; (2) the diversity of signs and symptoms that may appear; (3) failure to secure reliable identification of species eaten; or (4) inability to identify the toxin present in blood or excreta of the victim. While intermediate or mixed syndromes are sometimes seen, three major syndromes may be distinguished clinically (mycetismus cerebralis, mycetismus choleriformis, and mycetismus nervosa. **mycetismus cerebralis**. a hallucinosis due to the ingestion of any of several species or types of mushrooms. **mycetismus gastrointestinalis**. a mild mycetismus characterized by gastrointestinal distress. It is caused by ingestion of *Omphalotus olearius* (orange Jack o' lantern) and a variety of other mushrooms. The symptoms are those of gastrointestinal distress that may include nausea, vomiting, and diarrhea. **mycetismus nervosus.** neurotoxic poisoning due to ingestion of *Amanita pantherina* and *A. muscaria*. The chief toxicant, muscarine, is a parasympathetic stimulant. Symptoms include salivation, sweating, tearing, persistent peristalsis, vomiting and retching, miosis, intense excitement, delirium, and coma. **mycetismus sanguinarius**. the syndrome caused by ingestion of *Helvella* spp. Symptoms include severe abdominal distress, hemoglobinuria, and jaundice.

mycetogenic. mycetogenetic.

mycoderma. mucosa.

mycodermatitis. any inflammation or disease of the skin caused by a fungus.

mycogenetic (mycogenic; mycetogenic). caused by fungi.

mycogenic. mycogenetic.

mycoin C_3. See patulin.

mycophenolic acid. a crystalline fungistatic and bacteriostatic antibiotic, $C_{17}H_{20}O_6$, from *Penicillium brevi-compactum* and related species. It is moderately toxic to laboratory mice.

mycorrhiza. a symbiotic association between a fungus and the roots of a plant.

mycosis (plural, mycoses). any disease caused by a fungus.

mycostasis. suppression of fungal growth.

mycostat. fungistat.

mycostatic. fungistatic.

mycotic. relating to or caused by a fungus, relating to a mycosis.

mycotoxic lupinosis. lupinosis.

mycotoxicosis. **1:** poisoning by fungal or bacterial toxins, usually in contaminated food or animal feeds. **2:** poisoning due to ingestion of fungi (e.g., *Claviceps* spp., poisonous mushrooms. **mushroom mycotoxicosis**. mycetismus.

mycotoxin. **1:** any toxic chemical that is produced naturally by fungi. Numerous mycotoxins have been identified. They are active at very low concentrations, sometimes making detection difficult, and mechanisms of action are remarkably varied. See, for example, ochratoxin, patulin, penicillic acid, *Penicillium*, toxin (e.g., T-2 toxin). **2:** sometimes restricted to any poisonous substance produced by fungi that infest food or animal feed. **anthraquinoid mycotoxin**. toxic derivatives of anthraquinone (*q.v.*) that occur in fungi. See, for example, luteoskyrin, rubroskyrin. **bisfuranoid mycotoxin**. any of a class of mycotoxins that contain the dihydrofurobenzofuran system. All metabolites of these mycotoxins characteristically include either the unsaturated 7,8-dihydrofurano(2,3-b)furan or the more reduced 2,3,-7,8-tetrahydrofuro(2,3-b)furan. They are hepatotoxic and tumorigenic to various animals. Examples are sterigmatocystin and aflatoxins.

mycotoxin T-2. T-2 toxin.

mycotoxinization. inoculation with a fungal toxin.

Mycteroperca venenosa (sea bass). a marine bony fish (Family Serranidae) of the western tropical region of the Atlantic Ocean. See ciguatera.

mydaleine. a poisonous ptomaine in putrefying viscera. Symptoms of intoxication include salivation, miosis, intermittent fever, and arrest of the heart in diastole.

mydatoxin (mydatoxine). a deadly ptomaine, $C_6H_{13}NO_2$, from putrefying viscera and flesh or from human intestines maintained at low temperatures for a long time.

mydatoxine. mydatoxin.

mydriasis (pupillary dilation; dilation of pupils). prolonged, excessive, or abnormal dilation of the pupil of the vertebrate eye. Causes are varied and include fear, hysteria, sudden changes in emotional state, coma, anesthesia, drugs and certain other xenobiotic substances, botulism, and irritation of the cervical sympathetic nerve.

mydriatic. **1:** causing mydriasis (pupillary dilation). **2:** a drug that causes mydriasis (pupillary dilation).

mydriatic effect. dilation of the pupil of the vertebrate eye by certain drugs such as the tropane alkaloids.

mydriatic rigidity. tonic pupil of the eye.

myel-, myelo-. a prefix referring to or indicating **a:** bone marrow, **b:** spinal cord.

myelinotoxic. demyelinating; able to damage myelin with consequent effects on the CNS.

myelinotoxicity. the quality of being myelinotoxic.

myelo-opticoneuropathy, subacute. See iodochlorhydroxyquin.

myelopathy. any disease, disturbance, pathological change in the spinal cord. *Cf.* neuropathy, encephalopathy.

myelosuppressive. **1:** able to suppress or inhibit bone marrow activity with a consequent decrease in the production of blood cells and platelets. **2:** an agent that is able to suppress bone marrow activity.

myelotoxic. **1:** poisonous to bone marrow or any of its components. **2:** of, pertaining to, or arising from diseased bone marrow.

[illegible]

myelotoxin. a cytotoxin that destroys bone marrow cells.

Mygale. a genus of very large tropical spiders that reputedly attack and kill small birds.

mylabris (Chinese cantharides; Chinese blistering flies). dried insects, *Mylabris sidae*, of China and East India. It is a potent source of cantharidin. See also cantharides, cantharidin poisoning.

myliobatid. common name of eagle rays or bat rays, *Myliobatis*, *Aetobatis*, *Rhinopter* (Family Myliobatidae). They are venomous, often with a large, well-developed sting that lies close to the base of the tail. See stingray.

Myliobatidae (eagle rays; bat rays). a family of venomous cartilaginous fish, the eagle rays or bat rays. See also stingray.

myocarditis. inflammation of the myocardium. **toxic myocarditis**. that due to poisoning from any of numerous sources such as drugs; toxins elaborated by any of numerous infectious organisms, as in diphtheria; carbon monoxide. In this condition in humans the apex beat is extremely weak and rapid. *Cf.* myolysis.

myoclonia. a condition of intermittent clonic spasms or twitching of a muscle or group of muscles.

myoclonus. a twitching or clonic spasm of a muscle or muscles.

myoclonus multiplex (pramyoclonus multiplex). a condition marked by continuous muscular spasms in unrelated muscles.

myoglobinuria. the presence of myoglobin in the urine due, for example, to muscle phosphorylase deficiency, following protract vigorous exercise in susceptible individuals, or in crushing injuries.

myohematin. cytochromes c.

myolysis. disintegration or fatty degeneration of muscle tissue. **cardiotoxic myolysis** (myolysis cardiotoxica). myolysis due to systemic poisoning, often due to infectious diseases.

myolysis cardiotoxica. cardiotoxic myolysis (See myolysis).

myoparesis. muscle weakness.

myopathia. myopathy.

myopathic. **1:** of or pertaining to a myopathy. **2:** a person suffering from myopathy.

myopathy. any disease or abnormal condition of striated muscle. **alcoholic myopathy**. myopathy that affects alcoholics. It is usually marked by acute myoglobulinuria, sometimes accompanied by proximal limb weakness. **cortisone myopathy**. myopathy especially of the limbs resulting from repeated high dosages of corticosteroid preparations. The subject generally recovers if the dosage is lowered or administration of the drug discontinued. **thyrotoxic myopathy**. a chronic effect of hyperthyroidism marked by progressive weakness and muscle atrophy.

myosis. miosis.

myospasm. a muscle spasm.

myotic. miotic (def 2).

myriapod. common name of any member of the arthropod classes Chilopoda, Diplopoda, Symphyla, or Pauropoda.

Myrica cerifera (bayberry). a fragrant, perennial, evergreen shrub (Family Myricaceae) suspected of being carcinogenic.

Myripristis murdjan (squirrelfish). a marine bony fish (Family Holocentridae) of the Indo-Pacific region. See ciguatera.

Myristica fragrans (nutmeg). a tropical tree (Family Myristicaceae) with entire leaves and flowers that produce a fleshy fruit. The seeds are toxic and narcotic in effect. The chief active principle is myristicin. See also nutmeg; oil of nutmeg.

Myristicaceae (nutmeg family). a family of trees with unisexual flowers and monadelphous stamens; the seeds are arillate.

myristicin. a toxic, crystalline phenolic ether, $C_{11}H_{12}O_3$, with a strong odor. It is the chief toxic substance in the ripe seeds of this plant. Ingestion of large amounts of the spice (5-15 g) causes symptoms similar to those of atropine poisoning. It reduces the levels of cytochrome P-450 and thus inhibits monooxygenations catalyzed by this enzyme. Myristicin occurs also in a number of essential oils.

myrmicacin. a fungicide secreted by the leaf-cutting, fungus-growing ant, *Atta sexdens*.

myronate potassium. sinigrin.

Mytelase. a parasympathomimetic agent given orally.

mytilotoxin. a former name for paralytic shellfish poison.

mytilotoxism. poisoning due to ingestion of mussels that contain paralytic shellfish poison, formerly termed mytilotoxin.

Mytilus. a large genus of marine mussels (Class Pelecypoda also known as Bivalvia). They are wedge-shaped bivalve mollusks that occur, often abundantly, on various substrates near the reach of low tide. Some species are edible. ***M. californianus*** (common mussel). a species found from Alaska and the Aleutian Islands to Socorro Island. It is often a transvector of paralytic shellfish poison to humans. ***M. edulis*** (bay mussel). a species found from the Arctic Ocean to South Carolina and from Alaska to Baja, California; it is nearly cosmopolitan in temperate waters. The bay mussel is a frequent transvector of paralytic shellfish poison to humans.

mywe howk. *Naja naja*.

N. the symbol for nitrogen.

n. nano-.

Na^+-K^+ ATPase. a plasma membrane protein with ATPase activity. See sodium-potassium pump.

Na. symbol for sodium.

NAA. 1-naphthaleneacetic acid.

nabam (1,2-ethanediylbascarbamodithioic acid disodium salt; ethylenebis[dithiocarbamic acid] disodium salt; disodium ethylenebis[dithiocarbamate]). the sodium salt of dimethyldithiocarbamate, $NaSSCNHCH_2CH_2NHCSSNA$. When pure, it is a colorless, crystalline, highly water-soluble substance. Nabam is a very toxic agricultural fungicide. It is irritant to the skin and mucous membranes and has narcotic effects at high concentrations. Its toxicity to animals is low. See dithiocarbamate fungicides. Exposure together with ethanol can cause violent vomiting. See also dithiocarbamate fungicide.

nachiku. *Naja nigricollis*.

nAChR nicotinic acetylcholine receptor. See nicotinic cholinergic receptor.

NAD (nicotinamide adenine dinucleotide). a coenzyme that acts as a hydrogen carrier in oxidation-reduction reactions as in the electron transport chain in aerobic respiration. NAD and NADP are derivatives of nicotinic acid. They are reduced to NADH and NADPH, respectively, when hydrogen atoms are transferred to them from the substrates in a reaction catalyzed by a substrate-specific dehydrogenase.

NAD^+. the oxidized form of NAD.

NAD-nucleotidase. an enzyme known from at least 9 species of venomous snakes. It catalyzes the hydrolysis of nicotinamide N-ribosidic linkage of NAD to produce nicotinamide and adenosine diphosphate riboside. It is heat labile with an optimal pH of 6.5-8.5. Its toxicity in snake venom is uncertain.

NADH. the reduced form of NAD.

nadide (adenosine 5′-(trihydrogen diphosphate) 5′→5′-ester with 3-(aminocarbonyl)-1-β-D-ribofuranosylpyridinium, hydroxide, inner salt; 3-carbamoyl-1-β-D-ribofuranosylpyridinium hydroxide 5′-ester with adenosine 5′-pyrophosphate inner salt; diphosphopyridine nucleotide; nicotinamide-adenine dinucleotide (NAD); codehydrogenase; coenzyme I; Co I; cozymase; DPN; adenine-D-ribosephosphate-phosphate-D-ribose-nicotinamide; ARPPRN). a hygroscopic white powder, $C_{21}H_{27}N_7O_{14}P_2$, nadide is a component of fresh baker's yeast. It is the naturally occurring form of nicotinamide-adenine dinucleotide. It is the coenzyme of apozymase, which is the essential catalyst of the alcoholic fermentation of glucose and is used therapeutically as a narcotic and alcohol antagonist. See also NAD.

NADP (nicotinamide adenine dinucleotide phosphate). a coenzyme with an action similar to that of NAD, *q.v.*

NADPH. the reduced form of NADP.

NAEL. no-adverse-effect level.

Naematoloma fasciculare (sulfur tuft). a bitter-tasting poisonous mushroom that occurs throughout much of North America on stumps and logs or on soil over buried wood. Whereas some individuals in North America have suffered gastric distress for a few hours with complete recovery, fatalities have been reported in Asia and Europe.

nag. *Naja naja*.

naga pambu. *Naja naja*.

nahuyaca. *Bothrops nasutus*, *B. schlegelii*.

Naja (cobras). a genus of highly venomous snakes (Family Elapidae), all of which, except *Naja naja*, are indigenous to Africa. The head is depressed, with a distinct canthus and rounded snout. It is rather broad, but barely distinct from the neck. The body is slightly depressed, tapered, and moderately slender with a tail of moderate length. The neck can spread, forming a hood. The eyes are of moderate in size with round pupils. The crown has 9 scales; the rostral is rounded and the frontal is short. There is no loreal scale, which is distinctive of cobras and cobra-like elapids. Each nasal abuts one or two preoculars. The dorsal body scales are smooth and are arranged in 17-25 oblique rows at midbody, fewer posteriorly. There are 159-232 ventral scales; 42-88 subcaudal scales, most of which are paired; the anal plate is entire. There are two rather large tubular maxillary fangs with external grooves followed by a gap and 0-3 small teeth. When agitated cobras usually spread the hood and rear up with about ⅓ of the body off the ground. On occasion cobras will bite without rearing up and sometimes do not spread the hood. Many nonvenomous snakes also spread the neck and forebody. Signs and symptoms of cobra envenomation may include pain, localized edema, numbness of the affected area, weakness, drowsiness, slurred speech, difficulty in swallowing, ptosis, blurred vision, headache, changes in breathing, excessive salivation, nausea, vomiting, abdominal pain, pain in regional lymph nodes, weak pulse, hypotension, shock, localized vesicles, localized necrosis, paresis or paralysis, muscle fasciculations, convulsions. See also *Ophiophagus hannah*. ***N. haje*** (Egyptian cobra; rock cobra; banded cobra; black cobra; brown cobra; asigirikolongo; balor; brillenschlange; Brillenschlange; camamala; de Capello cobra; hontipeh; iloyi; keautia; kiparaa nunga; mpili; mpiri; MWe bwe; mywe howk; nachiku; nag; nalla pambu; nchuweira; oraj sinduk; sakamala; tadioko; tedong naga; ular bedul; ular biludak; ular tedong sendok; [illegible]; Urauschlange). a North African species that occurs in a wide variety of habitats including flatland with bunch grass or scrubby bushes, irrigated fields, rocky hillsides, old ruins, and near villages. It is a rather retiring snake, often making little effort to defend itself. The venom is about as toxic as that of *N. naja*. The Egyptian cobra is distinguished from other African cobras by the presence of small subocular scales that separate the eye from the upper labials. The body is cobra-like but moderately stout. The scales are smooth with a dull luster; scale rows are strongly oblique; the anal plate is entire; subcaudals are paired. The ground color is extremely varied, ranging from brownish yellow to very dark brown. The head and neck are usually somewhat darker than the rest of the body. There are dark cervical crossbars. The young are yellowish with a black head and neck and the body is crossed by wide dark bands. ***N. melanoleuca*** (forest cobra; black cobra; black and white-lipped cobra; black-lipped cobra; fumbe; liteo; muyirima; pshissapa; wakabi). a large, dark, terrestrial cobra that occurs in tropical rain forest, subtropical forested areas, and throughout most of west and central Africa, southward to Angola and Natal. It is most easily recognized by the highly polished dorsal scales and the creamy white labial scutes which are edged in black. It is sometimes mistaken for *Dendroaspis polylepis*, which does not occur within the rain forest region. The hood is long and wedge-shaped as in *N. nigricollis* and is often mistaken for the dark phase of the latter species. The forest cobra is rarely aggressive and seldom bites humans. ***N. naja*** (Asian cobra; Indian cobra; common cobra; spectacled cobra; bespectacled snake; Central Asiatic cobra; Borneo cobra; de Capello cobra; monocellate cobra; monocled cobra; oxus cobra; Pakistanian cobra; aguason; alupong). all central and south Asian members of this genus are considered subspecies of *N. naja*. While color patterns vary among the races, this species is characterized by the presence of conspicuous dark bars or spots on the ventral side of the neck near the level of the hood. The venom is produced in large volume and is highly lethal. These snakes prey chiefly on small terrestrial vertebrates, perhaps especially rats. They are thus common in areas densely populated by humans, sometimes entering homes (usually at twilight). Adults may reach a length of nearly 2 m although the average length of adults is much less. It kills large numbers of humans annually (probably tens of thousands). ***N. n. atra*** (Chinese cobra). a race indigenous to Thailand and south China east to North Viet Nam, Hainan, and Taiwan. Adults reach a maximum length of about 1.7 m; they are grayish brown, olive, or nearly black with widely spaced, narrow, light bands.

Marks on the hood are usually similar to those of the *N. n. kaouthia*. The belly is pale, sometimes mottled in brown. The young are black with well-defined whitish crossbands. ***N. n. kaouthia*** (monocellate cobra). a brown or black cobra found chiefly in the lowlands of East Pakistan, the Indian state of West Bengal, Assam and the Union of Myanmar (formerly Burma), Thailand, Malaya and southeast China. It often has alternate wide and narrow transverse dark bands, and is typically speckled or variegated with white or pale yellow. Adults are paler and have less intense crossbands than do the young. The hood is marked dorsally by a pale circle edged with black that encloses 1-3 dark spots; ventrally it is marked by a pair of dark spots or a wide dark band. ***N. n. miolepis*** (Borneo cobra). a very dark brown or black snake above, with a yellow to dark gray belly. The hood lacks a dorsal mark. Young specimens have widely spaced white or yellow crossbands and a chevron-shaped light mark behind the head. It is very similar in appearance to *Naja n. samarensis*. Adults may reach a length of up to 1.4 m. It occurs in Borneo and Palawan (Philippines). ***N. n. naja***. a race indigenous to the Indian subcontinent with the exception of the extreme northwestern portion and areas east of the Ganges delta; it also occurs in Ceylon. Adults may be uniformly brown or black, or variegated with rows of dappled or bi-colored scales. The dorsum of the hood bears a "spectacle" type of marking, except in adult melanistic specimens or populations. The belly may be light anteriorly becoming clouded posteriorly, or it may be generally dark with light areas on the neck. The young are paler and more variegated than adults. This race is responsible for tens of thousands of human deaths per year. ***N. n. oxiana*** (oxus cobra). a brown cobra with dark bars on the neck and occasionally traces of wide, dark crossbands on the body; the belly is pale; the hood is narrow with no dorsal markings. This snake occurs in mountainous terrain up to an elevation of about 2300 m. from northwest Pakistan and Afghanistan into eastern Iran and southern parts of Russia. The young are tan or buff with dark crossbands. ***N. n. philippinensis*** (Philippine Cobra). a race indigenous to Luzon and Mindoro, Philippines. The ground color of adults is light brown or olive dorsally, with no hood marking, and cream-colored to light brown ventrally. The young are darker with a reticulate pattern of light lines on the body. Adults average about 1.2 m in length. ***N. n. samarensis***. a cobra of the Visayan Islands of the Philippines. It is very similar in size and appearance to *N. n. miolepis*. ***N. n. sputatrix*** (Malay cobra). a brown, gray, or black cobra that occurs on the Malay peninsula and many of the larger Indonesian islands. It lacks a definite body pattern. Dorsal markings on the hood are absent in many individuals, but when present, are very similar to those of *N. n. kaouthia*. The belly is dark; the throat may be marked by white blotches. The discharge orifice of the fang is small and positioned well below the tip, a condition associated with the spraying or "spitting" venom, as sometimes observed in this snake. ***N. nigricolis*** (black-collared cobra; black-necked cobra; black-necked spitting cobra; cracheur; dab kwingu; disi; dundugu; imfezi; achil; ipHimpi; kawosia; khuppur; kigau; kikanga; komourtiou; phakhuphakhu; nachiku; saaman; spuugslang; swartnek cobra; swartnek koperkapel; sweela). an extremely aggressive snake that occurs throughout Africa. It rarely bites, but can spit a stream of venom into the eyes of its prey from a distance of more than 2 m. The venom is harmless to the intact skin, it is damaging to the eyes. Blindness usually results if the venom is not immediately washed away. *N. nivea* (yellow cobra; Cape cobra; bruinkapel; geelkapel; isikhotsholo; kaapse geelslang; koperkapel; udlezinya; umdlezinye). a large, very dangerous cobra of southern Africa, with the most toxic venom of any cobra. Adults may reach a length of about 2.1 m. The bite causes rapid onset of edema, dyspnea, and cardiac arrhythmia. The case mortality rate is high in the absence of treatment. antivenin is available.

naled. a colorless, liquid, organophosphate insecticide, $C_4H_7Br_2C_{12}O_4P$, that is slightly soluble in aliphatic hydrocarbons and very soluble in aromatic hydrocarbons. It is poisonous by ingestion and inhalation and moderately toxic by skin contact. A cholinesterase inhibitor, it is dangerous to both production workers and applicators. It produces multiple symptoms, ranging from minor eye irritation, cramps, and weakness to paralysis, convulsions, and hypotension.

nalla pambu. *Naja naja*.

nalorphine (*N*-allylnormorphine; allorphine;

antorphine). the allyl (—CH_2—CH=CH_2) derivative of morphine, $C_{19}H_{21}NO_3$. It is an effective and specific antagonist against morphine and related narcotics and is able to neutralize most effects of such drugs. It is also an antidote for acute morphine poisoning.

nalorphine dinicotinate (7,8-didehydro-4,5-epoxy-17-(2-propenyl)morphinan-3,6-diol bis-(3-pyridinecarboxylate) (ester); *N*-allylnormorphine dinicotinate; bis(nicotinic acid) diester of *N*-allylnormorphine; *N*-allylnormorphine bis(pyridine-3-carboxylic acid) ester; nalorphine bis(nicotinate)). a narcotic antagonist, $C_{31}H_{27}N_3O_5$. See nalorphine.

nalorphine hydrochloride [USP], the hydrochloride of nalorphine, $C_{19}H_{21}NO_3 \cdot HCL$, is a white, or nearly white, crystalline powder; used as a narcotic antagonist. Administered intravenously, its main effect is to relieve respiratory depression following narcotic overdosage. Addicts of narcotic drugs undergo withdrawal symptoms following its subcutaneous administration to addicts. Nalorphine is thus useful in the diagnosis of narcotic addiction. In some cases, the symptoms of withdrawal are severe and violent.

naloxone (1-*N*-allyl-14-hydroxynordihydromorphinone; 1-*N*-allyl-7,8-dihydro-14-hydroxynormorphinone; 12-allyl-7,7a,8,9-tetrahydro-3,7a-dihydroxy-4aH-8,9c-iminoethanophenanthro-[4,5-bcd]furan-5(6H)-one; 4,5-epoxy-3,14-dihydroxy-17-(2-propenyl)morphinan-6-one; *N*-allylnoroxymorphone). a crystalline substance, $C_{19}H_{21}NO_4$, that is soluble in chloroform but nearly insoluble in petroleum ether. It is a specific opiate antagonist used in human and veterinary medical practice. See naloxone hydrochloride.

naloxone hydrochloride (17-allyl-4,5α-epoxy-3,14-dihydroxymorphinan-6-one hydrochloride). a white or off-white crystalline substance or powder, $C_{19}H_{21}NO_4 \cdot HCL$, that is soluble in water and ethanol, and nearly insoluble in ether. It is a narcotic antagonist that is moderately or very toxic when administered parenterally; less toxic *per os* (oral doses in excess of 1 gm have been tolerated by humans, without serious effects). It is administered parenterally in the diagnosis of morphinism, as an antidote to narcotic overdosage, and as an antagonist for pentazocine overdosage. See naloxone.

nalukonge. *Atheris squamigera*.

namahamba. See *Dendroaspis*, *Dispholidus*; also applied to a variety of nonvenomous snakes.

name. see brand name, generic name, trade name, U.S.N.

nano- (n) a prefix to units of measurement that indicates a quantity of 1 millionth (10^{-9}) the value indicated by the root. See International System of Units.

nanogram (millimicrogram; micromilligram; milligamma; symbols, mγ; symbols, mμg, ng). a unit of mass equivalent to 10^{-9} gram or 10^{-3} microgram.

nanometer (nm). 10^{-9} meter or 10 Ångstrom units.

nap-at-noon. *Ornithogalum umbellatum*.

naphtha. **1:** benzin; coal tar naphtha; petroleum distillates (naphtha); a colorless, flammable liquid used as a dry-cleaning solvent and in the production of insecticides, rubber, paint, varnish, and plastics. Injurious effects of exposure are usually mild and may include chapping of the skin, eye irritation, and irritation of the upper respiratory tract. Severe effects include CNS depression, indicated by dizziness, convulsions, and loss of consciousness. **2:** petroleum ether; any refined, partly refined, or unrefined petroleum products and liquid products of natural gas. They are comprised chiefly of mixtures of predominantly C_5 to C_{13} aliphatic hydrocarbons that boil over a range of 30-238°C. Petroleum naphtha can thus be said to include any cut (fraction) more volatile than kerosene. All fractions in this range are moderately toxic. **3:** sometimes used as a synonym for petroleum ether.

1-naphthalenamine. 1-naphthylamine.

2-naphthalenamine. 2-naphthylamine.

naphthalene (naphthalin; naphthaline; moth flakes; naphthene; tar camphor; white tar). a moderately to very toxic, white, crystalline, solid hydrocarbon with volatile flakes, $C_{10}H_8$, and a strong coal tar odor; it is the most abundant constituent of coal tar. It is insoluble in water, but soluble in benzene, absolute alcohol, and ether. Usually prepared from coal tar, naphthalene is used mostly as a moth repellent (See mothball) and insecticide; also used in dyes, and disinfectants. In addition to systemic poisoning of humans by ingestion or exposure to the fumes, the liquid, vapor, and dust are irritants of the eyes and skin. Naphthalene is questionably carcinogenic. *Cf.* naphthene. See also naphthalene poisoning, Ah receptor. **chlorinated naphthalene** (chloronaphthalene). any of a class of compounds produced by the chlorination of naphthalene. They vary physically from oily liquids to crystalline solids, depending on the degree of chlorination. They are used, for example, in the manufacture of cables, computer circuit boards, condensers, and storage batteries. Exposure to most of these compounds may result in skin eruptions, headaches, fatigue, and inappetence. Liver damage is a common finding. Trichloronaphthalene and tetrachloronaphthalene, for example, are strong irritants and are toxic by ingestion, inhalation, and cutaneous absorption. See hyperkeratosis.

naphthalene poisoning (mothball poisoning). poisoning by ingestion of naphthalene or prolonged inhalation of fumes. Symptoms of human intoxication may include nausea, vomiting, abdominal pain, diaphoresis, hematuria, headache, pyrexia, disorientation, anemia, and coma with or without convulsions. Lesions include clumping and destruction of erythrocytes with extrusion of hemoglobin and kidney damage; liver necrosis (in rare instances); acute hemolytic anemia in certain individuals. Prolonged exposure to the vapor can cause cataracts. Delayed acute intravascular hemolytic crisis with hematuria is a characteristic and dramatic feature in sensitive individuals. About 2% of the earth's human population, including 20% of black Americans and about 50% of North American Jews, have an enzyme deficiency that results in anemia if they are exposed to naphthalene; such individuals and certain others are sensitive to naphthalene.

α-naphthaleneacetic acid. 1-naphthaleneacetic acid.

1-naphthaleneacetic acid (NAA; α-naphthaleneacetic acid; naphthylacetic acid). a synthetic auxin-type plant growth regulator and selective herbicide related to auxin 1AA. It is moderately toxic to laboratory rats. It is, for example, lethal to *Brassica kaber* (= *B. arvensis*) (charlock) and *Beta vulgaris* (sugarbeet), but is essentially nontoxic to cereal crops. NAA is moderately toxic to laboratory rats.

1-naphthalenol. 1-naphthol.

2-[(1-naphthalenylamino)carbonyl]benzoic acid. naptalam.

naphthalidine. 1-naphthylamine.

naphthalin. naphthalene.

naphthaline. naphthalene.

naphthanthracene. 1,2-benzanthracene.

naphthene. used variously in the literature, it use in toxicology is not recommended. **1:** any petroleum hydrocarbon (also termed naphthenic oil) that has physical and chemical properties similar to those of alkanes. Naphthenes, so defined, occur in both crude oils and refined petroleum products. At low concentrations, naphthenic oils usually have anesthetic effects, becoming dangerously neurotoxic at higher concentrations. **2:** any saturated or nearly saturated cycloalkane that occurs in petroleum. As with other cycloalkanes, they are CNS depressants and are potentially lethal. **3:** sometimes used as a synonym for naphthalene, but generally not in the petroleum industry.

naphthenic oil. naphthene.

naphthol (naphtholum). a compound from coal tar with monoclinic crystals, $C_{10}H_7OH$, prepared by oxidation of naphthalene. It is soluble in water and chloroform and very soluble in ethanol and benzene. Naphthol is used as an antiseptic and in some dyes. It occurs in two forms, 1-naphthol and 2-naphthol, *q.v.*

α-naphthol. 1-naphthol.

β-naphthol. 2-naphthol.

1-naphthol (α-naphthol; 1-hydroxynaphthalene; α-hydroxynaphthalene; 1-naphthalenol; C.I.

76605; C.I. Oxidation Base; 1-naphthol). a colorless or yellow powder or prisms, $C_{10}H_7OH$, with a disagreeable taste. It is a severe irritant of the skin and eyes, injuring the lens and cornea, and is moderately toxic by ingestion or cutaneous absorption. Ingestion of large amounts can prove fatal. 1-naphthol is teratogenic and a reproductive poison to laboratory animals. It is used in dyes, as a reagent in organic syntheses, and the production of perfumes. See naphthol poisoning.

2-naphthol (β-naphthol; 2-hydroxynaphthalene; β-hydroxynaphthalene; C.I. azoic coupling component; C.I. Developer 5; 2-naphthol). a white powder or lustrous crystalline substance, $C_{10}H_7OH$, becomes dark with age; it has a slight phenolic odor. It is very toxic by ingestion, cutaneous absorption, and probably other routes. It is a skin and eye irritant and a mutagen. It is used in dyes, pigments, insecticides, as an antioxidant, and in the synthesis of fungicides, antiseptics, pharmaceuticals, and perfumes. See naphthol poisoning.

naphthol poisoning (naphtholism). signs and symptoms of acute poisoning by 1-naphthol and 2-naphthol are similar, although the 1-isomer is thought to be more toxic. Symptoms of poisoning usually include nausea, vomiting, abdominal pain, and diarrhea; sometimes also convulsions, circulatory collapse, and death. Percutaneous or intestinal absorption may produce a severe nephritis, liver damage, and acute hemolytic anemia.

naphtholism. naphthol poisoning.

β-naphtholsulfonic acid. a powerful poison, $OH\text{-}C_{10}H_6SO_2OH$, that can induce a deep narcosis similar to diabetic coma. It is a sulfonated aromatic acid derived from β-naphthol and used as an azo dye intermediate.

naphtholum. naphthol.

naphthyl. the radical of naphthalene, $C_{10}H_7$-.

naphthylacetic acid. 1-naphthaleneacetic acid.

α-naphthylamine. 1-naphthylamine.

1-naphthylamine (1-naphthalenamine; 1-aminonaphthalene; α-naphthylamine; naphthalidine). a toxic, carcinogenic solid, $C_{10}H_9N$, with an unpleasant odor, that becomes red when exposed to the atmosphere. It is slightly soluble in water, but freely soluble in ethanol and ether. The dust and vapor are hazardous.

2-naphthylamine (β-naphthylamine; 2-naphthalenamine; 2-aminonaphthalene). a compound, $C_{10}H_7NH_2$, with white to pinkish, lustrous, leaf-like crystals that is soluble in hot water, ethanol, and ether, b.p. 306°C. It is moderately toxic by all routes of exposure, is a confirmed human carcinogen, and probable mutagen. Prolonged exposure to even small amounts may cause bladder and pancreatic cancer. Formerly used chiefly in chemical, dye, rubber tire, coal gas, nickel, and copper industries; now used only in research. See also amine (aromatic amine), metabolic activation.

naphthylmercapturic acid. a substance found in the blood of laboratory rabbits that have ingested naphthalene.

2-(naphthyloxy)propionamide. napropamide.

α-naphthylphthalamic acid. naptalam.

***N*-1-naphthylphthalamic acid**. naptalam.

α-naphthylthiourea. 1-naphthylthiourea.

1-naphthylthiourea (ANTU; α-naphthylthiourea; 1(naphthyl)-2-thiourea). a derivative of thiourea, it is a nearly tasteless, odorless, gray powder, $C_{10}H_7NHCSNH_2$, that is insoluble in water and sparingly soluble in most organic solvents. It is a highly toxic antithyroid agent used as a rodenticide. Symptoms of intoxication include breathing impairment due to pulmonary edema and a decrease in body temperature; the principal lesion is fatty degeneration of the liver. Dogs are highly susceptible to ANTU poisoning. The rodent:human toxicity ratio is very high. The lethal dose to monkeys is about 4.0 g/kg body weight. Humans are similarly resistant (at least partly because of vomition).

1(naphthyl)-2-thiourea. 1-naphthylthiourea.

naphtol. naphthol.

1-naphtol. 1-naphthol.

Naples yellow. lead antimonate.

napropamide (*N,N*-diethyl-2-(1-naphthalenyloxy)propanamide; *N,N*-diethyl-2-(1-naphthyloxy)propionamide; 2-(naphthyloxy)propionamide). a herbicide of low toxicity to mammals.

Naprosyn. phenylbutazone.

naptalam (2-[(1-naphthalenylamino)carbonyl]-benzoic acid; *N*-1-naphthylphthalamic acid; α-naphthylphthalamic acid). a colorless, crystalline amide, $C_{10}H_7NHCSNH_2$, that is soluble in alkaline solutions and sparingly soluble in ethanol, acetone, and benzene. It is used as an analytical reagent. The sodium salt is used as a soil-applied preemergence herbicide that is absorbed through the roots. It may persist in the soil for as much as 2 months. Naptalam closely resembles a morphactin in its physiological action in that it promotes the elongation of isolated coleoptile segments but inhibits the elongation of intact coleoptiles. It inhibits root elongation. Morphactins, naptalam, and certain substituted benzoic acids interfere with geotropic and phototropic curvatures. *Cf.* alachlor, diphenamid, propachlor.

naramycin A. cycloheximide.

Narcissus (narcissus). a genus of bulbous plants (Family Amaryllidaceae). Many of these are favored as house and garden plants. Certain hardy forms are commonly referred to as daffodils. The use of common names leads to confusion (see daffodil, jonquil) and should be used professionally with care, if at all. Nomenclature is further complicated by the numerous wild and cultivated hybrids that are known. The bulb is toxic, producing symptoms of severe gastrointestinal distress, vomition, purging, and sometimes tremors and convulsions. Gastric lavage or emesis is recommended. The sap may produce a contact dermatitis. Common or garden species include *N. jonquilla* (jonquil), native to southern Europe and Algeria, it is the only species which may properly be called the jonquil; *N. poeticus* (poet's narcissus), a species native to southern Europe, especially Greece; *N. tazetta*, a species with especially large bulbs; *N. triandrus* (angel's tears), originally native to Spain and Portugal.

narco-. a prefix meaning numbness or stupor.

narcose. narcous.

narcosis. **1:** a general, nonspecific, reversible depression of neuronal excitability in the CNS that can be caused by any of a number of physical and chemical agents. This is manifested in a state of stupor or insensibility. **2:** sleep-like unconsciousness or stupor due to the influence of narcotics. Unlike sleep, the subject cannot be fully roused. **3:** an obsolete synonym for anesthesia.

narcostimulant. **1:** the quality of being both narcotic and stimulant in effect.

narcotic. **1:** of, pertaining to, producing, or inducing dullness, stupor, insensibility, or sleep; and relief of pain. **2:** a drug that produces narcosis. Included are natural (e.g., codeine, morphine), semisynthetic (modifications of natural narcotics such as ethylmorphine), or synthetic (e.g., meperidine, propoxyphene, ethadone) nitrogen-containing heterocyclic compounds that characteristically induce narcosis and analgesia. Certain of these drugs are used in medicine to relieve moderate or severe pain. Abuse can lead to tolerance and drug addiction or dependence. **2:** many drugs that do not fit def. 1 have narcotic effects (e.g., chloroform, barbiturates, benzene).

narcotic antagonist. a drug (e.g., nalorphine, naloxone) that is used chiefly to treat narcotic-induced respiratory depression.

narcotic drug. narcotic (def. 2).

narcotic poisoning. intoxication by a narcotic or by sleep-inducing poisons (e.g., opium and its derivatives, chloral, barbital). Symptoms are largely those of classic CNS depression, bradycardia, respiratory depression, stupor, sleep-like unconsciousness, followed by coma. In acute cases, death may ensue within a few hours. Narcotic poisonings are often due to accidental overdosage (e.g., in therapy or due to an addict's error in estimating his or her tolerance to a particular narcotic) or to suicide.

narcotico-acrid. having both narcotic and acrid qualities.

narcotico-irritant. having both narcotic and irritant qualities.

narcotism. **1:** narcosis; a state of stupor or insensibility due to the influence of narcotics. **2:** addiction to narcotics, defined as dependence on a narcotic such that discontinuance of usage causes symptoms that are rapidly relieved by a dose of the drug.

narcotize. to expose to or put under the influence of a narcotic.

narcous (narcose). stuporous, insensible; in a state of stupor.

Nardelzine. See phenelzine.

Nardil. See phenelzine.

narrow-banded sea snake. *Hydrophis spiralis*.

narrow-banded snake. *Brachyurophis fasciolatus*.

narrow-headed sea snake. *Microcephalophis gracilis*.

Narthecium ossifragum (bog asphodel). a European bog herb (Family Liliaceae) with linear leaves and greenish-yellow flowers. It causes hepatogenic photosensitization in livestock. There is a similar species in the United States, *N. americanum* (also known as the bog asphodel). See also phylloerythrin.

nashornviper. *Bitis nasicornis*.

nasogastric tube. a slender tube, passed through the nose into the stomach, and used to drain stomach secretions or to feed patients that are unable to eat normally.

natality (birth rate). **1a:** the ratio of the number of births in a target population during a stated period of time to the total number of individuals in the population. **1b:** operationally, the number of offspring produced per female or per individual in a specified population per unit time. **1c:** treated as an emergent property at the population level or organization, natality is the rate of production of new individuals by birth, germination or fission. **2**. the production of new individuals (by whatever means) in a specified population. See also mortality. In this definition, natality (defs. 1a, 1b, 1c) is substituted by the term "natality rate." This usage can be confusing and is not recommended.

natality rate. See natality (def. 2).

natamycin (pimaricin; antibiotic A 523; tennecetin). a polyene antifungal, antibacterial antibiotic isolated from the soil fungus *Streptomyces natalensis* and *S. chatanoogensis*. It is moderately toxic to laboratory rats.

natimortality. **1:** perinatal mortality. **2:** the ratio of fetal and neonatal deaths to overall natality.

National Formulary (NF). a formulary originally issued by the American Pharmaceutical Association, now published by the U.S. Pharmacopeial Convention, Inc. It is a book of standards for pharmaceuticals and preparations that have been established as useful, but which are not included in the USP. It is revised every five years, and is recognized by the U.S. Pure Food and Drugs Act of 1906 as a book of official standards.

National Toxicology Program (NTP). a project established by the U.S. government in November, 1978, to conduct tests and research on hazardous substances, and to protect the population of the United States from the many dangerous chemicals in use. Several thousand chemicals have thus far been tested.

natremia. hypernatremia.

natrite. See sodium carbonate.

natrium. sodium.

Natrix viperinus. See *Nerodia viperinus*.

natron. See sodium carbonate.

Natulan. procarbazine hydrochloride. See procarbazine.

natural gas. **1:** any gas or gaseous mixture formed naturally in the earth's crust including those generated by volcanic activity. Many of these (e.g., hydrogen sulfide) are dangerously toxic. **2:** a naturally occurring gaseous combustible mixture used as fuel. American natural gas as supplied for use is comprised of ca. 85% methane, 9% ethane, 3% propane, 2% nitrogen, 1% butane, and smaller amounts of higher hydrocarbons. Unprocessed natural gas typically contains natural gasoline that must be recovered and sometimes carbon dioxide, hydrogen sulfide and helium. An occasional wellhead will yield nearly pure methane. Nitrogen and other gases (e.g., helium) may also be present in amounts sufficient to warrant extraction prior to usage. Natural gas when present in ambient air at high concentrations is a deadly narcotic and asphyxiant. As with other hydrocarbon fuels, carbon monoxide results from incomplete combustion. See also carbon monoxide poisoning.

natural gasoline. a liquid alkyl hydrocarbon contained in raw natural gas.

natural order (in botany). a former name for family.

natural product. See toxic natural product.

nausea. a sickening or unpleasant sensation that usually precedes vomiting; a feeling of queasiness; a feeling of imminent vomiting. Nausea may be accompanied by sweating, pallor, and sometimes dizziness. It is common in humans in many types of poisonings.

nauseant. **1:** nauseating; disgusting; inducing nausea. **2:** a substance or phenomenon that causes nausea.

nauseate. to disgust; to sicken, to cause nausea.

nauseated. affected with nausea or revulsion.

nauseous. **1:** experiencing or exhibiting symptoms of nausea. **2:** manifesting nausea or revulsion. **3:** causing nausea.

nawama. *Bitis arietans*.

Nb. the symbol for niobium.

nchuweira. *Naja haje*.

Nd. the symbol for neodymium.

Ne. the symbol for neon.

Near East Viper. *Vipera xanthina*.

Nebraska fern. *Conium maculatum*.

nebularine. a toxic principle isolated from *Clitocybe nebularis* (cloudy clitocybe; Family Agaricaceae). Nebularine is extremely toxic to laboratory rats and guinea pigs. *C. nebularis* is a mushroom of the Rocky Mountains and Pacific Coast of North America. It has a disagreeable odor and is not considered edible.

nebulization. the reduction of a solid or liquid to a mist or spray made up of very fine particles.

nebulizer. a device used to administer drugs, as in the treatment of asthma and similar respiratory conditions. It converts medication into a fine mist that can be inhaled deeply into the lungs. See also nebulization.

neburon (u*N*-butyl-*N'*-(3,4-dichlorophenyl)-*N*-methylurea; 3-(3,4-dichlorophenyl)-1-methyl-1-*n*-butylurea). a herbicide used, for example, to control annual weeds in certain evergreens and to control *Stellaria media* (common chickweed) and *Cerastium vulgatum* (mouse-ear chickweed) in crops such as *Medicago sativa* (alfalfa), tomatoes, and strawberries. Neburon is only slightly toxic to laboratory rats.

NEC. no-effect concentration.

necine. any of a class of nitrogenous compounds produced by the hydrolysis of the hepatotoxic pyrrolizidine alkaloids of *Senecio*, *Crotalaria*, and certain other plants of the family Boragnaceae. Necines are not hepatotoxic, but they can be lethal if produced in sufficient amounts. See especially *Senecio*. See also retrorsine, necine base.

necine base. a 1-methylpyrrolizidine such as retronecine. Such compounds are esters with various stereochemical configurations and degrees of hydroxylation; they occur in alkaloids produced by plants of the genera *Senecio*, *Crotalaria*, and a number of plants of the family Boragnaceae. See also necine.

necklace snake. See *Elapsoidea sundevallii*.

necklace snake, Australian. *Brachyurophis australis*.

necklaceweed. *Actaea*.

necrocytosis. cell death, decomposition and/or dissolution.

necrocytotoxin. a toxin that destroys cells.

necrogenic. able to cause necrosis or death.

necrolysis. separation or exfoliation of necrotic tissue. **toxic epidermal necrolysis** (Lyell's syndrome, scalded skin syndrome; TEN). an exfoliative skin condition in which erythema spreads rapidly over the entire body with the development of large, flaccid bullae and subsequent sloughing off of the skin in large sheets as in a second degree burn. The level of the separation is subdermal. TEN is commonly due to drug sensitivity, but is sometimes of unknown etiology.

necropsy (necroscopy; ptomatopsia; ptomatopsy). autopsy, *q.v.*; examination of a corpse.

necroscopy. necropsy.

necrose. **1:** to cause necrosis. **2:** to undergo necrosis; to become necrotic.

necroses. plural of necrosis.

necrosis (plural, necroses). **1a:** tissue death; death of a circumscribed (focal) portion of tissue. Necrosis in animals may be due to disease, impaired circulation, to chemicals acting locally or it may be secondary to systemic insult. Necrosis in plants often results from fungal infection, and the shape of necrotic areas often indicates the specific cause. **1b:** the morphological changes that attend the death of cells due to the progressive degradative action of enzymes. This term is usually applied to small areas of dead tissue surrounded by healthy tissues. The causes are varied (e.g., trauma, insufficient blood supply, intense radiation (e.g., x-irradiation, infrared, ultraviolet radiation, electricity, poisonous materials acting locally or internally).

necrotic. **1:** of, pertaining to, or concerned with, or characterized by necrosis. **2:** concerned with death; pertaining to or characterized by death or necrosis.

necrotizing. causing or able to cause necrosis.

necrotizing factor. cobra bites are sometimes accompanied by extensive necrosis with few systemic effects. It is hypothesized that venoms of some individual cobras may have large amounts of a necrotizing principle relative to the lethal component.

necrotoxin. a substance secreted by certain staphylococci, which kills tissue cells.

NED. no-effect dose. See dose (no-effect dose).

neem, neem tree. *Melia azadirachta*.

neem oil (nim oil; margosa oil). an aromatic oil from the fruit and seeds of *Melia azadirachta*. It is an anthelmintic, a pesticide (in certain formulations), an insect repellent, antifeedant, and an ethanol denaturant.

negligence. **1a:** carelessness, inattentiveness, indifference. **1b:** failure to take prudent care (e.g., of a patient; in the use of equipment). *Cf.* malpractice.

NEL. no-effect level.

Nelson's milksnake. *Lampropeltis triangulum nelsoni*.

nematicide. nematocide.

nematocide (nematicide). **1:** destructive to nematodes. **2:** a substance that kills nematode worms. See anthelmintic.

nematocyst (cnidocyst; cnidocil; cnidocell; thread capsule; thread cell; thread tube). any of a type of minute, elaborate, thread-like or spherical, stinging capsule. They occur in representatives of all classes of coelenterates except the Ctenophora. They are distributed chiefly on the tentacles and housed in Most are imbedded in specialized epidermal cells (cnidoblasts). Nematocysts contain unusual toxins and function largely in defense and the paralysis and capture of prey. The nematocysts of some coelenterates can penetrate human skin and at least 80 species are known to injure humans.

nematocystic. of or pertaining to nematocysts.

neodymium (symbol, Nd)[Z = 60; A_r = 144.24]. a toxic, silvery-white element of the lanthanoid series of metals; it yellows on exposure to air. It is used as a catalyst and a component of various metal alloys.

neologism. a new term or new usage of an existing term. Such terms often meet resistance because of their novelty, lack of precision, failure to meet conventional standards of usage, and/or the confusion that a proliferation of terms can add to the existing vocabulary of a discipline or profession. The introduction of neologisms is most common in the applied (as opposed to basic) sciences because there may not be a well-established tradition for the formulation of new terms or for the redefinition of existing terms. In toxicology, new terms enter not only directly through the activity of scientists but frequently also through the thoughtless adoption of bureaucratic, legislative, commercial, and even popular language. Neologisms introduced from such sources may be used in contravention of existing terminology and definitions. They also tend to change rapidly in meaning and are often rapidly replaced by newer terms. Not infrequently, the sources named above may rapidly pervert the meaning of existing terms when they come to their attention (examples are "the ecology" and "biorhythms"). Governmental entities may even force inappropriate usage on contractors and other professionals (e.g., "parameter" used as a synonym for any variable). Persons entering a field sometimes invent new terms when suitable terms already exist; such may be propagated in the professional literature.

neomycin. a moderately toxic complex of antibacterial aminoglycosides secreted by a strain of *Streptomyces fradiae*. **neomycin B**. the most common neomycin currently found in commercial preparations. It is administered *per os* as a pre-surgical intestinal antiseptic and in the management of hepatic coma. It has also been administered parenterally in the management of severe, life-threatening infections, although its use in this way has been nearly abandoned because of the high risk of serious systemic poisoning. Neomycin is toxic by all routes of administration and has caused sensorineural deafness, respiratory depression, and renal failure. Even so, the main danger from topical application is hypersensitivity which may produce no more than a skin rash.

neon (symbol, Ne)[Z = 10; A_r = 20.18]. an inert monatomic gaseous element; one of the rare gasses. It is used in neon lights, gas lasers, and electrical equipment.

Neophoca cinerea (Australian sea lion). a species of sea lion (Family Otariidae) that occurs only on the coast of southern Australia. The meat is reportedly toxic.

neoplasm. **1:** the original and preferred definition: any new tissue growth. **2a:** a new growth of tissue that serves no known physiological function, at least where located. **2b:** any tumor. **3:** any new growth comprised of cells that proliferate in an uncontrolled and progressive manner. **benign neoplasm** (benign tumor). any noncancerous neoplasm; one that is neither invasive nor metastatic or is mildly so, with a low degree of anaplasia. **malignant neoplasm** (malignant tumor). any cancerous neoplasm; cancer. *Cf.* tumor.

neosaxitoxin. See paralytic shellfish poison.

Neostigmin™. neostigmine.

neostigmine (3-[[(dimethylamino)carbonyl]oxy]-*N,N,N*-trimethylbenzenaminium; (3-dimethylcarbamoxyphenyl)trimethylammonium; proserine; synstigmin; Neostigmin; Prostigmin). a deadly parasympathomimetic quarternary ammonium compound, $C_{12}H_{19}N_2O_2$, used as the bromide ($C_{12}H_{19}BrN_{20}O_2$) or as the methylsulfate ($C_{13}H_{22}N_2O_6S$) ester. It is deadly poisonous by all routes of exposure. It is sometimes used to counteract the toxic effects of tubocurarine chloride. See neostigmine methyl sulfate.

neostigmine methyl sulfate (*N,N*-dimethylcarbamic acid-3-dimethylaminophenyl ester methosulfate; dimethylcarbamic acid ester with (*m*-hydroxyphenyl)trimethylammonium methyl sulfate). the methyl sulfate ester of neostigmine, $C_{12}H_{19}N_2O_2 \cdot CH_3O_4S$. Both compounds have similar toxicities.

neotropical rattlesnake. *Crotalus durissus*.

Nepal aconitine. pseudoaconitine.

nepaline. pseudoaconitine.

nephralgia. pain centered in a kidney.

nephralgic. of or pertaining to nephralgia.

nephredema. renal congestion; nephremia.

nephremia. nephredema.

nephric. renal; pertaining to the kidney.

nephridium. **1a:** any tubule in a metazoan that has an external opening and performs an excretory or osmoregulatory function. **1b:** an excretory organ in certain invertebrates that corresponds functionally to the vertebrate kidney. **2:** the embryonic kidney tubule of vertebrates.

nephritic. **1:** of or pertaining to nephritis. **2:** pertaining to the kidneys. **3:** a drug or agent used in the treatment of renal disease.

nephritides. **1:** plural of nephritis. **2:** a collective term, used in reference to all types of nephritis.

nephritis (plural, nephritides). acute or chronic inflammation of the kidney that may be a focal, or a diffuse proliferative, or destructive process that may involve the glomeruli, tubules, or interstitial tissue. Causes vary but include bacteriotoxins, streptococcal infections, septicemia, drugs, heavy metals. **nephrotoxic nephritis**. that produced in a vertebrate animal by the injection of antikidney antibody from a different species that had earlier received a preparation of glomerular basement membrane from an animal of the same species as the one to be inoculated. **saturnine nephritis**. that seen in chronic lead poisoning. **tartrate nephritis**. an acute form of nephritis induced by subcutaneous injection of racemic tartaric acid.

nephritogenic. causing, or able to cause, nephritis.

nephrolysine. nephrotoxin.

nephron. the functional unit of the kidney of reptiles, birds, and mammals. While nephrons vary in structural detail, all contain a structure (the Malpighian corpuscle) that contains a knot of capillaries (a glomerulus) that receives a filtrate of circulating blood and an associated tubule that reabsorbs parts of the filtrate into the circulation. The remainder is excreted.

nephropathia. nephropathy.

nephropathic. of, pertaining to, characterized by, or productive of nephropathy.

nephropathic. of, pertaining to, characterized by, or productive of nephropathy.

nephropathy. any disease or morbid condition of the kidneys. *Cf.* nephrosis. **toxic nephropathy**. nephrotoxic disorder.

nephrosis. **1a:** any disease or lesion of the kidney that is accompanied by degenerative lesions of the renal tubules with attendant, noninflammatory edema, albuminuria, and decreased serum albumin. See also nephrotic syndrome. **1b:** a clinical syndrome characterized by edema, excess albumin in urine, and cholesterol in blood. **mercurial nephrosis**. that due to poisoning by mercuric chloride. **toxic nephrosis.** nephrosis caused by a toxicant, most often by bichloride of mercury. *Cf.* nephropathy.

nephrotic. of, pertaining to, resembling, or caused by nephrosis.

nephrotic syndrome. the set of characteristic signs and symptoms that accompany nephrosis, *q.v.*

nephrotoxic. toxic to kidney cells.

nephrotoxic agent. a toxicant that is harmful or damaging to the kidney of vertebrates. There are numerous such agents including many analgesics, antibiotics, anticancer, antiepileptics, biologicals, botanicals, diagnostic drugs, herbicides, immune complex inducers, oxalosis-inducing agents (e.g., oxalic acid), pesticides, prostaglandin synthetase inhibitors, solvents, all nonsteroidal antiinflammatory drugs, and most, if not all, heavy metals.

nephrotoxic disorder (toxic nephropathy). any functional or morphologic injury or abnormality of the vertebrate kidney due to exposure to a toxicant.

nephrotoxicity (renal toxicity). **1:** the quality, property or capacity of a particular toxicant to poison, whether directly or indirectly, the kidneys. **2:** adverse effects on the kidneys by the action of a toxicant. **3:** the degree of toxicity of a substance with respect to kidney tissue.

nephrotoxin. a toxin that specifically poisons kidney cells.

nephrotropic. said of an agent that has a special affinity for or exerts its main effect upon kidney tissue.

Neptal. Mersalyl with theophylline.

neptunium (symbol, Np)[Z = 93; most stable isotope = ^{237}Np, $T_{1/2}$ = 2.2 X 10^6 yr]. a toxic, silvery, radioactive element of the actinoid series of metals. It occurs naturally in minute quantities in uranium ore and is produced as a by-product of ^{239}Pu production.

nereistoxin. an insecticidal tertiary amine produced by certain venomous marine polychaete annelids that inject toxins through bites or by bristles.

neriifolin ((3β,5β)-3-[(6-deoxy-3-*O*-methyl-α-L-lucopyranosyl)oxy]-14-hydroxycard-20(22)-enolide) monoacetylneriifolin). a cardiotonic, cardiotoxic glycoside derived from *Cerbera odollam*, *Thevetia peruviana*, and *T. thevetioides* seeds.

neriolin. oleandrin.

nerioside. a toxic cardiac glycoside isolated from *Nerium oleander*.

Nerium (oleander). a genus of evergreen shrubs (Family Apocynaceae) native to southern Europe and Japan. ***N. indicum*** (oleander). a common species found in Hawaii. It contains several toxic glycosides, including the steroid glycoside, odoroside. Symptoms and toxicity are similar to those of *N. oleander*. ***N. oleander*** (oleander; rose-bay). an ornamental evergreen shrub or bush native to Europe that is widely cultivated as a house and garden plant. All parts of the plant are toxic; the leaves are extremely poisonous. They contain toxic cardiac glycosides, the most important of which are oleandroside and nerioside. Symptoms in humans and other animals include discoloration of the mouth, nausea and vomiting, weakness and sometimes dizziness, a rapid pulse, sweating, cold extremities, mydriasis, anorexia, abdominal pain, and bloody feces. In fatal cases, the heartbeat is weak and irregular; dyspnea and coma are common. See also oleandrin, oleandrin poisoning.

Nerodia viperinus (= *Natrix*; viperine snake). a small nonvenomous snake (Family Colubridae) of southern Europe and northern Africa. Its color pattern resembles that of the viper, *Vipera berus*.

nerve gas. any of several volatile, poisonous, odorless, colorless, and tasteless liquid warfare agents that are rapidly absorbed through the eyes, lungs, or skin, causing systemic effects, notably respiratory and nuromuscular paralysis. See also warfare agent, isopropylphosphonofluoride.

nerve tract. a bundle of neurons that form a transmission pathway through the brain and spinal cord.

nervous syndrome of cattle and sheep (rape blindness). a disease, reported from Canada and the United Kingdom, that results from ingestion of rape, *Brassica napus*. Affected animals initially stand alone or wander, often bumping into solid objects, signaling blindness. The disease often arrests at this point or the animal becomes aggressive. Specific lesions are absent and the eye exhibits no abnormality except the absence of a pupillary reflex. Blindness may be permanent, but with good feed and care, vision usually returns gradually.

nervous system. a highly organized, complex system that is present in all multicellular animals. All such systems are comprised of a highly organized receptor-conductor-effector system comprised of cells (nerve cells or neurons) that generate and transmit information in the form of electrical impulses by means of electrochemical mechanisms. Additional types of cells that support or provide nutrients to the neural tissue may also be present. The nervous system receives information in the form of both external and internal stimuli, interprets it, conveys it, and coordinates a rapid and suitable (adaptive) muscular and/or glandular response. In higher animals, the nervous system can be said to coordinate the adjustments and reactions to stimuli such that behavior is appropriate to both internal and environmental conditions. That of most, if not all, metazoans also stores information and has some capacity to modify its processing of information based on experience (learning). The vertebrate nervous system consists of a central nervous system (CNS) and a peripheral nervous system (PNS). See also autonomic nervous system, parasympathetic nervous system.

nervous tissue. electrochemically active tissue of a nervous system. The cells of such tissue nervous system, sympathetic nervous system. receive, transmit, and in some cases generate, electrical impulses.

nettle. (stinging nettle). **1:** any plant of the family Urticaceae, especially those of the genus *Urtica*. **2:** any of numerous unrelated plants with stinging hairs (e.g., *Hesperocnide*, *Laportea*, *Urtica*)

nettle rash. urticaria.

nettle syndrome. a disease of hunting dogs caused by the stinging hairs of *Urtica chamaedryoides*. It is characterized by excessive salivation, emesis, pawing or wiping at the muzzle, a slow irregular heart beat, and muscular weakness. See also syndrome.

neuralgia. severe paroxysmal pain that encompasses the course of one or more nerves. There are usually no morbid or morphological changes associated with neuralgia. Causes are many and varied including toxins and other poisonous substances, pressure on a nerve trunk, neuritis, syphilis, malaria, diabetes.

neuralgic. of the nature of, or pertaining to, neuralgia.

neurasthenia. a condition of vague functional fatigue that often accompanies or follows depression. See also neurotoxia.

neurine (trimethylvinylammonium hydroxide). a toxic amine or ptomaine with a fishy smell obtained mainly from brain, bile, and egg yolk, and which is formed by the dehydration of choline in putrefying meat. Neurine is also a toxic constituent of certain mushrooms.

neuritis. **1:** inflammation of a nerve or nerves that is often associated with a degenerative process. Neuritis is attended by pain and tenderness over the nerves; anesthesia and paresthesia; paralysis; wasting; and elimination of reflexes. Causes vary, but include compression or trauma; direct infection of the affected nerves; various disease states; metabolic disorders (e.g., toxemias of pregnancy; toxins; heavy metal poisoning; ethanol; carbon tetrachloride. **2a:** sometimes used in routine practice to denote noninflammatory lesions or disease of the peripheral nervous system. **2b:** often used interchangeably with neuropathy, *q.v.* **alcoholic neuritis**. alcoholic neuropathy. **disseminated neuritis**. polyneuritis. **lead neuritis**. neuritis saturnina. **multiple neuritis**. polyneuropathy. **radiation neuritis**. radioneuritis. **saturnine neuritis**. neuritis saturnina. **toxic neuritis**. that due to poisons such as arsenic, mercury, thallium, various hydrocarbons and organic solvents.

neuritis saturnina (lead neuritis, saturnine neuritis). neuritis due to plumbism.

neurobehavioral toxicology. See toxicology (neurobehavioral toxicology).

neuroclonic. marked by nervous spasms.

neurocyte. neuron.

neurodynia. pain in a nerve or nerves.

neurohypophysis. the posterior lobe of the pituitary gland. See pituitary gland.

neuroleptic (antipsychotic). **1:** of or pertaining to the action of, or a condition produced by, an antipsychotic drug. **2:** an antipsychotic or neuroleptic drug.

neuroleptic drug (antipsychotic drug**;** major tranquilizer**;** neuroleptic (def. 2)**;** antipsychotic (def. 2)). any drug that moderates or modifies psychotic symptoms (e.g., butyrophenones, phenothiazines, thioxanthines) that alter mood, thought, and behavior in helpful ways. Such drugs are used in the treatment of all psychoses (e.g., schizophrenia, delusions, hallucinations, mania, severe paranoia). Neuroleptics act chiefly as dopamine receptor antagonists, but may also have anticholinergic or antiadrenergic properties. The therapeutic index for these drugs used alone is very high. Fatal overdose is a rare problem. They do, however, interact with other depressant drugs. Toxic effects are primarily neurological and are commonly of types: (1) extrapyramidal effects such as pseudoparkinsonism or acute dystonia). (2) movement disorders (e.g., tardive dyskinesia) that appear following long-term treatment. They are usually of a permanent nature and are refractory to pharmacotherapy.

neuroleptic malignant syndrome (NMS). a toxic reaction to a powerful neuroleptic. Signs and symptoms include a combination of catatonic rigidity, stupor, unstable blood pressure, dyspnea, profuse sweating, hyperthermia, and incontinence. Symptoms may persist for 5-10 days following discontinuation of the drug(s). The case mortality rate may be as high as 20%.

neurologic. pertaining to neurology, to nervous tissue, the nervous system, or to a state or condition of nervous tissue or to the action of nerves.

neurology. **1:** the biological science that deals with the nervous system and its behavior. **2:** the field of medicine that deals with the diagnosis and treatment of disorders of the nervous system.

neurolysin. any toxin that attacks nerve cells; nearly synonymous with neurotoxin.

neurolytic. destructive of nerve cells.

neuromimetic. **1:** able to elicit a response in an effector organ very similar to that elicited by nervous impulses. **2:** an agent that elicits such responses. **3:** pertaining to neurotic or hysterical simulation of organic disease.

neuromotor. **1:** of or pertaining to nerves and skeletal muscles, especially to their combined action. **2:** pertaining to efferent nerve impulses that terminate at a myoneural junction.

neuromuscular (neuromyal). pertaining to muscles and nerves considered jointly. *Cf.* neuromotor.

neuromuscular blocking agent. See blocking agent.

neuromyal. neuromuscular.

neuromyopathic. pertaining to or causing neuropathy that affects the nervous system and muscle, including the heart.

neuron (neurone; neurocyte). a nerve cell, it is the structural and functional unit of a nervous system. Neurons are conductive cells that transmit an electrical impulse under certain conditions. A typical vertebrate neuron consists of a nerve cell body, which contains the nucleus and the surrounding cytoplasm (perikaryon); dendrites, which are short, radiating processes; and the axon, which is a long process that carries the impulse generated by the cell. The axon terminates in small branches (telodendrons) and may have collateral branches along its course. The axon together with a sheath constitutes the nerve fiber.

neuronal. of or pertaining to a neuron or neurons.

neurone. neuron.

neuropathic. of, pertaining to, or characterized by neuropathy.

neuropathology. the branch of biology and medicine that treats diseases of the nervous system, especially morbid changes in morphology and function.

neuropathy. **1:** any disease, inflammation, disturbance, damage, or pathological change in the peripheral nervous system, regardless of etiology (e.g., arsenical neuropathy, diabetic neuropathy, ischemic neuropathy, traumatic neuropathy). *Cf.* polyneuropathy, encephalopathy, myelopathy, neuritis. **2:** noninflammatory lesions in the peripheral nervous system, as opposed to neuritic inflammatory lesions. **alcoholic neuropathy** (polyneuritis potatorum; alcoholic neuritis). neuropathy due to thiamine deficiency in chronic alcoholism. **buckthorn polyneuropathy**. that due to ingestion of the fruit of shrubs of the genus *Rhamnus* (buckthorns). It is a symmetrical ascending neuropathy beginning in the lower limbs and terminating in the brain stem with consequent respiratory paralysis and death. Survivors recover slowly, but usually completely. There is no specific therapy. **delayed neuropathy**. organophosphorus-induced delayed neurotoxicity.

neuropathy target enzyme. neurotoxic esterase.

neuropharmacological. pertaining to neuropharmacology.

neuropharmacology. that branch of pharmacology concerned especially with the action of drugs upon various components of the nervous system.

neurospasm. a muscle spasm due to a nervous disorder.

neurotoxia. **1:** a toxic condition of the nervous system. **2:** a form of neurasthenia, *q.v.*, probably due to autointoxication.

neurotoxic. poisonous to, functionally harmful to, or damaging to nervous substance, tissue, or to the nervous system; applied to any toxicant that impairs neural function.

neurotoxic agent. a substance that impairs neural or nervous system function. Some such agents produce focal lesions or may act preferentially only on a particular class of cells.

neurotoxic esterase (NTE; neuropathy target enzyme). any of a class of membrane-bound hydrolases (mw 155-178,000 daltons) situated in the brain and spinal cord. They are generally accepted as the target enzyme of those organophosphates that elicit delayed neurotoxicity (OPIDN); the toxicity of some organophosphorous esters depends on their ability to combine with a specific neurotoxic esterase. Inhibition of NTE can produce a delayed form of neurotoxicity. Thus in organophosphate poisoning, if NTE is inhibited, signs of neurotoxicity appear from 8-21 days following exposure.

neurotoxicant. a toxicant that acts on the nervous system.

neurotoxicity. **1:** the capacity of a toxicant to poison, functionally impair, or damage neural tissue or the nervous system of an animal. **2:** any toxic effect on a nervous system function, whether behavioral, structural, physiological, chemical, or neurological.

neurotoxicological. pertaining to neurotoxicology

neurotoxicology. See toxicology (neurotoxicology).

neurotoxin (neurolysin). **1a:** neurolysin; any toxin that chiefly, but not necessarily exclusively, poisons nervous tissue. **1b:** a toxin that has more marked effects on nervous tissue than on other tissues. **1c:** sometimes applied to any poison that produces primary toxic effects on the nervous system. **2:** an exotoxin that has a marked affinity for nerve tissue, causing fatty degeneration of the myelin sheath of peripheral nerves, of the white matter of the brain and spinal cord, and of certain other tissues (e.g., cardiac muscle). The principal effects are inhibition of nerve conduction or of transmission at the synapse by linkage to a voltage-gated sodium channel protein.

neurotropic (polioclastic). having an affinity for nervous tissue, or exerting its principal effect upon the nervous system. See also encephalotropic, myelotropic.

neurotropic ergotism (convulsive ergotism; chronic ergotism). a type of ergotism that provokes muscle spasms, hallucinations, and sometimes paralysis; the victim's body feels on fire (hence, the name St. Anthony's fire), and the skin feels as though it is crawling. When cattle feed on pasturage or consume large amounts of hay infested with ergot (most often by *Claviceps paspali*), the affected animals display nervous symptoms (including convulsions) in addition to those of gastrointestinal distress. Early signs are hyperexcitability and sometimes belligerence. Trembling and incoordination ensue; running animals exhibit an exaggerated flexure of the forelimbs, and if startled or excited, animals will often fall. Finally, kicking may alternate with periods of tetanic rigidity and emprosthotonos. Death may intervene within 3 days to a month, although early removal of stock from affected vegetation may result in complete recovery. Neurotropic ergotism is rarely caused, in humans or cattle, by *Claviceps purpurea*. The disease is occasionally seen in animals grazing on *Hilaria mutica*. (tobosagrass) which is parasitized by *C. cinerea*. In the United States, neurotropic ergotism of cattle results usually from the ingestion of infected *Paspalum dilatatum* (Dallis grass) or *P. notatum* (Argentine bahia) that are parasitized by *C. paspali*

neutral lead acetate. lead acetate.

neutralization. **1:** the process of making neutral as in the neutralization of an acid or base such that the substance no longer exhibits either acidity or alkalinity. **2:** inactivation (e.g., of a toxicant, a virus, or other biologically active agent). **3:** many hydrolytic reactions are reversible. Neutralization is the reversed reaction. See hydrolysis.

neutropenia. a decrease in neutrophils in the circulating blood. It is sometimes inaccurately used as a synonym of agranulocytosis.

nevite. sodium carbonate, decahydrate.

New Guinea death adder. *Acanthophis antarcticus*.

New Guinea mulga snake. *Pseudechis papuanus*.

NF. National Formulary.

ng. nanogram.

ngu sam liem. *Bungarus fasciatus*.

Ni. the symbol for nickel.

nialamide (1-((2-benzylcarbamyl)ethyl-2-isonicotinoylhydrazine). a crystalline powder, C_5H_4-$NCO(NH)_2(CH_2)_2CONHCH_2C_6H_5$, it is white and is soluble in slightly acid aqueous solution. It is a monoamine oxidase inhibitor that is no longer available in the United States because of its toxicity. It is especially dangerous when used in combination with certain other drugs (barbiturates, antihistamines, and other antidepressants such as meperidine, morphine, and aminopyrine).

nickel (symbol, Ni)[Z = 28; A_r = 58.70]. a toxic, caustic, lustrous white, hard, ferromagnetic transition metal that occurs naturally as the sulfide and silicate. It closely resembles and is often associated with cobalt. Nickel causes dermatitis in sensitive individuals. The ingestion of soluble nickel salts causes gastrointestinal symptoms including nausea, vomiting, and diarrhea. It is used, for example, as a catalyst in the hydrogenation of alkenes, nickel plating, and in coinage alloys.

nickel carbonyl (nickel tetracarbonyl). a colorless, volatile, flammable liquid, $Ni(CO)_4$, that is soluble in many organic compounds and concentrated nitric acid, but insoluble in water. It is a zero-valent compound in which the nickel atom is joined to the carbonyl moieties by a covalent bond with each of the 4 carbon atoms. The carbonyl groups form a tetrahedron about the nickel atom. One of the most dangerous industrial chemicals, this compound is extremely toxic, carcinogenic, and widely used as an industrial chemical. See carbonyl compound. It is used industrially as a catalyst, in plating metals, and in the preparation of nickel. The fumes are extremely toxic; inhalation can cause serious pulmonary edema with dyspnea and focal hemorrhage. Nickel carbonyl oxidizes in air and explodes at ca. 60°C.

nickel cyanide. a toxic, water-insoluble apple-green powder, $Ni(CN)_2 \cdot 4H_2O$, used in metallurgy and electroplating.

Nicotiana (tobaccos, nicotiana). a genus of plants (Family Solanaceae) all, or nearly all, of which contain the highly toxic alkaloid nicotine, the alkaloid pyridine, and a number of carcinogenic hydrocarbons. Accidental or intentional use or misuse of tobacco produces frequent intoxication or other effects on humans, livestock and other animals. Species known to have poisoned livestock in the United States include *N. attenuata* (wild tobacco, coyote tobacco), *N. glauca* (tree tobacco), *N. tribonophylla* (wild tobacco, desert tobacco), and *N. tabacum*. All of these produce similar, mainly neurogenic effects. Symptoms may include muscular weakness, tremors which may be localized or involve the entire body, staggering, and eventual collapse. Violent heart action may be observed, but the pulse is weak and rapid and the extremities become cold. Dyspnea is a common feature. The time to death varies from a few minutes to several days. In at least one case, pregnant swine that ingested tobacco stalks developed arthrogryposis. *N. tabacum* (tobacco; cultivated tobacco). a major toxicant of humans. See also cigarette, nicotine.

nicotinamide adenine dinucleotide. NAD.

nicotinamide-adenine dinucleotide. See NAD. See also nadide.

nicotinamide adenine dinucleotide phosphate. NADP.

nicotine (3-(1-methyl-2-pyrrolidinyl)pyridine; β-pyridyl-α-n-methylpyrrolidine). an extremely toxic alkaloid. The pure compound is a thick, colorless, flammable, levorotatory oil, C_5H_4-$NC_4H_7NCH_3$, that is miscible with water; it turns brown on exposure to air. Nicotine occurs in all parts of tobacco plants (*Nicotiana spp.*), but is most concentrated in the leaves. It is nearly odorless, with a sharp burning taste and is hygroscopic and soluble in a number of organic solvents. Nicotine is one of the most toxic and addicting of all drugs and is toxic by all routes of exposure including the intact skin. The lethal dose for an adult humans is considered to be less than 5mg/kg body weight. Chronically heavy users (cigarette smokers), however, acquire a pronounced tolerance to nicotine. It is also used as a contact insecticidal fumigant in closed spaces and in mickey finns, *q.v.* It is apparently the addictive component of cigarettes. See nicotine poisoning, nicotinism, *Equisetum*.

nicotine sulfate (1,10methyl-2-(3-pyridyl)-pyrrolidine sulfate; (S)-3-(1-methyl-2-pyrrolidinyl)pyridine sulfate (2:1); 1,3-(1-methyl-2-pyrrolidyl)pyridine sulfate). a poisonous substance, $C_{20}H_{26}N_4 \cdot O_4S$, used as an insecticide. Animals poisoned by nicotine or by nicotine sulfate show tremors, incoordination, nausea, and disturbed respiration, and finally dark bloody hemorrhages in the heart and in the lungs, and congestion of the brain. Mildly affected animals recover rapidly and spontaneously. When heated to decomposition it emits toxic gases.

nicotinic. **1a:** of or pertaining to nicotine. **1b:** of or pertaining to the effects of nicotine and other drugs that, at high dosage, initially stimulate and later inhibit neuronal impulses at autonomic ganglia and neuromuscular junctions.

nicotinolytic. destructive or inhibitory to the toxic action of nicotine.

β-nicotyrine (3-(1-methyl-2-pyrryl)pyridine). an oily, liquid, insecticidal alkaloid, $C_{10}H_{10}N_2$, from tobacco; it has a characteristic odor.

nicouline. rotenone.

night adder. See *Causus*.

night adder, African. *Causus rhombeatus*.

night adder, rhombic. *Causus rhombeatus*.

nightshade. common name of any of a number of plants of the genus *Solanum* (Family Solanaceae) and of some other plants of the Family Solanaceae. See also *Atropa belladonna, Phytolacca americana*.

nightshade family. Solanaceae.

nim. *Melia azadirachta*.

niobium (formerly columbium, symbol, Nb). a rare, steel gray, lustrous, ductile and malleable metallic transition element, found mostly in association with tantalum. It is used primarily in ferrous metallurgy[$Z = 41$; $A_r = 92.91$].

nitarsone (4-nitrophenylarsanilic acid; *p*-nitrobenzenearsonic acid). a toxic substance used as an antihistomonad in domestic fowl.

niter. potassium nitrate.

nitogenin. diosgenin.

niton. obsolete synonym for radon.

nitracrine (*N,N*-dimethyl-*N'*-(1-nitro-9-acridinyl)-1,3-propanediamine;9-[[3-(dimethylamino)-propyl]amino]-1-nitroacridine). a cytotoxic and cytostatic antimetabolite derived from acridine. It is a crystalline compound, $C_{18}H_{20}N_4O_2$, that is nearly insoluble in water but soluble in most organic solvents. Nitracrine occurs also as a dihydrochloride monohydrate, which is soluble in water. The latter (commercial names: C-283, Ledakrin) used as an antineoplastic, is extremely toxic to laboratory rats and mice.

nitralin (4-(methylsulfonyl)-2,6-dinitro-*N,N*-dipropylbenzenamine; 4-(methylsulfonyl)-2,6-dinitro-*N,N*-dipropylanaline). a herbicide used to control annual grasses and weeds in a number of crops when incorporated into the top 2-5 to 5 cm of soil. Nitralin inhibits cell division in root meristems. It is, at most, moderately toxic to laboratory rats.

***m*-nitraniline**. *m*-nitroaniline.

***o*-nitraniline**. *o*-nitroaniline.

***p*-nitraniline**. *p*-nitroaniline.

nitraphen. nitrofen.

nitrate. a salt or ester of nitric acid. Nitrates are used in the treatment of angina pectoris. Nitrates as such are only moderately toxic, but if they are not promptly absorbed during digestion, they can be reduced to nitrites by microflora in the intestines. The latter are extremely toxic to humans, somewhat less so to most laboratory animals. They may also combine with amines forming nitrosamines, which are carcinogenic. As a consequence, nitrates are among the most dangerous food additives. In humans, the major source of nitrate poisoning is from foods that have been improperly preserved by nitrates. Nitrate intoxication of livestock and other animals has been by ingestion of plants that contain toxic levels of nitrates. These include certain crops such as corn, oat hay, and sorghum, and weedy pasture plants from a number of families (e.g., Amaranthaceae, Chenopodiaceae, Brassicaceae, and Solanaceae). Other sources of nitrate poisoning of animals are via the intake of pond water or well water with high nitrate content, and even the ingestion of fertilizer, machine oil, silage.

nitrate poisoning. poisoning by nitrates from various sources. The major source of human poisoning is from foods that have been improperly preserved by nitrates. Livestock and wildlife that graze on the above-ground portion of plants that contain toxic levels of nitrates, including crops such as corn, oat hay, and sorghum, and a variety of weedy pasture plants may be seriously poisoned; such poisoning occurs mainly under conditions and in seasons when plant growth has been especially rapid. Animals have also been poisoned by drinking pond water or well water with high levels of nitrate, and on occasion by ingesting fertilizer, machine oil, or silage. See also nitrite poisoning. Thus, the symptoms of nitrate poisoning are usually those of nitrite poisoning, or in some cases derive mostly from the non-nitrate moiety of the responsible compound. Cyanosis is a common feature of nitrate poisoning in human infants who have been poisoned by nitrate-containing well water.

nitrazepam (1,3-dihydro-7-nitro-5-phenyl-2*H*-1,4-benzodiazepin-2-one; Benzalin, Mogadan, Mogadon, Nelbon, Noctesed, Somnased, Somnibel, Somnite). a sedative/hypnotic that is moderately toxic to laboratory mice. It is a benzodiazepine used therapeutically as an anticonvulsant and hypnotic. It is a controlled substance.

nitre. potassium nitrate.

nitric acid (aqua fortis; azotic acid; engraver's acid). a colorless or slightly yellowish, hygroscopic, strongly oxidant, dangerously reactive, corrosive, suffocating, highly toxic liquid, HNO_3. It is miscible with water and may boil violently (b.p. 86°C). Nitric acid is a mineral acid with industrial applications such as the production of fertilizers and explosives and is widely employed in research. It is a fire hazard. Nitric acid is an important component of acid precipitation. **anhydrous nitric acid**. a liquid at standard temperature and pressure that soon becomes yellow due to nitric oxide production. It decomposes above the freezing point, releasing NO_2, H_2O, and O_2. Anhydrous nitric acid forms white monoclinic crystals below the freezing point (-41.59°C). **fuming nitric acid**. concentrated nitric acid that emits toxic fumes. It is a strong oxidizing agent, a fire hazard, and may explode on contact with organic material. It is highly toxic by inhalation and is corrosive to the skin, eyes, and mucous membranes. **white fuming nitric acid** (WFNA). a highly concentrated form of nitric acid that contains more than 97.5% nitric acid and less than 2% water. **red fuming nitric acid** (RFNA). concentrated nitric acid that contains more than 85% of the acid and usually 6-15% nitrogen dioxide.

nitric acid poisoning. nitric acid is poisonous by all routes of exposure. The symptoms produced by nitric acid empoisonment are very similar to those of sulfuric acid. Included are severe pain and burns on the skin with contact, an accelerated respiratory rate, decreased vital capacity, hypotension, an elevated blood platelet count, severe pain, vomiting, thirst, and shock. Pulmonary disease may result from inhalation of nitric acid fumes.

nitric oxide (nitrogen monoxide). a colorless, flammable gas, NO, with a sharp, sweet odor. It is used to produce nitric acid and certain plastics, paints, lacquers, and artificial fabrics. It finds additional uses in blasting, welding, electroplating, and metal cleaning. Nitric oxide is the principle oxide of smog, slowly oxidizing under suitable conditions in air to yield nitrogen dioxide. It is a powerful irritant of the lungs, eyes, and nose.

nitrile. any organic compound that contains trivalent nitrogen bound to carbon (—CN). Nitriles are colorless, pleasant-smelling liquids. See, for example, dichlobenil, bromoxynil.
nitrile glycoside. See glycoside.

nitrite. a salt or ester of nitrous acid. These compounds depress motor centers of the spinal cord; have an antispasmodic action; and as vasodilators, they lower blood pressure. Symptoms of intoxication may include headache, vertigo, palpitations, visual disturbances, flushed skin, nausea, vomiting, diarrhea, methemoglobinemia in infants, coma, and death. They are used in the treatment of angina pectoris. Both organic and inorganic nitrites are found in the home in medications, as are nitrates (*q.v.*). Those most commonly used as drugs are amyl, ethyl, potassium, and sodium nitrites. Nitrite is added to 60-65% of all pork produced in the United States, as well as some other meat, poultry, fish, and cheese. It is especially prevalent in processed meats such as bacon, sausage, luncheon meats, and hot dogs to preserve the pink color and inhibit the growth of bacteria that cause botulinum poisoning. Nevertheless, most known cases of nitrite poisoning in the United States are those in which infants have been given well water contaminated with nitrites. Nitrites are also of environmental significance because they react with some amines yielding carcinogenic nitrosamines. See also nitrite poisoning.

nitrite poisoning. nitrites are released during the digestion of food materials that contain nitrates. They are substantially more toxic than nitrates and are the most common proximate cause of intoxication from the ingestion of nitrates. Nitrites impair the oxygen-carrying capacity of the blood by oxidizing ferrous hemoglobin (Hb) to the ferric methemoglobin. Symptoms appear rapidly. They include cyanosis, trembling, weakness, and severe dyspnea. The blood becomes characteristically dark, chocolate-brown in nitrite poisoning. In some cases, the dominant symptoms are diuresis, pronounced vasodilation with concomitant reduction in blood pressure, increased cardiac activity, and coma. In these cases, vasoconstrictive drugs can be effective. In other cases, anoxia dominates. The anoxia is responsive to methylene blue. In active animals, acute intoxication results in death by asphyxiation when ca. 50% or more of the hemoglobin has been converted to methemoglobin. See also nitrate poisoning.

***m*-nitroaniline** (3-nitrobenzenamine; *m*-nitraniline). a toxic, yellow, crystalline compound, $C_{12}H_9NO_2$, used as a dye intermediate.

***o*-nitroaniline** (*o*-nitraniline). a toxic, orange-red, crystalline compound, $C_{12}H_9NO_2$, that is soluble in ethanol. It is used, for example, as a fungicide, a dye intermediate, plasticizer, wood preservative.

***p*-nitroaniline** (*p*-nitraniline; *p*-aminonitrobenzene; C.I. 37035; C.I. Azoic Diazo Component 37; C.I. Developer 17; Diazo Fast Red GG; PNA). a bright yellow, flammable powder or crystalline substance, $NO_2C_6H_4NH_2$, that is insoluble in water but soluble in ethanol and ether. It is toxic by ingestion, and by intravenous and intraperitoneal routes of exposure. Acute exposures may cause headache, nausea, vomiting, muscular weakness, stupor, methemoglobinemia, and cyanosis. Chronic exposure may cause liver damage.

3-nitrobenzenamine. *m*-nitroaniline.

nitrobenzene (nitrobenzol; essence of mirbane; oil of mirbane). a combustible, greenish-yellow, crystalline substance or yellow, oily liquid, $C_6H_5NO_2$, that is slightly soluble in water and soluble in benzene, ethanol, and ether. It is used as a solvent, in the manufacture of aniline, a constituent of metal and shoe polishes, and in the manufacture of benzidine, azobenzine, quinoline, and related compounds. It is toxic by all routes of absorption. Symptoms of intoxication include cyanosis, shallow breathing, vomiting, and death.

nitrobenzol. nitrobenzene.

***p*-nitrobiphenyl** (4-nitro-1,1'-biphenyl; 4-nitrobiphenyl; *p*-nitrodiphenyl; PNB). a water-insoluble, crystalline compound, $C_{12}H_9NO_2$, that is slightly soluble in cold ethanol, more soluble in hot ethanol, soluble in chloroform and ether. It is a known human carcinogen.

nitrochloroform. chloropicrin.

nitrofen (2,4-dichloro-1-(4-nitrophenoxy)benzene; 2,4-dichlorophenyl *p*-nitrophenyl ether;

nitraphen; nitrophen; nitrophene). a moderately toxic, selective, pre- and postemergent herbicide that is similar in use and action to trifluralin. It should, however, be placed on the soil surface and not incorporated into soil. Nitrofen is a probable carcinogen.

nitrofene. nitrofen.

nitrogen (symbol, N)[Z = 7; A_r = 14.007]. a very electronegative, odorless, diatomic, relatively inert, gaseous element that condenses to a liquid at -195.79°C (77.36° K). It comprises about 78% by volume of the atmosphere. Nitrogen is an important component of living matter. It is a simple asphyxiant in high concentrations. Many of its compounds are very toxic (e.g., many alkaloids, nitroso compounds, nitrites, nitrogen oxides, ammonia, toxalbumins, amines, nitric acid). Pharmaceutical-grade nitrogen containing not less than 99.0% by volume of N_2 is used as a diluent for medicinal gases, and to replace air in pharmaceutical preparations. See also bends.

nitrogen dioxide. a heavy, yellow-brown or reddish-brown, toxic, oxidant gas that normally exists in equilibrium with other nitrogen oxides. It converts at low temperatures to the colorless tetroxide. NO_2 is a common industrial pollutant, a component of smog, and is the chief component of nitrous fumes (See nitrogen oxide). It is also produced during the fermentation of fodder to form silage (See silo-filler's disease). See also nitrogen oxide red.

nitrogen dioxide poisoning. nitrogen dioxide intoxication of humans may result from exposure to silage (See silo-filler's disease) or to a number of industrial processes. Symptoms vary as a function of concentration and duration of exposure. Acute exposures may have a fatal outcome within a few hours to a few days as a result of pulmonary edema. Presenting symptoms often include muscular weakness, dyspnea, and coughing which may persist for 2-3 weeks. Following this period, the victim may experience fever (with or without chills), severe dyspnea, heavy ineffective coughing, and cyanosis. Treatment may be ineffective and subjects often die within 3½-6 weeks following exposure. Victims occasionally recover spontaneously. Lesions in such cases include discrete pulmonary nodules and pronounced neutrophilic leukocytosis. Symptoms in milder cases are similar to those of bronchopneumonia.

nitrogen fluoride (nitrogen trifluoride). a colorless gas, NF_3, used as an oxidizer of high-energy fuels. Chronic cutaneous exposure can cause mottling of the teeth, and skeletal changes.

nitrogen monoxide. nitric oxide.

nitrogen oxide (NO_X). **1:** any of the gaseous oxides of nitrogen. These include nitric oxide (NO); nitrogen dioxide (the monomer, NO_2, and the dimer, N_2O_4); nitrogen oxide red; nitrogen pentoxide; nitrogen trioxide; nitrous oxide. All of these gases are pulmonary irritants and can cause acute pulmonary reactions when inhaled. The most hazardous of these oxides are NO and NO_2. **2:** nitrous fumes. **3:** nitrous oxide.

nitrogen oxide poisoning. poisoning by inhalation of NO_2 or its dimer, N_2O_4. It is severely irritating to the pulmonary epithelium and associated tissues. Unless the vapor concentration is very high, there may be no symptoms near the time of exposure with the possible exception of a transient cough, nausea, and fatigue; symptoms usually present within 5-72 hours. Acute exposure may cause pulmonary edema and fatal bronchiolitis fibrosa obliterans. Inhalation for even very brief periods of time of air containing 200-700 ppm of $N0_2$ can be fatal. $N0_2$ disrupts certain enzyme systems such as lactic dehydrogenase and may also cause lipid peroxidation. These two gases are usually the chief component of most fumes that contain nitrogen oxides and cause most of the lung damage in such cases. See also nitrogen oxide.

nitrogen oxide red. a nitrogen dioxide, N_2O_4, it is often produced together with nitrogen dioxide during the fermentation of fodder to form silage (see silo-filler's disease).

nitrogen sesquioxide. nitroglycerin.

nitrogen trifluoride. nitrogen fluoride.

nitroglycerin (synonyms, def. 2: 1,2,3-propanetriol trinitrate; glycerol nitric acid triester; nitroglycerol; glonoin; blasting gelatin; blasting oil; glyceryl trinitrate; nitrogen sesquioxide;

trinitroglycerin; trinitroglycerol; trinitrin). **1:** any nitrate of glycerol, especially the trinitrate. **2:** an extremely hazardous, highly unstable, flammable, explosive, colorless to pale yellow liquid, $CH_2NO_3CHNO_3CH_2NO_3$, produced by the reaction of nitric acid (with sulfuric acid as a catalyst) and glycerine. It is highly explosive on concussion or heating and is a severe explosion hazard. It is used in the manufacture of dynamite, other explosives, and rocket propellants. Its chief role in medicine is that of a vasodilator, especially in the prophylaxis and treatment of angina pectoris. Nitroglycerin is toxic by all routes of exposure. See also nitroglycerin poisoning.

nitroglycerin poisoning. poisoning by nitroglycerine. The condition is marked by an almost immediate fall in blood pressure due to peripheral vasodilation; an intense, throbbing headache; faintness; dizziness; excessive muscular relaxation; nausea; vomiting; skin flushed then cyanotic; methemoglobinemia; postural hypertension; paralysis; convulsions due to anoxia induced by methemoglobinemia; and death. Lesions include vasodilatation, methemoglobinemia, and stagnation of venous and capillary blood with resultant anoxia.

nitroglycerol. nitroglycerin.

nitrohydrochloric acid (*aqua regia*; nitromuriatic acid; chloronitrous acid; chloroazotic acid). an extremely caustic, fuming yellow liquid mixture of concentrated nitric and hydrochloric acids (usually about 18 to 82 parts, respectively). In addition, it may also contain some nitrosyl chloride, NOCl, and free chlorine. It is used chiefly to dissolve metals (e.g., gold or platinum) and in etching. It dissolves all metals, hence the name "aqua regia." The concentrated acid is extremely corrosive and irritant.

nitrophen. nitrofen.

nitrophene. nitrofen.

***N*-(4-nitrophenyl)-*N*′-(3-pyridinylmethyl)urea**. pyriminyl.

β-nitropropionic acid. See *Aspergillus fumigatus*.

1-nitropyrene. 3-nitropyrene.

3-nitropyrene (1-nitropyrene). a probable human carcinogen.

nitrosamine. any of a group of *N*-nitroso derivatives of secondary amines, $R_2{=}N{-}N{=}O$ (where R is an alkyl or aryl group), formed by the combination of nitrates with amines or amides present in the organism. They occur in many foods, whiskey, cosmetics, herbicides, rubber factories, tanneries, iron foundries, and a number of other industrial settings. Some are strongly carcinogenic to experimental animals. See nitrate. Nitrosamines are the most studied of the *N*-nitroso compounds.

nitroso-. a prefix indicating the presence of the nitroso radical.

***N*-nitroso compound** (NOC). any of a class of compounds that contain the nitroso radical. They are carcinogens that can cause tumors in many different organs in experimental animals of various species.

nitroso radical. a reduction product, -N:O, of nitrite (NO_2).

***N*-nitrosodiethylamine** (*N*-ethyl-*N*-nitrosoethanamine; diethylnitrosamine; DEN; DENA; NDEA). a toxic, slightly yellow liquid, $(C_2H_5)NNO$, that is soluble in water, alcohol and ether. It is an antioxidant, stabilizer, and an additive to gasoline and lubricants. A known animal carcinogen and probable human carcinogen, DENA is also present in trace amounts in tobacco smoke. See also nitrosamine.

***N*-nitrosodimethylamine** (dimethylnitrosamine; DMN; DMNA). an oily, flammable, yellow, water- and ethanol-soluble liquid, $(CH_3)_2NNO$. It is an extremely toxic, carcinogenic nitrosamine and is toxic by all routes of exposure, including skin contact. Reaction time is usually a few hours. Early symptoms include nausea, vomiting, headache, abdominal cramps, diarrhea, muscular weakness, and fever. The major lesion is liver damage which may cause death. [illegible] liver, lung, and kidney cancer. When added as a preservative to stale herring meal, sodium nitrite is converted into a compound which is very toxic to mink, causing hepatic degeneration, ascites, and extensive internal hemorrhage. DMN is a probable human carcinogen. It is used as an industrial solvent, a rubber accelerator, an intermediate in the production of rocket propellants, a gasoline additive, antioxidant, and pesticide.

nitrosohemoglobin. a form of hemoglobin in which the nitroso radical binds to the heme moiety of hemoglobin. It is toxicologically similar to carbon monoxide hemoglobin. See nitrosohemoglobin poisoning.

nitrosohemoglobin poisoning. a condition similar to that of carbon monoxide poisoning. It has caused extensive mortality among cattle especially in the central United States. Symptoms are also similar to nitrite poisoning, but the blood has a normal coloration and contains no methemoglobin.

nitrous fumes. gaseous mixtures of oxides of nitrogen; nitrogen oxide (def. 2).

nitrous oxide (dinitrogen monoxide; laughing gas; hyponitrous acid anhydride; factitious air). a colorless, sweet-tasting, ethanol-soluble, slightly water-soluble gas, N_2O. It is used as an inhalation anesthetic in dentistry and surgery. See also nitrogen oxide.

nitroxyl chloride. nitryl chloride.

nitryl chloride (nitroxyl chloride). a toxic, corrosive, colorless gas, NO_2Cl, with a chlorine-like odor. Either as a gas or liquid (m.p. -145°C), this compound may explode violently on contact with organic matter. It is used as a nitrating and chlorinating agent in organic syntheses.

Nitsche's tree viper. *Atheris nitschei*.

nivalenol (12,13-epoxy-3,4,7,15-tetrahydroxy-trichothec-9-en-8-one; 3α,4β,7α,15-tetrahydroxyscirp-9-en-8-one). a crystalline substance, $C_{15}H_{20}O_7$, that is slightly soluble in water and soluble in most polar organic solvents. It is an extremely toxic tricothecene mycotoxin isolated from *Fusarium nivale*. Nivalenol is a powerful hemorrhagenic; symptoms of poisoning, in addition to hemorrhage, may include blisters, tissue necrosis, dizziness, nausea, vomiting, diarrhea, and death. It is extremely poisonous in minute amounts and has possibly been used as a warfare agent (used together with T-2 toxin) in Southeast Asia.

nm. nanometer.

NMS. neuroleptic malignant syndrome.

No. the symbol for nobelium.

no-effect concentration (NEC). **1:** the highest concentration of a drug or other biologically active substance that produces no observable effects in a given type of organism or population under known specified conditions (as in a test or controlled experiment). Operationally, a determination of "no effect" simply means that the results of the test or experiment were not statistically significantly different from those of controls. Usually only adverse effects are considered. *Cf.* no-effect level.

no-effect level (NEL; no-observed-effect level). **1:** the highest concentration or dosage of a drug or other biologically active agent that produces no observable effects in a given type of organism or population under known specified conditions (as in a test or controlled experiment). Operationally, a determination of "no effect" simply means that the results of the test or experiment were not statistically significantly different from that of controls. Usually only adverse effects are considered. Toxicologists often use the NEL interchangeably with the no-adverse-effect level (NAEL), the no-observed-effect level (NOEL), and the no-observed-adverse-effect level (NOAEL). See also dose (minimum effective dose). *Cf.* no-effect concentration. **2:** "The maximum dose used in a test which produces no adverse effects." the Organization for Economic Cooperation and Development (1981). So defined, the NEL of a test substance may be expressed, for example, as g of test substance per kg body weight of test animal or as mg of test substance per kg of food, as mg/liter of water, or as ppm in food or water. **no-effect level, food additives**. "the level of a substance that can be included in the diet of a group of animals without toxic effects" {the Food and Agriculture Organization/World Health Organization (FAO/WHO) Expert Committee on Food Additives (1974)}.

no-observed-adverse-effect concentration. a no-effect concentration; only adverse effect(s) are specified for observation or measurement. *Cf.* no-observed-adverse-effect level.

no-observed-adverse-effect level (NOEL). a no-effect level in which only adverse effect(s) are specified for observation or measurement. *Cf.* no-observed-effect concentration. The NOEL derived from toxicity testing that measures the most sensitive end point in the most sensitive organism is the one that is usually used for setting standards and for regulatory purposes.

no-observed-effect concentration (NOEC). a no-effect concentration that causes no observable adverse effects.

no-observed-effect level (NOEL). a no-effect level that causes no observable adverse effects.

NOAEL. no-observed-adverse-effect level.

nobeliumy (symbol, No)[Z = 102]. an unstable, artificial, transuranium element of the actinoid series. Several very short-lived isotopes have been prepared by bombardment of curium with carbon nuclei and similar heavy ions on other transuranium elements.

nodus atrioventricularis. atrioventricular node.

NOEC. no-observed-effect concentration.

NOEL. no-observed-effect level.

Nolina texana (sacahuista; sacahuiste; beargrass). a perennial herb (Family Liliaceae) of the foothills and rangelands of western Texas, Arizona, and Mexico. Grazing animals normally feed on the buds, blooms and mature fruit which contain no chlorophyll. If the animals also graze on other plants or plant parts containing chlorophyll, photosensitization will usually develop. The buds, blossoms, and fruit, which contain hepatotoxic and nephrotoxic principles. The minimum lethal dose and minimum toxic dose are nearly equivalent. Symptoms initially include apathy, anorexia, followed within a day or two by generalized and progressively severe jaundice. Affected animals remain, as possible, near water and in the shade. The urine may be red, due to hemoglobinuria, or yellow. Discharges from nostrils and orbit are common. Cachexia and increasing debility leading to death within a week following onset of symptoms is usual. See also phylloerythrin.

non-proliferative dust. See dust.

nonanoic acid. pelargonic acid.

nonantigenic. **1a:** having no antigenic capacity; not provoking an antibody production or cell-mediated immunity. **1b:** pertaining to properties of an antigen other than its antigenicity.

nonemé. *bitis nasicornis*.

nonoic acid. pelargonic acid.

nononcogenic. not productive of tumors.

nonoxinol. nonoxynol.

nonoxynol (α-(4-nonylphenyl)-ω-hydroxypoly-(oxy-1,2-ethanediyl); polyethyleneglycols mono(nonylphenyl) ether; macrogol nonylphenyl ether; nonoxinol; polyoxyethylene(*n*)nonylphenyl ether; nonylphenyl polyethyleneglycol ether; nonylphenoxypolyethoxyethanol). an alkyl phenoxy polyethoxy ethanol used in its various forms as spermicidal agents (nonoxynol-9, 11), pharmaceutic aids (nonoxynol-4, 15, 30), nonionic detergents, emulsifiers, wetting agents, dispersants, and as intermediates in the synthesis of anionic surfactants. They appear to be no more than moderately toxic to higher animals, *per os*. Animals poisoned orally by this compound give evidence of gastrointestinal irritation (e.g., bloating, diarrhea). See nonoxynol-9. **nonoxynol-9**. the *p*-nonylphenyl ether of an ethylene glycol polymer with 9 units. Spermicidal contraceptives may contain 2-12.5% of this compound. Users often experience a transient burning irritation of the penis or vagina. See also Ortho-creme. **nonoxinol 11**. used as a spermicide. See Duragel, Duracreme. See also contraceptive.

nonpolar. of or pertaining to a molecule or to material that does not have a strong net electrical charge or polarity. Nonpolar molecules have low solubilities in water and higher solubilities in lipids.

nonsmoker. see cigarette and side stream smoke.

nonspecific. **1:** not known to be caused by a particular agent (i.e., xenobiotic, stimulus, pathogen). **2:** not dependent on the nature of the causative agent. **3:** not directed against a specific agent as in nonspecific therapy.

nonspecific agent. an agent (e.g., a toxicant) that affects many targets and fails to act selectively on any one.

nonspecific change. **1a:** a change that can be elicited by any of many agents. **1b:** a change or response of a cell, organ or tissue, the nature of which is independent of the stimulus or triggering agent.

nonspecific cholinesterase. pseudocholinesterase.

nonsteroidal anti-inflammatory drug (NSAID). a class of weakly acid, nonsteroidal, nonionizable, protein-bound antiinflammatory agents that include fenamates, indenes, oxicams, propionates, pyrazolones, and salicylates. They interrupt inflammatory intercellular communication by inhibiting prostaglandin synthesis. Unintended effects may include headache, nausea, gastritis, peptic ulcer, nephritic syndrome, platelet dysfunction, confusion, allergic rhinosinusitis, and aggravation of asthma and allergic dermatoses.

nontoxic. there is no commonly accepted usage for this term. A determination of nontoxic status (defs. 1 and 2) must, in principle, and in fact, be determined for each population of organisms, taking into account such factors as age or developmental stage and sex, and prevailing environmental variables. As a consequence of the difficulties in doing so, and because many factors can modify toxicity, the use of this term is rarely justified. A determination of nontoxic cannot be based on lethality. **1:** practically nontoxic; atoxic; not toxic at concentrations or in amounts to which a living system might conceivably be exposed. **2:** not toxic to a given biological system at concentrations or in amounts routinely encountered by the system in its natural environment (or in a specified environment). **3:** when appearing on the label of a consumer product, this term probably means nontoxic (as in def. 1) if used according to instructions on the label. The product in question may be significantly toxic under some conditions of exposure or to some individuals. At best, this usage of "nontoxic" implies that a product produced limited toxicity in certain laboratory animals during short-term toxicity tests. Repeated, long-term exposure to the so-called "nontoxic" dose is rarely made. In the case of skin or eye contact, the label term "nontoxic" can only be taken to mean that no serious damage resulted from tests involving short term eye or skin contact. **4:** not productive of poisons. **essentially nontoxic**. where a substance produces no demonstrable harm under conditions of normal usage or exposure over a long period of time. **practically nontoxic**. **1a:** nontoxic (defs. 1, 2); of or pertaining to an agent that produces adverse effects on a designated living system only when present in large amounts or at very high concentrations. **1b:** a substance that has a probable lethal oral dose for humans of more than 15g/kg body weight is usually considered practically nontoxic. See also toxicity, toxicity rating.

nontoxic substance. a substance that is practically nontoxic.

nontropical sprue. See gluten-induced enteropathy.

nonvenomous. **1:** not productive of venoms. **2:** of or pertaining to an animal or group of animals that is not venomous.

nonviable. incapable of living.

nonylic acid. pelargonic acid.

19-nor-17-α-pregna-1,3,5(1)-trien-20-yne-3,17-diol. ethinylestradiol.

norbormide (3a,4,7,7a-tetrahydro-5-(hydroxyphenyl-2-pyridinylmethyl)-8-(phenyl-2-pyridinylmethylene)-4,7-methano-1H-isoindole-1,3-(2H)-dione; 5-(α-hydroxy-α-2-pyridylbenzyl)-7-(α-2-pyridylbenzylidene)-5-norbornene-2,3-dicarboximide). a toxic, white, water-insoluble solid that is soluble in dilute acid used as a rodenticide.

norepinephrine (4-(2-amino-1-hydroxyethyl)-1,-2-benzenediol;α-(aminomethyl)-3,4-dihydroxybenzyl alcohol; 2-amino-1-(3,4-dihydroxyphenyl)ethanol; 1-(3,4-dihydroxyphenyl)-2-aminoethanol; noradrenaline). an adrenergic hormone produced by the chromaffin cells of the adrenal medulla of vertebrates. It is a vasoconstrictor (but with little effect on cardiac output) and mediates transmission of sympathetic nerve impulses. It is used therapeutically to maintain blood pressure in acute hypotension resulting from trauma, heart disease, or cardiovascular collapse; it has similar usage in veterinary practice. Norepinephrine (usually given as the bitartrate) may cause local tissue necrosis, headache, bradycardia, and hypertension. Some individuals develop hypersensitivity to this substance and it may not be used in conjunction with halothane or cyclopropane anesthesia. *Cf.* epinephrine.

norhyoscyamine (1-tropic acid 3α-nortropanyl ester; pseudohyscyamine; solandrine). an alkaloid, $C_{16}H_{21}NO_3$, from plants of the family Solanaceae, toxicological properties similar to those of hyoscyamine.

Norisodrine™. isoproteronol hydrochloride.

normal. ordinary, usual, typical; standard; average, common;

normal lead acetate. lead acetate.

normo- a prefix meaning normal, usual; conforming to the rule.

normotonia. normal tone or tension.

normotonic. of, pertaining to, or characterized by normotonia.

normoxia (normoxicity). the condition of having a normal level of molecular oxygen (a level that fully supports aerobic life) present in the surroundings; used especially in reference to a habitat. *Cf.* anoxia, oligoxia.

normoxic. of, pertaining to, or characteristic of normoxia.

normoxicity. normoxia.

nornicotine (3-(2-pyrrolidinyl)pyridine; 2-(3-pyridyl)pyrrolidine). a toxic, rather viscous, hygroscopic liquid, $C_9H_{12}N_2$, with a slightly amine odor. It is found in plants of the genus *Nicotiana*; it occurs in smoking tobacco and other tobacco products. It is used as an insecticide on farms and horticultural plants. It is highly toxic to laboratory mammals and is about one-third as toxic as nicotine to humans. Symptoms of intoxication include light-headedness, muscular weakness, severe nausea, vomiting, diarrhea, prostration, and collapse.

norsympatol. octopamine.

norsynephrine. octopamine.

North African horned viper. *Cerastes cerastes*.

North Carolina pine. *Pinus taeda*.

northern bearded milky. *Lactarius representaneous*.

northern copperhead. *Agkistrodon contortrix mokeson*.

northern horse mussel. *Modiolus modiolus*.

northern mole viper. *Atractaspis microlepidota*.

Northern Pacific rattlesnake. *Crotalus viridis oreganus*.

Norway rat. *Rattus norvegicus*. See also laboratory rat.

nose-horned viper. *Bitis nasicornis*, *Vipera ammodytes*.

noso-. a prefix indicating relationship to disease.

nosocomial. pertaining to or originating in a hospital, as a disease. *Cf.* iatrogenic.

nosogenic. pathogenic.

nosointoxication. poisoning by the toxic products of disease.

nosopoietic. pathogenic.

nosotoxic. productive of nosotoxicosis.

nosotoxicity. the quality of being nosotoxic.

nosotoxicosis. any disease caused by or associated with poisoning.

nosotoxin. any toxin that causes or is associated with disease.

Notechis (tiger snakes). a genus of extremely venomous snakes of Australia (Family Elapidae). ***N. scutatis*** (Australian tiger snake; black tiger snake; brown-banded snake; Krefft's tiger snake). possibly the most deadly of all terrestrial snakes, it inhabits relatively wet, brushy, rocky situations in Tasmania and southern Australia from the border of Queensland to the coast of South Australia; it inhabits some of the offshore islands as well. The body is commonly brown or creamy yellow with a large number of dark bands, but ground color varies among individuals from yellowish, greenish-gray, orange, and brown to black. Dark individuals may show little if any banding. Adults are 1.2-1.5 m long in most parts of the range, but may reach 2 m or more in Victoria and Tasmania. Dorsal scales have pointed tips. This snake is undoubtedly the most dangerous snake in southern Australia. It is nocturnal and is not aggressive unless molested. Most strikes seem to occur when the snake is accidentally stepped on in the dark. There are usually few focal effects, but the venom is the most virulent known among terrestrial snakes. Systemic effects appear rapidly, are severe, and prove fatal in well over half of untreated human victims. The victim usually experiences pain, edema, numbness, mental disturbances, and rapidly developing paralysis. Treatment with antivenin is usually quite effective.

Notesthes robusta (bullrout). a venomous fish (Family Scorpaenidae) found in the waters of New South Wales and Queensland, Australia. The venom apparatus is of the *Scorpaena* type (See Scorpaenidae).

Notholaena sinuata (Jimmy fern; cloak fern). a neurotoxic perennial, evergreen, upright fern that occurs on dry, rocky hills especially those with limestone soils in western Texas, Arizona, New Mexico, and Mexico. It is responsible for "jimmies" of range livestock (especially sheep).

nourishing. providing or able to provide nourishment.

nourishment. **1a:** nutrition. **1b:** food, foodstuffs, edibles; any substance that nourishes and supports life. **2:** the act or process of nourishing or of being nourished.

nourodiou. *Atractaspis microlepidota*.

Novocain™. procaine hydrochloride.

NO_X. nitrogen oxide.

noxa. any agent that is harmful to health.

NOXA. naphth-2yl-oxyacetic acid.

noxious. **1a:** offensive. **1b:** harmful, injurious, hurtful, malignant, pernicious, poisonous, damaging to tissue.

Np. the symbol for neptunium.

NPN. nonprotein nitrogen.

NSAID. nonsteroidal anti-inflammatory agent.

NSC 138780. T-2 toxin.

NTE. neurotoxic esterase.

NTEL. no-toxic-effect level.

NTP. See National Toxicology Program.

nucin. juglandic acid.

nuclear reactor (atomic pile; pile; reactor). the fuel element in a nuclear reactor. It contains sufficient, extremely toxic, fissionable material arranged in such a way that, when properly managed, it can maintain a controlled, self-sustained nuclear fission chain reaction.

nuclear waste. radioactive waste.

nucleotoxin. **1:** a toxin extracted from cell nuclei. **2:** a toxin that acts on cell nuclei.

nuclide (nuclear species). a species of atomic nucleus (or atom) defined by the number of protons, neutrons and energy content. Thus all of the following are different nuclides. The nucleus of ^{12}C has 6 protons and 6 neutrons, ^{23}Na has 11 protons and 12 neutrons, and ^{24}Na has 11 protons and 13 neutrons, and ^{24}Mg has 12 protons and 12 neutrons. A nuclide is alternatively identified by the proton number (atomic number), mass number, and atomic mass. To be considered a nuclide, an atom must have a measurable lifetime, generally more than 10^{-10} second.

numb. **1:** having diminished sensation; unfeeling; anesthetized; **2:** to diminish sensation; to deaden, desensitize, anesthetize, paralyze.

number. any cigarette made from marijuana.

numbness. absence of or diminished sensation of a body part.

nutmeg. **1:** ***Myristica fragrans.*** **2:** the ground, dried kernels of the seeds of *M. fragrans*, used as a spice. Ingestion of large amounts (5-15 g) produces symptoms similar to that of atropine: CNS excitation, euphoria, hallucinations, delirium, flushing of the skin, tachycardia, and lack of salivation. This substance is sometimes abused because of the narcotic effects. The chief toxic principle is myristicin.

nutmeg family. Myristicaceae.

nutrasweet. See aspartame.

nutrient. **1a:** any substance that provides an organism with the elements that are essential to normal metabolism; those substances that promote growth, development, and support life processes. **1b:** often defined operationally as any substance assimilated by a living organism that promotes growth (nitrogen and phosphorus in wastewater can be considered nutrients under this definition, although they are sometimes environmentally damaging). **2:** imparting, serving as, or providing nourishment.

nutrient deficiency. See malnutrition.

nutrient excess. See malnutrition.

nutriment. a nutritious substance; that which nourishes.

nutrition (nourishment). **1:** all processes involved in the acquisition and utilization of foods that support growth, development, repair, and maintenance of an organism, including the provision of energy, essential elements, and nutrients that support metabolism, vital activities, and reproduction. Functions (normal and pathological) such as ingestion, storage, digestion, assimilation, are thus processes included under this term. The nutritional state of an animal can affect toxicity; good nutrition is protective against many toxicants. Thus, restriction of overall food intake in normally overfed laboratory rats and mice can cause a conspicuous reduction in the incidence of chemically induced tumors. A protein deficient diet in laboratory animals can increase or decrease toxicity of various toxicants. Serious injury, whether traumatic or due to empoisonment, is often accompanied by nutrient losses. Thus caustic or irritant poisons may damage the tissues of the alimentary canal to such an extent that ingestion; absorption, and digestion are severely impaired. Severe liver damage from hepatotoxicants can greatly alter the metabolism of nutrients. Poisons of many types affect nutrition, but the nature and seriousness of their effects vary as a function of nutrition before and/or after exposure. Thus, for example, the acute toxicities of rotenone and TNT are potentiated by a diet high in fat, whereas such diets reduce the toxicity of cyanide. See also starvation, malnutrition. Thiamine deficiency sensitizes an animal to the cardiotoxic action of arsenicals and of cobalt. **2:** the science that treats processes involved in nutrition as defined above. Included is the study of dietary deficiencies, malnutrition, the problems of overnutrition, the relationships between nutrition and disease processes. Included also is the study of the role and importance of nutrition in toxicology. Indeed the fields of nutrition and toxicology have generated the science of nutritional toxicology (See toxicology).

nutritional. pertaining to nutrition.

nutritional deficiency. **1a:** any suboptimal amount, concentration, or proportion of a particular nutrient in a particular diet can be regarded as a deficiency of that nutrient. The deficiency of a single nutrient may affect the toxicity of xenobiotics. Thus a deficiency of thiamine increases the sensitivity of an animal to the cardiotoxicity of cobalt and of arsenicals. **1b:** a nutritionally deficient diet is one that contains suboptimal amounts or concentrations of one or more essential nutrients. **1c:** operationally a diet is nutritionally deficient if the organism maintained on the diet fails to grow or thrive.

nutritious (nutritive). nourishing, wholesome, sustaining; affording nutriment.

nuttall death camas. *Zigadenus nuttallii*

nux vomica (quaker buttons; bachelor's buttons; poison nut; dog buttons; vomit nut). the dried ripe seed of *Strychnos nux-vomica*. It contains several toxic alkaloids, most importantly strychnine and brucine. It has been used in medicine as a bitter tonic and CNS stimulant (but See strychnine) and is used in veterinary medicine as a bitter tonic and to treat chronic indigestion, inappetence, and atony of the rumen.

O. the symbol for oxygen.

^{15}O. the symbol for oxygen-15.

^{16}O. the symbol for oxygen-16.

^{17}O. the symbol for oxygen-17.

^{18}O. the symbol for oxygen-18.

oak poisoning (quercus poisoning). a toxic syndrome to which many species of mammals are susceptible. It is usually acquired by ingestion of buds and immature leaves of most, if not all, species of oak in North America and Europe. The active principle is the gallic acid contained in the tannin of most, perhaps all, species of *Quercus*, *q.v.* Symptoms and lesions are chiefly those due to gastrointestinal distress and a distinctive renal dysfunction. Clinical signs include anorexia, nasal discharge, rumen stasis in ruminants, polydipsia, polyuria, constipation, and abdominal tenderness and pain. Poisoning has been prevented experimentally by adding calcium hydroxide to tannic acid. If added to the ration to make up 15% of the bulk, the syndrome may be prevented in livestock and may relieve symptoms if given early in the course of the disease. The syndrome is similar to that of acorn poisoning. Animals are usually poisoned from browsing on scrub oak or upon the foliage of felled trees.

oasis mole viper. *Atractaspis engaddensis*.

oats. See *Avena sativa*.

objective. **1:** detached, impartial, free from judgmental bias; having an awareness or based on an awareness of things as they really are; substantive; factual; concrete; not a matter of opinion. *Cf.* subjective. **2:** a goal; something strongly desired; what one intends to accomplish.

obligate. **1:** to compel, constrain, require. **2:** confined to performing a particular action or function. **3:** used in reference to a living organism that can survive only under a restricted set of environmental conditions. See also metallophyte, accumulator, selenophile. *Cf.* facultative.

obligate accumulator. See accumulator.

obs. obsolete.

obtund. **1a:** to render dull or less acute. **1b:** to soothe or deaden pain. **2:** to render insensitive to disturbing or painful stimuli by decreasing the level of consciousness (e.g., by anesthesia or the use of a narcotic analgesic).

obtundation. **1:** diminished alertness, usually accompanied by hypersomnia. **2:** the use of an agent that soothes or deadens pain by blocking transmission of pain stimuli at some level in the CNS as with an anesthetic or tranquilizer.

obtundent. **1:** having the capacity to dull pain or to diminish alertness. **2:** a soothing or partially anesthetic drug.

OC. oral contraceptive.

occupancy. applied to a body part, it is the period of time that elapses from the administration (by a specified route and method) of a substance until it is excreted or is decomposed.

occupational poisoning. poisoning associated with the pursuit of an occupation.

ochratoxin. any of a number of hepatotoxic mycotoxins (e.g., from *Aspergillus ochraceus*, *A. sulphureus*, *A. melleus*, and *Penicillium viridicatum*). They sometimes contaminate cottonseed, corn, other stored grains, and peanuts. **ochratoxin A**. the main ochratoxin, it is a chlorine-containing mycotoxin, $C_{20}H_{18}ClNO_6$, produced by a number of fungi including *Aspergillus ochraceus* and *Penicillium viridicatum* that contain an α,β-unsaturated lactone moiety. It is hepatotoxic and nephrotoxic and known to be extremely toxic to laboratory mice, rats, monkeys, dogs, domestic fowl, ducklings, and trout. **ochratoxin B**. a dichloro derivative of ochratoxin A, $C_{20},H_{19}NO_6$; it is very toxic, but is the least toxic of the ochratoxins. **ochratoxin C**. the amorphous ethyl ester of ochratoxin A, $C_{22}H_{22}ClNO_6$. It is hepatotoxic, nephrotoxic, and very toxic to a number of laboratory animals.

octachlorocamphene. toxaphene.

octadecanoic acid mercury salt. mercuric stearate.

3,3a,4,4a,7a,8,9,9a-octahydro-4-hydroxy-4a,8-dimethyl-3-methyleneazuleno[6,5-b]furan-2,5-dione. helenalin.

[1*R-trans*]-octahydro-2*H*-quinolizine-1-methanol. lupinine.

octamethyl pyrophosphoramide. schradan.

octamethyldiphosphoramide. schradan.

octane, *n*-octane. a clear, colorless, flammable liquid, $CH_3(CH_2)_6CH_3$, with an odor like gasoline. It is a lower alkane (See alkane) used chiefly as a solvent and fuel. It is a dangerous fire and explosion hazard. It is a lower alkane, and is an irritant and simple asphyxiant. Large numbers of persons are exposed daily to dangerous levels of octane. Symptoms of intoxication include irritation of the eyes and nose, drowsiness, dermatitis, and perhaps pneumonia.

octanoic acid (caprylic acid; octylic acid; octoic acid). a colorless, oily, malodorous liquid, $CH_3(CH_2)_6COOH$, with a rancid taste; it emits vapors that induce coughing and is also a skin irritant. It is mildly toxic by ingestion and moderately toxic by intravenous exposure.

1-*n*-octanol (*n*-octyl alcohol, primary; alcohol C-8; heptyl carbinol). a colorless liquid, CH_3-$(CH_2)_6CH_2OH$, with a penetrating aromatic odor. It is miscible with ethanol, chloroform, and mineral oil but immiscible with water and glycerol. This substance is well known in toxicology (See partition coefficient).

octopamine (α-(aminomethyl)-4-hydroxybenzenemethanol; α-(aminomethyl)-*p*-hydroxybenzyl alcohol; 1-(*p*-hydroxyphenyl)-2-aminoethanol; norsympatol; norsynephrine; *p*-hydroxyphenylethanolamine; D-*p*-hydroxyphenylethanolamine). an ethanolamine, octopamine is a probable invertebrate neurotransmitter; it is adrenergic and neurotoxic. Found in the salivary glands of octopods, *Octopus vulgaris*, *O. macropus*, and *Eledone moschata*. See also catecholamines, eledoisin.

Octopoda. one of two suborders of the molluscan Order Dibranchia that includes octopi. The body is spherical or sac-like, without fins; head large; eight arms that bear sessile suckers; shell usually absent.

octopus. common name of decapod molluscs of the genus *Octopus*.

Octopus. a genus of more than 50 species of venomous cephalopod molluscs (Suborder Octopoda; Order Dibranchia) characterized by a nearly spherical or sac-like body, lacking fins; a large head with eight similar, outspread arms (0.3-6.7 m long) that bear sessile suckers. They are usually shy, nocturnal or crepuscular, marine animals that may occur in shallow to deep water. The venom apparatus consists of paired posterior and paired anterior salivary glands, each connected via short ducts to a common salivary duct, the buccal mass, and mandibles (beak). Envenomation by bites paralyzes the small crustacea on which it feeds. They are not usually dangerous to humans, but fatalities are known. The beak is comprised of dorsal and ventral chitinous mandibles that can close with great force, usually producing two small puncture wounds. Symptoms of envenomation in humans usually include a burning sensation, swelling, redness, heat, and often profuse bleeding from the wound. Most envenomations result from improper handling of small specimens. The bites of North American octopi are rarely serious. ***O. lunulatus*** (= *Hapalochaena maculosa*; Australian blue-ringed octopus). a venomous octopus of the Indian Pacific region, especially along the southern Australian coast. Mature specimens are about 20 cm long and the tentacles have a span of about 10 cm. The venom is very toxic, containing a high concentration of hyaluronidase. It is neurotoxic, hypotensive, convulsant, and paralytogenic. The bite, which can be lethal to humans, is often painless initially, with pallor, edema, and hemorrhage appear within a few minutes. A local stinging or burning sensation soon follows. Early symptoms include tingling or a sensation of numbness of the neck, mouth and head. Nausea and vomition may follow. In severe cases, one suffers respiratory distress, ocular paralysis, dizziness, incoordination, difficult speech, swallowing, muscular weakness, and even complete paralysis lasting ca. 4-12 hours in nonfatal cases. Coma and death may result from respiratory paralysis. See also maculotoxin.

octylene glycol. ethohexadiol.

odoroside. a toxic steroid glycoside produced by *Nerium indicum, q.v.*

ODTS. organic dust toxic syndrome.

Oenanthe crocata (hemlock water dropwort; water dropwort). highly toxic. The convulsant toxin, enanthotoxin, is very similar to that of water hemlock, *Cicuta spp.*, *q.v.* It is generally considered to be the most toxic plant in England; it has been introduced into the United States.

oenanthotoxin. enanthotoxin.

oesophagus. esophagus.

oestrus. estrus.

oestrus cycle. estrus cycle.

Ogmodon vitianus (Fiji snake). a small, terrestrial, burrowing snake, usually less than 0.5 m long (Family Elapidae). It occurs on Vita Levu, the largest island, and perhaps others, of the Fiji group in the western South Pacific Ocean. The head is small, with a pointed snout, and is indistinct from the neck; the third upper labial is often separated, forming a preocular; there is no canthus rostralis. The body is rather slender and cylindrical with a short tail. This snake is not believed to be dangerous to humans.

Ohio buckeye. *Aesculus glabra*.

oidium (arthrospore). an asexual spore produced by the fragmentation of a fungal hypha into individual cells of a fungal hypha.

Oignon sauvage. *Arisaema atrórubens*.

oil. See petroleum or specific type of oil (e.g., mustard oil).

oil beetle. a common name of beetles of the family Meloidae.

oil gland. sebaceous gland (See skin).

oil of bitter almond. bitter almond oil.

oil of calamus. a highly toxic, carcinogenic oil. Toxicity probably resides in its asarone or safrole content. See also *Acorus calamus*, *Sassafras albidum*.

oil of chenopodium. See wormseed oil, *Chenopodium*.

oil of cherry laurel (cherry laurel oil). a volatile, highly toxic and hazardous, yellow oil from the leaves of *Prunus laurocerasus*. It contains hydrogen cyanide, benzaldehyde, benzaldehyde cyanhydrin, and benzyl alcohol. The toxicity is due chiefly to the cyanide content. This material must be kept sealed and protected from light when stored.

oil of Euphorbia. mercurialine.

oil of mirbane. nitrobenzene.

oil of myristica. oil of nutmeg, volatile.

oil of nutmeg, volatile (oil of myristica). a steam-distilled oil from the kernels of ripe seeds of *Myristica fragrans*. Ingestion of large amounts are narcotic, causing delirium and death. The toxic principle is myristicin.

oil of rue. a toxic, volatile oil from *Ruta graveolens*. A pale yellow liquid with an unpleasant odor. The principal components are methyl nonyl ketone and methyl anthranilate. Frequent exposure by dermal contact causes vesication and erythema. Ingestion of large amounts may cause gastrointestinal distress with abdominal pain, nausea, vomiting, confusion, convulsions, and death.

oil of tansy. a poisonous, volatile, yellow oil that becomes brown on exposure to the atmosphere. It is obtained from the leaves and tops of *Tanacetum vulgare*. Constituents include thujone, borneol, camphor. See tansy poisoning.

oil of turpentine, sulfurated. sulfurated oil of turpentine.

oil of vitriol. sulfuric acid.

oil of wintergreen. See methyl salicylate.

oil slick. See slick.

Okinawa pit viper. *Trimeresurus okinavensis*.

Okinawan habu. *Trimeresurus okinavensis*.

Old World viper. See Viperidae.

oldfield pine. *Pinus taeda*.

oleander. *Nerium oleander*.

oleandrin(16β-(acetyloxy)-3β-[(2,6-dideoxy-3-*o*-methyl-L-arabino-hexopyranosyl)oxy]-14-hydroxycard-20-(22)-enolide; neriolin). **1:** a cardiac glycoside, $C_{30}H_{46}O_8$, from oleander, composed of digitalose and digitaligenin. **2:** an alkaloid, $C_{32}H_{48}O_9$, produced by *Nerium oleander*. It is a powerful diuretic, formerly used in the management of cardiac insufficiency.

oleandrin poisoning. poisoning by ingestion of any part of the bush, *Nerium oleander*, q,v. Symptoms may include intense abdominal pain, nausea, vomition, hypotension, bloody diarrhea, hypothermia, cardiac arrhythmias, cyanosis, convulsions, coma, and death.

oleandrism. poisoning by oleandrin or by *Nerium oleander*.

oleate of mercury. mercuric oleate.

olefiant gas. ethylene.

olefin. See alkene.

oleic acid. See fatty acid.

oleic acid lead salt. lead oleate.

oleum. fuming sulfuric acid (See sulfuric acid).

olig-, oligo-. a prefix meaning few, very little, or deficient.

oligemia. hypovolemia.

oligemic. hypovolemic.

oligohemia. hypovolemia.

oligohemic. hypovolemic.

oligopnea. hypoventilation.

oligoxia (oligoxicity). the condition of having a level of molecular oxygen present in the surroundings; used especially in reference to a habitat. *Cf.* anoxia, normoxia.

oliguresis. oliguria.

oliguria (oliguresis). diminished production of urine in relation to fluid intake, a condition that may be observed following bleeding, diarrhea, or profuse sweating. The condition may also result from renal failure due to any cause. Urine may be retained in cases of CNS disorders, shock, poisoning, deep coma, and hypertrophy of the prostate gland. *Cf.* anuria. See also shock.

oliguric. of, pertaining to, or characterized by oliguria.

olive brown sea snake. *Aipysurus laevis*.

olive-green viper. *Causus lichtensteinii*.

olwero. *Bitis arietans*.

omnivore. an animal that eats a broad range of food materials of both plant and animal origin.

omnivorous. pertaining to an animal that is able to subsist on or has the ability to consume a wide variety of foods, including both plants and animals or their products. Whereas *Homo sapiens* is omnivorous, many individuals (strict vegetarians for personal, religious, or social reasons) and populations (e.g., those in the far north that subsist on a few types of animal prey) are not omnivorous

OMPA. schradan.

Omphalotus illudens. See *Omphalotus olearius*.

Omphalotus olearius (formerly *Clitocybe illudens*

in part; Jack o' Lantern, orange Jack o' lantern; false chanterelle). an orange to yellowish-orange, mild to moderately poisonous mushroom (Family Tricholomataceae) with sharp-edged gills that usually occurs in clusters at the base of stumps or buried roots of deciduous trees in eastern North America and California. Some experts consider this part of a species complex, in which case the eastern North American form, should be designated *O. illudens* and its western counterpart, *O. olivascens*. All parts are poisonous. If ingested, it typically causes nausea, vomiting and/or diarrhea, gastric distress, sensory disturbances, and muscle relaxation that persist for a few hours to 2 days. See mycetismus gastrointestinalis.

Omphalotus olivascens. See *Omphalotus olearius*.

On-scene Coordinator (OSC). in the United States, a person responsible for responses to spills of hazardous substances.

onco-. a prefix indicating relationship to a tumor, swelling, or mass.

oncogene (oncogenic gene). **1a:** a gene that contains cancer-inducing genetic sequences. When activated it can convert a normal cell into a neoplastic cell. **1b:** operationally, a gene that brings about the formation and development of a neoplasm. *Cf.* protooncogene.

oncogenic (oncogenous). giving rise to, or producing tumors, especially malignant tumors; suitable for the formation and development of a neoplasm.

oncogenous. oncogenic.

oncology. the health science that deals with the pathogenesis, structure, biological and chemical properties, and treatment of neoplasms. **toxicological oncology** (cancer toxicology). the branch of oncology (or of toxicology, depending upon the main thrust and tools applied) that is directed toward determining the potential for and role of toxicants and the induction, development, and treatment of neoplasms.

oncotropic. having a special affinity for neoplasms or neoplastic cells.

oncovirus. a virus (Subfamily Oncovirinae) that can cause an RNA tumor.

one hundred pace snake. *Agkistrodon acutus*.

onion. See *Allium cepa*.

onion-stalked lepiota. *Lepiota cepaestipes*.

onobaio. a powerful arrow poison, *q.v.*, from Obok, Africa; it has a cardioinhibitory action.

Onoclea sensibilis (sensitive fern). a neurotoxic perennial fern found in open woods, meadows, and old fields throughout the eastern United States, west to Texas, and north into Canada. Toxicity and symptoms in horses that graze on this plant vary but are neurological in character.

Oonopsis. (goldenweeds). a genus of plants (Family Compositae), all of which are obligate accumulators of selenium from soil.

OP. organophosphate, organophosphate ester.

opàrcâ. *Vipera ammodytes*.

ophi-, ophio-. a prefix that **1:** refers to or denotes snakes. **2a:** indicates a thing that suggests a snake or is snake-like. **2b:** denotes the quality of being or resembling a snake; used in relation to a specified structure or quality.

ophic. ophidic.

Ophidia. Serpentes.

ophidian. **1:** snakelike. **2:** a snake.

ophidiasis. ophidism.

ophidic. of, pertaining to, caused by, or obtained from snakes.

ophidism (ophidiasis; ophiotoxemia; ophitotoxemia). **1:** any condition that results from envenomation by a snake. **2:** snake venom poisoning.

ophiobolin A. the toxic principle of *Helminthosporium oryzae*.

Ophiophagus hannah (king cobra; hamadrayad; belalang; Königskobra; oraj totok; raj nag;

tedong selar; ular anang; ular kunyett terus). a very large, diurnally active cobra (Family elapidae), it occurs throughout much of southeast Asia, including the Philippines and the larger islands of Indonesia. It is larger than any Asian snake except the pythons and exceptional specimens of the nonvenomous *Zaocys* (keeled rat snake) which sometimes reaches a length of about 3.7 m. Adults king cobras average about 2-4 m throughout most of the range and occasional individuals may approach 5 m in length. This snake is easily recognized by its great size and the presence of large occipital shields, which are unique to this species. The hood is relatively narrower than that of the *Naja*. The body is slender, tapering, with a long tail, and the neck can expand into a small hood. The head is rather short, depressed, with a broad snout, a rounded indistinct canthus, and is somewhat distinct from neck. The eyes are of moderate size with round pupils. The ground color of adults is olive, brown, or greenish yellow; darker on the tail; the scales on the head are edged in black; the throat is yellow or orange, sometimes with black markings. Young individuals are black with buff, white or yellow chevron-shaped narrow crossbands. This pattern is usually retained by the adults in the Indian state of East Bengal, in The Union of Myanmar (formerly Burma) and in Thailand. It is one of the most dangerous snakes in the world. Effects of envenomation include rapid swelling, dizziness, loss of consciousness, dyspnea; cardiopulmonary effects, including cardiac arrhythmia, may be pronounced and life-threatening. There are usually no gross lesions of internal organs nor about the bite (except for the wound itself). Mortality varies sharply with the amount of venom involved, but nonlethal amounts are injected in most cases. This snake infrequently bites humans, although unprovoked attacks are recorded.

ophiotoxemia. ophidism.

ophiotoxicology. See toxicology.

Ophiuroidea. the largest class of the phylum Echinodermata, containing the brittle stars (e.g., *Ophiothrix)*. The body, covered with articulating skeletal plates, consists of a small central disc from which long, fragile, radiating arms are sharply delimited. The mouth is on the ventral surface. Locomotion is effected by the arms, the tube feet being used only for feeding.

ophthalmia. severe inflammation of the eye.

ophthalmotoxin. **1:** a toxin formed from an emulsion of the ciliary body. **2:** a toxin that acts on the eye. See also *Hemachatus*, *Naja naja sputatrix*, *Naja nigricollis*.

opiate. any drug that contains or is derived from opium.

opiate poisoning. poisoning due to overdosage of an opiate. Symptoms of acute opiate poisoning include euphoria, flushing, itching of the skin, miosis, drowsiness, decreased rate and depth of breathing, hypotension, bradycardia, and a fall in body temperature. Death may result in the absence of suitable treatment.

OPIDN. acronym for organophosphorus-induced delayed neurotoxicity.

opioid (e.g., heroin; morphine; codeine; meperidine; methadone; hydromorphone; opium; pentazocine; propoxyphene; fentanyl; sufentanil). **1:** a synthetic narcotic; one not derived from opium. Acute poisoning or overdosage may decrease body temperature, depress respiration, and reduce blood pressure, sometimes with symptoms of shock. Depending upon dosage (the specific opioid involved) and tolerance of the victim, the following may be observed: euphoria or stupor, miosis, diminished or absent reflexes, pulmonary edema, constipation, convulsions, cardiac arrhythmias; coma. Withdrawal from usage results in a syndrome marked by dilated pupils, erection of skin papillae as seen also in cold or shock (goose flesh), tachycardia, lacrimation, abdominal cramps, muscle twitches. One may also observe vomiting, diarrhea, shakiness, yawning, and apprehension or anxiety. **2:** of, pertaining to, or indicating a substance (e.g., an enkephalin or endorphin) that occurs naturally in the body and acts on the CNS to relieve pain. **3:** having the characteristics or action of opium or of an opioid (def. 1).

Opisthoglypha. a group of genera of snakes (Family Colubridae) that have one or more pairs of grooved teeth in the posterior part of upper jaw. They are usually venomous, but are rarely harmful to humans.

opisthotonoid. resembling opisthotonos. See emprosthotonos.

opisthotonos. emprosthotonos.

opisthotonus. emprosthotonos.

opium. **1:** gum opium; crude opium. a highly toxic, addictive substance, it is a mixture that contains morphine, codeine, thebaine, papaverine, and narcotine. The crude resin is a gummy substance obtained from the unripe seed capsules of the opium poppy, *Papaver somniferum*; it also occurs in the fruit and sap of this plant. It is usually smoked or chewed illicitly, but the liquid form, which is usually thick, syrupy, and very sweet is sometime drunk. Combined with other drugs, it has been used in medications such as laudanum and paregorics. **2:** *Papaver somniferum*

opium poisoning. poisoning due to a large overdose of opium. The effects are very similar to that of morphine poisoning, *q.v.*

opium poppy. *Papaver somniferum.*

opiumism. **1:** addiction to opium; the opium habit. **2:** the condition that results from continuing overuse of opium. See also opium poisoning.

Opsanus. a genus of toadfish (Family Batrachoididae). ***O. tau*** (common toadfish; oysterfish). a toadfish that occurs along the Atlantic Coast of the United States from Massachusetts to the West Indies.

optical activity. that property of a molecule that causes rotation of the plane of polarization of polarized light that is passed through a solution containing the compound.

optimal. **1a:** of or pertaining to the optimum; most favorable; ideal (as opposed to pessimal). **1b:** pertaining to the intensity (temperature, concentration) or amount (heat, dosage) of an environmental factor that most favors growth and reproduction of a living system. Some essential substances have an optimum range of concentration or dosage above which toxic effects occur and below which deficiencies may occur. **1c:** the most efficient; the most cost-effective (e.g., use of a mechanism that optimizes the efficiency or effectiveness of energy acquisition; use of a mechanism that minimizes energy expenditure of a life process). **2:** of or pertaining to any property or condition of a living system (e.g., a behavioral trait such as optimal foraging; optimal reproductive strategy) or its environment that favors production of the largest possible number of viable offspring.

optimum. **1a:** that condition of an environment (e.g., surroundings, habitat, locale, region) that is conducive to an organism's or population's best or ideal (See optimal) functioning, including the realization of optimal reproductive function and the closest approach to biotic potential. **1b:** that degree (intensity) or amount of any environmental factor or set of factors that supports or allows full development of a living system.

ora. plural of *os*.

oraj boengka laoet. *Trimeresurus gramineus, T. popeorum.*

oraj bungka. *Trimeresurus gramineus.*

oraj hedjo. *Trimeresurus gramineus*

oraj kalakai. *Trimeresurus puniceus.*

oraj lemah. *Agkistrodon rhodostoma, Trimeresurus puniceus.*

oraj sinduk. *Naja naja.*

oraj totok. *Ophiophagus hannah.*

oraj welang. *Bungarus fasciatus.*

oraj weling. *Bungarus fasciatus.*

oral. **1:** of, pertaining to, or concerning the mouth. **2:** in anatomy, denotes or describes the surface of a tooth that faces the oral cavity or tongue, as opposed to the buccal or facial surface of a tooth.

orange fire worm. *Eurythoe complanata*

orange lead. lead oxide red.

orange root. *Hydrastis canadensis.*

orange-bellied snake. *Pseudechis australis*.

ordeal bean. common name of physostigma and *Physostigma venosum*.

order. **1a:** a group of related families of plants or animals. **1b:** a taxonomic category subordinate to a class that contains one or more families that share certain characteristics and are deemed to be closely related. Names of orders typically end in *-ales* in plants and *-a* in animals. See also taxon.

Oregon water hemlock. *Cicuta vagans*.

orellanin, orellanine. together with muscarine, orellanin is the most common poison of the cort family (Cortinariaceae). It is the chief toxic principle of the mushroom, *Cortinarius orellanus*. It causes silent damage to the liver and kidneys, with symptoms that may not appear for several days. See also *Cortinarius, Inocybe*.

organ (organum). a structure within the body of most multicellular organisms that is comprised of one or more tissues that interact to perform a single, usually specialized, coherent, function or set of functions. **critical organ**. the organ that first reaches the critical concentration of a toxicant under specified conditions of exposure for a given species or population. See also critical concentration, effect (critical effect).

organic. **1:** integral to, central to. **2:** having an organized structure. **3a:** living, alive; biotic; biotic in origin. **3b:** arising from an organism. **4:** pertaining to substances derived from living organisms. **5:** of or pertaining to an organ or organs. **6:** of or pertaining to chemical substances that contain carbon. **7:** of, pertaining to, or cultivated by, the use of fertilizers from animal or vegetable sources as opposed to the use of synthetic chemicals.

organic decay. See decay.

organic disulfide. See disulfide.

organicidia. the presence of an organic acid, as in the stomach.

organism. any individual living being including humans.

organochlorine. **1:** of or pertaining to chlorinated hydrocarbons. **2:** an organochlorine compound.

organochlorine compound (organochlorine). chlorinated hydrocarbon (See hydrocarbon).

organochlorine insecticide (chlorinated organic insecticide). chlorinated organic insecticide. Any chlorinated hydrocarbon that is prominent in the formulation of an insecticide. Symptoms of intoxication in humans may include immediate or delayed vomiting; paresthesia and itching of lips, tongue, and face; headache; sore throat; fatigue; tremors; ataxia; weakness; confusion; convulsions; coma; and death from respiratory failure. There are usually no pathological findings.

organolead compound. an organic compound that contains one or more lead atoms.

organolead poisoning. poisoning by organolead compounds such as triethyllead, presents a syndrome that includes hyperexcitability, hyperactivity, disorientation, psychosis (which may be transient) with hallucinations and paranoid ideation, anorexia, nausea, vomiting, diarrhea, impotence, and impairment of memory.

organomercury. **1:** of or pertaining to organomercury compounds. **2:** an organomercury compound.

organomercury compound. an organic compound that contains one or more mercury atoms.

organometal. organometallic compound.

organometallic. pertaining to organic substances that contain one or more carbon-bound metal moieties.

organometallic compound (organometal). **1:** an organic compound that contains a carbon-metal covalent (or coordinate-covalent) bond (e.g., organotin, organolead, alkyl lead, methyl mercury). Metals are often more [illegible] organic compound rather than an inorganic compound; many organometallic compounds are immunotoxic. **2:** sometimes loosely used to include organometalloid compounds.

organometalloid. **1:** an organometalloid compound. **2:** pertaining to, or having the properties of an organometalloid compound.

organometalloid compound (organometalloid). a compound in which a metalloid element is covalently bonded to at least one carbon atom in an organic group. *Cf.* organometallic compound.

organophosphate ester. See organophosphate.

organophosphate warfare agent. See warfare agent.

organophosphorus. See organophosphorus compound.

organophosphorus compound (organophosphorus). an organic compound that contains bound phosphorus. Many of these compounds are powerful acetylcholinesterase inhibitors used as insecticides. Those developed for use as a nerve gas, *q.v.*, are among the most toxic synthetic compounds ever produced; they are lethal to humans in minute amounts. There are many types of organophosphorus compounds, most notably phospholipids, which are widely distributed in nature; esters of phosphinic and phosphonic acids which are economic poisons; pyrophosphates which are extremely toxic cholinesterase inhibitors used as insecticides; phosphoric esters of glycerol, glycol, etc., which serve as components of fertilizers. See also organophosphorus ester. See also pesticide, organophosphorus poisoning.

organophosphorus ester-induced delayed neurotoxicity. organophosphorus-induced delayed neurotoxicity.

organophosphorus-induced delayed neurotoxicity (OPIDN; organophosphorus ester-induced delayed neurotoxicity; organophosphorus-induced delayed neuropathy). a neurologic condition in humans; certain other mammals, especially cattle; (but not laboratory rats and mice), domestic fowl, and turkeys (but not Japanese Quail) due to exposure to certain organophosphate esters. The condition is unknown in poikilothermic vertebrates and invertebrates. The disease is progressive, signs of ataxia appearing after 8-14 days; the victim becomes weak and unsteady, with eventual flaccid paralysis of the extremities. The characteristic lesions are those of a Wallerian-like axonal degeneration which appears at about the time of onset of symptoms. Certain sensory and motor pathways of the spinal cord are affected as are large fibers of the peripheral nerves (often only the sciatic nerve). See also neurotoxic esterase, tri-*o*-cresyl phosphate.

organophosphorus insecticide. an organophosphate compound used as an insecticide. Such compounds act to inhibit a wide range of esterase enzymes, including acetylcholinesterse. Inihibition of the latter is chiefly responsible for organophosphate poisoning in both invertebrates and vertebrate animals. Toxicities of these compounds in higher vertebrates varies widely. Symptoms in humans include bronchoconstriction with wheezing and complaints of tightness in the chest; nausea, vomiting, and diarrhea due to stimulation of muscles in the walls of the gastrointestinal tract; twitching; cramps; restlessness, anxiety, and emotional instability; headache and insomnia. Severe poisoning may further cause respiratory and circulatory depression; convulsions; coma; and death due to respiratory paralysis. Some of organophosphate insecticides also produce lesions in birds and mammals that are unrelated to their anticholinesterase activity. In these cases, symptoms appear 1-2 weeks following exposure. The hind limbs, and in severe cases the fore limbs also, become paralyzed. Originally referred to as demyelination, this condition is due largely to damage of the axon itself, not to the myelin sheath.

organophosphorus military poison. organophosphorus compounds developed for use as military poisons or nerve gases. They are among the most toxic synthetic compounds known. Examples are Tabun (*O*-ethyl *N,N*-dimethylphosphoramidocyanidate), Soman (*o*-pinacolyl methylphosphonofluoridate), and "DF" (methylphosphonyldifluoride). See also organophosphorus compound. See also nerve gas.

organophosphorus poisoning. acute poisoning by an organophosphorus compound, many of which are highly toxic to humans and other animals. They are neurotoxic producing their effects largely by inhibition of acetylcholinesterase. Precise symptoms of organophosphorus poisoning vary considerably among taxa. Signs and symptoms in man may include bradycardia, salivation, sweating, increased bronchial secretion, miosis, hyperglycemia, muscular weakness, headache, hypotension, disorientation, slurred speech, anxiety, neurosis, convulsions; respiratory failure and death due to bronchoconstriction and paralysis of the respiratory centers of the brain stem.

organoselenium. organoselenium compound.

organoselenium compound (organoselenium). an an organic compound that contains one or more selenium atoms. Such are produced synthetically (e.g., 1,4-diselenane) or by the action of various organisms (e.g., dimethylselenide, dimethyldiselenide and dimethylselenone). All are of considerable environmental and toxicological importance. Some organisms convert inorganic selenium to dimethylselenide. This type of biomethylation is efficiently carried out by several genera of fungi; this activity is accompanied by the very strong "ultragarlic" odor of the product. The biotransformation of organic and inorganic selenium to dimethylselenide and dimethyldiselenide occurs in animals (e.g., laboratory rats) and the volatile compounds are exhaled. Dimethylselenone is produced by bacteria. See also selenium.

organosulfur compound. any organic compound that contains sulfur. Examples are thiols (R-SH) and thioethers (R-S-R). Sulfur that is bonded directly to hydrocarbon moieties can also bond to oxygen, producing a diversity of organosulfur compounds. Many, but not all, are toxic amd many have strong, disagreeable odors provide a warning of their presence, thereby reducing the risk of accidental poisoning.

organotellurium compound. an organic compound of tellurium. Even though tellurium is a rather rare element, these compounds are of some environmental and toxicological concern. They can be produced synthetically and by microorganisms. Dimethyltelluride, for example, can be produced by fungi from inorganic tellurium compounds. Tellurium compounds are toxic, but less so than their selenium analogs.

organotin compound, organotin. any organic compound that contains tin. Many have the general formula, RnSnX4-n, where n = number, R = a hydrocarbon group, and X = the organic or inorganic moiety linked to a tin atom by a carbon atom. In general, the toxicity of a series of organotins is maximal where n = 3. In most cases, toxicity is conferred largely by the R group. Most are readily absorbed through the skin. The principal mode of action, especially of R3SnX molecules involves bonding to proteins, presumably by the sulfur on the cysteine and histidine residues. Mitochondrial function is disrupted by any of several biochemical processes. Annual worldwide production of organotins may exceed 40,000 metric tons. Applications are numerous and include use as heat stabilizers, catalytic agents, antifouling paints, fungicides, acaricides, and disinfectants. See also alkylated organotin compound, arylated organotin compound, dialkylated organotin compound, immunotoxicant, organometallic compound, trialkylated organotin compound, triarylated tin compound.

organotrophic. heterotrophic. Do not confuse with organotropic.

organotropic. pertaining to, or characterized by, organotropism. Do not confuse with organotrophic.

organotropism. the state or condition of having a special affinity for particular tissues or organs. Said of chemicals and pathogenic agents.

organozinc compound, organozinc. any organic compound that contains one or more zinc atoms. Certain of these compounds are highly flammable and some (e.g., dimethyl and diethyl zinc compounds) produce fumes that contain very fine particles of zinc oxide. They are chemically and toxicologically similar in many ways to organomagnesium compound. They do not, however, react with carbon dioxide, as do some organomagnesium compounds. Infants may contract an acute fatal pneumonitis from inhaling the fumes of zinc soap. The condition is characterized by pulmonary lesions similar to, but more serious than, those caused by talc. See also organometallic compound, zinc oxide, zinc metal fume fever, zinc pyridine thione.

organum. organ.

oriental catfish. *Plotosus lineatus*.

oriental coral snake. See *Caliophis*, *Hemibungarus*, *Maticora*.

oriental poppy. See *Papaver*.

orizaba jalap root. See *Ipomoea*.

ornamental snake. *Denisonia maculata*.

ornithine, ornithine conjugation. See conjugation.

ornithine carbamyl transferase (ornithine transcarbamoylase). an enzyme of the urea cycle that catalyzes the transfer of carbamyl groups from carbamyl phosphate to ornithine with the formation of citrulline. It is a normal constituent of mitochondria in liver cells (hepatocytes), but does not normally occur in significant amounts in serum. Its presence in serum is a specific indicator of damage to liver cells.

Ornithodorus (pajaroello). a genus of soft ticks that parasitize a variety of mammals; many are disease vectors. They occur in Mexico and the southwestern United States. The bite of some species causes a local vesiculation, pustulation, rupture, ulceration, and eschar, with varying degrees of local swelling and pain. Such reactions sometimes occur from the bites of other ticks. See tick, Ixodoidea.

Ornithogalum (star-of-Bethlehem, nap-at-noon). a genus of bulbous plants (Family Liliaceae) that is common to roadsides, meadows, and grasslands. It is an Old World genus with 2 North American species, *O. umbellatum* and *O. nutans*. *O. umbellatum* is one of the most dangerous lilies in the United States. The bulbs are toxic and have caused deaths of children. Livestock have been killed by bulbs of wild-growing plants brought to the surface by plowing, frost heaving or rooting by swine. ***O. umbellatum*** (star of Bethlehem; nap-at-noon; snowdrop). an Old World onion-like perennial herb and garden plant that was introduced to North America from the Mediterranean region; it now occurs in grasslands, thickets, and along roadsides from Newfoundland to Ontario and from New England to North Carolina, Kansas, and Nebraska. All parts of the plant contain toxic alkaloids. The bulbs are especially toxic and have caused losses of sheep and cattle. Children in various countries have been poisoned by eating the bulbs of related species. Symptoms are chiefly those of severe gastrointestinal distress. See also convallatoxin.

Ornithogalum umbellatum (nap-at-noon; Star-of-Bethlehem; snowdrop). an Old World onionlike perennial herb and garden plant (Family Lilaceae) introduced to North America from the Mediterranean region and naturalized, occurring in grasslands, thickets and along roadsides from Newfoundland to Ontario, and New England to North Carolina, Kansas and Nebraska. The bulbs are toxic and have caused losses of sheep and cattle. Children have been poisoned by eating the bulbs of related species.

Ornithorhynchus anatinus (duck-billed platypus). the common duck-billed platypus of eastern Australia and Tasmania (Order Monotremata). It is a primitive (prototherian), aquatic, burrowing mammal with small eyes, a duck-like bill, webbed feet, a beaver-like tail. Adults may attain a length of up to 0.5 m. The male has a sharp spur on the heel that communicates with a poison gland. This mammal feeds on small fresh-water invertebrates and occasional vegetation. The female lays 1-3 small, fertilized eggs in the burrow; newly hatched young feed on milk secreted by scattered ventral mammary glands. See also Monotremata.

ornithuric acid (dibenzoylornithin). a substance found in the urine of birds that have ingested benzoic acid.

oro otto. *Maticora intestinalis*.

Orontium aquaticum. an aquatic perennial herb (Family Araceae) native to North America. The roots and seeds are toxic.

oropel. *Bothrops schlegelii*.

orosomucoid. α-acid glycoprotein.

orpiment. the naturally occurring mineral form of arsenic trisulfide.

Orthene™. acephate.

Ortho 12420™. acephate.

Ortho-creme. the brand name of a product that contains nonoxynol-9.

ortho-phthalic acid. See phthalic acid.

***ortho*-veratric acid** (2,3-dimethoxy-benzoic acid). an isomer of veratric acid.

orthoboric acid. boric acid.

orthocrasia. a state in which the response of the body to ingested or injected xenobiotics is normal.

orthocresol. *o*-cresol.

orthophosphate. See phosphate.

orthostatic. pertaining to or caused by standing erect, as in orthostatic syncope.

oryzalin (4-(dipropylamino)-3,5-dinitrobenzenesulfonamide; 3,5-dinitro-N^4,N^4-dipropylsulfanilamide). a preemergence herbicide that is particularly effective in orchards and vineyards. It is carcinogenic to some laboratory animals.

os. **1:** mouth, plural, ora. **2:** bone, plural, ossa.

Os. the symbol for osmium.

Osage copperhead. See *Agkistrodon contortrix*.

OSC. On-scene Coordinator.

osmic acid. osmium tetroxide.

osmic acid anhydride. osmic tetroxide.

osmic tetroxide. osmium tetroxide.

osmium (symbol, Os)[Z = 76; A_r = 190.2]. a transition metal element of the platinum group.

osmium oxide. osmium tetroxide.

osmium tetroxide (osmic acid; osmic tetroxide; osmium oxide; osmic acid anhydride; perosmic acid anhydride; perosmium tetroxide anhydride; perosmic oxide). a corrosive and toxic, pale yellow crystalline or amorphous compound, OsO_4, with a pungent, disagreeable odor that is soluble in water, ethanol, and ethyl ether. It is a strong oxidizing agent and irritant to the eyes and mucous membranes. Precautions must be taken when opening a container of the poison as the fumes are extremely toxic and damaging to the eyes, skin and respiratory tract. The vapor can cause blindness by coating the eyeball with osmium. Osmic acid is used in photography, as a histological stain, and an oxidation catalyst in organic chemistry.

osmoregulation. the maintenance of osmolarity of a cell or organism with respect to the surrounding medium.

osmose. to pass through a semipermeable membrane by osmosis.

osmosis. the diffusion molecules of a pure solvent through a semipermeable membrane (a membrane that selectively prevents the passage of solute molecules but is permeable to the solvent) that separates two solutions of differing solute concentrations. The transfer is from a region of relatively low solute concentration to one of relatively high solute concentration. In living systems, the solvent is water and the migration tends to equalize solute concentrations on either side of a cell membrane. The net direction of osmosis is determined by the relative osmotic pressures of solutions within the cell and those without (See endosmosis, exosmosis).

osmotic. of or pertaining to osmosis.

ossa. plural of os (bone).

Osteichthyes. the class of vertebrate animals that contains the bony fishes, characterized by the presence of an endoskeleton chiefly of bone. They are the dominant fishes, and considered the most highly evolved, invading all types of waters. The single external gill opening is covered by an operculum, and in modern forms the spiracle is greatly reduced or absent. Most have an air bladder (swim bladder) and a covering (operculum) over the gills.

osteomalacia (osteomalacosis). a disease marked by increasing softness of bones due to loss of calcium salts. The condition is accompanied by rheumatic pains in the limbs, vertebral column, and chest; anemia; and progressive muscular weakness. See fluorosis, itai, itai disease, renal failure.

osteomalacic. pertaining to or characterized by osteomalacia.

osteomalacosis. osteomalacia.

osteopetrosis. a disease characterized by excessive calcification of bone and spontaneous fractures. It is considered inherited in humans, but in the case of certain herbivores the disease can result from prolonged calcium intake or poisoning by vitamin D_3 analogs that are present in certain plants such as *Trisetum flavescens*, and *Cestrum diurnum*. See also enzootic calcinosis.

osteosarcoma. See sarcoma.

ostracitoxin. crude ichthyocrinotoxin isolated from secretions of the skin of the trunkfish, *Ostracion lentiginosis*. It is extremely toxic to other fish, is heat stable, nondialyzable, soluble in water, methanol, ethanol, acetone, and chloroform, but insoluble in benzene and diethyl ether. Other reef fishes, exposed to ostracitoxin that is free in the water or injected *i.v.*, may react by "gasping," a decrease in rate of opercular movements, loss of equilibrium and decreased locomotor activity, followed by intermittent convulsions and death. It is very toxic to laboratory mice. See also ichthyocrinotoxin. *Cf.* pahutoxin.

ostreotoxismus. poisoning by the ingestion of contaminated oysters.

Otariidae (eared seals; sea lions; fur seals). a family of marine mammals (Suborder Pinnipedia) of the Pacific Ocean. They have a long neck, are larger than true seals, and have small external ears. The underfur is absent and the hind limbs can be rotated forward. See *Neophoca cinerea*.

OTBE. tributyltin.

ototoxic. **1:** poisonous to the organs of hearing. **2a:** in the case of vertebrate animals, poisonous to the organs of hearing and balance. **2b:** having a deleterious effect upon the eighth cranial nerve, or upon the organs of hearing and equilibrium in vertebrate animals.

ototoxicity. the quality of being ototoxic.

otter, puff. *Bitis arietans*.

Ottoman viper. *Vipera xanthina*.

ouabain (3β-[(6-deoxy-α-L-mannopyranosyl)-oxy]-1β,5β,11α,19-pentahydroxycard-20(22)-enolide octahydrate; G-strophanthin; Gratus strophanthin; strophanthin-G; acocantherin). a white crystalline substance or white powder that is water- and ethanol-soluble; it melts with decomposition at 190°C. It is an extremely toxic, cardiotonic, digitalis glycoside derived from certain plants of the family Apocynaceae: *Strophanthus spp.*, especially *S. gratus*, and the *Acocanthera* group of trees, especially *A. ouabaio*, *q.v.* Ouabain is a specific inhibitor of Na^+/K^+ ATPase, *q.v.* Ouabain has the same actions as digitalis, but causes digitalization more rapidily. Its digitalis-like effects on the heart are due to inhibition of Na^+/K^+ ATPase. It is used therapeutically, *i.v.*, to treat acute congestive heart failure, nodal paroxysmal tachycardia and atrial flutter. It has a specific binding site that is pharmacologically similar to Na^+/K^+ ATPase. The primitive use of ouabain is on weapons, but it has sometimes been used in trial by ordeal. See also strophanthin and sodium-potassium pump. Ouabain is especially toxic to laboratory rats; males are more sensitive than females; the *i.v.* LD_{50} is 14 mg/kg body weight.

"ouch, ouch." itai, itai disease (See disease, itai, itai).

ourari. curare.

ova. plural of ovum.

oven cleaner. See cleaner.

overdosage. **1:** the administration of an excessive dose. **2:** the condition resulting from an excessive dose. **absolute overdosage**. that due to an error in the dosage administered to or taken by an individual, whether due to an error in prescription or failure to follow the prescription. **relative overdosage**. that in which the normal dosage administered to or taken by a specific individual is toxic due, for example, to liver or kidney disease or a metabolic disorder in which the drug is not normally metabolized or excreted.

overdose. **1:** to administer (or take) a dose of a drug that is excessive in relation to the therapeutic requirements, the therapeutic index, or toxicity. **2:** an excessive dose in the sense of def. 1.

overexposure. exposure beyond specified limits.

overventilation. hyperventilation.

ovine. of, from, or pertaining to sheep (genus *Ovis*).

ovotoxicity. See toxicity (ovotoxicity).

ovum (plural, ova; egg). the mature female gamete or germ cell.

oxalate. any salt of oxalic acid that contains the (COO_2 radical (e.g., ammonium oxalate, ethyl oxalate, sodium oxalate). Numerous green plants and some fungi contain oxalates, but only a few contain concentrations high enough to be considered dangerous. The toxicity of oxalic acid and its salts resides in the reactivity of the oxalate ion. See oxalic acid, oxalate poisoning.

oxalate poisoning (oxalic acid poisoning). **1:** systemic oxalate poisoning is most commonly due to ingesting plant material with high concentrations of oxalic acid, *q.v.* Symptoms, which appear within 2-6 hours, may include depression, respiratory distress, hemorrhage, weak pulse, cold skin, cyanosis, convulsions, coma, collapse, and death. The most toxic oxalate-containing plants occur in desert or semidesert situations or dry grasslands. A number of food plants (e.g., rhubarb, spinach, and sometimes beets, purselane, Russian thistle, sorrels, etc.) contain enough oxalate to be regarded as dangerous. Systemic poisoning may also result from accidental ingestion of oxalic acid or inhalation of the vapor. Calcium gluconate is given *per os* as an antidote to oxalate poisoning. **2:** poisoning due to ingestion of plants containing crystals of calcium oxalate. Most of these plants are members of the family Araceae, which includes Jack-in-the-pulpit, wild calla, skunk cabbage, and calamus. The crystals penetrate the tongue and tissues of the mouth and pharynx, producing edema and an intense burning sensation. The victim may be unable to talk for a while and the tongue may obstruct breathing. See oxalate, oxalic acid.

oxalemia. the presence of an excess of oxalates in circulating blood.

oxalic acid (ethanedioic acid). a colorless, transparent, crystalline powder that is soluble in water, ethanol, and ether. It is a highly toxic dibasic mineral acid, $(COOH)_2$. Numerous plants, including some consumed by humans and browsing or grazing mammals (e.g., beet tops, spinach) contain soluble (sodium, potassium) and insoluble (calcium) salts of oxalic acid. They occur in especially high concentrations in some desert plants. Oxalic acid is corrosive when ingested. It is often used in the home as a bleach, stain remover, rust remover, and in polishes. See oxalate, oxalate poisoning.

oxalic acid poisoning. oxalate poisoning.

Oxalidaceae. the wood-sorrel family. See *Oxalis*.

Oxalis (sorrel). a genus of small, rather delicate herbs (Family Oxalidaceae). Some species sometimes contain lethal concentrations of soluble oxalates. Species of the genus *Rumex* are also commonly referred to as "sorrel," thus causing occasional confusion in reports of livestock losses in the United States. ***Oxalis corniculata*** (creeping lady's sorrel). a hardy species that can contain lethal amounts of soluble oxalates. ***O. pes-caprae*** (*O. cernua*; Bermuda buttercup; Bermuda oxalis; soursob; sorrel). a species that occurs as an introduced weed in California and Florida. It sometimes contains lethal concentrations of soluble oxalates. It has caused sheep fatalities in Australia. Symptoms and lesions are those of oxalate poisoning, *q.v.*

oxalism. poisoning by oxalic acid or by an oxalate. See oxalate, oxalate poisoning.

oxalosis. a generalized deposition of calcium oxalate in renal and extrarenal tissues, including the formation of urinary calculi. Oxalosis-inducing agents include oxalic acid, methoxyflurane, ethylene glycol, antirust agents, all of which are nephrotoxic.

oxamyl (2-(dimethylamino)-*N*-[[(methylamino)-carbonyl]oxy]-2-oxoethanimidothioic acid methyl ester; *N',N'*-dimethyl-*N*-[(methylcarbamoyl)oxy]-1-thiooxamimidic acid methyl ester; *N,N*-dimethyl-α-methylcarbamoyloxy-imino-α-(methylthio)acetamide; methyl 1-(dimethylcarbamoyl)-*N*-(methylcarbamoyloxy)thioformimidate; thioxamyl). a white crystalline compound used as a pesticide with tobacco, fruits, other crops, and ornamental plants.

oxepinac. isoxepac.

oxidant. the electron acceptor in an oxidation-reduction (redox) reaction.

oxidase (oxidizing enzyme). any of a class of (metalloprotein) enzymes that catalyze the metabolic oxidation of chemical compounds with the reduction of molecular oxygen. Specific oxidases are usually named for the group oxidized (e.g., amine oxidase, amine oxidase). See also oxidation, reduction, dehydrogenation.

oxidation. **1:** formerly, the chemical combination of oxygen with another substance. **2a:** any reaction in which electrons are transferred from one substance (oxidation) to another. The substance that gains electrons is termed the oxidizing agent, whereas the substance that loses electrons is termed the reducing agent. See also reaction (detoxification reaction, oxidation reaction, redox reaction), oxidase. **2b:** partial loss of electrons as when electrons are displaced within a molecule is also considered oxidation. **3:** dehydrogenation.

oxidation reaction. a chemical reaction in which (1) electrons are removed from an atom or ion; (2); the chemical addition of oxygen to a compound, or (3) the loss of hydrogen from a compound.

oxidative. **1a:** having the capacity to oxidize. **1b:** pertaining to a process that involves oxidation.

oxidative injury. oxidative lesion.

oxidative lesion (oxidative injury). molecular oxygen is reduced in the body to active species that can damage tissue. Tissue injuries caused by these highly reactive oxygen-containing molecules and radicals are sometimes referred to as oxidative lesions. A number of enzymes (e.g., peroxidase, superoxide dismutase, and catalase) scavenge these radicals from the system and repair of some tissue is possible through the agency of DNA-repair enzymes. See oxygen (molecular oxygen, active oxygen).

oxidizing enzyme. oxidase.

oxidizing substance. any compound that spontaneously evolves oxygen under mild conditions (i.e., at or slightly above about 20°C). Examples are chlorates, nitrates, perchlorates, permanganates, and peroxides. Such compounds react vigorously when in contact with reducing substances or materials (e.g., organic compounds or living tissue). Many are fire and explosion hazards.

oxidosis acidosis

oxirane. ethylene oxide.

22-oxovincaleukoblastine sulfate. vincristin sulfate (See vincristine).

oxus cobra. *Naja naja*.

oxybiotic. aerobic.

1,1′-oxybis(2-chloro)ethane. *sym*-dichloroethyl ether.

1,1′-oxybisethane. ethyl ether.

oxybutyric acid. hydroxybutyric acid.

oxycanthine. a white alkaloid, $C_{37}H_{40}O_6N_2$, isolated from the root of *Berberis vulgaris*. It is a CNS poison.

oxygen (symbol, O)[Z = 8; A_r = 15.9994]. a colorless, odorless, tasteless, neutral, nonmetallic, diatomic gas that supports combustion, and the most abundant and widely distributed chemical element on Earth. It combines with most other elements to form oxides and is essential to animal and plant life. Pharmaceutical-grade oxygen contains not less than 99.0%, by volume, of O_2. It is used in carbon monoxide poisoning; to relieve hypoxia; in cryosurgery; and at hyperbaric pressures to conduct cardiac and other surgery. Therapeutic overdosage can cause convulsions. **active oxygen**. highly reactive species of oxygen such as superoxide, hydroxyperoxyl radical, hydrogen peroxide, and the hydroxyl radical. They destroy microsomal cytochromes by initiating autoxidation. Active oxygen species are formed in the animal body by successive additions of an electron (e-) and a hydrogen ion (H+) to a molecule of molecular oxygen, yielding $H0_2$-.

oxyhemoglobin (oxyhaemoglobin). the form of hemoglobin in which molecular oxygen is reversibly bound to the heme moiety; it is the form normally found in systemic arterial blood. About 97% of the oxygen in blood is transported in this form. See also hemoglobin.

oxymethylene. formaldehyde.

oxyosis. acidosis.

oxyphyte. a plant that grows under acidic conditions.

oxytetracycline. an antibiotic with actions and applications similar to those of tetracycline.

Oxytropis (locoweeds; point locoweeds; point-vetches). a genus of perennial herbs and small shrubby plants (Family Fabaceae, formerly Leguminosae) that occur chiefly in mountainous areas of Europe, Asia, and North America. Examples are *O. lambertii* (white loco, white point loco, stemless loco), *O. saximontana*, and *O. sericea* (white loco, white point loco). There is a intergradation of characteristics, which is probably due to hybridization, among the locoweed species of this genus. No doubt many of the specific names have been used falsely in the toxicological literature. Toxic species contain indolizidine-1,2,8-triol. See also *Astragalus*.

Oxyuranus scutellatus (taipan; brown canesnake; giant brownsnake; diriora; gobori). a highly venomous, aggressive, extremely dangerous snake (Family Elapidae) that inhabits grasslands and savannahs of northern Australia and southeastern New Guinea. The average length of an adult is about 2.0-2.2 m; rare individuals may reach a length of more than 3 m. The ground color is coppery or dark brown in Australia and grayish-black with a posterior reddish-orange stripe in specimens from New Guinea. The fangs (1.3 cm or more in a large adult) are unusually long for an elapid; copious amounts of a very toxic venom can be delivered. When irritated, the head is depressed, the neck compressed, and the body expands revealing the white skin shows between the scales. The taipan attacks swiftly, without warning, usually delivering a succession of bites before the victim is able to react. Very few humans survive an attack by this snake unless treated with "Taipan" antivenin. Paralysis with severe dyspnea develops rapidly and most victims die within minutes. See also venom (taipan venom).

oxyuricide. a chemical that destroys oxyurids (pinworms).

oxyurid. **1a:** of or pertaining to a nematode of the superfamily Oxyuroidea. **1b:** a pinworm, seatworm, threadworm, or any nematode of the superfamily Oxyuroidea.

oysterfish. *Opsanus tau*.

oz. ounce.

ozone (triatomic oxygen). **1:** an allotropic form of oxygen, O_3, it is a bluish, highly unstable gas (or dark blue liquid) with a pungent odor. It is a stronger oxidizing agent than oxygen, a fire and explosion hazard, and is toxic by inhalation. Ozone is used as a bactericide; an oxidizing agent in various chemical processes and the oxidation of cyanides and phenols; in the purification of water; industrial waste treatment; deodorization of air and sewage gases; the bleaching of waxes, oils, textiles, wet paper, paper pulp; the aging of liquor and wood; the drying of varnishes and printing inks. Ozone is toxic to humans and other terrestrial vertebrates by inhalation. It is a strong irritant of skin, eye, lungs, and mucous membranes of the nose and throat. It impairs breathing, causes changes in cell membranes and lung tissue that can be fatal. Exposure of humans by inhalation is usually marked by coughing, dyspnea, or even choking; lacrimation; and a sore throat. Systemic effects may include changes in the visual field, discomfort, and profound fatigue. Persons who are chronically exposed to ozone also experience a greater susceptibility to infection. Teratogenicity and reproductive effects have been demonstrated experimentally. Ozone is a major air pollutant, massive amounts being produced by the action of the sun on hydrocarbons and atmospheric nitrogen oxides, by the combustion of fossil fuels, by lightning, and in the stratosphere by ultraviolet radiation. Ozone is also produced by X-ray machines, mercury vapor lamps, electric arcs, and by electrolysis of alkaline perchlorate solutions. It is the chief component of photochemical smog, *q.v.* Many trees throughout the world are damaged annually by ozone. **2:** air that contains a perceptible amount of ozone (def. 1).

P. the symbol for phosphorus.

P-450 (P_{450}). cytochrome P-450.

P-450 monooxygenase enzyme system. cytochrome P-450-dependent monooxygenase system.

P-C. prophlogistic corticoid.

^{32}P. the symbol for phosphorus-32. See radiophosphorus.

^{33}P. the symbol for phosphorus-33. See radiophosphorus.

p.o. per os.

p-. chemical abbreviation for *para*-.

P2S. pralidoxime mesylate.

P_{450}. cytochrome P-450.

P_{450} monooxygenase system. cytochrome P-450-dependent monooxygenase system.

Pa. the symbol for protoactinium.

PA. parathion

PAB. *p*-aminobenzoic acid.

PABA. *p*-aminobenzoic acid.

pacemaker. sinoatrial node.

pachi virian. *Trimeresurus trigonocephalus*.

Pachlioptera aristolociae. a butterfly that contains aristolochic acid, a toxic substance, which the caterpillar acquires from the plants on which it feeds.

pachy-. prefix meaning thick.

Pachycephalidae (thickheads; whistlers). a family of perching birds (Order Passeriformes); formerly a subfamily of the Muscicapidae. Most species are insectivorous although some such as the poisonous ***Pitohui dichrous***, *q.v.*, subsist chiefly on berries. The body is robust; the sexes usually look alike.

pachyderma. abnormal thickening of the skin.

Pachylomerides. a common genus of trapdoor spiders (Family Ctenizidae).

Pachysandra procumbens (box). a hedge plant (Family Buxaceae) that sometimes escapes cultivation. It contains alkaloids and certain other biologically active compounds. Numerous animals have been killed by eating clippings from box hedges. The leaves are very toxic; a small amount can be lethal. Symptoms are those of severe gastrointestinal irritation, sometimes producing bloody stools.

Pacific coral snake. See *Micropechis*.

Pacific poison oak. *Rhus diversiloba*.

Pacific ratfish. See *Hydrolagus colliei*.

Pacific rattlesnake. *Crotalus viridis*.

packed cell volume (PCV).

paederine. pederin.

PAGE. polyacrylamide gel electrophoresis q.v.

Pagellus erythrinus (porgy; porgie). a deep-bodied, inshore bony fish (Family Sparidae) of the Black Sea, Mediterranean Sea, and the eastern Atlantic Ocean from Scandinavia and the British Isles south to the Azores, Canary Islands, and Fernando Po. Some populations are transvectors of ciguatera.

pagha wubré. *Atractaspis microlepidota*.

Pagrus pagrus (red porgy). a reddish, deep-bodied, bony fish (Family Sparidae) of the western Atlantic Ocean from New York to Argentina, the eastern Atlantic Ocean and the Mediterranean Sea. It inhabits coastal waters over vegetated sandy bottoms or coral to a depth of ca. 175 m. See ciguatera.

PAH. **1:** polycyclic aromatic hydrocarbon (See hydrocarbon).

pahutoxin. a choline chloride ester of 3-acetoxyhexadecanoic acid, it is the purified form of ostracitoxin, *q.v.*

pain (dolor). distress, suffering, or agony experienced at times by nearly all humans, by nearly all higher animals. Pain is usually a more or less localized unpleasant or even agonizing sensation that results from stimulation of sensory nerve terminals located in almost every part of the vertebrate body and presumably of that of many other multicellular animals. Numerous agents, indeed nearly any injury or strong stimulus, can cause pain. If the stimulus is external, the affected organism normally withdraws from the source, thereby usually limiting damage and relieving pain. The kinds of external stimuli that produce pain and aversive behavior in humans generally produce aversive behavior in lower animals also. The physiological mechanisms, however, usually differ greatly from that of vertebrates.

paint brush. See *Castilleja*. hydroxymethylbenzeneacetic acid

painted cup. *Castilleja*.

painter's colic. lead colic.

pajaroello. *Ornithodorus*.

Pakistanian cobra. *Naja naja*.

pala polonga. *Trimeresurus trigonocephalus*.

pale laurel. a common name of *Kalmia polifolia*.

pale-mouthed mamba. *Dendroaspis augusticeps*.

Palestinian viper. *Vipera xanthina palaestinae*.

palidoxime mesylate. See pralidoxime chloride.

palladium (symbol, Pd)[Z = 46; A_r = 106.4]. a silver-white metallic element resembling platinum. It also occurs as a black powder and soft compressible masses.

Pallas' pit viper. *Agkistrodon halys*.

Pallas' viper. *Agkistrodon halys*.

pallor. lack of color, paleness, especially of the skin.

palm crab. *Birgus latro*.

palm family. Palmaceae (= Arecaceae, Palmae).

palm viper. any of various arboreal species of tropical American (e.g., *Bothrops nigroventris*) or Asian (*Trimeresurus spp.*) crotalid snakes. They are often green in color with prehensile tails.

palma christi. *Ricinus communis*.

Palmaceae (Arecaceae; Palmae). some 2000 species of mainly tropical, woody trees and some vines and shrubs (Family Palmaceae). They usually have a tall columnar trunk that lacks cambium and a crown of large feathery leaves. See *Areca catechu*.

Palmae. Palmaceae.

palmotoxin B_0. a mycotoxin from *Aspergillus flavus*. It is extremely toxic (*i.p.*) to rats.

palpitations. an irregular rapid heartbeat that is noticeable to the affected individual.

palsy. **1a:** a condition of uncontrollable tremor of the body or of one or more body parts. **1b:** paralysis. **2:** a weak or enfeebled condition. **3:** to become palsied or paralyzed. **lead palsy**. a distal, usually painless, polyneuropathy that chiefly affects adults with chronic lead poisoning. The wrist and hand are mainly affected, showing weakness, pain, and paresthesia with wrist drop.

palustrine. an alkaloid of *Equisetum*, *q.v.*

Palythoa (death seaweed of Hana; limu-mak-o-hana). a genus of venomous, sometimes poisonous, zoanthid corals (Order Alcyonacea) that occur both in tidal pools or other protected shallow bodies of water in the Caribbean Sea and the tropical Pacific, including the Hawaiian Islands. The tissues of some species are extremely poisonous. Humans swimming in pools containing *Palythoa* may develop numbness or a tingling sensation about the lips and mouth. If the slime comes into contact with an open wound, the subject may experience malaise, abdominal pains, and muscle cramps. Such contact is potentially lethal. These animals should be avoided or treated with extreme caution. The toxic principle is palytoxin, q.v. ***P. toxica*** (limu-mak-o-hana). extremely toxic; the original source of palytoxin, q.v. ***P. tuberculosa***. extremely toxic; palytoxin isolated from this species differs somewhat from that of *P. toxica*.

palytoxin (palytoxin (C5-55 hemiacetal)**;** PTX). the most toxic nonproteinaceous substance known. It is a water-soluble toxin produced by soft corals of the genus *Palythoa*, especially *P. toxica*. Palytoxin is neurotoxic, myotoxic, and a powerful vasoconstrictor. It is easily and rapidly destroyed by heat or acids, probably including gastric juice. It is lethal in extremely small amounts to a diversity of experimental animals (e.g., crabs, dogs, mice). Doses >0.06 μg/kg body weight, *i.v.*, are rapidly fatal to dogs; LD_{50}, *i.v.*, in the laboratory mouse is reportedly ca. 0.15 to 0.45 μg/kg body weight. Palytoxin from *P. toxica* ("Limu-make-o-Hana"), an Hawaiian species, and *P. tuberculosa*, an Okanawan species, differ somewhat structurally.

palytoxin (C51-55 hemiacetal). palytoxin.

2-PAM. pralidoxime.

L-PAM. melphalan.

2-PAM chloride. pralidoxime chloride.

2-PAM iodide. pralidoxime iodide.

2-PAM mesylate. pralidoxime mesylate.

pama. *Bungarus fasciata*.

pamaquine (6-methoxy-8-(1-methyl-4 diethylamino)butylaminoquinoline; Fourneau 694; Fourneau 710). a toxic substance, $C_{19}H_{29}N_3O_4$, derived from 8-aminoquinoline. It is an antimalarial agent that is active against avian malaria and against the exoerythrocytic forms (gametocytes) of all human malarial parasites. Pamaquine is more toxic than chloroquine or primaquine and has thus been replaced by primaquine.

Pamphobeteus. a genus of tarantulas that are not native to the United States although individuals sometimes enter on shipments of produce or other materials. They are dangerous to humans.

panacene. an aromatic bromoallene antifeedant, isolated from *Aplysia brasiliana*.

pancreas. a vertebrate organ located near the stomach that produces and secretes digestive enzymes into the duodenum. It also helps to regulate concentrations of sugar and certain other nutrients by producing and secreting insulin into the circulating blood. The pancreas is a target of a number of poisons (e.g., Alloxan, carbon tetrachloride, ethanol); some can cause hyperglycemia by attacking the β-cells of the pancreas.

pancreatitis. inflammation of the pancreas.

pancuronium (Pavulon). a highly toxic neuromuscular blocking agent.

pandemic. **1a:** widespread throughout a given nation, geographic region, or larger geographic area. **1b:** cosmopolitan; worldwide in distribution.

pang. a sudden, acute pain.

panic grasses. *Panicum*.

Panicum. (panic grasses) a widely distributed genus of grasses (Family Gramineae) that occupy a diversity of habitats. They can produce hepatogenic photosensitivity in grazing animals. See also phylloerythrin. ***P. capillare*** (witchgrass). toxic concentrations of nitrate have also been reported.

panther mushroom. *Amanita pantherina*.

papala. *Trimeresurus wagleri*.

papapa bean. *Dolichos lablab*.

Papaver. (poppies). a genus of about 50 plant species (Family Papaveraceae) that are nearly worldwide in distribution, although few species occur naturally in North America. Most contain a variety of toxic alkaloids, sometimes in high concentrations. ***P. nudicaule*** (Iceland poppy). a species, sometimes cultivated, that has been implicated in the poisoning of livestock. ***P. orientale*** (oriental poppy). a common garden poppy that contains some alkaloids, but is generally considered nontoxic. ***P. rhoeas*** (corn poppy; red poppy). occasionally cultivated, it has been implicated in the poisoning of livestock. ***P. somniferum*** (opium poppy; gum opium; poppy seed; poppy; opium). an herb introduced to North America from Europe, it has escaped from cultivation, often occurring near dwellings, along roadsides, and abandoned fields from Newfoundland to North Dakota, Puget Sound, and southward. It contains a large number of isoquinoline alkaloids including codeine, morphine and papaverine. The unripe seed capsule is very toxic and seeds used as animal feed following removal of the oil have proven lethal. See also nontoxic, opium.

Papaveraceae (poppy family). a family of herbs, or sometimes shrubs, many of which are common garden plants and nearly all of which produce toxic alkaloids or other poisonous principles. See, for example, *Sanguinaria canadensis*, *Chelidonium majus*, *Papaver spp.*, *Argemone mexicana*, *Dicentra*, and *Corydalis*.

papaverine (1-[(3,4-dimethoxyphenyl)methyl]-6,7dimethoxyisoquinoline; 6,7-dimethoxy-l-veratrylisoquinoline). a toxic narcotic alkaloid, $(CH_3O)_2C_6H_3CH_2C_9H_4N(OCH)_2$ (not an opioid), extracted from crude opium. It is a white crystalline powder, soluble in chloroform, hot benzene, glacial acetic acid, aniline, and acetone; slightly soluble in ethanol and ether; insoluble in water. It is a smooth muscle relaxant used chiefly (as the water-soluble hydrochloride) in the treatment of hypertension. Patients may experience drowsiness, gastrointestinal distress, tachycardia, facial flushing, and possibly liver damage. Administration of high doses can induce cardiac arrhythmias.

paper chromatography. See chromatography.

paper viper. *Trimeresurus gramineus*.

paper wasp. any member of the family Polistidae.

paperflower. See *Psilostrophe*.

papilla. a small protuberance from the surface of a tissue or organ.

papillary tumor. papilloma.

papilledema. a swelling of the optic disc that is visible by use of an ophthalmoscope. The proximate cause is increased intracranial pressure. This is possible because the sheaths that encase the optic nerves are continuous with the meninges such that increased intracranial pressure is transmitted from the brain to the optic disc.

papilloma (papillary tumor). a benign, branching or lobular, epithelial neoplasm. See also xeriderma pigmentosum.

papoose root. *Caulophyllum thalictroides*.

PAPP. *p*-aminopropiophenone.

PAPS. 3′-phosphoadenosine-5′-phosphosulfate.

Papuan blacksnake. *Pseudechis papuanus*.

papule. any small, raised, red, brown, yellow, white or skin-colored skin lesion. The tip may be pointed, dome-shaped, or flat.

PAR. population attributable risk (See risk).

para- (*p*-). **1:** prefix meaning positioned beside, situated near, surrounding, accessory to, beyond, apart from, etc. **2:** in chemistry, *para-* or its abbreviation, *p*-, indicates the substitution of 2 atoms, each bonded to a carbon atom or situated on opposite sides of a benzene ring.

parabendazole ((5-butyl-1*H*-benzimidazol-2-yl)-carbamic acid methyl ester; methyl-5-butyl-2-benzimidazole carbamate; 5-butyl-2-(carbomethoxyamino)benzimidizole). a veterinary anthelmintic with nematocidal action, $C_{13}H_{17}N_3O_2$. It is moderately toxic to laboratory rats and mice.

Parabolin. diethylpropion.

Paracentrotus lividus (European sea urchin). a species of sea urchin (Phylum Echinodermata,

Class Echinoidea) that occurs along the Atlantic coast of Europe and the coastal waters of the Mediterranean Sea. It has been implicated in the poisoning of humans by ingestion. Symptoms are those of acute gastroenteritis, including nausea, vomiting, diarrhea, abdominal pain, and severe headaches.

paracetaldehyde. paraldehyde.

paracetamol. acetaminophen.

paracresol. *p*-cresol.

paradichlorobenzene (PDB; *p*-dichlorobenzene; 1,4-dichlorobenzene; Dichloricide). a toxic, carcinogenic, volatile chemical that can cause severe irritation to the nose, throat, and lungs; depression. Chronic or intermittent exposure over a long period of time can cause hepatic and renal injury. See also mothball.

paraesthesia. paresthesia.

paraffin. **1:** alkane. **2:** a white translucent mixture of hydrocarbons that is obtained from petroleum; it is solid at room temperature. Mild exposure causes chronic dermatitis and boils. Intense exposure is associated with cancer of the scrotum.

paraganglioma. chromaffinoma.

paraldehyde (2,4,6-trimethyl-1,3,5-trioxane; paracetaldehyde). a cyclic polymer (a trimer) of acetaldehyde, $C_6H_{12}O_3$. It is a rapid-acting CNS depressant and hypnotic that may be addictive. It is used largely on institutionalized patients. Therapeutic doses given orally can irritate the throat and stomach; administration *i.m.* can result in necrosis and nerve damage Large doses have anticonvulsant and antidelirium properties. The effective doses, however, can cause respiratory depression and hypotension.

paraldehyde poisoning (paraldehydism). poisoning by overdosage of paraldehyde. Symptoms are similar to those of poisoning by chloral hydrate. Symptoms may initially include rapid and labored respiration with subsequent cardiac and respiratory depression; acidosis; leukocytosis; hepatitis and nephrosis; pulmonary hemorrhages and edema; dizziness; collapse with partial or complete anesthesia. The odor of the breath is a distinctive sign. Repeated use of paraldehyde can result in tolerance and dependence.

paraldehydism. paraldehyde poisoning.

paralgia. unusual or abnormal pain.

paralyses. plural of paralysis.

paralysis (plural, paralyses). **1:** palsy (in part); a temporary or permanent impairment, inhibition, or loss of voluntary motor function, usually accompanied by a loss of sensation in part of the body. The proximate cause can be impairment of sensory or motor function. Paralyses may be named or classified functionally (ascending, motor, sensory), in terms of the causative agent (e.g., arsenical, toxic, traumatic), on the basis of the affected nerve, muscles, or other body parts (e.g., obturator, ulnar), or in various other ways. See also palsy, paraplegia, paresis. **2:** loss, impairment, or reduced ability to perform any bodily function (e.g., secretion, sensation). **3:** the masking of an immune response by the presence of excessive amounts of antigen which gives the appearance of immunologic tolerance. **acute ascending paralysis** (Landry's paralysis; Kussmaul-Landry paralysis). a paralysis that rapidly expands progressively from the legs to the trunk, arms, and neck in humans, sometimes terminating in death within 1-3 weeks. **alcoholic paralysis**. that due to habitual drunkenness. **ascending paralysis**. a paralysis that progresses from the lower to the upper (or posterior to anterior) portions of the body in humans and other vertebrate animals or centrad from the periphery of the body. **anesthesia paralysis**. that following anesthesia. **arsenical paralysis**. that due to arsenical poisoning. **Chastek paralysis**. paralysis of foxes or mink due to thiamine deficiency caused by feeding on raw fish that contain an enzyme destructive of thiamine. Effects include loss of appetite, emaciation, and eventual paralysis and death. **diphtheritic paralysis** (postdiphtheric paralysis). paralysis of the muscles of the palate, eyes, intercostal muscles, and limbs as a complication of diphtheria caused by diphtheria toxin. **ginger paralysis**. Jamaica ginger paralysis. **global paralysis**. paralysis that involves the entire body. It is usually rapidly fatal. **immunological paralysis**. lack of specific antibody production following exposure to large doses of the antigen. Elimination of antigen relieves the condition. **jake paralysis**. Jamaica ginger paralysis. **Jamaica ginger paralysis** (jake paralysis; jake neuritis; ginger paralysis; Jamaica ginger polyneuritis). paralysis of the extremities, especially of the legs, due to imbibing Jamaica ginger as a beverage. **Kussmaul-Landry paralysis**. acute ascending paralysis. **Landry's paralysis**. acute ascending paralysis. **lead paralysis**. that caused by lead poisoning. It is due to a peripheral neuritis, and marked by wristdrop. **mixed paralysis**. combined motor and sensory paralysis. **motor paralysis**. loss or impairment of muscular contraction. **narcosis paralysis**. that which occurs during anesthesia. **postdiphtheric paralysis**. diphtheric paralysis. **sensory paralysis**. that due to loss or impairment of sensory function. **tick-bite paralysis**. a progressive, ascending, flaccid motor paralysis observed in humans (mostly children) and domestic animals. It is due to the bites of certain ticks, notably those of the genera *Dermacentor* and *Ixodes*. The condition is presumably due to a venomous saliva. The victim usually recovers following removal of the tick. **toxic paralysis**. that due to exposure to a toxic substance. **traumatic paralysis**. that due to trauma.

paralysis time. a test usually conducted by challenging laboratory mice with zoxazolamine, which produces a transient flaccid motor paralysis. The xenobiotic tested may shorten or prolong the paralysis. Thus, inducers of cytochrome P-450 shorten and inhibitors prolong the duration of paralysis.

paralytic. **1:** pertaining to paralysis, the capacity to cause paralysis, or to suffering from paralysis. **2:** a chemical agent that paralyzes. *Cf.* paralytogenic.

paralytic dementia. dementia paralytica.

paralytic ileus. inhibition of peristalsis as by some toxicants. This also slows the absorption of the toxicant.

paralytic poisoning. paralytic shellfish poisoning (See shellfish poisoning).

paralytogenic. causing paralysis. *Cf.* paralytic.

paralyzant. **1:** causing paralysis. **2:** any substance that paralyzes.

paralyze. to effect a state of paralysis; to render an organism or part of an organism unable to move.

paralyzed tongue. a condition of domestic cattle that results from long-term consumption of *Descurainia pinnata*, *q.v.* Partial or complete blindness is the first indication of poisoning. Subsequently, the affected animal loses the use of its tongue and cannot swallow. The animal becomes progressively thinner and weaker, but the symptoms gradually disappear with improved feed and the daily administration of 2-3 gallons of water.

paralyzer. **1:** that which causes paralysis. **2:** any substance that inhibits a chemical reaction.

paramecin. a water-soluble toxin produced by kappa particles, *q.v.*, of certain strains of the ciliate, *Paramecium*. It is released by diffusion into the surrounding medium where it may kill other sensitive strains. The terms "kappa particle" and "paramecin" are sometimes erroneously used interchangeably. See also *P*-particle, kappa particle.

Paramecium (slipper animalcule; paramecium). a genus of ciliate Protozoa that occur commonly in puddles, stagnant ponds, and hay infusions. Paramecia are often used in biological research. **killer paramecium**. any strain of *Paramecium* that releases paramecin into the surrounding medium in sufficient quantity to destroy sensitive strains of *Paramecium*.

paramorphine, *para*-morphine. thebaine.

paramyoclonus multiplex. myoclonus multiplex (See myoclonus).

Paranaja multifasciata (burrowing cobra). a rare, small (usually less 0.6 m long), cylindrical, and moderately slender, but probably dangerous cobra (Family Elapidae) of western central Africa. The head is short, depressed, and slightly distinct from body. The fangs are relatively large; pupils round; the hood is lacking.

paranaphthalene. anthracene.

Paranemertes peregrina. a venomous marine nemertine worm (Phylum Nemertea, Order Hoplonemertea). The venom contains a paralytic substance, anabascin.

paranoia. a chronic mental disorder in which a person suffers delusions, usually of persecution; the victim generally believes that he or she is being discussed or plotted against.

paraoxon (*O,O*-diethyl-*O*-(p-nitrophenyl) phosphate; phosphoric acid diethyl 4-nitrophenyl ester; diethyl *p*-nitrophenyl phosphate; phosphacol; E-600). an extremely toxic activation product of parathion. It is formed by the desulfuration by microsomal monooxygenases of parathion. Paraoxon is a potent anticholinesterase and other serine esterases and as such, finds wide usage as a pharmacologic tool.

paraparesis. a mild paralysis of the lower extremities.

paraphasia. jargon.

Parapistocalamus hedegeri (Hediger's snake) known only from Bougainville Island, Solomons group. It is a small burrowing snake; the largest known specimen is about 0.5 m in length. It does not appear to be dangerous to humans.

paraplegia. paralysis of the lower body and legs in humans. **alcoholic paraplegia**. that due to chronic alcoholism. **toxic paraplegia**. that due to poisons in circulating blood.

paraquat (1,1′-dimethyl-4,4′-bipyridinium; *N,*-*N*′-dimethyl-γ,γ'-dipyridylium; methyl viologen (2+)). a yellow, solid, water-soluble, bipyridylium quaternary ammonium salt, used as a rapid-acting, nonspecific contact herbicide. It is better known than other members of this group (e.g., diquat, chlormequat, and difenzoquat). It disrupts plant membranes causing wilting and desiccation. It is highly toxic and has caused large numbers of human deaths. Intoxication of humans and other animals may occur by all routes of absorption (suicidal hypodermic injections are known). Systemic effects in such cases are independent of the route of exposure and the formulation encountered. Paraquat is, nevertheless generally considered safe to use if suitable precautions are taken. Contamination of food and drinking water are serious concerns that must be monitored and guarded against. Poisoning by paraquat is complex and devastating to humans and other animals. Chronic effects due to long-continued exposure to low concentrations are not well known; paraquat is carcinogenic to some laboratory animals. Acute exposure of test animals to aerosols causes pulmonary fibrosis, and the lungs are affected even when exposure is through nonpulmonary routes. Acute exposure may cause focal erosion of tissue and extensive systemic effects including pulmonary fibrosis; variations in the levels of catecholamine, glucose, and insulin; and severe damage to a number of organs. Presenting symptoms include burning sensations in the mouth and throat, vomiting, dyspnea, and sometimes diarrhea, followed within a few days by dyspnea, cyanosis, and signs, impairment of renal, heart, and liver function. Progressive interstitial pneumonia may develop. Long-term nonlethal exposure can cause skin irritation, nosebleeds, and severe eye injury. Death is usually due to pulmonary insufficiency. A constant lesion in fatal cases is pulmonary fibrosis, often accompanied by pulmonary edema and hemorrhaging.

parasite. an organism that lives on or within, and at the expense of another organism (host).

parasiticidal. lethal to parasites.

parasiticide. 1: lethal to parasites. **2:** an agent that kills parasites.

parasol mushroom. a mushroom of the genus *Lepiota*.

parasorbic acid. a carcinogenic lactone.

parasuicide. apparent attempted suicide where death is not the intended outcome, often involving self-poisoning or self-mutilation.

parasympathetic. **1:** of, pertaining to, functioning like, or identifying a division of the autonomic nervous system of vertebrate animals. **2:** cholinergic; a drug or other agent that produces effects similar to those of the parasympathetic nervous system. *Cf.* parasympatholytic, parasympathomimetic. See also cholinergic, muscarinic.

parasympathetic nervous system (cholinergic nervous system). the craniosacral division of the autonomic nervous system of vertebrates, it is in turn comprised of ocular, bulbar and sacral divisions. It is distinguished by long preganglionic fibers with ganglia that are located close to the target organ. the postganglionic fibers are cholinergic and fewer in number than the preganglionic fibers. The parasympathetic nervous system generally promotes activities associated with normal bodily functions. It thus controls contraction of the smooth muscles of the blood vessels and gastrointestinal tract gut as well as glandular secretions (e.g., from the salivary glands, bladder, genital organs, and rectum). Via the vagus nerve, this system serves the viscera as a whole. Parasympathetic stimulation is broadly antagonistic to that of the sympathetic system (the other division of the autonomic nervous system). It acts, for example, to constrict pupils, lower blood pressure, to slow heart rate, constrict bronchi, contract the urinary bladder, stimulate digestion, etc. *Cf.* sympathetic nervous system.

parasympatholytic. **1:** producing effects that resembling those that result from interruption of the parasympathetic nerve supply to a part. **2:** anticholinergic; an agent that opposes or blocks effects of the parasympathetic nervous system. Symptoms of acute poisoning include tremor, marked peristalsis with involuntary defecation and urination, pinpoint pupils, vomiting, cold extremities, hypotension, bronchial constriction, dyspnea, wheezing, twitching of muscles, fainting, slow pulse, convulsions, death from asphyxia or cardiac insufficiency. Lesions on autopsy include congestion of brain, lungs, and gastrointestinal tract and sometimes pulmonary edema. Repeated small doses may reproduce the symptoms of acute poisoning. Atropine can bring about immediate recovery. *Cf.* parasympathetic. See also anticholinergic, antimuscarinic.

parasympathomimetic. **1:** having an action or producing effects similar to that produced by stimulation of parasympathetic innervation of a body part. **2:** parasympathomimetic agent (e.g., acethylcholine).

parasympathomimetic agent (parasympathomimetic (def. 2); cholinergic (def. 2), cholinergic agent). an agent such as acetylcholine that produces cholinergic effects, i.e., effects similar to those produced by stimulation of parasympathetic nerves. The basic feature of poisoning by these agents is dyspnea. Symptoms may also include tremor, marked peristalsis, involuntary defecation and urination, miosis, vomiting, hypotension, coolness of extremities, bronchconstriction, wheezing, trembling, fainting, slow pulse, convulsions, death from asphyxia or cardiac arrest. Lesions seen at autopsy include cerebral, pulmonary, and gastrointestinal tract congestion, often with pulmonary edema. These are cumulative poisons and repeated small doses may produce symptoms identical to that of acute poisoning. Administration of atropine may bring about immediate recovery.

parathion (phosporothioic acid *O,O*-diethyl *O*-(4-nitrophenyl; PA) ester; *O,O*-diethyl *O*-*p*-nitrophenyl phosphorothioate; diethyl-*p*-nitrophenyl monothiophosphate; DNPT). a highly toxic, highly dangerous, phosphorothioate organophosphate insecticide and acaracide, $(C_2H_5O)_2$-$P(S)OC_6H_4NO_2$. It is a dark brown to yellow liquid, that is soluble in alcohols, aromatic hydrocarbons, esters, ketones, ethers, and oils; and nearly insoluble in water, kerosene, and petroleum ether. Parathion is a phosphorothionate ester first licensed for use in 1944. It is applied as an emulsion in water, dust, wettable powder, or aerosol. Parathion is highly toxic to a wide range of insects, but should not be used in the home or any shelter that houses animals because it is toxic also to mammals. Parathion is supertoxic (as little as 120 mg is known to have killed an adult human; and 2 mg a child). Most accidental poisonings of humans are due to cutaneous absorption; some are due to inhalation. The proximate poison is paraoxon, the oxygen analog of parathion, which is produced metabolically from parathion by the microsomal cytochrome P-450-dependent monooxygenase system. It is a powerful, irreversible inhibitor of acetylcholinesterase. Because this conversion is required for parathion to have a toxic effect, symptoms develop several hours after exposure, whereas the toxic effects of TEPP or paraoxon develop much more rapidly. Symptoms of parathion poisoning in humans first appear several hours following exposure. All are referable to the inhibition of acetylcholinesterase with consequent accumulation of acetylcholine and nerve terminals.

Symptoms may include tearing, nausea, vomiting, muscular twitching, diarrhea, abdominal cramps, salivation, headache, vertigo, runny nose, miosis, respiratory distress, generalized and profound muscular weakness, confusion, jerky movements, convulsions, coma. Death may result from respiratory failure due to CNS paralysis. See also fenthion, formothion.

parathion-methyl. methyl parathion.

parathyroid. **1:** situated near the thyroid gland. **2:** the parathyroid gland.

parathyroid gland (parathyroid). two pairs of small glands that control calcium and phosphorus metabolism in vertebrate animals and regulate concentrations in circulating blood and bones. They are located at the base of the neck within or near the thyroid gland. See also gland.

parathyrotoxicosis. acute parathyroid intoxication; a morbid condition due to acute parathyroid intoxication.

paratoid. of, pertaining to, one of a number of poison glands that extend in a double row along the dorsum of certain amphibians.

Pardachirus marmoratus (Red Sea flatfish). a bony fish with 212-235 glands along the dorsal and anal fins, the secretions of which are toxic to other fishes. The toxic principle is pardaxin. See crinotoxic fish.

pardaxin. a protein crinotoxin with a single chain and four sulfide bridges, MW ca. 15,000, isolated from the Red Sea flatfish, *Pardachirus marmoratus*. It inhibits Na^+-K^+-ATPase; stimulates esterase activity; hemolytic in the domestic dog.

paregoric. a camphorated tincture of opium used as an analgesic and in the treatment of diarrhea. Some individuals are hypersensitive. Paregoric should not be used when diarrhea is due to the effects of a poisonous substance. Used properly, however, adverse effects are uncommon and usually mild (e.g., GI disturbances, constipation).

parent compound. a chemically or metabolically unmodified compound. **active parent compound**. a compound such as a toxicant that is absorbed and conveyed to receptors without metabolic modification. **inactive parent compound**. a compound that is inactive unless metabolically converted to an active form.

parenteral. **1a:** of or pertaining to any route of exposure to, or administration of, a drug or xenobiotic other than via the gastrointestinal route. **1b:** definition 1a plus the further exclusion of percutaneous and pulmonary routes (inhalation). 1a is the preferred definition.

paresis. **1:** partial or complete muscle paralysis; sometimes reported as muscular weakness. **2:** dementia paralytica; a disease of the brain that is syphilitic in origin. It is also called paralytic dementia and Boyle's disease.

paresthesia (paraesthesia). **1a:** an abnormal, morbid, or perverted sensation, or the condition of having such sensations (e.g., numbness, prickliness, burning, tickling, or tingling), with or without additional symptoms. **1b:** numbness and tingling in an extremity which often occur with the return of the blood flow to a nerve following temporary blockage (e.g., by the application of pressure) of, or mild injury to a blood vessel. **2:** beyond (normal) sensation.

paresthetic. pertaining to or affected by paresthesia, usually denoting numbness and tingling in an extremity similar to that which usually occurs on the resumption of the blood flow to a nerve in a temporarily blocked or mildly injured blood vessel.

paretic. pertaining to, suffering from, or characterized by paresis.

pari mortem. at or near the time of death.

parietal scute (parietal). either of the large, paired, posterior scutes of a snake's crown; they lie directly behind the frontal and supraocular scutes.

Paris blue. See ferric ferrocyanide.

Paris green. cupric acetoarsenite.

Paris red. lead oxide red.

Paris yellow. lead chromate.

Paris yellow [C.I. 77600]. chrome yellow.

park lily. *Convallaria majalis.*

Parkinson's disease. See parkinsonism.

Parkinsonia aculeata (horsebean). often used as a hedge plant, it is a small evergreen tree (Family Fabaceae, formerly Leguminosae) with clusters of fragrant yellow flowers and feathery pendulous branches. Toxic concentrations of nitrate have been reported in the foliage.

parkinsonism (Parkinson's disease). a neurologic disorder that affects movement. It is characterized by tremor, muscle rigidity, hypokinesia, disturbance of posture, a slow shuffling gait, difficulty in chewing, swallowing, and speaking. The manifestations of the disease are nearly indistinguishable from those of idiopathic Parkinson's disease. It is caused by lesions in the extrapyramidal system that destroy neurones (especially in the striatum) and disrupt dopamine neurotransmission. Parkinsonism may develop during or following acute encephalitis, syphilis, malaria, poliomyelitis, and carbon monoxide poisoning. Parkinsonian symptoms (pseudoparkinsonism) also often occur in patients treated with antipsychotic drugs (e.g., reserpine, amitriptyline, chlorpromazine, fluphenazine, loxapine, thioridazine and other phenothiazine derivatives). See also procyclidine.

Parmelia molliuscula (ground lichen). a lichen that contains usnic acid. It is at most moderately toxic to cattle and sheep, but is potentially lethal. Most poisonings occur during winter when forage is scarce. Most cases are mild, symptoms include ataxia especially of the hind legs. Total paralysis beginning in the hind legs may occur in severe cases and the animal is unable to rise. Affected animals usually remain alert and appetite is normal.

Parnate™. tranylcypromine sulfate (See tranylcypromine).

Parnitine™. tranylcypromine.

paroxysm. **1:** a sudden recurrence or increase in severity of the symptoms of a disease. **2:** a sudden spasm or convulsion.

paroxysmal. pertaining to or occurring during paroxysms.

parrot green. cupric acetoarsenite.

parrot snakes. See *Leptophis*.

parrotfish. fish of the family Scaridae, especially those of the genus *Scarus*.

Parthenocissus quinquefolia (Virginia creeper, woodbine, American ivy). a deciduous high-climbing vine (Family Vitaceae) native from New England southward. The berries are toxic.

parti-colored sea snake. *Pelamis platurus*.

***P*-particle**. a kappa particle that has been released into the medium; it secretes paramecin which kills sensitive strains of *Paramecium*.

particulate. of, or pertaining to particulate matter or particulates.

partition. screen, separation; to devide or separte; the tendency of a substance to exhibit an affinity for one material over another. See partition coefficient.

partition coefficient (PC). a measure of the relative solubility of a chemical in either of two non-miscible liquids when placed in a container holding both liquids. The PC of a substance is the ratio of its concentration in one of the liquids to its concentration in the other. **octanol/water partition coefficient**. the ratio of the solubility of a test substance in a 1-*n*-octanol (the lipid phase) to that of an aqueous phase. It is a measure of the ease with which a substance is transferred from water to lipids. This value is very helpful in estimating the ability of a xenobiotic to cross cell membranes. In practice, it helps toxicologists to infer the bioavailability or uptake (especially via the gill tissue of fish) of a given toxicant since those with high octanol/water PCs are lipophilic and are usually much more readily absorbed than those with low coefficients.

parts per billion (PPB). the number of parts of a given substance in one billion parts of another substance, usually on the basis of weight.

parts per million (PPM). the number of parts of a given substance in one million parts of another substance, usually on the basis of weight.

parturient apoplexy. See milk fever.

parturient fever. See milk fever.

parturient paralysis. See milk fever.

parturient paresis. milk fever. *Cf.* paralysis.

parturition. **1:** passage and delivery of a newborn organism through the terminal portion of the female reproductive tract or birth canal. **2:** childbirth.

Parupeneus chryserydros (surmullet; goatfish). a marine bony fish (Family Mullidae) found from Polynesia west to eastern Africa. Certain populations are transvectors of ciguatera.

parvoline. a toxic, amber-colored liquid, $C_9H_{13}N$, in decaying fish or horse meat.

PAS. 4-aminosalicylic acid, a therapeutic agent.

paspalism. paspalum staggers.

Paspalum. a large, widely distributed genus of mostly perennial grasses (Family Gramineae). All species have one or more flattened racemes with hard, papery bracts that surround the flowers. The seed heads are millet-like. Some species, when infested with the fungus *Claviceps paspali*, are responsible for ergot poisoning. Most important in this regard are *P. dilatatum* (dallis grass) originally from South America, but now growing wild in the United States and elsewhere; *P. notatum* (Argentine Bahia); and *P. scrobiculatum* (kodo millet), a species of India. See paspalum staggers, ergotism (neurotropic ergotism).

paspalum staggers (paspalism). poisoning of cattle grazing on the millet-like seed heads of grasses of the genus *Paspalum*, *q.v.* when infested by the fungus *Claviceps paspali*. The condition, usually marked by tremors, incoordination, a spastic pelvic limb gait, and falling is due to one or more alkaloidal tremorigenic mycotoxins produced by the fungus. A large, single dose can produce signs that persist for several days. Animals display continuous trembling of the large muscle groups; walking is jerky; limb movements are uncoordinated; and they fall in awkward attitudes if they attempt to run. Condition is lost after prolonged exposure and complete paralysis can occur. The condition most often affects cattle, but horses and sheep are also susceptible. Guinea pigs have been poisoned experimentally. Early signs have been shown to appear in cattle after about 100 g sclerotia per day have been administered for >2 days. Recovery follows removal of the animals to feed not contaminated with sclerotia of *C. paspali*. Topping of the pasture to remove affected seed heads has been effective in controlling this condition.

pasque flower. a common name of plants of the genus *Anemone*.

passion flower. See *Adenia*.

passive. **1:** inactive, acquiescent, not spontaneous; not alert or spontaneous. **2:** not produced by active efforts; able to proceed without expenditure of metabolic energy.

passive diffusion. See diffusion.

passive immunity. See immunity.

passive process. a system or process that proceeds without the expenditure of metabolic energy. *Cf.* active process.

passive smoke. See cigarette smoke.

passive transport. passive diffusion (See diffusion).

passive uptake. the absorption of ions by a passive process, q.v., such as diffusion. See also active uptake, transport (passive transport).

Patagonian pit viper. *Bothrops ammodytoides*.

patency. **1:** said of blood vessels or of any hollow organ that is open (not blocked or obstructed). Blood vessels or any hollow organs that clog or become blocked are said to lose their patency. **2:** of or pertaining to an argument or observation that is obvious.

patent. **1:** open, unobstructed. **2:** clear, obvious, apparent, evident.

paternoster pea. *Abrus precatorius*.

pathoamine. **1:** a toxic amine that causes or results from a disease or pathological condition. **2:** a ptomaine.

pathobiological phenomenon. any of three types of effects produced by toxic chemicals: acute lethal injury, autoxidative injury, and immunological injury.

pathobiology. a science that is closely allied to the science of pathology considered broadly inclusive of subdisciplines such as plant pathology (plant pathobiology) and animal pathology (animal pathobiology) with emphasis on biological processes and mechanisms as opposed to medical or legal aspects.

pathobolism. a condition of deviant metabolism of a morbid or diseased nature. *Cf.* pathometabolism.

pathogen. **1:** any disease-producing agent; any virus, organism, or substance (e.g., a toxin) that can cause disease. **2:** in common usage, this term is often restricted to a disease-causing living organism usually a microorganism.

pathogenesis (pathogeny; pathogenesy). **1:** the course of development of a disease or morbid state especially with respect to cellular and tissue-level processes. **2:** the pathologic, physiologic, or biochemical mechanisms resulting in the development of a disease or morbid process. *Cf.* etiology. **drug pathogenesis**. the production of morbid symptoms by exposure to drugs.

pathogenesy. pathogenesis.

pathogenetic. pathogenic.

pathogenic (pathogenetic; morbific; morbigenous; nosogenic; nosopoietic). of, pertaining to, or causing pathogenesis; giving rise to disease, pathologic conditions, or morbid symptoms.

pathogenicity. the state or quality of causing or being able to cause disease.

pathogeny. pathogenesis.

pathognomonic (pathognostic). **1a:** distinctive or characteristic of a disease or pathologic condition. **1b:** pertaining to any effect or that is distinctive or characteristic of a particular disease or morbid state or to any sign or symptom on which a diagnosis can be made.

pathognomy. the science of diagnosis by studying the typical signs and symptoms of a disease or pathological state.

pathognostic. pathognomonic.

pathography. a descriptive study or treatise on pathology or disease.

pathologic, pathological. **1a:** morbid or diseased. **1b:** indicative of or caused by a morbid condition. **2:** pertaining to pathology.

pathologic histology (histopathology). the histology of morbid or diseased tissues. See also pathology (anatomical pathology), histopathology.

pathological anatomy. anatomical pathology (See pathology).

pathological examination. a laboratory study of abnormal tissue to establish or confirm a diagnosis.

pathologist. a specialist in the practice or study of pathology.

pathology. **1:** the largely descriptive study of the causes and nature of disease and dysfunctional states, especially the study of morphological and functional changes in tissues as a consequence of disease whether caused by pathogenic organisms or by exogenous or endogenous chemical or physical agents. Tools used range from those of microscopy, to gross and subgross anatomy; and from those of biochemistry to physiology. While largely descriptive in nature, both quantitative and experimental studies are also important, as is the study of toxic substances. **2:** the structural and functional characteristics of disease, especially at the tissue and organ level. **anatomical pathology** (pathological anatomy). the gross, subgross, and microscopic study of diseased organs and tissues. **cellular pathology**. **1:** the study and interpretation of cellular lesions and changes of diseased tissues. **2:** cytopathology. **clinical pathology**. **1:** the study of pathological changes during a disease process that may benefit the care of patients. **2:** the science that treats and applies, as possible, the theoretical aspects and procedures of the natural sciences (toxicology, biochemistry, physiology, microbiology, parasitology, immunology, hematology) to the diagnosis, prevention, and treatment of disease, or to the care of patients. **comparative pathology**. the study of animal pathology across taxonomic categories with a view to better understanding disease processes, especially in relation to humans. **molecular pathology**. the study of biochemical and biophysical mechanisms that underlie cellular changes in disease. **oral pathology** (dental pathology). a branch of dentistry that treats the all aspects of disease processes that affect the teeth, bone, salivary glands, and soft tissues of the oropharynx. **plant pathology**. the study of the causes and nature of disease and dysfunctional states in green plants.

pathomaine. any of the pathogenic cadaveric alkaloids.

pathometabolism. metabolism in diseased or morbid states. *Cf.* pathobolism.

pathonomia. pathonomy.

pathonomy (pathonomia). the science that deals with the principles of morbid transformations.

pathophysiologic, pathophysiological. pertaining to pathophysiology or to pathophysiological states.

pathophysiology. **1a:** the study of the physiology of disordered or pathological processes. **1b:** the physiological manifestations of pathological processes.

pathosis. a morbid or diseased condition.

patient. **1:** a person who is suffering, ill, diseased, or facing the prospect of such condition and is under treatment by a health care professional. **2:** frequently defined as any client of a health care professional, as in the case of a normal, healthy, pregnant woman.

patoca. *Bothrops lansbergii*.

patulin (4-hydroxy-4*H*-furo[3,2-c]pyran-2(6H)-one; clavacin; clavatin; claviformin; expansine; penicidin). a highly toxic, crystalline antimicrobial mycotoxin isolated from numerous species of *Aspergillus* and *Penicillium*, including *A. clavatus* and *P. patulum*. It contains an α,β-unsaturated lactone moiety. Patulin is extremely toxic to laboratory mice; very toxic (*p.o.*) to chicks and cockerels of the domestic fowl. Patulin inhibits $K+$ uptake into erythrocytes and inhibits various forms of $Na+/K+$-ATPase. It is a potent carcinogen, although apparently not carcinogenic to laboratory mice and rats.

Pavulon. pancuronium.

pawpaw. *Asimina triloba*.

paxilline. An indole mycotoxin produced by *Penicillium paxilli*.

Paxillus involutus (poison paxillus). a reputedly immunotoxic mushroom (Family Paxillaceae) which gradually sensitizes a subject following several exposures, ultimately causing kidney failure. They occur on or near wood in mixed deciduous evergreen forests throughout much of North America.

Pb. the symbol for lead (plumbum).

PB. phenobarbital.

PBB. polybrominated biphenyl(s).

PCB. polychlorinated biphenyl; procarbazine.

PCDD. polychlorinated dibenzodioxin.

PCDF. polychlorinated dibenzofuran. See TCDF.

pCi/L. picocuries per liter.

pCi. picocurie.

PCNB. See pentachloronitrobenzene.

PCP. **1:** pentachlorophenol. **2:** phencyclidine.

PCPA. chlorophenoxyacetic acid.

Pd. the symbol for palladium.

PDB. paradichlorobenzene.

PDC. 3-pentadecylcatechol.

3-PDC. 3-pentadecylcatechol.

PDE. phosphodiesterase.

pea. See *Lathyrus*.

PEA. phenethylamine.

pea family. See Fabaceae.

pea flower locust. *Robinia pseudoacacia*.

peach. *Prunus persica*.

pear oil. amyl acetate.

pearly gates. *Pomoea violacea*.

peat scours. See scours.

peccant. unhealthy, ill, diseased; causing illness or disease.

pectenine. a toxic alkaloid isolated from *Cereus pecten*, a Mexican cactus.

pectoralgia. pain in the chest.

Pecusanol™. lindane.

pederin (*N*-[[6-(2,3-dimethoxypropyl)tetrahydro-4-hydroxy-5,5-dimethyl-2*H*-pyran-2-yl]methoxymethyl]tetrahydro-α-hydroxy-2-methoxy-5,6-dimethyl-4-methylene-2*H*-pyran-2-acetamide; *N*-[[6-(2,3-dimethoxypropyl)tetrahydro-4-hydroxy-5,5-dimethyl-2*H*-pyran-2-yl]methoxymethyl]tetrahydro-α-hydroxy-2-methoxy-5,6-dimethyl-4-methylene-2*H*-pyran-2-glycolamide; pederine; paederine). a crystalline toxin, $C_{25}H_{45}NO_9$, isolated from blister beetles of the genus *Paederus* (Family Staphylinidae). It is a blistering agent and a potent inhibitor of protein biosynthesis and mitosis.

pederine. pederin.

pedicellariae (singular, pedicellaria; tube feet). the pedicellariae are the principal venom organs of the Echinoidea (q.v.). They are minute pincer-like structures that stud the surface of some echinoderms (e.g., starfish). They usually seize, kill, and discard small animals that touch them and are also used to capture prey and to groom. The pincer-like "head" of a pedicellaria is made up of 3 calcareous jaws or valves, each with a rounded tooth-like fang. Starfish and sea urchins have pedicellariae that contain venom glands in the concavity of the valves. Humans who have been stung by the pedicellariae of sea urchins generally report an immediate, intense, sometimes prolonged, pain; focal edema; and sometimes muscle weakness. There are few if any systemic manifestations, however. See venom apparatus, sea urchin.

pediculicide (lousicide). **1:** lethal to lice. **2:** an agent that kills body lice; usually applied to the skin.

pee-un. *Bungarus caeruleus*.

peeling. a sloughing off or loss of epidermis (e.g., in sunburn and toxic epidermal necrolysis).

pegali. *Bitis nasicornis*.

Peganum harmala (African rue). a bushy perennial herb (Family Zygophallaceae) native to the deserts of Africa, southern Asia, and common also to dry rangelands of the southwestern United States. It contains toxic alkaloids, three of which contain the indole configuration (see harmine, harmaline). The seeds and young leaves are highly toxic. Symptoms of intoxication in guinea pigs include paralysis of the posterior extremities, weakness of back muscles and usually death when fed 10 - 15 mg of plant materials (young leaves and ground seed) / kg of body weight on a dry weight basis. See also harmine, harmaline.

PEL. permissible exposure limit, a maximum (legally enforceable) allowable level for a chemical in workplace air.

pelagic sea snake. *Pelamis platurus*

Pelamis platurus (pelagic sea snake; parti-colored sea snake; yellow-bellied sea snake; Plattchen Seeschlange; umi hebi). the most widely distributed sea snake, and although it is the most common sea snake in the shallow waters over continental shelves, it is the only truly pelagic sea snake, often seen hundreds of miles from land. It is not known to enter brackish or freshwater bodies. The venom is weak with a toxicity about ¼ that of *Enhydrina* and is thought usually to be injected only in minute amounts. Validated accounts of human fatalities are unknown. The head is usually dark on top and yellow on the sides; it is depressed, elongate, and rather narrow, although the neck is distinct. It is a rather slender snake with a very strongly compressed body and is eel-like in appearance. The body is usually black or dark brown above and dark yellow to brown below with a pale yellow lateral stripe. Yellow individuals with a brown or black dorsal stripe are also fairly common and other variants occur occasionally. The tail is whitish and barred or mottled with black. The average adult is about 70 cm; rare individuals may reach 1.1 m.

pelargonic acid (nonanoic acid; nonylic acid; nonoic acid). a colorless or yellowish, very toxic, strongly irritant, oily liquid, $CH_3(CH_2)_7CO_2H$. It is soluble in ethanol and ethyl ether, but insoluble in water. Pelargonic acid is used as a chemical intermediate in organic syntheses, as a flotation agent and in the manufacture of hydropic salts, pharmaceuticals, artificial food flavorings and aromas, lacquers, and plastics.

pelecypod. any bivalve mollusk.

Pelecypoda (Bivalvia; formerly Lamellibranchiata). a class of about 15,000 species of mollusks that includes all bivalve mollusks (e.g., clams, oysters, mussels, scallops, cockles, teredo). The shell is comprised of two lateral, nearly symmetrical, calcium carbonate valves that are fastened together by a dorsal hinge, a dorsal ligament, and one or two adductor muscles. A mantle lines the valves of shell and forms posterior siphons which control the flow of water through the mantle cavity. There is no head, jaws, or radula; the gills are plate-like; the foot is singularly hatchet shaped and often protrudes between valves during locomotion. Most marine species have trochophore and veliger larval stages, while freshwater species typically have glochidium larvae.

Peltandra virginica (arrow arum). a perennial herb (Family Araceae) with dark green leaves shaped like arrow heads. They are native to swamps, bogs, and borders of ponds and slow streams in eastern North America. All parts of the plant are toxic.

penchlorol. pentachlorophenol.

pencil lead. graphite; usually not toxic if ingested.

penetrant. **1:** any agent that increases the speed and ease with which a liquid penetrates or permeates a membrane, tissue, or other material. Penetrants may be used to facilitate absorption of toxicants or other pharmacologically active agents by cells or tissues. See, for example, dimethyl sulfoxide (DMSO). **2:** See nematocyst.

penetrate. to pierce, enter; invade; infiltrate, permeate; to pass deeply into (e.g., cells, tissues, a body cavity).

penetrating. **1a:** invasive, piercing, cutting, or entering deeply, as in a penetrating wound. **1b:** able to pass deeply, e.g., into tissue. **2:** sharp, e.g., a penetrating odor.

penetration. **1a:** piercing, entering, ingress. **1b:** invasion. **1c:** absorption, *q.v.*, infiltration, permeation **2:** intelligence, insight, mental acumen. **3:** depth of focus of a lens.

penetration, route of. See exposure (route of exposure).

penicidin. patulin.

penicillamine (U.S.N.; d,3-mercaptovaline; 3-mercapto-D-valine). a hydrolytic degradation product of penicillin, $(CH_3)_2C(SH)CH(NH_2)$-COOH. It is a chelating agent used in the treatment of heavy metal poisoning, especially that of lead, copper, mercury, or zinc. This compound promotes the urinary excretion of these metals. It is nephrotoxic and hemotoxic, affecting the hemopoietic system.

penicillia. common name of ascomycete fungi of the genus *Penicillium*.

penicillic acid (3-methoxy-5-methyl-4-oxo-2,5-hexadenoic acid; γ-keto-β-methoxy-δ-methyleneΔ-$^{\alpha}$-hexenoic acid). a very toxic mycotoxin antibiotic, $C_8H_{10}O_4$, produced by *Penicillium puberulum*, a mold found on maize; also from *P. baarnense*, *P. cyclopium*, *P. suaveolens*, *P. thomii*, *Aspergillus melleus*, and *A. ochraceus*. It contains an α,β-unsaturated lactone moiety. Penicillic acid is toxic to both

Gram-positive and Gram-negative bacteria and to animal tissues; it and is also carcinogenic.

penicillin. any of a large class of natural or semisynthetic bacteriostatic antibiotics based on a β-lactam ring structure. They disrupt cell-wall synthesis of many bacteria and some fungi with consequent osmotic lysis; peptidoglycan synthesis in bacteria is inhibited by penicillins. They are produced by *Penicillium* spp. and other soil-inhabiting fungi and are generally relatively nontoxic to the host; they can also be synthesized from penicillic acid. Penicillins are especially active against Gram-positive pathogens (e.g., staphylococci and gonococci) but are usually only mildly toxic to animals. See also penicillinase.

penicillinase. any of a class of enzymes secreted by penicillin-resistant bacteria (e.g., staphylococci) that inactivates penicillins by hydrolysing the β-lactam ring.

Penicillium. a genus of imperfect fungi (Family Moniliaceae, order Moniliales) the fruiting bodies of which resemble a broom. Most species of *Penicillium* produce mycotoxins that are extremely toxic. Thus *P. brevicompactum*, produces the mycotoxins brevianamide (an indole) and mycophenolic acid. *P. chrysogenum* produces a number of penicillins. *P. citreoviride* produces the mycotoxin citreoviridin. *P. citrinum* produces the antibiotic mycotoxin citrinin. *P. crustosum*. produces the indole mycotoxins penitrem A and C. *P. cyclopium* produces the indole mycotoxins cyclopiazonic acid and penitrem A and B. *P. decumbens* produces the mycotoxin decumbin. *P. expansum* produces the mycotoxin curvularin. *P. fellatum* produces citreoviridin. *P. griseofulvum* produces the mycotoxin griseofulvin. *P. islandicum* is a causal fungus of yellowed rice; it produces the mycotoxins cyclochlorotine, erythroskyrine, and luteoskyrin, and (-)rugulosin. The most toxic and carcinogenic of these is (-)luteoskyrin. *P. janczewski* (*P. nigricans*). produces the antimycotic griseofulvin. *P. nigricans*, *P. janczewskii*. *P. notatum* produces several penicillins. *P. ochrasalmoneum* produces the mycotoxin citreoviridin. *P. oxalicum* produces the mycotoxins oxaline (an indole) and secalonic acid. *P. palitans* produces the mycotoxin penitrem A. *P. patulum* produces the antimycotics patulin and griseofulvin. *P. paxilli* produces the indole mycotoxin paxilline. *P. puberulum* produces the mycotoxin penicillic acid. *P. pulvillorum* produces citreoviridin. *P. purpurogenum* produces the mycotoxins rubratoxin A and rubratoxin B. *P. roqueforti* produces the mycotoxins PR-toxin and roquefortine (an indole). *P. rubrum* produces the mycotoxins rubratoxin A and rubratoxin B. *P. rugulosum* produces the mycotoxin (+)rugulosin. *P. uticale* (*P. patulum*) produces patulin. *P. terlikowskii* produces gliotoxin. *P. verruculosum* produces the indole mycotoxin verruculogen (TR-1 toxin). *P. viridicatum* produces ochratoxins A, B and C.

penitrem A. an indole mycotoxin produced by *Penicillium cyclopium, P. crustosum and P. palitans*. It is extremely toxic and tremorgenic at low doses to laboratory mice; at high doses, tremors progress to clonic or tetanic convulsions and death. Neurotoxic, nephrotoxic and diuretic to a number of laboratory mammals and to the domestic fowl.

penitrem B (dechloro-penitrem A). an indole mycotoxin produced by *Penicillium cyclopium*. It is extremely toxic (*i.p.*) to laboratory mice.

penitrem C. an indole mycotoxin produced by *Penicillium crustosum*. It is very toxic (*i.v.*, *s.c.*) or moderately toxic (*p.o.*) to laboratory mice.

pennyroyal. a common name for various plants of the family Labiatae, especially those of the genus *Mentha*. Examples are *Hedeoma pulegoides* (pennyroyal of America) and *Mentha pulegium* (true pennyroyal).

pennyroyal of America. *Hedeoma pulegoides*.

Penstemon (beard tongue). a large genus of North American herbs (Family Scrophulariaceae) that are facultative accumulators of selenium from soil.

penta. pentachlorophenol.

penta-. a prefix indicating five, five of, grouped in fives, etc.

pentaborane. a stable neurotoxic compound,

$B_{10}H_{14}$, with colorless crystals that causes autonomic and behavioral dysfunction similar to that caused by reserpine or decaborane. Uses include those of a catalyst, corrosion inhibitor, fuel additive, oxygen scavenger, mothproofing agent, reducing agent, and propellant. It lowers brain norepinephrine and serotonin in laboratory animals. Acute exposures may cause visual disturbances, tremors, generalized muscle spasms, rhabdomyolysis and coma; liver and kidneys may also be affected. In humans, symptoms may develop from inhaling concentrations as low as 25 ppm. Low concentrations in air are associated with drowsiness, headache, nausea and vertigo. See also borane.

pentacarbonyl iron. iron pentacarbonyl.

pentachlorin. DDT.

pentachloroethane (pentalin). a dense, colorless, flammable, water-insoluble liquid, $CHCl_2CCl_3$. It is toxic by ingestion or inhalation and is carcinogenic to some laboratory animals.

pentachloronitrobenzene (PCNB). a colorless crystalline compound, $C_6Cl_5NO_2$, with a musty odor. It is a pesticide, soil fungicide, and has been used as a seed dressing. It is an irritant and is known to cause cancer in certain strains of laboratory mice.

pentachlorophenol (penta; penchlorol; PCP). a toxic, fungicidal, bactericidal, light-brown, solid, chlorinated hydrocarbon, C_6Cl_5OH, prepared by catalytic chlorination of phenol. It is used as a wood preservative, protecting it from bacteria, fungi, and slime. It is also used for termite control and as a general herbicide and defoliant. Pentachlorophenol is a component of shampoos, paints, and laundry starches and is found as a contaminant in numerous foods that have been stored in containers treated with PCP or that have been sprayed with PCP. It is very toxic to laboratory rats and is carcinogenic to some laboratory animals. Toxic exposures may result from ingestion, inhalation of pentachlorophenol-containing dusts, and by absorption through the skin. Some effects in humans (e.g., chloracne) may be due to contamination by dibenzodioxins. Effects of exposure can include any of a variety of symptoms resulting from irritation to the skin and mucous membranes of the respiratory tract and upper gastrointestinal tract, and from damage to the liver, kidneys, and lungs. Observed effects may thus include a contact dermatitis, chest pain, dizziness, sweating, anorexia and weight loss, pyrexia, and dyspnea. Pentachlorophenol is known to have caused the death of human infants (when used as an antibacterial soap). Cattle have become ill from licking the wood or breathing the fumes from fences made from PCP-treated wood. Pentachlorophenol is also fetotoxic and teratogenic. See pentachlorophenol poisoning.

pentachlorophenol poisoning (PCP poisoning). poisoning marked by signs that may include nervousness, weakness, rapid pulse, rapid respiratory rate, muscle tremors, fever, convulsions, and death. Chronic poisoning may result in CNS depression, lightheadedness, sleepiness, tremors, nausea, loss of appetite and weight loss, fatty liver, nephrosis, and cancer. There is no known antidote. Livestock fed in troughs made of PCP-treated lumber may salivate excessively and display irritation of the oral mucosa.

3-pentadecyl-1,2-benzenediol. 3-pentadecylcatechol.

3-pentadecylcatechol (3-pentadecyl-1,2-benzenediol; 3-pentadecylpyrocatechol; tetrahydrourushiol; hydrourushiol; dihydrohengol; 3-PDC). a toxic white powder, C_6Cl_5OH, that is soluble in acetone, benzene, ethanal, and ethyl ether. It is used as a chemical intermediate, a diagnostic aid, and as an algicide, bactericide and herbicide. It is a constituent of urushiol.

3-*n*-pentadecylcatechol. a contact allergen; one of the toxic principles of *Rhus radicans*.

3-pentadecylpyrocatechol. 3-pentadecylcatechol.

(2β,3β,5β,22R)-2,3,14,20,22-pentahydroxycholest-7-en-6-one. ponasterone A.

2,3,14,22,25-pentahydroxycholest-7-en-6-one. α-ecdysone.

3,5,7,3′,4′-pentahydroxyflavone. quercetin.

4,9,12-β-13,20-pentahydroxy-1,6-tibliadiene-3-ol. tetradecanoylphorbol 13-acetate.

pentalin. pentachloroethane.

pentamethylenediamine. cadaverine.

4,4′-(pentamethylenedioxy)dibenzamidine. pentamidine.

pentamidine (4,4′-[1,5-pentanediylbis(oxy)]bis-benzenecarboximidamide; 4,4′-(pentamethyl-enedioxy)dibenzamidine; 4,4′-diamidino-α,ω-diphenoxypentane). a colorless, crystalline, highly toxic compound, $C_{19}H_{24}N_4O_2$, used therapeutically as an antiprotozoal against *Babesia*, *Leishmania*, and *Pneumocystis*.

1,5-pentanediamine. cadaverine.

4,4′-[1,5-pentanediylbis(oxy)]bis-benzenecarboximidamide. pentamidine.

1-pentanethiol (*n*-amyl mercaptan; amyl mercaptan; amyl hydrosulfide; amyl sulfhydrate; amyl thioalcohol; pentyl mercaptan). a clear, colorless to yellow, flammable, liquid odorant, b.p. 124°C. Dangerous to humans, it is moderately toxic when inhaled. It is a weak sensitizer and allergen, able to cause a localized contact dermatitis. See thiol.

pentanochlor. a postemergent anilide herbicide that rapidly degrades in the soil. It is used mainly to weed tomatoes and umbelliferous crops such as carrots, celery, and parsley.

pentanol-2. 2-pentanol.

1-pentanol. *n*-amyl alcohol, primary.

2-pentanol (*sec*-amyl alcohol, active; *sec-n*-amyl alcohol; *sec*-amyl alcohol; 1-methylbutyl alcohol; methylpropylcarbinol; pentanol-2; *sec*-pentyl alcohol). a commercially available straight chain secondary alcohol and isomer of amyl alcohol. It is a colorless, toxic, flammable liquid, $CH_3(CH_2)_4OH$, that is slightly soluble in water and miscible with ethanol and ether. A narcotic and irritant to the eyes, nose, and throat; it is moderately toxic by all routes of exposure. It is also a moderate fire and severe explosion hazard. *Cf.* 3-pentanol.

3-pentanol (*sec-n*-amyl alcohol; 1-ethyl-1-propanol; diethyl carbinol). a commercially available straight chain secondary alcohol and isomer of amyl alcohol. It is an irritant, colorless liquid, used as a solvent, flotation agent, and in pharmaceuticals. *Cf.* 2-pentanol.

***n*-pentanol**. *n*-amyl alcohol, primary (See alcohol).

***tert*-pentanol**. *tert*-pentyl alcohol (See alcohol).

2-pentanone. (methyl propyl ketone; ethyl acetone; MPK). a clear flammable liquid with a strong odor, $CH_3(CH_2)_2COCH_3$, used as a solvent and synthetic food flavoring. Technical grade pentanone is usually a 3:1 mixture of 1-pentanone and 2-pentanone. It is an irritant to the eyes and nasal passages, has a narcotic effect, and at high concentrations can cause coma and death.

pentavalent sodium stibogluconate. See stibogluconate sodium, definition 1.

pentazocine (2-dimethylallyl-5,9-dimethyl-2′-hydroxy benzomorphan). a synthetic noncumulative opioid, $C_{19}H_{27}NO$, It has been claimed to be as effective as morphine, but not addictive.

pentobarbital (5-ethyl-5-(1-methylbutyl)2,-4,6(1H,3H,5H)-pyrimidinetrione monosodium salt; sodium 5-ethyl-5(1- methylbutyl)barbiturate; pentobarbital sodium; pentobarbitone sodium; Nembutal). a sedative-hypnotic barbiturate, $C_{11}H_{18}O_3N_2$, that is sometimes used to treat insomnia and as a premedication drug prior to surgery, but has been largely replaced by the benzodiazepines which are much safer. Adverse effects are those typical of barbiturates. Dependence and tolerance are known to occur. In veterinary practice, pentobarbital is used as an anesthetic, an anticonvulsant, and for euthanasia.

pentothal. thiopental sodium.

pentothal sodium. thiopental sodium.

pentyl. amyl.

pentyl mercaptan. 1-pentanethiol.

3-pentyl-6,6,9-trimethyl-6a,7,8,10a-t etrahydro-6*H*-dibenzo(b,d)pyran-1-ol. (—)-Δ^1-3,4-*trans*-tetrahydrocannabinol.

pentylenetetrazole (6,7,8,9-tetrahydro-5*H*-tetrazolo[1,5-a]azepine; Metrazol). a CNS stimulant, and analeptic agent. It can induce seizures at dosages in the therapeutic range and is now used only as a diagnostic tool in screening for latent epileptogenic foci and in basic research.

pentymal. amobarbital.

pepper. **1:** any of a number of products of plants of the genus *Piper*. **2:** to apply pepper; to sprinkle as one sprinkles pepper; to make spicy. **black pepper**. **1:** *Piper nigrum*. **2:** a widely used condiment from *Piper nigrum*. It is usually nontoxic, unless inhaled in mass. **Java pepper**. *Piper cubeba*. **tailed pepper**. *Piper cubeba*. **true pepper**. any plant of the genus *Piper*.

pepper coral. *Millepora*.

peppery milky. *Lactarius piperatus*.

pepsin. a protein-digesting enzyme secreted by the gastric glands of vertebrates.

peptic ulcer. ulcer of the stomach.

peptid. peptide.

peptidase. any proteolytic enzyme that catalyzes the hydrolysis of peptide bonds in polypeptides or proteins. See snake venom peptidases.

peptide (peptid). any of a class of low molecular-weight compounds comprised of two or more amino acids in which the alpha carboxyl group of one bonds to the alpha amino group of a neighbor. These bonds, -CO-NH, called peptide bonds are formed with the loss of a molecule of water from the NH_2 and COOH groups of adjacent amino acids. Depending on the number of amino acids in the molecule, peptides are known as di-, tri-, tetra-, etc. peptides or polypeptides. Peptides are the component parts of proteins. **toxic peptide**. any poisonous peptide. Such are not uncommon among poisonous and venomous animals; snake venoms are usually complex mixtures chiefly of proteins and peptides. See also *Anemonia sulcata*, anaphylatoxin, apamin, bombesin, *Bombina*, *Buthus occitanus*, cardiotoxin, cobratoxin, *Conus*, eledoisin, mellitin.

peptide bond. the bond, -CO-NH, that joins two amino acids within a peptide chain. The alpha-carboxyl group of one amino acid binds to the alpha-amino group of another, with loss of a water molecule.

peptinotoxin. a poisonous intestinal substance formed by partial or defective gastric digestion.

peptone. a protein or mixture derived by partial hydrolysis of native protein (e.g., in egg albumin, meat, milk) by an acid or enzyme. All are soluble in water, insoluble in ether and ethanol. They are not precipitated by heat, alkalis, or saturation with ammonium sulfate. Peptones are used mainly to make nutrient media in microbiology. **venom peptome**. a peptone in snake venom. See peptotoxin (def. 2).

peptotoxin. used variously in the literature. **1:** a poisonous alkaloid or ptomaine that occurs in certain peptones. **2:** a poisonous derivative of a peptone. **3:** a poisonous intestinal substance resulting from partial or imperfect digestion in the stomach. **4:** putrefying proteins.

per os (*p.o.*; *po*). by mouth; orally. Used in reference to the route of administration of a drug or toxicant. *Cf.* peroral.

peracute. said of a disease: very acute.

perc. perchloroethylene.

perchlorate. any salt of the chlorooxyacids. Most are not very toxic and toxicity depends upon the cation in the compound. Most are skin irritants. Ammonium perchlorate, however, is a highly reactive, and powerful oxidizing agent; it is toxic and a fire and explosion hazard. See perchloric acid.

perchloric acid. a colorless fuming liquid, $HClO_4$, having the highest oxygen content of the chlorine acids. It reacts explosively with organic matter or other reducible materials. It is a strong irritant and is toxic by ingestion and inhalation.

perchloride (hyperchloride) a chloride that contains the largest possible amount of chlorine.

perchloropentacyclo(5.2.1.0^{2,6}.0^{3,9}.0^{5,8})-decane. mirex.

perchlorobenzene. hexachlorobenzene.

perchlorocyclopentadiene. hexachlorocyclopentadiene.

perchloroethylene (tetrachloroethylene; TCE; ethylene tetrachloride; tetrachloroethene; carbon bichloride; carbon dichloride; perc). a clear, colorless, nonflammable, extremely stable, organic liquid, Cl_2-C=C-Cl_2, with an ether-like odor. It is used as a dry cleaning solvent (chiefly in coin-operated dry-cleaning facilities), in spot removers, in rug, carpet, and upholstery cleaners, and sometimes as a vapor-degreasing agent, fumigant, and vermifuge. It is effective in treating hookworms and intestinal trematode infections. It is an irritant, especially of the skin, eyes, and nose; in high concentrations, and contact can lead to dermatitis through defatting of the skin. It is mildly narcotic. Immediate effects of exposure may include light-headedness, dizziness, drowsiness, nausea, tremors, loss of appetite, and disorientation. Long-term exposure may damage the liver, heart, kidneys, or CNS, sometimes ending in death. Acute exposures to high concentrations can be deadly. It is carcinogenic to some laboratory animals and is a known human carcinogen. The fumes are toxic and carcinogenic.

perchloromethane. carbon tetrachloride.

perchloropentacyclo[5.2.1.$0^{2,6}$.$0^{3,9}$.$0^{5,8}$.]-decane. mirex.

perchloryl fluoride. a colorless, noncorrosive gas or liquid, $ClFO_3$, with a sweetish odor. It is used as an oxidant in rocket propellants; as an oxidizing and fluorinating agent in chemical reactions; and as an insulating gas in high-voltage electrical systems. It is a lung and skin irritant and may damage blood cells. A strong oxidizing agent, it should not be allowed to come into contact with organic material.

percolate. **1:** to strain or filter a liquid through porous material. **2:** to trickle slowly through a porous substance. **3:** a liquid that percolates.

percolation. **1a:** the slow movement of a liquid (usually water or a solvent) through a porous material (e.g., soil). **1b:** the extraction of soluble materials from the materialo through which a solvent percolates. In toxicology, it is usually used to extract soluble portions of a drug or other suitable xenobiotic. **2a:** the movement of water (e.g., rain water) downward and radially through subsurface soil layers. The excess of water over that amount which is sufficient to saturate the soil, will usually continue downward into the ground water. Percolation may play an important role in the transport of environmental chemicals (e.g., pollutants and toxic chemicals) to new locations. In some cases (as in the case of improperly sited land fills or hazardous waste dumps), percolation can extend pollution to nearby streams and in extreme cases may cause irreversible pollution of groundwater aquifers. **2b:** the flow of groundwater due to gravity through pore spaces in rock or soil.

percutaneous (diadermic; transcutaneous; transdermic). of or pertaining to the passage of a substance through unbroken skin of a given type of animal. The permeability of the skin to a given substance varies with the site on the body of the animal and the species concerned (compare frog, human, rhinoceros).

percutaneous absorption. absorption of chemicals through the skin. both epidermis and dermis. Although this does not occur at appreciable rates for water-soluble chemicals, it is an important route of exposure for lipophilic toxicants such as parathion. See also exposure (route of exposure).

Percy Island snake. *Demansia psammophis*.

perennial pea. *Lathyrus latifolius*.

perennial snakeweed. *Gutierrezia microcephala*.

perforation. an abnormal hole or opening; puncture; tear; slit. **intestinal perforation**. a perforation of the intestinal wall, usually due to trauma or a complication of an existing condition such as ulcers or carcinomas. Such can also result from the erosion of the intestinal wall by a corrosive poison. Aside from the trauma itself, which may be quite severe, intestinal contents may pass into the abdominal cavity, causing severe inflammation.

perforin. a protein produced by cytotoxic T cells and natural killer cells that lyse target cells by forming pores in the cell membrane.

perfume. an artificial fragrance usually with a high ethanol content. A small child may become intoxicated by ingesting as little as a full tablespoon of perfume. This may cause a reduction in blood sugar resulting in unconsciousness. More serious consequences including coma and death might result from ingestion of this or a slightly greater amount.

perfusion. **1:** the pouring of a liquid over or through a material. **2:** artificial passage of a fluid through blood vessels of an organ or whole animal. *Cf.* infiltration, suffusion. **organ perfusion**. the maintenance, by perfusion, of an organ isolated from any vascular supply either *in situ* or isolated from the body. The organ is perfused with a suitable fluid driven mechanically through its own vascular system. Such preparations offer excellent opportunities to test the behavior and effects of a xenobiotic substance and its interactions with the test organ.

perianth. a collective term for the sepals and petals of a flower.

pericarditic. pertaining to pericarditis.

pericarditis. inflammation of the pericardium.

pericardium. **1:** pericardial; around the heart. **2:** the fibroserous sac that surrounds the heart and roots of the great vessels in vertebrates. It is made up of an outer layer of fibrous tissue (the pericardium fibrosum) and an inner serous layer (the pericardium serosum). The base of the pericardium in mammals attaches to the central tendon of the diaphragm.

perinatal. of or pertaining to the time near birth. See perinatal period.

perinatal mortality. See mortality.

perinatal period. **1a:** near the time of birth. **1b:** the time when viability *ex utero* is considered possible. **1c:** the last third of the gestation period. In *Homo sapiens* the perinatal period is usually taken to begin at the 20th week of gestation. For some purposes the perinatal period in humans may include the first postnatal week. See mortality (perinatal mortality).

Peringuey's adder. *Bitis peringueyi*.

period. **1a:** a cycle; an interval or division of time between two regularly repeated occurrences. **1b:** the time elapsed between regular recurrence of a phenomenon. **latent period**. an interval of seeming inactivity, such as the time interval between exposure to a poison, carcinogen, or other injurious agent and the appearance of an effect (e.g., a sign, or symptom). **silent period**. an interval during the course of a disease in which symptoms ameliorate or disappear for a while.

periodic. recurring at regular intervals in time or space.

periodontitis. inflammation of the gums.

peripheral. **1:** superficial, marginal; pertaining to or situated at or near the margin or periphery. **2a:** outlying; located away from the center of a structure or away from a central structure. **2b:** nearer to the periphery than some designated point or structure.

peripheral nervous system (PNS). **1:** that portion of the nervous system of an animal that lies outside the central nervous system (that part which includes a "brain"). **2:** in vertebrate animals the PNS includes the nerves and ganglia that lie outside of the brain and spinal cord.

peripheral neuropathy. any disease of the peripheral nervous system that may affect specifically or collectively, somatic motor, sensory, and autonomic neurons. Symptoms vary according to the parts of the CNS affected. Different parts of neurones may be preferentially involved: neuronal cell bodies (neuronopathy); proximal axon (proximal axonopathy); distal axon (distal axonopathy; Schwann cell or myelin sheath (myelinopathy). Etiologies vary functionally and may arise from toxic, metabolic, hereditary, autoimmune, ischemic, traumatic, neoplastic, infectious or idiopathic proximate or ultimate causes according to which the disease may be designated (e.g., toxic peripheral neuropathy).

peripheral vascular system (PVS). a network of arteries, veins and lymphatic channels that supply the head, arms, and legs.

periphyton. a complex matrix of aquatic algae and heterotrophic microbes attached to submerged surfaces. It is a major source of food for numerous aquatic invertebrates and herbivorous fishes. Periphyton can bioconcentrate environmental chemicals such as mercury and cadmium by several orders of magnitude over concentrations that occur in the surrounding water.

perirectal. of or pertaining to the skin and underlying tissue about the rectum.

perish. to die or be destroyed, especially violently or from intense ambient conditions.

peristalsis. a series of involuntary, usually rhythmic contractions that move the contents along in a tubular organ with muscular walls (e.g., a digestive tract). In portions of a digestive tract it may also produce a churning action that mixes the contents and aids digestion.

peritoneal dropsy. ascites.

peritonitis. an inflammation of the peritoneum.

peritubular capillary. a capillary that surrounds a nephron and functions in reabsorption during urine formation.

permeability. **1:** the quality of being permeable. **2a:** the relative ability of a membrane, cell, soil, or other potential barrier to permit passage of fluids or dissolved materials. Permeability of membrane or other semipermeable structure to a toxic substance is a function of properties of the structure, the substance, and the environment in which contact occurs (e.g., temperature, Ph). **2b:** the rate or relative rate of passage of liquids or dissolved materials through such barriers.

permeable. pervious, penetrable; permitting the passage of fluids or dissolved materials (said of a membrane or other potential barrier). **selectively permeable**. a property of a living cell membrane whereby permeability is regulated.

permeant. able to pass through a specific semipermeable membrane. *Cf.* permeant.

permease. a generic term for a stereospecific membrane transport carrier. The suffix emphasizes the functional similarities of transport carriers with enzymes (e.g., specificity).

permeate. **1:** to penetrate, infiltrate, saturate, or pass through, as through a filter or semipermeable membrane. **2:** any component of a solution or suspension that can permeate (def. 1). *Cf.* permeant.

permeation. The process of spreading through, infiltrating, saturating, or penetrating, e.g., a substance, cell, tissue or organ.

permissible exposure limit (PEL). primarily health-based limits on exposure to a chemical that are developed as enforceable standards in the United States. by the Occupational Safety and Health Administration

pernicious. **1:** noxious, destructive, deadly, fatal, tending to be fatal **2:** of or pertaining to a severe, often fatal disease.

pernicious vomiting. continuing uncontrollable vomiting.

Pernox™. haloperidol.

Peron's sea snake. *Acalyptophis peronii*.

peroral. *per os*; through the mouth. Passed through, administered through, exposed by way of, or performed through the mouth. Denoting usually a method of administration or passage of a material into the body.

perosmic acid anhydride. osmium tetroxide.

perosmic oxide. osmium tetroxide.

perosmium tetroxide anhydride. osmium tetroxide.

peroxide. **1:** any compound that contains a bivalent O—O group. All such compounds are strong oxidizing agents and a fire hazard when stored in contact with flammable organic materials or under conditions of high temperature. Their primary uses are as bleaches, oxidizers, and initiators of polymerization. Inorganic peroxides are contact poisons that may injure the skin or mucous membranes. Toxicities vary considerably. They are strong oxidizing agents and moderate to very dangerous fire and explosion hazards if in contact with reducing agents. Organic peroxides are strong oxidizers. Many are highly toxic and are strong irritants to the skin, eyes, and mucous membranes. They are very dangerous fire hazards when in contact with reducing agents or exposed to high temperatures. They are severe explosion hazards if shocked or exposed to heat; they may also explode spontaneously. Many are very unstable chemically. **2:** often used in speech to mean hydrogen peroxide.

peroxide dye. See color (HC dye).

1,4-peroxido-*p*-menthene-2. ascaridole.

peroxisomal enzyme. See peroxisome.

peroxisome. a small organelle, about 0.5-1.0 μm in diameter and bounded by a single membrane. They may occur in all types of cells, but are prominant in hepatocytes and erythrocytes. They contain catalase, peroxidases, other oxidative enzymes. They appear to play a role in lipid metabolism and are presumed to play an important role in detoxication reactions (e.g., of ethanol). See also glyoxysome.

peroxisome proliferation. certain xenobiotics (e.g., clofibrate, di(2-ethylhexyl) phthalate) produce a discernible proliferation of peroxisomes in liver cells. Such compounds are often induce liver tumors and are thought to be promoters of carcinogenesis.

Persea americana (avocado). a tree (Family Lauraceae), the leaves, fruit, bark, and seeds of which are toxic. Livestock, various mammals and birds, and probably fish have been poisoned under natural conditions. In cattle, horses and goats milk flow is reduced and a noninfectious, sometimes fatal mastitis develops. The mastitis can be controlled, but milk flow never returns to normal. The ripe fruit is toxic to canaries. Varieties of avocado range in toxicity from weakly to highly toxic.

Persian horned viper. *Pseudocerastes persicus*.

Persian insect powder. pyrethrin, pyrethrum flowers.

persic oil. See apricot-kernel oil.

persistence. **1a:** stability, endurance, tenacity; durability; resistance to degradation. **1b:** durability or stability of a characteristic structure or behavior in the face of opposition or adverse environmental conditions. **2a:** chemical stability of a substance, usually under a stated set of conditions. **2b:** of or pertaining to the length of time a chemical compound will remain unaltered under specified conditions or in designated situations or environments. **absolute environmental persistence**. a term that applies to those chemicals (e.g., heavy metals) that never undergo dissolution in the environment. Burdens of such chemicals are reduced at a given location only via transport mechanisms. **aquatic persistence**. chemical stability of a substance in a water body. **environmental persistence**. stability, or the tendency to stability, of a substance following introduction into the environment. The most hazardous toxicants in the environment are generally those that persist; they are not necessarily the most toxic.

persistent. **1a:** having the quality of persistence, *q.v.* **1b:** lasting, enduring, durable, long-lived, resistant to degradative forces or to decomposition, remaining. **1c:** in toxicology, said of a chemical substance that is stable or durable under specified conditions or in specified situations; one that will maintain its structure (does not decompose) and activity in the face of degradative forces.

persistent environmental chemical. any chemical, usually of anthropogenic origin, that decomposes slowly, if at all, in the environment (or a specified environment or portion of an environment).

persistent pesticide. a pesticide such as DDT that decomposes slowly, if at all, or remains active following one growing season. While such pesticides may give longer protection against the targeted pests than nonpersistent pesticides, they may contaminate the harvest or may bioaccumulate, poisoning wildlife, livestock, pets, and even humans (examples are DDT and related organochlorine pesticides, MCPA, and PCB).

personal sampler. See sampler, personal.

person (plural, people or persons). a human of any age and of either sex.

perspiration. **1:** sweat. **2:** sweating; the secretion of sweat.

perturbation. **1:** a disturbance; any departure of a living system from a dynamic steady state. **2a:** distress; **2b:** uneasiness of mind; the state of being greatly disturbed, agitated, or stressed.

pertussigen. pertussis toxin. See toxin.

Peruvian bark. cinchona.

Peruvian coca. *Erythroxylon novagranatense*.

Peruvian tarantula. *Glyptocranium gasteracanthoides*.

pervious. **1:** permeable, porous, penetrable, capable of being penetrated. **2:** penetrating.

pessimal. least favorable (as opposed to optimal), as in the case of environmental variables that are near the limits of tolerance of a living system.

pessimum. those conditions that are suboptimal for maintenance, growth and development of a living system.

pest. **1a:** any organism considered to be disruptive to human activities, especially one that interferes with human well-being, comfort, the value or well-being of pets or domestic animals and plants, or other human resources. **1b:** most commonly applied to any organism that has a negative economic impact, such as one that damages crops or reduces their yield or marketability; one that irritates or injures pets or livestock; one that damages or reduces the value or pleasure one takes in house or garden; one that spreads disease. An organism that, furthermore, has no appreciable redeeming features in the context in which it is considered a pest. **emergent pest**. a special case of faunal imbalance whereby alternative species become pests by virtue of some environmental change such as an imbalance in, or even extinction of, the population of an another pest (e.g., DDT control of the codling moth has resulted in a burgeoning of mite populations to the point where they became serious pests).

pesticidal synergist. a compound (e.g., piperonyl butoxide) added to an insecticide or other pesticide to increase its effectiveness. See also synergist.

pesticide. any compound or formulation (e.g., herbicide, insecticide, miticide, rodenticide) used to destroy plant and animal pests or to inhibit their activity. Virtually all are toxic to humans, some seriously so. **contact pesticide**. a chemical that kills pests through contact with the body surface, i.e., ingestion or inhalation is not required. **household pesticides**. any pesticide used in the home (e.g., fly spray, ant and roach baits, fly spray, insect repellents). In the United States alone, about 2.5 million humans are thought to be affected annually by such substances; 70% of those affected are children under the age of five. Symptoms of pesticide exposure in the home vary but often include nausea, coughing, dyspnea, eye irritation, dizziness, weakness, blurred vision, muscle spasms, and convulsions. Most are lipid-soluble cumulative poisons.

pesticide-resistant. pertaining to pests that are not easily killed or damaged by pesticides, but continue to multiply when exposed to pesticides.

pesticide tolerance. the amount of pesticide residue permitted under law to remain in or on a harvested food crop. In the U.S., the Environmental Protection Agency attempts to set these at levels well below the point where the chemicals might be harmful to consumers.

pestilent. epidemic, widespread and destructive; devastating; epidemic.

petechia (plural, petachiae). a flat, perfectly round, pinpoint, purplish red spot caused by intradermal or submucosal hemorrhage. *Cf.* ecchymosis.

petechiae. plural of petechia.

petechial. characterized by, or of the nature of, petechiae.

petechiation. the formation of petechiae.

Peters sea snake. *Hydrophis bituberculatus*.

pella. *Hemachatus haemachatus*.

petiole. the stalk of a leaf.

petit gui. *Arceuthobium pusillum*.

Petit Prêcheur. *Arisaema atrórubens*.

petrohol. 2-propanol.

petrol. gasoline (in the United Kingdom).

petroleum (crude oil). a naturally occurring complex mixture of tens of thousands of paraffinic, cycloparaffinic (naphthenic) and aromatic liquid hydrocarbons, usually with small amounts of sulfur, varying amounts of heavy metals, and traces of nitrogen and oxygen compounds. Some crude oils contain as much as 10% sulfur. Full chemical characterization of any crude oil has proven impossible. Petroleum is toxic and a skin irritant. Crude oils vary considerably in physical properties and toxicity. Oil pollution of the marine environment produces heavy mortality of fish and wildlife worldwide. In addition, large spills have caused environmental disasters of major proportions. The oiling of feathers or fur of marine animals so impairs mobility that death by drowning or starvation may follow. Ingestion of oil (as by preening) may cause death by dehydration due to interference with ion transport and water balance in the alimentary canal. Some dissolved petroleum fractions are highly toxic and damage or kill biota through ingestion or absorption through the body surface. Minute amounts of crude and refined petroleum are toxic to the embryos of aquatic birds when the oil contacts the egg shell.

petroleum benzine. petroleum ether.

petroleum distillate. any of a wide variety of generally neurotoxic materials of varying toxicity distilled from petroleum (e.g., kerosene, gasoline, paint thinner). All are common constituents of cleaning fluids and household polishes. These chemicals dissolve fat, causing reddened and calloused skin. Symptoms of intoxication may include depression, coma, and occasionally convulsions. Mild heart attacks may follow ingestion or inhalation. Other effects often associated with poisoning by these chemicals are a burning sensation in mouth, throat, and/or stomach; nausea with vomiting and diarrhea; drowsiness; restlessness; disorientation. Pulmonary involvement usually signals impending fulminating hemorrhagic bronchopneumonia. Petroleum distillates cause numerous deaths from accidental poisoning, especially of very young children. In the United States nearly one fourth of all deaths from accidental poisoning result from exposure to these substances. See hydrocarbon poisoning.

petroleum distillates (naphtha). See naphtha. Do not confuse with the term petroleum distillate.

petroleum distillate poisoning. poisoning due to ingestion or inhalation of a petroleum distillate. Symptoms may include nausea, vomiting, chest pain, dizziness, and severe CNS depression. Lesions may include a fatal pneumonitis is the distillate is aspirated. See also petroleum distillate, kerosene poisoning.

petroleum ether (petroleum naphtha; petroleum benzine; benzine). **1:** a petroleum distillate comprised chiefly of n-pentane and n-hexane, with a boiling point range of 35-80°C. It is moderately toxic and may cause CNS depression. Because of its high volatility, it does not readily induce pneumonia when aspirated. When aspirated by laboratory rats, petroleum ether displaces oxygen from the lungs rapidly causing profound anoxia; cardiac arrest and/or brain damage may soon follow. **2:** sometimes used as a synonym for naphtha and light ligroin.

petroleum fuel. any of a wide range of petroleum products used as liquid fuels for heating equipment or engines. They may be broadly classified as distillate fuels (See petroleum distillate) and residual oils. Residual oil is the heavy oil that remains following removal of distillates in a refinery. Included are some diesel fuels and fuel oils. All are polluting, some to a high degree, and most will emit carbon monoxide fumes if burned under less than optimal or prescribed conditions.

petroleum naphtha. petroleum ether.

petroleum naphthene. See naphthene (def. 3).

petroleum pitch. asphalt.

petty splurge. *Euphorbia peplus*. See *Euphorbia*.

Peucetia viridans (green lynx spider). a bright leaf-green, ivory or tan spider with yellowish legs that bear many black spines. It occurs in fields and woods on tall grasses and flowerheads, especially of wild buckwheat, in the southern United States and Mexico. The female, 14-16 mm long, is dangerous to humans.

peyote (peyoti). **1:** any of several Mexican cacti (Family Cactaceae) of the genus *Lophophora,* especially *L. williamsii, q.v.* **2:** the active principle (mescaline) from the dried flowering heads. It is used by native Americans of North America to produced altered states of consciousness. In some tribes this is done as part of religious ceremonies. The members of such tribes are permitted to use this drug, even though it is classed as a narcotic and its use is otherwise restricted to research.

pg. picogram.

pH. symbol for the negative logarithm (base 10) of the hydrogen ion concentration of an aqueous solution. It is a measure of the acidity or alkalinity of the solution. A solution with pH of 7.00 is neutral; one with a pH of 7.0-14.0 is alkaline; one with a pH lower than 7.00 is acidic. This value is of great significance in toxicology. The pH of a toxicant can, for example, greatly influence its absorption and thus its toxicity. Poisons with extremely high or extremely low pH are extremely irritant and corrosive to tissue. The pH of biological fluids, cells or tissues, on the one hand, and environmental media (soil, sediments, surface water, groundwater) can strongly affect or control the direction, speed, and extent of chemical reactions including those that involve toxicants.

Ph. phenyl.

PHA. phytohemagglutinin. See lectin.

phacoid. pertaining to or resembling the lens of the eye.

phacotoxic. exerting a toxic (injurious) effect upon the crystalline lens.

Phaeolepiota aurea (golden false pholiota). a mushroom (Family Agaricaceae) that is toxic to some individuals but not others. It generally grows on leaf litter or in compost piles in Alaska and the Pacific Northwest region of the United States.

Phaeophyta (brown algae). a plant phylum containing the largest of the algae. They are multicellular, almost exclusively marine, and are filamentous or composed of leaf-like structures (thalli). They sometimes provide major forage for livestock. A few, such as the giant kelp, *Macrocystis pyrifera*, contain moderately toxic substances. See also alga.

phagocyte. any cell (e.g., a leukocyte) capable of phagocytosis, especially of other cells or microorganisms.

phagocytosis. a process of ingestion by phagocytes whereby the cell membrane of the phagocyte engulfs bacteria, other cells, and debris, taking them into the protoplasm of the cell where digestion takes place. Phagocytosis sometimes facilitates the absorption of toxicants into the body, especially in the gastrointestinal tract and alveoli of the lungs. *Cf.* pinocytosis.

phakhuphakhu. *Naja nigricollis*.

Phalaris arundinacea (reed canarygrass). a wild North American grass (Family Gramineae) that may become heavily ergotized. See *Claviceps*.

phallacidin. See phallotoxin.

phallisacin. See phallotoxin.

phalloidin (phalloidine). a polypeptide phallotoxin, $C_{30}H_{39}O_9N_7S$, isolated from the mushroom, *Amanita phalloides*. A component of amanitatoxin, it binds to actin filaments and prevents cell movement. The LD_{50} in albino mice has been reported at 3.3 mu-g/g *i.m.* It is nephrotoxic, hepatotoxic and less toxic, but more rapid-acting than amanitin. See also phallotoxin.

phalloidine. See phallotoxin.

phallolysin. a glycoprotein, it is a heat-sensitive toxin of the *Amanita phalloides*. It is destroyed by cooking.

phallotoxin (phallacidin; phalloidin; phallisacin). any of at least 6 related, thermostable, cyclic polypeptide toxins that have the same heptapeptide skeleton. They occur in several species of wild mushrooms of the genus *Amanita*. They are deadly, but much less toxic than amatoxins. Phallotoxins attack the plasma membranes of hepatocytes and the epithelial cells of the proximal convoluted tubules of the kidney.

Death of laboratory animals is due to hepatic and/or renal failure. See also amatoxin.

pharmacal. pertaining to pharmacy.

pharmaceutic. pertaining to pharmacy, pharmaceuticals (drugs), or pharmaceutics.

pharmaceutical. **1:** pertaining to pharmacy or drugs. **2:** a medicinal drug.

pharmaceutics. **1:** pharmacy (def. 1). **2:** pharmaceutical preparations.

pharmacist (apothecary; druggist; chemist (in Great Britain)). a person licensed to prepare and dispense medicinal drugs and to fill prescriptions.

pharmaco-. an affix denoting relationship to a drug or medication.

pharmacodynamic. pertaining to pharmacodynamics or to the biological effects or action of biologically active substances. See pharmacodynamics.

pharmacodynamic (drug) interaction. any drug interaction in which the concurrent administration of two or more drugs produce the same (or opposing) pharmacologic effects, or in which one drug alters the responsiveness or sensitivity of target tissues to another. See also drug interaction, pharmacodynamic, pharmacokinetic (drug) interaction.

pharmacodynamics. **1a:** the action of drugs within an organism, all aspects of behavior, and mechanisms of pharmacodynamic interactions. **2:** the study of the biochemical and physiological effects and mechanisms of the action of drugs, medicines, and other biologically active chemicals, including venoms and other toxic agents. See pharmacology, pharmacokinetics.

pharmacogenetic. of or pertaining to genetic differences in sensitivity or mechanisms of response to a toxicant within and among species of living organisms.

pharmacogenetics. the study of genetically influenced responses to drugs. This discipline treats the genetic basis of variations in drug responses among individual organisms within a given population or species.

pharmacognosy. the study of the properties of crude drugs.

pharmacokinetic. pertaining to pharmacokinetics (def. 2).

pharmacokinetic (drug) interaction. any drug interaction in which the administration of two drugs alters the amount and duration of drug availability at receptor sites mainly by altering the rate of absorption, distribution, metabolism, or excretion. In this type of interaction, the magnitude and duration of the effect of one or both of the drugs are altered, but the type of effect remains unaffected. See also drug interaction, pharmacokinetic, pharmacodynamic (drug) interaction.

pharmacokinetics. **1:** the science that deals with and describes, often through the use of mathematical models, the kinetic processes involved in the absorption, distribution, metabolism, and elimination of drugs and toxic substances. **2:** the behavior and time course of movements of drugs and toxic substances within an organism, including absorption, distribution, localization, metabolism, and elimination. See pharmacology, pharmacodynamics.

pharmacologic, pharmacological. **1:** pertaining to pharmacology or to the properties and actions of drugs. **2:** sometimes used in reference to a dose of a biologically active agent that mimics that of a naturally occurring substance (e.g., a hormone) that is so much greater or more potent than would be experienced naturally that it might have qualitatively different effects. *Cf.* autopharmacologic.

pharmacologic treatment. the treatment of patients by the use of drugs. *Cf.* physiological treatment, psychological treatment.

pharmacologist. a specialist in pharmacology; one who studies the biological actions of drugs, mostly therapeutic drugs. *Cf.* toxicologist.

pharmacology. the study of biologically active chemicals, especially of medicinal drugs, their origins, nature, chemistry, action, applications. This field embraces pharmacognosy, pharmacokinetics, pharmacodynamics, pharmacotherapeutics, and portions of toxicology. **biochemical pharmacology**. the subdiscipline that deals with the biochemical processes and mechanisms responsible for drug actions. **clinical pharmacology**. that division concerned with the use of therapeutic agents in the prevention, treatment, and control human disorders. **veterinary pharmacology**. the study and application of the principles and practices of pharmacology to the concerns of veterinary medicine.

pharmacomania. an uncontrollable craving to take or administer medicines.

pharmacopeia (pharmacopoeia). an authorized compendium on drugs that contains information on their preparation, effects, dosages, standards, and legal requirements of purity, strength, and quality. **U.S. Pharmacopeia** (U.S.P.). the legally recognized pharmacopeia published by the United States Pharmacopeial Convention, Inc. See also unit (USP unit).

pharmacophore. that part of a molecule that is responsible for causing the specific pharmacologic effects of a drug.

pharmacopoeia. pharmacopeia.

pharmacopsychosis. a toxic response to a drug manifested as a psychosis.

pharmacotherapeutic. **1:** of or pertaining to a drug or drugs used in the treatment of disease. **2:** referring to the efficacy of a drug in the treatment of disease.

pharmacy. **1:** the health science that deals with the preparation, dispensing, and proper use of drugs. **2:** a dispensary, store, or other location where drugs are compounded and/or dispensed.

pharyngitis. inflammation of the throat.

phase I reaction. See metabolic reaction.

phase II reaction. See metabolic reaction.

phase of toxicity. the action of a toxic substance usually proceeds through two major phases, a kinetic phase and a dynamic phase. **kinetic phase of toxicity**. the initial phase; it includes absorption, metabolism, temporary storage, distribution, and to an extent, excretion of the prototoxicant or toxicant. If the toxicant fails to reach a target tissue due to the action of one or more of the above processes, no injury or damage is sustained and the dynamic phase does not occur. On the other hand, if a prototoxicant that is itself nontoxic is absorbed and transformed into a toxicant (during the kinetic phase), and is not then detoxified or excreted, but is transported to target organs and detrimental effects are to be expected. **dynamic phase of toxicity**. the phase during which the toxicant reacts with a receptor to initiate, in a primary reaction step, one or more biochemical reactions that produce physiological and/or behavioral manifestations of poisoning.

phasein. lectin.

phaseolunatin. linamarin.

Phaseolus (kidney beans; bush beans; lima beans; etc.). a genus of warm-climate, cyanogenic, twining, annual or perennial herbs (Family Fabaceae, formerly Leguminosae). A number are grown for their edible pods and seeds. ***P. coccineus*** (scarlet runner bean). an ornamental vine with pods that are considered edible, has a record of toxicity in Europe. ***P. limensis*** (= *P. lunatus* in part) (lima bean; Java bean; Burma bean; Sieva bean; etc.). there are numerous varieties of this bean, all of which contain the cyanogenic glycoside, linamarin (= phaseolunatin). The lima beans commonly cultivated in the United States and other countries in the temperate zone are essentially nontoxic. They are small, light green to nearly white in color, easily recognized, and essentially non-toxic. An exception is the black lima bean of Puerto Rico, which has proven toxic to humans. Tropical varieties are often quite toxic. They are usually small, plump beans that are solid-colored or sometimes spotted. Many instances of cyanide poisoning are known from New Guinea due to eating raw lima beans (cooking is preventive). Lima beans imported into Canada or the United States for use as food or feed must by law be analyzed for cyanide. Specimens from the United states contain less than 0.01% hydrocyanic acid (HCN), whereas tropical varieties contain more than 0.01% HCN, frequently much more. Some tropical varieties contain as much as 0.3% HCN. Plants containing 0.02% HCN are potentially hazardous to livestock.

phasin. lectin.

pheasant's eye. a common name for plants of the genus *Adonis*.

phenacemide (*N*-(aminocarbonyl)benzeneacetamide; (phenylacetyl)urea; phenacetylurea; Phenurone). a white, or nearly white, crystalline substance or fine crystalline powder, $C_9H_{10}N_2O_2$, that is nearly insoluble in water and slightly soluble in ethanol, benzene, chloroform, and ether. It is an anticonvulsant used as an adjunct in the management of psychomotor, grand mal, and petit mal epilepsy and in mixed seizures. Because of its toxicity, it should be limited and effects closely monitored. Adverse effects of overdosage may include gastrointestinal distress, dermatitis, behavioral effects, and aplastic anemia. Lesions are those of hepatitis and nephritis. Toxicity to humans is uncertain.

phenacetin. acetophenetidin.

phenacyl chloride. ω-chloroacetophenone.

phenalzine. phenelzine.

phenanthrene. a shining, colorless, crystalline, tricyclic, water-insoluble, flammable, aromatic hydrocarbon, $C_{14}H_{10}$, present in coal tar that is isomeric with anthracene. It is soluble in organic solvents. Phenanthrene is moderately toxic to laboratory mice when ingested. It has tumorigenic activity, is a human carcinogen and probable mutagen, and can cause photosensitization of the skin in humans.

2-phenanthrylamine. a carcinogenic polycyclic arylamine. See also aromatic amine.

phenarsazine chloride (10-chloro-5,10-dihydrophenarsazine; 5-aza-10-arsenaanthracene chloride; 10-chloro-5,10-dihydroarsacridine; diphenylaminechlorarsine; diphenylaminearsine chloride; phenazarsine chloride; adamsite; DM). a toxic war gas, $NH(C_6H_4)_2AsCl$, dispersed in the atmosphere as fine particulates (smoke). Also used in combination with chloroacetophenone (tear gas) to control riots and in wood-treating formulations to control marine borers and similar pests. It is a skin and respiratory irritant. Effects on humans include severe sinus, nasal and chest pain with sneezing, coughing and a copious watery nasal discharge. Further symptoms include nausea, vomiting, weakness, and pronounced depression. Sensory disturbances may eventually appear.

phenazarsine chloride. phenarsazine chloride.

phenazone. antipyrine.

phenazopyridine (3,9-phenylazo-2,6-pyridinediamine). a brick-red azo dye used as a urinary tract analgesic. It is carcinogenic to some laboratory animals. Acute poisoning of humans is marked by methemoglobinemia and Heinz body anemia, which is more severe in cases of glucose-6-phosphate dehydrogenase deficiency.

phencyclidine (1-(1-phencyclohexyl)piperidine; angel dust; HOG; PCP; CI-395; Sernyl). a very toxic arylcyclohexylamine, usually prepared as the hydrochloride, $C_{17}H_{26}ClN$. A substance of abuse, used mainly as an hallucinogen, phencyclidine is a controlled substance listed in the U.S. Code of Federal Regulations. Although classed as a central stimulant, it also has depressant, anesthetic, analgesic, and hallucinogenic activity and even small doses may induce delirium and a variety of other signs and symptoms to include hyperactivity; slurred speech; sweating, muscular rigidity; numbness in fingers and toes; visual, auditory, and tactile illusions; delusions; eyes crossing; incoordination; lack of sensation; hyperextension; ataxic gait; facial grimaces; anxiety; hostility; feelings of inebriation; disorientation; prominent body image distortions; transient amnesia; lack of pain perception; and wild movements. Larger doses cause tachycardia, convulsions, stupor or coma, and sometimes high fever and seizures. Fatal or near fatal intoxications may further be marked by high blood pressure, convulsions, grand mal seizures, suppressed or absent reflexes, renal failure. Respiratory arrest is a major cause of death. Laboratory animals and humans can develop tolerance.

1-(1-phencyclohexyl)piperidine. phencyclidine.

phenelzine ((2-phenethyl)hydrazine; β-phenylethylhydrazine; phenalzine). a white, water-soluble powder, $C_6H_5CH_2CH_2NHNH_2$, it is a very toxic monoamine oxidase inhibitor. The acid sulfate (e.g., Estinerval, Kalgan, Nardelzine, Nardil) is used therapeutically as an antidepressant.

phenesterine (cholesteryl-*p*-bis(2-chloroethyl)-amino phenylacetate; (*p*-(bis(2-chloroethyl)-amino)phenyl)acetic acid cholesterol ester; (4-(bis(2-chloroethyl)amino)phenyl)acetic acid cholesteryl ester; 5-cholesten-3-*β*-ol 3-(*p*-(bis-(2-chloroethyl)amino)phenyl)acetate;fenesterin; fenestrin; phenestrin). a substance that is carcinogenic to laboratory mice and rats and is a suspected human carcinogen. Upon decomposition by heating, it emits very toxic fumes (Cl^- and NO_x).

phenestrin. fenesterine.

phenethyl diguanide. phenformin.

phenethylamine (*β*-phenethylamine; benzene-ethanamine; *β*-phenylethylamine; 2-phenylethylamine; 1-amino-2-phenylethane; *β*-aminoethylbenzene; PEA). an endogenous primary amine, related structurally and pharmacologically to amphetamine. It is a metabolic product of the decarboxylation of phenylalanine and can be synthesized chemically from phenylethyl alcohol and ammonia under pressure. PEA is used in organic synthesis, as a reagent, and as a carbon dioxide absorber in scintillation counting. It is a combustible liquid, $C_6H_5C_2NH_2$, with a fishy odor, that is soluble in water, ethanol, and ether. It is a strong base and is toxic on contact or by ingestion. PEA, like amphetamine has anorectic properties and induces hyperactivity in mice and, at high doses, stereotypical behavior in rats. It is a skin irritant and possible sensitizer. It is also associated with migraine and may play a role in the development of schizophrenia. It is a toxic principle in *Phoradendron flavescens* (American mistletoe), *q.v.* The LD_{50} in laboratory mice is approximately 470 mg/kg body weight, *s.c.*

β-phenethylamine. phenethylamine.

1-phenethylbiguanide. phenformin.

***N'*-β-phenethylformamidinyliminourea**. phenformin.

(2-phenethyl)hydrazine. phenelzine.

***p*-phenetol carbamide**. dulcin.

***p*-phenetylurea**. dulcin.

phenformin (*N*-(2-phenylethyl)imidodicarbonimidic diamide; 1-phenethylbiguanide; phenethyldiguanide; *N'*-*β*-phenethylformamidinyliminourea; fenformin; fenormin). a very toxic, oral hypoglycemic substance, $C_{19}H_{15}N_5$, associated with a high incidence of fatal lactic acidosis. An antidiabetic, usually given as the hydrochloride. It has been declared an imminent hazard to public health by the Food and Drug Administration and its use is no longer permitted in the United States.

phenformin hydrochloride. (1-phenylbiguanide monohydrochloride). a substance, $C_{19}H_{16}ClN_5$, with a toxic action essentially similar to that of phenformin, *q.v.*

phenic acid. phenol (def. 1).

phenindione (2-phenyl-1*H*-indene-1,3(2H)-dione; 2-phenyl-1,3-indandione; 2-phenyl-1,3-diketohydrindene). a creamy white to pale yellow crystalline or powdery compound, $C_{15}H_{10}O_2$, that is soluble in a number of organic solvents. It is an indanedione anticoagulant, administered orally. It is marked by rapid onset and short duration of action.

pheniprazine ((1-methyl-2-phenylethyl)hydrazine; (*α*-methylphenethyl)hydrazine; (1-benzylethyl)hydrazine; 1-phenyl-2-hydrazinopropane; *β*-phenylisopropylhydrazine; PIH). a crystalline substance, $C_6H_5CH_2CH(CH_3)NHNH_2$, it is a monoamine oxidase inhibitor. Formerly used as an antihypertensive antidepressant. It is especially dangerous when used in combination with certain other drugs such as barbiturates, antihistamines, and other antidepressants such as meperidine, morphine, and aminopyrine.

phenmedipham (3-methylphenyl)carbamic acid 3[(methoxycarbonyl)amino]phenyl ester; *m*-hydroxycarbanilic acid methyl ester m-methylcarbanilate; methyl-3-(*m*-tolylcarbamoyloxy)phenylcarbamate). a selective carbamate herbicide. It is slightly toxic to laboratory rats.

phenmetrazine(3-methyl-2-phenylmorpholine;3-methyl-2-phenyltetrahydro-2*H*-1,4-oxazine; 2-phenyl-3-tetramethylhydro-1,4-oxazine; A 66;

Preludin). a white, water-soluble, crystalline powder. It is a highly toxic sympathomimetic drug used as an appetite suppressant in managing the diet of obese individuals; it is usually given as the hydrochloride. Not only is its effectiveness unpredictable in a given case, but numerous additional effects may appear even at therapeutic dosages. Including an evaluation of blood pressure, disturbances of heart rhythm, tachycardia, restlessness, excessive activity, dizziness, incoordination, impaired judgment, confusion, insomnia, euphoria, tremor, headache, dryness of mouth, unpleasant tastes, diarrhea, upset stomach, changes in sex drive, and impotence, aggressiveness, and psychosis. Addiction is frequent. Overdosage may cause hallucinations, aggressiveness, panic, and hypertension or hypotension. The victim may become comatose and death from circulatory collapse may follow. Users become tolerant after a few weeks of treatment. If treatment is terminated abruptly, the patient may suffer irritability, extreme depression, hyperactivity, and behavioral changes.

phenobarbital (5-ethyl-5-phenyl-2,4,6(1H, 3H,5H)pyrimidinetrione; 5-ethyl-5-phenylbarbituric acid; PB; phenobarbitone; phenylbarbital; phenylethylbarbituric acid; phenylethylmalonylurea; Sulfoton). a white, shining, odorless, somewhat bitter-tasting, crystalline powder, $C_{12}H_{12}N_2O_3$, that is soluble in water and a variety of organic solvents. It is a barbiturate used primarily as a hypnotic, a long-acting sedative, and as an anticonvulsant. Phenobarbital was the first effective antiepileptic developed. It raises the seizure threshold and also limits the spread of seizures. It is especially effective against grand mal epilepsy and cortical focal seizures. When used properly as a sedative-hypnotic, the incidence of serious side effects is low. It may, however, lower body temperature, retard development, and cause neonatal bleeding and hyperbilirubinemia. Apparently noncarcinogenic in humans, it induces liver tumors in laboratory mice. Withdrawal in a phenobarbital-dependent individual is characterized by convulsions. **sodium phenobarbital** (luminal sodium). a water-soluble salt of phenobarbital which is more rapidly absorbed than the parent compound.

phenobarbitone. phenobarbital.

Phenoclor. polychlorinated biphenyl(s).

phenol. 1: (carbolic acid; benzophenol; hydroxybenzene; phenic acid; phenyl alcohol; phenylic acid). a highly toxic, colorless, pink or red, highly corrosive, crystalline solid, C_6H_5OH, with a distinctive odor. It is produced by the distillation of coal tar. Phenol is a strong irritant to the skin, is toxic when ingested, inhaled, or absorbed by the skin, and must be handled cautiously; absorption through the skin alone can prove fatal. Phenol may be eliminated from the body unchanged, but most is first converted metabolically by Phase II reactions into water-soluble sulfates and glucuronides. See carbolic acid poisoning, phenol poisoning. **2:** (phenolic; phenolic compound) a class of aromatic derivatives - aryl alcohols of benzene that have 1 or more hydroxy groups attached directly to the benzene ring. They are byproducts of petroleum refining, tanning, and textile, dye, and resin manufacturing. Included are phenol itself (def.1), cresols, flavonoids, xylenols, cinnamic acid, coumarin, resorcinol, and naphthols. Their behavior is distinctly different from other alcohols. Low concentrations cause taste and odor problems in water; higher concentrations can kill aquatic life and humans. Some are immunotoxic; several have been shown to suppress antibody production by spleen cells in mice. **chlorinated phenol** (chlorinated phenolic). a class of phenols that importantly includes pentachlorophenol and the trichlorophenol isomers which are fungicidal and insecticidal and have thus been widely used as wood preservatives. **plant phenol**. any of numerous, extremely diverse phenols, including the flavinoids, that are produced by plants. Some are allelopathic.

phenol carbinol. benzyl alcohol.

phenol poisoning. 1: severe poisoning caused by swallowing phenolic compounds and mixtures (e.g., carbolic acid, creosote, naphthol, pentachlorophenol). Victims exhibit burns of the mouth and pharynx, often with stricture of the esophagus; seizures; and collapse of all major systems. **2:** carbolic acid poisoning.

phenol red. phenolsulfophthalein.

phenolemia. the presence of phenols in circulating blood.

phenolic. **1:** phenol (def. 2). **2:** any phenol-formaldehyde resin or plastic. **chlorinated phenolic**. chlorinated phenol (See phenol).

phenolphthalein (3.3-bis(4-hydroxyphenyl)-1-(3H)-isobenzofuranone). a white or faintly white crystalline powder, $C_{20}H_{14}O_4$. It is an orally administered cathartic. It is used chiefly in the chemical and biological sciences as a pH indicator. The glucuronide has been used in studies of hepatobiliary dysfunction caused by toxic chemicals.

phenolsulfophthalein (phenol red). a bright to dark red, crystalline powder, $C_{19}H_{14}O_5S$, that is soluble in alkali hydroxides and carbonates; slightly soluble in water, ethanol, and acetone; and nearly insoluble in chloroform and ether. It differs chemically from phenolphthalein by the presence of an SO_2 group rather than a CO group. Phenolsulfophthalein is a pH indicator used in the pH 6.8-8.4 range. It is also used clinically to assess kidney function and experimentally to assess tubular activity in studies of nephrotoxic agents.

phenomenology. phenology, *q.v.*

phenomenon (plural, phenomena). **1a:** event, happening, occurrence. **2:** a marvel, wonder, sensation, curiosity; any unusual fact or occurrence. **3a:** a symptom, occurrence, or event of any kind that derives from or bears on a disease. **3b:** a sign that is often associated with a specific disease or condition, one that is thus diagnostically important.

phenothiazine. **1:** 10*H*-phenothiazine; thiodiphenylamine; dibenzothiazine. a photocontact allergen and antiemetic, $C_{12}H_9NS$, used as an insecticide and anthelmintic. It also lowers body temperature. **2:** any of a class of antipsychotic drugs such as chlorpromazine, prochlorperazine, and trifluoperazine. They are derivatives of the phenothiazine nucleus. They slow and regulate mental system activity and are usually used to treat anxiety and other mental conditions; they are also used as antihistaminics and sometimes used to induce sleep. Symptoms of overdosage may include symptoms such as dry mouth, incoordination, muscular rigidity, tremors, uncontrollable facial grimacing, lowered body temperature, decreased respiratory rate, cardiac arrhythmias, convulsions, somnolence, stupor, and coma. In addition to central nervous (extrapyramidal) effects, phenothiazines are toxic to marrow and can induce granulocytopenia. In addition to central nervous (extrapyramidal) effects, phenothiazines are toxic to marrow and can induce granulocytopenia. Lesions may include CNS depression, extrapyramidal seizures, liver damage; neonatal bleeding, developmental retardation. See also phenothiazine, phenothiazine sudden death, phosphodiesterase.

10*H*-phenothiazine. phenothiazine.

phenothiazine drug. See phenothiazine (def. 2).

phenothiazine sudden death. sudden death from phenothiazine-induced respiratory or vasomotor collapse. Victims are usually psychiatric patients who have received large doses of chlorpromazine or other phenothiazines. Patients are thought to die from asphyxiation during a convulsive seizure or from cardiovascular collapse during a hypotensive crisis. Overdosage may be signaled by extrapyramidal effects as indicated in the definition of phenothiazine (def. 2), *q.v.*

phenotype. **1:** any individual discernable or measurable trait (structural or functional) of an organism (or type of organism) as opposed to genotype. A phenotype is the result of interactions between genotype and environment. **2:** the sum of all such traits of an individual organism or population.

phenotypic. pertaining to phenotype.

phenoxyacetic acid. **1:** a light tan, flammable powder, $C_6H_5OCH_2COOH$, used, for example, as an intermediate for dyes, pesticides, and pharmaceuticals, and a precursor in antibiotic fermentations, especially penicillin V. **2:** any of a class of substituted compounds of phenoxyacetic acid (def. 1), some of which show auxin-like activity. Various chlorine-substituted species disrupt plant metabolism especially at the meristems and have been widely used as selective herbicides. These compounds are especially effective as weedkillers because they are immune from the plant's endogenous IAA oxidizing system, which usually inactivates excess auxins. 2,4-D, 2.4.5-T, and MCPA are phenoxyacetic acids.

phenoxybenzamine (*N*-(2-chloroethyl)-*N*-(1-methyl-2-phenoxyethyl)benzenemethanamine; *N*-(2-chloroethyl)-*N*-(1-methyl-2-phenoxyethyl) benzylamine; *N*-phenoxyisopropyl-*N*-benzyl-β-chloroethylamine; bensylyt; 688A). a crystalline, benzene-soluble compound, $C_{18}H_{22}ClNO$. The hydrochloride (phenoxybenzamine hydrochloride; Dibenyline; Dibenzyline; Dibenzyran), a hypertensive, has been established as carcinogenic to some laboratory animals.

phenoxybenzamine hydrochloride. See phenoxybenzamine.

***N*-phenoxyisopropyl-*N*-benzyl-β-chloroethylamine**. phenoxybenzamine.

phentanyl. fentanyl citrate. See fentanyl.

phentermine (α,α-dimethylbenzeneethanamine; Fastin; Ionamin). a colorless, oily liquid, $C_{10}H_{15}N$, an adrenergic isomer of amphetamine used as an appetite suppressant. It also elevates blood pressure, disturbs heart rhythm, and may produce any of a number of symptoms at prescribed doses such as restlessness, hyperactivity, insomnia, euphoria, tremor, headache, dryness of mouth, unpleasant tastes, upset stomach, diarrhea, reduced libido, impotence, and sometimes depression and psychosis. Overdoses may result in "accidental" death from hallucinations, aggressiveness, and panic. It should not be prescribed for persons with heart disease. *Cf.* benzphetamine, diethylpropion, phenmetrazine.

phenthion. fenthion.

Phenuron. See phenacemide.

phenvalerate. fenvalerate.

phenyl (Ph). the univalent radical, C_6H_5, of benzene.

phenyl alcohol. phenol.

phenyl hydroxide. phenol.

***N*-phenylacetamide**. acetanilide.

3-α-phenyl-β-acetylethyl-4-hydroxycoumarin. warfarin.

***p*-phenylalanine**. *p*-biphenylamine.

phenylalanine mustard. melphalan.

1-phenylalanine mustard. melphalan.

L-phenylalanine mustard. See melphalan.

phenylalaninenitrogenmustard. melphalan.

phenylamine. aniline.

4-(phenylamino)benzenesulfonic acid. *N*-phenylsulfanilic acid.

1-phenyl-2-aminopropane. amphetamine.

phenylaniline. diphenylamine.

***N*-phenylaniline**. diphenylamine.

(*p*-phenylazo)aniline. aminoazobenzene.

1-[[4-(phenylazo)phenyl]azo]-2-naphthalenol. sudan III.

1-(*p*-phenylazophenylazo)-2-naphthol. sudan III.

3,9-phenylazo-2,6-pyridinediamine. phenazopyridine.

(phenylazo)thioformic acid 2-phenylhydrazide. dithizone.

phenylbarbital. phenobarbital.

phenylbenzene. biphenyl.

***N*-phenylbenzeneamine**. diphenylamine.

phenylbromide. bromobenzene.

phenylbutazone (4-*n*-butyl-1,2-diphenyl-3,-5-pyrazolidinedione; Butazolidin; diphebenzol; Naprosyn). a white or very light yellow powder, $C_{19}H_{20}N_2O_2$, that is slightly water soluble and freely soluble in acetone, ether, and ethyl acetate. The taste is somewhat bitter and it has a very slight aromatic odor. A pyrazolone derivative, phenylbutazone is synthesized and occurs also in certain herbs. It is a strong antiinflammatory, used also as a mild analgesic. Phenylbutazone is nephrotoxic, and can also cause acute myelogenous leukemia. This drug and other pyrazolone derivatives (e.g., aminopyrine, antipyrine, apazone) can cause serious blood disorders such as a fatal agranulocytosis in sensitive individuals. Phenylbutazone may potentiate the action of certain other drugs or may have toxic effects through drug interactions. *Cf.* aminopyrine, apazone, antipyrine, salicylate, pyrazolone derivative. See also analgesic, antiinflammatory drug, hypoglycemia.

phenylcarbamic acid ethyl ester. phenylurethane.

phenylcarbamic acid 1-methylethyl ester. propham.

phenylcarbamide. 1-phenyl-2-thiourea.

phenylcarbinol. benzyl alcohol.

2-phenylcinchoninic acid. cinchophen.

1-(1-phenylcyclohexyl)piperidine. phencyclidine.

***trans*-(±)-2-phenylcyclopropanamine**. tranylcypromine.

phenyldiazinecarbothioic acid 2-phenylhydrazide. dithizone.

2-phenyl-1,3-diketohydrindene. phenindione.

phenyldimethylpyrazolone(e). antipyrine.

***N*-phenyl-*N'*,*N'*-dimethylurea**. fenuron.

phenylethanolamine *N*-methyltransferase. See *N*-methylation.

β-phenylethylamine. phenethylamine.

2-phenylethylamine. phenethylamine.

phenylethylbarbituric acid. phenobarbital.

phenylethylene. styrene.

β-phenylethylhydrazine. phenelzine.

***N*-(2-phenylethyl)imidodicarbonimidic diamide**. phenformin.

phenylethylmalonylurea. phenobarbital.

***N*-(1-phenylethyl-4-piperidinyl)-*N*-phenylpropionamide**. fentanyl.

***N*-(1-phenylethyl-4-piperidyl)propionanilide**. fentanyl.

phenylformic acid. benzoic acid.

phenylglycuronic acid. an organic compound, ($C_6H_5{\cdot}O{\cdot}CO{\cdot}(CHOH)_4{\cdot}CHO$), that occurs in the urine following ingestion of phenol.

phenylhydrazine (PHZ). a colorless liquid, $C_6H_5NHNH_2$, that becomes red-brown when exposed to air. A hemolytic agent, toxic by all routes of exposure. It is a potent redox-active agent that denatures erythrocyte components, especially hemoglobin, causing anemia in many animals. PHZ converts oxyhemoglobin to methemoglobin with the release of benzene and nitrogen.

(2-phenyl)hydrazine. phenelzine.

1-phenyl-2-hydrazinopropane. pheniprazine.

1-phenyl-1-hydroxy-2-methylaminopropane. ephedrine.

2-phenyl-1,3-indandione. phenindione.

2-phenyl-1*H*-indene-1,3(2*H*)-dione. phenindione.

(phenylisopropyl)amine. amphetamine.

DL-β-phenylisopropylamine hydrochloride. amphetamine hydrochloride (See amphetamine).

***N*-phenyl isopropyl carbamate**. propham.

β-phenylisopropylhydrazine. pheniprazine.

***d*-phenylisopropylmethylamine**. methamphetamine.

phenyl isothiocyanate. phenyl mustard oil.

phenyl mercaptan. benzenethiol.

phenylmercuric acetate ((acetato)phenylmercury; acetoxyphenylmercury; phenylmercury acetate; PMA; PMAC; PMAS). a white to cream-colored, crystalline compound, $C_6H_5HgOCOCH_3$, that is slightly soluble in water and soluble in benzene, ethanol, and glacial acetic acid. It is used as a bacteriostatic preservative, antiseptic, fungicide, and herbicide (especially for crabgrass), a mildewcide for paints, and a slimicide in paper mills. Phenylmercuric acetate is a strong irritant and is toxic by ingestion, inhalation, and skin absorption.

phenyl mercury. See monophenyl mercury, diphenylmercury.

phenylmethane. toluene.

phenylmethanol. benzyl alcohol.

1-phenyl-2-methylaminopropane. methamphetamine.

1-phenyl-2-methylaminopropanol. ephedrine.

1-phenyl-3-methyl-5-pyrazolyl dimethylcarbamate. Pyrolan®.

phenyl mustard oil (thiocarbanil; phenyl isothiocyanate; phenylthiocarbonimide). a colorless or pale yellow, penetrating liquid with an irritating odor. It is an irritant and is toxic by ingestion and inhalation.

***N*-phenyl-β-naphthylamine**. a light gray, flammable powder, $C_{10}H_7NHC_6H_5$, used in the manufacture of rubber and lubricants. It is a carcinogen linked to bladder cancer.

***N*-phenyl-*N*-[1-(2-phenylethyl)-4-piperidinyl]-propanamide**. fentanyl.

phenylphosphine (phosphaniline). a colorless, reactive, moderately flammable liquid, $C_6H_5PH_2$, b.p. 160°C. The liquid, if ingested, and the fumes are highly toxic. See phosphine.

phenylphosphonothioic acid *O*-ethyl *O*-(4-nitrophenyl) ester. EPN.

phenylpropanolamine. a CNS stimulant (See stimulant). See also drug abuse.

3-phenyl-2-propenylanthranilate. cinnamyl anthranilate.

3-phenyl-2-propen-1-yl anthranylate. cinnamyl anthranilate.

phenyl-2-pyridylmethyl-β-*N,N*-dimethylaminoethyl ether. doxylamine.

2-phenylquinoline-4-carboxylic acid. cinchophen.

phenyl-4-quinolinecarboxylic acid. cinchophen.

***N*-phenylsulfanilic acid** (4-(phenylamino)benzenesulfonic acid; 4-diphenylaminesulfonic acid; *p*-anilinobenzenesulfonic acid). a solid, $C_{12}H_{11}NO_3S$, with white plates that turn blue in light; soluble in water and ethanol; decomposes above 200°C. It is used as a colorimetric indicator in the determination of nitrates.

2-phenyl-3-tetramethylhydro-1,4-oxazine. phenmetrazine.

phenylic acid. phenol (def. 1).

phenylisopropylamine. amphetamine (def. 1).

phenylmercuric nitrate, basic ((nitrato-*O*)-phenylmercury; merphenyl nitrate). a fine, white, crystalline substance or light gray powder, $C_6H_5HgNO_3 \cdot C_6H_5HgOH$, that is insoluble in ether, slightly soluble in water and ethanol, and moderately soluble in glycerol. It is used as a germicide, a fungicide, and is sometimes used as a preservative in drug or toxicant formulations. It is probably toxic to humans by all routes of exposure.

phenylmercury acetate. phenylmercuric acetate.

phenylthiocarbonimide. phenyl mustard oil.

phenylthiourea. See 1-phenyl-2-thiourea.

1-phenylthiourea. 1-phenyl-2-thiourea.

1-phenyl-2-thiourea (1-phenylthiourea; phenylcarbamide; phenylthiourea; PTU). a compound with bitter, needle-like crystals, $C_7H_8N_2S$, that is soluble in water, ethanol, and aqueous ether. It is a selective toxicant and rodenticide. While much more toxic to rodents than to humans, it is extremely hazardous to humans and is poisonous by ingestion and by intraperitoneal routes. Toxicity to other animals is largely unknown; there is no evidence of carcinogenicity. Phenylthiourea is thoroughly metabolized, and some of the sulfur is excreted as sulfate in urine.

phenylurethan. phenylurethane.

phenylurethane. (phenylurethan; phenylcarbamic acid ethyl ester; ethyl phenylcarbamate; ethyl phenylurethane; ethyl carbanilate). a white crystalline aryl carbamate, $C_6H_5NHCOOC_2H_5$, that is soluble in common organic solvents and hot water. It has an aromatic odor and clove-like taste. Phenylurethane is a herbicide that interferes with photosynthesis. It inhibits the growth of barley at concentrations that do not affect the spindle. It delays the germination of oats and wheat with ensuing abnormal growth.

phenylvinyl ketone. a carcinogenic lactone.

phenytoin (5,5-diphenyl-2,4-imidizolidinedione; Dilantin; etc.). formerly diphenylhydantoin, a non-sedative, fetotoxic, carcinogenic anticonvulsant powder that has proven useful in the treatment of all types of convulsive disorders except absence seizures. It is also a cardiac depressant used as an antiarrhythmic agent and in the treatment of all forms of epilepsy except petit mal. Unintended effects include slurred speech, visual disturbances, gastrointestinal distress and drug reactions. An excessive rate of *i.v.* administration may cause cardiac arrhythmias, with or without hypotension and may also cause CNS depression. Phenytoin can cause congenital abnormalities such as cleft palate. Chronic administration can cause gingival hyperplasia.

PHI. phosphine.

Phidippus (jumping spiders). tenacious, venomous, jumping spiders, the bite of which causes a light, sharp pain; the area of the wound becomes transiently tender and painful in humans and an erythematous wheal up to 5 cm in diameter develops; diffuse edema; some pruritus; and mild lymphadenitis. Symptoms usually abate within 2 days.

Philippine cobra. *Naja naja philippinensis*.

Philippine long-glanded snake. *Maticora intestinalis*.

philodendron. *Philodendron*.

Philodendron (philodendron). a genus of mostly shrubby, climbing, tropical plants (Family Araceae) of the moist American tropics (Family Araceae). Some are popular as house and garden plants. Many are shrubby, climbing plants; all parts of the plant are toxic. When plant material is bitten into or chewed, it produces symptoms similar to those of *Dieffenbachia* due to the presence of calcium oxalate crystals in the plant tissue. The needle-sharp crystals are released by ruptured plant cells and penetrate the tissues of the mouth, tongue, and throat, quickly producing a severe burning sensation and irritation of the mucous membranes with swelling of the affected tissues; nausea; vomiting; diarrhea; salivation. The swollen tongue may become immobile, interfering with speech, swallowing, and breathing; obstruction of the airway is a rare complication.

philosopher's wool. zinc oxide.

phlebarteriectasia. vasodilatation.

phlebitis. inflammation of a vein.

phlebotomy. the removal of blood from blood vessels. This was done in earlier times to "cure" many diseases. Its purpose today is to acquire blood samples as a diagnostic tool.

phlogistic agent. a phlogogenic agent such as formalin or mustard powder.

phlogogenic. causing inflammation.

phlogotic. inflammatory.

phlorhizin. a glycoside that blocks the tubular reabsorption of glucose by the kidney.

phlyctenule. a small ulcerated nodule or vesicle of the corneal or conjunctival epithelium that is characteristic of phlyctenulosis.

phlyctenulosis. a nodular hypersensitive condition of corneal and conjunctival epithelium caused by an endogenous toxin.

phobia. fear of a kind that cannot be overcome by reason.

Phocidae (earless seals, harbor seals or hair seals, sea elephants). a family of marine mammals (Suborder Pinnipedia) with hind limbs that are larger than the forelimbs and which cannot rotate forward. The external ear is absent. See *Erignathus barbatus*.

phocomelia. a developmental anomaly in which the proximal portion of limb or limbs is missing and the distal portions (hands and/or feet) are attached to the skeleton of the trunk by a single, small irregularly shaped bone. See thalidomide.

Phodopus. See hamster.

Pholiota. a mushroom genus of brown-spored agarics of North America and Europe (Family Strophariaceae). They have an annulus and grow on open ground or decaying wood. ***P. autumnalis***. former scientific name of *Galerina autumnalis*. ***P. squarrosa*** (scaley pholiota). a mushroom that grows in clusters at the base of trees, most commonly aspen and birch, and on logs and stumps. It causes gastric distress in some individuals.

Phoma. a genus of fungi, the various species of which synthesize a number of toxic metabolites including cytochalasins, and phomenone.

phomenone. a substance secreted by fungi of the genus *Phoma* that inhibits seed germination.

phomin. **1:** a type of mycotoxin produced by a wide variety of fungi. They are structurally and toxicologically related to the cytochalasins and chaetoglobosins. See also phomin, macrolide mycotoxin. **2:** a former name of cytochalasin B.

Phomopsis paspalli. a fungus that produces the toxic macrolide mycotoxins kodo-cytochalasin-1 and -2. It commonly infests kodo millet, *Paspalum scrobiculatum*

phonocardiograph. device to record heart sounds.

phoorsa. *Echis carinatus*.

Phoradendron serotinum (formerly *P. flavescens*; American mistletoe; false mistletoe). a yellowish-green shrubby epiphyte that grows on various deciduous trees in eastern North America, from Florida to eastern Texas, north to southern Ontario, southern Indiana, southern Illinois, southern Missouri, and southeastern Kansas. The leaves are broad, thick, greenish; the fruit is a globose, pulpy drupe. All parts, but especially the small white berries are toxic by ingestion, producing gastrointestinal symptoms and bradycardia. Leaves have been used for tea and a coffee substitute, apparently with no ill effects. The active principle is phenethylamine. Tyramine has sometimes been erroneously cited as the toxic principle; toxic proteins have also been implicated. See also *Arceuthobium pusillum* and *Viscum album*.

phorate (phosphorodithioic acid *O,o*-diethyl *S*-[(ethylthio)methyl] ester; *O,o*-diethyl *S*-(ethylthio)methyl phosphorodithioate; *O,o*-diethyl *S*-ethylmercaptomethyl dithiophosphate; thimet; Agrimet; Geomet; Granutox; Rampart; Thimenox). a clear, slightly water-soluble liquid used as a broad-spectrum, systemic organophosphate insecticide. The oral LD_{50} in male laboratory rats is 2-4 mg/kg body weight; the dermal LD_{50} in guinea pigs is 20-30 mg/kg body weight. It is a neurotoxicant that acts via acetylcholinesterase inhibition. It is highly dangerous and toxic by all routes of exposure, but is used in some household products to eradicate pests.

phoratoxin. a toxin that inhibits synthesis of proteins in the intestinal wall.

phorbol (4,9,12-β-13,20-pentahydroxy-1,6-tib liadiene-3-ol; 1,1a,1b,4,4a,7a,7b,8,9,9a-decahydro-4a,7b,9,9a-tetrahydroxy-3-(hydroxymethyl)-1,1,6,8-tetramethyl-5H-cyclopropa-[3,4]benz[1,2-e]azulen-5-one). a compound used in biochemical and toxicological research, it is the parent compound of the tumor-producing esters in croton oil, the most important of which is tetradecanoylphorbol 13-acetate, *q.v.*

phorbol myristate acetate. tetradecanoylphorbol 13-acetate.

phosalone (phosphorodithioic acid *S*-[(6-chloro-2-oxo-3(2H)-benzoxazolyl)methyl] *O,o*-diethyl ester; phosphorodithioic acid, *O,o*-diethyl ester, *S*-ester with 6-chloro-3-(mercaptomethyl)-2-benzoxazolinone; 3-(*O,o*-diethyldithiophosphorylmethyl)-6-chlorobenzoxazolinone; 6-chloro-3-(*O,o*-diethyldithiophosphorylmethyl)-benzoxazolone; *S*-(6-chloro-2-oxobenzoxazolin-3-yl)methyl diethyl phosphorothiolothionate). a toxic, organic compound, $C_{12}H_{15}ClNO_4PS_2$, with colorless crystals that is soluble in alcohols and aromatic solvents. It is used mainly as an insecticide and molluscicide. It is a component of some household products used to eradicate pests; such products should be considered dangerous and handled with care.

Phosdrin®. a mixture of organophosphates that contains more than 60% mevinphos and less than 40% chemically related insecticides. See also mevinphos.

phosethyl Al. fosetyl aluminum.

phosethyl aluminum. fosetyl aluminum.

phosgene (carbonic dichloride; carbonic acid dichloride; carbonic dichloride; carbonyl chloride; chloroformyl chloride; carbon oxychloride). a colorless to light yellow, noncombustible, extremely toxic, volatile liquid (below 8° C), or gas, $COCl_2$, with a sweet, oppressive, sometimes stifling odor. It is used, e.g., as an asphyxiant war gas; in the manufacture of dyes and metal products; and in a variety of organic syntheses including the production of insecticides and pharmaceuticals. It is a strong irritant of the eyes and mucous membranes. Inhalation may produce no symptoms for an hour or two; serious symptoms may not appear until 6-24 hours after exposure. In acute exposures by inhalation, symptoms of intoxication in humans may include weakness, vertigo, choking, vomiting, a sensation of coldness, a tight feeling in the chest, pain in the region of the diaphragm with painful breathing, coughing with bloody sputum, peribronchial edema, severe pulmonary congestion, bronchitis, and alveolar edema or pneumonia which may rapidly bring about convulsions, coma, and death from respiratory or cardiac failure. Pulmonary damage results from the hydrolysis of phosgene which yields HCl and $C0_2$. Chronic exposure may produce dizziness, chills, coughing, and a choking sensation. It can also cause bronchitis, pneumonia, as well as irreversible emphysema and pulmonary fibrosis.

phosgene oxime (dichloroformoxime). an asphyxiant war gas, $CCl_2:N\cdot OH$.

phosmet (phosphorodithioic acid S-[(1,3-dihydro-1,3-dioxo-2*H*-isoindol-2-yl)methyl] *O,o*-dimethyl ester; phosphorodithioic acid *O,o*-dimethyl ester *S*-ester with *N*-(mercaptomethyl)phthalimide; *O,o*-dimethyl *S*-phthalimidomethyl phosphorothionate; *N*-(mercaptomethyl)phthalimide *S*-(*O,o*-dimethyl phosphorodithioate); phthalophos). a colorless, crystalline, partially water-soluble, organophosphate, $C_{11}H_{12}NO_4PS_2$. It is a dimethyl ester of phosphorodithioic acid. It is an anticholinesterase used as an insecticide and acaricide.

phosphacol. paraoxon.

phosphamidon (2-chloro-2-diethylcarbamoyl-1-methylvinyldimethylphosphate). a highly toxic, colorless, liquid, organophosphate insecticide, $(CH_3O)_2P(O)OC(CH_3):C(Cl)C(O)N(C_2H_5)_2$, that is soluble in water and organic solvents. It is more toxic than malathion, but less toxic than parathion. Effects are due to cholinesterase inhibition. It is a component of some household products designed to eradicate pests. Phosphamidon is toxic to humans by all routes of exposure.

phosphaniline. phenylphosphine.

phosphatase. See phosphomonoesterase.

phosphate. any of the three series of salts of phosphoric (V) acid. **metaphosphates**: those with the anion HPO^{2-}; all are neutral. **orthophosphates**: all of the naturally occurring phosphates, the anion of which is HPO^-; all are acidic. **pyrophosphates**: those with the anion PO^{3-}; all are alkaline. Phosphates vary enormously in toxicity, some being virtually harmless and others deadly in trace amounts. Toxic phosphates kill primarily by the reversible inhibition of acetyl cholinesterase, although they affect many other enzyme systems also. In general, an exposed animal dies or recovers completely from the action of phosphates.

phosphate conjugation. phosphate conjugation of xenobiotics is rare among animals and is known chiefly from insects. A conjugating enzyme from cockroach gut utilizes, for example, ATP in the phosphorylation of α-naphthol and *p*-nitrophenol. The process requires Mg^2+.

phosphate ester. See organophosphate ester.

phosphate ester insecticide. an organophosphate insecticide such as paraoxon and mevinphos, based on the activity of phosphate ester(s). Such compounds do not contain sulfur and are highly toxic.

phosphine. **1:** hydrogen phosphide; phosphoretted hydrogen; PHI. A colorless, very toxic, hydride of phosphorus, H_3P, with a characteristic foul garlicky (or dead-fish) odor that is soluble in ethanol, ether, and cuprous chloride solution. It is produced in small amounts during putrefaction of phosphorus containing organic matter and is synthesized from white phosphorus and aqueous alkali hydroxide. PHI is used as a war gas; in organic reactions (e.g., in the synthesis of organophosphate compounds); and as an active agent in some rodenticides. It is readily flammable and ignites spontaneously in air at 100° (because of the presence of diphosphine as a contaminant) with the release of phosphorus oxide. It is a dangerous fire and explosion hazard. Phosphine is a very strong pulmonary tract irritant and central nervous system depressant. Inhalation affects the eyes and causes dilation of the pupils. A high concentration in the air is rapidly fatal, producing fainting, lowered blood pressure, pulmonary damage, convulsions, paralysis, and coma. Somewhat less serious exposures in humans may cause fatigue, vertigo, vomiting, a sensation of coolness, bronchitis, edema, dyspnea, and associated pain in the region of the diaphragm. It is a potential hazard in industrial processes and in the laboratory because of its inadvertent production in chemical syntheses involving other phosphorus compounds. Phosphine is produced in small amounts during putrefaction of phosphorus-containing organic matter. *Cf.* dimethylphosphine, methylphosphine; See also zinc phosphide. **2:** any of a class of organophosphorus compounds derived from phosphine by substitution of organic groups for H atoms. They are analogous to amines, but are weaker bases. See, for example, methylphosphine, dimethylphosphine, trimethylphosphine, triethylphosphine, phenylphosphine. **3:** an orange-yellow basic dye that is essentially a nitrate of chrysaniline. Its assigned name in the Colour Index is orange 15, but a common commercial name is phosphine. It is extremely toxic to infusoria. See also chrysaniline yellow.

1,1′,1″-phosphinothio ylidynetrisaziridine. triethylenethiophosphoramide.

1,1′,1″-phosphin ylidynetrisaziridine. triethylenephosphoramide.

3′-phosphoadenosine-5′-phosphosulfate (PAPS). a compound formed from ATP in a sequence of reactions catalyzed by ATP sulfurylase and adenosine-5-phosphosulfate kinase. It yields sulfate conjugates of xenobiotics through the agency of sulfotransferases. See also sulfate conjugation.

phosphodiesterase (PDE; cyclic nucleotide phosphodiesterase). any of a number of orthophosphoric diester phosphohydrolases that occur in the venoms of all families of snakes. They are exonucleotidases that attack DNA and RNA by hydrolyzing cyclic nucleotides with the release of 5-mononucleotide from the polynucleotide chain. They also lyse derivatives of arabinose. Methylxanthines and phenothiazines inhibit phosphodiesterase activity *in vitro* but their action is probably insignificant *in vivo*.

phospholan. phosfolan.

phospholine. a topically administered parasympathomimetic agent.

phospholipase. any of four classes of lytic enzymes that catalyze the hydrolysis of phospholipid, two of which (phospholipases A and D) are of particular interest in toxicology. **phospholipase A** (formerly lecithinase A). **1:** a class of enzymes that occurs in the liver, pancreas, kidney, muscle, heart, and adrenal glands and in the venoms of most families of venomous snakes. They catalyze the hydrolysis of one of the fatty ester linkages in diacyl phosphatides, yielding lysophosphatides and fatty acids. **2:** sometimes used specifically for a hemolytic constituent of cobra venom. Synonyms are cobra phospholipase; formerly cobra lecithinase, lecithinase A. **phospholipase B** (formerly lecithinase B). its distribution is the same as that of phospholipase A (def. 1). It catalyzes the hydrolysis of lysolecithin and lysocephalin to yield a fatty acid. **phospholipase C** (formerly lecithinase C). a hydrolytic enzyme of the pancreas, liver, intestinal mucosa, and brain. It catalyzes the hydrolysis of glycerylphosphophorylcholine into a diglyceride and choline phosphate. **phospholipase D**. (lecithinase). a type of enzyme in the toxin of *Clostridium welchii*. It catalyzes the hydrolysis of phosphatidylcholine into choline and phosphatidate.

phospholipid. any lipid that contains phosphorus. Sometimes called structural lipids, they are especially important in the formation of cell membranes.

phosphomonoesterase (phosphatase). any of a class of enzymes that occur widely among snake venoms except for those of the Colubridae. The activity is that of an orthophosphoric monoester phosphohydrolase. Many snake venoms contain both acid (optimal pH, 5.0) and alkaline (optimal pH, 8.5) phosphatases; others contain only one type. See also enzymes of snake venoms.

phosphonate. any phosphonic acid ester, *q.v.*

phosphonecrosis. necrosis of the bony tissue of the jaw due to long-continued inhalation of phosphorus fumes; an occupational disease. See also phossy jaw.

phosphonic acid (phosphorous acid; orthophosphorous acid). a colorless, deliquescent solid, H_3PO_3. It is a reducing agent and a dibasic acid, yielding the anions $H_2PO_3^-$ and HPO_3^{--} in aqueous solution. On warming, it decomposes to yield phosphine and phosphoric acid. See phosphonic acid ester.

phosphonic acid monoethyl ester aluminum salt. fosetyl aluminum.

phosphonodithioimidocarbonic acid cyclic ethylene *P,P*-diethyl ester. phosfolan.

***N*-(phosphonomethyl)glycine**. glyphosphate.

phosphonuclease. See nucleotidase.

phosphooxythirane ring. a hypothetical intermediate in the oxidative desulfuration of certain organophosphorus insecticides that contain a thionosulfur (P=S) group. The process is catalyzed by the cytochrome P-450-dependent monooxygenase system.

phosphoptomaine. any toxic compound found in the blood as a result of phosphorus poisoning.

phosphoramide mustard cyclohexylamine salt (*N,N*-bis(2-chloroethyl)phosphorodiamidic acid, cyclohexyl ammonium salt). a moderately toxic, bifunctional, alkylating agent. It is teratogenic and is a probable human mutagen. Very toxic fumes, PO_x and NO_x, are released upon decomposition by heating.

phosphordiamidic acid cyclic ester monohydrate. cyclophosphamide.

phosphoretted hydrogen. phosphine.

phosphoric acid. a colorless, highly toxic, corrosive liquid. It is used, for example, in the manufacture of fertilizers, detergents, animal feed, drugs, and soft drinks; it is used also in sugar refining, metal cleaning, engraving, and lithography. It can severely damage eyes and skin on contact and is moderately toxic if ingested. Under certain conditions it can cause respiratory difficulties, especially as an aerosol of mist.

phosphoric acid, anhydrous. phosphorus pentoxide.

phosphoric acid 7-chlorobicyclo-[3.2.0]hepta-2,6-dien-6-yl dimethyl ester. heptenophos.

phosphoric acid 2-chloro-1-(2,4-dichlor ophenyl diethyl ester. chlorfenvinphos.

phosphoric acid 2-chloro-1-(2,4,5-trichlor ophenyl)ethenyl dimethyl ester. stirofos.

phosphoric acid 2,2-dichloroethenyl dimethyl ester. dichlorvos.

phosphoric acid diethyl 4-nitrophenyl ester. paraoxon.

phosphoric acid 3-(dimethylamino)-methyl-3-oxo-1-propenyl dimethyl ester. dicrotophos.

phosphoric acid dimethyl ester, ester with *cis*-3-hydroxy-*N,N*-dimethylcrotonamide. dicrotophos.

phosphoric acid triethyleneimide. triethylenephosphoramide.

phosphoric acid tris(methylphenyl) ester. tritolyl phosphate.

phosphoric bromide. phosphorus pentabromide.

phosphoric chloride. phosphorus pentachloride.

phosphoric oxide. phosphorus pentoxide.

phosphoric perbromide. phosphorus pentabromide.

phosphorism. chronic phosphorus poisoning. See phosphorus.

phosphorodithioic acid *S*-[2-chloro-1-(1,3-dihydro-1,3-dioxo-2*H*-isoindol-2-yl)ethyl] *O,O*-diethyl ester. dialifor.

phosphorodithioic acid *S*-[(6-chloro-2-oxo-3(2*H*)-benzoxazolyl)methyl] *O,O*-diethyl ester. phosalone.

phosphorodithioic acid S-[[(4-chlorophenyl)-thio]methyl] *O,O*-diethyl ester. carbophenothion.

phosphorodithioic acid, *O,O*-diethyl ester, *S*-ester with 6-chloro-3-(mercaptomethyl)-2-benzoxazolinone. phosalone.

phosphorodithioic acid *O,O*-diethyl ester *S*-ester with *N*-(2-chloro-1-mercaptoethyl)-phthalimide. dialifor.

phosphorodithioic acid *O,O*-diethyl *S*-[2-(ethylthio)ethyl] ester. disulfoton.

phosphorodithioic acid *O,O*-diethyl *S*-[(ethylthio)methyl] ester. phorate.

phosphorodithioic acid *S*-[(1,3-dihydro-1,-3-dioxo-2*H*-isoindol-2-yl)methyl] *O,O*-dimethyl ester. phosmet.

phosphorodithioic acid *O,O*-dimethyl ester *S*-ester with *N*-(mercaptomethyl)phthalimide. phosmet.

phosphorodithioic acid *O,O*-dimethyl ester, ester with 2-mercapto-*N*-methylacetamide. dimethoate.

phosphorodithioic acid *O,O*-dimethyl ester *S*-ester with *N*-formyl-2-mercapto-*N*-methylacetamide. formothion.

phosphorodithioic acid *O,O*-dimethyl *S*-[2-(methylamino)-2-oxoethyl] ester. dimethoate.

phosphorodithioic acid *O,O*-dimethyl ester, ester with 2-mercapto-*N*-methylacetamide. dimethoate.

phosphorodithioic acid *S,S'*-1,4-dioxane-2,3-diyl *O,O,O',O'*-tetraethyl ester. dioxathion.

phosphorodithioic acid *S*-[2-(formylmethylamino)-2-oxyethyl] *O,O*-dimethyl ester. formothion.

phosphorofluoridic acid bis(1-methylethyl) ester. diisopropyl fluorophosphate.

phosphorofluoridic acid diisopropyl ester. diisopropyl fluorophosphate.

phosphorothioic acid. Ronnel.

phosphorothioic acid *O,O*-diethyl *O*-[2-(ethylthio)ethyl] ester mixture with *O,O*-diethyl *S*-[2-(ethylthio)ethyl]phosphorothioate. demeton.

phosphorothioic acid *O,O*-diethyl *O*-(4-nitrophenyl) ester. parathion.

phosphorothioic acid *O,O*-diethyl *O*-(3,5,6-trichloro-2-pyridinyl) ester. chlorpyrifos.

phosphorothioic acid *O,O*-dimethyl *O*-[3-methyl-4-(methylthio)phenyl] ester. fenthion.

phosphorothioic acid *O,O*-dimethyl *O*-(4-nitrophenyl) ester. methyl parathion.

phosphorothioic acid *O*-[2-(ethylthio)ethyl] *O,O*-dimethylester mixture with *S*-[2-(ethylthio)ethyl]. demeton-*S*-methyl.

phosphorothionate ester. a class of esters in which the oxygen of the phosphate group has been replaced by sulfur. These compounds which include the insecticides chlorothion, diazenon, parathion, and tributylphosphorothionate, thus contain an $O_3P{:}S$ group. These compounds are often more toxic to insects relative to mammals than their oxygen-containing analogs.

phosphorothionate insecticide. See phosphorothionate ester.

phosphorous. relating to or containing phosphorus.

phosphorous acid. phosphonic acid.

phosphorus (symbol, P)[Z = 15; A_r = 30.975]. a reactive, solid, highly combustible, nonmetallic allotropic element that is widely distributed in nature, always in a combined state. There are no stable isotopes, but several radioactive isotopes. It occurs as a phosphate in all living cells. Elemental phosphorus and its fumes have a characteristic fishy odor and are extremely poisonous. It occurs in three main forms: black, red, and white. Elemental phosphorus causes severe burns or intense inflammation and fatty degeneration of the liver and other viscera; repeated inhalation of the fumes often causes a necrosis of the lower jaw (phosphonecrosis). The largest single use of phosphorus compounds, many of which are caustic or toxic, is in fertilizers. Cupric sulfate is used therapeutically as an antidote. See also phosphoptomaine, radiophosphorus. **black phosphorus**. a black, lustrous crystalline substance that resembles graphite or coal. It is stable in air, does not ignite spontaneously, is electrically conductive, and is insoluble in most solvents. **ordinary phosphorus**. white phosphorus. **red phosphorus** (amorphous phosphorus). a dark violet-red amorphous powder that is infusible, insoluble in most solvents including carbon disulfide, and much less reactive than white phosphorus. Pure red phosphorus is much less toxic than white phosphorus. It may explode from friction or on contact with an oxidizing agent; large masses may ignite spontaneously. It reacts with oxygen and water vapor to emit phosphine fumes. Red phosphorus is used in pyrotechnics, in the manufacture of fertilizers, pesticides, and various organic syntheses, and in the manufacture of incendiary shells, cartridges, and smoke bombs. **white phosphorus** (ordinary phosphorus; yellow phosphorus). an extremely poisonous white or transparent crystalline, wax-like solid that darkens and yellows on exposure to light, and ignites spontaneously in moist air at about 30°C. It must thus be stored under water away from heat or dissolved in carbon disulfide. White phosphorus is toxic by ingestion or inhalation and contact with the skin can cause severe burns. Ingestion of as little as ca. 50-100 mg can cause severe gastrointestinal irritation with bloody diarrhea, skin eruptions, liver damage, oliguria, cardiovascular collapse, convulsions, coma, and death. The effects of chronic systemic intoxication may include phosphonecrosis, weight loss, spontaneous fractures, and anemia. It is used in the manufacture of rodenticides, smoke screens, and analytical chemistry. **yellow phosphorus**. white phosphorus.

phosphorus-32. See radiophosphorus.

phosphorus-33. See radiophosphorus.

phosphorus anhydride. phosphorus pentoxide.

phosphorus chloride. phosphorus trichloride.

phosphorus halide. See halide.

phosphorus oxychloride (phosphoryl chloride). a colorless or faintly yellow, fuming liquid, $POCl_3$, with a pungent odor. It decomposes exothermically in water or ethanol. It is toxic by inhalation and ingestion and is strongly irritant to the skin and mucous membranes. It is used as a catalyst, a chlorinating agent, and in the manufacture of various anhydrides.

phosphorus oxyhalide. See halide.

phosphorus pentabromide (phosphoric bromide; phosphoric perbromide). a yellow solid, PBr_5, soluble in water (decomposes), carbon disulfide, carbon tetrachloride, and benzene. It decomposes at approximately 100°C. It is corrosive to skin, eyes and mucous membranes.

phosphorus pentachloride (phosphoric chloride; phosphorus perchloride). a toxic, flammable, yellow, crystalline substance, PCl_5, that sublimates on heating, but melts at 281°C. It is soluble in carbon disulfide and water, but decomposes in water with which it reacts strongly. PCl_5 is used as a catalyst in organic synthesis, as a chlorinating agent, and as a raw material in the synthesis of phosphorus oxychloride. It is corrosive to the skin, eyes, and mucous membranes.

phosphorus pentafluoride. a colorless, nonflammable gas, PF_5, that decomposes when wet. It is corrosive to the eyes and skin.

phosphorus pentoxide (phosphorus anhydride; phosphoric oxide; phosphoric acid, anhydrous). a soft white, flammable, hygroscopic powder, P_2O_5, that forms meta-, pyro-, or orthophosphoric acid in moist air. P_2O_5 is the oxide most commonly formed by the combustion of elemental white phosphorus and many phosphorus compounds. It is a fire hazard and reacts violently with water. Phosphorus pentoxide is a desiccant and corrosive irritant to the skin, eyes and mucous membranes.

phosphorus perbromide. phosphorus pentabromide.

phosphorus perchloride. phosphorus pentachloride.

phosphorus trichloride (phosphorous chloride). a clear, colorless fuming liquid, PCl_3, that is soluble in a number of organic solvents but decomposes in water. It is used as a solvent for phosphorus, as a chlorinating agent, and in the manufacture of saccharin. PCl_3 is corrosive to skin and other tissues and reacts with water (and thus body fluids) to yield hydrochloric acid.

phosphorus trifluoride. a poisonous, colorless gas, PF_3.

phosphorus trioxide (diphosphorus trioxide). a very poisonous, transparent, crystalline substance or colorless liquid (P_2O_3); exists also as P_4O_6), that is soluble in benzene and carbon disulfide. It is irritant and corrosive to the eyes, skin, and mucous membranes. It reacts with cold water to yield red phosphorus, phosphine, and phosphoric acid. The liquid is highly combustible.

phosphoryl chloride. phosphorus oxychloride.

***O*-phosphoryl-4-hydroxy-*N,N*-dimethyltryptamine**. psilocybin.

phosphorylated. pertaining to a substance that has undergone phosphorylation.

phosphorylation. any reaction in which phosphorus combines with an organic compound, usually as a trivalent phosphoryl group. In cellular metabolism this may occur by addition of an inorganic phosphate group (e.g., from ATP) or of an organophosphorus moiety (e.g., from an organophosphorus insecticide or a nerve gas) to a macromolecule. Phosphorylation (and dephosphorylation) plays a critical role in the energy metabolism of an organism. It is also plays an important role in enzyme formation and vitamin activity.

phosphothion. malathion.

phosphureted hydrogen. phosphine.

phosporothioic acid *O,O*-diethyl*O*-(4-nitrophenyl) ester. parathion.

phossy jaw. phosphonecrosis with fracture of the lower mandible in *Homo sapiens*. It is associated with brittleness of other bones with deterioration and loss of teeth due to chronic exposure to white phosphorus.

Phosvel. See leptophos.

photo-. a prefix indicating a response, sensitivity, or relationship to light.

photoactive. chemically reactive to light in the visible or ultraviolet portions of the spectrum.

photoallergic. of, pertaining to, or characterized by, an intensified, delayed contact-type sensitivity to light, especially sunlight.

photoallergic dermatitis. photoallergic photodermatitis.

photoallergy. **1a:** an allergic sensitivity to light. **1b:** a type of delayed contact dermatitis produced by the interaction of light and certain chemicals. See also dermatitis, photosensitivity, photocontact allergen. Light produces the sensitivity reaction.

photoallergy test. See test (phototoxicity test).

photoautotrophic organism. See autotrophic organism.

photochemical. **1a:** pertaining to the chemical action or activity of light. **1b:** chemically reactive in the presence of light or other radiation. **1c:** influenced or initiated by light (said of a chemical or chemical reaction). See also pollutant (photochemical air pollutant), photochemical oxidant.

photochemical oxidant. any air pollutant produced by the action of sunlight on oxides of nitrogen and hydrocarbons. See photochemical air pollutant.

photochemical reaction. a chemical reaction induced by electrical excitation initiated by the action of light. Such reactions can be initiated only if the light is absorbed by the reactive system. Such reactions are usually effected by the visible, or more often, by the ultraviolet portion of the electromagnetic spectrum. Such reactions are not only responsible for the production of photochemical smog, but photolytic reactions are often responsible for chemical decomposition in the environment.

photodermatitis. a rapidly developing dermatitis due to exposure to ultraviolet light. It results from the action of u.v. light (usually that in sunlight) following exposure of a subject to a phototoxic substance or photoallergen. *Cf.* photodermatosis, photosensitivity. **photoallergic photodermatitis** (photoallergic dermatitis). a rapidly developing allergic dermatitis of sun-exposed areas of the skin known mainly from livestock and humans. **phototoxic photodermatitis** (phototoxic dermatitis). a rapidly developing dermatitis with hyperpigmentation of sun-exposed areas of the skin, following exposure to agents containing photosensitizing substances (e.g., coal tar, some perfumes, drugs, or plants that contain psoralens), followed by exposure to sunlight. See photodermatitis.

photodermatosis. a morbid condition of the skin resulting from exposure to light.

photoinhibition. inhibition of a chemical reaction, reaction sequence, or of a biological process (e.g., germination) by light.

photolethal. of or pertaining to the lethal capacity of visible or ultraviolet light.

photolysis. light-induced chemical decomposition.

photonosus (photopathy). a disease or pathological effect due to exposure to light. Such effects may result from phototoxis or photoallergy.

photopathy. photonosus.

photopharmacology. the science that treats the effects of light and other radiations on drugs and on their pharmacologic action.

photosensitive. **1:** highly sensitive to or reactive to light. See photosensitivity, photosensitization. **2:** highly responsive to photoperiod or changes in photoperiod.

photosensitivity. **1:** sensitivity of an organism or a chemical (e.g., a pigment) to light. **2:** the property or condition of a biological clock such that it is reset daily by the photoperiod. **3a:** any abnormal response to sunlight (usually a skin reaction) due to the presence of a sensitizing agent in the body. **3b:** hypersensitivity of an animal to light due to the presence of any of a number of photodynamic pigments in the peripheral circulation. In such cases, exposure to sunlight produces erythema and pruritis of lightly pigmented, unprotected skin followed by edematous suffusions and usually necrotic skin lesions. Thus cattle are most affected on white or unpigmented areas of the skin and white sheep are most affected about the head. See also photoallergy, phototoxicity. **hepatogenic (hepatogenous) photosensitivity**. a condition in which the photodynamic pigment is a normal breakdown product of digestion that is eliminated by the healthy liver. Thus, the condition is the immediate result of liver dysfunction which is usually due to hepatotoxic principles present in the photosensitizing plant. The condition may be accompanied by pronounced general jaundice. See also *Agave lecheguilla*, *Brassica napus*, *Lantana spp.*, *Nolina texana*, *Panicum spp.*, *Tetradymia spp.*, and *Tribulus terrestris*, phylloerythrin. **ictrogenic photosensitivity**. a term sometimes applied to hepatogenic photosensitivity. The generalized jaundice and intensely pigmented livers result from an inability of the affected animal to excrete pigment molecules. **primary photosensitivity.** the case in which the photodynamic pigment, unaltered by digestion, enters the general metabolism. Only two North American plants, *Fagopyrum sagittatum* and *Hypericum perforatum*, are known to cause primary photosensitivity.

photosensitization. a condition in which the skin reacts abnormally to light, especially to ultraviolet rays or sunlight, due to the presence of a phototoxic pigment in the skin. Light energy is absorbed by the pigment inaugurating a signal to receptor molecules that rapidly initiates chemical reactions that are injurious to the skin. Included are the production of reactive oxygen intermediates or changes in cell membrane permeability.

photosensitizer (phototoxic agent). any of a class of systemic poisons (mostly pigments of plant origin) that cause photosensitization. Known photosensitizers include coal tar derivatives, oil of bergamot, furocoumarins (psoralens), calcium cyclamate, griseofulvin, nalidixic acid, and tetracycline.

photosynthesis. a process in which the energy of sunlight is converted to chemical energy. It occurs in green (chlorophyll-bearing) plants and phototrophic bacteria that contain a similar pigment, bacteriochlorophyll. In a sequence of reactions, carbohydrates are manufactured from oxygen, carbon dioxide, and water in the presence of sunlight. The net reaction can be summarized as:

$$6CO_2 + 6H_2O + \text{light energy} \rightarrow C_6H_{12}O_6 + 6O_2$$

net photosynthesis. total or gross photosynthesis less respiration. It is usually determined by measuring the net uptake of carbon dioxide by the leaf.

phototoxic. **1a:** of, pertaining to, characterized by, or causing phototoxis. **1b:** of or pertaining to the capacity of a chemical to (nonimmunologically) sensitize the skin to a harmful, rapidly developing, light-induced reaction (e.g., sunburn).

phototoxic agent. photosensitizer. See also dermatitis (phototoxic contact dermatitis).

phototoxic agent. photosentsitizer.

phototoxic contact dermatitis. See dermatitis (phototoxic contact dermatitis).

phototoxic dermatitis. See dermatitis (phototoxic dermatitis).

phototoxic photodermatitis. See photodermatitis.

phototoxicity. the quality of being phototoxic.

phototoxin. a phototoxic substance of biological origin; usually a sensitizing plant pigment. See phototoxis, phylloerythrin, photosensitization.

phototoxis. a disorder produced by overexposure to light in the visible or ultraviolet range in concert with a sensitizing phototoxic substance. See also photosensitization, phototoxic.

phreatic. **1a:** pertaining to a well. **1b:** pertaining to groundwater that is reachable or probably reachable by drilling.

phreaticole. an organism that inhabits phreatic groundwater.

phreaticolous. inhabiting groundwater (phreatic) habitats.

phreatobiology. the study of groundwater organisms.

phreatophyte. a deep-rooted plant that absorbs water from the permanent water table or from the soil layer immediately above it. See also phreatic.

phrynin. a poison with digitalin-like action secreted by the skin of various toads. See also *Bombinator igneous*.

phrynolysin. a lytic toxin secreted by *Bombinator igneous*, the fire-toad.

phthalate. phthalic acid ester.

phthalate ester. phthalic acid ester.

phthalic acid. **1:** *o*-phthalic acid. It is a colorless, crystalline substance that is soluble in alcohol, water, and ether. It yields the anhydride when heated above its melting point of 213°C. **2:** any of the three isomeric benzene carboxylic acids derived from phthalic anhydrides. The ortho acid is also named 1,2-benzenedicarboxylic acid and benzene orthocarboxylic acid; the para isomer is termed terephthalic acid; and the meta form is isophthalic acid.

***m*-phthalic acid**. isophthalic acid.

***o*-phthalic acid**. See phthalic acid.

***p*-phthalic acid**. terephthalic acid.

phthalic acid dibutyl ester. *n*-butyl phthalate.

phthalic acid dimethyl ester. dimethyl phthalate.

phthalic acid ester (phthalate ester; phthalate). any of a family of stable, irritant, environmentally persistent compounds produced by the direct action of alcohol on phthalic anhydride (e.g., di-*n*-butyl phthalate (DBP) and di(2-ethylhexyl) phthalate (DEHP)). They are chiefly used as primary plasticizers of polyvinyl chloride products. The fumes of volatile (low molecular weight) phthalates are irritating to the eyes, mucous membranes, and usually the skin. Chronic exposure to fumes may cause asthma, bronchitis, and emphysema and can cause permanent damage to the eyes. The acute toxicities of most phthalic acid esters are slight to moderate. Continued exposure to low doses may, however, cause pulmonary or hepatic damage; they also induce peroxisome proliferation and may thus promote tumors. Reproduction may be impaired in some aquatic organisms. Phthalate esters are widely distributed environmental pollutants; they are dispersed in all environmental media. Phthalates such as *n*-butyl- and bis(2-ethylhexyl) phthalate are especially common water pollutants.

phthalic anhydride. a white, alcohol-soluble, crystalline acid anhydride, $C_6H_4(CO)_2O$, prepared by the oxidation in air of naphthalene using a catalyst such as vanadium oxide. It is used in the synthesis of phthalic acid esters, various polyester resins, pharmaceutical intermediates, and the formulation of insect repellents and insecticides.

α-phthalimidoglutarimide. thalidomide.

3-phthalimidoglutarimide. thalidomide.

phthalophos. phosmet.

***N*-phthaloylglutamimide**. thalidomide.

***N*-phthalylglutamic acid imide**. thalidomide.

phthisis. **1**. wasting; any wasting or atrophic disease. **2:** pulmonary tuberculosis.

phyco-. prefix indicating a relationship with algae.

Phyllanthus abnormis (spurge). a small annual or perennial herb (Family Euphorbaceae) of western and southwestern Texas. It is nephrotoxic and hepatotoxic to most types of livestock. Toxicity varies geographically.

Phyllobates. a genus of very small tree frogs (Family Dendrobatidae). One adult Cuban species has a head and body length of only 9 mm. A native arrow poison, batracin, is prepared from the skin of the South American species, *P. chocoensis*. The following toxins have been isolated from frogs of this genus: batrachotoxin, batrachotoxin A, homobatrachotoxin, dihydrobatrachotoxin, and 3-*O*-methylbatrachotoxin, *q.v.*

phylloerythrin. a normal product of the digestion of chlorophyll by vertebrate animals, it is a photosensitizing porphyrin pigment. It is an anaerobic breakdown product of chlorophyll in the forestomachs of ruminants. Normally it is rapidly excreted via the bile. In certain cases of liver dysfunction or bile duct occlusion, it may reach the peripheral circulation where it absorbs sunlight and initiates a phototoxic reaction. It is nearly always the photosensitizing pigment in hepatogenous photosensitization. This has been verified in many cases as in copper poisoning or poisoning by *Lippia rehmanni*, *Lantana camara*, *Myoporum laetum* (ngaio), *Narthecium ossifragum* (bog asphodel), *Panicum* spp., *Tribulus terrestris* (puncture vine). *Myoporum laetum* (ngaio).

Phylloxera (plant louse). a genus of plant lice (Family Phylloxeridae) that are destructive to many commercially important plants. See also mercuric nitrate.

phylum (plural, phyla). **1:** a group of related classes of animals, representing the highest (and broadest) category within the animal kingdom; roughly equivalent to the "division" in botany. **2:** used also by some plant taxonomists in preference to "division."

phyone. somatotrophin.

Physalia pelagica (Portuguese man-of-war; Portuguese man-o'-war; blue bottle). a large, pelagic, venomous, colonial hydroid that is often mistaken for a jellyfish. It occurs in the tropical and subtropical waters of the Atlantic Ocean as far north as the Bay of Fundy, the Hebrides; also the Mediterranean Sea; and the Indo-Pacific region. It floats by means of a large, air-filled sac that projects above the ocean's surface. The extensive colony of polymorphic hydroids is suspended below the float, forming the so-called tentacles. Whereas the float may reach a length of 35 cm or more, the "tentacles" may extend as much as 50 feet below the surface of the water. The crude venom is extremely toxic and rich in glutamic acid.

Contact with human skin causes immedi-ate, sometimes severe localized pain which may spread to the larger muscle masses; severe chest and abdominal rigidity; numbness and pain of the extremities, and dysphagia. Urticarial wheals appear, quickly increase in size, and may undergo vesiculation, pustulation and desquamation. Systemic manifestations such as weakness, nausea, anxiety, headache, muscle spasms, dyspnea, cyanosis, shock, renal failure, and death may ensue. The venom contains physalitoxin, a protein.

Physalis (ground cherry, husk tomato). a genus of annual or perennial herbs (Family Solanaceae). The seeds and unripe fruit are toxic. The toxic principle is an extremely toxic protein, physalitoxin.

physalitoxin. an extremely toxic hemolytic protein, lethal to humans, isolated from the Portuguese man-of-war, *Physalia* spp. It is inactivated by concanavalin A.

physic nut. See Jatropha.

physiogenic. **1:** originating from physical or environmental causes. **2:** caused by the physical activity of an organ or part. **3:** pertaining to embryology (not in common usage).

physiograph. an instrument used to record a myogram.

physiologic, physiological. **1:** pertaining to physiology or the phenomena of physiology. **2:** of or pertaining to normal (not pathologic) vital processes. **3:** of or pertaining to a process or biologic entity that is identified by its functional activity or effects rather than its anatomic structure (e.g., a physiologic sphincter). **4:** of or pertaining to a dosage (or the effects) of a xenobiotic, at a concentration, amount, or potency that falls within the normal range of the naturally occurring endogenous substance (e.g., a hormone or neurotransmitter) that it mimics or resembles.

physiological, physiologic. pertaining to physiology or the phenomena of physiology.

physiological chemistry. a term formerly applied to biochemistry.

physiological saline solution. physiological saline. See saline.

physiological salt solution. physiological saline. See saline.

physiological treatment. a term sometimes used in psychiatry in reference to such measures as the treatment of patients with drugs as opposed to psychological treatment, *q.v.* This term is poorly chosen, as treatment with artificial drugs is an unnatural form of treatment and does not truly represent a physiologic approach. It is best termed pharmacologic. Only in the case of treatment with hormones in order to improve hormonal balance is this approach justifiably termed "physiologic," to distinguish it from such totally unnatural means of treatment or therapy with artificial drugs. The consequences of "physiological treatment" is a proper concern for the science of toxicology.

physiology. the biological science that treats the chemical and physical processes in the functioning of cells, tissues, organs, organ systems, and whole organisms. While this discipline does not treat structure per se, it is strongly rooted in histology and anatomy. **cell physiology**. the study of the physiology of cells. **comparative physiology**. the study and comparison of the physiology of different taxa. **general physiology** the study of the general principles and scientific bases of physiology. **pathologic physiology**. the study of pathological states and processes. **special physiology**. the physiology of the special senses.

physiopathologic. **1:** pertaining to physiology and pathology. **2:** pertaining to a pathological change in a normal function.

Physobrachia douglasi. a sea anemone (Class Anthozoa) that is poisonous when raw, but is cooked and apparently safely eaten by Samoans. Early symptoms of intoxication are those of acute gastritis. The victim may soon become stuporous, lacking superficial reflexes, and may eventually pass into sustained shock and die with pulmonary edema. There is no known antidote. See Anthozoa.

physometra. air or gas in the womb.

physostigma (calabar bean; ordeal bean; chop nut; split nut). the highly poisonous, dried ripe seed of *Physostigma venenosum*. It contains the alkaloids physostigmine, eseridine, eseramine, physovenine, and a mixture of phytosterols.

Physostigma venenosum (calabar bean; ordeal bean; chop nut; split nut). a forest vine (Family Fabaceae, formerly Leguminosae) that bears large, extremely poisonous, kidney-shaped beans. It is indigenous to West Africa near the mouths of the Niger and Old Calabar rivers and has been introduced into India and Brazil. The principal toxicant is physostigmine (eserine). It also contains eseramine, eseridine (geneserine) and physovenine. Symptoms of intoxication include vomiting, colic, salivation, diarrhea, sweating, dyspnea. vertigo, slow pulse, miosis, and extreme prostration. See also physostigma.

physostigmanism (physostigmine poisoning). poisoning due to ingestion of physostigma, q.v. See also physostigmine.

physostigmine ($3\alpha S$-cis)-1,2,3,3α,8,8α-hexahydro-1,3α,8-trimethylpyrro[2,3-b]indol-5-ol methylcarbamate (ester); eserine; physostol). an alkaloid of physostigma, it is an extremely toxic parasympathomimetic alkaloid, derived from the Calabar bean (See *Physostigma venenosum*) that reversibly inhibits cholinesterases (with little if any effect on other serine esterases), thus preventing the destruction of acetylcholine. It is supertoxic to laboratory mice and probably to humans when delivered i.v. It is generally used in medicine as the salicylate in the treatment of poisoning by substances with anticholinergic effects (e.g., atropine, scopolamine, and imipramine and other tricyclic antidepressants). It is usually administered *i.m.* or *i.v.* at a usual adult dosage of 0.5 to 2.0 mg (and *i.v.* administration should not exceed a dose rate of 1 mg/min). Physostigmine salicylate is used also to produce miosis and to decrease intraocular pressure in the treatment of glaucoma. It induces a cholinergic syndrome in rats that may progress to status epilepticus.

physostigmine aminoxide. eseridine.

physostigmine poisoning. physostigmanism.

physostigmine salicylate. See physostigmine.

physostol. physostigmine.

phytagglutinin. lectin.

phytar. cacodylic acid.

phytoalexin. any chemical constituent of a plant that repels, inhibits the growth of, or destroys insects or pathogens, especially fungi. They are usually formed in response to chemical stimuli from the pathogen. Examples are pisatin and ipomeamarone.

phytoanaphylactogen (phytosensitinogen). an antigen of plant origin that can induce anaphylaxis.

phytochemistry. the biochemical science concerned with chemical phenomena in plants, especially the identification, biosynthesis, and metabolism of chemical constituents of plants.

phytocidal. destructive or lethal to plants.

phytocide. a chemical used to destroy and control plants. *Cf.* herbicide.

phytodermatitis. dermatitis caused by exposure to plants. The condition can be due to any of a number of mechanisms and modes of exposure (e.g., mechanical or chemical irritation, toxicity, allergy, or photosensitization). See also phytophotodermatitis, *Urtica*.

phytoecdysone. any plant substance with ecdysone-like activity. Some are extremely potent. The principal ecdysone of plant extracts is β-ecdysone. See also ponasterone.

phytogenous. of plant origin, produced by plants.

phytohemagglutinin (PHA). a lectin of plant origin that agglutinates erythrocytes.

phytohormone (plant hormone). any substance normally produced by a plant at a location removed from the site of action within the plant and is active in minute amounts. Phytohormones control plant functions such as growth. There are three types: auxins, cytokinins, and gibberellins.

phytokinin. cytokinin.

Phytolacca (pokeweed; pokeberry; scoke; garget; pigeonberry). a genus of perennial poisonous herbs, shrubs and trees (Family Phytolaccaceae) of tropical and warm areas. Glycoproteins in species of pokeweed in Africa produce lymphocytes that strongly resemble those of Burkitt's lymphoma. ***P. americana*** (synonym, P. decandra L.) (pokeweed, scoke, poke, garget, pigeonberry). a tall, perennial herb with smooth stems and leaves, and an unpleasant odor. It is a common plant that grows on low ground often in barnyards, clearings and along roadsides from Florida to Texas, northward to New England, southern Quebec, New York, and southern Ontario. The ovaries are green and the berries, when ripe in autumn, are dark purple. All parts, and the seeds and roots are extremely toxic and deadly. Young growth (not including any of the root) is sometimes eaten as a substitute for asparagus. Narcotic effects have been observed and children have died from eating the berry. Symptoms in humans include an immediate burning sensation in the throat, followed in about 2 hours by vomiting, gastrointestinal cramps and diarrhea. The pulse is weak and rapid and respiration is shallow. Animals are rarely poisoned. Swine, however, have been poisoned by rooting out and eating the roots. Even now people are sometimes killed when they eat the extremely toxic, dark purple berry or the root which may be mistaken for wild parsnip or horseradish. Gloves should be worn when handling this plant. The poisonous principles are a variety of saponins and glycoproteins such as phytolaccine.

Phytolaccaceae (pokeweed family). a family of plants with alternate entire leaves and perfect flowers, and similar in general characteristics to plants of the family Chenopodiaceae. See *Phytolacca*.

phytolaccine. a toxicologically active principle of *Phytolacca americana*, q.v.

phytonosis. any morbid condition caused by a plant.

phytopathogenic. pathogenic to plants; of or pertaining to a substance or an organism that is pathogenic to or productive of disease in plants.

phytopathology. the study of pathological states and processes in plants. See pathology.

phytophagus. herbivorous.

phytopharmacology. the study of the effects of biologically active substances on plants, especially on growth. See also pharmacology.

phytophlyctodermatitis. meadow grass dermatitis.

phytophotodermatitis. phototoxic dermatitis due to the ingestion of parts of plants such as certain species of *Brassica*, *Erodium*, *Euphorbia*, *Medicago*, *Poa*, and *Trifolium* with subsequent exposure to light. See also phytodermatitis.

phytophysiology. the science of plant physiology.

phytoplankton. that portion of the plankton community comprised of microscopic or nearly microscopic plants (e.g., algae, diatoms).

phytosensitinogen. phytoanaphylactogen.

phytosterol. any sterol isolated from higher plants. See sterol.

phytoteratology (plant teratology). the study of abnormal development, malformations, and monstrosities of plants.

phytotoxic. **1:** used in reference to a phytotoxin. **2:** toxic to, injurious to, damaging to, or inhibiting the growth of plants.

phytotoxicity. the capacity to poison plants.

phytotoxicologist. a biotoxicologist concerned with poisonous plants; one who contributes to the body of phytotoxicology or one who applies the findings of phytotoxicology to problem solving or to the needs of other disciplines or professions. *Cf.* biotoxicologist, zoototoxicologist.

phytotoxicology. the science of poisonous plants, and of toxicologically active plant substances and products and their effects, usually upon domestic animals and humans, by plants and plant substances. One serious difficulty in this field is the frequent failure of victims and even professionals to accurately identify the plant or plant materials involved in a poisoning.

phytotoxin (plant toxin). **1:** toxalbumin; any substance with properties similar to those of a bacterial exotoxin that is produced by certain species of higher plants (e.g., abrin, convallatoxin, crotin, ricin, and robin). They are complex proteins that are produced by a relatively small number of plants and are structurally similar to bacterial toxins. They have antigenic properties, causing *in vitro* agglutination of erythrocytes. Their extreme toxicity, however is due to their proteolytic activity. They are heat labile, nondialyzable, resist proteolytic digestion, and may be identified by antibody-specific precipitin tests. They are effective orally, being rather easily absorbed through the wall of the gastrointestinal tract, though there is usually some delay between ingestion and the onset of symptoms. Symptoms are predominantly hemorrhagic inflammation of the gastrointestinal tract and edema in a number of organs. Only a few native North America plants produce phytotoxins. An exception is *Abrus precatorius*, q.v. **2:** sometimes applied to any poisonous substance of plant origin. By this definition, most nitrogen-containing phytotoxins are alkaloids. Cyanides, and a small number of polypeptides and amines are exceptions. When present, alkaloids are often distributed throughout the plant, thus any part may be dangerous to wildlife, to livestock and even to humans. **3:** sometimes used by other than toxicologists in reference to substances toxic to plants.

PHZ. phenylhydrazine.

pi electron. π electron. See hydrocarbon (aromatic hydrocarbon).

π electron (pi electron). See hydrocarbon (aro matic hydrocarbon).

pica. perverted appetite whereby an individual ingests material (e.g., dirt, clay, plaster, ashes, flakes of paint) that is unfit as food. Pica is sometimes seen in pregnancy, hysteria, chlorosis, helminthiasis, certain psychoses, and in children it is sometimes the result of iron-deficiency anemia. Such individuals may have a higher than normal rate of exposure to certain poisons.

picloram (4-amino-3,5,6-trichloro-2-pyridinecarboxylic acid; 4-amino-3,5,6-trichloropicolinic acid). a crystalline solid or colorless powder, $C_6H_3Cl_3N_2O_2$, with the odor of chlorine. A pyridine herbicide and defoliant, it is toxic also to higher animals and humans by ingestion and inhalation and is a suspected occupational carcinogen. Picloram and its salts are absorbed by leaves and roots and translocated to new growth. It is used to control broadleaved weeds (except Cruciferae) in grasses. Even at low application rates it alters leaf morphology and inhibits the growth of leaves. With normal usage, terminal growth ceases. Following high application rates, it may persist in the soil for nearly a year.

pico- (p) a prefix to units of measurement that indicates a quantity of 10^{-12} or one trillionth of the value indicated by the root. See International System of Units.

picocurie, picoCurie (pCi). a measure of radioactivity, a pCi is one million millionth, or one trillionth, of a curie, and represents 2.22 radioactive particle disintegrations per minute (0.037 disintegrations per second).

picocuries per liter (pCi/L). a unit of measure used to chiefly express levels of radon gas. (See picocurie.)

picogram ($\mu\gamma$; pg; microgamma; micromicrogram). a unit of mass, equal to 10^{-12} or one-trillionth of a gram.

α-picoline (2-methylpyridine). a pyridine derivative, C_6H_7N, and component of coal tar and bone oil that is used as an industrial solvent and chemical intermediate in the manufacture of dyes and resins. It is a colorless liquid with a strongly disagreeable odor. It is a respiratory irritant and is moderately toxic to laboratory rats. See also *Nicotiana*.

picosclerotine. a toxic alkaloid of ergot.

picramic acid (2-amino-4,6-dinitrophenol; 4,6-dinitro-2-aminophenol). a crystalline, acidic substance, $C_6H_2(NO_2)_2(NH_2)OH$, that forms dark red needles from ethanol solution. It is a product of the partial reduction of picric acid. Uses include that of a reagent in tests for albumin and in the manufacture of azo dyes. Picramic acid sometimes occurs in the blood of persons poisoned by picric acid.

picratonitine. a toxic alkaloid of common occurrence in various species of *Aconitum*. It initially stimulates and then depresses the central nervous system. See also aconitine.

picric acid (2,4,6-trinitrophenol; picronitric acid; carbazotic acid; nitroxanthic acid). a toxic, explosive, highly oxidative, bitter-tasting yellow, crystalline substance, $C_6H_2(NO_2)_3OH$, that is soluble in water and a number of organic solvents. Picric acid is a severe explosion hazard when shocked or heated and is especially reactive with metals or metallic salts. It is used in explosives, to etch copper, and in the manufacture of batteries, dyes, and matches.

picronitric acid. picric acid.

picrotin ([1aR-(1aα,2aβ,3β,6β,6aβ,8aS*,8bβ,-9S*)]-hexahydro-2a-hydroxy-9-(1-hydroxy-1-methylethyl)-8b-methyl-3,6-methano-8H-1,5,7-trioxacyclopenta[ij]cycloprop[a]azulene-4,8-(3*H*)-dione). the generally nontoxic moiety of picrotoxin. It is nevertheless very toxic to laboratory mice. See also picrotoxin, picrotoxinin.

picrotoxin (cocculin). a highly toxic, extremely bitter, neutral, shiny, crystalline alkaloid or microcrystalline powder, $C_{30}H_{34}O_{13}$, found primarily in the leaves and seeds (cocculus indicus) of the shrub, *Anamirta cocculus*, but also occurs in *Tinomiscium philippinense*. It is comprised of equimolar parts of picrotoxinin and picrotin. It is a CNS and respiratory stimulant that acts by blocking γ-aminobutyric acid. It is extremely toxic to fish, laboratory animals, and humans. Symptoms of intoxication include salivation, emesis, convulsions, and hypertension. Diazepam is an effective antidote for picrotoxinism. Picrotoxin was formerly used as an antidote for acute poisoning by barbiturates and certain other CNS depressants. See also cocculus, picrotoxinin.

picrotoxin poisoning. picrotoxinism.

picrotoxinin ([1aR-(1aα,2aβ,3β,6β,6aβ,8aS*,-8bβ,9R*)]-hexahydro-2a-hydroxy-8b-methyl-9-(1-methylethenyl)-3,6-methano-8*H*-1,5,7-trioxacyclopenta[*ij*]cycloprop[*a*]azulene-4,8-(3H)-dione). a lactone, it is the toxic moiety of picrotoxin. It is extremely toxic to fish, laboratory animals, and humans.

picrotoxinism (picrotoxin poisoning). poisoning by picrotoxin, q.v. See also Diazepam.

picry. *Rhus radicans*, *Rhus toxicodendron*

Pictet-Spengler condensation reaction. Pictet-Spengler isoquinoline synthesis.

Pictet-Spengler isoquinoline synthesis (Pictet-Spengler condensation reaction). a type of nonenzymatic reaction that yields tetrahydroisoquinoline derivatives from the condensation of β-arylethylamines such as serotonin with carbonyl compounds and cyclization of the resulting Schiff bases.

Pictou disease (Molten's disease; Winton disease). a disease of domestic horses and cattle in Nova Scotia marked by cirrhosis of the liver. It results from ingestion of *Senecio jacobeus*,

pieris. common name of plants of the genus *Pieris*.

Pieris (pieris). **1:** a genus of butterflies (Family Pieridae). The best known and most widespread species of which is *P. rapae*, the cabbage butterfly; the caterpillar feeds on cabbage and other Cruciferae. See sinigrin, allyl isothiocyanate. **2:** a genus of evergreen shrubs or small trees of eastern North America and Asia (Family Ericaceae). They contain grayanotoxins. See also sinigrin. ***P. floribunda*** (fetter bush). a shrub of moist hillsides in the Allegheny Mountains from Virginia to Georgia. ***P. japonica*** (formerly *Andromeda japonica*; Japanese pieris). a woody shrub or small tree to 10 m tall, native to Japan but grown also as an ornamental. Specimens contain toxic resinoids.

pig. a container, usually lead, used to ship or store radioactive materials.

pigbel. a necrotizing enteritis endemic to the Papua New Guinea highlands. It is caused by type B toxin of *Clostridium* perfringens type C. It occurs mainly in children because of their limited immunity to B toxin and the low level of intestinal proteases. The latter are not induced because of a diet that is low in protein and high in sweet potatoes.

pigeonberry. **1:** *Duranta repens*. **2:** *Phytolacca americana*.

pigment. **1:** any colorant, natural or synthetic; a dye, stain, tint, coloring, or paint. See also color. **2:** any of numerous substances or formulations in the form of a solid (a crystal or powder), that impart a specified positive color to another substance or material. By this definition, talc, whiting, barytes, and clays are not pigments, but carbon black can be considered a pigment. Most are insoluble in water or organic solvents. Whereas a pigment is usually dispersed in a vehicle or substrate, it is physically and chemically unaffected by the vehicle. Pigments are classified in any of numerous ways not treated here. But See color. In some classifications, white is considered a color. Toxicologists should be aware that in the United States, only pigments that exhibit an acute oral toxicity of more than 5000 mg/kg body weight in tests of laboratory rats are defined as toxic under the Federal Hazardous Substances Act. Nearly all pigments used in the United States are defined as nontoxic under this act. **3:** a tincture; any medicinal preparation that is painted onto the skin (tincture of iodine). *Cf.* color, colorant, dye. **animal pigment**. any pigment that occurs naturally in animals. Examples are rhodopsin, melanin. **biological pigment**. any pigment that occurs in a living organism. Pigments play many diverse roles in life processes (photoreception, diverse photochemical reactions; photosensitization; the provision of cryptic and warning coloration; smoke screens; internal respiration; enzyme activity; detoxication). Examples of biological pigments are hemoglobin, chlorophylls, cytochromes, ferrichromes, gossypol. All of these examples contain iron. See also Lepidoptera. **chromate pigment**. chromates used as pigments. All (e.g., zinc chromate, lead chromate) are toxic. See also chromate. **iron pigment**. pigments that contain iron. Many play critical biological roles; many are toxic. Examples are hemoglobin, cytochromes, ferrichromes, iron oxides. **lead pigment**. any pigment that contains lead compounds. All are potentially toxic if ingested. Examples are lead carbonate, basic, lead molybdate, lead antimonite, lead nitrate. **photosensitizing pigment**. any pigment, usually a plant pigment, that causes photosensitization in animals (usually herbivores) when ingested and the animal is subsequently exposed to sunlight. See, for example, hypericin, phylloerythrin, See also photodermatitis, photosensitivity, photosensitization. **phytotoxic pigment**. a toxic plant pigment. Photosensitizing pigments are phytotoxic. **plant pigment** (vegetable pigment). any of a large variety of organic pigments (after def. 2, above) that are produced by plants. Examples are carotenoids such as carotene and xanthophyll; flavanoids such as catechins; flavones. Some can be prepared synthetically. **respiratory pigment**. a pigment that reversibly binds oxygen and carbon dioxide. See hemoglobin. **synthetic pigment**. any pigment that is synthesized and not found in nature. Examples are phthalocyanine, lithos, toluidine, para red, toners, lakes.

pigment red. lead oxide red.

pigmy rattler. pigmy rattlesnake (See *Sisturus*).

pigmy rattlesnake. **1:** any snake of the genus *Sistrurus*. **2:** *Sistrurus miliarius*.

pigskin poison puffball. *Scleroderma citrinum*.

pigweed. common name for species of *Amaranthus* and *Chenopodium*.

PIH. pheniprazine.

pile. **1:** a little used synonym for nuclear reactor, *q.v.* **2:** a mound or heap of waste; a waste pile.

pill. **1:** capsule, lozenge, pastille; tablet; a medicine (or a placebo) in the form of a small solid mass or pellet (usually oval or globular in shape) to be swallowed or chewed. In addition to the active component, a pill contains a diluent or filler and an excipient to provide adhesiveness, pliability, and firmness to allow shaping of the pill. **2:** the pill; a term applied to a hormonal contraceptive that is taken orally.

pilocarpine. an extremely toxic parasympathomimetic agent (an alkaloid), $C_{11}H_{16}N_2O_2$, obtained from leaflets of *Pilocarpus microphyllus or P. jaborandi*. It is used therapeutically in eyedrops to treat glaucoma. It is also used as a diaphoretic, sialogogue, and stimulant of intestinal motility. It is used as the hydrochloride ($C_{11}H_{16}N_2O_2 \cdot HCl$) and the nitrate. Pilocarpine induces a cholinergic syndrome in rats that may progress to status epilepticus.

Pilocarpus microphyllus and ***P. jaborandi***. shrubs (Family Rutaceae), of the West Indies and tropical America; the source of pilocarpine.

piloerection. the erection or apparent stiffening of body hair due to stimulation and contraction of the erector pili muscles. It is usually accompanied by dilation of the pupils. It is often due to excitement, especially fear.

pimaricin. natamycin.

pimelic ketone. cyclohexanone.

pimeluria. lipid in the urine.

Pimephales promelas (fathead minnow). a common minnow (Family Cyprinidae) in the United States east of the Rocky Mountains. It is a standard species used in aquatic toxicity.

pinacoloxymethylphosphoryl fluoride. soman.

pinacolyl methylphosphonofluoridate. soman.

***O*-pinacolyl methylphosphonofluoridate**. soman.

Pindione™. phenindione.

pine. *Pinus*.

pineapple snake. *Lachesis mutus*.

pingue. *Hymenoxys richardsonii*.

pink allamanda. *Cryptostegia grandiflora*.

pink disease. See acrodynia.

pink family. Caryophyllaceae.

pink jellyfish. *Cyanea capillata*.

pink-fringed milky. *Lactarius torminosus*.

pink-tipped anemone. *Condylactis gigantea*.

pinna. **1:** the outer, funnel-like structure of the ear, as in humans, that gathers sound waves. **2:** a leaflet of a fern frond or of any compound pinnate leaf.

pinnate. situated along a central axis as in a compound leaf with leaflets on either side of a common petiole.

pinocytosis. a process whereby cells ingest or absorb nutrients and fluid by engulfment and pinching off of a small, fluid-filled portion of the cell membrane with the formation of small vacuole within the cell's cytoplasm. The nutrient, now inside is subject to metabolism. It is not generally considered to be important in the uptake of toxicants. *Cf.* phagocytosis.

pinpoint pupil. the extreme or ultimate state of miosis in which the pupil of the eye is extremely small.

Pinus (pine). a genus of about 80 species of coniferous trees (Family Pinaceae), mostly of the North Temperate Zone; some occur on high mountains in the tropics. ***P. ponderosa*** (ponderosa pine; western yellow pine; blackjack pine). a large to very large dominant coniferous tree that grows to about 60 m under favorable conditions. It is widely distributed at moderate elevations in western North America from southern California, New Mexico, and trans-Pecos Texas to southern British Columbia to North Dakota. The needles and buds are embryotoxic and abortifacient to cattle.

Pipa (common Surinam toad; Suriman toad). a genus of toads that contains only the poisonous Surinam toad (Family Pipidae, q.v.).

Piper (true peppers). a large genus of tropical plants (Family Piperaceae), the true peppers. ***P. cubeba*** (tailed pepper; Java pepper). a tropical shrub of southern Asia. See cubeb, cubebism. ***P. nigrum*** (black pepper). a true pepper (Family Piperaceae) and source of a widely used condiment. The fruit contains safrole. See piperism.

piperazine (diethylenediamine). a toxic cyclic liquid amine of low volatility; inhalation is unlikely except in dust.

piperidic acid. γ-aminobutyric acid.

piperidine (hexahydropyridine). a highly toxic, soapy feeling, liquid alkaloid with the characteristic odor of pepper; it is a completely saturated six-membered ring that contains 1 nitrogen atom, $CH_2CH_2CH_2CH_2CH_2NH$, used chiefly as an insecticide. It occurs in some species of *Lupinus* and in small amounts in *Piper nigrum*. See also piperine, piperism, pyridine.

piperidine alkaloid. any of a group of alkaloids

that include the piperidine ring. Most are derived from lysine. Examples are nicotine, piperidine.

3-(2-piperidinyl)pyridine. anabasine.

piperine (1-[5-(1,3-benzodioxol-5-yl)-1-oxo-2,4-pentadienyl]piperidine; 1-piperoylpiperidine). a crystalline, slightly soluble alkaloid from *Piper nigrum* (black pepper) and related species such as *P. clusii*, *P. geniculatum*, *P. longum*, and *P. retrofractum*. It yields piperidine and piperic acid on hydrolysis. Piperine provides the pungent taste of brandy; it is used as an insecticide and is more toxic to *Musca domestica* (the house fly) than is pyrethrum. See piperism.

piperism. poisoning by black pepper.

piperonyl butoxide (5-[[2-(2-butoxyethoxy) ethoxy]methyl]-6-propyl-1,3-benzodioxole; α-[2-(2-butoxyethoxy)ethoxy]-4,5-methylenedioxy-2-propyltoluene; [3,4-(methylenedioxy)-6-propylbenzyl] butyl diethyleneglycol ether). a light brown, combustible liquid with a mild odor that is insoluble in water, but soluble in ethanol, benzene, and petroleum hydrocarbons. It is a methylenedioxyphenyl synergist of pyrethrins and related pesticides, in oil solutions, emulsions, aerosols, or powders. It acts by inhibiting cytochrome P-450-dependent detoxication reactions (monooxygenations). It is highly toxic to humans, is noncarcinogenic, and only slightly toxic to laboratory rats and rabbits. The estimated safe dose for chronic ingestion is 42 mg/kg body weight/day in the diet.

piperonyl sulfoxide. carcinogenic in some laboratory animals.

1-piperoylpiperidine. piperine.

Pipidae. a small family of northern South American toads (Order Opisthocoela), individuals of which have a star-shaped cluster of dermal papillae on the tip of each digit and lack both tongue or eyelids. The developing embryos are borne in individual dermal pockets on the back of the female. See *Pipa*, Anura.

Piptadenia peregrina (cohoba). a tropical American tree (Family Fabaceae, formerly Leguminosae). Its seeds contain bufotenin and are used to prepare a narcotic snuff, cohoba. Natives of what are now Peru and Columbia have used this plant for many centuries (inhaled as a powder) to become intoxicated.

piri. *Bitis arietans*.

pirimicarb (dimethylcarbamic acid 2-(dimethylamino)-5,6-dimethyl-4-pyrimidinyl ester; 2-(dimethylamino)-5,6-dimethyl-4-pyrimidinyl dimethylcarbamate; 5,6-dimethyl-2-dimethylamino-4-dimethylcarbamoyloxypyrimidine). a carbamate insecticide used in agriculture largely as an aphicide that is highly toxic, systemically, to mammals. It binds strongly to soil and is more environmentally persistent than many carbamates. It is, however, poorly absorbed through the skin.

pirimiphos-ethyl (o-[2-(diethylamino)-6-methyl-4-pyrimidinyl]phosphorothioic acid *O,o*-diethyl ester; *O,o*-diethyl *o*-[2-(diethylamino)-6-methyl-4-pyrimidinyl]-phosphorothioate). a straw-colored liquid that decomposes at 130°C. It is used to control soil insects in crops.

pisatin. a phenolic phytoalexin that accumulates in tissues of *Lathyrus* spp. in the presence of various fungi.

Pisces. one of two superclasses of the subphylum Vertebrata (Phylum Chordata) comprised of all true fish (Classes: Chondrichthyes and Osteichthyes) including those of toxicological significance. Fish often accumulate lead, mercury and PCB to concentrations well above that of the medium in which they live. Species sensitivities to PCB are quite variable.

piscicidal. poisonous to fish.

piscicide. a substance poisonous to fish.

piscine. pertaining to fish, especially bony fish; fishlike.

pistil. the central structure(s) of a flower which develops into the fruit following fertilization.

pit. See loreal pit.

pit viper. any snake of the family Crotalidae, *q.v.*, having a depression or pit (loreal pit) between the nostril and the eye. The Old World vipers (Viperidae) lack a loreal pit.

pit viper venom poisoning. See crotalid venom poisoning.

pitch poisoning (coal tar poisoning; coal tar pitch poisoning; clay pigeon poisoning). **1a:** poisoning by any of a variety of coal tar derivatives (e.g., anthracene cresols, crude creosote, heavy oils) that induce acute to chronic disease in livestock and pets. Poisoning usually occurs in animals that chewing on or consume products made of coal tar distillates (e.g., bitumen-based flooring, clay pigeons, creosote-treated wood, and tar paper). Clinical effects include acute to chronic hepatic damage with attendant signs of icterus, ascites, anemia, and death. It is known chiefly from Canada, Germany, Ireland, Poland, and the United States. **1b:** a sometimes fatal condition of livestock that have eaten expended clay pigeons. The affected animals weaken and become ataxic and icteric with an enlarged liver; hepatic failure and death may result.

Pithecolobium lobatum (velvet bean; djenkol bean). a tropical shrub (Family Fabaceae, formerly Leguminosae) with bipinnate leaves and globose flower heads. It is the chief source of djenkol poisoning, *q.v.* Because the velvet bean is high in vitamin B, it continues to be used for food in spite of its toxic properties. See also djenkolic acid.

pithomycotoxicosis. facial eczema.

pito. *Maticora bivirgata*.

pitohui. See *Pitohui dicrous*.

Pitohui dichrous (hooded pitohui). a brightly colored, jay-like perching bird (Order Passeriformes; Family Pachycephalidae) and the only bird known to be poisonous. It is a true "thick-head" or "whistler" (See Pachycephalidae). The hooded pitohui is a heavy-bodied, thick-headed, jay-sized, fruit-eating bird of the tropical rain forest and scrub in New Guinea. It is easily recognized by its size and coloration: the hood, wings, and tail are black and the remaining body plumage is an orangish rufous brown. The sexes look alike. The muscle, skin and feathers contain homobatrachotoxin, an extremely toxic steroidal alkaloid earlier known only from Amazonian frogs of the family Dendrobatidae. Handling these birds with bare hands could lead to serious poisoning.

pituitary. **1:** pituitary gland. **2:** pertaining to the pituitary gland.

pituitary gland, pituitary body (hypophysis cerebri; hypophysis; pituitary). a small round, unpaired, compound endocrine gland of vertebrate animals. It is sometimes called the master gland. The pituitary is attached to the brain by a stalk (the infundibular stalk) originating in the floor of the hypothalamus. There are 2 major divisions of this gland, the anterior and posterior lobes. The hormones secreted by this gland regulate major aspects of metabolism and play a crucial role in the regulation of growth and reproduction. See also hormone, pituitary gland toxicity. **anterior lobe of the pituitary gland** (adenohypophysis; anterior pituitary gland; anterior pituitary). the anterior lobe of the pituitary gland. It secretes several tropic hormones in response to stimulation by releasing factors (hormones) from the hypothalamus. This hypothalamic-pituitary system (often termed the hypothalamic-hypophyseal axis) thus controls the activity of several peripheral endocrine glands. The hormones secreted by the anterior pituitary are: ACTH, FSH, LH, somatotrophin, and TSH, *q.v.* Each of these hormones stimulates or activates a specific endocrine gland and triggers the release of its hormones. **intermediate lobe of the pituitary gland**. a region of the pituitary gland of poikilotherms ("cold-blooded" animals) that secretes intermedin, a hormone that controls the activity of chromatophores (pigment cells). **posterior lobe of the pituitary gland** (neurohypophysis; posterior pituitary gland; posterior pituitary). the portion of the pituitary gland that is connected by a stalk to the hypothalamus. It secretes the hormones oxytocin and vasopressin (antidiuretic hormone). The former controls uterine contractions and milk let-down; the latter controls vasoconstriction and water reabsorption in the kidney.

pituitary growth hormone. somatotrophin.

pityriasis rubra. exfoliative dermatitis (See dermatitis).

pivalolactone. a poisonous compound by ingestion, $C_5H_8O_2$. It is carcinogenic to some laboratory animalls and is a questionable human carcinogen; some studies indicate mutagenicity.

pl. plural.

placebo. a drug or treatment that has no demonstrated pharmacological effect on disease but may provide relief in those individuals who believe in its efficacy.

placenta. a specialized organ of eutherian mammals, formed from the chorion and uterine tissue following implantation of the fertilized egg in the uterus. It is the organ of metabolic exchange between the mother and the embryo or fetus. The placenta is also the organ of transport of toxicants from maternal blood to that of the fetus. See also barrier (placental barrier).

placental barrier. See barrier.

placental membrane. See barrier (placental barrier).

Placidyl. ethchlorvynol.

Plagiobothrys (popcorn flower). a genus of North American herbs (Family Boriginaceae) that sometimes contain toxic concentrations of nitrate.

plains whorled milkweed. *Asclepias pumila*.

plankton. microscopic plants and animals that live in water bodies, fresh-water, brackish, and marine. See also phytoplankton, zooplankton.

plant. any member of the kingdom Plantae. The responses of plant populations to a xenobiotic chemical vary as a function of phenological stage and effects of environmental variables (past and current) such as nutrient availability, moisture, and interactions with other plants. In most cases, plant tissue concentrations of toxic chemicals do not bear a quantitative relationship with exposure concentrations. Effects can be indirect in the sense of impact on soils or plant species found in association with the affected plant. **accumulator plant**. **1a:** historically any plantspecies, the individuals of which accumulate, from the environment, relatively high concentrations of a specific chemical element (e.g., a heavy metal) in their tissues. **1b:** any plant that is an accumulator (def. 1). **African milk plant**. *Euphorbia lactea*. **annual plant**. a plant that completes its life cycle in one growing season; usually applied only to herbaceous plants. **bee plant**. *Cleome serrulata*. **biennial plant**. a plant that matures and dies after two seasons; usually applied only to herbaceous plants. See also annual plant, perennial plant. **California soap plant**. *Chlorogalum pomeridianum*. **castor bean plant**. *Ricinus communis*. **castor oil plant**. *Ricinus communis*. **dog button plant**. *Strychnos nux-vomica*. **food plant**. any plant, some part of which is edible by wild or domestic animals or by humans (with or without preparation). Many food plants have poisonous parts, poisonous life stages, or require special preparation to assure safety. Avocado leaves; apple or pear seeds; rhubarb leaves; apricot, cherry, peach, and plum pits; tomato leaves; and the outer green husks of walnuts can all have toxic effects if enough is ingested. Even a small amount of green potato sprouts can be very toxic. See also garden plant, houseplant. **garden plant**. any plant, other than an agricultural crop, that is typically cultivated in a garden or out-of-doors as an ornamental, for use in landscaping, or occasionally for food. Some or all parts of numerous garden plants are pharmacologically active or poisonous when ingested or in some cases touched. Many are deadly. **green plant**. any plant that contains chlorophyll; nearly all are autotrophic. **hedge plant**. **1:** a tree, shrub, and sometimes even a rapidly growing annual plant that can form a continuous close growth or hedge. A number of these plants are poisonous. See, for example, *Taxus* (yew), *Pachysandra procumbens* (box), *Hydrangea* (hydrangeas), and *Ligustrum* (privet). **2:** *Ligustrum vulgare*. *Dieffenbachia seguine*. **house plant**. See houseplant. **perennial plant**. any plant that lives more than 2 years. See also annual plant, biennial plant; usually applied only to herbaceous plants. Some tend to live indefinitely, others just a few years. **pie plant**. *Rheum rhabarbarum* (rhubarb). **poisonous plant**. any plant whose tissues, in whole or in part contain poisons in sufficient quantity or concentration to injure or kill specified organisms that might feed upon it or might otherwise conceivably be exposed to the plant such that sufficient poison is absorbed to cause injury or death. Nearly all major classes of plants contain poisonous species or variants. Only mosses, some groups of algae, and some fungi include no toxic species. See also atoxic, phytotoxin. **weather plant**. *Abrus precatorius*.

plant amine oxidase. an amine oxidase (*q.v.*) of plant origin. *Cf.* monoamine oxidase.

plant hormone. phytohormone.

plant louse. See *Phylloxera*.

plant teratology. phytoteratology.

Plantae. the plant kingdom; usually defined to include all green plants (i.e., all organisms that contain chlorophyll). In many classifications the fungi and bacteria are also included. Organisms included in this kingdom are usually easily distinguished from members of the animal kingdom (Animalia). Most are autotrophic; they are typically attached to a substrate and unable to move freely about (exceptions are fungi and certain flagellate algae); in addition to a semipermeable cell membrane, most plant cells are surrounded by a cell wall of cellulose; starch is a common storage polysaccharide.

plaque. **1a:** a small raised area of abnormal material on a surface such as the skin or lining of a blood vessel. **1b:** a soft mass of fatty material, notably cholesterol, beneath the inner linings of arteries. **2:** a mixture of bacteria and calcium deposited on the teeth that causes cavities, gum diseases, and finally bone lesions.

Plaquenil. quinine.

plasma. See blood plasma.

plasma cell. any of a type of specialized antibody-forming cell responsible for humoral immunity and derived from a B lymphocyte. They produce and secrete massive numbers of antigen-specific antibodies. See also antigen-antibody reaction, immune response. *Cf.* B lymphocyte.

plasma membrane. plasmalemma.

plasmalemma (cell membrane; cytoplasmic membrane; plasma membrane). a semipermeable membrane that surrounds the cytoplasm of a cell and regulates the passage of molecules into and out of the cell. It plays a critical role in determining the extent to which a poisonous substance penetrates the cell. The membrane itself can be poisoned, with changes in membrane permeability with a consequent change in the rate of passive transport through the membrane. Carriers involved in active transport may also be poisoned, affecting cellular penetration.

plasmid. any small, autonomously replicating DNA segment that exists in bacteria apart from the chromosome and replicates independently of it. Plasmids carry information that renders the bacteria resistant to antibiotics. Plasmids are often used in genetic engineering to carry desired genes into organisms. See also factor (R factor).

plasmodicidal. lethal to plasmodia.

plasmodicide. an agent that is lethal to plasmodia.

Plasmodium. a genus of sporozoa (Family Plasmodiidae) that are parasitic in the red blood cells of lizards, birds, and mammals; the malarial parasite.

plasmolysis. contraction or shrinking of the cytoplasm away from the cell wall, as of a plant cell, due to the loss of water through exosmosis.

plasmolytic. of, pertaining to, characterized by, subject to, or tending toward plasmolysis.

plasticizer. a substance (usually an organic liquid) which, when added to a natural or synthetic rubber, plastic, or resin, confers plasticity (flexibility). A plasticizer accomplishes this by weakening the attractive forces between the macromolecules by increasing the degree of separation of macromolecules through chemical reaction or by physically fitting between the molecules. This permits sliding or displacement of the macromolecules over one another. There is an enormous variety of plasticizers. Examples are camphor, TOCP, triethylphosphate, waxes, and phthalic acid esters (e.g., bis(2-ethylhexyl)phthalate, diphenylphthalate). The latter are used in large quantities as plasticizers of polyvinyl chloride (PVC) plastics.

platelet (blood platelet; thrombocyte). a round or oval disk-like fragment, about 2-4 micra in diameter, of a megakaryocyte that normally initiates the process of blood clotting in vertebrates.

platinosis (platinum poisoning). a morbid condition of the skin and upper respiratory tract due to exposure to soluble platinum salts.

platinum (symbol, Pt)[Z = 78; A_r = 195.09]. a silver-gray, lustrous, malleable, and ductile metallic transition element that is resistant to oxidation and to attack by acids (except aqua regia). Inhalation of dust containing platinum salts can cause sneezing, rhinitis, coughing, wheezing, dyspnea with tightness of chest, wheezing, and cyanosis. Platinum black, a finely powdered form, is an important catalyst in hydrogenation. Some platinum salts have been used to treat syphilis.

***cis*-platinum II**. cisplatin.

platinum poisoning. platinosis.

plattchen Seeschlange. *Pelamis platurus*.

platterful mushroom. *Tricholompsis platyphylla*.

plattnerite. See lead dioxide.

platyphylline. an alkaloid, $C_{18}H_{27}O_5N$, isolated from *Senecio platyphyllus* and other species of *Senecio* (Family Compositae).

platypus. duck-billed platypus. See Ornithorhynchus.

Plectropomus (seabass). a genus of marine bony fish (Family Serranidae) of Indonesia, the Philippine, Caroline and Marshall islands. Certain populations (e.g., of *P. obligacanthus* and *P. truncatus* are transvectors of ciguatera.

pleura (pleural membrane). the thin serous tissue that invests the lungs and lines the thoracic cavity of vertebrates. See pleurisy.

pleural. pertaining to the pleura.

pleural effusion (pleural fluid effusion). fluid that collects around the lungs, usually due to inflammation of the lungs and pleura or congestive heart failure.

pleural fluid effusion. pleural effusion.

pleurisy. inflammation of the pleura. Acute pleurisy is marked by initial reddening of the pleura with secretion of an exudate of lymph, fibrin, and cellular elements onto the surface of the pleura and into the cavity. The disease may progress with copious effusion of exudate, adhesions (usually permanent) form between pleural surfaces chills, a pain in the side, followed by fever and a dry cough. Eventually, pain recedes and the victim becomes dyspneic.

pleurisy root. *Asclepias tuberosa*.

pleurodynia. pain in the thoracic wall.

pleurotin. a toxic antibiotic, $C_{20}H_{22}O_5$, active against the tubercle bacillus and staphylococcus of boils. It is produced by the mushroom, *Pleurotis griseus*.

plexus. a network of nerves, veins, or arteries. The solar plexus is a nexus of nerves that lie behind the stomach.

plotolysin. the hemotoxic fraction of plototoxin.

plotospasmin. the neurotoxic fraction of plototoxin.

Plotosus lineatus (oriental catfish). a venomous catfish (Family Plotosidae), the sting of which causes intense pain which may persist for several days. It commonly occurs in the Indo-Pacific region at the mouths of rivers. The toxic principle is plototoxin. See catfish.

plototoxin. a venom isolated from the catfish, *Plotosus lineatus*. It is composed of a hemotoxic fraction (plotolysin) and a neurotoxic fraction (plotospasmin) fraction.

plumbagin (5-hydroxy-2-methyl-1,4-naphthoquinone). a yellow, crystalline, irritant, antibacterial, and abortifacient compound, $CH_3 \cdot C_{10}H_4(:O)_2 \cdot OH$, isolated from roots and shrubs of the genus *Plumbago*.

Plumbago. a genus of widely distributed, warm-climate herbs, shrubs, and woody vines (Family Plumbaginaceae). They have alternate sessile leaves and spicate white, blue, or rosy-red flowers. The roots of some species contain plumbagin.

plumbic. pertaining to or containing lead.

plumbism. **1:** chronic lead poisoning. **2:** lead poisoning.

plumboplumbic oxide. lead oxide red.

plumbous fluoride. lead fluoride.

plumbous oxide. lead monoxide.

plumbum. lead.

plume. **1:** a visible or measurable discharge of a contaminant (thermal or chemical) from a single point of origin in water or air (e.g., a plume of smoke). **2:** the area of measurable and potentially harmful radiation from a damaged reactor. **3:** the distance from a release of toxic fumes considered hazardous for persons who might be exposed.

plumeless thistle. *Carduus*.

Plummer's disease. hyperthyroidism due to a nodular toxic goiter, usually unaccompanied by exophthalmos.

pluriresistant. resistant to more than one pharmacologically active xenobiotic.

plutonism. poisoning by plutonium, as demonstrated in experimental mammals when exposed to plutonium present in atomic piles. Common symptoms and lesions include hepatic damage, changes in bone structure, and the graying of hair. Severe acute exposure of dogs by inhalation quickly results in death, as a rule, from radiation pneumonitis. Dogs that survive such exposures for more than 1000 days develop pulmonary fibrosis and usually die from neoplasms.

plutonium (symbol, Pu)[Z = 94; most stable isotope = 244Pu, T½ = 8.2 X 107 yr]. a highly toxic, highly reactive, silvery-white, synthetic transuranium radioactive element of the actinoid series that is formed in nuclear reactors. Its chemical behavior is similar to that of uranium. Of the isotopes, the best known α-emitter is ^{239}Pu (T½ = 24,000 yr) which, like ^{235}Pu, is fissionable and used as a nuclear fuel and in nuclear explosives; ^{238}Pu (T½ = 86 yr) is an energy source in pacemakers. Plutonium compounds are highly toxic and should be considered extremely hazardous. They vary in toxicity depending mainly upon the atoms that bind to plutonium. Soluble salts of plutonium concentrate in bone, liver, and body fluids and may cause bone cancer and occasionally liver cancer. Symptoms of plutonium intoxication (plutonism) are dose dependent (See plutonism).

plutonium poisoning. plutonism.

Pm. the symbol for promethium.

PMA. phenylmercuric acetate.

PMAC. phenylmercuric acetate.

PMAS. phenylmercuric acetate.

PMN. premanufacture notification.

PNA. paranitroanaline. See *p*-nitroanaline.

PNB. *p*-nitrobiphenyl.

pneumatic. pertaining to breath.

pneumatotaxis. pneumotaxis.

pneumocardial. cardiopulmonary.

pneumoconiosis (pneumokoniosis; pneumonoconiosis; anthracotic tuberculosis). an inflammation of the lungs and upper respiratory tract, leading ultimately to pulmonary fibrosis. Such a condition results from long-continued inhalation and permanent accumulation of relatively large amounts of particulate matter in the lungs and the reaction of lung tissue to its presence. Most cases are occupational (e.g., coal mining, stone cutting) or environmental in origin. Symptoms usually include chest pain, coughing with little or no expectoration, dyspnea, reduced thoracic excursion, fatigue following light exertion, and sometimes cyanosis. The degree of disability depends on the type of particles inhaled, and the degree of exposure to them. There are three main functional types of pneumonoconiosis: simple, irritant, and allergic, *q.v.* **allergic pneumoconiosis**. that due to organic dusts that induce a kind of allergic reaction (e.g., byssinosis, bagossis). **bauxite pneumoconiosis** (Shaver's disease; bauxite worker's disease). a rapidly progressing condition that results in an exceedingly severe pulmonary emphysema. It is caused by occupational exposure to bauxite fumes given off during the manufacture of alumina abrasives. Symptoms typically include coughing, fatigue, and labored breathing with a combined obstructive and restrictive breathing pattern, and impaired diffusing capacity. **coal worker's pneumoconiosis** (coal miner's lung; collier's lung; miner's lung; black phthisis; miner's phthisis). a pneumoconiosis caused by the deposition of large amounts of coal dust in the lungs and characterized by centrilobar emphysema. *Cf.* anthracosis, black lung. **collagenous pneumoconiosis**. an irritant pneumoconiosis marked by a permanent scarring of pulmonary tissue, due to inhalation of fibrogenic dust (e.g., asbestos, silica) or to a modified response to a normally non-fibrogenic dust. **irritant pneumoconiosis**. that due to the inhalation of irritant dusts such as silica and asbestos. This type causes scarring and gradual destruction of lung tissue. **simple pneumoconiosis**. that which results from the deposition of inert dust (e.g., iron, tin, and carbon dusts) in the lungs without significant adverse effects. **talc pneumoconiosis**. that due to inhalation of talc. It is characterized by coughing, shortness of breath, fatigue, weakness, and weight loss. Prolonged exposure to large quantities of talc may result in pulmonary fibrosis.

pneumoconiosis siderotica. siderosis, def. 1.

pneumokoniosis. pneumoconiosis.

pneumomelanosis. the blackening of lung tissue by the deposition of inhaled coal dust in the lungs. *Cf.* black lung.

pneumonectasis. pulmonary emphysema due to overdistension of the lungs.

pneumonia. inflammation of the lung parenchyma with consolidation of the affected part, the alveoli becoming filled with exudate. Pneumonia is usually caused by bacteria or viruses, and much less commonly by rickettsias, fungi, yeasts. Pneumonia sometimes results from inhalation of irritants or trauma to the thoracic wall. Pneumonia may or may not affect the entire lung, hence the terms lobar pneumonia, lobular pneumonia, and segmental pneumonia. **aspiration pneumonia**. pneumonia that develops following inhalation of foreign matter into the lungs. **bronchopneumonia**. pneumonia accompanied by bronchitis (inflammation of the bronchae). **chemical pneumonia**. that caused by inhalation of fumes or a toxic gas such as phosgene or chlorine. In such cases, the lungs may become hemorrhagic and edematous. The air passages may become filled with large amounts of fluid, literally drowning the victim. In survivors, permanent lung damage remains, and recurring pulmonary infections are common.

pneumonitis (pulmonitis). inflammation of the lungs.

pneumonoconiosis. pneumoconiosis.

pneumorrhagia. pulmonary hemorrhage. See hemorrhage.

pneumotactic. pertaining to pneumotaxis.

pneumotaxic. pertaining to the regulation of breathing.

pneumotaxic center. the region of the respiratory control center, located in the pons of the vertebrate brain.

pneumotaxis (pneumatotaxis). the directed response of a motile organism towards (positive pneumotaxis) or away from (negative pneumotaxis) the stimulus of dissolved carbon dioxide or other gas.

pneumotropic. pertaining to pneumotropism.

pneumotropism. the orientation of an organism to the stimulus of dissolved carbon dioxide or other gas.

pneusis. breathing.

PNS. peripheral nervous system.

po, p.o. *per os*.

Po. the symbol for polonium.

Poa (meadow grasses, speargrasses, bluegrasses). a genus of grasses (Family Gramineae) native to North America that may become heavily ergotized. See *Claviceps*.

podalgia. foot pain.

podophyllin (podophyllum resin; podophyllin resin). a resinous, violently cathartic extract of *Podophyllum peltatum*. It is a variable mixture of at least 16 physiologically active compounds, including some which have pronounced cytotoxic effects. It is an abortifacient. It is still used by some as a laxative, but is unsafe. Overdosage may prove lethal. See also podophyllinic acid and podophyllotoxin.

podophyllin resin. podophyllin.

podophyllinic acid ($C_{24}H_{30}N_2O_8$; 2-ethylhydrazide). an antineoplastic agent produced by *Podophyllum.*

podophyllotoxin (5,8,8a,9-tetrahydro-9-hydroxy-5-(3,4,5-trimethoxyphenyl)furo-[3′,-4′:6,7] naphthol [2,3-d]-1,3-dioxol-6 (5aH)-one). a highly toxic antineoplastic glucoside, $C_{22}H_{22}O_8$, it is the chief active principle isolated from the roots and rhizomes of *Podophyllum peltatum*. It is a metaphase poison and mutagen that binds specifically to the spindle protein tubulin and inhibits its polymerization. This inhibits mitosis at metaphase, resulting in polyploidy, a usually lethal mutation in animals. Podophyllotoxin also poisons the hematopoietic and lymphoid systems by inhibiting nucleoside transport, and mitochondrial electron transport. Common symptoms of intoxication include nausea, vomiting, and alopecia.

podophyllum (mandrake root; mayapple; Indian apple; vegetable calomel). the dried rhizome and roots of *Podophyllum peltatum*, the chief active principle of which is podophyllotoxin. See also podophyllin and podophyllinic acid.

Podophyllum. a genus of perennial North American herbs (Family Berberidaceae). ***P. peltatum*** (mayapple; mandrake; Indian apple; vegetable calomel). a simple, 1- or 2-leafed toxic herb (Family Berberidaceae) rising from a horizontal root. It contains cathartic and cytotoxic principles. The green fruit, foliage, and roots are toxic. Poisoning is rare as animals do not commonly consume mayapple. Poisoning in humans usually occurs from misuse of preparations made from mandrake. American Indians once ingested the young shoots of mayapple to commit suicide. Symptoms are those of a severe, purging gastroenteritis with vomiting. Contact produces a focal dermatitis. See also podophyllum, podophyllin, podophyllinic acid.

podophyllum resin. podophyllin.

POF. thioctic acid.

pofadder. *Bitis arietans*.

Pogonomyrmex (harvester ants). a genus of herbivorous harvester or agricultural ants (Family Formicidae). Some species are mound builders. The bite, which introduces enzymes and formic acid, causes a sharp stinging pain. The skin about the bite in humans itches and becomes whitish. Ulceration and fever may ensue in severe cases. ***P. barbatus***. a large stinging species that is distributed widely throughout the deserts of the southwestern United States. The venom is a cholinergic. The sting (usually on an extremity) produces immediate severe pain and progresses to a deep aching. A wheal and flare quickly develop at the site of the sting. Severe inflammation, sweating, and piloerection appear over the eruption, persisting for up to 2 days. The affected area remains sensitive to touch for a few more days. Systemic reactions and even death in children may result from multiple stings.

poinciana (bird of paradise). **1:** prickly tropical shrubs of the genus *Caesalpinia* (formerly *Poinciana*). **2:** royal poinciana refers only to *Delonix regia* (formerly *Poinciana regia*).

poinsettia. **1:** common name of *Euphorbia pulcherrima*. **2:** common name of any plant of the subgenus *Poinsettia*.

Poinsettia. a subgenus of spurges (Genus *Euphorbia*; Family Euphorbaceae), some variants of which are used as ornamental house plants. Ingestion of small amounts of these plants can produce severe digestive upset and has proven fatal to children. See poisoning (poinsettia poisoning).

poinsettia poisoning. poisoning by ingestion of any part of *Euphorbia pulcherrima* (poinsettia). Symptoms include abdominal pain with vomiting and diarrhea. Fatalities are rare.

point locoweed. *Oxytropis*.

point source. See source.

pointed-scales pit viper. *Trimeresurus mucrosquamatus*.

pointvetches. *Oxytropis*.

pois vivace. *Lathyrus latifolius*

poison. **1a:** any biologically active agent that injures, impairs, or kills a living system, a system component, or the normal maintenance and perpetuation of the system by virtue of its chemical or physical interaction with the metabolic processes of the system. **1b:** a substance that causes alterations or disturbances in the functioning of an exposed organism or its progeny or the functioning of current or future generations of a plant or animal population. **2:** any substance which, in small amounts, is injurious to health or endangers the life of a living organism when in contact with, absorbed by, or developed within the body of the organism. All chemicals are poisonous at some level of exposure. **3:** any substance that endangers the fitness of a species by virtue of its chemical or physical interactions with the metabolic processes of individual organisms. **4:** in chemistry, any substance that destroys the activity of a catalyst; a catalyst poison. *Cf.* chemical poison. **5:** in law, a substance is generally a poison if, when taken into the body, it is inherently capable of destroying life by directly impairing vital functions or preventing the continuance of life by rendering body fluids and solids incapable of supporting the continuance of life. *Cf.* bane, toxicant, toxic substance, toxin, venom. **acrid poison**. a poison that can produce a damaging local irritation or inflammation as well as systemic effects (e.g., mineral acids, strong bases, caustic alkaloids, phosphates). *Cf.* corrosive poison. **acronarcotic poison**. any poison (chiefly plant poisons such as belladonna, stramonium) that sometimes causes irritation and sometimes narcoticism, or both concurrently. **acrosedative poison**. any poison (usually a plant poison such as aconite) that sometimes causes irritation and sometimes sedation, or both concurrently. **anticoagulant poison**. an anticoagulant used generally to control small mammal pests, especially rodents of the general *Mus* and *Rattus*. They are the most widely used agents for this purpose. See coumarin, warfarin. **arrow poison**. **1:** any of a number of toxic substances, usually alkaloids, prepared from various plant and/or animal extracts or mixtures by primitive peoples in various parts of the world and used to prepare arrows with poison coated tips. Examples are abric acid, batracin, curare, tanghin, ukambin. See also antiar, batrachotoxin, calotropin, strophanthin, *Cerbera tanghin*, *Dendrobates*, *Derris*, *Euphorbia lactea*, *Phyllobates*, onobaio. **2:** a common name for curare. **African arrow poison**. any poisonous preparation, e.g., calotropin, used by natives of Africa to coat the tip of arrowheads for use in killing game or humans. See also *Acocanthera*. **autonomic poison**. **1:** a poisonous autonomic agent. **2:** a poison that acts entirely or in part on the autonomic system. Such poisons may produce effects such as miosis, bronchoconstriction, salivation, lacrimation, respiratory failure, cardiovascular collapse, and death. **barbasco poison**. a substance prepared from barbasco that is sometimes used by the natives of eastern Peru to paralyze piranhas before catching them for food. See also *Jacquinia*. **beaver poison**. *Cicuta maculata*. **biogenic poison**. any poison produced metabolically by a living organism. See toxin. **black poison**. *Metopium brownei*. **cardiac poison**. any poison that damages the heart or causes it to malfunction (e.g., squill, digitoxin). **catalytic poison**. catalyst poison. **catalyst poison** (catalytic poison). a poison that prevents or interferes with adsorption of the reactants of the chemical reaction that would otherwise by catalyzed. **cellular poison**. a substance that damages or destroys living cells. **chemical poison**. a substance that inhibits or impairs the activity of another substance or one that inhibits or disrupts the progress of a chemical reaction or process. **chronic poison**. a poison that produces toxic effects following long-continued, or numerous brief exposures. There is very little trustworthy information on the toxicity of such poisons. **ciguatera poison**. a mixture of toxins that includes ciguatoxin and sometimes maitotoxin, scaritoxin, and perhaps others. Of these, ciguatoxin is the principle toxin that causes ciguatera. **clam mussel poison**. paralytic shellfish poison. **clam poison**. paralytic shellfish poison. **class A poison**. a term applied by the U.S. Department of Transportation to extremely hazardous gas or liquid toxicants that are dangerous to life, even in very small amounts, if mixed with the atmosphere (e.g., hydrocyanic acid, nitrogen peroxide, phosgene). **class B poison**. a term applied by the U.S. Department of Transportation to liquid, solid, paste, or semisolid substances, other than class A poisons, that are established as toxic to humans and considered hazardous to human health during transport. **CNS (central nervous system) poison**. a poison that affects the CNS. **cone shell poison**. any poison produced by venomous marine snails of the genus *Conus*. These poisons cause temporary paralysis of the limbs and prolonged difficulty in breathing at the very least. A numbness in the lips progresses to include the whole body and is followed by dizziness, tightness in the throat, and pain on breathing; the pulse is weak and rapid. Death due to respiratory arrest may intervene in 1-8 hours following the initial symptoms. If the victim survives for 10 hours, the prognosis is good. **contact poison**. any poison that acts on contact with the skin or outer surface of a living organism and does not require another exposure route to have its full effect. See also insecticide (contact insecticide). **convulsant poison**. a toxicant such as camphor, strychnine, and picrotoxin that acts directly on the CNS with resulting convulsions, increased respiration, cardiac edema, vomiting and death. See convulsant. **corrosive poison**. any extremely irritant, caustic, or corrosive substance (e.g., mineral acids and strong alkalis) that destroys tissues on contact, when sufficiently dilute, they are referred to as irritants. *Cf.* acrid poison. **cow-poison**. *Amianthium muscitoxicum*. **crow poison**. See *Zigadenus densus*. **cumulative poison**. a toxicant that is not normally eliminated from the body or is so slowly eliminated that continued or repeated exposure to nontoxic amounts or concentrations may result in a build-up to toxic levels. In some cases, there may be cumulative subclinical effects that eventually result in damage, even though the poison is normally eliminated. **cutaneous poison**. any chemical that affects the skin whether by irritation, erosion, defatting, or allergic reactions. **dart poison**. any poisonous preparation (e.g., antiar) used by certain natives (e.g., of Africa, South America, the southwest Pacific) to coat the tip of darts for use in blowguns to kill game or humans. **depressant poison**. a poison (e.g., carbon tetrachloride, chloral hydrate, chloroform) that depresses CNS function. The term may also be applied to poisons that depress cardiac or respiratory function (e.g., aconite). **dinoflagellate poison**. paralytic shellfish poison. **ecosystem poison**. any toxicant such as certain herbicides and pesticides (e.g., chlorinated hydrocarbons) that can damage or impair ecosystems or key ecosystem components. Such substances often affect a broad spectrum of organisms. **environmental poison**. any toxicant, usually of anthropogenic origin, that becomes dispersed in the environment in quantities at least large enough and dispersed enough to poison or burden large or significant components of the environment such as an air shed, watershed, or one or more water bodies (e.g., an aquifer, a lake, or a stream). Some environmental poisons are global in distribution. An environmental poison is a pollutant, but not all pollutants are environmental poisons. Many environmental poisons threaten humans by entering groundwater or surface water bodies, or by accumulating in food webs that include *Homo sapiens*. **enzyme poison**. a catalyst poison in which the catalyst is an enzyme. **fatigue poison**. fatigue toxin. **fish poison**. **1:** any poison used to kill fish (e.g., certain saponins). **2:** ichthyotoxicon (fish poison). **3:** tetrodotoxin (fish poison). **fugu poison**. tetrodotoxin. **gonyaulax poison**. paralytic shellfish poison. **hemolytic poison**. a poison that lyses erythrocytes; one that causes anemia. See toxanemia, toxic anemia. **hemotropic poison**. a poison that has a specific affinity for erythrocytes. **household poison**. any chemical poison commonly found in the home. These typically include any of an array of medicines, cleaning agents, disinfectants, polishes, bleaches, volatile organic solvents, pesticides, toiletries, and cosmetics. In most instances, a person's greatest exposure to toxic substances is in the home. Among pesticides that find their way into the home, sometimes via household products, probably the most hazardous are aldicarb, aldrin, bufencarb, carbofuran, carbophenothion, chlorpyrifos, DDVP, demeton, dichlorvos, dicrotophos, dieldrin, dinitrocresol, dioxation, disulfoton, DNOC, endrin, EPN, fensulfothion, fonophos, methamidophos, methylparathion, mevinphos, mexacarbate, monocrotophos, nicotine, paraquat, parathion, phorate, phosalone, phosphamidon, pentachlorophenol, propoxur, schradan, and TEPP. See also household poisoning. **irritant poison**. See irritant. See also corrosive poison. **microbial poison**. a poison produced by a microorganism. **military poison** (war poison; warfare poison). a toxicant whose sole or primary application is in military combat (e.g., mustard gas, soman). **mineral poison**. a poison chemically derived from minerals. These include inorganic acids, bases, salts, and various compounds or solutions that are generally caustic or corrosive. **mitotic poison**. a poison that interrupts or disrupts cell division in plants or animals. **mouse poison** (mouse killer). any of a large variety of poisons used to poison *Mus musculus* (the house mouse). Mouse and rat poisons are among the most harmful pesticides available for use in the home.

They may contain arsenic, strychnine, or phosphorus, all of which are rapid acting and lethal when ingested. See also *Nerium oleander*. **muscle poison**. a poison (e.g., barium salts, benzene, digitalis, potassium salts) that interferes with the normal functioning of muscle; some are deadly. **mussel poison**. paralytic shellfish poison. **mutagenic poison**. any poison that causes damaging mutagenic effects. **narcotic poison**. any poison (e.g., hyoscyamus, opium) that causes stupor or delirium. **nerve poison**. any toxicant that acts directly on nervous tissue. **organophosphorus military poison**. any of a number of organophosphates used in warfare. They are powerful inhibitors of acetylcholinesterase. **paralytic shellfish poison** (PSP; gonyaulax poison; gonyaulax toxin; dinoflagellate poison; mussel poison; mytilotoxin). an extremely toxic complex of an unknown and variable number of toxins found in some molluscs, arthropods, and echinoderms that have ingested toxic protistans, most often *Gonyaulax* spp. Included in this complex are saxitoxin, gonyautotoxin II (GTX_2), gonyautotoxin III (GTX_3), and neosaxitoxin. The most common transvectors of PSP are the pelecypod genera *Mya*, *Mytilus*, *Modiolus*, *Protothaca*, *Spisula*, and *Saxidomus*. Common types of shellfish poisoning are allergic, gastrointestinal, and paralytic. See also red tide, batrachotoxin, saxitoxin. **pesticidal poison**. a poison whose toxic properties are commercially exploited to kill pests. **plant poison**. any poison of plant origin. **protoplasmic poison**. a poison that damages or destroys cells with resultant inflammation of the involved tissue. Such poisons (e.g., arsenic, cyanide, formaldehyde, phenol) alter cell membranes and inhibit enzyme activity. **puffer poison**. tetrodotoxin. **rat poison**. any of a large variety of poisons and special concoctions (e.g., warfarin) that are used to poison the domestic or Norway rat, *Rattus norvegicus*. They may contain arsenic, strychnine, or phosphorus, all of which are rapid-acting and lethal when ingested. Rat and mouse poisons are among the most dangerous pesticides available for home use. **reproductive poison**. any poison that damages the reproductive system or otherwise impairs reproductive processes. **sedative poison**. a poison (e.g., hydrocyanic acid, hydrogen sulfide, potassium cyanide) that directly depresses the vital centers of the CNS. **spindle poison** (antimitotic agent). a poison (e.g., cholchicine, nitrous oxide, nonphysiological temperature, taxol, vinblastine, vincristine), usually chemical, that disrupts the structure and/or function of the mitotic or meiotic spindle. **squill-toad poison.** bufanolide. **stomach poison**. a poison (usually a pesticide) that is active only via the oral route (e.g., stomach insecticide). **systemic poison**. **1a:** a poison that produces generalized or whole-body effects. **1b:** a poison that produces toxic effects in major parts of an organism that are remote from the site of contact between the organism and the toxicant. *Cf.* toxicosis. **teratogenic poison**. any poison that produces teratogenic effects. **toot poison**. a poison isolated from the seeds of *Coriaria* spp., especially *C. ruscifolia* and *C. sarmentosa* of New Zealand. **vascular poison**. a poison that acts on vessels of the blood vascular system. **war poison**. military poison. **warfare poison**. military poison.

POISON (with skull and crossbones). a signal word which, when displayed on a product label, indicates that ingestion of a tiny pinch can kill an adult human. The signal word "DANGER" used in this context is equivalent.

poison, mitotic. See mitotic poison.

poison arrow frog. See Dendrobatidae.

poison ash. *Rhus vernix*.

poison bean. See *Sesbania*.

poison bulb. *Crinum asiaticum*.

poison camas, poison camass. See *Zigadenus nuttallii*.

poison camass. *Zigadenus nuttallii*.

poison center. See Poison Control Center, Poison Information Center.

Poison Control Center. a facility that meets the staffing and equipment standards of the American Association of Poison Control Centers and can provide information on, or treatment of, victims of poisoning. Most are affiliated with a large hospital, medical school, or school of pharmacy. They are available by telephone for emergency instructions, usually on a 24-hr basis. *Cf.* Poison Information Center.

poison creeper. *Rhus toxicodendron.*

poison darnel. *Lolium temulentum.*

poison dart frog. See Dendrobatidae.

poison dogwood. *Rhus vernix.*

poison elder. *Rhus vernix.*

poison flag. *Iris versicolor.*

poison fool's parsley. *Conium maculatum.*

poison gland. See gland.

poison hemlock. *Conium maculatum.* See also hemlock.

Poison Information Center. a facility comprised solely of a reference library and no treatment capability. See also Poison Control Center.

poison ivy. **1:** the common name of *Rhus radicans*. **2:** correctly applied only to *R. radicans*, but often used interchangeably for *R. diversiloba* (Pacific poison oak), *R. radicans* (poison ivy), and *R. toxicodendron* (poison oak). See *Rhus*. **3:** *Rhus* dermatitis (poison ivy dermatitis).

poison ivy dermatitis. *Rhus* dermatitis.

poison lady. See *Latrodectus*.

poison laurel. *Kalmia latifolia.*

poison nut. nux vomica.

poison oak. **1:** the common name of *Rhus toxicodendron*. **2:** often used interchangeably for *R. diversiloba* (Pacific poison oak), *R. radicans* (poison ivy) and *R. toxicodendron* (poison oak). See *Rhus*. **3:** *Rhus* dermatitis (poison oak dermatitis).

poison oak dermatitis. *Rhus* dermatitis.

poison parsley. *Conium maculatum.*

poison paxillus. *Paxillus involutus.*

poison pie. *Hebeloma crustuliniforme.*

poison sego. *Zigadenus venenosus.*

poison suckleya. *Suckleya suckleyana.*

poison sumac. **1:** the common name of *Rhus vernix*. **2:** *Rhus* dermatitis (poison sumac dermatitis).

poison sumac dermatitis. *Rhus* dermatitis.

poison tobacco. *Hyoscyamus niger.*

poison tree. *Metopium taxiferum.*

poison vine. *Rhus radicans.*

poisoning. **1a:** the administration of, or the act of administering, a poison. **1b:** accidental or intentional contact of an organism with a poison. **2:** to make a thing poisonous by adding poison. **3:** intoxication; toxication; toxicosis; the state or condition of a living organism resulting from the intake of, introduction to, or exposure to, a poisonous substance or other toxic agent. **acute poisoning**. **1a:** poisoning due to a single dose of a poison or to an exposure of short duration, usually of a relatively large amount or at a relatively high concentration. **1b:** poisoning that runs a relatively short, severe course, with or without recovery. **chronic poisoning**. **1a:** poisoning due to repeated or continuous exposure to a poison or poisons for a protracted period of time. **1b:** poisoning of long duration and (usually) moderate severity. **2:** the persistence of the signs or symptoms of poisoning for a protracted period; a condition of long duration. **subacute poisoning**. **1:** poisoning at a dosage or concentration between that considered chronic and that considered acute for a particular poison. **2:** poisoning of moderate duration or severity. **subclinical poisoning**. poisoning in which the symptoms or manifestations are of slight degree or in early stages such as to remain below that detectable, diagnosable, or treatable in a clinical setting.

poisonous. **1a:** denoting, containing, having the properties of, able to act in whole or in part as a poison. **1b:** able to convey a poison. Many plants are poisonous. Poisonous species of animals occur widely throughout the animal kingdom (Animalia) and are represented among all classes of vertebrates. **2a:** descriptive of any illness that is caused by poisonous substances. **2b:** mephitic, toxic, toxical, toxicant, toxiferous, venomous.

poisonous plant. See plant (poisonous plant).

poisonous snake. a term very much subject to error in usage. **1:** any snake that contains poison in any organs or tissues; not necessarily a venomous snake. All venomous snakes are, *de facto* poisonous. **2:** a venomous snake. This usage of the term "poisonous snake" can be confusing and should normally be avoided.

poisonous trich. *Tricholoma pardinum*.

poisonvetch. See *Astragalus*.

poisonwood. *Metopium taxiferum*.

poke. *Phytolacca americana*.

pokeberry. *Phytolacca americana*.

pokeroot. the dried root of *Veratrum viride*, the most poisonous part of the plant. It finds use as an antihypertensive agent.

pokeroot poisoning. poisoning due to ingestion of pokeroot. Symptoms may include nausea, vomiting, abdominal pain, drowsiness, vertigo, convulsions, depressed heart action, respiratory paralysis, coma, and sometimes death.

pokeweed. *Phytolacca americana*. See also Phytolaccaceae.

pokeweed family. Phytolaccaceae

polar. descriptive of a molecule with a strong net electric charge or polarity. Such is soluble in water but not in lipids (fats).

polar bear. See *Thalarctos*.

polar body. a nonfunctioning daughter cell having little cytoplasm, which is formed during oogenesis.

polar translocation (polar transport). the movement of materials through tissues or cells in one direction only

polar transport. polar translocation.

polenosis. hay fever.

Polillo pit viper. *Trimeresurus flavomaculatus*.

polioclastic. neurotropic.

polioencephalitis. an inflammation of the gray substance of the brain.

polish. See floor polish, furniture polish.

Polistes. a large cosmopolitan genus of wasps (Family Polistidae). They secrete an alkaline venom containing neurotoxic and hemolytic agents. The venom is neutralized by lemon juice or vinegar. The sting produces symptoms similar to those of bees (See, for example, *Apis*). See also Polistidae.

polistid. a wasp of the family Polistidae.

Polistidae (paper wasps). a family of hymenopterans. They are characteristically large slender wasps similar to vespids. See *Polistes*.

pollen. microspores produced in vast numbers in the anthers of flowering plants. They are essential to sexual reproduction. They form a fine dust in the atmosphere and are considered a natural or background pollutant. They are the cause of pollinoses.

pollinosis (pollenosis). an allergic reaction to inhaled airborne pollen resulting in the seasonal type of hay fever.

pollutant. any substance or form of energy, of natural or anthropogenic origin, that is present in the environment in greater than natural concentrations or amounts due to human activity, and which has a net detrimental or undesirable effect on humans, the environment, or upon something of value to humans in the environment. Most are a matter of concern because of their toxicity. **air pollutant**. any pollutant in any form or chemical phase that occurs in air. **conventional pollutant**. any pollutant listed under statute and which is understood well by scientists. **photochemical air pollutant**. any of a large number of trace compounds (e.g., acrolein, formaldehyde, ozone, peroxyacetyl nitrate, higher oxides of nitrogen, aldehydes and ketones, and a number of inorganic and organic acids) that are a product of the action of sunlight on certain primary air pollutants (e.g., NO_X, hydrocarbons). See photochemical oxidant. **primary air pollutant. 1a:** a pollutant emitted into the atmosphere from an identifiable source. **1b:** the form or species of a chemical pollutant that is released into the environment from a polluting source. **secondary air pollutant. 1a:** a pollutant formed by chemical reaction in the atmosphere. **1b:** a transformation product of a primary pollutant following its release into the environment. **toxic pollutant**. a pollutant, natural or anthropogenic, that causes disease or toxic effects in organisms that ingest or absorb them. **water pollutant**. any pollutant that occurs in fresh water or marine water bodies (lakes, streams, reservoirs, oceans) regardless of the route or mode of entry.

polon thelissa. *Agkistrodon hyphnale*.

polonium (symbol, Po)[Z = 84; most stable isotope = ^{209}Po, $T_{1/2}$ = 103 yr]. a hazardous, radioactive, metallic element that occurs in minute amounts in uranium ores. More than 30 radioactive isotopes are known. Inhalation of insoluble polonium salts can damage lungs; ingestion of soluble compounds can cause systemic damage, especially to kidneys and spleen. It is an alpha emitter.

poly-. prefix meaning many.

polyalcoholism. intoxication or poisoning due to the combined effects of a variety of alcohols.

polyamine. a compound that that contains two or more amino groups. **alkyl polyamine**. a polyamine that contains two or more amino groups bonded to alkane moieties. The more familiar alkyl polyamines are used commercially, for example, as solvents, emulsifiers, epoxy resin hardeners, stabilizers, starting materials for dye synthesis, and as chelating agents. Partly because they are strongly alkaline, many of these compounds, especially the smaller, are irritants to the skin, eyes, and respiratory tract.

polyarthritis. the inflammation of many joints.

polybasic acid. an acid than contains more than two ionizable atoms of hydrogen per molecule.

polybrominated biphenyl (PBB; polybromobiphenyl). any of a class of brominated hydrocarbon compounds used as fire retardants and components of heat resistant plastics. They are chemically related to polychlorinated biphenyls. All are immunotoxic in humans and laboratory animals. They can cause atrophy of lymphoid tissue, reduced size of the spleen, lymphopenia, and suppression of humoral immunity in several species of laboratory animals. In laboratory rats and mice, for example, PBBs produce thymus atrophy and consequent impairment of thymic-dependent immune functions. Accidental human exposures have also altered immune functions, and have produced acneform skin lesions, pigmentation of skin and nails, and liver damage. Massive livestock poisonings occurred in Michigan in 1973 due to the accidental addition of about two tons of PBB flame retardant to livestock feed during its formulation. Some 500 cattle farms and 150 dairy farms were placed under quarantine. As a precautionary measure, 30,000 cattle and swine were killed. Signs exhibited by cattle exposed to the contaminated feedstock usually included salivation, lacrimation, diarrhea, inappetence, dehydration, weight loss, alopecia, abnormal growth of hooves, decreased milk production, and teratologies; gravid cows aborted. Autopsy revealed lesions of the kidney, liver, gallbladder and thymus. Failing and moribund animals suffered extensive subcutaneous edema and hemorrhage.

polybrominated biphenyl mixture (Firemaster FF-1). this mixture has been shown to be carcinogenic in some laboratory animals.

polybromobiphenyl. polybrominated biphenyl.

polycardia. tachycardia.

Polychaeta. a class (formerly an order of the annelid class Chaetopoda) of mostly burrowing or tube-dwelling annelid worms having separate sexes; well-defined segmentation, with paired setate parapodia ("setae") on most segments; the head region has a prostomium and peristomium and bears tentacles and palps. The larva is a trochophore, but some species reproduce by budding. The parapodia of some species form a stinging mechanism (e.g., *Eurythöe complanata, Hermodice carunculata*). Certain other species have hard chitinous biting jaws (e.g., *Glycera dibranchiata, Eunice aphroditois*). See Annelida.

polychaete (polychete; polychaete worm). **1:** polychaetous; of or pertaining to annelid worms of the class Polychaeta. **2:** an annelid worm of the class Polychaeta.

polychaete worm. See polychaete.

polychaetous. polychaete (def. 1).

polychete. polychaete.

polychlorinated biphenyl (chlorobiphenyl; chlorinated diphenyl; chlorinated diphenylene; polychlorobiphenyl; PCB; Aroclor; Clophen; Fenclor; Phenoclor; Pyralene; Santotherm). any of a series of about 200 nonflammable, highly persistent organic liquids derived from biphenyl (a compound with two linked benzene rings), $C_6H_5C_6H_5$, in which some or all of the hydrogen atoms attached to the benzene ring have been replaced by chlorine atoms. Differences in physical and toxicological properties of the various isomers are due chiefly to the degree of chlorination. Thus technical mixtures comprised of various isomers and compounds range from mobile oily liquids to white crystalline solids or hard resins. Acute toxicity is generally low, but PCBs are of concern because of their persistence in the environment and in body fat. They are very stable, have become widespread in the environment, and considerable concern has been expressed over their effects on human health, wildlife, and the environment. Effects on animals due to ingestion of various PCBs include hepatic neoplasia, hepatic porphyria, alterations in mixed function oxidase systems, liver hypertrophy, increased bile flow, reproductive dysfunction (in birds), and reproductive failure (in mink, probably in fish). PCBs cause eggshell thinning in birds and have thereby contributed to the extinction or near extinction of certain birds. Effects on humans include gastrointestinal distress, chloracne, excessive discharge from the eyes, edematous eyelids, pigmentation of nails and skin, and characteristic hair follicles, and birth defects. PCBs are immunotoxic, are known tumorigens and are carcinogenic to some laboratory animals and are suspected carcinogens. Most commercially produced PCBs (Aroclors) are comprised of several isomers and sometimes contain polychlorinated terphenyls also. PCB are used in the manufacture of certain polymers (mostly plastics), as insulating materials in transformers and capacitors and as a lubricant in gas pipeline systems. They have also been used in a variety of other ways (e.g., as hydraulic fluids, plasticizers, and flame retardants). Further sale or new use was banned in the United States in 1979. See Aroclor.

polychlorinated dibenzodioxin (PCDD). any of a class of immunotoxic compounds that have the same basic structure as TCDD, but different numbers and arrangements of chlorine atoms on the ring structure. They exhibit varying degrees of toxicity.

polychlorobiphenyl. polychlorinated biphenyl.

polychlorocamphene. toxaphene.

polychrome. esculin.

polycondensed hexahydrotricarboxytriphenylmethane. trigentisic acid.

polycyclic. **1:** having many cycles, circuits, circles, whorls, rounds, or rings, etc. **2:** in organic chemistry, having two or more ring structures in a molecule. **3:** in botany, having **a:** many whorls (of flowers); or **b:** a vascular system comprised of several concentric cylinders.

polycyclic musk. acetylethyltetramethyltetralin.

polycythemia. a condition characterized by a higher than normal erythrocyte mass in circulating blood, or a large mass relative to plasma volume. Tissue hypoxia, induced directly or indirectly by drugs or toxicants, is one of a number of causes of polycythemia. See cobalt.

polydipsia. **1:** excessive thirst. Such is characteristic of a number of conditions (e.g., diabetes mellitus, diabetes insipidus). The condition may also result from any number of forms of renal dysfunction, including that caused by empoisonment by various xenobiotics. **2:** informally applied to alcoholism.

polyene. an alkene with more than two double bonds per molecule.

polyethylene. a thermosetting, high-temperature-resistant, flammable, white solid. Burning releases toxic fumes. It is a suspected carcinogen. It is used to make food containers and as a packaging film for food products; high density polyethylene is used to make plastic bottles. Other products made from polyethylene include chewing gum, coffee stirrers, drinking glasses, carpet fibers safety covers for electrical outlets, heat-sealed plastic packaging, kitchenware, paper coating, plastic bags, plastic pails and garbage cans, swizzle sticks, and toys. **high-density polyethylene**. a polymer used to make plastic bottles and other products. It produces toxic fumes on combustion.

polyethyleneglycols mono(nonylphenyl) ether. nonoxynol.

Polygonaceae (buckwheat family). a family of widely distributed shrubs, trees, and erect or climbing herbs. A number of species are cultivated as ornamentals. See *Fagopyrum sagittatum*, *Polygonum*.

Polygonum (smartweeds). a large, widely distributed (rare in the tropics) genus of annual or perennial herbs and small shrubs (Family Polygonaceae). At least some species are photosensitizing. Toxic concentrations of nitrate have been reported for some.

polyhalogenated hydrocarbon. See hydrocarbon.

polyhidrosis. hyperhidrosis.

polymer. **1a:** a large molecule made of at least five identical, chemically bound units (monomers). **1b:** a macromolecule, natural or synthetic, comprised of a chain of similar, covalently bonded molecules.

polymer of *N,N,N′,N′*-tetramethylhexamethylenediamine and trimethylene bromide. hexadimethrine bromide.

polymerase. an enzyme that catalyzes the synthesis of DNA or RNA. An important mode of action of many xenobiotics is the inhibition of polymerases.

polymorphic. exhibiting polymorphism; occurring in more than one form; said of a group or population comprised of two or more recognizably different sorts (morphs or phases) of individuals.

polymorphism. **1:** the capacity to exist in more than one form (morph or phase). Said of a group or population. **2:** the state or condition of population or species comprised of two or more recognizably different sorts (morphs or phases) of individuals occurring within a single population. Polymorphism can extend to differing sensitivities or responses to a given toxicant.

polyneuritic. pertaining to or afflicted by polyneuritis.

polyneuritic psychosis. Korsakoff's syndrome.

polyneuritis (multiple neuritis; disseminated neuritis). the concurrent inflammation of a large number of nerves as in lead palsy. It is marked by paralysis, pain, and wasting. See also polyneuropathy (nutritional polyneuropathy). **Jamaica ginger polyneuritis.** Jamaica ginger paralysis. See also neuropathy (alcoholic neuropathy). **metabolic polyneuritis**. that resulting from metabolic malfunction, often as a result of nutritional deficiency (especially of thiamine), toxemias of pregnancy, and sometimes xenobiotics. **toxic polyneuritis**. that resulting from toxicants such as heavy metals, ethanol, carbon monoxide, and various organic compounds.

polyneuritis potatorum. alcoholic neuropathy. See neuropathy.

polyneuropathy (multiple neuritis; polyneuritis). **1:** concurrent neuropathy affecting a number of peripheral nerves, especially those of a noninflammatory nature. **2:** multiple disorders of the nervous system. **ascending polyneuropathy**. a polyneuropathy that progresses from the lower extremities. **buckthorn polyneuropathy**. an ascending polyneuropathy due to ingestion of the fruit of *Karwinskia humboldtiana* (buckthorn). **distal polyneuropathy**. peripheral polyneuropathy. **nutritional polyneuropathy**. that due to thiamine deficiency, e.g., as in chronic alcoholism. **peripheral polyneuropathy** (distal polyneuropathy). polyneuropathy that affects only the peripheral nervous system.

polyoxyethylene(*n*)nonylphenyl ether. nonoxynol.

polypeptide. a peptide that contains a large number of amino acid residues and a molecular weight that is usually greater than 5000. Polypeptides are the building blocks of proteins. The polypeptides of snake venoms are best described as nonenzymatic, low-molecular-weight proteins. They are the most toxic components of some snake venoms. More than 60 pharmacologically active polypeptides have been isolated from snake venoms. See crotoxin, crotactin, crotamine, erabutoxins, venom.

polypeptides of snake venoms. nonenzymatic polypeptides that are best described as low-molecular weight proteins. More than 60 pharmacologically active polypeptides have been isolated from snake venoms. They are the most toxic components of some snake venoms and are often improper designated as neurotoxins. See also crotoxin, crotactin, crotamine, erabutoxins, venom.

polypharmacy. **1:** the practice of concurrent administration of multiple drugs. **2:** the excessive administration of medication. In both cases (defs. 1, 2), the risk of drug interactions, immune reactions, poisoning, and other unintended, possibly adverse effects is enhanced. In addition the effects from a therapeutic perspective may differ from that intended.

polypnea. tachypnea.

poly(*N,N,N',N'*-tetramethyl-*N*-trimethylenehexamethylenediammonium dibromide). hexadimethrine bromide.

polysialia. ptyalism.

polystyrene. a common plastic made from styrene monomer, *q.v.*

polysulfide. See sulfide.

polyurethane. a polymer. The cured polymer appears to be biologically inert, but may release toluene-2,4-diisocyanate, *q.v.*, which is a powerful irritant to the eyes, skin, and respiratory tract. It can produce severe pulmonary effects and sensitization. Frequent contact with polyurethanes can cause bronchitis, coughing, and skin and eye problems. Studies conducted by the World Health Organization (WHO) evidenced a high frequency of cancer in mice given daily insertions of a polyurethane sponge, the same material from which contraceptive sponges are made. Polyurethane foam is used to make cushions, mattresses, and pillows.

polyuria. **1:** abnormally frequent discharge of urine or an abnormally sharp increase in urination; excessive secretion and excretion of (a usually colorless) urine, with or without abnormal constituents. The condition may be due to any of a number of causes including chronic nephritis, diabetes insipidus, diabetes mellitus, following edematous states, and excessive intake of fluid, especially diuretics. **2:** an abnormally sharp increase in the amount of urine excreted.

Polyvalent Crotalid Antivenin. See antivenin.

polyvinyl chloride (PVC). any of a number of solid, non-flammable, water-insoluble, vinyl chloride polymers of low chemical reactivity. They have an extensive and varied usage as a rubber substitute in the manufacture, e.g., of electrical insulation, raincoats, phonograph records, and floor coverings. Because PVC releases vinyl chloride, especially when new and because it is environmentally persistent and nearly indestructible, it is one of the most dangerous of plastics. At high temperature or on burning, PVC may release toxic gases such as hydrochloric acid and ethyl chloride. See vinyl chloride. Long-term production workers risk development of angiosarcoma. See vinyl chloride.

polyvinylpyrrolidone (PVP; polyvinylpyrrolidone plastic). a toxic, white, amorphous powder or aqueous solution, $(C_6H_9NO)_n$, that is soluble in water and organic solvents. When heated to decomposition, it releases toxic fumes (NO_x). PVP has a variety of applications including its use in pharmaceuticals; as a complexing agent; in the detoxification of many chemicals (e.g., iodine, phenol, dyes, certain drugs); in hair sprays, shampoos, lipsticks, lotions, hand creams; and detergents. It is a suspected human carcinogen and is responsible for a pulmonary disease (a form of thesaurosis) that affects some users of PVP-containing products such as hairspray, styling mousse, shampoo, and lipstick. The disease is marked by enlarged lymph nodes, lung masses, and changes in blood cells. It is reversible if the offending product is avoided.

pome. a fleshy fruit such as an apple or pear with a fleshy outer portion and a papery core with seeds.

pomnica. *Acanthropis antarcticus*.

ponasterone. any of four polyhydroxylated steroids isolated from *Podocarpus nakaii* (Family

Podocarpaceae). They exert potent ecdysone activity on houseflies, silkworms, and other insects. Ingested plant material containing as little as 1 μg/kg ponasterone can be lethal to caterpillars of the cecropia moth (*Samia cecropia*). See also ponasterone A, phytoecdysone.

ponasterone A ((2β,3β,5β,22R)-2,3,14,20,22-pentahydroxycholest-7-en-6-one; 25-deoxyecdysterone). the first of four ponasterones (A, B, C, D) to be isolated. It is highly toxic to caterpillars of the cecropia moth (*Samia cecropia*). See ponasterone.

ponderosa pine. *Pinus ponderosa*.

pooled. composite.

pooled sample. See composite.

poor Annie. *Veratrum viride*.

poor-man's-weatherglass. *Anagallis arvensis*.

popcorn flower. *Plagiobothrys* spp.

Pope's pit viper. *Trimeresurus popeorum*.

Pope's tree viper. *Trimeresurus popeorum*.

popolo. *Solanum sodomeum*.

Popperian method. a partially deductive scientific method in which theories or hypotheses are formulated from which simple, testable hypotheses or predictions are inferred. A theory or hypothesis may thus be falsified or sustained but not proven.

poppy. See *Papaver*.

poppy family. Papaveraceae.

population. **1:** a group of interbreeding organisms that occupy a particular space. Populations are of particular interest to toxicologists because there are emergent properties at this level or organization that are not due simply to nature or number of the individuals that make up a population. These include properties such as immigration rate, emigration rate, natality, and mortality, all of which may be sensitive to a particular toxicant. **2:** all individuals of one or more species within a prescribed area. **3:** any set of items or individuals that come under investigation. **sample population**. a subset of a population of interest that is manageable for the purpose intended (e.g., a toxicity test) and is thought to be representative of the population of interest. Whereas a sample may be representative in some sense, it is questionable that it can be representative of any large population as a whole. See also sample.

porcupinefish. a common name of fish of the family Diondontidae.

porgie. **1:** *Pagellus erythrinus*. **2:** porgy. See also Sparidae.

porgy. common name of certain deep-bodied food fishes (e.g., *Pagellus erythrinus* and *Pagrus pagrus*) mainly of the Atlantic Ocean and Mediterranean Sea. See *Pagellus*.

Porichthys. a genus of toadfish (Family Batrachoididae).

Porifera (sponges). an animal phylum of some 5000 species of animals. Most are marine. The skeleton may consist of calcareous spicules, silicon dioxide spicules, an organic material (spongin), or spongin and silicon dioxide spicules. There is no body cavity and no organs, just a differentiation of cells into tissues. Some species release toxins into the medium that can be lethal to other animals. Many have sharp spicules that can simultaneously injure human skin and expose it to poison.

porphobilinogen synthase (aminolevulinic acid dehydratase; ALAD). an enzyme that catalyzes the production of porphobilinogen (protoporphyrin precursor). See protoporhyrin (free erythrocyte protoporphyrin).

porphyria. any of a group of predominantly heritable disorders of porphyrin metabolism which are defined by the an abnormal increase in the production and excretion of porphyrins or their precursors. Porphyria may also be environmental in origin (e.g., through the agency of drugs or other xenobiotics). Authors generally consider two main types of porphyria (erythropoietic and hepatic), although four types have been described. While causes and effects differ, the various porphyrias usually share certain symptoms: abdominal pain, photosensitivity, and neuropathy. **acute intermittent porphyria**.

a rare hepatic porphyria characterized by episodes of porphyrinuria, acute abdominal pain, and neuropathies. It is also marked by a deficiency of uroporphyrinogen synthetase with an increase in the production of δ-aminolevulinic acid synthetase, and consequent overproduction of δ-aminolevulinic acid and porphobilinogen. Urinary excretion of these substances is greatly increased. Common symptoms are intermittent hypertension, abdominal colic, psychosis, and neuropathy. While not the proximate cause of this form of porphyria, episodes can be precipitated by barbiturates, sulfonamides, and certain other drugs. **congenital erythropoietic porphyria**. porphyria in which large amounts of porphyrins are produced and excreted by the erythroid cells of bone marrow. The condition is often accompanied by hemolytic anemia and persistent cutaneous photosensitivity. **erythropoietic porphyria**. **1:** an obsolete synonym for congenital erythropoietic porphyria. **2:** a usually mild heritable porphyria with increased production and increased fecal excretion of protoporphyrin III. **porphyria cutanea tarda** (symptomatic porphyria). porphyria with onset of symptoms between the ages of 10 and 30. The skin is photosensitive. Symptoms include blistering of the fragile skin, damage to the eyes, excessive hair growth, wasting of skeletal muscles, sometimes anorexia, and weight loss. Some symptoms may possibly be due to impurities (e.g., polychlorinated dibenzodioxins) in products used by the victim. **porphyria hepatica**. that due to disturbances of liver metabolism as a result of hepatitis or poisoning (e.g., by heavy metals, benzene hexachloride). **symptomatic porphyria**. porphyria cutanea tarda.

porphyrin. any of a group of iron- or magnesium-free substituted pyrroles that occur in protoplasm. They are widely distributed in nature (e.g., bile pigments, cytochromes, heme) and are essential moieties in animal and plant respiratory pigments. Porphyrins consist of four substituted pyrroles joined in a ring (porphin) by methylene bridges. A metal (e.g., copper, iron, magnesium) is central to the molecule and is bound to the nitrogen atoms of the pyrrole rings. Heme, for example, is the prosthetic group of the toxicologically important hemoglobins and cytochromes.

porphyrinuria (porphyruria). the excretion of excessive amounts of porphyrins and related compounds in the urine. See also porphyria.

porphyruria. See porphyrinuria.

Port Darwin brownsnake. *Pseudechis australis*.

Port Darwin sea snake. *Hydrelaps darwiniensis*.

portal. a point of entry; gate; gateway; avenue. See especially portal of entry.

portal circulatory system. the network of veins that drain blood from the gastrointestinal tract; also called the portal-vein system. The smaller veins empty into the portal vein, which transports blood into the liver; blood exits via the hepatic vein.

portal cirrhosis. Laennec's cirrhosis (See cirrhosis).

portal of entry. **1:** the point or site of entry whereby a pathogen enters the body. **2:** sometimes used in reference to the site of entry of a xenobiotic into the body. Use of the terms "route of exposure", q.v., or "route of administration" are preferred in most cases.

portal vein. See portal circulatory system.

Portuguese man-o'-war. Portuguese man-of-war.

Portuguese man-of-war (Portuguese man-o'-war). *Physalia pelagica* and related species.

Portulaca oleracea (purslane; pusley). an annual, low-growing, herbaceous weed (Family Portulacaceae) of the eastern United States. It contains toxic concentrations of soluble oxalates.

positive (+) **1:** having a value greater than zero. **2:** affirmative; definite; not negative. **3:** of or pertaining to the presence, existence, or evocation of a response, reaction, condition, or entity. **false positive**. a result of a test or observation which appears to possess the attribute of concern, but in fact does not.

positive taxis or tropism. a tendency to move or grow toward the source of the stimulus.

post-. a prefix indicating location behind, or occurring after.

post mortem, post-mortem. after death; occurring or performed after death.

postembryonic. occurring after the embryonic stage.

posterior pituitary gland, posterior pituitary. posterior lobe of the pituitary gland. See pituitary gland.

postestrus (post). metestrus (See estrus cycle).

postganglionic axon (postganglionic fiber). in the autonomic nervous system, a nonmyelenated axon that conducts impulses away from the ganglion, not the myelinated axon arising in the brain that synapses with the ganglion. *Cf.* preganglionic axon.

postganglionic fiber. postganglionic axon.

postganglionic neuron. a nonmyelenated neuron that conducts impulses away from a ganglion.

posthumous. occurring after death.

postmortal. after death.

postmortally. occurring after death.

postmortem. **1:** of or pertaining to the period following death; after death; post-mortem. **2:** pertaining to or denoting that which occurs or is performed after death. **3:** pertaining to or used in a postmortem examination. **4:** post mortem; post-mortem; a postmortem examination (See autopsy). **5:** to perform a postmortem on (See autopsy, def. 2). **6:** an examination or evaluation following an event.

postmortem examination. autopsy.

postmortem lividity. livor mortis.

postnatal. after birth. *Cf.* antenatal.

postnatal period. the period occurring after birth. An ambiguous term, it is usually regarded as the early neonatal period of development, but in principle extends throughout the period of development, which in humans, for example, might easily be interpreted as the first 25 years of life, as the period during which development is completed. However defined, the adult years and senescence are usually excluded. In toxicology, this term is often applied to exposures or effects that may have occurred *in utero*, but in which at least some significant effects are not apparent until sometime following birth. As an example, effects on reproductive potential might not be manifested until late adolescent or early adult years of life.

postoestrus. postestrus.

postparalytic. subsequent to a paralytic attack.

postparturient hemoglobinemia. See hemoglobinuria.

postsynaptic membrane. a membrane that is part of a synapse and receives a neurotransmitter substance.

posture. the position of the body with respect to surrounding space (as in erect, prone, supine) and the gravitational field. Unusual postures or changes in posture are sometimes indicative of disease or poisoning. Furthermore, posture (as in many venomous snakes) can signal prospective attack. *Cf.* carriage. See, for example, trembles.

pot curare. See curare.

potable. drinkable; safe to drink; pertaining to water that is safe to drink; water that is free of harmful substances.

potable water. water that is safe for drinking and cooking.

potai pambu. *Lapemis curtus*.

Potamotrygon motoro (South American freshwater stingray). a venomous, extremely dangerous, freshwater stingray (Family Potamotrygonidae) found in the Amazon River as far south as Rio de Janeiro, Brazil, and in freshwater rivers of Paraguay.

Potamotrygonidae (freshwater stingrays; river rays). a family of extremely dangerous, venomous freshwater stingrays. They occur in freshwater rivers from Paraguay and the Amazon River south to Rio de Janeiro, Brazil. The sting is of the dasyatid type and the wound can be extremely painful. See *Potamotrygon motoro*, stingrays, dasyatid.

potassa. potassium hydroxide.

potassic. pertaining to or containing potassium.

potassiocupric. pertaining to or containing potassium and copper.

potassiomercuric. pertaining to or containing potassium and mercury.

potassium (symbol, K; kalium)[Z = 19; A_r = 39.09]. a soft, highly reactive, alkali metal that occurs abundantly in nature, always in combination. It reacts vigorously with oxygen, water, acids, and halogens. It must be kept under liquid containing no oxygen. Many of its salts are toxic; many are used medicinally. **potassium-40** (^{40}K). a naturally occurring radioactive isotope of potassium. It is a pure beta emitter with a half-life of 1.3 billion years and is the chief source of natural radioactivity of living tissue. **potassium-42** (^{42}K). an artificial radioactive isotope of potassium; it is a pure beta emitter with half-life of 12.47 hours.

potassium acid arsenate. potassium arsenate.

potassium acid fluoride. potassium bifluoride.

potassium acid oxalate. potassium binoxalate.

potassium antimonyltartrate. antimony potassium tartrate.

potassium arsenate (potassium acid arsenate; potassium dihydrogen arsenate; Macquer's salt). a toxic, colorless, water-soluble, ethanol-insoluble crystalline compound, KH_2AsO_4. It is used as an insecticide, an analytic reagent, in textile printing, and in the preservation of hides. It is a strong irritant and is toxic by ingestion or inhalation.

potassium arsenite (potassium metarsenite). a very toxic, hygroscopic, ethanol-soluble, white powder. The commercial product is rather variable, but corresponds approximately to the formula, $KH(AsO_2)_2$. It is a strong irritant and is toxic by all routes of exposure. It is used as a reducing agent in silvering mirrors and as an analytic reagent.

potassium arsenite solution (Fowler's solution; arsenical solution). a very toxic solution prepared in the proportions of 10 g arsenic trioxide; 7.6 g potassium bicarbonate; 30 ml ethanol; and water to 1 liter. It is a clear, colorless liquid used therapeutically in medicine as an antineoplastic and dermatologic; and in veterinary medicine as a tonic in pulmonary emphysema, chronic cough, chronic skin diseases, and anemias.

potassium atractylate. atractyloside.

potassium bichromate. potassium dichromate.

potassium bifluoride (potassium acid fluoride; potassium hydrogen fluoride; Fremy's salt). a colorless, toxic, corrosive crystalline compound, KHF_2, used to etch glass and as a metallurgical flux. It is corrosive to tissue.

potassium binoxalate (potassium acid oxalate; sorel salt; salt of sorrel; sal acetosella). a toxic, white, odorless, crystalline compound, $KHC_2O_4 \cdot H_2O$, used to remove ink stains, to clean wood, as a mordant in dyeing and in photography. It is toxic by ingestion.

potassium bis(cyano-C)argentate(1—). potassium silver cyanide.

potassium bromate. a white, crystalline or powdery substance, $KBrO_3$. It is a strong oxidizing agent; an irritant to the skin, eyes, and mucous membranes; and can ignite spontaneously if in contact with organic material. It is used as an analytic reagent and a constituent of permanent-wave solutions. It is a strong irritant. Symptoms of poisoning by ingestion are initially those of acute gastritis or gastroenteritis. The reaction time is 5-20 minutes. Symptoms may include vomiting, abdominal pain, diarrhea, lethargy, anuria or oliguria, deafness, hypotension, rapid pulse, coma, and convulsions. Lesions include kidney damage and minuscule, red spots that appear late and persist after death. See also bromate, sodium bromate. Potassium bromate is possibly carcinogenic.

potassium bromide. a white, hygroscopic, bitter-tasting, compound, KBr, consisting of crystalline granules or powder that are soluble in water and glycerine. It is toxic by inhalation and ingestion.

potassium chlorate (potcrate). a transparent, colorless, crystalline compound or white powder, transparent, soluble in boiling water, in ethanol- and alkali-soluble crystalline compound, $KClO_3$. It forms explosive mixtures with flammable materials such as sulfur and sugar. It is a strong oxidizing agent and is used in explosives and matches; in the manufacture of paper; and in textile dyeing.

potassium chloride (potassium muriate). a salty tasting, colorless, water-soluble, ethanol-insoluble, crystalline compound, KCl. It is used in pharmaceuticals, as a fertilizer, in photography, and as a food additive. Large oral doses can cause gastrointestinal distress, weakness, purging, and cardiovascular problems. Near-fatal poisonings by this salt are known.

potassium cyanate. a very toxic, white, water-soluble, crystalline powder, KOCN. It is used as a herbicide, in the manufacture of organic chemicals and drugs, and the treatment of sickle cell anemia.

potassium cyanide. an extremely poisonous, water-soluble, white, ionic, amorphous or crystalline solid, KCN, with a faint odor of bitter almonds. It is a violent poison; extremely toxic to laboratory rats. Intoxication can occur by ingestion, absorption through the skin or by inhalation of hydrogen cyanide fumes which can be released by the action of acids or slowly by atmospheric CO_2. Symptoms of empoisonment are essentially those of cyanide poisoning. It is used as a commercial fumigant, in the extraction of ores, electroplating and in various manufacturing processes.

potassium dichromate (potassium bichromate; red potassium chromate). a poisonous, bright yellowish red, metallic-tasting, water-soluble crystalline substance, $K_2Cr_2O_7$, with anhydrous, nondeliquescent triclinic crystals. It is used as an analytical reagent and oxidizing agent, and in electroplating, explosives, and matches. It is a strong oxidizing agent and is toxic by ingestion or inhalation.

potassium dichromate, zinc chromate, and zinc hydroxide (1:3:1). zinc chromate, potassium dichromate, and zinc hydroxide (3:1:1).

potassium dicyanoargentate. potassium silver cyanide.

potassium dicyanoargentate(I). potassium silver cyanide.

potassium dihydrogen arsenate. potassium arsenate.

potassium ferrocyanide (tetrapotassium hexakis-(cyano-C)ferrate(4—); potassium hexacyanoferrate(II); yellow prussiate of potash; yellow potassium prussiate). a soft, slightly efflorescent, lemon-yellow, crystalline substance or powder; a trihydrate, $K_4Fe(CN)_6·3H_20$, it is used in the preparation of various cyanides and is an antidote to copper sulfate poisoning. A compound of relatively low toxicity, it evolves highly toxic fumes when heated to red heat.

potassium fluoride. a toxic, strongly irritant, white, saline-tasting, deliquescent, crystalline substance, KF or $KF·2H_2O$, that is soluble in water and hydrofluoric acid, but insoluble in ethanol. It is used as a preservative, in etching glass, and as an insecticide.

potassium hexacyanoferrate(II). potassium ferrocyanide.

potassium hydrate. potassium hydroxide.

potassium hydrogen fluoride. potassium bifluoride.

potassium hydroxide (potassium hydrate; caustic potash; potassa). an extremely caustic, water-soluble, white solid, KOH. It is extremely toxic and destructive to living tissue. It is used as an analytical reagent, a chemical intermediate, in the manufacture of soap and matches, and is a component of many cleaning solutions, drain cleaners, aquarium products, cuticle remover, and some small batteries. It destroys tissues on contact. Ingestion by humans is characterized by a soapy taste, nausea, difficulty in swallowing, a severe burning pain in the mouth, throat, and epigastrium with hematemesis, abdominal cramps, a weak rapid pulse, bloody purging, collapse and death. Death may result from shock, but if death is not immediate, the victim may die from asphyxia due to glottic edema. Lesions may include laryngeal or glottic edema; erosion with or without perforation of the upper gastrointestinal tract; and late esophageal stenosis.

potassium iodide. white, water- and ethanol-soluble, saline-tasting crystals, granules, or powder, KI, with a bitter, saline taste. It is used in photography, as an analytic reagent, in feed additives for livestock, spectroscopy, scintillation. It is also used as a dietary supplement at a level of up to 0.01% in table salt. KI is teratogenic, producing congenital goiter.

potassium mercuriiodide solution. potassium triiodomercurate(II) solution.

potassium metarsenite. potassium arsenite.

potassium muriate. potassium chloride.

potassium myronate. sinigrin.

potassium nitrate (**niter; nitre; saltpeter**). a colorless or white, slightly hygroscopic, crystalline powder, KNO_3, with a pungent salty taste. It is a strong oxidizing agent and there is considerable risk of fire or explosion if it is shocked, heated, or comes into contact with organic matter. Potassium nitrate is used in gunpowder and other explosives, pyrotechnics, matches, fertilizers, preservatives, as a reagent, and as a color fixative in cured meat products; it is also used in pickling brine and in chopped meat. Potassium nitrate is a suspected human carcinogen and may have mutagenic and reproductive effects.

potassium nitrite. a very toxic, white to yellowish compound with monoclinic crystals, KNO_2. It is used therapeutically as an antidote to cyanide poisoning and as a vasodilator. Additional applications are as a curing agent and color fixative in meats and as an oxidizing agent. It is a dangerous fire and explosion hazard. In the body of humans and higher animals, nitrite combines with other chemicals in the gastrointestinal tract, forming powerful carcinogens such as nitrosamines and nitrosamides. Its carcinogenic and mutagenic properties and possible further effects have not been fully explored.

potassium oxalate. a colorless, odorless, efflorescent, water-soluble crystalline substance, $K_2C_2O_4 \cdot H_2O$. It is used as a bleach, an analytic reagent, in photography, and as a source of oxalic acid. It is toxic if inhaled.

potassium permanganate. a toxic, water-soluble compound, $KMnO_4$, with dark purple crystals that have a blue metallic sheen. It is a powerful oxidizing agent and serious fire and explosion hazard when brought into proximity to organic material. It is used, for example, in radioactive decontamination of skin, in air and water purification, and as an oxidizer, deodorizer, disinfectant, antiseptic, bleach, and dye. It is also used in solution as a gastric lavage in poisoning from morphine, strychnine, aconite, and picrotoxin. The use of potassium permanganate for any purpose should, however, be considered hazardous. Topical application to the vagina or urethra will cause severe burning, hemorrhages, and collapse of the blood vessels. This compound has sometimes been used unwisely or criminally to induce abortion by topical application to the vaginal wall. Any amount or concentration that might be sufficient to induce abortion would almost undoubtedly prove fatal. See potassium permanganate poisoning.

potassium permanganate poisoning. poisoning by potassium permanganate by contact or ingestion. **1a: potassium permanganate poisoning by ingestion**. the chief clinical signs of permanganate poisoning by ingestion of the crystals or strong solutions are, in humans, those of erosion with brown discoloration and edema of the mucous membranes of the mouth and pharynx including the larynx; acute gastrointestinal distress; bradycardia; and hypotension. Death may be nearly immediate due to trauma-induced shock or delayed. Unless death intervenes promptly, liver and kidney failure with icterus and oliguria or anuria may result and ultimately causing death. Lesions revealed on autopsy usually include erosion, hemorrhage, and necrosis of mucous membranes that came in contact with the potassium permanganate; liver and kidney damage. **1b: potassium permanganate poisoning by vaginal application**. a small amount of potassium permanganate crystals or a concentrated solution when applied to the mucous membranes of the vagina or urethra may cause severe burning, hemorrhage, and vascular collapse. Perforation of the vaginal wall may occur, resulting in peritonitis, fever, and abdominal pain. Examination reveals a characteristic chemical burn that is stained brown. Lesions observed at autopsy include necrosis and hemorrhage of the vaginal wall and usually liver and kidney damage.

potassium rhodanide. potassium thiocyanate.

potassium salt. any of a number of compounds of potassium with a nonmetal (e.g., potassium chloride, potassium nitrate, potassium phosphate, and potassium sulfate); most are moderately toxic. They are normal constituents of animal tissues and fluids and are not toxic at normal physiologic levels. Acute poisoning by ingestion is rare because the amount that is sufficient to poison will normally induce vomiting in humans and most other terrestrial vertebrates. In individuals with normal kidney function potassium is rapidly excreted. Near fatal poisonings are known, however.

potassium silver cyanide (potassium bis(cyano-C)argentate(1—); potassium dicyanoargentate(1—); silver potassium cyanide). a poisonous, white, water-soluble, light-sensitive, crystalline compound, KAgCN; a bactericide.

potassium sulfate. a colorless or white, crystalline or powdery compound, K_2S0_4, that is water soluble. It is used, for example, as an analytical reagent, a cathartic, in the manufacture of fertilizers for chloride sensitive crops (such as citrus and tobacco), alums, and mineral water. It is also a water corrective used in the brewing industry. Potassium sulfate is poorly absorbed from the gastrointestinal tract of vertebrates. Nevertheless, exposure to large amounts can cause purging and gastrointestinal distress with severe bleeding. Poisoning is due largely to the potassium cation.

potassium sulfocyanate. potassium thiocyanate.

potassium sulfocyanide. potassium thiocyanate.

potassium telluride. See potassium tellurite.

potassium tellurite. a granular, hygroscopic, water-soluble, white powder, K_2TeO_3. It is hemolytic, although the active species is probably the reduced form, potassium telluride.

potassium thiocyanate (potassium rhodanide; potassium sulfocyanate; potassium sulfocyanide). a colorless, transparent, deliquescent, odorless, crystalline substance, KCNS, that is soluble in water, ethanol, and acetone. It is very toxic to humans *per os*, the probable lethal single dose for adults appears to lie between 15 and 50 gm, with time to death ranging from ca. 10-48 hours. Symptoms of acute intoxication may include nausea, vomiting, extreme cerebral excitement, convulsions, delirium, and spasticity of the extensor muscles. This compound is considered more toxic than sodium thiocyanate, because of possible potassium intoxication.

potassium triiodomercurate(II) solution (mercuric potassium iodide solution; potassium mercuriiodide solution; solution potassium iodohydrargyrate; Channing's solution; Thoulet's solution). a poisonous aqueous mixture of 1.0 g HgI_2 and 0.8 g KI in 100 g water. It is a topical antiseptic and disinfectant.

potato. *Solanum tuberosum*.

potato family. See Solanaceae.

potato poisoning. poisoning due to the ingestion of potatoes (*Solanum tuberosum*) or potato parts (the skin or green sprouts) that contain high levels of solanine, q.v. Potatoes normally contain about 7 mg of solanine per 100 g of potato; 20-25 mg/100g is toxic. Symptoms of intoxication may include headache, abdominal pain, nausea, vomiting, diarrhea, pyrexia, and neurologic disturbances such as restlessness, drowsiness, apathy, confusion, stupor, visual disturbances and hallucinations.

potcrate. potassium chlorate.

potency. **1:** power, force, intensity, or strength; the condition or quality of being potent. **2:** the power of a drug, toxicant, or other biologically active agent to produce a desired or specified effect. **2a:** toxicity; a measure of the amount (or concentration) of a substance required to produce a given effect. **2b:** pharmacological activity or therapeutic strength of a substance. **3:** the ability of a presumptive embryonic tissue or organ to develop normally and fully. **sexual potency**. a term usually applied to a male, indicating the ability to fertilize an egg.

potent. **1a:** powerful, vigorous, intense (with respect to a specified quality), strong. **1b:** having the quality of, or being potent. **1c:** intensely active, capable, or powerful in a specified way. **2:** having sexual potency, *q.v.*

3: indicating the ability of a primitive cell to differentiate normally. **4:** said of a toxicant: being highly toxic, or having a substantially higher toxicity than one with which it is compared. The utility of this term depends upon the stated boundary conditions: relative potency may vary with changing environmental conditions, with dosage, and such factors as stability of the toxicants, and with the age, sex, species, strain, etc., of the target organism.

potentiate. to increase or enhance the strength, intensity, or degree of activity of a substance or other entity.

potentiation. the interaction of two or more substances where one is not biologically active by itself (or is weakly active) on a specific living system, but when added to a biologically active agent (e.g., a toxicant), the effect of the latter is enhanced. **Note:** synergism and potentiation are sometimes treated as equivalent terms in the literature of chemistry and toxicology. *Cf.* synergism.

potion. a concoction, elixir, tonic; a drink, draft, swig, or large dose of liquid medication.

poultry hemorrhagic syndrome. See aflatoxicosis.

powder. **talcum powder** (face powder). the greatest danger from face powder is the presence of talc which may be contaminated with asbestos. During application, talc-containing respirable dust is raised. Possibly harmful also are artificial fragrances contained in many talcum powders; they are a common cause of allergic reactions to cosmetics.

Poznan cort. *Cortinarius orellanus*.

ppb. parts per billion.

PPB. parts per billion.

PP_i. pyrophosphate.

ppm. parts per million.

PPM. parts per million.

ppt. **1:** parts per trillion. **2:** parts per thousand.

Pr. the symbol for praseodymium.

PR-toxin. a highly toxic mycotoxin (*i.p.* or *s.c.*; somewhat less toxic *p.o.*) from *Penicillium roqueforti*.

practically nontoxic. See nontoxic.

practolol. an antiarrhytmic drug, $C_{14}H_{22}N_2O_3$, it is an immunotoxic β-adrenergic blocker.

prairie rattlesnake. *Crotalus v. viridis*.

prajmaline (17,21-dihydroxy-4-propylajmalanium; N^4-propylajmalinium; prajmalium; *N*-propylajmaline). a very toxic antiarrhythmic, $[C_{23}H_{33}N_2O_2]^+$, used therapeutically usually as the tartrate, *q.v.*

prajmaline bitartrate. prajmaline tartrate.

prajmaline tartrate (prajmaline bitartrate; 17R,21-α-dihydroxy-4-propylajmalaniumhydrogen tartrate). an antiarrhythmic compound, $C_{23}H_{32}N_2O_2 \cdot C_4H_6O_6$, that is poisonous to laboratory animals and humans *i.v.* and orally. Human systemic effects of overexposure by ingestion include hallucinations and distorted perceptions.

prajmalium. prajmaline.

pralidoxime (2-[(hydroxy-imino)methyl]-1-methylpyridinium; *N*-methylpyridinium-2-aldoxime; 2-PAM). a cholinesterase reactivator, $C_7H_9N_2O^+$, that inhibits the action of certain anticholinesterases by reversing the organophosphate acetylcholinesterase inhibition by dephosphorylating the phosphorylated enzyme. It is moderately toxic orally and is an antidote to nerve gases and to cholinesterase-inhibiting insecticides. It is used as the chloride, the iodide (pralidoxime iodide; 2-pyridine aldoxime methiodide; 2-PAM; Protopam iodide), or the mesylate (palidoxime mesylate; Protopam methanesulfonate; P2S), all of which act in similar fashion and with similar potency.

pralidoxime chloride (2-[(hydroxyimino)methyl]-1-methylpyridinium chloride; 2-formyl-1-methylpyridinium chloride oxime; 1-methyl-2-formylpyridinium chloride oxime; *N*-methylpyridinium-2-aldoxime chloride; 2-pyridine aldoxime methyl chloride; 2-PAM chloride; Protopam chloride). a white to pale yellow, crystalline powder, $C_7H_9ClN_2O$. It is a chlorinated oxime that is only moderately toxic to humans, *per os*. It is used as an antidote for nerve gases and cholinesterase-inhibiting pesticides of the parathion type. It reactivates cholinesterase by removing phosphoryl groups.

pralidoxime iodide (pralidoxime methiodide; 2-pyridine aldoxime methiodide; 2-PAM iodide; protopam iodide). the iodide salt of pralidoxime, $C_7H_9IN_2O$. It is probably somewhat less toxic than pralidoxime chloride, and like the latter compound, it is a reactivator of acetylcholinesterase, and is used to reverse the effects of anticholinesterases. See pralidoxime.

pralidoxime mesylate (protopam methanesulfonate; Contrathion; 2-PAM mesylate; P2S). the methanesulfonate of pralidoxime, it is a water-soluble, crystalline compound, $C_8H_{12}N_2O_4S$. It is moderately toxic *per os*, with functions and applications similar to those of pralidoxime chloride and pralidoxime iodide.

pralidoxime methiodide. pralidoxime iodide.

prandial. relating to a meal.

praseodymium (symbol, Pr)[Z = 59; A_r = 140.91]. a soft, ductile, malleable, silvery element of the lanthanoid series of metals. It is used as a catalyst in several alloys, and in the glass and ceramic industries.

Pratt's snake. *Apistocalamus pratti*.

prayer bead. *Abrus precatorius.*

pre-. a prefix meaning situated before or occurring before.

preadaptation. an inherited attribute that enables the adaptation of an organism to environmental factors to which it has not yet been exposed. Thus, synthetic poisons that resemble natural poisons may be less toxic to those organisms that are adapted to the natural substances.

preagonal. immediately preceding death.

precancerous. characteristic of a growth that has the potential to become cancerous.

precarcinogen. procarcinogen.

precatory bean. *Abrus precatorius*.

prechronic. sometimes used as a synonym for subchronic (the preferred term).

precipitate. **1a:** to arouse, trigger or initiate abruptly (e.g., to precipitate a crisis). **1b:** abrupt, sudden. **2:** to separate out from solution; to form a precipitate (def. 3); to settle, to sink. **3:** a solid that separates from a solution because of some chemical or physical change.

precipité blanc. mercurous chloride.

precision. 1a: in science and statistics the degree of agreement among replicated measurements of the same quantity under specified conditions. **1b:** the closeness of agreement between results obtained by applying an experimental procedure several times under prescribed conditions. The smaller the random uncertainties that influence the results, the greater the precision. Precision, however, does not have a numerical value. Consequently, the term imprecision may be preferable where a rigorous treatment is desirable. **2:** the exactness with which a quantity is stated.

preclinical. pertaining to a disease state before it becomes clinically recognizable.

preconvulsant. pertaining to occurrence prior to convulsions. *Cf.* preconvulsive.

preconvulsive. preconvulsant; pertaining to a stage or period of time that precedes the convulsive stage. *Cf.* preconvulsant.

precordial. referring to the precordium.

precordialgia. pain in the precordium.

precordium. the region over the heart and lower thorax of mammals, especially humans.

precursor. **1:** any substance involved in the formation of another. Examples of usage: **a:** a protein from which an enzyme, hormone, or other active principle may be produced by further chemical modification. **b:** in the photochemistry of pollutants, precursors are compounds (e.g., a volatile organic compound) that react in sunlight to form ozone or other photochemical oxidants. **2:** a cell from which later cells can develop.

predisposing. in the health and environmental

sciences, conferring a tendency toward a particular disease, condition, or outcome.

preeclampsia. See toxemia (toxemia of pregnancy).

Prefamone. diethylpropion.

preganglionic axon (preganglionic fiber). in the autonomic nervous system, the myelinated axon that synapses at a ganglion, not the numerous unmyelinated neurons that conduct impulses away from the ganglion. *Cf.* preganglionic neuron, postganglionic axon.

preganglionic fiber. preganglionic axon.

preganglionic neuron. a myelinated neuron of the autonomic system that arises in the brain and synapses at a ganglion with numerous postganglionic unmyelinated neurones that terminate at the effectors. *Cf.* preganglionic axon.

pregnancy. **1:** gestation. **2a:** the period of development within the uterus from conception to birth of the baby. **2b:** the condition of a woman from conception to the birth of the baby that arises from the growth and development within a woman of the new individual. Pregnancy (or gestation) is a period of heightened sensitivity of the female and/or the developing embryo or fetus to certain poisons. See woman.

pregnant. **1:** a synonym of gravid, but applied most often to humans and sometimes domestic mammals. **2:** of or pertaining to a woman, or to the condition of a woman, who carries within her a developing embryo or fetus.

Δ^4-pregnen-3,20-dione. progesterone.

pregn-4-ene-3,20-dione. progesterone.

4-pregnene-11β,17α,21-triol-3,20-dione. hydrocortisone.

4-pregnen-21-ol-3,20-dione. desoxycorticosterone.

Δ^4-pregnen-21-ol-3,11,20-trione. dehydrocorticosterone.

preicteric. of or pertaining to the phase of hepatic disease prior to the appearance of jaundice (icterus).

Preludin. phenmetrazine.

Premanufacture Notification (PMN). most industrially advanced nations have legal systems that require manufacturers to report drugs, industrial chemicals, and other potentially hazardous chemicals prior to manufacture. For example, with certain exemptions, the U.S. Toxic Substances Control Act (TSCA) requires manufacturers to notify the U.S. Environmental Protection Agency 90 days prior to production of a new chemical or an existing chemical product when new applications might increase human and environmental exposure (e.g., by increased production and more widespread distribution). The notification must include information on the product, including toxicological properties, the method and amount of production, and distribution. Assurances that production workers and consumers have been fully notified of the properties of the chemical must be included in the notification.

premature. **1:** underdeveloped, embryonic, vestigial. **2:** early, hasty, precipitate. **3:** immature.

premature birth. birth of a viable fetus prior to the normal end of gestation.

premature infant. a viable infant that is born prior to the normal end of gestation; an infant born prematurely.

premature labor. **1a:** labor beginning before that expected for a given mammalian species; **1b:** usually taken as labor in humans that occurs prior to 39 weeks of pregnancy.

premolar. the tooth anterior to a molar.

premonitory. **1:** forewarning, presaging, signaling; giving a warning, as an early sign or symptom. **2:** in toxicology, pertaining to a premonitory sign or symptom, especially one that presages the major effects of a poisoning. See sign, symptom.

premortal. occurring immediately before death.

prenatal (antenatal). occurring or existing before birth; before birth; pertaining to the period of development of the offspring prior to birth (i.e., during pregnancy).

prenatal period. the period of development of a new individual prior to birth. Prenatal exposure to toxicants can cause embryonic or fetal death or teratogenesis.

preparation. 1: readiness, making ready, especially of a drug, for use. **2:** concoction, product, a drug that is ready for use. **3:** training, education, background. **4:** a specimen made ready for demonstration (e.g., of anatomy, histology, pathology).

preparation K. dixanthogen.

presbyopia. form of nearsightedness that normally accompanies advancing age.

pressor (hypertensor; pressor substance). a substance capable of increasing blood pressure.

pressor substance. pressor.

pressure filtration. a process whereby small molecules leave a capillary due to blood pressure.

presymptom. an indication which is a forerunner of the actual symptoms of a condition. *Cf.* symptom (premonitory).

presymptomatic. **1a:** existing prior to the appearance of symptoms. **1b:** descriptive of the state of health prior to clinical appearance of signs and symptoms of a disease state. See also symptom (premonitory).

presynaptic. **1:** pertaining to location near, but above a synapse (i.e., above the point where a nerve terminal ends at a synapse). **2:** occuring prior to (usually implying just prior to) transmission across a synapse.

presynaptic membrane. a membrane (neurolemma) that is part of a synapse and releases a neurotransmitter substance.

prevention. **1:** deterrence, avoidance. **2:** any action taken to prevent illness or empoisonment.

preventive. **1:** deterrent, prophylactic, protective. **2a:** tending to prevent, slow, stop, or interrupt the course of a disease. **2b:** at a community or population level, tending to reduce the incidence of a disease.

prickle. a short, woody, pointed, epidermal outgrowth of a plant. *Cf.* spine (def. 3).

prickly lettuce. *Lactuca scariola*.

prickly poppy. See *Argemone*.

pride of India. *Melia azedarach*.

prim. *Ligustrum vulgare*.

prima facie. a fact in law that is presumed to be true unless there is sufficient evidence to the contrary to disprove it.

primaquine phosphate. very toxic 8-aminoquinoline that has replaced the more highly toxic pamaquine in controlling the exo-erythrocytic forms of malaria. It is especially effective in terminating relapsing vivax malaria. This compound is hemolytic; normal therapeutic doses sometimes cause extravascular hemolysis in sensitive non-Caucasian individuals. On rare occasions, this drug may trigger methemoglobinemia in individuals that are deficient in methemoglobin reductase activity.

primary carcinogen. See carcinogen.

primary disorder. a basic disease from which complications may stem. Thus diabetes mellitus is a primary disorder often with secondary disorders (complications) of the kidneys, blood vessels, and eyes.

primary photosensitivity. See photosensitivity.

primary radiation barrier. See barrier.

primary reaction step. See phase of toxicity.

primate. any mammal assigned to the order Primates. Included are tree shrews, lemurs, monkeys, apes, and humans. Most species are arboreal with limbs adapted to climbing, leaping, or brachiating; they have a highly developed, large brain relative to body size; a short snout; an elaborate visual apparatus, often with stereoscopic vision.

primin.(dimethyl-5-(1-isopropyl-3-methylpyrazolyl)carbamate; dimethylcarbamic acid 3-methyl-1-(1-methylethyl)-1H-pyrazol-5-yl ester). an insecticide and the active principle of *Primula*, $C_{10}H_{17}N_3O_2$. It is toxic to humans by ingestion, skin contact, and possibly other routes. It is tumorigenic, possibly carcinogenic. See *Primula*.

primrose. *Primula*.

Primula (primrose). a large genus of perennial herbs (Family Primulaceae). They are common greenhouse and garden plants. The stems and leaves of several species are toxic, causing a contact dermatitis similar to that produced by *Rhus radicans*. The active principle is primin.

prince's plume. See *Stanleya*.

principle. **1:** a chemical component. **2a:** the essential or characteristic ingredient in a substance. **2b:** an often ill-defined substance in a secretion, poison, venom, toxic organism, etc., in which the biological activity resides; sometimes an extract, fraction, or mixture, sometimes a specific substance. **3:** a general or fundamental postulate, precept, or tenet. **active principle**. any constituent of a mixture, biological material or product, organism, or drug in which the activity (e.g., toxicity) resides, or on which the activity strongly depends. **toxic principle**. any constituent of a biological material, organism, or drug that confers toxicity.

Pristis (sawfish). a genus of shark-like rays (sawfish) up to 6.6 m long. The snout is elongate and bears large tooth-like projections along margins that disable prey.

privet. See *Ligustrum*.

proactinium. protactinium.

proactivator. the inactive precursor or storage form of an activator. See, for example, factor (properdin factor).

probability. **1:** in statistics (and the usual meaning in the natural sciences), a formal numerical statement of the likelihood of a given event or outcome. The probability of an impossible event is zero; certainty of outcome is 1.0. **2:** often used in its vernacular sense meaning chance, likelihood, or prospect. **3:** an empirical assertion of likelihood about the world or of outcomes based on the relative frequencies or limiting frequencies of a kind event. **4:** an expression of a logical relationship (quantitative or nonquantitative) between a proposition and a body of knowledge in which degrees of belief are objectively inferred from the given evidence by a formal, logical process. **5:** the subjective (often unique) degree of belief one may hold about the likelihood of an event or outcome.

probiosis. *Cf.* antibiosis.

procaine. See procaine hydrochloride.

procaine hydrochloride (procaine; Novocain). a synthetic alkaloid that is soluble in water and ethanol; slightly soluble in chloroform; and nearly insoluble in ether. It occurs as an odorless, white powder or colorless crystals, $C_6H_4NH_2COOCH_2CH_2N(C_2H_5)_2 \cdot HCl$. It is less toxic than cocaine and is used as a generally safe local anesthetic by injection and infiltration, but not by topical application. It is also used for epidural, caudal, and other regional anesthetic procedures. The anesthetic effect is prolonged by concurrent injection of epinephrine. It can cause potentially serious neurologic and cardiovascular reactions if administered *i.v.*; allergic reactions may also occur. Its use is contraindicated for individuals with known hypersensitivity to related anesthetics and should not be introduced into inflamed or infected tissues. Large doses may not be given to patients with heart block.

procalmadiol. meprobamate.

procarbazine (*N*-(1-methylethyl)-4-[(2-methylhydrazino)methyl]benzamide; *N*-isopropyl-α-(2-methylhydrazino)-*p*-toluamide; *N*-4-isopropylcarbamoylbenzyl-*N'*-methylhydrazine; *p*-(N^1-methylhydrazinomethyl)-*N*-isopropylbenzamide; ibenzmethyzin; MIH; PCB). a dialkylhydrazine (*q.v.*) used as an antineoplastic. It is poisonous *i.p.* and *i.v.* and is moderately toxic (as the hydrochloride) orally to laboratory rats. Procarbazine is a confirmed carcinogen, teratogen, and a probable mutagen. Toxic fumes, NO_x, are released on decomposition by heating. **procarbazine hydrochloride** (Matulane; Natulan). the hydrochloride, $C_{12}H_{19}N_3O \cdot HCl$, of procarbazine. **procarbazine hydrobromide**. the hydrobromide, $C_{12}H_{19}N_3O \cdot HBr$, of procarbazine.

procarcinogen (precarcinogen; proximate carcinocinogen). a substance that becomes carcinogenic following metabolic transformation. It is metabolically activated by monooxygenases and other enzymes. See also bioactivation, promutagen.

procaryote. prokaryote.

procatarxis. **1a:** predisposition. **1b:** a predisposing cause. **2:** the production of a disease or of intoxication partially due to predisposition.

procedure. a particular methodical way, usually consisting of several steps, whereby a desired result may be accomplished.

process. **1:** a mechanism, means, method, procedure, or mode of action used to achieve a certain result. **2:** the progress of a disease. **3:** an anatomical projection or outgrowth as from an organ or tissue such as bone.

prochlorperazine. a phenothiazine derivative. See phenothiazine.

procyclidine (α-cyclohexyl-α-phenyl-1-pyrrolidinepropanol; 1-cyclohexyl-1-phenyl-3-(1-pyrrolidinyl)-1-propanol; 1-cyclohexyl-1-phenyl-3-pyrrolidino-1-propanol). a crystalline anticholinergic, $C_{19}H_{29}NO$, often given as the hydrochloride. **procyclidine hydrochloride** (1-cyclohexyl-1-phenyl-3-pyrrolidino hydrochloride). a white, crystalline powder $C_{19}H_{30}ClNO$, which is water soluble, soluble in ethanol and chloroform, and only slightly soluble in ether. It is a synthetic anticholinergic used as a muscle relaxant in the management of parkinsonism. See anticholinergic.

α-prodine. alphaprodine.

prodromal (plural, prodromata; prodromic). premonitory; presaging; pertaining to any early, often minor symptom or sign (a prodrome) of disease or pathology that presages or indicates the onset of the actual condition.

prodromal symptom. a premonitory symptom; one that indicates the onset of a disease. **prodromal stage**. the initial stage of acute radiation syndrome.

prodromata. plural of prodroma.

prodrome. an early or premonitory sign, symptom, or precursor of a morbid condition or disease. See prodromal.

prodromic. prodromal.

producer. See autotrophic organism.

producer gas (blow gas). an inefficient, but inexpensive, industrial fuel gas with high levels of nitrogen (ca. 75%), carbon monoxide (ca. 10%), and carbon dioxide (14%); it also contains about 1% argon. It is a hazardous asphyxiant gas, produced by burning solid fuel in an oxygen-deficient atmosphere or by conducting a stream of air and steam through a bed of incandescent coke. Producer gas is the nitrogen source in the manufacture of ammonia. See especially carbon monoxide.

product. See household product.

proemial. **1:** introductory; an introduction or indication; prodromal. **2:** potentially hazardous or unsafe.

proestrus. See estrus cycle.

profound. **1:** deep, penetrating. **2:** acute, extreme, intense. **3:** enigmatic, mysterious.

progestational hormone. progesterone.

progesterone (pregn-4-ene-3,20-dione; Δ^4-pregnen-3,20-dione; corpus luteum hormone; luteohormone; progestational hormone; progestin). a steroid hormone produced naturally by the corpus luteum, the adrenal gland, the maternal placenta, and also synthetically. In humans, it stimulates the changes that occur in the uterine endometrium during the second half of the menstrual cycle prior to implantation of the blastocyst. It functions in the maintenance of pregnancy, suppresses ovulation, and stimulates development of the mammary glands and that of the maternal placenta following implantation. Progesterone may at times also contribute to fatigue, depression, and changes in mood. Progesterone is also used, for example, in the treatment of abnormal uterine bleeding, premenstrual tension, endometriosis, suppression of lactation following birth, suppression of endometrial carcinoma, and inhibition of testicular function in males. Alone or in combination with estrogens, progesterone is used in oral contraceptives and the treatment of amenorrhea, dysmenorrhea, and impending abortion in humans. In veterinary practice, it has been used to control habitual abortion and to suppress or synchronize estrus. Its use of progesterone and other progestogens in therapy is contraindicated in individuals with thrombophlebitis, breast cancer, missed abortion, liver dysfunction, or hypersensitivity to the drug. Adverse effects of therapy include disturbances of electrolytes and catabolic effects. One may observe edema, weight gain, headache, loss of appetite, rash, irregular menstrual cycles, breast tenderness, and ovarian cysts. See also contraceptive (oral contraceptive), progestogen, progestomimetic.

progestin. progesterone.

progestogen. **1:** any substance that is able to produce effects similar to those of progesterone, *q.v.* See oral contraceptives. **2:** a synthetic derivative of testosterone that has some of the pharmacologic and physiological action of progesterone.

progestomimetic. pertaining to or having a physiologic action similar to that of progesterone.

prognose. prognosticate.

prognosis. a forecast, especially of the probable future course of a disease including the prospect for recovery or other outcome. A realistic prognosis in many poisonings requires reliable information on the source or even the specific toxicant and depends further on the amount encountered, the route of exposure, and often other factors such as the general health and medical history of the victim. Critical information is often lacking or in error.

prognostic. pertaining to prognosis or to a sign or symptom that is indicative of the course or outcome of a disease process generally or in a specific case.

prognosticate (prognose). to formulate and make a prognosis.

progress. **1:** advancement; development; course of a disease. **2:** to advance, improve. **3:** to continue along an unfavorable course (said of a disease).

progression. **1:** advancing, moving forward. **2:** advancement of a neoplasm toward a less differentiated state or more aggressively malignant behavior.

progressive. **1a:** advancing, developing, as the course of a disease. **1b:** gradual, incremental. **1c:** continuing, worsening, spreading. **2:** advancing; the continuation of a disease along an unfavorable course.

proinvasin. a precursor of hyaluronidase (invasin). See anti-invasin II.

proinvasin I. See anti-invasin.

projectile vomiting. vomiting with great force.

prokaryote (procaryote; prokaryotic cell). a simple unicellular organism (protist) in which there is no true nucleus; the genetic material is thus not separated from the cytoplasm by a membrane. They have very few subcellular structures and are further characterized by a high rate of reproduction (cell division). Included are bacteria and cyanobacteria. They are used in a variety of mutagenicity tests. See also eukaryote.

prokaryotic. of, pertaining to, or resembling prokaryotes; lacking the organelles found in more complex cells (eukaryotes).

prokaryotic cell. prokaryote.

proliferative dust. See dust.

promecarb (3-methyl-5-(1-methylethyl)phenol methylcarbamate; methylcarbamic acid *m*-cym-5-yl ester; *m*-cym-5-yl methylcarbamate; 3-methyl-5-isopropyl *N*-methylcarbamate; Carbamult). a poisonous, colorless, crystalline, water-insoluble substance, $C_{12}H_{17}NO_3$. It is used as a nonsystemic carbamate insecticide for potato, fruit, and corn crops. It is a reversible anticholinesterase with very high acute oral toxicity; it is toxic also by skin contact and absorption. Atropine is antidotal.

promethium (symbol, Pm)[Z = 61]. a radioactive element of the lanthanoid series of metals, promethium lacks a characteristic terrestrial isotopid composition.

prometon. an atrazine used only as a nonselective herbicide for the control of most annual and perennial broadleaved and grass weeds on non-crop areas. See triazine herbicide.

prometryne. an atrazine herbicide, similar in biological properties to atrazine and simazine, but is absorbed through foliage as well as the roots. See triazine herbicide.

promoter. **1:** a chemical that potentiates the activity of a catalyst when added to it in relatively small amounts. **2a:** a substance which, when administered after an initiator has been given, fosters the change of an initiated cell into a cancerous one. **2b:** operationally a promoter is a substance that increases the incidence of response to a carcinogen that has been previously encountered or administered or that shortens the latency period during which no response is seen. Numerous substances can act as promoters, inluding hormones and various drugs, plant products, ethanol, and phorbol esters. See also initiator.

promotion. a process whereby previously initiated cells (See initiation) are stimulated to grow and develop into neoplasms. See also initiation.

promutagen. a substance that becomes mutagenic only when modified or activated by metabolic processes. Promutagens are metabolically activated by monooxygenases and other xenobiotic-metabolizing enzymes. See also procarcinogen.

prone. the position of the body when lying face downward. *Cf.* supine.

proof. **1:** compelling acceptance by the mind of the truth or existence of a structure or phenomenon. **2:** the establishment of certainty or validity of phenomenon or the establishment of something as fact. So defined, proof has no operational standing in the natural sciences including those subdivisions of toxicology that are based on research and toxicity testing; it is strictly a layman's or dilettante's term. Scientific research can, by its nature, only rule out (disprove) possibilities (e.g., hypotheses), thereby narrowing the number of plausible possibilities. One sometimes speaks of proof when new possibilities are, at least for the time being, no longer conceived. **3:** in applied fields of toxicology (e.g., forensic toxicology, clinical toxicology), proof of a thing may be accepted as sufficiently established if the preponderance of evidence justifies a decision or action (e.g., a course of therapy). "Proof" in such cases is still qualified to a degree, however, and is presumptive. **4:** a measure of the ethanol content of intoxicating beverages. In the United States, proof spirits (100 proof) contain 50% ethanol by volume; in the United Kingdom, they contain 42.28% ethanol by weight (= 57.1% by volume at 16°C). Spirits with lower amounts of ethanol are labeled as parts of proof (e.g., 60 proof).

proof spirits. 100 proof spirits. See proof (def. 4).

propachlor (2-chloro-*N*-(1-methylethyl)-*N*-phenylacetamide; 2-chloro-*N*-isopropylacetanilide; *N*-isopropyl-α-chloroacetanilide). a tan powder, $C_{11}H_{14}NOCl$, that is soluble in ethanol and benzene. It is a preemergence selective anilide herbicide that is absorbed through the roots and inhibits root elongation by inhibiting cell division by influencing RNA and protein synthesis. It is moderately toxic to laboratory rats. Other soil-applied amides with similar action include alachlor, diphenamid and naptalam.

propane. a hydrocarbon, C_3H_8, it is a gas at room temperature and natural atmospheric pressure. Under normal circumstances, propane is a simple asphyxiant. At high concentrations, however, propane is a neurotoxicant that affects the CNS. **chlorinated propane**. a derivative of propane in which one or more hydrogen atoms are substituted by chlorine. See 1,2,2,3-tetrachloropropane.

propanenitrile. propionitrile.

propanethiol. 1-propanethiol.

1-propanethiol (*n*-propyl mercaptan; 3-mercaptopropanol; propyl mercaptan; propanethiol). a highly flammable oil, C_3H_8S, with a disagreeable odor and a boiling range of 67-73°C, used as a herbicide. It is moderately toxic, orally, to humans, mildly toxic when inhaled, and is a severe eye irritant. Toxic fumes, SO_x, are released on heating to decomposition.

1,2,3-propanetriol diacetate. diacetin.

1,2,3-propanetriol trinitrate. nitroglycerin.

propanil (*N*-(3,4-dichlorophenyl)propanamide; 3′,4′-dichloropropionanilide; *N*-(3,4-dichlorophenyl)propionamide). a post-emergent contact anilide herbicide, applied to the foliage, it rapidly degrades in the soil. It is used mainly to weed umbelliferous crops such as celery, carrots, and parsley and also in rice and potatoes. Damage is in the form of localized or general necrosis, depending on the dose applied. It also inhibits growth but this is probably a secondary effect. Propanil is moderately toxic to laboratory rats.

propanoic acid. propionic acid.

1-propanol (*n*-propyl alcohol; propylic alcohol). an industrial solvent with an alcoholic and somewhat intoxicating odor, $CH_3CH_2CH_2OH$. It is a mild irritant of the eyes and mucous membranes and a moderately toxic CNS depressant, with effects similar to that of ethanol.

2-propanol (isopropyl alcohol; isopropanol; secondary propyl alcohol; dimethyl carbinol; petrohol). a slightly bitter, toxic, flammable liquid, $CH_3CHOHCH_3$. It is used as an industrial solvent for gums, essential oils, shellac, the extraction of alkaloids, and the manufacture of acetone, glycerol, and isopropyl acetate. It is a component of antifreeze formulations and of a number of other commercial preparations such as inks, hand lotions, and after shave lotions; it is a denaturant of ethyl alcohol; and is often the active constituent of rubbing alcohol. It also used as an antiseptic and in veterinary medicine as a rubefacient. It is moderately toxic to humans who can be seriously affected by inhalation or ingestion of fairly large quantities. Symptoms of intoxication include headache, flushing, dizziness, nausea, vomiting, headache, mental depression, narcosis, anesthesia, coma, and even death; about 100 ml can be fatal. It is only slightly toxic to laboratory rats.

propanone. See acetone.

2-propanone. acetone.

propanone 1-oxine. isonitrosoacetone.

propantheline bromide (*N*-methyl-*N*-(1-methylethyl)-*N*-[2-[(9*H*-xanthen-9-ylcarbonyl)oxy]ethyl]-2-propanaminium bromide; (2-hydroxyethyl)diisopropylmethylammonium bromide xanthene-9-carboxylate). a white or off-white crystalline substance, $C_{23}H_{30}BrNO_3$, that is soluble in water, ethanol, chloroform; nearly insoluble in ether, benzene. It is an anticholinergic with an action similar to that of belladonna. It inhibits gastrointestinal hypermotility and hyperacidity. It is used especially as an adjunct peptic ulcer therapy. Overdosage can cause tachycardia, dry mouth, and decreased sweating. Some individuals experience hypersensitivity reactions.

2-propenyl isovalerate. allylisovalerate.

2-propanyl 3-methylbutanoate. allylisovalerate.

propazine (2,4-bis(isopropylamino)-6-chloro-*s*-triazine). a triazine derivative, $C_9H_{16}ClN_7O_2$, used as a pre-emergence herbacide to control weeds in sorghum culture. It is moderately irritant to the eye and is moderately toxic to humans by ingestion. Propazine is a possible carcinogen.

2-propenal. acrolein.

propenamide. acrylamide.

2-propenamide. acrylamide.

2-propen-1-amine. allylamine.

propene. propylene. See also chlorinated propene.

propene amide. acrylamide.

propene nitrile. acrylonitrile.

propene oxide. propylene oxide.

propene-1,2,3-tricarboxylic acid. aconitic acid.

2-propenenitrile. acrylonitrile.

2-propene-1-ol. allyl alcohol.

2-propene-1-thiol (allyl mercaptan). a highly toxic, volatile liquid (b.p. 60°C) alkenyl mercaptan with a strong garlic odor. It is strongly irritating to mucous membranes when inhaled or ingested.

1-propenol-3. allyl alcohol (See alcohol).

2-propenylamine. allylamine.

5-(2-propenyl-1,3-benzodioxole). safrole.

2-propenyl isothiocyanate. allyl isothiocyanate.

properdin. a heat-labile serum protein involved in the lysis of erythrocytes as well as Gram-negative bacteria and viruses. It is not an antibody, but can lyse cells in the presence of magnesium and complement. See properdin system.

properdin factor. See factor.

properdin factor B. See factor.

properdin factor E. See factor.

properdin system. an alternative pathway of complement fixation in which the C3 step in antibody formation proceeds without the activation of C1, C4, and C2. The process may be initiated, for example, by endotoxin, bacterial capsule, or IgA. See properdin. See also factor (cobra venom factor, properdin factor B, properdin factor E).

property. any attribute, feature, character, quality, or trait.

propham (phenylcarbamic acid 1-methylethyl ester; carbanilic acid isopropyl ester; *N*-phenyl isopropyl carbamate; isopropyl carbanilate; *O*-isopropyl *N*-phenyl carbamate; INPC; IPC; Iso-PPC). an aryl carbamate herbicide applied to soil by spraying. It is generally effective against monocotyledons only, and does not control crabgrass (*Digitaria spp.*). It is moderately toxic to laboratory rats. It causes contraction of the plant chromosomes, is an inhibitor of gibberellin-induced α-amylase synthesis in barley endosperm, and thus probably inhibits gene derepression (i.e., the onset of metabolic changes) but not processes that are already in progress. It causes mitotic aberrations (e.g., anaphase bridges, blocked metaphases, nuclear fragments, giant vesiculate nuclei, increased numbers of chromosomes) in certain root and shoot cells of *Avena* and *Allium*. See also chlorpropham.

prophlogistic corticoid (P-C). any corticoid that is especially effective in facilitating inflammation (e.g., desoxycorticosterone acetate, desoxocortisone). *Cf.* prophlogistic hormone.

prophlogistic hormone. any hormone that is especially effective in facilitating inflammation (e.g., somatotrophic hormone, desoxycorticosterone acetate, desoxycortisone). *Cf.* prophlogistic corticoid.

prophylaxis. prevention of or preventive treatment, as of an illness; protection; prevention of the spread of a disease. *Cf.* synteresis.

β-propiolactone. a carcinogenic α,β-unsaturated lactone.

propionate. a salt or ester of propionic acid, a number of which are used as nonsteroidal anti-inflammatory drugs.

propionic acid (propanoic acid; methylacetic acid; ethylformic acid; carboxyethane; ethane carboxylic acid; metacetonic acid). a transparent, colorless, highly flammable, oily liquid, CH_3CH_2COOH, with a pungent odor that is miscible with water, ethanol, chloroform, and ether. It is strongly corrosive to the skin, eyes, and mucous membranes and is moderately toxic by ingestion, skin contact, and intravenous injection. It occurs in sweat and is a component of certain ant venoms. When heated to decomposition, it releases irritant smoke and fumes. Propionic acid is used in nickel-electroplating solutions; in the synthesis of various propionates, artificial flavors, and perfume esters; and as a cellulosic solvent.

propionic nitrile. propionitrile.

propionitrile (propanenitrile; ethyl cyanide; cyanoethane; ether cyanatus; hydrocyanic ether; propionic nitrile). a colorless, flammable liquid, C_2H_5CN, with an ethereal odor that is miscible with ethanol and ether; and boils at 97.1°C. It is poisonous by most or all routes of exposure. It is moderately toxic by inhalation. It is an eye irritant and is teratogenic to laboratory animals. When heated to decomposition, it releases toxic fumes (NO_X and CN_-). Propionitrile is used as a solvent in petroleum refining and as a raw material in drug manufacture.

propoxur (2-(1-methylethoxy)phenol methylcarbamate; *o*-isopropoxyphenyl-*N*-methylcarbamate; 2-isopropoxyphenyl-*N*-methylcarbamate; methyl-2-isopropoxyphenylcarbamate; aprocarb; isocarb). a toxic, white, crystalline powder, $(CH_3)_2CHOC_6H_4OOCNHCH_3$, that is soluble in water and most polar solvents. It is a dangerous carbamate insecticide, molluscicide, and is a component of some household products to eradicate pests. It is toxic by ingestion and inhalation. It inhibits cholinesterase, but injection of atropine sulfate readily reverses the inhibition. Symptoms of empoisonment are similar to those of organophosphate poisoning. Oral toxicity may vary considerably. The oral LD_{50} in laboratory rats has been reported as 95 mg/kg body wt, but >800 mg/kg in domestic goats. See also insecticide (carbamate insecticide).

propoxyphene (Darvon; dextropropoxyphene; α-(+)-4-dimethylamino-1,2-diphenyl-3-methyl-2-butanolpropionate ester; α-d-4-dimethylamino-3-methyl-1,2-diphenyl-2-butanol propionate; α-[2-(dimethylamino)-1methylethyl]-α-phenylbenzeneethanol propanoate; [S-(R*,S*)]-α-[2-(dimethylamino)-1-methylethyl]-α-phenylbenzene-ethanol propanoate (ester); (+)-4-dimethylamino-1,2-diphenyl-3-methyl-2-propionyloxybutane; (+)-1,2-diphenyl-2-propionoxy-3-methyl-4-dimethylaminobutane; D-propoxyphene). an addictive opioid, analgesic, $C_{22}H_{29}NO_2$. It is similar chemically to methadone and is very toxic, but is much less potent than morphine, having about the same analgesic effect as codeine. This drug also lacks the antipyretic and anti-inflammatory action of morphine. The α-diastereoisomers are optically active and provide a greater analgesic effect. Overdosage can prove fatal. In the United States, usage is restricted by the Federal Drug Administration (FDA).

***d*-propoxyphene**. propoxyphene.

propoxyphene hydrochloride. a narcotic analgesic. See propoxyphene.

propylacetone. methyl *n*-butyl ketone.

***N*-propylajmaline**. prajmaline.

N^4-propylajmalinium. prajmaline.

propyl carbinol. 1-butanol.

propyl mercaptan. 1-propanethiol.

***n*-propyl mercaptan**. 1-propanethiol.

propylene (propene). a colorless, simple asphyxiant gas, $CH_3CH{:}CH_2$, with chemical, physical and toxicological properties that are very similar to those of ethylene. It is used in the manufacture of polypropylene polymer. Propylene is highly flammable and a dangerous fire hazard.

propylene oxide (1,2-propylene oxide; methyloxirane; 1,2-epoxypropane; propene oxide). a colorless, reactive, irritant, moderately toxic, volatile, highly flammable, liquid epoxide, CH_2OCHCH_3. Its uses and toxic effects are similar to those of ethylene oxide, although the effects are less severe. Propylene oxide is carcinogenic to some laboratory animals.

1,2-propylene oxide. propylene oxide.

2-propylpiperidine. coniine.

6-propyl-2-thiouracil. propylthiouracil.

propylthiouracil (6-propyl-2-thiouracil; 2,3-dihydro-6-propyl-2-thioxo-4(1H)-pyridinone).

proscillaridin (proscillaridin A; 3-[(6-deoxy-α-L-mannopyranosyl)oxy]-14-hydroxybufa-4,20,22-trienolide; 14-hydroxy-3β-(rhamnosyloxy)bufa-4,20,22-trienolide; desglucotransvaaline; scillarenin 3β-rhamnoside; 3β-rhamnosido-14β-hydroxy-$\Delta^{4,20,22}$-bufatrienolide; cillarenin; urgilan). a white, powdery, very toxic, bitter, crystalline, cardiac glycoside, $C_{30}H_{42}O_8$, isolated from *Urginea burkei* and *U. (Scilla) maritima*, *q.v.* It can be prepared by acid cleavage of scillarin A or by enzymatic decomposition of glucoscillaren A by strophanthobiase. It is used therapeutically as a thyroid inhibitor. It is hepatotoxic and can cause cholestasis.

proscillaridin A. proscillaridin. See also scillaren A.

proserine. neostigmine.

prostigmin. an oral parasympathnomimetic agent.

Prostigmin™. neostigmine.

prostrate. pertaining to a condition of prostration.

prostration. a condition in which a human or animal assumes a recumbent position due to a pronounced loss of strength or exhaustion. A subject in such a state may show intermittent uncoordinated movements.

protactinium (symbol, Pa; protactinium; protoactinium)[Z = 91; most stable isotope = ^{231}Pa, $T_{1/2}$ = 3.2 x 10^4 yr]. a bright, lustrous, toxic, radioactive element (an alpha emitter) of the actinoid series of metals; it is an alpha particle emitter. The isotopes ^{231}Pa and ^{234}Pa occur naturally in minute quantities in uranium ores (e.g., pitchblende, carnotite) as a radioactive decay product of actinium. Protactinium is an inhalation hazard and can also produce serious systemic effects if ingested. It is a highly radiotoxic element and a confirmed human carcinogen.

protamine. any of a class of simple basic proteins that occur naturally combined with nucleic acid in the sperm of certain fish (e.g., salmon) and that yield basic amino acids on hydrolysis. They are soluble in water, dilute acids, and ammonia water. Protamines contain few amino acid residues, but all contain alanine, arginine, and serine. Examples are clupeine (from herring sperm), iridine (from trout sperm), salmine (from salmon sperm), and scombrine (from mackerel sperm). See also protamine sulfate.

protamine sulfate. a purified, white or whitish, finely powdered, or crystalline form of protamine used as an antidote to heparin.

protapam methanesulfonate. pralidoxime mesylate.

protease. proteinase.

protectant (protective agent). any agents, regardless of mechanism, that can protect an organism, partially or completely, from the detrimental effects of a toxicant. Protectants range from mechanical barriers to exposure such as protective clothing or protective equipment such as air filters and gas masks to chemicals that may be used to inhibit the toxic effects of one or more toxicants. Selenium, for example, is a natural protectant against a variety of toxic chemicals.

protectin. lectin.

protective. **1:** preventive as in protective against infection or a toxicant; or in providing immunity. **2:** an agent that shields or protects mechanically (as a bandage or plaster cast) or chemically (an antidote, a chemical prophylactic).

protective agent. See protectant.

protective barrier. See barrier.

protein. any of a large variety of complex nitrogenous macromolecules, essential for life, that occur in all living cells and in biological fluids such as blood plasma. They are composed of one or several long polypeptides composed of α-amino acids, R—CH(NH_2—COOH), that are connected by peptide linkages (—CO·NH—) which are formed by elimination of H_2O between the NH_2 group and COOH group of successive amino acid residues. R is hydrogen or a ring or chain of carbon atoms. There are two major classes (globular and fibrous). Most enzymes are globular proteins, whereas structural and contractile proteins are usually fibrous. Proteins have complex configurations, the contributing components of which fall into several structural categories: primary, secondary, tertiary, and quaternary. The primary structure is defined as the sequence of amino acids in the polypeptide chain which determines the higher order, three-dimensional shape of the molecule. The secondary structure, which depends on hydrogen bonds, is the coiling (α-helix) or pleating (β-pleated sheets) of the chain and the tertiary structure, seen in globular proteins, is an additional folding of the coiled or pleated fibers into a three-dimensional shape that is held together by weakly hydrophobic and polar interactions among the amino acid residues. Quaternary structure is the oligomeric protein molecule formed by the linkage of multiple polypeptide chains. **oligomeric protein**. a protein formed by two or more polypeptide chains.

protein poisoning. grass tetany.

proteinase (protease; proteolytic enzyme). any protein-digesting enzyme, especially one that acts on native protein molecules, converting them into polypeptides; proteolytic enzyme.

proteinuria. the appearance of protein (usually albumin, hence albuminuria) in the urine. This condition may result from poisoning by substances that affect renal function (e.g., arsenic, mercurials).

protemetine. an alkaloid constituent of ipecac.

proteolysin. any enzyme that catalyzes the hydrolysis of proteins into proteoses, peptones, and other substances. Snake venoms contain proteolysins.

proteolysis. enzymatic (via proteolysins) or nonenzymatic hydrolysis of proteins into proteoses, peptones, and other substances. Snake venoms contain proteolysins.

proteolytic. of or pertaining to proteolysis; having the ability to break down proteins into smaller molecules.

proteolytic enzyme. proteinase.

proteotoxin. **1a:** a toxic protein (e.g., anaphylatoxin) formed by the interaction of a bacterial protein with the host's serum. **1b:** endotoxin.

proteum. hydrogen.

prothoracic gland. a gland in the prothorax (the first segment of the thorax) of insects that secretes ecdysone and sometimes poisonous materials. See, for example, *Dysticus*.

Prothromadin™. warfarin sodium. See warfarin.

prothrombin. plasma protein made by liver that must be present in blood before clotting can occur.

prothrombin time (PT). a one-stage test in which thromboplastin and calcium are added simultaneously to a sample of a patient's plasma and to that of a normal (control) individual. The length of time to clot formation is measured. If thrombin formation (from prothrombin in the presence of thromboplastin and calcium) is normal, clotting time will generally be normal. The test is used to detect and diagnose clotting defects and to control anticoagulation in some diseases of the heart and blood vessels. An extend PT indicates a deficiency in factor V, VII, or X (as in liver disease), vitamin K deficiency, or the anticoagulant effects of coumarin.

protista. a general category or taxon of life forms which includes all unicellular algae, protozoans, bacteria, yeasts, etc. **marine protista**. more than 80 species, mostly of dinoflagellates, are known to be toxic to animals.

proto-oncogene. a normal cellular gene that can become an oncogene. Such genes seem to readily form cancer-inducing genetic sequences. Proto-oncogenes are designated as such in written communications by use of the prefix c- followed by the name of the oncogene.

protoactinium. protactinium.

protoanemonin (5-methylene-2(5*H*)-furanone**;** 4-hydroxy-2,4-pentadienoic acid γ-lactone**;** 5-methylene-2-oxodihydrofuran). an antibacterial, moderately toxic, vesicant oil, $C_5H_4O_2$, isolated from the ranunculaceous herb, *Anemone pulsatilla*, by enzymatic maceration and is thus derived from the original ranunculin present in the intact plant. Protoanemonin is a constituent of all herbs of the genus *Ranunculus*, and in at least some plants of other ranunculaceous genera (e.g., *Anemone*, *Helleborus*) of the family Ranunculaceae. It is a volatile, unstable, potent irritant and vesicant oil; a breakdown product of ranunculin, it is also the precursor of anemonin which occurs in *Caltha palustris*. When ingested, it produces a burning sensation in the oropharynx, profuse salivation, emesis, colicky gastroenteritis, diarrhea and sometimes hematuria, polyuria, and painful urination. See also ranunculin.

protogen A. thioctic acid.

proton number (symbol, Z**;** atomic number). the number of protons in the nucleus of an atom. The electron structure, which determines the nature of chemical bonding, depends upon the degree of electrostatic attraction of the nucleus. The latter is a function of the proton number which is thus the prime determinant of an element's chemical behavior.

protopam chloride. pralidoxime chloride.

protopam iodide. pralidoxime iodide.

protopam methanesulfonate. pralidoxime mesylate.

protopine (7-methyl-2,3:9,10-bis(methylenedioxy)-7,13a-secoberbin-13a-one; 4,6,7,14-tetrahydro-5-methyl-bis-[1,3]-benzodioxolo[4,5-c;5′,6′g]azecin-13(5*H*)-one; biflorine; fumarine, macleyine). **1:** a crystalline alkaloid, $C_{20}H_{19}NO_5$, found in small amounts in opium, and in a number of herbs of the families Papaveraceae and Fumaraceae (e.g., *Eschscholtzia californica*, *Argemone mexicana*, *Chelidonium majus*, *Papaver*, *Fumaria officinalis*, *Chelidonium majus*). It is an anodyne and hypnotic. Protopine is poisonous by ingestion and intraperitoneal routes and it releases NO_X when heated to decomposition. **2:** a toxic alkaloid from various perennial herbs of the genus *Dicentra*.

protoporphyrin. any of a set of naturally occurring precursors of blood and plant pigments that contain four pyrrole nuclei, $C_{34}H_{34}N_4O_4$. **free erythrocyte protoporphyrin** (FEP; erythrocyte protoporphyrin; EPP). an indirect measure of lead concentrations in circulating blood.

protoporphyrin IX. heme.

Prototheria. one of three infraclasses of the class Mammalia. It is comprised of a single order: Monotremata, *q.v.*

prototoxicant. a substance that is nontoxic or nearly so, but when absorbed by an organism is converted to a toxic metabolite and transported to receptors where it has a toxic effect. See also procarcinogen, phase of toxicity. See also procarcinogen, phase of toxicity.

prototoxin. an obsolete term for one of three groups of toxins that differ in their affinity for antitoxin. They are, in order of increasing affinity for antitoxin: tritoxin, deuterotoxin, prototoxin.

prototoxoid (protoxoid). an obsolete term for a hypothetical nontoxic substance (a toxoid) in a bacterial culture that has a stronger affinity for antitoxin than has the toxin. See toxoid.

protoveratrine. a mixture of the ester alkaloids, protoveratrine A and B obtained from the rhizome of *Veratrum album* and *V. viride*. Both are cardioactive vasodilators that act chiefly via the carotid sinus receptors to lower blood pressure, and are used therapeutically, usually in combination as antihypertensive agents to control certain forms of hypertension.

protoveratrine B. See protoveratrine.

protoxoid. prototoxoid.

protozoan (plural, protozoa). common name of any member of the phylum of unicellular animals that includes ameba and paramecia.

protozoicide. **1:** lethal to protozoa. **2:** an agent that is lethal to protozoa.

proximal. describes a point or part of the body or an organism that is closer to a central point or other reference point than some other part. The part that is further away is referred to as distal. *Cf.* distal.

proximate. immediate or nearest.

proximate carcinogen. procarcinogen.

prunasin. the d-form of mandelonitrile glucoside, *q.v.*

pruning spider. *Glyptocranium gasteracanthoides*.

Prunus. a genus of mostly deciduous, small trees and shrubs (Family Rosaceae) of the North Temperate Zone and the Andean Mountains of South America. Most species have edible fruits and many are cultivated. ***P. amygdalus*** (synonym, *Amygdalus communus*; common almond). an edible almond, but see *P. amygdalus amara*. ***P. amygdalus amara*** (synonym, *Amygdalus communis amara*, bitter almond). a variety of common almond that has a very bitter seed; it contains amygdalin and benzaldehyde. The latter compound confers a characteristic odor to the nut. See amygdalin, cyanide poisoning. ***P. armeniaca*** (apricot). a tree grown for its edible fruit. The pits contain amygdalin. ***P. caroliniana*** (cherry laurel). a lovely, shiny-leaved, poisonous wild cherry which may grow to a height of more than 40 feet. The leaves, seeds, and bark contain amygdalin. ***P. laurocerasus*** (European cherry laurel). a species that is very similar to the American species. The leaves, seeds, and bark contain amygdalin. ***P. persica*** (peach tree). a widely cultivated fruit tree native to China. The roots contain large amounts of amygdalin. ***P. serotina*** (black cherry, wild cherry). the bark, leaves, and especially the seed are toxic. Symptoms of intoxication from chewing the seeds include stupor, vocal cord paralysis, twitching, convulsions, and coma. Amygdalin is the source of the hydrocyanic acid that produces these effects. This species is also known to be a cause of arthrogryposis in terrestrial vertebrates.

pruritic. of or pertaining to pruritus.

pruritus. **1:** severe itching as a result of a disease process, allergic response, or psychic factors. **2:** itch.

Prussian blue. See ferric ferrocyanide.

prussiate. **1:** a cyanide; a salt of hydrocyanic acid. **2:** a ferricyanide or ferrocyanide.

prussic acid. hydrogen cyanide or hydrocyanic acid. See also cyanide.

pseud-, pseudo-. prefixes that indicate falseness or imitation; similarity; or a close relationship.

pseudo-BTX. pseudobatrachotoxin.

pseudoaconitine ((1α,3α,6α,14α,16β)-20-ethyl-1,6,16-trimethoxy-4-(methoxymethyl)aconitane-3,8,13,14-tetrol 8-acetate 14-(3,4-dimethoxybenzoate); acraconitine; feraconitine; nepaline; veratroylaconine; English aconitine; Nepal aconitine). a very poisonous substance isolated from aconite (the dried roots of *Aconitum napellus* and *A. ferox*). It is a white, crystalline substance or syrupy mass, $C_{36}H_{51}NO_{12}$, that is insoluble in water, but soluble in ethanol and ether. It is poisonous by skin contact, ingestion, and inhalation. When heated to decomposition, it releases highly toxic fumes (NO_X).

pseudoacrorhagus. a tubercle near the margin of certain sea anemones that contains ordinary epidermal nematocysts. *Cf.* acrorghagus.

pseudoanaphylactic. anaphylactoid.

pseudobatrachotoxin (pseudo-BTX). one of four major steroidal alkaloids isolated from the skin of poison-dart (poison-arrow) frogs of the genus *Phyllobates*. See batrachotoxin.

pseudocholinesterase (nonspecific cholinesterase). an esterase that catalyzes the hydrolysis of other esters, in addition to choline esters. It occurs in blood serum, pancreas, and liver, and can be prepared from horse serum.

pseudocyesis. false pregnancy.

pseudodigitoxin. gitoxin.

***d*-pseudoephedrine**. a plant alkaloid similar to ephedrine.

Pseudohaje (tree cobras). a genus of dangerous venomous snakes (Family Elapidae) comprised of two arboreal species that inhabit the tropical rain forests of central and western Africa. Mambas (*Dendroaspis*) are the only other arboreal elapids. The snout is broad and rounded and the head is foreshortened, narrow, and only slightly distinct from the neck; the canthus is distinct; the eyes are very large with round pupils. The two maxillary fangs are short with external grooves followed by a gap and 2-4 small teeth. The body is slender and tapered. Unlike other African cobras, there is merely the hint of a hood; there are just a few scale rows (13-15) at midbody; and the tail (more than 20% of body length) is unusually long. Adults occasionally reach 2.4 m in length. ***P. goldii*** (Gold's tree cobra; black cobra; black forest cobra; tchissapa). a rarely seen, long-tailed, mamba-like cobra of the tropical rain forest region of Africa from Nigeria eastward to Uganda and southward to southwest Africa. It is a shiny black snake with large eyes. It has a single preocular (as opposed to 3 in mambas) that either abuts the nasal or is separated from it by a loreal. There are 15 rows of dorsals on the neck and at midbody. ***P. nigra***. this species is even less well known than *P. goldii*; it occurs as far west as Sierra Leone.

pseudohyscyamine. norhyoscyamine.

pseudojervine ((3β,23β)-17,23-epoxy-3-(β-D-glucopyranosyloxy)veratraman-11-one). the glucoside, $C_{33}H_{49}NO_8$, of jervine, isolated from *Veratrum album*, it consists of shiny leaflets that are soluble in benzene and chloroform, slightly soluble in ethanol, and nearly insoluble in ether.

pseudometallophyte (facultative metallophyte). a plant that can tolerate high concentrations of heavy metals but one not confined to soils, sediments, or other media that contain heavy metals. See also metallophyte.

Pseudomonas tabaci. a bacterium (Family Pseudomonadaceae), that contains wildfire toxin and is the causative organism of wildfire disease in tobacco plants. See toxin (wildfire toxin).

pseudonarcotic. inducing sleep by means of a sedative effect that is not directly narcotic.

pseudoparkinsonism. one of several neurological sequalae of disturbances of the basal ganglia of the brain, commonly caused by antipsychotic drug treatment. See parkinsonism. See also akathesia, dystonia, extrapyramidal side effect.

Pseudopolydesmus serratus. a species of millipede (Phylum Arthropoda; Class Diplopoda) that releases hydrogen cyanide when threatened.

pseudoscorpion. any member of the arachnid order Pseudoscorpionida, q.v.

Pseudoscorpionida (pseudoscorpions). an order of small (up to 5 mm long), flat arachnids. They superficially resemble scorpions but have a short, oval abdomen that lacks both tail and stinger. The pincer-like pedipalps of some species bear venom glands and are used to immobilize prey. Pseudoscorpions occur in diverse habitats. See Chernetidae.

pseudostratified. descriptive of the layered appearance of some epithelial cells where each cell actually meets a base line and true layers do not exist.

pseudotoxin. a poisonous substance isolated from belladonna leaves.

pshissapa. *Naja melanoleuca*.

psilocin(3-[2-(dimethylamino)ethyl]-1*H*-indole-4-ol; 4-hydroxy-*N,N*-dimethyltryptamine; psilocyn). a minor hallucinogenic alkaloid mycotoxin produced by various *Psilocybe* species and others such as *Conocybe smithii*, *Gymnopilus spectabilis*, and *Panaeolus subbalteatus*. It is an indole derivative related to psilocybin.

psilocin phosphate ester. psilocybin.

psilocybe. part of the common name of mushrooms of the genus *Psilocybe*.

Psilocybe. a large genus of mostly small, brownish mushrooms (Family Strophariaceae), a few species of which are known to be hallucinogenic. This genus shares characteristics of *Stropharia* and *Naematoloma*. Some taxonomists believe that all three genera should be combined. ***P. baeocystis*** (potent psilocybe). occurs on decayed wood, wood chips, and decaying moss in the Pacific Northwest of the United States. It is strongly hallucinogenic, sometimes with serious side effects that are not yet well established. It contains several active substances. ***P. caerulipes*** (blue-foot psilocybe). found on decayed deciduous wood and wood mulch from Maine to North Carolina west to Michigan. A small, hallucinogenic brown mushroom that turns blue when handled for several minutes. ***P. coprophila*** (dung-loving psilocybe). occurs throughout much of North America on horse and cow dung. It is mildly hallucinogenic. ***P. cubensis*** (common large psilocybe). found commonly on horse and cow dung in pastures along the Gulf Coast of North America. It is hallucinogenic. ***P. cyanescens*** (bluing psilocybe). found predominantly in coniferous mulch from British Columbia to San Francisco. It is strongly hallucinogenic if large amounts are consumed. ***P. mexicana*** (God's chair, Teonanacatl, sacred mushroom of Mexico). the best known of the hallucinogenic species of *Psilocybe*, used for centuries by Aztec priests. Active principles of this species and certain others are psilocin and psilocybin. ***P. pelliculosa*** (conifer psilocybe). occurs individually or in clusters on conifer mulch in woodlands from British Columbia to northern California. It is weakly hallucinogenic and often mistaken for *P. semilanceata*. ***P. semilanceata*** (liberty cap). occurs, sometimes in large numbers, in tall grass or grass hummocks in cow pastures in the Pacific Northwest of the United States and has been reported in Quebec. It is perhaps the best known hallucinogen of the Oregon coast. ***P. stuntzii*** (Stunz's blue legs, Washington blue veil). it occurs in coniferous wood-chip mulch and sometimes on lawns in the Pacific Northwest of the United States. It is hallucinogenic and extremely dangerous to use because it is easily confused with the *Galerina autumnalis*. See also psilocybe poisoning.

psilocybe poisoning. a condition of children marked by sometimes persistent convulsions and sometimes by death due to the ingestion of *Psilocybe mexicana*, q.v., and a few other members of this genus. These mushrooms as normally used by adult humans are hallucinogenic, usually without further serious consequences.

psilocybin (3-[2-(dimethylamino)ethyl]-1*H*-indol-4-ol dihydrogen phosphate ester; *o*-phosphoryl-4-hydroxy-*N,N*-dimethyltryptamine; psilocin phosphate ester; Indocybin). an alkaloid mycotoxin, $C_{12}H_{17}N_2O_4P$. Together with its unphosphorylated analog, psilocin, it occurs in the fruiting bodies of mushrooms such as *Psilocybe mexicana* (Mexican "magic" mushroom), and certain others of the genera *Panaeolus, Conocybe, Gymnopilus*, *Psathyrella*, and *Stropharia*. It is the *N'N'*-dimethyl derivative of 4-hydroxytryptamine and is chemically related to psilocin, q.v. Psilocybin is a hallucinogenic that is much less potent than LSD, but much more potent than mescaline. It is moderately toxic to humans by ingestion and intraperitoneal routes and is probably toxic by all routes of exposure. It may accelerate the onset of psychoses in predisposed individuals. Accidental poisonings of adults resemble alcohol intoxication. Systemic effects of large doses include euphoria, hallucinations, toxic psychosis, nausea, vomiting, muscular weakness, hyperthermia, effects on the visual field, unconsciousness and, in children, tonic clonic convulsions. When heated to decomposition it emits very toxic fumes (NO_X and PO_X).

psilocyn. psilocin.

psilosis (sprue). falling out of hair.

psilostrophe. common name of a plant of the genus *Psilostrophe*.

Psilostrophe (paperflowers). a genus of small, toxic, bushy, wooly herbs with a perennating taproot (Family Compositaceae) that occur in open dry rangelands of the western United States. They have caused substantial losses of sheep. Affected sheep begin to show some incoordination when running. They become sluggish and anoretic and often cough violently, with or without vomiting. The animals become progressively depressed and emaciated and eventually die. Lesions are generally limited to some albuminous degeneration of renal tubules.

psoralen (7*H*-furo[3,2-g][1]benzopyran-7-one; 6-hydroxy-5-benzofuranacrylic acid δ-lactone; furo[3,2-g]-coumarin; ficusin). one of a class of furocoumarin phytoalexins, $C_{11}H_6O_3$, that occur in numerous species of plants of the families Rutaceae (e.g., bergamot, cloves, limes), Apiaceae (e.g., celery, parsnips), Fabaceae (e.g., *Psoralen coryfolia*), and Moraceae (e.g., figs). Psoralins are defensive against attacks by insects and fungi. They are photosensitizing, and phototoxic not only to insects, but to laboratory animals and humans also. They have been used in photochemotherapy to treat psoriasis, vitiligo, and mycosis fungoides.

PSP. paralytic shellfish poison. See poison.

psychedelic. **1:** hallucinogenic. **2:** of or pertaining to certain drugs, said to be mind-expanding, that affect CNS function such that it seems to the user that there is a heightening of consciousness. Included are LSD, hashish, mescaline. **3:** a hallucinogen.

psychedelic agent. hallucinogen.

psychological treatment. the treatment of patients and clients by psychologists and psychiatrists using the tools and principles of psychology (e.g., psychotherapy) as opposed to the use of drugs. *Cf.* physiological treatment, pharmacological treatment.

psychopharmacology. the science that treats the effects of drugs on the mind.

psychosis (plural, psychoses). **1:** obsolete: any mental disorder or severe emotional illness, or insanity of any kind. **2:** a mental disorder of a quality and magnitude that it reflects a severe loss of mental capacity, derangement of personality, and loss of contact with reality. The individual cannot relate to or communicate with others well enough to meet the ordinary demands of life. Signs of psychosis in laboratory animals are similar to those of primary sensory or motor disturbances, and may include ataxia, paralysis, or somnolence. Clinical manifestations of psychosis in humans include delusions, hallucinations, or illusions that may be reflected in inadequate, abnormal, or antisocial behavior. Psychoses may arise from organic brain damage or emotional origins. **alcoholic psychosis**. that caused chiefly by excessive alcoholic consumption. Clinical manifestations may include delirium tremens, acute hallucinosis (See hallucinosis), Korsakoff's psychosis, and pathological intoxication. **amnestic psychosis**. Korsakoff's syndrome. **drug psychosis**. that due to or provoked by exposure to a drug. **organic psychosis.** psychosis due to a lesion of the CNS. **polyneuritic psychosis**. Korsakoff's syndrome. **toxic psychosis**. that due to exposure to a toxic agent. See also, for example, anorectic (def. 2), bemegride, DMT, haloperidol, organolead poisoning, pharmacopsychosis, porphyria, psilocybin, LSD, manganese poisoning, morphinomania, scopolamine.

psychosomatic. of both mind and body, or of one caused by the other.

psychosomimetic. psychotomimetic.

psychotic. **1:** of, pertaining to, affected by, or caused by psychosis. **2:** a person who exhibits psychosis.

psychotogen. a substance that produces psychotic manifestations.

psychotogenic. able to induce psychosis or manifestations of psychosis; used especially in reference to psychedelic drugs such as LSD.

psychotomimetic (psychosomimetic). **1:** able to induce psychological changes that resemble psychosis. **2:** a psychomimetic drug.

psychomimetic agent. See psychomimetic drug.

psychotomimetic drug (psychotomimetic agent; psychotomimetic). a drug that has a psychotomimetic action.

psychotropic. **1a:** able to affect behavior, emotional state, or mental activity. **1b:** of or pertaining to drugs used in the treatment of mental illness.

psychotropic agent. psychotropic drug.

psychotropic drug (psychotropic agent; tranquilizer). a psychotropic therapeutic agent.

pt. **1:** pint. **2:** point.

Pt. the symbol for platinum.

PT. prothrombin time.

***cis*-Pt II**. cisplatin.

Pteridium (bracken; bracken fern). a genus of perennial, moderately toxic ferns that occur chiefly in open woods, upland pastures or abandoned fields on dry, sandy soils. They are global in distribution. The toxic principle is thiaminase. Poisoning in livestock occurs only if large amounts of bracken are ingested. Ingestion of hay that contains about 20% bracken usually produces symptoms in horses in about 1 month. The poisoned animal initially loses weight, suffers loss of condition and incoordination that becomes progressively severe. Affected livestock later become lethargic and stand with widespread legs. Severe tremors later develop and the animal is unable to stand and convulsions, collapse, and death may ultimately result. ***P. acquilinum*** (bracken; bracken fern). an attractive, highly variable species that occurs widely in tropical and temperate regions; it is common in woodlands, hillsides, and pastures. Ingestion by nonruminants may result in thiamine deficiency and ultimately death. This plant does not produce thiamine deficiency in cattle and other ruminants because the ruminal bacteria produce a sufficient internal supply. Bracken, however, can be lethal to cattle because it contains a toxin that inhibits the production of leucocytes by bone marrow cells. Symptoms in the latter case include internal bleeding and bloody feces, with occasional hemorrhage from the eye sockets and/or nostrils. Fever and infections due to activity of intestinal microflora are common. Either infection or bleeding alone can prove fatal.

Pterois (zebrafish; turkeyfish). a genus of ornate, venomous, coral reef fishes (Family Scorpaenidae). They are unafraid of humans and are dangerous to individuals working or bathing in shallow waters of coral reefs. The venom apparatus consists of small, sharp, dorsal, anal, and pelvic spines with small, well-developed venom glands. ***P. volitans*** (lionfish). a species that frequents shoals near beaches in Barbados. Bathers are sometimes injured by the sharp spines of this fish. The site of the sting is very painful, edematous, and inflamed. Additional effects are usually severe and include profuse perspiration, dyspnea, tachycardia, vomiting, diarrhea, and intense abdominal pain.

ptomaine (ptomatine). any nitrogenous organic base, usually toxic, formed by putrefaction of proteins and amino acids of animal or sometimes plant origin. Examples are toxic amines, choline, putrescine, cadaverine, muscarine.

ptomaine poisoning (ptomainotoxism). food poisoning caused by a ptomaine. One of the most common ptomaine poisonings of humans is that caused by *Salmonella typhimurium*. Symptoms are flu-like; death sometimes intervenes.

ptomainemia. **1a:** the presence of ptomaines in circulating blood. **1b:** the toxic condition caused by ptomaines in circulating blood.

ptomainotoxism (ptomaine poisoning). poisoning by a ptomaine.

ptomatine. ptomaine.

ptomatopsia. necropsy.

ptomatopsy. necropsy.

ptomatropine. a ptomaine from putrid sausages and from the viscera of corpses killed by typhoid fever. It is produced by the action of bacteria that decarboxylate amino acids and have toxic effects similar to those of atropine.

ptosed. exhibiting ptosis. *Cf.* ptotic.

ptosis. **1a:** a dropping, drooping, sinking, or prolapse of an organ or structure. **1b:** optic ptosis. **optic ptosis**. a drooping of one or both upper eyelids in paralysis, presumably due to impaired conduction of the third cranial nerve.

ptotic. pertaining to, characterized by or exhibiting ptosis. *Cf.* ptosed.

PTU. 1-phenyl-2-thiourea.

PTX. palytoxin.

ptyal-. ptyalo.

ptyalagogic. promoting the flow of saliva.

ptyalagogue (sialagogue; sialogogue). **1:** stimulating salivation. **2:** any agent that stimulates salivation.

ptyalism (hyperptyalism; hygrostomia; polysialia; sialism; sialismus; sialorrhea; sialosis). excessive or profuse salivation. It sometimes occurs in humans in the early months of gestation. Ptyalism is induced by poisons as in muscarinism, aconite poisoning, alcoholism, and that induced by various poisons or poison-containing materials (e.g., amanitin, aplysin, barium, *Dieffenbachia*, iodides, mercury, and pilocarpine). See also saliva, salivation.

ptyalo-, ptyal-. prefixes denoting or pertaining to saliva or to salivary glands. See also sialo-.

Ptychodiscus brevis. *Gymnodinium brevis*.

ptychotis oil. ajowan oil.

Pu. the symbol for plutonium.

puerperal hemoglobinuria. See hemoglobinuria.

puff adder. *Bitis arietans*. See also Viperidae.

puff otter. *Bitis arietans*.

puffer (puffer fish; pufferfish). the preferred common name of fish of the family Tetraodontidae. Any of some 100 species of fish of the family Tetraodontidae. Tetrodotoxin occurs in the ovaries, roe, liver, and skin of numerous species examined. All should be regarded as poisonous. Known as fugu, these fish are a delicacy in Japan but unless expertly prepared, an often fatal poisoning may follow ingestion. Effects may include incoordination, paresthesia, diarrhea, and paralysis. See also *Arothron*, *Tetraodon*, tetrodonic acid, tetrodotoxin.

puffer fish, pufferfish. puffer.

puffer poison. tetrodotoxin.

puffer poisoning. tetrodotoxism.

pul surattai. *Echis carinatus*.

pulicicide (pulicide). a chemical agent destructive to fleas.

pulicide. pulicicide.

pulmonary. of or pertaining to lungs and breathing.

pulmonary alveolar proteinosis. a rare disease of unknown etiology; the condition occurs mainly in healthy, mature individuals, 20-60 years of age. Lesions are limited to the lungs. While the alveolar lining of the lung and interstitial cells are usually normal, the alveoli are plugged with amorphous PAS-positive granules that contain plasma lipoproteins, plasma proteins, and other components of blood. Interstitial fibrosis is a rare occurrence. Clinical findings are variable, some patients being asymptomatic while others suffer severe respiratory insufficiency. *Cf.* silicoproteinosis.

pulmonary artery. a blood vessel in which blood flows from the heart to the lungs.

pulmonary diffusion capacity. an important measure of respiratory function in toxicology, it is the rate of diffusive exchange of oxygen and carbon dioxide across the lumen of the alveoli and the capillaries. It is partly determined by the length of the diffusion path which decreases, with a reduction in diffusive capacity, with the development of pulmonary fibrosis. This is a common effect in respiratory toxicology.

pulmonary edema. See edema.

pulmonary excretion. See excretion.

pulmonary hypertension. increased pressure in the blood vessels of the lungs.

pulmonary resistance (respiratory resistance). resistance of the respiratory tree to the passage of air. Contributions to this resistance are made by the elastic recoil of the chest and lungs, and by frictional resistance to the flow of air through the airways. Factors that alter the diameter or elasticity of the airways or the activity of the muscles will alter pulmonary resistance and the ease and even the effectiveness of breathing.

pulmonary vein. a vein through which blood passes from the lungs to the heart.

pulmonitis. Pneumonitis.

pulsatilla camphor. anemonin.

pulse. **1:** the heartbeat as transmitted through an artery. It is the rate or rhythm of arterial expansion that follows ventricular contraction and the consequent increased volume of blood that enters the arterial system. The pulse may be felt as a vibration with the finger held on the skin over an artery that lies near the body surface (e.g., at the wrist, or neck). The frequency is essentially the same as that of the heart beat. A pulse may sometimes occur in a vein or a highly vascular organ, such as the liver. A large variety of toxicants affect the pulse rate. A given toxicant may not affect the pulse, may decrease the rate (bradycardia), increase the rate (tachyardia), and/or produce irregularities in the rate (arrhythmia). Alcohols cause bradycardia or tachycardia; amphetamines, some belladonna alkaloids, cocaine, and tricyclic antidepressants may cause either tachycardia or arrhythmia; and toxic doses of digitalis may cause bradycardia or arrhythmia. Arrhythmias are also caused by arsenic, caffeine, some belladonna alkaloids, certain organic solvents, phenothiazine, and theophylline. Barbiturates, carbamates, clonidine, muscarinics, opiates, organophosphates, and local anesthetics generally cause bradycardia. **2:** any brief surge as in an electrical current. **3:** a common name of plants of the bean family (Fabaceae, formerly Leguminosae).

pulse family. See Fabaceae.

pulse rate. the number of pulsations of an artery per minute.

pulses. See Fabaceae.

Pultox™. lindane.

pumiliotoxin. any of three extremely toxic alkaloids (pumiliotoxin A, B, and C) isolated from the skin secretions of frogs of the families Atelopidae and Dendrobatidae. Pumiliotoxin A can produce ataxia, clonic convulsions, and death within minutes. Pumiliotoxin B potentiates evoked contractions of skeletal muscle. See batrachotoxin, *Dendrobates pumilio*, and *D. auratus*.

pump. **1:** a device that transfers fluids from one location to another by creating a pressure differential between the two locations (e.g., by suction). **2:** to force a fluid along a particular pathway by creating a pressure differential along the pathway. **Na^+-K^+ pump**. sodium-potassium pump. **sodium pump**. sodium-potassium pump. **sodium-potassium pump** (sodium pump; Na^+/K^+-pump). the system that simultaneously exchanges Na^+ and K^+ ions across the membrane (neurolemma) of a neuron (nerve cell). Because the neurolemma is semipermeable, sodium ions continuously enter the neurone and tend to accumulate; permeability increases during the passage of an impulse. The pump is an active transport carrier, with ATPase enzymatic activity, that acts to move the Na^+ ions out of the cell in the face of a concentration gradient in exchange for K^+ ions, thereby maintaining a gradient of these ions across the cell membrane that is responsible for the resting potential of the neurone. Such a pump operates in other excitable cells (e.g., muscle cells) in which the maintenance of an electrochemical gradient is functionally important. The responsible enzyme is a plasma membrane protein with ATPase activity, Na^+-K^+ ATPase. **stomach pump**. a device that removes stomach contents.

puncture vine. *tribulus terrestris*.

pungent. sharp; acrid; strong; harsh. Generally said of the taste or odor of a substance.

pupil (pupilla). **1:** the central (circular) aperture of the iris of the vertebrate eye, through which light enters. The size varies with the state of contraction of the ciliary body around the iris. Pupillary size is often a good indicator of poisoning. Many xenobiotics dilate the pupils (e.g., depressants, atropine and related alkaloids, cocaine, gelsemine, nicotine, propantheline, methanol); others are associated with contracted pupils (miosis) (e.g., antihistamines, morphine, physostigmine, phosphate ester insecticides, ciguatera, betel nuts, darnel seeds). **2:** the central spot of an eye-spot (ocellus).

pupilla. plural of pupil.

pupillary. pertaining to the pupil of the eye.

pupillary dilation. mydriasis.

pure. **1a:** unadulterated, undiluted, unmixed; **1b:** free from admixture or contamination by other materials. A toxicant or other biologically active agent is functionally pure if it contains no other chemicals that measurably alter its action.

purgation. catharsis.

purgative. **1:** cathartic, cathartic. **2:** an agent that stimulates bowel movement. See also catharsis, cathartic, purge.

purge. **1:** to evacuate the bowels by use of a cathartic. **2:** a cathartic; a drug that purges.

purge nut. *Jatropha curcas*.

purging croton. *Croton tiglium*.

purified talc. talc.

purine ribonucleoside. nebularine.

purine. any of a type of nitrogenous base found in DNA and RNA that have two interlocking rings.

6-purinethiol. 6-mercaptopurine.

purity. **1a:** the state of being pure, pureness, homogeneity. **1b:** the degree of purity (def. 1a).

Purkinje fiber. a specialized type of cardiac muscle fiber that conducts electrical impulses from the AV bundle into the ventricular wall.

puromycin. a cytotoxic, nucleosidic antibiotic, $C_{22}H_{229}N_7O_9$, produced by *Streptomyces albo-niger*. It inhibits protein synthesis and is toxic to all types of cells. The action is due to incorporation of puromycin into a polypeptide chain during its formation with the release of the incomplete chain from the ribosome.

purple allamanda. *Cryptostegia grandiflora*.

purple cockle. See *Agrostemma*.

purple cudweed. *Gnaphalium purpureum*.

purple foxglove. *Digitalis purpurea*.

purple loco. *Astragalus mollissimus*. a true locoweed.

purple snail. a snail of the genus *Thais*.

purple spotted pit viper. *Trimeresurus purpureomaculatus*.

purpurin. **1:** 1,2,4-trihydroxyanthraquinone, $C_6H_4(CO)_2C_6H(OH)_3$, a reddish, crystalline glycoside from the roots of *Rubia tinctorum* (madder) and synthesized by oxidation of alizarin. It is used as a nuclear stain, as a dye for cotton, in the manufacture of acid and chrome dyes, and as a reagent in the analysis of boron. **2:** uroerythrin (a pink or reddish pigment found in many pathologic urines and often in very small quantities in normal urine). **3:** purpurine.

purpurine. **1:** a glycoside, isolated from the median zone of the hypobranchial gland (purple gland) of gastropods of the genus *Purpura*. It appears to be an ester or mixture of esters of choline and is identical or nearly identical to murexine. **2:** See purpurin.

purse crab. *Birgus latro*.

purslane. *Portulaca oleracea*.

purulence (purulency). the state or condition of containing or producing pus.

purulency. purulence.

purulent. containing, comprised of, producing pus.

pus. a thick, usually yellowish or sometimes green fluid that forms at the site of a local infection (usually with inflammation), often appearing within an enclosed sac (an abscess), which may eventually burst as pus accumulates. Composition varies but usually includes at least degenerate and necrotic cells including various phagocytes and cells of local tissue, much of which has been liquefied by the action of proteolytic and histolytic enzymes elaborated by polymorphonuclear leukocytes. See also inflammation (purulent inflammation).

pusley. *Portulaca oleracea*.

puss-moth caterpillar. See *Egalopyge opercularis*.

pustulant. **1a:** having or characterized by pus; **1b:** causing a pustular eruption. **2:** an agent able to produce pustules.

pustular. pertaining to or marked by pustules.

pustulation. the formation or presence of pustules.

pustule. a small discrete elevation of the skin that contains purulent material.

putrefaction. See decay (def. 2).

putrefy. decay (def. 1).

putrescent. decaying; in a state of decay.

putrescine (1,4-diaminobutane). a poisonous polyamine ptomaine, $NH_2(CH_2)_4NH_2$, formed from arginine produced during the putrefaction of animal and some plant tissue. It is also a constituent of ergot. *Cf.* cadaverine.

putrid. decayed, decaying and odiferous.

putromaine. any poison produced by the decomposition of food within a living organism.

PVC. polyvinyl chloride.

PVP. polyvinylpyrrolidone.

PVP plastic. polyvinylpyrrolidone plastic (See plastic). See also shampoo (dandruff shampoo).

PVS. peripheral vascular system.

pwéré. *Causus rhombeatus*, *Echis carinatus*.

pycnosis. contraction of the nuclear material of dying cells into a compact, densely staining mass.

pydrin. fenvalerate.

pyelitis. **1:** inflammation of the renal pelvis. **2:** formerly a synonym for pyelonephritis.

pyemia (metastatic infection; pyohemia). a generalized septicemia with multiple secondary foci of suppuration and the formation of abscesses in various parts of the body, especially the lungs. Symptoms include recurrent chills, high intermittent fever, sweating. May progress to include septic pneumonia, empyema, and possibly death.

pygidial gland. See *Dysticus*.

pygmy rattlesnake, pygmy rattler. **1:** a snake of the genus *Sistrurus*. **2:** *Sistrurus miliarius*.

pyogenic. pus-producing.

pyohemia. pyemia.

pyoktanin. See auramine hydrochloride.

pyoktanin, yellow. auramine hydrochloride.

pyoluteorin. a chlorine-containing antibacterial and antiprotozoal fungal metabolite.

pyracantha. *Pyracantha coccinea*.

Pyracantha coccinea (scarlet firethorn; pyracantha; firethorn). a shrub that bears clusters of small bright red and orange, toxic, berry-like fruit. Ingestion of large numbers of berries by small birds, mammals, and children is not known to have had serious effects.

Pyralene. See polychlorinated biphenyl.

pyramidal tract. a descending neuronal pathway in the white matter of the spinal cord that conducts impulses from the brain to the anterior horn cells on the opposite side of the body. The nerve cell bodies reside in the precentral cortex and via this tract control voluntary and reflex activity of the muscles. See also extrapyramidal tract, extrapyramidal system.

pyrazolone derivative (pyrazolone). any of a number of potent nonsteroidal antiinflammatory drugs (e.g., aminopyrine, antipyrine, apazone, and phenylbutazone). Most of these drugs are also mild antipyretics and analgesics. They sometimes cause serious, sometimes fatal, blood dyscrasias (e.g., agranulocytosis). They are generally more toxic than the other mild analgesics, but may be indicated for patients who are hypersensitive to salicylates.

pyrazolone. See pyrazolon derivative.

pyrazon. a pyridazine herbicide, it is a plant growth inhibitor with a tendency to produce abnormal plants. The most prominent biochemical effect is a significant reduction in the rate of photosynthesis.

pyrectic. **1:** febrile; of, of the nature of, or pertaining to fever. **2:** a fever-inducing agent.

pyrene (benzo[def]phenanthrene; β-pyrine). a colorless, often yellow (when contaminated with tetracene), water-insoluble, condensed, solid, tetracyclic, aromatic hydrocarbon. It is fairly soluble in common organic solvents; solutions are somewhat bluish. Pyrene is poisonous if inhaled and is moderately toxic with oral and intraperitoneal administration. It is an irritant of the skin and mucous membranes. It is tumorigenic and is a debatably carcinogenic.

pyrethrin (pyrethrum (in part); pyrethrum insecticide; pyrethrum flowers, pyrethrum extract, Dalmatian insect powder, Persian insect powder). **1:** a botanical comprised of several esters extracted from the flowers of three species of chrysanthemum (Family Asteraceae) using hot carbon dioxide under high pressure. Pyrethrin is rapidly degraded photochemically in the environment. The most intense reactions in humans and other terrestrial vertebrates result from inhalation, but poisoning may occur through ingestion or absorption through the skin. In serious acute poisoning, the first sign is dermatitis, which may appear within half an hour of exposure; the skin becomes red and burns and itches at the site of contact or more generally; the cheeks soon swell. The initial effects may be followed by progressive hyperexcitability, loss of coordination, tremors, nausea, vomiting, diarrhea, convulsions, and muscular paralysis. Death from respiratory paralysis may intervene. Less intense exposure may result in headache, gastrointestinal distress, and paresthesia of lips and tongue. When removed from exposure, the victim's symptoms may require 2 days to 2 weeks to subside. Commercial sprays used as a plant insecticide usually contain about 0.1% pyrethrin and may include other active agents. Though seldom used now, pyrethrin has been used in concentrated form as a dust to kill fleas on animals. **2:** any of six insecticidal allethrin analogs extracted from pyrethrum flowers, q.v.: cinerins I and II, jasmolins I and II, and pyrethrins I and II, q.v. Pyrethrins I and II comprise 71% of the extract. All of these compounds are highly toxic (to the target organisms), fast-acting paralytic, broad-spectrum contact insecticides. They are only moderately toxic to laboratory rats, *per os*. They act primarily by slowing the opening and closing of sodium ion channels in nerve membranes with a consequent increase in Na^+ permeability and depolarization. Symptoms of acute intoxication in humans are those of gastroenteritis and CNS disturbance and may include diarrhea, nausea, vomition, tinnitis, headache, numbness of tongue and lips, syncope, hyperexcitability, incoordination, convulsions, prostration, and death due to respiratory paralysis. These substances are also hepatotoxic and nephrotoxic and can cause severe allergic dermatitis systemic allergic reactions. *Cf.* pyrethrum.

pyrethrin I (2,2-dimethyl-3-(2-methyl-1-propenyl)cyclopropanecarboxylic acid 2-methyl-4-oxo-3-(2,4-pentadienyl)-2-cyclopenten-1-yl ester; chrysanthemummonocarboxylic acid pyrethrolone ester). See pyrethrin (def. 2).

pyrethrin II. (3-(3-methoxy-2-methyl-3-oxo-1-propenyl)2,2-dimethylcyclopropanecarboxylic acid 2-methyl-4-oxo-3-(2,4-pentadienyl-2-cyclopenten-1-yl ester; chrysanthemumdicarboxylic acid monomethyl ester pyrethrolone ester). See pyrethrin (def. 2).

pyrethroid. **1:** pyrethrin-like. **2:** a pyrethroid insecticide, q.v.

pyrethroid insecticide. any of a class of synthetic, neurotoxic, pyrethrin-like insecticides with a mechanism of action similar to that of the naturally occurring pyrethrins. Examples are fenvalerate, resmethrin, bioresmethrin and permethrin.

pyrethrolone (2-methyl-4-oxo-3-(2,4-pentanedienyl)-2-cyclopentenol). a constituent of pyrethrins, it is an analog of allethrolone.

pyrethrosine. a naturally occurring epoxide auxin.

pyrethrum (Spanish chamomile). **1:** Spanish chamomile, the root of *Anacyclus pyrethrum*. **2:** an extract of pyrethrum flowers, q.v. **3:** pyrethrin (def. 2).

Pyrethrum. a former genus (Family Asteriaceae), the species of which are now included in the genus *Chrysanthemum*. See pyrethrum flowers.

pyrethrum extract. pyrethrin.

pyrethrum flowers (Dalmatian insect powder, Persian insect powder). **1:** the dried flowers of *Chrysanthemum (Pyrethrum) cinariifolium* and *C. coccineum* (= *C. roseum*), both indigenous to Dalmatia, Montenegro, and western Asia; and from those of *Anacyclus pyrethrum* which is native to Morocco. The powdered flowers or the extract (pyrethrum) are used in insecticidal preparations, as a scabicide, and in veterinary practice as an ectoparaciticide. **2: pyrethrin**. See also pyrethrum.

pyrethrum insecticide. pyrethrin.

pyretic. **1:** febrile; of or pertaining to a fever; having the characteristics of a fever. **2:** capable of causing a rise in body temperature.

pyreticosis. any febrile affection.

pyretogen. a rarely used synonym for the term pyrogen.

pyretogenetic, pyretogenic. pyrogenic.

pyretogenous. **1:** productive of fever. **2:** caused by fever.

pyrexia (pyrexy). fever.

pyrexial. of or pertaining to fever.

pyrexiogenic. pyrogenic.

Pyribenzamine Citrate™. tripelennamine citrate.

pyridine. a colorless or slightly yellow, slightly alkaline, water-soluble, flammable, liquid, $N(CH)_4CH$, with a penetrating offensive odor and burning taste. An alkaloid, pyridine is a single-ring aromatic amine found in *Conium*, *Lobelia*, and *Nicotiana* (tobacco). It is also recovered from coal tar or synthesized from acetaldehyde and ammonia. A nitrogen atom is included as part of the ring. It is a fire and explosion hazard and is toxic by all routes of exposure. It is used, for example, in chocolate flavorings for beverages, ice cream, ices, candy, and baked goods; as a denaturant for ethanol and antifreeze mixtures; as an initiator in vulcanization of rubber; in the synthesis of drugs and vitamins; and as a fungicide. Although only moderately toxic, pyridine has caused human fatalities. In addition to narcotic effects and mental depression, symptoms of intoxication include anorexia, nausea, and fatigue. Large quantities if inhaled or ingested, can cause headaches, vertigo, and tremor. It is an irritant to the eyes and mucous membranes.

2-pyridine aldoxime methiodide. pralidoxime iodide.

2-pyridine aldoxime methyl chloride. pralidoxime chloride.

4-pyridinecarboxylic acid hydrazide. isoniazid.

1-(3-pyridinyl)ethanone. methyl pyridyl ketone.

pyridostigmine bromide (3-[[(dimethylamino)-carbonyl]oxy]-1-methylpyridinium bromide; 3-hydroxy-1-methylpyridinium bromide dimethylcarbamate; 1-methyl-3-hydroxypyridinium bromide dimethylcarbamate; 3-(dimethylcarbamyloxy)-1-methylpyridinium bromide; Ro 1-5130; Mestonin Bromide; Kalymin; Regonol). a cholinergic drug, $C_9H_{13}BrN_2O_2$, used as an anti-nerve gas agent.

pyridoxine (5-hydroxy-6-methyl-3,4-pyridinedimethanol). a moderately toxic, water-soluble substance, it is one form of vitamin B_6, q.v. A number of drugs and toxic chemicals interfere with pyridoxine activity. The administration of supplemental pyridoxine may be beneficial in such cases. See also *Gyromitra esculenta*.

pyridoxine hydrochloride (5-hydroxy-6-methyl-3,4-pyridinedimethanol hydrochloride; pyridoxol hydrochloride; vitamin B_6 hydrochloride; pyridoxinium chloride; adermine hydrochloride; 2-methyl-3-hydroxy-4,5-bis(hydroxymethyl)pyridine hydrochloride; 3-hydroxy-4,5-dimethylol-α-picoline hydrochloride). the hydrochloride salt, $C_8H_{11}NO_3{\cdot}HCl$, of pyridoxine, in the form of colorless or white, moderately toxic crystals or powder. It occurs in many foodstuffs and is the form of pyridoxine given as a prophylactic against or in treatment of vitamin B_6 deficiency. It has also been used in the management of dermatoses, irradiation sickness, certain neuromuscular and neurological diseases, and treatment of nausea and vomiting in pregnancy. Pyridoxine hydrochloride is moderately toxic. Excess amounts are rapidly excreted. Acute oral overdosage can cause tonic convulsions and death in laboratory rats; lesions revealed on autopsy include cerebral cortical hemorrhage and adrenal gland enlargement. In one study, i.v. injections of 650 mg/kg caused instant tonic and clonic convulsions with complete recovery or death within five minutes.

pyridoxinium chloride. pyridoxine hydrochloride.

pyridoxol hydrochloride. pyridoxine hydrochloride.

***N*-3-pyridylmethyl-*N'*-*p*-nitrophenylurea**. pyriminyl.

β-pyridyl-α-*n*-methylpyrrolidine. nicotine.

2-(3-pyridyl)piperidine. anabasine.

2-(3-pyridyl)pyrrolidine. nornicotine.

pyrimidinetrione. barbituric acid.

2,4,5,6(1H,3H)-pyrimidinetetrone. alloxan.

2,4,6-(1*H*,3*H*,5*H*)-pyrimidinetrione. barbituric acid.

pyriminyl (N-(4-nitrophenyl)-*N'*-(3-pyridinylmethyl)urea; *N*-3-pyridylmethyl-*N'*-*p*-nitrophenylurea; pyrinuron). an extremely hazardous substance, $C_{13}H_{12}N_4O_3$, used as a rodenticide. It is extremely toxic to humans, laboratory rats, and feral Norway rats by ingestion. Systemic effects include distorted perceptions, hallucinations, muscular weakness, nausea and vomiting. When heated to decomposition, toxic NO_X fumes are released.

pyrinuron. pyriminil.

pyroacetic ether. acetone.

pyrocatechin. pyrocatechol.

pyrocatechol (1,2-benzenediol; 2-hydroxyphenol; pyrocatechin; catechol; 1,2-dihydroxybenzene; *o*-dihydroxybenzene). do not confuse with catechin which is sometimes called catechol (flavin). It is an organic compound used in electroplating, photography, the dyeing of fur, as a reagent, and in antioxidants, inks, and light stabilizers. It is also used as a topical antiseptic. It is very toxic, orally and *i.p.*, to laboratory mice. Skin contact can cause eczematous dermatitis in humans. Ingestion can produce hypertension and phenol-like symptoms; convulsions are more severe. Death from respiratory failure may result. Pyrocatechol is also carcinogenic.

Pyrodinium phoneus. a toxic dinoflagellate implicated in paralytic shellfish poisoning of humans in The Netherlands.

pyrogallic acid. pyrogallol.

pyrogallol (1,2,3-trihydroxybenzene; 1,2,3-benzenetriol; pyrogallic acid; C.I. 76515; C.I. Oxidation Base 32). a toxic acid, $C_6H_6O_3$, derived from gallic acid. It is used externally as an antimicrobial and irritant; also as a reagent.

pyrogen. pyretogen is a rarely used synonym. **1a:** any fever-inducing substance. **1b:** any chemical agent that produces a febrile response when administered parenterally to humans and certain animals. **1c:** any of a group of pyrogenic substances (chiefly polysaccharides) that are microbial in origin.

pyrogenetic. pyrogenic.

pyrogenic (pyrogenetic; pyretogenic; pyretogenetic; pyretogenous; pyrexiogenic). inducing fever; fever-inducing; productive of fever.

pyrexy. pyrexia.

pyrolan® (dimethylcarbamic acid 3-methyl-1-phenyl-1*H*-pyrazol-5-yl ester; 3-methyl-1-phenyl-5-pyrazolyl dimethylcarbamate; 1-phenyl-3-methyl-5-pyrazolyl dimethylcarbamate; 3-methyl-1-phenyl-5-pyrazolyl dimethylcarbamate). a crystalline substance, $C_{13}H_{15}N_3O_2$, formerly used as an insecticide. It is a reversible cholinesterase inhibitor, and is very to extremely toxic by ingestion. Though more toxic, the effects on pyrolan on humans are presumably similar to those of carbaryl.

pyrolysis. the decomposition of a chemical by the application of extreme heat.

pyrophosphate. See phosphate.

pyrophosphoric acid tetraethyl ester. tetraethyl pyrophosphate.

pyrosulfuric acid. fuming sulfuric acid.

pyrotic. caustic, burning.

pyrotoxin. **1:** a hypothetical toxin produced during a febrile condition. **2:** a toxic principle obtained from many normally nonpathogenic bacteria which produces fever and wasting when injected into laboratory animals or humans.

3-(2-pyrrolidinyl)pyridine. nornicotine.

pyrrolizidine. a highly toxic alkaloid that occurs in *Crotalaria* (rattlebox), *Echium* (viper's bugloss), *Heliotropium* (heliotrope), and *Senecio* (groundsel). See pyrrolizidine alkaloid.

pyrrolizidine alkaloid (senecio alkaloid). any of a class of highly toxic hepatotoxic alkaloids that cause exceptionally uniform and severe liver damage and are suspected human carcinogens. They are metabolically activated to electrophilic intermediates by cytochrome P-450. They occur, for example, in plants of the genera *Senecio*, *Crotolaria*, *Echium*, *Heliotropium*. and *Trichodesma*. See also monocrotaline, pyrrolizidine, *Callimorpha jacobea*.

Pyrus (pear trees). a genus of mostly deciduous trees (Family Rosaceae) native to cool-temperate regions; included are a number of valuable fruit trees and ornamentals. The leaves of some species contain arbutin. ***P. malus*** (apple). sometimes given as the scientific name for the common apple. The presumed ancestral species of the modern common apple tree are *Malus communis*, *M. pumila*, and *M. domestica*. *Malus* is the botanical name for all apples. Scientifically, in this author's view, *Malus* should be treated as a subgenus of *Pyrus*. Furthermore, because of the long history of domestication and hybridization, a species designation for the common (domesticated) apple is not warranted. The seeds are cyanogenic.

pyruvaldoxine. isonitrosoacetone.

q.v. (Latin for *quo vide*). which see.

Q_{10}. See temperature coefficient.

QCD 84924. (—)-Δ^1-3,4-*trans* tetrahydrocannabinol.

QNB. the benzilate ester of quinuclidinol, *q.v.*

QSAR. quantitative structure-activity relationship.

qt. quart.

quack grass. *Agropyron* spp. See Claviceps.

quaker buttons. nux vomica.

qualitative. descriptive; non-numerical; not measurable; not able to be counted. Qualitative differences are those that represent or are based on two or more discrete values or states of a system that cannot be represented numerically. Thus, the dose, amount, or concentration of a substance given in general terms is qualitative (e.g., a large overdose, highly concentrated). See also quantitative.

quantal data (incidence data). in toxicology, quantal data are those that indicate the number of individuals (e.g., test animals) that are affected by a toxicant under a specified set of conditions, but not the degree of harm that is done.

quantitative. an observation, measurement, estimate, or inference based on precise numbers; measurable or based on measurements; or based on ratios or other numerical values. This term is generally used in reference to a property that can be characterized or defined in terms of measurements, counts, ratios, or other numerical values. The amount or concentration of a substance given in precise or exact numerical terms is quantitative. See also qualitative.

quantitative structure-activity relationship (QSAR). the study of quantitative relationships between chemical structure and biological activity. Such relationships are either often unclear or restricted in application. There are strong relationships between the chemical properties and molecular structural properties of various substances and their toxicities. Assessing such relationships in a quantitative and reproducible way is often exceedingly difficult. One hope is that general principles may eventually be discovered that will allow the faithful prediction of chemical toxicities.

quartz dust. See dust.

quaternary ammonium compound (quaternary ammonium salt). **1a:** any of a type of salt formed from an amine by the addition of a proton with the production of a positive ion. The simplest of these salts are formed by the reaction of ammonia with an acid as in the formation of ammonium chloride (the fumes of which are toxic):

$$NH_3 + HCl \rightarrow NH_4{}^+Cl^-$$

All other amines, such as methylamine (itself, a strong irritant) can add protons:

$$CH_3NH_2 + HX \rightarrow [CH_3NH_3]^+ + X^-$$

where X^- is an acid radical. These salts are biocidal; they are often highly toxic to animals and are generally also phytocidal. Such reactions can also occur with heterogeneous nitrogen compounds such as the nitrogen bases, adenine, cytosine, thymine, and guanine. **1b:** any organic nitrogen compound with a molecular structure that includes a central nitrogen atom that is coordinately bonded to four positively charged organic groups and a negatively charged acid radical. These compounds are surface active agents that tend to adsorb to surfaces. They are generally used as disinfectants, sterilizers, cleansers, biocides, and also have applications in cosmetics. See, for example, octadecyldimethylbenzyl ammonium chloride and hexamethonium chloride. See also protein, paraquat, tetraethylammonium.

quaternary ammonium salt. See quaternary ammonium compound.

quebrachine. an alkaloid isolated from the dried bark (quebracho, def. 2) of *Aspidosperma quebrachoblanco* (Family Apocynaceae). It is an adrenergic blocking agent that is identical to yohimbine. It was formerly used therapeutically in cardiac dyspnea. *Cf.* yohimbine.

quebracho (aspidosperma). **1:** a common name for any of a number of Central and South American trees of differing genera, including *Aspidosperma*, the source of quebrachine and adiposamine. **2:** the dried bark of a genus of trees, *Aspidosperma* (Family Apocynaceae), especially *A. quebracho-blanco* (white quebracho); also called quebracho tannin and schinopsis lorentzii tannin). It contains a number of alkaloids including adiposamine, aspidospermine, and quebrachine. Quebracho tannin is toxic *i.p.* and *i.v.*, tumorigenic, and is a possible carcinogen. When decomposed by heating it emits acrid, irritating smoke and fumes. It has been used as a respiratory stimulant in emphysema, dyspnea, and chronic bronchitis. Chlorinated quebracho can be used to control nematodes and other parasitic worms in soil. **3:** the wood of quebracho. **white quebracho**. quebracho (def. 2).

quebracho tannin. See quebracho (def. 2).

Queen Anne's lace. *Daucus carota*.

"queen mother of poisons". aconite.

queen sago, queen-sago. *Cycas circinalis*.

queen's delight. *Stillingia treculeana*.

Quellada™. lindane.

quercetin (C.I. 75670; C.I. Natural Red 1; C.I. Natural Yellow 10; meletin; 3,5,7,3′,4′-pentahydroxyflavone; quercetine; quercetol; quercitin; 3′,4′,5,7-tetrahydroxyflavan-3-ol). the aglycon of quercitrin, rutin, and certain other glycosides. It occurs usually as the 3-rhamnoside and used therapeutically in the treatment of abnormal capillary fragility. It is poisonous at least by oral, subcutaneous, and intravenous routes, with teratogenic and reproductive effects. Tumorigenic, neoplastic, and carcinogenic, and mutagenic data are available; it is a questionable human carcinogen.

quercetine. quercetin.

quercetol. quercetin.

quercitin. quercetin.

Quercus (oak). a genus of perennial, woody, deciduous trees and shrubs (Family Fagaceae). Most, if not all species of which have similar toxicities and produce comparable symptoms and lesions. Losses of livestock to oak poisoning in the southwestern United States alone is economically very substantial. Toxicity is low, but oak buds and immature leaves are highly favored by ruminants and are browsed preferentially. Symptoms may appear within a week when oak is freely available and usually run a course terminating fatally in 3 - 10 days in untreated animals. Symptoms commonly include anorexia, rumen stasis, abdominal pain, constipation, rough coat and dry muzzle, intense thirst and frequent urination. Typical lesions are those of nephritis and gastroenteritis. The kidney is pale, enlarged and turgid; often with petechial hemorrhages. Some tubules contain precipitated albumin. Many of the proximal and ascending renal tubules lack epithelial cells and contain a mass that contains evenly distributed degenerated cell nuclei and stains red with eosin-hematoxylin. The abomasum and small intestines are inflamed and hemorrhagic and petechial hemorrhages occur in organs and membranes throughout the body. Edema and copious amounts of fluid in the body cavities are common findings. See also tannin, gallic acid, oak poisoning, and acorn poisoning.

quercus poisoning. oak poisoning.

quicklime. calcium oxide.

quicksilver. elemental mercury. See mercury.

quicksilver vermillion. red mercuric sulfide (See mercuric sulfide).

quina. cinchona.

Quinacrine. quinine.

quinacrine mustard (9-(4-(bis-β-chloroethylamino)-1-methylbutylamino)-6-chloro-2-methoxyacridine). a deadly poison, $C_{23}H_{28}Cl_3N_3O$, *i.v.* and *i.p.* and probably a human mutagen. It releases very toxic fumes (Cl^- and NO_x) when heated to decomposition.

quinaquina. cinchona.

quinidine. a very toxic sterioisomer of quinine, it is a cardiac depressant and is also an effective antimalarial at doses similar to that of quinine, *q.v.*

α-quinidine. cinchonidine.

quinine (Quinacrine, chloroquine, Atabrine, Plaquenil). a white, odorless, very bitter, amorphous powder or crystalline substance, $C_{20}H_{24}N_2O_2{\cdot}3H_2O$, that is very slightly soluble in water but soluble in ethanol, ether, carbon disulfide, chloroform, glycerol, alkalies, and acids (with which it forms salts). It is the most important alkaloid isolated from cinchona. It is used most notably in medicine to treat cerebral malaria and other severe cases of malignant tertian malaria; it is also used to treat cases caused by chloroquine-resistant strains of *Plasmodium falciparum*. Quinidine, a stereoisomer of quinine is also an effective antimalarial. Quinine is used also as an antipyretic, analgesic, sclerosing agent, stomachic. It is sometimes used as an oxytocic and in the treatment of atrial fibrillation, and certain myopathies. Quinine is a very toxic cardiac depressant. Overdosage can cause blurred vision, a progressive ringing in the ears, muscular weakness, hypotension, hemoglobinuria, oliguria, and cardiac irregularities. Injection or ingestion of large doses causes sudden onset of cardiac depression which may be accompanied by convulsions; death due to heart failure or respiratory arrest may ensue. Quinine is a skin irritant and, by ingestion, is a cause of toxic amaurosis. See also amaurosis, amblyopia.

quininism. cinchonism.

quinoform. iodochlorhydroxyquin.

quinol. hydroquinone.

quinone (chinone). **1a: *p*-quinone** (2,5-cyclohexadiene-1,4-dione; 1,4-cyclohexadienedione; 1,4-benzoquinone; *p*-benzoquinone; 1,4-dioxybenzene). a yellow substance, $CO(CHCH)_2CO$, with monoclinic crystals, an irritant vapor and a penetrating chlorine-like odor. It is soluble in ethanol, ethyl ether, and alkalies. It is sometimes simply referred to as benzoquinone. This substance and related benzoquinones, *q.v.*, are secreted by a wide variety of insects, sometimes mixed with hydrocarbons that may promote penetration of the toxicant through the cuticle of the victim. *p*-quinone is toxic or very toxic to humans and laboratory animals by any route of exposure. It causes severe damage to the skin and mucous membranes on contact, whether in the solid, liquid, or vapor state. It can discolor and damage the skin; prolonged exposure causes swelling, ulceration, and even necrosis. Even brief exposures to high concentrations of vapor can produce serious disturbances of vision. **1b: *o*-quinone**. a colorless or red isomer of *p*-quinone with essentially similar toxicologic properties. It is used in the manufacture of dyes. **2:** benzoquinone.

***o*-quinone**. See quinone.

***p*-quinone**. See quinone.

quinoneimine. a metabolically active, hepatotoxic form of paracetamol.

***N′*-(2-quinoxalinyl)sulfanilamide**. sulfaquinoxaline.

quinquina. cinchona.

3-quinuclidinol (1-azabicyclo[2.2.2]octan-3-ol; 3-hydroxyquinuclidine). a very water-soluble compound used therapeutically as a hypotensive agent. The benzylate ester (3-quinuclidinyl benzilate; BZ; QNB) is an incapacitating chemical warfare agent; and the acetate ester (aceclidine; 3-acetoxyquinuclidine; 3-quinuclidinyl acetate) is cholinergic. See warfare agent.

3-quinuclidinyl acetate. the acetate ester of quinuclidinol, *q.v.*

3-quinuclidinyl benzilate. the benzilate ester of quinuclidinol, *q.v.*

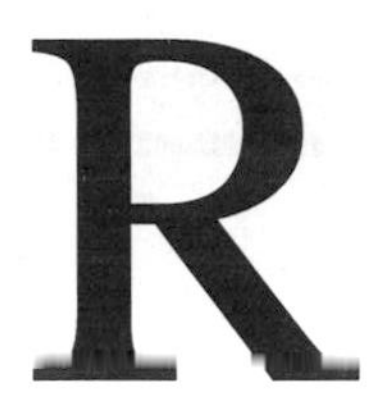

®. Registered Trademark.

r. abbreviation for racemic, Roentgen.

R. abbreviation for gas constant; electrical resistance; empiric risk (See risk); radical (usually an alkyl or aryl group, thus ROH is an alcohol); respiration; respiratory exchange ratio; Roentgen; remainder of a chemical formula; the unit of resistance in the cardiovascular system.

R-1132. diphenoxylate.

Ra. the symbol for radium.

rabbit brush. *Tetradymia*.

rabbit's pea. *Tephrosia virginiana*.

rabbitfishes. common name of fishes of the family Siganidae.

rabo de chuncha. *Bothrops punctatus*.

rabo de raton. *Bothrops neglectus*.

race (subspecies; variety). a subspecies or a geographic subdivision of a species. A geographically defined group of breeding populations that differs from other similar groups in the frequency of one or more heritable traits.

racemate. a racemic compound or the salt or ester of such a compound. See also racemic.

raceme. a rather elongated and slender inflorescence in which the pedicels attach to a simple central axis.

racemethorphan ((±)-3-methoxy-17-methylmorphinan; *dl-cis*-1,3,4,9,10, 10a-hexahydro-6-methoxy-11-methyl-2*H*-10,4a-iminoethanophenanthrene; *dl-cis*-1,2,3,9,10,10a-hexahydro-6-methoxy-11-methyl-4*H*-10,4a-iminoethanophenanthrene;deoxydihydrothebacodine; methorphan). commonly called methorphan, it is a widely used, slightly toxic, cough suppressant that acts by elevating the threshold of the CNS cough reflex. It is the d-isomer of levorphanol (a codeine analog). It does not appear to be addictive. It is a weak CNS depressant.

racemic (r). of or pertaining to an optically inactive mixture, due to the presence of an equal number of separable dextro- and levorotatory substances.

racers, West Indian. snakes of the genus *Alsophis*.

rad. See RAD.

RAD (rad). an acronym for radiation absorbed dose (See dose). 1 RAD is equivalent to the absorption by tissue of 10^{-2} joules of energy by 1 kg of matter. The biological effect of 1 RAD varies with the type of ionizing radiation. The RAD has been superseded by the gray, an SI unit. 1 RAD is equivalent to 0.01 grays.

Radianthus paumotensis. a poisonous sea anemone (Anthozoa) found in the tropics of the Indo-Pacific region. It is unsafe to eat even if cooked. There is no known antidote. See Anthozoa.

radiation. **1:** divergence in all directions from a center or focal point. **2a:** the emission and propagation (transmission) of waves transmitting energy through space or through some medium; for example, the emission and propagation of electromagnetic, sound, or elastic waves. When unqualified, the term usually refers to electromagnetic radiation. **2b:** the energy itself; radiant energy. **3:** corpuscular radiation, including charged (e.g., alpha particles, electrons, protons, deuterons) and uncharged (e.g., neutrons) particles; a stream of particles, such as electrons, neutrons, protons, α-particles, or high-energy photons, or a mix ture of these. **4:** a ray (e.g., of light). **5:** the emission or use of electromagnetic radiation (e.g., light, short radio waves, ultraviolet light, X-rays) for any purpose, as in chemical analysis, diagnosis, or therapy. **6:** the process of rapid evolutionary diversification and dispersal of plant or animal populations, species and higher taxonomic categories into new niches and habitats. **background radiation**. **1:** the sum of all natural radioactive radiation (cosmic and terrestrial) at a given location, including elementary particles from outer space (cosmic radiation) and terrestrial radiation from naturally occurring isotopes. **2:** the level of radioactive radiation (natural and anthropogenic) that would occur at a given location if the level were not influenced by a specified source. **beta radiation**. radiant energy from a source of beta rays. **internal radiation**. ionizing radiation emanating from radionuclides deposited within the body as by ingestion, inhalation, absorption (e.g., through the surface of the body or external gills), or injection. **ionizing radiation**. **1:** particles (e.g., beta particles, neutrons, alpha particles) or electromagnetic radiation (X rays, gamma rays) of sufficient energy to directly produce ionization in their passage through, for example, living cells and tissue. Such radiation can damage or destroy living cells and tissues by altering their chemical makeup. Ionizing radiation can originate from natural or man-made sources. **2:** particles that are capable of interacting with atomic nuclei such that sufficient energy is released to produce ionization. **nonionizing radiation**. ultraviolet light is the most widespread type of nonionizing radiation of toxicological interest. It is a component of sunlight, but much is absorbed by the atmosphere. While ultraviolet light can penetrate only superficial cells and tissues of the animal body, it damages RNA and DNA, one effect of which is the development of skin cancer. Radio-frequency electromagnetic radiation (microwaves) heats tissues and can be very damaging. Because microwave ovens, though used widely, are well-shielded with safety cutoffs that stop radiation emission when the door is open, microwaves do not pose much risk. Other types of nonionizing radiation to which humans may be exposed are magnetic fields and ultrasound, both of which can be damaging.

radiation absorbed dose. See dose.

radiation barrier. See barrier.

radiation biology. actinobiology.

radiation burn. See burn.

radiation dermatitis. radiodermatitis.

radiation ecology. the science that deals with the effects of radiation on populations, communities, and ecosystems; on the relationships between living organisms and environment; and on the fates and effects of radioactive substances in the environment.

radiation equivalent man. See REM.

radiation standard. a regulation that sets maximum exposure limits for protection of the public from radioactive materials.

radiation syndrome. See acute radiation syndrome.

radical. **1:** a plant part that arises from a root close to the ground such as a basal leaf or the stem of a flower. **2:** an ionic group with one or more positive or negative charges (e.g., OH^-, NH_4^+, SO_4^{--}). Such groups do not exist in the free state but as a unit in a compound. **free radical**. a usually short-lived, highly reactive fragment formed by the splitting of a molecular bond. It has one or more unpaired electrons. Such a species is symbolized in chemical formulas by a dot (e.g., $(C_2H_5)^{\bullet}$, $Cl^{\bullet}$), which denotes that each has an unpaired electron. Even though short-lived, free radicals initiate many kinds of chain reactions. Such a reaction occurs when free radicals attack a substrate and are regenerated during the progress of the reaction. Free radicals form, for example, in tissues exposed to ionizing radiation and are thus a proximate cause of radiation damage that may occur. See also active oxygen, superoxide.

radicicol (monorden). a naturally occurring epoxide antibiotic mycotoxin produced by *Nectria radicicola*.

radio-labeled. pertaining to a substance to which a radioactive tracer has been added.

radioaction. radioactivity.

radioactive. of, pertaining to, or having the quality of radioactivity. **radioactive substance**. any element, compound, or mixture that exhibits radioactivity. Many of these substances are extremely toxic or otherwise hazardous. Those who generally incur the greatest risks from exposure to radioactive substances are uranium miners, radiologists, and workers in the nuclear industry. Other potential sources of serious or even catastrophic exposure are nuclear accidents (e.g., core meltdown at a nuclear power plant), radioactive wastes, and nuclear warfare.

radioactive waste. any waste that contains radioactive materials. Sources of such waste include (1) high level waste produced during the reprocessing of radioactive fuel following usage in a nuclear reactor, (2) transuranic waste resulting from the reprocessing and fabrication of plutonium to make nuclear weapons, and (3) low level wastes contained in the solutions, rags, clothes, and tools used to clean and decontaminate nuclear reactors. Nuclear wastes remain hazardous for extremely long periods of time. For example, plutonium 239, the fuel of breeder reactors, remains extremely hazardous for 24,000 years. Secure, long-term storage of radioactive wastes is a problem that has yet to be fully resolved.

radioactive fallout. See fallout.

radioactive half-life. See half-life.

radioactive isotope. an isotope of an element that is radioactive; a radionuclide.

radioactive label. See label.

radioactive tag. radioactive label (See label).

radioactivity. **1:** the property of some atomic nuclei of spontaneously disintegrating with emission of electromagnetic radiation (gamma rays) and subatomic ionizing particles (alpha and beta rays). Radioactivity is a property of all chemical elements of atomic number above 83, and can be induced in all other known elements. **2:** a property of radioactive waves or particles generated from atomic nuclei (See def. 1) or from other natural sources (e.g., cosmic rays). Particularly important from a toxicological standpoint is radioactivity released into human space or into the environment from human activities, i.e., that released by nuclear reactors, nuclear-weapon tests, radioactive wastes, and color TV. The presence of radioactive material in the atmosphere is a serious pollution problem.

radiobiology. the study of radiation effects, especially the harmful effects, on living systems; included is the study of the mechanisms of action and of system response to radiation.

radiocesium. cesium-137.

radiochemical. of or pertaining to radioactive chemicals or to radiochemistry.

radiochemistry. the branch of chemistry that deals with radioactive substances.

radiodermatitis (radiation dermatitis; Roentgen ray dermatitis; X-ray dermatitis). a dermatitis with destructive changes in the dermis due to exposure to ionizing radiation (e.g., gamma rays, Roentgen rays). *Cf.* radiodermatitis, radioepidermitis, radioepitheliitis).

radioepidermitis. destructive changes in the epidermis due to exposure to ionizing radiation.

radiogenesis (actinogenesis). the production or formation of rays (= actinogenesis) or of radiation, especially ionizing radiation.

radiogenic (actinogenic). characteristic of or pertaining to the production or formation of rays or of radiation, especially ionizing radiation.

radioimmunoassay (RIA). a very sensitive competitive binding assay used to detect and measure substances by the use of radioactively labelled specific antibodies or antigens.

radioisotope. a radioactive isotope.

radiomimetic. resembling radiation or radiation effects; frequently applied to mutagenic chemicals.

radiomimetic drug. an immunosuppresive drug that mimics the action of ionizing radiation.

radionecrosis. necrosis produced by ionizing radiation.

radioneuritis. neuritis caused by prolonged or repeated exposure to ionizing radiation.

radionuclide. a radioactive nuclide. Such are physically characterized by their atomic mass and atomic number. Many can cause radiation sickness and some can cause serious health or biological effects due to their chemical nature. Some would be highly toxic even if they were not radioactive.

radiopathology. the science, a branch of pathology, that treats the effects of radioactive substances on cells and tissues. *Cf.* radiobiology.

radiophosphorus (labeled phosphorus; phosphorus-32; phosphorus-33). either of two radioactive phosphorus isotopes (^{32}P, ^{33}P) with atomic weights, respectively, of 32 and 33. ^{32}P is a pure beta-emitter with a half-life of 14.3 days. It is used in solution or in colloidal form as a metabolic and environmental tracer and as a therapeutic agent in certain diseases of the hematopoietic and osseus systems.

radiopoison. a radioactive poison.

radioresistant. **1a:** resistant to the effects of radiation. **1b:** of or pertaining to cells or tissues that have a relatively high resistance to the effects of radiation, especially ionizing radiation. **1c:** of or pertaining to cells or tissues that are not destroyed by ionizing radiation in the usual therapeutic range of dosage.

radiosensitive. of or pertaining to cells or tissues that are readily affected by radiation, especially ionizing radiation.

radiosensitivity. the condition of being readily affected by radiation, especially ionizing radiation.

radiotoxemia. an obsolete term for radiation sickness.

radiotoxicity. the capacity of a radionuclide to produce local or systemic toxic effects in an exposed organism by virtue of the ionizing radiation released. In practice, radiotoxicity usually indicates the relative hazard of radionuclides introduced into the human body.

radiotropic. sensitive to radiation.

radish. *Raphanus sativus*.

radium (symbol, Ra)[Z = 88; most stable isotope = ^{226}Ra; $T_{1/2}$ = 1602 yr]. a radioactive, brilliant white, luminescent, alkaline earth metal that occurs in uranium ores. It undergoes spontaneous disintegration to yield radon. Exposure to radium via contact, inhalation, or ingestion may cause injury to the skin, blood dyscrasias, pulmonary carcinoma, osteitis, and osteogenic sarcoma. It is used in radiography of metals, as a source of radon, to some degree therapeutically as an antineoplastic radiation source, and was formerly used in luminous paints.

radium bromide. a poisonous, radioactive compound, $RaBr_2$, with white crystals that become pink or yellow on standing. It is soluble in water and ethanol and melts at 728°C. Radium bromide is used as a luminous paint, in cancer therapy, and in physical research. It is hazardous chiefly due to the radium content, but is also corrosive to tissue.

radium carbonate. an amorphous, nearly water-insoluble, poisonous, white (when pure) radioactive powder, $RaCO_3$. It is often yellow, orange, or pink due to the presence of impurities. Radium carbonate is hazardous chiefly due to the radium content.

radium chloride. a poisonous, corrosive, irritant, radioactive compound, $RaCl_2$, with yellowish-white crystals that become pink or yellow on standing; it melts at 1000°C. It is hazardous chiefly due to the radium content. It is used in cancer therapy and physical research.

radium emanation (emanation). obsolete synonym for radon.

Radius of Vulnerable Zone. the maximum distance from the point of release of a toxic or otherwise hazardous substance in which the airborne concentration could reach the level of concern under specified weather conditions.

radon (symbol, Rn)[Z = 86; most stable isotope = ^{222}Rn, $T_{1/2}$ = 30.0 s]. a colorless, odorless, inert, monatomic radioactive element of the rare gas group. In some locations, it is an important indoor pollutant. The maximum permissible concentration of ^{222}Rn in air in the United States is 10^{-8} μ-Curie/cc. (National Bureau of Standards Handbook 69:79, 1959). Radium emanation is an obsolete synonym for radon. Radon is used therapeutically as an antineoplastic radiation source.

radon daughter. See cigarette smoke.

radon decay products. the immediate products of the radon decay chain. These include Po-218, Pb-214, Bi-214, and Po-214, which have an average combined half-life of about 30 minutes.

radula. a flexible, rasp-like or tongue-like structure in the anterior part of the digestive tract of all mollusks except the Pelecypoda. It bears a series of transverse rows of teeth that are moved rapidly back and forth against the upper surface of the pharynx, thus macerating ingested material. In some groups it can be extended from the mouth and used to scrape or drill.

ragwort. *Senecio aureus*.

rain lily. *Zephyranthes atamasco*.

rainbow trout. *Salmo gairdneri*.

raj nag. *Ophiophagus hannah*.

Raja. common genus of skates of the suborder Batoidea.

Raji. Batoidea.

Rajida. Batoidea.

rale. any abnormal respiratory sound detected by auscultation.

Ramaria (corals). a genus of mushrooms (Family Clavariaceae). Most species are edible, but the two listed here cause gastric upset in humans. They are among the few corals considered toxic. *R. formosa* (formerly *Clavaria formosa*; yellow-tipped coral). It is widely distributed in North America, growing mainly under conifers, especially Douglas-fir and western hemlock. *R. gelatinosa* (jellied-base coral). It grows under western hemlock in the Pacific Northwest region of the United States.

Rampart. phorate.

ramycin. fusidic acid.

random error. See error.

random process (stochastic process). a process controlled in part by a mechanism that is comprised of, or is responsive to, information which is generated or received at random (or nearly random) intervals. A process guided by chance events (internal or external).

random sample. See sample.

random variable. a variable in statistics that can assume any of a given set of values with an assigned probability.

Ranidae. one of three families of true frogs. They are nearly cosmopolitan in distribution, but are absent from the Australian and Neotropical regions. The upper jaw is toothed, the tongue is forked posteriorly.

Ranunculaceae (crowfoot or buttercup family). a family of herbs, at least some members of which contain a glycoside (ranunculin) that produces protoanemonin on enzymatic hydrolysis. Species associated with intoxications include baneberry (*Actaea pachypoda*), buttercups (*Ranunculus* spp.), marsh marigold (*Caltha palustris*), pasque flowers (*Anemone* spp.), and clematis (*Clematis* spp.). All parts are toxic. Symptoms following ingestion may include paresthesia, a burning sensation of the mouth and skin, nausea, vomiting, hypotension, weak pulse, convulsions. See also Ranunculus, anemenol, ranunculin.

ranunculin. a nontoxic glycoside found in a number of members of the Ranuculaceae (buttercups, cowslips). Ranunculin readily undergoes hydrolysis, releasing a toxic, volatile aglycone (protoanemonin) which produces severe irritation of the mouth and digestive system in livestock; the sap of some species can burn the skin. See also Ranunculaceae, *Anemone*, *Ranunculus*.

Ranunculus. (buttercup, crowfoot). a very large genus of toxic herbs, the type genus of the family Ranunculaceae. All species contain a moderately toxic vesicant oil, protoanemonin, in varying amounts. All parts of the plant are toxic. Ingestion may produce a severe gastroenteritis in which salivation, diarrhea, and signs of abdominal pain are typical. Oral tissues may be inflamed. All forms of livestock are susceptible. Buttercups are distasteful to most livestock and poisoning is rare. Protoanemonin is unstable and buttercup hay is therefore nontoxic. See also anemonin, dermatitis (primary contact dermatitis), Ranunculaceae.

rape. **1:** *Brassica napus*. **2:** a usually violent crime associated with heterosexual or homosexual sex, with or without penetration. In some cases alcohol or other drugs are used to reduce the ability of the victim to resist.

rape blindness. a nervous syndrome of cattle and sheep. See syndrome.

rape poisoning. poisoning by ingestion of *Brassica napus*, q.v. Symptoms in grazing ruminants include ruminal atony, adipsia, aphagia, and death unless suitable food and care are given.

rape scald. a phototoxic syndrome of livestock. Some species of *Brassica*, *Erodium*, *Medicago*, and *Trifolium* have been implicated.

rapeseed meal. a feed concentrate used extensively during World War II. It is toxic to livestock, but its effects can be partly eliminated by hot water extraction of the meal or supplementation with iodine. See *Brassica napus*.

rapeseed oil. the oil expressed from the seeds of *Brassica campestris*. It is used in the manufacture of margarine, soaps, and lubricants.

raphania (rhaphania). a spasmodic disease due to ingestion of the seeds of *Raphanus sativus*. The condition is allied with ergotism.

Raphanus (radishes). a genus of herbs that includes both wild and cultivated radishes. ***R. raphanistrum*** (wild radish; jointed charlock). an annual plant (Family Brassicaceae) that is dangerous to livestock in North America and Europe. It has been implicated in poisonings of cattle and sheep, including lambs. The animals may develop severe gastroenteritis, apparently because of the mustard oil content. See also raphania. ***R. rusticanus***. an outdated scientific name for *Armoracia rusticana* (horseradish). ***Raphanus sativus*** (radish; wild radish). an herb (Family Brassicaceae) from which the cultivated radish was developed. The seeds of this plant are the cause of raphania. Toxic concentrations of nitrate have also been reported.

rapture of the deep. See nitrogen narcosis.

rare earths, rare earth metals. See lanthanoid.

rascasse. *Scorpaena porcus*.

rash. a vernacular term for any temporary cutaneous eruption. See, for example, urticaria, nettle. **brown-tail rash**. See dermatitis (brown-tail moth dermatitis).

rat. a common name that includes a wide variety of medium-sized rodents, such as members of the following genera: *Rattus, Sigmodon,* and *Dipodomys*. See also Cricetidae, Muridae. **albino rat**. **1:** a phenotype that sometimes occurs in various populations of *Rattus*. The fur is white and the eyes are pink, due to a lack of pigment. **2:** any strain of laboratory rat inbred for albinism. **laboratory rat**. any rat bred and reared in captivity, originally derived from a small population of *Rattus norvegicus*, the Norway rat, for use in laboratory studies. There are many strains and lines of such rats, all of which are now genetically far removed from the wild type. While they can be considered congeneric with *Rattus*, they should not be characterized as *R. norvegicus*, nor referred to as the Norway rat. Laboratory rats no longer have any counterparts in nature. Laboratory rats are often used as test species in toxicology, but often respond differently to a specific chemical than other mammals including humans and other rodents. Thus, for example, the adult laboratory rat is highly resistant to the action of cardiac glycosides and to the fetotoxic effects of thalidomide. **Wistar rat**. an inbred strain of laboratory rat that is homozygous at most loci. This strain is the result of strict brother-sister inbreeding over many generations to develop animals of closely simililar genetic composition that are suitable for research.

rat killer. rat poison. See poison (rat poison).

ratfish. See Chimaeridae, Holocephali.

raticide. an agent that destroys rats.

ratsbane. elemental arsenic.

rattle. **1:** a sound or rale heard on auscultation. **2:** the jointed series of cornified structures at the tip of the tail of most rattlesnakes. When the snake is disturbed the tip of the tail is upturned and the rattle vibrates rapidly, making a distinctive buzzing sound that warns one that a rattlesnake is near and may be about to strike. See rattlesnake, *Crotalus*, *Sistrurus miliarius*. **3:** the distinctive buzzing sound produced by arattle (def. 2) **death rattle**. a gurgling or crack-gurgling or crackling sound heard in the trachea of a dying person.

rattlebox. See *Crotalaria sagittalis*, *Sesbania*.

rattlebrush. See *Sesbania*.

rattler. a vernacular term applied especially to those rattlesnakes that normally have a rattle. See rattlesnake, rattle, *Crotalus*, *Sistrurus*.

rattlesnake (rattler). the common name of some 25 species of venomous American snakes (Family Crotalidae) of the genera *Crotalus* and *Sistrurus*. Most occur in the southwestern United States and northern Mexico, but the range of *C. durissus* extends into southern South America, while *C. unicolor* occurs only on the island of Aruba in the Caribbean Archipelago near the coast of Venezuela. Most species are easily identified by the distinctive horny, segmented rattle at the tip of the tail (but See *Sistrurus miliarius*). All species are live-bearing, have a facial (loreal) pit, keeled scales and undivided subcaudal scales. About 70% of venomous snakebites in the United States and nearly all deaths are caused by rattlesnake bites.

rattlesnake poisoning. poisoning by any species of rattlesnake. In moderate to severe envenomations, ecchymosis usually appears about the bite area within 3-6 hours. Another early effect is a sharp rise in packed cell volume, although in severe cases, the hematocrit may fall. In North America it is most severe following bites by *Crotalus adamanteus*, *C. atrox*, *C. v. viridis*, and *C. v. oreganus*. Necrosis about the bite is common in untreated cases. The surrounding superficial blood vessels may be thrombosed. Effects may include hemorrhaging from the gums, hematemesis, melena, and hematuria, prolonged clotting times, low platelet counts. Effects of the venom usually peak by the 4th day.

rattleweed. **1:** *Astragalus diphysus* (= *A. lentiginosus* var *diphysus*), a true locoweed. **2:** *Baptisia tinctoria*.

Rattus. a genus of omnivorous rats (Family Muridae). ***R. norvegicus*** (Norway rat; house rat; barn rat; wharf rat; domestic rat; brown rat; sewer rat). the most common and widely distributed rat. It is a pest and disease carrier in many human communities; strains originating more than a century ago from poorly defined populations of *R. norvegicus* stock are the most important subjects used in many fields of biological research (e.g., in genetics, immunology, neurobehavioral sciences, oncology, pharmacology, physiology, transplantation). Such are also extremely important in toxicological research and testing, especially to predict the effects of chemicals on humans. See also rat (laboratory rat). ***R. rattus*** (black rat). it is smaller and darker in color than the Norwegian, sewer, or brown rat *(Rattus norvegicus)* and has longer ears and tail. It is much less common and appears to be driven out wherever it is in contact with *R. norvegicus*. The black rat is the species that ordinarily transmits plague to humans by means of the flea, *Xenopsylla cheopis*.

Rauhschuppige bush viper. *Atheris squamigera*.

Rauwolfia alkaloid. any indole alkaloid of *Rauwolfia serpentina*, *q.v.*

Rauwolfia serpentina. a small Asiatic shrub (Family Apocynaceae) found from India to Sumatra. The powdered whole root contains many indole alkaloids (e.g., ajmaline, reserpine, yohimbine, serpentine, and serpentinine); approximately 50% of the total activity is due to reserpine. They are potent antihypertensives and sedatives of low conventional toxicity.

ray fungi. a common name of Actinomyces.

rayless goldenrod. See *Haplopappus*.

razor clam. *Ensis directus*.

Rb. the symbol for rubidium.

rbc, RBC (red blood cell). erythrocyte.

RBC. erythrocyte.

RBE. relative biological effectiveness.

Rc-toxin. deoxynivalenolmonoacetate.

RCL I. See ricin mixture, ricin agglutinin.

RCL II. See ricin mixture, ricin agglutinin.

RCL III. See ricin mixture, ricin.

RCL IV. See ricin mixture, ricin.

Rd-toxin. deoxynivalenol.

Re. the symbol for rhenium.

reactant. **1:** ready to or capable of reacting; that which reacts. **2:** a substance that enters into or undergoes a reaction.

reaction. **1:** the response of an organism or a component part, such as a muscle or nerve, to a stimulus. **2:** chemical reaction; a process whereby a chemical change takes place, such that one or more substances are transformed into one or more other substances. **3:** an opposing or countervailing action. **adverse drug reaction**. an unintended adverse reaction to a therapeutic drug. **adverse reaction**. a reaction (e.g., to a drug or other xenobiotic) which is harmful, injurious, damaging, or even life-threatening to the recipient organism. **allergic reaction**. (allergic response; sensitization reaction). **1a:** a reaction to a second or subsequent exposure to an antigen or allergen that is more intense than that experienced by most individuals. **1b:** an immune response characterized by an intense (usually detrimental) physiologic reaction to a substance that causes no symptoms in nonsensitive individuals. Examples include systemic lupus erythematosus (SLE), serum sickness, tuberculosis and hemolytic anemia. **2:** those hypersensitivity (anaphylactic) reactions produced by IgE antibodies. Included are common reactions such as rhinitis or asthma produced by pollens and danders. Not included in this definition are delayed hypersensitivity reactions such as contact dermatitis caused by poison ivy, drugs, cosmetics and certain metals, and reactions not mediated by IgE. Note: an allergic reaction is not considered a poisoning in the narrow sense because it depends on an individual, preexisting sensitivity. *Cf.* anaphylactic shock, antibiotic, drug allergy, hypersensitivity. **anamnestic reaction**. anamnestic immune response. **anaphylactic reaction**. anaphylaxis. **antigen-antibody reaction**. any reaction or combination of an antibody with its specific antigen. Such reactions may result in agglutination, precipitation, neutralization (of exotoxin), complement-fixation, or increased susceptibility of the antigen to phagocytosis. This phenomenon occurs *in vivo* and *in vitro* and provides the basis for immunity. See also complement fixation, immune response. **chemical reaction**. reaction (def. 2). **complement-fixation reaction**. the process of complement fixation. **conjugation reaction** (phase 2 reaction). a reaction in which a conjugating agent unites with a metabolite (e.g., an amino acid, glutathione, a sugar, a sulfate) to yield a new compound (the conjugate). Two types of conjugation reactions are recognized: **type I conjugation reaction** (as in the formation of a glycoside or sulfate) is the union of an activated conjugating agent with a substrate to yield a conjugated product. **type II conjugation reaction** is the union of an activated substrate with an amino acid to yield a conjugated product. The product in the latter case is almost always more polar and thus less lipid soluble and more water soluble than the parent compound and is typically less toxic and more easily excreted. See also activation, metabolic reactions, glucuronides, conjugation (amino acid conjugation), glycoside conjugation, sulfate conjugation. **cross reaction**. a reaction between an antiserum and an antigen or antigen complex that did not evoke the specific antibodies of the antiserum. This occurs because each antigen has one or more antigenic determinants that are equivalent to cerrtain of those of the other antigen. **cytotoxic reaction** (cytotoxic hypersensitivity; class I hypersensitivity). a complement-dependent, rapid-acting, humoral immunologic (allergic) reaction in which noncytotropic IgG or IgM antibody combines with foreign cells or with cell surface antigens. The antigen-antibody complex so formed activates complement, which rapidly damages the cells. In the absence of complement, T lymphocyte activity may be enhanced with increased vulnerability of the foreign cells to phagocytosis. See hypersensitivity, immune response. **detoxication reaction**. **1:** detoxication. **2:** any reaction in a detoxication sequence. See detoxication. **detoxification reaction**. any metabolic reaction involved in eliminating the toxic properties or effects of a substance. There are four basic types of detoxification reactions: conjugation, hydrolysis, oxidation, and reduction. **early reaction**. immediate reaction. **false-negative reaction**. the mistaken assumption, inference, or conclusion that a response or a test result is negative, when in fact it is positive. **false-positive reaction,** an erroneous or mistakenly positive response; a result inferred to be positive when in fact it is negative. **fixation reaction**. See complement fixation. **focal reaction** (local reaction). any reaction which occurs at or near the point of entrance of an injection, an infecting organism, or at the site of an envenomation. **immediate reaction** (early reaction). either a focal or systemic reaction that presents within about an hour following exposure to an antigen to which an individual has been sensitized. **immune reaction**. a reaction that indicates the presence of antibodies in the blood, a condition that is indicative of a high degree of immunity or resistance to a specific antigen. **irreversible reaction**. **1:** any reaction in which the reacting system cannot (at least under normal circumstances) return to the original state. **2:** a chemical reaction that results in a permanent chemical or physicochemical change in the reacting system. **3:** a reaction or response of a living organism to a xenobiotic or pathogen that results in permanent injury or morbidity. **local reaction**. focal reaction. **metabolic reduction reaction**. See reduction reaction. **oxidation reaction**. a chemical reaction in which (1) electrons are removed from an atom or ion; (2) the chemical addition of oxygen to a compound; or (3) the loss of hydrogen from a compound. *Cf.* reduction reaction. **phase 2 reaction**. conjugation reaction. **redox reaction**. a chemical reaction in which electrons are transferred from one reactant (oxidation) to another (reduction). See also oxidation, reduction. **reduction reaction**. a chemical reaction that is the opposite of an oxidation reaction. Such reactions may involve (1) acceptance of one or more electrons by an atom or ion; (2) the elimination of oxygen from a compound; or (3) the introduction of hydrogen to a compound. See also redox reaction. A number of functional groups of xenobiotics that are of toxicological importance are reduced metabolically (e.g., aldehydes, alkenes, arsenates, disulfides, ketones, nitrogen oxides, sulfoxides). Such reductions are catalyzed by reductases. See reductase. **toxification reaction**. **1a:** a reaction that converts a substance into a more toxic substance. **1b:** a reaction in a metabolic toxification sequence. *Cf.* detoxication reaction. See also toxication, toxification, detoxification.

reaction time. the time interval between the application of a stimulus and a response. See latent period.

reactivity. **1a:** the capacity to react or respond (e.g., to a stimulus). **1b:** the capacity of a substance to react chemically, or in any other sense, with another substance. High reactivity is usually a basis for classifying a material as toxic or otherwise hazardous. **2a:** a measure of how readily, or how violently, a chemical reacts with a specified substance or type of substance under a given set of circumstances. **2b:** a measure of how readily a substance detonates or explodes under more or less normal conditions of handling. **3:** the process of reacting.

reacton. the smallest possible biologic target. A primary, subcellular component of a living system that exhibits selective reactivity to biologically active agents. Examples include organelles, subcellular chemical systems, and the smallest intercellular receptors that are able to react selectively.

reactor. See nuclear reactor.

reagent. any substance that is used in a chemical reaction for purposes of measuring, indicating, detecting, examining, or analyzing other substances. High purity is generally required of such substances when used in a laboratory or research situation.

real-time measurement. a measurement that is made nearly simultaneously with the event that is measured. In air pollution science, for example, real-time measurement can be contrasted with integrating or time-averaging methods in which the average concentration of a pollutant over a given period of time is determined. Real-time measurements are of particular importance when the pollutants involved are hazardous to health.

realgar. arsenic disulfide.

rear-fanged snake. See Opisthoglypha.

recalcitrant compound. See recalcitrant substance.

recalcitrant substance. any substance (recalcitrant compound; recalcitrant molecule) that is resistant to degradative processes in the environment whether physical (e.g., light, ultraviolet light), chemical (e.g., hydrolysis, oxidation/reduction), or biological (e.g., enzyme-catalyzed hydrolysis) in nature (usually applied to organic, man-made compounds). Such materials may persist in the environment for very long periods of time.

receiving water concentration (RWC). the concentration of a toxicant (often determined by measuring toxicity) in the receiving water after mixing has occurred; formerly "instream waste concentration."

receptor. **1:** a structure (e.g., a sensory nerve terminal, eye, inner ear) that receives information or responds to stimuli from the environment. **2a:** as proposed by Paul Ehrlich, it is a specialized side-chain of a molecule (or molecular configuration) within the cytoplasm or on the surface of a cell that binds to foreign substances, certain of which are involved in the initial (and sometimes intermediate) steps of a toxic response. **2b:** any site on or in a living organism that is acted upon by a xenobiotic substance or its metabolites. Such receptors are almost always proteins or portions of proteins, usually enzymes. Examples of nonenzyme receptors are opiate (nerve) receptors, gonads, and the uterus. **3:** a hypothetical moiety in a cell (or released from the cell into the blood serum) that can bind a haptophore group of a toxicant or other chemical. **4:** any cell, cell component, tissue, or organ of a living organism that detects or otherwise interacts directly with a substance or with internal or external stimuli. Receptors are often specialized tissue or cellular elements with which a substance (e.g., a pharmacologically active agent or toxicant) interacts to produce a characteristic effect or effects. Some receptors show a high degree of specificity for one type of chemical or stimulus or even to a particular chemical species or stimulus. Organisms have several major receptors (enzymes, cell membranes) that are affected by toxicants. In general, receptor molecules are thought generally to be stereospecific macromolecules. **5:** any person, plant, animal or object that is subjected to a contaminating substance, such as a pollutant or toxic substance. See also toxicant-receptor complex. **acetylcholine receptor**. cholinergic receptor. **adrenergic receptor** (adrenoceptor; adrenoreceptor). either of two types of receptors (α-adrenergic and β-adrenergic) postulated to reside on effector organs that are innervated by postganglionic adrenergic nerve fibers. See also adrenergic. ***α*-adrenergic receptor**. one of two types of adrenergic receptors, it responds to noradrenalin and to blocking agents such as phenoxybenzamine and phentolamine. ***β*-adrenergic receptor**. one of two types of adrenergic receptors, it responds to adrenalin and to blocking agents such as propanolol. There are two subtypes, β_1- and β_2-adrenergic receptors. **β_1-adrenergic receptor**. one of two types of β-adrenergic receptors, it is involved in lipolysis and cardiostimulation. **β_2-adrenergic receptor**. one of two types of β-adrenergic receptors, it is involved in bronchodilation and vasodilation. **adrenoceptor**. adrenergic receptor. **adrenoreceptor**. adrenergic receptor. **Ah receptor** (TCDD-binding protein). a cytosolic receptor protein, presumably determined by the Ah locus, that is present in many laboratory mammals and in humans. Variations in the Ah gene may produce variations in (or even the absence of) this receptor among strains of laboratory mice and may account in part for strain differences in susceptibility to certain toxicants. The binding of aromatic hydrocarbons (Ah) to this receptor is essential to induction of a number of xenobiotic-metabolizing enzymes. Thus, for example, the differing toxicities to various strains of mice of TCDD are strongly associated with modifications of the Ah receptor as is susceptibility to the induction of cataracts by naphthalene and paracetamol. See also Ah locus. **cholinergic receptor** (cholinoceptor; acetylcholine receptor). a cell-surface protein on an effector organ that is innervated by cholinergic neurons that mediate the response of the organ to the acetylcholine secreted by these neurons. There are two types of such receptors, nicotinic and muscarinic, q.v., based on their binding affinity for nicotine or muscarine. These receptors can be blocked, for example, by atropine, decamethonium, scopolamine, and d-tubocurarine. **cholinoceptor**. cholinergic receptor. **ligand receptor**. a receptor site that has a high affinity for a particular ligand. Those that interact with endogenous ligands are the most important as they facilitate intracellular or extracellular communication. See also ligand-receptor complex, muscarinic acetylcholine receptor. muscarinic cholinergic receptor. **muscarinic cholin-**

ergic receptor (muscarinic receptor; muscarinic acetylcholine receptor; mAChR). any of a type of receptor that occurs at autonomic effector cells and is stimulated by muscarine. These receptors mediate the effects of postganglionic parasympathetic neurons. Some also occur in autonomic ganglia and in cortical and subcortical neurons in the brain. Cholinergic agonists for muscarinic receptors include bethanechol and methacholine. Atropine is an effective antagonist. **muscarinic receptor**. muscarinic cholinergic receptor. **nicotinic acetylcholine receptor**. nicotinic cholinergic receptor. **nicotinic cholinergic receptor**. (nicotinic receptor; nicotinic acetylcholine receptor; nAChR). an ion channel that occurs in autonomic ganglia in the brain, in the postsynaptic terminals of neurons, and at the endplates of skeletal muscle. They are stimulated by the alkaloid nicotine. Cholinergic agonists of nicotinic receptors include hexamethonium, nicotine, certain snake toxins, and d-tubocurarine. **nicotinic receptor**. nicotinic cholinergic receptor. **opiate receptor** (opioid receptor). any site in the brain that is able to bind morphine and related opioids. Some are located in areas related to pain; others are not. **opioid receptor**. See opiate receptor.

recidivation. **1a:** relapse, retrogression, recurrence; reversal. **1b:** relapse or recurrence of a symptom, syndrome, disease, activity, state of behavior, or other condition.

recidivism. **1:** a tendency to revert to, or relapse into, a previous condition, pattern, or behavioral mode. **2:** in law, to repeat or revert to criminal behavior; to repeatedly or habitually act in a criminal fashion.

recommended exposure limit (REL). primarily health-based limits on exposure to a chemical that are developed as guidelines by the U.S. Occupational Safety and Health Administration (OSHA).

Recommended Maximum Contaminant Level (RMCL). the maximum level of a contaminant in drinking water at which no known or anticipated adverse effect on human health would occur, and including an adequate margin of safety. RMCLs are nonenforceable health goals.

recrudescence. **1a:** assuming renewed activity following a period of inactivity or dormancy. **1b:** the regrowth or reactivation of tissue following a period of inactivity or involution. **2:** the recurrence or reemergence (within a few days or weeks) of a morbid process or of symptoms following abatement or remission.

recrudescent. of, pertaining to, or exhibiting recrudescence; developing, becoming active, or breaking out again.

rectum. the terminal portion of the intestine in vertebrate animals.

recuperate. to recover from an illness; to undergo a process of recovery.

recuperation. recovery from or the process of recovery from an illness; restoration to a normal state of health following illness.

recycling. See enterohepatic circulation.

red algae. See *Laurencia*.

red arsenic. arsenic disulfide.

red arsenic glass. arsenic disulfide.

red arsenic sulfate. arsenic disulfide.

red bead vine. *Abrus precatorius*.

Red 3. a food coloring used in candy, desserts, and baked goods. It can cause thyroid tumors and chromosomal damage.

Red 40. a food coloring used in candy, desserts, and baked goods that causes lymphatic tumors. Its use is banned in several European countries.

red-bellied Black Snake. *Pseudechis porphyriacus*.

red-bellied blacksnake. *Pseudechis porphyriacus*.

red-bellied long-glanded snake. *Maticora bivirgata*.

red-bellied snake. *Aspidomorphus squamulosus*.

red blood cell. erythrocyte.

red blood corpuscle. erythrocyte.

red-brown trich. *Tricholoma pessundatum*.

red clover. *Trifolium pratense*.

red cohosh. a common name of *Actaea (= Actea) rubra* and sometimes of *A. spicata*.

red copper oxide. cuprous oxide.

red corpuscle. erythrocyte.

red diamond rattlesnake. *Crotalus ruber*.

red dinoflagellate bloom. red tide.

Red Dye No. 2. amaranth.

Red No. 2. amaranth.

red-headed krait. *Bungarus flaviceps*.

red-hot milky, red hot milky. *Lactarius rufus*.

red lead. lead oxide red.

red lead oxide. lead oxide red.

red-legged widow spider. *Latrodectus bishopi*.

red mercuric iodide. mercuric iodide.

red mercuric sulfide. See mercuric sulfide.

red mercury sulfide. red mercuric sulfide (See mercuric sulfide).

red mercury sulfuret. red mercuric sulfide (See mercuric sulfide).

red-mouth bolete. *Boletus subvelutipes*.

red-naped snake. *Aspidomorphus diadema*, *Pseudelaps diadema*.

red oil (turkey-red oil, sulfated castor oil). usually not toxic if ingested.

red orpiment. arsenic disulfide.

red porgy. *Pagrus pagrus*.

red potassium chromate. potassium dichromate (VI).

red precipitate. See mercuric oxide, red.

red puccoon. *Sanguinaria canadensis*.

red rattlesnake. *Crotalus ruber*.

red-ringed snake. *Calliophis macclellandii*.

red sage. *Lantana camara*.

Red Sea flatfish. *Pardachirus marmoratus*.

red snake, striped. *Calliophis sauteri*.

red snapper. a common name of *Lutjanus spp*. See snapper.

red spotted toad. a common name of *Bufo punctatus*.

red squill. *Urginea maritima*.

red squill powder. a widely used, neurotoxic rodenticide prepared from the plant *Urginea maritima*. It contains cardiac glycosides and is readily eaten by pest mice and rats. It is fairly specific in its effectiveness due to the inability of rats and mice to vomit.

red tail snake. *Trimeresurus stejnegeri*.

red tide (red water; dinoflagellate bloom; Gonyaulax bloom; Gymnodinium bloom; red dinoflagellate bloom). a red, marine, water bloom composed of vast numbers of dinoflagellates (notably *Gonyaulax* and *Gymnodinium*) that appears at times in nearshore waters in various parts of the world. The incidence of such tides and the number and extent of areas affected may be increasing. In North America they occur most often along the shores of the south Atlantic and Gulf states. They have been responsible for innumerable instances of mass mortality of fish and other marine organisms. Toxins contained in these microorganisms become concentrated in shellfish that feed on them; animals, including humans, that eat these shellfish may be fatally poisoned. Fish sometimes die within minutes of swimming into an area of high *Gymnodinium* density; the principal toxin is virulent and extremely fast acting. Paralytic shellfish poisoning is associated with red tides due to large numbers of *Gonyaulix*. See also saxitoxin, brevetoxin, shellfish poisoning, *Saxidomus*, *Schizothaerus*.

red water. **1:** red tide. **2:** any of several diseases of certain domestic animals indicated by reddish urine.

redroot. plants of the genus *Amaranthus*.

redskin. allyl isothiocyanate.

redtop. *Agrostis alba*.

reduced arsenic. arsenite; trivalent arsenic; any arsenical in which the arsenic is in a trivalent state.

reducing enzyme. reductase.

reductase (reducing enzyme). any of a class of enzymes that catalyze the metabolic reduction of chemical compounds. Specific reductases are usually named for the group reduced, thus enzymes that catalyze the reduction of nitro groups are called nitroreductases. Reductases occur chiefly in the liver, and to some extent in kidneys, lungs, and certain other organs. The intestinal microflora contain reductases and thus the reduction of xenobiotics can also take place in the intestinal tract. See also reduction reaction.

reductase enzymes. enzymes found largely in the liver and to a certain extent in other organs, such as the kidney and lung. They also occur in intestinal bacteria and reduction of xenobiotics can occur in the intestinal tract. Nitroreductase enzyme, for example, catalyzes the reduction of the nitro group.

reduction. reduction is the addition of electrons to a compound undergoing reaction. See also reaction (detoxification reaction, oxidation reaction, redox reaction).

reductive dehalogenation. See dehalogenation.

redwater disease. urinary syndrome of cattle and sheep.

reed bentgrass. *Calamagrostis* spp.

reed canarygrass. Phalaris arundinacea.

reef. See coral reef.

reef crab. See *Atergatus floridus*, *Demania toxica*.

reef sea snake. *Hydrophis ornatus*.

reefer. a cigarette made from marijuana.

reentry interval. the period of time following application of a pesticide at a given location during which workers are not permitted to reenter the area.

reference material. in chemical analysis, a substance that contains known quantities of target analytes in solution or in a homogeneous matrix. Such is used to test the bias of an analytical process and thus the accuracy of results.

reflex (reflex action). a usually rapid, involuntary muscular response to a stimulus. It is an automatic response in the sense that it is triggered by a relatively short nerve pathway that does not involve the brain. **conditioned reflex** (CR). either a modification of an inborn reflex or a completely new reflex that is developed, usually gradually, as the result of an animal's experience. Such reflexes are often the product of training and association by frequent repetition of a definite stimulus. CRs were first demonstrated by Ivan Pavlov in the 19th century. He conditioned a dog to salivate at the sound of a bell by first training the dog to associate the ringing of a bell with the appearance of food. **conjunctival reflex**. closing of the eyes in response to irritation of the conjunctiva. **cough reflex**. reflex coughing in response to irritation of the larynx or upper respiratory tree in terrestrial vertebrates and some (perhaps many) aquatic vertebrates such as triggerfish (Family Balistidae). **gag reflex**. gagging or retching that may be triggered by a noxious gas or by pressure exerted by a foreign body on the mucous membrane of the fauces, or in some cases on the upper respiratory tree. **nasal reflex**. sneezing caused by mechanical or chemical irritation of the nasal mucous membrane. **nociceptive reflex**. any reflex caused by a painful stimulus. **pharyngeal reflex**. **1:** vomiting reflex. **2:** swallowing reflex. **pupillary reflex**. reflex variation in the size of the pupil due to a change in illumination of the retina; the pupil contracts when the intensity of light increases, dilating in dim light. **righting reflex** (righting response). the ability of an animal, when placed on its back, to right itself. **swallowing reflex** (pharyngeal reflex, def. 2). swallowing induced by stimulation of the palate, fauces, or dorsal wall of the pharynx. **unconditioned reflex. 1a:** generally defined as an inborn reflex that is not dependent on previous learning or experience. Research in comparative psychology, however, has shown that a number of so-called unconditioned reflexes involve experiences during very early development. In the strictest sense, it is probably impossible to demonstrate that any reflex is unconditional. **1b:** operationally, an unconditioned reflex is one that is characteristic of a given species and appears at birth or an approximately equivalent stage of development (e.g., at hatching, at metamorphosis) in the absence of obvious training. Note that whereas the pecking reflex of domestic fowl fits this definition, there is excellent evidence that it develops by conditioning prior to hatching. **vomiting reflex** (pharyngeal reflex, def. 1). vomiting that is induced by any of a number of stimuli (e.g., application of pressure to the wall of the fauces, many xenobiotics, certain diseases). **wink reflex**. reflex closing of the eyelids regardless of the stimulus (e.g., touch, motion toward the eye).

reflex action. reflex.

reflux. **1:** a backward flow. *Cf.* regurgitation. **2:** in a distillation process, during which the vapor from a boiling liquid passes through a condenser that returns it to its original liquid state and returns it to the still (refluxing). The condensate returned to the still is reflux.

refractory. **1a:** unresponsive to or responding slowly to a stimulus or to a biologically active agent. **1b:** unresponsive to treatment or therapy.

Regenon. diethylpropion.

regianin. juglone.

registered trademark (symbol, ®). a trademark that is filed in the U.S. Patent and Trademark office, with the required description and other statements required by law, and duly recorded, thus securing exclusive use of the trademark to the person prompting its registration. *Cf.* trademark.

®. the symbol for registered trademark.

Registered Trademark. ®.

Reglone. diquat.

Regonol. pyridostigmine bromide.

regressive. retrogressive; pertaining to or characterized by regression.

regulation. **1:** the control or maintenance of structural, and functional integrity in the face of often changing environmental conditions. **2:** a rule or requirement promulgated by the government department or agency that is responsible for implementing a particular law. The purpose of a regulation and the requirements stated therein is to bring those subject to the law into compliance. All regulations are published in the Code of Federal Regulations.

regurgitant. **1:** flowing back or in the opposite direction from normal. **2:** material that is regurgitated.

regurgitate. **1:** to flow backward, as in the case of blood flowing through an incompetent heart valve. **2:** to expel small amounts of the contents of the stomach short of vomiting as seen, for example, in some animals when feeding newly born or recently hatched young. **3:** to vomit.

regurgitation. **1:** a backward flow, as of blood through an incompetent valve of the heart. *Cf.* reflux. **2:** the return of gas or small amounts of food from the stomach.

rehmannic acid (lantadene A). a highly toxic, hepatotoxic, polycyclic triterpene found in plants of the genera *Lippia* and *Lantana*. It can be lethal to animals grazing on plants of this genus.

Reichstein's Substance H. corticosterone.

rejecta. those components of ingested materials that are not utilized by the body. Included are egesta and excreta, q.v.

relapse. **1a:** to recur; to regress; to experience a recurrence of a disease or symptoms of a disease following apparent partial or complete recovery. **1b:** the recurrence of a disease or the return of its symptoms following apparent recovery. **2:** a stage of illness in which the patient gets worse after having improved for a period of time usually taken to be a matter of weeks or months. *Cf.* recrudescence.

relapsing. said of a disease that recurs following a period of improvement or apparent recovery.

relative atomic mass (A_r). formerly termed atomic weight, it is the ratio of the average mass per atom of a naturally occurring element to 1/12 of the mass of the ^{12}C nuclide, q.v. The relative atomic mass of many elements varies with the isotopic makeup of the material assayed. See carbon-12.

relative biologic effectiveness (RBE). the effectiveness or strength of a particular kind of radiation in causing a specified biological effect relative to 250 keV of X-rays or gamma rays. See also REM.

relative density. specific gravity.

relative molecular mass (M_r). formerly termed molecular weight, it is the ratio of the average mass per molecule of the naturally occurring form of an element or compound to 1/12 of the mass of the ^{12}C nuclide.

relative risk. See risk.

relative sensitivity. See sensitivity.

relaxant. **1:** lessening or reducing tension. **2:** an agent that reduces tension. **muscle relaxant**. an agent that helps to reduce muscle tension. Some (e.g., meprobamate) act at the polysynaptic neurons of motor nerves; others (e.g., curare) at the myoneural junction.

release. **1:** the escape of a substance from containment or confinement as by leaking or accidental discharge (e.g., an oil release or spill). **2:** a substance that has been so released into the environment. See also airborne release.

relief. the alleviation, reduction, or removal of pain, distress, or suffering.

relieve. to remove or allay pain, distress, or suffering.

REM, rem (Roentgen Equivalent Man; radiation equivalent man). the energy of a source of radiation multiplied by its RBE (relative biological effectiveness, q.v.). It is a measure of radioactive dose to humans. One REM is equivalent to the absorption of one rad of high-voltage X-rays or gamma rays by an average adult human male. 1 rem = 1 rad X RBE, where RBE is the relative biological effectiveness, which varies according to the type of radiation. See also RAD.

remediable. curable, correctable.

remedial. acting as a remedy; helpful, corrective, curative, or acting as a remedy.

remedy. any agent or activity that relieves, cures, or prevents the appearance of, the symptoms of disease.

remission. **1:** an abatement or diminution in the severity of the symptoms of an illness. **2:** the stage of a chronic illness during which the symptoms become less severe or even disappear completely.

remit. to become temporarily less severe.

remittance. a temporary reduction in severity (as of symptoms).

remittent. having periods of diminution or amelioration (as of symptoms) and of exacerbation.

Remsed. promethazine.

renal (nephric). of, pertaining to, or related to the kidney; having the characteristics of the kidney; having the shape of a kidney.

renal clearance. See clearance (renal).

renal excretion. See excretion (renal).

renal failure (Bright's disease; renal dropsy). failure of the kidney to maintain water and electrolyte balance, with consequent progressive systemic acidosis. The causes of acute renal failure include trauma, vascular insults, uncontrolled hypertension, or any condition that impairs blood flow to the kidneys; poisoning (e.g., by heavy metals, bacterial toxins, by carbon tetrachloride, ethylene glycol, any of a number of drugs, glomerulonephritis, acute obstruction of the urinary tract). Unless reversed by treatment, acute renal failure progresses to chronic or end-stage renal failure. This stage is marked by acidosis and the toxicity of specific electrolytes, which leads to volume overload, hypertension, congestive heart failure, osteomalacia, secondary hyperparathyroidism and death.

renal insufficiency. reduced ability of the kidney of vertebrates to maintain water and electrolyte balance. See renal failure.

Renard's viper. *Vipera ursinii*.

renopathy. nephropathy.

rep. Roentgen-equivalent-physical.

REP. Roentgen-equivalent-physical.

repellent. **1:** able to drive off or repel; repulsive. **2:** an agent that repels noxious organisms (e.g., insect pests, ticks, mites). **3:** an agent that reduces swelling.

repolarization. the reestablishment of the resting membrane potential following depolarization of a neuron.

repression. restraint, subjugation, **enzyme repression**. inhibition of enzyme synthesis by a product (or products) of the metabolic pathway in which the enzyme is active. It is a metabolic control mechanism.

repressor. a substance, usually a small molecule, that inhibits the production of a specific enzyme by an organism. See also inducing agent, enzyme repression.

reproductive toxicity. the ability of a toxicant to injure or damage the reproductive system or to alter normal reproductive performance of an organism. Other systems and functions may or may not remain normal. Thus the life cycle and behavior of adult predatory and piscivorous birds, including reproductive behavior, usually appear normal with DDT burdens that cause egg shell thinning and often disaster for the futurity of the affected population.

reptilase, reptilase R. batroxobin.

Reptilia. a class of amniote, poikilothermic, air-breathing, predominantly terrestrial, usually pentadactyl tetrapod vertebrates. They are characterized by a bony skeleton, a cranium that bears a single occipital condyle, a three-chambered heart (four-chambered in the case of crocodilians), a dry horny skin that bears epidermal scales, plates, or scutes, and internal fertilization. The lungs are functional throughout life. Most are oviparous (egg-laying), but some are ovoviviparous in the sense that the eggs hatch internally and the young are born live. The egg shell in most cases is leathery. Included in this class are lizards (Order Squamata, Suborder Sauria) and snakes (Order Squamata, Suborder Serpentes).

RER. rough endoplasmic reticulum that has attached ribosomes. See endoplasmic reticulum.

reregistration. the reevaluation and relicensing of existing pesticides in the United States that were originally registered prior to current scientific and regulatory standards.

rescue grass. *Bromus catharticus*.

reserpine ((3β,16β,17α,18β,20α)-11,17-dimethoxy-18-[(3,4,5-trimethoxybenzoyl)-oxy]yohimban-16-carboxylic acid methyl ester; 3,4,5-trimethoxybenzoyl methyl reserpate; Sandril; Serpasil). an odorless, white or pale buff to slightly yellow powder, $C_{33}H_{40}N_2O_9$, that is insoluble in water, but soluble in benzene and chloroform. It darkens slowly on exposure to light or more rapidly in solution. Reserpine, an alkaloid, is obtained from the root of certain species of *Rauwolfia* and has been used as a sedative, antihypertensive, antipsychotic, and antimanic. These actions are due in part to the depletion of 5-hydroxytryptamine and catecholamine concentrations by reserpine in both the CNS and PNS. It lacks specificity, has a number of adverse effects which may include, in some individuals, abdominal cramps, diarrhea, hypotension, and severe depression; it is also a probable human carcinogen. It is thus in little use today.

reserve. **1:** that which is retained and available for future use. **2:** self-control; control of one's feelings and thoughts. See also alkali reserve.

resibufogenin ((3β,5β,15β)-14,15-epoxy-3-hydroxy-5-bufa-20,22-dienolide). a cardiotoxic, cytotoxic component, $C_{24}H_{32}O_4$, of toad venom and secretions.

residence time. **1a**: the length of time that a given substance remains within a given system. **1b:** the length of time that a given molecule remains within a given system. Thus, in air pollution studies, it is the length of time during which a given molecule of an air pollutant remains in the atmosphere. Because it may be replaced by another molecule by the end of that time, the temporal contribution of the substance in question to the air pollution cannot be predicted from the residence time.

residua. plural of residuum.

residual. **1a:** remainder, remnant, excess; the portion that remains. **1b:** pertaining to or of the nature of a residue. **2a:** that portion of a substance that remains following a process that removes the bulk of the substance. **2b:** the amount or portion of a pollutant that remains in the environment after a natural or technological process has removed most of it (e.g., the sludge that remains following initial wastewater treatment). **3:** a by-product; material that is unintentionally produced during the production, processing, or treatment of other materials.

residue (residuum). **1:** that which remains following removal of one or more substances. **2a:** a compound (e.g., an amino acid, monosaccharide, nucleotide) that is part of a larger molecule. **2b:** a free compound that remains following partial decomposition of the parent compound.

residuum (plural, residua). residue.

resin. **1:** any of a chemically diverse group of complex nitrogenous compounds that occur in some plants and share certain physical properties. They are solid or semisolid at room temperature, brittle, easily melted and flammable. They are insoluble in water. Some toxic resinoids are among the most virulent of poisons to humans and other animals. They are direct irritants of the nervous system or of muscles. **2:** rosin. **3:** any of a variety of organic monomers (e.g., acrylic resin, autopolymer resin) that are insoluble in water. *Cf.* resinoid.

resinoid. any substance that contains or resembles a resin.

resinous. pertaining to or derived from a resin.

resmethrin. a semi-synthetic mixture of pyrethrin-like pyrethroid isomers. Resmethrin (under brand names such as Bioresmethrin, Chryson, Cismethrin, and Synthrin) is used largely in homes, gardens, and greenhouses to control a wide variety of insects. It is moderately toxic to mammals including humans. Acute dermal LD_{50} in laboratory rats is about 2.0 g/kg; acute oral toxicity varies among rat strains from about 1 to 4 g/kg body weight. The mode of action is via an effect on the sodium channels, leading to hyperexcitability. The various brands, however, vary considerably in toxicity. Bioresmethrin is essentially nontoxic when administered orally to laboratory rats, whereas Cismethrin (due to the *cis*-component) is very toxic, 90 to 170 mg/kg body weight.

resorcin. resorcinol.

resorcine. resorcinol.

resorcinism. chronic resorcinol poisoning. Effects include methemoglobinemia, paralysis, and damage to the capillaries, kidneys, heart, and nervous system. See also resorcinol poisoning.

resorcinol (1,3-benzenediol; *m*-benzenediol; resorcin; resorcine; resorcinum; *m*-dihydroxybenzene; 3-hydroxyphenol; C.I. 76505; C.I. developer 4; C.I. oxidation base 31; resorcin; resorcine; resorcinum). a flammable bacterial, fungicidal, keratolytic, exfoliative, and antipruritic, $C_6H_6O_2$, occurring as white, or nearly white, monoclinic crystals or powder. It is used as a topical keratolytic in the treatment of acne and other dermatoses, such as seborrheic dermatitis. It is also used in hair dyes, some drugs, and as an adhesive in rubber products. Resorcinol is toxic *per os*, by cutaneous absorption, and other routes. It is an irritant of the skin, eyes and mucous membranes. It can cause chronic health problems as well as hematologic and neurologic effects. This substance may also be carcinogenic. See phenol, resorcinism, resorcinol poisoning.

resorcinol poisoning. cutaneous absorption of resorcinol can cause local effects such as hyperemia, dermatitis, edema, and corrosion of the skin with associated enlargement of regional lymph nodes. Whether ingested or absorbed through the skin, resorcinol can cause systemic poisoning that involves both the cardiovascular and nervous systems. Such poisoning may be characterized by restlessness, methemoglobinemia, cyanosis, convulsions, tachycardia, dyspnea, and death. Chronic resorcinol poisoning is termed resorcinism, q.v.

resorcinum. resorcinol.

resource. **1a:** anything needed or used by a population or individual organism to better adapt it to the environment or to maintain or improve the quality of life. **1b:** a reserve source of supply of a material or of energy.

respirable. **1:** suitable for respiration, or able to be respired. **2:** inhalable; able to be breathed (e.g., respirable dust).

respirable dust. See dust (respirable dust).

respirable particle. See dust (respirable dust).

respiration (R). **1a:** the taking in of oxygen by an organism, its metabolism, and the release of carbon dioxide. **1b:** the interchange of oxygen and carbon dioxide between an aerobic organism and the environment or between individual living cells and the surrounding medium. **2:** ventilation or breathing as by gills or lungs. **audible respiration**. abnormal or abnormally discernable respiratory sounds of any lung-breathing animal, particularly mammals or birds (e.g., wheezing, rales). **Cheyne-Stokes respiration**. the breathing pattern characteristically observed in coma due to affection of the nervous centers of respiration. It is marked by a gradual increase in depth breathing, and sometimes in rate, to a maximum, followed by a decrease ending in apnea. The cycles normally range from 30 seconds to 2 minutes in duration, with 5 to 30 seconds of apnea.

respiratory. **1a:** of or pertaining to respiration. **1b:** pertaining to the lungs and other organs of breathing.

respiratory arrest. the cessation of spontaneous breathing.

respiratory burn. See burn.

respiratory center (respiratory control center). the structure or portion of the medulla oblongata of brain stem that regulates the depth and rate of breathing. In humans it is comprised of an inspiratory center situated in the rostral half of the reticular formation that overlies the olivary nuclei, and an expiratory center that lies dorsal to the inspiratory center. A portion of the respiratory center, the pneumotaxic center is situated in the pons of the vertebrate brain.

respiratory chain. See electron transport chain.

respiratory control center. respiratory center.

respiratory failure. **1a:** a condition in which the organs of breathing deliver atmospheric air (lungs) or dissolved oxygen (gill systems) to the respiratory surfaces. This crisis is often associated with cardiovascular failure. Proximate causes include certain diseases that affect lung tissue (emphysema), paralysis or weakness of the respiratory muscles, hemorrhage, shock. The ultimate cause may be any of various poisonings. **1b:** a complete inability to exchange respiratory gases with the environment. **acute respiratory failure** (ARF). severe, critical respiratory failure of rapid onset.

respiratory membrane. a thin, moistened membrane within the vertebrate lungs, or other breathing organ (e.g., gills, skin) through which gas exchange takes place. The respiratory membrane of the vertebrate lung is composed of an alveolar portion and a capillary portion through which gaseous exchange occurs.

respiratory rate (breathing rate). the number of breathing cycles as inspirations (or expirations) per unit length of time. Numerous toxicants affect respiratory rate. Thus, toxic doses of cocaine, amphetamines, fluoroacetate, nitrites, methanol, salicylates and hexachlorobenzene increase respiratory rate; carbon monoxide and cyanide may either increase or decrease respiratory rate; alcohols (except methanol), analgesics, narcotics, sedatives, phenothiazines, and opiates decrease respiratory rate.

respiratory sensitization. a type I (anaphylactic) allergic reaction involving the respiratory system. Such a reaction can be chemically induced as in asthma.

respiratory syndrome, of cattle. toxic pulmonary emphysema, of cattle. See emphysema.

respiratory tree. See upper respiratory system.

respire. **1:** to breathe. **2:** the exchange, by a cell, tissue, or organism, of oxygen and carbon dioxide with the environment as a result of respiration.

response. **1a:** any activity (behavioral, metabolic, etc.) of a cell or organism that occurs in response to a stimulus. **1b:** classically, any observed or measured action or movement due to the application of a stimulus. **2:** an action or actions taken to minimize the impacts of a release of contaminating materials to the environment. Also termed response action. **immune response**. the response of the immune system following exposure to a specific antigen. Such a response may be manifested in the production of antibodies, cell-mediated immunity, or immunological tolerance; immune reaction. **toxic response**. any deleterious response of a cell or organism due to exposure to a toxicant.

resting potential. the measured voltage within a neuron when it is not conducting nerve impulses.

restorative (anastatic). **1:** reviving, renewing, rejuvenating, promoting a return to health or to consciousness. **2:** an agent that helps to restore health, consciousness, or vitality.

Restricted Use. when a pesticide is registered in the United States, some or all of its uses may be restricted if the pesticide requires special handling because of its toxicity. Restricted-use pesticides may be applied only by trained, certified applicators or those under their direct supervision.

retch. **1a:** an involuntary attempt to vomit. **1b:** to attempt, involuntarily, to vomit. *Cf.* retching.

retching (dry heaves; dry vomiting; vomiturition). an involuntary attempt to vomit, sometimes repeatedly.

retention. **1:** memory, recall. **2:** the amount of a substance deposited in the lungs less the amount cleared (e.g., by ciliary action, coughing, phagocytosis) from the respiratory tract.

reticuloendothelial system (RES; mononuclear phagocyte system; MPS). a fine fibrillar meshwork of phagocytic cells within a connective tissue matrix within the lymphoid organs (e.g., spleen, lymph nodes) and scattered elsewhere throughout the body of vertebrates (e.g., bone marrow, adrenal cortex, liver, kidneys, hypophysis, microglia of the CNS). This system plays a key role in removing foreign particulate matter (e.g., microorganisms, colloids) and worn out or dead cells (most importantly erythrocytes) from the blood. It also has a role in repairing tissue and a role in the immune system. Antigenic xenobiotics taken up by the phagocytes of the reticuloendothelial system in lymphoid organs, encounter T lymphocytes and B lymphocytes which then initiate a specific immune response.

reticulum. a network. See endoplasmic reticulum.

retina. the light-sensitive innermost layer (3rd tunic) of the eye of vertebrates. It receives images formed by the lens, transforms and conveys the information via rods and/or cones via the optic nerve, to the CNS.

retinal. of or pertaining to the retina.

retinitis. inflammation of the retina.

retinoic acid (3,7-dimethyl-9-(2,6,6-trimethyl-1-cyclohexen-1-yl)-2,4,6,8-nonatetraenoic acid; vitamin A acid; tretinoin). all-*trans* retinoic acid is the naturally occurring form; it is moderately toxic. See also especially isotretinoin.

retinoid. **1:** resembling the retina; retina-like. **2:** resinous; resembling a resin.

retinol. a form of vitamin A found in various animal tissues.

retinopathy. any non-inflammatory disease or damage to the retina associated with changes in retinal blood vessels. Retinopathies can be produced by a number of xenobiotics, including chloroquine and indomethacin, as well as certain phenothiazines. High concentrations of oxygen (hyperoxia) are also retinotoxic as evidenced by the development of retrolental fibroplasia in preterm infants in hyperoxic incubators. Retinopathies are common secondary complications of diabetes mellitus. See also retinotoxic.

retinotoxic. toxic to the retina; able to produce a toxic or injurious effect upon the retina.

retiolental fibroplasia. a bilateral retinopathy marked by vascular dilatation, proliferation, and tortuosity, edema, and retinal detachment with ultimate replacement of the retina by a fibrous mass appearing as a dense membrane behind the lens. It occurs chiefly in infants placed in a high oxygen environment. The eye usually fails to grow properly resulting in microophthalmia and possible blindness. See toxicity (oxygen toxicity).

retrobulbar. behind the eyeball.

retrocollis. spasmodic torticollis in which the head is drawn directly to the back.

retronecine ((1*R*-trans)-2,3,5,7a-tetrahydro-1-hydroxy-1*H*-pyrrolizine-7-methanol). the necine base, $C_8H_{13}NO_2$, of monocrotaline, senecionine, seneciphylline, retrorsine, and numerous other hepatotoxic pyrrolizidine alkaloids that occur in *Crotalaria*, *Senecio*, and various genera of the family Boraginaceae. It is a moderately toxic 1-methylpyrrolizidine, but the intact alkaloids which it forms are very toxic. *Cf.* retrorsine. See also monocrotaline.

retrorsine (12,18-dihydroxysenecionan-11,16-dione; β-longilobine). a common hepatotoxic, carcinogenic, pyrrolizidine alkaloid, $C_{18}H_{25}NO_6$, of *Senecio* spp. Originally isolated from *S. retrorsus* and *Crotalaria usaramoensis*, it is composed of retronecine and isatinecic acid. Retrorsine is responsible for a fatal cirrhosis of the liver in cattle and horses. Retrorsine is poisonous by most or all routes of exposure. It is neoplastigenic and tumorigenic to laboratory animals and a possible mutagen; human carcinogenicity is questionable. When heated to decomposition it releases toxic NO_X fumes. See also necine base.

retrovirus. a virus (Family Retroviridae) capable of integrating its genes into those of the host by utilizing RNA to DNA transcription. They uniquely contain a reverse transcriptase. The virion RNA, multiple copies of which may be present, is single stranded. Included in this family are both oncogenic and non-oncogenic viruses.

reversal. **1:** a turning, change in direction, or setback as in the progress of a condition, disease, condition, or symptom.

reversal of succession. See succession.

reversible. **1a:** able to change back and forth (fully reversible). **1b:** able to reverse or to be reversed. This term is applied, for example, in toxicology to processes (e.g., chemical reactions, diseases, the effects of a toxicant) if they do not persist long following exposure, or if they are fully alleviated by medication or other treatment.

reversible toxic effect (reversible effect). **1a:** any toxic effect that does not persist when a subject is not longer exposed to the toxicant. **1b:** any lesion or pathological effect that is reversible upon removal of the causative agent. Such a capacity depends upon the ability of the target cells or tissue to regenerate or for the essential metabolic mechanisms that support normal cellular activity to remain intact or to be resupplied following pathological injury. Reversibility of pathological injury may be relative and depends also upon the type and extent of injury.

revertant. **1a:** an organism that has reverted to a previous phenotype by means of back mutation. **1b:** a gene that undergoes back mutation (i.e., to a previous sequence of bases). **2:** a microbial mutant that has reverted to its former genotype (true reversion) or to the original phenotype by means of a suppressor mutation.

revivification (vivification). **1a:** an attempt to restore to life. **1b:** restoration to life or conciousness, including the restoration of body parts (e.g., a frozen limb, a severed appendage). **2:** refreshing and repairing the edges of a wound or of diseased surfaces to promote healing (e.g., of a wound).

Reye's syndrome. a striking, often fatal disease of children mainly in southeast Asia, India, and Africa. The etiology is by no means well established. All of the following have been implicated: consumption of aflatoxin-contaminated food; increased use of aspirin; viral infections such as chicken pox and influenza in association with aspirin use. Symptoms include fever, vomiting, diarrhea, convulsions, coma, and death. Lesions include fatty degeneration of heart, liver and kidneys; cerebral edema; neural degeneration; and lymphocytolysis. See aflatoxin, aspirin.

***Rf* value**. in chromatography, the ratio of the distance traveled by the solute from the origin to the distance traveled by the solvent front from the origin under standard or specified conditions.

RFD, RfD. reference dose.

RFNA. red fuming nitric acid. See nitric acid.

RH 6201. scifluorofen.

Rh. the symbol for rhodium.

rhabarberone. aloe emodin.

rhabdomyolysis. an acute, sometimes fatal condition that is sometimes induced by convulsions that caused empoisonment by water hemlock (*Cicuta spp.*, *q.v.*). The disease is characterized by destruction of skeletal muscle.

Rhamnaceae. the buckthorn family. See *Rhamnus*.

rhamnoside. a glycoside that yields rhamnose on hydrolysis.

3β-rhamnosido-14β-hydroxy-$\Delta^{4,20,22}$-bufatrienolide. proscillaridin.

Rhamnus (buckthorn). a genus of evergreen or deciduous shrubs and small trees (Family Rhamnaceae). Most occur in the temperate regions of the Northern Hemisphere. The leaves and fruit are toxic. Species that occur in North America include *R. caroliniana* (Carolina buckthorn, Indian cherry), *R. cathartica* (common buckthorn), *R. frangula* (buckthorn; alder buckthorn), and *R. purshiana* (cascara buckthorn, bearberry, bearwood, coffeeberry). The last named species is a small tree of the Pacific Coast of North America. The dried bark (cascara sagrada) contains emodin and is used as a mild laxative. See also polyneuropathy (buckthorn polyneuropathy).

rhaphania. raphania.

rhenium (symbol, Re)[Z = 20; A_r = 186.21]. a very rare, black to silver-gray, crystalline transition metal that usually occurs in nature in association with molybdenum. No toxic effects have been reported.

Rheum. a genus of Asiatic perennial plants (Family Polygonaceae). ***R. rhabarbarum*** (*R. rhaponticum*; rhubarb; pie plant). a widely grown garden vegetable with a leaf stalk of rhubarb that has a sour but pleasant taste, due chiefly to the presence of nearly nontoxic malic acid. The leaf also has a sour taste due to the presence of citric and oxalic acids. The content of oxalic acid and soluble oxalates in the leaf is usually sufficient to produce severe effects and even death in humans. Symptoms of intoxication of humans include nausea, vomiting, abdominal pain, anuria, hemorrhage. emodin, *q.v.*, also occurs chiefly in the roots. ***R. rhaponticum***. the former scientific name of *R. rhabarbarum*.

rheum emodin. emodin.

rheumatism weed. *Apocynum androsaemifolium*.

rhinitis. inflammation of the nasal mucosa. **allergic rhinitis**. any allergic reaction of the nasal mucosa. **anaphylactic rhinitis**. allergic rhinitis. **nonseasonal allergic rhinitis**. any perennial rhinitis. **seasonal allergic rhinitis**. hay fever.

rhinoceros horned viper. *Bitis nasicornis*.

rhinoceros viper. *Bitis nasicornis*.

Rhinocrichus lethifer. a millipede found in Haiti that can spray an extremely toxic and irritant venom a distance of up to 100 cm. The venom can apparently cause phlyctaena and desquamation of human skin. In addition, small animals may be blinded and sometimes die of starvation because they are unable to feed; presumably some may be lost to predation as a result of blinding.

Rhinoplocephalus bicolor (Muller's snake). the sole member of this genus (Family Elapidae), it is a small, venomous snake of southwestern Australia. Adults reach a length of no more than about 36 cm. The head is small and nearly indistinct from the neck; the snout is broad and depressed; the eyes are small with round pupils; the body is rather slender and cylindrical with a small tail. Although its behavior remains largely unknown, it is not thought to be dangerous.

Rhinoptera bonasus. a large venomous ray (Family Rhinopteridae) that has a snout thought to resemble that of a cow. It occurs in the Atlantic Ocean from Cape Cod, Massachusetts to Florida.

Rhinopteridae. a family of venomous cartilaginous fish, the cow-nosed rays. See also stingray.

rhizome. an underground, often horizontal, stem of plants such as *Acorus calamus*, *Iris*, and *Sanguinaria*.

Rhodactis howesi. a sea anemone that is poisonous by ingestion. See Anthozoa.

Rhodamine B (*N*-[9-(2-carboxyphenyl)-6-(diethylamino)-3*H*-xanthen-3-ylidene]-*N*-ethylethanaminium chloride; tetraethylrhodamine; D & C Red No. 19; Basic Violet 10; C.I. Basic Violet 10; C.I. 45170; C.I. Food Red 15; Food Red 15). a coloring agent, $C_{28}H_{31}N_2O_3 \cdot Cl$, used as a dye (mainly for paper) and as a biological stain, and reagent. It is poisonous, *i.v.* and *i.p.*, and is moderately toxic by ingestion. While it has been listed as a possible component of drugs and cosmetics, it is carcinogenic to some laboratory animals.

rhodanate. thiocyanate.

rhodanic acid. thiocyanic acid.

Rhodenwasserstoffsäure. the German term for thiocyanic acid.

rhodium (symbol, Rh)[Z = 45; A_r = 102.91]. a silvery-white, soft, ductile, malleable, corrosion-resistant transition metal. No toxic effects have been reported, although rhodium chloride, $RhCl_3$, is moderately toxic to laboratory rats.

rhododendron. See *Rhododendron*.

Rhododendron (rhododendron; rosebay; azalea). a large genus of deciduous and evergreen shrubs, small trees and a few epiphytic plants (Family Ericaceae) of the Northern Hemisphere (one Australian species). All parts of most species and varieties are toxic. Symptoms following ingestion may include salivation, lacrimation, nasal discharge, vomiting, bradycardia, hypotension, convulsions, and paralysis. The active principle is andromedotoxin. ***R. albiflorum*** (white-flowered rhododendron). a woody deciduous shrub up to 2 m tall, found in wet sites at higher elevations in the mountains from Oregon to British Columbia and Montana. ***R. macrophyllum*** (California rose bay). an evergreen shrub or small tree, 1-5 m tall found in woodlands and thickets at mid elevations of the Pacific Coast of North America from California to Washington. ***R. maximum*** (great laurel; rose bay). a shrub or small evergreen tree up to 3½ m tall usually found in damp woods or swamps in eastern North America from Nova Scotia and Ontario to northwestern Georgia and Alabama. ***R. occidentale*** (western azalea). a slender, deciduous shrub, 1-3 m tall found in thickets with moist soil to mid elevations on the coastal slopes of the mountains of California and Oregon.

Rhodotypos scandens (**jet berry bush**). a deciduous upright, spreading shrub (Family rosaceae), the berries of which contain amygdaline and other cyanogenetic glycosides. Symptoms following ingestion are those of cyanide poisoning and include nausea, vomiting, abdominal pain and rigidity, fever, convulsions and coma.

Rhoicissus cuneifolia (wild grape). a shrubby vine with a watery acid sap (Family Vitaceae). The root is toxic, causing gastrointestinal distress and respiratory depression.

rhombic night adder. *Causus rhombeatus*.

Rhopdactis howesi. a sea anemone (Anthozoa) that is poisonous when eaten raw, but is cooked and apparently safely eaten by Samoans. Early symptoms of intoxication are those of acute gastritis. The victim may shortly become stuporous, suffer a loss of superficial reflexes, and may eventually pass into sustained shock and die with pulmonary edema. There is no known antidote. See Anthozoa.

Rhothane. See DDD.

rhubarb. *Rheum rhaponticum*.

Rhus. a genus of toxic deciduous and evergreen shrubs, vines and small trees (Family Anacardiaceae) that are widely distributed in the temperate zone and subtropics. Most have compound leaves which turn a brilliant scarlet color in the autumn. The flowers are usually small and occur in large panicles. Some species are virulently poisonous and may prove fatal to humans. All North American species produce a severe allergic dermatitis (*Rhus* dermatitis) in sensitive individuals that is marked by itching and burning of the skin and edema. On rare occasions, serious cases can be life-threatening; fatalities are known. Poisonous members of the genus with smooth fruits and leaves are sometimes placed in the genus *Toxicodendron*, q.v. ***R. diversiloba*** (syn. *Toxicodendron diversilobum*; Pacific poison oak; Western poison oak). a shrub or vine, with three or five toothed or lobed leaflets and growing to 2.5 m. Closely related to *R. radicans*, q.v., it occurs in western North America from southern British Columbia to northern Baja California. The toxic principle is urushiol. ***R. quercifolia***. *Rhus toxicodendron*. ***R. radicans*** (syn. *Toxicodendron radicans*; formerly *R. toxicodendron*; poison ivy; poison vine; cowitch; "poison oak"; poison creeper; picry; mercury; markry). a climbing vine, small shrub, or rarely a small tree with three-lobed leaflets; most common in shaded woodlands, woodland glades or borders, or along roadsides. It is toxic to humans and closely related primates throughout the year. Birds eat the fruit in winter with impunity. The active principle is urushio, q.v., *R. radicans* and *R. diversiloba* cause more cases of delayed contact sensitivity than all other known sensitizers combined. Their effects are quite similar: dermal contact produces a mild to severe self-limiting dermatitis which develops within 12-14 hours following reexposure of a sensitized individual. The condition is characterized by a cutaneous erythema with edema, itching, and burning of the skin. Blistering (with discharge of serum) and pruritus with exudation may develop. Ingestion can seriously injure the mucous linings of the mouth and gastrointestinal tract. Except in hypersensitive individuals, healing is complete in about 10 days. Sensitivity slowly diminishes in the absence of further exposures. Contrary to common belief, immunity is not conferred by chewing the leaves or fruit of the plant. Perhaps 30% of North Americans have an antigen-specific tolerance to these plants. ***R. succedanea*** (wax tree). a poisonous tree of Japan that grows to about 10 m. It has compound leaves with 9-15 lustrous leaflets. The whitish fruit contains a wax-like substance used in candlemaking. ***R. toxicodendron*** (syn. *Toxicodendron quercifolium*; *R. quercifolia*; poison oak; Eastern poison oak). **1:** a shrub of the scrub oak-pine savannahs of the Atlantic and Gulf coastal plains of North America. It grows to a height of 2 m, with leaves that are three-lobed or rounded and hairy on both surfaces. It is easily confused with *R. radicans*, and its effects on humans are identical. The active principle is urushiol. **2:** earlier literature treated *R. radicans* as a variant of *R. toxicodendron*. ***R. verniciflua*** (varnish tree; lacquer tree). a poisonous, ornamental tree that is indigenous to China and Japan. It grows to a height of about 20 m. The leaves are compound, with 11 to 15 large leaflets and long, hanging clusters of whitish flowers. It yields the lacquer used in Japan to protect and to put a beautiful high gloss on woodenware. ***R. vernix*** (syn. *Toxicodendron vernix*; poison sumac; poison dogwood; poison elder; poison ash; swamp sumac; thunderwood). a shrub or small tree up to 5 m high. It occurs in damp, marshy situations from Quebec to central Florida. All parts of the plant are toxic, producing an allergic contact dermatitis (*Rhus* dermatitis). The active principle is the resin urushiol.

Rhynchoelaps (desert banded snakes). a genus of two species of small venomous snakes (Family Elapidae). Neither species is considered dangerous to humans. The head is small, compressed, has a prominent snout, and is indistinct from the neck; the rostral lacks a sharp edge and the nasal abuts the first two labials only. They inhabit arid regions of Australia. The tail is short and the maximum length attained is about 0.4 m. ***R. approximans*** (desert banded snake). a species restricted to arid regions of northwestern Australia. ***R. bertholdi*** (bandy-bandy; desert banded snake; Jan's banded snake; short-tailed snake). a species restricted to arid regions of southwestern Australia.

rhythm. any cyclical variation in the intensity of a process or phenomenon (e.g., behavioral, biochemical, physiological), including responses to xenobiotic chemicals.

rhythmic. periodic; displaying rhythmicity.

RIA. radioimmunoassay.

rib. one of a set of bones that are hinged to the vertebral column and sternum and which, along with muscle, define the top and sides of the chest cavity.

ribbed-stalked cup. *Helvella acetabulum*.

ribonuclease. See RNase.

ribosome. any of numerous minute particles attached to endoplasmic reticulum or loose in the cytoplasm of a cell. They are the site of protein synthesis.

ribosylpurine. nebularine.

richweed. *Eupatorium rugosum*.

ricidine. ricinine.

ricin (ricine). **1:** either of two extremely poisonous glycoprotein toxins (RCL III and RCL IV) found in the castor bean plant, *Ricinus communis*. The two toxins are present in highest concentration in the seeds. They are violent irritants and hemorrhagins. They act by inhibiting ribosomal protein synthesis. Each of these two nearly identical lectins is composed of two polypeptide chains connected by a disulfide bridge. When ingested, one chain binds to cells of the intestinal wall, allowing entry of the other chain into the cytoplasm of a cell. A single molecule of ricin is sufficient to kill the cell. They can induce labor in pregnant women and very small amounts are lethal to humans. **2:** ricin mixture. **3:** the individual lectins. This usage is confusing and with current knowledge unacceptible. *Cf.* ricinine. See also ricin agglutinin, ricin mixture.

ricin agglutinin. either of two components of ricin mixture (the active principle of the castor bean, *Ricinus communis*, q.v. They are two nearly identical lectins (RCL I and RCL II). Each is composed of four polypeptide subunits. These components of ricin mixture are inactive orally and affect erythrocytes only when given intravenously. Thus, contrary to much of the literature, hemolysis is not a major effect of castor bean poisoning. *Cf.* ricin.

ricin mixture. the toxic principle of *Ricinus communis*, the castor bean plant, originally defined as a single, extremely toxic glycoprotein present in highest concentration in the seeds. It is actually a mixture of several different proteins of which two agglutinins (RCL I and RCL II) and two toxins (RCL III and RCL IV) have been isolated. Toxicity resides largely in the two toxins. The minimum lethal oral dose in some laboratory animals may be as low as 0.01 mg/kg body weight. When injected, the minimum lethal dose is about 0.1 ng/kg. Even so, castor bean press cake, a livestock feed, can be detoxified simply by heating. Individual variation in sensitivity is high, but on the average horses are most vulnerable and poultry are most resistant to ricin poisoning. Very low doses have antitumor activity. See also ricin, ricin agglutinin.

ricin poisoning. ricinism.

ricine. ricin.

ricinine (1,2-dihydro-4-methoxy-1-methyl-2-oxo-3-pyridinecarbonitrile; 1,2-dihydro-4-methoxy-1-methyl-2-oxonicotinonitrile; ricidine). a white, crystalline alkaloid, $C_8H_8N_2O_2$, that is slightly soluble in water, chloroform, and ether. It is isolated from the leaves and seeds of *Ricinus*. Symptoms of poisoning of humans by ingestion may include nausea, vomiting, hemorrhagic gastroenteritis, convulsions, hypotension, coma, and death usually by respiratory failure. Lesions seen on autopsy are renal and hepatic damage.

ricinism (ricin poisoning). poisoning by ingestion of ricin or ricin mixture from seeds or leaves of the *Ricinus communis* (the castor bean). Ricinism is nearly synonymous with castor bean poisoning.

ricinoleic acid. ([R-(Z)]-12-hydroxy-9-octadecanoic acid). a cathartic fatty acid isolated from the seeds of *Ricinus communis*. It is the chief fatty acid component of the triglyceride fraction of castor oil.

Ricinus. a genus of large herbaceous plants native to tropical Africa (Family Euphorbiaceae). Because of the attractive foliage, some species such as *R. communis* are cultivated as annuals. ***R. communis*** (castor Bean; gourd; African coffee tree; castor-oil plant; palma Christi; koli). a plant of 2 m or so in height, grown in Africa and India as an ornamental and for commercial use and as an ornamental in North America and most tropical areas. The seed if chewed is toxic, producing severe gastrointestinal distress, convulsions, and uremia. Symptoms may include burning in the mouth, nausea, bloody vomit, a bloody stool, convulsions, coma, and death. Ricin, a toxic lectin, hemolyzes erythrocytes even at extremely low concentrations, causing severe hemorrhaging. It can also induce labor in pregnant women. Initial symptoms may appear within 2 hours to 2 days. Death may occur up to 12 days following ingestion. If a seed with an intact coat is swallowed, absorption and poisoning does not occur. The active principle is ricin.

Riddell's groundsel. *Senecio riddellii*.

ridge-nosed rattlesnake. *Crotalus willardi*.

Riesen Korallenotter. See *Micrurus spixii*.

rifampicin. a semi-synthetic derivative of the antibiotic rifamycin.

rifamycin. an antibiotic from a *Streptomyces* that specifically inhibits the initiation of RNA synthesis in bacterial cells.

righting response. righting reflex.

rigidity. **1a:** stiffness or inflexibility; tenseness. **1b:** rigor; abnormal or morbid stiffness or inflexibility. **cadaveric rigidity**. rigor mortis. **postmortem rigidity**. rigor mortis.

rigor. **1a:** severity **1b:** a state of rigidity, hardness, or stiffness. **2:** a sudden paroxysmal chill with high body temperature (the cold stage) followed by profuse sweating and a sensation of heat (the hot stage). **acid rigor**. the coagulation of muscle protein by acid. **calcium rigor**. cardiac arrest with the heart in a fully contracted state due to calcium poisoning.

rigor mortis (cadaveric rigidity; postmortem rigidity). a rigidity or stiffness of the muscles and joints of a dead body associated with the depletion of adenosine triphosphate in muscle tissue. It appears usually within a few hours following death and persists for some time. It is a recognized test of death. *Cf.* livor mortis.

ring. a thin, loose tissue ring about the stalk of a mushroom, often called the veil or annulus.

ring-necked spitting cobra. *Hemachatus haemachatus*.

ringed sea snake. *Emydocephalus annulatus*.

ringed snake. *Vermicella annulata*.

ringhals, ringhals cobra. *Hemachatus haemachatus*.

ringless honey mushroom. See *Armillariella*.

rinkals. *Hemachatus haemachatus*.

Rio de Janiero coral snake. *Micrurus corallinus*.

riot control agent. See harassing agent.

risk. **1:** the probability of occurrence of a specified undesirable or adverse effect (e.g., loss, harm, injury, death) as a function of hazard and exposure. As the probability of experiencing the event approaches zero, so does risk.

Risk is based on the probability of the undesired event occurring, and the severity of the outcome if it does occur. **1b:** the probability that harm will result to a specified population under specified conditions. Risk is expressed in various terms [e.g., per person, per unit of time (incidence), per event]. **1c:** in toxicology, risk is the probability that a substance will cause harm (usually injury, damage, or death). **2:** the probability of occurrence of an adverse effect combined with the probability that the effect, if it does occur, will be at an unacceptable (or specified) level of severity. **acceptable risk**. a value judgment determined by society usually through designated government representatives; sometimes by litigation. It is usually based upon considerations of costs, benefits and disbenefits to society (or strata of society that might be affected) in relation to perceived or estimated risk. In the United States today, one fatality per 100,000 humans or one per 1,000,000 is generally considered acceptable risk for exposure to a hazardous substance. The level of risk that is acceptable is a societal decision that varies as a function of perception, the nature and potential intensity of the risk, and other factors. **empiric risk**. an estimate of risk based on empirical evidence alone, without appeal to formal theory, inference, or analytical framework. **population attributable risk** (PAR). the proportion of a disease among individuals in a given population that is attributable to a specific etiologic agent. It is a measure of the contribution of chemical exposure to a population's disease burden. **relative risk** (RR). the relationship between exposure to a hazard and occurrence of a specified impact (e.g., disease, poisoning, injury, damage) or deleterious effect compared to the incidence of such an impact in unexposed individuals. It is often expressed as the ratio of the incidence of the selected effect in exposed versus unexposed individuals. **significant risk**. a risk of moderate likelihood of causing an unacceptable impact (e.g., to human health). **toxicological risk**. the probability that an adverse effect (e.g., injury, cancer) will result from a given exposure to a xenobiotic.

risk agent. an entity (toxicant, biological organism, radioactive material), process, activity, or phenomenon that is potentially hazardous.

risk analysis. **1:** application of the entire range of scientific and policy evaluation to risk assessment, risk management, and risk communication processes. **2:** risk assessment.

risk assessment (risk analysis). the process of estimating the magnitudes and probabilities of occurrence of adverse effects of specified processes, activities, accidents, or catastrophes, either natural or as a consequence of human activities. While risk assessments can be helpful in identifying risks, even the most sophisticated assessments suffer from irreducible uncertainties. As a consequence (1) a large array of diverse, ever-changing methodologies have been used to estimate the probability of undesirable events; (2) risk assessments offer only one element in making decisions (e.g., regulatory decisions) regarding risk in any given case. **toxicological risk assessment**. the determination of the probabilities and magnitudes of potential toxic effects due to exposure to xenobiotics or to ionizing radiation in a specified situation. Both the assessment of toxicity and probability of exposure are involved. Four major components, q.v., are usually involved: **1:** hazard identification; **2:** dose-response assessment. **3:** exposure assessment; **4:** risk characterization.

risk avoidance. the reduction of risk by altering or curtailing activities.

risk characterization. the summary output of a risk assessment, it is a projection of ultimate health and environmental consequences of existing conditions; qualitative and quantitative evaluation of uncertainties; characterization of the degree of risk reduction associated with different degrees of source reduction or contaminant removal.

risk communication. any purposeful exchange, formal or informal, of information concerning health, environmental, or other risks among interested parties.

risk control. societal control of risk whether through social, legal, scientific, or engineering methods and activities.

risk estimate. in toxicology, a description of the probability that organisms exposed to a specified dose of a xenobiotic will experience an adverse effect.

risk extrapolation. See extrapolation.

risk management. the process of judgment and evaluation that uses the results of risk assessment to evaluate alternative regulatory and nonregulatory options to reduce risk to an acceptable level and selecting among them. In addition to scientific/technical input, this process necessarily takes into account legal, economic, political, and social factors.

risus. laughter, a laugh.

risus caninus. risus sardonicus.

risus sardonicus (risus caninus; canine spasm; cynic spasm; sardonic grin). a grotesque grinning expression caused by acute spasm of the facial muscles in tetanus, as in the tetanic grimace of one dead from strychnine poisoning. The name stems from a plant of Sardinia, probably a ranunculaceous plant, that was believed to cause it.

river jack. *Bitis nasicornis*.

river ray. a stingray, q.v., of the family Potamotrygonidae.

river snake, Clarence. See *Tropidechis carinatus*.

Rn. the symbol for radon.

RNase (ribonuclease). an enzyme that lyses ribonucleic acids. It occurs in small amounts in some snake venoms.

Ro 1-5130. pyridostigmine bromide.

Roaccutane™. the 13-*cis*-form of retinoic acid (isotretinoin). See also retinoic acid, Accutane.

road asphalt. asphalt.

road tar. asphalt.

robin. a potent phytotoxin found in the roots, bark, leaves and seeds of *Robinia pseudoacacia*. *Cf.* antirobin.

Robinia pseudoacacia (black locust; bastard acacia; black acacia; false acacia; pea flower locust). a large, common, rough-barked, native tree (sometimes a large shrub; Family Fabaceae, formerly Leguminosae) of eastern North America, especially from Pennsylvania through Georgia and the Ozark Mountains. It sometimes grows to 25 m. The branches bear spines. The bark, foliage, seeds, and sprouts are very toxic due to the presence of a powerful phytotoxin, robin. A number of other toxic compounds have been isolated from the tissues of this plant. Ingestion of relatively small amounts of the toxic plant material can produce symptoms that include vomiting, diarrhea, dullness, depression, a weak pulse, and coldness of the extremities. Toxicity, at least of the leaves, may be seasonal. Horses are much more sensitive than cattle. The latter may be poisoned when grazing on cut-over woods. A mass poisoning of humans is known from 1887 when 32 boys from the Brooklyn Orphan Asylum ate bark from fence posts of black locust. All recovered.

rock cobra. *Naja haje*.

rock poppy. See *Chelidonium*.

rock rattlesnake. *Crotalus lepidus*.

rock salt. sodium chloride.

rock shell. See *Murex*.

rockfish. **1:** a common name of any of a great number of fishes that inhabit rocky sea bottoms. **2:** a common name of many fish of the family Scorpaenidae (e.g., *Sebastodes*, *Sebastes*). See also *Pholis*.

Rocky Mountain bee plant. *Cleome serrulata*.

Rocky Mountain wood tick. *Dermacentor andersoni*.

rodent. see Rodentia.

Rodentia (rodents). the largest order of placental mammals (class Eutheria), containing some 6400 known species and subspecies. All have one pair of continuously growing, rootless, chisel-like upper incisors used in gnawing; no canines; and flat-crowned premolars and molars used in grinding. The lower jaw is quite mobile and is able to move foreward and backward, laterally, as well as up and down. Otherwise, rodents are extremely diverse behaviorally, physiologically, and nutritionally. They range in size from a few grams to ca. 60 kg, are nearly world-wide in distribution (absent from the Australian region); and occupy a wide range of habitats. While most are omnivorous or herbivorous, a few are carnivorous. Included, for example, are various mice, voles, rats, cavies, guinea pigs, hamsters, nutria, pacas, porcupines, squirrels, and beavers. Most have not been studied toxicologically and almost all that is known of rodent toxicology is based on tests or studies of a few varieties of small rodents that were born, reared, and studied under laboratory conditions and whose genome is not represented in nature. See also Cricetidae, Muridae, rat, *Rattus*, *Sigmodon*.

rodenticide 1: any substance that is lethal to rodents (e.g., 6-mercaptopurine, malachite green, rubratoxin, scilliroside). **2:** any chemical compound or formulation produced for the purpose of killing or repelling rodents. Most are applied to the control of the house mouse, *Mus musculus* or *Rattus norvegicus*, the Norway rat. Rodenticides are of considerable toxicological concern because their use frequently results in direct or secondary poisoning of nontarget animals (e.g., of pets, foxes, swine, certain omnivorous birds, birds of prey; (See, for example, sodium hexafluorosilicate, sodium fluoroacetate, TEPP, zinc phosphide); they are also responsible for accidental and suicidal poisonings of humans. See also anticoagulant poison, ANTU (1-naphthylthiourea), ANTU poisoning, aconitase, aminopterin, arsenic disulfide, barium chloride, barium nitrate, brodifacoum, bromadiolone, bromethalin, calcium cyanide, coumafuryl, coumarin, 2-diphenyl-1,3-indanedione, 2,4-dithiobiuret, fluoroacetamide, fluoroacetate, hydrogen cyanide, 1-phenyl-2-thiourea, phosphine, phosphorus (white phosphorus), 2,4,6-(1*H*,3*H*,5*H*H)-pyrimidinetrione, red squill powder, strychnine, thallium, thallium sulfate, thiourea, *Urginea maritima*, vitamin D_2, vitamin K antagonist, warfarin.

rodentine. pertaining to or characteristic of rodents; rodent-like.

roe. **1:** eggs of fishes. The roe of many bony fishes (Class Osteichthyes) are edible, but that of most marine species of bony fishes is at times poisonous; sometimes extremely so. **2:** the swollen ovaries or extruded eggs of decapod crustaceans. **3:** Old World deer of the genus *Capreolus* (roe, roe deer).

Roentgen (röntgen; symbol, R formerly, r). a unit of measurement of an exposure dose of gamma- or X-ray radiation defined by the ability to ionize air. It corresponds to that amount of ionizing radiation that, in dry air at STP, is sufficient to liberate electrons and positrons carrying a total charge of 2.58 x10^{-4} coulombs/kg of air, or to produce two ionizations per cubic μm of water or living tissue.

Roentgen-equivalent-physical (REP, rep). an obsolete unit of measurement; that amount of ionizing radiation which, if absorbed by living tissue, increases the energy per gram of tissue by that produced by 1 rad of X-rays or gamma-rays. See RAD.

Roentgen intoxication. radiation sickness.

Roentgen ray. X-ray.

Roentgen ray dermatitis. radiodermatitis.

Roentgenism (Roentgenization). an obsolete term. **1:** the use of X-rays in the diagnosis and treatment of disease. **2:**. injurious or deleterious effects of X-rays on tissues.

Roentgenization. See Roentgenism.

Rokitansky's disease. acute yellow atrophy (See atrophy).

Roman vitriol. the pentahydrate of cupric sulfate.

Ronnel (phosphorothioic acid**;** O,O-dimethyl O-(2,4,5-trichlorophenyl)ester**;** dimethyltrichlorophenyl thiophosphate**;** fenchlorphos). a systemic organophosphate insecticide and parasiticide, $C_8H_8Cl_3O_3PS$, used in the home and on farms. Ronnel is a cholinesterase inhibitor; it is an eye irritant and can cause liver and kidney damage; it is a possible human carcinogen and teratogen. When therapeutic doses were given orally to pregnant *Alopex lagopus* (blue foxes), for example, the average litter size fell from 9.5 per vixen in controls to 1.2 in test animals. Effects on the fetus included incomplete ossification of cranial bones, palatoschisis, and hydrocephalus.

röntgen. Roentgen.

roof rat. *Rattus rattus alexandrinus*.

root. **1:** the underground portion of a plant. **2:** the proximal end of a nerve. **3:** a portion of an organ imbedded in tissue.

roquefortine. an indole mycotoxin produced by Penicillium roqueforti.

roridin A. a trichothecene mycotoxin, similar in structure to the verrucarins. It was first isolated from *Myrothecium roridum*.

roridin C. trichodermol.

Rosaceae. (the rose family). a family of herbs, shrubs, and trees of various growth forms, that are widely distributed throughout the Northern Temperate Zone. Many are cultivated as ornamental garden or house plants, for their fruit, or (in a few cases) for their medicinal value.

rosary bead. *Abrus precatorius*.

rosary pea. *Abrus precatorius*.

rose. See Rosaceae.

rose bay. *Rhododendron*.

rose-bay. *Nerium oleander*.

rose daphne. *Daphne cneorum*.

rosebay. See *Rhododendron*.

Rosellina necatrix. a fungus that produces cytotoxic cytochalasins.

rosemary pine. *Pinus taeda*.

Rosen's snake. *Denisonia fasciata*.

rosin. the solid, amber, translucent, resin from various pine trees (*Pinus* spp.). It is used chiefly to prepare plasters and ointments.

rostral. **1:** resembling a beak. **2:** cephalad; toward the cephalic end of the body. **3:** rostral plate.

rostral plate (rostral). the single large plate at the tip of a snake's snout. It is modified or even absent in some snakes.

rotenone ([2R-(2α,6aα,12aα)]-1,2,12,12a-tetrahydro-8,9-dimethoxy-2-(1-methylethenyl)-[1]benzopyrano-[3,4-b]furo[2,3-h][1]benzopyran-6(6aH)-one; Derrin; nicouline; tubatoxin). a dangerously toxic, white crystalline compound, $C_{23}H_{22}O_6$, soluble in ethyl ether and acetone, but insoluble in water. It decomposes on exposure to light. It has been used as an insecticide, a scabicide, a fish poison (currently the most important application), in fly sprays, flea collars, and in moth-proofing agents. Fatalities from overexposure have occurred, although it is safe for most mammals as normally used. It is poisonous to humans and other higher vertebrates by ingestion and possibly other routes of exposure; swine are especially susceptible; it is very toxic to fish. It is teratogenic, may be mutagenic, and affects the reproductive system of laboratory animals. Rotenone is a questionable human carcinogen. Symptoms of acute poisoning of humans include nausea, numbness, and tremors. Lesions resulting from chronic exposure include liver and kidney damage. Rotenone is the most active of several rotenoids synthesized by at least 70 species of legumes, most notably those of the genera *Derris* and *Lonchocarpus*. See also derris.

rough endoplasmic reticulum (RER). endoplasmic reticulum that has attached ribosomes. See endoplasmic reticulum.

rough fish. those fish, not prized for eating, such as gar and suckers. Most are more tolerant of pollution and changing environmental conditions than game species.

rough-scaled snake. *Tropidechis carinatus*.

round stingray. a stingray, q.v., of the family Urolophidae, q.v.

Roundup™. glyphosphate.

roundworm. any animal of the phylum Nematoda.

route of absorption. See absorption (route of absorption).

route of administration. See administration (route of administration).

route of entry. route of exposure. See exposure (route of exposure).

route of exposure. See exposure.

route of penetration. See exposure (route of exposure).

Roxarsone. a trademark name for 4-hydroxy-3-nitrophenylarsonic acid. It is marketed primarily as a growth promoter added to animal feed and as an antibacterial for domestic fowl. It is a pentavalent organic arsenical with a relatively high acute toxicity to rats and dogs. Among the effects observed in these animals are internal hemorrhage, kidney congestion, and gastroenteritis. Rats fed fatal doses of about 400 ppm in the diet exhibited progressive weakness prior to death.

RQ. respiratory quotient, *q.v.*

RR. relative risk. See risk.

Ru. the symbol for ruthenium.

RU-486™. mifepristone.

RU-38486™. mifepristone.

rubber vine. Cryptostegia grandiflora.

rubber weed. *Actaea (= Actea) richardsoni.*

rubberweed. *Hymenoxys odorata* (bitter rubberweed), *Hymenoxys richardsonii* (Colorado rubberweed).

rubbing alcohol. isopropanol. See alcohol.

rubefacient. **1:** reddening the skin; able to redden the skin. **2:** a counterirritant that increases blood flow to the area that it contacts, usually on the skin (with reddening) but sometimes internally.

rubellin. a cardiac glycoside with a digitalis-like action isolated from *Urginea rubella* (Family Liliaceae).

rubescent. flushing, growing red; exhibiting redness.

rubidium (symbol, Rb)[Z = 37; A_r = 85.48]. a soft, lustrous, silvery, highly reactive element of the alkali-metal group. Its chemical behavior is similar to that of potassium. It reacts vigorously with oxygen, water, halogens, and mercury; and it ignites spontaneously in an oxygen atmosphere. It must be stored under benzene, petroleum, or other liquids that are free of oxygen. Its salts are used in medicine, largely for the same purposes as the corresponding sodium and potassium salts.

rubidomycin. daunorubicin.

rubigervine. rubijervine.

rubijervine (solanid-5-ene-3β,12α-diol; Δ^5-3β,12α-dihydroxysolanidene; rubigervine). a very toxic, crystalline alkaloid, $C_{27}H_{43}NO_2$, isolated from various species of *Veratrum.* It is slightly soluble in water and soluble in organic solvents such as ethanol, benzene, chloroform, and methanol. It has been used therapeutically as an antifungal.

Rubomycin. daunorubicin.

rubor. the redness or discoloration of inflammation; together with calor, dolor and tumor, it is one of the principal signs of inflammation asserted by Celsus.

rubratoxin. either of two mycotoxins (rubratoxins A and B) produced by *Penicillium rubrum* and *P. purpurogenum* which often infest cereal grains. Both appear to be extremely toxic and are known from outbreaks of toxicoses in the United States. Rubritoxin B is extremely toxic to laboratory rodents, rabbits, cats, and dogs. It is hepatotoxic, embryotoxic, and teratogenic. Effects typically include anorexia, dehydration, somnolence, diarrhea, and jaundice; internal hemorrhage is common.

rubroskyrin. one of a group of highly toxic, hepatotoxic, anthraquinoid mycotoxins synthesized by *Penicillia* (e.g., *P. brunneum*, *P. islandicum*, *P. rubrum*, and *P. tardum*). See also luteoskyrin, rugulosin.

ruby arsenic. arsenic disulfide.

rudimentary. **1:** imperfectly formed or developed; undeveloped, immature, incomplete. **2:** arrested, blocked, or inhibited at an early stage of development or evolution. **3:** it is not a synonym of vestigial as is sometimes encountered in the literature.

rug and carpet cleaner. See cleaner.

rug cleaner. See upholstery cleaner.

rug shampoo. See upholstery cleaner.

rugando. *Atheris nitschei.*

rugulosin. either of two hepatotoxic hydroxyquinones produced by *Penicillia* (e.g., *P. brunneum*, *P. islandicum*, *P. rubrum*, and *P. tardum*). See also luteoskyrin. **(+)rugulosin**. produced by *Penicillium rugulosum*, it is very toxic to laboratory mice and extremely toxic to laboratory rats, HeLa cells, the protozoan, *Tetrahymena pyriformis*, and mutant *Escherichia coli*. Responses of exposed mice to (+) rugulosin generally include ruffled fur, loss of skin tone, inactivity, shallow respiration, prolonged coma, and death. The liver of mice subjected to acute intoxication exhibits centrolobar necrosis with coagulation necrosis of the cells, karyolysis, karyorrhexis, hyperchromatosis of the nuclear membrane, and fatty metamorphosis. Atrophy of the thymus, spleen, and lipid tissue accompany the liver necrosis. **(-)rugulosin**. produced by *P. islandicum* and *Myrothecium verrucaria*, it is extremely toxic to HeLa cells and *Escherichia coli* mutants. In various tests, (-)rugulosin has proven to be ca. 4-10 times more toxic than (+)rugulosin. See also cyclochlorotine, erythroskyrine, luteoskyrin.

rum cherry. *Prunus serotina.*

rumenitis. inflammation of the rumen of ruminant animals.

Rumex (dock; "sorrel"). a nearly cosmopolitan genus of hardy, mostly biennial and perennial, often weedy herbs and shrubs (Family Polygonaceae). Most species are native to, and widely distributed within, north temperate regions. Many are common to pastures, meadows, roadsides, and other situations typified by dry, acidic, or poor soil. Some species are frequently referred to in the literature as sorrel. This causes some confusion in reports of livestock losses in the United States. Many species are toxic due to the presence of sometimes lethal amounts of soluble oxalate in the plant tissues. Examples of species that have caused losses among livestock in various countries: *R. crispus* (yellow dock), *R. acetosella* (sheep sorrel, common sorrel), *R. acetosa* (garden sorrel). Those species with fleshy leaves are often sour but edible, and used especially in salads. Toxic concentrations of nitrate have been reported for various species under conditions of rapid growth. See also emodin.

ruminant. an animal that regurgitates and rechews food (the cud) from the first stomach (the rumen); examples are antelope, cow, deer, sheep.

run-off. runoff.

running spiders. See *Chiricanthium.*

runoff (run-off; environmental runoff). that part of precipitation, snow melt, or irrigation water that is not retained by the soil, but runs freely off the land into streams or other surface waters. It can carry dissolved or suspended pollutants from the air and land (e.g., pesticides and other agrochemicals, industrial wastes) into the receiving waters. The water bodies thus contaminated can be toxic to humans and other animals directly or via food-chain contamination. See also transport.

rush. *Equisetum.*

Russell's sea snake. *Hydrophis obscurus.*

Russell's viper. *Vipera russelii.*

Russian fly. See *Lytta.*

Russian thistle. *Salsola pestifer.*

Russula emetica (emetic russula). a poisonous, strongly emetic mushroom (Family Russulaceae) that usually grows in sphagnum moss or occasionally on rotten wood in wet, boggy, mixed or coniferous woods throughout much of North America.

rustyleaf. See *Menziesia.*

Ruta graveolens (common rue, herb of grace). a fragrant, poisonous, shrubby, evergreen plant

(Family Rutaceae) that was formerly cultivated for its therapeutic and aromatic properties. It is now locally common in old fields and along roadsides in the northeastern United States from Vermont to Virginia and west to Missouri. It is an acronarcotic poison and a photosensitizing agent, such that exposure to sunlight after ingestion can cause severe sunburn. Contact with the foliage, flowers, or fruit can produce erythema and vesication resembling that of poison ivy. The oil is very dangerous and may be abortifacient. Large doses can cause nausea, vomiting, confusion, gastrointestinal pain, convulsions, and prostration, but are rarely lethal.

rutabaga. *Brassica napus* var. *napobrassica*.

ruthenium (symbol, Ru)[Z = 44; A_r = 101.1]. a lustrous, hard, transition metal of the platinum group of elements. It catalyzes the synthesis of long-chain hydrocarbons.

Ruvettus pretiosus (castor oil fish; escolar). a large, deepwater, mackerel-like marine bony fish (Family Gempylidae) of the tropical Atlantic Ocean, Mediterranean Sea, and Indo-Pacific region. It usually occurs at depths greater than 700 m. The flesh and bones are oily with a mild purgative effect. Treatment is not required.

RWC. receiving water concentration.

ryania. a naturally occurring insecticide.

rye. *Secale cereale* (cultivated rye). *Elymus* (wild rye).

rye ergot. See ergot.

rye smut. See ergot.

S

σ. See sigma. For terms beginning with this prefix, see the root word.

Σ. See sigma. For terms beginning with this prefix, see the root word.

S. the symbol for sulfur.

s.-K. strophanthin.

S-9 fraction. an enzyme preparation used *in vitro* to test a mutagenic chemical to ascertain whether metabolic activation is essential to the expression of mutagenicity. It is usually the supernatant or cytosolic fraction of a rat liver homogenate centrifuged at about 9000 G for 10 to 20 minutes. Included in this fraction are enzymes of the cytochrome P-450-dependent monooxygenase system that have usually been induced by treating the rats with Aroclor 1254. S-9 fractions from uninduced rats and from other species and organs are sometimes used. See test (Ames test), cytochrome P-450, microsome.

SA. sinoatrial node.

SA node. sinoatrial node.

saaman. *Naja nigricollis*.

sabadilla (cevadilla; caustic barley). **1:** common name of the plant, *Schoenocaulon officinale*. **2:** dried ripe seeds of *S. officinale*, used as a pedicullicide and as an insecticide and ectoparasiticide for cattle. They contain 3-6% total alkaloids by weight, chiefly cevadine and veratridine, with smaller amounts of sabadilline, sabadine, and sabadinine. Sabadillic and veratric acids are also present.

sabadine ((3β,4α,16β)-cevane-3,4,12,14,16,-17,20-heptol 3-acetate; sabatine). a toxic, crystalline alkaloid isolated from the dried seeds of *Schoenocaulon officinale*; a component of veratrine mixture. See veratrine, sabadilla.

sabatine. sabadine.

sabinism (savin poisoning). poisoning by savin.

sac. a bag or pouch-like structure.

sacahuista. See *Yucca* and *Nolina texana*.

saccharin (1,2-benzisothiazol-3(2H)-one 1,1-dioxide). a formerly popular nonnutritive sweetener, discovered in 1879. At relatively high concentrations it increases the rate of bladder cancer in some laboratory animals and is a suspected human carcinogen. It promotes bladder carcinogens such as *N*-methylnitrosourea.

saccharin sorghum. *Sorghum vulgare*.

Saccharomyces cerevisiae. a species of yeast frequently used to test, *in vitro*, the mutagenic potential of a chemical.

sacred mushroom of Mexico. *Psilocybe mexicana*.

sacrifice (kill). a term commonly used to indicate that a test organism is killed in the course of, or at the termination of a test or experiment, usually in order to recover tissue or material to measure or otherwise evaluate the necessary end point(s) for interpretation of the test or experiment.

saddle-shaped false morel. *Gyromitra (Helvella) infula*.

safe. **1:** secure, sheltered, protected; unharmed. **2:** harmless, innocuous. **3:** reasonably free from danger, as applied, for example, to the proper use of a food, chemical substance, drug, material, or process. **4:** a strongbox or secure vault.

safety. **1:** security, protection. (note: safety is always relative or uncertain). **2a:** the reciprocal of risk, or the probability that harm will not occur under a set of specified conditions. **2b:** the approximate opposite of hazard; the improbability of exposure to possible loss or injury. See also margin of safety. **3:** the expectation that a biologically active agent will cause relatively little or no harm under a given set of circumstances. See also margin of safety.

saffron (meadow saffron). See *Colchicum*.

safrole (5-(2-propenyl-1,3-benzodioxole); 4-allyl-1,2-methylenedioxybenzene; allyl catechol methylene ether; allyldioxybenzene methylene ether). a pale yellow, volatile, oily constituent, $C_3H_5C_6H_3O_2CH_2$, of the fruit of *Piper nigrum* (black pepper) and *Sassafras albidum*. It is a methylenedioxyphenyl compound that is responsible for the odor of sassafras, camphorwood, and certain other oils. Safrole finds commercial use in the manufacture of heliotropin, soaps, and in perfumery. Safrole is hepatogenic and carcinogenic to some laboratory animals; it induces hepatic tumors in the laboratory rat at dietary concentrations as low as 0.5%. The active metabolite is probably the sulfate ester of the 1′-hydroxy derivative of safrole. It inhibits cytochrome P-450-dependent monooxygenations. Safrole was formerly used in root beer and medically as an anodyne.

sage. See *Artemisia*, *Salvia californica*. **annual sage**. *Salvia reflexa*. **hop sage**. *Grayia*. **sand sage**. *Artemisia filifolia*.

sage sickness. a rapidly developing disease of horses unaccustomed to foraging on sage. The animals become nervous and tend to fall when forced to move due to a partial paralysis of the forelimbs. Within a week or two the animals recover and are able to consume large quantities of sage without recurrence of symptoms. The species responsible for the disease is *Artemisia filifolia* and sometimes *A. spinescens*.

sagebrush. See *Artemisia*.

sago palm. *Cycas*.

Sahara rock viper. *Vipera mauritanica*.

Sahara sand viper. *Cerastes vipera*.

Sahara viper. *Vipera lebetina*.

saindé. a colloquial name of *Causus rhombeatus*.

Saint Anthony's fire. ergotism.

Saint Lucia viper. *Bothrops caribbaeus*.

sakamala. *Naja haje*.

Sakashima habu. *Trimeresurus elegans*.

sal acetosella. potassium binoxalate.

sal ammoniac. ammonium chloride.

sal ammoniasalivation. the secretion of saliva. *Cf.* ptyalism.

sal chalybis. ferrous sulfate.

sal soda (washing soda). the technical grade of sodium carbonate, decahydrate, q.v. It is used in the treatment of scaly skin diseases, but is otherwise rarely used therapeutically because of its irritant action. See also sodium chloride, clorox 2.

salamander. See Urodela.

salicylate. any salt of salicylic acid. They form a class of nonsteroidal anti-inflammatory (analgesic) drugs, the most familiar of which is acetylsalicylic acid (aspirin). Sodium salicylate is sometimes used as an alternative to aspirin. Toxic doses of salicylates initially stimulate the CNS including the respiratory center, with consequent hyperpnea leading to CO_2 loss. Effects of acute intoxication include a life-threatening respiratory alkalosis, sometimes a subsequent very serious metabolic alkalosis which occurs most often in children, hyperkalemia, hyperthermia, dehydration, convulsions, and shock. The initially acute CNS stimulation is followed by depression. Extremely high doses may produce a respiratory paralysis leading to respiratory acidosis for which the body may not be able to compensate because bicarbonate ions were excreted during the respiratory alkalosis that occurs in the initial stages of intoxication. Lesions are those of disturbed acid-base balance. Death may result from respiratory collapse. Salicylates are also fetotoxic, and may cause tremors, convulsions and bleeding that result in neonatal death. Chronic use of salicylates may cause tinnitis, nausea, vomiting, deafness, and disturbance of equilibrium.

salicylate poisoning. salicylism.

salicylic acid (*o*-hydroxybenzoic acid). a derivative of salicin, it is a toxic, white, acrid powder, $C_6H_4(OH)(COOH)$, the dust of which forms an explosive mixture in air. It is used in the manufacture of aspirin and salicylates and as a topical keratolytic agent, antiseptic, and fungicide. See also salicylate, aspirin.

salicylic acid acetate. aspirin.

salicylism (salicylate poisoning). chronic poisoning by salicylic acid or by salicylates. Symptoms usually include tinnitis, nausea and vomiting, and may include deafness and disturbances of equilibrium. See salicylates.

Salientia. a synonym in some classifications for the amphibian subclass Anura (frogs and toads). See also Batrachia.

saline (salty). **1:** salty; salt-like, of the nature of a salt; of, pertaining to, or containing salts or a salt, especially sodium chloride. **2:** saline solution; a salt solution. **3:** pertaining to water or soil that is rich in soluble salts. **physiological saline** (physiological saline solution; physiological salt solution). an isotonic aqueous solution of salts at a concentration that is isotonic with the cells or body fluids of a particular type of organism. Such solutions are used chiefly to bathe and temporarily maintain living cells *in vitro*, or as the carrier of drugs or other substances that are administered by injection. Human physiological saline solutions contain 0.9% sodium chloride. Physiological saline is a physiological salt solution, q.v.

saline solution. See saline.

salinity. **1a:** saltiness; a measure of the total concentration of dissolved sodium chloride in water. **1b:** a measure of the total weight of dissolved salts per unit volume of solvent. Thus, the salinity of sea water is taken as the total amount by weight (in parts per thousand, ppt) of dissolved solids following the conversion of all bromide and iodide to chloride; of all carbonate to oxide; and the complete oxidation of all organic matter. The salinity of ocean water averages 35 ppt. See also chlorinity.

salinization. **1:** the process whereby soluble salts accumulate in soil or other environmental matrices. This often results in a reduction in arable land due to the toxicity of the salinized soil to plants. **2:** excessive salinity, especially of soil.

saliva. the secretion of salivary glands. Such glands are present and open into or near the mouth in many vertebrate and invertebrate animals. The usual function of saliva is the moistening and lubrication of food; in some cases it may also serve an excretory role (See excretion). The saliva of humans and most mammals is a clear, viscous fluid (secreted also by mucous glands of the mouth) that contains water, mucin, organic salts, and ptyalin. In some tetrapods and insects, saliva contains amylases (e.g., ptyalin in mammals) which may initiate the digestion of starch. The saliva of insects contains various digestive enzymes depending upon the nature of the diet. In blood-sucking animals such as leeches and mosquitos, saliva contains anticoagulants. The saliva of many animals is poisonous, often serving as a venom (See, for example, Acanthina, *Blarina brevicauda*, *Bufo*, Cephalopoda, *Dermacentor*, gland (proboscis gland), ledoisin). The rate of secretion of saliva and/or its composition may be affected by a many toxicants (See ptyalism).

salivary (sialine; sialic). of or pertaining to saliva or to the secretion of saliva.

salivation. the secretion of saliva, including ptyalism.

salivatory. pertaining to the secretion of saliva.

salmin. a toxic substance isolated from the milt of salmonid fish (Class Osteichthyes, Family Salmonidae).

Salmo gairdneri (rainbow trout). a standard species of freshwater bony fish (Family Salmonidae) used in aquatic toxicology tests. It is native to the Pacific Coast of North America, but has been widely introduced throughout the United States and elsewhere.

Salmonella. a genus of pathogenic, toxin-secreting, aerobic and facultatively aerobic Gram-negative bacteria (Family Enterobacteriaceae). ***S. cholerae-suis***. a species that infects swine; it sometimes also causes an acute gastroenteritis and enteric fever in humans. ***S. enteritidis***.

(Gärtner's bacillus). a cause of gastroenteritis in humans. ***S. hirschfeldii*** (Hirschfeld's bacillus). a cause of enteric fever in humans. ***S. schotmülleri*** (Schotmüller's bacillus). a cause of enteric fever in humans. ***S. typhi*** (=*S. typhosa*; typhoid bacillus). the cause of typhoid fever; exposure of humans is usually via contaminated water and food. ***S. typhimurium***. this species secretes a ptomaine, which causes the most common form of food poisoning in humans. Victims are afflicted with flu-like symptoms and sometimes die from the effects of the toxin. See also *Salmonella*/microsome test. ***S. typhosa***. *S. typhi*.

salmonellosis. any disease caused by bacteria of the genus *Salmonella*.

Salmonidae. a family of bony fishes that includes trout and salmon. The mouth is relatively large with well-developed dentition; the scales are small; and there is a small, dorsal, adipose fin anterior to the base of the tail. Most species inhabit cool water (freshwater and/or marine) in the northern hemisphere. See salmin.

salpetersäure (German). nitric acid.

Salsola pestifer (Russian thistle). a bushy annual herb (Family Chenopodiaceae) that is most commonly seen in the prairie regions of the United States and in waste areas. It may contain hazardous concentrations of soluble oxalates and probably salsoline (which has been isolated from *S. richteri*). Concentrations of nitrate that are toxic to livestock have also been reported.

salsoline(1,2,3,4-tetrahydro-7-methoxy-1-methyl-6-isoquinolinol; 6-hydroxy-7-methoxy-1-methyl-1,2,3,4-tetrahydroisoquinoline). a moderately toxic crystalline substance isolated from *Salsola richteri*, $C_{11}H_{15}NO_2$.

salt. **1:** sodium chloride. **2a:** a compound produced by the reaction of an acid with a base. Such a compound is usually composed of a metal with a nonmetal and may react chemically as a metal or a nonmetal. **2b:** a compound resulting from the replacement of one or more hydrogen atoms of an acid by metal ions or electropositive radicals (e.g., NH_4^+). Salts are generally crystalline at ordinary temperatures, and are strongly ionic in aqueous solution. Examples of salts are chlorides, nitrates, carbonates, sulfates, silicates, and phosphates. **3:** a substance (e.g., Epsom salt) used as a purgative. **common salt, common table salt, table salt**. sodium chloride.

salt gland (superciliary gland). one of a pair of glands near the eyes of seabirds that removes excess salt (accumulated through intake of sea water) from the body fluids. This allows substantial intake of seawater (e.g., in association with feeding) without suffering toxic effects.

salt poisoning. a lethal disease of animals, notably pigs, fed on garbage, due largely to excessive intake of sodium chloride (table salt). If sufficient quantities of fresh drinking water are available, poisoning does not usually result.

salt of Saturn. lead acetate.

salt of sorrel. potassium binoxalate.

salt water intrusion. the invasion of fresh surface- or groundwater by salt water. If the salt water comes from the ocean it may be called sea water intrusion. Such invasion may, if the rate of recharge by fresh water is low, create a semipermanent or permanent toxic environment, even a virtual desert at and about the site.

saltatory conduction. the mode of transmission of nervous impulses in myelinated nerves, whereby the myelin sheath provides insulation that permits efficient passive transmission of the impulse by local currents along the neuronal axis in the internode region. The signal is amplified at the nodes of Ranvier, where the absence of myelin permits the generation of an impulse. This type of conduction is much more rapid and efficient than continuous conduction.

saltbush. See *Atriplex*.

saltbush snake. *Demansia psammophis*.

saltpeter. potassium nitrate.

salvarsan (dihydroxydiaminoarsenobenzene dihydrochloride; arsphenamine; 606). a derivative of arsenic and benzene, this compound, $C_{12}H_{14}O_2N_2Cl_2As_2 \cdot 2H_2O$, was synthesized by Dr. Paul Ehrlich in 1907 and widely used as a specific in the treatment of syphilis. Many consider this to mark the beginning of modern chemotherapy. It is no longer used.

Salvia leucophylla (sage). this shrub (Family Labiatae) and certain others such as *Artemisia californica*, that comprise the soft chaparral communities of southern California, produce volatile terpenes that are apparently adsorbed by the soil and inhibit the growth of grasses (Family Gramineae), sometimes producing bare areas of one or a few meters wide about these communities. ***Salvia reflexa*** (syn. *S. laciniata*)(annual sage, mintweed). an annual, bushy weed (Family Labiatae) that is relatively common in dry, open areas from Wisconsin to Montana, south to Mexico; uncommon in the eastern United States. It has caused severe losses among livestock in Australia and cattle in the United States. Australian material is high in nitrate and the symptoms are those of nitrate poisoning. Symptoms differ in the United States, however, where the poisoned cattle quickly develop muscular weakness and symptoms of gastrointestinal distress.

salycilated mercury. mercuric salycilate.

Salzburg vitriol. the pentahydrate salt of cupric sulfate, q.v.

samandaridin. a neurotoxin, it is an extremely toxic, steroid alkaloid secreted by glands in the skin of various salamanders. It has neurotoxic, hypertensive, and anesthetic action. See Urodela.

samandaridine. an alkaloid, $C_{21}H_{31}O_3N$, secreted by glands in the skin of various salamanders. It is less toxic than samandarine.

samandarin. a powerful, steroid, alkaloid toxin from salamanders. It has neurotoxic, pressive, and anesthetic activity.

samandarine (1α,4α-epoxy-3-aza-A-homoandrostan-16β-ol). a toxic, convulsant, steroid alkaloid, $C_{19}H_{31}O_2N$, secreted by glands in the skin of *Salamandra atra* (the Alpine salamander) and *S. maculosa* (the European fire salamander). See also Urodela.

samandarone. a toxic, alkaloid ketone corresponding to the secondary alcohol of semandarine. It is secreted by skin glands of *Salamandra atra* (Alpine salamander) and *S. maculosa* (European fire salamander).

samandenon. a steroid, alkaloid toxin secreted by the skin glands of salamanders. See Urodela.

samanin. a steroid, alkaloid toxin secreted by the skin glands of salamanders. See Urodela.

samarium (symbol, Sm)[Z = 62; A_r = 150.35]. a silvery yellow element of the lanthanoid series of metals. It occurs in nature in association with other lanthanoids. The trichloride of samarium, $SmCl_3$, is very toxic to laboratory mice.

Sambucus (elders). a genus of large, deciduous shrubs or small trees that have compound leaves (Family Caprifoliaceae). Documented poisonings of humans by members of this genus are rare. Nevertheless, most species should be considered dangerous, especially to children. ***S. canadensis***. (common elder; elderberry; black elder; red-berried elder; scarlet elder). a shrub 1-4 m tall that occurs in the northern United States and Canada. The stems have a white pith and are only slightly woody. The twigs bear 5-11 elliptic to lanceolate, sharply serrated leaflets. All parts of the plant are poisonous, including the raw berries. The leaves, shoots, bark, roots, should be considered very toxic. Neverthless, limited intake of raw berries is not likely to cause harmful effects. Cooked elderberries are often used to make pies and jams and as such are safely eaten. The active principle is sambunigrin, a cyanogenic glycoside. Symptoms of intoxication, which usually appear several hours following ingestion of the plant material, include dizziness, headache, nausea, vomiting, tachycardia, and possible death. ***S. pubens*** (American red elder; red-berried elder; stinking elder). a large shrub up to eight m tall, with warty bark and brown pith, pubescent branches; 5-7 ovate-lanceolate leaflets which are downy or glabrous beneath. The ripe, cooked berries are edible. The roots, stems, leaves and possibly the flowers are toxic and cathartic. The leaves and roots contain cyanogenic glucosides, and high concentrations of nitrate that may affect livestock have been reported. ***S. racemosa*** (European red-berried elder). a hardy species that grows to a height of about 4 m. Toxicity and effects are similar to those of *S. canadensis*.

Samoan sea snake. *Laticauda semifasciata*.

sample. **1a:** any subset of a population. **1b:** a representative subset of a population. **1c:** a subset of a population collected with a view to estimating some characteristic of the population by observation or measurement of the sample. The specimens that form the sample are selected and collected in such a way that the sample can be regarded as representative of the population with respect to the characteristic(s) to be investigated. *Cf.* specimen. **grab sample**. a sample, usually of a single specimen, presumably taken randomly (often not). In air pollution science, a grab sample of air is considered a type of short-period sample. **random sample**. a sample in which every member of the population from which it is drawn has an equal and independent chance of being included. **representative sample**. **1a:** a representative part of a larger unit used to study the properties of the whole. **1b:** a true random sample of sufficient size such that the distribution of values about a suitable measure of central tendency (e.g., average or mean value) is identical (or sufficiently similar) so that sample observations or measurements can be used to make confident predictions about the population as a whole. See also population. **short-period sample** (spot sample). a sample of air collected over a short period of time and usually taken to a central laboratory for analysis. The WHO Expert Committee on Atmospheric Pollutants (1963) recommended that for purposes of international comparison of routine measurements a short-period air sample be defined as one taken over a period of 30 minutes. See also grab sample. **spot sample**. short-period sample.

sampler, personal. a device attached to a person that samples air in his or her immediate vicinity so that the individual's exposure to toxic or otherwise hazardous materials may be determined.

sand briar. *Solanum carolinense*.

sand clam. *Mya arenaria*.

sand dollar. any of a type of echinoderm (Class Echinoidea) that has a flat, thin, nearly circular test that bears a thick coat of spines. They aggregate on sandy bottoms. Common genera are *Clypeaster* and *Dendraster*.

sand filtration. See slow sand filtration.

sand natter. *Vipera ammodytes*.

sand viper. **1:** *Vipera ammodytes*. **2:** any of various small, usually greenish, arboreal pit vipers of the genera *Bothrops* and *Trimeresurus*.

sandcorn. *Zigadenus paniculatus*.

Sandimmun(e). cyclosporin A.

sandrasselotter, sandrassel otter. *Echis carinatus*.

Sandril™. reserpine.

sangchul. a colloquial name of *Bungarus caeruleus*.

sangrel. serpentaria.

sanguinaria canadensis (red puccoon; bloodroot). a striking perennial herb (Family Papaveraceae) of eastern North America from southern Canada to Florida and west to Texas. Throughout its range it occurs most often in rich woodlands and among bushes along fence rows. The flowers, which appear early in the spring, are shiny, white, and poppy-like with 8 petals, and a yellow stamen. It has thick roots, unusual leaves, and blood-red sap. All parts of the plant are moderately toxic. Symptoms of intoxication by ingestion usually appear within 1-2 hours and may include a burning sensation in the mouth, gastritis, intense thirst, nausea and violent vomiting, a depressed heart rate, faintness, paralysis, vertigo, dimness of vision, prostration, and collapse. The active principle is sanguinarine, q.v. Natural poisonings are rare. The sap may cause a contact dermatitis.

sanguinarine(13-methyl[1,3]benzodioxolo[5,6c]-1,3-dioxolo[4,5-i]phenanthridinium; pseudochelerythrine, ψ-chelerythrine). a toxic isoquinoline alkaloid, closely related to chelerythrine, isolated from the roots of *Sanguinaria canadensis*, *Chelidonium majus*, *Papaver spp.*, and other members of the family Papaveraceae and also from many of the Fumariaceae. It is also a constituent of argemone oil obtained from the poppies of the genus *Argemone* (prickly poppies), especially *A. mexicana*. Sanguinarine is moderately toxic to laboratory rats, *p.o.* but is highly toxic to laboratory mice and rats, *i.v.* or *s.c.* It depresses heart function, skeletal muscle action, and that of the nervous system. Death may follow violent vomiting, extreme thirst, gastritis, abdominal pain, epigastric soreness, heaviness of the chest, dyspnea, miosis, faintness, coldness of skin, and cardiac paralysis. The reaction time of humans is 1-2 hr.

sanguis. blood; the fluid that circulates through the cardiovascular system.

sansert. *Claviceps purpurea*.

santonica. See α-santonin.

α-santonin. a white powder, $C_{15}H_{18}O_3$, that turns yellow on exposure to light; it is odorless and initially tasteless, but turns bitter. It is only slightly soluble in water, but is soluble in ethanol, chloroform, and in most fatty and volatile oils. Alpha-santonin has a tricyclic structure and is the inner anhydride or lactone of santoninic acid. It is extracted from santonica, the unexpanded flower heads of various species of *Artemisia*, *q.v.* It has been used as a nonirritating, powerful anthelmintic, especially against *Ascarus lumbricoides*, and in the treatment of urinary incontinence, but because of its toxicity it is generally no longer used. It causes disturbances of vision, especially affecting color vision in humans and sometimes causes disorders of taste, hearing, and smell. Large doses stimulate the CNS causing nausea, vomiting, headache, epigastric pain, muscle twitching, confusion, diarrhea, and tonic-clonic convulsions. One can expect ensuing or intermittent periods of depression, and even coma, depressed body temperature, and death from cardiopulmonary collapse.

Santotherm. polychlorinated biphenyl(s).

Santox. See EPN.

sap. **1:** any naturally occurring juice of a living organism or tissue. **2:** the sugary fluid transported through the phloem of a vascular plant. **3:** the cytoplasm (ground substance) of a cell.

Sapindaceae (the soapberry family). a family of tropical trees and shrubs that bear leathery fruits that contain a saponin that is used for cleaning by natives. The most notorious of these plants is *Blighia sapida*, *q.v.*

sapo. *Thalassophryne maculosa*, *Thalassophryne reticulata*.

sapogenin. the aglycone of a saponic glycoside; all are steroidal alcohols. Many are known and are widely distributed throughout the plant kingdom. See glycoside.

Saponaria (soapworts). a genus of hardy annual and perennial European herbs (Family Caryophyllaceae) with white, red, or pink flowers. They contain toxic saponins. ***S. officinalis*** (bouncing bet; cow cockle). a stout perennial plant of Asia, but widely naturalized and common in many places as a roadside plant. The flowers are large and rose colored. It contains saponins and is toxic to livestock, although actual instances of poisoning are rare. ***S. vaccaria***. a coarse annual plant with pale red flowers. It contains saponins and is toxic to livestock but poisonings are rare.

saponic glycoside. See glycoside.

saponin (saponic glycoside). any of a class of bitter-tasting, steroidal glycosides of plant origin. The aglycone, a sapogenin, is a steroid alcohol. Saponins form nonalkaline colloids in water that characteristically produce a froth or foam when shaken. Many are toxic to animals and have been used as fish poisons. Some (e.g., diosgen) are used in the commercial production of steroid hormones. They may function in the plant as a deterrent to herbivores. Toxic saponins are gastrointestinal irritants and when present in sufficient concentration in plant tissue may produce severe effects. They also have hemolytic effects. Saponins are not readily absorbed through the intact gastrointestinal tract and the most serious effects occur in association with other irritant substances that are able to injure the wall of the digestive tract. See *Phytolacca americana*, asterosaponin.

sapotoxin. a toxic saponin; it is a phytotoxin. See especially *Aleurites*.

sapremia. septicemia.

saprobe. **1:** a saprophytic organism. **2:** an organism thriving in water rich in organic matter. See saprobic system.

saprobic (saprobiont). of or pertaining to water with a high content of decaying organic matter.

saprobic system. an ecological classification of polluted water bodies that are undergoing self-purification. The system is based on relative levels of pollution, oxygen concentration, and types of indicator microorganisms present.

saprobiont. saprobic.

saprogenic (saprogenous). causing, or caused by, the decay of organic matter.

saprogenous. saprogenic.

saprophyte. a heterotrophic organism such as a bacterium or fungus that externally breaks down dead organic matter before absorbing the products.

saprozoic. of or pertaining to organisms that feed on solutions of organic material rather than on solid organic matter. Saprozoic organisms include the flagellate protozoan *Polytoma uvella* and many gastrointestinal parasites.

sapwood (alburnum). See wood.

SAR. structure-activity relationship. See quantitative structure-activity relationship.

Sarcobatus (greasewood). a genus of oxalate-containing plants (Family Chenopodaceae) of the southwestern United States. ***S. vermiculatus*** (black greasewood). a low, spiny, woody, deciduous shrub that contains lethal amounts of soluble oxalates. Mass mortality of sheep has occurred in a number of locations, but cattle are rarely poisoned. Symptoms are those of oxalate poisoning. It is considered useful forage by most ranchers assuming that moderate amounts of other forage are available.

sarcolemma. the membrane that surrounds striated muscle cells.

L-sarcolysine. melphalan.

sarcoma. a usually highly malignant connective tissue neoplasm comprised of cells, mesodermal in origin, that are similar to embryonic connective tissue cells. It is a tumor of fibrous tissue (fibrosarcoma) or of bone (osteosarcoma).

sarcoplasmic reticulum. the form of endoplasmic reticulum found in muscle cells, it is a membranous network of channels and tubules within the muscle fiber. This structure conveys nerve signals to the contractile filaments of the fiber.

sardonic. of or pertaining to a kind of spasmodic or tetanic grin or involuntary smile as in *risus sardonicus*.

sardonic grin. risus sardonicus.

sarin (isopropyl methanefluorophosphonate; methylphosphonofluoride acid, isopropyl ester; methylphosphonofluoridic acid 1-methylethyl ester; isopropoxymethylphosphoryl fluoride; isopropylmethylphosphonofluoridate; GB). a colorless, volatile liquid (f.p. -58°C, b.p. 147°C), irreversible anticholinesterase, $[(CH_3)_2CHO](CH_3)FPO$. It is a synthetic supertoxic poison that is rapidly lethal to humans and laboratory animals by inhalation, skin absorption, and all other routes of exposure. It is produced by *Homo sapiens*, q.v., for use chiefly as a military nerve gas. Sarin is more toxic to humans (LD_{50} = 14 μg/kg) than tabun or soman. Systemic effects in humans from inhalation or skin absorption may include bronchiolar constriction (with asthma), muscular weakness, nausea and vomitimg, flaccid paralysis, and miosis. The dominant lesion is irreversible cholinesterase inhibition. When heated to decomposition or exposed to steam, it releases very toxic fumes (F^- and PO_X).

Sarothamnus scoparius. *Sparteum junceum*.

sarsa. sarsaparilla.

sarsaparilla (sarsa). the dried root of *Smilax aristolochiaefolia* and other congeners. It is used as a flavoring agent in beverages and in the treatment of psoriasis. It contains smilagenin which is used in the synthesis of hormones in the pregnane series. It also contains a poisonous glycoside, smilacin.

sarsasapogenin. See smilagenin.

sassafras. the root bark of *Sassafras spp*. used in folk medicine as a muscle relaxant and in the treatment of rheumatism, skin diseases, and typhus. It contains safrole, *q.v.* Sassafras is a suspected carcinogen in laboratory mice and rats and their use as flavors or food additives has been banned in the United States. One cup of sassafras tea made from *S. albidum* contains 200 mg of safrole, four times the amount considered hazardous to humans. See also *Sassafras albidum*.

Sassafras albidum (white sassafras). a tree (Family Lauraceae) up to 40 m tall with mitten-shaped leaves and aromatic bark. It occurs in woods and thickets from New England, Michigan, and Illinois, south to Virginia and Arkansas. Varieties occur locally also in the south, southwest, and midwest. Sassafras root bark was at one time used to make a pleasant-tasting tea and provided the distinctive flavoring of root beer. Sassafras oil and an active principle, safrole, are now prohibited for use in food flavorings. Sassafras is very toxic. It is carcinogenic, hepatotoxic, nephrotoxic, and cardiotoxic. Two other species of sassafrass occur in eastern Asia. See also sassafras, safrole.

sassafras oil. a volatile oil obtained by distillation from the bark of *Sassafras albidum* and *S. vanifolium*. It has been used as a topical antiseptic, carminative, pediculicide, and flavoring agent.

satellite DNA. a fraction of DNA that separates from the bulk of the genomic DNA of an organism during centrifugation. DNA of mitochondrial, chloroplast, centromeric, and ribosomal types may all be recovered from centrifuged samples as distinct satellites, especially when the DNA is run through gradients of sucrose or caesium chloride.

satratoxin H. a trichothecene mycotoxin isolated from *Stachybotrys atra*.

saturated compound. an organic compound that contains double or triple bonds. Such a compound may undergo substitution reactions, but not addition reactions since the molecule already contains the maximum possible number of single bonds. *Cf.* unsaturated compounds.

saturated solution. a solution that is in equilibrium with its solute. Such a solution contains the greatest equilibrium amount of solute that is possible at a given temperature.

saturation. **1:** the presence of a factor at a concentration or level equal to, or in excess of, that required for maximum response or activity. **2:** the equilibrium condition of a community in which immigration is balanced by extinction. **3:** the state or condition of containing the maximum amount (number, concentration, etc.) of a substance, structure, relationship, or entity.

Saturn red. lead oxide red.

saturnine. **1:** of, pertaining to, or caused by, lead or lead poisoning. **2:** having the properties of lead.

saturnine colic. lead colic.

saturnine tremor. a tremor due to chronic lead poisoning.

saturnism. chronic lead poisoning. See lead poisoning.

Sauria. a suborder of the order Squamata (Class Reptilia), comprised of lizards. They are mostly terrestrial, pentadactyl quadrupeds; with long, slender bodies and a long tail. A few lizards, e.g., *Anguis* (slow worm), are limbless. A movable quadrate bone provides a flexible joint between the upper jaw and the skull, increasing the gape of the jaw, the teeth are fused to the jaws. Most have moveable eyelids, are oviparous, and insectivorous; some are herbivorous. Only two living species are venomous. See also Helodermatidae, Squamata.

saurine (scombrotoxin). the toxic principle in scombroid poisoning, *q.v.* It is a heat-stable, histamine-like toxin formed by bacterial action within the skeletal musculature of improperly preserved or stored scombroid fish (e.g., albacore, bonito, mackerel, tuna). It causes severe allergy-like symptoms. A significant portion of the literature names histamine as the toxic component. This is not true, although histamine may be involved. See poisoning (scombroid poisoning).

sauroid. reptile-like; resembling a reptile.

sausage poisoning (allantiasis). botulism acquired from eating improperly preserved sausage; an obsolete term.

savin (savin oil). a volatile oil responsible for the toxicity of *Juniperus sabina*, q.v. In small doses, it facilitates water loss and onset of menstruation. It is a highly toxic irritant, vesicant, and convulsant. Skin contact will produce blisters and sometimes focal necrosis. If ingested, the oil can cause hemorrhagic gastroenteritis with vomiting of greenish masses having an ether-like odor. Polyuria with bloody urine may occur, followed by oliguria and anuria, convulsive coma, and acute renal damage. The renal damage is generally more severe than that produced by turpentine. In serious cases, death from respiratory arrest may occur in 10 hours to several days.

savin oil. savin.

savin poisoning. sabinism. See savin.

saw-scaled viper. See *Echis*.

sawah tadung. *Trimeresurus puniceus*.

sawfish. See *Pristis*.

Saxidomus (butter clams). a genus of bivalve mollusks (Class Bivalvia = Pelecypoda). ***S. giganteus*** (Alaskan butter clam; butter clam; smooth Washington clam; Washington clam). a clam, similar to *S. nuttalli*, but somewhat larger and lacking the concentric sculpture and rust-stain color of that species. It occurs from San Francisco Bay, California to Sitka, Alaska. It is generally highly regarded as food, but is often involved in paralytic shellfish poisoning of humans. ***S. nuttalli*** (butter clam; common Washington clam; Washington clam; money-shell). a large, oblong-oval, somewhat convex, tan or grayish edible clam found on America's west coast from Humbolt Bay, California to San Quenton Bay, Baja California. It is often involved in paralytic shellfish poisoning of humans.

Saxifragaceae (the saxifrage family). herbs or shrubs of varied aspects, but distinguishable from the Rosaceae by the production of seeds that contain large amounts of albumen. The leaves are opposite or alternate and lack stipules. See *Bergenia crassifolia*, *Hydrangea*.

saxifrage family. Saxifragaceae.

saxitoxin (STX). **1:** a powerful neurotoxin and myotoxin, $[C_{10}H_{17}N_7O_4]^{2+}$, found in bivalve mollusks, it was originally isolated (and named for) *Saxidomus giganteus*, the Alaskan butter clam; it is, however, identical to Gonyaulax toxin, the toxin produced by *Gonyaulax catenella* and *G. tamarensis*, dinoflagellates that are commonly ingested by shellfish. **2:** often used as a synonym for paralytic shellfish poison and related terms, all of which refer rather ambiguously to mixtures of toxins. In North America, California sea mussels *Mytilus californianus*, Alaskan butter clams *Saxidomus giganteus* and scallops may ingest *Gonyaulax*, thus serving as secondary sources of this and other toxins. See also paralytic shellfish poison, red tides, batrachotoxin.

Sb. the symbol for antimony (stibium).

***s.c.*, S.C., SC**. subcutaneous.

Sc. the symbol for scandium.

scabby. pertaining to grain (e.g., oats, wheat, corn, barley) that is infected with fungi.

scabby barley. barley infested with the fungi, *Gibberella* or *Hordeum vulgare*.

scabby barley poisoning. a disease of livestock caused by ingestion of barley infested with the fungus, *Gibberella* or *Hordeum vulgare*, q.v. Ingestion of small amounts of scabby barley induces vomiting in swine, but cattle are unaffected even following ingestion of large quantities.

scabicide. any of a large number of drugs used topically that destroy *Sarcoptes scabiei*, the itch mite. They should be used with caution, especially in treating children.

scad. a fish of the family Carangidae.

scalded skin syndrome. toxic epidermal necrolysis (See necrolysis).

scaley pholiota. *Pholiota squarrosa*.

scandium (symbol, Sc)[Z = 21; A_r = 44.96]. a lightweight, silvery transition element, often associated in nature with lanthanoids. The chloride is moderately toxic to laboratory mice. Scandium is used in high intensity lights and electronic devices.

Scaridae (parrotfish). a family of brightly colored, herbivorous, bony fish (Class Osteichthyes) with large, cycloid scales and a single, continuous dorsal fin (Family Scaridae). The teeth fuse to form beak-like plates in either jaw. They are confined to rather shallow waters and are very common on coral reefs. See *Scarus*.

scaritoxin. a toxin sometimes found in ciguatera poison (See poison).

scarlet elder. *Sambucus canadensis*, *S. pubens*.

scarlet firethorn. *Pyracantha coccinea*.

scarlet pimpernel. *Anagallis arvensis*.

scarlet runner bean. *Phaseolus coccineus*.

Scarus. a genus of parrotfish (Family Scaridae). ***S. caeruleus*** (parrotfish). a moderately deep-bodied marine bony fish of coral reefs. It is light to deep (large adults) blue and occurs from Maryland to Brazil, including the West Indies, Bermuda, and the Caribbean Sea. See ciguatera. ***S. microrhinos*** (parrotfish). a moderately deep-bodied marine bony fish of the Indo-Pacific region. See ciguatera.

scat (dropping; fecal dropping). the fecal dropping of an animal. Scats are fairly easy to collect for some animals, whereas the animals themselves may be difficult to observe or capture. They often offer the opportunity to identify some of the animals present in an area and of gaining information regarding the recent diet of an individual or population of animals and the hazards encountered thereby (e.g., deficiencies, toxicants).

scatemia. intestinal autointoxication.

scavenger. **1:** an organism that feeds predominantly on carrion or offal. **2:** a substance added to another substance, mixture, or other system to remove pollutants or other impurities.

scavenging process. a process or mechanism (chemical, physical, or biological) that results in the removal of pollutants from the atmosphere (e.g., the removal of suspended particulate matter by rain; the removal of sulfur dioxide by living plants).

Scheele's Green. cupric arsenite.

Schilbeodes (madtoms). a genus of small bony fishes (Family Ameiuridae) that are common in streams of the eastern half of the United States. The head is depressed, mouth large, pectoral spine with a basal poison gland.

schinopsis lorentzii tannin. quebracho (def. 2).

Schistosoma. a genus of three species of trematode worms (Order Digenea) that are parasitic in the blood vessels of mammals. The body is elongate and the sexes are separate. See also antimony potassium tartrate, antimony sodium tartrate, dermatitis (schistosome dermatitis).

schistosome dermatitis. See dermatitis.

schitomordnik. a colloquial name of *Agkistrodon halys*.

Schizothaerus nuttalli (gaper; summer clam). a clam found on the west coast of North America from Prince William Sound, Alaska, south to Baja California, and in northern Japan. It is often a source of paralytic shellfish poisoning of humans.

Schlank Korallenottern. See *Leptomicrurus*.

Schlegel's palm viper. *Bothrops schlegelii*.

Schlegel's snake. *Aspidomorphus schlegelii*.

Schmidt's sea snake. *Praescutata viperina*.

Schmiedeberg's digitalin. digitalin.

schmuckottern. *Calliophis*.

Schneider's sea krait. *laticauda colubrina*.

Schoenocaulon officinale (*Asagraea officinalis*; sabadilla; cevadilla; caustic barley). a toxic plant indigenous to the Andes of Mexico, Guatemala, and Venezuela (Family Liliaceae). It occurs also in coastal areas of thc Gulf of

Mexico and the Caribbean Sea. The seeds contain a number of toxic alkaloids, chiefly cevadine and veratridine. See also sabadilla, sabadine.

schoolmaster. See snapper.

schradan (octamethyldiphosphoramide; octamethyl pyrophosphoramide; bis[bisdimethylaminophosphonous]anhydride; bis-*N,N*,N′,N′-tetramethylphosphorodiamidic anhydride; OMPA). a viscous liquid that is miscible with water and soluble in most organic solvents $[(CH_3)_2N]_2P(O)OP(O)[N(CH_3)_2]_2$. A cholinesterase inhibitor, it is toxic by ingestion and inhalation. Schradan is used as a systemic insecticide. See also poison (household poison).

Schultze's pit viper. *Trimeresurus flavomaculatus*.

schwarzen mamba. *Dendroaspis polylepis*.

Schweinfurt green. cupric acetoarsenite.

scifluorofen. the sodium salt of acifluorfen, q.v.

scilla. **1:** squill (def. 1). **2:** any plant of the genus *Scilla*.

Scilla (squill; bluebell; wild hyacinth; scilla). a genus of small, Old World bulbous plants (Family Lilaceae) that have been introduced into North America where they are often found along roadsides, vacant lots, pastures, and old fields. They are closely related to plants of the genus *Urginea*, some of which, including *S. nonscripta* and *S. peruviana,* are toxic. See *U. maritima*. ***S. autumnalis*** (autumn squill; starry hyacinth). a small (up to 15 cm tall), toxic herb with white or blue flowers that is sometimes erroneously called autumn crocus (See *Colchicum*). ***S. maritima***. *Urginea maritima*. ***S. nonscripta*** (English bluebell; harebell). a widely cultivated toxic species that has escaped cultivation in the United States where it occurs in vacant lots, along roadsides, and in old fields. ***S. peruviana*** (Cuban lily). a toxic species of southern Europe and a common greenhouse plant. It has clusters of white, reddish, blue, or purple blossoms and long, broadleaves. *Cf. Colchicum*. See also scillaren, scillarenin, scilliroside.

scillaren. an extremely poisonous mixture of the cardiotoxic glycosides, scillarin A and B, in the same proportions (2:1) in which they occur in *Urginea maritima* (squill). **scillaren A** (glucoproscillaridin A, transvaalin). a bitter-tasting, extremely toxic, crystalline , steroidal glycoside. It is extracted from *Urginea* bulbs and can be hydrolyzed to glucose and proscillaridin A, and the latter then hydrolyzed to rhamnose and the aglycone, scillaridin A. **scillaren B**. an amorphous, water-soluble, glycosidal mixture of the cardiotoxic compounds that remain following extraction of scillaren A from *Urginea*. These include glucoscillaren A, scillipheoside, glucoscillipheoside, scillicryptoside, scilliglaucoside, scillicyanoside, and scillazuroside.

scillarenin. **1:** proscillaridin. **2:** an extremely poisonous, cardiotoxic glycoside produced by the enzymatic decomposition of proscillaridin A from *Urginea burkei*.

scillarenin 3β-rhamnoside. proscillaridin.

scillaridin A. the aglycone of scillaren A.

scillazuroside. a cardiotoxic glycoside. It is a component of scillaren B.

scillicryptoside. a cardiotoxic glycoside. It is a component of scillaren B.

scillicyanoside. a cardiotoxic glycoside. It is a component of scillaren B.

scilliglaucoside. a cardiotoxic glycoside. It is a component of scillaren B.

scillipheoside. a cardiotoxic glycoside. It is a component of scillaren B.

scilliroside. an extremely toxic cardiac glycoside isolated from the bulbs of *Urginea maritima*. It is used as a rodenticide.

scillism. poisoning by *Urginea maritima* (squill).

scillitic. pertaining to *Urginea maritima* (squill).

scirpentriol. a trichothecene produced by *Fusarium spp.*

scissors snake. *Vipera russelii*.

sclera. the white fibrous outer layer of the eyeball in vertebrate animals.

Scleroderma citrinum (= *S. aurantium*; pigskin poison puffball). the most common species of the fungus *Scleroderma* (Family Sclerodermataceae), it is a poisonous, warty, thick-skinned ball with an apical pore. It induces vomiting, and reportedly tetany and paresthesias. It grows in woodlands near trees on sandy soil, or on dead wood throughout much of North America.

sclerotia. plural of sclerotium.

sclerotic acid (sclerotinic acid). an active principle of ergot.

Sclerotinia. a large genus of ascomycetous fungi (Family Sclerotiniacea), many of which grow on fallen rotting fruit (e.g., pears, peaches); many are virulent plant pathogens. ***S. sclerotiorum***. a mushroom that secretes the mycotoxins 8-methoxypsoralen and 4,5′,8-trimethylpsoralen.

sclerotinic acid. sclerotic acid.

sclerotium (plural, sclerotia). a dormant stage of various true fungi in the form of a usually dark, compact, hardened mass of mycelium. When mature it detaches from the host and begins to grow, sending out hyphae or producing fruiting bodies, when conditions become favorable. *Claviceps purpurea*, an ascomycete fungus parasitic on rye, produces sclerotia known as ergots in place of the ovary of the host grass flower. These ergots contain alkaloids that are responsible for outbreaks of the disease ergotism; however, they are also a source of the drug ergotamine, which is used to assist childbirth, etc. Ergots are overwintering structures that germinate the following spring to produce stalked stroma in which the perithecia are formed.

Sclerotium. a genus of fungi that may infest corn or sweet potatoes. Ingestion of infested feed produces severe respiratory distress and anorexia. The course of the disease is rapid, terminating fatally within a week.

scoke. *Phytolacca americana*.

Scolopendra. a common, widely distributed genus of large centipedes. The venom is extremely virulent. The bite is painful, but seldom lethal to humans. See also Chilopoda, Myriapoda. ***S. galapagensis***. a large species that reaches ca. 30 cm in length. The venom is extremely virulent, but of poorly known composition. ***S. morsitans***. a species indigenous to the southern United States. Its bite causes localized erythema, inflammation, and edema. Purpura and systemic symptoms sometimes occur, but usually disappear within a few hours. ***S. vircanis***. the LD_{50} of the venom in laboratory mice is ca. 1/300th of a venom gland.

Scomberoides sanctipetri (leatherback; lae). a venomous fish (Family Carangidae) that occurs throughout the Indo-Pacific region. The venom apparatus is comprised of seven dorsal spines and two anal spines with the associated venom glands, integumentary sheaths, and musculature. The puncture wound with envenomation causes intense pain, redness, and edema which may persist for several hours.

Scombridae. a family of large marine percomorph fishes (suborder Scombroidea) that includes numerous commercially important species such as mackerels (*Scomberomorus* and *Scomber*), bonito or skipjack (*Sarda*), and tuna (*Thunnus*). See also scombroid poisoning.

scombroid. **1a:** common name of the mackeral-like bony fish of the suborder Scombroidea (e.g., albacore, bonito, mackeral, and tuna). **1b:** of or pertaining to fish of the suborder Scombroidea. See also Scombridae, scombroid poisoning.

scombroid poisoning (scombrotoxism). a form of ichthyosarcotoxism due to ingestion of raw or inadequately cooked, and poorly preserved scombroid (mackeral-like) fish. Initial symptoms appear within 2 hours following ingestion of such fish. These initially include epigastric pain, nausea, vomiting, diarrhea, headache, dysphagia, burning of the throat, followed by thirst, numbness, generalized urticaria, and pruritis. Symptoms usually disappear within a day. Some scombroids have also been implicated in ciguatera poisoning, *q.v.* See saurine.

scombrotoxic. of, pertaining to, or caused by saurine (scombrotoxin). See also poisoning (scombroid poisoning).

scombrotoxin. saurine.

scombrotoxism. scombroid poisoning.

scoparius. the dried tops of *Cytisus scoparius*, q.v., and sometimes (though not authorized) including seeds and flowers. It is used as a diuretic and cathartic.

scopine (6,7-epoxytropine; 6,7-epoxy-3-hydroxytropane). scopolamine without the tropic acid side chain.

scopine tropate. scopolamine.

scopolamine(6β,7β-epoxy-1α-H,5αH-tropan-3α-ol (—)-tropate; α-(hydroxymethyl)benzeneacetic acid-9-methyl-3-oxa-9-azatricyclo[3.3.1.$0^{2,4}$]non-7-yl ester; 6β,7β-epoxy-3α-tropanyl S-(-)-tropate; 6,7-epoxytropine tropate;6a,7a-epoxy-1αH,5αH-tropan-3α-ol(-)-tropate; hyoscine; scopine tropate; tropic acid ester with scopine; *l*-scopolamine). a highly toxic, anticholinergic, tropane alkaloid isolated from the leaves and seeds of *Atropa belladonna*, *Datura metel*, *Duboisia myoporoides*, *Scopola carniolica*, *S. japonica*, and other solanaceous plants. It is also a constituent of duboisine. Scopolamine is used therapeutically in the management of motion sickness, parkinsonian tremor, as a preanesthetic medication that reduces salivary and branchial secretions, and as an antidote to organophosphates and other cholinesterase inhibitors. In certain hyperactive or spastic conditions, it relaxes the organs of the digestive system. This drug is, however, extremely poisonous (usually considered to be more toxic than atropine) and has been so used for centuries. Fatalities from clinical doses are, however, rare. It paralyzes the parasympathetic nervous system functions by blocking the action of acetylcholine at nerve terminals. Scopolamine crosses the blood-brain barrier, producing CNS effects such as drowsiness, fatigue, euphoria, loss of REM sleep, and amnesia even at therapeutic doses. Large overdoses may cause excitement, delirium, and psychotic behavior, as in the case of atropine. Scopolamine psychosis is followed by pronounced and prolonged depression.

***l*-scopolamine**. scopolamine.

scopolamine hydrobromide (hyoscine hydrobromide). a moderately toxic derivative of scopolamine with an anticholinergic action similar to that of atropine. See also scopolamine.

scopolamine methylbromide (methscopolamine bromide). a quaternary ammonium derivative of scopolamine with spasmolytic and antisecretory action. It lacks the CNS effects produced by scopolamine and is thus much less toxic.

scopoletin (7-hydroxy-6-methoxy-2*H*-1-benzopyran-2-one, 7-hydroxy-6-methoxycoumarin, 6-methoxyumbelliferone, β-methylesculetin, chrysatropic acid, gelsemic acid, geleminic acid). the aglucone of scopolin, it is a coumarin derivative and plant growth inhibitor stored in the roots of *Avena sativa*, *Atropa belladona*, *Convolvulus scammonia*, *Scopolia japonica*, *S. carniolica*, and also occurs in *Helianthus annus*.

scopolia. the dried rhizome and roots of *Scopolia carniolica*.

Scopolia. a genus of belladonna-like herbs (Family Solanaceae). ***S. carniolica***. a herb of Austria and neighboring countries of Europe; it resembles belladonna in pharmacologic action and contains scopolin. ***S. japonica*** (Japanese belladonna). a species, the leaves, root, and seeds of which contain scopolamine and scopolin.

scopolin (7-(β-D-glucopyranosyloxy)-6-methoxy-2H-1-benzopyran-2-one). the mono-β-glucopyranoside of scopoletin, it occurs in the roots of *Scopolia japonica* and *S. carniolica* (Family Solanaceae) and *Nerium odorum* (Family Apocynaceae).

scopoline. a decomposition product of scopolamine, it is an isomer of scopine, differing only in the location of the epoxy and hydroxyl groups.

scopomorphinism. a combined addiction to scopolamine and morphine.

Scorpaena (scorpionfishes; lionfishes). a genus of shallow water, venomous fish (Family Scorpaenidae) that usually occur along sandy beaches, bays, and rocky coastlines or coral reefs. Most are bottom dwellers of the intertidal zone to depths of about 100 m or more. The venom apparatus consists of moderately long and heavy, dorsal (often anal and pelvic) spines with well-developed venom glands and no evident venom duct. The area about the wound commonly becomes inflamed, edematous, and very painful. The pain from envenomation by tropical and subtropical species can be excruciating and often radiates within minutes to all parts of an affected limb. Symptoms of envenomation may include nausea, vomiting, diarrhea, pallor, weakness, dyspnea, tachycardia, profuse perspiration, diarrhea, intense abdominal pain, syncope, and shock. Symptoms may subside within several hours. In severe cases, primary shock associated with bradycardia, hypothermia, and respiratory distress may present. Examples are *S. guttata*, which occurs from central California southward into the Gulf of California; *S. plumieri* which occurs along the Atlantic Coast from Massachusetts to the West Indies and Brazil; *S. porcus* (scorpionfish, rascasse, sea pig) which occurs along the Atlantic Coast of Europe from the English Channel to the Canary Islands, French Morocco, the Mediterranean Sea, and the Black Sea. ***S. guttata.*** (scorpion fish). a fish of the temperate and tropic zones of all seas. Spines of the opercula (gill covers) can penetrate skin, causing intense local pain and swelling of the entire extremity. See also Scorpaenidae.

Scorpaenidae (ocean perch or rosefish, scorpionfishes, lionfishes, sculpins, zebrafishes, stonefishes, bullrout, and waspfish). marine bony fishes that are widely distributed throughout temperate and tropical waters; the range of a few species extends into arctic waters. Several genera within this family are venomous. The venom apparatus of most species consists of several dorsal and anal spines and two pelvic spines, and a glandular complex that secretes the venom. All of these are encased by an integumental sheath. The glandular component sits in anterolateral spinal grooves within the sheath. The venom apparatus varies substantially among genera but falls into three structural types: *Scorpaena*, *Pterois*, and *Synanceja* types, *q.v.* Pharmacology and potency of the venom vary among and within genera, but symptoms, for the most part, differ chiefly in degree. Envenomation by *Synanceja spp.* (stonefishes) may be more serious than those of any other venomous fishes. Symptoms are similar to those described below, but are more severe and necrosis with sloughing of tissues at the site of a sting is common. Envenomation by tropical or subtropical *Scorpaena spp.* (scorpion fishes and lionfishes) may produce immediate, sometimes excruciating pain in the area of the wound, but this often radiates within minutes and may persist for a few hours or for many days. The area about the wound becomes ischemic and then cyanotic. The surrounding area becomes swollen and inflamed. Tissues about the wound may later slough off. Symptoms of systemic effects commonly include nausea, vomiting, pallor, weakness, syncope, and shock. Symptoms often subside within a 8-12 hours. Primary shock accompanied by bradycardia, hypothermia, and respiratory distress may be observed in severe cases.

Scorpaenopsis (scorpionfishes). a genus of venomous fish (Family Scorpaenidae). An example is *S. diabolus* which occurs in The Netherlands Indies, Melanesia, Polynesia, and Australia.

scorpamine. See scorpion venom.

Scorpio. an Old World genus of scorpions.

scorpion. any member of the arachnid order Scorpionida. All are venomous, but the stings of relatively few require medical attention. Scorpions are not usually aggressive toward humans and attack only when suddenly disturbed or aggravated. See *Centuroides*, *Androctonus australis*, *Buthus occitanus*.

scorpion fly. any fly of the insect order Mecoptera.

scorpion venom. See venom.

Scorpiones. See Scorpionida.

scorpionfish, scorpion fish. any member of the family Scorpaenidae, but especially reserved for the genera *Scorpaena* (scorpionfish proper) and *Scorpaenopsis*.

Scorpionida (scorpions). an order (also termed Scorpiones) of medium to large (40-127 mm) nocturnal, viviparous, arachnids comprised of 1500-2000 species, all of which are venomous. *Centrurus*, *Hadrurus*, and *Vejovis* are representative genera. Most species prey upon spiders and large insects. They are largely tropical and subtropical in distribution, although some live in cool climates as in Mongolia and parts of Europe. Scorpions are missing from Antarctica, New Zealand, and southern portions of Chile and Argentina. More than 70 species occur in North America. The body is elongate, with a compact cephalothorax that is broadly joined to a long abdomen, with a postabdomen of 6 segments that terminates in a bulbous venom gland and sting. There is a pair of ventral, comblike sensory structures (pectines) on the second abdominal segment; small chelicerae; large chelate pedipalps; and book lungs. There are two eyes in the center and two to five eyes on either side of the cephalothorax. A layer of wax in the cuticle affords protection against excessive water loss, which is an especially valuable adaptation in hot, desiccating environments. Only a few scorpions are dangerous to large animals and humans; most do not sting unless disturbed. The sting of most North American species is painful, but all except *Centruroides exilicauda (sculpturatus)* are relatively harmless; the sting usually causing no more than minimal tenderness, edema, and increased skin temperature about the wound; lymphangitis with regional lymph gland swelling; and an occasional localized tissue reaction. See also venom (scorpion venom), Arachnida.

scorpionism. poisoning from scorpion stings.

Scotch broom. *Sparteum junceum* (*Cytisus scoparius*).

scouring. See scours.

scouring powder. any household composition used to scrub and whiten hard, scratch-resistant, difficult-to-clean surfaces. Almost all such powders contain chlorine bleach as a whitener and stain remover. The powder releases chlorine fumes when wet. The fumes can be highly irritating to the eyes, nose, throat, and lungs. Headaches, fatigue, and difficult breathing are sometimes reported. Detergents and talc (*q.v.*) are also ingredients of scouring powders. Talc, though practically nontoxic itself, may be contaminated with asbestos (*q.v.*), thus posing a hazard. See chloramine. When scouring powder is used, some fumes almost always enter the air and possibly into the user's lungs. Scouring powders should be considered hazardous and those containing talc should probably be considered unsafe to use as there is no safe level of asbestos exposure.

scours (scouring). diarrhea or dysentery in livestock, notably in the newborn. **peat scours. 1a:** that due to molybdenum poisoning of grazing cattle. **1b:** molybdenum poisoning of cattle.

screen grip. See grip strength.

screening-level assessment. qualitative assessment of expected effects based upon preliminary monitoring data, available toxicity data, and standard assessment models.

Scrophulariaceae (the figwort family; the snapdragon family). a family chiefly of temperate herbs and occasional shrubs or trees. Many species (most importantly *Digitalis*) are toxic or have medicinal value. The sap of most contains cucurbitacin, *q.v.*.

scrotal cancer (chimney-sweeps' cancer; soot wart). an epidermoid neoplasm of the scrotum, appearing initially as a small sore that may ulcerate. It occurs most often in elderly men, especially those exposed to soot, crude oils, mineral oils, polycyclic hydrocarbons, arsenic fumes from copper smelting operations. Scrotal cancer among chimney sweeps in London was described by Sir Percival Pott, Surgeon General of Britain, during the reign of King George III.

scrotal sac. See scrotum.

scrotum (scrotal sac) the pouch of skin that houses the testes and part of the spermatic cord of male mammals that have descended testes. It does not appear to be a common portal of entry for xenobiotics (but See scrotal cancer).

sculptured centroides. *Centruroides exilicauda (sculpturatus)*.

scups. See Sparidae.

Scyphozoa (true jellyfish). true medusae or jellyfish comprise one of the three classes of coelenterates. The medusa stage is the dominant or only stage. It is free-swimming and bell- or umbrella-shaped, being typically convex above and concave below. The polyp, when present, is restricted to a small larval stage. The medusa bears tentaculocysts but lacks a velum and stomodaeum. The body cavity (coelenteron) is divided into 4 pouches and contains a gastrovascular system (a canal system) that distributes food. The diameter of medusae (in the horizontal plane) among the Scyphozoa ranges from about 70 mm (e.g., in *Aurelia*) to about 2 m (in *Cyanea*) Typical genera are *Aurelia*, *Cassiopeia*, *Cyanea*. See also Cnidaria.

scyphozoan. common name for a true jellyfish. See Scyphoa.

SDMH. 1,2-dimethylhydrazine.

SDR. standardized disease ratio.

Se. the symbol for selenium.

Se-methyl-selenocysteine. a storage form of selenium in plants such as *Astragalus pectinatus* that accumulate this element.

sea anemone. **1a:** a vernacular term sometimes applied to any anthozoan coelenterate that produces no skeleton. **1b:** commonly reserved for the larger, more brightly colored species of anthozoans that belong to the Order Actiniara. They are usually solitary, with a cylindrical body, and often occur anchored to rocks; the free end has a mouth-like opening surrounded by tentacles that are often arranged in multiples of six. They are venomous and poisonous although some may be safely eaten when cooked (e.g., *Physobrachia douglasi, Rhodactis howesi*). The toxic principles of the venom are diverse; a substantial number of cytolysins are included. See Anthozoa.

sea bass. fishes of the family Serranidae. See especially *Cephalopholis argus*. See also seabass.

sea bream. See Sparidae.

sea catfish. See *Bagre marinus*.

sea catfishes. See Ariidae.

sea cucumber. a common name of echinoderms of the class Holothuroidea.

sea hare. See *Aplysia*.

sea krait. any snake of the family Laticaudae.

sea krait, yellow-lipped. *laticauda colubrina*.

sea lilly. See Echinodermata.

sea lion. any of a group of mostly marine pinniped mammals of the Pacific Region (Family Otariidae). They usually occur on or near shore, although migrations may take them far out to sea. Sea lions are larger than true seals and lack the underfur. Genera include *Zalophus, Eumetopias,* and *Otaria*. See *Neophoca cinerea*, Otariidae.

sea nettle. a large jellyfish, *Chrysaore quinquecirrha*.

sea onion. *Urginea maritima*.

sea pig. *Scorpaena porcus*.

sea salt. sodium chloride.

sea serpent, common. *Hydrophis cyanocinctus*, *Laticauda semifasciata*.

sea slug. any gastropod mollusk of the suborder Nudibranchia.

sea snake. any snake of the family Hydrophidae, q.v.

sea snake bite. See snakebite.

sea snake venom poisoning (hydrophid venom poisoning). poisoning by the bite of a sea snake (Family Hydrophidae). The wounds characteristically consist of multiple pinhead-sized punctures with little or no pain about the bite, although some tenderness and pain (exacerbated by movement) may develop in the skeletal muscle masses including the neck. Common effects include paresthesia about the mouth, excessive sweating, thirst, pain on swallowing. Some victims experience generalized weakness, extraocular muscle weakness, trismus, dilation of the pupils, and ptosis. Features of severe cases may include severe hemorrhage, respiratory distress, a variety of muscle pains, and paralysis. Mortality is usually low, partly because venom is rather infrequently injected. Lesions include myoglobinuria, which is diagnostic. Some antivenins are available.

sea star. any echinoderm of the class Asteroidea.

sea urchin. the common name for certain members of the Class Echinoidea, Phylum Echinodermata. They do not have arms or rays and are characterized by a globular, subglobular, or hemispherical test or skeleton. The radial symmetry of many species is nearly perfect. Some species of sea urchins are poisonous (e.g., *Paracentrotus lividus*, *Tripneustes ventricosus*) and perhaps lethal if ingested by humans. Others are venomous. Envenomation by venomous sea urchins is accomplished via spines in some species and by the pedicellariae in others. Penetration of the skin by sea urchin spines may be painful, producing erythema, edema, an aching sensation, and sometimes numbness and muscular paralysis. Envenomation by globiferous pedicellarae may produce an immediate intense, radiating pain. Symptoms commonly include light headedness, numbness, generalized muscular paralysis, loss of speech, respiratory distress, and sometimes death. The pain usually disappears in a few minutes, but paralysis may persist for 6 hours or more. *Diadema setosum* (black long-spined sea urchin) is the most venomous species. The symptoms produced by red sea urchins, e.g., *Toxneustes elegans* and *Asthenosoma jimoni*, are milder.

sea wasp. See cubomedusae.

sea water intrusion. See salt water intrusion.

sea whip. any of several marine hydrozoans that form long, slender, whip-like colonies. See lophotoxin.

seabass. correctly used as the common name of a wide variety of marine bony fishes other than the family Serranidae (sea basses). See, for example, *Plectropomus obligacanthus*, *Variola louti*.

seabather's eruption. schistosome dermatitis. See dermatitis.

seal. See Pinnipedia.

seasonal allergic rhinitis. hay fever.

seaweed. any of a variety of macroscopic algae that occur free floating in the sea or found along rocky shores. Most belong either to the family Phaeophyta, q.v., or Rhodophyta.

Sebastes (= *Sebastodes*; rockfishes). a genus of viviparous fish (Family Scorpaenidae) that occur along rocky shores in cool and temperate seas. The head is armored and some species have venomous spines of the scorpionfish type (See Scorpaenidae). Examples are *S. paucispinis* (the brown to red bococcio of the Pacific Coast), *S. goodei* (chillipepper), and *S. mystinus* (black rockfish) all of which are important in commercial and sport fishing.

Sebastodes. *Sebastes*.

seborrhea. excessive secretion of fatty substances from the skin.

sebum. a lipoidal secretion of the sebaceous glands of the skin of mammals. Sebum from the ears is wax-like and called cerumen; that from the foreskin of the penis is called smegma. See skin.

Secale cereale (cultivated rye; rye). a hardy annual grass (Family Grammineae) and agricultural crop grown for grain and straw. It has been heavily ergotized at times (See *Claviceps*, ergotism). Toxic concentrations of nitrate have been reported at times, especially during periods of unusually rapid growth.

secale cornutum. ergot.

secalinotoxin. a principle isolated from ergot.

secalonic acid. a mycotoxin from *Penicillium oxalicum*. It is extremely toxic to laboratory mice.

secobarbital (5-(1-methylbutyl)-5-(2-propenyl)-2,4,6(1H,3H,5H)-pyrimidinetrione monosodium salt; sodium 5-allyl-5-(1-ethylbutyl)barbiturate; Seconal). a moderately toxic barbiturate when taken in the absence of potentiating substances; it has a very long physiological half-life, commonly used as a sedative-hypnotic. Abuse can lead to dependence and tolerance. Death from secobarbital poisoning usually follows a coma that lasts for several days, unless it was used in combination with another drug that increase its toxicity. Alcohol may potentiate the toxicity of secobarbital.

secobarbital sodium. the monosodium salt of secobarbital; its uses and actions are the same as those of the base.

(1α,3β,5Z,7E)-9,10-secocholesta-5,7,10(19)-triene1,3,25-triol. calcitriol.

9,10-secoergosta-5,7,10(19),22-tetraen-3-ol. vitamin D_2.

Seconal™. secobarbital.

Seconal Sodium™. secobarbital sodium.

second messenger. any of a variety of compounds (cyclic AMP, cyclic GMP, phosphoinositols) the biosynthesis of which is stimulated or inhibited by the binding of first messenger (a neurotransmitter or hormone) to a receptor. Second messengers may bind to one or more receptor sites, sometimes producing a variety of cellular metabolic alterations. The action of many toxicants (e.g., choleragen, pertussis toxin) depends upon their ability to alter the concentrations of second messengers. Thus a primary action of choleragen, q.v., is stimulation of adenylate cyclase activity in the intestinal mucosa. See especially choleragen, adenylate cyclase.

secondary poisoning. intoxication of one organism by contact with another that has been poisoned. Exposure usually occurs via ingestion of dead or dying animals, the major source being unabsorbed toxicants in the digestive tract. Other common sources of secondary poisoning are contaminated milk, ectoparasites, and transfer of toxicants to the fetus across the placenta. The most common victims of secondary poisoning are carrion feeders and carnivorous animals. A broad spectrum of animals fall victim to secondary poisoning by pesticides.

secondary radiation barrier. See barrier.

secondary sex ratio. See sex ratio.

secreta. secretion (def. 2).

secrete. **1:** to discharge or release (through cell membranes) a liquid product from a cell or tissue. **1b:** to synthesize and release (through cell membranes) a liquid product; said of a gland or living cell.

secretion. **1a:** the process of discharge or release (through the cell membrane) by a cell or gland of a liquid product (e.g., tears, saliva, a hormone). **1b:** the process of synthesis and release (through the cell membrane) by a cell or gland of a liquid product. **1c:** the process of release of any intracellular molecules (e.g., ions, enzymes, hormones, inorganic molecules) from a cell or gland even though it was not produced by the cell or gland (e.g., the excreta from the salt gland of seabirds). See exocytosis. **2:** secreta; the product of secretion (def. 1).

secretitious. produced by secretion.

secretory. of or pertaining to the process of secretion or to cells and tissues that secrete substances.

Sectral. a beta-adrenergic blocking agent.

sedate. to calm, to allay excitement, or facilitate sleep.

sedation. **1:** an induced state of calmness, reduced activity, or sleep. **2:** the process of calming or of producing a sedative effect.

sedative. **1:** quieting; having the ability to calm, or to allay excitement and activity. **2:** of or pertaining to a process, measure, or substance that sedates. **3:** an agent that calms, diminishes irritability, and allays excitement, and reduces activity. Some sedatives have a general effect on the body; some affect chiefly cardiac activity, and that of the stomach, intestines, respiratory system, nerve trunks, or vasomotor system. Sedatives vary in their usage and may be to reduce pain, induce sleep, allay anxiety states, facilitate the induction of anesthesia, or to treat convulsive conditions or irritable bowel syndrome. Toxic doses decrease respiratory rate. Most sedatives are toxic, some extremely so. Overdosage or prolonged usage can lead to serious problems, even death. *Cf.* analeptic, depressant. See, for example, acetyl carbromal, barbiturate, cannabis, chloral hydrate, doxepin, glutethimide, hexobarbital, lithium bromide, pentobarbital, phenobarbital, thalidomide. **cardiac sedative**. a sedative that decreases the force of the heart. **nervous sedative**. a sedative that affects the nervous system.

sedative-hypnotic (sleeping pill). drugs that help relieve anxiety and promote sleep. Included are barbiturates, the much safer benzodiazepines, chloral derivatives, tertiary alcohols, and gluthimide. These are drugs of abuse; abrupt withdrawal can be a problem.

sedge viper. *Atheris nitschei*.

sediment. particulate matter that has been transported by wind, water or ice and later deposited in an area or has been precipitated from water. Sediments have been repeatedly shown to accumulate toxic chemicals. Under favorable conditions sediments provide useful records of pollution chronology and can demonstrate the sources of specific pollutants. **bottom sediment**. that which has settled onto the bottom of a water body such as a lake, pond, or stream. Sediments that accumulate beneath bodies of standing or slow-moving water play a very important role in freshwater ecology. They contain nutrients and often a rich biota as well as toxic chemicals. Organic compounds in sediments may be taken up by humans and a wide variety of organisms, sometimes with disastrous consequences (e.g., methyl mercury). **suspended sediment**. particulate matter that remains in air or water either because of its small size or due to resuspension by wind or water currents. Resuspension of sediments is often a continuing source of environmental poisons. *Cf.* seston.

sedimentation. the process of forming sediment.

sedimentation rate. a test that measures the rate at which blood settles in a test tube. Results offer an indication of infection, inflammation or tissue damage.

seed. the ripened ovule of a plant following fertilization of the egg or, in some plants, the seed is produced asexually (agamospermy); the embryonic plant within a protective coat.

seed dressing. an antifungal or insecticidal substance applied to protect seeds from infestations. Seeds so treated are toxic to animals and cannot be used in feeds. Seeds so treated have, at times, devastated wildlife (especially bird) populations. See also organomercurial fungicide.

Seeschlangen. Hydrophidae.

segmental demyelination. See demyelination.

segmented worm. See Annelida.

seismonasty. nastic movement in response to shock. See haptonasty.

seizure. **1:**. an attack, fit, paroxysm; sudden onset of the symptoms of a disease. **2:** a convulsion.

Selachii. an order of fishes of the vertebrate class, Chondrichthyes. It is comprised chiefly of marine, cartilaginous, predatory fishes, some bottom-feeders, and a few fresh-water species. Some classifications include only the sharks; others include also the skates and rays. The body is streamlined with a well-developed heterocercal tail; the mouth has a wide gape and is usually lined with numerous sharp teeth; pharyngeal gills; a spiracle behind each eye and separate gill slits open laterally (sharks) or ventrally (skates and rays); a cloaca is present.

Selaginellales. See Lycopsida.

selection. See natural selection.

selective. **1:** discerning; discriminating; exhibiting a high degree of specificity with respect to activity, action, or choice. **2:** descriptive of herbicides that kill some plants and not others. Broadleaved plants are usually unaffected by such herbicides. Such herbicides are thus valuable for applications on cereal crops and lawns. Some selective herbicides (e.g., Simazine) kill germinating plants only. **3:** descriptive of the degree to which a drug produces the desired effect in relation to adverse effects that may occur; pertaining to drug selectivity. See also selectivity, selective reabsorption, toxicity (selective toxicity).

selective advantage. the greater fitness of one genotype relative to a competing genotype.

selective reabsorption. a process involved in the formation of urine; pertaining to the differential reabsorption of nutrient molecules from the renal (kidney) tubule into the blood in relation to the reabsorption of other molecules, especially nutrients in relation to waste products.

selective toxicity. See toxicity.

selectivity. specificity; for example, toxicants and other biologically active substances (agonists) are selective (specific) in that they react only with specific receptors. See also specificity, selective.

selenide. a compound of selenium, with another element or radical.

seleniferous. containing selenium, as in seleniferous soils.

seleniferous soil. a soil with a high selenium content; such soils are encountered in various parts of the world. Farming on such soils has often resulted in the production of highly toxic crops and animal feedstocks. Occasionally, when animal feeds contain only moderate levels of selenium, the addition of protein supplements (e.g., linseed meal), sulfur compounds, or inorganic arsenicals to the diets may counter some or all of the toxic effects of selenium. The efficacy of food additives is, however, extremely undependable.

seleninyl fluoride. selenium oxyfluoride.

selenium (symbol, Se)[Z = 34; Ar = 78.96]. an allotropic nonmetallic (metalloid) element that is chemically similar to sulfur. It is used industrially in the manufacture of photoelectric cells and semiconductors, in certain photocopying processes, in the manufacture of glass, and for many other purposes. As a component of glutathione peroxidase, selenium is an essential trace element in the diet. It is, nevertheless, highly poisonous to domestic animals, wildlife, and humans at concentrations in the diet somewhat above that required (ca. 50-70 μg/day). Chronic selenium poisoning (selenosis) has occurred in humans who consumed about 40 μg/kg body weight on a daily basis. The median lethal dose ranges from about 1.5-6 mg/kg body weight, depending upon the species and compound tested. Selenium is probably toxic by all routes of exposure. It produces reproductive effects, causes liver and kidney damage, and is tumorigenic and teratogenic in experimental animals. Many soils contain selenate and selenite compounds. In some cases, soil concentrations are such that selenophilic plants (e.g., *Astragalus pectinatus*) can accumulate selenium to toxic concentrations. The storage form in such cases is often Se-methyl-selenocysteine. Alkali disease or blind staggers is an acute or subacute condition of livestock due to the consumption of grain or accumulator plants that contain high concentrations of selenium (usually above 5mg/kg of total mass). Elemental selenium is a questionable human carcinogen. Selenium sulfide is carcinogenic to some laboratory animals. Inorganic selenium compounds probably attach to protein sulfhydryl groups, as does inorganic arsenic. In general, organoselenium compounds are regarded as being less toxic than inorganic selenium compounds. selenium may protect against or allay the toxic effects of heavy metals such as mercury, lead, cadmium and arsenic; these, in turn, may protect against the toxic effects of selenium.

selenium disulfide. selenium sulfide.

selenium hydride. hydrogen selenide.

selenium oxychloride (seleninyl chloride). a nearly colorless or yellowish, fuming, corrosive liquid, Cl_2OSe, that is miscible with benzene, carbon disulfide, carbon tetrachloride, chloroform, and toluene. It decomposes in water with the release of HCl and selenious acid. It is an extremely toxic irritant and severe vesicant.

selenium oxyfluoride (seleninyl fluoride; selenyl fluoride). a corrosive, fuming liquid, F_2SeO, that rapidly attacks glass. It is a strong irritant and vesicant; acute poisoning can cause a fatal pulmonary edema.

selenium poisoning (selenosis). chronic poisoning of horses, cattle. and swine, caused by ingesting grains and forage raised on soils high in selenium. It occurs only in arid regions from eating certain plants which are selenium accumulators. **acute or subacute selenium poisoning** (in animals). blind staggers.

selenium sulfide (selenium disulfide). a bright orange powder, SeS_2, that is nearly insoluble in water and organic solvents. It is a strong irritant to the skin and eyes and is toxic by ingestion. It is carcinogenic to some laboratory animals. Selenium sulfide is used therapeutically in the treatment of seborrhea or dandruff; it is usually applied to the scalp as a mixture of crystalline selenium sulfide and solid solutions of selenium and sulfur in an amorphous form. The selenium content of such a suspension is 52-55.5% It is also used in medicated shampoos. See also shampoo (dandruff shampoo).

selenocysteine. cysteine that contains selenium rather than sulfur. It occurs in nature and is partly responsible for certain curative effects of cysteine.

selenomethionine. methionine that contains selenium rather than sulfur.

selenophile. any plant that accumulates selenium from the surrounding medium (usually soil). **facultative selenophile**. any selenophile that grows on soils that do not contain selenium as well as soils that do. **obligate selenophile.** any plant that grows only on seleniferous soils and contains selenium at some stage of growth, e.g., *Astragalus*.

selenosis. selenium poisoning.

selenyl fluoride. selenium oxyfluoride.

self-empoisonment. See autointoxication.

self-limited. pertaining to a process that is limited by its own traits or peculiarities as opposed to the action of an outside influence; said of a disease that runs a definite, limited course.

self-poisoning. autointoxication.

semantic. pertaining to the meaning or significance of words.

semeiotic, **semiotic**. of or pertaining to symptoms.

semeiotics (semiotics; symptomatology). **1:** the branch of biomedical science that treats the signs and symptoms of disease. **2:** the symptoms of a particular disease considered as a whole.

semelo. *Trimeresurus gramineus*.

semen. a fluid (spermatozoa and seminal fluid) secreted by the male, predominantly in animals in which fertilization is internal. It consists of spermatozoa (elaborated by the testes) and secretions from various accessory sex glands (in mammals, these include the prostate gland and seminal vesicles). Semen is typically transferred from the male to the female as an ejaculate during copulation.

semicircular canal. a tubular structure within the inner ear that contains the receptors that are essential to the sense of dynamic equilibrium.

semicoma. a coma from which an individual can be aroused.

semicomatose. of, pertaining to, or descriptive of a state of semicoma; in a state of semicoma.

semiconscious. partially conscious; stuporous.

semiconsciousness. partial consciousness or stupor. *Cf.* semiunconsciousness; See also unconsciousness, stupor.

semilunar valve. a type of heart valve in vertebrates that resembles a half moon. Such valves are situated between the ventricles and the vessels that convey blood from the heart.

seminal fluid. the fluid component of semen produced by various glands (accessory glands) situated along the male reproductive tract.

seminal vesicles. convoluted sac-like structures attached to the ductus deferens near the base of the bladder in male vertebrates.

seminatural. said of ecological communities modified by human activities.

seminiferous tubules. highly coiled ducts within the testes of vertebrate animals that produce and transport sperm.

seminole bead. *Abrus precatorius*.

semiochemical. **1:** any substance produced by a living organism that conveys information to other organisms, especially that which stimulates or helps to regulate behaviors (e.g., social behavior, aggregation, sexual stimulation, trail following). **2:** of or pertaining to semiochemicals or to semiochemical interactions.

semiochemical interaction. any interaction that involves the transmission of information between organisms by means of chemicals elaborated by some or all of those that interact. See allelochemical, semiochemical.

semiotics. semeiotics.

semipermeable membrane. a selectively permeable membrane, one with pores of a size that permits the passage of solvent and some solute molecules but restricts the passage of other solute molecules.

semiunconscious. partially unconscious or stuporous.

semiunconsciousness. partial unconsciousness or stupor. *Cf.* semiconsciousness; unconsciousness.

Semper's sea snake. *Hydrophis semperi*.

Sencopelma communis (black tarantula). a large, venomous, black tarantula of Panama. Effects in humans are almost always localized.

senecio. *Senecio*.

Senecio (groundsel; ragwort; squaw weed; life root; senecio). a semicosmopolitan genus of more than 1200 species of herbs with many-flowered, chiefly yellow heads (Family Compositaceae), of which more than two dozen are known to be poisonous. They contain hepatotoxic pyrrolizidine alkaloids that cause hepatogenic photosensitization. These yield a nitrogen-containing fraction (necine) and a mono- or dicarboxylic necic acid on hydrolysis, which cause hepatogenic photosensitization. Nitrogen oxides also occur in *Senecio*. They are hepatotoxic and produce liver lesions that differ somewhat from those produced by the pyrrolizidine alkaloids. Most cases of senecio poisoning involve cattle and horses. Sheep and goats are more resistant, but have been poisoned experimentally. Symptoms of chronic poisoning are similar in cattle and horses. Onset of symptoms is usually abrupt following a latent period of several months. The animal is initially sluggish, uneasy and anorectic, becoming rapidly weak and emaciated with signs of abdominal pain and reduced sensibility. Death may intervene in as little as a week. The mucous membranes assume a yellow or muddy coloration and the skin (and milk of cattle) develops a distinctive and unpleasant odor. In the later stages the animal may become progressively weak, dying quietly. Some animals become restless and walk aimlessly, into or through structures such as fences. Death is thus sometimes accidental. In cattle the coat becomes roughened and the muzzle, dry and scaly and tenesmus with constipation or diarrhea is common. Symptoms of chronic senecio poisoning of humans (based on cases in Africa, the West Indies and elsewhere) are similar to venous occlusive disease and include ascites, headache, hepatomegaly, nausea, vomition, abdominal pain, apathy and emaciation. Both vomitus and stool may be bloody. Human intoxications in Africa usually stem from ingestion of bread made from contaminated flour. In the West Indies, poisoning may result from the native custom of brewing a medicinal tea from various indigenous plants. Acute poisoning is rare in humans and livestock. Specific liver lesions may be observed on post mortem examination of affected animals. The exact nature of these lesions depends mainly upon the duration of the condition. Thus, necrosis and severe hemorrhage into the lobules are characteristic of acute intoxications. Advanced chronic cases present fibrotic or cirrhotic livers. In cattle, additional lesions occur; the gall bladder may be greatly enlarged, with an accumulation of bile and nephropathies observed. Examples of poisonous North American species are *S. glabellus* (bitterweed), *S. integerrimus*, *S. longilobus* (thread-leaf groundsel), *S. plattensis*, *S. riddellii* (Riddell's groundsel), *S. spartioides* (broom groundsel), and *S. vulgaris* (common groundsel). ***S. aureus*** (life root; squaw weed; ragwort). a common weed of the eastern United States, formerly used in the treatment of amenorrhea and other menstrual irregularities. It is hepatotoxic and unsafe to use. ***S. jacobea*** (stinking Willie). a monopolistic, weedy annual (rarely perennial) herb (Family Compositae) common to roadsides, fields, and pastures. It contains the toxic alkaloid, jacobine, and is avoided by grazing animals. ***S. platyphyllus***. this and other species of *Senecio* produce the alkaloid platyphylline. See necine, phylloerythrin, senecionine, seneciophylline, senecifoline, senecine, necic acids, retrorsine, and retronecine. See also senecio poisoning.

senecio alkaloid. See pyrrolizidine alkaloid.

senecio poisoning. a sometimes fatal poisoning of herbivores (usually cattle and horses) and sometimes of humans due to ingestion of herbs of the genus *Senecio*. In livestock, onset of the disease is usually abrupt, typically following a latent period of several months. The animal is initially sluggish, uneasy, and anorectic, becoming rapidly weak and emaciated with signs of abdominal pain and reduced sensibility. Death may intervene in as little as a week. The mucous membranes take on a yellow or muddy coloration and the skin (and milk of cattle) develops a distinctive sweetish and unpleasant odor. In the later stages, the animal may become progressively weak, dying quietly. Some animals become restless and walk aimlessly into or through structures such as fences. Death is thus sometimes accidental. In cattle, the coat roughens and the muzzle becomes dry and scaly; tenesmus with constipation or diarrhea is common. **acute senecio poisoning**. Such is rare in humans and livestock. Post mortem examination of affected animals may reveal specific liver lesions, the exact nature of which depends chiefly on the duration of the condition. Thus, necrosis and severe hemorrhage into the lobules are characteristic of acute intoxications. See *Senecio*. **senecio poisoning, chronic**. human intoxications in Africa usually stem from ingestion of bread made from contaminated flour. In the West Indies, poisoning may result from the native custom of brewing a medicinal tea from various indigenous plants. Signs and symptoms in humans are similar to those of venous occlusive disease and include ascites, headache, hepatomegaly, nausea, vomiting, abdominal pain, apathy, and emaciation. The vomitus and stool may be bloody. Lesions include those of hepatic fibrosis or cirrhosis. In cattle, the gall bladder may also be greatly enlarged with an accumulation of bile; nephropathies may be observed.

senecioic acid (3-methyl-2-butenoic acid; 3,3-dimethylacrylic acid; methylcrotonic acid). the acid component of binapacryl, q.v. It is used as a fungicide and miticide and is a precursor of isoprenoid and terpene compounds.

seneciolylcholine. a choline ester isolated from the hypobranchial glands of the marine gastropod *Thais floridiana*.

senecionine (12-hydroxysenecionan-11,16-dione; aureine). a highly toxic hepatotoxic pyrrolizidine alkaloid from *Senecio vulgaris* and other species of this genus.

seneciosis. liver degeneration and necrosis in senecio poisoning.

senecioylcholine. a toxic choline ester associated with the hypobranchial gland of the gastropod mollusc, *Thais floridana*. Activity resembles that of muscarine and nicotine. It is cardiotoxic, hypotensive, stimulates gastric motility and secretion, and increases the rate of respiration.

seneciphylline (13,19-didehydro-12-hydroxysenecionan-11,16-dione; jacodine; α-longilobine). a hepatotoxic pyrrolizidine alkaloid, $C_{18}H_{25}O_6N$, isolated from the herb, *Senecio jacobea*.

senecoic acid (2-*sec*-butyl-4,6-dinitrophenyl ester). binapacryl.

senescence. **1:** the process of growing old or the state of the organism during the process. **2:** the period between maturity and natural death of a whole organism or part.

senescent. growing old; being old.

sensitinogen. any antigen that has a sensitizing effect or produces hypersusceptibility (e.g., an anaphylactogen, an allergen).

sensitive. **1:** responsive; susceptible; able to receive and respond to external stimuli. **2a:** unusually or abnormally receptive or responsive to external stimuli. **2b:** able to respond quickly and intensely to external stimuli.

sensitivity. **1a:** receptiveness; the state or quality of being sensitive. **1b:** irritability; the capacity of an organism to respond to stimuli. **2:** in sensory analysis, the ability to perceive, identify, and/or differentiate, qualitatively and/or quantitatively, one or more stimuli by means of the sense organs. **3:** in analytical chemistry, the sensitivity of a procedure. For a simple procedure, sensitivity is the slope of the calibration curve which is the differential, dx/dc, where x

is the measurement and *c* is the actual concentration. The does not represent the smallest amount, or the lowest concentration that the procedure can detect (i.e., it is not the same as the limit of detection). **delayed contact sensitivity**. allergic contact dermatitis. See dermatitis. **normal sensitivity**. **1a:** the sensitivity of those individuals that lie in mid-range of the dose-response curve for a particular toxicant. **1b:** the sensitivity of those individuals within a given test population that survive exposure to a particular toxicant at the dosage corresponding LD_5, but are killed at LD_{50}. See also hypersensitivity, hyposensitivity.

sensitization reaction. allergic reaction.

sensitize. **1:** to render sensitive. **2:** to render an individual sensitive to a chemical. **3:** to immunize. See also sensitized antigen.

sensitized. rendered sensitive.

sensitizer. **1:** antibody. **2:** a substance that causes a substantial proportion of exposed animals to develop an allergic response following additional exposure to the same or to closely related substances.

sensitizing substance. See substance.

Senso. See Ch'an Su.

sensory. **1a:** of or pertaining to the ability to feel or experience sensations such as sound, light, touch, or pain. **1b:** of or pertaining to structures or processes that enable an organism to experience sensations as in def. 1a.

sensory adaptation. See adaptation.

sensory nerve. a nerve that contains only dendrites of sensory neurons.

sensory neuron. an afferent neuron; one that transmits impulses from the periphery of the body to the CNS.

sensu lato. Latin, meaning in the broad sense. Abbreviated *s.l.*

sensu stricto. Latin: in the narrow sense, in the strict sense. Abbreviated *s. str.* or *s.s.*

sepal. a unit of the outer whorl of sterile leaf-like parts of a flower. Such are often green but are sometimes colored.

sepsin. a toxic crystalline substance isolated from decaying yeast and from animal matter.

sepsis. **1a:** the presence in circulating blood and other tissues, of disease-producing microorganisms (bacteria or viruses) or their toxins. **1b:** the condition resulting from 1a, which may include abscesses throughout the body or septicemia.

septemia. septicemia.

septic. **1:** infected; any condition that produces pus, and is caused by sepsis. **2:** heavily polluted; said of a habitat or freshwater zone that is rich in decomposing organic matter, high in carbon dioxide, and very low in dissolved oxygen.

septic fever. septicemia.

septic intoxication. septicemia.

septic shock. the life-threatening stage of septicemia. It is an extremely serious condition with tissue damage and a dramatic drop in blood pressure due to septicemia and toxemia. Symptoms vary with the site and extent of tissue damage, but are chiefly those of septicemia. Additional symptoms may include a weak, rapid pulse; marked hypotension; cold extremities with or without cyanosis; and sometimes vomiting and diarrhea. Low urine output may signal kidney damage.

septicemia (septemia; sapremia; septic fever; hematosepsis; septic intoxication). **1a:** sepsis; the rapid multiplication or presence of pathogenic microbes or their toxins in circulating blood. **1b:** a condition marked by localized or systemic tissue destruction due to disease-causing bacteria or their toxins absorbed from the bloodstream. Such a condition may become life-threatening if suitable treatment is delayed. The affected individual becomes suddenly and seriously ill, with chills, a high fever, hyperpnea, headache, and often clouding of consciousness and short term memory loss. Skin rashes may occur, the hands may be unusually warm, and the individual may become jaundiced. *Cf.* blood poisoning. See also sepsis, septic shock, toxemia. **2:** bacteremia. **3:** pyemia. **4:** toxemia.

septicemic. pertaining to, or denoting septicemia or its consequences.

septicine. a ptomaine produced from hexylamine and amylamine during the putrefaction of flesh.

septicopyemia. pyemia and septicemia occurring at the same time.

septicopyemic. pertaining to septicopyemia.

septis. poisoning due to products of putrefaction.

sequela (plural sequalae). any lesion or affection that persists (e.g., scar tissue, a paralyzed limb, a behavioral deficit) following recovery from a diseased state or condition (e.g., as from poisoning).

sequestering agent. a substance that removes a metal ion from solution by forming a complex that has properties differing from those of the metal ion.

sequestrant. an agent that binds or complexes with a chemical substance rendering it chemically or biologically less active. *Cf.* chelating agent.

SER. endoplasmic reticulum without attached ribosomes.

sera. plural of serum.

sere. a completed succession of plants and animals, ending in the establishment of a climax community. It is made up of a series of seral communities (or stages). A primary sere is one that has become established in an area never before inhabited; a secondary sere occurs in a region previously occupied by established communities, such as an abandoned cropland. A secondary sere usually takes a much shorter time to be completed than a primary sere, because the nutrients are already available and conditions are more favorable to colonization. Seres developing in different environments can be specified; for example, a hydrosere develops in an aquatic environment, a halosere in a salt marsh, a xerosere in a desert, etc.

Sernyl. phencyclidine.

serologic, serological. pertaining to serology.

serologist. a scientist or practitioner who is an expert in serology.

serology. the study, *in vitro*, of reactions (e.g., agglutination reactions, complement fixation, precipitation reactions) between antigens and antibodies. A major aim is to establish the specificity of antigenic substances from various sources in relation to the establishment and nature of immunity. As a result, certain important tests have been developed (e.g., complement fixation and precipitin tests) and antisera (including antivenins) have been developed for therapeutic use.

serolysin. a bactericidal substance in blood serum.

seropositive. a condition in which antibodies specific to a poison or other disease-producing agent occur in blood serum.

serosa. a serous epithelial membrane that covers the surface of a visceral organ; examples are the pericardium, peritoneum, and pleura.

serotherapy. treatment by injection of serum.

serotonin (5-hydroxytryptamine; 5-HT; 5-hydroxy-(2-aminoethyl)indole; 3-(2-aminoethyl)-1*H*-indol-5-ol; 3-(β-aminoethyl)-5-hydroxyindole; 5-hydroxy-3-(β-aminoethyl)indole; enteramine; thrombocytin; thrombotonin). a catecholamine with neurotransmitter properties. In vertebrate animals it is concentrated in the hypothalamus within the granular synaptic vesicles of nerve terminals. Serotonin nerve terminals also occur in nuclei of the parasympathetic and sympathetic nervous systems and in the pineal gland. It is synthesized in the brain from tryptophan. Serotonin is associated with the control of wakefulness, mood, and pain sensation (e.g., in mammals); it is a chiefly inhibitory CNS neurotransmitter and a potent constrictor of smooth muscle, especially of the larger blood vessels, thus elevating blood pressure in vertebrate animals. It increases vascular permeability, inhibits gastric secretion, and stimulates smooth muscle contraction (e.g., of blood vessels). Serotonin occurs widely in (especially invertebrate animal) tissues and fluids (e.g., blood platelets, brain, carcinoid tumors, and mast cells of some species). In domestic rabbits, and several other mammalian species, but not in humans, serotonin participates in anaphylaxis. It occurs also in nettles and in some animal toxins and venoms. Its action peripherally is similar to that of adrenalin, but more pronounced. See *Bombina*, *Murex*.

serotoxin. **1:** a toxin formed in and from blood serum. **2:** anaphylatoxin.

serous. **1:** pertaining to, resembling, or in the nature of a serum. **2:** producing or containing serum or a serous secretion, as in the case of a serous gland or cyst.

serous membrane. serosa.

Serpasil™. reserpine.

serpent. snake, q.v.

serpent du bananier. *Dendroaspis jamesoni*, *D. viridis*.

serpentaria (bitter tonic; snakeroot; snakeweed; sangrel; birthwort). an astringent, aromatic, bitter material comprised of the dried rhizome and roots of *Aristolochia serpentaria*, and *A. reticulata* (Family Aristolochiaceae). It contains a volatile oil which contains borneol, aristolochin, and aristolochic acid. No longer used in medicine because of its toxicity.

Serpentes (Ophidia). a suborder (Class Reptilia, Order Squamata) comprised of the true snakes. Snakes are characterized by the absence of limbs, external ear openings, an ear drum, sternum, and urinary bladder. The eyes are covered by an immovable, transparent eyelid, the body is elongate, and the tongue is long, slender, protrusile, and deeply forked (bifid). The gape of the jaw is extremely wide due to the loose articulation between the maxillae and the skull; the prey is swallowed whole. The skin is dry, scaly, and may be molted several times per year.

Serranidae (sea basses). a large family of percomorph bony fishes (Class Osteichthyes) with a single, slightly notched dorsal fin. One species (*Serranus subligarius*) is the only known hermaphroditic vertebrate. See *Cephalopholis argus*, *Epinephelus fuscoguttatus*, *Mycteroperca venenosa*, *Plectropomus* spp., *Variola louti*, ciguatera. See also especially seabass.

serum. **1a:** any thin fluid, especially that which moistens the surfaces of serous membranes. **1b:** the clear portion of any animal liquid separated from its more solid elements. **2:** blood serum; blood plasma less the fibrinogen; the thin, watery portion of vertebrate blood that remains after coagulation. **3:** immune serum. **ACS serum**. antireticular cytotoxic serum. **active serum.** a serum that contains complement. **allergenic serum**, **allergic serum**. a serum that induces hypersensitivity to antigen. **anallergenic serum**, **anallergic serum**. a serum that does not produce hypersensitivity. **antibothropic serum**. a serum effective against envenomation by snakes of the genus *Bothrops* and that by some related crotalids. *Cf.* bothropic antitoxin (See antitoxin). **anticomplementary serum**. a serum that destroys or inhibits the activity of complement. **anticrotalus serum**. an antivenomous serum which is protective against envenomation by crotalids. *Cf.* crotalus antitoxin (See antitoxin). **antiophidic serum**. antivenomous serum. **antiphalloidian serum**. a serum that is effective against poisoning by *Amanita phalloides* if administered soon after symptoms appear. **antireticular cytotoxic serum** (ACS serum; Bogomolet's serum). serum produced by inoculating horses with an extract of spleen and bone marrow. Small doses may stimulate the reticuloendothelial system; large doses are cytotoxic to reticuloendothelial cells. **antisnakebite serum**. snake antivenin (See antivenin). **antitoxic serum**. serum that contains antitoxin. *Cf.* antitoxin. **antivenomous serum**. snake antivenin (See antivenin). **blood serum**. the thin, watery fraction of blood that remains following removal of fibrin, coagulation factors, and formed elements (corpuscles) of whole blood. See also serum. Blood serum is more suitable for use in many types of tests and the preparation and use of antisera where clotting might interfere with the process or results. **Bogomolet's serum**. antireticular cytotoxic serum. **Calmette's serum.** snake antivenin (See antivenin). **immune serum**. blood serum that contains specific antibodies. Such serum, when introduced into the body, produces passive immunization by virtue of the antibodies that it contains. **leukotoxic serum.** a serum that destroys leukocytes. **lymphatolytic serum**. a serum that is destructive to lymphatic tissues. **monovalent serum**. antiserum that contains antibody to only one strain or species of microorganism or to one type of antigen. **nephrolytic serum, nephrotoxic serum**. a serum that is specifically toxic to the kidney. Such a serum may be produced by immunizing an animal with a brei or emulsion of kidney tissue. **neurolytic serum, neurotoxic serum**. a serum that is selectively toxic to the brain and spinal cord. Such a system may be produced by immunizing an animal with a brei or an emulsion of nervous tissue. **polyvalent serum**. antiserum that contains antibody to more than one kind of antigen. Such sera are produced by mixing monovalent sera or by immunizing an animal with multiple antigens. *Cf.* polyvalent antivenin (See antivenin). **thymotoxic serum**. a serum that is selectively toxic to thymus tissue. **thyrolytic serum, thyrotoxic serum**. a serum that is selectively toxic to the thyroid gland.

serum albumin. See albumin.

serum disease. serum sickness.

serum reaction. serum sickness

serum sickness. a pronounced anaphylactic reaction that sometimes results from injection of foreign serums, especially horse serum. See anaphylaxis, anaphylactic shock, antivenin.

sesbane. See *Sesbania*.

Sesbania (coffeeweed; coffeebean; rattlebox; bagpod; rattlebrush; sesbane; poison bean). a genus of warm climate, annual or perennial herbs, shrubs, and small trees (Family Fabaceae, formerly Leguminosae). The seeds contain toxic saponins. The three closely related species named below occur in the Gulf Coastal Plain of the southern United States and are toxic to livestock. Symptoms vary somewhat among types of stock, but are similar for all three species of *Sesbania*. The acute condition is characterized by gastroenteritis that may produce necrotic lesions, diarrhea (sometimes constipation in cattle) and may be rapidly fatal. ***S. drummondii*** (= *Daubentonia drummondi*, *D. longifolia* (in part); coffeebean; rattlebrush; rattlebox, etc.). ***S. platycarpa***. *Sesbania vesicaria*. ***S. punicea*** (= *Daubentonia punicea*, *D. longifolia*, in part)) (purple sesbane, purple rattlebox, etc.). ***S. vesicaria*** (= *S. platycarpa*, *Glottidium vesicarium*) (bagpod, bladderpod, coffeebean, etc.).

sesquimustard (1,2-bis(2-chloroethylthio)-ethane).

seston. the total particulate matter suspended in a natural body of water. *Cf.* sediment.

seta (plural, setae; chaeta). a chitinous bristle, or hair-like structure that arises from the epidermis of many invertebrates. See, for example, *Eurythoe*.

setae. plural of seta.

setterworth. *Helleborus foetidus*.

Seven. See carbaryl.

seven bark. *Hydrangea arborescens*.

seven pacer. *Vipera russelii*.

Seveso, Italy. in 1976, near Seveso, Italy, a massive industrial accident resulted in the release of a huge cloud of vaporized chemicals dominated by TCDD, into the atmosphere. Several tens of thousands of people were exposed to these emissions which spread over an area of approximately 3 square miles. Many persons in the vicinity suffered at least some of the following effects: cutaneous burns and sores, personality changes (many became agitated or moody, irritable), inappetence, excessive tiredness. Ninety women had medical abortions and 51 aborted spontaneously. Innumerable pets, livestock, and wildlife suffered and large numbers were killed; severe liver damage was a common finding at necropsy. See also TCDD. A comprehensive investigation, conducted by investigators at Catholic University (Rome), revealed no evidence of an increase in malformations among children born in the area of impact during the following six years over that observed in a control population. The same conclusion applies also to children born within 9 months of the accident, whose mothers were in the exposure area. No major malformations were found in children born to mothers living in the most highly contaminated area.

Sevin™. carbaryl.

Sevin Carbaryl™. an agricultural pesticide. See also Bhopal.

sewer gas. a toxic and potentially explosive gas, of variable composition, but usually mostly methane, produced by the decomposition of organic matter in sewers.

sewers. See combined sewers.

sex. the sum of all structural, functional and behavioral characteristics that distinguishes among males, females, asexual forms, intersexes, and hermaphrodites; sexual.

sex ratio. the ratio of one sex relative to the other in a population of animals, often expressed as the number of males per 100 females or as the proportion of male births. **secondary sex ratio**. the ratio of male/female live births in a population or community.

sex-linked gene. a gene located on a sex chromosome that controls somatic traits.

sexual. of or pertaining to sex, sex-related characteristics, or to the process of sexual reproduction.

SFS. sodium formaldehydesulfoxylate.

shad. clupeoid fish; those of the family Clupeidae.

shaggy parasol. *Lepiota rachodes*.

shaggy-stalked lepiota. *Lepiota clypeolaria*.

shallow breathing. breathing with an abnormally low tidal volume.

shamrock. *Trifolium repens*.

shark. the common name for all cartilaginous fish (Class Chondrichthyes) of the suborder Squali (Order Selachii). Most are strong, almost torpedo shaped, fast moving, pelagic, aggressive swimmers. Both small bottom-dwelling sharks such as *Scyliorhinus* and *Squalus* (dogfishes) and larger pelagic sharks such as *Cetorhinus* (basking shark) are included. See also elasmobranch poisoning.

sharp-nosed pit viper. *Agkistrodon acutus*.

shaushawane. *Bitis caudalis*.

Shaver's disease. bauxite pneumoconiosis.

Shaw's sea snake. *Lapemis curtus*.

shchitomordnik. *Agkistrodon halys*.

sheep laurel. a common name of *Kalmia augustifolia*.

sheep loco. *Astragalus nothoxys*, a true locoweed.

sheep sorrel. *Rumex acetosella*.

sheepkill. *Kalmia angustifolia*.

sheepshead. See Sparidae.

sheepshead minnow. *Cyprinodon variegatus*.

Shelford's law of tolerance. any environmental variable, value, quantity or factor (e.g., temperature, concentration of a particular chemical) below a certain minimum and/or above a certain maximum will exclude certain organisms from living in that area. Adapted to toxic environmental chemicals, one may say that (as a rule) too much or too little of a substance to which an organism is naturally exposed is potentially troublesome.

shellfish. common name of shelled mollusks and crustaceans, especially those used as human food; sometimes restricted to mollusks that have shells.

shellfish poison. paralytic shellfish poison.

shellfish poisoning. any of three types of poisoning (allergic, paralytic, and venerupin (or gastrointestinal) poisoning in which shellfish act as transvectors to humans and other animals that feed on them. Death or serious impairment may result. **allergic shellfish poisoning**. it is caused by specific substances secreted by toxic dinoflagellates that are sometimes present in shellfish. The dinoflagellates are digested and the toxins, to which the shellfish are insensitive, accumulate in their tissues. Symptoms appear in humans within a few minutes to three hours following ingestion of a toxic dose. The victim may suffer slight topical anesthesia about the lips and mouth on contact with the shellfish, rapidly progressing to numbness about the lips, face and fingertips, nausea, ataxia, headache, thick speech and a deepening general paralysis. Death usually results from respiratory collapse. The prognosis is excellent, however, if the victim survives for 24 hours. **gastrointestinal shellfish poisoning**. venerupin poisoning. **paralytic shellfish poisoning** (mussel poisoning; clam poisoning; paralytic poisoning). a type of shellfish poisoning marked by a servere neurologic reaction in higher animals and humans that may terminate in death. Mussels are the most common transvectors of this form of shellfish poisoning. See also paralytic shellfish poison, gonyaulax toxin, saxitoxin, mussel poisoning, red tide. *Cardium edule*, *Donax serra*, *Ensis directus*, *M. modiolus*, *Mya arenaria*, *Mytilus californianus*. *M. edulis*, *Saxidomus*, *Spisula solidissima*,

sheushewane. *Bitis caudalis*.

shfifon. *Cerastes cerastes*, *Pseudocerastes fieldi*.

shiddil. *Kerilia jerdonii*.

shield-nosed pit viper. *Agkistrodon acutus*.

shield-nosed snake. See *Aspidelaps*.

shield snake. *Aspidelaps scutatus*.

Shigella. a genus of pathogenic Gram-negative bacteria that cause gastroenteritis and bacterial dysentery. ***S. dysenteriae***. a species, certain populations or strains of which cause a severe form of dysentery due to a toxin they secrete, which causes intestinal hemorrhage and gastrointestinal tract paralysis.

shiner. See Cyprinidae.

shivers. a disease of Australian and African livestock (cattle, sheep and horses) probably caused by consumption of large amounts of *Malva parviflora*. It is a nervous syndrome characterized by severe muscular tremors which are intensified by exercise. The appetite remains normal and mild cases recover quickly if the animals are kept quiet. *M. nicaeensis* (= *M. borealis*) has been implicated in a similar syndrome in livestock in the United States. *Malva spp*., when ingested by laying fowl, produce a discoloration of yolk in stored eggs. Malvalic and sterculic acids are responsible for this effect.

shock. **1:** any sudden emotional or physical disturbance. **1b:** a condition of profound mental and physical depression or anguish due to severe physical or psychological trauma or any of a variety of toxicants. **2:** a condition arising from an acute, uncontrolled fall in peripheral blood pressure, which in some cases becomes irreversible with a fatal outcome. It is marked by hypotension, weakness, pallor, dry mouth, a cold sweat, coldness of the skin, usually tachycardia, oliguria, and often anxiety, even feelings of doom. It is most often due to cardiac insufficiency, changes in peripheral resistance to blood flow, or tissue damage. Contributing causes may include hemorrhage, diarrhea, vomiting, inadequate fluid intake, or excessive renal loss with ensuing hypovolemia. Any of a large variety of poisons can cause shock. A few examples are corrosive acids and alkalis, CNS depressants, certain mycotoxins, various venoms, many endotoxins, arsenates, ascaron, caffeine, histamine (endogenous or introduced), iodine, protoanemonin. See also disulfiram, *Aleurites*, blister beetle poisoning, fluoride toxicosis. **3:** the impact of an accentuated heart sound sensed by a hand on the chest wall. **allergic shock**. anaphylactic shock. **anaphylactic shock** (allergic shock). a severe, often fatal reaction to a second injection of serum or protein (e.g., a toxin or venom). It is due to anaphylaxis and is marked by smooth muscle contraction and capillary dilation initiated by cytotropic antibodies. It does not occur in a type IV allergic reaction. See *anaphylaxis*. **anaphylactoid shock** (anaphylactoid crisis; pseudoanaphylactic shock). a reaction similar to that of anaphylactic shock, but which does not depend upon anaphylaxis, q.v., nor is it a type of antigen-antibody reaction. The condition may result from intravenous injection of serum that is pretreated with any of a number ofmaterials such as kaolin or starch, trypsin, organic colloids, peptone. It may occur in cinchonism and hymenopterism. **anesthesia shock**. that induced by an anesthetic, usually an overdosage. **cardiac shock**. shock due to failure of blood supply due to insufficient cardiac output. **endotoxin shock**. that caused by bacterial endotoxins, especially of *Escherichia coli*. **erethistic shock** (delirious shock). traumatic or toxic delirium that may occur following shock. **hemorrhagic shock**. that resulting from acute hemorrhage. It is marked by hypotension; tachycardia; a weak, rapid pulse; pale, cold, clammy skin; and oliguria. **hypovolemic shock**. that due to reduced blood volume as a result, for example, of hemorrhage or dehydration. **insulin shock**. circulatory insufficiency due to severe hypoglycemia induced by overdosage of insulin. Symptoms include sweating, vertigo, anxiety, diplopia, and tremor; ultimately with delirium, convulsions, and collapse. **nitroid shock.** a syndrome similar to that of acute nitrite poisoning. It is sometimes caused by *i.v.* administration of a drug such as arsphenamine. **primary shock**. shock marked by hypotension and loss of consciousness. The proximate cause is usually cerebral anoxia which may result from physical trauma, psychological disturbance, or any of a variety of toxicants. The condition is usually reversible. **secondary shock**. shock due to peripheral blood flow that is insufficient to return enough blood to the heart to provide normal levels of oxygen to the tissues. Coma and death may ensue. Any of a wide variety of toxicants can cause secondary shock. **serum shock**. shock due to a reaction to injection of serum. **thyrotoxin shock**. symptoms of thyrotoxicosis due to overdosage of thyrotoxin. **toxic shock**. See syndrome (toxic shock).

shock organ. a specific organ involved in an antigen-antibody reaction.

shocking dose. the smallest amount of antigen that will elicit a particular clinical response or syndrome.

shootur sun. *Hydrophis obscurus*.

shore pit viper. *Trimeresurus purpureomaculatus*.

short-fanged snake. *Ultrocalamus preussi*.

short-tailed shrew. *Blarina brevicauda*.

short-term exposure limit value (STEL). See limit value.

shot. See lead shot.

showy milkweed. *Asclepias speciosa*.

shrews. See Insectivora.

Si. the symbol for silicon.

SI. International System of Units.

SI Units. See International System of Units.

sial-. sialo-.

sialagogic. ptyalagogic.

sialagogue. ptyalagogue.

sialemesia. sialemesis.

sialemesis (sialemesia). **1a:** the vomiting of saliva or of vomitus that contains primarily saliva. **1b:** vomiting accompanied by excessive secretion of saliva.

sialic. salivary.

sialine. salivary.

sialismus. ptyalism.

sialitis. inflammation of a salivary gland.

sialo-, sial-. prefixes denoting or pertaining to saliva or to salivary glands. See also ptyalo-.

sialogen. an agent that induces salivation.

sialogenous. producing saliva.

sialogogic. ptyalagogic.

sialogogue. ptyalagogue.

sialorrhea. ptyalism.

sialoschesis. suppression of the secretion of saliva. See, for example, atropinic poisoning.

sialosis. **1:** the flow of saliva. **2:** ptyalism.

siana. *Dispholidus typus*.

sick. **1:** unwell; ill. **2:** the state of feeling unwell; nauseous, nauseated. **3:** disturbed or mentally ill.

sickness. **1a:** any condition marked by pronounced deviation from a healthy state; illness. an ailment, disease, disorder, or malady. **1b:** a feeling of being ill. **compressed-air sickness**. bends. **decompression sickness**. bends. **green tobacco sickness**. a brief, recurrent, occupational illness of tobacco harvesters. Symptoms include headache, dizziness, vomiting, and prostration. **Jamaican vomiting sickness,** ackee poisoning. **lambing sickness**. a disease of domestic ewes that is nearly identical to milk fever of cattle. **milk sickness** (lactimorbus). a disease of humans caused by ingesting contaminated milk from cows suffering from trembles. Symptoms include severe vomiting, dyspnea, delirium, convulsions, coma, and death. Recovery, if not fatal, is slow. **radiation sickness** (actinotoxemia; Roentgen intoxication; radiotoxemia (obsolete)). **1a:** the original definition prior to the development and deployment of the atomic bomb, it is the illness produced by whole-body x-irradiation of humans and laboratory animals at a level exceeding about 100 rem. The severity and character of the disease are dose dependent. In mild cases, the usual symptoms are nausea, vomiting, anorexia, debility, and leukopenia. More serious cases are marked by a reduction or total loss of platelets with bleeding; anemia due to reduced capacity to produce new erythrocytes; and an increased risk of intercurrent infection due to severe loss or disappearance of leucocytes. **1b:** illness produced by exposure to any type or source of ionizing radiation. It is essentially indistinguishable from def. 1a, above. In its acute form it is usually referred to as acute radiation syndrome. **serum sickness** (serum disease; serum reaction). an immune complex disease that presents several days following administration of a foreign protein. It is marked by both focal and systemic effects that may include urticaria, edema, pyrexia, general lymphadenopathy, joint pains, and sometimes albuminuria.

side-strean smoke. See cigarette smoke.

sideromycin. a metal-containing antibiotic such as grisein.

siderosilicosis (silicosiderosis). a pneumoconiosis due to inhalation of dust containing iron and silica.

siderosis. **1:** a pneumoconiosis due to inhalation of iron dust. **2:** hemosiderosis. **3:** an excess of iron in circulating blood. **4:** a disease of the eye due to the deposition of iron within the eye. The condition is marked by degeneration of the retina, lens, and uvea.

siderotic. pertaining to or characterized by siderosis.

sidewinder. *Crotalus cerastes* and certain other snakes that occur in sandy deserts.

Sieva bean. *Phaseolus limensis*.

sievert (symbol, Sv). the SI unit of equivalent absorbed dose of ionizing radiation. This unit has largely replaced the becqueral in toxicology and related fields. One sievert is the absorbed tissue dose of ionizing radiation that produces the same biological effect as one gray of gamma rays or X-rays. The sievert is equivalent to 100 REMs.

Siganidae (rabbitfishes). a family of venomous spiny-rayed fishes of moderate size. The first and last rays of the pelvic fins are spinous, a diagnostic trait. They inhabit reefs and rocks from the Red Sea to Polynesia. See also *Siganus*.

Siganus (rabbitfish). a genus of venomous fish (Family Siganidae). The venom apparatus is comprised of 13 dorsal, 4 pelvic, and 7 anal spines, and associated venom glands. Each spine has a deep groove on either side of the midline that conveys venom to the wound. *S. fuscescens* occurs throughout the Indo-Pacific region; *S. lineatus* inhabits the Philippine Islands, the Santa Cruz Islands, New Guinea, the Solomon Islands, Australia, Okinawa, and the Ryukyu Islands; *S. puellus* occurs in the East Indies, the Philippine Islands, Palau, and the Gilbert, Marshall, and Solomon Islands.

sigma (symbols, σ, Σ). **1:** the 18th letter of the Greek alphabet. **2:** symbol for 0.001 sec. **3:** σ, the symbol denoting the standard deviation of a population. **4.** Σ, the symbol for summation.

Sigmodon (cotton rat). a genus of medium-sized herbivorous rats with a rough coat. They are sometimes pests on farms or ranches. Cotton rats occur in grasslands and brushy areas in the southern United States, Central America, and northern South America.

sigmoid curve. an S-shaped curve in graphics.

sign. **1:** evidence of disease in a living system that can be observed and measured, in contrast to a symptom in the case of humans, which is an expression of the victim's subjective experience. For example, blood pressure or red tonsils are signs; headache or nausea are symptoms. Signs of toxicity often differ among laboratory animals and between laboratory animals and humans or even more closely related wild animals. Thus, observations or tests of animals may overpredict or underpredict effects of a particular agent on humans or wildlife. *Cf.* symptom. **2:** an objective symptom of disease. Thus (as in this dictionary), the term "symptom" is often used in practice to include the term "sign." **accessory sign** (assident sign). a sign that is often, but not dependably present and thus not characteristic of a disease. **assident sign**. accessory sign. **characteristic sign**. a sign that is almost always associated with a particular disease or condition. **clinical sign**. sign; any sign discoverable by observation or examination. **non-pathognomonic sign**. accessory sign. **pathognomonic sign**. any sign that is specific to a particular disorder or disease thus serving as a basis for an accurate diagnosis. **premonitory sign**. an early, usually minor, sign that precedes a major disease or health problem. **vital sign**. a sign indicative of life. The presence of all vital signs is a measure of health or vitality and as such are indicators of toxic exposure and severity of intoxication. Vital signs in higher animals and humans are breathing, pulse or heart beat, and sustained blood pressure, and usually a characteristic body temperature.

signal. **1:** sign, indication, warning. **2:** any behavior, color, structural, or chemical feature (e.g., exudation, emanation, taste, odor, irritant) that conveys information from one organism to another. Thus warning coloration and warning substances often signal the toxic, venomous or otherwise dangerous nature of a plant or animal. See also allelochemical, allomone, kairomone.

signal word. See warning.

significance. **1:** importance, relevance, meaning. **2:** a measure of the statistical reliability of a finding. It is the probability that a finding (e.g., results of a toxicity test) is the result of chance. See test of significance.

sikimi. *Illicium religiosum.*

sikimin (sikimitotoxin). a poisonous hydrocarbon, $C_{10}H_{16}$, found in the leaves of *Illicium religiosum.*

sikimitotoxin. See sikimin.

silage. feed for livestock (fodder) prepared by storing and fermenting green forage plants (often chopped stalks and leaves of corn) in a silo. See silo-filler's disease.

silane (silane compound). **1:** silicon tetrahydride. **2:** any compound comprised of silicon and hydrogen with a general formula of $Si2_nH_{2n+2}$. Examples are silyl, SiH_3, and disilanyl (a cyclic silane, Si_2H_5). All are dangerous fire and explosion risks, are poisonous, and emit toxic fumes on decomposition. They are irritants to the eyes, skin and mucous membranes. See also disilane. **chlorosilane**. any substituted silane with the formula SiH_nCl_{4-n} as in dichlorosilane, SiH_4Cl_2. All yield HCl on hydrolysis and are thereby violent irritants to the skin, eyes, and mucous membranes and are poisonous by ingestion and inhalation. They are highly reactive, may ignite spontaneously in air, and emit Cl^- on decomposition by heating. **dichlorosilane**. a flammable, poisonous gas, SiH_2Cl_2, that ignites spontaneously in air. See chlorosilane.

silent. **1:** quiet, still, calm; mute, unspoken. **2:** certain disease states or morbid conditions that produce no detectable signs or symptoms are said to be silent.

silica. silicon dioxide.

silicane. silicon tetrahydride.

silicate. any of a large class of complex metal-silicon-oxygen compounds, the negative ions of which are SiO_4^{-4} and $Si_2O_7^{-7}$. Many of these compounds polymerize, however, with SiO_4 units bound into long chains, sheets, rings, lattices, or any of a large number of cross-linked, three-dimensional arrangements.

silicatosis. silicosis.

siliceous (silicious). containing silicon dioxide.

silicic. pertaining to silicon or silicon dioxide.

silicic anhydride. silicon dioxide.

silicious. siliceous.

silicoanthracosis. silicosis combined with coal worker's pneumoconiosis.

silicochloroform. trichlorosilane.

silicoethane. disilane.

silicofluoric acid. fluorosilic acid.

silicofluoride. a compound of silicon and fluorine with another element.

silicon (symbol, Si)[Z = 14; A_r = 28.086]. a hard, brittle, gray to black, crystalline metalloid that occurs in nature as silicon dioxide (silica) or in silicates, q.v. It is the second most abundant element on earth, comprising about 27.6% of the crust. Silicon forms a vast number of silicates and organic derivatives. Pure silicon is used as a semiconductor and in solar batteries.

silicon dioxide (silica; silicic anhydride; silicon oxide). a hard crystalline compound, SiO_2. It occurs in nature as sand, quartz, tridymite, and crysobalite, which are usually not toxic if ingested. It has many applications as in the manufacture of abrasives, glass, and ceramics. Inhalation of silica dust over a period of years can cause silicosis. Exposure to high atmospheric concentrations of silica dust for much shorter periods of time can cause silicoproteinosis, q.v. See also silicon.

silicon oxide. silicon dioxide.

silicon proteinosis. silicoproteinosis.

silicon tetrachloride. a fuming liquid with a suffocating odor, used in the manufacture of fumed silica (finely divided SiO_2). It reacts with water to emit HCl vapor.

silicon tetrahydride (monosilane; silicane; silane). a gaseous silane, q.v., with a disgusting odor, that slowly decomposes in water. It readily ignites in air (and may explode), spontaneously ignites in oxygen, explodes or ignites on contact with halogens or covalent halides, and burns or explodes when heated. It is an irritant to the eyes, skin, and mucous membranes and is mildly poisonous by inhalation.

silicoproteinosis (silicon proteinosis). an acute, always fatal, pulmonary disorder, due to relatively short exposure to high atmospheric concentrations of silica dust. It is marked by rapid onset and is radiographically and histologically similar to pulmonary alveolar proteinosis.

silicosiderosis. siderosilicosis.

silicosis (silicatosis; stone masons' disease). an irritant type of pneumoconiosis, q.v., caused by chronic inhalation of silica dust (silicon dioxide). It is one of the most common disabling occupational diseases caused by a hazardous substance. It poses a hazard for anyone who is regularly exposed to silica dust (e.g., coal miners, quarry workers, and stone workers). The inhaled dust causes a slowly progressive fibrosis and scarring of the lungs. Lung capacity is lowered with increased shortness of breath, predisposing the victim to pulmonary diseases such as pneumonia and tuberculosis. Severe cases can cause death due to oxygen insufficiency or heart failure. Shortness of breath and frequent attacks of chronic bronchitis over a number of years may lead to emphysema, an increased incidence of pulmonary tuberculosis and spontaneous pneumothorax (an accumulation of air in the pleural cavity).

silicotic. pertaining to or characterized by silicosis.

silicotuberculosis. silicosis in association with tuberculous pulmonary lesions.

silivicide. a herbicide used to destroy woody shrubs and trees.

silkweed. See *Asclepias*.

silo-filler's disease. a rare, acute poisoning of agricultural workers due to the inhalation of nitrogen oxides (usually nitrogen dioxide and nitrogen tetroxide) by workers collecting silage in poorly ventilated areas such as a silo or a room at the base of a silo. The gases, which are heavier than air, tend to collect in the bottom of the silo during fermentation, sometimes exposing workers to high concentrations of these gases. Concentrations of nearly 60,000 ppm of nitrogen dioxide have been recorded from silos housing fermenting silage. A breath or two under such conditions can produce acute intoxication. Symptoms may not appear until several hours following exposure and typically include respiratory distress, pulmonary edema, sometimes loss of consciousness, and rarely death. The acute disease may cause death due to pulmonary edema; subacute or chronic exposures may produce a chronic proliferative pulmonary disease with edema, emphysema, and in more protracted cases, adenomatosis. Except in severe exposures, death is uncommon, although eventual invalidism may result. The elaboration of nitrogen oxides by silage often reduces the possibility that the consuming livestock will suffer nitrate poisoning. See also nitrate.

silver (symbol, Ag; argentum)[Z = 47; A_r = 107.873]. a white, lustrous, transition metal. Elemental silver is not especially toxic, but prolonged exposure to silver compounds can cause a blue-gray skin discoloration (argyria or argyrosis). Many silver salts are irritants and most are light sensitive.

silver, liquid. mercury.

silver manganate (VII). silver permanganate.

silver nitrate. an odorless, bitter-tasting, water-soluble, corrosive compound, $AgNO_3$, with colorless, transparent, tabular, rhombic crystals that become gray to grayish black when exposed to light in the presence of organic matter. It is a powerful oxidizing agent and a strong irritant to skin and other tissue. Silver nitrate is used as a light-sensitive photographic film, in indelible inks, silver plating, the silvering of mirrors, as a hair dye, a disinfectant and reagent. It is a very toxic, lethal poison as the salt or in solution, damaging both the kidney and liver. Symptoms of severe poisoning following ingestion include pain and burning in the mouth; a blackening of the skin, mucous membranes, throat, and abdomen; salivation; nausea; a black vomitus; diarrhea; anuria; shock; convulsions or coma; collapse; and death. The lethal dose varies greatly among test animals. Sodium chloride is an antidote to silver nitrate poisoning. Repeated application or ingestion causes argyria.

silver permanganate (silver manganate(VII)). a light-sensitive, violet, crystalline powder, $AgMnO_4$, used in gas masks and as an antiseptic. It decomposes in ethanol and may explode when shocked or heated.

silver poisoning (argeria). poisoning due to ingestion of silver compounds in which the symptoms are mainly due to the silver ion. It is marked by the blue-gray discoloration of the skin (argyria). See, for example, silver nitrate poisoning.

silver polish. See cleaner (metal cleaner).

silver potassium cyanide. potassium silver cyanide.

silver salt. any salt of silver. Most are strong irritants, are light sensitive, and are used in photography. Ingestion may cause intense pain in the mouth, throat, and gastrointestinal tract; bloody stools; vertigo; convulsions; coma; and death. Lesions may include discoloring due to deposits of metallic silver and severe erosion of the gastrointestinal mucosa in acute cases.

silverleaf nightshade. *Solanum eleagnifolium*.

silvex (2-(2,4,5-trichlorophenoxy)propionic acid; fenoprop; 2,4,5-TC; 2,4,5-T/silvex). the α-methyl derivative of 2,4,5-T, it is a selective pre- and postemergence herbicide and is superior to 2,4,5-T on certain woody species such as *Quercus* (oak). Silvex is generally somewhat less toxic than 2,4-D. It is moderately toxic to laboratory rats and is irritating to skin, eyes, and mucous membranes. In mammals, most is excreted unchanged in the urine. The use of this herbicide on orchards, rice fields, sugarcane, rangeland, and other noncrop sites was terminated in the United States by the EPA in March, 1985. Silvex is carcinogenic to some laboratory animals.

Silybum marianum (variegated thistle). a thistle-like herb (Family Asteraceae) with spiny, white-spotted leaves. Toxic concentrations of nitrate have been reported.

silyl. See silane.

simulation. the imitation or mimicking of the behavior of a system as by a physical or mathematical model.

singlet oxygen. active oxygen.

sinigrin (1-thio-β-D-glucopyranose 1-[*N*-(sulfooxy)-3-buteninidate] monopotassium salt; sinigroside; myronate potassium; potassium myronate; allyl glucosinolate). a crystalline β-glucopyranoside and glycoside of allyl isothiocyanate, $C_{10}H_{16}KNO_9S_2$, isolated from various species of Cruciferae (mustard family, crucifers), most notably from the seeds of *Brassica nigra* and the root of *Alliaria officinalis*. It decomposes in a reaction catalyzed by myrosinase when a plant is damaged, releasing the mustard oil, allyl isothiocyanate, and potassium hydrogen sulfate. The isothiocyanate is a potent vesicant and an irritant to many species of animals, thus tending to inhibit grazing. This substance, however, attracts butterflies of the genus *Pieris*, *q.v.*

sinigroside. sinigrin.

sinistral. on or to the left.

sink. **1:** a basin or depression in a flat surface. **2a:** a basin, pool, or pit used to deposit waste or sewage. **2b:** any repository used for long-term storage of matter or energy. **2c:** any reservoir large enough receive or absorb energy or matter without undergoing significant change. Thus, body fat is a sink for lipophilic xenobiotics (e.g., organochlorine compounds). **3a:** any component of a living system that removes and retains, for some time, wastes or other materials that are inimical to other system components or to the system as a whole. Thus metallotheionein may act as a sink in the removal of toxic metals from the systemic circulation. **3b:** any living system or system component (metabolite, cell, organ, tissue) that is a net importer, storage depot, and end-user of a metabolite or other resource. The root of a plant is a sink that receives sugars which were synthesized in the leaves and converts them to a storage form (polysaccharides). **4:** in air pollution science, any area or environmental component that, by whatever process removes one or more pollutants from the air. Moist ground and green foliage, for example, form important sulfur dioxide sinks. See also storage, metallotheionein.

sinkaline. choline.

sinker. a chemical substance that is heavier than water and has low solubility in water.

sinking. a means of controlling toxic and other effects of oil spills by using an agent that traps the oil and sinks it to the bottom of the water body where the agent and the oil are (albeit slowly) biodegraded.

sinoatrial node (SA node; SA; pacemaker). a small volume of specialized cardiac tissue in the wall of the right atrium that initiates and maintains the rhythm of the cardiac cycle.

Sinox™. dinitrocresol (used in Europe).

sinsemilla. See *Cannabis sativa*.

sintomycetin. See chloramphenicol.

sinus. **1:** a cavity; usually one filled with air or blood. There are many sinuses throughout the human body, but the term is most commonly usually in reference to the cavities in the bone behind the nose. **2:** a drainage channel formed from an abscess to the skin's surface or to an internal organ.

siphon. a bent, flexible tube used to remove fluid from a cavity or vessel by atmospheric pressure.

siphonage. the removal of fluid from a body cavity such as the stomach by use of a siphon.

Siphonophora. an order of pelagic marine hydrozoans that form often massive floating or swimming colonies composed of polymorphic hydroids (polyps) and medusoids. Some individuals of the colony are modified as a float that aids movement along the water surface, as in the Portuguese Man o' War. The nematocysts are numerous and powerful, and the medusae remain attached to the parent hydroid by a stem or disc. Siphonophores are often mistakenly called jellyfish. Notable examples are *Physalia* and *Velella*. See also physalitoxin.

sirocucu. *Lachesis mutus*.

Sistrurus (pigmy rattlesnakes; ground rattlesnakes; pygmy rattlers). a genus of small North American rattlesnakes (Family Crotalidae); they are the least dangerous of the rattlers. with nine large scutes (scales) or crown shields as opposed to the rather extensively fragmented shields of rattlesnakes of the genus *Crotalus*. Three species are known, none of which are generally considered dangerous to adult humans, although the bites of *S. catenatus* are occasionally fatal to children. The head is broad and distinct from the narrow neck; the canthus rostralis is obtuse to acute. The pupils are moderate in size and elliptical. The body is slender to moderately stout, cylindrical and tapered with a short tail that bears a small terminal, horny, segmented rattle. The rattle, *q.v.*, is too small to be used as a means of field identification. ***S. catenatus*** (massasauga rattlesnake). one of two rattlesnakes (See *S. miliarius*) in the United States with large crown shields. It has a well-developed rattle and has a shorter tail than *S. miliarius*. The ground color of the body is usually gray, buff, tan or yellowish with dark gray, black, or brown spots in rows along the body; there are crossbars on the tail. Occasional specimens from the northeastern portion of the range are uniformly black. The range extends from the region of the Laurentian Great Lakes southwestward to extreme southeastern Arizona, southern Texas and adjacent areas in northern Mexico. The massasauga is secretive and is usually not aggressive unless angered. While not usually considered dangerous to adult humans, the venom is highly toxic and the bite is sometimes fatal to children. ***S. miliarius*** (pigmy rattlesnake; Southern rattlesnake; Western rattlesnake). the smallest crotaline snake in the United States, usually reaching a length of less than 0.5 m; the maximum length attained is about 0.8 m. The ground color of the body is reddish-orange, tan, or gray with an orange or rusty, dorsal midline stripe. There are five rows of short crossbars or sooty spots along the length of the body; the tail is barred. The belly is white and densely clouded, with black spots. The tail is long and slender, bearing a tiny terminal rattle which can only be heard in close proximity to the snake. It occurs mostly in dry areas with low vegetation from eastern North Carolina south to the Florida keys; west to southeastern Missouri, eastern Colorado, eastern Oklahoma, and eastern Texas. This snake is usually alert and relatively ill-tempered. Even the bite of small specimens can produce intense pain and extensive swelling. The bite is rarely, if ever, fatal to humans. ***S. ravus*** (Mexican pigmy rattlesnake). a small, rather stout snake that may reach a length of about .6 m. The body is gray or brownish above and marked by 25-35 dark, irregular spots that run along the back and small lateral spots that sometimes merge with the dorsal blotches, forming irregular crossbands. The lower surface of the body is yellowish, with brown spots. The tail is marked by 6-8 dark crossbands. It occurs in southcentral Mexico on the southern part of the Mexican plateau. It does not seem to be dangerous to humans.

sitotoxin. any food poison, especially if produced by bacteria or fungi in grain. See also sitotoxism.

sitotoxism. food poisoning, especially that caused by the presence of bacteria or fungi in grain.

606. salvarsan.

688A. phenoxybenzamine.

six-gilled shark. *Hexanchus grisseus*.

skate. the common name of any of several genera of bottom-feeding cartilaginous fish of the family Rajidae that have a greatly depressed, disc-like body, a short tail with no spine, large wing-like pectoral fins used in locomotion, and much reduced (or absent) dorsal, anal, and caudal fins. (e.g., *Raja*). *Cf.* batoid, Batoidea.

skatole (3-methyl-1*H*-indole; *β*-methylindole). a white, moderately toxic, crystalline compound, $D_6H_4{\cdot}C(CH_3){:}CH{\cdot}NH$, with a strong, characteristic, feces-like odor that is soluble in hot water. It occurs together with indole in the feces and intestines of the African civet, in human feces, in coal tar, nectandra wood, beetroot, and a number of other plants, such as the Javanese tree, *Celtis reticulosa*. It is used in perfumery, chiefly as a fixative. It is moderately toxic to frogs (*Rana*). See also skatoxyl, skatoxylglycuronic acid.

skatoxyl. an oxidation product, $CH_3{\cdot}C_8{\text -}H_6NO$, of skatole that occurs in urine in certain diseases. See also skatoxylglycuronic acid. See also skatoxylglycuronic acid.

skatoxylglycuronic acid. a conjugated, detoxicated form of skatoxyl.

skeletal muscle. striated muscle.

skeletonweed. *Lygodesmia juncea*.

skeptophylaxis. **1:** the situation whereby a minute dose of a poisonous substance will produce immediate temporary immunity to the action of the poison. **2:** allergic desensitization by prior injection of a small amount of the allergen as is often done before injecting an antiserum.

SKF-525A. a powerful inhibitor of cytochrome P-450-dependent monooxygenation that has been widely used to investigate the effects of monooxygenation on detoxication, activation, and interactions among xenobiotics.

skin (cutis). **1:** the outer covering or integument of an animal, especially of a vertebrate animal. It is an important route of entry into the body for many chemicals and classes of chemicals. See absorption (route of absorption), absorption of xenobiotics. The vertebrate skin is comprised of two layers: (1) the epidermis, comprised of epithelium and (2) the dermis, comprised of connective tissue with associated structures. The skin of fishes and some reptiles is covered by dermal or epidermal scales or dermal plates. The outermost layer of the epidermis (stratum corneum) of terrestrial vertebrates is cornified and variously modified as spines, nails, claws, beaks, and other structures. These epidermal modifications and also the sebum or oils secreted by terrestrial vertebrates and the mucus secreted by fishes and amphibians variously impede or potentiate the absorption of xenobiotics. The skin of many animals secretes poisons (See, for example, Amphibia, isobatrachotoxin, spiropiperidine alkaloid). The skin is an organ of breathing in many animals, especially those that are aquatic or semiaquatic (e.g., Amphibia). In so-called warm-blooded animals (homeotherms), the skin is important in temperature control; feathers in birds serve this function as do hair and sweat glands in mammals. The skin of mammals has follicles that produce hair, sebaceous (oil) glands, and usually sweat glands. Xenobiotics that enter the mammalian body via the skin do so mostly through epidermal cells. While individual cells of the sebaceous glands and those in the wall of hair follicles, are usually much more permeable than epidermal cells, they present a much smaller surface area than the latter. The skin is often a good indicator of exposure to toxic substances, usually by a change in color (e.g., pallor, flushing, jaundice, the bluish-black discoloration of argyria), the amount of moisture present, unusual sensations experienced by the subject (e.g., hot, cold, burning, tingly). Lipid-soluble substances generally penetrate the skin more readily than those that are water soluble. Thus, aconitine, dimethylsulfoxide (DMSO), hormones, vitamins D and K, and Sarin are readily absorbed by the skin. The skin of most vertebrates is easily damaged by corrosive or irritant poisons such as alkanes, mineral acids, strong alkalis, and oxidizing agents such as permanganate or dichromate (See also aerosol). Percutaneous absorption of systemic poisons may, however, cause little or no injury to the skin itself. Some xenobiotics are carcinogenic to the skin (e.g., allylisovalelrate, mineral oil). See also stratum corneum, percutaneous.

skipjack tuna. *Euthynnus pelamis*.

SKULL and CROSS BONES. a figure that usually signifies deadliness. When displayed on a product label, it is equivalent to the signal word, DANGER, indicating the product is highly poisonous or otherwise poses a serious hazard to life; its absence from a product label bears no significance. See also warning, label.

skunk cabbage. common name for *Symplocarpus foetidus* and *Lysichitum americanum*. *Veratrum californicum* is locally referred to as "skunk cabbage."

skunkweed. See *Croton*.

skyflower. *Duranta repens*.

s.l. abbreviation of the Latin *sensu lato*.

SLE. systemic lupus erythematosus.

sleeping gough. *Bothrops schlegelii*.

sleeping nightshade. *Atropa belladonna*.

sleeping pill. a sedative-hypnotic in pill or capsule form.

sleepy stage (forage poisoning; stomach stage). a disease of horses of unknown etiology, usually associated with the feeding on moldy hay and grain.

slender coral snake. *Calliophis maculiceps*, *Leptomicrurus*.

slick. an aggregation of floating debris resulting in reduced wave activity and a shiny water surface. **oil slick**. a slick comprised of petroleum hydrocarbons, as from an oil spill. Such a slick can be extremely hazardous to wildlife, due to toxicity and to fouling of fur or feathers.

slinkweed. *Gutierrezia microcephala*.

slipper animalcule. a common name for *Paramecium* that is descriptive of its shape.

slippery Jack. *Suillus luteus*.

sloe. the fruit (e.g., plums, cherries) of trees of the genus *Prunus*. Whereas the fruit of most species is edible, the pit usually contains amygdalin.

slough. **1a:** to shed dead tissue cells (as of the endometrium, which is shed during menstruation). **1b:** to be shed (said of necrotic cells or tissue). **2:** dead tissue that has been shed or separated from a living structure.

slow-reacting substance. former designation for three leukotrienes (C4, D4, and E4) produced during the antigenic degranulation of basophils.

sludge. any solid, semi-solid, or liquid waste generated from a municipal, commercial, or industrial wastewater treatment plant or from an air pollution control facility (exclusive of the treated effluent from a wastewater treatment plant).

Sm. the symbol for samarium.

small intestine. the portion of the alimentary canal between the stomach and the cecum that functions in the absorption of food and nutrients.

small jack-in-the-pulpit. *Arisaema triphyllum*.

small-spotted coral snake. *Calliophis maculiceps*.

smartweed. See *Polygonum*.

smilacin. a poisonous glycoside, $C_{18}H_{36}O_6$, from sarsaparilla.

smilagenin (sarsasapogenin). a steroid precursor, $C_{27}H_{44}O_3$, from several species of *Smilax*. It is an intermediate in the manufacture of hormones of the pregnane series.

Smilax (greenbriar). a genus of widely distributed, mostly woody, climbing plants (Family Lilaceae), which are sources of sarsaparilla. The starchy root of several species was used as food by American Indians. Several species contain a steroid precursor, smilagenin.

Smith's small-headed sea snake. *Hydrophis parviceps*.

smog. **1a:** originally defined as a mixture of smoke and fog dispersed in the atmosphere. **1b:** a mixture of smoke, fog, and other air pollutants dispersed in the atmosphere. **2:** air pollution associated with oxidants; photochemical smog. **oxidant smog**. photochemical smog. **photochemical smog** (smog; oxidant smog). a highly irritating, usually dense, lacrimatory haze produced by photochemical reactions that occur under the influence of strong sunlight in air polluted by various sources, but most importantly by automobile exhaust gases under temperature inversion conditions.

Smog Index. a mathematical relationship by which the presence or absence of photochemical smog can be determined. Such indices are based on the relationship between smog and the meteorological conditions that are conducive to its formation and take into account factors such as temperature, relative humidity, wind speed, and degree of temperature inversion.

smoke. **1:** in chemistry, an aerosol of solid (e.g., magnesium oxide smoke) or liquid (e.g., tobacco smoke) particles generated by combustion, thermal decomposition, or evaporation. **2a:** a suspension in the atmosphere of small particles produced by combustion. **2b:** visible emissions (other than vapor) from a smoke stack or chimney. **2c:** particles in the atmosphere resulting from the incomplete combustion of fossil fuels or other organic material. **3a:** a cigarette. **3b:** a cigarette made from marijuana. **secondary smoke**. tobacco smoke that is inhaled by a nonsmoker who is in a location where smoking occurs. **side-stream smoke**. smoke produced by a cigarette and released at the distal tip. Studies indicate that such smoke poses a risk to nonsmokers (and presumably smokers). Only about 4% of the total smoke produced by a cigarette is inhaled by the smoker. The remainder is side-stream waste in which the concentration of pollutants is about twice that inhaled by the smoker.

smoker. See cigarette smoke.

smoker's respiratory syndrome. a common syndrome in habitual smokers, marked by chronic pharyngitis, wheezing and dyspnea, with or without coughing, and susceptibility to respiratory infections. The affected individual may also have pain in the chest and a hoarse voice. Lesions include small lymphoid nodules in the pharynx with subsequent development of an edematous fibroma on the vocal cords.

smoking, passive. a term that is applied to the exposure of a person who is not smoking to inhale smoke from those who are. See tobacco, See smoke (side-stream smoke).

SMON. subacute myelo-optico neuropathy. See iodochlorhydroxyquin.

smooth brome grass. *Bromus inermis*. See *Claviceps*.

smooth endoplasmic reticulum (SER). endoplasmic reticulum that appears smooth rather than granular due to the lack of ribosomes.

smooth hammerhead. *Sphryna zygaena*.

smooth lepiota. *Lepiota naucina*.

smooth Washington clam. *Saxidomus giganteus*.

smoothing. the averaging of data with respect to time or space in order to compensate for random errors or small variations that are considered insignificant to a stated problem.

SMR. standardized mortality ratio.

smut. **1:** dirt, grime, soot. **2:** any of a group of basidiomycete fungi (Order Ustilaginales). They are normally facultative parasites, often causing substantial damage to agricultural crops. Infested grains may poison livestock. See, for example, smutty oat disease.

smutty oat disease. smutty oat poisoning.

smutty oat poisoning (smutty oat disease). a disease of cattle and horses known from a few cases in the 19th Century. Mortality was high among horses feeding on oat grain infested with smut. The disease in cattle was associated with hay that had been cut before the (already-infested) grain was ripe. Symptoms in horses were paralysis, terminal convulsions and death within 4-48 hours following ingestion of the smutty grain. In cattle, death occurred in about 18 hours due chiefly to acute gastroenteritis.

Sn. the symbol for tin (stannum).

snails. See Gastropoda.

snake (serpent; ophidian). common name of any of about 3,000 species of limbless reptiles (Order Squamata, suborder Serpentes). The blood of most is mildly toxic. See Squamata. **nonvenomous snake**. any snake that lacks fangs and a venom gland. **sea snake** (true sea snake). common name of any of about 50 species of snakes of the family Hydrophidae. The venom of all species is extremely toxic. Examples are *Enhydrina schistosa* and *Hydrophis nigrocinctus*. **venomous snake**. any snake that has fangs and venom glands, which are the only anatomic attributes that distinguish venomous from nonvenomous snakes. There are more than 400 species of venomous snakes worldwide; about 360 are terrestrial; less than 200 are considered dangerous to humans. Most fatalities from snakebite (about ⅔) occur in Asia. Collectively, venomous snakes are called thanatophidia or toxicophidia. **viperine snake**. any snake of the family Viperidae, the true vipers.

snake of hundred design. *Agkistrodon acutus*.

snake venom. See venom. See also enzymes of snake venoms.

snake venom enzymes. See enzymes of snake venoms.

snake venom poisoning. poisoning due to the venom received from a snakebite. The amount and potency of the venom delivered varies widely and not all bites of venomous snakes introduce venom. Just as dangerous, in many cases is the danger of tetanus (as does almost any animal bite) or other infectious disease. Anyone bitten by any snake (venomous or nonvenomous) should seek medical treatment. See also cobra venom poisoning, colubrid venom poisoning, crotalid venom poisoning, elapid venom poisoning, mamba venom poisoning, sea snake venom poisoning, viperid venom poisoning. Each of these definitions of poisoning by different taxa gives a brief summary of the effects of a more or less typical empoisonment of humans in a moderately severe case. *Cf.* ophidism.

snake venom polypeptides. See polypeptides of snake venoms.

snakebit, snakebitten. bitten (or struck) by a snake. See snakebite.

snakebite. **1:** the bite or penetration of teeth, especially fangs, with or without envenomation, when a snake strikes an animal. Throughout much of the earth, most snakebites inflicted on humans are by nonvenomous species (such may become seriously infected). Most snakes strike or bite an extremity. The bite of many types of snakes are distinctive or nearly so and should be noted before treatment, especially with antivenin. Thus, for example, the bite of a viper is usually marked by one or two rather large penetrating wounds; additional tooth marks are not often seen. Elapids typically produce 1 or 2 puncture marks, sometimes 3 or 4. Sea snakes may produce multiple (2-20) pinhead-size puncture marks and some teeth may be broken, remaining in the wound. Envenomation does not always result from the bite of a poisonous snake. If envenomation occurs, symptoms may rapidly appear. See poisoning (snake venom poisoning). **2:** the condition resulting from the bite or strike of a snake, especially in the region of the wound. Symptoms vary according to the family, genus, or species of snake involves, with the depth and location of the wound, the amount of venom delivered, and other factors including the condition of the victim. Nevertheless, the bite of a venomous snake usually soon produces unambiguous symptoms of envenomation. See snake venom poisoning.

snakeroot. **1:** common name of various plants (See *Aristolochia*, *Eupatorium rugosum*; *Asarum*). **2:** serpentaria.

snakeroot poisoning. milk sickness, trembles.

snakeroot, white. See *Eupatorium urticae folium*.

snakeweed. **1:** See *Gutierrezia*. **2:** serpentaria.

snapdragon family. Scrophulariaceae.

snapper (e.g., red snapper; gray snapper; yellowtail snapper). a common name applied to many marine, mostly warm-water, bony fish (e.g., *Aprion virescens*, *Gnathodentex*, *Lethrinus miniatus*, and *Monotaxis grandoculis*), mostly of the family Lutjanidae.

SNARL. Suggested-no-adverse-response-level.

sneezeweed. *Helenium hoopsii*.

sneezeweed poisoning (spewing sickness). poisoning of sheep, and perhaps other grazing livestock, by certain plants of the genus *Helenium*, *q.v.* Symptoms include dullness and depression; sometimes nausea and vomiting which may be accompanied by sialism with frothing, and belching; a rapid and irregular pulse; and rapid breathing. Postmortem examination may reveal a variety of lesions that include gastrointestinal irritation, necrosis, and sometimes perforation of the intestinal wall where it comes in contact with ingested plant material; hepatic and renal damage, and usually necrotic lesions of the lungs.

sneezing gas. diphenylchlorarsine.

snouted night adder. *Causus defilippii*.

snow-on-the-mountain. *Euphorbia marginata*.

snowball. *Hydrangia macrophylla*.

snowbank false morel. *Gyromitra (Helvella) gigas*.

snowdrop. a common name for herbs of the genus *Galanthus*, especially *Galanthus nivalis*, and *Ornithogalum umbellatum*.

snowflake. a common name for herbs of the genus *Leucojum*.

snub-nosed adder. *Vipera latasti*.

snub-nosed viper. *Vipera latasti*.

snuff. **1:** a form of shredded, smokeless tobacco. It is generally used either by inhaling (snuffing) a small amount into the nose or by holding an amount in the mouth (usually between the gum and cheek, occasionally under the tongue). Snuff has been used as a medicinal powder (by snuffing) to promote nasal discharge. It is carcinogenic, sometimes causing cancer of the mouth, pharynx, and esophagus that sometimes results in death. See nicotine, tobacco. **2:** street argot, meaning to kill or to murder; to snuff out life.

soap. **1:** soap compound; any alkali metal salt of a fatty acid. The alkalis most often used are sodium and potassium hydroxides. the fatty acids are derived from animal fats and various vegetable oils. A soap is a type of detergent in which the water-solubilizing group is a carboxylate ion, —COO^-, and the positive ion is usually sodium, Na^+, or potassium, K^+. Soaps are usually made by saponifying a vegetable oil with caustic soda. Soap is an antidote or counteragent in mineral acid and heavy metal poisoning. In veterinary practice, soap is also used as a counterirritant in liniments, as an antacid, and a laxative. **2:** commercially produced soaps usually contain a soap compound together with fragrances and other materials and molded into a bar. **deodorant soap**. any of a type of soap considered (in the United States) to be an over-the-counter drug and not cosmetics. Deodorant soaps work by killing bacteria that supposedly cause body odor. The most commonly used an- [illegible] bromosalicylanilide. Most formulations probably contain additional potentially toxic substances (e.g., phenol, borax, sodium perborate).

soap plant. *Zigadenus venenosus*.

soapberry family. Sapindaceae.

soapstone. talc.

soapwort. See *Saponaria officinalis*.

SOD. superoxide dismutase.

soda. See sodium carbonate. Soda is most often used in reference to the decahydrate.

soda ash (anhydrous sodium carbonate, technical; calcined soda). the technical and common commercial grade of sodium carbonate, anhydrous. It is a nonflammable, grayish-white powder or lumps, Na_2CO_3, that are soluble in water, insoluble in ethanol. It contains as much as 99% sodium carbonate. In the past soda ash was largely produced synthetically, but because of pollution and high energy costs, it is now obtained in the United States from natural deposits, mostly in the northcentral Great Plains and western states. See also sodium carbonate.

soda crystals. sodium carbonate, monohydrate.

soda lime. a toxic mixture of sodium or potassium hydroxide with lime (calcium oxide) used to absorb water vapor and gaseous carbon dioxide. It absorbs 25-35% of its weight of CO_2. Soda lime is a powerful corrosive and irritant to the eyes, skin, and mucous membranes. Ingestion can cause severe gastroenteritis with extensive erosion of tissue and death.

soda lye. sodium hydroxide.

soda monohydrate. sodium carbonate, monohydrate.

soda niter. sodium nitrate.

sodium (natrium; symbol, Na)[$Z = 11$; $A_r = 22.99$]. a soft, reactive, alkali metal, *q.v.*, that oxidizes readily in air or water. It occurs in nature as NaCl in seawater and deposits (halite) [illegible] water soluble and many are toxic or otherwise hazardous (e.g., azo dyes, sodium hypochlorite, mustard gas, soda lime, sodium azide, sodium borate, sodium cyanate, sodium cyanide, sodium selenate, sodium selenite, sodium thiocyanate). Some are extremely hazardous. Sodium is an essential element that is critical to a number of processes in living systems (See, for example, alkali reserve, bile acids, electrolyte, sodium-potassium pump). These processes are directly or indirectly affected by certain poisons (e.g., aconitine, aldosterone, barium, brevetoxin, ciguatoxin, cypermethrin, DDT, deltamethrin, digitoxin, emicholinium-3, pyrethrin. See also water intoxication. Sodium salts are used extensively used in medicine, in the home, and industry (e.g., carboxymethylcellulose sodium, delead, ditiocarb sodium, edetate sodium, erythrosine, hexobarbital sodium, household bleach, sodium chloride, sodium hypochlorite solution, soap, socium nitrite, sodium salicylate, sodium thiosulfate, stibogluconate sodium, thiopental sodium, trichloroacetic acid, trisodium phosphate).

sodium acetylsalicylate. aspirin.

sodium acid carbonate. sodium bicarbonate.

sodium 5-allyl-5-(1-ethylbutyl)barbiturate. secobarbital.

sodium aminoarsonate. sodium arsanilate.

sodiumaminophenylarsonate. sodiumarsanilate.

sodium *p*-aminophenylarsonate. sodium arsanilate.

sodium anilinearsonate. sodium arsanilate.

sodium antimonyl tartrate. antimony sodium tartrate.

sodium antimonylgluconate. See stibogluconate sodium, def. 2.

sodium arsanilate ((4-aminophenyl)arsonic acid sodium salt; sodium aminoarsenate; sodium *p*-aminophenylarsonate; sodium aminophenylarsonate; sodium anilinearsonate). a derivative of arsanilic acid, it is a white, odorless, crystalline powder, $C_6H_4NH_2(AsO{\cdot}OH{\cdot}ONa)$, that is often hydrated with one or more water groups; it is soluble in water and slightly soluble in ethanol. It was formerly used as an antisyphilitic and to treat sleeping sickness. Sodium arsanilate has also been used as an anthelmintic in swine and poultry. It is highly toxic by ingestion, inhalation, and by subcutaneous route. It can cause blindness.

sodium arsenate (sodium metaarsenate). a clear, colorless, water-soluble, crystalline compound, $Na_3AsO_4{\cdot}12H_2O$. It is slightly soluble in ethanol and glycerol, and insoluble in ether. Sodium arsenate is extremely toxic by ingestion and inhalation. See also arsenic.

sodium arsenite (sodium metaarsenite). a whitish or light grayish powder that is soluble in water, slightly soluble in ethanol, and absorbs atmospheric carbon dioxide. It is a derivative of arsenic trioxide and is used, for example, as an antiseptic, insecticide, a contact herbicide (it causes rapid wilting), and by taxidermists in arsenical soaps. It is a confirmed human carcinogen. It appears to be deadly poisonous by all routes of exposure, although most human poisoning is by ingestion. It is teratogenic and has reproductive effects in laboratory animals. Sodium arsenite poisoning in humans is essentially that of arsenic poisoning, q.v. This salt has been shown to inhibit respiration and growth, and to upset normal mitosis in affected plants. It emits toxic fumes, As and Na_2O, when heated to decomposition. **sodium arsenite, liquid**. sodium arsenite in solution. It is deadly poisonous with activity similar to that of sodium arsenite. See also arsenic.

sodium aurothiosulfate. gold sodium thiosulfate.

sodium azide. an extremely toxic, unstable, explosive, white crystalline compound, N_3NA, that is soluble in water and liquid ammonia. It is used to make lead azide explosives, as a propellant used to inflate automotive safety bags, as a herbicide, agricultural pesticide, and to control fruit rot. Sodium azide is a potent vasodilator and has been used therapeutically as an antihypertensive. Symptoms of intoxication can include hypotension, tachycardia, palpitations, tachypnea, severe headache, hypothermia, and unconsciousness, usually followed by rapid recovery. Overdosage sometimes, however, causes convulsions and death by respiratory arrest or heart failure. It is a mutagen in plants and bacteria and a suspected carcinogen.

sodium benzoate. a white, combustible, water- and ethanol-soluble crystalline substance or granular powder, C_6H_5COONa. It is moderately toxic and can cause intestinal upsets and allergic reactions. It is used in chronic and acute rheumatism and as a liver function test.

While generally considered relatively safe for most individuals, it can cause adverse reactions in people who are sensitive to petrochemical derivatives. It is sometimes used as a preservative in food (the content in food is currently limited to 0.1% in the United States) and certain pharmaceuticals. *Cf.* benzyl alcohol.

sodium bicarbonate (baking soda; sodium acid carbonate). a white, water-soluble powder or lumps, with a slightly alkaline taste. It is soluble in water and ethanol. It slowly decomposes in moist air. It is often used to relieve the discomfort and edema of bee stings. It is used therapeutically in the home to relieve heartburn, indigestion, and the pain of peptic ulcer. If used frequently over a long period of time, it may cause cramps, weakness, fatigue, and vomiting. Its use is contraindicated for individuals with heart failure or a history of kidney disease. Sodium bicarbonate is used also as an antacid, a mouthwash, in cleaners, in the manufacture of effervescent beverages, baking powder, pharmaceuticals, sponge rubber, in fire extinguishers, and in the prevention of timber mold.

sodium bromate (bromic acid, sodium salt). a white, odorless, water-soluble crystalline substance, $NaBrO_3$, that is soluble in ethanol. It is an oxidizing agent and dangerous fire hazard when near organic material; and is toxic by ingestion. In addition to organic matter, it reacts violently with Al, As, C, Cu, F_2, P, S, oils, and metal sulfides. When heated to decomposition it releases toxic fumes of Na_2O and Br_-. Sodium bromate is used as an analytical reagent.

sodium bufalin-3-sulfate. See bufalin.

sodium cacodylate ([(dimethylarsino)oxy]sodium As-oxide; sodium dimethylarsonate). the sodium salt of cacodylic acid, it is a white, deliquescent, powdery or amorphous crystalline substance, $(CH_3)_2AsOONa \cdot 3H_2O$, that is soluble in water and ethanol. It is used as a herbicide and is toxic to terrestrial vertebrates by ingestion and inhalation. It was once used medicinally (e.g., in chronic skin diseases, leukemia) and is still used in veterinary practice as a hematinic. Of note is the fact that the safe parenteral dose (300 mg) in humans is substantially larger than the safe oral dose (60 mg). This is due to the rapid acidification of this salt in the stomach with the liberation of inorganic arsenic, probably as arsenic acid, which is then reduced to arsenous oxide (As_2O_3).

sodium calciumedetate. edetate calcium disodium.

sodium carbonate. **1:** a compound, Na_2CO_3, that occurs in nature only as the hydrate (thermonatrite) and the decahydrate (natron, natrite). It is an irritant to the skin, eyes, and mucous membranes. In addition to severe gastroenteritis, concentrated solutions may cause local necrosis of mucous membranes. It is a highly corrosive alkali found in many cleaning agents. In its various forms, it is used, for example, in the manufacture of sodium salts, soap, and glass; as a general cleanser; as a bleach especially of linen and cotton fabrics; in water softeners; and as a chemical analytic reagent. It has also been used therapeutically as an emetic, and as a skin cleanser. It neutralizes bee venom and can thus relieve the discomfort and edema of bee stings. Soda ash is the crude sodium carbonate of commerce. See also sodium carbonate, monohydrate; sodium carbonate, decahydrate; solvay soda; sal soda; sodium bicarbonate; sodium sesquicarbonate; clorox 2. **2:** a term used, sometimes casually in the literature, for any of a number of related compounds that should be more precisely designated. **anhydrous sodium carbonate**. solvay soda. **anhydrous sodium carbonate, technical**. soda ash.

sodium carbonate, decahydrate (nevite; soda). transparent crystals, $Na_2CO_3 \cdot 10H_2O$, soluble in water, insoluble in ethanol, that effloresce on exposure to air. It readily loses water at temperatures at or above its melting point, 32.5-34.5°C. It is used as a general cleaner, to bleach linen and cotton fabrics, and to wash textiles. The aqueous solution is strongly alkaline. Store in a tightly closed container in a cool location. See also sodium carbonate.

sodium carbonate, monohydrate (carbonate crystal; soda monohydrate; soda crystals). a nonflammable, odorless, crystalline powder or small crystals, $Na_2CO_3 \cdot H_2O$, with an alkaline taste, that are soluble in water and glycerol, insoluble in ethanol. It is used chiefly as a food additive, in cleaners, to clean and bleach textiles, in photography, as an intermediate in thermochemical reactions, and as an analytical reagent. See sodium carbonate, solvay soda.

sodium carboxymethylcellulose. carboxymethylcellulose sodium.

sodium cellulose glycolate. carboxymethylcellulose sodium.

sodium chlorate. a colorless, odorless, flammable, crystalline substance, $NaClO_3$, with a saline taste; it is a strong oxidant and bleach. It is a widely used herbicide and defoliant and has numerous other applications, being used, for example, in matches, explosives, flares, and pyrotechnics, and in the production of perchlorates. Plants treated with sodium chlorate and contaminated clothing worn by the applicator are highly combustible and constitute fire hazards. Poisoning of livestock by sodium chlorate has occurred both from the ingestion of treated plants and from accidental consumption of feed to which it was mistakenly added as sodium chloride. Cattle are sometimes attracted to foliage treated with sodium chlorate. Considerable quantities must be consumed before signs of toxicity appear. The minimum lethal dose is 1.1 g/kg body wt for cattle, 1.54-2.86 g/kg for sheep, and 5.06 g/kg for poultry. Ingestion results in the conversion of normal hemoglobin to methemoglobin.

sodium chloride (salt; common salt; table salt; sea salt; rock salt; halite). colorless, transparent, somewhat hygroscopic, salty tasting, crystals or white crystalline powder, NaCl, that is soluble in water and glycerol and slightly soluble in ethanol. An essential nutrient, it is generally considered moderately toxic. It is used as a condiment, in curing foods, and as an analytical reagent. Overindulgence may cause stomach irritation and vomiting; the most common systemic effect from ingestion of table salt is hypertension and existing hypertension can be exacerbated with renal injury. Acute poisoning by ingestion may also cause vomiting and dehydration. Intraperitoneal administration can terminate pregnancy. Teratogenic and reproductive effects have been induced in laboratory animals and there is some indication of mutagenicity.

sodium chloride aerosol. an aerosol is produced by the breaking of waves at sea and on shore. It is a common and seriously corrosive air pollutant in maritime districts.

sodium cyanate (cyanic acid sodium salt). a very toxic, white, crystalline powder or colorless, monoclinic crystals, NaOCN, soluble in water, ethanol, and ethyl ether. It is used as a nonselective herbicide, a chemical intermediate, in the study of sickle cell anemia, the preparation of certain medications, and heat treatment of steel.

sodium cyanide. a white, deliquescent, crystalline powder, NaCN, that is soluble in water and slightly soluble in ethanol. The aqueous solution is strongly alkaline, but decomposes rapidly on standing. It is extremely toxic by ingestion, inhalation, and percutaneous absorption. Symptoms of empoisonment are essentially those of cyanide poisoning. As with other soluble cyanides, it is among the fastest-acting poisons known. To be effective, treatment must begin promptly. Sodium cyanide is used chiefly in the extraction of ores, electroplating and in various manufacturing processes.

sodium cyclamate. a white, crystalline, nearly odorless powder, $C_6H_{12}NO_3S{\cdot}Na$, used in many countries as a nonnutritive sweetener, it is about 30 times sweeter than cane sugar. It is slightly toxic to laboratory rats and mice, the oral LD_{50} being 15.25 and 17.0 g/kg, respectively for these animals. It appears to be a promoter in the induction of bladder cancer in rodents. It is a questionable human carcinogen and possible mutagen. Teratogenic and male reproductive effects have also been seen experimentally in laboratory rats and mice.

sodium 5-(1-cyclohexenyl)-1,5-dimethylbarbiturate. hexobarbital sodium.

sodium dichromate (bichromate of soda; sodium acid chromate; sodium bichromate). a poisonous, red to orange, water-soluble, deliquescent, crystalline substance, $Na_2Cr_2O_7$. It is used as a chemical reagent, corrosion inhibitor, tanning agent, in electroplating, and as an oxidizing agent in the manufacture of pigments (e.g., dyes and inks) and many other organic materials. Sodium dichromate is a caustic and irritant compound; it is used therapeutically as a topical antiinfective agent. See also dichromate.

sodium dimethylarsonate. sodium cacodylate.

sodium edetate. edetate sodium.

sodium 5-ethyldihydro-5-(1-methylbutyl)-2-thiobarbiturate. thiopental sodium.

sodium fluoride. a white, crystalline powder, NaF, with a saline taste that is soluble in water and very slightly soluble in ethanol. When used as an insecticide, it is often dyed blue. It is toxic to humans and other terrestrial vertebrates by ingestion and inhalation; it is very toxic to laboratory rats. See also sodium fluoride poisoning, fluorides.

sodium fluoride poisoning. over exposure to sodium fluoride may cause conjunctivitis, nausea, retching or vomiting, eventual weakening of the heart, disturbances of renal function, and poor coagulability of blood.

sodium fluoroacetate (sodium monofluoroacetate; fluoroacetic acid, sodium salt; compound 1080; 1080; Fratol). a fine, white, odorless, tasteless, hygroscopic, nonvolatile, highly water-soluble, extremely toxic powder, $CH_2FCOONa$. It is nonspecific in its action and is apparently highly toxic to all terrestrial vertebrates, causing convulsions and ventricular fibrillation. Formerly used as a rodenticide in rural situations and as a repellant against rodents and predatory mammals such as the coyote (*Canis latrans*) and Arctic fox (*Alopex algopus*). It has often accidentally killed wildlife. Poisoning may result from inhalation or ingestion of the powder or ingestion of a solution of this compound. It blocks cellular metabolism, affecting all body cells, especially those of the central nervous system. Symptoms of intoxication may include vomiting, irregular breathing, auditory hallucinations, anxiety, paresthesia of the face, facial tics, irregular pulse, convulsions, and coma. Death may intervene due to cardiorespiratory collapse as a result of pulmonary edema or ventricular fibrillation.

sodium fluorosilicate. sodium hexafluorosilicate.

sodium fluosilicate. sodium hexafluorosilicate.

sodium formaldehydesulfoxylate (hydroxymethanesulfinic acid sodium salt; formaldehyde sodium sulfoxylate; formaldehydesulfoxylic acid sodium salt; sodium methanalsulfoxylate). a moderately toxic, water-soluble crystalline substance, CH_3NaO_3S, used in the treatment of mercury poisoning.

sodium hexacyanoferate(II). sodium ferrocyanide.

sodium hexafluorosilicate (sodium fluosilicate; sodium fluorosilicate; sodium silicofluoride). a very toxic, white, granular powder, Na_2SiF_6, used as an insecticide, rodenticide, mothproof of woolen fabrics, and as a pediculicide in veterinary practice. It yields 2NaF and SiF_4 in aqueous solution. Humans have died from exposure to this compound. It is sometimes used as an insecticide.

sodium hydrate. sodium hydroxide.

sodium hydroxide (caustic soda; sodium hydrate; soda lye; lye; white caustic). a white deliquescent solid, NaOH, usually in the form of beads or pellets, but sometimes ground or as flakes. It absorbs water and carbon dioxide from the atmosphere. NaOH is commercially the most important caustic. It has numerous uses in science and industry; it is used in many cleaning agents and some detergents, even in the home where it should be treated as very dangerous, especially to children. It is highly corrosive to moist tissue; a strong irritant to eyes, skin, and mucous membranes. Symptoms following ingestion are those of severe gastrointestinal distress, including intense pain in the mouth; difficulty in swallowing; gastrointestinal pain; emesis; and a weak, rapid pulse. Death may ensue due to shock or to asphyxia from glottic edema. Lesions may include laryngeal or glottic edema; corrosion, possibly with perforation of the upper gastrointestinal tract; late esophageal stenosis.

sodium hydroxide with lime. See soda lime.

sodium o-[(3-hydroxymercuri-2-methoxypropyl)carbamoyl]phenoxyacetate. Mersalyl.

sodium 4-(hydroxymercuri)-2-nitrophenolate. mercurophen.

sodium hydroxymethanesulfinate. sodium formaldehydesulfoxylate.

sodium hypochlorite. a compound with the formula NaOCl. **anhydrous sodium hypochlorite**. a hazardous, heat-sensitive, friction-sensitive, very explosive compound, NaOCl. It explodes on contact with formic acid at or above 55°C. It reacts violently with phenyl acetonitrile, cellulose, and ethylene imine. It also reacts with various amines and ammonium salts to yield explosive products (e.g., ammonium acetate, ammonium oxalate, aziridine, methanol). **aqueous hypochlorite** (clorox; eau de labarraque). **sodium hypochlorite, pentahydrate**. a highly unstable, pale greenish substance with a disagreeable, sweetish odor, $NaOCl{\cdot}5H_2O$. It is soluble in cold water; decomposes in hot water. It is a strong oxidizing agent and fire risk when in contact with organic material; it is also unstable in air unless mixed with sodium hydroxide. It is usually used and stored in solution. Sodium hypochlorite is a strong irritant. Ingestion produces acute gastrointestinal distress with corrosion of the mucous membranes, laryngeal edema, and perforation of the esophageal or gastric wall. Similarly, inhalation of the fumes may cause severe bronchial irritation with pulmonary edema. It can damage the eyes and skin as well. It is used, for example, as a bleach (e.g., of paper pulp, textiles), in water purification, as a fungicide and swimming pool disinfectant, germicide, and analytical reagent. See also calcium hypochlorite, hypochlorite poisoning.

sodium hypochlorite, pentahydrate. See sodium hypochlorite.

sodium hypochlorite solution, alkaline (antiformin). a clear, yellowish, strongly alkaline, sodium hypochlorite solution that contains 5.68 g of active chloride, 7.8 g of NaOH, and 32 g of Na_2CO_3. It is used as a deodorizer, disinfectant, and germicide. It is strongly corrosive and irritant to the skin, eye, and mucous membranes; See sodium hypochlorite.

sodium hypochlorite solution, diluted (modified Dakin's solution; Carrel-Dakin solution; surgical chlorinated soda solution. a colorless or slightly yellow solution with a faint chlorine-like odor that contains 0.45-0.50 g of sodium hypochlorite per 100 ml of water. Toxicity and action are that of sodium hypochlorite. It is used in medical practice as a topical anti-infective and in veterinary practice as an antiseptic for wound irrigation.

sodium hyposulfite. sodium thiosulfate.

sodium iodide. white, somewhat bitter-tasting, cubical crystals or powder, NaI or $NaI{\cdot}2H_2O$, that slowly turns brown in the atmosphere. It is soluble in water, ethanol, and glycerol. It is used, for example, as a solvent for iodine, in photography, as a feed additive, and an expectorant. Sodium iodide is nephrotoxic.

sodium lead hyposulfate. lead sodium thiosulfate.

sodium metaarsenate. sodium arsenate.

sodium metaarsenite. sodium arsenite.

sodium metavanadate. a nonflammable salt of vanadium with colorless, water-soluble, monoclinic, prismatic crystals or a pale green, crystalline powder, $NaVO_3$ or $NaVO_3{\cdot}4H_2O$. It is used in photography, in inks, fur dyeing, mordants, and a corrosion inhibitor in gas-scrubbing systems. Sodium metavanadate is highly poisonous by ingestion.

sodium methanalsulfoxylate. sodium formaldehydesulfoxylate.

sodium fusidate. an antibacterial, it is the sodium salt, $C_{31}H_{47}NaO_6$, of fusidic acid. It is very toxic (*i.v.*) to laboratory mice.

sodium monofluoroacetate. sodium fluoroacetate.

sodium nitrate (soda niter). colorless, transparent, odorless crystals, granular powder, or white or nearly white, opaque fused masses or sticks, $NaNO_2$. It is an oxidizing agent, is a fire hazard near organic material, and may explode when shocked or at high temperature (537°C). The natural, impure form is Chile saltpeter and the impure industrial form is called caliche. Sodium nitrate is used, for example, in solid rocket propellants, fertilizer, soldering flux, pyrotechnics, a color fixative, fertilizer, and as a reagent. It is used to enhance the taste of cured and various processed meats (e.g., bacon, bologna, frank-furters) and fish and as a preservative to prevent the growth of *Claustridium botulinum* in meats. It is used also as an antidote for cyanide poisoning; in the relief of angina pectoris; Raynaud's disease; asthma; and lead colic, spastic colitis, and similar conditions. Sodium nitrate is toxic by ingestion and, together with other nitrates, it can combine with natural stomach and food substances and produce carcinogenic nitrosamines. Nitrosamines have been found in nitrate-treated fish.

sodium nitrite. a white to slightly yellowish flammable powder, needles, or pellets, $NaNO_2$, formed by the thermal decomposition of sodium nitrate. It is soluble in water and slightly soluble in ethanol. It oxidizes in air and is a dangerous fire and explosion risk. Sodium nitrite has a bacteriostatic action and finds wide usage in curing fish and meat products. Sodium nitrite residues in foods have caused human fatalities. Note that meat tenderizer is nearly pure sodium nitrite. It is also used as an industrial corrosion inhibitor, and in bleaching and dyeing. It is carcinogenic to some laboratory animals; in some foods, especially (e.g., fried bacon), it is thought to be carcinogenic to humans. Sodium nitrite is an antidote for cyanide poisoning.

sodium oxalate (ethanedioic acid disodium salt). a water-soluble salt of oxalic acid that is insoluble in ethanol. It is a white, crystalline powder, $Na_2C_2O_4$, used in leather and textile finishing, pyrotechnics, blueprinting, and as a reagent. Sodium oxalate is toxic by ingestion; toxicity is essentially the same as that of oxalic acid. Ingestion of concentrated solutions may cause severe gastrointestinal distress with hematemesis, cardiac and CNS depression, and may be fatal. Ingestion of dilute solutions or of plant parts that contain this compound may cause muscular weakness and twitching and, in rare instances, may cause convulsions, coma, and death may See oxalate. Chronic poisoning is marked by urinary calculi and hypocalcemic tetany.

sodium prussiate yellow. sodium ferrocyanide.

sodium pump. See pump.

sodium pyroborate. sodium borate (See borate).

sodium rhodanate. sodium thiocyanate.

sodium selenate. a toxic, very water-soluble, decahydrate, white crystalline substance, $Na_2SeO_4{\cdot}10H_2O$, used as a horticultural insecticide on nonedible plants. Anhydrous sodium selenate is extremely poisonous.

sodium selenite. white, tetragonal crystals, Na_2SeO_3 or $Na_2SeO_3{\cdot}5H_2O$, that are freely soluble in water, insoluble in ethanol. Sodium selenite is extremely toxic by ingestion. It is used as a reagent in microbiology, in glass manufacture, to decorate porcelain, and in testing the germination of seeds.

sodium sesquicarbonate. a white, nonflammable, mildly alkaline substance with monoclinic crystals, $Na_2CO_3{\cdot}NaHCO_3{\cdot}2H_2O$, that is less alkaline than sodium carbonate, is soluble in water, less so in ethanol. It is an irritant to the skin, eyes, and mucous membranes. It is used as a food additive, detergent, bath crystals, a general purpose cleaning agent, a water softener, and in the tanning of leather.

sodium subsulfite. sodium thiosulfate.

sodium sulfocyanate. sodium thiocyanate.

sodium sulfocyanide. sodium thiocyanate.

sodium TCA. sodium trichloroacetate.

sodium tellurate (IV) (sodium tellurite). 8641 Merck.

sodium tellurite. sodium tellurate (IV).

sodium tetraborate. sodium borate (See borate).

sodium thiocyanate (sodium sulfocyanate; sodium sulfocyanide; sodium rhodanate; sodium rhodanide). a hygroscopic compound with colorless, deliquescent crystals or white powder that is soluble in water or ethanol; it is affected by light. This compound was formerly given orally as a hypertensive agent but is now rarely used because it is very toxic. It is a general metabolic depressant, but has its most prominent effects on the brain and heart. Symptoms of intoxication may include disorientation, weakness, low blood pressure, confusion, psychotic behavior, muscular spasms, convulsions, and death.

sodium thiosulfate (sodium subsulfite; hypo; sodium hyposulfite). a powder or white, translucent crystals, $Na_2S_2O_3 \cdot 5H_2O$ (also available as the anhydride), that is soluble in water and oil of turpentine, nearly insoluble in ethanol; it is deliquiscent in humid air. It has a cool taste and a bitter aftertaste. Sodium thiosulfate is an antidote to cyanide poisoning in human and veterinary medicine; it acts by combining enzymatically with cyanide to yield thiocyanate. It also used in veterinary practice as a "general detoxifier." Among many additional uses are the removal of chlorine from solution, the bleaching of wood pulp (as antichlor), photography, leather making, extraction of silver from ores, and in dyes. Toxicity is low and it is poorly absorbed from the gastrointestinal tract.

sodium trichloroacetate (sodium TCA). the sodium salt of trichloroacetic acid, it is a water-soluble, very corrosive compound, CCl_3COONa, used as a herbicide and pesticide. It is absorbed by the roots of plants and is used chiefly to control grasses (Family Gramineae) such as Agropyron repens (couch grass) and *Avena fatua* (wild oats) that infest crops such as *Beta vulgaris* (sugarbeet), peas and *Brassica oleracea* (kale). It is used also as a protein precipitant and a histological decalcifier and fixative.

sodium-potassium pump. See pump.

soft coral. any anthozoan of the Order Alcyonacea, q.v. See also *Palythoa*, palytoxin.

soft detergent. any cleaning agent that readily breaks down in nature.

soft palate. the soft, boneless, posterior (or dorsal) portion of the roof of the mouth in mammals. It is made up entirely of mucous membrane and fibrous connective tissue.

soft-shelled clam. *Mya arenaria*.

soil. a mixture, usually stratified, of mineral matter from weathered rock together with organic components due to the decomposition or partial decomposition of vegetation. Both classes may form a soil *in situ* or, more commonly may be transported by wind or water to a particular site. Soil often exhibits extreme variability in composition, chemical and physical properties and gross structure even within a small area. A simple description or classification of soil to be sampled, e.g., based on soil profile analysis is insufficient to ascertain its toxicologically relevant properties. Variations in properties such as Ph within soil types can decisively alter their behavior with respect to persistence, mobility, extractability, and bioavailability of the environmental chemicals contained therein. In nearly all studies that involve chemical analyses of soils, the variability among results is high. This generally reflects true variability or inadequate sampling rather than chemical analytical errors or limitations. Furthermore, whereas biota usually make up well-defined statistical units, soils rarely do. Beyond this, concentrations of toxic chemicals in soils may bear little immediate relationship to their ecotoxicological significance or bioavailability. Thus, knowledge of a chemical's presence or concentration in soil alone is unlikely to provide useful information regarding any aspect of environmental toxicology, risk assessment, or management.

soil cover. a layer of uncontaminated soil that overlies contaminated soil.

soil sterilant. a pesticidal agent used to destroy all biota in soil for an extended period of time.

soil-to-plant transfer factor. the ratio of radionuclide activity per unit mass (usually on the basis of wet weight) of plant tissue to activity per unit mass of soil.

solan. an amide herbicide applied to foliage. It produces a localized or general necrosis, depending upon the dosage. It also inhibits growth.

Solanaceae (nightshade or potato family). a family of herbs and shrubs that includes about 85 genera and nearly 2000 species. A number of genera such as *Aconitum* (monkshood), *Atropa* (nightshades), *Hyoscyamus* (henbanes), *Mandragora* (mandrakes), *Aconitum* (mandrake), *Datura* (jimsonweeds), and *Nicotiana* (tobaccos), *Solanum* (nightshades, potato), are dangerously toxic.

solanaceous. of or pertaining to plants of the family Solanaceae or to the pharmacologics derived from them.

solanaceous alkaloid. any of a number of mostly toxic alkaloids found in plants of the family Solanaceae. Examples are aconitine, scopolamine, hyoscyamine. See also Solanaceae.

solandra. any plant of the genus *Solandra*.

Solandra (trumpet flower; chalice vine; solandra). a small genus of poisonous, tropical American shrubs (Family Solanaceae). They are tall with shiny leathery leaves and single large, white to yellow, showy flowers.

solandrine. pseudohyoscyamine.

solanid-5-ene-3β,12α-diol. rubijervine.

solanid-5-en-3β-ol. solanidine.

solanidine (solanid-5-en-3β-ol; solatubine). an extremely poisonous, neurotoxic, steroid alkamine that occurs in plants of the genera *Lycopersicon* (tomato) and *Solanum* (potato, nightshades). It is the biologically active aglycone of solanine, q.v.

solanine (solatunine). an extremely toxic alkaloid and steroid glycoside, $C_{45}H_{73}NO_{15}$, found in a number of species of *Solanum* and *Lycopersicon*. Solanine, which is not readily absorbed, yields a neurotoxic steroid alkamine (solanidine), the biologically active aglycone, and a sugar (solanose) on hydrolysis. Solanidine is considered the most toxic moiety. Symptoms and lesions of poisoning by solanine are similar among various mammals, but vary specifically with the relative amounts of the irritant glycoalkaloids and the released alkamines that reach the animal's system. Livestock have been killed by grazing on vines of solanine-containing plants and humans have died from ingesting potatoes (i.e., the tubers) with "sun burned" spots or with sprouts. Spoiled potatoes may contain more solanine than normal. Solanine has been used in agriculture as an insecticide and was formerly used to treat bronchitis, epilepsy, and asthma. See solanum poisoning.

solanine alkaloid. any of a class of solanine-like steroidal alkaloids that occurs in nightshades (*Solanum* spp.), the unripe fruit of some groundcherries (*Physalis* spp.), jessamines of the genus *Cestrum*, and various other plants.

solanine poisoning. poisoning by ingestion of plant materials that contain solanine. The chief proximate poison is solanidine, a product of the hydrolysis of solanine. See also solanine, solanum poisoning. Symptoms vary with source (as other toxicants may be involved). See especially potato poisoning, solanum poisoning.

Solanum (nightshades; potato; etc.). a largely tropical genus of herbs, shrubs, trees, and vines (Family Solanaceae); some species are highly toxic. These plants commonly have prickly veined leaves and white, purple, or yellow flowers. Many are ornamental, some are vegetables (potato, tomato, eggplant, Jerusalem cherry), others have medicinal value. Toxic species contain the steroid alkaloid, solanidine; the steroid glycoside, solanine; and a variety of other glycoalkaloids. Solanine, which is not readily absorbed, releases on hydrolysis an easily absorbed neurotoxic steroid alkamine aglycone (solanidine). Both fresh and dry plant material are toxic, thus hay containing poisonous *Solanum* spp. is hazardous to livestock. Symptoms and lesions produced by the various species are similar, but vary with the relative amounts of the particular irritant glycoalkaloids and alkamines that reach the animal's system. Concentrations of nitrate that are toxic to livestock have also been reported. See also solanum poisoning. ***S. americanum*** (American nightshade; "black nightshade"). indigenous to North America, it shows only minor differences from the introduced *S. nigrum*, both of which are common throughout the eastern United States, extending as far west as Texas and North Dakota. *S. americanum* commonly occurs in rocky or dry open woods, thickets, or shorelands. *S. nigrum*, however, most often occurs on disturbed soils and waste areas. ***S. carolinense*** (horse nettle; wild tomato; bull nettle; sand briar). all parts are toxic; it is most poisonous at maturity. Ingestion of the fruit can be fatal. Symptoms include pyrexia, sometimes headache, a prodromal scratchy sensation in the oropharynx followed by nausea, vomiting, anorexia, and diarrhea. The green fruit, if ingested, produces gastrointestinal distress with abdominal pain, and circulatory, and respiratory depression. The active principle is solanine. ***S. dulcamara*** (European bittersweet; climbing nightshade). a shrubby vine native to Europe; introduced and widely naturalized in North America. Leaves and berries are toxic.

Ingestion produces a prodromal scratchy sensation in the oropharynx followed by abdominal pain, sometimes headache, anorexia, nausea, vomiting, diarrhea, shock, and respiratory depression. Cattle, horses, and sheep have been poisoned by ingestion of this plant. The berries, red when ripe, are toxic and have caused fatalities in children. *S.* ***eleagnifolium*** (silverleaf nightshade; white horse nettle). a plant of dry, open woods, prairies, and disturbed areas, mostly in the southwestern United States. Cattle have been severely poisoned by ingestion of as little as 1 g/kg body weight of this plant and it has, at times, caused substantial losses among cattle. Sheep are more resistant; goats are even more resistant. ***S. esuriale***. ingestion of this plant causes calcinosis in herbivores. See enzootic calcinosis. ***S. intrusum*** (garden huckleberry; wonderberry). a cultivated species that is very closely related to *S. nigrum*. The fruit appears to be nontoxic. ***S. malacoxylon*** (glaucophyllum). a plant indigenous to Argentina and Brazil, it causes calcinosis in herbivores. ***S. melongena***. a species that bears lavender flowers and purple fruit. ***S. m. esculentum*** (eggplant). a food plant, the vines contain solanaceous alkaloids and may be toxic to livestock. ***S. nigrum*** (black nightshade; deadly nightshade; common nightshade; garden huckleberry). an annual weak-stemmed plant that often lies on the ground, but sometimes grows erect to 2/3 m. The leaves are highly toxic as are the berries, especially when green. Symptoms of intoxication following ingestion are chiefly those of gastrointestinal poisoning and include abdominal pain, nausea, vomiting, diarrhea, and circulatory and respiratory depression. Numerous poisonings have been reported though some, perhaps many, of these poisonings in the United States were due to *S. americanum*, q.v. ***S. pseudocapsicum*** (jerusalem cherry). a small, common, shrubby, potted plant with bright red berries. The fruit is toxic, causing gastrointestinal symptoms, convulsions, and respiratory and central nervous system depression. ***S. sodomeum*** (apple of Sodom; popolo). a shrubby, spiny, toxic weed native to the Mediterranean region and now common in Hawaii. ***S. torvum***. contains substances that can induce calcinosis in herbivores. See enzootic calcinosis. ***S. triflorum*** (three-flowered nightshade; cutleaf nightshade). an annual herb, native to the prairies of North America, now a weed common to disturbed soils and waste areas of British Columbia and most of the United States. It contains solanaceous alkaloids, and is known to be toxic to livestock. ***S. tuberosum*** (potato; Irish potato). a tropical and subtropical perennial herb now cultivated widely in temperate regions as a vegetable. The vines, sprouts, peelings, and sunburned or spoiled potatoes have poisoned livestock on many occasions; human poisoning has also been validated. These plants contain solanine. The highest concentrations occur in the green parts, especially sprouts and the sun-greened skin of the tuber. Symptoms and lesions are generally those of acute solanum poisoning, severe GI symptoms; headache; cold, clammy skin; circulatory and respiratory depression. ***S. villosum*** (hairy nightshade). a toxic annual weed that is common in the eastern United States and locally in the west.

solanum alkaloid. solanidine or any solanidine-like steroid alkaloid from *Solanum* species.

solanum poisoning. poisoning by any of various species of plants of the genus *Solanum*, q.v. It is essentially a form of solanine poisoning. Poisoning is almost always acute, but not always fatal; chronic poisoning is known, however. **solanum poisoning, acute**. symptoms and lesions are similar for the various species of plants and include the effects of the gastroirritant glycoalkaloids and the neurotoxic effects of one or more alamines such as solanidine. Symptoms of gastrointestinal irritation range from slight to severe. They include nausea, vomition, anorexia, anorexia, abdominal pain, constipation or diarrhea, sometimes with bloody feces. Neurotoxic effects may include apathy, drowsiness, and even stupefaction; tremor; dyspnea; salivation; and progressive debilitation (weakness, paralysis, prostration, and coma). Death (from paralysis) or recovery occurs within a few hours. **solanum poisoning, chronic**. the effects of chronic poisoning of cattle (established for *Solanum carolinense*) include anorexia, emaciation, constipation and ascites.

solatubine. solanidine.

solatunine. solanine.

solder. a metallic compound used to seal the joints between pipes. Until recently, most solder contained 50 percent lead.

soldierfishes. common name for fish of the family Holocentridae.

solenocyte. See flame cell.

solenopsin A. one of several alkaloids that occur in the venom of fire ants of the genus *Solenopsis*, q.v.

Solenopsis (fire ants, in part). a genus of large, bright red biting ants (Family Formicidae). A number of "fire ants" that aggressively attack humans are included in this genus. The bites are multiple, clustered, and burn intensely, forming a local pustule; the venom has neurotoxic, hemolytic, insecticidal, and antibiotic properties, producing systemic effects similar to those of *Latrodectus matans* (black widow). An allergic reaction sometimes occurs and a very few bites can be lethal to a sensitive individual. A number of species are indigenous to North America, *S. geminata* having the widest range. Chlordecone has been used effectively against fire ants; its use in the United States is, however, severely restricted. ***S. geminata***. an omnivorous ant that feeds on other small arthropods, seeds, flowers, and vegetables. There are 2 or more worker castes,. The body length rangies from about 1-6 mm. Individuals are dull yellow to red or black. The head is large, the jaws (usually lacking teeth) are incurved; the waist (pedicel) has two segments; the legs are long; and fine "hair" occurs chiefly on the head and abdomen. This ant may be found in fields, woodlands, and open areas with moist soil from Florida and the Gulf states to the Pacific Coast, northward to British Columbia. While these ants sometimes damage young plants, they are rarely a threat to established crops. Indeed, as a predator of other insects, their presence may at times be beneficial to crops. ***S. saevissima richteri***. a South American form that has been introduced into the southern United States where it is an established and iserious pest.

solenostele. See stele.

solid aerosol. See aerosol.

solid surf clam. *Spisula solidissima*.

solid-state carcinogen. See carcinogen.

solid waste. non-liquid, non-soluble materials in sources such as household garbage; municipal, commercial, and industrial wastes; sewage sludge, agricultural refuse, demolition wastes, mining residues. Such waste is usually chemically complex, and sometimes toxic or otherwise hazardous. Technically, waste liquids and gases in containers are solid wastes.

Solidago (goldenrod). a genus of erect herbs (Family Asteraceae), usually with golden (occasionally white) composite flowers. There are many species among which hybridization is common. Some species are considered toxic to livestock, due at times to the presence of toxic concentrations of nitrates. ***S. virgaurea***. an aromatic species that occurs in Europe and North America. It has a diuretic action.

soln. abbreviation for solution.

soln potassium iodohydrargyrate. potassium triiodomercurate(II) solution.

Solomons copperhead. *Denisonia par*.

Solubacter™. triclocarban.

solubility. **1a:** the capacity or tendency, of one substance (the solute) to mix or blend uniformly with another (the solvent). **1b:** the extent of such mixing under specified conditions. The solubility of a given solute varies among solvents and and is affected by factors such as temperature and pH. Solubility is the main factor that affects the rate of absorption of a xenobiotic by a living organism and hence (in part) its toxicity. Insoluble salts and ionized compounds are poorly absorbed, while lipid-soluble substances are generally readily absorbed, even through intact skin. Unless a xenobiotic is soluble in body fluids or is converted to a soluble form in the body, it will not be poisonous.

solubility limit. the maximum concentration of a chemical that will dissolve in water or another specified liquid.

solubilize. to render soluble.

soluble. capable of being dissolved.

soluble toxin. exotoxin (See toxin).

solute. the substance or substances dissolved in

another substance (the solvent). The solute is uniformly dispersed in the solvent, either as molecules or ions. The solvent together with the solutes comprise a mixture termed a solution.

solute potential. osmotic potential.

solution. **1:** a uniformly dispersed or homogenous mixture (at the molecular or ionic level) of two or more substances regardless of phase (gas, liquid, solid). One or more solutes dispersed as molecules or ions in a solvent. Such a system is usually a liquid, but may be a gas, liquid, or solid. The solutes may be gas, liquid, or solid and although the solvent is usually liquid, it may be solid. **2:** the process of dissolving. **3:** a loosening or separation.

solution potassium iodohydrargyrate. potassium triiodomercurate(II) solution.

solvay soda. pure, anhydrous sodium carbonate. It is a white, odorless, hygroscopic powder or small crystals with an alkaline taste. It gradually absorbs 1 molecule of water (about 15% of total mass) on exposure to air. It is soluble in glycerol, insoluble in ethanol. See sodium carbonate. See sodium carbonate. See soda ash.

solvent. a substance (usually a liquid) capable of dissolving one or more other substances (solutes) with the formation of a uniformly dispersed mixture (solution). **industrial solvent**. a solvent used in bulk in an industrial process. Such solvents are of major toxicological concern, as many are irritant or toxic to humans, especially to workers who are repeatedly exposed to a given solvent or its fumes. Examples are hexane, methanol, methyl isobutyl ketone, methylene chloride, naphtha, *N*-nitrosodimethylamine, propylene, propylene glycol, toluene. **organic solvent**. an organic compound or mixture (usually liquid) used as a solvent. Included are aliphatic and aromatic hydrocarbons, alcohols, aldehydes, ketones, chlorinated hydrocarbons and carbon disulfide. The vapors of organic solvents may be toxic. Exposure can occur by contact with the liquid or, in most cases with the fumes or vapor. Contact may cause skin irritation, defatting, or dermatosis — especially with long-continued usage. Exposure to solvent vapors is usually by inhalation, although skin absorption may also occur. Most organic solvent vapors have an anesthetic effect on the CNS; some are nephrotoxic and/or hepatotoxic; and some can damage the erythropoietic system. Additional specific effects are known.

solvent blue 8. methylene blue.

solvent partitioning. a method whereby a solute is distributed between two immiscible liquids (e.g., octanol and water). At equilibrium, the solute will be apportioned between the two liquids according to its relative solubility in the two liquids. The ratio of distribution of solvent between the two liquid layers is constant for any two immiscible solvents and is termed the partition coefficient. Thus a partition coefficient of 1 for a given system indicates that the solute is equally distributed between the two solvents.

soman (GD; methylphosphonofluoridic acid-1,2,2-trimethylpropylester; O-1,2,2-trimethylpropyl methylphosphonofluoridate; *o*-pinacolyl methylphosphonofluoridate). a colorless liquid, $(CH_3)_3CCH(CH_3)OPF(O)CH_3$, with colorless fumes. It is an organophosphate that is highly toxic by all routes of exposure. Even brief exposure to the gas can be fatal; the lethal dose for humans is 0.01mg/kg body weight. It is one of a class of volatile, liquid anticholinesterases used as warfare agents. Soman reacts irreversibly with the cholinesterase (ChE), permitting an accumulation of acetylcholine at nerve endings that can be rapidly fatal. Symptoms of intoxication may include miosis, salivation, lacrimation, twitching and fasciculations, diarrhea, frequent urination, repetitive convulsions, coma, respiratory failure and death. Pretreatment of laboratory rats with muscarinic antagonists such as atropine or benzactyzine or anticonvulsants such as diazepam modifies or inhibits these convulsions.

somasthenia. somatasthenia.

somatasthenia (somasthenia). chronic physical weakness, weariness, and fatigability.

somatic. pertaining to the body or to any nongerminal organ, tissue, cell, structure, or process.

somatic nervous system (voluntary nervous system). that part of the vertebrate peripheral nervous system comprised of motor neurons that control skeletal (voluntary) muscles.

somatotrophin, somatotropin (adenohypophyseal growth hormone; anterior pituitary growth hormone; growth hormone; GH; hypophyseal growth hormone; phyone; pituitary growth hormone; somatotropic hormone; STH). a species-specific anabolic protein hormone produced by the acidophile cells of the anterior lobe of the pituitary gland. It promotes body growth; stimulates protein synthesis; promotes lipid mobilization; inhibition of glucose utilization; and stimulates protein synthesis. See also hormone, pituitary gland; somatotropic hormone.

somatotropic hormone. somatotrophin.

Somnibel™. nitrazepam.

somnifacient. **1:** sleep inducing; hypnotic. **2:** an agent that induces sleep.

Somniosus microcephalus (Greenland shark). a poisonous shark of the Arctic portion portions of the Atlantic Ocean, the North Sea east to the White Sea and west to the Gulf of St. Lawrence, Greenland. Ingestion of liver or musculature can produce severe effects. See elasmobranch poisoning.

Somnite™. nitrazepam.

Sonchus (sow thistles). a genus of plants (Family Compositae) in which toxic concentrations of nitrate have been reported.

songo. *Dendroaspis angusticeps*, *D. polylepis*.

songwe. *Dendroaspis polylepis*.

sonitus (tinnitus aurium). subjective noises (e.g., ringing, tinkling, cricket sounds) in the ear.

Sonoran coral snake. *Micruroides euryxanthus*.

soot. any brown to black dust or powder with a high carbon content produced by the incomplete combustion of any carbonaceous material. It is a mixture containing variable amounts of organic tars (sometimes more than 50% by wt.), colloidal carbon, and refractory inorganic materials depending upon the composition of the combusted material and the conditions of combustion. Soot has physical and chemical characteristics that distinguish it from carbon black. Long exposure to soot via direct contact or by inhalation can result in pulmonary tumors and cancer of the skin or scrotum.

soot wart. scrotal cancer.

sopa fish. a species of globefish (Family Diodontidae). Tobriand Islanders are known to commit suicide by eating the raw gall bladder.

Sophora (frijolito). a genus of ornamental deciduous or sometimes evergreen trees (Family Fabaceae, formerly Leguminosae) that contain the alkaloid quinolizidine. Among the more commonly encountered are *S. japonica* (Japanese pagoda tree), *S. secundifolia*, *S. tetraptera*, and *S. microphylla*. ***S. secundiflora*** (burning bean; mescal bean). an evergreen shrub or tree that grows to a height of about 12 m. It contains neurotoxic quinolizidine alkaloids. Poisoning impairs skeletal muscle tone; principal symptoms are nausea, emesis, dizziness, a sensation of sweating, and sometimes clonic convulsions. Death is by paralysis of respiratory muscles.

sophorine. cytisine.

soporiferous. inducing deep sleep.

soporific. **1:** causing or inducing deep sleep. **2:** any agent that causes drowsiness or induces sleep.

soporous. associated with or affected with coma or profound sleep.

sorbefacient. **1:** promoting absorption. **2:** an agent that promotes absorption.

sorb. to absorb or adsorb.

sorbent. **1:** having the capacity to absorb or adsorb. **2a:** a substance or material that absorbs or adsorbs. **2b:** a substance that can absorb and retain a contaminating substance; includes absorbents and adsorbents.

sorghum. See *Sorghum*.

Sorghum (sorghum; Johnson grass; Columbus grass; sudan grass). a genus of grasses (Family Gramineae) that have been cultivated in Africa and Asia since antiquity. Hundreds of varieties are known, but are not well characterized genetically. They are thus usually divided into two categories: (1) the cultivated (grain) sorghums (See *S. vulgare*) and sudan grass which are annuals, and (2) Johnson grass, a weedy perennial. Sudan grasses produce hydrocyanic acid in varying concentration, depending upon the strain and growing conditions. Conditions that do not support normal growth usually increase cyanide formation. Soils high in nitrogen may yield a plant with highly toxic levels of cyanide. Livestock pastured in fields with stunted or immature sudan grass, especially if it is dark green in color, may be at risk. ***S. almum***. a perennial sorghum originating in South America. A forage crop, it has sometimes caused serious mortality in livestock. This species hybridizes with *S. halepense*, producing very toxic seeds. ***S. halepense*** (Johnson grass; Egyptian millet). native to Eurasia, it is a locally common noxious weed in old fields and waste areas in many parts of the United States. It has sometimes produced serious mortality of cattle, especially in a number of southern states. Elsewhere it is sometimes used as a forage and hay crop. ***S. vulgare*** (cultivated (grain) sorghums). a species that includes a large variety of cultivated sorghums, many of which are unnamed. Included are the saccharin (forage) sorghums (sweet or cane sorghums, sorgo), grain sorghums (including hegari, kafir corn, and milo), and broom corn. Many varieties are unnamed. Sorghum toxicity of the seeds and foliage is due to the cyanogenetic glycoside, dhurrin (q.v.), and symptoms of poisoning in livestock are those of cyanide poisoning. Cattle suffer the most fatalities among livestock. The saccharin sorghums may also accumulate toxic levels of nitrates, and sudan grass has been implicated in photosensitization of sheep. ***S. v. sudanense*** (sudan grass). a variety that is sometimes toxic; a photosensitizer.

sorghum forage poisoning. poisoning of livestock by certain species and varieties of *Sorghum*, q.v.

sorgo. *Sorghum vulgare*.

sorption. **1a:** the process or state of being sorbed. **1b:** absorption or adsorption. **2:** the process of soaking up (absorbing) or attracting substances; a process used in many pollution control systems.

sorrel. a common name most appropriately applied to *Oxalis pes-caprae*, but often used in reference to dock and certain other species of *Rumex*. The latter usage has caused some confusion in reports of livestock losses in the United States.

sorrel salt. potassium binoxalate.

soursob. *Oxalis pes-caprae*.

South African mamba. *Dendroaspis augusticeps*.

South African spitting cobra. *Hemachatus haemachatus*.

South American freshwater stingray. *Potamotrygon motoro*. See also *Potamotrygonidae*.

South American rattlesnake. *Crotalus durissus*.

Southeast Asian banded krait. *Bungarus multicinctus*.

southern copperhead, Southern copperhead. *Agkistrodon contortrix*.

southern coral snake, Southern coral snake. *Micrurus frontalis*.

Southern mole viper. *Atractaspis bibronii*.

Southern Pacific rattlesnake. *Crotalus viridis helleri*.

Southern rattlesnake. *Sistrurus miliarius*.

sow thistle. See *Sonchus*.

soya leaf necrosis. See *Rhizobium japonicum*, rhizobotoxin.

soybean. *Glycine max*.

sp. abbreviation for species, singular.

sp. gr. specific gravity.

SP104. (—)-Δ^1-3,4-*trans*-tetrahydrocannabinol.

Spanish broom. a common name of *Sparteum junceum* (*Cytisus scoparius*) and *Genista hispanica*.

Spanish chamomile. pyrethrum.

Spanish fly. **1:** *Lytta (Cantharis) vesicatoria*, a European beetle which is an important source of cantharidin. **2a:** cantharides. **2b:** processed or unprocessed extract of *L. (Cantharis) vesicatoria*. **3:** cantharis.

Spanish tea. *Chenopodium ambrosioides*.

Spanish vetch. *Lathyrus clymenum*.

spanon. chlordimeform.

Sparidae (porgies, sheepshead, sea bream, scups). a family of deep-bodied, mostly tropical and subtropical percomorph fishes (Class Osteichthyes). The mouth is small, terminal, and bears human-like incisors or canines in the front and molars in the sides of the jaws. They are mostly marine fish typically confined to inshore waters; some occur in brackish or freshwater. See *P. pagrus*, *Pagellus erythrinus*, ciguatera.

sparteine ([7*S*-(7α,7aα,14aα,14aβ)]-dodecahydro-7,14-methano-2*H*,6*H*-dipyrido[1,2-a:1′,2′-e][1,5]diazocine; *l*-sparteine; lupinidine). a toxic, viscous liquid, tetracyclic alkaloid extracted especially from the tops of *Sparteum Junceum* (*Cytisus scoparius*) but also from *Anagyris foetida*, *Lupinus luteus*, and *L. niger*. It is cardiotoxic with a digitalis-like action. **sparteine sulfate**. a crystalline compound used therapeutically as an oxytocic and formerly as a cardiotonic drug.

***l*-sparteine**. sparteine.

Sparteum junceum (*Cytisus scoparius; Sarothamnus scoparius*; weavers' broom; Spanish broom; Scotch broom). a leguminous shrub, the seeds and leaves of which are toxic and contain two toxic principles, sparteine and cytisine. Symptoms of intoxication include a weak pulse, hypotension, intestinal paralysis, and muscular weakness. The active principle is sparteine.

spasm (spasmus; muscle spasm). **1:** a sudden (sometimes violent) uncontrolled muscular con traction, either of a muscle, or group of muscles, usually accompanied by pain and awkward involuntary movement. Spasms may be caused by various factors, such as muscular fatigue or emotional stress. **2:** increased muscular tone and shortening which cannot be relieved voluntarily. **3:** a sudden, transitory constriction of a duct, canal, or orifice. **canine spasm**. risus sardonicus. **clonic spasm**. alternate involuntary rigidity and relaxation of muscles. **cynic spasm**. risus sardonicus. **fixed spasm**. permanent rigidity of a muscle or set of muscles. **intention spasm**. a spasm that occurs when voluntary movement is attempted. **massive spasm**. a spasm involving most of the body musculature. **respiratory spasm**. a spasm that involves the muscles of respiration. **tetanic spasm** (tetanus spasmus). **1:** a spasm that occurs in tetanus. **2:** tonic spasm. **tonic spasm** (entasia; entasis). a continuous muscle spasm. **tonoclonic spasm**. a convulsive spasm. **torsion spasm**. a spasmodic twisting of the trunk and pelvis. **toxic spasm**. a spasm due to the action of a poison.

spasmatic. spasmodic.

spasmo-. a prefix denoting spasm.

spasmodic (spasmatic). **1:** of, or pertaining to, spasm. **2:** pertaining to a sudden intermittent symptom.

spasmogen. a substance that induces spasms.

spasmogenic. **1:** causing or inducing spasms. **2:** pertaining to the production of spasms.

spasmolysant. **1:** arresting or relieving spasms or convulsions. **2:** an agent that relieves spasms or convulsions.

spasmolysis. the arrest, relief, or elimination of spasm or convulsion.

spasmolytic. **1:** pertaining to spasmolysis. **2:** antispasmodic; of or pertaining to a substance that relieves or eliminates smooth muscle spasms.

spasmotin. a poisonous ecbolic and acid principle, $C_{20}H_{21}O_9$, from ergot, which has oxytocic activity.

spasmus. spasm.

spastic. **1:** referring to, of the nature of, or characterized by spasm(s) or spasticity; **2:** a person or animal exhibiting spasticity.

spasticity. **1:** a condition marked by spasms. **2:** a condition marked by hypertonicity (excessive muscle tension; stiffness; rigidity) of muscles with heightened deep tendon reflexes and rigid limbs; awkward movements. It is often produced by an upper motor neuron injury.

Spatangoida (heart urchins). an order of echinoderms (Class Echinoidea). The test is elongate, and somewhat heart-shaped; the mouth near the narrow end; the anus, at the broad end. Aristotle's lantern is absent. *Enchinocardium* is a common, burrowing, North American genus.

speargrasses. *Poa* spp. See Claviceps.

specialization. **1a:** the evolution of a special adaptation that adapts an organism to a particular environment. This can result in extinction, locally or generally, of the species if the environment changes considerably. **1b:** a term applied to individuals or classes of individuals within a population that serve different functions as seen often in social insects, for example. **2:** physiological (biological) specialization. The existence in a population of forms or phases (physiological forms) that, although identical in appearance, differ biochemically from one another. It is especially important in plant pathology, when a parasite attacks only one form of a species or variety. *Cf.* specificity, selectivity.

species-dependent. of or pertaining to a characteristic, behavior, or response that is specific to a given species, although it may not be invariably present or observed in the species. See also species-typical, specific (species-specific).

species-specific. See specific.

species-specific behavior. **1a:** any behavior pattern that is performed by all members of a given species under the same conditions and which is not modified by learning. **1b:** behavior that is specific to and characteristic of a given species. See also species-typical.

species specificity. See specificity.

species-typical. of or pertaining to a characteristic, behavior, or response that is typical of a given species; one that is neither restricted solely to the species nor invariably observed in the species. *Cf.* specific (species specific). See also species-dependent, specific (species-specific).

specific activity. the ratio of radioactive atoms to total atoms of the same element.

specific cholinesterase. acetylcholinesterase.

specific gravity (sp. gr.; relative density). the ratio of the density of a material to the density of some standard material such as water at a specified temperature (e.g., 4°C), or for gases or air at standard conditions of temperature and pressure.

speckled loco. *Astragalus lentiginosus*, a true locoweed.

speckled pit viper. *Trimeresurus wagleri*.

speckled rattlesnake. *Crotalus mitchellii*.

spectacled cobra. a common name applied to *Naja naha* or to *Naha n. naha*.

spectral binding constant. See K_S.

"speed." **1:** dextroamphetamine sulfate. **2:** methamphetamine.

sperm. **1:** a spermatozoan; any male gamete produced by an animal. **2:** semen.

sperm duct. See vas deferens.

spermaticide (spermicide). an agent that destroys sperm. Chemical spermicides are used as vaginal foams, jellies, creams, suppositories, or in sponges. Spermaticides do not always kill sperm, and sometimes, sperm that have been damaged by a spermatocide are able to fertilize eggs, resulting in deformed offspring. See also intrauterine contraceptive devices.

spermatocidal (spermatocidal). **1:** destructive of spermatozoa. **2:** pertaining to spermaticides.

spermatocyte. the parent cell of a spermatid. **primary spermatocyte**. a sperm mother cell that gives rise, by the first meiotic division, to two secondary spermatocytes. **secondary spermatocyte**. a cell that gives rise, at the second meiotic division, to four haploid spermatids.

spermatogenesis. the production of sperm in males by the processes of meiosis and maturation.

spermatogonium (plural, spermatogonia). a cell that gives rise to primary spermatocytes by mitosis.

spermatotoxin. spermotoxin.

spermatoxin. spermotoxin.

spermatozoa. plural of spermatozoon.

spermatozoon (plural, spermatozoa). sperm (def. 1); the mature, motile, male gamete of animals; it typically consists of a head, which contains the nucleus, and a tail comprised of a single flagellum.

spermicidal. spermatocidal.

spermicide. spermaticide.

spermine oxidase. a tetramine oxidase. *Cf.* monoamine oxidase.

spermolysin. spermotoxin.

spermotoxic. pertaining to a spermotoxin.

spermotoxin (spermatotoxin; spermolysin; spermotoxin). a toxin or specific antiserum that destroys spermatozoa.

spewing sickness. sneezeweed poisoning. See also *Helenium hoopesii*.

sphacelate. to become or to make gangrenous or necrotic.

sphacelation. **1:** the process of sloughing; of becoming gangrenous or necrotic. **2:** gangrene or necrosis.

sphacelinic acid. a toxic sphacelating principle of ergot.

sphacelotoxin. **1:** spasmotin. **2:** a poisonous, yellow resin isolated from ergot.

Sphaeroides annulatus (gulf puffer). a tropical marine fish (Family tetraodontidae) that ranges from California to Peru, and the Galapagos Islands. It is one of the more poisonous tetraodontoid fishes. See tetrodotoxism.

sphagnum-bog galerina. *Galerina tibicystis*.

spheroidine. tetrodotoxin.

sphincter. a ring of muscle that constricts a passage or an opening when the muscle contracts.

sphingolipid. any of a group of lipids that contains one fatty acid residue, one sphingosine residue, and one base residue per molecule. The most abundant type of sphingolipid is sphingomyelin.

sphingomyelin. any of a number of diaminophosphatides found chiefly in nervous tissue. They are the most abundant of sphingolipids and contain choline, sphingosine, a fatty acid, and phosphoric acid. All are insoluble in water and acetone, but are soluble in hot absolute alcohol.

sphingosine (1,3-dihydroxy-2-amino-4-octadecene). a solid, waxy, lipid alcohol, CH_3-$(CH_2)_{12}CH{:}CHCH_2OCH(NH_2)CH_2OH$, that is soluble in ether. It is comprised of a long hydrocarbon tail and a polar head. Sphingosine is a component of phosphatides such as sphingomyelins, gangliosides, and cerebrosides.

Sphryna zygaena (smooth hammerhead). a species of hammerhead shark (Family Sphrynidae) found in marine waters from Nova Scotia south to Florida, and off the southern California coast. Adults reach a length of about 4 m. The liver is sometimes a cause of elasmobranch poisoning, *q.v.*

Sphyraena barracuda (great barracuda). a torpedo-shaped, robust bony fish (Family Sphyraenidae) that occurs in inshore waters of the Indo-Pacific region and the western Atlantic Ocean from Massachusetts, Florida, Bermuda, the Gulf of Mexico, and the Caribbean southward to Brazil. This fish occasionally attacks humans, with serious consequences. Together with the moray eel, it is the most common cause of ciguatera in the Caribbean region. See ciguatera.

Sphyraenidae (barracudas). a family of carnivorous, voracious, mainly piscivorous, elongate, pike-like, marine percomorph fishes (Class Osteichthyes) of warm seas. Some species are dangerous to humans. See *Sphyraena barracuda*, ciguatera.

spider. with the exception of 2 small groups, all spiders are venomous. Fortunately, the fangs of most species are too short or fragile to penetrate human skin. Nevertheless, at least 60 species in the United States alone have been implicated in bites to humans. Species that are dangerous include the widow spiders, *Latrodectus mactans*, and related species; the brown or violin spider, *Loxosceles reclusa* (sometimes called the brown recluse), and related species; the jumping spiders, *Phidippus* species; the tarantulas, *Aphonopelma* and *Pamphobeteus* species; the trap-door spiders, *Bothriocyrtum* and *Ummidia* species; the so-called banana spiders, *Phoneutria* and *Cupiennius sallei, Lycosa* (Wolf spider), and *Heteropoda*; the crab spider, *Misumenoides aleatorius*; the running spiders, *Liocranoides* and *Chiracanthium*; the orbweavers, *Neoscona vertebrata, Araneus* species, and *Argiope aurantia* (orange argiope); the running or gnaphosid spiders, *Drassodes*; the green lynx spider, *Peucetia viridans*; and the comb-footed or false black widow, *Steatoda grossa*. *Pamphobeteus, Cupiennius,* and *Phoneutria* are not native to the United States, but may be brought into the country on produce or other materials.

spider lily. a common name usually applied to bulbous plants of the genus *Hymenocallis* (*q.v.*) and locally applied to plants of the genus *Lycoris*. See *Hymenocallis*.

spider venom. See venom.

spider wasp. any of various species of solitary wasps that place spiders, which they have killed or paralyzed by stinging, in their brood cells. The spiders are used as food by the young.

spike, temperature. a high, but brief, febrile episode.

spilanthol. affinin.

spill. the unintentional or accidental release or discharge (in small or large amounts) of a sub stance from a container, generally of short duration. **chemical spill**. a spill involving one or more chemicals, usually of a hazardous nature (e.g., toxic, flammable, explosive). **oil spill**. a spill of crude oil or other petroleum-based liquid into the environment, either at the well head, during storage or processing, during transportation (whether via truck, rail, or tanker), or during delivery. Oil spills of greatest concern are generally those from oil tankers or offshore drilling rigs that have at times severely impacted marine, estuarine, and shoreline environments. Some of the more massive spills have destroyed innumerable animals and plants directly by poisoning or coating and impairing mobility, or indirectly by extensive destruction of habitat. A far greater volume of oil, and far more damage in both marine and freshwater ecosystems is due to frequent small spills that take place during the processing, storage and transport of oil. The potential long-term chronic or cumulative effects of petroleum in the environment through toxic effects, carcinogenesis, chemotactic responses, and impoverishment of habitat are poorly documented, but are a matter of great concern.

spinach. *Spinacia oleracea*.

Spinacia oleracea (spinach). a common short-season garden herb. It contains moderate amounts of oxalic acid. Has produced death from calcium deficiency in experimental animals maintained on diets that just meet minimum calcium requirements.

spinal. of or pertaining to the spine.

spinal cord. a characteristic organ of vertebrate animals, it is a dorsal column of nervous tissue that extends from the brain throughout the length of the body. Together with the brain it forms the CNS.

spinal nerve. a nerve that arises from the spinal cord.

spinant. an agent that acts directly upon the spinal cord to increase its reflex activity.

spindle. an apparatus in the cell nucleus composed of microtubules to which the chromosomes are attached during cell division.

spindle poison. an antimitotic agent that acts on the spindle.

spindle tree. See *Euonymus*.

spine. **1a:** in animals, any stiff, hard, rigid or semi-rigid pointed process, projection or appendage (usually external); spines of vertebrate animals, are usually of bone or horn. Spines are organs of envenomation in many venomous fish. **1b:** a pointed projection on a bone. **2:** common name for the vertebral column. **3:** in plants, a sharp, pointed, thorn-like, rigid projection formed from a modified leaf, part of a leaf, or stipule. *Cf.* prickle, thorn.

spine-tailed sea snake. *Aipysurus eydouxi*.

spinneret. any of several (usually six) cylindrical or conical organs on the abdomen of spiders perforated at the tip by numerous tiny spinning tubes through which liquid silk, that hardens on contact with the air, is extruded.

spinners. See Loxoscelidae.

spiny anteater. See Monotremata.

spiny dogfish. See *Squalus*.

spiny shark. See *Squalus*.

spiracle. **1:** a small gill slit in certain fish, situated between the mandibular and hyoid arches. It is a channel through which water flows inward; is not patent in most teleosts. **2:** the external opening of the tracheae in insects. The entry and exit of air is regulated by a valve.

spirilla. plural of spirillum.

spirillicidal. destructive to spirilla.

spirillicide. **1:** destroying spirilla. **2:** an agent that destroys spirilla.

spirillum (plural, spirilla). **1:** a relatively rigid, spiral-shaped bacterium. **2:** any microorganism of the genus *Spirillum*.

Spirillum. a genus of spiral-shaped, motile, freshwater and saltwater, microorganisms (Family Spirillaceae, Order Pseudomonadales). Included are: *S. itersonii, S. kutscheri, S. lipoferum, S. minus, S. serpens, S. tenue, S. undula, S. virginianum,* and *S. volutans*.

spirit. **1:** any volatile or distilled liquid. **2:** a solution of a volatile material in ethanol.

spirit (spirits) of turpentine. oil of turpentine.

spirits of wine. ethanol (alcohol).

spirituous. alcoholic; having a high alcohol content.

Spirochaeta. a genus of slender, motile, spiral microorganisms (Family Spirochaetaceae). *S. pallida* is the causative agent of syphilis.

spirochete. **1:** any microorganism of the order Spirochaetales. **2:** more strictly, a microorganism of the genus *Spirochaeta*.

spirocheticidal. destructive to spirochetes.

spirocheticide. an agent that destroys spirochetes.

spirochetolysin. a substance that lyses spirochetes.

spirochetolysis. the lytic destruction of spirochetes.

spirochetolytic. of, pertaining to, able to cause spirochetolysis.

spirometric. pertaining to spirometry or the spirometer.

spirometry. the measurement of the air capacity of the lungs; a test of pulmonary function.

spiropiperidin. a neurotoxic alkaloid secreted by the skin of *Dendrobates histrionicus* (Family Dendrobatidae). Two branches of the spiropiperidin molecule have acetylenic groups. It produces a number of additional toxic steroids from spiropiperidin. See histrionicotoxin.

spiropiperidine alkaloid. See alkaloid.

(25R)-5α-spirostan-2α,3β,15β-triol. digitogenin.

(5R)-spirost-5-en-3β-ol. diosgenin.

(25-R)-spirost-5-en-3β-yl *O*-6-deoxy-α-L-mannopyranosyl-(1→2)-*O*-6-[deoxy-α-L-mannopyranosyl-(1→4)]-β-D-glucopyranoside. dioscin.

Spisula solidissima (solid surf clam). occurs from North Carolina to Labrador. It is often involved in paralytic shellfish poisoning of humans.

spitting cobra. See *Hemachatus*, *Naja naja sputatrix*, *Naja nigricollis*.

spitting snake. *Hemachatus haemachatus*.

Spitzkopfotter. *Vipera ursinii*.

spleen. a large, blood-filled, lymphoid organ located in the intestinal mesentery of most vertebrate animals. It consists of a network of venous sinuses that lie in a loose meshwork of connective tissue, which is interspersed with globular masses of lymphocytes. In humans it is located near the left side of the stomach. The spleen generally filters foreign bodies; produces lymphocytes; destroys old, damaged, or defective erythrocytes in adults; manufactures erythrocytes in the fetus and newborn; and is an important storage organ of blood cells.

splenitis. inflammation of the spleen.

splenomegaly. enlargement of the spleen.

splenotoxin. a cytotoxin specific destroys cells of the spleen.

split nut. common name of physostigma and *Physostigma venosum*.

spondylalgia. pain in the vertebra.

sponge. **1:** common name of any member of the Phylum Porifera. **2:** a commercial product comprised of the spongin skeleton of certain sponges. See Porifera

sponge fisherman's disease. a condition due to handling the actinarian *Sarortia elegans*, wherein small spicules of the coral sometimes break off in the hand and become imbedded in the skin, sometimes resulting in infection.

spontaneous mutation. See mutation.

spoonwood. *Kalmia latifolia*.

spore. **1:** a minute structure other than a seed that is capable of developing into a new individual. **2:** a reproductive body in nonseed plants.

sporicidal. able to destroy spores.

sporicide. an agent that kills spores.

sporidesmin. a mixture of chlorine-containing, epipolythiadioxopiperazine mycotoxins (A-H and J) secreted by *Pithomyces chartarum* (= *Sporidesmium bakeri*) (Class Deuteromycetes), a saprophyte of decaying Grammineae. It is very toxic to laboratory mice and extremely toxic to laboratory rats, guinea pigs, rabbits, and sheep. It is known to cause facial eczema and liver damage in sheep and cattle in New Zealand and Australia. It is the causative agent of facial eczema in domestic sheep. **sporidesmin G**. a possible cause of facial eczema in humans.

spot remover. any dry-cleaning formulation used to clean small soiled areas of clothing. They often contain perchloroethylene, the fumes of which are highly toxic and carcinogenic to humans.

spotted blacksnake. *Pseudechis colletti*, *P. guttatus*.

spotted brownsnake. *Demansia textilis*.

spotted coral snake. *Calliophis gracilis*.

spotted cow bane. *Conium macalatum*.

spotted dumb cane. *Dieffenbachia maculata*.

spotted parsley. *Conium macalatum*.

spotted pit viper. *Trimeresurus monticola*.

spotted ratfish. *Hydrolagus colliei*.

spotted sea snake. *Hydrophis ornatus*

spotted snake. *Denisonia punctata*.

spotted spurge. See *Euphorbia lactea*.

spotted-headed snake. *Demansia olivacea*, *Denisonia psammophis*.

spotted-stalked milky. *Lactarius scrobiculatus*.

spp. abbreviation for species, plural.

spring adonis. Adonis vernalis.

spring parsley. *Cymopterus watsonii*.

spring vetch. *Vicia sativa*.

springhalt. stringhalt.

sprue. psilosis.

spurge. certain plants of the family Euphorbaceae, especially those of the genus *Euphorbia*, q.v.; sometimes applied also to other unrelated plants such as those of the family Thymelaeaceae. See *Daphne*, *Phyllanthus abnormis*.

spurge family. Euphorbiaceae.

spurge laurel. a common name of plants of the genus *Daphne*.

spurge nettle. *Cnidoscolus stimulosus*.

spurge olive. a common name of plants of the genus *Daphne*, especially *D. mezereum*.

sputum. a secretion of the lungs, coughed up in large amounts in some lung diseases.

spuugslang. *Hemachatus haemachatus*, *Naja nigricollis*.

Squali. a suborder of cartilaginous fishes (order Selachii) that includes true sharks. The body is torpedo- or spindle-shaped; tail heterocercal; 5-5-7 pairs of gill slits at least partly lateral; upper eyelids are free; small to medium-sized pectoral fins. Common genera are *Chlamydoselachus, Mustelus, Carcharias, Rhineodon, Squalus, Pristiophorus,* and *Pliotrema*.

Squalus acanthias (spiny dogfish, dogfish shark, grayfish). a species of horned or spiny sharks, it is a temperate and subarctic marine species that occurs chiefly in shallow bays on either side of the North Atlantic and North Pacific oceans. The length rarely exceeds 1 m. It bears two dorsal spines (dorsal stings), one in front of each of the two dorsal fins. It occasionally stings humans. The venom apparatus consists of the dorsal sting and a venom gland that lies in a shallow groove on the back of the upper part of each spine. When the spine pierces the skin, the venom gland is ruptured and releases the venom into the wound. Symptoms in humans include immediate intense pain which may persist for several hours. Tenderness about the wound and erythema and severe edema of the affected part may develop early and persist for several days. The sting may prove fatal on occasion.

Squamata (lizards, snakes, and amphisbaenians). an order (Class Reptilia) primarily of terrestrial vertebrates that includes the lizards (Suborder Sauria), snakes (Suborder Serpentes), and the amphisbaenians or worm lizards (Family Amphisbaenidae).

squaretail. See *Tetragonurus cuvieri*.

squash. *Cucurbita maxima*.

squaw root. *Caulophyllum thalictroides*.

squaw weed. *Senecio aureus*.

squill (scilla). **1:** *Scilla spp*. **2:** *Urginea maritima*. **3:** the dried, fleshy, inner scales of the bulb of the white variety of *U. maritima* or *U. indica*, excluding the central portion of the bulb. This preparation contains the cardiac glycosides, scillaren-A and scillaren-B. **4:** a poisonous powder prepared from the dried bulbs of *U. maritima* and to a lesser extent, *U. indica*.

squint. See strabismus.

squirrel corn. **1:** *Dicentra canadensis*. **2:** corydalis.

squirrel food. *Zigadenus venenosus*.

squirrelfish. common name of marine bony fishes of the family Holocentridae. See *Myripristis murdjan*.

Sr. the symbol for strontium.

s.s. See *s. str*.

ssp. "subspecies q.v."

s.str. (*s.s.*). abbreviation of *sensu stricto*.

St. Anthony's fire (mal des ardents; feu de Saint-Antoine). **1:** still used occasionally today, it was the original name (feu de Saint-Antoine) for ergotism, based on the first recognized case in Dauphiné, France, in 1039 AD, where St. Anthony's remains were housed. The cause of the condition was unknown until 1676. **2:** convulsive ergotism. **3:** *Claviceps purpurea*.

St. Ignatius' bean. the seed of *Strychnos ignatii*.

St. John's lily. *Crinum asiaticum*.

St. John's wort, St. Johnswort. plants of the genus *Hypericum*, especially *H. perforatum*.

St. Lucia serpent. *Bothrops caribbaeus*.

stability. **1:** resistance to change; the tendency to remain in or return to a state of equilibrium during or following a disturbance. **2:** the ability of a population of living organisms to withstand perturbations (e.g., those due to extremes of weather or pollution) in their environment without marked changes in composition.

stachbotryotoxicosis. a hemorrhagic mycotoxicosis of horses and cattle following ingestion of oats, hay, or fodder that has been overgrown by the fungus *Stachbotrys atra* (= *S. alternans*). Humans may be poisoned if exposed to infested forage either by inhalation or percutaneous absorption of the toxins. Typical effects in humans are skin rash, pharyngitis, and mild leukopenia. See also trichothecene.

Stachybotrys atra. the cause of stachbotryotoxicosis. The active principle is the mycotoxin satratoxin.

stachydrine (*N*-methylproline methylbetaine; 2-carboxy-1,1-dimethylpyrrolidinium hydroxide inner salt; methyl hygrate betaine; hygric acid; methylbetaine). an alkaloid, $C_{17}H_{13}NO_2$, it is the betaine of proline found in numerous species of plants (e.g., alfalfa, chrysanthemum, citrus plants, hedge nettles).

staff-tree family. Celastraceae.

stafftree. See Celastraceae.

stage. **1:** state, level, phase, step, interval. **2:** any discernable phase of growth or development of a living system.

stagger bush, staggerbush. *Lyonia mariana*.

staggergrass. *Amianthium muscitoxicum*.

staggers. **1:** a common term for any of a number of mostly unrelated neurotoxic diseases of livestock, e.g., atamasco poisoning, bracken staggers, grass staggers (See grass tetany), blind staggers (subacute selenium poisoning), *Lamium amplexicaule* poisoning, annual ryegrass staggers, paspalum staggers, perennial ryegrass staggers, choking staggers (See moldy corn poisoning), stomach staggers (See moldy corn poisoning). **2:** a type of decompression sickness characterized by mental confusion, vertigo, and muscular weakness. **3:**. gid, a disease in sheep, marked by swaying and uncertain gait, caused by the presence of the larva of *Taenia multiceps*. **annual ryegrass staggers**. an often fatal disease of livestock of any age or sex that results from grazing on *Lolium rigidum* (annual ryegrass). It is prevalent at times in western and southern Australia and South Africa from November through March. **ryegrass staggers**. either of two related neurotoxic diseases.

staggerweed. **1:** *Delphinium barbeyi* and related species. **2:** *Dicentra eximia*.

stagnant hypoxia. See hypoxia.

stalk. a stem-like structure that forms the (usually supportive) base of a flower or leaf (used in this dictionary in place of pedicel, peduncle, and petiole).

stamen. that part of the flower that produces pollen; the pollen-bearing organ of a flower composed of the anther (pollen sac) and filament (stalk).

standard. **1a:** that which is accepted (e.g., by a profession, by an authority) as a criterion, model, pattern, rule, norm, or ideal. **1b:** an accepted criterion (or set of criteria) used to guide or direct a process or activity as in a standard treatment or a standard protocol for the treatment of a specific type of empoisonment. **1c:** an entity, substance, or preparation of known value, quality, composition, or strength. **2:** an evaluation (e.g., a measurement) of a substance, process, or phenomenon that serves as an accepted basis of comparison in the evaluation of similar substances, processes, or phenomena. **3:** an accepted reference sample used to establish a unit of measurement for a physical or chemical quantity. **4a:** a mixture or solution of known composition or concentration that can be used as a reference in the preparation or measurement of similar preparations. **4b:** a chemical (e.g., a pharmaceutical) preparation of known composition, quantity, strength, and/or biological activity that can be used to determine the composition of another such preparation. **5:** the maximum safe concentration of a hazardous substance, usually expressed as mg/m_3, ppm, or ppb.

standard deviation. the square root of the sum of the deviations from the mean squared, divided by one less than the number of observations. A measure of the variation within a population of individuals. See also coefficient of variation.

standard error. the standard deviation divided by the square root of the number of observations. A measure of the variation of a population of means.

standard reference material. a reference material certified and distributed by a certified vendor such as the National Institute of Standards and Technology (NIST) in the United States.

standard solution. a solution of specifically known concentration. See also standard.

standardized disease ratio (SDR). a ratio arrived at by a procedure that closely resembles that of the standardized mortality ratio. Disease incidence is substituted for mortality to determine standardized disease ratios. The SDR may be adjusted for variables other than age that correlate with the end point selected.

standardized mortality ratio (SMR). a measure of chemically related mortality in humans, SMR is the ratio of observed to expected deaths in a population. The expected number of deaths is determined from the schedule of age-specific mortality rates in the comparison population and is related to the distribution of person-years in the study population. In some cases, the SMR may offers an approximation of relative risk, but [illegible] mortality must be consistent across age bands and the age bands cannot be too large (10 years or less) for such.

standing crop biomass. See standing crop.

Stanleya (prince's plume). a genus of plants (Family Brassicacea). All species are obligate accumulators of selenium from soil.

stannane (tin hydride). an unstable, poisonous, colorless gas, SnH_4, prepared by the action of lithium aluminum hydride on stannous chloride. It is a reducing agent.

stannic. **1:** pertaining to, resembling, or containing tin. **2:** in chemistry, pertaining to tetravalent tin (Sn^{4+}) or to compounds of tin that contain tetravalent tin.

stannic acid. stannic oxide.

stannic anhydride. stannic oxide.

stannic oxide (stannic anhydride; tin peroxide; tin dioxide; stannic acid). a white (when pure) or gray, anhydrous powder that contains varying amount of water, SnO_2 or $SnO_2 \cdot H_2O$. It has many applications in industry. It causes stannosis, *q.v.*

stannosis. a benign pneumoconiosis of industrial workers due to the inhalation of stannic oxide. It lacks symptoms unless complicated by silicosis.

stannous. pertaining to divalent (Sn^{2+}) or to compounds of tin that contain divalent tin.

stannous chloride (fuming tin chloride; tin crystals; tin salt; tin dichloride; tin protochloride). a white, crystalline mass, $SnCl_2$ or $SnCl_2 \cdot H_2O$, that is soluble in water, alkalies, ethanol, and tartaric acid. It is a reducing agent that readily absorbs oxygen from the atmosphere, converting it into the insoluble oxychloride. Stannous chloride has many applications. It is, for example, used as a food preservative; in the manufacture of various chemicals, intermediates, and dyes; and is used to revive yeast. As an antioxidant it is used in processing soft drinks, canned asparagus, and other foods. It is an irritant to the skin, eyes, and mucous membranes. It may also have mutagenic and reproductive effects.

stannum. tin.

staphylinid. **1:** common name of beetles of the

family Staphylinidae. **2:** pertaining to, characterized, or due to staphylinid beetles.

Staphylinidae. a family of beetles (Order Coleoptera), some species secrete a vesicant or irritant that causes blistering of human skin within 1 or 2 days following contact.

staphylocide. staphylococcide.

staphylococci. plural of *staphylococcus*.

staphylococcide. an agent that destroys staphylococci.

staphylococcin. a bacteriocin produced by *Staphylococcus*.

staphylococcinogenic. producing staphylococcins (See bacteriocin).

staphylohemolysin. a mixture of hemolysins (alpha, beta, gamma, and delta), elaborated by staphylococcal; the alpha hemolysin has a pronounced effect on vascular muscle.

staphyloleukocidin. a toxin produced by staphylococcus that destroys leukocytes.

staphylolysin. staphylococcolysin. **1:** a principle with hemolytic activity that is elaborated by staphylococci. **2:** an antibody that lyses of staphylococci.

staphylotoxin. a toxin released *in vitro* by the various species of *Staphylococcus*, especially *S. aureus*. See also staphylohemolysin.

star anise. any tree of the genus *Illicium*.

star anise, Japanese. *Illicium anisatum.*

star of Bethlehem, star-of-Bethlehem. *Ornithogalum umbellatum.*

starfish (sea star). any echinoderm of the class Asteroidea; they usually have five radially disposed arms. See Asteroidea.

stargazer, star-gazer. common name of fishes of the family Uranoscopidae especially *Uranoscopus scaber*.

starry hyacinth. *Scilla autumnalis*.

starvation. **1:** deprivation or withholding of food to an organism for a relatively long period of time. **2:** the suffering or condition resulting from lack of food or the inability to take in, digest, absorb, or metabolize nutrient substances. Such a condition is marked by loss of biomass and many changes in behavior and metabolism. Unless remedied, death is the ultimate outcome. Poisons may cause starvation by interfering with any of these processes, by impairing an organism's mobility, or altering behavior. Thus, the oiling of feathers or fur of marine animals by petroleum often impairs mobility such that death by starvation may ensue (See corrosive poisoning, *Rhinocrichus lethifer*, trembles). See also acidosis, inanition, ketosis.

starwort. *Aster*.

stasis, pl. stases. **1:** the stagnation or cessation of the flow of blood or other body fluids. **2:** the retardation or cessation of growth or movement.

state. **1:** condition, situation, status; the condition of a system; the condition in which a system exists. **2:** a minimum set of numbers that collectively contain enough information about a system or a systems history to (1) fully describe its condition or behavior, or to (2) enable its future behavior to be calculated. **3:** a stage in the life cycle of a plant. **4:** a variant that arises in a bacterial culture and alters the gross appearance of the culture.

static. inert, inactive, passive, without change or motion; stagnant.

statistic. any measurable property or function of a sample, especially one thought to correspond to that of the population from which the sample was drawn.

statistical inference (inductive statistics). the process and methods of predicting or estimating population variables on the basis of sample data.

statocyst. a small organ in the ear of vertebrates. It is concerned with the perception of gravity and maintenance of balance. Structure varies. Similar organs occur in many invertebrates.

status. state, condition; situation; rank.

status epillepticus. prolonged, generalized epileptic seizures following one another in rapid succession, each seizure may be followed by brief periods of coma. If seizures are of the tonic-clonic type, the condition may be life-threatening.

steam. See water, vapor (water vapor)

steam-fitter's asthma. asthma resulting from exposure to asbestos insulation of heating and plumbing conduits and fixtures. It is accompanied by asbestosis.

steamer clam. *Mya arenaria*.

stearic acid. usually not toxic if ingested.

stearic acid lead salt. lead stearate.

steatite. talc.

Steatoda (cobweb spiders). a genus of small, Old World spiders introduced in the New World; often mistaken for black widow spiders. The bite produces localized pain and tissue damage. Little is known regarding the chemistry and pharmacology of the venom or of further effects of envenomations by these spiders. The symptoms produced may vary among species. ***S. grossa*** (comb-footed black widow; false black widow). one of the larger species, it is considered dangerous to humans.

Steinbuhl yellow. C.I. pigment yellow 31.

Stejneger's palm viper. *Trimeresurus stejnegeri*.

STEL. short-term exposure limit value. See limit value.

stele (vascular bundle; vascular cylinder). the central core of stems and roots of a vascular plant. It is comprised of vascular tissue, ground issue (e.g., pith, medullary rays), and the pericycle (the outermost layer of the stele). **atactostele**. a stele typical of monocotyledons. The vascular tissue is distributed more or less irregularly within the ground tissue. **solenostele**. a siphonostele that contains a core of pith internal to the xylem and also a cylinder of phloem internal to the xylem.

Stellaria media (chickweed). a perennial weed (Family Caryophyllaceae) that grows to about 0.6 m. It is endemic to Europe, but has become established in much of North America. Toxic concentrations of nitrate have been reported for this herb which is sometimes used in salads or steamed as a green. It is high in vitamin C.

stem turnip. *B. oleracea* var. *gongylodes*.

stemless loco. See *Oxytropis*.

stemphone. a mycotoxin from *Stemphylium sarcinaeforme*. It is extremely toxic to the embryo of the domestic fowl.

stenecious. stenoecious.

stenoecic. stenoecious.

stenoecious (stenoecic; stenecious; stenecious; stenoecious; stenokous). of or pertaining to a living organism that is restricted to or tolerant of a narrow range of habitats and environmental conditions; one characterized by stenoky. See also amphioecious, euryoecious.

stenoekous. stenoecious.

stenohaline. of or pertaining to an aquatic organism that tolerates only very small variations in the salinity of the water; true of certain jelly fish and numerous fishes. *Cf.* euryhaline.

stenoionic. tolerating or characterized by a narrow range of pH. See also euryionic.

stenoky. the quality or condition of being stenoecious, whereby a species or population of organisms tolerates a relatively narrow range of ecological conditions. See also euroky.

stenophagous. pertaining to an organism that subsists on only one or a few kinds of food.

stenopodium. See biramous appendage.

stenosis. a narrowing or constriction of a tube, duct, or vessel within an organism.

stenosubstratic. tolerant of only a narrow range of types of ecological substrata. *Cf.* euryosubstratic.

stenotropic. said of organisms that exhibit a limited response or fail to adapt to changing environmental conditions. *Cf.* eurytropic.

stenotropism. the state or condition whereby an organism responds little or fails to adapt to changing environmental conditions. *Cf.* eurytropism.

Stentz's blue legs. *Psilocybe stuntzii*.

Stephen's banded snake. *Hoplocephalus stephensii*.

Steppe adder. *Vipera ursinii*.

stereochemical formula (spatial formula). a chemical formula in which the spatial arrangement of the atoms or atomic groupings is shown.

stereoisomer. a compound with the same molecular formula as another, but with a uniquely different three-dimensional configuration.

stereoline glutinant. See nematocyst.

stereospecificity. the property of a chemical whereby it is specific for only one of possible receptors of a cell due to (often minor) differences in configuration of the active site of the otherwise potentially suitable receptors. Stereospecificity often plays a major role in toxicant (drug)-receptor activity. The spatial arrangement of a portion of the molecule can affect the ease with which the agonist (e.g., a toxicant) can approach the critical area of the receptor or its ability to bind to areas adjacent to the specific receptor site. See also antagonist (competitive antagonist).

sterigmatocystin. a bisfuranoid mycotoxin from *Aspergillus versicolor*. It is closely related to the aflatoxins, but contains a substituted xanthine ring. It is moderately toxic, *p.o.*, to laboratory mice; extremely toxic, *i.p.*, to laboratory rats; and has induced hepatomas in laboratory mice and pulmonary adenomas in laboratory rats.

sterilant. **1:** an agent that renders an animal sterile or incapable of reproduction. **2:** an agent that destroys microbial activity (e.g., toxicants, steam, ultraviolet light, high velocity electron bombardment). **chemical sterilant**. a chemical that eliminates the viability of a microbe. But See chemosterilant. **eugenic sterilant**. an agent used to render a person incapable of reproduction because of the opinion that the offspring would be in some way undesirable.

sterile. **1a:** of or pertaining to sterility. **1b:** unable to produce offspring; barren. **2:** aseptic; not producing microorganisms; free from living microorganisms. **3:** axenic; pertaining to cultures of a particular microorganism that are uncontaminated by other microorganisms.

sterility. **1:** infertility; in general, the inability to conceive or produce viable offspring (female sterility), to fertilize, or induce conception (male sterility). **2:** a condition of being aseptic, or free from living microorganisms and their spores.

sterilization. **1:** the use of chemical or other means (e.g., steam, UV light, high-velocity electron bombardment) to destroy all microorganisms in water or on surfaces. *Cf.* disinfection. **2:** the inability to reproduce. **3a:** any process (e.g., surgery, exposure to a xenobiotic) whereby the capacity of an individual organism to reproduce is eliminated. **3b:** in pest control, the use of radiation or chemicals to severely impair or kill body cells that are essential to reproduction. The destruction of all living organisms in water or on the surface of various materials.

sternalgia. pain in the sternum or region of the sternum.

steroid. **1:** pertaining to steroids (def. 2). **2:** any of a large family of polycyclic lipids that contain the tetracyclic cyclopenta[a]phenanthrene nucleus. Steroids are closely related to terpenes. Included are many hormones (e.g., adrenocortical hormones, gonadal steroids), precursors of certain lipid-soluble vitamins, certain active principles in toad venoms (e.g., bufanolide), and a variety of drugs (e.g., synthetic estrogens, certain digitalis derivatives). **3:** sometimes used for any of a number of substances that are structurally related to steroids (def. 2). Included are bile acids, sterols, cardiac glycosides, saponins, and a number of drugs. **4:** often used for any synthetic or semi-synthetic compound that has biological effects similar to those of a steroid hormone, whether the substance is chemically similar to a steroid (def. 2) or not. Such usage is not recommended.

steroid alkaloid. See alkaloid.

steroid glycoside. See glycoside.

steroidal (steroid, def. 1). pertaining to steroids.

sterol. any of a class of colorless (usually), crystalline, steroidal (def. 2) alcohols (e.g., cholesterol, ergosterol) with a cyclopentanoperhydrophenanthrene nucleus. They occur free or as fatty acid esters in higher plants (phytosterols) and in animals (zoosterols).

stertor. a noisy inspiration (snoring or laborious breathing) occurring in coma or deep sleep, sometimes due to obstruction of the upper air passages or larynx.

stertorous. pertaining to or characterized by stertor or snoring; provoking stertor.

stethoparalysis. paralysis of the muscles involved in breathing.

stethoscope. a device for listening to sounds in the chest (e.g., heart sounds).

Stevens-Johnson syndrome. a severe, and sometimes life-threatening form of erythema multiforme, *q.v.*, marked by lesions of, and sloughing of mucous membranes.

STH. **1a:** lyophylized anterior pituitary tissue. **1b:** somatotrophin.

stibialism. antimonial poisoning.

stibiated. containing or impregnated with antimony.

stibiation. impregnation with antimony.

stibine (antimony hydride). an extremely toxic, [illegible] greeable odor that is less stable than arsine. It is produced, for example, during the charging of storage batteries, or when welding, soldering, or etching. It is used to fumigate and fill hydrogen balloons. Stibine is a hepatotoxicant and CNS poison that can cause permanent blood and liver damage. Symptoms of acute intoxication are similar to those of arsine and may include headache, nausea, vomiting, weakness, abdominal pain, hemolysis, hematuria, bradypnea, a weak and irregular pulse, and in some cases icterus and death. A concentration of ca. 100 ppm is lethal to laboratory mice.

stibium. antimony.

stibogluconate sodium (antimony sodium gluconate). **1:** a toxic pentavalent antimony compound (antimony sodium gluconate; pentavalent sodium stibogluconate), $C_6H_9Na_2O_9Sb$, administered intravenously or intramuscularly as an antileishmanial. **2:** a toxic trivalent antimony compound (trivalent antimony sodium gluconate; trivalent sodium stibogluconate; sodium antimonylgluconate), $C_6H_8NaO_7Sb$, administered intravenously as an antischistosomal.

stilboestrol. diethylstilbestrol.

stillbirth **1:** the birth of a dead fetus. **2a:** in *Homo sapiens*, the death of a fetus during and after the 28th week is generally termed stillbirth. **2b:** the death of a human fetus that survives to the 20th week of gestation (the beginning of the perinatal period) is sometimes termed a stillbirth. *Cf.* abortion.

stimulant. **1:** stimulating, arousing, activating, exciting, exciting to action. **2:**. excitant; excitor; stimulator. **3:** a drug or other biologically active agent that increases the activity of an organism in whole or in part. See also stimulus. **3:** See allelochemical. **central nervous system stimulant**. See stimulant drug. **CNS stimulant**. See stimulant drug. **diffusible stimulant**. a stimulant that produces a rapid, transient effect. **general stimulant**. one that produces whole-body effects. **local stimulant**. one that acts only at the site of application.

stimulant drug. a drug that increases the activity of the CNS (e.g. cocaine, amphetamine, dex- [illegible] zine, phenylpropanolamine).

stimulate. **1a:** to provoke a response. **1b:** to excite an organ or structure to functional activity, or to bring about an increase in activity.

stimulation. **1:** the provoking of a response; arousal; inducing a higher level of activity (e.g., of a cell, tissue, organ, organism). **2:** the state of being provoked or aroused. **3:** the application of a stimulus.

stimulatory. having the capacity to stimulate.

stimuli. plural of stimulus.

sting. 1a:. a sharp, usually momentary prick, pain, or burning sensation; most commonly due to a puncturing of the skin by any of numerous species of arthropods, including insects, myriapods, and arachnids; many other invertebrates such as jellyfish, sea urchins, sponges, mollusks; and several species of venomous fish, such as stingray, toadfish, rabbitfish, and catfish also sting. *Cf.* bite. See also, for example, aculeate, *Anemonia sulcata*, *Centruroides*, *Megalopyge*, *Millepora*, mule killer, pedicellariae, *Plotosus lineatus*, *Pogonomyrmex*, *Polistes*, Polychaeta, Potamotrygonidae, Pseudoscorpionida, *Pterois*, Scorpaenidae, Scorpionida, *Squalus acanthias*, stinging hairs, *Synanceja*, tarantula, Theraphosidae, *Toxopneustes*, *Uranoscopus scaber*, Urolophidae, *Urtica*. **1b:** an injury due to the combined effects of the mechanical injury incurred and the venom injected into the wound. Except for the associated trauma, which may be slight to severe, stings are not usually serious. Nevertheless, poisoning, infection, or rarely anaphylactic shock must always be considered possible and preventive measures taken (See, for example, hymenoptera sting). **2:** a stinger, *q.v.*; venom apparatus; the organ used to inflict a sting (defs. 1a and 1b). The venom apparatus may consist of a chitinous spicule; bony spine; other hard, sharp, piercing organ; and a venom gland, sac, or scattered cells that secrete venom. **3:** to introduce a venom by stinging. **hymenoptera sting**. some hymenopterans are not venomous, while others can be deadly. In mammals and birds, the sting usually produces some tenderness, apparent pain and/or itching, erythema, and edema at the site of a sting; occasionally intense pain. Stings may also result in more generalized symptoms that include itching, urticaria, weakness and malaise, and sometimes anxiety, nausea, vomition, dizziness, edema, dyspnea, wheezing, and abdominal distress. Severe cases may also exhibit some of the following: hoarseness, thickened speech, dysphagia, confusion, and despondency. Hypersensitive individuals may develop deep, systemic (anaphylactic) shock with any or all of the following: hypotension, cyanosis, incontinence, collapse, and unconsciousness. Delayed anaphylactoid reactions may occur and are a risk in subsequent stings.

sting ray. See stingray.

stingaree. stingray.

stinger. a general type of venom apparatus, *q.v.*, it is a hard, piercing organ of envenomation of certain animals such as bees, wasps, and stingrays. **bee stinger**. the stinger of the worker bee is a sharp, barbed, modified ovipositor situated in the posterior wall of the abdomen. It is connected to the venom sac. The bee only stings once because when the bee drives the stinger into a victim, the barbs anchor it and the stinger is torn from the bee (who later dies) as it flies off. The stinger of the queen bee lacks barbs. Drones have no stinger. **hornet stinger.** the stinger is similar to that of a bee, but is retained following envenomation. **wasp stinger**. the stinger is similar to that of a bee, but is retained following envenomation.

stinging cell. nematocyst.

stinging hair. See *Hermodice*.

stinging nettle. *Urtica dioica*.

stinging spurge. *Cnidoscolus stimulosus*.

stingray (stingaree; sting ray;). **1:** whipray; whip-tailed sting ray; any of a group of cartilaginous fishes (Family Dasyatidae); whip-tailed sting rays; body thin and discoid, with a long, thin tail bearing one to several saw-edged spines on the upper surface; they produce ugly wounds that heal slowly; some produce deep, penetrating wounds and can introduce tetanus bacilli with resultant tetanus. See also *Dasyatis brevicauda*, *Urobatis halleri*. **2:** any of numerous species of selachians (Suborder Batoidea) in seven families. These are Dasyatidae (stingrays or whiprays), Potamotrygonidae (freshwater stingrays, river rays), Gymnuridae (butterfly rays), Myliobatidae (eagle rays or bat rays), Rhinopteridae (cow-nosed rays), Mobulidae (devil rays or mantas), and Urolophidae (round stingrays). Stingrays inhabit tropical, subtropical and warm temperate seas. All but the freshwater Trygonidae are primarily marine forms. Rays are generally nonmigratory and favor shallow waters with sandy bottoms, but occur also at moderate depths. They range in length from about 0.3-4.3 m. All are presumed venomous; some are extremely dangerous, others are essentially harmless to humans. The venom apparatus or sting (there are sometimes two stings) differs among taxa in structure and effectiveness, but is always part of the caudal appendage. The sting incorporates a bilaterally serrated, dorsocaudal spine composed of hard, bone-like material (vasodentine) and is enclosed in an integumental sheath. Shallow grooves lie along the length of the spine and deep grooves that contain a strip of grayish, spongy, venom-producing tissue throughout their length are situated ventrolaterally. Envenomation occurs during laceration or puncture of the skin by a spine usually propelled by the caudal appendage. The force of the strike alone is enough to cause release of the venom into the grooves. The wound is usually extremely painful and may become necrotic if unattended; most symptoms are restricted to the area of the wound. Systemic effects may include hypotension, diarrhea, vomiting, sweating, tachycardia, and paralysis. Chest wounds, which are rare, may prove fatal, otherwise human fatalities are rare. Removal of the serrated sting may cause extensive tissue damage. The four types of sting are the gymnurid, myliobatid, dasyatid, and urolophid, *q.v.*

stinkweed. *Datura stramonium.*

stinking clover. *Cleome serrulata.*

stinking elder. *Sambucus pubens.*

stinking hellebore. *Helleborus foetidus.*

stinking nightshade. *Hyoscyamus niger.*

stinking willie. *Senecio jacobaea.*

stinkweed. *Datura stramonium.*

stipule. one or two small bracts or leaves at the base of a leaf.

Stirifos. tetrachlorvinfos.

stochastic process. random process.

Stoddard solvent. a colorless, widely used dry-cleaning solvent, spot, and stain remover, degreasing solvent, and a paint thinner. It is a petroleum distillate related to the naphthas. It is a fire hazard and is toxic by inhalation, ingestion, or through the skin. Focal effects include irritation of the mucous membranes and dermatitis. Systemic effects include dizziness.

stoichiometry. **1:** the branch of chemistry that treats the quantities of substances that enter into and are produced by chemical reactions. **2:** the description of a chemical or metabolic reaction or set of reactions in terms of relative quantities of molecules of each reactant and each product.

Stoke's sea snake. *Astrotia stokesii.*

stoma (plural, stomata). **1a:** any of a large number of small apertures in the epidermis of portions of the aerial parts of green plants (e.g., leaves, stems, flowers), in some underground rhizomes, and in the leaves of some aquatic plants. The exchange of gases and loss of water by transpiration take place through these openings. Each stoma is bordered by a pair of guard cells, which control its opening and closing. **1b:** this term is sometimes taken to include the aperture and the guard cells combined.

stomach. **1:** the most anterior or superior organ of the vertebrate digestive tract (the gastrointestinal tract), it is usually situated between the esophagus and intestine. There is structural and functional variation among various taxa. Thus the stomach may be simple or compartmentalized. In birds the posterior end (the gizzard) is muscular and grinds food; in ruminant herbivores the stomach is comprised of several chambers, which facilitate the digestion of cellulose by commensal bacteria. The stomach typically functions in the temporary storage of food, digestion (mechanical and chemical), and absorption with translocation to other parts of the body. The food is broken up by the churning action of the muscular walls (muscularis mucosa) of the stomach. Glandular cells in the mucosa secrete digestive (gastric) juices that contain hydrochloric acid, which lowers the pH of food, and pepsin, which initiates protein digestion (a hydrolytic process); most protein digestion takes place in the stomach. Some substances that are ionic at a pH near 7 and above are neutral in the stomach and are thus readily absorbed. Stomach contents, other than HCl, may include food particles, gastric mucin, gastric lipase, and pepsin. A pyloric sphincter at the posterior or inferior end of the stomach regulates the passage of food into the duodenum. **2a:** any structure in an invertebrate that is analogous to the vertebrate stomach; such organs are common. **2b:** the entire digestive cavity of an invertebrate.

stomach ache. gastrodynia.

stomach insecticide. See insecticide (stomach insecticide).

stomach staggers. moldy corn poisoning.

stomachal. stomachic; relating to the stomach.

stomachalgia (stomachodynia). stomach ache (obsolete).

stomachic. pertaining to the stomach.

stomachodynia. stomachalgia.

stomata. plural of stoma.

stomatalgia (stomatodynia). pain in the mouth.

stomatitis. inflammation of oral mucosa. Symptoms vary, but usually include heat, pain, increased salivation, fetor, sometimes fever. Causes vary (e.g., pathogens, toxins, heavy metals, allergic responses to e.g., drugs, components of toothpaste, mouthwash). **corrosive stomatitis**. that caused by intentional or accidental exposure to strong irritants. **lead stomatitis**. that due to lead poisoning. It is characterized by the presence of a bluish-black line that follows the contours of the marginal gingiva where lead sulfide has precipitated because of the inflammation. **mercurial stomatitis**. that seen in chronic mercury poisoning of persons who work with mercury or following the administration of very large doses of mercury (or even small doses in highly susceptible people). Lesions may include mucosal erythema and edema of the mucosa, ulceration, and deposition of mercurial sulfide in the inflamed tissues. The resulting discoloration gives an appearance similar to that of lead stomatitis. **mycotic stomatitis**. that due to a fungus infection, seen especially in infants and young children. It is often accompanied by fever and gastrointestinal inflammation. **stomatitis medicamentosa**. a condition of the oral mucosa due to systemic drug allergy. The reaction produces lesions, usually erythema, angioneurotic edema, vesicles, bullae, or ulcerations. **stomatitis venenata**. localized stomatitis due to an allergic response to direct contact of the causative agent.

stomato-. a prefix that denotes a relationship to the mouth or to the mouth of the mammalian uterus (ostium uteri).

stomatopathy (stomatosis). any disease of the oral cavity.

stomatosis. stomatopathy.

stone masons' disease. silicosis.

stonefish. *Synanceja*. See also *Choridactylus multibarbis*.

stool. 1: defecation. **2:** a discharge of feces; the feces from a single bowel movement. **3:** feces.

stork's-bill. a common name of plants of the genus *Erodium*.

Storm's water cobra. *Boulengerina annulata*.

STP. 2,5-dimethoxy-4-methylamphetamine hydrochloride.

strabismic. displaying strabismus.

straight-stalked entoloma. *Entoloma strictus*.

strain. **1:** the force acting across a unit area of solid material that tends to resist external disruptive forces. **2a**. a stimulus or repetitive stimuli of such a nature and magnitude that they tend to disrupt homeostasis of an organism. **2b:** any physical or chemical change in a living organism produced by a stress, *q.v.* **3a:** a group of very similar individuals with common physiological traits and known or presumed common ancestry (e.g., a pure line, clone, physiological race, or mating strain). **3b:** an infraspecific group having characteristic properties and taxonomic status.

stramonium. the dried leaves and flowering or fruiting tops with branches of *Datura stramonium* or *D. tatula*. It is an antispasmodic and has been used in the treatment of asthma and parkinsonism. See also stramonium poisoning.

stramonium poisoning. poisoning, usually of children, due to accidental ingestion of stramonium or in adults due to an intentional overdose. Effects are similar to those of atropine poisoning, q.v.

stratification. layering; a separation into layers; separating into layers.

stratified. layered, as in stratified epithelium, which contains several layers of cells.

stratum corneum. the outer, horny layer of the epidermis of terrestrial vertebrates. It consists of clear, dead, denucleated epithelial cells whose cytoplasm has been replaced by keratin. The cells become increasingly flattened (squamous) as they approach the surface of the skin. It is the major barrier to dermal absorption of toxicants. The permeability of skin is inversely related to the thickness of this layer. Thus in humans, skin absorption is least on the soles and palms where this layer is thickest; greater on the skin of the abdomen, back, legs, and arms; and greatest in the genital (perineal) area. Injury to this layer (e.g., abrasions, lacerations, abrasion, irritation, and inflammation increase the permeability of the skin; the use of substances that hydrate the skin can further increase the rate of absorption of xenobiotics.

Straub-tail. a condition seen chiefly in laboratory mice, in which the tail is carried in a nearly vertical position. This sign is commonly associated with exposure to opiates such as morphine.

straw-colored fiber head. *Inocybe fastigiata*.

street drug. See drug.

Strelitzia. (**bird of paradise**). a genus of South African herbs (Family Cannaceae) that are often seen in flower shows and are sold in flower shops, but are rarely cultivated outside of California. The seeds and pods are toxic, producing gastrointestinal distress, vertigo, and drowsiness. See also Caesalpinia.

streptococcicide. an agent that destroys streptococci.

streptococcolysin. streptolysin.

Streptococcus. a genus of spherical nonmotile, non spore-forming, Gram-positive, facultatively anaerobic bacteria (Family Streptococcaceae) that occur in pairs or in long chains. Many are hemolytic and cause any of a variety of diseases, including throat infections, puerperal fever, rheumatic fever, and scarlet fever.

streptocolysin. streptolysin.

streptoleukocidin. a leukocytic toxin produced by streptococci cultures.

streptoline glutinant. See nematocyst.

streptolydigin. **1:** a naturally occurring epoxide antibiotic. **2:** any of a class of antibiotics that inhibit bacterial transcription by interacting with the β subunit of RNA polymerase.

streptolysin (streptocolysin; streptococcolysin). a filterable hemolysin produced by streptococci.

streptolysin O. a hemolysin, active only in the reduced state, produced by β-hemolytic streptococci. Antistreptolysin O, produced during the course of the infection, is diagnostic.

Streptomyces. a genus of some 150 species of microorganisms of the Order Actinomycetales (Family Streptomycetaceae). Most occur in soils; some are parasitic on plants or animals. They produce azaserine (an amino acid antagonist); many produce antibiotics, some of which are toxic to higher organisms (e.g., dactinomycin). *S. alboniger* produces puromycin; *S. aureofaciens* produces aureomycin; *S. caespitosus* produces the antineoplastic antibiotics mitomycins A, B, and C; *S. chromogenes* and *S. griseolus* produce anisomycin; *S. chrysomalus* and several other members of the genus produce dactinomycin (actinomycin D); *S. erythreus* produces erythromycin; *S. fardiae* produces neomycin; *S. griseus* is the source of streptomycin; *S. halstedii* produces carbomycin; *Streptomyces peucetius* var. *caesius* is the source of doxorubicin; and *S. venezuelae* is a source of chloramphenicol.

streptomycete. any of a type of filamentous, prokaryotic microorganisms of the family Streptomycetaceae, that occur widely in soil. They are characterized by the formation of a permanent mycelium and reproduction by means of conidia. some produce antibiotics (See, for example, *Streptomyces*).

streptomycin (*O*-2-deoxy-2-(methylamino)-α-L-glucopyranosyl-(1→2)-*O*-5-deoxy-3-C-formyl-α-L-lyxofuranosyl-(1→4)*N,N'*-bis(amino-iminomethyl)-D-streptamine; streptomycin A). a waste product and broad-spectrum trisaccharide antibacterial antibiotic secreted by the soil fungus, *Streptomyces griseus* (Family Actinomycetaceae); also produced commercially in large amounts. It inhibits growth by disrupting several bacterial metabolic pathways. It inhibits protein biosynthesis binding to ribosomal subunits, disrupting codon-anticodon recognition. *S. griseus* itself is not resistant to this antibiotic. Streptomycin is effective against spirochaetes, mycobacteria, Gram-negative and Gram-positive bacteria and many other microbes. It is poisonous to mink.

streptomycin A. streptomycin.

streptonigrin. an antibiotic that causes chromosome breakage. It is produced by the actinomycete *Streptomyces flocculus*.

streptothricin. the first antibiotic to be isolated, it is active against both Gram-negative and Gram-positive bacteria. It is too toxic to humans for systemic applications.

streptozocin (streptozotocin; 2-deoxy-2-[[(methylnitrosoamino)carbonyl]amino]-D-glucopyranose). an antibiotic substituted *N*-methylnitrosamine. It has been used as an antineoplastic and in the treatment of malignant pancreatic insulinoma. Symptoms of intoxication include nausea and vomiting. Lesions include liver damage and proximal tubular damage in the kidneys. Anemia, leukopenia, or thrombocytopenia also occur in about 20% of patients. Streptozotocin is an effective diabetogenic agent that selectively destroys pancreatic beta-cells; it causes chemically induced diabetes when administered to laboratory animals at a dosage of ca. 50 mg/kg, *i.v.* It acts in mammalian cells by inhibiting DNA synthesis and in microorganisms by alkylating the bases in DNA. The *i.v.* LD_{50} in rats is 138 mg/kg, whereas in dogs it is 25-50 mg/kg.

stress. **1:** any factor (external or internal) that disturbs the equilibrium of a system. The distribution of xenobiotic chemicals and their toxicities may be modified under conditions that produce stress. Thus, for example, crowding, increases the toxicity of CNS stimulants in rats. **2:** any stimulus or environmental factor that restricts growth and reproduction of an organism or population. See also strain.

stress avoidance. stress resistance, q.v., in which severe stress is avoided by means of a physical, chemical, or metabolic barrier that insulates living tissue from the stressor.

stress resistance. **1:** the ability of a living organism to survive exposure to an unfavorable environmental factor or factors (a strong stimulus, or repetitive stimuli). **2:** the amount of stress necessary to produce a specific strain (See strain, def. 2b).

stress tolerance. stress resistance q.v., through the ability of an organism to decrease or repair the strain (See strain, def. 2b) induced by the stress; See also stress avoidance.

stressor. an agent capable of producing stress. **systemic stressor**. in general, a nonspecific agent that both induces stress and stimulates defense mechanisms in a living system. At certain concentrations or amounts, virtually all substances become stressors. The resulting stress may go unnoticed in toxicity tests, for example, and the substance may be considered nontoxic at that level. **topical stressor**. an agent that elicits a localized stress response (e.g., inflammation).

Streunex™. lindane.

stringhalt (springhalt). a myoclonic affliction of one or both hindlimbs, chiefly of horses. It presents as spasmodic overflexion of the joints. Lesions of a peripheral neuropathy have been observed, involving the sciatic, peroneal, and tibial nerves. While of unknown etiology, se-

forms of the disease have been attributed by some to lathyrism (sweet pea poisoning) in the United States and to flat weed (dandelion) intoxication in Australia. Flexion may be mild, with only a spasmodic lifting and grounding of the foot. In severe cases, the foot may be drawn sharply up, touching the belly and then struck violently on the ground. Serious cases may involve atrophy of the lateral thigh muscles or even the quarters. The condition may be progressive, and euthanasia may become necessary.

stripe-back scorpion. *Centruroides vittatus*.

striped coral snake. *Calliophis japonicus*.

striped muscle. striated muscle.

striped red snake. *Calliophis sauteri*.

stripper's asthma. byssinosis.

Strobane® (terpene polychlorinates, Dichloricide Aerosol, Dichloride Mothproofer, Insecticide 3960-X14). an amber liquid mixture of polychlorinated terpenes used as an insecticide. It is a mild irritant and is carcinogenic to some laboratory animals.

strong sensitizer. See label (warning label).

strontium (symbol, Sr)[Z = 38;Ar = 87.62]. a soft, low-melting, silvery-white metallic element of the alkaline earth series. Its chemical and biological properties are similar to those of calcium and it is thus readily absorbed and stored in bone. The strontium molecule is large with a low ionization potential and is thus very electropositive and highly reactive. The anions of various strontium salts (e.g., bromide, iodide, lactate) are used therapeutically. The stable (nonradioactive) isotope is not very toxic. **strontium-89** (^{89}Sr). a radioactive isotope of strontium, a beta emitter, $T\frac{1}{2}$ = 51 days. It is used as a tracer in studies of strontium absorption by living tissues and systems (e.g., bone). **strontium-90** (^{90}Sr). a radioactive isotope of strontium, a beta emitter, $T\frac{1}{2}$ = 29 yr. It comprises about 5% of the products of uranium fission. ^{90}Sr is highly dangerous. Because of the amounts produced, its long half-life, and selective incorporation (with slow turnover) into bone tissue, ^{89}Sr poses a serious hazard to vertebrate animals. Relatively low doses cause bone cancer in many vertebrates, although sensitivity varies. Swine, for example are much more resistant than dogs. In one study, the channel catfish (*Ictalurus punctatus*) exposed to elevated levels of ^{90}Sr accumulated this isotope in their vertebrae.

strontium chromate(VI). strontium chromate.

strontium yellow. strontium chromate.

strophanthidin (3β,5,14-trihydroxy-19-oxo-5β-card-20(22)-enolide). a cardioactive glycoside, whose effect on the heart is mediated by inhibition of Na^+ -K^+ ATPase.

strophanthidin α-L-rhamnoside. convallatoxin.

strophanthin (K-strophanthin; K-s; s.-K.). a white or yellowish powder, it is a cardiotonic, cardiotoxic, neurotoxic glycoside or mixture of steroidal glycosides isolated from *Strophanthus kombé* and other plants of the family Apocynaceae. Strophanthin is given, *i.v.*, when a cardiotonic of rapid onset and short duration is needed. It is extremely toxic and also used as an arrow poison. The action is similar to that of ouabain, *q.v.* See *Strophanthus*.

G-strophanthin. ouabain.

K-strophanthin. strophanthin.

Strophanthus. a genus of shrubs, trees, and woody vines (Family Apocynaceae) that are confined largely to tropical Africa. Several species are known to be poisonous. The seeds of *S. gratus* are the principal source of ouabain. The dried ripe seeds of *S. kombe* and *S. hispidus* are poisonous and contain strophanthin, kombic acid, choline, and trigonelline; the seeds have been used to prepare arrow poison. *S. hispidus* also contains pseudostrophanthin and *S. kombé* yields also strophanthidin. *S. sarmentosus* yields sarmentocymarin, sarmentogenin, and **sarmentose**.

stropharia family. Strophariaceae.

Strophariaceae (stropharia family). a family of fungi that contains mushrooms, a few of which (as in the genus *Psilocybe*) are hallucinogenic. More importantly, the edibility of most has not been established. Most species grow on the ground in lawns, dung, humus, or on wood.

structural gene. any gene that determines protein structure.

structure-activity relationship. See quantitative structure-activity relationship.

strumitis. inflammation of the thyroid gland.

strychnia. strychnine.

strychnic. pertaining to or produced by strychnine; of or pertaining to poisoning by strychnine.

strychnic poisoning. poisoning by strychnine which characteristically heightens sensitivity to sensory stimuli and causes convulsions that impair breathing with resultant respiratory and metabolic acidosis. The convulsions are violent and take the form of general involuntary tetanus, with agonizing pain. The onset of symptoms is dramatically sudden. The victim may become irritable, uneasy or apprehensive, with muscle stiffness (especially in the neck), twitching of the face and arms, sweating, and a terrified countenance. To an observer there may be no warning; the normal acting victim may suddenly shudder and go into a convulsive seizure in which the head snaps back, the arms and legs extended, hands clenched, and the soles of the feet incurved. These seizures are accompanied by excruciating pain; the victim may shriek in agony. The height of a convulsion is marked by rigidity, hyperextension (arching of the body which is supported on the head and feet); the face and lips may become cyanotic, chest and abdomen stiff, eyes fixed and staring, the pulse slow and strong. Convulsions are usually followed by intervals of relaxation. The slightest sound or movement will induce convulsions, however, and the spasms usually worsen until the victim is almost continuously in hyperextension. Severe exposures may cause the face and lips to become cyanotic and may terminate in paralysis with contraction of the muscles of respiration and death in 1-2 hours from anoxia and exhaustion. Rigor mortis sets in immediately upon death, leaving the body in the convulsed position with eyes opened wide and an extreme facial grimace or cynic spasm. See also strychnine, *Homo sapiens*, risus sardonicus, spasm. **chronic strychnic poisoning**. strychninism.

strychnidin-10-one. strychnine.

strychnine (strychnidin-10-one; strychnia; dog button, mole-nots, mole death). **1:** a colorless, extremely bitter, nearly water-insoluble, crystalline powder, $C_{21}H_{22}N_2O_2$, that occurs in plants of the genus *Strychnos*. It is a fast-acting, convulsant alkaloid that is extremely toxic to mammals, acting chiefly by blocking the inhibition of nerve conduction by glycine in the spinal cord. Strychnine is used mainly as a rodenticide; because of this, it is a significant cause of poisoning of small animals, including pets. In medicine and allied fields it is used almost entirely as a research tool. Strychnine has no justifiable therapeutic applications, although it has been used (as the phosphate, sulfate, or hydrochloride) as a CNS stimulant or tonic, an antidote for depressant poisons, and in the treatment of myocarditis. See also strychnic poisoning, strychninism, spasm, *Homo sapiens*. **2:** to poison by strychnine.

strychnine poisoning. strychnic poisoning; strychninism.

strychninism (chronic strychnic poisoning; chronic strychnine poisoning). a toxic condition due to the misuse of strychnine. Early signs are apprehension, tremors, and twitching of the face and arms. The condition may advance to include severe convulsions and respiratory arrest. Symptoms are chiefly those of central nervous system stimulation and may also include increased acuity of hearing, vision, touch, taste, and smell, followed by tonic convulsions and vomiting. See also strychnine. See also strychnine, strychnic poisoning.

strychninization. **1:** the process of subjecting a human or other animal to the action of strychnine. **2:** the condition induced by large doses of strychnine.

strychninize (strychnize). to subject to, or expose to the action of strychnine; to bring under the influence, of strychnine.

strychninomania. a mania or mental aberration due to strychnic poisoning.

strychnism. poisoning from use of strychnine.

strychnize. strychninize.

Strychnos. a large genus of tropical trees and woody climbing vines (Family Loganaceae) with 3 to 5-nerved leaves, cymose flowers with a salvar-shaped corolla; the fruit is a berry with a thick rind. Most South American species contain chiefly quaternary neuromuscular blocking alkaloids; several species are used to make curare. The African, Asiatic, and Australian species contain tertiary strychnine-like alkaloids. See also curare, nux vomica, strychnine; brucine. ***S. nux-vomica*** (dog button plant). a tree of the Indian subcontinent, with fruit that resembles a mandarin or Chinese orange in shape and color; the seeds look like gray-velvet covered buttons as large a nickels. The fruits are attractive-looking, and many people are tempted to eat them despite their somewhat bitter taste. The entire tree including the bark contains strychnine but the highest concentrations occur in the seeds. It also contains curare and brucine. See also nux vomica. ***S. guianensis***. a source of curare. ***S. ignatii*** (St. Ignatius' bean). a tree of the Philippine jungles. The seeds are toxic and when ingested, produce symptoms similar to those of strychnic poisoning. This plant contains ignatia and brucine. ***S. toxifera***. a South American species used by natives in the Amazon and Orinoco valleys to make curare. It is the source of a number of extremely virulent curare alkaloids. See toxiferine.

stupefacient. **1:** stupefactive; stupor-inducing; able to induce stupor. **2:** an agent that induces stupor.

stupefactive. stupefacient (def. 1); able to induce narcosis or stupor.

stupor. **1:** semiunconsciousness; a state of partial or nearly complete unconsciousness such that the victim can be aroused only briefly by repeated, intense, stimulation. *Cf.* coma. **2:** a psychiatric disorder marked by diminished responsiveness.

stuporose. stuporous.

stuporous (stuporose). pertaining to, affected with, or marked by stupor.

stupp. poisonous soot that accumulates in the condensers of mercury smelters; it contains finely divided droplets of metallic mercury.

sturgeon. a fish of the family Acipenseridae. See *Acipenser*.

STX. saxitoxin.

stylopine (6,7,12b,13-tetrahydro-4*H*-bis[1,3]-benzodioxolo[5,6a:4′5′-g]quinolizine; tetrahydrocoptisine. a papaver alkaloid obtained from the roots of a number of plants of the family Papaveraceae, including *Chelidonium majus*, *Corydalis cava*, and *Stylophorum diphyllum*.

Stypven®. trademark for a preparation of Russell's viper venom used as a hemostatic agent.

styrene (cinnamene; cinnamol; ethenylbenzene; phenylethylene; styral; styrene monomer; styrol; vinylbenzene). an aromatic organic compound, it is an important monomer and intermediate in the production of plastics, synthetic rubber, and resins. It also serves (with divinylbenzene) as the basis for many synthetic ion exchangers. Generally regarded as moderately toxic, note that styrene is activated metabolically to the carcinogenic styrene epoxide. It is an irritant to the eyes and upper respiratory system, is nephrotoxic, hepatotoxic, and a CNS depressant. Symptoms of intoxication in humans may include cutaneous irritation, headache, fatigue, muscular weakness, cardiac arrhythmia, depression, dizziness, and unconsciousness. Lesions include pulmonary edema, renal and hepatic damage, glutathione depletion in the liver, reduced DNA and RNA in leukocytes, and amino acid depletion in the brain, and peripheral neuropathy. The latter results from disruption of AA transport across the blood-brain barrier. See also plastic (ABS plastic), styrene plastic, polystyrene. This family of effects is not uncommon among workers exposed to fumes or mist and is referred to as "styrene sickness."

styrene epoxide. the metabolically active form of styrene, *q.v.* It is carcinogenic.

styrene monomer. styrene.

styrene plastic. plastics such as ABS plastic and polystyrene that are made from styrene monomer. Products made with styrene plastics include air conditioners, automobile dashboards, building insulation panels, car and airplane models, cleaning brushes, clocks, coasters, floor polishes, flotation devices, ice-buckets, insulation on soft drink bottles, kitchen and bathroom wall tile, lighting fixtures, luggage tags, paints, poker chips, serving trays, sewing machine bobbins, silverware, throwaway hot drink cups, toys, telephones, and typewriter carrying cases.

styrene sickness. See styrene.

styrol. styrene.

subacute. **1a:** somewhat acute; between acute and chronic. **1b:** of or referring to the course of a disease of moderate duration or severity.

subacute myelo-optico neuropathy, subacute myelooptic neuropathy (SMON). a physiologic disorder that usually presents with abdominal pain or diarrhea with subsequent sensory and motor disturbances of the lower limbs, impaired vision, ataxia, and convulsions or coma. Fatalities are uncommon, but neurological debilities persist. Most cases are reported from Japan and Australia. Drugs of the halogenated oxyquinoline group are at least partially responsible. See iodochlorhydroxyquin.

subchronic. of intermediate duration; often applied in toxicological research or testing of vertebrate animals to exposures between 5-90 days (30-90 days in humans and longer-lived mammals). See also exposure, toxicity.

subchronic exposure. See exposure.

subchronic toxicity. See toxicity (subchronic toxicity).

subclinical. lacking clinical expression or manifestations as in the early stages or slight expression of an infection, poisoning, or disease state.

subcutaneous. **1:** beneath the skin. **2:** pertaining to a tissue layer (hypodermis) found in vertebrate skin that lies just beneath the dermis and tends to contain fat cells.

subcutaneous administration. See administration.

subcutaneous tissue (hypodermis). a layer of loose connective tissue that lies just beneath the dermis of vertebrate skin. It usually contains fat cells, large blood vessels, and nervous tissue.

subdelirium. a mild or partial state of delirium.

suberitin (suberitine). a toxin isolated from the siliceous marine sponge *Suberites domunculus*. It has hemorrhagic and respiratory effects.

suberitine. suberitin.

suberosis. a type of allergic alveolitis due to inhalation of moldy cork dust by individuals who are sensitive to the antigen which derives from a species of *Penicillium*.

suberyl arginine. a component of bufotoxins, $COOH(CH_2)_6CONHCO{:}NH(CH_2)_3CH(NH_2)COOH$, that is functionally analogous to the glucose in glucosides.

subintrance. recurrence of a paroxysm: **1a:** after a shorter than normal interval. **1b:** at decreasing intervals. **1c:** before the completion of a prior cycle; anticipation.

subjective. **1a:** perceived by or pertaining only to the affected individual. **1b:** perceived by or affecting individuals differently. **1b:** personal bias or opinion; personal, arbitrary, individual; partial, biased, prejudiced. **2:** biased; prejudiced, partial; not objective.

sublethal. less than lethal; pertaining to an agent or stimulus below the level of intensity, concentration, or amount that can cause death. A term used largely in reference to an agent or stimulus that is nearly, but not quite, lethal of and by itself; or to a quantity, intensity, level of dosage, concentration, or exposure which is nearly, but not quite, lethal.

subliminal. pertaining to a stimulus that is too weak to elicit a sensation; below the threshold of perception.

sublingual. below the tongue. See also route of administration.

subnarcotic. slightly or moderately narcotic.

subparalytic. partially paralytic.

substance. matter, stuff, material; including elements, compounds, and mixtures. See also subentries under substantia. **controlled substance**. a substance regulated under the U.S. Controlled Substances Act (1970), which regulates the manufacture, storage, prescription, sale, and distribution of such substances. They are allocated as appropriate to any of five schedules according to their 1) potential for or evidence of abuse, 2) potential for psychic or physiologic dependence, 3) contribution to public health risk, 4) harmful pharmacologic effects, or 5) potential or role as a precursor of other controlled substances. **sensitizing substance**. complement-fixing antibody. **toxic substance**. **1:** any poisonous substance. **2:** a legalistic term used in reference to a chemical element, compound, or mixture that presents an unreasonable risk of injury to human health or the environment.

substance abuse. drug abuse.

substrate. **1:** a specific substance or class of substances on which an enzyme of ferment acts catalytically. **2a:** any solid surface on which a layer or film of another substance is deposited. **2b:** substratum; a material that underlies and supports another (e.g., the soil on which a plant grows; a layer of rock beneath the soil surface. **suicide substrate**. a substrate that complexes with an enzyme to yield a reactive metabolite that inhibits further activity of the enzyme.

subterranean clover. *Trifolium subterraneum*.

subtetanic. slightly or mildly tetanic.

subtilin. an antibiotic secreted by *Bacillus subtilis*.

subtoxic. of or pertaining to a dose or concentration of a substance that is below that normally demonstrated as producing toxic effects for a given target (e.g, species, tissue, organ, sex, or age class).

succession. progression, sequence, series. **allogenic succession**. the replacement of one biotic community by another as a result of changes in the external environment. Reverse (allogenic) succession is a common result of chronic pollution. *Cf.* autogenic succession. **autogenic succession**. the replacement of one biotic community by another as a result of biological activities within the community. Toxic biological secretions often play a role in this form of succession. *Cf.* allogenic succession. **ecological succession**. the series of stages whereby a biotic community gradually changes in a particular area under a given set of environmental circumstances. This may be perceived as the appearance of a sequence of different communities over a period of time, usually with transitional stages, until a community (the climax community) results that can remain stable for a long period of time. Autogenic succession and reversed allogenic succession are of toxicological interest. **primary succession**. succession that begins with pioneer species (e.g., algae, lichens) that become established in an area devoid of life (e.g., a recent lava flow; an excavation to bare rock; a sand dune). **reversed allogenic succession**. allogenic succession in which the train of communities is reversed over that which naturally occurs; regression toward an earlier stage of succession. Reversal of succession is a fairly common effect of pollution at the community level. **reversed succession**. reversed allogenic succession. **secondary succession**. the establishment of a new community on a site previously occupied by an established community (e.g., a clearing in a forest, an abandoned field). See also allelopathy, sere.

succinamide (butanediamide). a crystalline compound, $H_2NOCCH_2CH_2CONH_2$, that is soluble in warm water, insoluble in ether. It is a nephrotoxic antiepileptic.

succinic acid bis[β-dimethylaminoethyl] ester dimethochloride. succinylcholine chloride.

succinic acid 2,2-dimethylhydrazide. daminozide.

succinic anhydride. a carcinogenic lactone.

succinylcholine chloride (2,2'-[(1,4-dioxo-1,4-butanediyl)bis(oxy)]bis[*N,N,N*-trimethylethanaminium] dichloride; bis[2-dimethylaminoethyl]succinate bis[methochloride]; 2-dimethylaminoethyl succinate dimethochloride; diacetylcholine dichloride; suxamethonium chloride; choline succinate dichloride; succinic acid [β-dimethylaminoethyl] ester dimethochloride; choline chloride succinate (2:1); listenon; Anectine chloride; Sucostrin chloride). a white, odorless, slightly bitter-tasting powder that is highly water-soluble. While relatively stable in acid solution, it is unstable in alkaline solution; it should be stored in a refrigerator. Succinylcholine chloride is an extremely toxic, synthetic, neuromuscular depolarizing blocking agent with effects that are superficially similar to those of tubocurarine. It is administered *i.m.* or *i.v.* as a muscle relaxant of short duration, an adjuvant in surgical anesthesia, to prevent trauma in electroconvulsive shock therapy, and in the management of homicidal poisonings. This drug is dangerous and should be administered only by physicians with extensive training in its use. Even then, equipment for respiratory and cardiovascular resuscitation must be immediately at hand. Overdosage or misuse may induce respiratory paralysis which requires immediate treatment. If a patient taking digitalis is given succinylcholine, he or she often experiences irregular heartbeats which may result in a heart attack.

Suckleya suckleyana (poison suckleya). a rare, but locally abundant cyanogenic herb (Family Polygonaceae) of the Great Plains from Montana to Colorado and New Mexico. Symptoms and lesions are those cyanide poisoning. At times suckleya is nontoxic. It has caused heavy losses among cattle in Colorado and New Mexico.

Sucostrin™. succinylcholine chloride.

Sucostrin™ chloride. succinylcholine chloride.

sudan grass. *Sorghum vulgare* var. *sudanense*.

sudor. sweat, perspiration.

sudorific (diaphoretic). **1:** productive of or promoting the flow of sweat. **2:** an agent that promotes or induces sweating.

suffocant. an suffocating agent (e.g., an asphyxiant gas).

suffocate. **1:** to smother, asphyxiate, to impair breathing or respiration. **2:** to suffer suffocation, to be asphyxiated. See suffocation.

suffocation. **1:** impaired or arrested breathing or respiration. **2:** the state, condition, or situation of being choked, smothered, or asphyxiated, as by inhalation of an asphyxiant gas, smothering, drowning, throttling, obstruction of an air passage. **3:** the act of causing suffocation (def. 2).

suffusion. **1:** diffusion, permeation, saturation. **2:** the spreading of a bodily fluid into surrounding tissues, as in the case of extravasation. *Cf.* perfusion, extravasation. **3:** the pouring of a liquid over the body, usually as a therapeutic measure.

sugar beet. *Beta vulgaris*.

sugar of lead. lead acetate.

suggested-no-adverse-response-level (SNARL). the level of a particular contaminant of drinking water below which no adverse effects are expected to occur following a specified period of exposure. A SNARL is not a legal standard for drinking water, but intended as a tool to help protect the public from contaminated drinking water on a temporary basis.

suicide. self-destruction; the intentional killing of oneself. In some jurisdictions, attempted suicide is a crime. Worldwide, poisoning is by far the leading cause of deaths by suicide. In the United States, suicide by poisoning is usually by overdosage of analgesics, soporifics, or by inhalation of automotive exhaust fumes. See *Homo sapiens*, Cleopatra's asp.

Suillus luteus (slippery Jack). a normally edible mushroom (Family Boletaceae), it is sometimes toxic, producing gastrointestinal distress with full recovery usually within a few hours.

sulfa drug. any sulfonamide.

sulfabenzpyrazine. sulfaquinoxaline.

sulfallate (diethylcarbamodithioic acid 2-chloro-2-propenyl ester; diethyldithiocarbamic acid 2-chloroallyl ester; 2-chloroallyl diethyldithiocarbamate; CDEC; CP 4742; Vegadex). an amber, liquid, herbicide, $C_8H_{14}ClNS$, that is soluble in water and most organic solvents. It is an irritant to the skin, eyes, and mucous membranes. It is carcinogenic in some laboratory animals and a suspected human carcinogen.

***p*-sulfamidoaniline**. sulfanilamide.

sulfanilamide (4-aminobenzenesulfonamide; *p*-anilinesulfonamide; *p*-sulfamidoaniline; *p*-aminobenzenesulfonamide). a sulfur-containing aromatic amide, $NH_2{\cdot}C_6H_4{\cdot}SO_2NH_2$, it was the first sulfonamide discovered and is chemically the simplest. Sulfanilamide is a powerful bactericide that suppresses bacterial growth by inhibiting purine synthesis. It is also an antimetabolite of the folic acid precursor *p*-aminobenzoic acid. Sulfanilamide is regarded as moderately toxic to humans and is now used mainly as an antimicrobial in veterinary practice. Toxic effects in humans include acidosis, cyanosis, hemolytic anemia, and agranulocytosis.

2-sulfanilamidoquinoxaline. sulfaquinoxaline.

sulfaquinoxaline (4-amino-*N*-2-quinoxalinylbenzenesulfonamide; *N'*-(2-quinoxalinyl)sulfanilamide; 2-sulfanilamidoquinoxaline; *N'*-(2-quinoxalyl)sulfanilamide; sulfabenzpyrazine). a sulfonamide, $C_{14}H_{12}N_4O_2S$, used chiefly as an antimicrobial and coccidiostat in veterinary practice. It upsets normal blood-clotting mechanisms of mink and causes extensive internal hemorrhage, which results in serious losses.

sulfate. a salt or ester of sulfuric acid. See also conjugation (amino acid conjugation).

sulfate conjugation. See conjugation.

sulfathiazole (4-amino-*N*-2-thiazolylbenzenesulfonamide; N^1-2-thiazolylsulfanilamide). a sulfonamide, $C_9H_9N_3O_2S_2$, the use of which is now confined to use in veterinary practice as an antimicrobial because of serious unintended effects on humans.

sulfation. the joining of SO_3H to a carbon or nitrogen atom. See also conjugation reactions.

sulfhemoglobin. sulfmethemoglobin.

sulfhemoglobinemia. the presence of sulfmethemoglobin in circulating blood.

sulfhydric acid. See hydrogen sulfide.

sulfhydryl. the univalent radical, SH; the thiol group.

sulfide. **1:** any compound of sulfur with a more electropositive element that contains one or more S^{2-} ions; disulfides contain two, S_2^{2-}; polysulfides may contain chains of sulfur atoms, S_x^{2-}. **2:** organic sulfides may contain bound sulfur, RSR'.

sulfinylbismethane. dimethyl sulfoxide.

sulfite. a salt or ester of sulfurous acid. Used as an additive in many types of food, primarily to reduce spoilage and discoloration, sulfites can induce severe allergic reactions in sensitive individuals. Effects may include dyspnea, nausea, vomiting, diarrhea, loss of consciousness, abdominal pain, cramps, and hives. In rare cases, anaphylactic shock can result in nearly instant death. Asthmatics, about 10% of whom are sulfite sensitive, form a special risk group. Sulfites are added to many foods such as dried fruits, clams and other shellfish, soups, wine vinegar, vegetables, packaged lemon juice, avocado dip, maraschino cherries, potatoes, sauces, salad dressings, gravies, and corn syrup. Food products in United States appear on container labels as sulfur dioxide, sodium sulfite, sodium and potassium bisulfite, and sodium and potassium metabisulfite. Sulfites are nearly always used in wine and beer production, but are not listed on the labels. Restaurants may also use sulfites to keep foods looking fresh. Unknown to the public, sulfites are often used in salad bars, on fresh fruits and vegetables, precut potatoes, seafood, cooked vegetable dishes, and bakery products. See also food additive.

sulfocarbamide. isothiocyanate.

sulfocarbonic anhydride. carbon disulfide.

sulfocyanate. thiocyanate.

sulfocyanic acid. thiocyanic acid.

sulfolane (1,1-dioxide tetrahydrothiofuran; tetramethylene sulfone). a toxic, flammable liquid, $(CH_2)_4SO_2$, that is miscible with a variety of organic and inorganic liquids. It is the most widely used sulfone. When ionic compounds are dissolved in sulfolane, the cations (but not the anions) are strongly solvated. Sulfolane has a number of commercial applications (e.g., as a curing agent for epoxy resins) and is used in medicine as an antibacterial. It is an irritant to the eye and skin, but its overall toxicity is rather low.

sulfonamide (sulfa drug; sulphonamide). any of a class of antimetabolites with the general formula, $R \cdot SO_2 \cdot NH_2$. Included are sulfanilamide and its derivatives (e.g., sulfapyridine, sulfathiazole, sulfadiazine). They are antagonists of *p*-aminobenzoic acids, and active against bacteria; some are herbicidal (e.g., oryzalin). Bacteria that do not require folic acid or that are able to use preformed folic acid are not affected by these drugs. All are toxic. Some tend to cause injury to the urinary tract by precipitating in the kidney.

sulfonic acid. a type of organic acid that contains the sulfonic group, $-SO_2 \cdot OH$; the simplest of these is benzenesulfonic acid, $C_6H_5SO_2 \cdot OH$. The sulfonic group can direct other groups, by electrophilic substitution, into the 3-position on the benzene ring. Most sulfonic acids are soluble in water and are strong acids in aqueous solution due to the virtually complete separation of ionizable H^+.

sulfonyl chloride. sulfuryl chloride.

4,4′-sulfonyldianiline. dapsone.

Sulfoton™. phenobarbital.

sulfoxidation. the formation of a sulfoxide by addition of an oxygen atom to a sulfur atom of an organic molecule. The most important metabolic sulfoxidations are catalyzed by the cytochrome P-450-dependent monooxygenase system or by FAD-containing monooxygenase.

sulfoxism (sulfuric acid poisoning). because it superficially resembles a syrup or glycerine, sulfuric acid is on occasion accidentally ingested. In such cases, there may be extensive tissue destruction (e.g., of skin and mucosa) about the lips, mouth and esophagus. Additional symptoms may include intense pain with swelling of affected tissues, hypersalivation, painful and difficult swallowing, a hoarse voice, dyspnea with gasping, repeated painful vomiting, and shock. See also sulfuric acid.

sulfur (symbol, S; brimstone; sulfur)[$Z = 16$; $A_r = 32.066$]. an irritant, usually yellow, low-melting, nonmetallic, water-insoluble, solid-allotropic element that occurs naturally in the free state or combined, chiefly as sulfides or sulfates. It combines with oxygen to form sulfur dioxide (SO_2) and sulfur trioxide (SO_3); these two gases combine with water to form strong acids. Sulfur reacts with most metals to yield sulfides. Most transition elements also form sulfides which are mostly covalent and often nonstoichiometric. Sulfur is mildly laxative and has been used therapeutically to treat bronchitis, gout, and rheumatism. Sulfur compounds (e.g., potassium bisulfate, sodium bisulfate, metabisulfite, sulfur dioxide, sulfuric acid) can cause fatal allergic anaphylactic shock and asthma attacks. Sulfur destroys vitamin B1 (thiamin); is mutagenic to viruses, bacteria, and yeast; and can act to potentiate carcinogens. ^{35}S is a radioactive isotope, a beta-emitter, $T_{1/2} = 87.1$ days.

sulfur dioxide (sulfurous anhydride; sulfurous oxide). a colorless, nonflammable, irritant gas or liquid, SO_2, with a strong, acrid, suffocating odor that is soluble in ethanol, water, and ether. It is a powerful reducing agent used to prevent oxidative decay of food and medicinal products; as a bleach; as an intermediate in the production of sulfuric acid, sodium sulfite, glass, disinfectants, fumigants, wine, and in brewing. It has numerous additional uses. It forms sulfurous acid, H_2SO_3, in aqueous solution. Sulfur dioxide is a by-product of coal and fuel oil combustion and is a global air pollutant of considerable significance and is chiefly responsible for acid rain. Atmospheric sulfur dioxide corrodes metals and contributes to the decay of some building materials. When inhaled at low atmospheric concentrations most sulfur dioxide is removed in the upper respiratory tract. It is nevertheless a powerful irritant of the eyes, upper respiratory tract and bronchi. It can cause rapid, acute irritation of the eyes with tearing and inflammation; its action on the respiratory tract causes coughing, dyspnea, and spasm of the larynx; it can also cause pulmonary edema and respiratory paralysis. Symptoms of exposure may include rhinitis, conjunctivitis, corneal erosion and opacity, an altered sense of smell, cough, bronchoconstriction, rales, fatigue, pneumonia, dyspnea, a thickening of the respiratory mucous layer, and inhibition of ciliary movement. High and lethal concentrations are asphyxiating. Sulfur dioxide potentiates the carcinogenicity of polycyclic aromatic hydrocarbons. Ocupational exposures may occur in sulfur mining or the mining of ores that contain sulfur or in smelters that roast sulfur-containing ore; in the manufacture of paper; and the pulp industry; in factories that manufacture sulfuric acid, and in some manufacturing plants where sulfuric acid or sulfur dioxide is used for organic synthesis. See especially sulfuric acid.

sulfur dust. an inorganic fungicide.

sulfur hexafluoride. a colorless gas, SF_6, which is which is essentially nontoxic when pure. It is often contaminated, however, with toxic lower fluorides.

sulfur monochloride. an oily, fuming, orange liquid, S_2Cl_2. It is a strong irritant to eyes, skin, and lungs.

sulfur monofluoride (S2F2). a colorless gas, S_2F_2, with a toxicity similar to that of hydrogen fluoride.

sulfur mustard. **1:** mustard gas (See gas). **2:** any of a number of highly toxic gases (e.g., mustard oil, sesquimustard, O-mustard) used as military poisons.

sulfur tetrachloride. a nonflammable, strongly irritant gas (b.p. -40°C, m.p. -124°C), SF_4, that decomposes in water. It is used as a fluorinating agent in the production of water and oil repellents and lubricity improvers.

sulfur trioxide (sulfuric oxide). a solid anhydride of sulfuric acid, SO_3, that reacts with water or steam to form sulfuric acid. The solid compound exists in at least three forms, only one of which is stable; all are extremely reactive, corrosive, and poisonous. It is an important product of the combustion of fossil fuels. It does not exist as such in the atmosphere, however, but forms sulfuric acid due to the moisture that is always present. Air pollution data given as atmospheric concentrations of sulfur trioxide should be regarded as concentrations of sulfuric acid expressed as sulfur trioxide for conversion.

sulfur tuft. *Naematoloma fasciculare*.

sulfurated oil of turpentine (Haarlem oil; Dutch oil; Dutch drops). a mixture of 1 part sulfurated linseed oil and 3 parts turpentine. Toxic effects are similar to those of turpentine, *q.v.*

sulfureted hydrogen. hydrogen sulfide.

sulfuric acid (hydrogen sulfate; oil of vitriol; battery acid; electrolyte acid). the most widely used industrial chemical, it is a dense, colorless, oily, highly corrosive, highly reactive, oxidizing, dehydrating liquid, H_2SO_4. It is an important laboratory reagent and is used in the synthesis/manufacture of many chemical compounds including many sulfates, acetic acid, citric acid, and phenol; also in the manufacture of fertilizers, nitrate explosives, artificial fibers, dyes, drugs, detergents, glue, paint, paper, furs, and food products. Additional uses include those of a dehydrating agent, an electrolyte in wet-cell batteries, the pickling of steel. Sulfuric acid dissolves most metals. It erodes, oxidizes, dehydrates, and/or sulfonates most organic materials. The dilute acid is an irritant to the skin, eyes, and mucous membranes. In concentrated form (commercial grade acid ranges from 93-98% H_2SO_4), however, it is exceedingly corrosive to these tissues causing tissue erosion and necrosis. It readily penetrates skin and mucous membranes to erode underlying tissues with charring and necrosis. Ingestion causes severe gastroenteritis; perforation of the esophagus or stomach may produce rapidly fatal results. Chronic, long-term exposure of industrial workers to sulfuric acid aerosols may cause erosion of teeth and chronic bronchitis. Concentrated sulfuric acid is also phytotoxic, readily and rapidly penetrating leaf tissue and other soft tissues. Although cell walls are apparently unaffected, cytoplasm is rapidly destroyed without cytolysis; and chlorophyll and chloroplasts are apparently destroyed. Sulfuric acid combines with the magnesium atom of chlorophyll *in vitro,* destroying this pigment. The primary action, otherwise, on plant tissue may be via dehydration, taking up bound water. See also sulfoxism. **fuming sulfuric acid** (pyrosulfuric acid; oleum). a solution of sulfur trioxide in sulfuric acid, $xH_2SO_4 \cdot SO_3$, it is a heavy, extremely hygroscopic, colorless to dark brown, oily liquid that emits suffocating fumes, especially in moist air. It is widely used in industry as a sulfating, sulfonating, and nitrating agent. Fuming sulfuric acid is an extremely hazardous material. It is strongly irritant and corrosive to living tissue (See sulfuric acid); it reacts violently with water, and is explosive.

sulfuric acid, cadmium salt, hydrate. cadmium sulfate (1:1) hydrate (3:8).

sulfuric acid, cadmium salt, tetrahydrate. cadmium sulfate tetrahydrate.

sulfuric acid diethyl ester. diethyl sulfate.

sulfuric acid dimethyl ester. dimethyl sulfate.

sulfuric acid ester. See ester sulfate.

sulfuric acid mist. the form of sulfuric acid that occurs in the atmosphere. It is an aerosol produced by the oxidation of atmospheric sulfur dioxide as well as by direct emissions from coal-fired powerplants and other coal-burning industrial operations. This mist is far more difficult to remove from the atmosphere than gaseous sulfur dioxide. The droplets are respirable and can reach the alveoli in the lungs without being absorbed in the nose, pharynx, or bronchial tree; they are thus potentially quite harmful.

sulfuric acid monomethyl ester. methyl sulfate.

sulfuric acid poisoning. sulfoxism.

sulfuric anhydride. sulfur trioxide.

sulfuric chloride. sulfuryl chloride.

sulfuric ether. ethyl ether.

sulfuric oxide. sulfur trioxide.

sulfuric oxychloride. sulfuryl chloride.

sulfurous anhydride. sulfur dioxide.

sulfurous oxide. sulfur dioxide.

sulfuryl chloride (chlorosulfuric acid; sulfonyl chloride; sulfuric chloride; sulfuric oxychloride). a colorless, toxic, corrosive liquid, SO_2Cl_2, m.p. -54°C, b.p. 69°C, used for organic synthesis.

sulky. *Apistus carinatus*.

sulphate. sulfate.

sulphonamide. sulfonamide.

sulphur. sulfur.

sumac. **1a:** a common name for nonpoisonous plants of the genus *Rhus*. **1b:** common name of *R. vernix* (poison sumac).

Sumatran pit viper. *Trimeresurus sumatranus*.

sumicidin. fenvalerate.

summation. **1a:** a cumulative process or effect. **1b:** the production of an effect by repeated exposure of a living system to subthreshold doses of a biologically active chemical. **1c:** the induction of a nerve impulse by a sequence of subthreshold stimuli. **1d:** increasingly stronger contraction of a muscle due to continuing or frequent stimulation that does not allow complete relaxation to occur.

summer clam. *Schizothaerus nuttalli*.

summer cypress. *Kochia scoparia*.

summer fescue toxicosis (summer toxicosis). a warm season condition of cattle, sheep, and horses marked by a reduction in weight gain or in milk production. Animals are afflicted during the summer when eating *Festuca arundinacea* (tall fescue) forage or seed. The severity of the condition varies from field to field and from year to year. It is caused by toxic alkaloids in forage or seed associated with an endophytic fungus, *Acreminium coenophialum*, thatinfests the fescue. Reduced performance may appear within 1-2 weeks after fescue feeding is started. Further signs of intoxication include elevated body temperature, increased breathing rate, rough hair coat, lower serum prolactin levels, and excessive salivation; the animals seek wet spots or shade. Lowered reproductive performance and agalactia have been reported for both horses and cattle. A thickened placenta and birth of weak foals have been reported in horses.

summer kill. a partial or complete kill of a fish population in a pond or lake during the warm season. Such kills may result from excessively warm water, depletion of dissolved oxygen, release of toxic substances from a decaying algal bloom, or by a combination of such factors.

summer toxicosis. summer fescue toxicosis.

sun spurge. *Euphorbia helioscopia*.

sunbeam snakes. See *Maticora*.

sunflower. See *Helianthus annus*.

sungahuni. *Thelotornis kirtlandii*.

sunscald. a form of photosensitization that affects swine in the southeastern United States during autumn and winter when pastured on *Avena sativa*. It is especially common on wet pasturage. Symptoms include intense itching with necrosis and sloughing of skin. The ears may be lost in acute cases.

superacid. excessively acid.

superacidity. excessive acidity; abnormally high acidity.

superactivity. hyperactivity (I); abnormally great activity.

superacute. **1:** markedly acute or severe. **2:** of or pertaining to a disease marked by rapid progress and extremely severe symptoms.

superalkalinity. greatly excessive alkalinity.

superb snake. *Denisonia superba*.

superciliary gland. See salt gland.

superexcitation. extreme excitement.

superfemale. a female that has three X chromosomes.

superglue. See cyanoacrylate adhesive.

superior hemorrhagic polioencephalitis. Wernicke's syndrome.

superovulation. the production of more ova than usual.

superoxide (hyperoxide; hyperperoxyl radical; superoxide anion). **1a:** the highly reactive superoxide anion, O_{2-}, produced when molecular oxygen loses a single electron or hydrogen peroxide gains an electron. It is formed when ionizing radiation passes through water or is produced metabolically by a number of enzymatic reactions (e.g., the conversion of hemoglobin to methemoglobin during phagocytosis of bacteria by granulocytes). Superoxide anion has a greater reducing capacity than oxidizing capacity and easily reduces the oxidized form of cytochrome C. **1b:** an inorganic compound that contains the O_{2-}ion. A number of superoxides are stable; all are paramagnetic and yellowish in color at room temperature. Such compounds may be formed metabolically by the action of flavoenzymes. See also dismutase (superoxide dismutase).

superoxide anion. See superoxide.

superoxide dismutase. See dismutase.

superphosphate. a fertilizer comprised mainly of calcium phosphates and calcium sulfate. It is manufactured by treating calcium phosphate ("phosphate rock"), $Ca_3(PO_4)_2$, with sulfuric acid. If the rock contains fluorides they release hydrogen fluoride and other gaseous fluorine compounds, which can be hazardous in the absence of efficient absorbing equipment.

supertoxic. poisonous to a given type of organism in extremely small amounts. Often taken to mean that even a drop of a liquid poison could be fatal. See toxicity rating.

supine. **1:** pertaining to the position in which a body is lying on its back with the face upward. **2:** the position of a hand or foot with the palm or foot facing upward. *Cf.* prone.

supperation. the formation and discharge of pus; the process of being converted into pus.

suppress. subdue, restrain, repress; conceal. See suppression.

suppressant. **1:** able to suppress, restrain, subdue; or induce suppression. **2a:** an agent that stops or greatly slows secretion, excretion, or discharge. **2b:** an agent that is able to suppress the external manifestations of a disease state (e.g., a cough suppressant). See also allomone.

suppression. **1:** restraint, repression. **2:** repression of the external manifestation of a morbid condition. **3:** complete failure of secretion as opposed to retention, in which normal secretion occurs but the discharge is retained within the body or organ. **4:** conscious inhibition of an idea or desire. See also immunosuppression.

suppressor T cell. suppressor T lymphocyte (See T lymphocyte).

suppurant, suppurative. **1a:** associated with or able to produce pus. **1b:** marked by suppuration. **2:** an agent that causes pus formation.

suppurantia. substances that cause supuration.

suppurative inflammation. purulent inflammation.

supraliminal. denoting stimuli that are above the threshold of sensation or perception.

supranormal. excessive; greater than normal; present or occurring in amounts or with values that are abnormally high.

suprapharmacologic. denoting a dosage or concentration of a drug that is in excess of that which is normally therapeutic.

supraphysiologic, supraphysiological. **1:** denoting a dosage or concentration of a physiologically active substance that is in excess of that which occurs naturally larger or more potent than would occur naturally. **2:** pertaining to the effects of a supraphysiologic dose or concentration.

suprarenal. superior to or above the kidney (in bipedal animals).

suprarenal gland. adrenal gland.

surattai pambu. *Echis carinatus*.

surface water. See water.

surfactant. **1:** a substance produced by the lungs that lowers the surface tension within the alveoli. **2:** a surface-active substance that lowers surface tension and thus has a wetting, emulsifying, foaming, and dispersive capacity. Surfactants are used in detergents, for example, to reduce the surface tension of water and to cause foaming.

surgeonfish. a common name of any fish of the family Acanthuridae, especially those of the genus *Acanthurus*.

surgical chlorinated soda solution. dilute sodium hypochlorite solution.

Surinam coral snake. *Micrurus surinamensis*.

Surinam toad. *Pipa*.

Surital. See thiamylal.

surma. a lead sulfide that was traditionally applied to the eyelids in India as a cosmetic. It was a significant, or at least an occasional, source of lead poisoning.

surmullet. See *Parupeneus chryserydros*, *Plectropomus obligacanthus*, and *Upeneus arge*. See also ciguatera.

surucucu patiabo. *Bothrops bilineatus*.

surugatoxin (SGTX). a toxin isolated from the carnivorous marine gastropod, *Babylonia japonica*. It is a water-soluble, toxic bromoindole joined to a pteridine derivative that is weakly soluble in methanol and ethanol, heat labile, dialyzable, ninydrin positive and biuret and Dragendorff negative. It is a potent mydriatic.

surveillance. continuous observation or monitoring, e.g., of the presence of a disease or a toxic substance in a community.

survival (survival rate). **1:** continuing existence; subsistence. **2:** the complement of mortality; survival = 1 - mortality. See also survivorship.

survival curve. survivorship curve.

survival rate. See survival, survivorship.

survivorship. the proportion of individuals from a given cohort, *q.v.*, surviving at a given age; the data are typically presented as a survivorship curve, *q.v.*

survivorship curve (survival curve). a graph that shows the relationship between age and the number of surviving individuals in a particular population or cohort, throughout the life cycle. There are three main forms that such a curve takes. Type 1, where probability of survival decreases with age; Type 2, where the probability of survival is constant with age; and Type 3 in which the probability of survival increases with age.

survivorship schedule. a set of demographic data that gives the number of individuals of a cohort that survive to each particular age in a population. See also survivorship curve.

susceptibility. **1a:** the state or condition of being sensitive, vulnerable; prone to influence by a stimulus, a poison, or to infection. **1b:** the degree of susceptibility to a stimulus, provocation, or influence. Biological systems (individuals, species, ecosystems) and also their components (e.g., organs, tissues, organelles, males, young, emergent vegetation) vary, sometimes substantially and even qualitatively in their susceptibility to a particular toxicant or constellation of toxicants. **2:** the likelihood that an organism will succumb to the action of a pathogen, toxicant, or other biologically active agent. Hypersensitivity, *q.v.*, is a state of extreme susceptibility. **species susceptibility**. the susceptibility of a given species or population, often judged in relation to one or more other species. In terms of toxicology, the concept of species susceptibility includes all those aspects of environment and behavior that influence the likelihood of exposure to a given poison. Thus a predator may be less susceptible to a particular plant poison that is detoxifed by a herbivorous prey, and it may be more susceptible than the prey in the case of toxicants that bioaccumulate. Even closely related organisms or those that occupy a similar niche, often differ substantially both in their susceptibility to a particular toxicant and the degree of exposure that will affect them adversely.

susceptible. sensitive, vulnerable; unprotected; prone to influence by a stimulus, a poison, or to infection.

suspended sediment. See sediment.

suspended solids. a mixture of fine, nonsettling particles on the surface of, or within sewage or other liquids. They resist removal by conventional means.

suspension. **1:** a fluid that includes fine, insoluble, particulate matter that is evenly dispersed throughout. **2:** a temporary stoppage, delay, or interruption as in suspension of the use of a substance (e.g., a pesticide) when deemed necessary by the responsible governmental authority in order prevent an imminent hazard to result from continued use.

suspension culture. individual cells or small clumps of cells growing in a liquid nutrient medium.

suxamethonium chloride. succinylcholine chloride.

swallow wort. See *Asclepias*, *Chelidonium*.

swamp. a type of wetland (fresh or salt water, tidal or non-tidal) that is dominated by trees or other woody vegetation.

swamp lily. *Crinum americum*.

swamp milkweed. *Asclepias incarnata*.

swamp sumac. *Rhus vernix*.

swartnek cobra. *Naja nigricollis*.

swartnek koperkapel. *Naja nigricollis*.

sweat gland. an exocrine gland in the skin of many mammals that secretes sweat.

sweat pea. See *Lathyrus*.

sweating. See sweat, diaphoresis.

swedish turnip. *Brassica napus* var. *napobrassica*.

sweela. *Naja nigricollis*.

sweet-birch oil. methyl salicylate.

sweet cane. calamus; *Acorus calamus*.

sweet clover (sweetclover). See *Melilotus*, sweet clover poisoning.

sweet clover hay. See sweet clover poisoning.

sweet clover poisoning. a pernicious hemorrhagic disease of animals that consume toxic quantities of spoiled (moldy) sweet clover hay or silage (See *Melilotus*). The natural coumarins in sweet clover are converted to dicumarol (bishydroxycoumarin) during spoilage. Hypothrombinemia results from ingestion of the spoiled feed. All herbivores are susceptible, but cattle are the most common victims of such poisoning, while sheep, swine, and horses are sometimes poisoned. All clinical signs are due to hemorrhage resulting from impaired blood coagulation. The appearance of clinical disease following consumption of toxic sweet clover varies greatly as a function of dicumarol content of the feed, the amount consumed, and the age of the animals affected. In some cases, clinical signs may appear only after several months of feeding on contaminated feed; onset of symptoms may be abrupt. As a consequence, death from massive hemorrhage may occur following trauma of any kind with little or no warning. Initial signs may include subcutaneous swellings due to subdermal hemorrhage, stiffness or lameness from bleeding into muscles and joints, epistaxis, the appearance of hematomas, and gastrointestinal bleeding. The blood has a reduced capacity to coagulate and bleeding may become massive, leading rapidly to death. Lesions revealed on autopsy typically include large subcutaneous and internal extravasations of blood. Neonates may be affected but rarely die unless the dam exhibits clinical signs of the disease.

sweet flag. See *Acorus calamus*.

sweet grass. See *Acorus calamus*.

sweet pea. *Lathyrus latifolius*, *L. odoratus*.

sweet pea poisoning. lathyrism.

sweet perennial pea. *Lathyrus latifolius*.

sweet potato. *Ipomoea batatas*.

sweet root. *Glycyrrhiza glabra*.

sweet sorghums. *Sorghum vulgare*.

sweetclovers. See *Melilotus*.

sweetener, nonnutritive. any of a number of natural or synthetic food additives that are usually much sweeter than sucrose. Included are saccharin, cyclamates (banned in the United States), and a number of newer sweeteners (the dihydrochalcone disaccharides, glycyrrhizin (licorice extract), dulcin(4-ethoxyphenylurea), a polypeptide extract from the "serendipity berry," and Aspartame (a combination of aspartic acid and L-phenylalanine. The latter has been approved by the US Federal Drug Administration.

swellfish. one of the many common names of bony fish of the family Tetraodontidae. See also puffer.

swellhead. lecheguilla poisoning.

Swift's disease. See acrodynia.

swimmer's itch. schistosome dermatitis (See dermatitis).

Swiss blue. methylene blue.

Swiss chard. *Beta vulgaris*.

swooning. syncope.

sword bean. *Canavalia gladiata*.

sylvatic. pertaining to, affecting, or occurring in feral animals.

sympathetic. **1:** pertaining to, caused by, or exhibiting sympathy, compassion, or understanding. **2a:** of or pertaining to the sympathetic nervous system of vertebrates or any of its components. **2b:** of or pertaining to segmental nerves that supply the spiracles of insects. **2c:** of or pertaining to coloration of an organism that imitates that of the surroundings.

sympathetic nervous system (thoracolumbar division of the autonomic nervous system). that part of the autonomic nervous system concerned with processes involving homeostasis, the utilization of energy, and stress or emergencies. The neurones emerge from the cranial, thoracic, and lumbar regions of the CNS. They have short preganglionic fibers and numerous, mainly adrenergic postganglionic fibers. The ganglia form a chain on either side of the spinal column that communicates with the CNS by a branch to a corresponding spinal nerve. Adrenalin, the chief neurotransmitter of this system, is liberated in heart, visceral muscle, glands and internal vessels. Acetylcholine also plays a role as a neurotransmitter at ganglionic synapses and at sympathetic terminals in skin, skeletal muscle, and the walls of blood vessels. The sympathetic system is concerned mainly with homeostasis. It controls heat loss which is accomplished in mammals, for example, by regulating blood vessel tone, the secretion of sweat, and erection of hair, etc. It is also plays a critical role in preparing an animal to respond to an emergency by, for example, inducing cardioacceleration, vasoconstriction of blood vessels other than those to the heart and skeletal muscles, mobilizing glucose (or glycogen) from the liver, and by stimulating the secretion of adrenalin. The latter prepares the animal for 'flight or fight', as appropriate. The actions of the sympathetic system tend to oppose those of the parasympathetic system. See also nervous system, parasympathetic nervous system, parasympathetic nervous system.

sympatheticomimetic. sympathomimetic.

sympatheticoparalytic. due to or affected by paralysis of the sympathetic nervous system.

sympathicolytic. antiadrenergic.

sympathicomimetic. sympathomimetic.

sympathin. adrenalin or noradrenalin in their role as neurotransmitters.

sympatholytic. antiadrenergic.

sympathomimetic (sympathicomimetic; sympatheticomimetic). **1:** mimicking the action or effects of the sympathetic nervous system. **2a:** an agent that produces the effects similar to those produced by the sympathetic nervous system, especially those produced by the action of the adrenergic fibers. **2b:** a sympathomimetic drug.

sympathomimetic drug. a drug similar to adrenalin in its action.

sympathoparalytic. antiadrenergic.

sympectothion. ergothioneine.

Symphytum (comfrey). a genus of perennial European herbs (Family Boraginaceae). Certain species were formerly thought to have medicinal value. Because they promote cell growth, they were used especially on boils, bruises, and other skin conditions. Some people still drink tea brewed from comfrey leaves and roots. Such internal use is ill-advised, and could be quite hazardous since allantoin, a major constituent, is carcinogenic to laboratory rats. Comfrey also contains pyrrolizidine alkaloids that are tumorigenic to laboratory animals. ***Symplocarpus foetidus*** (skunk cabbage). a toxic perennial herb (Family Araceae) with heavy roots and broadleaves that is native to low, swampy areas throughout the eastern United States. The leaves and rhizomes are toxic. The raw leaves can produce an intense burning sensation, irritation, and edema when chewed; and can cause severe gastroenteritis if swallowed. The active principle is calcium oxalate. See also *Dieffenbachia*. ***S. officinale*** (common comfrey; comfrey). this species (Var. *variegatum*) is an ornamental that has become feral in parts of North America. It grows to nearly 1 m.

symptom. **1a:** any bodily condition or feeling experienced (and noticed) by a person (e.g., pain, nausea, dizziness, anxiety, depression.) that is indicative of a disease, abnormality, or morbid phenomenon. Symptoms of poisoning may vary greatly as a function, not only concentration or dosage, but also with the route of exposure, the type and source of the poison, various environmental factors that affect toxicity, and individual variation in susceptibility. **1b:** subjective evidence of a morbid phenomenon or physical disturbance conveyed by the patient. *Cf.* sign. See also phenomenon. **2:** sometimes used as a collective term that includes the term sign. As a convenience it is frequently so used in the present work. **accessory symptom** (assident symptom). a minor symptom; one that is not pathognomonic. **accidental symptom**. one that occurs incidentally during the course of a disease, but has no relationship to the disease. **assident symptom**. accessory symptom. **cardinal symptom**. the symptom of greatest significance in the diagnosis of an illness or morbid condition. **cardinal symptoms**. often taken as those related to the pulse, temperature, and respiration. **characteristic symptom** (guiding symptom). a symptom that is almost always associated with a particular disease or condition. *Cf.* pathognomonic symptom. **constitutional symptom** (general symptom). a symptom that is due to, or indicative of, a disease or disorder that affects the entire body. **delayed symptom**. a symptom that appears some time after exposure to the precipitating cause of an illness or morbid condition. **direct symptom**. a symptom due to the direct effects of a disease. **equivocal symptom**. a symptom that is associated with a number of different diseases or morbid conditions. **focal symptom**. a symptom at a specific location (e.g., epigastric pain). **general symptom**. a constitutional symptom. **guiding symptom**. a characteristic symptom. **local symptom**. a symptom that indicates the location of the disease process (e.g., as in the case of pain about the site of a snakebite). **negative pathognomonic symptom**. a symptom that never occurs in a particular disease or condition. Its presence thus rules out that disease. **presenting symptom**. a symptom that prompts a person to seek medical attention. Presenting symptoms are not necessarily the first symptoms to appear.

symptomatic. **1:** indicative, characteristic; of or pertaining to the nature of a symptom. **2:** pertaining to or indicative of a particular disease or disorder. **3:** of or pertaining to the combination of symptoms typifying a disease. **4:** exhibiting the symptoms of a particular disease but having a different cause. **5:** directed to the relief of symptoms, as in symptomatic treatment or therapy.

symptomatology. See semeiotics.

symptomatolytic (symptomolytic). causing the elimination of symptoms.

symptomolytic. symptomatolytic.

syn. synonym.

Synanceja (stonefishes). a genus of venomous, well-camouflaged, brownish bony fish (Family Scorpaenidae) that occur throughout the south Pacific and Indian oceans. They usually occupy tide pools and shoal reef areas where they often lie motionless in crevices, holes, under rocks, or buried in sand. The venom apparatus consists of short, heavy, dorsal, anal, and pelvic spines with very large, well-developed venom glands. The sting, which is the most severe of the Scorpaenidae (and probably of any bony fish), causes excruciating pain that may persists for several days. The victim may scream, thrash about, and even lose consciousness. The wound becomes cyanotic, numb, and often necrotic. It may swell to such an extent that movement is impaired and sometimes the entire limb becomes rapidly paralyzed. In severe cases, symptoms (similar to those of *Scorpaena*) may include delirium, convulsions, and cardiac failure, nausea, vomiting, lymphangitis, swelling of lymph nodes, fever, aching joints, respiratory distress, convulsions, and occasionally death. Many months may be required for recovery. An antivenin is produced by hyperimmunizing horses with the stonefish venom. It was first developed by the Commonwealth Serum Laboratories, Melbourne, Australia. ***S. horrida*** (deadly stonefish). an extremely dangerous venomous fish (Family Scorpaenidae) that occurs in coastal waters of India, the East Indies, China, the Philippine Islands, and Australia.

synapsis. the attraction and pairing of homologous chromosomes during meiosis.

synaptic. of, pertaining to, or occurring at a synapse.

synaptic cleft. a small gap between presynaptic and postsynaptic membranes of a synapse.

synaptic ending. the swelling at the end of an axon in a synapse.

synchronous culture. a culture of microorganisms or of tissue cells in which cell division is synchronized, i.e., each cell is at the same stage of mitosis at any one time. This synchrony is achieved by various techniques, including varying the temperature and limiting the nutrients, but cannot be maintained after a few generations.

synclonus. **1:** muscular tremor, or the successive clonic contraction of several muscles. **2:** any disease marked by muscular tremors.

syncopal (syncopic). pertaining to or characterized by syncope.

syncope (fainting; swooning; a fainting spell; a swoon). a sudden, temporary loss of consciousness due to a generalized cerebral ischemia.

syncopic. syncopal.

syncytia. plural of syncytium.

syncytiolysin. syncytotoxin.

syncytiotoxin. syncytotoxin.

syncytium (plural syncytia). **1:** a multinucleate mass of protoplasm (e.g., skeletal muscle). **2:** a group of cells in which the protoplasm of one cell is continuous with that of adjoining cells (e.g., the mesenchyme cells of the embryo).

syncytotoxin (syncytiotoxin; syncytiolysin). a cytolytic serum that acts specifically on placental cells; it is produced by immunizing animals with placental cells.

syndrome (symptom complex). the complex of signs, symptoms, and indications that collectively characterize a given morbid process or disease state; a group of associated symptoms; any complex of symptoms. Examples are: acute radiation syndrome, alcohol amnestic syndrome, bovine hemorrhagic syndrome, cerebral syndrome, fetal alcohol syndrome, fetal trimethadione syndrome, fetal warfarin syndrome, gastrointestinal radiation syndrome, hemorrhagic disease (syndrome) of poultry, hemoglobinuria, hepatorenal syndrome, Korsakoff's syndrome, moldy corn poisoning, nephritic syndrome, nervous syndrome of cattle and sheep, nettle syndrome, smoker's respiratory syndrome, toxic shock, urinary syndrome of cattle and sheep, Wernicke's syndrome, withdrawal syndrome.

syndromic. pertaining to or denoting a syndrome.

synergetic. synergistic.

synergia. synergism.

synergic. synergistic.

synergism (synergia, synergy). **1:** the interaction of two or more structures, biologically active agents (e.g., certain combinations of auxins, hormones, toxicants) or other phenomena (e.g., stimuli, physiological processes), acting simultaneously or sequentially, such that their combined effect on a specific living system is greater than the sum of their individual effects. Synergism and potentiation are sometimes treated as equivalent terms. **2:** mutualism. **3:** the action of two microorganisms whereby an effect is produced that would not otherwise occur or would proceed at a slower rate in axenic culture. *Cf.* potentiation, summation, antagonism.

synergist. a substance, structure, agent, or process that facilitates or potentiates the action of another. Thus, for example, the effectiveness of some insecticides is enhanced by inclusion of a synergist, a compound which has little or no effect by itself. *Cf.* antagonist.

synergistic (synergic; synergetic; synergistical). **1:** of or pertaining to synergism. **2:** denoting a synergist, an agent that has the ability to act in synergism.

synergistical. synergistic.

synergy. synergism.

synocytotoxin. syncytotoxin.

synstigmin. neostigmine.

syntectic. characterized by or pertaining to syntexis.

synteresis. prophylaxis; preventative treatment.

syntexis. wasting or emaciation.

synthesis. **1:** the process of forming a complex substance from simpler substances (elements or simpler compounds), as in the synthesis of polypeptides and proteins from amino acids. **2:** the union of elements to yield compounds. **3:** a compound, combination, composite, blend, mixture; a creation; that prepared by synthesis.

synthesize. **1a:** to produce or manufacture by synthesis. **1b:** to combine, blend, or mix substances to create a new substance or material that is not necessarily a chemical compound.

synthetic. **1:** artificial, man-made, manufactured; pertaining to a substance, material, structure, or entity that is not produced naturally; that does not occur in nature. **2:** prepared by synthesis.

synthetic steroid. See steroid.

synthomycin. See chloramphenicol.

syntrophic. associated or mutually dependent on one another; said in reference to living organisms.

synxenic. pertaining to a culture comprised of two or more organisms under controlled conditions.

syphilis. chronic, contagious venereal disease caused by a spirochete bacterium. See also arsenoactivation, neuralgia, primary lesion, salvarsan.

syringe. an instrument with a hollow needle used to administer drugs by injection or for withdrawing blood or other fluids.

syrup of ipecac. a syrup sometimes used as an emetic to eliminate toxicants from the stomach. It is a mixture of ipecac fluid extract, glycerin, and syrup. See ipecac, emetism.

systematics. approximately synonymous with taxonomy, it is the science of analyzing the diversity of plants and animals and their genetic and evolutionary interrelationships including their identification, classification, and nomenclature.

Systéme International d'Unités. International System of Units.

systemic. **1:** of or pertaining to a system. **2a:** of, pertaining to, or involving the whole organism. *Cf.* somatic. **2b:** involving parts of the body of an organism that are remote from the site of exposure to a xenobiotic. **3:** translocated; descriptivc of a xcnobiotic chemical that is absorbed by, and transported throughout the tissues or a living organism. Examples are systemic fungicides and pesticides applied to a plant in order to render plant tissues toxic to fungi, insects, or other pests. **4:** describing that part of the circulatory system that serves body parts other than the gas-exchanging surfaces of lungs or gills.

systemic clearance. See clearance.

systemic fungicide. See fungicide (systemic fungicide).

systemic lupus erythematosus (SLE). a chronic, mild to severe or fulminating, inflammatory syndrome that affects connective tissue, including that of the skin, nervous system, and mucous membranes. The etiology remains obscure although it is considered an autoimmune, or immune complex disease. It most commonly affects women of child-bearing age. Symptoms vary and may implicate any organ. A specific diagnosis may be difficult. There is no cure, but existing symptomatic treatments may alleviate symptoms in many cases. Chloroquine, a highly toxic antimalarial may be used to treat the skin rash; corticosteroids may be employed to alleviate fever, pleurisy, and neurological symptoms. **drug-induced systemic lupus erythematosus**. a syndrome, triggered by prolonged use of various drugs, including most commonly various anticonvulsants, hydralazine, isoniazid, and procainamide. It closely resembles systemic lupus erythematosus.

systemic toxicity. the quality or capacity of a toxicant to produce effects in an organism that affect major parts of the body that are remote from the initial site of contact between the organism and the toxicant.

systole. the contractile phase of the beating of the vertebrate heart, especially the contraction of the ventricles (ventricular systole); contraction of the atria is usually referred to as atrial systole.

systolic pressure. maximum blood pressure; the arterial blood pressure during the systolic phase of the cardiac cycle.

Systox. demeton.

T

™. trademark.

T. tera-.

T_{c}'. cytotoxic T lymphocyte.

2,4,5-T (2,4,5-trichlorophenoxyacetic acid). a synthetic, light tan or gray solid chlorophenoxy auxin with three substituted chlorine atoms, $C_6H_2Cl_3OCH_2CO_2H$. Originally developed as a chemical warfare agent in World War II, it has been widely used as a selective, post-emergence herbicide and defoliant; also as a bactericide, fungicide. It is a component of agent orange. It is a highly toxic irritant of the skin and gastrointestinal tract; is hepatotoxic; a probable abortifacient; and is thought to be carcinogenic, teratogenic and fetotoxic. The acute oral LD_{50} in laboratory rats is 300 mg/kg and in dogs is 100 mg/kg. Test mammals usually show ataxia, skin irritations, mild spasticity, nephritis, hepatitis, enteritis, and bloody stools. Autopsied livestock (sheep) in areas of usage revealed nephritis, hepatitis, and enteritis. In the late 1970s where forests were sprayed in the United States, the fetuses of many local pregnant women miscarried, a number of mammals sickened and died, and many species of birds disappeared. Because of trace contamination with dioxin, its use on all food crops except rice was banned in the United States in 1970 by the Department of Agriculture. Because of its probable potency as an abortifacient, its use was further restricted in 1979, by the U. S. Environmental Protection Agency to use mainly as a herbicide on highway borders, rangelands, rice, and sugarcane fields. In 1985, its use on all noncrop sites was banned by the EPA. See warfare agent.

2,4,5-T/silvex. silvex.

2,4,6-T (2,4,6-trichlorophenol). a toxic, nonflammable chlorinated phenol. It is a solid that forms yellow flakes and is soluble in ethanol, acetone and ethyl ether. It is used as a fungicide, herbicide and defoliant. See trichlorophenol.

$T_{1/2}$. half-life.

T-496. a buckthorn toxin, *q.v.*

T-514. a buckthorn toxin, *q.v.*

T-516. a buckthorn toxin, *q.v.*

T-544. a buckthorn toxin, *q.v.*

T cell. T lymphocyte.

T lymphocyte (T-cell; thymus-dependent lymphocyte; thymus-derived lymphocyte). any of a type of lymphocyte that develops from pluripotent bone marrow stem cells that migrate to and develop in the thymus. The mature cell is transported to other lymphoid tissues (e.g., lymph nodes, spleen) where they interact with B cells and accessory cells to stimulate antibody production. T cells also interact directly with antigen-bearing cells, thereby providing cell-mediated immunity. There are three main subpopulations of T cells: cytotoxic, helper, and suppressor cells. **cytotoxic T-lymphocyte** (cytotoxic T-cell; T_c; CTL). a type of T-lymphocyte that interacts antigenically directly with and destroys virus-infected, parasite-infected, or otherwise antigenically altered or abnormal cells. It is the effector cell of cell-mediated immunity. **helper T lymphocyte** (helper T cell; TH cell; $CD4^+$ helper/inducer T lymphocyte; T_h). a type of T-lymphocyte that helps to stimulate the antibody production of B lymphocytes and certain other T lymphocytes by antigens. **suppressor T lymphocyte** (suppressor T cell; T_s). a T lymphocyte that inhibits cell division or the production of specific antigens in certain other T and B lymphocytes. *Cf.* B lymphocyte.

T-spider. See *Latrodectus*.

T^c. cytotoxic T-lymphocyte.

T_h. helper T lymphocyte (See T-lymphocyte).

T_s. suppressor T lymphocyte (See T lymphocyte).

T.S. test solution.

T.U. toxic unit; tuberculin unit.

Ta. the symbol for tantalum.

Tabac du diable. *Veratrum viride.*

tabacism. chronic tobacco poisoning. *Cf.* tabacosis.

tabacosis. 1: poisoning by tobacco, especially by inhalation of tobacco dust. **2:** tabacosis pulmonum; a type of pneumoconiosis associated with tobacco dust. *Cf.* tabacism, tabagism, nicotinism.

tabacosis pulmonum. See tabacosis (def. 2).

tabagism (nicotinism). a condition resulting from excessive use of tobacco. *Cf.* tabacism, tabacosis, nicotinism.

tabes. a progressive wasting away; any chronic disease with a gradual, progressive wasting. **tabes ergotica. 1a:** tabes resulting from the use of ergot. **1b:** the ataxia, amyotrophy, and neuralgic pain of ergotism.

tabescence. the condition of progressively wasting away.

tabescent. wasting; pertaining to a progressive wasting or withering; characteristic of tabes.

table salt. sodium chloride.

tabun (*O*-ethyl *N,N*-dimethylphosphoramidocyanidate; diisopropylfluorophosphate; dimethylaminoethodycyanophosphine oxide; dimethylphosphoramidocyanidic acid, ethyl ester; ethyl phosphorodimethylamidocyanidate; GA). a highly volatile liquid organophosphate war gas developed in Nazi Germany just prior to World War II and used extensively by that nation from 1941-1945. It is an extremely powerful, irreversible anticholinesterase that acts chiefly on the sympathetic nervous system. The action and symptoms of intoxication of tabun are similar to those of parathion. Symptoms may include a runny nose, dimness of vision, miosis, dyspnea, drooling, nausea, vomiting, headache, cramps, involuntary defecation and urination, excessive sweating, drowsiness, tremor, convulsions, coma, respiratory failure, and death. A lethal exposure (which, for humans may be as low as 0.01 mg/kg body weight) rapidly produces death by respiratory failure. See also anticholinesterases, G agent, organophosphate, *Homo sapiens.*

tachilett. *Echis carinatus.*

tachlit. *Cerastes cerastes.*

tachycardia (polycardia; tachyrhythmia; tachysystole). an abnormally rapid heart rate, usually taken as a rate greater than 100 beats per minute in resting humans. Tachycardia in small mammals and birds, though not precisely defined would be much higher than the value for humans and for larger homeothermic and most other animals would be substantially lower. *Cf.* bradycardia.

tachycrotic. of, pertaining to, characterized by, or causing a rapid pulse.

tachyphylaxis. a rapid, progressive decrease in response to repetitive exposure to a pharmacologically or physiologically active substance.

tachypnea (polypnea). an unusually rapid rate of breathing.

tachyrhythmia. tachycardia.

tachysystole. tachycardia.

tacrine (1,2,3,4-tetrahydro-9-acridinamine; 9-amino-1,2,3,4-tetrahydroacridine; 1,2,3,4-tetrahydro-5-aminoacridine). a bitter, water-soluble, yellow crystalline substance. A centrally active acetylcholinesterase used as a respiratory stimulant and antidote to curare.

tadioko. *Naja naja.*

[illegible]

taeniacide. teniacide.

tag. See label.

tailed pepper. *Piper cubeba.*

tailflower. *Anthurium.*

taipan. *Oxyranus scutellatus.*

Taiwan banded krait. *Bungarus multicinctus.*

Taiwan coral snake. *Calliophis macclellandii.*

Taiwan krait. *Bungarus multicinctus*

Taiwan-hai. *Calliophis sauteri.*

talc (talcum; soapstone; purified talc; steatite). a fine, native, light green to gray powder of hydrous magnesium silicate; contamination with asbestos is common. Talc is used as a dusting powder (talcum powder, face powder) to control rashes in infants and for use with cosmetics. Talc has many industrial applications and occurs in many commercial preparations including chewing gum, cosmetics and toiletries (many of which contain allergens and other toxic or potentially toxic materials), electrical insulation, lubricants, paints, rubber, and vitamin supplements. It is a lung irritant and infants have occasionally died within a few hours from inhaling massive amounts. In acute cases, the mucous membranes of the lungs and respiratory tree are desiccated, the broncioles are congested, and pulmonary edema may develop with pneumonia as a complication. The Japanese treat rice with talc, which has been suggested as a cause of stomach cancer. *Cf.* talcosis.

talcosis. an occupational disease (e.g., of talc miners). It is a pneumoconiosis marked by pulmonary fibrosis and pleural sclerosis, sometimes with involvement of the lymph nodes. **pulmonary talcosis**. that due to inhalation of talc dust.

talcum. talc.

talcum powder (face powder). a powder used, for example, as a cosmetic and to control diaper rash. It may contain talc contaminated with asbestos fibers. It is usually applied with a soft cloth or "powder puff." As one pats it on the face, a portion of the powder enters the air and some inevitably reaches the lungs. *Cf.* talc. See also asbestos.

tall fescue. *Festuca arundinaceae.*

tamaga. *Bothrops lansbergii.*

tanacetin. thujone.

tanacetol. thujone.

tanacetone. thujone.

Tanacetum (tansy). a genus of annual and perennial herbs (Family Asteraceae). All parts of the plant contain a toxic oil (cedar leaf oil, *q.v.*), the toxic component of which is thujone (formerly termed tanacetin from this source). Small amounts of flowers and young leaves are used as a seasoning, but can cause violent reactions and death if consumed in large quantities. The thujone content of these plants varies from none to ca. 95%. ***T. vulgare*** (common tansy). a highly toxic perennial herb, growing to a height of 1 m. It is indigenous to Europe, but has become naturalized elsewhere, occurring widely in the United States, especially in the eastern states and the Pacific Northwest. The leaves, flowers, and stem contain a toxic oil, tanacetin.

tandaruma. *vipera superciliaris.*

tang. a common name of any of the many species of marine bony fish of the family Acanthuridae.

tanghin. **1:** *Cerbera tanghin*, the ordeal tree of Madagascar. **2:** tanghinia; a virulently poisonous extract prepared from the seed of *C. tanghin*. It is used as an arrow poison. *Cf.* tanghinin.

tanghinia. tanghin.

Tanghinia (ordeal trees). a genus of evergreen Madigascan trees (Family Apocynaceae). ***T. madagascariensis***. a source of tanghinin. ***T. venenifera*** (Tanghin, ordeal bean of Madagascar). a species indigenous to Madagascar, but scattered specimens occur in Hawaii. It is a fragrant, very attractive plant with star-shaped flowers. The sap is milky and viscous. The seeds are notoriously poisonous and symptoms of intoxication may appear almost immediately following ingestion of the seed. See tanghin, tanghinin.

tanghinigenin. the extremely toxic genin of tanghinin.

tanghinin. an extremely toxic glucoside isolated from the seeds of *Tanghinia madagascariensis* and *T. venenifera*. It has digitalis-like activity, and almost immediately produces digitalis-like symptoms. The genin is tanghinigenin. *Cf.* tanghin.

tangs. See Acanthuridae.

tanier. a common name of any herb of the genus *Xanthosoma*.

tannate. any salt or ester of tannic acid.

tannic. of or pertaining to tan (tan-bark) or tannin.

tannic acid (gallotannic acid). **1:** a glucoside, $C_{76}H_{52}O_{46}$, that yields gallic acid and glucose on hydrolysis. It occurs in many plants especially in the bark of oak trees and related species (Family Fagaceae). Tannic acid has been used as a styptic and astringent and to treat diarrhea. It is toxic by ingestion and inhalation. **2:** tannin. See also oak poisoning.

tannin. **1:** any of a heterogeneous group of acidic materials derived from gallic acid. The taste is astringent. They occur in the leaves, vascular tissues, bark, unripe fruits, galls, and seed coats of many higher plants. They are found in vacuoles, cytoplasm, and sometimes cell walls or in specially enlarged tannin cells or systems of cells that appear as granular masses or yellow, red, or brown bodies. They are used to tan animal hides to produce leather, in photography, in the preparation of inks and dyes, and as clarifying agents for beer and wine. **2:** tannic acid. **3:** any of a number of substances that contain tannin. See also gallic acid, oak poisoning, acorn poisoning.

tansy. *Tanacetum*.

tansy aster. See *Gutierrezia*, *Machaeranthera*.

tansy mustard. *Descurainia pinnata*.

tansy oil. See thujone.

tansy poisoning. poisoning by herbs of the genus *Tanacetum*, *q.v.* Symptoms of tansy poisoning include a weak, rapid pulse; severe gastritis; miosis; foaming at the mouth; violent seizures; convulsions; and death. Kidney damage is revealed on autopsy. Touching the plant causes a contact dermatitis. Tansy is also an abortifacient. The toxic principle is thujone. "Oil of tansy" is used to kill intestinal worms, induce abortion, and encourage menstruation. Humans are often poisoned by taking overdoses of the oil or of tea made from the leaves. Small amounts of flowers and young leaves used as a seasoning are probably safe; consumption of large amounts can cause violent reactions and death.

tantalum (symbol, Ta)[Z = 73; A_r = 180.95]. a gray, heavy, malleable, ductile, noncorrosive transition metal.

Tantilla. a genus of small, slender, burrowing snakes (Family Colubridae) found in the southern half of United States southward into South America. At least some species, while not harmful to humans, secrete a mildly toxic venom. ***T. atriceps*** (Mexican black-headed snake). a brown or gray-brown snake, reddish-orange below, with a distinct black cap that extends downward to just above the corner of the mouth. This species is indigenous to southern Arizona and adjacent Sonora east to west Texas and Coahila. It is a secretive, nocturnal species that often hides under rocks, leaves, or loose boards. The average length of adults is about 19 cm. Two grooved fangs are borne in the rear of either side of the upper jaw. Secretion of a mildly toxic venom apparently helps to subdue invertebrate prey.

tapioca. *Manihot esculenta*.

tapopote. *Ephedra nevadensis*.

tar acne. chloracne.

tar camphor. naphthalene.

tara. *Colocasia esculenta*.

tarantula. **1:** a common name of any spider of the family Theraphosidae (also called bird spiders), *q.v.* Tarantulas are very large, hairy, and highly venomous, most are relatively inoffensive to humans. The bite in most cases is no more harmful than a bee sting. **American tarantula**. *Eurypelma hentzii*. **Arkansas tarantula**. *Eurypelma hentzii*. **black tarantula**. *Sencopelma communis*. **European tarantula**. *Lycosa tarantula*. **Peruvian tarantula** (pruning spider). *Glyptocranium gasteracanthoides*. See also Theraphosidae. **2:** a common genus of whipscorpions (Family Tarantulidae, Order Pedipalpi); not a spider. They occur in the southern United States. See also tarantula.

tardive. late; marked by lateness; usually applied to an illness in which the characteristic lesion is long-delayed in appearing.

tardive dyskinesia (dyskinesia tarda). a sometimes irreversible and often iatrogenic neurological syndrome usually resulting from long-term use of antipsychotic (neuroleptic) drugs. It occurs in about 10-20% of those undergoing treatment with antipsychotic drugs. Characteristic symptoms are pronounced involuntary repetitive movements of the facial, oral, buccal, and cervical musculature. The elderly are most often affected. There is no effective treatment.

target. **1a:** that which is acted upon or reacts to an agent. **1b:** that which is selectively affected or engaged (e.g., an organ) by a particular agent (e.g., a hormone, neurotransmitter, toxicant). In toxicology, the whole organism, a structure (e.g., heart), or a chemical (e.g., acetylcholinesterase) may be a target of a given toxicant in a particular instance. Some biologically active agents are relatively nonselective and have more than one target; others act selectively on one or few targets. **2a:** an entity or area to which something is selectively directed. **2b:** an aim, goal, or end. Thus the targets of one's attention, in an experiment, a test, or an assessment can, for example, can be directed to any structural or functional level from parts of molecules, to cells, organs, populations, communities, ecosystems, geographic areas, or regions.

target dose. the amount or concentration of a substance that reaches the site of action, producing a measurable effect.

target organ. **1:** an end-organ upon which a hormone or nerve acts. **2a:** any organ on which a chemical substance has a significant effect. **2b:** an organ intended to be affected by a toxicant or a therapeutic drug. **2c:** any organ, organ system or tissue for which a particular toxicant displays a strong affinity or a capacity to harm. Most systemic poisons are selective with respect to the organs (or tissues) that are affected. The organ or organs that are most affected by a toxicant are referred to as target organs. See target species, target system, target tissue.

target species. **1:** any susceptible species or population of organisms on which a biological agent acts, is intended to act, or to which such an agent (e.g., a toxicant) is applied. **2:** those species which react to a biologically active agent under the conditions obtaining when it is administered or applied. Most toxic chemicals are not species- or taxon-specific; there is usually much risk in assuming that a chemical intended for one species or one type of population will not affect others.

target system. 1: a generic term that applies to any target that can be conceived of or modelled as a holistic system (e.g., a cell, organ, population, species, etc.). **2:** a term that applies to an organ system (e.g., the cardiovascular system) that is a target of a specified agent.

target tissue. the basic definitions of "target organ" apply also to tissues. Most systemic poisons are selective with respect to the tissues (or organs) that are affected. The tissue or tissues most affected by a toxicant are usually referred to as target tissues. See target organ.

tarichatoxin (tarichotoxin). a potent neurotoxin isolated from newts of the genus *Taricha*. It is apparently identical with tetrodotoxin, *q.v.*

tarichotoxin. tarichatoxin.

taro. **1:** *Colocasia esculenta*. **2:** the rootstock of *C. esculenta*.

tarsomalacia. softening of the edge of an eyelid.

tartar emetic. antimony potassium tartrate.

tartarized antimony. antimony potassium tartrate.

tartrated antimony. antimony potassium tartrate.

tartrazine (4,5-dihydro-5-oxo-1-(4-sulfophenyl)-4-[(4-sulfophenyl)azo]-1*H*-pyrazole-3-carboxylic acid trisodium salt; C.I. acid yellow 23; 3-carboxy-5-hydroxy-1-*p*-sulfophenyl-4-*p*-sulfophenylpyrazole trisodium salt; 5-hydroxy-1-(*p*-sulfophenyl)-4-[(*p*-sulfophenyl)azo]pyrazole-3-

carboxylic acid trisodium salt; hydrazine yellow; C.I. 19140; FD&C Yellow No. 5; Food Yellow 4). a food coloring used in a variety of foods including catsup, juices, jams and jellies, dairy products, in pharmaceuticals, and as a dye for wool and silk. Sulfanilic acid is the major metabolite. Tartrazine is an allergen and is carcinogenic to some laboratory animals. The World Health Organization estimates that half the aspirin-sensitive people in the world are sensitive to this color. This substance can cause many different types of reactions including angioedema, rhinitis, urticaria, and sometimes life-threatening asthma attacks.

tarweed. *Amsinckia intermedia*.

taste bud. an organ containing the receptors associated with the sense of taste.

TAT, T.A.T. toxin-antitoxin.

taurine. a sulfonic acid-containing amino acid, $H_2NCH_2CH_2SO_20H$, that is a component of the bile salt taurocholate. It has a hyperpolarizing action on certain neurones of the central nervous system. See conjugation (amino acid conjugation).

taurine conjugation. See conjugation.

tautog. See Labridae.

taxa. plural of taxon.

Taxaceae (the yew family). a family of resinous evergreen trees and shrubs, the yews, distributed throughout the warmer regions of the earth. Some species are cultivated for their ornamental value. A number of forms are known to be toxic; See especially *Taxus*.

taxine, taxine alkaloids. a mixture of cardiotoxic alkaloids isolated from the berries and leaves of the yew, *Taxus baccata* (Family Taxaceae) and certain other members of the genus. Taxines are pharmacologically poorly understood cardiac depressants. In taxine poisoning, the heart is [illegible] deer seem impervious to these alkaloids at the concentrations present in at least some cultivars of *T. baccata* (personal observation by the author). See also *Taxus*.

taxine poisoning. poisoning by taxines. See taxine.

taxis (plural, taxes). a directed response or orientation by a motile organism toward the source of a weak or "positive" stimulus or away from a "negative" or strong stimulus (e.g., a strong irritant or other toxicant). See teleorganic, approach withdrawal, adience, appetitive.

taxon (plural, taxa). any defined taxonomic group of organisms of any rank (e.g., species, family, order, phylum).

taxonomic. of or pertaining to taxonomy.

taxonomy. approximately synonymous with systematics, it is the science of naming and classifying organisms. It is a complex and fundamental field of biological science that provides a sound and necessary basis for all biology-based sciences such as toxicology. It is the study of the theory, procedures, and rules of classification of organisms, usually according to phenotype and inferred evolutionary (genetic) relationships. **classical taxonomy**. this discipline identifies homologous morphological structures, and serological and biochemical data to describe and classify plants and animals. **cytotaxonomy**. a specialized subdiscipline in which description and classification is based on the size, shape, and number of the chromosomes in somatic cells. DNA analysis is employed if possible. **experimental taxonomy** (biosystematics). a subdiscipline of taxonomy that relies largely on experimental breeding to establish or clarify relationships among closely related species or genera and to define taxa. **numerical taxonomy**. a set of approaches based on sometimes sophisticated mathematical methods to classify organisms on the basis of similarities and differences.

Taxus (yew). a genus of poisonous evergreen trees and shrubs (Family Taxaceae) that are native throughout the Northern Hemisphere. A genus of plants (Family Taxaceae), the needles and the scarlet berry-like fruit contain a com- [illegible] and related alkaloids. While the outer fleshy part of the fruit (the aril) is generally edible, the seeds are poisonous. Human intoxication by most species is very rare, thus clinical experience and knowledge are limited. Nevertheless, symptoms usually appear within 1-3 hours following ingestion and may include nausea, vomiting, acute abdominal pain, shallow respiration or dyspnea, and disturbances of cardiac conduction similar to that of hyperkalemia; P waves may be absent from the EKG. Instances of subacute intoxication in humans may not appear until 2 days following ingestion but are generally fatal. The toxicology of *T. baccata* has been the most extensively studied. ***T. baccata*** (English yew). a poisonous woody plant that grows to a height of more than 12 m. There are a large number of cultivars; toxicity might vary among them. All parts are toxic, producing severe gastrointestinal distress, diarrhea, mydriasis, muscular weakness, coma, convulsions, and cardiac and respiratory depression. *T. baccata* is readily eaten by livestock and has been responsible for numerous fatalities among cattle. ***T. canadensis*** (American yew, Canadian yew, ground hemlock). a hardy, low, spreading bush that sometimes grows to a height of 1.4 m. It occurs in rich woodlands and thickets from Newfoundland, Nova Scotia and New England to portions of Virginia, Kentucky, Indiana, Illinois, and Iowa and is often used as a hedge plant. The pulp is edible, but seeds and wilted foliage contain taxine and are poisonous and notoriously lethal to livestock.

Tb. the symbol for terbium.

2,3,6-TBA (2,3,6-trichlorobenzoic acid). a substituted benzoic acid, TBA is a potent synthetic auxin. TBA and its aldehyde are all-purpose, non-selective, preemergence and early postemergence herbicides. It is used to control weeds in *Zea mays* (corn, maize) at doses of 0.5-1.0 kg/ha. It is absorbed by plant leaves and roots and can persist in soil for long periods with high residual toxicity.

TBII. hcxachlorocyclohexane.

TBT. tributyltin.

TBT paint. See tributyltin.

TBTO. tributyltin.

Tc. the symbol for technetium.

Tc cell. cytotoxic T lymphocyte.

TC. toxic concentration (See concentration). TC is usually followed by a number indicating the % of individuals of a test population estimated to be killed at the stated dose.

2,4,5-TC. silvex.

TC50. threshold concentration; that concentration which produces a specific toxic effect within a specified period of time.

TCA. **1:** thyrocalcitonin (See calcitonin). **2:** trichloroacetic acid. **3:** citric acid (1,2,3-tricarboxylic acid).

TCA cycle. tricarboxylic acid cycle.

TCC. triclocarban.

TCD_{50}. tissue culture infectious dose. See dose.

TCDD (2,3,7,8-tetrachlorodibenzo[b,e][1,4]-dioxin; 2,3,7,8-tetrachlorodibenzo-*p*-dioxin; 2,3,7,8-tetrachlorodioxin; "dioxin"). a compound formed as a by-product in the manufacture of certain chlorinated benzene compounds, notably trichlorophenol and its derivatives, some of which are widely used as herbicides; it may occur as a contaminant in such products. It is one of the most virulent poisons known, and is also teratogenic and a potent carcinogen to laboratory animals. Effects of exposure may include edema, hemorrhage, severe weight loss, thymic atrophy, kidney anomalies, lowered gestation index, decreased fetal weight, increased ratio of liver to body weight, and increased fatty deposition in the liver. There have been reports of illness and death resulting from occupational exposure to TCDD. TCDD is sometimes used erroneously as a synonym for dioxin; but it is only one of the dioxins; it is the most toxic. See also polychlorinated dibenzodioxin; receptor (Ah receptor); Seveso, Italy; TCDF; trichlorophenol.

TCDD-binding protein. Ah receptor. See receptor (Ah receptor).

TCDF (2,3,7,8-tetrachlorodibenzofuran). an extremely toxic compound. It is structurally and toxicologically similar to TCDD, *q.v.*

TCE. trichloroethane, tetrachloroethylene. See perchloroethylene.

tchissapa. *Pseudohaje goldii*.

$TCID_{50}$. tissue culture infectious dose (See dose).

TCM. chloroform.

TCP. **1:** tricresyl phosphate. **2:** tritolyl phosphate (TOCP).

***m*-TCPN**. chlorothalonil.

TD. toxic dose. See dose.

TD_{01}. toxic dose/1% response.

TD_{50}. median toxic dose (See dose).

TD_{99}. toxic dose/99% response.

TDE. 1,1-dichloro-2,2-bis(*p*-chlorophenyl)-ethane.

***p,p'*-TDE**. tetrachlorodiphenylethane (See 1,1-dichloro-2,2-bis(*p*-chlorophenyl)ethane.

TDI. toluene diisocyanate. See toluene-2,4-di-isocyanate.

TDS. See total dissolved solids.

Te. the symbol for tellurium.

TEA. tetraethylammonium.

teamsters' tea. *Ephedra nevadensis*.

tear. **1:** to rip, split, or pull apart by force. **2:** the liquid excreted into the eyes of a vertebrate animal by the lacrymal glands. See also bloody tears.

tear gas. **1:** chloropicrin. **2:** irritant smoke, gases or vapors (e.g., acetone, benzene bromide, xylol) that cause profuse lacrimation and are used chiefly as riot control agents. *Cf.* lacrimator.

teart. **1:** soil or plants that contain unusual concentrations of molybdenum. **2:** molybdenum poisoning of livestock that graze on vegetation grown in soil that contains unusually high concentrations of molybdenum (that graze on teart).

teartness. a synonym for molybdenum poisoning in the United Kingdom. See teart.

technetium (symbol, Tc)[Z = 43; most stable isotope = ^{98}Tc, $T_{1/2}$ = 4.2 X 10^6 yr]. an artificial radioactive transition metal.

technic. technique.

technical. **1:** pertaining to a formal technique or method; especially one that is complex, intricate, or detailed. **2:** pertaining to science, a specialized field of study, art, craft, or trade. **3:** of, designating, or pertaining to a substance that contains appreciable quantities of impurities. Often used in reference to a commercial product.

technique (technic). **1:** skill, proficiency, or the specific manner of performance that one applies to any process or activity. **2:** a means, method, or specific type of performance, or a detailed description of such (whether manual or mechanical) that is used to accomplish work or to elicit a response.

TECP. tetrachloropropane.

Tedania toxicalis. a toxic sponge (Phylum Porifera).

tedong naga. *Naja naja*.

tedong selar. *Ophiophagus hannah*.

teeth. modified dermal papillae present in the oral cavity of most vertebrate animals, the surfaces of which are sheathed by a layer of calcified tissue. They are used in biting, crushing, tearing, and masticating food, general manipulation, fighting and wounding, and in the delivery or injection of venom. See also fang.

Tegretol. See carbamazepine.

TEL. tetraethyl lead.

tel-karawala. *Bungarus ceylonicus*.

teleological. pertaining to teleology, *q.v.* The use of teleological expressions (See especially teleology, defs. 1a and 1b) in speech or text by academic or professional persons is common and, unfortunately, is often misinterpreted by persons who are untutored or unsophisticated regarding the legitimate content of scientific inference.

teleology. **1a:** the usually false assumption or belief that a process is directed toward a final goal or purpose. **1b:** the belief that ends are inherent in nature, that everything is directed toward some final purpose. **2:** the philosophical study of the evidence of purpose in nature. **3:** the fact or quality of being directed by, or shaped by purpose.

teleorganic. 1a: vital; necessary to organic life. **1b:** of or pertaining to processes or functions that are vital to the life of an organism (e.g., breathing, excretion). The full scope of toxicology is achieved only if teleorganic functions are considered fully in determining the toxicological status of a suspected toxicant for a particular type of organism. Thus, a chemical that temporarily induces sneezing in civilized humans would hardly be treated as a toxicant or hazard, but one can see possible uses for such a substance as a warfare agent. Egg shell thickness is teleorganic in birds, especially certain large species. DDT brought a number of bird populations to the point of extinction not because it injured or impaired life functions of adults, but because it caused the thinning of egg shells, many of which were broken before hatching, with fatal consequences for the progeny.

Teleostei. the largest subclass, Actinopterygii, of bony fishes. The endoskeleton is completely ossified; the body is covered by thin rounded bony scales; the tail is symmetrical (homocercal); the jaws are shortened with reduced cheek bones that allow a wide gape to the mouth. In most teleosts, the fins are supported by a few strong movable spines and the pelvic fins are at the anterior end of the body. The swim bladder normally has a hydrostatic function, conferring great maneuverability on these fishes.

telepathine. harmine.

telesthesia. sensation perceived as though from a distance.

telluric. **1:** relating to, characteristic of, or originating from the earth. **2:** pertaining to tellurium.

telluric acid. a toxic, unstable, dibasic acid, H_2TeO_4. It is reduced, liberating tellurium, during certain smelting operations. It may poison plants and animals in the vicinity of the operation.

telluric effluvium. miasma; alleged disease-causing emanations from the earth.

tellurism. a disease caused by emanations presumed to come from the earth or soil. See also telluric effluvium.

tellurium (Symbol, Te)[Z = 52, Ar = 127.6]. a rare toxic, grayish-white, lustrous, crystalline metalloid element related to sulfur. Tellurium and those of its compounds with valences of -2 (telluride), +4 (tellurite) and +6 (tellurate) are nauseants and CNS depressants. Inorganic tellurium is used as a coloring agent, as a pigment in some porcelain products and in certain specialized alloys (e.g., with lead in pipes and sheets used in chemical plants and nuclear shielding). Volatile species such as tellurium dioxide and hydrogen telluride are the most hazardous industrially. Commonly seen effects include hemolysis, weight loss, blue/black discoloration of the skin and a distinctive garlicky breath. Tellurium may produce specific neuropathies such as communicating hydrocephalus in the offspring of laboratory rats. Lipofuscinosis and peripheral neuropathies have also been demonstrated experimentally. See also dimethyltelluride, organotellurium compounds, potassium tellurite, telluric acid, tellurism.

Telone II. 1,3-dichloropropene.

telophase I. the stage of meiosis during which the nuclear envelope and nucleolus reappear while the spindle disappears.

telophase II. the stage of meiosis during which the spindle disappears as the nuclear envelope reappears and the cell membrane increases, splitting to yield 2 haploid cells.

TEM. triethylenemelamine.

temephos. abate, def. 3.

Temik™. aldicarb.

tempeh poisoning. bongkrek poisoning.

temperature coefficient. **1a:** the rate of change of a quantity or process with respect to temperature. **1b:** the proportional change in any physical property per degree rise in temperature. **1c:** the ratio by which the velocity of an enzyme-catalyzed reaction increases for a given rise in temperature. **1d:** a common temperature coefficient used in physiology and related sciences is Q10, which is the quotient of the two rates at which a reaction proceeds at temperatures that differ by 10°C. This is a convenient standard because the velocity of many biochemical reactions and physiological processes approximately doubles for each increase in temperature of 10°C (i.e., Q10≈2).

temperature spike. See spike, temperature.

template. a pattern that serves as a mold for the production of an oppositely shaped structure. Thus, for example, one strand of DNA is a template for synthesis of the complementary strand.

temple snake. *Trimeresurus wagleri*.

temporal lobe. area of the cerebrum responsible for hearing and smelling, the interpretation of sensory experience, and memory.

temulence. inebriation; intoxication.

temuline. a presumably toxic alkaloid contained in seeds of *Lolium temulentum*.

TEN. toxic epidermal necrolysis.

1080. sodium fluoroacetate.

tenderness. abnormal sensitivity, registering pain to light touch or pressure.

tendon. connective tissue that connects muscle to bone.

tenesmic. of or pertaining to tenesmus; exhibiting tenesmus.

tenesmus. straining, especially painful or ineffectual straining during urination or evacuation of the bowels.

teniacide (taeniacide; tenicide). **1:** lethal to tapeworms. **2:** an agent that is lethal to tapeworms.

tenicide. teniacide.

teniotoxin. a poison that occurs in tapeworms.

tennecetin. natamycin.

Tenormin. a beta adrenergic blocking agent.

tentacular bulb. the swollen base of a tentacle in many coelenterate medusae. It is the site of nematocyst formation and also of digestion.

tenthmeter. Ångstrom.

tentoxin. a cyclic tetrapeptide comprised of *N*-methylalanine, glycine, *N*-methyl dehydrophenylalanine and leucine. It has been isolated from various species of phytopathogenic *Alternaria*.

Tenuate. diethylpropion.

Tenuate Dospan. diethylpropion.

Teonanacatl. *Psilocybe mexicana*.

TEP. triethylphosphate.

TEPA. triethylenephosphoramide.

Tepanil. diethylpropion.

Tephrosia (hoary pea). a genus of mostly tropical herbs and small shrubs (Family Fabaceae, formerly Leguminosae) with odd-pinnate leaves and purplish or white flowers. They occur in Africa, America, and Australia. ***T. virginiana*** (catgut; devil's shoestring; goat's rue; rabbit's pea; turkey pea). the leaves are toxic and were earlier dried and used as an anthelmintic. ***T. vogelii***. a source of rotenone and tephrosin.

tephrosin (hydroxy-duguelin). an isomer of toxicarol obtained from the leaves of *Tephrosia vogelii* and the roots of *Derris* and *Lonchocarpus*.

TEPP (ethyl pyrophosphate; tetraethyl pyrophosphate). a clear, white to amber, hygroscopic, liquid organophosphate, $(C_2H_5)_4P_2O_7$, that quickly hydrolyzes on contact with water. It is miscible with water and all organic solvents except aliphatic hydrocarbons. It has been used as an insecticide and rodenticide. It is a potent irreversible anticholinesterase. TEPP is no longer in use as an insecticide because it is supertoxic to humans and extremely toxic also to other mammals; it appears to be a cumulative poison, is toxic by all routes of exposure, and is rapidly absorbed through the skin. One herd of 29 cattle, including calves and adults was accidently sprayed with 0.33% TEPP emulsion; all reportedly died within 40 minutes.

tera- (T). a prefix to units of measurement that indicates a quantity of 1 trillion, 10^{12}, times the value indicated by the root. See International System of Units.

teras (plural, terata). a malformed individual or morphologic anomaly (a monster) that arises during development and growth. Such monsters may be exemplified by the presence of redundant, defective, misplaced, or grossly malformed parts, or the absence of critical parts. In the case of mammals, a teras is a grossly malformed fetus.

terat. abbreviation for teratology.

terat- or **terato-**. **1:** a prefix implying portent, marvel, monster. **2:** denoting relationship to a teras. See also teratism.

teratism (teratosis). **1a:** an anomaly of development that produces a teras. **1b:** an anomaly of organic form and structure; a monster. *Cf.* teras.

teratogen. any dysmorphogenic xenobiotic chemical or environmental factor that can cause serious malformations during development and growth of animals or plants. In vertebrate animals teratogens are those agents that act from fertilization to hatching, birth, or during metamorphosis. Teratogens are mechanistically diverse and include, for example, malnutrition, various xenobiotics (e.g., thalidomide), ionizing radiation, and virus infections (e.g., rubella).

teratogenesis (teratogeny). **1a:** the production of monstrous growths or fetuses. **1b:** abnormal development or growth that produces monsters (teras), monstrous growths, or fetuses whether viable or nonviable. *Cf.* dysmorphogenesis. **2:** the set of processes involved in the production of a monster or teras.

teratogenetic. pertaining to teratogenesis. *Cf.* dysmorphogenic, teratogenic.

teratogenic. **1:** of, pertaining to, having the action of a teratogen. **2:** able to or tending to produce developmental malformations and monstrosities.

teratogenicity. the capacity of an agent to produce fetal malformations. See also teratism, teratogen.

teratogeny. teratogenesis.

teratoid. **1:** abnormal in formation (e.g., a tumor). **2:** characteristic of a teratoma. **3:** a teratoma.

teratologic. pertaining to teratology.

teratological specimen. an individual malformed embryo, foetus, or early developmental or growth stage of a plant or animal; a malformed organ or body part; a teratism, a monstrosity.

teratologist. a student of or specialist in teratology.

teratology. **1:** the study of malformations, monstrosities, or critical deviations from that which is morphologically and functionally normal in developing and growing organisms. **2:** the study of the origin and nature of terata and the processes involved in their initiation and development. Much of this field is weighted toward abnormalities caused by radiation, viruses, and xenobiotics, including drugs. Teratology generally combines knowledge from developmental biology, pathology, and specific knowledge of the physiology and biochemistry of the particular organisms studied. Branches of this science include phytoteratology and zooteratology.

teratoma (plural, teratomas or teratomata). a tumor derived from more than one embryonic layer and comprised of a heterogeneous mixture of tissues.

teratosis. teratism.

terbacil. (pyrimidine: a uracil.) a selective herbicide absorbed mainly through the roots. It is used to control many annual and some perennial weeds without damage to a number of crops.

terbium (symbol, Tb)[Z = 65; A_r = 158.93]. a soft, ductile, malleable, silver-gray metal of the lanthanoid series of elements that readily oxidizes in air. Terbium chloride hexahydrate, $TbCl_3 \cdot H_2O$, is moderately toxic to laboratory mice, *i.v.*, but is only slightly to moderately toxic, *per os*.

terebinthinate (terebinthine). **1:** impregnated with or containing turpentine. **2:** a preparation that contains turpentine.

terebinthine. terebinthinate.

terebinthinism. turpentine poisoning.

Terebra. a genus of slender, marine, gastropod mollusks (Family Terebridae), with a long shell that tapers to a sharp point. All are carnivorous and some are venomous. ***T. concava*** (concave auger). a small, slender, venomous, marine snail about 25 mm long and 6 mm wide that occurs along the North Carolina coast to both of the Florida coasts. It preys upon small invertebrates, especially polchaetes, which it immobilizes by introducing a mild toxin via fangs on the radula. See Toxoglossa.

Terebridae (augers; auger shells). a family of mostly tropical, carnivorous gastropods, related to conidae, that have a very long tapered shell with a small aperture. Some species are venomous. See *Terebra*.

Teriosal™. the 13-*cis*-form of retinoic acid. See also retinoic acid, Accutane.

terminal. **1:** denoting, referring to, or placed at the end. **2a:** descriptive of a condition or illness that is expected to terminate in death. See dying. **2b:** indicating or referring to a near-death condition. **3:** an on-line device through which an operator can send information to and receive information from a computer or information network.

terminal residue. that amount of a pesticide residue that remains in an agricultural product at the point of human ingestion. See also maximum residue limit.

terminated. **1:** complete, concluded, finished, ended (as of life). **2:** aborted, discontinued, stopped. **3:** killed. **4:** adjudicated.

termiticide. a substance that destroys termites. The most favored method until recently was to spray potent, environmentally persistent chemicals (e.g., chlordane, aldrin, dieldrin, heptachlor) onto or into the soil near a building's foundation, and in the crawl space or basement. The use of such potent and persistent pesticides has been associated with chronic or recurring health and environmental problems (e.g., poisoned fish ponds and well water, contamination of indoor air). The EPA determined that 90 percent of the homes properly treated with termiticides still had detectable residues in the air one year later. Chlordane can persist in the treated area for 30 years or more. The sale of the above-named compounds has been prohibited in the U.S. since April 15, 1988. Chlorpyrifos (sold as Dursban, Lorsban, or Pyrinex), *q.v.*, is one of the most popular termiticides now in use, even though it is extremely toxic to fish, birds, and aquatic invertebrates and is known to persist at toxic levels for nearly 20 years in some cases.

terpene. **1:** any of a large class of naturally occurring saturated hydrocarbons produced by plants that are formed completely by 5-carbon isopentyl (isoprene) C_5 units with the general formula, $(C_5H_8)_n$. The smaller terpene molecules occur in plants (e.g., geraniums, mints, etc.), giving them their characteristic odors. Examples of terpenes with larger molecules are carotenoids, lanosterol, phytol, squalene, vitamin A, and natural rubber. Some terpenes are allopathic agents. See also aplysin, aplysinol, roridin A, fusidic acid. strobane, Artemesia californica. Some are allopathogenic agents (See *Artemisia*, *Salvia*). **2:** a moderately toxic, unsaturated hydrocarbon, $C_{10}H_{16}$, found in plant oleoresins and essential oils.

terpene polychlorinates. strobane®.

terpenism. poisoning by a terpene; the condition is marked by nausea, vomiting, convulsions, unconsciousness, pulmonary edema, and tachycardia.

terpenoid. any compound with an isoprenoid structure similar to that of terpenes; many are [illegible], menthol, terpineo, borneol, and geraniol are terpenoids.

terphenyllin. a mycotoxin from *Aspergillus candidus*.

terramycin. an antibiotic isolated from *Streptomyces*. It is effective against a wide variety of disease-producing organisms.

terreic acid. a dermatotoxic antibiotic epoxide mycotoxin from *Aspergillus terreus*. It is very toxic to laboratory mice.

terrestrial. **1:** of or pertaining to the land. **2a:** living entirely or mainly on land. **2b:** living in a situation where land or soil is the dominant substrate. **2c:** of an animal living on or under the ground, not in air or water. **2d:** of a plant living or growing on land or in the soil. **3:** used in or conducted on land.

terrestrial ecosystem. an ecosystem in which terrestrial biota predominate. Thus a watershed is a terrestrial system, even though it includes surface water and aquatic life. The water bodies within a watershed considered alone constitute an aquatic ecosystem.

2,2′:5′2″-terthiophene. α-terthienyl.

tertiary butanol. tert-butanol.

test (See also toxicity test). **1a:** to examine, **1b:** an assay, trial, examination, or experiment. See also alternative test method, assay, auditory startle, BALB/3T3, bioassay, color (FD&C color, for example), concentration (lethal concentration, etc.), conditioning, cosmetic, dose (lethal dose, etc.), double blind, experiment, extrapolation, microbioassay, no-effect level, paralysis time, partition coefficient (octanol/water partition coefficient), prothrombin time, sodium benzoate, toxicity test. **1c:** a criterion, gage, standard, or norm. **2:** a chemical reaction or reagent that has clinical significance (See chemical test). **3a:** a usually hard, often rigid, protective covering (e.g., of certain protozoans, echinoderms of the class Echinoidea). **3b:** the exoskeleton of *Limulus*. **3c:** testa; the protective outer covering of a seed. **allergenicity test**. any direct test of the allergenicity of a substance (e.g., dermal sensitization tests). See also radioallergosorbent test. **Ames test** (Ames assay; Bruce Ames procedure). an assay of bacterial growth used to assess the capacity of a substance to cause mutations *in vitro*. In this test, one measures the rate of reversion (in the presence of the test substance) of mutant histidine-requiring strains of *Salmonella typhimurium* back to a form that can synthesize their own histidine. The bacteria are inoculated onto a medium that does not contain histidine, and those that mutate back to a form that can synthesize histidine establish colonies which are assayed on the growth medium, thus providing both a qualitative and quantitative indication of specific mutagenicity. The test is also used as a preliminary screen for possible carcinogenicity of a chemical. **Anstie's test**. a test for ethanol in the urine. Anstie's reagent is a solution of one part potassium dichromate in 300 parts concentrated sulfuric acid. The solution is added dropwise to the urine; appearance of emerald-green color signifies the presence of a toxic amount of alcohol in in the body. **ascorbate-cyanide test**. a test for glucose 6-phosphate-deficient erythrocytes. A subject's blood is incubated with sodium cyanide and ascorbate. Hydrogen peroxide is generated and is able to rapidly oxidize hemoglobin to methemoglobin in glucose 6-phosphate-deficient cells, yielding the typical brown color of methemoglogin. **Bartlett's test**. a test for departures from homogeneity of within-group variance in nonparametric statistics. If the error variances are homogeneous, the sampling distribution of the Bartlett's statistic is approximated by the χ^2 distribution with k^{-1} degrees of freedom, where *k* equals the number of groups. **Bettendorff's test**. a test for arsenic in which hydrochloric acid is mixed with a test fluid and stannous chloride is then added. If the resultant fluid contains arsenic, a piece of tin foil placed in the fluid will form a brown precipitate. **Buehler test**. a dermal sensitization test in which the test chemical is repeatedly applied to the skin under a protective patch followed in two weeks by a challenge dose applied in the same manner. **carcinogenicity test**. any test of the potential of a chemical to induce cancer. Such are usually long-term tests; most frequently tests of laboratory rats and mice that are conducted throughout most of the life of the test animal (2 years or more for rats; 1.5-2.0 yrs for mice). Depending on the nature of the test chemical and the probable route of exposure of humans, the test chemical may be delivered via food, gavage, drinking water, dermal application, or inhalation. **chemical test**. **1a:** the process of examining a substance, tissue, or other material to determine its chemical composition or structure (e.g., Bettendorff's test, Fleitmann's test, Reuss' test, Wormley's test, Yvon's test). **1b:** the reagent or substance (e.g., barium hydroxide, diphenylamine, picramic acid) used in such a test. **eukaryote mutagenicity test**. any short-term test that uses cultures of eukaryotic cells or lower metazoans to ascertain the mutagenic potential of chemicals (e.g., yeast mutation tests, the sex-linked recessive lethal test in *Drosophila*, the specific locus test in mice, tests of the HGPRT, ouabain, or TK loci in cultured mammalian cells). **eye irritation test**. a test of the ability of a chemical to irritate the eye when topically applied; albino rabbits are the test animals of choice. See also Draize test. The test chemical is introduced into the conjunctival sac of one eye of each test animal; the other eye serves as the control. Endpoints include edema of the eyelid; subjective estimates of the opacity of the cornea; appearance of the iris and its ability to respond to light; redness of the conjuctiva. **field test**. any test conducted out of doors or in a situation not under the control of the investigator. The natural environment (and even the human indoor environment) is usually far more complex and more variable than that which obtains in a laboratory situation. Even so, test results from field tests may be more realistic in some cases. **Fleitmann's test**. a test for arsenic in which hydrogen is generated in a test tube containing the test liquid. The material is heated and filter paper treated with silver nitrate solution is placed over the top of the test tube. Arsenic is present if the moist paper is blackened. **Gutzeit's test**. a test for arsenic in a liquid. A small amount of sulfuric acid and a small piece of zinc are added to the suspected liquid, which is then boiled. Filter paper treated with silver nitrate solution is placed in the vapor stream above the test tube. If paper turns yellow, arsenic is present. **immune function test**. any of a number of tests and methods used in conjunction with routine chronic or subacute toxicity tests to ascertain the immunotoxicity of a tested substance. **Katayama's test**. a qualitative colorimetric test of the presence of carboxyhemoglobin in the blood. **laboratory test**. any test conducted in a laboratory, under at least partially controlled environmental conditions. See, for example, electrolyte measurement. **mitochondrial function test**. any test of mitochondrial function. Such tests may include measurements of oxygen uptake, oxidative phosphorylation, dehydrogenase activity, activity of aminolevulinic acid synthetase (involved in heme biosynthesis) and other enzymes involved in mitochondrial protein synthesis. Measurements of enzymes of the cytochrome P-450-dependent monooxygenase system and of UDP glucuronosyltransferase also offer very useful endpoints. **organophosphorus-induced delayed neurotoxicity test** (delayed neuropathy test). the potential of an organophosphate to produce this condition is usually tested by observing the appearance of paralysis of the leg muscles of treated mature domestic fowl or examination, at autopsy, for motor nerve degeneration. The clinical response of these animals is similar to that of humans. **photoallergy test**. See phototoxicity test. **picrate test**. a sensitive test for cyanide that is suitable for use in the field. Filter paper dipped in a 5% aqueous solution of sodium carbonate and 0.5% picric acid is allowed to dry slightly and then suspended in a test tube over a biological sample (e.g., macerated plant material, minced liver, rumen contents) to which a few drops of chloroform or dilute acetic acid have been added. **radioallergosorbent test (RAST)**. a radioimmunoassay test to detect elevated levels of specific IgE-bound allergen. In this test, the allergen is bound to an insoluble material and the subject's serum is mixed with this conjugate. If the serum contains the specific antibody, it will complex with the allergen. A high level of the IgE complex is a predictor of an anaphylactic response to the substance. **Reinsch's test**. a test for arsenic in which a copper strip is placed in the test liquid, which is then acidified and boiled with hydrochloric acid. The presence of arsenic is indicated by the appearance of a gray deposit on the copper. **Reuss' test**. a test for the presence of atropine in a fluid. Oxidizing agents and sulfuric acid are added to a test liquid. If the odor of orange flowers and roses is generated, the liquid contains atropine. **Rimini's test**. a test for the presence of formaldehyde in liquids (e.g., milk, urine) using a dilute solution of phenylhydrazine hydrochloride, sodium hydroxide, and sodium nitroprusside. ***Salmonella*/microsome test**. a mutagenicity test in which the test organism is a mutant strain of *Salmonella typhimurium* that does not normally grow in a histidine-deficient medium. The test system includes a histidine-deficient medium, the bacterium, the test chemical, and rat liver microsomes. If growth occurs in the test system, then a back mutation to the wild type, induced by the test chemical, is inferred. **significance, test of**. a statistical test that provides a criterion to help determine whether a difference between the expected (hypothetical) and observed values is small enough to be reasonably due to chance. **skin irritancy test**. See test (dermal test). **skin sensitization test**. dermal sensitization test. **toxicokinetic test**. any test that measures rates of absorption, rates and pathways of distribution, biotransformations, and excretion rates in the test population under a given exposure regime. **Wormley's test**. a test for alkaloids in which the test solution is treated with picric acid or a dilute iodine/potassium-iodide solution. The presence of alkaloids is indicated by a color reaction. **Yvon's test**. **1:** a test for alkaloids in which the test solution is added to an aqueous mixture of bismuth subnitrate, potassium iodide, and hydrochloric acid. The presence of alkaloids is indicated by a red color reaction. **2:** a test for acetanilid in which a urine specimen is treated with chloroform and heated with yellow nitrate of mercury. The presence of acetanilid in the urine is indicated by a green color reaction.

testis (plural, **testes**). the male gonad or organ of generation; the primary sex organ that produces sperm. In vertebrates there is a pair of testes, which produce both spermatozoa and steroid hormones under the control of the pituitary gonadotrophins.

testoid. a compound that simulates the action of a male hormone or androgen.

testosterone (17-β-hydroxy-androst-4-en-3-one). a steroid, testosterone (the active form of which is dihydrotestosterone) is the most potent of the androgens and the principal androgen produced and secreted by the mammalian testis, $C_{19}H_{28}O_2$. It is also an intermediate in the biosynthesis of estrogens. Testosterone is secreted chiefly by the testicular Leydig cells, under the control of interstitial cell-stimulating hormone (ICSH). It can cause virilization of female fetuses and excess testosterone in adult women causes masculinization. Testosterone is very toxic; LD_{100}, *i.p.*, in female rats is 325 mg/kg body weight. Testosterone may be prescribed for androgen deficiency, breast cancer in females (but not in males), weight gain, stimulation of growth, and erythropoiesis. Adverse effects include fluid retention, acne, masculinization, and erythrocythemia. Some individuals are hypersensitive to exogenous testosterone.

tetanic. **1:** pertaining to or characterized by sustained muscular contraction without relaxation, as in tetanus. **2:** causing or inducing tetanus. **3:** an agent that induces tonic muscular spasm (e.g., strychnine).

tetanic contraction. sustained muscle contraction without relaxation.

tetanic convulsion. See convulsion (tonic convulsion).

tetaniform. tetanoid.

tetanigenous. able to cause or induce tetanus or tetaniform spasms.

tetanilla. See tetany.

tetanode. of or pertaining to the quiescent interval between recurring tonic spasms in tetanus.

tetanoid (tetaniform). **1:** tetanus-like, resembling tetanus; of the nature of tetanus. **2:** having the appearance of tetany.

tetanolysin. the hemolytic component of tetanus toxin (tetanotoxin). It is a hemolytic substance that appears to play no role in the etiology of tetanus.

tetanospasmin. the heat-labile, easily oxidized, neurotoxic component of tetanus toxin (tetanotoxin) which is responsible for the characteristic signs and symptoms of tetanus. It is a protein comprised of one heavy and one light peptide chain joined by a disulfide bond. Tetanospasmin acts chiefly on the anterior (ventral) horn cells of the spinal cord by binding to the presynaptic endings of motor neurons, blocking the release of amino acid neurotransmitters from inhibitory synapses with resulting tetanic spasms.

tetanotoxin. tetanus toxin.

tetanus. **1:** a smooth contraction of a muscle (as opposed to twitching). **2:** a state of sustained high tension, muscular contraction due to a series of stimuli repeated so rapidly that the individual muscular responses are fused. **3:** an acute disease marked by painful, sustained, tonic (tetanic) spasm and hyperreflexia of certain skeletal muscles. This rather quickly results in trismus (lockjaw) and ultimately generalized muscle spasm, emprosthotonos, glottal spasm, seizures, respiratory spasms and paralysis. Tetanus is caused by the neurotropic protein toxin, tetanospasmin, that acts on the CNS. It is secreted by an anaerobic bacterium, *Clostridium tetani*, growing at the site of an injury such as deep puncture wound. **4:** tetany. **apyretic tetanus**. tetany. **artificial tetanus**. drug tetanus. **benign tetanus**. tetany. **drug tetanus** (artificial tetanus). that caused by any tetanic drug. **generalized tetanus**. tetanus completus. **grass tetanus**. grass tetany. **intermittent tetanus**. tetany. **lactation tetanus**. grass tetany. **local tetanus**. an early stage of tetanus that affects muscles near the infected wound as a result of the action of the tetanospasmin on the anterior horn cells of the spinal cord at the level of the wound. **toxic tetanus**. that produced by an overdose of nux vomica or strychnine.

tetanus completus. generalized tetanus; that which involves most of the musculature of the body.

tetanus dorsalis. emprosthotonos.

tetanus posticus. emprosthotonos.

tetanus spasmus. tetanic spasm. See spasm.

tetanus toxin (tetanotoxin). a potent exotoxin secreted by *Clostridium tetani*. It is comprised of two toxins, tetanospasmin (a neurotoxin) and tetanolysin (a hemolysin), *q.v.*

tetany. 1: intermittent cramp or tetanus; tetanilla; apyretic tetanus; benign tetanus; neuromuscular hyperexcitability with intermittent cramp or tetanus. The condition in humans is marked by severe, intermittent, involuntary, tonic muscular contractions (twitching) with associated fibrillary tremors, paresthesia, and muscular pain. Spasms usually begin in the hands, followed by the face and trunk; the laryngeal muscles are sometimes affected. Both motor and sensory nerves exhibit heightened irritability to electrical and mechanical stimuli. The proximate cause of this disorder is a reduction of ionized calcium in blood plasma which may be the result, for example, of parathyroid insufficiency, vitamin D deficiency, alkalosis. **2:** tetanus (def. 3). See also grass tetany.

TETD. tetraethylthiuram disulfide. See disulfiram.

tethys. *Aplysia*.

tetracarbonyl nickel. nickel carbonyl.

tetracemate disodium. edetate disodium.

tetracemate tetrasodium. tetrasodium edetate.

tetracemin. edetate sodium.

2,4,5,6-tetrachloro-1,3-benzenedicarbonitrile. chlorothalonil.

2,3,5,6-tetrachloro-1,4-benzenedicarboxylic acid dimethyl ester. chlorthaldimethyl.

2,3,7,8-tetrachlorodibenzo[b,e][1,4]dioxin. TCDD.

2,4,5,6-tetrachloro-3-cyanobenzonitrile. chlorothalonil.

2,3,6,7-tetrachlorodibenzodioxin. TCDD.

2,3,7,8-tetrachlorodibenzo[B,E][1,4]dioxin. TCDD.

2,3,7,8-tetrachlorodibenzo-*p*-dioxin. TCDD.

2,3,7,8-tetrachlorodibenzofuran. TCDF.

2,4,5,6-tetrachloro-1,3-dicyanobenzene. chlorothalonil.

tetrachlorodiphenylethane. 1,1-dichloro-2,2-bis(*p*-chlorophenyl)ethane.

tetrachloroethane. See *sym*-tetrachloroethane.

***sym*-tetrachloroethane** (acetylene tetrachloride). a heavy, nonflammable, corrosive liquid, $Cl_2HC{-}CHCl_2$, with a chloroform-like odor. It is slightly soluble in water and soluble in ethanol and ether. It is used chiefly as a solvent (e.g., for fats, oils, waxes, resins); to degrease metals; extraction of fats and oils; in organic syntheses, the manufacture of paint and varnish removers, photographic films, lacquers, and insecticides; as an alcohol denaturant, a weed killer, a fumigant, and an intermediate in the manufacture of other hydrocarbons. It is highly toxic by inhalation, ingestion, and skin contact. It is more toxic than chloroform or carbon tetrachloride. Effects of intoxication include narcosis, liver damage, kidney damage, and gastroenteritis.

tetrachloroethene. perchloroethylene.

tetrachloroethylene. perchloroethylene.

***N*-(1,1,2,2-tetrachloroethylmercapto)-4-cyclohexene-1,2-dicarboximide**. captafol.

***N*-(1,1,2,2-tetrachloroethylsulfenyl)-*cis*-4-cyclohexene-1,2-dicarboximide**. captafol.

***N*-(1,1,2,2-tetrachloroethylthio)-4-cyclohexene-1,2-dicarboximide**. captafol.

***N*-(1,1,2,2-tetrachloroethylthio)-Δ^4-tetrahydrophthalimide**. captafol.

tetrachloroisophthalonitrile. chlorothalonil.

tetrachloromethane. carbon tetrachloride.

***m*-tetrachlorophthalodinitrile**. chlorothalonil.

1,2,2,3-tetrachloropropane (TECP). a hepatotoxic, chlorinated propane, $C_8H_4Cl_4$.

2,3,5,6-tetrachloroterephthalic acid dimethyl ester. chlorthal-dimethyl.

tetrachlorvinphos (2-chloro-1(2,4,5-trichlorophenyl)vinyl dimethyl phosphate; 2-chloro-1-(2,4,5-trichlorophenyl)ethenyl phosphoric acid dimethyl ester; Stirifos). a moderately toxic, organophosphate insecticide used to control pests on food crops and livestock. It is one of the least toxic to vertebrates of the cholinesterase inhibitors (due probably to its low solubility in water). No fatal cases of human exposure are known. This compound is sold as wettable powders and dusts, emulsifiable concentrates, and in a special formulation for cattle which is mixed with the feed. At high concentrations technical grade tetrachlorvinfos causes hyperplastic thyroid lesions and adrenal cortical adenomas in laboratory rats and is carcinogenic to laboratory rats and mice.

tetracycline [*Lederle*]. a yellow, odorless, crystalline powder, $C_{22}H_{24}N_2O_8 \cdot 0.6H_2O$, that is nearly insoluble in water, chloroform, and ether; slightly soluble in ethanol; and very soluble in dilute hydrochloric acid and solutions of alkali hydroxide. Tetracycline is a modified naphthacene. It is a broad-spectrum antibiotic obtained from certain bacteria of the genus *Streptomyces*; it can also be prepared from oxytetracycline. Tetracycline is often prescribed to treat urinary infections, streptococcal infections, pneumonia, brucellosis, and rickettsial diseases such as typhus. It may also be used to treat bacterial infections in persons who are sensitive to penicillin. Bacteria may become resistant to this antibiotic, but this does not usually occur with a short course of treatment. It is very toxic to laboratory rats. Prolonged use in humans may result in dental discoloration, fatty liver, and and renal damage. Tetracycline is teratogenic, and may cause severe damage to teeth and cataracts in the fetus if administered to the prospective mother.

tetradecahydro-3,9-dihydroxy-4,11b-dimethyl-8,11a-methano-11a*H*-cyclohepta[a]naphthalene-4,9-dimethanol. aphidicolin.

tetradecanoylphorbol 13-acetate (phorbol myristate acetate). an ester of phorbol, it is a potent, extremely lipophilic tumor promoter in croton oil. This and other phorbol esters are also immunotoxic and have important effects on leukocytes; they are toxic to T cells.

(3β)-1,2,6,7-tetradehydro-14,17-dihydromethoxy-16(15*H*)-oxaerythrinan-15-one. β-erythroidine.

6,7,8,14-tetradehydro-4,5-epoxy-3,6-dimethoxy-17-methylmorphinan. thebaine.

Tetradymia (horse brush; rabbit brush). a common hepatotoxic, photosensitizing bushy plant of the western United States. It causes more livestock losses from photosensitization than any other plant in the United States. Bighead appears in sheep within 16-24 hours following ingestion of a toxic dose. If a sufficient amount of *Tetradymia* forage is consumed, death can result from liver damage alone. See also phylloerythrin.

tetraethyl lead (TAL; lead tetraethyl; ethyl gas; tetraethyllead; tetraethylplumbane). a poisonous, immunotoxic and possibly carcinogenic, flammable, water-insoluble, colorless, oily liquid, $Pb(C_2H_5)_4$, with a characteristic odor. It is widely used as a solvent for fatty materials and as an antiknock agent added to fuels to prevent preignition in internal combustion engines. It is a common air contaminant. Human exposure to TAL is most often via inhalation or by absorption through the skin. Effects of intoxication are largely due to CNS effects and may include anorexia, nausea, vomiting, diarrhea, tremors, muscular weakness, irritability, nervousness, insomnia, anxiety, and death. Teratogenic and reproductive effects have been produced experimentally. The toxic action of tetraethyl lead is probably due to its metabolic conversion to triethyl lead, *q.v.*

tetraethyl pyrophosphate. the legal label name of TEPP.

tetraethylammonium (TEA). a quaternary ammonium compound that blocks the generation of nerve impulses by selectively blocking potassium conductance channels in neurones.

tetraethyllead. tetraethyl lead.

tetraethylplumbane. tetraethyl lead.

tetraethylrhodamine. rhodamine B.

tetraethylthioperoxydicarbonic diamide. disulfiram.

tetraethylthiuram disulfide (TTD). disulfiram.

tetrafluorohydrazine. a colorless, mobile liquid or colorless gas, F_2NNF_2. It is an oxidizing agent and irritant that explodes on contact with reducing agents and under high pressure. It is used as an organic reagent and an oxidizer for rocket fuels.

tetrafluoromethane. carbon tetrafluoride.

Tetragonurus cuvieri (squaretail). a marine bony fish found in temperate waters throughout the world. See ciguatera.

(3β,12β)-1,2,6,7-tetrahydro-12,17-dihydro-3-methoxy-16(15*H*)-oxaerythrinan-15-one. α-erythroidine.

1,2,3,4-tetrahydro-9-acridinamine. tacrine.

1,2,3,4-tetrahydro-5-aminoacridine. tacrine.

6,7,12b,13-tetrahydro-4*H*-bis[1,3]benzodioxolo[5,6a:4′5′-g]quinolizine. stylopine.

(1)-Δ¹-tetrahydrocannabinol. (—)-Δ¹-3,4-*trans*-tetrahydrocannabinol.

(—)-Δ⁶-3,4-*trans*-tetrahydrocannabinol. one of two isomers of tetrahydrocannabinol, *q.v.*

(—)-Δ⁸-*trans*-tetrahydrocannabinol. (—)-Δ⁶-3,-4-*trans*-tetrahydrocannabinol.

(—)-Δ⁹-3,4-*trans*-tetrahydrocannabinol. (—)-Δ¹-3,4-*trans*-tetrahydrocannabinol.

Δ¹-tetrahydrocannabinol. (—)-Δ¹-3,4-*trans*-tetrahydrocannabinol.

Δ⁹-tetrahydrocannabinol. (—)-Δ¹-3,4-*trans*-tetrahydrocannabinol.

tetrahydrocannabinol (tetrahydro-6,6,9-trimethyl-3-pentyl-6*H*-dibenzo[b,d]pyran-1-ol; THC). either of two psychoactive isomers, the Δ¹-3,4-*trans* and Δ⁶-3,4-*trans* forms, that occur in the crude resin of cannabis (marihuana). They have not been isolated from other sources. THC produces pharmacological effects that are indistinguishable from those of cannabis. **(—)-Δ¹-3,4-*trans*-tetrahydrocannabinol** (9-*trans*-tetrahydrocannabinol; 1-*trans*-Δ⁹-tetrahydrocannabinol; 3-pentyl-6,6,9-trimethyl-6a,7,8,10a-tetrahydro-6*H*-dibenzo(b,d)pyran-1-ol; 6,6,9-trimethyl-3-pentyl-7,8,9,10-tetrahydro-6*H*-dibenzo(b,d)pyran-1-ol; Δ¹-tetrahydrocannabinol; Δ¹-THC; (1)-Δ¹-tetrahydrocannabinol; Δ⁹-tetrahydrocannabinol; *trans*-Δ⁹-tetrahydrocannabinol; Δ⁹-THC; (—)-Δ⁹-3,4-*trans*-tetrahydrocannabinol; drocannabinol; dronabinol; QCD 84924; SP104). a moderately toxic, psychoactive drug with variable effects on the CNS, depending upon the dose and individual affected. This isomer probably accounts for about 95% of the psychomimetic activity of cannabis. It is moderately toxic to laboratory rats orally and extremely toxic by inhalation. Effects may include euphoria, heightened states of extero- and proprioceptive sensitivity, a distorted time sense, changes in short-term memory, dry mouth, tachycardia. THC is easily metabolized by liver microsomal enzymes, producing a number of intermediates, some of which are biologically active. THC and its metabolites are excreted in urine and feces, with a biological half-life of 24-33 hours. **(—)-Δ^6-3,4-*trans*-tetrahydrocannabinol** (6-*trans*-tetrahydrocannabinol; Δ^6-THC; 1-*trans*-Δ^8-tetrahydrocannabinol; (—)-Δ^8*trans*-tetrahydrocannabinol; Δ^8-THC). a hallucinatory drug that is toxic by all routes of exposure and is a possible human mutagen. It is extremely toxic to laboratory mice, *i.v.*, and moderately toxic *p.o.* It is used in some countries as an anti-emetic. **2:** any of a class of some 30 cannabinoid resins. All have three 6-membered rings, one phenolic hydroxyl group, and an amyl side chain. The Δ^9-*trans*-tetrahydrocannabinols and to some extent the Δ^8-*trans*-tetrahydrocannabinols, account for the psychoactive properties of cannabis; they are very toxic. No human fatalities are known. See also cannabinoid, cannabis, *Cannabis*.

1-*trans*-Δ^8-tetrahydrocannabinol. (—)-Δ^6-3,4-*trans*-tetrahydrocannabinol.

1-*trans*-Δ^9-tetrahydrocannabinol. (—)-Δ^1-3,4-*trans*-tetrahydrocannabinol.

([-])-Δ^9-*trans*-tetrahydrocannabinol (Δ^9-THC; Δ^1-THC; THC). a cannabinoid produced by *Cannabis sativa*, it is the major source of psychoactivity of cannabis. THC occurs at very low concentrations, if present at all, in the stems, roots or seeds, but concentrates selectively in the bracts, flowers and leaves. See also cannabinol, cannabinoid.

6-*trans*-tetrahydrocannabinol.(—)-Δ^6-3,4-*trans*-tetrahydrocannabinol.

9-*trans*-tetrahydrocannabinol.(—)-Δ^1-3,4-*trans*-tetrahydrocannabinol.

***trans*-Δ^9-tetrahydrocannabinol**. (—)-Δ^1-3,4-*trans*-tetrahydrocannabinol.

tetrahydrocoptisine. stylopine. See also *Chelidonium majus*.

[2R-(2α,6aα,12aα)]-1,2,12,12a-tetrahydro-8,9-dimethoxy-2-(1-methylethenyl)-[1]benzopyrano-[3,4-b]furo[2,3-h][1]benzopyran-6-(6aH)-one. rotenone.

2,3,7,8-tetrahydrofuro(2,3-*b*)furan). See mycotoxin (bisfuranoid mycotoxins).

2,3,6a,9a-tetrahydro-9a-hydroxy-4-methoxy-cyclopenta[c]furo[2,3-h][1]benzopyran-1,11-dione. aflatoxin M_1.

2,3,6a,9a-tetrahydro-9a-hydroxy-4-methoxy-cyclopenta[c]furo[3′,2′:4,5]furo[2,3-h][1]-benzopyran1,11-dione. aflatoxin M_1.

3a,4,7,7a-tetrahydro-5-(hydroxyphenyl-2-pyridinylmethyl)-8-(phenyl-2-pyridinylmethylene)-4,7-methano-1*H*-isoindole-1,3(2H)-dione. norbormide.

(1*R*-*trans*)-2,3,5,7a-tetrahydro-1-hydroxy-1*H*-pyrrolizine-7-methanol. retronecine.

tetrahydro-4-hydroxy-*N,N,N*,5-tetramethyl-2-furanmethanaminium. muscarine.

5,8,8a,9-tetrahydro-9-hydroxy-5-(3,4,5-trimethoxyphenyl)furo-[3′,-4′:6,7] naphthol [2,3-d]-1,3-dioxol-6 (5aH)-one. podophyllotoxin.

2,3,6aα,9aα-tetrahydro-4-methoxycyclo-penta-[c]furo[3′,2′:4,5]furo[2,3-h][1]benzopyran-1,11-dione. aflatoxin B_1.

3,4,7aα,10aα-tetrahydro-5-methoxy-1H,12*H*-furo[3′,2′:4,5]furo[2,3-h]pyrano[3,4-c]-[1]benzopyran-1,12-dione. aflatoxin G_1.

1,2,3,4-tetrahydro-7-methoxy-1-methyl-6-isoquinolinol. salsoline.

4,6,7,14-tetrahydro-5-methyl-bis[1,3]-benzodioxo[4,5-c:5′,6′-g]azecin-13(5H)-one. protopine.

5,6,6a,7-tetrahydro-6-methyl-4*H*-dibenzo-[d,e,g]quinoline-10,11-diol. apomorphine.

[R-(R*,S*)]-6-(5,6,7,8-tetrahydro-6-methyl-1,3-dioxolo[4,5-g]isoquinolin-5-yl)furo[3,4-e]-1,3-benzodioxol-8(6H)-one. bicuculline.

1,2,5,6-tetrahydro-1-methylnicotinic acid. arecain.

1,2,3,6-tetrahydro-1-methyl-4-phenylpyridine. MPTP.

1,2,5,6-tetrahydro-1-methyl-3-pyridine-carboxylic acid. arecain.

1,2,5,6-tetrahydro-1-methyl-3-pyridinecarboxylic acid methyl ester. arecoline.

1,2,3,4-tetrahydro-1-phenyl-1,4-naphthalenedicarboxylic acid bis(8-methyl-8-azabicyclo-[3.2.1]oct-3-yl) ester. bellkadonnine.

***N*-(5,6,7,9-tetrahydro-1,2,3,10-tetramethoxy-9-oxobenzo[a]heptalen-7-yl)acetamide**. colchicine.

6,7,8,9-tetrahydro-5*H*-tetrazolo[1.5-a]azepine. pentylenetetrazole.

tetrahydrothiophene (thiophane). the saturated analog of thiophene.

1,2,3,6-tetrahydro-*N*-(trichloromethylthio)-phthalimide. captan.

3a,4,7,7a-tetrahydro-2-[(trichloromethyl)thio]-1*H*-isoindole-1,3-(2*H*)-dione. captan.

tetrahydro-6,6,9-trimethyl-3-pentyl-6*H*-dibenzo[b,d]pyran-1-ol. tetrahydrocarbinol.

tetrahydrourushiol. 3-pentadecylcatechol.

1,3,4,5-tetrahydroxycyclohexanecarboxylic acid. chlorogenic acid.

3′,4′,5,7-tetrahydroxyflavan-3-ol. quercetin.

$\Delta^{20:22}$-3β,12β,14,21-tetrahydroxynorcholenic acid lactone. digoxigenin.

3α,4β,7α,15-tetrahydroxyscirp-9-en-8-one. nivalenol.

3,5,3′,5′-tetraiodothyronine. thyroxine (See thyroid hormone).

tetraiodothyronine. thyroxine (See thyroid hormone).

tetrakis hydroxyl-methyl phosphonium (THP). See flame retardant.

tetrakis hydroxyl-methyl phosphonium chloride (THPC). See flame retardant.

6′,7′10,11-tetramethoxyemetan. emetine.

6,6′,7′,12′-tetramethoxy-2,2,2′,2′-tetramethyltubocuraranium diiodide. metocurine iodide.

tetramethyl lead (TML; tetramethylplumbane). a colorless, toxic, flammable liquid, $Pb(CH_3)_4$, similar to, but more active than, tetraethyl lead as an antiknock agent. Toxicity and effects are similar to those of tetraethyl lead but it is more volatile and thus a greater fire hazard and presumably more hazardous than tetraethyl lead if used as a fuel additive. Such use has been banned in many countries.

tetramethyl thiurane disulfide. thiram.

tetramethylammoniumhydroxide (*N,N,N*-trimethylmethanaminium hydroxide; tetramine). a venom, $(CH_3)_4NOH$, with an ammonia-like odor that has a curare-like effect on vertebrates. It is produced by the salivary glands of the carnivorous marine gastropods, *Busycon* spp., *Buccinum* spp., *Neptunea* arthritica, and *N. antiqua*, forming the toxic fraction of the venom; it is also produced and by the sea anemone, *Actinia equina*.

tetramethylbiarsine. cacodyl.

tetramethylene sulfone. sulfolane.

***N,N,N′,N′*-tetramethyl-1,6-hexanediamine polymer with 1,3-dibromopropane**. hexadimethrine bromide.

tetramethylplumbane. tetramethyl lead.

tetramethylputrescine. an extremely poisonous crystalline base, $N(CH_3)_2(CH_2)_4N(CH_3)_2$, derived from putrescine. Its toxic action is similar to that of muscarine.

tetramethylthionine chloride. methylene blue.

tetramethylthiuram disulfide. thiram.

2,5,7,8-tetramethyl-2-(4′,8′,12′-trimethyltridecyl)-6-chromanol. vitamin E.

tetramine. tetramethylammonium hydroxide.

Tetraodon. a genus of marine bony fish (Family Tetraodontidae), more than half of which are known to be toxic; all should be regarded as poisonous. See also tetrodonic acid, tetrodotoxin.

tetraodon poisoning. tetrodotoxism.

Tetraodontidae (puffer; swellfish; fugu; globe fish; blowfish; balloonfish; toadfish; toado; botete; fahaka; tinga). a family of some 100 species of marine, bony fishes capable of rapidly inflating their body with water or air. The preferred common name in English is puffer, as they can inflate themselves by gulping large amounts of air or water. They are mostly tropical, although the ranges of many extend into the temperate zone. Most are dull colored. They are similar in behavior and morphology to the Diodontidae, but the skin bears numerous prickles. They have heavy jaw teeth and some species can inflict serious bites. More than 50 species are known to be poisonous and are among the most poisonous of all marine organisms. All are generally regarded as highly poisonous. The musculature is normally edible, but the integument, intestines, liver, and gonads are extremely toxic and contain the neurotoxicant, tetrodotoxin. Some of the most toxic species are *Arothron hispidus*, *A. meleagris*, *A. nigropunctatus*, and *Sphaeroides annulatus*. See also tetrodotoxism, *Tetraodon*, tetrodonic acid.

tetraodontoid. of or pertaining to bony marine fish of the suborder Tetraodontoidea (Order Tetraodontiformes). Many are poisonous. See tetrodotoxism.

tetraodontoid fish. any bony fish of the suborder Tetraodontoidea (Order Tetraodontiformes).

Tetraodontoidea. a suborder of bony marine, mostly tropical, fish (Order Tetraodontiformes) that includes puffers (Family Tetraodontidae) and allied fish such as parrotfish (Family Diodontidae) and ocean perch (Family Molidae). Many species are poisonous. See tetrodotoxin.

tetraodontoxin. tetrodotoxin.

tetraodontoxism. tetrodotoxism.

T-2 tetraol. a trichothecene mycotoxin produced by *Fusarium* spp.

2,4,5,6-tetraoxohexahydropyrimidine. alloxan.

tetraphene. 1,2-benzanthracene.

tetrapotassium hexakis-(cyano-C)ferrate(4—). potassium ferrocyanide.

tetrasodium ethylenebis(iminodiacetate). edetate sodium.

tetrasodium ethylenediaminetetraacetate. edetate sodium.

tetrasodium hexakis(cyano-C)ferrate(4—). sodium ferrocyanide.

tetrazobenzene-β-naphthol. sudan III.

tetrodonic acid. a toxic acid produced by various puffer fish of the genus *Tetraodon*. See also tetrodotoxin.

tetrodontoxin. tetrodotoxin.

tetrodotoxin (maculotoxin; spheroidine; tarichatoxin; tetraodontoxin; tetrodontoxin; fugu poison; puffer poison; TTX). an extremely toxic, highly lethal, crystalline neurotoxin, $C_{11}H_{17}N_3O_3$. It occurs in the ovaries and liver of the globe fish, *Spheroides rubripes*; numerous other fishes of the order Tetraodontoidea (especially those of the family Tetraodontidae); in the skin and egg clusters of the frog of the genus *Atelopus varius*; certain other frogs; in the eggs of salamandrids of the genus *Taricha* (known as tarichatoxin); and in *Octopus maculosus* (known as maculotoxin). Tetrodotoxin acts on both the central and peripheral nervous systems, causing nerve and skeletal muscle paralysis by selectively blocking the regenerative sodium conductance channel in axons of preganglionic cholinergic fibers, somatic motor nerves, and muscle fibers. Intoxication is characterized by rapid onset (a few minutes) of symptoms that include dizziness; pallor; muscular weakness; paresthesia of the lips, tongue and throat; diaphoresis, excessive salivation; hypotension; and bradycardia. Ataxia, shock, cyanosis, convulsions, and death from respiratory collapse due to paralysis of the diaphragm often follow. See also tetrodotoxism, *Atergatus floridus*, *Carpilius maculatus*, *Demania toxica*, batrachotoxin A., Tetraodontidae, maculotoxin, zetekitoxin.

tetrodotoxism (puffer poisoning; fugu poisoning; tetraodon poisoning; tetraodontoxism). a type of ichthyotoxism caused by eating fishes of the order Tetraodontoidea, including the families Tetraodontidae (puffers or swellfishes), Diodontidae (porcupine fishes), and Molidae (ocean sunfishes). It is one of the most violent types of poisoning known. Ingestion of portions of the integument, liver, or gonads can produce a swift, agonizing death. Central and peripheral neurological responses dominate symptoms. Included are paresthesias, beginning with a tingling sensation of the lips and tongue spreading to the extremities, ultimately resulting in a numbness of the entire body. Pronounced respiratory distress with eventual intense cyanosis is usual. Symptoms often also include ataxia, aphonia, paralysis, and convulsions; hypotension is a consistent feature. The victim typically remains conscious until near death. Occasional symptoms include blistering, extensive scaling of the skin, and petechial hemorrhages. Gastrointestinal distress may be evidenced. Mortality is about 60%, death occurring within 24 hours. There is no antidote or specific treatment. See tetrodotoxin. See also *Arothron*, tetrodotoxin.

teturamin. disulfiram.

Texas croton. *Croton texensis*.

Texas diamondback. *Crotalus atrox*.

Texas snakeroot. *Aristolochia reticulata*.

6-TG. 6-thioguanine.

Th. the symbol for thorium.

TH cell. helper T cell.

Thais (dogwinkles; dog whelks; purple snails). a cosmopolitan genus of carnivorous marine snails with thick, striated, nodular shells, usually about 2.5 cm long. ***T. floridana***. See senecioylcholine. ***T. haemastoma***. the venom of this species contains dihydroxymurexine. A poison (a vasodilator) has also been isolated from extracts of the salivary gland. It is a hypotensive agent, causes bradycardia, and appears to have behavioral effects.

Thalarctos maritimus (polar bear). a creamy white species of bear (Family Ursidae) that is circumarctic in distribution. It feeds mainly on fish and seals and is found chiefly in coastal areas and even at sea in Arctic America. The average weight of adult males is about 350 kg. Poisonings of humans have often resulted from consumption of the liver and kidneys of this animal. Symptoms typically include frontal headaches which may be intense, dizziness, nausea, vomition, diarrhea, drowsiness, abdominal pain, extensive scaling of the skin, irritability, photophobia, collapse, and convulsions; usually not fatal. The principal toxicant is vitamin A.

thalassin. a toxic substance derived from tentacles of the sea anemone, *Anemonia sulcata*, though it does not appear to reside in the nematocysts. When injected into dogs, it produces allergic signs.

Thalassophryne. a genus of toadfishes (Family Batrachoididae) ***T. reticulata*** (toadfish; bagre sapo; sapo). a toadfish that occurs along the Pacific Coast of Central America. ***T. maculosa*** (toadfish, sapo). a toadfish of the West Indies.

thalassotherapy. treatment of disease by living near the sea, sea voyages, or sea bathing.

thalidomide (2-(2,6-dioxo-3-piperidinyl)-1H-isoindole-1,3(2*H*)-dione; *N*-(2,6-dioxo-3-piperidyl)phthalimide; α-phthalimidoglutarimide; 3-phthalimidoglutarimide; 2,6-dioxo-3-phthalimidopiperidine; *N*-phthalylglutamic acid imide; *N*-phthaloylglutamimide). a piperidinedione sedative shown to be very effective in relieving severe pain associated with acute lepra (psoriasis-like) reactions. It is used to treat Hanson's disease and was formerly widely used as a sedative-hypnotic. It is a teratogen that produces gross morphological deformities in the human fetus during the first trimester of pregnancy. These may result in death of the fetus or infant and include especially amelia or phocomelia (absence or severe shortening of the limbs); cardiovascular, gastrointestinal and urogenital abnormalities; deafness; and a permanent peripheral neuritis.

thallitoxicosis. thallotoxicosis.

thallium (symbol, Tl)[Z = 81; A_r = 204.37]. a highly toxic, soft, malleable, bluish- or grayish-white, metallic element of group III of the periodic table. Thallium and certain of its salts occur in lead and cadmium ores as well as in flue dust and during zinc smelting. It readily forms toxic compounds on contact with water. Thallium is toxic by all routes of exposure; avoid contact. It was formerly used as an insecticide and rodenticide. Symptoms of acute poisoning by thallium and many of its salts commonly include nausea, vomiting, severe abdominal pain, tremors, delirium, convulsions, dyspnea, paralysis, coma, collapse, and death. Chronic poisoning may cause alopecia. Lesions include hemorrhagic gastroenteritis and encephalopathy.

thallium acetate (thallous acetate). a compound, $TlOCOCH_3$, with white, deliquescent crystals; soluble in water and ethanol; toxic behavior similar to that of thallium.

thallium bromide. a highly toxic salt, TlBr; the crystals are mixed with those of thallium iodide for use in infrared radiation transmitters. See also thallium.

thallium carbonate (thallous carbonate). a highly toxic salt, Tl_2CO_3, used in the manufacture of simulated diamonds. See thallium.

thallium chloride (thallous chloride). an extremely hazardous, highly toxic salt, TlCl, used as a catalyst in chlorination and in sun lamp monitors. It is poisonous by ingestion or by intraperitoneal routes. When heated to decomposition, it releases very toxic fumes (Tl and Cl^-). See thallium.

thallium hydroxide (thallous hydroxide). a highly water-soluble, highly toxic salt, TlOH·HOH, used as an indicator in chemical analysis. See thallium.

thallium iodide (thallous iodide). a highly toxic salt, TlI; the crystals are mixed with those of thallium bromide for use in infrared radiation transmitters. See thallium.

thallium monoxide (thallium oxide; thallous oxide). a toxic, black, water- and ethanol-soluble powder, Tl_2O, that decomposes in water; it melts at 300°C. It is used as an analytical reagent and in optical glass and artificial gems; it oxidizes on exposure to air and should be tightly stoppered when not immediately in use.

thallium nitrate (thallous nitrate). a highly toxic salt, $TlNO_3$, and strong oxidizing agent. When heated to decomposition it releases very toxic fumes (Tl and NO_x). It is used as a reagent in analytical chemistry and in pyrotechnics (green fire). See thallium.

thallium oxide. thallium monoxide.

thallium poisoning. thallotoxicosis.

thallium sulfate (thallous sulfate). an extremely toxic, colorless, water-soluble, crystalline substance, Tl_2SO_4. Thallium sulfate is used as a rodenticide; in ant baits; and as a reagent in analytical chemistry. It is similar toxicologically to thallium, *q.v.*

thallotoxicosis (thallitoxicosis; thallium poisoning). poisoning by thallium or thallium-containing substances. It usually occurs following accidental ingestion of a rodenticide. Symptoms in humans include vomiting, diarrhea, stomatitis, gastroenteritis, peripheral and retrobulbar neuritis, temporary alopecia, and endocrine disorders.

thallous acetate. thallium acetate.

thallous bromide. thallium bromide.

thallous carbonate. thallium carbonate.

thallous chloride. thallium chloride.

thallous hydroxide. thallium hydroxide.

thallous iodide. thallium iodide.

thallous nitrate. thallium nitrate.

thallous oxide. thallium monoxide.

thallous sulfate. thallium sulfate.

thallous sulfide. thallium sulfate.

thamaha. *Bitis arietans*.

thamaha-dinkotsane. *Bitis arietans*.

Thametopoea pityocampa. the pine tree caterpillar of Europe. It has stinging bristles that can cause urticaria in humans.

thanato-. a prefix indicating or pertaining to death.

thanatobiological. pertaining to the processes of life and death.

thanatognomonic. 1a: of or pertaining to fatal prognosis; prognostic of death. **1b:** symptomatic or indicative of approaching death.

thanatoid. **1:** resembling death. **2a:** deadly, lethal; mortal. **2b:** deadly poisonous.

thanatologist. a scientist or student of the processes and mechanisms of death and dying.

thanatology. the science of death and dying.

thanatophidia (toxicophidia). venomous snakes, collectively.

thanatophidial. pertaining to venomous snakes collectively or in general.

thanatophoric. lethal; deadly; causing death.

thanatopsy (thanotopsy). See autopsy (defs. 1 and 2).

thanotopsy. thanatopsyt (See autopsy defs. 1 and 2).

thanatosis. **1:** necrosis; gangrene. **2:** tonic immobility.

thatta pam. *Hydrophis mamillaris*.

THC. tetrahydrocannabinol

Δ^1-THC. (—)-Δ^1-3,4-*trans*-tetrahydrocannabinol.

Δ^9-THC. ([-])-Δ^9-*trans*-tetrahydrocannabinol.

The Colour Index. a publication that gives chemical details of commercially available dyes, pigments, and colorings.

"the pill." oral contraceptive pill.

theine. the alkaloid of tea, it is an isomer of caffeine.

theaism. theinism.

thebaine (dimethyl morphine; *para*-morphine; paramorphine). a crystalline, alkaloid, $C_{19}H_{21}NO_3$, obtained from opium. It is extremely toxic with an action similar to that of strychnine. Sometimes used as an anodyne; possibly addictive.

theinism (theism; theaism). chronic poisoning from drinking excessive quantities of tea. The condition is marked by headache, nervousness, dyspepsia, palpitation, and insomnia.

theism. theinism.

Thelotornis kirtlandii (bird snake; birit tiu; ehé; konkati; sungahuni; twig snake; ukhokhothi; vine snake; yangalukwe; likwétéma; lukukuru). a slender-snouted arboreal snake of tropical and southern Africa (Family Colubridae). It is one of only two colubrids that cause serious injury and even death to humans; the other is *Dispholpholidus typus* (the boomslang). It is characterized by an elongate head with a flattened crown and shallow lateral grooves that extend anteriorly from the eyes. The head is distinct from the neck with a distinct, projecting canthus rostralis that forms a shallow lateral groove below; the eyes are large with horizontally elliptical pupils; the body is slender, cylindrical, and elongate; the tail is long. The head bears 9 scales on the crown; internasals are large; 3 large scales abut the parietals posteriorly and 1-3 laterally; most individuals have two loreal scales which separate the nasal from the preocular. The dorsal scales on the body are narrow, weakly keeled, and have apical pits; 19 rows of oblique dorsals occur from midbody forward, with 11-13 posteriorly. The ventrals are rounded and the anal plate is divided and the subcaudals are paired. There is a series of 11-16 small maxillary teeth followed by a short gap and 3 long, grooved fangs. Even though this snake rarely bites humans, it should be considered dangerous. The venom is highly toxic; occasional bites are fatal. When irritated it assumes a threatening display with the neck greatly inflated with the presentation of a conspicuous pattern of black crossbands on a light background.

theobromine (3,7-dihydro-3,7-dimethyl-1*H*-purine-2,6-dione; 3,7-dimethylxanthine). a pseudoalkaloid, it is one of the methylxanthine central stimulants that occur in the cacao bean, cola nuts and tea. It may be prepared from the dried ripe seed of *Theobroma cacao* or produced synthetically. Its actions and effects are similar to those of caffeine and it is used therapeutically as a diuretic, myocardial stimulant, a smooth muscle relaxant, and a dilator of the coronary arteries. See also caffeine, methylxanthines, pseudoalkaloid.

theophylline (3,7-dihydro-1,3-dimethyl-1H-purine-2,6-dione; 1,3-dimethylxanthine; theocin). a white, crystalline, odorless, methylxanthine central stimulant with a bitter taste, $C_7H_8N_4O_2$, that occurs in small amounts in tea. It is pharmacologically similar to caffeine. It is used medically as a vasodilator, smooth muscle relaxant, diuretic, cardiac stimulant. It is also used in angina pectoris, peripheral vascular disease, and bronchial asthma. It may cause cardiac arrhythmias. Aminophylline is a complex of theophylline and ethylenediamine (ethane-1,2-dione).

theophylline ethylenediamine. See aminophylline.

theory. a scientifically accepted general principle or explanation of a class of phenomena by inferring that they are necessary consequences of other phenomena that are considered more primitive. Thus, organic evolution (according to Darwinian theory) is caused (in part) by natural selection. An acceptable theory must take into account, and be supported by a considerable body of coherent evidence, or several lines of evidence.

therapeusis. therapeutics.

therapeutic. **1:** curative. **2:** pertaining to therapeutics or the treatment of disease.

therapeutics (therapeusis; therapia; therapy). **1a:** the science and craft of treating disease. **1b:** the treatment of a disease. Therapy for poisoning usually depends on the nature of the poison, the type (or route) and intensity of exposure, the source of the poison, and the organism exposed. Such therapy may be specific (e.g., use of a specific antivenin) or nonspecific (e.g., use of an emetic). In practice, therapy for poisoning is often seriously complicated because the poison to which a victim was exposed may not be established. See also therapy. **2:** a technical description of the treatment of a disease.

therapeutic index (TI). variously defined and sometimes used synonymously with margin of safety, *q.v.* **1:** a measure of the relative safety of a drug expressed as **a:** the ratio of the lethal dose (LD) or toxic dose (TD) to the therapeutically effective dose (ED); most commonly $TI = LD_{50}/ED_{50}$). **b:** the difference between the lethal dose (LD) or toxic dose (TD) and the therapeutically effective dose (ED).

Theraphosidae (tarantulas; bird spiders). a family of large, hairy, burrowing American spiders. These spiders reach a length of 35 mm or more with a legspan of up to 150 mm. They have 8 closely set eyes; 3 are positioned on each side of the head and 2 large eyes are centered in between. The bite of some South American species can be deadly. The bite of any of the more than 30 North American species is, however, no more toxic than the sting of a bee or wasp, and most are reluctant to bite humans. The bite of some American species can be fatal. Common or well-known genera are *Dugesiella*, *Eurypelma*, *Glyptocranium*, *Lycosa*, *Sencopelma*, and *Theraphosa*.

therapia. therapeutics.

therapy. therapeutics (def. 1b). **nonspecific therapy**. therapy that is at least partially effective over a range of illnesses of differing etiology. Such therapy may be oriented, for example, toward the relief of symptoms, the maintenance of the patient, or prevention of further encroachment of the disease. In the case of poisoning, nonspecific therapy might include the maintenance of vital signs and prevention of further absorption of the toxicant. In many cases, if the nature of the poison and the conditions surrounding the poisoning are unknown or specific antidotes are unknown or unavailable, the application of nonspecific therapy, while the only recourse, may ultimately offer poor or inadequate results. **specific therapy**. therapy addressed to the mode of action of the poison or based upon specific antidotes (e.g., entivenin therapy).

theriac. **1:** theriaca. **2a:** cure-all **2b:** medicinal, antidotal.

theriaca (theriac). **1:** a complex mixture, usually of some 70 pharmacologically active substances, including opium, that were ground to powder and reduced with honey into a confectionery paste (electuary) that was believed to effect extraordinary cures, especially against the bites of snakes and other venomous animals. It was developed in the Middle Ages as an antidote to poisoning. **2:** treacle, molasses (British).

Theridiidae (comb-footed spiders). a family of small, dark spiders with long, slender legs that bear 3 claws on each tarsus and 6-10 inconspicuous comb-like bristles on the hind tarsi of most species. One of the largest families of spiders. There are more than 200 species in North America alone; included is *Latrodectus*.

thermalgia. a sensation of burning pain.

thermobiology. the study of the effects of heat on living organisms and biological processes.

thermolabile. easily altered or destroyed by high temperatures.

thermonatrite. See sodium carbonate.

thermopenetration. medical diathermy (See diathermy).

thermostabile. thermostable.

thermostable (thermostabile). not readily altered or destroyed by moderately high temperatures.

thermotoxin. a poison formed endogenously due to excessive heat.

thermotoxy. death or injury caused by high temperature.

thesaurosis. a condition resulting from the accumulation of foreign or endogenous substances in the body. See also polyvinylpyrrolidone.

thetic acid. 1-naphthaleneacetic acid

Thevetia neriifolia. synonymous with *Thevetia peruviana*.

Thevetia peruviana (synonym, *T. neriifolia*; yellow oleander; be-still tree; lucky nut; tiger apple; trumpet flower). a poisonous, tropical, American, broadleaved evergreen shrub or tree (Family Apocynaceae) that grows to 10 m. It is a common ornamental plant in Florida and Hawaii. All parts of the plant are toxic; even smoke from burning foliage and water in which the flowers have been placed are toxic. The chief toxic agent is the cardiac glycoside, thevetin. Seeds also contain the cardiotoxic glycoside, neriifolin (*q.v.*). Fatal poisonings of humans and various animals are known. On the Hawaiian island of Oahu this plant has been a frequent cause of human intoxication. Symptoms are similar to those of digitalis poisoning. They usually include nausea, vomiting, digestive upset, and a slow, irregular pulse. a source of neriifolin. ***T. thevetioides***. a source of cerberoside and neriifolin.

thevetin. a cardiac glycoside, found in *Thevetia peruviana*, that is chemically and physiologically related to the digitalis glycosides.

thevetin A. a cardiotoxic glycoside, $C_{42}H_{64}O_{19}$, isolated from *Thevetia peruviana*. **thevetin B**. cerberoside.

thiabendazole. (2-(4′-thiazoyl)benzimidazole; 4-[2-benzimidazolyl]thiazole). a white to tan crystalline substance, $C_{10}H_7N_3S$, used as a systemic fungicide on citrus fruits. Rats and dogs tolerate a relatively high dose.

thiaminase. an enzyme that catalyzes the hydrolysis of thiamine (vitamin B_1) into pyrimidine and a thiazole derivatives. It occurs in raw fish, certain bacteria, and an occasional plant (e.g., *Dryopteris felix-mas* (male fern), *Equisetum* (horsetails), and *Pteridium acquilinum* (bracken)). It is the only known plant enzyme that is toxic to animals. See, for example, bracken staggers, *Pteridium*, equisetosis, *Equisetum*.

thiamine (vitamin B_1; 3-(4-amino-2-methyl-pyrimidyl-5-methyl)-4-methyl-5,β-hydroxy-ethylthiazolium chloride). a water-soluble vitamin, $C_{12}H_{17}ClN_4OS$, also called the anti-neuritic vitamin. It is essential for growth and prevention of beriberi. It is a precursor of the coenzyme thiamine pyrophosphate, which plays a role in carbohydrate metabolism in the decarboxylation of α-keto acids. Many organisms are unable to synthesize thiamine and it therefore has to be supplied from an external source. Thiamine deficiency in the techno logically advanced nations is uncommon (but see thiamine shock). Thiamine occurs, for example, in wheat germ, whole grains, enriched breads and cereals, and pork. Thiaminase is responsible for thiamine deficiency in certain poisonings, especially of livestock (See, for example, bracken staggers, equisetosis). Some poisons can cause disturbances of thiamine metabolism (See, for example, herbicide (translocated herbicide)).

thiamine shock. challenge of humans by very high levels of thiamine as in the treatment of thiamine deficiency occasionally produces reactions termed "thiamine shock." The Recommended Daily Allowance for humans is 0.5 mg/kcal. Thiamine-deficient individuals may be treated with parenteral injection of ca. 500 mg/kcal. This amount can cause a mild to severe reaction in sensitive individuals. In mild cases of thiamine shock, symptoms may include muscular weakness, a burning sensation, and nausea. In more severe cases, clinical findings may include gastrointestinal hemorrhage, pulmonary edema, collapse, and sudden death.

thiamylal. (dihydro-5-(1-methylbutyl)-5-(2-propenyl)-2-thioxo-4,6(lH,5H)-pyrimidinedione; 5-allyl-5-(1-methylbutyl)-2-thiobarbituric acid; thioseconal; Surital). a depressant anesthetic that is given intravenously and used in conjunction with inhalant anesthetics. It acts via the barbiturate-binding site of the GABA receptor complex. Thiamylal can dangerously depress respiration or cause hypotension. It is a controlled substance and may be habit forming.

thiasine. ergothioneine.

2-thiazolamine. 2-aminothiazole.

2-(4′-thiazoyl)benzimidazole. thiabendazole.

N^1-2-thiazolylsulfanilamide. sulfathiazole.

thick-stalked morel. *Gyromitra fastigiata*.

thickheads. See Pachycephalidae.

thiéby. *Bitis arietans*.

5-(2-thienyl)-2,2′-bithiophene. α-terthienyl.

thimbleweed. common name of plants of the genus *Anemone*.

Thimenox. phorate.

thimet. phorate.

thioallyl ether. allyl sulfide.

1,1′-thiobis[2-chloroethane]. mustard gas (See gas).

2,2′-thiobis(4,6-dichlorophenol). bithionol.

***N,N′*-[thiobis[(methylimino)carbonyloxy]]bisethanimidothioic acid dimethyl ester**. thiodicarb.

3,3’-thiobis-1-propene. allyl sulfide.

thiocarbamide. thiourea.

thiocarbanil. phenyl mustard oil.

thiocarbonates. a compound that contains the thiocarbonate group, --CS.SS.CS--. See also herbicide (thiocarbamate h.).

thioctic acid (6,8-thioctic acid; dl-α-lipoic acid; 1,2-dithiolane-3-pentanoic acid; 1,2-dithiolane-3-valeric acid; 6,8-dithiooctanoic acid; 5-(1,2-dithiolan-3-yl)valeric acid; 5-(1,2-dithiolan-3-yl)pentanoic acid; acetate replacing factor; pyruvate oxidation factor; POF). a crystalline substance, $C_8H_{14}O_2S_2$, that is nearly insoluble in water but soluble in fat solvents. It is a growth factor in many bacteria and protozoa. It is used in the treatment of liver disease and *Amanita* poisoning; it is thought to be an antidote for *Amanita* poisoning. It occurs naturally in yeast and liver and is prepared synthetically.

6,8-thioctic acid. thioctic acid.

thiocyanate (rhodanate; sulfocyanate). **1a:** a salt of thiocyanic acid. **1b:** any compound that contains the radical —SCN; organic thiocyanates are derivatives of thiocyanic acid, HSCN, in which the hydrogens have been replaced by hydrocarbon moieties (e.g., CH_3). They were the first synthetic organic insecticides, killing on contact. Toxicities vary greatly among thiocyanates. Certain metabolic processes release HCN from thiocyanates; this can be fatal. The volatile, lower molecular mass thiocyanates (methyl, ethyl, isopropyl) are rapid-active potent poisons used as fumigants in the control of insects. Thiocyanates are thyrotoxic to livestock, causing thyroid hyperplasia and symptoms of hypothyroidism. Effects on lambs are similar to those of L-5-vinyl-2-thiooxazolidone (*q.v.*). Thiocyanate glycosides have been isolated from plants which have caused symptoms of hypothyroidism in livestock. See isothiocyanate.

thiocyanate radical. —SCN. See thiocyanate, thiocyanic acid.

thiocyanic acid (hydrogen thiocyanate; rhodanic acid; Rhodenwasserstoffsäure (German); sulfocyanic acid). a toxic substance obtained by distilling a thiocyanate salt with dilute sulfuric acid. It polymerizes readily and exists under ambient conditions as a colorless gas or white solid, depending upon the degree of polymerization. It is thought to be a tautomeric mixture of hydrogen thiocyanide (HSCN) and isothiocyanic acid (HNCS). Thiocyanic acid is soluble in some organic solvents and is freely water soluble, forming a very strong, unstable liquid acid with a strong odor.

thiodemeton. disulfoton.

2,2′-thiodiethanol. thiodiglycol.

thiodiethylene glycol. thiodiglycol.

thiodiglycol (2,2′-thiodiethanol; thiodiethylene glycol; β-bis-hydroxyethyl sulfide; dihydroxyethyl sulfide). a syrupy, colorless, flammable liquid, $HOCH_2CH_2SCH_2CH_2OH$, that is miscible with water and ethanol, slightly soluble in ether; not to be used with HCl. It is used as an intermediate in the production of elastomers and antioxidants; a solvent for dyes in textile printing; in photo development solutions; and ink for ballpoint pens. It is a precursor in the manufacture of mustard gas (See gas).

thiodiphenylamine. phenothiazine.

thioethanolamine. cysteamine.

thioether. any of a class of organosulfides (substituted hydrogen sulfides) with the general formula R-S-R.

(1-thio-*d*-glucopyranosato)gold. gold thioglucose.

1-thio-β-D-glucopyranose 1-[*N*-(sulfo-oxy)-3-butenimidate] monopotassium salt. sinigrin.

6-thioguanine (6-TG; 2-amino-6-mercaptopurine). an antineoplastic metabolic antagonist and immunosuppressant used to treat certain types of leukemia. It inhibits DNA synthesis by conversion to a metabolic intermediate, 6-thioGMP, that inhibits purine biosynthesis and can be phosphorylated and incorporated into DNA. It causes bone marrow depression and gastrointestinal effects in humans.

thioimidodicarbonic diamide. 2,4-dithiobiuret.

thiol (mercaptan) **1:** sulfhydryl. **2:** any organic compound that contains the —SH (sulfhydryl) group; they are substitution products of hydrogen sulfide; many have repulsive odors. Thiols are analogous to and resemble alcohols (R-OH). Many aliphatic thiols are toxic by inhalation and are flammable (e.g., methanethiol). The lighter alkyl thiols are characterized by their "ultragarlic" odor; some are rather fairly common air pollutants. Their action is similar to that of H_2S; they are precursors of cytochrome oxidase poisons. They cause nausea and headaches when inhaled even at very low concentration. At higher concentrations they can cause cardioacceleration, cyanosis, and coldness of hands and feet in humans. Extreme exposures may cause unconsciousness, coma, and death.

-thiol. a suffix indicating that a substance is a thiol.

thiolhistidine-betaine. ergothioneine.

thiolutin. (6-(acetamido)-4-methyl-1,2-dithiolo-[4,3-b]pyrrol-5,4H-one; 3-acetamido-5-methylpyrrolin-4-one[4,3-d]-1,2-dithiole; acetopyrothine). an extremely toxic antibiotic isolated from several strains of *Streptomyces albus*. It is a yellow substance with monoclinic crystals, is slightly soluble in water and more soluble in ethanol, methanol, chloroform, and other organic solvents. It is active against Gram-negative bacteria and is [illegible] and fungicidal.

thioneine. ergothioneine.

thiooxazolidone. **1:** L-5-vinyl-2-thiooxazolidone (*q.v.*). **2:** any thiooxazolidone alkaloid.

thiooxazolidone alkaloid. See L-5-vinyl-2-thio-oxazolidone.

thiopental sodium (pentothal sodium; sodium 5-ethyldihydro-5-(1-methylbutyl)-2-thiobarbiturate; pentothal). a barbiturate, it is widely used as an intravenous anesthetic of ultrashort action because it is rapidly absorbed by fat storage depots. It has also been used rectally for basal anesthesia, but is presumed inactive *per os*. It rapidly induces sleep and is thus suitable for induction of anesthesia. Thiopental sodium also reduces excitement and controls convulsions. It is detoxified by all body tissues. This substance is very toxic to laboratory mice, *i.v.*, and extremely toxic, *i.p.* Effects of acute exposure include hypotension and respiratory depression.

thioperoxydicarbonic acid diethyl ester. dixanthogen.

thiophane. tetrahydrothiophene.

thiophene. the most common cyclic organosulfide. It is a heat-stable liquid (b.p., 84°C) with a solvent action similar to that of benzene. It is used in the manufacture of pharmaceuticals, dyes, and even resins that also contain phenol or formaldehyde.

thiophosphorus insecticide. any sulfur-containing organophosphosphate insecticides (e.g., methylparathion, parathion). Metabolic activation requires desulfuration of these compounds to their oxon analogs (e.g., methylparaoxon and paraoxon).

β-thioquanidine deoxyriboside. carcinogenic in some laboratory animals.

thioseconal. thiamylal.

thiosulfate. See sodium thiosulfate.

thio-TEPA. thiotepa.

thiotepa (triethylenethiophosphoramide; tris(1-aziridinyl)phosphine sulfide; thio-TEPA). an extremely toxic cytotoxic alkylating agent related to triazone, *q.v.* It is used as an insect sterilant and antineoplastic agent. It causes delayed reproductive effects in human females and is a human carcinogen. Inhalation or contact with thiotepa particles is hazardous.

thiourea (thiocarbamide). a sulfur analog of urea with white, lustrous, bitter-tasting crystals, $(NH_2)_2CS$, that has been used as a rodenticide and is moderately to highly toxic to humans. It is a skin irritant and affects bone marrow, causing anemia. Thiourea is a known carcinogen and may not be used in food products in the United States, and is known to cause liver and thyroid cancer in experimental animals.

thioxamyl. oxamyl.

thioxanthene. **1:** a crystalline compound (thiaxanthene; dibenzothiopyran; dibenzopenthiophene; diphenylene methane sulfide), $C_{13}H_{10}S$, that sublimes readily and is moderately soluble in ethanol and ether. **2:** any of a major class of structurally related neuroleptic drugs (e.g., chlorprothixene, thiothixene) that are based on substituents of the compound, thioxanthene. Thioxanthenes appear to be antagonists of both the Dl and D2 classes of dopamine receptors.

thiram (tetramethyl thiurane disulfide; tetramethylthiuram disulfide; bis-(dimethylthiocarbamyl) disulfide; thiuram; TMTD). a white or yellow, water-insoluble, crystalline powder, $[(CH_3)_2NCH]_2S_2$, with a characteristic odor, that is soluble in a number of organic solvents. It is a dithiocarbamate fungicide that is used also as a disinfectant of seeds, nuts and fruit; to control bacteria in edible oils and fats; and as an ingredient in antiseptic sprays and soaps and in suntan lotion. It is toxic by ingestion and inhalation, and an irritant of skin, eyes, and mucous membranes, causing sneezing, coughing, and skin and eye irritations; the reaction is especially severe if a subject drinks alcohol while exposed to thiram. These properties led to the development of disulfiram, which is sold under the name Antabuse.

thirst. a sensation marking the desire for fluid, most importantly water. It may occur when mucous membranes, especially those of the pharynx, are dry or from lack of salivary secretion. It occurs also in dehydration (e.g., due to vomiting, profuse sweating, loss of fluid through excessive urination, or hemorrhage). Thirst may also accompany fevers and certain other illnesses. Thirst is absent in certain illnesses.

thistle. *Cirsium arvense* (Canada thistle), *Carduus* (plumeless thistle), *Salsola pestifer* (Russian thistle), *Silybum marianum* (variegated thistle).

thiuram. thiram.

Thlaspi arvense (fanweed). a toxic, widely distributed temperate zone weed (Family Brassicaceae) that causes severe gastroenteritis in livestock when ingested, presumably because of the mustard oil content.

THM. See chloroform.

thoracolumbar division of the autonomic nervous system. sympathetic nervous system.

Thorazine. chlorpromazine.

thoria. thorium dioxide.

thorium (symbol, Th)[Z = 90; A_r = 232.04]. a toxic, flammable, radioactive, soft, ductile, grayish-white, lustrous, powder. It is a metallic element of the actinoid series. ^{232}Th is the only naturally occurring isotope and is the most stable nuclide, $T_{1/2} = 1.4 \times 10^{10}$ yr. It is used in nuclear reactors and in some industrial processes. It is used, for example, in colloidal form as a stain for acid mucopolysaccharides in electron microscopy.

thorium anhydride. thorium dioxide.

thorium dioxide (thorium oxide; thorium anhydride; thoria). a white, heavy, infusible crystalline powder, ThO_2, used chiefly in the chemical, steel, ceramics, and incandescent lamp industries, with applications also in nuclear reactors and metal refineries. It causes liver cancer in humans; exposed workers are at risk.

thorn. a woody, spine-like structure with a sharp point, which is continuous with the vascular system of the plant.

thornapple. a common name of plants of the genus *Datura*, especially *D. stramonium*.

thoron. an isotope of *radon* with a mass number of 220. It is a decay product of thorium.

thoroughwort. See *Eupatorium*.

THP. tetrakis hydroxyl-methyl phosphonium. See flame retardant.

THPC. tetrakis hydroxyl-methyl phosphonium chloride. See flame retardant.

thread capsule. nematocyst.

thread cell. nematocyst.

thread-leaf groundsel. *Senecio longilobus*.

thread tube. See nematocyst.

three-flowered nightshade. *Solanum triflorum*.

D-threo-*N*-dichloroacetyl-1-*p*-nitrophenyl-2-amino-1,3-propanediol. chloramphenicol.

D(—)-threo-2-dichloroacetamido-1-*p*-nitrophenyl-1,3-propanediol. chloramphenicol.

D-threo-*N*-(1,1′-dihydroxy-1-*p*-nitrophenylisopropyl)dichloroacetamide. chloramphenicol.

threshold. **1a:** the intensity of a stimulus that is just sufficient to produce a sensation or to elicit a response in an irritable tissue, organ or, in some cases, a whole body response. **1b:** in physiology, the stimulus intensity that is just sufficient to elicit a motor response. **1c:** the membrane potential at which an impulse is initiated in a neurone. **2:** in toxicology, the dose or concentration of a toxicant that is just sufficient to produce a toxic effect. The concept of threshold has many limitations at best and has little credibility among many toxicologists as the "threshold" dose or concentration of a given toxicant may be extremely sensitive to slight changes in value of any of numerous variables, both endogenous and exogenous. Even where the toxicity of a given toxicant varies little within the range of variation in environmental or biological variables, it is often costly and difficult to accurately establish a threshold dose or concentration experimentally. **air pollutant threshold**. the minimum concentration of an air pollutant that will cause injury to a population under stipulated conditions during a specified length of time.

threshold limit value-biological (TLV-BLV). See limit value.

threshold limit value-ceiling (TLV-C). See limit value.

threshold limit value-short-term exposure limit (TLV-STEL). See limit value.

threshold limit value-time-weighted average (TLV-TWA). See limit value.

thriftiness. superior weight gain of an animal or human relative to that expected from the quantity and quality of the diet.

thrips. See Thysanoptera.

throb. **1a:** a pulsating or beating, usually more or less rhythmic, movement or sensation. **1b:** to pulsate, beat, palpitate, pound.

throbbing. palpitations; pulsating or beating, often rhythmically; said in reference to movement or sensation (as in throbbing pain).

thrombin. a proteolytic enzyme that catalyzes the complex conversion of fibrinogen to fibrin during blood clotting. It is formed from an inactive precursor (prothrombin) that is present in normal blood plasma. Thrombin is present only in blood removed from circulation and intravenous injection of thrombin causes immediate clotting. See venom (snake venom).

thrombin-like enzymes. See enzymes of snake venoms.

thrombocyte. obsolete name for a platelet, which was originally thought to be a true cell.

thrombocytin. serotonin.

thrombocytopenia. a condition of abnormally low platelet (thrombocyte) concentrations in circulating blood. Affected individuals exhibit excessive bruising and bleeding from wounds. This condition is often due to an autoimmune reaction due to the formation of antigens by the linkage of such drugs as antihistamines, aspirin, digoxin, or sulfonamides with platelet proteins. It also results when bone marrow function is depressed by chemicals or ionizing radiation; and following exposure to cytosine arabinoside.

thrombotonin. serotonin.

thrombus. a blood clot; usually a clot that remains in the blood vessel where it developed.

throughwort. See *Eupatorium perfoliatum*.

thuja (thuya). the fresh tops of *Thuja occidentalis*, it is a source of cedar leaf oil. See thujone.

Thuja occidentalis (American arborvitae; white cedar). an ornamental evergreen tree of eastern North America (Family Pinaceae), often growing to a height of 20 m or more. It is a source of cedar leaf oil, *q.v.*

thuja oil. See thujone.

3-thujanone. thujone.

thujol. thujone.

thujone (3-thujanone; thujol; thuyol; thuyone; absinthol; tanacetol; tanacetin; tanacetone). a moderately toxic, aromatic, terpene ketone, $C_{10}H_{16}O$, that occurs in many essential oils, such as thuja oil, oil of tansy, oils from sage, and wormwood. It is, for example, the principal component of cedar leaf oil, *q.v.* See also *Artemisia absinthium*, *Tanacetum*, *Thuja occidentalis*, tansy poisoning.

thulium (symbol, Tm)[Z = 69; A_r = 168.93]. a soft, malleable, ductile, silvery white metal of the lanthanoid series of elements. Thulium chloride heptahydrate, $TmCl_3 \cdot 7H_2O$, is toxic.

thunderwood. *Rhus vernix*.

Thurber loco. *Astragalus thurberi*, a true locoweed.

thuya. thuja.

thuyol. thujone.

thuyone. thujone.

thymic. of or pertaining to the thymus.

thymidine kinase locus. TK locus.

thymus gland. an unpaired, bilobed, lymphpoid organ of vertebrates that develops from the third pharyngeal pouch. It lies ventrally in the region of the lower neck and anterior to, or in bipeds, superior to the heart. The thymus is essential to the development of immunity and controls cell-mediated immunity, homograft rejection, and allergic aspects of the immune response in mammals. It plays an essential role in the development of T-cells. At sexual maturity the thymus atrophies in most vertebrates other than birds. It is active only if the pool of T cells is depleted.

thymus-dependent lymphocyte. T. lymphocyte.

thymus-derived lymphocyte. T. lymphocyte.

thyreotrophic hormone. TSH.

thyrocalcitonin. See calcitonin.

thyroglobulin. See thyroid hormone.

thyroid gland (*glandula thyroidea*). a paired endocrine gland of vertebrates that produces, stores, and secretes the thyroid hormones that regulate metabolic rate and contribute to the regulation of normal growth and development. It also secretes thyrocalcitonin. See also thyroid hormone.

thyroid hormone. either or both of two iodine-containing amino acid hormones secreted by the follicles of the vertebrate thyroid gland: thyroxine (thyroxin; 3,5,3′,5′-tetraiodothyronine), $HOC_6H_2I_2OC_6H_2I_2CH_2CH(NH_2)$-COOH, and 3,5,3′-triiodothyronine. They are derivatives of tyrosine. Thyroxine is converted into triiodothyronine, the apparent active form, within the target tissues. Both hormones increase the metabolic rate and oxygen consumption of tissues generally. In accomplishing this, they probably act synergistically with growth hormone, adrenocortical steroids, and adrenalin. Thyroid hormone has many additional effects on vertebrate tissues, all of which arise from stimulation of specific protein and enzyme syntheses by thyroid hormone; including stimulation of molt and metamorphosis. Excessive secretion of thyroid hormones in humans (most frequently women), causes a rare disorder, Graves' disease. This disease is characterized by an enlarged, pulsating goiter (toxic goiter); elevated basal metabolic rate, pronounced tachycardia, weight loss, nervousness (evidenced by fine muscle tremors, restlessness, and irritability), a tendency toward diaphoresis, emaciation, and often exophthalmos. The etiology of Graves' disease is unkown, although it is thought that hyperthyroidism in humans may be due to (1) excessive secretion of TSH from tumors, or (2) the elaboration of antibodies with TSH-like activity. See also thyrotoxicosis.

thyroid storm. thyroid crisis (See crisis).

thyroid-stimulating hormone. TSH.

thyroiditis. inflammation of the thyroid gland.

thyroidotoxin. a toxin that acts specifically on thyroid tissue.

thyrointoxication. thyrotoxicosis.

thyrotoxic. **1a:** of or pertaining to thyrotoxicosis or the ability to cause thyrotoxicosis. **1b:** toxic to the thyroid gland. **2:** pertaining to, characterized by, or denoting toxic activity of the thyroid gland.

thyrotoxicosis (thyrointoxication). poisoning of humans and other vertebrates due to excessive quantities of endogenous or exogenous thyroid hormones. It is marked by many of the symptoms of hyperthyroidism in humans (Graves' disease) such as nervousness, tachycardia, elevated levels of thyroid hormones in circulating blood, and thyroid gland hyperplasia or hypertrophy. See also encephalopathy (thyrotoxic encephalopathy), thyrotoxin, shock (thyrotoxin shock). **apathetic thyrotoxicosis** (chronic toxicosis). a condition that presents as cardiac disease or as a wasting syndrome. The affected individual exhibits few of the common symptoms of thyrotoxicosis.

thyrotoxin. **1:** a complement-fixing antigenic factor seen in certain diseases of the thyroid gland. **2:** a hypothetical toxin produced by hyperplastic thyroid glands and thought to be the cause of Graves' disease. **3:** any substance that is toxic to thyroidal tissue. This use is not recommended.

thyrotrophic hormone. TSH.

thyrotrophin. TSH.

thyrotropic hormone. TSH.

thyrotropin. TSH.

thyroxine, thyroxin. See thyroid hormone.

Ti. the symbol for titanium.

TI. **1:** therapeutic index; topical irritation. **2:** topical irritation.

TIA. topical irritation arthritis.

TIBA (2,3,5-triiodobenzoic acid). an anti-auxin herbicide that inhibits cell division in primary meristems. This compound and certain other substituted benzoic acids have similar but weaker antitropistic properties to those of morphactins.

tic. a brief, involuntary muscle spasm.

tic polonga. *Vipera russeli*.

tick. common name of any representative of the arachnid superfamily, Ixodoidea, *q.v.* Bites produce itching sensations and local skin irritation. Some species are disease vectors (e.g., of typhus, Rocky Mountain spotted fever). The bite of poisonous species produces pain, erythema, swelling, muscle cramps and, in some cases, tick paralysis, *q.v.* The application of gasoline or heat to the tick should cause it to loosen its jaws. It can then be removed with forceps (tweezers). Do not leave jaws embedded in the wound, which should be carefully washed. If there is any question of infection or complications, a physician should be notified.

tick paralysis. a condition caused by the venomous saliva of certain ticks of the genera, *Amblyomma* and *Dermacentor*. Symptoms and signs include anorexia, lethargy, muscle weakness, incoordination, nystagmus, and a flaccid ascending motor paralysis. Bulbar or respiratory paralysis may develop. Ticks generally present a greater hazard to humans as vectors of many serious diseases such as Rocky Mountain spotted fever and lyme disease.

tiger apple. *Thevetia peruviana*.

tiger moth. common name of moths of the family Arctiidae.

tiger ottern. See *Notechis*.

tiger rattlesnake. *Crotalus tigris*.

tiger snake, tigersnake. **1:** a common name of *Notechis* especially *N. scutatus*. **2:** common name of *Hoplocephalus stephensii*.

tight junction. a zipper-like union between two cells that prevents the passage of molecules.

tigra mariposa. *Bothrops venezuelae*.

timba. *Bothrops nummifer*.

timber rattlesnake. *Crotalus horridus*.

timbo. See *Lonchocarpus*.

time, lethal. lethal time.

time to observance. See time to tumor.

time to tumor (time to observance). the time at which a tumor is first detected by palpation or by gross or histologic examination of an animal at death.

time. **1:** duration; a measure of duration. **2a:** a dimension that represents, and serves as a measure of, sequential changes of matter in space. **2b:** the interval between the beginning and end (or between two designated points) of any process or activity. **3:** a term that is expressive of a precise, designated point (instant) or terminus in a time sequence (e.g., the time of day). **4:** a term that expresses an interval between two points in time (def. 3a) (e.g., an hour, a calendar year). **5:** a mode of perception that can be altered by any of a large variety of drugs or toxicants. See also time to tumor, time-averaged measurement, time-weighted average, real-time measurement. **lethal time**. the time lapsing from exposure to a toxicant and time of death. **lethal time 50** (LT_{50}). the time elapsing from exposure to a toxicant and the time when half of a test population is dead.

time-averaged measurement. See real-time measurement.

time-weighted average. the average value of a function that varies with time, weighted for the time duration of the sample taken.

Timolide. a beta-adrenergic blocking agent.

tin (symbol, Sn; stannum)[Z = 50; A_r = 118.69]. a soft, malleable, silver-white, lustrous, low-melting, metallic element with many uses. It is used to plate other metals and in the production of alloys, soft solders, and tinfoil. It is an occupational poison, affecting steel workers, miners, welders, refinery workers, smelters, solderers, and welders. Mild irritation to skin and mucous membranes are early and sometimes the only symptoms, but prolonged exposure to the fumes or dust may produce a pneumoconiosis. ^{113}Sn is a radioisotope with $T_{1/2}$ = 115 days.

tin crystals. stannous chloride.

tin dichloride. stannous chloride.

tin dioxide. stannic oxide.

tin hydride. stannane.

tin peroxide. stannic oxide.

tin protochloride. stannous chloride.

tin salt. stannous chloride.

tincal. sodium borate (See borate).

tincture. See pigment (def. 3).

tinga. one of numerous common names of fishes of the family Tetraodontidae.

tingling. a light prickly sensation of short or long-continued duration. The former may be caused by low temperature, or by striking a nerve. Prolonged, persistent, or intermittent tingling can result from drug reactions or from any of various diseases of the central nervous system.

tinnitus. a usually persistent ringing sound or similar low-level noise that is unrelated to external sounds. Although tinnitis occasionally is seen in poisoning or chronic use of certain drugs, it is not a reliable symptom of such.

tinnitus aurium. sonitus.

Tinomiscium philippinense. a toxic plant that contains picrotoxin.

tissue. an assemblage of specialized, usually adherent cells that are structurally similar and perform similar functions. Organs are comprised of spatially arranged combinations of tissues. The cells of a tissue may be of a single specialized type (e.g., neurons) of mixed types (e.g., connective tissue).

tissue culture (explantation). the growth in a suitable medium of tissue specimens taken from a living organism.

tissue death. See necrosis.

tissue specificity. See specificity (target organ specificity).

titanium (symbol, Ti)[Z = 22; A_r = 49.90]. a dark gray, lustrous transition metal that is reactive at high temperatures.

titer (titre). **1:** the amount or concentration of a substance in solution as determined by titration. **2a:** the concentration of a substance in blood or other body fluids. **2b:** a measure of the amount of antibody present in serum. The titer is estimated by assessing the concentration in the highest dilution of the serum at which agglutination by antigen-antibody reaction is detected; this value is expressed as the reciprocal of this dilution.

titre. titer.

Tityus. a genus of scorpions, two species of which (*T. bahiensis* and *T. serrulatus*) are venomous. Both occur in Mexico and Brazil.

TK locus (thymidine kinase locus). the locus that allows cultured mammalian cells to incorporate pyrimidines from the medium so that these pyrimidines may be converted into nucleic acids.

Tl. symbol for thallium.

TL_{50}. median tolerance limit.

TLC. **1:** thin layer chromatography. **2:** threshold limit concentration (See limit value).

TLC-C. threshold limit value ceiling. See limit value.

TLV. threshold limit value. See limit value.

TLV-BLV. threshold limit value-biological. See limit value.

TLV--C. threshold limit value, ceiling. See limit value.

TLV--STEL. threshold limit value, short-term exposure limit.

TLV--TWA. threshold limit value, time-weighted average.

Tm. symbol for thulium.

TML. tetramethyl lead.

TMTD. thiram.

TNA. See trinitrotoluene.

TNF-α. tumor necrosis factor α. See factor.

TNF-β. See lymphotoxin.

TNT. 2,4,6-trinitrotoluene.

TNT poisoning. a sometimes fatal poisoning by 2,4,6-trinitrotoluene (TNT), chiefly of munitions workers. It is marked by gastric and intestinal disturbances and dermatitis. Symptoms of intoxication may include sneezing and coughing due to inflammation of the upper respiratory tract, dyspnea, cataracts, weakness, muscle pain, neuritis, and cardiac arrhythmias. Lesions may include damage to bone marrow, kidney, and liver. Toxic hepatitis has developed in some systemically exposed workers and some older workers have developed aplastic anemia. Twenty-two fatal TNT poisonings were documented in the United States during the second world war.

toad. any of a large number of tailless amphibians of the subclass Anura that usually have dry, rough or warty skin. They are stouter and more sluggish than frogs with a parotid gland that forms a conspicuous raised area behind and above the tympanum; eggs are laid in water. *Bufo* is a well-known genus. Toads typically occur in cool, moist terrestrial habitats.

toadfishes. See Batrachoididae, Haplodoci, Tetraodontidae.

toadflax. See *Comandra pallida*.

toado. one of numerous common names of fishes of the family Tetraodontidae.

toadstool. **1:** any of numerous fungi with an umbrella-like cap (pileus). This term is essentially synonymous with mushroom in both the narrow and broad senses, but is more often used for inedible species. **2:** todesstuhl; death's stool; a vernacular term applied to any unappealing, unpalatable, or poisonous mushroom. *Cf.* mushroom. See also mycetismus.

toadstool poisoning. a common term referring to intoxication by poisonous mushrooms. See also toadstool, mycetismus.

tobacco. **1:** the dried, usually processed, leaves of *Nicotiana* spp., as that in cigars, cigarettes, snuff, pipe tobacco, or chewing tobacco. It is carcinogenic to humans in all forms. Tobacco is teratogenic and a cause of arthrogryposis in livestock. See also cigarette, cigarette smoke. **2:** *Nicotiana* spp., *q.v.* **chewing tobacco**. a form of smokeless tobacco. **smokeless tobacco**. snuff and chewing tobacco. These types of tobacco are generally held in the mouth for some time during use. Individuals who use these products risk consequences such as mouth sores, gum disease, tooth loss, and cancer of the mouth, pharynx, and esophagus. The cancer is occasionally fatal. See snuff.

tobacco heart. a sometimes painful condition of cardiac irritability, with irregular action and palpitations, due to excessive use of tobacco.

tobacco smoke. See cigarette smoke.

toboba chingu. *Bothrops nummifer*.

toboba de pestana. *Bothrops schlegelii*.

tobosagrass. *Hilaria mutica*. See also *Claviceps*.

α-tocopherol. vitamin E.

TOCP (tri-*o*-cresyl phosphate; tricresylphosphate; TCP; tri-*o*-tolyl phosphate; TOTP). an almost colorless, odorless, liquid mixture of phosphate esters, $(CH_3C_6H_4O)_3PO$, in which the hydrocarbon moieties are *meta* and *para* cresyl substituents. It is a very poisonous, neurotoxic industrial chemical with numerous applications, e.g., as a lubricant, gasoline additive, flame retardant, a plasticizer in lacquers and varnishes, solvent for nitrocellulose, plasticizer, and a coolant for machine guns. It has relatively low acute toxicity and fatalities among humans are rare. Acute poisoning of laboratory mice, rats, guinea pigs, and rabbits is characterized by hyperexcitability, spastic muscular incoordination, flaccid paralysis, dyspnea. Death usually results from asphyxiation that is partly due to neuromuscular failure and partly to depressed CNS activity. Pure TOCP is a colorless liquid (f.p. -27°C, b.p. 410°C). Modern commercial preparations, however, contain less than 1% TOCP, contamination of earlier products (with up to 20% TOCP) have caused severe poisonings. Oral exposure of humans may cause nausea, vomiting, diarrhea, or even progressive polyneuritis leading to paralysis of the extremities. Presenting symptoms of oral exposure in humans include nausea, vomiting, and diarrhea with severe abdominal pain. These symptoms typically subside for a period of 1-3 weeks to be followed by signs of peripheral paralysis evidenced by "wrist drop" and "foot drop." This late neuropathy follows activation to 0-2-tolyl *O,O*-saligenin cyclic phosphate and is associated with degeneration of central and peripheral neurons. In some cases, a progressive polyneuritis leads to paralysis of the extremities. The effects can be devastating, but recovery is often complete. See also ginger jake; organophorus-induced delayed neurotoxicity (OPIDN).

toddaline. chelerythrine.

todesstuhl. toadstool.

tofoni. *Echis carinatus*.

Tofranil™. imipramine hydrochloride.

TOK. nitrofen.

toka. *Dendroaspis jamesoni*.

tolectin. See tolmetin.

tolerance (toleration). **1:** the ability to endure pain or hardship. **2:** the ability to endure exposure to a large amount (or a high concentration) of a substance (e.g., food, drug, toxicant) without adverse effects. This term may apply to an individual organism, a type of organism, a population or species, or even to an ecological community. **3:** the capacity of an organism to show decreasing sensitivity to a particular substance with subsequent exposure. See also intolerance. **drug tolerance**. showing a progressive decrease in sensitivity to a drug with subsequent doses. **immunologic tolerance**. a state of immunologic responsiveness to an antigen such that an amount that would ordinarily induce an immune reaction does not do so. **law of tolerance**. that the distribution of an organ ism is limited by its tolerance to the fluctuations of a single factor. This is actually a working hypothesis that may hold in some instances and not others. Nevertheless, microdistribution of some species is limited by toxic plant secretions. For example, we may be facing reductions and/or redistributions of certain species of certain populations of organisms due to toxic chemicals in the environment. **pesticide tolerance**. the permissible residue levels for pesticides in raw agricultural produce and processed foods. In the United States, when a pesticide is registered for use on a food or a crop, a tolerance (or exemption from the tolerance requirement) must be established. These are established by EPA and are enforced by the Food and Drug Administration and the Department of Agriculture. **self tolerance**. See horror (horror autotoxicus).

tolerance level. that concentration of a chemical residue in food or feed above which adverse health effects are possible and above which corrective action should be taken.

tolerance limit. **1a:** the greatest amount, concentration, or intensity of a biologically active agent or stimulus to which a living system can be exposed for a specified period time and (1) not give evidence of distress, injury, or damage, or (2) still survive. See also limit, limit value, exposure limit, maximum allowable concentration, ceiling level, ceiling value. **1b:** in toxicology this term usually refers to the concentration or dose of a substance that a living organism can endure without ill effects. **median tolerance limit** (TL_{50}). the concentration of a substance in a suitable diluent that 50% of a test population can survive for a specified period of exposure.

tolerant. having the capacity for tolerance; exhibiting tolerance; having the quality of tolerance. In toxicology, said of a living system who is able to endure exposure to high concentrations or large doses of a specific toxicant without ill effects.

toleration. tolerance.

tolerogen. an antigen that induces a state of specific immunological tolerance or unresponsiveness to subsequent exposure to the antigen.

tolerogenesis. the induction of immunologic tolerance.

tolerogenic. able to induce immunologic tolerance.

tolguacha. See *Datura*.

***o*-tolidine** (3,3′-dimethylbenzidine; diaminoditolyl). a white to reddish, flammable substance, $[C_6H_3(CH_3)NH_2]_2$, with glossy plates that is soluble in ethanol and sparingly soluble in water. It is used in dyes and as a reagent in the detection of gold and free chlorine. This compound is carcinogenic to some laboratory animals; *o*-tolidine-based dyes are also carcinogenic to humans.

***O*-tolidine**. carcinogenic in some laboratory animals.

***O*-tolidine-based dyes**. carcinogenic in humans and some laboratory animals.

tolmetin (1-methyl-5-(4-methylbenzoyl)-lH-pyrole-2-acetic acid; l-methyl-*S*-*p*-toluoylpyrrole-2-acetic acid; Tolectin). a crystalline, antiinflammatory, analgesic, antipyretic substance, $C_{15}H_{15}NO_3$. It is effective in moderate doses and is better tolerated by most individuals than aspirin. Nevertheless it may cause discomfort and problems in the form of dyspepsia, epigastric pain, nausea and vomiting.

toluene (methylbenzene; phenylmethane). a colorless, flammable, liquid hydrocarbon, $C_6H_5CH_3$, with a benzene-like odor, derived from coal tar, that boils at 101.4°C. Among the many uses of toluene are its use in aviation gasoline; as a solvent for paints and various resins, gums, oils, rubber, and plastics; as a component of fuels; and in the synthesis of organic compounds such as toluene diisocyanate, phenol, saccharin, nitrotoluenes, and numerous benzyl, benzoyl and benzoic acid derivatives. It is moderately toxic to humans by inhalation or ingestion but only slightly toxic by dermal exposure. Exposure of humans to atmospheric concentrations of about 500 ppm may cause headache, nausea, lassitude, and impaired coordination without detectable physiological effects. At extremely high concentrations toluene has a narcotic effect that can lead to coma. Poisoning by ingestion causes a burning sensation in the mouth and stomach, nausea, vomiting, coughing, chest pains, headache, silliness, dizziness, ataxia, confusion, stupor, restless coma, and late severe blood dyscrasias with severe bone marrow damage. In severe cases, death from respiratory failure from CNS depression or from ventricular fibrillation may intervene. Pathology: respiratory failure from CNS depression or ventricular fibrillation.

***m*-toluenediamine**. toluene-2,4-diamine.

2,4-toluenediamine. toluene-2,4-diamine.

toluene-2,4-diamine (2,4-toluenediamine; 3-amino-*p*-toluidine; 5-amino-*o*-toluidine; *m*-tolylenediamine; MTD; C.I. 76035; C.I. Oxidation Base; 1,3-diamino-4-methuylbenzene; 2,4-diamino-1-methylbenzene; diamino toluene; 2,4-diaminotoluene; 4-methyl-1,3-benzenediamine; 4-methyl-*m*-phenylenediamine; *m*-toluenediamine). a colorless, crystalline substance, $CH_3C_6H_3(NH_2)_2$, used in the manufacture of various dyes in the textile, leather, fur, silk, wood, paper, and cotton industries. Formerly used in hair dyes. It is an irritant to the eyes and skin. Symptoms of exposure include nausea, vomiting, jaundice, and anemia. Major lesions are those of the CNS and liver. It is also carcinogenic.

2,6-toluene diisocyanate. toluene-2,6-diisocyanate.

toluene-2,4-diisocyanate (2,4-tolylene diisocyanate; *m*-tolylene diisocyanate; TDI). a clear, colorless to yellow liquid, that solidifies below about 2°C, $CH_3C_6H_3(NCO)_2$, with a sharp, pungent odor. It is obtained chiefly from coal tar and petroleum; it is soluble in many organic solvents. TDI reacts with water, yielding CO_2, and often reacts violently with hydrogen-containing compounds. It is mostly used to make polyurethane foams. TDI is considered hazardous at any concentration. It is a strong irritant, damaging tissue on contact. Most poisonings are, however, by inhalation of the vapor with resultant nausea, vomiting, abdominal pain, and dyspnea; sometimes also, temporary headaches, insomnia, and paranoid depression. Severe poisoning by inhalation of fumes causes pulmonary edema. Long-term exposure (weeks to years) sensitizes some people to TDI, altering protein in lung tissues and producing an allergic reaction with asthmatic attacks and some loss of lung function. TDI is carcinogenic to some laboratory animals. Long-term effects may include anemia and blood cell damage. *Cf.* toluene-2,6-diisocyanate, diphenylmethane-4,-4′-diisocyanate.

toluene-2,6-diisocyanate. this isomer has properties essentially similar to those of 2,4-toluene diisocyanate; they are most commonly available as a mixture. It is carcinogenic in some laboratory animals.

***p*-toluenesulfonic acid**. a strong irritant to skin, eyes, and mucous membranes.

α-toluenol. benzyl alcohol.

***o*-toluidine** (*o*-aminotoluine). a flammable, light yellow, liquid amine, $CH_3C_6H_4NH_2$, that turns reddish brown on exposure to the atmosphere and light. It is slightly soluble in water and soluble in ethanol and ether. It is toxic by ingestion, inhalation, and percutaneous absorption; it is carcinogenic.

***p*-toluidine** (*p*-aminotoluine). a flammable, slightly water-soluble amine, $CH_3C_6H_4NH_2$, with white lustrous plates or leaflets. It is toxic by ingestion, inhalation, and percutaneously in humans. It is carcinogenic to some laboratory animals and a contaminant of D&C Green No. 6.

***o*-toluidine hydrochloride**. carcinogenic in some laboratory animals.

0-2-tolyl *O,O*-saligenin cyclic phosphate. See TOCP.

2,4-tolylene diisocyanate. toluene-2,4-diisocyanate.

***m*-tolylene diisocyanate**. toluene-2,4-diisocyanate.

***m*-tolylenediamine**. toluene-2,4-diamine.

Tolypocladium inflatum. See cyclosporin.

tomato. *Lycopersicon esculentum.*

tomigoff. *Bothrops atrox.*

tommygoff. *Bothrops nummifer.*

tone. **1:** a state of continuous partial contraction of muscle, maintained by reflex activity; sometimes defined as resistance to elongation or stretch. Tonic reflexes form the basic mechanism of posture control. **2:** tonus; a healthy state of an organ or other part of an organism. **3:** a particular quality of a sound.

tongue. a protrusible muscular organ of the mouth. The tongue of vertebrates is attached to the floor of the oral cavity. It is typically an organ of chemoreception (e.g., taste) and touch, although one of these functions may be the paramount or even sole function in some organisms. In some forms (e.g., woodpeckers, anteaters, geckos) it is used to seize prey.

tonic. **1a:** of or pertaining to a state of persistent action, especially of muscular contraction. **1b:** producing tension. **2:** rigidity of the body that is characteristic of generalized tonic-clonic seizures. **3:** refreshing, stimulating, especially with regard to mental state or strength. **4:** a remedy claimed to be a general restorative; able to correct frailties and promote vigor of the body as a whole or of a specific organ or system such as the heart or digestive system. **5:** pertaining to a quality of sound, tone.

tonic contraction. sustained contraction of a muscle, as employed in the maintenance of posture. See also tone.

tonic convulsion. See convulsion.

tonic immobility. See thanatosis.

tonicity. **1a:** tonus; the quality of having tone, especially muscular tone. **1b:** a state of normal tension of the tissues, especially muscle fibers while at rest. In the case of muscle, tonicity is a state of continuous activity or tension beyond that related to the physical properties that account for resistance to stretch; in skeletal muscle it is dependent upon the efferent innervation. **2:** the osmotic pressure or tension of a solution, usually relative for example, to that of the cytoplasm of a cell or the blood of a higher animal, the root of a green plant. See also isotonicity.

tonicoclonic. tonoclonic.

tonka bean camphor. coumarin.

tonoclonic (tonicoclonic). of or pertaining to a condition in which both tonic and clonic muscular spasms occur.

tonus. **1a:** tonicity; a condition of persistent partial stimulation as in muscle tonus, with a consequent state of partial contraction. **1b:** in certain nerve centers, the state whereby motor impulses are continuously emitted without any sensory input.

topical. **1a:** of local occurrence; applied to or confined to a small area. **1b:** having a limited distribution. **2:** pertaining to drugs applied locally to relatively small surface areas of an organism (e.g., in humans, those applied to the skin, conjunctiva, mucous membranes of the mouth, nose, vagina, or rectum).

topical irritation arthritis (TIA). a condition caused by injection of a phlogistic agent such as formalin or mustard powder into the periarticular tissues.

topochemotaxis. a response of a motile organism toward the source of a chemical stimulus.

torciopelo. *Bothrops atrox*.

torsion. postural incoordination with a rolling gait that is usually associated with disturbances of the vestibular (ear canal) system.

torticollis (wryneck). a continual or intermittent twisting of the neck conferring an unnatural position of the head due to spasm of the cervical muscles.

total dissolved solids (TDS). a measure of the total amount of inorganic salts and other substances dissolved in a water sample.

total dose. See dose

TOTP. See TOCP.

toulou. *Bitis nasicornis*.

toumou. *Bitis nasicornis*.

tox-, toxi-, toxico-, toxo-. prefixes that indicate or denote toxicity, poisonousness, poisoning, or relationship to a toxicant or poison.

toxaemia. toxemia.

toxalbumic. obsolete; pertaining to or caused by toxalbumin.

toxalbumin. **1:** obsolete; used in reference to any poisonous albumin (e.g., abrin, phallin, risin, snake venoms). See also *Ricinus communis, Abrus precatorius*. **2:** sometimes used as a synonym for abrin. **3:** a phytotoxin, def. 1.

toxanaemia. toxanemia.

toxanemia (toxanaemia). anemia due to the effects of a hemolytic toxin.

toxaphene (chlorinated camphene; octachlorocamphene; polychlorocamphene). the generic name for a complex insecticidal mixture of chlorinated camphenes, chiefly octachlorocamphene, which has the approximate empirical formula, $C_{10}H_{10}Cl_8$. Toxaphene is an extremely toxic, amber, waxy, solid mixture of more than 170 compounds with a mild odor of chlorine and camphor. The individual compounds vary widely in toxicity. The insecticidal activity, however, resides in just a few of these compounds. Symptoms of intoxication in exposed mammals include nausea, confusion, agitation, tremors, epileptiform convulsions, and unconsciousness. The oral LD_{50} in laboratory rats is 69 mg/kg. Toxaphene is carcinogenic to some laboratory animals and is a suspected human carcinogen. All registrations for toxaphene in the United States have been cancelled.

toxemia (toxaemia; toxicemia; toxicohemia; toxinemia; blood poisoning). **1:** any of a variety of intoxications, pathological conditions, or disease syndromes caused by the presence of toxicants, especially bacterial toxins, in circulating blood. **1a:** septicemia; a generalized systemic intoxication due to dissemination of bacterial toxins via the circulatory system from a focal infection. **1b:** clinical manifestations of 1a; a functional or organic disturbance or pathological condition due to dissemination of toxicants derived from protein catabolism. See, for example, ophidism. **2:** injury to a plant by insect or other poisons. **alimentary toxemia**. a form of autointoxication due to the absorption of poisonous substances from the alimentary canal that were produced therein. **toxemia of pregnancy** (eclampsia and preeclampsia). a poorly defined term used in reference to metabolic disorders of pregnancy in humans that are marked by edema and serious disturbances of renal function (with albuminuria), central nervous system function, and blood pressure. **eclampsia**. a metabolic disorder of late pregnancy that may occur from the 20th week of pregnancy until a week after delivery. It is characterized by an intensification of the symptoms described for preeclampsia, above, plus the development of muscle spasms, seizures, and coma. The etiology is unknown. Both of these toxemias usually diminish and disappear within a week following childbirth, but are sometimes fatal. See toxemia (def. 3). **preeclampsia**. a metabolic disorder of late pregnancy that may occur from the 20th until a week after delivery. It is marked by edema and serious disturbances of renal function (with albuminuria), central nervous system function, and blood pressure. It is sometimes fatal. If convulsions and coma also occur, the condition is termed eclampsia. The condition may be mild (mild preeclampsia), with a significant rise in blood-pressure, a puffiness of the face, hands and feet, proteinuria, and excessive weight gain during the last trimester. In severe cases (severe preeclampsia), the above symptoms are more pronounced and the victim may have headache, blurred vision, abdominal pain, and irritability.

-toxemia, -toxaemia. suffix indicating a particular type of toxicant in circulating blood, e.g., [illegible]

toxemic. of, pertaining to, characteristic of, caused by, or affected with toxemia.

toxemic jaundice. copper poisoning.

toxenzyme. any poisonous enzyme.

toxi-. See toxico-.

toxic. **1a:** harmful to living systems. **2a:** poisonous; noxious; toxical, toxicant, toxiferous, venomous; having the capacity to poison; of the nature of a poison; of or pertaining to a poison. **2b:** pertaining to the quality of being poisonous. **2c:** pertaining to a poisonous substance, or to an effect or reaction of a living system to such a substance especially if the potential for harm is substantial. **2d:** produced by or resulting from the action of a poison; induced by a poison or toxin. **2e:** producing or containing a poison ortoxin. **3:** pertaining to a toxin. **4:** in chemistry, able to destroy the activity of a catalyst. **acutely toxic**. **1a:** able to produce toxic effects following a brief exposure. **1b:** pertaining to a xenobiotic or to radiation that can cause toxic effects (usually severe) following a single brief exposure. **chronically toxic**. pertaining to a toxicant that can produce toxic effects, at a dosage or concentration below that which is acutely toxic, following long-continued, or numerous brief exposures. **highly toxic**. impossible to define without using a number of qualifiers (e.g., species, age, sex, race or strain, size, individual or population history, duration of exposure, environmental conditions, route of exposure or administration). **1:** pertaining to a toxicant with a median lethal dose (LD_{50}) or median lethal concentration (LC_{50}) that is fatal to laboratory animals whether exposure is by contact, ingestion, or inhalation. **2:** pertaining to a substance with an LD_{50} of 50 mg/kg body weight or less when administered orally to albino laboratory rats weighing between 200-300 g each. **3:** pertaining to a toxicant with a median lethal dose (LD_{50}) of 200 mg/kg body weight or less when administered by continuous contact for 24 hours with the bare skin of albino laboratory rabbits weighing between 200-300 each (less if death occurs within 24 hours). **4:** pertaining to a toxicant that has a median lethal concentration (LC_{50}) in air of [illegible] vapor, or 2 mg/liter referred to mist, fume, or dust when administered to albino laboratory rats weighing between 200-300 grams each by continuous inhalation for 1 hour (less if death occurs within 1 hour). See also toxicity.

toxic agent. a toxicant or poison.

toxic amblyopia. a reduction in visual acuity believed to be due to a toxic reaction in the orbital portion of the optic nerve.

toxic amine. See amine.

toxic anemia. See anemia.

toxic bloom. See bloom.

toxic cirrhosis. cirrhosis of the liver due to chronic poisoning (e.g., by lead or carbon tetrachloride).

toxic cloud. an airborne mass of gases, vapors, fumes, or aerosols that contains toxic materials.

toxic conjugate. See conjugate.

toxic environmental chemical. See environmental chemical (toxic environmental chemical).

toxic fat syndrome (water belly). a condition occurring in 3-10 week old domestic fowl (*Gallus*) that have been maintained on fat-supplemented feeds. It is marked by edema of the pericardium and abdomen, a waddling gait, and sudden death.

toxic glycoside. See glycoside, aglycone.

toxic goiter. Graves' disease; the enlarged goiter of Graves' disease.

toxic natural product. any poison produced by a living organism. Included are an enormous variety of substances, including toxic by-products of metabolism as well as poisons produced by special mechanisms (e.g., venoms) which may have an adaptive function.

toxic peanut meal. a disease of livestock caused by ingestion of fungus-infected peanut (*Arachis*) meal.

toxic psychosis. any psychosis induced by exposure to a toxic substance (e.g., lead, ethanol).

toxic shock (toxic shock syndrome). a form of blood poisoning caused by an endotoxin released into the blood by staphylococci. Signs and symptoms include sudden high fever of at least 101°F (38.3°C), vomiting, diarrhea, and a scarlatiniform rash followed by desquamation. Complicating symptoms can include thirst, headache, sore throat, renal failure, rapid pulse, hypotension, and severe shock, congestive heart failure, extreme fatigue and weakness, mental confusion and feelings of impending doom. Fatalities occur, but most victims recover with early diagnosis and prompt treatment. Most serious cases have come from staphylococci in the vagina of women using supplemental tampons, although toxic shock syndrome can also arise from wounds or infections in the throat. See also tampon, toxic shock toxin; *Staphylococcus aureus*.

toxic shock syndrome (TSS). See toxic shock.

toxic substance. **1:** a poison; any substance that can poison a living system under conditions that such a system can be expected to encounter during its normal existence. **2a**. a compound or chemical mixture that presents an unreasonable risk of injury to human health or to the environment. **2b:** a substance declared to be toxic or registered by the appropriate governmental authority as a toxic substance.

Toxic Substances Control Act (TSCA). a U.S. federal law passed in October 1976. This act authorizes the EPA to develop and implement means to minimize dangers from toxic substances. Pesticides, drugs, tobacco products, nuclear materials, cosmetics, foods, and food additives are excluded from compliance with this act or the regulations promulgated therefrom.

toxic tetanus. See tetanus (toxic tetanus).

toxic waste. **1:** any waste material that is toxic. **2:** any waste material that contains poisonous substance(s) as designated under law such that their use, transportation, and disposal is regulated. Several laws apply in the United States. These are The Resource Conservation and Recovery Act of 1976, as amended; The Toxic Substances Control Act of 1976, as amended; the Comprehensive Environmental Response Compensation and Liability Act of 1980 (CERCLA). Many governmental jurisdictions have similar laws that regulate the handling and disposal of such wastes.

toxic-allergic syndrome. a syndrome due to ingestion of rapeseed oil containing acetanilide and contaminated with aniline. Symptoms include allergic pneumonopathy, respiratory distress, headache, nausea, abdominal pain, fever, a rash, and myalgia. Lesions include hepatomegaly and eosinophilia.

toxical. toxic; poisonous; containing poison.

toxicant. **1:** poisonous. **2a:** a toxic agent or poison; any substance that can poison or toxicate. **2b:** an alcoholic poison; any poisonous substance (e.g., ethanol, an alcoholic beverage) that causes intoxication in the popular sense. *Cf.* poisonous, toxic, toxical, toxiferous, venomous.

toxicant-receptor complex. the action of many toxicants involves initial binding of the toxicant to one or more populations of sites (receptors) that have a relatively high (often specific) affinity for the toxicant; this unit may be referred to as a toxicant-receptor complex. *Cf.* ligand-receptor complex.

toxicants, types of. toxicants may be grouped in any of numerous ways, some of which are more useful than others and all of which have weaknesses and limitations. In particular, none of these classifications includes all toxicants, nor are the categories mutually exclusive. They may, for example, be classified according to **1**: origin or source (e.g., botanicals, mycotoxins). **2**: mode of action (e.g., carcinogens, mutagens, hepatotoxics). **3**: level of organization acted upon (e.g., cytotoxicants, ecosystem poisons). **4**: chemical class (e.g., thiols, alcohols, halides, etc.). **5**: use or application (e.g., insecticide, industrial solvent, dye). **6**: distribution (e.g., air pollutants, water pollutants, soil contaminants). **7**: persistence (stability). toxicants vary greatly in their stability or persistence within a given milieu (e.g., liver tissue, stomach, a white-water river). In many cases, the persistence of a toxicant is more important than solubility, toxicity, etc. in assessing hazard or risk. Thus cyanide is extremely toxic, but is not an appreciable hazard in an oxidizing environment. **8**: affinity for a particular tissue, for example, as in the case of lipohilic compounds can be extremely important in determining the hazard of a substance that may be chronically present only in small concentrations. Numerous other bases for classification can be, and have been, employed. A useful classification system is usually tailored to the particular problem at hand.

toxicate. to poison. Used in reference to the action of the poison itself, not to the act of administering a poison.

toxication. the toxic action of a poison, poisoning (See toxicate).

toxicemic. toxemic.

toxicide (toxolysin). **1:** a toxolysin; a substance that is destructive to toxins. **2:** a chemical antidote; any substance able to render a toxic agent harmless. *Cf.* antitoxin.

toxicity. **1:** the state or quality of being poisonous; the capacity to poison living systems; having a toxic or poisonous quality. This is sometimes erroneously taken to be an inherent property of a substance. The toxicity of a substance is always due, to a greater or lesser extent, to certain interactive properties of the substance, the specific target (e.g., species, strain, sex, age, organ, prior exposure to the toxicant), and the environment. **2:** potency, virulence, deadliness, lethality, or the degree of virulence of a poison or of a toxic microbe. The degree to which a toxicant is injurious, lethal, or poisonous to plant or animal life. **3:** toxicity is often operationally defined in terms of the amount (or concentration) of a particular toxicant required to produce a stated detrimental effect in a given type or species of living organism under specified conditions of exposure for a given end point (e.g., mortality, morbidity, eggshell thinning, loss of fecundity); the toxic dose is often given in terms of unit body weight of the organism of concern (e.g., mg of toxicant per kg body weight). See also factor (modifying factor), nontoxic, toxic, toxicity rating. **absolute toxicity**. the toxicity of a substance without consideration of dilution. **absolute effluent toxicity**. the toxicity of a waste stream or effluent without consideration of dilution. **acute toxicity**. **1:** the capacity of a biologically active agent to cause serious harm, injury, or death to a living system following a single dose or a single brief exposure. **2:** harm to a living system that is manifested within a relatively short time following exposure to the toxicant. Dependent on the test species and other aspects of the test situation, LC_{50} (96 hours) and EC_{50} (48 hours) usually satisfy this definition. **3:** the amount of pesticide that will seriously affect or destroy a test animal in a single dose. See also toxic. **aquatic toxicity**. the capacity of a substance in a body of water (usually taken to be a freshwater system - a stream, lake, reservoir, or groundwater body) to poison those organisms that normally live in such a body. *Cf.* marine aquatic toxicity. **biogenic toxicity**. the chemical inhibition of one organism by another (allelopathy); sometimes taken to also include autotoxicity. **chronic toxicity**. **1:** the capacity of a biologically active agent to produce toxic effects following long-continued, or numerous brief exposures. Trustworthy quantitative data on chronic toxicity are rarely available for a drug or poison. Because acute toxicity testing is faster and less expensive than chronic testing, chronic toxicity is often inferred on the basis of results from acute toxicity tests. See also extrapolation. **2:** any poisonous effect that results from long-continued or numerous brief exposures. **3:** occasionally used synonymously with delayed toxicity. This usage is not recommended. See also toxic. **comparative toxicity**. **1:** the capacity of a toxicant to poison a particular taxon (taxonomic group, e.g., a species, genus, family, order, class, or even phylum) relative to its toxicity to other taxa of the same rank. If comparisons are made among a set of species, all should be taxonomically related (e.g., all are members of the same genus, a set of closely related genera, or the same family). If one is comparing higher taxonomic categories (e.g. birds vs. mammals) serious problems may arise, especially that of selecting species or populations that are representative of the higher taxonomic category to which each belongs. **2:** relative toxicity. This is confusing and inappropriate usage. *Cf.* relative toxicity, below. See also toxicology (comparative toxicology). **delayed toxicity** (latent toxicity). the appearance of clinical toxic effects following a relatively extended latent period (e.g., tumor development and accompanying symptoms; poisonings by certain mushrooms). This term is not synonymous with chronic toxicity. **dermal toxicity**. the ability of a toxicant to poison people or other vertebrates by contact with the skin. **direct toxicity**. **1a:** the capacity of a substance to poison an exposed organism without metabolic transformation. **1b:** the capacity of an organism to poison other organisms by virtue of a poison or poisons produced by the organism as opposed to poisons accumulated from the environment. **hepatic toxicity**. hepatotoxicity. **hepatotoxicity**. a main entry, *q.v.* **indirect toxicity**. the capacity of a substance or organism to exert toxic effects (1) only following metabolic transformation to another chemical species (the proximate poison) or (2) by virtue of contamination by or accumulation of a poison or poisons. *Cf.* secondary toxicity. **latent toxicity**. delayed toxicity. **marine aquatic toxicity**. the toxicity of a substance in a saltwater body to those organisms that normally live in such a body. *Cf.* aquatic toxicity. **nephrotoxicity**. a main entry, *q.v.* **O_2 toxicity**. oxygen toxicity. **organ toxicity**. **1a:** the quality, property or capacity of a particular toxicant to poison a particular organ via direct or indirect action. **1b:** the virulence of a toxicant with respect to a specified organ. **2:** the effects of a toxicant upon a particular organ. **3:** any adverse effect on an organ due to the direct or indirect (e.g., through altered metabolic processes, circulatory changes) action of a toxicant. **overdosage toxicity**. the toxic effect that predictably occurs when the dosage of a drug that is administered to a subject exceeds the therapeutic range. Overdosage toxicity is most often encountered with the use of drugs with a low therapeutic index. Such effects are usually unintended. **ovotoxicity**. the quality, property, or capacity of a particular toxicant to directly or indirectly poison ova. **oxygen toxicity** (O_2 toxicity). **1:** the toxicity of oxygen. **2:** oxygen poisoning, *q.v.* **pulmonary toxicity**. **1:** the quality, property or capacity of a particular toxicant to poison, whether directly or indirectly, the lungs of vertebrate animals. Because of the extremely high respiratory surface area of the lung, the high volume of air passing over the respiratory surfaces during the course of time, and the relative fragility and lack of protection of the respiratory epithelium, many respirable toxicants, perhaps especially irritants, can damage the lung. Chronic exposure to various respirable dusts (as in the case of coal miners, stone cutters, textile workers, and many other workers) also produces diseases of the lung (See pneumoconiosis) that may not present clinically for many months or years. The lung is, of course, also a major route for airborne systemic poisons. It is also a major excretory route for certain poisons or their metabolites and is thus exposed by this route. Agents that are notoriously pneumotoxic include allergens such as molds and fungal spores, ammonia, asbestos, coal dust, chlorine gas, nitrogen oxides, ozone, phosgene, and silica dust. **relative toxicity. 1:** the toxicity of one toxicant or class of toxicants relative to another for a specified target under specified conditions. This term is often confused with comparative toxicity, *q.v.* Factors that affect relative toxicity in addition to the inherent properties of the toxicants under consideration at a given time are numerous and may include dosage, route of exposure, physical nature or phase of the toxicant, temperature of the toxicant, ambient temperature, humidity, health and condition of the subject (individual, population, system), and interactions with other chemicals (poisonous and nonpoisonous). **2a**: the toxicity of a chemical determined for a given effect relative to that for another effect of the same chemical. This is a special case of relative toxicity. **2b:** the toxicity of two or more chemicals with respect to one specified adverse effect. This is a special case of relative toxicity. **3a:** the toxicity of an effluent discharge (e.g., from a manufacturing plant) relative to that following mixture with the receiving waters. **3b:** the toxicity of an effluent relative to a dilution using water that has a composition that is similar to that of the receiving waters. **renal toxicity**. nephrotoxicity. **secondary toxicity**. the state, quality, or capacity (which may be temporary) of an ordinarily nonpoisonous organism, material, or substance to poison living systems. The situation may arise by accumulation or contamination with toxicants from the environment. Cadmium, lead, copper, and fluorine are examples of chemicals that may accumulate to dangerous levels in certain plants. *Cf.* indirect toxicity. **selective toxicity**. **1a:** the state whereby the action of a given toxicant is restricted to one or a few structures, processes, species, or higher taxon of organism. Only rarely do poisonous compounds used to control pests and weeds possess selective toxicity. **1b:** as applied to pesticides and herbicides, selective toxicity is the nature of a poison such that it injures one kind of living system (the uneconomic form) without injuring another (the economic form), even though the two systems are in intimate contact. Very few substances exhibit selective toxicity. **subchronic toxicity**. toxicity due to nearly chronic exposure to quantities of a toxicant that do not cause any evident acute toxicity, with repeated doses or continuous exposure (in food, water, or ambient air) for an extended period of time, but of shorter duration than in chronic exposure and substantially less than the life span of the organism exposed. In humans and most other mammals, 30-90 days of exposure is considered normal. *Cf.* chronic toxicity, above.

toxicity assessment. a formal evaluation of the toxicity of a chemical based on currently available field and laboratory data on humans, animals, or other living systems as appropriate to the goals of the assessment. See also risk assessment.

toxicity rating. toxicity has been rated variously, but all classifications, indices, and scales used for this purpose are limited in their applicability by at least one or more of the following: the types of poisons, the kinds of organisms, age, sex, physiological condition and state of health, environmental conditions, the conditions and routes of exposure, the time-course of exposure. See also toxic, toxicity.

toxicity rating, a generalized scheme. a scheme for rating chemical substances especially when comparative or quantitative information on the chemicals of interest is lacking and where environmental conditions and exposure regimes are within the range of those normally experienced by the target organism or system: **essentially nontoxic**. where a substance produces no harm under conditions of normal usage or exposure. **slightly toxic**. where acute or chronic exposure produces only slight effects on the organism (or system of concern) under normal circumstances of exposure. **moderately toxic**. where a substance produces moderate effects on the organism (or system of concern). **very toxic**. where acute or chronic exposure to a substance can produce severe irreversible impairment. **extremely toxic** (highly toxic). where acute or chronic exposure to a substance can be life threatening or otherwise threatening to the integrity of the system.

toxicity rating, a widely used scheme. a scheme based upon oral lethal dose for a 70 kg (154 lb) human (after Gosselin et al., 1984, See "Clinical Toxicology of Commercial Products," Williams & Wilkins Co.: Baltimore). Toxicity is rated from 1-6:

Rating/class	Dose
6. Super toxic.	< 5 mg/kg (a taste, < 7 drops)
5. Extremely toxic.	5-50 mg/kg (7 drops - 1 tsp)
4. Very toxic.	50-500 mg/kg (1 tsp - 1 oz)
3. Moderately toxic.	0.5-5 gm/kg [1 oz. - 1 pt (1 lb)]
2. Slightly toxic.	5-15 gm/kg (1pt - 1 qt)
1. Practically nontoxic	>15 gm/kg More than one qt (2.2 lb)

The above rating system is sometimes applied loosely to laboratory animals. See also nontoxic.

toxicity test (toxicity assay). any of extremely numerous and varied tests used to establish the toxicity of a substance to any of a wide variety of living systems (e.g., cells, organisms, populations, communities) tested under various exposure regimes and for various purposes, often to help establish regulations. Unlike most research protocols, toxicity tests may come into general use and the protocol rapidly becomes standardized. All tests, to the extent that they are effective, help to determine either the relative or comparative toxicity of a toxicant, potential toxicant, or a set of toxicants under highly prescribed and controlled test conditions. The test system is purposely as simple as possible to assure reproducibility, and usually quantitative results. Thus, the requirements of a test are paramount over many aspects of applicability to real world exposures. Applications of toxicity tests to human or animal health or to the environment invariably involve extrapolation; such is by no means always justified (See extrapolation). More realistic testing and less extrapolation are highly desirable and have been repeatedly sought, but all tests ultimately confront complexities, system and enviromental variation, costs, and time that frustrate the purpose. A principal application of the results of toxicity testing is in the promulgation of health and environmental regulation of toxic chemicals including manufacture, transport, usage, and disposal. A catalogue listing toxicity tests that are or have been in use would more than fill a text the size of this dictionary. A selected set is included below. See also test. **acute toxicity test**. any test designed to yield a quantitative measure of acute toxicity (usually lethality, e.g., LD_{50} or LC_{50}) of the test chemical. Such tests may also provide information on the clinical manifestations of acute toxicity and provide data to help guide the selection of dosages for other types of tests. **behavioral toxicity test**. any test of a xenobiotic on behavior. Such tests are not often used for humans and are not prescribed by regulation at this time in most nations. Most behavioral tests are conducted on feral animals or surrogates for feral animals, especially aquatic species. Some behavioral tests are based upon conditioning, (1) operant conditioning in which the test animals learn to perform a task to obtain a reward or to avoid a punishment, or (2) simple pavlovian conditioning, in which the subject learns to associate a conditioning stimulus with a simple behavior normal to the animal, usually a reflex. Other tests take advantage of a normal unconditioned response of an animal to a particular stimulus (e.g., an unconditioned approach-withdrawal response, predator-prey interactions). Natural endogenously generated behavior can also be employed in such tests (e.g., foraging behavior, diurnal rhythms). In all such tests the behavior of the test group in response to challenge by a toxicant is compared with that of untreated controls. **biochemical toxicity test**. any toxicity test that measures biochemical endpoints. While the target system may be a tissue, organ, organism, or occasionally a population, the test system is comprised of cells or fractions of cells. **chronic toxicity test**. a type of test that is similar to a subchronic test, but of longer duration. These tests may extend over the life of the animal and are often used to determine the carcinogenic potential of the test chemical. **ecological effects toxicity test**. any of a wide variety of tests that may employ a combination of field tests, laboratory tests, microcosms (constructed or *in situ*), and physical or computer models. All are designed to test the toxicity and effects of particular chemicals on natural (sometimes urban or agricultural) populations, communities, or ecosystems. Such tests often include studies of environmental transport and fate of the test chemical and effects of intermediates formed during decomposition. ***in vitro* toxicity test**. **1a:** in general, a test using living material other than the intact living organism, usually one that employs tissues, cells, organelles, or enzymes maintained under suitable conditions usually on natural or synthetic culture media. **1b:** any test conducted outside the organism's body. ***in vivo* toxicity test**. any toxicity test of intact organisms. **inhalation toxicity test**. any toxicity test in which the test substance is inhaled by the subject. **phototoxicity test**. any test that evaluates the dermal effects of light, especially ultraviolet light, on the test organism subsequent to treatment with the test chemical. Such tests may be employed to assess phototoxicity and photoallergenicity of a test chemical. In general, the chemical is applied to a portion of the skin of controls and experimentals and the area then irradiated with ultraviolet light in the experimental groups. **reproductive toxicity test**. any of a wide variety of tests of reproductive function in an intact organism or a mating system (e.g., a pair). Many of these tests follow complex protocols and may not be suitably standardized or easily reproducible. Such tests applied to laboratory mammals often also include information on embryos if conception and development occur under at least some treatments. See also viability index, weaning index. **subacute toxicity test**. a test designed to assess the toxicity of a substance due to repeated exposure and usually to help determine doses to be used in subchronic tests. **subchronic toxicity test**. usually a type of pilot study used to characterize dose-response relationships of the test chemical and to help determine doses for chronic tests. It is most often 90 days in duration, employing dogs and laboratory rats as test animals. Any of a wide variety of endpoints may be used, but often include body weight, food consumption, mortality, as well as pharmacologic, hematologic, and toxicologic signs and lesions.

toxico-. a prefix indicting the quality of being poisonous or relationship to a poison or poisoning.

Toxicocalamus (elongate snakes). a genus of two species of venomous snakes (Family Elapidae), *T. longissimus* (elongate snake) and *T. stanleyanus* (Mount Stanley snake). Both are indigenous to New Guinea and the Fergusson, Misima, and Woodlark Islands. The body of each bears a long terminal spine that is keeled above; the internasals are distinct. Adults may reach a length of about 0.7 m. The fangs are short and neither snake appears to be dangerous.

toxicodendric acid. a supposedly toxic compound isolated from *Rhus toxicodendron* (poison ivy).

Toxicodendron. **1:** See *Rhus*. **1b:** poisonous members of the genus *Rhus* with smooth fruits and leaves are sometimes placed in this genus (e.g., *R. diversiloba*, *R. quercifolia*, *R. radicans*, *R. vernix*). See *Rhus*.

toxicoderma (toxicodermatosis). any disorder or disease of the skin caused by a poison or by a toxin-secreting microorganism.

toxicodermatitis. any inflammation of the skin due to the action of a toxicant. *Cf.* toxicoderma.

toxicodermatosis. toxicoderma.

toxicodynamic. pertaining to toxicodynamics or to the effects or action of toxic chemicals. See also pharmacodynamic.

toxicodynamics. **1:** the dynamics of toxicants in living systems and the responsible mechanisms; the pharmacodynamics of toxic agents. **2:** the study of uptake, transport, transformations, binding, and the physiological, biochemical, and molecular interactions of toxicologically active molecules at their loci of action within living systems. See pharmacology, pharmacokinetics, toxicodynamics. *Cf.* pharmacodynamics. **3:** toxicometrics.

toxicogenic. toxigenic.

toxicogenic conjunctivitis. conjunctivitis produced by topical application of a microbial toxin.

toxicogenicity. toxigenicity.

toxicognath. either of a pair of fangs on the anterior segment of a centipede. They are modified legs.

toxicohemia. toxemia.

toxicoid. **1:** having the nature of or an action resembling that of a poison. **2:** temporarily poisonous.

toxicokinetic. pertaining to toxicokinetics.

toxicokinetics. **1a:** the pharmacokinetics of toxic chemicals. **1b:** the dynamic behavior and movements of toxic substances and their metabolites within a living system. **1c:** the (usually quantitative) study of the dynamic behavior of toxic substances and their metabolites within an organism including uptake, binding, distribution, biotransformations, and elimination. Included also is the study of mechanisms affecting these phenomena.

toxicologic, toxicological. pertaining to or having to do with toxicology.

toxicological risk. See risk.

toxicologist. **1a:** a professional person whose primary activity is the study of the behavior and adverse effects of potentially toxic agents on living systems or the application of such knowledge (e.g., in law, the health professions, the management of toxic waste). **1b:** a scientist or other professional who specializes in toxicology. Such individuals specialize to one degree or another (e.g., clinical toxicologist, forensic toxicologist, ecotoxicologist). Some examples are given here, but to gain a better grasp of the diversity of specialties among toxicologists, see toxicology. *Cf.* pharmacologist. **descriptive toxicologist**. one concerned directly and chiefly with toxicity testing. **mechanistic toxicologist**. one that elucidates the mechanisms by which chemicals exert their toxic effects on living systems. **regulatory toxicologist**. one that evaluates existing (health or environmental) data and determines whether a drug or other chemical is of sufficiently low risk to be marketed and used for stated purposes. See also venomologist. See also toxinologist, venomologist.

toxicology. the organized study of the nature, effects, and mechanisms of action of toxic substances on living systems as well as the application of the acquired body of knowledge to human interests. Included is the quantitative assessment of the severity and frequency of toxic effects in relation to exposure regimes and types of systems affected. Toxicology is an extremely broad and diverse field of study, as evidenced by the large number of subspecialties and the increasingly complex systems of poisons and living systems with which toxicology deals. Classically, this field dealt only with individual organisms, primarily with a view of protecting health, especially that of humans. Today, this science in its various branches integrates scientific information from a diversity of fields. The range of inquiry and application of toxicology includes the detection and recognition of poisons; their physical and chemical effects and the severity of effects; mechanisms of action; diagnostics; therapeutics; antidotes; and all of their interactions with and effects on individual living organisms and on populations, communities, and ecosystems. Advancement from the study of single organisms to that of ecosystems has brought many complexities that are not yet fully appreciated or understood. Included in the field is the study of adaptation (both physiological and evolutional) and natural selection. Branches of toxicology also deal specifically with clinical, industrial, legal, and other types of issues that are, in part, intimately concerned with poisons. **aquatic toxicology**. a subdiscipline that treats the toxicology of individual aquatic organisms, populations, and/or aquatic ecosystems largely from the perspective either of environmental toxicology or of ecotoxicology. Professionals in this field sometimes conduct studies, tests, or assessments that bear upon the secondary effects of chemicals in the aquatic environment on humans, livestock, and other economically or aesthetically important organisms. This discipline is allied with wildlife toxicology; each draws upon the other for relevant information. **behavioral toxicology**. the study of the behavioral effects of poisons and of behaviors related to poisons and poisoning (e.g., adaptive behavior, predator-prey relationships, feeding behavior, avoidance behavior, warning behavior, specific behaviors related to sources of poisoning). *Cf.* neurobehavioral toxicology. **biochemical toxicology** (molecular toxicology). the branch that deals with biochemical and molecular aspects of any phase of the complex web of events that make up the set of interactions between a toxicant and a living system. Included is the study of mechanisms of absorption; metabolic transformations, transport, and fate of a toxicant, including the generation of reactive and other significant intermediates and end-products and their interactions with cellular macromolecules; the nature and function of the enzyme systems involved in the metabolism of xenobiotics. Because it deals to a high degree with basic mechanisms, this field is fundamental to and contributes significantly to other branches of toxicology. **biotoxicology**. the science of poisons and venoms contained in, produced by, secreted by, or transmitted by living organisms. This branch is centered about any or all of the following: toxicity, symptomology, the treatment of poisoning or envenomation; and natural history and adaptive biology of poisonous and venomous organisms and their prey or otherwise affected organisms. Ecotoxicology subsumes biotoxicology. **cancer toxicology**. See oncology (toxicological oncology). **cardiac toxicology**. cardiotoxicology. **cardiotoxicology** (cardiac toxicology). the study of cardiotoxic substances, their effects and mechanisms of action; included is the study of indirect effects (e.g., effects on the vagus nerve that affect heart function) of poisons on the heart. **cellular toxicology**. cytotoxicology. **clinical toxicology**. an applied branch that deals with the diagnosis and treatment of clinically expressed disease caused by exposure to toxic substances. It is practiced either as a specialty of medicine or of veterinary medicine. **comparative toxicology**. the study of the toxicities, effects, intermediary metabolism, and mechanisms of action of xenobiotics in one or more taxa or genetic strains of animals or plants compared to others that are, at some genetic or taxonomic level, related. In the broadest sense, it also includes the study of differing biological and behavioral mechanisms among taxa that affect the action of the toxicants investigated and of differing susceptibilities and responses. Although any aspect of toxicology can be studied from a comparative point of view, this field has, until now, dealt predominantly with experimental studies of acute toxicity or the metabolism of toxicants, with a view to developing hypotheses regarding the poisoning of humans and of selecting animal models for tests oriented toward elucidating toxic responses in humans. Other areas of concern include the development of selective toxicants (e.g., pesticides), the demonstration or elucidation of physiological, behavioral, and [illegible] specific toxicants, and the study of the phenology of xenobiotics. This discipline also has much to offer the burgeoning field of ecotoxicology. **cytotoxicology**. the branch of toxicology that focuses on cellular responses (adaptive and pathological) to toxic challenge. This discipline employs both physiological and biochemical knowledge, theory, and methods. It is closely related to and overlaps the field of biochemical toxicology. **developmental toxicology**. the science that deals with the effects of chemicals on the developing organism from fertilization to birth or hatching. The scope of this field in humans extends postnatally, for some purposes, to the attainment of sexual maturation. **economic toxicology**. the subdivision that deals chiefly with the toxicology of and establishment of standards for economic poisons such as food additives, cosmetics, pesticides, and drugs. **ecotoxicology**. the branch of toxicology that deals with naturally occurring and anthropogenic toxic chemicals in the environment and their interactions with ecological systems. The focus of ecotoxicology is on populations and communities. Ecotoxicology also considers harm to species, populations or biotic communities that may not be reflected in obvious harm to individuals (e.g., reproductive impairment, changes in competitive relationships, population decline). In practice, this major subdiscipline is an interdisciplinary field drawing on scientists whose advanced education has been primarily in ecology, biogeography, various related fields of biology, chemistry, or toxicology. It includes the study of the direct and indirect harmful effects on humans (e.g., food-chain transfer). Ecotoxicology subsumes a portion of biotoxicology. **environmental toxicology**. **1:** the applied science that deals with the toxic properties and effects of chemicals that enter the environment with or without modification and without regard to source, or that are transported through the environment by environmental media (air, water, soil, or sediment) or through the agency of living organisms. This science is concerned with sources, transport, and transformations that take place in the environment; environmental conditions that modify toxicity and other aspects of toxicant behavior; fate; effects; hazard and risk to humans and other biota; safety limits of incidental exposure of organisms to environmental chemicals. This discipline is closely [illegible] hand and ecotoxicology and wildlife toxicology on the other. It is also closely allied to, and interactive with, the field of environmental chemistry. **2:** a branch of toxicology that centers on the direct and indirect effects of chemicals in the environment, e.g., through effects on food organisms and other organisms of economic or aesthetic interest. **forensic toxicology**. a science that is a combination of the application of analytic chemistry and fundamental toxicological principles to the study of the medicolegal aspects of toxic substance, especially those that cause severe injury or death to humans. This field is devoted mainly to establishing the cause and circumstances of death due to the criminal use of toxicants as well as to accidental poisoning or the inept usage of poisonous substances and the legal issues arising therefrom. Forensic toxicology is heavily concerned with medicolegal aspects of clinical, industrial, and occupational toxicology. It is also extensively involved with environmental toxicology and ecotoxicology as they relate to sources, behavior, and impacts of pollutants and hazardous wastes. To some extent, this field also deals with the diagnosis and treatment of the effects of poisons. **general toxicology**. the science that attempts to establish the basic principles and mechanisms of toxic action and biological response. This draws heavily upon and contributes very substantially to all other branches of toxicology. **genetic toxicology** (genotoxicology). the science that explores alterations of DNA by toxicants and all mechanisms whereby genetic effects are produced. Also studied are genetic differences in metabolism, transport, fate, effects, and susceptibility or resistance to various toxicants. **genotoxicology**. genetic toxicology. **ichthyotoxicology**. the branch of biotoxicology that deals with poisons and venoms contained in or produced by fishes. This science is concerned with toxicity, symptomology, the treatment of poisoning or envenomation, the natural history and adaptive biology of poisonous/venomous fishes and their prey or otherwise affected organisms. **immunotoxicology**. the science that identifies and deals with the effects of toxicants on or via the immune system. **industrial toxicology**. that branch of occupational toxicology that deals with the toxicology of industrial settings. The study and practice of industrial toxicology involves aspects of forensic, environmental, and economic toxicology. **kinetic toxicology** (metabolic toxicology; pharmacologic toxicology). a subdiscipline of systemic toxicology that treats the transport and metabolism of systemic poisons within a living organism. **marine aquatic toxicology**. a rather new discipline devoted to the study of marine contaminants under realistic conditions and the development of suitable and sensitive methods for such study. This field arose chiefly from the needs of health and environmental assessment of toxic marine contaminants of human origin. Work was initially devoted largely to the development and application of acute toxicity tests on marine life forms. *Cf.* aquatic biology. **metabolic toxicology**. kinetic toxicology. **molecular toxicology**. biochemical toxicology. **neurobehavioral toxicology**. the study of behavioral responses to toxicants and the underlying nervous system effects and mechanisms. This science combines the application of knowledge and methods of the behavioral sciences (especially comparative and physiological psychology) with those of toxicology to detect, measure, diagnose, or evaluate the toxic properties of psychoactive agents on neurobehavioral performance (e.g., learning, memory, motor behavior); to determine the underlying causes of toxicity-induced behavioral changes; and to develop sensitive indicators of neurobehavioral toxicity. *Cf.* behavioral toxicology. **neurotoxicology**. the branch of toxicology that is most closely allied with neuropharmacology. It deals with toxicants that injure the central and/or the peripheral nervous tissues, including evaluation of the nature and mechanisms of neurotoxic insult and rebound or recovery. This science deals with the effects of toxicants on nervous tissue, and in higher animals, on the nervous system in part or in whole. The focus is largely on the sensitivities, effects, and mechanisms of response and recovery of nerves and other nervous system components to toxic insult. This field overlaps with biochemical and neurobehavioral toxicology; with cytotoxicology; immunotoxicology; and, at times, with most other branches of toxicology. **nonkinetic toxicology** (nonmetabolic toxicology; nonpharmacologic toxicology). the study of the toxicology of chemicals (e.g., irritants, caustics) that act on contact, are not transported to other sites, and do not produce direct systemic effects. This science is concerned with the generalized harmful effects produced at the point of contact or exposure (e.g., erythema, tissue destruction, edema) to a toxicant. Indirect effects (e.g., psychic; effects of pain) are dealt with as appropriate. **nonmetabolic toxicology**. nonkinetic toxicology. **nonpharmacologic toxicology**. nonkinetic toxicology. **nutritional toxicology**. the study of the effects of diet, dietary components, and deficiencies on the expression and mechanisms of toxic action. **occupational toxicology**. the study or practice of toxicology that is focused on the prevention and treatment of occupational exposures to toxic substances. A rapidly developing part of forensic toxicology is that of legal liability because of harmful effects from accidental exposure to toxicants in the workplace. **ophiotoxicology**. the study of snake venoms, their effects, and to some extent the associated natural history, physiology, and behavior of venomous snakes. **pharmacologic toxicology**. kinetic toxicology. **pulmonary toxicology**. the study of toxicants that affect pulmonary tissues, lung function, and the consequences of altered function. **regulatory toxicology**. the branch of toxicology that supports the formulation and administration of regulations that govern the manufacture, sale, use, and disposal of toxic or potentially toxic products. **renal toxicology**. the science that treats the effects and mechanisms of action of toxicants on the kidney tissue and function. **reproductive toxicology**. the study of the toxic effects of chemicals on the reproductive system to include the fertilized egg and developing embryo, as appropriate. The interests of this field often overlap with those of developmental toxicology. **systematic toxicology**. the science that applies the principles of pharmacology and general toxicology to the study of groups of structurally and functionally related drugs. **systemic toxicology**. the study of systemic poisons, their behavior, effects, and metabolic interactions. It includes the subdiscipline of kinetic toxicology. **veterinary toxicology.** the study, diagnosis, and treatment of poisonings of livestock, zoo animals, pets, and wildlife. The field deals chiefly with pets and animals of economic value. Secondary poisoning of humans by animals with which they come in contact or which form part of the diet is also a concern. This field broadly overlaps with wildlife toxicology, although methods and focus differ very substantially. **wildlife toxicology**. concerned with the study and treatment of feral animals, especially those of economic, esthetic, or recreational importance. This field draws upon theory and methods of nearly all of toxicology as well as those of zoology, ecology, and veterinary medicine. See also toxinology.

Toxicology Data Base. an on-line computer data base that contains information on toxic chemicals that is compiled from the peer-reviewed literature by the National Library of Medicine, Bethesda, MD, in the United States. Included is information on nomenclature, chemical and physical properties, threshold limit values, pharmacology, and toxicology. It is made available for on-line search. See also Toxline.

toxicomania. **1:** an abnormal craving for narcotics or toxicants. **2:** a condition affecting herbivorous mammals, including livestock, in which the animals forage preferentially on toxic plants.

toxicometrics. the study of the absorption, transport, transformations, receptor binding, interactions, elimination, and the kinetic variables of toxicologically active substances at their loci of action within living systems. Also included is the study of their interactions with other metabolically active substances and with the metabolic and kinetic variables that affect their distribution and dynamics. This term is partially synonymous with toxicodynamics, *q.v.*

toxicon. an obsolete term for toxin.

toxicopathic. toxipathic.

toxicopathy. toxipathy.

toxicopectic (toxicopexic). pertaining to, characterized by, or promoting toxicopexis.

toxicopexic. toxicopectic.

toxicopexis (toxicopexy). the fixing or neutralization of a poison within the organism.

toxicopexy. toxicopexis.

toxicophidia. thanatophidia.

toxicophobia (toxiphobia). an unfounded and morbid fear of being poisoned.

toxicophorous. toxiferous.

toxicoses. plural of toxicosis.

toxicosis (plural, toxicoses). poisoning, the state or quality of being poisoned; intoxication, especially systemic poisoning, usually chronic. **endogenic toxicosis**. autointoxication. **exogenous toxicosis**. any condition due to a poison not produced in the body. **retention toxicosis**. a diseased state due to toxicants that are normally excreted soon after their formation in the body. *Cf.* toxinosis.

toxics. **1:** jargon, usually legalistic or bureaucratic, for toxic chemicals; sometimes used also by professionals and the public, often without reference to precise meaning. To understand this term (and often, indeed, the term toxic chemical) one must know the context in which it is used and the source. **2:** jargon, under various legislation, for toxic chemicals other that have not been given another designation. **air toxics**. government jargon; a term used, for example, by the U.S. EPA in reference to toxic air pollutants other than those (criteria air pollutants) regulated by EPA based on national ambient air quality standards.

Toxics Release Inventory (TRI). a computerized data base, updated annually, initiated by the U.S. Environmental Protection Agency under the Emergency Planning and Community Right-to-Know Law of 1986. Information on toxic chemicals includes chemical identity, quantities released into the environment, types of waste treatment; effectiveness of industry in reducing waste generation.

toxicyst. a type of protozoan trichocyst which may induce paralysis or lysis of the prey on contact.

toxidermitis. toxicodermatitis.

toxiferine. any of a class of extremely virulent curare alkaloids; *Strychnos toxifera* is the chief source.

toxiferous (toxicophorous). **1:** producing, containing, or imparting a poison. **2:** toxic, poisonous; venomous.

toxification. the process of making a substance poisonous or more poisonous. Thus a xenobiotic substance may be made more toxic (toxified) by metabolic processes. **metabolic toxification**. the phenomenon whereby a substance (usually a xenobiotic) is metabolically transformed into a more toxic substance (e.g., acetaminophen, glutethimide, methanol, organophosphate insecticides). This process may be intensified by agents that induce hepatic enzymes. See metabolite (active metabolite). *Cf.* detoxification, toxication.

toxify. to make poisonous or more poisonous. *Cf.* detoxify, detoxification.

toxigenic (toxicogenic; toxinogenic). **1:** producing, capable of producing, or elaborating a toxin or other toxicant; said of a cell, tissue, organ or organism. **2:** caused by a poison.

toxigenicity (toxicogenicity; toxinogenicity). **1:** the capacity of an organism to produce or elaborate a toxin; applied chiefly to microorganisms. **2:** the virulence of a parasite that acts by elaborating a soluble toxin; the degree to which an organism can produce toxin.

Toxiglossa. a division of the gastropod Suborder ectinibranchia, comprised of carnivorous species (especially those of the familics Conidae and Terebridae) in which the radula has two long hollow teeth in each row that convey toxic salivary secretions of the proboscis gland. *Conus* is a typical example.

toxiglossate. having hollow radular teeth that convey venom secreted by the salivary glands into the bite, as in certain carnivorous marine gastropods (See Toxiglossa).

toxignomic. **1a:** specifically distinctive or peculiar to the toxic action or effects of a given poison. **1b:** having the toxic action specific to a poison. **2:** pertaining to a sign or symptom on which a diagnosis of poisoning can be made. *Cf.* pathognomonic.

toxin (toxinum; biotoxin; toxicon (obsolete)). variously defined by different specialists, sometimes without regard to existing usage. Definitions 1a and 1b are preferred. **1a:** true toxin; a biogenic poison; a usually proteinaceous poison, produced metabolically, that can induce the production of antitoxins when injected into the tissues of an animal, especially a vertebrate animal. The term biotoxin is sometimes used or, as appropriate, bacteriotoxin, microbial toxin, mycotoxin, phytotoxin, or zootoxin. Toxins are produced by certain bacteria, a few higher plants, and by certain venomous animals (e.g., stinging hymenoptera, venomous snakes). A number of toxins have been developed for use as biological warfare agents (See warfare agent). **1b:** in the original sense, a poison of high molecular weight, usually a polypeptide or protein, synthesized by a microorganism, especially a bacterium. See microbial toxin. **2:** in pathology and medical jurisprudence, a toxin is generally considered to include any diffusible alkaloidal substance (e.g., ptomaines, abrin, brucin); snake venoms and wasp venoms; microbial toxins. **3:** often used as a synonym for any poison including mixtures (e.g., snake venoms). This definition is not recommended. **acetyl T-2 toxin**. T-2 toxin. **animal toxin**. zootoxin. **anthrax toxin** (bacillus anthracis toxin). a culture filtrate of *bacillus anthracis* that contains at least three different substances: an edema factor, a lethal factor, and a protective antigen. **bacterial toxin**. bacteriotoxin. **bacteriotoxin** (bacterial toxin). a microbial toxin produced by a bacterium; any of a large number of poisonous substances produced by bacteria (endotoxins, exotoxins). Such toxins play important roles in a large number of disease processes and in food poisoning. See also warfare agent. **biotoxin**. a term that is redundant with toxin (definition 1a). any poison that is biogenic in origin. **botulinum toxin** (botulin; botulinus toxin; botulismotoxin). the generic name of a set of at least eight extremely toxic, antigenically distinct protein neurotoxins produced anaerobically by the bacterium, *Claustridium botulinum*. They sometimes occur in imperfectly preserved or canned meats and vegetables. These toxins act chiefly to block the release of acetylcholine at myoneural junctions, causing paralysis and associated symptoms. The probable lethal dose for an adult human is about two micrograms (one 1/14 millionth of an ounce). Humans are usually exposed by consuming inadequately sterilized canned or bottled food. The botulinum toxins resist decomposition by gastric and intestinal juices. See botulism. They are chemically similar to abrin. **botulinus toxin**. botulinum toxin. **botulismotoxin**. botulinum toxin. **cholera toxin**. choleragen. **cortinarius mushroom toxin**. a category that includes at least one unidentified hepatotoxin found in some species of *Cortinarius*. **cyclopeptide mushroom toxin**. any of the toxins produced by *Amanita phalloides* except phallolysin (a glycoprotein) and can be assigned to either of two subgroups: phallotoxins and amatoxins. See also amatoxin, phallotoxin, gyromitrin, coprine, 1-cyclopropanone, muscarine, ibotenic acid, muscimol, psilocybin, psilocin. **dermonecrotic toxin**. dermotoxin. **diagnostic diphtheria toxin**. Schick test toxin. **Dick toxin**. erythrogenic toxin. **dinoflagellate toxin**. a virulent neurotoxin thought to impair the synthesis or release of acetylcholine. **diphtheria toxin**. the causative agent of diphtheria. It is an antigenic protein toxin secreted by *Corynebacterium diphtheriae* that interferes with protein synthesis in eukaryotic cells; it is also destructive of tissue. **dysentery toxin**. the exotoxin of various species of *Shigella*. **esotoxin**. an obsolete synonym for endotoxin. **exotoxin** (ectotoxin (obsolete)**;** extracellular toxin**;** soluble toxin). any pathogenic proteinaceous toxin produced by bacteria, usually Gram-positive bacteria, that are released into the surrounding medium. Many are highly active enzymes, are more heat labile and generally more toxic than endotoxins, and cause more severe effects. Exotoxins are detoxified (but retain their antigenicity) by treatment with agents such as formaldehyde (formol toxoid) that do not affect endotoxins. Exotoxins (e.g., botulinum toxin) are the most toxic substances known and include cardiotoxins, hemolysins, cardiotoxins, diphtheria toxin, neurotoxins, plague toxin and various cell-disrupting enzymes such as lecithinase and collagenase). Ectotoxin is an obsolete synonym. See botulinum toxin, endotoxin, enterotoxin, toxemia, toxoid. **extracellular (bacterial) toxin**. exotoxin. **fatigue toxin**. kinotoxin. **gonyaulax toxin**. paralytic shellfish poison. **gyromitra toxin**. any toxin produced by fungi of the genera *Gyromitra* and *Helvella*. They produce systemic effects and can be lethal. The presenting symptoms may include severe headache, vomiting, diarrhea, and muscular incoordination. The actual toxins present vary according to species or even variety, location, time of year, and unknown environmental factors; they may include gyromitrin, helvellic acid, methyl hydrazine, muscarine, phalloidine, and perhaps others. *Cf.* gyromitrin. **HT-2 toxin**. a trichothecene first isolated from *Fusarium*. **intracellular (bacterial) toxin**. endotoxin. **microbial toxin**. an extremely toxic poison of high molecular mass synthesized metabolically by a microorganism. It is, by and large, the virulence of the toxins produced that determines the virulence of a particular microbe. Microorganisms also produce a variety of toxic substances (e.g., acetaldehyde, formaldehyde, hydrogen sulfide) that are usually not termed toxins. See also bacteriotoxin, endotoxin, exotoxin, toxinologist. **milk toxin**. any of the aflatoxins G. See aflatoxin. **mushroom toxin**. any toxin produced by the any of the numerous species of poisonous mushrooms. The known mushroom toxins have been variously named and classed such that casual examination of the relevant literature can be confusing and misleading. Chemically, known mushroom toxins fall into the following categories or classes: cyclic polypeptides (cyclopeptides), e.g., phallotoxins; gyromitrins (which decompose to release the active toxicant, monomethylhydrazine); coprine (which yields the active agent, cyclopropanone, on hydrolysis); muscarine; ibotenic acid and muscimol; tryptamine derivatives (especially psilocybin and psilocin). Two further classes, not now based on chemical characteristics, should be considered: *cortinarius* toxins; miscellaneous and unknown toxins. (Selection of the above categories is based in part on the work of G. Lincoff and D. H. Mitchel, 1977, *Toxic and Hallucinogenic Mushroom Poisoning*, Van Nostrand Reinhold Co. and in part on the work of R.E. Gosselin et al., 1984, *Clinical Toxicology of Commercial Products*, Williams & Wilkins). A fairly large variety of mushrooms contain mostly unidentified gastrointestinal irritants, some of which produce severe effects. Of the above classes, only the cyclopeptides and gyromitrins are lethal to humans as a rule. **normal toxin**. a toxin solution that holds exactly 100 lethal doses per ml of diluent. **pertussis toxin** (histamine sensitizing factor; pertussigen). the causative agent of whooping cough, produced by the Gram-negative bacterium *Bordetella pertussis*. **phytotoxin**. any toxin produced by a plant. Most are alkaloids and produced by fungi; toxins are less common in higher plants (examples are juglone, ricin, abrin). See also *Amanita phalloides*, *Claviceps purpurea*, *Aspergillus flavus*. **plant toxin**. phytotoxin. **scarlet fever erythrogenic toxin**. See erythrogenic toxin. **scarletinal toxin**. See erythrogenic toxin. **Schick test toxin** (diagnostic diphtheria toxin). *Corynebacterium diphtheriae* toxin diluted so that the *inoculated* dose (0.1 or 0.2 ml) will contain 1/50th of the guinea pig minimal lethal dose. See also Schick test. **soluble toxin**. exotoxin. **streptococcus erythrogenic toxin**. See erythrogenic toxin. **T-2 toxin** (acetyl T-2 toxin; 12,13-epoxytrichothec-9-ene-3,4,8,15-tetrol 4,15-diacetate 8-(3-methylbutanoate); 3α-hydroxy-4β,15-diacetoxy-8α-(3-methylbutyryloxy)-12,13-epoxy-Δ^9-tricothecene; 8α-(3-methylbutyryloxy)-4β,15-diacetoxyscirp-9-en-3α-ol; fusariotoxin T-2; insariotoxin, mycotoxin T-2, NSC 138780). an extremely toxic trichothecene mycotoxin first isolated from *Fusarium tricinctum*. It is caustic, neurotoxic, and immunosuppressive. Contact with skin or other tissues may produce blistering and necrosis, dizziness, nausea, vomiting, hemorrhage, and possibly death. **tetanus toxin** (tetanotoxin). **1:** the neurotropic, heat-labile exotoxin of *Clostridium tetani*; it is the cause of tetanus. It is a crystalline protein (molecular weight 67,000) that consists of two parts: a neurotoxin (tetanospasmin) and a hemolysin (tetanolysin). It is one of the most toxic substances known. The major action is apparently the blocking of inhibitory synaptic impulses. **true toxin**. toxin (See def. 1a, above). A term that encompasses exotoxins produced by certain bacteria and toxins produced by a few fungi, a few higher plants, and by venomous animals such as stinging Hymenoptera and venomous snakes. **wildfire toxin** (*N*-[2-amino-4-(3-hydroxy-2-oxo-3-azetidinyl)-1-oxobutyl]-L-threonine). a colorless, hygroscopic, fungal phytotoxin, $C_{11}H_{19}N_3O_6$, that is soluble in water and methanol and moderately soluble in ethanol. It is produced by *Pseudomonas tabaci*, the bacterium that causes wildfire disease of tobacco. It is highly toxic to diverse organisms including algae, bacteria, animals, and higher plants. **Note:** a few related terms of interest, mostly toxins, are: acetyladonitoxin; acetyldigitoxin; actinotoxin; adonitoxin; adrenotoxin; aflatoxin; algelasine; allelotoxin; allotoxin; amanitatoxin; amatoxin; antitoxin; apitoxin; atelopidtoxin; atroxin; autocytotoxin; autotoxin; batrachotoxin; batrachotoxinin A; biotoxication; botulinic acid; brevetoxin; bromatotoxin; buckthorn toxin; bufagin; bufotoxin; bungarotoxin; α-bungarotoxin; buxine; cardiotoxin; catenarin; cephalotoxin; chaetoglobosin; cicutoxin; ciguatoxin; convallatoxin; crotalotoxin; crotoxin; cytotoxin; decumbin; digitoxin; ectoantigen; enterotoxin; equinatoxin; erabutoxin; ergotoxine; flavoskyrin; galitoxin; gitoxin; grayanotoxin; heterocytotoxin; holothurin; hypnotoxin; ichthyocrinotoxin; ichthyootoxin; isotoxin; latrotoxin; mycotoxin; neurotoxin; ochratoxin; ostracatoxin; palytoxin; paralytic shellfish poison; paramecin; patulin; penitrems A; B; C; phallotoxin; physalitoxin; picrotoxin; plototoxin; podophyllotoxin; prototoxin; pumiliotoxin; rhizobotoxin; rubratoxin; sapotoxin; saurine; saxitoxin; surugatoxin; tetrodotoxin; toxoid; tarichatoxin.

toxin-antitoxin (TAT; T.A.T.). a neutral mixture of a toxin with a corresponding antitoxin, formerly used to immunize against diphtheria.

toxinemia. toxemia; poisoning of the blood.

toxinic. of or pertaining to a toxin.

toxinicide. anything that is able to destroy toxins.

toxinogenic. toxigenic.

toxinogenicity. toxigenicity.

toxinologist. **1:** a scientist or practitioner who deals with microbial toxins [(See toxin (microbial toxin)]. **2:** a scientist who studies proteinaceous toxins of plant, animal, or microbial origin.

toxinology. the science of toxins produced by pathogenic microbes, animals, and certain higher plants.

toxinosis (toxonosis). **1:** any disease, condition, or lesion produced by the action of a toxin. **2:** sometimes given as a synonym for toxicosis. This usage is not recommended.

toxinum. toxin.

toxipathic (toxicopathic). of or pertaining to any pathologic state caused by a poison.

toxipathy (toxicopathy). any disease due to poisoning, especially chronic poisoning.

toxiphobia. toxicophobia.

toxiphoby, index of. a qualitative scale that ranks species according to their tolerance of toxic or polluted habitats.

toxisterol. a toxic derivative produced by excessive irradiation of ergosterol or calciferol.

toxitabella (plural, toxitabellae). a poisonous tablet. Such tablets usually have an angular shape and/or are marked with a skull and cross-bones.

toxitherapy. the use of toxins in the treatment of disease. Thus botulinum toxin is used to treat spasmodic torticolis and local injections of the same toxin are used to treat certain types of eye muscle imbalance.

toxituberculid. a skin lesion due to the toxin of *Mycobacterium tuberculosis*, the tubercle bacillus.

Toxline. a comprehensive bibliographic computer data base that contains more than one million references from the toxicological literature. Toxline is compiled by the National Library of Medicine, Bethesda, MD in the United States, which has made it available to the public for on-line searches.

Toxneustes elegans. a moderately venomous red sea urchin (Class Echinoidea, Phylum Echinodermata).

toxo-. a prefix that denotes relationship to a toxin or poison.

toxoalexin. an alexin that neutralizes the activity of bacterial toxins.

toxogenin. a hypothetical substance in blood that results from the injection of antigens. It is presumed to be nontoxic by itself but causes anaphylaxis when antigen is added anew.

Toxoglossa. a group of families of marine carnivorous gastropod mollusks (especially Conidae and Terebridae), many of which are venomous. In the latter forms, the radula has a few large teeth (fangs) that convey the poison secreted by a large proboscis gland.

toxognath. either member of the first pair of limbs in centipedes. It is an organ of envenomation, bearing the external opening of the venom duct.

toxoid (anatoxin). a modified bacterial exotoxin that has been treated, usually with formalin, destroying its toxicity but not its antigenicity, i.e., it can still effectively stimulate the production of antibodies that are competent against the original toxin. Toxoids can be used to immunize against various diseases such as diphtheria and tetanus. See also epitoxoid.

toxolecithid. toxolecithin.

toxolecithin (toxolecithid). a toxic lecithin, as in cobra venom, in which lecithin is chemically bound to a toxin.

toxolysin. **1:** antitoxin. **2:** toxicide.

toxon (toxone). a hypothetical substance of feeble toxicity and weak affinity for antitoxin that is elaborated by bacteria.

toxone. toxon.

toxonosis. toxinosis.

toxopeptone. a protein derivative produced by the action of a toxin on a peptone.

toxophil, toxophile. **1a:** having a special affinity for toxins. **1b:** susceptible to the action of a poison.

toxophilic. pertaining to a toxophil; having the characteristics of a toxophil.

toxophore (toxophore group). **1a:** of or pertaining to the toxophore (def. 1b). **1b:** that part (atomic group) of the molecule of a toxin that confers its specific toxic properties following firm attachment to the haptophore.

toxophore group. toxophore.

toxophorous. pertaining to the toxophore of a toxin molecule or to its toxic properties.

toxophylaxin. a substance that neutralizes the action of bacterial toxins.

Toxopneustes. a genus of common tropical sea urchins. ***T. elegans***. a venomous sea urchin, dangerous to humans, that occurs in the coastal waters of Japan. The pedicellariae are globiferous, flower-like in appearance and extend slightly beyond the spines; they can inflict a painful and sometimes lethal sting.

toxoprotein. a toxic protein or a mixture of toxin and protein.

TR-1. verruculogen.

trace element. **1:** an element that occurs in living organisms in trace or very low concentrations. Most are essential to normal growth, development, and functioning of an organism, but are present in, and used by, the organism only in minute quantities. A total lack of certain of these eventually causes death, but concentrations that are higher than normal (e.g., of chromium, cobalt, copper, fluorine, iodine, selenium) can be toxic. **2:** an element present in a material only in minute or trace quantities (e.g., trace elements in coal); these sometimes have toxicological significance as in the burning of fossil fuels. **essential trace element**. a trace element that forms part of the structure of metabolically important molecules (e.g., enzymes), or are important cofactors or catalysts of metabolic processes. The total (or near total) lack of which eventually causes death; excessive amounts are toxic. The essential trace elements of humans and most vertebrates include Zn, Mg, Cu, Ni, Mn, Co, Cr, I, and Se; those of plants include Cu, Zn, Mg, Mn, and B.

trace metal. See metal.

tracer. an atom, molecule, or material that carries a chemical or radioactive marker or label such that it can be readily detected and identified even at very low concentrations. It can be added to another substance, the presence of which can then be detected or followed, for example, during a chemical reaction, intermediary metabolism, or during transport within an organism or in the environment. Thus, for example, fluorescein can be used to trace respirable dust emitted from a smoke stack without confusing it with dust from other sources. **isotopic tracer**. an isotope that is present in small amounts (or added to) a material often provides a signature that allows a substance to allow identification and tracking. **radioactive tracer**. a radioactive molecule or atom used as a tracer. Such tracers are widely used to investigate chemical change or to follow the movements of substances that are transported in small amounts through the body of an organism or through environmental media (air, water, soil, bottom sediment).

Trachinidae. See weever.

Trachinus (weeverfishes). a genus of rather small (less than 46 cm), venomous, temperate zone bottom fishes (Family Trachinidae). They are aggressive, have a well-developed venom apparatus, and remain concealed in bottom sediments most of the time. ***T. draco*** (greater weever). a species that occurs in the eastern Atlantic Ocean and Mediterranean Sea from Norway to the more northerly coasts of North Africa. ***T. vipera*** (lesser weever). its range includes the North Sea, southward in the coastal waters of Europe to and in the Mediterranean Sea. The venom apparatus is made up of the dorsal and opercular spines and venom glands that produce both a neurotoxin and hemotoxin. The tip of the spine is extremely sharp and inflicts a painful wound. The pain becomes progressively intense, peaking within about 30 minutes and usually subsiding with 24 hours. The pain can be nearly overwhelming and the patient may scream, eventually falling into unconsciousness. Symptoms may include chills, fever, nausea, vomition, dizziness, headache, aching joints, perspiration, loss of speech, palpitations, bradycardia, cyanosis, mental depression, convulsions, dyspnea, and death. Gangrene is an occasional complication. In serious nonlethal cases, recovery may take several months. See also weever.

Tracium. atracurium.

trade name (commercial name). **1:** the lawful title or name which is used by a given manufacturer, industry, or merchant that identifies and symbolizes the nature and/or reputation of the business. Do not confuse with trademark, *q.v.* A trade name does not apply to vendable commodities. See also brand name, generic name, trademark. **2:** the name used among traders for an article (e.g., a product, chemical compound). Thus blue vitriol is a common "trade name" for copper sulfate. **3:** the U.S. Department of Agriculture takes the term "trade name" to be any name of a food product, mixture, or compound that distinguishes it from any other food product, mixture, or compound. It is used in this context synonymously with the terms "arbitrary name" and "fancy name." *Cf.* brand name, trademark, registered trademark, U.S.N.

trademark (symbol, ™). in general, a distinctive mark, symbol, or group of words that authenticates the ownership and distinguishes the products of one manufacturer or merchant from those who manufacture, market, or distribute similar products. Do not confuse with "trade name," *q.v.* A trademark is only applied to vendable commodities. **trademark name**. a trademark in the form of a name. *Cf.* brand name, trade name, U.S.N.

Traill's green mamba. *Dendroaspis jamesoni.*

trait. a distinguishing feature or characteristic property; sometimes taken to be a heritable feature. **acquired trait**. a trait that is not inherited. **inherited trait**. a genetic or heritable trait transmitted through germ cells.

tramp's spurge. *Euphorbia corollata.*

tranquilizer. a psychotropic drug; one intended to reduce mental tension, to be quieting, calming, comforting, or soothing, without depression or interference with mental ability. Unintended effects (especially from the use of chlorpromazine and reserpine) may include icterus, nausea, dermatitis, and occasionally severe mental depression. Some tranquilizers may injure developing embryos.

trans-. prefix meaning across.

trans-Pecos copperhead. *Agkistrodon contortrix pictigaster.*

transaminase. an enzyme that catalyzes a transamination reaction. The coenzyme in all such reactions is pyridoxal phosphate.

transamination. the enzymic transfer of an amino group, NH_2, from an amino acid to an alpha-keto acid with the formation of a new keto acid and a new amino acid without the appearance of free ammonia. The amino group is transferred to the coenzyme (pyridoxal phosphate) forming pyridoxamine phosphate, which then transfers the amino group to the keto acid.

transcription. the process whereby a strand of mRNA that is complementary to a segment of DNA is formed.

transdermic. percutaneous.

transdifferentiation. the transformation of a living cell from one fully differentiated state to another; metaplasia, def. 1a.

transformation. **1:** the conversion of a chemical substance from one form to another within an organism or within environmental media, often with changes in activity. Used especially in reference to toxic and hazardous chemicals. See also biotransformation. **2:** a genetic change of form; metamorphosis. **3a:** a change due to the acquisition of foreign genetic material. **3b:** the genetic alteration of a bacterium by absorption of DNA that has been added to the culture medium. It also occurs naturally during mixed infections. **3c:** genetic recombination in bacteria due to the addition of DNA from a different strain to the culture. **4:** in statistics, a change in the statement of a variable such that necessary calculations are simplified; or, for example, to transform the distributions of random variables to a normal distribution which is interpreted more easily. **malignant transformation**. neoplastic transformation. **neoplastic transformation** (malignant transformation). the changes that take place in cultured cells following treatment with carcinogens or infection with tumor viruses, X-irradiation, etc. The transformed cells are characterized by the ability to divide indefinitely, and sometimes produce a daughter cell that can develop into a malignant tumor.

transformation rate. **1:** the rate at which the properties of an environmental chemical, on which its hazardous nature depends (e.g., toxicity, flammability, volatility, explosiveness), change chemically or physically in such a way as to pose a lesser or greater hazard to humans or to the environment. **2:** the rate at which a chemical substance changes from one form to another within an organism or within environmental media, often with changes in activity (e.g., toxicity). It is a term used especially in reference to toxic and hazardous chemicals.

transfuse. to transfer blood from one organism to another.

transfusion. transferring (blood) from one organism to another. This is often accomplished by introducing blood through a needle placed in a blood vessel (usually a vein in humans).

transfusion reaction. an undesirable effect or condition resulting from a blood transfusion.

transient. **1:** brief; fleeting; transitory; of short duration. **2:** a stage in the phylogeny of a species.

translation. a process involving mRNA, ribosomes, and tRNA whereby a polypeptide is synthesized that has an amino acid sequence dictated by the sequence of codons in mRNA.

translocation. **1:** the transport of materials in solution from one part of a multicellular organism to another, referring especially to transport in the conducting systems of plants (xylem and phloem). **2:** movement or transport to a different location or habitat. **3:** a chromosome mutation in which a chromosome segment detaches and becomes reattached to a different (nonhomologous) chromosome.

transplacental. a term referring to the passage of chemicals across the placenta. See barrier (placental barrier).

transport protein. a protein that binds to a specific type of molecule and facilitates its transport from one site in the organism to another. Transport proteins in vertebrate blood are usually albumins and lipoproteins. See also carrier protein.

transudation. the extravasation of fluid, electrolytes, and low molecular weight proteins as in inflammation. *Cf.* effusion, exudation.

transuranium element. an element that has an atomic number greater than 92; all are radioactive and produced by artificial nuclear changes; they are members of the actinide group (elements 89 through 103).

transvaalin. See scillarin.

transvector. **1:** to transfer or transmit a poison from one organism to another (e.g., via a food chain). **2:** an organism that transmits a poison (e.g., via a food chain) from one organism to another. See also shellfish poison, ciguatera.

tranylcypromine (trans-(±)-2-phenylcyclopropanamine; Parnitine™). a highly toxic, potent, monoamine oxidase inhibitor. The sulfate (e.g.,

Parnate™, Tylciprine™) is used as an antidepressant; the action is similar to those of dextroamphetamine and iproniazid. It depresses appetite, and at doses above the recommended therapeutic dose, it exhibits anticonvulsant activity. It potentiates the activity of certain other drugs such as Hexobarbital, probably by interfering with their detoxificaition. In laboratory animals, Parnate causes irritability, restlessness, hyperreflexia, hypertonia, lacrimation, salivation, mydriasis, and mild hypertension. Overdosage of humans in the range of 300 to 1400 mg may cause dizziness, confusion, somnolence, and hypotension.

trapdoor spider. any of a number of spiders that build silk-lined burrows that are closed to the outside by a trapdoor lid. Common genera (Family Ctenizidae) are *Bothriocyrtum*, *Pachylomerides*, and *Ummidia*. All are indigenous to the southern and western United States.

trauma. **1:** physical injury, wounding, or stress due to an external agent as via the bite, strike, or sting of a venomous animal. Trauma can be self-inflicted. **2:** emotional or psychological shock and any attendant disordered feelings or behavior. *Cf.* harm, damage, injury.

tread-softly. *Cnidoscolus stimulosus*.

tree cobra. any snake of the genera *Dendroaspis* and *Pseudohaje*.

tree frogs. See Hylidae.

tree toads. See Hylidae.

tree tobacco. See *Nicotiana glauca*.

tree viper. *Atheris squamigera*.

trefoil. a common name of *Lotus corniculatus* and plants of the genus *Trifolium*.

trefoil dermatitis. trifoliosis.

trembles (tremetol poisoning of cattle). a disease primarily of cattle caused by consuming tremetol-containing plants such as *Eupatorium rugosum*, *E. urticaefolium*, and *Haplopappus venetus*. Humans have been secondarily poisoned by drinking milk from, and occasionally eating the meat of an affected animal. The condition in humans is called milk sickness. Symptoms in cattle may include weakness and anorexia, conspicuous trembling, stiffness of the limbs, and frequent falling especially upon forced exertion, and gaseous abdominal distension. In acute cases, the animal may lose consciousness and die within a few days. Chronic cases are typified by periods of remission and exacerbation. Symptoms in humans are similar to those in cattle, but also include vomiting, complaints of vague pains, listlessness, abdominal distress, and constipation. The proximate cause of the weakened condition in humans and cattle is probably hypoglycemia; death results from ketosis and acidosis. Conspicuous lesions are fatty degeneration of the liver, kidneys, and sometimes muscle. Death in the chronic form of trembles usually results from starvation or intercurrent infection.

trembling. shivering, quivering, or quaking as of a tremor.

tremetol (2-isopropenyl-2,3-dihydro-5-acetylbenzofuran; tremetone). a lipophilic alcohol, it is a toxin produced by certain plants such as *Eupatorium rugosum*, *E. urticaefolium*, and *Haplopappus heterophyllus*. It causes muscular weakness and trembles, *q.v.*, in cattle and in humans who drink contaminated milk or who occasionally eat the meat of an affected animal. It causes milk sickness, *q.v.*

tremetol poisoning. tremetol poisoning of cattle is called trembles, q.v; that of humans is called milk sickness, *q.v.*

tremetone. tremetol.

tremor. **1:** fine, involuntary, oscillating, sometimes rhythmic, muscular movements (quivering, twitching) that involve one or more body parts; involuntary trembling. Tremors may be persistently present or may occur irregularly. Causes include acute anxiety; poisoning; a hyperactive thyroid gland; and failure of the kidneys or liver. **2:** a minute ocular movement or tremor that occurs during fixation on an object. **alcoholic tremor**. tremor potatorum. **arsenic tremor**. that due to chronic arsenic poisoning. **arsenical tremor**. arsenic tremor. **coarse tremor**. that in which the movements are slow. **fine tremor**. that in which the movements are rapid. **intention tremor**. that which occurs only when voluntary movements are attempted. Intention tremor in humans is symptomatic of cerebellar disease. **mercurial tremor** (tremor mercurialis). a slight muscular tremor observed in mercurialism and poisoning by other heavy metals. **metallic tremor**. that which occurs in poisoning by any of several metals. **opium tremor**. tremor opiophagorum. **postural tremor**. that which occurs while a subject is assuming a fixed posture, but does not occur at rest. **resting tremor**. that seen in resting or relaxed individuals. In humans it usually involves the extremities, is associated with Parkinson's disease, and is indicative of a basal ganglia disorder. The condition is aggravated by stress and inhibited by voluntary movement. **saturnine tremor**. that caused by chronic lead poisoning. **toxic tremor**. that seen in various types of chronic poisoning.

tremor mercurialis. mercury tremor. See tremor.

tremor opiophagorum. (opium tremor). that due to habitual usage of opium.

tremor potatorum (alcoholoc tremor). that which occurs in cases of chronic alcoholism.

tremorgenic. tremorigenic.

tremorigen (tremorigenic agent). any agent that causes tremor, *q.v.*

tremorigenic (tremorgenic). causing or able to cause tremor.

tremorigenic agent. tremorigen.

tremulous. trembling or shaking; exhibiting or characterized by tremor. *Cf.* trepidant.

trepidant. characterized by anxious fear or tremor. *Cf.* tremulous.

trepidation. **1:** tremor, trembling. **2:** uneasiness, anxiety, fearfulness, anxious fear.

tretamine. triethylenemelamine.

Tri-excel™. a commercial preparation of pyrethrin, rotenone and ryania.

tri-*o*-cresylphosphate. See TOCP.

tri-*o*-tolyl phosphate. TOCP.

Tri-6™. lindane.

TRI. Toxics Release Inventory.

Triacetin (glyceryl triacetate). usually not toxic if ingested.

Triactis producta. a small but dangerous sea anemone found in the Red Sea. It is ca. 4 cm in diameter and reaches a height when fully extended of ca. 8 cm. This anemone inflicts an extremely painful wound. See also Anthozoa.

trialkylated organotin compound (trialkyltin; trialkyltin compound). biocidal agents used, e.g., in wood and paper preservatives and antifouling paints. These compounds disturb mitochondrial functions by inhibiting oxidative phosphorylation. They cause swelling and uncouple mitochondria by inhibiting oxidation of α-ketoacids. See also alkylated organotin compounds, dialkylated organotin compounds.

trialkyltin, trialkyltin compound. any trialkylated organotin compound.

triallate (bis(1-methylethyl)carbamothioic acid S-(2,3,3-trichloro-2-propenyl) ester; diisopropylthiocarbamic acid S-(2,3,3-trichloroallyl) ester; 2,3,-trichloro-2-propen-1-thiol diisopropylcarbamate; S-(2,3,3-trichloro-2-propenyl) bis(1-methylethyl)carbamothioate). a thiocarbamate herbicide, $C_{10}H_{16}Cl_3NOS$, closely related to diallate. It is used chiefly to control wild oats and blackgrass in cereal crops. It is a mitotic poison in young shoots, but its major effect is suppression of cell elongation. In wild oat seedlings exposed to triallate, cell divisions are fewer and mitotic abnormalities can be observed.

triallylphosphate. a clear, white liquid, $(CH_2{:}CHCH_2O)_3PO$, it is the phosphate triester of allyl alcohol and contains unsaturated C=C bonds. This compound is a liquid (f.p. -50°C). It is highly toxic and causes abnormal tissue growth when administered subcutaneously. It sometimes explodes during distillation.

2,4,6-triamino-*s*-triazine. melamine.

triarylated organotin compound. an organotin compound with 3 aromatic groups. These compounds are biocidal, immunotoxic agents, used mainly as fungicides and insecticides.

triaryltins. triarylated organotin compounds.

triatomic oxygen. ozone.

triazine. **1:** any of three isomers, $C_3H_3N_3$, that contain a 3-membered carbon ring, three double bonds, and three nitrogen atoms. Triazine is used as a herbicide, but is very or even extremely toxic to sheep, cattle, and horses. These animals have been fatally poisoned when grazing on triazine-treated pastures as much as a week following spraying. Signs of intoxication may include anorexia, weight loss, tenseness and general depression, muscular weakness. **2:** any of various structurally related derivatives of these compounds. See atrazine, herbicide (triazine herbicide).

triazine compound. triazine derivative.

triazine derivative (triazine compound). **1:** any of a number of compounds that are structural derivatives of triazine; a number of which are used as herbicides. See atrazine, ametryn, herbicide (triazine herbicide). **2:** triazine (def. 2).

triaziquone (2,3,5-tris(1-aziridinyl)-*p*-benzoquinone). a highly toxic antineoplastic agent related to thiotepa, *q.v.* It is also active against *Entamoeba histolytica*. It is a probable carcinogen.

triazoic acid. hydrazoic acid.

1*H*-1,2,4-triazol-3-amine. 3-amino-1,2,4-triazole.

1,2,4-triazol-3-ylamine. 3-amino-1,2,4-triazole.

tribasic acid. an acid that contains three ionizable atoms of hydrogen per molecule. See carboxylic acid.

tribasic zinc phosphate. zinc phosphate.

tribe. a taxonomic unit of the plant kingdom. It is a subdivision of a subfamily that is used only in very large families. The name is derived from one species in the tribe; it ends in *-eae*. Thus, in the Compositae, for example, the large subfamily Carduoideae is made up of ten tribes, one of which is Astereae. It contains the genera *Aster, Bellis,* and *Erigeron*, etc.

Tribulus terrestris (puncture vine; caltrop). A prostrate annual weed (Family Zygophyllaceae), introduced to southern United States from southern Europe. Common on dry soils along roadsides and deserts mainly in the southern and southeastern states. It causes hepatogenic photosensitization in grazing animals. It is a causative agent of bighead in sheep. The hepatotoxin inhibits the conjugation of pigments which normally occurs prior to excretion via the bile; it also inhibits excretion. Concentrations of nitrate that are toxic to livestock have also been reported. See also phylloerythrin, bighead.

tributyl tin, tributyl tin compound (bis(tributyltin) oxide; TBT; TBTO; OTBE; Biomet TBTO; Butinox; hexabutyldistannoxane). an alkyltin compound, $[(C_4H_9)_3Sn]_2O$, with bactericidal, fungicidal, muolluscicidal, and insecticidal properties. TBTs are of particular environmental importance because of their growing use as industrial biocides, e.g., in the control of slime, fungi, and bacteria. A major use is that of TBT paints used to coat maritime ships to prevent the growth of fouling organisms. It is very or extremely toxic to laboratory rats by ingestion, and is considered very toxic to humans. Mechanisms of toxic action are unknown and the direct primary effects are poorly known. It appears to be neurotoxic and immunotoxic and is a suspected teratogen. Exposed humans may experience headaches, blurred vision, stomach aches, dizziness, fatigue, and chronic dermatitis.

tributylphosphorothionate. a phosphorothionate insecticide, it is a colorless liquid (b.p. 143°C). It is an anticholinesterase as are some of its metabolic products.

1,2,3-tricarboxylic acid. citric acid.

tricarboxylic acid cycle (citric acid cycle; Krebs cycle; TCA cycle). an important series of metabolic reactions in aerobic cellular respiration in plants and animals. In this cycle, acetyl CoA formed from pyruvate produced during glycolysis is oxidized completely to carbon dioxide and water through a sequence of interconversions of various carboxylic acids (oxaloacetate, citrate, ketoglutarate, succinate, fumarate, and malate). The indirect result of this process is the synthesis (reconstitution) of ATP, although some ATP is produced by direct coupling with cycle reactions. The two molecules of pyruvate that enter a cycle from glycolysis, produce 30 ATP molecules. See citric acid.

trichlofos. See chloral derivative.

trichloride of arsenic (arsenic chloride).

1,1,1-trichloro-2,2-bis(*p*-fluorophenyl)ethane. DFDT.

trichloroacetaldehyde (chloral; anhydrous chloral). a colorless, mobile, oily, bitter-tasting liquid aldehyde, CCl_3CHO, with a pungent odor, produced by chlorinating ethanol, then treating with sulfuric acid and distilling. Used in the manufacture of DDT and chloral hydrate. A hypnotic, it may be habit forming and is a controlled substance in the U.S. See also chloral hydrate.

trichloroacetaldehyde monohydrate. chloral hydrate.

trichloroacetic acid (TCA). a colorless, extremely toxic, hygroscopic solid, $C_2HCl_3O_2$, with colorless, deliquescent crystals. It is a strong acid and an oxidative metabolite of trichloroethylene (TCE). It is used as an herbicide (sodium trichloroacetate) and an intermediate in the manufacture of pesticides. As in the case of other haloaliphatic acids, TCA inhibits plant growth, causes leaf chlorosis and necrosis. It affects numerous additional plant processes. TCA is used in the laboratory to precipitate protein. It is corrosive to skin, eyes, and mucous membranes and is carcinogenic to some laboratory rodents.

trichloroacetonitrile (trichloromethylnitrile). an insecticide, C_2Cl_3N, and a strong irritant of the eyes and skin. It is very toxic to laboratory rats.

1,2,4-trichlorobenzene (*unsym*-trichlorobenzene). a colorless, flammable, stable, water-insoluble liquid, $C_6H_3Cl_3$, with a pleasant odor similar to that of *o*-dichlorobenzene; it is miscible with most oils and organic solvents. It is used chiefly as a solvent in the manufacture of chemicals, in dyes and intermediates, herbicides, and insecticides. It is toxic by contact, ingestion, and inhalation. The principle lesions found in experimental exposures of laboratory animals are those of the liver, kidney, and lung. Humans are exposed predominantly by runoff from precipitation and drinking water. It is hepatotoxic and a strong irritant of the skin, eye, and upper respiratory tract. Symptoms of intoxication may cause drowsiness, lack of coordination, and unconsciousness.

***unsym*-trichlorobenzene**. 1,2,4-trichlorobenzene

2,3,6-trichlorobenzeneacetic acid. chlorfenac.

2,3,6-trichlorobenzoic acid. 2,3,6-TBA.

1,1,1-trichloro-2,2-bis(*p*-chlorophenyl)ethane. DDT.

3,4,4′-trichlorocarbanilide. triclocarban.

2,4,5-trichloro-α-(chloromethylene)benzyl phosphate ester. stirofos.

1,1,1-trichloro-2,2-bis(*p*-chlorophenyl)ethane. DDT.

1,1,1-trichloroethane (methyl chloroform; TCE). a colorless, chlorinated, liquid solvent, CH_3CCl_3, that is widely used as an industrial degreasing agent. It is an irritant to the eyes and is taken into the body largely through the lungs and the skin; it is somewhat soluble in blood, but is strongly lipophilic. Effects of exposure can include corneal burns, narcosis, and liver damage. Exposure to large amounts depress the CNS, slow reaction time, and sometimes death.

1,1,2-trichloroethane (vinyl trichloride; β-trichloroethane; TCE). a clear, colorless solvent, $CHCl_2CH_2Cl$, with a sweet, characteristic odor. It is used as a solvent for fats and oily products and in organic synthesis. It is an irritant, is readily absorbed by the skin, and is carcinogenic to some laboratory animals.

β-trichloroethane. 1,1,2-trichloroethane.

2,2,2-trichloro-1,1-ethanediol. chloral hydrate.

2,2,2-trichloroethanol (trichloroethyl alcohol). a moderately to very toxic hypnotic and anesthetic, CCl_3CH_2OH, it is the active intermediate, *in vivo* of chloral hydrate and presumably all other chloral derivatives.

trichloroethene. trichloroethylene.

trichloroethylene (trichloroethene; ethinyl trichloride; TCE). a stable, low boiling, colorless, nonflammable, sweet-smelling, volatile liquid, $CCl_2=CHCl$, that has been used as an anesthetic; as a household and industrial solvent, especially to degrease machine parts; and in cleaning septic tanks. It is also used in dry cleaning and in the removal of caffeine from coffee. TCE is a widespread environmental contaminant. It is toxic by inhalation, acting on the central nervous system; moderate exposure can cause symptoms similar to alcohol intoxication; symptoms may include disturbed vision, headaches, nausea, paresthesia, and cardiac arrhythmias (which may have a fatal consequence). TCE also has a toxic effect on the liver and kidneys and may damage these organs and the central nervous system. High doses cause liver carcinoma in some laboratory animals and probably in humans. It has been largely replaced by methylene chloride for removing the caffeine from coffee. See also trichloroacetic acid.

1,1′-(2,2,2-trichloroethylidene)bis[4-chlorobenzene]. DDT.

1,1′-(2,2,2-trichloroethylidene)bis[4-fluorobenzene]. DFDT.

1,1′-(2,2,2-trichloroethylidene)bis[4-methoxybenzene]. methoxychlor.

trichloroethylidene glycol. chloral hydrate.

2,4,4′-trichloro-2′-hydroxydiphenyl ether. triclosan.

trichloromethane. chloroform.

trichloromethylchloroformate. diphosgene.

***N*-trichloromethylmercapto)-Δ^4-terahydrophthalimide**. captan.

***N*-trichloromethylmercapto-4-cyclohexene-1,2-dicarboximide**. captan.

trichloromethylnitrile. trichloroacetonitrile.

***N*-(trichloromethylthio)cyclohex-4-ene-1,2-dicarboximide**. captan.

***N*-(trichloromethylthio)-4-cyclohexene-1,2-dicarboximide**. captan.

***N*-trichloromethylthio-3a,4,7,7a-tetrahydrothalimide**. captan.

trichloromonosilane. trichlorosilane.

trichloronitromethane. chloropicrin.

trichlorophenol. either of two toxic nonflammable chlorinated phenols, 2,4,5-T and 2,4,6-T. They are toxic and may be carcinogenic. They have a strong phenolic odor. Effects of exposure on humans may include chloracne, muscle weakness, liver damage, and porphyria. 2,4,6-trichlorophenol may disrupt mitochondrial phosphorylation. TCDD (probably formed by the chemical combination at high temperature of two molecules of 2,4,5-T to form one molecule of TCDD) is a common contaminant of trichlorophenols. See also chlorophenol, pentachlorophenol. *Cf.* silvex.

2,4,5-trichlorophenoxyacetic acid. 2,4,5-T.

2-(2,4,5-trichlorophenoxy)ethyl-2,2-dichloropropionate. Erbon.

2-(2,4,5-trichlorophenoxy)propionic acid. silvex.

2,3,6-trichlorophenylacetic acid. chlorfenac.

1,2,3-trichloropropane (triCP). a colorless, flammable liquid, $CH_2ClCHClCH_2Cl$, that is slightly soluble in water and is a solvent of oils, fats, waxes, and many resins. It is used as a paint and varnish remover and degreasing agent. TriCP is a strong irritant and is hepatotoxic by inhalation and skin absorption.

2,3,3-trichloro-2-propen-1-thiol diisopropylcarbamate. triallate.

***S*-(2,3,3-trichloro-2-propenyl)bis(1 methylethyl)carbamothioate**. triallate.

trichlorosilane (trichloromonosilane; silicochloroform). a compound, $HSiCl_3$, used in the production of organotrichlorosilanes and elemental silicon for semiconductors. It is a fuming liquid with a suffocating odor. It reacts with water, releasing HCl vapor. See silane.

Trichoderma viride. produces the mycotoxins trichotoxin A, trichodermin, and trichodermol (roridin C).

trichodermin. a trichothecene mycotoxin first isolated from *Trichoderma viride*.

trichodermol (roridin C). a trichothecene first isolated from *Trichoderma viride*.

Tricholoma. a genus of agarics (Family Tricholomataceae). ***T. muscarium***. a poisonous agaric mushroom and a source of the isoxazole, tricholomic acid, *q.v.* ***T. pardinum*** (dirty trich, poisonous trich). a poisonous mushroom that grows under trees, especially fir, throughout much of northern North America. It causes severe gastric distress. ***T. pessundatum*** (red-brown trich). a species that grows mainly under conifers in eastern North America. Symptoms are those of gastrointestinal distress which rarely persists for more than a day or two.

tricholomic acid. an isoxazole, isolated from *Tricholoma muscarium*, that causes visual damage, mental confusion, spatiotemporal dislocation, and memory loss in humans. It is the most toxic of the isoxazoles. See isoxazole, muscazone, muscimol.

Tricholompsis platyphylla (platterful mushroom). a normally edible mushroom (Family Tricholomataceae), but sometimes toxic, producing gastrointestinal distress; the affected person usually recovers within a few hours.

1,1,1-trichol or-2,2-bis(*p*-methoxyphenyl)ethane. methoxychlor.

trichothecene, trichothecene mycotoxin. any of a class of more than 40 sesquiterpenoid mycotoxins produced by a variety of fungi that grow chiefly on grains (e.g., *Cephalosporium, Fusarium, Myrothecium, Stachybotrys, Trichoderma, Trichothecium*, and *Verticimonosporium*). Many of these compounds have bacteriocidal, fungicidal and insecticidal activity and are highly toxic to various plants and animals. They invade numerous plants and many agricultural products, the ingestion of which can cause serious poisonings of humans and other animals. An outbreak of trichothecene poisoning in Siberia in 1944 provided considerable information on the effects and toxicities of these substances on humans. Because of a shortage of food related to World War II, many people ate moldy grains such as barley, millet, and wheat. The skin of those who did became inflamed and they suffered gastrointestinal disorders, including vomiting and diarrhea; and multiple hemorrhage. About 10% of those affected died. All trichothecenes are dermatotoxic; some (e.g., T-2 toxin) are neurotoxic; the crude toxins may be immunosuppressive. Examples are roridin, verrucarins, deoxynivalenol (Rd-toxin), deoxynivalenol-monoacetate (Rc-toxin), diacetylnivalenol, diacetoxyscirpenol, monoacetylnivalenol (fusarenon-X), nivalenol, trichothecin, and T-2 toxin.

trichothecene poisoning. poisoning of humans or other animals by ingesting plants and agricultural products infested by fungi that secrete trichothecene mycotoxins. Vomiting is an early and characteristic sign of trichothecene poisoning by cats, dogs, swine, and birds. These mycotoxins assault the hematopoietic system of vertebrate animals, especially bone marrow. Leucopoiesis is depressed in acutely poisoned laboratory cats and mice. Subacute exposure of domestic cats causes pronounced leucopenia and cellular destruction in bone marrow; pulmonary and intestinal hemorrhage are also common.

trichothecin. structurally the simplest of the trichothecines, it is a naturally occurring epoxide trichothecene antibiotic from *Trichothecium roseum*.

trichotoxin. an antibody or cytotoxin that is injurious specifically to ciliated epithelium.

trichotoxin A. a mycotoxin from *Trichoderma viride*.

triclocarban (*N*-4-chlorophenyl)-*N*-(3,4-dichlorophenyl)urea; 3,4,4′-trichlorocarbanilide; 1-(3′,4′-dichlorophenyl)-3-(4′-chlorophenyl)urea; TCC; Solubacter). a bacteriostat and disinfectant often used in deodorant soaps. It is possibly carcinogenic.

triclosan (5-chloro-2-(2,4-dichlorophenoxy)phenol; 2,4,4′-trichloro-2′-hydroxydiphenyl ether). a white, crystalline powder, $C_{12}H_7Cl_3O_2$, with a faint, aromatic odor that is used chiefly as a bacteriostat and preservative for cosmetic and detergent formulations; also used as a disinfectant. See also deodorant.

triCP. 1,2,3-trichloropropane.

tricresol. See cresol.

tricresyl phosphate (tritolyl phosphate; TTP; TCP). a mixture of isomers, the *ortho* isomer of which is highly toxic. See TOCP.

tricresylphosphate. See TOCP.

1,4,5-trideoxy-1,4-imino-5-(4-methoxyphenyl)-D-xylo-pentitol 3-acetate. anisomycin.

trientine (*N,N'*-bis-(2-aminoethyl)-1,2-ethanediamine; Cuprid). a chelating agent used in copper chelation therapy in Wilson's disease for patients who are intolerant of penicillamine.

trietazine. an atrazine herbicide. First described by Gysin and Knusli (1958), it was introduced in 1972 as a component of herbicide formulations for weed control in potatoes and peas.

triethyl lead (TEL; triethyllead). a metabolite of tetraethyllead used as an antiknock additive in gasoline. It is an extremely toxic, lipophilic neurotoxicant in both laboratory animals and humans that causes extensive encephalopathy with accompanying behavioral, pathological and biochemical changes that are typical of organolead poisoning generally. It inhibits myelogenesis in the fetus.

triethylenethiophosphoramide. thiotepa.

2,4,6-triethylenimino-1,3,5-triazine. triethylenemelamine.

triethyllead. triethyl lead.

triethylphosphate (TEP). a moderately toxic, neurotoxic, colorless, high-boiling, water-insoluble liquid, $(C_2H_5)PO_4$, that is soluble in most organic liquids. It is used as a solvent and plasticizer for gums, resins, and plastics, and in the manufacture of pesticides. Like other phosphate esters, it inhibits cholinesterase, and may cause nerve damage, but less so than other anticholinesterases.

***N*-(3-trifluoromethylphenyl)-*N'*,*N'*-dimethylurea**. fluometuron.

1,1,1-trifluoro-2,6-dinitro-*N,N*-dipropyl-*p*-toluidine. trifluralin.

α,α,α-trifluoro-2,6-dinitro-*N,N*-dipropyl-*p*-toluidine. trifluralin.

3-trifluoromethyl-4-nitrophenol. TFM.

***N*-(3-trifluoromethylphenyl)-*N'*,*N'*-dimethylurea**. fluometuron.

α,α,α-trifluoro-4-nitro-*m*-cresol. TFM.

trifluralin (1,1,1-trifluoro-2,6-dinitro-N,N-dipropyl-p-toluidine; 2,6-dinitro-*N,N*-dipropyl-4-(trifluoromethyl)benzenamine; α,α,α-trifluoro-2,6-dinitro-N,N-dipropyl-p-toluidine; 2,6-dinitro-N,N-dipropyl-α,α,α-trifluoro-p-toluidine; 2,6-dinitro-N,N-dipropyl-4-trifluoromethylaniline; N,N-dipropyl-2,6-dinitro-4-trifluoromethylaniline). a yellowish-orange solid, $F_3C(NO_2)_2C_6H_2N(C_3H_7)_2$, that is insoluble in water and soluble in ethanol, acetone, and xylene. A herbicide used to control annual grass and weeds in a number of crops (chiefly cotton) by incorporation into the top 2-5 cm of soil. Trifluralin inhibits cell division in root meristems, arresting cells in prophase within a few hours of treatment. This is usually followed by other mitotic aberrations, the enlargement of some meristematic cells which have become multinucleate, and a progressive decrease in the amount of meristematic tissue. It may completely inhibit lateral root development, especially in cotton. It is very toxic to laboratory rats and is carcinogenic to some laboratory animals.

trifoliosis (dew poisoning; alsike poisoning; trefoil dermatitis; clover disease). alsike-induced photosensitization that occurs in livestock due to ingestion of any of several types of clover and alfalfa (See *Brassica*, *Erodium*, *Medicago*, and *Trifolium*). In sensitized horses, the skin reddens from sunlight and a dry necrosis of the skin, or edema with a serous discharge from the affected areas ensues. The most affected parts of the body are usually the muzzle and feet which are most exposed to moisture when on pasture (hence "dew poisoning"). Symptoms may also include excitement or depression; and sometimes colic, stomatitis, and diarrhea which may be severe, in which case the affected animal becomes emaciated.

Trifolium (clover; trefoil). a genus of annual, biennial, or perennial herbs (Family Fabaceae, formerly Leguminosae). Various species have been implicated in trefoil dermatitis, trifoliosis, and rape scald, as have species of *Brassica*, *Erodium*, and *Medicago*. ***T. hybridum*** (alsike clover). originating in the Alsike parish of Sweden, it is a tall, slender perennial with white flowers. Alsike poisoning has affected cattle, horses, sheep, and swine. Animals are photosensitized, and usually succumb on alsike pasturage during sunny weather. See also trifoliosis. ***T. pratense*** (red clover). a low-growing herb, often used as a forage crop. Hay containing substantial amounts of second cuttings or late-season cuttings of red clover may seriously affect cattle, horses, and sheep. The syndrome is initially characterized by slobbering, but other symptoms soon appear and become increasingly severe unless contaminated hay is removed from the diet. Symptoms include bloating, a stiff gait, reduced milk flow, diarrhea, emaciation, and abortion. Pasturage in fields with red clover may produce photosensitization within a few days, with consequent edema and necrosis of tissues surrounding the eye and sometimes consequent impaired vision or even blindness has been reported. ***T. repens*** (white clover, shamrock). a low, creeping perennial herb with small, round, white flower heads. Some strains are moderately cyanogenic. ***T. subterraneum*** (subterranean clover). a species that sometimes accumulates lethal concentrations of copper from the soil. The growing foliage contains an estrogen, genistein, which has caused extensive damage to flocks of sheep in Australia. Symptoms of poisoning associated with genistein include infertility, abnormal lactation, dystocia and prolapse of the uterus in affected sheep.

triggerfish. certain fish of the family Balistidae. They are characterized by 3 dorsal spines, and an oval, compressed body and large, thick, diamond-shaped scales; those immediately above the base of the pectoral fin are usually enlarged and separated, forming a flexible "tympanum." Some species are poisonous, others are edible. See *Balistoides conspicillum*.

Triglochin (arrowgrass). A semicosmopolitan genus of grass-like perennial plants (Family Juncaginaceae) that grow in clumps. They have long linear leaves which, unlike grasses, are thick. These plants contain widely variable amounts of HCN and have been responsible for intoxication of livestock. Symptoms and lesions are those of cyanide poisoning.

1,2,4-trihydroxyanthraquinone. purpurin.

2,3,5-triiodobenzoic acid. TIBA.

3,5,3′-triiodothyronine. See thyroid hormone.

trimercuric orthophosphate. mercuric phosphate.

Trimeresurus (Asian lance-headed vipers; Asian lanceheads; Asiatische Lanzenottern). a genus of venomous snakes (Family Crotalidae) that are closely related to the tropical American lance headed vipers, *Bothrops*. The Asian lanceheads are indigenous to southeast Asia and adjacent islands. The large species are dangerous to humans, but while the bite of the smaller species can be very painful, such is rarely if ever fatal. The head is large, depressed, and triangular in dorsal aspect; the neck is much narrower than the base of the head; the canthus is obtuse to sharp, and the eyes are small to moderate in size with vertically elliptical pupils. The facial pit and the absence of large plates on the crown distinguish the members of this genus from most other snakes of the region. The body is nearly cylindrical to moderately compressed; moderately slender to stout; the tail is short to moderately long; and the subcaudal scales may be divided or undivided. Signs and symptoms of envenomation by Asian lance-headed vipers usually include pain, swelling, and edema, discoloration of the skin, ecchymosis, weakness, a weak, rapid pulse, hypotension or shock, hemorrhage, and increased clotting time. Species not described below include *T. cantori*, a species restricted to the Andaman and Nicobar Islands; *T. chaseni* (Chasen's pit viper) and *T. convictus*, both of the island of Borneo; *T cornutus* (horned pit viper) of Vietnam; *T. elegans* (elegant pit viper; Sakashima habu), a species indigenous to the Ryuku Islands; *T. erythrurus*, a species that occurs in east Pakistan, northeastern India, eastern and southeastern Asia including Nepal, Myanmar, and northern Thailand; *T. flavomaculatus* (yellow-spotted pit viper; McGregor's pit viper; manda-dalag; Polilo pit viper; Schultze's pit viper), a species indigenous to the Phillipines; *T. hageni* (Hagen's pit viper), a Sumatran species; *T. puniceus* (ashy pit viper; flat-nosed pit viper; oraj kalakai; oraj lemah; sawah tadung; ular gebuk), indigenous to Malaya, Sumatra, Java, Lesser Sunda Island, and Borneo; *T. sumatranus* (Sumatran pit viper), a species that occurs in Thailand, Malaya, Sumatra, Lesser Sunda Island, and Borneo); and *T. trigonocephalus* (green pit viper; pachi virian; pala polonga), a lance-headed viper indigenous to Ceylon. ***T. albolabris*** (white-lipped tree viper). a small, chiefly nocturnal species of southeastern China, probably Taiwan and Hainan, south through the Sunda Archipelago to northeastern India. Adults average about 0.5 m in length, the female being much larger than the male, some individuals reaching a length of about 1 m. The upper lip is pale green, yellow, or white and the nasal shield is fused to the first upper lip shield; the iris is yellow. The ground color in most specimens is a noticeably paler green than that of *T stejnegeri* and the males have a white lateral stripe on either side of the body; the end of the tail is bright red. This lancehead typically inhabits brushy or wooded situations in low-lying, but often hilly areas; It seldom occurs above 500 m; it is often common near human habitation, especially in gardens. ***T. flavoviridis*** (Okinawa habu; habu). a slender, graceful, light brown or olive snake found only on the larger of theAmami and Okinawan islands. The habu is the largest of the Asian lanceheads, adult specimens averaging about 1.4 m in length, although individuals as long as 2.3 m are known. The body bears elongated green or brownish blotches edged in yellow; these often coalesce, forming wavy stripes. The venter is whitish with dark mottling laterally. The scales on the crown are small. The habu occupies crevices in stone walls, caves, old tombs, and other structures, usually in the ecotone between cultivated fields and palm forest. This is an active, irritable, rather aggressive, and mostly nocturnal snake. It strikes rapidly with little provocation, the incidence of bites in some locals being very high. The venom is not very toxic to humans, however, and the case fatality rate is only about 3%; permanent disabilities occur in 6-8% of affected persons. ***T. gramineus*** (Indian green tree viper; bamboo pit viper; esau; Jerdon's viper; paper viper; keisau; oraj hedjo; oraj boengka laoet; semelo). a so-called bamboo viper, it is a largely nocturnal, arboreal snake of southeast Asia including peninsular India, east Pakistan, Thailand, Taiwan, and the Andaman and Nicobar Islands. It typically inhabits wooded hilly country with dense underbrush. The ground color is green, usually with darker green flecks and a light, irregular line on either side. Adults average somewhat less than 1 m in length. The conformation of the head is that typical of the genus; the iris of the eye is yellow. This lancehead is distinguished from other tree vipers within its range by the smooth dorsal scales, only a few posterior rows of which bear keels. It remains quietly hidden on the ground under bark or debris. It is not especially aggressive during the day although it will strike if touched or otherwise aggravated. It is, nevertheless, the leading cause of snakebite in some areas; at greatest risk are persons who harvest tea, cut bamboo, or clear underbrush. The bite is not lethal to adult humans, but fatalities among children are knwon. ***T. monticola*** (Chinese mountain viper; Chinese pit viper; spotted pit viper; Arisan habu). a gray or olive snake that is stockier than *T. mucrosquamatus*. It occurs in Nepal, eastward through China, and southward through the Malay Peninsula. The crown is dark brown to black, sometimes bearing a light-colored Y-shaped mark. The body is speckled with black and has a linear series of large brown or reddish, more or less square blotches; the belly is white, flecked with dark brown spots; the tail is not prehensile. Adult specimens usually measure about 1 m to 1.2 m in length; occasional individuals may approach 1.4 m. This viper commonly inhabits forested mountains at elevations up to 1,900 m. It is normally sluggish and not especially irritable. A female coiled about her eggs is, however, unpredictable, easily aggravated, and should be considered dangerous. ***T. mucrosquamatus*** (Chinese habu; pointed scales pit viper; turtle-designed snake). a grayish-brown to buff or olive snake bearing three rows of darker gray or brown spots (edged in yellow) along the body; spots of the middle row are largest, and sometimes coalesce, forming a broken wavy stripe along the dorsum; the belly is whitish infiltrated by brown. Adult specimens average about 0.9 m in length; the maximum is about 1.2 m. The Chinese habu is indigenous to a region that includes Taiwan and southern China west to eastern Mayanmar. It occurs chiefly in low, grassy or sparsely wooded hill country but is not uncommon on agricultural lands and in suburban areas. This snake is similar to appearance *T. elegans* and remarkably similar to *T. flavoviridis, q.v.*, in its behavior, frequently striking humans. ***T. okinavensis*** (himehabu; Okinawan habu; Okinawa pit viper; kufah). a sluggish snake that is less aggressive, smaller and stockier than *T. popeorum*. Its distribution is nearly coincident with that of *T. flavoviridis*; it is smaller, stockier, and less aggressive than *T. flavoviridis* and is not considered especially dangerous to humans. ***T. popeorum*** (Pope's tree viper; Pope's pit viper; esau; oraj boengka laoet). a chiefly nocturnal species indigenous to Assam and Mayanmar eastward to Cambodia and southward through Malaysia and Indonesia. It resides mainly in hilly country, usually at elevations from about 1000 to 1600 m; it occurs commonly on tea plantations. This snake is similar in appearance to *T. stejnegeri*, differing most obviously in the structure of the male genitalia. Furthermore, the iris of live specimens is yellow (not reddish) and the stripe is indistinct in the adult. Mature specimens are larger than *T. stejnegeri*, often reaching a length of about 1 m. ***T. purpureomacula*** (mangrove viper; purple spotted pit viper; shore pit viper). a species largely restricted to coastal areas and islands from east Bengal, southern Mayanmar, the Malay Peninsula, Sumatra, and the Andaman islands. The average length of adult specimens is slightly less than 1 m. The Mangrove viper usually inhabits rocky sites or areas with low vegetation, especially in mangrove swamps. It is similar in appearance to the green tree vipers, although the ground color varies. It also has more scale rows at midbody (25-27 as opposed to 19-21). Common color phases include purplish brown (with or without a whitish lateral line) and olive or gray (irregularly spotted with brown); the venter is whitish, usually clouded by brown or spotted with gray and brown. It is a fairly common cause of snakebite in some areas (e.g., coastal Malaya). ***T. stejnegeri*** (Chinese green tree viper; Chinese bamboo viper; red tail snake; Stejneger's palm viper). a so-called bamboo viper, this snake commonly inhabits mountainous country in central and southeastern China and Taiwan. Favored habitats include wooded areas near streams, scrub, and partially cultivated areas. It is mostly nocturnal, has a prehensile tail, and is a more or less typical tree viper of tropical Asia. The ground color is bright green to chartreuse dorsally and yellow or pale green ventrally. A white or yellow lateral line with a red margin runs the length of the body; the terminus of the tail is reddish. It is a stout snake; the upper lip is not fused with the nasal shield; the iris is orange or coppery. The dorsal scales are keeled; the average length of an adult is about 0.5 m. ***T. wagleri*** (Wagler's pit viper; djalimoo; papala; speckled pit viper; temple snake; ular bakaw; ular nanti bulau; ular puckuck). an unusually stout arboreal snake with a prehensile tail, it is extremely sluggish, not easily aggravated, even said to be gentle, and is not considered dangerous to humans. It is indigenous to Thailand, Malaysia, Indonesia, Borneo, and the Phillipines where it commonly occurs in lowland jungle and on plantations. The head is unusually broad; the scales between the eyes and on the chin and throat are strongly keeled; the keeled scales on the throat are diagnostic throughout its range. The average length of adults is about 0.8 m, but an occasional individual may reach a length of slightly more than 1 m. Adults are green with scales edged in black speckled with green; it displays broad crossbands that are green above, shading laterally into yellow; the belly is greenish, mottled with yellow; the tail is black and the side of the head may be greenish or yellow. subadult specimens are green with a row of red and white spots; the tail is usually reddish.

trimethyl-β-acetoxypropylammoniumchloride. methacholine chloride.

2,6,8-trioxypurine. uric acid.

tripelennamine (*N,N*-dimethylamino-*N'*-(phenylmethyl)-*N'*-2-pyridinyl-1,2-ethanediamine; 2-[benzyl(2-dimethylaminoethyl)amino]pyridine). an antihistaminic with properties similar to those of diphenhydramine and related compounds. It is highly toxic by ingestion.

tripelennamine citrate (pyribenzamine citrate; 2-

[benzyl-(2-dimethylaminoethyl)amino]pyridine dihydrogen citrate). a white, bitter, crystalline powder that is soluble in ethnol and water; slightly soluble in ether; and forms acidic solutions. It is an antihistaminic

Trisetum flavescens (golden oats; yellow oats; yellow oat grass). a tall grass (Family Gramineae), growing to nearly 1 m in height, with flat leaves and a terminal, rudimentary flower. It is indigenous to middle Europe, but has been naturalized elsewhere (e.g., in North America). It contains substances that can induce calcinosis accompanied by osteopetrosis in herbivores. See enzootic calcinosis.

trismus (lockjaw). a functional disturbance of the motor branch of the trigeminal nerve, with tonic spasm especially of the muscles of mastication. As a consequence the jaws are firmly closed and can be opened only with great difficulty. Trismus is an early and characteristic sign of general tetanus.

trismus sardonicus. risus sardonicus.

trisodium phosphate (TSP). a cleaning agent used mainly to prepare wood for painting. It is produced by mixing soda ash with phosphoric acid forming disodium phosphate; caustic soda is then added to yield the final product. TSP is a minor skin irritant and is moderately toxic when ingested. It should be kept out of the reach of children and gloves worn while using it.

tropate. any salt or ester of tropic acid.

-trophic. a suffix denoting nutrition.

tropic acid ester with scopine. scopollamine.

-tropic. a suffix denoting: **1a:** affinity for or a turning toward. **1b:** a change in the visual axis. **2a:** a tendency to influence, as in the case of the pituitary hormones (See pituitary gland). **2b:** a tendency to be influenced by, as in the case of radiotropism (influenced by radiation).

true pennyroyal. *Mentha pulegium*.

true viper. See Viperidae.

trumpet flower. any plant of the genus *Solandra*.

TSH (thyrotropic hormone; thyrotrophic hormone; thyreotrophic hormone; thyroid-stimulating hormone; TTH; TSH). a glycoprotein composed of 2 large peptide subunits, it is a hormone elaborated by the pars distalis of the mammalian pituitary gland and homologous tissue of other vertebrates. TSH is essential to the normal functioning of the thyroid gland. It stimulates growth of the thyroid gland, the uptake of iodine, and the secretion of thyroxine. Secretion of TSH is inhibited by massive doses of vitamin A; it is inactivated by oxidants (e.g., potassium permanganate, elemental iodine), certain other chemicals, heat, and proteolysis. See also hormone, thyroxine, pituitary gland.

tsp. teaspoonful.

TSP. trisodium phosphate.

TSS. toxic shock syndrome. See toxic shock.

TTH. TSH.

tube feet. pedicellariae.

d-tubocurarine chloride (tubocurarine chloride). a highly toxic, white to light tan, odorless, crystalline alkaloid, $C_{38}H_{44}Cl_2N_2O_6{\cdot}5H_2O$. The active principle of curare, it is isolated from the stems of *Chondodendron*, especially *C. tomentosum*. The crude extract is used by certain South American natives as an arrow poison. Tubocurarine is a muscle relaxant, formerly used during surgery, but has been replaced by synthetic curarimemetics. It raises the threshold for acetylcholine at the myoneural junction by competitively binding to the receptors; it also blocks ganglionic transmission and releases histamine. Toxic doses thus cause paralysis of striated muscle. The LD_{50}, *i.p.*, in laboratory mice has been reported as 0.63 mg/kg.

tubulin (colchicine-binding protein). the spindle protein seen during the division of eukaryotic cells. It has the ability of self-assembly, forming the protein filaments or microtubules. As such, it plays an important role in the determination of form and dynamics of eukaryotic cells. It may be involved in phagocytic motility. They process of microtubule formation may be inhibited, or microtubules disassembled by the action of colchicine and by vinblastine.

tullidora. See *Karwinskia humboldtiana*.

tumefacient. causing or tending to cause swelling.

tumefaction. **1:** tumentia; a swelling. **2:** tumescence.

tumefy. to swell; to cause swelling.

tumentia. tumefaction, definition l.

tumescence. **1:** the state of being or becoming swollen or tumid. *Cf.* turgescence. **2:** a swelling. *Cf.* tumor, turgescence.

tumescent. swollen, tumid, turgescent; of or pertaining to tumescence. *Cf.* turgid.

tumid. turgid, swollen, tumescent.

tumor. **1:** a swelling or enlargement. **2:** a neoplasm; a new mass of abnormal tissue; an abnormal new growth or spontaneous enlargement in a tissue or organ due to the unusually abnormal or uncontrolled proliferation of cells. **benign tumor**. a benign neoplasm; a growth that does not invade or infiltrate other nearby tissues, and does not spread to other parts of the body. **malignant tumor**. a malignant neoplasm; a tumor that grows uncontrollably, invades or infiltrates nearby tissue, and spreads to other parts of the body that are remote from the tumor. **3:** swelling, one of the four signs of inflammation (calor, dolor, rubor, tumor). *Cf.* neoplasm.

tumor-inducing principle. a plasmid conveyed by the bacterium *Agrobacterium tumefaciens* that causes crown gall disease in plants. It transforms normal host tissue into tumor tissue. This is accomplished by incorporation of the plasmid into the plant genome.

tumor necrosis factor β. lymphotoxin.

tumor promoter. a substance which, though not carcinogenic itself, potentiates or accelerates the effects of a carcinogen.

tumoricidal. lethal to neoplastic cells.

tumorigen. tumorigenic, able to produce tumors; an agent that is able to produce tumors.

tumorigenesis. the production of tumors; the development of tumors.

tumorigenic. **1:** giving rise to or producing tumors, especially malignant tumors. **2:** of or pertaining to any agent that can cause a tumor (e.g., certain chemicals, a tumor virus). **2:** said of a cell that can give rise to a tumor.

tumorigenicity. the capacity or ability to give rise to or cause a tumor.

tumorous. tumor-like; swollen, protuberant.

tuna. a common name of marine fish of the genus *Thunnus*, and related genera (Family Scombridae). See Scombridae, scombroid poisoning.

tung meal. See *Aleurites*.

tung nut. See *Aleurites*.

tung oil (Chinawood oil). a yellow, flammable drying oil that is soluble in other oils, chloroform, ether, and carbon disulfide. It is extracted from the nuts of *Aleurites*, chiefly from *A. fordii*. It is used in the formulation of paints and varnishes, in linoleum, and the waterproofing of paper. The nuts and leaves also contain saponins. The ingestion of one nut is sufficient to produce severe illness in humans. Vomiting and severe abdominal pain may appear within 1/2 hour, followed somewhat later by diarrhea. Severe poisoning may end in exhaustion and collapse. Livestock that ingest prunings or fallen branches of *Aleurtes* present similar symptoms. Little, if any, tung oil is now produced in the United States.

tung-oil tree. See *Aleurites*.

tung tree. *Aleurites*.

tungstate white. barium tungstate.

tungsten (symbol, W; formerly wolfram; wolframium)[Z = 74; A_r = 183.85]. a somewhat toxic, steel-gray to nearly silver-white, transition metal that occurs in wolframite.

tunica mucosa. mucosa.

Turbinaria ornata. a brown, benthic seaweed the surface of which is sometimes inhabited by *Gambierdiscus toxicus*. It thus comprises part of the food web that transfers ciguatoxin to fish and ultimately to *Homo sapiens*.

turgescence. swelling or enlargement of a cell or body part. *Cf.* tumescence.

turgescent. swollen, inflated. *Cf.* tumescent.

turgid. swollen, bloated; in a state of turgor. This term is used chiefly for distention of plants, plant cells, and prokaryotes due to the admission of water into the cells. *Cf.* tumid.

turgor. **1a:** the normal tension in a cell. **1b:** distension, swelling; the state of being maximally swollen or distended due to the admission of water as in the case of a cell, tissue, or green plant. A state of full turgor is reached when a net intake of water is no longer possible by osmosis. A cell with a high degree of turgor is said to be turgid. **turgor vitalis**. the normal fullness of the capillaries and other blood vessels.

turgor pressure. the internal pressure (elastic and osmotic) developed by the fluid content within a turgid cell.

turkey corn. See *Dicentra*.

turkey pea. *Tephrosia virginiana*.

turkey X disease. a mycotoxicosis of domestic turkeys caused by aflatoxins B from *Aspergillus flavus*, *q.v.* See aflatoxicosis.

turkey-red oil. red oil.

turkeyfish. a common name of fish of the genus *Pterois*.

turkeyfishes. fish of the genus *Pterois*.

turnip. *Brassica rapa* var. *rapifera*.

turnover. the quantity of a substance that is metabolized or otherwise processed within a given period of time.

turpentine. **1:** a toxic oleoresin from various species of *Pinus*, especially *P. palustris* and other species of *Pinus*. It is used chiefly as a source of spirits of turpentine. It is an irritant of the skin and mucous lining of the gastrointestinal tract, and also affects the nervous system. It was formerly used in both human and veterinary medicine as a rubefacient and counterirritant. **2:** spirits of turpentine. **oil of turpentine**. spirits of turpentine. **spirits of turpentine** (oil of turpentine). **1:** a colorless, volatile, water-insoluble oil distilled from turpentine gum, $C_{10}H_{16}$. It is a toxic, colorless, flammable, liquid mixture of terpenes (chiefly pinene and diterpene) and other hydrocarbons (e.g., ethers, alcohols, esters, ketones). It is used as a solvent for paint, varnish, waxes, resins, and similar materials. It was formerly used in liniments and applied topically as a counterirritant. Exposure is usually by inhalation, but has often been used to attempt suicide. Mild or moderate exposures may irritate the eyes, nose, and throat. More severe intoxication from inhalation of vapor causes an initial CNS depression with eventual tissue damage and a disposition to pneumonia. *Cf.* sulfurated oil of turpentine. **2:** turpentine. See also turpentine poisoning. **turpentine gum** (gum terpentine). a gum secreted by the sapwood of pine trees and certain related conifers. See spirits of turpentine. **turpentine oil**. See turpentine.

turpentine poisoning (terebinthinism). poisoning by oil of turpentine. Exposure is usually by inhalation of spirits of turpentine, but it has been ingested in attempts to commit suicide. Serious intoxication from inhalation of vapor causes an initial CNS depression; with continued exposure, the subject may develop chronic nephritis and a disposition to pneumonia. Early symptoms of severe poisoning are a warm or burning sensation in the esophagus and stomach, with nausea, vomiting, headache, confusion, disturbed vision, cramps, diarrhea, and inflammation of the urinary tract. Pulse and respiration become slow, weak, and irregular. Symptoms resembling alcoholic intoxication may also appear. Turpentine poisoning is rarely fatal because ingestion of turpentine is too painful to breathe or swallow for one to take a lethal dose.

turpentine weed. *Gutierrezia microcephala*.

turpeth mineral. mercuric subsulfate.

turtle-designed snake. *Trimeresurus mucrosquamatus*.

turtle poisoning. sometimes very serious poisoning by ingestion of sea turtles. See also Chelonidae, *Chelonia mydas*, *Dermochelys coriacea*, and *Eretomochelys imbricata*.

Tussilago farfara (coltsfoot). a perennial herb (Family Compositae) that is native to Europe, Asia and North Africa, and introduced to North America. It has been used as an antitussive agent, but is probably carcinogenic and is now considered unsafe to use.

tutu. the Maori name for plants of the genus *Coriaria*.

TWA. time-weighted average.

twig snake. *Thelotornis kirtlandii*.

twin-spotted rattlesnake. *Crotalus pricei*.

twinge. a sudden, momentary, sharp pain.

twitch. 1: to jerk spasmodically or convulsively. **2:** a brief spasmodic contraction of a muscle.

twitching. repeated twitches of some part of the body. See fasciculation.

Tykenol. See acetaminophen.

Tylciprine™. tranylcypromine sulfate.

type I pyrethroids. See pyrethroids.

Tylenol. a trade brand of acetaminophen.

Tylinal. diethylpropion.

tympanitis. inflammation of the middle ear.

type II pyrethroids. See pyrethroids.

typhoid bacillus. *Salmonella typhi*.

typholysin. a hemolysin produced by *Salmonella typhosa*.

tyraminase. monoamine oxidase.

tyresin. a principle derived from snake venom and from certain mushrooms. It was formerly thought to be an antidote for snakebite. See also tyrosamine, tyramine.

tyrosine hydroxylase. an enzyme found in cells that synthesize catecholamine. It catalyzes the hydroxylation of L-tyrosine to L-3,4-dihydroxyphenylalanine (L-dopa), which is the initial step, and the rate-limiting step in the metabolic production of catecholamines (dopamine, norepinephrine and epinephrine). It also catalyzes the hydroxylation of phenylalanine to tyrosine.

tyrotoxism. poisoning by a milk product such as cheese.

U. **1:** the symbol for uranium. **2:** unit (def. 2).

U.S. United States of America.

U.S. Adopted Name (U.S.N.). a nonproprietary name approved by the American Pharmaceutical Association, the American Medical Association, and the U.S. Pharmacopeia. The application of such a name to a drug does not imply endorsement. Their use, however, in labeling and advertising is required by law.

U.S. certified color. See color (FD&C color).

U.S. National Institute for Occupational Safety and Health (NIOSH). an agency of the U.S. government that develops recommendations for (1) limits in the workplace of exposure to hazardous substances and conditions; and (2) preventive measures to reduce or eliminate the adverse effects on health of hazardous materials in the workplace.

U.S. Pharmacopeia (USP). See pharmacopeia.

U.S. Public Health Service (USPHS). a bureau of the U.S. Department of Health and Human Services, served by a corps of medical officers presided over by the Surgeon General; concerned with scientific research, domestic and insular quarantine, administration of government hospitals, publication of sanitary reports, and statistics; associated with it are the National Institutes of Health, Centers for Disease Control, and other units.

U.S.N. U.S. Adopted Name.

uao-uao. *Laticauda colubrina.*

ubiquitous. widespread; cosmopolitan; worldwide in distribution, effect or influence; pervasive; pandemic.

udlezinya. *Naja nivea.*

UDLH. immediately dangerous to life or health.

UDMH. 1,1-dimethylhydrazine.

UF. uncertainty factor (See factor).

ukambin. an African cardiotoxic arrow poison from plants of the family Apocynaceae. Its action is similar to that of digitalis or strophanthus.

ukhokhothi. *Thelotornis kirtlandii.*

ular anang. *Ophiophagus hannah.*

ular bakaw. *Trimeresurus wagleri.*

ular bandotan bedor. *Agkistrodon rhodostoma.*

ular bedul. *Naja naja.*

ular biludak. *Naja naja, Agkistrodon rhodostema.*

ular bisa. *Trimereserus gramineus, T. popeorum, T. sumatranus.*

ular chabe. *Maticora intestinalis.*

ular gebuk. *Agkistrodon rhodostoma, Trimeresurus puniceus.*

ular kapac daun. *Agkistrodon rhodostoma.*

ular kapala dua. *Maticora intestinalis.*

ular katam tabu. *Bungarus fasciatus.*

ular kunyett terus. *Ophiophagus hannah.*

ular matahari. *Maticora bivirgata.*

ular nanti bulau. *Trimeresurus wagleri.*

ular puckuk. *Trimeresurus wagleri.*

ular sina. *Maticora bivirgata*.

ular tanah. *Agkistrodon rhodostoma*.

ular tandjon api. *Bungarus flaviceps*.

ular tedong sendok. *Naja naja*.

ular tjabeh. *Maticora intestinalis*, *Maticora bivirgata*.

ular welang. *Bungarus fasciatus*.

ular weling. *Bungarus candidus*.

ulcer. a local defect, excavation, or open sore on the surface of an organ or tissue that is produced by sloughing of inflamed necrotic tissue. Simple ulcers may result from trauma, caustic chemicals, intense heat or cold, or arterial or venous stasis. **chrome ulcer** (tanner's ulcer). an ulcer caused by exposure to chromium or its salts. **indolent ulcer**. a chronic ulcer with hard, elevated edges that does not tend to heal. **tanner's ulcer**. chrome ulcer.

ulcerate. to become affected by or to produce an ulcer.

ulcerated. of the nature of, or afflicted with an ulcer.

ulceration. suppuration on a free surface or the erosion of the surface or lining of an organ that exposes underlying tissue. Thus ulceration of the stomach lining exposes blood vessels and may cause their erosion with bleeding. If ulceration proceeds to the point where the wall of an organ is breached (perforation), the consequences may be serious, even grave.

ulcerative. of, pertaining to, or characterized by an ulcer or ulcers.

ulcerogenic. producing or productive of ulcers. Beryllium, for example, is an ulcerogenic toxicant.

ulexine. the original name given to cytisine isolated from the leguminous shrub *Ulex europaeus*.

ultimate carcinogen. See carcinogen.

ultramarine yellow. C.I. pigment yellow 31.

ultraviolet. ultraviolet radiation.

ultraviolet light. ultraviolet radiation.

ultraviolet radiation (**ultraviolet light; UV radiation; UV light; UV; ultraviolet**). electromagnetic radiation with wavelengths from about 100-3900Å, which is below that of visible light. The sun is the source of natural UV radiation; other sources include mercury vapor lamps and welding arcs. UV radiation is toxic to all forms of life. Acute exposure can seriously damage the eyes and skin (sunburn); UV radiation can cause conjunctivitis, keratitis, and skin cancer, all of which are fairly common among those who work outdoors or are otherwise frequently exposed (e.g., arc welders, sunbathers) during many years. UV radiation is used as a means of sterilization in hospitals.

Ultrocalamus preussi (short-fanged snake). the sole species in its genus (Family Elapidae), found in New Guinea and Seleo, an offshore island. Adults may reach a length of about 75 cm. The head is small and indistinct from the neck; the eyes are small with round pupils. It is apparently not dangerous to humans.

ulupung. *Naja naja*.

um jenaib. *Echis carinatus*.

umbelatine. berberine.

umbellatine. berberine.

Umbelliferae. the former designation for Apiaceae (the carrot, celery, or parsnip family).

umbrella snake. *Bungarus multicinctus*.

umdlezinye. *Naja nivea*.

umi hebi. *Pelamis platurus*.

Ummidia. a common genus of trapdoor spiders (Family Ctenizidae). They are dangerous to humans.

uncomplemented. not united with complement and thus inactive. See complement.

unconscious. **1:** insensible, insensate; nonresponsive or nearly so to most or all sensory stimuli. Having little or no conscious experience or awareness. **2:** in freudian psychiatry that part of the personality comprised of feelings and drives of which one is unaware.

unconsciousness (exanimation). a state of insensibility, unawareness, or reduced awareness of one's environment. The depth of unconsciousness may vary from that of stupor to a state of coma. Normal states of unconsciousness include sleep, hibernation, and torpor (which may be natural or pathological). Abnormal or pathological states of unconsciousness include syncope (fainting), trance, deep states of traumatic or psychologically induced shock (anoxia), and coma. Unconsciousness may be part of the pattern of intoxication from exposure to any of numerous chemical agents, as in insulin shock, narcosis, carbon monoxide poisoning, and acute alcoholic intoxication. *Cf.* coma.

uncoupling agent (uncoupler). an uncoupler or oxidative phosphorylation — a poisonous substance (e.g., arsenate, dicumarol, and dinitrophenol) that allows mitochondrial oxidation (respiration) to proceed without phosphorylation and the production of ATP.

undernourishment. See malnutrition.

Uniform Substances Act. a U.S. federal law that provides the legal basis for the classification, regulation, control, distribution, sale, and use of controlled substances. See controlled substance.

unintentional effect. See effect (unintended effect).

unit. **1a:** one; a single item. **1b:** of, pertaining to, or equivalent to a unit (e.g., a unit measure). **2a:** U; a quantity taken as a standard of measurement. **2b:** a standard of physical quantity of a physical entity, material, energy, or quality. Examples are units of mass, length, electrical charge. **2c**. the quantity of a vaccine, drug, poison, serum, whole blood for transfusion, of vitamins, or other agents required to produce a specific effect. **3:** a group of items or individuals treated as a whole (e.g., a pod of whales, a squad of soldiers). See also International System of Units. **absolute unit**. a unit, the value of which does not change regardless of time, place, or conditions. **alexin unit**. complement unit. **amboceptor unit**. hemolysin unit. **antigen unit**. the smallest amount of antigen that will fix 1 unit of complement in the presence of specific antiserum. **antitoxic unit**. antitoxin unit. **antitoxin unit** (immunizing unit; standard antitoxin unit; antitoxic unit). a unit that expresses the strength or activity of an antitoxin. The various units were originally defined biologically, but are now compared to a weighed standard specified by the U.S. Public Health Service and the World Health Organization. The American antitoxin unit is that which is equivalent to a 1/6000 g of a reference standard maintained at the National Institutes of Health in Washington, D.C. This material is the a dried unconcentrated horse serum antitoxin preserved since 1905. Both the international and American antitoxin units are equivalent. **antivenene unit**. antivenin unit. **antivenin unit** (antivenene unit). a unit that expresses the strength or activity of an antivenin (an antitoxin or serum containing a mixture of antitoxins prepared from the blood of animals immunized against a particular venom). **base unit**. See SI base unit. **biological standard unit**. a specified quantity of a biologically active material (e.g., venom, antitoxin, antibiotic, enzyme, hormone) based on reference material that is usually dried and maintained at a designated organization in the dark, in an atmosphere of nitrogen at 10°C. **cat unit**. obsolete; that amount of digitalis per kilogram of body weight of a domestic cat that is just sufficient to kill upon slow, continuous injection into a vein. **centimeter-gram-second unit**. CGS unit. **C.G.S. unit**. CGS unit. **cgs unit**. CGS unit. **CGS unit** (centimeter-gram-second unit; cgs unit; C.G.S. unit). any unit of the centimeter-gram second system. **complement unit** (alexin unit). the smallest amount (highest dilution) of complement that will completely hemolyze a standard suspension of erythrocytes (1 ml of a 5% emulsion of washed erythrocytes) in the presence of a hemolysin unit. **digitalis unit**. any of several units formerly used in bioassay of digitalis preparations and named according to the animal in which it was determined, as cat unit (given above as an example). **diphtheria antitoxin unit**. approximately that amount of antitoxin which will spare the life of a 250 gm guinea pig for at least 96 hours following subcutaneous injection of a mixture that is 100 times the minimum lethal dose of diphtheria toxin. **enzyme unit**. that quantity of an enzyme which will catalyze the reaction of 1 micromole of substrate or 1 microequivalent of the substrate group (if more than one bond is involved in the reaction) per minute under standard conditions of temperature (30°C) and at optimal pH and concentration of substrate. **hemolysin unit** (hemolytic unit; amboceptor unit). the smallest amount or highest dilution of inactivated immune serum (hemolysin) that sensitizes a standard suspension of erythrocytes (1 ml of a 5% emulsion of washed erythrocytes) so that standard complement will cause complete hemolysis. **hemolytic unit**. hemolysin unit. **hemorrhagin unit**. the least amount of a snake venom that will cause hemorrhaging in the vascular network of a three-day-old chick embryo. See hemorrhagin. **hereditary unit**. a gene. **immunizing unit**. antitoxin unit. **international digitalis unit**. the activity of 0.1 g of the international reference standard of powdered digitalis. **international unit**. any unit defined and adopted by the International Conference for the Unification of Formulae. Used in the measurement of hormones, enzymes, and some vitamins. **SI base unit**. any of the fundamental units of mass (kilogram), length (meter), time (second), electricity current (ampere), temperature (Kelvin), amount of material (mole), and luminous intensity (candela) as defined in the International System of Units (SI). **SI unit**. any unit specified by the International System of Units. **standard antitoxin unit**. antitoxin unit. **tetanus antitoxin unit,** the antitoxin activity of 0.3094 Meq of standard tetanus antitoxin. This is approximately the amount of antitoxin which will spare the life of a 350 gm guinea pig for at least 96 hours following injection of a mixture that is 100 times the minimum lethal dose of tetanus toxin. **toxic unit** (toxin unit). **1:** obsolete; formerly synonymous with minimal lethal dose. **2:** the smallest dose of toxin that will kill a guinea pig weighing about 250 gm. in 3-4 days. **3:** current; the amount of standard antitoxin with which a specific toxin combines. **toxin unit**. toxic unit. **urotoxic unit**. the smallest amount of urotoxin that will kill an animal weighing 1 gm. **USP unit**. any unit defined and adopted by the U.S. Pharmacopeia.

univalve. **1:** any mollusk having a one-piece shell. **2:** pertaining to a one-piece shell.

unknown substances poisoning. a poisoning in which the nature of the poison is unknown and the signs and symptoms are not recognized as due to a particular substance. Treatment is limited in such a case and specific antidotes cannot be administered.

unmoonarbomma. *Pseudechis australis*.

unobhiya. *Hemachatus haemachatus*.

unobibi. *Hemachatus haemachatus*.

unofficial. **1:** unauthorized, unapproved, casual. **2:** denoting a drug that is not listed in the U.S. Pharmacopeia or the National Formulary.

unphysiologic. **1a:** pertaining to physiologically abnormal conditions with an organism. **1b:** physiologically abnormal amounts of a substance that is normally present in a living organism. **1c:** pertaining to the administration of substances to an organism in unphysiologic amounts.

unsaturated compound. an organic compound that contains at least one double or triple carbon to carbon bond. Such bonds are relatively weak and as a result, unsaturated compounds readily undergo additional reactions with the formation of single bonds. *Cf.* saturated compound.

unslaked lime. calcium oxide.

unsoundness. unhealthiness, shakiness. In livestock (especially horses), unsoundness is any departure from normal that renders the animal unfit for service.

unstable. erratic, liable to change, changeable, not constant; reactive, explosive.

unsteady gait. an erratic manner of walking.

***unsym*-dimethylhydrazine**. 1,1-dimethylhydrazine.

unthriftiness. poor weight gain of an animal or human relative to that expected from the quantity and quality of the diet.

upas. **1:** upas tree, *Antiaris toxicaria*. **2:** any of a number of poisonous plants of the genus *Strychnos* used to manufacture arrow poisons.

3: a poisonous mixture of tar-like consistency that is usually made from the boiled down sap of *A. toxicaria* and used primarily as an arrow and dart poison. See also antiar. **4:** used in a social context in reference to a poisonous or harmful situation or institution.

Upeneus arge (surmullet; goatfish). a marine bony fish (Family Mullidae) of Polynesia and Micronesia. See ciguatera.

upholstery cleaner. See cleaner (upholstery cleaner).

upholstery shampoo. See cleaner (upholstery cleaner).

upper respiratory system (upper respiratory tree). upper part of the breathing system of terrestrial and semiterrestrial vertebrates consisting of the nose, throat, larynx (in mammals), trachea, and bronchial tubes.

upper respiratory tree. upper respiratory system.

uptake. absorption of a substance (e.g., water, nutrients, a toxicant) as by a cell, tissue, organ, or compartment of a living system, with temporary or permanent retention. Sometimes applied to a compartment of the environment or to a whole organism, especially where absorption occurs through the integument.

uracil. a pyrimidine base found in RNA.

uracil mustard (5-[bis(2-chloroethyl)amino]-2,4-(1*H*,3*H*)-pyrimidinedione; 5-[bis(2-chloroethyl)amino]-uracil; uramustine; desmethyldopan). an alkylating type of cytotoxic drug used to treat certain types of malignant tumors. This super toxic drug must be handled with extreme care. One should wear protective gloves, a protective mask, protective glasses, and handle only in a well-ventilated hood. The work area must be thoroughly cleaned after handling and the hands should be thoroughly washed with soap and water for several minutes.

uramustine. uracil mustard.

uranium (symbol, U)[Z = 92; A_r = 238.03]. a highly toxic, weakly radioactive, silvery white, lustrous element of the actinoid series of metals. There are three naturally occurring isotopes, ^{238}U, ^{235}U, and ^{234}U, with respective abundances of 99.283, 0.711, and 0.005%, found mainly in pitchblende. ^{235}U is readily fissionable and was the first substance demonstrated to support a self-sustaining chain reaction and is used in atomic and hydrogen bombs and as a fuel in nuclear reactors. ^{234}U is a fuel in nuclear reactors. ^{238}U is a source of fissionable ^{239}Pu. Uranium and its salts are highly toxic. Exposure may cause dermatoses, renal damage, acute necrotic arterial lesions, and death. Inhalation of fine particles (ca. 1μ diam.) poses a carcinogenic risk.

uranium nephritis. a nephritis produced experimentally by the administration of uranium nitrate.

Uranoscopidae (stargazers). a family of tropical, bottom-dwelling, marine fish, often buried in mud or sand with only the eyes and mouth protruding. The head is cuboidal, with eyes on the upper surface of the head and a nearly vertical mouth. Some, perhaps all, are venomous. See *Uranoscopus scaber*.

Uranoscopus scaber (stargazer). a bottom-dwelling fish of the eastern Atlantic and Mediterranean Sea (Family Uranoscopidae). The venom apparatus consists of two cleithral spines and their associated venom glands. The sting may be fatal.

urari. curare.

uratemia. the presence of urates, especially sodium urate, in circulating blood.

uratosis. any morbid condition due to the presence of urates in the blood or other tissues.

Urauschlange. *Naja haje*.

urchin. sea urchin.

urea. the diamide of carbonic acid, urea is a crystalline solid, $CO(NH_2)_2$, it is a white crystalline substance or powder. It is the primary nitrogenous waste of elasmobranchs, amphibians, and mammals where it occurs in blood, lymph, and urine. Urea and carbon dioxide are the end products of protein metabolism. It is produced in the liver chiefly from ammonia released during the deamination of amino acids; it is also formed from ammonia compounds in the body and can be formed directly from arginine. Urea normally comprises 80-90% of total urinary nitrogen in humans. Urea was the first organic compound to be synthesized in the laboratory. Aside from its metabolic importance, urea is produced industrially high volume. Industrial applications include its use in the manufacture of plastics and resins, herbicides, fertilizers, animal feeds, pharmaceuticals, adhesives, dentifrices, cosmetics, flame retardants, as a stabilizer in explosives, and in the preparation of biuret. See also uremia, ureotelic, uric acid.

urea nitrogen. See blood urea nitrogen.

Urechites suberecta (yellow nightshade; wild allamanda). an herb (Family Solanaceae), the fruit of which is toxic, causing a burning sensation in the mouth and throat, drowsiness, paralysis, convulsions, and death due to respiratory collapse. See urechitin, urechitoxin.

urechitin. a toxic glycoside, $C_{28}H_{42}O_8$, produced by *Urechites subercta*, *q*.v.

urechitoxin. a toxic glycoside, $C_{13}H_{20}O_5$, produced by *Urechites subercta*, *q*.v.

Urecholine®. bethanechol chloride.

uremia. **1a:** the presence of abnormally high amounts of urine and other nitrogenous wastes in blood and other tissues. **1b:** uremic poisoning, the complex of symptoms associated with uremia (def. 1a). Uremia is a potentially fatal toxicosis. **renal uremia**. that due to renal insufficiency or failure.

uremic. of or pertaining to uremia.

uremic poisoning. **1:** uremia (def. 1b). **2:** any poisoning that causes uremia.

uremigenic. **1:** uremic in origin or caused by uremia. **2:** causing uremia.

ureotelic. **1a:** of or pertaining to the excretion of nitrogen in the form of urea. **1b:** pertaining to or designating animals that excrete most excess nitrogen in the form of urea. Examples are elasmobranchs, amphibians, and mammals. *Cf.* uricotelic.

urethan. urethane.

urethane (carbamic acid ethyl ester; ethyl carbamate; ethyl urethane; ethyl urethan; urethan). a moderately toxic, synthetic white powder or crystalline ester amide, $CO(NH_2)OC_2H_5$. It is odorless, tastes like saltpeter, and forms solutions that are neutral to litmus. It is the repeating unit of polyurethane resins. It is used as antineoplastic agent, a solubilizer and cosolvent for fumigants and pesticides, and as a veterinary anesthetic. Its use is incompatible with alkalies, acids, antipyrine, camphor, chloral hydrate, menthol, salol, or thymol. Urethane is moderately toxic to laboratory mice. Urethane is a hepatotoxicant, causing liver necrosis; an immunosuppressive agent; and a direct-acting carcinogen.

urethylane. *N*-methyl carbamate.

urgilan. proscillaridin.

urginea. the bulbs of *Urginea indica* (Indian squill) and *U. maritima*. It is the source of squill.

Urginea. a genus of Old World bulbous herbs (Family Lilaceae), chiefly of the Mediterranean region; discovered in Algeria. ***U. burkei***. toxic properties and component glycosides are similar to those of *U. maritima*, *q.v.* See also scillarenin. ***U. indica*** (Indian squill). a species found in the Orient. It contains one or more cardiac glycosides and is used as an expectorant, a diuretic and cardiac stimulant. The young dried bulbs are used to prepare squill. See urginea. ***U. maritima*** (= *Scilla maritima*) (Mediterranean squill; red squill; medicinal squill; sea onion). a bulbous herb (Family Lilaceae), of southern Europe and northern Africa. The bulbs are cardiotoxic and widely used in commercial rodenticides (as red squill powder) because they are readily consumed (and assimilated) by rats, whereas they are distasteful to humans and species other than rats. The bulbs contain several cardiotoxic glycosides, including proscillaridin, scillaren A, scillaren B, and scilliroside. See also proscillaridin, red squill powder, rodenticide, *Scilla*, scilliroside, scillism. ***U. rubella***. a species that contains rubellin, *q.v.*

uric acid (2,6,8-trioxypurine; lithic acid; uric oxide). an odorless, tasteless, white, crystalline compound, CHNO, that is nearly insoluble in water; insoluble in ethanol and ether; soluble in hot, concentrated sulfuric acid; and soluble in glycerol, aqueous solutions of alkali hydroxides, sodium acetate, and sodium phosphate. When heated, it emits toxic fumes of hydrogen cyanide. Uric acid is the primary nitrogenous waste of birds, reptiles, and certain invertebrates, especially insects. It is the endproduct of purine metabolism in primates including humans, some dogs, and certain other mammals. If not excreted, uric acid remains in the body unchanged and may form renal calculi and gouty concretions. Decreased elimination of uric acid may indicate lead poisoning; but may also be due to nephritis, chlorosis, or a protein-free diet. See also uricotelic.

uric oxide. uric acid.

uricotelic. **1a:** of or pertaining to the excretion of nitrogen in the form of uric acid. **1b:** pertaining to animals that excrete most excess nitrogen in the form of uric acid. Birds are uricotelic. *Cf.* ureotelic.

urinalysis. the determination of the composition of urine.

urinary syndrome of cattle and sheep (redwater disease). this disease is a hemolytic anemia resulting from consumption of rape, *Brassica napus*. The symptoms and lesions are almost indistinguishable from copper poisoning. Affected animals stand apart and do not feed. The urine is dark colored and foams on striking the ground. Visible mucous membranes and the carcass are pale or icteric. Blood hemoglobin levels are very low; the liver is unusual in appearance and necrotic; the kidney is congested and dark appearing. In recently postpartum dairy cattle, this syndrome is indistinguishable from puerperal hemoglobinuria (See hemoglobinuria).

urine. a fluid (e.g., in mammals, amphibians), solid or semisolid (e.g., in birds, reptiles) excretion from the kidneys of most vertebrate animals. It contains the endproducts of protei n metabolism. See urea, uric acid. Urine is often a key indicator of disease or poisoning (e.g., via changes in color, clarity, pH, odor, volume discharged, presence of erythrocytes or other cellular material, chemical makeup). The following materials, when present in urine (or present in abnormal amounts) are indicators of pathological states: albumin, bile, blood casts, hemoglobin, cystine, epithelial cells, glucose, ketone bodies, pus, bacteria, fat, pus, mucous casts, proteins, proteoses, sulfanilamide derivatives. For examples indications of poisoning see acetylethyltetramethyltetralin, albuminuria, cacodylic acid, carbamate, diazepam, diuretic, hemoglobinuria (toxic hemoglobinuria), iodine poisoning, jaundice (toxic jaundice), *Juniperus virginiana*, ketosis, poisoning (acute mercury poisoning), ornithuric acid, phenylglycuronic acid, purpurin (def. 2), red water, savin, septic shock, silvex, test (Anstie's test), tetrahydrocannabinol, urinary syndrome of cattle and sheep, venom (hydrophid venom).

Urobatis. a nearly cosmopolitan genus of stingrays (Family Dasyatidae). All large freshwater varieties are dangerous. Penetration of the skin by the barb in the tail usually causes intense local pain, dizziness, weakness, generalized cramps, sweating, and hypotension. Symptoms may persist for 24-48 hr. Penetration of the chest or abdomen is uncommon, but may be fatal. In all cases, the condition of the affected individual may be complicated by septicemia. Thus, in general, the wound should be thoroughly irrigated to remove foreign matter and soaked in water as hot as can be tolerated. Early removal to a medical facility is highly desirable, to secure suitable treatment which usually includes surgical debridement and closure of the wound; tetanus prophylaxis; antibiotics.

urocanylcholine. murexine.

Urodela (Caudata; salamanders, newts, etc.). a subclass of Amphibia (sometimes treated as an order of the subclass Lissamphibia) that includes newts and salamanders, which are almost entirely restricted to the Northern Hemisphere. The body is long, tailed, and with limbs that are short, weak, and of similar size. Most are largely or wholly aquatic but some (e.g., *Triturus*) are terrestrial as adults, although confined to moist situations. Eggs are deposited in water or in moist, protected environments. Adults have lungs; the gilled larvae resemble the adults, but in some cases the larval gills are retained and the lungs tend to atrophy in the adults. Certain salamanders of the genus *Salamandra* are poisonous as are newts of the genera *Taricha* and *Triturus*. The steroid alkaloid toxins secreted by the epidermal glands of salamanders include samanin, samandaridine, samandarine, samandenon, cycloneosamandaridin, cycloneosamandion, samandarin, and samandaridin, samandarone. See also tarichatoxin.

urofolitrophin. FSH.

urolophid. any of the round stingrays (Family Urolophidae), especially those of the genus *Urolophis*. All are venomous and are dangerous to humans. The sting is similar to the dasyatid type and the caudal appendage is short, muscular, and capable of delivering a well-directed sting. See stingrays.

Urolophidae (round stingrays). a family of venomous, stinging, cartilaginous fish, the round stingrays. They are dangerous to humans; have a short muscular caudal appendage that bears a sting similar to the dasyatid type. They are able to deliver a forceful, well-directed sting. See also stingray, urolophid.

Urolophis. a genus of round sting rays (Family Urolophidae). They are venomous and dangerous to humans.

urophile. an organism that thrives in habitats rich in ammonia.

urophilic. able to thrive in habitats rich in ammonia.

urophily. the quality of an organism that allows it to thrive in habitats rich in ammonia.

urotoxia. the toxicity of urine.

urotoxicity. the toxicity or toxic nature of urine.

urotoxin. a toxin in the urine.

ursin. arbutin.

ursini's viper. *Vipera ursinii*.

urtica. **1:** a wheal or pomphus. **2:** any herb of the genus *Urtica*.

Urtica (nettle). a widely distributed genus of annual and perennial herbs (Family Urticariaceae) introduced to North America from Eurasia. They bear stinging hairs that penetrate the skin and inject irritant chemicals that include formic acid and histamine. This produces an almost immediate nonimmunologic contact urticaria, *q.v.*, with a burning sensation and pruritis. These plants are technically venomous since the tips of the hairs break off on contact and inject the chemicals into the skin. ***U. chamaedryoides*** (nettle). a slender nettle with stinging bristles, it occurs in rich woods, bottomlands, and waste areas from Florida to Texas and Mexico and northward to West Virginia, Kentucky, Missouri, and Oklahoma. It is the cause of nettle syndrome, *q.v.* The bristles contain toxicologically significant amounts of acetylcholine and histamine. ***U. dioica*** (stinging nettle). contact with the stiff bristles ("stinging hairs") on the leaf will sting exposed skin for a short while. The active principle is thought to be a mixture of acetylcholine, histamine, and serotonin. ***U. procera*** (nettle). occurs in thickets and along roadsides in much of eastern North America. The stem is only slightly, if at all, bristly. In North America, it occurs on roadsides, in waste areas, and other areas of second growth from Newfoundland to Manitoba, southward to Nova Scotia, New England, and Virgina, and west to Illinois. See also *Dendrocnide moroides*, *Hylesia*.

urticant. **1a:** producing or able to produce a wheal or an itching or stinging sensation, or both. **1b:** an urticarial reaction in the skin. **2:** an agent that produces an urticarial reaction in the skin. See urticaria.

urticaria (hives; nettle rash; urtication). a skin eruption of short duration (a few hours or a few days) that is often associated with allergy or contact with an irritant; proximate causes may include irritation of internal mucous membranes, emotions, menstruation, certain external allergens or irritants, foods, certain drugs, physical agents, insect bites, serum sickness, pollens, drugs, or neurogenic factors. The condition is characterized by the sudden appearance of smooth, red, slightly elevated, itching patches or wheals up to several inches across. They are reddish, but usually paler than the surrounding skin. See also lepidopterism, *Urtica*, *Thaumetopoea pityocampa*). **acute urticaria**. that marked by the abrupt appearance of extensive wheals of brief duration (e.g., from insect bites or the administration of drugs). **aquagenic urticaria**. that due simply to exposure to water. **chronic urticaria** (recurrent urticaria). a form marked by recurrent episodes during a period of weeks or months. **contact urticaria**. urticaria due to contact with an external agent such as a nettle (See *Urtica*) **endemic urticaria**. urticaria endemica. **epidemic urticaria**. urticaria endemica. **recurrent urticaria**. chronic urticaria. **urticaria endemica** (caterpillar poisoning; urticaria epidemica; endemic urticaria). a rare, severe form of urticaria caused by the hairs of certain caterpillars and occurring as an endemic or sometimes an epidemic. See also *Thametopoea pityocampa*. **urticaria epidemica**. urticaria endemica. **urticaria medicamentosa**. urticaria caused by certain drugs.

urticaria endemica. See urticaria.

urticaria epidemica. See urticaria.

urticaria medicamentosa. See urticaria.

urticarial, urticarious. pertaining to or marked by urticaria.

urticate. **1:** to produce an urticarial reaction in the skin. **2:** marked by the presence of urticaria-like skin eruptions.

urtication. **1:** urticaria (in part). **2:** flogging of a body part with nettles to produce counterirritation. **3:** a burning or itching sensation resembling that of urticaria. **4:** an eruption of itching wheals.

uruta. *Bothrops alternatus*, *Bothrops neuwiedi*.

urutu. *Bothrops alternatus*.

USDA. U.S. Department of Agriculture.

Usnic acid (usninic acid). a yellow, crystalline solid, $C_{18}H_{16}O_7$, that is insoluble in water, but slightly soluble in ethanol and ether. It is a tricyclic compound with antibiotic properties that occurs in many lichen species and was isolated early from *Usnea barbata*, an epiphytic lichen. See also *Parmelia molliuscula*.

usninic acid. usnic acid.

USP. U.S. Pharmacopeia. See pharmacopeia.

uterine. of or pertaining to the uterus.

uterine cycle. the regularly occurring changes that take place in the uterine lining of female mammals during the reproductive cycle.

uteritis. inflammation of the uterus.

uterus. **1:** the organ in female mammals in which the fetus develops, is maintained, and nourished prior to birth. **2:** an enlarged, modified portion of the oviduct (e.g., as in birds) of an animal in which the eggs or young develop prior to birth or laying.

utricle. an enlarged cavity comprising part of the membranous labyrinth of the inner ear.

UV light. ultraviolet radiation.

UV radiation. ultraviolet radiation.

UV. See ultraviolet radiation.

Uvasol™. arbutin.

V. the symbol for vanadium.

V_d. volume of distribution.

v-onc. symbol for a viral oncogene.

vaccinate. to confer active immunity by vaccination.

vaccination. the administration of antigens to an animal in order to induce active immunity by stimulating the production of adequate levels of specific antibody and lymphocytes carrying cell-bound antibody to be protective in the event of future natural exposures to the antigen. Antigens are ingested or injected in the form of a vaccine.

vaccine. a suspension of antigen (e.g., dead or weakened pathogenic bacteria or viruses; proteinaceous toxins from pathogenic bacteria) that will, when introduced into animal tissues, provoke the formation of antigen-specific antibodies. *Cf.* antivenin. See also vaccination.

Vaccinium (blueberry; cowberry; cranberry). a genus of widely distributed deciduous, evergreen shrubs (Family Ericaceae). The leaves of a number of species contain arbutin [e.g., *V. angustifolium* (low bush blueberry), *V. corymbosum* (high bush blueberry), *V. macrocarpon* (cranberry) *V. vitis-idaea* (cowberry)].

vagina. a sheath-like passage of the female reproductive tract in mammals that extends from the vulva to the cervix (neck of the uterus). Vaginal absorption is an uncommon, but sometimes important route of entry of poisons. See potassium permanganate.

vaginitis. inflammation of the vagina.

vagus nerve. a long cranial nerve that arises in the base of the brain and passes to the chest and abdomen. It contributes to the regulation of heart rate, breathing, swallowing, digestion, and many other bodily functions in vertebrates.

valakachiyan. *Enhydrina schistosa.*

valakadyen. *Enhydrina schistosa.*

valid. authentic, legitimate, sound, effective; verifiably correct.

validate. to verify, authenticate, certify, substantiate, confirm.

validation. **1a:** the process of establishing validity; authentication; the process of verifying. **1b:** the process of making valid. **2:** as part of the risk management process, validation is the evaluation of the chosen management strategy by means of post-decision monitoring or epidemiological investigation. This is to assure both legal and regulatory compliance.

validity. **1a:** the property of being valid. **1b:** the degree to which information or a data set is true and correct. **2:** the extent to which an observation reflects the true situation. **3:** an index of the effectiveness of a test or procedure in measuring what it claims to measure.

Valium™. diazepam.

vanadic acid (HVO_3; metavanadic acid). an oxidation product of vanadium which may cause chronic poisoning. See vanadium.

vanadic acid anhydride. vanadium pentoxide.

vanadic anhydride. vanadium pentoxide.

vanadium (symbol, V)[$Z = 23$; $A_r = 50.9414$]. a light gray or white transition element present in small amounts in complex ores. It exists in compounds in the +3, +4, and +5 oxidation states (most often +5). It is tumorigenic and questionably carcinogenic. Vanadium and its compounds (some of which are highly toxic) are of environmental concern because of their high levels in residual fuel oils and subsequent emission as small particulate matter from the combustion of these oils in urban areas. Vanadium occurs as chelates of the porphyrin type in crude oil and it concentrates in the higher boiling fractions during the refining process. The potential for industrial exposure of workers is significant.

vanadium pentoxide (vanadic acid anhydride; vanadic anhydride). a toxic, yellow to red, crystalline powder, V_2O_5, that is insoluble in water, but soluble in acids and alkalies. It is an irritant to the skin, eyes, and lungs; and is poisonous by all routes of exposure, although exposure by inhalation is most common. Exposure by inhalation commonly causes bronchospasm, a greenish-black tongue, chest pain, palpitation, bronchitis, and bronchial pneumonia; it may also cause asthma, coughing, and dyspnea. Severe acute exposure may also affect the gastrointestinal tract, kidneys, and nervous system. Teratogenic and reproductive effects have been demonstrated experimentally.

vancomycin. an antibiotic that blocks cell wall formation in bacteria.

Vapo-Iso™. isoproteronol hydrochloride.

vapor. **1a:** any diffuse or gaseous matter (e.g., mist, haze, smoke, steam) dispersed in air, reducing its transparency. **1b:** a liquid that has become gaseous. **2:** any substance in a gaseous state. **3:** a gasified liquid or solid. **water vapor**. the vapor phase of water; water droplets dispersed in air; steam.

vapor dispersion. the movement of vapor clouds in air due to wind, gravity, spreading, and mixing.

vapor capture system. any combination of hoods and ventilation systems that capture or retain organic fumes in order that they may be conveyed to an abatement or recovery device.

vapor plume. a flue gas that is visible because it contains water droplets.

vaporability. the capacity to vaporize or be vaporized.

vaporable (vaporizable). pertaining to a substance that can be vaporized.

vaporizable. vaporable.

vaporization. the process of vaporizing or of being vaporized.

vaporize. **1:** to cause something to become vapor. **2:** to become vapor.

variability. the state of being variable in form or quality; mutability.

variable. **1a:** changeable, alterable, mutable; apt to vary or change; subject to variation. **1b:** that which is characterized by variation or variability. **1c:** having the characteristics of a variable (e.g., of a numerical variable, a measurement). **2a:** not conforming to a single type or value. **2b:** aberrant, inconstant (as is sometimes the case with biological variables or characteristics of species, population, etc.). **3:** that which is inconstant. **4:** a property with respect to which individuals within a sample differ in some discernible way. **5a:** any term to which a number of different numerical values may be assigned. See also factor, accuracy, precision. **5b:** a datum that can assume any of a set of values. **confounding variable**. **1a:** a variable that is puzzling or confusing; one that can obscure an issue or the relationships among a set of variables or that leads to errors in inferences based upon a set of variables thought to be relevant to a given problem. **1b:** a variable that is inconsistent with variables to which it is thought to be related. **continuous variable**. a variable that can theoretically take on any value intermediate between any two specified values. **demographic variable** (population variable). a variable that is descriptive of a population (e.g., mortality, natality, emigration rate, immigration rate). **discontinuous variable**. a variable that can take on only discrete values. Thus, for example, a variable that can assume only whole numbers is discontinuous. Variables such as rank and gender are discontinuous. **environmental variable**. any variable or factor in a stated environment (e.g., temperature, humidity, pollution index, concentration of a toxic chemical). **population variable**. demographic variable.

variance. **1:** the quality or condition of being varied. **2:** the mean squared deviation from the mean; the square of the standard deviation. **3:** government permission for a given party to delay implementation or to depart from the literal requirements of a given law, ordinance, or regulation due to a unique hardship or special circumstances. A variance does not free a party from substantial compliance in accord with the spirit and purpose of the law.

variant. an individual, item, process, or group that shows a pronounced deviation or divergence from type in form, quality, or behavior.

variation. **1a:** change, modification **1b:** variety, diversity. **1c:** differences among individuals. Such differences may be heritable or may be largely environmental in origin. **2:** a variant, variety, type. **3a:** divergence from a type or norm with respect to certain characteristics. **3b:**deviation from the mean. **age variation**. variation as a function of age. Variation in the production, distribution, and effectiveness of xenobiotic-metabolizing enzymes as a function of age. **continuous variation**. genetic variation due to the expression of many genes (polygenes), such that each gene has a small effect. Variation in height and weight are examples, although environmental factors also play an important role in determining the variation in these characteristics. *Cf.* quantitative variation. **discontinuous variation**. genetic variation due to the expression of one or a few major genes, which may exist in the population in two or more allelic forms. These alleles give rise to alternative phenotypes that differ sharply (often qualitatively) from each other. *Cf.* qualitative variation. **environmental variation**. that due largely to the environment to which an individual organism, population, or community is exposed. Such variation exhibits low heritability and is thus said to be phenotypically plastic. It may result from any of a constellation of factors (biotic or abiotic) and may or may not be dependent on population density. In the chemical sphere environmental variation may relate to nutrient status of the environment or the presence of biologically available toxic chemicals. **genetic variation**. heritable variation; it is ultimately due to mutation, but is generally a more immediate result of gene recombination in sexually reproducing organisms. **individual variation**. differences in characteristics (e.g., degree of response to a poison) among individuals within a species, population, or community. Such variation may derive largely from the genetic constitution of the individual (genotype), from the environment, or to interactions between them. Humans are characterized by great individual variation in many characteristics. Thus, the response of one individual to a poison may be extreme, while the response of another may be quite limited. The causes of such variation are sometimes known, as in the case of hypersensitivity, but in many cases variation is idiosyncratic and inaccessible to investigation. **interspecies variation** (interspecific variation; species variation). variation between species. Differences in the toxicity of a chemical among even closely related species are sometimes dramatic; while usually quantitative, such variation is sometimes qualitative. Variations in susceptibility and specific responses to toxicants, for example, among related species can be striking, even extreme. Such variations my be due to specific differences in physiology, behavior, rates and routes of absorption of particular substances, differences in composition of intestinal microflora, and quantitative or qualitative differences in the metabolism of xenobiotics. See also specificity, susceptibility. **intraspecies variation** (intraspecific variation). variation within a species. Variation within a population (e.g., as a function of sex, age, strain or race, morph, dominance, health, nutritional status) with respect to the toxicity and effects of a given xenobiotic can be very substantial. **qualitative variation**. that due to differences in kind; differences that cannot be expressed numerically (e.g., red and gray color phases in a population of owls). **quantitative variation**. that due to differences in degree; differences that can be measured or expressed numerically (e.g., body weight, length). **sexual variation**. variation between male and female individuals (and sometimes sexual variants). Sexual differences in susceptibility and response to xenobiotics are common. **species variation**. interspecies variation.

variegated thistle. *Silybum marianum.*

variety. a subdivision of a species. The term is usually applied loosely to a number of a groups within the species. It is most often applied to a cultivated form of a plant and occasionally to a breed of animal. Cultivated varieties of horticultural plants are termed cultivars. Cultivars do not have latinized names. Some varieties of plants have, for example, been developed for their resistence to disease.

Variola louti (seabass). a tropical, marine, bony fish of the Indo-Pacific region. See ciguatera, seabass.

vas deferens (sperm duct). the tube that carries the sperm from the epididymis of the testes to the seminal vesicles (alongside the prostate gland) where sperm is stored in mammals; or to the exterior; or to the urogenital canal or cloaca of animals other than mammals.

vascular bundle. stele.

vascular cylinder. stele.

vasculitis. inflammation of blood vessels.

vasoconstriction. decreased cross-sectional area of blood vessels (usually arterioles or capillaries) due to the contraction of the smooth muscles in their walls. It can be induced by the stimulation of vasoconstrictor nerve fibers of vasomotor nerves that serve the blood vessel walls, or by injection locally of vasopressors such as epinephrine. Reflex vasoconstriction occurs in response to pain, loud noises, fear, deep breathing, or a fall in blood pressure. See also vasomotor nerve, angiotensin II, scopolamine, cocaine hydrochloride, hypertension, nitrite poisoning, palytoxin, pituitary gland (posterior lobe of the pituitary gland), sympathetic nervous system, vasodilatation, vasohypertonic.

vasoconstrictor. **1:** causing vasoconstriction, *q.v.* **2:** any agent (e.g., a drug or nervous stimulation) that causes vasoconstriction. See vasoconstriction.

vasoconstrictor drugs. medications that cause vasoconstriction, *q.v.*

vasodentine. a hard, bone-like material. See stingray.

vasodepression. reduced tone of the smooth muscles in blood vessels with vasodilation and a resultant drop in blood pressure. See vasodilatation.

vasodepressor. **1:** causing vasodepression, *q.v.* **2:** an agent that causes vasodepression.

vasodilatation (vasodilation; phlebarteriectasia). increased cross-sectional area of small blood vessels, especially arterioles, due to reduced tone of the smooth muscles in their walls; the process of effecting such an increase. Stimulation of cholinergic vasodilator nerve fibers can induce vasodilatiation, as can reduction of the sympathetic vasoconstrictor innervation, injection of acetylcholine, histamine, or kinins (See bradykinin), hypertension, exercise, high ambient temperature, temperature, certain types of stress. It can also be induced by any of numerous poisons (See, for example, disulfiram, nitrite, nitroglycerine, protoveratrine, sodium azide).

vasodilative. **1:** of a chemical with the ability to dilate blood vessels (especially arterioles and capillaries. **2:** vasodilator.

vasodilator. **1:** vasodilative; causing vasodilatation. **2:** an agent that causes vasodilatation.

vasoinhibitor. vasodilator.

vasoinhibitory. See vasodilator.

vasomotor. pertaining to the contraction or relaxation of the smooth muscles of blood vessels, especially of arterioles or capillaries, with concomitant constriction or dilation of blood vessels. See vasoconstriction, vasodilatation.

vasomotor nerve. an autonomic neuron that induces an alteration in the diameter of a blood vessel. Included are vasoconstrictor neurons, which are sympathetic in origin and vasodilator neurons which are far less common; these may be sympathetic or parasympathetic in origin. See vasoconstriction, vasodilatation.

vasopressive. having the action of a vasopressor.

vasopressor. **1:** a drug or other agent that raises blood pressure by causing vasoconstriction, especially of capillaries and arterioles. **2:** producing or having the capacity to produce vasoconstriction with a concomitant rise in systemic, usually arterial, blood pressure.

vasostimulant. **1:** able to stimulate a vasomotor response. **2:** an agent that excites or increases the actiion of vasomotor nerves. **3:** vasotonic.

vasotonia (angiotonia). the tension or tone of blood vessels, especially of arterioles.

vasotonic. **1:** angiotonic, of or pertaining to vascular tone. **2:** vasostimulant; an agent that increases vascular tension or tone.

VC. vinyl chloride.

VCR. vincristine.

vector. **1:** an organism, often an insect or rodent, that transmits disease. See also transvector. **2:** an object that transports genes into a host cell (a plasmid, virus, or other bacteria). A gene is placed in the vector which then "infects" the host.

Vegadex. sulfallate.

vegetable calomel. *Podophyllum peltatum.*

vegetal. *Blighia sapida.*

vein. **1:** a blood vessel that, in vertebrate animals, conveys blood from the capillary bed to the heart, lungs, or gills. Veins have a larger lumen and thinner walls than arteries of comparable size because they lack the muscular and elastic tissues that present in arterial walls. Most veins contain valves that prevent backflow, thus ensuring blood flow only toward the heart. Blood pressure is much lower in veins than in arteries. **2:** one of the vascular bundles of a leaf. **3:** one of the chitinous tubes that strengthen and support the wing of an insect.

Vejovis (devil scorpions). a genus of common crab-like scorpions of the western and southwestern United States. They have a long telson that terminates in a bulbous sac and stinger. Envenomation can produce local or systemic effects. Allergic reactions are possible. See Scorpionida.

velvet bean. *Pithecolobium lobatum.*

velvet grass. *Holcus lanatus.*

velvet tail. *Crotalus horridus*

velvet-tail rattlesnake. *Crotalus horridus.*

velvety green night adder. *Causus resimus.*

vena cava. one of two large veins that convey deoxygenated blood to the right atrium of the vertebrate heart.

venenation. poisoning, as by envenomation; the state of being poisoned.

venene. a mixture of snake venoms.

veneniferin. cerberin.

veneniferous. **1:** able to convey poison, as by a sting or bite. **2:** bearing or containing poison.

venenific. producing or able to produce poison (said of a venomous organism).

venenosa. formerly a collective name for venomous snakes.

venenosalivary (venomosalivary). pertaining to the secreting of a poisonous saliva or venom, used especially in reference to venomous reptiles.

venenosity. **1a:** the state of being venomous. **1b:** the state of being poisonous, a term applied only to poisonous or venomous animals or their secretions.

venenous. poisonous; venomous. Applied especially to poisonous or venomous animals or to their secretions or venoms.

venenum. a poison, usually of animal origin.

venerupin. a poison isolated from the oyster, *Crassostrea gigas.* See venerupin poisoning.

venerupin poisoning (gastrointestinal shellfish poisoning). a frequently fatal poisoning of humans following ingestion of certain oysters and clams (formerly placed in the genus *Venerupis*) in Japanese coastal waters. The responsible shellfish contain venerupin, a toxicant that causes impaired liver functioning, gastrointestinal distress, and leukocytosis in humans. About one third of these poisonings are fatal. Venerupin poisoning has an incubation period of a day or more. Symptoms include headache, gastric pain, anorexia, nausea, vomiting, halitosis, constipation, and weakness. In more severe cases, effects may also include nervousness, hematemesis, and hemorrhage from mouth, nose, mucous membranes, and gums. Anemia, leukocytosis, and prolonged blood coagulation time are sometimes observed. In severe cases, icterus may occur, together with petechial hemorrhages and ecchymosis on the chest, neck, and arms. See also venerupin, *Crassostrea gigas*, shellfish poisoning.

venesection. cutting open a vein.

venin. any specific poison found in animal venoms. Used most frequently in reference to snake venoms.

venin de crapaud. a poison used in medieval times. It was produced by distilling the juices of toads and other animals which had been fed arsenic.

venipuncture. the puncturing of a vein with a needle.

venom. **1:** any of numerous, usually complex, chemically and pharmacologically diverse toxic secretions of animals that are transmitted to other animals by means of a bite, sting, puncture, or in rare instances by "spitting" (See envenomation). Crude venoms often have properties that are not explained by the behavior of individual components. They were historically classified on the basis of the type of the tissue presumed to be affected (e.g., hemotoxin, neurotoxin). Most venoms, however, appear to affect nearly any cell or tissue which they contact and the specific effects observed are due chiefly to the degree of accumulation at a particular locus in the victim's body. **2:** unacceptable usage: a poison of animal origin. See also allomone. **ant venom**. a venom produced by an ant. Constituents vary but usually include enzymes and one or more organic acids including formic acid. Formic acid predominates in the venoms of red ants, *Formica spp.*, whereas the venom of *Myrmicaria natalensis* contains approximately 35% acetic acid, 31% isovaleric acid, and 22% propionic acid. Envenomation may be painful. **arthropod venom**. any venom secreted by an arthropod. Most spiders and many insects are venomous. Both the venom and the apparatus of envenomation vary considerably among arthropods. **bee venom**. the venom of bees (Order Hymenoptera). It is more complex and contains a greater variety of proteinaceous materials than do wasp and hornet venoms. These venoms produce neurotoxic, hemolytic, and strong histamine-like effects. Honey bee venom, for example, contains histamine, small peptides, mellitin, apamin, phospholipase, proteins (five of which are antigenic to humans), free amino acids, a histopeptide, simple sugars, and several phospholipids. See also apamin, mellitin, mast cell degranulating peptide. **black widow venom**. See latrodectus venom. **bungarus venom**. venoms from snakes of the genus *Bungarus*, *q.v.* They are highly toxic and contain bungarotoxins. **cobra venom**. the venom of true cobras, those of the genus *Naja*. It is a complex mixture of toxins and enzymes, the most important of which are cobra venom factor, cobra hemotoxin, and cobrotoxin. The latter is a powerful neurotoxin and is usually the dominant toxin of cobra venom. In some populations or individual snakes, the venom is high in necrotizing factor but low in neurotoxin. See cobraism. **cone shell venom** (conus venom). venoms produced by marine gastropods of the genus *Conus*. Some (e.g., that of *Conus geographus*) are lethal to humans. They contain a complex mixture of biologically active amines, peptides, and proteins and also *N*-methylpyridinium, homarin, and τ-butyrobetain. Each of the last three substances produce curare-like effects. **conus venom**. See cone shell venom. **crotalid venom**. the venom of a crotalid snake. They are generally rich in proteolytic enzymes. The venoms of New World crotalids alter peripheral resistances and often damage blood vessels; alter blood cells and blood coagulation mechanisms, cardiac and pulmonary dynamics, and nervous system functions. The venom is extremely toxic but intensity of effects varies as a function of the amount and kind of venom injected and other factors. Symptoms in humans may include local pain, inflammation, hemorrhage and necrosis; dizziness, motor and sensory depression, hypotension (may be severe), collapse. Death may occur, usually 18-32 hours following the bite. **elapid venom**. any venom produced by a snake of the family Elapidae. These venoms exhibit little or no proteolytic activity and cause less tissue damage than viperid or crotalid venoms. Elapid venoms, however, produce serious effects on sensory and motor functions as well as cardiac and respiratory effects. **heloderma venom**. sometimes called lizard venom; the venom of the two lizards of the genus *Heloderma*, *H. suspectum* and *H. horridum*. It contains serotonin, amine oxidase, phospholipase A, hyaluronidase, and proteolytic activity. Injection of a large dose into a laboratory animal lowers systemic arterial pressure with a concomitant decrease in circulating blood volume, respiratory distress, and tachycardia. A lethal dose reduces ventricular contractility. **hemotoxic snake venom** (e.g., of *Crotalus*, *Agkistrodon*). any snake venom that contains cytolysins that enzymatically destroy cell membranes and tissues, with consequent hemorrhage from various organs. Such venoms may cause considerable pain, swelling, discoloration, hemorrhage, and necrosis that spreads rapidly from the area of the bite. Hematuria usually occurs and hemorrhaging from the mouth, nose, eyes, and gastrointestinal tract is common. Nausea, vomition, circulatory collapse, and death may ensue. Hemorrhage is due to one of three types of hypofibrinogenemia: **1:** without fibrinolysis or thrombocytopenia; **2:** with fibrinolysis but no thrombocytopenia; **3:** with thrombocytopenia and no fibrinolysis. **honey bee venom**. See bee venom. **hornet venom**. wasp and hornet venoms are distinguished from bee venoms by their lower content of peptides. They do contain kinin peptide, which may cause smooth muscle contraction and lowered blood pressure. Two biogenic amines in wasp or hornet venoms (serotonin and acetylcholine) lower blood pressure and cause pain. Acetylcholine may cause malfunction of heart and skeletal muscles. **hydrophid venom** (sea snake venom). any venom secreted by a sea snake (Family Hydrophidae). The venoms of hydrophids are the most toxic of snake venoms; all are highly toxic, even supertoxic. All contain erabutoxin and at least 61 amino acids. They are hemotoxic, neurotoxic, and have direct myotoxic effects. Some species can inject as much as 0.25 ml in a single bite, but approximately 0.01 ml can kill an adult human. The LD_{50} of *Enhydrina schistosa* has been reported at 0.01 mg/kg, *i.v.*, in laboratory rats). Hydrophid venoms exhibit little or no proteolytic activity. The venom causes skeletal muscle pain although the wound itself is not painful, thus the victim may not associate the bite with the ensuing symptoms. Early symptoms, the first of which are weakness and soreness of skeletal muscles, usually appear within one to a few hours following envenomation and vary from mild euphoria to generalized aching and anxiety; ptosis is an early and characteristic sign. An ascending paralysis beginning with the legs follows, and the trunk, arms and neck muscles are usually affected within two hours. The pulse becomes weak and irregular; the pupils dilate. Vision is blurred, the tongue and mouth feel paralyzed and speaking and swallowing become difficult leading to trismus, a conspicuous symptom. Nausea and vomition are common; tremor and spasms may also be seen. Weakness increases until the patient can barely move. The venom acts directly on the muscles, releasing myoglobin and potassium such that the urine is stained red, the kidneys are injured, and cardiac irregularities develop. In cases of severe intoxication, symptoms intensify, the skin becomes cold, clammy, and cyanotic. Convulsions and severe respiratory distress lead to unconsciousness and death. **hymenoptera venom**. hymenopteran venom. **hymenopteran venom** (hymenoptera venom). any venom produced by a hymenopteran. They are typically composed of water-soluble, nitrogen-containing, chemical species in concentrated mixtures. Although they contain certain chemical compounds in common, the venoms from different species vary. See, for example, bee venom. The major types of chemical species present are biologically synthesized (biogenic) amines; peptides and small, nonenzymatic proteins; and enzymes. The most common of the biogenic amines is histamine, found in the venoms of bees, wasps, and hornets (See also 5-hydroxytryptamine). Wasp and hornet venoms contain serotonin, and hornet venom contains the biogenic amine acetylcholine. Among the peptides and low-molecular-mass proteins in hymenopteran venoms are apamin, mellitin, and mast cell degranulating peptide in bee venom; wasp kinin, and hornet kinin. Enzymes contained in bee, wasp, and hornet venom include phospholipase A and hyaluronidase. Phospholipase B occurs in wasp and hornet venom. **insect venom**. any venom produced by an insect. The composition of insect venoms varies greatly among taxa and within the class. **krait venom**. the venom and its effects are similar to that described for cobra venom, but while the bite is usually less painful than that of a cobra, the clinical manifestations may be more severe. Systemic effects include shock, often with conspicuous respiratory depression and coma; abdominal pain may be intense; and chest pain is sometimes reported. **latrodectus venom**. extremely toxic, neurotoxic venom, the composition of which remains unclear. It is comprised of about 5 or 6 proteins (some are enzymes, including hyaluronidase) and a neurotoxic principle that appears to be a polypeptide with a MW of perhaps 130,000. It acts on nerve endings or myoneuronal junctions, destroying peripheral nerve endings or producing an ascending paralysis. The known pharmacological properties of the crude venom are inconsistent with the effects of this venom on humans. Early effects of envenomation in humans usually include muscle fasciculations; muscular weakness; sweating; tender, painful, and often enlarged lymph nodes; lymphadenitis; and hypertension. Severe cases often exhibit lower back and abdominal pain with muscular (especially abdominal) rigidity, and sometimes severe paroxysmal muscle cramps and arthralgia. Additional symptoms that may occur include nausea, vomition, conjunctivitis, edema of the eyelids, hyperemia, skin rash, and pruritis. See also black widow venom poisoning. **lizard venom**. See *Heloderma* venom. **loxosceles venom**. a largely cytotoxic, hemotoxic, necrotizing, and coagulant venom produced by spiders of the genus *Loxosceles*, most notably that of *L. reclusa* (brown recluse). The bite produces an unusual and characteristic necrotic lesion in humans; systemic effects may include disseminated intravascular coagulation. The venom is comprised of 10-12 proteins with enzyme activity that is greater than that of latrodectus venom. Two toxins (mw ca. 24,000) have been isolated from the protein fraction of *L. reclusa* venom. This venom contains a number of enzymes, is necrotizing and lethal to mice and rabbits. No principle has been isolated that produces the lesion that characteristic of *Loxosceles* bites. **mamba venom**. that produced by snakes of the genus *Dendroaspis*; it is similar in most respects to krait venom. **neurotoxic snake venom** (e.g., of *Micrurus*, *Naja*). any snake venom that contains one or more neurotoxins that dominate the effects produced by envenomation. Type A neurotoxins act primarily on the higher cardiorespiratory centers of the CNS; type B neurotoxins generally act on the myoneural junctions. The human victim of a neurotoxic snake venom may experience muscular weakness, paralysis of the mouth and throat, and ultimately of the respiratory muscles. Additional effects seen in serious cases of envenomation may include drowsiness, optic ptosis, muscular weakness, difficulty in swallowing, trismus, occasionally convulsions (e.g., from cobra venom), bradycardia, a weak pulse, dyspnea, with eventual cardiac and respiratory collapse and death. **rattlesnake venom**. a venom produced by the venom glands of snake of the family Crotalidae, especially those of the genera *Crotalus* and *Sistrurus*. The venom is complex and contains many poisonous constituents including hemorrhagins. The lung appears to be the major target organ, where increased vascular permeability causes pulmonary congestion and hemorrhage. The venom may cause local bleeding, severe pain, superficial edema of rapid onset, and hemorrhagic bleb formation. **Russell's viper venom**. the venom of Russell's viper (*Vipera russelii*). This venom acts *in vitro* as an intrinsic thromboplastin and is used to investigate defects in blood coagulation. Serious effects of envenomation often include pain, swelling and edema, discoloration, prolonged clotting time, and internal hemorrhage. Also commonly seen are ecchymosis, necrosis, nausea, vomiting, thirst, weakness, diarrhea, abdominal pain, pupillary dilation, a weak pulse, hypotension, anemia, albuminuria, proteinuria, hemorrhage about the wound and sometimes from the gums, gastrointestinal and urinary tracts. The lethal dose of the venom for *Homo sapiens* is ca. 40-70 mg, yet a large snake yields 150-250 mg when milked. The venom, in a 1:10,000 solution, is used as a local coagulant in the management of hemorrhage in hemophilia. See also viperid venom poisoning. **scorpion venom** (scorpamine). any venom produced by a scorpion. In the amount delivered by a sting, although usually painful, the venom of relatively few species produces effects that require medical attention. The most toxic venoms are produced by scorpions of the family Buthidae. The buthid venoms contain extremely toxic neurotoxins which are small basic proteins with a mw of about 7000. They also contain 5-hydroxytryptamine which probably does not significantly contribute to toxicity. The buthid venom [illegible] release of acetylcholine by motor neurons and by postganglionic autonomic neurons with increased axonal Na^+ permeability. Symptoms in humans may include focal pain and inflammation, nervousness, debility, cardiac arrhythmia, and respiratory distress. All cases of severe poisoning must be treated with anti-scorpion serum as soon as possible. Small children are especially at risk from scorpion stings. **sea snake venom**. hydrophid venom. **sea urchin venom**. a venom produced by sea urchins and introduced via spines or by globiferous pedicellariae. The pedicellarial venoms are more potent than those within spines. See also venom apparatus. **snake venom**. a venom produced by a snake. Such venoms are complex mixtures that do not lend themselves to complete analysis. Included are polypeptides; ezymes (e.g., phospholipases, peptidases, proteases, collogenases, phosphoesterases, acetylcholinesterase) and other toxic proteins; biogenic amines; carbohydrates; and metal ions. Among the most important components are the smaller (nonenzymatic) polypeptides (those containing 60-70 amino acid residues), many of which are neurotoxic; some are more toxic than the crude venom. Proteolytic enzymes are also prominent constituents, often causing widespread tissue damage and necrosis. In terms of effects, snake venoms are comprised of variable amounts of blood coagulants, anticoagulants, agglutinins, cytolysins, proteolysins, antibactericidin, and neurotoxins. Most components appear to occupy specific chemical and physiologic receptor sites. Envenomation by a snake may be complicated by the release of autopharmacologic substances (e.g., histamine, serotonin) that can make diagnosis and treatment difficult. Individual snake venoms produce multiple effects, yet they are often classified on the basis of what is considered by some as the primary toxic effect (e.g., cardiotoxic, hemotoxic, myotoxic, neurotoxic). The use of such classification has sometimes caused serious errors in the treatment of snakebite. See also enzymes of snake venom, snake venom poisoning. Snake venoms sometimes cause severe anaphylaxis and nearly instantaneous death in humans or animals (especially livestock) that have experienced snakebite in the past. **spider venom**. that secreted by a venomous spider. **stingray venom**. any of a number of extremely toxic venoms produced by sting rays. Deaths of striken humans are, however, rare and symptoms may be largely restricted to the area about the sting. They are generally cardiotoxic to mammals. Low concentrations administered to laboratory mammals produce vasodilation or vasoconstriction, nominal bradycardia, and a lengthened P-R interval in the electrocardiogram; intermediate concentrations may produce cardiac ischemia and occasionally, myocardial injury; high concentrations produce substantial changes in heart rate and amplitude of systole, often leading to complete, irreversible cardiac arrest. Other symptoms in laboratory animals given high doses include depressed respiration and behavioral effects; lethal doses may bring about hyperkinesis, pronounced dyspnea, prostration followed by cyanosis, atonia, gasping, coma and death. **taipan venom**. the venom of *Oxyuranus scutellatus*, *q.v.* One of the most toxic of snake venoms, it is otherwise similar in most respects to krait venom. It is hepatotoxic and contains a clotting factor that usually brings death to a human victim within minutes unless treated with "Taipan" antivenin. **tarantula venom**. any venom produced by spiders of the family Theraposidae. Such venoms contain a number of proteins, at least one of which is cardiotoxic. The bite of a single tarantula is rarely seriously harmful to a human. Some tarantulas are dangerous to humans, but those native to the United States are not considered dangerous. See also *Pamphobeteus*, tarantula. **toad venom**. the venom of toads of the genus *Bufo*; it contains bufotenin. **viper venom**. viperid venom. **viperid venom** (viper venom). any venom produced by snakes of the family Viperidae. These venoms may cause extensive focal and systemic tissue damage, defective coagulation, hemotoxic effects, blood vessel damage. Damage to the heart, kidneys, and lungs may occur in some cases. **yellow jacket venom**. a hymenopteran toxin that may induce a possibly fatal (type I hypersensitivity) anaphylactic shock. The liberation of vasoactive amines may cause urticaria, chills, fever, a tightness in the chest, dyspnea, and even cardiovascular collapse and death.

venom apparatus. the structural components that produce, transport, and deliver the venom. In snakes, it is usually composed of two venom glands, two venom ducts, and two or more teeth or fangs. *Cf.* stinger.

venom apparatus of sea urchins. sea urchins are generally equipped with one of two types of venom organs: **1:** venomous spines or **2:** globiferous pedicellariae. The spines of most sea urchins are solid with rounded or blunt tips. the spines of some, however, are long, slender, hollow, sharp, and brittle. Such species are extremely dangerous to handle as the spines can easily penetrate deeply into flesh; they readily break off in the wound and are extremely difficult to remove. The spines may contain (or possibly even secrete) venom. There are several different types of pedicelariae in sea urchins. In the globiferous type, the head is the venom organ. It is equipped with calcareous pincer-like valves and is nearly globe-shaped. The inner surface of each valve contains thin sensory hairs that trigger the valve to close immediately upon contact with a foreign body. The outer surface of the valve is overlain by a large gland that delivers venom to the terminal fang of the valve. A sensory bristle situated within each valve causes the small muscles at the base of the valve to contract, thereby closing the valves and injecting venom into the integument of the victim.

venom extract therapy. the prophylactic administration of antivenin against the toxic effects of future envenomations by a specific animal. *Cf.* antivenin therapy.

venom hemolysin. See hemolysin.

venom immunotherapy. hyposensitization to the bite of a venomous animal by the serial administration of gradually increasing doses of the specific antigenic substance secreted by the animal.

venomization. treatment of a substance with snake venom.

venomologist. a toxicologist who studies venoms, their properties, mechanisms of action, and/or effects.

venomology. the branch of toxicology that treats venoms, their properties, mechanisms of action, and effects. See also toxicology.

venomosalivary. venenosalivary.

venomous. **1a:** venom-producing or secreting; noxious; of or pertaining to an animal that has poison glands and is able to inflict a poisonous wound by biting, stinging, or sometimes by "spitting." **1b:** having venom gland(s) that communicate with an organ of envenomation (e.g., a fang, stinger, or spine. **2:** poisonous, toxiferous, noxious; virulent. Such usage is not recommended in toxicology. *Cf.* poisonous, toxiferous.

venomous animal. a term reserved for animals that produce toxicants (venom) in a highly developed secretory organ and convey them to other animals by biting, stinging or in a few instances spraying. *Cf.* poisonous animal.

venomous fish. a term reserved for fish that produce toxicants (venom) in a highly developed secretory organ and convey them to other animals by biting or stinging (e.g., with a spine). There are numerous species of venomous fish, mostly marine. There are more than 200 species of venomous marine species. Most are non-migratory, slow-moving bottom-dwellers or inhabitants of reef communities, or other protected situations. The chemical, pharmacologic, and toxicologic properties of their toxins differ substantially from those of poisonous species. They are, in particular, unstable at room temperature and toxicity is easily lost or reduced. Consequently, the basic structures of fish venoms are as yet unknown. See also venomous animal.

venomous insect. any insect that has one or more venom glands or salivary glands that secrete caustic or otherwise poisonous substances together with fangs, a stinger, or other apparatus that can introduce the venom into another animal. Relatively few insect species produce enough venom to endanger humans. Nevertheless, insects probably cause more fatal poisonings than all other venomous animals combined. Most venomous insects belong to the order Hymenoptera.

venous tension (intravenous pressure). the blood pressure within a vein.

veratralbine. a component of veratrine; also found in *Veratrum viride*.

veratramine. a steroid alkaloid that occurs in *Amianthium* (staggergrass), *Veratrum* (false hellebore), and *Zigadenus* (death camas). See *Veratrum*, veratrine.

veratrate. a salt or ester of veratric acid.

veratria. veratrine.

veratric acid (3,4-dimethoxybenzoic acid; dimethylprotocatechuic acid). **1:** a crystalline acid prepared from the seeds of *Schoenocaulon officinale* (*Sabadilla officinarum*); also formed by the decomposition of cevadine (veratrine) and related alkaloids. **2:** ortho-veratric acid, *q.v.*

veratridine (4,9-epoxycevane-3,4,12,14,16,17,-20-heptol 3-(3,4-dimethoxybenzoate; 3-veratroylveracevine; veratrum alkaloid). an amorphous, slightly water-soluble, neurotoxic nitrate alkaloid that occurs in the rhizome of some species of *Veratrum* and the seeds of *Schoenocaulon officinale*. It is the most toxic of the veratrum alkaloids. Its action is physiologically indistinguishable from that of aconitine. Veratridine is a component of veratrine mixture (See veratrine).

veratrin. See *Helleborus*.

veratrine (veratria). **1:** cevadine. Cevadine that has been isolated from *Veratrum* is commonly referred to as veratrine. **2:** veratrine mixture. **3:** veratramine.

veratrine alkaloid (veratrum alkaloid; veratrum viride alkaloid). any of numerous and complex alkaloids (glycoalkaloids and ester alkaloids) contained in plants of the genus *Veratrum* (especially *V. viride*) and certain other plants such as *Amianthium*, *Helleborus*, and *Zigadenus*. They have been used medicinally, but may cause myotonia, muscular spasms, and neuropathy. The alkamine moiety of these compounds has a steroidal configuration. There is some confusion of terms in the literature on the nomenclature of veratrine alkaloids and various alkaloidal extracts of these plants (See, for example, veratrine). Examples of veratrine alkaloids are cevadilline, cevadine (veratrine), cevine, germidine, germitrine, jervine, pseudojervine, rubijervine, sabadine, veratralbine, veratramine, veratridine, veratrin, veratroidine, veratroidine, veratrosine, and veratrosine. See also veratrine mixture, veriloid.

veratrine mixture. a toxic mixture of alkaloids from the seeds of *Schoenocaulon officinale* (sabadilla). It is an acrid-tasting powder that contains cevadine (veratrine), veratridine, cevadilline, cevine, and sabadine. It is myotoxic, neurotoxic, and an intense local irritant; it causes violent sneezing when inhaled. It has been used therapeutically as a counterirritant in neuralgia and arthritis. See especially veratrine, veratridine.

veratroidine. a toxic, crystallizable base, $C_{32}H_{53}NO_9$, that occurs in *Veratrum album* and *V. viride*. It is a potent cardiac inhibitor and neurotoxicant.

veratrosine. a toxic glycoalkaloid of *Veratrum*, the sugar moiety is D-glucose and the alkamine is veratramine.

veratroylaconine. pseudoaconitine.

Veratrum (false hellebore). a genus of nearly 50 species of hardy and showy perennial toxic herbs (Family Lilaceae), with neurotoxic, hypotensive, and cathartic activity; some species are teratogenic to sheep. They occur in moist pastures and woodlands throughout North America and are cultivated in wild gardens around ponds or streams. They contain several complex neurotoxic and cardiotoxic alkaloids that cause dilation of the small arteries and constriction of the small veins. They also reduce the rate and force of cardiac contractions. All parts of the plant are toxic. The roots, which are especially toxic, are sometimes used to make an insecticide. Poisoning of humans is rare, however, and grazing animals generally avoid this plant (but, See monkeyface). *V. album* (European white hellebore), *V. californicus*, and *V. viride* are sources of antihypertensive alkaloids. ***V. album*** (European false hellebore; European white hellebore). a species with greenish-white flowers, growing to a height of ca. 1.3 m. Its toxicity is similar to that of *V. viride*. See jervine. ***V. californicum*** (false hellebore; cornlily; "skunk cabbage"). an attractive herb of the western United States that has white flowers marked with green and grows to a height of ca. 2 m. The toxic principles and effects are similar to those of *V. viride*. If this plant is ingested during the second and third weeks of pregnancy by ewes, craniofacial malformations form in the newborn lambs. Affected lambs cannot breathe or feed properly and usually die within a few hours of birth. See monkeyface. ***V. nigrum***. a plant that grows to a meter or more and bears crowded purple flowers. ***V. viride*** (American hellebore; American veratrum; white hellebore; American white hellebore; bear corn; earth gall; Indian poke; itchweed; devil's bit; devil's tobacco; Tabac du diable;

duck ratten; poor Annie). a coarse, erect, sparingly leafy herb of swamps and low, moist situations that grows to a height of 1 to 2½ m or more. The flowers are white or yellowish-green flowers. It occurs in low or swampy grounds from Quebec, Minnesota, New Brunswick, and New England, south to Maryland. It occurs also in uplands as far south as Georgia and Tennessee. *V. viride* is toxicologically typical of the genus. All parts of the plant are toxic. It is a violently irritant and hypotensive agent. Symptoms of intoxication may include burning in the throat, nausea, abdominal pain, impaired vision, depressed heart action, dyspnea, paralysis, coma, and sometimes death. Human intoxication often results from confusion of *V. viride* with other herbs. It contains numerous and complex alkaloids (glycoalkaloids and ester alkaloids) in which the alkamine moiety has a steroidal configuration. These include jervine, pseudojervine, rubijervine, cevadine (veratrine), germitrine, germidine, veratralbine, veratramine, veratrosine, veratroidine, veratridine. ***V. woodii*** (western false hellebore). a stout leafy plant that grows to a height of about 1.3 m or more. It occurs mainly in woodlands or on bluffs and hillsides in the western United States from Oregon to Iowa, south to Missouri and Oklahoma.

veratrum alkaloid. veratrine alkaloid.

veratrum therapy (veriloid therapy). See veriloid.

veratrum viride alkaloid. veratrine alkaloid.

Verbesina encelioides (crownbeard). a branching, many-flowered, annual herb (Family Compositae) with alternate leaves that grows to as much as 0.6 m in height. It occurs from Montana to Arizona, eastward to Kansas and Texas, and occurs casually in the northeastern United States. It sometimes contains toxic concentrations of nitrate.

veriloid. a partially purified alkaloidal extract of *Veratrum viride*. It was used for many centuries in the treatment of hypertensive cardiovascular disease. Even well controlled conventional doses of this drug, however, often causes vomiting; and some patients experience reversible myotonic-like syndromes. Atropine may reverse or alleviate symptoms of intoxication. Veratrum (veriloid) therapy is no longer used.

veriloid therapy. veratrum therapy (See veriloid).

Vermaxan™. lindane.

vermicidal. lethal to worms, especially to intestinal parasitic worms.

vermicide. **1:** able to destroy worms. **2a:** an agent that can kill intestinal worms. **2b:** any drug used to treat an infection of parasitic worms.

vermifugal. anthelmintic.

vermifuge. an agent that destroys or eliminates intestinal worms. *Cf.* anthelmintic.

vermilion. red mercuric sulfide (See mercuric sulfide).

vermin. a noun in common usage that includes all noxious, parasitic, and nuisance animals such as lice, Norway rats, house mice, flies, and bedbugs. The eye is in the beholder, however, and animals such as crows, snakes, and prairie dogs may be considered to be vermin by some people and at some locations, but not by others or at other locations.

vernal pheasant's eye. *Adonis vernalis*.

Verpa bohemica. an edible mushroom that sometimes produces a mild gastroenteritis. Motor incoordination has been reported for some persons.

verrucarin. any of a class of extremely toxic and caustic macrocyclic trichothecene mycotoxins produced by *Myrothecium verrucaria*. They are antibiotic, antifungal and cytostatic agents.

verrucarin A. a dermotoxic, trichothecene mycotoxin from *Myrothecium verrucaria*. It is extremely toxic to laboratory mice.

verrucarol. a trichothecene mycotoxin produced by *Fusarium spp.*

verruculogen (TR-1). a neurotoxic, tremorgenic indole mycotoxin produced by *Aspergillus caespitosus* and *Penicillium verruculosum*. It is extremely toxic to laboratory mice and domestic fowl, eliciting hypersensitivity to sound, ataxia, and tetanus spasmus.

verrugosa. *Lachesis mutus*.

Vertebrata (Craniata). the principal subphylum of the phylum Chordata. Included are cartilaginous and bony fishes, amphibians, reptiles, birds, and mammals. Representatives typically have an endoskeleton of cartilage or bone, a dorsal vertebral column (backbone) that encloses a tubular nerve cord (spinal cord), and a complex nervous system with a well-developed brain, in the cranium (skull). The blood and digestive systems are situated ventral to the vertebral column. All living vertebrates except the Agnatha have jaws, formed from the anterior pair of visceral arches. There are nine classes, including those known only from the fossil record: Agnatha, Placodermi, Acanthodii, Chondrichthyes, Osteichthyes, Amphibia, Reptilia, Aves, and Mammalia.

vertebrate. **1:** having vertebrae. **2:** a vertebrate animal; an animal of the subphylum Vertebrata, *q.v.*

Verticimonosporium. a genus of fungi that secrete epoxide mycotoxins. ***V. diffractum***. a species that produces the trichothecene mycotoxin, vertisporin.

vertigo. **1:** a disorder of equilibrium that gives a person the illusion of movement when at rest. This usually takes one of two forms, objective vertigo or subjective vertigo. Vertigo may result from any disease or disturbance of the inner ear, the vestibular apparatus, or nerve tracts and centers within the CNS concerned with equilibrium. It may be induced, for example, by toxic chemicals, sudden disturbances of eye function, and psychic disturbances. **2:** the terms dizziness and vertigo are erroneously used as synonyms by some toxicologists. *Cf.* dizziness. **objective vertigo**. the sensation of external objects spinning around the subject. **subjective vertigo**. the sensation that the subject is revolving in, or moving around in space when at rest. **toxemic vertigo**. toxic vertigo. **toxic vertigo** (toxemic vertigo). a form of vertigo due to poisoning (e.g., alcoholism, food poisoning, uremia, lithemia). **true vertigo**. a disturbance of the organs of equilibrium (i.e., of the inner ear, the vestibular centers, or tracts within the CNS) that the affected individual suffers an illusion of movement, either objective vertigo or subjective vertigo, *q.v.*

vertisporin. a trichothecene mycotoxin isolated from *Verticimonosporium diffractum*.

vervain family. Verbenaceae.

vesicant. **1:** vesicatory; causing or forming blisters or vesicles; blistering. **2:** an agent (e.g., cantharidin, mustard gas) that blisters the skin on contact. They may also cause tissue necrosis and additional effects upon extravasation, or when fumes are inhaled. See gas (vesicant gas).

vesicating gas. vesicant gas (See gas).

vesicatory. vesicant (def. 1).

vesiculation. the formation of blisters.

Vespa. a genus of wasps (Family Vespidae) that secrete an alkaline venom containing neurotoxic and hemolytic agents. The venom is neutralized by lemon juice or vinegar. The sting produces symptoms similar to those of bees. See *Apis*, venom (hornet venom). ***V. diabolica*** (yellow jacket). a species that nests in or on the ground. See venom (yellow jacket venom).

vespid. of or pertaining to a vespid wasp; a vespid wasp; any hymenopteran of the family Vespidae.

Vespidae. a family of medium to large colonial wasps (Order Hymenoptera) with powerful stingers. They are yellow or red with black or brown markings. Included are paper wasps, yellow jackets, and hornets. See venom (hornet venom, yellow jacket venom, hymenopteran venom). The nest is often quite large and made of material (chewed wood) that has a papery consistency. Each new colony originates from a single over-wintering queen. Vespula and Vespa, *q.v.*, are typical genera.

vespoid. wasp-like.

Vespula. a common genus of hornets (Family Vespidae) that secrete an alkaline venom containing neurotoxic and hemolytic agents. The venom is neutralized by lemon juice or vinegar. The sting produces symptoms similar to those of bees (See, for example, *Apis*). See also Vespidae, venom (hornet venom).

vestibule. an enlarged space or cavity at the entrance of a canal.

vestigial. the nonfunctional remnant of a structure that was functional in some ancestral species.

vetch See *Vicia*.

vetch seed. See *Vicia sativa*.

vetchling. See *Lathyrus*.

veterinary medicine. the medical science or practice that treats the diseases of animals including domestic species, wildlife, and zoo and aquarium animals.

viability. depending on context: **1a:** vitality, capacity to live, ability to continue living, ability to survive. **1b:** ability to live, grow, and develop. **1c:** ability to grow and propagate or reproduce.

viability index. **1:** a measure of the survival of offspring in reproductive toxicity tests. It is usually expressed as the percentage of live progeny that survive to a specified age or stage. **2:** a relative measure of the number of surviving individuals of any given phenotypic or genotypic class under a specified set of conditions.

viable. vital, alive, capable of living, having the capacity to be normally active, live, grow, and develop.

viable count. a count of the proportion of cells in a sample (e.g., of bacteria, yeasts, or cells in a tissue culture) that are able to grow and reproduce.

Vibrio cholerae (*Vibrio comma*). a Gram-negative bacterium (Family Spirallaceae) that is the etiologic agent of Asiatic cholera in humans. The proximate cause of the disease is the exotoxin, choleragen, *q.v.*

Vibrio comma. *Vibrio cholerae*.

viburnum. *Lantana*.

Vicia (vetches). a genus of mostly weedy herbs (Family Fabaceae, formerly Leguminosae). Some species contain vicine, *q.v.* ***V. faba*** (fava bean; broad bean; horse bean). the only species of *Vicia* normally cultivated for human food. ***V. sativa*** (spring vetch). the seed is cyanogenic and the foliage is photosensitizing to livestock.

vicine (2,5-diamino-4,6-diketopyrimidine-3-β-D-glucoside). a white, crystalline, mononucleoside glycoside, $C_{10}H_{16}N_4O_8$, that occurs in *Vicia sativa* and related vetches. *V. sativa* are often present among *Lathyrus sativus* seeds used in livestock feeds and may be responsible for some of the symptoms of lathyrism.

Viet Nam sea snake. *Lapemis hardwickii*.

vikane. sulfuryl fluoride.

vinblastine (VBL; vincaleukoblastine). an antimitotic and neurotoxic alkaloid, derived from *Vinca spp.*, that inhibits axonal transport. The sulfate is used as an antineoplastic, especially in the treatment of acute leukemia. VBL inhibits microtubule formation, killing rapidly dividing cells by disrupting the mitotic spindle. It binds to the microtubule subunit, tubulin, at a site distinct from that of colchicine and podophyllotoxin. This binding is temperature-dependent, rapid, and reversible. It results in large tubulin aggregates of highly ordered structure (vinblastine paracrystals). It also causes alopecia. See also antineoplstic drug, colchicine, poison (spindle poison), tubulin, vincristine.

vinblastine sulfate. See vinblastine.

Vinca. a genus of erect or trailing herbs or small shrubs (Family Apocynaceae). *V. rosea* (= *Catharanthus roseus*) is believed to cause defects in grazing calves similar to those of crooked calf disease.

vincaleukoblastine. vinblastine.

vincaleukoblastine sulfate (1:1)(salt). vincablastine sulfate. See vincablastine.

vincristine (VCR). an antimitotic and neurotoxic aldehyde derivative of vinblastine. The sulfate is used as an antineoplastic agent, especially in the treatment of acute lukemia. It inhibits axonal transport in peripheral nerves; it also causes delayed alopecia in humans. The toxicity and effects of VCR are very similar to those of vinblastine, *q.v.*

vincristine salt. vincristine sulfate (See vincristine).

vine snake. *Thelotornis kirtlandii*.

vinegar acid. acetic acid.

vinyl (ethenyl). the hydrocarbon radical, $CH_2=CH$-.

vinyl chloride (chlorethylene; chloroethylene; chlorethene; chloroethene; ethylene monochloride; monochloroethene; monochloroethylene; vinyl chloride monomer; vinyl monomer; VC). a colorless, flammable liquid, C_2H_3Cl, or gas (at room temperature) with a faint ether-like odor. Vinyl chloride is a very dangerous fire hazard. It is an irritant and is moderately toxic if ingested or inhaled. While it is a gas at room temperature (b.p. -13.9^o), it is usually handled as a refrigerated liquid. Occupational exposures are almost exclusively via inhalation or contact. The public may be exposed to vinyl chloride via drinking water and in consumer products. The central nervous system, respiratory system, liver, and blood and lymph systems are all affected by exposure to vinyl chloride. Among the symptoms of poisoning are fatigue, weakness, cardiac arrhythmias, and abdominal pain. Cyanosis may occur. Chronic poisoning with vinyl chloride produces angiosarcoma of the liver and probably lung cancer in humans. It causes liver, lung, kidney, and mammary cancers and cancers of the blood vessels in laboratory mice and rats. It is also mutagenic and has reproductive effects (e.g., on spermatogenesis). It is used in large quantities as the monomer in the manufacture of polyvinyl chloride. It has also been used as an intermediate in other synthetic reactions, as a solvent, and as a propellent in household aerosol products. Many of them are being or have been restricted, and the last mentioned has been discontinued. Acute poisoning causes CNS depression with symptoms similar to those of alcoholic intoxication, and hepatotoxicity. The carcinogenicity and mutagenicity of vinyl chloride is apparently due to its oxidation by the cytochrome P-45 monooxygenase enzyme system in the liver. The chloroethylene oxide so formed has a strong tendency to bond covalently to macromolecules (protein, DNA, and RNA) and to undergo transformation to chloroacetaldehyde, an established mutagen. Both of these intermediates can be inactivated and eliminated from the body by conjugation with glutathione. During the above oxidation, the activity of cytochrome P-45 is destroyed.

vinyl chloride monomer. vinyl chloride.

vinyl compound. a compound that contains the vinyl group, $CH_2=CH—$. Important examples are vinyl halides, vinyl acetate and related esters, all of which are toxic and highly reactive. They readily form polymers in which the vinyl group is part of the repeating unit. These polymers, and a number of important plastics, are also classed as vinyl compounds.

vinyl cyanide. acrylonitrile.

vinyl halide. See halide.

vinyl methyl ketone (3-buten-2-one). a colorless, flammable liquid, $CH_3CO=CH_2$, that is soluble in water and alcohols. An alkylating agent, it is an irritant to the skin and eye. *Cf.* methyl vinyl ketone.

L-5-vinyl-2-thiooxazolidone. a thyrotoxic substance that causes hyperplastic enlargement of the thyroid and symptoms of hypothyroidism. Thiooxazolidone glycosides have been isolated from plants which have caused livestock mortality especially among lambs from ewes fed on plants or plant products that contain these substances; many lambs are stillborn. Those that live have enlarged thyroids, are poorly developed, apathetic, and may fail to nurse. The ewes usually show few signs, although some enlargement of the thyroid may be observed on autopsy. Thiooxazolidone is the goitrogenic factor in various species of *Brassica*, including broccoli (*B. oleracea* var. *botrytis*), Brussels sprouts (*B. o.* var. *gemmifera*), kale (*B. o.* var. *acephala*), and kohlrabi (*B. o.* var. *gongylodes*, synonym, *B. o. caulorapa*), and rape (*B. napus*).

vinyl trichloride. 1,1,2-trichloroethane.

vinylbenzene. styrene.

17α-vinyl-5(10)-estren-17β-ol-3-one. norgesterone.

vinylestrenolone. norgesterone.

vinylethylene. 1,3-butadiene.

vinylformic acid. acrylamide.

violin spider. See *Loxosceles*.

viosterol. vitamin D_2.

viper. **1a:** a true viper is any snake of the family Viperidae. **1b:** any snake of the families Viperidae and Crotalidae. **2:** any snake of the genus *Vipera*. **3a:** in Europe, often used as the common name of *Vipera berus*. **3b:** colloquially, any venomous snake.

viper bite. See snakebite.

viper venom poisoning. viperid venom poisoning.

viper's bugloss. See *Echium*.

Vipera (true vipers). a small, but diverse, genus of venomous snakes (Family Viperidae) found in northern Europe and Asia southward into northern Africa. Some species are small and relatively harmless (e.g., *V. berus*), a few are larger and extremely dangerous (e.g., *V. lebetina*, *V. russeli*). A zigzag line along the dorsum is often present. The head is broad and distinct from the narrow neck; the canthus rostralis is distinct; the eyes are small to moderately large with vertically elliptical pupils; except for crown scutes, scales on the head are small and 1-4 rows separate the eye from the supralabials; the nasal touches the rostral or is separated by a single, large, nasorostral scalel; the body is cylindrical, relatively slender to stout, with a short tail; dorsal body scales are keeled with apical pits and form 19 - 31 nonoblique rows at midbody; ventrals are rounded; the subcaudals are paired. See also adder, viper, viperidae. ***V. ammodytes*** (long-nosed viper; nose-horned viper; sand viper; Armenian sand viper; eastern sand adder; western sand adder; long-nosed viper; nose-horned viper; opàrcâ; sand natter; sand viper; vipera cu corn). a species found from southeastern Europe to Asia Minor, chiefly in dry hilly country, especially on rocky limestone slopes, from an elevation above sea level of about 600 to 1700 m. The body is ash-gray, yellow, pale orange, coppery or brownish, with a prominent zig-zag dorsal line. Adults average 64-76 cm in length. The male is larger than the female and the color pattern is more intense. *V. ammodytes* is further characterized by the sharply upturned, horn-like protuberance on the snout, which is used in burrowing. The head bears no distinct dorsal markings and is covered by small scales that are irregular in size and disposition. While it appears to be sedentary and retiring, this snake is easily irritated and quick to strike. It is generally thought to be the most dangerous of the European vipers. The venom is very toxic, and of varied composition throughout the range. ***V. aspis*** (asp viper; European asp; Hugy's viper; Jura viper). a rather sluggish viper found chiefly in hilly or mountainous regions of the western parts of southern Europe. The head is more triangular than *V. berus*; snout slightly upturned; crown shields fragmented and only 2 or 3 are enlarged as a rule. The ground color is similar to that of *V. berus*, but often reddish or brown; the dark spots may coalesce, sometimes forming a zigzag band; the belly is dark gray with light spots; the dark head mark is poorly defined; the tip of tail is yellow or orange beneath. The body length of adults averages about ½ m; males are larger than females. ***V. berus*** (European viper; common adder, European adder, common European adder, viper; common viper; cross adder; Balkan cross adder; Iberian cross adder). a small, stout, relatively harmless, viperine snake of Europe (to locations above the Arctic Circle in Norway), North Africa, the near East. It is the only venomous snake of northern Europe; in central and southern Europe it is generally confined to mountains, generally above 3000 m. Some individuals reach a length of about 1 m. The ground color may be gray, brown, red, or nearly black. Darker individuals are usually females. There is a dark brown or black zigzag band that runs the length of the back with black spots along the sides. The head is ovoid rather than triangular and is distinct from neck; the snout is flat, blunt, and is not upturned; there is a chevron mark or dark "X" on the crown behind the eyes; the top of the head is covered by 5 large smooth shields. The bite may cause hemorrhage and tissue damage, but is rarely fatal to humans. Antivenin is available. Viper and adder are common colloquial names. ***V. hindii*** (montane viper). a rare viper of central Kenya. ***V. kaznakovi*** (caucasus adder; Kaukasus Otter). a species that occurs only in southwestern European Russia. ***V. latasti*** (snub-nosed viper; snub-nosed adder). a species usually of rocky or open sandy situations in lowlands and at moderate elevations on the Iberian Peninsula (Spain and Portugal) and in northwest Africa. It is about the size of *V. aspis* (the other European viper) and similar in appearance to *V. mauritanica*. The shields on the crown are fragmented and usually asymmetrical. It is not generally considered dangerous to humans. In contrast to *V. aspis*, its snout is pointed and more upturned. ***V. lebetina*** (levantine viper; Atlas adder; koufi; desert adder; Levantine adder; gjurza; Sahara viper; vipére lébétine; coffin snake; haia amia; Levante viper; mountain viper). an extremely dangerous, ovoviviparous snake that occurs in barren rocky regions, at elevations from about 900-2100 m in northern Eurasia, North Africa, and on Cyprus and the Cyclades Islands in the eastern Mediterranean Sea. It is a terrestrial snake, although it sometimes climbs into bushes. It is active chiefly at night and is slow moving and very difficult to irritate during the day. Nevertheless, it can strike rapidly and savagely at any time. The head is triangular, fairly long, and distinct from the neck; the nostrils are lateral, and the crown is covered by small keeled scales; the supraocular is comprised of 3 small shields; and one normally finds 3 scale rows between the eye and upper labials; The body is stout with a tail that begins to taper abruptly behind the vent. The average length of adults is about 1 m, with occasional specimens reaching 1.5 m or so. The ground color is light gray, khaki, or buff dorsally with minute darker punctation usually presenting a dusty appearance. There is also a series of small, rectangular brown, reddish or gray blotches above. The belly is buff, often clouded with gray, and the tail is pinkish brown. The dorsal body scales are keeled and occur in 23 to 27 rows at midbody; the ventrals are not keeled; the subcaudals are divided. The amount and toxicity of the venom is similar to that of *V. russelii*. Envenomation may produce severe nausea, vomiting and diarrhea. Other signs and symptoms of envenomation by this viper may include pain, discoloration of the skin, ecchymosis, swelling and edema about the wound, facial edema, pupillary dilation, weak pulse, albuminuria, proteinuria, hypotension, shock, hemorrhage in the area of the wound, and prolonged clotting time. ***V. mauritanica*** (Sahara rock viper). a viper, closely related to *V. lebetina*, found on stony hillsides with scrubby vegetation in the northwestern part of the Sahara to northwest Libya. It is most active during twilight hours and is not often encountered during the daytime as it remains hidden in crevices and tunnels. The average length of adults is about 1 m or so. It is dangerous to humans. The ground color is grayish, reddish, or brown with a series of oval or rectangular dark blotches that tend to fuse forming a dorsal zigzag stripe; the belly is pale and heavily clouded with dark gray. Its pattern is very similar to that of *V. x. palaestinae*. Average length is 35 to 45 inches. It is similar in appearance to *Cerastes*, but is distinguished from it by the absence of keels on the ventral scales and the lack of serrated keels on the lateral scales. The presence of paired subcaudals and lack of serrated keels on the lateral scales distinguish it from *Echis*. The lateral position of the nostrils, more slender body and fewer than 27 scale rows at midbody distinguish it from *Bitis arietans*. *Cf. V. latasti*. ***V. russelii*** (Russell's viper; chain viper; chain snake; chandra bora; daboia; daboya; glass viper; Ketten viper; kannadi virian; kunuadi vyrien; kusari hebi; kuturee pamhoo; scissors snake; seven pacer; tic polonga). a deadly, thick-bodied, brightly marked, chiefly nocturnal, terrestrial viper of southern and southeastern Asia from India to Sumatra, Java and Taiwan. It is easily distinguished from most other Asian snakes due to the wide, rather long head; the absence of large crown plates; lack of a loreal pit; and a bright coloration with distinctive markings. The ground color is light brown, tan, or deep yellow (some populations are more grayish or olive) marked by 3 rows of large, oval, dark, black-ringed spots, sometimes with a thin white edging; spots of the middle row on the posterior half of the body often fuse; the belly is pinkish brown to white, bearing black spots. The crown is marked by a light V or X. Adults average about 1.5 m in length, less in some populations; males are larger than females. Russell's viper is primarily a snake of open grassy or bushy country and is often seen in or near cultivated fields and villages. Some populations are found in hilly or even mountainous terrain. It typically moves slowly and appears sluggish. It is, however, the deadliest snake throughout its range, hissing loudly when disturbed and striking with great force and speed. Even though the case mortality rate is lower than that of bites by kraits, cobras and saw-scaled vipers, bites by Russell's viper are more frequent and cause more human deaths than any of these snakes, especially where it occurs in open country. Serious effects of envenomation often include pain, swelling, and edema, discoloration, prolonged clotting time, and internal hemorrhage. Also commonly seen are ecchymosis, necrosis, nausea, vomiting, thirst, weakness, diarrhea, abdominal pain, pupillary dilation, a weak pulse, hypotension, anemia, albuminuria, proteinuria, hemorrhage about the wound and sometimes from the gums, gastrointestinal and urinary tracts. The lethal dose of the venom for *Homo sapiens* is estimated at 40-70 mg, yet a large snake yields 150-250 mg when milked. The venom, in a 1:10,000 solution is used as a local coagulant in the management of hemor-rhage in hemophilia. ***V. superciliaris*** (domino viper; African lowland viper; tandaruma). a rare viper, known from the lowlands of Tanzania in central Africa and Mozambique. It is not generally considered dangerous to humans. ***V. ursinii*** (Karst adder; Renard's viper; Spitzen-kopfotter; Steppe adder; Ursini's viper; Wiesenotter). a species found in southern Europe through southeastern France to European Russia; east in the northern Mediteranean region into the Near and Middle East and Southeast Asia. It is the one species that has all 9 crown scutes. ***V. xanthina*** (Ottoman viper; Near East viper; Armenian mountain adder; bergotter). a nocturnal snake, the range of one subspecies, *V. x. bungarus* (Near East viper; Ottoman viper) extends into European Turkey. Otherwise the range extends from the Caucasus Mountains, northwestern Iran, and western Turkey, south to Israel and Jordan. It occurs in cultivated areas, often near human habitation as well as semiarid regions, where it occupies sites that support vegetation (e.g., river valleys). The ground color is sandy yellow, golden brown, gray or reddish brown with a series of oval or round spots with lighter centers and pale edging. These spots often fuse, forming a zigzag band. The average length of an adult is about 0.9 m maximum; rare specimens may attain a length of about 1.2 m. The head is large, but shorter than that of *V. lebetina*. The crown has a conspicuous dark, V-shaped mark or pair of dark, elongate spots and there is a conspicuous dark stripe behind the eye. The supraocular is undivided and there are usually 1 or 2 scale rows between the eye and upper labials; dorsal scales usually lie in 23 rows (occasionally 25) rows at midbody. ***V. x. palestinae***. a snake that bites more humans than any other in Israel and adjoining lands. The case of fatality rate is about 5%. Color pattern is very similar to that of *V. mauritanica*.

viperan. viperine.

vipère à cornes. *Bitis gabonica, B. nasicornis, Cerastes cerastes*.

vipère des pyramides. *Echis carinatus*.

vipère lèbètine. *Vipera lebetina*.

vipère rhinocèros. *Bitis nasicornis*.

vipère de l'Erg. *Cerastes vipera*.

vipère du Cap. *Bitis arietans*.

vipère hèbraique. *Bitis arietans*.

vipère heurtante. *Bitis arietans*.

vipère nasicornis. *Bitis nasicornis*.

viperian. viperine.

viperid. **1:** any snake of the family Viperidae. **2:** of or pertaining to Viperidae or to a snake of the family Viperidae. See also viperine.

viperid venom poisoning (viper venom poisoning). the result of envenomation by a viperid. The bite generally produces intense, burning pain with rapid onset of moderate to severe edema and associated swelling, ecchymosis, and patchy skin discoloration about the bite. Clotting time of blood is greatly extended and is a very helpful diagnostic feature. Common symptoms in viperid venom poisoning also include moderate to severe thirst, weakness, nausea, and vomiting, hemorrhage, and fairly protracted bleeding time. Extravasation of blood from the wound is a common feature of strikes by *Echis carinatus* and *Vipera russelii*. Bleeding (often severe) from the gums, GI, and urinary tracts occurs commonly also in these species. See also snake venom poisoning.

vipera cu corn. *Vipera ammodytes*.

Viperidae (true vipers; Old World vipers; Afri can vipers; adders; puff adders; viperids). a family of about 50 species of widely distributed venomous snakes, the Old World vipers and adders. All species are dangerous; some are deadly. Though widely distributed, the majority of species are indigenous to Africa; the largest in size occur in tropical and southern Africa. The paired erectile, caniculated fangs are long, recurved, and situated in the front of the upper jaw. They usually remain folded (recessed within a fold of the palate), but are joined to movable bones that permit erection of the fangs when the mouth opens during a strike. Viperids differ from crotalids by lack of a loreal pit. Some species (e.g., those of *Atheris*, *Bitis*, and *Vipera*) have broad heads that are distinct from the neck; the pupils of the eyes are vertically elliptical. Viperids of the genera *Atractaspis* and *Causus* have no external characteristics that distinguish them from nonvenomous snakes.

viperiform. viperine; resembling a viper.

viperine (viperan; viperian). **1:** viper-like; viperiform; resembling a viper. **2:** pertaining to or denoting any snake of the genus *Vipera* or to venomous snakes of similar appearance or behavior. **3:** sometimes used synonymously with viperid. **4:** venomous.

viperine sea snake. *Praescutata viperina*.

viperine snake. **1:** any snake of the family Viperidae. **2:** *Natrix viperinus*.

viperling. a young viper.

vipers, true. Viperidae.

Virginia creeper. *Parthenocissus quinquefolia*.

Virginia snakeroot. *Aristolochia serpentaria*.

viricidal. able to destroy or neutralize a virus.

viricide (virucide). destructive of viruses.

viriditoxin. a mycotoxin from *Aspergillus viridinutans*. It is extremely toxic (*p.o.*) to laboratory mice.

viridobufagin. a cardiac poison, $C_{23}H_{34}O_5$, from the skin glands of *Bufo viridis*.

virology. the natural science that treats viruses, their structure and behavior.

virtiginous. of or pertaining to vertigo; afflicted with vertigo.

virucide. viricide.

virulence. **1a:** deadliness, lethality, toxicity; the capacity to be virulent. **1b:** the ability of any pathogenic, noxious or poisonous agent to produce deleterious biological effects, especially as measured by the intensity, frequency, and/or injuriousness of effects. **1c:** the quality of being toxic; the degree of toxicity. **2a:** the ability of a microorganism or virus to overcome host resistance. **2b:** the capacity of a organism (especially a microorganism) or a virus to cause disease in a given host, as indicated by the median lethal dose, LD_{50}, or the median infective dose, ID_{50}. See, for example, toxin (microbial toxin). **2c:** the degree of destructiveness of a pathogenic organism (especially a microorganism) or virus, as indicated (usually) by the median lethal dose, LD_{50}, or the median infective dose, ID_{50}.

virulent. **1:** severe, malignant, or rapidly progressing; referring especially to any infection marked by rapid onset of severe symptoms due to the toxic effects produced by the pathogen. **2:** relating to or characterized by virulence; extremely toxic, pathogenic, or harmful. **3:** pathogenic, or able to overcome defenses of the host and cause disease. **4:** pertaining to any infectious disease or poisoning marked by rapid onset of severe symptoms. **5:** pertaining to or denoting an extremely pathogenic microorganism.

virulicidal. capable of destroying the potency of a pathogenic, noxious, or poisonous agent.

virus. a minute, obligate, intracellular, infectious, cytopathic agent. It is characterized by its simple structure and a mechanism of replication that employs genetic information in a single strand of DNA or RNA. The genetic material is covered by a protein, or sometimes a lipid, coat. A virus is metabolically inert when outside the host cell and is not usually considered to be a living organism. Because they pass through filters that retain bacteria, viruses are often termed filtrable. Viruses were formerly thought of as intracellular parasites.

viscera. the internal organs of an animal collectively.

visceral. of, relating to, or pertaining to the viscera.

visceral arch. **1:** one of a series of skeletal arches and associated soft tissue in the lateral walls of the pharynx of fish and in tetrapod embryos. These arches develop in relation to the mouth and pharynx and include the gill arches. They separate the mouth from the spiracle, the spiracle from the first gill slit, and the gill slits from each other. In fish they support the gill apparatus structurally and include the gill arches and gill bars. **2:** the bony or cartilaginous skeleton that supports a visceral arch. It is usually comprised of four components on each side, the most ventral of which articulate medially. The first, or mandibular, arch forms the jaw; the second forms elements that support the jaw; and the third and subsequent arches, termed branchial arches, support the gills.

visceral nervous system (involuntary nervous system). **1:** the autonomic nervous system of vertebrate animals, *q.v.* **2:** components of invertebrate systems that are functionally analogous to the autonomic system of vertebrate animals.

viscin. a glutinous principle obtained from the berries of *Viscum album* and from certain other plants.

viscum. **1:** *Viscum album* or its berries. **2:** the foliage of *Phoradendron flavescens*.

Viscum album (European mistletoe). a poisonous parasitic shrub (Family Loranthaceae) that grows on various trees (e.g., apple, pear); the true Old World mistletoe. It has been used as an oxytocic. Toxic principles include amines and phoratoxins. Fatalities have resulted from eating the berries. The leaves contain a toxic lectin. Large amounts are cardiotoxic and hepatitis-B-like symptoms have been reported for individuals using *V. album* remedies. See also *Phoradendron*.

viscumin. a phytotoxin of *Viscum album*.

visha. *Aconitum ferox*.

Visken. a beta-adrenergic blocking agent.

vital. **1:** of, pertaining to, or characteristic of life. **2:** essential to the maintenance of life, critical, integral.

vital capacity. the volume of air that can be expelled following maximum inspiration or that which can be inspired following a maximum exhalation. Loss of vital capacity may result, for example, from pulmonary fibrosis as seen in pneumoconioses, or to pneumonia. Either of which result from exposure to any of a large number of xenobiotics. See, for example, acrolein, aluminosis, ammonium hydroxide, anthracosilicosis, *Bayleya multiradiata*, asbestosis, bagassosis, baritosis, bituminosis, black lung, byssinosis, dust, hematite, nitrogen dioxide poisoning, pneumoconiosis, octane, paraquat, petroleum distillate, siderosis, turpentine, vanadium pentoxide.

vital sign. **1a:** any of the traditional signs of life such as breathing, heart beat, body temperature, pulse. **1b:** any indication that an organism is still alive (used especially in reference to vertebrates). In humans, for example, these include body temperature, the presence of a pulse (at the neck or wrist), breathing as evidenced by sensible exhalations or movements of the chest that are typical of breathing, blood pressure, and constriction of the eye when exposed to a bright light. Human subjects suffering from acute poisoning usually show changes in the vital signs. The latter may thus, in appropriate circumstances, indicate toxic exposure and the severity of intoxication.

vitality. **1:** a property peculiar to living organisms, that of being fully alive and vigorous. **2:** that quality of an organism that enables it to face and survive unfavorable environmental conditions.

vitals. the body parts and organs that are essential to life.

vitamin. usually a coenzyme that an animal does not synthesize and is thus acquired in the diet. Numerous studies have indicated that the minimum daily requirements (established many years ago) may not provide the full benefits of certain vitamins. Lipid soluble vitamins can accumulate in the body, sometimes causing serious problems. See, for example, vitamin A, vitamin A hypervitaminosis, vitamin K, vitamin D hypervitaminosis.

vitamin A (retinol). a moderately toxic, fat-soluble vitamin, $C_{20}H_{30}O$, produced by carotinoids in the liver and intestinal tract of vertebrates and stored in the liver. Vitamin A is essential to growth, development, normal skin, and the regeneration of visual purple in the retina. Symptoms of deficiency include night blindness and xerophthalmia. It is abundant in fish liver oils, eggs; and milk. Some mammals accumulate enough vitamin A in the liver to seriously poison a human being. See hypervitaminosis A, *Thalarctos maritimus*, *Erignathus barbatus*.

vitamin A vitaminosis. See hypervitaminosis A.

vitamin antagonist. antivitamin.

vitamin B_1. thiamine.

vitamin B_6. a water-soluble complex of vitamins concerned with amino acid metabolism, the degradation of tryptophan, and the breakdown of glycogen to glucose-1-phosphate. Pyradoxine, pyridoxine hydrochloride, pyridoxal, and pyridoxamine all possess vitamin B_6 activity. They occur in most foodstuffs, are readily absorbed in the gastrointestinal tract and converted to pyridoxal phosphate; excess amounts are rapidly excreted. Acute oral overdosage of these vitamins can cause tonic convulsions and death in laboratory rats; lesions revealed on autopsy include cerebral cortical hemorrhage and adrenal gland enlargement. A number of drugs and toxicants interfere with vitamin B_6 activity. See pyridoxine, pyridoxine hydrochloride.

vitamin B_6 hydrochloride. pyridoxine hydrochloride.

vitamin D_2 (9,10-secoergosta-5,7,10(19),22 tetraen-3-ol; calciferol, ergocalciferol, oleovitamin D_2, activated ergosterol, viosterol). an antirachitic vitamin and rodenticide.

vitamin deficiency. the chronic dietary intake of a suboptimal amount of one or more vitamins. Such a deficiency usually has little direct bearing on the behavior of toxic chemicals. The toxicity of some xenobiotics may be affected, however, due to alterations in metabolism. Certain poisons can act as vitamin antagonists, causing a virtual deficiency of the vitamin. See vitamin K antagonist, malnutrition, nutrition.

vitamin E. a nutritional factor present in many foods; deficiencies are rare. It is an antioxidant that is protective against ozone and nitrogen oxide exposure, at least *in vitro*. It is known as the antisterility vitamin.

vitamin K. a fat-soluble antihemorrhagic factor essential to the hepatic clotting factors such as prothrombin. A deficiency of vitamin K prolongs blood coagulation time, resulting in hemorrhage. Excessive vitamin K can cause hyperbilirubinemia. See vitamin K antagonist, warfarin.

vitamin K antagonist. any antimetabolite of vitamin K (e.g., hydroxycoumarins). Because they inhibit prothrombin synthesis, such substances are often used as rodenticides. Each new exposure to such a substance (e.g., warfarin) used as a bait, progressively reduces blood prothrombin levels. The affected rodent eventually dies of hemorrhagic shock. Accidental poisonings of humans and other animals have occurred. The administration of vitamin K can effectively reverse the process.

vitaminosis. See hypervitaminosis.

Viton™. lindane.

vitreous humor. the substance that occupies the space between the iris and retina of the eye.

vivification. revivification.

VLB. vinblastine.

VLDLP. very-low-density lipoprotein.

volatile. **1:** of or pertaining to a substance that evaporates readily. **2:** unstable; explosive; readily becoming violent or explosive; rapidly changing. **highly volatile**. a liquid that rapidly enters the gas phase when unconfined at room temperature (20° C).

volatile oil. protoanemonin.

volatile oil of mustard. allyl isothiocyanate.

volatilization. evaporation.

volatilize. evaporate.

volume of distribution. **1:** a method of determining the volume of fluids in a body fluid compartment (blood or blood plasma). In this method a solute is injected into the target compartment (e.g., inulin). When the solute is evenly distributed within the compartment, a sample is removed. The quantity of solute removed from the compartment (e.g., by metabolism, transport, excretion, etc.) relative to that which was administered can be determined by measuring a series of subsequent samples (as can the half-life of the solute in the compartment). **2:** a proportionality "constant" (V_d) that relates the amount of a solute in the body to the concentration in a body fluid such as blood or blood plasma. See also Michaelis-Menton kinetics, clearance (systemic clearance).

voluntary muscle. striated muscle.

voluntary nervous system. somatic nervous system.

volvent. See nematocyst.

vomit. **1:** to voluntarily or involuntarily expel, partially or completely, stomach contents through the mouth. See also vomiting, regurgitate. **2:** vomitus, the matter so expelled. **3:** an emetic.

vomit nut. nux vomica.

vomiting (emesis; vomition; vomitus). the voluntary or involuntary forcible ejection of partial or complete stomach contents through the esophagus and mouth. Vomiting provides protection against numerous noxious agents. Prolonged vomiting, however, may result in dehydration or the loss of essential electrolytes and can be fatal. Humans vomit readily when noxious agents are ingested or even tasted or smelled. Many species of livestock (e.g., cattle, horses, goats, and sheep) and wildlife do not readily vomit and are therefore more susceptible to many toxicants than are animals that vomit readily. **pernicious vomiting**. continuing, uncontrollable vomiting. **projectile vomiting**. vomiting with great force.

vomiting gas. **1:** a gas (e.g., chloropicrin) that causes vomiting and gastrointestinal disorders. **2:** chloropicrin.

vomiting sickness. See Jamaican vomiting sickness.

vomition. vomiting.

vomitive. emetic.

vomitory. **1:** emetic. **2:** an emetic.

vomiturition. retching.

vomitus. 1: vomit. **2:** vomiting.

vomitus cruentes. hematemesis.

VSD. virtual safe dose. See dose.

vulnerability analysis. assessment of components of a community that are susceptible to damage should a release of toxic or other hazardous materials occur.

vulnerable zone. the area over which the airborne concentration of a toxic or otherwise hazardous chemical involved in an accidental release could reach the level of concern.

vulva. the external genitalia of the female mammal that lie near the opening of the vagina. See also vulvovaginitis.

vulvovaginitis. inflammation of the vulva and vagina or the vulvovaginal glands. In livestock, this is sometimes caused by ingestion of fungus-infested corn (*Zea*). See fusarium estrogenism.

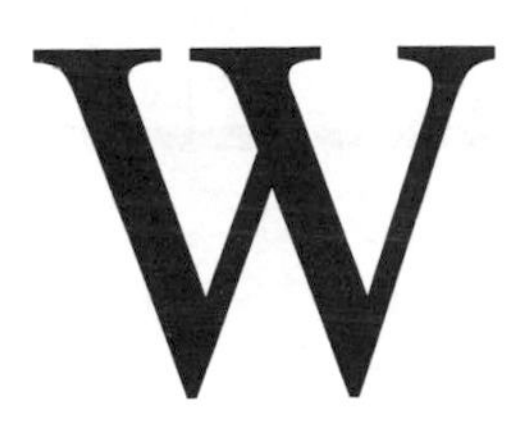

W. the symbol for tungsten.

Wagler's pit viper. *Trimeresurus wagleri*.

wahoo. *Euonymus atropurpurea*.

wakabi. *Naja melanoleuca*.

walking disease. a disease of domestic horses associated with frequent ingestion of rather small amounts of *Amsinckia* seeds (nutlets) over an extended period of time.

Wall's krait. *Bungarus walli*.

Wallerian degeneration (secondary degeneration; orthograde degeneration). the degeneration and ultimate destruction of an axon distal to the point of cutting or crushing. The characteristic pattern of degeneration includes a primary degeneration of the axon with secondary degeneration of the myelin sheath (if present) accompanied by axonal swelling, disintegration, tumefaction, inflammation and fibrosis. The phenomenon is due to separation of the axon from its trophic center.

wallflower. See *Erysimum cheiranthoides*

walnut. See *Juglans nigra*.

walnut stain. manganese(III) oxide.

walo-walo. *Lapemis hardwickii*.

Walter Innes's snake. *Walterinnesia aegyptia*.

Walterinnesia aegyptia (desert black snake; Walter Innes's snake; desert cobra; Egyptian blacksnake; Morgan's cobra; Walter Innes's snake; Wüstenkobra; achipeten; bargil; haia soda). a relatively rare, venomous snake (Family Elapidae) of desert areas of Egypt and the Near and Middle East. The average length of adults is about 1.2 m; the body is slender to moderately stout, cylindrical, and tapered, with a short tail. The head is compressed, and rather distinct from the neck; snout broad; and there is a distinct canthus. The eyes are of moderate size with round pupils. The crown is covered by the usually 9 scales; the rostral is broad. The nasal abuts the single elongate preocular. Adults are uniformly black, very dark brown, or gray above; somewhat lighter beneath. In some populations, young specimens may have light, narrow crossbands. This snake may be distinguished from other dark species (e.g., the Egyptian cobra) within its range by the lustrous scales. It can usually be distinguished from cobras and nonvenomous snakes by a set of features that includes the absence of a loreal plate; dorsal scales that are smooth anteriorly, but weakly keeled posteriorly; the anal plate is divided and most subcaudals are paired. While this species occupies rocky hillsides and sandy desert with sparse vegetation, human encounters occur most often in oases, gardens, and irrigated areas. It should be considered dangerous. It neither rears up nor spreads its hood before striking, but when disturbed may strike more than half its length. The toxicity of the venom is similar to that of the Indian cobra but the amount delivered is substantially less.

wamon-beni-hebi. *Calliophis macclellandii*.

wandering sea snake. *Laticauda laticauda*.

war gas. a chemical warfare agent delivered as a gas. See warfare agent.

war poison. a chemical warfare agent (See warfare agent).

Warburg's respiratory enzyme. cytochrome oxidase.

warfare agent. any material used for the purpose of killing or incapacitating enemy troops or depriving them of food or water. See also teleorganic (def. 1b). **biological warfare agent**. any of a large number of warfare agents. Included are a large number of microbial toxins and viral, bacterial, and fungal pathogens that are effective against plants or animals, including humans. Examples are toxins of staphylococci that cause severe food poisoning and pathogens of such human diseases as anthrax, plague, tularemia, Venezuelan equine encephalomyelitis, and coccidiomycosis. In addition to agents that directly attack troops, are those that cause plant or animal diseases such as foot-and-mouth disease, fowl plague, black stemrust of cereal grasses, potato blight, and rice blast disease. **chemical warfare agent** (military poison; military gas; war poison; warfare poison; war gas; chemical warfare gas). any toxic substance (other than a toxin) developed to kill or incapacitate enemy troops or to deprive them of meat, crops, or plant cover. Such agents have been used on civilian populations. Included are nerve agents, incapacitating agents, psychological agents designed to disorient troops, defoliants, and others. See, for example, nerve gas, Agent Orange, 2,4,5-T. **organophosphate warfare agent**. an organophosphate compound developed for use as a chemical warfare agent. These "nerve gases" are among the most toxic of synthetic compounds. They are potent inhibitors of acetylcholinesterase. Included are Sarin, Tabun (O-ethyl N,N-dimethylphosphoramidocyanidate), Soman (o-pinacolyl methylphosphonofluoridate), and "DF" (methylphosphonyldifluoride).

warfare poison. a chemical warfare agent (See warfare agent).

warfarin (4-hydroxy-3-(3-oxo-1-phenylbutyl)-2*H*-1-benzopyran-2-one; 3-(α-acetonylbenzyl)-4-hydroxycoumarin; 1-(4′-hydroxy-3′-coumarinyl)-1-phenyl-3-butanone; 3-α-phenyl-β-acetylethyl-4-hydroxycoumarin; compound 42). a synthetic coumarin anticoagulant, $C_{19}H_{16}O_4$. A vitamin K antagonist, it depresses the formation of prothrombin and increases capillary fragility, with increased hemorrhaging. It is used primarily as a rodenticide, usually applied as the sodium salt (warfarin sodium) or the potassium salt (warfarin potassium). Warfarin is named after the Wisconsin Alumni Research Foundation, where it was developed. **warfarin potassium**. a white, crystalline powder, $C_{19}H_{15}KO_4$, with anticoagulant activity and uses similar to those of warfarin sodium. **warfarin sodium**. a white, amorphous or crystalline powder, $C_{19}H_{15}NaO_4$. It is a cumulative anticoagulant with an action of intermediate duration, It is used as an oral rodenticide, but can also be effectively administered *i.v.* or *i.m.* It has the same action as dicumarol and is extremely toxic to female laboratory rats, very toxic to male laboratory rats and mice, but less toxic to laboratory rabbits. Symptoms of intoxication include hematuria, bloody feces, shock, coma and death.

warning coloration (aposematic coloration). the brilliant and striking colors and color patterns seen in some animals, especially insects. Such species are often distasteful or poisonous to predators. Because of the striking colors (and patterns) predators more easily learn to avoid the species. Some harmless species mimic the warning coloration and patterns of noxious species.

warning label. See label.

warning substance. Some organisms secrete substances that signal toxicity or danger to potential predators. The effects are similar to those of a warning coloration.

wartweed. *Euphorbia helioscopia*.

washing soda. sal soda.

washington blue veil. *Psilocybe stuntzii*.

Washington clam. *Saxidomus nuttalli*.

wasp. any of a very large number of insects (Order Hymenoptera), all having a slender body and an abdomen that is narrowly attached to the thorax. They have biting mouth parts, and a sting in the females and workers which can be used repeatedly for envenomation. Some species are social, other are solitary. Adults feed chiefly on insects and nectar. The nests are either constructed of paper-like material (masticated wood fiber), mud or other materials; some live in burrows. Vespidae is a prominent family. See also venom (hymenoptera venom, hornet venom), *Vespa*.

wasp venom. See hornet venom.

waspfish. *Centropogon australis*.

Wasserkobra. any snake of the genus *Boulengerina*.

wassermokassinschlange. *Agkistrodon piscivorus*.

waste. hazardous waste. toxic waste.

wasting. **1a:** enfeebling. **1b:** emaciating; loss of strength or excessive loss of body mass (especially of muscle and vital tissue). **1c:** to cause wasting (defs. 1a, 1b). See also phthisis.

water. **1a:** a clear, tasteless, odorless, polar-covalent liquid, H_2O; f.p. = 0°C (32°F), b.p. = 100°C (212°F). It is a universal component of living systems and is critical to life. It is often referred as the "universal solvent" because of its solvent power. Exogenous water is a vital nutrient to living systems, but water from natural water bodies, reservoirs, and other sources are often and variously contaminated with toxic chemicals. Concentrations of such chemicals within a given water body may fluctuate widely as a function of rate and turbulence of flow, turbidity, and other factors. Furthermore, nearly any major human disturbance (road-building, waste disposal, deforestation, mining, etc.) will alter the hydrologic cycle and the flow and concentrations of environmental chemicals. Water and water bodies can be important in transferring chemicals directly and indirectly (via food chains) to biota including humans. In addition to its critical importance to the survival of individual organisms, water and water flow (influx, efflux, and internal cycling) are vital to the continuing existence and structure of ecosystems. It is a major vehicle whereby chemical pollutants are transported to and within ecosystems. It is a major medium of chemical transport and the major medium in which chemical reactions take place. Uncontaminated fresh water is usually thought of as essentially nontoxic; but See water intoxication. See also poison (environmental poison), toxicity (aquatic toxicity). **1b:** the liquid phase of H_2O; the other phases are solid (ice) and vapor (steam). **2:** a euphemism for urine. **drinking water**. water from any source, treated (i.e., that from a municipal water supply) or untreated (e.g., that from a well) that is (1) considered sufficiently clean or pure to drink without harming one's health, or (2) normally taken in by humans or other animals, especially livestock. Contamination of drinking water is a major concern. Sources of contamination by xenobiotics include waste dumps, pollutants discharged into the atmosphere or directly into surface waters, including reservoirs that serve municipalities. See Recommended Maximum Contaminant Level. See also acid mine drainage, cadmium, carbofuran, chlorination, cyanobacteria, exposure limit, fluoridation, groundwater, salt poisoning, Suggested-no-adverse-response level, 1,2,4-trichlorobenzene, vinyl chloride. **groundwater, ground water**. subsurface water; all water beneath the soil surface that has no direct connection to the surface. Groundwater bodies lie over impervious strata. In many areas groundwater is major or even sole source of drinking water. Some groundwater bodies can be contaminated by chemicals from the surface by leaching or percolation through the soil and other strata above or by leakage or flow from waste dumps or mines. Because the rate of water flow within groundwater bodies is often negligible, contamination by persistent chemicals can be permanent and extremely difficult or impossible to remediate. See also phreatic and related terms, agricultural ecosystem, alachlor, cadmium. **surface water**. water in the hydrological cycle that is retained for a time, or moves on the surface of the ground (e.g., ponds, lakes, rivers, streams, impoundments, swamps, estuaries, oceans). Such water is readily contaminated from direct exposure to air, precipitation, dust, materials in surface run-off, or the activities and decomposition of living organisms. Also included under this term are water bodies that are partially underground (e.g., springs, wells, collectors) that contain fresh water and are directly influenced by above-ground water.

water arum. *Calla palustris*.

water beetle. a common name applied to any of a large variety of fresh-water beetles. See *Dysticus*.

water belly. toxic fat syndrome.

water bloom. See bloom.

water cobra. See *Boulengerina*.

water dropwort. *Oenanthe crocata*.

water hemlock. **1:** *Cicuta spp*., especially *C. maculata*. **2:** *Conium macalatum*.

water intoxication. **1a:** an excessive volume of free water in the body, due to dilutional hyponatremia. The condition may result from excessive intake of water; excessive use of tapwater enemas; the administration, *i.v.*, of hypotonic solutions; cerebral concussion; excess secretion of antidiuretic hormone (ADH); or hypothalamic tumors. **1b:** severe hyponatremia, a condition resulting from excessive retention of water with sodium depletion. Symptoms include nausea, vomiting, lethargy, confusion, excitability of muscle, convulsions, and coma. Recovery requires the restoration of fluid and electrolyte balance.

water lily. *Zigadenus venenosus*.

water moccasin. *Agkistrodon piscivorus*.

water quality. the quality of surface water with respect to its intended usage (e.g., human consumption, recreational use), based on a number of properties including the degree of pollution and factors that affect solubilities, transport rates, and toxicities of xenobiotics that may be present. A number of properties and constituents of water that influence the solubilities, transport rates and toxicities of xenobiotics (e.g., alkalinity, conductivity, hardness).

water solubility. the maximum concentration of a substance that can result when it is dissolved in water. If a substance is water soluble it can very readily disperse in the environment.

weaning index. an index of survival to weaning of offspring in a reproductive toxicity test. In the laboratory rat, it is the number of offspring surviving to weaning (taken as 21 days) as a percentage of those alive at four days.

weavers' broom. *Sparteum junceum* (*Cytisus scoparius*).

weedkiller. herbicide.

weever (weever fish; weeverfish; wivre). any of several small, edible, bony marine fishes (Family Trichinidae) that are restricted to the littoral zones of the eastern Atlantic Ocean and the Mediterranean Sea. They are especially common in certain shallow waters with sandy bottoms. The head is broad, eyes dorsal, and dorsal fins bear many long, sharp venomous spines. The venom apparatus consists of two opercular spines, five to eight dorsal spines, and secretory tissues, all of which are encased by integumental sheaths. The venom is secreted into, and retained within, various spinal grooves. Envenomation may be via single or multiple puncture wounds. The pain from such wounds is intense, lasting from 24-48 hours. Low-grade infections from the sting and necrosis about the wound are not uncommon. Stinging of humans is common in some areas, but fatalities are rare. The name is apparently a corruption of the Anglo-Saxon "wivre," which means viper. See *Trachinus*.

weever fish, weeverfish. weever.

Weid's lancehead. *Bothrops neuwiedi*.

welder's lung. a relatively benign pneumoconiosis caused by the deposition of fine metallic particles in the lung of welders.

Wernicke's disease. Wernicke's syndrome.

Wernicke's encephalopathy. Wernicke's syndrome.

Wernicke's syndrome (Wernicke's disease; Wernicke's encephalopathy; superior hemorrhagic polioencephalitis). a condition due chiefly to thiamine deficiency that is often seen in chronic alcoholics. It is characterized by pupillary alterations, disturbances in ocular motility, nystagmus, and ataxia with tremors; an organic-toxic psychosis is often an associated finding. It is often accompanied by Korsakoff's syndrome and may then be termed Wernicke-Korsakoff syndrome.

Wernicke-Korsakoff syndrome. combined or coexisting Wernicke's and Korsakoff's syndromes.

werr. *Denisonia coronata*.

West African mamba. *Dendroaspis viridis*.

West African tree viper. *Atheris chlorechis*.

West Indian racer. any snake of the genus *Alsophis*.

western azalea. *Rhododendron occidentale.*

western diamondback rattlesnake. *Crotalus atrox.*

western duck sickness. See alkali disease.

western false hellebore. *Veratrum woodii.*

western flat-topped. *Agaricus meleagris.*

western green mamba. *Dendroaspis viridis.*

western mole viper. *Atractaspis corpulenta.*

western monkshood. *Aconitum columbianum.*

western poison oak. See *toxicodendron diversilobum.*

western sand adder. *Vipera ammodytes.*

western stinging nettle. *Hesperocnide.*

western (water) hemlock. *Cicuta occidentalis.*

western yellow pine. See *Pinus ponderosa* (or sometimes *P. jeffreyi*).

Western diamond rattlesnake. *Crotalus atrox.*

Western diamondback. *Crotalus atrox.*

Western hog-nosed viper. *Bothrops lansbergii.*

Western mole viper. *Atractaspis corpulenta.*

Western pygmy rattlesnake. *Sistrurus miliarius.*

Western rattlesnake. *Crotalus viridis.*

Western yellow pine. See *Pinus ponderosa*; (sometimes refers to *P. jeffreyi*).

wet deposition. See acid deposition.

wet lung. an edematous lung.

WFNA. white fuming nitric acid. See nitric acid.

wharf rat. *Rattus norvegicus*. See also laboratory rat.

wheal. a welt; a smooth, circumscribed, slightly elevated, local, transitory area of edema of the skin, which appears as a slightly reddened urticarial lesion. It may change in size and shape and spread to adjacent areas and is typically accompanied by intense itching. Such lesions are often the result of exposure to allergenic substances in susceptible persons and are evidence of allergy.

wheat. *Triticum aestivum*

wheat grasses. *Agropyron* spp. *See Claviceps.*

wheat pasture poisoning. grass tetany.

wheeze. **1a:** a high-pitched whistling or sighing sound marked by a high-velocity flow of air, during inspiration and/or expiration, through a narrowed airway. Asthma, chronic bronchitis, croup, hay fever, for example, may cause a wheeze on both expiration and inspiration. An obstruction in an airway, inflammatory stenosing lesions, and bronchogenic carcinoma may, for example, cause a unilateral wheeze. **1b:** to breathe with a wheeze.

wheezing. the process of breathing with a wheeze.

whelk. See *Busycon.*

whip snake. See *Demansia, Denisonia.*

whip-tailed sting ray. a stingray, *q.v.*, of the family Dasyatidae.

whipray. a stingray, *q.v.*, of the family Dasyatidae.

whisper of death. curare.

Whistlers. See Pachycephalidae.

white arsenic. arsenic trioxide.

white baneberry. *Actaea (= Actea) alba.*

white bryony. *Bryonia alba.*

white camas. *Zigadenus elegans.*

white caustic. sodium hydroxide.

white cedar. *Thuja occidentalis, Melia azedarach.*

white clover. *Trifolium repens*.

white cohosh. *Actaea (= Actea) alba*.

white damp. See damp.

white-flowered rhododendron. *Rhododendron albiflorum*.

white hellebore. *Veratrum viride*.

white horse nettle. *Solanum eleagnifolium*.

white lead. basic lead carbonate.

white-lipped bamboo viper. *Trimeresurus albolabris*.

white-lipped pit viper. *Trimeresurus albolabris*.

white-lipped snake. *Denisonia coronoides*.

white-lipped tree viper. *Trimeresurus albolabris*.

white loco. See *Oxytropis*.

white mercuric precipitate. ammoniated mercuric chloride (See mercuric chloride).

white mineral oil. See mineral oil.

white mussel. *Donax serra*.

white mustard. *Brassica hirta*.

white-nosed snake. *Pseudoechis australis*.

white point loco. See *Oxytropis*.

white precipitate. ammoniated mercuric chloride (See mercuric chloride). Do not confuse with the French precipité blanc ("white precipitate") which is mercurous chloride.

white quebracho. *Aspidosperma quebracho-blanco*. See also quebracho.

white ragweed. *Franseria discolor*.

β-methylindole. skatole.

white sanicle. *Eupatorium rugosum*.

white sassafras. *Sassafras albidium*.

white sea urchin. *Tripneustes ventricosus*.

white snakeroot. *Eupatorium rugosum*.

white-spotted puffer. *Arothron meleagris*.

white sweet clover. *Melilotus officinalis*.

white tar. naphthalene.

whorehouse tea. *Ephedra nevadensis*.

whorled milkweed. *Asclepias subverticillata*.

wicky. *Kalmia angustifolia*.

widow spider. See *Latrodectus*.

Wiesenotter. *Vipera ursinii*.

Wijs' chloride. iodine monochloride.

wild allamanda. *Urechites suberecta*.

wild balsam-apple. *Momordica charantia*.

wild black cherry. *Prunus serotina*.

wild cabbage. *Brassica oleracea*.

wild calla. *Calla palustris*. Note that the common name, calla, usually refers to the calla lily of the genus *Zantedeschia*.

wild carrot. *Daucus carota*.

wild celery. See *Angelica*.

wild cherry. *Prunus serotina*.

wild fire toxin. a fungal phytotoxin, secreted by *Pseudomonas tabici*, which infects tobacco. This toxin is a dipeptide with a threonine and a hydroxy-diamino-diacid moiety. It is toxic to chloroplasts and causes leaf chlorosis, possibly by acting on glutamine synthetase.

wild ginger. a common name of plants of the genus *Asarum*.

wild grape. See *Rhoicissus cuneifolia*.

wild hippo. *Euphorbia corollata*.

wild hyacinth. *Scilla*.

wild hydrangea. *Hydrangea arborescens*.

wild indigo. *Baptisia tinctoria*.

wild ipecac. **1:** *Apocynum androsaemifolium*. **2:** *Euphorbia*.

wild jasmine. *Cestrum diurnum*.

wild licorice. *Abrus precatorius*.

wild monkshood. See *Aconitum uncinatum*.

wild mustard. *Brassica kaber*.

wild onion. *Zigadenus venenosus*.

wild parsley. *Aethusa cynapium*.

wild pea. See *Lathyrus*.

wild radish. *Raphanus raphanistrum, Raphanus sativus*.

wild rodent. See rodentia.

wild rosemary. *Ledum palustre*.

wild rye. *Elymus spp*. See *Claviceps*.

wild sage. *Lantana camara*.

wild sunflower. *Helianthus annuus*.

wild tobacco. See *Nicotiana glauca*.

wild tomato. *Solanum carolinense*.

wildfire disease. a disease of tobacco caused by wildfire toxin produced by *Pseudomonas tabaci* (See toxin).

willow-leaved jessamine. *Cestrum parqui*.

Wilson's disease. **1:** a heritable disease characterized by tremor, muscular rigidity, involuntary movements, spastic contractions, psychic disturbances, dysphagia, and progressive weakness and emaciation. A pigmented ring (Kayser-Fleischer ring) about the margin of the cornea is pathognomonic. The physiologic basis for this disease is the loss of ceruloplasmin in circulating blood and the resulting accumulation of copper in organs such as the brain, cornea, liver, and kidneys. Lesions include degenerative changes in brain tissue, hepatic cirrhosis, splenomegaly. See also ceruloplasmin, trientine, dermatitis (exfoliative dermatitis). **2:** dermatitis exfoliativa. See exfoliative dermatitis (See dermatitis).

windflower. *Anemone* spp.

winter aconite. *Helleborus viridis*.

winter fern. *Conium macalatum*.

wintergreen oil. methyl salicylate.

Wintersteiner's compound F. cortisone.

wintersweet. Chimonanthus.

Winton disease. Pictou disease.

wisteria. See *Wisteria*.

Wisteria (wisteria). a genus of deciduous, twining vines (Family Fabaceae, formerly Leguminosae) that are native to Asia and North America. The seeds are very toxic. ***W. sinensis*** (Chinese wisteria; wisteria). a vigorous climbing vine, growing to nearly 10 m. The pods are toxic and can cause severe gastrointestinal symptoms and collapse. The active principles are resin and the glucoside, wisterin.

witchgrass. *Panicum capillare*

witchweed. *Veratrum viride*.

withdrawal. removal, departure, retreat, revocation; detachment. See approach-withdrawal. See also withdrawal syndrome.

withdrawal syndrome. a condition characterized by untoward physiologic changes that occur when the use of a drug is discontinued or when its effect is counteracted by a specific antagonist.

wivre. See weever.

WL. Working Level.

WLM. Working Level Month.

wobbles. a disorder of cattle in South America produced by grazing on *Zamia integrifolia* and related members of the genus. The condition is characterized by atactic movements of the hind legs and peculiarities of stance. Symptoms appear several weeks following ingestion of toxic amounts of *Zamia* spp. The injury is permanent.

wolf's milk. *Euphorbia esula*.

wolfbane, wolfsbane. See *Aconitum*.

wolfram. former name for the element tungsten.

wolfram white. barium tungstate.

wolframium. a former name of tungsten.

wooly-pod milkweed. *Asclepias eriocarpa*.

woman (plural, women). **1a:** a female human, especially an adult. **1b:** the female member of any species of the family Hominidae as distinguished from the male. Female and male *Homo sapiens* have a number of toxicological distinctions relating chiefly to differences in endocrinology and reproductive biology. These differences have by no means been adequately studied or clarified. The poisoning of an adult male usually directly affects only him, whereas the poisoning of a female before or during pregnancy, or during lactation, may harm her offspring, with or without harm to the woman. As one example, while intake of alcoholic beverages or drugs during pregnancy may not have serious or lasting effects on the woman, it can cause death or irreversible impairment of the fetus or newborn. Thalidomide is a mild sedative that rarely causes problems for the woman, but causes drastic malformations in the fetus if taken during the first trimester of pregnancy. See also human, hominid, man, thalidomide, ethanol.

womb. See uterus.

women. plural of woman.

wonderberry. *Solanum intrusum*.

wood alcohol. methanol.

wood naphtha. methanol.

wood preservative. any compound (e.g., creosote, pentachlorophenol, and arsenicals) used to protect wood from decay or pests. In 1985 the U.S. Environmental Protection Agency banned public sale of the above named preservatives because they are demonstrated carcinogens and mutagens in animals. Only certified individuals may now purchase these substances.

woodbine. *Parthenocissus quinquefolia*.

woody aster. a common name of herbs of the genus *Xanthosoma* and *Xylorrhiza* (= *Aster*).

Woolf equation. See Woolf plot.

Woolf plot. one of a number of methods (e.g., Eadie-Hofstee plot, Lineweaver-Burk plot), based on transformations of the Michaelis-Menton equation, that are used to linearize enzyme or receptor-binding data. For receptor-binding, the Woolf equation is:

$$[F]/B = (l/B_{max})[F] + K_d/B_{max}$$

where [F] is the concentration of free ligand, B is the amount bound, K_d is the dissociation constant, and B_{max} is the theoretical number of sites. A plot of [F]/B versus [F] produces a straight line with a Y intercept of Kd/B_{max} and a slope of l/B_{max}.

wooly croton. *Croton capitatus*.

wooly loco. *Astragalus mollissimus*.

woorari. curare.

Wooton loco. *Agave wootonii*, a true locoweed.

Working Level (WL). a unit of measure used to document exposure to radon decay products. One working level is equal to approximately 200 picocuries per liter.

Working Level Month (WLM). a unit of measure used to determine cumulative exposure to radon. See Working Level.

workplace. See industrial chemicals.

worm-eating viper. *Adenorhinos barbouri*.

worm lizard. any reptile of the family Amphisbaenidae.

wormseed. *Chenopodium ambrosioides*.

wormseed mustard. *Erysimum cheiranthoides*.

wormseed oil. a dangerous anthelmintic oil expressed from the seeds of *Chenopodium ambrosioides*. It contains ascaridole, a toxic terpene. Symptoms of intoxication are those of acute gastroenteritis; an overdose can be fatal.

wormwood. *Artemisia absinthium*.

wort weed. See *Chelidonium*.

wourara. curare.

wrasse. common name of many fishes of the Family Labridae. See *Coris gaimardi*, *Epibulus insidiator*.

wryneck. torticollis.

Wuchstoff. auxin.

Wüstenkobra. *Walterinnesia aegyptia*.

wutu. *Bothrops alternatus*.

wyree. *Demansia psammophis*.

X-radiation. See X-ray.

X-ray (Roentgen ray; X-radiation). electromagnetic radiation of extremely short wavelength (0.06-120Å) that is emitted when electron transitions occur in the inner orbits of heavy atoms in response to heavy bombardment by cathode rays in a vacuum tube. X-rays can be extremely dangerous to living things. Radiologists and others in the health fields who work with X-rays are at significantly greater risk from skin cancer and leukemia than the general population, as are patients who are subjected to X-ray examination; X-rays are mutagenic and cytotoxic. Extreme exposures can be lethal. See also mutagen, ozone, Roentgenism. **hard X-ray**. any X-radiation of the shortest wavelengths. Such are the most intense and dangerous to living organisms.

X-ray dermatitis. radiodermatitis.

xanthine. **1:** a yellowish-white water- and acid-insoluble powder, $C_5H_4N_4O_2$, that is soluble in potassium hydroxide. It sublimates in air with partial decomposition. It is a toxic purine base that occurs, for example, in vertebrate muscle tissue, liver, spleen, pancreas, blood, and urine. It also occurs in the tissues of certain plants. It is produced during the metabolism of nucleoproteins and can be produced synthetically by the action of nitrous acid on guanine. **2:** any derivative of xanthine. Examples are methylxanthine; theobromine and theophylline which are dimethylxanthines; and caffeine, a trimethylxanthine. See also xanthine oxidase.

xanthine oxidase. a metalloflavoprotein, closely related to aldehyde oxidase, that occurs in animal tissues. Both have a mass of about 300 kd and broadly overlapping substrate specificities. Xanthine oxidase catalyzes, for example, the hydroxylation of hypoxanthine, xanthine, aldehydes, and reduced coenzyme I, with transfer of electrons from the substrate to a variety of acceptors. Both enzymes appear to play a role in two types of detoxication: (1) the hydroxylation of exogenous aldehydes, purines, pyrimidines and other heterocyclic compounds; and (2) the transfer of electrons to exogenous compounds (e.g., reduction of nitrogen oxides to the free base).

Xanthium (cocklebur; clotbur). a genus of toxic, semicosmopolitan, coarse, weedy, annual herbs (Family Compositae). All species have similar toxicities and in many reports of poisoning it is difficult to determine which of the approximately 20 species were involved, although *X. spinosum* and *X. strumarium* have often been implicated in North America. One or the other occurs in fields and wastelands throughout most of North America. Poisoning of livestock is common throughout much of the United States. The seeds are very toxic due to the presence of hydroquinone, but are rarely eaten because they are contained in the burrs. The hydroquinone disperses throughout the plant, however, when the seed germinates. Thus, while the very young plants have a pleasant taste, they are toxic. All known cases of poisoning by this plant are due to the ingestion of seedlings in the cotyledonary stage of growth. Symptoms are similar in all classes of livestock and usually include anorexia, depression, nausea, vomiting, muscular weakness, weakened cardiac contractions, dyspnea, and prostration. Emprosthotonos and convulsions sometimes occur in severe cases. Vomiting may not occur in cattle. Swine are frequent victims.

xanthoascin. a cardiotoxic mycotoxin from *Aspergillus candidus*. It causes ventricular dilatation and accumulation of myeloid filaments in the myocardium.

xanthocillin-X. a mycotoxin from *Aspergillus chevalieri*. It is extremely toxic to laboratory mice and rats, rather less toxic to guinea pigs.

xanthopsia (yellow vision). a condition in which objects appear yellow. It can result from picric acid poisoning, santonin poisoning, icterus, or digitalis intoxication.

Xanthosoma (woody aster; yautia; tanier; malanga; caladium). a genus of tropical American herbs (Family Araceae) with thick tubers. Some species are easily confused with and have been grouped with the closely related *Caladium*. Both groups share "caladium" as a common name. If material from these plants is chewed, an intense burning sensation, irritation and edema rapidly follow. This is probably due to the presence of calcium oxalate crystals. All plants of this genus are also obligate accumulators of selenium from soil. ***X. violaceum*** (malanga). a species cultivated in South America for its edible, root-like, underground tubers. Some parts of this plant are toxic as described above. *Cf. Dieffenbachia, q.v.*

xanthostrumarin. the glycosidic fraction of *Xanthium* (*q.v.*) seeds. It is toxic, but is not the cause of livestock poisoning by these plants as was formerly thought.

xanthothricin. toxoflavin.

xanthotoxin. methoxsalen.

Xe. the symbol for xenon.

xeno-. a prefix meaning strange, foreign.

xenobiotic. **1a:** foreign to a living system. **1b:** a foreign compound; any natural or synthetic compound that is foreign to a living system. **1c:** any substance that enters and interacts with an organism and is neither endogenously produced or part of a normal metabolic pathway of the organism. **1d:** in toxicology and allied fields, a pharmacologically, endocrinologically, or toxicologically active substance that is exogenous in origin and does not occur in the normal intermediary metabolism of a living organism. **2:** a foreign (allochthonous) organic chemical. **3:** xenogeneic.

xenobiotic agent. See xenobiotic (def. 1c).

xenobiotic substance. any natural or synthetic substance that is foreign to a living system; an exogenous substance that is not produced endogenously and is not part of a normal metabolic pathway. See also xenobiotic.

xenobiotic transformation. See metabolism (xenobiotic metabolism).

xenogeneic. xenobiotic; xenogenic; heterologous with respect to a tissue graft taken from another species, especially one not closely related to the recipient.

xenogenic (xenogenous). **1a:** exogenous; allochthonous; originating from outside the organism or system that is the subject of concern. **1b:** produced by a foreign compound within the organism or system of concern. **2:** xenogeneic.

xenogenous. xenogenic.

xenon (symbol, Xe)[Z = 54; A_r = 131.3]. a colorless, odorless, monatomic gas of the rare-gas series of elements. It is stable and unreactive under most natural conditions.

xenylamine. *p*-biphenylamine.

***p*-xenylamine**. *p*-aminodiphenyl.

xeroderma pigmentosum (xerodermia pigmentosum). a rare, often fatal skin disease in humans that is due to an hereditary sensitivity to the carcinogenic effects of UV light. The disease is progressive from childhood. Skin and eyes become very sensitive to light. Eventually ocular damage is severe; tumors of the eyelid and cornea appear; and the skin develops keratoses, papillomas, carcinomas, and melanomas. Total protection from sunlight, if instituted early, completely reverses these effects.

xerodermia pigmentosum. xeroderma pigmentosum.

xerostomia. dryness of the mouth due to an abnormal reduction in the flow of saliva. There are a variety of causes (e.g., fear or hysteria, diabetes, acute infections, certain neuroses). It is a common adverse reaction to drugs or toxic substances (e.g., atropine, nicotine).

Xiphosura. See Merostomata.

xylene (dimethylbenzene; xylol). a clear, colorless, toxic, anesthetic, flammable liquid, $C_6H_4(CH_3)_2$, sold commercially as a mixture of three isomers. It is used widely as an industrial solvent for paints, varnishes, inks, dyes, cements, and cleaning fluids. It is used also in the manufacture of plastics, synthetic textiles, perfumes, and drugs. It is toxic by contact, inhalation, and ingestion. Exposure to small amounts irritates the eyes, nose, throat, and skin. Ingestion or inhalation may cause a burning sensation in the mouth and stomach, nausea, vomiting, chest pains, coughing, headache, dizziness, ataxia, confusion, stupor, restless coma, and death from respiratory failure or ventricular fibrillation. Lesions produced by large doses include CNS depression, liver and kidney damage, severe and possibly fatal bone marrow damage, and late severe blood dyscrasias. The toxicities of the three isomers (*m*-xylene, *o*-xylene, *p*-xylene) are similar.

Xylocopa (carpenter bees). a genus of robust, mostly tropical bees (Family Xylocopidae) that bore into and nest in dry, solid wood. The venom is acidic, containing hemolytic enzymes. The sting causes local pain and a burning sensation, with swelling, erythema, whitishness about the wound, respiratory distress, and shock. Multiple stings may require hospitalization. Death from anaphylactic shock occurs, but is rare.

L_S-xylokainic acid. kainic acid.

xylol. xylene.

Xylorrhiza (= *Aster*; = *Machaeranthera*, in part, woody asters). a genus of plants (Family Compositaceae) closely allied with *Aster*; some authorities do not consider woody asters sufficiently distinct to rank as a separate genus. All species of woody aster are selenium accumulators and most are reliable indicators of selenium-rich soils. The selenium content of woody asters varies with soil content and perhaps with species. Thus consumption of these plants by livestock may produce chronic or acute selenium poisoning depending upon the selenium content and the proportion of forage comprised by *Xylorrhiza*. Even though plants of this genus are generally unpalatable to sheep, extensive losses have occurred as a result of consuming woody aster, especially *X. parryi*, when other forage is limited.

Y. the symbol for yttrium.

yageine. harmine.

yamuhando. *Dispholidus typus*.

yangalukwe. **1:** *Dispholidus typus*. **2:** *Thelotornis kirtlandii*.

yarara. *Bothrops jajaraca*.

yarara nata. *Bothrops ammodytoides*.

yararaguassu. *Bothrops jararacussu*.

yautia. a common name of herbs of the genus *Xanthosoma*.

Yb. the symbol for ytterbium.

yellow arsenic sulfide. arsenic trisulfide.

yellow atrophy. See acute yellow atrophy.

yellow-banded snake. *Hoplocephalus stephensii*.

yellow-bellied sea snake. *Pelamis platurus*.

yellow cobra. *Naja nivea*.

yellow cross. mustard gas (See gas).

yellow cross liquid. mustard gas (See gas).

yellow enzyme. flavoprotein.

yellow-faced whip snake. *Demansia psammophis*.

yellow-foot agaricus. *Agaricus xanthodermus*.

yellow-headed krait. *Bungarus flaviceps*.

yellow-headed snake. *Bungarus flaviceps*.

yellow jacket. See Vespidae.

yellow jasmine. *Gelsemium sempervirens*.

yellow jessamine. *Gelsemium sempervirens*.

yellow latex milky. *Lactarius vinaceorufescens*.

yellow-lined palm viper. *Bothrops lateralis*.

yellow-lipped sea krait. *Laticauda colubrina*.

yellow lupin bean. *Lupinus luteus*. See sparteine.

yellow-naped snake. *Aspidomorphus christicanus*.

yellow nightshade. *Urechites suberecta*.

yellow oat grass, yellow oats. *Trisetum flavescens*.

yellow oleander. *Thevetia peruviana*.

yellow-orange fly agaric. *Amanita muscaria* var. *formosa*.

yellow parilla. *Menispermum canadense*.

yellow pine flax. *Linum neomexicanum*.

yellow potassium prussiate. potassium ferrocyanide.

yellow precipitate. See mercuric oxide, yellow.

yellow prussiate of potash. potassium ferrocyanide.

yellow prussiate of soda. sodium ferrocyanide.

yellow pyoktanin. auramine hydrochloride.

yellow sea snake. *Hydrophis spiralis*.

yellow sorrel. *Rumex crispus*.

yellow-spotted palm viper. *Bothrops nigroviridis*.

yellow-spotted pit viper. *Trimeresurus flavomaculatus*.

yellow-spotted snake. *Hoplocephalus bungaroides*.

yellow sweet clover. *Melilotus officinalis*.

yellow-tipped coral. *Ramaria formosa*, a poisonous mushroom.

yellow turnip. *Brassica napus* var. *napobrassica*.

yellow vision. xanthopsia.

yellowed rice toxins. any of several hepatotoxic hydroxyquinones (e.g., luteoskyrin, rubroskyrin, rugulosin) produced by various species of *Penicillia*.

yennai viriyan. *Bungarus caeruleus*.

yettadi viriyan. *Bungarus caeruleus*.

yew. plants of the family Taxaceae. See also *Taxus*.

yew poisoning. poisoning, usually of livestock, by ingesting the needles or berries of certain taxine-containing, woody, evergreen shrubs and trees of the genus *Taxus*. Symptoms of acute yew poisoning in humans are those of gastroenteritis. Death from cardiac or respiratory failure may intervene without further symptoms. See especially *Taxus baccata*. See also taxine.

Yodoxin™. iododquinol.

Yperite. mustard gas (See gas).

ytterbium (symbol, Yb)[Z = 70; A_r = 173.04]. a soft, malleable, ductile, silvery element of the lanthanoid series of metals; it darkens on exposure to light. Some salts such as ytterbium fluoride, YbF_3, and ytterbium nitrate, $Yb(NO_3)_3$, are toxic. The oxide (yttria; yttrium oxide), O_3Y_2, is very toxic to laboratory rats.

Z

Z. the symbol for proton number.

zakra. *Causus rhombeatus*.

Zamia. a genus of perennial tropical and subtropical plants (Family Zamiaceae). They superficially resemble both palms and ferns, but are not closely related to either. ***Z. integrifolia*** (coontie; Florida arrowroot). a woody, fern-like, plant with palm-like leaves that arise from a thick, underground stem. They occur in dry, sandy soils throughout peninsular Florida. The leaves are commonly used as decorative foliage by florists. This and other species of *Zamia* cause wobbles in cattle.

Zantedeschia aethiopica (calla; calla lily). a sturdy, bulbous plant that grows to a height of about 0.8 m. The leaves are smooth and arrow-shaped; the creamy white spathe is lily-like, flaring out but coming to a pointed tip. Biting into any part of the plant produces effects similar to those produced by *Dieffenbachia*, *q.v.* It is unrelated to plants of the genus *Calla*.

Zea mays (maize; corn; Indian corn). an important grain, forage, and fodder plant (Family Cyperaceae) that presumably originated in Mexico and now consists of countless varieties. It is known to cause a number of illnesses in livestock. Problems may result, for example, from consumption of moldy corn or of forage or silage that contains unusually high concentrations of nitrate; *Zea mays* also has a cyanogenic potential. Poisoning of cattle by forage with high levels of nitrate can be fatal. Corn stalks may contain nitrate burdens, KNO_3, as high as 8%. Silage may contain high concentrations of nitrogen dioxide and nitrogen tetroxide. See also silo-filler's disease.

zearalenone (F_2 toxin; 3,4,5,6,9,10-hexahydro-14,16-dihydroxy-3-methyl-1*H*-2-benzoxacyclotetradecin-1,7(8H)-dione; 6-(10-hydroxy-6-oxo-*trans*-1-undecenyl)-β-resorcylic acid lactone; Compound F-2; FES). a mycotoxin with the general formula, $C_{18}H_{21}O_5$, produced by various strains of *Fusarium* that infest corn (e.g., *F. roseum*, *F. graminearum*). It is the only known mycotoxin whose effects are primarily estrogenic. It causes estrogenation in swine: females feeding on infested corn suffer symptoms of hyperfolliculinia with a high abortion rate and males develop female sexual characteristics. Zearalenone has been detected in corn, oats, barley, wheat, and sorghum (both fresh and stored); in compounded cattle and swine rations; in corn ensiled at the green stage; and in hay. Samples of the above-named plants from pastures in temperate climates occasionally contain zearalenone at levels thought to be sufficient to cause reproductive failure of grazing herbivores. It also occurs widely in animal tissues. It is carcinogenic to some laboratory animals.

zearalenone hyperestrogenism. *Fusarium* estrogenism.

zebrafish. a common name of fish of the genus *Pterois*.

zebu. *Bos indicus*.

zeilen-seeschlange. *Laticauda laticauda*.

zeitgeber. See circadian rhythm, diurnal rhythm.

zephyr lily. See *Zephyranthes*.

Zephyranthes (zephyr lily). a genus of bulbous, usually attractive plants (Family Amaryllidaceae, *q.v.*) most of which are native to the temperate and tropical areas of the Americas. ***Z. atamasco*** (atamasco lily; rain lily). the common white, spring-blooming species found in wooded areas of the southern United States from Virginia to Florida, west to the coastal plains of Mississippi. It causes "staggers" in horses. The bulb is the most toxic part of the plant. It is extremely toxic to livestock. Symptoms appear within 2 days of foraging on the bulbs or leaves of this plant. They include staggering, soft feces with bloody mucus, collapse, and death. Most poisonings occur during the spring when other forage is in short supply.

zero tolerance. See tolerance.

zetekitoxin. any of a class of extremely toxic principles from frogs of the genus *Atelopus*. An example is tetrodotoxin, *q.v.*

Zigadenus (*Zygadenus*; black snake root; death camas). a genus of North American and Asian perennial herbs (Family Liliaceae) with rhizomes or bulbs, grass-like leaves and whitish to yellow, greenish, or bronze flowers in clusters. The seeds, leaves, and bulbs of most species contain several toxic veratrum alkaloids and are among the most important poisonous plants of western North America. They have caused extensive loss of life among sheep on spring range. Toxicity varies considerably among species, but symptoms are remarkably similar. The minimum lethal dose of Z. *nuttallii* in sheep is about 0.8% that of Z. *elegans*. Individual species are difficult to identify, and all species should be considered hazardous. Initial symptoms in sheep include persistent excessive salivation, nausea, and sometimes emesis. Muscular weakness, bradycardia, hypotension, ataxia, tremors, and prostration follow. The pulse is weak and rapid, and heart action is depressed. Dyspnea is common, the animals may become cyanotic and in severe cases, the animal may become comatose and die. There are no characteristic lesions. Cattle, horses and domestic fowl may also be poisoned. Swine are susceptible, but natural cases of swine poisoning do not occur because swine promptly expel the ingested material by vomiting. Humans have been poisoned when they mistook the bulb for that of an edible species. Symptoms in humans are those of gastrointestinal irritation and cardiovascular collapse. ***Z. coloradensis***. a synonym for Z. *elegans*. ***Z. densus*** (black snake root; crow poison). a bulbous plant that occurs in damp pinelands and bogs in the southeastern United States. It is very toxic and is just as poisonous as, and strongly resembles, *Amianthium muscitoxicum*. ***Z. elegans*** (Z. *coloradensis*; white camas; alkali grass). a highly toxic species with very narrow leaves and greenish flowers in clusters up to 0.3 m long. It is much more toxic than Z. *nuttallii*. ***Z. nuttallii*** (nuttall death camas; death camas; death-camass; poison camas; merryhearts). a species with rather large flowers, nearly 1.3 cm across in loose clusters. It is extremely dangerous to humans and livestock. It contains several complex alkaloids. Symptoms include hypotensive effects, including reduced force and rate of cardiac contractions, accompanied by digestive upset, vomiting, diarrhea, and abdominal pain. Death camas probably causes greater mortality among sheep on the spring ranges of the western United States than any other plant. ***Z. paniculatus*** (foothill death camas; sandcorn). a species found in the western United States. It has yellow flowers and an extremely toxic bulb. ***Z. venenosus***. a synonym for Z. *venosus*. ***Z. venosus***. (Z. *venenosus*; death camas; alkali grass; black snake root; soap plant; poison sego; water lily; wild onion; squirrel food; hog's potato). an herb that is widely distributed in North America, excluding the extreme southwest of Canada, and Alaska. All parts of the plant are poisonous, but the seeds are especially so; ingestion of the bulb can be fatal. The leaves are long, narrow, grass-like, and gathered about the base of the stem. The stem bears a terminal branched cluster of greenish-white to yellow-white flowers. The bulb is onion-like but with a dark coat and lacking the onion odor. Ingestion may cause increased salivation, muscular weakness with staggering, or even complete prostration, dyspnea, coma, and death.

zinc (symbol, Zn; C.I. 77945; C.I. Pigment Black 16; C.I. Pigment Metal 6)[$Z = 30$; $A_r = 65.38$]. a bluish-white, lustrous transition metal that occurs naturally as the sulfide in zinc blende, or as the carbonate in smithsonite. It is a heavy metal, *q.v.*, of relatively low oral toxicity. Zinc is an essential trace element that normally occurs in animals in metalloenzymes, peptidases (e.g., carbonic anhydrase, alkaline phosphatase and dehydrogenase) or as cofactors (e.g., arginase and histamine diaminase) that catalyze the synthesis of DNA, proteins and insulin. Zinc poisoning is uncommon and is important only in the industrial setting. Because zinc is represented in every known chemical pathway in the human body, toxic effects can be widespread and serious if homeostatic clearance is exceeded. Systemic effects by ingestion on humans may include coughing, dyspnea, and sweating. Chronic exposure to zinc produces anemia (with lowered serum levels of iron, ferritin, and hemoglobin, with reduced cytochrome C and catalase activities), and liver and pancreatic fibrosis, and fetal death. Acute exposure to Zn fumes damages the lungs and may result in death. The most important industrial zinc poisoning is metal fume fever which is commonly associated with the inhalation of zinc oxide. Symptoms include delayed onset of chills, fever, sweating and muscular weakness.

zinc chloride. a white, granular or powdery crystalline solid, $ZnCl_2$, that is soluble in water, ethanol, glycerol, and ether. It is used, for example, as a catalyst; as a dehydrating and condensing agent in organic syntheses; in fireproofing; as a wood preservative; as a food preservative, in soldering fluxes, in dry cell batteries; and in oil refining. It is one of the most dangerous and toxic compounds of zinc. It is a potent irritant that can damage the eyes, skin, and mucous membranes on contact. Both the solid and fumes are toxic; inhalation of fumes may cause severe pain and damage to the eyes and upper respiratory tract, accompanied by pulmonary edema.

zinc chromate (basic zinc chromate; chromium zinc oxide; chromic acid, zinc salt; C.I. 77955; C.I. pigment yellow 36; zinc chromate hydroxide). a toxic, water-insoluble, acid-soluble, yellow powder, $ZnCrO_4$ or $Zn{\cdot}CrH_2O_4$ (the hydroxide), used as a pigment in paints and varnishes, linoleum, and epoxy laminates. It is a confirmed human carcinogen and probable mutagen.

zinc chromate hydroxide. zinc chromate.

zinc chromate, potassium dichromate, and zinc hydroxide (3:1:1) (potassium dichromate, zinc chromate, and zinc hydroxide (1:3:1)). a confirmed human carcinogen. See also zinc chromate, potassium dichromate, zinc hydroxide.

zinc colic. colic resulting from chronic zinc poisoning.

zinc diethyl. diethylzinc.

zinc dimethyl. dimethylzinc.

zinc dioxide. zinc peroxide.

zinc diphenyl. diphenylzinc.

zinc ethide. diethylzinc.

zinc ethylenebis(dithiocarbamate). zineb.

zinc hydroxide. a colorless, nearly water-insoluble, crystalline compound, $Zn(OH)_2$. It is a strong alkali that readily forms zinc salts and zincates. It is used as an absorbent in surgical dressings, in rubber compounding, and as a chemical intermediate in certain syntheses.

zinc metal fume fever. zinc fumes comprised of pure zinc powder or dust are relatively nontoxic to humans by inhalation. The oxidation of such fumes, however, or their contamination with other elements such as cadmium, antimony, arsenic, or lead can cause serious systemic effects, usually referred to as zinc metal fume fever. There may be a sweetish taste and dryness of the throat, coughing, weakness, generalized aches, chills and fever, nausea, and vomiting, tightness of the chest, dyspnea, and other pulmonary effects. See also zinc oxide.

zinc omadine. zinc pyridine thione.

zinc orthophosphate. zinc phosphate.

zinc oxide (Chinese white; C.I. 77947; C.I. pigment white 4; zinc white). an odorless, white, grayish, or yellowish powder, ZnO, with a bitter taste. It has a variety of uses as a healing agent, a dietary supplement, a feed additive, in ointments, a colorant in artists' colors, a pigment and mold-growth inhibitor in paints, a seed dressing, a pigment and reinforcing agent in rubber, etc. It is an irritant of the skin and eye and is poisonous, *i.v.* It is a teratogen and probable mutagen. Inhalation of the powder or dust can cause zinc metal fume fever, *q.v.*

zinc peroxide (zinc dioxide; zinc superoxide). an odorless, flammable, explosive, very dangerous, white or yellowish-white powder, O_2Zn. It is poisonous with effects similar to those of zinc oxide. It is a powerful oxidizing agent and severe fire and explosion risk; it should be kept secure from flammable materials; it reacts violently with Zn and Al; undergoes exothermic reactions with water or steam; reacts vigorously with reducing agents; and explodes on heating to a temperature of 212°C, releasing ZnO.

zinc phosphide. a toxic, ethanol-insoluble, dark gray, granular powder, Zn_3P_2, used as a rodenticide. It reacts violently with oxidizing agents and decomposes in water. Oral administration can produce rapidly fatal hepatic and pulmonary lesions in laboratory rats, rabbits and various avian species; dogs and cats are more resistant, the former may tolerate doses as high as 1g/kg body weight. Zinc phosphide itself is not very toxic, but it releases highly poisonous phosphine gas when it comes into contact with mineral acids such as HCl. Susceptible species release gastric HCl continuously, whereas the more resistant species secrete HCl intermittently.

zinc pyridine thione (zinc 2-pyridinethiol-1-oxide; zinc pyrithione; zinc omadine; ZPT). a compound that is very toxic orally to laboratory rats, especially females, but appears to be only moderately toxic to dogs, rabbits, and guinea pigs. Emesis induced by this substance has been shown to cause retinal detachment and blindness in dogs. This effect is apparently species-specific and does not occur in monkeys and laboratory rodents at dosages much higher than that given dogs.

zinc 2-pyridinethiol-1-oxide. zinc pyridine thione.

zinc pyrithione. zinc pyridine thione.

zinc salts. zinc salts are typically toxic, commonly causing increased salivation; violent vomiting and purging, followed by prostration. Lesions may include stricture of esophagus and pylorus, and destruction of glandular structure of stomach as well as ulceration and/or perforation of the stomach.

zinc soaps. the inhalation of fumes from zinc soaps by infants has been known to cause acute fatal pneumonitis characterized by lung lesions similar to, but more serious than, those caused by talc.

zinc white. zinc oxide.

zincophyte. a plant adapted to, or tolerating, high levels of zinc in the soil. Such plants are often used as indicators of soils high in zinc.

zineb. the zinc salt of dimethyldithiocarbamate, it is a fungicide of low toxicity to animals. *Cf.* maneb. See also dithiocarbamate fungicide.

ziram. the zinc salt of ethylenebisdithiocarbamate. Its toxicity to animals is low. It is, however, carcinogenic in some laboratory animals. See dithiocarbamate fungicide.

zirconium (symbol, Zr)[Z = 40; A_r = 91.22]. a transition element that occurs in the gemstone, zircon. Zirconium and its salts are perhaps slightly toxic, The low toxicity is due to low solubility. A granulomatous skin disease has been reported for users of a deodorant containing sodium zirconium lactate. Pulmonary effects have been reported also due to the use of hair sprays that contain zirconium salts.

ZMA. zinc meta-arsenite.

Zn. the symbol for zinc.

zoa. plural of zoon.

ZOA. zinc ortho-arsenate.

Zoacys (keeled rat snake). a large, Asian, nonvenomous snake. See *Ophiophagus hannah*.

Zoantharia. a subclass of Anthozoa. Included are the solitary sea anemones (e.g., *Actinia)* and the colonial stony (or true) corals (e.g., *Astrangia)*, which secrete a calcareous skeleton. Members of this subclass typically have more than eight tentacles. *Cf.* Alcyonaria.

zokalugwagu. *Dispholidus typus*.

zooecdysones. ecdysones that occur in insects and crustaceans; the main ecdysone of insect extracts is α-ecdysone.

zoogenous. originating, derived, or acquired from animals.

zoon. an animal; singular of zoa.

zoonosis (plural, zoonoses). a disease that is transferrable from animals to humans under natural conditions.

zoopathogenic. pathogenic to animals; of or pertaining to a substance or an organism that is productive of disease in animals.

zoopathology. the science of pathology applied to animals.

zoophobic. pertaining to plants that are avoided by animals because they are potentially injurious, harmful, or repugnant (e.g., because of spines, hairs, toxicants).

zooplankton. that portion of a plankton community comprised of microscopic or nearly microscopic animals.

zoosterol. See sterol.

zooteratology. the science of teratology, *q.v.*, as applied to animals.

zootogenous. originating, derived, or acquired from animals.

zootoxic. toxic to, or able to disrupt the iife cycle of animals.

zootoxicologist. a biotoxicologist who is concerned with poisonous and venomous animals; one who contributes to the body of zootoxicology or one who applies the findings of zootoxicology to problem solving or to needs of other disciplines or professions. *Cf.* phytotoxicologist, biotoxicologist.

zootoxicology. a subdiscipline of biotoxicology that deals with poisonous and venomous animals.

zootoxin (animal toxin). a toxin of animal origin as found, for example, in the venoms of snakes and stinging hymenopterous insects. All animal phyla contain species that produce toxins. Animal toxins are often complex mixtures that may include toxicologically active enzymes, other proteins, peptides, and small organic compounds such as alkaloids, biogenic amines, glycosides, and terpenes.

zootrophic. heterotrophic.

Zovirax. *See* ACYCLOVIR.

zoxazolamine (2-amino-5-chlorobenzoxazole). a compound, $C_7H_{15}ClN_2O$, sometimes used as a skeletal muscle relaxant and to promote the excretion (in urine) of uric acid in gout. It is a paralytic drug, the effects of which are closely associated with its concentration in circulating blood which is, in turn, related to its rate of metabolism by monooxygenases. See paralysis time.

ZPP. zinc protoporphyrin (See protoporphyrin).

ZPT. zinc pyridine thione.

Zr. the symbol for zirconium.

Zwerg-klaperschlangen. See *Sistrurus*.

zygacine. an ester alkaloid of *Zigadenus*.

zygadenine. a crystalline alkaloid that occurs with germine in various species of *Zigadenus* and *Veratrum*, q.v.

Zygaena. a genus of small brightly colored moths (Family Zygaenidae) with rounded wings and a well-developed proboscis. These moths can kill mice by injection. The venomous principles in *Z. lenicerae*, are histamine and hydrogen cyanide.

zymogen. an inactive precursor of an enzyme.